HANDBOOK OF ANIMAL MODELS OF INFECTION

HANDBOOK OF ANIMAL MODELS OF INFECTION

Experimental Models in Antimicrobial Chemotherapy

Edited by

Oto Zak
Formerly Head of Infectious Diseases
Research Department
Pharmaceuticals Division
CIBA-GEIGY Limited
Basel, Switzerland

Merle A. Sande
Professor and Chairman
Department of Internal Medicine
School of Medicine
Salt Lake City, Utah
USA

Academic Press
San Diego London Boston
New York Sydney Tokyo Toronto

This book is printed on acid-free paper.

Academic Press
24–28 Oval Road, London NW1 7DX, UK
http://www.hbuk.co.uk/ap/

Academic Press
a division of Harcourt Brace & Company
525 B Street, Suite 1900, San Diego, California 92101-4495, USA
http://www.apnet.com

ISBN 0-12-775390-7

Library of Congress Catalog Card Number: 98-89455

A catalogue record for this book is available from the British Library

Typeset by Phoenix Photosetting, Chatham, Kent, UK
Printed in Great Britain by The Bath Press, Bath, Avon, UK

99 00 01 02 03 BP 9 8 7 6 5 4 3 2 1

Contents

An eight page colour plate section appears between pages 576 and 577.

Section Editors

Oto Zak
Rheinparkstrasse 1/14,
CH-4127 Birsfelden,
Switzerland

Claude Carbon
Hopital Bichat,
Henri Huchard,
F75018 Paris, Cedex 18,
France

Bruno Fantin
Unite de Medicine Interne,
Hôpital Beaujon,
100 bd du General-Leclerc,
92118 CHICHY Cedex,
France

Terence O'Reilly
Novartis Pharma Research,
K125.1.05,
Basel, CH 4002,
Switzerland

Ronald Kaminsky
Swiss Tropical Institute,
Socinstrasse 57,
Postfach,
Basel CH 4002,
Switzerland

Present address:
Novartis,
Centre de Recherche Santé Animale SA,
CH 1566, St-Aubin,
Switzerland

Earl Kern
Dept of Paediatrics & Division of Clinical Virology,
University of Alabama at Birmingham,
309 Bevill Biomedical Research Building,
845 19th Street South,
Birmingham,
Alabama 35294-2170,
USA

List of Contributors

G. K. Abruzzo
Merck & Co Inc, Rahway, NJ 07065-0900, USA

N. L. Aguirre
California Regional Primate Research Center, Davis, CA 95616-8542, USA

J. D. Alder
Scriptgen Pharmaceuticals, Waltham, MA 02451, USA

T. Allen
Animal Welfare Information Center, Beltsville, MD 20705, USA

A. M. Arvin
Stanford University School of Medicine, Stanford, CA 94305-5208, USA

E. Azoulay-Dupuis
Hopital Bichat-Claude Bernard, 75877 Paris Cedex 18, France

I. A. J. M. Bakker-Woudenberg
Erasmus Medical Centre, 3015 GD Rotterdam, The Netherlands

S. Baqar
Naval Medical Research Center, Bethesda, MD 20889-5607, USA

B. Barry
Hôpital Bichat-Claude Bernard, 75877 Paris Cedex 18, France

K. Bartizal
Merck & Co Inc, Rahway, NJ 07065-0900, USA

A. S. Bayer
Harbor-UCLA Research and Education Institute, Torrance, CA 90509, USA

D. I. Bernstein
Children's Hospital Medical Center, Cincinnati, OH 45229-3039, USA

W. Behrens-Baumann
Universitätsklinik für Augenheilkunde, D-39120 Magdeburg, Germany

D. J. Bidanset
University of Alabama at Birmingham, Birmingham, AL 35294-2170, USA

B. Bilger
Universität Hohenheim (Parasitologie), D-70599 Stuttgart, Germany

B. L. Blagburn
College of Veterinary Medicine, Alabama 36849-5519, USA

N. Bourne
Children's Hospital Medical Center, Cincinnati, OH 45229-3039, USA

T. Bremell
Sahlgren University Hospital, S-413 45 Gothenburg, Sweden

I. Brook
Armed Forces Radiobiology Research Institute, Maryland, MD 20889-5603, USA

M. R. W. Brown
Aston University, Birmingham B4 7ET, UK

R. Brun
Swiss Tropical Institute, CH-4002 Basel, Switzerland

A. G. Buret
University of Calgary, Alberta, Canada T2N 1N4

E. F. Burg
Naval Medical Research Center, Bethesda, MD 20889-5607, USA

M. J. Burkhard
College of Veterinary Medicine, North Carolina State University, Raleigh, NC, USA

E. Buschman
McGill University Health Centre, Montreal, Canada

S. M. Callister
Gundersen Luther Medical Center, LaCrosse, WI 54601, USA

A. Casadevall
Albert Einstein College of Medicine, Bronx, NY 10461, USA

A. Cassone
Istituto Superiore di Sanita, Rome 00161, Italy

H. Ceri
University of Calgary, Alberta, Canada T2N 1N4

W. G. Cheadle
Veterans Affairs Medical Center and University of Louisville School of Medicine, Kentucky 40292, USA

N. D. Christensen
The Jake Gittlen Cancer Research Institute, Hershey, PA 17033-0850, USA

J. Cohen
Hammersmith Hospital, London W12 0NN, UK

G. C. Coles
University of Bristol, Bristol BS40 5DU, UK

C. H. Contag
Stanford University School of Medicine, Stanford, CA 94305 5208, USA

P. R. Contag
Xenogen Corporation, Alameda, CA 94501, USA

A. Contrepois
INSERM U158, hôpital Necker, 75015 Paris, France

R. M. Cozens
Novartis Pharma AG, CH-4002 Basel, Switzerland

R. Crawford
Animal Welfare Information Center, Beltsville, MD 20705, USA

S. L. Croft
London School of Hygiene & Tropical Medicine, London WC1E 7HT, UK

M. Cynamon
Veterans Affairs Medical Center, New York, NY 13210 2716, USA

F. de Bernardis
Istituto Superiore di Sanita, Rome, 00161 Italy

P. de Greeve
Inspectorate for Health Protection, 2500 BC, The Hague, The Netherlands

W. de Leeuw
Inspectorate for Health Protection, 2500 BC, The Hague, The Netherlands

D. De Shazer
University of Calgary Health Sciences Centre, Calgary, Canada T2N 4N1

M. DeStefano
SUNY Health Science Center, New York, NY 13210, USA

M. del Cerro
University of Rochester School of Medicine, Rochester, NY 14642, USA

G. Del Giudice
IRIS Chiron Vaccines, 53100 Siena, Italy

M. M. do Canto Cavalheiro
Rua Geraldo Martins 249 ap. 802 Icarai, Niteroi, Brazil

E. A. Deitch
University of Medicine and Dentistry, Newark 07103-2714, USA

K. G. Delgaty
University of Ottawa, Ottawa, Canada K1H 8M5

M. Duchet-Suchaux
Institut National de la Recherche Agronomique, 37380 Nouzilly, France

G. M. Duke
Aviron, Mountain View, CA 94043, USA

H. Dupont
Hopital Bichat, 75877, Paris, France

P. Dussault
Centre de pneumologie de l'Hôpital Laval, Université Laval, Quebec City, Canada G1V 4G5

P. Edelstein
University of Pennsylvania Medical Center, Philadelphia, PA 19104-4283, USA

T. B. Elliott
Armed Forces Radiobiology Research Institute, Maryland, MD 20889-5603, USA

L. S. Engel
Louisiana State University Medical Center, New Orleans, LA 70112-1393, USA

H. Erlendsdóttir
Landspitalinn (University Hospital), 101 Reykjavic, Iceland

F. Espersen
Statens Serum Institut, DK 2300 Copenhagen, Denmark

S. A. Estrem
University of Missouri, Columbia, Missouri 55212, USA

M. E. Etheridge
Johns Hopkins University, Baltimore, MD 21205, USA

B. Fantin
Hopital Beaujon, 92118 Clichy Cedex, France

O. M. Faye-Petersen
University of Alabama at Birmingham, Birmingham, AL 35294-2170, USA

R. J. Fenton
Glaxo-Wellcome Research & Development, Stevenage, UK

P. L. Fidel Jr
Louisiana State University Medical Center, New Orleans, LA 70112-1393, USA

G. Findon
University of Auckland, Auckland, New Zealand

A. M. Flattery
Merck & Co Inc, Rahway, NJ 07065-0900, USA

J. I. Flock
Karolinska Institute, S-141 86 Huddinge, Sweden

I. Foley
University of Warwick, Coventry, CV4 7AL, UK

J. G. Fox
Massachusetts Institute of Technology, Cambridge, MA 02139-4307, USA

N. Frimodt-Möller
Statens Serum Institut, DK-2300 Copenhagen, Denmark

M. Frosch
Institut für Hygiene & Mikrobiologie, 97080 Würzburg, Germany

P. N. Fultz
University of Alabama at Birmingham, Birmingham, AL 35294-2170, USA

P. M. Furr
Imperial College School of Medicine, London W2, UK

G. E. Garber
University of Ottawa, Ottawa, Canada K1G 8LG

C. J. Gauntt
University of Texas Health Science Center, San Antonio, TX 78284-7758, USA

T. G. Geary
Pharmacia and Upjohn Co, Kalamazoo, MI-49001, USA

B. M. Gebhardt
LSU Eye Center, Louisiana State University Medical Center School of Medicine, New Orleans, LA 70112-2234, USA

M. J. Gentry
Section of Infectious Diseases, Veterans Affairs Medical Center, Omaha, NE 68105, USA

P. Ghiara
IRIS Chiron Vaccines, 53100 Siena, Italy

G. S. Giebink
University of Minnesota, Minneapolis, MN 55455, USA

C. Gichuki
Kenya Trypanosomiasis Research Institute, PO Box 362, Kikuyu, Kenya

C. G. Gill
Merck & Co Inc, Rahway, NJ 07065-0900, USA

D. L. Goldman
Albert Einstein College of Medicine, Bronx 10461, USA

G. L. Gorby
Section of Infectious Diseases, Veterans Affairs Medical Center, Omaha, NE 68105, USA

G. Greif
Bayer AG, D-51368 Leverkusen, Germany

S. Gudmundsson
Landspitalinn (University Hospital) and Directorate of Health, 101 Reykjavic, Iceland

A. Haberkorn
Hindenburgstrasse 168, D-42117 Wuppertal, Germany

J. Haddad Jr
College of Physicians and Surgeons, Columbia University, New York, NY 10032-3784, USA

D. M. Hajek
University of Minnesota, Minneapolis, MN 55455, USA

A. K. Hansen
Department of Pharmacology and Pathobiology, Royal Veterinary and Agricultural University, Copenhagen, Denmark

S. F. Hayward-McClelland
University of Ottawa, Ottawa, Canada K1G 8M5

L. D. Hazlett
Wayne State University School of Medicine, Detroit, MI 48201, USA

A. Heimdahl
Karolinska Institute, S-141 86 Huddinge, Sweden

S. A. Hienz
Karolinska Institute, S-141 86 Huddinge, Sweden

J. M. Hill
LSU Eye Center, Louisiana State University Medical Center School of Medicine, New Orleans, LA 70112-1393, USA

J. A. Hobden
Wayne State University School of Medicine, Detroit, MI 48201, USA

N. Høiby
Rigshospitalet, DK-2200 Copenhagen, Denmark

W. J. Hopkins
University of Wisconsin Medical School, Madison, WI 53792, USA

E. A. Hoover
College of Veterinary Medicine, Colorado State University, Fort Collins, Colorado, USA

K. Ishida
Institute of Medical Science, Kawasaki 216, Japan

A. A. Izzo
University of New Mexico School of Medicine, Albuquerque, New Mexico 87131, USA

K. Janitschke
Robert Koche Institut Berlin, D-13352 Berlin, Germany

C. Jantos
Medizinische Mikrobiologie, Klinikum der Justus Liebig Universität, D-35392 Giessen, Germany

B. Ji
Faculte de Medecine Pitie-Salpetriere, 75634 Paris Cedex 13, France

H. K. Johansen
Rigshospitalet, DK 2200, Copenhagen, Denmark

D. E. Johnson
Dept of Veterans Affairs Medical Center, Baltimore, MD 21201, USA

S. S. Johnson
Pharmacia and Upjohn Co, Kalamazoo, MI-49001, USA

V. Joly
Hopital Bichat-Claude Bernard, 75877 Paris Cedex 18, France

A. B. Kaiser
Vanderbilt University Medical Center, Nashville, TN 37232-2358, USA

M. Kaku
Institute of Medical Science, Kawasaki 216, Japan

R. Kaminsky
Swiss Tropical Institute, CH-4002 Basel, Switzerland

H. E. Kaufman
LSU Eye Center, Louisiana State University Medical Center School of Medicine, New Orleans, LA 70112-2234, USA

G. W. Kemble
Aviron, Mountain View, CA 94043, USA

E. R. Kern
University of Alabama at Birmingham, Birmingham, AL 35294-2170, USA

K. A. Kernacki
Wayne State University School of Medicine, Detroit, MI 48201 USA

D. S. Kernodle
Vanderbilt University School of Medicine, Nashville, TN 37232-2358, USA

J. D. Knudsen
Statens Serum Institut, DK 2300 Copenhagen, Denmark

B. A. Korba
Georgetown University, Rockville, MD USA

J. W. Kreider
The Jake Gittlen Cancer Research Institute, Hershey, PA 17033-0850, USA

M. Laviolette
Centre de pneumologie de l'Hôpital Laval, Universite Laval, Quebec City, Canada G1V 4G5

E. S. Lazar
University of Rochester School of Medicine, Rochester, NY 14642, USA

G. D. Ledney
Armed Forces Radiobiology Research Institute, Maryland, MD 20889-5603, USA

J. Lee
Harbor-UCLA Research and Education Institute, Torrance, CA 90509, USA

A. C. A. P. Leenders
Bosch Medical Centre, 's-Hertogenbosch, The Netherlands

A. Lefort
Hopital Beaujon, 92118 Clichy Cedex, France

J. Leggett
Providence Portland Medical Center, Portland, OR 97213-2967, USA

L. L. Leon
Fundaçao Oswaldo Cruz-Instituto Oswaldo Cruz, Rio de Janeiro, Brazil

L. Levy
Hadassah University Hospital, Jerusalem, Israel

D. S. Lindsay
Virginia-Maryland Regional College of Veterinary Medicine, Virginia 24061-0442, USA

M. F. Lipscomb
University of New Mexico School of Medicine, Albuquerque, New Mexico 87131, USA

S. A. Locarnini
Victorian Infectious Diseases Reference Laboratory, North Melbourne, Australia

C. V. Lockatell
University of Maryland School of Medicine, Baltimore, MD 21201, USA

R. Lorenzini
Istituto Superiore di Sanita, Rome 00161, Italy

V. Lorian
The Bronx-Lebanon Hospital Center, Bronx, NY 10456, USA

J. Lovchik
University of New Mexico School of Medicine, Albuquerque, New Mexico 87131, USA

C. A. Luscombe
University of New Mexico Cancer Research & Treatment Center, Albuquerque, NM 87131-5266, USA

C. R. Lyons
University of New Mexico School of Medicine, Albuquerque, New Mexico 87131, USA

J. T. Mader
University of Texas Medical Branch, Galveston, TX 77555-0792, USA

L. Magnotti
University of Medicine and Dentistry, Newark 07103-2714, USA

R. P. Marini
Massachusetts Institute of Technology, Cambridge, MA 02139-4307, USA

T. Matsumoto
University of Occupational & Environmental Health, Kitakyushu, 807-8555, Japan

M. B. McKeough
University of Utah, Salt Lake City, UT 84132, USA

M. M. McNeal
Children's Hospital Medical Center, Cincinnati, OH 45229-3039, USA

T. E. Miller
University of Auckland, Auckland, New Zealand

L. Mizen
Smithkline Beecham Pharmaceuticals, Collegeville, PA 19426-0989, USA

E. S. Mocarski
Stanford University, Stanford, CA 94305-5124, USA

J. F. Moffat
Stanford University School of Medicine, Stanford, CA 94305-5208, USA

P. Moine
Centre Hopitalier de Bicetre, 94274 Le Kremlin-Bicêtre Cedex, France

P. Montravers
Centre Hospitaliar, Universitaire d'Amiens, 80054 Amiens Cedex 01, France

D. W. Morck
University of Calgary, Alberta, Canada T2N 1N4

J. D. Morrey
Institute for Antiviral Research, Logan, UT 84322-5600, USA

J. G. Morris Jr
University of Maryland School of Medicine, Baltimore, MD 21201-1192, USA

D. B. Morton
University of Birmingham, Edgbaston, Birmingham B15 2TT, UK

D. E. Mosier
Scripps Research Institute, La Jolla, CA 92037, USA

M. Motschmann
Universitätsklinik für Augenheilkunde, D-39120 Magdeburg, Germany

A.-M. Moulin
INSERM U 158, hôpital Necker, 75015 Paris, France

M. Muffat-Joly
Hôpital Bichat-Claude Bernard, 75877 Paris Cedex 18, France

J. R. Murphy
University of Texas, Houston, USA

J. M. Nazzaro
Boston University Medical Center, Boston, MA 02118, USA

E. A. Neuwelt
Oregon Health Sciences University, Portland, OR 97201, USA

T. Nishikawa
Kanazawa University, Kanazawa 921, Japan

C. E. Nord
Karolinska Institute, S-141 86 Huddinge, Sweden

R. J. O'Callaghan
Louisiana State University Medical Center, New Orleans, LA 70112-1393, USA

M. Ohkawa
Kanazawa Municipal Hospital, Kanazawa 921, Japan

A. Olomu
Stanford University School of Medicine, Stanford, CA 94305 5208, USA

M. E. Olson
University of Calgary, Alberta, Canada T2N 1N4

T. O'Reilly
Novartis Pharma Ltd, Basel, Switzerland CH-4002

I. M. Orme
Colorado State University, Fort Collins, CO 80523, USA

D. J. Ormrod
University of Auckland, Auckland, New Zealand

P. Panigrahi
University of Maryland School of Medicine, Baltimore, MD 21201-1192, USA

J. R. Perfect
Duke University Medical Center, Durham, NC 27710, USA

W. Peters
CABI Bioscience, St Albans, Herts AL4 0XU, UK

J. C. Peyton
Veterans Affairs Medical Center, Kentucky, 40292, USA

J. D. Pietsch
Price Institute of Surgical Research, Department of Surgery, University of Louisville School of Medicine, Kentucky 40292, USA

K. Piper
Mayo Graduate School of Medicine, Rochester, MN 55905, USA

A. Polak-Wyss
Spitzenrainweg 45, 4147 Aesch, Switzerland

H. C. Polk, Jr
Price Institute of Surgical Research, Department of Surgery, University of Louisville School of Medicine, Kentucky 40292, USA

L. C. Preheim
Section of Infectious Diseases, Veterans Affairs Medical Center, Omaha, NE 68105, USA

G. E. Price
University of Birmingham, Edgbaston, Birmingham B15 2TT, UK

G. A. Prince
Virion Systems Inc., Rockville, USA

M. K. Quartey
University of Minnesota, Minneapolis, MN 55455, USA

F. Quimby
College of Veterinary Medicine, Cornell University, USA

R. Rappuoli
IRIS Chiron Vaccines, 53100 Siena, Italy

R. R. Read
University of Calgary, Alberta, Canada T2N 1N4

F. P. Reinholt
Karolinska Institute, S-141 86 Huddinge, Sweden

J. Renneberg
Department of Medicine, Laegemiddelstyrelsen, 2700 Bronshoj, Denmark

J. Y. Richmond
Center for Disease Control, Atlanta, GA 30333, USA

B. L. Robinson
CABI Bioscience, St Albans, Herts AL4 0XU, UK

T. Romig
Universität Hohenheim (Parasitologie), D-70599 Stuttgart, Germany

M. S. Rouse
Mayo Clinic, Rochester, MN 55905, USA

M. A. Sande
Utah School of Medicine, Salt Lake City, USA

W. M. Scheld
University of Virginia, Charlottesville, VA 22908, USA

R. Schell
University of Wisconsin, Madison, WI 53706, USA

H. G. Schiefer
Medizinische Mikrobiologie, Klinikum der Justus Liebig Universität, D-35392 Giessen, Germany

A. Schmidt
Bayer AG, D-42096 Wuppertal, Germany

D. M. Scollard
GWL Hansen's Disease Center, Baton Rouge 70894, USA

K. B. Seydel
Washington University School of Medicine, St Louis, MO 63110-1093 USA

T. Shaw
Victorian Infectious Diseases Reference Laboratory, North Melbourne, Australia

J. Shimada
Institute of Medical Science, Kawasaki 216, Japan

M. E. Shirtliff
University of Texas Medical Branch, Galveston, TX 77555-1115, USA

M. Shroti
Veterans Affairs Medical Center and University of Louisville School of Medicine, Kentucky 40292, USA

R. W. Sidwell
Institute for Antiviral Research, Logan, UT 84322-5600, USA

E. Skamene
McGill University Health Centre, Montreal, Canada

J. G. Smith
Merck & Co Inc, Rahway, NJ 07065-0900, USA

J. D. Sobel
Wayne State University School of Medicine, Detroit, MI, USA

N. Søe-Nielsen
University of Copenhagen, 2900 Hellerup, Denmark

K. F. Soike
Tulane Regional Primate Research, Covington, LA 70433, USA

S. L. Spruance
University of Utah, Salt Lake City, UT 84132, USA

S. Sriskandan
Hammersmith Hospital, London W12 0NN, UK

F. S. Stals
St Laurentius Zickenhuis, 6043 CV Roermond, The Netherlands

L. R. Stanberry
Children's Hospital Medical Center, Cincinnati, OH 45229-3039, USA

S. L. Stanley Jr
Washington University School of Medicine, St Louis, NO 63110-1093, USA

J. M. Steckelberg
Mayo Clinic, Rochester, MN 55905, USA

C. A. Stoddart
Gladstone Institute of Virology & Immunology, San Francisco, CA 94141-9100, USA

C. Sweet
University of Birmingham, Edgbaston, Birmingham B15 2TT, UK

M. Takashima
Kanazawa University, Kanazawa 921, Japan

T. G. Takoudes
College of Physicians and Surgeons, Columbia University, New York, NY 10032-3784, USA

D. Taylor-Robinson
Imperial College School of Medicine, London W2, UK

B. C. Tennant
College of Veterinary Medicine, Ithaca, NY 14853-6401, USA

E. M. Thomas
Pharmacia and Upjohn Co, Kalamazoo, MI-49001 USA

R. C. A. Thompson
Murdoch University, Murdoch WA 6150, Australia

L. Tissi
University of Perugia, 06100 Perugia, Italy

S. Tokunaga
Maizuru Kyosai Hospital, Maizuru, Japan

G. C. Townsend
University of Virginia, Charlottesville, VA 22908, USA

G. M. Tremblay
Centre de pneumologie de l'Hôpital Laval, Universite Laval, Quebec City, Canada G1V 4G5

R. W. Truman
GWL Hansen's Disease Center, Baton Rouge 70894, USA

E. Tuomanen
St Jude Children's Hospital, Memphis, TN 38105-2794, USA

J. Tureen
San Francisco General Hospital, San Francisco, CA 94110, USA

S. J. Upton
Kansas State University, Manhattan, Kansas 665056, USA

E. W. M. van Etten
Erasmus Medical Centre, 3015 GD Rotterdam, The Netherlands

K. K. A. Van Rompay
California Regional Primate Research Center, Davis, CA 95616-8542, USA

E. D. Varnell
LSU Eye Center, Louisiana State University Eye Center School of Medicine, New Orleans, LA 70112-2234, USA

U. Vogel
Institut für Hygiene & Mikrobiologie, 97080 Würzburg, Germany

R. L. Ward
Children's Hospital Medical Center, Cincinnati, OH 45229-3039, USA

K. Wicher
David Axelrod Institute, Albany, NY 12201-0509, USA

V. Wicher
David Axelrod Institute, Albany, NY 12201-0509, USA

J. A. Wilder
University of New Mexico School of Medicine, Albuquerque, New Mexico 87131, USA

W. R. Wilson
Mayo Clinic, Rochester, MN 55905, USA

D. E. Woods
University of Calgary Health Sciences Centre, Calgary, T2N 4N1 Canada

D. Z. Xu
University of Medicine and Dentistry, Newark 07103-2714, USA

V. Yardley
London School of Hygiene & Tropical Medicine, London WC1E 7HT, UK

M. R. Yeaman
St John's Cardiovascular Research Centre, Torrance, CA 90509, USA

P. Yeni
Hopital Bichat-Claude Bernard, 75877 Paris Cedex 18, France

Z. Yuan
University of Minnesota, Minneapolis, MN 55455, USA

O. Zak
Rheinparkstrasse, CH 4127 Birsfelden, Switzerland

T. Zhang
Washington University School of Medicine, St Louis, MO 63110-1093, USA

W. Zimmerli
University Hospital, CH-4031 Basel, Switzerland

Preface

The development and testing of new antimicrobial agents has taken on new importance and urgency as we began to lose our currently available drugs for treating infections caused by highly drug-resistant microorganisms. Today, we have no antibiotics active *in vivo* for the treatment of infections due to some strains of *Enterococcus faecium*. The spread of highly resistant clones of *Streptococcus pneumoniae* around the world is dramatically changing the therapeutic approach to the treatment of infections caused by this organism. The appearance in Japan and then in the USA of vancomycin-resistant strains of *Staphylococcus aureus* has shaken the medical world. It is clearly apparent that, without new therapeutic strategies and drugs, the number of untreatable infections will increase markedly during the next 10 years. It is indeed conceivable that we will witness the end of the antibiotic era as we know it today. This problem is not restricted to more common bacterial infections, but also includes multi-drug-resistant *Mycobacterium tuberculosis* and *Plasmodium falciparum*. Furthermore, emerging infections present new therapeutic needs: adequate treatments for chlamydial infection and for Hantavirus and Filovirus infections are lacking.

However, developing new therapeutic strategies for the treatment of microbial infections is a complicated and expensive activity. The standard approach has been to screen large numbers of compounds for their *in vitro* activity against a series of reference organisms followed by determination of activity against a large number of clinical isolates. Next, candidate compounds possessing an appreciable antimicrobial activity are selected and their properties are delineated using many different *in vitro* tests (e.g. stability, protein binding, solubility, etc.), and their pharmacokinetic properties and toxicity profile using *in vivo* systems. Today, with the genomic sequences of many organisms available, screening based on molecular techniques (e.g. determination of novel genes expressed *in vivo* by genomic or proteomic techniques) should facilitate identification of new targets. Then, with genetic approaches to determine the suitability of the target (e.g. determination of the virulence mutant organisms) to provide a "proof of concept", followed by production and characterization of recombinant protein(s), rational drug design based upon *in vitro* screening of compounds against isolated target proteins may facilitate the discovery of new inhibitors by the use of a molecularly defined target methodology. This approach has been beautifully illustrated by the discovery of the protease inhibitors of the human immunodeficiency virus. However, no matter how sophisticated drug screening and development may be, the final essential step in development of new antimicrobial therapies prior to testing in man is evaluation of the drug for its antimicrobial efficacy and toxicity in animal models of infection.

Thirteen years ago we published a three-volume text elucidating in detail a series of experimental models for evaluating antimicrobial chemotherapy *in vivo*. We felt that having this information in one series would save investigators time and effort in formulating an approach to test a new potential therapeutic agent or combination of agents for *in vivo* efficacy and to position the therapy for specific infections where it might have a therapeutic promise. The series has accomplished these goals and, while it has a limited readership, those active in the field of research and development of antimicrobials have found it extremely useful.

Therefore, we have decided to update the series by publishing a new edition. For that we have chosen a one-volume strategy, a handbook format. This book is divided into five parts, the first covering general methodology and followed by parts describing experimental bacterial, mycotic, parasitic and viral infections. We have attempted to include many new models that have been developed within the last decade or that were omitted for various reasons in the first edition. For most part they are sophisticated models that allow for well-targeted use of new agents. We hope this update will be as useful as the last edition proved to be.

Lastly, we would like to extend our appreciation to the staff at Academic Press who have been involved in this project: Dr Lilian Leung, Emma White and Dr Tessa Picknett; their enthusiasm and dedication to this project, and patience with us, is most appreciated.

Oto Zak and Merle A. Sande, Editors

Claude Carbon, Bruno Fantin, Ronald Kaminsky,
Earl R. Kern and Terence O'Reilly, Section Editors

Introduction: The Role of Animal Models in the Evaluation of New Antibiotics

Oto Zak, Merle Sande and Terence O'Reilly

Two milestones in the history of antimicrobial chemotherapy are directly attributable to experiments conducted in animal models of infection. The first resulted from the surprising observation that a dyestuff, sulphonamido-crysoidine (Prontosil), although devoid of antibacterial activity *in vitro*, was effective against a pneumococcal infection in mice (Domagk, 1935). This finding furnished the very first proof that (1) systemic bacterial infections could be cured by drugs and (2) *in vivo* testing was an indispensable component of antimicrobial drug research. The second was penicillin. In retrospect, there seems to have been a fair chance of this antibiotic becoming available to patients much earlier than it did: Fleming only used it for differentiating cultures and in a few cases as a local antiseptic; he apparently gave no thought to the possibility that it might also be systemically effective. Not until the 1940s, at the suggestion of Florey and Chain, was the substance partially purified and shown to exert potent therapeutic activity, first in the mouse and afterward in humans (Abraham, 1980).

It is hardly conceivable that any new antibiotic could have been developed since then without thorough verification at an early stage of its antimicrobial efficacy (and toxicological innocuity) in models of infection in laboratory animals. Testing *in vivo* has come to be recognized as the essential link between *in vitro* sensitivity testing and clinical studies. Indeed, it is essential that new agents shown to be of interest following *in vitro* evaluation should exhibit sufficient activity *in vivo* to justify their continued clinical development. Guidelines for the clinical evaluation of anti-infective drugs specifically place experimental evaluation of new compounds (or novel combinations or therapeutic modalities) in animals as prerequisites for clinical trials (Beam *et al.*, 1992, 1993). Specifically, indications of pharmacokinetic properties of new molecules, including possible metabolism, pharmacodynamic attributes (i.e. determination of the interactions of drug, host and infecting microbe, e.g. determination of possible postantibiotic effects) and determination of efficacy in animal models mimicking human disease are required, and indeed may assist in the planning of clinical trials of new antibiotics or combinations (Beam *et al.*, 1992, 1993). However, it must be emphasized that a direct extrapolation of data obtained from animal models of infection may not be possible and instead the data obtained should guide the initiation of clinical research. Given this, the researcher using experimental models should bear in mind the nature of clinical infection and treatment when designing studies for the evaluation of anti-infectives.

Since 1960 far more than 1000 different animal models for experimental chemotherapy have been described. Depending on their purpose, nature and predictive value, they can be assigned to various categories (Zak, 1980); each has its justification at some point in the development of an antibiotic or therapeutic strategy.

The basic antimicrobial screening models are the *in-vivo* test systems most frequently used in the early evaluation of new antibiotics. The best known model in this category is that of acute septicemia in mice, and it is, so far, the simplest and most inexpensive test, allowing a rough estimate of whether or not an antibiotic is likely to be effective *in vivo*. The relevance of the results obtained in this model to clinical chemotherapy is, however, limited by its many inherent drawbacks (Miller, 1971; Bergeron, 1978; Barza, 1978; Zak, 1980; Zak and Sande, 1982), e.g. the too rapidly fatal course of infection in mice not being a characteristic of human disease, the sensitivity of the test to the size of the inoculum, the prophylactic rather than therapeutic regimens used in this test, etc.

Ex-vivo models appear to simulate human infections somewhat more closely than do screening tests. This category comprises models in which foreign bodies, e.g. fibrin clots or dialysis sacks, are implanted into animals and infected. Other variants use perforated rubber or plastic tubing, small spiral springs or perforated balls, which are infected after they have been surrounded by granulation tissue and filled with edematous ("tissue") fluid. The effects of antibiotics in these tests are then evaluated *in vitro*, e.g. by determination of the antibiotic concentrations reached or the bacterial counts in samples of the foreign bodies or the fluid.

A number of models used in the evaluation of antibiotics belong to the category described as monoparametric/polyparametric models. They differ from the screening models in that, instead of waiting for the ultimate therapeutic effects to appear (cure or death), only one single, or preferably many, parameter(s) is examined during the experiment. The determination of serial bacterial counts in the blood or tissues during treatment, or studies of the influence of antibiotics on the morphology of bacteria in the body cavities, can be included in this category.

Both *ex-vivo* and monoparametric models are frequently used during the more advanced secondary phase of the evaluation of a new antibiotic. In general they are very helpful in differentiating the properties of various antibiotics, although their main merit is in providing data on the effects of antibiotics that would be impossible to study in humans, whether for ethical or technical reasons, e.g. the effects of subtherapeutic doses or very short treatment regimens. Many of the drawbacks of screening models, however, also apply to the *ex-vivo* and monoparametric test systems and especially to their use in small rodents.

Experimental infections belonging to the category of discriminative models are designed to simulate human infection as closely as possible. They permit the potential therapeutic effects of new or already established drugs and drug combinations to be differentiated and delimit the indications in which they might be effective in humans. These systems are also suitable for investigating the interactions between the host and the drug or microorganism. The ideal model of this type should exhibit the features listed in Table 1.

Table 1 Features of the ideal discriminative animal model (adapted from Zak, 1980; modified from Harter and Petersdorf, 1960)

Technique of infection	*Simple*
Causative organism Route of entry Spread in the body	Identical or at least similar to situation in humans
Tissue involvement Severity, course, and duration of disease	Predictable Reproducible Amenable to analysis
Susceptibility to chemotherapy	Measurable Reproducible

Several models seem to satisfy these criteria adequately, a few examples of which include endocarditis in rabbits or rats, pneumonia in immunosuppressed guinea-pigs or dogs, osteomyelitis in rabbits or rats, various models of urinary tract infection in rats and pigs, meningitis in rabbits, eye infections, oral candidiasis in rats, and cryptococcal infection in guinea-pigs or rabbits. Investigations using these or other discriminative models have undoubtedly made a significant contribution to our knowledge of the pathogenesis of various types of meningitis (Scheld, 1981, 1986) or the value of specific antibiotics in specific infections, and they have been most helpful in establishing the guiding principles for the use of antimicrobial agents. As was recently noted by Sande (1981), many questions can be more clearly answered with the aid of these models than in humans, e.g.:

- When is a bacteriostatic drug as good as a bactericidal drug?
- When is the rate of bacterial killing important?
- How important is a high bactericidal titer in serum?
- What is the optimal dosing interval?
- How important are postantibiotic effects?

Results obtained in discriminative models of infection have often provided the impetus for the initiation of clinical trials or have confirmed and explained observations made in humans. For example, studies in animal models of endocarditis due to *Streptococcus viridans*, sensitive, relatively resistant or tolerant to penicillin, demonstrated that penicillin given alone was less efficacious against the resistant and tolerant strains than against the susceptible strains. A combination with aminoglycosides (e.g. streptomycin), on the other hand, was equally effective against all streptococci tested and markedly accelerated the rate at which the cardiac vegetations were cleared of bacteria (Zak *et al.*, 1982).

Clinical trials conducted by Wilson *et al.* (1978, 1981) showed that 2 weeks of therapy with this regimen was as effective as 4 weeks with penicillin alone. Recommendations for changes of therapeutic regimens have also resulted from findings made in animal models, e.g. the use of vancomycin combined with rifampicin, with or without gentamicin, in the treatment of endocarditis due to methicillin-resistant *Staphylococcus epidermidis* (Drake and Sande, 1986; Vazquez and Archer, 1980).

Finally, it is worth bearing in mind that had the model of bacterial meningitis been used to test the efficacy of cephalothin in this infection, therapeutic failures might not have been needed to reveal the agent's inferior activity in patients' cerebrospinal fluid (CSF). Cephalothin penetrates poorly into the CSF of rabbits with meningitis, and much of the drug is in the desacetyl form, which is not effective against meningococci or pneumococci (Ernest and Sande, 1982).

In short, animal models of infection are the best means presently available of estimating the efficacy of and tolerability to an antibiotic before its administration to humans and of finding new approaches to the treatment of infections.

The technical advantages of using animal models are no less important: studies can be performed in groups of animals large enough to bear statistical analysis; they provide reproducible results; and they permit individual effects to be examined separately through variation of the parameters (Zak, 1982). On the other hand, it cannot be denied that even the best animal model is not free of limitations, and investigators have to be as cognizant of these as of the merits of the model.

Perhaps the greatest limitation in the use of animal models for studying chemotherapy, and perhaps the single most ignored parameter, is the difference between the pharmacokinetic characteristics of antibiotics in most animals used for infectious disease models and those occurring in humans. The predominant differences lie in the faster elimination of drugs by small animals as compared to humans (e.g. O'Reilly *et al.*, 1996; Craig, 1998). Although the *in-vitro* potency of an antimicrobial remains a critical predictor of eventual efficacy, the *in-vivo* efficacy of a particular agent can be dramatically affected by the pharmacokinetics (metabolism and elimination of antimicrobials from the body) and pharmacodynamics (antimicrobial action of drugs *in vivo*) of the agent. A general background to pharmacokinetics/pharmacodynamics can be found in several sources, and the background to particular issues concerning antimicrobials has been recently reviewed (e.g. Craig, 1998; O'Reilly *et al.*, 1996). Consideration of these pharmacokinetic differences has led some researchers to obtain antibiotic pharmacokinetic profiles more similar to those expressed by humans by adapting the antimicrobial dosing regimens to compensate for the faster elimination of compounds by animals. Impairment of renal function to reduce antibiotic elimination rates, fractional administration of decreasing doses of antibiotics at various time intervals, and continuous infusion have all been used to mimic human pharmacokinetics of antimicrobial agents in small animals. However, the use of such protocols is limited in experimental evaluations of antibiotics as they require knowledge of the pharmacokinetic profiles of antibiotics in humans. They may retain value in the evaluation of new compounds with unknown human pharmacokinetics when the agents are administered in regimens similar to the regimens of standard compounds with established human pharmacokinetic profiles.

The focus of this book is to combine the collected experience of the world's leading investigators in the area of experimental anti-infective chemotherapy. Many previous reviews of the use of animal models for the evaluation of infection have appeared (for a list, see O'Reilly *et al.*, 1996). We have selected topics ranging from the ethical considerations of using animals for experimentation to the practical aspects of utilizing the various discriminative models. We have attempted to encompass models of most of the agents that infect humans where chemotherapy has been employed. Each section emphasizes not only the contributions of the model but also the pitfalls in design and in the interpretation of results. It should be noted, however, that the models presented in this book may not be suitable for the exploration of the mechanisms of pathogenesis during infection. Although the use of transgenic mice for such studies has proved valuable (for a review, see Kaufmann, 1994; Kaufmann and Ladel, 1994), the use of infected transgenic mice for the evaluation of host response during the antimicrobial treatment of infection has not received much attention.

It is hoped that this contribution will assist the new investigator in better planning experiments by avoiding many of the mistakes already made by others.

References

Abraham, E. P. (1980). Fleming's discovery. *Rev. Infect. Dis.*, **2**, 140–141.

Barza, M. (1978). A critique of animal models in antibiotic research. *Scand. J. Infect. Dis.*, **14** (Suppl.), 109–117.

Beam, T. R., Gilbert, D. N., Kunin, C. M. (1992). General guidelines for the clinical evaluation of anti-infective drug products. *Clin. Infect. Dis.*, **15** (Suppl. 1), S5–S32.

Beam, T. R., Gilbert, D. N., Kunin, C. M., the European Working Party (eds) (1993). *European Guidelines for the Clinical Evaluation of Anti-Infective Drug Products*. European Society of Clinical Microbiology and Infectious Diseases.

Bergeron, M. G. (1978). A review of models for the therapy of experimental infections. *Scand. J. Infect. Dis.*, **14** (Suppl.), 189–206.

Craig, W. A. (1998). Pharmacokinetic/pharmacodynamic parameters: rationale for antibacterial dosing of mice and men. *Clin. Infect. Dis.*, **26**, 1–10.

Domagk, G. (1935). Ein Beitrag zur Chemotherapie der bakteriellen Infektionen. *Deutsch. Med. Wschr.*, **61**, 250–153.

Ernst, J. D., Sande, M. A. (1982). Selected examples of failure of *in vitro* testing to predict *in vivo* response to antibiotics. In: *Action of Antibiotics in Patients* (ed. Sabath, L. D.), pp. 68–73. Hans Huber, Berne.

Harter, D. H., Petersdorf, R. G. (1960). A consideration of the pathogenesis of bacterial meningitis: review of experimental and clinical studies. *Yale J. Biol. Med.*, **32**, 280–309.

Kaufmann, S. H. (1994). Bacterial and protozoal infections in genetically disrupted mice. *Curr. Opin. Immunol.*, **6**, 518–525.

Kaufmann, S. H., Ladel, C. H. (1994). Application of knockout mice to the experimental analysis of infections with bacteria and protozoa. *Trends Microbiol.*, **2**, 235–242.

Miller, A. K. (1971). *In vivo* evaluation of antibacterial chemotherapeutic substances. *Adv. Appl. Microbiol.*, **14**, 151–183.

O'Reilly, T., Cleeland, R., Squires, E. (1996). Evaluation of antimicrobials in experimental animal infections. In: *Antibiotics in Laboratory Medicine*, 4th edn (ed. Lorian, V.), pp. 599–759. Williams & Wilkins, Baltimore.

Sande, M. A. (1981). Animal models in the evaluation of antimicrobial agents (guest editorial). *Infect. Dis.*, **4**(11), 4–20.

Scheld, W. M. (1981). Pathophysiological correlates in bacterial meningitis. *J. Infect.*, **3** (Suppl. 1), 5–19.

Scheld, W. M. (1986). Experimental animal models of bacterial meningitis. In: *Experimental Models in Antimicrobial Chemotherapy*, vol. 1 (eds Zak. O., Sande, M. A.). Academic Press, London.

Vazquez, G. J., Archer, G. L. (1980). Antibiotic therapy of experimental *Staphylococcus epidermidis* endocarditis. *Antimicrob. Agents Chemother.*, **17**, 280–285.

Wilson, W. R., Geraci, J. E., Wilkowski, C. J., Washington, J. A. II (1978). Short-term intramuscular therapy with procaine penicillin plus streptomycin for endocarditis due to viridans streptococci. *Circulation*, **57**, 1158–1161.

Wilson, W. R., Thompson, R. L., Wilkowski, C. J., Washington, J. A. II, Giulliani, E. R., Geraci, J. E. (1981). Short term therapy for streptococcal infective endocarditis: combined intramuscular administration of penicillin and streptomycin. *J. A. M. A.*, **245**, 360–363.

Zak, O. (1980). Scope and limitations of experimental chemotherapy. *Experientia*, **36**, 479–483.

Zak, O. (1982). Usefulness and limitations of animal models in the study of opportunistic nonbacterial infections. In: *Infections in Cancer Patients* (ed. Klastersky, J.), pp. 25–45. Raven Press, New York.

Zak, O., Sande, M. A. (1982). Correlation of *in vitro* antimicrobial activity of antibiotics with results of treatment in experimental animal models and human infection. In: *Action of Antibiotics in Patients* (ed. Sabath, L. D.), pp. 55–67. Hans Huber, Berne.

Zak, O., Sande, M. A., Wilson, W. R. (1982). Penicillin or penicillin plus streptomycin therapy of viridans streptococcal experimental endocarditis. 22nd Interscience Conference on Antimicrobial Agents and Chemotherapy, Abstract No 839.

Section I

Introductory Background to Animal Models of Infection

Chapter 1

Early History of Animal Models of Infection

A. Contrepois and A.-M. Moulin

The formation of the concept of specific infection has been a complex process dependent on a series of discoveries and the identification of anatomopathological and microbiological mechanisms, as well as clinical and epidemiological phenomena. In the development of a medical discipline such as infectious pathology, various steps can be outlined, among which one of the most crucial is the quest for animal models. These models were established after a long historical development in the understanding of human infections.

Clinical investigation and necropsy findings

Some nosological entities still acknowledged today were identified very early, on the sole basis of clinical signs, such as smallpox and measles; they were differentiated by the Arab physician Razi, as early as the 10th century. But at the turn of the 19th century in western Europe, the anatomoclinical method enabled a new system of correlations between lesions discerned during autopsy and symptoms observed in the live patient (Ackerknecht, 1967; Foucault, 1976; Gelfand, 1980; Maulitz, 1987).

One of the first founders of the doctrine of anatomoclinical specificity was Pierre Bretonneau in France (Foster, 1970). Bretonneau's grasp of the concept of specificity derived from his detailed study of two diseases, diphtheria and typhoid fever. He showed that the various sorts of diphtheria in fact constituted the same entity, quite distinct from the various other inflammatory and ulcerative conditions of the throat. Similarly, he separated typhoid from the undifferentiated mass of fevers by pointing out the characteristic lesions in the small intestine. He recognized both diseases to be contagious. During that period, the research on heart valve diseases provides a fascinating example of the construction of new entities (Contrepois, 1996). In 1806, Nicolas Corvisart in France described, as well as clinical signs and symptoms, outgrowths or soft "vegetations" observed on the cardiac valves. The Englishman Allan Burns wrote in 1809 about a "concretion of an irregular form, friable and of granular structure" in the heart. According to Friedrich Kreysig of Germany, cardiac valve alterations were linked to "blood clots" in the heart cavity; coagulation occurred before the clots adhered to the membrane.

In 1819 in Paris, Théophile Laënnec greatly improved clinical observation techniques, and information gained through percussion and auscultation enabled him to analyse new cardiac symptoms which he correlated to lesions observed during autopsy. In 1835, Jean-Baptiste Bouillaud named "endocardium" the internal membrane of the heart (described by Bichat in 1800) and called the inflammation of this membrane "endocarditis". In 1841, Bouillaud isolated a category of "typhic endocarditis". At the time, the generic term "typhus" or "typhoid fever" covered several febrile syndromes, usually accompanied by stupor and delirium.

In 1852, the Englishman William Senhouse Kirkes described in cases of endocarditis phenomena which were related to the consequences of the release of fragmented cardiac vegetations of varying size into the blood stream, and which blocked a vessel far away from the heart. This local lesion was the morbid phenomenon referred to by Rudolph Virchow as "embolism" in 1858. According to some authors in the 1860s, the symptoms (temperature peak, shivering, spleen enlargement) were due to the "intoxication" induced by a "morbid poison" present in the blood and continuously produced in the abnormal endocardium.

Around 1860, the microscopic observations of the thrombi gradually revealed the existence of numerous "small granulations" in vegetations on the valve. At that time, it was difficult for doctors to say anything more about the nature of these granulations.

From "miasmatic diseases" to the germ theory

Until microscopes were notably improved in the 1830s, microscopic observations were hardly relevant to substantiate new nosological entities. Muscaridine was investigated at that time by the Italian Agostino Bassi, who showed by inoculation experiments into healthy silkworms that a "parasitic fungus" was the cause of the disease (Foster, 1970). By 1840 the theory that some human diseases might be caused by "microparasites" was to some extent "in the air".

Handbook of Animal Models of Infection
ISBN 0-12-775390-7

Although difficult to understand today, a basic distinction was made between contagious diseases, transmissible through close contact between one individual and another, and infection carried by miasmas disseminated through the air from a focus (Delaporte, 1986). Syphilis illustrated the former, malaria the latter. In fact, most diseases associated the two modes of propagation.

The germ theory unified the understanding of infection by focusing on a single phenomenon: the actual presence of living germs, which are responsible for the disease. The germ theory rested on the assumption that the active agent of the diseases was a *contagium vivum*, reminiscent of Bassi's parasitic fungi. According to Jacob Henle in 1840, a human epidemic would start with germ dissemination in the air (Foster, 1970). The germs could infect a host, multiply, and in turn produce further germs that could spread either by direct contact between persons or by indirect transfer through the air or another means. The germ theory received a major impetus from the study of "fungi" or "parasites", easily visible under the microscope. Thus various "granules", "virus", "ferments", "bacteria" were described (Bulloch, 1938). But, as the new microscopes suddenly enhanced the power of the senses, this knowledge acquired a new meaning.

In 1869, during an endocarditis autopsy, the Norwegian Emmanuel Winge noticed "on the aortic valves, grey felt-like masses, resembling pebbles, which were easily detached". Under the microscope, these masses were made up of a fine network of "fibrinous filament". With greater magnifying power, these filaments "appeared to be entangled with microorganisms". Winge admitted that they were "parasitic" organisms which, having entered the bloodstream through a plantar excoriation, were transported to the heart through the veins (Contrepois, 1996). Although not questioning the existence of "parasites", some authors expressed reservations and were inclined to believe that these microorganisms were the *result* and not the *cause* of endocarditis — a general debate at the time on the meaning of microorganism findings. How could one prove that microorganisms were crucial for the emergence of the disease? Animal models were decisive in that respect.

Returning to the endocarditis, the problem was posed in the following way: in order to give a solid base to Winge's hypothesis, it was necessary to prove that the "germs" had grown in the vegetations on the valves during the patient's lifetime, that they were not just secondary putrefaction agents. It was also important to find out if the infectious agent was the same in all cases. Although all attempts had been inconclusive thus far, in order to prove the case, doctors tried very hard to reproduce the disease (Contrepois, 1995a).

Experimental physiology and experimental pathology

Physiology, in the time of François Magendie in France and Johannes Müller in Germany, in the early 19th century, evolved from being a study concerned with organs and other parts serving the animal soul — a discipline closely associated with anatomy — to a study of the processes of living bodies (Holmes, 1974, Albury, 1977; Maulitz, 1978; Pickstone, 1981; Lenoir, 1982; Lesch, 1984; Fye, 1987; Geison, 1987; Jacyna, 1988). One consequence of this new concern with living processes and experimental control was that most of the new life sciences were based to an unprecedented extent on experiments on living animals (Rupke, 1987; Tansey, 1989; Cunnigham and Williams, 1992). During the 19th century, Claude Bernard's and Karl Wunderlich's physiological medicine was based on the notion that diseases are the result of a disturbance in the normal functioning of an organ or system of organs, with a wide range from normal to pathological (Wunderlich, 1845; Bernard, 1865). They were generally opposed to the defenders of specific disease entities and a specific cause of disease.

First experimental "infections"

The supporters of animal experimentation assumed that there were phenomena which were common to both humans and animals (Bynum, 1990). Experimental pathology was present throughout the 17th and 18th centuries. The Italian Giorgio Baglivi, for example, tried to develop different kinds of "fevers" in order to "better know the essential cause of these fevers". He experimented on dogs and injected in veins different "liquors" until symptoms appeared. The Frenchman Antoine Deidier tried to reproduce the symptoms of plague by injecting animals with the bile of plague victims (Lancereaux, 1872).

In the last three decades of the 19th century, experimental pathologists tried to further their knowledge of "spontaneous diseases", by comparing "induced pathological phenomena" with "natural morbid phenomena" (Lancereaux, 1872). Between 1865 and 1868, Jean Antoine Villemin in Paris inoculated tuberculous matter from human cases into rabbits, and showed that the rabbits developed the characteristic tubercles and that the disease, once produced, could be transmitted to other rabbits (Villemin, 1866). But the question of the transmission of tuberculosis in humans was not resolved before the bacilli were discovered by Koch in 1882.

Other researchers tried to transmit human infective diseases to animals, with varying levels of success. Confusion often arose because an animal might die from an intoxication of the putrid material rather than from the infection. Between 1863 and 1870, Joseph Davaine in France produced an infectious septicaemia and established the bacilli of anthrax as the cause of the disease. At the same period, Johannes Orth in Germany claimed that he had transmitted human erysipelas to rabbits. Edwin Klebs went further and induced septicaemia with artificial cultures of micrococci from septic infections. Pasteur's demonstration that a pure culture of the anthrax bacillus in an artificial medium would cause anthrax in an experimental animal put the finishing touches to the work of Davaine and Koch on the aetiology of that disease (Foster, 1970).

Medical bacteriology and infectious pathology

Pathogenic germs, particularly bacteria, gradually began to be better known in the 1880s thanks to the work of different scientists and doctors across Europe, particularly Robert Koch and others in Germany, and Louis Pasteur and his team in France (Foster, 1970; Baldry, 1976). The microscopic observation of necroptic lesions was no longer at the forefront; what was crucial was to track the microorganisms responsible for the disease in the "live" patient. This new attitude to the disease gave rise to a new clinical practice—a different kind of interaction with the patient's body, "inhabited" by microscopic germs which circulated and grew inside it. These had to be tracked down by taking samples of blood, urine and pus, and adding them to the culture media. Close collaboration started between laboratory work and medical diagnosis.

The affirmation of the presence of pathogenic germs in the blood and their culture in an appropriate medium marked the birth of "medical bacteriology". It was important to detect microorganisms in the blood because the blood was "the main canal of the body", and could distribute germs throughout the organism far from the original focus. Besides, the idea of "cultivating" blood was closely related to the concept of autonomous, live, pathogenic germs which can therefore multiply and be cultivated. The development of the medical practice of sterile sampling evolved as much from new technical and material possibilities as from an increased knowledge of human pathology and medical bacteriology. This was the real birth of "infectious pathology". For example, the invention of an easily sterilizable syringe was a great improvement (Contrepois, 1995b). Sampling for bacterial cultures was gradually standardized in the 1890s and was a good example of the match between a theoretical "revolution" (i.e. the germ theory) and technical innovation (Löwy, 1993).

Koch introduced the technique of making thin smears of bacteria containing fluid on glass slides, fixing and staining them, and examining them under high-powered oil-immersion lenses (Brock, 1995). From 1880 onwards, new germs were sought and new diseases identified. Koch considered that to prove that a bacterium was the specific cause of a disease, it was necessary to isolate it in pure culture (in solid medium) and show that it produced the disease in question when injected in animals. In this way, animal models became crucial for the assessment of the germ as a cause of disease.

Specific and non-specific infections

The idea that the same germ did not always provoke the same disease was gradually accepted in the 1880s–1890s. Similarly, different types of bacteria could produce the same disease. This departed somewhat from the theory that a specific germ existed for each infectious disease. Unlike tuberculosis, or cholera (bacilli discovered by Koch in 1884), certain infectious diseases could be caused by a number of different bacteria. This was first recognized in the case of pneumonia, pleurisy, meningitis, angina, septicaemia and endocarditis. This fact encouraged clinicians to distinguish between specific and non-specific infectious diseases.

When Koch in 1879 became confident that "the comma bacilli are the constant companions of the choleraic process, and that they are present nowhere else", he aimed to demonstrate that the bacillus was the cause of cholera. Koch recognized the importance of transmitting the disease to experimental animals using a pure culture, as he had done with the tubercle bacillus, but no animals were naturally susceptible to cholera and he had not been able to produce the disease in experimental animals (as was also the case with the leprosy bacillus, discovered by Hansen in 1874, and the typhoid bacillus, discovered by Eberth in 1880).

The example of the experimental animal model of infective endocarditis

In 1876, Edwin Klebs in Germany developed experimental surgical techniques with the main objective of studying the circulatory consequences of valvular insufficiency (Klebs, 1876). Two years later, Ottomar Rosenbach, a student of Julius Cohnheim, examined the possibility of reproducing endocarditis in animals as well as its likely after-effects (Rosenbach, 1878). He established the first endocarditis animal model: he mechanically induced an injury in a rabbit's aortic valves by pushing a stylet into the carotid artery straight through the left ventricle. When the instrument was carefully sterilized, at autopsy the perforation was either still visible or healed but there was no trace of infective endocarditis. However, when the instrument was covered with septic matter, vegetations and fibrinous deposits formed around the wound.

Experimental endocarditis, therefore, was a result of the combination of the physiological model elaborated by Klebs, himself inspired by the Bernardian animal model of experimental physiology, and the infectious animal model of Henle, Koch and Pasteur, which was injected with a pure microorganism in order to observe how an infectious disease was transmitted. Separately the methods were ineffective but together they produced remarkable results. Indeed, the injection of pathogenic germs into the blood was not sufficient to produce the disease. The valves, for their part, had to be damaged, at least slightly, for the bacterial graft to be possible. On the other hand, it seemed that only certain bacteria could be grafted on to the valves.

Other doctors, such as Vladimir Wyssokowitsch (1886), Arnold Netter (1886), or Anton Weichselbaum (1885), considerably improved the model. In his experimental model, once the valvular lesion was produced following

Rosenbach's technique, Wyssokowitsch injected cultures of specific bacteria into the vein of a rabbit's ear (following Winge's theory). This model, standing squarely at the crossroads of physiological and bacteriological experimental approaches, was very close to the one which is still in use today.

Other non-specific experimental infections

As was the case for infective endocarditis, the aetiological agents of bacterial meningitis were isolated in the last decade of the 19th century. At that time, lumbar puncture in humans was first used both by Heinrich Quincke (Germany) and Walter Wynter (England) in 1891. This technique allowed investigators to establish the bacterial aetiology of human meningitis: pneumococci (cultivated in 1886 by Fraenkel), meningococci (by Weichselbaum in 1887) and *Haemophilus influenzae* (by Richard Pfeiffer in 1892). Pure germ cultures of various species were inoculated through different routes in animals — mostly rabbits — at the end of the 19th and at the beginning of the 20th centuries. Anton Weichselbaum in 1887 used the subdural inoculation of purulent "meningitic fluid" from deceased patients into animals to provoke experimental meningitis (Weichselbaum, 1887). Experimental meningitis was not always observed after intranasal inoculation (von Lingelsheim and Leuchs, 1906), via the intraperitoneum (Cantani, 1896; von Lingelsheim and Leuchs, 1906), subdural (Bettencourt and Franca, 1904) or intrathecal route (Councilman *et al.*, 1898; Flexner, 1906, 1907) and, in general, meningitis did not occur after injection of bacteria into the intravascular compartment (Wollstein, 1911; Austrian, 1918; Amoss and Eberson, 1919; Idzumi, 1920). Ultimately, the intracisternal route of inoculation, first utilized in 1913 (Dixon and Halliburton), appeared to be the most reliable method for producing experimental meningitis in animals.

In the same period, Haymann (1912–1913) demonstrated that the introduction of various bacteria into the middle-ear cavity of guinea-pigs resulted in middle-ear effusion. This effusion was produced experimentally by Beck (1919) in dogs by mechanically obstructing the eustachian tube. Subsequently, a number of investigators reported methods for creating mechanical tubal obstruction in dogs, cats and squirrel monkeys in order to study different aspects of the otitis inflammatory process.

In the early 20th century, experimental pneumonias were almost exclusively pneumococcal and were created using direct tracheal or bronchial instillations (Nungester and Jourdonais, 1936).

Concerning the early experimental infections of the urinary tract, a mechanical factor of obstruction was acknowledged as also playing a major role in predisposing the infected kidney to severe damage, as demonstrated by Guyon and Albarran in 1890 (Brewer, 1911), and later by Lepper (1921).

The first successful infection of rabbits with the yeast *Candida albicans* from vaginal thrush was reported by Winckel (1866). But when the fungus was injected into the blood, no vaginal lesion was observed. Colpe (1894) isolated a yeast from the vagina of a patient and directly inoculated it into the rabbit's genital tract.

Attempts have also been made to develop a reliable animal model for acute and chronic osteomyelitis. Rodet (1885) and Lexer (1894, 1896) produced experimental bone abscesses in rabbits by intravenous injection of *Staphylococcus aureus*.

Specific experimental infections

Syphilis was the leading venereal disease of the 19th century. Geoffroy Saint Hilaire, a member of the famous zoologist's family, repeatedly tried to inoculate monkeys in order to obtain an experimental disease and observe the results of various modes of treatment and prophylaxis. The many attempts at self-inoculation were provoked by the failure of inoculation in animals (Moulin, 1991). In 1875, Klebs observed a spiral-shaped microorganism in the exudate of a human chancre and for the first time transmitted syphilis to the monkey. In 1881, Haensell transferred syphilis to the rabbit; eye infection resulted in acute keratitis. In 1903, Élie Metchnikoff and Émile Roux successfully inoculated syphilis in both monkeys and apes. Finally, in 1905 Schaudinn and Hoffmann isolated and identified the causative agent, *Treponema pallidum*, from a human chancre. Rabbit infection inoculated into the testicles imperfectly mimics primary and secondary stages of human syphilis, but much of our current knowledge has emanated from this experimental model (Brown and Pearce, 1920, 1921).

The aetiological agent of leprosy was first described in 1874 by Armauer Hansen. *Mycobacterium leprae* is almost the only bacillus causing disease in humans that has not been successfully cultured. Hansen attempted — without success — to transmit leprosy to a variety of animal species (Hansen, 1880). Other investigators attempted to establish the disease in animals, but their efforts were also unsuccessful (Neisser, 1879). Only recently the tattoo inoculated into the plantar sole has become an interesting model, but the immunological peculiarities of the animal (close to immunological deficiencies) once again provided a model that was very far from the human disease.

It is well-known that Pasteur reported the successful transmission of rabies to rabbits in May 1881 and demonstrated by this transmission that the "virus" (never seen by Pasteur) existed in the spinal cord as well as the saliva without one being able to cultivate it. During the next months he enriched the knowledge of the disease with new techniques. He maintained the "virus" by serial passages in rabbits in the laboratory. He developed the technique of intracerebral inoculation which, unlike subcutaneous inoculation, invariably caused infection. He also demonstrated that human brain tissue from rabies cases also contained the "virus" and took the first step towards developing an effective vaccine (Moulin, 1991).

Conclusions

Up to this very day, all infectious diseases affecting humans are far from having appropriate animal models and, even in those cases where such infections are possible, the symptoms observed in animals and the course of the disease are often different from those encountered in humans (Löwy, 1992, 1995). The bacterial shifts required to make infectious diseases subject to laboratory investigation have reduced the messy clinical reality of an infectious disease, with its wide range of symptoms and individual manifestations, with its regional, seasonal and environmental variations, and often unpredictable outcomes, to the relatively predictable model diseases of laboratory animals infected with specified amounts of standardized and pure bacterial cultures grown on ever more precisely composed substrates under well-controlled conditions (Amsterdamska, 1998).

By manipulating bacteria in test tubes and on Petri dishes and by injecting them into animals, bacteriologists and experimental pathologists at the turn of the century were able to study various properties of pathogenic microorganisms, the host's induced immunity and some immunological reactions and pathological changes accompanying individual infection, and the therapeutic efficacy of serum therapy and, later, antimicrobial agents. In all these studies, infectious disease was an artificially transmitted individual event — the uniform result of an interaction between an increasingly well-characterized microorganism and an increasingly well-controlled animal host (Amsterdamska, 1998). The relationship between human infectious disease and its animal model and the similarities and dissimilarities in symptoms, provide an insight into the remote past of species and their divergences from each other and may give some clues to the history of infectious disease throughout evolution. The study of human immunodeficiency virus (HIV) animal models today provides us with a good example of these speculations and the difficulties involved.

References

Ackerknecht, E. H. (1967). *Medicine at the Paris Hospital 1794–1848.* John Hopkins Press, Baltimore.

Albury, W. R. (1977). Experiment and explanation in the physiology of Bichat and Magendie. *Studies Hist. Biol.*, **I**, 47–131.

Amoss H. L., Eberson, F. (1919). Experiments on the mode of infection in epidemic meningitis. *J. Exp. Med.*, **29**, 605–618.

Amsterdamska, O. (1998). Standardizing epidemics: infection, inheritance, and environment in prewar experimental epidemiology. In: *Transmission: Human Pathologies between Heredity and Infection* (eds Gaudillière, J. P., Löwy, I.) Harvard Amsterdam (in press).

Austrian, C. R. (1918). Experimental meningococcus meningitis. *Bull. Johns Hopkins Hosp.*, **29**, 183–185.

Baldry, P. (1976). *The Battle against Bacteria.* University Press, Cambridge.

Beck, K. (1919). Ueber mittelohrveränderungen bei experimenteller läsion der tube. *Z. Ohrenheilk.*, **78**, 83-108.

Bernard, C. (1865). *Introduction to the Experimental Medicine.* Translated by Copley Greene, H. New York. New York, Dover Books, 1957 (original French edition Paris, 1865).

Bettencourt, A., Franca, C. (1904). Ueber die meningitis cerebrospinalis epidemica und ihren specifischen erreger. *Z. Hyg. Infektionskr.*, **46**, 463–516.

Brewer, G. E. (1911). The present state of our knowledge of acute renal infections. *J.A.M.A.*, **LVII**, 179–187.

Brock, T. D. (1995). *Robert Koch: A Life in Medicine*, p. 94. Sc. Tech. Publishers, Madison, Wisconsin.

Brown, W. H., Pearce, L. (1920). Experimental syphilis in the rabbit. Primary infection in the testicle. *J. Exp. Med.*, **31**, 475–498.

Brown, W. H., Pearce, L. (1921). Experimental syphilis in the rabbit. VI. Affections of bone, cartilage, tendons, and synovial membranes. *J. Exp. Med.*, **33**, 495–514.

Bulloch, W. (1938). *The History of Bacteriology*, pp. 178–201. University Press, Oxford.

Bynum, W. F. (1990). "C'est un malade": animal models and concepts of human diseases. *J. Hist. Med. Allied Sci.*, **45**, 397–413.

Cantani, A. Jr. (1896). Wirkung der influenzabacillen auf das centralnerven-system. *Z. Hyg. Infektionskr.*, **23**, 265–282.

Colpe, J. (1894). Hefezellen als krankheitserreger im weiblichen genitalcanal. *Arch. Gynaekol.*, **47**, 635–645.

Contrepois, A. (1995a). Notes on the early history of infective endocarditis and the development of an experimental model. *Clin. Infect. Dis.*, **20**, 461–466.

Contrepois, A. (1995b). The birth of blood culture. *Rev. Praticien*, **45**, 942–947.

Contrepois, A. (1996). Towards a history of infective endocarditis. *Med. Hist.*, **40**, 25–54.

Councilman, W. T., Mallory, F. B., Wright, J. H. (1898). *Epidemic Cerebrospinal Meningitis and its Relation to other Forms of Meningitis. A Report of the State Board of Health of Massachusetts.* Wright and Potter, Boston.

Cunnigham, A., Williams, P. (1992). *The Laboratory Revolution in Medicine*, pp. 1–13. University Press, Cambridge.

Delaporte, F. (1986). *Disease and Civilization. The Cholera in Paris, 1832.* MIT Press, Cambridge, Massachusetts.

Dixon, W. E. and Halliburton, W. D. (1913). The cerebrospinal fluid I. Secretion of the fluid. *J. Physiol. (Lond.)*, **47**, 215–242.

Flexner, S. (1906). Experimental cerebrospinal meningitis and its serum treatment. *J.A.M.A.*, **47**, 560–566.

Flexner, S. (1907). Concerning a serum-therapy for experimental infection with *Diplococcus intracellularis*. *J. Exp. Med.*, **9**, 168–185.

Foster, W. D. (1970). *A History of Medical Bacteriology and Immunology*, pp. 5–8. Heinemann, London.

Foucault, M. (1976). *The Birth of the Clinic: An Archaeology of Medical Perception.* Translated by Sheridan, A. M. Tavistock, London (original French edition, Paris, PUF, 1963).

Fye, W. B. (1987). *The Development of American Physiology: Scientific Medicine in the 19th Century.* Baltimore.

Geison, G. L. (1987). *Physiology in the American Context 1850–1940.* Bethesda, Maryland.

Gelfand, T. (1980). *Professionalizing Modern Medicine: Paris Surgeons and Medical Science and Institutions in the 18th Century.* Greenwood Press, Westport, Connecticut.

Hansen, G. A. (1874). Undersøgelser angaende spedalskhedens arsager. *Norsk Magazin Laegevidenskaben*, **3**, 1–88.

Hansen, G. A. (1880). Studien über *Bacillus leprae*. *Virchow's Arch. (Cell Pathol.)*, **79**, 32–42.

Haymann, L. (1912–13). *Arch. Ohrenheilk*, **90**, 267. In: Friedmann, I. (1955). The comparative pathology of otitis media: experimental and human. I. Experimental otitis of the guinea pig. *J. Laryngol. Otol.*, **69**, 27–50.

Holmes, F. L. (1974). *Claude Bernard and Animal Chemistry: The Emergence of a Scientist*. Cambridge, Massachusetts.

Idzumi, G. (1920). Experimental pneumococcus meningitis in rabbits and dogs. *J. Infect. Dis.*, **26**, 373–387.

Jacyna, L. S. (1988). The laboratory and the clinic: the impact of pathology on surgical diagnosis in the Glasgow Western Infirmary, 1875–1910. *Bull. Hist. Med.*, **62**, 384–406.

Klebs, E. (1876). Ueber operative Verletzungen der Herzklappen und deren Folgen. *Prager Med. Wochenschr.*, **1**, 29–36.

Lancereaux, E. (1872). *De la maladie expérimentale comparée à la maladie spontanée*, pp. 5–6. Parent A., Paris.

Lenoir, T. (1982). *The Strategy of Life: Teleology and Mechanism in 19th Century German Biology*. Dordrecht, reprinted Chicago (1989).

Lepper, E. (1921). The production of coliform infection in the urinary tract of rabbits. *J. Pathol Bacteriol.*, **24**, 192–204.

Lesch, J. (1984). Science and medicine in France: the emergence of experimental physiology, 1790–1855. Cambridge, Massachusetts.

Lexer, E. (1894). Zur experimentellen erzeugung osteomyelitischer herde. *Arch. Klin. Chir.*, **48**, 181–200.

Lexer, E. (1896). Experimente über osteomyelitis. *Arch. Klin. Chir.*, **53**, 266–277.

Löwy, I. (1992). From guinea pigs to man: the development of Haffkine's anticholera vaccine. *J. Hist. Med. Allied Sci.*, **47**, 270–309.

Löwy, I. (1993). Medicine and change. In *Medicine and Change: Historical and Sociological Studies of Medical Innovation* (ed. Löwy, I.), pp. 1–20. INSERM, Paris.

Löwy, I. (1995). Whose body? The experimental body and 20th century medicine. Workshop on the body in twentieth century medicine, Manchester, 1995.

Maulitz, R. (1978). Rudolph Virchow, Julius Cohnheim and the program of pathology. *Bull. Hist. Med.*, **52**, 162–182.

Maulitz, R. (1987). *Morbid Appearances: the Anatomy of Pathology in the early 19th century*. University Press, Cambridge.

Moulin, A.-M. (1991). *Le dernier langage de la médecine. Histoire de l'immunologie de Pasteur au SIDA*, pp. 37–38. PUF, Paris.

Neisser, A. (1879). Zur aetiology der lepra. *Breslauer Artz. Z.*, **1**, 200–202.

Netter, A. (1886). De l'endocardite végétante d'origine pneumonique. *Arch. Physiol.*, 106–161.

Nungester, W. J., Jourdonais, L. F. (1936). Mucin as an aid in the experimental production of lobar pneumonia. *J. Infect. Dis.*, **59**, 258–265.

Pickstone, J. V. (1981). Bureaucracy, liberalism and the body in post-revolutionary France: Bichat's physiology and the Paris School of Medicine. *Hist. Sci.*, **19**, 115–142.

Rodet, A. (1885). Étude physiologico-pathologique expérimentale sur l'ostéomyélite infectieuse. *C. R. Acad. Sci.*, **99**, 569–571.

Rosenbach, O. (1878). Ueber artificielle Herzklappenfehler. *Arch. Exp. Pathol. Pharmak.*, **9**, 1–30.

Rupke, N. A. (1987). *Vivisection in Historical Perspective*. London.

Tansey, E. M. (1989). The Wellcome physiological research laboratories 1894–1904: the Home Office, pharmaceutical firms, and animal experiments. *Med. Hist.*, **33**, 1–41.

Villemin, J. A. (1866–1867). *Etude sur la tuberculose: preuves rationnelles et expérimentales de sa spécificité et de son inoculabilité*, Bulletin de l'Académie de médecine, f. 32, p. 152, Paris.

von Lingelsheim, J. M. and Leuchs, S. A. (1906). Tierversuche mit dem *Diplococcus intracellularis* (meningococcus). *Klin. Jahrb.*, **15**, 489–506.

Weichselbaum, A. (1885). Zur Aetiologie der akuten Endokarditis. *Wien. Med. Wochenschr.*, **35**, 1241–1246.

Weichselbaum, A. (1887). Ueber die aetiologie der akuten meningitis cerebrospinalis. *Fortschr. Med.*, **5**, 573–583.

Winckel, F. (1866). Ueber die bedeutung pflanzlicher parasiten der scheide bei schwangeren. *Berl. Klin. Wochenschr.*, **3**, 237–239.

Wollstein, M. (1911). Influenzal meningitis and its experimental production. *Am. J. Dis. Child.*, **1**, 42–58.

Wunderlich, K. (1845). Das Verhältniss der physiologischen Medicin zur ärtzlichen Praxis. *Arch. Physiol. Heilkunde*, **4**, 1–13.

Wyssokowitsch, V. (1886). Beiträge zur Lehre von der Endocarditis. *Arch. Pathol. Anat. Phys.*, **103**, 301–332.

Chapter 2

General Methodologies for Animal Models

M. S. Rouse and W. R. Wilson

The ideas in this chapter are provided as a framework upon which to build a detailed protocol to be used to conduct investigations in infectious diseases using animal models. The resources we have listed were chosen because they provide basic information that we believe is a good foundation for experimental design and yet encourage adaptation to different investigators' interests or needs.

When designing an experiment using an animal model, one may focus on the similarities of the animal model to the human disease; however, one must also be aware of the inherent differences between animal models and making observations in humans. We want to design experiments that capitalize on these similarities and avoid unknowingly having differences influence the experimental data or its interpretation.

One must be aware of the many limitations of using laboratory animals to model infectious processes. Laboratory animals often have a limited genetic diversity; one must recognize that observations made in a group of nearly isogenic hosts with a single or few strains of a specific pathogen may not represent the actual disease process seen in genetically diverse hosts (patients) with a genetically diverse group of pathogens (naturally acquired). This may be further complicated by clinical designations of groups of closely related bacterial species such as viridans streptococci or coagulase-negative staphylococci. Some laboratory animals do not tolerate antimicrobial therapy well and the host response to a particular antimicrobial or carrier agent may influence or obscure the results of the experiment. The reader is referred to Chapter 9 for a more detailed discussion of this topic. Many animals must be surgically or medically altered to reproduce a disease state—this is both good and bad. We can closely control the alteration so that all laboratory animals are altered equally, but one must acknowledge and understand how any such alterations may influence the disease process being studied.

Using animals as research subjects also offers several advantages over similar studies in humans. Small mammals have accelerated life cycles. This can be used to reduce the time and cost of compiling experiments. We can observe large numbers of identical subjects easily; this allows us to generate adequate data to perform meaningful statistical analysis. We can closely control many important variables in each experiment, such as the timing of onset of infection and treatment; this allows us to eliminate confounding influences that are part of making experimental observations in humans. The pathophysiology of many infectious processes is very similar to the human disease; we can observe infectious processes associated with significant morbidity and mortality in animals that would be unacceptably dangerous to human subjects. Animal models may be the only effective method of study for conditions which are uncommon in human beings.

Models of infection in animals are often used to compare the efficacy of novel antimicrobial compounds to that of currently available antibiotics, to study pathogens with novel patterns of antimicrobial resistance, or to study the pathophysiology or prevention of specific infection. These data may then be used to design further studies, often as a prelude to human studies. Animal models of infection are important tools used to study different aspects of infectious diseases because they provide all the factors involved in the complex host–pathogen interactions. The host immune response, the pharmacokinetics of antimicrobial exposure and pharmacodynamic response, and the influences on the host of any bacterial toxins are important aspects of infections that are not currently part of any *in vitro* testing systems.

Key to the success of any research endeavor is organization. The scientific method requires that prior to an experiment, a hypothesis is clearly articulated (with the hypothesis in the form of the null hypothesis), an experimental test of the hypothesis is designed, the experiment is performed (with accurate record-keeping), and then the experimental results are compared with the hypothesis. This basic idea is worth repeating here because it has been our observation that many challenges experienced doing an animal experiment are related to departing from this basic concept.

Animal experiments start out as a question. Information to consider when developing a hypothesis includes the current literature on the topic, and *in vitro* activity, and *in vivo* pharmacokinetics (if known) of antimicrobials studied. Comprehensive literature searches can be performed in little time and are a prerequisite to designing an animal model of infection experiment. Published information on a specific animal model is an efficient way of becoming aware of specific problems, issues, and concerns other investigators have experienced doing similar experiments.

Handbook of Animal Models of Infection
ISBN 0-12-775390-7

The electronic environment available in most research institutions provides ready access to current literature on the topic being investigated. Electronic mail is an efficient low-cost mode of communication among the global community of scientists and is a valuable resource for almost any research laboratory. The internet may also be a source of specific information. Bear in mind that information published on web sites is often not subject to peer review, but none the less these are sources of information that may be helpful.

Knowledge of the *in vitro* activity of antimicrobials being studied, the mechanism of antimicrobial activity, and any mechanisms of resistance among the bacteria being studied is needed to design a hypothesis-driven protocol. If the study is of community-acquired pathogens, I prefer to perform susceptibility testing in our laboratory with a collection of clinical isolates from our institution. This is useful if one is studying an organism with a specific antimicrobial susceptibility to determine the frequency of the specific antibiogram. If the focus of the experiment is a particular species, knowing the *in vitro* activity of the antibiotics to be studied will allow one to pick an organism with a susceptibility pattern representative of the collection for study *in vivo*.

Issues that need to be considered when designing an experiment to test the hypothesis should include the ethical considerations regarding the use of laboratory animals, proper care and use of animal subjects, an understanding of the pathophysiology of the infection in humans, pathophysiology of the specific organism in the specific animal host used, pharmacology of any medications used in human and animal subjects, technical expertise in any medical or surgical procedures on the animal subjects, biosafety responsibilities of working with infectious agents, availability of any support services that may be needed, and costs of materials and services needed.

The lay press regularly debates ethical considerations regarding the use of animals for medical research. Primary ethical considerations regarding the use of laboratory animals for medical research are a sense of stewardship for the animals being used and respecting institutional and legal regulation of the care and use of laboratory animals. Is each experiment designed to test the hypothesis? Is the statistical analysis strong enough to justify the use of animal subjects? Are the physical needs of the animals included in the protocol? Virtually every modern country in the world has legal regulation of the care and use of animals used for medical research. It is important that all investigators using laboratory animal subjects work within the guidelines set forth by our governments and scientific community. These guidelines not only contribute to the scientific merit of the research, but also promote good will and trust among fellow researchers.

One needs to understand the clinical features of an infection in order to model it. Few animal models are exact duplicates of human infection. Investigators need to know what features of the human infection are or are not present in the model used and how this affects the interpretation of the results. Studies to document similarities and differences between the pathophysiology of an infection in humans and in the model being used should be part of any animal model of infection protocol. This may be histological studies to compare features of the modeled infection with the human infection, blood chemistry studies, radiological studies, or microbiological studies.

Knowledge of the pathophysiology of the specific organism in the specific animal host will help determine which animal species to use in the infection model. In general, we prefer to use the smallest practical species that models the important features of the infection being studied. The smallest species available is however not always the most appropriate. We previously used mice to model group A streptococcal necrotizing fasciitis; this was attractive for many reasons, one of which was the availability of effective isolation boxes for housing the mice. There is now evidence suggesting that mice are less susceptible to the pathology associated with group A streptococcal pyrogenic exotoxin than rabbits (Dinges and Schlivert, 1997). Rabbits may be more appropriate to model this infection. Conversely, we have performed clinical trials with a mouse model of *Haemophilus influenzae* type b pneumonia, comparing the efficacy of clarithromycin with other antimicrobials. One of the clarithromycin metabolites produced in humans acts synergistically with clarithromycin against *H. influenzae in vitro*. This metabolite is not produced in rodents or lagomorphs. We chose to co-administer the metabolite in our trials in mice, resulting in ratios of the antibiotic and the metabolite in mice serum similar to that in humans. The decision of which species to use depends on how well the selected species used models the pathophysiology or the infection being modeled.

The pharmacology in human and animal subjects of any medications used in the study needs serious consideration when performing clinical trials of antimicrobials in animals. Small animals metabolize most pharmaceuticals more rapidly than humans. This may influence the outcome of treatment that depends on the concentration of antibiotics in serum. If the aim of the study is the influence of pharmacokinetics on outcome of an infection, the pharmacokinetics can be altered, for example using computer-controlled infusion pumps. If the aim of the study is to compare the antimicrobial efficacy of two or more antimicrobials in an animal with an accelerated metabolic rate, a more frequent dosing schedule may better imitate treatment in humans and may be more appropriate. Pharmacokinetic differences may be further complicated by the use of antimicrobial formulations of two compounds where optimal antimicrobial activity depends on the ratio of the two compounds *in situ*. Another difficulty can be choosing a route of administration of medication. Repeated intravenous access can be a challenge, as can be oral administration with some animal species. We approach this dilemma by using an alternative route of administration that produces similar pharmacokinetics. Serum concentration of the compounds is often the priority in most models and needs to be monitored. Pharmacokinetic differences between human and animal models may be the most perplexing aspect of a study design and data interpretation. Chapters 10 and 11 provide a detailed discussion of pharmacokinetics in animal models of infection.

Expertise at working with animals (handling and surgical manipulations) can be gained by experience, most easily from others who have that experience. There are many investigators actively using animal models of infectious diseases who are willing to share their experience with other scientists. Many of these people have kindly contributed to this text. The European Network for the Study of Experimental Infections and the European Society for Clinical Microbiology and Infectious Diseases hosts the International Symposium on Infection Models in Antimicrobial Chemotherapy. This organization is a rich source of information for anyone designing, performing or interpreting the results of experiments using animal models of infection. Read their publications, participate in scientific conferences where they are presenting, ask them questions.

Scientists using animal models of infections need to understand the basic care of any laboratory animal being utilized. Laboratory animals may become ill from a variety of veterinary maladies. Preventing and recognizing illnesses if they arise is important because they may influence the results of an experiment or they may spread throughout the animal housing facility, possibly wasting valuable laboratory resources. Information on the care and use of laboratory animals can be found in this handbook.

Each investigator is responsible for the biological containment of pathogenic bacteria and of infected animals. These responsibilities need to be addressed specifically in each protocol. Biological containment can be a challenge when infected animals are housed in the same room with animals from several different laboratories.

The availability of any support services utilized in an experiment needs to be considered in the design of the experiment. Rarely are patient care facilities available for animal studies, especially infected animal studies. Expertise in several areas besides infectious diseases is required to complete an experiment modeling an infectious disease. Colleagues in other medical and surgical specialties can also be a valuable source of information on the design and interpretation of animal models of infection. I urge you to consult with them if their expertise may be helpful to you. Communication skills to convey ideas to funding sources, the scientific community, and regulatory agencies are needed. Technical expertise in handling animals, performing laboratory analysis, and performing statistical analysis are important. A team approach is the obvious answer, but it requires management skills to coordinate the efforts of the research team.

Management of costs of materials and services used for each experiment needs to be addressed in the protocol. Two aspects of laboratory finances that continue to astound are the high cost of biomedical supplies and the variability of the cost of some research supplies. Imagination and regular assessment of how laboratory resources are utilized are the only recommendations I can offer to deal with this issue. Ideas such as autoclaving non-sterile pipette tips for less than the cost of sterile pipette tips, or reducing polymerase chain reaction (PCR) mixture volumes to save on the cost of expensive enzymes need to be solicited from the lab staff. The variability of the cost of graphics presentations is often greater than 10-fold between different vendors. We value the work of graphics professionals depicting our scientific data in an attractive manner, but excellent presentations can also be prepared with a desktop computer.

Statistical analysis must be designed prior to collection of data to avoid bias in the decision of which statistical methods to employ. The experimental protocol needs to specify what type of descriptive statistics will be used and what statistical tests of hypothesis will be used to allow the investigator to choose statistical methods based on statistical principles without knowledge of any experimental results.

How many animals do I need to use in the study? What endpoint (mortality, infection site sterility, or actual quantitation of bacteria at the infection site) do I study? The answers are a balance of what type of data best addresses the hypothesis, what differences (if they exist) are clinically significant enough to detect, and what resources one intends to dedicate to addressing the hypothesis. One obvious pitfall to avoid is choosing to measure continuous data when nominal data best address the hypothesis in order to conserve resources. The investigator must also decide on what differences are clinically significant. Some issues I consider when deciding what is clinically significant are the extent to which resolving the infection is influenced by factors other than the independent variable and resources available. When several factors contribute to the resolution of an infection, small differences may not be significant enough to warrant use of lab resources to detect. As examples, an immune response significantly augments antimicrobial activity *in situ* in pneumococcal pneumonia, or some bone infections are treated with aggressive surgical intervention along with appropriate antimicrobials. Conversely, if resolution of the infection is primarily a function of antimicrobial activity, small differences may have more clinical significance, i.e., as in experimental endocarditis, where an immune response may be less important than bactericidal antimicrobial therapy in resolving the infection. It is also important to discuss what the investigators feel is clinically significant and why, in any reports or manuscripts.

Determining the actual number of subjects needed is not complicated after one has resolved the aforementioned concerns. The reference I use most often is the CRC handbook of tables for probability and statistics (Cochrane and Cox, 1957). Tables of sample sizes needed to provide specific power to reject the null hypothesis given the variability of the data of different types of data are listed. But how can we design the statistical analysis when we need to know the variability of the data? You must estimate the variability. Review prior experience—yours and others—and estimate the variability. There is a certain joy in being asked by a statistician for your best guess of variability upon which statistical analysis will be based.

When the data are collected, tests of hypothesis can be done. One method of comparing our experimental observations with our hypothesis is by statistical analysis. We regularly rely on current literature and the services of a professional statistician for information on statistical analysis.

Prospective clinical trials in animals are designed to address a specific hypothesis with the lowest number of specimens needed to test the hypothesis adequately: the sample size is limited. Continuous data is the type of data often generated in animal models of infections for chemotherapy and is compared by hypothesis testing. I will offer some considerations on the analysis of this type of data that are generalizations intended to be a starting point for designing statistical analysis. Detailed analysis needs to be individualized for each protocol.

Descriptive statistics for relatively small amounts of data are some measure of central tendency and variability. The mean or median is often used to describe data. I prefer the median as it is less influenced by data extremes (outliers) and biological data is frequently non-gaussian. There are several measures of variability commonly used — range, standard deviation, or estimate of percentiles. The range gives no information on the distribution of data in the sample; the standard deviation is designed to give a measure of distribution, but only if the data is symmetrically distributed within the range and that is rare in biological systems. I prefer the estimate of percentiles as a measure of variability of continuous non-gaussian data as it provides a measure of variability that reflects distribution, whether gaussian or non-gaussian. Another valuable descriptive statistic is the "eyeball statistic," a scattergram of the raw data. Simply looking at a graphical depiction of the data to give a different perspective than just knowing the numbers can provide valuable insight. We often use scattergrams along with tables in presentations to display the data.

Tests of the hypothesis for continuous, non-gaussian data among two samples with a single independent variable are limited. Wilcoxon's signed-rank test is appropriate. This non-parametric analysis offers the advantage of accommodating continuous measurements from sterile cultures to death from infection, with little power of detection lost if the data are evenly distributed. If bacterial quantification is an endpoint, sterile cultures can be accommodated in this test by assigning them all an equal value less than the lowest positive culture result.

Subject mortality is not uncommon when modeling infectious diseases: if mortality is not associated with an independent variable, it is probably not statistically significant. If mortality is associated with an independent variable, it may be due to lack of clinical efficacy or secondary to toxicity (physiological or an alteration of normal flora in the host) or of the compound or its solvent. These observations need to be addressed when they occur. Another dilemma often encountered is extreme variability of a few data points (outliers). This is understandable when a test system involves one biological system living within another biological system, but data cannot be excluded from analysis except for sound scientific reasons. Designing monitors into a protocol such as surveillance cultures, therapeutic monitoring of antimicrobial levels, radiological exams, or other appropriate test results can be very helpful when interpreting data.

Another test of hypothesis stressed by professional statisticians is worth mentioning: consider the data and conclusions and decide "does this make sense?" If your conclusions do not pass the "non-sense test", an inappropriate test (laboratory or statistical) may have been used or you may have a very novel observation.

There is no right design of any animal model of infection experiment. All approaches have strengths and limitations that we need to understand to insure that our data and data analysis specifically address the query in an unbiased manner.

References

Cochran, W. G., Cox, G. M. (1957). *Experimental Designs*, 2nd edn, pp. 24–25. John Wiley, New York, NY.

Dinges, M. M., Schlivert, P. M. (1997). Comparison of mouse and rabbit models of human toxic shock syndrome. In *Abstracts of the 97th Annual Meeting of the American Society for Microbiology*, abstract B-406. American Society for Microbiology.

Chapter 3

Ethics Committees in Europe — An Overview

P. de Greeve and W. de Leeuw

Introduction

In Europe, over the last few decades, the public attitude towards the use of animals in procedures has evolved. As a result, legislation aimed to protect laboratory animals has been issued on a national as well as on an international level.

The first European legislative document aimed to protect laboratory animals, the Convention for the Protection of Vertebrate Animals used for Experimental or other Scientific Purposes (ETS 123), was issued in 1986 by the Council of Europe.

The Council of Europe was set up in 1949 to enable the governments of European states to co-operate: to achieve a greater unity between its members for the purpose of safeguarding and realizing the ideals and principles which are their common heritage and facilitating their economic and social progress.

Amongst these *ideals and principles,* respect for animals is undoubtedly one of the obligations on which the dignity of the European citizen is based. Even in the 1960s the Committee of Ministers realized that the promotion of human rights and fundamental freedom advocated in the organization's statute was inextricably linked to recognition of the intrinsic value of animals. At present, in most member states public concern for animal welfare is so strong that governments are under pressure to take legislative action.

In 1986, after several years of consultation, the then 26 member states of the Council of Europe agreed on the Convention ETS 123.

The fundamental principle of this Convention reads as follows:

> *while accepting the need to use animals for scientific and other purposes, everything possible should be done to limit the use with the ultimate aim of replacing experiments, in particular by alternative methods.*

The Convention applies to any animal intended for use in any experimental or scientific procedure where that procedure may cause pain, suffering, distress or lasting harm. It does not apply to any non-experimental agricultural or clinical veterinary practice.

The Convention contains several provisions aimed at replacing, reducing and refining animal use.

The Council of Europe cannot impose rules on its member states, but it can achieve its objectives backed by the following legal instruments: *recommendations* — often referred to as soft law — and *conventions* — treaties concluded between states. The member states are not legally obliged to sign a convention. However, once a state has signed and ratified — its parliament has approved the instrument — and the convention has become effective, the State will be legally bound under international law to implement that convention.

Although the Convention has not added much to the national legislation that already existed in some member states in 1986, adoption of the Convention was a great step forward. Indeed, this Convention represents a series of principles and regulations, which after eight years of research and negotiations were finally found acceptable by governmental experts, delegates from animal welfare organizations, scientific researchers and representatives of directly concerned industries. These principles and regulations constitute a European-wide accepted basis for improvements.

The second European document is the *Directive for the Protection of Vertebrate Animals used for Experimental and other Scientific Purposes* (86/609/EEC), which was inspired by the Convention. However, the spirit of the two texts appears to be different in that the Convention gives priority to important principles such as animal rights and human needs whereas the main concern of the Directive is to harmonize national laws in order to avoid any distortion in the internal market.

According to the Convention and the Directive, member states have established procedures whereby experiments themselves or the details of persons conducting them are notified in advance to the authority. In addition, each animal experiment must be especially declared and justified to, or specifically authorised by, the responsible authority, if the animal will or may experience severe pain which is likely to endure.

Member states have implemented these provisions in various ways. In many institutions an animal care and use committee, an animal care committee or an animal procedure committee has to advise the authority on the admissibility of the animal experiment. Such a committee primarily considers the design of animal experiments themselves and their

Handbook of Animal Models of Infection
ISBN 0-12-775390-7

execution, including the care and accommodation of animals before, during and immediately after the experiment. In addition, some committees will also provide advice on ethical aspects of the use of laboratory animals.

Current situation

With respect to this overview, an enquiry has been made into the current situation in the member states of the Council of Europe by posing the following questions:

- Is animal experimentation to be justified by a (scientific) committee?
- Is animal experimentation to be justified by an ethical review committee?
- If so, is there a legal basis for such a committee and is any detailed information on the organization of such a committee available?

Belgium

In Belgium (Ödberg, personal communication) in 1994, the Deontological Committee was created. This committee comprises 19 people, appointed on the basis of their expertise in the field of animal experiments, alternative methods, animal welfare and protection or ethical assessment. On request or on its own initiative, this national committee could provide the Minister with recommendations on animal experiments or the use of alternatives. Currently, a provision is in preparation providing local ethical review committees with a legal base. Animal experiments must be approved by these committees before experiments can be undertaken.

Czech Republic

In the Czech Republic (Sovjak, personal communication), experiments on live animals have to be approved by a professional (scientific) committee. A legal basis for approval is provided by the Animal Protection Act 246/1992 and it is further elaborated in the proposed regulations concerning protection of vertebrate animals in experiments.

The provisions of the European Convention ETS 123 as well as of the European Directive 86/609/EEC will be implemented. It is expected that these regulations will enter into force on 5 February 1997. From that time, animal experimentation must be justified by an ethical review committee.

Denmark

In Denmark (Kjaersgard, personal communication), a licence to perform experiments on animals is given by the Animal Experiments Inspectorate to specific types of experiments and to specific species and numbers of animals. This authority, set up according to the Animal Experiments Act, is headed by a board of 11 members. This board consists of representatives of research organizations, the government, the pharmaceutical industry, animal protection organizations and representatives of organizations combating diseases. A member of the Animal Ethics Committee set up by the Animal Welfare Act is also a member of the Board of the Animal Experiments Inspectorate. This Animal Ethics Committee has stated that in principle basic research must be of real or essential benefit.

The Board can refuse to give a licence to perform experiments that are not—in the wording of the law—considered to be of real or essential benefit. This, obviously, is a rather vague statement, but it is a legal basis for taking into account ethical considerations when deciding on an application for a licence to perform animal experiments.

Finland

In Finland (Rantamäki, personal communication), according to the Statute on Animal Experimentation, establishments performing animal experiments must form a committee of animal experimentation. A prime function of the Committee is to handle all experimental plans and to decide unanimously on the classification of the animal experiment. There are two classes of animal experiments depending on the degree of suffering, pain or distress which might be imposed on the animal. The local committee has the right to give the experimenter the licence to perform experiments that cause minor suffering, distress or pain. If an experiment might cause severe distress, suffering or pain, then the committee sends the application to the Provincial State Office, which has the authority to grant licences for these types of experiments.

There is also special legislation regulating experiments with transgenic animals: the Gene Technology Act and, in particular, the Gene Technology Decree. According to this legislation, the Board for Gene Technology and the Advisory Board for Biotechnology were set up, both focusing particularly on the ethical considerations of biotechnology.

France

In France (Mahouy, personal communication) authorization is given to those responsible for experiments and who perform experiments on animals. There is no legal obligation to submit research projects and protocols for experiments on animals to a scientific or ethical committee. However, in practice, most experiments must be assessed and endorsed by a scientific committee, since most are financed by ministries or major research bodies. Most experiments performed in the industry must either fulfil the regulatory requirements or be carried out in accordance with good laboratory practice. Even though they are not legal instruments, ethical committees are increasingly being set up on the initiative of scientists. In general, they include one or two persons who are not in the employ of the establishment and who are not scientists (lay persons).

Germany

In Germany (Wille, personal communication), the authorities responsible under the regulations are responsible for implementing the Animal Protection Act and any regulations adopted on the basis thereof. This Animal Protection Act demands ethical justification of proposed experiments. Accordingly, each authority has appointed one or more committees to assist the authority responsible in deciding whether to authorize experiments on animals. These committees include participants with a knowledge of veterinary medicine, medicine or any other natural science. They also include members proposed by animal protection organizations. At committee meetings, the scientific and ethical aspects of animal experiments are discussed.

The results of a questionnaire which was completed in 1994 showed that most committees work well together with little resentment between scientists and animal welfare representatives. Ethical aspects have increased in the judgement of applications in recent years; 0–5% of applications are rejected; 2–68% were returned to the scientists for improvement; 2–38% were recommended to be allowed with restrictions, and 18–96% are recommended for approval without any restrictions (Gruber and Kolar, 1997).

Greece

In Greece (Kostomitsopoulos, personal communication), the national legislation aiming to protect laboratory animals is in accordance with the Convention ETS 123 and the European Directive 86/609/EEC. Animal experiment committees are not required in Greek law. In each prefecture, the Veterinary Directorate is responsible for the scientific and mainly the ethical consideration of experimental procedures. If necessary, the Veterinary Directorate can ask for the opinion of other scientists who are experts in the field concerned.

In some cases (for example, in hospitals or institutions), where scientific committees exist because of intramural regulations, these are not specifically for animal experiments.

Italy

In Italy (Guaitani, personal communication), all projects involving experiments on live vertebrate animals must be punctually justified by the scientists responsible for submitting the proposals. Detailed information on the scientific background of each project, its objectives, treatment and procedures, animal species, number of staff involved and their experience and education and veterinarian in charge have to be supplied to the Veterinary Department of the Italian Ministry of Health. It must also be documented that valid alternatives to the use of living animals are not available. A special Commission of the Ministry of Health revises all experimental projects.

The approval is in force for 3 years. One year before the end, the scientist responsible for the project has to submit a request for renewal.

Research projects involving cats, dogs or primates require special authorization. All projects involving painful procedures or stress to animals (e.g. experiments without anaesthesia) need a special permit. A Special Commission at the National Institute of Health (Istituto Superiore di Santa, Roma) is in charge of the critical revision of such projects, their approval or rejection. Following this approval the Italian Ministry of Health authorizes the exemption (*in deroga*) of such projects from the respective Articles of the Law (*Decrato Legge* 116).

At present, Institutional Ethical Committees are neither defined nor requested by law. However, many statements in law suggest the need for ethical consideration in study planning and in issuing protocols. In many institutes, on a voluntary basis, Institutional Animal Care and Use Committees and Ethical Committees provide consultation and advice on all ethical aspects of proposed projects and procedures, including suggestions for revision, before submitting to the Ministry of Health.

The above-mentioned Special Commission may be considered to be the official National Ethical Committee. It was constituted by special decree of the Ministry. Representatives of animal protection leagues also participate in this committee. This commission is responsible for review and official approval of these special projects and for their acceptance from an ethical point of view.

The Netherlands

In The Netherlands, research plans must be recommended by an ethical review committee, which has to consider the potential benefit from experiments and whether this justifies the distress caused to the animals which are to be used. If an ethical review committee has given a negative recommendation to the licensee, that animal experiment may not be conducted unless the central animal experiments committee has reversed a previous negative recommendation.

The structure of an ethical review committee is defined in the Experiments on Animals Act 1996, which requires there to be at least seven members, including equal numbers of experts in animal experiments, alternative methods, animal welfare and ethical assessment. Three participants, including the chairperson, should not be employed by any licence-holder applying to the committee. Ethical review committees must be recognized by the Minister, who will be advised by the Central Committee on Animal Experimentation. All committees are required to submit an annual report to the Minister. For several years now, training courses have been organized for members of ethical review committees. In 1996 an association of ethical review committees was founded. In 2000, the functioning of the ethical review committees will be evaluated by the Inspectorate.

If any biotechnological techniques are to be used, a second piece of legislation applies. Based on the assumption that biotechnology applications in animals yield problems, irrespective of how inconvenient their purpose may be, the Animal Health and Welfare Act prescribes an ethical review to assess the acceptability to society of the application. No animal experiment involving biotechnology may be conducted without a licence issued by the Minister of Agriculture, Nature Management and Fisheries. Before the Minister decides, he or she will be advised on each licence application by the national Committee on Animal Biotechnology. A licence will only be issued if the Minister is of the opinion that the biotechnological techniques concerned do not have unacceptable consequences for the health or welfare of animals and there are no ethical objections to the actions. The licensing procedure is public. The committee's recommendations contribute to public opinion and give greater structure to the public debate that may take place in the context of the licensing.

Norway

In Norway (Tore Wie, personal communication) all animal experiments are approved by the responsible local laboratory animal science specialist under the surveillance of the Norwegian Animal Research Authority (NARA). NARA consists of seven members, representing the scientific community and the Norwegian animal welfare organization as well as the judiciary. NARA is more or less in itself an ethical review committee.

Poland

In Poland (Pisula, personal communication) experiments to be carried out on animals are discussed by local scientific committees. In some scientific institutions, ethical committees were appointed in order to assess experiments carried out on animals. At present there is no legal basis for ethical committees, but a proposal for such legislation has been already submitted to the Polish parliament.

Portugal

In Portugal (Gouveia da Veiga, personal communication), in accordance with the Act for the protection of animals used for experimental or other scientific purposes, each project which includes animal experimentation must be justified by the national Advisory Committee to the General Director of Veterinary Services. There is not an ethical review committee on a national level.

Slovenia

In Slovenia (Pogacnik, personal communication), the Law on Animal Health contains a chapter dealing with animal experiments. The provisions are elaborated in detail in the Directive on the Conditions for Issuing Licences for Experiments on Animals for Scientific and Research Purposes. In one of the institutions which is most involved in animal experiments, an Animal Experiments Board operates. This board includes an ethics committee whose members are recognized domestic and foreign experts. The Committee considers every request of the Board for research work and later controls all research papers and postgraduate theses which use animals or parts of animals in experiments. A research project can only be approved with the consent of the ethics committee.

New legislation is in preparation which will contain the provisions of the European Convention ETS 123. This legislation will include the founding of a National Ethics Committee to deal with all projects which include the use of animals in experiments. Research projects will then only be approved with the consent of this committee.

Sweden

In Sweden (Stagh, personal communication), according to the Animal Protection Act (1988), all experiments using animals are subject to ethical review. The care and accommodation before, during and immediately after experiments is to be considered. This review is made by local committees. These include laypeople, research workers and representatives of the personnel who handle laboratory animals. Decisions made by a local ethical committee approving or recommending the rejection of animal experiments may not be appealed against. Decisions to approve experiments are valid for not more than 3 years.

Local ethical committees report to the National Board for Laboratory Animals decisions relating to animal experiments in which animals may suffer severe pain which is likely to endure, or to animal experiments which are of fundamental importance from an ethical point of view.

Switzerland

In Switzerland (Bernard-Summermatter, personal communication) the Swiss Academy of Medical Sciences and the Swiss Academy of Sciences formulated *Ethical Principles and Guidelines for Scientific Experiments on Animals* which serves as a code of conduct for all scientists and members of allied professions practising in

Switzerland. In addition, there are academic bodies at the different universities. They have to approve experiments on animals at their institutes before a cantonal authorization can be requested. The academic body will not support experiments on animals which violate the ethical principles and guidelines of the Swiss Academy of Science.

The Swiss parliament has also decided to create in the near future a new ethical committee to evaluate and to inform the Federal Council and the general public about the developments in gene technology, especially about genetically modified animals and plants, biodiversity and sustainable development. This committee will adopt a position on special projects related to gene technology.

United Kingdom

In the United Kingdom (Richmond, personal communication), the Animals for Scientific Procedures Act 1986 does not formally require that research institutions establish Ethical Review Committees.

The Animals of Scientific Procedures Inspectorate reviews project licences applications, and performs a cost–benefit analysis before advising the Secretary of State whether and under what conditions programmes of work should be licensed.

In April 1995, designated establishments were invited to consider the benefit of one of the local review processes outlined by the department. Five methods were identified:

1. ethics committees and wider consultation.
2. animal care and use committees.
3. named veterinary surgeon providing an ethical perspective.
4. project refinement reviews.
5. awareness-raising activities to encourage researchers to consider more thoroughly the 3Rs and other welfare and ethical aspects.

As mentioned above, an ethical review process is nót mandatory but correct management structures and communications are required to ensure the welfare of laboratory animals, and licence applicants must consider the justification for using animals in their projects.

The Animals Procedures Committee (a national advisory body) has been reviewing the working of the domestic legislation in the 10th year of its operation. Any improvements in practice are likely to be achieved without new legislation being drafted.

Concluding remarks

Based on the replies, it can be concluded that in each of the countries mentioned above, animal experimentation needs to be justified by a scientific committee.

Furthermore, in several countries, the debate on the *ethical* justification of animal experimentation has begun. In about half of these countries there is a legal basis for some form of ethical review. In some others legislation is in preparation.

Ethical review committees have to decide whether an experiment using animals may be performed only after a decision has been made that this experiment is justified, weighing the social, scientific or educational value against the potential effect on the welfare of the animal. These committees should not be made up of scientists alone, as their judgement on the ethical justification might be biased.

It is important that in the near future much effort is put into training members of ethical committees. In addition, more effort must be invested in the development of practical decision models for the ethical evaluation of animal experiments by ethical review committees. This will improve the quality and transparency of the ethical review procedures.

In Europe, the ethical review of animal experimentation has become an essential part of education and training courses for future scientists.

Transgenic animals

In the context of this textbook, the development of genetically modified animals needs to be addressed. Transgenic technology has evolved rapidly and has given rise to public concerns about the ethics of transgenic animal development.

Scientists stress the advantages of biotechnology techniques — the creation of better animal models which eventually may lead to a decrease in the number of animals used. However, the general public is very concerned about these techniques, because of their potential to alter nature radically.

Due to the ethical, economical and legal aspects involved and to the heightened public concern, the question of whether genetic modification in general requires specific regulation has been given great consideration by the Council of Europe and the European Union.

At the Council of Europe, the Parliamentary Assembly in its Recommendation 1213 on developments in biotechnology and the consequences for agriculture called for the elaboration of a new legal instrument covering the field of biotechnology, including transgenic animals. Replying to this recommendation and in the light of the opinions communicated by two committees competent respectively in the field of animal protection and bioethics, the Committee of Ministers on 7–11 September 1995 agreed to envisage holding a conference with representatives of all relevant professions and interest groups, in particular representatives of the world of science and ecology, law, industry, consumers, the health field and protection of animals. The committee also wishes to associate the European Union and the European Patent Office with such a conference. The conclusion of this conference, which would be preceded by studies involving various aspects, could prepare the way for determining the need to draw up a new Convention covering bioethical aspects of biotechnology.

At the moment, *INTERNATIONAL CONFERENCE ON ETHICAL ISSUES ARISING FROM THE APPLICATION OF BIOTECHNOLOGY*, is scheduled to take place in Oviedo (Spain) on 16–19 May 1999.

The European Union has issued a number of regulations and directives that are of importance for the creation or use of genetically modified animals. The main goal for the directives is to harmonize national legislations; this is of great importance for the realization of a true internal market.

On 21 November 1991, the European Commission set up a Group of Advisers on the Ethical Implications of Biotechnology (GAEIB).

The GAEIB's terms of reference are:

- to identify and define the ethical issues raised by biotechnology
- to assess, from an ethical point of view, the impact of the Community's activities in the field of biotechnology
- to advise the Commission, in the exercise of its power, on the ethical aspects of biotechnology and to ensure that the general public is kept properly informed

The GAEIB is strictly speaking an advisory body, but can also decide to investigate an issue on its own initiative.

On 21 June 1995, the GAEIB met in Brussels. According to the minutes of this meeting, some of the participants considered that it would be sufficient to adapt existing directives to cover transgenics. Others felt that the significant changes brought about by the genetic modification of animals generated a need for special legislation.

In May 1996, on a request from the European Commission, the GAEIB adopted an opinion concerning the ethical implications of the genetic modification of animals.

Extract:
Genetic modification of animals may contribute to human well-being and welfare, but is acceptable only when the aims are ethically justified and when it is carried out under ethical conditions (. .). In view of the consequences this technology may have for the health of humans and animals, for the environment and society, a policy of great prudence is required (. .). The scope of this policy should apply to:

- the making of genetically modified animals
- the use and care of these animals
- the release of these animals
- putting genetically modified animals and their products onto the market (including import/export) (. .).

Licensing bodies in all member states should have the task of assessing research projects and applications in the light of the ethical principle of prudence. There should be appropriate and understandable information for the public about genetic modification of animals and their products.

References

Ausems, E. J. (1994). Development and application of European legislation: the European Convention for the protection of vertebrate animals used for experimental and other scientific purposes. In *Welfare and Science, Proceedings of the Fifth FELASA Symposium 8–11 June 1993, Brighton UK 1994* (ed. Bunyan, J.), pp. 234–242. Royal Society of Medicine Press, London.

Council of Europe. (1986). *European Convention for the Protection of Vertebrate Animals Used for Experimental and other Scientific Purposes* (ETS 123) Strasbourg.

European Commission. (1986). *Directive for the Protection of Vertebrate Animals used for Experimental and other Scientific Purposes* (86/609/EEC). Off J C Eur Comm 1986; L 358: 1–29.

European Commission. (1991). Group of Advisers to the European Commission. On the Ethical Implications of Biotechnology (GAEIB). Luxembourg: Office for Official Publications of the European Communities, 1996 ISBN 92–827-7350-7.

Gruber, Fr. P., Kolar, R. (1997). Animal test advisory commissions: ethics committees in Germany. In *Animal Alternatives, Welfare and Ethics* (eds van Zutphen, L. F. M., Balls, M.), pp. 373–376. Elsevier Science, Amsterdam.

Chapter 4

Animal Care and Use Committees: An American Perspective

T. Allen and R. Crawford

The views expressed by the authors do not necessarily represent positions or policies of the US Department of Agriculture or any agency thereof and should not be interpreted as such.

Introduction

The Animal Welfare Act (AWA) (1966) is the primary federal law that governs the use of animals in research, testing, and teaching in the USA. Originally passed by Congress in 1966 and amended in 1970, 1976, 1985, and 1990, the AWA provides the basis for the regulatory authority given to the United States Department of Agriculture (USDA) to insure the welfare of covered animals used in regulated activities. The Act includes all warm-blooded vertebrates, as defined by the Secretary of Agriculture, but specifically exempts all farm animals used in food or fiber research or production. Rats of the genus *Rattus,* mice of the genus *Mus,* and all birds are administratively exempted at this time by the Secretary of Agriculture. With the passage of the 1985 Improved Standards for Laboratory Animals Act, sponsored by Senator Robert Dole and Representative George Brown, the provisions of the AWA were greatly expanded. The primary purpose of the new law was succinctly stated by Senator Dole in his remarks introducing the amendment. He said:

> Mr President, the farm bill contains legislation dealing with the humane treatment of animals. The main thrust of the bill is to minimize pain and distress suffered by animals used for experiments and tests. In so doing, biomedical research will gain in accuracy and humanity. We owe much to laboratory animals and that debt can best be repaid by good treatment and keeping painful experiments to a minimum. (Congressional Record, 1985).

The new law redefined humane care to include such factors as sanitation, ventilation, and housing. The USDA was directed to establish regulations to give dogs the opportunity for exercise and to set standards relating to a physical environment adequate to promote the psychological well-being of non-human primates. Not unexpectedly, the regulations greatly expanded the powers of the laboratory animal veterinarian and stressed the need to minimize pain and distress through adequate veterinary care and the proper use of anesthetics, analgesics, tranquilizers, and euthanasia. The principal investigator was also obligated to consider alternatives to any procedure *likely* to cause pain or distress. USDA considers alternatives in the spirit of the 3Rs of Russell and Burch (1959)—reduction in the numbers of animals used, refinement of techniques to minimize pain or distress, or replacement with non-animal techniques. To insure that the new regulations were being followed, the law also called for the establishment of Institutional Animal Care and Use Committees (IACUC). The IACUC was given broad powers to oversee all aspects of an institution's animal care and use program, including approval or disapproval of animal use protocols, development of training programs for animal care and use personnel, inspection of all animal areas, review of the animal care program, and investigation of any alleged problems. The development of the IACUC also allowed USDA to begin "enforced self-regulation." Under this concept, the institution is responsible for ensuring compliance with the AWA regulations and reporting any problems to USDA. To insure that the IACUC is performing its duties, USDA inspects all facilities at least once each year. The new law also established an information service (the Animal Welfare Information Center) at USDA's National Agricultural Library to assist both researchers and IACUCs in complying with provisions of the regulations. Finally, the law provided for severe penalties for any IACUC member who released proprietary information or other trade secrets garnered in the course of IACUC activities.

Although the AWA regulations are the only federal regulations governing the welfare of animals in research, the *Guide for the Care and Use of Laboratory Animals* (NRC, 1996) is a widely used reference on animal care and use. However, researchers receiving funding from the US Public Health Service (PHS) are obliged to follow the animal care standards found in the *Guide* and must assure the PHS that they are doing so. Unlike the AWA, the *Guide* covers all species of animals used in biomedical research. By and large, the standards found in the *Guide* and the AWA regulations have been harmonized.

Because of the legal burden placed upon IACUCs, it is important to understand the organization and make-up of

Handbook of Animal Models of Infection
ISBN 0-12-775390-7

the committee, the regulations they are required to follow, the processes involved in review and inspection of the institution's policies and physical plant, the process of protocol review, common IACUC problems, and finally, a look at special issues in infectious disease research.

Organization

Under the 1985 amendments, each institution must designate an Institutional Official (IO), who has the authority legally to commit on behalf of the research facility that the regulations of the AWA will be followed. The IACUC, which is appointed by the chief executive officer, reports directly to the IO. Although the IACUC is responsible for evaluating the animal care program, it is the IO who has ultimate legal responsibility for insuring compliance with the regulations and proper functioning of the IACUC. Under both the AWA and the *Guide*, the IACUC has final authority to disapprove any activity involving the care and use of animals. However, activities approved by the IACUC may be subject to further scrutiny by the institution (USDA, 1995b; NRC, 1996).

Membership

The AWA provides that, at a minimum, each IACUC shall be composed of the chairman, a veterinarian, and an unaffiliated member. The unaffiliated member, as the name implies, has no affiliation with the institution and can receive only minimal compensation (e.g., travel expenses, meals) from the institution. If the committee is composed of more than three members, not more than three members can be from the same administrative unit.

By contrast, the *Guide* requires that an IACUC consist of at least five members. Of these five people, one must be a practicing scientist experienced in research involving animals, one must be a non-scientist, and one must not be affiliated with the institution. A survey of 477 research facilities in 1995 found that most IACUCs consisted of 7 members with a range from 3 to 50 members (Borkowski, 1996).

The veterinarian

The veterinarian must be trained and experienced in laboratory animal science and medicine, and must have direct or delegated responsibility for the animal care program and activities involving animals. The veterinarian should also help determine the institution's goals for its animal care program, develop training programs to insure humane treatment of animals, and promote the use of alternatives to animals whenever possible (Van Hoosier, 1987; Schwindaman, 1994).

The non-affiliated member

The role of the non-affiliated member (NAM) has recently been the subject of much study in the USA. The inclusion of NAMs on animal care committees was a hard-fought victory for advocates of laboratory animals (Stevens, 1986). Under both the AWA and the *Guide*, the role of the NAM is to "provide representation for general community interests in the care and treatment of animals" (USDA, 1995b; NRC, 1996). Members of animal protection groups wanted the NAM to have a background in animal protection efforts and thought that the NAM would provide greater accountability from researchers. However, the NAM is appointed by the chief executive officer with little input from the community and oftentimes has no background in animal welfare or animal protection. One of the reasons for this hesitancy to appoint animal protectionists is the underlying fear that s/he may be obstructive or damaging to the actions of the committee (Levin and Stephens, 1995). In the industrial sector, the whole issue of the NAM is clouded by the possibility that trade secrets or other proprietary information could be made public. However, Congress recognized this possibility and provided for serious criminal penalties for release of trade secrets by anyone on the IACUC (USDA, 1995b).

Many observers agree that the NAM brings an unbiased view to the IACUC, and insures that animal use is essential. By providing public accountability, the NAM serves as "the built-in integrity factor to counter negative public perception or prevent the real potential for conflict of interest" (Theran, 1997).

Federal requirements for IACUC activities

The primary functions of an IACUC are to review and inspect all aspects of an institution's animal care and use program, including all animal facilities, review animal use protocols, review and investigate complaints about animal use, and make recommendations to the IO (USDA, 1995b; NRC, 1996). The purpose of these reviews and inspections is to provide a mechanism that insures compliance with all regulations and policies and allows for interaction between the IACUC and institutional staff members. The IACUC becomes a group of individuals rather than a faceless in-house regulatory body, which serves to lessen the sometimes adversarial nature of the review process.

Review of the animal care program

At least once every 6 months, the IACUC must conduct a thorough review of the institution's program for humane care and use of animals using the AWA regulations and the *Guide* as the basis. Evaluation of the program should concern itself with how these activities are administered, implemented, and documented. It will necessarily focus on record-keeping and review of written procedures and policies. The programs that should receive a thorough evaluation include all IACUC procedures and policies, methods for protection of personnel that may report deficiencies in animal care or treatment, procedures for filing of semi-annual and annual reports to USDA, the facility's program of veterinary care, the occupational health program for animal care personnel, and finally, the facilities training program for all personnel involved in the use of animals (McLaughlin, 1993).

Facility inspections

As with the program review, facility inspections must be completed once every 6 months by the IACUC. This is the opportunity for the IACUC to see first-hand not only the structures and equipment of the animal facility, but also the physical manifestation of the written policies of the animal care program. By observing the animals in their daily quarters, the IACUC can most readily determine if the institutional animal care policies are promoting the welfare of the animals. Another compelling reason is that the inspection insures that the facility is complying with all federal regulations and guidelines.

Members of the IACUC tour the facility's animal rooms, feed and bedding preparation areas, necropsy rooms, cage wash areas, and any other rooms used in the animal program. If animals are routinely transported to laboratories or are maintained in satellite facilities, these areas must also be inspected. In addition to examining the animals, animal care personnel should be questioned about the daily and weekly animal husbandry routine. Facilities housing dogs and non-human primates have to meet special requirements concerning the well-being of these animals. It is imperative that the IACUC pay special attention to the implementation of these requirements for exercise in dogs and environmental enrichment for non-human primates. During the inspection process, the IACUC member should take detailed notes documenting, as needed, minor and significant deficiencies, or outstanding innovations that have improved animal welfare. These notes will be used in the preparation of the report to the IO.

After the program review and facility inspection, a detailed report is generated noting any significant or minor deficiencies, the probable reason for the deficiency, and plans for corrective actions, including a timetable for completion of these actions. The report should also note outstanding aspects of the program and facilities. Significant deficiencies—those that pose an immediate threat to the health or safety of the animals—must be corrected within a reasonable time frame. Any failure to adhere to the plan that results in a significant deficiency remaining uncorrected must be reported to the USDA and any federal agency that has provided funding for a project. The final report must be approved by a majority of the IACUC and must include any dissenting viewpoints.

IACUC review of protocols

Both the AWA and the PHS *Guide* mandate the review of animal research protocols by the animal care and use committee before any research may begin. The AWA also requires the IACUC to review all approved protocols on an annual basis. The IACUC must review and approve, require modifications to a proposal in order to secure approval, or disapprove any protocol which it receives. The institution is given leeway in determining the most appropriate means of complying with these requirements (Dresser, 1989). The regulations and guidelines do not specify the frequency of meetings for IACUCs, leaving this to the needs of each institution. Animal care committees at large institutions may meet every week while smaller institutions may be able to function with bimonthly meetings.

The AWA mandates very specific criteria that must be met before an IACUC may grant approval to new proposals or changes in existing protocols (Schwindaman, 1994; USDA, 1995b). Those criteria include:

- procedures involving animals will avoid or minimize pain or distress to the animals
- an investigator must consider using alternatives to procedures that might cause pain or distress to animals. Further, the investigator must provide, at a minimum, a literature review that focuses on alternative methods, and discuss, in a written narrative, why alternatives can or cannot be used. The narrative must also include a list of the databases searched and the keywords used in the search strategy, and any other sources consulted
- the investigator must provide written assurance that the proposed activities are not unnecessarily duplicative
- if painful or distressful procedures are unavoidable, the procedures must:
  - be performed with appropriate anesthesia, analgesia, or sedatives, unless withholding them can be scientifically justified
  - involve a veterinarian in the planning (review by a veterinarian on the IACUC after the protocol has been submitted is not acceptable; the veterinarian should be consulted before the protocol is submitted)
  - never use paralytics without anesthesia
- animals that experience severe or chronic pain that cannot be relieved will be euthanized at the end of the experiment or, if appropriate, during the experiment
- the animals' living conditions must be appropriate for the species and must contribute to their health and comfort
- any medical care required by the animals will be provided by a qualified veterinarian, i.e. a veterinarian trained in laboratory animal medicine

- people who will be performing any procedures on animals will be qualified and trained to perform the procedures
- any procedures that involve surgery will include appropriate pre- and postoperative care. All survival surgeries must use aseptic technique, including the use of sterile gloves and masks. Operative procedures that penetrate or expose a body cavity or procedures that result in permanent impairment of physical or physiological functions (major procedures) must be performed in a dedicated surgical facility that is maintained in aseptic condition. Rodents are exempted from this requirement. Minor surgical procedures must be performed aseptically but do not require a dedicated site
- use of an animal in more than one major operative procedure, from which it is allowed to recover, is prohibited unless it can be justified scientifically, is necessary for the health of the animal, or unless special permission is obtained from the USDA
- use of professionally recognized methods of euthanasia, unless a different method can be scientifically justified. The guidelines developed by the American Veterinary Medical Association in 1993 are generally recognized as the currently accepted standard
- identification of the species to be used and the number of animals requested
- the scientific rationale for using animals, and the reasons for using the requested species and the number of animals
- a complete description of the procedures involving animals
- a complete description of the methods that will be used to minimize pain and discomfort to that which is unavoidable and necessary to the collection of *scientifically* valuable data

The animal use protocol review form

While federal regulations give rather specific requirements for what an IACUC must consider for approval or disapproval of animal use protocols, the actual method for collecting that information has been left to the scientific community. However, the protocol review form is the key to the entire process for it provides the IACUC members with the information necessary for them to perform their jobs (Prentice *et al.*, 1991). A successful review form should be regarded as a dynamic document that can change with the institution's experiences and evolving regulatory, professional, and societal standards (Prentice *et al.*, 1991). A well-designed review form challenges the scientist to examine and justify, both scientifically and ethically, all aspects of a procedure that affects the well-being of the animals (Russow, 1995). A sample protocol review form is shown in Figure 4.1 at the end of this chapter.

To assist the scientist and the members of the IACUC, each institution should develop policies or standard operating procedures on common painful experimental procedures that carry a "high ethical cost," so that everyone involved in the review process has a common point of reference and consistent decisions are rendered. This also leads to a more efficient IACUC as time is not spent resolving the same conflicts time and again. Some of these procedures would include the use of complete Freund's adjuvant, death as an endpoint, tumor burdens, food and water deprivation, and LD_{50} studies for both toxicology and infectious disease studies (Dresser, 1987). Institutional guidelines may require that alternative methods be used for particularly painful procedures, may list criteria for euthanizing animals during a painful procedure, or may provide guidance as to monitoring animals for symptoms of pain or distress.

The protocol review process

Contrary to public perception, the animal care and use committee is not an animal welfare committee. If IACUCs were required to approve protocols based solely on animal welfare issues, most protocols would be rejected. The primary purpose of protocol review is to promote the welfare of animals without compromising valid scientific objectives that might benefit other animals and humankind. Protocol review is a moral and ethical evaluation that necessarily requires the evaluation of the science involved (Prentice *et al.*, 1990, 1992). Without addressing the validity of the proposed scientific objectives and methods, the IACUC can't decide if the ethical cost weighed against the potential benefits is morally justifiable (Russow, 1995). It would appear that the *Public Health Service Policy on Humane Care and Use of Laboratory Animals* (PHS, 1996) provides the IACUC, as an appropriate institutional review board, with a legal basis for considering scientific merit of a proposed research activity. According to this document, "procedures involving animals should be designed and performed with due consideration of their relevance to human or animal health, the advance of knowledge, or the good of society" (OPRR, 1991; Prentice *et al.*, 1992).

The review process will be different at different institutions depending upon the needs of the facility. A committee may delegate review of protocols to a review team or may require all members of the IACUC to review the form. Regardless of the process used, the AWA and PHS *Guide* require that all members of the committee have full access to the protocol review form. Federal regulations and guidelines allow the designated reviewers to approve, require modifications to, or disapprove a protocol. But any member of the IACUC may request full committee review. Approval by the full committee requires that a quorum be present and that a majority vote to approve (USDA, 1995b; NRC, 1996). Because IACUCs are asked to review a broad range of activities, many of which may be outside the expertise of the committee, the use of consultants is allowed. However, the consultants do not have voting privileges unless they are members of the IACUC.

During the review process, the members of the IACUC should carefully assess the information provided by the scientist on the animal protocol form. If for any reason a reviewer is not satisfied with the information provided by the scientist, s/he may submit questions to the scientist asking for clarification or additional information. Typical reasons for a protocol being sent back to a scientist include:

- lack of justification for the species or number of animals requested
- criteria for alleviation of pain or distress, or use of euthanasia during an experiment inadequately addressed
- consideration of alternatives not addressed
- lack of assurance that the proposed procedure is not unnecessarily duplicative
- incomplete description of proposed activities, post-procedural care, or endpoints not clearly defined

Consideration of alternatives

Animal welfare regulations require that an investigator performing procedures that are painful or distressful to the animal provide assurance that no alternatives exist to the painful procedure. To provide this assurance the investigator must provide, at a minimum, a written narrative that describes the literature databases searched (e.g., Medline, EmBase, Biosis Previews, AGRICOLA, PREX), the keywords or strategy used to retrieve information, and a brief description of why alternatives are or are not available. The IACUC must satisfy itself that alternatives were adequately considered, must discuss the use of alternatives during meetings and note the discussion in its minutes. These must be made available to USDA inspectors (USDA, 1989). Stokes and Jensen (1995) have developed guidelines to assist IACUCs in reviewing protocols for alternatives. Some institutions have developed animal alternatives committees (James *et al.*, 1995; Holden, 1997) or appointed librarians familiar with this type of searching (Keefer and Westbrook, 1996) to help the IACUC with this aspect of protocol review. Smith (1994) has written a method paper on searching for alternatives that is an overview of this type of searching.

To assist both investigators and IACUCs, the 1985 amendment to the Animal Welfare Act established the Animal Welfare Information Center (AWIC) within USDA's National Agricultural Library. AWIC provides literature-searching services and produces numerous publications and workshops on animal welfare and the use of alternatives in research, testing, and teaching.

As mentioned previously, the law provides for an annual review of protocols by the IACUC. The investigator should use this review as an opportunity to reexamine the literature for alternatives that may have been developed since the prior review.

Expedited review of animal use protocols

Under the AWA the principal investigator is required to provide the animal care and use committee with a written description of all activities that involve the care and use of animals that are covered by the regulations. If a full committee review is not apparently necessary or is not requested, then an expedited review of the protocol may be made. For an expedited review, the committee chairman designates at least one member of the IACUC who is qualified to conduct the review to review the protocol. This designated individual is to review the protocol and has the authority to: approve the protocol, require modifications to the protocol, or request a full committee review of the protocol. If a full committee review is requested, approval may be granted only after review, at a convened meeting of a quorum of the IACUC, and with approval vote of a majority of the quorum present. No member of the IACUC may participate in a protocol review or approval, or be part of a quorum, if that member has a conflicting interest in the protocol, except to provide requested information to the IACUC. The IACUC member making the expedited review does not have the authority to disapprove a protocol. Disapproval or suspension of a protocol may only be done by a majority vote of a quorum at a convened meeting of the IACUC.

Problems with IACUC review

In a 1995 audit of 26 research facilities, the Inspector General of the USDA found that IACUCs, in general, were not meeting the standards of the Animal Welfare regulations. This was attributed to the fact that committee members are not always aware of their responsibilities and duties. Consequently, the audited institutions could not insure that pain or distress were minimized or that repetitive experiments were not being performed. The audit also showed that facilities were not adequately addressing the use of alternatives to painful procedures and were allowing research activities to begin before the proposals had been reviewed by the IACUC (USDA, 1995a).

Part of the problem was the lack of federal oversight of committee activities. While USDA inspectors routinely visited the institutions, the inspections focused on the conditions of the animal facilities and not on the policies and programs of animal care and review. The Inspector General recommended that USDA send official notices to animal care committees informing them of their legally mandated responsibilities (USDA, 1995a).

I. Non-technical synopsis
A brief, narrative description of the proposal or idea that is easily understood by non-scientists

II. Background
A. Background
This should include a brief statement of the requirement or need for the information being sought

B. Literature search
This search must be performed to prevent unnecessary duplication of previous experiments. A search of the scientific literature (MEDLINE, EmBase, AGRICOLA, AWIC, etc.) is highly recommended.
1. Literature source(s) searched
2. Date and number of search
3. Key words of search
4. Results of search: provide a narrative description of the result of the literature search

III. Objective or hypothesis
In non-technical terms, state the objective of this protocol, or the hypothesis to be accepted or rejected

IV. Materials and methods
A. Experimental design and general procedures
Provide a "complete description of the proposed use of animals." This section should succinctly outline the formal scientific plan and direction for experimentation. It is critical that reviewers of this protocol are able to follow your reasoning and calculations for the number of animals required, and can verify that the experimental design clearly supports the number of animals requested

B. Laboratory animals required and justification
1. Non-animal alternatives considered: were alternatives to animal use considered?
2. Animal model and species justification
3. Laboratory animals:
   a. Genus and species
   b. Strain/stock
   c. Source/vendor
   d. Age
   e. Weight
   f. Sex
   g. Special considerations: specialized requirements for the animals, e.g. SPF
4. Total number of animals required
5. Refinement, reduction, replacement: does this protocol have any provisions that would qualify it to be identified as one that refines, reduces, or replaces (3 Rs) the use of animals?

C. Technical methods
These should be presented in sufficient detail, documented or referenced, so that the IACUC can adequately review the procedure and obtain a clear understanding of what is to be done and how the animal will be handled, and make a reasonable determination as to whether this proposed use of laboratory animals is in compliance with federal regulations, guidelines, and law
1. Prolonged restraint: describe and justify in detail any prolonged restraint (greater than 3 hours) intended for use during the study. Also describe habituation procedures for the prolonged restraint
2. Surgery:
   a. Procedure: describe in detail any planned surgical procedures
   b. Pre- and postoperative provisions: detail the provisions for both pre- and postoperative care, including provisions for post-surgical observations
3. Animal manipulations: any injections, sampling procedures, or other manipulations of the animals necessary for the execution of the study must be described
   a. Injections
   b. Biosamples: cerebral taps, blood sampling, etc. List amounts taken and method for sampling
   c. Animal identification: microchip, tattoo, ear tags, cage cards, etc
   d. Behavioral studies: fully describe any intent to use aversive stimuli, food or water deprivation, etc., that would impact upon the animals in this study
   e. Other procedures: EKGs, radiology, aerosol exposure, etc.
4. Adjuvants: list any adjuvants and your plan for their use. Provide dosages and route
5. Study endpoint: what is the projected endpoint or termination of the study for the animals? You should insure that unnecessary pain or distress is prevented by carefully considering "When is the experimental question answered?" so that the animals can be removed from the study as soon as feasible. Explain the plan for the disposition of surviving animals. You must specifically address and justify any proposed use of death as an endpoint.

6. Euthanasia: explain the plan for euthanasia of the animals at the completion of the study and who will perform the procedure. The current AVMA guidelines for euthanasia must be followed
7. Pain: if a procedure involves pain or distress, the PI must consult with the attending veterinarian
   a. USDA (form 18-3) pain category: this information is reported by the organization to the USDA on USDA form VS 18-23. The PI or primary user should estimate the number of animals that will be counted in each pain category.
      i. no pain – number of animals: % (column C on USDA form)
      ii. alleviated pain – number of animals: % (column D)
      iii. unalleviated pain or distress – number of animals: % (column E)
   b. Pain alleviation: the attending veterinarian should be able to provide assistance in completing this section of the proposal
      i. anesthesia/analgesia/tranquilization: describe the methods or strategies planned to alleviate pain or distress. Provide agent, dosage, route and site, indication, etc.
      ii. paralytics: no use of paralytic agents without anesthesia is allowed.
   c. Alternatives to painful procedures:
      i. source(s) searched: e.g., AGRICOLA, MEDLINE, PREX, etc.
      ii. date of search
      iii. key words of search: e.g. pain, surgery, simulation, cell line
      iv. Results of search: provide a narrative description of the results of the alternative literature search
   d. Painful procedure justification: procedures causing more than transient or slight pain that are unalleviated must be justified on a scientific basis in writing by the PI. The pain must continue for only the necessary period of time dictated by the experiment, and then be alleviated, or the animal humanely euthanized. The PI must consult with the attending veterinarian or his or her designee in the planning of both alleviated and unalleviated painful procedures, and state it here

D. Veterinary care

Attending veterinary care of lab animals receives particular emphasis in the AWA.

1. Attending veterinary care: will the animals be observed daily or more frequently, and by whom? What is the plan if the animal becomes ill or debilitated during the study and requires supportive therapy? Will the animal be euthanized if it becomes critically ill or comatose, and by whom (study endpoint adjustment)? Justification for not providing supportive care for clinically ill animals is necessary
2. Enrichment strategy: written justification for restricting enrichment programs or activity programs of dogs, cats, or nonhuman primates must be provided

E. Data analysis

List the statistical test(s) planned or the strategy intended to evaluate the data

V. Biohazard/safety

Provide a list of any potential biohazards associated with this proposal, e.g., viral agents, toxins, radioisotopes, oncogenic viruses, chemical carcinogens, etc. Explain any safety precautions or programs designed to protect personnel from biohazards, and any surveillance procedures in place to monitor potential exposures

Figure 4.1 Sample animal protocol form to be submitted to an institutional animal care and use committee. Compiled following a form used by the US Department of Defense.

Special issues in infectious disease research

Appropriate animal numbers

One of the primary responsibilities of the IACUC is to insure that the fewest animals possible that will yield scientifically valid data are used in an experiment. Too many animals are an unethical waste of animals and an improper use of research funds. The same argument holds true for using too few animals. However, with proper planning and consultation with a biostatistician, calculating the proper number of animals can be attained. It should be noted, however, that other factors, such as ethical considerations, may also influence the number of animals in a sample size.

The scientist must provide adequate justification for the number of animals proposed for the protocol. The appropriate number of animals for a study will be determined by the variability of the parameter being studied and the statistical tests to be used in analyzing the experimental results (Festing, 1992). To use statistical formulas successfully for determining sample size, the scientist must have some idea of several parameters: the probability of accepting a false positive (alpha error) or a false negative (beta error), the smallest difference worth detecting (effect size), and the variability of experimental groups (Mann *et al.*, 1991; Festing, 1995). Proper scrutiny of the relevant scientific literature may provide information on effect size and variability, allowing the investigator to assign precise values to these variables for use in power analysis or other methods for estimating sample size.

Because of the profound effect of variability on the response of animals to an experimental challenge, it is necessary to understand this source of error if animal numbers are to be minimized (Festing, 1992). To minimize within-group variability it is imperative that animals be free of clinical or subclinical disease and not subject to environmental or dietary stress (except as part of a protocol). The use of inbred strains of laboratory animals will further reduce variability and the number of animals needed because their high degree of uniformity increases the statistical power (the probability that a statistical test will detect a difference when the difference actually exists; Festing, 1995).

Erb (1996) has provided an excellent review of the actual elements of the sample size calculations, issues that determine which sample size formula to use, and methods to decrease the needed sample size when the calculated sample size is impractical to use.

Minimizing pain and distress

US animal welfare regulations define a painful procedure as one that "would reasonably be expected to cause more than slight or momentary pain or distress in a human being to which that procedure was applied, that is, pain in excess of that caused by injections or other minor procedures" (USDA, 1995b). Investigators are required to consult with a veterinarian to insure that pain or distress is minimized. Because many infectious disease studies have the potential for causing pain in animals, the IACUC should establish guidelines for periodic monitoring of animals, set criteria for veterinary intervention to alleviate the pain through use of analgesics or euthanasia, and require training for investigators and animal care personnel to ensure that all can recognize symptoms of pain and distress. Morton and Griffiths (1985) and other authors (Soma, 1987; NRC, 1992) provide comprehensive reviews on the recognition of pain and distress in laboratory animals. The IACUC should make these available to animal use personnel.

Alternatives to death as an endpoint

Because many infectious disease studies may cause irreversible pain or distress before death ensues, the investigator should strive to incorporate earlier endpoints that will satisfy the requirements of the study and spare the animal a painful death (Siems and Allen, 1989; Olfert, 1995). Also, an earlier endpoint may result in better scientific data as the death of the animal may result in postmortem changes to tissue or body fluids (Amyx, 1987a; Siems and Allen, 1989). Because these earlier endpoints may be best addressed as part of the experimental design, IACUC review of the experimental design becomes very important in infectious disease studies (Amyx, 1987b).

Hamm (1995) has proposed guidelines for IACUC acceptance of death as an endpoint. By establishing clear guidelines or criteria for early euthanasia of animal subjects, the IACUC can clearly minimize the pain or distress that the animal will experience. Numerous authors have outlined these criteria (Morton and Griffiths, 1985; Tomasovic *et al.*, 1988; Hamm, 1995; Olfert, 1995). Frequently mentioned variables that should be observed include body weight, physical appearance, clinical signs such as temperature, heart rate, or bleeding, unprovoked behavior of the animal, and responses to external stimuli. In addition, the investigator should use information available from the infectious disease literature on progression of symptoms, time course of the disease and other unique features of the disease being studied to establish earlier endpoints (Soothill *et al.*, 1992). In some studies there may be a scientifically valid reason for allowing the progression to death. However, inconvenience to the investigator or cost of alternatives are not acceptable reasons (Tomasovic *et al.*, 1988).

Alternatives in antibody production

This has emerged as a very controversial topic in the USA. In 1997, the USDA and the National Institutes of Health were petitioned by the American Anti-Vivisection Society to ban the use of animals in the production of monoclonal antibodies (MAb) via the ascites method. Although several European countries have regulations limiting or prohibiting the use of animals for MAb production (McArdle, 1997), at this time USDA has decided not to prohibit the use of animals for MAb production, citing the need for further development and evaluation of *in vitro* methods.

Production of antibodies in animals usually involves the use of an adjuvant or priming agent, such as Freund's complete or incomplete adjuvant or pristane, in conjunction with a selected antigen to stimulate the immune system of an animal to produce titers of antibodies. But it should be remembered that the purpose of the adjuvant is to induce antibodies, not pain (Amyx, 1987b). Both types of Freund's adjuvants are known to produce serious inflammatory reactions that may result in abscesses, granulomas or tissue necrosis. Consequently, IACUCs should always question the use of these adjuvants and should urge investigators to use alternative adjuvants such as Montanide ISA or Ribi's. These alternative adjuvants may provide immunpotentiation similar to Freund's, but without the severe pain associated with the use of Freund's (Hanly and Bennett, 1997).

The production of MAbs via induction of ascites fluid has been an important tool in immunological and infectious disease research. However, unless the mouse or other animal is carefully monitored, the potential for severe pain is very real. In 1974 Kohler and Milstein showed that MAbs could be produced with *in vitro* methods. With Niels Jerne they won the 1984 Nobel Prize for their *in vivo* and *in vitro* work on MAb production (McArdle, 1997). Current *in vitro* methods widely used in Europe and the USA include modular bioreactors, static and agitated suspension cultures, and membrane-based and matrix-based culture systems (Marx *et al.*, 1997; Petrie, 1997). Modular bioreactors used in the USA have been found to rival the efficiency of the ascites method (Petrie, 1997). The use of these *in vitro* methods, besides eliminating the use of animals, has the added advantage of being free of contaminating antibodies, cytokines and similar biologically active materials. Because the primary purpose of the IACUC is to minimize pain in animals, it should always encourage the use of alternative methods of MAb production at both the institutional and laboratory level. To assist IACUCs, a comprehensive bibliography on adjuvants and antibody production is available (Smith *et al.*, 1997).

Concluding remarks

The 1985 amendments to the US AWA have had a profound effect on how scientific research is conducted in the USA. With the establishment of institutional animal care and use committees, scientists wishing to conduct research using animals must receive approval from a committee composed of their peers and a representative of the general public. This accountability, for the privilege of using live animals in scientific endeavors, is enforced by periodic, unannounced inspections by veterinarians from the USDA. Although no system of oversight is without problems, the animal care and use committee process in the USA seems to be doing its job of facilitating science while allowing for the welfare of those animals that must be used in research. However, with continuing societal concern over the use of animals in research, these committees should be advocates for alternative methods that implement the 3Rs of Russell and Burch (1959) whenever possible.

References

Amyx, H. L. (1987a). Alternatives to the LD-50 in infectious disease studies. *Sci. Center Animal Welfare Newslett.*, **9**, 1–2.

Amyx, H. L. (1987b). Control of animal pain and distress in antibody production and infectious disease studies. *J.A.V.M.A.*, **191**, 1287–1289.

Animal Welfare Act. (1966 as amended). 7 USC 2131–2157. U.S. Government Printing Office, Washington, D.C.

Borkowski, G. (1996). *IACUCs: Celebrating 10 Years of Experience.* http://vs247.cas.psu.edu/iacuc/IACUCRES.HTM.

Congressional Record. (1985). December 17. U.S. Government Printing Office, Washington, D.C.

Dresser, J. D. (1987). Refining the ACUC process: policies and procedures. *Sci. Center Animal Welfare Newslett.*, **9**, 3–5.

Dresser, J. D. (1989). Developing standards in animal research review. *J.A.V.M.A.*, **194**, 1185–1191.

Erb, H. (1996). A non-statistical approach for calculating the optimum number of animals needed in research. *Lab Anim.*, **25**, 45–49.

Festing, M. F. W. (1992). The scope for improving the design of laboratory animal experiments. *Lab. Anim.*, **26**, 256–267.

Festing, M. F. W. (1995). Variation and experimental design. *Sci. Center Animal Welfare Newslett.*, **18**, 3–9.

Hamm, T. E. (1995). Proposed institutional animal care and use committee guidelines for death as an endpoint in rodent studies. *Contemp. Topics Lab. Anim. Sci.*, **34**, 69.

Hanly, W. C., Bennett, B. T. (1997). Overview of adjuvants. In *Information Resources for Adjuvants and Antibody Production: Comparisons and Alternative Technologies 1990–1997* (eds Smith, C. P., Jensen, D., Allen, T., Kreger, M.), pp. 1–7. Animal Welfare Information Center, Beltsville, Maryland.

Holden, F. (1997). Alternatives committee established at Indiana. *Johns Hopkins Center Altern. Animal Test.*, **14**, 6–7.

James, M. L., Mininni, L. A., Anderson, L. C. (1995). Establishment of an animal alternatives committee. *Contemp. Topics Lab. Anim. Sci.*, **34**, 61–64.

Keefer, E., Westbrook, F. (1996). The role of the librarian in the work of the institutional animal care and use committee. *Anim. Welfare Inf. Center Newslett.*, **6**, 9–11.

Levin, L. H., Stephens, M. L. (1995). Appointing animal protectionists to institutional animal care and use committees. *Anim. Welfare Inf. Center Newslett.*, **5**, 1–2, 8–10.

Mann, M. D., Crouse, D. A., Prentice, E. D. (1991). Appropriate animal numbers in biomedical research in light of animal welfare considerations. *Lab. Anim. Sci.*, **41**, 6–14.

Marx, U., Embleton, M. J., Fischer, R. *et al.* (1997). Monoclonal antibody production: the report and recommendations of ECVAM workshop. *A.T.L.A.*, **25**, 121–137.

McArdle, J. (1997). Alternatives to ascites production of monoclonal antibody production. *Anim. Welfare Inf. Center Newslett.*, **8**, 1–2, 15–18.

McLaughlin, R. M. (1993). Institutional animal care and use committee review of animal care and use programs. *Contemp. Topics Lab. Anim. Sci.*, **32**, 12–15.

Morton, D. B., Griffiths, P. H. M. (1985). Guidelines on the recognition of pain, distress, and discomfort in experimental animals and an hypothesis for assessment. *Vet. Rec.*, **116**, 431–436.

NRC (National Research Council) (1992). *Recognition and Alleviation of Pain and Distress in Laboratory Animals.* National Academy Press, Washington, D.C.

NRC (National Research Council) (1996). *Guide for the Care and Use of Laboratory Animals.* National Academy Press, Washington, D.C.

Olfert, E. D. (1995). Defining an acceptable endpoint in invasive experiments. *Anim. Welfare Inf. Center Newslett.*, **6**, 3–7.

Olfert, E. D., Cross, B. M., McWilliams, A. A. (eds) (1993). *Guide to the Care and Use of Experimental Animals*, vol. 1. Canadian Council on Animal Care, Ottawa, Ontario, Canada.

OPRR (Office for Protection from Research Risks, NIH) (1991). The Public Health Service responds to commonly asked questions. *ILAR News*, **33**, 68.

Petrie, H. (1997). Modular bioreactors as an alternative to ascitic fluid production of monoclonal antibodies. *A.N.Z.C.C.A.R.T. News*, **10**, 6.

PHS (Public Health Service) (1992). *Institutional Animal Care and Use Committee Guidebook.* Publication no. 92-3415. National Institutes of Health, Bethesda, Maryland.

Prentice, E. D., Crouse, D. A., Rings, R. W. (1990). Approaches to increasing the ethical consistency of prior review of animal research. *Invest. Radiol.*, **25**, 271–274.

Prentice, E. D., Crouse, D. A., Mann, M. D. (1991). The IACUC protocol review form: one of the keys to a successful review. *A.A.L.A.S. Bull.*, **30**, 10–16.

Prentice, E. D., Crouse, D. A., Mann, M. D. (1992). Scientific merit review: the role of the IACUC. *I.L.A.R. News*, **34**, 15–19.

Russell, W. M. S., Burch, R. (1959). *The Principles of Humane Experimental Technique*. Special edition reprinted 1992. Universities Federation for Animal Welfare, Hertfordshire, UK.

Russow, L-M. (1995). Protocol review: too much paperwork? In *Current Issues and New Frontiers in Animal Research* (eds Bayne, K.A.L., Greene, M., Prentice, E.D.), pp. 15–18. Scientists Center for Animal Welfare, Greenbelt, Maryland.

Schwindaman, D. (1994). Federal regulation of experimental animal use in the United States of America. *Rev. Sci. Tech. Off. Int. Épiz.*, **13**, 247–260.

Siems, J. J., Allen, S. D. (1989). Early euthanasia as an alternative to death in chronic infectious disease studies using a systemic *Candida albicans* model. *Abstr. Annu. Meet. Am. Soc. Microbiol.*, **89**, 81.

Smith, C. (1994). AWIC tips for searching for alternatives to animal research and testing. *Lab. Anim.*, **23**, 46–48.

Smith, C. P., Jensen, D. J. B., Allen, T., Kreger, M. D. (1997). *Information Resources for Adjuvants and Antibody Production: Comparisons and Alternative Technologies 1990–1997*. Animal Welfare Information Center, Beltsville, Maryland.

Soma, L. R. (1987). Assessment of animal pain in experimental animals. In *Effective Animal Care and Use Committees* (eds Orlans, F. B., Simmonds, R. C., Dodds, W. J.), pp. 71–74. Special issue of *Lab. Anim. Sci*. Scientists Center for Animal Welfare, Bethesda, Maryland.

Soothill, J. S., Morton, D. B., Ahmad, A. (1992). The HID-50 (hypothermia inducing dose 50): an alternative to the LD-50 for measurement of bacterial virulence. *Int. J. Exp. Pathol.*, **73**, 95–98.

Stevens, C. (1986). Institutional animal care and use committees: new advocates for laboratory animals. *Advocate*, **4**, 16–18.

Stokes, W. S., Jensen, D. J. B. (1995). Guidelines for institutional animal care and use committees: consideration of alternatives. *Contemp. Topics Lab. Anim. Sci.*, **34**, 51–60.

Theran, P. (1997). The SCAW IACUC survey part II: the unaffiliated member. *Lab Anim.*, **26**, 31–32.

Tomasovic, S. P., Coghlan, L. G., Gray, K. N., Mastromarino, A. J., Travis, E. L. (1988). IACUC evaluation of experiments requiring death as an endpoint: a cancer center's recommendations. *Lab. Anim.*, **17**, 31–34.

USDA (U.S. Department of Agriculture) (1989). Animal welfare: final rules — institutional animal care and use committees. *Fed. Reg.*, **54**, 36125–36131.

USDA (U.S. Department of Agriculture) (1995a). Enforcement of the Animal Welfare Act. Audit report no. 33600-1-Ch., pp. 24–26. Office of Inspector General, Washington, D.C.

USDA (U.S. Department of Agriculture) (1995b). Title 9 CFR Chapter 1, Subchapter A — *Animal Welfare*. U.S. Government Printing Office, Washington, D.C.

Van Hoosier, G. L. (1987). Role of the veterinarian. In *Effective Animal Care and Use Committees* (eds Orlans, F. B., Simmonds, R. C., Dodds, W. J.), pp. 101–102. Special issue of *Lab. Anim. Sci*. Scientists Center for Animal Welfare, Bethesda, Maryland.

Further reading

PHS (Public Health Service) (1996). *Public Health Service Policy on Humane Care and Use of Laboratory Animals*. National Institutes of Health, Bethesda, Maryland.

Chapter 5

Ethical Aspects of the Use of Animal Models of Infection

D. B. Morton

In this chapter I will first give an outline of the main theoretical and applied ethical positions on our use of animals and then focus on the theoretical and practical aspects of animal "suffering". I will then go on to illustrate where research in infection could be refined to cause less animal suffering, and so be more ethically acceptable.

Status of animals in society

All uses of animals by society are coming under increasing scrutiny through the activities of animal rights and animal welfare organizations. Some associations focus on animal research (antivivisectionist groups), others on intensive farming, sports and spectacles, companion animals and wildlife. Most people would agree that animals are important and, furthermore, that humans have a moral responsibility to care for domesticated and captive (and sometimes wild) creatures—the increase in the number and membership of animal protection organizations reflects this concern (see Rupke, 1987, for a review specifically on the rise of the antivivisection movement). The advent of television has further heightened public concern as animal cruelty can now be seen for what it is—gratuitous violence, abuse or misuse—and is less a matter of verbal interpretation of interested parties in law courts. Partly as a result of all this, vegetarianism is increasing (from 2–3% 10 years ago to an estimated 7% today), but this is also motivated through health and financial concerns. Research into animal behaviour has played an important role in revealing the capacity of animals to experience pain, distress and fear and this too has raised our awareness of animal suffering. Consequently, our view of what is acceptable to do to animals is continually challenged and the argument that it is legal to treat animals in a particular way is no longer adequate. By the same token, the fact that animal research is legal does not mean that scientists can do anything to animals in the name of science: it depends on why the research is being done and how well it is done. I shall return to this point later.

Animal rights and animal welfare

Any action that humans carry out, as moral agents, can be described as being morally right or wrong either in terms of the outcome of that action (its ends or goal) or in terms of the action itself—it is right to tell the truth, it is wrong to kill. One can argue that killing is always wrong no matter what the outcome (the pacifist view), or decide that it is right depending on the circumstances—a "just war". The two philosophical theories commonly used in such analyses are *utilitarianism* and *deontology*, and they are examples of *consequentialist* and *non-consequentialist* theories respectively, that is, the consequences of actions are or are not taken into account. An *animal rightist* will hold the view that it is always wrong to kill animals no matter what the circumstances or the consequences. They consider that animals, like humans, have an intrinsic value regardless of their utility, and so should be the object of similar, if not equal, respect in terms of taking their lives. Note that they may not deny that humans and animals have different inherent values. The argument is often turned round to state that animals have a right not to be killed. Animal rightists also believe that it is wrong to cause animals to suffer in any way and so animals have another right—not to be caused pain or distress. Those who believe that animals have rights can never condone the use of animals in research as this involves killing animals as well as causing them to suffer and such people are, therefore, antivivisectionists. They might, however, find permissible the use of animals that are not sentient (i.e. not possessing the ability to experience pain) providing they are not killed. The point at which a believer in animal rights permits the use of medicines (herbal, homeopathic or allopathic) to kill other animals such as microorganisms or parasites that are causing disease is unclear, but patently they would protect all sentient living vertebrates. Regan (1983) is a modern philosopher who espouses such rights-based views.

Utilitarians, on the other hand, would make a judgement solely on the outcome of any action. They believe in the greatest good for the greatest number with the least amount of harm. To return to our example again, leaders of a country might argue that a war was justified if there was an increased gain in land or wealth which would provide an

Handbook of Animal Models of Infection
ISBN 0-12-775390-7

increased quality of life for their people. Utilitarianism could also justify killing one person to save the lives of many others, for example, using a healthy person to provide organs for transplantation to save the lives of five or six others. This would represent a greater good for a greater number of people. In terms of animal use, a utilitarian might argue that the act of killing was not morally important as the consequential benefits, say from animal research in terms of quantity and quality of life for humans (new antibiotics, vaccines, beautifying cosmetics) would be sufficient justification. Sentience (the ability to experience pain and pleasure) is a property of animals and is a key value to be considered. Consequently, causing animals to suffer without a justifiable outcome is unacceptable. Singer (1975, 1991) takes this line in the renaissance of philosophical concern for animals, following earlier utilitarian philosophers, e.g., Jeremy Bentham (1780) who said, referring to animals: "The question is not can they reason, nor can they talk, but can they suffer?"

It can be seen that neither deontological nor utilitarian theories, if applied strictly, accord with modern practices. In fact, a mixture of the two is often used. We are likely to take the view that some actions are indeed inherently wrong and should be used only under extreme circumstances, such as killing of humans, but in those circumstances some calculus has to be done to ensure the benefits are significant and commensurate (as in some wars). As regards animals, there is an underlying deontological assumption that we have a duty of care to them regardless of their utility. As a result, in many countries laws determine that it is inherently wrong to cause animals to suffer. Animals deserve protection, although killing animals is usually acceptable under certain circumstances (compare killing sheep and pigs with killing whales and elephants). In practice, whilst the theme of judging the outcome is maintained, the argument is somewhat modified along the lines that one should seek to minimize any suffering: if it is acceptable to kill animals, then it should be done humanely. This approach would be typical of an *animal welfarist* who would find it acceptable to eat meat, keep pets, and carry out research providing the minimum amount of suffering was inflicted. Moreover, the use of animals in research to determine the safety of a new antibiotic for a human or animal disease or to protect the environment might be considered justified outcomes, whereas the use of animals to produce a safer cigarette, a less addictive drug, or a beautifying cosmetic might not. The animal welfarist, therefore, has not only to make a judgement as to whether there is a greater good, but how much greater that good is, and whether it is offset by or commensurate with the harms done to the animals. This is called a cost– or harm–benefit analysis and is the mandate of ethics committees for animal research.

Singer (1975, 1991) looks at what is meant by *harms* and *interests* in detail and argues that humans should give equal consideration to equal interests in animals. Thus, as a general rule, both humans and animals would wish to avoid suffering, but what we mean by harm and measuring it can be difficult. Furthermore, it has to be perceived in a similar way to be comparable in a cost–benefit analysis. For example, a horse and a baby would feel pain differently if given an identical slap. Not being able to eat roughage will cause more misery to a calf than to a cat. The innate senses of the animal species concerned, its natural abilities and its capacity to suffer in response to a stimulus or deprivation have all to be taken into account. This becomes extremely relevant when trying to predict the harm caused by an experiment and *critical anthropomorphism* has been proposed as a useful method in such an assessment (Morton *et al.*, 1990).

Another dilemma arises in animal research which has been termed a "tragic choice" or "necessary evil" (Smith and Boyd, 1991) and this reflects a commonly held view. If we experiment on animals we cause them to suffer (a wrong action), but if we do not, then we fail to relieve human and animal suffering (a wrong outcome; Cohen, 1986). The animal welfarist sees the force of that argument but may reserve judgement depending on the details of the research and a harm–benefit analysis, whereas an animal rightist could never agree with any animal research.

How do scientists address the question: What are the morally relevant differences between humans and animals that make it acceptable for research to be carried out on animals that we would not permit on humans? Justifying it by stating that we are human and not animals is termed a *speciesist* argument (Ryder, 1975). Proponents of such an argument would then be challenged to say what characteristics of being human provide the *moral* justification. Anatomical differences such as having a tail, or humans being stronger, would not seem morally relevant. But what about our intelligence, our ability to create, to make plans for the future, to be aware of ourselves in relation to others and over time, our ability to communicate and even to hold this debate? Here we are faced by another problem, that of marginal cases. Some animals may be more intelligent, more creative and more communicative than some humans. Compare the chimpanzee with a mentally handicapped child. Or, to pose a similar dilemma, would we be causing a greater harm by using a normal chimpanzee or a human in a permanent vegetative state (from whom it had been decided to withdraw life support and who had made an advance directive consenting to such a procedure) for an experiment such as a pharmacokinetic study, or an investigation into rejection using a transgenic animal organ graft using extracorporeal perfusion? There is the added benefit of more accurate information by experimenting on humans as it removes the need to extrapolate between species. We may logically be forced into accepting that it is wrong to experiment on either or, conversely, that it is right to experiment on both (Singer, 1975; Frey, 1987, 1988). As a result of such dilemmas it has been argued that there are no absolute *morally relevant* differences that can separate *all* humans from *all* animals. Rather there is a continuum of differences which vary more in degree than kind — a view reflected to varying extents in different religions.

Speciesism and different species of animals

Speciesism is an anthropocentric view of human beings in relation to other animals, but it also describes differing human attitudes to different species. When we talk about animals we usually only think of mammals, but on occasions it may be possible to use the lower species in research. The issue clearly is about sentience, rather than phylogenetic order, and we should take care always to use the least sentient species. In fact, our moral concern increases the closer in terms of looks and in zoological similarity animals are to ourselves, thus primates rate higher than rats. Clearly, the ability of animals to experience pain, discomfort and distress is important and it is generally agreed that all vertebrates have this ability (Smith and Boyd, 1991) and so are normally protected by laws in most western countries. Domesticated mammals are also of concern and dogs, cats and equidae are given special consideration in UK law. Conversely, animals that are not high in public esteem may be of less concern: in the USA there is no legal requirement to keep records of the number of rats and mice used in experiments.

The fact that some animals are seen as less valuable than others, or that some animals suffer less than others, is probably a matter of human perception rather than any physiological or psychological reality. Only for the higher primates (such as chimpanzees) is there good scientific evidence supporting the notion that they possess self-awareness and self-consciousness, which may make them able to suffer more than other mammals but even their suffering remains unlikely to be as great as that of normal humans. However, we should be careful not to assume that non-primate mammals do not suffer in this way, as clearly there is considerable anecdotal evidence they do.

Do animals make good models for humans or other animal species?

It should be remembered that experiments are carried out on humans as well as animals, though not in such an invasive manner. We do not intend to take their lives or cause them harm. Any new medicine tried in patients for the first time is an experiment, no matter how many animal tests have been carried out beforehand. Despite the remarkable similarity between animals and humans, animal tests may get it wrong through unpredictable side-effects which may cause medicines to be withdrawn from use (e.g. thalidomide; see the rather provocative book by Sharpe, 1988). Animal tests can only give an indication that the chemical has the desired action and is not obviously toxic, but when many more humans than experimental animals are exposed to the substance, as in a clinical trial, one may start to pick up side-effects not measurable in animals, such as headaches or transient nausea, or side effects which are so infrequent that they will not be detected in the small number of animals used in preliminary safety tests.

Second, animal tests are inherently limited (some would say flawed) by virtue of having to extrapolate from one species to another when there are many species differences in clearance rates and metabolic routes. However, many medicines researched and developed in animals have been shown to work extremely effectively in humans and other animals — consider the medicines used in veterinary practice.

Third, non-human mammals share many aspects of their physiology and biology with humans and that is why they make good animal models. But, the more the experimental animals are like humans and the closer they are genetically, the greater is our moral concern for their use in research. The Darwinian theory of evolution would support such a concept of a continuum in abilities rather than differences reflected by DNA homogeneity, thus adding provenance to our concerns based on natural attributes and mutual abilities to suffer.

Fourth, and more specifically, we cannot lightly deny that they will not suffer like us in experiments, particularly as they make such good models for emotional states of suffering like pain, depression and anxiety. It is a two-way street. Animals are similar to humans in their responses to and interactions between stress and disease. For example, stress (such as transport, poor handling, poor husbandry) can predispose to disease. New medicines developed as a result of such studies — anaesthetics, analgesics, antidepressives and some anxiolytics — work well in both humans and animal species other than the one in which the drug was developed.

Finally, our ability to create more accurate models through the advent of new genetic engineering techniques (e.g. transgenic animals bearing defective human genes) cannot fail but to heighten such concerns.

In summary, humans have a duty of care to animals used in research not to cause them avoidable suffering. There must also be sufficient benefit to outweigh and so justify any animal suffering but, as judged by whom?

Applied ethics and responsible science

Having decided that it is acceptable to use animals in research, that it is the only way to achieve the scientific objective, and that the scientific objective is worth achieving, one then has to decide how to carry out the work. One practical ethical framework for evaluating the conduct of experiments in animals is that put forward by Russell and Burch (1959): the Three Rs. Its principles are that one should *replace* animals whenever possible by using non-sentient means such as *in vitro* techniques and computer modelling; *refine* the experimental design so as to minimize any suffering; and *reduce* the number of animals to the minimum needed to achieve an acceptable statistical standard.

The use of the word "alternative" can be applied to all three Rs, but it is commonly accepted that the term applies when no animals are needed at all. A true replacement would be the use of computer modelling and permanent cell culture lines. Killing an animal for tissue would not be a replacement as an animal could still suffer as a result of its husbandry or its manner of death (it could experience fear on being removed from its cage to be killed humanely) and this would be classified as a refinement; the same would be true for procuring tissues under terminal anaesthesia. Applied ethics in animal research can be summed up in part by asking the question: is it possible to carry out this scientific procedure and cause less suffering to animals? If so, it would be unethical to cause that *avoidable suffering* and this means that *best practice* should always be the goal. Animal researchers should refine their care and use of experimental animals for moral, legal, economic and scientific reasons. Avoidable animal suffering leads to poor science, as poor animal well-being caused by poor animal health, poor care or husbandry or clumsy experimental technique will lead to animal suffering which may well modulate an animal's response to the scientific variable being investigated, thus adding to the biological variation. At best it may skew the research data minimally; at worst it may totally alter the scientific interpretation of that data. The results obtained are then not only misleading but in the long run may lead to more animals having to be used as the work may have to be repeated in order for it to be validated — a resource issue in its own right. Paying attention to animal well-being makes sense for precise, humane and economic science.

Refinement can be defined as:

> Those methods which avoid, alleviate or minimize the potential pain, distress or other adverse effects suffered by the animals involved, or which enhance animal well-being. (Morton, 1995a)

Twenty years or so ago, animal welfare was considered mainly in the light of health — absence of disease — and scientists would have concentrated on the effects of microorganisms on research (Pakes *et al.*, 1984). Whilst disease is still important, it is less so than before thanks to the high health standards maintained by the breeders and nowadays other areas are emerging as important to animals, particularly psychological well-being. I wish to illustrate some practical ways in which animals can suffer, how these adverse states can be recognized, and how 'As best practice' can be determined and put into effect.

Animal suffering

Animal suffering is a generic term used rather loosely to refer to an animal's adverse reactions when exposed to some environmental or internal physiological or psychological disturbance. It should be possible to use physiological and psychological indices to recognize when animals are suffering and this is a prerequisite to be able to assess, measure and even separate these emotional conditions. Several general references are worth reading (Duncan and Molony, 1986; A.V.M.A., 1987; Dawkins, 1990, 1993; Smith and Boyd, 1991; I.L.A.R./N.R.C., 1992; Townsend, 1993; Morton and Townsend, 1995; Morton, 1997).

Each adverse state and its physiology will now be briefly described. These adverse states are often not clearly separated from each other: for example, animals in chronic pain or which are disabled may also be anxious and frightened (Figure 5.1). Animals are able to suffer psychologically. This is seen as abnormal behaviour, e.g. some mammals in captivity show signs of boredom and frustration (mental distress). Such signs can often be reversed through enriching the environment (or rather, making it less barren!). Refinement includes enrichment (Reese, 1991) and if we are under a moral obligation to reduce all avoidable suffering to the minimum necessary through utilizing best practice, then we must pay attention to the husbandry of animals as well as experimental procedures.

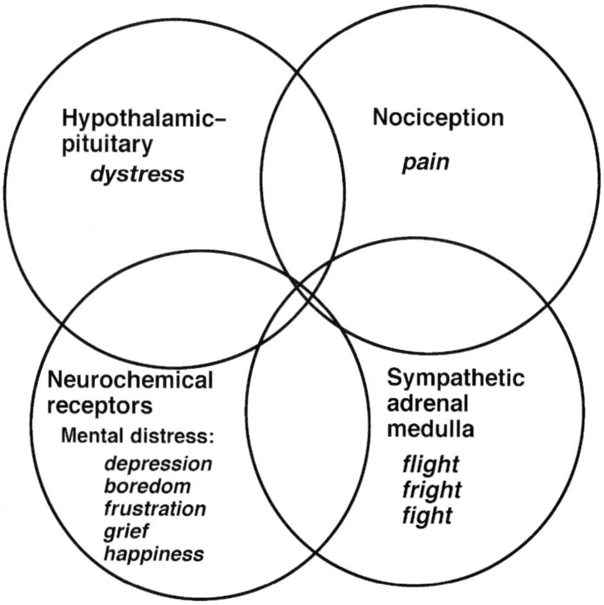

Figure 5.1 Possible adverse physiological and psychological states of vertebrates can occur independently as well as concurrently. It is likely that on many occasions they will occur at the same time and the circles should therefore overlap more closely.

Pain

Some people deny that animals can feel pain but this view is becoming discredited as research has shown on numerous occasions that animals and humans respond to painful stimuli very similarly both physiologically and psychologically.

In animals, pain is used to describe a physical, rather than a mental effect (as in humans) and involves specialized receptors (nociceptors) in the periphery of the body (e.g. skin) which respond to excessive pressure, temperature or chemical stimuli (Melzack and Wall, 1982). Nerve impulses pass from these nociceptors in afferent nerve pathways to the cerebral cortex where they are translated into the emotional feeling of pain. There are also descending nerve pathways which can modify the passage of impulses passing up the spinal cord and in this way pain tolerance and thresholds may be increased or decreased. These develop shortly after birth in some species and so, paradoxically, fetuses and neonates may experience more pain than adults (Fitzgerald, 1994).

When pain does occur in experimental research it is usually an unwanted side-effect and is a consequence — rather than an objective — of the scientific procedure. For example, pain associated with various infections is incidental to the evaluation of a novel antibiotic drug. Similarly, in the production of an animal model of a human disease (e.g. arthritis), the pain is inevitable but not intended. Pain can also be produced as a result of the safety testing of new chemicals and may cause a spectrum of physical signs according to the chemical and site concerned, e.g. skin irritation, colic, cancer. On some occasions it may be the scientific objective to study pain itself but in this area of research the amount of pain is carefully controlled and is usually kept to the minimum needed to obtain scientific data (International Association for the Study of Pain, 1979; Zimmermann, 1983).

Dystress

Dystress (coined from the Greek *dus*, having a connotation of bad — bad stress) is caused when an animal can no longer adapt to the stressors imposed upon it, such as excessive (high or low) temperatures or humidity, transport for long distances, restraint, inadequate or irregular food supply. The body is maintained within a narrow range of physiological parameters (homeostasis) and adapts to, or copes with, its new environment. Some environmental changes or stimuli are good (those stressors an animal can cope with: so-called eustress) and probably prevent other adverse states such as boredom and frustration. However, stressors that are severe or prolonged or repeated may eventually erode an animal's ability to cope and so cause dystress. Dystress can be increased when the animal is unable to predict or control the stressor (Wiepkemea and Koolhaus, 1993). Dystress will primarily activate the hypothalamic–pituitary–adrenal axis, ultimately leading to an increase in corticosteroid levels. Non-steroidal hormones may also be raised and these include prolactin, tumour necrosis factor and interleukins which in turn can affect other body systems and predispose animals to disease through reduced effectiveness of the immune system (Dantzer and Kelley, 1989). Dystress may lead to poor growth and to a reduction in breeding capacity (even short-term stressors such as noise and vibration are known to affect breeding). Novel and repetitive scientific procedures may stress animals but a certain level of stress is unavoidable in research. The important question is whether the experimenter can help an animal adapt to the stressors beforehand, in which case any stress is minimized and the experimental results will reflect more accurately the scientific variable under study.

Fear

Fear is probably the commonest adverse effect experienced by laboratory animals and physiologically involves the sympathetic nervous system and the adrenal medulla. It results in the release of catecholamines, which are the classic hormones released in the fight and flight (and fright) reactions preparing the body for defence or aggression. These hormones are responsible for diverting blood from gut to muscle as well as increasing those processes which can help an animal's immediate physiological needs: heart rate rises, the heart pumps more efficiently, blood glucose rises.

Lasting harm

Lasting harm is a term used to cover those adverse affects which may not be painful or cause dystress or indeed overt physical suffering. It refers to procedures which may damage the animal in a non-painful way, such as permanent paralysis of the nervous system, or a scientific procedure that may cause blindness, deafness or diabetes. Such adverse effects have the potential to cause fear, anxiety and chronic illness and may induce a state of mental distress.

Mental distress

Distress is used to describe adverse psychological (mental) states measurable by abnormal behavioural parameters — so-called stereotypic behaviours. Distress is often associated with the other adverse states described above but can occur in a more restricted sense when animals elaborate abnormal behaviours because they are unable to meet their mental (as opposed to physiological) needs. Such behaviours are usually repetitive, have seemingly no obvious function, and may be novel to that animal species or be an exaggerated form of a natural behaviour (in terms of the amount of effort or time spent carrying it out). Stereotypic behaviours vary in style and with species according to the environmental conditions an animal has previously experienced. Broadly, these behaviours reflect mental states such as boredom, frustration, torment, social isolation. Mental distress may involve serotonin, noradrenaline and dopamine receptors and drugs classified as reuptake inhibitors have been found to reverse such stereotypies in both humans and animals.

It may be important to animals for us to separate their behavioural *needs* from *desires* and freedom for them to carry out many behaviours. It seems intuitively obvious that some behaviour patterns, such as those that damage the ability of that animal to survive (*fitness* in the Darwinian sense) and which involve self-mutilation, are particularly unacceptable (chewing digits in monkeys, repeated gnawing leading to damage to the mouth and teeth—often called vices in the past, as if they were the fault of the animal!). Other behaviours seem less self-destructive in a physical sense but may be mentally destructive (weaving, inactivity, pacing). Such behaviours are seen in many mammals in zoos and laboratories and occur in nearly all species. Any behaviour not seen in the wild counterpart of a species should be treated with suspicion. It is not easy to determine the behavioural needs of animals but Dawkins and others have attempted to address these issues in two ways (Dawkins, 1990, 1992; Poole, 1992). Behaviours can be measured through choice experiments to determine what animals prefer by giving them a choice of two environments (e.g. to compare types of cages or bedding). These experiments have limitations in that an animal may like or dislike both but is forced to choose one; it may also not know what is good for it in the long or short term; and it may change its mind with time, experience and age. Another method is to measure how hard animals will work (e.g. lever-pressing or pecking) in order to gain access to a particular environment or enrichment. Such quantitative estimates enable one to rank environmental behaviours. Perhaps we will then be obliged to meet those that appear to be important for that species in so far as is practicable in the laboratory and within the scientific objective. In my view we should strive to meet any such needs in its normal husbandry, except for when the scientific protocol demands it.

Routine husbandry

The routine husbandry of animals and their environment, because of the stressors inevitably involved, may well affect the quality of the scientific data generated (Rose, 1994). It is important, therefore, that all such details should be reported when writing up the work (Ellery *et al.*, 1985; Morton, 1992). Animals should be given time (4–7 days) to acclimatize to their new conditions of husbandry (e.g. new diet, indigenous diseases, handlers, noises, odours) before being used in an experiment, particularly if they have been brought in from an outside supplier. Even transporting animals very short distances, say from another building on campus can be stressful (Toth and January, 1990; Tuli *et al.*, 1995). These stressors raise the levels of corticosteroids, prolactin and other hormones which may affect many areas of research. For example, acclimatizing animals to a metabolic cage for 21 days before a lethality test reduced the LD_{50} dose by 60 times compared with animals that had not been acclimatized (Damon *et al.*, 1986). There are also scientific consequences of keeping animals like rabbits in excessively confined conditions as their bones do not develop normally compared with those kept in more spacious accommodation. For example, in caged rabbits, the compact bone of the femur was half its normal thickness and other bones (vertebrae and ribs) were deformed in some way (Stauffacher, 1992). Such animals show osteopenia and osteoporosis and yet are being used in orthopaedic research to study normal bone healing!

Recognition and assessment of animal suffering

The types of adverse effects that animals may suffer in a laboratory from physiological and psychological (behavioural) viewpoints have been described and we now have to address the more difficult task of recognizing when animals are showing these effects and trying to quantify them. How can one tell when animals are in pain or are suffering in any way? A good animal technician may think it is obvious but how exactly does he or she know? What are we looking for? Clearly the ability of humans to recognize adverse effects in another animal depends partly on the species and, by and large, it is easier to recognize adverse effects in larger species, particularly if they are mammals. Perhaps it is because we are more familiar with these species (e.g. pet dogs and cats) than others. (Whilst this chapter concentrates mainly on mammals, it should not be forgotten that birds, fish, reptiles and amphibia are also considered to be able to suffer and consequently are also protected by legislation.) Whilst humans too are animals, we cannot simply assume that animals suffer like humans and it is important not to become too anthropomorphic because this will be misleading. A critical anthropomorphic approach, however, takes into account the animal's biology and biography (life experience), and provides a useful starting point of what adverse effects to look for (Morton *et al.*, 1990). It can sometimes help to imagine what it would be like to be that animal in that given experimental situation. For example, after a knee operation one might feel pain and therefore limp, and so in a similar (experimental) operation on an animal one might look for limping as a postoperative sign of pain. Such an anthropomorphic approach has limitations, and there are sometimes important biological differences between humans and animals which make it risky to extrapolate simplistically from one to the another. On the other hand, sometimes these differences can be very helpful, e.g. the position and action of the tail: compare sick with happy dogs and pigs! One must also be aware that many laboratory animals have reversed circadian rhythms compared with humans and that their main activity period is during the night. Therefore any interference with their normal activity may be picked up better at night than during our normal human working day. We are using this factor as a test of provoked behaviour to see if inactive animals are able to respond during the day by turning off the white light and observing them under red light. If the animal's nocturnal behaviour of increased activity has not started within 5–10 minutes then that animal is not normal.

Recognition of the normal is of fundamental importance and the scientist, animal technician and veterinarian should all know what is normal for a healthy individual animal of the particular species, strain, sex, age and background being used. Recognition of an animal's normal behaviour in the environment or circumstances in which it is being kept is as important as other physiological aspects, such as type and volume of faeces. Methods of husbandry can profoundly affect an animal's condition and these factors should be understood. If animals are poorly socialized or housed when young they may develop *abnormal* behaviour patterns which become accepted as normal. Whilst it is possible to learn what is normal from books, this must always be supplemented by direct observation of animals. Pre-experiment assessment and observation are helpful, and guidance from experienced animal care staff is invaluable. To develop an understanding of what is normal researchers should gain direct experience of normal animals by working with them in an animal unit for a few days, performing routine husbandry tasks such as feeding and watering, handling the animals and cleaning cages.

Difficulties in assessment with groups of animals and with small animals

As mentioned earlier, many, if not most, pet owners instinctively become accustomed to their pet animal's natural behaviours and foibles and so come to know their animals as individuals. This is not so in an experiment that could involve hundreds of laboratory rodents, but the fact that there may be many of them does not relieve us of our moral obligation to reduce suffering to a minimum in every individual experimental animal. In practice, however, it does seem difficult to observe behaviours and to recognize suffering in small animals like mice, and it is even more difficult to monitor individual animals when kept in groups. It is not easy to monitor an individual animal's dietary intake or faecal and urine output. On the other hand, it is possible to observe interactions between animals, especially when they stand out from others in the group by the very nature of their behaviour, posture or appearance.

An approach to the recognition and assessment of animal suffering using score sheets

One way to assess adverse states is to examine in detail the scientific procedures involved and to score those in some way. Wallace *et al.* (1990) tried to do this prospectively, but this approach assumes that all researchers will be equally competent on all occasions, and that all animals will react the same. An alternative approach is to look at it from the point of view of what the animal is telling us no matter what the procedure or species of animal, or who is carrying out the procedure. Many laboratories are approaching this difficult topic of the recognition and assessment of suffering using score sheets which provide a list of cardinal clinical signs encountered in that particular scientific procedure (Morton, 1990; 1995b; 1998; Morton and Townsend, 1995; Olfert, 1995). These are developed through experience and, by and large, are unique to the system of husbandry, to the specific experiment, as well as to the species, and even the breed or strain of animal being used. It is not possible to make a general score sheet for all experiments and for all species (or for all users!). One only has to consider the different potential adverse effects of a skin transplant compared with a kidney or heart transplant to appreciate the different clinical signs that might be seen (e.g. rejection of skin compared with a dependent kidney or an accessory heterotopic heart in the abdomen).

Practically, it is important to develop a disciplined approach and strategy to the recognition of adverse effects in animals (Morton and Griffiths, 1985). At the beginning of an assessment, the animal should be viewed from a distance and its natural undisturbed behaviour and appearance noted. Next, as the observer approaches the cage or pen, the animal will inevitably start to take notice and interact with the observer and that interaction can be used to determine whether it is responding normally (an animal may be inquisitive or show signs of fear). Finally, a detailed clinical examination can be carried out by restraining the animal in some way and observing its appearance carefully and then making clinical measurements of body weight, body condition and temperature in addition to its behaviour, as it may have become more aggressive or fearful, and may even vocalize.

Lists of clinical signs are developed by observing the first few animals undergoing a novel scientific procedure very closely. The list can then be modified with experience until a set of cardinal signs that most animals will show during that experiment and that are relevant to the assessment of suffering is determined. These key clinical signs are set out against time in a score sheet (Table 5.1). On the left-hand side are listed clinical and behavioural signs and along the top the days and time of the recorded observations. The method of scoring is that clinical signs can only be recorded as being present or absent. This is indicated by a plus or minus sign (or sometimes +/– if the observer is unsure). The convention is that negative signs indicate normality or within the normal range, and positive signs indicate compromised animal well-being. In this way, it is possible visually to scan a score sheet to gain an impression of an animal's well-being: the more plus signs, the more that animal has deviated from normality, with the inference that it is suffering more than it was earlier. Clinical treatments and other observations are also recorded. It is important to note that

Table 5.1 Animal score sheet (blank) for streptozotocin diabetes model

Rat No.		*Animal issue no:*					
Date of starving:		*Pre-starved weight:*					
Date							
Day							
Time							
From a distance							
Inactive							
Isolated							
Walking on tiptoe							
Hunched posture							
Starey coat							
Type of breathing*							
Grooming							
On handling							
Not inquisitive and alert							
Not eating							
Not drinking/average amount drunk (ml)							
Body weight (g)							
% Change from start							
Body temperature (°C)							
Pale or sunken eyes							
Dehydration							
Diarrhoea: 0 to 3 (+m or +b)†							
Distended abdomen/swollen							
Vocalization on gentle palpation							
Nothing abnormal detected (NAD)							
Given 5 ml saline s/c or p.o.							
Other signs noted: Loss of body condition							
Signature							

Special husbandry requirements
Animals should be kept on a grid cage with tray and cleaned twice daily, and mouse box for enrichment.
Two bottles should be provided for each cage and filled twice daily.
Deprivation of water overnight may be sufficient to cause death by dehydration.
Autoclaved diet must be provided.

Scoring details
* Breathing: R = rapid; S = shallow; L = laboured; N = normal.
†0 = normal; 1 = loose faeces on floor; 2 = pools of faeces on floor; 3 = running out on handling; +m = faeces contain mucus; +b = faeces contain blood.

Humane end points and actions
1. Any animals showing signs of coma within the first 24–48 hours will be killed.
2. Any animals weighing less than the starting weight after 7 days will be killed.
3. Any animal showing tiptoe or slow ponderous gait will be killed.

Inform scientist, named veterinary surgeon and animal technician in day-to-day care if any of 1–3 above is seen.

Scientific measures
Animals that have to be killed should have their kidneys placed in formal saline and the pots clearly labelled.

animals can be scored at any time and would certainly be scored more than once daily during critical periods when an animal's condition could predictably give rise to concern (e.g. in the immediate postoperative period; in a study on infection at the time of bacteraemia or septicaemia).

At the bottom of the sheet there are guidance notes for the animal technicians about what they should provide in terms of husbandry and care for animals on that scientific procedure. There are also guidelines on how to score qualitative clinical signs such as diarrhoea and respiration, as well as criteria by which to judge humane endpoints. If an animal has to be killed, there is guidance about what other actions should be taken, such as tissues to be retrieved and kept in formal saline; this helps ensure that the maximum information is always obtained from any animal in the study.

While these sheets take time to fill in, it is not difficult for an experienced person, such as an animal caretaker, to see if an animal is unwell so the time taken can be reduced by simply scoring that the animal is normal by ticking the NAD box (nothing abnormal detected). However, if an animal is not normal, it does take time to check it and to make judgements over what actions to be taken, but is that not the price to be paid for practising humane science?

In order to promote good care and good continuity of care we allocate an animal technician to be responsible for liaising with the scientists and other technical staff, and also to maintain and update the score sheets. The roles of the technician in charge are:

- to check that the appropriate licences are in order and marry up with what the scientist intends to do that day to the animal
- to check that the score sheet is appropriate before the experiment begins
- to know the purpose of the experiment and its scientific objectives, and to become familiar with the scientific procedures to be carried out on the animals and the clinical signs that may occur
- to ensure all personnel (technicians, scientists) know how to use score sheets and can recognize the clinical signs and interpret them clearly into humane endpoints
- to check that technicians not familiar with that experiment, say doing a weekend or holiday rota, are informed about animals
- to liaise with licensees over the experiment, e.g. timing, numbers of animals, equipment, endpoints
- to update the score sheets based on new signs or combinations of signs observed
- to report to the responsible persons any concerns over the animals or personnel involved

Table 5.2 shows a completed score sheet. At a glance it can be seen that there are more plus signs to the right than to the left. Several other points can be noted: first, along the top, as the animal became unwell, it was scored more frequently. During day 0 (the day of the operation) it scored abnormal in one or two predictable signs as it was recovering from the anaesthetic and the surgery (low body temperature and hunched) and so the NAD box was ticked. The next day (21 June) basic observations were made of the amount of food eaten, temperature and body weight, and again the NAD box was checked. However, towards the end of that day, the coat became starey (ruffled), the body temperature rose, and its breathing became more rapid. By the next morning, there was a significant body weight loss (12%) which increased during the day to 18%—a strong indication that the animal had not eaten or drunk much, if anything, or that it had diarrhoea. In fact there were so many abnormal clinical signs that it was decided to kill the animal on humane grounds before the end of the experiment. The sudden appearance of diarrhoea and the concomitant rapid weight loss and dehydration, laboured breathing, posture and lack of a red-light response all confirmed that the animal was becoming severely physiologically compromised and consequently—and an important criterion—it was not going to yield valid scientific results. Even more significantly, its temperature was now at 35.5°C—a very poor sign, and its extremities (feet and ears) were blue. In our experience, this animal would have died that night, if not sooner.

This scheme of scoring clinical signs for the recognition and assessment of adverse effects on animals during scientific procedures has been shown to have several advantages:

- Closer observation of animals is now carried out by all staff at critical times in the experiment as the sheets have indicated those times that are critical for the animal, and when the animals find their circumstances most aversive
- Subjective assessments of suffering by staff and scientists are avoided, thereby promoting more fruitful dialogue, as evidence-based opinion becomes possible based on the clinical signs. In a sense they empower the animal technicians and help them demonstrate to less experienced persons why an animal is "not right"
- Consistency of scoring is increased as the guidance is clear and the scoring options are limited
- Single signs or combination of signs can be used to indicate the overall severity of the procedure, as well as alleviative therapies or scientific procedures at set points in the experiment (e.g. blood sampling)
- The score sheets help to determine the effectiveness of any therapy intended to relieve adverse effects
- The score sheets help to determine which experimental models cause least pain, distress and dystress by comparing alternative animal models and thus helping to refine scientific procedures
- The score sheets help to train those inexperienced in the assessment of adverse effects

As mentioned before, the score sheets are constantly being updated with further experience; it is surprising how the process never seems to stop as new staff pick up new signs, or new signs develop as the experimental model is slightly modified, or staff start to perceive patterns of adverse effects that, when taken as a whole, indicate early death or early deterioration sufficient to warrant the animal being killed on scientific grounds alone. Such information has led to better animal care as well as providing useful scientific information such as the recognition of neurological deficits, times of epileptic fits or weight loss, as well as unexpected findings such as urinary retention in a model of renal failure. Furthermore, by picking up signs of poor animal well-being early, we can implement humane

Table 5.2 Animal score sheet (completed) for heterotopic kidney transplant

Rat No. HN1		*Issue no:* 234					
Date of operation: 20 JUNE at 11.00 h		*Preoperation weight:* 250 g					
Date	20	20	21	21	22	22	22
Day > surgery	0	0	1	1	2	2	2
Time	13.30	17.30	8.00	4.00	8.00	11.00	14.00
From a distance							
Inactive	–			–	–/+	+	+
Inactive? Try red light response ^					–	+	+
Isolated	–			–	+	+	+
Hunched posture	+			+/–	–/+	+	+
Starey coat	–			+	+	+	+
Rate of breathing	54			60	64	70	40
Type of breathing*					R	R	L
On handling							
Not inquisitive and alert	–			–	–	–	+
Eating/jelly mash? amount eaten†	–		50%	–	?	?	?
Not drinking	–			–	?	?	?
Bodyweight (g)	254		260	250	221	215	205
% Change from start	–		6%	0%	–12%	–14%	–18%
Body temperature (°C)	35	36,5	37	38	38	36.5	35.5
Crusty red eyes/nose	–			–	–	+	+
Excessive wetness on lower body	–			–	–	–	+
Sunken eyes ^	–			–	–	–	–
Dehydration ^	–			–	–	+	+
Coat/wet soiled	–			–	–	–	+
Pale eyes and ears	–			–	–	–	–
Blue extremities ^	–			–	–	–	+
Stitches OK?/date removed	–			–	–	–	–
Swelling of graft‡ ^	–			–	–	–	–/+
NAD		✓	✓	–	–	–	–
Dosing							
Other DIARRHOEA ^						+	+
Signature							

Special husbandry requirements
Animals should be put on a cage liner with tissue paper and a small piece of VetBed.
Scoring details
* Breathing: R = rapid; S = shallow; L = laboured; N = normal.
†Eaten/jelly mash–amount? Record as 0/25/50/75/100%.
‡Swelling score: 0 = normal; 4+ = rejection and large swelling.
Humane end-points and actions
1. Weight loss of 15% or more; inform the investigator, veterinarian and technician in charge.
2. Pre-moribund state (indicating a failing graft).
3. Any major clinical sign recurs after 24 hours (marked ^, less than 35 °C).

Scientific measures
Take 1 ml of blood and urine, if possible; place at 4°C.
Place transplanted kidney into 10 ml formal saline.

endpoints sooner rather than later, which avoids animals being inadvertently lost from an experiment. In the UK, where severity limits are imposed on each scientific procedure, the score sheet can be used to indicate when such limits have been reached, or are about to be breached, or may have to be reviewed, by a precise observation of clinical signs. The score sheet system provides a visual aid, opens up discussion between interested parties, and helps focus attention on to the animals' condition throughout the procedures.

Disagreement can still occur, however, but from a scientific viewpoint a skewing of data may arise due to culling animals breaching severity limits—survivors may be a separate subgroup for some reason. In any event, the question that the scientist must ask is whether an animal deviating from normality to that degree is really going to provide sound scientific data in terms of answering the specific scientific objective? If it is not, the animal should be killed.

An analysis of the score sheets can reveal patterns of recovery or deterioration and so gives a better picture of the overall effect of a procedure on the animals from start to finish. The sheet encourages all involved to observe the behaviour of animals, to recognize normal and abnormal behaviours, which will help in determining animals' responses to various procedures, and this will help to devise ways of refining experimental technique by highlighting the type and timing of any adverse effects. The scoring system has proved to be especially useful with new procedures, or when users are not always sure of what effects a procedure will have. In my experience the literature rarely records adverse effects on the animals, or how to avoid them or measure them, and I believe scientists have a moral obligation to do so (Morton, 1992). We now look more closely at ways of improving our perioperative care and in some experiments we have found that recovery is slower than it could be if we used different anaesthetics or analgesics, or intraoperative procedures such as maintaining body temperature or giving warm saline (Wadham, 1996). In the light of all this, we now have better systems of care, and we try to operate earlier in the day so that animals have maximum time under close observation when they can be given more support such as fluid therapy or special diets (e.g. jelly, fruit, vegetables). This has proven to save animals' lives as well as improving the speed of recovery, yielding scientific benefit, good science and good animal welfare.

Devising and validating humane endpoints

Where signs can be objectively measured (such as body weight, heart rate, blood glucose level and body temperature), they can be fitted into bands, e.g. respiration rate raised between 5 and 25%, or 25 and 50%, or greater than 50% from normal. Body weight loss can also be banded in this way, e.g. a loss of less than 10%, or between 10 and 20%, or greater than 20%, for an individual from the start of the experiment or compared with non-experimental cohorts or compared with controls. Subjective signs, such as eye closure and pain response, are recorded to obtain an idea of the degree of abnormality. For example, the eye may be normal, half-closed or fully closed and/or an eye may be discharging or ulcerated. The overall suffering for an animal or severity of a procedure can then be assessed from the range of abnormalities as well as their magnitude. Humane endpoints can then be assigned according to the parameters observed or measured.

Another way of assessing the overall state of the animal from these observations is to score each sign or band category, then add them up to indicate some action depending on the final figure. Thus, one could allocate 0 for normal body weight, 1 for a loss of 5–10%, 2 for 11–20% and 3 for greater than 20%. Similarly, for the eye: normal would score 0, half-closed 1, and fully closed 2. An eye which may be discharging or ulcerated might score 3. In this way the animal would be given an overall score which can be used to assess the overall degree of deviation from normal and place it in a pain severity band. The score could also dictate action to be taken, such as seek veterinary advice, give analgesia, and, as in the recording only system, this could be linked even more closely with a particular sign, for example, whimpering or squeaking when touching the abdomen, or ulceration of an eye.

Death as an endpoint

It is worth considering why animals die during experiments and why death is such a crude, inhumane and unscientific endpoint. Animals that die in coma are not suffering but what happened before that state is of concern. Ultimately death ensues because the heart stops or the brain stem is irreversibly destroyed so that respiratory and cardiac control centres fail, and is normally not a direct result of the infectious agent or whatever experimental variable is being studied. Animals may be unable to reach the water bottle or unable to eat, and this will lead to dehydration, increased viscosity of the blood and eventually to heart failure (which is what happens in the rabies potency test described below). Potentially, and probably on most occasions, death is preceded by considerable animal suffering, and so should only be used as a last resort and requires substantial justification (Tomasovic *et al.*, 1988; Kuijpers and Walvoort, 1991; Workman *et al.*, 1998) At the World Congress on Alternatives the following statement was agreed at the workshop (Mellor and Morton, 1997).

> In principle death should not be an endpoint in any experiment or test, but if death is proposed as an endpoint, it must be justified with increasingly strong arguments as the anticipated overall pre-death suffering increases.

There is also the possibility of losing valuable scientific data through the death of an animal as postmortem findings, particularly histological analysis and culture counts, will be severely compromised. Even in infection studies, alternatives such as bacterial colony counts or plaque-forming units before death at set time points after infection may be more valuable. Such quantitative data can be analysed on a variation of a dose–response curve and should achieve good scientific data without wasting animal lives. Little work seems to have been done in this area and there is scope for research in order to refine and determine humane endpoints.

I will now describe three examples to the implementation of humane endpoints in vaccine potency testing, in research into virulence, and in a trial of potential therapeutic agents using these score sheets.

Worked examples

The first concerns rabies vaccine potency testing. This work was carried out by Klaus Cussler in order to find humane endpoints to replace that of death (Cussler, 1997). Here the object of the challenge test is to test various dilutions of the batch of rabies vaccine against a standard challenge of live rabies virus. The clinical signs for this test are well-known, as are the times of their appearance, and the data were analysed during an ongoing challenge test. The signs were grouped chronologically in five stages as shown in Table 5.3. The data were then analysed from the viewpoint of what clinical signs could be used to predict death: from what stage could an animal recover, and at what stage did death always ensue after a particular group of clinical signs (Table 5.4)? From these preliminary data it can be observed that one mouse (number 6) recovered from stage 1, but no mouse that showed clinical signs of stage 2 or beyond ever recovered. It can also be seen that if mice had been killed at stages 2 or 3, then it would have been possible to reduce the animals' suffering by approximately 3 days. The animals should possibly have been examined more frequently as it can be seen that some stages were not recorded as they probably occurred between observation periods.

Table 5.3 Chronological appearance of clinical signs during rabies challenge tests

Clinical signs	*Stage*
Ruffled fur	1
Hunched back	
Slow movements	2
Circular movements	
Trembling, shaky movements	3
Convulsions	
Lameness	4
Paralysis	
Prostration	5
Death	

After Cussler *et al.* (1998).

The second example is that described by Soothill and co-workers in 1992. In this work they were investigating the virulence of *Pseudomonas aeroginosa, Staphylococcus aureus*, and *S. epidermidis* in mice, again using death as an endpoint. They found that the animals that became moribund or died always had a body temperature below 35°C, and those that survived always had a temperature above 35°C. It was possible then to use that temperature as a cut-off point. The mice also showed other clinical signs, such as reduced

Table 5.4 Actual appearance of clinical signs during a rabies vaccine potency challenge test

Day	*Mouse* 1	2	3	4	5	6	7	8	9	10
1	0	0	0	0	0	0	0	0	0	0
2	0	0	0	0	0	0	0	0	0	0
3	0	0	0	0	0	0	0	0	0	0
4	0	0	0	0	0	0	0	0	0	0
5	0	0	0	0	0	0	0	0	0	0
6	0	0	0	1	1	1	2	1	0	0
7	2	0	0	1	2	0	2	3	1	0
8	5	0	0	1	4	0	4	4	1	0
9	–	0	0	2	5	0	5	5	1	0
10	–	0	0	4	–	0	–	–	3	0
11	–	0	0	5	–	0	–	–	–	0
12	–	0	0	–	–	0	–	–	–	0
13	–	0	0	–	–	0	–	–	–	0
14	–	0	0	–	–	0	–	–	–	0
15	–	0	0	–	–	0	–	–	–	0

0 = animal normal, no clinical sign observed.
1–5 = recognized stages of clinical signs set out in Table 5.3.
– = the animal has died.
After Cussler *et al.* (1998).

mobility or hunched or starey coat, and some of these recovered; the appearance of these signs indicated which mice should start to have their temperatures taken every 2 hours. This approach is even more practical today with the introduction of temperature transponders along with individual identifiers, as they would be far preferable to manual temperature-taking with a thermistor, which takes time and makes the animals sore.

Finally, Townsend and Morton (1994) looked at ways of refining an experiment aimed at preventing toxic shock caused by bacterial endotoxins by the use of antibodies. The experiment was first altered by giving a dose of endotoxin that would kill the mice during the working day, instead of the traditional way of using a lower dose and seeing how many animals were alive after 4 days. Next the mice were observed carefully and scored on the basis of their appearance (hunched, starey coat and closed eyes), their behaviour (decreased activity) and clinical signs (diarrhoea, haematuria and body temperature drop). These data were treated numerically and a score of 1 was given to all the signs apart from body temperature, which was given 1 point for every degree centigrade drop below normal. The idea was that the higher the score, the more the animal had deviated from normality and that subtle partial cure effects could be recorded (Table 5.5 gives the results). It can be seen that in no case were any of the treatments successful, but there was a clear and statistically significant difference between those mice left unprotected and the control untreated group (as well as all other treatment groups). This way of handling data to try and quantify adverse effects may help in some experimental work; it should not necessarily be taken as indicating that an animal may be in more pain or distress, although this is likely to be the case (Morton and Griffiths, 1985).

Table 5.5 Mean scores for mice given endotoxin and various treatments

	Mean scores	*n*
Controls		
No endotoxin	1.53 ± 0.108*	8
50 μg endotoxin only	9.86 ± 0.57	8
Prophylaxis against endotoxin with:		
Non-immune serum	9.24 ± 0.36	4
Anti-endotoxin antibody	7.22 ± 0.27	4
Pre-incubation of endotoxin with:		
Normal serum	10.49 ± 1.20	5
Anti-endotoxin antibody	9.11 ± 0.59	5
Post-endotoxin treatment with:		
Normal serum	8.90 ± 0.64	6
Anti-endotoxin antibody	6.89 ± 0.79	6

* Significant differences at $P < 0.05$ when control untreated group was compared with all other groups.
n = number of animals.
After Townsend and Morton (1994).

Who monitors and when?

It is important that a researcher knows how to recognize adverse states and the various methods used to assess them and when to institute humane endpoints. Researchers should visit their animals on a regular basis at a frequency which is commensurate with the particular experimental conditions. For example, if animals have gut fistulae and are cared for by experienced stockpersons, they may need visiting only on a weekly basis. However, if the animal is on a virulence or protection study and expected to show adverse effects in a few hours or days, it may be necessary to attend hourly and for long periods. After surgery, care should be continuous until the animal has fully recovered from the acute effects of the procedure. After the initial postoperative period (say, 3–4 days depending on the operation and experience of those concerned), checking may be reduced to three times a day with one scoring session unless there is any cause for particular concern. This regime should be incorporated into any postoperative treatment, e.g. checking cannulae, dressings, administering antibiotics or analgesic drugs, taking samples.

The researcher is not alone in his or her responsibility to the animal, as this may be shared with others, such as the veterinarian, animal technicians and other co-workers. Animal technicians by virtue of their professional calling have a responsibility for the welfare of the animals in their care. Thus the user should take advantage of all the resources at his or her disposal to ensure that the animal is given the highest level of care and attention at all times.

Ways in which animal suffering can be avoided, reduced or alleviated

Some suffering may be inevitable in an experiment and no amount of skill or following best practice can avoid it being incurred except, of course, by not carrying out the experiment. This suffering can be viewed as necessary or unavoidable to achieve that scientific objective and is the cost to the animal that is taken into account when a harm–benefit analysis is carried out. However, as far as the scientist is concerned there are also a number of factors within his or her control which can contribute to the quality of life of the animal on experiment as well as improving the science (Table 5.6).

The first is careful planning of experiments to maximize the amount of valid and reliable information which can be obtained from any one animal, thus reducing the need for repetition of studies and the use of large numbers of animals. This would include allowing sufficient time to allow animals to settle into husbandry routines, their new accommodation and the experimental protocols. Second, using score sheets with clear endpoints, care protocols and standard operating procedures enables those concerned

Table 5.6 Refinement considerations

Consider how environmental aspects might affect the animals and the experiment
Husbandry, e.g. enrichment:
- Complexity of environment
- Interaction with humans and other animals (avoiding single caging or single penning of animals, stable groups, in sight of others)
- Novelty (e.g. diet, enrichment objects)

Consider how scientific procedures can be improved
- Controls (necessity or use historical data)
- Staging of experiments (critical experiments, relationship *in vitro* to *in vivo* experiments)
- Surgery and anaesthesia (asepsis, modern anaesthetics, analgesics)
- Euthanasia (humane, skilled persons)
- Statistics and experimental design (too many/too few? right design — obtain professional advice before starting work)
- Endpoints (humane and scientific; avoid painful endpoints where possible)
- Monitoring adverse effects (score sheets)
- Education, training and competence
- Question tradition and obtain data to support scientific and humane suppositions
- Database for animal models

with the care of animals to know exactly what is allowed and at what point action is taken to prevent further suffering. Third, scientific techniques and procedures should be refined to cause the least suffering whilst ensuring that the experimental variable being measured is not adversely affected (eg. will giving a sedative to bleed an animal affect the parameter being determined?). Most importantly, all those involved must have an open mind and be prepared to discuss how best to satisfy the scientific requirements whilst maintaining the animal and its welfare as a primary focus.

There are times when suffering, pain or distress cannot be prevented but must be relieved by the sensible use of analgesic drugs (opiates, non-steroidal anti-inflammatories, long-acting local anaesthetics). Sometimes it is not possible to avoid potentially confounding results through the administration of pain-relieving agents such as analgesics and anaesthetics due to the actions of these drugs in the body. In practice, however, the mechanism of action of such drugs is usually so well-known that appropriate allowances can be made. In any event, confirmation of any potential confounding effect of a drug could be investigated at an early stage and verified. Drugs such as antibiotics and analgesics should never be used to facilitate bad or sloppy practices, nor should pain be used as a means of keeping an animal immobile to compensate for poor surgical technique.

Ensuring competent researchers

Considerable animal suffering can be caused by persons not skilled in the scientific techniques they carry out. Incompetent administration of substances or removal of body fluids or surgical procedures carried out without due care and attention to asepsis can all cause more animal suffering than necessary to achieve the scientific objective. Therefore, best practice should be encouraged through training programmes together with an assessment of competence. This should be continually reviewed for all persons and all husbandry and scientific procedures, as advances are continually being made in technical methodology, e.g. physical restraint, anaesthesia, analgesia.

Ensuring good experimental design

The design of an experiment is crucial to its success but as well as obtaining good statistical advice much can be done strategically to reduce animal suffering (Morton, 1998), for example, thorough literature review and justification of the choice of model (Tables 5.7 and 5.8). The purpose of an experiment has also to be clearly defined as well as any application for which the results may be used. An important question is what is going to be done with the results. If there is a clear difference in consequent actions based on the difference in probability estimates, then providing that can be justified, it is acceptable. But if there is to be no difference in action between different statistical precisions, then the lowest acceptable precision that will lead to the use of the fewest animals should be used.

Another refinement of experimental design is when it has been decided that, even if one animal in a group fails a test, then a specific action will take place. For example, if a vaccine is going to be labelled as unacceptable on the basis of one animal in a group of six reacting adversely, and if the first animal reacts adversely, then there is no point in continuing with the experiment on the remaining five animals. It would have provided valid scientific information, of course, but the practical outcome has already been decided and so there is no point in continuing the study and causing pain and suffering to the remaining animals.

Protocols evaluating new treatments, such as the rodent protection test (RPT), help determine effectiveness and therapeutic levels of antibiotics for use in humans (also see Acred *et al.*, 1994, for some ideas on humane endpoints). In the RPT, the effectiveness of a novel compound to prevent or treat infection is investigated over a range of doses using standard microorganisms in an animal model (mouse or rat). There is little need for high precision in the animal model species; as opposed to the target species, and yet the approach of accepting a lower statistical probability ($P < 0.05$ versus $P < 0.01$) may mean the difference between 5 and 50 animals in a group. This is especially important in the area of infection where considerable suffering may

Table 5.7 Animal model data

1. Model of:

Purpose of the research (benefit)

2. Purpose that model can be used for (e.g. human disease, drug evaluation)
3. Advantages and disadvantages (scientifically and in its relevance and application to human or animal disease or treatment)
4. Clinical trials or medicine (drug) evaluation indicating relevance or predictability of the model to its intended purpose (see 2 above)

Animal details

5. Species of animal and experimental details (e.g. strain, inbred/outbred, sex, age, weight, health status, acclimatization period, other)
6. Husbandry special or critical requirements (e.g. caging or pen type, animal kept singly or in groups, diet, bedding, isolator or filtration boxes, breeding details, other)
7. Methodological details (e.g. equipment required, manual skills required, dosing and timings, timings of measurements, where advice can be obtained)
8. Refinement aspects (e.g. humane endpoints, score sheet giving cardinal signs, useful tests, and any other information)
9. Scoring of signs and severity grading (severity grading: give average and maximum expected)

Alternative models and staging strategies

10. Replacement (any alternative methods for all or part of the model available, e.g. pre-screening)
11. What information is necessary or would be useful before *in vivo* work begins
12. Are pilot studies necessary for dose sighting? Are the doses chosen the minimum necessary to obtain an indication of effectiveness before proceeding to more detailed trials, e.g. dose–response studies
13. Success rate of the model (give mean and range, and methods used to determine success)
14. Statistics (how is the data to be handled? What are the best means of analysis, data transformation, etc.)
15. If lethality is an endpoint please justify. If a LD_{50} dose is being obtained, justify the scientific necessity for a precise estimation in relation to the subsequent extrapolation to human or animal therapies
16. Ethical commentary
17. References
18. Keywords for literature searches

occur in animals which are not protected from the disease. Moreover, if a fudge factor, such as a further safety margin of 100× the toxic dose found in animals, is going to be used to calculate the hazard involved in any exposure for humans, then the futility of such precise experimentation is highlighted even further.

Pilot studies can be useful to give some idea of the variance, the likely adverse effects that may be encountered, and any practical local problems that are likely to be encountered during the main study. They help, therefore, to determine the appropriate number of animals that should be used in a group, as well as the avoidance and alleviation of any adverse effects. The use of pilot studies need not be a waste of animals as the results can be used for the main study, unless there are good reasons for not doing so. Furthermore, key pilot experiments before any controls or further experiments may provide crucial information on whether it is even worth progressing down certain experimental pathways.

Historical data is also very useful, particularly when considering both positive (e.g. a standard challenge of a known substance) and negative (e.g. giving vehicle alone) controls. The use of background data can influence the number of animals needed for a given study. In some instances it may not even be necessary to have a control group as the control "fact" is so well-known and accepted, e.g. total pancreatectomy leads to diabetes mellitus with a consequential rise in blood sugar and death in less than 10 days. However, it may be advisable to carry out pilot studies when using a new model in one's own animal facility, as differences in strain of animal, diet, staff, bedding, cleaning materials, husbandry and environment have all been shown to affect animals' responses in various ways (Claassen, 1994; Morton, 1995; Wadham, 1996). This was exemplified by a multicentre trial on the fixed dose procedure carried out by van den Heuvel and co-workers (1990; see also Schlede *et al.*, 1992). They found that while the LD_{50} of various chemicals was similarly ranked between laboratories, the actual LD_{50} dose varied considerably.

Order of work

Scientists should not only have a clear idea of how the work is to be carried out but also the order of work, and the interdependence between the proposed experiments, including any collaborative work with other research groups.

In vitro to *in vivo*

Research is sometimes aimed at investigating the interactions of substances, such as new drugs, in the whole body. If the *in vitro* work will yield information that shows that the effectiveness of the substance or viability of the cells or organisms has been compromised in some way, it may not be worth proceeding with the *in vivo* work. There must be confidence that the *in vitro* work can be extrapolated to the *in vivo* situation, but if that is true then the *in vitro* work should precede in *in vivo* experiments. For example, genetic modification of viruses may reduce their virulence in that they are no longer able to infect cells *in vitro* and, based on past experience, it is unlikely that they will infect cells *in vivo*. In cases of doubt, small pilot projects involving only two or three animals can be used (depending on the precision needed) to confirm such findings and so validate a

Table 5.8 Animal model data (example)

Purpose of the research (benefit)	*Reference*
1. Model of: Serum rhubarb disease in humans.	
2. *Purpose that model can be used for (e.g. human disease, drug evaluation).*	
This model has been used to study pathophysiological processes and the testing of replacement enzyme therapies, as well as other treatments.	1–10
It is the required European Pharmacopoeia standard test for uptake of enzyme from the lungs by inhalation.	11
3. *Adavantages and disadvantages (scientifically and in its relevance to human or animal disease or treatment).*	
As the model involves transgenesis of the defective gene in humans, it is therefore reproducible between animals, and appears to have a high level of fidelity.	12
4. *Clinical trials or medicine evaluation indicating relevance or predictability of the model to its intended purpose (2 above).*	
Validation studies in humans and animals show that there is a good degree of correlation on the following:	
Similar onset of disease, clinical appearance and signs	2, 6
Gross postmortem findings	1, 5, 10
Histological appearance	7, 9, 10
Expression of the defective protein	11
Responds to treatments known to be effective in humans in a similar timeframe and dosages (given pharmacokinetic data)	3, 4, 8
Animal details	
5. *Species of animal and experimental details (e.g. strain, inbred/outbred, sex, age, weight, health status, acclimatization period, other).*	
The transgenic model is available from Harlan Rivers on a female C57 Bl10/J inbred background.	
The animal develops the disease at 12 weeks of age (around 25 g).	5
Animal exposed to Sendai virus will show abnormal disease development (signs of bronchopneumonia as well as mucus build-up).	13
6. *Husbandry special or critical requirements (e.g. caging or pen type, animal kept singly or in groups, diet, bedding, isolator or filtration boxes, breeding details, other).*	
Because of the reduced immunocompetence it is recommended that these animals are kept in filter-topped boxes at a minimum, or in positive pressure isolators. They should be maintained in small groups of the same sex.	14
Colonies can be maintained by conventional breeding techniques, although the reproductive index will be low (around 0.3).	14
In situ hybridization probes are available to confirm transgenic breeding stock, which should be selected from homozygous males and heterozygous females.	15
7. *Methodological details (e.g. equipment and skills required, dose, timings, where advice can be obtained).*	
(This section is not relevant here but could be used when surgical models are produced, e.g. for renal hypertension).	
8. *Refinement aspects (e.g. humane endpoints, score sheet giving cardinal signs, useful tests, and any other information).*	
Animals will die if not humanely killed after showing signs of the disease for 4 weeks. The first indications are hypoxia and epileptic fits during the night. This can be tested during the day by exposing the animals to a 1 second flash of light. Weight loss begins at this time and animals will show some 15% loss per week compared with healthy non-transgenic cohorts of similar age and sex. Animals should be culled when body weight loss reaches 20% or fits are prolonged, i.e. for more than 1 minute, or are particularly severe, involving the animal damaging itself.	16–20

Table 5.8 *Continued*

	Reference
9. *Scoring of signs and severity grading (severity grading: give average and maximum expected).*	
(Again, this is more relevant for surgical and similar models, but in this case body weight loss and the number and duration of fits could be used as indicators).	
A model list of signs or a score sheet could be attached.	
This model has been scored as being of substantial severity in the Netherlands, moderate in the UK, Switzerland and France, and has not been classified elsewhere.	20, 21
Alternative models and staging strategies	
10. *Replacement (any alternative methods for all or part of the model available, e.g. pre-screening).*	
At the present time there are no validated alternative models but *in vitro* culture of lung alveolar macrophage cells taken from these animals can be useful experimental adjuncts and may be used for the development of new therapies and presentation and packaging of exogenous enzyme.	22
11. *What information is necessary or would be useful before* in vivo *work begins?*	
It could be that some *in vitro* toxicity work or sensitivity studies could help sight doses.	
12. *Are pilot studies necessary for dose sighting? Are the doses chosen the minimum necessary to obtain an indication of effectiveness before proceeding to more detailed trials e.g. dose–response studies?*	
If a substance is ineffective at a near-toxic dose, is it worth proceeding with the study?	
13. *Success rate of the model (give mean and range, and methods used to determine success).*	
The model has high fidelity: 100% of transgenic animals expressing the gene will go on to show the disease but only 10% of transgenic animals may express the gene.	12
14. *Statistics (how is the data to be handled? What are the best means of analysis and data transformation?).*	
The data should be recorded as colony-forming units and transformed to $\log_{10}$. A two tailed t-test should then be used to analyse for significance.	
15. *If lethality is an endpoint, please justify. If a LD_{50} dose is being obtained, justify the scientific necessity for a precise estimation in relation to the subsequent extrapolation to human or animal therapies.*	
Sometimes it appears that LD_{50} tests are done to satisfy a mathematical rather than a biological rationale.	
16. *Ethical commentary.*	
The occurrence of fitting in animals combined with the weight loss may cause severe suffering in animals. Animals bred surplus to requirements should be killed at the earliest opportunity and certainly before any signs of the disease are shown.	
Any breeding stock showing signs should be immediately killed.	
17. *References.*	1–22
(obvious)	
18. *Keywords for literature searches.*	
Serum rhubarb disease; transgenic; mouse; somatic cell therapy.	

predicted negative, especially when there is less likelihood of causing animals pain and suffering. When a positive harmful effect needs to be confirmed *in vivo* then only one animal may be needed as *any toxicity* will be likely to provide sufficient evidence, e.g. standard infective doses of microorganisms to check for viability. Furthermore, in cases where an adverse effect is predicted and needs to be confirmed, it may be possible to prevent the animal feeling any pain by carrying out the work under general or local anaesthesia.

Mild to high severity

Several examples can be given where the harm done to an animal can first be shown to be effective or ineffective at a low level before going to higher levels of animal pain and suffering. Thus, if an antibiotic does not protect at low challenge doses, it won't at higher doses. This stepwise approach can be a valuable approach to the refinement of many drugs.

Reporting experiments

Full and adequate reporting of scientific procedures is an ethical mandate as it prevents others making the same mistakes and helps avoid duplication of research (Morton, 1992). Furthermore, scientific publications in the UK (and probably elsewhere) are being increasingly read and scrutinized by non-scientists (after all, much research is supported by public funds). It may be appropriate, therefore, to insert a sentence or two in the introduction explaining why the work was carried out in terms of its potential benefit and not just the narrow scientific issue that is being studied. A separate subsection on Animals in the Materials and Methods section giving details of the animal model, signs of adverse effects, and their alleviation or avoidance, and any humane endpoints will all help to reduce animal suffering, as others can learn from earlier mistakes! Table 5.6 gives some idea of the scope for detailed reporting.

Conclusions

The ethics of animal experimentation rests at two levels. The first is whether it should ever be carried out and if so, on what occasions. The second is then to ensure that any suffering is minimized and that the number of animals used is statistically sound. In terms of refinement, the avoidance or relief of pain, fear, dystress, mental distress and lasting harm, in addition to boredom, frustration and fear, are moral imperatives for all those persons involved in research. Any adverse effect experienced by animals should be reduced to the minimum consistent with attaining a justifiable scientific benefit. The first step in achieving this end is to be able to recognize when animals are suffering and the circumstances in which this may happen. One can then avoid those conditions, or if they are necessarily inherent, at least try to minimize their impact through the use of alleviative treatments.

Score sheets provide an easy and reproducible way of recognizing and assessing the health and well-being of an animal. They have enabled us to develop improved husbandry systems as well as early humane endpoints to be instituted. Importantly, score sheets allow for reliable and reproducible scoring by different members of staff as the clinical signs are so specific and allow little room for individual interpretation. This gives some confidence to the scientists that animals are not going to be killed prematurely, and to all staff and scientists that animals are being treated humanely. Furthermore, it has to be appreciated that during the research stage we have to let some animals go further than we would normally accept, in order to validate potentially humane endpoints based on clinical signs. This is why research into refinement is so important as it can directly contribute to the relief of animal suffering in the short term, unlike replacement alternatives.

Acknowledgements

I am grateful to Diane Fleury, Jean Kelly, Sue Moseley, Shirley Peach and Kim Shimell of the Biomedical Services Unit at the University of Birmingham for their help, suggestions and efforts in developing the score sheets.

References

Acred, P., Hennessey, T. D., MacArthur-Clarke, J. A. *et al.* (1994). Guidelines for the welfare of animals in rodent protection tests. A report from the Rodent Protection Test Working Party. *Lab. Anim.*, **28**, 1–8.

A.V.M.A. (1987). Colloquium on recognition and alleviation of animal pain and distress. *J.A.V.M.A.*, **191**, 1184–1294.

Bentham, J. (1780). *Introduction to the Principles of Morals and Legislations* 1789. Reprinted by Anchor Books, Garden City, New York 1973.

Claassen, V. (1994). *Neglected Factors in Pharmacology and Neuroscience Research. Biopharmaceutics, Animal Characteristics, Maintenance, Testing Conditions.* p. 486. Elsevier, Amsterdam.

Cohen, C. (1986). The case for the use of experimental animals in biomedical research. *N. Engl. J. Med.*, **315**, 865–870; **316**, 551–553.

Cussler, K., Morton, D. B., Hendriksen, C. F. M. (1998) Klinische Endpunkte als Ersatz fur die Berstimmung der Letalitatsrate bei Tollwutinfektionsversuchen zur Impfstoffprufung. *Altex (Alternativen Tierexp.* (Suppl 98), 40–42.

Damon, E. G., Eidson, A. F., Hobbs, C. H., Hahn, F. F. (1986). Effect of acclimation to caging on nephrotoxic response of rats to uranium. *Lab. Anim. Sci.*, **36**, 24–27.

Dantzer, R., Kelley, K. W. (1989). Stress and immunity: an integrated view of relationships between the brain and the immune system. *Life Sci.*, **44**, 1995–2008.

Dawkins, M. S. (1990) From an animal's point of view: motivation, fitness and animal welfare. *Behav. Brain Sci.*, **13**, 1–61.

Dawkins, M. S. (1992). *Animal Suffering: The Science of Animal Welfare*, 2nd edn. Chapman & Hall, London.

Dawkins, M. S. (1993). *Through Our Eyes Only. The Search for Animal Consciousness*. W. H. Freeman Spektrum, Oxford.

Duncan, I. J. H., Molony, V. (1986). *Assessing Pain in Farm Animals*. Commission of the European Communities, Luxembourg.

Ellery, A. W. *et al.* (1985). Chairman of a Working Committee for the biological characterization of laboratory animals/GV-SOLAS. Guidelines for specification of animals and husbandry methods when reporting the results of experiments. *Lab. Anim.*, **19**, 106–108.

Fitzgerald M. (1994). Neurobiology of foetal and neonatal pain. In *Textbook of Pain*, 3rd edn (eds Wall, P., Melzack, R.) pp. 153–163. Churchill Livingstone, London.

Frey, R. G. (1987). Autonomy and the value of animal life. *Monist* **70**, 50–63.

Frey, R. G. (1988). *Moral Standing, The Value of Lives and Speciesism. Between the Species*, **4**, 191–201. c/o Schweizer Center, San Francisco Bay Institute, PO Box 254, Berkeley, USA, CA 94707.

I.L.A.R./N.R.C. (Institute of Laboratory Animal Resources and National Research Council) (1992). *Committee on Pain and Distress in Laboratory Animals: Recognition and Alleviation of Pain and Distress in Laboratory Animals*. National Academy Press, Washington, USA.

International Association for the Study of Pain (1979). Guidelines on painful experiments. Report of the International Association for the Study of Pain Subcommittee on Taxonomy. *Pain*, **6**, 249–252.

Kuijpers, M. H. M., Walvoort, H. C. (1991). Discomfort and distress in rodents during chronic studies. In *Animals in Biomedical Research* (eds Hendriksen, C. F. M., Koeter, H. W. B. M.), pp. 247–263. Elsevier, Amsterdam.

Mellor, D. J., Morton, D. B. (1997). Humane endpoints in research and testing. Synopsis of the workshop. In *Animal Alternatives, Welfare and Ethics* (eds van Zutphen, L.F.M., Balls, M.) pp. 297–299. Elsevier Science, Amsterdam, The Netherlands.

Melzack, R., Wall, P. (1982). *The Challenge of Pain*. Penguin Books, Harmondsworth, UK.

Morton, D. B. (1990). Adverse effects in animals and their relevance to refining scientific procedures. *A.T.L.A.*, **18**, 29–39.

Morton, D. B. (1992). A fair press for animals. *New Sci.*, **1816**, 28–30.

Morton, D. B. (1995a). Practical ideas for refinement in animal experiment. In: *Proceedings of Animals in Science Conference Perspectives on their Use, Care and Welfare* (ed. Johnston N. E.), pp. 82–87. Monash University, Australia.

Morton, D. B. (1995b). The post-operative care of small experimental animals and the assessment of pain by score sheets. In: *Proceedings of Animals in Science Conference Perspectives on their Use, Care and Welfare* (ed. Johnston N. E.), pp. 82–87. Monash University, Australia.

Morton, D. B. (1997). Ethical and refinement aspects of animal experimentation. In *Veterinary Vaccinology* (eds Pastoret, P.-P., Blancou, J., Vannier, P., Verscheuren, C.), pp. 763–785. Elsevier Science, Amsterdam.

Morton, D. B. (1998). The use of score sheets in the implementation of humane endpoints. Paper presented at 1997 ANZCCART Meeting in Auckland, NZ, on ethical approaches to animal-based science (in press).

Morton, D. B. (1999). The importance of non-statistical design in refining animal experimentation. ANZCCART Facts Sheet. ANZCCART News 11, No. 2 June 1998. Insert p. 12. ANZCCART, PO Box 19, Glen Osmond, SA5064, Australia.

Morton, D. B., Griffiths, P. H. M. (1985). Guidelines on the recognition of pain, distress and discomfort in experimental animals and an hypothesis for assessment. *Vet. Rec.* **116**, 431.

Morton, D. B., Townsend, P. (1995). Dealing with adverse effects and suffering during animal research. In: *Laboratory Animals—An Introduction for Experimenters* (ed. Tuffery, A. A.), pp. 215–231. John Wiley, Chichester, UK.

Morton, D. B., Burghardt, G., Smith, J. A. (1990). Critical anthropomorphism, animal suffering and the ecological context. *Hasting's Center Rep. Anim. Sci. Ethics.*, **20**, 13–19.

Olfert, E. D. (1995). Defining an acceptable endpoint in invasive experiments. *Anim. Welfare Inf. Center Newsletter*, **6**, 3–7.

Pakes, S. P., Yue-Shoung, L., Meunier, P. C. (1984). Factors that complicate animal research. In *Laboratory Animal Medicine* (eds Fox, J. G., Cohen, B. J., Loew, F. M), pp. 649–665. Academic Press, Orlando, FL.

Poole, T. B. (1992). The nature and evolution of behavioural needs in mammals. *Anim. Welfare*, **1**, 203–220.

Reese, E. P. (1991). The role of husbandry in promoting the welfare of animals. In *Animals in Biomedical Research* (eds Hendriksen, C. F. M., Köeter, H. B. W. M.), pp. 155–192. Elsevier, Amsterdam.

Regan, T. (1983). *The Case for Animal Rights*. University of California Press, Berkeley, California.

Rose, M. (1994). Environmental factors likely to impact on an animal's well-being—an overview. In *Improving the Well-being of Animals in the Research Environment* (eds Baker, R. M., Jenkin, G., Mellor, D. J.), pp. 99–116. Available from A.N.Z.C.C.A.R.T., PO Box 19, Glen Osmond, South Australia 5064, Australia.

Rupke, N. A. (1987). *Vivisection in Historical Perspective*. Croom Helm, London.

Russell, W. M. S., Burch, K. L. (1959). *The Principles of Humane Experimental Technique*. UFAW, London.

Ryder, R. D. (1975). *Victims of Science—The Use of Animals in Research* (2nd edn). National Anti-vivisection Society, London.

Schlede, E., Mischke, U., Roll, R., Kayser, D. (1992). A national validation study of the acute-toxic-class method—an alternative to the LD_{50} test. *Archi. Toxicol*, **66**, 455–470.

Sharpe, R. (1988) *The Cruel Deception—The Use of Animals in Medical Research*, p. 288. Thorsons, Wellingborough, Northants, UK.

Singer, P. (1975). *Animal Liberation*. Jonathan Cape, Thorsons, Wellingborough, Northamptonshire.

Singer, P. (1991). *Animal Liberation*, 2nd edn. HarperCollins, Glasgow.

Smith, J. A., Boyd, K. (1991). (eds) *Lives in the Balance*. Oxford University Press, Oxford.

Soothill, J. S., Morton, D. B., Ahmad, A. (1992). The HID_{50} (hypothermia-inducing dose 50): an alternative to the LD_{50} for measurement of bacterial virulence. *Int. J. Exp. Pathol.*, **73**, 95–98.

Stauffacher, M. (1992). Group housing and enrichment cages for breeding and fattening and laboratory rabbits. *Anim. Welfare*, **1**, 105–125.

Tomasovic, S. P., Coghlan, L. G., Gray, K. N., Mastromarino, A. J., Travis, E. L. (1988). IACUC evaluation of experiments requiring death as an end point: a cancer center's recommendations. *Lab. Anim.*, **17**, 31–34.

Toth, L. A., January, B. (1990). Physiological stabilisation of rabbits after shipping. *Lab. Anim. Sci.*, **40**, 384–387.

Townsend, P. (1993). Control of pain and distress in small laboratory animals. *Anim. Technol.*, **44**, 215–226.

Townsend, P., Morton, D. B. (1994). Practical assessment of

adverse effects and its use in determining humane end-points. In *Proceedings of 1993 5th FELASA Symposium: Welfare & Science* (ed Bunyan, J.) pp. 19–23. Royal Society of Medicine Press, London.

Tuli, J., Smith, J. A., Morton, D. B. (1995). Stress measurements in mice after transportation. *Lab. Anim.,* **29**, 132–138.

van den Heuvel, M. J., Clark, D. G., Fielder, R. J. *et al.* (1990). The international validation of a fixed dose procedure as an alternative to the classical LD_{50} test. *Food Chem. Toxicol.*, **28**: 469–482.

Wadham, J. J. B. (1996). Recognition and reduction of adverse effects in research on rodents. *PhD thesis.* University of Birmingham.

Wallace, J., Sanford, J., Smith, M. W., Spencer, K. V. (1990). The assessment and control of the severity of scientific procedures on laboratory animals. Report of the Laboratory Animal Science Association Working Party (assessment and control of severity). *Lab. Anim.*, **24**, 97–130.

Wiepkemea, P. R., Koolhaus, J. M. (1993). Stress and animal welfare. *Anim. Welfare,* **2**, 195–218.

Workman, P., Twentyman, P., Balkwill, F. *et al.* (1998). U.K.C.C.C.R. (United Kingdom Co-ordinating Committee on Cancer Research) Guidelines for the welfare of animals in experimental neoplasia (revised July 1997, 2nd edn). *Cancer*, **77**, 1–10.

Zimmermann, M. (guest editorial) (1983). Ethical guidelines for investigations of experimental pain in conscious animals. *Pain*, **16**, 109–110.

Chapter 6

The Impact of General Laboratory Animal Health on Experimental Models in Antimicrobial Chemotherapy

A. K. Hansen

Introduction

Animal health is important for all types of animal experimentation. The elimination of infectious agents which act as undefined experimental factors has attracted much attention over the last decades. There is no difference between the use of animal models for antimicrobial research and other kinds of animal research. However, it should be obvious that if animals are to be used as reliable tools for experimental infection, spontaneous infections should not be allowed to interfere. Therefore, microbiologists should feel a special obligation to define the health of their laboratory animals. In this chapter some examples will be given of how poor health status may interfere with animal models for antimicrobial research, and precautions to avoid such interference will be described.

The impact of animal health on experiments

Concerns for the health of laboratory animals have mostly focused on spontaneous infections, although disease in laboratory animals may also be caused by genetic or environmental determinants. Examples of infectious agents which are of importance for antimicrobial animal models are listed in Tables 6.1–6.3. Some of these may cause disease in laboratory animals, and such disease might interfere with research. However, if animals are clinically ill, they are seldom used for experiments, and therefore research interference is more to be feared from those complications which are not clinically observable. Some microorganisms only have the ability to influence the animal temporarily, while others act through a long period of the animal's life, maybe lifelong.

Table 6.1 Common virus infections occurring spontaneously in laboratory animals

Genera	*Strain (species affected)*
Adenoviruses	Various rodent-specific strains, infectious canine hepatitis
Herpesviruses	Cytomegaloviruses (many), Aujeszky (pig), Rhinotracheitis (cat)
Parvoviruses	Minute virus of mice, mouse orphan parvovirus, Kilham rat virus, Toolan's H1 virus (rat), rat orphan parvovirus, canine parvovirus, panleukopenia (cat)
Arenaviruses	Lymphocytic choriomeningitus (mouse, hamster)
Caliciviruses	Rabbit haemorrhagic disease
Coronaviruses	Mouse hepatitis virus, rat coronavirus, sialodacryoadenitis (rat), guinea-pig coronavirus, rabbit coronavirus, haemagglutinating encephalomyelitis (pig), transmissible gastroenteritis (pig), feline infectious peritonitis
Paramyxoviruses	Sendaivirus (mouse, rat, hamster, guinea-pig), pneumoniavirus (mouse, rat, hamster, guinea-pig), simian virus 5-like (hamster, guinea-pig), parainfluenzavirus 3 (guinea-pig), distemper (dog), canine parainfluenzavirus
Orthomyxoviruses	H1N1, H3N2 (pig)
Poxviruses	Ectromelia (mouse), myxomavirus (rabbit), rabbit poxvirus, feline poxvirus
Picornaviruses	Theiler's murine encephalomyelitis virus (mouse, rat), guinea-pig poliovirus, stillbirth mummification embryonic death infection (pig)
Reoviruses	Type 3 (all mammals), rotaviruses (mouse, rat, rabbit, pig)
Retroviruses	Feline leukaemia, feline immunodeficiency virus
Togaviruses	Lactic dehydrogenase virus (mouse)

The impact on animal models in antimicrobial chemotherapy differs between different viruses. Clinically apparent viral disease is seldom seen in laboratory animals, while all viruses may have an impact on the immune system and may contaminate biological products sampled from animals. The strains listed are examples. Several viruses which may infect the species included are not given here as they are less common. A comprehensive list may be found in Hansen *et al.* (1994). The table does not include primates, as different primate species harbour a wide range of viruses. A comprehensive list referring to primates may be found in *Working Committee for SPF and Gnotobiotic Laboratory Animals* (1980).

Handbook of Animal Models of Infection
ISBN 0-12-775390-7

Table 6.2 Common bacterial infections occurring spontaneously in laboratory animals

	Bacteria (species affected)
Gram-positive cocci	*Staphylococcus aureus* (many), *S. hyicus* (pig) Streptococci β-haemolytic type A/B/D/G (many), *Streptococcus zooepidemicus* (guinea-pig, pig), *S. pneumoniae* (rodents, rabbit), *S. suis* (pig)
Gram-positive rods	*Corynebacterium kutscheri* (mouse, rat) *Erysipelothrix rhusiopathiae* (pig) *Eubacterium suis* (pig) *Clostridium piliforme* (rodents, rabbit) *Clostridium difficile* (hamster, guinea-pig) *Clostridium perfringens* (pig)
Enterobacteriaceae	*Citrobacter freundii* Type 4280 (mouse) *Klebsiella pneumoniae* (many) Salmonellae, subgenus I (all mammals) *Yersinia* spp. *Yersinia pseudotuberculosis* (guinea-pig), *Y. enterocolitica* (pig, dog, cat)
Pasteurellaceae	*Actinobacillus pleuropneumoniae* (pig) *Pasteurella pneumotropica* *Haemophilus parasuis* (pig)
Spiral bacteria	*Treponema cuniculi* (rabbit) *Campylobacter coli/jejuni* (many) *Helicobacter* spp. *H. hepaticus/bilis* (mouse), *H. suis* (pig), *H. felis* (cat)
Other Gram-negative bacteria	*Bordetella bronchiseptica* (mouse, guinea-pig, pig, dog, cat) Cilia-associated respiratory (CAR) bacillus (rat, rabbit)
Mycoplasma	*Mycoplasma pulmonis* (mouse, rat), *M. hyopneumoniae* (pig)

The bacteria listed are examples and under certain circumstances may all cause disease. A few bacteria, such as streptococci and salmonellae, may have an impact on the immune system. Most spontaneous bacterial infections possess the potential for competing with spontaneous infections and become activated by immunosuppression. *Mycoplasma* are common contaminants of biological products. Also bacteria from the normal flora, such as *Escherichia coli* and *Pseudomonas aeruginosa*, may act as opportunistic pathogens. The table does not include primates, as different primate species harbour a wide range of bacteria. A comprehensive list for primates may be found in *Working Committee for SPF and Gnotobiotic Laboratory Animals* (1980) and a more detailed list for the other species may be found in Hansen *et al.* (1994).

This division is mainly connected with the ability of the agent to persist in the organism. However, that it is not always so can be illustrated by the very simple example given by those infections which introduce resistance against reinfection. It cannot be overemphasized that even good experimental designs cannot eliminate all kinds of microbial interference because infections may inhibit the induction of a certain animal model, may make it difficult to interpret the final results, may show a dose-related response or, last, but not least, increase variations within the experiment.

The impact of spontaneous infections on the animal model

Pathological changes, clinical disease and mortality

Clinically apparent disease due to infections with specific pathogens is rare in laboratory animals, as the most pathogenic agents, such as *Ectromelia* virus in mice, have been eliminated from most colonies. However, the presence of various less pathogenic microorganisms may cause changes in the organs, resulting in difficulties in the interpretation of the pathological diagnosis included in the evaluation of many microbiological models—a phenomenon often referred to as background noise. Furthermore, the pathology of experimental infections may be changed by spontaneous infections.

Example. Rodents used as models for acute pneumonia may be naturally infected with agents altering the pathology of other infectious agents. For example, rats infected with either *Mycoplasma pulmonis* alone, Sendai virus alone or both together show three different pictures of respiratory pathology (Schoeb *et al.*, 1985).

Immunomodulation

The immune system may be modulated by spontaneous infections in the absence of clinical disease. This effect may be either *suppressive* or *activating* or both at the same time, but on different parts of the immune system. As a general rule of thumb all viruses should be regarded as immunosuppressive. One of the reasons for this is the viraemic phase in the pathogenesis of many virus infections: during this phase cells of the immune system may be infected.

Table 6.3 Common parasite infestations occurring spontaneously in laboratory animals

Genera	*Parasite species*	*Animal species commonly infected*
Pinworms	*Syphacia obvelata/muris*	Rodents
	Passalurus ambiguus	Rabbit
	Ascaris suum	Pig
	Toxocara canis/cati	Dog, cat
	Toxascaris leonina	Dog
Cestodes	*Dipylidium caninum*	Dog, cat
Flagellates	*Giardia muris*	Rodents
	Spironucleus muris	Rodents
	Tritrichomonas spp.	Rodents
Microspora	*Encephalitozoon cuniculi*	Rodents, rabbit
Coccidia	*Eimeria* spp.	Rabbit, pig
	Isospora spp.	Pig
	Toxoplasma gondii	Cat
Hair follicle mites	*Demodex* spp.	Dog
Body mange mites	*Notoedres cuniculi (cati)*	Rabbit, cat
	Sarcoptes scabiei	Pig
Ear mange mites	*Otodectes cynotis*	Cat, dog
	Notoedres cati	Cat
Fur mites	*Chirodiscoides caviae*	Guinea-pig
	Cheyletiella parasitovorax	Dog, cat
Ticks	*Ctenocephalides* spp.	Dog, cat

The parasites listed under each species are examples. Several other infestations — although less common — may occur. Except for coccidia and the mange mites, most parasites only cause mild disease or no disease at all, but they have an impact on the immune system and may affect the absorption of compounds from the gastrointestinal tract. Other parasitic infestations than those listed in this table may occur. A comprehensive list may be found in Hansen *et al.* (1994) or *Working Committee for SPF and Gnotobiotic Laboratory Animals* (1980).

Examples. Infection with lactic dehydrogenase virus in mice, which clinically is totally inapparent, seems to influence the function of the macrophages (Stevenson *et al.*, 1980), e.g. an impaired antigen presentation has been described (Isakov *et al.*, 1982). This may lead to an increase in the severity of the symptoms shown by experimental infections (Bonventre *et al.*, 1980). Also non-viral microorganisms are known to influence the immune system, e.g. *Mycoplasma pulmonis* infection in mice and rats may change the symptoms observed after experimental infection (Howard *et al.*, 1978; Cassell *et al.*, 1986). Another mycoplasma, *M. arthritidis*, has been reported to increase susceptibility to experimental pyelonephritis (Thomsen and Rosendal, 1974).

Physiological modulation

Some microorganisms have a specific effect on enzymatic, haematological and other parameters which are monitored in the animal during an experiment. Organic function disturbances may change the outcome of the experiment without the knowledge of the scientist. For example, the altered function of the liver caused by some spontaneous hepatic infections may influence the pharmacokinetics of antimicrobial drugs.

Example. Acute infection with *Clostridium piliforme* in mice prolongs the half-life of trimethoprim (Friis and Ladefoged, 1979).

Competition between microorganisms within the animal

Spontaneous infections in an experimental infection model may compete with the experimental infection, which in worst cases may fail.

Example. Helicobacter pylori mouse models have been difficult to develop. Lately, it has been reported that the "cleaner" the mouse, the more successful the colonization rate of *H. pylori* in mice (Fox and Lee, 1997). This is equivalent to the fact that gnotobiotic or microbiologically defined (see below) pigs may be experimentally infected with *H. pylori*, while conventional pigs may not (Krakowka *et al.*, 1987). Such differences between different microbiological categories of animals may be explained by natural infection with related agents, and today species-specific *Helicobacter* spp. have been isolated from both mice (Fox *et al.*, 1994, 1995) and pigs (Queiroz *et al.*, 1990). Infections with species other than *Helicobacter* may play a role as well.

Contamination of biological products

Microorganisms present in the animal may contaminate samples and tissue specimens, such as cells and sera. This may complicate the *in vitro* maintenance of cell lines, and may interfere with experiments performed on cell cultures or isolated organs. Further, the reintroduction of such products into animal laboratories will impose a risk to the animals kept in that laboratory. Nicklas *et al.* (1993) found that 3% of a high number of cell lines and monoclonal antibodies were contaminated with murine viruses, the most important ones being lactic dehydrogenase virus, reovirus type 3 and minute virus of mice.

Example. An unidentified virus culture received at an American virus centre from a Romanian institute was inoculated on mice. Before the virus was identified as the poxvirus, mouse ectromelia, the virus had spread to the sarcoma cells used for the induction of ascites. Mice inoculated with the sarcoma cells now died earlier than previously. Afterwards, 28% of the institute's stored ascitic fluids were found to be contaminated with ectromelia, and so were 15 virus strains, which the institute had deposited at the American Tissue Culture Collection. The virus centre had to stop all experimental work for half a year, be totally depopulated of mice, disinfected and restocked with caesarian-derived mice (Shope, 1986).

The impact of the environment, genetics and the experiment on the health of the animals

Susceptibility to the development of spontaneous infectious diseases is under the control of genetics, sex, sexual cycle, age, and other characteristics of the host. It is obvious that immune-deficient rodents such as the nude mouse, the severe combined immunodeficiency (SCID) mouse, and the nude rat are more susceptible, e.g. in nude mice abscesses caused by bacterial species such as *Staphylococcus aureus*, *Pasteurella pneumotropica*, *Morganella morganii*, *Citrobacter freundii* and *Streptococci* (Rygaard, 1973; Custer *et al.*, 1973; Fortmeyer, 1981) are common. Variation in susceptibility to the development of infectious disease between inbred strains of rats and mice is often connected with the histocompatibility type of the animal (Brownstein, 1983; Hansen *et al.*, 1990). Infectious disease symptoms are most common in young animals (Fujiwara *et al.*, 1973; Onodera and Fujiwara, 1973; Lai *et al.*, 1976; Zurcher *et al.*, 1977), but if for some reason disease is developed at a later age, there is a tendency for the disease to be worse than in younger animals (Jersey *et al.*, 1973). Differences between sexes in susceptibility to the development of infectious disease may be observed, e.g. colitis and rectal prolapse caused by *C. freundii* in mice are more common in males (Fortmeyer, 1981).

Animals should be transported in specifically designed containers being sufficiently supplied with food and water. Road transport in purpose-equipped vehicles by trained staff directly from vendor to user is preferable. A British set of guidelines may be followed (LABA and LASA, 1993). Any kind of transportation will stress the animals, and therefore a period of acclimatization in the experimental facilities is a must. One of the most common errors in transportation is the use of vehicles without proper ventilation, cooling or heating, which may be the cause of various grades of respiratory infection, often with opportunistic bacteria, such as *Staphylococcus* spp. and *Pseudomonas* spp., involved. Primates are often transported long-distance by air, during which journey they are handled by untrained staff. This may result in wounds and scratches, which if necessary should be treated immediately upon arrival. In all cases of such improper transportation any trauma should be reported to the transporting agent as well as to the vendor in order to improve future transportations.

Principles for housing have been dealt with in another chapter. Improper housing may change different aspects of an animal model. For example, mycoplasmosis is more severe in rats exposed to ammonia than in unexposed rats (Pinson *et al.*, 1986) — a situation which may be caused by infrequent cage-changing and bad ventilation.

Subclinical respiratory disease caused by *Mycoplasma*, respiratory viruses and some bacteria may be responsible for increased mortality during anaesthesia.

Post-surgical infections are more common in large animals than in rodents. Infections are mostly caused by non-specific members of the normal flora, such as *Staphylococcus aureus, Bacillus fragilis, Pseudomonas aeruginosa* and *Escherichia coli*. Rats seem to be more resistant than mice and hamsters (Donelly and Stark, 1985).

Immunosuppression of laboratory animals, a common tool in infection research, may activate latent infections. Some examples from mice are given in Table 6.4.

Long-term treatment of laboratory animals with antibiotics may change the normal flora of the animals from Gram-positive to Gram-negative dominated flora. Some of the propagated bacteria, e.g. *Klebsiella pneumoniae*, may be opportunistic pathogens, and may, as with humans, be multiresistant to a wide range of antibiotics (Hansen, 1995). Application of certain antibiotics, e.g. all β-lactams (Young *et al.*, 1987), in hamsters and guinea-pigs may often lead to fatal *Clostridium difficile* enterotoxaemia. The effect may also be observed with other antibiotics, e.g. tetracyclines and macrolides. In rabbits, the enterotoxaemia is mostly observed after oral dosing (Olfert, 1981), and in general, rabbits should not be given antibiotics *per os*, if disturbance in the intestinal flora is to be avoided.

Zoonoses caused by laboratory animals

Zoonoses are infections which may spread from one animal species to another. In this connection a zoonosis will be regarded as an infection with the potential of spreading from animals to humans. Selected zoonoses are listed in Table 6.5.

Zoonoses from rodents, rabbits, dogs and cats used for experimental purposes are rare. Several agents with a potential of infecting both humans and animals exist, but nowadays all rodents for research can and should be purchased from vendors who efficiently keep their animals free of zoonoses. The most important zoonosis from rodents is lymphocytic choriomeningitis virus, an arenavirus, which

Table 6.4 Examples of latent infections in mice which may be activated by immunosuppression

Protection level needed to avoid infection	*Agent*	*Reference*
Barrier	*Citrobacter rodentium*	Juhr (1988)
	Clostridium piliforme	Fujiwara *et al.* (1964)
	Corynebacterium kutscheri	Takagaki *et al.* (1967)
	Streptobacillus moniliformis	Juhr (1988)
	Cytomegaloviruses	Sekizawa and Openshaw (1984)
	Mouse hepatitis virus	Dupuy *et al.* (1975)
	Sendai virus	Anderson *et al.* (1980)
Isolators	*Enterobacter cloacae*	Matsumoto (1980)
	Klebsiella pneumoniae	Matsumoto (1982)
	Pseudomonas aeruginosa	Taffs (1974)
	Staphylococcus spp.	Detmer *et al.* (1990)

Those infections which can be kept absent by barrier protection will normally be absent in mice from commercial breeders, as all of these keep their colonies behind barriers. However, if problems are to be expected from those infections which can only be avoided in isolators, one will have to specify the need for gnotobiotic animals (see Table 6.6).

may occasionally be found as a spontaneous infection in hamsters, and on rare occasions also in mice and gerbils. The portals of entry are probably the mucous membranes and broken skin. Most reported cases of infection in humans derive from hamsters (Baum *et al.*, 1966). Human symptoms are mostly influenza-like, but meningitis, abortions and, in rare cases, fatalities may occur (Shek, 1994). The microspore, *Encephalitozoon cuniculi*, is widespread in laboratory rabbits and guinea-pigs. Therefore, it should be noted that it has recently emerged as an opportunistic parasite in human immunodeficiency virus (HIV)-infected humans (Desplazes *et al.*, 1996). Dermatophycoses, i.e. skin infections with *Trichophyton* and *Microsporum* spp., may occur in a wide range of laboratory animals. Cats, in which the infection is often asymptomatic, impose the highest risk for humans. The coccidium *Toxoplasma gondii* may also occur in a wide range of species, but its natural reservoir is in the cat, and care should be taken only to purchase cats free of this infection. In pigs the major risk seems to be development of erysipeloid after contact with pigs infected with *Erysipelothrix rhusiopathiae*, a Gram-positive rod found in most agricultural pig herds (Takahashi *et al.*, 1987).

Non-human primates share a range of infections with humans. Although most research primates of the western world today are purpose-bred, they are in general not produced under hygienic regimes as strict as those which are applied to rodents. Therefore, primates should be regarded as a risk for human health. The most serious problem is herpesvirus type B, which in macaques produces symptoms equivalent to herpes simplex in humans, while in humans it may cause fatal encephalitis (Weigler, 1992). The virus is present in several commercial breeding colonies. Humans bitten by seropositive monkeys should as a routine be treated with acyclovir. Hepatitis viruses of various types, especially type A, are of greatest concern, when working with chimpanzees (Friedmann *et al.*, 1971), although it should be noted that antibodies may be found in other primate species as well. It is, however, not clear whether these are due to viruses capable of infecting humans or just equivalent monkey strains without such a potential. Fatal filoviruses, i.e. Ebola and Marburg, are not present in commercial colonies and can simply be avoided by not purchasing wild-caught monkeys.

When dealing with primates the opposite transmission of infections, i.e. from humans to monkeys, must be regarded as a serious risk for ruining projects. For example, infection with measles virus can easily eradicate a colony of the common marmoset, *Callithrix jacchus* (Working Committee for SPF and Gnotobiotic Laboratory Animals, 1980). Also herpes simplex may produce devastating disease in marmosets (Hendrickson, 1972). Therefore, people with symptoms of these or other kinds of viral disease should not be allowed to handle monkeys. For the same reasons, staff from primate facilites are commonly screened for tuberculosis in those parts of the world where this disease is prevalent in the human population.

Prevention of infections in laboratory animals

Production of animals free of research-interfering infections

To avoid disease, microbial interference and zoonoses laboratory animals are produced according to a three-step principle: *rederivation, barrier protection* and *health monitoring*.

Rederivation means the production of animals to initiate breeding colonies by either caesarian section (Foster, 1959) or embryo transfer (Bur, 1995). Such animals are *germ-free* at delivery, except in rare cases for transplacental infections. For both procedures foster mothers are used: these are

Table 6.5 Important zoonoses in laboratory animals

Disease or disease agent	*High-risk animal species*	*Transmission*	*Reference*
Primate zoonoses			
Filoviruses			
Marburg	African green monkeys	Contact with contaminated excretions	Held *et al.* (1968)
Ebola (African strains)	Macaques	Contact with contaminated excretions	Pattyn (1978)
Hepatitis viruses, types A and E	Mostly chimpanzees (type A)	Enterically	Gust and Feinstone (1988); Ticehirst *et al.* (1992)
Herpesvirus type B	Macaques	Invasive or mucosal contact	Hendrickson (1972)
Monkeypox	Various	Probably by contact	Ivker (1997)
Mycobacterium tuberculosis	All old-world monkeys	Aerogenously	Goss (1970)
Salmonella spp.	All species	Enterically	Williamson *et al.* (1963)
Shigella spp.	All species	Enterically	Williamson *et al.* (1963)
Zoonoses of other animal species			
Lymphocytic choriomeningitis virus	Hamster, mice, gerbils	Invasive or mucosal contact	Baum *et al.* (1966)
Hantaviruses	Rats	Invasive or mucosal contact	LeDuc (1987)
Erysipelothrix rhusiopathiae	Pigs	Skin contact	Takahashi *et al.* (1987)
Yersinia spp.	Guinea-pigs, pigs	Ingestion of contaminated faeces	Aleksic and Bockemuhl (1990)
Leptospira	Rats	Invasive or mucosal contact	Kiktenko (1985)
Pasteurella multocida	Cats	Bites	Griego *et al.* (1990)
Salmonella	All species	Enterically	Morris (1996)
Campylobacter	All species	Enterically	Morris (1996)
Rat bite fever			
Streptobacillus moniliformis	Rats, mice	Invasive or mucosal contact	Wullenweber (1995)
Spirillum minus	Rats, mice	Invasive or mucosal contact	Bhatt and Mirza (1992)
Ringworm			
Trichophyton	Guinea-pigs, rabbits	Skin contact	Pier *et al.* (1994)
Microsporum spp.	Cats	Skin contact	Pier *et al.* (1994)
Encephalitozoon cuniculi	Guinea-pigs, rabbits	Enterically	Desplazes *et al.* (1996)
Toxoplasma gondii	Cats	Ingestion of spores	Dubey (1996)
Hymeonolepis spp.	Rats, mice	Ingestion of eggs/insects	Schantz (1996)
Visceral larva migrans (*Toxocara* spp.)	Dogs, cats	Ingestion of eggs	Fenoy *et al.* (1997)

either germ-free or *gnotobiotic*, i.e. they have only a well-defined microflora. The foster mothers and the rederived offspring are housed in isolators (Figure 6.1). Approximately 8 weeks after birth or section, the foster mothers and some of the offspring are sampled for a microbiological screening.

Animals kept in isolators can be kept totally germ-free or gnotobiotic. Larger-scale production in isolators is, however, expensive, and therefore most laboratory animals are bred in a *barrier unit*, i.e. a unit where materials are decontaminated before introduction and the staff showers on the way in (Figure 6.1). Barrier-bred animals do not have a fully known flora, but they can be kept free of some specified agents. Before being moved from isolators into the barrier unit, the animals are often given a starter flora consisting of, for example, lactobacilli and some anaerobic rods. Furthermore, they catch microorganisms of human origin from the caretakers. All this becomes "the normal flora" (Hansen, 1992). Members of this flora can also interfere with some kinds of research, e.g. in immunosuppressed animals (Table 6.4).

To secure the absence of specific microorganisms in breeding and preferably also experimental colonies of laboratory animals, a number of animals from the colony are

Protection level	*Designation of animal units*	*Designation of animals*
Ventilation No filters *Staff* Unrestricted entrance No quarantine *Materials* No decontamination *Protection obtained* Microbiological status is insecure	 Conventional	Conventional
Ventilation Filters in inlet and outlet *Staff* Entrance through a three-room shower Quarantine after contact with other animals *Materials* Autoclave Chemical disinfection lock *Protection obtained* Animals can be kept free of certain agents, but typically harbour microorganisms of human origin	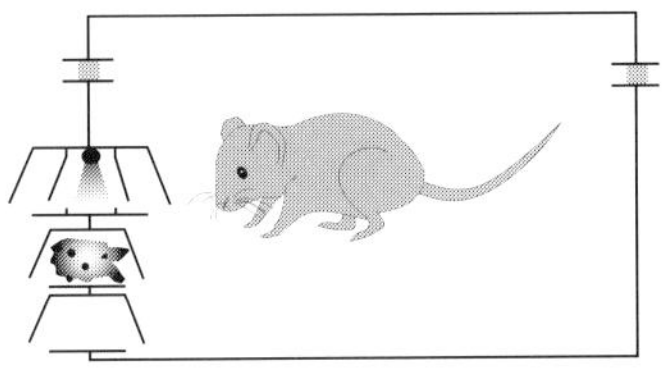 Barrier unit	*Preferred term* Microbiologically defined *Other terms* Specific pathogen-free (SPF) Virus Antibody-Free (VAF) Caesarian-Originated Barrier-Sustained (COBS)
Ventilation Filters in inlet and oulet *Staff* Animals are only handled from the outside through gloves *Materials* Disinfected, e.g. by irradiation and passed through a chemical disinfection lock *Protection obtained* Animals can be kept gnotobiotic (axenic), i.e. totally germ-free or with a fully defined microflora	 Isolator	*Preferred terms* Germ-free Gnotobiotic Axenic

Figure 6.1 Different hygienic levels for housing laboratory animals. Transition forms between the different levels exist, e.g. animal units in which materials are decontaminated but the staff do not shower on their way in. The so-called ventilated rack, in which each animal cage can be single-ventilated, is someway between a barrier unit and an isolator.

Table 6.6 An example of a health-monitoring report from a breeding unit for laboratory mice

	FELASA-approved health monitoring report
Name and address of the breeder:	Dept. Exp. Med., University of Copenhagen, Panum Institute
Date of issue: 5 May 1997	Unit no: 10.2 Latest test date: April 14 1997 Rederivation: 1987
Species: Mice	Strains: Pan:NMRI, DBA1/J/Pan

	Historical results	*Latest test results*	*Laboratory*	*Method*
Viral infections				
Minute virus of mice	Negative	0/8	Panum	ELISA
Mouse hepatitis virus	Negative	0/8	Panum	ELISA
Pneumonia virus of mice	Negative	0/8	Panum	ELISA
Reovirus type 3	Negative	0/8	Panum	ELISA
Sendai virus	Negative	0/8	Panum	ELISA
Theilers encephalomyelitis virus	Negative	0/8	Panum	ELISA
Ectromelia virus	Negative	NT	Panum	ELISA
Hantaviruses	Negative	NT	Panum	IFA
Lymphocytic choriomeningitis virus	Negative	0/8	Panum	ELISA
Lactic dehydrogenase virus	Negative	NT	Panum	Enzymatic
Bacterial and fungal infections				
Bordetella bronchiseptica	Negative	0/10	Panum	Culture
Citrobacter freundii (4280)	Negative	0/10	Panum	Culture
Clostridium piliforme	Negative	0/8	Panum	ELISA
Corynebacterium kutscheri	Negative	0/10	Panum	Culture
Leptospira spp.	Negative	NT	Panum	Mic. agg.
Mycoplasma spp.	Negative	0/8	Panum	ELISA
Pasteurella spp.				
Pasteurella pneumotropica	Positive	3/10	Panum	Culture
Other *Pasteurella* spp.	Negative	0/10	Panum	Culture
Salmonella	Negative	0/10	Panum	Culture
Streptobacillus moniliformis	Negative	0/10	Panum	Culture
β-Haemolytic streptococci				
Group G	Positive	0/10	Panum	Culture
Other	Negative	0/10	Panum	Culture
Streptococcus pneumoniae	Negative	0/10	Panum	Culture
Other species associated with lesions: None				
Parasitological infections				
Arthropods	Negative	0/10	Panum	Inspection
Helminths	Negative	0/10	Panum	Flotation
Eimeria spp.	Negative	0/10	Panum	Flotation
Giardia spp.	Negative	0/10	Panum	Microscopy
Spironucleus spp.	Negative	0/10	Panum	Microscopy
Other flagellates	Negative	0/10	Panum	Microscopy
Klossiella spp.	NT	NT		
Encephalitozoon cuniculi	NT	NT		
Toxoplasma gondii	NT	NT		

Pathological lesions observed
Stock: Pan:NMRI Lesions: None

Abbreviations for laboratories

Panum	Dept. Exp. Med., Panum Institute, Univ. Copenhagen, Blegdamsvej 3, DK-2200 Copenhagen N		
Positive	Positive results previously observed	0/10	No positives out of 10 samples
Negative	Positive results never observed	NT	Not tested

The report form has been standardized according to guidelines issued by the Federation of European Laboratory Animal Science Associations (FELASA; Kraft *et al.*, 1994).

currently sampled and subjected to a range of bacteriological, parasitological and serological investigations. This procedure is referred to as *health monitoring*. All commercial vendors issue reports on the health monitoring performed in their colonies (Table 6.6), and animals should never be purchased without first having studied such a report. Typical intervals between screening of colonies are 6–12 weeks for rodents and 6 months for larger animals such as rabbits, pigs, dogs and cats. Results of health monitoring are historical, i.e. animals may become infected in the period between two samplings, and therefore infected animals may have been used for research, before the infection has been discovered. Although the number of animals to sample for health monitoring can be judged from statistical principles (Hansen, 1993), such statistical principles are not commonly applied in health monitoring, and therefore infections listed as "not found" in health-monitoring reports may be present but not found due to the use of too few animals. The typical sample size ranges from 6 to 10 animals. If this sample size is used to screen for *Clostridium piliforme* in rat colonies by the means of serology it would be statistically valid, while it would not be if pathology is used due to the lower sensitivity of the histopathological methods (Hansen *et al.*, 1994). To minimize such concerns about the microbial status of laboratory animal guidelines relating to a number of species have been published in Europe (Hem *et al.*, 1994; Kraft *et al.*, 1994). These guidelines list what to test for, how often to do it and how many animals to sample. In general, it is a sound principle to ask commercial vendors for health reports issued according to these guidelines, but it should be noted that the guidelines do not ensure statistical validity (Hansen, 1996).

Animals bred behind a barrier and being currently health-monitored are sold under different terms, among which *microbiologically defined* seems to be the most precise (Figure 6.1). Rodents, rabbits, pigs, cats and dogs can be purchased as microbiologically defined at several commercial vendors all over the world.

Also, animal units housing animals in long-term studies should run a health-monitoring system. For larger animals, such as dogs, cats and rabbits, this may be done by sampling directly from the animals. When using rodents in long-term studies the experimental animals will normally be supplemented by some equivalent animals—so-called *sentinels*, i.e. animals kept in the facilities only for health-monitoring purposes. The sentinels should be left in the unit for at least 30 days before examination. It is normal practice to add some bedding from the other animals in the unit to the fresh bedding of the sentinels (Hansen and Skovgaard-Jensen, 1995).

Quarantine housing

All introductions of new animals into an animal unit run the risk of introducing infections. Therefore, animals may be housed in a quarantine facility before introduction. In general, all animals not coming from suppliers with a reliable health-monitoring system should as a minimum be housed in a quarantine facility for at least 2 weeks and thereafter screened for those agents from which they must be free. For quarantine housing a *ventilated rack* may be used. This is a cage system where each cage can be single-ventilated. In this system different deliveries may be isolated from one another; this is not possible in a common quarantine facility. Alternatively, animals may be quarantine-housed in isolators. All primates, no matter what their origin, should be quarantined for at least 30 days before entering a running primate facility. During this period the animals should be inspected for signs of clinical disease and screened for unwanted infectious agents, especially those with a zoonotic potential, such as herpesvirus type B, *Salmonella* spp. and *Shigella* spp. During quarantine experiments should not be performed on the animals.

Screening biological materials for the absence of infections

To avoid biological materials of animal origin—sera, cells, tissue preparations—carrying infectious agents when used in an animal facility, such materials should either have their origin in an animal department with a reliable health-monitoring system or be screened directly for the presence of infectious agents. This has traditionally been done by the mouse antibody production (MAP) test (Collins and Parker, 1972): Mice are injected with the material intraperitoneally and nasally, and kept in isolators for 4 weeks, whereafter serum is sampled and screened for the infectious agents by serology. In recent years polymerase chain reaction has been used as a supplement to the MAP test (Yagami *et al.*, 1995). If biological materials of uncertain microbial status are to be used, inoculation and maintenance of the inoculated animals should only be performed in facilities efficiently separating these animals from all other animals, e.g. in negative-pressure isolators.

References

Aleksic, S., Bockemuhl, J. (1990). Mikrobiologie unde Epidemiologie von *Yersinia* Infektionen [Microbiology and epidemiology of *Yersinia* infections]. *Immun. Infekt.*, **18**, 178–185.

Anderson, M. J., Pattison, J. R., Cureton, R. J., Argent, S., Heath, R. B. (1980). The role of host responses in the recovery of mice from Sendai virus infection. *J. Gen. Virol.*, **46**, 373– 379.

Baum, S. G., Lewis, A. M., Rowe, W. P., Huebner, R. J. (1966). Epidemic non-meningitic lymphocytic-choriomeningitis-virus infection: an outbreak in a population of laboratory personel. *N. Engl. J. Med.*, **274**, 934–936.

Bhatt, K. M., Mirza, N. B. (1992). Rat bite fever: a case report of a Kenyan. *East Afr. Med. J.*, **69**, 542–543

Bonventre, P. F., Bubel, H. C., Michael, J. G., Nickol, A. D. (1980). Impaired resistance to bacterial infection after tumor implant is traced to lactic dehydrogenase virus. *Infect. Immun.*, **30**, 316–319.

Brownstein, D. G. (1983). Genetics of natural resistance to Sendai virus infection in mice. *Infect. Immun.*, **41**, 308–312.

Bur, M. (1995). Embryo technology in laboratory animals. *Scand. J. Lab. Anim. Sci.*, **22**, 75–85.

Cassell, G. H., Davis, J. K., Simecka, J. W. *et al.* (1986). Mycoplasmal infections: disease pathogenesis, implications for biomedical research, and control. In *Viral and Mycoplasmal Infections of Laboratory Rodents. Effects on Biomedical Research* (eds Bhatt, P. N., Jacoby, R. O., Morse III, H. C., New, A. E.), Chapter 8. Academic Press, Orlando.

Collins M. J., Parker, J. C. (1972). Murine virus contaminants of leukemia viruses and transplantable tumors. *J. Natl. Cancer Inst.*, **49**, 1139–1143.

Custer, R. P., Outzen, H. C., Eaton, G. J., Prehn, R. T. (1973). Does the absence of immunological surveillance affect the tumour incidence in "nude" mice? First recorded spontaneous lymphoma in a nude mouse. *J. Natl. Cancer Inst.*, **51**, 707–711.

Desplazes, P., Mathis, A., Baumgartner, R., Tanner, I., Weber, R. (1996). Immunologic and molecular characterization of *E. cuniculi*-like microsporidia from humans and rabbits indicates the *E. cuniculi* is a zoonotic parasite. *Clin. Inf. Dis.*, **22**, 557–559.

Detmer, A., Hansen, A. K., Dieperink, H., Svendsen, P. (1990). Xylose-positive staphylococci as a cause of respiratory disease in immunosuppressed rats. *Scand. J. Lab. Anim. Sci.*, **18**, 13–18.

Donnelly, T. M., Stark, D. M. (1985). Susceptibility of laboratory rats, hamsters, and mice to wound infection with *Staphylococcus aureus*. *Am. J. Vet. Res.*, **46**, 2634–2638.

Dubey, J. P. (1996). Strategies to reduce transmission of *Toxoplasma gondii* to animals and humans. *Vet. Parasitol.*, **64**, 65–70.

Dupuy, J. M., Levey-Le-Blond, E., LeProvost, C. (1975). Immunopathology of mouse hepatitis virus type 3. II. Effect of immunosuppression in resistant mice. *J. Immunol.*, **114**, 226–230.

Fenoy, S., Cuellar, C., Guillen, J. L. (1997). Serological evidence of toxocariasis in patients from Spain with a clinical suspicion of visceral larva migrans. *J. Helminthol.*, **71**, 9–12.

Fortmeyer, H. P. (1981). *Thymusaplastische Maus (nu/nu) Thymusaplastiche Ratte (rnu/rnu) Haltung, Zucht, Versuchsmodelle*. Paul Parey, Berlin.

Foster, H. L. (1959). A procedure for obtaining nucleus-stock for a pathogen-free animal colony. *Proc. Anim. Care*, **9**, 135–142.

Fox, J. G., Lee, A. (1997). The role of *Helicobacter* species in newly recognized gastrointestinal tract diseases of animals. *Lab. Anim. Sci.*, **47**, 222–255.

Fox, J. G., Dewhirst, F. E., Tully, J. G. *et al.* (1994). *Helicobacter hepaticus* sp. nov., a microaerophilic bacteria isolated from livers and intestinal mucosa scrapings from mice. *J. Clin. Microbiol.*, **32**, 1238–1245.

Fox J. G., Yan F. E., Dewhirst F. E. *et al.* (1995). *Helicobacter bilis* sp. nov., a novel *Helicobacter* isolated from bile, livers and intestines of aged inbred mouse strains. *J. Clin. Microbiol.*, **33**, 445–454.

Friedmann, C. T. H., Dinnes, M. R., Bernstein, J. F., Heibreder, G. A. (1971). Chimpanzee associated infections, hepatitis, among personnel at an animal hospital. *J. Am. Vet. Med. Assoc.*, **159**, 541–545.

Friis, A. S., Ladefoged, O. (1979). The influence of *Bacillus piliformis* (Tyzzer) infections on the reliability of pharmacokinetic experiments in mice. *Lab. Anim.*, **13**, 257–261.

Fujiwara, K., Takagaki, Y., Maejima, K., Tajima, Y. (1964). Tyzzer's disease in mice. Effects of corticosteroids on the formation of liver lesions and the level of blood transaminases in experimentally infected animals. *Jpn. J. Exp. Med.*, **34**, 59–75.

Fujiwara, K., Hirano, N., Takenaka, S., Sato, K. (1973). Peroral infection in Tyzzer's disease in mice. *Jpn. J. Exp. Med.*, **43**, 33–42.

Griego, R. D., Rosen, T., Orengo, I. F., Wolf, J. E. (1990). Dog, cat, and human bites: a review. *J. Am. Acad. Dermatol.* **33**, 1019–1029.

Goss, L. J. (1970). Summary of zoo TB survey. *J. Zoo. Anim. Med.*, **1**, 16.

Gruber, F. (1975). *Immunologie der Versuchstiere*. Verlag Paul Parey, Berlin.

Gust, I. D., Feinstone, S. M. (1988). *Hepatitis A*. CRC Press, Boca Raton.

Hansen, A. K. (1992). The aerobic bacterial flora of laboratory rats from a Danish breeding centre. *Scand. J. Lab. Anim. Sci.*, **19**, 59–68.

Hansen, A. K. (1993). Statistical aspects of health monitoring of laboratory animal colonies. *Scand. J. Lab. Anim. Sci.*, **20**, 11–14.

Hansen, A. K. (1995). Antibiotic treatment of nude rats and its impact on the aerobic bacterial flora. *Lab. Anim.*, **29**, 37–44.

Hansen, A. K. (1996). Improvement of health monitoring and the microbiological quality of laboratory rats. *Scand. J. Lab. Anim. Sci.*, **23**, (suppl. 2), 1–70.

Hansen, A. K., Skovgaard-Jensen, H. J. (1995). Experience from sentinel health monitoring in units containing rats and mice in experiments. *Scand. J. Lab. Anim. Sci.*, **22**, 1–9.

Hansen, A. K., Svendsen, O. and Møllegaard-Hansen, K. E. (1990). Epidemiological studies of *Bacillus piliformis* infection and Tyzzer's disease in laboratory rats. *Z. Versuchstierkd.* **33**, 163–169.

Hansen, A. K., Andersen, H. V., Svendsen, O. (1994). Studies on the diagnosis of Tyzzer's disease in laboratory rat colonies with antibodies against *Bacillus piliformis* (*Clostridium piliforme*). *Lab. Anim. Sci.*, **44**, 424–429.

Held, J. R., Richardson, J. H., Mosley, J. W. (1968). A zoonosis associated with African green monkeys. *J. Am. Vet. Med. Assoc.*, **153**, 881–884.

Hem, A., Hansen, A. K., Rehbinder, C., Vopio, H. M., Engh, E. (1994). Recommendations for health monitoring of pig, cat, dog and gerbil breeding colonies. *Scand. J. Lab. Anim. Sci.*, **21**, 97–126.

Hendrickson, R. (1972). Herpesvirus infections of primates. *J. Zoo Anim. Med.*, **3**, 37–41.

Howard, C. J., Stott, E. J., Taylor, G. (1978). The effect of pneumonia induced in mice with *Mycoplasma pulmonis* on resistance to subsequent bacterial infection and the effect of a respiratory infection with Sendai virus on the resistance of mice to *Mycoplasma pulmonis*. *J. Gen. Microbiol.*, **109**, 79–87.

Isakov, N., Feldman, M., Segal, S. (1982). Acute infections of mice with lactic dehydrogenase virus (LDV) impairs the antigen-presenting capacity of their macrophages. *Cell. Immun.*, **66**, 317–332.

Ivker, R. (1997). Human monkeypox hits beleaguered Zaire [news]. *Lancet*, **8**, 349, 709.

Jersey, G. C., Whitehair, C. K., Carter, G. R. (1973). *Mycoplasma pulmonis* as the primary cause of chronic respiratory disease in rats. *J. Am. Vet. Med. Assoc.*, **163**, 599–604.

Juhr, N. C. (1988). Provocation of latent infections. In *New Developments in Biosciences: Their Implications for Laboratory Animal Science* (eds Beynen, A. C., Solleveld, H. A.), pp. 127–131. Martinus Nijhoff Publishers, Dordrecht.

Kiktenko, V. S. (1985). Influence of the mechanism of transmission of the infective agent on the aetiological structure of leptospiral foci. *J. Hyg. Epidemiol. Microbiol. Immunol.*, **29**, 77–82.

Kraft, V., Deeney, A. A., Blanchet, H. M. *et al.* (1994). Recommendations for the health monitoring of mouse, rat, hamster, guineapig and rabbit breeding colonies. *Lab. Anim.*, **28**, 1–12.

Krakowka, S., Morgan, D. R., Kraft, W. G., Leunk, R. D. (1987). Establishment of gastric *Campylobacter pylori* infection in the neonatal gnotobiotic piglet. *Infect. Immun.*, **55**, 2789–2796.

LABA (Laboratory Animal Breeders Association of Great Britain Limited), Laboratory Animal Science Association (LASA) (1993). Guidelines for the care of laboratory animals in transit. *Lab. Anim.*, **27**, 93–107.

Lai, Y. L., Jacoby, R. O., Bhatt, P. N., Jonas, A. M. (1976). Keratoconjunctivitis associated with sialodacryadenitis in rats. *Invest. Ophthalmol.*, **15**, 538–541.

LeDuc, J. W. (1987). Epidemiology of hantaan and related viruses. *Lab. Anim. Sci.*, **37**, 413–418.

Matsumoto, T. (1980). Early deaths after irradiation of mice contaminated by *Enterobacter cloacae*. *Lab. Anim.*, **14**, 247–249.

Matsumoto, T. (1982). Influence of *Escherichia coli*, *Klebsiella pneumoniae* and *Proteus vulgaris* on the mortality pattern of mice after lethal irradiation with X-rays. *Lab. Anim.*, **16**, 36–39.

Morris Jr, J. G. (1996). Current trends in human diseases associated with foods of animal origin. *J. Am. Vet. Med. Assoc.*, **209**, 2045–2047.

Nicklas, W., Kraft, V., Meyer, B. (1993). Contamination of transplantable tumors, cell lines and monoclonal antibodies with rodent viruses. *Lab. Anim. Sci.*, **43**, 296–300.

Olfert, E. D. (1981). Ampicillin toxicity in rabbits. *Can. Vet. J.*, **22**, 217, 1981.

Onodera, T., Fujiwara, K. (1973). Nasoencephalopathy in suckling mice inoculated intranasally with the Tyzzer's organism. *Jpn. J. Exp. Med.*, **43**, 509–522.

Pattyn, S. R. (1978). *Ebola Virus Haemorrhagic Fever*. Elsevier, Amsterdam.

Pier, A. C., Smith, J. M., Alexiou, H., Ellis D. H., Lund, A., Pritchard, R. C. (1994). Animal ringworm — its aetiology, public health significance and control. *J. Med. Vet. Mycol.*, **32** (suppl. 1), 133–150.

Pinson, D. M., Schoeb, T. R., Lindsey, J. R., Davis, J. K. (1986). Evaluation by scoring and computerized morphometry of lesions of early *Mycoplasma pulmonis* infection and ammonia exposure in F344/N rats. *Vet. Pathol.*, **23**, 550–555.

Queiroz, D. M. M., Rocha, G. A., Mendes, E. N., Lage, A. P., Carvalho, A .C., Barbose, A. J. (1990). A spiral organism in the stomachs of pigs. *Vet. Microbiol.*, **24**, 199–204.

Rygaard, J. (1973). *Thymus and Self. Immunobiology of the Mouse Mutant Nude*. John Wiley, London.

Schantz, P. M. (1996). Tapeworms (cestodiasis). *Gastroenterol. Clin. North. Am.*, **25**, 637–653.

Schoeb, T. R., Kervin, K. C., Lindsey, J. R. (1985). Exacerbation of murine respiratory mycoplasmosis in gnotobiotic F344/N rats by Sendai virus infection. *Vet. Pathol.*, **22**, 272–282.

Sekizawa, T., Openshaw, H. (1984). Encephalitis resulting from reactivation of latent herpes simplex virus in mice. *J. Virol.*, **50**, 263–266.

Shek, W. R. (1994). Lymphocytic choriomeningitis virus. In: *Microbiologic Monitoring of Laboratory Animals*, 2nd edn (eds Waggie, K., Kagiyama, N., Allen, A. M., Nomura, T.), pp. 35–42. US Department of Health and Human Services, National Institutes of Health, Bethesda.

Shope, R. E. (1986). Infectious disease research. In: *Viral and Mycoplasmal Infections of Laboratory Rodents. Effects on Biomedical Research* (eds Bhatt, P. N., Jacoby, R. O., Morse III, H. C., New, A. E.), Chapter 30. Academic Press Inc., Orlando.

Stevenson, M. M., Rees, J. C., Meltzer, M. S. (1980). Macrophage function in tumor-bearing mice: evidence for lactic dehydrogenase-elevating virus-associated changes. *J. Immun.*, **124**, 2892–2899.

Taffs, L. F. (1974). Some diseases in normal and immunosuppressed animals. *Lab. Anim.*, **8**, 149–154.

Takagaki, Y., Naiki, M., Ito, M., Noguchi, G., Fujiwara, K. (1967). Checking of infections due to *Corynebacterium kutscheri* and Tyzzer's organism among mouse breeding colonies by cortisone injection. *Exp. Anim.*, **16**, 12–19.

Takahashi, T., Sawada, T., Muramatsu, M. *et al.* (1987). Serotype, antimicrobial susceptibility and pathogenicity of *Erysipelothrix rhusiopathiae* isolates from tonsils of apparently healthy slaughter pigs. *J. Clin. Microbiol.*, **25**, 536–539.

Thomsen, A. C., Rosendal, S. (1974). Mycoplasmosis — experimental pyelonephritis in rats. *Acta Pathol. Microbiol. Scand.*, **82**, 94–98.

Ticehirst, J., Rhodes Jr., L. L., Krawczynski, K. *et al.* (1992). Infection of owl monkeys (*Aotus trivigatus)* and cynomolgous monkeys (*Macaca fascicularis)* with hepatitis E virus from Mexico. *J. Infect. Dis.*, **165**, 835–845.

Weigler, B. J. (1992). Biology of B virus in macaques and human hosts: a review. *Clin. Infect. Dis.*, **14**, 555–567.

Williamson, W. M., Tilden, E. B., Getty, R. (1963). Enteric infections occurring during an eight year period at the Chicago Zoological Park, Brookfield, Illinois. *Bijdragen Dierkde*, **33**, 86–88.

Working Committee for SPF and Gnotobiotic Laboratory Animals (1980). *Obligative or Facultative Pathogenic Germs of Primates, with Supplement: Classification of Primates*. GV-SOLAS, Basel.

Wullenweber, M. (1995). *Streptobacillus moniliformis*—a zoonotic pathogen. Taxonomic considerations, host species, diagnosis, therapy, geographical distribution. *Lab. Anim.*, **29**, 1–15.

Yagami, K. I., Goto, Y., Ishida, J., Ueno, Y., Kajiwara, N., Sugiyama, F. (1995). Polymerase chain reaction for detection of rodent parvoviral contamination in cell lines and transplantable tumors. *Lab. Anim. Sci.*, **45**, 326–328.

Young, T. D., Hurst, W. J., White, W. J., Lang, C. M. (1987). An evaluation of ampicillin pharmacokinetics and toxicity in guinea pigs. *Lab. Anim. Sci.*, **37**, 652–656.

Zurcher, C., Burek, J. D., van Nunen, M. C. J., Meihuizen, S. P. (1977). A naturally occurring epizootic caused by Sendai virus in breeding and aging rodent colonies. I. Infection in the mouse. *Lab. Anim. Sci.*, **27**, 955–962.

Chapter 7

Non-invasive Monitoring of Infection and Gene Expression in Living Animal Models

P. R. Contag, A. B. Olomu and C. H. Contag

Background

There is a need to accelerate data analyses and enhance the predictive value of animal models for infectious diseases, cancer and genetic disorders. A rapid method to monitor etiologic agents, pathological events, and target metabolic pathways non-invasively in living laboratory animals would accomplish these goals. The *ex vivo* methods by which these events are typically monitored require removal of tissue and consequently the loss of contextual influences of the living animal. For *in vivo* analyses, MRI (magnetic resonance imaging), PET (positron emission tomography) and X-ray (CT; computed tomography) can be used in many circumstances to monitor the effects of disease and response to therapy, but these methods require relatively expensive devices that are not practical for use with most animal models. Moreover, in the absence of significant pathological changes leading to increased density, selective uptake of contrast dyes, incorporation of radioactive tracers or other distinguishable features, these methods are not applicable and investigators are left with labor-intensive, time-consuming *ex vivo* methods of enumerating pathogens or nodules, analyzing nucleic acids or quantifying proteins to determine the efficacy of potential therapeutics in animal models.

The rapidly evolving paradigms in drug development, including combinatorial chemistries, genomics, high-throughput *in vitro* screening and chip technologies are generating, for a given disease, hundreds or thousands of potential lead compounds that require efficacy testing in animal models. Drug screening in animals remains a significant bottleneck in the development of therapeutics and can be severely limiting. Non-invasive *in vivo* assays would accelerate the data acquisition, require fewer animals for drug testing, and produce significantly more information per protocol. The increased depth of understanding that could be obtained through non-invasive assays for pathogenic mechanisms and potential interventions would yield improved preclinical data and ultimately preservation of time, effort and investments.

To address the unmet need of real-time access to data regarding infection and gene expression in living animals, we have developed a broadly applicable technology for monitoring biological events in living animals (Contag *et al.*, 1998). This technology is based on the observation that light is transmitted through living tissue with an efficiency suitable for use in monitoring both structure and function. The application of light as a biological monitor currently has both routine clinical uses (e.g. pulse oximetry) and novel experimental applications in medicine (Benaron *et al.*, 1997). In contrast to those methods which use external sources of light, the approach that we discuss here employs internally emitted biological sources of light to tag biological functions such as infection and gene expression (Contag and Contag, 1996; Contag *et al.*, 1995, 1996, 1997, 1998).

Luciferases have been used for decades as biological tags for gene expression and for determination of adenosine triphosphate concentrations *in vitro*. Based on this widespread use and the observation that light is transmitted through tissues, we reasoned that luciferases as internal sources of light could be monitored externally, thus serving as real-time indicators of biological functions *in vivo*. To detect the few photons that escape the scattering and absorbing environment of mammalian tissues after being emitted internally, ultra-sensitive cameras are required. Fortunately, photonic detection technologies have advanced in the areas of night vision and quality control in the silicon chip industry. These technologies are now being broadly applied to biological questions (Millar *et al.*, 1992, 1995; Brandes *et al.*, 1996); however, use in living mammals had not been described prior to 1995 (Contag *et al.*, 1995). *In vivo* monitoring of luciferase was initially pioneered with an infectious disease model—gastrointestinal infection by *Salmonella typhimurium*. This study indicated that the technique is not only feasible, but that it can be quantitative and its use, as predicted, results in more information in less time than conventional assays. Since a large variety of bacterial species can be labeled with luciferase, this approach can be generally applied in the field of microbiology. Moreover, mammalian cells can similarly be labeled and thus this method may also be applicable to the evaluation of genetic- and cell-based therapies (Contag *et al.*, 1997), anticancer drugs, as well as the study of gene expression and development (Olomu *et al.*, 1998; Spilman *et al.*, 1998; Sweeney *et al.*, 1998).

The depth of tissue that light penetrates is wavelength-dependent which is largely a function of absorbance (Jobsis, 1977; Wilson *et al.*, 1985). Shorter wavelengths of light

Handbook of Animal Models of Infection
ISBN 0-12-775390-7

(ultraviolet to blue) are absorbed by mammalian tissue to a greater extent than are longer wavelengths of light (red to near infrared). At wavelengths at, or longer than, 700 nm, absorbance is minimal, yet scattering continues to affect the path that a given photon travels in tissue. Wavelengths of bioluminescent light have been described from 490 nm to 700 nm (Widder *et al.*, 1984; Campbell, 1988; Hastings, 1996), although the best characterized bioluminescent reporters — those most often used in biological assays are blue to yellow-green (490–560 nm, bacterial, jellyfish and firefly luciferases). Since the ability to detect light through tissues is influenced by the tissue depth and optical features or opacity of the tissues, detecting light emitted from tissues deep within a mammal or through fairly opaque organs such as liver is less efficient than detection through nearly translucent tissues (skin and bone) at more superficial sites.

Some knowledge of the depth of the signal and nature of the tissue between the signal and the detector is useful in the analysis of optically based reporters in living tissues. In addition, the consistency and intensity (as determined, in part, by the promoter strength) of the bioluminescent signal contribute to its detectability and quantitation. Quantifying an internal signal is possible provided that the parameters of depth, intensity, and opacity are held constant in the model system. Thus, signals at specific tissue sites with a relatively constant physiology are quantifiable (Contag *et al.*, 1995, 1997).

A significant advantage to *in vivo* luciferase monitoring is that bioluminescent tags allow for an integrated approach whereby the same label can be used *in vitro* for quantitation and in cell culture correlates of *in vivo* processes, and then used *in vivo* to test the predictions made *in vitro*. Moreover, once studies in the animal model are completed the bioluminescent tag can then be used to quantify the tagged organism or process in an *ex vivo* biochemical assay. The results of the *in vivo* monitoring can thus be verified using more standard analyses to detect proteins, enzymatic activity or nucleic acid. In the bacterial models the infecting pathogen can be distinguished from normal flora using the bioluminescent tag. For gastrointestinal models this feature is a tremendous asset given the numbers of microorganisms that comprise the normal flora of the gut. Moreover, since pathogens tend to adhere to tissues, *in vivo* monitoring of bioluminescence may provide more accurate quantitation than aspirates, washes or homogenized tissues where adherent organisms pellet with debris or are not released from the tissues to form a homogeneous suspension and thus cannot be easily recovered and quantified. Examples of how *in vivo* luciferase monitoring can be used for studies in gene expression and infection are discussed below.

Methodologies

Labeling bacterial cells

For labeling bacterial pathogens *lux* operons from bioluminescent bacteria are ideal since these operons not only encode the luciferase genes (two genes for the heterodimeric enzyme, gene locus: *lux* and enzyme: Lux) but also encode the genes for the biosynthetic enzymes (fatty aldehyde synthases) for the substrate (Frackman *et al.*, 1990). Use of these operons obviates the need for an exogenous supply of substrate. Moreover, no deleterious effects of substrate biosynthesis on bacterial metabolism or virulence have been observed (Contag *et al.*, 1996). Among the *lux* operons from bioluminescent bacteria, that from *Photorhabdus luminescens* (previously known as *Xenorhabdus luminescens* (Boemare *et al.*, 1993; Rainey *et al.*, 1995) appears to be ideally suited for use in mammalian animal models given that mammalian body temperatures lie within the optimum temperature range for this enzyme (Frackman *et al.*, 1990; Szittner and Meighen, 1990; Meighen, 1993). This is in contrast to the low optimal temperature for beetle luciferases (Luc) and other characterized bacterial luciferases (those from *Vibrio* spp.).

Well-characterized vectors have been used for the expression of the *lux* operons in Gram-negative organisms. The original pUC-based plasmid used to clone the *lux* operon from *P. luminescens*, including the promoter region (vector designated pCGLS1; Frackman *et al.*, 1990) is well-suited for expressing luciferase from Gram-negative organisms such as *Escherichia coli* and *Salmonella* (Frackman *et al.*, 1990; Szittner and Meighen, 1990; Xi *et al.*, 1991). Introduction of the Lux encoding vector into cells of these bacterial species can be carried out using standard methods of bacterial cell transformation for Gram-negative organisms (Sambrook *et al.*, 1989).

Three *Salmonella* strains have been evaluated using this approach (Contag *et al.*, 1995). These include a mouse virulent strain (SL1344) (Hoiseth and Stocker, 1981; Sanderson and Stocker, 1987), a mutant of SL1344 (Tn5 *lac-* insertion) that was selected for decreased attachment to epithelial cells in culture (Jones and Falkow, 1994), and a laboratory strain with reduced virulence properties in culture and low virulence in mice (LB5000, LT2 strain). SL1344 and LB5000 were obtained from B.A.D. Stocker (Stanford University), and BJ66 was obtained from B. Jones (University of Iowa). These strains were transformed with the pCGLS1 plasmid using electroporation and standard methods (Sambrook *et al.*, 1989). Brightly bioluminescent colonies were selected for evaluation in culture and in mice (Contag *et al.*, 1995). The sensitivity of detection for these labeled strains is demonstrated in Figure 7.1 [all figures in this chapter are in the colour plate] where approximately eight bacterial cells (LB5000*lux*) yield a signal over background, in culture, both in the presence or absence of an absorbing and scattering biological medium — blood. This ability to detect small numbers of viable bacterial cells is useful for *ex vivo* analyses following excision of tissues from animal models.

Cell culture correlates

Cell culture correlates for pathogenesis and expression can provide a rapid means of evaluating tagged events prior to introduction into living animals. For some systems such correlates have not been established (e.g. propagation of human papillomavirus), but, if available, they can accelerate the development of *in vivo* assays, and in combination with luciferase tags, the same reporter can be used *in vitro*, *in vivo* and finally as an *ex vivo* assay to confirm *in vivo* observations. Use of bioluminescent reporters in these correlative assays provides a rapid means of quantitation that is amenable to high-throughput drug assays (Figure 7.2). We describe an additional functional assay where bioluminescence has been used in living cells, *Salmonella* adherence and entry (Figure 7.3) (Finlay and Falkow, 1989; Contag *et al.*, 1995). In the case of bacterial infections, such assays have been described for many organisms and some adopted for use with bioluminescent reporters, including for microscopic detection (Petterson *et al.*, 1996). In the case of mycobacterium, use of luciferase as a reporter can significantly accelerate *in vitro* assays and such labeled strains have begun to be used *in vivo* (Hickey *et al.*, 1996).

Bioluminescent *Salmonella* adherence and invasion assays

A human cervical epithelial cell line HEp-2 was obtained from American Type Culture Collection (ATCC) and propagated in RPMI with added glycine (in the absence of antibiotics) and 5% calf serum in an atmosphere of 5% CO_2. HEp-2 cells were seeded 12–18 hours in advance of the assay at $1x10^5$ cells per well in a 24-well tissue culture plate. *Salmonella* labeled with luciferase were grown in Luria Broth (LB) with ampicillin (50 µg/ml), to an optical density 600 of 0.08, in static culture conditions started from a 1:300 dilution of an overnight culture (Lee *et al.*, 1992). The statically grown *Salmonella* $1x10^7$ were added to the wells of HEp-2 cells (multiplicity of infection of 100) and co-cultured. Infected HEp-2 cells were washed (3×) with phosphate-buffered saline (PBS) to remove non-adherent *Salmonella*, and the cell cultures were imaged for 5 minutes with a Hamamatsu-intensified charge-coupled device (ICCD) camera. Parallel HEp-2 cultures inoculated with *Salmonella* were either treated with gentamicin (100 µg/ml for 30 minutes) to kill extracellular bacteria or left untreated. HEp-2 cultures were again imaged to observe adherent and intracellular bacteria or, in the antibiotic-treated cultures, the intracellular organisms only (Figure 7.3). Epithelial cells were lysed (lysis buffer) and bacteria plated and counted to correlate relative bioluminescence and colony-forming units (cfu).

Transient transfection of mammalian cells

Mammalian vectors can be propagated in *Escherichia coli* and used to transfect cell lines, or transduce primary mammalian cells in culture for the purpose of monitoring adoptive transfer of immune or other cells, or to tag cancer cells for tracking *in vivo* (Olomu *et al.*, 1998; Sweeney *et al.*, 1998). Alternatively these vectors may be used to transfect cells directly *in vivo*, as in DNA-based therapeutics and vaccines, and monitored non-invasively, or injected into embryos to generate transgenic animals where each of its cells has the potential of responding with light production to specific signals (Contag *et al.*, 1997).

Although a wide variety of methods have been described for transfection of cell lines in culture and cells *in vivo*, we have used cationic liposome-mediated gene transfer for each of these applications, using procedures recommended by the manufacturer (Gibco-BRL, Gaithersberg, MD). A wide variety of mammalian expression vectors that express firefly luciferase (gene locus *luc*) have been described. Mammalian promoters may contain cryptic bacterial promoters or the expression vector may contain upstream sequences that function as promoters in the bacterial cells such that *E. coli* containing an intact *luc* reading frame can be selected by growing the transformed cells on bacterial plates containing the substrate luciferin (0.5 mmol/l, Molecular Probes, American Luminescent Laboratories). Note, however, that even though this permits selection of transformed bacteria with intact *luc* reading frames, the intensity of bioluminescence in these bacterial cells does not predict levels of expression in mammalian cells.

Cell lines: Transient and stable expression

DNA delivery was optimized with respect to DNA and Lipofectin concentrations for each cell line. For cell transfections DNA ranges of 0.5–5 µg and Lipofectin amounts of 5–20 µl (in 5-µl increments) with each DNA concentration were used in the optimization, as recommended by the manufacturer (Gibco-BRL). Optimal conditions were determined by monitoring luciferase expression in living cells by adding luciferin to the transiently expressing cells at 24 hours post-transfection and monitoring expression (light production) with an ICCD camera (Hamamatsu model C2400-32) at 24-hour intervals for 72 hours. For transient expression assays in several cell lines, 2 µg of DNA and 6 µl of Lipofectin were optimal in 1 ml of medium.

Stable cell lines were generated using optimal conditions as determined above. To facilitate selection of stable lines the luciferase expression vector (pGL3) was co-transfected with a vector encoding a Zeocin resistance gene (pZeo, Invitrogen, San Diego, CA) at a ratio of either 10:1 or 5:1. After 24 hours Zeocin (a bleomycin derivative, Invitrogen, San Diego, CA) was added to the wells at a predetermined optimal concentration between 100 and 500 µg/ml. Such optima were determined prior to transfections by testing the susceptibility of cells to Zeocin for each cell line. Drug concentrations of 5–1000 µg/ml (5, 10, 25, 50 100, 250, 500 and 1000 µg/ml) were evaluated in triplicate for each cell line. The lowest drug concentration that resulted in significant killing of cultures in 3 days was used for selection of stable transfectants.

Optimal conditions for transfection were determined for NIH 3T3 and HeLa cells, and stable bioluminescent lines were generated by co-transfection with Luc and Zeocin expression vectors in 1 ml culture volume for 3 hours (in six-well dishes). Cultures were expanded to 6 ml in a 100 mm dish, and after 24 hours Zeocin was added to a final concentration of 500 μg/ml (Figure 7.4). Zeocin (250 μg/ml) has been used for the selection of stable human T-cell lines derived from Jurkat cells and CEM (human T-cell lines). When colonies were visible, luciferin was added to the cultures at a concentration of 0.5 mmol/l and cultures screened for bioluminescent colonies with the Hamamatsu ICCD camera (Figure 7.4). Bioluminescent colonies were selected by pipetting cells off the culture dish directly, or using a cloning ring, and cells were transferred to a 48-well dish. Cultures were expanded over a period of 15–21 days. Subsequently, samples were frozen in medium containing fetal bovine serum, growth medium and dimethyl sulfoxide (in a ratio of 5:4:1).

Transient transfections *in vivo*

Since the preparation of suitable transgenic mouse lines for target validation and other applications is time consuming and expensive, we have used transient transfections of tissues in living animals followed by *in vivo* luciferase monitoring to screen expression in living animals (Figure 7.5). This rapid evaluation of vector constructs in the living animal can be used as a prescreen for tissue- or condition-specific expression prior to generating transgenic animals with a particular construct.

In Figure 7.5, a commercially available *luc* expression vector (pGL3control vector, Promega, Madison, WI) was delivered to lung cells of 7-day-old Wistar rats (obtained from Simonsen Laboratories, Gilroy, CA) using liposome mediated gene transfer (Liu *et al.*, 1995; Contag *et al.*, 1997). Luciferase expression in this vector is under control of the SV-40 promoter/enhancer, and the coding sequence of the reporter has been modified. These modifications include removal of the peroxisome targeting site and optimization of codon usage for mammalian cells. These modifications lead to higher levels of expression in some mammalian cells. DNA–liposome complexes were injected directly into the lungs of 7-day-old rats in groups of three animals (100 μl of a cationic liposome solution, Lipofectin (Gibco-BRL), mixed with 100 μg of plasmid DNA). Treated animals were then injected with an aqueous solution of luciferin (150 mg/kg body weight; stock solution is 15 mg/ml), and imaged at 24, 48 (Figure 7.5) and 72 hours. Images were obtained with an integration time of 30 minutes (Contag *et al.*, 1997). For *in vivo* transfections the amounts of DNA and liposome can be prohibitive and we used 50, 100 and 200 μg of DNA in 100 μl of liposome preparation to determine optimal concentrations.

Assessing the extent of infection and efficacy of antibiotic therapy in living animals

The patterns of disease caused by three strains of salmonella in a BALB/c mouse model were determined by monitoring infection over an 8-day period using the light transmitted through the mouse tissues (Figure 7.6). BALB/c mice were infected orally — the natural route by which mice or humans become infected with *Salmonella* — and images were obtained daily. Groups of mice were infected with SL1344*lux*, the less invasive mutant, BJ66*lux* or the less-virulent strain, LB5000*lux*. At 1–2 days, the bioluminescent signal localized to a single focus in all infected animals. The distribution of bioluminescence did not spread in mice inoculated with the BJ66*lux*. This was in contrast to mice infected with the less virulent LB5000*lux* in that bioluminescence was not detected in any animal at 7 days. In BALB/c mice infected with the wild-type SL1344*lux*, bioluminescence was detected throughout the study period, with multiple foci of transmitted photons at 8 days. In one-third of the animals infected with SL1344*lux*, transmitted photons were apparent over much of the abdominal area at 8 days, resembling the distribution of photons in a systemic infection following an i.p. inoculation (Contag *et al.*, 1995).

The genetic locus called Nramp-1 (previously called Bcg/Ity/Lsh) determines the sensitivity of mice infection by intracellular pathogens. Resistance (*r*) is dominant and, after infection of mice that are heterozygous at the Nramp-1 locus (BALB/c × 129) with wild-type SL1344*lux*, the bioluminescent signal remained localized and persistent in a group of 10 mice throughout the study period (Contag *et al.*, 1995). This result was in contrast to the disseminated bioluminescence observed in SL1344*lux*-infected susceptible mice (*Nramp-1* s/s), but resembled the persistent infection of susceptible BALB/c mice with the less invasive BJ66*lux* (Figure 7.6). *Salmonella* were cultured from persistently infected resistant BALB/c × 129 mice and 80–90% of the colonies recovered after 8 days were amp^r; of these, greater than 90% were bioluminescent, suggesting that observed differences were not due to significant loss of the *lux* plasmid, but rather were due to real differences in pathogenicity of the bacterial strains.

To demonstrate the utility of *in vivo* imaging for drug screening and development, infected animals were treated with the antibiotic ciprofloxacin, known to be effective against systemic *Salmonella* infections (Magallanes *et al.*, 1993), while a second group of infected animals was left untreated as a control (Figure 7.7). Bioluminescence was then monitored over 5.5 hours. Bioluminescence over the abdomen of the ciprofloxacin-treated animal was reduced to undetectable levels during this period of time, while bioluminescence in the control increased 7.5-fold (Figure 7.7). Therefore, monitoring of optical signal intensity permitted a non-invasive assessment of the efficacy and kinetics of antibiotic action *in vivo*.

Monitoring expression in transgenic mice

Transgenic mice, containing the HIV LTR-*luc* fusion (mouse line 333 generated by Dr. John Morrey and coworkers (1991), Utah State University), were initially obtained from Jackson Laboratories (Bar Harbor, ME). The transgene in these mice includes the active part of the human immunodeficiency virus-1 (HIV-1) promoter fused to the coding sequence of the wild-type firefly luciferase gene. This construct has also been used in cell culture to monitor HIV replication over time (Aguilar-Cordova *et al.*, 1994; Contag *et al.*, 1997). A breeding nucleus was subsequently obtained directly from Dr Morrey and then bred and maintained in the Research Animal Facility at Stanford University under strict adherence to institutional guidelines. This transgenic mouse line was initially selected because of the documented expression of luciferase in the skin and the ability to activate the gene chemically and with ultraviolet light (Morrey *et al.*, 1991, 1992, 1993, 1994).

To activate LTR-*luc* expression, dimethyl sulfoxide (DMSO; 100%, Baker) (Morrey *et al.*, 1993) was applied topically twice to regions of skin at 24 and 16 hours prior to imaging (Figure 7.8A,B). This method of promoter induction was found to be optimal for expression of luciferase in adult LTR-*luc* Tg mice (Contag *et al.*, 1997). If necessary, skin was shaved to remove hair just prior to application of DMSO (Figure 7.8A,B). The HIV LTR could be specifically activated in small regions, single ears, or large areas, over half of the back. In addition expression could be activated internally and monitored externally. The substrate for the firefly luciferase was delivered systemically to these mice with peak light emission in the ears at 20 minutes following luciferin injection (i.p.), or transdermally in 100% DMSO with peak expression at 1 minutes post-application of substrate (Figure 7.8A,B).

In neonatal LTR-*luc* Tg mice, developmental effect on expression of the LTR was followed and in neonatal mice the LTR was not induced chemically or by any other known means (Figure 7.8C). Thus, increases in expression due to developmental activation of the promoter and basal levels of expression at other tissue sites were evaluated in neonatal LTR-*luc* Tg mice (Figure 7.8C and unpublished results). These data indicated that developmental gene expression could be monitored externally and temporal and spatial fluctuations in gene expression could be monitored in a single group of mice over time.

Animal handling

Anesthesia

For imaging, mice and rats were anesthetized and imaged as previously described (Contag *et al.*, 1997) using pentobarbital (35–70 mg/kg body weight). Low doses (35 mg/kg) were used with neonatal mice and to anesthetize rats: both mice and rats are sensitive to the effects of pentobarbital. While under anesthetic animals were observed closely, when not in the dark box, and respiration was monitored. Following the imaging animals were kept warm and allowed to recover.

Luciferase assays

Bacterial luciferase *ex vivo* and *in vitro*

Ex vivo analyses of bacterial luciferase levels in labeled pathogens were performed by removing the tissue and assessing the expression in tissue pieces or after homogenization in 10 ml of PBS using a stomacher. Since the bacterial cells synthesize substrate as well as enzyme, and viable bacteria have adequate levels of nicotinamide adenine dinucleotide (NADPH), bioluminescent signals can be observed in aerobic environments following removal of the tissue. This is also the case for labeled bacteria in culture.

The luminescent measurements could be made in a luminometer or, alternatively, imaged using the ICCD camera. In the case of *ex vivo* tissue analyses, where the spatial resolution provided by an image was not required, samples of tissue homogenates were placed in a cuvette and read in a luminometer (LKB-Wallac). In contrast, for selecting transformed bacterial colonies that were expressing luciferase or for culture correlates of virulence, spatial resolution is useful and the ICCD camera was used.

In vitro assays adopted for use with bioluminescent reporters in our laboratories have included sensitivity of detection determinations (Figure 7.1), determination of minimal inhibitory concentrations of antibiotic for *E. coli*, *Salmonella* (Figure 7.2) and *Pseudomonas*, and adherence and invasion assays as virulence correlates for *Salmonella* (Figure 7.3). In these *in vitro* assays bioluminescent signals correlated with cfu, except in the case of invasion of macrophages by *Salmonella*, where the bioluminescent signal per cfu appeared to be reduced relative to *Salmonella* that had entered epithelial cells in culture (Contag *et al.*, 1995). This may in part be due to differences in the intracellular environment for *Salmonella* in each cell type.

Firefly luciferase *ex vivo* and *in vitro*

Luciferase activity for the firefly enzyme expressed in mammalian cells was determined for tissue homogenates with a commercially available assay kit, that included acetyl coenzyme A to prolong light emission (Promega, Madison, WI). Tissues were removed and homogenized with a tissue disrupter in minimal volumes. Lysis buffer containing luciferin, adenosine triphosphate and acetyl coenzyme A was added to samples and read in a standard luminometer (LKB/Wallac, Stockholm, Sweden) according to the manufacturer's recommendations (data not shown).

For cells in culture, 10 µl of a 15 mg/ml stock solution of luciferin (per 1 ml of medium) was added to cells for a final concentration of 150 µg/ml. Immediately after substrate addition the cells were imaged, and bioluminescent signals were apparent in these cultures for hours after substrate addition. Titration of substrate in live HeLa cells expressing luciferase from the SV-40 promoter–enhancer (stable integration of pGL-3) indicated that increased bioluminescent signals could be obtained with increases in final substrate concentrations to 1 mmol/l. Bioluminescent signals were measured in living cells using the ICCD camera. If signals were too weak to be detected with this camera, cells were lysed and luciferase activity measured in a luminometer with the standard assay procedure.

In vivo: substrate and delivery to tissues

An aqueous solution of the substrate, luciferin (50 mmol/l), was injected into the peritoneal cavity 20 minutes prior to imaging for systemic delivery at a dose of 150 mg/kg. Alternatively, substrate was applied topically (50 mmol/l) in 100% DMSO for transdermal delivery to monitor luciferase expression in the skin. Luciferin has also been electroporated (Ionophoresis) into skin of the ear using a caliper electrode (Gentronics/BTX, San Diego, CA) connected to an electroporator (Electro Cell Manipulator, model 600, Gentronics/BTX) under the following conditions: 120 V, 100 μF, 72 Ω, resulting in a time constant of 6–7 ms (Contag *et al.*, 1997). The duration of detectable luciferase activity in tissues was variable with different modes of delivery. With systemic delivery peak activity was detected at 20 minutes in the skin and at 5 minutes from a source of luciferase expressed in the peritoneal cavity.

Imaging

At the time of imaging, anesthetized animals were placed in a light-tight chamber, and a gray-scale body surface reference image was collected in low room light provided by leaving the chamber door slightly open. Subsequently, the door to the chamber was closed, to exclude the room light that would obscure the relatively dimmer luciferase bioluminescence. Photons emitted from luciferase within the animal, and then transmitted through the tissue, were collected and integrated for a period of time between 5 and 30 minutes. A pseudocolor image representing light intensity (generally blue representing least intense and red most intense light) was generated on an Argus 20 image processor (Hamamatsu, Japan); images were transferred using a plug-in module (Hamamatsu, Japan) for Adobe Photoshop (Adobe, Mountain View, CA) to a computer (Macintosh 8100/100, Apple Computer, Cupertino, CA). Gray-scale reference images and pseudocolor images were superimposed using the image-processing software, and saved as PICT files. Superimposed images were opened within another graphics software package (Canvas, Deneba, Miami FL) and composites made and annotations added. An intensified CCD camera (C2400-32, Hamamatsu), fitted with a 50 mm f1.2 Nikkor lens (Nikon, Japan) and an image processor (Argus 20, Hamamatsu), has been used in most of our research. The capabilities of alternative low-light imaging devices employing different technologies have been assessed below.

Light-tight boxes are commercially available (e.g. Hamamatsu, Japan) or can be fabricated. Even the smallest light leaks can present significant background when using sensitive imaging devices and long integration times. The potential sites for light leaks are the door closure, camera mount (if camera is mounted externally) and entry point for the cables (if camera is mounted internally). Commercially available boxes typically have external camera mounts. We have used boxes with cameras mounted internally and externally, and each has advantages and disadvantages.

Charge-coupled devices

Low-light imaging systems employ technologies that either reduce the background noise or increase the signal. By cooling the CCD chip in a video camera to temperatures of about –30°C, the background due to infrared irradiation (so-called thermal noise) can be reduced, permitting detection of low-light signals. An alternative is specifically to increase the signal using an image intensifier. Intensifiers are designed around a high-voltage microchannel plate technology that amplifies signals in the form of electrons. A photocathode is used to convert photons to electrons, the numbers of which are then amplified over the microchannel plate; the electrons then contact a phosphor screen, generating photons which are then focused on a video camera (CCD). The materials used in the photocathode determine the spectral sensitivity of the intensifier, and to reduce thermal noise blue-sensitive materials can be employed. Alternatively, thermal noise on the intensifier can be reduced by cooling intensifiers with photocathodes made of materials that are sensitive to longer wavelengths of light (cooled-intensified cameras).

In a comparison between a cooled camera (Hamamatsu model C4880) and an intensified camera (Hamamatsu model C2400-32), essentially equivalent signals were obtained from neonatal Tg mice expressing luciferase in their eyes and skin (Contag and Oshiro, unpublished results). In adult mice with bioluminescence originating from deeper tissues, differences have been apparent between cooled intensified systems (sensitive to light ≤850 nm) and the C2400–32 ICCD camera (sensitive to light ≤600 nm). This is likely due to a suspected filter effect of the tissues where shorter wavelengths of light are more readily absorbed by mammalian tissues, and the longer wavelengths of light (luciferases have rather broad spectral peaks of up to 50 mm bands) may preferentially pass through tissues, and more red-sensitive instruments may be well-suited for collecting this light.

Summary

Non-invasive monitoring of light emitted from within a living mammal, where the light is constitutively expressed or reporting fluctuations in gene expression, is a platform technology that results in increased information about specific physiological processes and integrated, whole biological systems. The monitoring of biological light requires less time and fewer animals, reducing the cost of analyses. This is especially relevant to the costly processes of drug discovery where significant advances have been made in generating large numbers of potentially useful therapeutic compounds (e.g. combinatorial chemistries, genomics and high-throughput screening). In the absence of significant advances in *in vivo* analyses, the bottleneck in drug development is at the animal model step and the extraordinary developments in the earlier steps accentuate this bottleneck. *In vivo* luciferase monitoring addresses this limitation (Contag *et al.*, 1998) with broad applications in the areas of infectious disease (viral, bacterial and fungal), tumor progression and metastasis, gene therapy, mammalian development, and many other areas in which animal models are used as predictors for the human response to therapy.

References

Aguilar-Cordova, E., Chinen, J., Donehower, L., Lewis, D. E., Belmont, J. W. (1994). A sensitive reporter cell line for HIV-1 tat activity, HIV-1 inhibitors, and T cell activation effects. *AIDS Res. Hum. Retroviruses*, **10**, 295–301.

Benaron, D., Cheong, W.-F., Stevenson, D. (1997). Tissue Optics. *Science*, **276**, 2002–2003.

Boemare, N., Akhurst, R., Mourant, R. (1993). DNA relatedness between *Xenorhabdus* spp. (Enterobacteriaceae), symbiotic bacteria of entomopathogenic nematodes, and a proposal to transfer *X. luminescens* to a new genus, *Photorhabdus*. *Int. J. Syst. Bacteriol.*, **43**, 249–255.

Brandes, C., Plautz, J. D., Stanewsky, R. *et al.* (1996). Novel features of drosophila period transcription revealed by real-time luciferase reporting. *Neuron*, **16**, 687–692.

Campbell, A. K. (1988). *Chemiluminescence. Principles and Applications in Biology and Medicine*. Ellis Horwood and VCH Verlagsgesellschaft, Chichester, England.

Contag, C. H., Contag, P. R. (1996). Viewing disease progression through a bioluminescent window. *Optics Photonics News*, **7**, 22–23.

Contag, C. H., Contag, P. R., Mullins, J. I., Spilman, S. D., Stevenson, D. K., Benaron, D. A. (1995). Photonic detection of bacterial pathogens in living hosts. *Mol. Microbiol.*, **18**, 593–603.

Contag, C. H., Contag, P. R., Spilman, S. D., Stevenson, D. K., Benaron, D. A. (1996). Photonic monitoring of infectious disease and gene regulation. In *OSA TOPS on Biomedical Optical Spectroscopy and Diagnostics*, vol. 3. (eds Sevick-Muraca, E., Benaron, D.), pp. 220–224. Optical Society of America, Washington, DC.

Contag, C., Spilman, S., Contag, P. *et al.* (1997). Visualizing gene expression in living mammals using a bioluminescent reporter. *Photochem. Photobiol.* **66**, 523–531.

Contag, P., Olomu, I., Stevenson, D., Contag, C. (1998). Bioluminescent reporters in living mammals. *Nature Med.*, **4**, 245–247.

Finlay, B. B., Falkow, S. (1989). Salmonella as an intracellular parasite. *Mol. Microbiol.*, **3**, 1833–1841.

Frackman, S., Anhalt, M., Nealson, K. H. (1990). Cloning, organization, and expression of the bioluminescence genes of *Xenorhabdus luminescens*. *J. Bact.*, **172**, 5767–5773.

Hastings, J. W. (1996). Chemistries and colors of bioluminescent reactions: a review. *Gene*, **173**, 5–11.

Hickey, M. J., Arain, T. M., Shawar, R. M. *et al.* (1996). Luciferase *in vivo* expression technology: use of recombinant mycobacterial reporter strains to evaluate antimycobacterial activity in mice. *Antimicrob. Agents Chemother.*, **40**, 400–407.

Hoiseth, S. K., Stocker, B. A. D. (1981). Aromatic-dependent *Salmonella typhimurium* are non-virulent and effective as live vaccines. *Nature*, **291**, 238–239.

Jobsis, F. F. (1977). Noninvasive, infrared monitoring of cerebral and myocardial oxygen sufficiency and circulatory parameters. *Science*, **198**, 1264.

Jones, B. D., Falkow, S. (1994). Identification and characterization of a *Salmonella typhimurium* oxygen-regulated gene required for bacterial internalization. *Infect. Immun.*, **62**, 3745–3752.

Lee, C. A., Jones, B. D., Falkow, S. (1992). Identification of a *Salmonella typhimurium* invasion locus by selection for hyper-invasive mutants. *Proc. Natl Acad. Sci. USA*, **89**, 1847–1851.

Liu, Y., Liggitt, D., Zhong, W., Tu, G., Gaensler, K., Debs, R. (1995). Cationic liposome-mediated intravenous gene delivery. *J. Biol. Chem.*, **270**, 24864–24870.

Magallanes, M., Dijkstra, J., Fierer, J. (1993). Liposome-incorporated ciprofloxacin in treatment of murine salmonellosis. *Antimicrob. Agents Chemother.*, **37**, 2293.

Meighen, E. (1993). Bacterial bioluminescence: organization, regulation, and application of the lux genes. *FASEB J.*, **7**, 1016–1022.

Millar, A. J., Short, S. R., Chua, N. H., Kay, S. A. (1992). A novel circadian phenotype based on firefly luciferase expression in transgenic plants. *Plant Cell.*, **4**, 1075–1087.

Millar, A. J., Carre, I. A., Strayer, C. A., Chua, N. H., Kay, S. A. (1995). Circadian clock mutants in *Arabidopsis* identified by luciferase imaging. *Science*, **267**, 1161–1163.

Morrey, J. D., Bourn, S. M., Bunch, T. D. *et al.* (1991). *In vivo* activation of human immunodeficiency virus type 1 long terminal repeat by UV type A (UV-A) light plus psoralen and UV-B light in the skin of transgenic mice. *J. Virol.*, **65**, 5045–5051.

Morrey, J. D., Bourn, S. M., Bunch, T. D., Sidwell, R. W., Rosen, C. A. (1992). HIV-1 LTR activation model: evaluation of various agents in skin of transgenic mice. *J. AIDS*, **5**, 1195–1203.

Morrey, J. D., Jackson, M. K., Bunch, T. D., Sidwell, R. W. (1993). Activation of the human immunodeficiency virus type 1 long terminal repeat by skin-sensitizing chemicals in transgenic mice. *Intervirology*, **36**, 65–71.

Morrey, J. D., Bourn, S. M., Morris, J. L., Bunch, T. D., Sidwell, R. W. (1994). Activation of the human immunodeficiency virus long terminal repeat by abrasion of the skin in transgenic mice. *Intervirology*, **37**, 315-320.

Olomu, A., Sweeney, T., Mailander, V. *et al.* (1999). Visualizing tumor progression in living mammals. *In preparation*.

Petterson, J., Nordfelth, R., Dubinina, E. *et al.* (1996). Modulation of virulence factor expression by pathogen target cell contact. *Science*, **273**, 1231–1233.

Rainey, F., Ehlers, R.-U., Stackebrandt, E. (1995). Inability of the polyphasic approach to systematics to determine the relatedness of the genera *Xenorhabdus* and *Photorhabdus*. *Int. J. Syst. Bacteriol.*, **45**, 379–381.

Sambrook, J., Fritsch, E. F., Maniatis, T. (1989). *Molecular Cloning: A Laboratory Manual*. Cold Spring Harbor Laboratory Press, Cold Spring Harbor.

Sanderson, K. E., Stocker, B. A. D. (1987). *Salmonella typhimurium* strains used in genetic analysis. In *Escherichia coli* and *Salmonella typhimurium*. *Cellular and Molecular Biology*, vol. 2. (eds Neidhardt, F. C., Ingraham, J. L., Low, K. B., Magasanik, B., Schaechter, M., Umbarger E.), pp. 1220–1224, American Society of Microbiology, Washington, DC.

Benaron, D. A., Contag, C. H., Olomu, I. N., Spilman, S. D., Stevenson, D. K. (1999). Probing developmental gene regulation noninvasively in transgenic mice. *In preparation*.

Sweeney, T., Mailander, V., Olomu, A., *et al*. (1999). Visualizing the kinetics of tumor cell growth in living animals. *In preparation*.

Szittner, R., Meighen, E. (1990). Nucleotide sequence, expression and properties of luciferase coded by lux genes from terrestrial bacterium. *J. Biol. Chem.*, **265**, 16581–16587.

Widder, E., Latz, M., Herring, P. (1984). Far red bioluminescence of two deep-sea fishes. *Science*, **255**, 512–514.

Wilson, B., Jeeves, W., Lowe, D. (1985). *In vivo* and postmortem measurements of the attenuation spectra of light in mammalian tissues. *Photochem. Photobiol.*, **42**, 153–162.

Xi, L., Cho, K. W., Tu, S. C. (1991). Cloning and nucleotide sequences of lux genes and characterization of luciferase of *Xenorhabdus luminescens* from a human wound. *J. Bact.*, **173**, 1399–1405.

Chapter 8

Considerations for Working Safely with Infectious Disease Agents in Research Animals

J. Y. Richmond and F. Quimby

Animal facilities used for research with infectious diseases need to meet criteria that *exceed* the standards for normal experimental animal research because four criteria need to be met. We need to protect the workers from exposure to the infectious experimental agent; we need to prevent the spread of the infectious agent to other animals; we need to protect the integrity of the research project itself; and we need to protect the environment. This chapter will provide a framework for decision-making regarding choice of facilities needed to meet these goals based on a risk assessment approach that considers engineering design, administrative controls, and worker practices and procedures. Furthermore, emphasis is placed on issues related to regulated medical waste, with illustrative examples drawn primarily from the new animal research facilities at the College of Veterinary Medicine, Cornell University.

General principles for safe use of animals in infectious disease studies

Biosafety levels for animal studies

Guidelines for establishing appropriate combinations of facilities, containment, practices and procedures have been published for working with vertebrate animals used in infectious disease research. Published in the CDC/NIH publication, *Biosafety in Microbiological and Biomedical Laboratories* (Richmond and McKinney, 1999), are Agent Summary Statements concerning recommendations for the safe handling of those pathogens known to have caused laboratory-acquired infections. Recent recommendations have been published for containment facilities used for the safe housing and handling of livestock and poultry infected with pathogenic agents (Barbieto *et al.*, 1995). Special requirements for work involving recombinant DNA molecules have also been published (National Institutes of Health, 1995).

Factors to consider in risk assessment

Understanding the chain of infection concept may be helpful in evaluating both the risks and the intervention strategies (Richmond, 1991a). Consider the infected animal as the reservoir of the pathogen. How will the pathogen be transmitted from the animal (through its breathing, excretions, biting or in its blood or tissues)? In what quantities? Will the pathogen survive in the environment? For how long? How will the agent enter the new host? Is the host susceptible? Will the agent cause disease? If we can apply appropriate intervention strategies at one or more steps along this chain of infection, we will effectively prevent transmission of this pathogen from the reservoir to the new susceptible host.

What is known about the animal selected for the work will strongly influence the outcome of the risk assessment. Does the species of animal selected for the work have a history of harboring intrinsic infectious agents that may pose a particular health risk to humans or that may in some manner compromise the nature of the experimental work?

The animal's size will dictate the facility required, the type of physical containment needed, and the extent of specialized training required by the animal care-givers. If the animal is a natural host for the infectious agent, the expectation is that the agent will grow to high titers and become a potential source for infecting other susceptible animals. This generally will require containment and isolation of infected animals. Specific work practices will be required to minimize fomite transfer by care-givers. If the animal is known not to be susceptible to the infectious agent and the agent is non-infectious to humans, less stringent caging and housing requirements are needed. If the susceptibility of the host animal is unknown, intermediate level containment may be adequate.

The route of transmission is a salient factor in evaluating the risk (Barkley and Richardson, 1984). Agents spread by the aerosol route dictate more stringent facility containment requirements than agents spread by blood-to-blood contact. Isolation caging must be used, care must be taken while transferring animals from cage to cage, and respiratory-protective equipment must be worn by animal care-givers to protect against inhalation of infectious aerosols. When the agent is spread in urine or feces, special waste-handling procedures or treatment may be required to minimize exposure of other animals, humans, or the environment.

Handbook of Animal Models of Infection
ISBN 0-12-775390-7

With the information gathered during the risk assessment process, a job hazard analysis can then be performed (Richmond *et al.*, 1993). Determine what specific interventions are required that will provide sufficient engineering controls, personal protective equipment, and work practices to minimize exposure of the animal care-givers to the specific hazards. Engineering controls will include the building's air handling equipment, the physical integrity of the facility, selection of caging/containment devices, accessibility to autoclaves, cage washers, and other support equipment (Ruys, 1991; Richmond *et al.*, 1997). Work practices range from operations of clean/dirty corridors, requirements for clothing change/showering, and means for waste management, to assessment and conduct of needed training. Personal protective equipment will include those items worn as part of daily standard operating procedures (gloves, gowns, boots/shoe covers, etc.) and specialized items needed for the particular work (respirators, face shields/goggles, etc.).

Risk reduction

Veterinary interventions can be extremely helpful in reducing the risks associated with infectious disease research. Many experimental laboratory animals come from quite clean purpose-bred colonies, and it is often possible to minimize intrinsic risks even further by procuring specific pathogen-free animals. Even then, quarantine procedures must be followed when introducing new animals into closed colonies, and appropriate monitoring of all animals (often using sentinels) is necessary. When field-source animals are to be used, field and pasture management, preventive veterinary medical techniques, herd surveillance, and vermin control practices must be used to insure the health of the animals and minimize the chance of extraneous infections.

Protocol review by committees offers additional opportunities for identifying risk reduction strategies; the institutional animal care and use committee, the biohazard safety review committee, and the occupational health and safety committee all have unique perspectives to offer in risk assessment and reduction. The principal investigator proposing the work and the veterinarian-in-charge of the facility should take the leadership role in these discussions. The intent is to identify the type of caging required, any necessary restraint strategies, or other unique handling requirements. The level of training of the personnel who will be involved with animal care and treatment and the need for ongoing training must be defined at this stage.

A program for occupational health needs to be established (National Research Council, 1997). This may include employee immunizations, evaluation for capacity to wear respirators, exposure monitoring, incident/accident reporting, and emergency response (on-site and within the occupational health clinic). The strategy is to encourage employee awareness to possible sequelae of exposure to the infectious agent so that appropriate medical interventions can be initiated in a timely manner.

The contributions of the safety professionals (Richmond, 1991b; Fleming *et al.*, 1995) should also be to insure that proper waste disposal programs are used (National Research Council, 1989). With small animals, on-site autoclaving of used bedding and/or carcasses may be all that is needed. With larger animals, other technologies may need to be employed, such as incineration, rendering, or tissue digestion. Final treatment may be done in the animal facility, elsewhere on-site, or after transportation to another location. Local ordinances for disposal and/or transport of potentially hazardous materials need to be consulted when developing the waste management plan.

Application of the basic principles during animal research

Permitting requirements associated with regulated medical waste

All states have developed a program for monitoring the safe treatment and disposal of regulated medical waste (RMW). New York state provides the following definition: "Regulated medical waste is any medical waste that is solid waste, generated in the diagnosis, treatment or immunization of human beings or animals, in research pertaining thereto, or in the production or testing of biologicals" (New York State Department of Health, 1993).

For the purposes of this chapter, RMW includes "contaminated animal carcasses, body parts, and bedding of animals that were known to have been exposed to infectious agents (infectious to humans) during research, production of biologics or testing pharmaceuticals". It includes all cultures and stocks of infectious agents, contaminated culture dishes, and devices used to handle transfer or inoculate infectious agents. It includes all culture dishes and devices contaminated by impure cell lines and sharps used in animal care and treatment.

New York and other states have developed comprehensive regulations governing the treatment of regulated medical waste. Most recently New York state included all biosafety level 3 (BSL-3, Richmond and McKinney, 1999) facilities as RMW treatment facilities which require permits to construct and operate (New York State Department of Environmental Conservation, 1996). The permit requires institutions to plan carefully, in advance, for the safe operation of biosafety facilities. Since many of the general principles previously mentioned are incorporated in this application, and since the recommended biosafety and operation manual includes much of this information (Richmond and McKinney, 1999), it seems prudent to review the highlights of the process here.

The application consists of 13 sections including:

(a) an engineering report which describes the overall process and function of the facility and all equipment to be used in treatment of RMW;
(b) engineering plans and specifications, including site plans, construction specifications, equipment performance specifications, etc.;
(c) the treatment, destruction and disposal facility description — name, address and location of all facilities involved in these three operations;
(d) validation testing program — details on a program to validate all treatment operations;
(e) surety — proof of liability insurance sufficient to meet all responsibilities in the event that RMW from the facility causes bodily injury or property damage;
(f) final facility plans;
(g) operating plan — comprehensive procedures for the operation of every major component of the facility;
(h) maintenance and monitoring plan, which details how monitoring and inspections of equipment will be made, the spare parts inventory necessary and the details on challenge testing with an approved biological indicator;
(i) personnel staffing and training plan;
(j) waste control plan — an insurance that the facility will only receive, store, treat and destroy RMW specifically authorized by the state;
(k) contingency plan, describing procedures that will be used to minimize hazards to human health and the environment resulting from equipment failure, utility failure and/or other disasters;
(l) closure and financial assurance plans, including the steps necessary to close the facility;
(m) security plan, restricting the presence of and minimizing the possibility for any unauthorized entry on to the facility site and limiting contact with RMW.

While the specific details of this application go beyond the scope of this chapter, it is clear that the specific requirements listed in the operating plan, maintenance and monitoring plan, personnel training plan, contingency plan and security plan are important considerations for all institutions conducting infectious disease research. The details of this regulation will assist institutions in the preparation of the biosafety manuals.

Treatment of large-animal RMW

Special considerations are required when conducting studies with human pathogens in horses and cattle. The primary enclosure is often a room rather than equipment. Contaminated excrement loads may exceed 10 gallons per day per animal, and contaminated carcasses are quite large and difficult to manipulate. In an effort to address these points, Cornell University has constructed an animal biocontainment facility (ABSL-3) which incorporates the recommended performance standards (Richmond and McKinney, 1999) and also includes several unique features of construction and equipment.

Animal rooms are designed to be washed down on a daily basis into a dedicated sewerage system and manure sterilization plant. Rooms are constructed such that gaseous sterilants (vaporized hydrogen peroxide or chlorine dioxide) may be applied from a remote site, or contact foam sterilants may be applied directly to all room surfaces. Double-lined stainless steel plumbing carries liquid and solid wastes to a remote processing plant, itself designed for operation under ABSL-3 conditions. Internal sensors between the two layers of stainless steel plumbing monitor leaks and activate auxiliary pumps which guarantee that all contaminated liquids reach the treatment plant. Inside the manure processing plant, wastes are pumped, 250 gallons at a time, into a kill tank, brought to 275°F and 30 psig and held for sufficient time to kill all infectious agents before discharge into the municipal sewer.

If disposal of an infected animal carcass is necessary, special provisions have been made to avoid off-site treatment. Within the confines of the animal holding space we are installing a carcass digester which digests tissue based on the principle of alkaline hydrolysis. The alkaline hydrolysis digester is an insulated steam-jacketed stainless steel vessel with clamp-on lid into which infected animal remains are hydraulically lowered and sealed. Once the seals are tested, pumps circulate 1–2 N sodium hydroxide solution throughout the inner chamber, making contact with all animal remains. The steam-jacketed chamber elevates the internal temperature to 120°C; pressure is maintained at 15 psig. During a typical 18-h cycle, protein, lipids, RNA and DNA are converted to a soluble hydrosylate containing small peptides, amino acids, fatty acid salts, sugars and nucleic acids respectively. The process liberates the 70–80% water content of the carcass. The destruction of peptide bonds during hydrolysis reduces the original pH from 14 to 11. Finally, when temperature and pressure are reduced, the liquid contents are pumped to a municipal sewer, leaving sterile bone meal (about 3% original weight), a residue for solid waste disposal (or use as a fertilizer for institutional plantings). The digestion unit is manufactured with internal capacities varying from 400 to 3300 lb of animal remains.

Other equipment or disposables exit the facility via either a double-door autoclave or pass through a gas sterilization chamber which can accommodate formaldehyde, vaporized hydrogen peroxide or chlorine dioxide gas.

Treatment of small-animal RMW

As previously mentioned, contaminated small animal carcasses usually can be treated on-site in autoclaves and safely taken to an incinerator (or other appropriate device) for final disposal. Although sterile, the final disposal of intact animal carcasses may be limited by local ordinances and state regulations. Infected animal excrement is absorbed on to appropriate bedding which is also autoclaved on site before final disposal. However, the management of certain other contaminated waste products may be more difficult.

We have recently constructed a facility in which sheep will be studied during gestation. Because our institutional policy is to treat all sheep as though infected with *Coxiella burnetii*, special provisions were necessary for their care and use. Pregnant catheterized animals are maintained under ABSL-2 housing. Animals are not allowed to deliver in this facility and thus we developed a mechanism to transport these animals from ABSL-2 to ABSL-3 for surgical intervention and necropsy. Specific transport cages transfer the animal to a special chamber with electronically interlocking double doors. Once the animal (enclosed in a conveyance vehicle) is introduced into the chamber, all air within the chamber is evacuated before a second door (leading to ABSL-3) can be opened.

In surgery or necropsy all exhaust air from the room is diverted through downdraft tables. Contaminated liquids are isolated and diverted through dedicated plumbing to a 4000 gallon kill tank which treats contaminated liquids by chlorination. In the event of system failure, all waste water can be diverted manually to mobile stainless steel tanks (located under the tables) which can be sealed and transported to the autoclave for sterilization.

The necropsy facility is equipped with a through-the-wall autoclave which can operate using steam or provide cold sterilization with vaporized hydrogen peroxide. The latter provision allows investigators to remove fresh tissues, sealed in plastic containers, from necropsy to laboratories without contaminating the passageways.

Why develop an operations protocol before construction of the facility?

The Part 360 solid waste-permitting process of New York state requires the institution to propose an operating plan for all final RMW treatment modalities in the facility. This plan is essential to recognize those practices, equipment needs and construction details unique to ABSL-3 operations. We decided to expand on this operating plan and establish standard operating procedures for all aspects of animal care and use. Some oversights in our design (which became obvious during the production of the operating plan) are briefly described below.

The dirty side-cage wash area only receives cages and bottles after they have been sterilized in an autoclave; however, the central location of this area and limitations placed on personnel traffic between ABSL-2 rooms and this area could not preclude contamination of the floor with biosafety level 2 agents. This being the case, the entire dirty side of this area must be amenable to sanitation (preferable with sterilant solutions in foam). However, containers of detergents and other drummed consumable supplies posed a problem regarding treatment and removal. To rectify this problem, a reservoir tank for detergents is being constructed inside the clean loading dock with detergents plumbed over ceilings into the cage washer. This configuration avoids having to autoclave drums out and allows delivery of detergent without having the drums enter the facility.

Carbon dioxide gas for euthanasia in necropsy and medical oxygen to support anesthesia during surgery (both at ABSL-3) also had to be plumbed from the clean loading dock over ceilings to the appropriate location. The prospect of sterilizing empty compressed gas tanks was not considered in the original plans.

While remote sensing and the capability of making adjustments in ventilation from laptop computers was included in the original design, no thought had been given to the removal of maintenance equipment once contaminated on the ABSL-3 side. Currently steam sterilization and gas sterilization using vaporized H_2O_2 or chlorine dioxide are possible, but some sensitive electronic monitoring devices may need to remain within the facility at a secured location.

Great attention to detail is necessary in containment facilities and for some items; rehearsing each step in the research protocol (not just the animal care protocol) and the maintenance protocol is essential to insure safe operation of this facility.

Security

Biocontainment facilities pose four different security concerns: protection of the unknowing individual from inadvertent contamination, protection of the facility from contamination due to inappropriate actions by employees, alerting the employees to system failure, and protecting the facility and its contents from terrorism. Emergency lighting, alarms and signage are all important considerations in the security plan.

In developing a maintenance and monitoring plan, it became obvious that many individuals other than the animal care staff and investigators may need to enter the facility. Individual access cards can be issued to employees, but special provisions had to be made for access by life safety and maintenance personnel. Limiting access by limiting the number of entry points is essential. We have incorporated internal loading docks which allow trucks to be offloaded or loaded while at the same time denying access through the loading dock to the facility. Unfortunately in some instances, convenience is sacrificed for security.

Personnel training, contingency plans and validation

While each of these activities are quite different they each have one factor in common: a knowledgeable staff.

We mentioned previously the importance of operator knowledge for the proper performance of equipment, for appropriate use of personnel protective equipment, and for the proper conduct of procedures — each necessary for risk reduction. A comprehensive safety manual should always be available for training new employees and as a reference to trained employees. While certain aspects of facility operation will be standardized, e.g. operation of the autoclave, other aspects will vary according to the scientific protocol. For this reason it is essential that animal care staff and research staff develop these operating plans together.

Contingency plans deal with procedures to be employed to minimize hazards to human and environmental health resulting from equipment failure, utility failure or natural disasters. Appropriate input from maintenance, life safety, and environmental compliance personnel is essential. In the absence of an emergency, these plans should be reviewed and rehearsed at least annually by all staff members.

Validation under the Part 360 permit refers to the initial testing and analysis procedures used to evaluate efficacy on all equipment involved with the treatment of RMW. The tests must be performed on loads simulating worse-case operating conditions and make use of actual RMW expected to be treated. The testing program and results must be reviewed by the state before the facility is approved for operation. At Cornell a specialist is employed to develop and implement this validation plan. For the facilities discussed in this chapter the plan includes: the manure treatment plant, autoclaves and alkaline hydrolysis digester in the large animal containment facility, and the autoclaves, waste water chlorination system and incinerator associated with the small animal facility.

Once approved for operation the institution is obliged to conduct challenge testing, using an appropriate biological indicator, on each treatment unit once during every 40 hours of use. These tests measure the ability of each unit completely and consistently to kill the approved indicator. A detailed log is required for each unit, recording the dates and results of each challenge test and visual inspection. The operating requirements for each unit will be determined originally by validation and challenged testing. This plan may be modified at any time by the state based on the results of continuing challenge testing. For this reason it is essential that all personnel involved with equipment operation work closely with the biosafety and compliance officers of the institution to insure that measures have been taken to treat all RMW appropriately.

Finally, we have developed a three-tier training program for animal care personnel. All staff members must enroll in classes leading to certification by the American Association for Laboratory Animal Science (AALAS) at the Assistant Technicians level. In addition, classes leading to Technicians certification and assisted home study leading to Technologist certification are offered at the workplace. Paid time away from work is given to any employee attending one of these courses, and bonus pay is given upon certification. However, despite the level of certification, additional courses in biosafety must be successfully completed (with examination) before working at ABSL-2 or ABSL-3. These are also taught in-house and successful completion is necessary for promotion to a higher pay band. These courses include microbiology, biosafety practices, risk assessment and risk reduction (as previously detailed). Finally, work-site training for personnel working at ABSL-2 or -3 is provided by animal care supervisors, biosafety officers, equipment manufacturer representatives and research investigators to include all facets of the operating plan at a given facility site.

References

Barbieto, M. S., Abraham, G., Best, M. *et al.* (1995). Recommended biocontainment features for research and diagnostic facilities where animal pathogens are used, *Rev. Sci. Tech. Office Int. Epiz.*, **14**, 873–887.

Barkley, E. W., Richardson, J. H. (1984). *Control of Biohazards Associated with the Use of Experimental Animals. Laboratory Animal Medicine*, pp. 596–603. Academic Press.

Fleming, D. O., Richardson, J., Tulis, J. J., Vesley, O. (eds) (1995). *Laboratory Safety*, 2nd edn. ASM Press, Washington, DC.

National Institutes of Health (1995). *Guidelines for Research Involving Recombinant DNA Molecules*. Federal Register (60 CFR 20726), April 27. National Institutes of Health, Washington, DC.

National Research Council (1997). *Occupational Health and Safety in the Care and Use of Laboratory Animals*. P. National Academy Press, Washington, DC.

National Research Council (1989). *Biosafety in the Laboratory — Prudent Practices for Handling and Disposal of Infectious Materials*. National Academy Press, Washington.

New York State Department of Health (1993). *Regulated Medical Waste*. Title 10, NYCRR Part 70. New York State Department of Health, Albany, New York.

New York State Department of Environmental Conservation (1996). *Regulated Medical Waste Treatment, Storage, Containment, Transport and Disposal*. Title 6, NYCRR Part 360. New York State Department of Environmental Conservation, Albany, New York.

Richmond, J. Y. (1991a). Hazard reduction in animal research facilities. *Lab Animal*, **20**, 23–29.

Richmond, J. Y. (1991b). Responsibilities in animal research. *Lab Animal*, **20**, 41–46.

Richmond, J. Y., McKinney, R. W. (eds) (1999). *Biosafety in Microbiological and Biomedical Laboratories*, 3rd edn. CDC/NIH, HHS publication (CDC) 93–83, 95. US Government Printing Office, Washington, DC.

Richmond, J. Y., Arambulo, P., Ruiz, A. (1993). Protection from risk in animal studies. *Lab Animal*, **22**, 36–40.

Richmond, J. Y., Ruble, D. L., Brown, B., Jaax, G. P. (1997). Working safely at animal biosafety level 3 and 4: facility design implications. *Lab Animal*, **26**, 28–35.

Ruys, T. (1991). *Handbook of Facilities Planning*, vol. 2. *Laboratory Animal Facilities*. Van Norand Reinhold, New York.

Chapter 9

Analysis of Genetic Susceptibility to Infection in Mice

E. Buschman and E. Skamene

Introduction

The identification of human genes predisposing to infections is hardly ever straightforward, even in those cases where there exists compelling evidence that susceptibility to an infection is genetically regulated, such as in tuberculosis and leprosy (Schurr and Skamene, 1996). The complications arise from polygenic control, genetic heterogeneity, variable expression, epistasis, low penetrance and ambiguous clinical phenotypes. Although some of these difficulties can be alleviated by the utilization of a whole genome-scanning approach in family-based studies, for this approach to be useful, adequate numbers of individuals have to be identified and readily available for study. In addition, such genome scans can lead to chromosomal regions where there are no obvious candidate genes.

The rational answer to tackling many of the above problems has been to investigate mouse models of infection (Erickson, 1997; Skamene *et al.*, 1998). There are numerous justifications for studying mouse models of human diseases. The most central of these are controlled inbreeding and the extensive mouse–man homology at the genetic level (Nadeau *et al.*, 1995; Ehrlich *et al.*, 1997). This homology confers a high likelihood that a disease gene mapped in the mouse will be a candidate disease gene in humans. Such an approach has worked in several instances, for example, the interferon gamma receptor gene identified in disseminated mycobacterial infection (Newport *et al.*, 1996) and the *NRAMP1* gene linked with susceptibility to leprosy (Abel *et al.*, 1998). In light of these recent discoveries, it may be worthwhile pointing out that often research investigators do not follow exclusively either a mouse or human approach to genetic studies. Rather, as in the case of *Nramp1*, human linkage analysis had been proceeding for years in tandem with the positional cloning of mouse *Nramp*, in order to test for linkage as soon as the appropriate candidate genes or markers became available (Schurr *et al.*, 1991).

The basic structure–function analysis of mouse–man conserved genes or gene families is also likely to yield biological insight, even though the allelic variants and modifier genes may differ between species (Erickson, 1997). The development of mouse models for human diseases also provides a living resource in which genetically altered mice, or naturally mutant mice, can be treated by gene therapy techniques. Mouse models also provide a unique system to test for genetic interactions and gene epistasis which would be otherwise impossible.

The purpose of this chapter is to provide direction towards research articles and website resources for those interested in the genetic analysis of infection in mice. This chapter is not intended to be a review of specific mouse models of infection, but rather to point out the various mouse models which are available to study the genetics of infection. The article is organized according to three themes. First, we discuss the innate and acquired response to infection and the mouse models which have been used to dissect these responses. Second, we discuss strategies to analyze the genetic variation of infection. Last, we discuss the use of knockout breeding techniques to analyze gene function in resistance to infection.

Mouse models of innate and acquired resistance to infection

Exposure of inbred mice to infectious pathogens results in activation of a broad network of innate (non-adaptive) and immune (adaptive) responses (Bendelac and Fearon, 1997). In general, experimental analysis of these responses can take two approaches. In the first, a specific inbred (such as A/J), or mutant inbred strain of mice (such as *beige*) has been used to characterize the host defense pathways which are incurred following infectious challenge. In a second type of approach, combinations of inbred strains, and segregating crosses thereof (such as between A/J and C57BL/6J), have been used to analyze the genetic variations in host responses. Often, however, the first approach leads to the second, so that an observed difference in pathologic phenomena between two strains leads to an interest in gene identification. In fact, this is how the *Bcg* gene studies were initiated some 20 years ago, when it was observed during studies of humoral immunity that *Mycobacterium bovis* grew poorly in the A/J strain but rapidly in the C57BL/6J strain. Below we discuss mouse strains and models which

Handbook of Animal Models of Infection
ISBN 0-12-775390-7

have been used in the analysis of innate and acquired resistance to infection.

First line of host defense — the innate system

The innate system acts as the first component of host resistance by controlling pathogen growth during the acute phase of infection (see *Current Opinion in Immunology*, 1997, volume 9, "Innate Immunity"). A basic difference between the innate and immune systems is the recognition of pathogen-associated molecular patterns (PAMPs) by the innate host defense system (Medzhitov and Janeway, 1997). PAMPs are invariant ligands on foreign invaders and include lipopolysaccharides (LPS), yeast mannans and double-stranded RNA. The receptors for PAMPs differ from T- and B-cell receptors in that they are non-clonal and germline-encoded. The host receptors which recognize PAMPs include C-type lectins on natural killer (NK) cells and scavenger receptors on macrophages (Pearson, 1996). Below, certain mouse models investigating the functions of three cells of key importance in innate resistance (macrophages, NK cells and T gamma/delta cells) are discussed.

Macrophages

Scavenger receptors. Scavenger receptors (SRs) are proteins expressed on the macrophage surface membrane. There are three known SR classes, with the CD36 SR thought to play a role in clearance of *Plasmodium falciparum* oxidatively damaged red blood cells (Pearson, 1996). The SR-AI/II class is believed to function in clearance of Gram-positive bacteria, and an SR-AI/II knockout mouse is reportedly being investigated (Lougheed *et al.*, 1997).

The Lps gene. The ability of the macrophage to respond to LPS of bacteria is known to be regulated by the *Lps* gene, located on murine chromosome 4 (Sultzer *et al.*, 1993). The *Lps* gene is expressed either as responsiveness to LPS (Lps^n), or as LPS non-responsiveness (Lps^d, defective) found in the C3H/HeJ mouse strain. Interestingly, a gene closely linked to or identical with *Lps* also determines the susceptibility to *Salmonella typhimurium* in the C3H/HeJ mouse (O'Brien *et al.*, 1980). The *Lps* gene is currently being cloned by a positional cloning strategy (Qureshi *et al.*, 1996).

Bcg (Nramp1) gene. The *Nramp1* gene has two alternative alleles in inbred mouse strains that encode a single, non-conservative glycine to aspartic acid substitution at position 169 of the Nramp1 protein (Malo *et al.*, 1994) which controls resistance and susceptibility, respectively, to several intracellular pathogens, including mycobacteria, salmonella and leishmania (Vidal *et al.*, 1995a). The identity of the *Bcg* gene with *Nramp1* was established by experiments in $Nramp1^{-/-}$ gene-knockout mice (Vidal *et al.*, 1995b). The *Nramp1* gene does not encode a surface receptor protein or a PAMP receptor, but rather a transmembrane protein expressed in the macrophage phagosomal/late-endosomal compartment (Gruenheid *et al.*, 1997). Recent functional cloning studies support the role of Nramp proteins in the transport of charged ions (Fleming *et al.*, 1997).

NK cells

Beige mutation. Although there is no true NK knockout mouse, the *beige* mutant mouse, together with normal C57BL/6 controls, is used to assess the relative importance of NK cells in resistance to infection, as the most marked deficiency in *beige* mice is in NK cell cytolysis (Roder and Duwe, 1979). A candidate gene for *beige* and the related human disorder Chédiak–Higashi syndrome has been designated *Lyst*, for lysosomal trafficking regulator (Barbosa *et al.*, 1996). It has been shown that in mouse *beige* mutants, lysosomal fusion of NK and other cells is disrupted, which causes increased susceptibility to certain viral and bacterial infections (Scharton-Kersten and Sher, 1997). For example, *beige* mice have increased susceptibility to *Mycobacterium avium* (Florido *et al.*, 1997). However, it is recognized that NK cells of *beige* mice still retain the ability to secrete cytokines, and thus results from *beige* mice must be carefully interpreted (Scharton-Kersten and Sher, 1997).

Cmv1. The NK gene complex (NKC) on the distal portion of mouse chromosome 6 has also been linked with NK-cell mediated defense against viral infections. Two genes, termed *Rmp1* (Delano and Brownstein, 1995) and *Cmv1* (Scalzo *et al.*, 1992), have been linked to the NKC and control resistance against ectromelia virus and cytomegalovirus infections, respectively. The *Cmv1* gene is currently being cloned by a positional cloning approach (Brown *et al.*, 1997; Depatie *et al.*, 1997).

T gamma/delta cells

Although acquired immunity to intracellular infections is primarily mediated by alpha/beta T cells, there is considerable evidence that T cells expressing gamma/delta receptors play a role in protective innate resistance (Boismenu and Havran, 1997). T gamma/delta cells are thought to recognize non-peptide moities of pathogens and are believed to mediate resistance against mycobacteria (Tanaka *et al.*, 1994) and possibly malaria (Rzepczyk *et al.*, 1997). Further studies with T gamma/delta knockout mice should clarify their role in natural resistance to intracellular infections.

Co-stimulatory molecules involved in innate resistance

There are many proteins derived from macrophages, NK cells and T cells that indirectly or directly inhibit growth of microbes and are therefore important in innate immunity. These include interferon (IFN)γ, tumor necrosis factor (TNF), interleukin (IL-4), IL-12, INFγR, nuclear factor (NF)-IL6 and nitric oxide. Several mouse strains with targeted mutations in the genes encoding these proteins are commercially available (see the Jackson website). Studies using mouse knockouts of genes encoding these proteins have greatly contributed to our knowledge regarding innate mechanisms involved in resistance to infections. For example, TNF knockout mice are highly susceptible to *Listeria* and *Candida* infections (Marino *et al.*, 1997). IL-12 knockout mice are susceptible to *M. tuberculosis* (Cooper *et al.*, 1997) and *Leishmania major* infection (Mattner *et al.*, 1996). In addition, mice with a targeted mutation in the nitric oxide synthase NOS2 gene, and therefore unable to produce nitric oxide, are susceptible to *M. tuberculosis* infection (MacMicking *et al.*, 1997). Such gene knockout studies in mice are also a means of identifying candidate infectious disease susceptibility genes in humans (Newport *et al.*, 1996; Abel *et al.*, 1998).

The second line of host defense — the immune system

There are several immunological models which have been used to investigate the role of adaptive immune function on resistance to infection in mice. The primary objective of the mouse models dicussed below is to define the functional role of T and B lymphocytes, and the cytokines they produce, in immunity against various infections, rather than to identify resistance genes *per se*.

H-2 congenic mouse strains. Many host resistance genes have been mapped to the major histocompatibility complex (MHC) on mouse chromosome 17. Through the use of congenic lines bearing H-2 recombinant haplotypes, a particular trait can be shown to be controlled by one of the H-2 regions, or by the TNF locus. Examples of phenotypes which have been linked to the H-2 region are the development of immunity to murine visceral leishmaniasis caused by *L. donovani*, granuloma formation following *M. lepraemurium* infection, resistance to mouse-acquired immunodeficiency syndrome (MAIDS), recovery from Friend virus infection, expulsion of the intestinal helminth *Trichinella spiralis*, and the genetic control of encephalitis following *Toxoplasma gondii* infection (for general review see McLeod *et al.*, 1995).

Nude (nu) mouse. The athymic, hairless nude mouse has been a long-standing tool to analyze the T-cell arm of the cellular immune response to several infections, including BCG (Gros *et al.*, 1983). A *nu* candidate gene, termed Hfh11 on chromosome 11, has been isolated (Nehls *et al.*, 1994). Hfh11 encodes a winged helix protein gene, a DNA-binding domain which is mutated in the nude mouse. The nude mouse manifests a severely defective T-cell response, but retains a relatively intact B-cell population as well as normal macrophage phagocytic response. The nude mouse model has recently been used to demonstrate the pathogenic role of CD4+ T cells in *Leishmania amazonis* infection (Soong *et al.*, 1997).

SCID (severe combined immune deficiency). The mouse SCID mutant (in the C.B-17 strain) carries a nonsense mutation in the gene encoding the catalytic subunit of DNA-activated protein kinase (prkdc) which repairs breaks in double stranded DNA (Blunt *et al.*, 1996). SCID mice are deficient in both B- and T-cell function, and thus tolerate foreign grafts, including injected human mononuclear cells, called the human (hu) PBL/SCID mouse model. The SCID mouse model has been used recently to study human immune responses to influenza virus (Albert *et al.*, 1997) and *M. leprae* (Converse *et al.*, 1995). The human PBL/SCID mouse model is considered a useful model in which specific aspects of immune responsiveness and potential vaccine strategies can be studied in an *in vivo* context.

xid. The *xid* mutation, in the CBA/N mouse, causes a recessive B-cell immunodeficiency that is expressed in B lymphocytes by a variety of abnormal responses including expression of surface immunoglobulin receptors (Wicker and Scher, 1986). The genetic basis for *xid* is a mutation in the Bruton's tyrosine kinase [*btk*] gene, and is homologous to the human X-linked agammaglobulinemia (XLA) mutation (Rawlings *et al.*, 1993; Thomas *et al.*, 1993). Although several mouse strains with targeted immunoglobulin deficiencies have been produced recently (Bot, 1996), the *xid* mouse is still used to analyze the role of B cells and antibody in resistance to infections, such as in *Mycoplasma* infections (Sandstedt *et al.*, 1997), as well as in the response to BCG vaccination (Nikonenko *et al.*, 1996).

Analysis of genetic variation in resistance to infection

In this section, we review strategies to analyze genetic resistance to infection. The overall genetic control of host resistance to infection (encompassing both innate and acquired phases) is generally regarded as a complex trait. However, careful selection of mouse strain and phenotype can allow the complex trait to be dissected into traits under single gene control. Together, these individually mapped (or cloned) genes may be seen as quantitative trait loci (QTL) that may be useful in cross-species comparative gene mapping (Georges, 1997).

Mapping and cloning of host resistance loci (single genes)

Initially, an analysis of inbred mouse strains will show whether or not a phenotype is discontinuous. Should phenotypes of either susceptibility or resistance appear among inbred strains, a single locus with two alternative alleles may control the trait (for reviews see Malo and Skamene, 1994; Nadeau *et al.*, 1995). F1, F2 and backcross progeny from resistant and susceptible strains are then used to determine the mode of inheritance involved. From this point on, the impetus of the investigation will shift to gene cloning, usually by a candidate gene or positional approach. The chromosomal location of the gene is established through strain distribution pattern (SDP) analysis in recombinant inbred strains (RIS) of mice. This approach has been used to identify distinct chromosomal loci controlling resistance to many viral, bacterial and parasitic infections (Malo and Skamene, 1994; Nadeau *et al.*, 1995). Moreover, there are several studies ongoing at present to positionally clone the single genes *Lps* and *Cmv1* (discussed above) as well as *Lgn1,* the gene responsible for resistance to *Legionella pneumophila* (Beckers *et al.*, 1995).

Mapping and cloning of host resistance loci (multiple genes)

If an infectious phenotype in a panel of inbred strains reveals a pattern of continuous or quantitative variation, this may be due to the additive effects of QTL. Identification of the chromosomal regions that likely harbor the susceptibility QTL will proceed with a genome-wide scan of polymorphic markers spaced at 10–20 cM intervals. The analysis will detect maximum lod scores on each chromosome, and QTL are mapped by a non-parametric QTL analysis (Lander and Krugylak, 1995). This approach in experimental mouse crosses has recently been used to detect multiple gene control of variation in susceptibility to trypanosomiasis (Kemp *et al.*, 1997) and *Leishmania major* infection (Beebe *et al.*, 1997; Roberts *et al.*, 1997).

Congenic mapping. In addition to mouse strain intercrosses, the use of congenic strains can also be an effective mapping strategy, as has recently been demonstrated for the mapping of a new locus, *mtv-7*, which controls susceptibility to severe anemia and high parasitemia caused by *Plasmodium yoelli* infection in BALB/c mice (Swardson *et al.*, 1997). In addition, the use of congenic strains for gene mapping has been used to reveal epistatic interactions for the insulin-dependent diabetes (*Idd*) genes (Podolin *et al.*, 1997).

Wild-derived mice. The use of wild-derived mice in the analysis of genetic resistance to infection has had many applications. First, the use of interspecific backcross panels, such as between *Mus spretus* and C57Bl/6J, is used to identify informative polymorphic DNA markers required for fine-resolution genetic mapping (Qureshi *et al.*, 1996; Depatie *et al.*, 1997). Also, a recent study has mapped 146 new genetic loci found mostly in wild-derived strains, thus providing an improved method of genome-scanning for complex genetic trait loci (Elango *et al.*, 1996). Finally, because wild mice and laboratory mice have been subjected to different selective pressures, wild mice are an important model in which to assess population genetics. Recent studies have used wild mice compared to laboratory mice to study resistance to worm infections (Derothe *et al.*, 1997) and to murine leukemia viruses (Lyu and Kozak, 1996).

RCS (recombinant congenic strains). Recombinant congenic strains are created by backcrossing F1 offspring of two progenitor strains already known to show large phenotypic differences (Nadeau *et al.*, 1995). The individual progeny of the second backcross generation each carry approximately 12.5% of the alien genome on 87.5% of the background genome. Therefore, genes controlling any complex multigenic trait become separated (in a relatively narrow interval) in different recombinant congenic strains, allowing the mapping, isolation and study of the individual genes. The major advantage of an RCS panel is that each strain, once inbred, and especially once genotyped with informative chromosomal markers, makes possible the QTL mapping of a large variety of multigenic traits which show genetic variation.

QTL cloning

Once chromosomal regions harboring susceptibility QTL are identified, a strategy is then needed to target candidate genes, or in the absence of suitable candidate genes in the mapped chromosomal region, to use a positional cloning approach. Either approach is critically dependent on further fine-resolution mapping of the genetic interval, which is subject to constant modification, considering the rapidly evolving transcript map of the genome (Weissenbach, 1996). In a candidate approach, comparative QTL mapping can be used to directly test genes mapped in one species for their identity as QTL in another species. This strategy has been successfully used for the identification of *Lps* and *Nramp1* as QTL susceptibility genes for salmonellosis in chicken, genes that were originally identified in mice (Georges, 1997; Hu *et al.*, 1997). For the positional cloning approach, the RCS can be useful since each individual RCS out of the entire panel will usually contain only one donor strain gene influencing the studied trait. Therefore, RCS strains of interest can be used immediately to breed recombinants in the region of interest and to refine the genetic interval.

Analysis of gene function in mice

We have already discussed several targeted mutant mice which have been used in the study of infections. However, it is now generally recognized that in order maximally to utilize the information from these knockout mouse models, the function of the targeted genes should be studied in the context of different background (i.e. modifier) genes (Erickson, 1997). The targeted gene in question is usually transferred to the desired genetic background by standard backcrossing followed by interbreeding, but microsatellite technology has also been developed to speed up the construction of such strains (Matouk *et al.*, 1996).

Breeding special knockouts

Among the first models to study the effect of modifier genes was the creation of cystic fibrosis transmembrane conductance regulator (CFTR)-deficient mice on different inbred backgrounds (Rozmahel *et al.*, 1996). The results showed the existence of major modifier genes which clearly changed the disease severity of the cystic fibrosis phenotype. Several researchers have since extended this observation to the study of infections, considering that the background strain of most targeted genes (usually 129 or Friend virus B (FVB)) is known to influence susceptibility to most infectious pathogens. Gene knockouts of IL-12 and NOS2 have been bred on to different genetic backgrounds in the study of susceptiblity to *M. tuberculosis* (Cooper *et al.*, 1997; MacMicking *et al.*, 1997). In addition, there are commercially available at Jackson Laboratories gene-targeted mutants on several inbred strain backgrounds. For example, mice with a targeted deletion in the IFNγ gene are now available on three different backgrounds (C57BL/6J, BALB/C and DBA/1). Finally, by intercrossing two previously generated null mutant mouse lines, several double-gene knockout mice strains have been created (Tourne *et al.*, 1997; Weih *et al.*, 1997). There are numerous double mutant strains listed at Jackson's Induced Mutant Resource.

In conclusion, it is evident that mouse models continue to provide important insight into the genetics of infectious disease, and, in many instances, provide the best method of searching for candidate susceptibility genes in human diseases with complex phenotypes. We refer readers to the websites listed below for the continually unfolding database of genetic information.

Websites

Jackson Laboratories website

- http://www.jax.org

The Jackson Laboratories Website contains the listing of all the mouse genetic resources, such as the Mouse Mutant Resource and the Induced Mutant Resource (Transgenic and Targeted MutantMice). In addition, the Mouse Genome Informatics Project is a comprehensive, integrated genetics database, and includes the Mouse Genome Database (MGD), the Encyclopedia of the Mouse Genome (a genetic map-viewing application) and the Gene Expression Database Project.

Center for Genome Research, Whitehead Institute and MIT

- http://www.genome.wi.mit.edu/

This World-Wide Web server run by the Center for Genome Research at the Whitehead Institute for Biomedical Research in Cambridge, Massachusetts contains resources and databases on genome sequencing, genome maps, the Human Physical Mapping Project, the Mouse Genetic and Physical Mapping Project and the Rat Genetic Mapping Project.

National Center for Biotechnology Information and the National Library of Medicine Bethesda, MD, USA

- http://www.ncbi.nlm.nih.gov/

NCBI resources include the GenBank Sequence Database, the Gene Map of the Human Genome, OMIM: Online Mendelian Inheritance in Man, the Database of Expressed SequenceTags and Human/Mouse Homology Maps. Database services include PubMed MEDLINE, Entrez Search System and BLAST Sequence Similarity.

References

Abel, L., Sanchez, F. O., Oberti, J. *et al.* (1998). Susceptibility to leprosy is linked to the human *Nramp1* gene. *J. Inf. Dis.*, **177**, 133–145.

Albert, S. E., McKerlie, C., Pester, A., Edgell, B. J., Carlyle, J., Petric, M. (1997). Time-dependent induction of protective anti-influenza immuneresponses in human peripheral blood lymphocyte/SCID mice. *J. Immunol.*, **159**, 1393–1403.

Barbosa, M. D., Nguyen, Q. A., Tchernev, V. T. *et al.* (1996). Identification of the homologous beige and Chediak–Higashi syndrome genes. *Nature*, **382**, 262–265.

Beckers, M-C., Yoshida, S-I., Morgan, K., Skamene, E., Gros, P. (1995). Natural resistance to infection with *Legionella pneumophila*: chromosomal localization of the *Lgn1* susceptibility gene. *Mamm. Genome*, **6**, 540–545.

Beebe, A. M., Mauze, S., Schork, N. J., Coffman, R. L. (1997). Serial backcross mapping of multiple loci associated with resistance to *Leishmania major* in mice. *Immunity*, **6**, 551–557.

Bendelac, A., Fearon, D. T. (1997). Innate pathways that control acquired immunity. *Curr. Op. Immunol.*, **9**, 1–3.

Blunt, T., Gell, D., Fox, M. *et al.* (1996). Identification of a nonsense mutation in the carboxyl-terminal region of DNA-dependent protein kinase catalytic subunit in the scid mouse. *Proc. Natl. Acad. Sci. USA*, **93**, 10285–10290.

Boismenu, R., Havran, W. L. (1997). An innate view of gamma delta T cells. *Curr. Op. Immunol.*, **9**, 57–63.

Bot, A. (1996). Immunoglobulin deficient mice generated by gene targeting as models for studying the immune response. *Int. Rev. Immunol.*, **13**, 327–340.

Brown, M. G., Fulmek, S., Matsumoto, K. *et al.* (1997). A 2-Mb YAC contig and physical map of the natural killer gene complex on mouse chromosome 6. *Genomics*, **42**, 16–25.

Converse, P. J., Haines, V. L., Wondimu, A., Craig, L. E., Meyers, W.M. (1995). Infection of SCID mice with *Mycobacterium leprae* and control with antigen-activated "immune" human peripheral blood mononuclear cells. *Infect. Immun.*, **63**, 1047–1054.

Cooper, A. M., Magram, J., Ferrante, J., Orme, I. M. (1997). Interleukin 12 (IL-12) is crucial to the development of protective immunity in mice intravenously infected with *Mycobacterium tuberculosis*. *J. Exp. Med.*, **186**, 39–45.

Delano, M. L., Brownstein, D. G. (1995). Innate resistance to lethal mousepox is genetically linked to the NK gene complex on chromosome 6 and correlates with early restriction of virus replication by cells with an NK phenotype. *J. Virol.*, **69**, 5875–5877.

Depatie, C., Muise, E., Lepage, P., Gros, P., Vidal, S. M. (1997). High-resolution linkage map in the proximity of the host resistance locus Cmv1. *Genomics*, **39**, 154–163.

Derothe, J. M., Loubes, C., Orth, A., Renaud, F., Moulia, C. (1997). Comparison between patterns of pinworm infection (*Aspiculuris tetraptera*) in wild and laboratory strains of mice, *Mus musculus*. *Int. J. Parasitol.*, **27**, 645–651.

Ehrlich, J., Sankoff, D., Nadeau, J. H. (1997). Synteny conservation and chromosome rearrangements during mammalian evolution. *Genetics*, **147**, 289–296.

Elango, R., Riba, L., Housman, D., Hunter, K. (1996). Generation and mapping of *Mus spretus* strain-specific markers for rapid genomic scanning. *Mamm. Genome*, **7**, 340–334.

Erickson, R. P. (1997). Pigment platelets and Hermansky-Pudlak in human and mouse. *Proc. Natl. Acad. Sci. USA* **94**, 8924–8925.

Fleming, M. D., Trenor, C. C. III, Su, M. A. *et al.* (1997). Microcytic anaemia mice have a mutation in Nramp2, a candidate iron transporter gene. *Nat. Genet.*, **16**, 383–387.

Florido, M., Appelberg, R., Orme, I.-M., Cooper, A. M. (1997). Evidence for a reduced chemokine response in the lungs of beige mice infected with *Mycobacterium avium*. *Immunology*, **90**, 600–606.

Georges, M. (1997). QTL mapping to QTL cloning: mice to the rescue. *Genome Res.*, **7**, 665–667.

Gros, P., Skamene, E., Forget, A. (1983). Cellular mechanisms of genetically controlled host resistance to *Mycobacterium bovis* (BCG). *J. Immunol.*, **131**, 1966–1972.

Gruenheid, S., Pinner, E., Desjardins, M., Gros, P. (1997). Natural resistance to infection with intracellular pathogens: the Nramp1 protein is recruited to the membrane of the phagosome. *J. Exp. Med.*, **185**, 717–730.

Hu, J., Bumstead, N., Barrow, P. *et al.* (1997). Resistance to salmonellosis in the chicken is linked to NRAMP1 and TNC. *Genome Res.*, **7**, 693–705.

Kemp, S. J., Iraqi, F., Darvasi, A., Soller, M., Teale, A. J. (1997). Localization of genes controlling resistance to trypanosomiasis in mice. *Nat. Genet.*, **16**, 194–196.

Lander, E., Kruglyak, L. (1995). Genetic dissection of complex traits: guidelines for interpreting and reporting linkage results. *Nat. Genet.*, **11**, 241–247.

Lougheed, M., Lum, C M., Ling, W., Suzuki, H., Kodama, T., Steinbrecher, U. (1997). High affinity saturable uptake of oxidized low density lipoprotein by macrophages from mice lacking the scavenger receptor class A type I/II. *J. Biol. Chem.*, **272**, 12938–12944.

Lyu, M. S., Kozak, C. A. (1996). Genetic basis for resistance to polytropic murine leukemia viruses in the wild mouse species *Mus castaneus*. *J. Virol.*, **70**, 830–833.

MacMicking, J. D., North, R. J., LaCourse, R., Mudgett, J. S., Shah, S. K., Nathan, C. F. (1997). Identification of nitric oxide synthase as a protective locus against tuberculosis. *Proc. Natl. Acad. Sci. USA*, **94**, 5243–5248.

Malo, D., Skamene, E. (1994). Genetic control of host resistance to infection. *Trends Genet.*, **10**, 365–371.

Malo, D., Vogan, K., Vidal, S. *et al.* (1994). Haplotype mapping and sequence analysis of the mouse Nramp gene predict susceptibility to infection with intracellular parasites. *Genomics*, **23**, 51–61.

Marino, M. W., Dunn, A., Grail, D. *et al.* (1997). Characterization of tumor necrosis factor-deficient mice. *Proc. Natl. Acad. Sci. USA*, **94**, 8093–8098.

Matouk, C., Gosselin, D., Malo, D., Skamene, E., Radzioch, D. (1996). PCR-analyzed microsatellites for the inbred mouse strain 129/Sv, the strain most commonly used in gene knockout technology. *Mamm. Genome*, **7**, 603–605.

Mattner, F., Magram, J., Ferrante, J. *et al.* (1996). Genetically resistant mice lacking interleukin-12 are susceptible to infection with *Leishmania major* and mount a polarized Th2 cell response. *Eur. J. Immunol.*, **26**, 1553–1559.

McLeod, R., Buschman, E., Arbuckle, L. D., Skamene, E. (1995). Immunogenetics in the analysis of resistance to intracellular pathogens. *Curr. Op. Immunol.*, **7**, 539–552.

Medzhitov, R., Janeway, C. A. Jr. (1997). Innate immunity: impact on the adaptive immune response. *Curr. Op. Immunol.*, **9**, 4–9.

Nadeau, J. H., Arbuckle, L. D., Skamene, E. (1995). Genetic dissection of inflammatory diseases. *J. Inflamm.*, **45**, 27–48.

Nehls, M., Pfeifer, D., Schorpp, M., Hedrich, H., Boehm, T. (1994). New member of the winged-helix protein family disrupted in mouse and rat nude mutations. *Nature*, **372**, 103–107.

Newport, M. J., Huxley, C. M., Huston, S. *et al.* (1996). A mutation in the interferon-gamma-receptor gene and susceptibility to mycobacterial infection. *N. Engl. J. Med.*, **335**, 1941–1949.

Nikonenko, B. V., Apt, A. S., Mezhlumova, M. B., Avdienko, V. G., Yeremeev, V. V., Moroz, A. M. (1996). Influence of the mouse Bcg, Tbc-1 and xid genes on resistance and immune responses to tuberculosis infection and efficacy of bacille Calmette-Guerin (BCG) vaccination. *Clin. Exp. Immunol.*, **104**, 37–43.

O'Brien, A. D., Rosenstreich, D. L., Taylor, B. A. (1980). Control of natural resistance to *Salmonella typhimurium* and *Leishmania donovani* in mice by closely linked but distinct genetic loci. *Nature*, **287**, 440–442.

Pearson, A. M. (1996). Scavenger receptors in innate immunity. *Curr. Op. Immunol.*, **8**, 20–28.

Podolin, P. L., Denny, P., Lord, C. J. *et al.* (1997). Congenic mapping of the insulin-dependent diabetes (Idd) gene, Idd10, localizes two genes mediating the Idd10 effect and eliminates the candidate Fcgr1. *J. Immunol.*, **159**, 1835–1843.

Qureshi, S. T., Larivière, L., Sebastiani, G. *et al.* (1996). A high-resolution map in the chromosomal region surrounding the *Lps* locus. *Genomics*, **31**, 283–294.

Rawlings, D. J., Saffran, D. C., Tsukada, S. *et al.* (1993). Mutation of unique region of Bruton's tyrosine kinase in immunodeficient XID mice. *Science*, **261**, 358–361.

Roberts, L. J., Baldwin, T. M., Curtis, J. M., Handman, E., Foote, S. J. (1997). Resistance to *Leishmania major* is linked to the H2 region on chromosome 17 and to chromosome 9. *J. Exp. Med.*, **185**, 1705–1710.

Roder, J., Duwe, A. (1979). The beige mutation in the mouse selectively impairs natural killer cell function. *Nature*, **278**, 451–453.

Rozmahel, R., Wilschanski, M., Matin, A. *et al.* (1996). Modulation of disease severity in cystic fibrosis transmembrane conductance regulator deficient mice by a secondary genetic factor. *Nat. Genet.*, **12**, 280–287.

Rzepczyk, C. M., Anderson, K., Stamatiou, S. *et al.* (1997). Gamma delta T cells: their immunobiology and role in malaria infections. *Int. J. Parasitol.*, **27**, 191–200.

Sandstedt, K., Berglof, A., Feinstein, R., Bolske, G., Evengard, B., Smith, C. I. (1997). Differential susceptibility to *Mycoplasma pulmonis* intranasal infection in X-linked immunodeficient (xid), severe combined immunodeficient (scid), and immunocompetent mice. *Clin. Exp. Immunol.*, **108**, 490–496.

Scalzo, A. A., Fitzgerald, N. A., Wallace, C. R. *et al.* (1992). The effect of the Cmv-1 resistance gene, which is linked to the natural killer cell gene complex, is mediated by natural killer cells. *J. Immunol.*, **149**, 581–589.

Scharton-Kersten, T. M., Sher, A. (1997). Role of natural killer cells in innate resistance to protozoan infections. *Curr. Op. Immunol.*, **9**, 44–51.

Schurr, E., Skamene, E. (1996). The role of the Bcg gene in mycobacterial infections. In: *Tuberculosis* (eds Rom, W. N., Garay, S. M.) pp. 247–258. Little, Brown, New York.

Schurr, E., Malo, D., Radzioch, D. *et al.* (1991). Genetic control of innate resistance to mycobacterial infections. *Immunol. Today*, **12**, A42–A45.

Skamene, E, Schurr, E., Gros, P. (1998). Infection genomics: *Nramp1* as a major determinant of natural resistance to intracellular infections. *Annu. Rev. Med.*, **49**, 275–287.

Soong, L., Chang, C. H., Sun, J. *et al.* (1997). Role of CD4+ T cells in pathogenesis associated with *Leishmania amazonensis* infection. *J. Immunol.*, **158**, 5374–5383.

Sultzer, B. M., Castagna, R., Bandekar, J., Wong, P. (1993). Lipopolysaccharide nonresponder cells: the C3H/HeJ defect. *Immunobiology*, **187**, 257–271.

Swardson, C. J., Wassom, D. L., Avery, A. C. (1997). *Plasmodium yoelii*: resistance to disease is linked to the mtv-7 locus in BALB/c mice. *Exp. Parasitol.*, **86**, 102–109.

Tanaka, Y., Sano, S., Nieves, E. *et al.* (1994). Nonpeptide ligands for human gamma delta T cells. *Proc. Natl. Acad. Sci. USA*, **91**, 8175–8179.

Thomas, J. D., Sideras, P., Smith, C. I., Vorechovsky, I., Chapman, V., Paul, W. E. (1993). Colocalization of X-linked agammaglobulinemia and X-linked immunodeficiency genes. *Science*, **261**, 355–358.

Tourne, S., Miyazaki, T., Wolf, P., Ploegh, H., Benoist, C., Mathis, D. (1997). Functionality of major histocompatibility complex class II molecules in mice doubly deficient for invariant chain and H-2M complexes. *Proc. Natl. Acad. Sci. USA*, **94**, 9255–9260.

Vidal, S., Gros, P., Skamene, E. (1995a). Natural resistance to infection with intracellular parasites: molecular genetics identifies *Nramp1* as the *Bcg/Ity/Lsh* locus. *J. Leuk. Biol.*, **58**, 382–390.

Vidal, S., Tremblay, M., Govoni, G. *et al.* (1995b). The *Ity/Lsh/Bcg* locus: natural resistance to infection with intracellular parasites is abrogated by disruption of the *Nramp1* gene. *J. Exp. Med.*, **182**, 655–666.

Weih, F., Durham S. K., Barton D. S., Sha, W. C., Baltimore, D., Bravo, R. (1997). p50-NF-B complexes partially compensate for the absence of RelB: severely increased pathology in p50/relB/ double-knockout mice *J. Exp. Med.*, **185**, 1359–1370.

Weissenbach, J. (1996). Landing on the genome. *Science*, **27**, 479.

Wicker, L. S., Scher, I. (1986). X-linked immune deficiency (xid) of CBA/N mice. *Curr. Top. Microbiol. Immunol.*, **124**, 87–101.

Chapter 10

Formulation of Compounds and Determination of Pharmacokinetic Parameters

R. M. Cozens

Introduction

The ideal animal model of infection will provide data that allows the outcome of therapy in humans to be predicted. For a variety of reasons this ideal is difficult to attain; possible reasons include, among others, the different pharmacokinetics of compounds in animals and humans. It is well-established that pharmacokinetics of therapeutic agents is usually dependent on the species. Not only are there differences between animals and humans but there are also significant differences between animals — a fact made use of in making extrapolations from animals to humans, as will be described later.

Pharmacokinetics in any species is governed by the physico-chemical properties of the drug. The processes of absorption and disposition can be influenced by the physiology and anatomy of the animal and also metabolism, which is often unpredictable (Williams, 1974; Dedrick and Bischoff, 1984). Although not often a concern in animal experiments, it is worth remembering that the age of the subject can also have a profound influence an the pharmacokinetic properties of a compound not only in humans but also probably in animals. The differences often observed between the pharmacokinetics of anti-infective agents in animals and humans means that efficacy data obtained in model infections is really only interpretable as a predictor of the clinical situation if the pharmacokinetics of the drug is taken into account. A number of studies have shown that with a knowledge of the pharmacokinetics in both humans and animals it may be possible to design treatment regimens in animals which result in at least an approximation of the pharmacokinetic profiles in humans (Gerber *et al.*, 1986, 1991; Flückiger *et al.*, 1991; Lister and Sanders, 1995). There are a number of publications that deal with some of the issues surrounding the pharmacokinetics of anti-infectives in animals and the usefulness of pharmacokinetic data obtained in animals (Craig *et al.*, 1988; Drusano, 1988; Mizen and Woodnutt, 1988; Dalhoff and Ullmann, 1990; Barza, 1993). The aim of this chapter is not to add to this debate but rather to describe some of the ways pharmacokinetic data can be obtained from animals and give some hints on how the data can be used to aid the design, interpretation and predictive value of efficacy experiments in animals. However, it is up to the experimenter to establish how useful the pharmacokinetic information is and how it can best be used in a particular situation.

The determination of pharmacokinetic parameters and comparison between compounds and species only makes sense if the form in which the drug is presented is taken into account. The vehicle used to administer a drug, particularly via the oral route, can have a profound influence on the rate and extent of absorption. In animal studies, especially during the early evaluation of a new drug when optimized formulations are unavailable, it is often expedient to use formulations which may be quite different to those ultimately used during the clinical evaluation and use of the compound. This chapter will, therefore, also consider some of the formulations that can be used for the administration of anti-infective agents to animals.

Formulation of anti-infectives for administration to animals

If animals are to receive a compound already available for human use, it may be possible to use the compound in its clinical formulation; indeed, this is probably desirable. However, care must be taken to ensure that excipients are suitable for administration to animals and it must be remembered that it is seldom possible to give the dosage form directly to animals without some further preparation. For example, material presented in a capsule for human use must, in most cases, be removed from the capsule and dissolved or suspended in a vehicle for administration.

In this section I have limited myself to discussing preparation of formulations from the pure active compound and to those procedures which can be carried out in a laboratory equipped with apparatus available to most microbiology laboratories. More advanced formulations, such as liposomes (Bakker-Woudenberg *et al.*, 1993) or nanoparticles (Leroux *et al.*, 1995, 1996), may certainly have a role to play in the evaluation of anti-infectives but are best prepared by specialists. Only injection and oral administration will be considered. Other routes of administration may be considered for particular situations. For example, lung infections

Handbook of Animal Models of Infection
ISBN 0-12-775390-7

may be amenable to therapy administered in an aerosol or skin infections may be treatable by a topical formulation. Such formulations require specialist knowledge and equipment to manufacture and use, and are beyond the scope of this chapter. There are obviously some formulations for oral administration and injection which also require specialist equipment such as milling apparatus. The production of these is probably best left to specialist formulation laboratories. Although those administering compounds to animals will occasionally need to call on specialist expertise to solve intractable formulation problems, much can be done using simple, readily available materials and equipment. Obviously, if the aim is to attempt to correlate pharmacokinetics with efficacy then the same formulations need to be used in both circumstances. This may seem trivial but as efficacy studies and pharmacokinetic studies are often performed in different laboratories or even different departments in an institution, some degree of agreement needs to be reached. It may be that a proposed excipient, while eminently suitable for efficacy experiments, may interfere with the analytical method employed. In what follows, a number of excipients are mentioned. Those mentioned are by no means exhaustive and information, such as composition, toxicity data, etc., on these and others can be found in the *Handbook of Pharmaceutical Excipients* (Wade and Weller, 1994). For parenteral administration a useful handbook exists which gives details, including excipients, of injectable drugs (Trissell, 1992). This can provide useful starting points, particularly if compounds are congeners of those described in the handbook. It should be remembered that whatever excipients are included, they may not be pharmacologically inert. They may not only have direct effects on the animal but may also alter the basic pharmacokinetics of the compound (e.g. Sparreboom *et al.*, 1996).

Parenteral administration

If the compound is sufficiently soluble in aqueous solvents, then a simple solution in a simple vehicle will be the most acceptable formulation. The formulation can be easily sterilized by filtration and will also be suitable for oral administration. Possible binding of the compound to the filtration equipment should be checked. If necessary the concentration in the filtrate should be determined to ensure accurate dosing. The best vehicle is physiologically neutral (neutral pH and isotonic); saline or phosphate buffered saline is the most common. Deviations from isotonicity or neutrality may lead to irritation at the site of injection and should therefore be kept to a minimum.

Compounds which have poor aqueous solubility can often be administered as solutions in mixed-solvent systems. The compound may first be dissolved in a suitable organic solvent and then diluted with saline or phosphate-buffered saline (PBS). The amount of organic solvent in the final formulation should be kept to an absolute minimum. For example, dimethyl sulphoxide (DMSO) should be present at concentrations no higher than 10% v/v; ethanol, 10% v/v and benzyl alcohol, 6% v/v.

Cyclodextrins (Duchêne and Wouessidjewe, 1993; Loftsson and Brewster, 1996) have been used to solubilize compounds with poor water solubility. Cyclodextrins are a group of cyclical oligosaccharides consisting of 6 (α-cyclodextrins), 7 (β-cyclodextrins) or 8 (γ-cyclodextrins) glucose residues. The glucose residues may be derivatized to improve water solubility. The molecules form toroidal structures in aqueous solution. The interior of the torus is hydrophobic and can accommodate molecules of the correct size and hydrophobicity, forming a complex that is soluble in water. The choice of cyclodextrin will depend on the properties of the molecule of interest. On administration the compound is released from the complex. The potential problem with these molecules is that the rate of release of the active molecules may not be immediate, which may lead to atypical pharmacokinetic behaviour. These formulations can also be administered orally; however, it has been reported that cyclodextrins can influence the uptake of compounds and so may not play a simple carrier role.

Any solution intended for intravenous formulation should be checked for the possibility that on addition to plasma the compound does not remain in solution. Precipitation of the compound in the circulation could result in misleading pharmacokinetic properties or even cause unnecessary suffering to the animal. Precipitation of the active ingredient is unlikely to occur if the formulation contains only water or aqueous solutions but may be a problem when organic solvents are employed.

Suspensions should only be administered intravenously when they are well-characterized. This usually involves incorporation into liposomes or nanoparticles (Leroux *et al.*, 1995), although simple suspensions can be employed. In general, particle sizes should be 100 µm or smaller. Suspensions can be contemplated for other parenteral routes (e.g. subcutaneous and intramuscular) but formulations used for these routes of administration must be well-tolerated at the site of injection. These routes may lead to slower release of the compound into the circulation, although this cannot be assumed. For slow release, special formulations are required and these are best developed and produced by specialist formulation scientists and will not be considered further here.

Oral administration

Formulations for oral administration of compounds can vary from, for example, simple powder in gelatin capsules to complex mixtures designed to release the active ingredient in a particular portion of the GI tract (Leroux *et al.*, 1996).

Oral formulations must be designed to ensure accurate and reproducible dosing and to result in optimum absorption of the compound. The latter is easier said than done and optimization of a formulation will often require specialist assistance. However, whenever a compound is administered orally and results in relatively poor plasma concentrations, which can be shown to be the result of poor absorption, it may be worthwhile trying another simple formulation before having recourse to complex, specialist formulations.

As with intravenous administration, the easiest formulation is that in an aqueous vehicle. If the compound is soluble then often a simple solution in water is sufficient. Aqueous vehicles for oral administration need not be precisely physiological; however, if relatively large volumes are to be administered then extremes should be avoided.

Suspensions can be readily administered by the oral route. There are two important factors to be considered if suspensions are used. First, the suspension should be homogeneous to ensure accurate and reproducible dosing—it is impossible to administer an accurate and reproducible dose from a suspension which contains large clumps of material. Second, the particle size should be as small as possible and the particle size and distribution should be known if possible. As a minimum, microscopic examination of the suspension should be undertaken and particle size and distribution estimated. Generally compounds must be dissolved before they can be absorbed from the gastrointestinal tract. Particle size of poorly soluble compounds can markedly affect the rate of dissolution which, if too slow, can mean that a significant proportion of the compound will pass through the gastrointestinal tract before it has time to be absorbed. Micronized material (median particle size less than 10 μm) can be obtained by wet or dry milling. Milling requires specialist equipment and often results in a loss of quite large amounts of compound. The nature of the solid must also be known; whether the compound is crystalline (and the form of the crystals) or amorphous can also affect dissolution rates and hence pharmacokinetic properties.

The simplest suspensions are prepared in water containing an agent to increase the viscosity and thereby the physical stability of the suspension. Suitable agents are the chemically modified celluloses. Klucel HF (hydroxypropyl cellulose) at a concentration of 0.5% w/v is widely employed; the concentration can be increased if larger particles have to be kept in suspension. For some compounds it may be necessary to incorporate a wetting agent. Tween 80 at a concentration of 0.1–1% is most suitable. Other wetting agents include the poloxamers (e.g. Pluronic), polyoxyethylene alkyl ethers (e.g. Brij), the polyoxyethylene castor oil derivatives (e.g. Cremphor), etc. The final step is sonication of the suspension to ensure homogeneity. It should be remembered that too powerful or too long exposure to ultrasound can actually result in the formation of large aggregates. The use of low-energy ultrasound baths is frequently sufficient to produce a finely divided, homogeneous suspension. In some cases the wetting agents may render the compound soluble.

A procedure used frequently in our laboratories and found suitable for many compounds that are insoluble in water is detailed here.

Compounds are dissolved in DMSO to a concentration of 20× the final concentration. The DMSO solution is slowly diluted 1 : 20 with 1% Tween 80 in water. Thorough vortexing and possibly light sonication (low-energy bath) produces a finely divided suspension of amorphous active ingredient. Subsequent crystal formation in this formulation cannot be excluded and it should therefore be prepared just before use. The universal solvent DMSO is very useful for preliminary formulations; however, it is recognized that DMSO may have a positive effect on oral absorption of drugs and this must be taken into account when comparing treatment with different agents which may be formulated differently. In some cases ethanol or other organic solvents can be used in place of DMSO in this formulation.

Gelucires have proved to be useful for some compounds. Gelucires are polyglycerides which are manufactured in various grades having different melting points and dispersibility in water. The most useful is Gelucire 44/14 (the first number indicates the melting point and the second the dispersibility in water). The Gelucire is warmed to its melting point and the compound added. The compound may dissolve or be dispersed as a finely divided suspension. The melted Gelucire is then diluted with water and sonicated to produce a fine suspension. Alternatively, after the addition of the compound the Gelucire can be allowed to solidify and stored. Portions prepared in this way can be taken, water added and the suspension prepared as required.

In the smaller species used in the majority of experiments in anti-infective research, it is not possible to administer the compound of interest in a solid form, although this can be easily achieved in larger species. However, using a suitable applicator, small hard-gelatin capsules with dimensions ca. 3×5 mm can be administered by gavage to rats. Following administration of the capsules it is important to administer water (ca. 1 ml) to ensure efficient disintegration of the capsule.

Whichever formulation is chosen, it is usually preferable to prepare the material just before administration. If the chemical stability of the active ingredient and the physical stability (sedimentation and resuspension, growth of crystals, etc.) of the formulation are known then it may be possible to prepare larger batches of the formulation in advance. However, these properties are seldom known and can be difficult to determine and it is therefore preferable to use freshly prepared formulations.

Pharmacokinetics

General considerations

The need for the evaluation of pharmacokinetics in the use of anti-infectives is apparent in such things as dose monitoring, dose adjustment in patients with underlying disease or when receiving therapy for underlying diseases when drug interactions may be an issue. Dose monitoring

(Horrevorts and Mouton, 1993) is usually indicated when a drug has a narrow therapeutic window or shows large interindividual variation. In most instances these situations are not relevant to those investigating the effects of anti-infectives in animals. The need for pharmacokinetics in the laboratory situation is rather to aid interpretation of results and possibly to allow prediction to humans.

Pharmacokinetic determinations may also be useful as a preliminary screen employed before investigations of efficacy are undertaken. If an oral therapy is proposed then a small pilot experiment can be used to demonstrate oral bioavailability before a time-consuming efficacy experiment is undertaken. There are four main species used for anti-infective research and it is expedient to consider only the determination of pharmacokinetic parameters in these, namely, the mouse, the rat, the rabbit and the guinea pig. As the methods for administration of the compounds to these species are well-known, only the collection of samples for analysis will be considered here. At this point it is sensible to consider experimental design.

Before embarking on a pharmacokinetic experiment the method of analysis must be considered and some questions answered. Is whole blood or plasma or even serum to be analyzed? What anticoagulant should be used? Are any excipients likely to interfere with the analysis? How much sample is required? Once these questions have been answered then the next steps in the experiment can be planned. For how long will samples be collected? How many samples will be collected? The accuracy of the determination of the pharmacokinetic parameters is influenced by the number and frequency of samples. For example, an accurate terminal half-life can only be obtained when there are at least three (and preferably more) points on the linear portion of the curve. If a compound is rapidly absorbed and then excreted after oral administration then more early time points should be included. Conversely, for a compound which is absorbed slowly and has a long elimination half-life then more samples at later times are preferable. Given the limitations of the number of samples which can be taken from one animal it may be necessary to design an experiment in which not all animals are sampled at all time points. If a compound is being examined for which no pharmacokinetic information is available then a smaller pilot study may be useful to determine the times of sampling and the number of samples that should be taken. Food can have an influence on the pharmacokinetics of orally administered compounds. Therefore, it must be decided whether the animals should have free access to food and water or whether they should be fasted. Clearly, if the intention is to use pharmacokinetics to aid planning and interpretation of efficacy experiments then the pharmacokinetic experiments will be done in animals held under the same conditions as those employed in the efficacy experiments. With rodents it is often difficult to ensure that all animals employed have the same nutritional status. This means that if the oral bioavailability of a compound is markedly affected by the presence of food then in animals that have free access to food and water there is often a larger degree of variation than would be found if the animals had been fasted.

Sampling from animals

Mouse

Due to their small size and animal welfare considerations it is usually only possible to take one terminal blood sample from a mouse. Although it may be possible to take multiple samples via the tail vein, in most cases the volumes which can be taken are barely enough for one analysis. Similar and additional constraints apply to sampling from veins following surgical exposure. This means that experiments over a suitable time course require a number of animals and this can lead to data which can be rather variable. For this reason at each time point a suitable number of animals should be taken if variability is to be countered. A minimum of four mice at each time has been found to be suitable in most cases.

Some compounds may themselves have intrinsic variability due, for example, to having an absorption window or to being administered in poorly prepared formulations (see above). In the former case an increase in the number of animals at each time point should be considered, while in the latter the best solution would be to seek a better formulation. Samples are best collected under anesthesia from which the animals are not allowed to recover. Well-trained operators can retrieve at least 1 ml of blood from puncture of the inferior vena cava or directly from the heart. These two sites are probably the easiest from which to obtain blood which is uncontaminated with tissue fluid or other extraneous matter. The amount of blood or plasma required will of course depend on the method used for analysis. In some cases as little as 10 µl will be sufficient. With less sensitive methods then up to 1 ml (the maximum which can be guaranteed) may be required. Mice are frequently the first animal in which the efficacy of a new agent is investigated. As already mentioned above, in this situation it may be useful to obtain some basic knowledge of the pharmacokinetics of the compound before an efficacy experiment is undertaken. The small size of mice means that with very little compound some idea of oral bioavailability can be gauged.

In the search for orally active human immunodeficiency virus 1 (HIV-1) protease inhibitors, we have routinely used mice in a pharmacokinetic screen (Alteri *et al.*, 1993; Capraro *et al.*, 1996; Cozens *et al.*, 1996). Mice receive a standard oral dose of compound in a standard formulation. At 30, 60, 90 and 120 minutes after administration four mice are killed and plasma concentrations of the compound determined by high-performance liquid chromatography (HPLC). These experiments can be used to select compounds suitable for further evaluations. The time points chosen have been shown in most cases to encompass the maximum plasma concentration. When starting with a new class of compounds it is worthwhile investigating tissue concentrations and trying to correlate plasma concentration with efficacy. Compounds which are quickly and extensively distributed may still be effective but have relatively low concentrations in the circulation. One example of this would be the situation of azithromycin (Girard *et al.*, 1987, 1990), which is extensively and rapidly distributed to tissues. Once correlations have been obtained then it may be possible to select compounds for efficacy testing *in vivo* based on plasma concentrations attained in this simple model.

Rat

For pharmacokinetic purposes, rats can be used in the same way as mice, however, their larger size does open some other possibilities. Sampling from the tail vein of a rat, particularly in older animals, is very difficult, if not impossible. There are two additional veins from which blood can be collected, namely the sublingual and penile veins (these veins are also found in mice but are usually so small that sampling is not practicable). While it is possible to remove several samples from a single animal via these routes the technique requires an anesthetic to be given to the animal and this may be avoided if the trouble is taken to implant catheters in suitable blood vessels. Implantation of a catheter in an artery is usually preferable to one in a vein as the sampling is then much more secure and the catheters are much less likely to become blocked. Standard surgical techniques can be employed to implant catheters in the jugular vein or carotid artery. However, in this position there can be interference with administration of the compounds by the oral route and a preferred vessel is the femoral vein or artery. This also has the advantage that the vessels are rather bigger than those in the neck and therefore the implantation is quicker and easier.

If it is intended to administer the compound by the intravenous route then it is worth considering catheterization of both the femoral vein and artery: the latter is used to obtain the sample and the former to administer the compound. The use of suitable tethers, such as the Harvard swivel tether (Harvard Bioscience), will protect the catheter from the unwanted attentions of the rat and allow the possibility of sampling from an unanesthetized animal while allowing the rat almost full freedom of movement. With careful rinsing of the catheter with saline containing an anticoagulant it is easily possible to keep the rats for 48 hours after implantation of the catheter. As most single dose experiments will not last longer than 24 hours after administration of the drug, this length of time is adequate for most purposes.

If it is intended to examine the pharmacokinetics after multiple administration, then it is probably better to perform the surgery on rats which have already received treatment rather than attempt to keep animals for several days with implanted catheters. Using these techniques it is possible to take up to 4 ml of blood in a 24-hour period if the blood volume is replaced. Replacement of the blood volume can cause some small errors in the concentration in the circulation. Usually these are small; 4 ml of blood taken from a rat weighing 300 grams represents approximately 12% of the blood volume. The blood can be replaced with physiological saline in most cases. In special circumstances, for example when drug may partition strongly to erythrocytes, it may be desirable to use blood from donor animals. The need for this and the influence of blood volume replacement can be easily investigated in small pilot experiments *in vivo* or even *in vitro*.

Not only the circulation of rats can be sampled—recently the pharmacokinetics and distribution of the antiviral penciclovir to the brain of rats has been studied (Borg and Ståhle, 1997). The authors used microdialysis to sample the unbound extracellular fraction of the drug in the brain. The basic technique of microdialysis has been described by Ståhle (1991, 1993). The technique is performed in anesthetized animals but has the advantage that samples can be obtained without altering the volume of fluid in the sampling compartment and frequent samples can be obtained. The samples are also essentially free of protein and can be analyzed directly by HPLC. The technique is obviously not limited to rats or to the brain and could be adapted to other species and tissues.

The oral administration of compounds can result in low plasma levels due not to poor absorption but rather to efficient first-pass extraction by the liver. An indication of the relative contributions of first-pass effects and absorption to plasma levels can be gleaned if samples are taken from the hepatic portal vein and the peripheral circulation. In anesthetized rats a cannula can be inserted into the hepatic portal vein and secured in place with veterinary cyanoacrylic glue—a procedure which ensures a maintained blood flow through the vein. A catheter implanted in the carotid or femoral arteries allows sampling of the peripheral circulation. As the animals are anesthetized it is not possible to give the compound orally as stomach emptying and intestinal motility are invariably reduced by the anesthesia. Instead, the drug is given by intraduodenal injection, which has the added advantage that, if the concentrations in the peripheral circulation in this model are compared with those after administration by gavage, the possible effects of gastric instability can also be determined.

A comparison of the concentrations of compound in the hepatic portal vein and in the peripheral circulation allows the relative contributions of absorption and first-pass effects to be assessed. The effects of first-pass extraction alone could be assessed if the compound were to be injected into the hepatic portal vein.

Guinea pigs

Guinea pigs can be used in the same way as rats. This species has no easily accessible veins from which to remove blood, although marginal veins on the inside of the thigh, on the foot and ankle and the penile vein can be used by skilled operatives and other veins can be sampled after surgical exposure. This means that the use of indwelling catheters should be considered as the method of choice for this species.

Rabbits

The presence of the large blood vessels on the ears of rabbits means that blood collection from this species is relatively easy even in conscious animals. No special preparation is required. In some experiments it may be possible to insert a catheter into the vessel which remains in place throughout, but this is probably not necessary.

Other species

Although other species are not so often employed for efficacy studies a knowledge of the pharmacokinetics in larger animals such as dogs or monkeys can be useful to aid extrapolations of animal data to the situation in humans (see below). The larger species present no particular problems. Experienced operatives using trained dogs can administer compounds and obtain many blood samples from an individual animal with the minimum of stress to the animal.

The dog opens up several opportunities which are not possible in smaller animals. Recently, Sagara and co-workers (Sagara *et al.*, 1994), used pentagastrin and atropine sulphate to regulate the gastrointestinal physiology in dogs. Much of the differences in oral absorption between species depends on the rate of movement of drugs through the gut (intestinal motility and gastric emptying) and differences in gastric pH. These can be regulated by the co-administration of pentagastrin and atropine sulphate. A cautionary note is necessary; the co-administration of any compound can affect the pharmacokinetics of the compound under study and the extent of any influence should be determined in control experiments.

Another possibility is to investigate regiospecific absorption of the drug. This information can be useful to optimize formulations, particularly if slow-release formulations are contemplated. Kwei *et al.* (1995) have used beagle dogs fitted with chronic cannulas allowing direct administration of drug solutions to specific regions of the gastrointestinal tract to investigate the differences in absorption in different portions of the gut of an HIV-protease inhibitor.

Dogs and other larger species have the advantage that they can be reused. Interindividual variation can be considerable for some compounds. In dogs cross-over experimental designs can be contemplated. The only prerequisite is that sufficient time is left to ensure that essentially all the compound administered in the first use is eliminated (including metabolites) before the animals are reused. Such experiments get around the problem of interindividual variation and are recommended whenever possible.

Methods of analysis

For the investigation of new agents the choice falls between microbiological or other biological assay and HPLC. Other techniques, such as radioimmunoassay, which may be considered are compound-specific, would only be used on a case-by-case basis and usually will only be available for compounds already in clinical use or those in an advanced stage of development. Microbiological assays for antibacterials are well-defined and can quite quickly be established for a new compound. Other biological assays for other anti-infectives can be foreseen but are usually only developed on a case-by-case basis. Therefore for anti-infectives as a whole HPLC is probably the method of first choice.

There are two main weaknesses of HPLC: cost and the need for experienced operators to establish suitable methods. Microbiological assay has the advantage that it is relatively cheap and should be able to be established quickly in any microbiology laboratory. The two main strengths of HPLC is that it is specific, and can often be adapted to measure both parent compound and any potential metabolite, and has the potential to be more sensitive. The main weakness of microbiological assay is that it lacks specificity. If a compound were metabolized to an active metabolite then it would not be possible accurately to measure the concentration of the parent in a microbiological assay. The method is also relatively insensitive. Here a word of caution is needed. Often analysts strive for increased sensitivity. While in some cases this may be desirable, allowing smaller sample volumes to obtain the same data, in many cases it may not be necessary. If it is only required to know how long a compound is present at a level above the minimal inhibitory compound (MIC) then it is not necessary to strive to obtain a method which can detect and quantitate amounts several-fold less than the MIC.

Recent developments in mass spectrometry allied to HPLC mean that mass selective detectors are now available to all; hitherto liquid chromatography/mass spectrometry was a technique which required extensive practical and theoretical knowledge. HPLC allied to mass spectrometry has the possibility of improving sensitivity and selectivity, allowing quicker method development. Mass selective detection may also allow the simultaneous determination of compounds in a mixture (Berman *et al.*, 1997; Olah *et al.*, 1997). This opens up the possibility of administration of a carefully selected mixture to an animal and determination of individual compounds in the circulation. The savings in time and animals could be considerable and the method has much to commend it as a screening system. However, interaction of compounds can cause profound mutual changes in pharmacokinetic properties and therefore results should be interpreted with caution, and certainly the pharmacokinetics of interesting compounds repeated in subsequent experiments in which the compound is given alone.

The most difficult part of analysis by HPLC is the development of suitable methods. As already stated, the use of mass selective detectors may help in this regard. For screening purposes it is often possible to establish generic methods which can be used by all compounds of a particular class. The aim here is not necessarily to obtain the highest sensitivity; however, the method should be adequately sensitive, selective and the work-up of samples relatively easy. For more advanced pharmacokinetic experiments then the method used for screening may need to be altered to increase sensitivity. There are any number of different methods for sample preparation and chromatography. Even for one compound a review of the literature will reveal many methods, all of which could be used. The choice of method will often be dictated by one's own experience and knowledge as well as access to particular equipment. Klassen and Edberg (1996) have recently reviewed some of the methods for the analysis of antibacterials in body fluids and the reader is referred to this chapter and to the extensive literature for details of methods for individual compounds.

Pharmacokinetic parameters

A full treatise on pharmacokinetics is well beyond the scope and space limitations of this chapter. Pharmacokinetics is a science in itself and several monographs serve to introduce the reader to the basic and more advanced concepts. The book by Rowland and Tozer (1995) has become a standard reference work for those involved in pharmacokinetics, although other volumes are also useful for a less comprehensive coverage (for example, Krishna and Klotz, 1990). Here I will only discuss some of the key pharmacokinetic parameters which pertain to studies in animals (there are several pharmacokinetic parameters which are relevant in the clinical situation but which have little use in experimental infection models). Knowledge of this will aid in planning of experiments, interpretation of data (particularly where compounds are to be compared) and may allow some degree of prediction to the clinical situation. There are five key parameters: elimination half-life, apparent volume of distribution, total plasma clearance, absolute bioavailability (relative bioavailability can also be useful) and free fraction of drug. All of these, with the exception of free fraction of drug, can be determined by measurement of plasma concentrations after administration of the drug.

At this point it is worth considering in what other compartments, beyond the circulation, concentrations of an anti-infective should be measured. There is still much debate about the relative merits of measuring concentrations in organs or tissue fluid as opposed to the circulation. Some points in the debate can be gleaned from a number of reviews which have appeared over the past few years (Carbon, 1990; Barza, 1993). Clearly, to exert its effect an anti-infective needs to be present in the same space as the target organism. Whether concentration in the circulation is sufficiently predictive of the concentration in tissue fluid probably needs to be determined for each compound, although it is probable that at equilibrium mean concentrations of unbound drug in plasma and tissue fluids are equal (Barza, 1993). There are several ways in which antibiotic concentrations can be determined in tissue fluids and these have been comprehensively reviewed by Bergeron and Brousseau (1986) in the first edition of this handbook.

Although there are a few well-known exceptions, most bacterial infections are extracellular. Therefore in most cases intracellular concentrations of drug are not relevant. However, the observation that drug taken up into phagocytic cells can be targeted to the site of infection (Gladue *et al.*, 1989) means that even here there may be a rationale to measure intracellular concentrations. In the case of antivirals there is clearly a need in many cases for the drug to enter the cell. Here then intracellular concentrations may be the most relevant. The measurement of concentrations in organs is most easily accomplished in homogenates of the organ. Contamination of the organ with blood will lead to some error unless a correction is made for the blood content of the organ. However, in those circumstances in which the drug concentration in the organ equals or is higher than that in the circulation then clearly the content of the organ *per se* must be higher than the concentration in the circulation, indicating uptake into the organ from the circulation.

Elimination half-life

The elimination half-life is the time it takes the drug concentration during the elimination phase to decrease by 50%. Half-life can be estimated from the linear terminal portion of the log plasma concentration/time curve or it can be calculated:

$$t_{½} = \frac{0.693 \times Vd}{Cl} = \frac{0.693}{\mathrm{K}}$$

where, Vd is the apparent volume of distribution (see below), Cl is the total plasma clearance (see below) and K is the elimination rate constant.

This parameter is probably the most readily understood of the pharmacokinetic parameters. It can be used to predict suitable dosing intervals; as a starting point a drug could be administered once every half-life. It can also be used to predict the time required for a steady-state situation to be attained during multiple dosing or even continuous intravenous infusion and will also allow the differences between peak and trough concentrations during the steady state to be estimated. It can also be used to calculate a so-called accumulation factor:

$$R = \frac{1}{1 - e^{-\mathrm{K}\tau}}$$

where τ is the dosing interval.

This accumulation factor can in turn be used to calculate, in conjunction with a known maintenance dose, the loading dose required to bring plasma concentration to the required level immediately. However, loading doses are, it seems, seldom used in animal models.

Apparent volume of distribution

This is a proportionality constant that indicates the volume into which the given dose of a drug distributes, assuming its concentration in that volume to be the same as that in the plasma.

$$Vd = \frac{D}{C}$$

where D is the dose and C the concentration in plasma at zero time (determined by extrapolation).

In animal studies this parameter has only one use — to calculate a loading dose to attain a target steady-state concentration.

$$\text{Dose} = Vd \times Css$$

where Css is the target steady state concentration.

The volume of distribution can be useful to indicate which compounds may have better tissue distribution. All other factors being equal, then a compound with the greater tissue distribution will have the greater volume of distribution. However, as the volume of distribution also reflects distribution of fat-soluble compounds into fat deposits and binding to tissue components, not much about tissue distribution can be concluded in the absence of additional information.

Total plasma clearance

Clearance is defined as the portion of the distribution volume which is totally cleared of drug per unit time:

$$Cl = \frac{D}{AUC}$$

where D is the dose and AUC is the area under the whole of the concentration/time curve (calculated by the trapezoidal rule).

The area under the curve represents an important parameter which is easily derived from the concentration–time curve. Total clearance is the sum of all the routes of clearance; renal clearance and hepatic clearance are usually the main routes of elimination but other minor routes are recognized and may in some cases be significant. Clearance can be used to determine the average steady-state concentration of a drug during multiple dosing and similarly, when a steady-state concentration is required, then it can be used to calculate the rate of dosing. While probably not of relevance to the subject of this book, a knowledge of the two major components of total clearance may also provide insight into the processes by which a drug is eliminated in an animal, and which may be significantly different in humans.

Absolute bioavailability

The absolute bioavailability is most often applied to oral administration where it is defined as follows:

$$F = \frac{AUC_{\text{oral}}}{AUC_{\text{i.v.}}} \times \frac{D_{\text{i.v.}}}{D_{\text{oral}}}$$

By definition, intravenous administration results in 100% bioavailability. Similarly, bioavailability from other routes of administration can be calculated. The relative bioavailability is used to compare compounds given by the same route or possibly to compare the same compound given in different formulations.

The use of computer programs can expedite the calculation of these parameters. There are several programs which can calculate the parameters based on a non-compartmental procedure or data can be used to construct a compartmental model from which the parameters are calculated. However, all the parameters are easily calculated with the aid of graph paper and a hand-held calculator.

Free fraction of drug

Generally only drug which exists free in the plasma is able to exert its pharmacological effects or even to be metabolized or excreted; it is therefore important to know the amount of drug which is free to exert its pharmacological action. Usually determination of plasma concentrations is designed to measure the total drug in the circulation, i.e. bound and unbound; this is mainly because it is the easiest way to proceed. However, the determination of the concentration of only the free fraction may provide more meaningful information. It is important to recognize that the relative amounts of proteins in the circulation, and therefore the degree of protein binding, can change under certain disease states (Zini *et al*., 1990). Particularly if protein binding is high, drugs administered simultaneously can, by competition for binding sites, cause mutual changes in the unbound fraction.

Interspecies scaling

Although the pharmacokinetic differences between animals do not make life easy for those involved in the *in vivo* evaluation of anti-infectives, the differences can in some cases be used to extrapolate to the human situation. It has been noted that certain physiological processes can change in proportion to the size of the animal. These changes can be reflected in the dependence of certain pharmacokinetic parameters on the size of the species, a fact that can be used for predicting human pharmacokinetics from pharmacokinetic data collected in small animals.

Prediction of pharmacokinetics in humans from data from animals relies on general allometric relationships that link physiology and anatomy with body weight (Boxenbaum, 1982; Boxenbaum and Ronfield, 1983). In some cases pharmacokinetic parameters are better related to other variables such as brain weight or maximum life span potential (Mahmood, 1996; Mahmood and Balian, 1996). Unfortunately, application of these principles is limited to compounds for which the route of elimination and route and extent of metabolism (no metabolism is preferable) are the same in all species. In addition, first-order pharmacokinetics is an additional prerequisite. Usually, the relationships do not hold for absorption of drugs from the gastrointestinal tract; due to variations in gastrointestinal physiology (pH, gastrointestinal tract motility, etc.) rather than differences in the movement of compound across the gut wall. Thus a potentially useful method is limited by a variety of preconditions and it is up to the experimenter to determine experimentally if interspecies scaling could be of use in a particular situation. Lastly, even if the preconditions are met there is no guarantee that the predictions obtained will survive clinical scrutiny.

In conclusion, pharmacokinetics of anti-infectives in animals can be used as an aid to the design and interpretation of efficacy experiments, can allow prediction to the human situation, particularly if the human pharmacokinetics is known, and is essential for the development of suitable formulations. Pharmacokinetic data should always be interpreted with due regard to the formulation in which the compound is administered. Which pharmacokinetic parameters can or should be determined will depend on the particular situation: the compound under study, the route of administration, the site of infection, the species under study, etc. Much useful information can be obtained if, at a minimum, concentrations in the circulation are determined and the basic pharmacokinetic parameters are calculated.

References

Alteri, E., Bold, G., Cozens, R. *et al.* (1993). CGP 53437, an orally bioavailable inhibitor of human immunodeficiency virus type 1 with potent antiviral activity. *Antimicrob. Agents Chemother.*, **37**, 2087–2092.

Bakker-Woudenberg, I. A. J. M., Lokerse, A. F., ten Kate, M. T., Melissen, P. M. B., van Vianen, W., van Etten, E. W. M. (1993). Liposomes as carriers of antimicrobial agents or immunomodulatory agents in the treatment of infections. *Eur. J. Clin. Microbiol. Infect. Dis.*, **12**, S61–S67.

Barza, M. (1993). Pharmacokinetics of antibiotics in shallow and deep compartments. *J. Antimicrob. Chemother.*, **31**, 17–28.

Bergeron, M. G., Brousseau, L. (1986). Tissue fluid pharmacokinetic models in humans and animals. In: *Experimental Models in Antimicrobial Chemotherapy*, vol. 1 (eds Zak, O., Sande, M. A.), pp. 71–107. Academic Press, London.

Berman, J., Halm, K., Adkison, K., Shaffer, J. (1997). Simultaneous pharmacokinetic screening of a mixture of compounds in the dog using API LC/MS/MS analysis for increased throughput. *J. Med. Chem.*, **40**, 827–829.

Borg, N., Ståhle, L. (1997). Penciclovir pharmacokinetics and distribution to the brain and muscle of rats, studied by microdialysis. *Antiviral Chem. Chemother.*, **8**, 275–279.

Boxenbaum, H. (1982). Interspecies scaling, allometry, physiological time, and the ground plan of pharmacokinetics. *J. Pharmacokin. Biopharm.*, **10**, 201–227.

Boxenbaum, H., Ronfield, R. (1983). Interspecies scaling and the Dedrick plots. *Am. J. Physiol.*, **245**, R768–R774.

Capraro, H.-G., Bold, G., Fässler, A. *et al.* (1996). Synthesis of potent orally active HIV-protease inhibitors. *Arch. Pharmacy — Pharm. Med. Chem.*, **329**, 273–278.

Carbon, C. (1990). Significance of tissue levels for prediction of antibiotic efficacy and determination of dosage. *Eur. J. Clin. Microbiol. Infect. Dis.*, **9**, 510–516.

Cozens, R. M., Bold, G., Capraro, H.-G. *et al.* (1996). Synthesis and pharmacological evaluation of CGP 57813 and CGP 61755, HIV-1 protease inhibitors from the Phe-c-Phe peptidomimetic class. *Antiviral Chem. Chemother.*, **7**, 294–299.

Craig, W. A., Legget, J., Totsuka, K., Vogelman, B. (1988). Key pharmacokinetic parameters of antibiotic efficacy in experimental animal infections. *J. Drug Dev.*, **1** (suppl. 3), 7–15.

Dalhoff, A., Ullmann, U. (1990). Correlation between pharmacokinetics, pharmacodynamics and efficacy of antibacterial agents in animal models. *Eur. J. Clin. Microbiol. Infect. Dis.*, **9**, 479–487.

Dedrick, R. L., Bischoff, K. B. (1984). Species similarities in pharmacokinetics. *Fed. Proc.*, **39**, 54–59.

Drusano, G. L. (1988). Role of pharmacokinetics in the outcome of infections. *Antimicrob. Agents Chemother.*, **32**, 289–297.

Duchêne, D., Wouessidjewe, D. (1993). New possibilities for the pharmaceutical use of cyclodextrins and their derivatives. *Chimicaoggi*, **10**, 17–24.

Flückiger, U., Segessenmann, C., Gerber, A. U. (1991). Integration of pharmacokinetics and pharmacodynamics of imipenem in a human-adapted mouse model. *Antimicrob. Agents Chemother.*, **35**, 1905–1910.

Gerber, A. U., Brugger, H.-P., Feller, C., Stritzko, T., Stalder, B. (1986). Antibiotic therapy of infections due to *Pseudomonas aeruginosa* in normal and granulocytopenic mice: comparison of murine and human pharmacokinetics. *J. Infect. Dis.*, **153**, 90–97.

Gerber, A. U., Stritzko, T., Segessenmann, C., Stalder, B. (1991). Simulation of human pharmacokinetic profiles in mice, and impact on antimicrobial efficacy of netilmicin, ticarcillin and ceftazidime in the peritonitis–septicemia model. *Scand. J. Infect. Dis.* (suppl. 74), 195–203.

Girard, A. E., Girard, D., English, A. R. *et al.* (1987). Pharmacokinetic and *in vivo* studies with azithromycin (CP-62,993), a new macrolide with an extended half-life and excellent tissue distribution. *Antimicrob. Agents Chemother.*, **31**, 1948–1954.

Girard, A. E., Girard, D., Retsema, J. A. (1990). Correlation of extravascular pharmacokinetics of azithromycin with *in vivo* efficacy in models of localized infection. *J. Antimicrob. Chemother.*, **25** (suppl. A), 61–71.

Gladue, R. P., Bright, G. M., Isaacson, R. I., Newborg, M. F. (1989). *In vitro* and *in vivo* uptake of azithromycin (CP-62,993) by phagocytic cells: possible mechanism of delivery and release at sites of infection. *Antimicrob. Agents Chemother.*, **33**, 277–282.

Horrevorts, A. M., Mouton, J. W. (1993). Abnormal pharmacokinetics: the need for monitoring. *Eur. J. Clin. Microbiol. Infect. Dis.* (suppl. 1), 58–60.

Klassen, M., Edberg S. C. (1996). Measurement of antibiotics in human body fluids: techniques and significance. In: *Antibiotics in Laboratory Medicine* (ed. Lorian, V.), pp. 230–295. Williams & Wilkins, Baltimore.

Krishna, D. R., Klotz, U. (1990). *Clinical Pharmacokinetics: A Short Introduction*. Springer Verlag, Berlin.

Kwei, G. Y., Novak, L. B., Hettrick, L. A. *et al.* (1995). Regiospecific absorption of the HIV protease inhibitor L-735,524 in beagle dogs. *Pharm. Res.*, **12**, 884–888.

Leroux, J.-C., Cozens, R., Roesel, J. L. *et al.* (1995). Pharmacokinetics of a novel HIV-1 protease inhibitor incorporated into biodegradable or enteric nanoparticles following intravenous and oral administration to mice. *J. Pharm. Sci.*, **84**, 1387–1391.

Leroux, J.-C. Cozens, R. M., Roesel, J. L., Galli, B., Doelker, E., Gurny, R. (1996). pH-sensitive nanopartcles: an effective means to improve the oral delivery of HIV-1 protease inhibitors in dogs. *Pharm. Res.*, **13**, 485–487.

Lister, P. D., Sanders, C. C. (1995). Comparison of ampicillin-sulbactam regimens simulating 1.5 and 3.0 gram doses to humans in treatment of *Escherichia coli* bacteremia in mice. *Antimicrob. Agents Chemother.*, **39**, 930–936.

Loftsson, T., Brewster, M. E. (1996). Pharmaceutical applications of cyclodextrins. 1. Drug solubilization and stabilization. *J. Pharm. Sci.*, **85**, 1017–1025.

Mahmood, I. (1996). Interspecies scaling: predicting clearance of anticancer drugs in humans. A comparative study of three different approaches using body weight or body surface area. *Eur. J. Drug Metab. Pharm.*, **21**, 275–278.

Mahmood, I., Balian, J. D. (1996). Interspecies scaling: predicting clearance of drugs in humans. Three different approaches. *Xenobiotica*, **26**, 887–895.

Mizen, L., Woodnutt, G. A. (1988). A critique of animal pharmacokinetics. *J. Antimicrob. Chemother.*, **21**, 273–280.

Olah, T. V., McLoughlin, D. A., Gilbert, J. D. (1997). The simultaneous determination of mixtures of drug candidates by liquid chromatography atmospheric pressure chemical ionization mass spectrometry as an *in vivo* drug screening procedure. *Rapid Commun. Mass Spectrometry*, **11**, 17–23.

Rowland, M., Tozer, T. N. (1995). *Clinical Pharmacokinetics: Concepts and Applications*, 3rd edn. Williams & Wilkins, Philadelphia, USA.

Sagara, K., Yamada, I., Kawazoe, Y., Mizuta, H., Shibata, M. (1994). Gastrointestinal physiology-regulated dogs: utilization of a bioavailability study of a new thieno[3,2-f][1,2,4]triazolo[4,3-a][1,4]-diazepine, an antagonist of platelet-activating factor, and its preparations. *Biol. Pharm. Bull.*, **17**, 117–120.

Sparreboom, A., van Tellingen, O., Nooijen, W. J., Beijen, J. H. (1996). Non linear pharmacokinetics of paclitaxel in mice results from the pharmaceutical vehicle Cremophor EL. *Cancer Res.*, **56**, 2112–2115.

Ståhle, L. (1991). Use of microdialysis in pharmacokinetics and pharmacodynamics. In: *Methods in Neuroscience* (eds Robinson, T., Justice, J.), pp. 155–174. Elsevier, Amsterdam.

Ståhle, L. (1993). Microdialysis in pharmacokinetics. *Eur. J. Drug Metab. Pharm.*, **18**, 89–96.

Trissell, L. A. (1992). *Handbook on Injectable Drugs*. American Society of Hospital Pharmacists, Bethesda, MA, USA.

Wade, A., Weller, P. J. (1994). *Handbook of Pharmaceutical Excipients*. The Pharmaceutical Press, London.

Williams, R. T. (1974). Inter-species variations in the metabolism of xenobiotics. *Biochem. Soc. Trans.*, **2**, 359–377.

Zini, R., Riant, P., Barre, J., Tillement, P. (1990). Disease-indiced variations in plasma protein levels. Implications for drug dosage regimens (part 1). *Clin. Pharm.*, **19**, 147–159.

Chapter 11

Methods for Obtaining Human-like Pharmacokinetic Patterns in Experimental Animals

L. Mizen

Introduction

The pharmacokinetics of antibiotics in experimental animals can often differ markedly from those in humans. This is a consequence of species differences in absorption, distribution, metabolism and excretion (Mizen and Woodnutt, 1988; Dalhoff and Ullmann, 1990). These are critical parameters to consider not only for safety evaluation but also when studying the potential clinical efficacy of novel agents or their application in alternative clinical indications. In general, antibiotics are eliminated more rapidly in laboratory animals than in humans (Chung *et al.*, 1985; Mordenti, 1986; Mizen and Woodnutt, 1988; Zak and O'Reilly, 1991). The excellent reviews and studies of Gerber *et al.* (1984), Drusano (1988), Frimodt-Moller (1988), Vogelman *et al.* (1988), Dalhoff and Ullmann (1990), and Leggett *et al.* (1991) clearly support the existence of a relationship between dosing schedules and efficacy in experimental infections. It is a logical progression, therefore, to consider compensating for the more rapid elimination of antibiotics in animals compared with humans by adjusting dosing schedules to simulate human pharmacokinetic parameters in laboratory animal serum.

The techniques available to modify the concentration–time curve in serum fall into three main categories. These are first, modifying dosing schedules; second, reducing the rate of elimination with drugs coadministered with the antibiotic, and third, modifying infusion rates. All three methods have been used successfully, but this approach may not be suitable for all antibacterial agents; for example, metabolism and the metabolites produced can vary between species (Williams, 1974), as can binding to serum and tissue protein (Frimodt-Moller, 1988; Woodnutt *et al.*, 1988).

These techniques are generally complex and require more resources and more compound than the standard screening methods used to select antibiotics. However, because of the requirement for knowing the human pharmacokinetics prior to study, they are likely to be used for compounds in development or those marketed. The techniques need to be applied to experimental infections which should be of sufficient duration to allow for predictable and reproducible outcomes. In this way, human kinetic simulation should lead to more meaningful comparisons not only between antibacterial agents with regard to their relative efficacies against experimental microbial infections, but also between different dosing schedules. Overall, the objective of these studies is to contribute to a more informed view of clinical potential relative to appropriate dosing schedules and cost of goods.

Modification of dosing schedules

Eagle *et al.* (1950, 1953) and others demonstrated, in the mouse and rat, that the successful eradication of bacteria from infected sites and survival depended heavily upon appropriate dosing schedules. In the 1970s Hunter *et al.* (1976), demonstrated the importance of selecting the right dosing interval in a *Pseudomonas aeruginosa* thigh lesion model in mice. Gerber *et al.* (1986) developed these studies further, working with *P. aeruginosa* in the same model and in an acute intraperitoneal infection, comparing the effects of murine and simulated human pharmacokinetics on the efficacy of selected beta-lactam and aminoglycoside antibiotics. To obtain "human-adapted" pharmacokinetics the antibiotics were administered as a series of fractionated and decreasing doses to each mouse.The doses were administered subcutaneously at 15–20-minute intervals depending on the murine pharmacokinetics and for comparison the same total dose was administered to other groups of mice as a single bolus dose (murine kinetics). Control mice received only diluent. The studies lasted 8–12 hours and numbers of injections per mouse in the human simulation group varied; for example, mice designated for the 8-hour sample group would have received as many as 23 subcutaneous injections each over that period, whereas mice designated for the 2-hour sample group could have received 6 injections (Gerber *et al.*, 1986).

The "human-adapted" pharmacokinetics proved to be superior in activity to the single bolus dose for ticarcillin, ceftriaxone and ceftazidime, whereas for the aminoglycosides gentamicin and netilmicin, there was little difference between the therapeutic outcomes for the two dosing schedules. The single bolus dose of the aminoglycoside tended to be more effective initially, particularly for the *P. aeruginosa*

Handbook of Animal Models of Infection
ISBN 0-12-775390-7

infection, but the differences were less in neutropenic mice (Gerber *et al.*, 1986). These observations were confirmed in later studies by Gerber *et al.* (1991) and it was concluded that when used against *P. aeruginosa* experimental infections, the differences between the two classes of antibiotic were likely to reflect differences in rate of kill, extent of bactericidal activity relative to concentration and post-antibiotic effects (PAE)—features which are fundamental to all investigative studies with aminoglycosides and beta-lactams. Aminoglycoside activity was predominantly concentration-dependent, followed by a PAE, whereas that of the beta-lactams tested was primarily time-dependent with no PAE.

Fluckiger *et al.* (1991) continued to use the multiple injection approach in a comparison of proposed clinical intramuscular and intravenous formulations of imipenem in the mouse thigh lesion model, with *Escherichia coli* and *P. aeruginosa* as the infecting organisms. The outcome of the studies agreed with previous data for beta-lactam antibiotics, and imipenem was more effective the longer concentrations in serum remained above the minimum inhibitory concentration (MIC). Hishikawa *et al.* (1990, 1991) adopted the fractionated dose approach of Gerber *et al.* (1986) to simulate concentrations in human serum in the mouse for two cephalosporins, cefazolin and cefmenoxime. He compared the efficacy of single and multiple dosing schedules against experimental pneumonia caused by *Klebsiella pneumoniae*. The fractionated doses of cefazolin and cefmenoxime were administered by the subcutaneous route as series of decreasing doses over a period of four hours. The mice were dosed at 0 hours, 20 minutes, 40 minutes and 1 hour and thereafter either hourly or half-hourly depending on the compound. To mimic the concentrations in human serum for the multiple dose studies, these series were administered to mice every eight or twelve hours for a period of seven days. The therapeutic endpoint was bacterial enumeration in lung tissue. The authors found it was not possible to predict the therapeutic outcome of multiple dosing schedules from the single dose studies (Hishikawa *et al.*, 1990), and this was thought to be due to host dependent factors, including both cellular and humoral responses and the subsequent histopathological changes that would change with time during the multiple dosing studies.

Hatano and colleagues (1994) designed a mathematical multiple dosing formula to derive multiple dosing schedules which ensured that the serum concentrations in mouse serum were no more than twice those observed in human serum at each time interval, and also that the area under the curve values in mice were equal to those in humans. The simulated curves were for drip infusions of 1 g doses of four cephalosporins, FK037, ceftazidime and cefmenoxime and flomoxef, and efficacy was compared against either a respiratory tract infection with *Staphylococcus aureus* or *P. aeruginosa*. The studies continued until 24 hours after dosing, at which time lungs were removed for bacterial enumeration. Dose-response studies supported previous observations, including those of Schentag (1991) and indicated that the duration above MIC was critical for all, and although there were similarities between *in vitro* activities, the pharmacokinetics of the individual agents influenced the outcome.

Lister and Saunders (1995) took an alternative approach to the simulation of human pharmacokinetics in mice in their studies comparing the efficacies of ampicillin–sulbactam clinical dose regimens (1.5 and 3.0 g i.v. doses). Three *E. coli* strains of varying susceptibilities *in vitro* were used to induce intraperitoneal infections in mice. Only two subcutaneous injections were administered an hour apart to simulate the human kinetics for either a single 1.5 or 3 g dose. Although two peak concentrations were obtained in serum, it was assumed on the basis of previous work by Gerber *et al.* (1986) and others that there would be no dose relationship between rate of killing and amounts in serum for beta-lactam antibiotics. The most critical factor was seen to be the duration of time above the breakpoint concentration. Level of bacteremia was measured over 6 hours and survival over 48 hours. Differences in bacteremia and survival between the two dose levels were only observed for the highly resistant strain and the authors went on to conclude that, despite differences in peak to MIC ratio, the lower dose (1.5 g) of ampicillin–sulbactam would be sufficient for the treatment of susceptible organisms. The results for the highly resistant infection were of further interest in that, although the lower dose did not reduce bacterial numbers in blood, 60% of the mice survived compared with none in the untreated control groups. These studies were undertaken in immunocompetent mice and the latter observation would presumably indicate some contribution of host response to the outcome.

The use of granulocytopenic animals has been shown to increase the precision of the estimate of antibacterial effects *in vivo* by avoiding possible synergy that may occur between the action of the antibiotic and professional phagocytes (Gerber *et al.*, 1984; van Ogtrop *et al.*, 1990; Leggett *et al.*, 1991). The findings therefore may be particularly relevant to the treatment of immunocompromised patients who are only dependent upon the action of the antibiotic. The retrospective analysis of Drusano (1988) lends support to this view by providing evidence of similarities between clinical and experimental outcomes. The relative merits of inducing granulocytopenia will depend entirely upon the questions being asked of the studies.

Reduction in elimination rate

An alternative to the methods described in the first section is the co-administration of agents known to influence the elimination rate of antibiotics. Craig *et al.* (1991a, b) and Fantin *et al.* (1990) have undertaken a number of studies using the nephrotoxic agent uranyl nitrate (Giacomini *et al.*, 1981) to produce renal dysfunction, which in rats was apparent after 1 day. Necrosis of the distal portion of the proximal renal tubule was clearly seen at 4 days. The influence of a single s.c. dose of 10 mg/kg uranyl nitrate in mice, 3 days before the experiment, on the pharmacokinetics of an aminoglycoside and cephalosporin is shown in Table 11.1. Other agents capable of altering pharmacokinetics have included for example probenecid (Kunst and Mattie, 1978) and the non-steroidal antiinflammatory drug diclofonec (Joly *et al.*, 1988; Table 11.1). The elimination half-life of ceftriaxone increased significantly in rabbits after co-administration of diclofenac and this was reflected in therapeutic responses, however, the half-life did not equate to that in humans. Of further interest here (Joly *et al.*, 1988) is the similar elimination half-lives between rabbits and humans for cefotiam and cefmenoxime: this is not always the case for beta-lactams in this species (Woodnutt *et al.*, 1995). In mice, probenecid (oral dose: 0.5 ml of 20 mg/ml in 0.5 N sodium bicarbonate just prior to dosing) increased AUC values of cephradine and nafcillin and this paralleled increased therapeutic potency against a murine thigh infection of *E. coli* and urinary tract infection of *S. aureus*. However elimination half-lives were not reported (Kunst and Mattie, 1978; Mattie and Kunst, 1978). Thauvin *et al.* (1987) reported an increase in elimination half life for ampicillin in rats after administration of probenecid (Table 11.1). It is possible therefore to influence the rate at which compounds are eliminated from the body; but the method cannot be relied upon in all cases to yield pharmacokinetic parameters similar to those in humans.

The use of infusion techniques

Techniques in this section are concerned with modifying the rates of administration of the compounds to maintain antibiotic concentrations in the body and thus compensate for species differences in elimination mechanisms. Animals can be anaesthetized throughout the infusion process or conscious: when conscious, swivel systems connecting intravascular cannulae to the infusion systems have been used to allow free movement during the study. The flow rates of intravenous infusion should be within physiological limits for the particular species used and consideration needs to be given to the stability of the antibiotic within the infusion diluent. It is also important to determine whether the infusion process will influence the therapeutic outcome.

Continuous infusion and prolonged dosing

The continuous infusion of antibiotics is an administration route that is more frequently used for the treatment of severe infections, particularly in immunocompromised patients and for infections caused by Gram-negative bacteria (Craig and Ebert, 1992). This route is considered by some to be the route of choice for beta-lactam antibiotics in this group of patients (Craig and Ebert, 1992; Vondracek, 1995; Benko *et al.*, 1996), as most of these agents do not exhibit concentration-dependent killing. Other antibiotics, such as vancomycin and aminoglycosides, can also be infused in this patient group (Daenen and deVries-Hospers, 1988; James *et al.*, 1996). The need to determine not only the most effective dosing route but also the least expensive in terms of compound amounts and hospital staff requirements supports the use of animal studies for the development of appropriate clinical dosing schedules. Examples of the experimental studies undertaken are listed in Table 11.2.

Table 11.1. The effect of uranyl nitrate, diclofenac and probenecid on the pharmacokinetics of selected beta-lactams and aminoglycosides in laboratory animals and a comparison with elimination half lives in humans

	Antibiotic pharmacokinetic parameters in animal and humans			
Compound	*Normal* $t_{1/2}$	*+ Uranyl nitrate* $t_{1/2}$	*Human* $t_{1/2}$	*Reference (Animal/human)*
Mouse				
Cefpirome	14–17 min	1.8–2.2 h	2.3 h	Craig and Ebert (1992)/Kavi *et al.* (1988)
Amikacin	18.5–32.5 min	1.6–2.0 h	2.0 h	Craig *et al.* (1991b)/Walker *et al.* (1979)
Gentamicin	17 min	NR	2.1 h	Fantin *et al.* (1990)/Walker *et al.* (1979)
Rabbit	Normal $t_{1/2}$	+Diclofenac $t_{1/2}$	Human $t_{1/2}$	
Cefotiam	1.4 h	1.64 h	1.1 h	Joly *et al.* (1988)
Cefmenoxime	1.3 h	1.2 h	1.6 h	
Ceftriaxone	2.8 h	3.5 h*	5.8 h	
Rat	Normal	+ Probenecid 50 mg/kg per day	Human	
Ampicillin	0.3 h	1.0 h	~ 1.0 h	Thauvin *et al.* (1987)/Bergan (1978)

NR, Not reported.
*Significantly higher than normal; $P < 0.05$.

Table 11.2 Use of continuous infusion of undiluted and continuously diluted infusates to simulate human antibiotic serum concentration and concentration–time profiles in animal sera

Administration	*Reference*	*Species*	*Dose*	*Concentration–time profile*	*Flow rate (ml/h)*	*Infection model*	*Duration*
Intravenous infusion: undiluted	Thauvin *et al.* (1987)	Rat	Ampicillin 450 mg/kg per day = ~ 7–8 mg/kg per day human equivalent	Steady-state	0.8 ml/h	*Enterococcus faecalis* Endocarditis	5 days
	Thauvin-Eliopoulos *et al.* (1997)	Rat	Intermittent i.v. versus continuous infusion cefotaxime: repeated doses 400 mg/kg per day	Steady-state vs rat kinetics i.v.		*Klebsiella pneumoniae* Abdominal sepsis	3–4 days
Tissue infusion via tissue cage	Bakker-Woudenberg *et al.* (1985)	Rat	Cefoxitin 5.25 mg/kg per h Cefazolin 1.59, 2.3 mg/kg per h	Steady-state reached in 3 h	0.12 ml/h	*Klebsiella pneumoniae* Pneumonia	2–3 days
Prolonged infusion: osmotic pump i.p.	Naziri *et al.* (1995)	Mouse	Cefazolin 180 mg/kg per day for 3 days versus intermittent dosing	Steady-state for 3 days		*Klebsiella pneumoniae* Suture/wound model	2, 4, 7 days
Infusion of continuously diluted infusate (i.v.)	Woodnutt *et al.* (1988)	Rabbit (anaesthetized)	Simulated human kinetics	2 g i.v. temocillin vs bolus (rabbit)	2.0 ml/h	*Klebsiella pneumoniae* Meningitis	12 h
	Mizen *et al.* (1989)	Rabbit (anaesthetized)	Simulated human kinetics	TIC/CA† 3.1 g × 4, ceftazidime 2 g × 1.5 — all i.v.	2.0 ml/h	*Klebsiella pneumoniae* Meningitis	12 h
	Woodnutt *et al.* (1992a)	Rat	Simulated human kinetics	Cefazolin (1 g), piperacillin (2 g) ± BRL 42715 (100 mg) all i.v.	0.6 ml	i.p. infections *Serratia marcescens* *Escherichia coli*	13 h

† TIC/CA ticarcillin 3 g plus clavulanic acid 100 mg.

Bakker-Woudenberg *et al.* (1985) used an experimental *Klebsiella pneumoniae* respiratory infection in rats which required infusion of cephalosporins over 65 hours. They adopted an alternative approach to intravenous infusion as it was felt that infusion in animals over long periods of time, particularly when the infusion rate is slow (≤0.5 ml/h), could lead to cannulae losing patency (Table 11.2). These workers used a modification of the tissue cage method of Thonus *et al.* (1979). The tissue cages, made of a rigid material (Teflon) to avoid any movement of the animal distorting the cage and thus inadvertently altering the flow rate, were implanted subcutaneously at least 6–8 weeks prior to the study to allow a consistent encapsulation of the cages by host tissue. The catheter running from the cage was exteriorized in the neck region, passed through a coiled spring kept in place by a jacket around the chest of the rat, to an infusion pump via a swivel system to allow free movement. Steady-state serum levels were usually achieved after 3 hours of infusion.

Naziri *et al.* (1995) used a slightly different approach for infusion and employed Alzet peritoneal micro-osmotic pumps (Alza Corp., Palo Alto, CA) to compare continuous infusion of cefazolin with intermittent administration of an equal total amount. The pumps were implanted intraperitoneally and primed to deliver a dose of 180 mg/kg per day of cefazolin for 3 days. The mice had also received a loading dose 60 mg/kg i.p. 30 minutes before infection and implanting of the pumps. These groups were compared with those receiving 60 mg/kg 30 minutes before infection and thereafter 180 mg/kg per day for 3 days in three equally divided doses. The antibiotic concentrations in mouse serum in animals with implanted osmotic pumps were similar for a period of 3 days (20.3 ± 1.4 and 17.2 ± 5.1 μg/ml, days 1 and 2 respectively); by day 4, however, no levels were detected. After a normal bolus dose no levels were detectable by 8 hours. The therapeutic outcome (survival over 10 days and reduction in bacterial numbers in blood and tissues by 4 days) was superior for 'continuous infusion'.

Continuous vascular infusion versus intermittent i.m. injection of ampicillin was investigated by Thauvin *et al.* (1987) in the enterococcal endocarditis model in rats. At 24 hours after infection a Silastic tubing was inserted into the superior vena cava via the jugular vein, exteriorized and connected via a swivel system to an infusion pump (Sage pump 352). The flow rate was 0.8 ml/h. The total daily doses for infusion were 450 mg/kg and 4500 mg/kg. Repeated i.m. injections were dosed at 450 mg/kg (in three divided doses every 8 hours). Treatment for all groups lasted for 5 days or until death. The authors determined previously that continuous infusion of saline did not influence the course of infection. Differences between the therapeutic outcomes for the two routes of administration were significant and sterile vegetations were only observed in the infused groups and not in the group receiving intermittent intramuscular injections, however, there were no significant differences between the high and low dose infusions. These authors gave no indication of difficulties with the infusion procedures and have continued to use intravascular infusion to investigate efficacy in rats against experimental intra-abdominal abscesses caused by Gram-negative bacteria (Thauvin-Eliopoulus *et al.*, 1997).

Simulation of human serum concentration time curves in laboratory animals

Infusion of continuously diluted infusates

In order to simulate human serum–concentration profiles, pharmacokinetic analysis for the agents under study are required in both humans and animals. It has been shown that the continuously diluted intravenous infusion system is adaptable to single or multiple compartmental analysis. A mathematical model was devised on the basis of human and rabbit or rat pharmacokinetics to determine the initial concentrations of the infusate and the volumes for the reservoirs which contained the infusate as it was being diluted during infusion (Woodnutt *et al.*, 1988, 1992; Mizen *et al.*, 1989). A summary of the parameters for pharmacokinetics and for infusion is shown in Table 11.3.

In the studies simulating the human concentration profile for a 2 g i.v. bolus dose of temocillin in rabbits (Woodnutt *et al.*, 1988), the efficacy of temocillin against *Klebsiella pneumoniae* meningitis was compared for the simulated dose and an i.v. bolus dose (rabbit kinetics). The results are shown in Table 11.4 for therapeutic outcome and distribution 12 hours after the start of infusion, and demonstrate clearly the advantage of the more prolonged elimination half-life that is observed in humans.

The use of a continuously diluted infusate was also adopted for studies with ticarcillin/clavulanate and ceftazidime in the rabbit meningitis model (Mizen *et al.*, 1989). The technique was slightly modified as a 1 hour infusion of ticarcillin/clavulanate was simulated rather than that for a bolus dose. This required two infusates/reservoirs; the additional reservoir was to aid in the simulation of the concentrations occurring during the 1-hour infusion. In both studies rabbits were anaesthetized throughout for humane reasons. The activity of the beta-lactams piperacillin and cefazolin in the presence or absence of a beta-lactamase inhibitor BRL 42715 against intraperitoneal infections caused by *Serratia marcescens* or *Escherichia coli* was assessed in conscious rats (Woodnutt *et al.*, 1992a). There were marked differences in humans between the pharmacokinetics of BRL 42715 and either piperacillin or cefazolin: the inhibitor was eliminated far more rapidly than either of the antibiotics. Thus, in this series of studies different reservoirs and reservoir volumes were required for each compound to compensate for this. For administration of the antibiotic with the inhibitor the cannulae from each reservoir were joined just before entry into the external jugular vein. The rats were housed singly, and the cannulae were exteriorized dorsally, taken through a flexible metal sheath to the top of the cage and attached to a free-moving swivel system. A single dose only was simulated, and overall the

Table 11.3 Pharmacokinetics and dosing parameters for simulating a human serum concentration–time curve of temocillin (2 g i.v. dose) in rabbits and cefazolin, piperacillin and BRL 42715* in rat plasma (Woodnutt *et al.*, 1988, 1992a)

Parameter	*Temocillin*	*Cefazolin*	*BRL 42715*	*Piperacillin*
Animal	Rabbit	Rat	Rat	Rat
Kel (h^{-1})	2.6	1.4	2.3	1.9
V (l/kg)	0.35	0.3	2.2	0.48
Human				
β (h^{-1})	0.15	0.44	1.2	0.8
B (μg/ml)	41	108	0.9	53
Co (μg/ml)	243	300	8.0	184
Dosing				
V_{res} (ml)	15	1.5	0.48	0.75
C_{res} (mg/ml)	80	36.0	2.71	33.5
F (ml/h)	2	0.6	0.6	0.6
Bolus dose (mg/kg)†	82	90.0	16.8	87.5

Kel, Elimination rate constant; *V*, volume of distribution; β, elimination phase; *B*, zero time intercept for β; *Co*, initial plasma concentration; V_{res}, reservoir volume; C_{res}, initial concentration in reservoir; F, infusion flow rate.
* Following bolus administration of 1 g cefazolin, 2 g piperacillin, 1 g cefazolin plus 100 mg BRL 42715 or 2 g piperacillin plus 100 mg BRL 42715.
† Initial bolus dose to simulate distribution phase.

Table 11.4 Pharmacodynamics of temocillin efficacy in an experimental *Klebsiella pneumoniae* meningitis in rabbits: temocillin administered as either a bolus dose or by infusion to simulate results for a 2 g dose in humans (Woodnutt, *et al.*, 1988)

Dose	*No. of rabbits*	*Sample*	*AUC (μg/h per ml)*	$t_{1/2}h$	*Penetration (%)*	*cfu log_{10}/ml CSF at 12 h*
Bolus	6	Plasma	90.9 ± 20.2	0.3 ± 0.11		
		CSF	8.1 ± 7.5	0.83 ± 0.24	10 ± 11	6.4 ± 0.5
Infusion	10	Plasma	894.8 ± 329.6	5.1 ± 1.3		
		CSF	103.6 ± 34.9	6.3 ± 1.6	11.8 ± 2.9	< 2.0

AUC, Area under the curve; CSF, cerebrospinal fluid.

results supported the use of the inhibitor, despite the disparity in pharmacokinetics.

Infusion at variable flow rates

In the studies described above, reservoirs were required to be of fixed volumes, therefore new reservoirs had to be constructed for each series of studies. The introduction of computer-controlled, programmable infusion pumps to vary the flow of the infusate therefore heralded an approach to the simulation of human serum pharmacokinetics in animals, which could be more efficient and require less pre-test preparation of apparatus. Woodnutt *et al.* (1992b, 1993, 1995b–e, 1996) developed an alternative method using these pumps (Figure 11.1) to simulate in the conscious rat either oral or parenteral antibiotic concentration–time curves observed in human serum. Deconvolution equations (Veng Pedersen, 1980) were used to derive the various infusion rates that were required. The flow rates were calculated on a PC-based microcomputer linked serially to the infusion pumps (Woodnutt *et al.*, 1995b). A number of different infection models were used with this system (Table 11.5). The aims of these studies were first, to compare the efficacies of different dose regimens of amoxycillin/clavulanate and of amoxycillin with a view to selecting the most appropriate regimen relative to susceptibilities and duration of antibiotic concentrations over MIC, and second, to contribute to the determination of meaningful susceptibility breakpoints. Dosing continued for 2–3 days and the therapeutic endpoints were based on bacterial enumeration at the site of infection 14–16 hours after the cessation of therapy.

Blatter *et al.* (1993), Fluckiger *et al.* (1994), Entenza *et al.* (1994a,b, 1997) and used a computer-controlled infusion system with a programmable infusion pump (pump 44, Harvard apparatus, South Natick, MA) to simulate human serum pharmacokinetics of antibiotics in the rat endocarditis model (Table 11.5). Studies were directed towards determining whether or not co-administration of amino-

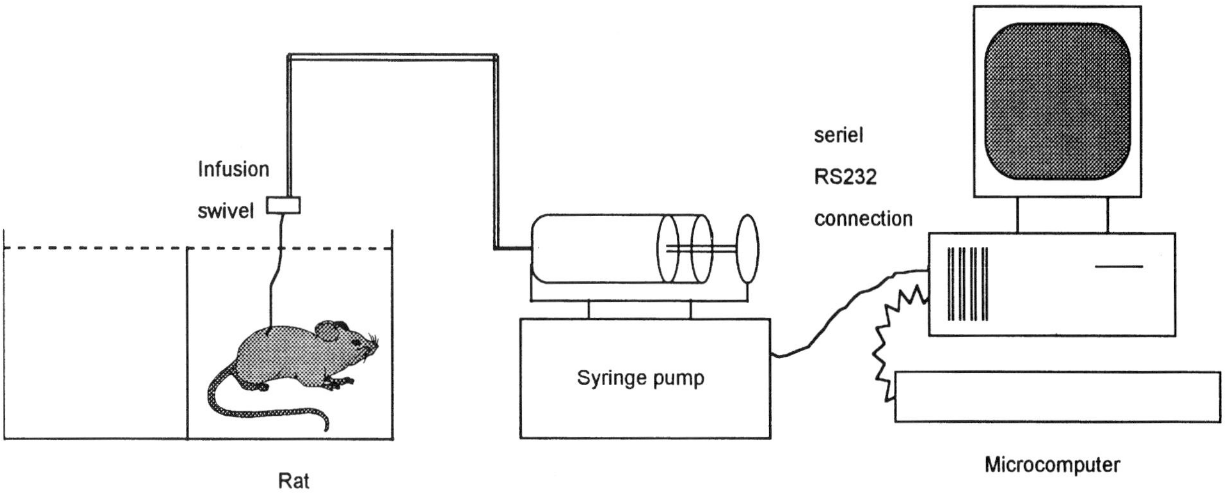

Figure 11.1 Apparatus for simulating human serum pharmacokinetics for antibiotics in animals using intravenous infusion at variable flow rates.

glycosides would enhance the therapeutic effects of ceftriaxone (Blatter *et al.*, 1993), defining a suitable prophylactic dose regimen with amoxycillin (Fluckiger *et al.*, 1994), exploring the clinical potential of new quinolones (Entenza *et al.*, 1994b, 1997) and investigating a possible use for Augmentin in the treatment of infections caused by methicillin-resistant *Staphylococcus epidermidis* strains (Entenza *et al.*, 1994a).

The use of simulated human serum profiles has also been applied to antibacterial studies in experimental endocarditis infections in rabbits (Gavalda *et al.*, 1996). These authors used an experimental *Enterococcus faecalis* endocarditis model to explore the use of a once-daily dose of gentamicin with ampicillin because of the potentially toxic effects of prolonged aminoglycoside therapy and the need for less toxic but equally effective dosing schedules. The mathematical model used to simulate the human serum concentration–time profile for a 2 g intravenous dose of ampicillin was based on the premise that at any time the human serum concentration was assumed to be higher than in rabbit serum.

The pharmacokinetic parameters for ampicillin in human and rabbits were applied to the equations. The authors felt this to be an amenable open one-compartment pharmacokinetic mathematical model to use with a computer-controlled infusion pump system. The model was not applied to gentamicin, as human pharmacokinetics were defined as an open three-compartment model. It was felt, however, that the half-life of 1.5–1.7 hours was close enough to that found in humans (2–3 hours). Gentamicin was administered by the subcutaneous route 0.5 hours before the appearance of the peak concentration of ampicillin. Details of the mathematical formulas used are described fully as an appendix to the paper. The therapeutic outcome of these studies was expressed in terms of bacterial numbers in cardiac vegetations. The results suggested that, as there was no significant difference in efficacy between the two dosing schedules of gentamicin, but both were significantly different to ampicillin alone, it should be possible to reduce the frequency of dosing of gentamicin in the susceptible patient population.

The work of Bugnon *et al.* (1996), similarly to that of Gavalda *et al.* (1996), was aimed at attempting to define an optimal therapeutic regimen for aminoglycosides relative to toxicity. In these studies an experimental *Serratia marcescens* endocarditis model in rabbits was used to determine whether or not once-a-day dosing of the aminoglycoside amikacin or a fractionated dosing schedule provided the best clinical therapeutic regimen. A computer-controlled infusion with pre-programmed variable flow rates was used. The amikacin protocol was established on the basis of a one-compartment model in the rabbit in order to obtain a monoexponential decay with an apparent elimination half-life of 2 hours, which was comparable to that in humans. These authors observed considerable interindividual variability in pharmacokinetic parameters, including AUC, volume of distribution and $t_{1/2}$h, and some animals showed totally unexpected profiles. It was thought that this large degree of variation could have been due in part to the septic status of the animals.

It is apparent therefore that not all experimental infections would be amenable to the simulation of human serum concentration time curves, particularly severe infections which could lead to septic shock and vascular collapse. In addition, to obtain human concentration profiles, high doses often need to be administered to the animals to compensate for the more rapid elimination in animals: this may lead to toxic effects with certain compounds.

Table 11.5 The use of continuous infusion at variable flow rates to simulate antibiotic serum–concentration time curves in conscious rats and rabbits

Species	*Source*	*Mathematical model for infusion*	*Infection model*	*Therapeutic outcome*	*Human dose regimens simulated*	*Duration of therapy*
Rat	Woodnutt *et al.* (1992b)	See Woodnutt *et al.* (1995a)	*Staphylococcus aureus* suture/wound	cfu/g wound	AMX/CA 250/125 mg p.o. t.i.d vs AMX 250 mg p.o. t.i.d.	3 days
	Woodnutt *et al.* (1993)	See Woodnutt *et al.* (1995a)	*Haemophilus influenzae* RTI	cfu/g lung	t.i.d. vs b.i.d. p.o. AMX/CA 625 mg and 375 mg cefuroxime axetil 250 mg b.i.d. cefaclor 250 mg t.i.d.	2.5 days
	Woodnutt *et al.* (1995c)	See Woodnutt *et al.* (1995a)	*Streptococcus pneumoniae* sensitive and resistant: RTI	cfu/lung	AMP q.i.d. 500 mg p.o. vs AMX 500 mg p.o. t.i.d.	2.5 days
	Woodnutt *et al.* (1995d)	See Woodnutt *et al.* (1995a)	*Escherichia coli*	cfu/g abscess	1.2 g AMX/CA t.i.d. relevant to clinical breakpoints	3 days
	Woodnutt *et al.* (1995e)	See Woodnutt *et al.* (1995a)	*Escherichia coli*	cfu/g abscess	3.1 g TIC/CA q.i.d. relevant to clinical breakpoints	3 days
	Woodnutt *et al.* (1996)	See Woodnutt *et al.* (1995a)	*Streptococcus pneumoniae* sensitive and resistant: RTI	cfu/lung	AMX 500 mg p.o. t.i.d. vs 875 mg p.o. b.i.d.	2.5 days
	Blatter *et al.* (1993)	NR	*Streptococcus sanguis* and *S. mitis* endocarditis	cfu/g vegetation	Ceftriaxone 2 g i.v. simulated with and without netilmicin s.c.: daily dose for both	3 days
	Fluckiger *et al.* (1994)	NR	*Streptococcus sanguis* and *S. intermedius* endocarditis	cfu/g vegetation	AMX 3 g p.o. prophylaxis 1 h prior to infection vs normal i.v. bolus dose	2–3 days
	Entenza *et al.* (1994a)	NR	*Streptococcus mitis* and *S. sanguis* endocarditis	cfu/g vegetation	Ceftriaxone 2 g i.v. o.d. simulated vs Sparfloxacin given s.c. q.i.d. to simulate 400 mg i.v. o.d.	3 or 5 days
	Entenza *et al.* (1994b)	NR	*Staphlococcus epidermidis*: MRSE endocarditis	cfu/g vegetation	AMX/CA 1.2 g i.v. q.i.d. vs VAN 1 g b.i.d.	3 days
Rabbit	Entenza *et al.* (1997)	NR	*Staphlyococcus aureus* MRSA, MSSA endocarditis	cfu/g vegetation	Levofloxacin 350 mg p.o., o.d. vs ciprofloxacin 750 mg p.o., b.i.d., vs flucloxacin 1 g i.v. q.i.d. vs VAN 1 g i.v. b.i.d.	3 days
Rabbit	Gavalda *et al.* (1996)		*Enterococcus faecalis* endocarditis	cfu/g vegetation	Simulated AMP 2 g i.v. every 4 h with or without gentamicin s.c. (two dose levels) 0.5 h before AMP	3 days
Rabbit	Bugnon *et al.* (1996)	NR	*Serratia marcescens* endocarditis	cfu/g vegetation	Amikacin 15 or 30 mg/kg per day i.v. simulated: o.d. vs t.i.d.	1 day

NR, not reported; AMX, amoxycillin; AMX/CA, amoxycillin/clavulanic acid (Augmentin); AMP, ampicillin; VAN, vancomycin; MRSA, MSSA, methicillin-resistant or sensitive *Staphylococcus aureus*; MRSE, methicillin-resistant *Staphlococcus epidermidis*; TIC/CA, ticarcillin/clavulanate; RTI, respiratory tract infection.

Summary

The development of models for obtaining human-like pharmacokinetic patterns in laboratory animals—in particular the use of computer-controlled infusion rates—is a major step forward in our efforts to find the right dose, the right dosing schedule and to obtain a better indication of how to find appropriate susceptibility breakpoints. For these studies to be meaningful, however, the *in vitro* antimicrobial properties also need to be clarified, for example, the rate of kill relative to concentration and test conditions, and whether or not there are PAEs. It is not possible to correlate directly experimental therapeutic outcomes with clinical efficacy but retrospective analyses have shown correlations between experimental and clinical studies. The use of investigative preclinical studies should be encouraged, but only when there is no alternative, as a number of these studies involve surgery and/or many repeated injections, both of which have ethical considerations with regard to the use of animals.

References

Bakker-Woudenberg, I. A., van den Berg, J. C., Vree, T. B., Baars, A. M., Michel, M. F. (1985). Relevance of serum protein binding of cefoxitin and cefazolin to their activities against *Klebsiella pneumoniae* pneumonia in rats. *Antimicrob. Agents Chemother.*, **28**, 654–659.

Benko, A. S., Cappelletty, D. M., Kruse, J. A., Rybak, M. J. (1996). Continuous infusion versus intermittent administration of ceftazidime in critically ill patients with suspected Gram-negative infections. *Antimicrob. Agents Chemother.*, **40**, 691–695.

Bergan, T. (1978). Pharmacokinetic comparison of oral bacampicillin and parenteral ampicillin. *Antimicrob. Agents Chemother.*, **13**, 971–974.

Blatter, M., Fluckiger, U., Entenza, J., Glauser, M. P., Francioli, P. (1993). Simulated human serum profiles of one daily dose of ceftriaxone plus netilmicin in treatment of experimental streptococcal endocarditis. *Antimicrob. Agents Chemother.*, **37**, 1971–1976.

Bugnon, D., Potel, G., Xiong, Y. Q. *et al.* (1996). *In vivo* antibacterial effects of simulated human serum profiles of once-daily versus thrice-daily dosing of amikacin in a *Serratia marcescens* endocarditis experimental model. *Antimicrob. Agents Chemother.*, **40**, 1164–1169.

Chung, M., Radwanski, E., Loebenberg, D. *et al.* (1985). Interspecies pharmacokinetic scaling of Sch 34343. *J. Antimicrob. Chemother.*, **15** (suppl. C), 227–233.

Craig, W. A., Ebert, S. C. (1992). Continuous infusion of β-lactam antibiotics. *Antimicrob. Agents Chemother.*, **36**, 2577–2583.

Craig, W. A., Moffat, J., Redington, J. (1991a). Effect of human kinetics on activity of cefpirome in murine thigh model. Abstract 1192, 31st Interscience Conference on Antimicrobial Agents and Chemotherapy, Chicago USA 29 Sept–2 Oct. American Society for Microbiology, Washington, DC.

Craig, W. A., Redington, J., Ebert, S. C. (1991b). Pharmacodynamics of amikacin *in vitro* and in mouse thigh and lung infections. *J. Antimicrob. Chemother.* (suppl. C), 29–40.

Daenen, S., deVries-Hospers, H. (1988). Cure of *Pseudomonas aeruginosa* infection in neuropenic patients by continuous infusion of ceftazidime. *Lancet* **1**, 937.

Dalhoff, A., Ullmann, U. (1990). Correlation between pharmacokinetics, pharmacodynamics and efficacy of antibacterial agents in animal models. *Eur. J. Clin. Microbiol. Infect. Dis.*, **9**, 479–487.

Drusano, G. L. (1988). Role of pharmacokinetics in the outcome of infections. *Antimicrob. Agents Chemother.*, **32**, 289–297.

Eagle, H., Fleischman, R., Musselman, A. D. (1950). Effect of schedule of administration on the therapeutic efficacy of penicillin: importance of the aggregate time penicillin remains at effectively bactericidal levels. *Am. J. Med.* **9**, 280–299.

Eagle, H., Fleischman, R., Levy, M. (1953). Continuous vs discontinuous therapy with penicillin: the effect of the interval between injections on therapeutic efficacy. *N. Engl. J. Med.* **248**, 481–488.

Entenza, J. M., Blatter , M., Glauser, M. P., Moreillon, P. (1994a). Parenteral sparfloxacin compared with ceftriaxone in treatment of experimental endocarditis due to penicillin-susceptible and -resistant streptococci. *Antimicrob. Agents Chemother.* **38**, 2683–2688.

Entenza, J. M., Fluckiger, U., Glauser, M. P., Moreillon, P. (1994b). Antibiotic treatment of experimental endocarditis due to methicillin-resistant *Staphylococcus epidermidis*. *J. Infect. Dis.*, **170**, 100–109.

Entenza, J. M., Vouillamoz, J., Glauser, M. P., Moreillon, P. (1997). Levofloxacin versus ciprofloxacin, flucloxacillin, or vancomycin for treatment of experimental endocarditis due to methicillin-susceptible or -resistant *Staphylococcus aureus*. *Antimicrob. Agents Chemother.*, **41**, 1662–1667.

Fantin, B., Ebert, S., Legget, J., Vogelman, B., Craig, W. A. (1990). Factors effecting duration of *in-vivo* post-antibiotic effect for aminoglycosides against Gram-negative bacilli. *J. Antimicrob, Chemother.*, **27**, 829–836.

Fluckiger, U., Segessenmann, C., Gerber, A. U. (1991). Integration of pharmacokinetics and pharmacodynamics of imipenem in a human-adapted mouse model. *Antimicrob. Agents Chemother.*, **35**, 1905–1910.

Fluckiger, U., Moreillon, P., Blaser, J., Bickle, M., Glauser, M. P., Francioli, P. (1994). Simulation of amoxicillin pharmacokinetics in humans for the prevention of streptococcal endocarditis in rats. *Antimicrob. Agents Chemother.*, **38**, 2846–2849.

Frimodt-Moller, N. (1988). Correlation of *in vitro* activity and pharmacokinetic parameters with effect *in vivo* for antibiotics. Observations from experimental pneumococcus infection. *Dan. Med. Bull.*, **35**, 422–437.

Gavalda, J., Cardona, P. J., Almirante, B. *et al.* (1996). Treatment of experimental endocarditis due to *Enterococcus faecalis* using once-daily dosing regimen of gentamicin plus simulated profiles of ampicillin in human serum. *Antimicrob. Agents Chemother.*, **40**, 173–178.

Gerber, A. U., Feller, C., Brugger, H. P. (1984). Time course of the pharmacological response to beta-lactam antibiotics *in vitro* and *in vivo*. *Eur. J. Clin. Microbiol.*, **3**, 592–597.

Gerber, A. U., Brugger, H. P., Feller, C., Stritzko, T., Stalder, B. (1986). Antibiotic therapy of infections due to *Pseudomonas aeruginosa* in normal and granulocytopenic mice: comparison of murine and human pharmacokinetics. *J. Infect. Dis.*, **153**, 90–98.

Gerber, A. U., Stritzko, T., Segessenmann, C., Stalder, B. (1991). Simulation of human pharmacokinetic profiles in mice, and impact on antimicrobial efficacy of netilmicin, ticarcillin and ceftazidime in the mouse peritonitis septicaemia model. *Scand. J. Infect. Dis.* (suppl. 74), 195–203.

Giacomini, K. M., Roberts, S. M., Levy, G. (1981). Evaluation of methods for producing renal dysfunction in rats. *J. Pharm Sci.*, **70**, 117–121.

Hatano, K., Wakai, Y., Watanabe, Y., Mine, Y. (1994). Simulation of human plasma levels of beta-lactams in mice by multiple dosing and the relationship between the therapeutic efficacy and pharmacodynamic parameters. *Chemotherapy*, **40**, 1–7.

Hishikawa, T., Kusunoki, T., Tsuchiya, K. *et al.* (1990). Application of mathematical model to experimental chemotherapy of fatal murine pneumonia. *Antimicrob. Agents Chemother.*, **34**, 326–331.

Hishikawa, T., Kusunoki, T., Tsuchiya, K. *et al.* (1991). Application of mathematical model to multiple-dose experimental chemotherapy for fatal murine pneumonia. *Antimicrob. Agents Chemother.*, **35**, 1066–1069.

Hunter, P. A., Rolinson, G. N., Witting, D. A. (1976). Effect of carbenicillin on *Pseudomonas* infection. In *Chemotherapy*, ed. JD Williams, AM Geddes, Vol. 2, pp. 289–293.

James, J. K., Palmer, S. M., Levine, D. P., Rybak, M. J. (1996). Comparison of conventional dosing versus continuous-infusion vancomycin therapy for patients with suspected or documented gram-positive infections. *Antimicrob. Agents Chemother.*, **40**, 696–700.

Joly, V., Pangon, B., Brion, N., Vallois, J. M., Carbon, C. (1988). Enhancement of the therapeutic effect of cephalosporins in experimental endocarditis by altering the pharmacokinetics with diclofenac. *J. Pharm. Exp. Ther.*, **246**, 695–700.

Kavi, J., Andrews, J. M., Ashby, J. P., Hillman, G., Wise, R. (1988). Pharmacokinetics and tissue penetration of cefpirome, a new cephalosporin. *J. Antimicrob. Chemother.*, **22**, 911–916.

Kunst, M. W., Mattie, H. (1978). Probenecid and the antibacterial activity of cephradine *in vivo*. *J. Infect. Dis.*, **137**, 830–834.

Leggett, J. E., Ebert, S., Fantin, B., Craig, W. A. (1991). Comparative dose–effect relations at several dosing intervals for beta-lactam, aminoglycoside and quinolone antibiotics against Gram-negative bacilli in murine thigh infection and pneumonitis models. *Scand. J. Infect. Dis.* (suppl. 74), 179–184.

Lister, P. D., Saunders, C. C. (1995). Comparison of ampicillin-sulbactam regimens simulating 1.5- and 3.0-gram doses to humans in treatment of *Escherichia coli* bacteremia in mice. *Antimicrob. Agents Chemother.*, **39**, 930–936.

Mattie, H. (1984). Animal models in antibacterial drug research. *J. Antimicrob. Chemother.*, **14**, 101–102.

Mattie, H., Kunst M. W. (1978). Animal models for the assessment of potentiation of antibiotics by probenecid and by host-response. *Infection*, **6** (suppl. 1), S36–S37.

Mizen, L., Woodnutt, G. (1988). A critique of animal pharmacokinetics. *J. Antimicrob. Chemother.*, **21**, 273–278.

Mizen, L., Woodnutt, G., Kernutt, I., Catherall, E. J. (1989). Simulation of human serum pharmacokinetics of ticarcillin-clavulanic acid and ceftazidime in rabbits, and efficacy against experimental *Klebsiella pneumoniae* meningitis. *Antimicrob. Agents Chemother.*, **33**, 693–699.

Mordenti, J. (1986). Man versus beast: pharmacokinetic scaling in mammals. *J. Pharmaceut. Sci.*, **75**, 1028–1040.

Naziri, W., Cheadle, W. G., Trachtenberg, L. S., Montgomery, W. D., Polk, H. C. Jr. (1995). Second place winner of the Conrad Jobst Award in the gold medal paper competition. Increased antibiotic effectiveness in a model of surgical infection through continuous infusion. *Am. Surg.*, **61**, 11–15.

Schentag, J. (1991). Correlation of pharmacokinetic parameters to efficacy of antibiotics: relationships between serum concentrations, MIC values, and the bacterial eradication in patients with Gram-negative pneumonia. *Scand. J. Infect. Dis.* (suppl. 74), 218–234.

Thauvin, C., Eliopoulos, G. M., Willey, S., Wennersten, C., Moellering, R. C. Jr. (1987). Continuous-infusion ampicillin therapy of enterococcal endocarditis in rats. *Antimicrob. Agents Chemother.*, **31**, 139–143.

Thauvin-Eliopoulos, C., Tripodi, M. F., Moellering, R. C. Jr, Eliopoulos G. M. (1997). Efficacies of piperacillin-tazobactam and cefepime in rats with experimental intra-abdominal abscesses due to an extended-spectrum beta-lactamase-producing strain of *Klebsiella pneumoniae*. *Antimicrob. Agents Chemother.*, **41**, 1053–1057.

Thonus, I. P., deLange-Macdaniel A. V., Otte, C. J., Michel, M. F. (1979). Tissue cage infusion: a technique for the achievement of prolonged steady state inexperimental animals. *J. Pharm. Methods*, **2**, 63–69.

van Ogtrop, M. L., Mattie, H., Guiot, H. F. L., van Strijen, E., Hazekamp-van Dokkum, A-M., vanFurth, R. (1990). Comparative study of the effects of four cephalosporins against *Escherichia coli in vitro* and *in vivo*. *Antimicrob. Agents Chemother.*, **34**, 1932–1937.

Veng-Pedersen, P. (1980). Model-dependent method of analysing input in linear pharmacokinetics systems having polyexponential impulse response I and II. *J. Pharm. Sci.*, **69**, 298 311.

Vogelman, B., Gudmundsson, S., Leggett, J., Turnidge, J., Ebert, S., Craig, W. A. (1988). Correlation of antimicrobial pharmacokinetic parameters with therapeutic efficacy in an animal model. *J. Infect. Dis.*, **158**, 831–847.

Vondracek, T. G. (1995). Beta-lactam antibiotics: is continuous infusion the preferred method of administration? *Ann. Pharmacother.*, **29**, 415–424.

Walker, J. M., Wise, R., Mitchard, M. (1979). The pharmacokinetics of amikacin and gentamicin in volunteers: a comparison of individual differences. *J. Antimicrob. Chemother.*, **5**, 95–99.

Williams, R. T. (1974). Inter-species variations in the metabolism of xenobiotics. *Biochem. Soc. Trans.*, **2**, 359–377.

Woodnutt, G., Catherall, E. J., Kernutt, I., Mizen, L. (1988). Temocillin efficacy in experimental *Klebsiella pneumoniae* meningitis after infusion into rabbit plasma to simulate antibiotic concentrations in human serum. *Antimicrob. Agents Chemother.*, **32**, 1705–1709.

Woodnutt, G., Berry, V., Mizen, L . (1992a). Simulation of human serum pharmacokinetics of cefazolin, piperacillin, and BRL 42715 in rats and efficacy against experimental intraperitoneal infections. *Antimicrob. Agents Chemother.*, **36**, 1427–1431.

Woodnutt, G., Berry, V., Catherall, E. (1992b). Simulation of human serum concentrations of amoxicillin/clavulanate in the rat using a computer controlled model — efficacy in an experimental wound infection. Abstract 29. p 46 Int. Symp. Bacterial Infection models in Antimicrobial Chemotherapy., Helsinore, Denmark Oct. 26–28.

Woodnutt, G., Larque, E., Berry, V., Wells, J. (1993). Efficacy of simulated human serum concentrations of amoxycillin and clavulanic acid against an *Haemophilus influenzae* respiratory tract infection — comparison with cefaclor and cefuroxime.

Abstract 371, 18th Int. Congress Chemother., Stockholm, Sweden, June 27–July 2., American Society for Microbiology, Washington, DC.

Woodnutt, G., Berry, V., Mizen, L. (1995a). Effect of protein binding on penetration of β-lactams into rabbit peripheral lymph. *Antimicrob. Agents Chemother.*, **39**, 2678–2683.

Woodnutt, G., Berry, V., Bryant, J., Gisby, J., Slocombe, B. (1995b). Efficacité de l'association amoxicilline-acide clavulanique dans un modèle d'abces sous-cutane a *E. coli* chez le rat après simulation de l'administration chez homme de 1 g/200 m (IVD) ou de 2 g/200 mg (perfusion): pp 23–26. In Augmentin® Actualites et mode d'emploi, La Lettre de L'Infectiologue, Numero hors, Fevrier, Edimark, France.

Woodnutt, G., Berry, V., Bryant, J., Gisby, J. (1995c). Relationship between *in vitro* activity of ticracillin/clavulanate and *in vivo* efficacy. Abstract 1199, 19th Int. Congress of Chemother., Montreal: In *Canadian J. Inf. Disease,* **6** Suppl C., Pulsus Group Inc., Canada.

Woodnutt, G., Berry, V., Bryant, J., Gisby, J. (1995d). *In vivo* efficacy of amoxycillin /clavulanate against strains of *E. coli* with differing *in vitro* susceptibility. Abstract 1200 19th Int. Congress of Chemother., Montreal: In *Canadian J Inf. Disease*, **6** Suppl C, Pulsus Group Inc., Canada.

Woodnutt, G., Berry, V., Bryant, J. (1995e). Comparative efficacy of amoxycillin and ampicillin against *Streptococcus pneumoniae* experimental respiratory tract infections using simulated human serum concentrations. Abstract A80., 35th Interscience Conference on Antimicrobial Agents and Chemotherapy, San Francisco, USA 17–20 Sept. American Society for Microbiology, Washington, DC.

Woodnutt, G., Berry, V., Bryant, J. (1996). Evaluation of reduced dosing schedule of amoxycillin using two pharmacodynamic models. Abstract A40 36th Interscience Conference on Antimicrobial Agents and Chemotherapy, New Orleans, USA, 15–18 Sept. American Society for Microbiology, Washington, DC.

Zak, O., O'Reilly, T. (1991) Animal models in the evaluation of antimicrobial agents. *Antimicrob. Agents Chemother.*, **35**, 1527–1531.

Chapter 12

Modes of Action of Antibiotics and Bacterial Structure: Bacterial Mass Versus their Numbers

V. Lorian

Introduction

During the last decade, the molecular mode of action of antibacterial agents has been extensively elucidated. The correlation of modes of action and morphologic-ultrastructural alterations of bacteria has been exemplified in several reports. This review presents antibiotics and their schematic mode of action together with their respective structural alterations produced on bacteria as observed *in vitro*, in animal models as well as after therapy in humans. Moreover, this review documents that mode of action changes with antibiotic concentrations, that at concentrations lower than the minimum inhibitory concentration (MIC), the antibacterial effect is not lower—the effect is different. These low antibiotic concentrations produce live, giant forms of bacteria, which are a real challenge for phagocytosis; therefore, the classical equation of the number of bacteria and their pathogenic activity is not applicable. Correct clinical interpretation requires the determination of their mass rather than their numbers.

Antibiotic-induced alteration of bacterial structure

The modes of action of antibiotics on the bacterial cell can be classified according to the target site and, to some extent, to the structural alterations produced. As depicted schematically in Figure 12.1, the largest group of agents is

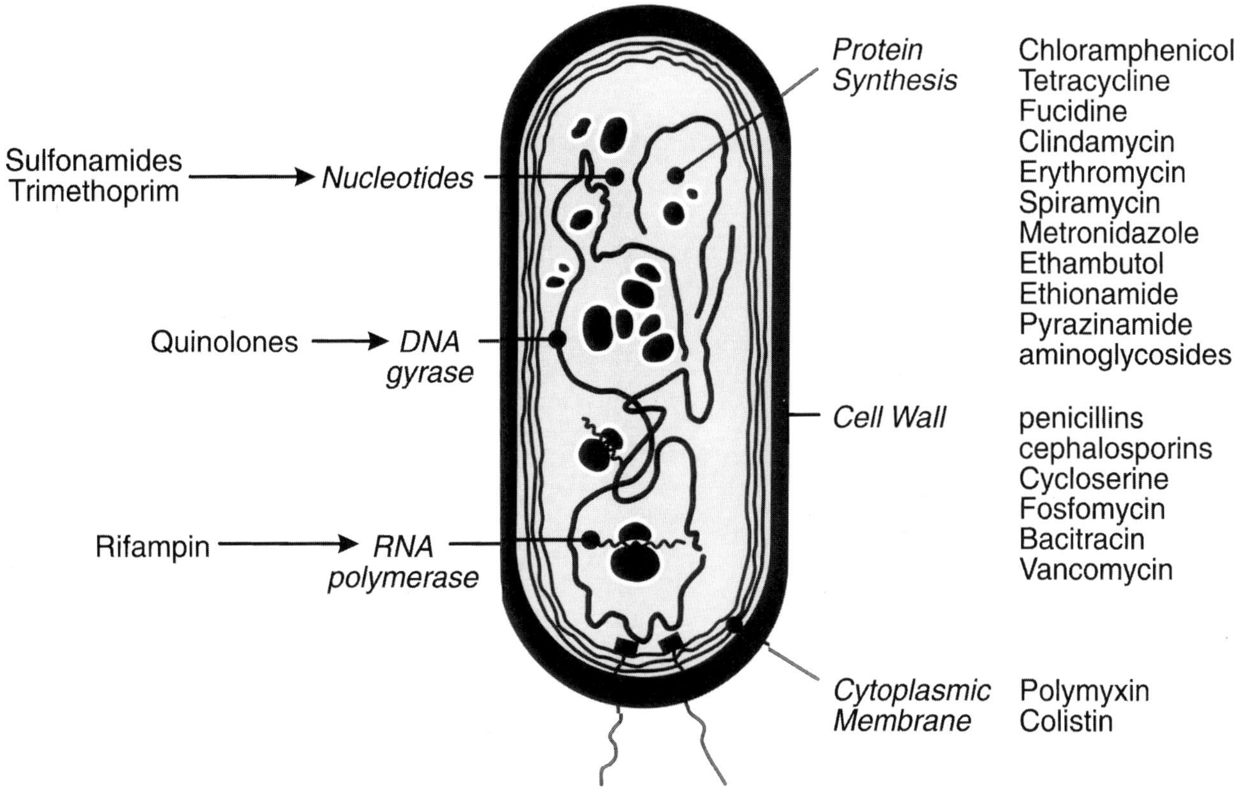

Figure 12.1 Schematic mode of action of antibiotics on bacteria, according to the target site of each group of agents.

Handbook of Animal Models of Infection
ISBN 0-12-775390-7

characterized by their action on the ribosome, resulting in the alteration of protein synthesis. The most frequently utilized antibiotics in clinical practice are found within the second largest group which consists of agents that interfere with the integrity of the cell wall.

Other modes of action include: disruption of the cytoplasmic membranes (polymyxin), alteration of RNA synthesis by inhibition of RNA polymerase (rifampin), inhibition of DNA gyrase (quinolones), and competitive inhibition of coenzyme synthesis (sulfonamides).

A summary of their modes of action and a description of the respective structural alterations, including electron micrographs, follows.

Cell wall modifiers

Maintenance of cellular shape in growing and dividing bacteria depends on the integrity of the cell wall murein (peptidoglycan). The most important step in the construction of the murein is the formation of peptide cross-bridges. The transpeptidase enzymes catalyzing this process are highly sensitive to inhibition by β-lactam antibiotics. Sensitive organisms possess specific high molecular weight proteins known as the penicillin-binding proteins (PBPs; Spratt, 1975; Spratt *et al.*, 1980; Bryan and Godfrey, 1991).

Before the discovery of the PBPs and their function, it had been observed that Gram-negative bacilli (normally 2–4 μm long) when exposed to certain β-lactam antibiotics produce filaments up to 93 μm long while others, such as cephaloridine, aminopenicillanic acid or aminocephalosporanic acid, produced spheroid cells (Lorian and Sabath, 1972).

The PBPs are now classified in accordance with the structural alteration produced by β-lactams on Gram-negative bacilli. Inhibition of PBP 1 A and B results ultimately in cell lysis; inhibition of PBP 2 results in the formation of spherical cells and inhibition of PBP 3 produces filaments by preventing septum formation. Other PBPs, 2A, 4, 5 and 6, are not directly responsible for bacterial shape. Antibiotic sensitivity is dependent on the availability of PBPs, which varies among the Gram-negative bacilli; susceptibility is also related to the PBP-binding affinity which differs for each β-lactam antibiotic. Cephaloridine and cefsulodin have a higher affinity for PBP 1 than for PBP 3 but minimal affinity for PBP 2; therefore, *Escherichia coli* exposed to these drugs will transform into filaments that lyse rapidly (Figure 12.2A). Mecillinam and imipenem have a high affinity for PBP 2 and, therefore, will change the bacilli into spheroid cells (Figure 12.2B). Even at concentrations that do not inhibit replication, exposure of *E. coli* to mecillinam results in production of spherical forms. In contrast, mezlocillin and aztreonam prefer PBP 3 and, therefore, will produce long and quite stable filaments of *E. coli*

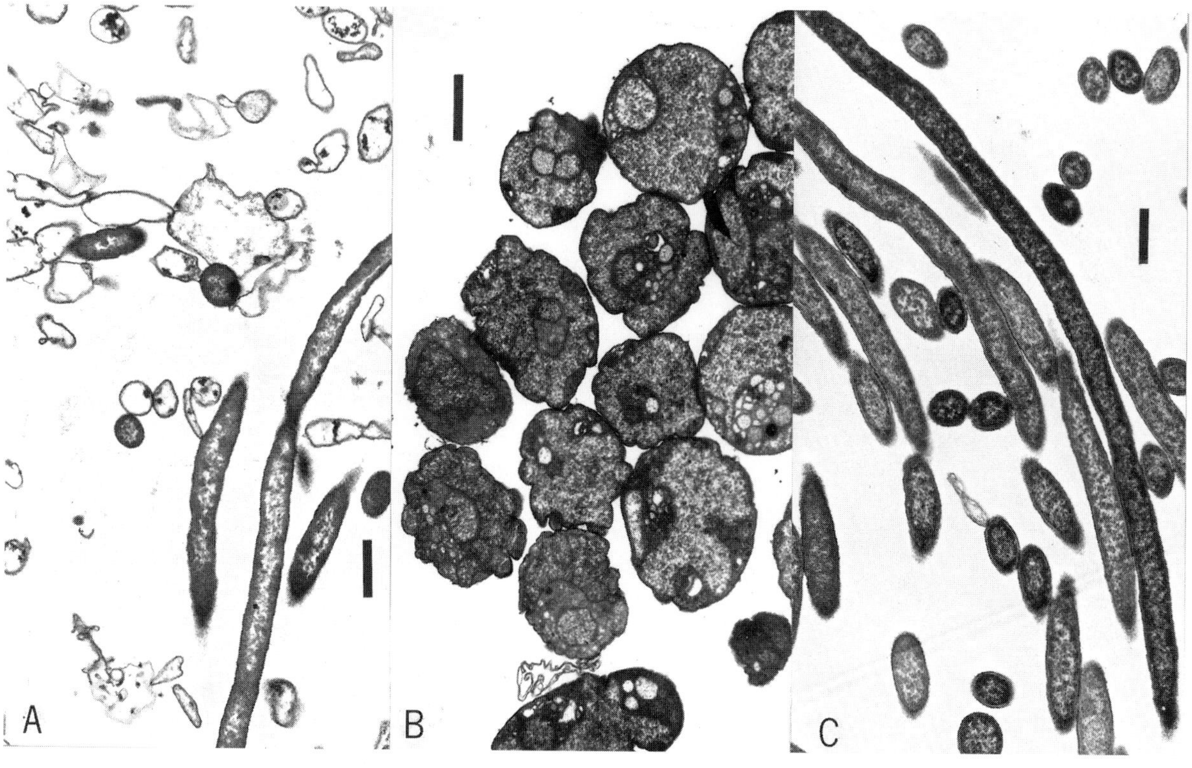

Figure 12.2 Binding affinities and structural effects. (A) Intense binding to PBP 1 results in extensive lysis; (B) binding to PBP 2 results in spherical forms; (C) intense binding to PBP 3 results in filament formation. Original magnification = 14 000×. Bar = 1 μm.

(Figure 12.2C). It has been suggested that the characteristics of filamentation can predict *in vivo* efficacy (Ryan and Monsey, 1981).

The relative binding affinity and binding stability of the antibiotic to each PBP will determine the multitude and diversity of aberrant forms. Exposure to subinhibitory concentrations (sub-MIC) will best identify the morphologic differences related to the blocking of each PBP. The bactericidal rate at concentrations above the MIC for most β-lactams is concentration-independent. Lysis is a consequence of saturation of the preferred target, leaving sufficient molecules to block PBP 1 also.

The quinolones, oxolonic acid, novobiocin, nitrofurantoin, sulfonamides and trimethoprim also produce filaments with a morphology similar to that of the filaments produced with β-lactam antibiotics. These drugs, however, do not bind to the PBPs.

In contrast to Gram-negative bacilli, there is little variation in the affinities of β-lactam antibiotics for the various PBPs of staphylococci, other Gram-positive cocci and *Neisseria*. Thus, all β-lactam antibiotics have similar morphologic effects on these bacteria. The β-lactam antibiotics at concentrations two to four times the MIC trigger the lysis and the total disintegration of staphylococci.

Untreated staphylococci have a diameter of 0.9–1.1 μm. When these staphylococci are grown for 4 hours with 1/3 of their MIC of any β-lactam antibiotic, large cells (1.5–3.0 μm in diameter) with multiple cross walls develop (Figure 12.3; Lorian, 1975).

At and above the MIC, antibiotic activity follows basically the same molecular mode of action. At concentrations lower than the MIC, usually at 1/2 to 1/8 MIC, antibiotics act by different mechanisms; the antibacterial effect is not proportionally lower—the effect is different.

During exposure to sub-MICs of β-lactam antibiotics, the number of colony-forming units (CFUs) increases only slightly; the bacteria divide, as shown by numerous genomes, but the new divided cells cannot separate.

Thus, each cell of the abnormal 'large staphylococci' consists of several generations derived from a single staphylococcus, and is in fact a cluster of staphylococci, prevented from separating by the failure to lyse adjoining cross walls. The inhibition of cross wall separation is reversible: when incubated on drug-free media, clusters of staphylococci separate into smaller clusters and eventually into many cells of normal shape and growth rate. The very same phenomenon was observed with enterococci and other streptococci when exposed to sub-MIC of β-lactam antibiotics. *Enterococcus faecium*, resistant to vancomcyin 400 μg/ml, when exposed to sub-MIC of vancomcyin (200 μg/ml), developed into large cells by the same mechanism (Lorian and Fernandes, 1997; Figure 12.4).

Cycloserine, fosfomycin, and vancomycin, which are not β-lactam antibiotics, inhibit the early stages of peptidoglycan assembly. These agents produce morphologic and ultrastructural alterations in Gram-positive cocci similar to those produced by β-lactam antibiotics, apparently without

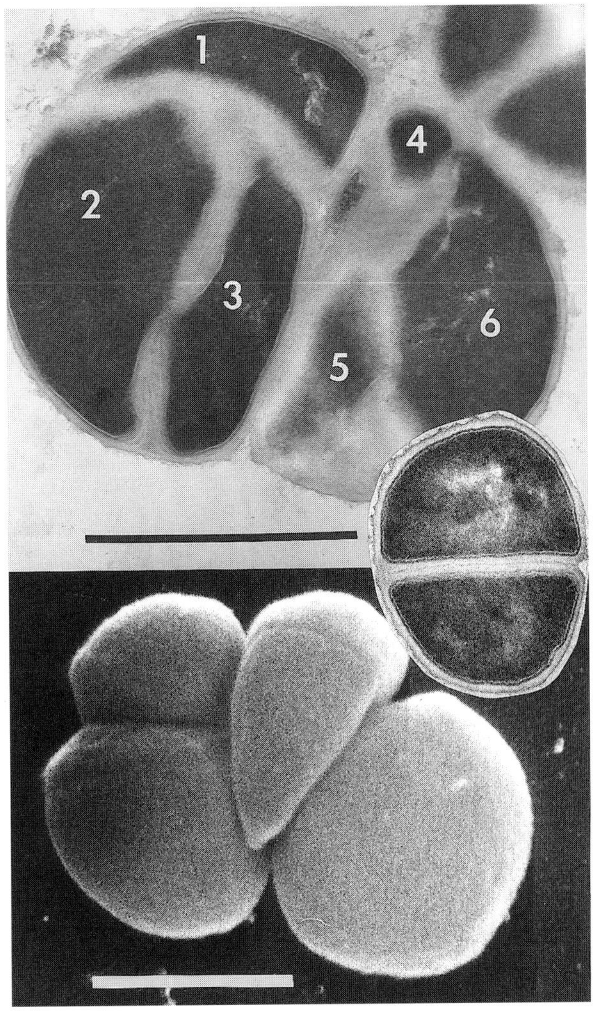

Figure 12.3 (Top) *Staphylococcus aureus* grown on agar containing 1/3 minimum inhibitory concentration oxacillin. Note the multiple thick cross walls holding together at least six divided staphylococci (transmission electron microscopy). (Bottom) Scanning electron microscopy of a comparable clump of *S. aureus*. The inset is a normal *S. aureus* control. Original magnification = 75 000×. Bar = 1 μm.

interfering with the PBPs. Exposure to fosfomycin results in an explosive lysis of staphylococci (Figure 12.5).

In brief, filamentation of Gram-negative bacilli is the result of multiplication without separation due to inhibition of septum initiation. In contrast, the large Gram-positive cocci with multiple thick cross walls are due to an increase in septum initiation and the failure of cross wall lysis.

The aminoglycoside antibiotics

Uptake of aminoglycosides depends on drug concentration and medium salt components; it could take *in vitro* up to 40 minutes for completion. This is the longest known time required for an antibiotic to reach its molecular target. The delay could influence total antibacterial activity when aminoglycoside synergism in combination with other

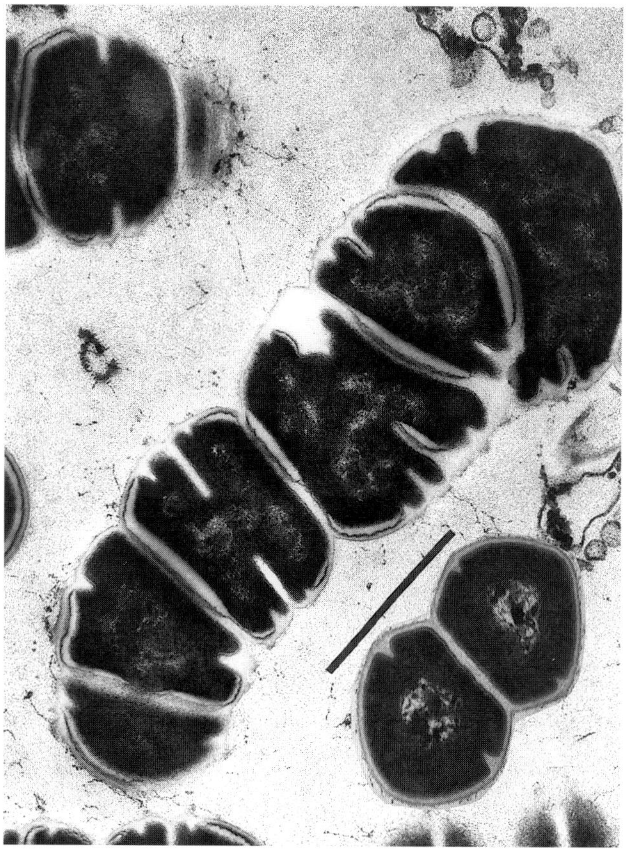

Figure 12.4 Vancomycin-resistant *Enterococcus faecium* exposed to 1/2 minimum inhibitory concentration vancomycin. (Bottom right) Control cell grown in medium alone. Original magnification = 50 000×. Bar = 1 μm.

antibacterial agents is expected (Davies, 1991). The classic description of their mode of action is as follows. The aminoglycoside is transported across the cytoplasmic membrane by enzymes requiring oxidative energy, therefore, antibacterial activity is minimal under anaerobic conditions.

The final target is the ribosome, and specific irreversible binding of the drug to the ribosomal RNA (not to the proteins) disrupts protein synthesis and kills the bacteria. The irreversibility of binding to the ribsome and the cidal action appear to distinguish the aminoglycosides from the other protein synthesis inhibitors with static action such as tetracycline and chloramphenicol. A recent view considers that these mechanisms may not be the most important mechanisms of the antimicrobial action; in fact, most susceptible Gram-negative bacilli are probably dead long before the drug arrives at the 30S ribosome. It is now recognized that aminoglycosides competitively displace cell biofilm-associated Mg^{2+} and Ca^{2+} that link the polysaccharides of adjacent lipopolysaccharide molecules (Kadurugamuwa *et al.*, 1993).

Ultrastructure alterations occurred in *E. coli* exposed to amikacin and gentamicin, consisting of a reduction in the number of ribosomes in the center of the cell. The cells showed a distinct pattern in the center of almost all cells after 2 hours contact with amikacin at levels twice their MIC. This was called the "tornado image" (Lorian and Atkinson, 1986) because it consisted of just a few ribosomes and an aggregation of nuclear material in a peculiar concentric pattern (Figure 12.6). While aminoglycosides could produce other images of ribosome deprivation, the "tornado image" has not been observed with any other antibiotic and can be considered specific for aminoglycosides.

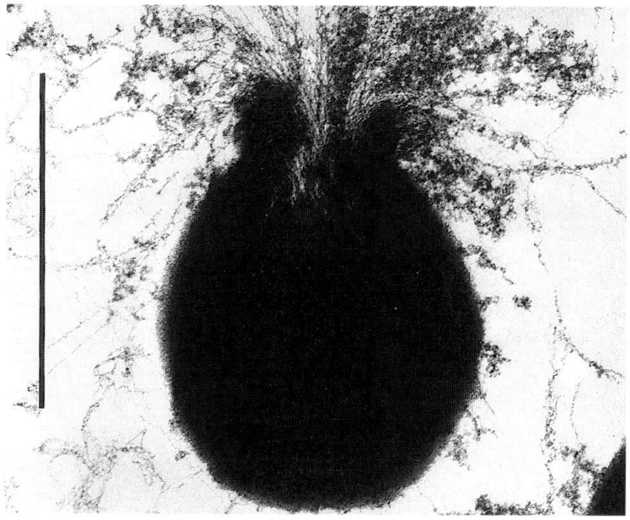

Figure 12.5 *Staphylococcus aureus* exposed to fosfomycin explodes. Original magnification = 125 000×. Bar = 1 μm. This photograph is now part of the permanent collection at the *Metropolitan Museum of Art* in New York City.

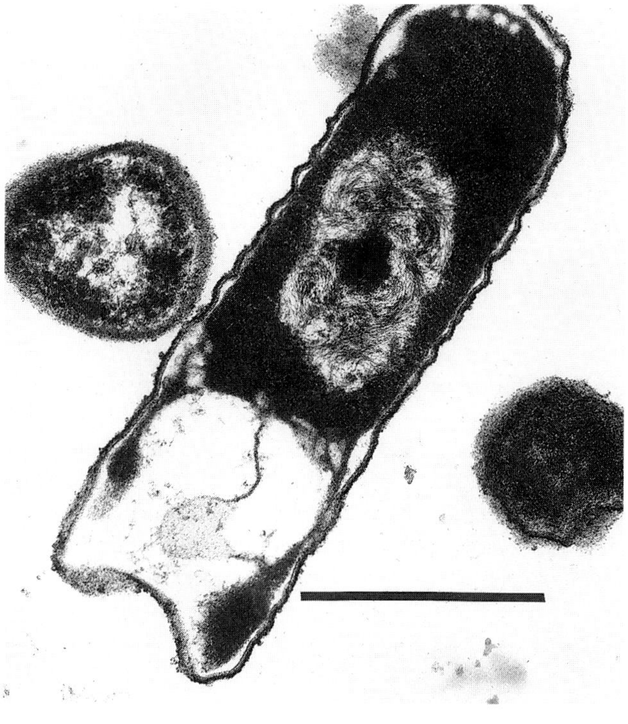

Figure 12.6 *Escherichia coli* exposed to amikacin shows the "tornado image" as well as cell wall and cytoplasmic membrane alterations. Original magnification = 60 000×. Bar = 1 μm. From Lorian and Atkinson (1986), with permission.

When the *E. coli* were exposed to higher concentrations of amikacin the ultrastructure confirms the "biofilm" mode of action; the cytoplasmic membranes disassociated from the cell walls, and some bacilli had damaged cell walls, with complete loss of cellular shape.

Tetracycline

There are a number of different tetracyclines which differ in lipophilicity but all have a similar mode of action. Tetracycline inhibits the binding of aminoacyl-tRNAs to the A site on the 30S ribosome subunit to prevent protein synthesis (Day, 1966). One molecule of drug per ribosome inhibits translation.

Low concentrations of chlortetracycline cause progressive breakdown (up to 50%) of polyribosome in metabolically active *Bacillus megaterium*. Tetracycline is bacteriostatic even at concentrations significantly higher than the MICs.

E. coli and other species of Gram-negative bacilli exposed to tetracycline exhibit many well-defined areas with nuclear material but few or no ribosomes. These areas are surrounded by ribosomes at a normal density. This structural pattern can be produced by many antibiotics, including β-lactams at low concentrations; it is therefore non-specific but it is always present after exposure to tetracycline.

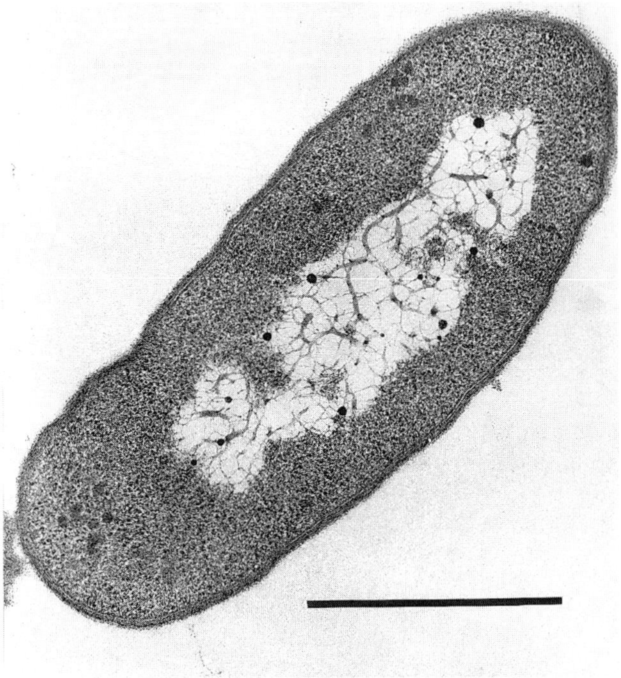

Figure 12.7 Salmonella exposed to chloramphenicol shows an extensive area devoid of ribosomes located along the longitudinal axis of the bacillus. Original magnification = 75 000×. Bar = 1 μm.

Chloramphenicol

Chloramphenicol inhibits protein synthesis (Goldberg, 1965). It binds to a histidine in the active site of ribosomal peptidyl transferase on the 50S subunit, thereby interfering with amino acid acetyl group migration. Chloramphenicol inhibits peptide chain elongation and, with it, the movement of ribosomes along the messenger RNA. Chloramphenicol is bacteriostatic for staphylococci, streptococci, and Enterobacteriaceae; it is bactericidal for *Haemophilus* and *Neisseria* species, and *Streptococcus pneumoniae*, by an unknown mechanism.

E. coli and *Salmonella* exposed to chloramphenicol exhibit an extensive elongated and centrally located area of nuclear material with few or no ribosomes (Morgan *et al.*, 1967). A wide layer of ribosomes is present in a normal density all along the inner part of the cytoplasmic membrane (Figure 12.7). The evidence suggests that chloramphenicol produces inside the bacillus a tubular area that is deprived of ribosomes and extends in some cases the entire length of the cell.

While chloramphenicol may produce only dispersed zones deprived of ribosomes (similar to tetracycline), the extensive, tubular-like areas of nuclear material surrounded with ribosomes along the inside of the cytoplasmic membrane are not produced by any other antibiotic, and remain specific for chloramphenicol.

Exposure of *Staphylococcus aureus* to chloramphenicol can produce thickening of the cell wall, and *S. aureus* with two or three cell wall layers have been observed after prolonged exposure to chloramphenicol (Giesbrecht and Ruska, 1968).

Clindamycin

Clindamycin and lincomycin inhibit protein synthesis in Gram-positive cocci (Reusser, 1975). The peptidyl transferase reaction as well as the binding to ribosomes of both acetyl-lencyl and lencyl pentanucleotides is inhibited. It also causes rapid and extensive degradation of polyribosomes.

S. aureus exposed to lincomycin showed an increase in the cell wall and cross wall thickness; however, the cross walls that are larger than normal exhibit a normal texture and the typical dense line common to staphylococci which are ready to separate (Lorian and Atkinson, 1976). Some staphylococci exhibit 2 to 4 cross walls. Such alterations can be caused by other drugs as well as changes in growth conditions and, therefore, are not specific for the action of lincomycin-clindamycin on staphylococci.

Pristinamycins

The combination of semisynthetic water-soluble derivatives of pristinamycin, quinupristin/dalfopristin, target the ribosome. These compounds close or narrow the channel through which proteins are extruded, leading to their accumulation on the ribosome. In addition, it is likely that this accumulation disturbs peptidyl-tRNA hydrolase activity, thereby depleting free tRNAs within the cells and inhibiting protein synthesis (Aumercier *et al.*, 1992). They are very active against all Gram-positive cocci as well as against *Moraxella* sp., *Legionella* sp. and *Mycoplasma* sp. (Bouanchaud, 1997).

Exposure of *S. aureus* to the pristinamycin combination results in two main structural alterations: enlargement of the cell due to the enlargement of the cytoplasmic mass and a marked thickening of the cell wall, including cells with several cell wall layers (Lorian *et al.*, 1994). Increase in cell wall thickness is produced by other antibiotics. However, only the pristinamycin combination produced staphylococci with four, five, and as many as six distinct layers of cell wall (Figure 12.8). Only this antibiotic produced increases in the cytoplasmic mass. Such structural alterations could be specific for pristinamycin activity on *S. aureus*.

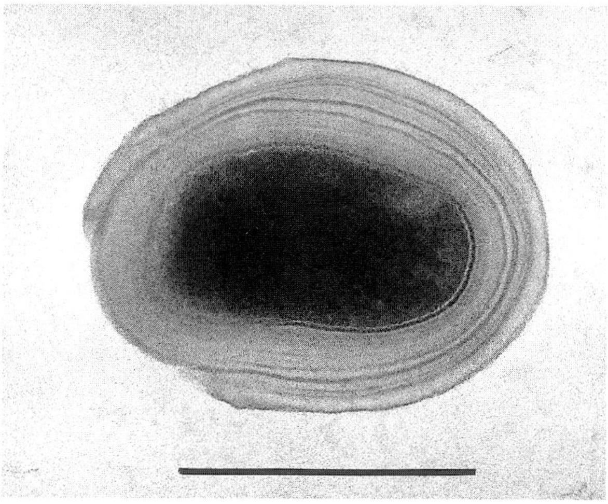

Figure 12.8 *Staphylococcus aureus* exposed to pristinamycin I and II, in combination, shows six layers of cell wall. Original magnification = 100 000×. Bar = 1 μm.

Rifampin

Rifampin is a semisynthetic derivative of the rifamycin class of antibiotics. Rifampin inhibits the RNA polymerase that employs a DNA template to produce ribosomal and messenger RNA. By binding to RNA polymerase, rifampin sterically blocks the translocation of pppApU and thereby inhibits further elongation of the RNA chain. One molecule of rifampin is bound to each enzyme monomer. Importantly, DNA can protect RNA polymerase against inactivation by rifampin, and RNA synthesis is not affected once chain elongation has started (Wehrli, 1983).

S. aureus exposed to rifampin exhibits an increase in thickness of both the cell wall and the cross wall. The outer part of the thick cell wall has a serrated appearance (Lorian *et al.*, 1983). The very thick cross wall, in contrast to similar abnormal cross walls produced with β-lactams, contains a well-defined dense line seen in normal cross walls of staphylococci ready to separate. Since thick cell walls and cross walls can be the result of exposure to other antibiotics or even the effect of prolonged incubation on a solid surface, the image is not specific for rifampin but is always evident after exposure to rifampin.

The quinolones

DNA gyrase, the A subunit, is the primary target of quinolones. The effects of ciprofloxacin are extensive and include reduction in negative supertwisting, decreased decatenation of interlocked DNA circles, antagonism of DNA replication (initiation and fork propagation), impaired segregation of replicated chromosomes into daughter bacterial cells, and interference with aspects of transcription, DNA repair, and recombination.

The quinolones are potent inducers of the SOS DNA repair–recombination systems which are responsible for filamentation of Gram-negative bacilli (Hooper and Wolfson, 1991). Drug exposure also causes inhibition of protein synthesis and membrane damage, usually resulting in rapid cell death.

After exposure to ciprofloxacin, 1/2 to 10× MIC, *E. coli* grows into unseptated filaments. The filaments produced by exposure of *E. coli* to ciprofloxacin are comparable to filaments produced by a large number of other antibacterial agents. There is, however, one main difference which was observed with ciprofloxacin; the filaments are riddled with small roundish areas, which appear on electron microscopy as spots containing a dense substance, or, depending on the fixation method, simply as holes (Figure 12.9). The presence of these little holes was confirmed by freeze-cracking electron microscopy.

In vitro *compared to* in vivo *activity*

The question arises whether the mode of action of antibiotics on bacteria with the respective ultrastructural expression observed *in vitro* is comparable to the action of antibiotics on bacteria *in vivo*. The criterion of comparison is the ultrastructure of bacteria exposed to antibiotics on a solid medium with that observed at the site of infection in experimental infections in animals or in human disease treated with antibiotics.

Since most characteristic morphology-ultrastructure alterations were produced with sub-MIC rather than with higher antibiotic concentrations, sub-MICs were used both in the *in vitro* and animal studies and in human antibiotic administration.

The first animal experimentation investigating therapeutic effects with an antibiotic dosage to result in sub-MIC at the site of infection was done in 1979 by Zak and Kradolfer. Rabbits were infected intraperitoneally with *E. coli* or *Proteus mirabilis*.

Sub-MICs of ampicillin or gentamicin were maintained for 6–10 hours by intravenous infusion. In infections with *E. coli* and *P. mirabilis*, sub-MICs of ampicillin or genta-

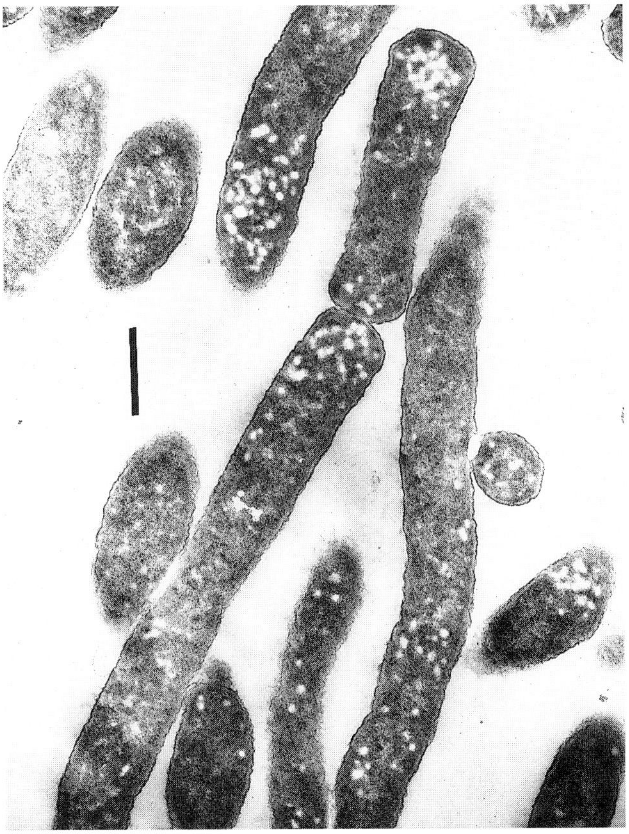

Figure 12.9 *Escherichia coli* exposed to ciprofloxacin shows filaments with small roundish holes. Original magnification = 25 000×. Bar = 1 μm.

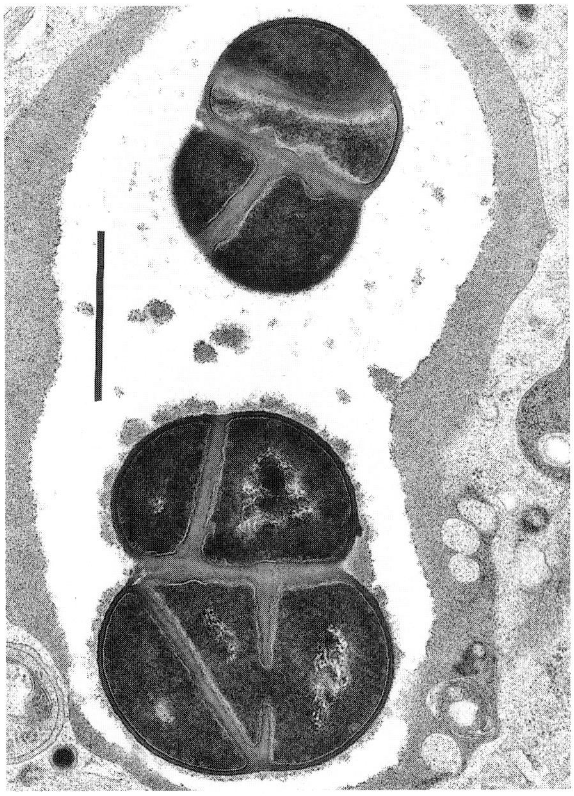

Figure 12.10 *Staphylococcus aureus* from the peritoneal exudate of a mice with peritonitis treated with cloxacillin. Compare with Figure 12.3; size and multiple cross walls are comparable. Original magnification = 50 000×. Bar = 1 μm.

micin prolonged the survival rates for infected animals beyond those for control animals (control, 17%>MIC, 87%; <MIC ampicillin, 51%; <MIC gentamicin, 57%). Filaments of *E. coli* or *P. mirabilis* were observed in the peritoneal exudate of animals treated with ampicillin.

Mice, with peritonitis induced experimentally with *S. aureus* Smith received oxacillin. Subsequent examination of the peritoneal fluid showed phagocytized staphylococci of normal size (phagocytosis protects staphylococci from oxacillin which cannot enter the polymorphonuclear cells) and free staphylococci that were 1.8–2.5 μm in diameter with multiple thick cross walls. When the experiment was repeated using pretreatment with cyclophosphamide to suppress phagocytosis, all organisms were observed to be up to 2–3 times the normal size with multiple thick cross walls, comparable to staphylococci exposed to 1/4 MIC of oxacillin *in vitro* on a membrane (Lorian *et al.*, 1982a).

In a comparable experiment, cloxacillin was used to treat the peritonitis induced in mice. The peritoneal fluid and the spleens were examined by transmission and scanning electron microscopy. The staphylococci in the spleens and the peritoneal exudate of treated mice were large and contained multiple cross walls (Figure 12.10); these cells are comparable to *S. aureus* grown on a membrane placed on blood agar containing sub-MIC of oxacillin (compare to Figure 12.3).

Rabbits with established staphylococcal (*S. aureus* MIC to cloxacillin = 0.25 μg/ml) endocarditis were injected twice at an interval of 2 hours with either 0.5 mg cloxacillin per kg or saline and were sacrificed 2.5 hours after the second injection. Vegetations were excised and ultrathin sections were prepared and examined by light, transmission, and scanning electron microscopy. Several valves were examined histologically. Concentrations of cloxacillin in serum were a mean of 0.166 μg/ml. Vegetations of treated rabbits contained staphylococci of normal size and form as well as organisms two to five times larger than normal.

Larger bacterial cells were usually located in areas close to blood; cells of normal size were usually embedded in fibrin (Figure 12.11). The large cells were about 5× larger than the control (Figure 12.12) and contained several cross walls (Figure 12.13). The size and structures of these staphylococci and those grown *in vitro* on a membrane in the presence of 0.09 μg cloxacillin per ml (1/3 MIC) were comparable (Lorian *et al.*, 1984).

A patient with a pleural effusion containing *S. aureus* was treated with nafcillin. Another patient with bilateral upper lobe infiltrates which grew *S. aureus* from a catheter inserted through a fiberoptic bronchoscope into the lesion was also treated with nafcillin. A third patient with a

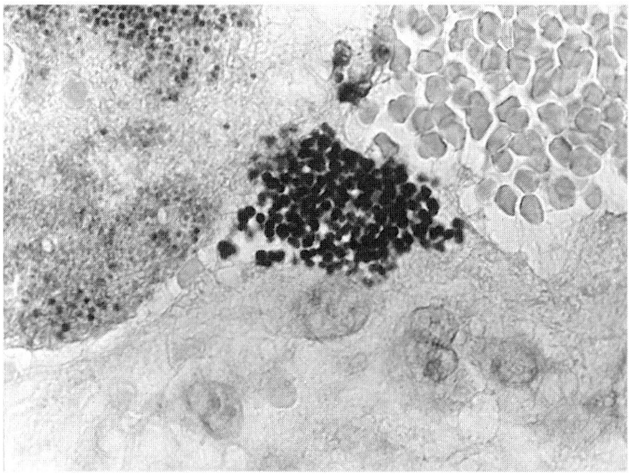

Figure 12.11 Histological section sowing the valve and the vegetation from a rabbit with endocarditis treated with cloxacillin. The microcolony of large staphylococci which developed on the valve is adjacent to blood. On the left side of the colony is fibrin containing normal-size staphylococci.

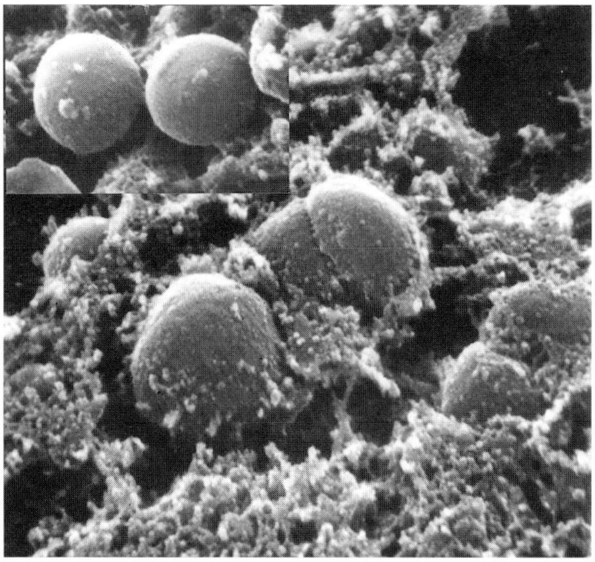

Figure 12.12 A large *Staphylococcus aureus* on a valve from a rabbit with endocarditis treated with cloxacillin. (Inset) Staphylococci from a rabbit control which was not treated. Original magnification = 43 000×. Bar = 1 μm.

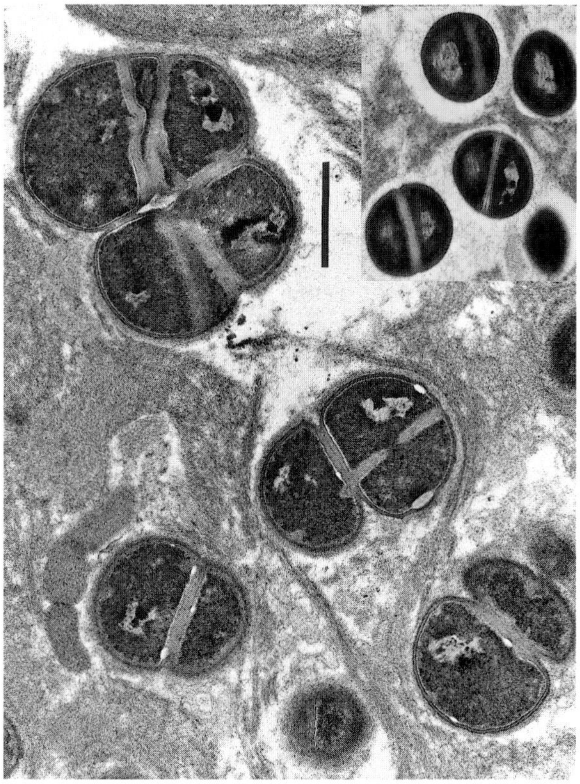

Figure 12.13 Large bacterial cells in a vegetation from a rabbit treated with cloxacillin, showing multiple cross walls. (Inset) Staphylococci from the vegetation of a control rabbit. Original magnification = 30 000×. Bar = 1 μm.

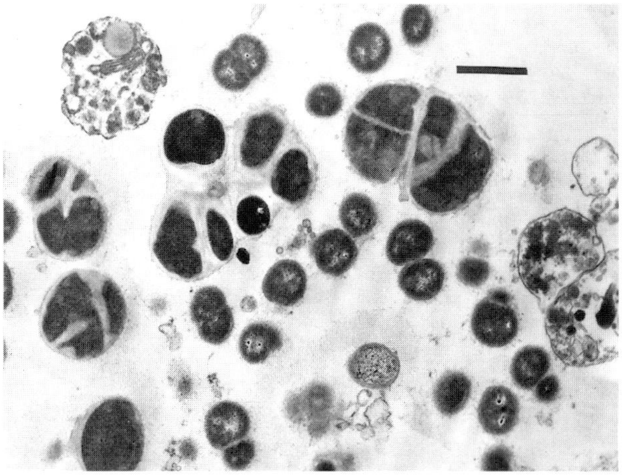

Figure 12.14 Staphylococci in the sputum of a patient with lung congestion treated with ampicillin. There are both normal-size staphylococci as well as larger cells showing several cross walls. One cell contains at least 6 staphylococci embedded in a cross wall net. Original magnification = 20 000×. Bar = 1 μm.

congestive lung image was treated with ampicillin and produced abundant sputum. Electron microscopy of organisms obtained from the pleural effusion, from the catheter and from the sputum, while the patients were receiving nafcillin or ampicillin, showed staphylococci which were much larger than normal and contained multiple cross walls and normal peripheral cell walls (Figure 12.14) (Lorian *et al.*, 1982b), (Ernst *et al.*, 1985) and (Lorian, 1985a).

A total of 20 patients with symptomatic urinary tract infections (>10^5 colony-forming units of *E. coli* per ml of urine) received 10 mg of ampicillin (1/200 of the usual dose) and 2 liters of fluid daily for 3 days. The MIC of *E. coli* ranged from 1 to 4 μg/ml of ampicillin; most were 4 μg/ml. The concentration of ampicillin in the urine ranged from 1/5 to 1/2 of the MIC of the respective *E. coli* strains producing the urinary infection. After 12 hours of therapy, the urine sediment contained filaments of *E. coli* 10–50 μm long. After 2 days, 16 patients had culture-negative urine without pyuria. A total of 18 similar patients received only

2 liters of fluid for 3 days. On day 4, all had $>10^5$ CFU of *E. coli* per ml of urine and $>10^4$ leukocytes per mm^3 of urine.

The data show that sub-MIC of ampicillin produce *in vivo E. coli* filaments comparable to those produced *in vitro*. It shows also that sub-MIC at the site of some infection could eradicate an infection (Redjeb *et al.*, 1982). In serious infections, however, a high dosage of a drug combination, resulting in >MIC at the site of infection, remains the indispensable choice (O'Reilly *et al.*, 1992).

It appears, therefore, that the morphology-ultrastructure and probably the mode of action of antibiotics on bacteria *in vivo* are comparable with the same phenomenon occurring *in vitro* if the bacteria are grown on a solid surface rather than in a liquid medium (Lorian, 1989).

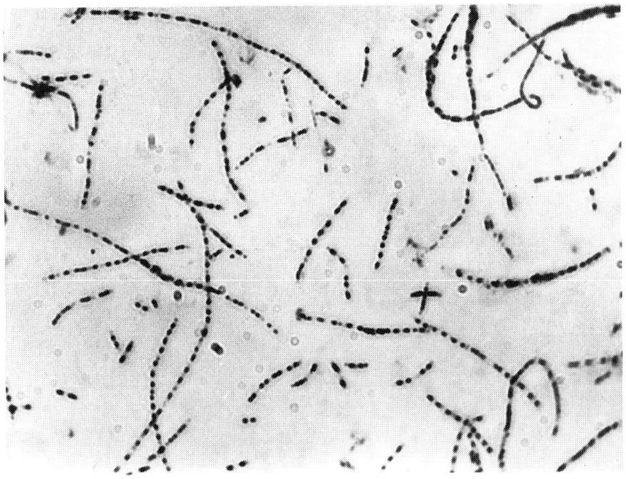

Figure 12.15 The dots are Giemsa-stained DNA in filaments of *Escherichia coli* after hydrochloric acid hydrolysis (the filaments were produced by exposure to ampicillin). From Lorian *et al.* (1985), with permission.

Clinical interpretation of bacterial mass versus bacterial numbers

An important consequence of morphologic alteration of bacteria exposed to antibiotics is that the equation of the number of bacteria and their pathogenic activity is no longer true. Standard relations between numbers and pathogenic activities were determined with normal-size organisms, therefore the standard unit in the pathogenicity equation for giant multinucleated bacteria is their total mass rather than their number. Bacterial virulence may be quantified; it can be defined as that minimal bacterial mass capable of producing injury to a given host; the smaller the mass, the higher the virulence. Bacterial mass can be measured by electron microscopy (Bahr *et al.*, 1976). However, when the mass of different bacterial sizes must be determined, measurement of direct weight is the more accurate method (Scullard and Meynell, 1966). In clinical laboratories quantification of bacteria is usually obtained by counting the CFUs. One bacterium on the surface of an agar medium will give rise to one CFU. This determination is accurate only if a one-to-one relationship of a single genome bacterium to a single CFU is strictly maintained.

As shown, exposure to antibiotics frequently alters the morphology of bacteria. Gram-negative bacilli elongate and become filaments and Gram-positive cocci develop clusters of cells held together by thick cross walls. The elongated bacilli and the unseparated clusters of Gram-positive cocci contain a variable number of genomes within a common cytoplasm (Figure 12.15); the number of genomes represents the number of replications that occurred while under exposure to the antibiotic. A normal *E. coli* has a mass of 2–4 picograms (pg) (dry weight), whereas a filament that results from exposure to ampicillin averages 45 pg. A single *S. aureus* weighs about 2 pg, whereas a cluster resulting from exposure to oxacillin may weigh as much as 18 pg.

When such a filament or cluster of cocci is removed from the drug-containing medium and is placed in drug-free medium, or *in vivo*, when the antibiotic concentration decreases to inactive levels, the organism may separate into individual bacilli or cocci corresponding in number to the number of viable genomes contained. In the CFU agar system, one filament or one cluster will give rise to one CFU (Figure 12.16) regardless of the size or the structure and the number of viable genomes contained. The utilization of the CFU system as a criterion for measuring bacteria will result in an erroneous interpretation because the size or mass and the number of viable genomes contained within each cell will not be measured. Therefore, the correct criterion for comparison and of these aberrant forms of bacteria produced by antibiotics with their untreated counterparts and the respective pathogenic effects should be their respective mass rather than their number (Lorian *et al.*, 1985). Since bacterial mass determination *in vivo* is not practical, comparable information can be obtained by calculating their volumes based on measurements of organisms on scanning electron micrograph photographs.

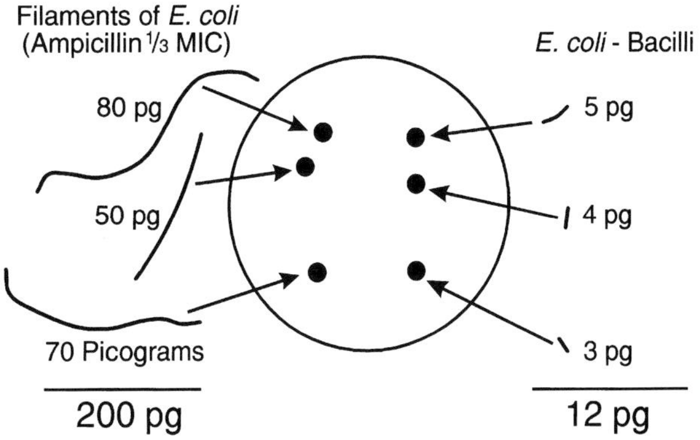

Figure 12.16 A single filament or a single bacillus planted on agar generates a single colony. (Left) Three filaments totaling 200 pg produced three colonies. (Right) Three bacilli totaling 12 pg also produced three colonies.

Using cellular mass to evaluate phagocytosis

The morphological effects of antibiotics also cause problems of interpretation in studies that utilize CFUs as a means of determining the phagocytic efficiency of polymorphonuclear cells against bacteria that were exposed to antibiotics. In a phagocytic system, either a filament or a cluster of bacterial genomes will yield only one CFU on drug-free agar; however, each filament or cluster of cocci that was engulfed was equivalent to as many bacteria as genomes in the respective aberrant form.

Since the sizes or masses of antibiotic-altered bacteria are so much larger than those of untreated organisms, a new basis for the quantification of bacteria replaced the CFU count. The use of weight or mass as a basis for the quantification of bacteria led to conclusions that differ considerably from those based on the number of CFUs (Lorian and Atkinson, 1984).

Figure 12.17 compares the number of CFUs phagocytized by human PMNs with the corresponding mass phagocytized. If the CFU system is employed as the criterion, it appears that the phagocytosis of filaments is equally effective as that of bacilli. However, when the mass criterion is employed, it becomes obvious that the mass of phagocytized bacteria, and hence the respective number of bacterial genomes, is more than 10 times greater for the filaments than for the untreated bacilli. Phagocytosis of filaments is, therefore, very efficient (Figure 12.18). These data demonstrate, in contrast to what was concluded from CFU numbers, that the phagocytic process is considerably more effective against filaments than against untreated bacilli. Similar results were obtained with antibiotic-induced clus-

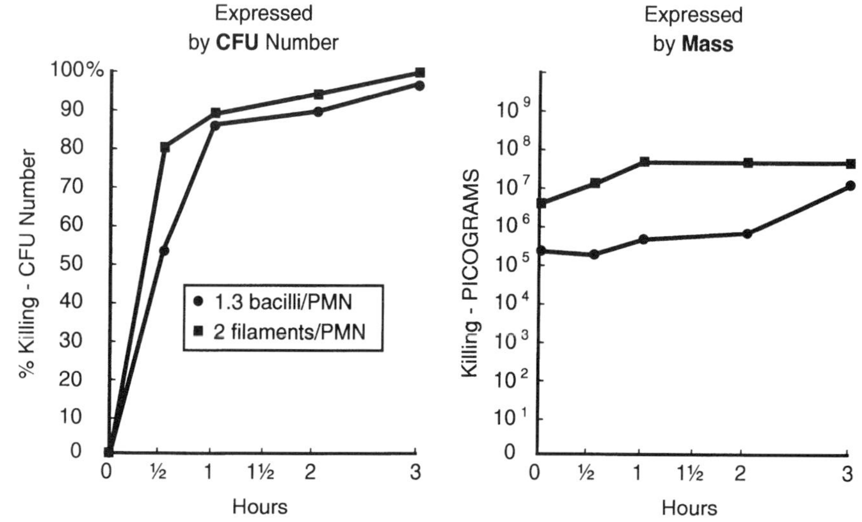

Figure 12.17 Phagocytosis of *Escherichia coli* bacilli or filaments by polymorphonuclear cells. When data are presented according to the number of organisms (colony-forming units: CFUs) phagocytized, the curves are comparable. When data are presented according to picograms of bacilli or filaments phagocytized, the number of picograms of filaments phagocytized is one $\log_{10}$ higher than the number of picograms of phagocytized bacilli. From Lorian and Atkinson (1986) with permission.

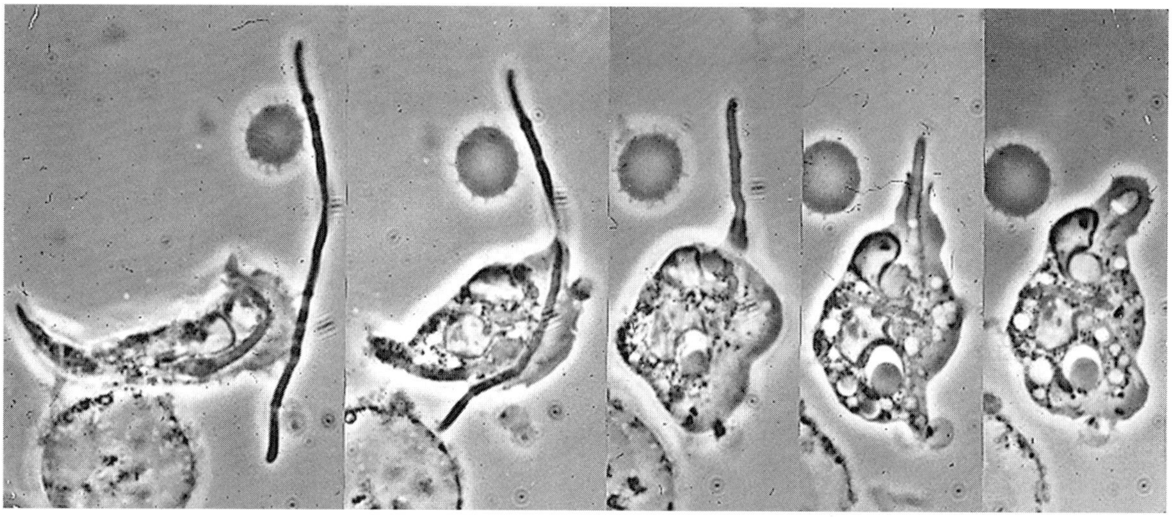

Figure 12.18 Phagocytosis of a filament by a polymorphonuclear cell.

ters of *S. aureus* (Lorian, 1985b). Thus, in a phagocytic system, where the targets are antibiotic-treated bacteria, the mass, and not the CFU count of phagocytized organisms, is the correct criterion for the quantitative evaluation of phagocytosis.

Although the MIC is desirable for optimizing the pharmacological activity of these agents, a sub-MIC may be unavoidable as the concentration falls between doses. These sub-MICs may cause morphological changes which include an increase in cell mass containing multiple genomes. These large targets make the phagocytic process more efficient by reducing the time and energy of the PMN which would otherwise be necessary to pursue many smaller, dispersed targets.

References

Aumercier, M., Bouhallab, S., Capmau, M. L., Le Goffic, F. (1992). RP59500: a proposed mechanism for its bactericidal activity. *J. Antimicrob. Chemother.*, **30** (suppl. A), 9–14.

Bahr, C. F., Engler, W. F., Mazzone, H. M. (1976) Determination of the mass of viruses by quantitative electron microscopy. *Q. Rev. Biophys.*, **9**, 459–489.

Bouanchaud, D. H. (1997). *In vitro* and *in vivo* antibacterial activity of quinupristin/dalfopristin. *J. Antimicrob. Chemother.* (suppl. A39), 15–21.

Bryan, L. E., Godfrey, A. J. (1991). β-Lactam antibiotics: mode of action and bacterial resistance. In: *Antibiotics in Laboratory Medicine* (ed Lorian, V.), pp. 599–604. Williams & Wilkins, Baltimore, Md.

Davies, J. E. (1991). Aminoglycoside-aminocyclitol antibiotics and their modifying enzymes. In: *Antibiotics in Laboratory Medicine* (ed Lorian, V.), pp. 691–713. Williams & Wilkins, Baltimore, Md.

Day, L. E. (1966). Tetracycline binding to ribosome. *J. Bacteriol.*, **91**, 1917–1920.

Ernst, J., Sy, E., Lorian, V., Kim, Y. (1985). Ultrastructure of staphylococci in respiratory infections treated with nafcillin. *Drugs Exp. Clin. Res.*, **6**, 357–360.

Giesbrecht, P., Ruska, H. (1968). Uber Veranderungen der Feinstrukturen von Bakteriien unter der Einwirking von Chloramphenicol. *Klin. Wochenschr.*, **11**, 575–579.

Goldberg, I. H. (1965). Mode of action of antibiotics II. Drugs affecting nucleic acid and protein synthesis. *Am. J. Med.*, **39**, 722–725.

Hooper, D. C., Wolfson, J. S. (1991). The quinolones: mode of action and bacterial resistance. In: *Antibiotics in Laboratory Medicine* (ed Lorian, V.), pp. 665–690. Williams & Wilkins, Baltimore, Md.

Kadurugamuwa, J. L., Lam, J. S., Beveridge, T. J. (1993). Interaction of gentamicin with the A bank and B bank lipopolysaccharides of *Pseudomonas aeruginosa* and its possible lethal effect. *Antimicrob. Agents Chemother.*, **37**, 715–721.

Lorian, V. (1975). Some effects of subinhibitory concentrations of penicillin on the structure and division of staphylococci. *Antimicrob. Agents Chemother.*, **7**, 864–870.

Lorian, V. (1985a). Phagocytosis of staphylococci after exposure to antibiotics. *J. Antimicrob Chemother.*, **16**, 129–133.

Lorian, V. (1985b). Low concentrations of antibiotics. *J. Antimicrob. Chemother.* (Suppl A), 15–26.

Lorian, V. (1989). *In vitro* simulation of *in vivo* conditions: physical state of the culture medium. *J. Clin. Microbiol.*, **27**, 2403–2406.

Lorian, V., Sabath, L. D. (1972). Penicillins and cephalosporins: differences in morphologic effects on *Proteus mirabilis*. *J. Infect. Dis.*, **125**, 560–564.

Lorian, V., Atkinson, B. (1976). Effects of subinhibitory concentrations of antibiotics on cross walls of cocci. *Antimicrob. Agents Chemother.*, **9**, 1043–1055.

Lorian, V., Atkinson, B. (1984). Bactericidal effect of polymorphonuclear neutrophils on antibiotic-induced filaments of Gram-negative bacteria. *J. Infect. Dis.*, **149**, 719–727.

Lorian, V., Atkinson, B. A. (1986). Amikacin-induced alterations in the structure of Gram-negative bacilli. *Diagn. Microbiol. Infect. Dis.* **5**, 93–97.

Lorian, V., Fernandes, F. (1997). The effect of vancomycin on the structure of vancomycin-susceptible and resistant *Enterococcus faecium* strains. *Antimicrob. Agents Chemother.*, **41**, 1410–1411.

Lorian, V., Atkinson, B., Waluschka, A., Kim, Y. (1982a). Ultrastructure, *in vitro* and *in vivo*, of staphylococci exposed to antibiotics. *Curr. Microbiol.*, **7**, 301–304.

Lorian, V., Waluschka, A., Kim, R. Y. (1982b). Abnormal morphology of bacteria in the sputa of patients treated with antibiotics. *J. Clin. Microbiol.*, **16**, 382–386.

Lorian, V., Atkinson, B., Kim, Y. (1983). Effect of rifampin and oxacillin on the ultrastructure and growth of staphylococci. *Rev. Infect. Dis.*, **5**, S418–S427.

Lorian, V., Zak, O., Kunz, S., Vaxelaire, J. (1984). Staphylococcal endocarditis in rabbits treated with a low dose of cloxacillin. *Antimicrob. Agents Chemother.*, **25**, 311–315.

Lorian, V., Tosch, W., Joyce, D. (1985). Weight and morphology of bacteria exposed to antibiotics. In: *Influence of Antibiotics on the Host–Parasite Relationship* (eds Gillissen, G., Hahn, H., Opferkuch, W.), pp. 65–72. Springer-Verlag, Berlin.

Lorian, V., Esanu, Y., Amaral, L. (1994). Ultrastructure alterations of *Staphylococcus aureus* exposed to RP59500. *J. Antimicrob. Chemother.*, **33**, 625–628.

Morgan, C., Rosenkranz, H. S., Carr, H. S., Rose, H. M. (1967). Electron microscopy of chloramphenicol-treated *Escherichia coli*. *J. Bacteriol.*, **93**, 1987–1990.

O' Reilly, T., Cleeland, R., Squires, E. L. *et al.* (1992). Relationship between antibiotic concentration in bone and efficacy of treatment of staphylococcal osteomyelitis in rats: azithromycin compared with clindamycin and rifampin. *Antimicrob. Agents Chemother.*, **36**, 2693–2697.

Redjeb, B. S., Slim, A., Horchani, A. *et al.* (1982). Effects of ten milligrams of ampicillin per day on urinary tract infections. *Antimicrob. Agents Chemother.*, **22**, 1084–1086.

Reusser, F. (1975). Effect of lincomycin and clindamycin on peptide chain initiation. *Antimicrob. Agents Chemother.*, **7**, 32–35.

Ryan, D. M., Monsey, D. (1981). Bacterial filamentation and *in vivo* efficacy; a comparison of several cephalosporins. *J. Antimicrob. Chemother.*, **7**, 57–63.

Scullard, G., Meynell, E. (1966). Bacterial mass measured with the MRC greywedge photometer. *J. Pathol. Bacteriol.*, **91**, 608–612.

Spratt, B. G. (1975). Distinct penicillin-binding proteins involved in the division, elongation and shape of *Escherichia coli*. *Proc. Natl Acad. Sci. USA*, **72**, 2999–3003.

Spratt, B. G., Boyd, A., Stoker, N. (1980). Defective and plaque-forming A transducing bacteriophage carrying penicillin-binding protein-cell shape genes: genetic and physical mapping and identification of gene products from the lip-lacA-rodA-pbpA-leuS region of the *Escherichia coli* chromosome. *J. Bacteriol.*, **143**, 569–581.

Wehrli, W. (1983). Rifampin: mechanisms of action and resistance. *Rev. Infect. Dis.*, **5**, 407–417.

Zak, O., Kradolfer, F. (1979). Effects of subminimal inhibitory concentrations of antibiotics in experimental infections. *Rev. Infect. Dis.*, **1**, 862–879.

Chapter 13

Activity of Antibiotics Against Adherent/Slow-growing Bacteria Reflecting the Situation *in vivo*

I. Foley and M. R. W. Brown

In chronic infections the causative organisms commonly exist in often complex, surface-associated microbial communities called biofilms. Such biofilm-associated infections are difficult to model *in vitro*. In the typically nutrient-rich, hospitable environment of laboratory growth media, growth as a biofilm provides little competitive advantage to the cells and is suppressed. In contrast, the growth of microbial cells as biofilms within hostile nutrient environments, including infected hosts, confers many advantages upon the component cells and is promoted (Allison and Gilbert, 1992). These advantages include:

1. Increased protection from the activity of phagocytes and other host immune defences and from antibiotics
2. Improved growth prospects due to a concentration of nutrients and extracellular products both within the biofilm and at its interface with the bulk phase. In mixed cultures, the co-operative mobilization of nutrients by different species within the consortia enables a wide range of nutrients to be utilized
3. Modulation of the physicochemical environment of the cells through the establishment and maintenance of pH and electropotential gradients across the biofilm
4. Increased genetic exchange brought about through the close proximity of cells within the biofilm.

Biofilms that develop in humans may occasionally be of benefit to the host, e.g. dense biofilms in the gut often competitively inhibit the development of pathogenic organisms. Alternatively, the development of pathogenic organisms within biofilms or indeed the establishment of "harmless" commensal biofilms at inappropriate sites, i.e. catheter blockage, are of clinical importance. Of particular concern is the failure of biofilm bacteria to respond poorly, if at all, to even aggressive antibiotic prescribing. Explanations for such therapeutic failure have been suggested, including first, the expression of a biofilm/surface-associated resistant phenotype influenced by the nutrient environment generated within the biofilm; second, acquisition of antimicrobial resistance, by the biofilm-promoted transfer of resistance genes (plasmids/transposons) within or between another species or genus; and third, direct modification of antibiotic action through the presence of extracellular polymers and antibiotic-modifying enzymes. However, before these are discussed it is appropriate to give a brief description of biofilm architecture and formation and the role they play in antibiotic resistance.

Biofilm formation

The initial stages in the formation of a biofilm involve the adhesion of bacteria originating from a fluid phase to a surface with subsequent colonization. Bacteria translocate from the bulk phase, usually a fluid, to a surface by a variety of processes which include diffusion, convective transport and active motility. Once at the surface, adhesion is dependent upon the physicochemical properties of both the bacterial cell envelope and of the substratum. The attachment of bacteria to solid substrata may also be promoted by the preconditioning of surfaces by the adsorption of biological molecules (Costerton *et al.*, 1987). This conditioning film is composed of products of the planktonic cells and/or constituents of the surrounding aqueous phase. Common examples of these conditioning molecules include serum and tissue proteins that coat biomaterials following their insertion into body cavities.

The ability of specific envelope components to bind to any of these molecules will contribute to the irreversible attachment of the microorganism by helping to overcome the repulsive forces that exist between the cell and the substratum. Microorganisms may also synthesize extracellular polysaccharides (EPS), such as soluble glucans and fructans and insoluble glucans. Although EPS may aid the primary attachment of bacteria to a surface, extensive EPS production generally only occurs following bacterial attachment to a surface and the deregulation of the *algD* gene in Pseudomonads (Allison, 1993). EPS may consequently have an important role as a secondary cement binding bacteria to the surface. Large amounts of exopolymers are eventually produced by the adhered cells to the extent that it entirely envelops them. *In vivo* this exopolymer matrix combines with bacterial products and host factors to form the bacterial glycocalyx (Costerton *et al.*, 1981). The biofilm will grow due to the internal replication of cells within the glycocalyx and sequestration of cells from the bulk phase.

Handbook of Animal Models of Infection
ISBN 0-12-775390-7

The net result of these processes is the formation of a biofilm that occludes the surface and consists of both microcolonies and single cells embedded within a highly hydrated glycocalyx. The replication and release, or mechanical sloughing of cells from biofilm will promote the dispersal of cells into the bulk phase. *In vivo* biofilms may act as sources of secondary infections in which the release of daughter cells generate bacteraemia and septicaemia that produce infections at sites remote from the parent biofilm. Indeed, the detection of bacteraemia within a patient is usually the first indicator of otherwise clinically indolent biofilm-associated infections such as infective endocarditis. Although bacteraemia may respond well to antibiotic therapy, the bacteria within the biofilm are often resistant. Cessation of treatment following an improvement in a patient's condition is therefore often followed by relapse.

Antibiotic resistance

In the majority of natural habitats, biofilms are composed of a variety of species and genera, but in biomedical situations associated with soft-tissue infection or infections of indwelling devices, monocultures are more usual. The structural organization of the glycocalyx appears to vary according to the prevailing physicochemical environment. Within the biofilm matrix diffusion of nutrients, and metabolic byproducts, is more restricted than it would be in planktonic culture. This leads to spatial organization of mixed-species biofilm communities, with associated cross-feeding, and functional interspecies dependence. In monospecies biofilms, nutrient and gaseous gradients, generated by metabolism, will cause growth rates of the enveloped cells, and nutrient availability, to vary with location relative to the substratum and biofilm–liquid phase interface.

The susceptibility of bacterial cells to antibiotics and other antimicrobial compounds is significantly affected by their nutrient status and growth rate, as well as by temperature, pH and exposure to subeffective concentrations of antibiotics (Brown and Williams, 1985; Williams, 1988; Brown *et al.,* 1990). Such environmental stimuli may induce the expression of genes within the bacterial genome, leading to a change in cell phenotype. In well-mixed suspension cultures all members of the community experience a common environment at any particular time. Single phenotypes therefore dominate such cultures which might, as a consequence, demonstrate a singular response towards antimicrobial treatment. Within biofilm communities, however, a plethora of phenotypes is represented for each component species, indicative of the heterogeneity of the environments within the biofilm (Gilbert *et al.*, 1990). The outcome of any antimicrobial treatments of the biofilm community will therefore reflect the susceptibility of the most resistant phenotype represented within it.

As the biofilms mature, and exopolymer deposition increases, then the magnitude of the nutrient and gaseous gradients will become increased and the net growth rate of the community will become further reduced, possibly with an onset of dormancy and triggering of stringent response genes (Zambrano and Kolter, 1995), contributing to the observations that aged, mature biofilms are more recalcitrant to antibiotic and biocide treatment than young ones (Anwar *et al.,* 1989). The presence of subinhibitory levels of antibiotic agents within the depths of the biofilm, whether caused by failure to penetrate adequately the glycocalyx or by decreases in susceptibility of the enveloped cells, may provide selective pressures for the development of more resistant phenotypes, and the selection and expression of resistance plasmids.

Other workers have used perfused biofilm fermenters (Gilbert *et al.,* 1989) directly to control and study the effects of growth rate *per se* within biofilms on antibiotic susceptibility. Using planktonic controls grown in chemostats the separate contributions of growth rate, and association within a biofilm could be evaluated. In this respect it was noted that the decreased susceptibility of biofilm cultures of both *Staphylococcus epidermidis* (Duguid *et al*., 1992a, b) and *Escherichia coli* (Evans *et al.,* 1990, 1991) could be explained largely in terms of growth rate. Cells resuspended from growth rate-controlled biofilms and planktonic cells of the same growth rate possessed virtually identical susceptibilities to this agent. When intact biofilms were treated, however, susceptibility was decreased somewhat from that of planktonic and resuspended biofilm, cells indicating some benefit to the cells of organization within a glycocalyx (see below). Similar studies with ciprofloxacin, conducted with the same pair of organisms (Evans *et al.,* 1991; Duguid *et al.,* 1992a), suggested that whilst growth rate played a crucial role in the susceptibility to this agent of both planktonic and biofilm cells, slow-growing ($\mu < 0.15$ hours) biofilms were especially sensitive. These observations, together with those of Ashby *et al.* (1994), suggest that some physiological properties, with a potential as antimicrobial targets, may be unique to biofilm-grown cells.

It has been predicted by Stewart (1994) that the reduced susceptibility of thicker biofilms was attributable to oxygen depletion. The exhaustion of oxygen within the biofilm is thought to lead to reduced growth rates and associated resistance. Oxygen gradients within the biofilm may also directly influence the activity of antibacterials themselves (Zabinski *et al*., 1995). Tresse *et al.* (1995) examined the susceptibility of *E. coli* biofilms entrapped within agar gel layers to aminoglycoside (tobramycin) and β-lactam (latamoxef) antibiotics under various states of oxygenation. This study showed that the increased resistance of sessile cells compared to free organisms to β-lactam antibiotics was growth rate-dependent, brought about by oxygen and nutrient depletion. However, the increased resistance of biofilms to aminoglycosides is probably growth rate-independent but rather may be due to the low uptake of antibiotics by oxygen-deprived cells. Since the nutrient and gaseous gradients within biofilms will increase with biofilm age and thickness, then growth rate effects such as these will become particularly noticeable in mature films (Anwar *et al.*, 1989; Anwar and Costerton, 1990). In this respect antibiotic therapies associated with infected indwelling devices should be initiated at the earliest possible opportunity.

Such observations support the prophylactic use of antibiotics in patients undergoing implant surgery where there is a high risk of infection. Since such chemotherapy may be protracted, there is an increased risk, particularly associated with attached cells, that subinhibitory concentrations of the antibiotics may select for resistant mutants. Clinicians should therefore be aware of the increased resistance of biofilm-associated infections and initiate aggressive therapies that target these cells rather than more susceptible, fast-growing bacteraemic cells. However, the significance of oxygen depletion within the biofilm population may need to be re-examined following the elucidation of biofilm architecture.

The development of sophisticated, non-invasive techniques such as confocal scanning laser microscopy (CFLM) in conjunction with fluorescent techniques has shown that some biofilm structures are punctuated by a semi-regular arrangement of channels and pores at approximately 20–40 μm intervals (Costerton *et al.*, 1995). These channels have been suggested to represent a rudimentary circulatory system, facilitating the transport of nutrients and oxygen to the base of the biofilm. This means that nutrients do not have to diffuse through the entire thickness of the biofilm. This latter finding reduces the significance of the dual protective role of EPS within the biofilm; it is believed to act as either a diffusion barrier against the penetration of antibiotics and/or react with and neutralize the antimicrobial agent.

Recently, there is evidence that an aspect of the stationary phase stress response known as the general stress response plays a role in chronic infection (Foley *et al.*, 1999).

Role of the glycocalyx and extracellular enzymes

The glycocalyx protects the underlying cells against desiccation, predation by phagocytic entities (white blood cells/protozoa), and may form an ionic and steric barrier to diffusion of agents from the surrounding medium. In this last respect it has been suggested that the glycocalyx can reduce access of antibiotics to the biofilm population (Gordon *et al.*, 1988; Nichols *et al.*, 1989). However, Brown *et al.* (1995) have examined planktonic and biofilm *Pseudomonas aeruginosa* populations and have shown that, although the quantities of EPS per cell were not significantly different, its distribution was. Within the biofilm glycocalyx, material was more or less evenly distributed among the cells and extended around them in a protective sheath. In contrast, planktonic cells were incompletely surrounded by a compressed glycocalyx. Therefore, planktonic cells would be more likely to have exposed surface-associated antibiotic/therapeutic target sites, increasing their susceptibilty to therapy. This study also highlighted the point that, once the majority of the biofilm had reached stationary phase, the increasing physical thickness of the biofilm was due to the accumulation of disrupted (non-viable) cells and other cell debris. Depending upon its location this detritus may act as an effective physical barrier to antibiotic penetration. As a consequence, antibiotics may either be occluded from deep-seated cells or only achieve submaximal inhibitory complex levels, leading to the selection of resistant phenotypes.

The synthesis of some matrix polymers is regulated by a variety of factors, of which surface attachment appears to be of particular importance. In this respect Davies *et al.* (1993) showed that EPS (alginate) production is upregulated in biofilm cells compared to planktonic cells in the liquid phase, and Evans *et al.* (1994) showed exopolysaccharide production to be increased at low growth rates and substantially higher for biofilm than planktonic cells. There is now speculation that, in some Gram-negative organisms, production of exopolysaccharides such as alginate may be under the control of signal substances such as homoserine lactone (HSL), which may also act as global regulators of biofilm-specific physiology (Williams *et al.,* 1992; Gambello *et al.,* 1993; Davies and Geesey, 1995). Within biofilms the latter effects would provide for increased exopolymer production in the slow-growing heart of a thick microcolony, where signal substances such as HSL might be concentrated. This would alter the distribution and density of cells throughout the matrix, and once again confer some structural organization upon the community to provide customized microniches at various points within the biofilm (Costerton *et al.*, 1994). The extent and nature of exopolymer production is also dependent upon physiological factors such as the relative availability of carbon and nitrogen (Sutherland, 1985). The presence of HSL may possibly also trigger rpoS (Latifi *et al.*, 1996) and the general stress response. Although potentially important, the role of the general stress response in chronic biofilm infections has received little study as yet.

The presence of an organism within such an exopolymer matrix will undoubtedly influence profoundly the access of molecules and ions, including protons (Costerton *et al.*, 1981), to the cell wall and membrane. It is not surprising therefore that many groups of workers have suggested that the glycocalyx physically prevents access of antimicrobials to the cell surface and that the recalcitrance of biofilms is purely and simply a matter of exclusion (Slack and Nichols, 1981, 1982; Costerton *et al.,* 1987; Suci *et al.*, 1994). Opposition to this view (Gordon *et al.*, 1988; Nichols *et al.*, 1988, 1989) is on the basis that reductions in diffusion coefficient for antibiotics such as tobramycin and cefsulodin within biofilms or microcolonies are insufficient to account for the observed changes in susceptibility of the contained cells. In this light, Gristina *et al.* (1989) examined susceptibilities towards antibiotics, of biofilms of slime-producing and non-slime-producing strains of *S. epidermidis* and saw no difference, suggesting that the slime was not significant in antibiotic penetration, and various antibiotic agents have been shown to perfuse biofilms readily (Dunne *et al.,* 1993) and attain concentrations within the matrix that exceed the MIC/MBC observed for planktonic organisms (Darouicue *et al.*, 1994).

Clearly, whether or not the glycocalyx constitutes a physical barrier to antibiotic penetration depends greatly upon the nature of the antibiotic, the binding capacity of the glycocalyx towards it, the levels of agent used therapeutically and the rate of growth of the microcolony relative to antibiotic diffusion rate (Kumon *et al.,* 1994). For antibiotics such as tobramycin and cefsoludin, therefore, such effects are suggested to be minimal (Nichols *et al.*, 1988, 1989), but for positively charged antibiotics such as the aminoglycosides, binding to the polyanionic matrix polymers is high and their access to cells is impeded (Nichols *et al.,* 1988). Curiously, macrolide antibiotics, which are also positively charged, but also very hydrophobic, are relatively unaffected by the presence of the exopolymers (Ichimiya *et al.,* 1994). Poor penetration through anionic matrices might therefore be a phenomenon restricted to the more hydrophilic, positively charged agents.

Evans *et al.* (1991) assessed susceptibility towards the quinolone antimicrobial ciprofloxacin of mucoid and non-mucoid strains of *P. aeruginosa* grown as biofilms. In this study possession of the mucoid phenotype was clearly associated with decreases in susceptibility. In a similar vein, activities of chemically highly reactive biocides such as iodine and iodine–polyvinylpyrollidone complexes (Favero *et al.*, 1983) are substantially reduced by the presence of protective exopolymers. In such instances not only will the polymers act as adsorption sites but they will also react chemically with, and neutralize, biocides. Similarly, the relatively large accumulation of drug-inactivating enzymes such as β-lactamases (Giwercman *et al.*, 1991) within the glycocalyx will create marked concentration gradients of the antibiotic across them and protect, to some extent, the underlying cells.

The ionic composition of the bulk fluids within which the biofilms form will affect the properties and integrity of the biofilm (Gordon *et al.,* 1991). This might have profound effects upon the susceptibility of biofilms growing at different sites within the body or in patients with some predisposing clinical condition (i.e., cystic fibrosis, chronic renal failure, hypercalcemia). Such conditions create abnormal body chemistries which include markedly elevated calcium ion levels. The presence of adsorbed ions within the biofilm–fluid interface may affect the net charge of the matrix and thereby the diffusion of antibiotic. Hoyle *et al.* (1992) have shown *in vitro* that the tobramycin-binding capacity of bacterial exopolysaccharides is less important in terms of reduced susceptibility than is the reduction in diffusivity of the matrix caused through steric hindrance imposed by Ca^{2+} condensation of the polymer. Such effects are dependent upon the type of ion and are not seen, for example, with Mg^{2+}. This suggests that a greater understanding of the biofilm penetration by antibiotics is required before extrapolating planktonic susceptibility data to biofilms. A potential application of these observations might be the use of adjuvants to alter the charge characteristics of the polymer matrix (Lee *et al.,* 1995), or to employ novel agents with non-lethal effects on exopolymer synthesis in combination with conventional antibiotics (Gagnon *et al.*, 1994; Pascual *et al.*, 1994). In this respect it is notable that clindamycin has been suggested to suppress biofilm formation by *P. aeruginosa*, at concentrations below its MIC (Ichimiya *et al.,* 1994).

In summary, reductions in diffusion coefficients across the glycocalyx, relative to liquid media, are alone insufficient to account for antibiotic recalcitrance of biofilms since, at equilibrium, concentrations at the cell surfaces will be the same as those in the bathing medium. With losses either through chemical or enzymic inactivation of the agent occurring in the biofilm, the reduction in diffusion coefficient, particularly if exacerbated by the presence of bound cations, might facilitate resistance to the agent. Such protection might also apply where the adsorptive capacity of the glycocalyx is very high with respect to antibiotic levels in the bulk phase.

Attachment-specific physiologies

Surface-induced stimulation of bacterial activity has often been described in the literature (Fletcher, 1984, 1986; Zobell, 1943). Dagostino *et al.* (1991) have attempted to identify such activity with the induction of particular genes. There is still much debate as to whether surface-induced stimulation of metabolic activity reflects derepression/induction of specific operons/genes or is purely a manifestation of the physicochemical presence of the surface on the surroundings of the cell (van Loosdrecht *et al.*, 1990). Degradation of the substrate nitrilotriacetate, which does not itself adsorb to surfaces, is enhanced when the degradative organisms are attached to inert surfaces (McFeters *et al.*, 1990). This suggests increased production of degradative enzymes by attached cells. Similarly, gliding bacteria lack extracellular polymer biosynthesis when grown in suspension culture (Humphrey *et al.*, 1979; Abbanat *et al.*, 1988), but rapidly initiate/increase such synthesis following irreversible adhesion to a surface.

In this respect, work on the regulation of lateral flagella gene transcription in *Vibrio parahaemolyticus* showed it to produce a single polar flagellum in liquid, and numerous lateral, unsheathed flagella on solid culture media (Belas *et al.*, 1984, 1986; McCarter *et al.*, 1988). These workers recognized that changes in flagellation reflected increased viscosity experienced by the organism at a surface, restricting movement of the polar flagellum. The net effect however was a switching on of the *laf* genes through contact with a surface. In a similar vein, Lee and Falkow (1990) recognized that reduced oxygen tensions, such as might be experienced by cells enveloped within a biofilm or in association with a surface, cause the triggering of expression of *Salmonella* invasins.

The balance of evidence is now in favour of the notion that bacteria are able to sense the close proximity of a surface, and through cell density transcriptional activation the close proximity of other cells. Indeed, in this respect Davies and Geesey (1995) have demonstrated, using a *LacZ* reporter fused to *AlgC* of *P. aeruginosa*, that alginate synthesis is upregulated within minutes of the cell attaching to a surface, and many other groups of workers are now postulating a general involvement of quorum-sensing transcriptional activation in the regulation of biofilm-specific physiologies (Davies *et al.*, 1998). Whilst it is likely that some of the physiological responses invoked will also confer some change in antibiotic susceptibility, this has not, to date, been proven. It has, however, been shown that for the action of chemical biocides such as the polymeric biguanides, quaternary ammonium compounds and peracetic acid, susceptibility of target cells is altered within 15 minutes of attachment to a polystyrene surface. Such changes are noted before there has been any significant growth of the organisms on the surface and before exopolymers have accumulated (Das *et al.*, 1995).

Genetic exchange within biofilms

As discussed, the biofilm mode of growth provides many unique benefits to the associated bacteria directly related to the structural and physiological composition of the biofilm that are not encountered by planktonic bacteria. However, the biofilm may also act as a catalyst, speeding up processes that occur within the planktonic population, in particular, the exchange of mobile genetic elements encoding antibiotic resistance genes (Belas *et al.*, 1984). This may occur through a variety of processes such as transposition, transduction and conjugation. However, in all cases it requires intimate contact between the donor and recipient strains. Planktonic cells such as those encountered during bacteraemia generally rely on the purely random event of colliding with a potential donor/recipient. Within the biofilm, often composed of a heterogeneous population in a state of flux, exhibiting various novel antibiotic resistance mechanisms, the cells are in intimate contact with one another at numbers of up to 10^{12}. These conditions would theoretically provide for the optimum acquisition and dissemination of resistance determinants throughout the biofilm.

A large number of studies have examined the conjugal transfer of plasmids in natural environments where biofilms represent the dominant mode of growth. Thus, *in situ*, and in laboratory models of soil (Stotzky, 1989) and aquatic systems (Fry and Day, 1990) plasmid exchange, selection and mobilization between planktonic donor cells and biofilms has been demonstrated. Whilst much of this work has been concerned with the transfer of resistance and degradation potential with respect to environmental pollutants, the results may be equally applicable to the transfer of antibiotic resistance or virulence determinants within medical situations.

The worldwide recognition of "superbugs" exhibiting resistance to a wide variety of originally therapeutically effective antibiotics has added fresh impetus to the understanding of genetic transfer within the natural environment. Indeed, in 1990, with the isolation of a strain of *Enterococcus faecium* that was resistant to all known antibiotics, the headlines pronouncing the end of the antibiotic era no longer remained a figment of the tabloid press but became a serious health care threat. The rich genetic diversity and highly specialized pheromone-mediated plasmid exchange mechanisms (Dunny *et al.*, 1978) of *E. faecalis* in particular led to the realization that mobile genetic elements may jump what were originally believed to be intraspecies boundaries between Gram-positive and Gram-negative organisms (Trieu-Cuot *et al.*, 1987). Such evidence should stimulate research into this area, particularly since the days of methicillin-resistant *Staphylococcus aureus* or other pathogens such as tubercle bacilli becoming untreatable are approaching.

The transfer of antibiotic resistance within biofilms is likely to be a multifactorial process, with the age, type, composition and location of the biofilm being as important a consideration as the properties of the individual plasmid, such as host range, copy number and stability. It is hoped that, as our understanding of biofilm physiology increases, then accurate, reproducible model biofilms may be developed which allow for the study and evaluation of the kinetics of plasmid acquisition and expression within natural communities.

References

Abbanat, D. R., Godchaux, W., Leadbeter, E. R. (1988). Surface-induced synthesis of new sulphonolipids in the gliding bacterium *Cytophaga johnsonae*. *Arch. Microbiol.*, **149**, 358–364.

Allison, D. G. (1993). Exopolysaccharide synthesis in bacterial biofilms. In: *Bacterial Biofilms and their Control in Medicine and Industry* (eds Wimpenny, J., Nichols, W., Stickler, D., Lappin-Scott, H.), pp. 25–30.

Allison, D. G., Gilbert, P. (1992). Bacterial biofilms. *Sci. Progress Oxford*, **76**, 305–321.

Anwar, H., Costerton, J. W. (1990). Enhanced activity of combination of tobramycin and piperacillin for eradication of sessile biofilm cells of *Pseudomonas aeruginosa*. *Antimicrob. Agents Chemother.*, **34**, 1666–1671.

Anwar, H., Dasgupta, M., Lam, K., Costerton, J. W. (1989). Tobramycin resistance of mucoid *Pseudomonas aeruginosa* biofilm under iron limitation. *J. Antimicrob. Chemother.*, **24**, 647–655.

Ashby, M. J., Neale, J. E., Knott, S. J., Critchley, I. A. (1994). Effect of antibiotics on non-growing cells and biofilms of *Escherichia coli*. *J. Antimicrob. Chemother.*, **33**, 443–452.

Belas, R., Mileham, A., Simon, M., Silverman, M. (1984). Transposon mutagenesis of marine *Vibrio* spp. *J. Bacteriol.*, **158**, 890–896.

Belas, R., Simon, M., Silverman, M. (1986). Regulation of lateral flagellar gene transcription in *Vibrio parahaemolyticus*. *J. Bacteriol.*, **167**, 210–218.

Brown, M. R. W., Williams, P. (1985). The influence of the environment on envelope properties affecting survival of bacteria in infections. *Ann. Rev. Microbiol.*, **39**, 527–556.

Brown, M. R. W., Collier, P. J., Gilbert, P. (1990). Influence of growth rate on the susceptibility to antimicrobial agents: modification of the cell envelope and batch and continuous culture. *Antimicrob. Agents Chemother.*, **34**, 1623–1628.

Brown, M. L., Aldrich, H. C., Gauthier J. L. (1995). Relationship between glycocalyx and povidone-iodine resistance in *Pseudomonas aeruginosa* (ATCC 27853) biofilms. *Appl. Environ. Microbiol.*, **61**, 187–193.

Costerton, J. W., Irvin, R. T., Cheng, K. J. (1981). The bacterial glycocalyx in nature and disease. *Annu. Rev. Microbiol.*, **35**, 399–424.

Costerton, J. W., Cheng, K. J., Geesey, G. G. *et al.* (1987). Bacterial biofilms in nature and disease. *Annu. Rev. Microbiol.*, **41**, 435–464.

Costerton, J. W., Ellis, B., Lam, K., Johnson, F., Khoury, A. E. (1994). Mechanism of electrical enhancement of efficacy of antibiotics in killing biofilm bacteria. *Antimicrob. Agents Chemother.*, **38**, 2803–2809.

Costerton, J. W., Lewandowski, Z., Caldwell, D. E., Korber, D. R., Lappin-Scott, H. M. (1995). Microbial biofilms. *Annu. Rev. Microbiol.*, **49**, 711–745.

Dagostino, L., Googman, A. E., Marshall, K. C. (1991). Physiological responses induced in bacteria adhering to surfaces. *Biofouling*, **4**, 113–119.

Darouicue, R. O., Dhir, A., Miller, A. J., Landon, G. C., Raad, I. I., Muscher, D. M. (1994). Vancomycin penetration into biofilm covering infected prostheses and effect on bacteria. *J. Infect. Dis.*, **170**, 720–723.

Das, J., Jones, M., Bhakoo, M., Gilbert, P. (1995). An automated method for the assessment of biofilm formation and antimicrobial screening. *Abstract 95th General Meeting of the American Society for Microbiology*, p. 460. ASM Press, Washington.

Davies, D. G., Geesey, G. G. (1995). Regulation of the alginate biosynthesis gene algC in *Pseudomonas aeruginosa* during biofilm development in continuous culture. *Appl. Environ. Microbiol.*, **61**, 860–867.

Davies, D. G., Chakrabarty, A. M., Geesey, G. G. (1993). Exopolysaccharide production in biofilms: substratum activation of alginate gene expression by *Pseudomonas aeruginosa*. *Appl. Environ. Microbiol.*, **59**, 1181–1186.

Davies, D. G., Parsek, M. R., Pearson, J. P., Iglewski, B. H., Costerton, J. W., Greenberg, E. P. (1998). The involvement of cell-to-cell signals in the development of a bacterial biofilm. *Science*, **280**, 295–298.

Duguid, I. G., Evans, E., Brown, M. R. W., Gilbert, P. (1992a). Growth-rate-independent killing by ciprofloxacin of biofilm-derived *Staphylococcus epidermidis*: evidence of cell-cycle dependency. *J. Antimicrob. Chemother.*, **30**, 791–802.

Duguid, I. G., Evans, E., Brown, M. R. W., Gilbert, P. (1992b). Effect of biofilm culture upon the susceptibility of *Staphylococcus epidermidis* to tobramycin. *J. Antimicrob. Chemother.*, **30**, 791–802.

Dunne, W. M. Jr., Mason, E. O., Kaplan, S. L. (1993). Diffusion of rifampicin and vancomycin through a *Staphylococcus epidermidis* biofilm. *Antimicrob. Agents Chemother.*, **37**, 2522–2526.

Dunny, G. M., Brown, B. L., Clewell, D. B. (1978). Induced cell aggregation and mating in *Streptococcus faecalis*: evidence for a bacterial sex pheromone. *Proc. Natl Acad. Sci. USA*, **75**, 3479–3483.

Evans, D. J., Brown, M. R. W., Allison, D. G., Gilbert, P. (1990). Susceptibility of bacterial biofilms to tobramycin: role of specific growth rate and phase in the division cycle. *J. Antimicrob. Chemother.*, **25**, 585–591.

Evans, D. J., Brown, M. R. W., Allison, D. G., Gilbert, P. (1991). Susceptibility of *Escherichia coli* and *Pseudomonas aeruginosa* biofilm to ciprofloxacin: effect of specific growth rate. *J. Antimicrob. Chemother.*, **27**, 177–184.

Evans, E., Brown, M. R. W., Gilbert, P. (1994). Iron chelator, exopolysaccharide and protease production of *Staphylococcus epidermidis*: a comparative study of the effects of specific growth rate in biofilm and planktonic culture. Microbiology, **140**, 153–157.

Favero, M. S., Bond, W. W., Peterson, N. J., Cook, E. H. (1983). Scanning electron microscope observations of bacteria resistant to iodophor solutions. In: *Proceedings of the International Symposium on Povidone*, pp. 158–160. University of Kentucky, Lexington, USA.

Fletcher, M. (1984). Comparative physiology of attached and free-living bacteria. In *Microbial Adhesion and Aggregation* (ed. Marshall, K. C.), pp. 223–232. Springer Verlag, Berlin.

Fletcher, M. (1986). Measurement of glucose utilisation by *Pseudomonas fluorescens* that are free living and that are attached to surfaces. *Appl. Environ. Microbiol.*, **52**, 672–676.

Foley, I., Marsh, P., Wellington, E. M. H., Smith, A. W. and Brown, M. R. W. (1999). General stress response master regulator *rpoS* is expressed in human infection: a possible role in chronicity. *Journal of Antimicrobial Chemotherapy*, **43**, 164–165.

Fry, J. C., Day, M. J. (1990) Plasmid transfer in the epithilon. In *Bacterial Genetics in Natural Environments* (eds Fry, J. C., Day, M. J.), pp. 55–80. Chapman and Hall, London.

Gagnon, R. F., Richards, G. K., Kostiner, G. B. (1994). Time-kill efficacy of antibiotics in combination with rifampicin against *Staphylococcus epidermidis* biofilms. *Adv. Peritoneal Dialysis*, **10**, 189–192.

Gambello, M. J., Kaye, S., Inglewski, B. H. (1993). *LasR* of *Pseudomonas aeruginosa* is a transcriptional activator of the line protease gene (*apr*) and an enhancer of exotoxin A expression. *Infect. Immun.*, **61**, 1180–1184.

Gilbert, P., Allison, D. G., Evans, D. J., Handley, P. S., Brown, M. R. W. (1989). Growth rate control of adherent bacterial populations. *Appl. Environ. Microbiol.*, **55**, 1308–1311.

Gilbert, P., Collier, P. J., Brown, M. R. W. (1990). Influence of growth rate on susceptibility to antimicrobial agents: biofilms, cell cycle and dormancy. *Antimicrob. Agents Chemother.*, **34**, 1865–1868.

Giwercman, B., Jensen, E. T., Hoiby, N., Kharazmi, A., Costerton, J. W. (1991). Induction of β-lactamase production in *Pseudomonas aeruginosa* biofilms. *Antimicrob. Agents Chemother.*, **35**, 1008–1010.

Gordon, C. A., Hodges, N. A., Marriot, C. (1988). Antibiotic interaction and diffusion through alginate and exopolysaccharide of cystic fibrosis derived *Pseudomonas aeruginosa*. *J. Antimicrob. Chemother.*, **22**, 667–674.

Gordon, C. A., Hodges, N. A., Marriot, C. (1991). Use of slime dispersants to promote antibiotic through the extracellular polysaccharide of mucoid *Pseudomonas aeruginosa*. *Antimicrob. Agents Chemother.*, **35**, 1258–1260.

Gristina, A. G., Jennings, R. A., Naylor, P. T., Myrvik, Q. N., Barth, E., Webb, L. X. (1989). Comparative *in vitro* antibiotic resistance of surface colonising coagulase negative staphylococci. *Antimicrob. Agents Chemother.*, **33**, 813–824.

Hoyle, B. D., Wong, C. K., Costerton, J. W. (1992). Disparate efficacy of tobramycin on Ca^{2+}-, Mg^{2+}-, and HEPES-treated *Pseudomonas aeruginosa* biofilms. *Can. J. Microbiol.*, **38**, 1214–1218.

Humphrey, B. A., Dixon, M. R., Marshall, K. C. (1979). Physiological and *in-situ* observations on the adhesion of gliding bacteria to surfaces. *Arch. Microbiol.*, **120**, 231–238.

Ichimiya, T., Yamaski, T., Nasu, M. (1994). *In-vitro* effects of antimicrobial agents on *Pseudomonas aeruginosa* biofilm formation. *J. Antimicrob. Chemother.*, **34**, 331–341.

Kumon, H., Tomochika, K.-I., Matunaga, T., Ogawa, M., Ohmori, H. (1994). A sandwich cup method for the penetration assay of antimicrobial agents through *Pseudomonas* exopolysaccharides. *Microbiol. Immunol.*, **38**, 615–619.

Latifi, A., Foglino, M., Tanaka, K., Williams, P., Lazdunski, A. (1996). A hierarchical quorum-sensing cascade in *Pseudomonas aeruginosa* links the transcriptional activators LasR and RhlR (VsmR) to expression of the stationary-phase sigma factor RpoS. *Mol. Microbiol.*, **21**, 1137–1146.

Lee, C. A., Falkow, S. (1990). The ability of *Salmonella* to enter mammalian cells is affected by bacterial growth state. *Proc. Natl Acad. Sci. USA*, **87**, 4304–4308.

Lee, C. K., Rubin, L. G., Moldwin, R. M. (1995) Synergy between protamine and vancomycin in the treatment of *Staphylococcus epidermidis* biofilms. *Urology*, **45**, 720–724.

McCarter, L., Hilmen, M., Silverman, M. (1988). Flagellar dynamometer controls swarmer cell differentiation of *Vibrio parahaemolyticus*. *Cell*, **54**, 345–351.

McFeters, G. A., Egil, T., Wilberg, E. *et al.* (1990). Activity and adaptation of nitrilo(NTA)-degrading bacteria: Field and laboratory studies. *Water Res.*, **24**, 875–881.

Nichols, W. W., Dorrington, S. M., Slack, M. P. E., Walmsley, H. L. (1988). Inhibition of tobramycin diffusion by binding to alginate. *Antimicrob. Agents Chemother.*, **32**, 518–523.

Nichols, W. W., Evans, M. J., Slack, M. P. E., Walmsley, H. L. (1989). The penetration of antibiotics into aggregates of mucoid and non-mucoid *Pseudomonas aeruginosa*. *J. Gen. Microbiol.*, **135**, 1291–1303.

Pascual, A., Ramirez de Arellano, E., Pera, E. J. (1994). Activity of glycopeptides in combination with amikacin or rifampin against *Staphylococcus epidermidis* biofilms on plastic catheters. *Eur. J. Clin. Microbiol. Infect. Dis.*, **13**, 515–517.

Slack, M. P. E., Nichols, W. W. (1981). The penetration of antibiotics through sodium alginate and through the exopolysaccharide of a mucoid strain of *Pseudomonas aeruginosa*. *Lancet*, **11**, 502–503.

Slack, M. P. E., Nichols, W. W. (1982). Antibiotic penetration through bacterial capsules and exopolysaccharides. *J. Antimicrob. Chemother.*, **10**, 368–372.

Stewart, P. S. (1994). Biofilm accumulation model that predicts antibiotic resistance of *Pseudomonas aeruginosa* biofilms. *Antimicrob. Agents Chemother.*, **38**, 1052–1058.

Stotzky, G. (1989). Gene transfer among bacteria in soil. In *Gene Transfer in the Environment* (eds Levy, S. B., Miller, R. V.), pp. 165–222. McGraw-Hill, New York.

Suci, P. A., Mittelman, M. W., Yu, F. U., Geesey, G. G. (1994). Investigation of ciprofloxacin penetration into *Pseudomonas aeruginosa* biofilms. *Antimicrob. Agents Chemother.*, **38**, 2125–2133.

Sutherland, I. W. (1985). Biosynthesis and composition of Gram-negative bacterial extracellular and wall polysaccharides. *Annu. Rev. Microbiol.*, **39**, 243–270.

Tresse, O., Jouenne, T., Junter, G. -A. (1995). The role of oxygen limitation in the resistance of agar-entrapped, sessile-like *Escherichia coli* to aminoglycoside and β-lactam antibiotics. *J. Antimicrob. Chemother.*, **36**, 521–526.

Trieu-Cuot, P., Arthur, M., Courvalin, P. (1987) Transfer of genetic information between Gram-positive and Gram-negative bacteria. In *Streptococcal Genetics* (eds Curtiss, R., Feretti, J. J.), pp. 65–68. American Society for Microbiology Press, Washington, DC.

van Loosdrecht, M. C. M., Lyklema, J., Norde, W., Zehnder, A. J. B. (1990). Influence of interfaces on microbial activity. *Microbiol. Rev.*, **54**, 75–87.

Williams, P. (1988). Role of the cell envelope in bacterial adaption to growth *in vivo* in infections. *Biochimie*, **70**, 987–1011.

Williams, P., Bainton, N. J., Swift, S. *et al.* (1992). Small molecule-mediated density dependent control of gene expression in prokaryotes: bioluminescence and the biosynthesis of carbapenem antibiotics. *FEMS Microbiol. Lett.*, **100**, 161–168.

Zabinski, R. A., Walker, K. J., Larsson, A. J., Moody, J. A., Kaatz, G. W., Rotschafer, J. C. (1995). Effect of aerobic and anaerobic environments on antistaphylococcal activities of five fluoroquinolones. *Antimicrob. Agents Chemother.*, **39**, 507–512.

Zambrano, M. M., Kolter, R. (1995). Changes in bacterial cell properties on going from exponential growth to stationary phase. In *Microbial Quality Assurance: A Guide Towards Relevance and Reproducibility of Inocula* (eds Brown, M. R. W., Gilbert, P.), pp. 21–30. CRC Press, Boca Raton, US.

Zobell, C. E. (1943). The effect of solid surfaces upon bacterial activity. *J. Bacteriol.*, **46**, 39–56.

Section II

Bacterial Infection Models

Chapter 14

The Mouse Peritonitis/Sepsis Model

N. Frimodt-Møller, J.D. Knudsen and F. Espersen

Background

As early as 1905 Kindborg (1905) demonstrated the lethal effect of pneumococci injected into the peritoneum of mice and the effect of enhancing virulence by passage through mice. The concept of studying the effect *in vivo* of an antibacterial compound by treating intraperitoneally inoculated mice was perhaps introduced in 1911 for optochin against pneumococci (Morgenroth and Levy, 1911). Another well-known early example of the use of the mouse peritonitis model was Domagk's demonstration of the activity of Prontosil, the first sulphonamide, against *Streptococcus pyogenes* (Domagk, 1935). Intraperitoneal inoculation of mice with various microorganisms to produce peritonitis or sepsis has since been used for such varying purposes as studying virulence, pathogenesis, course of infection, immunology and other types of host-response to bacteria or fungi (Browder *et al.*, 1987; Teale and Atkinson, 1992; Evans *et al.*, 1994; Briggs *et al.*, 1995; Echtenacher *et al.*, 1995–96; Ayala and Chaudry, 1996; O'Reilly *et al.*, 1996; Byrum *et al.*, 1997; Takashima *et al.*, 1997).

In the setting of this book we shall concentrate upon the use of the mouse peritonitis model for the study of antimicrobial chemotherapy against bacterial infections. The popularity of the model comes from the ease of its use incorporating small and cheap animals, which are easy to keep and handle, short-duration experiments with reproducible infections and simple end-points (death or survival), and it has therefore evolved as an important early screening method to study *in-vivo* effects of antibacterial compounds. For examples of diverse issues concerning aspects of antimicrobial chemotherapy that the mouse peritonitis model has been used for, see Table 14.1.

For a further introduction to the mouse peritonitis or peritonitis-protection model the reader is referred to recent reviews (Frimodt-Møller, 1988, 1993; O'Reilly *et al.*, 1996).

Animal species

Usually, outbred strains of mice are used for reasons of economy and to ensure a heterogenous population, which again ensures the relevance of a possible antibacterial effect and allows for further immunological or other host factor variations to be studied. Specialized strains of mice can be used to study particular facets of the bacterium–host relationship (e.g. transgenic NSE-hIL-6 mice with altered neuronal IL-6 expression (Santo *et al.*, 1996) or WBB6F1-W/Wv mice, which are mast-cell-deficient (Malaviya *et al.*, 1994)).

Mice of any age can be used, but usually those aged 3–4 weeks upwards (20–30 g) are preferred. Even at a similar age mice can vary in weight and for experiments incorporating antimicrobial chemotherapy variation of more than 2–3 g should be avoided. Female mice are easier to keep together five to six to a cage, while the aggressiveness of male mice can cause stress to the animals and even substantial loss of animals due to scratching or fighting.

Preparation of animals

No specialized housing or care nor specific pretreatment is required.

Details of surgery

Overview

Briefly, the unanaesthetized mice are inoculated by transcutaneous intraperitoneal injection. Anaesthesia is not necessary for subsequent antibiotic treatment, unless it must be administered intravenously usually in the tail. When blood or intraperitoneal contents are sampled in living animals by surgical procedures, the animals must be anaesthetized and some degree of anatomical and surgical expertise is needed.

Materials required

Anaesthetic, skin disinfectant, syringes (1 or 5 ml) and needles (26 g), scalpel handle plus blades, forceps, scissors, flat-plated material (either soft wood, hardboard or disposable plastic material), pins or tape, pipettes (up to 2 ml).

Handbook of Animal Models of Infection
ISBN 0-12-775390-7

Table 14.1 Examples from the literature of the use of the mouse peritonitis model in antimicrobial research for purposes other than screening of effect of new antibiotics

Reference	*Mouse strain*	*Bacteria*	*Purpose of study*	*Main results*
Beskid *et al*., 1981	Swiss albino	Various Gram-positive and Gram-negative human pathogens	Evaluate activity *in vivo* of 2′,3′-dideoxyadenosine (DDA), a nucleotide analogue *Effect parameter*: mortality	DDA effective by oral route against *Salmonella* and other enteric pathogens
Pechere *et al*., 1986	Swiss Webster, female	Various enteric pathogens and *S. aureus*	Evaluation of emergence of resistance after therapy with antibiotics alone or combined *Effect parameter*: colony counts in peritoneum	Combinations of *β*-lactams with aminoglycoside or quinolone significantly reduced development of resistance *in vivo*
Browder *et al*., 1987	ICR/HSD	*E. coli*	Evaluation of effect of glucan (immunostimulator) alone or together with gentamicin *Effect parameter*: IL-1, peripheral neutrophil counts, bone-marrow proliferation, blood bacterial counts, survival	Combined glucan–gentamicin significantly enhanced bone marrow proliferation and neutrophil counts, which correlated with survival from peritonitis
Kristiansen *et al*., 1988	CF1, female	*S. pneumoniae* type 3	Evaluation of effect of transclo-penthixol (TP; a weak neuroleptic) alone or together with penicillin *Effect parameter*: survival	TP had activity alone *in vitro* but not *in vivo*; however, *in-vivo* synergy in combination with penicillin
Swenson *et al*.,1990	CD-1, male	Various enteric pathogens	Pharmacokinetics and effect of liposome-encapsulated gentamicin *Effect parameter*: survival/50% protective dose (PD_{50})	Liposome-encapsulated gentamicin significantly better than gentamicin
Matsumoto *et al*., 1991	ddY, male made neutropenic with cyclophosphamide	*S. aureus*, *P. aeruginosa*, *E. coli*, *S. marcescens* and *C. albicans*	Evaluation of rG-CSF in combination with antibiotics *in vivo* *Effect parameter*: survival	rG-CSF in combination significantly improved survival as compared to antibiotics alone
Tokunaga *et al*.,1997	ddY, male	*E. coli* KC-14, virulent to mice; *E. coli* TT-48, mutant unable to utilize exogenous thymidine	Effect of trimethoprim on infection caused by thymidine-deficient *E. coli* in mice with thymidine concentrations *Effect parameter*: survival/ 50% effective dose (ED_{50})	Effect of trimethoprim higher with mutant than parent strain, i.e. host thymidine concentrations antagonize trimethoprim

Anaesthesia

The animals usually need short-duration anaesthesia only (10 minutes). The easiest and most rapid methods are small chambers (glass or metal jar) with ether or CO_2. The mice are placed in a closed chamber, which should preferably be transparent so that the animal can be monitored. Complete anaesthesia ensues in 1–2 minutes and lasts for about 5–10 minutes. The use of ether demands good ventilation in order to minimize ether inhalation for the persons involved in experimentation. Intraperitoneal (i.p.) injection of propanidid (Sombrevin®, Chemical Works of Gedeon Richter Ltd., Budapest, Hungary), 50 mg/ml, in doses of 0.1 mg/10 g mouse, or similar, can be used, but it takes longer for full anaesthesia to ensue. Further i.p. installation of anaesthetic may be needed, but usually one dose is sufficient for surgery.

Inoculation

Injection of inoculum can easily be performed by one person. The mouse is grabbed by the tip of its tail between the thumb and forefinger of one hand, and the tail is folded around the base of the little finger, which is then flexed to hold the tail tight. The grip on the tail tip by thumb and forefinger is then relaxed and the same fingers grab the neck of the mouse just behind the head. By pronating the hand the mouse is now held in one hand, thumb and forefinger tightly behind the head, so that the mouse is unable to bite the fingers, and the tail under the flexed little finger, with its abdomen exposed. The other hand is now free to handle the syringe and inject the inoculum intraperitoneally. The same hold on the animal can be used when injecting antibiotic, either in the hind quarters, by twisting the mouse slightly, or in the neck region between (and not into!) your thumb and forefinger.

Intraperitoneal injection is performed in either of the lateral lower quadrants of the abdomen in order to avoid damaging such organs as the liver or the spleen, which are both relatively large in mice. The needle should be directed cephally and penetrate about 1 cm into the peritoneum. Very seldom will the intestines be penetrated, but injection of the whole inoculum into the intestines is one reason for a mouse to survive a lethal inoculum, while bigger lesions of the intestine may cause unwanted bacterial contamination of the peritoneum. If necessary, this can be tested by subsequent opening of the peritoneum, visual inspection of the abdominal contents and bacterial culture of peritoneal fluid. For antibiotic treatment volumes larger than 0.5 ml should be avoided, since they can be painful and may cause both slow absorption of drug and fluid overload to the animals.

Sampling of fluids or organs

Blood

Sampling of blood from mice can be performed in several ways depending upon the purpose and volume of the blood sample. Several samples, but of small volumes (1–200 μl), can be obtained either by periorbital cutdown or by puncture of the tail vein. Larger volumes (up to 1–1.5 ml) involving complete exsanguination of the mouse can be obtained by periorbital cutdown, cardial puncture or axillary cutdown. We use axillary cutdown for bacterial counts in blood, and periorbital cutdown for blood antibiotic concentration determination. The axillary cutdown is shown in Figure 14.1. With the anaesthetized mouse lying on its back — in ventral position — with legs fixed, a 5 mm skin incision is performed over the front–medial part of the axilla parallel to the thorax, the axillary artery and/or vein is cut, and the blood can be sampled (Figure 14.1). For periorbital cutdown, a deep 5 mm incision around the periphery of one eye is performed with a scalpel in the anaesthetized mouse and blood is dripped into a container. For both of these methods, the animals are sacrificed immediately after sampling.

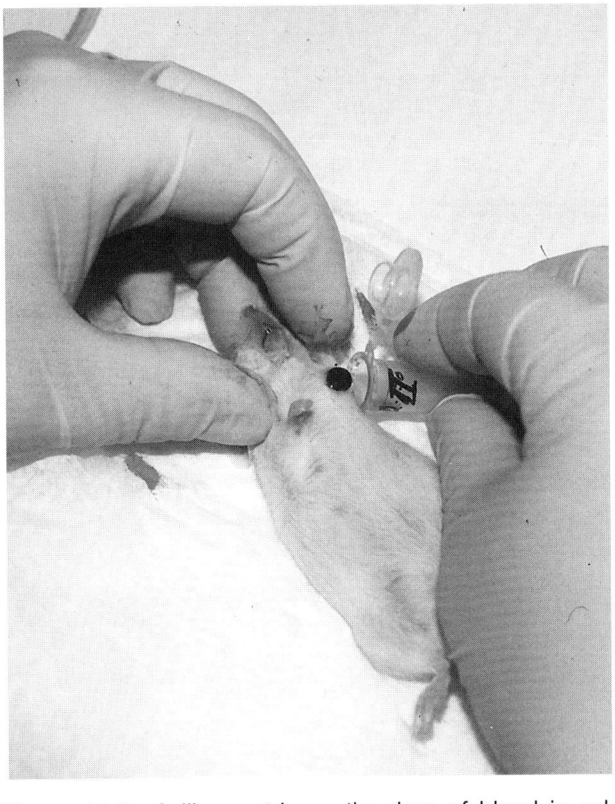

Figure 14.1 Axillary cutdown: the drop of blood is collected in a container and immediately processed in order to avoid coagulation.

Peritoneal fluid and intra-abdominal organs

For bacterial counts in the peritoneum the mouse is anaesthetized and secured in the ventral posistion as above. Two ml of sterile saline is injected i.p. and the abdomen is gently massaged between two fingers for 1 minute (Figure 14.2). After disinfection of the skin the abdomen is opened via a 1–1.5 cm incision with forceps and scissors (Figures 14.3 and 14.4). While one side of the peritoneal fold and skin is held with forceps, the pipette tip is introduced into the peritoneum and after gently pushing the intestines aside at least 1 ml of fluid can be sampled. For antibiotic concentration determination paper disks (e.g. 6 mm diameter) can be

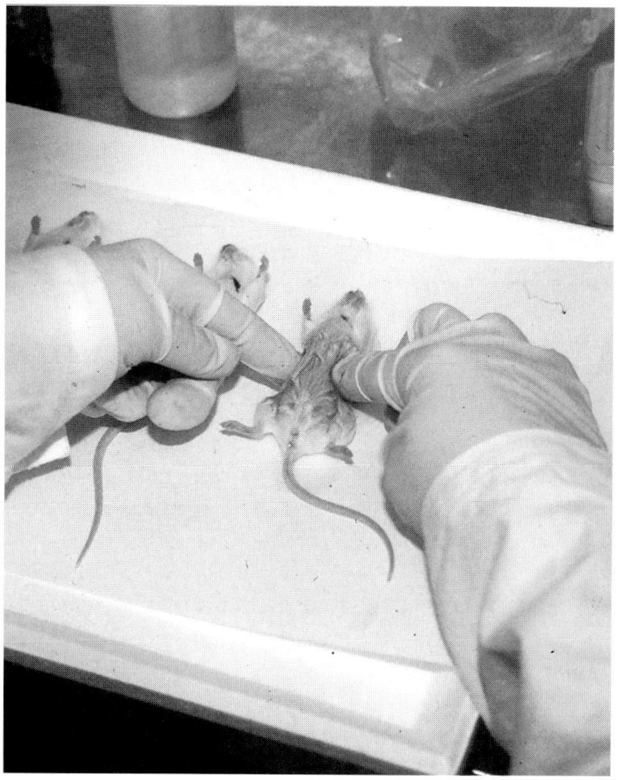

Figure 14.2 Bacterial counts in peritoneum. After injecting 2 ml of sterile saline into the peritoneum, the peritoneal contents are mixed by massaging the abdomen with two fingers.

placed in the peritoneum and retrieved after visual inspection for fluid absorption. For sampling of liver and spleen the incision is further elongated and the organs can be easily identified in the upper right and left quadrants respectively and removed.

Postoperative care

After inoculation mice should be kept in their cages under daily observation for development of signs of infection or death. Infected mice will become quiet, with less moving around, and the fur will become rough and appear less shining than usual. Shaking of the animal may ensue, and

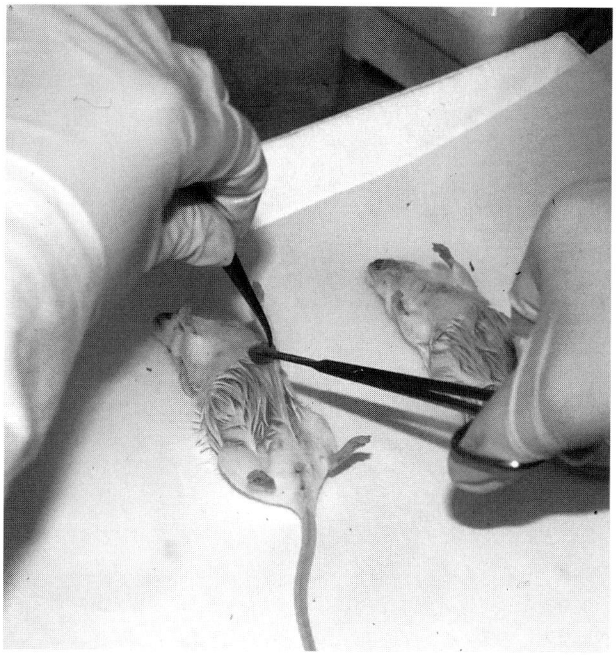

Figure 14.3 Bacterial counts in peritoneum. After mixing, an incision 1 cm long is performed on the abdominal wall and the peritoneum is opened as well.

respiration may appear more restrained or superficial. In some countries, for ethical reasons, the animals are not allowed to die from infection and must at this stage be sacrificed or processed for sampling. Dead animals should be removed from the cages as soon as possible.

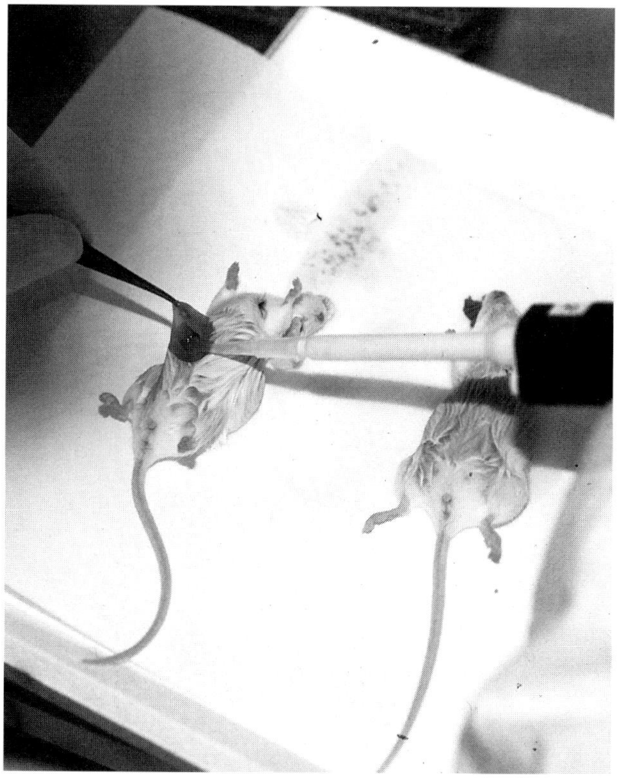

Figure 14.4 Bacterial counts in peritoneum. Peritoneal fluid is sampled with a pipette.

Storage and preparation of inocula

The model is amenable to using any species of bacterium, but because of differences in their virulence to mice some consideration should be given prior to choice of pathogen, adjuvant and inoculum. Among bacteria normally virulent to mice without adjuvant are many serotypes of capsulated *Streptococcus pneumoniae* (Kindborg, 1905; Mørch, 1943), some strains of *Streptococcus pyogenes* (Eagle, 1949), selected strains of *Salmonella* types (especially among *S. typhi*, *S. paratyphi* and *S. typhimurium*), *Yersinia pestis*, *Brucella* (Bowden *et al.*, 1995) and others. For *S. pneumoniae*, the penicillin-intermediate or -resistant serotypes are usually not virulent to mice — or only in very large numbers (Knudsen *et al.*, 1995). Most other common human pathogens, including staphylococci, other streptococci than those mentioned above, enterococci, Enterobactericeae and *Pseudomonas* spp. are not naturally virulent and demand some kind of adjuvant inhibiting the acute immunological reaction of the mice, or for the mice to be made immunoincompetent prior to inoculation. The most commonly used adjuvant has been mucin (Comber *et al.*, 1975), which inhibits local macrophage function for 2–3 hours (Nungester *et al.*, 1932). A solution with 5% wt/vol. mucin is usually sufficient to render most human pathogens virulent to mice with inocula around 10^5–10^7 cfu. In our hands it has worked well with pencillin-resistant pneumococci, leading to immediate growth of the bacteria in peritoneum and blood (Figure 14.5). For *Staphylococcus aureus* human plasma increases the virulence of the bacteria, resulting in 100% lethality with inocula of 10^{6-8} cfu , while more than 10^{9-10} cfu are otherwise necessary (Espersen *et al.*, 1984a). Other adjuvants include talcum powder (Pechere *et al.*, 1986), bakers yeast (Joó and Zsidai, 1982) or gelatine capsules, which dissolve after a while *in vivo* (Matsushita *et al.*, 1989).

Systemic inhibition of the immune system or parts of the immune defence is widely used in many animal models, particularly in the mouse thigh model, but is not so common in the mouse peritonitis model. Neutropenia can be induced reproducibly by pretreating the mice with cyclophosphamide (100 mg/kg once a day for 3 days). The complement system, which is crucial for defence against capsulated bacteria such as pneumococci, can be inhibited by administration of cobra-venom-factor (Bakker-Woudenberg *et al.*, 1984). Other methods that induce immunoincompetence are splenectomy (Brown *et al.*, 1981) or high-voltage radiation.

Even with naturally virulent bacteria such as pneumococcus type 3 the choice of vehicle is important for the fate of the bacteria *in vivo*. Inoculation of the pneumococci in saline leads to a 6–8 hours lag phase before the bacteria start to divide, which has great impact upon the effect of ß-lactam antibiotics (Frimodt-Møller *et al.*, 1983). Use of broth (beef, tryptic soy or Mueller–Hinton) as a vehicle will protect the bacteria and result in immediate growth of the bacteria *in vivo* (Frimodt-Møller and Thomsen, 1986).

For preparation of the inoculum either of two methods are used. A broth culture prepared the day before is diluted in fresh broth to the desired concentration, or the inoculum is prepared from a fresh agar plate culture, from which the colonies are picked and diluted in broth — or washed in saline by centrifugation — to the desired concentration by the use of a spectrophotometer or colorimeter. For both methods sufficient inoculum for several experiments can be prepared, divided into smaller volumes and stored (–80°C, frozen nitrogen or freeze-dried). We prefer the plate method to the broth method since it is easier to ensure on a fresh plate that there is no contamination, that the bacteria are capsulated, etc. The broth method is used if bacteria in the log phase are needed for the inoculum. Whether this has any advantage over the other methods mentioned has not been studied in detail.

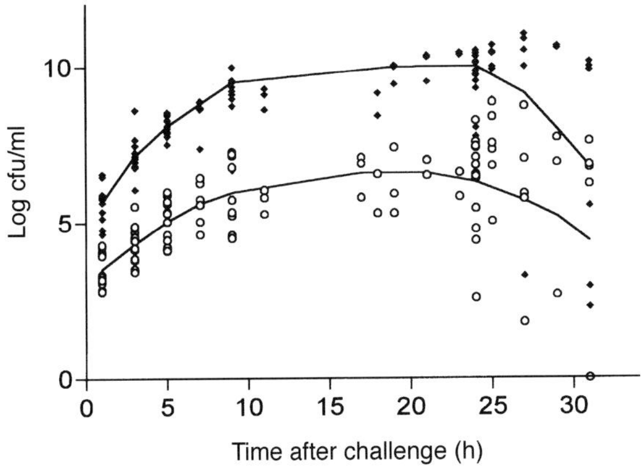

Figure 14.5 Bacterial counts in the mouse peritonitis model with a penicillin-resistant pneumococcus in 5% mucin. Closed circles, counts in peritoneum; open circles, counts in blood.

The infection process

After intraperitoneal inoculation the local immune response immediately comes into play, unless intentionally inhibited as described above. Local macrophages together with neutrophil granulocytes increase in number and activity; even mast cells have been shown to participate in the early defence (Malaviya *et al.*, 1994; Ayala and Chaudry, 1996). Bacteria are transported by the lymphatic system via the ductus thoracicus to the blood stream and bacteraemia at 1–2 log bacterial counts below the peritoneal counts is present within 30 minutes after inoculation (Figure 14.5; Frimodt-Møller *et al.*, 1983; Espersen *et al.*, 1984b; Frimodt-Møller and Thomsen, 1986, 1987a). This bacteraemic state is not necessarily indicative of the severity of the infection, since it may disappear after some hours without obvious harm to the mice (Espersen *et al.*, 1984b). The interleukine system is rapidly induced, and the immune system responds by inducing the complement cascades and increasing the concentration of acute-phase immunoglobulins (Brown *et al.*, 1983; Echtenacher *et al.*, 1995–96; Ayala and Chaudry, 1996; Takashima *et al.*, 1997). Specific IgG will be produced but usually not in order to cure the present infection (Brown *et al.*, 1983). However, for this reason surviving mice should not be used for new experiments. Most bacteria causing lethal infection will demonstrate growth curves *in vivo* similar to those found in broth culture *in vitro* (Figure 14.5). After an exponential phase, which is usually highly reproducible, a stationary phase ensues, ending in a declining phase during which the animals succumb. Death of the animal is due to overwhelming infection, but some degree of toxin production may also play a role. Thus, it has been shown for pneumococci that the main lethal effect probably stems from production of pneumolysin; Dick and Gemmell (1971) showed that penicillin treatment in late stages of infection, although reducing bacterial numbers, had no effect upon mortality of the mice.

It is usually a good idea to determine a 50% lethal dose (LD_{50}) for each strain used, i.e. groups of five to 10 mice are inoculated with increasing numbers (e.g. 10^2, 10^4, 10^6, 10^8 cfu) of bacteria together with the adjuvant. Most protection experiments are performed with inocula leading to 100% mortality; one should be aware, however, that the effect of antibiotics *in vivo* is directly correlated to the size of the inoculum (Eagle *et al.*, 1950a).

Key parameters to monitor infection and response to treatment

Two parameters for monitoring the course of the infection and the effect of treatment are commonly used: death or survival of the animals, or bacterial counts in peritoneum and/or blood. Death or survival of the animal has the obvious advantages of being easy to monitor and survival of the host is the ultimate goal in any treatment experiment. Monitoring the time of death of the mice may be important for study of the effect of treatment, which may be incorporated in survival curves. For short-duration experiments with death of mice within 24 hours observation of mice during the first 12–24 hours may be necessary. When treatment is part of the experiment death could also be due to some toxic side-effect; a non-inoculated control group of mice should be treated with the highest dose of drug used in order to monitor side-effects.

Number of deaths per treatment group can be incorporated into estimating a 50% survival end-point 50% protective dose (PD_{50}) or 50% effective dose (ED_{50}; parameters similar to LD_{50}, which has been used as an excellent effect parameter that generally shows good correlation with activity *in vitro* (Eagle *et al.*, 1950a; Tsuchiya *et al.*, 1981; Frimodt-Møller *et al.*, 1986; Knudsen *et al.*, 1995; O'Reilly *et al.*, 1996).

Intraperitoneal bacterial counts and blood counts are important and effective parameters for illustrating the course of infection as well as effect *in vivo* of antibiotics and for relating effect *in vitro* with effect *in vivo* (Eagle *et al.*, 1950a,b). The knowledge of the bacterial growth curve in peritoneum *in vivo* is crucial for optimal antibiotic therapy because of the following factors: for ß-lactam antibiotics activity depends upon growing or dividing bacteria, which means that there is no effect during the lag phase (Frimodt-Møller *et al.*, 1983). In contrast, antibiotics that are protein synthesis inhibitors show activity during all growth phases, but the effect is better during lag phase than during the exponential growth phase (Frimodt-Møller *et al.*, 1983). Therefore, comparison of antibiotics with different mechanisms of activity can favour some drugs in one growth phase and some in another. Peritoneal and blood bacterial counts are somewhat more tedious to collect than mortality data, but give important information on the effect of the drugs used such as bactericidal activity *in vivo*, development of antibiotic resistance or loss of virulence properties. The peritoneal counts are not necessarily indicative of the survival of the animal (Frimodt-Møller and Thomsen, 1986; Knudsen *et al.*, 1998). If bacterial counts *in vivo* are used, they should be performed at several time points in order to describe bacterial killing or growth.

Other parameters can be used to monitor the course of infection but have rarely been used to monitor the effect of antibiotics. The usual indicators of progressing infections are also applicable to the mouse, e.g. leucocyte counts in blood and in peritoneal fluid, various interleukins, liver enzymes, renal function parameters, IgM and IgG antibodies, etc. (Ayala and Chaudry, 1996; Echtenacher *et al.*, 1996; O'Reilly *et al.*, 1996). The small size of the animal, however, together with the short course of the infection and its usually short treatment, most often preclude the use of such paraclinical parameters in this model.

Antimicrobial therapy

A variety of antimicrobial agents have been used in the model, as well as a number of non-antibiotic drugs and compounds (Miraglia *et al.*, 1973; Rolinson, 1974; Comber *et al.*, 1975; Davis, 1975; Beskid *et al.*, 1981; Tsuchiya *et al.*, 1981; Fung-Tomc *et. al.*, 1989; Grappel *et al.*, 1985; Fernandes *et al.*, 1986; Frimodt-Møller *et al.*, 1986; Cacciapuoti *et al.*, 1987; Frimodt-Møller and Thomsen, 1987a; Imada *et al.*, 1992; Iwahi *et al.*, 1992; Alder *et al.*, 1995). Almost any type of drug administration has been used, including intravenous, subcutaneous, oral, intraperitoneal and local application. The intravenous route is best performed via a tail vein, which needs technical skill. The mouse may be anaesthetized, although it is not necessary for the skilled technician, and the vein is exposed by heating the tail, e.g. under a lamp. It is quite easy to see whether the drug stays in the vein or is injected subcutaneously.

The oral route can be employed in several ways. Feeding each mouse individually with a syringe carrying a blunt cannula, which is introduced into the mouth, has the advantage of making sure that all animals are treated; it is difficult to be sure, however, whether the drug stays in the mouse thereafter. Mixing drug in the drinking water or foodstuff for the animals is another means of oral administration, and it might be considered easier than the individual syringe method, especially if many animals are to be treated, but exact timing of the treatment is difficult and large variation in the amount of drug delivered to each animal makes this method unsatisfactory. Intraperitoneal administration is easy and can be considered to be relevant systemic treatment since most of the drug will be absorbed into the blood stream, but considering the intraperitoneal route of inoculation it will be difficult later to evaluate whether the effect of the antibiotic was due to local activity in the peritoneum or to a systemic effect.

The easiest and most practical method of antibiotic treatment is by the subcutaneous or intramuscular route. This results in very reproducible antibiotic concentration profiles in serum or other compartments with little variation among animals of the same age and weight (Frimodt-Møller *et al.*, 1986; Knudsen *et al.*, 1995). Since timing of antibiotic treatment in relation to bacterial growth curves *in vivo* can be crucial, this route of drug administration is recommended.

Dose calculation in mice must take into account that they are small animals with a relatively larger surface area relative to weight than larger animals, including man. Furthermore, they have a rapid turnover resulting in much shorter serum elimination half-lives of drugs than in man. Doses can be calculated from estimated surface area determinations: i.e. a full-grown mouse of 25–30 g weight has a surface area of 60–65 cm^2 (Diack, 1930). Working from the normal human surface area of 17 200 cm^2 similar doses as in humans can be calculated. Such calculations have worked well in our studies (Frimodt-Møller *et al.*, 1986).

The antibiotic should be diluted according to the manufacturer's instructions, but for dosing of mice volumes larger than 0.5 ml per dose should be avoided in order to prevent fluid overload. If several doses are given, the volume of each dose should be further reduced.

Determination of serum or peritoneal fluid antibiotic concentrations in mice used for protection studies are mandatory in order to correlate effect with antimicrobial activity. For optimal evaluation complete pharmacokinetic profiles (C_{max}, $t_{1/2}$, AUC, time above MIC for the infecting strain) should be determined at doses relevant to the doses showing effect *in vivo*. Optimally, infected animals should be used for this purpose, but because of the rapid course of infection one may have to rely upon non-infected mice for this purpose.

Antibiotic concentrations can be measured in peritoneal fluid as mentioned above, but generally drug concentrations in peritoneum will closely follow those in serum.

Pitfalls (advantages/disadvantages) of the model

Apart from the usual considerations regarding extrapolation from an animal as small as the mouse to humans the mouse peritonitis model has some obvious disadvantages. The infection induced is a peritonitis, but seldom with bacteria relevant for this type of infection, and results are usually extrapolated to other types of infection without considering the influence of factors peculiar to the peritoneum. On the other hand, local antibiotic concentrations in peritoneal fluid have usually correlated so well with serum antibiotic concentrations that sampling of blood alone has been sufficient to demonstrate relationships between pharmacokinetics and effect *in vivo*. Furthermore, the overwhelming infection with high inocula, leading to death, can be an advantage when survival is used as parameter of effect but is not always relevant to the clinical situations. Treatment is usually initiated shortly after inoculation according to knowledge of the bacterial growth curve *in vivo*, whereas in the clinical situation we do not know at which point in the bacterial growth phase we first treat the patient. The early treatment start might also prevent a number of the defence mechanisms that the host would otherwise respond with. The impairment of host defence mechanisms that is necessary in many cases in order to enhance virulence of many bacteria should always be borne in mind when evaluating the results of treatment regimens. It may, however, actually be relevant to the clinical situation, since infection often occurs in connection with pathological or iatrogenic impairment of host defences.

Further disadvantages of the small animal with a rapid course of infection are that the animals can only be used once for parameters such as organ or body fluid bacterial colony counts, and several animals are needed to monitor these parameters over time. Therefore the variation in the infection and its treatment must be standardized to a high degree.

Some drugs are more toxic to rodents than to humans, e.g. macrolides and fusidic acid; the interpretation of experiments with such drugs in the mouse model may therefore underestimate the potential of the drugs.

The mouse peritonitis model is excellent for evaluating interaction between drugs *in vivo* such as synergism or antagonism between different antibiotics or between antibiotics and other types of drug (Schaad *et al.*, 1982; Browder *et al.*, 1987; Frimodt-Møller and Thomsen, 1987b; Kristiansen *et al.*, 1988; Matsumoto *et al.*, 1991; Lang *et al.*, 1993).

Contribution of the model to infectious disease therapy

The mouse peritonitis model has been instrumental in early demonstration of the *in-vivo* effect of numerous antimicrobial compounds against common human pathogens, especially where *in-vitro* experiments had not predicted such an effect (Domagk, 1935). Similarly, the model has been important in testing for the effect of antibiotics for treatment of special pathogens such as *Brucella* spp. or group B streptococci (Baltimore *et al.*, 1979; Lang *et al.*, 1993). The model has had an important role in demonstrating fundamental issues in the relationship between infecting pathogen and antibiotic therapy such as the correlation between activity *in vitro* (MIC) and effect *in vivo* (Eagle *et al.*, 1950a; Frimodt-Møller *et al.*, 1986; Sullivan *et al.*, 1993; Knudsen *et al.*, 1995). For examples of the excellent correlations between MIC (or logMIC) and ED_{50} for β-lactam antibiotics which have been demonstrated in the model, see Figure 14.6. Furthermore, the use of the model has demonstrated: the relationship between the size of inoculum and antibiotic dose (Eagle, 1949, 1952; Eagle *et al.*, 1950b; Frimodt-Møller and Thomsen, 1986); the timing of treatment in relation to the *in-vivo* growth curve and the size of the dose (Eagle, 1949; Frimodt-Møller *et al.*, 1983; Knudsen *et al.*, 1998) the importance of bactericidal activity (Eagle *et al.*, 1950a; Davis, 1975; Frimodt-Møller and Thomsen, 1987c); the importance of different phases of the growth curve and effect of different classes of antibiotics (Frimodt-Møller *et al.*, 1983; Knudsen *et al.*, 1998); and the importance of protein binding for antibiotic activity *in vivo* (Mattie *et al.*, 1973; Rolinson, 1980).

Similarly, via this model fundamental principles were demonstrated in the area of pharmacodynamics, i.e. the relationship between *in-vitro* activity, pharmacokinetic parameters and effect *in vivo* (Eagle *et al.*, 1950b; Eagle, 1952; Frimodt-Møller *et al.*, 1986, 1987; Onyeji *et al.*, 1994; Knudsen *et al.*, 1995; Soriano *et al.*, 1996; den Hollander *et al.*, 1998). Thus, such issues as the importance of the time the antibiotic concentrations remain above the MIC for ß-lactam antibiotics and the importance of the peak antibiotic concentration for antibiotics such as aminoglycosides and quinolones were first demonstrated in the mouse peritonitis model (Eagle *et al.*, 1950b; Davis, 1975).

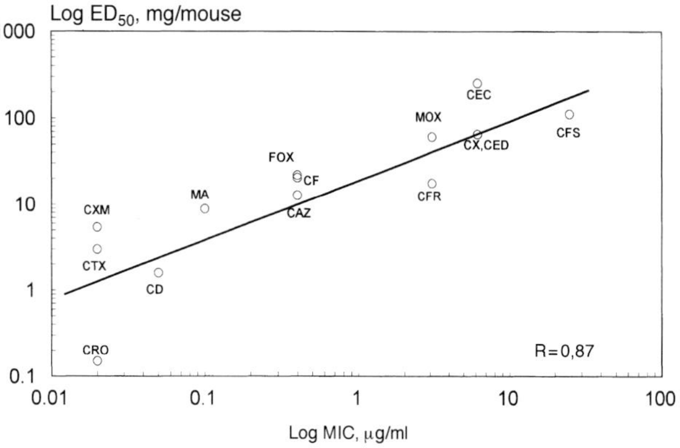

Figure 14.6 Correlation between $\log_{MIC}$ and $\log ED_{50}$ for 14 cephalosporins against a pneumococcus type 3 in the mouse peritonitis model. CRO, ceftriaxone; CTX, cefotaxime; CAZ, ceftazidime; FOX, cefoxitin; MOX, moxalactam; CXM, cefuroxime; CX, cephalexin; CD, cephaloridine; MA, cefamandole; CF, cephalothin; CED, cephradine; CEC, cefaclor; CFR, cefadroxil; CFS, cefsulodin. Adapted from Frimodt-Møller *et al.*, 1986 with permission.

Other areas where the mouse peritonitis model has contributed to our basic knowledge of the behaviour of antibiotic compounds are: interaction between drugs *in vivo* (Schaad *et al.*, 1982; Frimodt-Møller and Thomsen, 1987b; Kristiansen *et al.*, 1988); development of bacterial resistance to antibiotics during treatment (Pechere *et al.*, 1986; Marchou *et al.*, 1987; Pechere and Vladoianu, 1992); the relationship between various aspects of the immune system and antibiotics (Browder *et al.*, 1987; Matsumoto *et al.*, 1991; Teale and Atkinson, 1992); and testing of new preparations of old drugs (Swenson *et al.*, 1990) or new applications of old drugs (Kristiansen *et al.*, 1988).

Therefore, bearing in mind what the mouse peritonitis model has been able to achieve in regards of predicting relationships between effect of drugs *in vivo* and various pharmacological issues it will continue to be an important first-step model for these purposes in the future.

References

Alder, J., Clement, J., Meulbroek, J. *et al.* (1995). Efficacies of ABT-719 and related 2-pyridones, members of a new class of antibacterial agents, against experimental bacterial infections. *Antimicrob. Agents Chemother.*, **39**, 971–975.

Ayala, A., Chaudry, I. H. (1996). Immune dysfunction in murine polymicrobial sepsis: mediators, macrophages, lymphocytes and apoptosis. *Shock*, **6** (Suppl. 1), S27–S38.

Bakker-Woudenberg, I. A. J. M., van den Berg, J. C., Fontijne, P., Michel, M. F. (1984). Efficacy of continuous versus intermittent administration of penicillin G in *Streptococcus pneumoniae* pneumonia in normal and immunodeficient rats. *Eur. J. Clin. Microbiol.*, **3**, 131–135.

Baltimore, R. S., Kasper, D. L., Vecchitto, J. (1979). Mouse protection test for group B Streptococcus type III. *J. Infect. Dis.*, **140**, 81–88.

Beskid, G., Eskin, B., Cleeland, R. *et al.* (1981). Antibacterial activity of 2′, 3′-dideoxyadenosine *in vivo* and *in vitro*. *Antimicrob. Agents Chemother.*, **19**, 424–428.

Bowden, R. A., Cloeckaert, A., Zygmunt, M. S., Dubray, G. (1995). Outer-membrane protein- and rough lipopolysaccharide-specific monoclonal antibodies protect mice against *Brucella ovis*. *J. Med. Microbiol.*, **43**, 344–347.

Briggs, J. B., Oda, Y., Gilbert, J. H., Schaefer, M. E., Macher, B. A. (1995). Peptides inhibit selectin-mediated cell adhesion *in vitro*, and neutrophil influx into inflammatory sites *in vivo*. *Glycobiology*, **5**, 583–588.

Browder, W., Williams, D., Sherwood, E., McNamee, R., Jones, E., DiLuzio, N. (1987). Synergistic effect of nonspecific immunostimulation and antibiotics in experimental peritonitis. *Surgery*, **102**, 206–214.

Brown, E. J., Hosea, S. W., Frank, M. M. (1981). The role of the spleen in experimental pneumococcal bacteremia. *J. Clin. Invest.*, **67**, 975–982.

Brown, E. J., Hosea, S. W., Frank, M. M. (1983). The role of antibody and complement in the reticuloendothelial clearance of pneumococci from the blood stream. *Rev. Infect. Dis.*, **4** (Suppl. 4), 797–805.

Byrum, R. S., Goulet, J. L., Griffiths, R. J., Koller, B. H. (1997). Role of 5-lipoxygenase-activating protein (FLAP) in murine acute inflammatory responses. *J. Exp. Med.*, **185**, 1065–1075.

Cacciapuoti, A., Moss, E. L., Menzel, F. *et al.* (1987). *In vitro* and *in vivo* characterization of novel 8-methoxy derivatives of chlortetracycline. *J. Antibiot. (Tokyo)*, **40**, 1426–1430.

Comber, K. R., Osborne, C. D., Sutherland, R. (1975). Comparative effects of amoxicillin and ampicillin in the treatment of experimental mouse infection. *Antimicrob. Agents Chemother.*, **7**, 179–185.

Davis, S. D. (1975). Activity of gentamicin, tobramycin, polymyxin B, and colistimethate in mouse protection tests with *Pseudomonas aeruginosa*. *Antimicrob. Agents Chemother.*, **8**, 50–53.

Den Hollander, J. G., Knudsen, J. D., Mouton, J. W. *et al.* (1998). Comparison of pharmacodynamics of azithromycin and erythromycin *in vitro* and *in vivo*. *Antimicrob. Agents Chemother.*, **42**, 377–382.

Diack, S. L. (1930). The determination of the surface area of the white rat. *J. Nutr.*, **3**, 289–296.

Dick, T. B., Gemmell, C. G. (1971). The pathogenesis of pneumoccal infection in mice. *J. Med. Microbiol.*, **4**, 153–163.

Domagk, G. (1935). Ein Beitrag zur Chemotherapie der bakteriellen Infectionen. *Deut. Med. Wschr.*, **7**, 250–253.

Eagle, H. (1949). The effect of the size of the inoculum and the age of the infection on the curative dose of penicillin in experimental infections with streptococci, pneumococci, and *Treponema pallidum*. *J. Exp. Med.*, **90**, 595–607.

Eagle, H. (1952). Experimental approach to the problem of treatment failure with penicillin. *Am. J. Med.*, **11**, 389–399.

Eagle, H., Fleischman, R., Musselman, A. D. (1950a). The effective concentrations of penicillin *in vitro* and *in vivo* for streptococci, pneumococci, and *Treponema pallidum*. *J. Bacteriol.*, **59**, 625–643.

Eagle, H., Fleischman, R., Musselman, A. D. (1950b). Effect of schedule of administration on the therapeutic efficacy of penicillin. Importance of the aggregate time penicillin remains at effectively bactericidal levels. *Am. J. Med.*, **9**, 280–299.

Echtenacher, B., Hlultner, L., Míannel, D.N. (1995–96). Cellular and molecular mechanisms of TNF protection in septic peritonitis. *J. Inflamm.*, **47**, 85–89.

Espersen, F., Clemmensen, I., Rhodes, J. M., Jensen, K. (1984a). Human serum and plasma increase mouse mortality in *Staphylococcus aureus* intraperitoneal infection. *Acta Pathol. Microbiol. Immunol. Scand. Sect. B.*, **92**, 305–310.

Espersen, F., Clemmensen, I., Frimodt-Møller, N., Jensen, K. (1984b). Effect of human IgG and fibrinogen on *Staphylococcus aureus* intraperitoneal infection in mice. *Acta Pathol. Immunol. Scand. Sect. B.*, **92**, 311–317.

Evans, T., Carpenter, A., Silva, A., Cohen, J. (1994). Inhibition of nitric oxide synthase in experimental gram-negative sepsis. *J. Infect. Dis.*, **169**, 343–349.

Fernandes, P. B., Bailer, R., Swanson, R. *et al.* (1986). *In vitro* and *in vivo* evaluation of A-56268 (TE-031), a new macrolide. *Antimicrob. Agents Chemother.*, **30**, 865–873.

Frimodt-Møller, N. (1988). Correlation of *in vitro* activity and pharmacokinetic parameters with effect *in vivo* for antibiotics: observations from experimental pneumococcus infection. *Dan. Med. Bull.*, **35**, 422–437.

Frimodt-Møller, N. (1993). The mouse peritonitis model: present and future use. *J. Antimicrob. Chemother.*, **31** (Suppl. D), 55–60.

Frimodt-Møller, N., Thomsen, V. F. (1986). The pneumococcus and the mouse protection test: inoculum, dosage and timing. *Acta Pathol. Microbiol. Immunol. Scand. Sect. B.*, **94**, 33–37.

Frimodt-Møller, N., Thomsen, V. F. (1987a). The pneumococcus and the mouse-protection test: Correlation of *in vitro* and *in vivo* activity for beta-lactam antibiotics, vancomycin, erythromycin and gentamicin. *Acta Pathol. Microbiol. Immunol. Scand. Sect. B.*, **95**, 159–165.

Frimodt-Møller, N., Thomsen, V. F. (1987b). Interaction between beta-lactam antibiotics and gentamicin against *Streptococcus pneumoniae in vitro* and *in vivo*. *Acta Pathol. Microbiol. Immunol. Scand. Sect. B.*, **95**, 269–275.

Frimodt-Møller, N., Sebbesen, O., Thomsen, V. F. (1983). The pneumococcus and the mouse protection test: importance of the lag phase *in vivo*. *Chemotherapy*, **29**, 128–134.

Frimodt-Møller, N., Bentzon, M. W., Thomsen, V. F. (1986). Experimental infection with *Streptococcus pneumoniae* in mice: correlation of *in vitro* activity and pharmacokinetic parameters with *in vivo* effect for 14 cephalosporins. *J. Infect. Dis.*, **154**, 511–517.

Frimodt-Møller, N., Bentzon, M. W., Thomsen, V. F. (1987). Experimental pneumococcus infection in mice: comparative *in vitro* and *in vivo* effect of cefuroxime, cefotaxime and ceftriaxone. *Acta Pathol. Microbiol. Immunol. Scand. Sect. B.*, **95**, 261–267.

Fung-Tomc, J., Desiderio, J. V., Tsai, Y. H., Warr, G., Kessler, R. E. (1989). *In vitro* and *in vivo* antibacterial activities of BMY 40062, a new fluoronaphthyridine. *Antimicrob. Agents Chemother.*, **33**, 906–914.

Grappel, S. F., Giovenella, A. J., Phillips, L. (1985). Antimicrobial activity of aridicins, novel glycopeptide antibiotics with high and prolonged levels in blood. *Antimicrob. Agents Chemother.*, **28**, 660–662.

Imada, T., Miyazaki, S., Nishida, M., Yamaguchi, K., Goto, S. (1992). *In vitro* and *in vivo* antibacterial activities of a new quinolone, OPC-17116. *Antimicrob. Agents Chemother.*, **36**, 573–579.

Iwahi, T., Okonogi, K., Yamazaki, T. *et al.* (1992). *In vitro* and *in vivo* activities of SCE-2787, a new parenteral cephalosporin with a broad antibacterial spectrum. *Antimicrob. Agents Chemother.*, **36**, 1358–1366.

Joó, I., Zsidai, J. (1982). A comparative study of gastric mucin and commercial bakers yeast as virulence-enhancing agents in the typhoid mouse protection test. *J. Biol. Stand.*, **10**, 17–24.

Kindborg, A. (1905). Die Pneumokokken. Vergleichende Untersuchungen mit besonderer Berucksichtigung der Agglutination. *Zschr. Hygiene*, **51**, 197–232.

Knudsen, J. D., Frimodt-Møller, N., Espersen, F. (1995). Experimental *Streptococcus pneumoniae* infection in mice for studying correlation of *in vitro* and *in vivo* activities of penicillin against pneumococci with various susceptibilities to penicillin. *Antimicrob. Agents Chemother.*, **39**, 1253–1258.

Knudsen, J. D., Frimodt-Møller, N., Espersen, F. (1998). Pharmacodynamics of penicillin in relation to bacterial growth phases of *Streptococcus pneumoniae* in the mouse peritonitis model. *J. Antimicrob. Chemother.*, **41**, 451–459.

Kristiansen, J. E., Sebbesen, O., Frimodt-Møller, N., Jørgensen, T. A., Hvidberg, E. (1988). Antimicrobial action of trans-E-clopenthixol in mice including synergistic effect with penicillin. *Acta Pathol. Microbiol. Immunol. Scand. Sect. B.*, **96**, 1079–1084.

Lang, R., Shasha, B., Rubinstein, E. (1993). Therapy of experimental murine brucellosis with streptomycin alone and in combination with ciprofloxacin, doxycycline and rifampin. *Antimicrob. Agents Chemother.*, **37**, 2333–2336.

Malaviya, R., Ross, E., Jakschik, B. A., Abraham, S. N. (1994). Mast cell degranulation by type 1 fimbriated *Escherichia coli* in mice. *J. Clin. Invest.*, **93**, 1645–1653.

Marchou, B., Michea-Hamzehpour, M., Lucain, C., Pechere, J.-C. (1987). Development of β-lactam resistant *Enterobacter cloacae* in mice. *J. Infect. Dis.*, **156**, 369–373.

Matsumoto, M., Matsubara, S., Yokota, T. (1991). Effect of combination therapy with recombinant human granulocyte colony-stimulating factor (rG-CSF) and antibiotics in neutropenic mice unresponsive to antibiotics alone. *J. Antimicrob. Chemother.*, **28**, 447–453.

Matsushita, T., Okuno, S., Maezawa, I. *et al.* (1989). A new model of bacterial peritonitis in mice for evaluation of antibiotics. Effects of aspoxicillin and piperacillin. *Jpn J. Antibiot.*, **42**, 887–895.

Mattie, H., Goslings, R. O., Noach, E. L. (1973). Cloxacillin and nafcillin: serum binding and its relationship to antibacterial effect in mice. *J. Infect. Dis.*, **128**, 170–177.

Miraglia, G. J., Renz, K. J., Gadebush, H. H. (1973). Comparison of the chemotherapeutic and pharmacodynamic activities of cephradine, cephalothin, and cephaloridine in mice. *Antimicrob. Agents Chemother.*, **3**, 270–273.

Mørch, E. (1943). Serologic studies on the pneumococci. Einar Munksgaard, Copenhagen/Humphrey Milford, Oxford University Press, London.

Morgenroth, J., Levy, R. (1911). Chemotherapie der Pneumokokkeninfection. *Berlin. Klin. Wschr.*, **34**, 1560–1561.

Nungester, W. J., Wolf, A. A., Jourdonais, L. F. (1932). Effect of gastric mucin on virulence of bacteria in intraperitoneal injections in the mouse. *Proc. Soc. Exp. Biol. Med.*, **30**, 120–121.

Onyeji, C. O., Nicolau, D. P., Nightingale, C. H., Quintiliani, R. (1994). Optimal times above MIC of ceftibuten and cefaclor in experimental intra-abdominal infections. *Antimicrob. Agents Chemother.*, **38**, 1112–1117.

O'Reilly, T., Cleeland, R., Squires, E. L. (1996). Evaluation of antimicrobials in experimental animal infections. In *Antibiotics in Laboratory Medicine*, 4th edn (ed. Lorian, V.), pp. 604–765. Williams and Wilkins, Baltimore.

Pechere, J. C., Vladoianu, I. R. (1992). Development of resistance during ceftazidime and cefepime therapy in a murine peritonitis model. *J. Antimicrob. Chemother.*, **29**, 563–573.

Pechere, J. C., Marchou, B., Michea-Hamzehpour, M., Auckenthaler, R. (1986). Emergence of resistance after therapy with antibiotics used alone or combined in a murine model. *J. Antimicrob. Chemother.*, **17** (Suppl. A), 11–18.

Rolinson, G. N. (1974). Laboratory evaluation of amoxicillin. *J. Infect. Dis.*, **129** (Suppl.), S139–S145.

Rolinson, G. N. (1980). The significance of protein binding of antibiotics in antimicrobial chemotherapy. *J. Antimicrob. Chemother.*, **6**, 311–317.

Santo, E. D., Alonzi, T., Fattori, E. *et al.* (1996). Overexpression of interleukin-6 in the central nervous system of transgenic mice increases central but not systemic proinflammatory cytokine production. *Brain Res.*, **740**, 239–244.

Schaad, U. B., Grimm, L. M., Beskid, G., Cleeland, R., Nelson, J. D., McCracken, G. H. (1982). Mecillinam alone and in combination with ampicillin or moxalactam in experimental *Escherichia coli* meningitis. *Infection*, **10**, 90–96.

Soriano, F., Garcia-Corbeira, P., Ponte, C., Fernandez-Roblas, R., Gadea, I. (1996). Correlation of pharmacodynamic parameters of five β-lactam antibiotics with therapeutic efficacies in an animal model. *Antimicrob. Agents Chemother.*, **40**, 2686–2690.

Sullivan, M. C., Cooper, B. W., Nightingale, C. H., Quintiliani, R., Lawlor, M. T. (1993). Evaluation of the efficacy of ciprofloxacin against *Streptococcus pneumoniae* by using a mouse protection model. *Antimicrob. Agents Chemother.*, **37**, 234–239.

Swenson, C. E., Stewart, K. A., Hammett, J. L., Fitzsimmons, W. E., Ginsberg, R. S. (1990). Pharmacokinetics and *in vivo* activity of liposome-encapsulated gentamicin. *Antimicrob. Agents Chemother.*, **34**, 235–240.

Takashima, K., Tateda, K., Matsumoto, T., Izawa, Y., Nakao, M., Yamaguchi, K. (1997). Role of tumor necrosis factor alpha in pathogenesis of pneumococcal pneumonia in mice. *Antimicrob. Agents Chemother.*, **65**, 257–260.

Teale, D. M., Atkinson, A.M. (1992). Inhibition of nitric oxide synthesis improves survival in a murine peritonitis model of sepsis that is not cured by antibiotics alone. *J. Antimicrob. Chemother.*, **30**, 839–842.

Tokunaga, T., Oka, K., Takemoto, A., Ohtsubo, Y., Gotoh, N., Nishino, T. (1997). Efficacy of trimethoprim in murine experimental infection with a thymidine kinase-deficient mutant of *Escherichia coli*. *Antimicrob. Agents Chemother.*, **41**, 1042–1045.

Tsuchiya, K., Kondo, M., Keda, M. *et al.* (1981). Cefmenoxime (SCE-1365), a novel broad-spectrum cephalosporin; *in vitro* and *in vivo* antibacterial activities. *Antimicrob. Agents Chemother.*, **19**, 56–65.

Chapter 15

Murine Thigh Infection Model

S. Gudmundsson and H. Erlendsdóttir

Background of model

While several different animal models have been used to examine pharmacokinetic–pharmacodynamic interactions against a wide variety of micro-organisms (O'Reilly *et al.*, 1996), considerable experience has been accumulated with the versatile and simple murine thigh infection model. The model provides a sensitive experimental infection for evaluation of antimicrobial efficacy; it allows for easily reproducible determination of antimicrobial–microbe interactions and the simultaneous measurement of antimicrobial pharmacokinetics in the serum and/or tissue of the infected animal (Craig and Gudmundsson, 1996; O'Reilly *et al.*, 1996).

The model was first described by Eagle and co-workers (Eagle *et al.*, 1950) who in the late 1940s used a murine thigh infection to examine the effect of penicillin G in a single dose on the growth of various streptococci. Selbie and O'Grady (1954) used a similar model a few years later for measurable tuberculosis lesions in the mouse thigh. Rolinson and co-workers (Hunter *et al.*, 1973; Rolinson, 1973) used the same model to study the bactericidal efficacy and postantibiotic effect (PAE) of broad-spectrum penicillins against various Gram-negative bacilli. Normal mice were used in these early studies.

The model was modified by Craig and associates in the 1970s by rendering the mice neutropenic (Gerber *et al.*, 1983; Vogelman *et al.*, 1988a). It has since been used by several investigators for further study of the *in-vivo* PAE and to examine various pharmacokinetic and -dynamic parameters and their correlation with efficacy against a number of different organisms, both when used singly and in combinations for 24 hours (Kunst and Mattie, 1978; Haller, 1987; Vogelman *et al.*, 1988b; Leggett *et al.*, 1989; Valdimarsdóttir *et al.*, 1994).

Animal species

Most investigators have used Swiss albino mice, 23–27 g female ICR (Gerber *et al.*, 1983; Gudmundsson *et al.*, 1986; Vogelman *et al.*, 1988a,b), 25–30 g female NMRI (Valdimarsdóttir *et al.*, 1994), 20–25 g male ddy (Oshida *et al.*, 1990), 23–26 g male CF1 (Haller, 1987) or 23–27 g female BALB/c (Minguez *et al.*, 1992). Any strain of Swiss mice can probably be used in the neutropenic model, but a pilot study to determine the growth rate of untreated organisms and the optimal inoculum is advisable.

Preparation of animals

No specialized housing or care are needed. No special preparation or surgical procedure is required prior to inoculation.

While early studies in this model and occasional studies performed recently were done in normal (non-immunocompromised) animals (Craig and Gudmundsson, 1996), most of the experience with this model has been obtained in neutropenic mice. The mice are rendered neutropenic by 150 mg/kg and 100 mg/kg of cyclophosphamide injected intraperitoneally on days 0 and 3 respectively. This produces severe neutropenia by day 4 (the day of the experiment), which lasts for 2–3 days (Figure 15.1).

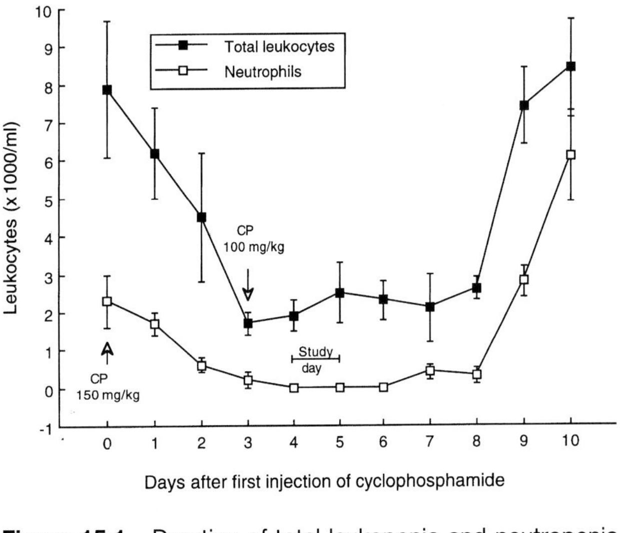

Figure 15.1 Duration of total leukopenia and neutropenia in female ICR mice following i.p. injection of 150 and 100 mg/kg of cyclophosphamide on days 0 and 3 respectively. Mean values ± SD for six to 10 mice are represented by points and bars.

Handbook of Animal Models of Infection
ISBN 0-12-775390-7

Total leukocyte counts generally remain at 1000/mm^3 and neutrophils below 50–100/mm^3 (Gerber *et al.*, 1983; Craig and Gudmundsson, 1996). It is advisable while using the model for the first time in a new laboratory to perform total white blood cell and differential counts daily on six to eight mice over a 10-day period following cyclophosphamide injection, and subsequently on day 4 (day of experiment) in every fourth or fifth experiment.

In order to simulate human pharmacokinetics in the mouse, the half-life of some antimicrobials can be prolonged by inducing transient renal impairment in the mice administering 10 mg/kg subcutaneously of uranyl nitrite 72 hours prior to the experiment (Giacomini *et al.*, 1981; Fantin *et al.*, 1990; Craig *et al.*, 1991; Craig and Gudmundsson, 1996).

Preparation of inocula

Several different species of micro-organism have been studied in this model, summarized in Table 15.1.

Table 15.1 Micro-organisms studied in the neutropenic murine thigh model

Staphylococcus aureus	*Escherichia coli*
Group A streptococci	*Klebsiella pneumoniae*
Streptococcus pneumoniae	*Enterobacter cloacae*
Haemophilus influenzae	*Serratia marcescens*
Mycobacterium tuberculosis	*Pseudomonas aeruginosa*

Both standard ATCC or NTCC and clinical strains from various sources have been employed.

MICs/MBCs are determined by standard methods, either tube dilution, microtitre dilution or E-test®. Mueller–Hinton broth can be used for *Staphylococcus*, Enterobacteriaceae and *Pseudomonas* and Todd–Hewitt or Schaedler's broth for *Streptococcus*. For pneumococci we have used hearth infusion broth supplemented with 10% horse serum. Mueller–Hinton broth is supplemented with Ca^{2+} (50 mg/l) and Mg^{2+} (20 mg/l) when aminoglycosides are used against Enterobacteriaceae and *Pseudomonas* (Stratton and Reller, 1977). Study organisms are grown in an appropriate broth (generally for 5–7 hours) into log phase at an OD_{580} of 0.30 or to a 0.5 McFarland standard.

Infection process

After 1:10 dilution into warm broth, 0.1 ml of the log growth suspension (~ 10^5–10^6 cfu) is injected into each thigh, while the mice are under light anaesthesia with ether. Thus, this part of the procedure needs to be performed under a ventilation hood. While the animals are asleep, they are held upside down by the paw with the forefinger and thumb, steadied by the palm of the hand and the rest of the fingers and the injection is made obliquely from above posterolaterally into the muscles of the thigh close to the femur. The injection is best performed with a 1 ml syringe with a 25G or 27G needle. Both thighs can be used. In certain experiments two different organisms can simultaneously be injected into opposite thighs.

Antimicrobial therapy

All antimicrobial classes in current use have been evaluated in the neutropenic murine thigh infection model, including penicillins, cephalosporins, carbapenems, glycopeptides, aminoglycosides, fluoroquinolones, macrolides, tetracyclines, lincosamines, rifamycins and streptogramins (Gerber *et al.*, 1983, 1986; Gudmundsson *et al.*, 1986; Haller, 1987; Vogelman *et al.*, 1988a,b; Leggett *et al.*, 1989; Fantin *et al.*, 1990; Oshida *et al.*, 1990; Minguez *et al.*, 1992; Caminero *et al.*, 1993; Craig and Ebert, 1993; Gudmundsson *et al.*, 1993). Most studies in this model have focused either on determination of the *in vivo* PAE following a single dose of antibiotic or on determination of various pharmacokinetic and pharmacodynamic parameters and their correlation with efficacy against a number of different organisms, both when used singly and in combinations for 24 hours. Thus, method descriptions for both types of experiments will follow in some detail.

Drug kinetics and microbiological assay

Single-dose antimicrobial pharmacokinetics need to be determined either before or simultaneously with treatment. Up to four groups of four mice each are sampled at intervals of 5–30 minutes over a 5–8 hour period. Animals are anaesthetized with ether, and blood is drawn from the retro-orbital sinus into 100 µl heparinized microhaematocrit tubes. After centrifugation, the tubes can be stored at –70°C until drug levels are determined. We have used standard microbiological assays, but other methods can be employed as well (Klassen and Edberg, 1991). The β- and γ-elimination rate constants, half-lives, the interval that serum levels exceed the MIC and the area under the concentration–time curve (AUC) are calculated from serum concentrations assuming a one-compartment model. Examples of single-dose pharmacokinetics of several antimicrobial agents are provided in Table 15.2.

Antimicrobial administration in determination of *in vivo* PAE

After allowing the bacteria to grow for 2 hours in the thigh, a single injection of the antimicrobial under study in a volume of 0.2 ml is administered subcutaneously to

Table 15.2 Single-dose pharmacokinetic parameters for antimicrobial doses used in the neutropenic murine thigh model (adapted from Vogelman *et al.*, 1988a; Leggett *et al.*, 1989; Gudmundsson *et al.*, 1993)

Antimicrobial agent	*Dose (mg/kg)*	*Peak level/dose (μg ml⁻¹/mg kg⁻¹)*	*Half-life (minutes)*	*AUC/dose (μg hours ml⁻¹/mg kg⁻¹)*
β-lactam				
Penicillin G	1–100	0.6–1.1	12.1–16.1	0.3–0.5
Ampicillin	2–25	0.9–0.9	12.0	0.3–0.4
Cefazolin	3.125–600	0.8–1.4	14.1–15.5	0.6–0.8
Ceftriaxone	1–100	2.4–4.1	52.2–65.3	3.0–8.4
Ceftazidime	37.5	1.3	20.9	0.8
Cefoperazone	40–1200	0.7–1.0	10.6–22.6	0.4–0.9
Ticarcillin	200–1200	0.7–1.1	17.0–17.5	0.5–0.7
	2400–3200	0.8–1.0	22.4–23.0	0.7–0.9
Imipenem	50–200	1.0–1.1	12.6–13.1	0.4–0.4
Aminoglycoside				
Gentamicin	4–8	1.2–1.3	14.9–17.3	0.7–0.7
Tobramycin	2–32	1.3–1.4	15.6	0.7–0.8
Netilmicin	6–24	0.9–1.4	18.4–18.8	0.5–0.8
Other				
Erythromycin	4–8	0.3–0.4	21.6–22.6	0.2–0.2
	100	0.2	44.5	0.3
Tetracycline	1–8	0.9	30.6	1.0
Vancomycin	24	1.4	26.5	1.2
Rifampicin	1.0	0.8	232	0.4
	25–75	0.6–0.7	335-378	0.2–0.3
Ciprofloxacin	3.12–50	1.0–1.5	37.5–41.5	0.8–0.9

approximately 60–70% of the mice in each experiment. The remaining mice serve as untreated controls. The time of injection is defined as time zero of the experiment. The dose of drug to be selected should, based on the previously performed pharmacokinetic studies, provide serum concentrations above the MIC of the test organism for 1–2 hours. Drug levels may also be determined from serum samples obtained at the time of sacrifice of each mouse during the experiment.

Antimicrobial administration in dosing studies

For studies of the effect of different dosing regimens of antimicrobials and the correlation of various pharmacokinetic and -dynamic parameters with maximum efficacy, mice have been treated with multiple dosing regimes over a 24-hour period (Vogelman *et al.*, 1988b; Leggett *et al.*, 1989; Valdimarsdóttir *et al.*, 1994). The regimens may be selected as follows. For each drug–organism combination four to eight 24 hour total doses of the antimicrobial(s) in question are selected and subsequently divided into individual doses to be administered at 1-, 2-, 3-, 4-, 6-, 8-, 12- and 24-hour intervals. The range of individual doses used should be wide. Examples of range of total 24 hour doses of some antimicrobials studied as single agents in this model include: penicillin, 0.5–384 mg/kg; erythromycin, 0.5–96 mg/kg; cefazolin, 3–4800 mg/kg; ticarcillin, 25–4800 mg/kg; ceftazidime, 0.5–1200 mg/kg; imipenem, 0.5–600 mg/kg; gentamicin, 0.25–256 mg/kg; tobramycin, 1–192 mg/kg; and netilmycin, 0.5–96 mg/kg (Vogelman *et al.*, 1988b; Leggett *et al.*, 1989).

This method of selecting dosing regimens allows one to minimize the interdependence between the pharmacokinetic parameters studied and to describe the complete dose–response relationship by achieving serum drug concentrations similar to, above and below those seen in humans. Generally 20–50 dosing regimens are studied for each drug–organism combination. This can both be done with combinations of drugs as well (Valdimarsdóttir *et al.*, 1994). Table 15.3 illustrates a possible experimental design for dosing schedules of two drugs in combination against a particular organism.

The agents are diluted to the desired concentrations by using the manufacturer's recommendations. They are administered subcutaneously in 0.2 ml volumes starting 2 hours after organism inoculation; administration continued for the entire 24 hour treatment period.

Key parameters to monitor infection and response to treatment

In the neutropenic animal, infection in the muscle is diffuse and organisms on Brown–Brenn tissue staining are distributed relatively evenly between muscle bundles without

Table 15.3 Example of experimental design for administration of same total dose of antimicrobial agents singly or in combination by different individual dosages and dosage intervals for 24 hours in the neutropenic murine thigh infection model (TD, total 24 hour dose of each drug; $TD_{A1} = TD_{A3} = TD_{A6} = TD_{A12} = TD_{A24}$ and $TD_{B1} = TD_{B3} = TD_{B6} = TD_{B12} = TD_{B24}$; *q1*, *q3*, *q6*, *q12*, *q24*, drug administered every 1, 3, 6, 12 or 24 hours; * represents two mice – four thighs)

		Drug A						
		o	***q1***	***q3***	***q6***	***q12***	***q24***	
	o	*	*	*	*	*	*	No drug B
	q3	*	*	*	*	*	*	TD_{B3}
Drug B	***q6***	*	*	*	*	*	*	TD_{B6}
	q12	*	*	*	*	*	*	TD_{B12}
	q24	*	*	*	*	*	*	TD_{B24}
		No drug A	TD_{A1}	TD_{A3}	TD_{A6}	TD_{A12}	TD_{A24}	

forming microcolonies (Erlendsdóttir and Gudmundsson, unpublished observations, 1997). Most organisms have attained a logarithmic growth 2 hours after inoculation (Figure 15.2), whereupon treatment is initiated. In non-neutropenic animals, growth may be slower and less predictable. A growth curve should be determined for each organism to be studied prior to planning and initiation of experiments.

The neutropenic animal becomes septic rather quickly and control mice generally die from sepsis within 18–36 hours, depending on organism. Thus, survival is a suitable end-point for treatment studies, but in this model the almost universal parameter used to follow the infection process has been viable bacterial counts per thigh (Craig and Gudmundsson, 1996). The results are reproducible and accurate, and mean viable counts for each group of four thighs usually have had consistently low SD values (0.1–0.5 log cfu per thigh).

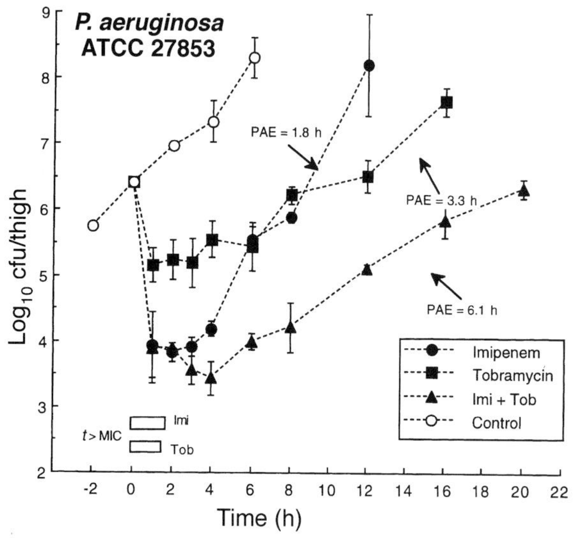

Figure 15.2 Growth curves of *P. aeruginosa* ATCC 27853 in the thighs of mice treated with a single dose of imipenem (Imi, 200 mg/kg) and tobramycin (Tob, 8.0 mg/kg), singly and in combination, and in untreated control mice. The open horizontal boxes denote the time that serum drug levels exceeded the MIC ($t > MIC$). Data points are mean ± SD (vertical error bars) values from four thighs.

Earlier, some investigators observed a correlation between the thigh diameter and the number of bacteria in the thigh (Selbie and O'Grady, 1954; Hunter *et al.*, 1973). However, we have not found a good correlation in neutropenic mice (Craig and Gudmundsson, 1996). Other parameters, including appearance, body weight, histopathology, blood cell responses, cytokine response, etc., have not been investigated in this model.

Animal sacrifice

In studies of regrowth after single doses of antibiotics (*in-vivo* PAE), treated mice (two to four at each time period) are sacrificed by cervical dislocation as frequently as every hour for the first 4 hours and every 2–4 hours until 16 hours (and occasionally 24 hours) after antimicrobial dosing. Untreated control animals should be sacrificed at time of inoculation, 2 hours later at time of initiation of therapy, and at intervals of 2–4 hours thereafter up to 6–8 hours.

In studies of the effect of 24 hour antimicrobial treatment, two control mice are sacrificed for organism quantification just after organism inoculation; 2 hours later, just before drug treatment; and 24 hours after the onset of therapy. Two mice for each dosing regimen are sacrificed at the end of 24 hours of therapy.

Processing of specimens

The thighs are removed at each sampling time, and the muscle is cut from the femur and pelvic bones. The muscle is immediately homogenized in ice-cold 0.85% NaCl in a tissue homogenizer in a volume up to 10 ml. In experiments with β-lactams 10–50 μl of penicillinase (10×10^6 U/ml) is added to the homogenate. Viable counts are determined on Müller–Hinton agar for staphylococci, Enterobacteriaceae and *Pseudomonas* and horse-blood agar for streptococci by plating in duplicate 10 µl of serially 10-fold diluted samples of homogenate.

Expression of results

The counts of cfu are generally expressed per thigh (Hunter *et al.*, 1973; Gerber *et al.*, 1983; Vogelman *et al.*, 1988a). For determination of PAE (Craig and Gudmundsson, 1996) the counts of cfu/thigh (mean of three to four mice) at each time point are then graphed and the duration of PAE is calculated by the following equation:

$$PAE = T - C - M$$

where M represents the time serum levels exceed the MIC; T is the time required for the counts of cfu in thighs of treated mice to increase 1 $\log_{10}$ above the count closest to but not less than time M; and C is the time required for the counts of cfu in thighs of untreated control mice to increase 1 $\log_{10}$ above the count at time zero. An example of a typical experiment involving PAE following combinations of two agents is shown in Figure 15.2.

In this model efficacy of an antimicrobial agent can be defined as the change in $\log_{10}$ number of cfu per thigh over the 24 hour treatment period and is calculated by subtracting the mean $\log_{10}$ number of cfu per thigh from control mice just before therapy from the mean number of cfu from the two thighs of each treated mouse at the end of therapy (Vogelman *et al.*, 1988b; Leggett *et al.*, 1989; Valdimarsdóttir *et al.*, 1994). The pharmacokinetic parameters studied (including duration of time that serum levels exceeded the MIC or MBC expressed as a percentage of the 24 hour treatment period, $\log_{10}$ peak serum level, $\log_{10}$ AUC, etc) can then be correlated with efficacy by using univariate and stepwise multilinear regression analysis (Vogelman *et al.*, 1988b). Other indicators of efficacy can be calculated, such as E_{max}, a measure of relative efficacy indicated by the maximum antimicrobial effect attributable to the drug, and P_{50}, a measure of potency indicated by the dose producing 50% of E_{max} (Leggett *et al.*, 1989; Craig and Ebert, 1993). *In-vivo* synergy/antagonism has been studied in this model (Gerber *et al.*, 1983; Fantin and Carbon, 1992; Gudmundsson *et al.*, 1993) and induction of resistant variants of bacteria following antimicrobial exposure (Gerber *et al.*, 1982).

Advantages and disadvantages of the model

The major disadvantages of this model would include the facts that the mice need to be anaesthetized for inoculation and that the model does not represent a 'natural' infection: pyomyositis is a very rare infection indeed (Adams *et al.*, 1985).

On the other hand, the model has several advantages (Craig and Gudmundsson, 1996). It is very simple and versatile. Mice are easy to handle and are relatively inexpensive. The model is very reproducible, the inoculum is accurate with low variability and, therefore, relatively few mice are needed for each data point.

Marked granulocytopenia is reliably induced and thus permits study of the interactions between micro-organism and antimicrobial agent without the interference of neutrophils or an abscess. However, the model can also be used in non-neutropenic mice.

Thigh muscles are easily inoculated and removed from the femur and pelvis. As an end-point most investigators have used viable bacterial counts; the number of bacteria is easily quantitated in thigh homogenates, and the results are generally easily reproducible. Other end-points, including survival, can furthermore be included.

The rapid elimination of antimicrobial agents in mice permits short exposure periods after a single dose when determining *in-vivo* PAE in this model. Human pharmacokinetics can be simulated in this model, both by repeated injections (Gerber *et al.*, 1986) and by inducing transient renal impairment in the mice (Fantin *et al.*, 1990; Craig *et al.*, 1991).

Although antimicrobial concentrations can be determined in thigh homogenates, it is unlikely that they represent the exact concentration at the site of infection. Ryan and Cars (1980) compared tissue concentrations in whole muscle with those in muscle tissue fluid obtained by implantation of cotton threads under muscle fascia. The muscle concentrations of two β-lactams were only 18–22% of the corresponding tissue fluid level. Moreover, the drug concentrations in muscle interstitial fluid closely approximated those observed in serum with only a short time lag. In their early studies, Eagle *et al.* (1950) used different doses of procaine penicillin to determine the serum concentration having the minimum bactericidal effect in muscle tissue. The value obtained for several different streptococci was virtually identical to the concentration that was minimally lethal *in vitro*. Thus, it appears that serum concentrations can be used to estimate relatively accurately the antimicrobial concentration in the interstitial fluid of muscle.

Contribution of the model to infectious disease therapy

The value of animal models in evaluating optimal antimicrobial therapy is unquestioned (O'Reilly *et al.*, 1996), but perhaps often overlooked. Animal models, particularly simple ones like the neutropenic mouse thigh infection model, offer the possibility of standardizing both the host factors that may affect outcome of infection and treatment as well as the offending organism and the therapy rendered. Animal models are suitable for investigation of complex situations that are difficult to mimic in test tubes and impossible to control and consistently reproduce in clinical trials (Calandra and Glauser, 1986). Recent guidelines on the clinical evaluation of antimicrobials issued by the Infectious Disease Society of America in conjunction with the US Food and Drug Administration emphasize the need for animal model evaluation of new compounds or new therapeutic modalities as a prerequisite for clinical trials (Beam *et al.*, 1992).

Dosing of antimicrobials

The major studies performed in the neutropenic murine thigh model during the past decade have involved determination and standardization of the PAE *in vivo* and determination of the association of various pharmacological variables with maximum efficacy of the drug under study (Gudmundsson *et al.*, 1986; Vogelman *et al.*, 1988a,b; Gudmundsson *et al.*, 1993; Craig and Gudmundsson, 1996).

These studies have confirmed the *in-vitro* data for the PAE, i.e. that cell-wall-active antibiotics (all four classes of β-lactams, vancomycin) generally induce a significant PAE against Gram-positive organisms, but none or very short PAE against Gram-negative bacilli, with the exception of penicillin failing to induce a PAE against pneumococci in this model. Drugs that inhibit protein or nucleic acid synthesis (aminoglycosides, trimethoprim, tetracyclines, clindamycin, erythromycin, chloramphenicol, 6-fluoroquinolones) induce a significant PAE against both Gram-positive cocci and Gram-negative bacilli. A notable exception to the general absence of a PAE after exposure of Gram-negative bacilli to β-lactams is the induction of a clinically significant PAE by imipenem and other carbapenems against *P. aeruginosa* (Vogelman *et al.*, 1988a; Craig and Gudmundsson, 1996).

These data on the PAE provide theoretical rationale for either intermittent or more continuous dosing of antimicrobials, depending on the infecting organism and the drug to be used. Intermittent dosing regimens would apply primarily to drug–organism combinations that exhibit a prolonged PAE. On the other hand, more continuous dosing would be necessary for drug–organism combinations that lack a PAE.

This hypothesis has been confirmed by employing the neutropenic mouse thigh model with the methodology described above (Vogelman *et al.*, 1988b; Leggett *et al.*, 1989; Craig and Gudmundsson, 1996). Thus, it was demonstrated that, for drugs inducing a significant PAE, serum levels needed not be maintained constantly above the MIC to achieve maximal efficacy. The major pharmacokinetic parameter predicting efficacy for such agents, e.g. aminoglycosides against Gram-negative bacilli, is the area under the concentration–time curve (AUC) and the optimal dosage interval is primarily dependent on the duration of the PAE, thus indicating that the optimal dosing regimens in terms of efficacy, cost and toxicity for these agents could include long intervals.

On the other hand, the β-lactams do not induce a PAE against Gram-negative bacilli and furthermore exhibit a bactericidal activity that is time- but not concentration-dependent. For such drugs the major pharmacokinetic parameter predicting efficacy and determining the optimal dosage interval was demonstrated to be the time that drug concentration exceeds the MIC, implying that the most efficacious drug administration in this case involves short dosage intervals or even by continuous administration.

Implication for therapy in humans

These observations have been of significant importance for studies on once-daily dosing of aminoglycosides. There are now dozens of clinical trials showing equal efficacy for once-daily versus two- or three-times-a-day dosing of aminoglycosides, either alone or in combination with a β-lactam antibiotic (Gilbert, 1991; Hatala *et al.*, 1996; Munckhof *et al.*, 1996). Although most studies have also exhibited similar toxicity, some trials have suggested a lower frequency of adverse reactions with a single daily dose of an aminoglycoside than with three-times-a-day dosing of the drug (Prins *et al.*, 1993). These observations have furthermore been extended to febrile neutropenic patients (International Therapy Cooperative Group of the EORTC, 1993).

In contrast to the current extensive information on once-daily dosing of aminoglycosides, very limited information is available on the clinical efficacy of continuous infusion of antimicrobials (particularly β-lactams) vs. intermittent dosing or the value of maintaining serum levels above the MIC for the majority of the dosing interval (Craig and Ebert, 1992; Mouton and Vinks, 1996). However, in a study performed approximately 20 years ago, Bodey and associates (1979) compared the efficacy of intermittent vs. continuous administration of cefamandole in combination with intermittent dosing of carbenicillin in cancer patients. A significant benefit for continuous administration of the β-lactam antibiotic could be demonstrated in patient subgroups, such as those infected with cefamandole-sensitive organisms and those with marked neutropenia that did not improve during therapy. Similarly, Shentag and co-workers (1984) demonstrated the association between efficacy of cefmenoxime in lower respiratory tract infections in intubated patients and the time serum drug levels remained above the MIC of the offending organism.

Antimicrobial combinations

In the neutropenic murine thigh model, β-lactam–aminoglycoside combinations induce a prolonged PAE *in vivo* against Gram-negative bacilli as compared to the PAEs of the agents alone (Gudmundsson *et al.*, 1993). Preliminary dosing experiments have been conducted in this model (Valdimarsdóttir *et al.*, 1994), where ceftazidime and imipenem were administered at different dosing frequencies, singly or in combination with tobramycin, against a strain of *P. aeruginosa*. Dosing regimens of the two ß-lactam–aminoglycoside combination that produced comparable efficacy allowed for longer dosing interval for imipenem (q12–q24 hours) than for ceftazidime (q3–q6 hours). This is probably associated with the enhancement of PAE by imipenem–tobramycin (both PAE-inducing agents) not observed with ceftazidime–tobramycin.

These observations indicate that different dosing schedules may be appropriate in drug combinations as compared to the drugs as single agents. Dosage interval may possibly be prolonged and total dose reduced without loss of efficacy. However, human studies to this effect have not been performed.

Concluding remarks

The model is very suitable for screening of new antimicrobial agents, particularly in regards to the association of antibacterial effect with the pharmacological variables prior to clinical trials (Craig and Ebert, 1993). Experience with aminoglycosides attests to the predictive ability of this model. Furthermore, it is simple, accurate and cheap. It needs neither special housing (except for standard animal facilities) nor special equipment, and is thus suitable for practically every researcher in the field of microbiology and infectious diseases.

References

Adams, E., Gudmundsson, S., Yocum, D. E., Haselby, R. C., Craig, W. A., Sundstrom, W. R. (1985). Streptococcal myositis. *Arch. Intern. Med.*, **145**, 1020–1023.

Beam, T. R., Gilbert, D. N., Kunin C. M. (1992). General guidelines for the clinical evaluation of anti-infective drug products. *Clin. Infect. Dis.*, **15** (Suppl. 1), S5–S32.

Bodey, G. P., Ketchel, S. J., Rodriguez, V. (1979). A randomized study of carbenicillin plus cefamandole or tobramycin in the treatment of febrile episodes in cancer patients. *Am. J. Med.*, **67**, 608–616.

Calandra, T., Glauser, M.P. (1986). Immunocompromised animal models for the study of antibiotic combinations. *Am. J. Med.*, **80** (Suppl. 5C), 45–52.

Caminero, M. M., Martinez, F. F., Izquierdo, J.I., Gómez-Lus Centelles, M. L., Prieto, J. (1993). *In-vivo* and *in-vitro* study of the postantibiotic effect of meropenem. *J. Antimicrob. Chemother.*, **32**, 917–918.

Craig, W. A., Ebert, S. C. (1992). Continuous infusion of ß-lactam antibiotics. *Antimicrob. Agents Chemother.*, **36**, 2577–2583.

Craig, W., Ebert, S. (1993). Pharmacodynamic activities of RP59500 in an animal infection model. Abstract 205. In *Programs and Abstracts of the Thirty-Third Interscience Conference on Antimicrobial Agents and Chemotherapy, New Orleans 1993*. American Society for Microbiology, Washington, DC.

Craig, W. A., Gudmundsson, S. (1996). The postantibiotic effect. In *Antibiotics in Laboratory Medicine*, 4th edn, (ed. Lorian, V.), pp. 296–329. Williams and Wilkins, Baltimore, MD.

Craig, W. A., Redington, J., Ebert, S. (1991). Pharmacodynamics of amikacin *in vitro* and in mouse thigh and lung infections. *J. Antimicrob. Chemother.*, **27** (Suppl. C), 29–40.

Eagle, H., Fleischman, R., Musselman, A. D. (1950). The bactericidal action of penicillin *in vivo:* the participation of the host, and the slow recovery of the surviving organisms. *Ann. Intern. Med.*, **33**, 544–571.

Fantin, B., Carbon, C. (1992). *In vivo* antibiotic synergism: contribution of animal models. *Antimicrob. Agents Chemother.*, **36**, 907–912.

Fantin, B., Ebert, S., Leggett, J., Vogelman, B., Craig, W.A. (1990). Factors affecting the duration of *in-vivo* postantibiotic effect for aminoglycosides against Gram-negative bacilli. *J. Antimicrob. Chemother.*, **27**, 829–836.

Gerber, A. U., Wiprachtiger, P., Stettler-Spichiger, U., Lebek, G. (1982). Constant infusions vs. intermittent doses of gentamicin against *Pseudomonas aeruginosa in vitro. J. Infect. Dis.*, **145**, 554–560.

Gerber, A. U., Craig, W. A., Brugger, H.–P., Feller, C., Vastola, A. P., Brandel, J. (1983). Impact of dosing intervals on activity of gentamicin and ticarcillin against *Pseudomonas aeruginosa* in granulocytopenic mice. *J. Infect. Dis.*, **147**, 910–917.

Gerber, A. U., Brugger, H. P., Feller, C., Stritzko, T., Stalder, B. (1986). Antibiotic therapy of infections due to *Pseudomonas aeruginosa* in normal and granulocytopenic mice: comparison of murine and human pharmacokinetics. *J. Infect. Dis.*, **153**, 90–97.

Giacomini, K. M., Roberts, S. M., Levy, G. (1981). Evaluation of methods for producing renal dysfunction in rats. *J. Pharmaceut. Sci.*, **70**, 117–121.

Gilbert, D. N. (1991). Once-daily aminoglycoside therapy. *Antimicrob. Agents Chemother.*, **35**, 399–405.

Gudmundsson, S., Vogelman, B., Craig, W. A. (1986). The *in vivo* postantibiotic effect of imipenem and other new antimicrobials. *J. Antimicrob. Chemother.*, **18** (Suppl. E), 67–73.

Gudmundsson, S., Einarsson, S., Erlendsdóttir, H., Moffat, J., Bayer, W., Craig, W. A. (1993). The postantibiotic effect of antimicrobial combinations in a neutropenic murine thigh infection model. *J. Antimicrob. Chemother.*, **31** (Suppl. D), 177–191.

Haller, I. (1987). Evaluation of ciprofloxacin alone and in combination with other antibiotics in a murine model of thigh muscle infection. *Am. J. Med.*, **82** (Suppl 4A), 76–79.

Hatala, R., Dinh T., Cook, D. J. (1996). Once-daily aminoglycoside dosing in immunocompetent adults: a meta-analysis. *Ann. Int. Med.*, **124**, 717–725.

Hunter, P. A., Rolinson, G. N., Witting, D. A. (1973). Comparative activity of amoxycillin and ampicillin in an experimental bacterial infection in mice. *Antimicrob. Agents Chemother.*, **4**, 285–293.

International Therapy Cooperative Group of the EORTC (1993). Efficacy and toxicity of single daily doses of amikacin and ceftriaxone versus multiple daily doses of amikacin and ceftazidime for infection in patients with cancer and granulocytopenia. *Ann. Intern. Med.*, **119**, 584–593.

Klassen, M., Edberg, S. C. (1991). Measurement of antibiotics in human body fluids: techniques and significance. In *Antibiotics in Laboratory Medicine*, 4th edn, (ed. Lorian, V.), pp. 230–295. Williams and Wilkins, Baltimore, MD.

Kunst, M. W., Mattie, H. (1978). Cefazolin and cephradine: relationship between antibacterial activity *in vitro* and in mice experimentally infected with *Escherichia coli*. *J. Infect. Dis.*, **137**, 391–402.

Leggett, J., Fantin, B., Ebert, S. *et al.* (1989). Comparative antibiotic dose–effect relations at several dosing intervals in murine pneumonitis and thigh-infection models. *J. Infect. Dis.*, **159**, 281–292.

Minguez, F. M., Izquierdo, J. I., Caminero, M. M., Martinez, F. F., Prieto, J. (1992). *In vivo* postantibiotic effect of isepamicin and other aminoglycosides in a thigh infection model in neutropenic mice. *Chemotherapy*, **38**, 179–184.

Mouton, J. W., Vinks, A. A. T. M M. (1996). Is continuous infusion of ß-lactam antibiotics worthwhile? —efficacy and pharmacokinetic considerations. *J. Antimicrob. Chemother.*, **38**, 5–15.

Munckhof, W. J., Grayson, M. L., Turnidge, J. D. (1996). A meta-analysis of studies on the safety and efficacy of aminoglycosides given either once-daily or as divided doses. *J. Antimicrob. Chemother.*, **37**, 645–663.

O'Reilly, T., Cleeland, R., Squires, E. L. (1996). Evaluation of antimicrobials in experimental animal infection. In *Antibiotics in Laboratory Medicine*, 4th edn. (ed. Lorian, V.), pp. 604–765. Williams and Wilkins, Baltimore, MD.

Oshida, T., Onta, T., Nakanishi, N., Matsushita, T., Yamaguchi, T. (1990). Activity of sub-minimal inhibitory concentrations of aspoxicillin in prolonging the postantibiotic effect against *Staphylococcus aureus*. *J. Antimicrob. Chemother.*, **26**, 29–38.

Prins, J. M., Büller, H. R., Kuijper, E. J., Tange, R. A., Speelman, P. (1993). Once versus thrice daily gentamicin in patients with serious infections. *Lancet*, **341**, 335–339.

Rolinson, G. N. (1973). Plasma concentrations of penicillin in relation to the antibacterial effect. In *Biological Effects of Drugs in Relation to Their Plasma Concentrations*, (eds Davies, D. S., Prichard, B. N. C.), pp. 183–189. University Park Press, Baltimore, MD.

Ryan, M., Cars, O. (1980). Antibiotic assays in muscle: are conventional tissue levels misleading as indicator of the antibacterial activity? *Scand. J. Infect. Dis.*, **12**, 307–309.

Schentag, J. J., Smith, I. L., Swanson, D. J. *et al.* (1984). Role for dual individualization with cefmenoxime. *Am. J. Med.*, **77** (Suppl. 6A), 43–50.

Selbie, F. R., O'Grady, F. (1954). A measurable tuberculosis lesion in the thigh of the mouse. *Br. J. Exp. Pathol.*, **35**, 556–565.

Stratton, C. W., Reller, L. B. (1977). Serum dilution test for bactericidal activity. I. Selection of a physiologic diluent. *J. Infect. Dis.*, **136**, 187–195.

Valdimarsdóttir, M., Einarsson, S., Erlendsdóttir, H., Gudmundsson, S. (1994). Dosing studies of antimicrobial combinations against *P. aeruginosa* in an animal model. In *Recent Advances in Chemotherapy, Proceedings of the 18th International Congress of Chemotherapy, Stockholm, 1993*, (eds. Einhorn, J., Nord, C. E., Norrby, S. R.), pp. 475–476. American Society for Microbiology, Washington, DC.

Vogelman, B., Gudmundsson, S., Turnidge, J., Leggett, J., Craig, W. A. (1988a). *In vivo* postantibiotic effect in a thigh infection in neutropenic mice. *J. Infect. Dis.*, **157**, 287–298.

Vogelman, B., Gudmundsson, S., Leggett, J., Turnidge, J., Ebert, S., Craig, W. A. (1988b). Correlation of antimicrobial pharmacokinetic parameters with therapeutic efficacy in an animal model. *J. Infect. Dis.*, **158**, 831–847.

Chapter 16

Mouse Subcutaneous Cotton Thread Model

J. Renneberg

Background of the model

Since the first antimicrobial agents became available it has been interesting to test if the drugs are active against bacteria in an almost clinical situation. Therefore attempts have been made to develop both *in-vitro* and *in-vivo* models capable of simulating the distribution of antibiotics in blood and tissue. By the early 1940s it was already recognized that penicillin was unevenly distributed in different body fluids and tissues. Thus Florey and Turton (1946) found elimination of penicillin from infected wound exudate to be slower than from blood, a finding confirmed by Schachter (1948) using a dog lymph model and by Ungar (1950) using a turpentine injector model in rabbits. It was also found that treatment results could not be solely explained by the serum concentration of the drug used (Eagle, 1948; Eagle and Musselman, 1948; Schmidt and Walley, 1951; Eagle, 1952; Eagle et al., 1953). Such findings as these prompted the development of a number of animal and human pharmacokinetic models.

As bacterial infections usually start in the interstitial fluid of soft tissues, increasing attention was paid to interstitial fluid models. Raeburn (1971) evolved the dermabrasion or skin window technique for sampling interstitial fluid and Tan *et al.* (1972) introduced the use of chambers placed over the dermabrasions to obtain larger amounts of fluid. Simon *et al.* (1973) introduced the cantharidine blister technique for antibiotics research. In 1977, the suction skin blister technique, previously developed by Kiistala and Mustakallio (1967), was adapted by Harding and Eilon (1977) for studying the pharmacokinetics of antimicrobial agents. Ryan (1979) introduced the cotton thread technique. A 20–25 cm long cotton thread is applied in surgically created subcutaneous or muscular pockets of 1–2 cm in animals or humans. After administration of antibiotic the thread is pulled through the tissue at different time intervals and cut off in 1 cm pieces, which are placed in phosphate-buffered saline (PBS). The A/V ratio for this model is more than 100, peak levels are reached with no remarkable delay and the peak levels are above 70% of the serum. Ryan primarily used the model in human volunteers, whereas Hoffstedt and Walder used an improved version of the technique in a number of studies (Hoffstedt and Walder 1981a,b; Hoffstedt *et al.*, 1982).

At the same time the surgical wound infection model was developed by Russel *et al.* (1968) and later modified by McRipley and Whitney (1976) to test local treatment of wound infections. The back of the mouse was shaved and a superficial surgical infection was produced by a midline incision 2.3 cm long. The skin was retracted and the wound infected, usually by placing a contaminated suture diagonally in the wound. The experiments performed did not involve examination of the sutures after treatment, but only tested blood and tissue specimens for eventual bacterial growth.

The mouse subcutaneous cotton thread model, which is described here, is based upon a combination of the above mentioned models.

Animal species

The mouse strain BALB-c, female mice 20–25 g, have been used in this model. Other strains have been tested but have not been found to be as suitable as the BALB-c — similarly, females are preferred. Preliminary experiments should be performed to determine the optimal bacterial inoculum to ensure infection without producing significant mortality.

Preparation of animals

The animals require routine husbandry and care with no specific pretreatment. The animals should be obtained 2 weeks prior to receiving anaesthesia. After the operative procedure the animals should be kept warm for at least 4 hours.

Details of surgery

Overview

Under light ether anaesthesia, a small incision (4–6 mm) is made in the back of the mouse. Through this incision the

Handbook of Animal Models of Infection
ISBN 0-12-775390-7

cotton strip, after immersion in the PBS suspension of bacteria, is implanted subcutaneously; the incision is left unsutured. Previous to the treatment series the model was tested with and without bacteria. It was found that, on day zero, drug penetration into the threads was independent of the presence or absence of bacteria. The pharmacokinetic profile (against time) of the inoculated strips resembled that of the fibrin clot model on day 2 (acute infection site) and that of an abscess model on day 6. The thread beds were examined macroscopically and histologically on different occasions after challenge (Figures 16.1 and 16.2).

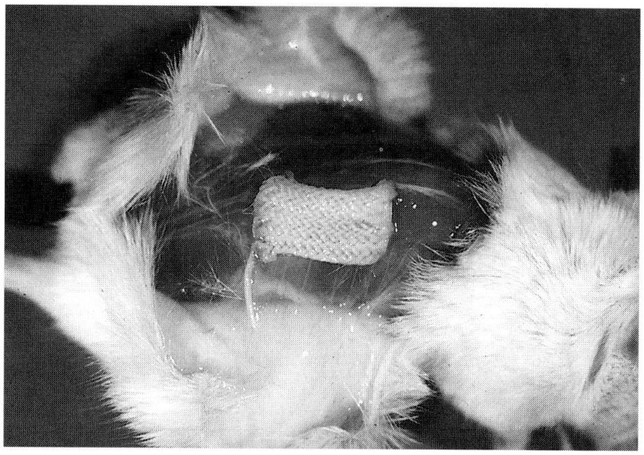

A

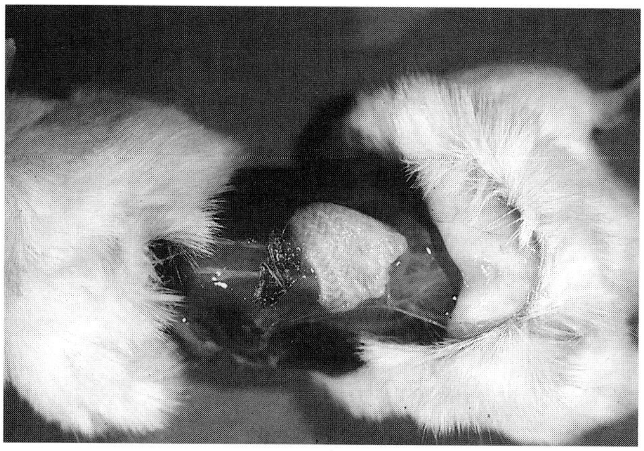

B

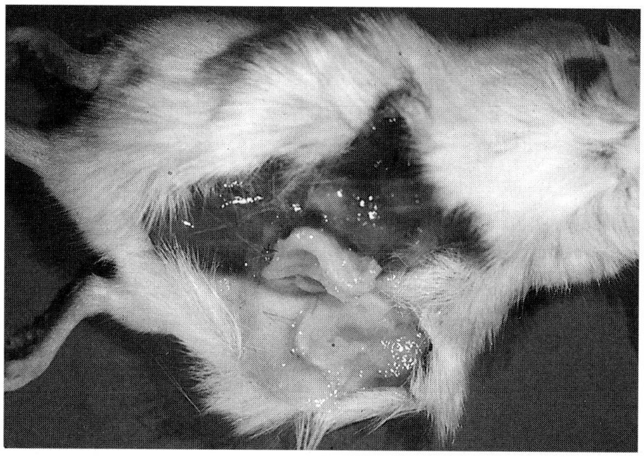

C

Figure 16.1 Macroscopic appearance of the thread bed. A: day 0, B: day 2, and C: day 6.

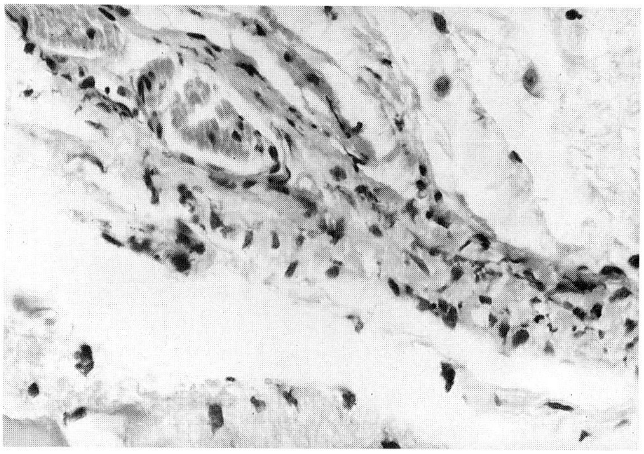

A

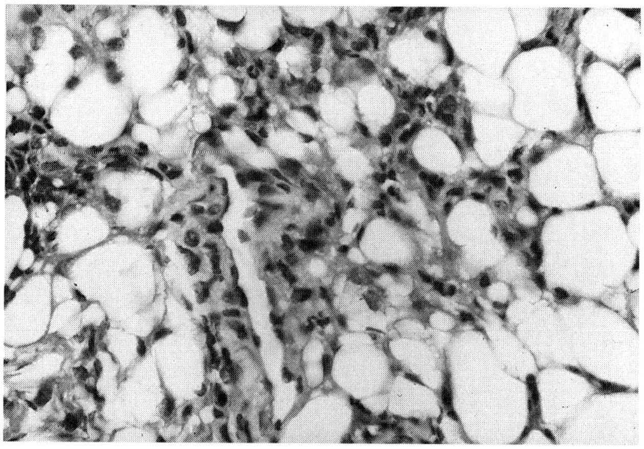

B

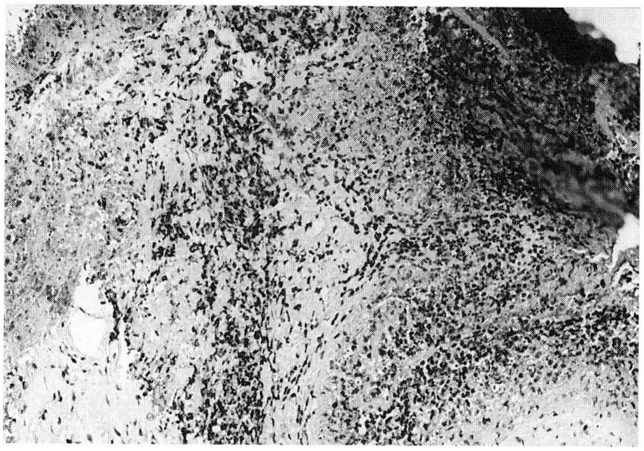

C

Figure 16.2 Microscopic appearance of the thread bed. A: day 0, B: day 2, and C: day 6.

Materials required

Pure cotton thread was used in 8×10 mm strips. Each strip absorbed 100±5μl fluid, as calculated from the difference between the wet and dry weights. Phosphate-buffered saline 0.9% and scissors are also required.

Anaesthesia

Ether inhalation (we used 2% for 30 seconds) for just long enough to ensure that the animal will not move and will not feel anything for the 1–2 minutes necessary to place the strip.

Surgical and infection procedure

Following the light ether anaesthesia, a 4–6 mm incision is made aseptically in the skin of the unshaved back of the mouse. A cotton strip previously immersed in a PBS suspension of bacteria is implanted through the incision under the skin upon the fascia. After implantation the bacteria in the thread should be allowed to multiply for at least 2 hours. This approach is used because all bacterial species used in the model until now showed a lag phase of around 2 hours. The incision is left open.

Postoperative care

The animals should be kept in their usual surroundings after the surgical/infection procedure—no special treatment is necessary so long as the mice do not bite each others' open wounds. The antibiotic treatment is initiated after 2 hours, 2 days or 6 days depending what infection state is needed for the experiment in question. The treatment is usually continued for a period of 6 hours (at most 8 hours). During this period, blood is drawn from the retro-orbital vein and thread strips are removed at precise intervals selected according to the drug(s) to be tested.

Storage and preparation of inocula

Both clinical isolates and reference strains have been used. The clinical isolates were identified by conventional methods. The reference strains were *Staphylococcus aureus* ATCC 25923, *Streptococcus faecalis* ATCC 28272, *Escherichia coli* ATCC 25922 and *Pseudomonas aeruginosa* ATCC 27853.

All strains used for inoculation of the threads were grown overnight in MHB. MIC and MBC values were determined in the same medium, except in the case of netilmicin, where Ca^{2+} and Mg^{2+} were added to the broth (Stratton and Reller, 1977). Viable counts were made on standard blood agar plates, except in the case of *Haemophilus influenzae*, which was grown on haematin agar.

To ascertain an appropriate strength of inoculum, different amounts of bacteria were tested in neutropenic and normal mice, cyclophosphamide being used to induce neutropenia (Vogelmann *et al.*, 1988). It was found that an inoculum of 10^{7-8} colony-forming units (cfu) per strip induced infection in normal mice whereas only 10^5 cfu per strip was required in neutropenic mice. For the preparation of inocula, broth cultures of the bacteria were grown overnight; the bacteria were then resuspended in PBS and adjusted to 10^{8-9} cfu/ml. The cotton strips were immersed in this suspension immediately before implantation in the mice. Where a combination of two pathogens was under study, a pool of separately grown strains was prepared and then resuspended in PBS. The inoculum was checked in each experiment by plating and counting.

Infection process

The infection process is described in Figures 16.1–16.2, which show both macroscopically and microscopically that the inoculated thread piece produces an inflammation leading first to acute infection and then to chronic infection after 6 days.

Key parameters to monitor infection

The infection should only be monitored by killing mice at selected intervals to count the numbers of bacteria left in the piece of thread with or without treatment. If this is done every 60 minutes (or 30 minutes) a killing curve can be produced.

Antimicrobial therapy

Several antibiotics have been used in the model. In each experiment the drugs were obtained from the manufacturers in dry powder form. The antibiotics tested include ampicillin and isoxazolyl penicillins (Renneberg and Walder, 1988; Renneberg and Forsgren, 1989), imipenem (Renneberg and Walder, 1989a), the aminoglycosides amikacin and netilmicin (Renneberg and Walder, 1989a), vancomycin (Renneberg *et al.*, 1993), clindamicin (Renneberg *et al.*, 1993), rifampicin (Renneberg *et al.*, 1993) and the quinolone norfloxacin (Renneberg and Walder, 1989). From a pharmacokinetic point of view, this panel of drugs is representative of most types of antibiotics.

Bacterial density in the threads

The thread strips are gently removed from the mice and immersed individually in 0.9 ml PBS (pH 7.4) for immediate homogenization. The strip bed and its surroundings should be checked for bacteria using sterile cotton swabs. Viable counts of bacteria for *H. influenzae* were determined on haematin agar, whereas all other bacteria were quantified by plating six 10-fold dilutions of 100 µl of the solutions obtained from the homogenized strips on blood agar. After 15–18 hours of incubation at 37°C, the cfu was counted, the detection limit for the viable count being 10 cfu/100 µl.

Antibiotic penetration into the threads

After removal from the mice, the thread strips were placed in 0.9 ml PBS and immediately homogenized. pH was determined in the strip bed before removal of the strips and after the strip was removed. *In-vitro* assays were then performed on serum and on the thread specimens. An agar well technique (Mattie *et al.*, 1973) was used. The plates measured 12 × 12 cm and contained 125 ml diagnostic sensitivity test (DST) agar. After prediffusion for 1 hour at room temperature, 8 mm diameter wells were punched in the agar and filled either with 70 µl standard solution prepared in pooled mice serum, PBS, serum specimen or subcutaneous tissue fluid (STF). Samples and standards were tested at least in duplicate, usually in triplicate. Homogenized strips previously immersed in fluid containing known concentrations of antibiotics in PBS were used as standards for the STF samples, since identical results were obtained in previous experiments where standards were prepared in non-infected STF, infected STF and in PBS. The indicator strains used were *Micrococcus luteus* CXX, *M. luteus* ATCC, and *M. luteus* ATCC 9341, *P. mirabilis* ATCC 21100, a *S. aureus* SBL 40 and a *Staphylococcus xylosus* 3329. The assay ranges for each experiment are given in the respective reports.

Measurement of protein binding to mouse serum and subcutaneous tissue fluid

The binding of antibiotic to serum proteins and to infected and non-infected subcutaneous tissue fluid was quantified by means of an ultracentrifugation method (Peterson *et al.*, 1977). The samples were centrifuged at 14 000*g* for 16 hours at 3°C. Then antibiotic activity was measured in starting samples and in the ultracentrifuge supernatant fluid as already described. The percentage of protein binding was then calculated as follows: 100% – [(ultracentrifuge supernatant activity/starting activity) × 100%].

Pharmacokinetic methods

The half-life of drugs was measured during the elimination phase using a two-compartment model (Barza, 1981; Bergeron *et al.*, 1981). The area under the curve (AUC) was measured with a compensating polar planimeter (Keuffel and Esser, USA). The forecasting of doses of β-lactam antibiotics followed the heartbeat theory of Mordenti (1985).

Key parameters to monitor response to treatment

The mice will only very seldom present systemic signs of infection. Therefore, the only measurable parameter of infection is the bacterial number in the thread pieces at certain time-points from placement of the inoculated threads and calculation of the difference in counts between animals having treatment and animals without treatment, and the histological signs of infection in the thread bed, which are required if the stage of infection is to be documented.

Pitfalls (advantages/disadvantages) of the model

Advantages

- It is a very easy model to work with.
- It is cheap.
- Results may be presented from day to day.
- It is easy to test large numbers of animals.
- It is possible to use the same model to evaluate antibiotics at different stages in the development of infection.
- It is possible to monitor concentrations of exoenzymes from the bacteria at the site of infection.
- It is possible to measure antibiotics and active metabolites at the site of infection.

Disadvantages

- The mouse infection is very distant from the clinical treatment situation.
- The thread could be seen as an *in-vivo* '*in-vitro* model'.
- Antibiotic pharmacokinetics in mice are completely different from those in humans.
- The animal immune system as far as infection is concerned cannot be compared to the human system.

Contributions of the model to infectious disease therapy

- The model has been used to show different *in-vivo* activity of the different isoxazolylpenicillins (Renneberg and Forsgren, 1989).

- The importance of free β-lactamase activity at the site of infection of a synergistic infection with streptococci and β-lactamase producers, when the drug of choice is penicillin G, has been shown (Renneberg and Walder, 1989b).
- The postantibiotic effect *in vivo* and synergistic effect *in vivo* have been described (Renneberg and Walder, 1989; Renneberg *et al.*, 1993).
- The model makes it possible to monitor antibiotic pharmacokinetics as well as pharmacodynamics at the site of infection (Renneberg and Walder, 1988a,b).

References

Barza, M. (1981). Principles of tissue penetration of antibiotics. *J. Antimicrob. Chemother.,* **8**, (Suppl. C), 7–28.

Bergeron, M. G., Beauchamp, D., Poirier, A., Bastille, A. (1981). Continuous versus intermittent administration of antimicrobial agents: tissue penetration and efficacy *in vivo*. *Rev. Infect. Dis.*, **3**, 84–97.

Eagle, H. (1948). Speculations as to the therapeutic significance of the penicillin blood level. *Ann. Intern. Med.*, **28**, 260–278.

Eagle, H. (1952). Experimental approach to the problem of treatment failure with penicillin. *Am. J. Med.*, **13**, 389–399.

Eagle, H., Musselman, A. D. (1948). The rate of bactericidal action of penicillin *in vitro* as function of its concentration, and its paradoxically reduced activity at high concentrations against certain organisms. *J. Exper. Med.*, **88**, 99–131.

Eagle, H., Fleischmann, R., Levy, M. (1953) "Continuass" vs "discontinous" therapy with penicillin. *N. Eng. J. Med.* **248**, 481–488.

Florey, M. E., Turton, E. C. (1946). Penicillin in wound exudates. *Lancet*, **Sept. 21**, 405–409.

Harding, S. M., Eilon, L. A. (1977). Clincal pharmacology of cefuroxime. In *The Early Evaluation of Cefuroxime*, (eds. Snell, E. S. *et al.*), pp. 57–68. Glaxo Research, Greenford, Middlesex.

Hoffstedt, B., Walder, M. (1981a). Influence of serum protein binding and mode of administration on penetration of five cephalosporins into subcutaneous tissue fluid in humans. *Antimicrob. Agents Chemother.*, **20**, 783–786.

Hoffstedt, B., Walder, M. (1981b). Penetration of ampicillin, doxycycline and gentamicin into interstitial fluid in rabbits and of penicillin V and pivampicillin in humans measured with subcutaneously implanted cotton threads. *Infection*, **9**, 7–11.

Hoffstedt, B., Walder, M., Forsgren, A. (1982). Comparison of skin blisters and implanted cotton threads for the evaluation of antibiotic tissue concentrations. *Eur. J. Clin. Microbiol.*, **1**, 33–37.

Kiistala, U., Mustakallio, K. K. (1967). Dermo-epidermal separation with suction. *J. Invest. Dermatol.*, **48**, 466–477.

Mattie, H., Goslings, W. R. O., Noach, E. L. (1973). Cloxacillin and nafcillin: serum binding and its relationship to antibacterial effect in mice. *J. Infect. Dis.*, **128**, 170–177.

McRipley, R. J., Whitney, R. R. (1976). Characterization and quantitation of experimental surgical-wound infections used to evaluate topical antibacterial agents. *Antimicrob. Agents Chemother.*, **10**, 38–44.

Mordenti, J. (1985). Forecasting cephalosporin and monobactam antibiotic halflives in humans from data collected in laboratory animals. *Antimicrob. Agents and Chemother.*, **27**, 887–891.

Peterson, L. R., Hall, W. H., Zinneman, H. H., Gerding, D. N. (1977). Standardization of a preparative ultracentrifuge method for quantitative determination of protein binding of seven antibiotics. *J. Infect. Dis.* **136**, 778–783.

Raeburn, J. A. (1971). A method for studying antibiotic concentrations in inflammatory exudate. *J. Clin. Pathol.*, **24**, 633–635.

Renneberg, J., Forsgren, A. (1989). The activity of isoxazolyl penicillins in experimental staphylococcal infection. *J. Infect. Dis.*, **159**, 1128–1132.

Renneberg, J., Walder, M. (1988a). A mouse model for simultaneous pharmacokinetic and efficacy studies of antibiotics at sites of infection. *J. Antimicrob. Chemother.*, **22**, 51–60.

Renneberg, J., Walder, M. (1988b). Mouse model for evaluation of antibiotic treatment of acute and chronic infections. *Eur. J. Clin. Microbiol. Infect. Dis.*, **7**, 753–757.

Renneberg, J., Walder, M. (1989a). Post-antibiotic effects of Imipenem, Norfloxacin and Amikacin *in vitro* and *in vivo*. *Antimicrob. Agents Chemother.*, **33**, 1714–1720.

Renneberg, J., Walder, M. (1989b). The role of beta-lactamase in mixed infections in relation to treatment with ampicillin. *J. Infect. Dis.*, **160**, 337–341.

Renneberg, J., Karlsson, E., Nilsson, B., Walden, M. (1993). Interactions of drugs acting against *Staphylococcus aureus in vitro* and in a mouse model. *J. Infection*, **26**, 265–277.

Ryan, D. M. (1979). Implanted cotton threads; a novel method for measuring concentrations of antibiotics in tissue fluid. *J. Antimicrob. Chemother.*, **5**, 735–736.

Russell, H. E., Gutekunst, D. P., Chamberlain, R. E. (1968). Evaluation of furazolium chloride in topical treatment of model infections in laboratory animals. *Antimicrob. Agents Chemother.*, **2**, 497–501.

Schachter, R. J. (1948) Fate and distribution of penicillin in the body. I. Circulation of penicillin in the lymph. *Proc. Soc. Exp. Biol. Med.*, **68**, 29–34.

Schmidt, L. H., Walley, A. (1951). The influence of the dosage regimen on the therapeutic effectiveness of penicillin G in experimental lobal pneumonia. *J. Pharm. Exp. Ther.* **103**, 479–488.

Simon, C., Malerczyk, V., Brahmstaedt, E., Toeller, W. (1973). Cefalozin, ein neues Breitspektrum-Antibiotikum. *Deutsche Med. Wschr.*, **51**, 2448–2450.

Stratton, C. W., Reller, L. B. (1977). Serum dilution test for bactericidal activity. I: Selection of a physiologic diluent. *J. Infect. Dis.*, **126**, 492–497.

Tan, J. S., Troot, A., Phair, J. P., Watanakunakorn, W. (1972). A method for measurement of antibiotics in human interstitial fluid. *J. Infect. Dis.*, **126**, 492–497.

Ungar, J. (1950). Penicillin in tissue exudates after injection. *Lancet*, **Jan. 14**, 56–59.

Vogelman, B., Gudmundsson, S., Turnidge, J., Legget, J., Craig, W. A. (1988). *In vivo* postantibiotic effect in a thigh infection in neutropenic mice. *J. Infect. Dis.*, **157**, 287–298.

Weinstein, L, Daikos, G. K., Perrin, T. S. (1951). Studies on the relationship of tissue fluid and blood levels of penicillin. *J. Lab. Clin. Med.*, **38**, 712–718.

Chapter 17

Infection after Ionizing Radiation

I. Brook, T. B. Elliott and G. D. Ledney

Background of human infection

Ionizing irradiation depresses normal host defenses (Mathe, 1965) and enhances the susceptibility of the immunocompromised host to local and systemic bacterial infections (Leroy, 1947; Clapper *et al.*, 1954; Kaplan *et al.*, 1958). There is a practical need to develop effective therapeutic modalities for infections following radiation injury. Lymphatic and other tissues from Japanese patients dying from the effects of the atomic blasts at Hiroshima and Nagasaki in 1945 frequently revealed microscopic bacterial colonies of both Gram-positive and Gram-negative bacteria in tissues (Keller, 1946). In some cases of accidental whole-body exposure, infections with enteric organisms occurred and presumably added to the radiation syndrome (Miller *et al.*, 1951). Mortality following exposure to irradiation during the Chernobyl nuclear accident in 1986 was also associated with septicemia due to Gram-negative and Gram-positive organisms (Gale, 1987). When a combination of trauma, exposure to toxic substances and other injuries occurs in conjunction with irradiation, the risk of developing a serious infection is increased.

The development of therapeutic modalities to cope with these infections is necessary, and animal models that explore the utility of these therapies are very important and useful as human experience is limited.

Background of the model

Generalized sepsis model

The source of most infections after a lethal dose of radiation in experimental animals is the gastrointestinal tract, which is normally colonized by aerobic and anaerobic bacteria (Brook *et al.*, 1984). Following sublethal to lethal γ or mixed-field irradiation of B6D2F1/J mice, a decrease occurs in the number of the aerobic and facultative anaerobic bacteria in the gastrointestinal tract of mice. That decrease starts 2 days after irradiation and is maximal at the ninth day following exposure to Cobalt-60 γ-photon irradiation (Brook *et al.*, 1988). However, following the ninth day, the number of Enterobacteriaceae increased (>10^9 organisms per gram of stool) while the number of anaerobes stayed low (10^3–10^4; Figure 17.1). The increase in the number of Enterobacteriaceae coincides with the appearance of endogenous *Escherichia coli* and *Proteus mirabilis* in the blood, spleen and liver of the animals and the beginning of mortality (Brook *et al.*, 1984, 1988; Figure 17.2). As the amount of radiation increases, anaerobic bacteria (mostly *Peptostreptococcus* spp. and *Bacteroides* spp.) are also recov-

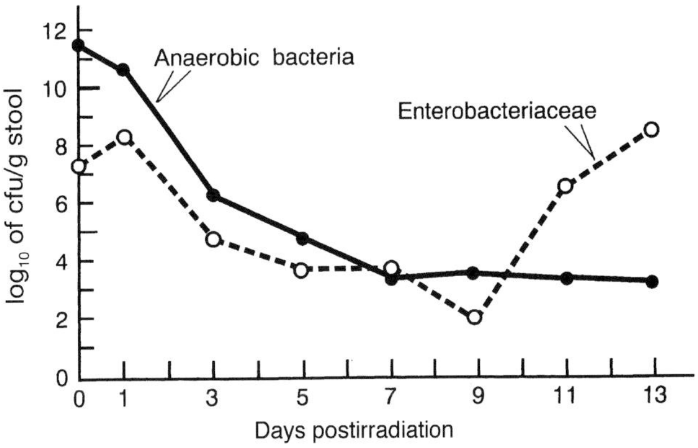

Figure 17.1 Changes in aerobic and facultative flora in ileum of mice after lethal irradiation (10.0 Gy). (Brook and Ledney, 1994a,b)

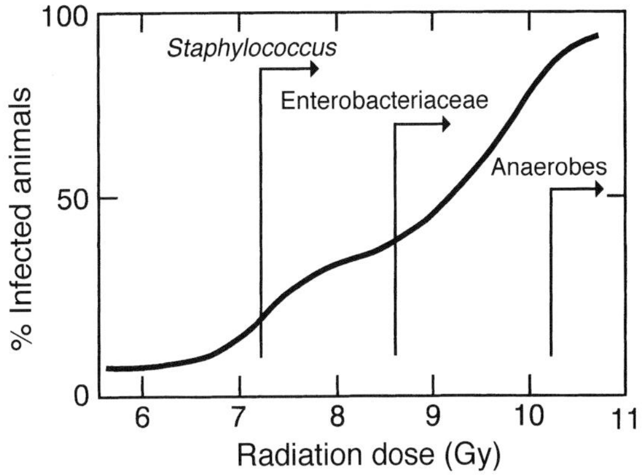

Figure 17.2 Micro-organisms recovered from irradiated mice in relation to radiation dose. (Brook *et al.*, 1986)

Handbook of Animal Models of Infection
ISBN 0-12-775390-7

ered from the blood (Figure 17.2). Other potential sources of serious infection in the irradiated host are exogenous organisms such as *Pseudomonas aeruginosa* and *Klebsiella pneumoniae*. Animals given γ-irradiation from Cobalt-60 become susceptible to colonization and systemic infections from these organisms starting 2 days after irradiation. This is associated with leukopenia and thrombocytopenia that occurs after irradiation (Figure 17.3).

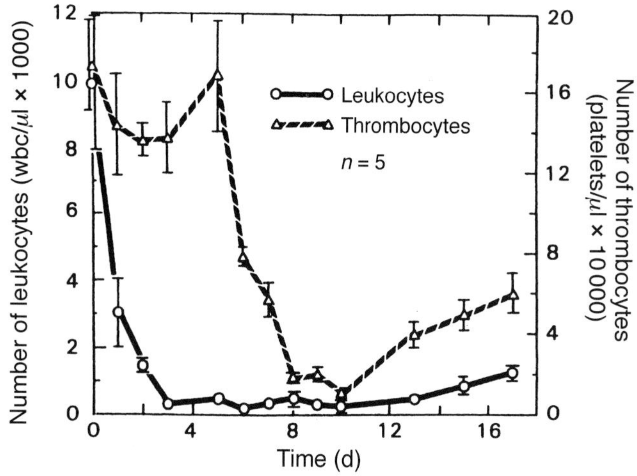

Figure 17.3 Numbers of leukocytes and thrombocytes in blood in mice exposed to 6.5 Gy ^{60}Co γ-rays. Blood was drawn from five 15-week-old female B6D2F1/J mice. Reproduced from Elliott *et al.*, 1990 with permission

Therapy for severe systemic infection due to Gram-negative bacteria in the immunocompromised host generally involves the use of aminoglycosides in combination with β-lactam antibiotics or single-agent therapy with ceftazidime or a carbapenem (Klastersky, 1998). An additional approach to the prevention of such infection is the use of selective decontamination of the gastrointestinal tract by using agents given after radiation that spare the anaerobic gut flora while inhibiting the Enterobacteriaceae (Brook and Ledney, 1992b).

Intra-abdominal model

Polymicrobial intra-abdominal infection involving aerobic and anaerobic bacteria such as *E. coli* and *Bacteroides fragilis* can occur following irradiation associated with intra-abdominal trauma. Studies illustrated the importance of administration of antimicrobials effective against both of these organisms in successful management of such infections (Solomkin *et al.*, 1984). However, treatment of mixed aerobic–anaerobic infections in irradiated hosts may be complicated by the adverse effects observed following the use of antimicrobials, which are capable of reducing the number of the anaerobic gut flora (Brook *et al.*, 1988). Such therapy enhanced the colonization of the intestinal tract with Enterobacteriaceae and contributed to mortality due to sepsis by these organisms.

Subcutaneous model

Susceptibility to infection was demonstrated in lethally irradiated mice by inoculating known doses of single species of bacteria to determine the $LD_{50/30}$, usually 4 days after irradiation (Madonna *et al.*, 1989). That time interval was selected because the number of circulating neutrophilic granulocytes decrease to barely detectable numbers by 3 days after 7.0 Gy γ- or 3.5 Gy mixed-field-irradiation (Figure 17.3). Micro-organisms are injected into the skin between the shoulders, that is pinched to form a pocket, in mice that are lightly anesthetized with methoxyflurane.

Wound model

Exposure to ionizing irradiation can be associated with concomitant wound infection, following either a nuclear accident or therapeutic irradiation and surgery to manage malignancies. The potential virulence of micro-organisms that cause wound infections in irradiated hosts is enhanced by the depressed defenses of the host (Kaplan *et al.*, 1958; Mathe, 1965; Brook, 1988; Elliott *et al.*, 1990) (Figure 17.4). Such infections are therefore difficult to control by local or systemic therapy (Miller *et al.*, 1952; Hammond *et al.*, 1955a; Brook, 1988). The organisms that cause most of the wound infections are skin flora organisms (staphylococci and streptococci) or environmentally acquired bacteria.

Animal species

Several strains of mice have been used in infection models: B6D2F1/J (20–25 g, female, Brook and Elliott, 1989; Brook and Ledney, 1991a, 1994a); C3H/HeN (20–25 g female, Brook and Ledney, 1994b); C3HeB/FeJ (20–25 g female, Brook *et al.*, 1986, 1988); CF#1 (20–25 g female, Hammond

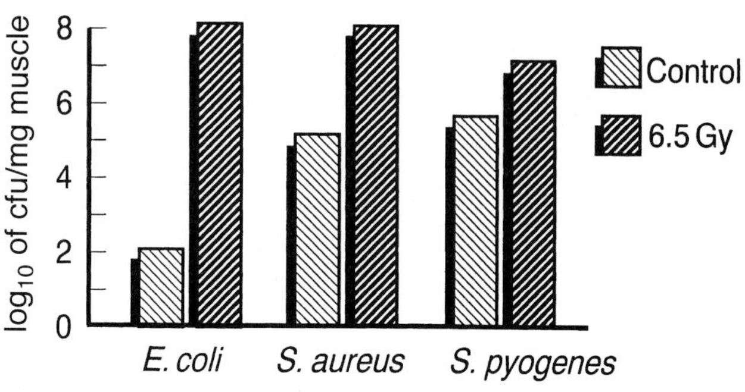

Figure 17.4 Number of organisms recovered in infected wounds in non-irradiated and irradiated animals 5 days after irradiation. (Elliott *et al.*, 1990)

et al., 1955b); and CBA (18–26 g, Wensinck and Renaud 1957; Klemparskaya *et al.*, 1987).

We used B6D2F1/J because it is a random-bred hybrid and C3H/HeN because of the presence of adequate numbers of Enterobacteriaceae in the intestines.

Rats have also been used in several studies: Sprague-Dawley (250–325 g, male, Taketa, 1962; 120 g female, Klaines *et al.*, 1967).

Most of our studies were done in animals aged 10–20 weeks old. It is preferable to use animals of the same age, since oral and gastrointestinal flora may vary with age. The unique endogenous bacterial flora of the mice influences the type of bacteria that causes endogenous sepsis – because the source of the sepsis is their own flora. It is therefore important to be aware of the established characteristic flora of the animals used for experimental models. The endogenous microbiology may vary between animal batches, and may influence the outcome of infection. There may also be differences in sensitivity to radiation among different animal batches and shipments. Because of this, a control untreated group is necessary in each experiment. Indeed, some commercial sources of mice have rid their animals of many opportunistic pathogens (i.e. *Pseudomonas* spp.).

Preparation of animals

Because of the neutropenia and immune suppression induced by radiation, extraordinary precautions should be taken to insure the preirradiation health of the animals and the absence of any subclinical colonization or infection by potential pathogens while maintained in the facility. To insure this, the mice are held in quarantine for 2 weeks after arriving in the facility. Representative samples are examined by microbiology, serology, and histopathology to assure the absence of specific bacteria, particularly *P. aeruginosa*, and common murine diseases (Elliott *et al.*, 1995).

Up to 10 mice can be housed in sanitized 46 × 24 × 15 cm polycarbonate boxes with a filter cover (MicroIsolator, Lab Products, Inc., Maywood, NJ) on hardwood-chip bedding. They should be maintained and irradiated according to guidelines set by the USDA and public health regulations. The animal holding rooms are maintained at approximately 70 °F and 50% (± 10%) relative humidity with a 12 hours light/dark, full-spectrum lighting cycle. Conditioned fresh air is changed at least 10 times per hour. The mice are given feed (Wayne Lab Blox, Continental Grain Co., Chicago, IL) and acidified (pH 2.5) water freely. (Acidified water is used to suppress bacterial growth in the water bottle.) It is important to note that the bedding material (which may contain micro-organisms and allergens), and the type of diet and nutrients, should be maintained throughout the experimental procedure.

Details of irradiation

The irradiation source (X-ray, γ, or neutron) and dose may vary according to the experimental design and animal susceptibility. Details regarding dosemetry can be obtained from Elliott *et al.*, 1995. End-points include: the LD_{50} of the specific mouse strain, the slope of the probits, and established measurement of morbidity, which are periodically repeated to assure their accuracy, as they may vary seasonally and among animal shipments.

Mixed-field irradiation

Studies of infection in neutron irradiated animals are important because of the apparent increase in sensitivity of the gastrointestinal tract to this type of radiation compared to gamma radiation.

Mice are exposed to mixed fisson-neutron and γ-photon radiation in an exposure room of a Training Research Isotope General Atomic (TRIGA) reactor. Mice are placed in well-ventilated aluminum holders that rotate at a speed no greater than 1.5 rpm. Mice should be arranged in an arc around the reactor wall lobe and a standard length above the exposure room floor. Rotation assures that the depth–dose distribution is as uniform as possible to allow optimal dosing of critical organs. All doses are at midline tissue (MLT). At the position of irradiation, the measured ratio of the neutron dose to the total dose (neutron plus γ) measured $n/(n + \gamma) = 0.67$, delivered at a nominal dose rate of 40 cGy/min MLT. The uniformity of the radiation field should be within 3% of the dose at the center of the array. The TRIGA reactor was used as a source of radiation because the proportion of neutrons and gamma photons provided to the animals can be adjusted to meet the needs of the investigators experiment.

Gamma irradiation

Mice should be placed in ventilated acrylic plastic boxes and exposed bilaterally to γ-photon radiation in a Cobalt-60 whole-body irradiation facility (Carter and Verrelli, 1973). Boxes are designed to minimize the air gap between the mouse and the plastic and are placed on a stationary platform. The MLT dose rate for mice is measured before irradiation of animals under conditions that are identical to those for irradiating mice by placing a 0.5 cm^3 tissue-equivalent ionization chamber in the center of a 2.5 cm diameter cylindrical acrylic mouse phantom at a known distance from the Cobalt-60 source. The ratio of the dose-rate measured in the phantom to the dose-rate in free air, for this array, should be determined (Elliott *et al.*, 1995). Exposure time is adjusted so that animals receive the desired dose at a nominal dose-rate of 40 cGy/minute. Variation of dose within the exposure field should be less than 3%. The average energy of γ photons is 1.25 MeV. Measurements should be in accordance with the protocol of the American Association of Physicists in Medicine for the determination of absorbed dose from high-energy photon and electron beams (American Association of Physicists in Medicine, 1983).

Radiation dose-responses

Mice are given graduated doses of mixed-field or γ-photon radiation in order to determine a current relative biological effectiveness and biologically comparable, or isoeffective, doses of the two qualities of radiation. This is done for the purpose of comparing experimental results in animals given either quality of radiation.

Details of surgery (for animal wounding)

Material required

Anesthetic, hair clippers or razor, skin disinfectant, scalpel handle plus blades, forceps, a steel punch, syringes and needles.

Anesthesia

Anesthesia is needed when animals undergo wounding or subcutaneous inoculation. Induction of intra-abdominal abscesses is described in Brooks' chapter on abdominal abscesses. Animals are sedated with methoxyflurane (Elliott *et al.*, 1990). A complete anesthesia lasts approximately 3 minutes.

Surgical procedure (wounding)

Punch wound

Ten minutes or 1 hour after irradiation the fully anesthetized mouse is subjected to various sizes of wound trauma. The folded panniculus carnosus muscle (twitch muscle) and overlying skin of the anterior dorsal surface between the shoulder blades (Ledney *et al.*, 1982) is removed by striking a steel punch with a hammer. The procedure is done on a clean disinfected Teflon-covered operating board. Various wound sizes are made according to the size of the mouse and the desired size of the wound. Wound sizes are generally 5%, 10%, 15% and 30% of the total skin surface. Immediately after trauma, the mice are placed on a clean paper towel and then transferred on to freshly autoclaved hardwood-chip bedding in clean cages.

Incision muscle wound infection

Wounding and inoculation of bacteria is done on the third day after irradiation, to fully anesthetized mice (Elliott *et al.*, 1990). The shaved skin over the inoculation site is wiped with gauze sponge that is moistened with 70% ethanol. A 0.1 ml volume of a suspension of bacteria is injected subcutaneously to form a bleb over the right medial gluteus muscle. When the liquid is absorbed after 20–30 minutes, the skin and muscle of the anesthetized mice are incised with a sterile no. 15 scalpel. The incision is approximately 12 mm long and approximately 30° to the spine caudally. The exposed medial gluteus is then incised three times about 3 mm deep with the blade at different angles along the same stroke as the initial incision.

Postoperative care

Animals are placed in freshly autoclaved hardwood-chip bedding and cages. They are housed in groups of four per cage.

Storage and preparation of inocula

The models of sepsis and wound are amenable to endogenous (animal's own flora) or exogenous infection. The endogenous skin infection is generally caused by streptococci, staphylococci, and Gram-negative enteric bacteria (from fecal or other sources) (Brook, 1995). Exogenous bacteria used for wound infection were *Staphylococcus aureus* (Brook and Elliott, 1989; Elliott *et al.*, 1990; Brook, 1995). *Escherichia coli, Klebsiella pneumoniae* and *Streptococcus pyogenes* (Elliott *et al.*, 1990).

Endogenous-induced sepsis is generally due to aerobic organisms and facultative anaerobic organisms (Brook *et al.*, 1986; Kurishita *et al.*, 1993; *E. coli, Proteus* spp., *Staphylococcus* spp., and *Streptococcus* spp.) and anaerobic organisms (*Peptostreptococcus* spp. and *Bacteroides* spp.; Brook *et al.*, 1988).

Exogenous bacteria used to cause sepsis were *P. aeruginosa* (Brook and Ledney, 1991a; Geraci *et al.*, 1985), and *K. pneumoniae* (Brook and Ledney, 1991a; Brook *et al.*, 1990).

The organisms used for systemic exogenous and wound infection should be susceptible to the antimicrobials used in the experiments. The organisms selected for inoculation are transferred from stock suspension on to Columbia sheep blood (5%) agar and incubated for 18–24 hours. Selected colonies are placed in brain heart infusion broth (BHIB) incubated at 35°C for approximately 2.5–3 hours, harvested in the logarithmic phase of growth and washed in 0.9% saline. The desired concentration of bacteria is prepared for inoculation by the desired route.

Infection process

Wound infection is initiated by direct injection of the bacterial suspension into the desired location (i.e. subcutaneous area over muscle). Colonization of the gastrointestinal tract and subsequent sepsis is induced by oral feeding of the bacterial suspension in the desired concentration in a volume of 0.1–0.2 ml. This is given by lavage using a 20 G animal feeding tube fitted to a 1.0 ml syringe (Brook and Ledney, 1990). Subcutaneous challenge can be performed by injecting 0.2 ml in the dorsal skin fold made by pinching the skin over the shoulder and lifting the skin.

The optimal time to induce these infections (wound or sepsis) is about 72–96 hours after irradiation, the time when the immune system is depressed (Figure 17.3) and the number of cfu of normal gastrointestinal flora is depleted (Figure 17.1; Brook and Ledney, 1994b).

Key parameters to monitor infection

Mortality is the final outcome of generalized sepsis. The effect of therapy in reducing death is monitored twice daily over a period of 30 days. Culture of heart blood or liver will document the bacterial cause of sepsis (Brook *et al.*, 1986). Quantitative microbiology of the wounded and infected muscle is an indicator for the efficacy of therapy. This is done by removing the gluteus medius muscle aseptically, weighing and homogenizing it in 0.9% NaCl solution, and performing serial dilution and plating. This should be done at intervals from 3–7 days after bacterial challenge with or without therapy. Monitoring wound closure after the punch is another indicator for efficacy.

Morbidity, particularly after low sublethal irradiation, can be assessed during the course of infection by several whole-animal, cellular, and molecular assays. Body mass (weight), water and food consumption (both reduced after irradiation and illness), and assessment of performance show the whole-body response to treatments. Assessment of performance includes testing the locomotor activity, grip strength, and motor co-ordination (Toth, 1997).

Hematological parameters such as WBC, differential counts and thrombocytes provide evidence for hematological recovery. Measuring of proinflammatory cytokines, including TNF-α, IL-1β, and IL-6, in plasma or tissue homogenates provides insight into the host response to infection at the molecular level (Neta and Oppenheim, 1988; Peterson *et al.*, 1994). These cytokines can be quantitatively monitored by enzyme-linked immunosorbent assay (EIA).

Antimicrobial therapy

A variety of antimicrobial agents have been evaluated in the wound, sepsis, and intra-abdominal abscess models (Tables 17.1–17.3). Antibiotics can be administered by a variety of routes (p.o., i.m., i.v., i.p.). However, the parenteral route may repeatedly expose the postirradiation immunodepressed animal to a potential source of exogenous infection (local or systemic) and the possibility of local bleeding (due to thrombocytopenia). For this reason, the oral route of administration of antimicrobial agents is preferred. Addition of the antibiotics to the drinking water has been tried (van der Waaij and van der Waaij, 1985; Geraci *et al.*, 1985); however, the altered taste of the water may reduce water consumption by the animal, which must be monitored. It is difficult to quantify the dose of antimicrobials as consumption varies among animals.

Antimicrobial therapy for endogenous infection is generally begun 3 days after irradiation, generally for 10–21 days. In the case of local inoculation or feeding of bacteria therapy is started 10–72 hours afterwards (Brook and Ledney, 1991a) depending on the experimental design to allow the bacteria sufficient time to establish a true infection. The optimal delay in initiation of therapy depends on the virulence of the organisms (Brook and Ledney, 1991a). Virulent organisms (such as *P. aeruginosa*) require earlier initiation of therapy within 10 hours after oral, and 24 hours after parenteral inoculation. Earlier initiation of therapy for endogenous therapy is not necessary, as the changes (depletion) in the gastrointestinal flora, and immunosuppression, appear only at that time (Brook and Ledney, 1991a).

The length of therapy can also effect survival. Longer therapy may be more optimal in some instances (Brook and Ledney, 1992a). The survival of animals inoculated with *K. pneumoniae* and treated with ofloxacin for 7 days was 55% (11 of 20), as compared to 90% (18 of 20) of those treated for 21 days. Experimental design should consider varying the length of therapy to cover the period of immunodeficiency.

The dosing schedule and choice of optimal antimicrobial dose should be given careful consideration, because of the different pharmacodynamics and pharmacodynetics of antimicrobials between humans and rodents, the possible effects of irradiation on the pharmacokinetics, and possible toxic side-effects of antimicrobials (Patchen *et al.*, 1993).

Measurement of peak and trough serum concentration can assist in the selection of the proper dose. The method used is generally by bioassay using agar diffusion (e.g. for gentamicin) or high-pressure chromatography (for metronidazole).

Advantages/disadvantages of the model

The irradiated animal model of infection satisfies the goal of simulating the difficulty in reproducing a complex scenario, bacteriological and physiological, that may occur after exposure to ionizing radiation. It produces the physiological, immunological and bacteriological events that lead to endogenous sepsis after lethal exposure to high levels of radiation. It permits infection of the nonlethally irradiated animal with pathogenic bacteria. It permits the investigator to determine or monitor

Table 17.1 Examples of antimicrobial therapy of generalized infection in mice for irradiation

Reference	*Bacteria*	*Irradiation dose/source*	*Antibiotics (dose)*	*When started/ route/dosing schedule/duration*	*Peak serum conc. (mg/l)*	*Comments*
Hammond *et al.*, 1955b	Endogenous flora	20 rep of mixed neutron and γ rays	Streptomycin (5–6 mg/day)	1–3 d/s.c./q.d./5–15 days	Not done	Streptomycin reduced mortality
Brook *et al.* 1988	Endogenous flora	10 Gy Cobalt-60	Gentamicin (7.5 mg/kg/day) Metronidazole (50 mg/kg/day)	3 d/i.m./q.d./21 days	6.2 ± 2.1 23.8 ± 6.6	Metronidazole decreased no. anaerobes in gut flora and enhanced sepsis gentamicin delayed mortality
Brook *et al.* 1990	*K. pneumoniae*	8 Gy Cobalt-60	Ofloxacin (40 mg/kg/day) Pefloxacin (50 mg/kg/day) Ciprofloxacin (50 mg/kg/day)	3 d/p.o./b.i.d. or q.i.d./7 d	2.6 ± 0.4 2.4 ± 0.3 2.8 ± 0.5	All quinolones effective in reducing mortality
Brook and Ledney, 1990	*P. aeruginosa*	7.0 Gy Cobalt-60	Ofloxacin (40 mg/kg/day)	3 d/p.o. or q.d./15 days	2.6 ± 0.4	Ofloxacin increased survival
Brook and Ledney, 1991a	*P. aeruginosa* *K. pneumoniae*	7.5 Gy Cobalt-6.0	Gentamicin (15 mg/kg/day) Gentamicin (7.5 mg/kg/day) Ofloxacin (40 mg/kg/day)	Variable/p.o./15 days Variable/i.m./15 days Variable/p.o./15 days	Not detected 5.5 ± 1.8 2.6 ± 0.4	Early gentamicin p.o. therapy (10 hours for *P. aeruginosa* and 24 hours for *K. pneumoniae*) prevented colonization and infection
Brook and Ledney 1991b	Endogenous flora	8.2 Gy Cobalt-60	Ofloxacin (40 mg/kg/day) Penicillin (250 mg/kg/day)	3 d/p.o./q.d./10 days 3 d/i.m./q.d./10 days	2.4 ± 0.3 38.5 ± 4.6	Combination of ofloxacin and penicillin more effective than either alone because of reduction of Enterobacteriaceae and streptococcal infection
Brook and Elliott 1991	Endogenous flora	9.5 Gy Cobalt-60	Ofloxacin (40 mg/kg/day) Pefloxacin (50 mg/kg/day) Ciprofloxacin (50 mg/kg/day)	4 d/p.o./b.i.d. or q.i.d./7 days	2.8 ± 0.4 2.6 ± 0.4 2.8 ± 0.5	All quinolones reduced mortality
Madonna *et al.*, 1991	Endogenous flora	8.0 Gy Cobalt-60	Gentamicin (7.5 mg/kg/day) Ofloxacin (40 mg/kg/day) Oxacillin (150 mg/kg/day)	1 d/s.c./q.d./10 days	Not done	Punch-wound-induced. Therapy with 5-TDC (immunomodulator) combined with ceftriaxone increased survival. Topical antimicrobials were also used
Brook and Ledney, 1992a	*K. pneumoniae*	8.0 Gy Cobalt-60	Ofloxacin (40 mg/kg/day)	5 d/p.o./q.d./7 or 21 days	2.4 ± 0.3	21 days more effective than 7 days in reducing mortality

Kurishita *et al.*, 1993	Endogenous flora	9.5 Gy	Aztreonam (100 mg/kg/day)	1 d/i.m./10 days	Not done	Immunomodulator (OK-432) was also used. Azteonam reduced infection with *Proteus mirabilis* but not with *Streptococcus faecalis*. OK-432 effective against infection by both organisms. Combination of both aztreonam and OK-432 more effective than either alone
Brook *et al.*, 1993a	Endogenous flora	8.3 Gy Cobalt 60	Ofloxacin (20 mg/kg/day) Vancomycin (50 mg/kg/day) Teicoplanim (50 mg/kg/day)	3 d/p.o./q.d./21 days 3 d/p.o. or i.m./q.d./21 days 3 d/p.o./q.d./21 days	2.1 ± 0.5 21.2 ± 2.8 Not done	Ofloxacin reduced Enterobacteriaceae, and glycopeptides streptococcal isolation. Animals treated with glycopeptides developed *Aspergillus* sp. and glycopeptide *Streptococcus bovis* infection
Brook *et al.*, 1993b	Endogenous flora	4.25, 4.50, 4.75, 5.00, and 5.25 mixed field (N + γ) irradiation	L-ofloxacin (20 mg/kg/day) Vancomycin (50 mg/kg/day) Teicoplanim (50 mg/kg/day)	3 d/p.o./q.d./21 days 3 d/i.m./q.d./21 days 3 d/i.m./q.d./21 days	2.4 ± 0.5 57.9 ± 6.4 49.6 ± 4.3	Ofloxacin reduced mortality after exposure to up to 4.5 Gy. Glycopeptide failed to reduce mortality due to enterococci because of development of resistance
Brook and Ledney, 1994a	Endogenous flora, *P. aeruginosa*	7.0 Gy and 8.2 Gy Cobalt-60	l-ofloxacin (50 mg/kg/day) Ofloxacin (50 mg/kg/day) Temafloxacin (50 mg/kg/day) Sparafloxacin (50 mg/kg/day) CI-960 (50 mg/kg/day) CI-990 (50 mg/kg/day)	3 d/p.o./b.i.d. or q.d./21 days	3.6 ± 0.5 3.0 ± 0.4 3.9 ± 0.5 4.2 ± 0.6 4.8 ± 0.4 5.3 ± 0.5	All quinolones effective in preventing endogenous and exogenous infection. Those effective also against anaerobes failed in preventing endogenous infection
Brook and Ledney, 1994b	Endogenous flora	7.0, 8.0 and 8.5 Gy Cobalt-60	Gentamicin (7.5 mg/kg/day) Metronidazole (50 mg/kg/day) Imipenem (50 mg/kg/day) Procaine-penicillin (250 mg/kg/day) Ofloxacin (40 mg/kg/day)	2 d/p.o.or i.m./b.i.d. or q.d./21 days	6.4 ± 1.8 24.6 ± 5.8 24.2 ± 4.8 18.9 ± 3.4 2.8 ± 0.5	Metronidazole and penicillin reduced anaerobic flora and increased mortality. Ofloxacin and gentamicin reduced mortality

Table 17.2 Antimicrobial therapy of intra-abdominal abscess in animals exposed to ionizing irradiation

Reference	*Bacteria*	*Irradiation dose/source*	*Antibiotics (dose)*	*When started/ route/dosing schedule/duration*	*Peak serum conc. (mg/l)*	*Comments*
Brook and Ledney, 1994c	*E. coli* *B. fragilis*	8.0 Gy Cobalt-60	Ofloxacin (40 mg/kg/day) Metronidazole (50 mg/kg/day) Imipenem (50 mg/kg/day)	12 hours/p.o. or i.m./b.i.d. or q.d./14 days	3.4 ± 0.6 26.6 ± 4.6 28.8 ± 6.2	Imipenem or ofloxacin plus metronidazole effective in prevention of abscesses. Metronidazole alone increased mortality

Table 17.3 Antimicrobial therapy of wound infection in irradiated mice

Reference	*Bacteria*	*Irradiation dose/source*	*Antibiotics (dose)*	*When started/ route/dosing schedule/duration*	*Peak serum conc. (mg/l)*	*Comments*
Hoogeterp *et al.*, 1987	*S. aureus*	6 Gy 5 MV linear accelerator	Rifampicin (0.5 mg/kg) Benzylpenicillin (16 mg/kg) Erythromycin (40 mg/kg)	1 hour/s.c./a single dose	0.229 ± 0.051 5.89 ± 0.81 4.3 ± 0.3	A higher dose of the erythromycin (bacteriostatic agent) was effective. Penicillin and rifampicin were effective
Brook and Elliott, 1989	*S. aureus*	6.5 Gy Cobalt-60	Penicillin G (62.5 mg/kg/day) Penicillin G (250 mg/kg/day)	1 d/i.m./10 days	>0.6 3.18 ± 0.48	Penicillin reduced the number of organisms in the infected muscle
Hoogeterp *et al.*, 1993	*S. aureus*	6 Gy 5 Mv linear accelerator	Erythromycin (10, 20 and 40 mg/dose) Clindamycin (10, 20 and 40 mg/day)	1 hour/s.c./a single dose	Not available	Both agents less effective *in vivo* after inoculation than *in vitro*

whole-animal, cellular, and molecular parameters of morbidity during the course of infection and responses to treatment.

Animal models allow for controlling the experimental conditions and for the testing of different antimicrobial therapies to prevent (prophylaxis) or treat the potential infection. The model enables the selection of the adequate duration (Brook and Ledney, 1992a) and proper route of treatment. It also allows for observing the effect of the antimicrobial therapy on the normal host gastrointestinal flora. For example, metronidazole was effective only against the anaerobic component of that flora, and was found to enhance mortality by reducing the number of these organisms. Metronidazole facilitated the growth of Enterobacteriaceae and emergence of earlier sepsis and mortality (Brook *et al.*, 1988).

The complex treatment of intra-abdominal abscess in irradiated mice, using agents effective against anaerobes, permitted other beneficial observations. Use of these models enabled the observation of the potential bone-marrow suppressive effect of certain antimicrobials (e.g. pefloxacin), which was only apparent following exposure to irradiation (Patchen *et al.*, 1993). Emergence of bacterial resistance to antimicrobials and appearance of new pathogens (e.g. fungi) was also observed in this model (Brook *et al.*, 1993a).

The drawbacks of irradiation animal models include the unpredictability of the bacteria that may cause endogenous generalized or wound sepsis. This effect may vary between mouse strains and even be different in the same strains at different times as a consequence of factors that are difficult to control. These factors include changes in the endogenous flora effected by shipment conditions, dietary changes, general health, and housing conditions. The exogenous infection models do not have these drawbacks, because the primary pathogen is introduced into the animal. However, changes in an animal's flora or health can affect the outcome.

There are limitations to the conclusions derived from the model of exogenous infection. Models of infectious diseases mimic many, but not necessarily all, signs and symptoms of disease or specific pathogens. Because all the animals receive a single or a set number of bacterial strains in a determined amount, the conclusions can be applied only to the species included and not to the many other species that can occur in patients. However, variable conditions can be controlled better in the experimental model than in the clinical setting and basic principles can be established. The variability in clinical infections is much more complex than these models in combinations of bacterial species and their relative concentrations at their introduction. The parameters used to evaluate the outcome of clinical cases are difficult to quantify. Similar difficulties exist in clinical trials of antimicrobials, where patient variables can interfere with fine distinctions between antimicrobials.

Contributions of the model to infectious disease therapy

Under controlled conditions, basic principles of the infections and their therapy can be established. Irradiated animal models have been used to understand the pathophysiological, immunological, physiological, and bacteriological events following exposure to ionizing irradiation and responses to infectious agents.

In the absence of an opportunity to perform clinical trials in patients, this type of animal model is the only way to test therapeutic modalities in treating infection after irradiation (Brook, 1988). The testing of antimicrobials includes the use of antibiotics alone (Brook and Ledney, 1992b) or in combination with non-specific immunomodulators (Madonna *et al.*, 1991). The work done in these models provided guidelines to the physicians who took care of the irradiation victims at Chernobyl and other incidences of exposure to ionizing irradiation (Gale, 1987; Korschunov *et al.*, 1996). It also demonstrated and allowed for the understanding of pitfalls of using therapies that will suppress the anaerobic gastrointestinal flora and promote mortality (Brook, *et al.*, 1988). The risk of emergence of new pathogens (Brook and Ledney 1991b), bacterial resistance (Brook *et al.*, 1993b) and untoward effects of therapy (Patchen *et al.*, 1993) were also demonstrated in these models.

References

American Association of Physicists in Medicine, Task Group 21, Radiation Therapy Committee. (1983). A protocol for the determination of absorbed dose from high-energy photon and electron beams. *Med. Phys.*, **10**, 741–771.

Brook, I. (1988). Use of antibiotics in the management of postirradiation wound infection and sepsis. *Rad. Res.*, **115**, 1–25.

Brook, I. (1995). Indigenous microbial flora. In *Surgical Infectious Diseases*, (eds Howard, R. J., Simons, R. L.), 3rd edn, pp. 37–46. Appleton and Lange, Norwalk, CT.

Brook, I., Elliott, T. B. (1989). Treatment of wound sepsis in irradiated mice. *Int. J. Radiat. Biol.*, **56**, 75–82.

Brook, I., Elliott, T. B. (1991). Quinolone therapy in the prevention of mortality after irradiation. *Radiat. Res.*, **128**, 100–103.

Brook, I., Ledney, G. D. (1990). Oral ofloxacin therapy of *Pseudomonas aeruginosa* sepsis in mice after irradiation. *Antimicrob. Agents Chemother.*, **34**, 1987–1989.

Brook, I., Ledney, G. D. (1991a). Oral aminoglycoside and ofloxacin therapy in the prevention of Gram-negative sepsis after irradiation. *J. Infect. Dis.*, **164**, 917–921.

Brook I., Ledney, G. D. (1991b). Ofloxacin and penicillin-G in the prevention of bacterial translocation and animal mortality after irradiation. *Antimicrob. Agents Chemother.*, **35**, 1685–1686.

Brook, I., Ledney, G. D. (1992a). Short and long courses of ofloxacin therapy of *Klebsiella pneumoniae* sepsis following irradiation. *Radiat. Res.*, **130**, 61–64.

Brook, I., Ledney, G. D. (1992b). Quinolone therapy in the management of infection after irradiation. *Crit. Rev. Microbiol.*, **18**, 235–246.

Brook, I., Ledney, G. (1994a). Quinolone therapy in the prevention of endogenous and exogenous infection after irradiation. *J. Antimicrob. Chemother.*, **33**, 777–784.

Brook, I., Ledney, G. (1994b). Effect of antimicrobial therapy on the gastrointestinal bacterial flora, infection and mortality in mice exposed to different doses of irradiation. *J. Antimicrob. Chemother.*, **33**, 63–72.

Brook, I., Ledney, G. (1994c). The treatment of irradiated mice with polymicrobial infection caused by *Bacteroides fragilis* and *Escherichia coli*. *J. Antimicrob. Chemother.*, **33**, 243–252.

Brook, I., MacVittie, T. J., Walker, R. I. (1984). Recovery of aerobic and anaerobic bacteria from irradiated mice. *Infect. Immun.*, **46**, 270–271.

Brook, I., MacVittie, T. J., Walker, R. I. (1986). Effect of radiation dose on the recovery of aerobic and anaerobic bacteria from mice. *Can. J. Microbiol.*, **32**, 719–721.

Brook, I., Walker, R. I., MacVittie, T. J. (1988). Effect of antimicrobial therapy on the gut flora bacterial infection in irradiated mice. *Int. J. Radiat. Biol.*, **53**, 709–716.

Brook, I., Elliott, T. B., Ledney, G. D. (1990). Quinolone therapy of *Klebsiella pneumoniae* sepsis following irradiation: comparison of pefloxacin, ciprofloxacin, and ofloxacin. *Radiat. Res.*, **122**, 215–217.

Brook, I., Tom, P., Ledney, G. (1993a). Development of infection with *Streptococcus bovis* and *Aspergillus* sp. in irradiated mice after glycopeptide therapy. *J. Antimicrob. Chemother.*, **32**, 705–713.

Brook, I., Tom, S. P., Ledney, G. D. (1993b). Quinolone and glycopeptide therapy for infection in mouse following exposure to mixed-field neutron-γ-photon radiation. *Int. J. Radiat. Biol.*, **64**, 771–777.

Carter, R. F., Verrelli, D. M. (1973). AFRRI cobalt whole-body irradiator. *Technical Note 73-3*. Armed Forces Radiobiology Research Institute, Bethesda, MD.

Clapper, W. E., Roberts, J. E., Meade, G. H. (1954). Radiation effects on pneumococcal infection produced by subcutaneous injections into white mice. *Proc. Soc. Exp. Biol. Med.*, **86**, 420–422.

Elliott, T. B., Brook, I., Stiefel, S. M. (1990). Quantitative study of wound infection in irradiated mice. *Int. I. Radiat. Biol.*, **58**, 341–350.

Elliott, T. B., Ledney, G. D., Harding, R. A. *et al.* (1995). Mixed-field neutrons and γ photons induce different changes in ileal bacteria and correlated sepsis in mice. *Int. J. Radiat. Biol.*, **68**, 311–320.

Gale, R. P. (1987). Immediate medical consequences of nuclear accidents: lessons from Chernobyl. *J.A.M.A.*, **285**, 625–628.

Geraci, J. P., Jackson, K. L., Mariano, M. S. (1985). Effect of *Pseudomonas* contamination or antibiotic decontamination of the GI tract on acute radiation lethality after neutron or γ irradiation. *Radiat. Res.*, **104**, 395–405.

Hammond, C. W., Ruml, D., Cooper, D. E., Miller, C. P. (1955a). Studies on susceptibility to infection following ionizing radiation. III. Susceptibility of the intestinal tract to oral inoculation with *Pseudomonas aeruginosa*. *J. Exp. Med.*, **102**, 403–411.

Hammond, C. W., Vogel, H. H. Jr, Clark, J. W., Cooper, L. B., Miller, C. P. (1955b). The effect of streptomycin therapy on mice irradiated with fast neutrons. *Radiat. Res.*, **2**, 354–360.

Hoogeterp, J. A. A. P. J., Mattie, H., Krul, A. M., vanFurth, R. (1987). Quantitative effect of granulocytes on antibiotic treatment of experimental staphylococcal infection. *Antimicrob. Agents Chemother.*, **31**, 930–934.

Hoogeterp, J. A. A. P. J., Mattie, H., vanFurth, R. (1993). Activity of erythromycin and clindamycin in an experimental *Staphylococcus aureus* infection in normal and granulocytopenic mice. *Scand. J. Infect. Dis.*, **25**, 123–132.

Kaplan, H. W., Speck, R. S., Jawetz, F. (1958). Impairment of antimicrobial defenses following total body irradiation of mice. *J. Lab. Clin. Med.*, **40**, 682–691.

Keller, P. D. (1946). A clinical syndrome following exposure to atomic bomb explosions. *J.A.M.A.*, **131**, 504–508.

Klastersky J. (1998). Science and pragmatism in the treatment of neutropenic infection. *J. Antimc. Chemother.* **41** (Suppl D), 13–24.

Klaines, A. S., Gorbach, S., Weinstein, L. (1967). Studies of intestinal microflora. Effect of irradiation on the fecal microflora of the rat. *J. Bacteriol.*, **91**, 378–382.

Klemparskaya, N. N., Pinegin, B. V., Shal'nova, G. A. *et al.* (1987). Normalizing effect of immunoglobulins in the treatment of endogenous infection and intestinal dysbacteriosis in irradiated mice. *J. Hyg. Epidemiol. Microbiol. Immunol.*, **31**, 91–98.

Korschunov, V. M., Smeyanov, V. V., Efimov, B. A., Tarabrina, N. P., Ivanov, A. A., Baranov, A. E. (1996). Therapeutic use of an antibiotic-resistant *Bifidobacterium* preparation in men exposed to high-dose γ-irradiation. *J. Med. Microbiol.*, **44**, 70–74.

Kurishita, A., Ono, T., Uchida, A. (1993). Prevention of radiation-induced bacteraemia by post-treatment with OK-432 and aztreonam. *Int. J. Radiat. Biol.*, **63**, 413–417.

Ledney, G. D., Exum, E. D., Stewart, D. A., Gelston, H. M. Jr., Weinberg, S. R. (1982). Survival and hematopoietic recovery in mice after wound trauma and whole-body irradiation. *Exp. Hematol.*, **10**, 263–278.

Leroy, G. V. (1947). The medical sequelae of the atomic bomb explosion. *J.A.M.A.*, **134**, 1143–1148.

Madonna, G. S., Ledney, G. D., Elliott, T. B. *et al.* (1989). Trehalose dimycolate enhances resistance to infection in neutropenic animals. *Infect. Immun.*, **57**, 2495–2501.

Madonna, G. S., Ledney, G. D., Moore, M. M., Elliott, T. B., Brook, I. (1991). Treatment of mice with sepsis following irradiation and trauma with antibiotics and synthetic trehalose dicorynomycolate (S-TDCM). *J. Trauma*, **31**, 316–325.

Mathe, G. (1965). Total body irradiation injury: a review of the disorders of the blood and hematopoietic tissues and their therapy. In *Nuclear Hematology*, (ed. Eszirami, R.), pp. 275–338. Academic Press, New York.

Miller, C. P., Hammond, C. W., Tompkins, M. (1951). The role of infection in radiation injury. *J. Lab. Clin. Med.*, **38**, 331–343.

Miller, C. P., Hammond, C. W., Tompkins, M. J., Shorter, G. (1952). The treatment of post-irradiation infection with antibiotics, an experimental study in mice. *J. Lab. Clin. Med.*, **39**, 462–479.

Neta, R., Oppenheim, J. J. (1988). Cytokines in therapy of radiation injury. *Blood*, **72**, 1093–1099.

Patchen, M. L., Brook, I., Elliott, T. B., Jackson, W. E. (1993). Adverse effects of pefloxacin in irradiated C3H/HeN mice: correction with Glucan therapy. *Antimicrob. Agents Chemother.*, **37**, 1882–1889.

Peterson, V. M., Adamovicz, J. J., Elliott, T. B. *et al.* (1994). Gene expression of hematoregulatory cytokines is elevated endogenously after sublethal gamma irradiation and is differentially enhanced by therapeutic administration of biologic response modifiers. *J. Immunol.*, **153**, 2321–2330.

Solomkin, J. S., Meakins, J. L. Jr, Allo, M. D., Dellinger, E. P., Simmons, F. L. (1984). Antibiotic trials in intra-abdominal infections: a critical evaluation of study design and outcome reporting. *Ann. Surg.*, **20**, 29–35.

Toth, L. A. (1997). The moribund state as an experimental endpoint. *Contemp. Topics.*, **36**, 44–48.

Taketa, S. T. (1962). Water electrolyte and antibiotic therapy against acute (3 to 5 days) intestinal radiation death in rats. *Radiat. Res.*, **16**, 312–326.

Van der Waaij, D., van der Waaij, J. M. (1985). Spread of multiresistant Gram-negative bacilli among severely immuno-compromised mice during prophylactic treatment with different oral antimicrobial drugs. In *Germfree Research: microflora control and its application to the biomedical sciences*, (ed. Wostmann, B. S.) pp. 245–250. Alan R. Liss, New York.

Wensinck, F., Renaud, H. (1957). Bacteremia in irradiated mice. *Br. J. Exp. Pathol.*, **38**, 483–488.

Chapter 18

Intra-abdominal Abscess

I. Brook

Background of human infections

Intra-abdominal sepsis is generally caused by the polymicrobial aerobic–anaerobic gastrointestinal flora. Most of these infections occur after breach of the barrier defenses usually as a result of perforation of an abdominal viscus, where the infecting organisms originate from the gut flora present at the perforation site. The number of bacterial species increases gradually in the distal portion of the gastrointestinal tract. The gastrointestinal flora at the colon is made of over 400 microbial species, the concentration of organisms approaches 10^{12} bacteria/g of stool, and the ratio of anaerobic vs aerobic bacteria is 1000 : 1 (Brook, 1995).

Intra-abdominal sepsis is a biphasic illness. Following the breach of the barrier defenses the patient generally suffers from generalized peritonitis, often associated with bacteremia due to Enterobacteriaceae (mostly *Escherichia coli*), and the secondary stage of the illness is characterized by the emergence of intra-abdominal abscesses, where anaerobic bacteria predominate (mostly *Bacteroides fragilis* group). The average number of isolates from the infected sites (peritoneum or abscess) that is recovered by clinical microbiology laboratories is five (three anaerobic and two aerobic bacteria; Brook, 1989; Swenson *et al.*, 1974).

Selection of antimicrobials for the treatment of intra-abdominal sepsis presents several problems. These include difficulty in knowing toward which of the isolates should therapy be directed and how effective certain antimicrobials are in clinical situations. Human clinical trials cannot answer these and other questions because of the variability in patients' clinical presentations, the need for drainage and corrective surgery, and other human variables (i.e. immunity, age, genetics, other medical problems). Because of these reasons an animal model of intra-abdominal sepsis is useful (Weinstein *et al.*, 1974; Bartlett and Gorbach, 1979; Onderdonk *et al.*, 1979; Gorbach, 1982; Bartlett, 1984).

Background of the model

The development of intra-abdominal abscess began by the group of Weinstein *et al.* (1974). The model simulated septic complication of colonic perforation by inoculation of stool into the peritoneal cavity of rats. The studies attempted first to identify the infections' dynamics and the organisms responsible for the infection (Weinstein *et al.*, 1974; Bartlett *et al.*, 1978; Onderdonk *et al.*, 1979), and subsequently the model was used to evaluate and compare the efficacy of antimicrobials in the management of the infection (Bartlett *et al.*, 1981).

The studies that established the model (Weinstein *et al.*, 1974; Onderdonk *et al.*, 1976) used pooled cecal contents obtained from 25–50 meat-fed rats (fed for 2 weeks prior to sacrifice). The colonic flora of meat-fed rats more closely simulates that of humans than does the flora of grain-fed rats. After the addition of peptone yeast and barium sulfate, and filtration (see below), a gelatin capsule containing 0.5 ml of the inoculum was inserted intraperitoneally.

The initial studies established the pathological and microbiological events and identified the true pathogens (Weinstein *et al.*, 1974; Onderdonk *et al.*, 1976). The infection was found to have two stages. During the initial 5 days after implantation, an acute generalized peritonitis developed, where 43% of the animals died within 4 days. *E. coli* was recovered from the blood of most of these animals. The second stage of the illness, between days 7 and 14 after challenge, was characterized by the formation of multiple intra-abdominal abscesses, where the predominate isolates were *B. fragilis, Fusobacterium varium, E. coli* and *Enterococcus*.

Animal species

Several strains of rat have been used in the model: Wistar (160–180 g male, Weinstein *et al.*, 1974; Louie *et al.*, 1977; Bartlett *et al.*, 1981; Bansal *et al.*, 1985; Wells *et al.*, 1985; Hau *et al.*, 1986; Thadepalli *et al.*, 1986; Muhvich *et al.*, 1988; McRitchie *et al.*, 1989; Willey *et al.*, 1989; Schwartz *et al.*, 1992; Pefanis *et al.*, 1993), Sprague–Dowley (150–400 g, Montravers *et al.*, 1994). Mice strains were also used: Balb/c (female, Nulsen *et al.*, 1983; Sawyer *et al.*, 1994); CF-1 (20 g female, Fu *et al.*, 1985), Swiss Webster (25–35 g male, Cheadle *et al.*, 1989); CH3/HEJ (20–25 g male, Brook, 1994); Swiss albino (20–25 g male, Brook and Gillmore, 1993).

Handbook of Animal Models of Infection
ISBN 0-12-775390-7

A key element in achieving reproducibility was to inoculate the animals with cecal contents (pooled from meat-fed rats obtained from multiple animals if endogenous rat flora is used), insure initial concentration of over 10^7g of anaerobes (usually *B. fragilis*) and 10^5g of facultatives or aerobes (usually *E. coli*); and add an adjuvant such as barium sulfate, which is necessary to consistently produce abscesses. Adjuvants other than barium sulfate include hemoglobin (Yull *et al.*, 1962), fibrin clot (Wells *et al.*, 1985), gum tragacanth, hog gastric mucin (Olitzki, 1948), dead tissue or foreign body such as sutures (Elek and Conen, 1957), a piece of potato or sponge (Tudor *et al.*, 1988), crystalline cellulose (Fu *et al.*, 1985), or fecal–agar pellet (Muhvich *et al.*, 1988). However, the role of the non-bacterial component has not been extensively studied. These products are believed to inhibit phagocytosis.

It is known that the injection of broth cultures of bacteria intraperitoneally in animals will cause septic shock and death due to rapid absorption from the peritoneal cavity (Scott and Robson, 1976), and little evidence of peritonitis is found on pathological examination. This type of experiment is not considered appropriate as a model of intra-abdominal sepsis in humans. The gelatin capsule that is used to inoculate the bacteria plays no role in the generation of the abscess.

Preparation of animals

No special housing or care nor specific pretreatment is required. Animals can be housed in groups of up to 10 before surgery. They are maintained on chow and water *ad lib*.

Details of surgery

Overview

Briefly, the anesthetized animals are surgically exposed and the inoculum is implanted intraperitoneally inside or not inside a capsule. It is designed to insure a uniform inoculum of the organisms for all animals.

Materials required

Anesthetic (preferably injectable), hair clippers or razor, skin disinfectant, scalpel handle plus blades, forceps, syringes and needles (23 G), suture materials.

Anesthesia

Animals are sedated by intramuscular (i.m.) or intraperitoneal (i.p.) injection of ketamine sulfate (Parke Davis, Morris Plains, NJ; 30 mg/kg of body weight). Complete anesthesia occurs in about 5 minutes and lasts about 15 minutes. Alternative anesthetics include phenobarbital sodium (Thadepalli *et al.*, 1986), ether (Wells *et al.*, 1985), a halothane–nitrous-oxide–oxygen mixture (McRitchie *et al.*, 1989), ketamine hydrochloride plus xylazine hydrochloride (Schwartz *et al.*, 1992), and sodium thiopental (Nichols *et al.*, 1979).

Surgical procedure

Following complete induction of anesthesia, the animal is placed on its back and the abdomen is shaved and prepared with topical skin disinfectant (i.e. iodine, betadine). A 1 cm midline incision is made that cuts through the abdominal muscle and the peritoneum, and the capsule containing the inoculum or fibrin clot is implanted preferably (for the gelatin capsule) into the pelvic region of the abdominal cavity. The wound is closed with musculoperitoneal layer and skin layer using interrupted nylon or silk sutures. Clips can also be used to close the incision (Wells *et al.*, 1985).

For experiments using combination of micro-organisms, the inocula are made by adding equal volumes of the individually prepared bacterial strain. Diluted broth is combined with pooled autoclaved rat colonic contents (50% vol./vol.) and barium sulfate (10% wt/vol.). Aliquots of 0.5 ml of the final product are placed using a 1 ml syringe into no. 0 gelatin capsule for intraperitoneal implantation. It is desirable to implant the capsules as soon as they are prepared, desirably within 5 minutes of preparation. However, in case of delay, the capsule should be kept refrigerated, to preserve the relative numbers of organisms. Quantitative bacteriological cultures of each inoculum are performed just before implantation.

Postoperative care

Animals should be kept warm after the surgery by placing under an infra-red lamp until they begin to recover from the anesthetic. Up to 5% acute mortality (within 4 hours) can be observed, secondary to surgery or anesthesia. Animals should be housed individually following surgery to facilitate wound healing and prevent wound infection. Food and water consumption can be low in the first 10 days.

Storage and preparation of inocula

The model is amenable to using several different bacterial species, alone or in combination. In the original work (Weinstein *et al.*, 1974) the rats were implanted with gelatin capsule containing an inoculum of barium sulfate and pooled cecal contents of meat-fed rats. However, the types of bacterial species and their numbers were difficult to control or replicate. For this reason it is preferable to choose the bacterial strains, grow them *in vitro*, and set their desired combinations and concentration in the inoculum (Onderdonk *et al.*, 1979a).

To obtain rat cecal flora inoculum the following procedures are done. The cecal contents are obtained by probing the cecal and large bowel of 15 rats. The abdomen of each rat is aseptically opened, the cecum and proximal large intestine are then clumped, excised and immediately entered into an anaerobic glove box where the processing is done. The contents of the cecum and bowel are extracted into a sterile beaker and the tissue is macerated. An equal volume of prereduced peptone–yeast–glucose broth is added to this material and vigorously mixed. This slurry is filtered twice through sterile gauze to remove particulate matter, sterile barium sulfate (10% wt/vol.) is added, and the inoculum is aliquoted into small portions (5 ml) and frozen at –80°C until use. Quantitative and qualitative bacteriology of this inoculum is performed to insure that appropriate microbial species are present. All procedures are carried out in an anaerobic chamber. The desired organisms in such an inoculum should include members of the *B. fragilis* group, *Clostridium* spp., Enterobacteriaceae (especially *E. coli*), and *Enterococcus* spp. The initial inocula must be repeatedly passed intraperitoneally in either mice or rats before they can be used in experiments. Stored inocula lose their virulence. Bacteria acquire capsule when passed and become virulent. This is especially important for *B. fragilis* (Brook *et al.*, 1984). Insuring encapsulation and maintenance of virulence is especially important in infection induced by one or two organisms (see below).

In experiments using chosen organisms, all bacterial strains are grown in prereduced peptone–yeast broth in an anaerobic chamber. Aliquotes of each culture are placed into sterile glass vials after incubation for 24–48 hours at 35°C. They are then frozen in liquid nitrogen and stored at –40°C until use. The desired inoculum can vary and has generally been made of organisms of the *Bacteroides fragilis* group (*B. fragilis, B. thetaiotaomicron, B. vulgatus, B. distasonis*), *Prevotella* species (*P. melaninogenica, P. intermedia*), *Porphyromonas asaccharolytica, Clostridium perfringens, Escherichia coli, Klebsiella pneumoniae*, and *Enterococcus* spp. *In-vitro* susceptibility of the organisms to the antimicrobials used in the experiment should be established to insure efficacy. The inoculum is prepared after thawing the vials of the test organisms inside an anaerobic chamber. They are diluted in a prereduced peptone–yeast broth. The number of anaerobic organisms per inocula is between 10^7 and 10^8, and the number of *E. coli* or other Enterobacteriaceae is generally 10^{5-6}. Higher inocula of *E. coli* can induce rapid fatal septicemia.

A different type of inoculum, made of fecal pellets, has also been developed (Nakatani *et al.*, 1984). A 0.6 ml mixture of powdered feces (160 mg), tripticase soy agar (19.2 mg), and water (0.54 ml) was placed into a mold made from a plastic syringe, and then autoclaved for 30 minutes at 121°C. After cooling the fecal gel to 45–50°C, 0.1 ml of sterile salve and 0.1 ml of live bacterial suspension is added and mixed well. The final pellet contains 20% (wt/vol.) dried feces, 2.4% (wt/vol.) agar, and the desired organism(s). The number of living organisms is 1 log less than the final concentration used. The syringe is capped with sterile aluminum foil and kept at 4°C until used (within 1 hour).

Infection process

The infection is initiated either by implantation of the organisms within the gelatin capsule (which dissolves shortly after insertion), fibrin clot, fecal pellets, or by direct inoculation. The inoculum is initially localized in the pelvic area. Within 24 hours a suppurative infection and ileus develops, and 0.2–0.5 ml of infected peritoneal fluid accumulates (Weinstein *et al.*, 1974). The fur appears ruffled, and the animals are lethargic and cold. At 48 hours, peritoneal adhesions appear anteriorly, and loose attached collections of purulent material can be noted. By day 3, up to half the untreated animals die, and the organism isolated from their blood cultures is generally *E. coli*. By 1 week, a well-formed abscess(es) can be palpated inferiorly along the anterior abdominal wall. There is rarely any excess peritoneal fluid, and the surgical wound is healed. The rats begin to gain weight and do not appear ill. The abscesses continue to enlarge and can have up to 0.3 ml of pus by 14 days. Sterile barium granulomas can be observed within the peritoneum and the viscera. Some abscesses may perforate spontaneously at 3–4 weeks, causing generalized peritonitis. However, others may persist for months. Such animals may have multiple intraperitoneal and or visceral abscesses. Abscesses generally contain the inoculated organisms—predominately the *B. fragilis* group.

Key parameters to monitor infection

The infection is initially in the peritoneum, but a sepsis stage rapidly develops and may last up to 1 week. Following that, the infection is confined to the abscess(es).

Death due to bacteremia generally occurs within the first week in up to half the untreated animals, and is a good indication for the efficacy of therapy directed at the eradication of Enterobacteriaceae. Confirmation of the organism causing mortality can be achieved by obtaining blood cultures, through percutaneous transthoracic cardiac puncture from additional living animals, or at necropsy. Appearance, body weight gain, food and water intake are also affected by the systemic infection. Changes in these parameters are especially apparent in the sepsis stage (first week) of the infection, and indicate the presence of systemic infection.

Pathological studies of the abscess cavity can be done to confirm the presence of the histopathological process. However, gross observation of these abscesses is generally sufficient. The criterion for defining an abscess is a localized collection of grossly purulent exudate, which shows bacteria and a predominance of polymorphonuclear leukocytes on Gram stain. The incidence of intra-abdominal abscess is also determined at a time that is set 12–16 days after implantation (Figure 18.1).

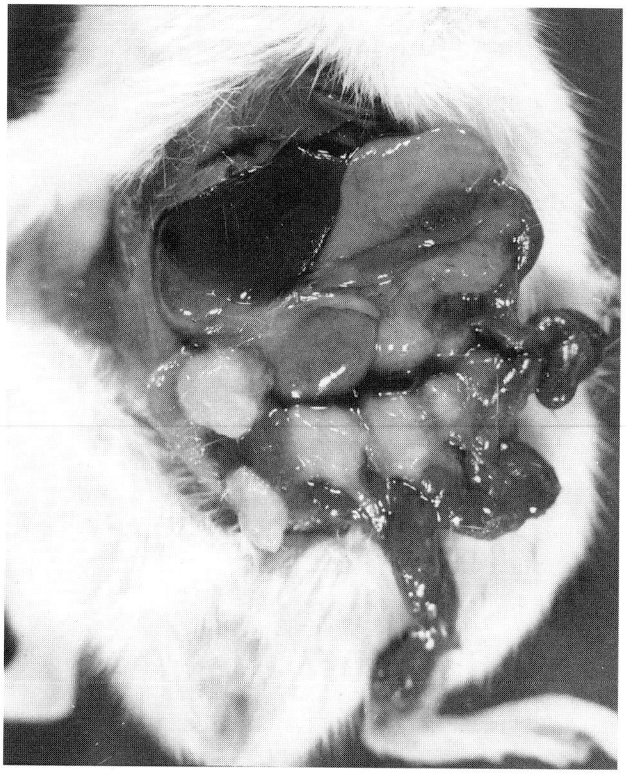

Figure 18.1 Intra-abdominal abscesses in a mouse.

Quantitative and qualitative analysis of the abscesses is also important. It can provide evidence regarding the efficacy of any therapy. The abscess(es) size should be measured and then aseptically removed, weighed (if knowledge of the total number of organisms is desired), and minced. The homogenate should be serially diluted and cultured qualitatively and quantitatively in an anaerobic chamber for aerobic and anaerobic bacteria.

Determining the bacterial cfu/g within the abscess requires performing serial dilution in prereduced saline, with proper and vigorous homogenization.

Determination of antibiotic concentration in serum and abscess

Antibiotic concentration in the serum and inside the abscess can be determined by a variety of methods, although bioassay appears to be the most common method used. Studies determined the antimicrobial concentration in the serum (Louie *et al.*, 1977; Bartlett *et al.*, 1981; Bansal *et al.*, 1985; Brook and Gillmore, 1993), the peritoneal cavity (Hau *et al.*, 1986; Thadepalli *et al.*, 1986) and abscess cavity (Tudor *et al.*, 1988; Cisneros *et al.*, 1990; Thadepalli *et al.*, 1997).

Determination of additional markers of infection and response to treatment

Tumor necrosis factor (TNF) α serum levels were measured by Thadepalli *et al.* (1997) in treated and untreated animals, and were used to monitor efficacy. The method used was the enzyme-linked immunosorbent essay (BioSource International, Camarillo, CA).

Antimicrobial therapy

A variety of antimicrobial agents have been evaluated in the intra-abdominal abscess model in rodents. They are summarized in Tables 18.1 and 18.2. The main lesson from the initial work in this model (Weinstein *et al.*, 1975) that applies to therapy of humans is that the proper therapy of all phases (both the peritonitis/sepsis and abscess phases) of the infection includes therapy effective against both the aerobic (Enterobacteriaceae) and anaerobic (*B. fragilis* group) component of the infection.

In the original studies (Weinstein *et al.*, 1975) the antimicrobials elected were gentamicin and clindamycin because of their unique spectra of activity, which allows for the distinction between the roles of coliforms and anaerobes. Gentamicin was effective only against Enterobacteriaceae and clindamycin against anaerobes. These drugs have opposite effects on the two phases of the infection. Gentamicin reduced the mortality seen in the first week from 100% to 4%, but more than half the surviving animals had intra-abdominal abscess. In contrast, clindamycin had an opposite effect, it did not reduce the initial mortality, but prevented abscess formation. The combination of both drugs produced the optimal results, reducing both mortality and morbidity. Most all other agents in single or combination therapies used in this model attempt to achieve a similar outcome.

For optimal benefits of therapy, antimicrobials should be started no later than 4 hours after bacterial challenge. Therapy delayed for 3 days after inoculation often fails. Prophylactic therapy should be given 1 hour prior to inoculation of bacteria (Brook, 1994). Therapy should be continual for 5–15 days, at intervals determined by the antimicrobial bioavailability. Prophylactic therapy should be shorter. Rats can be administered antibodies by a variety of routes (s.c., p.o., i.m., i.p.). Duration of treatment can affect the outcome of therapy, and experimental design should consider varying the length of treatment. Longer therapies can reduce the number of organisms within an abscess, and reduce their size and total number. The dosing schedule and choice of dose should be given careful consid-

Table 18.1 Examples of chemotherapy of abdominal infection in rats

Reference	*Bacteria*	*Therapy started/ route/duration (days)*	*Antibiotics (dose/frequency)*	*Peak serum content (mg/l)*	*Comments*
Louie *et al.*, 1977	Cecal contents	4 hours/i.m./10 days	Cephalothin (115 mg/kg/d, q.8 hours) Cefazolin (115 mg/kg/d, q.8 hours) Cefoxitin (115 mg/kg/d, q.8 hours) Cefamandole (115 mg/kg/d, q.8 hours) Clindamycin (115 mg/kg/d, q.8 hours) Gentamicin (11 mg/kg/d, q.8 hours) Tobramycin (11 mg/kg/d, q.8 hours)	35 107 130 63 6 7 9	All cephalosporins and aminoglycosides reduced mortality. Abscess formation was reduced by clindamycin, and was variable with cephalosporins — cefoxitin being the most effective
Nichols *et al.*, 1979	Human feces	2 hours/i.m./q.6 hours for 48 hours, then q. 8 hours for 7 days	Clindamycin (15 mg/dose) Tobramycin (2 mg/dose) Cephalothin (60 mg/dose) Cefamandole (60 mg/dose)	5.6 ± 1.5 6.6 ± 1.0 60.1 ± 32.9 133.6 ± 44.1	Clindamycin plus tobramycin most effective
Onderdonk *et al.*, 1979	*B. fragilis* group *E. coli* *Clostridium* spp.	4 hours/i.m./10 days	Chloramphenicol (16 mg/dose, q.8 hours) Metronidazole (16 mg/dose, q.8 hours)	22.0 30.0	Metronidazole reduces *E.-coli*-associated mortality.Chloramphenicol only slightly effective because of its inactivation by *B. fragilis* Antibiotics
Bartlett *et al.*, 1981	Cecal contents	4 hours/i.m./10 days	Amikacin (7 mg/dose) Gentamicin (2 mg/dose) Tobramycin (2 mg/dose) Cephalothin (20 mg/dose) Cefazolin (20 mg/dose) Cefamandole (20 mg/dose) Cefoxitin (20 mg/dose) Cefotaxime (20 mg/dose) Moxalactam (20 mg/dose) Penicillin G (60 mg/dose) Carbenicillin (90 mg/dose) Doxycycline (4 mg/dose) Chloramphenicol (16 mg/dose) Clindamycin (16 mg/dose) Erythromycin (16 mg/dose) Metronidazole (15 mg/dose) Trimethoprim (2 mg/dose) } q. 8 hours	40 6.9 7.1 6.0 60 23 18 24 36 50 290 2.5 19 3.1 3.2 33 3.2	effective against coliforms prevented mortality (aminoglycosides most effective), while those effective against *B. fragilis* group reduced occurrence of abscesses Exceptions: metronidazole also reduced mortality; chloramphenicol only modestly effective in reducing abscesses
Bansal *et al.*, 1985	*B. fragilis* *E. coli*	6 hours i.m./10 days	Penicillin (75000 UI/dose, q.8 hours) Penicillin and clavulanate (75000 UI/dose, q.8 hours) Ticarcillin (150 mg, q.8 hours) Ticarcillin and clavulanate (150 mg, q. 8 hours)	28.0 31.0	Highest success rate with ticarcillin–clavulanate

Table 18.1 Continued

Reference	*Bacteria*	*Therapy started/ route/duration (days)*	*Antibiotics (dose/frequency)*	*Peak serum content (mg/l)*	*Comments*
Onderdonk and Cisneros, 1985	*B. fragilis* *S. intermedius*	2 hours/i.m./10 days	Clindamycin (15 mg/dose, q.8 hours) Metronidazole (20 mg/dose, q.8 hours)	5.5 26	Metronidazole not effective against *S. intermedius*, while clindamycin was effective
Thadepalli *et al.*, 1986	*B. fragillis* *E. coli*	4 hours/i.m./13 days	Difloxacin (20 & 40 mg/kg/dose) A-56620 (20 & 40 mg/kg/dose) Ciprofloxacin (20 & 40 mg/kg/dose) Cefoxitin (130 mg/kg/dose) Clindamycin (75 mg/kg/dose) plus Gentamicin (20 mg/kg/dose) } q.8 hours	14.3 ± 0.8 1.98 ± 0.2	Difloxacin as effective as clindamycin–gentamicin. A-56620 equal to cefoxitin
Tudor *et al.*, 1988	*B. fragilis* *C. perfringens* *E. coli* *P. vulgaris* *S. faecalis*	1 dose oral/i.v. 4 days after inoculation	Fleroxacin (20 mg/kg i.v.) Fleroxacin (20 mg/kg oral) Ciprofloxacin (20 mg/kg i.v.) } q.8 hours	1.75 ± 0.33 2.86 ± 0.57 2.35 ± 0.4	In pus 12.7 ± 3.69 at 4 hours In pus 13.39 ± 3.13 at 2 hours In pus 2.25 ± 0.42 at 4 hours
Cisneros and Onderdonk, 1989	Pooled cecal contents	4 hours/i.m./7 days	Ciprofloxacin (9 mg/dose and 15 mg mg/dose, q.8 hours) Clindamycin (15 mg/dose, q.8 hours) Gentamicin (2 mg/dose, q.8 hours)	0.83 (for 9)– 3.19 (for 15) 5.5 6.9	Ciprofloxacin plus clindamycin equal to gentamicin plus clindamycin
Willey *et al.*, 1989	*E. faecalis*	4 hours/continuous infusion/5 days	Ampicillin (400 mg/kg) Clindamycin (400 mg/kg) Gentamicin (30 mg/kg) Metronidazole (300 mg/kg) Aztreonam (400 mg/kg)	15.4 ± 5.12 8.12 ± 1.86 3.10 ± 1.3 53.7 ± 12.5 19.3 ± 6.03	Ampicillin with or without gentamicin most effective. Clindamycin or metronidazole plus gentamicin also effective
Cisneros *et al.*, 1990	Pooled cecal content	4 hours/i.m./7 days	Ampicillin–sulbactam (72 mg/dose, q.6h)	154.5 ± 34.4	Similar levels to serum also in abscess Ampicillin–sulbactam effective in preventing mortality and abscesses
Rice *et al.*, 1991	*K. pneumoniae*	3–4 hours/continuous infusion/3 days	Cefoperazone (600 mg/kg/d) Cefoperazone (600 mg/kg/d) + sulbactam (300 mg/kg/d) Cefotaxime (400 mg/kg/d) Cefpirone (400 mg/kg/d) Ceftazidine (400 mg/kg/d) Imipenem–cilastalin (300 mg/kg/d)	13.5 ± 4.72 8.9 ± 3.22 17.7 ± 8.42 28.3 ± 2.06 19.4 ± 3.09 7.1 ± 2.08	Imipenem and cefoperazone were the most effective *in vivo*
Rice *et al.*, 1993	*E. coli*	2 hours/continuous infusion/3 days	Ampicillin (500 mg/kg/d) Sulbactam (250 mg/kg/d) Cefoxitin (500 mg/kg/d)	22 18.4 22.7	Cefoxitin superior to ampicillin and sulbactam
Pefanis *et al.*, 1993	*B. fragilis*	5 hours/i.m./3.5 days	Cefotaxime (180 mg/q.4 hours) Ceftriaxone (60 mg/15 hours) Ceftriaxone + tazobactam (40 mg/q.6 hours)	Not measured 192 ± 29.7 171 ± 35.9	All therapies effective in reducing number of *B. fragilis*

Table 18.1 Continued

Reference	Bacteria	Therapy started/ route/duration (d)	Antibiotics (dose/frequency)	Peak serum content (mg/l)	Comments
Pefanis *et al.*, 1994	*B. fragilis* group *Clostridium* spp. *E. coli* *Enterococcus* spp.	2 hours/i.v./8.5 days	Fleroxacin (10 mg/kg/q.6 hours) Clindamycin (80 mg/kg/q.3 hours) Gentamicin (8 mg/kg/q.8 hours) Metronidazole (250 mg/kg/24 hours)	6.9 ± 0.62 14.1 ± 1.46 21.5 ± 2.24 36.5 ± 2.12	Fleroxacin alone or in combinations effective against *E. coli*. Combination of fleroxacin with metronidazole or clindamycin effective against anaerobes
Thadepalli *et al.*, 1997	*B. fragilis* *E. coli*	4 hours/i.m./10 days	Trovafloxacin (40 mg/dose, q.8 hours) Clindamycin (75 mg/kg/dose, q.8 hours) Gentamicin (20 mg/kg/dose, q.8 hours)	2.7 ± 0.3	Trovafloxacin abscess concentration twice the serum concentration. Trovafloxacin equal to clindamycin plus gentamicin in efficacy

Table 18.2 Examples of chemotherapy of intra-abdominal infection in mice

Reference	Bacteria	Therapy started/ route/units	Antibiotics (dose/frequency)	Peak serum content (mg/l)	Comments
Fu *et al.*, 1985	*B. fragilis* *P. aeruginosa*	3.5 hours/s.c.: q.d./ 3 days	Rifampicin (10 mg/kg/dose, q.d.) Cefsulodin (100 mg/kg/dose, q.d.)	11.5 ± 1.6 102 ± 11.6	Combination of rifampicin and cefsulodin most effective. Single therapies prevented mortality
Brook and Gillmore, 1993	*E. coli* *B. fragilis* group	6 hours/i.m./10 days	Aztreonam (400 mg/kg/d, q.12 hours) Ciprofloxacin (60 mg/kg/d, q.d. × 10 d) Ofloxacin (60 mg/kg/d, q.d.) Gentamicin (7.5 mg/kg/d, q.d.) Clindamycin (160 mg/kg/d, q.d.) Cefoxitin (400 mg/kg/d, q.12 hours) Cefotetan (400 mg/kg/d, q.12 hours) Cefmetazole (400 mg/kg/d, q.12 hours) Ceftizoxime (400 mg/kg/d, q.12 hours) Ampicillin + sulbactam (400 mg/kg/d, q.12 hours) Imipenem + cilastatin (50 mg/kg/d, q.d.)	45.6 ± 4.0 3.2 ± 0.5 3.4 ± 0.6 5.6 ± 0.8 8.6 ± 2.0 62.0 ± 9.1 66.3 ± 8.4 60.4 ± 8.6 76.4 ± 6.0 135.2 ± 17.6 26.3 ± 5.2	No abscesses with clindamycin or imipenem only. Combination of clindamycin with either quinolones, aztreonam or gentamicin; or single therapy with imipenem, cefoxin, cefotetan, cefmetazole and ampicillin sulbactam reduced number of *E. coli* and *B. fragilis*
Brook, 1994	*E. coli* *B. fragilis* group	1 hour before/i.m./ 10 days	Cefoxitin (600 mg/kg/d, q.12 hours) Cefotetan (600 mg/kg/d, q.12 hours) Ampicillin–sulbactam (400 mg/kg/d, q. 12 hours)	62.0 ± 9.1 66.3 ± 8.4 138.5 ± 16.3	Mortality reduced. Cefotetan > cefoxitin > ampicillin–sulbactam. Cephalosporins more effective in reducing *E. coli* number in abscesses
Sawyer *et al.*, 1994	*B. fragilis* *E. coli*	1 hour before/i.m./ 3 days	Gentamicin (9 mg/kg/dose, q.8 hours) Aztreonam (200 mg/kg/dose, q.8 hours) Clindamycin (85 mg/kg/dose, q.8 hours)	Not measured	Aztreonam plus clindamycin superior to gentamicin plus clindamycin in preventing abscess formation and eliminating *E. coli*

eration, given the different pharmacokinetics of antibiotics between humans and rodents.

Pitfalls (advantages/disadvantages) of the model

The model satisfies the goal of simulating clinically encountered intra-abdominal infections, both bacteriologically and pathologically.

The initial model was not designed to examine the relative efficacy of antimicrobial agents (Weinstein *et al.*, 1974; Onderdonk *et al.*, 1976), but rather used them as probes to investigate their relative importance (Weinstein *et al.*, 1975). However, the model became very useful in evaluating and comparing agents capable of controlling the biphasic illness. This model allows for testing whether an antimicrobial or an antimicrobial combination is capable of eliminating both *E. coli* and the *B. fragilis* group.

However, there are limitations to the conclusions that can be drawn from the model. Since all animals receive a set number of strains in a set concentration, the conclusions drawn can be applied only to that isolate(s) and not to the many others that can occur in patients. The variability in clinical medicine is much more complex in combinations of isolates and bacterial species and their relative concentration. The parameters used to evaluate the outcome are difficult to quantify. Similar difficulties exist in clinical trials of antimicrobials, where patients' individual variability can interfere with finer distinction between antimicrobials.

The mouse model is apparently less used than the rat model. Some advantages of the rat model over the mouse is their tolerance to prolonged antibiotic administration, easier collection of blood and specimens, and generation of larger abscesses. The mouse model has the advantage of easier maintenance and handling, cheaper cost of acquisition and upcoming, smaller housing space, and the need for smaller doses of antimicrobials and inocula.

This model is not well designed to examine antimicrobial pharmacokinetic properties in an abscess, examine *in-vivo* efficacy against *B. fragilis*, and determine critical intervals for antibiotic effects. All these features can be easier to achieve in the subcutaneous abscess model.

Contribution of the model to infectious disease therapy

This model has enabled conclusions to be drawn regarding the pathogenic role of various organisms in mixed aerobic–anaerobic infections using techniques that cannot be used in patients for ethical reasons.

The correlation of results obtained in an animal model with clinical results is the ultimate test of the usefulness of the model. However, the assessment of such a correlation is often difficult because of inadequacies in comparative clinical studies, including a lack of stratification of patients, the exclusion of seriously ill patients, and the inclusion of inadequate numbers of patients. Because of these limitations, reliable clinical data are often unavailable, even after an antimicrobial agent has already been licensed. In contrast to clinical studies, those in animals are relatively uniform, creating a similar situation in numerous hosts that is devoid of the complicating factors seen in patients and measuring more precisely changes in bacterial populations after antimicrobial therapy. Studies in animals, therefore, may predict the clinical usefulness of antimicrobial drugs and may highlight the potential pitfalls of their use.

An example of the utility of *in-vivo* correlation is the demonstrated failure of chloramphenicol to eradicate anaerobic infection and the subsequent success of clindamycin (Gorbach and Thadepalli, 1974; Thadepalli *et al.*, 1977), despite the generally high level of *in vitro* effectiveness of both agents against anaerobic bacteria (Cuchural *et al.*, 1988). Such a failure of chloramphenicol was also demonstrated by Onderdonk *et al.* (1979) in an animal model and was attributable to the drug's inactivation by *Bacteroides fragilis* and *Clostridium perfringens* (Kanazawa *et al*, 1969).

The use of the intra-abdominal abscess model increased the understanding of the role of the various organisms in the infectious process, and established the biphasic nature of aerobic and anaerobic intra-abdominal infection. This was an initial peritonitis and bacteremia due to Enterobacteriaceae and the subsequent polymicrobial intra-abdominal abscess where the *B. fragilis* group predominates. This model allows the evaluation of various microbiological agents and their combinations in the treatment of the infectious process.

This animal model serves as an intermediary stage in drug development between the *in-vitro* laboratory antimicrobial susceptibility testing and clinical trials. This *in-vivo* evaluation can be done under carefully controlled conditions. It clearly illustrated, as was shown in clinical trials in patients, that numerous antimicrobial regimes are equally effective. The choice among them in clinical medicine is thereafter based on issues such as convenience of administration, bioavailability, half-life, side effects, toxicity, and cost considerations.

Acknowledgment

The author is grateful to Dr H. Thadepalli for his useful advice and review of the chapter.

References

Bansal, M. B., Chuah, S. K., Thadepalli, H. (1985) *In vitro* activity and *in vivo* evaluation of ticarcillin plus clavulanic acid against aerobic and anaerobic bacteria. *Am. J. Med.*, **79**(5B) 33–38.

Bartlett, J. G., Gorbach, S. L. (1979). An animal model of intra-abdominal sepsis. *Scand. J. Infect. Dis. Suppl.* **19**, 26–29.

Bartlett, J. G. (1984). Experimental aspects of intraabdominal abscess. *Am. J. Med.* **76**(5A), 91–98.

Bartlett, J. G., Onderdonk, A. B., Louie, T. J., Gorbach, S. L. (1978). Lessons from an animal model of intraabdominal sepsis. *Arch. Surg.*, **113**, 853–857.

Bartlett, J. G., Louie, T. J., Gorbach, S. L., Onderdonk, A. B. (1981). Therapeutic efficacy of 29 antimicrobial regimens in experimental intraabdominal sepsis. *Rev. Infect. Dis.*, **3**, 535–542.

Brook, I. (1989). A 12-year study of aerobic and anaerobic bacteria in intra-abdominal and postsurgical abdominal wound infections. *Surg. Gynecol. Obstet.*, **169**, 387–392.

Brook, I. (1994). Comparison of cefoxitin, cefotetan and the combination of ampicillin with sulbactam in the therapy of polymicrobial infection in mice. *J. Antimicrob. Chemother.*, **34**, 791–796.

Brook, I. (1995). Indigenous microbial flora in humans. In *Howard & Simons Surgical Infectious Diseases*, 3rd edn, pp. 37–46. Appleton & Lange, Norwalk, CT.

Brook, I., Gillmore, J. D. (1993). *In-vitro* susceptibility and *in-vivo* efficacy of antimicrobials in the treatment of intraabdominal sepsis in mice. *J. Antimicrob. Chemother.*, **31**, 393–401.

Brook, I. Coolbaugh, J. C., Walker, R. I. (1984). Pathogenicity of piliated and encapsulated *Bacteroides fragilis*. *Eur. J. Clin. Microbiol.*, **3**, 207–209.

Cheadle, W. G., Hershman, M. J., Mays, B., Melton, L., Polk, H. C. Jr (1989). Enhancement of survival from murine polymicrobial peritonitis with increased abdominal abscess formation. *J. Surg. Res.*, **47**, 120–123.

Cisneros, R. L., Onderdonk, A. B. (1989). Efficacy of a combination of ciprofloxacin and clindamycin for the treatment of experimental intraabdominal sepsis. *Curr. Ther. Res.*, **46**, 959–965.

Cisneros, R. L., Bawdon, R. E., Onderdonk, A. B. (1990). Efficacy of ampicillin/sulbactam for the treatment of experimental intra-abdominal sepsis. *Curr. Ther. Res.*, **48**, 1021–1029.

Cuchural, G. J. Jr, Tally, F. P., Jacobus, N. V. *et al.* (1988). Susceptibility of the *Bacteroides fragilis* group in the United States: analysis by site of isolation. *Antimicrob. Agents Chemother.*, **32**, 717–722.

Elek, S. D., Conen, P. E. (1957). The virulence of *Staphylococcus pyogenes* for man. A study of the problems of wound infection. *Br. J. Exp. Pathol.*, **38**, 573–579.

Fu, K. P., Lasinski, E. R., Zoganas, H. C., Kimble, E. F., Konopka, E. A. (1985). Efficacy of rifampicin in experimental *Bacteroides fragilis* and *Pseudomonas aeruginosa* mixed infections. *J. Antimicrob. Chemother.*, **15**, 579–585.

Gorbach, S. L. (1982). Interactions between aerobic and anaerobic bacteria. *Scand. J. Infect. Dis. Suppl.*, **31**, 61–67.

Gorbach, S. L., Thadepalli, H. (1974). Clindamycin in pure and mixed anaerobic infections. *Arch. Intern. Med.*, **134**, 87–92.

Hau, T., Jacobs, D. E., Hawkins, N. L. (1986). Antibiotics fail to prevent abscess formation secondary to bacteria trapped in fibrin clots. *Arch. Surg.*, **121**, 163–168.

Louie, T. J., Onderdonk, A. B., Gorbach, S. L., Bartlett, J. G. (1977). Therapy for experimental intraabdominal sepsis: comparison of four cephalosporins with clindamycin plus gentamicin. *J. Infect. Dis.*, **135**, 518–522.

Kanazawa, Y. Kuramata, T., Miyamura, S. (1969). Inactivation of chemotherapeutic agents by clostridia. *Jpn J. Bacteriol.*, **24**, 281–289.

McRitchie, D. L., Cummings, D., Rotstein, O. D. (1989). Delayed administration of tissue plasminogen activator reduces intra-abdominal abscess formation. *Arch. Surg.*, **124**, 1406–1410.

Montravers, P., Andremont, A., Massias, L., Carbon, C. (1994). Investigation of potential role of *Enterococcus faecalis* in the pathophysiology of experimental peritonitis. *J. Infect. Dis.*, **169**, 821–830.

Muhvich, K. H., Myers, R. A. M., Marzella, L. (1988). Effect of hyperbaric oxygenation, combined with antimicrobial agents and surgery in rat model of intraabdominal infection. *J. Infect. Dis.*, **157**, 1058–1061.

Nakatani, T., Sato, T., Marzella, L., Hiral, F., Trump, B. F., Siegel, J. H. (1984). Hepatic and systemic metabolic responses to aerobic and anaerobic intra-abdominal abscesses in a highly reproducible chronic rat model. *Circul. Shock.*, **13**, 27–294.

Nichols, R. L., Smith, J. W., Fossedal, E. N., Condon, R. E. (1979). Efficacy of parenteral antibiotics in the treatment of experimentally induced intraabdominal sepsis. *Rev. Infect. Dis.*, **1**, 302–309.

Nulsen, M. F., Finlay-Jones, J. J., Skinner, J. M., McDonald, P. J. (1983). Intra-abdominal abscess formation in mice: quantitative studies on bacteria and abscess-potentiating agents. *Br. J. Exp. Pathol.*, **64**, 345–353.

Olitzki, L. (1948). Mucin as a resistance-lowering substance. *Bacteriol. Rev.*, **12**, 149–162.

Onderdonk, A. B., Bartlett, J. G., Louie, T. J., Sullivan-Seigler, N., Gorbach, S. L. (1976). Microbial synergy in experimental intra-abdominal abscess. *Infect. Immun.*, **13**, 22–26.

Onderdonk, A. B., Louie, T. J., Taily, F. P., Bartlett, J. G. (1979a). Activity of metronidazole against *Escherichia coli* in experimental intra-abdominal sepsis. *J. Antimicrob. Chemother.*, **5**, 201–210.

Onderdonk, A. B., Kasper, D. L., Mansheim, B. I., Louie, T. J., Gorbach, S. L., Bartlett, J. G. (1979b). Experimental animal models for anaerobic infections. *Rev. Infect. Dis.*, **1**, 291–301.

Onderdonk, A. B., Cisneros, R. (1985). Comparison of clindamycin and metronidazole for the treatment of experimental intraabdominal sepsis produced by *Bacteroides fragilis* and *Streptococcus intermedius*. *Curr. Ther. Res.*, **38**, 893–898.

Pefanis, A., Thauvin-Eliopoulos, C., Eliopoulos, G. M., Moellering, R. C. Jr (1993). Efficacy of ceftriaxone plus tazobactam in a rat model of intraabdominal abscess due to *Bacteroides fragilis*. *J. Antimicrob. Chemother.*, **32**, 307–312.

Pefanis, A., Thauvin-Eliopoulos, C., Holden, J., Eliopoulos, G. M., Ferraro, M. J., Moellering, R. C. Jr. (1994). Activity of fleroxacin alone and in combination with clindamycin or metronidazole in experimental intra-abdominal abscesses. *Antimicrob. Agents Chemother.*, **38**, 252–255.

Rice, L. B., Yao, J. D. C., Klimm, K., Eliopoulos, G. M., Moellering, R. C. Jr (1991). Efficacy of different beta-lactams against an extended-spectrum beta-lactamase-producing *Klebsiella pneumoniae* strain in the rat intra-abdominal abscess model. *Antimicrob. Agents Chemother.*, **35**, 1243–1244.

Rice, L. B., Carias, L. L., Shlaes, D. M. (1993). Efficacy of ampicillin–sulbactam versus that of cefoxitin for treatment of *Escherichia coli* infections in a rat intra-abdominal abscess model. *Antimicrob. Agents Chemother.*, **37**, 610–612.

Sawyer, R. G., Adams, R. B., Spengler, M. D., Bruett, T. L. (1994). Aztreonam vs. gentamicin in experimental peritonitis and intra-abdominal abscess formation. *Am. Surgeon*, **60**, 849–853.

Schwartz, R. J., Dubrow, T. J., Rival, R. A., Wilson, S. E.,

Williams, R. A. (1992). The effect of fibrin glue on intraperitoneal contamination in rats treated with systemic antibiotics. *J. Surg. Res.*, **52**, 123–126.

Scott, R. E., Robson, H. G. (1976). Synergistic activity of carbenicillin and gentamicin in experimental *Pseudomonas* bacteremia in neutropenic rats. *Antimicrob. Agents Chemother.*, **10**, 646–672.

Swenson, R. M., Lorber, B., Michaelson, T. C., Spaulding, E. H. (1974). The bacteriology of intraabdominal infections. *Arch. Surg.*, **109**, 389–398.

Thadepalli, H., Gorbach, S. L., Bartlett, J. G. (1977). Apparent failure of chloramphenicol in the treatment of anaerobic infections. *Curr. Ther. Res.*, **22**, 421–426.

Thadepalli, H., Gollapudi, S. V. S., Chuah, S. K. (1986). Therapeutic evaluation of difloxacin (A-56619) and A-56620 for experimentally induced *Bacteroides fragilis*-associated intra-abdominal abscess. *Antimicrob. Agents Chemother.*, **30**, 574–576.

Thadepalli, H., Reddy, U., Chuah, S. K. *et al.* (1997). *In vivo* efficacy of trovafloxacin (CP-99,217), and new quinolone, in experimental intra-abdominal abscesses caused by *Bacteroides fragilis* and *Escherichia coli*. *Antimicrob. Agents Chemother.*, **41**, 583–586.

Tudor, R. G., Youngs, D. J., Yoshioka, K., Burdon, D. W., Keighley, M. R. B. (1988). A comparison of the penetration of two quinolones into intra-abdominal abscess. *Arch. Surg.*, **123**, 1487–1490.

Weinstein, W. M., Onderdonk, A. B., Bartlett, J. G., Gorbach, S. L. (1974). Experimental intraabdominal abscess in rats: development of an experimental model. *Infect. Immun.*, **10**, 1250–1255.

Weinstein, W. M., Onderdonk, A. B., Bartlett, J. G., Gorbach, S. L. (1975). Antimicrobial therapy of experimental intraabdominal sepsis. *J. Infect. Dis.*, **132**, 282–286.

Wells, C. L., Arland, L. A., Simmons, R. L., Rotstein, O. D. (1985). *In-vivo* bactericidal activity of Sch 34343 in *Bacteroides fragilis* abscesses and in *Bacteroides fragilis–Escherichia coli* abscesses. *J. Antimicrob. Chemother.*, **15**(Suppl C), 199–206.

Willey, S. H., Hindes, R. G., Elipoulos, G. M., Moellering, R. C. Jr (1989). Effects of clindamycin and gentamicin and other antimicrobial combinations against enterococci in an experimental model of intra-abdominal abscess. *Surg. Gynecol. Obstet.*, **169**, 199–202.

Yull, A. B., Abrans, J. S., Davis, J. H. (1962). The peritoneal fluid in strangulation obstruction. The role of red blood cell and *E. coli* bacteria in producing toxicity. *J. Surg. Res.*, **2**, 223–233.

Chapter 19

Mouse Peritonitis Model Using Cecal Ligation and Puncture

M. Shrotri, J. C. Peyton and W. G. Cheadle

Background of human infection

Incidence

Bacterial peritonitis, with its attendant local and distant effects, continues to be a common source of high morbidity and mortality in surgical patients. Despite the use of new antibiotic regimens and adequate surgical treatment of the focus of contamination, a large number of these patients with peritonitis will develop multiple organ failure (MOF) (Eiseman *et al.*, 1977; Fry *et al.*, 1980; Norton, 1985), require prolonged treatment in the intensive care unit, and frequently succumb in the end. The most commonly affected remote organ is the lung. This alone predisposes to the prolonged intensive care and total hospital stay, while it increases mortality with concomitant failure of other organs (Wickel *et al.*, 1997a). The current mortality rates (40–50%) from MOF have not improved over the last 10 years, as reported by McLauchlan *et al.* (1995).

Etiology

Perforated peptic ulcer and perforation of the appendix or the large bowel are the usual causes of peritonitis. Peptic perforations are mostly chemical in nature. Appendicular perforations are often localized and lead to the formation of an abscess or local peritonitis, rather than generalized peritonitis. Large-bowel perforation is the most common cause of severe intra-abdominal sepsis or bacterial peritonitis and can cause MOF when timely intervention does not follow (Polk, 1979).

The causes of large bowel perforation include secondary perforation of a grossly distended large bowel (commonly cecum) due to distal obstruction (benign or malignant), perforation of an inflamed bowel (commonly diverticular), and traumatic or less commonly iatrogenic (endoscopic manipulation) perforation. Often, perforations secondary to malignant obstruction and diverticular disease are insidious in nature, occur in the elderly, and thus commonly present to the clinician at a stage when the systemic inflammatory response syndrome (SIRS) has already set in. These patients invariably require intensive care, and mortality in this group is extremely high.

Pathology

The terms intra-abdominal sepsis and Gram-negative sepsis often are used interchangeably, although intra-abdominal sepsis usually results from polymicrobial infection. There is evidence that anaerobic bacteremia potentiates the effects of aerobic Gram-negative sepsis (Rotstein and Kao, 1988). Clinical studies have failed to demonstrate significant levels of endotoxin in more than half of patients with Gram-negative sepsis (Stumacher *et al.*, 1973). Thus, it is likely that bacteremia and the resultant cytokine response in the local and peripheral tissues are responsible for the pathological state seen in intra-abdominal sepsis.

Background of model

Historical review

Wichterman *et al.* (1980) have presented an eloquent review of various laboratory models for the study of sepsis. The importance of having a good laboratory model for sepsis cannot be overemphasized, considering that the diversity of diseases, the polymicrobial nature of commonly seen sepsis situations, and other variables in patients make it difficult to perform controlled studies. The common models of intra-abdominal sepsis include: (1) intravascular injection of bacterial endotoxin or Gram-negative bacteria; (2) intraperitoneal administration of bacteria (either Gram-negative or polymicrobial); and (3) peritoneal contamination with the animal's own intestinal flora (e.g. cecal ligation and puncture (CLP), as is described in this chapter).

Baker *et al.* (1983) first described the CLP model in mice based on the rat model developed by Wichterman *et al.* (1980). Unlike models based on intravascular or intraperitoneal administration of endotoxin–bacteria or bacteria with adjuvant, CLP represents a peritonitis model with clinical features of the infection comparable with those of

Handbook of Animal Models of Infection
ISBN 0-12-775390-7

peritonitis in humans. The former represent models of endotoxic shock, which is not representative of the true clinical situation. The CLP model has come to be an accepted and preferred one for intra-abdominal sepsis, and has been our standard model for bacterial peritonitis for the last 10 years.

Animal species

Mouse strain

Any variety of mouse strains can be used. We and others (Baker *et al.*, 1983) have used Swiss Webster (Taconic, Germantown, NY), C3H/HeN (Jackson Laboratory, Bar Harbor, ME), C3H/HeJ-endotoxin-resistant and C3H/HeOuJ or C3H/HeSnJ endotoxin-sensitive (Jackson Laboratory), WBB6F$_1$/J-W/W-(MCD) mastcell-deficient and WBB6F$_1$/J-W/W+/+(MCN) normal littermates (Jackson Laboratory), CD-1 (Charles River Laboratories, Wilmington, MA), CBA/J (Jackson Laboratories; Baker *et al.*, 1983), and C57/BL6 (Jackson Laboratories) mice.

Age and weight

Adult mice 6 to 8 weeks old, weighing 25–30 g, are used.

Sex

We use male mice in our CLP model, but mice of either sex may be used. The male and female data can be pooled, if required.

Preparation of animals

Animals are housed in a facility approved by the American Association for Accreditation of Laboratory Animal Care (AAALAC) and under the supervision of a veterinarian. All the mice remain housed in our facility for at least 2 weeks before the experiment to allow equilibration of intestinal flora. The mice are fed standard rodent chow and water *ad libitum*. No other specialized housing or care nor specific pretreatment is required. Studies are carried out in strict accordance with the National Institutes of Health guidelines for the treatment of laboratory animals.

Details of surgery

Preoperative care

No premedication or analgesic is necessary. The animals are fasted for 8 hours prior to surgery.

Anesthesia

A mixture of ketamine (80 mg/kg/body weight) and xylazine (16 mg/kg/body weight) is used. The mouse is weighed, and the appropriate amount of the anesthetic is drawn in an insulin syringe. The anesthetic is injected intramuscularly. Complete anesthesia ensues in about 5–7 minutes and lasts for 20–30 minutes. This is confirmed by the lack of response from the animal to pain, and paralysis. This combination of drugs provides adequate muscular relaxation, analgesia, and also some postoperative pain relief.

Instruments

The required instruments include a scalpel with a no. 15 blade, a pair of sharp dissecting forceps, blunt dissecting forceps, sharp dissecting scissors, blunt dissecting scissors, fine needle holder, and suture-cutting scissors.

Sutures

We use 3-0 silk, a 4-0 or 5-0 braided absorbable suture, and 4-0 nylon.

Intraoperative management

Respiration is observed carefully but not monitored mechanically nor supported. Body temperature is maintained. A further dose of ketamine–xylazine may be administered when needed to maintain surgical anesthesia (however, this is not usually necessary).

Surgical procedure

Once the animal is assessed to be completely anesthetized, it is placed supine on an adequately prepared surgical platform. The anterior abdominal wall is shaved, and the area is cleaned with isopropyl alcohol. A 7–10 mm vertical abdominal incision is made in the left lower quadrant using the scalpel. The muscular layer is opened using a pair of blunt scissors to split it. Once the peritoneal cavity is opened, the cecum is brought out. A 3-0 silk tie is placed around the cecum, immediately below the ileocecal valve, taking care not to obstruct the small or large intestine. The intensity of the systemic inflammatory response to cecal ligation can be varied by altering the position of the ligature from the ileocecal valve to the tip of the cecum. The cecum, which is isolated with the ligature, is then punctured once or twice using a hypodermic needle (Figure 19.1). The gauge of the needle can vary from 18G to 26G, depending on the desired mortality to be induced. Gentle pressure is then

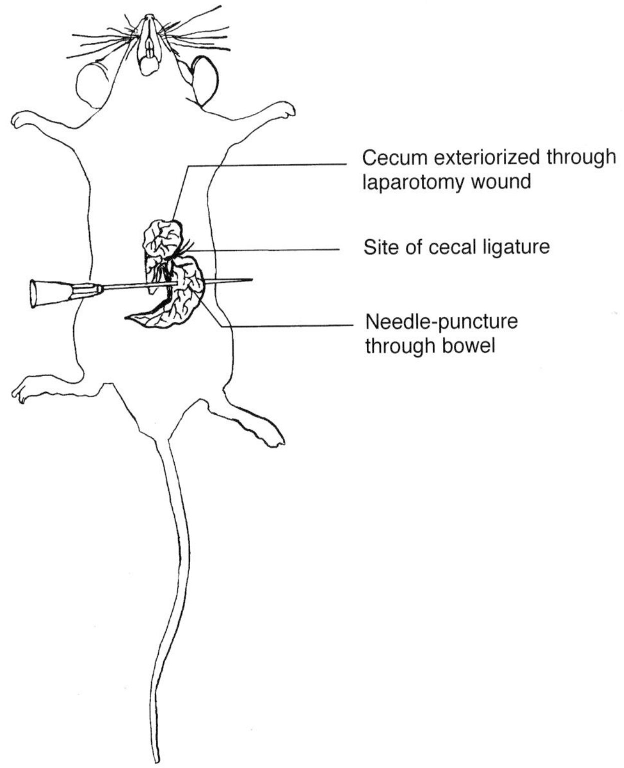

Figure 19.1 Diagrammatic representation of cecal ligation and puncture.

given on this segment to ensure that the puncture sites have not closed off and that a small amount of feces is extruded on to the surface of the bowel. The cecum then is returned to the peritoneal cavity. The wound is closed in two running layers using 4-0 braided absorbable suture for the muscle layer and 4-0 nylon for skin. In sham-operated mice, the cecum is returned in the peritoneum without ligation or puncture.

Postoperative care

Postoperatively, the animals are allowed to recover. Food and water is provided *ad libitum*. Routine use of analgesics (e.g. flunixin) is the rule, unless these agents alter the particular parameters to be studied. All animals are routinely examined every 4–6 hours in the postoperative period to confirm their well-being. Animals exhibiting undue stress or signs of imminent death such as inordinate weight loss, loss of righting reflex, or panting are euthanized promptly. Distress due to peritonitis is usually transient, and euthanasia will only be necessary within the first 48 hours after the induction of peritonitis. Postoperative care and euthanasia are managed by expert laboratory technicians who are accustomed to managing this model routinely. Our method of euthanasia is to open the chest while the animal is under standard ketamine–xylazine anesthesia and to exsanguinate by cardiac puncture. Cessation of cardiac function is observed by open pneumothorax.

Key parameters to monitor infection

Death

After CLP is performed as described, the majority of animals die within 2–3 days. Deaths are rare after day 5, and therefore it is acceptable to describe total mortality as 1 week mortality (not earlier). It is possible to increase survival time by limiting the amount of the cecum ligated and reducing the diameter of the needle used for puncture. A single puncture also reduces mortality from the standard double puncture, which provides a more rapidly developing lethal model. This has been nicely shown by Baker *et al.* (1983).

We have observed similar results with regards to number of punctures, including the effect of antibiotics on the mortality after CLP (Table 19.1). Single-puncture CLP with an 18G needle was associated with a 60% mortality rate, while double-puncture CLP with an 18G needle was associated with much higher mortality rate (95%). Another group of animals were subjected to double-puncture 18G CLP but received a continuous infusion of cefoxitin subcutaneously for 72 hours; the latter group of animals fared significantly better.

Blood culture

Blood samples obtained by cardiocentesis are plated on tryptic soy agar for aerobic or brain–heart infusion agar for anaerobic culture. Colony-forming units (cfu) are determined by manual counting of the plates, and results are expressed in cfu/ml blood. Our studies indicate that the initial bacteremia is anaerobic, noted at the earliest at 1 hour after CLP, with all mice having a positive culture by the end of 6 hours. The first indication of aerobic bacteremia observed is at 3 hours, with all the mice showing positive cultures at 24 hours (Figure 19.2).

Table 19.1 Mortality from cecal ligation and puncture (CLP; $n = 20$)

Infectious challenge	1 × CLP	2 × CLP	2 × CLP + cefoxitin
Mortality (%)	60*	95	55*

*Significantly less than double-puncture CLP (Fisher's exact test)

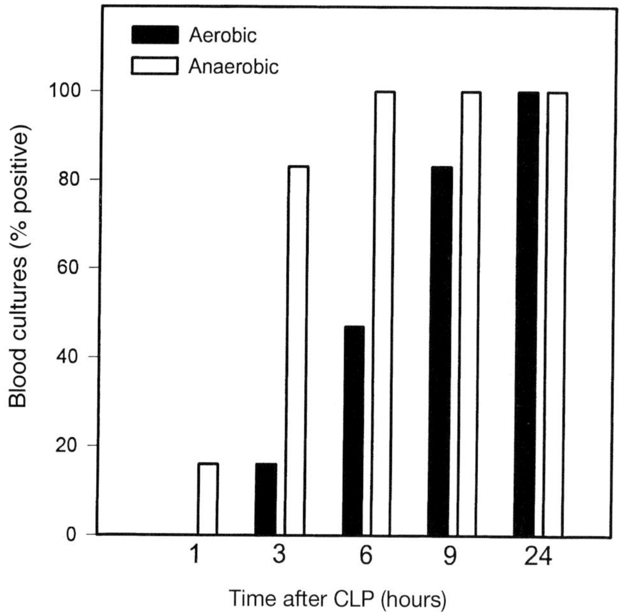

Figure 19.2 Pattern of bacteremia after cecal ligation and puncture.

Appearance

The signs of illness after CLP occur in stages (early sepsis to late) until reversal of these signs indicates either recovery or death. These signs are fairly easy to recognize: (1) reduced movement; (2) no attempt at congregation with other mice for group warmth; (3) tremors; (4) dirty or unkempt fur; (5) decrease in natural curiosity; (6) progressive lethargy; (7) piloerection; (8) reduced appetite and thirst; (9) tachypnea; (10) glazed eyes and crusting exudate; (11) concentrated, and sometimes frankly purulent, thick urine; (12) diarrhea; (13) animals laid supine without usual resistance; and (14) obtundation with subsequent rapid onset of death. Those mice that recover from peritonitis have localization of the infective or septic process on the left side of the abdomen in the form of an abscess that may discharge to the outside, with subsequent resolution of the septic focus.

Blood cell response

There are various ways of measuring or monitoring blood cell responses. The studies and techniques that we use are as follows.

Total/differential cell counts

Blood obtained from cardiocentesis (heparinized) is stained with Turk's solution. Total leukocytic count is performed manually by hemocytometer. Thin smears of whole blood are prepared and stained with Wright's and Giemsa stains. A differential count then is performed manually.

Peritoneal macrophage isolation

Peritoneal macrophages are harvested by peritoneal lavage after the animals are killed by cervical dislocation. Ice-cold RPMI 1640 medium (GIBCO/BRL, Bethesda, MD) is used for the lavage. Resident peritoneal macrophages from normal animals are used as controls. After centrifugation at 200*g*, the cells are resuspended at a concentration of 10^6 cells/ml in RPMI 1640 medium supplemented with penicillin (100 IU/ml), streptomycin (100 μg/ml), and glutamine (2 mmol/l). Ten ml of the cell suspension is plated into 100 mm tissue culture plates (Falcon, Lincoln Park, NJ). The macrophages are then allowed to adhere for 1 hour in a humidified CO_2 (5%) incubator at 37°C. The plates are washed three times with warm media to remove non-adherent cells. This protocol has resulted in routinely obtaining macrophage cultures containing more than 95% non-specific esterase-positive cells. Extraction of total RNA is then performed immediately after washing the plate cells. This technique has been well described by McMasters *et al.* (1994).

Peritoneal cell counts

Peritoneal exudate cellular response is measured and monitored with peritoneal lavage. Cell counts are performed manually as described for leukocytic counts.

Tissue neutrophil monitoring by myeloperoxidase (MPO) assay

This is performed on lung and liver tissue to accurately quantitate the number of neutrophils within the respective tissue (Schierwagen *et al.*, 1990).

Cytokines

Total tissue RNA extraction

This is performed according to the method described by Chomczynski and Sacchi (1987) using one whole lung or a partial lobe of the liver. Based on this extraction, any immunological mediator can be studied. Originally, we studied IL-1β and TNF-α mRNA.

Semiquantitation of IL-1β and TNF-α mRNA

This is a semiquantitative competitive differential polymerase chain reaction (PCR)-based protocol adapted to quantitate murine TNF-α and IL-β mRNAs (Chelly *et al.*, 1988; Hadjiminas *et al.*, 1994a). Total cellular RNA is reversely transcribed (Perkin-Elmer, Norwalk, CT), and the cDNA is amplified using specific primers (TNF-α, IL-β, β-actin, or GAPDH). To ensure that the reaction is in the exponential phase of the PCR amplification, the optimum number of PCR cycles is determined for each primer pair (Hadjiminas *et al.*, 1994a,d). DNA gives a larger product size than expected because all primer pairs span at least one intron. The ratios for cytokine to β-actin or GAPDH are calculated from densitometric analysis of negative films following electrophoresis and ethidium bromide staining of the PCR products.

Serum TNF-α assay

This was originally determined by WEHI 164 subclone 13 cell line bioassay, as described by Eskandari *et al.* (1990). Cell culture assays are now replaced by commercially available ELISA kits, which make estimation of various proinflammatory cytokines more sensitive and easier to perform. Serum IL-1 levels are extremely low, and TNF levels peak by 6 hours and decline to normal by 48 hours.

Other assays

Serum endotoxin assay is determined by 1:10 dilutions of mouse serum by the chromogenic limulus amoebocyte assay (QCL-1000, BioWhittaker, Walkersville, MD), according to the manufacturer's protocol.

Antimicrobial therapy

Broad-spectrum antibiotics (aerobic and anaerobic Gram-negative coverage) are used as in human polymicrobial infection. The various antibiotics that have been studied include cefoxitin, aztreonam, metronidazole, and gentamicin (Baker *et al.*, 1983). We have shown that use of continuous administration of antibiotics is superior compared with the intermittent dosage schedule (Hadjiminas *et al.*, 1995; Naziri *et al.*, 1995; Mercer-Jones *et al.*,1998). There are other reports on the advantage of using the continuous method of administration, mainly for cephalosporin antibiotics (Bodey *et al.*, 1979; Roosendaal *et al.*, 1989; Livingston and Wang, 1993).

We have also studied other agents that may affect survival after CLP, including pentoxifylline, heparin-binding protein (HBP), macrophage inflammatory protein and KC, anti-CD18 antibody, muramyl dipeptide, and anti-TNF and anti-IL-1 monoclonal antibodies.

Pentoxifylline

This drug has been shown to improve survival in endotoxemia by inhibiting the synthesis of endotoxin-induced TNF-α at the mRNA level. Anti-TNF treatment does not improve survival in the CLP model, and in fact worsens it. In contrast, IL-1ra treatment improved survival in the CLP model. We showed (Hadjiminas *et al.*, 1994c) that low-dose pentoxifylline improves survival in the CLP model without reduction in the tissue TNF mRNA. Pentoxifylline also did not affect the CD11b/CD18 expression on the leukocytes (which can mediate neutrophil sequestration and IL-1β mRNA) but did reduce lung IL-1β mRNA. It is likely that low-dose pentoxifylline may reduce the expression of some other adhesion molecules, mediated perhaps by reduction in IL-1 synthesis by virtue of reduced IL-1β mRNA.

Heparin-binding protein

This is a neutrophil azurophilic granule isolate that has mediator properties for *in-vivo* monocyte accumulation. Recombinant HBP has been shown to be a powerful chemoattractant and to increase cultured monocyte survival time (Ostergaard *et al.*, 1992).

Macrophage inflammatory protein-2 and KC

These are murine C-X-C chemokines, which are induced by lipopolysaccharide (LPS), TNF-α, and IL-1β. They upregulate neutrophil integrin (CD11b/CD18), stimulate neutrophil respiratory burst, and are neutrophil chemoattractants.

Anti-CD18 antibody

The neutrophil has been suggested to be an important component in the development of remote organ failure (Alexander *et al.*, 1992) and polymorphonuclear (PMN) leukocytosis is a common occurrence in this clinical situation. The key event appears to be the migration of the neutrophils across vascular endothelium into peripheral tissues, which is dependent on binding of the integrin complex (CD11b/CD18) on these cells to the ICAM-1 (intercellular adhesion molecule-1) receptor on the endothelium. We have studied the effect of anti-CD18 antibody in the CLP model and found that this antibody inhibits PMN migration into the peritoneal cavity, which would be potentially deleterious.

Muramyl dipeptide (MDP)

This is a small fragment of mycobacterial cell wall capable of adjuvant activity; in short, a non-specific immunomodulator. We have previously shown the beneficial effect of MDP in mice with peritonitis using CLP and other sepsis models (Brown *et al.*, 1986; Cheadle *et al.*, 1986, 1989; McMasters and Cheadle, 1993; Gallinaro *et al.*, 1994).

Anti-TNF and anti-IL-1 monoclonal antibody

TNF-α is one of the pro-inflammatory cytokines that is produced in response to bacterial endotoxin. The antibody to this cytokine was used to negate the actions of TNF-α (Hadjiminas *et al.*, 1994c). Similarly, IL-1 is a proinflammatory cytokine and is produced by activated macrophages and sequestered neutrophils. The antibodies to this cytokine are used to block its activity.

Advantages of the model

The general advantages of using mice as models are their inexpensiveness, their availability in large numbers, and the fact that inbred strains are genetically identical, matched for age and sex, and same controlled diet. These factors allow the use of a large number of animals for thorough statistical evaluation of results and for minimizing biologic variables.

Specific advantages (Baker *et al.*, 1983) in choosing the mouse peritonitis model with CLP are numerous.

- This model seems to be an appropriate analog of clinical peritonitis, a scenario in which mortality rates are decreased by surgical intervention or by antibiotics.
- The critically studied hemodynamic and the metabolic effects of CLP in the rat model by Wichterman *et al.* (1980) have shown good parallels in the human clinical situation.
- The immunogenic variables are better understood and modulated in mice as compared to rats.
- The clinical course of events in this model is closer to the patients who develop sepsis in a more gradual fashion, and thus seems more relevant than bolus intravascular or intraperitoneal injection models of sepsis.
- The polymicrobial nature of infection in peritonitis is better reflected by this model, as compared to the intravascular injection of endotoxin or intravascular/intraperitoneal injection of bacterial suspension.
- The technique of performing CLP is fairly simple.
- The severity of the model can be altered to suit the study protocols by altering the technique (i.e. the site of the ligature on the cecum, the diameter of the needle used for puncture, and the number of punctures performed).

Disadvantages of the model

CLP is an acute model of intra-abdominal infection rather than the more chronic form that is seen in humans. It is not a model of chronic organ failure *per se*, but one of acute remote organ injury in response to peritonitis. The natural history in survivors is for the infection to resolve spontaneously, often by extrusion of the abscess through the flank. This is not the case in humans, who require drainage and often bowel resection for cure.

The mortality rate can vary, particularly if inexperienced personnel perform the procedure. Even though the technique is fairly straightforward, close attention and adherence to certain technical details like standard laparotomy, gentle handling of tissues, constant site of cecal ligation, standardization of the needle diameter, and the number of punctures is vital in reproducing uniform results. Thus, this procedure and its results are technician-dependent. Close monitoring of technical performance is essential to achieve consistent results.

Philosophy of model

All clinical sepsis trials to date have been based on relatively high-dose endotoxemia and bacteremia models. Endotoxins were isolated about 62 years ago by Andre Biovin, but it was half a decade later that it was suggested that human Gram-negative shock was caused by endotoxins. Over the next 25 years, laboratory models were based on this suggestion, and many of the observations resulting from these studies were incorporated in elucidating the pathophysiology and modifying the treatment of sepsis and septic shock in the clinical situation. Unfortunately, later clinical studies failed to demonstrate significant levels of endotoxin in more than half the documented Gram-negative septic cases. Thus, the search for a more clinically relevant model was attempted, and this forms the basis for the excellent review article by Wichterman *et al.* (1980), who also proposed the CLP model. The elaborate studies detailing the hemodynamic and metabolic effects indicated the clear parallels to the human clinical situation. Thereafter, this model was successfully adapted by Baker *et al.* (1983) in mice with several advantages (see above).

Contributions of the model to infectious disease therapy

Indications of efficacy of antimicrobial agents

Our studies using the CLP model indicate that bacteremia (anaerobic and aerobic) rather than serum endotoxin is the determinant of effective proinflammatory cytokine response (Mercer-Jones *et al.*, 1998). Thus, it would seem that broad-spectrum antibiotics (including those for anaerobes) are likely to be efficacious in controlling the cytokine response to a significant extent, if the timing is right. Anaerobes are not only important in the context of potentiating the pathogenicity of Gram-negative aerobic bacteria, but also have a direct effect on the production of cytokines, and this too is evident from our study (Mercer-Jones *et al.*, 1998).

Development of novel therapeutic approaches

Newer agents

Pentoxifylline

Low-dose pentoxifylline in our study (Hadjiminas *et al.*, 1994c) has been shown to be beneficial in reduction of the lung injury, secondary to bacterial peritonitis by CLP. The challenge in further studies lies in identifying the therapeutic range of pentoxifylline and affecting this same benefit in the clinical setting.

Heparin-binding protein (HBP)

We have shown increased survival in mice undergoing CLP with prior treatment with intraperitoneal HBP, as compared to controls (Mercer-Jones *et al.*, 1996a). Subsequent studies have indicated that this survival advantage may be related to monocyte recruitment in the peritoneal cavity or systemic immune augmentation (Wickel *et al.*, 1997b).

Interleukin-1 receptor antagonist (IL-1ra)

IL-1ra has been shown to protect rats from the lethal effects of CLP (Alexander *et al.*, 1992). Furthermore, clinical trials (Fisher *et al.*, 1994a,b) have shown beneficial effects, in the form of reducing 28-day mortality. It is likely that a certain subset of patients, with a high propensity to progress to MOF would perhaps benefit with such targeted treatment.

Increase in understanding of pathology and pathogenesis, host inflammatory (cytokine, cellular), and immune response

There has been significant contribution to the understanding of the pathology–pathogenesis of bacterial peritonitis, along with host inflammatory and immune response. The various important findings that can be extrapolated from studies using the CLP model for mouse peritonitis are as follows:

- CLP causes polymicrobial peritonitis, which is similar to the human situation.
- Anaerobic bacteremia is the first to occur followed by aerobic bacteremia, and this is evident from Figure 19.2 (Mercer-Jones *et al.*, 1996b).
- Anaerobic component of the polymicrobial infection has a direct effect on the organ inflammatory cytokine production and does not just potentiate the aerobic Gram-negative component (Mercer-Jones *et al.*, 1996b).
- The elevation of TNF-α and IL-1β mRNA is only modest, as compared with that shown by direct effect of LPS on the respective circulating cytokines, and is similar to that observed in human subjects of sepsis (McMasters and Cheadle, 1993).
- Mortality from CLP did not differ in the endotoxin-resistant group of mice compared with the standard endotoxin-sensitive mice (the latter also showing a standard TNF-α response), indicating that factors other than endotoxin and TNF-α may be responsible for this (McMasters *et al.*, 1994).
- Lung injury is mediated by neutrophil sequestration, and we have shown that administration of anti-TNF and anti-IL-1 antibodies does not prevent this phenomenon (Hadjiminas *et al.*, 1994b).
- Septic challenge to the immune system by peritonitis is more insidious, prolonged, and progressive as compared with endotoxin injection. This reflects similarly in the tissue TNF-α mRNA. Also, TNF production varies in different tissues, and this makes serum TNF concentration an unreliable index of TNF activity in some tissues (Hadjiminas *et al.*, 1994a).
- Lung injury secondary to neutrophil sequestration and subsequent increase in lung C-X-C chemokines (macrophage inflammatory protein or MIP-2 and KC) mRNA is similar in endotoxin-resistant and endotoxin-sensitive strains of mice, indicating that the pulmonary inflammatory response is not endotoxin-dependent (Mercer-Jones *et al.*, 1997a).
- Our experiments using MIP-2 (Mercer-Jones *et al.*, 1996a) indicate that after CLP there is an early increase in MIP-2 mRNA expression in peritoneal leukocytes, anti-MIP-2 antibody reduces neutrophil migration after CLP, intraperitoneal injection of recombinant murine MIP-2 (rmMIP-2) causes peritoneal neutrophil migration, and peritoneal leukocytic migration is reduced in mast-cell-deficient mice after CLP or after intraperitoneal rmMIP-2 injection.
- Treatment with anti-CD18 antibody after CLP (Mercer-Jones *et al.*, 1997b) resulted in significant reduction in peritoneal neutrophil migration and increase in blood neutrophils as compared to controls, indicating that peritoneal neutrophil migration in response to fecal challenge is CD18-dependent. The study also demonstrated a paradoxical increase in remote organ (fivefold in liver and twofold in lung) neutrophil sequestration, which may indicate that this peripheral sequestration is not totally CD18-dependent. Similar effects were observed with anti-CD11b antibody administration.
- Increased survival for mice undergoing CLP is seen with the use of HBP, possibly as a result of increased monocyte recruitment in the peritoneal cavity or systemic immune augmentation (Wickel *et al.*, 1997b).

Critical questions to address

Did evaluation of antimicrobial agents using this model lead to clinical trials?

Our studies using this model have shown benefit of using broad-spectrum antibiotics during peritonitis. Also, the beneficial effects of pentoxifylline and heparin-binding protein that have been indicated in our studies suggest that future studies and clinical trials might be worthwhile in evaluating these agents. The advantages of using continuous infusion of antibiotics as opposed to intermittent dosing, as seen in our studies, has led to a clinical trial at one of our affiliated hospitals.

What is the predictive ability of this model?

The advantage of the mouse peritonitis model using CLP is its predictive capability as compared to the endotoxin model. This model has close parallels to the clinical setting, including the polymicrobial nature of the peritonitis and the pattern of remote organ failure, mainly lung damage and death. Any of the agents reducing mortality from CLP in mice has the potential for some use in the human. It is the best available small animal model for inducing peritonitis.

References

Alexander, H. R., Doherty, G. M., Venzon, D. J., Merino, M., Fraker, D. L., Norton, J. A. (1992). Recombinant interleukin-1 receptor antagonist (IL-1ra): effective therapy against gram-negative sepsis in rats. *Surgery*, **112**, 188–193.

Baker, C. C., Chaudry, I. H., Gaines, H. O., Baue, A.E. (1983). Evaluation of factors affecting mortality rate after sepsis in a murine cecal ligation and puncture model. *Surgery*, **94**, 331–335.

Bodey, G. P., Ketchel, S. J., Rodriguez, V. (1979). A randomized study of carbenicillin plus cefamandole or tobramycin in the treatment of febrile episodes in cancer patients. *Am. J. Med.*, **67**, 608–616.

Brown, G. L., Foshee, H., Pietsch, J., Polk, H. C. Jr (1986). Muramyl dipeptide enhances survival from experimental peritonitis. *Arch. Surg.*, **121**, 47–49.

Cheadle, W. G., Brown, G. L., Lamont, P. M., Trachtenberg, L. S., Polk, H. C. Jr (1986). Effects of muramyl dipeptide and clindamycin in a murine abdominal abscess model. *J. Surg. Res.*, **41**, 319–325.

Cheadle, W. G., Hershman, M. J., Mays, B., Melton, L., Polk, H. C. Jr (1989). Enhancement of survival from murine polymicrobial peritonitis with increased abdominal abscess formation. *J. Surg. Res.*, **47**, 120–123.

Chelly, J., Kaplan, J. C., Maire, P., Gautron, S., Kahn, A. (1988). Transcription of the dystrophin gene in human muscle and non-muscle tissue. *Nature*, **333**, 858–860.

Chomczynski, P., Sacchi, N. (1987). Single-step method of RNA isolation by acid guanidinium thiocyanate–phenol–chloroform extraction. *Anal. Biochem.*, **162**, 156–159.

Eiseman, B., Beart, R., Norton, L. (1977). Multiple organ failure. *Surg. Gynecol. Obstet.*, **14**, 323–326.

Eskandari, M. K., Nguyen, D. T., Kunkel, S. L., Remick, D. G. (1990). WEHI 164 subclone 13 assay for TNF: sensitivity, specificity, and reliability. *Immunol. Invest.*, **19**, 69–79.

Fisher, C. J. Jr, Dhainaut, J. F., Opal, S. M. *et al.* (1994a). Recombinant human interleukin 1 receptor antagonist in the treatment of patients with sepsis syndrome: results from a randomized, double-blind, placebo-controlled trial. *J.A.M.A.*, **271**, 1836–1843.

Fisher, C. J. Jr, Slotman, G. J., Opal, S. M. *et al.* (1994b). Initial evaluation of human recombinant interleukin-1 receptor antagonist in the treatment of sepsis syndrome: a randomized, open-label, placebo-controlled trial. *Crit. Care Med.*, **22**, 12–21.

Fry, D. E., Pearlstein, L., Fulton, R. L., Polk, H. C. Jr. (1980). Multiple system organ failure: the role of uncontrolled infection. *Arch. Surg.*, **115**, 136–140.

Gallinaro, R. N., Naziri, W., McMasters, K. M., Peyton, J. C., Cheadle, W. G. (1994). Alteration of mononuclear cell immune-associated antigen expression, interleukin-1 expression, and antigen presentation during intra-abdominal infection. *Shock*, **1**, 130–134.

Hadjiminas, D. J., McMasters, K. M., Peyton, J. C., Cheadle, W. G. (1994a). Tissue tumor necrosis factor mRNA expression following cecal ligation and puncture or intraperitoneal injection of endotoxin. *J. Surg. Res.*, **56**, 549–555.

Hadjiminas, D. J., McMasters, K. M., Peyton, J. C., Cook, M., Cheadle, W. G. (1994b). Passive immunization against tumor necrosis factor and interleukin-1 fails to reduce lung neutrophil sequestration in chronic sepsis. *Shock*, **2**, 376–380.

Hadjiminas, D. J., McMasters, K. M., Robertson, S. E., Cheadle, W. G. (1994c). Enhanced survival from cecal ligation and puncture with pentoxyphylline is associated with altered neutrophil trafficking and reduced interleukin-1β expression but not inhibition of tumor necrosis factor synthesis. *Surgery*, **116**, 348–355.

Hadjiminas, D. J., Peyton, J. C., Cheadle, W. G. (1994d). Enhanced interleukin-1 mRNA expression in the peritoneum and lung during experimental peritonitis in mice (abstract). *Intensive Care Med.*, **20** (Suppl. 1), 68.

Hadjiminas, D. J., Cheadle, W. G., Polk, H. C. Jr. (1995). Continuous antibiotic infusion during peritonitis reduces bacteremia and lung IL-1 mRNA compared to intermittent dosing. *Curr. Opin. Surg. Infect.*, **3**, (Suppl. 1), 38–40.

Livingston, D. H., Wang, M. T. (1993). Continuous infusion of cefazolin is superior to intermittent dosing in decreasing infection after hemorrhagic shock. *Am. J. Surg.*, **165**, 203–207.

McLauchlan, G. J., Anderson, I. D., Grant, I. S., Fearon, K. C. (1995). Outcome of patients with abdominal sepsis treated in an intensive care unit. *Br. J. Surg.*, **82**, 524–529.

McMasters, K. M., Cheadle, W. G. (1993). Regulation of macrophage TNFα, IL-1β and Iα (I-Aa) mRNA expression during peritonitis is site dependent. *J. Surg. Res.*, **54**, 426–430.

McMasters, K. M., Peyton, J. C., Hadjiminas, D. J., Cheadle, W. G. (1994). Endotoxin and tumour necrosis factor do not cause mortality from caecal ligation and puncture. *Cytokine*, **6**, 530–536.

Mercer-Jones, M. A., Heinzelmann, M., Peyton, J. C., Cook, M., Flodgaard, H., Cheadle, W. G. (1996a). Monocyte recruitment increases survival in fecal peritonitis. *Surg. Forum*, **47**, 106–108.

Mercer-Jones, M. A., Heinzelmann, M., Peyton, J. C., Wickel, D. J., Polk, H. C. Jr, Cheadle, W. G. (1996b). Peritoneal neutrophil (PMN) migration after peritonitis: a role for macrophage inflammatory protein-2 and mast cells. Presented to the Society for Leukocyte Biology, October 11–14, Verona, Italy.

Mercer-Jones, M. A., Heinzelmann, M., Peyton, J. C., Wickel, D. J., Cook, M., Cheadle, W. G. (1997a). The pulmonary inflammatory response to experimental fecal peritonitis: relative roles of tumor necrosis factor-alpha and endotoxin. *Inflammation*, **21**, 401–417.

Mercer-Jones, M. A., Heinzelmann, M., Peyton, J. C., Wickel, D. J., Cook, M., Cheadle, W. G. (1997b). Inhibition of neutrophil migration at the site of infection increases remote organ neutrophil sequestration and injury. *Shock*, **8**, 193–199.

Mercer-Jones, M. A., Hadjiminas, D. J., Heinzelmann, M., Peyton, J. C., Cook, M., Cheadle, W. G. (1998). Continuous antibiotic treatment for experimental abdominal sepsis: effects on organ inflammatory cytokine expression and neutrophil sequestration. *Br. J. Surg.*, **85**, 385–389.

Naziri, W., Cheadle, W. G., Trachtenberg, L. S., Montgomery, W. D., Polk, H. C. Jr (1995). Increased antibiotic effectiveness in a model of surgical infection through continuous infusion. *Am. Surg.*, **61**, 11–15.

Norton, L. W. (1985). Does drainage of intraabdominal pus reverse multiple organ failure? *Am. J. Surg.*, **149**, 347–350.

Ostergaard, E., Nielsen, O. F., Flodgaard, H. (1992). Comparison of the effects of methoxysuccinyl-Ala-Ala-Pro-Val-chloromethyl ketone-inhibited neutrophil elastase with the effects of its naturally occurring mutationally inactivated homologue (HBP) on fibroblasts and monocytes in vitro. *APMIS*, **100**, 1073–1080.

Polk, H. C. Jr (1979). Generalized peritonitis: a continuing challenge. *Surgery*, **86**, 777–778.

Roosendaal, R., Bakker-Woudenberg, I. A., Van den Berghe-van Raffe, M., Vink-van den Berg, J. C., Michel, M. F. (1989). Impact of the dosage schedule on the efficacy of ceftazidime, gentamicin and ciprofloxacin in *Klebsiella pneumoniae* pneumonia and septicemia in leukopenic rats. *Eur. J. Clin. Microbiol. Infect. Dis.*, **8**, 878–887.

Rotstein, O. D., Kao, J. (1988). The spectrum of *Escherichia coli–Bacteroides fragilis* pathogenic synergy in an intraabdominal infection model. *Can. J. Microbiol.*, **34**, 352–357.

Schierwagen, C., Bylund-Fellenius, A. C., Lundberg, C. (1990). Improved method for quantification of tissue PMN accumulation measured by myeloperoxidase activity. *J. Pharmacol. Methods*, **23**, 179–186.

Stumacher, R. J., Kovnat, M. J., McCabe, W. R. (1973). Limitation of the usefulness of the Limulus assay for endotoxin. *N. Engl. J. Med.*, **288**, 1261–1264.

Wichterman, K. A., Baue, A. E., Chaudry, I. H. (1980). Sepsis and septic shock: a review of laboratory models and a proposal. *J. Surg. Res.*, **29**, 189–201.

Wickel, D. J., Cheadle, W. G., Mercer-Jones, M. A., Garrison, R. N. (1997a). Poor outcome from peritonitis is caused by disease acuity and organ failure, not recurrent peritoneal infection. *Ann. Surg.*, **225**, 744–756.

Wickel, D., Mercer-Jones, M., Heinzelmann, M. *et al.* (1997b). Heparin binding protein increases survival in murine peritonitis. In *4th International Congress on the Immune Consequences of Trauma, Shock and Sepsis: Mechanisms and Therapeutic Approaches*, (ed. Faist, E.), pp. 413–416. Monduzzi Editore, Bologna, Italy.

Chapter 20

Murine Model of Peritonitis Involving a Foreign Body

F. Espersen and N. Frimodt-Møller

Background of the model

Foreign bodies or implants have been increasingly used in modern medicine over the last decades, and they have provided a range of benefits for patients. The major problem in using these foreign bodies is the high rate of bacterial infections, which limits their use.

Foreign-body infections have emerged as the major hospital-acquired infection (Dougherty, 1988; Pfaller and Herwald, 1988). The most common organisms are staphylococci, both *Staphylococcus aureus* and coagulase-negative staphylococci, but a range of other organisms may be involved. Foreign-body infections are characterized by a low infective inoculum, slow growth of the organisms and persistence of the infection despite the use of appropriate antibiotics (Zimmerli *et al.*, 1982; Christensen *et al.*, 1985; Dougherty, 1986; Gristina, 1987). Furthermore, foreign-body infections are often difficult to diagnose without removal of the foreign body, which prohibits controlled studies on antibiotic prophylaxis and/or treatment.

Experimental foreign-body infection models

In 1957 Elek and Conen found that a foreign body in form of a skin suture lowered the infective inoculum in humans. This principle has been used also in animals (James and Maclead, 1961; Renneberg and Walter, 1988). Further investigators have used tissue cage subcutaneously implanted in guinea pigs, rabbits or rats (Gardner *et al.*, 1973; Carbon *et al.*, 1977; Zimmerli *et al.*, 1984). More sophisticated models often necessitate the use of larger animals, but may offer the possibility of a close similarity to the clinical situation (Blomgren *et al.*, 1981; Malone *et al.*, 1975; Serota *et al.*, 1981).

We earlier worked with a murine peritonitis model, and found that mice challenged by this route were highly resistant to *S. aureus* (Espersen *et al.*, 1984a,b). On this background we developed a murine intraperitoneal model involving a foreign body (Espersen *et al.*, 1993, 1994).

Animal species

We have used CF-1 female mice of the age of 10–12 weeks, with a body weight of about 25–40 g (Statens Serum Institut, Copenhagen, Denmark) for most of our experiments. We have in few experiments used CF-1 mice of different age (10–22 weeks) and of opposite sex (males), and also Balb-c mice (Statens Serum Institut) without major differences in the results. It seems that most mice strains can be used, but initial experiments to determine the exact infective inoculum are necessary.

Details of surgery

Overview

Figure 20.1 illustrates the procedures. The mice are operated on under ether anaesthesia. Through an opening (~5 mm) in the lateral abdominal wall a catheter is inserted and the opening is closed with one suture. The bacterial inoculum is administered in the opposite lateral abdominal wall approximately 1 hour later.

Materials required

Anaesthetic, skin disinfectant, scalpel handle plus blades, forceps, scissors, syringes and needles, tape, suture materials, plastic tube (internal diameter > 25 mm) and catheter segments (i.e. the foreign bodies). The catheter used is a silicone catheter (Hedima, Glostrup, Denmark) with an internal diameter of 2.8 mm and an external diameter of 4.9 mm. We have used catheter segments from 5 mm to 20 mm long, with almost identical frequencies of infection. The length of 15 mm has been selected as it is easily accepted by the mice and also allows aspiration from the catheter (see below).

Anaesthesia

Animals are anaesthetized with ether and a plastic tube with ether-soaked gauze is used for continuation of

Handbook of Animal Models of Infection
ISBN 0-12-775390-7

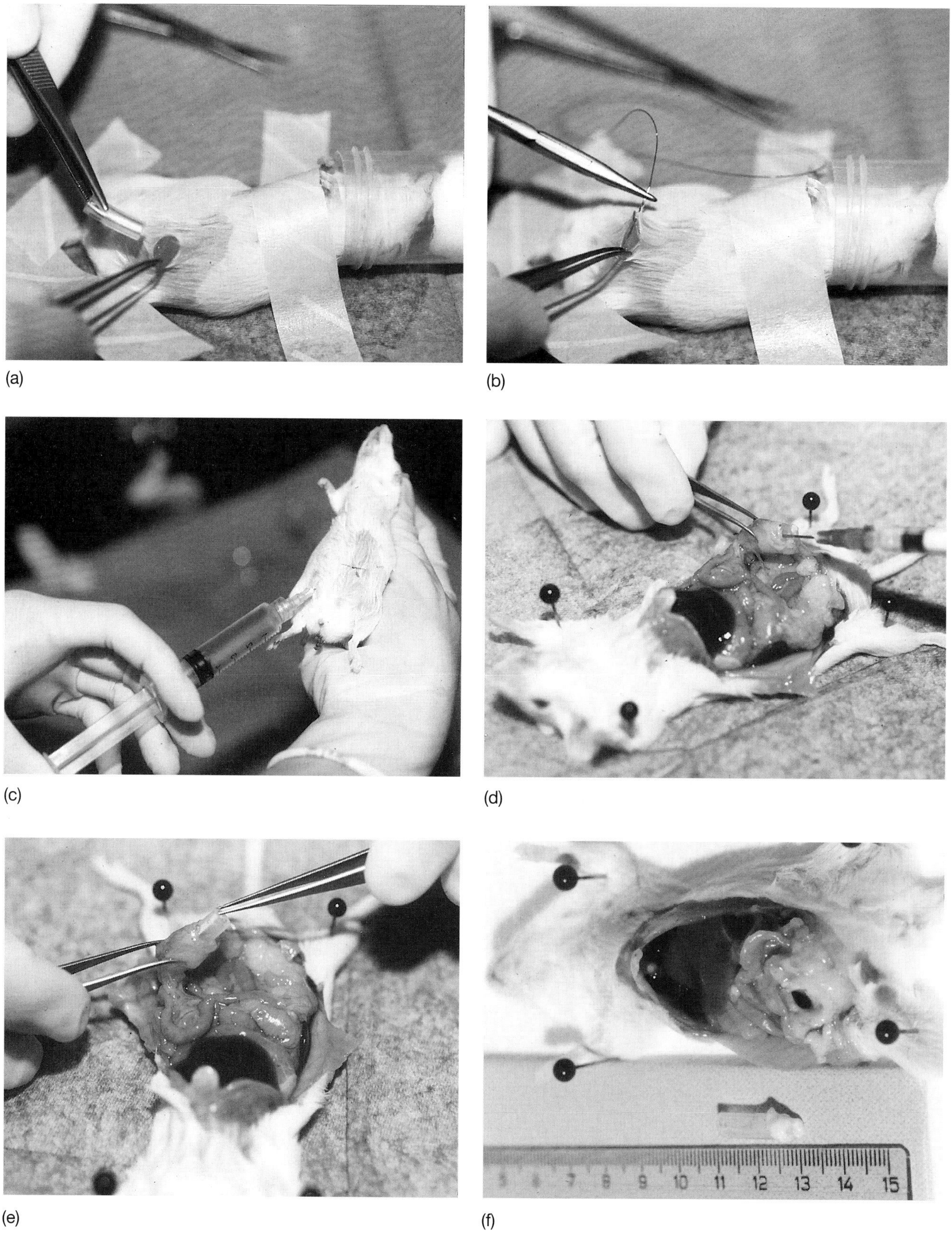

Figure 20.1 Procedures for establishing the murine model of peritonitis involving a foreign body. (a) Under ether anaesthesia the peritoneum is opened and the catheter inserted. (b) Suturing the abdominal wall. (c) Injection (i.p.) of *S. aureus* in saline. (d) The mouse is sacrificed and the peritoneum is opened. The contents of the catheter are aspirated. (e) The catheter is removed. (f) The empty abscess.

anaesthesia (Figure 20.1). The use of ether demands that the animals are handled in a fume cupboard.

As alternative anaesthesia we have used propanidid (Sombrevin, Chemical Works of Gedeon Richter Ltd., Budapest, Hungary). We have used intraperitoneal injections in the range of 20–40 mg. There are two problems with this type of anaesthesia. Firstly, the dose–response curve for anaesthesia is rather steep with a narrow range between effect and no-effect. Secondly, the animals seem to be more affected by the infective inoculum, resulting in a higher mortality, which necessitates an infective dose lower than 10^5 colony-forming units (cfu)/ml. This has, however, not been fully explored.

Surgical procedure

After induction of anaesthesia, the mouse is taped to the table as shown in Figure 20.1. An incision (~ 5 mm) is made in the lateral abdominal wall, and the peritoneum is opened after skin disinfection with 70% vol./vol. alcohol. Other disinfectant such as isopropanol- or iodine-based solution can also be used. The catheter is placed inside the peritoneum, and the skin and peritoneum are closed with one suture (Figure 20.1). The animals themselves remove the suture material in less than 6 hours following the procedure, without harm to the wound closure.

Infection procedure

Between 10 minutes and 1 hour after the surgical insertion, the mouse is infected with intraperitoneal injection of the infective inoculum in 1 ml of 0.9% saline in the opposite lateral abdominal wall (Figure 20.1). If the inoculum is introduced after 24 hours only approximately 10% of mice will develop a foreign body infection.

Postoperative care

The mice have been kept in cages of one to 10 each and have been allowed free access to food and water. The animals have normally returned to normal behaviour within a few hours. Postoperative analgesics have not been used, and the mice gain weight and have an unaffected food and water consumption for at least 2 weeks after infection. Later, the mice show signs of chronic infection (see below). Early mortality has been observed during the first 1–2 days.

Storage and preparation of inocula

Staphylococcus aureus, strain E 2371, isolated from a patient with bacteraemia, has been used mostly in this infection model, and as described earlier the bacteria were injected intraperitoneally in 1 ml of 0.9% saline. The bacteria were grown on solid medium, and a few colonies were inoculated into Mueller–Hinton broth (Statens Serum Institut) and incubated overnight at 37°C. Then the bacteria were harvested by centrifugation (1 600*g* for 10 minutes) and washed three times in saline under the same conditions. They were further adjusted to an optical density of 0.3 at 540 nm using a colorimeter and diluted 1:1 in saline (corresponding to approximately 10^8 cfu/ml).

The size of the inoculum has been adjusted based on experiments outlined in Figure 20.2. As can be seen, a challenge dose greater than 10^5 gives nearly 100% infection rate as defined by growth in catheter washings (see below) 3 days after challenge. Furthermore, if the inoculum is greater than 5×10^8 cfu an initial mortality is observed. With *S. aureus* strain E 2371, we have used infective dosages in the range of $0.5–1.0 \times 10^8$ cfu as routine.

We have used six other *S. aureus* strains, also isolated from blood cultures, with almost identical results. However, attempts to use four coagulase-negative staphylococci were less successful as at least 1.0×10^9 cfu have to be used to obtain 80–90% infection rates, but no mortality was observed.

At present no other species than staphylococci have been used in this model. Furthermore, no attempts have been made to store the inoculum using another suspension fluid than saline, or to perform the infection in immunosuppressed mice.

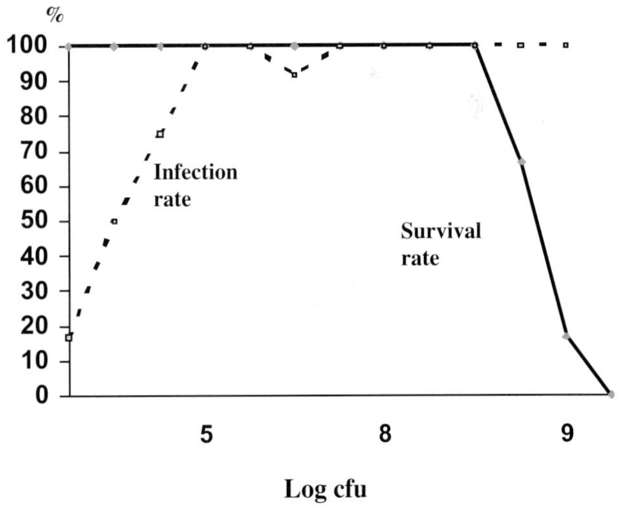

Figure 20.2 Effect of inoculum (log cfu) on infection and survival rate (% of infected animals in groups of six to 12 mice).

Characteristics of the model

We have followed the infection in animals for up to 30–40 days after challenge (Espersen *et al.*, 1994). Briefly, a few hours after infection the catheter is covered by the omentum, and 3 days later the infection becomes localized to an abscess around the catheter (Figure 20.1). The liver will enlarge after 10 days to approximately twice its size and the spleen to approximately three times. With an inoculum of 10^8 cfu the mice have peritoneal counts of 10^4–10^5/ml in few days, and then after 3–4 days the bacteria in the peritoneum disappear. In contrast, the bacteria will colonize the foreign body and have been found for up to 30 days. Simultaneously, blood cultures are positive (10–10^3/ml) for up to 3 days. Initially, we performed a similar operation and procedures in groups of mice without placement of a foreign body, and we obtained similar bacterial counts in both blood and peritoneum. Implantation of catheters without infection resulted in development of small sterile abscesses around the foreign body. If infected mice are followed for more than 30 days they start to show signs of wasting.

Key parameters to monitor infection and response to treatment

After a few days the infection is confined to the abscess around the foreign body and the initial diffuse peritonitis and bacteraemia have disappeared. The mice have slight blood neutrophilia, but haematological parameters have not been further studied. The antibody response has been investigated and compared to similarly infected mice without a foreign body. The ELISA with *S. aureus* strain E 1369 (protein A-free) was used, and antibody levels had already increased by day 3 and continued to increase for up to 2 weeks after challenge (Espersen, 1990, unpublished data). The ELISA was originally used to detect antibodies in peritonitis without a foreign body (Espersen and Clemmensen, 1985), and the presence of the foreign body and subsequent development of a chronic infection did not change the antibody levels; however, the duration of high titres was not further investigated (Espersen, 1990, unpublished data).

Bacteriological evaluation

Using this model we have developed the following principles for bacteriological evaluation:

1. Catheter aspirates

When the mice are killed, and the abdomen opened, it is possible to aspirate approximately 0.05–0.15 ml of fluid from the inside of the foreign body by puncturing the abscess (Figure 20.1). The cfu was obtained by serial dilutions in saline followed by plating. The lower limit of detection in aspirates is ~20 bacteria per millilitre of aspirate.

2. Catheter washings

The catheter is removed, placed in 1 ml of 0.9% saline and vortex mixed for 10 seconds, followed by quantitative cultures.

Intracellular bacteria

After treatment of mice, smears from aspirates have been evaluated by microscopy, and by means of counting the total number of bacteria per high-power field, and the rough number of intracellular and extracellular bacteria, the ratio of intracellular to extracellular bacteria can be followed during antibiotic treatment (Espersen *et al.*, 1994).

Antibiotic concentration

Antibiotic concentrations have been measured both in blood (orbital cut-down) or in aspirate from the foreign body. The concentrations have been measured by means of bioassay using 0.02 ml aspirate applied on discs. For gentamicin and methicillin the lower concentration limits have been determined to be 0.6 μg and 0.4 μg, respectively (Espersen *et al.*, 1994).

Antimicrobial therapy

Only few antimicrobial agents have yet been evaluated in the murine model of peritonitis involving a foreign body. The antibiotics can be administered by different routes in mice (s.c, i.p, p.o. and i.v.). We have only at present used s.c. administration of antibiotics in a volume of 0.1–0.5 ml of 0.9% saline (Espersen *et al.*, 1994).

The model has potential for investigations of antibiotic prophylaxis against foreign body infections, treatment of established foreign body infections and the evaluation of pathogenic parameters in these infections. So far the model has been used for antibiotic prophylaxis and for treatment using methicillin and gentamicin as model drugs.

In prophylaxis, it is important to determine the endpoints, and we have chosen to use the presence or absence of bacteria as measured by means of catheter washings. A problem in prophylaxis is to determine the time-schedule for the evaluation of the result after the challenge. We have, therefore, compared two different times, i.e. three days after challenge and 14 days after challenge, by giving a dose of either methicillin and gentamicin s.c. just before the mice received bacterial challenge i.p. The results indicate that three days is an optimal time for evaluation, since nearly identical results were obtained after three and 14 days (Table 20.1). Further studies of other antibiotics for prophylaxis need to be performed.

Table 20.1 Prophylaxis with methicillin or gentamicin; comparison of 3 days versus 14 days for evaluation of effect

Prophylaxis	*3 days*	*14 days*
None	1/56 (2%)*	1/20 (0)†
Methicillin (50 mg s.c.)	11/35 (31%)	6/19 (32%)†
Gentamicin (0.5 mg s.c.)	25/35 (71%)	14/20 (70%)†

*Cured (= no growth in catheter washings) mice/total mice (%).
$p < 0.05$ 3 days versus 14 days.

For treatment of established foreign body infections we chose to start treatment 3 days after bacterial challenge and continue the treatment, in the first study, for up to 6 days (Espersen *et al.*, 1994). The effect was evaluated by quantitative cultures from catheter aspirates and catheter washings 6 days after challenge (Espersen *et al.*, 1994). We also monitored the antibiotic concentrations both in serum and in catheter aspirates. Treatment gave a significant effect for both methicillin and gentamicin with a reduction of bacterial counts of about 2 log units, but no synergism was detected. The results of the treatment were relatively poor, when considering that the local concentration of methicillin was more than the MIC of the infective strain for at least 72 hours, and nine peaks (>13 μg/ml) was obtained for gentamicin (Espersen *et al.*, 1994). We investigated a range of parameters that may have caused this poor effect, but only identified an increased proportion of intracellular bacteria around the foreign body, that could have influenced the effect (Espersen *et al.*, 1994).

However, this model needs to be further explored for treatment with other antibiotics, longer treatment periods and treatment/prophylaxis experiments using other bacteria than *S. aureus*.

Pitfalls (advantages/disadvantages) of the model

Animal models have a range of shortcomings when compared to human infections; in mice especially, the different pharmacokinetics of antibiotics may play a role.

Except for the anatomical link with continuous ambulatury peritoneal dialysis this specific model has no direct similarities to any human infections, as the foreign body does not have a function. Another problem is the large inocula used in our present studies. However, it is possible to lower the inoculum to a range that may be more clinically relevant (Figure 20.2). On the other hand this model is easy to perform, highly reproducible, inexpensive and permits relatively short treatment periods. Also, the infection shows the same characteristics as foreign body infections in general, i.e. a low infective inoculum, a localization of the infection to the implant, development of a chronic infection and difficulty in eradicating an established infection.

The guinea-pig (and rat) model using a tissue cage (Zimmerli *et al.*, 1982) has some similarities with the present model, but the tissue cage allows repetitive punctures from the foreign body, which allows the kinetics and effect to be followed in individual animals. However, the present model is easier and less expensive and will be better for large scale (screening) use.

Contribution of the model to infectious disease therapy

The present model is new and has only been used in limited studies; consideration of its potential is, therefore, to some extent speculative.

Antimicrobial treatment and prophylaxis of foreign body infections

In these infections there is a need for cheap screening models, as clinical studies are difficult to perform on a large-scale basis, i.e. the diagnostic situation often necessitates the removal of the foreign body, which precludes the possibility of trials. The model can thus be an easy first line model to evaluate different treatment—or prophylactic regimens. The model also has potentials for studying the pharmacodynamics of such regimens, because of the possibility of determining of local antibiotic concentrations. Also, the high number of bacteria present in a focus and the difficulty of eradicating the infection gives potential for the study of the development of antibiotic resistance *in vivo*. An interesting future application could be to use the model to correlate the response with routine susceptibility testing methods, since the usual methods are not predictive of antibiotic activity in foreign-body infections (Blaser *et al.*, 1995).

Elucidation of pathophysiological parameters of foreign body infections

The model may be useful for evaluation of different foreign bodies, different polymers and different sizes of foreign body. The model will be able to establish the minimal and 50% infective dose. Furthermore, the model may have potential for elucidating the pathogenesis in foreign-body infections, the role of biofilm and leukocytes, and it could prove of value in investigation of antibiotic penetration into cells and studies of antibiotic synergy with specific reference to foreign-body infections.

References

Blaser J., Vergeres P., Widmer A. F., Zimmerli, W. (1995). *In vivo* verification of *in vitro* model of antibiotic treatment of device-related infection. *Antimicrob. Agents. Chemother.*, **39**, 1134–1139.

Blomgren, G., Lundquist, H., Nord, C. E., Lindgren, U. (1981). Late anaerobic haematogenous infection of experimental total joint replacement. *J. Bone Joint Surg.*, **62B**, 614–618.

Carbon, C., Contrepois, A., Brion, N., Lamotte-Barrillon, S. (1977). Penetration of cefazolin, cephaloridine and cephamandol into intestinal fluid in rabbits. *Antimicrob. Agents Chemother.*, **11**, 594–598.

Christensen, G. D., Simpson, W. A., Beachey, E. H. (1985). The adhesion of bacteria to animal tissue: complex mechanisms. In *Bacterial Adhesion: Mechanism and Physiological Significance*, (eds Savage, D. C., Fletcher, M.), pp. 279–305. Plenum Press, New York.

Dougherty, S. H. (1986). Implants infections. In *Handbook of Biomaterial Evaluation* (ed. von Recum, A. F.), pp. 276–289. Macmillan, New York.

Dougherty, S. H. (1988). Pathobiology of infection in prosthetic devices. *Rev. Infect. Dis.*, **10**, 1102–1117.

Elek, S. D., Conen, P. E. (1957). The virulence of *Staphylococcus pyogenes* for man. A study of the problem of wound infection. *Br. J. Exp. Pathol.*, **38**, 573–586.

Espersen, F., Clemmensen, I. (1985). Immunization of mice with the fibronectin-binding protein and clumping factor from *Staphylococcus aureus*: antibody response and resistance against peritoneal infection. *Acta Pathol. Immunol. Scand. Sect. C*, **93**, 53–58.

Espersen, F., Clemmensen, I., Rhodes, J. M., Jensen, K. (1984a) Human serum and plasma increase mouse mortality in *Staphylococcus aureus* intraperitoneal infection. *Acta Pathol. Immunol. Scand. Sect. B*, **92**, 305–310.

Espersen, F., Clemmensen, I., Frimodt-Møller, N., Jensen, K. (1984b). Effect of human IgG and fibrinogen on *Staphylococcus aureus* intraperitoneal infection in mice. *Acta Pathol. Immunol. Scand. Sect. B*, **92**, 311–317.

Espersen, F., Frimodt-Møller, N., Corneliussen, L., Thamdrup Rosdahl, V., Skinhøj, P. (1993). Experimental foreign body infection in mice. *J. Antimicrob. Chemother.*, **31** (Suppl. D), 103–111.

Espersen, F., Frimodt-Møller, N., Corneliussen, L., Riber, U., Thamdrup Rosdahl, V., Skinhøj, P. (1994). Effect of treatment with methicillin and gentamicin in a new experimental mouse model of foreign body infection. *Antimicrob. Agents Chemother.*, **38**, 2047–2053.

Gardner, W. G., Prior, R. B., Perkins, R. L. (1973). Fluid and pharmacological dynamics in a subcutaneous chamber implanted in rats. *Antimicrob. Agents Chemother.*, **4**, 196–197.

Gristina, A. G. (1987). Biomaterial-centered infection: microbial adhesion versus tissue integration. *Science*, **237**, 1588–1595.

James, R. C., Maclead, C. L. (1961). Induction of staphylococcal infections in mice with small inocula introduced on sutures. *Br. J. Exp. Pathol.*, **42**, 266–277.

Malone, J. M., Moore, W. S., Campagna, G. C., Bean, B. (1975) Bacteremic infectibility of vascular grafts: the influence of pseudointimal integrity and duration of graft function. *Surgery*, **78**, 211–216.

Pfaller, M. A., Herwaldt, L. A. (1988). Laboratory, clinical, and epidemiological aspects of coagulase-negative staphylococci. *Clin. Microb. Rev.*, **1**, 281–299.

Renneberg, J., Walder, M. (1988). A mouse model for simultaneous pharmacokinetic and efficacy studies of antibiotics at the site of infection. *J. Antimicrob. Chemother.*, **22**, 51–60.

Serota, A. I., Williams, R. A., Rose, J. G., Wilson, S. E. (1981). Uptake of radiolabeled leukocytes in prosthetic graft infection. *Surgery*, **90**, 35–40.

Zimmerli, W., Waldvogel, F. A., Vaudaux, P., Nydegger, U. E. (1982). Pathogenesis of foreign body infection: description and characteristics of an animal model. *J. Infect. Dis.*, **146**, 487–497.

Zimmerli, W., Lew, P. D., Waldvogel, F. A. (1984). Pathogenesis of foreign body infection. Evidence for a local granulocyte defect. *J. Clin. Invest.*, **73**, 1191–1200.

Chapter 21

Rat Polymicrobial Peritonitis Infection Model

H. Dupont and P. Montravers

Background of human infection

Peritonitis is the most frequent cause of intra-abdominal infection. Due to the large number of organisms isolated in the bowel, this infection is by nature polymicrobial. The standard care for this life-threatening situation consists of an emergency surgical intervention and empirical antibiotic therapy. Classification of the infection, diagnosis, antimicrobial therapy and adjuvant therapy have been reviewed previously (Gorbach, 1993; Nathens and Rotstein, 1994; Wittmann *et al.*, 1993, 1996).

Background of model

Several models of peritonitis using different animal species have been described, including dogs, pigs or sheep (Browne and Leslie, 1976; Wichterman *et al.*, 1980). The drawbacks with these large animals are linked to their cost and the limited number of individuals included in each experiment. As a consequence, the most common species used are mouse and rat (Browne and Leslie, 1976; Bartlett *et al.*, 1978; Wichterman *et al.*, 1980). Several rat models detailed below have been used from the beginning of the century, based upon the initial experiments of Powlowsky (1887) and Massart (1892).

Animal species

Sprague Dawley male rats are the most studied animals (180–490 g; Fry *et al.*, 1985; Matlow *et al.*, 1989; Andersson *et al.*, 1991; Ozmen *et al.*, 1993; Perdue *et al.*, 1994; Montravers *et al.*, 1994, 1997). The other strains used are Sprague Dawley female (180–300 g; Shyu *et al.*, 1987; Scarpace *et al.*, 1992; Kim *et al.*, 1996), Wistar male (190–250 g; Bartlett *et al.*, 1978; Almdahl and Osterud, 1987; Lorenz *et al.*, 1994; Jacobi *et al.*, 1997), Wistar female (180–280 g; Weinstein *et al.*, 1974; Bloechle *et al.*, 1995; Refsum *et al.*, 1996), Holtzman male (150–300 g; Wichterman *et al.*, 1980), Agouti female (Findon and Miller, 1995) and Norwegian female (300 g; Castillo *et al.*, 1991). Presumably, any strain could be used, but gender may be an important parameter to consider since Weinstein *et al.* (1974) reported an increased toxicity of the inoculum in female Wistar rats.

Preparation of the animals

No specialized housing or care is required. A model of prolonged starvation (water allowed *ad lib*) was evaluated by Wichterman *et al.* (1980) demonstrating an increased mortality in the animals that had been fasted for prolonged periods of time (≥5 days) when compared to fed animals. The bowel content can be altered by dietary manipulation. In the animals fed on a meat diet instead of the usual grain diet, a human-like flora with a high concentration of anaerobes is observed (Onderdonk *et al.*, 1974; Weinstein *et al.*, 1974). Pretreatments can be applied to obtain granulocytopenic animals (busulphan and cylophosphamide; Hau *et al.*, 1981) but these procedures do not have any specificity.

Details of surgery

Overview

The ideal model that reproduces human infection with a controlled inoculum does not exist. The models used for peritonitis can be roughly divided in three categories: direct intraperitoneal injection, bowel manipulations and peritoneal implantation. In every case, the purpose is to induce a massive and prolonged contamination of the peritoneal cavity.

Material required

Anaesthetic (preferably injectable), clippers or razor, skin disinfectant, surgical slippers, forceps, syringes and needles (20–23 g), suture material.

Handbook of Animal Models of Infection
ISBN 0-12-775390-7

Anaesthesia

The ideal anaesthetic agent is a rapidly acting drug (< 5–10 minutes), administered intramuscularly or subcutaneously, with a short duration of action (15–30 minutes), without respiratory depression and satisfactory muscular relaxation. Inhaled agents have been used: ether (Rotstein *et al.*, 1987; Bloechle *et al.*, 1995; Gürleyik *et al.*, 1996), or halothane 2% (Edmiston *et al.*, 1990). The intraperitoneal administration of an anaesthetic agent, although previously used (thiopental 15–60 mg/kg; Lally and Nichols, 1981; Nord *et al.*, 1986; Marshall *et al.*, 1988; Dunne *et al.*, 1996; Jacobi *et al.*, 1997), should not be recommended since the properties of the drug could affect the peritoneal response or the bacterial growth. Many drugs have been administered by intramuscular or subcutaneous route: droperidol subcutaneously at a dose of 0.2 ml (Hau *et al.*, 1983a,b); fentanyl combined with fluanisone (1 ml/kg Hypnorm®, Janssen Pharmaceuticals; Almdahl and Osterud, 1987; Olofsson *et al.*, 1986) or with haloperidol (10 mg/kg; Lorenz *et al.*, 1994), ketamine alone (30 mg/kg i.m.; Montravers *et al.*, 1994, 1997), or in association with xylazine (6–9 mg/kg; Terhar and Dunn, 1990; Ozmen *et al.*, 1993; Gürleyik *et al.*, 1996; Refsum *et al.*, 1996), or midazolam (10 mg/kg i.m.; Almdahl *et al.*, 1985; Olofsson *et al.*, 1986).

Surgical procedure

Intraperitoneal injection

The skin is disinfected with 70% vol./vol. ethanol, but other disinfectants are also applicable (isopropanolol or iodine-based solutions). Next, the puncture of the abdomen is performed with a needle of small diameter (25 G × 5/8″–0.5 × 16 mm) on the median abdominal line, 0.5–1 cm below the umbilicus, perpendicular to the muscular layer.

Surgical procedures

Following complete induction of anaesthesia, the animal is placed ventral side up. The abdominal skin is shaved with electric clippers. After skin disinfection (see above), a 1.5 cm midline incision is made at the median part of the abdomen. There are no major blood vessels in the region, but care should be taken to avoid any muscular damage, a source of subcutaneous haematoma. Exposure of the peritoneal cavity can be made by two incisions, first through the skin and second through the abdominal muscles. The skin incision can be made by a scalpel handle fitted with a curved-edge scalpel blade (e.g. Swann-Morton, Sheffield, UK). On the basis of our experience, we prefer to use sharp operating scissors (sharp, and sharp points, e.g. Collin, VM Tech SA, Paris, France) for the skin incision. A median 1–1.5 cm muscular incision is made at the linea alba level between abdominal rectis using delicate operating scissors (sharp, blunt points, e.g. Collin). Surgical exposure of the peritoneal cavity is obtained by a gentle removal of the peritoneo-muscular layer using thumb dressing forceps (e.g. Collin).

Infection procedures

In the caecal ligation procedure, the caecum is divided carefully avoiding all blood vessels (Wichterman *et al.*, 1980). The caecum is filled with faeces by milking stool back from the ascending colon. The caecum is then ligated just below the ileocaecal valve with a 3-0 silk ligature. Ligature at this limit allows bowel continuity to be maintained. The antimesenteric caecal surface is punctured with a needle (18–22 G) and the bowel is replaced into the peritoneal cavity.

In the peritoneal implantation model, the capsule previously prepared is inserted into the right lower quadrant of the pelvic cavity (Montravers *et al.*, 1994).

Wound closure

The wound is closed in two layers. The musculoperitoneal layer is closed using interrupted nylon sutures; two stitches are placed evenly over the wound. Care should be taken to avoid any visceral puncture. The skin layer is closed using interrupted nylon sutures, two stitches are placed evenly over the wound. Metal skin clips (11 mm, e.g. Michel, Collin) can also be used, but, in our experience, their use should be restricted to the procedures in which a prolonged survival is expected (> 48–72 hours) since sutures are easier for the rats to bite though risking removal or subcutaneous abscess before the wound is fully healed.

Postoperative care

Animals recover quickly from anaesthesia (< 30 minutes). Animals should be housed individually on wire supports following surgery. Postoperative analgesics have not been reported to be routinely used. The infection becomes apparent within a few hours after challenge.

Storage and preparation of inocula

Most studies have used high concentrations (10^7–10^9 cfu/ml) of *Escherichia coli*, *Enterococcus faecalis*, *Bacteroides fragilis*, corresponding to the organisms usually cultured from human infections (Altemeier, 1942; Fry *et al.*, 1985; Matlow *et al.*, 1989; Montravers *et al.*, 1994, 1997), or from rat infections (Onderdonk *et al.*, 1976). However, the model of peritonitis is amenable to using several different species of bacteria.

In the intraperitoneal injection models, the overnight

broth cultures of the organisms are sedimented by centrifugation and then resuspended at the chosen concentration in isotonic saline (Fry *et al.*, 1985) alone or associated with 10% barium sulphate (Matlow *et al.*, 1989).

In the model of intraperitoneal implantation, various inocula have been used. Faecal material from meat-fed rats (Weinstein *et al.*, 1975; Louie *et al.*, 1977; Barlett *et al.*, 1981) or from humans (Nichols *et al.*, 1979) have been used. The samples are pooled in an anaerobic chamber and mixed with an equal volume of prereduced peptone–yeast–glucose broth and sterile barium sulphate (10% wt/vol.). The pooled inoculum is filtered through surgical gauze and then frozen and stored at –40°C until used. At the time of challenge, the aliquots are thawed and prepared for inoculation without subculture (Weinstein *et al.*, 1975; Louie *et al.*, 1977; Nichols *et al.*, 1979; Barlett *et al.*, 1981). Aliquots of 0.5 ml of the mixture are placed into sterile gelatin capsules for intraperitoneal implantation. A double capsule obtained by the insertion of the 0.5 ml capsule (no. 1) into a 0.75 ml capsule (no. 0) permits ease in handling of the inoculum and slows its dissolution, while a single capsule dissolves immediately after peritoneal implantation (Weinstein *et al.*, 1974).

The implantation of a capsule with a calibrated inoculum has been used by our group (Montravers *et al.*, 1994, 1997) and by Onderdonk *et al.* (1976). Prereduced thioglycolate broth is used for culture of the anaerobes while brain–heart medium is used for the growth of the aerobes (Montravers *et al.*, 1994). The mixtures of strains are made when they are in log phase of growth; bacteria are diluted to obtain the number of micro-organisms for the challenge. Diluted broth cultures are combined with 2% (wt/vol.) agar (Montravers *et al.*, 1994) or with autoclaved rat colonic contents (50% vol./vol.; Onderdonk *et al.*, 1976) and barium sulphate (10% wt/vol.). Aliquots of 0.5 ml of the final product are placed into a double gelatin capsule (Weinstein *et al.*, 1974).

Infection processes

Many models simulating perforation have been proposed using perforation of the stomach, the caecum or the sigmoid colon (Weinstein *et al.*, 1974; Wichterman *et al.*, 1980; Almdahl *et al.*, 1985), while ischaemic models obtained by vascular ligation and ileal or caecal necrosis are scarce (Wichterman *et al.*, 1980; Nord *et al.*, 1986, Marshall *et al.*, 1988). The caecal ligation technique with perforation has become the reference because of its high reproducibility (Wichterman *et al.*, 1980). The severity of the disease and the initial mortality depends on the diameter of the needle and the number of caecal punctures. Puncturing the caecum once rather than twice with the same needle prolongs the survival of animals, while puncturing the caecum with an 18 G needle instead of a 22 G needle dramatically increases mortality (from 40–60% to 95% mortality within 48 hours).

Many compounds have been implanted within the peritoneum, including rat's faecal pellet (Wichterman *et al.*, 1980), human faeces (Nichols *et al.*, 1979), septic fibrin clots (Dunn *et al.*, 1984b; Rotstein *et al.*, 1987) or a gelatin capsule containing various inocula (faecal mixture, calibrated inoculum; Weinstein *et al.*, 1975; Lally *et al.*, 1983; Dunn *et al.*, 1989; Ozmen *et al.*, 1993; Montravers *et al.*, 1994, 1997; Dunne *et al.*, 1996). Various adjuvants have been combined with microorganisms in order to increase the severity of the disease and the local inflammatory response and to decrease the dissemination of the inoculum, such as mucin (Wichterman *et al.*, 1980), bile (Andersson *et al.*, 1989, 1991; Sonesson *et al.*, 1990), blood and haemoglobin (Weinstein *et al.*, 1974; Andersson *et al.*, 1989) but barium sulphate is the most used (Weinstein *et al.*, 1974, 1975; Onderdonk *et al.*, 1974, 1976; Montravers *et al.*, 1994, 1997). Various foreign bodies have been used to increase the severity of the disease: autoclaved tissue (Altemeier, 1942), sterile faeces (Onderdonk *et al.*, 1976), or agar (Montravers *et al.*, 1994, 1997).

Key parameters to monitor infection and response to treatment

A systemic dissemination of infection occurs within 1–3 hours after challenge. Bacteraemia is reported within 20 minutes after peritoneal injection (Dunn *et al.*, 1985) and within few hours after inoculation in the surgical models (Montravers *et al.*, 1994, 1997). The signs of sepsis are apparent within 6–12 hours (decreased mobility, piloerection). Then the animals become lethargic, stop food and water intake, their eyes appear glazed with crusting exudate, and some have diarrhoea associated with abdominal distension. Neutrophilia is common (Dunne *et al.*, 1996). A 10–25% loss of weight is observed within 1–2 days, which culminates on day 2–3 followed by a progressive improvement in survivors (Montravers *et al.*, 1994). The first deaths due to sepsis occur within 6–12 hours depending on the severity of the disease (Dunn *et al.*, 1985) and continue until day 4–5 (Weinstein *et al.*, 1974). The survivors are either completely cured or have intra-abdominal abscesses, depending on the inoculum (Bartlett *et al.*, 1978).

At the time of sacrifice, a midline laparotomy is performed after skin disinfection, and the peritoneal fluid is collected by direct aspiration or after lavage from all regions of the peritoneal cavity (Montravers *et al.*, 1994). Various counts of the organisms are reported within the peritoneal fluid depending on the concentration of the inoculum and the delay before sacrifice (Dunn *et al.*, 1985; Montravers *et al.*, 1997). An increased count in the peritoneal phagocytes is observed within the first 6 days of the disease that persists for at least 6 days (Montravers *et al.*, 1997). The cytokine production — TNF-α (bioassay LM cells) and IL-6 (bioassay B9 cells) — within the peritoneal fluid probably peaks between 1 and 3 hours after inoculation followed by a progressive decline until day 3 (Montravers *et al.*, 1997).

Antibiotic therapy

A variety of antibiotic regimens have been evaluated (Louie *et al.*, 1977; Nichols *et al.*, 1979; Barlett *et al.*, 1981; Montravers *et al.*, 1994). However, peritoneal enumeration of the remaining cfu and measurement of the peritoneal concentrations of the drugs have rarely been performed. Because of the ileus, oral therapy is unsuitable. Antibiotic therapy is usually begun early after inoculation (2–6 hours), administered intramuscularly every 6–8 hours and pursued for a period of 3–12 days. Duration of treatment can dramatically affect the number of remaining organisms (Montravers *et al.*, 1994) and the experimental design should consider varying the length of treatment. Determination of the peritoneal cfu should be done several hours after the cessation of treatment in order to allow the elimination of the antibiotic and reduce any effects of antibiotic carry-over. The dosing schedule and choice of dose should be given careful consideration, given the differing pharmacokinetics of antibiotics between humans and rats, especially for carbapenems and aminoglycosides. A gradient between plasma and peritoneal fluid is observed (Montravers *et al.*, 1994).

Pitfalls (advantages/disadvantages) of the model

The intraperitoneal injection is widely used because of the swiftness and the easiness of the procedure. A visceral puncture (liver, stomach, caecum, colon, bladder) during the inoculation is the most frequent complication of this technique. However, this model does not represent a model of peritonitis but rather a model of systemic infection (Dunn *et al.*, 1985) better suited to illustrating potentially pathological events in sepsis than to showing the therapeutic benefit of a particular therapy (Natanson *et al.*, 1994).

Many models simulating perforation have been proposed but the caecal ligation technique with perforation has progressively become the reference because it is highly reproducible. The size of the inoculum depends on the diameter of the needle and the number of punctures (see above; Wichterman *et al.*, 1980). The limitations of this model are linked to ignorance of the species and concentrations of bacteria in the caecum. In the normal faecal flora of rats, the concentration of aerobes outnumbers the concentration of anaerobes, an uncommon situation in human pathology (Weinstein *et al.*, 1974). Lactobacilli, *Staphylococcus epidermidis* and *Micrococcus* spp. are the predominant strains, followed at a lower concentration by Enterobacteriaceae and enterococci. The animals fed lean ground beef develop a human-like faecal flora, yelding anaerobes at a higher concentration than the Enterobacteriaceae associated to the persistence of the usual rat flora (Weinstein *et al.*, 1974). Since the various organisms cultured from these animals have rarely been identified and enumerated (Weinstein *et al.*, 1974, 1975), the reproducibility of the bacterial inoculum might be questionable.

The use of a gelatin capsule permits a slow dissolution and a progressive development of sepsis without overwhelming systemic toxicity (Weinstein *et al.*, 1974). The implantation of faecal content raises the limitations stated above. The use of calibrated inocula allows specific issues on pathophysiology or antibiotic therapy to be addressed. The use of adjuvants such as bile, haemoglobin and, to some extent, barium sulphate as adjuvants has some clinical relevance since these compounds are observed in human peritonitis and could interfere with peritoneal inflammation. However, these experimental conditions drift away from clinical practice. It is recommended that care be taken in the experimental design and subsequent interpretation of data on antimicrobial effectiveness. The evaluation of new agents, or novel combinations of agents, should be performed in comparison to reference agents. In addition, the dose levels of the antibiotics should be equated with the levels attainable in humans. The rat models are apparently less used than the mouse model of peritonitis. Some advantages of the rat models include the size of the animal, allowing repeated pharmacokinetic measurements, and an easy surgical procedure.

Contribution of the peritonitis models to infectious disease therapy

Pathophysiology of peritonitis

Most of the studies used the peritonitis models to study the pathophysiology of intra-abdominal sepsis. These models emphasize the pathogenic role of haemoglobin (Lee *et al.*, 1979; Hau *et al.*, 1983b), bile (Andersson *et al.*, 1989, 1991), lymphatic absorption (Dunn *et al.*, 1985; Dumont *et al.*, 1986; Olofsson *et al.*, 1986; Gürleyik *et al.*, 1996), fibrin (Dunn *et al.*, 1984b), immunosuppression (Marshall *et al.*, 1988), including surgical induced immunosuppression (Almdahl and Osterud, 1987), or inflammatory response (Scarpace *et al.*, 1992; Lorenz *et al.*, 1994; Dunne *et al.*, 1996; Refsum *et al.*, 1996). Various studies demonstrated the role of bacterial synergy, involving a co-operation between Enterobacteriaceae and anaerobes (Weinstein *et al.*, 1975; Onderdonk *et al.*, 1976; Rotstein *et al.*, 1987), and more recently of these organisms with enterococci (Fry *et al.*, 1985; Matlow *et al.*, 1989; Montravers *et al.*, 1994, 1997).

Evaluation of surgical techniques

Experimental models helped in clarifying the effects of surgical procedures—prolonged survival in using open management of the peritoneal cavity (Terhar and Dunn, 1990), delayed eradication of the pathogens induced by peritoneal dialysis (Findon and Miller, 1995). Various studies evaluated the effects of peritoneal lavage and demonstrated the superiority of low-dose povidione-iodide solutions and sterile normal saline (Lally and Nichols, 1981; McAvinchey *et al.*, 1983), the poor effects of local antibiotics when used alone (Hau *et al.*, 1983a; Lally and Nichols, 1981) and the need for combinations of antimicrobial therapy and peritoneal lavage (McAvinchey *et al.*, 1984).

Evaluation of antimicrobial chemotherapy

The need to treat Enterobacteriaceae and anaerobes initially demonstrated by Weinstein *et al.* (1975) has become a standard in human therapy and was confirmed in experimental (Louie *et al.*, 1977; Nichols *et al.*, 1979; Barlett *et al.*, 1981; Nord *et al.*, 1986) and in human studies (Gorbach, 1993; Nathens and Rotstein, 1994; Wittmann *et al.*, 1993, 1996), demonstrating the efficacy of penicillins, cephalosporins or pefloxacin. On the other hand, the need to treat enterococci remains controversial (Montravers *et al.*, 1994). Finally, polymicrobial peritonitis represents an interesting field in which to evaluate the effect of therapies aimed at modificating the immune system, such as granulocyte colony-stimulating factor (Dunne *et al.*, 1996; Lorenz *et al.*, 1994), TNF antibodies (Bagby *et al.*, 1991), pentoxyfilline (Refsum *et al.*, 1996), muramyl dipeptide (Dunn *et al.*, 1989) or free radical scavengers (Castillo *et al.*, 1991).

References

Almdahl, S. M., Osterud, B. (1987). Abdominal operations: effect on subsequent experimental faecal peritonitis. *Scand. J. Gastroenterol.*, **22**, 592–594.

Almdahl, S. M., Nordstrand, K., Melby, K., Osterud, B., Giercksky, K. E. (1985). Faecal peritonitis in the rat. An experimental model for evaluation of surgical and adjuvant therapies. *Acta Chir. Scand.*, **151**, 213–216.

Altemeier, W. A. (1942). The pathogenicity of the bacteria of appendicitis peritonitis. *Surgery*, **11**, 374–384.

Andersson, R., Tranberg, K. G., Alwmark, A., Bengmark, S. (1989). Factors influencing the outcome of *E. coli* peritonitis in rats. *Acta Chir. Scand.*, **155**, 155–157.

Andersson, R., Schalen, C., Tranberg, K. G. (1991). Effect of bile on growth, peritoneal absorption, and blood clearance of *Escherichia coli* in *E. coli* peritonitis. *Arch. Surg.*, **126**, 773–777.

Bagby, G. J., Plessala, K. J., Wilson, L. A., Thompson, J. J., Nelson, S. (1991). Divergent efficacy of antibody to tumor necrosis factor-alpha in intravascular and peritonitis models of sepsis. *J. Infect. Dis.*, **163**, 83–88.

Bartlett, J. G., Onderdonk, A. B., Louie, T., Kasper, D. L., Gorbach, S. L. (1978). Lessons from an animal model of intra-abdominal sepsis. *Arch. Surg.*, **113**, 853–857.

Barlett, J. G., Louie, T. J., Gorbach, S. L., Onderdonk, A. B. (1981). Therapeutic efficacy of 29 antimicrobial regimens in experimental intraabdominal sepsis. *Rev. Infect. Dis.*, **3**, 535–542.

Bloechle, C., Emmermann, A., Treu, H. *et al.* (1995). Effect of a pneumoperitoneum on the extent and severity of peritonitis induced by gastric ulcer perforation in the rat. *Surg. Endosc.*, **9**, 898–901.

Browne, M. K., Leslie, G. B. (1976). Animal models of peritonitis. *Surg. Gynecol. Obstet.*, **143**, 738–740.

Castillo, M., Toledo-Pereyra, L. H., Gutierrez, R., Prough, D., Shapiro, E. (1991). Peritonitis after cecal perforation. An experimental model to study the therapeutic role of antibiotics associated with allopurinol and catalase. *Am. Surgeon*, **57**, 313–316.

Dumont, A. E., Maas, W. K., Iliescu, H., Shin, R. D. (1986). Increased survival from peritonitis after blockade of transdiaphragmatic absorption of bacteria. *Surg. Gynecol. Obstet.*, **162**, 248–252.

Dunn, C. W., Horton, J. W., Walker, P. B. (1989). Additive effect of an immunomodulator and broad-spectrum antibiotic in fecal peritonitis. *Am. J. Surg.*, **157**, 548–551.

Dunn, D. L., Rotstein, O. D., Simmons, R. L. (1984). Fibrin in peritonitis. IV. Synergistic intraperitoneal infection caused by *Escherichia coli* and *Bacteroides fragilis* within fibrin clots. *Arch. Surg.*, **119**, 139–144.

Dunn, D. L., Barke, R. A., Knight, N. B., Humphrey, E. W., Simmons, R. L. (1985). Role of resident macrophages, peripheral neutrophils, and translymphatic absorption in bacterial clearance from the peritoneal cavity. *Infect. Immun.*, **49**, 257–264.

Dunne, J. R., Dunkin, B. J., Nelson, S., White, J. C. (1996). Effects of granulocyte colony stimulating factor in a nonneutropenic rodent model of *Escherichia coli* peritonitis. *J. Surg. Res.*, **61**, 348–354.

Edmiston, C. E. Jr., Goheen, M. P., Kornhall, S., Jones, F. E., Condon, R. E. (1990). Fecal peritonitis: microbial adherence to serosal mesothelium and resistance to peritoneal lavage. *World J. Surg.*, **14**, 176–183.

Findon, G., Miller, T. (1995). Bacterial peritonitis in continuous ambulatory peritoneal dialysis: effect on dialysis on host defense mechanisms. *Am. J. Kidney Dis.*, **26**, 765–773.

Fry, D. E., Berberich, S., Garrison, R. N. (1985). Bacterial synergism between the Enterococcus and *Escherichia coli*. *J. Surg. Res.*, **38**, 475–478.

Gorbach, S. L. (1993). Treatment of intra-abdominal infections. *J. Antimicrob. Chemother.*, **31** (Suppl. A), 67–78.

Gürleyik, E., Gürleyik, G., Unalmiser, S. (1996). Blockade of transdiaphragmatic lymphatic absorption reduced systemic inflammatory response syndrome during experimental peritonitis: evaluation with body oxygen kinetics in rats. *Eur. J. Surg.*, **162**, 729–734.

Hau, T., Lee, J. T. Jr, Simmons, R. L. (1981). Mechanisms of the adjuvant effect of hemoglobin in experimental peritonitis. IV. The adjuvant effect of hemoglobin in granulocytopenic rats. *Surgery*, **89**, 187–191.

Hau, T., Nishikawa, R., Phuangsab, A. (1983a). Irrigation of the peritoneal cavity and local antibiotics in the treatment of peritonitis. *Surg. Gynecol. Obstet.*, **156**, 25–30.

Hau, T., Nishikawa, R. A., Phuangsab, A. (1983b). The effect of bacterial trapping by fibrin on the efficacy of systemic antibiotics in experimental peritonitis. *Surg. Gynecol. Obstet.*, **157**, 252–256.

Jacobi, C. A., Ordemann, J., Böhm, B. *et al.* (1997). Does laparoscopy increase bacteremia and endotoxemia in a peritonitis model? *Surg. Endosc.*, **11**, 235–238.

Kim, Y. M., Hong, S. J., Billiar, T. R., Simmons, R. L. (1996).

Counterprotective effect of erythrocytes in experimental bacterial peritonitis is due to scavenging of nitric oxide and reactive oxygen intermediates. *Infect. Immun.*, **64**, 3074–3080.

Lally, K. P., Nichols, R. L. (1981). Various intraperitoneal irrigation solutions in treating experimental fecal peritonitis. *South. Med. J.*, **74**, 789–791, 798.

Lally, K. P., Trettin, J. C., Torma, M. J. (1983). Adjunctive antibiotic lavage in experimental peritonitis. *Surg. Gynecol. Obstet.*, **156**, 605–608.

Lee, J. T., Jr., Ahrenholz, D. H., Nelson, R. D., Simmons, R. L. (1979). Mechanisms of the adjuvant effect of hemoglobin in experimental peritonitis. V. The significance of the coordinated iron component. *Surgery*, **86**, 41–48.

Lorenz, W., Reimund, K. P., Weitzel, F. *et al.* (1994). Granulocyte colony-stimulating factor prophylaxis before operation protects against lethal consequences of postoperative peritonitis. *Surgery*, **116**, 925–934.

Louie, T. J., Onderdonk, A. B., Gorbach, S. L., Bartlett, J. G. (1977). Therapy for experimental intraabdominal sepsis: comparison of four cephalosporins with clindamycin plus gentamicin. *J. Infect. Dis.*, **135**, S18–S24.

Marshall, J. C., Christou, N. V., Meakins, J. L. (1988). Small-bowel bacterial overgrowth and systemic immunosuppression in experimental peritonitis. *Surgery*, **104**, 404–411.

Massart, J. (1892). Le chimiotactisme des leucocytes et l'immunité. *Ann. Inst. Pasteur (Paris)*, **6**, 221–272.

Matlow, A. G., Bohnen, J. M., Nohr, C., Christou, N., Meakins, J. (1989). Pathogenicity of enterococci in a rat model of fecal peritonitis. *J. Infect. Dis.*, **160**, 142–145.

McAvinchey, D. J., McCollum, P. T., McElearney, N. G., Mundinger, G., Jr., Lynch, G. (1983). Antiseptics in the treatment of bacterial peritonitis in rats. *Br. J. Surg.*, **70**, 158–160.

McAvinchey, D. J., McCollum, P. T., Lynch, G. (1984). Towards a rational approach to the treatment of peritonitis: an experimental study in rats. *Br. J. Surg.*, **71**, 715–717.

Montravers, P., Andremont, A., Massias, L., Carbon, C. (1994). Investigation of the potential role of *Enterococcus faecalis* in the pathophysiology of experimental peritonitis. *J. Infect. Dis.*, **169**, 821–830.

Montravers, P., Mohler, J., Saint Julien, L., Carbon, C. (1997). Evidence of the proinflammatory role of *Enterococcus faecalis* in polymicrobial peritonitis in rats. *Infect. Immun.*, **65**, 144–149.

Natanson, C., Hoffman, W. D., Suffredini, A. F., Eichacker, P. Q., Danner, R. L. (1994). Selected treatment strategies for septic shock based on proposed mechanisms of pathogenesis. *Ann. Intern. Med.*, **120**, 771–783.

Nathens, A. V., Rotstein, O. D. (1994). Therapeutic options in peritonitis. *Surg. Clin. North Am.*, **74**, 677–692.

Nichols, R. L., Smith, J. W., Fossedal, E. N., Condon, R. E. (1979). Efficacy of parenteral antibiotics in the treatment of experimentally induced intraabdominal sepsis. *Rev. Infect. Dis.*, **1**, 302–309.

Nord, C. E., Edlund, C., Lahnborg, G. (1986). The efficacy of pefloxacin in comparison to gentamicin in the treatment of experimentally induced peritonitis in rats. *J. Antimicrob. Chemother.*, **17** (Suppl. B), 59–63.

Olofsson, P., Nylander, G., Olsson, P. (1986). Endotoxin: routes of transport in experimental peritonitis. *Am. J. Surg.*, **151**, 443–446.

Onderdonk, A. B., Weinstein, W. M., Sullivan, N. M., Bartlett, J. G., Gorbach, S. L. (1974). Experimental intra-abdominal abscesses in rats: quantitative bacteriology of infected animals. *Infect. Immun.*, **10**, 1256–1259.

Onderdonk, A. B., Bartlett, J. G., Louie, T., Sullivan-Seigler, N., Gorbach, S. L. (1976). Microbial synergy in experimental intra-abdominal abcess. *Infect. Immun.*, **13**, 22–26.

Ozmen, V., Thomas, W. O., Healy, J. T. *et al.* (1993). Irrigation of the abdominal cavity in the treatment of experimentally induced microbial peritonitis: efficacy of ozonated saline. *Am. Surg.*, **59**, 297–303.

Perdue, P. W., Kazarian, K. K., Nevola, J., Law, W. R., Williams, T. (1994). The use of local and systemic antibiotics in rat fecal peritonitis. *J. Surg. Res.*, **57**, 360–365.

Powlowsky, A. D. (1887). Beitraege zur Aetiologie und Entstehungweise der akuten Peritonitis. *Zentralbl. Chir.*, **14**, 881–887.

Refsum, S. E., Halliday, M. I., Campbell, G., McCaigue, M., Rowlands, B. J., Boston, V. E. (1996). Modulation of TNF alpha and IL-6 in a peritonitis model using pentoxifylline. *J. Pediatr. Surg.*, **31**, 928–930.

Rotstein, O. D., Pruett, T. L., Wells, C. L., Simmons, R. L. (1987). The role of *Bacteroides* encapsulation in the lethal synergy between *Escherichia coli* and *Bacteroides* species studied in a rat fibrin clot peritonitis model. *J. Infect.*, **15**, 135–146.

Scarpace, P. J., Borst, S. E., Bender, B. S. (1992). The association of *E. coli* peritonitis with an impaired and delayed fever response in senescent rats. *J. Gerontol.*, **47**, B142–145.

Shyu, W. C., Nightingale, C. H., Quintiliani, R. (1987). Pseudomonas peritonitis in neutropenic rats treated with amikacin, ceftazidime and ticarcillin, alone and in combination. *J. Antimicrob. Chemother.*, **19**, 807–814.

Sonesson, A., Larsson, L., Andersson, R., Adner, N., Tranberg, K. G. (1990). Use of two-dimensional gas chromatography with electron-capture detection for the measurement of lipopolysaccharides in peritoneal fluid and plasma from rats with induced peritonitis. *J. Clin. Microbiol.*, **28**, 1163–1168.

Terhar, M. A., Dunn, M. M. (1990). Open peritoneal management in murine peritonitis. *Am. Surgeon.*, **56**, 451–454.

Weinstein, W. M., Onderdonk, A. B., Bartlett, J. G., Gorbach, S. L. (1974). Experimental intra-abdominal abscesses in rats: development of an experimental model. *Infect. Immun.*, **10**, 1250–1255.

Weinstein, W. M., Onderdonk, A. B., Bartlett, J. G., Louie, A. B., Gorbach, S. L. (1975). Antimicrobial therapy of experimental intraabdominal sepsis. *J. Infect. Dis.*, **132**, 282–286.

Wichterman, K. A., Baue, A. E., Chaudry, I. H. (1980). Sepsis and septic shock—a review of laboratory models and a proposal. *J. Surg. Res.*, **29**, 189–201.

Wittmann, D. H., Schein, M., Condon, R. E. (1996). Management of secondary peritonitis. *Ann. Surg.*, **224**, 10–18.

Wittmann, D. H., Walker, A. P., Condon, R. E. (1993). Peritonitis, intra-abdominal infection, and intra-abdominal abscess. In *Principles of Surgery* (eds Schwartz, S. I., Shires, G. T., Spencer, F. C.), pp. 1449–1484. McGraw Hill, New York.

Chapter 22

Murine Thigh Suture Model

J. D. Pietsch and H. C. Polk, Jr

Background of model

The thigh suture model was first developed 20 years ago by McCoy at the Price Institute of Surgical Research, University of Louisville, KY (Rouben *et al.*, 1977; Fagelman *et al.*, 1981). Adapted from work performed at the Lister Institute of Preventive Medicine in London (Polk and Miles, 1973), its purpose was to provide a simple, consistent, and reproducible model of surgical infection. The three essential components of surgical infection are surgical trauma, a foreign body, and defined bacterial contamination. It has been used as a model to assess the effects of non-specific immune-modulating agents such as bacillus Calmette–Guérin (Rouben *et al.*, 1977; Fagelman *et al.*, 1981), muramyl dipeptide (Galland *et al.*, 1981; Polk *et al.*, 1981; Ausobsky *et al.*, 1982; Lamont *et al.*, 1987), tumor necrosis factor (Hershman *et al.*, 1989), and γ-interferon (Hershman *et al.*, 1989), as well as the effects of single and multiple antibiotics on surgical infection (Galland *et al.*, 1982a). Studies on the effects of combined treatments such as muramyl dipeptide plus antibiotics (Polk *et al.*, 1982; Gaar *et al.*, 1994) and surgically induced infection on immunocompromised mice (Cobb *et al.*, 1986; Galland *et al.*, 1983) have also been performed. This model currently provides an accurate means to assess the systemic and local effectiveness of biological agents on the infection process.

Animal species

Both outbred and inbred strains of mice have been used effectively. Outbred strains provide an inexpensive survival model and allow bacterial clearance to be quantified. Inbred strains provide a more accurate measure of specific immunological changes that occur during the infection and recovery process.

Materials required

Suture

Sterile 3–0 or 2–0 cotton suture is commonly used, but other types of braided suture may also be used. Braided suture is required since the bacteria need a matrix to adhere to and grow on. The suture is cut into approximately 7 cm lengths. One length of suture is required for each animal. The size and density of the suture determine the amount of bacteria delivered at bacterial challenge.

Needles

Small, curved French-eye needles.

Surgical instruments

Needle holders, forceps, and small sharp scissors are needed. Curved tenotomy scissors are ideal to trim the suture close to the skin.

Miscellaneous

Sterile Petri dishes and alcohol swabs.

Fasting

This model is most effectively used in the fasted animal (Galland *et al.*, 1981, 1982b,c). Withholding food for 24 hours prior to the procedure will increase mortality. Fasting also tends to mimic the immunological insult that is observed clinically following major trauma, and it allows stratification of the groups so that significant differences can be observed. Mice have free access to food immediately following bacterial challenge, or up to 4 hours after challenge, depending on the effect desired.

Anesthesia

Anesthesia is not necessary since there is only slight and momentary pain. The amount of stress and discomfort is

Handbook of Animal Models of Infection
ISBN 0-12-775390-7

equivalent to that of a needle stick. If anesthesia is desired, the animals may be lightly anesthetized with an inhalation agent such as halothane, metofane, or ethyl ether. These agents can be administered by placing the animal into a small covered glass jar that contains the anesthetic.

Bacterial challenge

Many types of bacteria may be used with this model. Gram-negative bacteria have traditionally been used to mimic the clinical scenario. The bacterial challenge may be a single strain or a combined inoculation. We have found that a recently acquired hospital isolate works well. Virulence should be established, and the bacteria passaged every few weeks to ensure a consistent virulence. A baseline mortality rate of approximately 70% in untreated animals effectively assesses the positive or negative effect of treatment agents.

Preparation of inocula

The bacteria used for the challenge are inoculated into tryptic soy broth or other suitable media and grown overnight in an incubator. The sutures are added to the incubating media at least 1 hour prior to the procedure. If a dose-response challenge is desired, the incubated bacterial broth is diluted to the preferred concentrations before inserting the suture. The sutures are removed from the media at the time of the procedure and placed in a sterile Petri dish, and a small amount of broth is added to cover the sutures. Bacterial concentration of the challenge can be measured with a 1 cm strand of inoculated suture by homogenizing it in sterile saline and plating it on to sterile, nutrient-coated Petri dishes.

Surgical procedure

Two laboratory workers are needed to perform the procedure. Inoculated suture is attached to the French-eye needle, and the needle is secured in the needle holder. One person uses one hand to hold the mouse securely by the nape of its neck and the other hand to extend one of the legs of the mouse. Another person then has clear access to apply the suture to the thigh muscle. The area to be sutured is wiped with alcohol for two reasons: (1) it cleans the skin and (2) it mats the hair next to the skin, which helps when trimming the suture. In a smooth, fluid motion, the needle is inserted into and passed through the thigh muscle. The needle end of the suture is then cut off flush with the skin. The opposite side of the suture is also cut flush with the skin. The skin tissue around the entrance and exit sites of the suture is gently lifted to ensure that the suture is buried underneath the skin. If the suture is not buried, the animal may extract it with its teeth. The mouse is returned to a clean cage. A new strand of suture is used for each animal.

Postsurgical care

All infected animals are housed in isolation in a biohazard area in the animal care facility. They are given free access to food and water. Cages are covered with fabric bonnets or placed in laminar flow hoods to isolate them from any aerosol contamination.

Animals are monitored frequently to accurately plot the mortality curves. Mortality in this model occurs between 1 and 7 days after bacterial challenge. Depending on the strain of bacteria used, no further deaths usually occur after these 7 days.

Infection process

The thigh suture model can be used to measure survival and monitor the infection process. To accomplish both goals, parallel experiments should be conducted simultaneously. Two groups of animals are needed. One group is followed to record mortality and survival rates. The other group is sacrificed at specific time points to assess other parameters. Extra animals are needed for the second group to adjust for the animals that do not survive along the time course.

Recovery of bacteria

This model allows for the recovery and quantification of bacteria from three major sites: (1) systemic, (2) regional, and (3) local. It is best to collect these samples after the mouse is sacrificed by cervical dislocation. The bacterial recovery should be performed in the same chronological order as the three mentioned major sites of collection. (These samples may also be collected at the time of death from sepsis within the first 15 minutes after death.)

The animal is secured by all four legs to a dissection board. The chest is opened and the heart exposed. Using a heparinized syringe with a needle, blood is aspirated from the heart. Systemic blood can also be collected through the closed chest by cardiac puncture. This blood sample is used to determine systemic bacteremia. Inguinal lymph nodes can be dissected from the skin surrounding both the infected and uninfected legs. These samples provide a regional assessment of bacterial containment. Organs such as the liver and spleen can also be harvested at this time and analyzed for bacterial contamination as well as other immunological factors (Polk *et al.*, 1990). For local assessment, the infected leg is retracted and an incision is made through the skin, exposing the entire muscle mass. The muscle is dissected from the bone and placed in a sterile Petri dish. This can also be done with the other, non-infected leg. The muscle then is bluntly dissected to recover the suture. The suture length is measured and then returned to the Petri dish.

The blood is serially diluted and plated on to Petri dishes coated with nutrient agar. The muscle is weighed and then placed with the recovered suture in sterile saline and homogenized. The homogenate is then serially diluted and plated, as with the blood. The plates are incubated overnight and then counted for number of colony forming units.

References

Ausobsky, J. R., Trachtenberg, L. S., Polk, H. C. Jr (1982). Enhancement of nonspecific host defenses against combined local bacterial challenge. *Surg. Forum*, **33**, 46–47.

Cobb, J. P., Brown, C. M., Brown, G. L., Polk, H. C. Jr (1986). Muramyl dipeptide protects decomplemented mice from surgically-induced infection. *Int. J. Immunopharmacol.*, **8**, 799–803.

Fagelman, K. M., Flint, L. M., Jr, McCoy, M. T., Polk, H. C. Jr, Trachtenberg, L. S. (1981). Simulated surgical wound infection in mice: effect of stimulation on nonspecific host defense mechanisms. *Arch. Surg*, **116**, 761–764.

Gaar, E., Naziri, W., Cheadle, W. G., Pietsch, J. D., Johnson, M., Polk H. C. Jr (1994). Improved survival in simulated surgical infection with combined cytokine, antibiotic and immunostimulant therapy. *Br. J. Surg.* **81**, 1309–1311.

Galland, R. B., Trachtenberg, L. S., Polk, H. C. Jr (1981). Nonspecific enhancement of host defenses against infection in malnourished mice. *Surg. Forum*, **32**, 39–40.

Galland, R. B., Heine, K. J., Trachtenberg, L. S., Polk, H. C. Jr (1982a). Reduction of surgical wound infection rates in contaminated wounds treated with antiseptics combined with systemic antibiotics: an experimental study. *Surgery*, **91**, 329–332.

Galland, R. B., Trachtenberg, L. S., Rynerson, N., Polk, H. C. Jr (1982b). Nonspecific enhancement of resistance to local bacterial infection in starved mice. *Arch. Surg.*, **117**, 161–164.

Galland, R. B., Polk, H. C. Jr (1982c). Non-specific stimulation of host defenses against a bacterial challenge in malnourished hosts. *Br. J. Surg.*, **69**, 665–668.

Galland, R. B., Heine, K. J., Polk, H. C. Jr (1983). Nonspecific stimulation of host defenses against bacterial challenge in immunosuppressed mice. *Arch. Surg.*, **118**, 333–337.

Hershmann, M. J., Pietsch, J. D., Trachtenberg, L., Mooney, T. H. R., Shields, R. E., Sonnenfeld, G. (1989). Protective effects of recombinant human tumour necrosis factor α and interferon γ against surgically simulated wound infection in mice. *Br. J. Surg.*, **76**, 1282–1286.

Lamont, P. M., Maier, K. G., Melton, L., Polk, H. C. Jr (1987). Stimulatory effects of muramyl dipeptide upon neutrophils isolated from a local bacterial infection. *Br. J. Exp. Pathol.*, **68**, 655–661.

Polk, H. C., Jr, Miles, A. A. (1973). The decisive period in the primary infection of muscle by *Escherichia coli*. *Br. J. Exp. Pathol.*, **54**, 99–109.

Polk, H. C., Jr, Calhoun, J. H., Eng., M. *et al.* (1981). Nonspecific enhancement of host defenses against infection: experimental evidence of a new order of efficacy and safety. *Surgery*, **90**, 376–380.

Polk, H. C. Jr, Galland, R. B., Ausobsky, M. B. (1982). Nonspecific enhancement of resistance to bacterial infection: evidence of an effect supplemental to antibiotics. *Ann. Surg.*, **196**, 436–441.

Polk, H. C. Jr, Lamont, P. M., Galland, R. B. (1990). Containment as a mechanism of nonspecific enhancement of defenses against bacterial infection. *Infect. Immun.*, **58**, 1807–1811.

Rouben, D. P., Fagelman, K., McCoy, M. T., Polk, H. C. Jr (1977). Enhancement of nonspecific host defenses against local bacterial challenge. *Surg. Forum*, **28**, 44–45.

Chapter 23

Animal Models of Melioidosis

D. De Shazer and D. E. Woods

Background of human infection

Melioidosis, an infection of humans and animals, is caused by the Gram-negative bacterium *Burkholderia pseudomallei* (Leelarasamee and Bovornkitti, 1989; Dance, 1991, 1996; Yabuuchi and Arakawa, 1993; Kanai and Kondo, 1994). The disease is most common in south-east Asia and northern Australia, but sporadic cases have also been described in other regions of the world. The organism can be isolated from wet soil in endemic regions (Ellison *et al.*, 1969; Smith *et al.*, 1995; Wuthiekanun *et al.*, 1995). The route of infection is probably via inhalation of dust particles or direct inoculation of contaminated soil into cuts or abrasions. The outcome of a *B. pseudomallei* infection can vary from asymptomatic seroconversion to fulminant septicemic melioidosis and death. Septicemia is the most common clinical presentation, as 50–70% of melioidosis patients are bacteremic on admission to the hospital (Leelarasamee, 1986). Acute or chronic infection of any organ can occur, and lesions can form on any tissue but are most commonly found in the lungs, liver, spleen, lymph nodes, skin and soft tissues and urinary tract (Dance, 1996). Latent infections can also occur in which the organism can lie dormant for as many as 26 years before recrudescence into an active infection (Mays and Ricketts, 1975; Koponen *et al.*, 1991). *B. pseudomallei* appears to be an opportunistic pathogen as a relatively high number of melioidosis patients have underlying diseases such as diabetes mellitus and renal failure (Chaowagul *et al.*, 1989; Tanphaichitra, 1989; Currie *et al.*, 1993; Puthucheary *et al.*, 1992; Turner *et al.*, 1994).

Background of models

Hamsters are exquisitely sensitive to *B. pseudomallei* infection (Miller *et al.*, 1948a; Dannenberg and Scott, 1957; Ellison *et al.*, 1969; Brett *et al.*, 1997). The 50% lethal dose (LD_{50}) of *B. pseudomallei* in hamsters is approximately 10 bacteria. The LD_{50} is independent of the route of infection as similar values are obtained with intraperitoneal (i.p.), subcutaneous and respiratory infections. In addition, hamsters do not display individual variation in susceptibility to *B. pseudomallei*. The infection of hamsters with *B. pseudomallei* typically results in acute septicemic melioidosis and death within 2–3 days. In 1948 Miller *et al.* described the use of hamsters in the experimental chemotherapy of melioidosis (Miller *et al.*, 1948b). This animal model of melioidosis has also been used to assess the relative virulence of *B. pseudomallei* clinical and environmental isolates (Brett *et al.*, 1997) and isogenic mutants (DeShazer *et al.*, 1997).

Rats are relatively resistant to *B. pseudomallei* infection (Miller *et al.*, 1948a; Woods *et al.*, 1993). The LD_{50} of *B. pseudomallei* in infant rats can be lowered significantly by i.p. injection of the diabetogenic compound streptozotocin (STZ; Woods *et al.*, 1993). The LD_{50} of *B. pseudomallei* for nondiabetic infant rats is more than 1×10^8 bacteria while the LD_{50} in STZ-induced diabetic infant rats is approximately 1×10^4 bacteria (Woods *et al.*, 1993). STZ is thought to damage pancreatic ß-cell DNA through the formation of alkylating radicals (Okamoto, 1985). DNA damage activates poly (ADP-ribose) synthetase and cellular levels of NAD^+ are drastically reduced leading to the necrosis of insulin-producing β-cells (Okamoto, 1985). Recent evidence suggests that the GLUT2 transporter is expressed on rodent pancreatic β-cells and that this molecule mediates the specific uptake of STZ into these cells (Schnedl *et al.*, 1994). The infection of infant diabetic rats with *B. pseudomallei* typically results in acute septicemic melioidosis and death, usually within 7 days. This animal model of melioidosis has been used in passive immunoprophylaxis studies (Brett *et al.*, 1994; Bryan *et al.*, 1994; Brett and Woods, 1996) and has been useful in assessing the relative virulence of *B. pseudomallei* isogenic mutants (DeShazer *et al.*, 1997).

Animal species

We use female golden Syrian hamsters (*Mesocricetus auratus*) that are 6–8 weeks old and weigh 80–90 g. Male hamsters are also susceptible to *B. pseudomallei* infection and may be useful for those who wish to follow the development of the Straus reaction (Miller *et al.*, 1948a). We have arbitrarily chosen 6–8-week old animals to standardize this model of infection; but Ellison *et al.* (1969) described the use

Handbook of Animal Models of Infection
ISBN 0-12-775390-7

of weanling hamsters (3 weeks old), and we have found that older animals are also highly susceptible to *B. pseudomallei* infection. We also use female Sprague Dawley rats (*Rattus norvegicus*) that are approximately 3 weeks old and weigh 30–40 g. Approximately 10 pups (10–12 days) are delivered with one mother. The pups are weaned after they reach a weight of 30–40 g and injected with STZ as described below. Infant male rats may also be used in this model of infection (Woods *et al.*, 1993). It is important to note that adult rats should not be used because diabetic adult rats are not susceptible to i.p. infection with *B. pseudomallei*.

Preparation of animals

The average weight of 10–20 infant rats should be determined, and 80 mg of STZ/kg of body weight should be injected i.p. for 2 consecutive days in order to induce diabetes (Woods *et al.*, 1993). The STZ should be dissolved in 100 mM sodium citrate, pH 4.2, filter sterilized and put into sterile Monoject® blood collection tubes (or equivalent). We inject individual infant rats with the required amount of STZ in a volume of 100 µl using a 1 ml syringe fitted with a 25 G needle. The urine glucose levels of several random infant rats should be checked daily for diabetes progression. Diastix® (Bayer, Inc.) reagent strips for urinalysis (or equivalent) should be used to monitor urine glucose levels. This task can be accomplished more easily with two people as compared to one. One person should pick up the infant rat by the tail with one hand and with the forefinger and thumb at the base of the head with the other. Urination is common when rats are picked up in such a manner. The other person should hold the Diastix® strip and wet the reagent area of strip by holding it under the drops of urine. The infant rats are considered diabetic when the glucose in the urine reaches a concentration of 2 g/dl (111 mmol/l). It typically takes 2–3 days following the second STZ injection for infant rats to become diabetic. One week after the final STZ injection, the infant rats are infected with *B. pseudomallei*. No special preparation is required for hamsters.

Details of surgery

No surgical procedures are required in the hamster or infant diabetic rat models of melioidosis.

Storage and preparation of inocula

The strain we commonly use for animal studies is *B. pseudomallei* 1026b (Brett *et al.*, 1997; DeShazer *et al.*, 1997). This strain was isolated in Thailand from a human case of septicemic melioidosis with skin, soft tissue and spleen involvement. Other clinical and environmental isolates of *B. pseudomallei* yield similar results in animal studies (Woods *et al.*, 1993; Brett and Woods, 1996; Brett *et al.*, 1994, 1997). *B. pseudomallei* is grown for 2 days on a Luria–Bertani (LB) broth base (Becton Dickinson) agar plate in a 37°C incubator. A quantity of 2 ml of TSBDC broth (Brett *et al.*, 1997) is inoculated with an isolated colony and incubated in a 37°C rotary shaker (250 rpm) overnight; 100 µl of the saturated overnight culture is then transferred to 3 ml of fresh TSBDC broth, and the culture is incubated for 4 hours until it reaches the middle to late logarithmic phase of growth. The optical density at 600 nm is determined and compared with a standard growth curve to estimate the number of bacteria per milliliter of culture. The culture is serially diluted in sterile 0.85% NaCl, and aliquots of the appropriate dilutions are transferred to sterile blood collection tubes for subsequent injection of animals. The actual number of organisms in the starting culture is determined by spreading 100 µl portions of the appropriate dilutions in triplicate on to LB agar plates. In our hands, repeated subculture of *B. pseudomallei* does not affect its virulence capacity in hamsters. Infant diabetic rats, on the other hand, should be infected with a *B. pseudomallei* strain that has undergone at least one i.p. animal passage. The passage strain should be taken directly from a frozen stock and transferred to a LB agar plate as described above. It should also be noted that the media described here (LB agar and TSBDC broth) have been chosen in an attempt to standardize these animal models of infection, but other culture media have been used successfully in the past (Miller *et al.*, 1948a; Dannenberg and Scott, 1957).

Infection process

The diluted bacterial culture in the blood collection tube is drawn up into a 1 ml syringe fitted with a 25 G needle, and 100 µl aliquots are injected i.p. into hamsters and infant diabetic rats. We typically use five animals per bacterial dilution. The range of bacterial dilutions examined will vary depending on the experiment (chemotherapy, immunotherapy, or relative virulence studies) and the animal model (hamsters or infant diabetic rats).

Key parameters to monitor infection

Animal cages are checked daily, dead animals are removed and the day of death is noted. The LD_{50} is calculated using the method of Reed and Muench (Reed and Muench, 1938). The LD_{50} of *B. pseudomallei* for hamsters is approximately 10 bacteria at 2 days. There are rarely noticeable differences between infected and uninfected hamsters 24 hours following infection, even with infectious doses as high as 10^4 bacteria. There is profound morbidity and mortality 48 hours after infection, regardless of the infectious dose. Hamsters that are alive at 48 hours usually have a moribund appearance and exhibit a purulent ocular exudate. Hamsters rarely survive longer than 3 days after being infected with *B. pseudomallei*. Blood drawn from hamsters at the time of death typically contains more than 10^3 *B. pseudomallei*/ml (Brett *et al.*, 1997). Autopsies performed on hamsters 48 hours after infection reveal no gross lesions of the lungs, liver or spleen. All these tissues, however, are positive for *B. pseudomallei* when cultured. Peritonitis and splenomegaly are relatively common post-mortem observations in these animals. Splenomegaly is the result of hemorrhage in the spleen. No histopathological changes are observed in any other tissue including lungs, liver and kidney.

The LD_{50} of *B. pseudomallei* for STZ-induced diabetic infant rats is approximately 1×10^4 bacteria at 7 days. Infant diabetic rats that receive an infectious dose of more than 1×10^4 bacteria typically exhibit bacteremia and death within 7 days. Autopsies performed on dead animals reveal several enlarged organs, including the liver, spleen, and kidneys. The lungs are commonly filled with inflammatory exudate. No obvious lesions are present in any organs at the time of death.

Antimicrobial therapy

The mortality of septicemic melioidosis has decreased from approximately 80% to 40% with the use of the third-generation cephalosporin ceftazidime (White *et al.*, 1989). The current treatment of severe melioidosis involves 7–14 days of parenteral ceftazidime or amoxicillin/clavulanate followed by 12–20 weeks of oral amoxicillin/clavulanate or a combination of chloramphenicol, doxycycline and co-trimoxazole (White *et al.*, 1989; Suputtamongkol *et al.*, 1994; Rajchanuvong *et al.*, 1995). This treatment has reduced the overall mortality of severe melioidosis by half, but 40% mortality still represents a relatively high mortality rate. In addition, recent studies have demonstrated that ceftazidime is not bactericidal for *B. pseudomallei* (Sookpranee *et al.*, 1991; Smith *et al.*, 1994) and does not exhibit a postantibiotic effect (Walsh *et al.*, 1995). New and improved antimicrobial agents for the treatment of melioidosis are clearly needed.

There are no reports in the literature on the use of antimicrobial agents in the infant diabetic rat model of melioidosis. There is one published report describing experimental chemotherapy in the hamster model of melioidosis (Miller *et al.*, 1948b). Subcutaneous (s.c.) injections of sulfadiazine, streptomycin and penicillin were used to treat hamsters infected i.p. with $1–6 \times 10^3$ *B. pseudomallei*. Sulfadiazine treatment was effective in preventing death while streptomycin and penicillin were not (Miller *et al.*, 1948b). *B. pseudomallei* is relatively susceptible to sulfadiazine but is resistant to streptomycin and penicillin *in vitro* (Eickhoff *et al.*, 1970). Thus, there is a correlation of the activity of these antimicrobial agents *in vitro* and *in vivo* in the hamster model of melioidosis. There have been no studies with newer antimicrobial agents in experimental melioidosis. The carbapenems (imipenem, meropenem, and biapenem) and ciprofloxacin demonstrate low MICs and a significant postantibiotic effect for *B. pseudomallei in vitro* (Walsh *et al.*, 1995). These antibiotics are candidates for use in treatment of human melioidosis and in the hamster and infant diabetic rat models of melioidosis. Antimicrobial agents could be administered parenterally in hamsters (s.c. and i.p.) and infant diabetic rats (s.c., i.p. and intravenously via the tail vein). The dose, frequency, and duration of antimicrobial therapy will vary depending on the antibiotic used, and each parameter may have to be optimized empirically.

Key parameters to monitor response to treatment

When an infectious dose of *B. pseudomallei* equivalent to the LD_{50} is administered, effective antimicrobial agents should decrease or prevent the death of hamsters and infant diabetic rats in 2 days and 7 days respectively. As a result, such antibiotics should significantly increase the LD_{50} of *B. pseudomallei* for hamsters and infant diabetic rats. Treated animals may live longer than untreated animals and may even recover from *B. pseudomallei* infection (Miller *et al.*, 1948b). Morbidity and mortality of the treated animals can be monitored for weeks, as described above.

Pitfalls (advantages/disadvantages) of the models

The hamster and infant diabetic rat models of melioidosis most closely resemble acute septicemic melioidosis in humans. This form of the disease is the most common clinical presentation in humans and results in 40% mortality, even with vigorous antimicrobial therapy (Leelarasamee, 1986; White *et al.*, 1989; Suputtamongkol *et al.*, 1994). These animal models of *B. pseudomallei* infection may be useful for identifying new antimicrobial agents for use in human cases of acute septicemic melioidosis. On the other hand, the animal models described here do not resemble other clinical manifestations of human disease, including chronic or latent melioidosis. It is also important to note that no obvious lesions can be identified in the organs of hamsters or infant diabetic rats that are infected with *B. pseudomallei*. Visceral lesions are relatively common in human melioidosis cases and are typically found in the lungs, liver, spleen, lymph nodes, skin and soft tissues, and urinary tract (Dance, 1996).

Diabetes mellitus is a common underlying disease in melioidosis patients (Chaowagul *et al.*, 1989; Tanphaichitra, 1989; Puthucheary *et al.*, 1992; Currie *et al.*, 1993; Turner *et al.*, 1994). Similarly, diabetic infant rats are susceptible to infection with *B. pseudomallei*, but non-diabetic infant rats are not (Woods *et al.*, 1993). Streptozotocin-induced diabetes mellitus in rats is similar to insulin-dependent (juvenile- onset) diabetes mellitus in humans. Recent studies suggest that the majority of melioidosis patients with diabetes have non-insulin-dependent (adult- onset) diabetes mellitus rather than insulin-dependent diabetes mellitus (Currie *et al.*, 1993; Turner *et al.*, 1994; Currie, 1995). It is clear, however, that both types of diabetes mellitus are risk factors for *B. pseudomallei* infection. It is currently unclear why diabetics are highly susceptible to infection with *B. pseudomallei*, but the infant diabetic rat model of melioidosis should be useful in studying this interesting phenomenon.

Contributions of the models to infectious disease therapy

Several passive immunization studies have been conducted using the infant diabetic rat model of melioidosis (Brett *et al.*, 1994; Bryan *et al.*, 1994; Brett and Woods, 1996). *B. pseudomallei* surface antigens that are potential candidates for a melioidosis vaccine include the O-polysaccharide (O-PS) moiety of lipopolysaccharide and flagella. Rabbit antisera raised against purified O-PS, flagellin protein or an O-PS-flagellin conjugate passively protected diabetic rats from infection with *B. pseudomallei* (Brett *et al.*, 1994; Bryan *et al.*, 1994; Brett and Woods, 1996). The protective capacity of the specific antisera suggests that these antigens may be suitable vaccine candidates. It should be noted that this animal model is not amenable to active immunization studies, as active immunization would require a minimum of 4 weeks. At this time, the animals would be adults and would no longer be susceptible to i.p. challenge with *B. pseudomallei*.

We have used the hamster model of melioidosis to compare the relative virulence of clinical and environmental isolates of *B. pseudomallei* (Brett *et al.*, 1997). The exquisite sensitivity of hamsters to *B. pseudomallei* infection has allowed us to identify a new *B.-pseudomallei*-like organism that is relatively avirulent in hamsters (LD_{50} of $>1 \times 10^6$ bacteria). This *B.-pseudomallei*-like organism, recently named *B. thailandensis* (Brett *et al.*, in press), is antigenically, biochemically and morphologically similar to *B. pseudomallei* and can be isolated from the same environmental locations (Wuthiekanun *et al.*, 1996; Smith *et al.*, 1995; Brett *et al.*, 1997). A phylogenetic analysis based on 16S rDNA sequences confirms that *B. thailandensis* is closely related to *B. pseudomallei* but is clearly a separate species (Brett *et al.*, 1998). Further studies employing the hamster model of infection may aid in the identification of genetic factors responsible for the enhanced virulence of *B. pseudomallei* relative to *B. thailandensis*.

Finally, we have compared the relative virulence of *B. pseudomallei* 1026b and several isogenic Tn5-OT182 mutants in both animal models of melioidosis (DeShazer *et al.*, 1997). These studies have allowed us to identify and characterize those factors responsible for *B. pseudomallei* pathogenesis at the molecular level. For example, we have recently identified several *B. pseudomallei* Tn5-OT182 mutants that are susceptible to the bactericidal activity of 30% human serum. The serum-sensitive mutants are less virulent than the serum-resistant parental strain in both the hamster and infant diabetic rat models of melioidosis. In contrast, we found that there was no difference in the virulence of a Tn5-OT182 flagellin mutant and the wild-type parental strain in either animal model of *B. pseudomallei* infection (DeShazer *et al.*, 1997). We conclude from these studies that serum resistance is an important virulence determinant while flagella and/or motility are probably not significant virulence determinants in these animal models of melioidosis. Similar studies will allow investigators to assess the relative importance of specific genetic determinants of *B. pseudomallei* in the pathogenesis of melioidosis.

References

Brett, P. J., Woods, D. E. (1996). Structural and immunologic characterization of *Burkholderia pseudomallei* O-polysaccharide-flagellin protein conjugates. *Infect. Immun.*, **64**, 2824–2828.

Brett, P. J., Mah, D. C., Woods, D. E. (1994). Isolation and characterization of *Pseudomonas pseudomallei* flagellin proteins. *Infect. Immun.*, **62**, 1914–1919.

Brett, P. J., DeShazer, D., Woods, D. E. (1997). Characterization of *Burkholderia pseudomallei* and *Burkholderia pseudomallei*-like strains. *Epidemiol. Infect.*, **118**, 137–148.

Brett, P. J., DeShazer, D., Woods, D. E. (1998). *Burkholderia thailandensis* sp. nov., description of a *Burkholderia pseudomallei*-like species. *Int. J. Syst. Bacteriol.*, **48**, 317–320.

Bryan, L. E., Wong, S., Woods, D. E., Dance, D. A. B., Chaowagul, W. (1994). Passive protection of diabetic rats with anitsera specific for the polysaccharide portion of the lipopolysaccharide from *Pseudomonas pseudomallei*. *Can. J. Infect. Dis.*, **5**, 170–178.

Chaowagul, W., White, N. J., Dance, D. A. *et al.* (1989). Melioidosis: a major cause of community-acquired septicemia in northeastern Thailand. *J. Infect. Dis.*, **159**, 890–899.

Currie, B. (1995). *Pseudomonas pseudomallei*–insulin interaction. *Infect. Immun.*, **63**, 3745.

Currie, B., Howard, D., Hguyen, V. T., Withnall, K., Merianos, A. (1993). The 1990–1991 outbreak of melioidosis in the Northern Territory of Australia: clinical aspects. *Southeast Asian J. Trop. Med. Public Health*, **24**, 436–443.

Dance, D. A. (1991). Melioidosis: the tip of the iceberg? *Clin. Microbiol. Rev.*, **4**, 52–60.

Dance, D. A. B. (1996). Melioidosis. In *Manson's Tropical Diseases*, (ed. Cook, G. C.), pp. 925–930. W. B. Saunders, London:.

Dannenberg, A. M. Jr, Scott, E. M. (1957). Melioidosis: pathogenesis and immunity in mice and hamsters. I. Studies with virulent strains of *Malleomyces pseudomallei*. *J. Exp. Med.*, **107**, 153–187.

DeShazer, D., Brett, P. J., Carlyon, R., Woods, D. E. (1997). Mutagenesis of *Burkholderia pseudomallei* with Tn5-OT182: isolation of motility mutants and molecular characterization of the flagellin structural gene. *J. Bacteriol.*, **179**, 2116–2125.

Eickhoff, T. C., Bennett, J. V., Hayes, P. S., Feeley, J. (1970). *Pseudomonas pseudomallei*: susceptibility to chemotherapeutic agents. *J. Infect. Dis.*, **121**, 95–102.

Ellison, D. W., Baker, H. J., Mariappan, M. (1969). Melioidosis in Malaysia. I. A method for isolation of *Pseudomonas pseudomallei* from soil and surface water. *Am. J. Trop. Med. Hyg.*, **18**, 694–697.

Kanai, K., Kondo, E. (1994). Recent advances in biomedical sciences of *Burkholderia pseudomallei* (basonym: *Pseudomonas pseudomallei*). *Jpn. J. Med. Sci. Biol.*, **47**, 1–45.

Koponen, M. A., Zlock, D., Palmer, D. L., Merlin, T. L. (1991). Melioidosis. Forgotten, but not gone! *Arch. Intern. Med.*, **151**, 605–608.

Leelarasamee, A. (1986). Epidemiology of melioidosis. *Infect. Dis. Antimicrob. Agents*, **38**, 84–93.

Leelarasamee, A., Bovornkitti, S. (1989). Melioidosis: review and update. *Rev. Infect. Dis.*, **11**, 413–425.

Mays, E. E., Ricketts, E. A. (1975). Melioidosis: recrudescence associated with bronchogenic carcinoma twenty-six years following initial geographic exposure. *Chest*, **68**, 261–263.

Miller, W. R., Pannell, L., Cravitz, L., Tanner, W. A., Rosebury, T. (1948a). Studies on certain biological characteristics of *Malleomyces mallei* and *Malleomyces pseudomallei*. II. Virulence and infectivity for animals. *J. Bacteriol.*, **55**, 127–135.

Miller, W. R., Pannell, L., Ingalls, M. S. (1948b). Experimental chemotherapy in glanders and melioidosis. *Am. J. Hyg.*, **47**, 205–213.

Okamoto, H. (1985). Molecular basis of experimental diabetes: degeneration, oncogenesis and regeneration of pancreatic β-cells of islets of Langerhans. *BioEssays*, **2**, 15–21.

Puthucheary, S. D., Parasakthi, N., Lee, M. K. (1992). Septicaemic melioidosis: a review of 50 cases from Malaysia. *Trans. R. Soc. Trop. Med. Hyg.*, **86**, 683–685.

Rajchanuvong, A., Chaowagul, W., Suputtamongkol, Y., Smith, M. D., Dance, D. A. B., White, N. J. (1995). A prospective comparison of co-amoxiclav and the combination of chloramphenicol, doxycycline, and co-trimoxazole for the oral maintenance treatment of melioidosis. *Trans. R. Soc. Trop. Med. Hyg.*, **89**, 546–549.

Reed, L. J., Muench, H. (1938). A simple method of estimating fifty per cent endpoints. *Am. J. Hyg.*, **27**, 493–497.

Schnedl, W. J., Ferber, S., Johnson, J. H., Newgard, C. B. (1994). STZ transport and cytotoxicity. Specific enhancement in GLUT2-expressing cells. *Diabetes*, **43**, 1326–1333.

Smith, M. D., Wuthiekanun, V., Walsh, A. L., White, N. J. (1994). Susceptibility of *Pseudomonas pseudomallei* to some newer beta-lactam antibiotics and antibiotic combinations using time-kill studies. *J. Antimicrob. Chemother.*, **33**, 145–149.

Smith, M. D., Wuthiekanun, V., Walsh, A. L., White, N. J. (1995). Quantitative recovery of *Burkholderia pseudomallei* from soil in Thailand. *Trans. R. Soc. Trop. Med. Hyg.*, **89**, 488–490.

Sookpranee, T., Sookpranee, M., Mellencamp, M. A., Preheim, L. C. (1991). *Pseudomonas pseudomallei*, a common pathogen in Thailand that is resistant to the bactericidal effects of many antibiotics. *Antimicrob. Agents Chemother.*, **35**, 484–489.

Suputtamongkol, Y., Rajchanuwong, A., Chaowagul, W. *et al.* (1994). Ceftazidime vs. amoxicillin/clavulanate in the treatment of severe melioidosis. *Clin. Infect. Dis.*, **19**, 846–853.

Tanphaichitra, D. (1989). Tropical disease in the immunocompromised host: melioidosis and pythiosis. *Rev. Infect. Dis.*, **11** (Suppl. 7), S1629–S1643.

Turner, M. O., Lee, V. T., FitzGerald, J. M. (1994). Melioidosis in a diabetic sailor. *Chest*, **106**, 952–954.

Walsh, A. L., Smith, M. D., Wuthiekanun, V., White, N. J. (1995). Postantibiotic effects and *Burkholderia* (*Pseudomonas*) *pseudomallei*: evaluation of current treatment. *Antimicrob. Agents Chemother.*, **39**, 2356–2358.

White, N. J., Dance, D. A. B., Chaowagul, W., Wattanagoon, Y., Wuthiekanun, V., Pitakwatchara, N. (1989). Halving of mortality of severe melioidosis by ceftazidime. *Lancet*, **ii**, 697–701.

Woods, D. E., Jones, A. L., Hill, P. J. (1993). Interaction of insulin with *Pseudomonas pseudomallei*. *Infect. Immun.*, **61**, 4045–4050.

Wuthiekanun, V., Smith, M. D., Dance, D. A., White, N. J. (1995). Isolation of *Pseudomonas pseudomallei* from soil in north-eastern Thailand. *Trans. R. Soc. Trop. Med. Hyg.*, **89**, 41–43.

Wuthiekanun, V., Smith, M. D., Dance, D. A. B., Walsh, A. L., Pitt, T. L., White, N. J. (1996). Biochemical characteristics of clinical and environmental isolates of *Burkholderia pseudomallei*. *J. Med. Microbiol.*, **45**, 408–412.

Yabuuchi, E., Arakawa, M. (1993). *Burkholderia pseudomallei* and melioidosis: be aware in temperate area. *Microbiol. Immunol.*, **37**, 823–836.

Chapter 24

Low-Inoculum Model of Clean Wound Infection

A. B. Kaiser and D. S. Kernodle

Background of human infection

Wound infections associated with clean and clean-contaminated operative procedures present unique challenges to both clinicians and investigators. Unlike the operative categories of contaminated and infected, infection rates with clean and clean-contaminated procedures are low (< 5%), and placebo controlled studies have required the enrollment of large numbers of subjects in order to demonstrate advantages of prophylactic antimicrobials. Demonstrating differences among two or more active regimens has proved to be even more formidable; clinical trials involving 1000–2000 enrolees are often required to achieve meaningful results (Kaiser *et al.*, 1987; Townsend *et al.*, 1993). These categories of risk take on increasing importance in view of the fact that over 90% of all operative procedures are identified as either clean or clean contaminated (Culver *et al.*, 1991). In the USA alone, 400 000 infections occur annually despite the almost routine use of prophylactic antimicrobials (Haley *et al.*, 1985). Additionally, for the most part, clean and clean-contaminated surgical procedures are non-emergent, and patients and surgeons usually proceed with the understanding that wound infections are a rare and unexpected complication. When infections do occur, emotional and medical legal issues may dominate the care of the patient.

Elucidating the pathophysiology of these infections has also proved difficult. Infections that develop in clean and clean-contaminated procedures arise in near sterile operative environments where the level of contaminating bacteria is low. Based upon experimental data demonstrating that adjuvants are necessary to establish infection with low inocula of bacteria (Elek and Conen, 1958), it is likely that aspects of the operative procedure (devitalized tissues, hematomas, sutures, etc.) provide a mechanism enabling low numbers of contaminating organisms to initiate infection. Analysis of these operative variables has proved to be extremely challenging in the clinical arena. Efforts to understand the pathophysiology of such infection are further complicated by the fact that infecting pathogens (primarily *Staphylococcus aureus*) usually demonstrate *in vitro* susceptibility to antimicrobials used clinically in prophylaxis (Kernodle *et al.*, 1990).

In short, clinical trials have proved far too unwieldy to provide a systematic exploration of optimal prophylactic regimens or underlying mechanisms of infection.

Background of the model

In the absence of a foreign body or tissue devitalization, an inoculum of 10^6 or more colony-forming units (cfu) is needed to establish a subcutaneous *S. aureus* abscess in man (Elek and Conen, 1958). Early guinea-pig models of staphylococcal skin and soft tissue infection employed similarly high inocula. However instead of abscess formation, the area of intradermal induration at the injection site 24 hours following bacterial inoculation was used as the endpoint (Miles and Niven, 1950). With this model it was demonstrated that dehydration, shock, and the local administration of adrenaline produced larger inflammatory lesions. Antibiotics were also shown to reduce lesion induration, but only if they were given prior to or within the first 2 hours following bacterial inoculation (Miles *et al.*, 1957; Burke, 1961). Inhalation of low (12%) and high (45%) concentrations of oxygen by guinea-pigs increased and reduced lesion size respectively (Knighton *et al.*, 1984).

Although the intradermal guinea pig model has yielded important observations, its requirement for large inocula and the questionable validity of using induration size as an endpoint have made its relevance to clean surgical wound infection in man unclear. Using the finding by Ford *et al.* (1989) that the number of bacteria needed to establish infection in mice is reduced by the use of dextran or gelatin microbeads as an adjuvant, we have modified the guinea-pig model to more closely simulate the conditions associated with clean and clean-contaminated surgery, e.g.:

- low inocula of pathogenic bacteria reliably induce infection;
- the endpoint is consistent with the clinical definition of infection (i.e., purulent material yielding viable bacteria develops within previously healthy tissue);
- the pharmacokinetics of prophylactic agents may be adjusted to parallel the perioperative clinical environment; and

Handbook of Animal Models of Infection
ISBN 0-12-775390-7

- infections may be prevented with prophylactic antibiotics if care is taken to emulate important clinical parameters (antibiotic serum levels, inoculum size, and antimicrobial susceptibility of the bacterial pathogens).

Animal species

Albino Hartley guinea-pigs of either sex, weighing 500±50 g, have been used in all experiments involving this model.

Preparation of animals

Day prior to inoculation

Specialized housing is not required. Guinea pigs frequently exhibit severe colitis, Gram-negative septicemia, and death occurring 2–3 days after a single dose of some antimicrobial agents (Farrar and Kent, 1965), Therefore, two poorly absorbable antibiotics, gentamicin (80 μg/ml) and polymyxin B (50 μg/ml), are added to the drinking water of the guinea-pigs 24 hours before the procedure if the study involves antimicrobial prophylaxis. This reduces mortality to less than 10%. (Antibiotic-containing water may be prepared by adding 10 vials of 80 mg gentamicin and four vials of 500 000 units polymyxin B to 10 liters of autoclaved water.)

Day of inoculation

1. Sedate with i.p. sodium pentobarbital, 25 mg/kg (usually about 0.15 ml for 500 mg animal), initially; then administer one third to one half doses as needed to achieve and maintain sedation.
2. Clip dorsal hair with clippers using No. 10 blade.
3. Apply depilatory to residual, clipped hair. After 5 minutes, wipe off using paper towels and warm water. Dry guinea-pigs with towel.
4. Using felt-tipped permanent marker, number animals on ears and dorsum. Draw grid on back. In our experience, well-spaced lesions can be created by designating four columns that are parallel to the longitudinal axis of the animal and three rows that are perpendicular (see Figure 24.3 below).
5. Weigh and record on worksheet.

Following inoculation

Transport guinea-pigs back to cage (continue antibiotic-containing water, if appropriate). Observe daily, prior to the intended time of sacrifice. The usual time of sacrifice and harvesting of guinea pigs is 4 days following inoculation.

Storage and preparation of inocula

Overview

To date, only staphylococci have been evaluated in this model although it should be suitable for other bacterial species. *Staphylococcus aureus* strains are maintained at –70°C in tryptic soy broth containing 10% glycerol. After subculture to an overnight growth on tryptic soy agar, fresh bacterial colonies are suspended in phosphate-buffered saline (PBS) to achieve a standard turbidity (A_{600} reading of 0.4). Serial 10-fold dilutions and twofold dilutions are used to make a range of inocula that, when combined with the adjuvant, produce an abscess from 0% to 100% of the time (determined by preliminary *in-vivo* studies). Each dilution is mixed in a 1:1 volume:volume ratio with dextran microbeads (Cytodex®, Sigma Chemicals, St Louis, MO) that have previously been allowed to swell in PBS and have been sterilized by autoclaving.

Prior to the day of inoculation

Prepare microbeads as follows: add 250 ml of phosphate buffered saline, pH 6.0, to 500 ml flask. Gently tap contents of a 5 g vial of Cytodex® (Sigma C-0646) into the flask, while swirling gently. Allow to soak for 1–2 hours, with gentle swirling every 10–15 minutes. Autoclave. When cool, decant excess buffer until settled beads comprise approximately 50% of the total volume and transfer to sterile 50 ml Falcon tubes. Store in refrigerator until use. (Note: it is usually best to prepare two or three 5 g Cytodex® vials simultaneously.)

Make sterile phosphate buffered saline, pH 6.0, and disperse into glass (or plastic) tubes; prepare about 50 9 ml tubes and 80 4 ml tubes. Also prepare an extra 500 ml of PBS. PBS may be autoclaved if glass tubes are used.

Subculture organisms (stored in cryotubes at –70°C) to tryptic soy agar (TSA) at noon on the day prior to preforming inoculations. Incubate at 35–37°C.

Day of inoculation

Pick colonies from TSA plate and suspend in 8 ml of 0.1 M PBS, pH 6.0, to achieve A600 = 0.4 using any spectrophotometer.

Do three 10-fold dilutions and four or more twofold dilutions as needed to achieve desired inoculum sizes (Table 24.1).

Decant excess buffer from the top of the gravity-packed microbeads into the 50 ml Falcon tubes. Vortex gently and mix 1:1 with dilutions A–G (Table 24.1) or as needed to

Table 24.1 To prepare the bacterial inocula

1. Pick colonies from TSA plate and suspend in 8 ml of 0.1 M PBS, pH 6.0, to achieve a cell density of approximately A600 = 0.4. Label this dilution No. 1.
2. Do three 10-fold dilutions as follows:
   Transfer 1 ml of dilution No. 1 into 9 ml PBS, mix and label No. 2.
   Transfer 1 ml of dilution No. 2 into 9 ml PBS, mix and label No. 3.
   Transfer 1 ml of dilution No. 3 into 9 ml PBS, mix and label A.
3. Do four or more twofold dilutions as follows:
   Transfer 4 ml of dilution A into 4 ml PBS, mix and label B.
   Transfer 4 ml of dilution B into 4 ml PBS, mix and label C.
   Transfer 4 ml of dilution C into 4 ml PBS, mix and label D.
   Transfer 4 ml of dilution D into 4 ml PBS, mix and label E.
   Transfer 4 ml of dilution E into 4 ml PBS, mix and label F.
   Transfer 4 ml of dilution F into 4 ml PBS, mix and label G.
   – etc., as needed for the day's experiment

Table 24.2 Approximate backcounts of dilutions and suspensions

	Dilution	*Mix 1:1 with microbeads*	*Actually inoculated*
No. 1	2×10^8	1×10^8	2×10^7 CFU
No. 2	2×10^7	1×10^7	2×10^6 CFU
No. 3	2×10^6	1×10^6	2×10^5 CFU
A	2×10^5	1×10^5	20 000 CFU
B	1×10^5	5×10^4	10 000 CFU
C	5×10^4	25 000	5000 CFU
D	25 000	12 500	2500 CFU
E	12 500	6250	1250 CFU
F	6250	3125	625 CFU
G	3125	1562	312 CFU
H	1562	781	156 CFU
I	781	390	78 CFU
J	390	195	39 CFU
K	195	98	20 CFU
L	98	49	10 CFU
M	49	25	5 CFU
N	24	12	2.4 CFU
O	12	6	1.2 CFU
P	6	3	0.6 CFU

achieve desired inoculum sizes. The Falcon tubes are maintained on ice throughout the inoculation process. (Note: to determine the total volume of bacterial dilution:microbead mix to prepare, multiply the number of times the suspension will be inoculated during the day's experiments by 0.4. For example, if suspension *B* will be used 10 times, then a final volume of 4.0 ml is needed. Therefore, mix 2.0 ml of dilution *B* with 2.0 ml of microbeads.)

We have established a blinded method of inoculation by color-number coding each inoculation tube on the morning of inoculation and maintaining the color-number code at a site separate from the animal laboratory. The designation site for each inoculation is outlined on the daily worksheet. The code is not broken until after lesions have been harvested.

Calculation of backcount

To determine the number of bacteria inoculated with each preparation, backcounts are performed by using the flame loop method and spreading 50 µl of one of the twofold dilutions (prior to mixing with dextran microbeads) to each of three blood agar plates. (An ethanol-flamed glass rod is used for bacterial dispersal upon sheep blood agar.) Incubate at 35–37°C for 18–24 hours and count colonies.

The dilution to use for backcount is based upon the previously determined correlation between the number of cfu for that bacterial strain and the A_{600} reading that would result in approximately 200 colonies (i.e., within a range that facilitates easy counting; Table 24.2). For most strains of *S. aureus*, 50 µl of dilution G yields 150–200 colonies. The precise backcount of the particular dilution is determined from the mean of triplicate plating. The number of bacteria in other inocula preparations is determined by doubling/halving the directly determined backcount in a fashion that duplicates the derivation of the various preparations by twofold dilutions.

Infection process (inoculation procedure)

Overview

The guinea pig possesses a well-developed panniculus carnosus (skin muscle) that overlies the superficial trunk musculature (Langworthy, 1925). The infection is initiated by injecting the bacterial-strain–microbead suspension into the potential space between the fascia surrounding these two muscle groups (Figure 24.1B).

The interval between bacterial inoculation and placebo or anti-infective agent administration is determined by protocol. The use of timing clocks will facilitate the process when multiple animals are inoculated on a given day, particularly in circumstances when intervals between the administration of an anti-infective agent and the inoculation of organisms exceeds 10–15 minutes.

Materials required

Materials used on the day of bacterial inoculation include: anesthetic, hair clippers, depilatory agent, disinfectant, tuberculin syringes with 23 G needles, markers, towels, and sterile gauze. A 23 G needle is the smallest size that permits the microbeads to pass easily.

Procedure

Administer antibiotic or placebo into the subcutaneous fat pad on the dorsal neck and between the shoulder blades and set alarm clock for desired time of inoculation. (Inoculation may precede anti-infective agent administration if so determined by the experimental design.) Internal jugular vein and intraperitoneal sites have been used to administer prophylactic agents as required by protocol. When alarm sounds, inoculate at each of the 12 sites according to the previously prepared key. Use vortex genie to resuspend microbeads prior to drawing up each 0.2 ml aliquot into a tuberculin syringe with a 23 G needle. The intermuscular space can be easily entered by advancing the needle approximately 5 mm beneath the skin (Figure 24.2). A successful inoculation results in a 0.5–1.5 cm nodule palpable or observable beneath the skin (Figure 24.3).

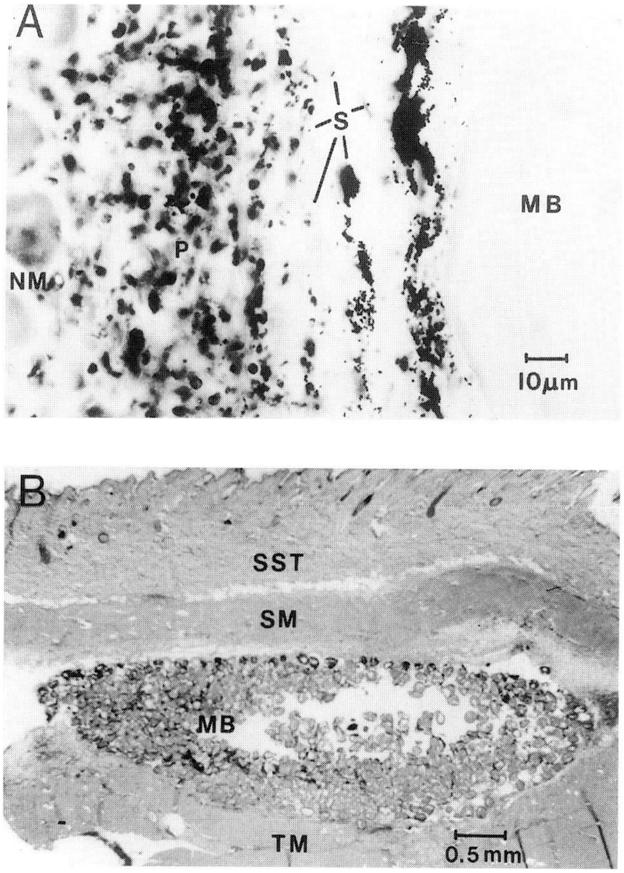

Figure 24.1 Photomicrographs of mature intermuscular lesions 72 hours after inoculation of a *Staphylococcus aureus* clinical isolate recovered from a deep wound infection. (**A**) Brown–Brenn preparation of high-inoculum lesion ($100\times ID_{50}$ staphylococci, ~ 300 organisms). One wall of the developing abscess cavity is pictured. A single dextran microbead (MB) is seen as well as numerous clusters and individual coccal forms of staphylococci (S), dense infiltration of fascial tissues with polymorphonuclear leukocytes (P), and associated necrotizing myositis (NM). (**B**) Hematoxylin–eosin preparation of low-inoculum lesion ($2\times ID_{50}$ staphylococci, ~ 6 organisms). The developing abscess cavity is located beneath skin and subcutaneous tissues (SST) and within the fascial plane that separates panniculus carnosus (skin muscle, SM) and the underlying trunk musculature (TM). Central cluster of dextran microbeads (MB) is clearly identified. Reproduced from Kaiser *et al*., 1992, with permission.

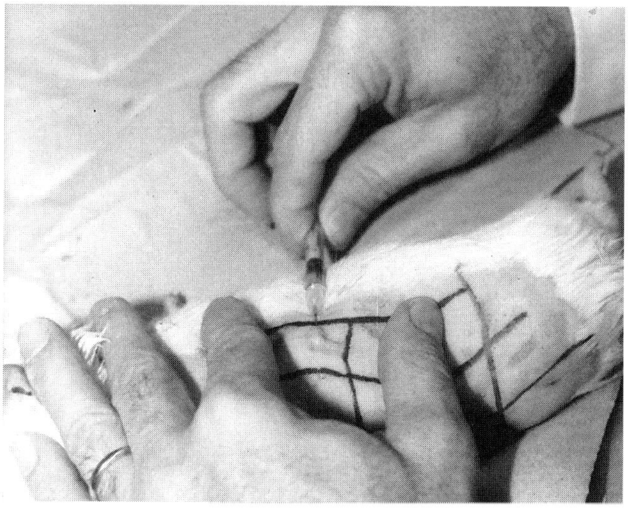

Figure 24.2 Intermuscular inoculation into one of 12 predetermined sites on dorsum of an anesthesized guinea pig. Bevel should be up on entering skin, turned over when dispensing the suspension. If the inoculation is too shallow, the injection will encounter resistance as the inoculant is forced into the dermis; if the inoculation is too deep, a nodule will not be present.

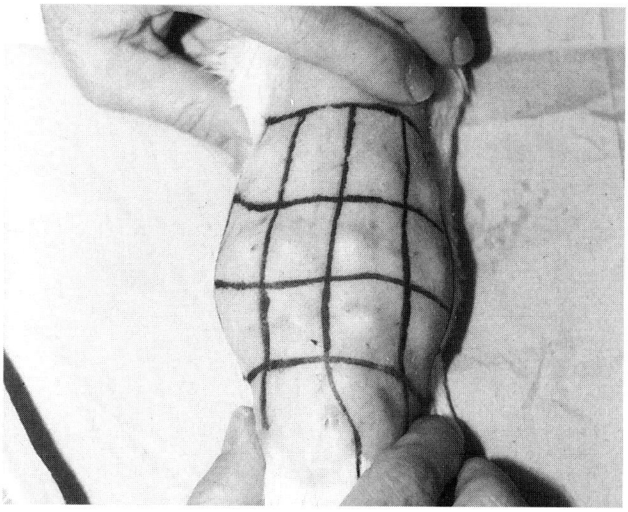

Figure 24.3 Conclusion of the successful inoculation of 12 sites on the dorsum of an anesthesized guinea-pig. The palpable nodules can be observed at several of the skin sites.

Key parameters to monitor infection and response to treatment

Overview

Three to 4 days following inoculation, the guinea-pigs are sacrificed and the presence or absence of viable bacteria on material recovered from each of the inoculation sites is determined. Logistic regression is used to plot the infection rate as a function of the number of bacteria inoculated and to determine the ID_{50} values and statistical significance of differences between bacterial-strain–prophylactic-regimen combinations.

Histopathology (Figure 24.1A)

For investigations into pathogenetic mechanisms, histologic sections of lesions may be prepared to evaluate enabling or preventing variables of the infecting process.

Materials for harvesting lesions

Alcohol burners and lighter, towel, 95% ethanol, gauze, depilatory, gloves, 6 mm biopsy punches, pentobarbital, forceps/scissors, 25 g needle affixed to tuberculin syringes, beakers, markers, blood agar plates (three per guinea-pig), alcohol swabs.

Procedure

Sacrifice with 75 mg/kg i.p. pentobarbital (usually about 0.5 ml). Identify guinea-pigs by number; renumber ears for clarity of identification. Depilate. Apply liberal amount of 95% ethanol to dorsum. Using a biopsy punch and sterile technique, remove a sample of material from each of the 12 inoculation sites (Figure 24.4) and streak across blood agar plates. Care is taken to insure that microbeads are clearly visible within the material transferred to the blood agar plate from each area of inoculation. Blood agar plates are incubated at 35°C for 24 hours and the presence or absence of bacterial growth is recorded (Figure 24.5).

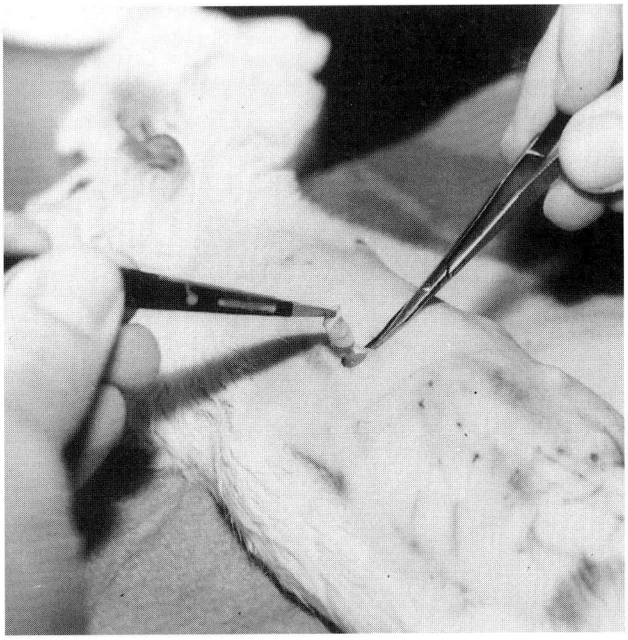

Figure 24.4 Removal of indurated abscess material 72 hours after intermuscular inoculation using sterile technique. The animal has been sacrificed and recurrent hair growth removed by depiliation.

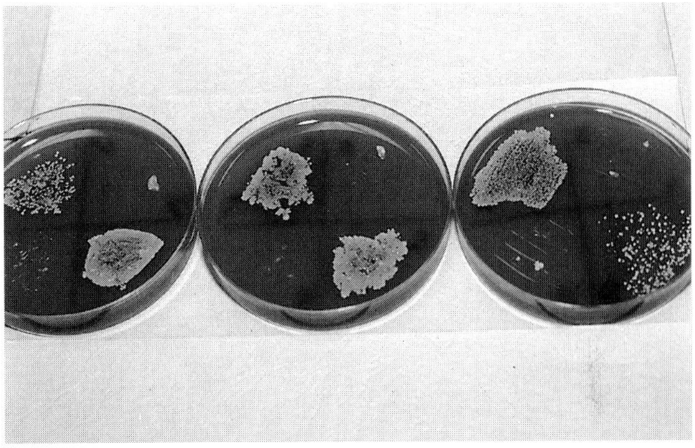

Figure 24.5 Blood agar plates after overnight incubation at 35°C. The plates had been inoculated with material resected from each of the 12 intermuscular inoculation sites from one guinea-pig. Four sites were inoculated per plate, resulting in a heavy growth of *Staphylococcus aureus* in two of the four sites on each plate. Collections of dextran microbeads can be observed on the remaining sites which were not overgrown by staphylococci.

Antimicrobial therapy

Overview

A variety of antimicrobial and immune-modulating agents may be employed in this model. Because pharmacokinetics of these agents in the guinea pig may differ substantially from the clinical model, and because pharmacokinetics of various agents may differ from each other, care must be taken to understand these differences and adjust the administration of anti-infective agents if necessary. For example, the cefazolin half-life in the guinea-pig is considerably shorter than that of vancomycin. Accordingly, in a protocol comparing the prophylactic efficacy of these two agents, cefazolin was redosed at 2 hours to produce a half-life virtually identical with that of vancomycin (Kernodle and Kaiser, 1993b).

Determination of serum antibiotic levels

During separate *in-vivo* investigations, 0.5 ml of intracardiac blood may be obtained at various intervals (e.g., 15,

30, 60, 120, 180 and 240 minutes) following the subcutaneous, intramuscular, intraperitoneal, or intravenous administration of the agent to be evaluated. Serum concentrations may be determined using appropriate techniques (e.g., agar diffusion bioassay).

Statistical analysis

Multiple logistic regression is used to assess the effect of different strains of staphylococcus and other microorganisms and different prophylactic regimens on the probability of infection (Figure 24.6). Adjustment is made for the amount of the inoculum in these regression models, using the $\log_{10}$ of the backcount. The ID_{50} (the estimated inoculum that has a 50% probability of producing an infection) is calculated as exp(–intercept/slope of $\log_{10}$ backcount) from the logistic regression using data for each combination of strain and prophylactic regimen. All analyses used either PC-SAS, release 6.04 (SAS Institute, Cary, NC) or JMP® statistical discovery software (SAS Institute, Cary, NC).

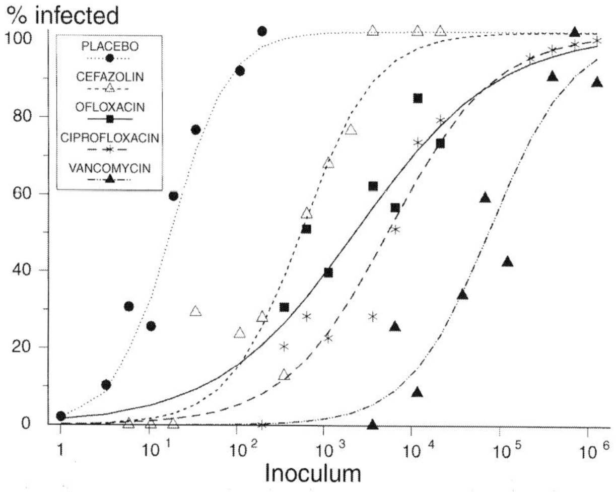

Figure 24.6 Semilogarithmic graph of infection rate versus inoculum size (cfu) for animals inoculated with a methicillin-resistant strain of *Staphylococcus aureus.* For each prophylactic regimen, the symbols show the observed infection rate at various inoculum sizes. The curves are idealized constructs derived by using the logistic model. Reproduced from Kernodle and Kaiser, 1994, with permission.

Pitfalls (advantages/disadvantages) of the model

The model's primary advantage is that, using the microbeads as a readily available and easily administered adjuvant, homogeneous intramuscular abscesses are reliably reproduced. ID_{50} values for stock staphylococcal isolates have remained remarkably constant over the 5 years of experience we have had with this model. This feature also relates to the primary disadvantage of the model in that trauma to tissues is minimal and may under-represent the impact of tissue devitalization and injury to the development of clinical infection.

Contributions of the model to infectious disease therapy

Profiling of antimicrobial prophylaxis

As depicted in Table 24.3, a wide variety of antimicrobial agents have been evaluated in this model. In general, for a given staphylococcal isolate, there was good correlation between the *in-vitro* activity of the antibiotics and their *in-vivo* efficacy. Unexpected results, such as the relatively high ID_{50} of ampicillin/sulbactam against MRSA (isolate 8001), could suggest *in-vitro* versus *in-vivo* discrepancies with the model. We believe it is more likely, however, that this result reflects the efficient binding of ampicillin to the PBP-2a of staphylococci. Binding affinity for penicillin-binding protein 2a correlates with the activity of β-lactam antibiotics against methicillin-resistant *S. aureus* in other *in vivo* models (Chambers *et al.*, 1990), and probably represents a biologic phenomenon which may have clinical relevance in the area of prophylaxis.

Elucidation of the pathophysiology of wound infection

As noted in Table 24.3, with placebo prophylaxis, the ID_{50} values of different strains of staphylococcus differ considerably, suggesting that certain strains (e.g., *S. haemolyticus*) possess less pathogenicity. The model lends itself to systematic evaluation of potential virulence factors of staphylococci as well as other clinical pathogens and will be particularly important in clinical settings where infection is thought to develop following low-level contamination of tissues.

We have recently demonstrated the effectiveness of immune-modulating agents in preventing infection (Kernodle *et al.*, 1998). Variations in the host risk factors (e.g., malnutrition) and local wound conditions (e.g., the presence of red blood cells) may be readily explored in this model.

Table 24.3 ID_{50} values associated with various prophylactic antibiotic regimens*; results reported as colony-forming units. Data are from Kernodle and Kaiser 1997.

	*Clinical isolates/resistance characteristics**						
	5001	*3177*	*3094*	*8001*	*5252*	*9021*	*9011*
Antibiotic regimen†	*PSSA*	*MSSA*	*BSSA*	*MRSA*	*MQRSA*	*MRSE*	*MRSH*
Placebo	5	3	2	7	8	20	6040
Ampicillin	119	—	381	75	—	—	—
Ampicillin/sulbactam	40 960	—	2629	2017	—	—	—
Cefazolin	227	—	335	26	140	8690	$> 4 \times 10^6$
Cefazolin/sulbactam	1978	—	1529	7	—	—	
Cefazolin redosed	1180	23 974	1690	30	—	$> 4 \times 10^6$	$> 4 \times 10^6$
Vancomycin	332 230	168 717	356 810	621 915	319 400	$> 4 \times 10^6$	$> 4 \times 10^6$
Ofloxacin	42 020	—	7630	—	29	$> 4 \times 10^6$	$> 4 \times 10^6$

* PSSA, penicillin-susceptible *S. aureus*; MSSA, methicillin-susceptible *S. aureus* – strain 3177 produces large amounts of type C staphylococcal β-lactamase; BSSA, borderline-susceptible *S. aureus* – strain 3094 belongs to phage group 94/96 and produces large amounts of type A staphylococcal β-lactamase (Kernodle and Kaiser, 1993a); MRSA, methicillin-resistant *S. aureus*; MSRSA, methicillin- and quinolone-resistant *S. aureus*; MRSE, methicillin-resistant *S. epidermidis*; MRSH, methicillin-resistant *S. haemolyticus*

† β-lactam, β-lactamase inhibitor, and quinolone antibiotics were given subcutaneously 15 minutes prior to the inoculation of bacteria; vancomycin was given intraperitoneally 1 hour prior to bacterial inoculation

References

Burke J. F. (1961). The effective period of preventive antibiotic action in experimental incisions and dermal lesions. *Surgery*, **50**, 161–168.

Chambers, H. F., Sachdeva, M., Kennedy, S. (1990). Binding affinity for penicillin-binding protein 2a correlates with the activity of beta-lactam antibiotics against methicillin-resistant *S. aureus* in other *in vivo* models. *J. Infect. Dis.*, **162**, 705–710.

Culver, D. H., Horan, T. C., Gaynes, R. P. *et al.* (1991). Surgical wound infection rates by wound class, operative procedure, and patient risk index. *Am. J. Med.*, **91**, 152S-157S.

Elek, S. D., Conen, P. E. (1958). The virulence of *Staphylococcus pyogenes* for man: a study of the problems of wound infection. *Br. J. Exp. Pathol.*, **38**, 573–586.

Farrar, E. W., Kent, T. H. (1965). Enteritis and coliform bacteremia in guinea pigs given penicillin. *Am. J. Pathol.*, **47**, 629–642.

Ford, C. W., Hamel, J. C., Stapert, D., Yancey, R. J. (1989). Establishment of an experimental model of a *Staphylococcus aureus* abscess in mice by use of dextran and gelatin microcarriers. *J. Med. Microbiol.*, **28**, 259–266.

Haley, R. W., Culver, D. H., White, J. W. *et al.* (1985). The nationwide nosocomial infection rate: a new need for vital statistics. *Am. J. Epidemiol.*, **121**, 159–167.

Kaiser, A. B., Petracek, M. R., Lea, J. W. IV *et al.* (1987). Efficacy of cefazolin, cefamandole, and gentamicin as prophylactic agents in cardiac surgery. Results of a prospective, randomized, double-blind trial in 1030 patients. *Ann. Surg.*, **206**, 791–797.

Kaiser, A. B., Kernodle, D. S., Parker, R. A. (1992). Low-inoculum model of surgical wound infection. *J. Infect. Dis.*, **166**, 393–399.

Kernodle, D. S., Kaiser, A. B. (1993a). Efficacy of prophylaxis with β-lactams and β-lactam/β-lactamase inhibitor combinations against wound infection by methicillin-resistant and borderline-susceptible *Staphylococcus aureus* in a guinea pig model. *Antimicrob. Agents Chemother.*, **37**, 702–707.

Kernodle, D. S., Kaiser, A. B. (1993b). Comparative prophylactic efficacy of cefazolin and vancomycin in a guinea pig model of *Staphylococcus aureus* wound infection. *J. Infect. Dis.*, **168**, 152–157.

Kernodle, D. S., Kaiser, A. B. (1994). Comparative prophylactic efficacy of ciprofloxacin, ofloxacin, cefazolin, and vancomycin in an experimental model of staphylococcal wound infection. *Antimicrob. Agents Chemother.*, **38**, 1325–1330.

Kernodle, D.S , Kaiser, A. B. (1997). Wound infections and surgical prophylaxis. In *The Staphylococci in Human Disease*, pp. 355–377 (eds Crossley, K. B., Archer, G. L.), Churchill Livingstone, New York.

Kernodle, D. S., Classen, D. C., Burke, J. P., Kaiser, A. B. (1990). Failure of cephalosporins to prevent *Staphylococcus aureus* wound infections. *J.A.M.A.*, **263**, 961–966.

Kernodle, D. S., Gates, H., Kaiser, A. B. (1998). Prophylactic anti-infective activity of PGG-glucan in a guinea pig nodel of staphylococcal wound infection. *Antimicrob. Agents Chemother.*, 42(3), 545–549.

Knighton, D. R., Halliday, B., Hunt, T. K. (1984). Oxygen as an antibiotic. *Arch. Surg.*, **119**, 199–204.

Langworthy, O. R. (1925). A morphological study of the panniculus carnosus and its genetical relationship to the pectoral musculature in rodents. *Am. J. Anat.*, **35**, 283–302.

Miles, A. A., Miles, E. M., Burke, J. (1957). The value and duration of defence reactions of the skin to the primary lodgement of bacteria. *Br. J. Exp. Pathol.*, **38**, 79–96.

Miles, A. A., Niven, J. S. F. (1950). The enhancement of infection during shock produced by bacterial toxins and other agents. *Br. J. Exp. Pathol.*, **31**, 73–95.

Townsend, T. R., Reitz, B. A., Bilker, W. B., Bartlett, J. G. (1993). A clinical trial of cefamandole, cefazolin, and cefuroxime for antibiotic prophylaxis in cardiac surgery. *J. Thorac. Cardiothorac. Surg.*, **106**, 664–667.

Chapter 25

Translocation of Gut Bacteria During Trauma

L. Magnotti and E. A. Deitch

Background of models

Clinical studies and experimental work from several laboratories using a variety of models have provided compelling evidence that injury-induced gut dysfunction may play a role in the pathogenesis of infection, systemic inflammation, and multiple organ failure (MOF; Deitch, 1990b). In fact, gut barrier failure and the subsequent translocation of bacteria and endotoxin from the gut has been proposed as a major contributor to the development of both systemic infection and MOF.

Normally, the intestinal mucosa functions as a major local defense barrier that helps prevent bacteria and endotoxin contained within the intestinal lumen from escaping and spreading to the extraluminal tissues and organs. However, under certain experimental and clinical circumstances, this intestinal barrier function becomes overwhelmed or impaired, resulting in the movement of bacteria or endotoxin to the mesenteric lymph nodes (MLN) and systemic tissues. Wolochow *et al.* (1966) and Fuller and Jayne-Williams (1970) were the first to use the term translocation for this passage of bacteria from the gastrointestinal (GI) tract. The term "bacterial translocation" does not indicate a mechanism but instead is used to describe the phenomenon of bacteria crossing the mucosal barrier. In this context, then, bacterial translocation is defined as the passage of viable indigenous bacteria from the GI tract to extraintestinal sites, such as the MLN complex, liver, spleen, and bloodstream. Three major mechanisms promote bacterial translocation: (1) intestinal bacterial overgrowth; (2) deficiencies in host immune defenses; and (3) increased permeability or damage to the intestinal mucosal barrier (Deitch, 1990b). Investigations of bacterial translocation with a variety of animal models, such as rodents subjected to thermal injury (Maejima *et al.*, 1984a,b; Deitch *et al.*, 1985), hemorrhagic shock (Baker *et al.*, 1988; Deitch *et al.*, 1988, 1990a,b), endotoxemia (Deitch *et al.*, 1989b; Spaeth *et al.*, 1990; Xu *et al.*, 1993) or zymosan challenge (Mainous *et al.*, 1991, 1993; Deitch *et al.*, 1992) have been described.

Animal species

Several different animal species have been used in these models: specific-pathogen-free (SPF) female, Crl:CDH(SD)BR Holtzman rats (Charles River Breeding Laboratories, Wilmington, MA; Maejima *et al.*, 1984a), adult male Sprague–Dawley rats (Harlan Sprague–Dawley, Indianapolis, IN; Mainous *et al.*, 1993; Deitch *et al.*, 1990a; Grotz *et al.*, 1995; Deitch *et al.*, 1988), SPF female Crl:CD-1(ICR)BR mice, outbred CD1 (ICR) mice, inbred Balb/c and inbred C57/Bl SPF mice (Harlan Sprague– Dawley Breeding Laboratories, Indianapolis, IN) with a stable indigenous GI tract flora (Maejima *et al.*, 1984b; Deitch *et al.*, 1989a), C3H/HeJ (macrophage-hyporesponsive), DBA/2 (congenitally complement-deficient) and W/W^v (congenitally mast-cell-deficient) mice, aged 8–12 weeks (Jackson Laboratories, Bar Harbor, ME; Deitch *et al.*, 1989a). These different strains of mice become important when factoring genetic susceptibility/resistance to injury into any model of bacterial translocation. However, any species of rat or mouse can presumably be used.

Bacterial translocation has also been described in animal models using dogs (Bibbo *et al.*, 1996), pigs (Gelfand *et al.*, 1991; Herndon and Zeigler, 1993), and sheep (Herndon and Zeigler, 1993).

Preparation of animals

In order to maintain a stable indigenous gut flora, all SPF animals are housed under barrier-sustained conditions with controlled temperature (22°C), humidity, and lighting (12 hour light–dark cycles). They are kept in autoclaved polystyrene cages (Maryland Plastics, New York, NY) with stainless steel lids covered with individual Fiberglas filter tops (Econ-filter cover, Scientific Products, Grand Prairie, TX), fed standard laboratory chow (Ralston Purina Co., St Louis, MO) and given acidified water (0.001 N HCl) *ad libitum*.

Handbook of Animal Models of Infection
ISBN 0-12-775390-7

Additional measures are necessary for germ-free and gnotobiotic animals in order to maintain their specific gut flora. These animals require autoclaved polystyrene cages with stainless steel wire lids inside Trexler-type flexible vinyl isolators (Germ free Supply Division, Standard Safety Equipment Co., Palatine, IL) sterilized with 2% peracetic acid (FMC Corp., Buffalo, NY) containing 0.1% Bio-Soft-N-300 (TEA linear alkylate sulfonate, 60% active, Stepan Chemical Co., Northfield, IL). The food, water and bedding for these animals is vacuum sterilized in a bulk sterilizer chamber (Hoeltge, Cincinnati, OH) with a 28 in (71 mm) vacuum cycle in an AMSCO automatic sterilizer adapted with a vacuum pump (Berg and Garlington, 1979).

Details of procedure

Overview

Briefly, anesthetized animals are exposed to various traumatic insults (burn, hemorrhagic shock, endotoxin, or zymosan), which subsequently result in gut mucosal injury (either directly or indirectly). After specific time periods, the animals are sacrificed and the degree of bacterial translocation is determined by culturing the MLN (primary site of bacterial translocation) and other organs, such as the spleen, liver, lung, and kidney, as well as the portal and systemic blood. The exact organ and tissues to culture may vary based on the experimental design or model used. However, the MLN is always cultured, as it is the initial tissue involved in the process of bacterial translocation.

Materials required

Anesthetic agent, Oster animal hair clipper, skin disinfectant, scissors, fine-tipped and curved forceps, polyethylene tubing, suture material, blood pressure recording device, syringes and needles, burn template (plastic holder).

Anesthesia

Mice are anesthetized with an intraperitoneal (i.p.) injection of thiopental sodium at a dose of 0.5 mg/10 g body weight (Deitch *et al.*, 1989a; Ma *et al.*, 1989). Additionally, 50 mg/kg of either pentobarbital sodium (Abbott Laboratories, Chicago, IL; Meyer *et al.*, 1995) or ketamine (Aveco Co., Fort Dodge, IA) plus xylazine (Phoenix Pharmaceuticals, St Joseph, MO), given i.p., can be used. Rats are also anesthetized with an i.p. injection of either pentobarbital sodium at 50 or 25 mg/kg body weight (Deitch *et al.*, 1985; Maejima *et al.*, 1984a) or ketamine (80 mg/kg) plus xylazine (8 mg/kg). Complete anesthesia ensues in approximately 5 minutes and lasts between 30 minutes and 1 hour. Supplemental anesthesia can be given by repeated i.p. injections of the original agent at a fraction of the induction dose depending on the desired level. Alternatively, a nose cone containing methoxyflurane (Mallinckrodt Veterinary, Mundelein, IL) can be placed over the animal to maintain an adequate level of anesthesia.

Thermal injury (burn) model

The technique used to burn the animals is adapted from Walker and Mason (1968), in which the burn size is based on the weight of the animal. Mice and rats are anesthetized as described above. The hair is then removed from both the back and the abdomen with an Oster animal clipper and the animal is placed in a protective template (plastic holder) containing an opening to allow exposure of about 20% of the body surface area to thermal injury. The skin of the back is immersed in boiling water (100°C) for 10 seconds to produce a third degree burn. Mice receiving a 20–25% total body surface area (TBSA) burn are administered 1 ml of sterile normal saline solution (NSS) i.p. immediately postburn to prevent shock (Deitch *et al.*, 1989a; Ma *et al.*, 1989). Larger burns can be produced by burning the animal's abdominal region in addition to the back. Since the skin of the abdomen is thinner than the back, the exposed abdomen is placed in contact with boiling water for 3 seconds to produce a third-degree burn. Since rats tolerate larger burns better than mice, we have examined 40% burns in a rat model. Rats receiving 40% burns are injected i.p. with 10 ml of sterile NSS to prevent shock (Deitch *et al.*, 1985; Maejima *et al.*, 1984a). The dimensions of the opening in the plastic burn holder can be varied and are selected to provide the size of burn desired in animals of known weight. Sham burn controls from both groups are anesthetized, placed in the protective template but not scalded.

An alternative method for inducing thermal injury is described by Hansbrough *et al.* (1996). A convex, body-conforming stainless steel template is heated to 250°C using a thermistor for calibration. The template is applied to the depilated dorsum of the anesthetized animal for 7 seconds. The animal is immediately resuscitated postburn with 20 ml of NSS given i.p. Template size is varied to produce burns of desired uniform percentage of total body surface area.

The animal's surface area and therefore the total percentage burn in any burn model is calculated based on Mech's formula (Walker and Mason 1965):

$A = kW^{2/3}$, where

A = surface area in cm^2, W = body weight in g, k = Mech coefficient; k = 9.0 (mouse), 9.1 (rat).

After the burn injury, the rodents are placed under a heat lamp to prevent hypothermia until they have recovered from the anesthesia. Food and water is made immediately available to the animals.

Hemorrhagic shock model

The rats are anesthetized as above. The skin over the femoral area is shaved with an Oster animal clipper and cleaned with povidone iodine solution (70% vol./vol. ethanol may also be used). The animals are placed supine on a Styrofoam board and allowed to breathe spontaneously. The skin over the femoral area is then excised with sterile scissors and the femoral artery is isolated aseptically using a combination of blunt and sharp dissection. Once isolated, the femoral artery is cannulated with polyethylene (PE-50) tubing (0.023 in i.d. × 0.038 in o.d.) containing 0.1 ml of 10 units/ml of heparinized saline. A three-way stopcock is attached in line for withdrawing blood and a Hewlett Packard Blood Pressure Module (No. 78205B) and Trend Recorder (No. 7825A) are used to monitor the animals' blood pressure during the shock period (Deitch *et al*., 1994). Blood is then withdrawn into a syringe containing 100 units of heparin in 0.3 ml of 0.9% NSS to prevent clotting. The mean arterial pressure is reduced to 30 mmHg and maintained at this level for 30, 60 or 90 minutes by either withdrawing or infusing shed blood (kept at 37°C) as needed (Deitch *et al*., 1990b). By using different periods of shock, the magnitude of the injury can be varied, as can the extent of bacterial translocation. Rectal temperature is monitored throughout the shock period. A heat lamp positioned over the animal is used to prevent hypothermia. At the end of the shock period, the animals are resuscitated by reinfusing all of the shed blood. The femoral catheter is removed and the artery is ligated. The incision is closed in two layers with 3-0 vicryl. The control (sham-shock) rats are anesthetized and their femoral arteries cannulated as described above, however no blood is withdrawn or infused. All animals are allowed immediate access to food and water *ad libitum* upon recovery from anesthesia. The rats are sacrificed immediately after (0 hour) or at 1, 3, 8 or 24 hours after the hemorrhagic shock or sham shock period depending on the experimental design (Baker *et al*., 1988; Guo *et al*., 1995).

Superior mesenteric artery (SMA) occlusion model

The rats are anesthetized as described above. The abdominal hair is then shaved with an Oster animal clipper. The abdomen is cleaned and prepped with povidone iodine solution. A midline celiotomy incision (approximately 5 cm in length) is made with sterile scissors and the SMA is found by deflecting the loops of intestine to the left of the animal with moist gauze swabs. Using blunt dissection (curved forceps), the SMA can be separated from its accompanying lymphatic trunk. It is temporarily occluded by placing a 2-0 silk ligature around its origin from the aorta. Immediate blanching of the small intestine and cecum verify that the blood supply to this region of the gut has been occluded. The abdomen is then covered with a sterile moist gauze pad. After 45 minutes, the ligature is removed and return of blood flow to the gut is confirmed visually. The celiotomy incision is then closed with a running 2-0 silk suture and the animals are allowed to awaken. Those animals subjected to sham SMA occlusion are anesthetized and undergo celiotomy with identification and isolation of the SMA as described above, however, the SMA is looped but not occluded by the 2-0 suture. The rats are sacrificed immediately after (0 hour) or at 1, 3, 8 or 24 hours after the actual or sham SMA occlusion period depending on the experimental design (Grotz *et al*., 1995).

Endotoxin model (non-lethal)

Animals (mice or rats) receive an i.p. injection of either sterile NSS (controls) or *E. coli* 0111:B4 endotoxin (List Biological Laboratories, Campbell, CA) at a dosage of 5 mg/kg (Xu *et al*., 1993). Generally, 24 hours after saline or endotoxin injection, animals are sacrificed (Spaeth *et al*., 1990); however, bacterial translocation can be detected as early as 4–6 hours after endotoxin challenge. The dose of endotoxin required to cause bacterial translocation may vary depending on the endotoxin preparation itself and the company that supplies it. Therefore, exact doses to be used should be determined experimentally.

Zymosan challenge model

Animals (mice or rats) receive an i.p. injection of either sterile NSS or zymosan (Sigma Chemical Co., St Louis, MO). Zymosan is suspended in sterile NSS at a concentration of 10 mg/ml or 50 mg/ml and injected i.p. at a dosage from 0.1 mg/g (mice and rats) to 1.0 mg/g (rats) body weight (Mainous *et al*., 1991) depending on the experiment. It is important to note that, at doses of 0.1 mg/g body weight, there is no mortality. However, as the dose is increased both mortality and the systemic spread of bacteria increase. Mice or rats are sacrificed at either 2, 4, 6, 10 or 24 hour following injection, depending on the experimental design (Deitch *et al*., 1992).

Modulation of gut flora

Since bacterial overgrowth can potentiate bacterial translocation and many of the therapeutic regimens used clinically (H_2-blockers, antacids, TPN, broad-spectrum antibiotics) as well as the development of ileus can lead to bacterial overgrowth, to better model the clinical state it may be necessary to test the effect of trauma, shock, or stress in animals with intestinal bacterial overgrowth. Likewise, the effect of gut flora on certain parameters after trauma, shock, or stress can be investigated by reducing as well as by increasing the gut bacterial population levels. Therefore, in order to further investigate these specific disease processes, the gut flora can be modified by a number of different techniques.

Antibiotic decontamination

The technique of antibiotic decontamination is used in order to eliminate the indigenous GI tract microflora (Figure 25.1). SPF animals are given 4 mg of streptomycin sulfate (Pfizer, New York, NY) and 4 mg of bacitracin (The Upjohn Co., Kalamazoo, MI) per milliliter of drinking water *ad libitum* for 4 days prior to any experimental procedure (Berg, 1980; Maejima *et al.*, 1984b). Alternatively, animals can be administered streptomycin (4 mg/kg) and penicillin (2 mg/kg) in their drinking water *ad libitum* for 4 days to achieve antibiotic decontamination (Guo *et al.*, 1995). Some animals may not drink the water because of the taste of the antibiotics. If this occurs, placing a small amount of saccharine in the water overcomes this problem. Fecal smears from these antibiotic-decontaminated animals are then Gram-stained and examined microscopically to verify visually that the indigenous gut flora has been eliminated or at least greatly reduced. These antibiotic regimens typically reduce the total aerobic cecal bacterial population levels from 10^6 to less than 10^2 colony-forming units (cfu) of bacteria per gram of cecum (Deitch *et al.*, 1985). Once the bacterial load has been sufficiently reduced (or eliminated), animals can then be colonized with a particular bacterial strain or directly tested.

Preparation of bacteria for monoassociation of animals

Steffen *et al.* (1988) demonstrated in monoassociated mice that indigenous Gram-negative enteric bacilli translocated in large numbers to the MLN when they reach high population levels in the gut ($> 10^{9-10}$ cfu/g), whereas Gram-positive bacteria translocated at intermediate levels and obligately anaerobic bacteria at only very low levels. Therefore, a streptomycin-resistant non-enteropathogenic strain of *E. coli* (*E. coli* C25) is used for monoassociation. This bacterial strain has certain inherent advantages: (1) it is resistant to streptomycin; (2) it is known to translocate *in vivo* without causing overt disease (non-pathogenic); and (3) it is part of the gut normal microflora. The *E. coli* are grown overnight in brain heart infusion medium (BHI, Difco Laboratories, Detroit, MI), washed and resuspended in sterile Dulbecco's modified Eagle's medium (DMEM; Sigma Chemical Co., St Louis, MO) to a concentration of approximately 4.2×10^{10} cfu/ml.

Monoassociation

Antibiotic-decontaminated animals are monoassociated (Figure 25.1) by oral gastric inoculation with 3×10^8 *E. coli* C25 resistant to streptomycin (Freter, 1962; Ozawa and Freter, 1964) using 6.35 cm 22 G stainless-steel feeding needles with 2 mm stainless steel bulbs on their tips (Popper

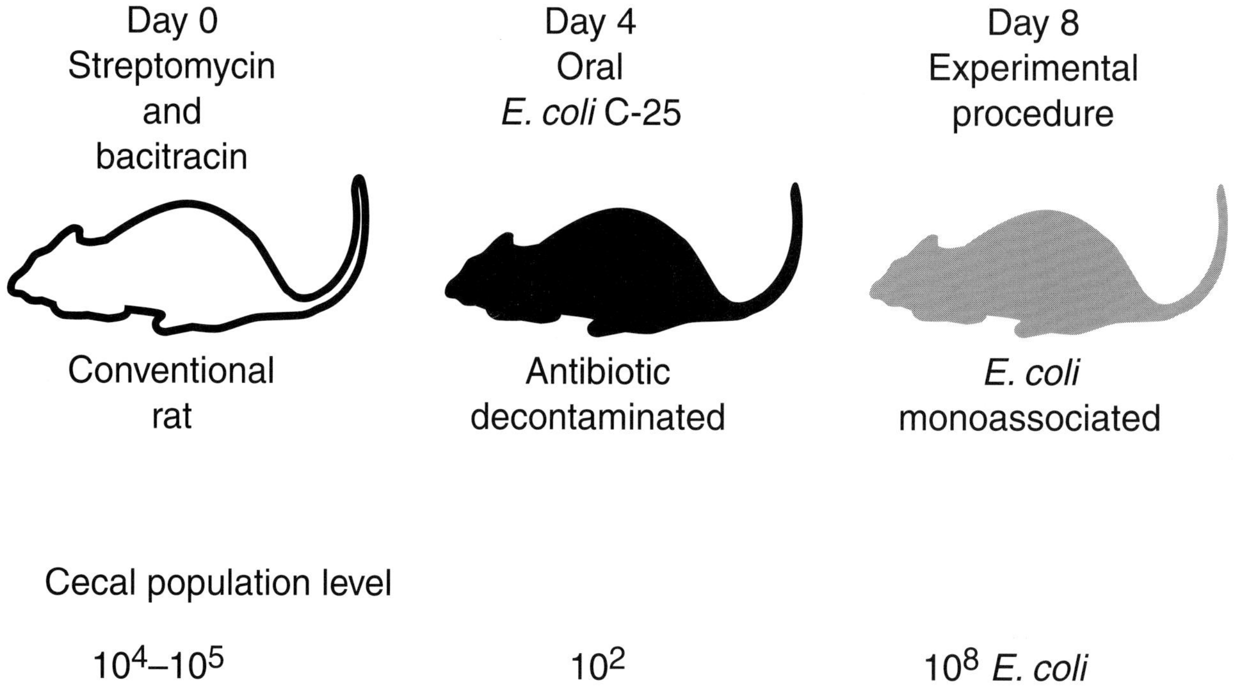

Figure 25.1 Schematic representation of the antibiotic decontamination–monoassociation process and the corresponding changes in the cecal bacterial population level.

& Sons, Hew Hyde Park, NY) as described by Berg (1978); the *E. coli* C25 can also be added to the drinking water, which is changed daily. The *E.-coli*-C25-monoassociated animals continue to receive streptomycin and occasionally bacitracin in their drinking water for a total of 4 days to prevent other bacteria from recolonizing their GI tracts prior to experimentation.

Antibiotic decontamination with or without monoassociation can be used in conjunction with any of the above models, depending on the specific experimental design.

Key parameters to monitor infection

Translocation of *E. coli* C25 in monoassociation models

The mice are sacrificed in a nitrogen atmosphere or via cervical dislocation at chosen times after being subjected to the experimental conditions of interest and their organs are cultured for translocating bacteria. Rats are anesthetized rather than killed prior to harvesting their organs. Next, the skin is cleaned thoroughly with 70% vol./vol. ethanol in order to avoid any accidental bacterial contamination while harvesting the intra-abdominal tissues and organs. Using sterile instruments, an incision is made through the skin and flaps are reflected laterally. The abdominal muscles are then also treated with 70% vol./vol. ethanol and the peritoneum is incised and reflected, thereby exposing the peritoneal cavity. Samples of portal venous and systemic blood (1 ml) are obtained if desired and transferred into tubes containing 5 ml of BHI (rats). These tubes are incubated for 48 hours at 37°C and examined for bacterial growth after 24 and 48 hours. The exposed viscera are swabbed with sterile, cotton-tipped applicator sticks. These swabs are then cultured in either tryptic soy broth (Difco) or BHI to detect any accidental bacterial contamination. The MLN complex draining the jejunum, ileum, and cecum is identified in the mesentery of the ascending colon (Figure 25.2) and the mesentery is sharply incised, exposing the MLN complex. It is important to remove the entire MLN complex as the rate of bacterial translocation differs with respect to the various segments of the MLN complex (Gautreaux *et al.*, 1994). The MLN, spleen, and liver (as well as other organs) are then removed, weighed, and placed in grinding tubes (Tri-R Instruments, Rockville Center, NY) containing 1 ml (MLN) or 5 ml (organs) of either sterile tryptic soy broth or BHI. Next, the cecum is removed and weighed. The ceca are placed in grinding tubes containing 5 ml of either phosphate-buffered saline (pH 7.3) or BHI. The organs are homogenized with teflon grinders (Tri-R Instruments) and portions (0.2 ml) of the MLN, spleen, and liver are cultured on either selective MacConkey's or Tergitol-7 agar plates. The plates are examined after 24 and 48 hours of aerobic incubation at 37°C. The cecal homogenate is serially diluted 1:100 and 0.1 ml of the serial dilutions are plated on to

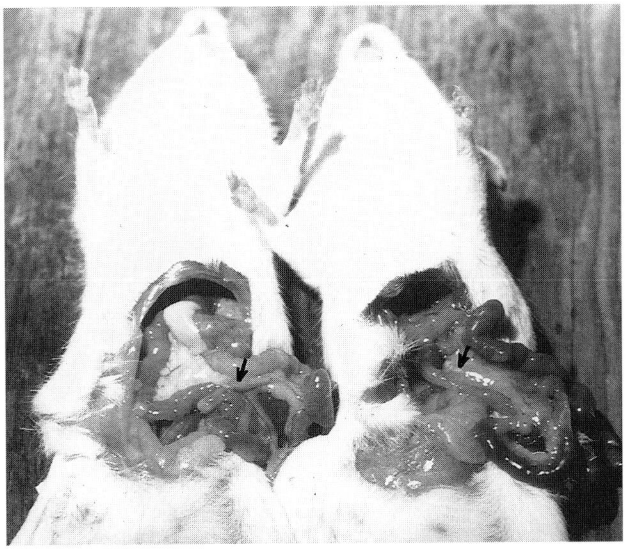

Figure 25.2 Exposure of the MLN complex located in the mesentery of the ascending colon. The main lymphatic chain is located at the tip of the arrow.

Tergitol-7 agar plates (Difco) containing streptomycin sulfate (Maejima *et al.*, 1984a) and incubated for 24 and 48 hours at 37°C. The numbers of viable *E. coli* C25 are determined per gram of organ or per whole organ:

CFU/g = (count × dilution)/sample weight.

Translocation of indigenous bacteria

The various organs are removed as described above and placed in grinding tubes containing sterile BHI to detect aerobic bacteria or sterile 10A broth to detect lactobacilli. The organs are homogenized as above and portions (0.2 ml) of the homogenates are then plated on blood agar (Difco) to detect total aerobic bacteria, MacConkey's agar to detect Gram-negative enterics or 10A agar plates to detect lactobacilli (Buchanan and Gibbons, 1974). The plates are then incubated and inspected at 24 and 48 hours. Gram-negative enteric bacteria that have translocated are subsequently identified with the API 20E system (Analytab Products, Plainview, NY; Smith *et al.*, 1972). Serial dilutions of the cecal homogenate (see above) are plated on blood and MacConkey's agar plates to quantitate total aerobic and Gram-negative enteric cecal bacterial populations (Mainous *et al.*, 1993). The organ homogenates are also Gram-stained to confirm that bacteria present in the homogenates grew on the agar plates.

Since obligate anaerobic bacteria are known to translocate rarely and when they do only at very low levels (Steffen *et al.*, 1988), we do not routinely perform anaerobic cultures. However, if anaerobic cultures are desired, the following special precautions (techniques) need to be observed. After removal, the organs and tissues are placed in grinding tubes containing prereduced, enriched tryptic soy broth with 0.05% dithiothreitol (Sigma Chemical Co., St Louis, MO) to reduce any oxygen contamination (Berg, 1978). The grinding tubes are transferred into an anaerobic environment: either an anaerobic glove box (Coy Manufacturing Co., Ann Arbor, MI) (Aranki and Freter, 1972) or an anaerobic isolation chamber described by Cox and Mangels (1976). The organs are homogenized within the anaerobic environment with Teflon grinders and portions (0.2 ml) are cultured on prereduced, enriched tryptic soy agar (TSA; Aranki and Freter, 1972) designed to support the growth of strictly anaerobic bacteria of the indigenous microflora (Deitch *et al.*, 1985). The enriched TSA contains 1 mg of polymyxin B (Pfizer, New York, NY) per milliliter to inhibit the growth of facultatively anaerobic bacteria. The plates are incubated at 37°C for at least 4 days inside the anaerobic environment. Obligate anaerobes are identified by procedures described by Holderman *et al.* (1977).

Morphologic analysis

Since insults such as burn injury, hemorrhagic shock and SMA occlusion are known to cause gut mucosal injury, which in turn leads to subsequent bacterial translocation, the morphology of the ileum and cecum are analyzed by light microscopy. Samples of the distal ileum and cecum are obtained at the time of sacrifice. The lumen of the gut is filled with fixative. Mucosal samples 2–4 mm in diameter are excised from both the ileum and cecum and immersed in Carson's modified Millonig's phosphate-buffered formalin solution (10 ml of 27–40% formaldehyde solution; 90 ml of tap water; 1.86 g of sodium phosphate monobasic; 0.42 g of sodium hydroxide) for at least 6 hours (Baker *et al.*, 1988; Deitch *et al.*, 1989a, 1990a; Ma *et al.*, 1989). The tissues are quickly dehydrated in 95% ethanol and embedded in glycol methacrylate (JB-4, Polyscience, Warrington, PA); 2 μm sections are then stained with 1% toluidine blue in 1% sodium borate buffer and examined with an Olympus Vanox microscope (Olympus Corporation of America, New Hyde Park, NY) for mucosal villi edema and/or necrosis.

Pitfalls (advantages/disadvantages) of the model

The major advantage of any model of bacterial translocation is that infection occurs via a natural route, which requires that the translocating bacteria colonize the host as well as invade and survive in systemic organs and tissues. Consequently, it represents a less complicated model with fewer confounding variables than most other infectious disease models.

A second advantage is the ability to model clinically common events, such as hemorrhagic shock, burn injury, or endotoxemia. Likewise, combinations of models can be used to more closely mimic what occurs clinically (e.g. burn injury plus endotoxemia or burn injury plus hypotension). In this context then, models of bacterial translocation allow investigation of what is known as the "two-hit phenomenon" of MOF. The two hit phenomenon is used to describe the clinical and biologic observation that an initial insult can prime the host such that a subsequent insult results in a much exaggerated host response than that subsequent insult would have induced if the host were not primed (Deitch, 1992). Experimentally, it has been shown that by combining physiologic insults (Figure 25.3), the degree of

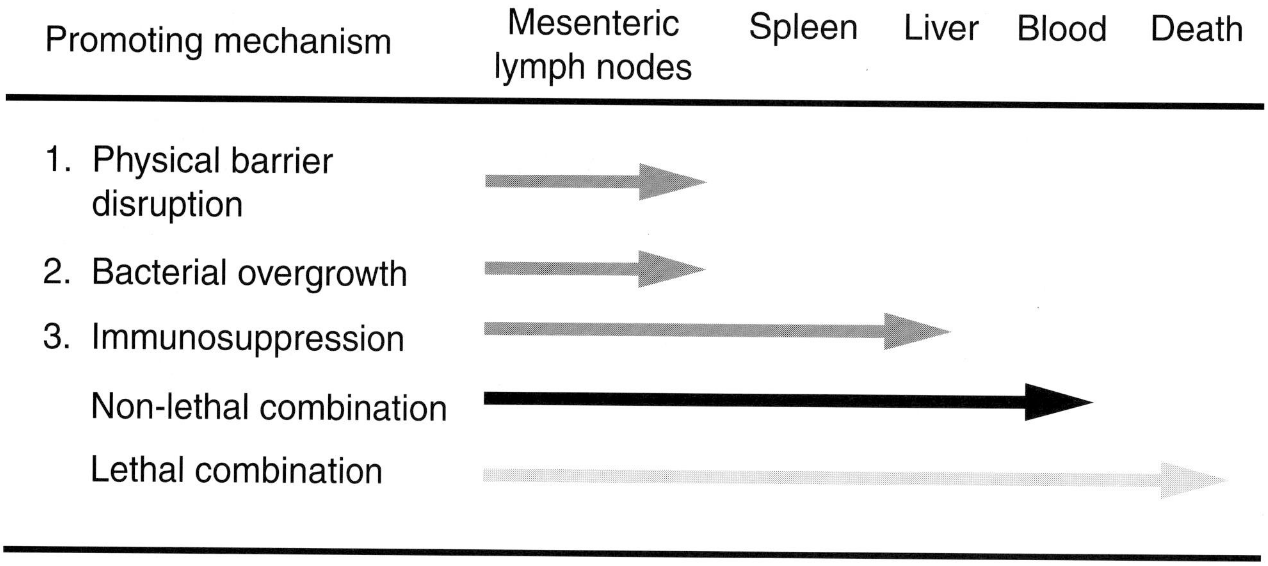

Figure 25.3 Graphical representation of the "two-hit" phenomenon showing that a combination of promoting mechanisms (insults) increases the degree of bacterial translocation and lethality versus single insults.

bacterial translocation increases as does mortality (Deitch and Berg, 1987; Mishima *et al.*, 1997). This is important clinically because most patients that ultimately develop gut origin sepsis or MOF do not experience a single insult but instead typically experience multiple insults. However, not all patients developing gut origin sepsis or MOF follow this paradigm, because clearly MOF and bacterial translocation can and do develop after a single, clinically definable insult if it is sufficiently severe.

The major disadvantage of any model of bacterial translocation is the limited sensitivity of the culture techniques used in assessing bacterial translocation. Alexander *et al.* (1991) documented that bacteriologic techniques that only quantitate viable bacteria grossly underestimate the magnitude of bacterial translocation by a factor of 10^3 as compared to radiolabeling techniques that quantitate both viable and non-viable translocated bacteria. Culture techniques therefore may fail to detect or underestimate the degree of bacterial translocation secondary to the fact that the translocating bacteria may have been killed by macrophages or, in the case of clinical studies, the administration of preoperative antibiotics. In this context then, the importance of culturing the entire MLN complex, especially the small segment (usually separate from the main body of the complex) is especially important in animal studies (Figure 25.4). Failure to culture the entire MLN complex could lead to erroneous conclusions concerning the degree of bacterial translocation from the GI tract. This is particularly important when culturing indigenous bacteria, which appear to be present in greatest numbers in the small segment of the MLN complex (Gautreaux *et al.*, 1994).

Finally, it is almost certain that the murine intestinal barrier is less stable than its human counterpart. Therefore, the ease with which bacterial translocation is induced in rodents means that results from rodent models must not be applied to humans without caution.

Contributions of the model to infectious disease therapy

Although bacterial translocation has been described in a variety of animal models, its clinical relevance still remains uncertain, since there is only limited direct evidence of bacterial translocation in humans. However, studies in animal models demonstrating that intestinal bacterial overgrowth directly promotes the translocation of certain indigenous bacteria from the GI tract to the MLN are probably relevant to the human clinical situation. For example, Berg (1981) showed that oral administration of penicillin, clindamycin or metronidazole to SPF mice for only 4 days allowed intestinal bacterial overgrowth by Gram-negative enterics and subsequent bacterial translocation. In fact, oral antibiotic treatment disrupts the GI ecology (loss of colonization resistance) to such a degree that bacterial transloca-

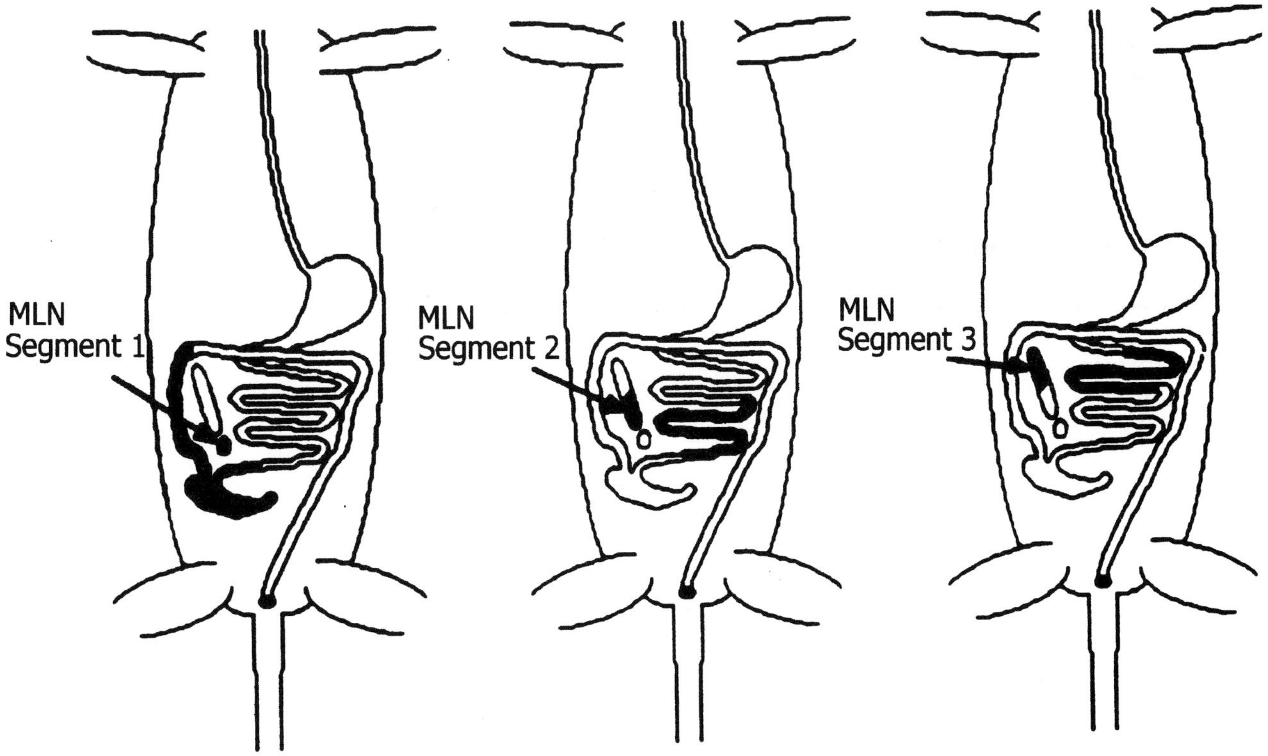

Figure 25.4 Schematic representation of the three segments of the MLN complex and the corresponding regions of the gut that each segment drains. MLN segment 1 (small segment) drains the distal ileum, cecum and ascending colon. MLN segment 2 drains the proximal ileum and segment 3 drains the jejunum.

tion continues after the antibiotic is discontinued (Berg, 1981). This phenomenon becomes important clinically when treating patients with certain systemic or oral antibiotics that kill the anaerobic gut flora because these patients may be at increased risk of bacterial overgrowth and subsequent translocation of Gram-negative enterics (antibiotic resistant) from their GI tract, not only during the treatment period but also for prolonged periods following cessation of treatment. In fact, routine fecal cultures from leukemic (or otherwise immunocompromised) patients reveal an association between the bacterial strain present in the highest concentration in the gut and that strain which is ultimately responsible for any subsequent septicemia (Tancrede and Andremont, 1985). Indigenous intestinal bacteria can also be cultured from: (1) the blood of patients with hemorrhagic shock when the systolic blood pressure has decreased to less than 80 mmHg (Rush *et al.*, 1988); and (2) directly from the MLN of patients with Crohn's disease (Ambrose *et al.*, 1984), colorectal cancer (Vincent *et al.*, 1988), and bowel obstruction (Deitch, 1989). These observations support the concept that bacterial translocation can and does occur routinely in humans. In fact, there is also evidence of increased gut permeability to lactulose and mannitol shortly after thermal (burn) injury (Deitch, 1990a), severe trauma and/or shock (Roumen *et al.*, 1993), and in healthy human volunteers receiving LPS challenge (O'Dwyer *et al.*, 1988). Finally, electron microscopic evaluation of MLNs of trauma patients demonstrated that the incidence of bacterial translocation was greater than that observed by culture technique alone (Brathwaite *et al.*, 1993). Therefore, it seems likely that bacterial translocation is occurring in humans to a much greater extent than previously thought, underscoring the importance of bacterial translocation as a clinically relevant and real phenomenon.

References

Alexander, J. W., Gianotti, L., Pyles, T., Carey, M. A., Babcock, G. F. (1991). Distribution and survival of *Escherichia coli* translocating from the intestine after thermal injury. *Ann. Surg.*, **213**, 558–567.

Ambrose, M. S., Johnson, M., Burdon, D. W., Keighley, M. R. B. (1984). Incidence of pathogenic bacteria from mesenteric lymph nodes and ileal serosa during Crohn's disease surgery. *Br. J. Surg.*, **71**, 623–625.

Aranki, A., Freter, R. (1972). Use of anaerobic glove box for the cultivation of strictly anaerobic bacteria. *Am. J. Clin. Nutr.*, **25**, 1329–1334.

Baker, J. W., Deitch, E. A., Li, M., Berg, R. D., Specian, R. D. (1988). Hemorrhagic shock induces bacterial translocation from the gut. *J. Trauma*, **28**, 896–906.

Berg, R. D. (1978). Antagonism among the normal anaerobic bacteria of the mouse gastrointestinal tract determined by immunofluorescence. *Appl. Environ. Microbiol.*, **35**, 1066–1073.

Berg, R. D. (1980). Inhibition of *Escherichia* coli translocation from the gastrointestinal tract by normal cecal flora in gnotobiotic or antibiotic decontaminated mice. *Infect. Immun.*, **29**, 1073–1081.

Berg, R. D. (1981). Promotion of the translocation of enteric bacteria from the gastrointestinal tracts of mice by oral treatment with penicillin, clindamycin or metronidazole. *Infect. Immun.*, **33**, 851–861.

Berg, R. D., Garlington, A. W. (1979). Translocation of certain indigenous bacteria from the gastrointestinal tract to the mesenteric lymph nodes and other organs in a gnotobiotic mouse model. *Infect. Immun.*, **23**, 403–411.

Bibbo, C., Petschenik, A. J., Reddell, M. T. *et al.* (1996). Bacterial translocation after mesenteric ligation in dogs. *J. Invest. Surg.*, **9**, 293–303.

Brathwaite, C. E. M., Ross, S. E., Nagele, R., Mure, A. J., O'Malley, K. F., Garcia-Perez, F. A. (1993). Bacterial translocation occurs in humans after traumatic injury: evidence using immunofluorescence. *J. Trauma*, **34**, 586–590.

Buchanan, R. E., Gibbons, N. E. (ed.) (1974). *Bergey's Manual of Determinative Bacteriology*, 8th edn, pp. 576–593. Williams & Wilkins, Baltimore, MD.

Cox, M. E., Mangels, J. J. (1976). Improved chamber for the isolation of anaerobic microorganisms. *J. Clin. Microbiol.*, **4**, 40–45.

Deitch, E. A. (1989). Simple intestinal obstruction causes bacterial translocation in man. *Arch. Surg.*, **124**, 699–701.

Deitch, E. A. (1990a). Intestinal permeability is increased in burn patients shortly after injury. *Surgery*, **107**, 411–416.

Deitch, E. A. (1990b). The role of intestinal barrier failure and bacterial translocation in the development of systemic infection and multiple organ failure. *Arch. Surg.*, **125**, 403–404.

Deitch, E. A. (1992). Multiple organ failure pathophysiology and potential future therapy. *Ann. Surg.*, **216**, 117–134.

Deitch, E. A., Berg, R. D. (1987). Endotoxin but not malnutrition promotes bacterial translocation of the gut flora in burned mice. *J. Trauma*, **27**, 161–166.

Deitch, E. A., Maejima, K., Berg, R. (1985). Effect of oral antibiotics and bacterial overgrowth on the translocation of the gastrointestinal tract microflora in burned rats. *J. Trauma*, **25**, 385–391.

Deitch, E. A., Bridges, W., Baker, J. *et al.* (1988). Hemorrhagic shock-induced bacterial translocation is reduced by xanthine oxidase inhibition or inactivation. *Surgery*, **104**, 191–198.

Deitch, E. A., Ma, L., Ma, J., Berg, R. D. (1989a). Lethal burn-induced bacterial translocation: role of genetic resistance. *J. Trauma*, **29**, 1480–1487.

Deitch, E. A., Taylor, M., Grisham, M., Ma, L., Bridges, W., Berg, R. (1989b). Endotoxin induces bacterial translocation and increases xanthine oxidase activity. *J. Trauma*, **29**, 1679–1683.

Deitch, E. A., Bridges, W., Ma, L., Berg, R., Specian, R. D., Granger, D. N. (1990a). Hemorrhagic shock-induced bacterial translocation: the role of neutrophils and hydroxyl radicals. *J. Trauma*, **30**, 942–952.

Deitch, E. A., Morrison, J., Berg, R., Specian, R. D. (1990b). Effect of hemorrhagic shock on bacterial translocation, intestinal morphology and intestinal permeability in conventional and antibiotic decontaminated rats. *Crit. Care Med.*, **18**, 529–536.

Deitch, E. A., Specian, R. D., Grisham, M. B., Berg, R. D. (1992). Zymosan-induced bacterial translocation — a study of mechanisms. *Crit. Care Med.*, **20**, 782–788.

Deitch, E. A., Xu, D., Franko, L., Ayala, A., Chaudry, I. H. (1994). Evidence favoring the role of the gut as a cytokine-generating organ in rats subjected to hemorrhagic shock. *Shock*, **1**, 141–146.

Freter, R. (1962). *In vivo* and *in vitro* antagonism of intestinal bacteria against *Shigella flexneri*. *J. Infect. Dis.*, **110**, 38–46.

Fuller, R., Jayne-Williams, D. J. (1970). Resistance of the fowl (*Gallus domesticus*) to invasion by its intestinal flora: II. Clearance of translocated intestinal bacteria. *Res. Vet. Sci.*, **11**, 368–374.

Gautreaux, M. D., Deitch, E. A., Berg, R. D. (1994). Bacterial translocation from the gastrointestinal tract to various segments of the mesenteric lymph node complex. *Infect. Immun.*, **62**, 2132–2134.

Gelfand, G. A., Morales, J., Jones, R. L., Kibsey, P., Grace, M., Hamilton, S. M. (1991). Hemorrhagic shock and bacterial translocation in a swine model. *J. Trauma*, **31**, 867–874.

Grotz, M. R. W., Ding, J., Guo, W., Huang, Q., Deitch, E. A. (1995). Comparison of plasma cytokine levels in rats subjected to superior mesenteric artery occlusion or hemorrhagic shock. *Shock*, **3**, 362–368.

Guo, W., Ding, J., Huang, Q., Jerrells, T., Deitch, E. A. (1995). Alterations in intestinal bacterial flora modulate the systemic cytokine response to hemorrhagic shock. *Am. J. Physiol.*, **269**, G827–G832.

Hansbrough, J. F., Wikstrom, T., Braide, M. *et al.* (1996). Neutrophil activation and tissue neutrophil sequestration in a rat model of thermal injury. *J. Surg. Res.*, **61**, 17–22.

Herndon, D. N., Zeigler, S. T. (1993). Bacterial translocation after thermal injury. *Crit. Care Med.*, **21**, S50–S54.

Holderman, L. V., Cato, E., Moore, W. E. C. (1977). *Anaerobic Laboratory Manual*, 4th edn, Virginia Polytechnic Institute and State University, Blacksburg, VA.

Ma, L., Ma, J., Deitch, E. A., Specian, R. D., Berg, R. D. (1989). Genetic susceptibility to mucosal damage leads to bacterial translocation in a murine burn model. *J. Trauma*, **29**, 1245–1251.

Maejima, K., Deitch, E. A., Berg, R. D. (1984a). Bacterial translocation from the gastrointestinal tracts of rats receiving thermal injury. *Infect. Immun.*, **43**, 6–10.

Maejima, K., Deitch, E. A., Berg, R. D. (1984b). Promotion by burn stress of the translocation of bacteria from the gastrointestinal tracts of mice. *Arch. Surg.*, **119**, 166–172.

Mainous, M. R., Tso, P., Berg, R. D., Deitch, E. A. (1991). Studies of the route, magnitude and time course of bacterial translocation in a model of systemic inflammation. *Arch. Surg.*, **126**, 33–37.

Mainous, M. R., Xu, D., Deitch, E. A. (1993). Role of xanthine oxidase and prostaglandins in inflammation-induced bacterial translocation. *Circ. Shock*, **40**, 99–104.

Meyer, T. A., Wang, J., Tiao, G. M., Ogle, C. K., Fischer, J. E., Hasselgren, P. (1995). Sepsis and endotoxemia stimulate intestinal IL-6 production. *Surgery*, **118**, 336–342.

Mishima, S., Yukioka, T., Matsuda, H., Shimazaki, S. (1997). Mild hypotension and body burns synergistically increase bacterial translocation in rats consistent with a "two-hit" phenomenon. *J. Burn Care Rehabil.*, **18**, 22–26.

O'Dwyer, S. T., Mitchie, H. R., Ziegler, T. R., Reuhaug, A., Smith, R. J., Wilmore, D. W. (1988). A single dose of endotoxin increases intestinal permeability in healthy humans. *Arch. Surg.*, **123**, 1459–1464.

Ozawa, A., Freter, R. (1964). Ecological mechanism controlling growth of *Escherichia coli* in continuous flow cultures and in the mouse intestine. *J. Infect. Dis.*, **114**, 235–242.

Roumen, R. M. H., Hendriks, T., Wevers, R. A., Goris, J. A. (1993). Intestinal permeability after severe trauma and hemorrhagic shock is increased without relation to septic complications. *Arch. Surg.*, **128**, 453–457.

Rush, B. F., Sori, A. J., Murphy, T. F., Smith, S., Flanagan, J. J., Machiedo, G. W. (1988). Endotoxemia and bacteremia during hemorrhagic shock. The link between trauma and sepsis? *Ann. Surg.*, **207**, 549–554.

Smith, P. B., Tomfohrd, K. M., Rhoden, D. L., Balows, A. (1972). API system: a multitube micromethod for identification of Enterobacteriaceae. *Appl. Microbiol.*, **24**, 449–452.

Spaeth, G., Specian, R. D., Berg, R. D., Deitch, E. A. (1990). Splenectomy influences endotoxin-induced bacterial translocation. *J. Trauma*, **30**, 1267–1272.

Steffen, E. K., Berg, R. D., Deitch, E. A. (1988). Comparison of translocation rates of various indigenous bacteria from the gastrointestinal tract to the mesenteric lymph node. *J. Infect. Dis.*, **157**, 1032–1038.

Tancrede, C. H., Andremont, A. O. (1985). Bacterial translocation and Gram negative bacteremia in patients with hematological malignancies. *J. Infect. Dis.*, **152**, 99–103.

Vincent, P., Colombel, J. F., Lescut, D. *et al.* (1988). Bacterial translocation in patients with colorectal cancer. *J. Infect. Dis.*, **158**, 1395–1396.

Walker, H. L., Mason, A. D. (1968). A standard animal burn. *J. Trauma*, **8**, 1049–1051.

Wolochow, G., Hildebrand, G. J., Lamanna, C. (1966). Translocation of microorganisms across the intestinal wall of the rat: effects of microbial size and concentration. *J. Infect. Dis.*, **116**, 523–528.

Xu, D., Lu, Q., Guillory, D., Cruz, N., Berg, R., Deitch, E. A. (1993). Mechanisms of endotoxin-induced intestinal injury in a hyperdynamic model of sepsis. *J. Trauma*, **34**, 676–682.

Chapter 26

Mouse Models of *Campylobacter jejuni* Infection

S. Baqar, E. F. Burg III and J. R. Murphy

Campylobacteriosis

Campylobacter jejuni, a non-spore-forming usually motile (amphitrichous flagella) pleiotropic Gram-negative rod (short and S-shaped, or longer spirals or rod-like; Figure 26.1; Vandamme and Ley, 1991) is one of the most frequent bacterial causes of enteritis and a less frequent cause of extraintestinal disease in humans (Allos and Blaser, 1995; Tauxe, 1992). Asymptomatic or mild symptomatic carriage of the organism by humans occurs with much greater frequency than colonizations causing disease. Immunoreactive complications may follow *C. jejuni* infection and include Guillain–Barré syndrome (Mishu and Blaser, 1993; Hartung *et al.*, 1995a,b), Reiter syndrome (Ponka *et al.*, 1981), reactive arthritis (Schaad, 1982; Ebright and Ryan, 1984) and erythema nodosum (Frohli *et al.*, 1990).

C. jejuni is distributed globally and has many zoonotic reservoirs, including the intestinal tracts of cattle, pigs, dogs, cats and most birds used as human food (Blaser *et al.*, 1983b; Smibert, 1984). Contamination of food is thought to be a main source of human infections. Data from the USA (Allos and Blaser, 1995) and the UK (Kendall and Tanner, 1982) show that about 1% of their populations get *Campylobacter* infection per year; highest rates are seen in the very young, very old, and immunocompromised (Riley and Finch, 1985).

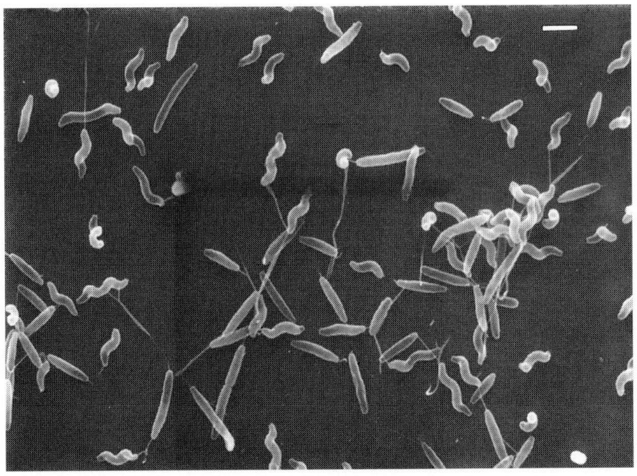

Figure 26.1 Scanning electron micrograph of *Campylobacter jejuni*; bar, 1.0 µm (broth phase of the biphasic cultures were used, for details see Preparation of *C. jejuni* for inoculation into mice, below).

In rank order of frequency, human colonization with *C. jejuni* results in asymptomatic carriage > watery diarrhea > invasive diarrhea > systemic infection; more than one manifestation may be concomitantly present. The bacterial and host factors that determine the outcome are not known. Evidence for heat-labile enterotoxin (Ruiz-Palacios *et al.*, 1983; Johnson and Lior, 1984), cytotoxin (Madden *et al.*, 1984; Johnson and Lior, 1986; Guerrant *et al.*, 1987; Mahajan and Rodgers, 1990), and cultured cell line invasion (Konkel *et al.*, 1990; Yao *et al.*, 1994) has been presented. However, direct evidence that these factors alone or in combination are responsible for clinical disease is lacking and not all isolates from ill individuals possess these putative virulence factors. Direct evidence that acquired immunity can moderate or prevent human *C. jejuni* disease comes from volunteer studies where acquired resistance to rechallenge was found concomitant with the development of anti-*C. jejuni* antibodies (Black *et al.*, 1988). Inferential evidence supporting human acquired immunity is found in the studies made in regions of high endemicity of *C. jejuni* that show a decrease in symptomatic infections with increasing age (Jones *et al.*, 1980; Kaldor *et al.*, 1983; Blaser *et al.*, 1986; Martin *et al.*, 1989; Taylor *et al.*, 1993) and that breast feeding protects from *C. jejuni* diarrhea (Ruiz-Palacios *et al.*, 1990; Torres and Cruz, 1993). However, the mechanisms of protection from disease and colonization are not known. Because this pathogen may both have toxic capacity and exist as a facultative intracellular parasite it is likely that both humoral and cellular arms of immune responses and non immunologic host responses contribute to protective immunity.

C. jejuni are almost uniformly susceptible to macrolides, quinolones, aminoglycosides, chloramphenicol, tetracycline and clindamycin, and resistant to cephalosporins, rifampin, penicillins, trimethoprim, and vancomycin (Taylor *et al.*, 1986; Sjogreen *et al.*, 1992). Reports of resistance of *C. jejuni* to erythromycin, ciprofloxacin, tetracycline, and ampicillin have come from differing geographic areas and are increasing (reviewed in Heresi and Cleary, 1997). However, except among immunocompromised patients (Allos and Blaser,

Handbook of Animal Models of Infection
ISBN 0-12-775390-7

1995), controversy exists on whether there is benefit from antibiotic treatment of patients with *Campylobacter* enteritis (Anders *et al.*, 1982; Pai *et al.*, 1983; Williams *et al.*, 1989). Patients with parenteral *C. jejuni* infection should be treated with antibiotics. Currently available procedures to prevent *C. jejuni* colonization center on establishment/maintenance of a hygienic food chain. Vaccines are being developed as attempts to limit the pathologic consequences of *Campylobacter* infection (Walker, 1994; Baqar *et al.*, 1995a,b; Scott *et al.*, 1997).

The frequency of *C. jejuni* infections, the cumulative morbidity caused by these and the unlikelihood that either *C. jejuni* will soon be eliminated from food chains or reservoirs of *C. jejuni* will soon be eliminated, combine to drive efforts to find other ways to limit the adverse impacts of this pathogen on man. In this context, vigorous efforts are ongoing to identify the bacterial and host factors responsible for pathogenesis and the immune components responsible for acquired resistance and formulation of protective vaccines. Animal models including mouse models are playing important roles in these investigations. Less attention has been directed to animal models of the postinfection immunoreactive complications and in the identification of new antibiotic therapy.

Models of campylobacteriosis

Table 26.1 lists some available animal models and schematically summarizes whether selected aspects of infection, signs of infection or immune responses to infection have been documented. The plethora speaks to no model meeting fully the needs of investigators; the large number of negative reports and characteristics not reported (NR in Table 26.1) speak to both the failure of many models to exhibit those characteristics investigators wanted and the incomplete nature of those studies that have been made of many mouse models.

Volunteer and monkey models

There are two series of studies using volunteers; in the first (Black *et al.*, 1988, 1992) up to 54% of volunteers had patterns of response to challenge consistent with confounding by previous exposure to *Campylobacter* or an antigenically similar agent; in the second (Tribble *et al.*, unpublished data) intensive screening for evidence of previous exposure to *Campylobacter* has reduced this confounding. Similarly, *Campylobacter* spp. is a common endemic agent in monkey colonies (Newell *et al.*, 1988). Thus, knowing the history of prior exposure with *Campylobacter* or antigenically similar organisms is important. The volunteer and monkey models, which best reproduce naturally acquired human campylobacteriosis, have drawbacks of genetic diversity, unknown histories of previous exposure to *C. jejuni* and cost.

Diarrhea

Diarrhea, a cardinal sign of clinically apparent infections in humans with enteropathogenic *C. jejuni* (Black *et al.*, 1988, 1992), is not universally found after infection of other mammalian species. For non-primates, diarrhea or surrogates of diarrhea — e.g. fluid accumulation in removable intestinal tie adult rabbit procedure (RITARD; Caldwell *et al.*, 1983) or rat ligated ileal loops (Saha and Sanyal, 1990) — occur mostly when young (Russell *et al.*, 1989; Bell and Manning, 1990b) or manipulated (McCardell *et al.*, 1986; Stanfield *et al.*, 1987) animals are infected. Where newborns or the very young are susceptible to *C.-jejuni*-induced diarrhea, susceptibility changes rapidly with age (Field *et al.*, 1981; Russell *et al.*, 1989), thus making the models unstable platforms for some experiments. Also, it has not been established that the mechanism(s) of diarrhea in non-primate mammalian newborns or manipulated non-primate animals is the same as that in humans. Thus, the relevance of findings regarding *C. jejuni* infections made in non-primate mammals to the human–*C.-jejuni* parasitism remains a concern to some.

Common findings from studies of primate and other mammalian models

These include:

- that different isolates of C. *jejuni* differ in pathogenic potential (Black *et al.*, 1988; Flores *et al.*, 1990; Baqar *et al.*, 1996);
- that the course of infection and to a degree the pathology caused is directly related to the number of live *C. jejuni* used as a challenge (Black *et al.*, 1988; Flores *et al.*, 1990; Baqar *et al.*, 1996);
- that infection by the oral route (including syringe delivery to the stomach) results in colonization of the intestine and often systemic dissemination (Blaser *et al.*, 1983c; Burr *et al.*, 1988; Vitovec *et al.*, 1989; Baqar *et al.*, 1995a);
- that the time courses of intestinal and extraintestinal infections (in the same animal) may differ (Blaser *et al.*, 1983c; Bell and Manning, 1990b; Baaqar *et al.*, 1996);
- that a first infection (or immunization with selected *C. jejuni* products) induces a degree of resistance to a second challenge (Baqar *et al.*, 1995a,b), which seems greatest if the second challenge is with the homologous strain (Abimiku and Dolby, 1988; Black *et al.*, 1988; Bell and Manning, 1991).

First infections and resistance to second challenges correlate with the development of or presence of anti-*C.-jejuni* antibodies (Abimiku and Dolby, 1987, 1988; Baqar *et al.*, 1995a), respectively; however, no direct evidence identifies an antibody specificity that protects from *C. jejuni*. There exists direct evidence from studies of rats and mice that the pathogenic potential of a *C. jejuni* isolate depends on the host animal's genome (Yrios and Balish, 1986a,b,c; O'Sullivan *et al.*, 1988; Baqar *et al.*, 1996). It is thought that serial passage

Table 26.1 Selected characteristics of some human and animal models of *Campylobacter jejuni* infection (+, positive; –, negative; NR, not reported; ND, not done; OG, ongoing)

			Infection		*Signs*			*Immunity to C. jejuni*		
Model (Reference)	*Route of infection*	*Manipulation*	*Intestinal colonization*	*Systemic infection*	*Watery diarrhea*	*Invasive enteritis*	*Immunoreactive complications*	*Resistance to 2nd challenge*	*Antibody*	*Cellular immunity*
Volunteers (1, 2)	Oral	Milk[a]	+	–	+	+	–	+*	+	NR
Volunteers (3)	Oral	Bicarbonate[b]	+	ND	+	+	–	+	OG	OG
Rhesus (4–6)	Oral	Fasting[c]	+	+	+	+	NR	+†	+	NR
Pig-tailed macaques (7, 8)	Oral	Bicarbonate	+	+	–	+	NR	+	+	NR
Calves (9–12)	Oral	Milk	+	+	–	+	–	NR	+	NR
Dog (13–15)	Oral	None	+	+	–	+‡	NR	NR	NR	NR
Rabbit (16)	Oral	Bicarbonate	+	+	–	–	NR	+	+	NR
Rabbit (16–19)	RITARD	Surgery[d]	+	+	+**	+	NR	+	+	NR
Hamster (20–22)	Oral	None	+	NR	+	–	–	NR	–	NR
Hamster (23)	Cecum	Surgery[e]	+	–	+	+	NR	NR	NR	NR
Pig (gnotobiotic) (24–26)	Oral	None	+	+	+	–	NR	NR	NR	NR
Pig (newborn) (27)	Oral	Fasting	+	NR	–	+	NR	NR	NR	NR
Ferret (28–31)	Oral	Bicarbonate	+	+	+	+	NR	+†	+	NR
Ferret (28)	Rectal	Enema[f]	+	–	+	+	NR	+	NR	NR
Guinea pig (32)	Intramuscular	None	+	+	–	–	NR	NR	+	NR
Guinea pig (33)	Intraperitoneal	None	NR	+	–	–	NR	NR	NR	NR
Chicken (33–36)	Oral	None	+	+	+	+	+	NR	NR	NR
Rat (37, 38)	Oral	Chemical[g]	+	–	–	–	NR	NR	NR	NR
Rat (39–41)	Intestine	Surgery[h]	NR	NR	+**	–	NR	NR	NR	NR
Mouse (newborn) (42–49)	Stomach	None	+	+	+	+	NR	+††	NR	NR
Mouse (newborn) (50)	Stomach	2nd infection[i]	+	–	–	–	NR	NR	NR	NR
Mouse (51, 52)	Intravenous	Chemical[j]	+	+	–	–	NR	NR	NR	NR
Mouse (53)	Intranasal	Anesthetic[k]	+	+	–	–	–	+	+	+
Mouse (54)	Intraperitoneal	None	+	+	–	+	NR	NR	NR	NR
Mouse (55–67)	Oral	None	+	+	–	–	–	+	+	+
Mouse (67–74)	Oral	Chemical[l]	+	+	–	+	–	+	+	NR
Mouse (Nu/Nu) (57–59, 67)	Oral	None	+	+	NR	+	NR	–	+	NR

Notes: a. *C. jejuni* was given in 150 ml of cow's whole milk, based on assumption that this might reduce stomach acidity and allow more live *C. jejuni* to pass the stomach; b. Bicarbonate solutions at varying doses were given on the assumption that this might reduce stomach acidity and allow more live *C. jejuni* to pass the stomach; c. Food withheld for 8 hours before challenge; d. Removable intestinal tie adult rabbit procedure (RITARD); e. Direct inoculation of *C. jejuni* into an exteriorized cecum followed by reinternalization and closure of wound; f. 3 ml of 1% formalin; g. Cadmium chloride; h. Surgical exposure of six to eight 5 cm intestinal loops starting at the ileocecal junction; i. Concomitant induced infection with *Cryptosporidium parvum*; j. Cobra venom factor; k. Methoxyflurane; l. Various chemicals, including antibiotics, iron dextran, ferric chloride, and colloidal carbon.

* Following homologous rechallenge, volunteers were protected from illness but not from colonization.

† Following homologous rechallenge, monkeys were protected from signs of infection and bacteremia, and had reduced fecal excretion of *C. jejuni*.

‡ Mild diarrhea; ** Fluid accumulation in loop; †† Infants of immunized dams were protected.

1. Black *et al.*, 1988; 2. Black *et al.*, 1992; 3. Tribble *et al.*, 1997; unpublished data; 4. Fitzgeorge *et al.*, 1981; 5. Baqar *et al.*, 1995b; 6. Anderson *et al.*, 1993; 7. Flores *et al.*, 1990; 8. Russell *et al.*, 1989; 9. Terzolo *et al.*, 1987; 10. Warner and Bryner, 1984; 11. Al-Mashat and Taylor, 1983; 12. Al-Mashat and Taylor, 1980; 13. Macartney *et al.*, 1988; 14. Olson and Sandstedt, 1987; 15. Prescott *et al.*, 1981; 16. Burr *et al.*, 1988; 17. Everest *et al.*, 1993; 18. Pang *et al.*, 1987; 19. Caldwell *et al.*, 1983; 20. Stills and Hook, 1989; 21. McOrist and Lawson, 1987; 22. Fox *et al.*, 1986; 23. Humphrey *et al.*, 1985; 24. Boosinger and Powe, 1988; 25. Tomancova *et al.*, 1989; 26. Vitovec *et al.*, 1989; 27. Babakhani *et al.*, 1993; 28. Bell and Manning, 1991; 29. Bell and Manning, 1990a; 30. Bell and Manning, 1990b; 31. Fox *et al.*, 1987; 32. Coid *et al.*, 1987; 33. SultanDosa *et al.*, 1983; 34. Li *et al.*, 1996; 35. Welkos, 1984; 36. Sanyal *et al.*, 1984; 37. Epoke *et al.*, 1992; 38. Epoke and Coker, 1991; 39. Chattopadhyay *et al.*, 1991; 40. Saha and Sanyal, 1990; 41. Saha *et al.*, 1988; 42. Van de Giessen *et al.*, 1996; 43. Takata *et al.*, 1992; 44. Field *et al.*, 1981; 45. Stewart-Tull *et al.*, 1984; 46. Hanninen, 1989; 47. Dolby and Newell, 1986; 48. Diker *et al.*, 1992; 49. Newell *et al.*, 1985; 50. Vitovec *et al.*, 1991; 51. O'Sullivan *et al.*, 1988; 52. Field *et al.*, 1991; 53. Baqar *et al.*, 1996; 54. McCardell *et al.*, 1986; 55. Blaser *et al.*, 1983c; 56. Yrios and Balish, 1986a; 57. Yrios and Balish, 1986b; 58. Yrios and Balish, 1986c; 59. Berndtson *et al.*, 1994; 60. Kita *et al.*, 1990; 61. Wu *et al.*, 1991; 62. Abimiku and Dolby, 1988; 63. Abimiku and Dolby, 1987; 64. Kita *et al.*, 1992; 65. Gao *et al.*, 1988; 66. Yrios and Balish, 1985; 67. Kita *et al.*, 1986; 68. Jesudason *et al.*, 1989; 69. Youssef *et al.*, 1987; 70. Fauchere *et al.*, 1985; 71. Stanfield *et al.*, 1987; 72. Dick *et al.*, 1989; 73. Merrell *et al.*, 1981; 74. Obi and Coker, 1989.

Table 26.2 Characteristics of selected mouse models of *Campylobacter jejuni*

Age	*Mice*	Campylobacter jejuni	*Pathogenesis and immunity*
Newborn	There are about 20 reports on responses to *C. jejuni* or products derived from *C. jejuni* in newborn and suckling mice (references 1–18). Most have employed animals from 1–9 days old and infections have been initiated either *per os* or by directly injecting bacteria into the stomach (milk-filled stomachs are easily visualized in neonatal mice). BALB/c mice were the most frequently used (six studies) followed by ddY (three studies) and NMRI (two studies). Some studies do not report the strain of mouse used. In some studies the mice were subjected to manipulations besides exposure to *C. jejuni* in attempts to enhance the virulence of the *C. jejuni* infection or to test the effect of the manipulation. These included initiation of second infections, fasting, injection of endotoxin, ferric chloride or iron dextran, bismuth subsalicylate, milky water and ferric ammonium citrate. Additional studies involving dam–newborn pairs will be reviewed separately below.	The *C. jejuni* used for these experiments varied widely. About 122 recent clinical isolates were tested (in four reports ranging from 15 to 50 isolates per report); at least 24 laboratory adapted strains were evaluated (81116 was the most frequent and the subject of three reports, with 158432 and 12650 tied for second rank with two reports each); a limited number of mutated campylobacters were tested, usually aflagellated or non-motile; in one instance a strain with a defect in chemiotaxis was tested. The remaining reports used unique isolates. The dose of *C. jejuni* administered ranged from undefined through $10–10^{10}$ cfu and included challenges with 'non-culturable' organisms. The procedures used to prepare campylobacters for challenges also varied widely and included passage in various animals, culture in different solid and liquid media at differing temperatures, and in differing environments. Most studies do not report the growth phase of the organisms used for challenge. Limited studies used cell-free culture supernates as challenges in attempts to detect 'toxins'.	The most frequently reported outcome was establishment of intestinal colonization following exposure to live *C. jejuni*. In general, colonization showed somewhat of a direct relationship to challenge dose; the duration of colonization differed for different isolates; colonization or duration of colonization was usually less for mutant organisms than for wild-type strains and non-culturable campylobacters did not colonize. Combined infection with *C. parvum* prolonged *C. jejuni* colonization. Treatments with endotoxin or bismuth reduced *Campylobacter* infection whereas treatments with iron enhanced. Rarely was diarrhea seen. Information on parenteral *C. jejuni* infection is lacking. Limited information is available on immunity in neonatal mice that were not co-housed with immune mothers (see below for summary of data on dam–newborn pairs). One report shows that IgM but not IgG antibodies with specificity for *C. jejuni* flagella reduced colonization.
Juveniles	Seven studies were reviewed that used mice of between 3 and 7 weeks old at the time of the start of the experiments (references 3, 19–23). Most were initiated by oral introduction of bacteria and two used intraperitoneal injection. Different strains of mouse were used. In some studies the mice were manipulated; manipulations included immunizations with formalin-killed *C. jejuni*, treatment with iron dextran, cobra venom factor, silica, colloidal carbon and withholding of food.	There was no pattern to the selection and preparation of *C. jejuni*. Nine clinical and at least five laboratory-adapted strains were evaluated. One aflagellated mutant was tested as an immunogen. Reported challenge doses range from $10^5–10^{10}$ cfu. Some studies did not report challenge dose. There is no apparent standardization to methods for preparation or characterization of challenge bacteria.	In all instances where tests were made for intestinal colonization, colonization was found. Diarrhea was found for iron-treated mice and infrequently in CD-1 mice. Some challenges caused the death of exposed mice and in some conditions lethality was enhanced by iron treatment. Occasionally, parenteral *C. jejuni* infection was shown as was persistence of *C. jejuni* in gall bladders for up to 60 days. Under certain conditions *C. jejuni* infection was associated with elevated liver enzymes. Following challenge or vaccination, *C. jejuni* specific antibodies and antibody-secreting cells at the local sites were shown. Compromise of the alternate complement pathway or macrophage function increased rates of death.

Adults	There are nearly 40 readily accessible reports of *C. jejuni* challenge of adult mice. The majority employed oral challenge (including syringe delivery of *C. jejuni* to the stomach). Fewer studies were made using other routes of infection, axenic, selectively microbially colonized, or genetically athymic mice. Additional studies involve dam–newborn pairs. These will be reviewed below. Thirteen studies where *C. jejuni* infection was initiated by the oral route were reviewed. The reported experiments were started using mice of 8–26 weeks old or reported as 'adult' (references 23–35). BALB/c animals were used most frequently but more than 15 strains, some of undefined origin, were used. Sometimes the source and strain of mouse was not provided. For some studies the mice were manipulated; manipulations included immunizations with *C. jejuni*-derived material, iron dextran, ferric chloride, delivery of selected cytokines *per os*, anesthesia, bicarbonate, various antibiotics, and withholding of food. Other studies (references 34, 36–43; see also Tables 26.3 and 26.4) have used intraperitoneal, intravenous, nasal or intestinal introduction of *C. jejuni* four, three, two, and one studies respectively). BALB/c was the most frequently used mouse strain. However, more than 10 strains were used among the experiments. For some studies the mice were manipulated; manipulations included immunizations with *C. jejuni*-derived material, iron dextran, ferric chloride, anesthesia, pregnancy, and laparotomy. Studies have been made of *C. jejuni* infection of axenic, monoxenic and holoxenic mice of C3H and BALB/c strains (references 44–46). All studies used oral challenge. Bicarbonate, antibiotic, withholding of food, and infection with *Clostridium perenne* were, at times, applied. Limited studies used genetically athymic BALB/c mice, some of which were axenic (references 47–50). Manipulations included immunization with formalin or heat-killed *C. jejuni*.	For studies that used oral administration of *C. jejuni*, more than 20 laboratory-adapted strains of the organism were used besides numerous recent human and animal isolates. There is no readily identified pattern to the selection or preparation of *C. jejuni*. Reported doses range from 10^{1}–10^{11} cfu; for some studies challenge doses were not reported. Many culture conditions including media, additives, temperature and duration were used to prepare *C. jejuni* inocula. Studies made using intraperitoneal, intravenous, nasal, or intestinal introduction of *C. jejuni* employed a variety of strains and isolates and used challenge doses from 10^{6}–10^{10} cfu. The methods to prepare the *C. jejuni* used for challenges are, for the most part, not comparable. Studies made of axenic, monoxenic and holoxenic mice have used at least 10 *C. jejuni* strains. No single *C. jejuni* was common across all studies. All studies of genetically athymic mice have used *C. jejuni* 45100 (a highly mouse-adapted strain) and in addition various other campylobacters. Most challenges were with 10^{8} cfu. For some reports, procedures used to prepare *C. jejuni* inocula are unclear.	In all instances where tests were made for intestinal colonization, such colonization was found for at least some challenged mice. The fraction that became colonized and the duration of colonization, which can be more than 6 months, seem to be functions of both mouse and *Campylobacter* strain. Pretreatment of mice with antibiotics results in more prolonged colonization. Reports of death or diarrhea occurring after oral *C. jejuni* challenge of non-manipulated adult mice are rare. Pretreatment with iron caused some mice to produce soft stools following *C. jejuni* challenge. Occasionally, dissemination of *C. jejuni* to liver, spleen, kidney, mesenteric lymph nodes, and blood are reported. Hepatitis and increases in liver enzymes are reported to follow *C. jejuni* challenge, as is nephritis and IgA immune complex deposition in kidneys. Again, the mouse and *Campylobacter* strain used in the experiment may be critical to the outcome obtained. *C.-jejuni*-specific antibody responses including antibodies in serum, urine, and intestinal lavage fluid were reported and *C. jejuni*-specific IgA, sIgA, IgM, and IgG were found, usually when they were searched for. Immunizations with heat-killed *C. jejuni* were shown to induce an antibody response to *C. jejuni* and an acquired resistance that reduced duration of intestinal colonization. Oral administrations of IL-5 or IL-6 both change the duration of intestinal colonization and *C. jejuni*-specific immune responses. Intraperitoneal, intravenous, nasal, or intestinal introduction of *C. jejuni* all can result in colonization of intestines with *C. jejuni*. Intraperitoneal, intravenous, and nasal *C. jejuni* challenges cause parenteral infection. Death of mice may result from intraperitoneal challenge. Diarrhea, including mucoid diarrhea, may follow intraperitoneal challenge. Comparative studies of axenic, monoxenic, and holoxenic mice show that intestinal colonization occurs for all and that a mild diarrhea may occur primarily in those animals harboring fewer non-*C. jejuni* bacteria. Additionally, the vigor of the *C. jejuni* infection (dissemination of bacteria, persistence of intestinal infection) was greater in animals with fewer concomitant bacterial colonizations. For all instances where it was evaluated, intestinal colonization with *C. jejuni* was demonstrated for athymic and euthymic mice. In most instances parenteral *C. jejuni* infection was shown. Diarrhea was commonly manifested in nu/nu mice.
Dam newborn pairs	Several studies (references 1, 2, 5, 10) investigated *C. jejuni* infection in dam–newborn pairs where either natural or foster mother designs were used. Manipulations included immunization with formalin-killed *C. jejuni* and flagella from *C. jejuni*. Most experiments used BALB/c mice.	*C. jejuni* 81116 was used in all experiments and other strains were also employed in some. An aflagellated mutant was used. Challenges were with 10^{7} cfu and were prepared on blood agar incubated at 37°C.	Intestinal colonization of newborns was found in all instances where the dams–newborns were *C. jejuni*-naive at *C. jejuni* challenge. Sometimes where the dam was *Campylobacter* experienced, newborns were protected from *C. jejuni* and protection was manifested as reduced intestinal colonization and was best against the homologous *Campylobacter* strain. Both *C. jejuni* infections of dams and some immunizations of dams with *C. jejuni*-derived products induced resistance that protected newborns. Some protection was manifested by suckling offspring of *C. jejuni*-naive mothers which were fostered fed on *C.-jejuni*-immune dams.

1. Abimiku and Dolby, 1987; 2. Abimiku and Dolby, 1988; 3. Coker and Obi, 1989; 4. Diker *et al.*, 1992; 5. Dolby and Newell, 1986; 6. Field *et al.*, 1981; 7. Hanninen, 1989; 8. Hanninen, 1990; 9. Morooka *et al.*, 1985; 10. Newell, 1986; 11. Newell and McBride, 1985; 12. Pang *et al.*, 1987; 13. Siddique and Akhtar, 1991; 14. Stewart-Tull *et al.*, 1984; 15. Takata *et al.*, 1992; 16. Ueki *et al.*, 1987; 17. Van De Giessen *et al.*, 1996; 18. Vitovec *et al.*, 1991; 19. Bar, 1988; 20. Coker and Obi, 1991; 21. Jian-Xin *et al.*, 1987; 22. Kita *et al.*, 1986; 23. Madge, 1980; 24. Baqar *et al.*, 1993; 25. Baqar *et al.*, 1995; 26. Berndtson *et al.*, 1994; 27. Blaser *et al.*, 1983; 28. Dick *et al.*, 1989; 29. Field *et al.*, 1984; 30. Gao *et al.*, 1988; 31. Gao *et al.*, 1991; 32. Kita *et al.*, 1990; 33. Kita *et al.*, 1992; 34. Stanfield *et al.*, 1987; 35. Wu *et al.*, 1991; 36. Baqar *et al.*, 1996; 37. Chaiyaroj *et al.*, 1995; 38. Field *et al.*, 1991; 39. McCardell *et al.*, 1986; 40. Merrell *et al.*, 1981; 41. O'Sullivan *et al.*, 1987; 42. O'Sullivan *et al.*, 1988; 43. Pancorbo *et al.*, 1994; 44. Fauchere *et al.*, 1985; 45. Lee *et al.*, 1986; 46. Youssef *et al.*, 1987; 47. Yrios and Balish, 1985; 48. Yrios and Balish, 1986a; 49. Yrios and Balish, 1986b; 50. Yrios and Balish, 1986c.

of *C. jejuni* within a host may result in bacteria that are more virulent for that host (Field *et al.*, 1981; Yrios and Balish, 1986a,b).

An avian model of campylobacteriosis (Table 26.1) has been employed for studies of pathogenesis and immunity (Sanyal *et al.*, 1984; Welkos, 1984), and the immunoreactive complication, Guillain–Barré syndrome (Li *et al.*, 1996).

Mouse models of C. *jejuni* infection

Table 26.2 presents selected results obtained using mouse models. In this table results are grouped for the purpose of the following discussion by age of mice and within some age groups by route of infection. The rationale for this grouping is that both factors may markedly influence the behavior of the model and the results obtained.

Outstanding are the demonstrations that *C. jejuni* has a capacity to colonize and that a first colonization engenders a resistance to a second. This immunity is also manifested in dam–neonate pairs where some acquired resistance seems to trace to antibodies. Also notable is the finding that procedures that disrupt the mouse's defense system's homeostasis may enhance or retard the consequences of *C. jejuni* challenge, including the numbers of bacteria present and the duration of the infection.

The table shows for some outcome measures that there is little influence of mouse strain or *Campylobacter* isolate. This observation may at first seem in conflict with the view expressed elsewhere in this review; that there are very significant differences that link to both mouse and *Campylobacter* strain. Both views are correct. Using low resolution outcome variables, such as a qualitative scoring of colonization, results are somewhat independent of mouse or *Campylobacter* strain. However, whenever attempts have been made to glean higher resolution, a critical dependence is found on both the mouse strain and *Campylobacter* strain.

Limited studies of histopathology have been made using the *C.-jejuni*–mouse model. These have shown that *C. jejuni* can be visualized in association with various sites along the intestine (time-, mouse-, and *Campylobacter*-strain-dependent) and inflammatory pathology may be seen in tissue sections of intestine (Gao *et al.*, 1988, 1991). At times this pathology can be sufficiently severe to rate comparison with inflammatory bowel disease (Gao *et al.*, 1991). Additionally, under some circumstances, long-term *C. jejuni* infection of mice may result in severe liver damage (Kita *et al.*, 1986, 1990, 1992).

Oral (Baqar, 1991) and nasal (Baqar, unpublished data) challenges of mice both trigger cytokine responses that include IL-1 and IL-6. However, significant levels of TNF-α are seen only after nasal challenge (Baqar, unpublished data). A most interesting finding is that oral delivery of IL-5 or IL-6 before *C. jejuni* challenges alter both course of infection and immune response to the *Campylobacter* (Baqar *et al.*, 1993). More investigations are required to allow a mechanistic understanding of the interactions of *C. jejuni* with these immunoregulatory components.

Mice are generally considered unsuitable for studies that focus on *C. jejuni*-caused watery diarrhea — which is known to occur only in newborn mice, invasive diarrhea — which occurs primarily in manipulated mice (reviewed in Table 26.1), and immunoreactive complications (existing studies are insufficient and most inappropriate in design to detect if these occur). These views are, for the most part, based on gross observations of *C.-jejuni*-exposed mice. The studies that have led to these negative views were not exhaustive and thus it is unclear whether further investigations might reveal macro- or microscopic equivalents of these pathologic processes.

It is established that: *C. jejuni* may replicate in mice; infection may either kill the animals or be controlled (Baqar *et al.*, 1996); immunizations with first *C. jejuni* infections or crude *C. jejuni*-derived products may protect against subsequent *C. jejuni* challenge (Baqar *et al.*, 1995a); protection is best against homologous strains (Abimiku and Dolby, 1988; see Table 26.4); infection can be successfully treated with antibiotics (Hof and Sticht-Groh, 1984; Lee *at al*, 1986); and, different isolates of *C. jejuni* differ in innate virulence for the same inbred mouse strain (Baqar *et al.*, 1996).

Thus, direct evidence shows that the *C. jejuni*–mouse parasitism is a robust platform for studies of mechanisms of colonization, distribution and clearance of *C. jejuni*, pathology, host defenses against *C. jejuni*, including non-specific and immunologic mediator and effector systems, antibiotic treatment, and bacterial factors. The availability of genetically defined yet diverse mouse strains, microbiologically defined (germ-free, gnotobiotic, specific-pathogen-free) mice and reagents and procedures should allow mechanistic studies to achieve high levels of resolution.

Selection and care of mice

Strain

Direct evidence establishes that lethality and aspects of *C. jejuni* infection differ following inoculation by the same route of identical doses of the same cloned *C. jejuni* into different inbred strains of mice (Baqar *et al.*, 1996). Because these fundamental aspects of *C. jejuni* parasitism are clearly mouse-strain-dependent, it is reasonable to assume that many outcomes of this parasitism may be functions of the mouse strain used. Selection of a mouse strain for an experiment should be based on citable information that the strain is suitable for the intended purpose or experiments should be made to establish the suitability of the selected mouse strain. Information developed in recent years make it no longer acceptable to randomly select a mouse strain, especially a random-bred strain, expose this to *C. jejuni* and from this draw wide-ranging conclusions on *C. jejuni* parasitism.

Age

Direct evidence obtained using BALB/c (Abimiku and Dolby, 1987), ddY (Morooka *et al.*, 1985; Takata *et al.*, 1992), and NMRI (Hanninen, 1989, 1990; Berndtson *et al.*, 1994) strains of mice shows an age dependence of the parasitism; newborn mice respond to challenge differently from adults. There is evidence from studies in humans that shows increased frequency of *C. jejuni* in immunosuppressed individuals (Leyes *et al.*, 1994). Also, there is an increase in frequency of infection in older individuals (Bokkenheuser, 1970), which may be a reflection of decreased immune capacity of the elderly (Scordamaglia *et al.*, 1991; Haeney, 1994; Caruso *et al.*, 1996). Therefore, it should be assumed, until proven otherwise, that the age of mice will influence results. Because of the rapid maturation of immune response capacity near birth and its decline at older ages, young adult mice are the preferred subjects of most experiments. This is based on the view that their physiologic status, including immune response status, is thought to be stable and to remain so for the interval of weeks or a few months that is the time line of most experiments. The validity of these assumptions for the *C. jejuni*–mouse parasitism has not been formally established by experiments but they seem reasonable based on composite results of available studies.

Presence of other microbes

Evidence has been presented showing that infection of mice with *Cryptosporidium parvum* (Vitovec *et al.*, 1991) or colonization with normal intestinal microbial flora (Field *et al.*, 1984; Yrios and Balish, 1986a,b,c) changes the consequences of concurrent or subsequent *C. jejuni* infection. It has also been shown that treatment of mice with selected antibiotics (Baqar *et al.*, 1993) before *C. jejuni* challenge, presumably by reducing gut colonization with non-*Campylobacter* microbes, exacerbates the consequences of subsequent *C. jejuni* infection. Changes in courses of infections and host responses to infections are well described for other facultative or obligate intracellular bacterial agents (Mackaness, 1969; Murphy 1981a,b; Crist *et al.*, 1984); the presence or absence and the timing of second infections with unrelated bacterial, viral or protozoal agents (Gupta and Pavri, 1987) markedly affect outcomes. These observations strongly suggest that steps should be taken to insure that the mouse's microbial environment is stable and free of unwanted know murine pathogens.

Selection and care of *C. jejuni*

Strain

Direct evidence shows that *C. jejuni* isolates differ in their virulence for mice (Baqar *et al.*, 1996) and that the organism may change in virulence with *in-vitro* passage (Caldwell *et al.*, 1985; Konkel *et al.*, 1990; Ketley, 1997). Inferential evidence (Field *et al.*, 1981) suggests that *in-vivo* passage may also cause changes in the bacterium (Cawthraw *et al.*, 1996). Thus, primary considerations in selection of *C. jejuni* strains are knowledge of whether the strain possesses the phenotypic characteristics needed for the proposed study and assurance that the strain will be available for future confirmatory experiments. Making experiments with severalisolates may be necessary (preferably including reference strains) to find one with a desired characteristic. Because of the potential for plasticity with passage of the bacterium, steps must be taken to limit this confounder. We have chosen the approach of cloning all isolates and preparing primary and secondary seed stocks from these clones (Baqar *et al.*, 1996). In all instances the cloned bacteria are stored frozen at $\leqslant -80$°C. Use of commercial 'bead' cryopreservation systems such as the Microbead® system (Pro-Lab Diagnostics, Austin, TX) markedly facilitates cryopreservation. The system reduces both the expansion of bacterial inoculum needed (and thus the number of *in-vitro* generations) and the volume required for storage.

If isolates from clinical materials are collected, steps should be taken to insure that they are processed similarly and frozen within five or fewer *in-vitro* passages. If established laboratory strains are used, it is important to know and report the passage history. Further, there exist differences among descendants of those isolates which, after more than a decade of investigational use, have become laboratory standards (Abimiku and Dolby, 1987; Cawthraw *et al.*, 1996); yet the strain designation of the original isolate is often retained. It is important, therefore, when reporting results obtained with a standard strain to identify its source and passage history.

Preparation of *C. jejuni* seed stocks

An isolate of *C. jejuni* of known source and passage history should be obtained and plated on to trypticase soy agar supplemented with 5% sheep blood (TSAB) to yield discrete colonies. Inoculated plates are incubated for 16 h under standard conditions (42°C in an atmosphere of 85% N_2, 10% CO_2, and 5% O_2). If a pure culture of *C. jejuni* is obtained, a single colony is transferred to TSAB, and after an additional 16 h growth, the resulting lawn is suspended in 0.5 ml of brain–heart infusion broth supplemented with 1% yeast extract (BHI+YE) and then the suspension is added to vials containing Micro-beads®. These are processed according to the manufacturer's instructions and then stored in liquid or vapor phase N_2 as a primary seed stock. For preparation of secondary 'working' seeds a single bead of the primary seed is placed into 0.2 ml of PBS and incubated for 5 minutes at room temperature. Then 0.1 ml of the resulting suspension is inoculated on to two TSAB plates that are then incubated for 16 hours. The resulting lawns are separately suspended in 0.5 ml of BHI, the suspensions combined and then prepared for freezing using Micro-beads®. These procedures are based on a *C. jejuni* doubling time of about 90 minutes and should be adjusted if a strain is used that has a significantly different growth rate. Primary seeds are used only for the making of secondary seeds. Secondary seeds are employed for all infections, antigen preparations, and other experiments. Vials of both primary and secondary seeds should be stored in two or more cryopreservation devices to protect from refrigeration failure.

Preparation of *C. jejuni* for inoculation into mice

A single bead of secondary seed is placed into 0.2 ml of PBS and incubated for 5 minutes at room temperature. Then 0.1 ml of the resulting suspension is inoculated on to a TSAB plate so that after incubation (see above) three-quarters of the plate shows a lawn and one-quarter yields isolated colonies. If microscopic examination of two isolated colonies confirms *Campylobacter*, the lawn is harvested in 15 ml of BHI+YE. The suspension is diluted until an OD_{625} of 0.08 is achieved (using 1.5 ml disposable cuvettes, Bio Rad, Hercules, CA) and 9 ml of this suspension is then overlaid on to the agar surface of a 25 cm^2 flask (Corning, Corning, NY) containing 8 ml of BHI + YE agar. After 18 hours incubation, the liquid phase of the culture is harvested and centrifuged (8000 *g*, 20 minutes, 4°C). The resulting pellet is suspended in spent medium (at one-fifth or one-20th of original volume for oral or nasal challenge suspensions respectively) and spectrophotometric means are used to estimate and dilutions with BHI+YE to adjust the concentration of bacteria. Serial dilutions of the inoculum are plated on to TSAB to detect the number of *C. jejuni* cfu. This method should yield, for laboratory adapted campylobacters, inocula where the number of culturable (cfu) as compared to nonculturable bacterial particles (as counted in a Petroff Hausser® chamber, Thomas Scientific, Swedesboro, NJ) is 1:2 or less. The ratio of live to dead campylobacters is thought to be important in studies of immunology and other host responses, because dead campylobacters are immunogens (Baqar *et al.*, 1995a,b) and irritants (Jian-Xin *et al.*, 1987).

C. jejuni, 81-176

The preceding methods and following exemplary data are based on the much studied 81-176 isolate of *C. jejuni*. The popularity of this isolate for experimental studies traces to its selection by Black *et al.* (1988) for inclusion into the first large volunteer study of campylobacteriosis. In those studies it was the more virulent of the studied strains. *C. jejuni* 81-176 (Penner serotype 23/36; Lior serogroup 5; see references Penner *et al.*, 1983 and Lior *et al.*, 1982 for details of these typing schemes) is motile and invasive in cell culture (Oelschlaeger *et al.*, 1993; Yao *et al.*, 1994). It was isolated in Minnesota in 1981 from the feces of a 9-year-old diarrheic female who became ill during a milk-borne outbreak (Korlath *et al.*, 1985). In 1984 a cryopreserved sample of the isolate was recovered, expanded in culture and used in volunteer studies at the Center for Vaccine Development (CVD), University of Maryland at Baltimore (Black *et al.*, 1988). The organism caused diarrhea in 18 of 39 (46%) volunteers given doses from 10^6 to 10^9 cfu; six also developed fever. The CVD-origin parent strain of the organisms used for the studies described in the next section was isolated on 12 October 1984 from a diarrheic stool of a challenged volunteer who had been shedding *C. jejuni* in stool for 3 days before the subject isolate was made (this isolate carries CVD designation CAINS 4-1006). The isolate was cryopreserved at –70°C at CVD until December 1990. At that time it was recovered from the freezer and subcultured. An aliquot of that subculture was sent to the Enteric Diseases Branch of the Naval Medical Research Institute (NMRI) where it was subcultured once and frozen. The *C. jejuni* used for the infections and antigen preparations detailed below are decendants of this frozen stock. Details of their handling within NMRI are presented by Baqar *et al.* (1996).

Details of selected adult BALB/c models of campylobacteriosis

Oral infection

Materials

10-week-old BALB/c female mice (Jackson Laboratories, Bar Harbor, ME; the use of female mice allows similarly treated groups to be housed in a single cage with a reduced risk of confounding from fighting); bacterial inoculum (desired number of cfu per 0.5 ml in BHI+YE; for the model described below the challenge dose is 8×10^9 cfu of *C. jejuni* strain 81-176); control inoculum (BHI+YE alone); 5% sodium bicarbonate solution; gloves; 3 ml syringe; feeding needle (stainless steel 21 G 2.5 cm needle with 2.25 mm ball-point end (Perfektum, Popper and Sons, New Hyde Park, NY)).

Anesthesia

None.

Procedure

Mice are restrained by firmly clasping the skin of the back of the neck between the thumb and first finger of a gloved hand, extending the animal's body across the palm and holding the base of the tail between the small finger and palm. The needle is inserted with the curve of the needle aligned with the curve of the mouse's body. For adult mice, the needle should be inserted fully — until the hub of the base reaches the opening of the mouth. A series of 0.5 ml injections is made as follows with the needle being removed between injections. At time –30 minutes and again at –15 minutes bicarbonate solution is injected (this is thought to improve survival of subsequently inoculated bacteria). At time 0, 0.5 ml of BHI+YE alone (control) or BHI+YE containing *C. jejuni* is inoculated.

Expected course of infection and host response

Oral introduction of *C. jejuni* does not result in any grossly observable signs of infection. There is no loss in body weight nor other signs (e.g., ruffled fur, hunched back, diarrhea) of illness. *C. jejuni* should be recovered from parenteral sites beginning a few minutes after challenge (Baqar *et al.*, 1995a) and should persist in feces of some mice for at least 3 weeks (Table 26.3). Humoral antibody, secretory antibody (Baqar *et al.*, 1995a), and cellular immune responses specific to *C. jejuni* antigens should develop because of the infection (Table 26.3). Neither *C. jejuni* nor immune responses to *C. jejuni* are expected in mice inoculated with BHI+YE only. Tables 26.3 and 26.4 present some expected values for distribution of *C. jejuni* and *C. jejuni* specific immune response that should follow the oral challenge described above.

Nasal infection

Materials

10-week-old BALB/c female mice (Jackson Laboratories, Bar Harbor, ME); bacterial inoculum (desired number of cfu per 0.030 ml of BHI+YE (see above); for the model described below the challenge dose is 8×10^9 CFU of *C. jejuni* strain 81-176); control inoculum (BHI + YE alone); gloves; micro-pipetman (Rainin Instruments, Emeryville, CA); methoxyflurane (Metofane®, Pitman-Moore, Mundelein, IL); an anesthesia jar (a 15 cm high × 13 cm diameter glass jar with a fine wire mesh placed in the bottom over 4 × 4 in gauze pads).

Anesthesia

Methoxyflurane, required.

Procedure

A quantity of 5 ml of methoxyflurane is added to the anesthesia jar. After a 5 minutes equilibration the subject mouse is placed into the jar and left in this environment for 1 minute. This produces a light anesthesia. Inocula are delivered as five 0.006 ml drops spaced at 3 s intervals.

Expected course of infection and host response

Nasal introduction of sufficient numbers of *C. jejuni* causes mice to appear ill (see below) and to lose weight. *C. jejuni* should be recovered from parenteral sites beginning a few minutes after challenge and should persist in feces of some mice for at least 3 weeks. Some challenged mice die of the *C. jejuni* colonization (Baqar *et al.*, 1996). Mice that survive possess humoral antibody, secretory antibody, and cellular immune responses specific to *C. jejuni* antigens (Baqar *et al.*, 1996, Table 26.3), will resist second challenges with homologous *C. jejuni* and may resist heterologous challenges to differing degrees. Resistance is manifested as a marked reduction in signs of illness, fecal colonization and rates of death. Tables 26.3 and 26.4 present selected expected values for distribution of *C. jejuni* and *C. jejuni* specific immune responses that should follow from the nasal challenge described above.

Considerations in monitoring *C. jejuni* infection and host responses to infection

Signs

Death and weight loss are the only grossly observable objective signs of *C. jejuni* infection of adult mice that have been repeatedly reported. These occur only after challenge of some mouse strains with high doses of selected campylobacters delivered by limited routes (Table 26.3). Other 'soft' signs that are often seen under conditions of *C. jejuni* challenge approximating those that result in death include a hunching of the back, ruffled fur, and a relative lethargy. Under those conditions where death results from *C. jejuni* challenge, essentially all animals that will die of this insult will have expired by day 5 after challenge (Baqar *et al.*, 1996). Thus, lethal dose data are based on counts of survivors made at 7 days after challenge. Weight loss following *C. jejuni* challenge, when and if it occurs, is rapid, with peak reductions seen at day 3 after infection (Table 26.3). Weight loss is usually expressed as percentage reduction as compared to weight immediately before challenge. In some situations it has been found advantageous to attempt to quantify signs to allow numeric comparison of the virulence of different *C. jejuni* or doses of *C. jejuni*. For this purpose an 'illness index' has been defined (Baqar *et al.*, 1996; and notes to Table 26.3).

Course of infection

Distribution of *C. jejuni* after introduction is monitored by enumeration of viable organisms in tissues and organs. *C. jejuni* in whole blood has been determined by collection of blood by cardiac puncture and plating on agar medium serial dilutions (Baqar *et al.*, 1996) . If dilutions are made soon after blood collection (using PBS as diluent) there is no need to employ an anticoagulant. If wanted, heparin may be used

Table 26.3 Expected outcomes after challenge of BALB/c mice with 8×10^9 cfu of *C. jejuni* 81-176 (ND, not done)

	Challenge	*Outcome measure*	*Day after infection* 1	3	4	7	21	42
Oral	Signs	Cumulative mortality		None				
		Illness*		None				
		Weight loss†				None		
	Infection	Fecal excretion‡	100%; 3.0, 1.2	100%; 2.6, 0.4	90%; 2.7, 0.2	90%; 2.0, 0.5	10%; 1.9, 0.1	
		Bacteremia	38%; 1.1, 0.5					
		Liver	69%; 2.0, 0.9		20%; 1.5, 0.8			
		Mesenteric lymph nodes	25%; 2.2, 0.3	50%; 2.4, 0.7	20%; 0.9, 0.2			
	Campylobacter-specific antibody	Serum IgA**					7.2, 0.6	
		Serum IgG					9.3, 0.2	
		Secretory IgA					6.8, 0.7	
	Campylobacter-specific antibody-secreting cells	Peyer's patches††				5.2, 0.8		
		Mesenteric lymph nodes				4.8, 0.7		
		Spleen				3.6, 0.2		
	Cellular immunity	Spleen cells‡‡						13 360 (0.2)
Nasal	Signs	Cumulative mortality				54%		
		Illness		100%; 0.9, 0.3				
		Weight loss		18%, 5%				
	Infection	Fecal excretion	100%; 3.4, 0.6	100%; 3.1, 0.1	100%; 3.4, 0.1	90%; 2.3, 0.4	50%; ND	
		Bacteremia	88%; 5.8, 1.2					
		Liver	80%; 3.3, 0.7		60%; 5.1, 0.4			
		Mesenteric lymph nodes	20%; 1.1, 0.1		40%; 1.4, 0.4			
	Campylobacter-specific antibody	Serum IgA					6.2, 0.9	
		Serum IgG					7.2, 0.5	
		Secretory IgA					10.3, 0.5	
	Campylobacter-specific antibody-secreting cells	Peyer's patches				3.9, 0.2		
		Mesenteric lymph nodes				1.6, 0.1		
		Spleen				3.9, 0.5		
	Cellular immunity	Spleen cells						8956, 0.4

* Data are: Percentage of mice showing signs; mean 'illness index', standard deviation of 'illness index'. Illness index is an estimate of severity of illness where the larger the number the more severe. This is derived from assigning values to signs that follow *C. jejuni* challenge as follows. Mice are observed once per day for 6 consecutive days following challenge, beginning at 24 hours after infection. At each observation interval each mouse is assigned a score of 0, 1 or 2 denoting apparently healthy, ill (a hunched back, ruffled fur, lethargic), or dead, respectively. For each observation day, the total score within each group is divided by the number of mice observed to yield the 'day index'. For similarly treated mice, the mean of these daily indices is the 'illness index'. Values are means of the indices for similarly treated individual mice and standard deviation.

† ((Weight at day 0 – weight at day 3)/weight at day 0) × 100.

‡ Percentage of animals from which the listed sample was positive for *C. jejuni*. Data are mean of $\log_{10}$ *C. jejuni* cfu, $\log_{10}$ standard deviation of cfu. The number of *C. jejuni* is expressed as per milligram of feces, per milliliter of whole blood, per liver or per pooled mesenteric lymph nodes.

** Anti-*Campylobacter* humoral responses were measured using an ELISA that employed a glycine extract of *C. jejuni* as antigen (Baqar *et al.*, 1995a). Data are geometric mean titers (log natural) and standard deviation.

†† Mean number of ASC per 10^6 mononuclear cells, which had been teased from the respective tissues or organ. See Measurement of correlates of acquired immunity (above) Baqar *et al.*, 1996 for details of methods.

‡‡ Spleen cells were isolated (Baqar *et al.*, 1996) and stimulated in triplicate in the presence of 10^5 formalin-inactivated *Campylobacter* cells for 7 days, at which time radioactive thymidine was added. After an additional 16 hours incubation the plates were harvested, and tritium-containing DNA was determined using standard liquid scintillation procedures (Murphy *et al.*, 1987). The counts per minute for each culture well were $\log_e$ transformed (Murphy *et al.*, 1989) and for similarly treated cultures, the mean $\log_e$ value was calculated. Data are back-transformed means ($\log_e$ standard deviation).

Table 26.4 Demonstration and measurement of acquired resistance of BALB/c mice to *C. jejuni* 81-176 (NA, not applicable)

Primary infection, 5 × 10⁷ cfu of:		*Second infection, 6 × 10⁹ cfu of* C. jejuni *81-176*		
			Outcome measure, (% protection)*	
Strain	*Route*	*Route*	*Illness index*	*% colonized at day 9‡*
None (BHI + YE)**	Oral††	Oral	NA	80
C. jejuni, 81-176	Oral	Oral	NA	11 (86)
C. jejuni, 81-176	Nasal‡‡	Oral	NA	13 (84)
None (BHI + YE)	Oral	Nasal	0.93; 0.36	75
C. jejuni, 81-176	Oral	Nasal	0.37; 0.31 (60)	29 (61)
C. jejuni, HC	Oral	Nasal	0.44; 0.36 (53)	62 (17)
None (BHI + YE)	Nasal	Nasal	0.92; 0.30	88
C. jejuni, 81-176	Nasal	Nasal	0.27; 0.39 (71)	8 (91)
C. jejuni, HC	Nasal	Nasal	0.28; 0.38 (70)	50 (43)
C. coli VC167	Nasal	Nasal	0.30; 0.40 (67)	8 (91)

*((Value for corresponding BHI+YE control) – (Value for treated group))/Value for corresponding control) × 100; †See footnotes to Table 26.3; ‡ Percentage of mice from whom *C. jejuni* was recovered from feces at 9 days after challenge; ** Brain–heart infusion broth plus yeast extract. This is the diluent in which challenges of campylobacters are delivered to mice.; †† Infection initiated by insertion of a blunted needle through the mouth as described in the text; ‡‡ Infection initiated by placing the bacterial suspension on the external nares as described in the text.

to prevent clotting. The number of viable *C. jejuni* in tissues or organs can be estimated by isolation of these using sterile techniques and subsequent application of procedures that disrupt the tissues without disrupting the campylobacters. The Omni-2000® (Omni International, Warrenton, VA) homogenization system has proved to be useful. Fecal excretion is estimated by determining the number of campylobacters in homogenates of stools. Fecal pellets are collected by placing an absorbent paper (Kaydry Lab Table Soaker®, VWR Scientific, New York, NY), absorbent side up, on the work surface of a laminar flow containment hood. A mouse is then placed on a section of the towel and the animal is restrained by placing a 400 ml beaker over it. After three or four pellets are deposited (usually in a few minutes) the mice are returned to their cages and deposited pellets are collected. For the interval between 3 and 5 days after some *C. jejuni* infections a marked reduction in the rate of feces production is found for some infected mice. Holding these animals in the collection device for a half hour or more may be necessary. All stool pellets from a mouse are placed into preweighed 12 × 75 mm plastic tubes and sufficient PBS is added to result in a 5% wt/vol. suspension. After 20–30 minutes, a disposable 10 μl plastic loop (Nalge Nunc Intl, Naperville, IL) is inserted into each tube and the tube containing the loop is vortexed at setting 7 of a Vortex Mixer® (Scientific Products, West Chester, PA) until the feces are dispersed. This procedure is carried out in a containment hood.

The vessels containing collected materials are immersed in a wet ice bath immediately after collection and all processing steps before plating on media are made at 4°C. The use of low temperature limits the overgrowth of other microorganisms that may be present in the samples (especially important for stool and intestinal samples).

Dilutions of the above suspensions (usually 10-fold dilutions) are plated in duplicate on CVA agar plates that are then incubated under standard conditions. Counts are made either of the number of colonies at the lowest dilution, if few campylobacters are present, or of the number at the highest dilution at which more than 10 cfu are found. Under some circumstances (stool samples, for example) confirming that colonies are *C. jejuni* may be necessary. For screening purposes this can be done by doing an oxidase test on suspect colonies and by determining the morphology of the organisms. For definitive purposes, subcultures are required with subsequent application of staining and chemical methods to confirm *C. jejuni*. Table 26.3 presents representative data obtained using the above procedures.

Measurement of correlates of acquired immunity

C. jejuni-specific antibodies and *C. jejuni*-specific cellular immune responses are seen after *C. jejuni* infection of mice. *C. jejuni*-specific antibodies of the IgM (Ueki *et al.*, 1987), IgG, and IgA (including secretory IgA) isotypes are found (Abimiku and Dolby, 1987; Baqar *et al.*, 1995a, 1996), as are antibody-secreting cells of these isotypes (Baqar *et al.*, 1996). Antibodies in blood have been successfully measured using whole blood (Baqar *et al.*, 1995a, 1996), serum (Wu *et al.*, 1991) and secretory antibodies using intestinal lavage fluid (Baqar *et al.*, 1993; Rollwagen *et al.*, 1993; Baqar *et al.*, 1995a), and antibody-secreting cells using isolated mononuclear cells from peripheral blood or cell suspensions from tissues and organs (Baqar *et al.*, 1996) or mononuclear cells from tissues (Jian-Xin *et al.*, 1987). Details of ASC and ELISA methods for measurements of murine antibodies specific to *C. jejuni* are found in Baqar *et al.*, 1996 and Rollwagen *et al.*, 1993 respectively. Levels of *C. jejuni* specific antibodies and ASC seen at selected intervals after challenge of BALB/c are presented in Table 26.3.

Most of the commonly used procedures, ELISA and ASC methods, for example, rely on very crude antigen preparations. Often these are prepared from a strain of *Campylobacter* that is readily available in the laboratory and which is further selected because it grows well under conditions suitable to yielding an antigen. Little effort has been directed to the biological or chemical standardization of these preparations. Thus, care should be exercised when interpreting between-experiment and between-laboratory results. The procedure used to make the antigen that yielded the results presented in Table 26.3 has been presented in detail (Baqar *et al.*, 1995a,b, 1996).

In-vitro antigen-driven mononuclear cell proliferation has been used to identify cellular immune recognition of *C. jejuni* (Table 26.3). Procedures that have been successfully employed have been summarized (Murphy *et al.*, 1987, 1989). We consider it important to employ the data processing procedures summarized in Murphy *et al.*, 1989. The antigens used in mononuclear cell proliferation assays are those described above for serologic procedures and suffer from the same standardization issues. Table 26.3 presents exemplary data on levels of *C. jejuni*-antigen-driven lymphocyte proliferation response expected after oral or nasal infection of BALB/c with *C. jejuni* 81-176.

Measurement of acquired resistance to *C. jejuni* challenge

Primary *C. jejuni* infection may provide a degree of protection from a second challenge with the same strain (Dolby and Newell, 1986; Abimiku and Dolby, 1988; Baqar *et al.*, 1993) and immunizations with *Campylobacter*-derived substances may induce resistance (Rollwagen *et al.*, 1993; Baqar *et al.*, 1995a,b). The way in which protection is manifested depends on the mouse model employed and may comprise: reduction in intestinal colonization measured either as *Campylobacter* present in the intestine or feces, or duration of colonization; reduction in dissemination of bacteria to parenteral sites (Baqar *et al.*, 1995a), reduction in signs of infection or rates of death (Table 26.4); or, in limited instances, a reduction in incidence or severity of diarrhea (Abimiku and Dolby, 1987).

The oral and nasal models presented above have been used to guide the development of the first generation *C. jejuni* vaccine. Candidate immunogens and candidate adjuvants (Baqar *et al.*, 1995a,b) have been evaluated. These studies are, in part, predicated on demonstrations that BALB/c have the capacity to develop effective antidisease and antimicrobial responses to *C. jejuni*. Table 26.4 presents representative results on acquired resistance as manifested in the above presented BALB/c-mouse–*C. jejuni*-81-176 parasitism. For the presented studies mice were exposed to either control material or one or another *Campylobacter* and 5–7 weeks later challenged with *C. jejuni* 81-176. Results from three separate experiments are summarized in the different frames of the table. The oral challenge of *Campylobacter*-naive mice, frame 1, resulted in 80% of mice having intestinal colonization with *C. jejuni* on day 9 after the challenge; nasal challenge of *Campylobacter*-naive mice resulted in 75–88% intestinal colonization. Nasal challenge of naive mice but not oral challenge caused animals to appear ill. For the two nasal challenge experiments, the degrees of illness as quantified as mean illness indices (Baqar *et al.*, 1996) were very similar, 0.92 and 0.93. The reproducibility of this model and of the measures used to quantify responses are sufficiently robust to support studies comparing quality of states of acquired resistance.

The results in the first section of Table 26.4 show that a first infection with *C. jejuni* 81-176, whether initiated by oral or nasal routes, causes generation of a resistance manifested as a reduction in percent of mice with campylobacters in stools at day 9 after a second challenge. Thus, a first infection engenders an animal with a capacity to restrict the homologous bacterium delivered as a second, subsequent challenge.

Data in the second frame of Table 26.4 address the question of whether infection with a different isolate of *C. jejuni* would engender protection against *C. jejuni* 81-176. For this experiment, nasal challenge was used because this route results in levels of intestinal colonization similar to those that follow oral challenge and because nasal challenge allows the measurement of *C. jejuni*-caused illness. The data show that first oral exposure to *C. jejuni* 81-176 or *C. jejuni* HC provides similar levels of protection from illness and that homologous first infection provides markedly better protection from colonization than does heterologous infection.

The third frame extends the preceding observations to demonstrations that nasal first infection protects from subsequent nasal challenge and that protection from illness can be induced by heterologous immunization. However, acquired resistance measured as protection from colonization is complex and mechanistically not yet understood. An example of this complexity is that little protection from colonization is afforded against *C. jejuni* 81-176 colonization by experience with a different isolate of the same species, *C. jejuni* HC. Whereas a remarkable level of protection against *C. jejuni* 81-176 colonization follows from previous exposure to an isolate of a different species, *C. coli* VC167.

The model systems summarized in Table 26.4 can also be used to measure resistance to hematogenous dissemination and resistance to organ/tissue colonization. However, insufficient data are currently available to establish reproducibility. The models can allow resolution of the mechanisms of the phenomenology.

Contributions of mouse models to knowledge of campylobacteriosis

C. jejuni naive mice if placed into a *C. jejuni*-contaminated environment will get infection and subsequently deployed mechanisms to control the growth and dissemination of the bacterium (Field *et al.*, 1981). Thus, *C. jejuni* infection of mice may be viewed as a natural host–parasite interaction. Manipulation of bacterial dose, strain and route of infection and mouse strain can be used to accentuate selected aspects of the parasitism, allowing the investigator to target interactions for analysis. Such manipulations have been used to glean information on the natural history of *C. jejuni* dissemination, pathology, immunoreactive responses, and immunoprotective responses (Tables and previous sections). To date most mouse-based studies have been descriptive. However, the extraordinary availability of mouse-based analytical reagents and procedures suggests that future studies should uncover the mechanistic foundations of the parasitism and resistance to *C. jejuni*.

Issues

A question frequently asked is whether results obtained with mouse models can be extrapolated to the human–*C.-jejuni* parasitism. A definitive answer is not available. However, given the limited number of categories of mechanisms of pathogenesis and immunity, it is reasonable to suggest that identification of a mechanistic function in a murine model would be a strong suggestion that it might be functional in another. Further, the lack of availability of volunteers and monkeys supports the use of an approach where candidate mechanisms are identified using a readily available and dissectable model and subsequently confirmed using the rarer primate resources. Currently ongoing investigations of *C. jejuni* infections of volunteers (Tribble, unpublished data) and of a candidate *C. jejuni* vaccine in volunteers are, in part, based on results from studies of *C. jejuni* infections and *C. jejuni* candidate vaccines and adjuvants made in mice (Baqar *et al.*, 1995a,b, 1996). The outcomes of these studies will yield important information on comparability of results obtained in the two experimental systems.

A second group of concerns turns on what could be summarized as 'inconsistencies' among mouse models of *C. jejuni*. Changes in *C. jejuni* isolate, passage history of a *C. jejuni* isolate, route of infection, mouse strain, mouse age, or mouse non-*Campylobacter* intestinal flora change the manifestations of the parasitism. Current knowledge suggests that this concern may be turned to advantage. Procedures are available to stabilize *C. jejuni* so that nearly identical bacteria can be used in repeated experiments made over years. Similarly, genetically defined and stable mice are readily available and genetically 'designed' mice may be created. Methods are available to limit the influence of non-*C jejuni* gut flora. Knowledge of the many variables that influence the *C. jejuni*–mouse host–parasite relationship is now sufficiently advanced to demand the inclusion of controls into future studies, i.e., mouse–*C. jejuni* strain pairs for which the pattern of the parasitism is known. In this way, it should be possible to develop data that are more easily compared between model systems and laboratories.

Human campylobacteriosis is fecal–oral-transmitted and most colonizations are asymptomatic (Kendall and Tanner, 1982; Tauxe, 1992; Allos and Blaser, 1995). The infectious dose for humans is thought to be 10^2–10^6 cfu (Black, 1988). This pattern of infection can be exactly mimicked in mice. Significant human-relevant data may be obtained through studies of mouse models. It might prove useful to view comparatively results obtained with lower-dose oral infection against those found with extreme doses, unnatural routes of introduction, and following surgical and chemical manipulations.

Acknowledgements

The opinions and assertions contained herein are not to be construed as official or as reflecting the views of the Navy Department or the naval service at large. These studies were supported by US Navy Research and Development Command Work Unit no. 62787A.001.01.EVX.1522. The authors are indebted to A. L. Bourgeois for critical review of the text. The technical assistance of Lisa Applebee, Jason Castro and Shaun Wright are appreciated.

References

Abimiku, A. G., Dolby, J. M. (1987). The mechanism of protection of infant mice from intestinal colonisation with *Campylobacter jejuni*. *J. Med. Microbiol.*, **23**, 339–344.

Abimiku, A. G., Dolby, J. M. (1988). Cross-protection of infant mice against intestinal colonisation by *Campylobacter jejuni*: importance of heat-labile serotyping (Lior) antigens. *J. Med. Microbiol.*, **26**, 265–268.

Allos, B. M., Blaser, M. J. (1995). *Campylobacter jejuni* and the expanding spectrum of related infections. *Clin. Infect. Dis.*, **20**, 1092–1099.

Al-Mashat, R. R., Taylor, D. J. (1980). Production of diarrhoea and dysentery in experimental calves by feeding pure cultures of *Campylobacter fetus* subspecies *jejuni*. *Vet. Rec.*, **107**, 459–464.

Al-Mashat, R. R., Taylor, D. J. (1983). Production of enteritis in calves by the oral inoculation of pure cultures of *Campylobacter fetus* subspecies *intestinalis*. *Vet. Rec.*, **112**, 54–58.

Anders, B. J., Lauer, B. A., Paisley, J. W., Reller, L. B. (1982). Double-blind placebo controlled trial of erythromycin for treatment of *Campylobacter* enteritis. *Lancet*, **i**, 131–132.

Anderson, K. F., Kiehlbauch, J. A., Anderson, D. C., McClure, H. M., Wachsmuth, I. K. (1993). *Arcobacter* (*Campylobacter*) *butzleri*-associated diarrheal illness in a nonhuman primate population. *Infect. Immun.*, **61**, 2220–2223.

Babakhani, F. K., Bradley, G. A., Joens, L. A. (1993). Newborn piglet model for campylobacteriosis. *Infect. Immun.*, **61**, 3466–3475.

Baqar, S. (1991). *Role of Cytokines in* Campylobacter jejuni *Infection and Immunity in Mice*. PhD thesis. University of Maryland, Baltimore, MD.

Baqar, S., Pacheco, N. D., Rollwagen, F. M. (1993). Modulation of mucosal immunity against *Campylobacter jejuni* by orally administered cytokines. *Antimicrob. Agents Chemother.*, **37**, 2688–2692.

Baqar, S., Applebee, L. A., Bourgeois, A. L. (1995a). Immunogenicity and protective efficacy of a prototype *Campylobacter* killed whole-cell vaccine in mice. *Infect. Immun.*, **63**, 3731–3735.

Baqar, S., Bourgeois, A. L., Schultheiss, P. J. *et al.* (1995b). Safety and immunogenicity of a prototype oral whole-cell killed *Campylobacter* vaccine administered with a mucosal adjuvant in non-human primates. *Vaccine*, **13**, 22–28.

Baqar, S., Bourgeois, A. L., Applebee, L. A. *et al.* (1996). Murine intranasal challenge model for the study of *Campylobacter* pathogenesis and immunity. *Infect. Immun.*, **64**, 4933–4939.

Bar, W. (1988). Role of murine macrophages and complement in experimental *Campylobacter* infection. *J. Med. Microbiol.*, **26**, 55–59.

Bell, J. A., Manning, D. D. (1990a). Reproductive failure in mink and ferrets after intravenous or oral inoculation of *Campylobacter jejuni*. *Can. J. Vet. Res.*, **54**, 432–437.

Bell, J. A., Manning, D. D. (1990b). A domestic ferret model of immunity to *Campylobacter jejuni*-induced enteric disease. *Infect. Immun.*, **58**, 1848–1852.

Bell, J. A., Manning, D. D. (1991). Evaluation of *Campylobacter jejuni* colonization of the domestic ferret intestine as a model of proliferative colitis. *Am. J. Vet. Res.*, **52**, 826–832.

Berndtson, E., Danielsson-Tham, M. L., Engvall, A. (1994). Experimental colonization of mice with *Campylobacter jejuni*. *Vet. Microbiol.*, **41**, 183–188.

Black, R. E., Levine, M. M., Clements, M. L., Hughes, T. P., Blaser, M. J. (1988). Experimental *Campylobacter jejuni* infection in humans. *J. Infect. Dis.*, **157**, 472–479.

Black, R. E., Perlman, D., Clements, M. L., Levine, M. M., Blaser, M. J. (1992). Human volunteer studies with *Campylobacter jejuni*. In Campylobacter jejuni, *Current Status and Future Trends* (eds Nachamkin, I., Blaser, M. J., Tompkins, L. S.), pp. 207–215. American Society for Microbiology, Washington, DC.

Blaser, M. J., Duncan, D. J., Osterholm, M. T., Islre, G. R., Wang, W. L. (1983a). Serologic study of two clusters of infection due to *Campylobacter jejuni*. *J. Infect. Dis.*, **147**, 820–823.

Blaser, M. J., Taylor, D. N., Feldman, R. A. (1983b). Epidemiology of *Campylobacter jejuni* infections. *Epidemiol. Rev.*, **5**, 157–176.

Blaser, M. J., Duncan, D. J., Warren, G. H., Wang, W. L. (1983c). Experimental *Campylobacter jejuni* infection of adult mice. *Infect. Immun.*, **39**, 908–916.

Blaser, M. J., Taylor, D. N., Echeverria, P. (1986). Immune response to *Campylobacter jejuni* in a rural community in Thailand. *J. Infect. Dis.*, **153**, 249–254.

Bokkenheuser, V. (1970). *Vibrio fetus* infection in man: I. Ten new cases and some epidemiologic observation. *Am. J. Epidemiol.*, **91**, 400–409.

Boosinger, T. R., Powe, T. A. (1988). *Campylobacter jejuni* infections in gnotobiotic pigs. *Am. J. Vet. Res.*, **49**, 456–458.

Burr, D. H., Caldwell, M. B., Bourgeois, A. L., Morgan, H. R., Wistar, R. Jr, Walker, R. I. (1988). Mucosal and systemic immunity to *Campylobacter jujuni* in rabbits after gastric inoculation. *Infect. Immun.*, **56**, 99–105.

Caldwell, M. B., Walker, R. I., Stewart, S. D., Rogers, J. E. (1983). Simple adult rabbit model for *Campylobacter jejuni* enteritis. *Infect. Immun.*, **42**, 1176–1182.

Caldwell, M. B., Guerry, P., Lee, E. C., Burans, J. P., Walker, R. I. (1985). Reversible expression of flagella in *Campylobacter jejuni*. *Infect. Immun.*, **50**, 941–943.

Caruso, C., Candore, G., Cigna, D. *et al.* (1996). Cytokine production pathway in the elderly. *Immunol. Res.*, **15**, 84–90.

Cawthraw, S. A., Wassenaar, T. M., Ayling, R., Newell, D. G. (1996). Increased colonization potential of *Campylobacter jejuni* strain 81116 after passage through chicken and its implication on the rate of transmission within flock. *Epidemiol. Infect.*, **117**, 213–225.

Chaiyaroj, S. C., Sirisereewan, T., Jiamwatanasuk, N., Sirisinha, S. (1995). Production of monoclonal antibody specific to *Campylobacter jejuni* and its potential in diagnosis of *Campylobacter* enteritis. *Asian Pacific J. Allergy Immunol.*, **13**, 55–61.

Chattopadhyay, U. K., Rathore, R. S., Pal, D., Das, M. S. (1991). Enterotoxigenicity of human and animal isolates of *Campylobacter jejuni* in ligated rat ileal loops. *J. Diarrhoeal Dis. Res.*, **9**, 20–22.

Coid, C. R., O'Sullivan, A. M., Dore, C. J. (1987). Variations in the virulence for pregnant guinea pigs, of campylobacters isolated from man. *J. Med. Microbiol.*, **23**, 187–189.

Coker, A. O., Obi, C. L. (1989). Endotoxin activity and enterotoxigenicity of human strains of *Campylobacter jejuni* isolated from patients in a Nigerian hospital. *Cent. Afr. J. Med.*, **35**, 524–527.

Coker, A. O., Obi, C. L. (1991). Effect of iron-dextran on lethality of Nigerian isolates of *Campylobacter jejuni* in mice. *Cent. Afr. J. Med.*, **37**, 20–23.

Crist, A. E. Jr., Wisseman, C. L., Jr, Murphy, J. R. (1984). Characteristics of *Rickettsia mooseri* infection of normal and immune mice. *Infect. Immun.*, **43**, 38–42.

Dick, E., Lee, A., Watson, G., O'Rourke, J. (1989). Use of the mouse for the isolation and investigation of stomach-associated, spiral-helical shaped bacteria from man and other animals. *J. Med. Microbiol.*, **29**, 55–62.

Diker, K. S., Hascelik, G., Diker, S. (1992). Colonization of infant mice with flagellar variants of *Campylobacter jejuni*. *Acta Microbiol. Hung.*, **39**, 133–136.

Dolby, J. M., Newell, D. G. (1986). The protection of infant mice from colonization with *Campylobacter jejuni* by vaccination of the dams. *J. Hyg. Lond.*, **96**, 143–151.

Ebright, J. R., Ryan, L. M. (1984). Acute erosive reactive arthritis associated with *Campylobacter jejuni*-induced colitis. *Am. J. Med.*, **76**, 321–323.

Epoke, J., Coker, A. O. (1991). Intestinal colonization of rats following experimental infection with *Campylobacter jejuni*. *E. Afr. Med. J.*, **68**, 348–351.

Epoke, J., Obi, C. L., Coker, A. O. (1992). *In vivo* effect of cadmium chloride on intestinal colonization of rats by *Campylobacter jejuni*. *E. Afr. Med. J.*, **69**, 609–610.

Everest, P. H., Goossens, H., Sibbons, P. *et al.* (1993). Pathological changes in the rabbit ileal loop model caused by *Campylobacter jejuni* from human colitis. *J. Med. Microbiol.*, **38**, 316–321.

Fauchere, J. L., Veron, M., Lellouch-Tubiana, A., Pfister, A. (1985). Experimental infection of gnotobiotic mice with *Campylobacter jejuni*: colonisation of intestine and spread to lymphoid and reticulo-endothelial organs. *J. Med. Microbiol.*, **20**, 215–224.

Field, L. H., Underwood, J. L., Pope, L. M., Berry, L. J. (1981). Intestinal colonization of neonatal animals by *Campylobacter fetus* subsp. *jejuni*. *Infect. Immun.*, **33**, 884–892.

Field, L. H., Underwood, J. L., Berry, L. J. (1984). The role of gut flora and animal passage in the colonisation of adult mice with *Campylobacter jejuni*. *J. Med. Microbiol.*, **17**, 59–66.

Field, L. H., Underwood, J. L., Payne, S. M., Berry, L. J. (1991). Virulence of *Campylobacter jejuni* for chicken embryos is associated with decreased bloodstream clearance and resistance to phagocytosis. *Infect. Immun.*, **59**, 1448–1456.

Fitzgeorge, R. B., Baskerville, A., Lander, K. P. (1981). Experimental infection of Rhesus monkeys with a human strain of *Campylobacter jejuni*. *J. Hyg. Lond.*, **86**, 343–351.

Flores, B. M., Fennell, C. L., Kuller, L., Bronsdon, M. A., Morton, W. R., Stamm, W. E. (1990). Experimental infection of pig-tailed macaques (*Macaca nemestrina*) with *Campylobacter cinaedi* and *Campylobacter fennelliae*. *Infect. Immun.*, **58**, 3947–3953.

Fox, J. G., Zanotti, S., Jordan, H. V., Murphy, J. C. (1986). Colonization of Syrian hamsters with streptomycin resistant *Campylobacter jejuni*. *Lab. Amin. Sci.*, **36**, 28–31.

Fox, J. G., Ackerman, J. I., Taylor, N., Claps, M., Murphy, J. C. (1987). *Campylobacter jejuni* infection in the ferret: an animal model of human campylobacteriosis. *Am. Vet. Res.*, **48**, 85–90.

Frohli, P., Hanselmann, R., Koelz, H. R. (1990). Erythema nodosum in *Campylobacter jejuni* colitis. *Schweiz. Med. Wschr. J. suisse med.*, **120**, 946–947.

Gao, J. X., Ma, B. L., Wang, W. C., Xie, Y. L., Yu, H. (1988). A mouse model of chronic *Campylobacter jejuni* infection. Post-infection nephritis. *Chin. Med. J. Engl.*, **10**, 623–630.

Gao, J. X., Ma, B. L., Xie, Y. L., Huang, D. S. (1991). Electron microscopic appearance of the chronic *Campylobacter jejuni* enteritis of mice. *Chin. Med. J.*, **104**, 1005–1010.

Guerrant, R. L., Wanke, C. A., Pennie, R. A., Barrett, L. J., Lima, A. A., O'Brien, A. D. (1987). Production of a unique cytotoxin by *Campylobacter jejuni*. *Infect. Immun.*, **55**, 2526–2530.

Gupta, A. K., Pavri, K. M. (1987). Alteration of immune response of mice with dual infection of *Toxocara canis* and Japanese encephalitis virus. *Trans. R. Soc. Trop. Med. Hyg.*, **81**, 835–840.

Haeney, M. (1994). Infection determinants at extremes of age. *J. Antimicrob. Chemother.*, **34** (Suppl. A), 1–9.

Hanninen, M. L. (1989). Effect of endotoxin on colonisation of *Campylobacter jejuni* in infant mice. *J. Med. Microbiol.*, **30**, 199–206.

Hanninen, M. L. (1990). Bismuth subsalicylate in the prevention of colonization of infant mice with *Campylobacter jejuni*. *Epidemiol. Infect.*, **104**, 397–404.

Hartung, H. P., Pollard, J. D., Harvey, G. K., Toyka, K. V. (1995a). Immunopathogenesis and treatment of the Guillain–Barré syndrome. Part I. *Muscle & Nerve*, **18**, 137–153.

Hartung, H. P., Pollard, J. D., Harvey, G. K., Toyka, K. V. (1995b). Immunopathogenesis and treatment of the Guillain-Barré syndrome. Part II. *Muscle & Nerve* **18**, 154–164.

Heresi, G., Cleary, T. (1997). *Campylobacter jejuni*. In *Textbook of Pediatric Infectious Diseases*, 4th edn (eds Feigin, R. D., Cherry, J. D.), pp. 1443–1451. W. B. Saunders, Philadelphia, PA.

Hof, H., Sticht-Groh, V. (1984). Antibacterial effects of niridazole: its efffects on microaerophilic *Campylobacter*. *Infection*, **12**, 36–39.

Humphrey, C. D., Montag, D. M., Pittman, F. E. (1985). Experimental infection of hamsters with *Campylobacter jejuni*. *J. Infect. Dis.*, **151**, 485–493.

Jesudason, M. V., Hentges, D. J., Pongpech, P. (1989). Colonization of mice by *Campylobacter jejuni*. *Infect. Immun.*, **57**, 2279–2282.

Jian-Xin, G., Huan-Niu, W., Gen-Fu, X. (1987). Studies on intestinal mucosal immunity in mice orally immunized with killed *Campylobacter jejuni*. *Adv. Exp. Med. Biol.*, **216B**, 1863–1875.

Johnson, W. M., Lior, H. (1984). Toxins produced by *Campylobacter jejuni* and *Campylobacter coli*. *Lancet*, **i**, 229–230.

Johnson, W. M., Lior, H. (1986). Cytotoxic and cytotonic factors produced by *Campylobacter jejuni*, *Campylobacter coli* and *Campylobacter laridis*. *J. Clin. Microbiol.*, **24**, 275–281.

Jones, D. M., Eldridge, J., Dale, B. (1980). Serological response to *Campylobacter jejuni/coli* infection. *J. Clin. Pathol.*, **33**, 767–769.

Kaldor, J., Pritchard, H., Serpell, A., Metcalf, W. (1983). Serum antibodies in *Campylobacter* enteritis. *J. Clin. Microbiol.*, **18**, 1–4.

Kendall, E. J., Tanner, E. I. (1982). *Campylobacter* enteritis in general practice. *J. Hyg. (Camb.)*, **88**, 155–163.

Ketley, J. M. (1997). Pathogenesis of enteric infection by *Campylobacter*. *Microbiology*, **143**, 5–21

Kita, E., Katsui, N., Nishi, K., Emoto, M., Yanagase, Y., Kashiba, S. (1986). Hepatic lesions in experimental *Campylobacter jejuni* infection of mice. *J. Gen. Microbiol.*, **132**, 3095–3103.

Kita, E., Oku, D., Hamuro, A. *et al.* (1990). Hepatotoxic activity of *Campylobacter jejuni*. *J. Med. Microbiol.*, **33**, 171–182.

Kita, E., Nishikawa, F., Kamikaidou, N., Nakano, A., Katsui, N., Kashiba, S. (1992). Mononuclear cell response in the liver of mice infected with hepatotoxigenic *Campylobacter jejuni*. *J. Med. Microbiol.*, **37**, 326–331.

Konkel, M. E., Babakhani, F., Jones, L. A. (1990). Invasion related antigens of *Campylobacter jejuni*. *J. Infect. Dis.*, **162**, 888–895.

Korlath, J. A., Osterholm, M. T., Judy, L. A., Forfang, J. C., Robinson, R. A. (1985). A point-source outbreak of campylobacteriosis associated with consumption of raw milk. *J. Infect. Dis.*, **152**, 592–596.

Lee, A., O'Rourke, J. L., Barrington, P. J., Trust, T. J. (1986). Mucus colonization as a determinant of pathogenicity in intestinal infection by *Campylobacter jejuni*: a mouse cecal model. *Infect. Immun.*, **51**, 536–546.

Leyes, M., Vara, F., Reina, J., Riera, M., Siquier, B., Villalonga, C. (1994). *Campylobacter* gastroenteritis in patients with human immunodeficiency virus infection. *Enferm. Inf. Microbiol. Clin.*, **12**, 332–336.

Li, C. Y., Xue, P., Tian, W. Q., Liu, R. C., Yang, C. (1996). Experimental *Campylobacter jejuni* infection in the chicken: an animal model of axonal Guillain–Barré syndrome. *J. Neurol. Neurosurg. Psychiat.*, **61**, 279–284.

Lior, H., Woodward, D. L., Edgar, J. A., Laroche, L. J., Gill, P. (1982). Serotyping of *Campylobacter jejuni* by slide agglutination based on heat-stable antigen factors. *J. Clin. Microbiol.*, **15**, 761–768.

Macartney, L., Al-Mashat, R. R., Taylor, D. J., McCandlish, I. A. (1988). Experimental infection of dogs with *Campylobacter jejuni*. *Vet. Rec.*, **122**, 245–249.

Mackaness, G. B. (1969). Influence of immunologically committed lymphoid cells on macrophage activation in vivo. *J. Exp. Med.*, **129**, 973–992.

Madden, J. M., McCardell, B. A., Shah, D. B. (1984). *Campylobacter jejuni* and *Campylobacter coli* cytotoxic toxin production by members of genus *Vibrios*. *Lancet*, **ii**, 1217–1218.

Madge, D. S. (1980). *Campylobacter* enteritis in young mice. *Digestion*, **20**, 389–394.

Mahajan, S., Rodgers, F. G. (1990). Isolation, characterization and host-cell binding properties of a cytotoxin from *Campylobacter jejuni*. *J. Clin. Microbiol.*, **28**, 1314–1320.

Martin, P. M., Mathiot, J., Ipero J., Kirimat, M., Georges, A. J., Georges-Courbot, M. C. (1989). Immune response to *Campylobacter jejuni* and *Campylobacter coli* in a cohort of children from birth to 2 years of age. *Infect. Immun.*, **57**, 2542–2546.

McCardell, B. A., Madden, J. M., Stanfield, J. T. (1986). A mouse model for the measurement of virulence of species of *Campylobacter* (letter). *J. Infect. Dis.*, **153**, 177.

McOrist, S., Lawson, G. H. (1987). Possible relationship of proliferative enteritis in pigs and hamsters. *Vet. Microbiol.*, **15**, 293–302.

Merrell, B. R., Walker, R. I., Coolbaugh, J. C. (1981). *Campylobacter fetus* ss. *jejuni*, a newly recognized enteric pathogen: morphology and intestinal colonization. *Scan. Electron Microsc.*, **4**, 125–131.

Mishu, B., Blaser, M. J. (1993). Role of infection due to *Campylobacter jejuni* in the initiation of Guillain–Barré syndrome. *Clin. Infect. Dis.*, **17**, 104–108.

Morooka, T., Umeda, A., Amako, K. (1985). Motility as an intestinal colonization factor for *Campylobacter jejuni*. *J. Gen. Microbiol.*, **131**, 1973–1980.

Murphy, J. R. (1981a). Host defenses in murine malaria: analysis of plasmodial infection-caused defects in macrophage microbicidal capacities. *Infect. Immun.*, **31**, 396–407.

Murphy, J. R. (1981b). Host defenses in murine malaria: nonspecific resistance to *Plasmodium berghei* generated in response to *Mycobacterium bovis* infection or *Corynebacterium parvum* stimulation. *Infect. Immun.*, **33**, 199–211.

Murphy, J. R., Baqar, S., Munoz, C. *et al.* (1987). Characteristics of humoral and cellular immunity to *Salmonella typhi* in residents of typhoid-endemic and typhoid-free regions. *J. Infect. Dis.*, **156**, 1005–1009.

Murphy, J. R., Wasserman, S. S., Baqar, S. *et al.* (1989). Immunity to *Sallmonella typhi*; considerations relevant to measurement of cellular immunity in typhoid-endemic regions. *Clin. Exp. Immunol.*, **75**, 228–233.

Newell, D. G. (1986). Monoclonal antibodies directed against the flagella of *Campylobacter jejuni*; production, characterization, and lack of effect on the colonization of infant mice. *J. Hyg. (Camb.)*, **96**, 131–141.

Newell, D. G., McBride, H., Dolby, J. M. (1985). Investigations on the role of flagella in the colonization of infant mice with *Campylobacter jejuni* and attachment of *Campylobacter jejuni* to human epithelial cell lines. *J. Hyg. (Camb.)*, **95**, 217–227.

Newell, D. G., Hudson, M. J., Baskerville, A. (1988). Isolation of a gastric *Campylobacter*-like organism from the stomach of four Rhesus monkeys, and identification as *Campylobacter pylori*. *J. Med. Microbiol.*, **27**, 41–44.

Obi, C. L., Coker, A. O. (1989). The role of macrophages in experimental *Campylobacter* infections of mice in Lagos, Nigeria. *Trop. Geogr. Med.*, **41**, 154–159.

Oelschlaeger, T. A., Guerry, P., Kopecko, D. J. (1993). Unusual microtubule-dependent endocytosis mechanisms triggered by *Campylobacter jejuni* and *Citrobacter freundii*. *Proc. Natl Acad. Sci. USA*, **90**, 6884–6888.

Olson, P., Sandstedt, K. (1987). *Campylobacter* in the dog: a clinical and experimental study. *Vet. Rec.*, **121**, 99–101.

O'Sullivan, A. M., Dore, C. J., Coid, C. R. (1987). Campylobacters and impaired fetal development in mice. *J. Med. Microbiol.*, **24**, 7–12.

O'Sullivan, A. M., Dore, C. J., Boyle, S., Coid, C. R., Johnson, A. P. (1988). The effect of campylobacter lipopolysaccharide on fetal development in the mouse. *J. Med. Microbiol.*, **26**, 101–105.

Pai, C. H., Gillis, F., Tuomanen, E., Marks, M. I. (1983). Erythromycin in treatment of *Campylobacter* enteritis in children. *Am. J. Dis. Child.*, **137**, 286–288.

Pancorbo, P. L., Gallego, A. M., Pablo, M., Alvarez, C., Ortega, E., Cienfuegos, G. A. (1994). Inflammatory and phagocytic response to experimental *Campylobacter jejuni* infection in mice. *Microbiol. Immunol.*, **38**, 89–95.

Pang, T., Wong, P. Y., Puthucheary, S. D., Sihotang, K., Chang, W. K. (1987). *In vitro* and *in vivo* studies of a cytotoxin from *Campylobacter jejuni*. *J. Med. Microbiol.*, **23**, 193–198.

Penner, J. L., Hennessy, J. N., Congi, R. V. (1983). Serotyping of *Campylobacter jejuni* and *Campylobacter coli* on the basis of thermostable antigens. *Eur. J. Clin. Microbiol.*, **2**, 378–383.

Ponka, A., Martio, J., Kosunen, T. U. (1981). Reiter's syndrome in association with enteritis due to *Campylobacter fetus* ss. *jejuni*. *Ann. Rheumat. Dis.*, **40**, 414–415.

Prescott, J. F., Barker, I. K., Manninen, K. I., Miniats, O. P. (1981). *Campylobacter jejuni* colitis in gnotobiotic dogs. *Can. J. Comp. Med.*, **45**, 377–383.

Riley, L. W., Finch, M. J. (1985). Results of the first year of national surveillance of *Campylobacter* infections in the United States. *J. Infect. Dis.*, **151**, 956–959.

Rollwagen, F. M., Pacheco, N. D., Clements, J. D., Pavlovskis, O., Rollins, D. M., Walker, R. (1993). Killed *Campylobacter* elicits immune response and protection when administered with an oral adjuvant. *Vaccine*, **11**, 1316–1320.

Ruiz-Palacios, G. M., Torres, J., Escamilla, N. I. (1983). Cholera-like enterotoxin produced by *Campylobacter jejuni*. Characterization and clinical significance. *Lancet*, **ii**, 250–253.

Ruiz-Palacios, G. M., Calva, J. J., Pickering, L. K. *et al.* (1990). Protection of breast fed infants against *Campylobacter* diarrhea by antibodies in human milk. *J. Pediatr.*, **116**, 707–713.

Russell, R. G., Blaser, M. J., Sarmiento, J. I., Fox, J. (1989). Experimental *Campylobacter jejuni* infection in *Macaca nemestrina*. *Infect. Immun.*, **57**, 1438–1444.

Saha, S. K., Sanyal, S. C. (1990). Production and characterisation of *Campylobacter jejuni* enterotoxin in a synthetic medium and its assay in rat ileal loops. *FEMS Microbiol. Lett.*, **67**, 333–338.

Saha, S. K., Singh, N. P., Sanyal, S. C. (1988). Enterotoxigenicity of chicken isolates of *Campylobacter jejuni* in ligated ileal loops of rats. *J. Med. Microbiol.*, **26**, 87–91.

Sanyal, S. C., Islam, K. M., Neogy, P. K., Islam, M., Speelman, P., Huq, M. I. (1984). *Campylobacter jejuni* diarrhea model in infant chickens. *Infect. Immun.*, **43**, 931–936.

Schaad, U. B. (1982). Reactive arthritis associated with *Campylobacter* enteritis. *Pediatr. Infect. Dis. J.*, **1**, 328–332.

Scordamaglia, A., Ciprandi, G., Indiveri, F., Canonica, G. W. (1991). The effect of aging on host defences. Implications for therapy. *Drugs Aging*, **1**, 303–316.

Scott, D. A., Baqar, S., Burr, D. H., Pazzaglia, G., Guerry, P. (1997). Vaccines against *Campylobacter jejuni*. In *New Generation Vaccines* (eds Levine, M. M, Woodrow, G. C., Kaper, J. B., Cobon, G. S.), pp. 885–896. Marcel Dekker, Basel.

Siddique, A. B., Akhtar, S. Q. (1991). Study on the pathogenicity of *Campylobacter jejuni* by modifying the medium. *J. Trop. Med. Hyg.*, **94**, 175–179.

Sjogreen, E., Kaijser, B., Werner, M. (1992). Antimicrobial susceptibilities of *Campylobacter jejuni* and *Campylobacter coli* isolated in Sweden: a 10-year follow-up report. *Antimicrob. Agents Chemother.*, **36**, 2847–2849.

Smibert, R. M. (1984). Genus *Campylobacter*. In *Bergey's Manual of Systematic Bacteriology* (eds Krieg, N. R., Holt, H. G.), vol. 1, p. 111. Williams & Wilkins, Baltimore, MD.

Stanfield, J. T., McCardell, B. A., Madden, J. M. (1987). *Campylobacter* diarrhea in an adult mouse model. *Microb. Pathog.*, **3**, 155–165.

Stewart-Tull, D. E., Ng, F. K. P., Wardlaw, A. C. (1984). Factors affecting the lethality of *Campylobacter fetus* subspecies *jejuni* in mice. *J. Med. Microbiol.*, **18**, 27–37.

Stills, H. F. Jr, Hook, R. R. Jr (1989). Experimental production of proliferative ileitis in Syrian hamsters (*Mesocricetus auratus*) by using an ileal homogenate free of *Campylobacter jejuni*. *Infect. Immun.*, **57**, 191–195.

SultanDosa, A. B., Bryner, J. H., Foley, J. W. (1983). Pathogenicity of *Campylobacter jejuni* and *Campylobacter coli* strains in the pregnant guinea pig model. *Am. J. Vet. Res.*, **44**, 2175–2178.

Takata, T., Fujimoto, S., Amako, K. (1992). Isolation of nonchemotactic mutants of *Campylobacter jejuni* and their colonization of the mouse intestinal tract. *Infect. Immun.*, **60**, 3596–3600.

Tauxe, R. V. (1992). Epidemiology of *Campylobacter jejuni* infections in the United States and other industrialized countries. In Campylobacter jejuni — *Current Strategy and Future Trends* (eds Nachamkin, I., Blaser, M. J., Tompkins, L. S.), pp. 9–19. American Society for Microbiology, Washington DC.

Taylor, D. E., Chang, N., Garner, R. S., Sherburne, R., Mueller, L. (1986). Incidence of antibiotic resistance and characterization of plasmids in *Campylobacter jejuni* strains isolated from clinical sources in Alberta, Canada. *Can. J. Microbiol.*, **32**, 28–32.

Taylor, D. N., Perlman, D. N., Echeverria, P. D., Lexomboon, U., Blaser, J. M. (1993). *Campylobacter* immunity and quantitative excretion rates in Thai children. *J. Infect. Dis.*, **168**, 754–758.

Terzolo, H. R., Lawson, G. H., Angus, K. W., Snodgrass, D. R. (1987). Enteric *Campylobacter* infection in gnotobiotic calves and lambs. *Res. Vet. Sci.*, **43**, 72–77.

Tomancova, I., Koudela, B., Matyas, Z., Vitovec, J. (1989). Dynamics of *Campylobacter jejuni* invasion in gnotobiotic piglets. *Vet. Med. Praha.*, **34**, 553–558.

Torres, O., Cruz, J. R. (1993). Protection against *Campylobacter* diarrhea: role of milk IgA antibodies against bacterial surface antigens. *Acta Paedriatr.*, **82**, 835–838.

Ueki, Y., Umeda, A., Fujimoto, S., Mitsuyama, M., Amako, K. (1987). Protection against *Campylobacter jejuni* infection in suckling mice by anti-flagellar antibody. *Microbiol. Immunol.*, **31**, 1161–1171.

Vandamme, P., Ley, J. D. (1991). Proposal for a new family, Campylobacteraceae. *Int. J. Sys. Bacteriol.*, **41**, 451–455.

Van de Giessen, A. W., Heuvelman, C. J., Abee, T., Hazeleger, W. C. (1996). Experimental studies on the infectivity of non-culturable forms of *Campylobacter* spp. in chicks and mice. *Epidemiol. Infect.*, **117**, 463–470.

Vitovec, J., Koudela, B., Sterba, J., Tomancova, I., Matyas, Z., Vladik, P. (1989). The gnotobiotic piglet as a model for the pathogenesis of *Campylobacter jejuni* infection. *Int. J. Med. Microbiol.*, **271**, 91–103.

Vitovec, J., Koudela, B., Vladik, P., Hausner, O. (1991). Interaction of *Cryptosporidium parvum* and *Campylobacter jejuni* in experimentally infected neonatal mice. *Zbl. Bakt.*, **274**, 548–559.

Walker, R. I. (1994). New strategies for using mucosal vaccination to achieve more effective immunization. *Vaccine*, **12**, 387–400.

Warner, D. P., Bryner, J. H. (1984). *Campylobacter jejuni* and *Campylobacter coli* inoculation of neonatal calves. *Am. J. Vet. Res.*, **45**, 1822–1824.

Welkos, S. L. (1984). Experimental gastroenteritis in newly-hatched chicks infected with *Campylobacter jejuni*. *J. Med. Microbiol.*, **18**, 233–248.

Williams, D., Schorling, J. B., Barrett, L. J. *et al.* (1989). Early treatment of *Campylobacter jejuni* enteritis. *Antimicrob. Agents. Chem.*, **33**, 248–250.

Wu, S. J., Pacheco, N. D., Oprandy, J. J., Rollwagen, F. M. (1991). Identification of *Campylobacter jejuni* and *Campylobacter coli* antigens with mucosal and systemic antibodies. *Infect. Immun.*, **59**, 2555–2559.

Yao, R., Burr, D. H., Doig, P., Trust. T. J., Niu, H., Guerry, P. (1994). Isolation of motile and nonmotile insertional mutants of *Campylobacter jejuni*: the role of motility in adherence and invasion of eukaryotic cells. *Molec. Microbiol.*, **14**, 883–893.

Youssef, M., Corthier, G., Goossens, H., Tancrede, C., Henry-Amar, M., Andremont, A. (1987). Comparative translocation of enteropathogenic *Campylobacter* spp. and *Escherichia coli* from the intestinal tract of gnotobiotic mice. *Infect. Immun.*, **55**, 1019–1021.

Yrios, J. W., Balish, E. (1985). Colonization and pathogenesis of *Campylobacter* spp. in athymic and euthymic germfree mice. In *Germfree Research: Microflora Control and Its Application to the Biomedical Sciences*, pp. 199–202. New York, Alan R. Liss.

Yrios, J. W., Balish, E. (1986a). Colonization and infection of athymic and euthymic germfree mice by *Campylobacter jejuni* and *Campylobacter fetus* subsp. *fetus*. *Infect. Immun.*, **53**, 378–383.

Yrios, J. W., Balish, E. (1986b). Pathogenesis of *Campylobacter* spp. in athymic and euthymic germfree mice. *Infect. Immun.*, **53**, 384–392.

Yrios, J. W., Balish, E. (1986c). Immune response of athymic and euthymic germfree mice to *Campylobacter* spp. *Infect. Immun.*, **54**, 339–346.

Chapter 27

Suckling Mouse Model of Enterotoxigenic *Escherichia coli* Infection

M. Duchet-Suchaux

Background of human infection

Incidence

Enterotoxigenic *Escherichia coli* (ETEC) is a major cause of diarrhoea in children of developing countries and in travellers to these regions (Black, 1990, 1993). In developed countries, infrequent sporadic outbreaks of ETEC diarrhoea have also been reported (Centers for Disease Control and Prevention, 1994, cited by Fleckenstein *et al.*, 1996). ETEC provokes watery diarrhoea and sometimes cholera-like syndromes with a risk of rapid dehydration, especially in children. In travellers, the symptoms induced by ETEC are less serious, and last only a few days in most cases, but result in discomfort and inconvenience (Black, 1990).

In children less than 5 years of age, the annual incidence of ETEC infections in the world is estimated to be 650 million cases of diarrhoea causing 800 000 deaths (Gaastra and Svennerholm, 1996). The proportion of diarrhoea originating from ETEC in children varies according to country (Black, 1990; Sommerfelt *et al.*, 1996; Paniagua *et al.*, 1997). ETEC has also been isolated from faeces of healthy children, and it was estimated that only one in three infections was associated with disease (Black, 1993). It was reported that 30–50% of travellers developed at least one diarrhoeal episode due to ETEC (Levine and Levine, 1994). More precisely, ETEC has been associated with 42% and 36% of diarrhoeal cases in Latin America and Africa respectively, and with only 16% of cases in Asia (Black, 1990). Seasonal variations have also been observed in the enteropathogens, and ETEC was most frequently isolated in autumn and summer (Mattila *et al.*, 1992). The model described here is more relevant to ETEC diarrhoea in children.

Aetiology

Diarrhoea caused by ETEC starts after contamination by the faecal–oral cycle or ingestion of contaminated water and food. ETEC strains are characterized by two types of virulence determinants, colonization factors (CFs), which mediate the colonization of the small intestine, and enterotoxins, which induce net secretion of fluid and diarrhoea. Colonization of the small intestine by ETEC and enterotoxic activities do not cause any histological damage in this organ (Garcia and Le Bouguénec, 1996; Sears and Kaper, 1996). However, recent observations suggest that the cell structure is subjected to some modifications. The organisms have been recently shown to invade epithelial cells (Elsinghorst and Kopecko, 1992; Fleckenstein *et al.*, 1996). Moreover, increases in intracellular levels of cAMP induced by LT-I enterotoxin affect cyotoskeletal proteins, including microtubules and the F actin (Sears and Kaper, 1996).

The knowledge of classical virulence factors of ETEC has increased considerably in the two last decades. Their properties have been reviewed recently (Gaastra and Svennerholm, 1996; Garcia and Le Bouguénec, 1996) and will be summarized very briefly here.

Colonization factors

A total of 20 antigenic colonization factors (CFs), with most frequently fimbrial morphology, have now been described in human ETEC (Gaastra and Svennerholm, 1996). They are designated colonization factor antigen (CFA), coli surface antigens (CS) or putative colonization factor (PCF; Gaastra and Svennerholm, 1996; Garcia and Le Bouguénec, 1996). Their classification is based on heterogeneous criteria, i.e. morphology, size of the major subunit, chemical composition, antigenicity or receptor recognition (Gaastra and Svennerholm, 1996). Some authors have distinguished them according to their antigenic types, composed either of one type of fimbria (CFA/I or CFA/III) or by two types of fimbria (CFA/II or CFA/IV; Garcia and Le Bouguénec, 1996; Table 27.1). Others have grouped human CFs according to their genetic relationship and similarity of amino-acid sequence (Gaastra and Svennerholm, 1996).

By using the first classification, some features of CFA/I, CFA/II and CFA/IV, which seem to occur most frequently (Gaastra and Svennerholm, 1996) and are the best characterized (Paniagua *et al.*, 1997), are summarized as examples (Table 27.1). Descriptions of other human CFs may be found in the recent reviews cited above. CFs are associated with a few serotypes of *E. coli*, some of them with only one or two serotypes (Table 27.1). The O:K:H serotypes of ETEC strains are different in domestic animals and man.

Handbook of Animal Models of Infection
ISBN 0-12-775390-7

Table 27.1 Some characteristics of human colonization factors CFA/I, CFA/II, CFA/IV (Gaastra and Svennerholm, 1996; Garcia and Le Bouguénec, 1996)

Name	*Fimbria diameter (nm)*	*Major subunit mol. mass (kDa)*	*Enterotoxins*	*O serogroups*	*Geographical repartition**
CFA/I					Brazil, Burma, Nicaragua†, Thailand
	7	15.0	STa + LT-I STa	4, 7, 15, 20, 25, 63 78, 110, 126, 128, 136, 153, 159	
CFA/II‡					Brazil, Chile, Egypt, Indonesia, Saudi Arabia
CS1	7	16.5	STa + LT-I	6, 139	
CS2	7	15.3–15.5	STa + LT-I	6	
CS3	2–3	15.0–15.1	STa + LT-I	8, 9, 78, 80, 115, 128, 139, 168	
CFA/IV**					Argentina, Bangladesh, Central African Republic, India††, Mexico, Peru, Saudi Arabia
CS4	6–7	17.0	STa + LT-I	25	
CS5	5–6	21.0	STa	6, 29, 92, 114, 115, 167	
CS6	A‡‡	14.5/16–18.5	STa + LT-I, STa, LT-I	25, 27, 92, 148, 153, 159	

* Geographical areas were determined for CFA/I, CFA/II and CFA/IV and not for the separate components CS1–CS6.
† Reported by Paniagua, *et al.*, 1997.
‡ Consists of CS3 alone or in combination with CS1 or CS2 (Cravioto, *et al.*, 1982; Smyth, 1982).
** Consists of CS6 alone or in combination with CS4 or CS5 (McConnell *et al.*, 1986; Thomas *et al.*, 1985).
†† Also reported by Sommerfelt *et al.* (1996).
‡‡ Afimbrial structure.

Human CFs are never found in ETEC strains from domestic animals (Orskov and Orskov, 1992). Studies have shown that the distribution of human CFs of ETEC varies according to geographical area (Table 27.1).

Most CFs have a fimbrial structure and are generally encoded by plasmids. The genes involved in their biosynthesis are organized in operons flanked by insertion sequences, which suggests that they were probably located on transposable elements that did not originate in *E. coli* (Gaastra and Svennerholm, 1996). These genes are fewer than those of ETEC from domestic animals. Some of them have been completely sequenced (CFA/I, CS1, CS3 and CS6), but the identification of gene products is not yet complete. Studies of the operons involved in the regulation of CF expression are in progress and show that they are sometimes located on a different plasmid than are the structural genes (Gaastra and Svennerholm, 1996). This last feature is another difference from ETEC of domestic animals, where the CF structural and regulatory genes are organized in operons close to each other on the same genetic element.

CFs are responsible for adhesion of ETEC to intestinal cells. Although few molecules in intestinal cells to which human ETEC adheres have been identified, they are most probably glycoconjugates, the diversity of which could explain the specificity of pathogens (Gaastra and Svennerholm, 1996). The ability of CFs to agglutinate red cells of different species is directly related to their adhesive properties and has long been used for their identification.

Enterotoxins

ETEC produce two types of enterotoxin, heat-stable (ST) and heat-labile (LT), which are among the best characterized of all enterotoxins (Sears and Kaper, 1996). STa is found in ETEC from man and domestic animals, whereas STb has mainly been associated with ETEC of porcine origin. However, the presence of the STb gene has been recently detected in human, bovine and chicken ETEC strains (Dubreuil, 1997). LT-I is produced by ETEC of both human and animal origin, while LT-II is only associated with animal ETEC strains. Human ETEC strains produce STa or LT-I only, or both toxins in combination. These two toxins only will be described here briefly. Production of either or both toxins is associated with expression of particular CFs (Table 27.1).

STa (Gyles, 1992; Sears and Kaper, 1996). STa is a short 18 or 19 amino-acid peptide with a molecular mass of approximatively 2 kDa. It is resistant to heating at 100°C for 15 minutes. The structure of STa has been elucidated.

Genes for STa are located on plasmids and can be associated with genes for drug resistance in human ETEC strains. The *estA* gene, which is a part of a transposon, encodes the toxin.

Protein receptors for ST are present on the brush border membrane of intestinal epithelial cells. In humans, they are found in decreasing numbers from the small intestine to colon, and their large number in infants becomes more and more reduced with age. Guanylate cyclase type C (GC-C), localized in the apical membrane of intestinal epithelial cells, has been identified as one STa receptor. High- and low-affinity receptors described in native intestinal tissues probably reflect the varied expression of this single protein and not the existence of multiple molecules acting as receptors.

Activation of GC-C by STa induces a rapid increase in intracellular cyclic GMP levels, which leads to the stimulation of chloride ions secretion in crypt cells and/or inhibition of NaCl absorption in villus tips. Consequently, net fluid secretion occurs in the intestine and its accumulation leads to diarrhoea. Another consequence of the chloride ion diffusion from the crypts is the stimulation of the $Cl^-HCO_3^-$ exchangers, enhancing HCO_3^- secretion, which may result in loss of HCO^-_3 in stools with the occurrence of systemic acidosis. Other mechanisms not cGMP-dependent but involving molecules such as prostaglandins, serotonin, phosphatidylinositol and/or diacylglycerol are controversial for the moment.

LT-I (Gyles, 1992; Spangler, 1992; Sears and Kaper, 1996). LT-I is a periplasmic protein, closely related to cholera toxin, as shown by their respective nucleotide sequences, showing approximately 80% homology. The molecular mass of LT-I is approximately 88 kDa. It consists of one subunit A associated with five smaller, identical B subunits which are arranged in a form of a ring. The polypeptide A subunit is subsequently cleaved into two peptides, A1 and A2, which remain connected by a disulphide bond, the reduction of which leads to the enzymatically active A1 peptide. Three-dimensional structures of the toxin and its subunits have now been produced.

The structural genes for LT-I are located on plasmids, which also frequently encode CFs (especially CFA/II) and sometimes drug resistance. In human ETEC strains producing STa and LT, the respective genes reside on the same plasmid. Genes for the A and B subunits, *eltA* and *eltB* respectively, were reported as being transcribed into a single mRNA. Overlapping of the *eltA* and *eltB* open reading frames has been demonstrated by sequencing.

Binding of LT-I to the membrane of intestinal epithelial cells is mediated by the B pentamer, the receptor of which consists of ganglioside GM_1. Subunit A is inserted into the cytosol, activated and becomes associated with one or more cytosolic factors. The activated subunit A_1 is able to bind NAD and to catalyse the ADP-ribolysation of $G_{s\alpha}$, which is a GTP-binding regulatory protein associated with adenylate-cyclase. Consequently, the intracellular level of cAMP dramatically increases. Another LT-I activity results in increases in prostaglandin synthesis. Both these cell events lead to modifications of chloride ion secretion and NaCl absorption described above for STa and responsible for fluid accumulation in the gut.

Management

Drug therapy in diarrhoea has to result in reduction of the volume of stools and duration of illness and in eradication of infectious agent. However, Rabbani (1996) reported that antibiotics are not useful in treatment of ETEC diarrhoea. Moreover, they are expensive and their use can lead to the emergence of bacterial resistance. Oral replacement of fluid and electrolytes can be an effective treatment, but prevention would be preferable (Black, 1993).

The reduction in ETEC transmission should be achieved by an improvement of sanitary structures and of hygiene levels (Bouchaud *et al.*, 1995). However, difficulties encountered in developing countries in attaining such an objective require other means of prevention. According to most authors, the use of vaccines is probably the most promising way for efficient prevention in the future. Advances in the understanding of ETEC and of the pathogenesis of illness have led to progress in the development of vaccines, which, however, are not ready yet to be commercialized.

Background of the model

In 1978 a scientific report of the World Health Organization concerning control programmes of diarrhoea required, among its recommendations, the development of an experimental model of neonatal diarrhoea in a laboratory animal. At the same time, neonatal diarrhoea was also widespread in domestic animals in Europe and North America, resulting in important economical losses in bovine, ovine and porcine rearing. As with human beings, prevention was preferred to treatment and vaccination appeared the best solution to manage the disease.

Most infectious agents involved in the aetiology of the disease were ETEC. Their characterization was in progress (Table 27.2). K88 was the first CF described by its antigenicity and fimbrial morphology. Present in porcine strains, it was also reported to be transmissible by Stirm *et al.* (1967). The production of enterotoxins ST and LT had been demonstrated in *E. coli* strains pathogenic for the intestine of human (Smith and Halls, 1967a; Smith and Linggood, 1971a) and animal origin (Smith and Linggood, 1972). These authors also demonstrated the transmissible nature of the genetic determinants of these toxins. Finally, other CFs such as K99 in bovine (Orskov *et al.*, 1975) and porcine (Moon *et al.*, 1977) strains, and 987P (Isaacson *et al.*, 1977; Nagy *et al.*, 1977); their genetic determinants located mostly on plasmids; and their association with few serotypes (Orskov and Orskov, 1992) and with one or both enterotoxins; had been also described (Table 27.2). In comparison, only CFA/I and CFA/II had been found in human ETEC strains (Evans *et al.*, 1975; Evans and Evans, 1978) and CFA/II was later characterized as CS3 alone or in combination with CS1 or CS2 (Cravioto *et al.*, 1982; Smyth, 1982).

The mechanisms involved in pathogenesis of ETEC diarrhoea and protection by vaccination were not completely understood. *In-vivo* studies may be performed in animals and had enabled the demonstration of the role of

Table 27.2 Characteristics of animal ETEC strains

Date	*Animal species*	*Colonization factor*	*Enterotoxin*	*Reference*
Before 1978	Bovine	K99	STa	1
	Porcine	K88	STa + LT	2
		K99	STa	3
		987P	STa	4,5
After 1978	Bovine	F41*	STa	6,7
		CS31A	STa	8
	Porcine	F41*	STa	9
		F42	STa	10
		8813	STa,STb, LT, STa + LT	11

* Alone or in combination with K99.
References: 1. Orskov *et al.*, 1975; 2. Stirm *et al.*, 1967; 3. Moon *et al.*, 1977; 4. Isaacson *et al.*, 1977; 5. Nagy *et al.*, 1977; 6. De Graaf and Roorda, 1982; 7. Morris *et al.*, 1982; 8. Girardeau *et al.*, 1988; 9. Morris *et al.*, 1983; 10. Yano *et al.*, 1986; 11. Salajka *et al.*, 1992.

some CFs and toxins in the pathogenesis of K88- and K99-positive strains in piglets, calves and lambs respectively (Jones and Rutter, 1972; Smith and Linggood, 1971b; Smith and Huggins, 1978). The efficacy of vaccinating dams with CFs to protect suckling neonates had been first demonstrated with K88 in swine by Rutter and Jones (1973) and was also demonstrated in the same species with K99 and 987P (Morgan *et al.*, 1978; Nagy *et al.*, 1978). The protective role of K99 as vaccine for calves was demonstrated slightly later (Acres *et al.*, 1979; Nagy, 1980). However, the existence of other important factors involved in ETEC virulence and protection by vaccination was strongly suspected.

Therefore, experimental infections of animal species naturally infected with ETEC had led to some interesting results and they very probably directed research in humans, such as inoculation of volunteers to demonstrate the role of CFA/I in pathogenesis of diarrhoea (Satterwhite *et al.*, 1978). However, there were obvious difficulties in repeating such experiments and experimental infections in domestic animals were easier, but not without some disadvantages. They are expensive and may be difficult to organize and carry out. The numbers of animals available may be limited in most cases and do not allow extensive screening. To continue to explore mechanisms of pathogenesis in animal ETEC diarrhoea and according to WHO recommendations, M. Plommet, the director of my laboratory, initiated the development of an infant mouse model for these illnesses. He had previously shown the usefulness of the murine models for understanding infectious diseases of the reproductive tract in domestic animals. Experimental infections of mice with *E. coli* pathogenic for intestinal disease in several species, including humans, had already been done, without obtaining complete reproduction of ETEC diarrhoeal syndrome (Mushin and Dubos, 1965; Davidson and Hirsch, 1975; Kétyi *et al.*, 1978).

The immaturity of the bacterial flora in intestine in the newborn animal allowed Plommet to consider that colonization of the gut by foreign bacteria might be possible. Furthermore, the susceptibility of infant mouse to STa enterotoxin (Dean *et al.*, 1972) had led him to the hypothesis that illness induced by ETEC could be reproduced by strains producing it in this species.

Animal species

The mouse strain most frequently used in this model has been the Swiss OF1, but several other outbred, inbred or crossed strains have also been tested (Table 27.3). The influence of the mouse strain in the susceptibility of infant mice to ETEC was very important in this model (Duchet-Suchaux *et al.*, 1987, 1990a) and was also reported by Newsome *et al.* (1987), with very similar results to ours. However, these authors attributed the differences observed in the susceptibility patterns to environmental factors. We have further provided evidence of the importance of the host genetic background by demonstrating a genetic origin of the susceptibility or the resistance of infant mice to ETEC (Duchet-Suchaux *et al.*, 1992a).

When they were 6–8 weeks old, virgin female mice were mated with males, generally of the same strain, except for genetic studies (Duchet-Suchaux *et al.*, 1992a; Table 27.3). Whereas infant mice showed susceptibility to ETEC diarrhoea up until 10 days of age, the model gave the most reproducible results after inoculation before 2 days of age (Duchet-Suchaux, 1980). This condition, defined as standard, has been applied in all the further studies.

Preparation of animals

Pregnant mice were placed in individual cages on day 15 of gestation. In order to randomize the dam effect, infant mice of the same strain were pooled together on the day after delivery and randomly redistributed to each dam. The litter size was reduced to eight animals per dam, which thus could feed each of them sufficiently and equally. It was necessary to wear gloves for handling, otherwise the dam

Table 27.3 Strains of mouse

Type	*Name*	*Reference*
Outbred	Swiss OF1	1–3, 5–10, 12–14
	Swiss CD-1	4, 5
Inbred	BALB/cBy	9, 10
	C3HB10	11
	C57BL/6	9, 10
	C57BL/10	11
	CBA	9, 10
	DBA/2	9, 10
Congenic	C57BL/10.D2	11
	C57BL/10.BR	11
Congenic	C57BL/10.G	11
	C57BL/10.P	11
	C3H.Q	11
	C3H.NB	11
Crosses F1	CBA × DBA/2	12
	DBA/2 × CBA	12
Backcrosses	F1* × CBA	12
	CBA × F1*	12
	F1* × DBA/2	12
	DBA/2 × F1*	12

* F1 = DBA/2 × CBA.
References: 1. Bertin, 1981; 2. Bertin, 1983; 3. Bertin, 1985; 4. De Rycke *et al.*, 1982; 5. Duchet-Suchaux, 1980; 6. Duchet-Suchaux, 1983; 7. Duchet-Suchaux, 1984; 8. Duchet-Suchaux, 1988; 9. Duchet-Suchaux, *et al.*, 1987; 10. Duchet-Suchaux, *et al.*, 1990a; 11. Duchet-Suchaux, *et al.*, 1990b; 12. Duchet-Suchaux *et al.*, 1992a; 13. Duchet-Suchaux *et al.*, 1992b; 14. Fourniat *et al.*, 1986.

was unable to recognize her offspring and ate them. Each experimental group consisted of at least three litters.

Storage and preparation of inocula

Several ETEC strains and derived ones (Tables 27.4 and 27.5), as well as transconjugant strains from mating ETEC strains with various *E. coli* recipients (Bertin, 1983, 1985), were used in the model. Non-enterotoxigenic *E. coli* strains, pathogenic (De Rycke *et al.*, 1982) or not (Table 27.6), were also tested. However, the ETEC reference strain B41 has been most frequently used (Table 27.4).

E. coli strains were kept in conservation medium (Institut Pasteur Production, France). The challenge strains were generally grown 18–24 hours at 37°C in Trypticase soy agar (TSA) or in Minca medium, which improves K99 and F41 expression in ETEC strains bearing either or both (Guinée *et al.*, 1977; De Graaf and Roorda, 1982). Use of either medium led to similar virulence of strain B41 for infant mice (unpublished). However, vaccine strains expressing K99 and/or F41 were cultured preferentially on Minca medium (Duchet-Suchaux, 1983; Duchet-Suchaux, 1988). The cultures were suspended into phosphate buffered saline (PBS), pH 6.8. Bacterial suspensions were adjusted to the appropriate concentration in PBS, after estimation of bacterial concentration by optical density at 600 nm. Inoculation of 10 colony-forming units

Table 27.4 ETEC strains (Ap, ampicillin; C, chloramphenicol; K, kanamycin; N, neomycin; P, penicillin; Sm, streptomycin; Su, sulfadiazine; Tc, tetracycline; ND, not done)

		Phenotype				
Origin	*Name*	*Serogroup*	*Colonization factor*	*Enterotoxin*	*Resistance to antibiotics*	*Reference*
Bovine	B41	O101:K99,F41	K99, F41	STa	Sm, Su, Tc	1–12, 14
	B41*†	O101:K99,F41	K99, F41	STa	Sm, Su, Tc	3
	B41M	O101:F41	F41	STa	ND	13
	H498	O101:K30, F41	F41	STa	ND	8
	B44	O9:K30,K99,F41	K99, F41	STa	C, K, N, P, Sm, Su, Tc	5, 8
	17C	ND	K99	STa	ND	5
	B80	O20:K17(?),K99	K99	STa	Sm, Su, Tc	8, 10
	B117	O8:K85,K99	K99	STa	Ap, K, P, Sm, Su, Tc	8
	559	O8:K25,K99	K99	STa	P, Sm, Su, Tc	8
	H926	O8:K25,K99	K99	STa	ND	Unpublished
Porcine	P2200	O149:K91,K88ac	K88	STa + LT	Sm, Su, Tc	2, 10
	P5148	O149:K91,K88ac	K88	STa + LT	Sm, Su, Tc	2
	987	O9:K103,987P	987P	STa	Sm, Su	10
	1676	O101:K30,F41	F41	STa	Sm, Su, Tc	10
	B2C	O9:K35, F41	F41	STa	ND	13
Human	H10407	O78:H11	CFA/I	STa + LT-I	Su	10
	E4833/76	O6:H16	CS2, CS3	STa + LT-I	None	10

References: 1. Bertin, 1981; 2. Bertin, 1983; 3. Bertin, 1985; 4. De Rycke *et al.*, 1982; 5. Duchet-Suchaux, 1980; 6. Duchet-Suchaux, 1983; 7. Duchet-Suchaux, 1984; 8. Duchet-Suchaux, 1988; 9. Duchet-Suchaux *et al.*, 1987; 10. Duchet-Suchaux *et al.*, 1990a; 11. Duchet-Suchaux *et al.*, 1990b; 12. Duchet-Suchaux *et al.*, 1992a; 13. Duchet-Suchaux *et al.*, 1992b; 14. Fourniat *et al.*, 1986.
† Supposed to be the same strain as B41, but from different source and possessing a single large plasmid that encoded K99, STa and drug resistance. In ETEC strain B41, virulence factors and drug resistance were coded by two different plasmids.

Table 27.5 *E. coli* strains derived from ETEC strains (Ap, ampicillin; K, kanamycin; Nal, nalidixic acid; P, penicillin; Sm, streptomycin; Su, sulfadiazine; Tc, tetracycline)

Strain		*Phenotype*			
Parental	*Derived*	*Colonization factor*	*Enterotoxin*	*Resistance to antibiotics*	*Reference*
B41†	B41A	F41	STa	Sm, Su, Tc	1, 2, 3
	B41B1	None	STa	Sm, Su, Tc	2, 3
	B41B2	None	STa	Sm, Su, Tc	2
B41*†	B41*C	F41	STa	Nal	2
B41*C	B41*CD	K99,F41	STa	Nal, Sm, Su, Tc	2
	B41*CB	None	None	Nal	2
B41*CB	B41*DB	K99	STa	Nal, Sm, Su, Tc	2
B44†	B44A	F41	STa	Sm, Su, Tc	3
B80†	B80A	None	STa	Sm, Su, Tc	3
B117†	B117A	None	STa	Ap, K, P, Sm, Su, Tc	3
559†	559A	None	STa	P, Sm, Su, Tc	3
P2200†	P2200 Raf–	None	STa	Sm, Su, Tc	1
P5148†	P5148 Raf–A	None	STa	Sm, Su, Tc	1
	P5148 Raf–B	None	None	Sm, Su	1

† ETEC strains described in Table 27.2.
References: 1. Bertin, 1983; 2. Bertin, 1985; 3. Duchet-Suchaux, 1988.

Table 27.6 Non-pathogenic *E. coli* strains (ND, not done)

		Name			
Use	*Origin*	*Parental*	*Mutant resistant to nalidixic acid*	*Serogroup*	*Reference*
Recipient	Faeces of healthy human	K12	K12 Nal^r	O–:K–	1
		C600 PK1046	ND	O–:K–	1,2
		K14a	K14 Nal^r	O9:K28	1,2
		H510a	H510 Nal^r	O101:K?:H33	1,2
	Human bile	G3404-41	3404 Nal^r	O8:K8:H4	1,2
	Faeces of healthy infant mice	C13	C13 Nal^r	ND	1
Vaccine	Faeces of healthy human	H510a	H510 Nal^r	O101:K?:H33	4
Control	Faeces of healthy infant mice	C11	ND	ND	3
	Faeces of healthy calf	C12	ND	ND	3

References: 1. Bertin, 1983; 2. Bertin, 1985; 3. Duchet-Suchaux, 1980; 4. Duchet-Suchaux, 1988.

(cfu) of the ETEC strain B41 were sufficient to cause diarrhoea and death in OF1 and CD-1 infant mice. However, the evolution of the disease and the reproducibility of the results were optimal after inoculation of 10 cfu of the same ETEC strain (Duchet-Suchaux, 1980). This latter dose was used as standard in most studies, but an inoculum size of 10^5 cfu (Bertin, 1983; 1985; Fourniat *et al.*, 1986; Duchet-Suchaux *et al.*, 1990a, 1992a) or 10^8 cfu (De Rycke *et al.*, 1982) were tested when *E. coli* strains were less or avirulent.

Infection process

Inoculation was done orally with a calibrated platinum loop for most studies, since newborn mice have to suckle in order to swallow. Inoculation with a tube or a pipette was rarely done (De Rycke *et al.*, 1982; Fourniat *et al.*, 1986), because it was very difficult to perform in infant mice less than 3 days of age.

Key parameters to monitor infection

Death

Death was the essential criterion used for monitoring infection. Indeed, susceptible infant mice developed severe diarrhoea after inoculation of ETEC strains virulent for this species. Illness led to dehydration and death in every affected animal (Duchet-Suchaux, 1980). In OF1 infant mice inoculated with ETEC strain B41 under standard conditions, death occurred early after the first symptoms, which appeared progressively. Maximum mortality rates were mostly attained on day 2 after challenge in these conditions. As strain B41 was one of the most virulent ETEC strains in the most susceptible infant mice (Duchet-Suchaux *et al.*, 1990a), animals were generally observed for 6 days after inoculation to estimate the virulence of ETEC strains, the susceptibility of mouse strains to ETEC (Bertin, 1983, 1985; Duchet-Suchaux *et al.*, 1990a, 1992a) or the activity of pharmacological treatments (Bertin, 1981). Fourniat *et al.* (1986) estimated the curative effect of a competitive bacterial treatment for 4 days postinoculation. Mortality rates were noted during 15 days postinoculation in the first vaccination experiments (Duchet-Suchaux, 1983), but this duration was reduced to 5 days postinoculation in later ones (Duchet-Suchaux, 1988), because results became too variable later (Duchet-Suchaux, 1983). Some observations were of a longer duration, e.g. during 12 days postinoculation to estimate the protection against ETEC strains less virulent than strain B41 in OFI infant mice (Duchet-Suchaux *et al.*, 1992b) or during 15 days to test the virulence of non-enterotoxigenic *E. coli* strains isolated from diarrhoeic calves (De Rycke *et al.*, 1982). Observation of death was not ambiguous, except in mouse inbred strains, where manipulated 1-day-old infant mice often died without any relation to diarrhoea. In this case, only data from litters that did not present dead animals within 24 hours after inoculation were retained (Duchet-Suchaux *et al.*, 1990a).

Bacterial enumeration in the intestine

Enumeration of the inoculated *E. coli* strain in the intestine of infant mice was not performed in every case. Indeed, not only bacterial counting was required, but serological identification of colonies also, since *E. coli* is commensal in the intestine. Moreover, the intestine could not be frozen, because the numbers of ETEC cfu decreased considerably. Thus, the numbers of samples were limited, since they had to be treated immediately after necropsy. This quantification was done to assess intestinal colonization in animals that developed diarrhoea (Duchet-Suchaux, 1984; Duchet-Suchaux *et al.*, 1990a) or did not do so (Bertin, 1983; Duchet-Suchaux *et al.*, 1990a) after a simple inoculation, or in infant mice that were protected against diarrhoea by passive or active immunization of their dams (Duchet-Suchaux, 1984; Duchet-Suchaux *et al.*, 1992b).

Pitfalls of the model

Advantages

The costs of experiments were obviously considerably reduced when infant mice were used rather than calves and piglets. Numbers of offspring per dam are also an advantage in this species, which is multiparous like pigs in contrast to cows or sheep. Mice are of course easier to manipulate than farm animals and the experiments in this species were much less difficult to perform. Sufficient numbers of infant mice could be easily obtained on the same day to perform synchronous inoculations. Inoculation of 25 litters could be done in 1 hour, if the inoculum was not modified during this time. The model is easy to use, because the mortality rate is the major criterion to monitor infection. This overall measure is nevertheless precise, because the mortality rates could be compared every day postinoculation and even several times a day. Differences between experimental groups could be observed on the first day postinoculation and not later (Duchet-Suchaux, 1988).

ETEC diarrhoea has been accurately reproduced by oral inoculation of relatively low doses of bacterial cfu of some strains pathogenic for infant mice. Intense multiplication of the challenge strain occurred in the intestine of diarrhoeic animals (Duchet-Suchaux, 1984; Duchet-Suchaux *et al.*, 1990a (Table 27.7), which is consistent with previous observations in experimental infections of calves (Smith and Halls, 1967b), lambs (Smith and Halls, 1967b; Morris *et al.*, 1980) and piglets (Smith and Halls, 1967b; Rutter and Anderson, 1972; Isaacson *et al.*, 1977; Moon *et al.*, 1977, 1983; Morgan *et al.*, 1978; Nagy *et al.*, 1976, 1977, 1978). However, avirulent *E. coli* strains colonized infant mouse intestine for at least 5 or 6 days postinoculation (Bertin, 1983; Duchet-Suchaux *et al.*, 1990a; Table 27.7). The reproducibility of results was quite good for estimation of the virulence of ETEC strains (Duchet-Suchaux, 1980). Illness induced by ETEC in infant mice was also determined by the production of STa and of some CFs, essentially K99 and F41, by these strains (Bertin, 1983; 1985; unpublished). As in domestic animals, newborns were protected against ETEC challenge by suckling dams vaccinated with homologous or hererologous strains presenting common antigens with the challenge one (Duchet-Suchaux, 1983, 1988). In murine species, maternal antibodies are transmitted to fetus and newborn via both placenta and milk, which in domestic animals occurs only via colostrum and milk. However, cross-fostering experiments showed that the protective effect of dam vaccination with ETEC was transmitted mostly via colostrum and milk in mice (Duchet-Suchaux, 1983). As in domestic animals, the protection of the newborn was due to the impaired intestinal colonization of the challenge strain in infant mice suckling vaccinated mothers but not to its elimination (Duchet-Suchaux, 1984; Table 27.7). Lastly, innate resistance to ETEC similar to that observed in piglets (Rutter *et al.*, 1975, Duchet-Suchaux *et al.*, 1991) has been observed in infant mice. Nevertheless, that resistance was not due to the immediate elimination of the challenge strain from the intestine (Table 27.7). All these observations enable the model to be considered as relevant to natural ETEC infections in the young. Genetic and immunological analyses may be performed in murine species especially, which gives unequalled possibilities for such studies.

Table 27.7 Intestinal colonization of infant mice with inoculated ETEC strains; unless otherwise stated, infant mice were of OF1 strain and inoculated with 10^3 bacterial cfu of ETEC strain B41

		Time after inoculation (day)		
Evolution	*Status of infant mice*	*1–2*	*5–6*	*Reference*
Increase*	Diarrhoeic	9.19 ± 0.47 (10/10)†	...	1
	Non diarrhoeic	7.80 ± 0.88 (8/8)	...	1
Decrease*	Vaccinated dams	7.13 ± 0.72 (15/15)	...	1
	Control	8.51 ± 0.97 (18/18)	...	1
Persistence	Inoculated with avirulent strain	...	5.50 ± 0.80 (8/8)‡	2
		...	6.74 ± 0.39 (5/5)**	3
	Resistant to ETEC strain	...	6.70 ± 0.99 (4/5)††	3
	Protected by MAb F41	...	7.90 ± 0.62 (4/5)‡‡	4

* Statistically significant ($p < 0.001$).
† Mean ± standard deviation of the numbers ($\log_{10}$) of lactose-fermenting colonies serologically identical to the challenge strain per gram of intestine (number of animals that harboured the challenge strain in the lactose-fermenting colonies tested/number of animals examined at necropsy).
‡ OF1 infant mice inoculated with 10^5 bacterial cfu of transconjugant strain B41 × H510Nal^r.
** BALB/c infant mice inoculated with ETEC strain 987.
†† DBA/2 infant mice inoculated with ETEC strain B80.
‡‡ OF1 infant mice inoculated with 10^3 bacterial cfu of ETEC strain B2C and protected by monoclonal antibody (MAb) F41-20 against epitope cluster 5.
References: 1. Duchet-Suchaux, 1984; 2. Bertin, 1983; 3. Duchet-Suchaux *et al.*, 1990a; Duchet-Suchaux, *et al.*, 1992b.

Disadvantages

Infant mice did not escape the restrictions of host specificity of ETEC strains described above in other species. They were generally susceptible to bovine, ovine and porcine ETEC strains bearing CFs K99 and/or F41 and secreting STa enterotoxin. In contrast, they were much more resistant to porcine ones expressing CFs K88 or 987P or human ones producing CFA/I, or CFA/II (CS2, CS3). However, all these strains were secreting STa enterotoxin (Table 27.4; Duchet-Suchaux *et al.*, 1990a). Newsome *et al.* (1987) had observed similar results for ETEC strains bearing K99 and/or F41, K88 and 987P, in their model similar to that presented here, but using a different mouse strain. These data were consistent with pathogenicity of ETEC expressing K99 and/or F41, which affect several animal species, whereas those expressing K88, 987P, CFA/I or CFA/II are very specific for one given species. Consequently, the possibilities of analyses in the model are limited. However, all existing mouse strains and ETEC strains, especially human ones, expressing other CFs than CFA/I and CFA/II have not yet been tested in our model. Moreover, each inoculated ETEC strain, except strain positive for 987P, induced weak mortality in at least one of the mouse strains tested (Bertin, 1983; Duchet-Suchaux *et al.*, 1990a) (Table 27.8). This result suggests that other mouse strains more susceptible to ETEC expressing K88, 987P, CFA/I and CFA/II might exist. For instance, Davidson and Hirsch (1975) described symptoms of diarrhoea after inoculation of one mouse strain (CF1), which we did not use, with an ETEC strain bearing CF K88.

Results were variable in experiments on vaccinal protection, especially from 5 days postinoculation and that is why we reduced the observation time to this duration (Duchet-Suchaux, 1983). If global protection by vaccination could be assessed in the model, immune effectors involved, like

Table 27.8 Susceptibility of five mouse strains to different ETEC (Duchet-Suchaux *et al.*, 1990a)

	% Mortality of infant mice inoculated with ETEC strain (colonization factor)*						
Mouse strain	*B41 (K99, F41) 10^3*	*B80 (K99) 10^3*	*1676 (F41) 10^3*	*P2200 (K88) 10^5*	*987 (987P) 10^5*	*H10407 (CFA/I) 10^5*	*E4833/76 (CS2, CS3) 10^5*
DBA/2	23	5	23	38	4	11	46
BALB/cBy	72	8	89	32	9	16	26
C57BL/6	80	0	96	0	9	21	11
CBA	80	92	62	23	10	42	3
OF1	94	62	100	12	0	0	0

* 6 days after inoculation of more than 20 animals, except in five cases (10–19).

classes of antibodies, are very probably different from one species to another. Innate resistance of the newborn to ETEC strains could be related to specific receptors in the intestinal cell and/or to bacterial competition with the indigenous intestinal flora. If specific receptors for a given CF are probably the same in every species, the intestinal flora is different according to the species (Ducluzeau and Raibaud, 1979; Savage, 1977). This latter is immature in newborns, but its implantation occurs soon after birth and could introduce factors of variability according to the animal species.

Contribution of the model to infectious disease therapy

Development of novel therapeutic approaches

Few tests have been done using the model for novel therapeutic approaches. However, treatment of ETEC-B41-inoculated infant mice with chlorpromazine or propranolol administrated orally or subcutaneously resulted in delayed mortality in every case (Bertin, 1981). The activity of chlorpromazine on the effects of enterotoxins had already been described, but data from experimental clinical diarrhoea induced by inoculation with an ETEC strain had been reported in only one case (Lönnroth *et al.*, 1979). A recent review, which reported successful treatment of human cholera with chlorpromazine associated with fluids for rehydration, did not mention additional data about treatment of ETEC diarrhoea with this drug (Rabbani, 1996).

Pathogenesis

The model was helpful in giving prominence to factors involved in the virulence of ETEC strains and different from those well-known until then, especially CF F41 (Table 27.2; Bertin, 1983; 1985; Table 27.9). Other data suggested that some other unknown virulence factors could be involved in some ETEC strains of porcine origin bearing K88 (Bertin, 1983) or of bovine origin expressing K99 alone (unpublished). However this work had to stop because of changed priorities in the laboratory. Innate genetic resistance to infectious diseases had been firstly described for diarrhoea induced in piglets by ETEC strains bearing K88 (Rutter *et al.*, 1975). The existence, in the host, of an innate resistance to ETEC strains bearing K99 and/or F41 (Duchet-Suchaux *et al.*, 1990a; Table 27.8) and the genetic origin of that related to K99-positive strains (Duchet-Suchaux *et al.*, 1992a) were demonstrated in the model. Data obtained in mice congenic for the *H2* locus suggested that this resistance was not related to the major histocompatibility complex (Duchet-Suchaux *et al.*, 1990b; Table 27.10).

Further studies showed differences in glycosyltransferase activities in the intestinal mucosa of infant mice,

Table 27.9 Role of K99 and F41 in the virulence of *E. coli* strains for infant mice (Bertin, 1983, 1985)

Colonization factor	E. coli *strain*†		*% mortality of infant mice*‡	*Virulence*
	Nature	*Name*		
K99 and F41	ETEC parental	B41	100	High
		B41*	100	High
		B41*CD clone 1	100	High
		B41*CD clone 2	100	High
F41 alone	ETEC-derived	B41A	100	High
K99 alone	ETEC derived	B41*DB clone 1	0	None
		B41*DB clone 2	0	None
	Transconjugant	B41 × C600 PK1046 (2)	0	None
		B41 × 3404 Nal^r(13)	4	None
		B41 × H510 Nal^r(13)	4	None
		B41 × C13 Nal^r(4)	8	Low
		B41 × K14 Nal^r(23)	21	Partial
None	ETEC derived	B41B1	0††	None
		B41B2	40††	Partial
	non ETEC recipient	C600 PK1046	4	None
		3404 Nal^r	0	
		H510 Nal^r	4	None
		C13 Nal^r	0	None
		K14 Nal^{r}PK1046	0	None

† All the strains secreted STa enterotoxin except non-ETEC recipients.
‡ 5 days after oral inoculation with 10^5 bacterial cfu. Each experimental group consisted of at least 24 animals, except for *E. coli* strain H510 Nal^r, which was inoculated to 16 animals.
†† Results pooled from two experiments.

Table 27.10 Susceptibility of mouse strains congenic for *H-2 locus* to ETEC strain B80 (Duchet-Suchaux *et al.*, 1990b)

	Genetic background			
	*B10**		*C3H*	
H-2 locus	*Name*	*Mortality†*	*Name*	*Mortality†*
b	B10	3/62 (6)	C3H	36/45 (80)
d	B10.D2	1/42 (2)	. . .	. . .
k	B10.BR	0/57 (0)	. . .	. . .
p	B10.P	1/27 (4)	C3H.NB	55/77 (71)
q	B10.G	2/51 (4)	C3H.Q	37/59 (63)

* = C57BL/10.
† Dead/challenged (%) 6 days after inoculation with 10^3 bacterial cfu.

susceptible and resistant to K99 ETEC infection (Grange and Mouricout, 1996). DBA/2 mouse strain, which was the most resistant to ETEC, showed reduced susceptibility to STa enterotoxin in comparison with more susceptible mouse strains (Bertin, 1992).

Immune protection

The model was used to evaluate the vaccinal value of different antigens present in ETEC strains. The importance of the presence of CF F41 in a vaccine strain for protection against challenge with an ETEC strain bearing both K99 and F41 was first demonstrated in infant mice (Duchet-Suchaux, 1988). In the same system, K99 and O antigens were shown to be involved to a lesser extent in protection (Table 27.11). The efficacy of F41 antigen as protective in vaccines was less well defined in cattle and swine (Moon and Bunn, 1993).

The model enabled differences to be shown in the protective effects of vaccinal ETEC strains of mostly widespread serotypes, expressing same antigens and used as vaccines in the field (Duchet-Suchaux, 1988; Table 27.11).

Protective epitopes of F41 antigen could be analysed in the model. Five epitope clusters of F41 antigen had been previously described (van Zijderveld *et al.*, 1989) and they induced equal protection, without any improvement of the protection after combination of them in groups of two or altogether. Specificity of the protection had been demonstrated by the inefficacy of a K99 monoclonal antibody injected into the dams in the same system. The protection had been clinical but not bacteriological, since the challenge

Table 27.11 Protection of OF1 infant mice challenged with ETEC strain B41 and suckling dams vaccinated with different *E. coli* strains (Duchet-Suchaux, 1988; pc, postchallenge with 10^3 bacterial cfu)

Vaccine strain						
Antigens			*% Survival* at*			
Colonization factor	*O*	*Name*	*1 day pc*	*2 days pc*	*5 days pc*	*Protection*
None	None		54†	12†	3†	None
K99, F41	101	B41	100	100	83	High
	9	B44	98	83	46	Partial
F41	101	B41A	100	100	85	High
	101	H498	100	96	54	High
	9	B44A	96	73	31	Partial
K99	8	B117	98	56	31	Partial
		559	100	34	19	Partial
	20	B80	98	52	25	Partial
None	101	B41B1	100	77	17	Partial
		H510a	98	54	19	Partial
	8	B117A	31	2	0	None
		559A	25	0	0	None
	20	B80A	47	19	12	None

* Each experimental group mostly consisted of six litters of eight animals, except those vaccinated with strains B80A and 559A, which were composed of four litters of eight animals.
† Mean for three experiments.

Table 27.12 Immune protection of OF1 infant mice induced by epitopes of K99 and/or F41 antigens (Duchet-Suchaux *et al.*, 1992b; unpublished data; MAb, monoclonal antibody)

	% survival* after challenge with ETEC strains bearing					
	K99 alone		F41 alone		K99 and F41	
Treatment of dams with	B80	H926	B2C	B41M	B44	B41
Saline	75	25	13	25	4	4
MAb anti-K99	100	50	4	33	21	12
F41†	75	33	100	96	87	29
K99 and F41†	ND	ND	ND	ND	100	75

* 5 days after the challenge of 24 animals.
† Mab F41-20 against epitope cluster 5.

strain had not been eliminated from the intestine (Duchet-Suchaux *et al.*, 1992b; Table 27.7). This latter observation had already been made in infant mice clinically protected against diarrhoea by vaccination of their dams (Duchet-Suchaux, 1984 and unpublished). Analysis of epitopes of K99 and F41 antigens responsible for protection was thereafter done in the same way against ETEC strains positive for K99 and/or F41. K99 epitopes were responsible for weak protection against ETEC strains bearing K99, associated with F41 or not. Considerable improvement of the protection induced by epitopes of a single antigen was observed after association of epitopes of both K99 and F41 antigens (Table 27.12).

Acknowledgements

Dr Michel Plommet should be thanked as the initiator of the model who has followed its development with interest and helpful advice. He is now retired and did not wish to participate in the production of this review. I would like to take this oppurtunity of wishing him a well-earned rest. I am grateful to Christiane Le Louedec for kindly updating the references and constructing the tables. I would like to thank Olivier Galland for his patience in translating encoded files and Dr Paul Barrow for carefully correcting the manuscript.

References

Acres, S. D., Isaacson, R. E., Babiuk, L. A., Kapitany, R. A. (1979). Immunization of calves against enterotoxigenic colibacillosis by vaccinating dams with purified K99 antigen and whole cell bacterins. *Infect. Immun.*, **25**, 121–126.

Bertin, A. (1981). Chlorpromazine and propranolol extend survival of infant mice inoculated with enterotoxigenic *Escherichia coli*. *Ann. Rech. Vét*, **12**, 137–141.

Bertin, A. (1983). Virulence factors of enterotoxigenic *E. coli* studied in the infant mouse model. *Ann. Rech. Vét*, **14**, 169–182.

Bertin, A. (1985). F41 antigen as a virulence factor in the infant mouse model of *Escherichia coli* diarrhoea. *J. Gen. Microbiol.*, **131**, 3037–3045.

Bertin, A. (1992). Comparison of susceptibility of inbred and outbred infant mice to *Escherichia coli* heat-stable enterotoxin STa. *Infect. Immun.*, **60**, 3117–3121.

Black, R. E. (1990). Epidemiology of travelers' diarrhea and relative importance of various pathogens. *Rev. Infect. Dis.*, **12** (suppl. 1), S73–S79.

Black, R. E. (1993). Epidemiology of diarrhoeal disease: implications for control by vaccines. *Vaccine*, **11**, 100–106.

Bouchaud, O., Cabié, A., Coulaud, J. P. (1995). Epidémiologie, physiopathologie et traitement de la diarrhée du voyageur. *Ann. Med. Interne*, **146**, 431–437.

Cravioto, A., Scotland, S. M., Rowe, B. (1982). Hemagglutination activity and colonization factor antigens I and II in enterotoxigenic and non-enterotoxigenic strains of *Escherichia coli* isolated from humans. *Infect. Immun.*, **36**, 189–197.

Davidson, J. N., Hirsch, D. C. (1975). Use of the K88 antigen for *in vivo* bacterial competition with porcine strains of enteropathogenic *Escherichia coli*. *Infect. Immun.*, **12**, 134–136.

De Graaf, F. K., Roorda, I. (1982). Production, purification and characterization of the fimbrial adhesive antigen F41 isolated from calf enteropathogenic strain B41M. *Infect. Immun.*, **36**, 751–758.

De Rycke, J., Boivin, R., Le Roux, P. (1982). Mise en évidence es souches de *Escherichia coli* à caractère septicémique dans les fèces de veaux atteints d'entérite mucoïde. *Ann. Rech. Vét.*, **13**, 385–397.

Dean, A. G., Ching, Y. C., Williams, R. G., Harden, L. B. (1972). Test for *Escherichia coli* enterotoxin using infant mice: application in a study of diarrhea in children in Honolulu. *J. Infect. Dis.*, **125**, 407–411.

Dubreuil, J. D. (1997). *Escherichia coli* STb enterotoxin. *Microbiology*, **143**, 1783–1795.

Duchet-Suchaux, M. (1980). Le souriceau, modèle expérimental de la diarrhée colibacillaire. *Ann. Microbiol. (Inst. Pasteur)*, **131B**, 239–250.

Duchet-Suchaux, M. (1983). Infant mouse model of *E. coli* diarrhoea: clinical protection induced by vaccination of the mothers. *Ann. Rech. Vét.*, **14**, 319–331.

Duchet-Suchaux, M. (1984). Modèle d'étude de la diarrhée colibacillaire chez le souriceau nouveau-né. In *La Diarrhée du jeune* (eds Desjeux, J.-F., Ducluzeau, R.) pp. 443–454. INSERM, Paris.

Duchet-Suchaux, M. (1988). Protective antigens against enterotoxigenic *Escherichia coli* O101:K99, F41 in the infant mouse model. *Infect. Immun.*, **56**, 1364–1370.

Duchet-Suchaux, M., Le Maitre, C., Bertin, A. (1987). Sensibilité de plusieurs lignées consanguines à différentes souches de colibacilles entérotoxinogènes (ETEC). *Ann. Rech. Vét.*, **18**, 336–337.

Duchet-Suchaux, M., Le Maitre, C., Bertin, A. (1990a). Differences in susceptibility of inbred and outbred infant mice to enterotoxigenic *Escherichia coli* of bovine, porcine and human origin. *J. Med. Microbiol.*, **31**, 185–190.

Duchet-Suchaux, M., Menanteau, P., Pla, M., Le Roux, H., Léchopier, P. (1990b). Contrôle génétique de la résistance à la colibacillose néonatale dans le modèle souriceau. *Ann. Rech. Vét.*, **21**, 294–295.

Duchet-Suchaux, M. F., Bertin, A. M., Menanteau, P. S. (1991). Susceptibility of Chinese Meishan and European Large White pigs to enterotoxigenic *Escherichia coli* strains bearing colonization factor K88, 987P, K99, or F41. *Am. J. Vet. Res.*, **52**, 40–44.

Duchet-Suchaux, M., Menanteau, P., Le Roux, H., Elsen, J. M., Léchopier, P. (1992a). Genetic control of resistance to enterotoxigenic *Escherichia coli* in infant mice. *Microb. Pathog.*, **13**, 157–160.

Duchet-Suchaux, M., Menanteau P., van Zijderveld, F. (1992b). Passive protection of suckling infant mice against F41-positive enterotoxigenic *Escherichia coli* strains by intravenous inoculation of the dams with monoclonal antibodies against F41. *Infect. Immun.*, **60**, 2828–2834.

Ducluzeau, R., Raibaud, P. (1979). *Ecologie microbienne du tube digestif*, p. 94. Masson, Paris.

Elsinghorst, E. A., Kopecko, D. J. (1992). Molecular cloning of epithelial cell invasion determinants from enterotoxigenic *Escherichia coli*. *Infect. Immun.*, **60**, 2409–2417.

Evans, D. G., Evans, D. J. (1978). New surface-associated heat-labile colonization factor antigen (CFA/II) produced by enterotoxigenic *Escherichia coli* of serogroups O6 and O8. *Infect. Immun.*, **21**, 638–647.

Evans, D. G., Silver, R. P., Evans, D. J., Chase D. G., Gorbach, S. L. (1975). Plasmid-controlled colonization factor associated with virulence in *Escherichia coli* enterotoxigenic for humans. *Infect. Immun.*, **12**, 656–667.

Fleckenstein, J. M., Kopecko, D. J., Warren R. L., Elsinghorst, E. A. (1996). Molecular characterization of the *tia* invasion locus from enterotoxigenic *Escherichia coli*. *Infect. Immun.*, **64**, 2256–2265.

Fourniat, J., Djaballi, Z., Maccario, J., Bourlioux, P., German, A. (1986). Influence de l'administration de *Lactobacillus acidophilus* tués sur la survie de souriceaux infectés avee une souche de *Escherichia coli* enterotoxinogène. *Ann. Rech. Vét.*, **17**, 401–407.

Gaastra, W., Svennerholm, A. M. (1996). Colonization factors of human enterotoxigenic *Escherichia coli* (ETEC). *Trends Microbiol.*, **11**, 444–452.

Garcia, M.-I., Le Bouguénec, C. (1996). Role of adhesion in pathogenicity of human uropathogenic and diarrhoeogenic *Escherichia coli*. *Bull. Inst. Pasteur*, **94**, 201–236.

Girardeau, J. P., Der Vartanian, M., Ollier, J. L., Contrepois M. (1988). CS31A, a new K88-related fimbrial antigen on bovine enterotoxigenic and septicemic *Escherichia coli* strains. *Infect. Immun.*, **56**, 2180–2188.

Grange, P. A., Mouricout, M. (1996). Susceptibility of infant mice to F5 (K99) *E. coli* infection: differences in glycosyl transferase activities in intestinal mucosa of inbred CBA and DBA/2 strains. *Glycoconjugate J.*, **13**, 45–52.

Guinée, P. A. M., Veldkamp, J., Jansen, W. H. (1977). Improved Minca medium for the detection of K99 antigen in calf enterotoxigenic strains of *Escherichia coli*. *Infect. Immun.*, **15**, 676–678.

Gyles, C. L. (1992). *Escherichia coli* cytotoxins and enterotoxins. *Can. J. Microbiol.*, **38**, 734–746.

Isaacson, R. E., Nagy, B., Moon, H. W. (1977). Colonization of porcine small intestine by *Escherichia coli*: colonization and adhesion factors of pig enteropathogens that lack K88. *J. Infect. Dis.*, **135**, 531–539.

Jones, G. W., Rutter, J. M. (1972). Role of the K88 antigen in the pathogenesis of neonatal diarrhea caused by *Escherichia coli* in piglets. *Infect. Immun.*, **6**, 918–927.

Kétyi, I., Kuch, B., Vertényi, A. (1978). Stability of *Escherichia coli* adhesive factors K88 and K99 in mice. *Acta Microbiol. Acad. Sci. Hung.*, **25**, 77–86.

Levine M. M., Levine, O. S. (1994). Changes in human ecology and behavior in relation to the emergence of diarrheal diseases, including cholera. *Proc. Natl Acad. Sci. USA.*, **91**, 2390–2394.

Lönnroth, I., Andren, B., Lange, S., Martinsson K., Holmgren, J. (1979). Chlorpromazine reverses diarrhea in piglets caused by enterotoxigenic *Escherichia coli*. *Infect. Immun.*, **24**, 900–905.

Mattila, L., Siitonen, A., Kyrönseppä, H. *et al.* (1992). Seasonal variation in etiology of travelers' diarrhea. *J. Infect. Dis.*, **165**, 385–388.

McConnell, M. M., Thomas, L. V., Scotland, S. M., Rowe, B. (1986). The possession of coli surface antigen CS6 by enterotoxigenic *Escherichia coli* of serogroups O25, O27, O148, and O159: a possible colonization factor? *Curr. Microbiol.*, **14**, 51–54.

Moon, H. W., Bunn, T. O. (1993). Vaccines for preventing enterotoxigenic *Escherichia coli* infections in farm animals. *Vaccine*, **11**, 213–220.

Moon, H. W., Nagy, B., Isaacson, R. E., Orskov, I. (1977). Occurrence of K99 antigen on *Escherichia coli* isolated from pigs and colonization of pig ileum by K99+ enterotoxigenic *E. coli* from calves and pigs. *Infect. Immun.*, **15**, 614–620.

Moon, H. W., Baetz, A. L., Gianella, R. A. (1983). Immunization of swine with enterotoxin coupled to a carrier protein does not protect suckling pigs against an *Escherichia coli* strain that produces heat-stable enterotoxin. *Infect. Immun.*, **39**, 990–992.

Morgan, R. L., Isaacson, R. E., Moon, H. W., Brinton, C. C., To, C. C. (1978). Immunization of suckling pigs against enterotoxigenic *Escherichia coli*-induced diarrheal disease by vaccinating dams with purified 987 or K99 pili: protection correlates with pilus homology of vaccine and challenge. *Infect. Immun.*, **22**, 771–777.

Morris, J. A., Wray, C., Sojka, W. J. (1980). Passive protection of lambs against enteropathogenic *Escherichia coli*: role of antibodies in the serum and colostrum of dams vaccinated with the K99 antigen. *J. Med. Microbiol.*, **13**, 265–271.

Morris, J. A., Thorns, C. J., Scott, A. C., Sojka, W. J., Wells, G. A. (1982). Adhesion *in vitro* and *in vivo* associated with an adhesive antigen (F41) produced by a K99 mutant of the reference strain *Escherichia coli* B41. *Infect. Immun.*, **36**, 1146–1153.

Morris, J. A., Thorns, C. J., Wells, G. A. H., Scott, A. C., Sojka, W. J. (1983). The production of F41 fimbriae by piglet strains of enterotoxigenic *Escherichia coli* that lack K88, K99 and 987P fimbriae. *J. Gen. Microbiol.*, **129**, 2753–2759.

Mushin, R., Dubos, R. (1965). Colonization of the mouse intestine with *Escherichia coli*. *J. Exp. Med.*, **122**, 745–757.

Nagy, B. (1980). Vaccination of cows with a K99 extract to protect newborn calves against experimental enterotoxic colibacillosis. *Infect. Immun.*, **27**, 21–24.

Nagy, B., Moon, H. W., Isaacson, R. E. (1976). Colonization of porcine small intestine by enterotoxigenic *Escherichia coli*: ileal colonization and adhesion by pig enteropathogens that lack K88 antigen and some acapsular mutants. *Infect. Immun.*, **13**, 1214–1220.

Nagy, B., Moon, H. W., Isaacson, R. E. (1977). Colonization of porcine small intestine by enterotoxigenic *Escherichia coli*: selection of piliated forms *in vivo*, adhesion of piliated forms to epithelial cells *in vitro*, and incidence of pilus antigen among porcine enteropathogenic *E. coli*. *Infect. Immun*, **16**, 344–352.

Nagy, B., Moon, H. W., Isaacson, R. E., To, C. C., Brinton, C. C. (1978). Immunization of suckling pigs against enteric enterotoxigenic *Escherichia coli* infection by vaccinating dams with purified pili. *Infect. Immun.*, **21**, 269–274.

Newsome, P. M., Burgess, M. N., Burgess, M. R., Coney, K. A., Goddard M. E., Morris, J. A. (1987). A model of acute infectious neonatal diarrhoea. *J. Med. Microbiol.*, **23**, 19–28.

Orskov, F., Orskov, I. (1992). *Escherichia coli* serotyping and disease in man and animals. *Can. J. Microbiol*, **38**, 699–704.

Orskov, I., Orskov, F., Smith, H. W., Sojka, W. J. (1975). The establishment of K99, a thermolabile, transmissible *Escherichia coli* K antigen called 'Kco', possessed by calf and lamb enteropathogenic strains. *Acta. Pathol. Microbiol. Scand. B*, **83**, 31–36.

Paniagua, M., Espinoza, F., Ringman, M., Reizenstein, E., Svennerholm, A.-M., Hallander, H. (1997). Analysis of infection with enterotoxigenic *Escherichia coli* in a prospective cohort of infant diarrhea in Nicaragua. *J. Clin. Microbiol.*, **35**, 1404–1410.

Rabbani, G. H. (1996). Mechanism and treatment of diarrhoea due to *Vibrio cholerae* and *Escherichia coli*: roles of drugs and prostaglandins. *Dan. Med. Bull.*, **43**, 173–185.

Rutter, J. M., Anderson, J. C. (1972). Experimental neonatal diarrhoea caused by an enteropathogenic strain of *Escherichia coli* in piglets: a study of the disease and the effect of vaccinating the dam. *J. Med. Microbiol.*, **5**, 197–210.

Rutter, J. M., Jones, G. W. (1973). Protection against enteric disease caused by *Escherichia coli*: a model for vaccination with a virulence determinant? *Nature*, **242**, 531–532.

Rutter, J. M., Burrows, M. R., Sellwood, R., Gibbons, R. A. (1975). A genetic basis for resistance to enteric disease caused by *E. coli*. *Nature*, **257**, 135–136.

Salajka, E., Salajkova, Z., Alexa, P., Hornich, M. (1992). Colonization factor different from K88, K99, F41 and 987P in enterotoxigenic *Escherichia coli* strains isolated from post-weaning diarrhea in pigs. *Vet. Microbiol.*, **32**, 163–175.

Satterwhite, T. K., Evans, D. G., Dupont, H. L., Evans, D. J. (1978). Role of *Escherichia coli* colonization factor antigen in acute diarrhoea. *Lancet*, **ii**, 181–184.

Savage, D. C. (1977). Interactions between the host and its microbes. In *Microbial Ecology of the Gut* (eds Clarke, R. T. J., Bauchop, T.), pp. 277–310. Academic Press, London.

Sears, C. L., Kaper, J. B. (1996). Enteric bacterial toxins: mechanisms of action and linkage to intestinal secretion. *Microbiol. Rev.*, **60**, 167–215.

Smith, H. W., Halls, S. (1967a). The transmissible nature of the genetic factor in *Escherichia coli* that controls enterotoxin production. *J. Gen. Microbiol*, **47**, 153–161.

Smith, H. W., Halls, S. (1967b). Observations by the ligated intestinal segment and oral inoculation methods on *Escherichia coli* infections in pigs, calves, lambs and rabbits. *J. Pathol. Bacteriol.*, **93**, 499–529.

Smith, H. W., Linggood, M. A. (1971a). The transmissible nature of enterotoxin production in a human enteropathogenic strain of *Escherichia coli*. *J. Med. Microbiol.*, **4**, 301–305.

Smith, H. W., Linggood, M. A. (1971b). Observations on the pathogenic properties of the K88, Hly and Ent plasmids of *Escherichia coli* with particular reference to porcine diarrhoea. *J. Med. Microbiol.*, **4**, 467–485.

Smith, H. W., Linggood, M. A. (1972). Further observations on *Escherichia coli* enterotoxins with particular regard with those produced by atypical piglet strains and by calf and lamb strains: the transmissible nature of these enterotoxins and K antigen possessed by calf and lamb strains. *J. Med. Microbiol.*, **5**, 243–249.

Smith, H. W., Huggins, M. B. (1978). The influence of plasmid-determined and other characteristics of enteropathogenic *Escherichia coli* on their ability to proliferate in the alimentary tract of piglets, calves and lambs. *J. Med. Microbiol.*, **11**, 471–491.

Smyth, C. J. (1982). Two mannose-resistant haemagglutinins on enterotoxigenic *Escherichia coli* of serotype O6:K15:H16 or H- isolated from travellers' and infantile diarrhoea. *J. Gen. Microbiol.*, **128**, 2081–2096.

Sommerfelt, H., Steinsland, H., Grewal, H. M. S. *et al.* (1996). Colonization factors of enterotoxigenic *Escherichia coli* isolated from children in North India. *J. Infect. Dis.*, **174**, 768–776.

Spangler, B. D. (1992). Structure and function of cholera toxin and the related *Escherichia coli* heat-labile enterotoxin. *Microbiol. Rev.*, **56**, 622–647.

Stirm, S., Orskov, F., Orskov, I., Birch-Anderson, A. (1967). Episome-carried surface antigen K88 of *Escherichia coli*. III Morphology. *J. Bacteriol.*, **93**, 740–748.

Thomas, L. V., McConnell, M. M., Rowe, B., Field, A. M. (1985). The possession of three novel coli surface antigens by enterotoxigenic *Escherichia coli* strains positive for the putative colonization factor PCF8775. *J. Gen. Microbiol.*, **131**, 2319–2326.

Van Zijderveld, F. G., Westenbrick, F., Anakotta, J., Brouwers, R. A. M., van Zijderveld, A. M. (1989). Characterization of the F41 fimbrial antigen of enterotoxigenic *Escherichia coli* by using monoclonal antibodies. *Infect. Immun.*, **57**, 1192–1199.

Yano, T., Leite, D. da S., de Camarguo, I. J. B., Pestano de Castro, A. F. (1986). A probable new adhesive factor (F42) produced by enterotoxigenic *Escherichia coli* isolated from pigs. *Microbiol. Immunol.*, **30**, 595–508.

Chapter 28

Rabbit Model of Shigellosis

M. E. Etheridge

Introduction

Shigella organisms are a group of Gram-negative, facultative anaerobes that each year cause millions of cases of diarrhea world wide. There is currently no licensed vaccine against shigellosis; therefore antibiotic therapy is the only means of controlling the often life-threatening infections that occur in young children in developing countries. This chapter describes the clinical and pathological features of a rabbit model for shigellosis, which may be useful in evaluating treatments or potential vaccines during this time when new, multiple-antibiotic-resistant strains are continually emerging. Emphasis is placed on the technical methods involved in using the model.

Background of human infection

Shigellosis, or bacillary dysentery, is a gastrointestinal disease caused by four species (or serogroups) of *Shigella* (*S. dysenteriae, S. flexneri, S. sonnei, S. boydii*). Infection is characterized by fever, tenesmus, abdominal cramps, and diarrhea, with or without blood and mucus (Samuelson and Von Lictenberg, 1994). In the USA, over 23 000 cases of shigellosis were reported to the Centers for Disease Control and Prevention in 1996, of which approximately 28% occurred in children less than 5 years of age (CDC, 1997). In developing countries, shigellosis is a serious problem and is responsible for 10% of the 5 million deaths that occur each year due to diarrheal diseases (WHO, 1991; Maurelli, 1992).

Background of model

Primates, both human and non-human, are the only animals that are naturally susceptible to shigellosis; and non-human primates, particularly Old World species from Africa and Asia (e.g. macaques), have been used as animal models of human infection (Rout *et al.*, 1975; Takeuchi, 1975, 1982; Banish *et al.*, 1993a,b). Although not naturally susceptible to enteric infection by *Shigella*, guinea-pigs and rabbits may develop a severe, purulent keratoconjunctivitis if drops of *Shigella* bacterial suspension are placed into the conjunctival sac. This procedure, called the Sereny test (Sereny, 1955), is sometimes used in conjunction with other microbiological methods to identify invasive forms of the organism (Sayeed *et al.*, 1992; Portnoy, 1994). Rabbits can be made susceptible prior to oral inoculation by a conditioning process (described below), which is patterned after an earlier described rabbit model for *Vibrio cholerae* infection (Cray *et al.*, 1983). Susceptibility to enteric infection can also be developed in guinea-pigs by initially exposing them to several days of fasting (Formal *et al.*, 1958, 1962).

Animal species

New Zealand White (NZW) rabbits (*Oryctolagus cunniculus*) of either sex weighing 2–2.5 kg (10–12 weeks old) are used in this model. All rabbits should be screened for the presence of diarrhea-causing organisms prior to inclusion in the study, since diarrhea following inoculation with *Shigella* is an important sign of infection, which would be confounded by the presence of other diarrheogenic organisms. Key bacterial agents to screen rabbits for include: *Salmonella* spp. (rectal swab culture) and *Clostridium (Bacillus) piliformis* (ELISA; DeLong and Manning, 1994); also screen for *Eimeria* spp. coccidia (direct fecal smear, fecal floatation, or concentration–floatation methods; Pakes and Gerrity, 1994). Only healthy, coccidia-free rabbits with normal fecal pellet consistency and rectal cultures free of enteric pathogens should be used.

Preparation of animals

Biosafety procedures — specialized care and housing

Because *Shigella* is highly infectious for humans, rabbits are housed in an isolated room in standard stainless-steel rabbit cages. The cage floors are constructed of flat,

Handbook of Animal Models of Infection
ISBN 0-12-775390-7

spaced bars, which allow all urine and contaminated feces to fall into pans containing a 1:256 dilution of a bactericidal reagent (LPH; Calgon Division of Merck, St Louis, MO). All materials used during the inoculation procedure (e.g. feeding tubes, oral speculum, gowns, etc.) are autoclaved before discarding or laundering, and table surfaces are disinfected with a bactericidal solution (LPH).

Preinoculation conditioning procedure

Rabbits are conditioned 36 hours before inoculation by withholding food and by the administration of a single 250 mg dose of tetracycline oral suspension (Sumycin; E. R. Squibb, Princeton, NJ) in the water bottle. The water bottle should contain only 250 ml of water to ensure that the rabbits drink all of the antibiotic solution. Once the solution has been consumed, the water bottle is refilled with fresh water. At all other times during the procedure rabbits are provided with fresh water and are fed *ad libitum* with standard pelleted rabbit food.

Details of inoculation

On the day of inoculation, rabbits are anesthetized with ketamine and xylazine (35 and 5 mg/kg respectively, intramuscularly). The hair over one marginal ear vein is clipped and the rabbit is given 75 mg of cimetidine hydrochloride intravenously at time 0 to inhibit gastric acid secretion. This is easily accomplished with a 25 G tuberculin needle and a 1 ml syringe. The rabbit is then positioned in dorsal recumbency with all four legs tied to a restraint board to facilitate gavage. Fifteen minutes after receiving i.v. cimetidine, the rabbit is gavaged with 15 ml of 5% sodium bicarbonate solution using a sterile plastic feeding tube (8F × 15 in/45 cm). Passage of the stomach tube is facilitated by a small oral speculum, which holds the mouth open slightly. Once the tube is correctly positioned in the lumen of the stomach, a gurgling sound can be ausculted when air from a 5 ml syringe is blown slowly into the catheter while a stethoscope is positioned over the stomach. Also, liquid stomach contents can be aspirated through the catheter into the syringe. (It is crucial to insure that the tube is correctly positioned to prevent inadvertently introducing inoculum into the lungs.) At time 30 minutes, a second 15 ml dose of 5% sodium bicarbonate is administered, followed immediately by 15 ml of *Shigella* bacterial suspension in brain–heart infusion broth. Also at this time, 2 ml of tincture of opium (containing 10 mg of morphine) is given intraperitoneally (i.p.) to reduce intestinal motility and thereby enhance colonization by *Shigella*. For i.p. injection, the skin caudal to the umbilicus is clipped and disinfected with betadine scrub followed by 70% isopropyl alcohol. The restraint board is repositioned so that the rabbit's hindquarters are elevated, allowing the viscera to fall cranially. A 1 in (2.5 cm) needle is inserted lateral to the midline and just caudal to the umbilicus, and is directed toward the spine. The syringe is aspirated before injecting to be certain the needle has not punctured a loop of bowel. After recovering from anesthesia, the rabbit is returned to its cage and is allowed to eat and drink *ad libitum*.

Storage and preparation of inoculum

The author has successfully used several species of pathogenic bacteria with this model, including *Shigella sonnei* and *Shigella flexneri*. Stock cultures were stored at –70°C on frozen Microbank beads (Pro-Lab Diagnostics, Round Rock, TX). To reconstitute frozen cultures, a frozen bead was streaked on to Congo red plates (Miller, 1991) and incubated for 18 hours at 37°C. Four to five Congo-red-positive, type-1 smooth colonies were transferred into 300 ml of BHI broth and incubated for 16 hours on a shaker (180 cycles/minutes) at 37°C. The cells were pelleted by centrifugation, washed once, then resuspended in sterile phosphate-buffered saline, pH 7.4 (Sigma Chemical Co., St Louis, MO). Each rabbit is given 1 ml of a bacterial suspension containing 10^{10} colony-forming units/ml in 14 ml of sterile BHI broth.

Animal- and bacteria-associated factors that impact the infection process

New Zealand White (NZW) rabbits weighing 2–2.5 kg are used in this model because larger, older rabbits seem to have a reduced susceptibility to infection. When larger animals are used, many do not develop diarrhea following oral inoculation. As stated earlier, rabbits are not naturally susceptible to enteric infection by *Shigella*. They can be made susceptible, however, by the administration of tetracycline, cimetidine, bicarbonate, and tincture of opium prior to oral inoculation with bacteria. Orally administered tetracycline is used to decrease competing intestinal flora, because preliminary studies indicated that rabbits given the identical challenge without antibiotics did not develop illness (Sack *et al.*, 1987). In addition, Cray *et al.* (1983) demonstrated, in their initial development of an oral inoculation technique using *Vibrio cholerae*, that intestinal colonization of rabbits was significantly increased when (1) gastric acid was neutralized with cimetidine and sodium bicarbonate, and (2) transient hypoperistalsis was induced using tincture of opium. The strain of bacterium selected should be tetracycline-resistant to avoid being affected by the oral tetracycline used in the initial conditioning procedure.

Key parameters to monitor infection and response to treatment

Infection with *Shigella* may be monitored by: (1) the rabbit's attitude; (2) the stool consistency; (3) culture results from stool and blood; and (4) pathology of the ileum (described below). Within 24 hours of inoculation, rabbits become very lethargic and reluctant to move. If a fairly high dose of a particularly pathogenic strain of *Shigella* is used, all the rabbits will become moribund, develop diarrhea and die within 24 hours. The stool will change from normal, formed pellets to watery diarrhea overnight. *Shigella* may be recovered from rectal swabs and from the blood using selective media and standard techniques for culture and isolation. We have not specifically investigated treatment of infected rabbits using this model.

Pathology

Rabbits that are necropsied 24 hours after inoculation have lesions confined to the terminal portion of the ileum. On gross examination, vessels on the serosal surface of the ileum are congested, and the ileum is distended with gas and pale yellow liquid. Cecal contents are also liquid and run out over the necropsy table when the lumen is opened. (This should be anticipated, and appropriate biosafety precautions taken to minimize contamination of the work area during necropsy.) The colon contains no formed feces. No gross lesions are observed in other organs examined (heart, lungs, spleen, liver, stomach, pancreas, bladder, kidneys, skeletal muscle).

Light-microscopic lesions are also confined to the ileum. The surface epithelium of the ileum is necrotic and sloughed into the lumen (Figure 28.1). The denuded lamina propria is distended with edema and numerous mixed inflammatory cells, including large clusters of heterophils. Capillaries in the lamina propria and submucosa are dilated and congested. (In Figure 28.2, notice the imminent destruction of capillary walls, leading to hemorrhage and the bloody diarrhea that is characteristic of *Shigella* dysentery.) A minority of rabbits may develop less severe histologic lesions, in which only the epithelial cells at the tips of the villi are affected (Figure 28.3). Cells in this area are

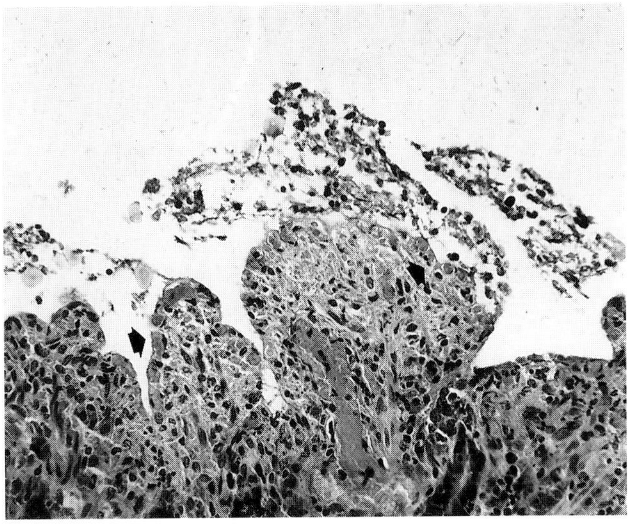

Figure 28.2 Close-up of Figure 28.1. Arrows illustrate the destruction of blood vessel walls, leading to characteristic intraluminal hemorrhage often associated with shigellosis. H&E stain; × 250.

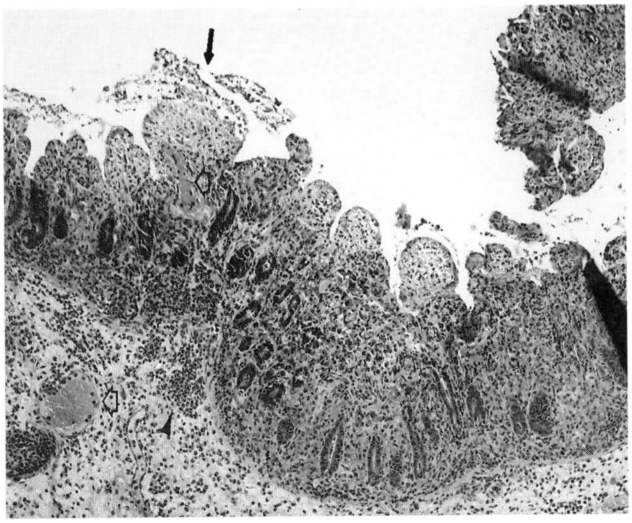

Figure 28.1 Photomicrograph of a section of the terminal portion of the ileum of a rabbit inoculated with *Shigella flexneri* 2a. Notice necrosis and sloughing of epithelial surface (long arrow), clusters of heterophils in the submucosa (closed arrowhead), and dilated blood vessels in the lamina propria and submucosa (open arrows). H&E stain; × 100.

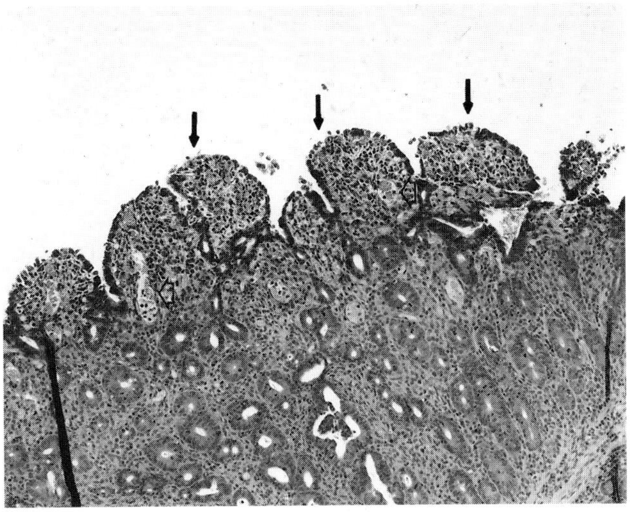

Figure 28.3 Photomicrograph of a section of the terminal portion of the ileum in a minimally affected rabbit inoculated with *Shigella flexneri* 2a. Notice the atrophic villi with dilated capillaries at the base (open arrows) and relatively intact epithelium with individual cell necrosis at the tips (closed arrows). H&E stain; × 100.

necrotic and sloughed into the lumen, whereas cells on the sides of the villi and in the crypts are noticeably spared. In these minimally affected rabbits, the villi are atrophic and blunt at the tips. Aside from a few scattered populations of heterophils, the cecum and colon are relatively unaffected.

Pitfalls (advantages/disadvantages) of the model

This model of shigellosis uses a commonly available and inexpensive small animal, the New Zealand White rabbit. Although rabbits are not naturally susceptible to shigellosis, as non-human primates are, rabbits have the advantage of being less expensive, more readily available, and easier to house than monkeys. A major disadvantage, however, is the rabbit's reduced susceptibility to *Shigella* infection, making a large inoculum dose (10^9) necessary. In humans, as few as 10–100 *Shigella* bacteria are sufficient to establish infection (Small, 1994).

The lesions caused by shigellosis in humans and in non-human primates are found in the colon and occasionally in the terminal portion of the ileum (Wassef *et al.*, 1989; Bennish, 1991), and a fibrinous, mucopurulent to muco-hemorrhagic exudate (pseudomembrane) often adheres to underlying mucosal ulcers (Jones and Hunt, 1983; Samuelson *et al.*, 1994). In the rabbit model, lesions are confined to the ileum and, although there is no obvious adherent pseudomembrane, the mucosal lesion is otherwise similar to the lesion that develops in non-human primates and humans.

Bacteremia as a complication of shigellosis is rare in developed countries; however, it is the most commonly reported lethal complication of *Shigella* infection in young children in developing countries. The rabbit model mimics human infection in this regard, in that five out of six rabbits inoculated with *S. flexneri* 2A (Etheridge, 1996) developed bacteremia.

Contributions of the model to infectious disease therapy

The rabbit model of shigellosis has been used extensively to elucidate the pathophysiology of shigella infection. Wassef *et al.* (1989) used an acute ileal loop preparation in rabbits to demonstrate that, upon entry into the intestinal lumen, pathogenic *Shigella* are initially phagocytized by, and then replicate within, the host's M-cells (specialized lymphoid-follicle-associated epithelial cells). Perdomo *et al.* (1994), also using the rabbit intestinal loop assay, demonstrated that the acute inflammatory response associated with infection by *Shigella* contributes more to the epithelial cell necrosis than does intracellular invasion by the organism. The usefulness of the rabbit model is further enhanced by the fact that the inoculation dose of *Shigella* organisms can be adjusted to result in colonization without illness. Thus rabbits can be inoculated with sublethal doses in order to study the immunologic events associated with the infection.

References

Banish, L. D., Sims, R., Sack, D., Montali, R. J., Phillips, L. Jr, Bush, M. (1993a). Prevalence of shigellosis and other enteric pathogens in a zoologic collection of primates. *J.A.V.M.A.*, **203**, 126–132.

Banish, L. D., Sims, R., Bush, M., Sack, D., Montali, R. J. (1993b). Clearance of *Shigella flexneri* carriers in a zoologic collection of primates. *J.A.V.M.A.*, **203**, 133–136.

Bennish, M. L. (1991). Potentially lethal complications of shigellosis. *Rev. Infect. Dis.*, **13** (Suppl. 4), S319–S324.

CDC (1997). *Summary of Notifiable Diseases, United States, 1996.* MMWR, **46**(54), 10. Centers for Disease Control and Prevention, Washington, DC.

Cray, W. C. Jr, Tokunaga, E., Pierce, N. F. (1983). Successful colonization and immunization of adult rabbits by oral inoculation with *Vibrio cholerae* 01. *Infect. Immun.*, **41**, 735–741.

DeLong, D., Manning, P. J. (1994). Bacterial diseases. In *The Biology of the Laboratory Rabbit* (eds Manning, P. J., Ringler, D. H., Newcomer, C. E.), pp. 143–149. Academic Press, San Diego, CA.

Etheridge, M. (1996). Pathologic study of a rabbit model for shigellosis. *Lab. Animal Sci.*, **46**, 61–66.

Formal, S. B., Dammin, G. J., LaBrec, E. H., Schneider, H. (1958). Experimental shigella infections: characteristics of a fatal infection produced in guinea pigs. *J. Bacteriol.*, **75**, 604–610.

Formal, S. B., Abrams, G. D., Schneider, H., Sprinz, H. (1962). Experimental shigella infections VI. Role of the small intestine in an experimental infection in guinea pigs. *J. Bacteriol.*, **85**, 119–125.

Jones, T. C., Hunt, R. D. (1983). Diseases due to simple bacteria. In *Veterinary Pathology*, p. 627. Lea & Febiger, Philadelphia, PA.

Maurelli, A. T. (1992). *Shigella* inside and out: lifestyles of the invasive and dysenteric. *ASM News*, **58**, 603–608.

Miller, M. J. (1991). Quality control of media, reagents, and stains. In *Manual of Clinical Microbiology* (ed Lennette, E.H.), p. 1240. ASM Press, Washington, DC.

Pakes, S. P., Gerrity, L. W. (1994). Protozoal diseases. In *The Biology of the Laboratory Rabbit* (eds Manning, P. J., Ringler, D. H., Newcomer, C. E.), p. 208. Academic Press, San Diego, CA.

Perdomo, O. J., Cavaillon, J. M., Huerre, M. *et. al.* (1994). Acute inflammation causes epithelial invasion and mucosal destruction in experimental shigellosis. *J. Exp. Med.*, **180**, 1307–1319.

Portnoy, D. A. (1994). Molecular and cellular biology of intracellular bacteria. In *Molecular Genetics of Bacterial Pathogens* (eds Miller, V. L., Kaper, J. B., Portnoy, D. A., Isberg, R. R.), p. 209. ASM Press, Washington, DC.

Rout, W. R., Formal, S. B., Giannella, M. D., Dammin, J. (1975). Pathophysiology of shigella diarrhea in the rhesus monkey: intestinal transport, morphological, and bacteriological studies. *Gastroenterology*, **68**, 270–278.

Sack, D. A., Cray, W. C., Alam, K. (1987). Comparison of prophylactic tetracycline and clioquinol in a rabbit model of intestinal infection with *Vibrio cholerae* and *E. coli*. *Chemotherapy*, **33**, 423–436.

Samuelson, J., Von Lictenberg, F. (1994). Infectious diseases. In *Pathologic Basis of Disease* (eds Cotran, R. S., Kumar, V., Robbins, S. L., Schoen, F. J.), p. 330. W. B. Saunders, Philadelphia, PA.

Sayeed, S., Sack, D. A., Qadri, F. (1992). Protection from *Shigella sonnei* infection by immunization of rabbits with *Plesiomonas shigelloides* (SVC 01). *J. Med. Microbiol.*, **37**, 382–384.

Sereny, B. (1955). *Shigella* keratoconjunctivitis: a preliminary report. *Acta Microbiol. Acad. Sci. Hung.*, **2**, 293–296.

Small, P. L. (1994). How many bacteria does it take to cause diarrhea and why? In *Molecular Genetics of Bacterial Pathogens* (eds Miller, V. L., Kaper, J. B., Portnoy, D. A., Isberg, R. R.), p. 479. ASM Press, Washington, DC.

Takeuchi, A. (1975). Bacillary dysentery, shigellosis. *Am. J. Pathol.*, **81**, 251–254.

Takeuchi, A. (1982). Early colonic lesions in experimental shigella infection in rhesus monkeys: revisited. *Vet. Pathol.*, **19** (Suppl. 7), 1–8.

Wassef, J. S., Keren, D. F., Mailloux, J. L. (1989). Role of M cells in initial antigen uptake and in ulcer formation in the rabbit intestinal loop model of shigellosis. *ASM News*, **57**, 858–863.

WHO (1991). *Bulletin of the World Health Organization*, **69**, 667–676.

Chapter 29

RITARD Rabbit Model for Studying *Vibrio cholerae* and Other Enteric Infections

P. Panigrahi and J. G. Morris Jr

Introduction

The removable intestinal tie adult rabbit diarrhea (RITARD) model was first proposed by Spira *et al.* (1981; Spira and Sack, 1982) in an attempt to create a simple animal model for studying the toxin-mediated diarrheal disease caused by *Vibrio cholerae* and enterotoxigenic *Escherichia coli* (ETEC). Prior to the description of this model, several animal species, including dogs, chinchillas, and mice, were being used with some success. Other models in use included the infant rabbit model of Dutta and Habbu and the adult rabbit ligated ileal loop model of De and Chatterjee. However, there were recognized difficulties with each of these models: species other than rabbits were difficult animals to work with because of size considerations, the infant rabbit assay was not optimally reproducible (in part because of the need to carefully standardize size and age of animals), and there were concerns about the "unphysiologic" permanent intestinal ties in the ligated loop model.

In this setting, Spira *et al.* modified the the idea of ligating the ileum originally suggested by De *et al.* to incorporate the use of a removable intestinal tie. The resultant model, dubbed RITARD, is now widely used in studying the pathogenesis of enteric pathogens. Key factors in its wide acceptance include the ease of the procedure, the availability of the animal species, and its "whole animal" characteristics (including the opportunity it provides to evaluate the intact immune system).

There is now an extensive literature on use of the model in work with *V. cholerae* and ETEC (Svennerholm *et al.*, 1990; Davis and Banks, 1991; Richardson, 1991; Russell *et al.*, 1992; Svennerholm *et al.*, 1992). Other species that have been studied in this model include *Campylobacter jejuni* (Pang *et al.*, 1987; Guerry *et al.*, 1994), enteroaggregative *E. coli* (Tickoo *et al.*, 1992), *Providencia alcalifaciens* (Mathan *et al.*, 1993), *Hafnia alvei* (Albert *et al.*, 1991), as well as non-specific organisms implicated in persistent diarrhea (Penny, 1992). We have used the RITARD model in a comparative study of different non-O1 *V. cholerae* strains and have addressed the role of toxins and capsule, as well as invasiveness, in the disease process (Russell *et al.*, 1992). This model has been of particular use in studying immunity against *V. cholerae* (Tokunaga *et al.*, 1984; Lopez-Vidal *et al.*, 1987; Pierce *et al.*, 1988; Dasgupta *et al.*, 1994; Ahmed *et al.*, 1994), ETEC (Sack *et al.*, 1988), and *C. jejuni* (Burr *et al.*, 1988; Pavlovskis *et al.*, 1991). Greenberg (1991) has reported the failure of this model (but success in piglet and mice models) in the study of somatostatin analogs in secretory diarrhea. However, other agents such as entericoated protease have been used quite successfully in the RITARD model of ETEC diarrhea (Mynott *et al.*, 1991). Some investigators have used this model in comparative studies along with the suckling mice and the rabbit ileal loop models to examine the effects of different phenotypes of *V. cholerae* (Richardson, 1991). Logan *et al.* (1989) have reported the utility of this model for studying phase variation in *Campylobacter coli*.

Preparation of animals

Outbred New Zealand White adult rabbits are used in this procedure. We have obtained consistent results and no mortality (from surgery) by using NZW rabbits weighing between 1.6 and 2.7 kg. The rabbits should be allowed at least 24 hours to acclimatize in the animal facility after shipment, although we routinely perform surgery 3–4 days after receiving the animals. Rabbits are fasted overnight and prepared for surgery by shaving the ventral part of the abdomen with an electric clipper after the animals are anesthetized. Iodine or other similar topical sterilizing agent is applied to the area and aseptic surgery is performed.

Anesthesia

A cocktail of ketamine (50 mg/kg body weight), acepromazine (1 mg/kg), and xylazine (8 mg/kg) is injected subcutaneously.

Surgical procedure

After induction of appropriate anesthesia, the animal is covered with a sterile surgical drape (with a small opening

Handbook of Animal Models of Infection
ISBN 0-12-775390-7

for conducting the surgery). A 4–5 cm midline incision is made using a scalpel through the skin and linea alba. The peritoneum is lifted up with toothed forceps and incised with a pair of scissors, taking care not to touch or injure the viscera. If the incision is made properly, the surgery turns out to be almost bloodless. The viscera is exposed and brought out gently and is left covered with warm, moist towels soaked in sterile saline. The ileocecal valve is identified and the cecum is tied off with a 0.25 in (6 mm) umbilical tape as close to the ileocecal valve as possible. Approximately 10 cm of terminal ileum is measured, and a slip knot is placed at that point using 0.047 in (1.2 mm) diameter silastic tubing. The proximal jejunum is mobilized and the test bacteria in saline suspension are injected using a 25 G or 26 G needle. Care must be taken not to inject around the vasculature. The puncture site should be gently pressed with moist gauze as the needle is withdrawn after injection of the bacterial inoculum. The parietal peritoneum and muscle layers are closed with 3.0 Dexon. The skin is closed with 3.0 nylon or stapled. The silastic tubing making the slip knot is exited using a large-eyed needle 2–3 cm lateral to the incision. Two hours after the inoculation of bacteria, the ileal tie is released by gentle tugs on the free end. Animals are then returned to the cage and allowed food and water *ad libitum* and observed for 1–7 days in a biohazard facility (if infectious agents are used).

Postoperative care of animals

After full recovery from anesthesia, the animals may be given bupernorphine for analgesia. All the animals do quite well the following day.

Evaluation of disease

Clinical observation

Animals should be monitored closely for signs of clinical disease. This could range from loose stool to signs of sepsis and intestinal necrosis, depending on the infectious agent and the dose used. Clinical judgment has to be made regarding conducting necropsy while dealing with new organisms. In spite of close observation, animals may die before presenting with overt signs of disease. In setting up a schedule of postoperative observation, it should also be kept in mind that animals may deteriorate rapidly and die overnight, in which case tissue samples at necropsy may have autolysis and yield minimal information on the disease process. Given these concerns, when dealing with a microorganism for which RITARD data are limited there needs to be judicious selection of the inoculating dose, starting with low dose and slowly increasing in increments.

Monitoring for diarrhea

There is no standardized grading system for evaluation of diarrhea in the RITARD model. Investigators from different laboratories set their own criteria, although watery liquid stool is almost always considered to be the highest grade. We have found it to be useful to score diarrheal stool as follows: grade 1, rabbit stool consisting of hard pellets; grade 2, soft stool; grade 3, catarrhal diarrhea where the stool takes the shape of the collection container; grade 4, watery liquid diarrhea. Starting the first postoperative day, rectal swabs should be collected, enriched in specific medium if available (e.g. in alkaline peptone water for *V. cholerae*) and then inoculated on to appropriate medium for incubation.

Necropsy examination

Rabbits should be anesthetized with ketamine for collection of blood from the heart and then sacrificed by administration of a lethal dose of pentobarbital (100 mg/kg). After opening of the skin and muscle layers, the peritoneum should be examined for signs of peritonitis and adhesions. Swab cultures should be taken from the peritoneum for culture. Different parts of the intestine should be observed for gross changes, including swelling, hemorrhage, exudation and perforation. Parts of jejunum, ileum, and colon should be fixed in 4F1G (4% formaldehyde + 1% glutaraldehyde). This preservative works well for conventional histopathologic as well as electron microscopic processing. Specimens for bacteriologic cultures should also be obtained from each intestinal site, as well as from liver, spleen, mesenteric lymph nodes and heart blood. Swabs should be taken from the permanent cecal tie.

Quantitation of bacterial colonization and adherence

Stool samples should be collected from ileum and colon, weighed, and plated for bacterial count. This provides a measure of total bacterial colonization in the intestine at the time of necropsy. For evaluation of bacterial adherence to mucosa, a portion of intestinal tissue should be weighed, ground in a homogenizer, and plated on appropriate media. The colony counts should be expressed per gram of tissue. Further qualitative analysis of adherence and invasion could be done during light and electron microscopic analysis of specimens from the same sites.

Histopathology

Tissues fixed in 4F1G should be embedded in paraffin, and the sections stained with hematoxylin and eosin. Light-microscopic examination of the sections should emphasize the condition of mucosa, crypts, lamina propria, and deeper layers. A range of abnormalities from mild submucosal edema to massive necrosis, hemorrhage, and collapse of the epithelium could be seen depending on the infecting organism used. Adherence of bacteria to epithelial cells, and invaded bacteria in deeper tissue, could also be visualized under high-power microscopy.

Electron microscopy

Transmission and scanning electron microscopy of the specimens from the RITARD rabbits provide many additional meaningful data on the pathogenesis of disease. Among many things, the condition of mucosal cells, tight junction, gap junction, exact mode of bacterial attachment (tight, loose, pedestal formation, etc.), path of invasion, and multiplication within membrane-bound vacuole should attract the attention of the investigator.

Discussion

As mentioned in the introduction, the RITARD model has been used to study a wide range of enteric bacteria. However, it should not be regarded as a generic model for diarrhea, nor should it be used as a general model system for examining the effect of drugs or other modifiers of diarrhea. In contrast to the rabbit ileal loops, this model provides an open and a more natural system to evaluate diarrheal response if the organisms used have the ability of causing such change (e.g. *V. cholerae*, ETEC). Since the animals can be monitored over a longer period, it also provides more insights into the pathogenesis and immune response of the host against the pathogen.

However, at times it may not be the best model, when a more precise evaluation of the organism and host response is required. An example would be the direct comparison of isogenic wild-type and mutant bacterial strains, looking specifically for a differential fluid production response. Because of the open nature of the system, small changes in fluid secretion may not be noted in this model, and the ileal loop model would be better in such a scenario. Another drawback of the RITARD model comes with the presence of indigenous rabbit flora in the intestine. While adult rabbits have been used very successfully to produce disease against human pathogens, concurrent presence of other organisms along with the experimental organism could sometimes pose a problem. For the same reason, it is also not possible to critically examine the bacteria (other than enumeration) for any *in-vivo* changes in the intestine. Although more artificial, the ileal loop model provides a closed *in-vivo* environment where the changes in the bacteria could be examined after the experiment.

Rabbits with ligated ileal loops have to be sacrificed within 12–48 hours, and provide a more acute injury model. In conditions where the organism takes more than 2 days to establish infection and produce disease, the ileal loop model may not provide enough time to produce the endpoint and elaboration of pathogenetic markers. Emphasizing this point, we have expanded the scope of the RITARD model and have used this system for examining different markers of pathogenesis (bacterial adherence, invasion, tissue pathology, septicemia, etc.) rather than diarrhea as an endpoint. This model could probably be tried with any enteric organism as long as the underlying mechanisms of disease (secretory diarrhea versus invasive disease, etc.) are kept in mind and appropriate markers are sought in the diseased animals after establishing the infection. There may be a need for changing the protocols with regard to inoculum size, time before removal of the intestinal tie, and duration of observation after surgery to critically evaluate all the changes. The nature of the expected injury, recovery, and factors such as local and systemic immunity will all depend on the characteristics of the infectious agent and the type of disease produced. Familiarity with these variables is of particular importance when drugs or other agents are evaluated in the face of this artificial infection.

No model is as good as a human model to study human disease. It is possible to use other models as long as we accept them as "models" only. It is often necessary to combine cell culture, explant culture, and, sometimes, more than one animal model to fully explore the complexity of enteric bacterial pathogenesis and the host responses. However, within these constraints, the RITARD model provides a useful tool for evaluating diarrheal pathogens.

References

Ahmed, Z. U., Hoque, M. M., Rahman, A. S., Sack, R. B. (1994). Thermal stability of an oral killed-cholera-whole-cell vaccine containing recombinant B-subunit of cholera toxin. *Microbiol. Immunol.*, **38**, 837–842.

Albert, M. J., Alam, K., Islam, M. *et al.* (1991). *Hafnia alvei*, a probable cause of diarrhea in humans. *Infect. Immun.*, **59**, 1507–1513.

Burr, D. H., Caldwell, M. B., Bourgeois, A. L., Morgan, H. R., Wistar, R., Jr., Walker, R. I. (1988). Mucosal and systemic immunity to *Campylobacter jejuni* in rabbits after gastric inoculation. *Infect. Immun.*, **56**, 99–105.

Dasgupta, U., Bhadra, R. K., Panda, D. K., Deb, A., Das, J. (1994). Recombinant derivative of a naturally occurring nontoxinogenic *Vibrio cholerae* 01 expressing the B subunit of cholera toxin: a potential oral vaccine strain. *Vaccine*, **12**, 359–364.

Davis, J. A., Banks, R. E. (1991). Modification to the RITARD surgical model. *J. Invest. Surg.*, **4**, 499–504.

Greenberg, R. N. (1991). Effects of somatostatin analog SMS 201-995 on enterotoxigenic diarrhea. *Dig. Dis. Sci.*, **36**, 1768–1773.

Guerry, P., Pope, P. M., Burr, D. H., Leifer, J., Joseph, S. W., Bourgeois, A. L. (1994). Development and characterization of recA mutants of *Campylobacter jejuni* for inclusion in attenuated vaccines. *Infect. Immun.*, **62**, 426–432.

Logan, S. M., Guerry, P., Rollins, D. M., Burr, D. H., Trust, T. J. (1989). *In vivo* antigenic variation of *Campylobacter flagellin*. *Infect. Immun.*, **57**, 2583–2585.

Lopez-Vidal, Y., Ahren, C., Svennerholm, A. M. (1987) Colonization, diarrhoea and protective immunogenicity of a CFA-deficient, enterotoxin-producing *Escherichia coli* mutant in a non-ligated intestine experimental model. *Acta Pathol. Microbiol. Immunol. Scand. B*, **95**, 123–130.

Mathan, M. M., Mathan, V. I., Albert, M. J. (1993). Electron microscopic study of the attachment and penetration of rabbit intestinal epithelium by *Providencia alcalifaciens*. *J. Pathol.*, **171**, 67–71.

Mynott, T. L., Chandler, D. S., Luke, R. K. (1991). Efficacy of enteric-coated protease in preventing attachment of enterotoxigenic *Escherichia coli* and diarrheal disease in the RITARD model. *Infect. Immun.*, **59**, 3708–3714.

Pang, T., Wong, P. Y., Puthucheary, S. D., Sihotang, K., Chang, W. K. (1987). *In-vitro* and *in-vivo* studies of a cytotoxin from *Campylobacter jejuni*. *J. Med. Microbiol.*, **23**, 193–198.

Pavlovskis, O. R., Rollins, D. M., Haberberger, R. L., Jr *et al.* (1991). Significance of flagella in colonization resistance of rabbits immunized with *Campylobacter* spp. *Infect. Immun.*, **59**, 2259–2264.

Penny, M. E. (1992). The role of the duodenal microflora as a determinant of persistent diarrhoea. *Acta Paediatr. Suppl.*, **381**, 114–120.

Pierce, N. F., Cray, W. C., Jr., Kaper, J. B., Mekalanos, J. J. (1988). Determinants of immunogenicity and mechanisms of protection by virulent and mutant *Vibrio cholerae* O1 in rabbits. *Infect. Immun.*, **56**, 142–148.

Richardson, K. (1991). Roles of motility and flagellar structure in pathogenicity of *Vibrio cholerae*: analysis of motility mutants in three animal models. *Infect. Immun.*, **59**, 2727–2736.

Russell, R. G., Tall, B. D., Morris, J. G., Jr (1992). Non-O1 *Vibrio cholerae* intestinal pathology and invasion in the removable intestinal tie adult rabbit diarrhea model. *Infect. Immun.*, **60**, 435–442.

Sack, R. B., Kline, R. L., Spira, W. M. (1988). Oral immunization of rabbits with enterotoxigenic *Escherichia coli* protects against intraintestinal challenge. *Infect. Immun.*, **56**, 387–394.

Spira, W. M., Sack, R. B. (1982). Kinetics of early cholera infection in the removable intestinal tie-adult rabbit diarrhea model. *Infect. Immun.*, **35**, 952–957.

Spira, W. M., Sack, R. B., Froehlich, J. L. (1981). Simple adult rabbit model for *Vibrio cholerae* and enterotoxigenic *Escherichia coli* diarrhea. *Infect. Immun.*, **32**, 739–747.

Svennerholm, A. M., Wenneras, C., Holmgren, J., McConnell, M. M., Rowe, B. (1990). Roles of different coli surface antigens of colonization factor antigen II in colonization by and protective immunogenicity of enterotoxigenic *Escherichia coli* in rabbits. *Infect. Immun.*, **58**, 341–346.

Svennerholm, A. M., McConnell, M. M., Wiklund, G. (1992). Roles of different putative colonization factor antigens in colonization of human enterotoxigenic *Escherichia coli* in rabbits. *Microb. Pathogen.*, **13**, 381–389.

Tickoo, S. K., Bhan, M. K., Srivastava, R. *et al.* (1992). Intestinal colonization and production of diarrhoea by enteroadherent-aggregative *Escherichia coli*. *Indian J. Med. Res.*, **95**, 278–283.

Tokunaga, E., Cray, W. C, Jr, Pierce, N. F. (1984). Compared colonizing and immunizing efficiency of toxinogenic (A+ B+) *Vibrio cholerae* and an A– B+ mutant (Texas Star-SR) studied in adult rabbits. *Infect. Immun.*, **44**, 364–369.

Chapter 30

The Mouse Model of *Helicobacter Pylori* Infection

P. Ghiara, G. del Giudice and R. Rappuoli

Background of model

Helicobacter pylori infects the human gastric mucosa and causes common diseases like chronic active gastritis and nearly all peptic ulcers, and is closely associated with the occurrence of gastric malignancies (Blaser and Parsonnet, 1994). Infection is very common and it is estimated that about 50% of the world population is infected, with a much higher prevalence in countries where hygienic and socio-economic conditions are lower. Severe symptomatic diseases (i.e. peptic ulcer and cancer) occur in only about 15% of the infected and are associated with the infection with a particular subset of *H. pylori* strains, called Type I, which expresses a potent toxin (VacA) that induces epithelial cell damage *in vitro* and *in vivo* (Telford *et al.*, 1994b) and is therefore thought to play an important role in ulcerogenesis (Figura *et al.*, 1989; Telford *et al.*, 1994a; Xiang *et al.*, 1995). Type I strains also harbour a pathogenicity island (PAI), called *cag*, which encodes for several disease-associated virulence factors, including an immunodominant surface-exposed antigen called CagA (Censini *et al.*, 1996; Covacci *et al.*, 1997). The strains that do not possess the *cag* PAI and lack toxic activity are called "Type II strains".

Chronic infection induces a strong immune response that is detectable both systemically and locally (reviewed in Telford *et al.*, 1997). However, this activation of the immune system is not efficacious in fighting off the colonizing bacteria.

Because of the severity of disease induced and to the possibility of gastric cancer as an outcome in long-term infections, the eradication of infection is now recommended. Several regimens of combined antibiotics and antisecretory drugs have proved to be efficacious in eradicating the bacteria from the gastric mucosa and therefore in modifying the clinical outcome of the infection (Tytgat, 1996). The increasing occurrence of strains resistant to the current chemotherapies is pushing many pharmaceutical companies to find a more effective "magic bullet" to control the infection. In this light a very attractive approach is certainly the use of either preventive or therapeutic vaccination against the infection (Telford and Ghiara, 1996). Whatever will be the optimal weapon to eradicate *H. pylori* infection, it is crucial for the investigators involved in this battle to have access to affordable, reproducible and relevant animal models for the preclinical screening of the efficacy of candidate molecules.

Several models of *H. pylori* infection have been described that, by using various host species, may allow various aspects of the infection to be studied (see Ghiara *et al.*, 1996 for a review). A very well characterized small animal model of *Helicobacter* infection is the *H. felis* mouse model described by Lee *et al.* (1990). However, a major limitation of this model is that *H. felis* does not express the virulence determinants (i.e. VacA and *cag* PAI) that are important in human infection (Xiang *et al.*, 1995). Moreover, *H. felis* is not a human pathogen. The only available animal model that involves a relatively affordable and well characterized small animal species and is relevant to the human disease is the mouse model of chronic *H. pylori* infection that we have developed (Marchetti *et al.*, 1995). The present chapter describes the technical aspects of this model and critically reviews its usefulness in the development of new strategies against this important human pathogen.

Mouse strains

Several mouse strains can be used for this animal model. We have initially described the infection in outbred CD1 mice (Marchetti *et al.*, 1995). Our data have have been recently confirmed by Elizalde *et al.* (1997). We have subsequently extended our studies to assess the feasibility of infection in several inbred strains, including BALB/c, C57BL/6, C3H/He and A/J mice. Other investigators have reported successful infection of DBA/2 and SJL mice (Sakagami *et al.*, 1996; Lee *et al.*, 1997). Both male and female 5–8-week-old mice can be used. As mentioned below, the outcome and persistence of the infection varies in the different mouse strains.

Housing and handling of mice

Mice are housed in a negative-pressure air-conditioned BL2 animal facility that allows a safe experimentation with

Handbook of Animal Models of Infection
ISBN 0-12-775390-7

pathogens like *H. pylori* and ensures that no contamination from the external environment may occur.

Mice are housed on a 12-hour light–dark schedule (illumination 7:00 am to 7:00 pm) and have *ad-libitum* access to autoclave-sterilized standard diet pellets and to filter-sterilized water. Animals are caged (maximum 10 per cage) onto sterile wood shavings in autoclave-sterilized plastic cages.

Infection procedure

Mouse-adapted *H. pylori* strains

Laboratory-passaged and frozen-type strains commonly available from bacterial cell collections have been reported to transiently colonize germ-free and athymic nude mice, but do not significantly colonize the xenobiotic mice (Karita *et al.*, 1991). Cellini *et al.* (1994) first reported that fresh clinical isolates were able to colonize transiently (for up to 4–8 weeks) xenobiotic mice. In 1995 we reported that bacteria freshly isolated from human biopsies were able to colonize the conventional xenobiotic mice, moreover we described a procedure to "adapt" the fresh isolate to stably colonize the mouse stomach (Marchetti *et al.*, 1995). Bacteria isolated from infected mice were then reinoculated into other mice. Several cycles of reisolation and inoculation, defined as "*in-vivo* passaging", have thus allowed selection of good colonizers that are adapted to the mouse gastric environment. Adapted strains can be frozen for at least 3 months without losing their ability to infect mice.

We have adapted to the mouse stomach several fresh isolates of both type I (virulent) and type II (non-virulent) *H. pylori*. Usually two to four passages *in vivo* are sufficient to adapt the strain to the mouse stomach. With type I strains the adaptation is usually much faster than with type II strains. Subsequently, this procedure has been followed successfully by other investigators (McColm *et al.*, 1995; Sakagami *et al.*, 1996; Hong *et al.*, 1996; Lee *et al.*, 1997). A more recent improvement in our laboratories consisted in obtaining streptomycin-resistant recombinant derivatives of these adapted strains, obtained by allelic exchange with a mutated *H. pylori* S12 gene sequence harboured by a naturally occurring streptomycin resistant strain of *H. pylori* (Marchetti *et al.*, manuscript in preparation). The introduction of a marker of selection will be useful to assess the colonization more easily.

Growth of bacteria for inoculation into mice

Bacteria to be inoculated into mice are grown on agar plates (Table 30.1, mixture A), which are incubated in microaerophilic conditions. Bacteria cultured either in liquid conditions or on agar plates incubated with 5% CO_2 instead of the microaerophilia fail to significantly colonize the mice.

Table 30.1 Composition of plates for the growth of *H. pylori*

***H. pylori* plates mixture A**	
Agar Columbia 44 g/l	
5% horse blood	
0.2% (w/v) cyclodextrin	
Vancomycin	5 μg/ml
Amphotericin	8 μg/ml
Trimethoprim	10 μg/ml
Cefsulodin	6 μg/ml
Cycloheximide	100 μg/ml
***H. pylori* plates mixture B (McColm mix)**	
Agar Columbia 44 g/l	
5% horse blood	
Vancomycin	100 μg/ml
Amphotericin	50 μg/ml
Polymixin B	3.3 μg/ml
Bacitracin	200 μg/ml
Nalidixic acid	10.7 μg/ml

Agar Columbia is from Difco, Detroit, MI. All the chemicals are from Sigma Co, St Louis, MD.

Horse blood used to prepare the plates deserves particular attention and its source has to be carefully selected for two reasons.

- We have observed that the blood of some of the horses to which we have had access in the past years have high titres of antibodies cross-reacting with *H. pylori* and that in the plates prepared with their blood the growth of *H. pylori* is inhibited (Rossi and Ghiara, unpublished observations).
- Horses may occasionally undergo antibiotic treatment. It is important to have access to an animal that has not recently received any antibiotic treatment, since the residual drug present in the circulation may affect the optimal growth of *H. pylori*.

One should make sure that the plates are freshly prepared (i.e. not older than 7–10 days) and spread, with a flamed 10 μl loop (blue loop; Nunc, Kamstrup, Denmark), a moderately heavy inoculum of bacteria as a 'lawn" over the plate. The plates are placed in the microaerophilic jar (Oxoid, Basingstoke, UK) and incubated at 37°C for 3–5 days. Bacteria grown from one plate will be passaged in the same way on to two or three new plates.

Preparation of bacterial suspension

Prepare the bacterial suspension immediately before administration to the mice. Collect the bacteria from each plate with a sterile cotton swab and resuspend them in 0.5 ml of sterile saline per plate. Keep the suspension on ice.

Prepare 1 ml of a 1:100 dilution of the obtained suspension and read the optical density (OD) at 530 nm. An $OD_{530} = 2.0$ corresponds to about 10^9 cfu/ml.

If necessary, dilute the bacterial suspension for the inoculation to mice such that 10^9 bacteria can be administered in 0.1–0.2 ml of sterile saline. It is important that the volume of saline in which the bacteria are collected is small, so that dilution instead of concentration is performed to obtain the proper bacterial concentration. In our experience, centrifugation of bacteria before inoculation to mice may decrease the efficiency of colonization, most probably as a result of damage to bacterial flagella, which are essential for colonization.

In our laboratory, bacteria recovered from one well-grown plate are usually sufficient to inoculate 10–15 mice. Keep the suspension on ice until it is administered to the mice. Bacteria has to be given to mice within 15–20 minutes from beginning to collect them from the plates.

Intragastric inoculation

Before each intragastric (i.g.) treatment mice are fasted for 24 hours, having *ad libitum* access to water only. For the i.g. inoculations a Luer lock stainless-steel (50 mm × 1 mm) gavage with a round tip is used (Figure 30.1A). This gavaging device can be autoclave-sterilized and we find that it is usually easier to use than flexible tubes. To perform i.g. inoculations, firmly hold the mouse by the back of the neck and keep it in a vertical position (Figure 30.1B) and pass the gavave gently down the oesophagus for about half of its length. If resistance is encountered while introducing the gavage, extract it and start again. The maximum volume that can be administered in a 5-week-old mouse is 0.5 ml, but smaller volumes are preferred.

To neutralize gastric acidity, fasted mice are given 0.2 ml of 0.2 M $NaHCO_3$ i.g. 10–20 minutes before receiving the bacterial inoculum. It is therefore a good practice to start this treatment while the bacterial suspension is being prepared.

A total of 10^9 bacteria in 0.1–0.2 ml of sterile saline are administered i.g. to mice. Mice are kept fasted for 2–3 hours after receiving the bacteria. The inoculation of bacteria is repeated as described three times every other day for a week.

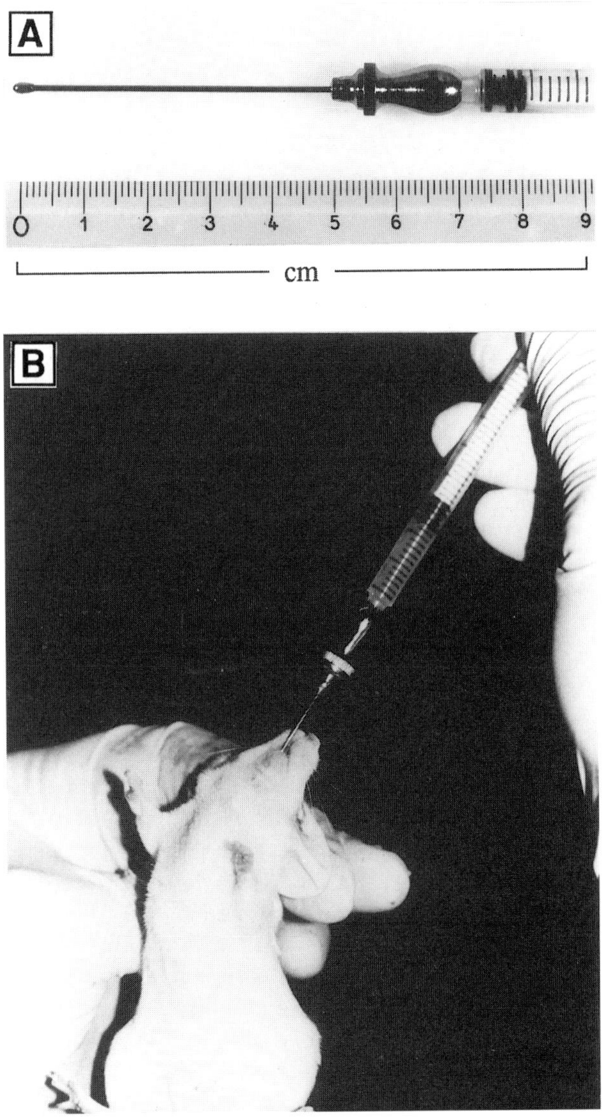

Figure 30.1 (A) The stainless-steel gavaging device used for i.g. inoculations. (B) The correct way to hold the mouse to receive the treatment.

Detection of infection

Before being sacrificed, mice are fasted for 24 hours with free access to water. During this fast, it is preferable also to prevent mice having access to wood shavings.

Surgical instruments used to manipulate the stomachs (forceps and scissors) are kept in ethanol before the collection of each organ. The mice are sacrificed by cervical dislocation and their stomachs are removed and placed on a piece of clean paper towelling. The stomach is cut along the lesser curvature (Figure 30.2, arrows) and flattened on the the towelling. The small amount of undigested food still present and any coarse material still present in the lumen are gently rolled out with the forceps. A specimen of the tissue including antral and oxyntic mucosa can be kept for standard histological examination.

The whole stomach is held firmly using the forceps, and gently laid on the plate, making sure that the side of the lumen is in contact with the agar surface of the plate (Figure 30.3A). With the aid of a flamed 10 μl sterile plastic loop (blue loop; Nunc), the mucosal surface of the tissue is then gently streaked on to the agar plate (Figure 30.3B).

We have found that the plate composition reported by McColm *et al.* (1995) is very effective to avoid overgrowth of contaminants (Table 30.1, mixture B), although the growth

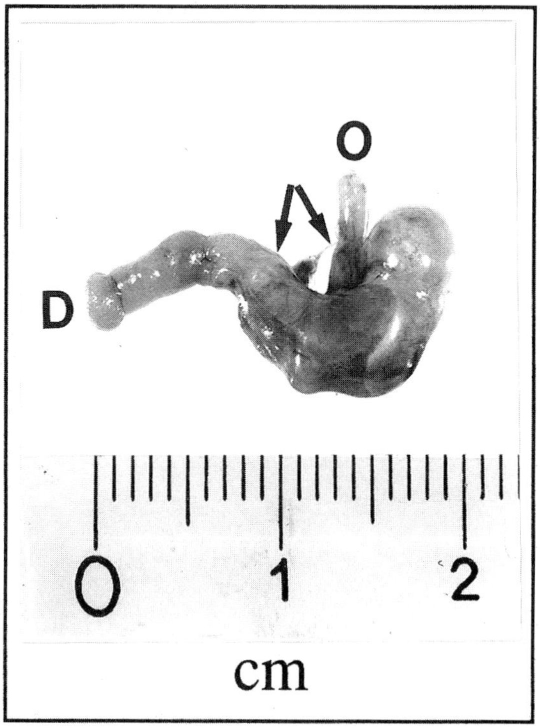

Figure 30.2 The mouse stomach. Arrows indicate the lesser curvature. O, oesophagus; D, duodenum.

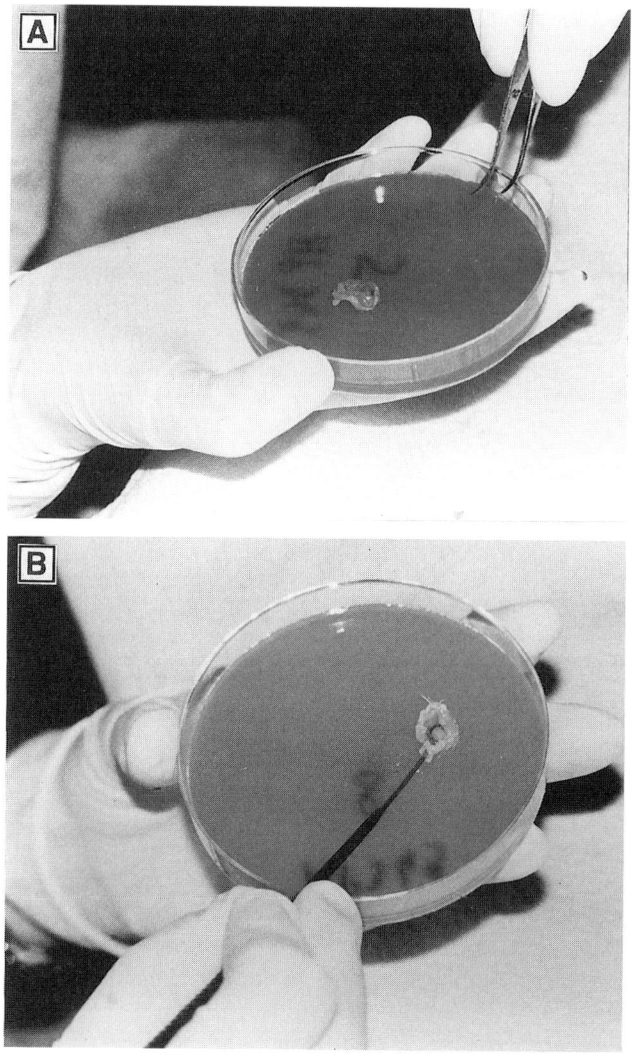

Figure 30.3 The collected stomach is placed on to the agar plate (A) and then gently streaked with the aid of a 10 µl sterile loop (B).

of *H. pylori* is much slower than in normal plates. If a streptomycin-resistant strain is used, then use normal plates (Table 30.1, mixture A) with the addition of streptomycin (400 µg/ml). The agar plates are subsequently incubated in microaerophilic conditions for 5–7 days. With plates prepared by the method of McColm *et al.* (1995), much longer times of incubation (up to 12 days) are needed.

The *H. pylori* colonies grown have a characteristic aspect (small, dome-shaped and translucent under incident light) and can be easily identified by visual inspection (Figure 30.4). Using the selective culture conditions described, and particularly when streptomycin-resistant strains are used for infection, contaminants are rare, but they may occur. Therefore, to gain confidence in the visual identification of *H. pylori* colonies grown on plates, it is useful to examine a few of them under a light microscope. To do this the colony is collected with a flamed 1µl plastic loop (clear loop; Nunc) and mixed with a 50µl drop of saline on a glass slide. The air-dried heat-fixed slide is then stained with carbolfuchsin (Difco, Detroit, MI) or with Gram's stain (Merck, Darmstadt, Germany) and examined under a microscope (1000× magnification) in comparison with control slides obtained from cultures of the same pure *H. pylori* strain before inoculation into mice.

Identity of colonies can be confirmed by the rapid urease test: the bacterial colony is collected with a 1 µl sterile plastic loop (clear loop, Nunc) and resuspended in a few drops of "urease reagent" (3 mM Na_2HPO_4 pH 6.8 containing 100 mM urea and 7 µg/ml phenol red).

In the presence of *H. pylori,* the orange colour of the pH indicator should turn to purple red within a few seconds, because of the enzymatic activity of urease, which rapidly causes a rise in the pH value as a consequence of the catalysed hydrolysis of urea into ammonia and CO_2.

Outcome of infection

We have reported that *H. pylori* strains freshly isolated from a patient biopsy and adapted to mouse gastric mucosa by *in-vivo* passaging, can establish persistent colonization in normal xenobiotic CD1 mice (Marchetti *et al.*, 1995). Infection by type I strains causes epithelial damage and inflammation that becomes evident after 8–12 weeks of infection. On the other hand, type II strains only cause a mild gastritis. These findings have subsequently been confirmed by others (Kleanthous *et al.*, 1995; Hong *et al.*, 1996; Sakagami *et al.*,

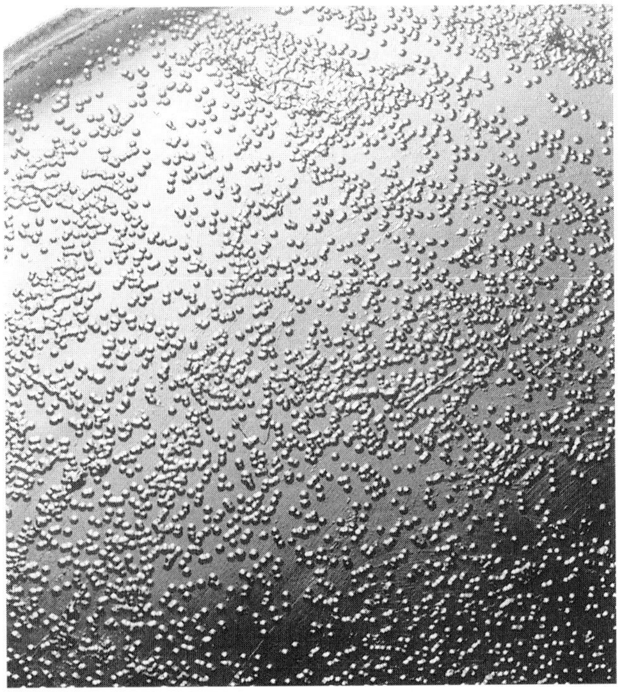

Figure 30.4 *H. pylori* colonies grown from an infected mouse stomach.

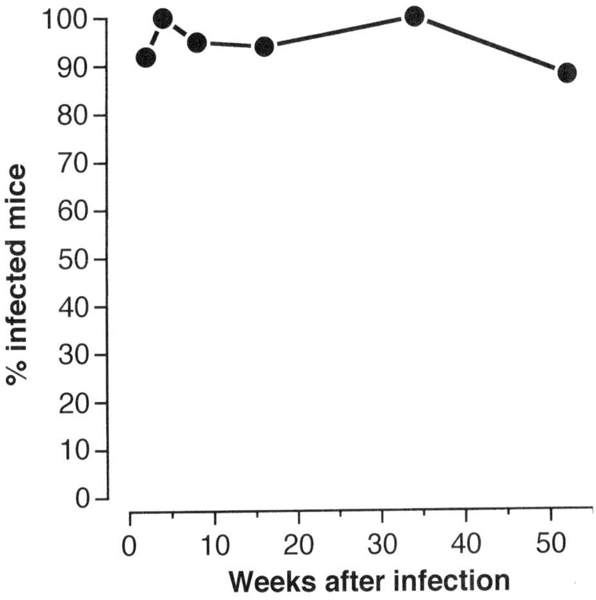

Figure 30.5 Persistence of infection by *H. pylori* in CD1 mice. A large group of 5-week-old male CD1/SPF mice was infected with a mouse-adapted *H. pylori* type I strain. At different time intervals, groups of mice were sacrificed and the gastric colonization was assessed as described.

1996; Lee *et al.*, 1997). As seen in Figure 30.5, inoculation of mice with an adapted strain of *H. pylori* results in a colonization of their gastric mucosa that persists for up to 1 year. The number of colonies that were recovered from the stomachs was also stable over the examined period of time and varied between 10^3 and 2×10^4 in the majority of infected mice (not shown).

The pathological changes induced by infection were followed by histological examination. Figure 30.6 shows pictures of representative histological fields of gastric mucosa of uninfected control mice (A) and mice after the onset of infection (B, C). At 8 weeks after infection, gastric lesions consist mainly of focal epithelial damage. Fields showing superficial epithelial erosive–reparative lesions with polymorphonuclear leukocytes infiltrating the lesions and the surrounding mucosa are present in most infected mice (B). These lesions, however, become less frequent in the mice

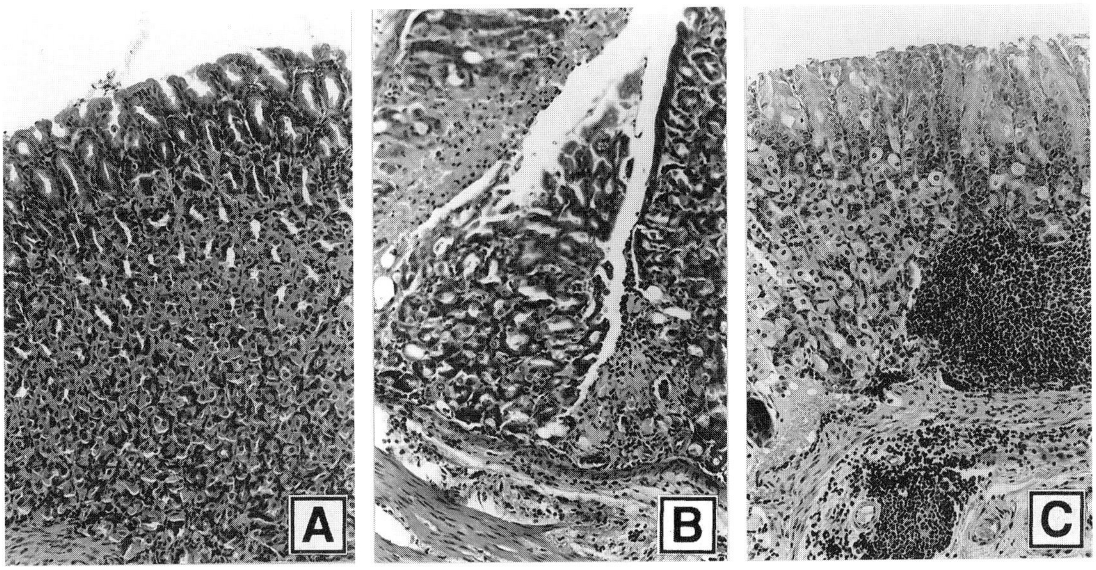

Figure 30.6 Gastric pathology induced by infection in CD1 mice. (A) The oxyntic mucosa on an uninfected control mouse. (B) The same region of the gastric mucosa in a mouse infected for 8 weeks; note the superficial erosions and the inflammatory cell infiltration. (C) A large lymphoid follicle in the oxyntic mucosa of a mouse infected for 1 year.

observed at subsequent times. At 16 and 34 weeks of infection, the most frequently observed lesions consist of inflammatory cells (mainly mononuclear cells) infiltrating lamina propria in the superficial layers of the mucosa (chronic superficial gastritis). Polymorphonuclear cells are also found, mainly associated with the surface epithelial layer. At 34 weeks of infection, small lymphoid aggregates appear in the lamina propria at the basis of the glands. In mice infected for 1 year or longer the main histopathological finding is a follicular gastritis with large lymphoid follicles present in the mucosa as well as in the submucosa (C).

The serum antibody response can be followed during infection by ELISA or Western blot analysis using *H. pylori* antigens. Using a whole *H. pylori* cell sonicate, specific antibodies are already detectable at 4–8 weeks after the onset of infection and increase slowly over time. The IgG response has been further assessed by determining the pattern of IgG_1 and IgG_{2a} isotypes during infection. A prevailing serum IgG_{2a} response is detectable in CD1 mice at all stages of infection, suggesting that a Th1-type of response is elicited by the infection. Figure 30.7 shows the *H.-pylori*-specific antibody titres in the sera of mice infected for 34 weeks. This observation is in agreement with what has been reported also in humans (see Telford *et al.*, 1997 for a review) and further indicates the relevance of this mouse model of *H. pylori* infection to human disease.

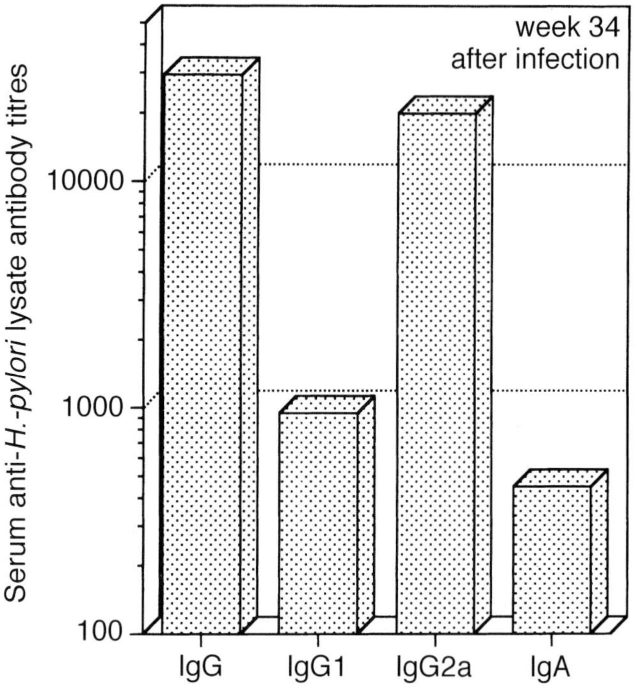

Figure 30.7 *H. pylori*-specific antibody titres in pooled sera from CD1 mice infected for 34 weeks.

The persistence and outcome of infection can vary in different mouse strains. In our experience, C57BL/6 mice are persistently colonized and develop a gastric pathology that is more evident than that of CD1 mice. On the other hand, BALB/c mice are less susceptible to infection than CD1 and C57BL/6. In fact they are able to clear the infection within 3–5 months. Moreover, BALB/c mice do not develop any significant pathology and mount a serum response against purified antigens characterized by a prevalence of ELISA titres of the IgG_1 isotype (Del Giudice and Ghiara, unpublished observations), which suggests a polarization of the $CD4^+$ T-cell response towards a Th2 phenotype.

These data further strengthen the hypothesis of a polarization of the immune response towards a Th1 phenotype in *H. pylori* infection, which is not effective in fighting off the infection and is also hypothesized to play a role in maintaining the chronic inflammatory status. Finally it is also tempting to speculate that a polarization towards a Th2 type of immune response is instead protective and might be helpful in conferring resistance to colonization of the gastric mucosa by *H. pylori*.

Usefulness of the model

Since its development in 1995, this mouse model has been used by several groups, mainly to assess the feasibility of vaccination (Kleanthous *et al.*, 1995; Marchetti *et al.*, 1995). This model is presently the only one that allows to test the potential, as vaccine candidates, of virulence factors that are peculiar to the more virulent (type I) *H. pylori* strains, such as VacA and those expressed by the *cag* PAI. This approach would not be feasible in animal models employing other *Helicobacter* spp. (for example *H. felis*) that do not express them (Xiang *et al.*, 1995). The feasibility of both the preventive and the therapeutic mode of vaccination has been assessed using oral vaccination with purified antigens together with mucosal adjuvants (Telford *et al.*, 1996).

A recent study has used this model to characterize some of the initial steps of the inflammatory response to infection (Elizalde *et al.*, 1997). Because of our excellent knowledge of the murine immune system, this model will be useful to better characterize several aspects of the immune response to infection, and also to define protective mechanisms responsible for the active immunity elicited by vaccination.

Characterization of pathogenesis has been also approached using this model. In our initial study, we have reported that type I (cytotoxic) strains only induce appreciable gastric disease in mice (Marchetti *et al.*, 1995). The model is currently being used in our and other laboratories to define the role of major putative players, such as the toxin VacA (Brzozowski *et al.*, 1996) and the proteins encoded by the PAI (Marchetti *et al.*, manuscript in preparation), in the disease.

Finally, this model has also been used for the rapid *in-vivo* screening of anti-*H. pylori* chemotherapeutics under development (McColm *et al.*, 1995; McColm and Bagshaw, 1996; Mellgard *et al.*, 1996).

It is clear that this model may not be ideal for all purposes. A major limitation is that, in order to detect the gastric colonization and pathology, mice have to be sacrificed. This makes the model unsuitable for long-term follow-up studies on the effects of infection on the same individual. Other limitations are the well-known differences in the gastric anatomy (absence of an acid-secreting fundus in mice) and physiology (i.e. the mouse gastric pH never goes below 4–4.5) between mouse and man. Also, as a result of the higher gastric pH and also of coprophagic behaviour, a commensal bacterial flora is commonly present in the stomach of mice. All these differences may make this model of infection not ideal for all investigators.

Conclusions

The development of this mouse model of persistent infection by *H. pylori* has represented a major breakthrough in the field because it has opened up the possibility of performing *in-vivo* studies for a larger number of laboratories, using an affordable host. Previously described models, which involved expensive and unpractical hosts such as germ-free or athymic mice (Karita *et al.*, 1991), monkeys (Dubois *et al.*, 1994) or germ-free piglets (Krakowska *et al.*, 1987), could only be performed in very specialized facilities.

Several aspects of the pathogenesis of this important human infection can now therefore be more easily investigated using this model.

Acknowledgements

We are grateful to Drs Marta Marchetti and Michela Rossi for their commitment to the development and optimization of this animal model. We thank Sonia Capecchi, Sara Bossini, Laura Pancotto, Luigi Villa, Alfio Ruspetti, and Fabrizio Zappalorto for their invaluable technical assistance. We also thank Giorgio Corsi for the artwork.

References

Blaser, M. J., Parsonnet, J. (1994). Parasitism by the 'slow' bacterium *Helicobacter pylori* leads to altered gastric homeostasis and neoplasia. *J. Clin. Invest.*, **94**, 4–8.

Brzozowski, T., Konturek, P. C., Karczewska, E., Ghiara, P., Hahn, E. G., Konturek, S. J. (1996). The mouse model of infection with cytotoxin expressing strain of *Helicobacter pylori* (Hp) in studying the pathogenesis of chronic gastric ulcer. *Gut*, **39** (Suppl. 2), A75.

Cellini, L., Allocati, N., Angelucci, D. *et al.* (1994). Coccoid *Helicobacter pylori* not culturable *in vitro* reverts in mice. *Microbiol. Immunol.*, **38**, 843–850.

Censini, S., Lange, C., Xiang, Z. Y. *et al.* (1996). Cag, a pathogenicity island of *Helicobacter pylori*, encodes type 1-specific and disease-associated virulence factors. *Proc. Natl Acad. Sci. USA*, **93**, 14648–14653.

Covacci, A., Falkow, S., Berg, D. E., Rappuoli, R. (1997). Did the inheritance of a pathogenicity island modify the virulence of *Helicobacter pylori*? *Trends Microbiol.*, **5**, 205–208.

Dubois, A., Fiala, N., Hemanackah, L. M. *et al.* (1994). Natural gastric infection with *Helicobacter pylori* in monkeys: a model for spiral bacteria infection in humans. *Gastroenterol.*, **106**, 1405–1417.

Elizalde, J. I., Gomez, J., Panes, J. *et al.* (1997). Platelet activation in mice and human *Helicobacter infection*. *J. Clin. Invest.*, **100**, 996–1005.

Figura, N., Guglielmetti, P., Rossolini, A. *et al.* (1989). Cytotoxin production by *Campylobacter pylori* strains isolated from patients with peptic ulcers and from patients with chronic gastritis only. *J. Clin. Microbiol.*, **27**, 225–226.

Ghiara, P., Covacci, A., Telford, J. L., and Rappuoli, R. (1996). *Helicobacter pylori*: pathogenic determinants and strategies for vaccine design. In: *Concepts in Vaccine Development* (ed. Kaufmann, S. H. E.), pp. 459–496. Walter de Gruyter, Berlin.

Hong, L., Mieziewska, K., Kalies, T., Berglindh, T., Mellgard B. (1996). Long term infection of *H. pylori* in mice: a study of infection, inflammation and immunological response. *Gut*, **39** (Suppl. 2), A76.

Karita, M., Kouchiyama, T., Okita, K., Nakazawa, T. (1991). New small animal model for human gastric *Helicobacter pylori* infection: success in both nude and euthymic mice. *Am. J. Gastroenterol.*, **86**, 1596–1603.

Kleanthous, H., Tibbits, T., Bakios, T. J. *et al.* (1995). *In vivo* selection of a highly adapted *H. pylori* isolate and the development of an *H. pylori* mouse model for studying vaccine efficiency and attenuating lesions. *Gut*, **37** (Suppl. 37), A94.

Krakowska, S., Morgan, D. R., Kraft, W. G., Leunk, R. D. (1987). Establishment of gastric *Campylobacter pylori* infection in the neonatal gnotobiotic piglet. *Infect. Immun.*, **55**, 2789–2796.

Lee, A., Fox, J. G., Otto, G., Murphy, J. (1990). A small animal model of human *Helicobacter pylori* infection. *Gastroenterology*, **99**, 1315–1323.

Lee, A., O'Rourke, J., Deungria, M. C., Robertson, B., Daskalopoulos, G., Dixon, M. F. (1997). A standardised mouse model of *Helicobacter pylori* infection: introducing the Sydney strain. *Gastroenterology*, **112**, 1386–1397.

Marchetti, M., Arico, B., Burroni, D., Figura, N., Rappuoli, R., Ghiara, P. (1995). Development of a mouse model of *Helicobacter pylori* that mimics human disease. *Science*, **267**, 1655–1658.

McColm, A., Bagshaw, J. (1996). Achlorhydric potentiation of antibiotic activity in mice colonised with *H. pylori*. *Gut*, **39** (Suppl. 2), A10.

McColm, A. A., Bagshaw, J., Wallis, J., McLaren, A. (1995). Screening of anti *Helicobacter* therapies in mice colonized with *H. pylori*. *Gut*, **37** (Suppl. 1), A92.

Mellgard, B., Varai, E., Larsson, H. (1996). Omeprazole enhances *H. pylori* eradication rate of amoxicillin/clarithromycin/-metronidazole combinations in a mouse model. *Gut*, **39** (Suppl. 2), A76.

Sakagami, T., Dixon, M., O'Rourke, J. *et al.* (1996). Atrophic gastritis changes in both *Helicobacter felis* and *Helicobacter pylori* infected mice are host dependent and separate from antral gastritis. *Gut*, **39**, 639–648.

Telford, J. L., Ghiara, P. (1996). Prospects for the development of a vaccine against *Helicobacter pylori*. *Drugs*, **52**, 799–804.

Telford, J. L., Covacci, A., Ghiara, P., Montecucco, C., Rappuoli, R. (1994a). Unravelling the pathogenic role of *Helicobacter pylori* in peptic ulcer: potential for new therapies and vaccines. *Trends Biotechnol.*, **12**, 420–426.

Telford, J. L., Ghiara, P., Dell'Orco, M. *et al.* (1994b). Gene structure of the *Helicobacter pylori* cytotoxin and evidence of its key role in gastric disease. *J. Exp. Med.*, **179**, 1653–1658.

Telford, J. L, Covacci, A., Rappuoli, R., Ghiara P. (1997). Immunology of *Helicobacter pylori* infection. *Curr. Op. Immunol.*, **9**, 498–503.

Tytgat, G. N. J. (1996). Current indications for *Helicobacter pylori* eradication therapy. *Scand J. Gastroenterol.*, **31** (Suppl. 215), 70–73.

Xiang, Z., Censini, S., Bayeli, P. F. *et al.* (1995). Analysis of expression of CagA and VacA virulence factors in 43 strains of *Helicobacter pylori* reveals that clinical isolates can be divided into two major types and that CagA is not necessary for expression of the vacuolating cytotoxin. *Infect. Immun.*, **63**, 94–98.

Chapter 31

Animal Models of *Helicobacter* (Ferrets)

R. P. Marini and J. G. Fox

Background of human infection

Helicobacter pylori is an important human pathogen that causes active, chronic gastritis and peptic ulcer disease and has been associated with gastric adenocarcinoma and lymphoma. It is the most common infection in the world, with levels of seroprevalence as high as 85% in Nigeria and 75% in Vietnam (Megraud *et al.*, 1989; Holcombe *et al.*, 1992). The infection is more prevalent in developing nations but is associated with lower socioeconomic status in both the industrialized and the developing world. Acquisition rate of *H. pylori* is lower in adulthood than in childhood and infection appears to be persistent (Vincent, 1995). Additional risk factors include large family size, crowding and ethnic group (Vincent, 1995).

Treatment of *H. pylori* typically consists of antibiotics, mucosal protectants, and antisecretory agents in various combinations. These regimens are variably successful in eradication of the bacterium, defined clinically in most trials as a negative *H. pylori* culture 4 weeks after termination of treatment (Chiba *et al.*, 1992; Marshall, 1993). Despite reasonable success, poor patient compliance, side effects, limited efficacy, and the emergence of bacterial resistance continue to motivate the search for simple, safe, and efficacious treatment options.

Background of the model

In 1985, a gastric spiral organism was isolated from a gastric ulcer of a ferret and subsequently identified as *Helicobacter mustelae* (Fox *et al.*, 1986). Since that time, *H. mustelae* infection of ferrets has been extensively studied as a model of *H. pylori* in people (Fox and Lee, 1997). The *H. mustelae* organism causes gastritis and has been associated with gastric adenocarcinoma and mucosal associated lymphoid tissue (MALT) lymphoma (Fox, 1994; Erdman *et al.*, 1997; Fox *et al.*, 1997). Ferrets become infected at around the time of weaning at 6 weeks of age; infection is life-long and gastritis severity increases with age (Table 31.1; Fox *et al.*, 1991a; Fox, 1994). Gastric and duodenal ulcer may occur and result in clinical signs of pallor, melena, vomiting, weight loss, and inappetance. Acute episodes of gastric bleeding occur uncommonly, but have been observed by the author in the setting of dexamethasone administration to ferrets subjected to long periods of anesthesia. Hypergastrinemia, a feature of *H. pylori* infection of people, has also been demonstrated in association with *H. mustelae* infection in the ferret (Perkins *et al.*, 1996).

Koch's postulates have been fulfilled for *H. mustelae* in the ferret (Fox *et al.*, 1991a). Administration of the organism to ferrets that are specific-pathogen-free (SPF) for *H. mustelae* results in gastric colonization, chronic gastritis, and elevation of serum anti-*H. mustelae* antibody titer (Fox *et al.*, 1991a). Histologic appearance corresponds to the presence and microscopic localization of *H. mustelae*. In the body, where *H. mustelae* colonizes the mucosal surface, superficial mononuclear gastritis is observed. This is in contrast to the diffuse mononuclear gastritis of the antrum, where the organism colonizes the surface, gastric pits, and

Table 31.1 Severity of gastritis related to ferret age — there was a statistically significant difference between groups ($p < 0.001$; $\chi^2 = 28$; Fox *et al.*, 1991b)

		No. (%) of ferrets in indicated category		
Age group (months)	*Total no. of animals*	*Minimal or no inflammation*	*Mild chronic gastritis*	*Moderate to severe chronic gastritis*
0–3	22	13 (59)	9 (41)	0 (0)
3–12	27	14 (52)	9 (33)	4 (15)
> 12	33	3 (9)	12 (36)	18 (55)

Handbook of Animal Models of Infection
ISBN 0-12-775390-7

the superficial portion of the glands (Fox *et al.*, 1990). The ferret model was originally described and proposed by Fox and has been used to explore such diverse areas of inquiry as *Helicobacter* pathogenesis, epizootiology, carcinogenesis, and antimicrobial therapy (Fox *et al.*, 1991a; Blanco *et al.*, 1993; Andrutis *et al.*, 1995, 1997; Batchelder *et al.*, 1996; Marini *et al.*, 1997).

Animals

Ferrets (*Mustela putorius furo*) may be purchased from a number of vendors. All ferrets used in studies by our laboratory are purchased from Marshall Farms (North Rose, NY). Ferrets of all ages may be used; animals aged from 3–4 days to 8 years have been used in the characterization of *H. mustelae* infection (McColm *et al.*, 1993; Fox and Lee, 1997). Gastritis severity appears to increase with age. Males are larger (approximate weight 900–1500 g) than females (700–1000 g) and, subjectively, appear to be more docile. In our experience, pregnant and periestrous females are more prone to mucosal bleeding from gastric endoscopic biopsy procedures; consequently, we commonly use males or spayed females in our *Helicobacter* studies. Natural infection with *H. mustelae* has been found in ferrets from the USA, UK, Canada, and Australia, but not from New Zealand (Fox, 1988; Fox *et al.*, 1991a; Fox and Lee, 1997).

Preparation of animals

Ferrets may be housed in a variety of primary enclosures and may be pair-housed with animals of the same sex. Stainless-steel (Research Equipment Co., Bryant, TX, 25 × 31 × 19 inches/63.5 × 78.75 × 48.25 cm) or plastic rabbit cages (MediCage Lock Solutions, Kenilworth, NJ) adapted for their use are commonly employed. Drop pans may be lined with absorbent litter or cageboard. For breeding programs, litter boxes should be provided 1 week prior to parturition and remain until weaning. Water should be available *ad libitum* and may be provided in bottles with sipper tubes or through automatic watering systems. Ferret chows are commercially available (Lab Diet No. 5L14; PMI Feeds, St Louis, MO and other ferret chows) and should be fed *ad libitum* in stainless-steel bowls.

Ferrets enjoy play devices manufactured for cats but are very curious and may ingest small objects, leading to gastrointestinal foreign body and obstruction. Fabric sleeping tubes (Marshall Farms, North Rose, NY) should be provided but must be laundered regularly. There are numerous catalogs available to provide environmental enrichment items; local pet stores will often carry a selection of ferret toys as well.

Details of surgery

Overview

Evaluation of mucosal biopsy samples obtained by gastric endoscopy is commonly used to follow the course of *H. mustelae* infection.

Materials required

Atropine, ketamine, xylazine, yohimbine, tuberculin syringes, 25G or 26G needles, Pentax pediatric 5 mm fiberoptic flexible bronchoscope (Pentax FB15-H, Orangeburg, NJ), or similar instrument, flexible pinch-biopsy forceps, monochromatic light source with variable illumination power (Wolf Electro Surgical Generator Model 2083.40, Rosemont, IL), or similar instrument, suction jar, and suitable vacuum. Stables *et al.* (1993) and McColm *et al.* (1993) report the use of a veterinary gastroscope with 6.3 mm insertion tube diameter (Scholt VFS2, Arnolds Veterinary Products, Shrewsbury, UK) to perform antral biopsies in female ferrets.

The suction/insufflation port of the endoscope is attached via sterile tubing to a three-way stopcock to which are also attached a vacuum source and a sphygmomanometer bulb or controlled air supply. Suction or insufflation can be achieved by choosing the appropriate stopcock setting and light pressure on the control port (Figure 31.1; Cabot and Fox, 1990).

Anesthesia

Food, but not water, is withheld overnight from ferrets. Ten minutes after atropine premedication (0.04 mg/kg i.m.), animals are anesthetized using ketamine (30 mg/kg i.m.: Ketaset®, Fort Dodge Laboratories, Fort Dodge, IO) and xylazine (3 mg/kg i.m.; Rompun®, Bayer Corp., Shawnee Mission, KS). Anesthesia may be prolonged by giving incremental dosages of ketamine at one-half the induction dose; xylazine may be dosed similarly if necessary. Atropine should be administered every 30–40 minutes to avoid xylazine-induced arrythmias (Green *et al.*, 1981; Moreland and Glaser, 1985). Ferrets are observed continuously for adequate respiration during the procedure. The α_2-antagonist yohimbine is administered (0.5 mg/kg i.m.; Lloyd Laboratories, Shenandoah, IA) at the completion of gastric endoscopic biopsy (Sylvina *et al.*, 1990).

Endoscopic procedure (Cabot and Fox, 1990)

Ferrets are maintained in left lateral recumbency. A mouth gag fashioned from a needle cap or syringe case is wedged between the left superior and inferior canines and holds the mouth in an open position. The scope is inserted into the

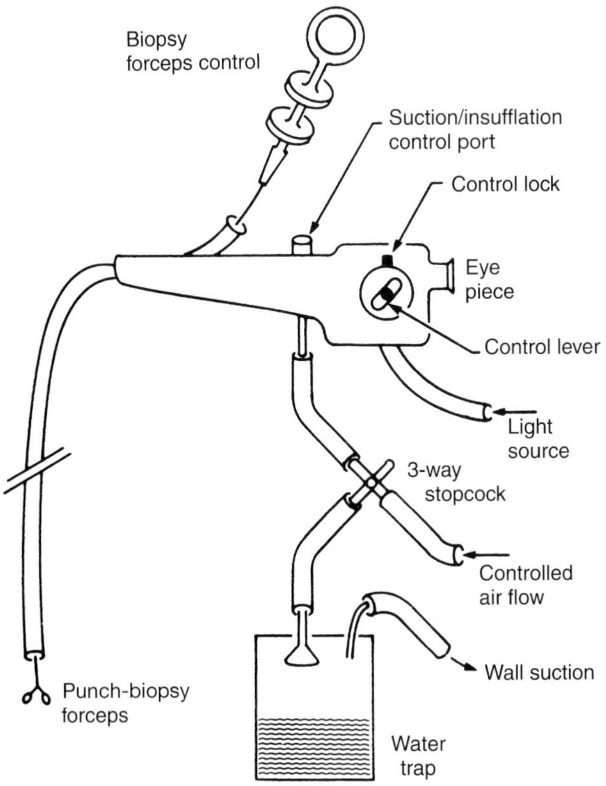

Figure 31.1 Modified esophagogastroscope for gastric mucosal endoscopic biopsies.

oropharynx and gently advanced to a length corresponding to the distance from the caudalmost limit of the costal arch to the rostralmost part of the mouth. Alternatively, the endoscopist may observe the scope's advance through the lower esophageal sphincter into the stomach. The location of the scope tip is verified by transillumination of the abdominal wall; the stomach may then be gently distended with room air via insufflation with the hand-held sphygmomanometer bulb. The scope tip is centralized, allowing examination of the mucosal surface. Mucus and other debris may be removed from the scope tip by irrigation with distilled water through the working channel. By advancing the scope and adjusting the tip angulation, the entire mucosal surface may be examined. The pylorus, *incisura gastrica*, fundus, and cardia may be examined and biopsied. Mucosal biopsy samples 2 mm in size are commonly obtained from the body and antrum; three to five samples are taken per site depending upon the study design. Upon completion of the procedure, air and fluid are aspirated and the scope is withdrawn with the scope tip straightened and the control knob unlocked.

Postprocedural care

Ferrets should be monitored frequently until they have recovered from anesthesia. Vomiting occurs commonly during emergence but is not associated with long-lasting sequelae. If animals are ambulatory, they may be offered their regular food and water on the evening of the procedure. Melena often occurs within the first 24 hours but should not persist beyond the morning of the day following the procedure. The most common serious untoward effect of gastric endoscopic biopsy in ferrets is mucosal bleeding; lethargy, pallor, and persistent melena are typical clinical signs of mucosal bleeding. Ferrets should be transfused if their hematocrits fall below 18–20%. The lack of blood groups makes prior cross-matching of donor and recipient blood unnecessary (Manning and Bell, 1990). Endoscopic examination for the purpose of mucosal cautery has not been necessary in our experience.

Storage and preparation of inocula

H. mustelae is a natural infection of ferrets, making experimental inoculation important only for specific-pathogen (*H.-mustelae*)-free animals. Available isolates include the North-American-type strain, ATCC 43772 (Rockville, MD) and the European-type strain NCTC 12032 (National Collection of Type Culture, London, UK). The organism may be stored at –70°C in a medium of brucella broth containing 10% glycerol (Fox *et al.*, 1988). Cultures may be grown from thawed aliquots of the organisms or from gastric endoscopic biopsies by plating on trypticase soy agar containing 5% sheep blood (TSAII-5%; BBL Microbiology Systems, Cockeysville, MD) at 37°C in vented jars containing 90% N_2, 5% H_2, and 5% CO_2. Cultures are checked every 3 days for 14 days for growth and kanamycin resistance (Andrutis *et al.*, 1995). Organisms are transferred from plates to *Brucella* broth or brain–heart infusion broth with 30% glycerol (Difco Laboratories, Detroit, MI) and kept on ice prior to administration to ferrets.

Phenotype

H. mustelae is a gram-negative, rod-shaped or slightly curved bacterium possessing four to eight sheathed flagella distributed both laterally and at each pole (Figure 31.2; Fox *et al.*, 1989; O'Rourke *et al.*, 1992). A summary of biochemical and other characteristics of *H. mustelae* and *H. pylori* is presented in Table 31.2 (Fox and Lee, 1997).

Infection procedure

A number of dosage regimens have been used for administration of *H. mustelae* to ferrets; all of them require an anesthetized animal. Food is withheld from ferrets overnight, cimetidine is administered (10–30 mg/kg, s.q.; Smith Kline Beecham, Philadelphia, PA; Fox *et al.*, 1991a) to induce transient hypochlorhydria, and ferrets are anesthetized 1 hour later with ketamine (30 mg/kg i.m.) and acepromazine

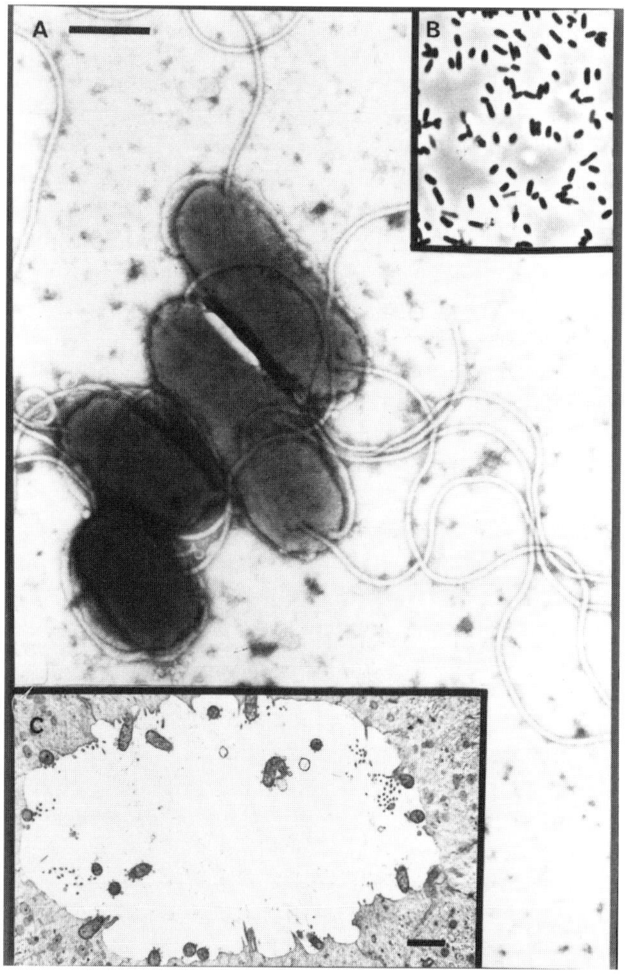

Figure 31.2 Electron micrographs of a pure culture of *H. mustelae* showing rod-shaped morphology and long polar and lateral flagella. (**A**) Transmission electron microscopy (bar = 0.2 μm). (**B**) Phase contrast microscopy. (**C**) Bacteria penetrating into epithelial cells (bar = 0.5 μm).

Table 31.2 Characteristics which differentiate *Helicobacter pylori* and *Helicobacter mustelae* (+, positive reaction; –, negative reaction; S, susceptible; R, resistant)

	H. pylori	*H. mustelae*
Catalase production	+	+
Nitrate reduction	–	+
Alkaline phosphatase hydrolysis	+	+
Urease	+	+
Indoxyl acetate hydrolysis	–	+
γ-glutamyl transpeptidase	+	+
Growth at 42°C	–	+
Growth with 1% glycine	–	–
Nalidixic acid (30 μg disc)	R	S
Cephalothin (30 μg disc)	S	R
Periplasmic fibers	–	–
No. of flagella	4–8	4–8
Distribution of flagella	Bipolar	Peritrichous
G + C content (mol %)	35–37	36

(0.25 mg/kg i.m.; Aveco Co., Fort Dodge, IO). After harvest of bacteria and suspension in brain–heart infusion broth with 30% glycerol (Fox *et al.*, 1991a), approximate concentration of *H. mustelae* is adjusted using optical density (Fox *et al.*, 1991a). Viability is assured by examination under phase microscopy (Fox *et al.*, 1991a). Microorganisms are administered via a premeasured orogastric tube inserted to the level of the costal arch (Feeding Tube, 8F, Professional Medical Products, Greenwood, SC). Inoculum doses vary: 1.5×10^8 cfu of *H. mustelae* in 3 ml of broth administered once daily for 2 days (Fox *et al.*, 1991a); 3 ml of 1.5×10^7 cfu/ml administered once (Batchelder *et al.*, 1996); two doses of 3×10^7 cfu per dose (Andrutis *et al.*, 1995); single 3 ml doses of 3.0×10^7 cfu/ml (Andrutis *et al.*, 1997). Typically, 1 ml of sterile broth and 1 ml of air is instilled into the feeding tube prior to its withdrawal from the ferret stomach.

Key parameters to monitor infection and response to treatment

Overview

The course of infection in both naturally and experimentally infected ferrets may be monitored by at least one of several manipulations of samples obtained by gastric endoscopic biopsy (Fox *et al.*, 1990; Figure 31.3), by evaluation of serum ELISA titers to *H. mustelae* (Fox *et al.*, 1990), by fecal culture (Fox *et al.*, 1992), or by urea breath test (McColm *et al.*, 1993).

Samples obtained by gastric endoscopic biopsy

Microbiology

Samples for culture are placed into 1 ml of sterile phosphate-buffered saline (PBS) and the tube is placed on to crushed ice pending transport to the laboratory. Samples are then homogenized using a sterile tissue grinder, and 0.25 ml of the homogenate and an equal volume of stock solution is plated on to blood agar plates and cultured as described above. Quantitative culture may be used to determine response to antimicrobial therapy in units of cfu per mg of tissue or per biopsy and is performed as follows: separate tubes containing 0.5 ml of PBS are pre-weighed, the biopsy is added, and the tube with biopsy sample is re-weighed. Tenfold serial dilutions of the homogenate are prepared in sterile PBS: from 10^1 to 10^5 for antrum; 10^1 to 10^3 for body. Concentrations are typically in the range of 10^4–10^9 cfu/biopsy in naturally infected, 9-month-old ferrets (unpublished).

Urease

Samples for urease are placed into 1 ml of sterile phosphate-buffered saline and the tube is placed on to crushed ice

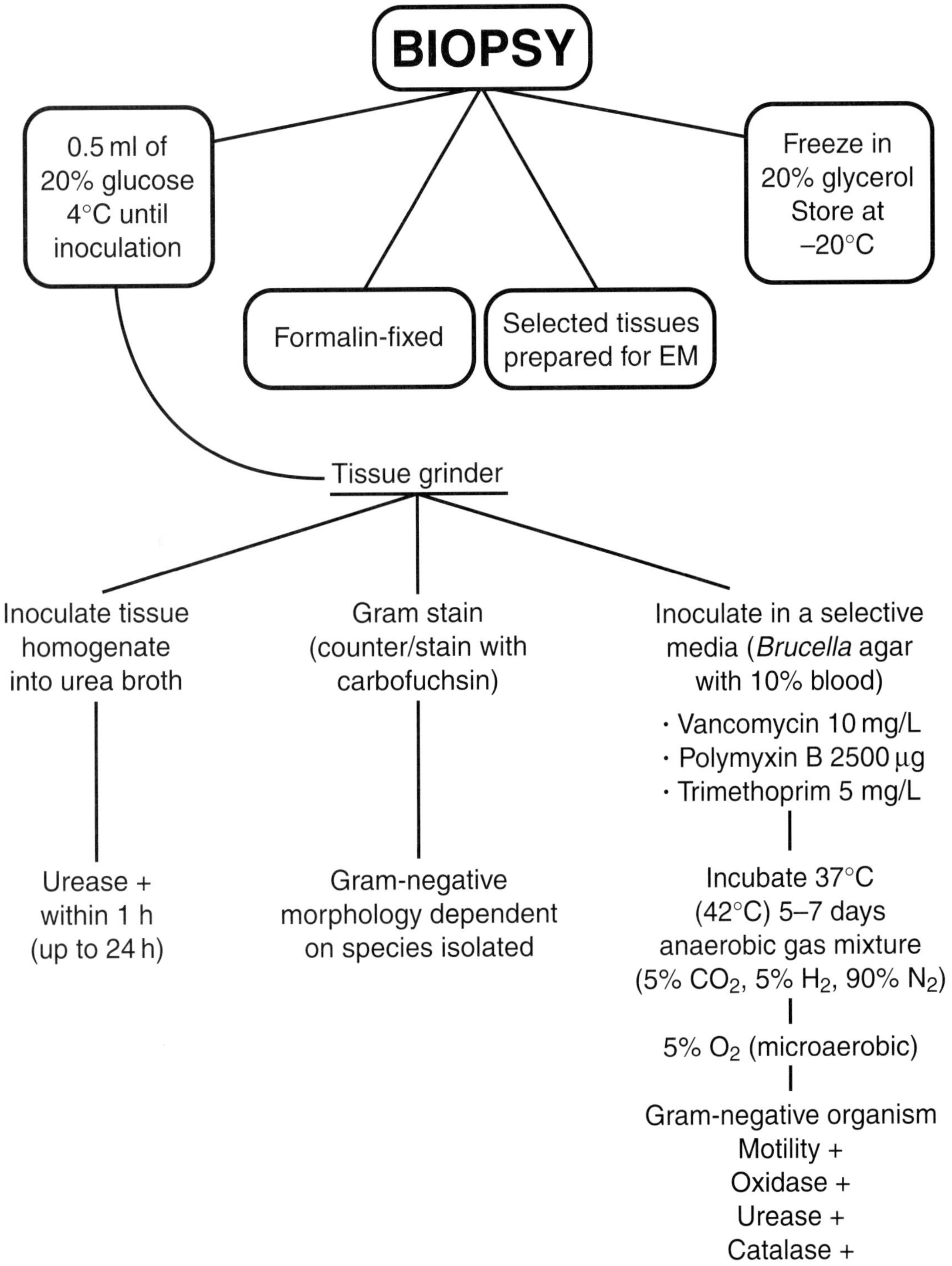

Figure 31.3 Isolation of *Helicobacter* spp. from gastric biopsies.

pending transport to the laboratory. The samples are tested for urease production according to the procedure of Hazell *et al.*, 1987. Briefly, each biopsy sample is placed into the well of a 96-well plate, 2 drops of urease reagent are added, and the plate is evaluated periodically for 24 hours. A color change indicates the presence of urease-positive organisms; most positive biopsy samples will effect a color change within 30 minutes. Although useful, this assay lacks specificity and sensitivity in our experience, as intestinal reflux and blood from the samples may cause false positive reactions and low-level infections may not produce a demonstrable color change (Andrutis *et al.*, 1995).

A modification of the urease assay provides semiquantitation of the number of organisms in each biopsy and is performed by incubation of 1–2 mm^3 of the biopsy sample in urea broth for 4 hours, centrifugation, and placement of duplicate 200 µl aliquots in microtiter plates. An ELISA reader set at 550 mm determines the amount of color change at 1, 6, 12, and 24 hours (Fox *et al.*, 1996).

Histopathology

Biopsy samples are placed into cassettes and oriented with 25 G needles so that the microtome blade cuts them perpendicular to the plane of the mucosal surface. Luminal and abluminal surfaces of these 2 mm biopsy samples may be identified by their differing appearances; the former appears dull and more textured while the latter is glistening and smooth. The samples also tend to contract and curl into comma shapes, with the luminal (mucosal) side comprising the convex surface. Cassettes are then placed into 10% neutral buffered formalin and the biopsy samples are processed routinely, embedded in paraffin and sectioned at 5 µm. The Warthin–Starry and hematoxylin and eosin stains are traditionally used for assessment of colonization and mucosal histopathologic morphology respectively (Figure 31.4).

Pathologic changes are scored as superficial gastritis, diffuse antral gastritis, and multifocal chronic atrophic gastritis according to the classification scheme of Correa (1988). Briefly, superficial gastritis refers to the presence of polymorphonuclear and/or mononuclear infiltration of the superficial layers of the mucosa. Mononuclear cell infiltration occupying the full thickness of the antral mucosa is referred to as diffuse antral gastritis; and multifocal glandular atrophy and inflammation of both antrum and body is referred to as multifocal chronic atrophic gastritis. Ferrets infected with *H. mustelae* typically have superficial gastritis of the oxyntic gastric mucosa, whereas the antrum and transitional mucosa may have diffuse mononuclear gastritis, focal glandular atrophy, and regeneration (Fox *et al.*, 1990). Ferrets that are not infected with *H. mustelae* do not have these lesions.

Serum ELISA

Immunogobulin to *H. mustelae* may be used to corroborate or monitor infection and correlates well with other diag-

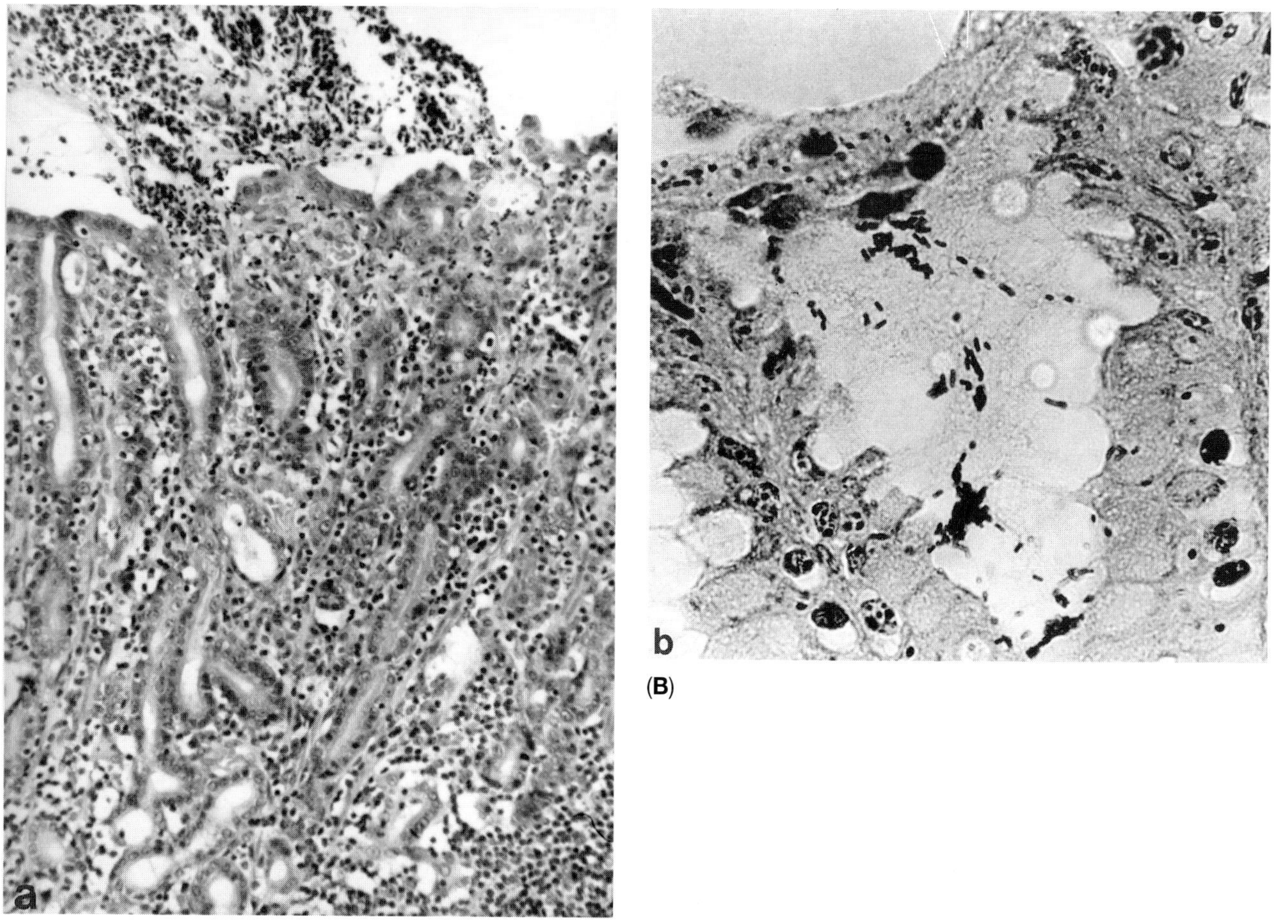

(A) (B)

Figure 31.4 (**A**) Diffuse mucosal inflammation, superficial erosion and luminal migration of leukocytes (H&E). (**B**) Warthin–Starry stain of ferret gastric mucosa.

nostic tests. We have used antigen from whole-cell sonic extracts of three isolates of *H. mustelae* (including ATCC 43772) using a protocol described for an *H. pylori* ELISA (Fox *et al.*, 1990). The titer of serum samples has been expressed as the dilution of serum that provides a reading that corresponds to the mean and 2 SD above the negative controls (Fox *et al.*, 1990, 1991a). The assay is specific for *H. mustelae* antibodies in ferret serum (Fox *et al.*, 1990; Otto *et al.*, 1990). Serum antibody to *H. mustelae* is not protective, as ferrets with high anti-*H.-mustelae* titers are colonized by the organism and have associated gastritis (Fox *et al.*, 1991a). ELISA titers also rise with age and associated chronicity of infection (Fox *et al.*, 1992). In contrast, ferrets that are *H.-mustelae*-free do not have serum ELISA titers to *H. mustelae*. We have demonstrated decreasing serum titers to *H. mustelae* in ferrets from which the bacterium had been eradicated; titers were below pretreatment baseline at approximately 4 months after termination of treatment (Marini *et al.*, 1997). Blood collection from ferrets may be performed by jugular venipuncture.

Fecal culture

H. mustelae may be cultured from infected ferrets by streaking fecal swabs on to CVA medium containing cefoperazone, vancomycin, and amphotericin B (Remel, Lenexa, KS) or TVP media containing trimethoprim, vancomycin, and polymyxin B (Remel). Plate incubation, urease testing, and identification of organisms are as described above.

In one study, 24 of 26 9-week-old (weanling) ferrets had *Campylobacter*-like organisms (CLOs) cultured from their feces; these same ferrets were fecal-culture-negative when retested at 20 weeks (Fox *et al.*, 1992). The feces of older ferrets tested in the same study were much less commonly culture positive.

When hypochlorhydria was induced in four adult ferrets by administration of omeprazole, 42% of sequential fecal samples were culture-positive for *H. mustelae* (Fox *et al.*, 1993a). Only 10% were positive in the same ferrets after cessation of omeprazole and consequent return to gastric acidity. In a separate experiment in the same study, a DNA species-specific probe demonstrated identical restriction enzyme patterns from organisms cultured from feces and stomach in four of five ferrets. These results support the possibility of fecal–oral spread in the epidemiology of *H. mustelae* and *H. pylori* infection and suggest that hypochlorhydria is an important factor in fecal shedding. Transient hypochlorhydria in association with *H. mustelae* infection has been documented in studies in which specific-pathogen-free ferrets were challenged with the organism. Hypochlorhydria occurred approximately 3–4 weeks after challenge; the corresponding age of hypochlorhydria in naturally infected ferrets would be 9–10 weeks (Dunn *et al.*, 1991). Eleven of the strains isolated had phenotypic and biochemical characteristics compatible with *H. mustelae*. Sixteen other strains were oxidase-positive catalase and urease-negative, and sensitive to cephalothin, nalidixic acid, and metronidazole. Sequence analysis of 16sRNA of these isolates demonstrated 97.2% similarity with *H. mustelae*, indicating that it represents a novel species of *Helicobacter*.

Urea breath test

The urea breath test, commonly used in people for diagnosis of gastric colonization of urease-positive organisms, has been adapted for use in the ferret by McColm *et al.* (1993). Urea labeled with the non-radioactive isotope, carbon-13, is administered to people *per os*; exhaled breath is subsequently analyzed for $^{13}CO_2$. In ferrets, adult and preweaning animals are administered labeled urea and placed in sealed glass metabolism chambers. Expired air is collected into sodium hydroxide by constant rate withdrawal of air from the chamber (250 ml/minute) and $^{14}CO_2$ is measured by beta scintillation counter. In McColm's study, administration of 370 kBq of [^{14}C]-urea to adult ferrets resulted in peak production of $^{14}CO_2$ 1 hour after administration (McColm *et al.*, 1993). Breath test activity was found to parallel quantitative cultures of *H. mustelae* in naive, neonatal (3–4 day old) ferret kits and correlated with *H. mustelae* status in adult animals subjected to various *Helicobacter* treatment strategies. Treatment with bismuth subcitrate, alone or in combination with amoxicillin and metronidazole, resulted in reductions both in $^{14}CO_2$ production and in colony forming units per biopsy.

Antimicrobial therapy

Many agents, including antibiotics, mucosal protectants, proton pump inhibitors, and H_2-receptor antagonists, have been used for eradication of *H. mustelae*, treatment of clinical signs of gastritis and gastric ulcer, and as a model to predict efficacy against *H. pylori* (Otto *et al.*, 1990; Fox and Lee, 1997). The most common regimen used for eradication of the organism in the ferret is triple therapy consisting of amoxicillin (30 mg/kg), Amoxi Drops®, Pfizer Animal Health, Exton, PA; metronidazole (20 mg/kg) and bismuth subsalicylate (17.5 mg/kg; Pepto-Bismol original formula; Procter & Gamble, Cincinnati, OH; Table 31.3). Suspensions of both amoxicillin and bismuth subsalicylate are used for oral dosing; the metronidazole is crushed and mixed with a nutritive paste favored by ferrets (Nutrical®, Evsco, Buena, NJ). All agents are given three times daily for 3 or 4 weeks. Suspensions may be administered by oral gavage with a dosing syringe. Ferrets are restrained by firmly grasping the scruff and suspending them by it (Figure 31.5). The syringe tip can then be placed in either diastema and the drug administered in small aliquots into the ferret's mouth. For administration of the medication-laced paste, a small volume of the paste (approximately 0.5 ml) is placed on a tongue depressor that has been split along its long axis.

Table 31.3 Comparison of eradication rates of *H. mustelae* and *H. pylori* using various chemotherapeutic regimens (many of the references are from review articles – please refer to originals for details of testing)

H. mustelae *(ferrets)*			H. pylori *(people)*		
Regimen		*Eradication rate*	*Regimen*		*Eradication rate*
Amoxicillin	30 mg/kg b.i.d.; 21–28 days	71% (Otto *et al.*, 1990)	Amoxicillin	500 mg t.d.s.; 7 days	
Metronidazole	20 mg/kg b.i.d.; 21–28 days	100% (Batchelder *et al.*, 1996)	Colloidal bismuth subcitrate	120 mg q.d.s.; 28 days	55–88% (Dixon, 1995)
Bismuth subsalicylate	17.5 mg/kg b.i.d.; 21–28 days		Metronidazole	400 mg t.d.s.; 7 days	
Enrofloxacin	8.5 mg/kg divided b.i.d.	100% (Blanco *et al.*, 1993)	Ciprofloxacin	500 mg b.i.d.; 14 days	36% (Stone *et al.*, 1988)
Bismuth subcitrate	12 mg/kg divided b.i.d.				
Clarithromycin	12.5 mg/kg b.i.d.; 14 days	83% (Marini *et al.*, 1997)	Clarithromycin	250 mg q.i.d.; 14 days	42% (Burette, 1995)
			Clarithromycin	500 mg q.i.d.	
			Clarithromycin	1 g/day	54% (Dixon, 1995)
			Clarithromycin	2 g/day	46%
Clarithromycin	12.5 mg/kg b.i.d.; 14 days	100% (Marini *et al.*, 1997)	Clarithromycin	500 mg b.i.d. first 14 days	96% (Axon *et al.*, 1997)
RBC (ranitidine	24 mg/kg b.i.d.; 14 days		RBC	400 mg b.i.d.; 28 days	
bismuth citrate)			Clarithromycin	250 mg q.i.d.; 14 days	92% (Axon *et al.*, 1997)
			RBC	400 mg b.i.d.; 28 days	

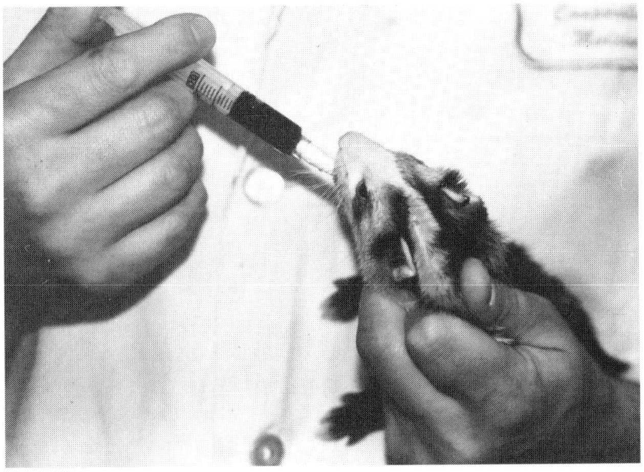

Figure 31.5 Scruff restraint and oral dosing of the ferret.

The paste is scraped off the stick by drawing it across the mucocutaneous junction of the upper lip. Alternatively, some ferrets lick the paste from the stick despite the presence of the medication. Ferrets quickly acclimate to both restraint and administration of medication. Initial studies with this regimen yielded eradication in 71% of ferrets treated and was used in pregnant jills for the purpose of obtaining specific-pathogen-free kits. No ill effects were observed in the kits (Otto *et al.*, 1990).

H. mustelae could not be cultured 12 weeks after termination of therapy from six ferrets treated with a combination of enrofloxacin (8.5 mg/kg/day divided b.i.d. *per os*; Baytril®, Bayer Corp., Shawnee Mission, KS) and colloidal bismuth subcitrate (12 mg/kg/day divided b.i.d. *per os*; DeNol®, Gist Brocades, West Byfleet, Surrey, UK; Blanco *et al.*, 1993). In humans, quinolone antibiotics have been evaluated as monotherapy but have not been used widely because of the emergence of antibiotic resistance. Ciprofloxacin in human clinical trials resulted in eradication in 36% of patients tested (Stone *et al.*, 1998).

Stables *et al.* (1993) described the use of the salt formed from the H_2-antagonist ranitidine and bismuth citrate in *H.-mustelae*-infected ferrets (ranitidine bismuth citrate). Doses of 12 mg/kg or 24 mg/kg administered b.i.d. *per os* for 28 days effected eradication in 50% and 69% of ferrets respectively. At the end of a 3-week drug-free period, however, all animals that had received the lower dose were recolonized; 10% of the ferrets in the higher dose group remained *H.-mustelae*-free at the end of 3 weeks. Parallel eradication and recolonization rates were observed in a subsequent study comparing ranitidine bismuth citrate with either tripotassium dicitrato bismuthate or bismuth citrate (Stables *et al.*, 1993).

In a study that identifed suppressive, non-eradicating doses of agents for use in synergistic combinations, the use of clarithromycin (12.5 mg/kg p.o. b.i.d. for 14 days, Biaxin®, Abbott Laboratories, North Chicago, IL) and a ranitidine/bismuth citrate combination drug (24 mg/kg b.i.d. *per os* for 14 days, GR122311X, Glaxo Wellcome, Research Triangle Park, NC) was 100% effective in eradicating *H. mustelae* from naturally infected ferrets (Marini *et al.*, 1997). Clarithromycin alone (12.5 mg/kg b.i.d. *per os* for 14 days) was also effective in eradicating the organism from five of six ferrets. These findings parallel eradication rates of 73–100% in people administered triple therapy combining bismuth, omeprazole and clarithromycin, or another antibiotic (Graham, 1995). In a recent study comparing the effects of ranitidine bismuth subcitrate (400 mg b.i.d. for 28 days) and clarithromycin at either 500 mg b.i.d. or 250 mg q.i.d. for 14 days, eradication rates of 96% and 92% respectively were demonstrated (Axon *et al.*, 1997).

Agents that are ineffective in eradicating *H. mustelae* include chloramphenicol palmitate (50 mg/kg b.i.d. *per os* for 21 days; Otto *et al.*, 1990), enrofloxacin (5 mg/kg/day divided b.i.d. *per os* for 14 days; Blanco *et al.*, 1993), amoxicillin (30 mg/kg/day *per os* divided b.i.d. for 4 weeks), omeprazole/amoxicillin (69 mg/kg/day divided b.i.d. *per os*; 30 mg/kg/day divided b.i.d.; Blanco *et al.*, 1993), and the triple therapy tetracycline hydrochloride (25 mg/kg t.i.d. for 4 weeks), metronidazole (20 mg/kg t.i.d. for 10 days), and bismuth subsalicylate (2.1 mg/kg t.i.d. for 4 weeks; Otto *et al.*, 1990). Oral omeprazole (0.7 mg/kg p.o. i.d.) may be used as an adjunct to antimicrobial therapy in cases of *H.-mustelae*-associated gastric or duodenal ulcers (Fox and Lee, 1997).

Pitfalls (advantages/disadvantages) of the model

As a laboratory animal used in biomedical research, the ferret has become an important model in the areas of gastrointestinal, cardiovascular, reproductive, and neurophysiology as well as infectious disease (Fox, 1998). Ferrets are of adequate size to allow manipulations that are impractical in rodents. They are readily available and may be purchased time-pregnant or in other life stages for epizootiological investigations. Vendors provide animals of consistent microbial status that are healthy and generally tractable. They are easily maintained in the modern laboratory animal facility.

H. mustelae infection in ferrets is a robust and faithful animal model of *H. pylori* infection in people. It is the only model of naturally acquired infection that has associated ulcer disease (Fox and Lee, 1997). From the perspective of comparative gastric biology, the ferret stomach has anatomic, physiologic, and histologic similarities with those of people. As a model of infection for experimental chemotherapy, efficacy trials in the ferret have paralleled those in people with regard to eradication rate and development of antibiotic resistance (Otto *et al.*, 1990; Marini *et al.*, 1997; Table 31.3). Ferrets are easily restrained, dosed, and bled; they may be anesthetized easily and reproducibly with widely available agents and recover well from anesthesia.

The ability to perform serial gastric endoscopic biopsy examinations within the same animal allows longitudinal investigation into the behavior of the bacterium and its response to therapeutic interventions. Additionally, ferrets that are specific-pathogen-free for *H. mustelae* may be derived to serve as negative controls in cofactor studies of gastric carcinogenesis, or for challenge experiments involving isogenic mutants.

There are also shortcomings of the ferret as a model of experimental chemotherapy against *H. pylori*. The first and most obvious is that *H. pylori* does not infect the stomach of the ferret. Although the responses of *H. pylori* and *H. mustelae* to antimicrobials are quantitatively similar, they are not identical (Table 31.3). This is evidenced by the enhanced eradication of *H. mustelae* by monotherapy with clarithromycin (Marini *et al.*, 1997), which has a more modest eradication rate against *H. pylori* in people (15–54%; Dixon, 1995; Graham, 1995). In some studies (Batchelder *et al.*, 1996), triple therapy with amoxicillin, metronidazole, and bismuth also has a 100% eradication rate, in excess of the 55–94% reported in the literature for *H. pylori* (McNulty *et al.*, 1989; Ruhl and Borsch, 1991; Dixon, 1995). Additionally, the histologic gastritis caused by the organisms is different, with *H. pylori* having an active (neutrophilic) component that is not commonly observed in ferrets. Nonetheless, animal models involving *H. pylori* as the gastric pathogen have other disadvantages that limit their use including cost (macaques, dogs and cats), need for specialized equipment and handling (gnotobiotic pigs and dogs; Krakowka *et al.*, 1987; Radin *et al.*, 1990), limitation in study duration (gnotobiotic pigs), potential presence of other gastric spiral organisms (macaques, cats and dogs), and size (rodents).

Contributions of the model to infectious disease therapy

Elucidation of the pathogenesis of *Helicobacter*-associated gastric disease

Isogenic mutants have been generated and used to gain insight into virulence determinants of *H. mustelae*. To date, urease-negative mutants as well as mutants to one or both of the flagellin proteins, Fla A or Fla B, have been constructed and used in infection studies in ferrets (Dunn *et al.*, 1991; Andrutis *et al.*, 1995, 1997). Urease negative isogenic mutants failed to colonize the ferret stomach (Andrutis *et al.*, 1995). For the flagellar mutant strains, the non-motile, aflagellar double mutant strain (*flaA flaB*) failed to colonize the ferret stomach while *flaA* (weakly motile, truncated flagella) and *flaB* (moderately motile, normal-appearing flagella) colonized initially at low levels. The single mutants established persistent infection with increasing numbers of organisms with time; gastritis severity correlated with numbers of organisms (Andrutis *et al.*, 1997).

Ferrets are an exciting new model for the study of *Helicobacter*-associated neoplasia (gastric MALT lymphoma and adenocarcinoma) because of the documented occurrence of these neoplasm and the ability to investigate dietary and microbial cofactors (Erdman *et al.*, 1997; Fox *et al.*, 1997). Yu *et al.* (1995) used proliferating cell nuclear antigen (PCNA/cyclin), a biomarker for cell proliferation, to determine that the labeling index of mucosa from both the gastric antrum and body of *H. mustelae* infected ferrets was significantly ($p < 0.001$) higher than that of SPF ferrets (Yu *et al.*, 1995). PCNA/cyclin positivity also correlated with histologic severity of gastritis. In another study, nine of 10 ferrets given a single dose of the nitrosamine N-methyl-N′-nitro-N-nitrosoguanidine (MNNG) developed gastric adenocarcinoma an average 63 months after administration (Fox *et al.*, 1993b). This rate of carcinogenesis following a single dose of MNNG is unprecedented (Hirono and Shibuya, 1972). These findings support the relationship, postulated to exist for *H. pylori* infection, between chronic gastritis, chronic multifocal atrophic gastritis, and the development of gastric neoplasia.

Transmission, reinfection and recrudescence

Ferrets have been used to investigate epizootiological phenomena, including age of first infection, mode of transmission, and susceptibility to natural or experimental reinfection following eradication (Fox *et al.*, 1992, 1993a; Batchelder *et al.*, 1996). The culture of *H. mustelae* from ferret feces (Fox *et al.*, 1992) was approximately coincident with that of *H. pylori* from human feces (Thomas *et al.*, 1992; Kelly *et al.*, 1994) and lends support to the concept of fecal–oral transmission. The age of infection (approximately 6 weeks), the age at which fecal culture was most often positive (9 weeks), and the enhanced recovery of *H. mustelae* from fecal cultures in adult hypochlorhydric ferrets may have parallels in *H. pylori* epidemiology in the developing world (Fox *et al.*, 1991a; Vincent, 1995).

In one study *H. mustelae* was eradicated from naturally infected ferrets by triple therapy; they remained negative as determined by culture of gastric endoscopic biopsy specimens until challenged 6 or 17 months later with a distinct isolate of *H. mustelae* (Batchelder *et al.*, 1996). Reinfection was successful, and restriction enzyme analysis of isolates demonstrated that recrudescence with the original strain had not occurred over a 3–12-month follow-up period. A second course of eradication was also successful in eliminating the new challenge organism from all ferrets. Animals were then cohoused with naturally infected animals harboring the original (natural) strain of *H. mustelae*. Five of seven ferrets that had been exposed to contact animals became infected with the original strain. An age association was observed in contact reinfection; ferrets that became reinfected after exposure to naturally infected contact controls were younger than those that were not reinfected. These findings suggest that prior infection with *H. mustelae* is not protective against subsequent challenge, that contact with infected cage mates may result in infection among adult animals, and that age of exposure may have an impact on exposure outcome. This and similar studies may provide insights into the epidemiology of *H. pylori* infection and help direct future efforts in public health, infection control, and vaccine development.

Novel therapeutic approaches

Existing therapeutic modalities for the treatment of *H. pylori* in people have, as shortcomings, the development of antibiotic resistance, side-effects, and poor patient compliance. A strategy to reduce side-effects, the number of medications taken, and the selection pressure for the development of resistance is to use non-eradicating, suppressive doses of agents that may act in synergy. Such a design strategy was used to determine synergistic doses of clarithromycin and the combination agent ranitidine bismuth citrate as described above (Marini *et al.*, 1997). A recent multicenter study demonstrated an *H. pylori* eradication rate of 96.2% in 111 patients 4 weeks after termination of treatment with clarithromycin (500 mg b.i.d. *per os* for 14 days) and ranitidine bismuth citrate (Pylorid; 400 mg b.i.d. *per os* for 25 days; Axon *et al.*, 1997).

References

Andrutis, K. A., Fox, J. G., Schauer, D. B. *et al.* (1995). Inability of an isogenic urease-negative mutant strain of *Helicobacter mustelae* to colonize the ferret stomach. *Infect. Immun.*, **63**, 3722–3725.

Andrutis, K. A., Fox, J. G., Schauer, D. B. *et al.* (1997). Infection of the ferret stomach by isogenic flagellar mutant strains of *H. mustelae*. *Infect. Immun.*, **65**, 1962–1966.

Axon, A. T. R., Ireland, A., Smith, M. J. L., Rooprams, P. D. (1997). Ranitidine bismuth citrate and clarithromycin twice daily in the eradication of *Helicobacter pylori*. *Aliment. Pharmacol. Ther.*, **11**, 81–87.

Batchelder, M., Fox, J. G., Hayward, A. *et al.* (1996). Natural and experimental *Helicobacter mustelae* reinfection following successful antimicrobial eradication in ferrets. *Helicobacter*, **1**, 34–42.

Blanco, M. C., Fox, J. G., Palley, L. S. (1993). Eradication of *Helicobacter mustelae* infection in the ferret: efficacy of multiple antibiotic therapies. Abstract. *Cont. Topics in Lab. Anim. Sci.*, **32**, 15.

Cabot, E. B., Fox, J. G. (1990). Bile reflux and the gastric mucosa: an experimental ferret model. *J. Invest. Surg.*, **3**, 177–189.

Chiba, N., Rao, B. V., Rademaker, J. W., Hunt, R. W. (1992). Meta-analysis of the efficacy of antibiotic therapy in eradicating *Helicobacter pylori*. *Am. J. Gastroenterol.*, **87**, 1716–1727.

Correa, P. (1988). Chronic gastritis: a clinicopathological classification. *Am. J. Gastroenterol.*, **83**, 504–509.

Dixon, J. S. (1995). *Helicobacter pylori* eradication: unravelling the facts. *Scand. J. Gastroenterol.*, **30**(Suppl. 212), 48–62.

Dunn, B. E., Sung, C. C., Taylor, N. S., Fox, J. G. (1991). Purification and characterization of *Helicobacter mustelae* urease. *Infect. Immun.*, **59**, 3343–3345.

Erdman, S. E., Correa, P., Li, X., Coleman, L. A., Fox, J. G. (1997). *Helicobacter mustelae*-associated gastric mucosa associated lymphoid tissue (MALT) lymphoma in ferrets. *Am. J. Pathol.*, **151**, 273–280.

Fox, J. G. (1988). *Biology and Diseases of the Ferret*. Lea & Febiger, Philadelphia, PA.

Fox, J. G. (1994). Gastric disease in ferrets: effects of *Helicobacter mustelae*, nitrosamines and reconstructive gastric surgery. *Eur. J. Gastroenterol. Hepatol.*, **6**(Suppl. 1), S57–S65.

Fox, J. G. (1998). *Biology and Diseases of the Ferret* (2nd edn). Williams & Wilkins, Baltimore, MD.

Fox, J. G., Lee, A. (1997). The role of *Helicobacter* species in newly recognized gastrointestinal tract diseases of animals. *Lab. Anim. Sci.*, **47**, 222–255.

Fox, J. G., Edrise, B. M., Cabot, E., Beaucage, C., Murphy, J. C., Prostak, K. S. (1986). Campylobacter-like organisms isolated from gastric mucosa of ferrets. *Am. J. Vet. Res.*, **47**, 236–239.

Fox, J. G., Cabot, E. B., Taylor, N. S., Laraway, R. (1988). Gastric colonization of *Campylobacter pylori* subsp. *mustelae* in ferrets. *Infect. Immun.*, **56**, 2994–2996.

Fox, J. G., Chilvers, T., Goodwin, C. S. *et al.* (1989). *Campylobacter mustelae*, a new species resulting from the elevation of *Campylobacter pylori* subsp. *mustelae* to species status. *Int. J. Syst. Bact.*, **39**, 301–303.

Fox, J. G., Correa, P., Taylor, N. S. *et al.* (1990). *Helicobacter mustelae*-associated gastritis in ferrets: an animal model of *Helicobacter pylori* gastritis in humans. *Gastroenterology*, **99**, 352–361.

Fox, J. G., Otto, G., Taylor, N. S., Rosenblad, W., Murphy, J. C. (1991a). *Helicobacter mustelae*-induced gastritis and elevated gastric pH in the ferret (*Mustela putorius furo*). *Infect. Immun.*, **59**, 1875–1880.

Fox, J. G., Otto, G., Murphy, J. C., Taylor, N. S., Lee, A. (1991b). Gastric colonization of the ferret with *Helicobacter* species: natural and experimental infections. *Rev. Infect. Dis.*, **13**(Suppl. 8), S671–S680.

Fox, J. G., Paster, B. J., Dewhirst, F. E. *et al.* (1992). *Helicobacter mustelae* isolation from feces of ferrets: evidence to support fecal–oral transmission of a gastric *Helicobacter*. *Infect. Immun.*, **60**, 606–611.

Fox, J. G., Blanco, M., Yan, L.-L. *et al.* (1993a). Role of gastric pH in isolation of *Helicobacter mustelae* from the feces of ferrets. *Gastroenterology*, **104**, 86–92.

Fox, J. G., Wishnok, J. S., Murphy, J. C., Tannenbaum, S., Correa, P. (1993b). MNNG-induced gastric carcinoma in ferrets infected with *Helicobacter mustelae*. *Carcinogenesis*, **14**, 1957–1961.

Fox, J. G., Perkins, S., Yan, L. *et al.* (1996). Local immune response in *Helicobacter pylori* infected cats and identification of *H. pylori* in saliva, gastric fluid and feces. *Immunology*, **88**, 400–406.

Fox, J. G., Dangler, C. A., Sager, W., Borkowski, R., Gliaffo, J. M. (1997). *Helicobacter mustelae* associated gastric adenocarcinoma in ferrets (*Mustela putorius furo*). *Vet. Pathol.*, **34**, 225–229.

Graham, D. Y. (1995). Clarithromycin for treatment of *Helicobacter pylori* infections. *Eur. J. Gastroenterol. Hepatol.*, **7**(Suppl. 1), S55–S58.

Green, C. J., Knight, J., Precious, S., Simpkin, S. (1981). Ketamine alone and combined with diazepam or xylazine in laboratory animals: a 10 year experience. *Lab. Anim.*, **15**, 163–170.

Hazell, S. L., Borody, T. J., Gal, A. (1987). *Campylobacter pyloridis* gastritis I: detection of urease as a marker of bacterial colonization and gastritis. *Am. J. Gastroenterol.*, **82**, 292–296.

Hirono, I., Shibuya, C. (1972). Induction of stomach cancer by a single dose of N-methyl-N′-nitro-N-nitrosoguanidine through a stomach tube. In *Topics in Chemical Carcinogenesis: Proceedings of the 2nd International Symposium of the Princess Takamatsu Cancer Research Fund* (eds Nakahara, W., *et al.*), pp. 121–132. University Park Press, Baltimore, MD.

Holcombe, C., Omotara, B. A., Eldridge, J., Jones, D. M. (1992). *H. pylori*, the most common bacterial infection in Africa: a random serological study. *Am. J. Gastroenterol.*, **87**, 28–30.

Kelly, S. M., Pitcher, M. C. L., Formery, S. M., Gibson, G. R. (1994). Isolation of *Helicobacter pylori* from feces of patients with dyspepsia in the United Kingdom. *Gastroenterology*, **107**, 1671–1674.

Krakowka, S., Morgan, D. R., Kraft, W. G. *et al.* (1987). Establishment of gastric *Campylobacter pylori* infection in the neonatal gnotobiotic piglet. *Infect. Immun.*, **55**, 2789–2796.

Manning, D. D., Bell, J. A. (1990). Lack of detectable blood groups in domestic ferrets: implications for transfusion. *J. A. V. M. A.*, **197**, 84–86.

Marini, R. P., Fox, J. G., Taylor, N. S., Yan, L., McColm, A., Williamson, R. Ranitidine–bismuth citrate and clarithromycin, alone or in combination, for eradication of *Helicobacter mustelae* from ferrets. *Am. J. Vet. Res.* (in press).

Marshall, B. J. (1993). Treatment strategies for *Helicobacter pylori* infection. *Gastroenterol. Clin. North Am.*, **22**, 183–198.

McColm, A. A., Bagshaw, J. A., O'Malley, C. F. (1993). Development of a ^{14}C-urea breath test in ferrets colonized with *Helicobacter mustelae*: effects of treatment with bismuth, antibiotics, and urease inhibitors. *Gut*, **34**, 181–186.

McNulty, C. A., Eyre-Brook, I. A., Uff, J. S., Dent, J. C., Wilkinson, S. P. (1989). Triple therapy is not always 95% effective. In *5th International Workshop on* Campylobacter *Infections*, sponsored by Insituto Nacional de la Nutricion, p. 73. Puerto Vallarta, Mexico.

Megraud, F., Brassens-Rabbe, M. P., Denis, F., Belbouri, A., Hoa, D. Q. (1989). Seroepidemiology of *Campylobacter pylori* infection in various populations. *J. Clin. Microbiol.*, **27**, 1870–1873.

Moreland, A. F., Glaser, C. (1985). Evaluation of ketamine, ketamine–xylazine and ketamine-diazepam anesthesia in the ferret. *Lab. Anim. Sci.* **35**, 287–290.

O'Rourke, J., Lee, A., Fox, J. G. (1992). An ultrastructural study of *Helicobacter mustelae*: evidence of a specific association with gastric mucosa. *J. Med. Microbiol.*, **36**, 420–427.

Otto, G., Fox, J. G., Wu, P.-Y., Taylor, N. S. (1990). Eradication of *Helicobacter mustelae* from the ferret stomach: an animal model of *Helicobacter (Campylobacter) pylori* chemotherapy. *Antimicrob. Agents Chemother.*, **34**, 1232–1236.

Perkins, S. E., Fox, J. G., Walsh, J. H. (1996). *Helicobacter mustelae* associated hypergastrinemia in the ferret (*Mustela putorius furo*). *Am. J. Vet. Res.*, **57**, 147–150.

Radin, J. M., Eaton, K. A., Krakowka, S. *et al.* (1990). *Helicobacter pylori* infection in gnotobiotic beagle dogs. *Infect. Immun.*, **58**, 2606–2612.

Ruhl, G. H., Borsch, G. (1991). Chronic active gastritis after eradication of *Campylobacter* (now: *Helicobacter*) *pylori*. *Pathol. Res. Pract.*, **187**(2–3), 226–234.

Stables, R., Campbell, C. J., Clayton, N. M. *et al.* (1993). Gastric anti-secretory, mucosal protective, anti-pepsin and anti-*Helicobacter* properties of ranitidine bismuth citrate. *Aliment. Pharmacol. Ther.*, **7**, 237–246.

Stone, J. W. *et al.* (1988). Failure of ciprofloxacin to eradicate *Campylobacter pylori* from the stomach. *J. Antimicrob. Chemother.*, **22**, 92–93.

Sylvina, T. J., Berman, N. G., Fox, J. G. (1990). Effects of yohimbine on bradycardia and duration of recumbancy in ketamine/xylazine-anesthetized ferrets. *Lab. Anim. Sci.*, **40**, 178–182.

Thomas, J. E., Gibson, G. R., Darboe, M. K., Dale, A., Weaver, L. T. (1992). Isolation of *Helicobacter pylori* from human feces. *Lancet*, **340**, 1194–1195.

Vincent, P. (1995). Transmission and acquisition of *Helicobacter pylori* infection: evidences and hypothesis. *Biomed. Pharmacother.*, **49**, 11–18.

Yu, J., Russell, R. M., Salomon, R. N., Murphy, J. C., Palley, L. S., Fox, J. G. (1995). Effect of *Helicobacter mustelae* infection on epithelial cell proliferation in ferret gastric tissues. *Carcinogenesis*, **16**, 1927–1931.

Chapter 32

The Hamster Model of Syphilis

J. D. Alder

Background of human infection

Syphilis is a sexually transmitted disease caused by the Gram-negative bacterial spirochete *Treponema pallidum* subsp. *pallidum*. Untreated syphilis has three distinct clinical stages with relatively long incubation periods between stages. The chancre or lesion at the site of infection is the most characteristic sign of the primary stage of syphilis. As the primary chancre forms, the infecting spirochetes migrate to lymph tissue and internal organs, including CNS tissue. Secondary syphilis involves generalized eruptions caused by secondary infections of mucocutaneous surfaces. If untreated, syphilis can be a lethal disease with tertiary neurologic involvement. The entire course of syphilis from primary to tertiary stages can last 30–40 years. There is a moderate to high probability of transmission during sexual contact between infectious and susceptible partners (Garnett *et al.*, 1997).

Syphilis has demonstrated both increases and decreases in incidence in the USA over the last several decades. There has been a recent decrease in the incidence of syphilis in adolescents, one of the groups with the highest rates of infection (Sells and Blum, 1996). The overall rate of syphilis infection increased steadily in the 1980s to a record high of 20.3 per 100 000 persons in 1990, but the rate has since decreased to 6.3 per 100 000 (Mushinski, 1997). The incidence of syphilis tends to be higher in the rural south and major cities.

Primary syphilis with the characteristic chancre is the clinical stage most often diagnosed and treated. Penicillin therapy at an early stage is usually effective in producing an apparent lifelong cure. However, it has been recognized that the standard single dose of penicillin G fails to achieve bactericidal concentrations in CNS fluid (Malone *et al.*, 1995). There have been treatment failures of penicillin therapy in AIDS patients, resulting in rapid progression to tertiary neurosyphilis (Gordon *et al.*, 1994). These treatment failures have raised doubts concerning the efficacy of penicillin therapy in immunosuppressed hosts. Multiple intramuscular injections of penicillin have been evaluated for treatment of syphilis in immunosuppressed patients (Crowe *et al.*, 1997). The therapy was effective, but compliance with multiple intramuscular injections has been an issue. Alternative treatments include oral doxocycline or erythromycin for prolonged periods, but a single intramuscular injection of penicillin is still the standard therapy in most cases. There have been no confirmed cases of penicillin-resistant isolates of *Treponema pallidum*.

History of the hamster model of syphilis

The hamster and rabbit are the laboratory rodents most commonly infected with *Treponema pallidum*. The hamster model of syphilis has been used primarily in immunologic studies of the role of antibody (Azadagen *et al.*, 1983), lymphocytes (Schell *et al.*, 1980; Liu *et al.*, 1990, 1991), and macrophages (Alder *et al.*, 1989) in treponemal infection. The rabbit model of syphilis was established first (Turner and Hollander, 1957) and remains the classical model of infection and treatment. These two animal models do not duplicate the complicated pathology of human syphilis, but primary chancre formation and invasion of lymph tissues is observed. The hamster has an advantage in terms of smaller size, allowing more animals to be housed in a smaller area, and the use of less drug per animal than in the rabbit. Inbred strains of the hamster are readily available. In the hamster, the infection rapidly centers in the regional lymph nodes, rather than in a skin lesion as in the rabbit. The resulting pathology in the hamster is characteristic of post-primary syphilis, while the rabbit best simulates the primary chancre of syphilis.

Three different subspecies of *Treponema pallidum* have been used to infect hamsters (Table 32.1). *Treponema pallidum* subspecies *pallidum*, the causative agent of syphilis in humans, produces an infection centered in lymph tissue with minimal skin lesions in hamsters. *T. pallidum* subsp. *pertenue*, the causative agent of yaws, and *T. pallidum* subsp. *endemicum*, the causative agent of endemic syphilis, have also been used. These two subspecies produce more extensive skin lesions than *T. pallidum* subsp. *pallidum*. All three of these subspecies of *T. pallidum* share apparent identical drug sensitivities and immunological cross resistance (Schell *et al.*, 1982a).

Handbook of Animal Models of Infection
ISBN 0-12-775390-7

Table 32.1 *T. pallidum* subspecies infection of hamsters

Subspecies	*Lesions*	*Duration*	*Treponemes*	*Human disease*
pallidum	Minimal	1 week	Moderate	Syphilis
pertenue	Moderate	3 weeks–3 months	Numerous	Yaws, frambesia
endemicum	Significant	9 months	Numerous	Endemic syphilis

Newer antibiotics are usually not targeted specifically against syphilis. The major reason is the perception of adequate therapy with penicillin G, and the lack of drug resistance in *Treponema pallidum*. For patients sensitive to penicillin, alternatives such as tetracycline, doxycycline, erythromycin, or newer macrolides such as clarithromycin or azithromycin are perceived as effective therapies. Adequate therapies with existing inexpensive generic drugs will probably limit extensive development of newer drugs against syphilis. Paradoxically, the smaller market size may cause animal models of syphilis to take on added importance. Newer drugs nearing approval for other indications could be tested against syphilis in animal models for a proof of principle in the absence of a clinical trial, stimulating potential off-label use.

Animal species

The outbred Syrian hamster supports the growth of all three subspecies of *T. pallidum* following intradermal infection. Of the inbred strains of hamster, the LSH/Se LAK and CB strains have been used to support *T. pallidum* infection but any hamster strain is probably also susceptible to infection (Schell *et al.*, 1979, 1980; Kajdacsy-Balla *et al.*, 1987). Either males or females can be used but, for prolonged group housing, females are preferred, as infighting can be a significant issue in male hamsters. Hamsters older than 6 weeks are typically used.

Preparation of animals

Hamsters infected as described below with *T. pallidum* subspecies *pertenue, endemicum*, or *pallidum* will develop skin lesions in the inguinal area. For this reason, wire flooring should be considered to avoid contamination of the infected site with bedding. If bedding is used, extra changes may be required to maintain sanitary conditions.

Storage and preparation of inoculate

Treponema pallidum can not be grown in quantity *in vitro*. Maintenance culture methods have been established (Norris, 1982), but *T. pallidum* must be maintained by repeated passage in animals. *T. pallidum* is a fragile pathogen and exposure to *in-vitro* heat and O_2 conditions must be minimized. In the inoculation procedure, hamsters infected 2–4 weeks previously are euthanized with CO_2 and the infected inguinal lymph tissue is removed into cold PBS (Schell *et al.*, 1979). Using sterile scissors, the lymph nodes are finely minced into cold RPMI 1640 supplemented with 10% fetal bovine serum, 10 mM N-2-hydroxyethylpiperazine-N′-2-ethanesulfonic acid (HEPES) and 0.63 nM dithiothreitol. A low-speed centrifugation at 1000*g* is used to pellet cellular debris, while the free treponemes remain in the supernatant. The concentration of treponemes is determined using dark-field microscopy and adjusted to 1×10^7 bacteria per milliliter. The inoculum can be used immediately, or may be quick-frozen in liquid nitrogen following addition of DMSO to 10% by volume as a cytopreservative. Frozen treponemes should not be used in a therapy trial, but instead should be passaged several times in animals at 2-week intervals to increase the pathogenicity of the strain. The inoculum is kept on ice while the animals are prepared. The entire preparation of the inoculum should be accomplished as rapidly as possible, without contaminating the solution.

Infection process

The hamsters are placed under light inhalation anesthesia such as sevoflurane or halothane. The *T. pallidum* inoculum prepared as described above is drawn into a 1 ml tuberculin syringe with a 27 G needle. By far the most common site injected is the intradermal inguinal region. The hamsters are shaved in the inguinal region and are injected intradermally with 0.1 ml of inoculum per site above both inguinal lymph nodes. A distinct white bleb should immediately form at the injection sites. The bleb will dissipate within 30 minutes and there should be minimal leakage from the injection site. Lack of a bleb indicates that the injection was too deep. This may result in a primary lymphatic infection, with little or no skin lesion. Infection of the inguinal site will lead to significant infection of the inguinal lymph nodes and in many cases, distinct skin lesions.

Parameters to monitor infection

In hamsters, *T. pallidum* infection can be monitored based on severity of skin lesions, quantity of treponemes in lymph node, and weight of lymph node. The three different subspecies of *T. pallidum* produce different levels of skin lesions. *T. pallidum* subsp. *pallidum* produces few or no skin lesions following intradermal inoculation (Schell *et al.*, 1982a). *T. pallidum* subsp. *pertenue* produces skin lesions that persist for up to 3 weeks in Syrian hamsters (Chan *et al.*, 1979), but can last for several months in CB hamsters (Schell *et al.*, 1979). *T. pallidum endemicum* produces extensive skin lesions that persist for up to 9 months and will involve a large surface area before resolving (Schell *et al.*, 1988).

Assay of lymph node tissue is a more sensitive technique to measure infection, but the method is terminal for the animal. In passive protection immune serum experiments, it has been demonstrated that hamsters inoculated with *T. pallidum pertenue* harbor treponemes in inguinal lymph nodes even though skin lesions do not develop (Schell *et al.*, 1978). The number of treponemes per lymph node can maximally reach approximately 10^7 organisms per node (Alder *et al.*, 1993). The treponeme burden in lymph tissue is determined by homogenizing the inguinal lymph node in 1 ml of PBS. The treponeme concentration is assessed by dark-field microscopy, and the overall burden is then calculated.

The inguinal lymph nodes of *T.-pallidum*-infected hamsters also demonstrate a significant increase in mass which initially parallels treponeme numbers (Schell *et al.*, 1979). Lymph nodes remain inflamed and enlarged for some period after treponemes are eradicated from the node. Therefore, treponeme number in the inguinal lymph nodes should be the primary method used to assess infection in the hamster, rather than lymph mass.

Antimicrobial therapy

Only a limited number of antimicrobial drugs have been tested in the hamster model of syphilis. Penicillins and macrolides have shown efficacy in the hamster model of syphilis (Table 32.2), and are also used with success clinically. A single intramuscular dose of benzathine penicillin at 25 000 U/kg is curative versus *T. pallidum* infection in the hamster (Schell *et al.*, 1979; Alder *et al.*, 1993). Intramuscular penicillin was demonstrated to be rapidly curative against active infection with *T. pallidum endemicum* (Schell *et al.*, 1982a) and *T. pallidum pertenue* (Chan *et al.*, 1982). Clarithromycin therapy at dosages of 10 mg/kg/day for 7 days was also demonstrated to be curative against both active and incubating infection with *T. pallidum pertenue* (Alder *et al.*, 1993). Clarithromycin was administered by the subcutaneous route, in contrast to the usual oral route used clinically. Great care must be taken when considering oral antibacterial therapy in hamsters. These animals are prone to colitis caused by gastrointestinal colonization with *Clostridium difficile*. Oral antibiotic therapy can alter normal gastrointestinal flora, causing an overgrowth of *C. difficile*. For therapy of *T. pallidum* infection in hamsters, orally active drugs can be administered by the subcutaneous route to avoid the colitis issue (Alder *et al.*, 1993).

Several additional antibiotics have been tested for syphilitic efficacy in the rabbit model. There are also issues with some antibiotics for prolonged oral therapy of rabbits causing alterations of local gut flora. Several of these antibiotics have also demonstrated clinical efficacy. Ceftriaxone and penicillin G demonstrated efficacy versus *T. pallidum pallidum* in the rabbit (Johnson *et al.*, 1982). Cefmetazole was determined to be as effective as benzathine penicillin in the rabbit (Baker-Zander and Lukehart, 1989). The macrolides roxithromycin (Lukehart and Baker-Zander, 1987) and azithromycin (Lukehart *et al.*, 1990) have both demonstrated efficacy. Additionally, rosaramicin phosphate (Baughn et al, 1981), Sch. 29482 (Baughn *et al.*, 1984), and cefetamet (Fitzgerald, 1992) demonstrated different degrees of efficacy versus *T. pallidum pallidum* infection in the rabbit. In contrast, aztreonam (Lukehart et al, 1984), and ofloxacin (Une *et al.*, 1987), failed to demonstrate acceptable efficacy in the rabbit model of syphilis.

Table 32.2 Burden of *T. pallidum* in lymph tissue of hamsters treated with clarithromycin or benzathine penicillin beginning 1 day after inoculation. (Adapted from Alder *et al.*, 1993; *n* = four hamsters per group; q.d., once daily; N/A, not applicable; count per node expressed as mean ± SD.)

Treatment	*Dose (mg or U/kg)/day*	*Schedule*	*Count × 10^5 per lymph node*
Penicillin	250	q.d. Day 1	13.5 ± 14.5
	2500		0.8 ± 1.0
	25 000		0.0 ± 0.0*
Clarithromycin	2.5	q.d. Days 1–7	0.5 ± 1.0*
	12.5		0.0 ± 0.0*
	25.0		0.0 ± 0.0*
Untreated	N/A	N/A	23.3 ± 15.1

* *p* = statistically significant pairwise comparison to untreated group.

Evaluation of therapy

Evaluation of treatment efficacy uses the same techniques as evaluation of infection. Lymph node weight, treponeme burden, and skin lesions can all be used to monitor response to therapy. A key consideration is the time interval between end of therapy and the assay for treponeme burden in lymph tissue. The response to therapy is determined by assay within 7 days after the end of therapy. The time frame for test of cure can be more problematic for infections with *T. pallidum pallidum* or *T. pallidum pertenue*, which do not produce as extensive skin lesions as *T. pallidum endemicum* in the hamster. A reasonable approach is to allow a sufficient period from the end of therapy to produce peak infection in an untreated hamster. A period of 3–4 weeks without therapy is usually sufficient to allow regrowth of surviving *T. pallidum* in the hamster.

Issues of antimicrobial therapy of *T. pallidum* infection in the hamster

T. pallidum infection of the hamster is a model of primary syphilis, with infection concentrated in the dermal tissue and proximal lymph tissue. The issue of latency for *T. pallidum* in the hamster is difficult to address due to the relative short lifespan of the species compared to humans. Because of the clinical efficacy of penicillins, the standard for therapy of syphilis is early cure. Therapy of secondary and tertiary syphilis is addressed with the same drugs (often at higher dosages) as used for primary syphilis (Crowe *et al.*, 1997; Garnett *et al.*, 1997). While neurosyphilis has not been studied in the hamster, the dermal and lymph infections provide a good simulation of primary syphilis. As such, the hamster model of *T. pallidum* infection provides a reasonable preclinical platform for therapy of syphilis.

Contributions of the hamster model of syphilis

The hamster model of syphilis has been used to demonstrate efficacy of antitreponemal drugs, leading to some clinical usage. No controlled clinical trials of newer syphilitic drugs have been performed recently, mainly because of the lack of penicillin resistance in *T. pallidum* and the efficacy of inexpensive drugs such as penicillins, tetracyclines, and macrolides. The scarcity of newer drug development is mainly due to the perceived lack of medical need for alternative therapies. Although not extensively tested, the predictive ability of the hamster model of syphilis is useful for assessment of potential clinical drug efficacy.

The hamster model of *T. pallidum* infection has been used primarily for immunological studies. The availability of inbred lines of animals and immunologic reagents has facilitated these immunological studies in the hamster. In the inbred strains of hamster, there is great variation in the level of IgG3 expression, but there is no known deficiency associated with a lack of IgG3 (Coe *et al.*, 1995). The effect of T-cell subpopulations and immune T cells on the response to *T. pallidum* infection has been studied in the hamster (Liu *et al.*, 1990, 1991; Schell *et al.*, 1982b). Also, the role of T suppressor cells in syphilis has been suggested by the hamster model (Tabor *et al.*, 1984). The effect of hamster immune serum and its fractions to confer complete protection from *T. pallidum* from animals challenge was shown (Azadegan *et al.*, 1983; 1986). Additionally, the role of hamster macrophages (Tabor *et al.*, 1985; Alder *et al.*, 1989) as well as the synergistic effect of IgG2 and macrophages (Azadegan *et al.*, 1988) against *T. pallidum* infection was demonstrated. These immunological studies have demonstrated the complexity of the protective immune response to *T. pallidum* infection and the great challenge in producing an effective vaccine.

There is a need to investigate *T. pallidum* latent infection, congenital syphilis, and neurosyphilis in normal and suppressed hosts. Animal models of these conditions are largely unexplored, and the pathology of *T. pallidum* infection in immunosuppressed inbred animals has not been extensively studied. There has been an attempt to model congenital syphilis in the hamster (Kajdacsy-Balla *et al.*, 1987), but many issues remain unexplored. The hamster does offer potential to approach multiple issues for study of the pathology and immunology of *T. pallidum* infection.

Summary

The hamster model of syphilis infection has been used for testing of drug efficacy and for immunological studies of *T. pallidum* infection. Because of the lack of new clinical studies, the hamster preclinical drug efficacy trials could take on added importance in stimulating clinical usage. Immunological studies in the hamster have demonstrated that multiple components are needed for an effective immune response to syphilis. The hamster also offers well defined endpoints for evaluation of *T. pallidum* infection and drug efficacy. The smaller size of the hamster allows testing on larger numbers of individuals with less drug. The hamster model of syphilitic infection has advanced both drug efficacy testing and investigation of immunological and pathological parameters of *T. pallidum* infection.

References

Alder, J. D., Daugherty, N., Harris, O. N., Liu, H., Steiner, B. M., Schell, R. F. (1989). Phagocytosis of *Treponema pallidum pertenue* by hamster macrophages on membrane filters. *J. Infect. Dis.*, **160**, 289–297.

Alder, J. D., Jarvis, K. J., Mitten, M., Shipkowitz, N. L., Gupta, P., Clement, J. (1993) Clarithromycin therapy of experimental *Treponema pallidum* infections in hamsters. *Antimicrob. Agents Chemother.*, **37**, 864–867.

Azadegan, A. A., Schell, R. F., LeFrock, J. L. (1983). Immune serum confers protection against syphilitic infection on hamsters. *Infect. Immun.*, **42**, 42–47.

Azadegan, A. A., Schell, R. F., Steiner, B. M., Coe, J. E., Chan, J. K. (1986). Effect of immune serum and its immunological fractions on hamsters challenged with *Treponema pallidum* ssp. *pertenue*. *J. Infect. Dis.*, **153**, 1007–1013.

Azadegan, A. A., Schell, R. F., Alder, J. D. *et al.* (1988). Synergistic effect of macrophage activation and immune serum, especially IgG2, on resistance to infection with *Treponema pallidum* ssp. *endemicum* in hamsters. *Reg. Immunol.*, **1**, 3–8.

Baker-Zander, S. A., Lukehart, S. A. (1989). Efficacy of cefmetazole in the treatment of active syphilis in the rabbit model. *Antimicrob. Agents Chemother.*, **33**, 465–469.

Baughn, R. E., Muscher, D. M., Adams, C. B., Knox, J. M. (1981). Evaluation of rosaramicin phosphatate in treatment of experimental syphilis in rabbits. *Antimicrob. Agents Chemother.*, **19**, 117–121.

Baughn, R. E., Adams, C. B., Muscher, D. M. (1984). Evaluation of Sch 29482 in experimental syphilis and comparison with penicillin G benzathine in disseminated disease and localized infection. *Antimicrob. Agents Chemother.*, **26**, 401–404.

Chan, J. K., Schell, R. F., LeFrock, J. L. (1979). Inability of immune cells treated with anti-thymocyte serum to confer on hamsters resistance to cutaneous infection with *Treponema pallidum*. *Infect. Immun.*, **25**, 208–212.

Chan, J. K., Schell, R. F., LeFrock, J. L. (1982). Mitogenic responses of hamsters infected with *Treponema pertenue*: lack of correlation with passive transfer of resistance. *Br. J. Infect. Dis.*, **58**, 292–297.

Coe, J. E., Schell, R. F., Ross, M. J. (1995) Immune response in the hamster: definition of a novel IgG not expressed in all hamster strains. *Immunology*, **86**, 141–148.

Crowe, G., Theodore, G., Forster, G. E., Goh, B. T. (1997) Acceptability and compliance with daily injections of procaine penicillin in the outpatient treatment of syphilis-treponemal infection. *Sex. Transm. Dis.*, **24**, 127–130.

Fitzgerald, T. J. (1992). Effects of cefetamet (Ro 15-8074) on *Treponema pallidum* and experimental syphilis. *Antimicrob. Agents Chemother.*, **36**, 598–602.

Garnett, G. P., Aral, S. O., Hoyle, D. V., Cates, W. Jr, Anderson, R. M. (1997). The natural history of syphilis. Implications for the transmission dynamics and control of infection. *Sex. Transm. Dis.*, **24**, 185–200.

Gordon, S. M., Eaton, M. E., George, R. (1994). The response of symptomatic neurosyphilis to high dose intravenous penicillin G in patients with human immunodeficiency virus infection. *N. Engl. J. Med.*, **331**, 1469–1473.

Johnson, R. C., Bey, R. F., Wolgamot, S. J. (1982). Comparison of the activities of ceftriaxone and penicillin G against experimentally induced syphilis in rabbits. *Antimicrob. Agents Chemother.*, **21**, 984–989.

Kajdacsy-Balla, A., Howeedy, A., Bagasra, O. (1987). Syphilis in the Syrian hamster. A model of human venereal and congenital syphilis. *Am. J. Pathol.*, **126**, 599–601.

Liu, H., Steiner, B. M., Alder, J. D., Baertschy, D. K., Schell, R. F. (1990). Immune T cells sorted by flow cytometry confer protection against infection with *Treponema pallidum* subsp. *pertenue* in hamsters. *Infect. Immun.*, **58**, 1685–1690.

Liu, H., Steiner, B. M., Stein-Streilein, J., Lim, L., Schell, R. F. (1991). Role of L3T4+ and 38+ T-cell subsets in resistance against infection with *Treponema pallidum* subsp. *pertenue* in hamsters. *Infect. Immun.*, **59**, 529–536.

Lukehart, S. A., Baker-Zander, S. A. (1987). Roxithromycin (RU 965): effective therapy for experimental syphilis infection in rabbits. *Antimicrob. Agents Chemother.*, **31**, 187–190.

Lukehart, S. A., Baker-Zander, S. A., Holmes, K. K. (1984). Efficacy of aztreonam in treatment of experimental syphilis in rabbits. *Antimicrob. Agents Chemother.*, **25**, 390–391.

Lukehart, S. A., Fohn, M. J., Baker-Zander, S. A. (1990). Efficacy of azithromycin for the therapy of active syphilis in the rabbit model. *J. Antimicrob. Chemother.*, **25**, 91–99.

Malone, J. L., Wallace, M. R., Hendrick, B. B. (1995). Syphilis and neurosyphilis in a human immunodeficiency virus type-I seropositive population: evidence for frequent serologic relapse after therapy. *Am. J. Med.*, **99**, 55–63.

Mushinski, M. (1997). Sexually transmitted diseases: United States, 1995. *Stat. Bull. Metrop. Insur. Co.*, **78**, 10–17.

Norris, S. J. (1982). *In vitro* cultivation of *Treponema pallidum*: independent confirmation. *Infect. Immun.*, **36**, 437–439.

Schell, R. F., LeFrock, J. L., Babu, J. P. (1978) Passive transfer of resistance to frambesial infection in hamsters. *Infect. Immun.*, **21**, 430–435.

Schell, R. F., LeFrock, J. L., Babu, J. P., Chan, J. K. (1979). Use of CB hamsters in the study of *Treponema pallidum*. *Br. J. Vener. Dis.*, **55**, 316–319.

Schell, R. F., LeFrock, J. L., Chan, J. K., Bagasra, O. (1980). LSH hamster model of syphilitic infection. *Infect. Immun.*, **28**, 909–913.

Schell, R. F., Azadagan, A. A., Nitskansky, S. G., LeFrock, J. L. (1982a) Acquired resistance of hamsters to challenge with homologous and heterologous virulent treponemes. *Infect. Immun.*, **37**, 617–621.

Schell, R. F., LeFrock, J. L., Chan, J. K. (1982b). Transfer of resistance with syphilitic immune cells: lack of correlation with mitogenic activity. *Infect. Immun.*, **35**, 187–192.

Schell, R. F., Steiner, B. M., Alder, J. D. (1988). Immunity to syphilitic infection. In *The Reticuloendothelial System* (eds Escobar, M. R., Utz, J. P.) Vol. 10, pp. 109–124. Academic Press, New York.

Sells, C. W., Blum, R. W. (1996). Morbidity and mortality among US adolescents: an overview of data and trends. *Am. J. Public Health*, **86**, 513–519.

Tabor, D. R., Azadegan, A. A., Schell, R. F., LeFrock, J. L. (1984). Inhibition of macrophage C3b-mediated ingestion by syphilitic hamster T cell-enriched fractions. *J. Immunol.*, **133**, 2698–2705.

Tabor, D. R., Azadegan, A. A., LeFrock, J. L. (1985). The participation of activated peritoneal macrophages in *Treponema pallidum* subspecies *pertenue* infection in Syrian hamsters. *J. Leukocyte Biol.*, **38**, 625–634.

Thomas, J. C., Kulik, A. L., Schoenbach, V. J. (1995) Syphilis in the south: rural rates surpass urban rates in North Carolina. *Am. J. Public Health*, **85**, 1119–1122.

Turner, T. B., Hollander, D. H. (1957). Biology of the treponematoses. *WHO, Monograph Series*, **35**, 214–233.

Une, T., Nakajima, R., Otani, T., Katami, K., Osada, Y., Otani, M. (1987) Lack of effectiveness of ofloxacin against experimental syphilis in rabbits. *Arzneimittelforschung*, **37**, 1048–1051.

Chapter 33

Guinea-pig Model of Acquired and Congenital Syphilis

V. Wicher and K. Wicher

Background of human infection

Although it was predicted that venereal syphilis would be eradicated by the early 1970s, syphilis is still a serious problem, ranking third among the sexually transmitted diseases reported to the Centers for Disease Control (1995), while approximately 3.5 million new cases per year occur worldwide. The disease is caused by the spirochete *Treponema pallidum* subsp. *pallidum* and humans are the only natural host. Congenital syphilis is the result of transplacental infection by a syphilitic mother. Early antibiotic treatment of acquired or congenital syphilis may effectively eradicate the pathogen. In untreated or insufficiently treated patients the disease becomes chronic. In a certain percentage of patients it may reactivate, causing pathological damage to almost any organ. Untreated congenital syphilis may cause stillbirths, death immediately after birth, or a series of hematological, tissue, and bone abnormalities. The clinical descriptions of syphilis, congenital syphilis, and the pathogenesis of treponemal diseases in humans are the subject of a number of chapters in books and reviews (Musher and Knox, 1983, Wicher and Wicher, 1983; Baughn, 1989, Holmes *et al.*, 1990; Lukehart, 1992).

Non-venereal treponematoses include: yaws, caused by *T. pallidum* subsp. *pertenue*, endemic syphilis, caused by *T. pallidum* subsp. *endemicum*; and pinta, caused by *T. carateum* (Musher and Knox, 1983; Wicher and Wicher, 1983).

Background of the model

One common characteristic of the pathogenic treponemes is their inability to propagate *in vitro*. Animal models must be used to maintain the treponemal strains and to delineate the mechanisms of pathogenicity and immune responses under controlled experimental conditions (Turner and Hollander, 1957b). Owing to its high susceptibility to *T. pallidum* infection (10–50 organisms, Turner and Hollander, 1957b), the rabbit is the animal of choice for propagation of *T. pallidum* and for detection of small numbers of virulent organisms in body fluids and tissues. The rabbit contributed a great deal of information on the kinetics and nature of the humoral response, histopathologic changes, and antibiotic treatment regimens. However, it has not fulfilled all expectations. There are not readily available the inbred strains required to delineate the cellular mechanisms of immunity and for studies of genetic differences in the immune response to *T. pallidum*. Additionally, the relatively short gestation period in the rabbit (~28 days), and the very slow growth rate (33 hours) of *T. pallidum* (Magnuson *et al.*, 1948) have been the main obstacles in its application for studies of congenital syphilis (Fitzgerald, 1985).

The guinea-pig model has been employed since 1907 by a number of investigators for studies of syphilis. For review of the literature see Wicher and Wicher, 1989. Unfortunately, the use of different sizes of inoculum, the unknown breeds of animal, and the indiscriminate use of males and females of different ages have provided conflicting results. Consequently, the use of the guinea-pig as experimental host of venereal syphilis was practically cast into oblivion for many years. Other animal models such as hamsters (Turner and Hollander, 1957b; Schell, 1983), rats (Schereschewsky, 1936), mice (Gueft and Rosahn, 1948; Klein *et al.*, 1980), and monkeys (Turner and Hollander, 1957b) have been used. Unfortunately, clinical manifestation of the venereal disease cannot be regularly induced in the three small animal models, and monkeys are prohibitively expensive and difficult to handle. The hamster has been successfully used for studies of endemic syphilis and yaws (Turner and Hollander, 1957b; Schell, 1983).

Animal species

The guinea-pig, *Cavia porcellus* (Wagner, 1976), although less susceptible than the rabbit to *T. pallidum* infection, has re-emerged as a useful model for studies of acquired (Pierce *et al.*, 1983; Pavia and Niederbuhl, 1985; Wicher and Wicher, 1989), neonatal (Wicher *et al.*, 1990, 1994), and congenital syphilis (Wicher *et al.*, 1992b, 1994). There are several inbred and outbred strains of guinea-pig with different susceptibilities to cutaneous infection with *T. pallidum* (Wicher *et al.*, 1985; Wicher and Wicher, 1989). The C4-deficient (C4D) guinea-pig genetically associated with

Handbook of Animal Models of Infection
ISBN 0-12-775390-7

inbred strain 13 is the most susceptible (ID_{50} = 10^2 organisms, Wicher and Wicher, 1991), whereas the Albany strain (a Hartley subline), haplotype-identical to inbred strain 2, is the most resistant (ID_{50} > 10^9 organisms). Inbred strains 2 and 13, and outbred strain Hartley have an intermediate susceptibility (ID_{50} ~ 10^5, Pierce *et al.*, 1983; Wicher *et al.*, 1985). Although C4D animals are genetically deficient in the C4 complement component (Ellman *et al.*, 1970) their humoral and cellular immunocompetences are similar to those of the complement-sufficient strain (Ellman *et al.*, 1971; Burger and Shevach, 1979). Owing to their remarkable different cutaneous response to *T. pallidum* infection, both C4D and Albany strains have been extensively used in our studies. All animals used in our laboratory are raised in our institutional farm and are free of *Treponema paraluiscuniculi* and *Encephalitozoon cuniculi*.

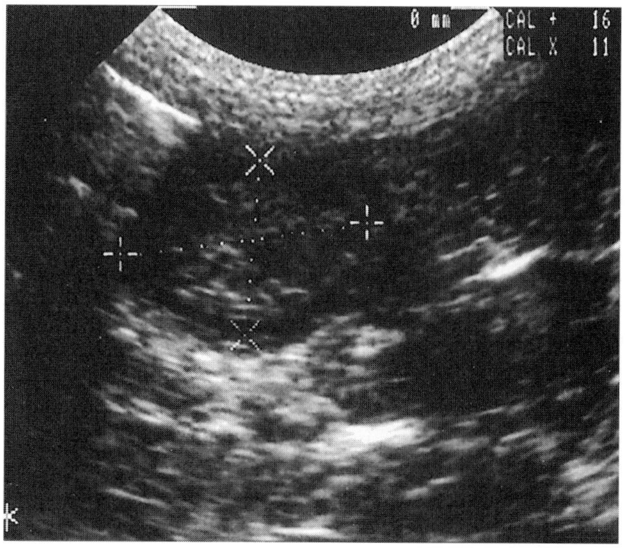

Figure 33.1 Ultrasound image of 23-day-old fetuses with caliper display (X) to measure the size of the fetuses. The conditions of the examination are listed on the left vertical axis, and the time, gain, and compensation (TGC) curve are displayed on the right vertical axis (unpublished observations).

Preparation of animals

Reproduction

Of all small laboratory mammals, the guinea-pig reproductive system most resembles humans in that it has a long cycle, ovulates spontaneously, and has an actively secreting corpus luteum (Reed and Hounslow, 1971). Like humans, the guinea-pig placenta is hemochorial (Sisk, 1976). In our experience the optimal conditions for mating include placing four to five dams together with a single male breeder in a large fiberglass run. Although females become sexually mature after 1 month of age, the best reproductive period is between 3 and 12 months of age. The gestation period in guinea-pigs is approximately 68 days and is inversely related to the number of fetuses (Goy *et al.*, 1957). Depending on the strain of guinea-pig, the litter size ranges from one to seven with an average of three. For more information on guinea-pig behavior see Harper, 1976, and on guinea-pig physiology see Sisk, 1976. In pregnant dams, guinea-pig fetuses are detectable by manual palpation as early as 25 days of gestation. Employing an ultrasound system (Ultramark™ 4, Advanced Technology Laboratories, Bothell, WA), we could detect fetuses as early as 15–18 days gestation (Figure 33.1). We used an optimal transducer energy power of 53%, a dynamic range (DRNG) of 58 dB, and a depth between 25 and 34 mm.

Housing

Neonates are housed together with their mothers until approximately 4 weeks of age, when they are weaned. Adult guinea-pigs are individually housed. Infected animals are kept in air-conditioned rooms at 15–16°C to enhance the spirochetal growth (Turner and Hollander, 1957c). The animals are fed with antibiotic-free Purina 5025 guinea-pig diet reinforced with vitamin C and water *ad libitum*. Guinea-pigs require a dietary source of vitamin C. For a 300 g guinea-pig a daily intake of 6 mg of vitamin C provides adequate protection against deficiencies (Nungester and Ames, 1948). This requirement increases to at least 20 mg per day for adult pregnant or lactating females (Bruce and Parker, 1947).

Anesthetics and tranquilizers

Guinea-pigs are anesthetized by an intramuscular (i.m.) injection of ketamine (40 mg/kg, Ketaset™, Fort Dodge Laboratories, Fort Dodge, IA) for treatments such as clipping, shaving, bleeding, and infection. The animal returns to normal behavior within 1 hour. For skin biopsy and tattooing the i.m. injection of ketamine is supplemented with a local injection of 0.05 ml of lidocaine 2% (The Butler Company, OH). This is followed by the analgesic buprenorphine (Buprenex, Reckitt & Colman Pharmaceuticals, VA) given i.m. 0.01 mg/kg twice a day for 3–5 days. For final sacrifice of the animal, an intracardiac (i.c.) injection of 0.5–1 ml/kg of Somlethol (American Hoechst, NJ) is given to the previously sedated guinea-pig.

Clipping and depilation

This is done prior to intradermal infection and then periodically to keep the infected area cool. This also facilitates the observation and follow-up of lesion development. Before infection the area is clipped and then shaved or depilated with Magic Shaving ™ powder as directed by the manufacturers. Before infection the skin area is disinfected with 70% vol./vol. alcohol.

Tattooing

An electric vibrating-needle tattoo machine (Spaulding & Rogers Manufacturing, Voorheesville, NY) and commercially available tattoo ink are used for this purpose. This procedure is employed in our laboratory when sites of intradermal inoculation must be identified for an extended period of several months. For shorter periods of observations the common laboratory permanent marker can be used to delineate the area of interest as guinea-pig skin is easily sensitized to various chemicals. Therefore, repeated depilation or demarcation of skin sites should be avoided when the local cutaneous immune response is the subject of the investigation. Appropriate controls consisting of guinea-pigs injected with non-pathogenic treponemes resuspended in the *T. pallidum*-free testicular supernatant (ITF) or ITF alone must always be included.

Bleeding and intravenous injection

Unlike rabbits, guinea-pigs do not have good-sized veins for bleeding or intravenous injection. Bleeding from the saphenous vein, marginal ear vein, cutting the nail bed, or puncture of the dorsal metatarsal vein yields only a very small volume of blood difficult to obtain under sterile conditions. (For references see Hoar, 1976). Moreover, small blood vessels are easily damaged after a few trials. Nonetheless, they may be used for blood cell counts or serological testing and injection of very small volumes.

Cardiac bleeding or injections are well tolerated by guinea-pigs. We have bled or infected intracardially (i.c.) hundreds of guinea-pigs with no loss to the procedure. Cardiac puncture is easy to perform in sedated and well-restrained animals. It allows injection of larger volumes (1–3 ml) and sterile bleedings of up to 5 ml every 3–4 weeks in adult animals or 0.5–1 ml every 3–4 weeks in young pups. Final cardiac bleeding of 20–40 ml may be procured from a single guinea-pig. It should be emphasized that, when the cardiac route has been employed several times, some structural changes in the heart must be expected and should be taken into consideration when evaluating the experimental results.

Preparation of inocula

T. pallidum is propagated in rabbit testes. The sedated rabbit is firmly held by an assistant ventral side up. The testicles are then exteriorized, the skin is disinfected with 70% alcohol, and with a syringe and a 25 G 22 mm needle each testis is slowly injected with 0.5–1 ml of infectious inoculum containing 10^7–10^8 organisms. The animal is observed every other day until orchitis develops, in approximately 9–13 days. At this point the rabbit is anesthetized with Ketaset (40 mg/kg) and sacrificed by i.v. injection of Somlethol. The testes are surgically removed under strictly sterile conditions and used for extraction of treponemes in antibiotic-free RPMI medium or sterile phosphate-buffered saline. After extraction the blood cells and tissue debris are removed by centrifugation at 200*g* for 10 minutes. The motile spirochetes in the supernatant are counted by darkfield microscopy and utilized for further propagation in rabbits or for infection of guinea-pigs.

Further purification of treponemes can be achieved by Percoll-density gradient centrifugation (Hanff *et al.*, 1984). Since we experienced significant loss of virulence by this procedure, the purified material is not used in our laboratory for infection of animals, although it contains the cleanest source of antigenic material for treponemal-DNA extraction, and for ELISA.

We have been able to maintain a guinea-pig-adapted strain of *T. pallidum* in C4D animals for a period of 2 years. Every 3–5 weeks extracts from infected lymph nodes were used for inoculation of the scrotal and hind limb areas of naive C4D (Wicher *et al.*, 1996c). Although the suspension contained a relatively small number of treponemes (~10^4 organisms/ml), the material was infectious for rabbits and its virulence for guinea-pigs seems to increase, as more severe lesions developed after the third passage. Also, the guinea-pig humoral response to *T. pallidum* (FTA-ABS) is higher when a guinea-pig- rather than rabbit-adapted strain of *T. pallidum* is used for infection (Wicher *et al.*, 1996c). Pregnant females injected with that small a number of treponemes are capable of intrauterine infection (Wicher and Wicher, unpublished data).

Routes of infection

For studies of acquired syphilis in guinea-pigs, cutaneous infection is the best route for examination of lesion development, particularly when done in the scrotal area or hind limb region (Turner and Hollander, 1957b; Pierce *et al.*, 1983; Pavia and Niederbuhl, 1985; Wicher *et al.*, 1988). Depending on the experiment, doses of 10^6–10^8 freshly extracted treponemes in 0.1 ml or in 0.5–1 ml are used for i.d. or i.v. infection respectively. Testicular infection of guinea-pigs is seldom done, since syphilitic orchitis in this animal is not as prominent as in the rabbit.

For studies of congenital syphilis, either the i.d. or i.c. route may be used to infect pregnant dams. Since guinea-pig litters range from one to seven, maternal infection is generally done with a large dose of treponemes (10^8 organisms).

Laboratory personnel involved in the extraction and inoculation of *T. pallidum* must wear surgical gown, gloves, mask or protector face shield to avoid infection.

Course of infection

Acquired syphilis

Neither the rabbit nor the guinea-pig can reproduce all stages of human syphilis. Both experimental models reproduce primary and latent syphilis but not the secondary stage which is characterized by a disseminated skin rash, or tertiary syphilis lesions (gummas, cardiovascular syphilis, or neurosyphilis). *T. pallidum* infection in guinea-pigs is affected by the factor of age, sex, and breed (Wicher *et al.*, 1988). Young animals are more susceptible than old and males are more susceptible than females. Likewise, the clinical and pathological consequences of syphilitic infection in humans vary among individuals (Gjestland, 1955) and are modulated by hormonal factors (Moore, 1922).

In the C4D guinea-pigs the onset of primary lesions at the site of inoculation is approximately 5–6 days after infection. It starts as an induration, which evolves into severe ulcerative lesions (chancre-like) of 8–18 mm diameter between 9 and 13 days (Figure 33.2A). The lesions are painless and disappear within 2–4 weeks, leaving no scars. Lesion aspirates contain large number of treponemes when examined by darkfield microscopy. The resistant Albany animals respond very mildly to i.d. infection, producing in approximately 7 days an induration of 5–7 mm in diameter lasting for 3–5 days (Figure 33.2B; Wicher *et al.*, 1985).

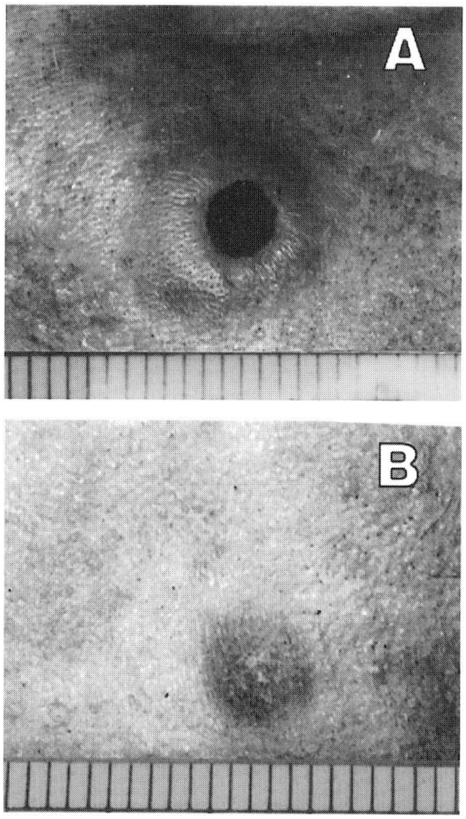

Figure 33.2 Dark-field positive lesions in C4D (**A**) and Albany (**B**) guinea-pigs. All animals were infected i.d. with 8×10^7 *T. pallidum* subsp. *pallidum*. Lesions at 23 days after infection.

Infection with *T. pallidum* does not cause the animals discomfort or increase of body temperature as the pattern of behavior and intake of food and water resemble those of normal animals. Like humans, untreated susceptible or resistant guinea-pigs may harbor virulent organisms for years in several organs, especially in draining lymph nodes, heart, and brain (Wicher *et al.*, 1996b). Intravenous infection does not produce obvious clinical symptoms in the guinea-pig.

Congenital syphilis

Transplacental infection in guinea-pigs occurs regardless of the maternal susceptibility to *T. pallidum* infection (Wicher *et al.*, 1992b). This is facilitated by the animal's relatively long gestation period (~ 68 days). Like a large percentage (> 50%) of infected human infants, congenitally infected guinea-pigs are asymptomatic (Wicher *et al.*, 1992b). Nonetheless the humoral response in the experimental host (Wicher *et al.*, 1992b, 1994) is identical to that of human infants (Dobson *et al.*, 1988; Sanchez *et al.*, 1989; Baughn, 1989). In the course of our studies on congenital syphilis we observed several cases of stillbirth, or death shortly after birth, underdeveloped, hemorrhagic and macerated fetuses, and even malformations (Figure 33.3).

Examination of several organs from underdeveloped or abnormal fetuses showed them to be *T.-pallidum*-DNA-positive. Notwithstanding, it was very difficult to correlate the congenital abnormalities with *in-utero* infection as most asymptomatic littermates were both serologically (IgM) and PCR positive (Wicher and Wicher, unpublished data).

Neonatal syphilis

Levels and antigenic specificity of natural antibodies and serum neutralizing activity present in mother and neonates of both strains of guinea-pig are very similar (Table 33.1; Jakubowski *et al.*, 1987). Nevertheless, the susceptibility to cutaneous infection in neonates is remarkably different from that of their mothers and adult controls. Table 33.2 summarizes the distinctive features in infected animals of both strains.

These findings, together with the late development of both cutaneous lesions and humoral response in C4D neonates, precluded the concept of an antibody-associated natural resistance. *T. pallidum* infection of neonates was not associated with depletion of any particular lymphoid tissue as reported for rabbits (Festenstein *et al.*, 1967). However, marked age- and strain-dependent histological differences were noted (Wicher *et al.*, 1990). This was further confirmed in studies of strain- and age-associated phenotypic markers of lymphoid cells in both strains of guinea-pig (Zhao *et al.*, 1992).

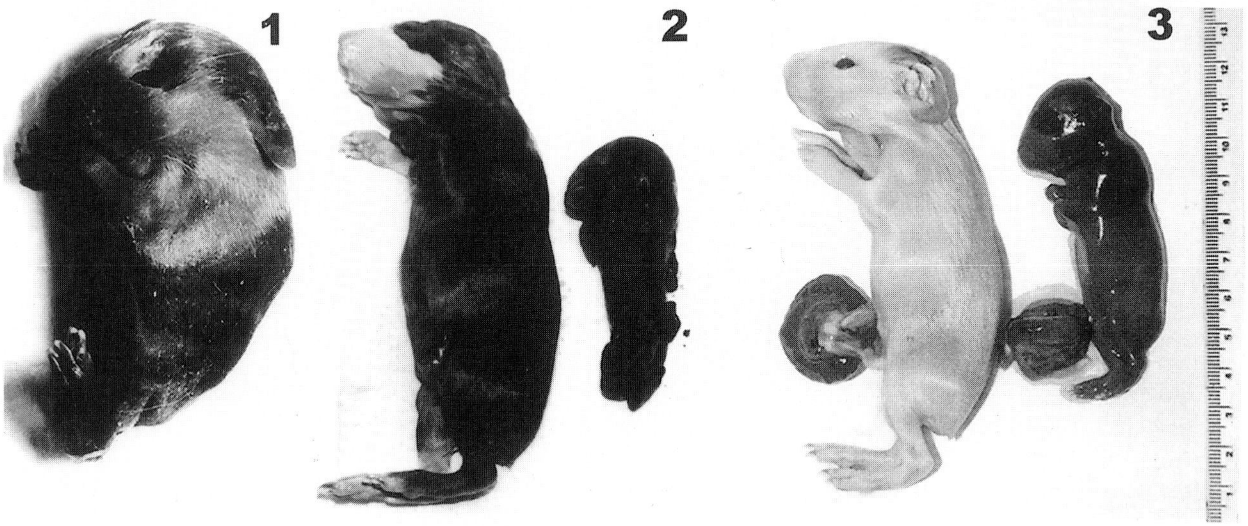

Figure 33.3 Congenital abnormalities in *T.-pallidum*-infected guinea-pigs. A stillborn Albany neonate underdeveloped and missing both eyes (**1**). The PCR-positive brain had a soupy consistency and was hemorrhagic. Two pairs of C4D (**2** and **3**), representing two litters of three and seven conceptuses respectively, were obtained from interrupted pregnancies at 60 days gestation. In both litters, the larger littermates approached normal size and the smaller fetuses were intensively hemorrhagic and dead. One to three organs from the living and dead fetuses were PCR-positive (unpublished observations).

Table 33.1 ELISA-Ig titres of natural treponemal antibodies in representative guinea-pig families (TP, *Treponema pallidum*; TR, *T. reiter*; TV, *T. vincentii*; reproduced from Jakubowski *et al*., 1987. *Immunology*, **60**, 281–285, with permission)

				Titre to:		
Strain	*Guinea-pig*	*Age*	*Sex*	*TP*	*TR*	*TV*
C4D	Mother	18 months		80	2560	640
	Offspring 1	1 day	F	80	2560	320
	2	1 day	M	80	2560	640
	Mother	12 months		160	2560	1280
	Offspring 1	7 days	M	160	2560	1280
	2	7 days	M	160	1280	640
Hartley A	Mother	27 months		10	640	80
(Albany)	Offspring 1	1 day	M	10	640	80
	2	1 day	M	10	640	80
	Mother	21 months		10	640	80
	Offspring 1	7 days	M	10	640	80
	2	7 days	M	10	640	80
	3	7 days	M	10	640	80
	4	7 days	F	5	640	80
2	Mother	24 months		40	1280	320
	Offspring 1	1 day	F	40	1280	320
	2	1 day	F	40	1280	320
	3	1 day	F	40	1280	160
	Mother	12 months		80	1280	80
	Offspring 1	7 days	M	80	1280	80
	2	7 days	M	80	1280	80
	3	7 days	M	40	1280	80
	4	7 days	F	40	1280	160

Table 33.2 Lesion development in neonates, their mothers, and young adult guinea-pigs infected with *T. pallidum* (reproduced from Wicher, V. *et al.*, 1990. *Clin. Immunol. Immunopathol.*, **55**, 23–40, with permission)

Strain	*Group*	*Age*	*Sex M/F*	*Animals with lesion/total*	*Percentage*	*Character**	*Incubation (days)*	*Maximum size (mm)*	*Duration (days)*
C4D	Neonates (*n* = 24)	1–5 days	12/12	4/24	17	Typical (early)	17.7 ± 2.6	14 ± 4.7	40 ± 5
				18/24	75	Typical (late)	77.5 ± 21.7†	9.18 ± 2	56 ± 11
	Mothers (*n* = 6)	13–27 months		5/6	83	Typical	14 ± 6.6	10.8 ± 1	43 ± 13
	Young adults (*n* = 12)	3–5 months	12/0	12/12	100	Typical	10 ± 1	13.4 ± 3	50 ± 7.4
Albany	Neonates (*n* = 41)	1–5 days	21/20	13/41†	32	Typical	15.2 ± 9.3	7.5 ± 1.5	14.3 ± 3.7
				15/41†	36	Atypical	11.2 ± 3.8	6.0 ± 2.2	7.33 ± 3.6
	Mothers (*n* = 12)	9–22 months		2/12	17	Atypical	14 ± 8.5	5.5 ± 0.7	13.5 ± 9
	Young adults (*n* = 17)	3–5 months	17/0	3/17	17	Atypical	9.6 ± 0.6	4.7 ± 0.6	5 ± 1

* Typical, progressive, ulcerative lesions; atypical, indurated, nonprogressive, nonulcerative lesions.
† $p < 0.01$ Mann–Whitney U test.

Mechanisms of immunity

Despite the early profuse humoral response to *T. pallidum* infection, passive transfer experiments with immune serum or its immunoglobulin IgG fraction, although encouraging, have not been totally successful. (For review of the literature see Wicher *et al.*, 1992a.) This may be explained in part by the extensive crossreactivity between pathogenic and non-pathogenic organisms (Wos and Wicher, 1986; Wicher *et al.*, 1991, 1992a). On the other hand, adoptive transfer of immunity to naive animals using large number of purified *T.-pallidum*-immune syngeneic lymphocytes, while effective in preventing infection in recipient animals (Wicher *et al.*, 1987), was unable to eliminate the pathogen from the donor. Indeed, untreated syphilis becomes a chronic disease characterized by the development of an incomplete immunity, also known as concomitant or infectious immunity. This is consistent with the slow clearance of the pathogen from the site of infection (Wicher *et al.*, 1996a) and its survival for months and years in several organs of the untreated host (Wicher *et al.*, 1996b). This may also account for the late manifestations of tertiary syphilis in humans affecting various organs (Sparling, 1990).

Key parameters to monitor infection and response to treatment

Serology, dark-field microscopy, immunofluorescence, silver staining, polymerase chain reaction (PCR), and rabbit infectivity test (RIT) are used to monitor infection and effectiveness of antibiotic treatment. In guinea-pig acquired and neonatal syphilis, IgM antitreponemal antibodies are elicited within 7–10 days, and IgG antibodies within 20–30 days after infection. The level of antibodies is lower in neonates compared with adults. These antibodies are detectable by fluorescence treponemal antibody absorption test (FTA-ABS), enzyme linked immunosorbent assay (ELISA), and immunoblot (Wicher *et al.*, 1994). As apposed to rabbits and humans, guinea-pigs do not produce antibodies to cardiolipin (Pierce *et al.*, 1983). By all *in-vitro* criteria used, transplacentally and neonatally infected hosts, regardless of their genetic background, respond immunologically in a very different manner as shown in Table 33.3 and Figure 33.4 (Wicher *et al.*, 1994). This is not surprising since, unlike neonatal syphilis, infection *in utero* involves not only the immature fetal immune system but also the fetal–maternal relationship (Wegman *et al.*, 1993).

Since *T. pallidum* cannot be visualized by light microscopy, dark-field microscopy is used to examine the morphology, motility, and number of spirochetes obtained from biological fluids or lesion aspirates. Localization of treponemes in tissues can be achieved by non-specific silver staining, or by immunofluorescence using specific antibodies. But none of these methods have the sensitivity or specificity of PCR or RIT (1–10 organisms, Turner *et al.*, 1969; Grimprel *et al.*, 1991). PCR has been extensively used in the guinea-pig model of syphilis. It was used to verify infection, dissemination, and efficacy of antibiotic treatment, as well as to monitor the clearance of virulent and dead treponemes from the site of infection in normal and immune animals (Wicher *et al.*, 1996a). It was used to identify target organs of infection in acquired and congen-

Table 33.3 Comparison of immune responses in congenitally and neonatally infected guinea-pigs (reproduced from Wicher, V. *et al.*, 1994. *Immunology*, **82**, 404–409, with permission).

Strain and no. of animals	*Bleeding after birth/infection (weeks)*	*Congenital syphilis*			*Neonatal syphilis*		
		*CIC**		*RF†*	*CIC*		*RF*
		IgM	*IgG*	*IgM*	*IgM*	*IgG*	*IgM*
C4D							
CS (n = 9)	1	4.8 ± 2.0	< 2.0	730 ± 359	< 2.0	< 2.0	< 100
NS (n = 15)	4	3.0 ± 2.0	< 2.0	386 ± 269	< 2.0	< 2.0	< 100
	8	< 2.0	< 2.0	< 100	< 2.0	< 2.0	< 100
	12	< 2.0	< 2.0	< 100	< 2.0	< 2.3 ± 1	< 100
	16	< 2.0	< 2.0	< 100	< 2.0	ND	< 100
Albany							
CS (n = 7)							
NS (n = 12)	1	6.3 ± 3.6	< 2.0	743 ± 204	< 2.0	< 2	< 100
	4	3.1 ± 2.1	< 2.0	377 ± 214	< 2.0	< 2	< 100
	8	< 2	< 2.0	115 ± 46	< 2.0	< 2	< 100
	12	< 2	< 2.0	< 100	< 2.0	< 2	< 100
	16	< 2	< 2.0	< 100	< 2.0	3.2 ± 1.5	< 100

* Circulating immune complexes expressed as numbers of SD above the mean of 30 normal guinea-pig sera; values of ≥ 2 SD indicate presence of CIC.
† Rheumatoid factor expressed as CPM, values of ≥ 100 (≥ 2 SD above 30 negative control sera) indicate presence of RF.

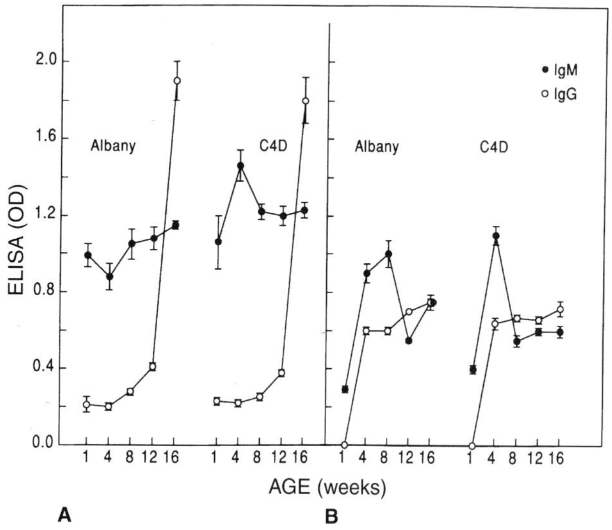

Figure 33.4 TP-ELISA. IgM (●) and IgG (○) antibodies reacting with 10% alcohol-treated *T. pallidum* produced during 4 months in transplacentally and neonatally infected guinea-pigs. Results express the mean OD ± SEM of nine C4D and seven Albany congenitally infected (**A**) and 15 C4D and 12 Albany neonatally-infected (**B**) with *T. pallidum*. (Reproduced from Wicher *et al.*, 1994. *Immunology*, **82**, 404–409, with permission.)

ital syphilis in a susceptible and a resistant strain of guinea-pig (Wicher *et al.*, 1996b). PCR is currently used to monitor vertical transmission of *T. pallidum* at various stages of gestation, from a single mother to several litters, and through several generations (Wicher and Wicher, submitted for publication). Since PCR cannot distinguish between dead and living organisms, in all these studies the rabbit infectivity test (RIT) was used to verify virulence or lack of it in PCR-positive samples. In spite of this shortcoming, PCR is becoming the method of choice in clinical and experimental settings, as the execution of the RIT is performed in very few laboratories and is very laborious and time-consuming.

Antibiotic treatment

The most effective antibiotic for treatment of human or rabbit syphilis is penicillin. Unfortunately, penicillin is toxic to and fatal for guinea-pigs (Hamre *et al.*, 1943), as it changes the intestinal flora from one that is predominantly Gram-positive to one predominantly Gram-negative (Formal *et al.*, 1963). Daily i.m. injections of a broad-spectrum antibiotic such as Cefotaxime sodium (Pavia and Niederbuhl, 1985) or chloramphenicol (Wicher *et al.*, 1987) in doses of 50 mg/kg for a period of 7 days has effectively eradicated *T. pallidum* from infected guinea-pigs.

Advantages and disadvantages of the experimental model

Advantages

On the basis of cumulative evidence provided by orthologically inherited genes (D'Erchia *et al.*, 1996), it has been suggested that the guinea-pig is not a rodent. This could explain why the guinea-pig is the only animal species with a well-defined complement system comparable to that of man. The similarity of bone marrow physiology and hematopoiesis appears to be greater between humans and guinea-pigs than between humans and rabbits, mice, or rats (Sisk, 1976). The observed strain- and age-associated susceptibility to *T. pallidum* infection makes this model suitable for exploration of the ontogeny of the immune response, genetic factors controlling presentation of antigen, and the nature of the immune effector mechanisms in syphilis.

The availability of inbred strains has been instrumental in the delineation of cellular immune mechanisms by adoptive transfer of immune-T cells (Wicher *et al.*, 1987). Guinea-pigs with C1, C2, C3, and C4 deficiencies of the classical complement system are available for analysis of the role of single complement components in humoral and cellular mechanisms of immune response (Bitter-Suermann and Burger, 1986). Unlike mouse, rat, or rabbit, the guinea-pig has a relatively long gestation period, which allows one to monitor transplacental infection at various stages of gestation and how it correlates with the ontogeny of the fetal immune response. In most recent experiments we could recognize that the guinea-pig is also a suitable model for cutaneous infection with *T. pallidum* subsp. *pertenue,* the causative agent of yaws. Intradermal infection of C4D and Albany guinea-pigs with *T. pallidum* subsp. *pertenue* produces irregular ulcerative skin lesions of similar size to those caused by *T. pallidum* subsp. *pallidum* (Figure 33.5).

Nonetheless, the degree of systemic dissemination, ability to cross the placenta, and humoral response differ between the strains of treponeme (Wicher and Wicher, unpublished observations). The guinea-pig has also been employed for studies of other spirochetal diseases such as Lyme disease caused by *Borrelia burgdorferi* (Sonnesyn *et al.*, 1993) and leptospirosis (Adler *et al.*, 1980). We have recently elaborated cDNA probes for detection of mRNA for guinea-pig cytokines IL-2, IL-10, IL-12, and TGFβ. These probes are presently used to examine the spontaneous expression of cytokines mRNA in various guinea-pig organs (Scarozza *et al.*, 1998) as well as the pattern of cytokines in *T. pallidum*-infected guinea-pigs (Wicher *et al.*, 1998).

Disadvantages

So far only the primary and latent stages of syphilis are reproduced in guinea-pigs. We have not explored the possibility of secondary disseminated infection by intravenous

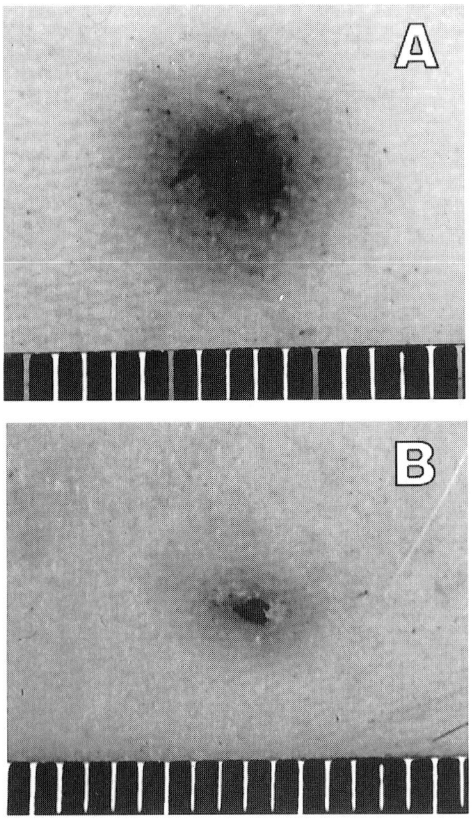

Figure 33.5 Dark-field positive lesions in C4D (**A**) and Albany (**B**) guinea-pigs infected with *T. pallidum* subsp. *pertenue.* The animals were i.d. injected with 2×10^7 virulent organisms. Lesions at 15 days after infection (unpublished observations).

infection and constant clipping of large areas of skin, as done in the rabbit model (Turner and Hollander, 1957c). Although lymph node, brain, and heart are the most common organs harboring *T. pallidum* for prolonged periods of time, no obvious signs of tertiary syphilis have been observed in the experimental model.

Another important limitation is the lack of readily available immunological reagents for this animal. There are very few molecular probes for detection of cytokines or monoclonal antibodies for examination of guinea-pigs' phenotypic markers, and only a few of them are commercially available. Although infected rabbit testes are still the major source of infectious organisms, further elaboration of the guinea-pig-adapted strain of *T. pallidum* is highly desirable.

Contribution of the animal model to treponemal infection

The guinea-pig model has been instrumental in the elucidation of the mechanisms of humoral and cellular response in acquired, congenital and neonatal syphilis. The animal is also suitable for exploration of experimental yaws. It has the potential to be employed for investigation of the different invasive properties displayed by *T. pallidum* subsp *pallidum* and *T. pallidum* subsp. *pertenue* (manuscript in preparation). The animal model has broadened our understanding of the immunopathology and natural history of syphilis in symptomatic and asymptomatic hosts.

Acknowledgment

The editorial help of Dr. M. King is greatly appreciated. This work was supported by grant No. AI21833 from the National Institute of Allergy and Infectious Disease, USA Public Services.

References

Adler, B., Faine, S., Muller, H. K., Green, D. E. (1980). Maturation of humoral immune response determines the susceptibility of guinea pigs to leptospirosis. *Pathology*, **12**, 529–538.

Baughn, R. E. (1989). Congenital syphilis: immunological challenges. *Pathol. Immunopathol. Res.*, **8**, 161–178.

Baughn, R. E., Wicher, V., Jakubowski, A., Wicher, K. (1987). Humoral response in *Treponema pallidum*-infected guinea pigs. II. Circulating immune complexes and autoimmune responses. *J. Immunol.*, **138**, 4435–4440.

Bitter-Suermann, D., Burger, R. (1986). Hereditary deficiencies in animals. Guinea pigs deficient in C2, C4, C3, or the C3a receptor. *Progr. Allergy.*, **39**, 134–158.

Bruce, H. M., Parker, A. S. (1947). Feeding and breeding of laboratory animals. III. Observations on the feeding of guinea pigs. *J. Hyg.*, **45**, 70–78.

Burger, R., Shevach, E. M. (1979). Evaluation of the role of C4D in the cellular and immune response *in vitro*. *J. Immunol.*, **122**, 2388–2394.

Centers for Disease Control (1995). *Sexually Transmitted Disease Surveillance 1995.* US Department of Health and Human Services, Washington, DC.

D'Erchia, A. M., Gissi, C., Pesole, G., Saccone, C., Arnason, U. (1996). The guinea pig is not a rodent. *Nature*, **381**, 597–600.

Dobson, S. R. M., Taber, L. H., Baughn, R. E. (1988). Recognition of *Treponema pallidum* antigens by IgM and IgG antibodies in congenitally infected newborns and their mothers. *J. Infect. Dis.*, **157**, 903–910.

Ellman, L., Green, I., Frank, M. (1970). Genetically controlled total deficiency of the fourth component of complement in the guinea pig. *Science*, **170**, 74–75.

Ellman, L., Green, I., Judge, F., Frank, M. M. (1971). *In vivo* studies in C4-deficient guinea pigs. *J. Exp. Med.*, **134**, 162–175.

Festenstein, H., Abrahams, C., Bokkenheuser, V. (1967). Runting syndrome in neonatal rabbits infected with *Treponema pallidum*. *Clin. Exp. Immunol.*, **2**, 311–320.

Fitzgerald, T. J. (1985). Experimental syphilis in rabbits. *Can. J. Microbiol.*, **31**, 757–762.

Formal, S. B., Abrams, G. D., Schneider, H., Sandy, R. (1963). Penicillin in germ-free guinea pigs. *Nature*, **198**, 712.

Gjestland, T. (1955). The Oslo study of untreated syphilis. An epidemiological investigation of the natural course of the syphilitic infection based upon a re-study of the Boeck–Bruusgaard material. *Acta Dermatol. Venereol. (Stockh.)*, **35**, 1–368.

Goy, R. W., Hoar, R. M., Young, W. C. (1957). Length of gestation in the guinea pig with data on the frequency and time of abortion and stillbirth. *Anat. Rec.*, **128**, 747–757.

Grimprel, E., Sanchez, P. J., Wendel, G. D. *et al.* (1991). Use of polymerase chain reaction and rabbit infectivity testing to detect *Treponema pallidum* in amniotic fluid, fetal and neonatal sera, and cerebrospinal fluid. *J. Clin. Microbiol.*, **29**, 1711–1718.

Gueft, B., Rosahn, P. D. (1948). Experimental mouse syphilis, a critical review of the literature. *Am. J. Syphilis Gonorrhea Vener. Dis.*, **32**, 59–88.

Hamre, D. M., Rake, G., McKee, C. M., MacPhillamy, H. B. (1943). Toxicity of penicillin as prepared for clinical use. *Am. J. Med. Sci.*, **206**, 642–652.

Hanff, P. A., Norris, S. J., Lovett, M. A., Miller, J. N. (1984). Purification of *Treponema pallidum* Nichols strain by Percoll density gradient centrifugation. *Sex. Transm. Dis.*, **11**, 275–286.

Harper, L. V. (1976). Behavior. In *The Biology of the Guinea Pig* (eds Wagner, J. E., Manning, P. J.), pp. 31–51. Academic Press, New York.

Hoar, R. M. (1976). Biomethodology. In *The Biology of the Guinea Pig* (eds Wagner, J. E., Manning, P. J.), p. 15. Academic Press, New York.

Holmes, K. K., Mardh, P. A., Sparling, P. F., Weisner, P. J. (eds.) (1990). *Sexually Transmitted Diseases*, 2nd edn, chs 18 23. McGraw-Hill, New York.

Jakubowski, A., Wicher, V., Gruhn, R., Wicher, K. (1987). Natural antibodies to treponemal antigens in four strains of guinea pigs. *Immunology*, **60**, 281–285.

Klein, J. R., Monjan, A. A., Hardy, P. H., Cole, G. A. (1980). Abrogation of genetically controlled resistance of mice to *Treponema pallidum* by irradiation. *Nature*, **283**, 572–574.

Lukehart, S. A. (1992). Immunology and pathogenesis of syphilis. In *Advances in Host Defense Mechanisms. Sexually Transmitted Diseases* (ed. Quinn, T. C.), pp. 141–163. Raven Press, New York.

Magnuson, H. J., Eagle, H., Fleischman, R. (1948). The minimal infectious inoculum of *Spirocheta pallida* (Nichols strain) and a consideration of its rate of multiplication *in vivo*. *Am. J. Syphilis Gonorrhea Vener. Dis.*, **32**, 1–18.

Moore, J. E. (1922). Studies in asymptomatic neurosyphilis: apparent influence of pregnancy on incidence of neurosyphilis in women. *Arch. Intern. Med.*, **30**, 548–554.

Musher, D. M., Knox, J. M. (1983). Syphilis and yaws. In *Pathogenesis and Immunology of Treponemal Infection* (eds Schell, R. F., Musher, D. M.), pp. 101–120. Marcel Dekker, New York.

Nungester, W. J., Ames, A. M. (1948). The relationship between ascorbic acid and phagocytic activity. *J. Infect. Dis.*, **83**, 50–54.

Pavia, C. S., Niederbuhl, C. J. (1985). Acquired resistance and expression of protective humoral immune response in guinea pigs infected with *Treponema pallidum* Nichols. *Infect. Immun.*, **50**, 66–72.

Pierce, C. S., Wicher, K., Nakeeb, S. (1983). Experimental syphilis: guinea pig model. *Br. J. Vener. Dis.*, **59**, 157–168.

Reed, M., Hounslow, W. F. (1971). Induction of ovulation in the guinea pig. *J. Endocrinol.*, **49**, 203–211.

Sanchez, P. J., McCracken, G. H., Wendel, G. D., Olsen, K., Threlkeld, N., Norgard, M. V. (1989). Molecular analysis of the fetal IgM response to *Treponema pallidum* antigens: implications for improved serodiagnosis of congenital syphilis. *J. Infect. Dis.*, **159**, 508–517.

Scarozza, A. M., Ramsingh, A. I., Wicher, V., Wicher, K. (1998). Spontaneous gene expression in normal guinea pig blood and tissues. *Cytokine*, **10**, 851–859.

Schell, R. F. (1983). Rabbit and hamster models of treponemal infection. In *Pathogenesis and Immunology of Treponemal Infection* (eds Schell, R. F., Musher, D. M.), pp 121–135. Marcel Dekker, New York.

Schereschewsky, J. (1936). Culture de spirochetes pales provenant de la rate de la souris blanche. *Bibl. Soc. Fr. Dermatol. Syph.*, **43**, 1063–1064.

Sisk, D. B. (1976). Physiology. In *Biology of the Guinea Pig* (eds Wagner, J. E., Manning, P. J.), pp 63–98. Academic Press, New York.

Sonnesyn, S. W., Manivel, J. C., Johnson, R. C., Goodman, J. L. (1993). A guinea pig model for Lyme disease. *Infect. Immun.*, **61**, 4777–4784.

Sparling, P. F. (1990). Natural history of syphilis. In *Sexually Transmitted Diseases*, 2nd edn (eds Holmes, K. K., Mardh, P. A., Sparling, P. F., Wiesner, P. J.), pp. 213–219. McGraw-Hill, New York.

Turner, T. B., Hollander, D. H. (1957a). Sources of strains studied. In *Biology of the Treponematoses*, Monograph Ser. No 35, pp. 15–30. WHO, Geneva.

Turner, T. B., Hollander, D. H. (1957b). The experimental disease in laboratory animals. In *Biology of the Treponematoses*, Monograph Ser. No 35, pp. 31–69. WHO, Geneva.

Turner, T. B., Hollander, D. H. (1957c). Factors affecting the evolution of experimental treponematosis. In *Biology of the Treponematoses*, Monograph Ser. No 35, pp. 70–94. WHO, Geneva.

Turner, T. B., Hardy, P. H., Newman, B. (1969). Infectivity tests in syphilis. *Br. J. Vener. Dis.*, **45**, 183–196.

Wagner, J. E. (1976). Introduction and taxonomy. In *The Biology of the Guinea Pig* (eds Wagner, J. E., Manning, P. J.), pp 1–4. Academic Press, New York.

Wegman, T. G., Lin, H., Guilber, L., Mosmann, T. R. (1993). Bidirectional cytokine interactions in the maternal–fetal relationship: is a successful pregnancy a Th2 phenomenon? *Immunol. Today*, **13**, 353–356.

Wicher, K., Wicher, V. (1983). Immunopathology of syphilis. In *Pathogenesis and Immunology of Treponemal Infection* (eds Schell, R. F., Musher, D. M.), pp. 148–159. Marcel Dekker, New York.

Wicher, K., Wicher, V. (1989). Experimental syphilis: guinea pig model. *CRC Crit. Rev. Microbiol.*, **16**, 181–234.

Wicher, K., Wicher, V. (1991). The median infective dose of *Treponema pallidum* determined in a highly susceptible guinea pig strain. *Infect. Immun.*, **59**, 453–456.

Wicher, K., Wicher, V., Gruhn, R. F. (1985). Differences among five strains of guinea pigs in susceptibility to infection with *T. pallidum*. *Genitourin. Med.*, **61**, 21–26.

Wicher, V., Wicher, K., Jakubowski, A., Nakeeb, S. M. (1987). Adoptive transfer of immunity to *T. pallidum* infection in inbred strain 2 and C4D guinea pigs. *Infect. Immun.*, **55**, 2502–2508.

Wicher, K., Wicher, V., Jakubowski, A., Gruhn, R. F. (1988). Factors affecting the course of *Treponema pallidum* infection in guinea pigs. *Int. Arch. Allergy Appl. Immunol.*, **85**, 252–256.

Wicher, V., Wicher, K., Rudosfky, U., Zabek, J., Jakubowski, A., Nakeeb, S. M. (1990). Experimental neonatal syphilis in susceptible (C4D) and resistant (Albany) strains of guinea pig. *Clin. Immunol. Immunopathol.*, **55**, 23–40.

Wicher, V., Zabek, J., Wicher, K. (1991). Pathogen-specific humoral response in *Treponema pallidum*-infected humans, rabbits and guinea pigs. *J. Infect. Dis.*, **163**, 830–836.

Wicher, K., Zabek, J., Wicher, V. (1992a). Effect of passive immunization with purified specific or cross-reacting immunoglobulin G antibodies against *Treponema pallidum* on the course of infection in guinea pigs. *Infect. Immun.*, **60**, 3217–3223.

Wicher, K., Baughn, R. E., Wicher, V., Nakeeb, S. (1992b). Experimental congenital syphilis: guinea pig model. *Infect. Immun.*, **60**, 271–277.

Wicher, V., Baughn, R. E., Wicher, K. (1994). Congenital and neonatal syphilis show a different pattern of immune response. *Immunology*, **82**, 404–409.

Wicher, K., Abbruscato, F., Wicher, V. (1996a). Persistence of virulent and heat-killed *Treponema pallidum* in rabbit and guinea pig as determined by PCR. *American Society for Microbiology 96 Annual Meeting. New Orleans, LA*, Abstract D51, p. 250.

Wicher, K., Abbruscato, F., Wicher, V., Baughn, R. E., Noordhoek, G. T. (1996b). Target organs of infection in guinea pigs with acquired and congenital syphilis. *Infect. Immun.*, **64**, 3174–3179.

Wicher, K., Wicher, V., Abbruscato, F., Baughn, R. E., Parsons, L. M. (1996c). The course of syphilis infection in C4D guinea pigs infected with gp-adapted strain of *T. pallidum*. *American Society for Microbiology 96th Annual Meeting, New Orleans, LA*. Abstract D50, p. 250.

Wicher, V., Scarozza, A. M., Ramsingh, I. A., Wicher, K. (1998). Cytokine gene expression in skin of susceptible guinea pig infected with *Treponema pallidum*. *Immunology*, **95**, 242–247.

Wos, S. M., Wicher, K. (1986). Extensive cross-reactivity between *Treponema pallidum* and cultivable treponemes demonstrated by sequential immunoadsorption. *Int. Arch. Allergy Appl. Immunol.*, **79**, 282–285.

Zhao, J., Wicher, V., Burger, R., Schafer, H., Wicher, K. (1992). Strain- and age-associated differences in lymphocyte phenotypes and immune responsiveness in C4-deficient and Albany strains of guinea pigs. *Immunology*, **77**, 165–170.

Chapter 34

The Guinea-pig Model of Legionnaires' Disease

P. H. Edelstein

Background of human disease

Legionnaires' disease is a type of acute bacterial pneumonia caused by bacteria of the genus *Legionella*, most commonly *L. pneumophila* (Edelstein *et al.*, 1984; Barbaree *et al.*, 1993; Edelstein, 1993). The disease was first recognized in 1976, when it caused a large epidemic of pneumonia amongst convention attenders in Philadelphia. About 10 000 cases of legionnaires' disease are estimated to occur in the USA annually, and it is estimated that the disease causes from 0.5–5% of all adult pneumonias requiring hospitalization (Marston *et al.*, 1997). *L. pneumophila* is an intracellular pathogen, which is transmitted to man via water containing the bacterium. The most common route of infection is by aerosol inhalation, although there is also some evidence that microaspiration of contaminated water may on occasion cause the disease.

The intracellular location of *L. pneumophila* protects the bacterium from many antimicrobial agents (Edelstein, 1995). Only antimicrobial agents active against intracellular *L. pneumophila* appear to be effective for treatment of the disease. Such agents includes tetracyclines, macrolide antibiotics, and fluoroquinolone antimicrobial agents. Aminoglycoside and β-lactam antimicrobial agents are inactive against the organism.

Prompt and specific therapy of legionnaires' disease usually cures from 70–99% of patients, depending on underlying illnesses and immunosuppression. Delays in therapy, severe immunosuppression, and very severe disease can lead to fatality rates as high as 80%. Some 80–98% of otherwise healthy patients may recover from legionnaires' disease without specific antimicrobial therapy, albeit after sometimes prolonged and debilitating illnesses.

Overview of animal models

Historical background and utility of various models

Intraperitoneal inoculation of clinical material into guinea-pigs was used for the isolation of *Legionella* species for several decades before the recognition of legionnaires' disease, when these bacteria were thought to be rickettsia-like bacteria (Tatlock, 1944, 1947; Bozeman *et al.*, 1968). After the recognition of legionnaires' disease, it was found that most rodents were completely or relatively resistant to infection by the bacterium, except for the AKR/J mouse (Hedlund *et al.*, 1979) and the guinea-pig. Guinea-pigs were found to be very susceptible to infection, and could be used in studies of pathogenesis and chemotherapy (Fraser *et al.*, 1978; Chandler *et al.*, 1979; Berendt *et al.*, 1980; Mycrowitz *et al.*, 1980). Winn, Davis, and colleagues demonstrated that both the aerosol and intratracheal routes of infection could be used to establish fatal infection in guinea-pigs and less severe infection in rats (Davis *et al.*, 1982, 1983; Winn *et al.*, 1982; Gump *et al.*, 1983). Monkeys were later found to be suitable models for studies of pathogenesis (Baskerville *et al.*, 1983a,b; Fitzgeorge *et al.*, 1983).

The mouse model appears to be most valuable for host defense studies because of the large body of knowledge about mouse genetics and immunology, as well as the ready availability of immunologic reagents for this animal (Klein *et al.*, 1993; Yamamoto *et al.*, 1993; Brieland *et al.*, 1994; Beckers *et al.*, 1995; Williams *et al.*, 1995; Miyamoto *et al.*, 1996; Smith *et al.*, 1997). The reason for the unique susceptibility of the A/J mouse strain is due to the permissiveness of its macrophages for bacterial multiplication, which has been attributed to the *Lgn1* gene (Beckers *et al.*, 1995). Recently the suckling CD1 mouse has also been shown to be susceptible to *L. pneumophila* infection (Pastoris *et al.*, 1997).

The guinea-pig model mimics legionnaires' disease as seen in immunocompromised humans, with high untreated fatality rates, and as such lends itself to studies of experimental chemotherapy (Edelstein *et al.*, 1984, 1996a; Pasculle *et al.*, 1985; Fernandes *et al.*, 1986; Saito *et al.*, 1986; Nowicki *et al.*, 1988). Guinea-pig infection is the method most often used to assess the relative virulences of different *Legionella* bacterial strains, and has also been used to study host immune defenses (Fitzgeorge *et al.*, 1983; Baskerville *et al.*, 1983c; Jepras *et al.*, 1985; Breiman and Horwitz, 1987; Blander *et al.*, 1989a; Cianciotto *et al.*, 1990; Marra *et al.*, 1992; Moffat *et al.*, 1994). This chapter describes only the guinea-pig model because of its primacy for the study of experimental chemotherapy.

Handbook of Animal Models of Infection
ISBN 0-12-775390-7

Guinea-pig source and types

Young Hartley strain guinea-pigs are the guinea-pigs most commonly used for models of *L. pneumophila* infection. Most work has been performed with 250–300 g animals. No comparative studies have been performed with older animals, so it is unknown if older animals are less susceptible to *L. pneumophila* infection. There appears to be no preference for a particular gender, although I have always used male animals. It is essential to obtain the healthiest animals possible. Source animal colonies should be screened by serologic or culture means for respiratory tract colonization with *Bordetella bronchicanis*, an exceptionally rare phenomenon with high-quality pathogen-free animals. Specific-pathogen-free guinea-pigs obtained from Charles River Laboratories are of very high quality and are especially recommended.

Preoperative care

Preoperative observation and conditioning of the animals for 4–7 days is required to ensure that all animals are healthy and have recovered from the stress of shipment. It is wise to monitor animal weights starting 1 day after receipt and every 2–3 days thereafter to ascertain that the animals are gaining weight. In addition, it is important to establish baseline body temperature before surgery. Body temperature is most easily measured using flexible rectal temperature probes designed for this purpose (Yellow Springs Instruments, Yellow Springs, OH). Supplementation of guinea-pig food with vitamin C is important to prevent naturally occurring pneumonia. Animals with diarrhea, weight loss, or failure to gain weight should not be used.

Details of surgery

Anesthesia

Adequate anesthesia is crucial to surgical success in the guinea-pig, making attention to this step crucial. A mixture of ketamine and xylazine is given intramuscularly, followed by local anesthesia using lidocaine. Mix ketamine (100 mg/ml) with xylazine (20 mg/ml) in a 16:5 volume ratio, e.g. 4 ml ketamine added to 1.25 ml xylazine. Each animal receives 1.05 ml/kg of this mixture. This will result in ketamine and xylazine doses of 80 and 5 mg/kg, respectively. I prefer the use of intramuscular rather than intraperitoneal ketamine/xylazine because of the greater speed of action and more reliable effect of the intramuscular route of administration. Withholding of food or water before anesthetic administration is not needed. The ketamine/xylazine can not be mixed together too soon before administration as older admixtures increase the perioperative mortality, by induction of pulmonary edema. Admixtures used within 2–4 hours appear safe, but older mixtures are not. The drugs are given in the hind thigh muscles, using a short 27 G needle; an insulin syringe is ideal. I use the mean weight of the animal group to decide on the ketamine/xylazine dose, as there appears to be a relatively broad range of tolerability and effectiveness; however animals with body weights significantly (perhaps more than 25% different) above or below the mean value may benefit from the use of individualized dosages. Intramuscular administration of ketamine/xylazine generally results in unconsciousness within 3–5 minutes; if after 10–15 minutes there has been only a minimal effect on consciousness, then an additional half-dose is given, which is almost uniformly successful. Ketamine/xylazine should be given about 20 minutes before the start of surgery, to allow adequate time for it to have its maximal effect, and to allow time for the lidocaine anesthesia to work. Animals generally awaken within 1 hour after administration of the ketamine/xylazine, although it may take up to another hour before they are normally active and resume eating and drinking. During this recovery phase, it is common to observe chorea-like movements. Vomiting or coughing are not observed during the recovery stage.

Use of local anesthesia allows for use of relatively low dosage of ketamine/xylazine and, more importantly, reduces animal movement. Lidocaine (1% without epinephrine) is administered in a dose of about 5–7 mg/kg once the animal is unconscious, and after the anterior neck has been disinfected with 10% povidone iodine; never administer more than 10 mg/kg. The lidocaine is given intradermally in the midline, as well as in the shallow subcutaneous tissues (about 50% of dose); deep drug injection should be avoided. Approximately 0.20 ml of lidocaine is sufficient for adequate local anesthesia, providing that sufficient time is allowed for drug diffusion in the tissues and for optimal drug effect. About 10–15 minutes is required for this optimal effect. The lidocaine is best administered using a tuberculin syringe and intradermal 26 G needle. Administration of up to 0.15 ml more does not appear to be harmful, but administration of this additional drug volume makes rapid and relatively bloodless dissection difficult because of tissue edema resulting from the injected drug. Epinephrine is not used because adequate hemostasis can be maintained without its use, and because its use may result in wound infections. Common errors are to inject too much lidocaine and to start surgery before the drug has had time to work effectively. Unlike the xylazine/ketamine, lidocaine is very stable and can be prepared in syringes the night before, as long as precautions are taken to prevent contamination of the syringes. On occasion, there may be a small dermal or subcutaneous localized area that is not completely anesthetic, usually discovered during the dissection. Additional lidocaine is usually not needed in these cases, as the affected areas are very small, and once dissection moves to a deeper layer the anesthesia is again sufficient.

Adequate ketamine/xylazine anesthesia causes unconsciousness but does not impair the rate or depth of respiration. Animals do not right themselves and make no voluntary movements. Mildly painful stimuli, such as injection of lidocaine, cause no reaction. However, more painful stimuli, such as skin incision, will result in protective voluntary movement. Administration of lidocaine results in complete local anesthesia, such that skin incision invokes no animal movement.

Skin disinfection

The anterior neck region is disinfected with 10% povidone iodine contained in individual ampules (Sepps™, Medi-Flex, Overland Park, KS), using the standard method of starting in the center and then moving outward in a circumferential manner. About three ampules are used per animal. A period of 3–5 minutes should be allowed for the iodine to work, although lidocaine injection can be started within a minute of the first iodine application. When the iodine dries completely, sometimes the fur becomes matted and difficult to cut through; in this case, another application of the disinfectant may be useful to make the dissection easier. Shaving or clipping the fur is neither helpful nor necessary.

Surgical instruments and other supplies

Sterile, autoclaved, instruments are used, with a different set used for each animal. The only exception is that the same disposable scalpel may be used to make the initial incision for every four to six animals. Surgical grade $4^1/_2''$ (11 cm) thumb forceps and $4^1/_2''$ (11 cm) straight iris scissors are used. Surgical stainless steel 9 mm clip applicators and removers are also needed (Totco Autoclip, Clay Adams, New York, NY). Wounds should not be closed with sutures instead of clips because of the much higher infection rate associated with suture closure of wounds. Sterile 5 × 5 cm gauze pads are needed to help control minor bleeding during dissection; one such gauze pad per animal is sufficient. Finally two to four pillows should be constructed of clean paper towels.

Preparation of inocula

Many different *L. pneumophila* strains have been used in animal models of *Legionella* pneumonia (Baskerville *et al.*, 1981; Winn *et al.*, 1982; Davis *et al.*, 1983; Meenhorst *et al.*, 1983; Edelstein *et al.*, 1984; Tartakovskii *et al.*, 1984; Jepras *et al.*, 1985; Plouffe *et al.*, 1986; Saito *et al.*, 1986; Twisk-Meijssen *et al.*, 1987; Catrenich and Johnson, 1988; Nowicki *et al.*, 1988; Cianciotto *et al.*, 1990; Marra *et al.*, 1992; Tully *et al.*, 1992; Blander and Horwitz, 1993; Fitzgeorge *et al.*, 1993; Mauchline *et al.*, 1994; Weeratna *et al.*, 1994; James *et al.*, 1995; Williams and Lever, 1995; Nikaido *et al.*, 1996). The most commonly used *L. pneumophila* strains are in serogroup 1. Strains F889, 130b, and AA100 (same as, or derived from, 130b) have been used in a large number of studies, as has the Philadelphia 1 strain. Of these strains, F889 is the most virulent. The type strain of *L. pneumophila* serogroup 1, ATCC 33152, is often avirulent, and should not be used unless it has been virulence-enhanced by passage through guinea-pigs, or perhaps by passage through cells or amoebae. *L. micdadei*, strain EK (ATCC 33204), was successfully used in one guinea-pig study (Pasculle *et al.*, 1985). Many of the Legionella species other than *L. pneumophila*, such as *L. dumoffii* and *L. bozemanii*, are relatively avirulent in guinea-pigs and require very high inoculum sizes to produce disease. Many, but not all, of the virulent *L. pneumophila* species used in animal models are salt-intolerant, which means that their suspension should be in distilled water rather than saline or PBS, unless they are shown to be salt-tolerant.

It can be exceptionally difficult to reproducibly make the same inoculum concentration of *L. pneumophila*, at least within a twofold range. The key to making this as reproducible as possible is to grow the bacteria to early log phase in BYEα broth (never more than $10^{8.4}$ cfu/ml). Since some frozen *L. pneumophila* strains may grow very poorly in BYEα broth, it may be necessary to make the first passage in BCYEα broth instead. Colonies picked from plates can have extremely variable viability; always pick individual colonies from the third or fourth quadrant if making the infectious inoculum from a plate. The correlation between viable organism concentration and broth turbidity or optical density is often strain-dependent, and is always growth-temperature- and growth-time-specific. Because of this variability, it is best to use a shaking water bath (for constancy of temperature and agitation), and to grow bacteria in broth for the same time each experiment. *L. pneumophila* may stick to polystyrene plastic tubes and pipets, so it is best to use glass or polypropylene tubes and pipets for all manipulations.

The dose–response curve of guinea-pigs to intrapulmonary *L. pneumophila* is sigmoidal, and quite steep in its slope (Figure 34.1). The dose–response curve is different for each bacterial strain, and for each route of inoculation. The LD_{50} for bacteria delivered by aerosol is about 1% of the corresponding value for the intratracheal route. However, this comparison is for retained bacteria (aerosol) versus delivered bacteria (intratracheal), so the total number of bacteria retained per lung is probably very similar for equivalent effects. Because of aerosol die-off and limited deposition of the aerosol into the lungs, about 10^{10} cfu must be aerosolized in a whole body chamber to achieve retained lung bacterial concentrations of 10^4–10^5 cfu. The approximate infectious doses of *L. pneumophila* strain F889 to use for intratracheal administration are 10^6–10^7 cfu per animal (delivered) and about 10^4–10^5 cfu per animal (retained).

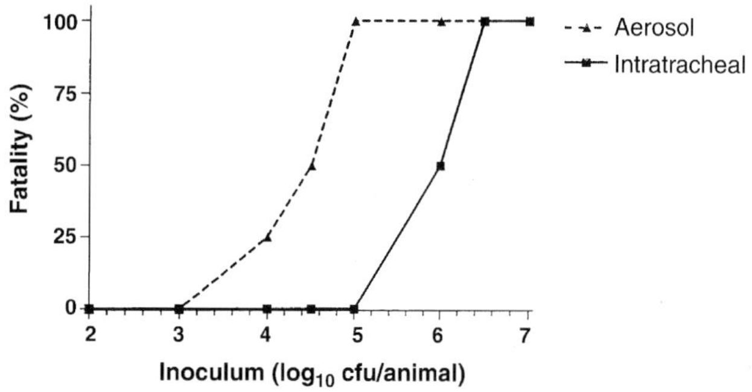

Figure 34.1 Typical fatality rates of guinea-pigs infected with different inocula of virulent *L. pneumophila* delivered by either the aerosol (retained dose) or intratracheal route (delivered dose). Fatality rate versus $\log_{10}$ inoculum.

Infection process

Infection route

Aerosol, intranasal, and intratracheal routes of infection have been successfully used to produce *Legionella* pneumonia in experimental animals. Intranasal infection can be used to induce respiratory infection in susceptible mice, but is ineffective in producing guinea-pig infection (Fitzgeorge *et al.*, 1983; Berendt and Jaax, 1985; Engleberg *et al.*, 1986). Intraperitoneal inoculation of guinea-pigs produces peritonitis, systemic infection, sepsis, and bacteremic pneumonia. While peritoneal infection is easy to produce, it is not a good model to use for legionnaires' disease, which is a primary pneumonia. Bacterial clearance, organ involvement, and organ histopathology differ between the peritonitis and pneumonia models (Fitzgeorge *et al.*, 1983; Eisenstein *et al.*, 1984). It is unclear if the peritonitis model can be used to successfully predict antimicrobial drug efficacy for legionnaires' disease (reviewed in Edelstein, 1995). Guinea-pig infection by either the aerosol or intratracheal route appears to lead to the same type of infection, although early pneumonia induced by the aerosol route is undoubtably more uniform in distribution. However, by 24–48 hours after infection, both types of inoculation result in bilateral pneumonia (Davis *et al.*, 1982; Winn *et al.*, 1982). Use of either route calls for special skills and equipment.

Aerosol infection can be administered via snout-only or whole-body exposure method (Baskerville *et al.*, 1981; Davis *et al.*, 1982; Watanabe *et al.*, 1985). In either case, special aerosol chambers must be used, both of which require special attention to limiting chamber leaks and protection of the user from aerosols generated. Improper chamber design or biosafety cabinet failure could result in extreme hazard to personnel. Whole body chambers require a large amount of biosafety hood space unless they are themselves self-contained. Throughput with a whole body chamber can be low, because of the need to first saturate the chamber with moisture, the 30 minutes animal exposure time, and to allow time for aerosol settling and removal before opening the chamber. In addition, terminal disinfection of the chamber may be time-consuming. The retained aerosol dose is dependent on a large number of factors, including relative humidity, strain type, culture phase and suspending medium, type of nebulizer used, and aerosol chamber construction. Comparison of strain virulences requires that a representative sample of animals be killed immediately after exposure for direct assay of retained dose, as aerosol stability may differ between strains (Hambleton *et al.*, 1983; Dennis and Lee, 1988). These various technical factors may outweigh the convenience of the aerosol method of delivery.

Intratracheal inoculation is technically demanding, requiring surgical skill and experience, as well as anesthesia. Unlike the aerosol method, from 1–10% of animals may suffer from perioperative mortality. The delivered dose is not dependent on a number of external factors. Strain variability of aerosol stability is not a complicating factor for the intratracheal route, although saline resistance must be accounted for in the intratracheal method (Catrenich and Johnson, 1989; Vogel *et al.*, 1996). An experienced operator can infect as many as 30–45 animals in an hour, in contrast to 10–20 animals per hour in an aerosol chamber. Surgical instruments are costly, and the time required to clean and sterilize them can be considerable.

Tracheal exposure and animal infection

The anesthetized and disinfected animal is placed supine with the neck extended and in neutral horizontal position. A 5 mm vertical incision through skin is made in the center of the anterior neck with a scalpel. Blunt dissection using scissors is performed into the cephalad and caudad subcutaneous tissues. This incision is then widened vertically by the scissors. Any fascia overlying the deeper neck structures is incised, always paying particular attention not to cut a blood vessel. The subcutaneous structures are then visualized, using both the forceps and scissors to gently move fat pads from the midline; if needed, the fat pads can be dissected to mobilize them away from the midline. Bleeding is usually very minor, but persistent oozing should be controlled by gentle pressure with sterile gauze pads. The midline, muscle-sheathed bundle containing the trachea should now be easily visualized. If it is not, check for positioning of the animal and the dissection itself, to be certain that the neck is not rotated and that you are exactly in the midline. Even minor neck rotation changes the visible anatomy and makes finding the trachea very difficult. There are two vertical strap muscles comprising the anterior sheath. The right strap muscle should be carefully transected, after which the membrane between that muscle and the underlying sheath is dissected cephalad about 5 mm. Great caution needs to be used, as immediately adjacent to these strap muscles are major blood vessels, which if cut will cause death from exsanguination. There should now be a relatively bloodless field, with the white tracheal rings easily visualized. The trachea is grasped gently with forceps to ascertain its mobility, then released.

While continuing to hold the forceps in one hand, the other hand is used to place the scissors in a discard pan and then to pick up the syringe containing the bacterial inoculum. The hand holding the forceps is used to uncap the needle. The trachea is then gently grasped with forceps as far cephalad as possible, and straightened. The needle is inserted into the trachea, trying to stay within the tracheal lumen. The needle is then partially withdrawn to ensure that it is in the lumen, and the forceps are removed from the trachea. The bacterial inoculum is slowly instilled into the trachea over a 10–15 second interval, after which the needle is removed. During instillation of the inoculum, fluid should be seen moving in the trachea with respirations. Look for any fluid that accumulates outside the trachea during the instillation of the inoculum; this is a very rare occurrence, and does not result in local infection. If the inoculum is inadvertently injected outside the trachea, simply mopping up the free fluid with a sterile gauze pad appears to be sufficient. If more than a very small amount of the inoculum has leaked outside the trachea, the animal may not be a valid experimental subject.

The wound is then closed in a single layer. The skin and subcutaneous tissues are grasped by the forceps to align the wound, and three to four sterile stainless steel clips are used to close the wound. The animal is then held in the head-up position and gently shaken, to help distribute the inoculum into the lungs. After this, the animal is placed supine with the head on a pillow made of paper towels, and allowed to recover from the immediate effects of the instillation. Immediately during or after instillation hyperpnea occurs, often followed by apnea or hypopnea. Cyanosis is sometimes observed, as demonstrated by eye or footpad observation; these respiratory abnormalities usually last less than a minute.

Postoperative course and complications

Animals generally awaken within 1 hour after administration of the ketamine/xylazine anesthesia, although it may take up to another hour before they are normally active, and resume eating and drinking. During this recovery phase, it is common to observe chorea-like movements. Vomiting and coughing are not observed during the recovery stage. The animals appear to have no postoperative pain or distress, unless there have been surgical complications. The most common postoperative complication is death without awakening from the anesthetic. These immediate postoperative deaths generally occur within 10–30 minutes after closing the wound, and certainly by 1 hour post-procedure. Necropsies of animals dying immediately postoperatively uniformly show pulmonary edema. The frequency of immediate postoperative deaths is about 1–2%; use of old mixtures of ketamine/xylazine increases this figure to as high as 10%. Surgeons unfamiliar with the surgical procedure may have complications of serious bleeding, but in experienced hands this is fortunately very rare (< 0.1%). Late bleeding into the wound rarely occurs, and is usually of no significance, but may be fatal. It is sometimes possible to successfully treat postoperative bleeding by wound exploration, but most serious wound bleeding is often obvious only relatively late, as the bleeding is confined to the subcutaneous tissues. Wound infection is extraordinarily rare and appears to be related most often to esophageal damage during surgery. These infections, which are not just of the superficial wound but also of the deep neck tissues, are invariably fatal because of the great difficulty in properly draining the infection.

Key parameters to monitor infection

Course of the *L. pneumophila* pneumonia

Guinea-pig illness from the *Legionella* infection is dependent on the bacterial dose administered and the virulence of the bacterium (Figure 34.1). Too low an infectious dose results in minor illness only, such as fever during the first two postoperative days, without death. Too high an inoculum can result in rapid death from sepsis and pneumonia within a day or two postinfection. Animals given intratracheal doses of approximately 10^6–10^7 cfu of *L. pneumophila* develop fever 1 day postinfection (Figure 34.2). Starting about 2 days postinfection, the animals appear to look ill, with ruffled fur and hyperpnea; they do not eat but appear normally active when handled; animal weight is lower than the preinfection weight (Figure 34.3). Three days after infection animals continue to lose weight and have persistent fever. They may be somewhat less active. Most animals die 4 days postinfection, at which point their body temperature starts to decline (Figures 34.2 and 34.4). They become lethargic, cyanotic (as judged by the change

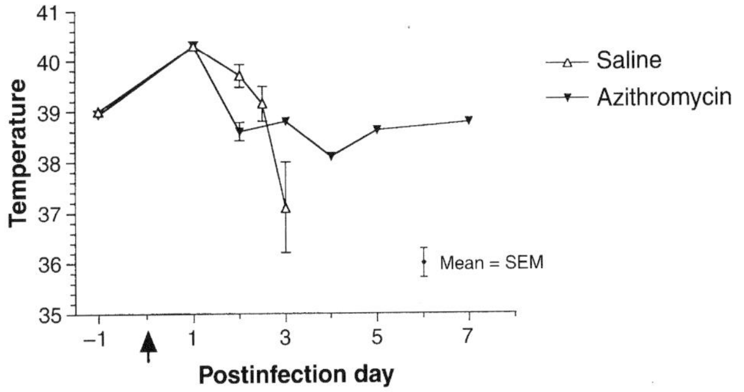

Figure 34.2 Temperature curves of guinea-pigs infected with *L. pneumophila* strain F889 that were treated either with saline (1 ml i.p. daily for 5 days) or azithromycin (10 mg/kg i.p. given once), starting 1 day postinfection. The infection day is shown by the arrow. Mean temperature (± SEM) versus postinfection day.

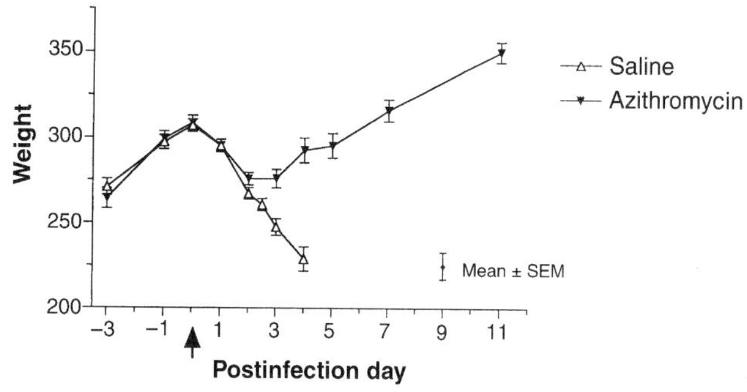

Figure 34.3 Guinea-pig weights (grams) after infection with *L. pneumophila* strain F889. Animals were treated either with saline (1 ml i.p. daily for 5 days) or azithromycin (10 mg/kg i.p. given once), starting 1 day postinfection. The infection day is shown by the arrow. Mean weight (± SEM) versus postinfection day.

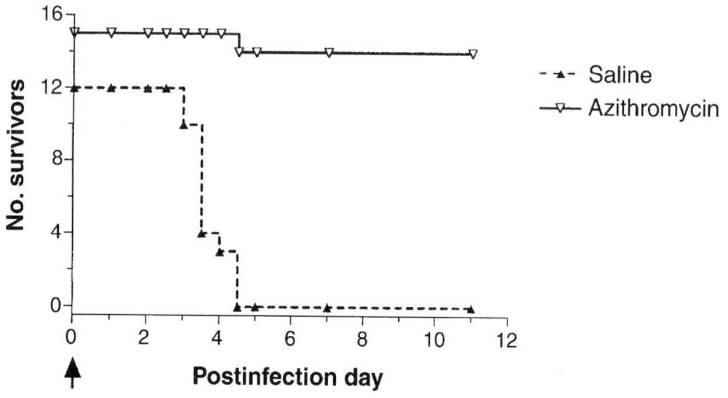

Figure 34.4 Guinea-pig survival after infection with *L. pneumophila* strain F889. Animals were treated either with saline (1 ml i.p. daily for 5 days) or azithromycin (10 mg/kg i.p. given once), starting 1 day postinfection. The infection day is shown by the arrow.

in eye color from bright red to dark red), and increasingly hyperpneic, with use of accessory muscles of respiration. They may not be normally active, and the very ill animals may position themselves semiupright in the cage. Even so, they may normally try to avoid handling. In some cases, apparently purulent conjunctivitis develops, as may rectal prolapse.

Death can be reliably predicted in animals with rectal temperatures at or above 41.5°C or at or below 37.5°C, especially when combined with weight loss of 25% or more of baseline, lethargy, and absence of eating or drinking. Animals meeting these criteria are humanely killed with an overdosage of pentobarbital, 150 mg/kg, given i.p. Carbon-dioxide euthanasia should not be used, as it results in pulmonary edema, which confounds postmortem studies. It is important to correlate clinical findings with the temperature determination as the rectal temperature probe is sometimes defective and fails to register a temperature.

By day 5 most animals start to recover from their infection, with weight gain, greater activity and less hyperpnea. Several days more is often required before the body temperature normalizes.

Frequency of observation

Animals should be observed twice daily during the first 4 days postinfection, and then daily thereafter. This monitoring schedule allows for the prompt detection of animal distress, and its amelioration if needed. Since body temperature is used for determination of endpoints, accuracy of the temperature probe should be ascertained at the start of each experiment using a reference thermometer. Body temperature can be increased by as much as 1°C if inexpertly taken, as a result of heat generated by an improperly restrained animal. Body temperatures and weights should be taken at approximately the same time each day.

Necropsy procedures

Animals must be disinfected before necropsy if organ bacteriology is performed. This can be done by dipping all but the head and neck of the dead animal in 2% Amphyl (National Laboratories, Montvale, NJ) for a few seconds. No disinfectant should be allowed to enter the mouth or nose because of the potential for disinfection of the bronchial tree. The neck and lower face can then be disinfected using 2% Amphyl in a wash bottle. Alternatively, a freshly prepared 10% bleach solution made from sodium hypochlorite, 5% wt/vol., can be used as a disinfectant; bleach is a better disinfectant, but it can be toxic to personnel and cause damage to instruments and hood surfaces. In either case, allow at least 2–5 minutes for full disinfectant action. The wet animal is then placed on an absorbent pad in the biosafety cabinet; the absorbent pad prevents gross hood soilage, contains spills, and helps to prevent aerosolization of droplets falling on to the hood surface.

The key to obtaining lung and spleen tissue that is not contaminated by skin flora is to use multiple sets of sterile instruments, which are used to perform dissection in layers. It is also very important not to enter the esophagus or bowel, something that is easily done if anything other than sharp dissection under direct vision is used. Sterile surgical instruments are used; $5\frac{1}{2}''$ (14 cm) straight operating scissors (sharp/sharp) are valuable for use in these dissections.

A vertical midline skin incision is made from the low neck to just above the xiphoid process. At midchest level, two lateral skin incisions are then made to expose the entire rib cage, dissecting the overlying soft tissues as needed, with sterile scissors and forceps. The scissors are then used to dissect down to the rib surface on the right chest wall. A new set of sterile scissors and forceps is used to open the chest cavity between two lower ribs, and to extend the incision so as to remove the entire anterior chest wall. If desired, up to 2 ml of heart blood can be aspirated at this juncture. The right lower lung lobe is then removed for culture, using a third pair of sterile scissors and forceps. Great care must be taken not to sever the esophagus if any of the left lung lobes are removed, as this will result in contaminated cultures. The entire remaining lung can then be removed *en bloc* if desired. It is very difficult to remove the entire lung *en bloc* without gross contamination of the lung by the esophageal bacteria.

Aseptic spleen removal is best accomplished by extending the midline skin incision caudad to the pelvis, cutting through the skin and fascia; this is done with a new set of scissors and forceps. A horizontal incision is then made that starts at the intersection of the midline incision with the imaginary line connecting the caudad end of the left rib cage. This incision is extended to the rib cage, and the resulting flap is transected and reflected upwards. The animal is then moved from the supine to the right lateral decubitus position. The spleen should then be easily visualized. The spleen is grasped with a new pair of sterile forceps and gently dissected from its vascular attachments. Great care should be used so as not to open the stomach, which is close to the spleen.

Antimicrobial therapy

Overview and prerequisite studies

Before embarking on an animal treatment study, it is important to perform prerequisite studies. These include confirmation of the *in-vitro* extracellular activity of the study drug and, more important, its intracellular activity against *L. pneumophila*. Determination of the pharmacokinetics of the study drug in the guinea-pig pneumonia model is essential for study interpretation, and may be important for choosing an animal dosage schedule. Methods for performing *in-vitro* susceptibility and pharmacokinetic studies are not given here, but appear in many publications (Edelstein, 1995; Edelstein *et al.*, 1996a).

Both intratracheal and aerosol routes of infection have been used to study the experimental chemotherapy of legionnaires' disease, and appear to give results predictive of outcome in humans (reviewed in Edelstein, 1995). The peritonitis/systemic infection model has also been used for the study of experimental therapy (Fraser *et al.*, 1978; Yoshida *et al.*, 1985), but may not be as reliable as the pneumonia model for the extrapolation of results. No parallel studies between the peritonitis and pneumonia models have been performed, but major differences in the specific inflammatory cell response, the organs involved, and probably in the mechanism of death make it desirable to use the pneumonia model to predict the response to antibiotic therapy of humans with legionnaires' disease. In addition, systemic and organ-specific drug pharmacokinetics may be different in the two different models, as well as the pharmacokinetics of intracellular drugs. For example, it is known that azithromycin and erythromcyin are handled differently in guinea-pigs with *L. pneumophila* pneumonia than in healthy animals, and that some of the reasons for these differences are due to the influx of drug-containing polymorphonuclear leukocytes and macrophages into the infected lungs (Stamler *et al.*, 1994). It is unknown if similar phenomena occur in the peritonitis model, and if so whether they can be used to model the response of the human pneumonia.

An intratracheal dose of about 2–3 LD_{50} should be given, as this (6×10^6 to 1×10^7 cfu per animal of strain F889) leads to death in about 4 days for most untreated control animals, but full recovery for animals treated with effective antimicrobial agents. Antibiotic therapy is started about 18–24 hours after infection, and given for 1–5 days, depending on drug pharmacokinetics and intracellular activity.

The antibiotic dosing regimen is dependent in part on the purpose of the study. If the study hypothesis is that the study drug is as effective as standard therapy (usually erythromycin), then to avoid unneeded experiments, it is best to be conservative and give a full 5 days of once- or twice-daily therapy at a dosage giving peak serum levels similar to those in man. However, if the experiment is designed to show superiority to standard therapy, then fewer days of study drug therapy, lower dosages, or less frequent dosing may be indicated (Fitzgeorge *et al.*, 1990, 1993; Edelstein *et al.*, 1992, 1996b).

Practical aspects of drug administration

Guinea-pigs can be very susceptible to gastrointestinal tract toxicity from some antibiotics, especially the β-lactam drugs, but also erythromycin, tetracyclines, and rifampin. This toxicity appears to be due to alteration of bowel flora, and is most prominent for those drugs likely to alter bowel flora. Late deaths due to drug-induced gastrointestinal tract toxicity may affect as many as 30% of animals. The toxicity does not appear to be due to *C. difficile* toxins, as stool cytotoxins are not detected in these animals, and no pseudomembranes are formed (unpublished data). Prophylactic oral administration of vancomycin or lactobacillus (Lactinex™) has not been helpful in my experience (Edelstein *et al.*, 1984), although it may have worked in another laboratory (Pasculle *et al.*, 1985). Poor animal quality appears to be a major factor in the risk of this disease, especially animals having diarrhea prior to antibiotic treatment. Fear of this β-lactam toxicity has led to use of a rat model by one group of investigators (Smith *et al.*, 1991, 1992). Luckily, most drugs causing this toxicity are inactive against intracellular *L. pneumophila*. Most fluoroquinolone drugs have not caused this problem, nor have newer macrolide or macrolide-like drugs. This toxicity does not necessarily prevent analysis of drug efficacy, as it tends to occur 5–7 days after the start of antibiotic therapy, after the time that the pneumonia responds to therapy. Use of additional measures of successful therapy, such as lung histopathology and culture, can distinguish between drug treatment failure and drug toxicity. Since animals with this type of drug toxicity lose weight, it is not possible to use that parameter to measure drug treatment success in animals with drug toxicity.

An additional problem related to drug administration is the route of administration. I prefer parenteral administration by i.p. injection because of its convenience and predictability; 1 ml volume injections are used. Some investigators have used oral administration and others intramuscular administration. Intramuscular administration of some drugs is painful, and in the case of some leads to injection site necrosis. Oral administration requires acid stability of the drug and good technique to avoid lung instillation. Parenteral administration of drugs does require that they be soluble in non-toxic solvents, something not to be overlooked until the day of administration! It is also useful to keep in mind that some prodrugs requiring metabolism for activation may not be metabolized by guinea-pigs. All drug solutions administered by the parenteral route should be sterile, non-pyrogenic, and preservative-free if possible.

Comparator drugs

Erythromycin is considered by many to be the standard of therapy, and the best comparator for standard therapy; it is generally given in a dose of 30 mg/kg i.p. twice daily for 5 days. However the drug is potentially toxic to guinea-pigs, especially if the animals are not of high quality. Several drugs are more active than erythromycin and less toxic for guinea-pigs. Azithromycin is much easier to administer to guinea-pigs than erythromycin, as a single dose of 10 mg/kg i.p. is curative and results in little or no toxicity to animals. In addition, azithromycin sterilizes the lung, in contrast to the persistent lung infection seen in erythromycin-treated animals. Another very active drug that can be given once daily for 3–5 days is levofloxacin (10 mg/kg once a day).

Every treatment study should include a placebo treatment group to ascertain that the infection is indeed fatal unless properly treated. Administration of saline once or twice daily (1 ml per dose) is generally used, although 5% dextrose or water for injection can also be used.

Numbers of animals

A minimum of 15 animals should be used in the each of standard and experimental therapy arms of the study. Since the fatality rate of the placebo-treated group is close to 100% in a properly conducted study, only about 12 animals are needed in that group. This should allow the detection of about 40–50% change in outcome with 80% power (α_2 = 0.05). Since perioperative fatalities occur, it is wise to start with about 45 animals in a three-arm study (placebo, standard therapy, and experimental therapy), and about 60 animals in a four-arm study. Animals surviving on the first postoperative day are then randomly allocated into the treatment groups. If the mean weights (use the infection day weights) in each group are significantly different after randomization, then it is best to manually adjust the randomization so that the weights are equivalent; this can usually be done by changing the allocations of two animals. Randomization is important because the guinea-pig population may not always be homogeneous in terms of weight, and because the time from infection to first drug dose will necessarily differ by several hours between the first and last animals infected.

Parameters used to gauge the success of chemotherapy

Animal survival, weight, body temperature, and time to death can all be used to determine the effectiveness of therapy (Figures 34.2 and 34.4). Survival rate and weight appear to be overall the most useful parameters to follow. Since most fatalities occur within the first 4–5 days of infection or not at all, time to death is usually not especially revealing. Monitoring body temperature is required, as it is an accurate predictor of eventual death (see above). Following animals for more than 5–7 days post-therapy (10–12 days postinfection) is not needed, as very late deaths as the result of relapsing infection are extremely rare. When very late deaths occur they are usually due to antimicrobial agent toxicity.

Quantitative lung culture is an especially powerful index of drug efficacy, and is also useful in distinguishing infectious from non-infectious death (Figure 34.5). A piece of the right lower lobe of known weight (≈ 1 g) is ground with a known volume (≈ 1 ml) of Mueller–Hinton broth in an all glass tissue grinder. Decimal dilutions of the resulting homogenate are made in Mueller–Hinton broth and plated on to BCYEα plates in 0.1 ml volumes. There is no need to use selective media for this purpose. Generally spleen culture adds little information. Lung and spleen both contain inhibitors to the growth of *L. pneumophila*, such that the

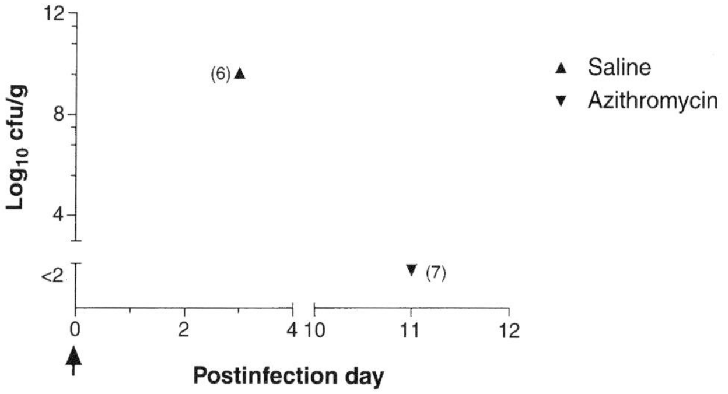

Figure 34.5 Guinea-pig lung concentrations of *L. pneumophila* strain F889. Animals were treated either with saline (1 ml i.p. daily for 5 days) or azithromycin (10 mg/kg i.p. given once), starting one day postinfection. The infection day is shown by the arrow.

lower limit of detection is about 100 cfu/g of tissue. Lungs of placebo-treated animals contain 10^9–10^{11} cfu/g. Lungs from animals treated with erythromycin usually contain 10^2–10^4 cfu/g 5 days post-therapy, whereas lungs from those animals treated with levofloxacin or azithromycin usually have rare or no detectable *L. pneumophila* bacteria present.

Histopathology of the lung may also aid in gauging the success of therapy. Lungs from placebo-treated animals reveal diffuse organizing pneumonia with microabscesses, neutrophilic and monocytic infiltrates, and alveolar–septal necrosis. Those from animals treated with erythromycin generally show focal areas of consolidation with monocytic infiltrates and fibrosis. Azithromycin- or levofloxacin-treated animals may have normal lung histopathology or rare scattered monocytic infiltrates. The best results are obtained when lungs are viewed without knowledge of treatment, and when a precise method is used to score histologic abnormalities.

Safety issues

Bacterial inoculum preparation, animal surgeries and necropsies must be performed in a recently certified and meticulously cleaned CDC/NIH class IIb biosafety cabinet. The hood should not be overloaded with animals or supplies, and the work surfaces should be covered with an adsorbent material in case of spills. During the surgical procedure, the surgeon and assistants should wear cap, surgical mask, gown, and sterile surgical gloves. The mask is used to prevent gross oral ingestion of infectious material only; use of a respirator (HEPA, PAPR, or N95) is not required.

Needles must not be recapped before disposal in an appropriate container. Infected animals are not infectious for uninfected animals (Katz *et al.*, 1982), and there is no evidence for animal-to-human spread (or human-to-human spread; Hoge and Brieman, 1991). No reports exist of laboratory-acquired infection from infected animals, which agrees with experience in my laboratory, with more than 60–100 person years of exposure to thousands of infected animals. The laboratory procedures most likely to transmit infection are the preparation and administration of the infectious inoculum and the handling of infected animal tissues. There is no need to disinfect animal fur before removal of the animal from the biosafety hood. All instruments used in the procedure must be placed into a covered container before removal from the hood, and then autoclaved in the container before being removed from it.

Animals can be safely housed in filter-top cages, which themselves are kept in a CDC/NIH Class II animal cubicle. Minor procedures such as weighing animals, giving injections, and taking temperatures can be performed in the cubicle without risk to the handler; there is no need to perform these operations in a biosafety cabinet. A surgical face mask, gown, shoe covers, cap, and gloves should be worn during these procedures, to comply with CDC/NIH guidelines. Use of respirators (HEPA, PAPR, or N95) is not required for the protection of personnel.

Necropsies are probably among the most potentially dangerous of the procedures used in these animal model studies. This is because lung may contain up to 10^{10} cfu/g of *L. pneumophila*, blood 10^4 cfu/g and spleen 10^6 cfu/g. All tissues must be placed in secure screw-cap containers before removing them from the hood, and all contaminated instruments handled as described above. It is important to know that the interior of large organ pieces placed into 10% formalin may contain viable *L. pneumophila* for several days.

All laboratory personnel working with *L. pneumophila* should be in good health. Specifically they should not be cigarette smokers, have chronic lung or heart disease, or be immunosuppressed because of disease or medications. No data exists regarding the proper action to be taken in case of serious biohazardous laboratory accidents. I have administered short-course antibiotic prophylaxis once when there was hood failure during grinding of infected guinea-pig lung; ofloxacin 500 mg given 12 hours apart for two doses was given, with the first dose administered within 4 hours of exposure.

Pitfalls of the model

In general there has been excellent correlation between the results of a properly conducted animal treatment study and what is known about the response of humans with legionnaires' disease to the same drugs (reviewed in Edelstein, 1995). The model system is a very severe test of antibiotic therapy, and some drugs showing adequate, but not exceptional, activity in this model system appear to be effective for the treatment of legionnaires' disease in humans. The animal model differs from the situation from man in several respects:

- Guinea-pig infection resembles infection in immunocompromised patients rather than the normally healthy host. As such it is a severe test of antimicrobial chemotherapy.
- The infection is treated promptly in the guinea-pig, which allows for recovery with active antimicrobial chemotherapy. Some humans are treated very late in the course of their disease, and have very poor response to what is otherwise active chemotherapy.
- Relatively few strains of *Legionella* bacteria have been tested in the model system, and *Legionella* species other than *L. pneumophila* and *L. micdadei* have not been tested at all. This may mean that extrapolation to the treatment of other species is not accurate, although there is no scientific reason to believe that this is so.
- Pharmacokinetics of antimicrobial agents given to guinea-pigs do not mimic the human situation. In most

instances drug half-lives are much shorter in guinea-pigs.

Mistaking drug toxicity for inactivity is an additional possible pitfall of this model system.

Contributions of the model to infectious disease therapy

Profiling of antimicrobial therapy

Screening of promising new antimicrobial agents in this animal model is a prerequisite to using the agents in the treatment of human legionnaires' disease. Animal screening only need be done after screening in cell models of infection show the agent to be active against intracellular *L. pneumophila*.

Evaluation of the virulence of genetic mutants

The guinea-pig pneumonia model has been used by a number of investigators to detect avirulent genetic mutants of *L. pneumophila* (Blander and Horwitz, 1989b; Cianciotto *et al.*, 1989, 1990; Weeratna *et al.*, 1994). While it is relatively easy to detect frankly avirulent genetic mutants in this model system, it is much more difficult to detect more subtle differences (Moffat *et al.*, 1994).

References

Barbaree, J. M., Breiman, R. F., Dufour, A. P. (1993). *Legionella*: current status and emerging perspectives, American Society for Microbiology, Washington, DC.

Baskerville, A., Fitzgeorge, R. B., Broster, M., Hambleton, P., Dennis, P. J. (1981). Experimental transmission of legionnaires' disease by exposure to aerosols of *Legionella pneumophila*. *Lancet*, **ii**, 1389–1390.

Baskerville, A., Dowsett, A. B., Fitzgeorge, R. B., Hambleton, P., Broster, M. (1983a). Ultrastructure of pulmonary alveoli and macrophages in experimental Legionnaires' disease. *J. Pathol.*, **140**, 77–90.

Baskerville, A., Fitzgeorge, R. B., Broster, M., Hambleton, P. (1983b). Histopathology of experimental Legionnaires' disease in guinea-pigs, rhesus monkeys and marmosets. *J. Pathol.*, **139**, 349–362.

Baskerville, A., Fitzgeorge, R. B., Conlan, J. W., Ashworth, L. A., Gibson, D. H., Morgan, C. P. (1983c). Studies on protective immunity to aerosol challenge with *Legionella pneumophila*. *Zentralbl. Bakteriol. Mikrobiol. Hyg. (A)*, **255**, 150–155.

Beckers, M. C., Yoshida, S., Morgan, K., Skamene, E., Gros, P. (1995). Natural resistance to infection with *Legionella pneumophila*: chromosomal localization of the *Lgn1* susceptibility gene. *Mamm. Genome*, **6**, 540–545.

Berendt, R. F., Jaax, N. K. (1985). Effect of prior influenza virus infections on susceptibility of AKR/J mice to respiratory challenge with *Legionella pneumophila*. *J. Lab. Clin. Med.*, **105**, 124–131.

Berendt, R. F., Young, H. W., Allen, R. G., Knutsen, G. L. (1980). Dose-response of guinea-pigs experimentally infected with aerosols of *Legionella pneumophila*. *J. Infect. Dis.*, **141**, 186–192.

Blander, S. J., Horwitz, M. A. (1989). Vaccination with the major secretory protein of *Legionella pneumophila* induces cell-mediated and protective immunity in a guinea-pig model of Legionnaires' disease. *J. Exp. Med.*, **169**, 691–705.

Blander, S. J., Horwitz, M. A. (1993). Major cytoplasmic membrane protein of *Legionella pneumophila*, a genus common antigen and member of the hsp 60 family of heat shock proteins, induces protective immunity in a guinea-pig model of Legionnaires' disease. *J. Clin. Invest.*, **91**, 717–723.

Blander, S. J., Breiman, R. F., Horwitz, M. A. (1989a). A live avirulent mutant *Legionella pneumophila* vaccine induces protective immunity against lethal aerosol challenge. *J. Clin. Invest.*, **83**, 810–815.

Bozeman, F. M., Humphries, J. W., Campbell, J. M. (1968). A new group of rickettsia-like agents recovered from guinea-pigs. *Acta Virol.*, **12**, 87–93.

Breiman, R. F., Horwitz, M. A. (1987). Guinea-pigs sublethally infected with aerosolized *Legionella pneumophila* develop humoral and cell-mediated immune responses and are protected against lethal aerosol challenge. A model for studying host defense against lung infections caused by intracellular pathogens. *J. Exp. Med.*, **165**, 799–811.

Brieland, J., Freeman, P., Kunkel, R. *et al.* (1994). Replicative *Legionella pneumophila* lung infection in intratracheally inoculated A/J mice. A murine model of human Legionnaires' disease. *Am. J. Pathol.*, **145**, 1537–1546.

Catrenich, C. E., Johnson, W. (1988). Virulence conversion of *Legionella pneumophila*: a one-way phenomenon. *Infect. Immun.*, **56**, 3121–3125.

Catrenich, C. E., Johnson, W. (1989). Characterization of the selective inhibition of growth of virulent *Legionella pneumophila* by supplemented Mueller–Hinton medium. *Infect. Immun.*, **57**, 1862–1864.

Chandler, F. W., McDade, J. E., Hicklin, M. D., Blackmon, J. A., Thomason, B. M., Ewing, E. P., Jr (1979). Pathologic findings in guinea-pigs inoculated intraperitoneally with the Legionnaires' disease bacterium. *Ann. Intern. Med.*, **90**, 671–675.

Cianciotto, N., Eisenstein, B. I., Engleberg, N. C., Shuman, H. (1989). Genetics and molecular pathogenesis of *Legionella pneumophila*, an intracellular parasite of macrophages. *Mol. Biol. Med.*, **6**, 409–424.

Cianciotto, N. P., Eisenstein, B. I., Mody, C. H., Engleberg, N. C. (1990). A mutation in the mip gene results in an attenuation of *Legionella pneumophila* virulence. *J. Infect. Dis.*, **162**, 121–126.

Davis, G. S., Winn, W. C. Jr., Gump, D. W., Craighead, J. E., Beaty, H. N. (1982). Legionnaires' pneumonia after aerosol exposure in guinea-pigs and rats. *Am. Rev. Respir. Dis.*, **126**, 1050–1057.

Davis, G. S., Winn, W. C. Jr., Gump, D. W., Beaty, H. N. (1983). The kinetics of early inflammatory events during experimental pneumonia due to *Legionella pneumophila* in guinea-pigs. *J. Infect. Dis.*, **148**, 823–835.

Dennis, P. J., Lee, J. V. (1988). Differences in aerosol survival between pathogenic and non-pathogenic strains of *Legionella pneumophila* serogroup 1. *J. Appl. Bacteriol.*, **65**, 135–141.

Edelstein, P. H. (1993). Legionnaires' disease. *Clin. Infect. Dis.*, **16**, 741–749.

Edelstein, P. H. (1995). Antimicrobial chemotherapy for legionnaires' disease: a review (Review). *Clin. Infect. Dis.*, **21**, S265–S276.

Edelstein, P. H., Meyer, R. D. (1984). Legionnaires' disease. A review. *Chest*, **85**, 114–120.

Edelstein, P. H., Calarco, K., Yasui, V. K. (1984). Antimicrobial therapy of experimentally induced Legionnaires' disease in guinea-pigs. *Am. Rev. Respir. Dis.*, **130**, 849–856.

Edelstein, P. H., Edelstein, M. A. C., Holzknecht, B. (1992). *In vitro* activities of fleroxacin against clinical isolates of *Legionella* spp., its pharmacokinetics in guinea-pigs, and use to treat guinea-pigs with *L. pneumophila* pneumonia. *Antimicrob. Agents Chemother.*, **36**, 2387–2391.

Edelstein, P. H., Edelstein, M. A., Lehr, K. H., Ren, J. (1996). *In-vitro* activity of levofloxacin against clinical isolates of *Legionella* spp, its pharmacokinetics in guinea-pigs, and use in experimental *Legionella pneumophila* pneumonia. *J. Antimicrob. Chemother.*, **37**, 117–126.

Edelstein, P. H., Edelstein, M. A. C., Ren, J. J., Polzer, R., Gladue, R. P. (1996b). Activity of trovafloxacin (cp-99,219) against *Legionella* isolates: *in-vitro* activity, intracellular accumulation and killing in macrophages, and pharmacokinetics and treatment of guinea-pig with *L. pneumophila* pneumonia. *Antimicrob. Agents Chemother.*, **40**, 314–319.

Eisenstein, T. K., Tamada, R., Meissler, J., Flesher, A., Oels, H. C. (1984). Vaccination against *Legionella pneumophila*: serum antibody correlates with protection induced by heat-killed or acetone-killed cells against intraperitoneal but not aerosol infection in guinea pigs. *Infect. Immun.*, **45**, 685–691.

Engleberg, N. C., Carter, C., Demarsh, P., Drutz, D. J., Eisenstein, B. I. (1986). A *Legionella*-specific DNA probe detects organisms in lung tissue homogenates from intranasally inoculated mice. *Isr. J. Med. Sci.*, **22**, 703–705.

Fernandes, P. B., Bailer, R., Swanson, R. *et al.* (1986). *In vitro* and *in vivo* evaluation of A-56268 (TE-031), a new macrolide. *Antimicrob. Agents Chemother.*, **30**, 865–873.

Fitzgeorge, R. B., Baskerville, A., Broster, M., Hambleton, P., Dennis, P. J. (1983). Aerosol infection of animals with strains of *Legionella pneumophila* of different virulence: comparison with intraperitoneal and intranasal routes of infection. *J. Hyg. (Lond.)*, **90**, 81–89.

Fitzgeorge, R. B., Featherstone, A. S., Baskerville, A. (1990). Efficacy of azithromycin in the treatment of guinea-pigs infected with *Legionella pneumophila* by aerosol. *J. Antimicrob. Chemother.*, **25** (Suppl. A), 101–108.

Fitzgeorge, R. B., Lever, S., Baskerville, A. (1993). A comparison of the efficacy of azithromycin and clarithromycin in oral therapy of experimental airborne Legionnaires' disease. *J. Antimicrob. Chemother.*, **31** (Suppl. E), 171–176.

Fraser, D. W., Wachsmuth, I. K., Bopp, C., Feeley, J. C., Tsai, T. F. (1978). Antibiotic treatment of guinea-pigs infected with agent of Legionnaires' disease. *Lancet*, **i**, 175–178.

Gump, D. W., Davis, G. S., Winn, W. C., Jr., Beaty, H. N. (1983). Protein and antibody in lavage fluid of guinea-pigs with *Legionella pneumophila* pneumonia. *Zentralbl. Bakteriol. Mikrobiol. Hyg. (A)*, **255**, 145–149.

Hambleton, P., Broster, M. G., Dennis, P. J., Henstridge, R., Fitzgeorge, R., Conlan, J. W. (1983). Survival of virulent *Legionella pneumophila* in aerosols. *J. Hyg. (Lond.)*, **90**, 451–460.

Hedlund, K. W., McGann, V. G., Copeland, D. S., Little, S. F., Allen, R. G. (1979). Immunologic protection against the Legionnaires' disease bacterium in the AKR/J mouse. *Ann. Intern. Med.*, **90**, 676–679.

Hoge, C. W., Brieman, R. F. (1991). Advances in the epidemiology and control of *Legionella* infections. *Epidemiol. Rev.*, **13**, 329–340.

James, B. W., Mauchline, W. S., Fitzgeorge, R. B., Dennis, P. J., Keevil, C. W. (1995). Influence of iron-limited continuous culture on physiology and virulence of *Legionella pneumophila*. *Infect. Immun.*, **63**, 4224–4230.

Jepras, R. I., Fitzgeorge, R. B., Baskerville, A. (1985). A comparison of virulence of two strains of *Legionella pneumophila* based on experimental aerosol infection of guinea-pigs. *J. Hyg. (Lond.)*, **95**, 29–38.

Katz, S. M., Habib, W. A., Hammel, J. M., Nash, P. (1982). Lack of airborne spread of infection by *Legionella pneumophila* among guinea-pigs. *Infect. Immun.*, **38**, 620–622.

Klein, T. W., Newton, C., Widen, R., Friedman, H. (1993). Delta 9-tetrahydrocannabinol injection induces cytokine-mediated mortality of mice infected with *Legionella pneumophila*. *J. Pharmacol. Exp. Ther.*, **267**, 635–640.

Marra, A., Blander, S. J., Horwitz, M. A., Shuman, H. A. (1992). Identification of a *Legionella pneumophila* locus required for intracellular multiplication in human macrophages. *Proc. Natl Acad. Sci. USA*, **89**, 9607–9611.

Marston, B. J., Plouffe, J. F., File, T. M. *et al.* (1997). Incidence of community-acquired pneumonia requiring hospitalization—results of a population-based active surveillance study in Ohio. *Arch. Intern. Med.*, **157**, 1709–1718.

Mauchline, W. S., James, B. W., Fitzgeorge, R. B., Dennis, P. J., Keevil, C. W. (1994). Growth temperature reversibly modulates the virulence of *Legionella pneumophila*. *Infect. Immun.*, **62**, 2995–2997.

Meenhorst, P. L., Reingold, A. L., Gorman, G. W. *et al.* (1983). *Legionella pneumonia* in guinea-pigs exposed to aerosols of concentrated potable water from a hospital with nosocomial Legionnaires' disease. *J. Infect. Dis.*, **147**, 129–132.

Miyamoto, H., Maruta, K., Ogawa, M., Beckers, M. C., Gros, P., Yoshida, S. (1996). Spectrum of *Legionella* species whose intracellular multiplication in murine macrophages is genetically controlled by *Lgn1*. *Infect. Immun.*, **64**, 1842–1845.

Moffat, J. F., Edelstein, P. H., Regula, D. P., Jr, Cirillo, J. D., Tompkins, L. S. (1994). Effects of an isogenic Zn-metalloprotease-deficient mutant of *Legionella pneumophila* in a guinea-pig pneumonia model. *Mol. Microbiol.*, **12**, 693–705.

Myerowitz, R. L., Dowling, J. N., Pasculle, A. W. (1980). Immunity to Pittsburgh pneumonia agent in guinea-pigs. In *20th Interscience Conference on Antimicrobial Agents and Chemotherapy* (Abstract).

Nikaido, Y., Nagata, N., Kido, M., Yoshida, S. (1996). Increased plasma adenosine-deaminase activity in the early phase of *Legionella pneumophila* infection in guinea-pigs. *Fems Immunol. Med. Microbiol.*, **14**, 39–43.

Nowicki, M., Paucod, J. C., Bornstein, N., Meugnier, H., Isoard, P., Fleurette, J. (1988). Comparative efficacy of five antibiotics on experimental airborne legionellosis in guinea-pigs. *J. Antimicrob. Chemother.*, **22**, 513–519.

Pasculle, A. W., Dowling, J. N., Frola, F. N., McDevitt, D. A., Levi, M. A. (1985). Antimicrobial therapy of experimental *Legionella micdadei* pneumonia in guinea-pigs. *Antimicrob. Agents Chemother.*, **28**, 730–734.

Pastoris, M. C., Proietti, E., Mauri, C., Chiani, P., Cassone, A. (1997). Suckling CD1 mice as an animal model for studies of *Legionella pneumophila* virulence. *J. Med. Microbiol.*, **46**, 647–655.

Plouffe, J. F., Para, M. F., Fuller, K. A. (1986). Intratracheal infection with *L. pneumophila* subtypes. *J. Clin. Lab. Immunol.*, **20**, 119–120.

Saito, A., Koga, H., Shigeno, H. *et al.* (1986). The antimicrobial activity of ciprofloxacin against *Legionella* species and the treatment of experimental *Legionella pneumonia* in guinea-pigs. *J. Antimicrob. Chemother.*, **18**, 251–260.

Smith, G. M., Abbott, K. H., Wilkinson, M. J., Beale, A. S., Sutherland, R. (1991). Bactericidal effects of amoxycillin/clavulanic acid against a *Legionella pneumophila* pneumonia in the weanling rat. *J. Antimicrob. Chemother.*, **27**, 127–136.

Smith, G. M., Abbott, K. H., Sutherland, R. (1992). Bactericidal effects of co-amoxiclav (amoxycillin clavulanic acid) against a *Legionella pneumophila* pneumonia in the immunocompromised weanling rat. *J. Antimicrob. Chemother.*, **30**, 525–534.

Smith, M. S., Yamamoto, Y., Newton, C., Friedman, H., Klein, T. (1997). Psychoactive cannabinoids increase mortality and alter acute phase cytokine responses in mice sublethally infected with *Legionella pneumophila*. *Proc. Soc. Exp. Biol. Med.*, **214**, 69–75.

Stamler, D. A., Edelstein, M. A. C., Edelstein, P. H. (1994). Azithromycin pharmacokinetics and intracellular concentrations in *Legionella pneumophila*-infected and uninfected guinea-pigs and their alveolar macrophages. *Antimicrob. Agents Chemother.*, **38**, 217–222.

Tartakovskii, I. S., Barkhatova, O. I., Prozorovskii, E. V. (1984). [Models for the study of the toxic action of *Legionella pneumophila*]. *Zh. Mikrobiol. Epidemiol. Immunobiol.*, 52–55.

Tatlock, H. (1944). A Rickettsia-like organism recovered from guinea-pigs. *Proc. Soc. Exp. Biol. Med.*, **57**, 95–99.

Tatlock, H. (1947). Studies on a virus from a patient with Fort Bragg fever (pretibial fever). *J. Clin. Invest.*, **26**, 287–297.

Tully, M., Williams, A., Fitzgeorge, R. B. (1992). Transposon mutagenesis in *Legionella pneumophila*. II. Mutants exhibiting impaired intracellular growth within cultured macrophages and reduced virulence in vivo. *Res. Microbiol.*, **143**, 481–488.

Twisk-Meijssen, M. J., Meenhorst, P. L., van Cronenburg, B. J., Mulder, J. D., Scheffer, E., van Furth, R. (1987). The course of *Legionella* pneumonia in guinea-pigs after inhalation of various quantities of *L. pneumophila*. *Immunobiology*, **176**, 108–124.

Vogel, J. P., Roy, C., Isberg, R. R. (1996). Use of salt to isolate *Legionella pneumophila* mutants unable to replicate in macrophages. *Ann. NY Acad. Sci.*, **797**, 271–272.

Watanabe, M., Nakahama, C., Hino, J., Soejima, R. (1985). [*Legionella pneumophila* infection. Studies on the host defense mechanism to aerosol challenge with *L. pneumophila*]. *Rinsho. Byori.*, **33**, 618–622.

Weeratna, R., Stamler, D. A., Edelstein, P. H. *et al.* (1994). Human and guinea-pig immune responses to *Legionella pneumophila* protein antigens OmpS and Hsp60. *Infect. Immun.*, **62**, 3454–3462.

Williams, A. and Lever, M. S. (1995). Characterisation of *Legionella pneumophila* antigen in urine of guinea-pigs and humans with Legionnaires' disease. *J. Infect.*, **30**, 13–16.

Williams, A., McBride, B. W., Hall, G., Fitzgeorge, R. B., Farrar, G. H. (1995). Experimental Legionnaires' disease in SCID-Beige mice reconstituted with human leucocytes. *J. Med. Microbiol.*, **42**, 433–441.

Winn, W. C., Jr., Davis, G. S., Gump, D. W., Craighead, J. E., Beaty, H. N. (1982). Legionnaires' pneumonia after intratracheal inoculation of guinea-pigs and rats. *Lab. Invest.*, **47**, 568–578.

Yamamoto, Y., Klein, T. W., Friedman, H. (1993). Ethanol affects macrophage production of IL-6 and susceptibility to infection by *Legionella pneumophila*. *Adv. Exp. Med. Biol.*, **335**, 169–173.

Yoshida, S., Mizuguchi, Y., Ohta, H., Ogawa, M. (1985). Effects of tetracyclines on experimental *Legionella pneumophila* infection in guinea-pigs. *J. Antimicrob. Chemother.*, **16**, 199–204.

Chapter 35

Murine Models of Tuberculosis

I. M. Orme

Introduction

The mouse provides a versatile and flexible model of mycobacterial infections, including *Mycobacterium tuberculosis* (Orme, 1995a). Animals are cheap to purchase and maintain, and the mouse has the additional attraction that it is a favored model for analysis of immunity; as a result, a vast number of antibodies and other reagents are widely available.

Mice are generally resistant to tuberculosis infection, being able to contain and control the infection in target organs and hence giving rise to a chronic disease state (Orme, 1997). This state may last for the lifetime of the host before recrudescence is seen when the mouse is very old (Orme 1995b), although more recent data suggest that certain inbred mouse strains are prone to reactivation disease before immunosenescence occurs (Orme, unpublished observations). In the more resistant strains, such as the C57BL/6 mouse, a gradual disorganization of the lung granulomas occurs as the mouse ages, and is accompanied by micronecrosis, cholesterol deposition, and extensive fibrosis (Rhoades *et al.*, 1997).

In contrast, a more acute form of disease process is seen in guinea-pigs exposed to tuberculosis by aerosol (McMurray, 1994). In these animals the infection is initially contained by a granulomatous response, but after 8–15 weeks the centers of certain lesions degenerate, leading either to mineralization of the lesion or extensive caseous necrosis or cavitation, eventually resulting in the death of the animal. This process appears to mirror the course of events seen in untreated human patients.

Despite this closeness to the human condition, animals larger than mice are rarely used to evaluate antimycobacterial therapies. The reason, of course, is cost. Guinea-pigs, for example, are about five times as expensive as mice, and about the same in terms of housing.

Mice can be productively infected by a variety of routes, including subcutaneous inoculation, intravenously, or by exposure to an aerosol of bacteria. Laboratories experienced in these procedures tend to find that the established course of infection is usually highly reproducible. Currently, the most widely used approach to drug screening involves high-dose intravenous inoculation, although new models involving a more realistic aerosol infection have recently been described (Kelly *et al.*, 1996).

Host response

Potential new drugs for treatment of tuberculosis can be tested in mice in which the capacity to express acquired immunity is intact or has been ablated in some way.

In the intact host, the cell-mediated immune response consists of two mechanisms: protective immunity and delayed-type hypersensitivity (DTH). The purpose of the former is to activate macrophages harboring living bacilli by means of exposure to the cytokine γ-interferon (IFN), and the purpose of the latter is to bring in large numbers of macrophages to form a granuloma, which prevents bacilli from disseminating away from the primary site of bacterial implantation (Orme *et al.*, 1993).

The former mechanism in tuberculosis infection is mediated by short-lived CD4 T cells that secrete large amounts of IFN and thus saturate the lung interstitium with this cytokine [Orme, 1994]. DTH is mediated by CD4 T cells also, but these by definition represent at least some specific cells that have entered the recirculating pool by leaving the primary infectious site. IFN secretion by these cells stimulates macrophages presenting mycobacterial antigens to secrete TNF, which in turn stimulates surrounding local tissue cells to produce a battery of chemokine molecules that are chemoattractant for monocytes passing through adjacent arterioles (Rhoades *et al.*, 1995). As a result, one can observe a characteristic perivascular cuffing (Rhoades *et al.*, 1997), which is ICAM-dependent (Johnson C. and Orme, I.M., unpublished observation).

A variety of immunodeficient mouse models are now available, although to date most have been used in the context of evaluating therapies against *M. avium* rather than *M. tuberculosis*. These include the beige mouse (*bg/bg*) in which mutations in the gene *Lyst* result in defective oxidative metabolism in neutrophils and macrophages, making the mouse more susceptible to *M. avium* but apparently not to *M. tuberculosis* (Figure 35.1). Reductions in CD4 T cells to mimic the effects of HIV have been described, either by

Handbook of Animal Models of Infection
ISBN 0-12-775390-7

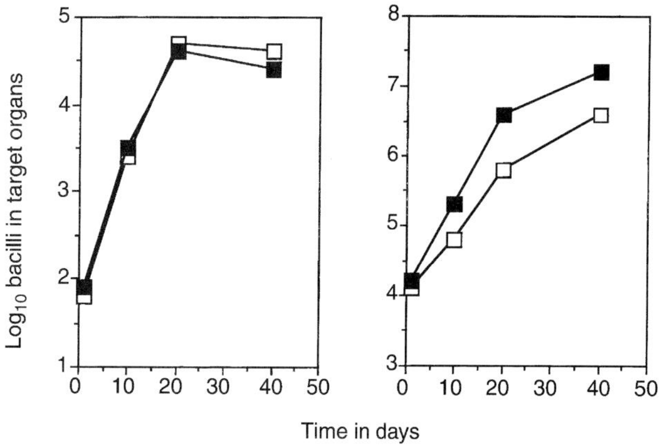

Figure 35.1 Growth of mycobacterial infections in the lungs of C57BL/6 control mice (open squares) and C57BL/6 beige mice (black squares) after aerosol exposure to *M. tuberculosis* (**A**) or intravenous inoculation with *M. avium* (**B**).

complete T-cell eradication by irradiation and thymectomy, followed by B-cell reconstitition, or by infusion of anti-CD4 monoclonal antibodies. An additional approach has been to selectively inhibit T-cell responses using the virus LP-BM5 (MAIDS virus; Orme *et al.*, 1992).

A newer approach, albeit rarely used so far, involves mice in which genes have been specifically targeted by homologous recombination (gene knock-out mice). In several models, such as IFN-KO (Cooper *et al.*, 1993) and β_2-microglobulin-KO (Flynn *et al.*, 1992), mutant mice undergo progressive disease and pulmonary necrosis.

Despite these models, however, most drug screening exercises *in vivo* use normal mice infected with a high intravenous bacterial inoculum. How closely this mimics reality is, of course, open to debate.

Screening of drugs *in vitro*

The simplest way to determine if a drug can inhibit *M. tuberculosis* is to titrate the drug against bacilli grown in a nutrient broth (such as 7H9 broth) to determine the concentration of drug at which the culture cannot grow, the minimal inhibitory concentration (MIC). Further, by plating each culture to see if bacilli are dead as opposed to a state of stasis, one can determine the minimal bactericidal concentration (MBC).

A more expensive way to determine drug susceptibility is by the use of automated equipment such the BACTEC device, which measures CO_2 production by bacterial cultures (either by radiometry, or more recently by infrared detection), and which can provide additional sophisticated analysis of bacterial growth. Other approaches include detection directly of bacterial growth, such as reduction of alamar blue dye (Yajko *et al.*, 1995), or the elegant luciferase assay pioneered by Jacobs and his colleagues (Sarkis *et al.*, 1995), in which a bacteriophage containing the luciferase gene infects the bacillus and produces light if the individual bacterium is still alive (add effective drug, and "out goes the light").

Our approach, however, looks at this question a little differently. In the MIC assay, for example, the bacillus is in a very favorable environment (a hippo in mud would be a good analogy) and the drug has only to penetrate the mycobacterial cell wall, albeit a formidable barrier (Figure 35.2). In reality, however, there are two membranes, the host microphage cell membrane and the phagosomal membrane, that must be crossed first. Moreover, the host cell is physiologically activated; it can potentially concentrate drugs internally, or it can pump them out. In addition, it responds to cytokine activation by forming proton pumps on the phagosomal membrane to acidify the phagosomal interior [Sturgill-Koszycki *et al.*, 1994]; the bacilli may respond by producing ammonia. All of this can potentially affect the properties of a given drug. (None of this happens inside the BACTEC machine!)

As a result, we have developed a macrophage screening assay to test new potential antimycobacterial drugs (Skinner *et al.*, 1994). In this assay a homogeneous monolayer of bone-marrow-derived macrophages is first generated in culture, and these cells are then infected with a small dose of *M. tuberculosis* bacteria.

After a lag phase of 1–2 days the bacilli begin to grow and can multiply by about 1 log in about 8 days. We curtail the experiment here; monolayers can last up to 20 days (some even claim 30), but after 10–12 days the cells begin to look disheveled and some can float off (if not accounted for

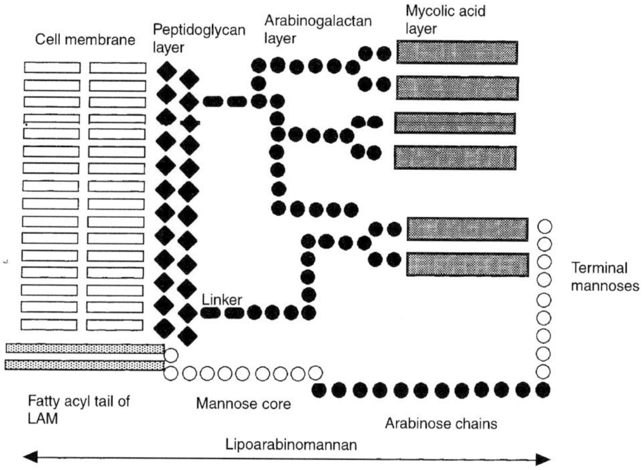

Figure 35.2 A generic cartoon of the cell wall of *M. tuberculosis*. The mycolic acid and arabinogalactan layers pose a substantial barrier to antimicrobial radicals and drugs. A rhamnose-containing linker connects this barrier to the peptidoglycan layer, and hence is a potential drug target. The lengthy lipoarabinomannan molecule attaches through the wall to the cell membrane, and extends out beyond the mycolic acid layer.

one will undercount bacilli). The monolayer supernatants are carefully removed and the cells are lysed, then plated on nutrient agar to determine bacterial numbers.

As shown in Figure 35.3, this assay can be used to determine an 'intracellular MIC". However, perhaps a more useful index is to use the assay to predict the serum concentration of drug needed to destroy 99% of the bacilli, a value we believe may be of more practical therapeutic value than the conventional MIC. This bactericidal concentration/ 99%, or BC_{99}, is determined by subtracting 2 logs from the control (no drug) culture cfu values at 8 days, and then reading off the drug concentration at which this occurs. An example is given in Figure 35.4; in this assay isoniazid repeatedly inhibits the growth of *M. tuberculosis* Erdman at a concentration of 0.7 μg/ml, and hence can be used as a positive control in each assay.

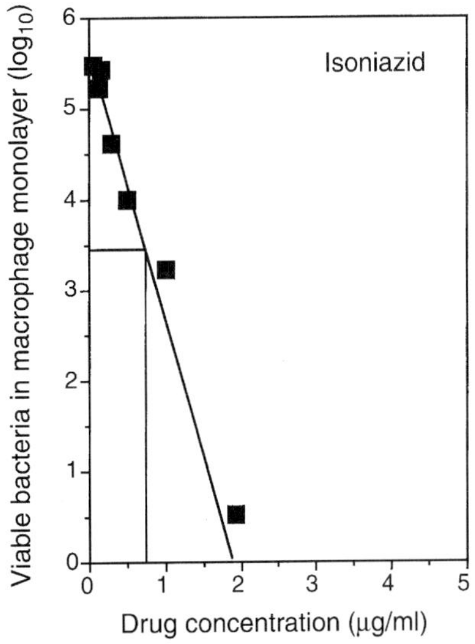

Figure 35.4 Application of the BC_{99} assay to the activity of isoniazid against the growth of *M. tuberculosis* Erdman.

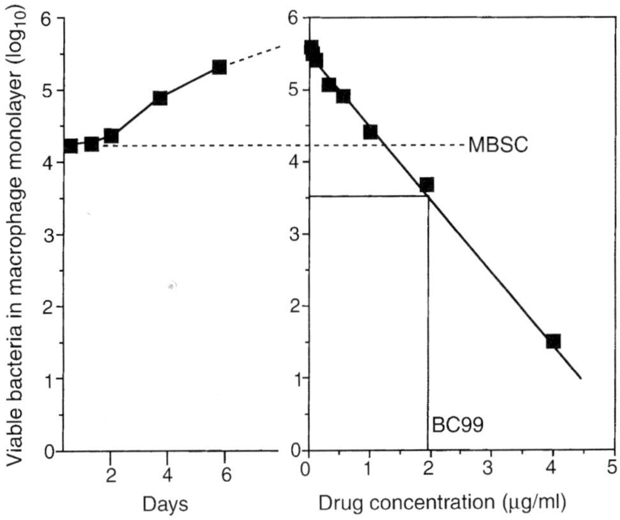

Figure 35.3 The basis of the macrophage drug screening model. By titrating drug dose versus bacterial growth one can make an estimate of the minimal serum concentration that is bacteriostatic (MBSC) and the concentration needed to establish a therapeutic effect; we arbitrarily define this as the BC_{99}, or bactericidal concentration of the drug that destroys 2 logs [99%] of *M. tuberculosis* over the 8-day culture period.

In-vivo screening

As described above, the mouse is the small animal most commonly used in the evaluation of new antimycobacterial therapies. Its advantages include cost-effectiveness and reproducibility, but there are obvious drawbacks as well, including the need for stringent level III biosafety containment facilities and personnel trained in handling virulent mycobacteria, as well as issues relating to the different pharmacokinetics of drugs between rodents and humans. Dosages apart, however, one can state with some certainty that drugs that work well in mice tend also to do so in humans.

In high-dose intravenous models most of the inoculum is taken up by phagocytes in the spleen and liver; only 1–2% can be detected in the lungs (if the lung uptake is much higher, this indicates that the inoculum was not dispersed correctly and is heavily clumped). This then grows progressively for 10–15 days in the spleen, and to a lesser extent in the liver, until acquired immunity comes into play, resulting in a "chronic" disease pattern. Thus, the drug under test can be given soon after inoculation (can it stop the progressive growth?) or after the disease has become chronic (can it decrease bacterial load at this stage?). The second model is of course more realistic; the patient is probably at this stage before diagnosis is first made.

Another approach is to give "standard chemotherapy," such as isoniazid, pyrazinamide, and rifampin, for a while to bring the load down, then replace this with the new drug under test (Truffot-Pernot *et al.*, 1991). A further model is to continue therapy until bacteria can no longer be detected, then give the new drug for a while, and then stimulate "persisting" bacilli to regrow by injecting the mouse with cortisol. This is called the Cornell model after its place of origin (McCune *et al.*, 1966). A current vogue is to test antianaerobic drugs such as metronidazole in this model, given the possibility that surviving bacilli switch into an anaerobic mode for a while (my own experiments in this regard have been negative to date). All of these approaches take advantage of the fact that mice tolerate long-term gavage well, if handled gently and correctly.

Our model of choice is to expose mice to a very low inoculum of bacilli using an aerosol generation chamber. After uptake of about 50 bacilli in the lungs the infection grows progressively at first and is then curtailed around 20 days. Laboratory strains such as Erdman attain 4–5 logs in the lungs by this time; more virulent strains such as CSU93

(Tennessee outbreak) and strain W (New York) can grow to between 6 and 7 logs in this time.

The standard regimen we use is to start the drug under test on day 20 (when the mouse first becomes DTH-positive) and to treat mice for 30 days, harvesting animals and determining bacterial loads at 35 and 50 days. Isoniazid (25 mg/kg/day) is given as the positive control in each assay.

Drug targets: new and old

Despite being the drug of choice for several decades, it is only recently that the mechanism of action of isoniazid has been determined. Early work pointed to the mycolic acid layer of the cell wall, but only recently was this narrowed down to enzymes that elongate the two fatty acid chains of the molecule. A gene encoding a major enzyme involved in this process has been designated *inhA*, the molecule has been synthesized, and its crystal structure has been elucidated (Banerjee *et al.*, 1994; Dessen *et al.*, 1995). Given that exposure to isoniazid is fatal to the bacillus, truncation of the mycolic acid layer probably allows the influx of protons, nitric oxide, oxygen radicals, peroxynitrite, and other noxious materials through the cell wall to damage and disrupt essential enzymatic processes.

Other drugs also target the cell wall, such as ethambutol, which disrupts the arabinogalactan layer, which in turn prevents mycolic acid attachment to the terminal arabinan motif (McNeil and Brennan, 1991). A potential new target may be a rhamnose-containing linker molecule that joins the peptidoglycan layer to the outer arabinogalactan layer; if this could be blocked then bacterial construction of the wall should cease almost at the source (McNeil *et al.*, 1990).

More conventional targets are the DNA gyrases (fluoroquinolones) and protein synthesis (rifamycins, macrolides). Unfortunately to date new drug development has been slow, a troubling thought in light of increasing drug resistance in tuberculosis isolates both here and throughout the world (Bloom and Murray, 1992).

Acknowledgments

This work was supported by NIH grant AI-45239 from the DMID, NIAID. I thank the many people who have contributed to the ideas expressed above, including Pam Hair, Syndy Furney, Brian Kelly, Jason Brooks, and Alan Roberts.

References

Banerjee, A., Dubnau, E., Quernard, A. *et al.* (1994). *InhA*, a gene encoding a target for isoniazid and ethionamide in *Mycobacterium tuberculosis*. *Science*, **263**, 227–230.

Bloom, B. R., Murray, C. J. L. (1992). Tuberculosis: commentary on a reemergent killer. *Science*, **257**, 1055–1064.

Cooper, A. M., Dalton, D. K., Stewart, T. A., Griffin, J. P., Russell, D. G., Orme, I. M. (1993). Disseminated tuberculosis in gamma interferon gene-disrupted mice. *J. Exp. Med.*, **178**, 2243–2247.

Dessen, A., Quemard, A., Blanchard, J. S., Jacobs, W. R. Jr, Sacchettini, J. C. (1995). Crystal structure and function of the isoniazid target of *Mycobacterium tuberculosis*. *Science*, **267**, 1638–1641.

Flynn, J. L., Goldstein, M. M., Triebold, K. J., Koller, B., Bloom, B. R. (1992). Major histocompatibility complex class I-restricted T cells are required for resistance to *Mycobacterium tuberculosis* infection. *Proc. Natl Acad. Sci. USA*, **89**, 12013–12017.

Kelly, B. P., Furney, S. K., Jessen, M. T., Orme, I. M. (1996). Low-dose aerosol infection model for testing drugs for efficacy against *Mycobacterium tuberculosis*. *Antimicrob. Agents Chemother.*, **40**, 2809–2812.

McCune, R. M., Feldmann, F. M., Lambert, H. P., McDermott, W. (1966). Microbial persistence. I. The capacity of tubercle bacilli to survive sterilization in mouse tissues. *J. Exp. Med.* **123**, 445–468.

McMurray, D. N. (1994). Guinea pig model of tuberculosis. In *Tuberculosis; Pathogenesis, Protection, and Control* (ed. Bloom, B. R.), pp. 135–147. ASM Press, Washington, DC.

McNeil, M., Brennan, P. J. (1991). Structure, function, and biogenesis of the cell envelope of mycobacteria in relation to bacterial physiology, pathogenesis, and drug resistance; some thoughts and possibilities arising from recent structural information. *Res. Microbiol.*, **142**, 451–463.

McNeil, M., Daffe, M., Brennan, P. J. (1990). Evidence for the nature of the link between the arabinogalactan and peptidoglycan components of *Mycobacterium* cell wall. *J. Biol. Chem.*, **265**, 18200–18206.

Orme, I. M. (1994). Protective and memory immunity in mice infected with *Mycobacterium tuberculosis*. *Immunobiology*, **191**, 503–508.

Orme, I. M. (1995a). *Immunity to Mycobacteria*. R. G. Landes, Austin, TX.

Orme, I. M. (1995b). Mechanisms underlying the increased susceptibility of aged mice to tuberculosis. *Nutr. Rev.*, **53**, S35–S40.

Orme, I. M. (1997). The host response to *Mycobacterium tuberculosis*. In *Host Response to Intracellular Pathogens* (ed. Kaufmann, S. H. E.), pp. 115–130. R. G. Landes, Austin, TX.

Orme, I. M., Furney, S. K., Roberts, A. D. (1992). Dissemination of enteric *Mycobacterium avium* infections in mice rendered immunodeficient by thymectomy and CD4 depletion or by prior infection with murine retroviruses. *Infect. Immun.*, **60**, 4747–4753.

Orme, I. M., Andersen, P., Boom, W. H. (1993). The T cell response to *Mycobacterium tuberculosis*. *J. Infect. Dis.*, **167**, 1481–1497.

Rhoades, E. R., Cooper, A. M., Orme, I. M. (1995). Chemokine response in mice infected with *Mycobacterium tuberculosis*. *Infect. Immun.*, **63**, 3871–3877.

Rhoades, E. R., Frank, A. A., Orme, I. M. (1997). Distinct phases of granuloma formation in mice aerogenically infected with *Mycobacterium tuberculosis*. *Tubercle Lung Dis.*, **78**, 57–66.

Sarkis, G. J., Jacobs, W. R. Jr, Hatfull, G. F. (1995). L5 luciferase reporter mycobacteriophages: a sensitive tool for the detection and assay of live mycobacteria. *Mol. Microbiol.*, **15**, 1055–1067.

Skinner, P. S., Furney, S. K., Jacobs, M. K., Klopman, G., Ellner, J. J., Orme, I. M. (1994). A bone marrow derived murine macrophage model for evaluating efficacy of anti-mycobacterial drugs under relevant physiological conditions. *Antimicrob. Agents Chemother.*, **38**, 2557–2563.

Sturgill-Koszycki, S., Schlesinger, P., Chakraborty, P. *et al.* (1994). Lack of acidification in *Mycobacterium* phagosomes produced by exclusion of the vesicular proton-ATPase. *Science*, **263**, 678–681.

Truffot-Pernot, C., Ji, B., Grosset, J. (1991). Activities of pefloxacin and ofloxacin against mycobacteria: *in vitro* and mouse experiments. *Tubercle*, **72**, 57–64.

Yajko, D. M., Madej, J. J., Lancaster, M. V. *et al.* (1995). Colorimetric method for determining MICs of antimicrobial agents for *Mycobacterium tuberculosis*. *J. Clin. Microbiol.*, **33**, 2324–2327.

Chapter 36

Beige Mouse Model of Disseminated *Mycobacterium avium* Complex Infection

M. H. Cynamon and M. S. DeStefano

Background of human infection

In the pre-AIDS era *Mycobacterium avium* complex (MAC) predominantly caused pulmonary mycobacteriosis, which was difficult to distinguish radiologically from tuberculosis and was rarely associated with extrapulmonary sites or disseminated disease (Wolinsky, 1979). During the AIDS era, disseminated MAC became a frequent diagnosis in late-stage HIV-infected patients, usually occurring in those patients with CD4$^+$ T-cell counts below 50/mm^3. The portal of entry is likely to be the gastrointestinal tract, with food-stuffs and water being the source of the organisms. The syndrome is characterized by fever, chills, sweats, malaise, weight loss, abdominal pain, and diarrhea. Patients may develop anemia, secondary to bone marrow involvement, and sometimes have a striking elevation of serum alkaline phosphatase (Burman and Cohn, 1995). Diagnosis is usually made by culture of blood; however, MAC is sometimes grown from another site (e.g. bone marrow, liver, or lymph nodes). Culture of MAC from sputum or bronchoalveolar lavage fluid in HIV-infected patients usually indicates colonization rather than local or disseminated disease. A stool specimen which is acid-fast bacillus smear-positive usually suggests disseminated MAC disease; however, we have occasionally seen patients with this presentation having disseminated tuberculosis. In HIV-infected patients with advanced disseminated MAC infection the bone marrow, lymph nodes, spleen, liver, and gastrointestinal tract are frequently infiltrated with masses of mycobacteria (Hawkins *et al.*, 1986). The central nervous system is rarely involved. It is now clear that disseminated MAC infection causes increased mortality as well as significant morbidity in HIV-infected patients (Chaisson *et al.*, 1994; Shafran *et al.*, 1996).

MAC is relatively resistant to most antituberculosis agents (Heifets, 1991). Pulmonary disease in the pre-AIDs era was usually treated with four to five drugs, including rifampin, ethambutol, and an aminoglycoside (streptomycin or amikacin; Iseman, 1995). HIV-infected patients with disseminated MAC disease have also been treated with multidrug regimens including rifampin, amikacin, clofazamine, and ethambutol (Parenti *et al.*, 1995; Shafran *et al.*, 1996). Azithromycin and clarithromycin represent significant improvements for the therapy of disseminated MAC infection in HIV-infected patients. These agents as well as rifabutin have been demonstrated to decrease the occurrence of disseminated MAC in HIV-infected patients with CD4$^+$ T-cell counts below 100/mm^3 (Nightingale *et al.*, 1993; Havlir *et al.*, 1996; Pierce *et al.*, 1996). Azithromycin and clarithromycin are the key agents in regimens for the treatment of MAC disease and probably for other nontuberculous mycobacterial infections (Dautzenberg *et al.*, 1991; Young *et al.*, 1991).

Background of model

During the past 30 years, various animal models of MAC infection have been explored. Pigs, sheep, fowl, rabbits and goats have been used in studies by Kleeberg and Nel (1973). The pathogenicity of MAC has been explored in hens (Shaefer *et al.*, 1970), mice and guinea-pigs (Engbaek, 1961). Several groups have worked with rabbits (Engbaek, 1961; Armstrong *et al.*, 1967; Meissner, 1981). Brown *et al.* (1993) developed a model using rats immunosuppressed with cyclosporin A. Much has been accomplished using the mouse as a model for MAC infection. Pioneer work with the mouse model was done by Collins *et al.* (1978) and Dunbar *et al.* (1968). The beige (C_{57}BL/6J bg^j/bg^j) mouse (Jackson Laboratory, Bar Harbor, ME) model of disseminated MAC was developed by Gangadharam *et al.* (1983). The beige mouse is considered to be the standard strain used for studying MAC infection (Bertram *et al.*, 1986; Gangadharam, 1986a, 1995; Reddy *et al.*, 1994). Other models of disseminated MAC infection in mice have been studied, such as: C57BL/6 (Collins, 1972; Gangadharam, 1986b; Cohen *et al.*, 1995); interferon-gamma knockout mice (Sacco *et al.*, 1996); Swiss Webster, AKR, DBA/2, BDF1, CBA/N, BALB/c (Gangadharam, 1986b); nude (Ji *et al.*, 1991); MAIDS (Orme *et al.*, 1992); and SCID (Appelberg *et al.*, 1994).

Storage and preparation of inocula

Primary cultures of MAC (*M. kansasii* or other mycobacteria) to be used for infection may be obtained from

Handbook of Animal Models of Infection
ISBN 0-12-775390-7

clinical isolates of HIV-infected patients with disseminated MAC infection, or the American Type Culture Collection (ATCC). We have primarily studied ATCC 49602 (serotype 1) strain LPR, a clinical isolate from our medical center, and MAC 101 (provided by Lowell Young, California Pacific Medical Center Research Institute, San Francisco, CA). All procedures involving the manipulation of these BSL-2 organisms should be carried out in a biological safety cabinet, type A. Guidelines for handling BSL-2 organisms are outlined in the CDC/NIH publication. *Biosafety in Microbiological and Biomedical Laboratories* (US Department of Health and Human Services, 1993). Organisms are grown in modified 7H10 broth (7H10 agar formulation with agar and malachite green omitted), pH 6.6, with 10% (vol./vol.) Middlebrook oleic-acid–albumin–dextrose–catalase (OADC) enrichment (Difco Laboratories, Detroit, MI) and 0.05% (vol./vol.) Tween 80 (Sigma, St Louis, MO). Broth cultures should be started from one transparent, smooth, flat colony (SmT) grown on an agar plate. The culture tube is then placed in an orbital shaker and incubated at 37°C for 3–5 days. Culture suspensions should be predominately (>95%) of the smooth, transparent, and flat (SmT) phenotype. This phenotype has been shown to be more virulent (Shaefer *et al.*, 1970; Inderlied *et al.*, 1993; Reddy *et al.*, 1994) and more resistant to antimycobacterial agents (Woodley and David, 1976; Rastogi *et al.*, 1981; Saito and Tomioka, 1988) than the smooth, domed, opaque (SmD) or rough phenotypes. After incubation, the culture is diluted in 7H10 broth to a concentration of 10 Klett units/ml (Manostat colorimeter, Manostat, New York, NY) or approximately 5×10^7 cfu/ml. The inoculum is then titrated in triplicate on 7H10 agar plates (Difco) supplemented with 5% (vol./vol.). Middlebrook OADC enrichment. Plates are taped with Blenderm® (3M, St Paul, MN), incubated for 2–3 weeks at 37°C, and then counted to determine the precise inoculum.

Animal species

The beige mouse is the model that we use in our laboratory. These mice are natural-killer-cell-deficient (Roder and Duwe, 1979). The mutation affects pigment granules of the optic cup, retina and neural crest. Lysosomal granules (Vassalli *et al.*, 1978), mast cells (Chi *et al.*, 1978) and Type II pneumocytes are also affected. The beige trait in mice is similar to the Chediak–Higashi syndrome of human beings and the Aleutian disease of mink and cattle. Beige mice can be infected with many of the nontuberculous mycobacteria: MAC, *M. kansasii, M. simiae, M. malmoense* and *M. genavense* (Klemens and Cynamon, 1994b and unpublished observations). Same-sex mice 5–7 weeks old are allowed to acclimate for 1 week in the facility before being used in an experiment. They are housed in microisolator units (lab products, Maywood, NJ) and are randomly distributed six to a group.

Infection process

The ability to infect beige mice using a variety of routes including intraperitoneal, intranasal, oral, intrarectal; and intravenous has been demonstrated (Gangadharam *et al.*, 1989). Intravenous infection has proved to yield a consistent disseminated disease model with viable organisms recovered from spleens, lungs, livers, blood, and lymph nodes. The i.v. infection route is the most commonly used to assess chemotherapeutic agents (Bertram *et al.*, 1986). Oral and rectal challenge results in lower levels of spleen and lung infection compared to i.v. challenge with the same number of organisms (Gangadharam *et al.*, 1989). As early as day 1 postinoculation cfu were present in the spleen, liver and lungs. Intranasal inoculation yielded viable counts only in the lungs (Gangadharam *et al.*, 1989). Other routes of inoculation resulted in counts in the above organs at week 2 (rectal) and week 6 (oral; Gangadharam *et al.*, 1989). In our model, the mice are infected i.v. via a caudal tail vein. The inoculum is injected in a 0.2 ml volume using a 0.5 ml syringe with an attached 28 G 0.5 in needle to deliver a total of approximately 10^7 cfu/mouse. Mice are placed in a restrainer (PlasLab, Lansing, MI), which is placed in a warming box for approximately 2–3 minutes to promote the dilation of the tail veins. The inoculation procedure takes place in a BSC, as noted above. When ready for injection, the restrainer is attached to a clamp on the holding ring stand, tail facing the injector. The tail vein injecting shield (Figure 36.1) is attached by inserting the base of the restrainer into the slot at the top of the shield. The tail is placed in the groove of the shield and threaded through the small hole at the bottom of the shield and grasped underneath with the non-dominant hand. The tail may be wetted with 70% ethanol for enhancement of the veins. The needle is gently inserted into the tail vein, bevel up, and the bolus is slowly injected. The needle is withdrawn and pressure is applied at the injection site with a 4 × 4 gauze. The mouse is released from the restrainer and returned to its isolator cage. Cages should have cards indicating the number and type of mouse, sex, date of birth, inoculum, isolate identification, and a red isolation or biohazard sticker. All items involved in the inoculating procedure need to be properly decontaminated for 15–30 minutes with an appropriate disinfectant such as Lysol® I.C.™ (National Laboratories, Montvale, NJ) (diluted to a final concentration of 0.5–5.0% phenolic compounds) or hypochlorite (diluted to a final concentration of 5000–10 000 ppm of available chlorine).

Each experiment consists of an early control (sacrificed 1 week postinfection at the initiation of therapy) and a late control (sacrificed at the end of therapy) group, neither of which receives any treatment. One treatment group consists of a drug known to have activity (e.g. azithromycin or clarithromycin). Treatment is started 7 days postinfection and is generally continued for 10 days in succession. In extended therapy experiments, drugs are given daily (Mon–Fri) for 4 or more weeks. All drugs are preweighed into individual tubes and reconstituted daily. Agents used

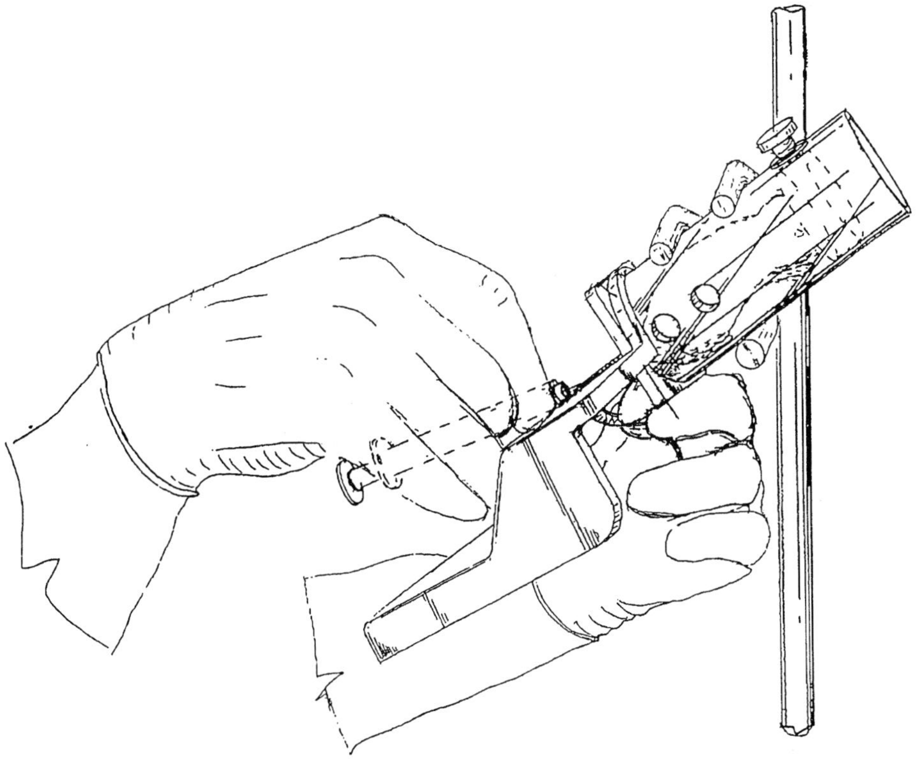

Figure 36.1 Technique for inoculating using injection shield (US patent no. 5 816 197).

for treatment are reconstituted to a solution or suspension not exceeding a final concentration of 20% (vol./vol.) solvent (e.g. ethanol, DMSO) and water. Solvents are added to the compounds, vortexed, and brought to final volume with water. Agents are administered in a 0.2 ml volume. Administration routes (p.o., s.c., i.v.) are determined by drug. Oral drugs (p.o.) are given using an animal feeding needle, 20 G × 40 mm, with a 0.25 mm ball at the tip attached to a 1 ml tuberculin syringe. Mice are handled using a Kevlar® glove (Lab Safety Supply, Janesville, WI) to avoid a biting hazard.

Mice are weighed at the beginning and end of each experiment, and averaged by group. Mice are evaluated daily and changes in appearance or behaviour are noted. In general, it has been our experience that infected mice appear outwardly well, and continue to gain weight with these infections. There have been some exceptions where the infected control animals have succumbed to the infection (Ji *et al.*, 1994; unpublished observations). Untreated mice develop splenomegaly, hepatomegaly (with visible lesions), and enlarged lungs. Enlargement of the spleen, although minimal, is evident at 1 week postinfection (average weight 0.12–0.14 g) and continues to increase for the duration of the experiment (average weight 0.65–0.80 g at 18–20 days postinfection). After 1 week of infection, the usual organ cell counts in the lungs of early control mice (4.5 log) is significantly less than that in the spleen (7 log). The late control mice (19 days postinfection) average 6 log and 7.8 log cfu in their lungs and spleens respectively.

At the completion of the experiment, approximately 2 days after the end of the treatment phase, mice are euthanized using CO_2 inhalation. Their spleens and right lungs are asceptically removed. Spleens are placed in preweighed tubes to determine their weight. Each organ is placed in a grinding assembly (Figure 36.2) containing saline with 0.05% (vol./vol.) Tween 80. The grinding assemblies have an attachable pestle that is secured to the motor-driven homogenizer. Each sample is ground and allowed to sit for 15 min to allow for settling of aerosols. The tops of the grinding assemblies are removed and an aliquot is removed. Dilutions are made to the appropriate concentrations for plating using tubes containing double-distilled H_2O (to promote red blood cell lysis) with 0.05% (vol./vol.) Tween 80. Spleen weights are used to estimate dilutions for the plating of the homogenate. In general, large spleens, such as those belonging to members of the late control group, are heavily infected and need to be diluted to 10^{-5}, 10^{-6}, and 10^{-7}. Agents with good activity, such as clarithromycin or azithromycin, result in smaller spleens (0.2–0.4 g) and need to be diluted 10^{-3}, 10^{-4}, and 10^{-5}. Although rifapentine and rifabutin show some activity, they produce misleadingly large spleens (0.35–0.5 g) and are diluted 10^{-3}, 10^{-4}, and 10^{-5}. (Note that the lungs are less infected than the spleens in this model and usually require 10–100 times less dilution.) Each organ is plated at three different dilutions on 7H10 agar (Difco) plates containing 5% OADC enrichment. Plates are incubated for 2–3 weeks and counted to determine viable cell counts, which are expressed as counts/organ.

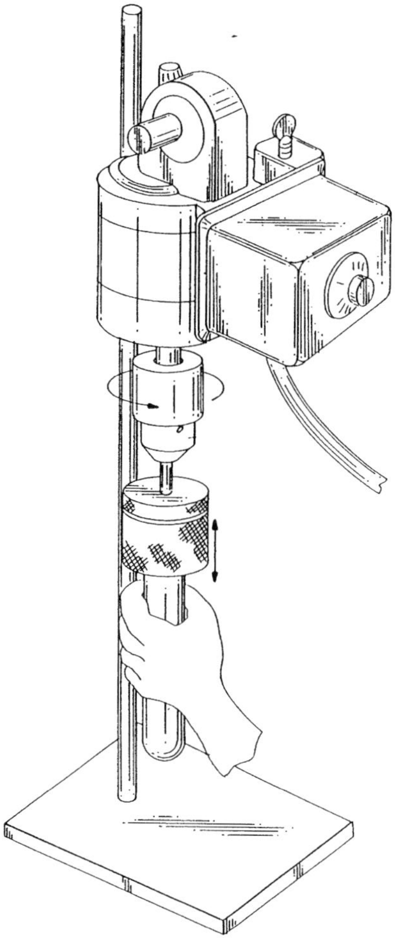

Figure 36.2 Aerosol-resistant grinding assembly (US patent no. 5 829 696).

Key parameters to monitor infection and response to treatment

Observing mice for signs of infection is an important evaluation tool. Because of the subjective nature of monitoring, one often has to rely on experience to ascertain the condition of the mice. Guidelines for evaluation and regulations for the welfare of the animals are available in *Guide for the Care and Use of Laboratory Animals* (National Research Council, 1996). Generally speaking, mice may show some or all of the following signs of infection. Often they exhibit a hunched posture with or without difficulty ambulating. They may isolate themselves from the rest of the group in their cage. Breathing may become labored, and shivering may be apparent. There may be evidence of lack of eating and/or drinking. Signs of diarrhea may be evident, such as areas of increased wetness in the bedding, especially in the corners of the cage.

It is necessary to evaluate the response of the mice to any therapy, particularly one not previously studied, when treatment is initiated. Upon introduction of a new agent, mice should be monitored initially and during the first hour for any indications of pain or discomfort, allergic reaction or swelling. During the first 3 days of therapy, the mice are monitored at least twice daily to note their general appearance. Any injection site is inspected for swelling or irritation. The general condition of the animals is carefully noted. Mice having an acute reaction should be euthanized immediately and the therapy should be re-evaluated.

Efficacy is evaluated relative to the reduction of viable organisms in the spleens and lungs of therapy groups in comparison to those in the spleens and lungs of the control (infected, but untreated) groups. Additional comparison should be made between the therapy groups and the group given a standard therapy such as clarithromycin or azithromycin. Spleen weights can be useful in evaluating the therapy's effectiveness (Ji *et al.*, 1994), although it should be noted that in our experience many of the rifamycins cause splenomegaly (unpublished observations).

Antimicrobial therapy

Antimicrobial agents that have been studied in the beige mouse model of disseminated MAC infection are summarized in Table 36.1. The antimicrobial agents have been administered by various routes (s.c., p.o., i.v.). Little attention has usually been given to formulation of the antimicrobial agent when the oral route is used. We have usually begun therapy 7 days postinfection; however, a variety of pretreatment periods have been used. Determination of the spleen and lung viable cell counts is usually done several days after the completion of therapy to reduce any effects of antimicrobial carry-over during the quantitative plating. This is more of a theoretical than a real issue, because of the high dilutions used for titration of viable cell counts, a reflection of the relatively limited activity of current agents. Although daily treatment for 10 days allows for differentiation of relative activities, longer treatment periods (4–12 weeks) are useful to characterize efficacy, particularly of combination regimens, and to study the development of resistance (Bermudez *et al.*, 1996b; Lounis *et al.*, 1995). Once-weekly dosing of azithromycin with rifapentine has been found to be effective for therapy of disseminated MAC (Klemens and Cynamon, 1994a). The selection of dose and interval of administration should be carefully considered given the differing pharmacokinetics of antimicrobial agents in humans and mice. Most investigators have not carefully defined the pharmacokinetics of their study drugs in the beige mouse. Furthermore, it is unclear whether the goal should be to mimic the peak serum level or the area under the curve for agents used in this model. When an agent is to be evaluated in mice prior to the study of its pharmacokinetics in humans, it is important to evaluate several doses and to define the maximally tolerated dose. When the

Table 36.1 Correlation of antimicrobial efficacy in the beige mouse model of MAC infection with clinical experience

	Experimental experience			*Clinical experience*			
Agents	*Dosing (mg/kg daily)*	*Outcome*	*Reference*	*Dosing*	*Outcome*	*Reference*	*Comments*
Prophylaxis							
Azithromycin (AZI)	200	Decreased incidence of bacteremia, decreased cfu in livers and spleens	Bermudez *et al.*, 1995	1200 mg/week	70% efficacy	Havlir *et al.*, 1996	Reasonably good correlation between experimental data and clinical experience for CLARI, AZI and RBT in the prophylaxis setting
Clarithromycin (CLARI)	200	Significant reduction of cfu in spleens and livers	Bermudez *et al.*, 1994b	500 mg bid.	70% efficacy	Pierce *et al.*, 1996	
Rifabutin	30	Higher dose decreases incidence of bacteremia	Bermudez *et al.*, 1995	300 mg/d	50% efficacy	Nightingale *et al.*, 1993	
	60	Did not decrease cfu in spleens and livers					
Therapy							
Azithromycin	100 200	Relatively good activity in spleens and lungs against eight clinical isolates	Cynamon *et al.*, 1994	600 mg/d	Good activity	Berry *et al.*, 1993	Good correlation between experimental data and clinical experience for CLARI and AZI
	200 mg/kg/week	Good activity alone and in combination with RPT 20 mg/kg/wk	Klemens and Cynamon, 1994a				
Clarithromycin	200	Relatively good activity in spleens and lungs against eight clinical isolates	Cynamon *et al.*, 1994	500 mg bid	Good activity	Chaisson *et al.*, 1994	
Rifabutin	20	Modest activity against 2/5 MAC isolates	Klemens *et al.*, 1994	300 mg/d	Good activity in CLARI/RBT/EMB combination	Shafran *et al.*, 1996	Role in therapy unclear (Is AZI or CLARI plus RBT better than AZI or CLARI plus EMB?) CLARI/RBT interaction, Wallace *et al.*, 1995. RBT uveitis, Shafran *et al.*, 1994
Rifampin (RIF)	20	Little activity as monotherapy in combination with CLARI did not enhance activity	Klemens and Cynamon, 1992 Klemens *et al.*, 1992	600 mg/d	Not able to reduce, bacteremia	Kemper *et al.*, 1994	Inactive in mice and in clinical experience
Rifapentine (RPT)	20	Modest activity against 2/5 MAC isolates	Klemens *et al.*, 1994		Not studied		

Table 36.1 (*Continued*)

	Experimental experience			*Clinical experience*			
Agents	*Dosing (mg/kg daily*	*Outcome*	*Reference*	*Dosing*	*Outcome*	*Reference*	*Comments*
KRM 1648	20	Modest activity	Tomioka *et al.*, 1992 Bermudez *et al.*, 1994a		Not studied		
Ethambutol (EMB)	100	Modest activity as monotherapy In combination with CLARI did not enhance activity	Young and Bermudez, 1995 Klemens *et al.*, 1992	15 mg/kg	Modest reduction in bacteremia	Kemper *et al.*, 1994	Good correlation between experimental data and clinical experience
Clofazimine (CFZ)	20	In combination with CLARI did not enhance activity	Klemens *et al.*, 1992	200 mg/d	Not active in reducing bacteremia	Kemper *et al.*, 1994	Good correlation between experimental data and clinical experience
Ciprofloxacin	40	Inactive	Inderlied *et al.*, 1989		Not adequately studied in monotherapy		Data not adequate to determine correlation of experimental and clinical experience for the quinolones
Levofloxacin	200	Inactive	Bermudez *et al.*, 1996a		Not adequately studied in monotherapy		Data not adequate to determine correlation of experimental and clinical experience for the quinolones
Sparfloxacin	50	Modest activity	Perronne *et al.*, 1992		Not adequately studied in monotherapy		
Amikacin	50–100	Modest activity In combination with CLARI did not enhance activity	Gangadharam *et al.*, 1988 Klemens *et al.*, 1992	1 g/d	Not adequately evaluated in monotherapy		
Streptomycin	150	Modest activity	Duzgunes *et al.*, 1991		Not adequately studied		
Paromomycin	200	Modest activity	Kanyok *et al.*, 1994				

human pharmocokinetics are known for an agent it is useful to evaluate this agent over a range of doses (usually seven to 10 times the maximally tolerated human dose for agents cleared by the kidneys). It is more difficult to define the appropriate murine dose for agents that are cleared by the liver. It is unclear whether the measurement of serum or tissue levels for various agents in mice provide useful information regarding therapeutic efficacy (e.g. azithromycin). Experience with antituberculosis therapy in mice and humans established that once-daily dosing is preferable to fractional dosing. The data are less clear for MAC therapy; however, once-daily dosing would seem most appropriate in light of the relatively long doubling time for this organism.

Pitfalls (advantages/disadvantages) of the model

Apart from the previously mentioned shortcoming of differing pharmacokinetics of antimicrobials in mice and humans, the murine model of disseminated MAC is suitable for the evaluation of antimicrobials for the treatment of disseminated MAC in HIV infected patients. The murine model enables one to rapidly evaluate multiple doses and combinations of agents relatively inexpensively. The i.v. model of disseminated MAC infection may be less useful for the study of immunology, since the beige mouse is immunocompromised differently from HIV infected patients. The i.v. model probably underestimates the potential efficacy of various agents for preventive therapy of disseminated MAC in HIV-infected humans. The oral infection model is better for this purpose; however, there is still the pitfall of the requirement for relatively large inocula, which are not likely to mimic the clinical situation (Bermudez *et al.*, 1994b, 1995). One can now use either clarithromycin or azithromycin as reference agents for the evaluation of new drugs alone or in combination to determine efficacy and the ability of the new agent to delay the emergence of resistance.

It is important to evaluate new agents against several MAC isolates in the murine model to insure that a less biased reflection of an agent's potential clinical activity is obtained. The rifamycins have better activity against ATCC 49601 (serotype 1) strain LPR than they do against MAC 101 (Klemens *et al.*, 1994). This difference led to somewhat different impressions of the potential clinical usefulness of these agents based on a myopic view of the universe of MAC organisms. Clofazamine has limited activity alone in the beige mouse model of disseminated MAC infection; however, it enhanced the activity of rifapentine when used in combination (Klemens and Cynamon, 1991). Clofazamine performed poorly when evaluated in humans with disseminated MAC infection (Kemper *et al.*, 1994).

Contributions of the model to infectious disease therapy

In the light of the potential problems with the beige mouse model of disseminated MAC infection as a predictor for the clinical efficacy of antimicrobial agents, it has proved to be useful for azithromycin (Cynamon *et al.*, 1994) and clarithromycin (Fernandes *et al.*, 1989). In murine studies both agents had activity against eight clinical MAC isolates (Cynamon *et al.*, 1994). The mean log reduction in cell counts/spleen was 1.2 log, 0.75 log and 0.61 log for mice treated for 10 days with azithromycin 200 mg/kg or 100 mg/kg, and clarithromycin 200 mg/kg, respectively. The mean reduction in cell counts/lung were 0.38 log for azithromycin 200 mg/kg and 0.67 log for clarithromycin 200 mg/kg. Once-weekly treatment with azithromycin 200 mg/kg alone or in combination with rifapentine 20 mg/kg for 8 weeks was effective for disseminated MAC infection in beige mice (Klemens and Cynamon, 1994a). These results are consistant with the efficacy of azithromycin in MAC prophylaxis (Havlir *et al.*, 1996).

The correlation of activity of rifamycins in the beige mouse model of disseminated MAC infection and human disease has been less clear. Rifampin has little activity against MAC *in vitro* or in mice (Klemens and Cynamon, 1992). Rifabutin, rifapentine, and KRM 1648 have been found to have promising activity against some isolates in the murine system (Tomioka *et al.*, 1992; Klemens *et al.*, 1994 and to have little activity against others (Bermudez *et al.*, 1994a; Klemens *et al.*, 1994). These discordant observations may be due in part to the use by the investigators of isolates with different intrinsic susceptibilities to rifamycins. Rifampin does not appear to be effective for the treatment of disseminated MAC infection based on a monotherapy study of bacteremia (Kemper *et al.*, 1994). The therapeutic efficacy of rifabutin, rifapentine, and KRM 1648 in human disease has not been adequately studied. Rifabutin has been demonstrated to decrease by one-half the occurrence of MAC bacteremia when given at 300 mg/d to HIV-infected patients with $CD4^+$ T-cell counts below $100/mm^3$ (Nightingale *et al.*, 1993).

Ethambutol had modest activity in both the beige mouse (Young and Bermudez, 1995) and in disseminated MAC infection in humans (Kemper *et al.*, 1994). Several investigators have evaluated the activities of antimicrobial agents in liposome preparations. This drug delivery system would seem to be attractive to treat disseminated MAC infection in mice and humans (Cynamon *et al.*, 1989; Klemens *et al.*, 1990; Duzgunes *et al.*, 1991). Liposome-encapsulated agents have not been adequately evaluated in clinical disseminated MAC infections.

References

Appelberg, R., Castro, A. G., Pedrosa, J., Silva, R. A., Orme, I. M., Minoprio, P. (1994). Role of gamma interferon and tumor necrosis factor during T-cell-independent and -dependent phases of *Mycobacterium avium* infection. *Infect. Immun.*, **62**, 3962–3971.

Armstrong, A. L., Dunbar, F. P., Cacciatore, R. (1967). Comparative pathogenicity of *Mycobacterium avium* and Battey bacilli. *Am. Rev. Respir. Dis.*, **95**, 20–32.

Bermudez, L. E., Kolonoski, P., Young, L. S., Inderlied, C. B. (1994a). Activity of KRM-1648 alone or in combination with ethambutol or clarithromycin against *Mycobacterium avium* complex in the beige mouse. *Antimicrob. Agents Chemother.*, **38**, 1844–1848.

Bermudez, L. E., Inderlied, C. B., Kolonoski, P., Petrofsky, M., Young, L. S. (1994b). Clarithromycin, dapsone, and a combination of both used to treat or prevent disseminated *Mycobacterium avium* infection in beige mice. *Antimicrob. Agents Chemother.*, **38**, 2717–2721.

Bermudez, L. E., Petrofsky, M., Inderlied, C. B., Young, L. S. (1995). Efficacy of azithromycin and rifabutin in preventing infection by *Mycobacterium avium* complex in beige mice. *Antimicrob. Agents Chemother.*, **36**, 641–646.

Bermudez, L. E., Inderlied, C. B., Kolonoski, P., Wu, M., Barbara-Burnham, L., Young, L. S. (1996a). Activities of Bay Y3118, levofloxacin, and ofloxacin alone or in combination with ethambutol against *Mycobacterium avium* complex *in vitro*, in human macrophages, and in beige mice. *Antimicrob. Agents Chemother.*, **40**, 546–551.

Bermudez, L. E., Nash, K. A., Petrofsky, M., Young, L. S., Inderlied, C. B. (1996b). Effect of ethambutol on emergence of clarithromycin-resistant *Mycobacterium avium* complex in the beige mouse model. *J. Infect. Dis.*, **174**, 1218–1222.

Berry, A., Koletar, S., Williams, D. (1993). Azithromycin therapy for disseminated *Mycobacterium avium–intracellulare* in AIDS patients (abstract). In *Program and Abstracts of the First National Conference on Human Retroviruses and Related Infection, Washington, DC, December 12–16, 1993*, p. 106. American Society for Microbiology, Washington, DC.

Bertram, M. A., Inderlied, C. B., Yadegar, S., Kolonoski, P., Yamada, J. K., Young, L. S. (1986). Confirmation of the beige mouse model for study of disseminated infection with *Mycobacterium avium* complex. *J. Infect. Dis.*, **154**, 194–195.

Brown, S. T., Edwards, F. F., Bernard, F. M., Tong, W., Armstrong, D. (1993). Azithromycin, rifabutin, and rifapentine for treatment and prophylaxis of *Mycobacterium avium* complex in rats treated with cyclosporine. *Antimicrob. Agents Chemother.*, **37**, 398–402.

Burman, W. J., Cohn, D. L. (1995). Clinical disease in human immunodeficiency virus-infected persons. In Mycobacterium avium–*Complex Infection: Progress in Research and Treatment* (eds Korvick, J. A., Benson, C. A.), pp. 79–108. Marcel Dekker, New York.

Chaisson, R. E., Benson, C. A., Dube, M. P. *et al.* (1994). Clarithromycin therapy for bacteremic *Mycobacterium avium* complex disease: a randomized, double-blind, dose-ranging study in patients with AIDS. *Ann. Intern. Med.*, **121**, 905–911.

Chi, E. Y., Ignacio, E., Langunoff, D. (1978). Mast cell granule formation in the beige mouse. *J. Histochem. Cytochem.*, **26**, 131–137.

Cohen, Y., Perronne, C., Lazard, T. *et al.* (1995). Use of normal C57BL/6 mice with established *Mycobacterium avium* infections as an alternative model for the evaluation of antibiotic activity. *Antimicrob. Agents Chemother.*, **39**, 735–738.

Collins, F. M. (1972). Acquired resistance to mycobacterial infections. *Adv. Tuberc. Res.*, **18**, 1–30.

Collins, F. M., Morrison, N. E., Montalbine, V. (1978). Immune response to persistent mycobacterial infection in mice. *Infect. Immun.*, **20**, 430–438.

Cynamon, M. H., Swenson, C. E., Palmer, G. S., Ginsberg, R. S. (1989). Liposome encapsulated amikacin therapy of *Mycobacterium avium* complex infection in beige mice. *Antimicrob. Agents Chemother.*, **33**, 1179–1184.

Cynamon, M., Klemens, S. P., Grossi, M. A. (1994). Comparative activities of azithromycin and clarithromycin against *Mycobacterium avium* infection in beige mice. *Antimicrob. Agents Chemother.*, **38**, 1452–1454.

Dautzenberg, B., Truffot, C., Legris, S. *et al.* (1991). Activity of clarithromycin against *Mycobacterium avium* infection in patients with acquired immune deficiency syndrome: a controlled clinical trial. *Am. Rev. Resp. Dis.*, **144**, 564–569.

Dunbar, F. P., Pejovic, I., Cacciatore, R., Peric-Golia, L., Runyon, E. H. (1968). *Mycobacterium intracellulare*. Maintenance of pathogenicity in relationship to lyophilization and colony form. *Scand. J. Resp. Dis.*, **49**, 153–162.

Duzgunes, N., Ashtekar, D. R., Flasher, D. L. *et al.* (1991). Treatment of *Mycobacterium avium–intracellulare* complex infection in beige mice with free and liposome-encapsulated streptomycin. *J. Infect. Dis.*, **164**, 143–151.

Engbaek, H. C. (1961). Pathogenicity and virulence of atypical mycobacteria for experimental animals (with particular reference to pathologic processes in joints and tendon sheaths of rabbits). *Acta. Tuberc. Scand.*, **40**, 35–50.

Fernandes, P. B., Hardy, D. J., McDaniel, D., Hanson, C. W., Swanson, R. N. (1989). *In vitro* and *in vivo* activities of clarithromycin against *Mycobacterium avium*. *Antimicrob. Agents Chemother.*, **33**, 1531–1534.

Gangadharam, P. R. J. (1986a). Animal models for nontuberculosis mycobacterial diseases. In *Experimental Models in Antimicrobial Chemotherapy 3* (eds Zak, O., Sande, M. A.), pp. 1–90. Academic Press, Orlando, FL.

Gangadharam, P. R. J. (1986b). Murine models for mycobacterioses. *Semin. Respir. Infect.*, **1**, 250–261.

Gangadharam, P. R. (1995). Beige mouse as a model for *Mycobacterium avium* complex disease. *Antimicrob. Agents Chemother.*, **39**, 1647–1654.

Gangadharam, P. R., Edwards, C. K. III, Murthy, P. S., Pratt P. F. (1983). An acute infection model for *Mycobacterium intracellulare* disease using beige mice: preliminary results. *Am. Rev. Resp. Dis.*, **127**, 648–649.

Gangadharam, P. R., Perumal, V. K., Podapati, N. R., Kesavalu, L., Iseman, M. D. (1988). *In vivo* activity of amikacin alone or in combination with clofazimine or rifabutin or both against acute experimental *Mycobacterium avium* complex infection in beige mice. *Antimicrob. Agents Chemother.*, **32**, 1400–1403.

Gangadharam, P. R. J., Perumal, V. K., Parikh, K. *et al.* (1989). Susceptibility of beige mice to *Mycobacterium avium* complex infections by different routes of challenge. *Am. Rev. Respir. Dis.*, **139**, 1098–1104.

Havlir, D. V., Dube, M. P., Sattler, F. R. *et al.* (1996). Prophylaxis against disseminated *Mycobacterium avium* complex with weekly azithromycin, daily rifabutin, or both. *N. Engl. J. Med.*, **335**, 392–398.

Hawkins, C. C., Gold, J. W. M., Whimbey, E. *et al.* (1986). *Mycobacterium avium* complex infections in patients with the acquired immunodeficiency syndrome. *Ann. Intern. Med.*, **105**, 184–188.

Heifets, L. B. (1991). Dilemmas and realities in drug susceptibility testing of *M. avium–intracellulare* and other slowly growing nontuberculous mycobacteria. In *Drug Susceptibility in the Chemotherapy of Mycobacterial Infection* (ed. Heifets, L. B.), pp. 123–146. CRC Press, Boca Raton, FL.

Inderlied, C. B., Kolonoski, P. T., Wu, M., Young, L. S. (1989). Amikacin, ciprofloxacin and imipenem treatment for disseminated *Mycobacterium avium* complex infection of beige mice. *Antimicrob. Agents Chemother.*, **33**, 176–180.

Inderlied, C., Kemper, C. A., Bermudez, L. E. (1993). The *Mycobacterium avium* complex. *Clin. Microbiol. Rev.*, **6**, 266–310.

Iseman, M. D. (1995). Pulmonary disease due to *Mycobacterium avium* complex. In Mycobacterium avium*–Complex Infection: Progress in Research and Treatment* (eds Korvick, J. A., Benson, C. A.), pp. 45–77. Marcel Dekker, New York.

Ji, B., Lounis, N., Tuffot-Pernot, C., Grosset, J. (1991). Susceptibility of immunocompetent beige and nude mice to *Mycobacterium avium* infection and response to clarithromycin. In *Program and Abstracts of 31st Interscience Conference on Antimicrobial Infections and Chemotherapy*, Washington DC, Abstract no. 291.

Ji, B., Lounis, N., Truffot-Pernot, C., Grosset, J. (1994). Effectiveness of various antimicrobial agents against *Mycobacterium avium* complex in the beige mouse model. *Antimicrob. Agents Chemother.*, **38**, 2521–2529.

Kanyok, T. P., Reddy, M. V., Chinnaswamy, J., Danziger, L. H., Gangadharam, P. R. J. (1994). *In vivo* activity of paromomycin against susceptible and multi-drug resistant *M. tuberculosis* and *M. avium* complex strains. *Antimicrob. Agents Chemother.*, **38**, 170–173.

Kemper, C. A., Havlir, D., Haghighat, D. *et al.* (1994). The individual microbiologic effect of three antimycobacterial agents, clofazamine, ethambutol, and rifampin, on *Mycobacterium avium* complex bacteremia in patients with AIDS. *J. Infect. Dis.*, **170**, 157–164.

Kleeberg, H. H., Nel, E. E. (1973). Occurrence of environmental atypical mycobacteria in South Africa. In *Proceedings of the Third International Colloquium on Mycobacteria*, pp. 201–214.

Klemens, S. P., Cynamon, M. H., Swenson, C. E., Ginsberg, R. S. (1990). Liposome encapsulated gentamicin therapy of *Mycobacterium avium* complex infection in beige mice. *Antimicrob. Agents Chemother.*, **34**, 967–970.

Klemens, S. P., Cynamon, M. H. (1991). *In vivo* activities of newer rifamycin analogs against *Mycobacterium avium* infection. *Antimicrob. Agents Chemother.*, **35**, 2026–2030.

Klemens, S. P., Cynamon, M. H. (1992). Activity of rifapentine against *Mycobacterium avium* infection in beige mice. *J. Antimicrob. Chemother.*, **29**, 555–561.

Klemens, S. P., Cynamon, M. H. (1994a). Intermittent azithromycin for treatment of *Mycobacterium avium* infection in beige mice. *Antimicrob. Agents Chemother.*, **38**, 1721–1725.

Klemens, S. P., Cynamon, M. H. (1994b). Activities of azithromycin and clarithromycin against nontuberculous mycobacteria in beige mice. *Antimicrob. Agents Chemother.*, **387**, 1455–1459.

Klemens, S. P., Destefano, M. S., Cynamon, M. H. (1992). Activity of clarithromycin against *Mycobacterium avium* complex infection in beige mice. *Antimicrob. Agents Chemother.*, **36**, 2413–2417.

Klemens, S. P., Grossi, M. A., Cynamon, M. H. (1994). Comparative *in vivo* activities of rifabutin and rifapentine against *Mycobacterium avium* infection. *Antimicrob. Agents Chemother.*, **38**, 234–237.

Lounis, N., Ji, B., Truffot-Pernot, C., Grosset, J. (1995). Selection of clarithromycin-resistant *Mycobacterium avium* complex during combined therapy using the beige mouse model. *Antimicrob. Agents Chemother.*, **39**, 608–612.

Meissner, G. (1981). The value of animal models for study of infection due to atypical mycobacteria. The Schaefer Memorial Lecture. *Rev. Infect. Dis.*, **3**, 953–959.

National Research Council (1996). *Guide for the Care and Use of Laboratory Animals*. National Academy Press, Washington, DC.

Nightingale, S. D., Cameron, D. W., Gordin, F. M. *et al.* (1993). Two controlled trials of rifabutin prophylaxis against *Mycobacterium avium* complex infection in AIDS. *N. Engl. J. Med.*, **329**, 828–833.

Orme, I. M., Furney, S. K., Roberts, A. D. (1992). Dissemination of enteric *Mycobacterium avium* infections in mice rendered immunodeficient by thymectomy in CD4 depletion or by prior infection with murine AIDS retroviruses. *Infect. Immun.*, **60**, 4747–4753.

Parenti, D., Ellner, J., Hafner, R. *et al.* (1995). A phase II/III trial of rifampin, ciprofloxacin, clofazimine, ethambutol, ± amikacin in the treatment of disseminated *Mycobacterium avium* infection in HIV-infected individuals (abstract). In *Abstracts of the Second National Conference on Human Retroviruses*, p. 56. American Society for Microbiology, Washington, DC.

Perronne, C., Cohen, Y., Truffot-Pernot, C., Grosset, J., Vilde, J. L., Pocidalo, J. J. (1992). Sparfloxacin, ethambutol, and cortisol receptor inhibitor RU-40 555 treatment for disseminated *Mycobacterium avium* complex infection of normal C57BL/6 mice. *Antimicrob. Agents Chemother.*, **11**, 2408–2412.

Pierce, M., Crampton, S., Henry, D. *et al.* (1996). A randomized trial of clarithromycin as prophylaxis against disseminated *Mycobacterium avium* complex infection in patients with advanced acquired immunodeficiency syndrome. *N. Engl. J. Med.*, **335**, 384–391.

Rastogi, N., Fresel, C., Ryter, A., Ohayon, H., Lesourd, M., David, H. L. (1981). Multiple drug resistance in *Mycobacterium avium*: is the wall architecture responsible for the exclusion of antimicrobial agents? *Antimicrob. Agents Chemother.*, **20**, 666–677.

Reddy, V. M., Parikh, K., Luna-Herrera, J., Falkinham, J. O. III, Brown, S., Gangadharam, P. R. (1994). Comparison of virulence of *Mycobacterium avium* complex (MAC) strains isolated from AIDS and non-AIDS patients. *Microbiol. Pathol.*, **16**, 121–130.

Roder, J., Duwe, J. (1979). The beige mutation in the mouse selectively impairs natural killer cell function. *Nature*, **278**, 451–453.

Sacco, R. E., Jensen, R. J., Thoen, C. O., Weinstock, J., Lynch, R. G., Dailey, M. D. (1996). Cytokine secretion and adhesion molecule expression by granuloma T lymphocytes in *Mycobacterium avium* infection. *Am. J. Pathol.*, **148**, 1935–1948.

Saito, H., Tomioka, H. (1988). Susceptibilities of transparent, opaque, and rough colonial variants of *Mycobacterium avium* complex to fatty acids. *Antimicrob. Agents Chemother.*, **32**, 400–402.

Shaefer, W. B., Davis, C. L., Cohn, M. L. (1970). Pathogenicity of transparent, opaque and rough variants of *Mycobacterium avium* in chickens and mice. *Am. Rev. Respir. Dis.*, **102**, 499–506.

Shafran, S. D., Deschenes, J., Miller, M., Phillips, P., Toma, E. (1994). Uveitis and pseudojaundice during a regimen of clarithromycin, rifabutin and ethambutol. *N. Engl. J. Med.*, **330**, 438–439.

Shafran, S. D., Singer, J., Zarowny, D. P. *et al.* (1996). A comparison of two regimens for the treatment of *Mycobacterium avium* complex bacteremia in AIDS: rifabutin, ethambutol, and clarithromycin versus rifampin, ethambutol, clofazamine, and ciprofloxacin. *N. Engl. J. Med.*, **335**, 377–383.

Tomioka, H., Sato, H., Sato, K. *et al.* (1992). Chemotherapeutic efficacy of a newly synthesized benzoxazinorifamycin, KRM-1648, against *Mycobacterium avium* complex induced in mice. *Antimicrob. Agents Chemother.*, **36**, 387–393.

US Department of Health and Human Services (1993). *Biosafety in Microbiological and Biomedical Laboratories* (eds Richmond, J. Y., McKinney, R. W.). Government Printing Office, Washington, DC.

Vassalli, J. D., Granelli-Piperno, A., Griscelli, E., Reich, E. (1978). Specific protease deficiency in polymorphonuclear leucocytes of Chediak–Higashi syndrome and beige mice. *J. Exp. Med.*, **147**, 1285–1290.

Wallace, R. J., Jr, Brown, B. A., Griffith, D. E., Girard, W., Tanaka, K. (1995). Reduced serum levels of clarithromycin in patients treated with multidrug regimens including rifampin or rifabutin for *Mycobacterium avium–M. intracellulare* infection. *J. Infect. Dis.*, **171**, 747–750.

Wolinsky, E. (1979). Nontuberculous mycobacteria and associated disease. *Am. Rev. Respir. Dis.*, **119**, 107–159.

Woodley, C. L., David, H. L. (1976). Effect of temperature on the rate of the transparent to opaque colony type transition in *Mycobacterium avium*. *Antimicrob. Agents Chemother.*, **9**, 113–119.

Young, L. S., Wiviott, L., Wu, M., Kolonoski, P., Bolan, R., Inderlied, C. B. (1991). Azithromycin for treatment of *Mycobacterium avium–intercellulare* complex infection in patients with AIDS. *Lancet*, **338**, 1107–1109.

Young, L. S., Bermudez, L. E. (1995). Animal models in anti-*Mycobacterium avium* complex drug development. In Mycobacterium avium*–Complex Infection: Progress in Research and Treatment* (eds Korvick, J. A., Benson, C. A., pp. 141–161. Marcel Dekker, New York.

Chapter 37

The Armadillo Leprosy Model with Particular Reference to Lepromatous Neuritis

D. M. Scollard and R. W. Truman

Introduction

Mycobacterium leprae is the only bacterial pathogen that infects peripheral nerves. The resulting sensory deficiencies and motor weakness are often what bring these patients to seek medical attention, at which time they present with a wide variety of skin lesions (Clements and Scollard, 1997). The mechanisms by which *M. leprae* localizes to peripheral nerve, and the mechanisms of subsequent nerve injury, are very poorly understood. A major reason for this is that nearly all information about nerve injury in leprosy has come from human biopsy material. The severe medical and ethical limitations on nerve biopsies have dictated that only selected sensory nerves have been studied systematically in patients with leprosy (primarily the sural and a sensory branch of the radial cutaneous nerve), sometimes with little or no clinical evidence that these particular nerves are affected in each individual.

The cell-mediated immune (CMI) response of the host to *M. leprae* determines the type of inflammatory lesion observed in tissues of infected individuals. The human response to *M. leprae* is uniquely broad and comprehensive, covering the entire spectrum of CMI (Skinsnes, 1964; Ridley and Jopling, 1966). Patients with strong delayed-hypersensitivity and CMI responses to *M. leprae* antigens exhibit dermal granulomatous lesions similar to those of tuberculosis, and are hence termed *tuberculoid*. At the opposite extreme are patients who have virtually no cellular immunity to *M. leprae*, termed *lepromatous*, whose skin lesions are filled with organisms growing in macrophages and cutaneous nerves. Most patients' responses fall into a wide *borderline* range of patterns between these two extremes, with variable features of both polar forms (Clements and Scollard, 1997). This broad spectrum of cellular immune dysfunction in leprosy has been described in increasing detail as it involves T-lymphocytes, macrophages, and cytokine production (Krahenbuhl and Adams, 1998; Modlin and Rea, 1994). At this time, however, no unifying hypothesis explains the extraordinarily wide range of CMI to *M. leprae*. Nerve involvement occurs in all types of disease, although the speed of progression of clinical neuropathy varies across the spectrum. Multiple mechanisms are likely to mediate nerve injury, including the CMI of the host.

Most evidence indicates that the most likely source of human infection is from another infected individual, but the exact means of transmission is not known (Nelson, 1998). Only approximately 5% of the human adult population appears to be susceptible to this organism. *M. leprae* appears to grow slowly, as determined from inoculation studies in mouse footpads (Shepard, 1960), and the incubation period in man usually ranges from 2–7 years, sometimes up to 20 years (Doull *et al.*, 1945). The infection is curable with multidrug antimicrobial therapy, and, although WHO programmes over the last 15 years have extended the reach of this treatment and resulted in the removal of many patients from active disease registries, the new-case prevalence rates have not fallen, and millions of people continue to suffer from deformities, ostracism, and other effects of the disease. In endemic countries in Asia, Africa, and Latin America, leprosy remains a major health problem and a major burden on patients and their families.

The armadillo as an experimental leprosy model

The nine-banded armadillo (*Dasypus novemcinctus*) is the only immunologically intact animal that exhibits a high frequency of natural susceptibility to infection with *M. leprae*, and its value as an immunopathologic model of leprosy and as an epidemiologic model have been reviewed previously (Storrs *et al.*, 1988b; Kirchheimer and Storrs, 1971; Truman and Sanchez, 1993; Truman, 1994). It is now emerging as a model for neurologic aspects of *M. leprae* infection. A primary focus of interest has been the immunologic nonresponsiveness of the animals that develop disseminated infection (Storrs, 1971), analogous to human lepromatous leprosy. Highly susceptible armadillos develop weak reactions to killed *M. leprae* (Job *et al.*, 1982, 1985a, 1987; Krotoski *et al.*, 1993), without granuloma formation, similar to those of human lepromatous patients. Strong lepromin reactions, with granulomatous inflammation, are associated with the development of high resistance to infection in armadillos and with tuberculoid leprosy in man. Lymphocytes from resistant animals proliferate *in vitro* in response to *M. leprae*

Handbook of Animal Models of Infection
ISBN 0-12-775390-7

antigens, while cells from susceptible animals show little or no proliferation, also comparable to leprosy in man (Kirchheimer *et al.*, 1978; Shannon *et al.*, 1984).

Efforts to further explore mechanisms of CMI (and its defects) in armadillos have been hampered by the dependence on outbred animals and the lack of reagents to identify T-cells and their subsets and cytokines in this species. The first T-cell marker for armadillo lymphocytes has recently been demonstrated by cross-reactivity of monoclonal antibody TCRδ_1 (recognizing a common determinant on the human δ chain) with a population of armadillo lymphocytes (Lathrop *et al.*, 1997). The distribution of this lymphocyte population was comparable to that of γ,δ cells in other species. However, although γ,δ cells have been implicated in immune responses to various mycobacterial antigens, no correlation has been documented between γ,δ-reactive lymphocytes and *M. leprae* infection in armadillos. Macrophage activating factor, present in supernatants of armadillo spleen cell cultures, has been shown to activate anti-*Toxoplasma* activity in armadillo macrophages by a nitrate-independent mechanism, but it had no anti-*M. leprae* activity (Adams and Krahenbuhl, 1996). This factor has not been purified or cloned.

Early studies identified armadillo immunoglobulin families and specific antibody to the phenolic glycolipid-1 of *M. leprae* in infected armadillos (Truman *et al.*, 1986a; Vadiee *et al.*, 1988; Job *et al.*, 1992), and documented the evolution of the antibody response in armadillos infected with *M. leprae* (Truman *et al.*, 1986a,b; Vadiee *et al.*, 1990; Santos-Argumedo *et al.*, 1995). Antibodies offer no protective effects against *M. leprae* infection in armadillos or in human leprosy, but in man their titers tend to rise with exacerbation or relapse of infection and fall with improvement, suggesting that they may provide useful markers for the effectiveness of antimicrobial treatment of experimental infection (Cho *et al.*, 1991).

The armadillo as an epidemiologic model of leprosy

Nine-banded armadillos in parts of Texas and Louisiana are recognized as highly endemic natural hosts for *M. leprae*. Archived serum samples collected on Louisiana armadillos during 1960 showed that sylvatic leprosy already existed among these animals before they were ever used in leprosy research, and it originated by natural means (Truman *et al.*, 1986b). The geographic distribution of *M. leprae* infection among wild armadillos roughly follows the same pattern for leprosy in man (Walsh *et al.*, 1986) and some reports suggest the possibility of zoonotic transmission (West *et al.*, 1988; Truman *et al.*, 1990). Enzootic leprosy in armadillos is also being studied as a population model in which to examine basic mechanisms of leprosy transmission and evaluate environmental risk factors for the infection (Truman *et al.*, 1991).

The armadillo as a model of leprosy neuropathy

A recent study of peripheral nerves in armadillos inoculated with *M. leprae* indicated that animals that developed systemic disease in less than 2 years all had substantial *M. leprae* infection in at least one major peripheral nerve trunk (Scollard *et al.*, 1996). Semi-quantitative assessment of the bacterial load along the course of the affected nerves indicated that infection was greatest distally and decreased steadily in more proximal segments, also comparable to involvement in man. Notably, this localization to peripheral nerves occurs in the natural course of experimental infection, without any additional experimental manipulation of the nerves themselves and, as with human nerve involvement, is asymmetrical (McDougall, 1997; Scollard and Truman, 1997). This model therefore appears to offer for the first time an animal model for nerve involvement in leprosy. Histopathologic evaluation revealed an interstitial neuritis, with an infiltrate composed entirely of mononuclear cells (Figure 37.1). *M. leprae* were found in macrophages and Schwann cells, but their greatest concentration was observed in epineural tissues.

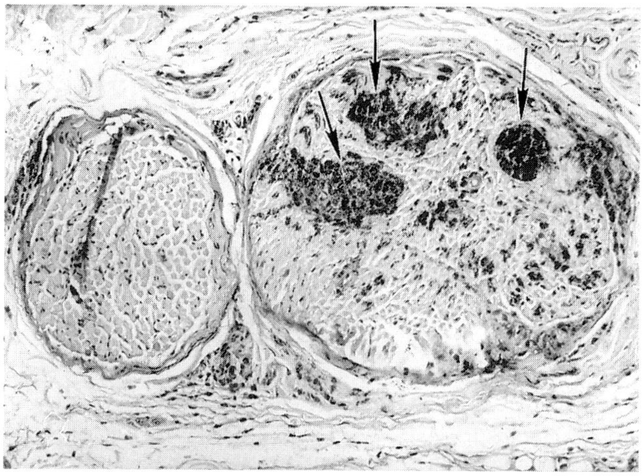

Figure 37.1 Interstitial neuritis in armadillos after experimental *M. leprae* infection. A cross-section of the ulnar nerve in an infected armadillo reveals large mononuclear cell infiltrates (arrows) containing abundant acid-fast organisms. (Fite & hematoxylin; original magnification × 400).

Animal species

Armadillos are small mammals of the superorder Edentata (Storrs, 1971). Lacking true teeth, they have only primitive molars and are unable to bite. They are equipped with short powerful legs and claws, which are well adapted to burrowing, and care must be taken when the animals are physically restrained in the laboratory. Armadillos have poor eyesight, but are especially sensitive to vibration. Amazingly agile, they can run rapidly. Adults weigh 3–5 kg, measure approximately 40 cm long and stand 20 cm from the floor. There are no established means to reliably estimate the age of individual animals, but they can be stratified into adult and subadult categories using a combination of weights and plasma progesterone concentrations (Truman *et al.*, 1991).

The armadillo's head, shoulders and pelvic area are covered by a hard carapace that is connected in the middle by nine movable bands interspaced by folds of skin (Talmage and Buchanen, 1954), from which it derives its name, the Spanish meaning 'little armored one' (Smith and Doughty, 1984). The carapace affords good protection but also lends to poor thermal regulation (Talmage and Buchanen, 1954). The armadillo's normal core body temperature ranges from 30–35 °C, fluctuating with ambient conditions. Armadillos pant at 39 °C and shiver at 16 °C (Johansen, 1961).

Preparation of animals

Armadillos exhibit polyembryony and delayed implantation. They regularly give birth to monozygotic quadruplets and delays of up to 3 years between natural mating and birth have been reported (Storrs *et al.*, 1988a). However, they reproduce only rarely in captivity and in most cases captive females cannibalize their young within a few days of parturition (Job *et al.*, 1984). There are no reliable commercial vendors and armadillos must be obtained from the wild for investigative purposes.

The principal maladies associated with housing and their medical needs have been reviewed elsewhere (Truman and Sanchez, 1993). They suffer immediate, severe captive stress reactions that seem to never entirely subside; these are the single greatest source of loss for the colony. Plasma corticosteroid concentrations among indoor-housed armadillos may equal or exceed those of free-ranging animals enduring a hard winter (Rideout *et al.*, 1985). Injections and bleeding are accomplished through the saphenous and subclavian veins. Armadillos respond well to most antibiotics and injectable or gas anesthetics in dosages appropriate for cats.

The principal requirement for containment is a sturdy unit capable of withstanding this strong animal's constant digging and clawing (Gilbert and Giametti, 1972; Burchfield *et al.*, 1976). They easily burrow to depths of 2 metres and are not contained by moats. Each requires at least 1 m^2 of cage space. Usually nocturnal, armadillos react to photo-period and other environmental cues (Johansen, 1961; Smith and Doughty, 1984). Insectivorous to omnivorous in nature, a number of diets have been described for suitable maintenance of the animals (Meritt, 1973; Ramsay *et al.*, 1981). Menadione requirements, however, are 10-fold greater than what is usually present in most commercial carnivore diets and a source of vitamin K must be added to armadillo feed in order to maintain the animal's blood clotting ability.

The principal use of armadillos has been for the *in-vivo* propagation of *M. leprae*. Wild-caught armadillos obtained from non-enzootic areas are conditioned to captivity over a 6-month period. Screening for sylvatic *M. leprae* infection includes histopathological examination of ear biopsies and buffy coat blood samples for acid fast bacteria (AFB), and serologic testing for *M.-leprae*-specific antibodies (Truman *et al.*, 1986a). For *in-vivo* propagation of *M. leprae* armadillos are maintained at 22 °C, to produce an internal temperature of approximately 32 °C.

Experimental infection with *M. leprae*

Experimental leprosy infections are initiated with bacilli derived from human biopsy material, armadillos, or nude mice. The incubation period is shortest when high doses (5×10^8 to 2×10^9) of *M. leprae* are delivered intravenously, although respiratory routes are also effective (Truman, unpublished observations). Bacilli rapidly become lodged in reticuloendothelial tissues, where they slowly proliferate and later disseminate predominantly to cooler regions of the body such as ears, nose, tongue, and footpads, as well as to the nasopharynx, bronchi, and lungs (Job *et al.*, 1985b).

The specific response and time course of experimental leprosy varies between individual animals; incubation periods range from 10 months to 4 years (Job *et al.*, 1985b). The disease produces few visual clues, and the earliest sign of successful infection is typically a sustained progressive evolution of IgM antibody to the major *M.-leprae*-specific antigen phenolic-glycolipid-1 (PGL-1; Truman *et al.*, 1986a). Usually detected 6–9 months after inoculation, its appearance corresponds with a proliferation to approximately 10^6 *M. leprae* per gram in either liver, spleen, or lymph node tissue (Job *et al.*, 1992). PGL-1 IgM antibody levels progressively rise with increased bacillary dissemination and the antibody continues to be detectable over the remaining course of infection. In successful lepromatous infections, AFB disseminate throughout the body and are usually detectable in ear biopsies within 1 year. Most develop heavy infections within 24 months of inoculation, with approximately 10^{12} recoverable bacilli in the liver and spleen (Truman *et al.*, 1986a). The late stages of infection last about 5 months and are marked by chronic hypochromic, microcytic anemia (Job *et al.*, 1992). Marked impairment of liver and renal function is observed (Table 37.1), and a characteristic lepromatous bronchopneumonia ensues.

Contributions of the model. In contrast to the mouse footpad model (Shepard, 1960), armadillos present different levels of infection, involvement of multiple organ systems and, because of their large size, permit evaluation of medications given orally — in feed or by capsule — as well as percutaneously or intravenously. The natural course of infection in armadillos offers unique opportunities for

Table 37.1 Representative clinical values for normal armadillos and animals experimentally infected with *M. leprae* manifesting fully disseminated late-stage disease (ALP-IFCC, alkaline phosphatase determined by International Federation of Clinical Chemists procedure; LDH-L, lactate dehydrogenase using l-lactate reaction

Test	*Fully disseminated**	*Normal†*
Chemistry‡		
Glucose	78 (± 19)	81 (± 21)
BUN	24 (± 7.6)	24 (± 6.5)
Creatinine	0.95 (± 0.13)	0.78 (± 0.19)
ALP (IFCC)	86 (± 13.7)	64 (± 22.6)
LDH-L	3394 (± 1220)	740 (± 45)
ALT	9.7 (± 5.5)	6.4 (± 3.2)
*Hematology***		
RBC (× 10^6/ml)	5.1 (± 0.8)	6.9 (± 0.3)
Hgb (g/dl)	9.8 (± 1.5)	13 (± 0.7)
HCT (%)	29 (± 5.6)	40 (± 3)
MCV (mu^3)	58.1 (± 2.2)	61 (± 2)
MCH (pg)	19.2 (± 0.7)	19.5 (± 0.5)

* Mean and standard deviation from sacrifice blood samples of seven animals that yielded more than 10^9 *M. leprae* per gram of liver tissue.
† Means and standard deviation from 45 naive armadillos held in captivity for 60 days.
‡ Chemistries determined on Ciba–Corning autoanalyzer with reagents and procedures according to manufacturers' recommendations.
** Performed on Coulter JT-3 gated for human cell types.

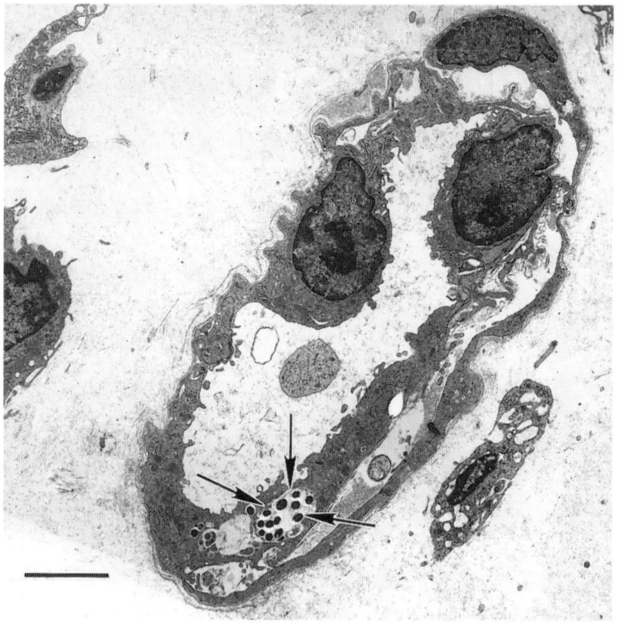

Figure 37.2 *M. leprae* in an epineural endothelial cell. This small vessel, closely approximated to the epineurium in a peripheral nerve of an experimentally infected armadillo contains several bacilli (arrows) in a large cytoplasmic vacuole. (Bar = 3 μm).

studies of relapse after chemotherapy and of long-term persistence of *M. leprae*. Preliminary studies on clearance of dapsone (DDS) and related compounds indicate that rates of clearance are similar to those in mice (Truman, unpublished observations).

Effective antimicrobial and anti-inflammatory treatment of neuritis is a high clinical priority in leprosy, and until now no model has been available in which this could be addressed experimentally. No other animal model has the capability to specifically examine issues related directly to nerve infection. Such issues, which can be examined in the armadillo model, include (1) how well each agent penetrates peripheral nerves; (2) how effective each of the agents is in killing *M. leprae* in nerve; and (3) whether such killing of *M. leprae* in nerves is entirely beneficial or whether it induces additional neuritis. Studies are under way to assess nerve conduction during the course of experimental leprosy in armadillos, possibly offering a non-invasive means of evaluating neurologic aspects of response to antimicrobial or anti-inflammatory treatment.

In addition, basic studies related to the mechanisms of localization and entry of *M. leprae* into peripheral nerve may offer new avenues for prevention, treatment, and assessment of neuritis in this disease. Studies in progress indicate that *M. leprae* is localized primarily in the vicinity of blood vessels in the epineurium (Figure 37.2; Scollard *et al.*, 1999). This is consistent with earlier reports indicating that cutaneous small blood vessels are major sites of bacillary localization in the skin (Balentine *et al.*, 1976; Coruh and McDougall, 1979; Boddingius, 1984; Burchard and Bierther, 1985; Mukherjee and Meyers, 1987), and suggest new avenues for investigation of the mechanisms by which *M. leprae* preferentially localizes to peripheral nerves. It will be of interest, for example, to determine whether or not armadillo nerves contain the same isoforms of laminin which have now been shown to bind *M. leprae* to the basement membrane of human Schwann cells (Rambukkana *et al.*, 1997).

The complex cellular and molecular mechanisms of the immune response to *M. leprae* have maximal impact on the slender peripheral nerves that are uniquely vulnerable to this pathogen, and the resulting nerve injuries are the primary cause of the debility, deformity, and stigma associated with this infection. The armadillo is the only animal model suitable for assessment of such mechanisms *in vivo*, and future studies may help to devise new strategies for the prevention and treatment of nerve injury in leprosy.

Acknowledgments

We are grateful to Ms Renee Painter for her assistance in preparing the manuscript. Portions of this work were supported by NIAID contract Y02-AI-0015-01.

References

Adams, L. B., Krahenbuhl, J. L. (1996). Responses of activated armadillo macrophages to challenge with *M. leprae* (Abstract). American Society of Microbiology, New Orleans, LA.

Balentine, J. D., Chang, S. C., Issar, S. L. (1976). Infection of armadillos with *M. leprae*: ultrastructural studies of peripheral nerve. *Arch. Pathol. Lab. Med.*, **100**, 175–181.

Boddingius, J. (1984). Ultrastructural and histophysiological studies on the blood–nerve barrier and perineural barrier in leprosy neuropathy. *Arch Neuropathol. (Berl.)*, **64**, 282–296.

Burchard, P. J., Bierther, M. (1985). An electron microscopic study of small cutaneous vessels in lepromatous leprosy. *Int. J. Lepr.*, **53**, 70–74.

Burchfield, H. P., Storrs, E. E., Walsh, G. P., Vidrine, M. F. (1976). Improved caging for nine-banded armadillos. *Lab. Anim. Sci.* **26**(2), 234–236.

Cho, S. N., Cellona, R. V., Fajardo, T. T. *et al.* (1991). Detection of phenolic glycolipid-1 antigen and antibody in sera from new and relapsed lepromatous patients treated with various drug regimens. *Int. J. Lepr.*, **59**, 25–31.

Clements, B. R., Scollard, D. M. (1997). Leprosy. In *Atlas of Infectious Diseases, vol VIII. External Manifestations of Systemic Infections* (eds Mandell, G. L., Fekety, R.), Churchill Livingstone, Philadelphia, PA.

Coruh, G., McDougall, A. C. (1979). Untreated lepromatous leprosy: histopathological findings in cutaneous blood vessels. *Int. J. Lepr.*, **47**, 500–501.

Doull, J. A., Guinto, R. C., Rodriguez, J. N., Bancroft, H. (1945). Risk of attack of leprosy in relation to age at exposure. *Am. J. Trop. Med.*, **25**, 435–439.

Gilbert, B. M., Giametti, L. (1972). An environmental caging system and maintenance program for the nine-banded armadillo *Dasypus novemcinctus*. *Lab Anim. Sci.*, **22**, 739–740.

Job, C. K., Kirchheimer, W. F., Sanchez, R. M. (1982). Tissue response to lepromin, an index of susceptibility of the armadillo to *M. leprae* infection — a preliminary report. *Int. J. Lepr.*, **50**, 177–182.

Job, C. K., Sanchez, R. M., Kirchheimer, W. F., Hastings, R. C. (1984). Attempts to breed the nine-banded armadillo *Dasypus novemcinctus* in captivity: a preliminary report. *Int. J. Lepr.*, **52**, 362–364.

Job, C. K., Sanchez, R., Hastings, R. C. (1985a). Effect of repeated lepromin testing on experimental nine-banded armadillo leprosy. *Indian J. Lepr*, **57**, 716–727.

Job, C. K., Sanchez, R. M., Hastings, R. C. (1985b). Manifestations of experimental leprosy in the armadillo *Dasypus novemcinctus*. *Am. J. Trop. Med. Hyg.*, **34**, 151–161.

Job, C. K., Sanchez, R., Hunt, R., Hastings, R. C. (1987). Prevalence and significance of positive Mitsuda reaction in the nine-banded armadillo (*Dasypus novemcinctus*). *Int. J. Lepr.*, **55**, 685–688.

Job, C. K., Drain, V., Truman, R. W., Deming, A. T., Sanchez, R. M., Hastings, R. C. (1992). The pathogenesis of leprosy in the nine-banded armadillo and the significance of IgM antibodies to PGL-1. *Indian J. Lepr.*, **64**, 137–151.

Johansen, K. (1961). Temperature regulation in the nine-banded armadillo (*Dasypus novemcinctus*). *Physiol. Zool.*, **34**, 126–144.

Kirchheimer, W. F., Storrs, E. E. (1971). Attempts to establish the armadillo (*Dasypus novemcinctus*) as a model for the study of leprosy, Part 1. Report of lepromatoid leprosy in an experimentally infected armadillo. *Int. J. Lepr.*, **39**, 693–702.

Kirchheimer, W. F., Sanchez, R. M., Shannon, E. J. (1978). Effect of specific vaccine on cell-mediated immunity of armadillos against *M. leprae*. *Int. J. Lepr.*, **46**, 353–357.

Krahenbuhl, J. L., Adams, L. B. (1999). *Mycobacterium leprae* as an opportunistic pathogen. In *Opportunistic Intracellular Pathogens and Immunity* (ed. Friedman, H.), Plenum Publishing, New York, pp. 75–90.

Krotoski, W. A., Mroczkowski, T. F. *et al.* (1993). Lepromin skin testing in the classification of Hansen's disease in the US. *Am. J. Med. Sci.*, **305**, 18–24.

Lathrop, G., Scollard, D. M., Dietrich, M. (1997). Reactivity of a population of armadillo lymphocytes with an antibody to human γ, δ cells. *Clin. Immunol. Immunopathol.*, **82**, 68–72.

McDougall, C. (1997). Bacillary and histopathological findings in the peripheral nerves of armadillos experimentally infected with *M. leprae* (letter). *Int. J. Lepr.*, **65**, 260.

Meritt, D. A. Jr (1973). Edentate diets. Part 1: Armadillos. *Lab. Anim. Sci.*, **23**, 540–542.

Modlin, R. L., Rea, T. H. (1994). Immunopathology of leprosy. In *Leprosy* (ed. Hastings, R. C.), pp. 225–234. Churchill Livingstone, New York.

Mukherjee, A., Meyers, W. M. (1987). Endothelial cell bacillation in lepromatous leprosy; a case report. *Lepr. Rev.*, **58**, 419–424.

Nelson, K. E. (1998). Leprosy. In *Textbook of Public Health*, 14th edn. (eds Maxcy, Rosenau and Last). Appleton-Century-Crofts, Norwalk, CT.

Rambukkana, A., Salzer, J. L., Yurchenco, P. D., Tuomanen, E. I. (1997). Neural targeting of *M. leprae* mediated by the G domain of the laminin-α_2 chain. *Cell*, **88**, 811–821.

Ramsey, P. R., Tyler, D. F. Jr., Waddill, J. R., Storrs, E. E. (1981). Blood chemistry and nutritional balance of wild and captive armadillos *Dasypus novemcinctus*. *Comp. Biochem. Physiol. A Comp. Physiol.*, **69**, 517–522.

Rideout, B. A., Gause, G. E., Benirschke, K., Lasley, B. L. (1985). Stress-induced adrenal changes and their relation to reproductive failure in captive nine-banded armadillos *Dasypus novemcinctus*. *Zoo Biol.*, **4**, 129–138.

Ridley, D. S., Jopling, W. H. (1966). Classification of leprosy according to immunity. A five-group system. *Int. J. Lepr.*, **34**, 255–273.

Santos-Argumedo, L., Guerra-Infante, F., Quesada-Pascual, F., Estrada-Parra, S. (1995). Identification and purification of armadillo immunoglobulins: preparation of specific antisera to evaluate the immune response in these animals. *Int. J. Lepr.*, **63**, 56–61.

Scollard, D. M., Truman, R. W. (1997). Reply to Dr McDougall (letter). *Int J. Lepr*, **65**, 261.

Scollard, D. M., Lathrop, G. W., Truman, R. W. (1996). Infection of distal peripheral nerves by *M. leprae* in infected armadillos: an experimental model of nerve involvement in leprosy. *Int. J. Lepr*, **64**, 146–151.

Scollard, D. M., McCormick, G. T., Allen, J. (1997). Early localization of *M. leprae* to epineural blood vessels (abstr) *J. Fed. Exp. Soc. Exp. Biol. Med.*, **A671**,

Scollard, D. M., McCormick, G. T., Allen, J. L. (1999) Localization of *M. leprae* to endothelial cells of epineurial and perineurial blood vessels and lymphatics. *Am. J. Pathol.* (In press.)

Shannon, E. J., Powell, M. D., Kirchheimer, W. F., Hastings, R. C. (1984). Effects of *M. leprae* antigens on the *in vitro* responsiveness of mononuclear cells from armadillos to Concanavalin-A. *Lepr. Rev.*, **55**, 19–31.

Shepard, C. C. (1960). The experimental disease that follows the injection of human leprosy bacilli into foot pads of mice. *J. Exp. Med.*, **112**, 445–454.

Skinsnes, O. K. (1964). Immunological spectrum of leprosy. In *Leprosy in Theory and Practice* (eds Cochrane, R. G., Davey, T. F.), pp. 156–182. John Wright, Bristol.

Smith, L. L., Doughty, R. W. (1984). *The Amazing Armadillos. Geography of a Folk Critter*. University of Texas Press, Austin, TX.

Storrs, E. E. (1971). The nine-banded armadillo, a model for leprosy and other biomedical research. *Int. J. Lepr.*, **39**, 703–714.

Storrs, E. E., Burchfield, H. P., Rees, R. J. W. (1988a). Super-delayed parturition in armadillos: a new mammalian survival strategy. *Lepr. Rev.*, **59**, 11–15.

Storrs, E. E., Walsh, G. P., Burchfield, H. P., Binford, C. H. (1988b). Leprosy in the armadillo: new model for biomedical research. *Science*, **183**, 851–852.

Talmage, R. V., Buchanen, C. D. (1954). The armadillo (*Dasypus novemcinctus*). A review of its natural history, ecology, anatomy, and reproductive physiology. *Rice Institute Pamphlet Monograph in Biology*, **XLI**(2), 1–135.

Truman, R. W. (1994). Environmental associations for *M. leprae*. In *Environmental Contaminants, Ecosystems, and Human Health* (eds Majumdar, S. K., Miller, E. W., Brenner, F. J.) pp. 437–449. Pennsylvania Academy of Sciences, Easton, PA.

Truman, R. W., Sanchez, R. M. (1993). Armadillos: models for leprosy. *Lab. Anim.*, **22**, 28–32.

Truman, R. W., Morales, M. J., Shannon, E. J., Hastings, R. C. (1986a). Evaluation of monitoring antibodies to PGL-1 in armadillos experimentally infected with *M. leprae*. *Int. J. Lepr.*, **54**, 556–559.

Truman, R. W., Shannon, E. J., Hagstad, H. V., Hugh-Jones, M. E., Wolff, A., Hastings, R.C. (1986b). Evaluation of the origin of *Mycobacterium leprae* infections in the wild armadillo *Dasypus-novemcinctus*. *Am. J. Trop. Med. Hyg.*, **35**, 588–593.

Truman, R. W., Job, C. K., Hastings, R. C. (1990). Antibodies to the phenolic glycolipid-1 antigen of *Mycobacterium leprae* for epidemiologic investigations of leprosy in armadillos (*Dasypus novemcinctus*). *Lep. Rev.*, **61**, 19–24.

Truman, R. W., Kumaresan, J. A., McDonough, C. M., Job, C. K. (1991). Seasonal and spatial trends in the detectability of leprosy in nine-banded armadillos. *Epidemiol. Infect.*, **106**, 549–560.

Vadiee, A. R., Shannon, E. J., Gillis, T. P., Mshana, R. N., Hastings, R. C. (1988). Armadillo IgG and IgM antibody responses to phenolic glycolipid-1 during experimental infection with *M. leprae*. *Int. J. Lepr.*, **56**, 422–427.

Vadiee, A. R., Harris, E., Shannon, E. J. (1990). The evolution of the antibody response in armadillos infected with *M. leprae*. *Lepr. Rev.*, **61**, 215–226.

Walsh, G. P., Meyers, W. M., Binford, C. H. (1986). Naturally acquired leprosy in the nine-banded armadillo, *Dasypus novemcinctus*, a decade of experience 1975–1985. *J. Leukocyte Biol.*, **40**, 645–656.

West, B. C., Todd, J. R., Lary, C. H. (1988). Leprosy in six isolated residents of northern Louisiana. *Arch. Intern. Med.*, **148**, 1987–1992.

Chapter 38

Models of Leprosy Infection in Mice

B. Ji and L. Levy

Introduction

Although *Mycobacterium leprae* was the first human bacterial pathogen to be recognized, by Hansen in 1873, it has yet to be cultivated on artificial media. In the absence of a method for cultivation *in vitro*, "cultivation" of the organisms in the animal has proved to be an important and most useful alternative for leprosy research.

Before 1960, countless efforts to develop an animal models of leprosy and to obtain multiplication of *M. leprae* in an animal had ended in failure. In 1960, Charles C. Shepard described (1960a, b) a method for cultivation of *M. leprae in vivo*. Application of the technique has proved of immense value in leprosy research and has made possible the screening of new drugs and new combined-drug regimens, and assessment of their efficacy in chemotherapy in a specific and quantitative manner, providing information of critical importance to efforts to control leprosy.

Two major factors contributed greatly to Shepard's success in establishing the mouse footpad technique. First, the fact that leprosy affects predominantly the skin, nasal mucosa and peripheral nerves suggested that the temperature at which *M. leprae* multiplies optimally is less than 37°C (Brand, 1959); based on the demonstration by Fenner (1956) of multiplication in the footpads of mice of *M. ulcerans* and *M. marinum*, for which the optimal temperature for multiplication *in vitro* is lower than 37°C, Shepard inoculated *M. leprae* recovered from skin biopsy specimens or from the sediment of nasal washings from untreated patients with lepromatous leprosy into the hind footpads of mice, in which the temperature is cooler, about 30°C at an ambient temperature of 20–25°C, as compared with the 37°C of the core temperature. Second, Shepard counted the organisms inoculated and later harvested by direct microscopy, and regularly sacrificed mice in order to examine the inoculated footpads histopathologically; this permitted detection of the limited and extraordinarily slow multiplication of *M. leprae* in the footpads of immunocompetent (hereafter "normal") mice, which occurred after inoculation of a small number of organisms, i.e. 10^4 per footpad or fewer, without gross lesions.

In chronological order, the subsequent important contributions to the animal models were the demonstration of the greater susceptibility to infection by *M. leprae* of thymectomized and irradiated mice (Rees, 1966), nine-banded armadillos (*Dasypus novemcinctus*; Kirchheimer and Storrs, 1971), the congenitally athymic "nude" mouse (Colston and Hilson, 1976; Kohsaka *et al.*, 1976), and the neonatally thymectomized rat (Fieldsteel and Levy, 1976).

Laboratory mice are convenient to work with, permit multiplication of *M. leprae* after inoculation of a few viable organisms, and the results of multiplication are remarkably consistent and reproducible. Although the armadillo is much more difficult to handle, it may harbor very much larger numbers of *M. leprae* than do normal mice; yields of 10^{12} *M. leprae* (approximately 300 mg wet weight of organisms) per armadillo, a million times the number of organisms yielded by a normal mouse, are not uncommon. Consequently, armadillos have primarily been used for the production of *M. leprae*. Because mice are widely used to screen new drugs, to test the drug susceptibility of the isolates of *M. leprae* and to assess the therapeutic efficacy of new drugs or regimens in clinical trials, this chapter is devoted mainly to a description of infection by *M. leprae* of both immunocompetent and immunodeficient mice, and the important applications of these experimental models.

Basic laboratory techniques for establishing the mouse model

Recovery of *M. leprae* from human biopsy specimens

The preferred sites for biopsy are the skin lesions with the largest bacterial indices in skin smear examination. The patient's wishes should be taken into account; in particular, biopsy from the face should be avoided. In general, a full-thickness specimen 4–5 mm in diameter, weighing 40–80 mg, is sufficient. The biopsy specimen is trimmed free of fat, weighed, placed in a sterile Petri dish to which one or two drops of Hanks's balanced salt solution (HBSS) have been added, and minced with sharp scissors. The minced tissue is than transferred on the blades of the scissors to the piston of a ten Broeck tissue grinder of 15 ml

Handbook of Animal Models of Infection
ISBN 0-12-775390-7

capacity, 2 ml HBSS is added and the tissue is ground until lumps of tissue can no longer be detected. The resulting tissue homogenate is transferred to a sterile test tube. An aliquot of the suspension is taken for preparation of smears for counting.

Counting acid-fast bacilli (AFB) in the tissue

Microscope slides with fused ceramic circles, each about 1 cm in diameter, on their surfaces are employed for counting. The diameter of the circles must be measured on slides selected at random from each new lot; this can be done to an accuracy of 0.01 cm with the microscope by means of the vernier on the mechanical stage.

To prepare smears for counting AFB, 10 μl of formol-milk is first applied to each of the circles. Then, 10 μl of the suspension to be counted is added to the circle, and the liquids are immediately mixed and carefully spread over the entire area of the circle with the aid of a platinum wire that has been bent to an angle of 60°. One circle is completed before the aliquot of suspension is added to the next; the wire is flamed between circles. The 10 μl volumes may be delivered with disposable capillary micropipettes or with an automatic micropipette that uses disposable, autoclavable tips. These operations are carried out on a levelling table on which the slides remain until dry. The smears are then fixed by exposure to formaldehyde fumes and controlling heating, and stained by the room-temperature acid-fast staining technique described by Shepard and McRae (1968).

The AFB are counted under optimal microscope conditions, employing Köhler illumination, an 100× apochromatic oil-immersion objective and 10–12.5× compensating oculars. The AFB are counted in fields selected every 0.5 mm across the equator of each of the three circles, employing the scale on the mechanical stage; thus, if the diameter of the circle is 1 cm, about 20 fields per circle or a total of 60 fields per slide are examined. When 10 μl of suspension has been spread over a circle of diameter D, and the diameter of the oil-immersion field is d, the following formula applies, expressing D and d in the same unit of measurement:

$$\frac{\text{No. AFB}}{\text{ml sample}} = \frac{\text{No. AFB counted}}{\text{No. fields examined}} \times \frac{1\ \text{ml}}{10\ \mu\text{l}} \times \frac{(D/2)^2}{(d/2)^2}$$

$$= \frac{\text{No. AFB counted}}{\text{No. fields examined}} \times 100 \times (D/d)^2$$

in which $(D/d)^2$ is the number of microscope fields per circle.

The results of the counting may be presented as the number of organisms per milliliter of tissue suspension, per footpad, or per milligram of tissue. In a far advanced, previously untreated lepromatous leprosy patient, as many as 10^8 *M. leprae* may be recovered from a skin biopsy specimen, or 10^6 organisms/mg of tissue. However, in most multibacillary patients, significantly fewer *M. leprae* are recovered.

Inoculation of mouse footpads

The number of *M. leprae* inoculated per footpad depends upon the immunological status of the mice. Normal mice are usually inoculated with 5×10^3 to 1×10^4 organisms per footpad. Although one might theoretically inoculate nude mice with any number of AFB, it is rarely possible to inoculate more than 5×10^6 per footpad because of its small size, which limits the volume of the inoculum to 0.03 ml per footpad. Moreover, the number of AFB that can be inoculated into the footpads of nude mice is also limited by the number of *M. leprae* that may be recovered from a skin biopsy.

Very often, both hind footpads of the mouse are inoculated. Employing a 0.5 ml or 1.0 ml syringe and a sharp 27 G or 30 G needle, the suspension is injected subcutaneously into the footpads. Instead of injecting a measured volume, which is difficult and time-consuming, injection of the suspension is continued until the tissues of the footpad are filled; the initial volume of the suspension in the syringe is recorded, as is the volume remaining after a group of mice has been inoculated, and the average volume per inoculum is calculated. In general, this volume is close to 0.03 ml.

Harvesting *M. leprae* from inoculated mouse footpads

Depending upon the objectives of the inoculation, 3–6 months after inoculation two to four mice are sacrificed by cervical dislocation, and the inoculated feet are cleaned with soap and water with the aid of gauze squares, rinsed with sterile water and dried with sterile gauze squares. The footpads are harvested individually or as a pool. Footpad tissues are removed in three layers—skin and subcutaneous tissue, tendon and muscle—with a sterile scalpel and haemostat. The tissues from an individual footpad, or the pooled tissues if the footpads are pooled, are minced and homogenized, as is done for skin biopsy specimens, smears are prepared from the resulting tissue suspension and the AFB are counted. When the total volume of a suspension prepared from the tissues of a single footpad is 2.0 ml, the diameter (D) of the circle is 1.13 cm and the diameter (d) of the oil-immersion field is 0.018 cm, the lower detectability level of the counting method, i.e. detection of a single bacillus in the course of examining 60 oil-immersion fields in three circles, is 1.3×10^4 AFB per footpad.

There are several ways to present the results of the harvests: the number of *M. leprae* from each harvested footpad; the mean number of organisms per footpad harvested from a given group at a specific interval after inoculation; and the number of footpads showing multiplication of *M. leprae* as a fraction of the number of footpads harvested. Multiplication of *M. leprae* is defined as a yield of 10^5 or more organisms per footpad if the inoculum was 10^4 per footpad or less, or a twofold or greater increase over the number inoculated if the inoculum was 10^5 per footpad or more (Ji *et al.*, 1996a).

Immunocompetent (normal) mouse model

Mice are readily available laboratory animals and easy to maintain. Although mice of inbred strains provide genetically uniform individuals, and although the maximum number of *M. leprae* harvested from BLAB/c, CBA as well as CFW mice may be greater than in A/J and C57BL mice in some experiments (Shepard and Habas, 1967), the overall influence of the inbred strain on multiplication of *M. leprae* in the footpads appears to have been marginal (Shepard, 1981), and most laboratories have reported satisfactory multiplication in outbred "Swiss" mice.

After inoculation of 5×10^3 to 1×10^4 *M. leprae* per footpad, AFB may be detectable in the inoculated footpads within 3–6 months; subsequently, the number of organisms increases gradually to a maximum of about 10^6 per footpad, after which the number remains constant or declines slightly (Figures 38.1, 38.2). One may thus identify the lag, logarithmic and stationary phases of the multiplication.

The "doubling time" or "generation time", i.e. the average interval required for each twofold increase in the number of organisms, of *M. leprae* during the logarithmic phase of multiplication is 11–13 days (Levy, 1976). This is an extraordinarily long doubling time compared to that of other microorganisms, including mycobacteria.

The maximum bacterial multiplication varies slightly among strains of mice and of *M. leprae*; the mean value is approximately 1×10^6 to 3×10^6 but never reaches 10^7 per footpad in normal mice. If the maximum is 10^7 or more, it is likely that either the counting procedure is in error or that the mice have become contaminated by mycobacteria other than *M. leprae*.

The stationary or "plateau" phase represents, in fact, the effects of the immune response of the mouse. Evidence for this is the failure of multiplication when normal mice are inoculated with 10^5 organisms per footpad or more (Shepard, 1960a, b), the resistance of *M. leprae*-infected mice to a second challenge with *M. leprae* (Levy, 1975; Shepard *et al.*, 1976) and the higher ceiling to multiplication in immunosuppressed rodents (Rees, 1966; Colston and Hilson, 1976; Fieldsteel and Levy, 1976; Kohsaka *et al.*, 1976). After the plateau is reached, the number of AFB in the footpad remains relatively constant, but the proportion of viable *M. leprae* decreases (Welch *et al.*, 1980). If one wishes to pass the strain to another batch of mice, subinoculation should be carried out as soon as the plateau has been reached.

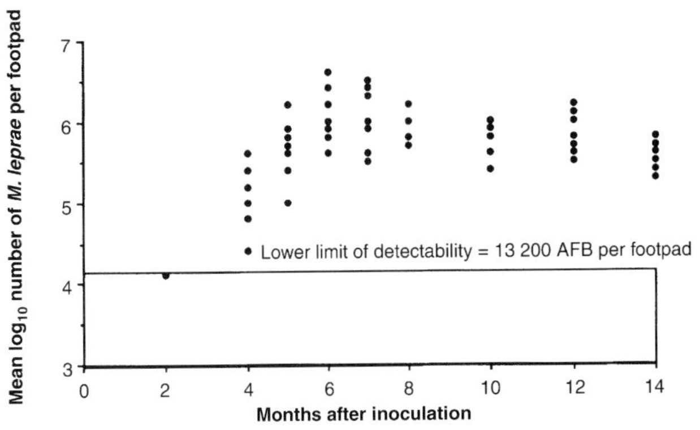

Figure 38.1 Results of harvests of *M. leprae* from the normal mice that had been inoculated with 5×10^3 AFB per footpad. Harvests were begun 2 months after inoculation and repeated at intervals of 1 or 2 months. At each interval, eight footpads were harvested individually. Each point represents the result of harvest from a single footpad. Because some results were identical, the numbers of points at each interval may be less than eight. The lower limit of detectability of the counting method is 1.3×10^4 organisms per footpad.

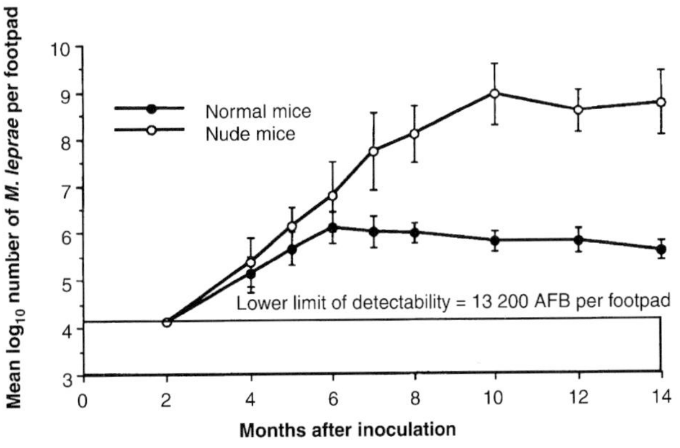

Figure 38.2 Growth curves of *M. leprae* in the footpads of normal and nude (*nu/nu*) mice that had been inoculated with 5×10^3 AFB per footpad.

The same maximum multiplication of *M. leprae* was encountered in the immunocompetent laboratory rat (Hilson, 1965), hamster (*Mesocricetus auratus*; Waters and Niven, 1966), gerbil (*Meriones unguiculatus*; Shepard, 1960b) and mystromys (*Mystromys mystromys*; Binford, 1968) as in normal mice, despite the much larger size of their footpads. Therefore, these animals have not been widely used for experimental work in leprosy.

Nude mouse model

In the footpads of thymectomized and whole-body-irradiated (TR) mice, *M. leprae* continues to multiply to a maximum of 10^7–10^8 per footpad in the course of 2 years following inoculation with 5×10^3 organisms, and swelling of the inoculated foot has been observed in about 10% of TR mice (Rees, 1966; Rees *et al.*, 1967; Shepard, 1985). Because they permit multiplication from a larger inoculum of *M. leprae* than do normal mice, TR mice were used to detect the persisting *M. leprae* in a large clinical trial, a procedure that requires inocula of 10^5 AFB per footpad or more (Subcommittee of THELEP, 1987). However, thymectomy and whole body irradiation resulted in poor survival of the mice (Shepard, 1985). *M. leprae* may also multiply after inoculation of larger numbers of the organisms in the footpads of neonatally thymectomized rats (NTR; Fieldsteel and Levy, 1976) and athymic (nude) rats (Colston *et al.*, 1979), but in none of them was the multiplication of *M. leprae* greater than was encountered in the TR mice. In addition, rats are much more difficult to work with than mice. Therefore, immunosuppressed or natively immunodeficient rats have rarely been used to assess the therapeutic effects in clinical trials (Gelber and Levy, 1987).

On the other hand, various lines of nude mouse are commercially available in large quantities. In addition, they permit the use of larger inocula of *M. leprae* than do TR mice (McDermott-Lancaster *et al.*, 1987). Consequently, since the beginning of the 1990s, the nude mouse has largely replaced the TR mouse in leprosy research.

Because the nude mouse is congenitally athymic, it lacks mature T lymphocytes and is extremely vulnerable to infections by many bacteria and viruses that are regarded as non-pathogenic for normal mice. Consequently, with few exceptions (Rungruang *et al.*, 1983), nude mice do not survive long under conventional animal-house conditions. For leprosy research, the lifespan of the nude mouse should be maintained to a similar length to the normal mouse, otherwise the advantages of using nude mice (see below) disappear. To prevent the death of nude mice from contamination by pathogens other than *M. leprae*, they must be bred and maintained under a germ-free or specific-pathogen-free isolator or laminar-flow rack, and everything that comes into contact with them, including the air, must be sterile. Therefore, the husbandry of nude mice is far more difficult and demanding than that of normal or other immunodeficient mice, and requires considerable investment in terms of equipment, manpower and operational costs. In one of our experiments involving 180 female Swiss nude mice, all but 5% of the animals survived the first year, and losses during the period 12–18 months after inoculation accounted for only an additional 10% (Ji *et al.*, 1996a); thus, under optimal conditions, the mortality of nude mice may be reduced to that of normal mice.

As shown in Figure 38.2, when mice are inoculated with 5×10^3 *M. leprae* per footpad, the organisms multiply to the level of 10^6 per footpad at the same rate in both normal and nude mice. Thereafter, however, multiplication ceases in normal mice, whereas it continues in nude mice. Although it was reported that, in nude mice, the maximum multiplication is of the order of 5×10^{10} AFB per footpad, achieved 12–18 months after inoculation (McDermott-Lancaster *et al.*, 1987), this is not always the case. In the experiment presented in Figure 38.2, multiplication slowed gradually after the number of organisms reached 10^8 per footpad and virtually ceased after the number had reached 10^9, suggesting that in nude mice mechanisms independent from T lymphocytes may also restrict the multiplication of *M. leprae*.

Some 10–12 months after inoculation, when the number of *M. leprae* exceeded 10^8 per footpad, more than 90% of the inoculated footpads were swollen (Ji *et al.*, 1996a); thereafter, the swelling gradually intensified, often ending in ulceration of the footpad. Another important feature of the process following inoculation of *M. leprae* into the footpads of nude mice is the dissemination of the organisms to uninoculated footpads, snout, testes, lymph nodes, tail-skin, ears, liver, spleen and tongue (Colston and Hilson, 1976; Kohsaka *et al.*, 1976; Chehl *et al.*, 1983).

Major applications of mouse models

Experimental applications

Drug screening

In the absence of a method for cultivating the organisms in cell-free media, *M. leprae* infection in the footpads of normal mice represents the only universally accepted system for screening drugs for activity against the organisms. Three methods have been employed in drug screening: (1) the continuous method; (2) the kinetic method; and (3) the proportional bactericidal method.

Continuous method. This was the first technique employed for demonstrating the anti-*M. leprae* activity of a drug in mice (Shepard and Chang, 1964). After inoculation with 5×10^3 or 1×10^4 *M. leprae* per footpad, the mice are divided among untreated control and treated group(s) with 10–20 mice in each. In the treated group, the drug is administered from the day of inoculation until the mice are sacrificed. If the compound to be screened is well absorbed after oral administration, it is administered either by gavage or by being incorporated into mouse chow (see below). If nothing is known of the activity of the drug against *M. leprae*, it is administered in the largest dosage tolerated by the mice.

Multiplication of *M. leprae* in the untreated control group is monitored by interval harvests. When evidence of multiplication is unequivocal, i.e. the mean number of AFB has reached 5×10^5 per footpad, harvests are performed from the footpads of both control and treated mice. An active treatment is one that inhibits multiplication of *M. leprae*. If, on two successive occasions approximately 1 month apart, the number of AFB harvested from treated mice falls below the range of the results of four simultaneous harvests from control mice, the treatment can be stated to have been active, with a probability of 0.04. Having established that a drug is active, one may, by the same method, establish the minimal effective dosage (MED)—the smallest dosage of the drug that produces an anti-*M. leprae* effect. The minimal inhibitory concentration (MIC) of the drug is its concentration in the serum or plasma of mice administered the drug in the MED.

This continuous method is the most sensitive method in detecting activity of treatment against *M. leprae*. However, it cannot distinguish between merely bacteriostatic and bactericidal activity. Many drugs exhibit bacteriostatic activity against *M. leprae*, whereas only a few are bactericidal (Shepard *et al.*, 1971); yet, only bactericidal drugs may be considered as potential components of multidrug regimens to be employed in treating patients and in leprosy control. Another disadvantage of the method is that it requires a rather large quantity, usually more than 10 g, of the drug. Unless the compounds are being produced commercially or are undergoing active development, the pharmaceutical firms are often unable to provide such a quantity.

Kinetic method. The method is identical to the continuous method, except that the compound to be screened is administered for only a limited period, usually beginning 60 days after inoculation, at which time the organisms are in early or mid-logarithmic phase, and continuing for 60–90 days (Shepard, 1967, 1969). The activity of the drug is assessed in terms of the "growth delay", determined by comparing the growth curves of the treated and untreated control mice. A purely bacteriostatic drug is one that inhibits multiplication of *M. leprae* only as long as the drug is administered; the growth curve of *M. leprae* in mice treated by such a drug is parallel to, but lags behind that in the control mice by a period of time no greater than that during which the drug was administered. The absence of bactericidal activity is reliably demonstrated by this approach. The failure of bacterial multiplication to resume immediately following cessation of drug administration suggests that *M. leprae* are killed during treatment, or that prolonged bacteriostasis (also termed "bacteriopause") has occurred. Prolonged bacteriostasis may result from persistence of the drug in the tissues or within the organisms, or it may reflect the recovery of organisms that have been reversibly damaged. Thus, the kinetic method can distinguish between purely bacteriostatic and so-called "bacteriocidal-type" activity (Shepard *et al.*, 1971), but it cannot distinguish between purely bacteriopausal and bacteriocidal activity, unless there is total failure of resumption of bacterial multiplication after cessation of drug administration.

Proportional bactericidal method. This method may clearly identify the bactericidal activity of the treatment. Groups of mice are inoculated with serially 10-fold diluted suspensions of *M. leprae*, ranging from 5×10^{-1} (0.5) or 1×10^{0} (1) organism to 5×10^{3} or 1×10^{4} organisms per footpad. Control mice are left untreated, whereas other groups of mice are treated for a period of time that varies, depending on the drug, from 1–60 days. After treatment, the mice are then held for 12 months, a period of time theoretically sufficient to permit a single surviving organism to multiply to a readily detectable level. Harvests of *M. leprae* are then performed from individual footpads, usually 10 per dilution of inoculum for each drug or drug concentration. The *M. leprae* are considered to have multiplied (or, at least one organism to have survived the treatment) in those footpads found to contain 10^5 AFB or more. The proportion of viable *M. leprae* remaining after the treatment and the significance of differences between the groups are calculated in terms of the median infectious dose by the method of Spearmen and Kärber (Shepard, 1982). If the largest inoculum was 5×10^3 AFB per footpad, a proportion of viable *M. leprae* as small as 0.00006 may be measured. One may then calculate the proportion of viable *M. leprae* killed by the treatment by comparing the proportions of viables in treated and control mice. Application of this method is demonstrated in an experiment in which the bactericidal effects against *M. leprae* of single doses of several combinations of drugs were measured (Table 38.1). The

Table 38.1 Testing the bactericidal effect of various treatments against *M. leprae* in mice by proportional bactericidal method

Group no. and regimen (mg/kg)	*Proportion of footpads showing multiplication* of* M. leprae *with inoculum* 5×10^3	5×10^2	5×10^1	5×10^0	5×10^{-1}	*Proportion of viable* M. leprae	*% of* M. leprae *killed*
1. Untreated control	10/10	10/10	10/10	3/10	2/10	13.77	–
2. RMP (10)†	10/10	9/10	2/10	0/10	–	0.55	96.0
3. One month MDT‡	2/10	0/10	0/10	0/10	–	0.007	99.95
4. OFLO (150) + MINO (25)†	10/10	10/10	10/10	1/10	–	5.48§ ~ 6.90**	49.9†† ~ 60.2‡‡
5. OFLO (300) + MINO (50)†	10/10	10/10	3/10	2/10	–	1.38§ ~ 2.18**	84.2†† ~ 90.0‡‡
6. CLARI (100) + OFLO (150) + MINO (25)†	10/10	10/10	9/10	0/10	–	3.46	74.9
7. RMP (10) + OFLO (150) + MINO (25)†	10/10	9/10	0/10	1/10	–	0.44	96.8
8. RMP (10) + OFLO (300) + MINO (50)†	10/10	6/10	2/10	0/10	–	0.28	98.0
9. RMP (10) + CLARI (100) + OFLO (150) + MINO (25)†	10/10	3/10	0/10	0/10	–	0.09	99.4

* ≥ 10^5 *M. leprae* harvested per footpad.
‡ A single dose of 10 mg RMP/kg plus 0.01% DDS and 0.005% CLO in the mouse diet for 30 days.
§ Minimum estimated value, assuming that no multiplication of *M. leprae* would have occurred in mice inoculated with 5×10^{-1}.
** Maximum estimated value, assuming that multiplication of *M. leprae* would have occurred in the same proportion of footpads in mice inoculated with 5×10^{-1} as those in mice inoculated with 5×10^0.
†† Minimum killing rate, calculated from the maximum estimated proportion of viable organisms.
‡‡ Maximum killing rate, calculated from the minimum estimated proportion of viable organisms.

activity of a single dose of the combination OFLO–MINO was found to be dosage-related: the combination of 300 mg OFLO plus 50 mg MINO per kilogram body weight was bactericidal, whereas the combination of the same drugs in half dosage was not.

Experimental chemotherapy of nude mice with established M. leprae *infection*

The recent discovery of three classes of drug—ofloxacin (OFLO, a fluoroquinolone), clarithromycin (CLARI, a macrolide) and minocycline (MINO, a tetracycline derivative)—are bactericidal against *M. leprae*, suggested the possibility of designing new generation of MDT regimens, which may be more effective and of shorter duration than the standard regimen (Ji *et al.*, 1996a,b). However, the many possible combinations of the new drugs with the components of the standard regimen—rifampicin (RMP), dapsone (DDS) and clofazimine (CLO)—cannot all the tested in clinical trials; there are simply not enough suitable patients or qualified institutes where such trials may be carried out. In addition, ethical considerations prohibit the inclusion in clinical trials of some important controls, such as RMP monotherapy or the combination of DDS–CLO, making it difficult to assess the role of each component in the tested combinations. Moreover, the long duration of the clinical trials would delay by many years the application of truly useful new regimens to the control of leprosy. Experimental chemotherapy of nude mice with established *M. leprae* infection provides a means of testing all possible combinations and identifying a few candidate regimens that could then be studied more efficiently in the clinical trials. A study carried out in nude mice illustrates this application (Ji *et al.*, 1996a).

Each hind footpad of 180 female Swiss nude mice was inoculated with 7.5×10^5 *M. leprae*; 12.5 months later, swollen footpads were observed in 155 (93.4%) of the 166 surviving mice, and the mean number of AFB per swollen footpad was found to be 4×10^7. The 155 mice with swollen footpads were randomly allocated to an untreated group and 12 treated groups, and treatment was begun. By 24 weeks after the beginning of treatment, the mean number of *M. leprae* per footpad had increased to 3×10^8 among the control mice, whereas it had remained virtually unchanged among the treated mice; the solid ratio of the bacilli, i.e. the ratio of bacilli uniformly, brightly and intensely stained by carbol–fuchsin in acid-fast staining, was declining in all treated groups, but did not differ significantly among them. These results indicated that, as often seen in clinical trials, the assessment of the therapeutic effect cannot rely upon the determination of the bacterial population size, and the solid ratio was not sensitive enough to distinguish the therapeutic effect of weak regimens from that of potent regimens in nude mice. The only reliable method that can measure precisely the bactericidal effect of the treatment is to determine the proportion of viable *M. leprae* in the bacterial population before and after treatment.

As is done in clinical trials (Grosset *et al.*, 1990; Ji *et al.*, 1993, 1994, 1996c), the proportion of viable *M. leprae* in each group of mice was measured at regular intervals during the experiment. AFB were harvested from four footpads (two nude mice) of each group, suspended in HBSS and counted, after which the suspensions derived from the four footpads were pooled and the numbers of AFB were counted again. The pooled suspension was then serially 10-fold diluted with HBSS for subinoculation; the AFB harvested from untreated mice and from those that had received only a single dose of any treatment were subinoculated only into normal mice, whereas the AFB harvested from nude mice that had been administered more than a single dose of treatment with RMP-containing regimens were subinoculated into both normal and nude mice. The maximal inoculum for normal mice was 10^4 AFB per footpad, and the inocula for nude mice were 10^5 AFB per footpad or the maximal available inoculum, which refers to the inoculation of 0.03 ml of undiluted suspension pooled from four footpads harvested simultaneously from the same group, and they always contained more than 10^6 AFB per 0.03 ml. In order to obtain results from 10 inoculated footpads for each of the serially 10-fold diluted suspensions, each dilution was subinoculated into both hind footpads of six normal or 10 nude mice. Harvests of *M. leprae* from the subinoculated footpads were carried out 12 months later. The organisms were considered to have multiplied (i.e. the inoculum contained viable organism) if 10^5 AFB per footpad or more were harvested from normal mice or, among the subinoculated nude mice, a twofold or greater increase over the number of organisms inoculated. The proportion of viable organisms in the suspensions and the significance of their differences were also calculated by the method of Spearman and Kärber (Shepard, 1982). If only normal mice are subinoculated with the maximal inoculum of 10^4 AFB per footpad, a proportion of viable *M. leprae* as small as 0.00003 may be measured; if both normal and nude mice are subinoculated with the maximal inoculum of 10^6 AFB per footpad, the smallest proportion of viable organisms that can be measured is 0.0000003, smaller by two orders of magnitude than the minimum proportion that can be measured by subinoculation of normal mice only. Typical results of such measurements are shown in Table 38.2.

In *M. leprae*-infected nude mice, both the mean numbers of organisms per footpad and the mean proportion of viable *M. leprae* are considerably larger than those usually found in human biopsy specimens; consequently, the maximum available inocula prepared from infected nude mice are at least 10 times larger than in the suspensions prepared from human biopsies, with less variation among the individual footpads than is usual among human biopsy specimens. As a result, it is possible to measure the bactericidal activity of the treatment in nude mice with greater sensitivity and greater precision than in clinical trials.

Table 38.2 Example of the results of determining the proportion of viable *M. leprae* in experimental chemotherapy in nude mice

Treatment (mg/kg body weight/dose)	No. footpads showing multiplication* of M. leprae/footpads harvested, at the inoculum of: Nude mice		Normal mice					Proportion of viable M. leprae
	Maximum†	10^5	10^4	10^3	10^2	10^1	10^0	
Untreated control			9/10	9/10	7/10	6/10	4/10	0.069
RMP (10), single dose			6/10	5/10	4/10	3/10	1/10	0.017
0.01% DDS & 0.005% CLO in mouse diet, 4 weeks			10/10	5/10	1/10	0/10	0/10	0.00087
WHO/MDT for 4 weeks			0/10	0/10	0/10	0/10	0/10	<0.00003
RMP (10) + CLARI (50) + MINO (25) daily, three doses	10/10	10/10	1/10	0/10				0.00003
RMP (10) + CLARI (50) + MINO (25) + OFLO (50) daily, three doses	10/10	9/10	9/10	2/10				0.00022
RMP (10) + CLARI (100) + MINO (50) once every 4 weeks, three doses	0/10	0/10	0/10	0/10				<0.0000003

* Multiplication was defined as harvests of ≥ 10^5 AFB per footpad for inocula ≤ 10^4 and a twofold or greater increase for inocula ≥ 10^5 AFB per footpad.
† Maximum available inoculum in 0.03 ml.

Clinical applications

Drug susceptibility testing

Measurement of the susceptibility to drugs of strains of *M. leprae* recovered from leprosy patients is one of the most important applications of the mouse footpad technique. At present, susceptibility of the organisms to the three components of the multidrug therapy (MDT) regimen — RMP, DDS and CLO — is of interest.

An active-appearing lesion is biopsied, and *M. leprae* are recovered from the biopsy specimen by the method already described. Normal mice are inoculated with 5×10^3 to 1×10^4 AFB per footpad and divided among a number of groups of 10–20 mice each. One group is held without treatment, and the mice of the remaining groups are treated, usually beginning immediately after inoculation and continued until sacrifice.

Because they are well absorbed from the gastrointestinal tract, all three drugs may be administered *per os*, either incorporated into the mouse diet or by gavage; the former method is generally preferred because of its greater convenience. DDS is administered in one of three concentrations — 0.0001, 0.001 and 0.01 g per 100 g diet — whereas CLO is usually administered in one of the two concentrations — 0.001 and 0.01 g per 100 g diet; some "wild" strains appear to be resistant to 0.0001 g CLO per 100 g diet.

A common source of error in testing drug susceptibility of *M. leprae* is inadequate mixing of drugs in the diet. Because the drug–diet mixture must be as uniform as possible, it is best to add the drug in solution and mix it in the diet by means of a liquid–solid blender. If only a solid–solid blender is available, a weighed portion of the powdered drug should first be diluted by a small quantity of some inert solid, such as lactose or the powdered diet, and one should then continue to dilute the drug gradually by adding weighed amounts of the powdered diet to the blender. If no blender is available, exhaustive mixing by hand is required. When preparing a series of drug-containing diets, one should always begin with the diet containing the smallest concentration of the drug, especially in the case of a drug such as DDS, the MIC of which is of the order of 3 ng/ml for susceptible strains of *M. leprae*. In this way, one may prepare diets of progressively greater concentration without stopping to wash the blender between diets.

Although RMP can be incorporated into the mouse diet, it appears to be unstable in this situation, and its potency may diminish with time. Therefore, in testing the susceptibility of *M. leprae* to RMP, freshly prepared suspensions of RMP are administered by an esophageal cannula (gavage) in most laboratories (Grosset *et al.*, 1989). Because RMP displays powerful bactericidal activity against *M. leprae*, in measuring drug susceptibility, RMP is administered once weekly in a dosage of 10 mg/1 kg body weight (Grosset *et al.*, 1989).

Approximately 6 months after inoculation, harvests of *M. leprae* are performed individually from the inoculated footpads of two to four untreated mice, and repeated at intervals of 2 months until the organisms are found to have multiplied to a mean of at least 5×10^5 AFB per footpad, at which time *M. leprae* are harvested immediately from the footpads of all treated groups. Harvests should also be performed from treated mice if, by 12 months after inoculation, the organisms have multiplied in the control mice, but to a mean of less than 5×10^5 AFB per footpad.

The criterion of multiplication of *M. leprae* has been carefully defined. As mentioned earlier, the lower limit of detectability of *M. leprae* in the mouse footpad, based on the observation of only a single AFB, is of the order of 1×10^4; in such a situation, one cannot be certain that the organisms observed are not those inoculated. To be confident that the organisms have, in fact, multiplied, an increase to at least 10^5 AFB per footpad is required. Employing this criterion of multiplication, if the organisms have multiplied only in

Table 38.3 DDS susceptibility tests of six strains of *M. leprae*

	No. footpads showing multiplication/no. harvested*				
	Concentration of DDS (g/100 g diet)				
Strain no.	*0*	*0.0001*	*0.001*	*0.01*	*Interpretation*
1	10/10	0/10	0/10	0/10	Susceptible
2	4/9	0/10	0/10	0/10	Inconclusive
3	8/8	7/9	0/10	0/10	Low resistance
4	10/10	9/9	7/7	0/10	Intermediate resistance
5	9/9	10/10	4/10	0/9	Intermediate resistance
6	8/8	10/10	10/10	10/10	Full resistance

* $\geq 10^5$ *M. leprae* harvested per footpad.

untreated mice but in none of the treated mice, the strain is considered susceptible; whereas the strain is considered resistant if the organisms are found to have multiplied in even one treated mouse. Finally, if the organisms have not multiplied in treated mice, but have multiplied in so small a proportion of the untreated mice that the proportion is not significantly different from zero by Fisher's exact probability caculation, the susceptibility test of the strain is inconclusive. In this event, one may wish to subinoculate the organisms into new groups of mice and to repeat the susceptibility test.

The degree of resistance to a drug is determined by the diet containing the largest concentration of drug that permits multiplication of the organisms. For DDS, the degree of resistance is defined as low, intermediate or high, depending on the ability of the organisms to multiply in mice given DDS in a concentration of, respectively, 0.0001, 0.001 or 0.01 g/100 g diet. Six examples of tests of the susceptibility of *M. leprae* to DDS are presented in Table 38.3.

Monitoring the efficacy of antimicrobial treatment in clinical trials

The most sensitive means of assessing therapeutic efficacy of an antimicrobial drug or regimen is to measure, by means of the mouse footpad technique, the rate at which the patients' *M. leprae* are killed during treatment. Before and at intervals during treatment; serial skin biopsies are obtained, preferably from the same lesion, and *M. leprae* are recovered from the biopsy specimens. Mice, usually 20 mice per specimen, are inoculated with 5×10^3 to 1×10^4 AFB per footpad (Shepard, 1981), and harvests of *M. leprae* are performed at intervals thereafter. The response to the treatment is often reported in terms of the duration of treatment, the time of the last specimen in which viable organisms were detected and that of the first specimen in which viable *M. leprae* could not be detected (Shepard, 1981). However, one may obtain more quantitative information, in terms of the decrease of the proportion of viable *M. leprae*, and also recognize bactericidal effect earlier by inoculating additional mice with serially diluted suspension of the organisms (Ji *et al.*, 1996 c).

The results of a clinical trial in which bactericidal effects of a treatment were measured in terms of the proportion of viable *M. leprae* before and after treatment are presented in Table 38.4. Although the treatment exhibited bactericidal activity in all patients, this activity was demonstrated in two patients (cases 5 and 38) only by measurement of the proportion of viable organisms. Had only a single inoculum of 5×10^3 AFB per footpad been employed, the bactericidal effect of treatment in these two patients would not have been recognized, because the proportion of footpads showing multiplication of *M. leprae* among footpads inoculated after treatment with 5×10^3 AFB per footpad did not differ significantly from those before treatment. These findings indicate that the titration of the proportion of viable *M. leprae* in the bacterial population is a sensitive and precise method for monitoring the therapeutic effect of the treatment.

Because the number of *M. leprae* that may be inoculated into the footpads of normal mice is limited to 5×10^3 to 1×10^4, inoculation of normal mice alone permits measurement of the bactericidal activity of only >99% to >99.9% of the viable organisms initially present. On the other hand, as many as 100 times more organisms may be inoculated into each footpad of nude mice, thus increases the sensitivity of the measurement by one or two orders of magnitude, permitting the demonstration of killing of 99.99–99.999% of the viable *M. leprae* initially present (Ji *et al.*, 1994). However, as compared with the population of viable *M. leprae* in untreated lepromatous leprosy patients, which may be as large as 10^{10}–10^{11} organisms, the employment of nude mice yields only a modest increase of sensitivity, and it is also very expensive. As a result, it is difficult to justify the use of nude mice to monitor therapeutic efficacy in a clinical trial, unless the additional information to be obtained is crucial. For example, to identify a drug or drug combination that can eliminate spontaneously occurring RMP-resistant *M. leprae* during the course of treatment, because the number of such resistant mutants present in a previously untreated lepromatous patient is estimated to be no greater than 10^4, it is important to measure the killing of more than 99.99% of the viable organisms initially present, in which case inoculation of nude mice may well be justified.

Table 38.4 Examples of results of determining the proportion of viable *M. leprae* in mouse footpads before and after treatment (Ji *et al.*, 1996c)

Case no.‡	*Date of biopsy*	*No. footpads showing multiplication* of M. leprae/no. footpads harvested at inoculum*				*Proportion of viable M. leprae*§	*% of viable bacilli killed by treatment***
		5×10^3	5×10^2	5×10^1	5×10^0		
7	D0	10/10	5/10	2/10	1/10	0.0028	
	D31	0/10	0/10	0/10	0/10	< 0.00006	> 97.9
6	D0	10/10	10/10	6/10	4/10	0.0435	
	D31	0/10	0/10	0/10	0/10	< 0.00006	> 99.9
38	D0	10/10	10/10	10/10	8/10	0.2745	
	D31	10/10	7/10	3/10	1/10	0.0055	98.0
5	D0	9/10	9/10	4/10	3/10	0.0138	
	D31	7/10	2/10	0/10	0/10	0.0004	97.5
21	D0	10/10	7/10	4/10	0/10	0.0055	
	D31	0/10	0/10	0/10	0/10	< 0.00006	> 98.9

* $\geq 10^5$ *M. leprae* harvested per footpad.
† Case nos 7, 6, 38, 5 and 21 were treated, respectively, with 1 month of the standard multidrug regimen for multibacillary leprosy; a single dose of 600 mg RMP; 30 days of DDS and CLO components in the standard multidrug regimen for multibacillary leprosy; a single dose of 2000 mg of CLARI plus 200 mg of MINO; and a single dose of 2000 mg of CLARI plus 200 mg of MINO plus 800 mg of OFLO.
‡ D0 and D31 refer, respectively, to before and after treatment.
§ Derived from the equation 0.69/50% infectious dose (Shepard, 1982).
** Comparison of the proportion of viable *M. leprae* before and after treatment.

It is essential to point out that failure to demonstrate viable *M. leprae* by inoculation of mice—even nude mice—at the end of treatment should not be taken as evidence that all viable *M. leprae* within the hosts have been killed. Many viable organisms may still survive after 99.999% have been killed. The only available means to evaluate the long-term efficacy of treatment is to measure the relapse rate after stopping treatment. One of the criteria in confirming the diagnosis of relapse is the demonstration of viable *M. leprae* by mouse footpad inoculation (Ji *et al.*, 1997).

References

Binford, C. H. (1968). The transmission of *M. leprae* to animals to find an experimental model. *Int. J. Lepr.*, **36**, 599.

Brand, P. W. (1959). Temperature variation and leprosy deformity. *Int. J. Lepr.*, **27**, 1–7.

Chehl, S. K., Shannon, E. J., Job, C. K., Hastings, R. C. (1983). The growth of *Mycobacterium leprae* in nude mice. *Lepr. Rev.*, **54**, 283–304.

Colston, M. J., Hilson, G. R. F. (1976). Growth of *Mycobacterium leprae* and *M. marinum* in congenitally athymic (nude) mice. *Nature (London)*, **262**, 399–401.

Colston, M. J., Fieldsteel, A. H., Lancaster, R. D., Dawson, P. J. (1979). The athymic rat: immunological status and infection with *Mycobacterium leprae*. *Int. J. Lepr.*, **47**, 679.

Fenner, F. (1956). The pathogenic behaviour of *Mycobacterium ulcerans* and *Mycobacterium balnei* in the mouse and the developing chick embryo. *Am. Rev. Tuberc.*, **73**, 650–673.

Fieldsteel, A. H., Levy, L. (1976). Neonatally thymectomized Lewis rats infected with *Mycobacterium leprae*: responses to primary infection, secondary challenge and large inocula. *Infect. Immun.*, **14**, 736–741.

Gelber, R. H., Levy, L. (1987). Detection of persisting *Mycobacterium leprae* by inoculation of the neonatally thymectomized rat. *Int. J. Lepr.*, **55**, 872–878.

Grosset, J. H., Guelpa-Lauras, C. C., Bobin, P. *et al.* (1989). Study of 39 documented relapses of multibacillary leprosy after treatment with rifampin. *Int. J. Lepr.*, **57**, 607–614.

Grosset, J. H., Ji, B., Guelpa-Lauras, C. C., Perani, E. G., N'Deli, L. N. (1990). Clinical trial of pefloxacin and ofloxacin in the treatment of lepromatous leprosy. *Int. J. Lepr.*, **58**, 281–295.

Hilson, G. R. F. (1965). Observations on the inoculation of *M. leprae* in the footpad of the white rat. *Int. J. Lepr*, **33**, 662–665.

Ji, B., Jamet, P., Perani, E. G., Bobin, P., Grosset, J. H. (1993). Powerful bactericidal activities of clarithromycin and minocycline against *Mycobacterium leprae* in lepromatous leprosy. *Antimicrob. Agents Chemother.*, **168**, 188–190.

Ji, B., Perani, E. G., Petinon, C., N'Deli, L., Grosset, J. H. (1994). Clinical trial of ofloxacin alone and in combination with dapsone plus clofazimine for treatment of lepromatous leprosy. *Antimicrob. Agents Chemother.*, **38**, 663–667.

Ji, B., Perani, E. G., Petinon, C., Grosset, J. H. (1996a). Bactericidal activities of combinations of new drugs against *Mycobacterium leprae* in nude mice. *Antimicrob. Agents Chemother.*, **40**, 393–399.

Ji, B., Levy, L., Grosset, J. H. (1996b). Chemotherapy of leprosy. Progress since the Orlando Congress and prospect for the future. *Int. J. Lepr.*, **64**, 580–588.

Ji, B., Jamet, P., Perani, E. G. *et al.* (1996c). Bactericidal activity of single dose of clarithromycin plus minocycline, with or without ofloxacin, against *Mycobacterium leprae* in patients. *Antimicrob. Agents Chemother.*, **40**, 2137–2141.

Ji, B., Jamet, P., Sow, S., Perani, E. G., Traore, I., Grosset, J. H. (1997). High relapse rate among lepromatous leprosy patients treated with rifampin plus ofloxacin daily for 4 weeks. *Antimicrob. Agents Chemother.*, **41**, 1953–1956.

Kirchheimer, W. F., Storrs, E. E. (1971). Attempts to establish the armadillo (*Dasypus novemcinctus* Linn.) as a model for the study of leprosy. I. Report of lepromatoid leprosy in an experimentally infected armadillo. *Int. J. Lepr.*, **39**, 693–702.

Kohsaka, K., Mori, T., Ito, T. (1976). Lepromatoid lesion developed in the nude mice inoculated with *Mycobacterium leprae*. *Lepro*, **45**, 177–187.

Levy, L. (1975). Superinfection in mice previously infected with *Mycobacterium leprae*. *Infect. Immun.*, **11**, 1094–1099.

Levy, L. (1976). Studies of the mouse foot pad technique for cultivation of *Mycobacterium leprae*. 3. Doubling time during logarithmic multiplication. *Lepr. Rev.*, **47**, 103–106.

McDermott-Lancaster, R. D., Ito, T., Kohsaka, K., Guelpa-Lauras, C. C., Grosset, J. H. (1987). Multiplication of *Mycobacterium leprae* in the nude mouse, and some applications of nude mice to experimental leprosy. *Int. J. Lepr.*, **55**, 889–895.

Rees, R. J. W. (1966). Enhanced susceptibility of thymectomized and irradiated mice to infection with *Mycobacterium leprae*. *Nature (London)*, **211**, 657–658.

Rees, R. J. W., Waters, M. F. R., Weddell, A. S. M., Palmer, E. (1967). Experimental lepromatous leprosy. *Nature (London)*, **215**, 599–602.

Rungruang, S., Ramasoota, T., Sampattavanich, S. (1983). Study of the use of nude mice in the cultivation of *M. leprae* in a normal, non-specific pathogen free room at a temperature of 30–35 °C without air conditioning. *Lepr. Rev.*, **54**, 305–308.

Shepard, C. C. (1960a). Acid fast bacilli in nasal excretions in leprosy, and results of inoculation of mice. *Am. J. Hyg.*, **71**, 147–157.

Shepard, C. C. (1960b). The experimental disease that follows the injection of human leprosy bacilli into foot pads of mice. *J. Exp. Med.*, **112**, 445–454.

Shepard, C. C. (1967). A kinetic method for the study of activity of drugs against *Mycobacterium leprae* in mice. *Int. J. Lepr.*, **35**, 429–435.

Shepard, C. C. (1969). Further experience with the kinetic method for the study of drugs against *Mycobacterium leprae* in mice. Activities of DDS, DFD, ethionamide, capreomycin and PAM 1392. *Int. J. Lepr.*, **37**, 389–397.

Shepard, C. C. (1981). A brief review of experiences with short-term clinical trials monitored by mouse-foot-pad inoculation. *Lepr. Rev.*, **52**, 299–308.

Shepard, C. C. (1985). Experimental leprosy. In *Leprosy* (ed. Hastings, R. C., pp. 269–286. Churchill Livingstone, Edinburgh.

Shepard, C. C., Chang, Y. T. (1964). Effect of several anti-leprosy drugs on multiplication of human leprosy bacillus in foot pads of mice. *Proc. Soc. Exp. Biol. Med.*, **109**, 636–638.

Shepard, C. C., Habas, J. A. (1967). Relation of infection to tissue temperature in mice infected with *Mycobacterium marinum* and *Mycobacterium leprae*. *J. Bacteriol.*, **93**, 790–796.

Shepard, C. C., McRae, D. H. (1968). A method for counting acid-fast bacteria. *Int. J. Lepr.*, **36**, 78–82.

Shepard, C. C., van Landingham, R. M., Walker, L. L. (1971). Recent studies of antileprosy drugs. *Int. J. Lepr.*, **39**, 340–349.

Shepard, C. C., van Landingham, R., Walker, L. L. (1976). Immunity to *Mycobacterium leprae* infections in mice stimulated by *M. leprae*, BCG and graft-versus-host reactions. *Infect. Immun.*, **14**, 919–928.

Subcommittee of THELEP (1987). Persisting *Mycobacterium leprae* among THELEP trial patients in Bamako and Chingleput. *Lepr. Rev.*, **58**, 325–337.

Waters, M. F. R., Niven, J. S. F. (1966). Experimental infection of the ear and foot pad of the golden hamster with *Mycobacterium leprae*. *Br. J. Exp. Pathol.*, **47**, 86–92.

Welch, T. M., Gelber, R. H., Murray, L. P., Ng, H., O'Neill, S. M., Levy, L. (1980). Viability of *Mycobacterium leprae* after multiplication in mice. *Infect. Immun.*, **30**, 325–328.

Chapter 39

Hamster Model of Lyme Arthritis

R. F. Schell and S. M. Callister

Background of human infection

Lyme borreliosis, caused by the spirochete *Borrelia burgdorferi sensu lato* (Barbour *et al.*, 1983a; Benach *et al.*, 1983; Johnson *et al.*, 1984) is the most frequently reported tick-borne illness in the USA (Steere, 1989), with over 13 000 cases reported in 1996 (Centers for Disease Control and Prevention 1997). Classically, Lyme borreliosis is characterized by an expanding skin lesion (erythema migrans) accompanied by constitutional symptoms that include fatigue, headache, mild stiff neck, bone and muscle aches, and fever (Steere *et al.*, 1983a, 1987). Other clinical manifestations include mild heart conduction system blockage, neurologic abnormalities, and polyarthropathies (Pachner and Steere, 1985; McAlister *et al.*, 1989; Preac-Mursic *et al.*, 1989). Arthritis is frequently the first recognized complication of Lyme borreliosis in the USA (Steere *et al.*, 1987), with intermittent episodes developing in approximately 60% of afflicted individuals (Steere *et al.*, 1987). In severe cases, chronic inflammatory Lyme arthritis can lead to cartilage and bone erosion (Steere *et al.*, 1979, 1987).

The diagnosis of Lyme borreliosis is based on characteristic clinical findings and exposure to ticks of the *Ixodes ricinus* complex (Barbour *et al.*, 1983b) in an endemic focus. It is well recognized that tests that detect antibody responses against *B. burgdorferi* serve only as an adjunct to other methods, especially the history and physical findings of the patient. Unfortunately, patients often present with indistinct clinical pictures and Lyme borreliosis endemic foci are common throughout the USA.

In addition, Lyme borreliosis symptoms may closely resemble other unrelated illnesses such as influenza, ehrlichiosis, multiple sclerosis, aseptic meningitis, and rheumatoid arthritis. For these reasons, clinicians have and will continue to rely heavily on the ability of the laboratory to detect accurately evidence of infection with *B. burgdorferi*. However, the sensitivity and specificity of the testing procedures need to be greatly improved (Bakken *et al.*, 1992, 1997). Accurate test results would greatly assist clinicians with the clinical diagnosis of Lyme borreliosis and prevent its over-diagnosis (Steere *et al.*, 1994). The development of the borreliacidal antibody test that detects antibodies in Lyme borreliosis sera that are lethal to *B. burgdorferi in vitro* (Callister *et al.*, 1996; Schell *et al.*, 1997) should be the most helpful of the current serodiagnostic tests.

Background of model

Animal models of Lyme borreliosis are extremely important for elucidating the mechanisms of pathogenesis and development of vaccines for human and animal usage (Schmitz *et al.*, 1991). These models increase our knowledge and understanding of the basic infectious process and assist in developing strategies to prevent infection and disease in humans. Unfortunately, no single animal model mimics all the pathological and clinical manifestations associated with Lyme borreliosis in humans. Skin lesions that resemble human erythema migrans develop in monkeys, rabbits and guinea pigs (Krinsky *et al.*, 1982; Burgdorfer, 1984; Kornblatt *et al.*, 1984; Philipp and Johnson, 1994). Moderate or chronic progressive synovitis develops in mice and young dogs (Barthold *et al.*, 1990; Appel *et al.*, 1993), while severe chronic arthritis can be induced in severe combined immunodeficiency mice and irradiated hamsters (Schmitz *et al.*, 1988; Schaible *et al.*, 1989) after infection with *B. burgdorferi*. Carditis, nephritis, and hepatitis can also be detected in some of these animal models infected with the Lyme spirochete (Barthold *et al.*, 1993; Philipp and Johnson, 1994; Munson *et al.*, 1996). Detection of neuroborreliosis in animal models has not been reported.

Animal species

The hamster, especially inbred LSH hamsters, is an excellent model to study the immune responses to infection or vaccination with *B. burgdorferi*. When adult inbred LSH hamsters are injected in the hind paws with 10^6 *B. burgdorferi sensu lato*, clinical manifestations of Lyme arthritis are induced. Inflammation or swelling of the hind paws can be detected 7 days after infection, peaks on day 10, and gradually decreases. At week 1, the tibiotarsal, intertarsal, and interphalangeal joints show evidence of acute

Handbook of Animal Models of Infection
ISBN 0-12-775390-7

inflammation. The synovial lining is hypertrophic and hyperplastic, and areas of ulceration are easily detected. Adherent fibrin protrudes into the joint spaces and is associated with inflammatory cells, especially neutrophils. The neutrophils also penetrate the subsynovial connective tissue and periarticular structures, including ligaments, tendons, tendon sheaths, fibrous capsule, and periosteum. Spirochetes (> 20 per high-power field) are readily observed in the subsynovial tissues. By week 3 after infection, the number of spirochetes has greatly decreased and the inflammatory response is resolving. A chronic synovitis characterized by hypertrophic villi, focal erosions of articular cartilage, and subsynovial mononuclear infiltrate persists for approximately 1 year.

Severe destructive Lyme arthritis can also be elicited in vaccinated hamsters after challenge with different isolates of *B. burgdorferi* (Lim *et al.*, 1994). Hamsters are vaccinated with a whole-cell preparation of formalin-inactivated *B. burgdorferi sensu stricto* isolate C-1-11 in adjuvant. A severe destructive arthritis is readily evoked in vaccinated hamsters challenged with the homologous *B. burgdorferi sensu stricto* isolate C-1-11 before high levels of protective borreliacidal antibody develop. Once high levels of C-1-11 borreliacidal antibody develop, hamsters are protected from homologous challenge and development of arthritis. Vaccinated hamsters, however, still develop severe destructive arthritis when challenged with other isolates of the three genomic groups of *B. burgdorferi sensu lato* (*B. burgdorferi sensu stricto* isolate 297, *Borrelia garinii* isolate LV4, and *Borrelia afzelii* isolate BV1) despite high levels of C-1-11-specific borreliacidal antibody. Vaccines that contain whole spirochetes in adjuvant induce destructive arthritis, but this effect is not dependent on the isolate of *B. burgdorferi sensu lato* or the type of adjuvant. The development of severe destructive arthritis is dependent on $CD4^+$ T lymphocytes. The hamster's propensity to develop arthritis, especially severe destructive arthritis, is an effective way to evaluate vaccines for adverse effects. In addition, the hamster mimics the vaccine response of humans to a field-trialed recombinant subunit vaccine (Padilla *et al.*, 1996).

Storage and preparation of inoculum

B. burgdorferi isolates are readily available from investigators. *B. burgdorferi* isolate 297 was obtained from Russell C. Johnson (University of Minnesota, Minneapolis, MN). The strain was originally isolated from human spinal fluid (Steere *et al.*, 1983b) and has been maintained by passage in modified Barbour–Stoenner–Kelly (BSK) medium (Callister *et al.*, 1990). The hamster-passed spirochete is grown in BSK at 32°C for 5 days. The suspension of *B. burgdorferi* is then adjusted with fresh BSK to contain approximately 10^7 spirochetes/ml. Samples of 1 ml are dispensed in vials, which are sealed and stored in liquid nitrogen until used.

Infection process

Hamsters are mildly anesthetized with ether and challenged subcutaneously in each hind paw with 0.2 ml of BSK containing 10^6 viable organisms of *B. burgdorferi*. Quantitation of spirochetes is performed by using dark-field microscopy. Hamsters can also be easily infected intradermally or intravenously via tongue veins. We have infected hamsters with 10^2–10^8 *B. burgdorferi* organisms and recovered spirochetes from bladders, spleens, kidneys, and hearts cultured in BSK medium. Recovery rates varied (≤ 66%) for challenge inocula of between 10^2 and 10^4 spirochetes. When hamsters are challenged with 10^6 spirochetes, spirochetes are recovered from all the above tissues.

Key parameters to monitor infection

Assessment of arthritis

Swelling of the hind paws of vaccinated or non-vaccinated hamsters challenged with isolates of *B. burgdorferi sensu lato* is used to evaluate the inflammatory response. The hind paws are measured every day for 21 days with a plethysmograph (Buxco Electronics, Sharon, CT). Measurements are obtained by mildly anesthetizing hamsters with ether, carefully dipping a hind paw into a column of mercury up to the ankle, and measuring the amount of mercury displaced (in milliliters). The mean plethysmograph value for three hamsters (six hind paws) per group is used as an index of severity of swelling from arthritis. Mercury displacement is standardized with a volume calibrator.

Recovery of spirochetes from tissues

Approximately 21 days after challenge, hamsters are killed by CO_2 inhalation. The urinary bladder, spleen, kidney, and heart are removed aseptically, homogenized through a 5 ml syringe, and inoculated into 5 ml of BSK supplemented with rifampin (100 μg/ml; Sigma, St Louis, MO) and phosphomycin (100 μg/ml; Sigma). Cultures are incubated at 32°C and examined weekly for 2 weeks by dark-field microscopy for motile spirochetes. If spirochetes are not detected, 0.5 ml of the culture is inoculated into 4 ml of BSK supplemented with antibiotics and examined weekly for another 3 weeks. If no spirochetes are detected, the culture is considered negative.

Preparation of tissues for histology

The hind legs of all vaccinated or non-vaccinated hamsters are amputated 21 days after challenge at the mid-femur, fixed in 10% neutral buffered formalin, placed in decalcifying solution (Lerner Laboratories, Pittsburgh, PA) for 18 hours, and stored in 10% formalin prior to processing. The knees and hind paws are dissected longitudinally, embedded in paraffin, cut into 6 μm sections, placed on glass slides, and stained with hematoxylin and eosin. Hind legs are randomly selected from each group of hamsters for histopathological examination.

Hamster sera

Sera are obtained from hamsters with or without vaccination at various intervals after vaccination or challenge. Concomitantly, sera are obtained from non-infected normal hamsters. Hamsters are anesthetized with ether and bled by intracardiac puncture. The blood is allowed to clot and serum is separated by centrifugation at 500*g*, pooled, divided into 1 ml aliquots, dispensed into 1.5 ml screw-cap tubes (Sarstedt), and frozen at –20 °C until used.

Borreliacidal assay

Sera from vaccinated or non-vaccinated hamsters are heat-inactivated at 56 °C for 30 minutes, diluted 1 : 10 with fresh BSK, and filter-sterilized through a 0.2 μm pore-size filter apparatus (Acrodisc, Gelman Sciences, Ann Arbor, MI). Frozen suspensions of *B. burgdorferi sensu lato* are thawed, inoculated into fresh BSK, and incubated at 32 °C for 72 hours. Spirochetes are enumerated by dark-field microscopy and with a Petroff–Hausser counting chamber, and the suspensions are adjusted to contain 10^5 spirochetes/ml with BSK. Samples of the spirochetal suspensions, 100 μl each, are added to round-bottomed wells of a 96-well microtiter plate (Gibco Laboratories, Grand Island, NY). Subsequently, 100 μl of sera or twofold dilutions of sera from vaccinated or non-vaccinated hamsters and 20 μl of sterile guinea-pig complement (hemolytic titer, 200 CH_{50} units per ml; Sigma) are added to each well of the microtiter plate. The plate is shaken gently and incubated at 32 °C for 16 hours. All assays are performed in duplicate.

Flow cytometry data acquisition and analysis

After incubation of assay samples, 100 μl is removed and diluted 1 : 5 with PBS (pH 7.4) and 50 μl of acridine orange (5.4 nmol/l; Sigma) is added. Controls include samples containing normal serum with viable or heat-killed (56 °C for 30 minutes) spirochetes in BSK and complement. The samples are then analyzed with a FACScan flow cytometry (Becton Dickinson Immunocytometry Systems, Mountain View, CA) with FACScan LYSYS II software for data acquisition. Initially, viable and heat-killed spirochetes are detected and differentiated from BSK, serum, and complement particles by using forward scatter, side scatter, and acridine orange fluorescence. Live gating is performed only on profiles of spirochetes during data acquisition to exclude all BSK, serum, and complement particles. Data are acquired for 1 minute. Assay samples are then analyzed by histogram profiles of acridine orange fluorescence with FACScan LYSYS II software. Gates are established for viable and heat-killed spirochetes on the basis of their incorporation of acridine orange. Three parameters are evaluated: events per minute (number of labeled spirochetes), percentage shift in fluorescence (number of dead spirochetes), and mean channel fluorescence (intensity of fluorescently labeled spirochetes). Borreliacidal activity is determined by a decrease in events per minute and increases in percentage shift in fluorescence and mean channel fluorescence compared with values obtained with normal serum. The borreliacidal titer is the highest dilution of immune serum that kills spirochetes compared with normal serum.

Antibody reagents

Hybridoma cell line 14-4-4s (ATCC HB-32) secreting murine monoclonal antibodies (MAb) recognizes a surface cell marker on hamster B lymphocytes. Hybridoma 14-4-4s is grown in Dulbecco's modified Eagle's medium at 37 °C in a humidified atmosphere of 7.5% CO_2. After 7 days, the culture supernatant is collected after centrifugation at 500 *g* for 10 minutes at 4 °C, dispensed into 12 ml aliquots, and frozen at –20 °C until used. Unconjugated goat anti-mouse immunoglobulin, heavy- and light-chain-specific (Organon Teknika Corp., Durham, NC) is used to coat tissue culture dishes for panning and isolation of T lymphocytes. In the flow cytometric analysis of T-lymphocyte preparations, a phycoerythrin-conjugated goat anti-hamster immunoglobulin, specific for both heavy and light chains (Boehringer Mannheim Biochemicals, Indianapolis, IN), is used for the detection of B-lymphocytes. $CD4^+$ T lymphocytes are detected with a phycoerythrin-conjugated rat anti-mouse CD4 (L3T4) antibody (Boehringer Mannheim Biochemicals). This antibody has specificity for the CD4 (L3T4) molecule on murine and hamster T lymphocytes. Phycoerythrin-conjugated goat and rat immunoglobulins are used as isotype controls.

Isolation of enriched populations of T lymphocytes

The isolation of enriched populations of hamster T lymphocytes by using MAb 14-4-4s has been described previously by Liu *et al.* (1991). Briefly, enriched populations of T lymphocytes are isolated from the inguinal lymph nodes of hamsters 14 days after vaccination with formalin-inactivated *B. burgdorferi* and from those of non-vaccinated hamsters. B lymphocytes are removed from the lymph node suspensions with immunoglobulin-coated tissue culture dishes. The dishes (100 × 20 mm; Corning Glass Works) are prepared by coating the surfaces with 100 μg of nunconjugated goat anti-mouse immunoglobulin in coupling buffer (15 mM Na_2CO_3, 35 mM $NaHCO_3$ [pH 9.6]) overnight at 4 °C and washing four times with PBS before use. Single-cell suspensions of 10^7 immune and non-immune lymph node cells per milliliter are incubated with MAb 14-4-4s in Dulbecco's modified Eagle's medium for 30 minutes at 4 °C. During this 30 minute incubation, the cell suspension is periodically mixed. The suspensions of cells are washed twice with PBS by centrifugation (500 *g* for 10 minutes at 4 °C), resuspended to 10^7 cells/ml, poured over the immunoglobulin-coated tissue culture dishes, and incubated for 60 minutes at 4 °C. Non-adherent cells are then collected by gently rinsing the tissue culture dishes with cold Dulbecco's modified Eagle's medium. Enriched T-lymphocyte suspensions obtained from several tissue culture dishes are aspirated, pooled, and centrifuged at 500 *g* for 10 minutes at 4 °C. After centrifugation, the supernatant is decanted and the pellet is resuspended in Dulbecco's modified Eagle's medium. The suspension is poured over another set of immunoglobulin-coated dishes and incubated for 60 minutes at 4 °C. This process is repeated three times. After the last panning cycle, the cells are washed twice with PBS by centrifugation (500 *g* for 10 minutes at 4 °C) and resuspended in PBS. Cell viability is determined by trypan blue (Sigma) exclusion. Enriched T-lymphocyte preparations obtained by this method were shown to contain less than 5% B-lymphocyte contamination by flow cytometric analysis. Giemsa-stained smears of B-lymphocyte-depleted lymph node cells obtained by this method showed a homogeneous population of lymphocytes with no other types of leukocyte visible.

Analysis of enriched T-lymphocyte preparations by flow cytometry

Samples 100 µl in size containing 10^5 lymph node cells obtained before and after panning with MAb 14-4-4s are stained for the presence of B lymphocytes or $CD4^+$ T lymphocytes. B lymphocytes are stained with a phycoerythrin-conjugated goat anti-hamster immunoglobulin, specific for both heavy and light chains (Boehringer Mannheim Biochemicals; 1:100), for 15 minutes at 4 °C. $CD4^+$ T lymphocytes are stained with a phycoerythrin-conjugated rat, anti-mouse CD4 (L3T4) antibody (Boehringer Mannheim Biochemicals; 1:100) for 15 minutes at 4 °C. This antibody has specificity for the CD4 (L3T4) molecule on murine and hamster T lymphocytes. Samples are then washed twice with PBS by centrifugation, fixed with 1% paraformaldehyde (Sigma), and kept in the dark until analyzed by flow cytometry. Phycoerythrin-conjugated goat and rat antibodies are used as isotype controls. Other controls include unstained suspensions of lymph node cells. All samples are analyzed by using a FACScan flow cytometer (Becton Dickinson Immunocytometry Systems, Mountain View, CA) with FACScan LYSYS II software for data acquisition. Cells are detected by forward scatter, side scatter, and phycoerythrin fluorescence. Data from 5000 cells are acquired. Cell samples are then analyzed by means of histogram profiles of phycoerythrin fluorescence evaluated with FACScan LYSYS II software. Gates are established by using unstained samples and samples stained with the isotype control antibodies. The percentages of B lymphocytes and $CD4^+$ T lymphocytes present in the cellular suspensions are determined by the percentage shifts in the phycoerythrin fluorescence of the stained cells.

Cell transfer and infection of hamsters

Three hamsters per group are mildly anesthetized and injected subcutaneously in each hind paw with 0.4 ml of PBS containing 10^6 viable macrophages and/or T lymphocytes. Within 13–14 hours after the cell transfer, recipient hamsters are mildly anesthetized and infected subcutaneously in each hind paw with 0.2 ml of BSK medium containing 10^5 viable *B. burgdorferi sensu stricto* isolate 297 spirochetes. This concentration of spirochetes can readily be detected in hamsters when their tissues are cultivated in BSK medium. The viability of the spirochetes is determined by motility and dark-field microscopy.

Antimicrobial therapy

Groups of *B.-burgdorferi*-infected hamsters are treated daily for 4 days with 2.5 mg of ceftriaxone, 2 mg of doxycycline or other antimicrobial agents at various weeks after infection. Controls include groups of hamsters infected with *B. burgdorferi* for 0, 1, 3, 5, 7, 9 weeks or later intervals after infection.

Key parameter to monitor response to treatment

Recovery of spirochetes from tissue(s) of treated and non-treated *B.-burgdorferi*-infected hamsters is compared to determine the efficacy of therapy.

Pitfalls (advantages/disadvantages) of the model

The compartmentalization of clinical Lyme borreliosis features that occur in animal models is beneficial for studying Lyme borreliosis. Each animal model contributes information that enhances the ability of investigators to define common pathological and immunological principles that are likely to occur in humans. Divergent findings are even more important and generally have a far greater impact on defining mechanisms of resistance or other pathological mechanisms induced by infection or vaccination. It is imperative, then, to define the uniqueness of each animal model to obtain a broad and comprehensive picture of pathological events that might or do occur in humans. No animal model is superior to another. The quality of answers is limited only by the imagination and inventiveness of the investigator. Hopefully, the answers apply to humans and not a single animal is tested without obtaining meaningful data.

The advantage of the hamster model of Lyme borreliosis is the ability to study the onset and resolution of Lyme arthritis in an immunocompetent mature animal. In addition, severe destructive Lyme arthritis can be induced in vaccinated hamsters after infection with *B. burgdorferi*. Most importantly, the ability of a Lyme borreliosis vaccine to induce adverse effects can be readily determined before field trials are begun in humans.

A major criticism of the hamster model of Lyme borreliosis is that the murine system is better defined for determining the mechanisms of pathogenesis. Generally, these statements reflect unfamiliarity with the immunologic reagents that can be used in hamsters to separate T lymphocytes. Hamster T cells can be separated by MAb GK 1.5 into two phenotypically and functionally distinct subsets (Witte *et al.*, 1985; Liu *et al.*, 1991). The T-cell subset recognized by MAb GK 1.5 (L3T4a) has characteristics of helper T cells ($CD4^+$), whereas the other cells have characteristics of cytotoxic/suppressor T cells ($CD8^+$). Unfortunately, the identification of modulatory cytokines induced by hamster macrophages and T lymphocytes has been complicated by the unavailability of species-specific antibody reagents.

Contribution of the model to infectious therapy

The hamster models of Lyme arthritis (immunocompetent mature hamsters) and severe destructive Lyme arthritis (vaccinated mature hamsters) may facilitate the development of new therapeutic options to circumvent the development of Lyme arthritis in individuals infected with *B. burgdorferi*.

References

Appel, M. J. G., Allan, S., Jacobson, R. H. *et al.* (1993). Experimental Lyme disease in dogs produces arthritis and persistent infection. *J. Infect. Dis.*, **167**, 651–664.

Bakken, L. L., Case, K. L., Callister, S. M., Bourdeau, N. J., Schell, R. F. (1992). Performance of 45 laboratories participating in a proficiency testing program for Lyme disease serology. *J.A.M.A.*, **268**, 891–895.

Bakken, L. L., Callister, S. M., Wand, P. J., Schell, R. F. (1997). Interlaboratory comparison of test results for detection of Lyme disease by 516 participants in the Wisconsin State Laboratory of Hygiene/College of American Pathologists proficiency test programming. *J. Clin. Microbiol.*, **35**, 537–543.

Barbour, A. G., Tessier, S. L., Todd, W. J. (1983a). Lyme disease spirochetes and ixodid tick spirochetes share a common surface antigenic determinant defined by a monoclonal antibody. *Infect. Immun.*, **41**, 795–804.

Barbour, A. G., Burgdorfer, W., Grunwaldt, E., Steere, A. C. (1983b). Antibodies of patients with Lyme disease to components of *Ixodes dammini* spirochete. *J. Clin. Invest.*, **72**, 504–515.

Barthold, S. W., Beck, D. S., Hansen, G. M., Terwilliger, G. A., Moody, K. D. (1990). Lyme borreliosis in selected strains and ages of laboratory mice. *J. Infect. Dis.*, **162**, 133–138.

Barthold, S. W., deSouza, M. S., Janotka, J. L., Smith, A. L., Persing, D. H. (1993). Chronic Lyme borreliosis in the laboratory mouse. *Am. J. Pathol.*, **143**, 959–971.

Benach, J. L., Bosler, E. M., Hanrahan, J. P. *et al.* (1983). Spirochetes isolated from the blood of two patients with Lyme disease. *N. Engl J. Med.*, **308**, 740–742.

Burgdorfer, W. (1984). The New Zealand white rabbit: an experimental host for infecting ticks with Lyme disease spirochete. *Yale J. Biol. Med.*, **57**, 609–612.

Callister, S. M., Case, K. L., Agger, W. A., Schell, R. F., Johnson, R. C., Ellingson, J. L. E. (1990). Effects of bovine serum albumin on the ability of Barbour–Stoenner–Kelly medium to detect *Borrelia burgdorferi*. *J. Clin. Microbiol.*, **28**, 363–365.

Callister, S. M., Jobe, D. A., Schell, R. F., Pavia, C. S., Lovrich, S. D. (1996). Sensitivity and specificity of the borreliacidal-antibody test during early Lyme disease: a 'gold standard'? *Clin Diag. Lab. Immunol.*, **3**, 399–402.

Centers for Disease Control and Prevention (1997). Lyme disease — United States, 1996. *Morbid. Mortal. Weekly Rep.*, **45**, 1132–1136.

Johnson, R. C., Hyde, F. W., Schmid, G. P., Brenner, D. J. (1984). *Borrelia burgdorferi* sp. nov.: etiologic agent of Lyme disease. *Int. J. Syst. Bacteriol.*, **34**, 496–497.

Kornblatt, A. N., Steere, A. C., Brownstein, D. G. (1984). Infection of rabbits with the Lyme disease spirochete. *Yale J. Biol. Med.*, **57**, 613–614.

Krinsky, W. L., Brown, S. J., Askenase, P. W. (1982). *Ixodes dammini*: induced skin lesions in guinea pigs and rabbits compared to erythema chronicum migrans in patients with arthritis. *Exp. Parasitol.*, **53**, 381–395.

Lim, L. C. L., England, D. M., DuChateau, B. K. *et al.* (1994). Development of destructive arthritis in vaccinated hamsters challenged with *Borrelia burgdorferi*. *Infect. Immun.*, **62**, 2825–2833.

Liu, H., Adler, J. D., Steiner, B. M., Stein-Streilein, J., Lim, L. C. L., Schell, R. F. (1991). Role of $L3T4^+$ and 38^+ T-cells subsets in resistance against infection with *Treponema pallidum* subsp., *pertenue* in hamsters. *Infect. Immun.*, **59**, 529–536.

McAlister, H. F., Klementowicz, P. T., Andrews, C., Fisher, J. D., Field, M., Furman, S. (1989). Lyme carditis: an important cause of reversible heart block. *Ann. Intern. Med.*, **110**, 339–345.

Munson, E. L., DuChateau, B. K., Jobe, D. A. *et al.* (1996). Hamster model of Lyme borreliosis. *J. Spirochet. Tick-borne Dis.*, **3**, 15–21.

Pachner, A. R., Steere, A. C. (1985). The triad of neurologic complications of Lyme disease: meningitis, cranial neuritis and radiculoneuritis. *Neurology*, **35**, 47–53.

Padilla, M. L., Callister, S. M., Schell, R. F. *et al.* (1996). Characterization of the protective borreliacidal antibody response in humans and hamsters after vaccination with a *Borrelia burgdorferi* outer surface protein A vaccine. *J. Infect. Dis.*, **174**, 739–746.

Philipp, M. T., Johnson, B. J. B. (1994). Animal models of Lyme disease pathogenesis and immunoprophylaxis. *Trends Microbiol.*, **2**, 431–437.

Preac-Mursic, V., Weber, K., Pfister, H. W. *et al.* (1989). Survival of *Borrelia burgdorferi* in antibiotically treated patients with Lyme borreliosis. *Infection*, **17**, 355–359.

Schaible, U. E., Kramer, M. D., Museteanu, C., Zimmer, G., Mossman, H., Simon, M. M. (1989). The severe combined immunodeficiency (SCID) mouse: a laboratory model for the analysis of Lyme arthritis and carditis. *J. Exp. Med.*, **170**, 1427–1432.

Schell, R. F., Callister, S. M., Jobe, D. A., DuChateau, B. K. (1997). The borreliacidal antibody test: an alternative approach for confirming Lyme borreliosis. *J. Spirochet. Tick-borne Dis.*, **4**, 4–6.

Schmitz, J. L., Schell, R. F., Hejka, A., England, D. M., Konick, L. (1988). Induction of Lyme arthritis in LSH hamsters. *Infect. Immun.*, **9**, 2336–2342.

Schmitz, J. L., Schell, R. F., Lovrich, S. D., Callister, S. M., Coe, J. E. (1991). Characterization of the protective antibody response to *Borrelia burgdorferi* in experimentally infected LSH hamsters. *Infect. Immun.*, **59**, 1916–1921.

Steere, A. C. (1989). Lyme disease. *N. Engl. J. Med.*, **321**, 586–596.

Steere, A. C., Gibofsky, A., Patarroyo, M. E., Winchester, R. J., Hardin, E. D., Malawista, S. E. (1979). Chronic Lyme arthritis: clinical and immunogenetic differentiation from rheumatoid arthritis. *Ann. Intern. Med.*, **90**, 896–901.

Steere, A. C., Bartenhagen, N. H., Craft, J. E. *et al.* (1983a). The early clinical manifestations of Lyme disease. *Ann. Intern. Med.*, **99**, 76–82.

Steere, A. C., Grodzicki, A. L., Kornblatt, A. N. *et al.* (1983b). The spirochetal etiology of Lyme disease. *N. Engl J. Med.*, **308**, 733–740.

Steere, A. C., Schoen, R. T., Taylor, E. (1987). The clinical evolution of Lyme arthritis. *Ann. Intern. Med.*, **107**, 725–731.

Steere, A. C., Taylor, E., McHugh, G. L., Logigian, E. L. (1994). The overdiagnosis of Lyme disease. *Journal of American Medical Association*, **269**, 1812–1816.

Witte, P. L., Stein-Streilein, J., Streilein, J. W. (1985). Description of phenotypically distinct T-lymphocyte subsets which mediate helper/DTH and cytotoxic functions in the Syrian hamster. *J. Immunol.*, **134**, 2908–2915.

Chapter 40

Rabbit Model of Bacterial Conjunctivitis

M. Motschmann and W. Behrens-Baumann

Background of human infection

Bacterial conjunctivitis is the most common infectious disease of the eye. Nearly every person is affected sometime during their lifetime. *Staphylococcus* spp., *Streptococcus* spp. and *Haemophilus* spp. are infectious agents in approximately 90% of cases (Fechner and Teichmann, 1991). In immunocompetent patients bacterial conjunctivitis is a self-limiting disease, normally healing within a few days (Gigliottti *et al.*, 1984; Fisher, 1987). However, in clinical practice local antibiotic therapy is usual to prevent complications. The fear of vision-threatening keratitis, in particular, is in most cases the reason for this type of prophylaxis. However, an uncritical application of local antibiotics may cause sensitivity to the substance administered. This may cause problems if it is later necessary to administer the substance systemically for another acute infection (Behrens-Baumann, 1991). A further danger exists in creating cross-reactions.

In addition, bacterial resistance against antibiotics is an increasing problem worldwide. For instance, gentamicin is ineffective against more than 30% of *Staphylococcus epidermidis* isolated from eyes with endophthalmitis (Puliafito *et al.* 1989). Methicillin-resistant *Staphylococcus aureus* also is resistant to gentamicin, tobramycin, amikacin, or erythromycin in 90% of all cases (Maple *et al.*, 1989). Nearly 25% of *Pseudomonas aeruginosa* strains are resistant to ciprofloxacin (O'Brien, 1991). A growing therapeutic concern involves the possibility of vancomycin-resistant *Staphylococcus aureus* species (Tomasz, 1994).

On the other hand, it is also possible to treat bacterial conjunctivitis with antiseptics. In these cases the negative side-effects of antibiotics can be avoided (Behrens-Baumann, 1991). However, as yet no *in-vivo* data are available that demonstrate the superiority of either substance under standardized conditions. *In-vitro* results are not sufficient. The efficacy of a drug depends on several factors that only operate under *in-vivo* conditions (Dolder and Skinner, 1983).

Background of the model

Since the first description by Leber (1891), published more than 100 years ago, several experiments have been performed to establish a model of bacterial conjunctivitis (Table 40.1).

Subconjunctival injection of a suspension of infectious agents may certainly initiate a local infection (Gasparrini, 1883; Fedukowicz, 1953). However, microbial suspension simply dropped into the conjunctival sac does not result in conjunctivitis in various species. Scarification with subsequent drops of microbial suspension does not provoke manifest infection either (Nöldeke, 1899; Römer, 1899; Lange, 1924; Sereny, 1955, 1956; Trabulisi, 1965; Howcraft *et al.*, 1978; Srivastava *et al.*, 1986). This is obviously because of the natural immune response of the animal. It is only the imitation of an open injury of the conjunctiva, e.g. by a twig, with the pathogens being transferred underneath the conjunctiva, that a fulminant conjunctivitis is provoked. This method was first described by Behrens-Baumann and Begall (1993a).

Animal species/preparation of animals

Pigmented rabbits weighing 1.8–2.7 kg were used in this model. They were given Altromin rabbit chow and water *ad libitum*.

Infection process/details of surgery

The animals were anesthetized using Ketavet and Rompun at 0.25 ml/kg body weight.

To imitate an infection, the agents were dropped into the lower conjunctival fornix in four eyes (method A).

To simulate a superficial scratch injury the conjunctiva

Handbook of Animal Models of Infection
ISBN 0-12-775390-7

Table 40.1 Animal models of conjunctivitis

	Animal	*Infectious agent*	*Method*	*Result*
Leber, 1891	Rabbit	*Staphylococcus aureus*	Instillation Sewing up eyelids	Negative
Römer, 1899	Rabbit	Pneumococcus	Instillation	Negative
Nöldeke, 1899	Guinea-pig	Pneumococcus	Instillation Scarification	Negative
Lange, 1924	White mouse	*Streptococcus* Pneumococcus	Instillation	Negative
Sereny, 1955/1956	Guinea-pig	*Enterobacter*	Instillation	Negative (keratitis)
Trabulisi, 1965	Guinea-pig	*Enterobacter*	Instillation	Negative (keratitis)
Howcraft *et al.*, 1978	Guinea-pig	*Staphylococcus aureus*	Instillation	Negative
Srivastava *et al.*, 1986	Mouse Rabbit	*Haemophilus*	Instillation	Blepharitis
Gasparrini, 1883	Rabbit	Pneumococcus	Subconjunctival injection	Infection
Fedukowicz, 1953	Rabbit	*Moraxella lacunata*	Subconjunctival injection	Infection

was scarified by a scalpel 6 mm in a radial direction from the limbus corneae before dropping the agent, in suspension, into four more eyes (method B).

To simulate a deeper wound, e.g. by a twig, a radial 6 mm incision of the conjunctiva was made with a pair of scissors in 12 eyes. The conjunctiva had previously been elevated by careful air-injection to prevent unintentional incision of subconjunctival tissue or the tenon capsule. A 25 G needle was inserted 3 mm beneath the edges of the wound and 25 μl of inoculum was injected (method C).

After instillation of the agents the eyelids were closed several times in order to achieve regular distribution of the suspension (methods A–C). Care was taken not to squeeze out any inoculum during this maneuver.

In addition to general anesthesia two drops of oxybuprocaine without preservative had previously been administered in the eye for methods B and C. Some animals were also treated according to methods B and C without delivering an infectious agent in order to control the effects of the artificial injury.

Storage and preparation of inocula

Haemophilus influenzae, *Staphylococcus epidermidis* and *Staphylococcus aureus* (ATCC 29213) were the challenge organisms used. The strains were transferred onto fresh chocolate agar (*H. influenzae*) or blood agar (*Staphylococcus* spp.) plates every two weeks, incubated at 36°C until colonies appeared, and then the plates stored at 4°C.

Inocula for experimental infections were prepared as follows: a colony was picked up from the agar plate and put into glucose or casein soybean peptone broth which was incubated for 6 hours at 36°C, then the bacteria collected by centrifugation for 10 minutes at 3000 rpm. After the supernatant was removed, the bacterial pellet was resuspended in sterile saline, and the suspension subsequently diluted to give a final cell concentration of 1.05×10^3/50 μl of *S. epidermidis* or *H. influenzae*, and 1.35×10^3/50 μl of *Staphylococcus* spp. as determined using McFarland barium-sulfate standard density scale.

Key parameters to monitor infection

A conjunctival swab was taken at the fourth and seventh day postinfection for microbiological examination. In addition, the clinical findings were monitored with regard to conjunctival hyperemia daily for 16 days. A staging of the conjunctival hyperemia was used (grade 0: white and not irritated; grade 1: slight; grade 2: moderate; and grade 3: severe conjunctival hyperemia (Figure 40.1 – see colour plate). The four quadrants of the conjunctiva were assessed according to this staging. The results of each quadrant were added. The lowest score was 0 (white and smooth conjunctiva in all quadrants), and the highest was 12 (severe conjunctival hyperemia in all quadrants).

Results of microbiological examination

All conjunctival swabs of eyes treated with method A were negative at the fourth day postinfection. Only one was

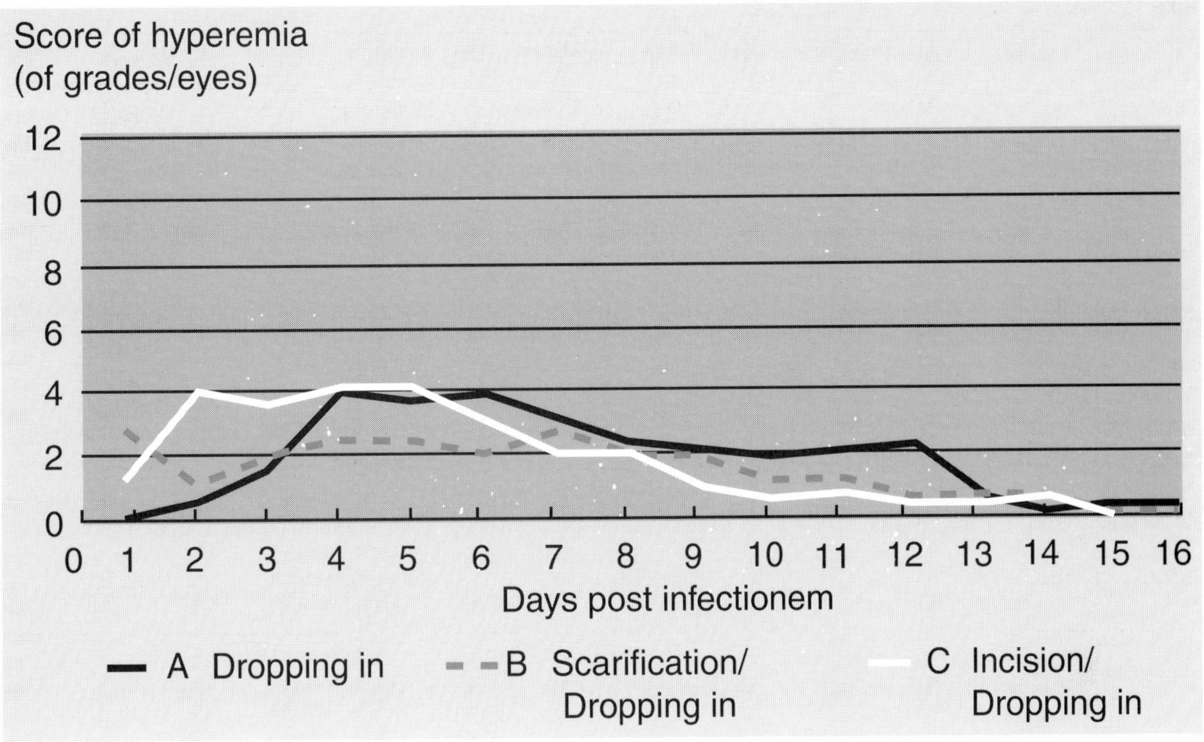

Figure 40.2 Total amount of hyperemia — grades of all eyes in method A–C using *Haemophilus influenzae*.

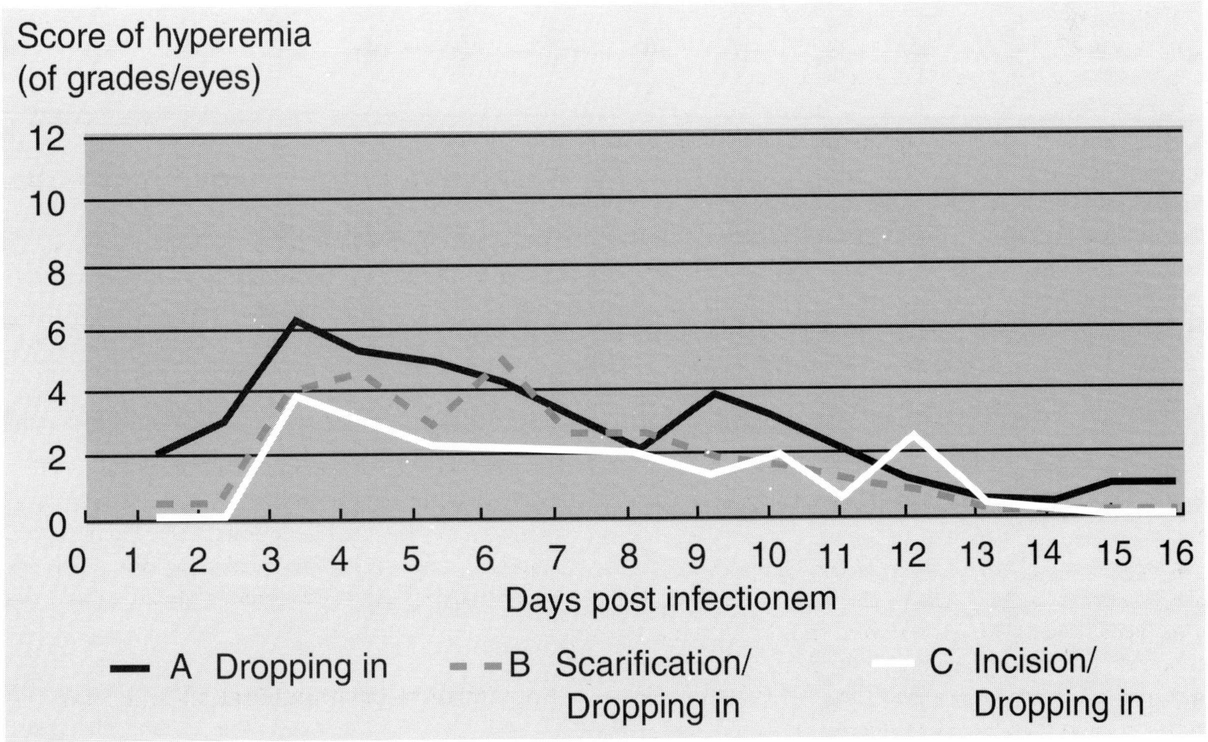

Figure 40.3 Total amount of hyperemia — grades of all eyes in method A–C using *Staphylococcus epidermidis*.

positive at the same time in method B. In all eyes treated using method C positive findings were proved at the fourth day. However, 7 days postinfection no more infectious agents were found in any eye.

Results of clinical findings

The hyperemia scores of the four conjunctiva quadrants are shown in Figures 40.2–40.4. *Haemophilus influenzae* provokes only a minimal irritation in all methods of agent inoculation, which has almost disappeared after 1 week (Figure 40.2). The results are hardly different when using *Staphylococcus epidermidis* (Figure 40.3).

In eyes prepared according to method A (instillation of inoculum) *Staphylococcus aureus* does not cause a marked reaction either. Using method B (scarification/instillation) the hyperemia is rather stronger. Only in eyes treated by method C (incision/injection) does a distinct infection arise that persists for more than a week and disappears gradually afterwards (Figure 40.4).

The eyes of the control group showed only a grade 2 hyperemia for about 2 days.

Antimicrobial therapy/key parameters to monitor response to treatment

Using this model, several anti-infectives have been studied for therapy of experimental conjunctivitis caused by *Staphylococcus aureus* (Behrens-Baumann and Begall, 1993b).

At 48 hours after infection, polyvinylpyrrolidone (PVP)-iodine, bibrocathol (Noviform, NF), ethacridine (Biseptol, BS) and bacitracin + polymyxin B + neomycin (Polyspectran, PS) were applied five times per day. A total of 16 eyes (one eye per animal) were used for each substance. An additional 16 animals were given NaCl drops and served as controls, as contralateral eyes should not be compared (Odenberger and Babicki, 1973; Immich, 1977). A conjunctival swab was taken every second day and a photograph every third day.

Conjunctival hyperemia was graded 1–3 for each quadrant. Regression of hyperaemia was registered and was controlled five times in a masked fashion using the photographs. The ranking order of the substances tested in achieving regression of conjunctival hyperemia over time is demonstrated in Table 40.2, column 1. The group treated with PVP-iodine was the first to become white and not inflamed, followed by the groups receiving the other antiseptics and the antibiotic. In the control group (NaCl) the hyperemia was prolonged. The rank order was established by summarizing the mean values obtained for each group of eyes using the scoring system. Differences between the various groups were evaluated using Wilcoxon's test (Table 40.3).

The ranking order of the substances in eliminating *Staphylococcus aureus* per time is presented in Table 40.2, column b. In the group treated with ethacridine, 12 of 16

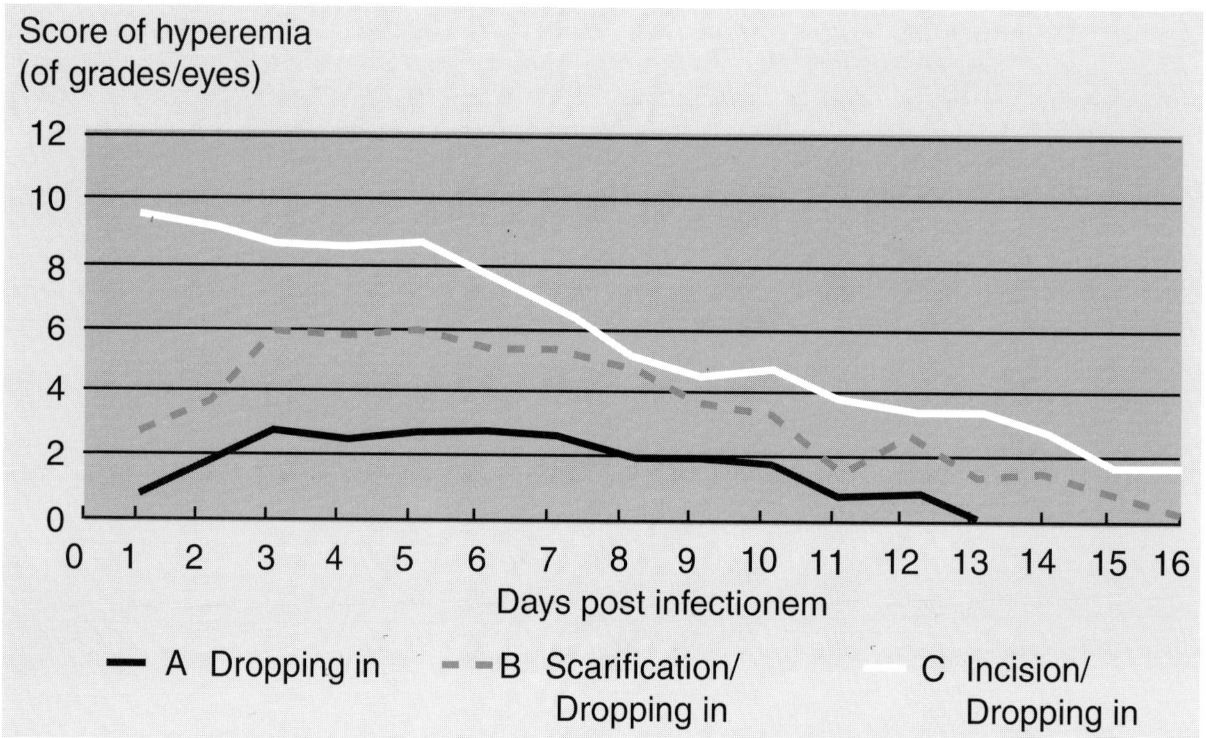

Figure 40.4 Total amount of hyperemia — grades of all eyes in method A–C using *Staphylococcus aureus*.

Table 40.2 Substances applied five times per day and their ranking order in achieving regression of conjunctival hyperemia (**A**), in eliminating *Staphyloccus aureus* (**B**), and in fulfilling both criteria (**C**).

Substance	*A*	*B*	*C*
Povidone-iodine eyedrops	1	2	1
Noviform ointment	2	5	4
Biseptol eyedrops	3	1	2
Polyspectran ointment	4	3	3
NaCl 0.9% eyedrops	5	4	5

Table 40.3 Comparison of the various experimental groups using Wilcoxon's test (level of significance, $p < 0.05$)

Group	*Significance (p)*
Polyspectran/Biseptol	0.933
Polyspectran/Noviform	0.939
Polyspectran/NaCl	0.314
Polyspectran/PVP-iodine	0.289
NaCl/Biseptol	0.394
NaCl/Noviform	0.556
NaCl/PVP-iodine	0.054
Biseptol/Noviform	0.819
Biseptol/PVP-iodine	0.342
Noviform/PVP-iodine	0.664

swabs were sterile at day 4, followed by the PVP-iodine group (11 of 16 swabs). No germ could be cultured at day 6 in the ethacridine group or at day 8 in all groups, including the controls (Table 40.4).

The overall rank order established by adding the values obtained for both criteria demonstrates that PVP-iodine was the most effective substance in achieving regression of conjunctival hyperemia as well as bacterial elimination (Table 40.2, column 3).

Pitfalls of the model

This rabbit model mimics a bacterial conjunctivitis under standardized conditions. A small injury is necessary to establish the model. However, this procedure is only a theoretical disadvantage, because under natural conditions small injuries, e.g. by a twig, can cause bacterial conjunctivitis. Therefore this method mimics a realistic situation. Research into optimal therapy is now possible for the first time under standardized conditions without immunosuppression of the animal.

Contributions of the model to infectious disease therapy

Like human conjunctivitis (Gigliotti *et al.*, 1984; Fisher, 1987) experimentally induced conjunctivitis in rabbits is a self-limiting disease. Our findings demonstrate that antibiotics are not superior to antiseptics in controlling bacterial conjunctivitis. In fact, PVP-iodine was the most effective substance in achieving bacterial elimination and reducing conjunctival hyperemia. This antiseptic agent acts via inhibition of the respiratory chain of the bacterium (Forth *et al.*, 1984). It has been successfully used in various studies (Hiti *et al.*, 1978; Neuhann and Sommer, 1980; Auerbach *et al.*, 1985), especially for preoperative disinfection and against Adenovirus (Apt *et al.*, 1985; Janthure *et al.*, 1985; Maeck *et al.*, 1990). It is well tolerated by the corneal epithelium (MacRae *et al.*, 1984).

In conclusion, this rabbit model study demonstrates that antiseptics are as effective as antibiotics in the treatment of this standardized experimental conjunctivitis induced by *Staphylococcus aureus* as well as imitating a traumatic genesis. Antiseptics may therefore be sufficient to control human bacterial conjunctivitis as well.

Table 40.4 Positive cultures of *Staphylococcus aureus* obtained at various days following infection

	Days							
Group	*0*	*2*	*4*	*6*	*8*	*10*	*12*	*14*
Polyspectran	16	15	7	2	0	0	0	0
Biseptol	16	11	4	0	0	0	0	0
Noviform	16	9	7	7	0	0	0	0
PVP-iodine	16	10	5	2	0	0	0	0
NaCl	16	10	8	5	0	0	0	0

In purulent conjunctivitis of an immunocompetent patient an antiseptic should be applied in the first instance. If there is no improvement a conjunctival swab should be taken and an antibiotic should be prescribed. If there is still no improvement a different antibiotic can be given according to the results of the sensitivity test. This recommendation seems to be a reasonable option that takes into consideration benefits, risks and side effects as well as health costs (Behrens-Baumann, 1997).

In addition, antiseptics may be of beneficial value for prophylactic purposes (Isenberg *et al.*, 1985; Apt *et al.*, 1994; Kramer and Behrens-Baumann, 1997).

References

Apt, L., Isenberg, S. J., Yoshimori, R. (1985). Antimicrobial preparation of the eye for surgery. *Hosp. Infect.* (Suppl.), **6**, 163–172

Apt, L., Isenberg S. J., Yoshimori, R. *et al.* (1994). The effect of povidone-iodine solution applied at the conclusion of ophthalmic surgery. *Am. J. Ophthalmol.*, **119**, 701–705.

Auerbach, B., Reich, E., Schuhmann, G. (1985). Polyvinyl-Jod in der Ophthalmochirurgie. *Klin. Monatsbl. Augenheilk.* **187**, 361–362.

Behrens-Baumann, W. (1991). Antibiotika und Antiseptika aus der Sicht des Ophthalmologen. *Augenärztl. Fortb.*, **14**, 27–31.

Behrens-Baumann, W. (1997). Efficacy of selected antiinfectives in the rabbit conjunctivitis model. *Hyg. Med.*, **22**, 73–76.

Behrens-Baumann, W., Begall, T. (1993a). Reproduzierbares Modell einer bakteriellen Konjunktivitis. *Ophthalmologica*, **206**, 69–75.

Behrens-Baumann, W., Begall T. (1993b). Antiseptics versus antibiotics in the treatment of the experimental conjunctivitis caused by *Staphylococcus aureus*. *Ger. J. Ophthalmol.*, **2**, 409–411.

Dolder, R., Skinner F. S. (1983). *Ophthalmika*, 3rd edn, pp. 68–70; 123–329. Wissenschaftliche Verlagsgesellschaft, Stuttgart.

Fechner, P. U., Teichmann, K. D. (1991). Medikamentöse Augentherapie. In *Bücherei des Augenarztes*, vol. 67, pp. 182–185. Enke, Stuttgart.

Fedukowicz, H. (1953). The Gram-negative Diplobacillus in hypopyon keratitis. *Arch. Ophthalmol.*, **49**, 202–211.

Forth, W., Henschler, H., Rummel, W. (1984). *Allgemeine und spezielle Pharmakologie und Toxikologie*, 4th edn, pp. 613–618. Wissenschaftsverlag, Mannheim.

Fisher, M. C. (1987). Conjunctivitis in children. *Pediatr. Clin. North Am.*, **34**, 1447–1456.

Gasparrini, E. (1883). Il diplococco di Fraenkel, in pathologia oculare. *Annali di Ottalmologia*, **22**, 6. (cited by Axenfeld, T. (1929). Infektionen der Conjunctiva. In *Handbuch der pathogenen Mikroorganismen*, 3rd edn (eds Kolle, W., Kraus, R., Uhlenhuth, P.), pp. 281–354. Urban & Schwarzenberg, Jena.

Gigliotti, F., Hendley, J. O., Morgan, J., Michaels, R., Dickens, M., Lohr, J. (1984). Efficacy of topical antibiotic therapy in acute conjunctivitis in children. *J. Pediatr.*, **104**, 623–626.

Hiti, H., Haselmayer H., Hofmann H. (1978). Erfahrungen in der Therapie und Prophylaxe der Keratokonjunktivitis epidemica. *Klin. Monatsbl. Augenheilk.*, **174**, 456–461.

Howcroft, M. J., Okumoto, M., Ostler, H. B., Schachter, J. (1978). Staphylococcal infection superimposed on guinea pig inclusion conjunctivitis. *Can. J. Ophthalmol.*, **13**, 39–44.

Isenberg, S., Apt, L., Yoshimori, R., Khwarg, S. (1985). Chemical preparation of the eye; comparison of povidone-iodine on the conjunctiva with a prophylactic antibiotic. *Arch. Ophthalmol.*, **103**, 1340–1342.

Immich, H. (1977). Versuchsplanung und Auswertung. In *Arzneimittelnebenwirkungen am Auge*, (eds Hockwin, O., Koch, H. R.), pp. 189–215. Gustav Fischer, Stuttgart.

Janthure, E., Blessing, J., Ehrich, W., Wigand, R. (1985). Polyvinyl-Pyrrolidon-Jod und Arginase: Einfluß auf Hornhautregeneration und antivirale Wirkung. *Klin. Monatsbl. Augenheilk.*, **186**, 25.

Kramer, A., Behrens-Baumann, W. (1997). Prophylactic use of topical sub-infectives. *Ophthalmologica* (Suppl. 1), 68–76.

Lange, B. (1924). Über die Infektion von weißen Mäusen auf den natürlichen Wegen durch die Haut, die Mund- und Darm-schleimhaut sowie die Augenbindehaut. *Z. Hyg. Infektionskr.*, **102**, 224–261.

Leber, T. (1891). *Die Entstehung der Entzündungen und die Wirkung der entzündungserregenden Schädlichkeiten*, pp. 134–136. Engelmann, Leipzig.

MacRae, S., Brown, B., Edelhauser, H. (1984). The corneal toxicity of presurgical skin antiseptics. *Am. J. Ophthalmol.*, **97**, 221–232.

Maeck, C., Eckardt, C., Höller, C. (1990). Präoperative Desinfektion der Konjunktiva mit PVP-Jod. *Fortschr. Ophthalmol.*, **87**, 320–323.

Maple, P., Hamilton-Miller, J., Brumfitt, W. (1989). Worldwide antibiotic resistance in methicillin-resistant *Staphylococcus aureus*. *Lancet*, 1 Mar., 537–540.

Neuhann, T., Sommer, G. (1980). Erfahrungen mit Jod-Povidon zur Behandlung der Keratokonjunktivitis epidemica. *Zeitschr. Prakt. Augenheilk.*, **1**, 65–68.

Nöldeke, E. (1899). Experimenteller Beitrag über die Bedeutung des Diplokokkus lanceolatus Fraenkel in der Pathologie des Auges. Medical dissertation, Kaiser-Wilhelm-Universität, Strassburg i.E.

O'Brien, T. P. (1991). Ciprofloxacin in der Behandlung der Keratitis. Paper presented at the International Symposium DOG 'Infektionskrankheiten des Auges', Münster.

Odenberger, J., Babicki, A. (1973). Alkali burns of the rabbit cornea. *Ophth. Res.*, **5**, 1–9.

Puliafito, C. A., Baker, A. S., Haaf, J., Foster, C. (1989). Infectious endophthalmitis. *Ophthalmology*, **8**, 921–929.

Römer, P. (1899). Experimentelle Untersuchungen über Infektionen vom Conjunctivalsack aus. *Hyg. Infektionskr.*, **32**, 295–326.

Sereny, B. (1955). Experimental shigella keratoconjunctivitis: a preliminary report. *Acta Microbiol. Hung.*, **2**, 293–296.

Sereny, B. (1956). Experimental keratoconiunctivitis shigellosa. *Acta Microbiol. Hung.*, **4**, 367–376.

Srivastava, K. K., Pick, J. R., Johnson, T. P. (1986). Charakterization of a *Haemophilus* sp. isolated from a rabbit with conjunctivitis. *Lab. Anim. Sci.*, **36**, 291–293.

Tomasz, A. (1994). Multiple-antibiotic-resistant pathogenic bacteria. *N. Engl. J. Med.*, **330**, 1247–1251.

Trabulisi, L. R. (1965). Experimental keratoconjunctivitis of the guinea pig by enterobacteria. *Rev. Inst. Med. Trop. Sao Paulo.*, **7**, 16–23.

Chapter 41

Murine Model of Bacterial Keratitis

K. A. Kernacki, J. A. Hobden and L. D. Hazlett

Introduction

Bacterial keratitis

Bacterial corneal infection is a highly destructive process that often leads to loss of vision. Because of the severity of the disease, initial therapy for bacterial ulcerative keratitis consists of an intensive regimen of broad-spectrum antibiotic chemotherapy (McLeod *et al.*, 1995). In many cases corneal transplantation is required to restore visual acuity. Approximately 30 000 bacterial corneal ulcers are treated annually in the USA (Pepose and Wilhelmus, 1992). One important risk factor in the development of bacterial ulcerative keratitis is the use of extended-wear contact lenses (Schein *et al.*, 1994). Other predisposing factors include corneal trauma, pre-existing ocular surface disorders, corneal surgery and immunosuppression. A variety of microorganisms have been reported to cause keratitis. These include *Pseudomonas aeruginosa, Staphylococcus aureus, Streptococcus pneumoniae, Neisseria gonorrhoeae, Moraxella catarrhalis, Haemophilus influenzae, Bacillus* spp., *Corynebacterium diphtheriae*, and *Serratia marcesens*, as well as several anaerobic bacteria.

Mouse model

Several experimental models of bacterial keratitis using mice, rabbits, rats, and guinea pigs have been described in which aspects of tissue pathology, host response, and course of infection are similar to those reported for human infections. Development and refinement of the murine-scratch model of bacterial keratitis can be attributed to the work of Gerke and Magliocco (1971), Hazlett *et al.* (1976), and Ohman *et al.* (1980). Essentially all of the studies in the mouse have used various strains of *P. aeruginosa* as the infectious agent. Because the organism is one of the most destructive of all the ocular bacterial pathogens, the focus of this review will be on the pseudomonal model of keratitis. In theory, methods described for the pseudomonal model should be applicable for other infectious agents as well.

Mice

Both inbred strains of mice and outbred mice have been used in this model (Table 41.1). Outbred mice are less expensive, generally more hardy and immunologically more accurately represent the heterogeneous human population. Inbred strains of mouse, which contain a homogeneous genetic background, should be used for any studies where a precise assessment of immunological parameters is of importance. Various inbred strains of mouse have been characterized as either susceptible or resistant to corneal infection on the basis of their ability to restore corneal clarity and ocular integrity after infection (Berk *et al.*, 1979, 1981). After infection, the cornea of a mouse with a susceptible phenotype will develop a purulent infection leading to corneal perforation and often endophthalmitis accompanied by atrophy of the whole eye (phthisis bulbi) within 7–14 days. In contrast, mice with a resistant phenotype will develop a similar purulent infection to susceptible animals; however, bacteria will be eliminated from ocular tissue, corneas will clear, and the near-normal architecture of the eye will be preserved. The phenotype of resistance and susceptibility can be attributed to many factors, including macrophage and T-cell function (Berk, 1993) and the kinetics of the inflammatory cell response (Hazlett *et al.*, 1992; Hobden *et al.*, 1995). In general, outbred mice exhibit a resistant phenotype;

Table 41.1 Mice used for bacterial keratitis models

Mouse	*Phenotype*	*Inbred strain*	*Ocular response*	*Source*
C57BL/6J	Black coat	Yes	Susceptible	Jackson Labs*
C3H/HeJ	Agouti	Yes	Susceptible	Jackson Labs
DBA/2J	Grey coat	Yes	Resistant	Jackson Labs
Balb/cJ	Albino	Yes	Resistant	Jackson Labs
Swiss Webster	Albino	No	Resistant	Harlan-Sprague†
Swiss ICR	Albino	No	Resistant	Harlan-Sprague

*The Jackson Laboratory, Bar Harbor, ME.
† Harlan-Sprague Dawley, Inc., Indianapolis, IN.

Handbook of Animal Models of Infection
ISBN 0-12-775390-7

however animals 1 year or more of age are usually susceptible to corneal infection (Hazlett *et al.*, 1990).

The choice of whether to use an outbred mouse or an inbred mouse with a resistant or susceptible phenotype will depend on the nature of the study. For example, a mouse with a resistant phenotype may be useful for long-term studies where the objective is to evaluate the effects of a chemotherapeutic regimen on minimizing stromal scarring. For acute studies, where the objective is to observe the efficacy of a particular antimicrobial agent with respect to bacterial clearing, a mouse with a susceptible phenotype may be desirable. An additional consideration in selection of which mouse strain to use is pigmentation of the animal. Ocular pathology, particularly in the anterior chamber, is easier to observe in an albino mouse.

Table 41.2 *Pseudomonas aeruginosa* strains used to establish bacterial keratitis in mice

Strain	*Source*	*Reference(s)*
PA103	American Type Culture Collection*	Ohman *et al.*, 1980; O'Callaghan *et al.*, 1996
PAO1	*Pseudomonas* Genetic Stock Center†	Ohman *et al.*, 1980; Preston *et al.*, 1997
27853	American Type Culture Collection	O'Callaghan *et al.*, 1996
19660	American Type Culture Collection	Hazlett *et al.*, 1976; Fleiszig *et al.*, 1994
PAK	American Type Culture Collection	Hazlett *et al.*, 1991; Fleiszig *et al.*, 1994

*American Type Culture Collection, Manassas, VA.
†Pseudomonas Genetic Stock Center, East Carolina University School of Medicine, Greenville, NC.

Animal care

Mice for these studies are generally obtained from vendors (Table 41.1) between the ages of 5 and 6 weeks. Female mice are generally preferred because they are able to be housed in multiples. Animals are housed in caging specified by the National Research Council's *Guide for the Care and Use of Laboratory Animals* (1996). Animals are fed commercial rodent chow, unless they are breeders and then they are fed mouse chow. Acidified water (3.1 ml of concentrated HCl added to 5 gallons of tap water, resulting pH 3.8) is provided *ad libitum* to prevent the animals from being colonized by *P. aeruginosa* in the drinking water. Animals are held for 1 week after arrival from the vendor before experimental use.

Preparation of bacterial inoculum

A variety of *P. aeruginosa* strains and clinical isolates (Preston *et al.*, 1995; Fleiszig *et al.*, 1994, 1996) have been used in experimental corneal infection models. Strains from culture collections are preferable to clinical isolates obtained from ocular infections. Culture collection strains are better characterized and described in the literature than clinical isolates and the virulence and viability of the organisms are assured. The more common strains and their respective sources are listed in Table 41. 2.

Cultures of the bacterial strains are kept frozen at –70°C in PTSB—0.25% trypticase soy broth (Sigma Chemical Co., St Louis, MO) and 5% peptone (Difco Laboratories, Detroit, MI)—mixed with 15% glycerol. Stock cultures of *P. aeruginosa* used to prepare an infecting inoculum are maintained on PTSB slants (PTSB solidified with 1.7% agar (Difco)) at 4°C. Fresh slants are prepared every 2 weeks using frozen stocks as an inoculum. Cultures for the infection process are grown in 25 ml of PTSB at 37°C on a rotary shaker at 150 rpm for 18 hours to an optical density of approximately 1.6 at 540 nm. Cultures are then centrifuged at 6 000 *g* for 10 minutes at 15°C. The bacterial pellet is resuspended in 10 ml of sterile saline (0.85 M NaCl, pH 7.2), centrifuged again for 10 minutes and resuspended in 5 ml of sterile saline to a concentration of approximately 2×10^{10} cfu/ml, using a standard curve relating viable counts to optical density at 540 nm. Depending on the infecting dose, further dilutions may be made from this stock culture. The infecting dose for ATCC strain 19660, routinely used in our studies, is 1×10^6 cfu in a total volume of 5 μl. Others have reported using infecting doses between 10^3 and 10^8 cfu depending on the bacterial strain and the level of anesthesia (Preston *et al.*, 1995; Fleiszig *et al.*, 1994; Ohman *et al.*, 1980; Gerke and Magliocco, 1971).

Infection of mice

In most published studies using the murine-scratch model of infection, mice are anesthetized with isoflurane (Aerrane; Anaquest, Madison, WI) or ethyl ether. The use of ethyl ether is not recommended because it is very volatile and the vapors are explosive. The anesthetic is administered by inhalation by placing a mouse within a jar containing the anesthetic. The degree of anesthesia can be monitored by the animal's loss of equilibrium and rapid but steady respiration. The length of time required to anesthetize a mouse varies with the construction of the anesthetizing jar and with the size and strain of mouse. In general, full anesthesia (defined as a lack of voluntary muscle tone) is accomplished in less than a minute. Mice generally recover from the effects of inhalant anesthesia in less than 2 minutes after removal from the anesthetizing jar.

Recently, Preston *et al.* (1995) described the use of an alternate method of anesthesia. They inject 0.2 ml of a mixture of 6.7 mg ketamine hydrochloride and 1.3 mg xylazine/ml i.p. into mice prior to corneal wounding. Using this method, the animals remain immobile for 15–30 - minutes depending on the murine strain. Using this alternate method mice could be infected with much lower doses of *P. aeruginosa*. These studies described the 50% infective dose of various strains to be between 3×10^2 and 1×10^5 cfu.

There are advantages and disadvantages to each of these methods of anesthesia. The advantages of using the inhalant method are that the animals recover quickly after anesthesia and that the ocular surface, i.e., the tear film, remains relatively intact. The disadvantages of using inhalation anesthetics are that a larger inoculum is required to initiate infection and that the investigator needs to work quickly to wound and topically apply the bacterial inoculum. Using the injection anesthesia method, a lower bacterial inoculum can initiate disease, but the normal tear film is altered by the lack of a blink reflex while the animals are anesthetized.

Following anesthesia, mice are placed beneath a stereoscopic microscope at 40× magnification and the central cornea of each mouse is scarified with three 1-mm incisions using a 26 G needle (Figure 41.1). The wounds are randomly examined histologically to ensure that they penetrate the epithelial basal lamina, but extend no deeper than the superficial corneal stroma (Figure 41.2). After wounding, 5 μl of the bacterial suspension is topically applied to the scarified cornea using a calibrated micropipette with a sterile disposible tip. After infection, mice are placed into a holding cage to recover from the anesthesia. After recovery, they are transferred to a cage containing fresh bedding.

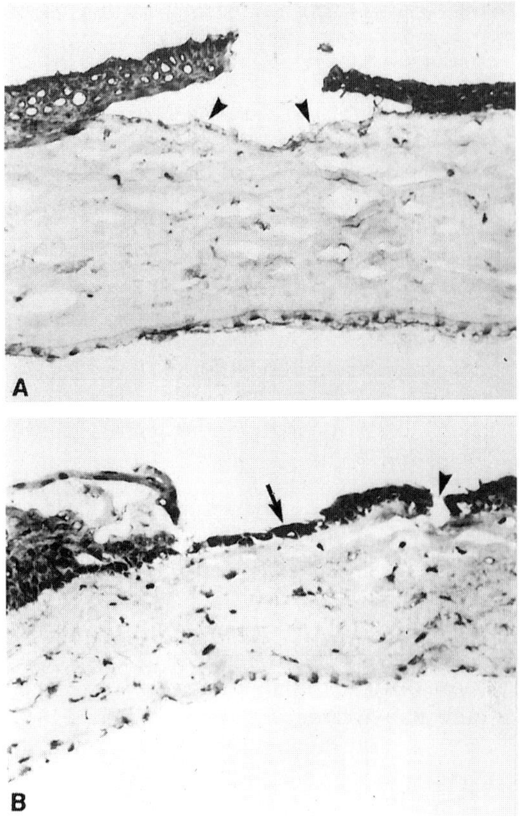

Figure 41.2 Corneal surface following scarification. (**A**) Corneal wound that penetrates epithelial basal lamina into superficial stroma (arrowheads). (**B**) Corneal wound that does not penetrate epithelial basal lamina (arrow).

Key parameters to monitor infection and response to treatment

Gross observation

At daily intervals after infection, animals are grossly observed to ensure that animals are infected and to grade the progression of the disease. Two different methods of grading ocular pathology in the mouse have been described and are shown in Tables 41.3 and 41.4 (Gerke and Magliocco, 1971; Hazlett *et al.*, 1987, 1991). To aid in grading, eyes may be examined using a 40 × dissecting microscope or with a slit-lamp microscope. In some cases, the eyelid may be sealed after infection. To facilitate grading of sealed eyes, the eyelid may be gently swabbed with a sterile cotton swab moistened with sterile saline and then gently opened sufficiently widely to grade the ocular disease.

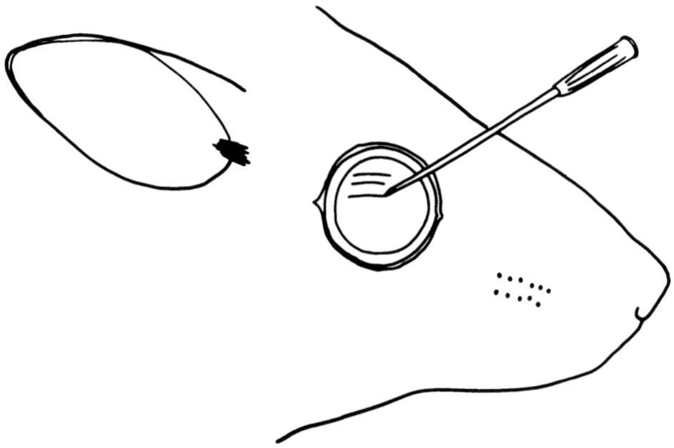

Figure 41.1 Diagram of the method used for scarification of the mouse cornea. Three 1-mm incisions are made to the upper right quadrant of the cornea using a sterile 26 G needle.

B. Quantitation of viable bacteria

Examination of numbers of viable bacteria in infected ocular tissues at different time points after infection additionally provides information on the progression of the infection as well as the efficacy of antimicrobial therapy (Kernacki and Berk, 1994; Hobden *et al.*, 1997). Following infection, mice are sacrificed by cervical dislocation and whole eyes or corneal tissue are collected. The ocular tissue is homogenized in a sterile tissue grinder containing 1.0 ml of sterile saline (0.85 M NaCl, pH 7.2) with 0.25% bovine

Table 41.3 +0 to +4 scoring system (Hazlett *et al.*, 1987)

Score	*Pathology*
+0	Clear or slight opacity, partially covering the pupil
+1	Slight opacity fully covering the anterior segment
+2	Dense opacity, partially or fully covering the pupil
+3	Dense opacity covering the entire anterior segment
+4	Corneal perforation or phthisis bulbi

Table 41.4 +0 to +10 scoring system (Gerke and Magliocco, 1971)

Score	*Pathology*
+0	No pathology
+1	Iritis
+2	Hazy opacity in wound or central cornea
+3	Hazy opacity over entire cornea
+4	Dense opacity in central cornea with remainder of cornea hazy
+5	Same as +4 but with prominent descemetocele
+6	Dense opacity of entire cornea
+7	Dense opacity of entire cornea with ulcer
+8	Same as +7 but cornea has perforated
+9	Dense opacity of most or all of cornea with iridial prolapse
+10	Any pathology not previously described, including phthisis bulbi

serum albumin (BSA). The samples are diluted serially in the same solution and plated in triplicate on *Pseudomonas* isolation agar plates (Difco, Detroit, MI). The number of viable bacteria in the infected ocular tissue is determined by counting the individual colonies on the plates from the different dilutions after incubating the plates at 37°C for 18–24 hours.

PMN quantitation

Within 18–24 hours after infection, PMNs are the predominant cell infiltrating the infected cornea. The gross observation of corneal opacity is a qualitative reflection of PMN infiltration. PMN in infected corneal tissues can be quantitated by assaying the activity of myeloperoxidase (MPO), an enzyme found in high concentrations in the azurophilic granules of PMNs (Williams *et al.*, 1983; Kernacki and Berk, 1995). After collection of either whole eyes or corneas, the tissue is homogenized in 2.0 ml of 0.5% hexadecyltrimethylammonium bromide (Sigma Chemical, St Louis, MO) in potassium phosphate buffer (50 mM, pH 6.0). Following homogenization, samples are sonicated for 10 seconds in an ice bath. The samples are then freeze-thawed three times, after which they are sonicated one more time on ice for 10 seconds. Samples are then centrifuged at 8000 *g* for 20 minutes. An aliquot of the resulting supernatant (0.1 ml) is mixed with 2.9 ml of potassium phosphate buffer containing O-dianisidine dihydrochloride (16.7 mg/100 ml; Sigma) and hydrogen peroxide (0.0005%). The change in extinction at 460 nm is monitored for 3 minutes. Determination of MPO activity can then be calculated. One unit of MPO activity degrades 1 mmol of peroxide/minute at 25°C. The degradation of 1 mmol of peroxide has been reported to give a change in extinction of 1.13×10^{-2} minutes.

The numbers of PMNs can be determined by preparing a standard curve relating PMN numbers to enzyme activity. A relatively pure population of PMNs can be collected from 2-day-infected corneal tissue by corneal disaggregation (Badenoch *et al.*, 1983). A pool of 10 infected corneas is placed in 10 ml of filtered sterilized phophate-buffered saline (PBS, 0.1 M, pH 7.4) containing 0.05% type IV collagenase (Sigma) and 0.25% pancreatin (Sigma). The disaggregation procedure is allowed to proceed for 45 minutes at 37°C with stirring. After the incubation procedure, the enzymes are inactivated by adding 2.0 ml of fetal calf serum. The sample is then centrifuged at 180 *g* for 10 minutes and resuspended in 2 ml of PBS. Viable cell counts are determined by trypan blue exclusion staining. Serial dilutions of the original sample are used to produce a standard curve relating units of MPO activity to numbers of PMNs.

Histopathology

The progression of corneal infection and the effects of antimicrobial therapy can be directly determined by histopathological evaluation. The procedures used for embedding and sectioning of ocular tissues has been described (Hazlett *et al.*, 1992). At various time points after infection, whole eyes are enucleated from the mice and briefly washed in 0.1 M sodium phosphate buffer, pH 7.4. Eyes are then fixed intact for 3 hours at 4°C in a 1:1:1 solution of 2% osmium tetroxide, 2.5% gluteraldehyde in 0.1 M sodium phosphate buffer, and 0.2 M Sorenson's buffer, pH 7.4. After fixation, eyes are rinsed briefly with 0.1 M sodium phosphate buffer, brought to room temperature, dehydrated with a series of ethanol solutions to 100%, and then transferred to propylene oxide. Specimens are infiltrated with epon-araldite resin and propylene oxide (1 : 1 mixture for 1 hour, followed by a 3 : 1 mixture for 24 hours) and infiltrated with fresh resin for 1 week before embedding in fresh resin containing the polymerizing agent DMP-30. Eyes should be oriented in the resin to facilitate cross-sectioning through the cornea. Sections (1.5 μm in thickness) from the area of the central cornea are cut with an ultramicrotome and stained for 3–5 minutes with a modified Richardson's stain. Sections may then be examined microscopically using standard bright-field optics.

Antimicrobial therapy

There are very few studies in the literature that have

examined the effect of antimicrobial agents using experimental murine models of bacterial keratitis. Most of these studies have employed the rabbit model of keratitis. One murine study by Tanaka (1981) described the efficacy of topical treatment every 3 hours for 24 hours with antibiotic (habekain, tobramycin, and gentamicin). Likewise, this study examined the effects of intramuscular injection of the same antibiotics on the corneal response to infection. The 50% effective dose was approximately 1 μg/mouse when the drug was topically applied every 3 hours after the infection, and about 0.2 mg/mouse when the antibiotic was intramuscularly injected 1 hour after the bacterial challenge. A second study introduced the antibiotic cepharolidine into the conjunctival sac four times daily for 7 days to control *P. aeruginosa* growth (Miyaji, 1977).

Advantages/disadvantages of the model

Advantages

There are several practical advantages of using mice as a model for bacterial keratitis. Unlike rabbits and guinea-pigs, mice are relatively inexpensive to purchase and house. Because of this relative inexpense, a larger number of mice can be incorporated into a given study, thereby strengthening any statistical analyses performed. Because of their size, mice are also easier to handle than larger animals. Another advantage related to the mouse's size is that topical antimicrobial therapy can be achieved with smaller drop volumes (5 μl) than with larger animals, which require drop volumes of 20–50 μl. Furthermore, a wide variety of immunological reagents are available for studies using mice. Finally, several of the *in-vitro* and *in-vivo* observations in the mouse have been demonstrated in organ-cultured human corneas and human corneal cells (Hazlett and Rudner, 1994; Wu *et al.*, 1995, 1996; Gupta *et al.*, 1997).

Disadvantages

The mouse's size also serves as a disadvantage. Unlike rabbits and guinea-pigs, it is difficult to observe any gross pathology in great detail in mice without the aid of a microscope. Unlike larger animals, mice must be handled with care during late stages of the infection (usually after 5 days post-infection) because the cornea is extremely fragile and excessive handling may lead to traumatic corneal perforation.

Contributions of the model to infectious disease therapy

The mouse model of *P. aeruginosa* keratitis has been used extensively to increase our understanding of host receptor/bacterial adhesion interactions (Hazlett *et al.*, 1987, 1991, 1993), host immune/inflammatory responses to infection (Hazlett *et al.*, 1990, 1992; Kernacki and Berk, 1994, 1995), and the role of bacterial virulence factors in pathogenesis (Ohman *et al.*, 1980; O'Callaghan *et al.*, 1996). Very few investigators have used the mouse model of bacterial keratitis to examine the efficacy of antimicrobial agents. Most studies evaluating new antimicrobial agents (Hobden *et al.*, 1992), chemotherapeutic regimens (Engel *et al.*, 1995), or drug delivery techniques (Rootman *et al.*, 1988) have relied on the rabbit model of bacterial keratitis. In spite of this dearth of previously published studies, the mouse model of bacterial keratitis should prove useful for evaluating the efficacy of novel antimicrobial agents or chemotherapeutic regimens, especially in circumstances where the experimental pharmaceuticals are in short supply. Finally, the genetic homology of inbred mice, the consistency of the murine immune response and its similarity to the human system, and the availability of immunological and molecular reagents make the mouse an attractive alternative to other animal species.

References

Badenoch, P. R., Finlay-Jones, J. J., Coster, D. J. (1983). Enzymatic disaggregation of the infected rat cornea. *Invest. Ophthalmol. Vis. Sci.*, **24**, 253–257.

Berk, R. S. (1993). Genetic regulation of the murine corneal and non-corneal response to *Pseudomonas aeruginosa*. In Pseudomonas aeruginosa *as an Opportunistic Pathogen* (eds Campa, M., Bendinelli, M., Frideman, H.), Plenum Press, New York.

Berk, R. S., Leon, M. A., Hazlett, L. D. (1979). Genetic control of the corneal response to *Pseudomonas aeruginosa*. *Infect. Immun.*, **26**, 1221–1223.

Berk, R. S., Beisel, K., Hazlett, L. D. (1981). Genetic studies on the murine corneal response to *Pseudomonas aeruginosa*. *Infect. Immun.*, **34**, 1–5.

Engel, L. S., Callegan, M. C., Hobden, J. A., Reidy, J. J., Hill, J. M., O'Callaghan, R. J. (1995). Effectiveness of specific antibiotic/steroid combinations for therapy of experimental *Pseudomonas aeruginosa* keratitis. *Curr. Eye Res.*, **14**, 229–234.

Fleiszig, S. M. J., Zaidi, T. S., Fletcher, E. L., Preston, M. J., Pier, G. B. (1994). *Pseudomonas aeruginosa* invade corneal epithelial cells during experimental infection. *Infect. Immun.*, **62**, 3485–3493.

Fleiszig, S. M. J., Preston, M. J., Grout, M., Evans, D. J., Pier, G. B. (1996). Relationship between cytotoxicity and corneal epithelial cell invasion by clinical isolates of *Pseudomonas aeruginosa*. *Infect. Immun.*, **64**, 2288–2294.

Gerke, J. R., Magliocco, M. V. (1971). Experimental *Pseudomonas aeruginosa* infection of the mouse cornea. *Infect. Immun.*, **3**, 209–216.

Gupta, S. K., Masinick, S., Garrett, M., Hazlett, L. D. (1997). *Pseudomonas aeruginosa* lipopolysaccharide binds galectin-3 and other human corneal epithelial proteins. *Infect. Immun.*, **65**, 2747–2753.

Hazlett, L. D., Rudner, X. L. (1994). Investigations on the role of flagella in adhesion of *Pseudomonas aeruginosa* to mouse and human corneal epithelial proteins. *Ophthalmic Res.*, **26**, 375–379.

Hazlett, L. D., Rosen, D. D., Berk, R. S. (1976). Experimental eye infections caused by *Pseudomonas aeruginosa*. *Ophthalmic Res.*, **8**, 311–318.

Hazlett, L. D., Moon, M. M., Strejc, M., Berk, R. S. (1987). Evidence for N-acetylmannosamine as an ocular receptor for *P. aeruginosa* adherence to scarified cornea. *Invest. Ophthalmol. Vis. Sci.*, **28**, 1978–1985.

Hazlett, L. D., Kreindler, F. B., Berk, R. S., Barrett, R. (1990). Aging alters the phagocytic capability of inflammatory cells induced into cornea. *Curr. Eye Res.*, **9**, 129–138.

Hazlett, L. D., Moon, M. M., Singh, A., Berk, R. S., Rudner, X. L. (1991). Analysis of adhesion, piliation, protease production and ocular infectivity of several *P. aeruginosa* strains. *Curr. Eye Res.*, **10**, 351–362.

Hazlett, L. D., Zucker, M., Berk, R. S. (1992). Distribution and kinetics of the inflammatory cell response to ocular challenge with *Pseudomonas aeruginosa* in susceptible versus resistant mice. *Ophthalmic Res.*, **24**, 32–39.

Hazlett, L. D., Masinick, S., Barrett, R., Rosol, K. (1993). Evidence for asialo GM1 as a corneal glycolipid receptor for *Pseudomonas aeruginosa* adhesion. *Infect. Immun.*, **61**, 5164–5173.

Hobden, J. A., O'Callaghan, R. J., Hill, J. M., Hagenah, M., Insler, M. S., Reidy, J. J. (1992). Ciprofloxacin and prednisolone therapy for experimental *Pseudomonas* keratitis. *Curr. Eye Res.*, **11**, 259–265.

Hobden, J. A., Masinick, S. A., Barrett, R. P., Hazlett, L. D. (1995). Aged mice fail to upregulate ICAM-1 after *Pseudomonas aeruginosa* corneal infection. *Invest. Ophthalmol. Vis. Sci.*, **36**, 1107–1114.

Hobden, J. A., Masinick, S. A., Barrett, R. P., Hazlett, L. D. (1997). Proinflammatory cytokine deficiency and pathogenesis of *Pseudomonas aeruginosa* keratitis in aged mice. *Infect. Immun.*, **65**, 2754–2758.

Kernacki, K. A., Berk, R. S. (1994). Characterization of the inflammatory response induced in mice by corneal infection with *Pseudomonas aeruginosa*. *J. Ocular Pharm.*, **10**, 281–288.

Kernacki, K. A., Berk, R. S. (1995). Characterization of arachidonic acid metabolism and the polymorphonuclear leukocyte response in mice infected with *Pseudomonas aeruginosa*. *Invest. Ophthalmol. Vis. Sci.*, **36**, 16–23.

McLeod, S. D., LaBree, L. D., Tayanipour, R., Flowers, C. W., Lee, P. P., McDonnell, P. J. (1995). The importance of initial management and treatment of severe infectious corneal ulcers. *Ophthalmology*, **102**, 1943–1948.

Miyaji, A. (1977). Experimental study on keratitis in mice. The influence of topical antibiotics on establishment of *Pseudomonas aeruginosa*. *Nippon Ganka Gakki Zasshi—Acta Soc. Ophthalmol. Jpn.*, **81**, 560–568.

Moon, M. M., Hazlett, L. D., Hancock, R. E. W., Berk, R. S., Barrett, R. (1988). Monoclonal antibodies provide protection against ocular *Pseudomonas aeruginosa* infection. *Invest. Ophthalmol. Vis. Sci.*, **29**, 1277–1284.

National Research Council (1996). *Guide for the Care and Use of Laboratory Animals*. National Academy Press, Washington, DC.

O'Callaghan, R. J., Engel, L. S., Hobden, J. A., Callegan, M. C., Green, L. C., Hill, J. M. (1996). *Pseudomonas* keratitis. The role of an uncharacterized exoprotein, protease IV, in corneal virulence. *Invest. Ophthalmol. Vis. Sci.*, **37**, 534–543.

Ohman, D. E., Burns, R. P., Iglewski, B. H. (1980). Corneal infections in mice with toxin A and elastase mutants of *Pseudomonas aeruginosa*. *J. Infect. Dis.*, **142**, 547–555.

Pepose, J. S., Wilhelmus, K. R. (1992). Divergent approaches to the management of corneal ulcers. *Am. J. Ophthalmol.*, **114**, 30–32.

Preston, M. J., Fleisiz, S. M. J., Zaidi, T. S. *et al.* (1995). Rapid and sensitive method for evaluating *Pseudomonas aeruginosa* virulence factors during corneal infection in mice. *Infect. Immun.*, **63**, 3497–3501.

Preston, M. J., Seed, P. C., Toder, D. S. *et al.* (1997). Contribution of proteases and LasR to the virulence of *Pseudomonas aeruginosa* during corneal infections. *Infect. Immun.*, **65**, 3086–3090.

Reidy, J. J., Hobden, J. A., Hill, J. M., Forman, K., O'Callaghan, R. J. (1991). The efficacy of topical ciprofloxacin and norfloxacin in the treatment of experimental *Pseudomonas* keratitis. *Cornea*, **10**, 25–28.

Rootman, D. S., Hobden, J. A., Jantzen, J. A., Gonzalez, J. R., O'Callaghan, R. J., Hill, J. M. (1988). Iontophoresis of tobramycin for the treatment of experimental *Pseudomonas* keratitis in the rabbit. *Arch. Ophthalmol.*, **106**, 262–265.

Schein, O. D., Buehler, P. O., Stamler, J. F., Verdier, D. D. (1994). The impact of overnight wear on the risk of contact lens-associated ulcerative keratitis. *Arch. Ophthalmol.*, **112**, 186–190.

Tanaka, Y. (1981). Effects of habekacin, a novel aminoglycoside antibiotic, on experimental corneal ulceration due to *Pseudomonas aeruginosa*. *J. Antibiot.*, **7**, 892–897.

Williams, R. N., Patterson, C. A., Eakins, K. E. (1983). Quantitation of ocular inflammation: evaluation of polymorphonuclear leukocyte inflammation by measuring myeloperoxidase activity. *Curr. Eye Res.*, **2**, 465–471.

Wu, X., Gupta, S. K., Hazlett, L. D. (1995). Characterization of *P. aeruginosa* pili binding human corneal epithelial proteins. *Curr. Eye Res.*, **14**, 969–977.

Wu, X., Kurpakus, M., Hazlett, L. D. (1996) Some *P. aeruginosa* pilus binding proteins of human corneal epithelium are cytokeratins. *Curr. Eye Res.*, **15**, 782–791.

Chapter 42

The Rabbit Intrastromal Injection Model of Bacterial Keratitis

R. J. O'Callaghan, L. S. Engel and J. M. Hill

Background

A model of *Pseudomonas* keratitis initiated by an intrastromal injection of bacteria into the rabbit cornea has been employed for many years to study ocular antimicrobial therapies (Table 42.1) and for determining mechanisms of pathogenesis (Table 42.2). The procedures used for the *Pseudomonas* model have been employed for developing models of pneumococcus (Johnson and Allen, 1971, 1975; Harrison *et al.*, 1983, Johnson *et al.*, 1991, 1992, 1995) and *Staphylococcus aureus* keratitis (Kupferman and Leibowitz, 1977; Leibowitz *et al.*, 1981; Mondino and Kowalski, 1982; Edwards and Schlech, 1985; Mondino *et al.*, 1987a,b; Kaufman *et al.*, 1991; Callegan *et al.*, 1992a,b, 1994a,b,c,d, 1995; Engel *et al.*, 1996; Moreau *et al.*, 1997; O'Callaghan *et al.*, 1997). There are a few reports describing the development of a rabbit intrastromal model for *Serratia* keratitis (Berstein and Maddox, 1973; Lyerly *et al.*, 1981; Hume *et al.*, 1998), but no quantitative chemotherapeutic studies have been described. A topical route of inoculation, employing

Table 42.1 Chemotherapy of *Pseudomonas* keratitis in the rabbit intrastromal injection model

Antibiotic class	*Studies*
β-lactams	Piatkowska, 1966; Galin *et al.*, 1968
Aminoglycosides	Fugiuele, 1968; Bohigian *et al.*, 1971; Smolin *et al.*, 1973, 1974; Belfort *et al.*, 1975; Kupferman and Leibowitz, 1976; Alpren *et al.*, 1979, Davis *et al.*, 1979; Hobden *et al.*, 1988a,b, 1989; Rootman *et al.*, 1988; Gritz *et al.*, 1992; Engel *et al.*, 1995
Quinolones	Galin *et al.*, 1968; Hobden *et al.*, 1990a, b, 1992, 1993a,b,c; O'Brien *et al.*, 1988; Reidy *et al.*, 1991; Gritz *et al.*, 1992; Engel *et al.*, 1996
Polymyxins	Fugiuele, 1968; Kupferman and Leibowitz, 1976

Table 42.2 Studies of pathogenic mechanisms for *Pseudomonas* keratitis in the rabbit intrastromal injection model

Pathogenic factor	*Studies*
Polymorphonuclear cell activity	Hobden *et al.*, 1993a
Pseudomonas elastase	Kreger and Gray, 1978; Gray and Kreger, 1975; Kessler and Spierer, 1984
Exotoxin A	O'Callaghan *et al.*, 1996
Protease IV	O'Callaghan *et al.*, 1996; Engel *et al.*, 1997

Handbook of Animal Models of Infection
ISBN 0-12-775390-7

bacteria adhering to contact lenses as an inoculum, has been described for producing *Pseudomonas* keratitis in rabbits (Brockman *et al.*, 1992), but this model has been used on a limited basis for chemotherapeutic (Brockman *et al.*, 1992) and pathogenic studies (O'Callaghan *et al.*, 1996). The mouse model of *Pseudomonas* keratitis has been employed extensively for the study of bacterial pathogenesis and for immunologic aspects of keratitis (Chapter 41); however, this model is not suited for quantitative measurements of antibiotic therapy.

Rabbits

Young New Zealand white rabbits, 2–3 kg, when intrastromally injected with *Pseudomonas*, produce a marked inflammatory response while older rabbits (5–6 kg) produce a less intense inflammatory reaction to infection (Hobden *et al.*, 1993a). Typical chemotherapy experiments employ rabbits weighing 2–3 kg.

Materials required for chemotherapeutic studies

Bacteriology laboratory

Storage of bacterial strains is best accomplished by freezing the organisms in an ultra-low freezer (–70 °C). Preparation of the bacterial inoculum requires both solid and liquid culture media. A shaking incubator is useful for the propagation of bacteria in broth culture. Agar plates are incubated in a standard incubator of a relatively large size to accommodate the large number of plates required for chemotherapeutic determinations. Specialized media for detection of proteases and hemolysins are recommended to ensure that the bacteria maintain their virulence throughout storage and throughout infection.

Ocular inoculation and evaluation

The corneal inoculation process requires an injection of 5–10 µl of the test sample using a microliter syringe (Hamilton Co., Reno, NV) with a 30 G needle. Locking forceps (e.g., Graefe-type fixation forceps, Roboz Surgical Instrument Company, Rockville, MD) are used for grasping the eye when making an intrastromal injection. A biomicroscope (e.g., Topcon, Kogakukai K.K., Tokyo, Japan) is required to perform slit lamp examination (SLE) of the eyes. Changes in ocular inflammation can be further documented by photography and therefore a camera is highly desirable. A tissue homogenizer is required for bacterial enumeration because the cornea is a particularly fibrous tissue and homogenous dispersal of the tissue allows accurate and precise determination of the number of bacteria. An Ultraturax Tissumizer (Tekmar, Cincinnati, OH) or its equivalent is recommended.

Anesthesia and medications

Ketamine (100 mg/ml; Aveco, Fort Dodge, IA) in 10 ml bottles, and xylazine (100 mg/ml; Miles, Shawnee, KS) in 50 ml bottles, can be obtained from the Butler Company, Columbus, OH. Proparacaine (0.5%), in 15 ml bottles, can be obtained from Bausch & Lomb, Tampa, FL. Topical antibiotic ointment in 3.5 g tubes, either Gentrasul (gentamicin 3 mg/ml, Bausch & Lomb) or Maxitrol (3.5 mg/ml neomycin, 10 000 units/ml polymyxin B, and 0.1% dexamethasone, Alcon, Fort Worth, TX), were employed. Pentobarbital powder (Sigma Chemical Co., St Louis, MO) was dissolved in sterile water at a concentration of 100 mg/ml.

Miscellaneous

For membranectomies, a hand-held electric cauterizer (high-temperature fine-tip cauterizer, Aaron Medical Industries, St Petersburg, FL) is essential for limiting bleeding during this surgical procedure. At the termination of the experiment, a scalpel with a no. 10 blade, surgical scissors, a hemostat, and forceps are needed for the removal of the corneas from sacrificed rabbits.

Procedures for anesthesia

For general anesthesia, rabbits are injected intramuscularly with a mixture of ketamine (50 mg/kg) and xylazine (10 mg/kg). Immediately prior to intrastromal injection or membranectomy, the eye is anesthetized by topical administration of two drops of proparacaine. Because the injection of ketamine and xylazine can effect iritis and possibly other types of ocular inflammation, slit-lamp examination is best performed prior to the administration of anesthesia.

Nictitating membranectomy (only for drug-delivery devices)

The nictitating membrane should be removed at least 3 days prior to the initiation of infection. This surgery involves the application of general and local anesthesia followed by the cutting of the membrane. The nictitaing membrane is extended with forceps and clamped with a hemostat at the base. The membrane extending beyond the hemostat is then cut with a scalpel. The membrane stump is cauterized with a fine tipped cauterizing unit to avoid bleeding following the release of the hemostat. The membrane stump is treated with antibiotic ointment to prevent postoperative infection.

Bacterial strains

Storage

The methods used for storage of bacterial cultures and the preparation of the bacterial inoculum are critical in achieving reproducible evaluations of antibiotic effectiveness. The bacterial strain used should be well characterized, preferably obtained from the American Type Culture Collection or from research laboratories employing the intrastromal keratitis model. Strains used previously in models of keratitis are preferred. *Pseudomonas aeruginosa* strain ATCC 27853 and *Staphylococcus aureus* strain ATCC 25923 have been used extensively in the study of antibiotic effectiveness in the rabbit intrastromal keratitis model.

Upon receipt of a new strain for use in the keratitis model, a large stock of the first passage of the bacteria is grown and stored. To prepare these stock samples, the bacteria are grown on solid medium to determine that only one colony type is present. From this growth on agar, three or more isolated colonies are subcultured into liquid medium (e.g., tryptic soy broth, TSB, about 75 ml) and grown to log phase. These bacteria are mixed with sterile glycerol to a final concentration of 15%, distributed to sterile tubes, labeled with the strain and passage date, and frozen at $-70\,^{\circ}$C. Multiple aliquots of the strain from the first passage ensures that reproducible infections can be generated over long periods of time. To avoid loss of cultures due to freezer failure, a few culture tubes are stored in an alternate freezer, separate from the remainder of the tubes. Strains are also preserved by growing in a semi-solid storage agar (e.g., 0.75% tryptic soy agar, TSA, Difco Laboratories, Detroit, MI) in screw-capped test tubes whose lids are subsequently sealed with paraffin.

Bacterial growth

To prepare a bacterial inoculum, one vial of frozen bacteria is used to streak an agar plate resulting in multiple isolated colonies. The production of bacterial products related to virulence (i.e., protease or hemolysins) is determined by zones of proteolysis or hemolysis around isolated colonies on specialized agar plates. *Pseudomonas* and *Serratia* strains are grown on skim milk agar to observe production of protease. *Staphylococcus aureus* is grown on sheep and rabbit blood agars to detect the action of various hemolysins. *Staphylococcus* ATCC 25923 often contains some non-hemolytic colonies that are avoided for typical chemotherapy experiments.

Bacterial characterization

Every strain of bacteria used for bacterial keratitis should be characterized in terms of its growth characteristics both *in vitro* and *in vivo*. The *in-vitro* analysis is needed to determine when 10^8cfu/ml of log phase bacteria are present. The *in-vivo* growth rates are needed to select treatment times for antibiotic evaluation; most efficient antibiotic killing of bacteria is generally obtained when bacteria are growing at their maximal rate.

For *in-vitro* growth studies, bacteria from isolated colonies are inoculated into TSB and incubated with shaking overnight. The resulting stationary-phase culture is then subcultured (a 1:100 dilution) in TSB with shaking. After 1 hour, and periodically thereafter, the optical density is determined at a wavelength of 650 nm and the number of cfu/ml is determined by plating aliquots of serial dilutions on TSA. From these data a curve of the OD_{650} versus bacterial cfu/ml can be defined. The curve will indicate the OD_{650} of the culture that corresponds to 10^8 cfu/ml. (For *Pseudomonas* strains this value is about 0.2 and for *Staphylococcus* the value is about 0.3.)

For *in vivo* studies of intracorneal growth rate, groups of rabbits (e.g., five groups of two rabbits per group) are injected with bacteria in both corneas. An inoculum (10 µl) of 10^2 cfu per cornea is used for a rapid-growing strain (e.g., *Staphylococcus*) or 10^3 cfu per cornea for a slower-growing strain (e.g., *Pseudomonas*). Eyes are evaluated every 5 or 6 hours by SLE and by determining the number of cfu per cornea in a group. The experiment is ended by sacrificing the rabbits when the SLE score reaches 20. The experiment provides data for two curves, a plot of cfu per cornea with time and a plot of SLE scores with time. The cfu data indicate the times when bacteria are growing at logarithmic rates and when this rapid growth declines. The SLE data indicate initiation of ocular inflammation and tissue damage and how they progress with time.

Preparation of the bacterial inoculum

A frozen culture is thawed, inoculated on to agar plates, and incubated overnight. Isolated colonies producing characteristic virulence factors (protease or hemolysins) are inoculated into TSB. After overnight incubation, the broth culture is subcultured (1:100 dilution) in TSB. Once the bacteria reach 10^8/ml, the culture is diluted in TSB to obtain 10^5 cfu/ml for *Pseudomonas* or 10^4 cfu/ml for *Staphylococcus*. A 10 µl injection of these dilutions will deliver precisely 10^3 cfu of *Pseudomonas* or 10^2cfu of *Staphylococcus* per cornea. Rabbits are randomly assigned to groups and anesthetized for injection.

Injection procedure

The superior rectus muscle is grasped with locking forceps and held to immobilize the eye. This muscle lies just above the cornea in the center of the eye and is held by pressing the forceps against the sclera. The bacterial inoculum in a microsyringe with a 30 G needle is held with the opposite hand. The needle, with the bevel up, is inserted into the cornea, just penetrating the epithelium, about 2–3 mm from the corneal center, on an angle such that the needle almost parallels the corneal surface. The needle is advanced into the center of the cornea until the bevel fully penetrates the corneal tissue and the fluid is injected. The liquid will form a bleb and sometimes causes a "cracked glass" appearance of the cornea. The needle is slowly withdrawn, completing the process. Rabbits are allowed to awaken after the injection without an analgesic. The infection process does not induce stress in the rabbit until the SLE score exceeds a value of 20; experiments must be terminated before the infections reach such extremes.

Parameters to monitor during infection and for quantifying the effectiveness of the treatment

SLE scoring

Other than the determination of cfu per cornea, the most important measurement to perform is SLE scoring. Rabbits should not be anesthetized for SLE scoring. The slit-lamp microscope is used to grade each of seven ocular parameters on a scale of 0 for normal to 4 for most severe in increments of 0.25. The seven parameter grades are then added to yield a single SLE score for the eye. Two or more individuals should independently score the same eyes in a masked fashion (without knowing which, if any, treatment the eyes received). The seven parameters graded are: conjunctival chemosis (edema), conjunctival injection (redness), corneal edema, corneal infiltrate, fibrin accumulation in the anterior chamber, hypopyon formation (polymorphonuclear cell — PMN — accumulation in the anterior chamber), and iritis (Table 42.3). The SLE for a maximally inflamed eye is 28, but this is a theoretical value and eyes should not be allowed to advance significantly beyond a score of 20. SLE scoring is typically performed, as a minimum, before starting an experiment, at the start of chemotherapy, and 1 hour after the last application of chemotherapy (at the time of sacrifice).

Cfu per cornea

This is the essential assay that determines the effectiveness of the chemotherapy being evaluated. Rabbits are sacrificed by anesthetizing with ketamine and xylazine, as described above for general anesthesia, and then by injecting pentobarbital (50 mg/kg) directly into the heart. After assuring the absence of heart sounds and reflex reactions, the corneas are harvested by cutting the tissue at the corneal–scleral junction. The corneas are transferred to a sterile Petri dish and cut into multiple pieces (10–15) using a sterile scalpel. The corneal pieces are placed into a sterile test tube containing 3.0 ml of sterile phosphate buffered saline (PBS, 0.02 M phosphate, pH 7.2, 0.15 M NaCl) at 4 °C. The tissue is then homogenized for 30 seconds and aliquots are serially diluted 10-fold (0.5 ml plus 4.5 ml) in sterile PBS. Samples from each dilution are plated in triplicate by pipetting 100 μl on to the surface of each agar plate (e.g., tryptic soy agar) and spread-

Table 42.3 Parameters for slit-lamp examination scoring (PMNs, polymorphonuclear cells)

Ocular site	*Parameter*	*Scoring**
Conjunctiva	Injection	"Redness" of tissue due to inflamed blood vessels†
	Chemosis	Edema due to capillary leakage
Cornea	Infiltrate	Migration of PMN into the stroma from the limbus
	Edema	Fluid accumulation expressed as a percentage of total area showing thickening
Anterior chamber	Fibrin	Size of fibrin deposit viewed by opacity of fibrinous clot
	Hypopyon	Size of accumulated PMN and debris
Iris	Iritis	Swelling of blood vessels and overall redness

*All scoring is based on a scale of 0 for normal to 4 for maximum change; values are assigned subjectively based on the degree of change observed.
† Mild inflammation of the conjunctiva results from grasping of the rectus muscle during the intrastromal injection process.

ing evenly over the surface using a sterile bent glass rod. Plates are incubated for 24 hours and the positions of all colonies are marked and counted. The plates are incubated for an additional 24–48 hours and examined again for the emergence of additional colonies; this second incubation is needed because antibiotic present in the corneal homogenate can slow the appearance of colonies. The colony counts are expressed as log values determined from the average of the three plates containing between 30 and 300 colonies.

Other assays of ocular inflammation

The influx of PMNs into the cornea can be assayed in direct relationship to the amount of myeloperoxidase activity in a corneal homogenate. The homogenate prepared for quantifying the cfu per cornea can be employed for this determination, as described previously (Hobden *et al.*, 1993a,b; O'Callaghan *et al.*, 1997).

Conditions for antibacterial therapy

Typically, one decides upon the course of therapy relative to the growth curve of the bacteria within the cornea and the increase in the SLE score during infection. The most effective therapy can be obtained while bacteria are replicating logarithmically, as determined from the intracorneal growth curve. Many antibiotics that are bactericidal (e.g., fluoroquinolones, vancomycin, or cephazolin) show far less effectiveness when applied while the bacteria are in stationary phase. However, therapy should *not* begin before the injection of bacteria into the eye or very shortly after.

Therapy started too soon after bacterial injection is unrealistic relative to the human condition and precedes the ocular changes associated with infection that could alter the effectiveness of a medication. Ideally, the time to start therapy is when the experimentally infected eye is most similar to the human eye when the patient first seeks medical attention. We have estimated that in the rabbit eye a SLE score between 3 and 6 is a preferred condition for initiating therapy.

The total treatment time for therapy is determined by a variety of factors. The concentration of drug used in experimental therapy should match that used clinically if the drug is licensed for clinical use. Newly developed compounds are tested initially at the highest practical concentrations that are not toxic to the eye. There is value in matching the dosing schedule to that used previously for other forms of therapy. By standardizing the dosing schedule, the effectiveness of various drugs can be compared. We have treated *Pseudomonas*-infected corneas from 16–26 hours post-infection (p.i.) and *Staphylococcus*-infected corneas from 4–9 hours p.i. (early therapy), from 10–15 hours p.i. (late therapy) or from 4–15 hours p.i. (extended therapy). If a drug formulation produces sterile corneas, the duration of therapy is decreased to obtain a quantitative measure of the drug's effectiveness.

The route of drug administration for keratitis is generally by topical drops. Iontophoresis, drug-delivery devices (e.g., collagen shields) or subconjunctival injection have also been tested using the rabbit model of bacterial keratitis (Unterman *et al.*, 1988; Hobden *et al.*, 1988a,b, 1989, 1990a,b; Reidy *et al.*, 1991; Clinch *et al.*, 1992; Hill *et al.*, 1992a,b; Callegan *et al.*, 1994a, 1995).

Therapy is ended about 1 hour before the animal is sacrificed. If tissue is harvested any sooner after the time of drug administration, drug on the corneal surface could possibly contaminate the tissue sample. Such contamination could alter bacterial growth when determining the cfu per cornea or lead to artificially high drug concentrations found in corneal tissue.

Determination of antibiotic in ocular tissues

The aqueous humor, the whole cornea or components of the cornea (corneal epithelium or stroma) can be assayed. Antibiotic can be determined by bioassay, HPLC, or immune assay (RIA or EMIT; Engel *et al.*, 1996; Green *et al.*, 1996). Whole corneal homogenates or stromal homogenates offer the disadvantage of high protein content and an inherent difficulty in achieving efficient drug extraction. Aqueous humor is often preferred because it can be used directly in bioassays and in immune assays. Aqueous humor is obtained by anesthetizing a rabbit and then inserting a 30 G needle through the cornea for fluid aspiration. Aqueous humor samples of 100 μl or more can be routinely obtained.

Assessments of the model

Disadvantages

- *Intrastromal injection requires technical skill*. The corneal injection procedure requires a steady hand, good vision, and depth perception. With experience, the injection process can be mastered. Injections that perforate the cornea alter the course of infection and could yield misleading information regarding the effectiveness and pharmacokinetics of the therapeutic formulation.
- *Precision of the model is dependent upon the number and quality of the bacteria in the inoculum*. The model requires the inoculum to consist of a precisely determined number of bacteria in the log phase and genetically stable organisms that express key virulence factors. Assessment is best accomplished by a team of researchers that includes a bacteriologist, a pharmacologist, and an ophthalmologist.

- *Atypical route of infection*. The intrastromal injection of bacteria is not analogous, in terms of the route of inoculation, to the most common forms of human bacterial keratitis. The invasion of tissue by bacteria from the corneal surface is the more natural means of initiating infection. Unfortunately, topical inoculations often fail to produce an infection or yield infections in an imprecise fashion (large variations in the number of bacteria in the tissue).

Advantages

- *Precision for an in-vivo model*. The experienced researcher can demonstrate a statistically significant decrease of less than 1 log in the cfu per cornea of a treated eye relative to an untreated eye. This is possible because the bacteria grow in a manner that is highly predictable from eye to eye and from experiment to experiment.
- *Versatility*. The model can be used for a variety of chemotherapeutic studies or for the analysis of pathogenic mechanisms. The killing of bacteria can be precisely determined so the effectiveness of various drugs (or drug combinations) can be compared to other drugs tested simultaneously, previously, or subsequently. Tissue can be obtained and assayed for pharmacokinetic determinations; this is of particular advantage when studying new formulations or when assessing the value of experimental drug-delivery systems. The effectiveness of anti-inflammatory agents can be determined because inflammatory changes associated with infection are reproducible and can be quantified by SLE and other means.
- *The data are applicable*. Because the rabbit cornea has been a standard model for numerous ocular studies, data can be extrapolated to the therapy of human keratitis. Measurements of a drug's effectiveness in non-human primates can be bypassed in many situations.

Contributions of the model to infectious disease therapy

The rabbit model of bacterial keratitis has been used extensively to determine the effectiveness of ocular antibiotics (Table 42.1). The effectiveness of experimental drug-delivery devices (Hobden *et al.*, 1988a,b, 1989, 1990a,b; Reidy *et al.*, 1991; Clinch *et al.*, 1992; Hill *et al.*, 1992a,b; Callegan *et al.*, 1994a, 1995) and of combinations of antibiotics and anti-inflammatory drugs have also been analyzed in these models (Bohigian and Foster, 1977; Fraser-Smith and Matthews, 1988; Hobden *et al.*, 1992, 1993b,c; Engel *et al.*, 1995). Experiments on ciprofloxacin formulations (Hobden *et al.*, 1990b, 1992, 1993b,c,d; Reidy *et al.*, 1991; Callegan *et al.*, 1992a,b, 1994b; Engel *et al.*, 1993, 1996) and on collagen shields for drug delivery (Hobden *et al.*, 1988a, 1990b; Clinch *et al.*, 1992; Hill *et al.*, 1992b; Callegan *et al.*, 1994a) are two examples of chemotherapeutic improvements that have been pioneered in the rabbit intrastromal model of bacterial keratitis.

Elucidation of pathophysiology

Studies of experimental bacterial keratitis have demonstrated that the tissue damage that causes loss of visual acuity and blindness is mediated by a combination of bacterial and host factors. Bacterial keratitis in leukopenic rabbits is significantly less severe than that of normal, immunocompetent rabbits (Harrison *et al.*, 1983; Hobden *et al.*, 1993a). PMNs, through the action of oxidative molecules and proteases, are a prime mediator of corneal damage.

Efforts to identify bacterial factors responsible for tissue damage have shown that each pathogen produces a few proteins that are very toxic to the eye (Kessler *et al.*, 1977; Kreger and Gray, 1978; Lyerly *et al.*, 1981; Kessler and Spierer, 1984; Spierer and Kessler, 1984; Johnson *et al.*, 1991, 1992, 1995; Callegan *et al.*, 1994c; O'Callaghan *et al.*, 1996, 1997; Engel *et al.*, 1997; Moreau *et al.*, 1997). Studies of *Pseudomonas* keratitis in the rabbit model have demonstrated that proteases are vital for ocular virulence (O'Callaghan *et al.*, 1996; Engel *et al.*, 1997, 1998a,b). Such studies have led to the recognition of *Pseudomonas* proteases IV as an important ocular virulence factor (O'Callaghan *et al.*, 1996; Engel *et al.*, 1997, 1998a,b). In the *Staphylococcus* model, studies to date have demonstrated that α-toxin, a hemolytic pore-forming toxin, mediates much of the ocular damage and inflammation (Callegan *et al.*, 1994c; O'Callaghan *et al.*, 1996; Moreau *et al.*, 1997; Sloop *et al.*, 1999). Other molecules toxic to the cornea may be produced by some strains, but these toxins have not yet been identified. In the pneumococcus model, virulence has been related to the production of pneumolysin, a pore-forming toxin that activates complement (Johnson *et al.*, 1991, 1992, 1995). This research is making possible a search for inhibitors that can abolish the activity of toxic bacterial proteins. Such inhibitors could be used in conjunction with antibiotics and antinflammatory drugs to prevent the corneal scarring and resulting blindness mediated by bacterial keratitis.

References

Alpren, T. V., Hyndiuk, R. A., Davis, S. D. (1979). Cryotherapy for experimental *Pseudomonas* keratitis. *Arch. Ophthalmol.*, **97**, 711–714.

Belfort, R. Jr, Smolin, G., Okumoto, M., Kim, H. B. (1975). Nebcin in the treatment of experimental *Pseudomonas* keratitis. *Br. J. Ophthalmol.*, **59**, 725–729.

Berstein, H. N., Maddox, Y. T. (1973). Corneal pathogenicity of *Serratia marcescens* in the rabbit. *Trans. Am. Acad. Ophthalmol.*, **77**, 432–440.

Bohigian, G. M., Foster, C. S. (1977). Treatment of *Pseudomonas* keratitis in the rabbit with antibiotic-steroid combinations. *Invest. Ophthalmol. Vis. Sci.*, **16**, 553–556.

Bohigian, G., Okumoto, M., Valenton, M. (1971). Experimental *Pseudomonas* keratitis. Treatment with and evaluation of carbenicillin and gentamicin in combination. *Arch. Ophthalmol.*, **86**, 432–437.

Brockman, E. B., Tarantino, P. A., Hobden, J. A. *et al.* (1992). Keratomy model of *Pseudomonas* keratitis: gentamicin chemotherapy. *Refract. Corneal Surg.*, **8**, 465–468.

Callegan, M. C., Hobden, J., Hill, J. M., Insler, M., O'Callaghan, R. J. (1992a). Methicillin-resistant *Staphylococus aureus* keratitis in the rabbit: therapy with ciprofloxacin, vancomycin, and cefazolin. *Curr. Eye Res.*, **11**, 1111–1119.

Callegan, M. C., Hobden, J., Hill, J. M., Insler, M., O'Callaghan, R. J. (1992b). Topical antibiotic therapy for the treatment of experimental *Staphylococcus aureus* keratitis. *Invest. Ophthalmol. Vis. Sci.*, **33**, 3017–3023.

Callegan, M. C., Clinch, T. F., Engel, L. S., Hill, J. M., Kaufman, H. E., O'Callaghan, R. J. (1994a). Tobramycin-hydrated collagen shield and drops applied to shields (DATS) for staphylococcal keratitis. *Curr. Eye Res.*, **13**, 875–878.

Callegan, M. C., Engel, L. S., Hill, J. M., O'Callaghan, R. J. (1994b). Tobramycin versus ciprofloxacin for the treatment of staphylococcal keratitis. *Invest. Ophthalmol. Vis. Sci.*, **35**, 1033–1037.

Callegan, M. C., Engel, L. S., Hill, J. M., O'Callaghan, R. J. (1994c). Corneal virulence of *Staphylococcus aureus*: The role of α-toxin and protein A in pathogenesis. *Infect. Immun.*, **62**, 2478–2482.

Callegan, M. C., O'Callaghan, R. J., Hill, J. M. (1994d). Pharmacokinetics in the treatment of bacterial keratitis. *Clin. Pharmacokinet.*, **27**, 129–149.

Callegan, M. C., O'Callaghan, R. J., Hill, J. M. (1995). Ocular drug delivery: a comparison of transcorneal iontophoresis to corneal collagen shields. *Int. J. Pharm.*, **123**, 173–179.

Clinch, T. F., Hobden, J., Hill, J. M., O'Callaghan, R. J., Kaufman, H. E. (1992). Collagen shields containing tobramycin for sustained therapy (24 hours) of experimental *Pseudomonas* keratitis. *CLAO*, **18**, 245–247.

Davis, S. D., Sarff, L. D., Hyndiuk, R. A. (1979). Comparison of therapeutic routes in experimental *Pseudomonas* keratitis. *Am. J. Ophthalmol.*, **87**, 710–716.

Edwards, J. G., Schlech, B. A. (1985). Efficacy of topical chloramphenicol and tobramycin ophthalmic solutions in preventing severe *Staphylococcus aureus* keratitis in rabbits. *Curr. Eye Res.*, **4**, 821–829.

Engel, L. S., Callegan, M. C., Hill, J. M., O'Callaghan, R. J. (1993). Bioassays for quantitating ciprofloxacin and tobramycin in aqueous humor. *J. Ocular Pharmacol.*, **9**, 311–320.

Engel, L. S., Callegan, M. C., Hobden, J. A., Reidy, J. J., Hill, J. M., O'Callaghan, R. J. (1995). Effectiveness of specific antibiotic/steroid combinations for therapy of experimental *Pseudomonas aeruginosa* keratitis. *Curr. Eye Res.*, **14**, 229–234.

Engel, L. S., Callegan, M. C., Hill, J. M., Folkens, A., O'Callaghan, R. J. (1996). The effectiveness of two ciprofloxacin formulations for experimental *Pseudomonas* and *Staphylococcus* keratitis. *Jpn. J. Ophthalmol.*, **40**, 212–219.

Engel, L. S., Hobden, J. A., Moreau, J. M., Callegan, M. C., Hill, J. M., O'Callaghan, R. J. (1997). *Pseudomonas* deficient in protease IV has significantly reduced corneal virulence. *Invest. Ophthalmol. Vis. Sci.*, **38**, 1535–1542.

Engel, L. S., Hill, J. M., Caballero, A. R., Green, L. C., O'Callaghan, R. J. (1998a) Protease IV, a unique extracellular protease and virulence factor from *Pseudomonas aeruginosa J. Biol. Chem.*, **273**, 16792–16797.

Engel, L. S., Hill, J. M., Moreau, J. M., Green, L. C., Hobden, J. A., O'Callaghan, R. J. (1998b). *Pseudomonas aeruginosa* protease IV produces corneal damage and contributes to bacterial virulence. *Invest. Ophthalmol. Vis. Sci.*, **39**, 662–665.

Fraser-Smith, E. B., Matthews, T. R. (1988). Effect of ketorolac on *Pseudomonas aeruginosa* ocular infection in rabbits. *J. Ocular Pharmacol.*, **4**, 101–109.

Fuguiele, F. P. (1968). Treatment of *Pseudomonas* infection of the rabbit cornea. Comparative study. *Am. J. Ophthalmol.*, **66**, 276–279.

Galin, M. A., Davidson, R. A., Harris, L. S. (1968). Experimental corneal infections. Evaluation of nalidixic acid in *Proteus* and *Pseudomonas* keratitis. *Am. J. Ophthalmol.*, **66**, 447–451.

Gray, L. D., Kreger, A. S. (1975). Rabbit corneal damage produced by *Pseudomonas aeruginosa* infection. *Infect. Immun.*, **12**, 419–432.

Green, L. C., Callegan, M. C., Engel, L. S. *et al.* (1996). Pharmacokinetics of topically applied ciprofloxacin in rabbit tears. *Jpn. J. Ophthalmol.*, **40**, 123–126.

Gritz, D. C., McDonnell, P. J., Lee, T. Y., Tang-Liu, D., Hubbard, B. B., Gwon, A. (1992). Topical olfoxacin in the treatment of *Pseudomonas* keratitis in a rabbit model. *Cornea.*, **11**, 143–147.

Harrison, J., Karcioglu, Z., Johnson, M. (1983). Response of leukopenic rabbits to pneumococcal toxin. *Curr. Eye Res.*, **2**, 705–710.

Hill, J. M., O'Callaghan, R. J., Hobden, J., Kaufman, H. E. (1992a). Collagen shields for ocular drug delivery. In *Ophthalmic Drug Delivery Systems* (ed. Mitra, A. K.), pp. 261–273. Marcel Dekker, New York.

Hill, J. M., O'Callaghan, R. J., Hobden, J. (1992b). Ocular iontophoresis. In *Ophthalmic Drug Delivery Systems* (ed. Mitra, A. K.), pp. 331–354. Marcel Dekker, New York.

Hobden, J., Reidy, J., O'Callaghan, R., Hill, J., Insler, M., Rootman D. (1988a). Treatment of experimental *Pseudomonas aeruginosa* using collagen shields containing tobramycin. *Arch. Ophthalmol.*, **106**, 1605–1607.

Hobden, J., Rootman, D. S., O'Callaghan, R. J., Hill, J. M. (1988b). Iontophoretic application of tobramycin to uninfected and *Pseudomonas aeruginosa*-infected rabbit corneas. *Antimicrob. Agents Chemother.*, **21**, 978–981.

Hobden, J. A., O'Callaghan, R. J., Hill, J. M., Reidy, J. J., Rootman, D. S., Thompson, H. W. (1989). Tobramycin iontophoresis into corneas infected with drug-resistant *Pseudomonas aeruginosa*. *Curr. Eye Res.*, **11**, 1163–1169.

Hobden, J. A., Reidy, J. J., O'Callaghan, R. J., Hill, J. M. (1990a). Ciprofloxacin iontophoresis for aminoglycoside-resistant *Pseudomonas* keratitis. *Invest. Ophthalmol. Vis. Sci.*, **31**, 1940–1944.

Hobden, J. A., Reidy, J. J., O'Callaghan, R. J., Insler, M. S., Hill, J. M. (1990b). Quinolones in collagen shields for therapy of drug-resistant *Pseudomonas* keratitis. *Invest. Ophthalmol. Vis. Sci.*, **31**, 2241–2243.

Hobden, J., O'Callaghan, R. J., Hill, J. M., Hagenah, M., Insler, M. S., Reidy, J. J. (1992). Ciprofloxacin and prednisolone therapy for experimental *Pseudomonas* keratitis. *Curr. Eye Res.*, **11**, 259–266.

Hobden, J. A., Engel, L. S., Callegan, M. C., Hill, J. M., Gebhardt,

B. M., O'Callaghan, R. J. (1993a). *Pseudomonas aeruginosa* in leukopenic rabbits. *Curr. Eye Res.*, **12**, 461–467.

Hobden, J. A., Engel, L. S., Hill, J. M., Callegan, M. C., O'Callaghan, R. J. (1993b). Prednisolone acetate or prednisolone phosphate concurrently administered with ciprofloxacin for the therapy of experimental *Pseudomonas aeruginosa* keratitis. *Curr. Eye Res.*, **12**, 469–473.

Hobden, J. A., Hill, J. M., Engel, L. S., O'Callaghan, R. J. (1993c). Age and therapeutic outcome of experimental *Pseudomonas aeruginosa* keratitis treated with ciprofloxacin, prednisolone and flurbiprofen. *Antimicrob. Agents Chemother.*, **37**, 1856–1859.

Hobden, J., O'Callaghan, R. J., Insler, M., Hill, J. M. (1993d). Ciprofloxacin ointment versus ciprofloxacin drops for the therapy of experimental. *Pseudomonas* keratitis. *Cornea*, **12**, 138–141.

Hume, E. B. H., Moreau, J. M., Conerly, L. L., Cannon, B. M., Engel, L. S., Stroman, D. W., Hill, J. M., O'Callaghan, R. J. (1998). *Serratia marcescens* Keratitis: strain-specific corneal pathogenesis in rabbits. *Exp. Eye Res*.

Johnson, M. K., Allen J. H. (1971). Ocular toxin of the pneumococcus. *Am. J. Ophthalmol.*, **72**, 175–180.

Johnson, M., Allen, J. (1975). The role of cytolysin in pneumococcal ocular infections. *Am. J. Ophthalmol.*, **80**, 518–521.

Johnson, M. K., Hobden, J. A., Hagenah, M., O'Callaghan, R. J., Hill, J. M., Chen, S. (1991). The role of pneumolysin in ocular infections with *Streptococcus pneumoniae*. *Curr. Eye Res.*, **9**, 1107–1114.

Johnson, M. K., Hobden, J., O'Callaghan, R. J., Hill, J. M. (1992). Confirmation of the role of pneumolysin in ocular infections with *Streptococcus pneumoniae*. *Curr. Eye Res.*, **11**, 1221–1225.

Johnson, M. K., Callegan, M. C., Engel, L. S., O'Callaghan, R. J., Hill, J. M. (1995). Growth and virulence of a complement-activation-negative mutant of *Streptococcus pneumoniae* in the rabbit cornea. *Curr. Eye Res.*, **14**, 281–284.

Kaufman, A. H., Darrell, R. W., Shieh, E. *et al.* (1991). Treatment of methicillin-resistant *Staphylococcus aureus* keratitis in rabbits with ciprofloxacin, norfloxacin, ofloxacin, vancomycin and cefazolin. *Invest. Ophthalmol. Vis. Sci.*, Abstract.

Kessler, E., Spierer, A. (1984). Inhibition by phosphoramidon of *Pseudomonas aeruginosa* elastase injected intracorneally in rabbit eyes. *Curr. Eye Res.*, **3**, 1075–1078.

Kessler, E., Kennan, H. E., Brown, S. I. (1977). *Pseudomonas* protease, purification, partial characterization, and its effect on collagen proteoglycan and rabbit corneas. *Invest. Ophthalmol. Vis. Sci.*, **16**, 488–497.

Kreger, A. S., Gray, L. D. (1978). Purification of *Pseudomonas aeruginosa* proteases and microscopic characterization of pseudomonal protease-induced rabbit corneal damage. *Infect. Immun.*, **19**, 630–648.

Kupferman, A., Leibowitz, H. M. (1976). Quantitation of bacterial infection and antibiotic effect in the cornea. *Arch. Ophthalmol.*, **94**, 1981–1984.

Kupferman, A., Leibowitz, H. M. (1977). Topical antibiotic therapy of staphylococcal keratitis. *Arch. Ophthalmol.*, **95**, 1634–1637.

Leibowitz, H. M., Ryan, W. J., Kupferman, A. (1981). Route of antibiotic administration in bacterial keratitis. *Arch. Ophthalmol.*, **99**, 1420–1423.

Lyerly, D., Gray, L., Kreger, A. (1981). Characterization of rabbit corneal damage produced by *Serratia* keratitis and by *Serratia* protease. *Infect. Immun.*, **33**, 927–932.

Mondino, B. J., Kowalski, R. P. (1982). Phlyctenulae and catarrhal infiltrates: occurrence in rabbits immunized with staphylococcal cell walls. *Arch. Ophthalmol.*, **100**, 1968–1971.

Mondino, B. J., Caster, A. L., Dethlefs, B. (1987a). A rabbit model of staphylococcal blepharitis. *Arch. Ophthalmol*, **105**, 409–412.

Mondino, B. J., Laheji, A. K., Adamu, S. A. (1987b). Ocular immunity to *Staphylococcus aureus*. *Invest. Ophthalmol. Vis. Sci.*, **28**, 560–564.

Moreau, J., Sloop, G., Engel, L., Hill, J., O'Callaghan, R. (1997). Histopathological studies of staphylococcal alpha-toxin effects on rabbit corneas. *Curr. Eye Res.*, **16**, 1221–1228.

O'Brien, T. P., Sawusch, M. R., Dick, J. D., Gottsch, J. D. (1988). Topical ciprofloxacin treatment of *Pseudomonas* keratitis in rabbits. *Arch. Ophthalmol.*, **106**, 1444–1446.

O'Callaghan, R. J., Engel, L. S., Hobden, J. H., Callegan, M. C., Green, L. C., Hill, J. M. (1996). *Pseudomonas* keratitis: the role of an uncharacterized exoprotein, protease IV, in corneal virulence. *Invest. Ophthalmol. Vis. Sci.*, **37**, 534–543.

O'Callaghan, R. J., Callegan, M. C., Foster, T. J. *et al.* (1997). The roles of alpha-toxin and beta-toxin during *Staphylococcus* corneal infection. *Infect. Immun.*, **65**, 1571–1578.

Piatkowska, B. (1966). Attempts at treatment of bacterial keratitis with low temperature. *Polish Med. J.*, **5**, 461–465.

Reidy, J. J., Hobden, J. A., Hill, J. M., Forman, K., O'Callaghan, R. J. (1991). The efficacy of topical ciprofloxacin and norfloxacin in the treatment of experimental *Pseudomonas aeruginosa*. *Cornea*, **10**, 25–28.

Rootman, D. S., Hobden, J. A., Jantzen, J., Gonzalez, J. R., O'Callaghan, R. J., Hill, J. M. (1988). Iontophoresis of tobramycin for the treatment of experimental *Pseudomonas* keratitis in the rabbit. *Arch. Ophthalmol.*, **106**, 262–265.

Sloop, G. D., Moreau, J. M., Conerly, L. L., Dajcs, J. J., O'Callaghan, R. J. (1999). Acute inflammation of the eyelid and cornea in *Staphylococcus* keratitis in the rabbit. *Invest. Ophthalmol. Vis. Sci.*, **40**, 385–391.

Smolin, G., Okumoto, M., Wilson, F. M. (1973). The effect of tobramycin on *Pseudomonas* keratitis. *Am. J. Ophthalmol.*, **76**, 555–560.

Smolin, G., Okumoto, M., Wilson, F. M. (1974). The effect of tobramycin on gentamicin-resistant strains in *Pseudomonas* keratitis. *Am. J. Ophthalmol.*, **77**, 583–588.

Spierer, A., Kessler, E. (1984). The effect of 2-mercaptoacetyl-L-phenylalanyl-L-leucine, a specific inhibitor of *Pseudomonas aeruginosa* elastase, on experimental *Pseudomonas* keratitis in rabbit eyes. *Curr. Eye Res.*, **3**, 645–650.

Unterman, S. R., Rootman, D. S., Hill, J. M., Parelman, J. J., Thompson, H. W., Kaufman, H. E. (1988). Collagen shield drug delivery: therapeutic concentrations of tobramycin in the rabbit cornea and aqueous humor. *J. Cataract Refract. Surg.*, **14**, 500–504.

Chapter 43

Gerbil Model of Acute Otitis Media

B. Barry and M. Muffat-Joly

Background of human infection

Acute otitis media (AOM) is one of the most frequent pediatric infections and the main indication for antimicrobial therapy in children. More than 90% of cases occur in children under 5 years and almost 70% of cases occur between the ages of 6 months and 3 years (Leizorovicz, 1995). The pathogenesis of AOM involves eustachian tube dysfunction and immune immaturity (Pelton, 1996). Upper respiratory tract viral infection probably triggers epithelial injury and bacterial colonization of the middle ear from the nasopharynx (Uhari *et al.*, 1996).

Pathogenic bacteria are cultured from 70% of cases of purulent AOM. *Haemophilus influenzae, Streptococcus pneumoniae* and *Moraxella catarrhalis* are the commonest isolates. *M. catarrhalis* accounts for 5–15% of cases. *H. influenzae* and *S. pneumoniae* account for most cases, the predominance of each pathogen varying from one country to another (Berche *et al.*, 1994; Jacobs *et al.*, 1996).

Strains of *H. influenzae* and *S. pneumoniae* responsible for AOM are those that colonize the nasopharynx, i.e., non-typable *H. influenzae* (85–90% of *H. influenzae* strains) and *S. pneumoniae* of non-invasive serotypes (6, 9, 14, 19 and 23). Epidemiological studies on antimicrobial resistance of these pathogens show large geographic differences. At time of writing, 25–35% of *H. influenzae* are beta-lactamase producers and the incidence of *S. pneumoniae* strains with decreased susceptibility to penicillin is still increasing (Berche *et al.*, 1997; Manninen *et al.*, 1997).

Acute otitis media is a self-limiting disease in up to 70–80% of cases (van Buchem *et al.*, 1981; Froom *et al.*, 1997) but antibiotic treatment is usually given to prevent local and meningeal complications (Klein, 1995).

Background of the gerbil model

Experimental models of bacterial or viral infection of the middle ear have been developed with rats (Daniel *et al.*, 1973; Grote and van Blitterswijk, 1984; Hermansson *et al.*, 1988; Melhus *et al.*, 1997), guinea pigs (Friedmann, 1955; Thore *et al.*, 1982, 1985; Ohashi *et al.*, 1991; Egusa *et al.*, 1995; Kawana, 1995), BALB/c mice (Krekorian *et al.*, 1991), ferrets (Buchman *et al.*, 1995) and monkeys (Fujiyoshi *et al.*, 1990), but the species most commonly used for acute otitis media are the chinchilla and gerbil.

The gerbil model is based on the chinchilla models developed by Giebink *et al.* (1976, 1978) and was first described by Fulghum *et al.* (1982). These two animal species share features that make them well adapted to experimental otitis media: (1) they have a hypertrophied middle ear bulla; (2) their nearly vertical eustachian tube does not favor spontaneous middle ear infection (Daniel *et al.*, 1982); (3) they do not harbor pathogenic strains in the external ear (Thompson *et al.*, 1981); (4) their middle-ear mucosa is histologically close to the human middle-ear mucosa (Chole and Chiu, 1985).

Various chinchilla models have been extensively used in pathophysiological studies on infectious, inflammatory and immune processes (see Juhn *et al.*, 1991 for a review). For example, the role of viral infection associated with bacterial infection has been reported (Giebink *et al.*, 1980; Abramson *et al.*, 1982; Giebink and Wright, 1983; Suzuki and Bakaletz, 1994), vaccine protection has been evaluated (Giebink *et al.*, 1978, 1979, 1993, 1995, 1996; Giebink, 1981; Barenkamp, 1986, 1996; De Maria *et al.*, 1996), and the roles of bacterial enzymes or components and mediators of inflammation in the pathogenesis of pneumococcal acute otitis media have been investigated (Carlsen *et al.*, 1992; Sato *et al.*, 1996). Chinchilla models have been also used to study the influence of antibiotics on middle ear infection (Lewis *et al.*, 1980; Supance *et al.*, 1982; Reilly *et al.*, 1983; Hotaling *et al.*, 1987; Kawana *et al.*, 1992; Rosenfeld *et al.*, 1992; Magit *et al.*, 1994; Bolduc *et al.*, 1995; Sato *et al.*, 1995; Alper *et al.*, 1996; Post *et al.*, 1996; Jauris-Heipke *et al.*, 1997) and drugs kinetics in the middle ear (Canafax *et al.*, 1989, 1995; Jossart *et al.*, 1994).

In experimental studies to evaluate antimicrobial efficacy on AOM pathogens, the gerbil model is often used mainly for economic reasons and because of animal husbandry/housing considerations.

Handbook of Animal Models of Infection
ISBN 0-12-775390-7

Animal species

Young adult (less than 6 months) Mongolian gerbils (*Meriones unguiculatus*) are always used. In our laboratory young adult females are purchased from Centre d'Elevage R. Janvier (Le Genest Saint Isle, France). At time of the experiments they are 8–9 weeks old (body weight 40–50 g).

Preparation of animals

On their arrival animals are placed in standardized transparent cages in a protected unit with slight negative pressure, filtered air and a 12–12 hours light–dark cycle (Iffa Credo, l'Arbresle, France). They are given free access to water and food (rodent diet enriched with vitamins A and D). An acclimatization period of at least 1 week is respected before experiments. All the animals are identified at the time of the experiments and are followed individually.

Gerbils are not subject to spontaneous AOM, but spontaneous aural cholesteatoma can develop in aging animals (Fulghum and Chole, 1985). Otoscopic examination is recommended before use, to confirm that the ears are healthy. None of the young females used in our laboratory have presented ear abnormalities.

In initial works, surgery was performed to enlarge the external auditory canal and facilitate tympanic access, 1 week before the experiments: the antero-inferior cutaneous and cartilageous part of the auricle, including the tragus, was excised and the wounds were washed with povidone iodine (Betadine®). However, simply pulling up the pinna allows most of the tympanic membrane to be examined with an operating microscope, rendering surgery unnecessary.

weeks (Fulghum *et al.*, 1985a); higher inocula result in a high rate of lethal systemic complications (Muffat-Joly *et al.*, 1994). When strains of non-invasive serotypes are used (types 19 and 23), inoculation of 10^{3-4} cfu/ear produces mild self-limiting disease (Fulghum *et al.*, 1985a; Barry *et al.*, 1993); the inoculum must be increased to 10^{6-7} cfu to obtain acute otitis media in over 90% of animals, but the course of the disease remains strain-dependent (Barry *et al.*, 1995). Figure 43.1 compares the time course of bacterial counts in middle ears infected with serotype 3, 19 and 23 strains.

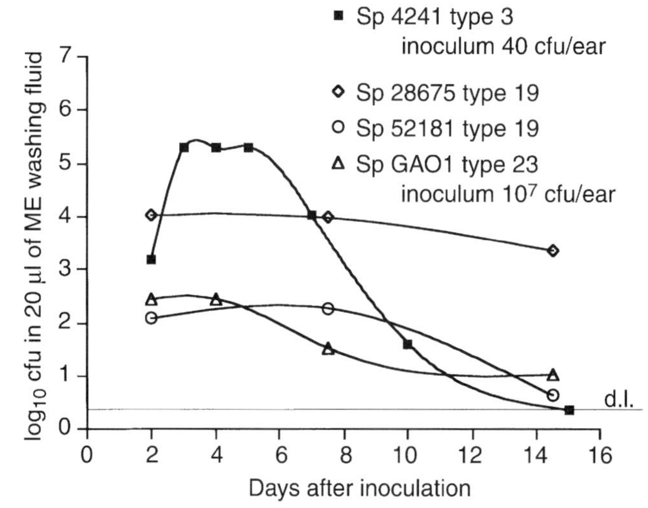

Figure 43.1 Comparative time course of bacterial counts in the middle ear (ME) during pneumococcal otitis induced by either 40° cfu of a serotype 3 strain or 10^7 cfu of serotypes 19 and 23 strains. Detection limit (d.l.) for bacterial counts was 0.34 log cfu in 20 μl of washing fluid.

Infecting agents — inocula

The gerbil model of middle ear infection has been studied by Fulghum and co-workers with *S. pneumoniae* (Fulghum *et al.*, 1982, 1985a, 1987), type b and non-typable *H. influenzae* (Fulghum *et al.*, 1982, 1985b), *M. catarrhalis* (Fulghum and Marrow, 1996), *E. coli* (Fulghum and Beamer, 1991) and various anaerobes, alone or in polymicrobial inoculates (Fulghum *et al.*, 1982; Fulghum and Beamer, 1991). With the three most common clinical pathogens, these authors obtained acute otitis media only with *S. pneumoniae* and *H. influenzae*; *M. catarrhalis* produced self-limiting inflammation with rapid bacterial clearance (Fulghum and Marrow, 1996). Similar differences in the middle-ear pathogenicity of bacterial species were observed in the chinchilla model (Fulghum *et al.*, 1982; Fulghum and Marrow, 1996).

With *S. pneumoniae* strains of the invasive serotype 3, a small inoculum (< 100 cfu/ear) is sufficient to consistently induce middle-ear otitis, with an acute phase lasting 2

We have only used *S. pneumoniae* strains isolated from human respiratory tract or systemic infections. Virulence is maintained by passage in mice. Aliquots of bacterial suspension are kept at –80°C. On each day of experimentation a freshly thawed pneumococcal aliquot is incubated for 6 hours at 37°C in brain–heart infusion broth (Bio-Merieux, Lyon, France) enriched with 5% horse serum (Diagnostics Pasteur). Bacterial density is determined by 10-fold serial dilutions of the culture in saline, and the pour-plate colony-counting method (Columbia agar with 5% sheep blood, Bio-Mérieux).

With non-typable *H. influenzae* strains, 10^{3-4} cfu/ear are required to induce otitis media in 100% of gerbils (Fulghum *et al.*, 1985b). Spontaneous resolution occurs during the second week and no bacteria are recovered after the third week (Fulghum *et al.*, 1985b). In later studies using this model, animals were inoculated with 10^{5-6} cfu/ear (Girard *et al.*, 1987; Clement *et al.*, 1990; Hardy *et al.*, 1990; Retsema *et al.*, 1990; Swanson *et al.*, 1991).

Details of the infection process

Overview

The bacterial suspension is inoculated percutaneously into the middle ear bulla of anesthetized animals.

Materials required

Skin disinfectant; Luer-lock 1 ml standard syringes and 50 μl Hamilton syringes; Luer-lock sterile disposable needles: 0.45 × 10 mm (26G ⅜″), 0.45 × 16 mm (26G ⅝″), 0.3 × 13 mm (30G ½″) with a transparent white lock (B-D® Microlance® 3); operating microscope providing magnifications up to 20 times; small, sharp forceps; sterile swabs.

Anesthesia

For all invasive procedures animals are anesthetized by intramuscular injection in the thigh of a mixture of 40 mg/kg ketamine (Ketalar®, Parke-Davis) and 13 mg/kg xylazine (Rompun®, Bayer Pharma) given in a volume of 150–200 μl. Anesthesia is complete within a few minutes, stillness is ensured for at least 15 minutes, and there is no anesthesia-related swelling of the tympanic membrane. A higher dose of ketamine (87 mg/kg) was used in the ketamine–xylazine mixture by initial authors (Fulghum *et al.*, 1982). Alternative anesthetics used in various reports include diethyl ether (Hardy *et al.*, 1990; Swanson *et al.*, 1991) and a ketamine (0.125 mg/g)/levopromazine (6.25 mg/g)/atropine (0.01 mg) mixture administered intraperitoneally (von Unge *et al.*, 1997).

Inoculation

The animal is placed on its side. The bulla corresponds to the cephalad portion of the middle-ear cavity. The postero-superior chamber is covered by a thin bony membrane (Daniel *et al.*, 1982). This anatomical area forms a discrete, easily palpable bump. Some authors perform surgical bone exposure before inoculation (von Unge *et al.*, 1997), but aseptic preparation of the skin is usually sufficient. The slight convexity of the bone behind the superior part of the auricle is located with the left hand of a right-handed experimenter. Light pressure of a 0.3 × 13 mm needle attached to a 50 μl Hamilton syringe held with the right hand pierces the thin bony membrane and penetrates perpendicular to the bone. The action must be precise to limit needle penetration into the middle-ear cavity and avoid injury, in particular perforation of a large vein that runs unprotected a short distance away. Slow injection of the inoculum in a volume of 20–30 μl is then possible without resistance. Proper inoculation causes epitympanic membrane swelling without rupture, and this is verified with the operating microscope. The procedure does not induce cutaneous abscesses or bone infection at the injection point, except when very large inocula of invasive serotypes of *S. pneumoniae* are used (Muffat-Joly *et al.*, 1994). Inoculation can be unilateral or bilateral.

Postoperative care

No particular housing is necessary after inoculation. Infected animals are caged in groups of four or five. They return to normal behaviour within a couple of hours after the anaesthesia has worn off. No postoperative analgesics are used.

Key parameters to monitor infection and response to treatment

Animals are followed up for behavior, body weight, tympanic features and bacterial counts in middle-ear washing fluid. Leukocyte counts can also be done on the samples. Meningeal involvement can be assessed by bacterial counts in cerebrospinal fluid (CSF).

Clinical parameters

Development of acute otitis media is not necessarily accompanied by behavioral signs. Ruffled fur, prostration, inability to stand, circling, and/or spine stiffness are mostly associated with systemic complications. Body-weight loss reflects the local or systemic nature of the infection.

Prior to middle-ear sampling, the presence and severity of otitis media is evaluated by otoscopic examination of the tympanum. A global scoring system ranging from 0 to 5 has been proposed by Fulghum *et al.* (1985b). We prefer to consider separately the presence of a retrotympanic exudate as the sign of actual infection, and examine the epitympanic membrane for thickening, inflammation and shape changes. We have opted for simple unambiguous grading, by attributing four grades to each membrane feature, as follows: thickening: 0 = none, 1 = opalescence, 2 = opacity, and 3 = opacity plus graining; inflammation: 0 = none, 1 = mild vasodilatation, 2 = frank hyperemia, and 3 = acute inflammation with intratympanic hemorrhage; shape: 0 = normally concave (or slightly retracted), 1 = flat, 2 = convex, and 3 = bulging (Barry *et al.*, 1996).

Bacterial density in middle-ear effusion

Middle-ear sampling is performed by transepitympanic washing. A quantity of 20 μl of saline is injected and withdrawn through the epitympanic membrane (0.3 × 13 mm needle attached to a 50 μl Hamilton syringe) under visual control with the operating microscope. With this procedure, the middle-ear effusion is diluted approximately 2.5-fold in the washing fluid. This sampling method, provided that it is well standardized by the experimenters, is satisfactory for infection/treatment monitoring and comparisons between

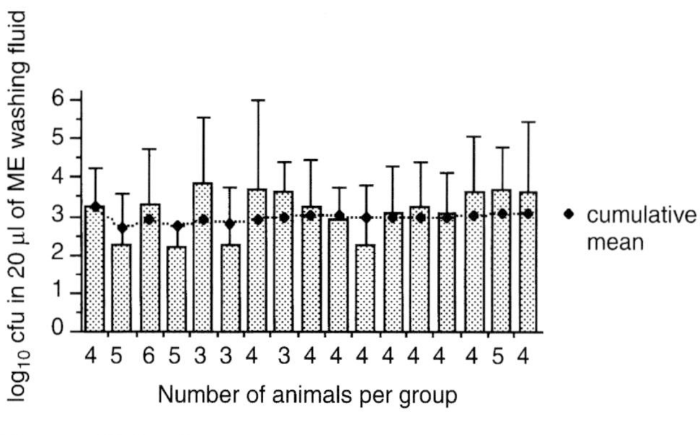

Figure 43.2 Overview of bacterial counts in 70 gerbil middle ear (ME) samples taken 2 days after transbullar inoculation of the same *S. pneumoniae* serotype 19 strain (Sp 15986, inoculum: $1.4 \times 10^7 \pm 0.5 \times 10^7$ cfu/ear). Chronologically ordered data for 17 sets of experiments over a 30-month period. Bars are mean ± SD per group. The final cumulated mean ± SD was 3.11 ± 1.30 cfu in 20 µl of washing fluid.

experimental groups. Figure 43.2 shows the reproducibility of mean bacterial load in a large number of experiments.

Determination of antibiotic penetration in middle ear and serum

Antibiotic penetration in middle-ear effusion can be determined by bioassay or HPLC in middle-ear samples. Pharmacokinetic data for serum and middle-ear effusion are available for erythromycin and analogs (Clement *et al.*, 1990), clarithromycin and its ^{14}OH metabolite (Hardy *et al.*, 1990), azithromycin (Girard *et al.*, 1996), amoxicillin (Barry *et al.*, 1993), ampicillin (Barry *et al.*, 1994), ceftriaxone (Barry *et al.*, 1996) and cefditoren (authors' unpublished data). Selected data are given in Table 43.1. The study by Girard *et al.* (1996) on azithromycin kinetics in middle ears infected by *H. influenzae* showed that 75% of azithromycin recovered in bulla washes was associated with local inflammatory cells. In the same conditions, the corresponding proportions for clarithromycin, roxythromycin, erythromycin and amoxicillin were 20%, 34%, 27% and 5%, respectively.

For pharmacokinetic determinations in middle-ear effusion, we inoculated the animals with 40 cfu of the same serotype 3 *S. pneumoniae* strain. Middle-ear and serum samples were taken 3 days after inoculation, i.e., at the onset of acute otitis media, when the effusion is abundant and tympanic inflammation is mild enough to avoid blood contamination of the samples. In these conditions, the middle-ear disease (inflammation and bacterial density) is close to that observed in children at the acute stage. Data on drug penetration in middle ear fluid are dependent on both the stage of infection/inflammation and the sampling method (whole-bulla washing after sacrifice or non-destructive transtympanic washing), which makes it difficult to compare published data. However, data on different compounds reported by a given team are usually comparable.

Determination of additional markers of infection and response to treatment

Pneumococcal otitis in gerbils is frequently followed by meningeal bacterial spread, the rate and outcome of which is strain- and inoculum-dependent (Muffat-Joly *et al.*, 1994; Barry *et al.*, 1995). No comparable studies have been reported for *H. influenzae* otitis. In the longitudinal study of *H. influenzae* otitis in gerbils by Fulghum *et al.* (1985b) only one animal died, of unidentified causes. Bacterial invasion of CSF can be used as an additional marker of infection. Samples can be obtained by percutaneous suboccipital intracisternal puncture. Animals are placed ventral side down. To present the cisternal region, the nape of the neck is exposed by keeping the head bent forward, and the fur is parted in the center with an alcohol swab. Shaving is not essential. Puncture is performed with a 0.3 × 13 mm needle attached to a 1 ml disposable syringe. The needle is slowly inserted perpendicular to the spine; slight depression is maintained to permit CSF to flow into the syringe as soon as the cisterna is reached. The transparent needle lock permits the sample to be controlled. Once again, the action has to be very precise to avoid spinal-cord injury and/or blood contamination of samples. In optimal conditions, 30–40 µl CSF can be withdrawn. Sampling is well tolerated.

Inflammatory cytokine production has not been studied in gerbil acute otitis media, but some studies have been performed in other pathological situations such as parasitic infections (Wang *et al.*, 1992; Campbell and Chadee, 1997) and cerebral ischaemia (Saito *et al.*, 1996). Other animal models have been used to explore local expression of inflammatory mediators/effectors during pneumococcal acute otitis media, e.g., cytokines in the chinchilla (Sato *et al.*, 1996) and rat (Zeevy, 1995), products of oxidative metabolism in the chinchilla (Kawana *et al.*, 1991) and guinea-pig (Parks *et al.*, 1994, 1995, 1996), and heat shock protein in the guinea-pig (Egusa *et al.*, 1995).

Antimicrobial therapy

Beta-lactams, macrolides and quinolones are the main antibiotic families so far evaluated in the gerbil model. Antibiotics can be administered perorally or subcutaneously. Oral administration must be cautious, because gastrointestinal disorders may occur, especially with group A penicillins. Protocols for treatment and control differ according to the objective, but only a short treatment course (1–2 days) can discriminate antibacterial activity from spontaneous bacterial clearance. Table 43.2 summarizes the results of studies by Clement *et al.* (1990), Girard *et al.*, (1987, 1993), Hardy *et al.* (1990, 1991), Retsema *et al.* (1990) and Swanson *et al.*, (1991) aimed at evaluating and comparing the activity of various antibiotics in the treatment of AOM induced by *H. influenzae* (beta-lactamase producers and non-producers) and penicillin-susceptible *S. pneumo-*

Table 43.1 Pharmacokinetic data in gerbils following a single dose of various antibiotics during acute otitis media (HPLC, high performance liquid chromotography; Bio, microbiological method assay)

				Serum or plasma				Middle ear effusion				
Compound	*Route*	*Dose*	*Method*	C_{max} (µg/ml)	T_{max} (h)	$T_{1/2}$ (h)	$AUC_{0-\infty}$ (µg/ml.h)	C_{max} (µg/ml)	T_{max} (h)	$T_{1/2}$ (h)	$AUC_{0-\infty}$ (µg/ml.h)	*Reference*
Erythromycin	PO	100	Bio	3.7	1.0	1.9	5.3	1.8	1.0	2.1	2.3	Clement *et al.*, 1990
A-69334		100	Bio	5.5	1.0	4.4	15	2.0	1.0	8.9	6.8	
Clarithromycin	PO	100	HPLC	2.1	3.0	1.5	9.1	1.2	3.0	1.7	5.7	Hardy *et al.*, 1990
14OH-Clari	PO	12	HPLC	0.8	3	1.1	3.3	0.25	3.0	1.4	1.3	
Azithromycin	PO	100	Bio*	4.0	2.0			0.58	24			Girard *et al.*, 1996
Amoxicillin	SC	100	Bio†	53	0.5	0.78	74	17	1–2	1.6	63	Barry *et al.*, 1993
Ampicillin	SC	100	Bio†	29	0.5	0.69	42	17	1	1.5	54	Barry *et al.*, 1994
Ceftriaxone	SC	50	Bio†	61				15				Barry *et al.*, 1996
		100	Bio†	95	0.5	0.88	173	40				
		50	HPLC	141				25				
		100	HPLC	268	0.5	0.77	488	43	2	2.3	176	
Cefditoren	SC	25	Bio†	10	0.5	0.73	12					Barry, unpublished data
		200	Bio†	60	0.5	0.56	103					
		25	HPLC	51	0.5	0.68	60	3.7	2	>12	18‡	
		200	HPLC	326	0.5	0.45	453	50	1	>12	205‡	

* Using standard curves in serum.
† Using protein-free standard curve (bioactive equivalent concentration).
‡ AUC values from 0–6 hours after injection (the uncertainty of $T_{1/2}$ values rules out $AUC_{0-\infty}$ calculation by the extrapolation method).

niae strains of invasive serotypes. Treatment is usually started within 24 hours after inoculation, and bacterial counts are performed on whole-bulla washing fluid after sacrifice, 18–24 hours after treatment cessation. Results are reported either as the daily dose yielding bacterial eradication, or as the effective dose for 50% of animals (ED_{50}).

In our studies conducted with non-invasive *S. pneumoniae* serotypes, the large inoculum, the short course of the disease and the risk of lethal meningeal complications demanded earlier treatment (2–4 hours p.i.). Animals were not killed, and bacterial counts in middle-ear washings were performed from day 2 to day 15 postinfection (Barry *et al.*, 1993). However, treatment efficacy is optimally observed on day 2. Our main results are summarized in Table 43.3.

Advantages/disadvantages of the model

With non-typeable *H. influenzae* and *S. pneumoniae* strains of invasive serotypes, the course of the infection is close to the human situation, as a small inoculum is sufficient to induce acute otitis media, and spontaneous resolution is achieved within 2 weeks.

Regarding pneumococcal otitis, investigations have focused on strains belonging to non-invasive serotypes, which are the commonest etiologic agents of middle ear infection and show cross-resistance to antimicrobials, a major health problem in clinical practice. When these strains are used in the gerbil model, a large amount of bacteria must be inoculated to obtain AOM. Bacterial density in the middle ear is maximal at the time of inoculation, which is not representative of the pathogenesis of AOM in humans. However, the model cannot be considered as a model of bacterial clearance: bacterial density first falls, then plateaus in middle ear effusion between day 2 and day 4 postinoculation, and complete clearance has not been achieved by days 7–8. In an attempt to render the animals more susceptible and therefore allow the use of a smaller bacterial challenge, preinfection leukopenia was induced from the normal count of 10 000 leukocytes/ml to 2000/ml with 100 mg/kg cyclophosphamide (Endoxan®) in an intramuscular injection 3 days before inoculation; a second 25 mg/kg dose given at time of inoculation maintained leukopenia at the same level for 4 days. This approach was unsuccessful, as 10^{4-5} cfu/ear were still necessary and the rate of lethal systemic complications was greatly increased (unpublished data).

The high rate of otogenic meningeal complications associated with use of large inocula, although necessitating early treatment, could be of interest. Prevention of bacterial CSF involvement can be an additional marker of antimicrobial activity, and the pathogenicity of strains in the middle ear can be defined in terms of both local bacterial persistence and meningeal tropism. Meningeal involvement subsequent to otitis media has also been exploited in a model of *Listeria-monocytogenes*-induced rhomboencephalitis in gerbils infected via the middle ear (Blanot *et al.*, 1997).

It is noteworthy that a large number of studies in auditory

Table 43.2 Evaluation of various antibiotics in the treatment of *H. influenzae* (Hi) and *S. pneumoniae* (Sp) otitis media in gerbils; data are from studies using similar protocols (treatment given perorally for 2 days) and similar criteria for treatment evaluation

Reference	*Pathogen*	*Treatment: start; frequency*	*Compound*	*MIC (μg/ml)*	*Treatment evaluation*	*cfu in untreated controls (log/ear)*	*Dose achieving bacterial eradication (mg/kg/day)*	*ED_{50} (mg/kg)*
Girard *et al.*, 1987	Hi 54A131	day 1 q.8 hours	Azithromycin	0.10	1 day	ND		37
			Erythromycin	1.56				>100
			Amoxicillin	>100				>100
Clement *et al.*, 1990	Hi ATCC 43095	17 hours q.12 hours	A-69334*	2–4	18 hours	7.1 ± 0.25	300	
			Erythromycin A	2–4			no effect	
Hardy *et al.*, 1990	Hi ATCC 43095	17 hours q.8 hours	^{14}OH-clari	4	18 hours	7.5 ± 0.8	300	
			Clarithromycin	8		7.5 ± 0.8	no effect	
			^{14}OH-clari + clari			5.3 ± 1.7	48 + 300	
						7.1 ± 0.3	96 + 300	
			Ampicillin	0.25		Either	96	
Retsema *et al.*, 1990	Hi	Day 1 q.8 hours	Azithromycin	0.78	1 day	ND		31
			Roxythromycin	3.12				>100
			Erythromycin	1.56				>100
Swanson *et al.*, 1991	Hi ATCC 43095	17 hours q.12 hours	Temafloxacin	0.015	18 hours	7.4 ± 0.1	10	
			Ciprofloxacin	0.015		7.4 ± 0.1	50	
			Ofloxacin	0.030		7.4 ± 0.1	10–50	
			Ampicillin	0.25		7.4 ± 0.1	50	
Hardy *et al.*, 1991	Hi ATCC 43095	17 hours q.12 hours	A-69334*	2	18 hours	7.1 ± 0.1	300	
			Erythromycin A	1		7.1 ± 0.1	no effect	
			Ampicillin	0.25		7.1 ± 0.1	100	
Girard *et al.*, 1987	Sp 02J025	Day 1 q.12 hours	Azithromycin	0.02	1 day	ND		4.1
			Erythromycin	0.01				6.8
			Amoxicillin	0.003				≤1.6
Swanson *et al.*, 1991	Sp ATCC 6303	17 hours q.12 hours	Temafloxacin	0.25	18 hours	5.0 ± 0.4	250	
			Ciprofloxacin	1		5.0 ± 0.4	50	
			Ofloxacin	0.5		5.0 ± 0.4	250	
			Ampicillin	0.03		5.0 ± 0.4	≤2	
Girard *et al.*, 1993	Sp 02J025	Day 1 q.12 hours	Azithromycin	0.19	1 day	7.9 (/ml bulla wash)	25	2.9
			Clarithromycin	0.02			25	6.3

* Erythromycin analog (9-12-epoxy erythromycin).

research are conducted in gerbils; some of them involved middle ear infection (von Unge *et al.*, 1997). In addition, the gerbil model of middle-ear infection could find useful links with *in-vitro* studies on transport in middle-ear epithelium, which are currently done on gerbil cells lines (Herman *et al.*, 1992, 1995, 1997; Furukawa, *et al.*, 1997).

Contributions of the model to infectious disease therapy

This gerbil model has proved suitable for evaluating new strategies in the treatment of pneumococcal AOM due to strains with decreased susceptibility to beta-lactams (Barry *et al.*, 1993, 1996). The outcome of experimental therapy correlates with the results of recent clinical trials. In the clinical study by Boucherat *et al.* (1997), a success rate of 100% (11/11) was obtained with a high oral dosage of amoxicillin (150 mg/kg per day) after failure of first-line treatment of AOM due to strains intermediate or resistant to penicillin. In two other studies of bacteriologically documented AOM, ceftriaxone 50 mg/kg once daily for 3 days yielded 87% (27/31) and 89% (24/27) cure rates in patients infected by intermediate or resistant pneumococci at inclusion (Géhanno *et al.*, 1997; Leibovitz *et al.*, 1997).

Experimental results and clinical outcome with macrolides are less consistent. The antibacterial activity of

Table 43.3 Beta-lactams in the treatment of pneumococcal otitis in gerbils induced by susceptible or resistant strains

	Untreated control group		*Treatment*				
Infecting strain (serotype)	*cfu counts in ME**	*Rate of culture-positive CSF (%)*	*Compound*	*MIC (μg/ml)*	*No. of inj.*	*Dose achieving ME bacterial eradication (mg/kg/inj.)*	*Remaining culture-positive CSF (%)*
GAO1 (23)	2.5 ± 1.5	14/46 (30)	Amoxicillin	≤0.016	2	≤2.5	2/12 (17)
			Ampicillin	≤0.06	2	≤2.5	ND
			Cefditoren	≤0.016		≤2.5	ND
54B (23)	3.0 ± 1.3	13/32 (41)	Amoxicillin	2	2	10	0/11
17346 (23)	2.1 ± 1.4	7/15 (47)	Ampicillin	2	2	10	ND
15986 (19)	3.1 ± 1.3	35/59 (59)	Amoxicillin	2–4	2	25	2/12 (17)
			Ampicillin	4	2	100	ND
			Cefditoren	0.5	2	25	ND
			Ceftriaxone	0.5	1	50	1/14
32870 (19)	2.4 ± 1.0	6/9 (67)	Ceftriaxone	4	1	100	2/10 (20)
40984 (14)	1.1 ± 0.8	8/14 (57)	Ceftriaxone	8	1	100	1/14 (7)

Data are from the authors' personal studies. Animals were inoculated with 10^7 cfu/ear. Treatment was started 2–4 hours after inoculation and given subcutaneously for 1 day. Counts were performed 1 day after the last injection.
* Data are mean ± SD in 20 μl of middle ear (ME) washing fluid.

azithromycin and clarithromycin in gerbil middle-ear infection due to *H. influenzae* (Hardy *et al.*, 1990; Retsema *et al.*, 1990; Alper *et al.*, 1996) does not tally with the lack of efficacy (70–80% rate of bacteriological failures) in the clinical setting, which was observed in studies involving bacteriological tests before and during treatment (Howie, 1992; Dagan *et al.*, 1997).

In our pharmacokinetic/pharmacodynamic studies of four beta-lactams (amoxicillin, ampicillin, cefditoren and ceftriaxone) against the same highly penicillin-resistant strain (MIC 4–8 μg/ml), we failed to identify any pharmacodynamic criteria determined in middle-ear effusion consistently predictive of treatment efficacy. In contrast, after 1 day of treatment, the four drugs were effective when serum concentrations of microbiologically active drug exceeded the MIC for 20–30% of the dosing interval (unpublished data). These results are consistent with a retrospective clinical analysis on pharmacodynamics in acute otitis media by Craig and Andes (1996).

References

Abramson, J. S., Giebink, G. S., Quie, P. G. (1982). Influenzae A virus-induced polymorphonuclear leukocyte dysfunction in the pathogenesis of experimental pneumococcal otitis media. *Infect. Immun.*, **36**, 289–296.

Alper, C. M., Doyle, W., Seroky, J. T., Bluestone, C. D. (1996). Efficacy of clarythromycin treatment of acute otitis media caused by infection with penicillin-susceptible, -intermediate, and -resistant *Streptococcus pneumoniae* in the chinchilla. *Antimicrob. Agents Chemother.*, **40**, 1889–1892.

Barenkamp, S. J. (1986). Protection by serum antibodies in experimental nontypable *Haemophilus influenzae* otitis media. *Infect. Immun.*, **52**, 572–578.

Barenkamp, S. J. (1996). Immunization with high-molecular-weight adhesion proteins of nontypeable *Haemophilus influenzae* modifies experimental otitis media in chinchillas. *Infect. Immun.*, **64**, 1246–1251.

Barry, B., Muffat-Joly, M., Géhanno, P., Pocidalo, J. J. (1993). Effect of increased dosages of amoxicillin in treatment of experimental middle ear otitis due to penicillin-resistant *Streptococcus pneumoniae*. *Antimicrob. Agents Chemother.*, **37**, 1599–1603.

Barry, B., Muffat-Joly, M., Géhanno, P., Pocidalo, J. J. (1994). Effect of increased dosages of ampicillin and amoxicillin in the treatment of experimental middle ear otitis induced by highly penicillin-resistant *Streptococcus pneumoniae*. Abstract 838. In: *Abstract Book of the 6th International Congress for Infectious Diseases*, p. 269. International Society for Infectious Diseases.

Barry, B., Muffat-Joly, M., Géhanno, P., Pocidalo, J. J. (1995). Pathogenicity and meningeal spread of *Streptococcus pneumoniae* serotype 19 and 23 with varying susceptibility to penicillin in experimental acute otitis media. In: *Recent Advances in Otitis Media – Proceedings of the Sixth International Symposium* (eds Lim, D., Bluestone, C., Casselbrant, M., Klein, J., Ogra, P), pp. 520–521. B. C. Decker, Hamilton, Ontario.

Barry, B., Muffat-Joly, M., Bauchet, J. *et al.* (1996). Efficacy of single-dose ceftriaxone in experimental otitis media induced by penicillin-resistant and cephalosporin-resistant *Streptococcus pneumoniae*. *Antimicrob. Agents Chemother.*, **40**, 1977–1982.

Berche, P., Géhanno, P., Duval, F., Lenoir, G. (1994). Epidémiologie des otites moyennes aiguës de l'enfant en France en 1993. *Lett l'Infectiologue*, **18** (suppl.), 11–19.

Berche, P., Nguyen, L., Ferroni, A. (1997). Epidémiologie des bactéries rencontrées au cours des otites moyennes aiguës de l'enfant. *Méd. Maladies Infect.*, **27**, 388–396.

Blanot, S., Muffat-Joly, M., Vilde, F. *et al.* (1997). A gerbil model for rhomboencephalitis due to *Listeria monocytogenes*. *Microb. Pathog.*, **23**, 39–48.

Bolduc, G. R., Tam, P. G., Pelton, S. I. (1995). Therapeutic approaches to experimental otitis media due to penicillin-resistant *Streptococcus pneumoniae*. In: *Recent Advances in Otitis Media — Proceedings of the Sixth International Symposium* (eds Lim, D., Bluestone, C., Casselbrant, M., Klein, J., Ogra, P.), pp. 518–519. B. C. Decker, Hamilton, Ontario.

Boucherat, M., Cohen, R., Geslin, P., Valentin, L. (1997). Experience clinique du traitement par amoxicilline des échecs thérapeutiques de l'otite à pneumococque de sensibité diminuée aux β-lactamines. *Méd. Maladies Infect.*, **27** (spécial), 68–72.

Buchman, C. A., Swarts, J. D., Seroky, J. T., Panagiotou, N., Hayden, F., Doyle, W. J. (1995). Otologic and systemic manifestations of experimental influenza A virus infection in the ferret. *Otolaryngol. Head Neck Surg.*, **112**, 572–578.

Campbell, D., Chadee, K. (1997). Interleukin (IL)-2, IL-4, and tumor necrosis factor-alpha responses during *Entamoeba histolytica* liver abscess development in gerbils. *J. Infect. Dis.*, **175**, 1176–1183.

Canafax, D. M., Nonomura, N., Erdmann, G. R., Le, C. T., Juhn, S. K., Giebink, G. S. (1989). Experimental animal models for studying antimicrobial pharmacokinetics in otitis media. *Pharm. Res.*, **6**, 279–285.

Canafax, D. M., Russlie, H. O., Lovdahl, M. J., Giebink, G. S. (1995). Ciprofloxacin middle ear and plasma pharmacokinetics in experimental acute otitis media. In: *Recent Advances in Otitis Media — Proceedings of the Sixth International Symposium* (eds Lim, D., Bluestone, C., Casselbrant, M., Klein, J., Ogra, P.), pp. 241–243. B. C. Decker, Hamilton, Ontario.

Carlsen, B. D., Kawana, M., Kawana, C., Tomasz, A., Giebink, G. S. (1992). Role of bacterial cell wall in middle ear inflammation caused by *Streptococcus pneumoniae*. *Infect. Immun.*, **60**, 2850–2854.

Chole, R. A., Chiu, M. (1985). Ultrastructure of middle ear mucosa in the mongolian gerbil, *Meriones unguiculatus*. *Acta Oto-Laryngol. (Stockh.)*, **100**, 273–288.

Clement, J. J., Shipkowitz, N. L., Swanson, R. L., Lartey, P. A., Alder, J. D. (1990). Efficacy of a 9–12-epoxy erythromycin derivative, A-69334, in *Haemophilus influenzae* induced otitis media in gerbils. Abstract 814. In: *Program and Abstracts of the 30th Interscience Conference on Antimicrobial Agents and Chemotherapy*, p. 222. American Society of Microbiology, Washington, DC.

Craig, W. A., Andes, D. (1996). Pharmacokinetics and pharmacodynamics of antibiotics in otitis media. *Pediatr. Infect. Dis. J.*, **15**, 266–269.

Dagan, R., Leibovitz, E., Jacobs, M., Fliss, D., Leiberman, A., Yagupsky, P. (1997). Bacteriologic response to acute otitis media cause by *Haemophilus influenzae* treated with azithromycin. Abstract K102. In: *Program and Abstracts of the 37th Interscience Conference on Antimicrobial Agents and Chemotherapy*, p. 345. American Society for Microbiology, Washington, DC.

Daniel, H. J., Means, L. W., Dressel, M. E., Koesche, P. J. (1973). Otitis media in laboratory rats. *Physiol. Psychol.*, **1**, 7–8.

Daniel, H. J., Fulghum, R. S., Brinn, J. E., Barrett, K. A. (1982). Comparative anatomy of eustachian tube and middle ear cavity in animal models for otitis media. *Ann. Otol. Rhinol. Laryngol.*, **91**, 82–89.

De Maria, D. F., Murwin, D. M., Leake, E. R. (1996). Immunization with outer membrane protein P6 from non-typeable *Haemophilus influenzae* induces bactericidal antibody and affords protection in the chinchilla models of otitis media. *Infect. Immun.*, **64**, 5187–5192.

Egusa, K., Huang, C. C., Haddad, J. J. (1995). Heat shock proteins in acute otitis media. *Laryngoscope*, **105**, 708–713.

Friedmann, I. (1955). The comparative pathology of otitis media — experimental and human. I. Experimental otitis of the guinea-pig. *J. Laryngol. Otol.*, **69**, 27–50.

Froom, J., Culpepper, L., Jacobs, M. *et al.* (1997). Antimicrobials for acute otitis media? A review from the International Primary Care network. *Br. Med. J.*, **315**, 98–102.

Fujiyoshi, T., Watanabe, T., Mogi, G. (1990). Three-dimensional images of the temporal bone and experimental otitis media in Japanese monkeys. *Arch. Otolaryngol. Head Neck Surg.*, **116**, 813–819.

Fulghum, R. S., Beamer, M. E. (1991). Experimental otitis media with anaerobic bacteria. *Ann. Otol. Rhinol. Laryngol. Suppl.*, **154**, 23–29.

Fulghum, R. S., Chole, R. A. (1985). Bacterial flora in spontaneously occurring aural cholesteatomas in Mongolian gerbils. *Infect. Immun.*, **50**, 678–681.

Fulghum, R. S., Marrow, H. D. (1996). Experimental otitis media with *Moraxella (Branhamella) catarrhalis*. *Ann. Otol. Rhinol. Laryngol.*, **105**, 234–241.

Fulghum, R. S., Brinn, J. E., Smith, A. M., Daniel, H. J., Loesche, P. J. (1982). Experimental otitis media in gerbils and chinchillas with *Streptococcus pneumoniae, Haemophilus influenzae* and other aerobic and anaerobic bacteria. *Infect. Immun.*, **36**, 802–810.

Fulghum, R. S., Hoogmoed, R. P., Brinn, J. E., Smith, M. A. (1985a). Experimental pneumococcal otitis media: longitudinal studies in the gerbil model. *Int. J. Pediatr. Otorhinolaryngol.*, **10**, 9–20.

Fulghum, R. S., Hoogmoed, R. P., Brinn, J. E. (1985b). Longitudinal studies of experimental otitis media with *Haemophilus influenzae* in the gerbil. *Int. J. Pediatr. Otorhinolaryngol.*, **9**, 101–114.

Fulghum, R. S., Chole, R. A., Brinn, J. E., Branigan, A. E. (1987). Mongolian gerbil tympanic membrane: normal and with induced otitis media. *Arch. Otolaryngol. Head Neck Surg.*, **113**, 521–525.

Furukawa, M., Ikeda, K., Yamaya, M., Oshima, T., Sasaki, H., Takasaka, T. (1997). Effects of extracellular ATP on ion transport and $[Ca^{2+}]i$ in Mongolian gerbil middle ear epithelium. *Am. J. Physiol.*, **272** (3), C827–C836.

Géhanno, P., Berche, P., N'Guyen, L., Derriennic, M., Pichon, F., Goehrs, J. M. (1997). Resolution of clinical failure in acute otitis media confirmed by in vivo bacterial eradication — efficacy and safety of ceftriaxone injected once daily, for 3 days. Abstract LM45. In: *Program and Abstracts of the 37th Interscience Conference on Antimicrobial Agents and Chemotherapy*, p. 372. American Society for Microbiology, Washington, DC.

Giebink, G. S. (1981). The pathogenesis of pneumococcal otitis media in chinchillas and the efficacy of vaccination in prophylaxis. *Rev. Infect. Dis.*, **3**, 342–352.

Giebink, G. S., Wright, P. F. (1983). Different virulence of influenza A virus strains and susceptibility to pneumococcal otitis media in chinchillas. *Infect. Immun.*, **41**, 913–920.

Giebink, G. S., Payne, E. E., Mills, E. L., Juhn, S. K., Quie, P. G. (1976). Experimental otitis media due to *Streptococcus pneumoniae*: immunopathogenic response in the Chinchilla. *J. Infect. Dis.*, **134**, 595–604.

Giebink, G. S., Schiffman, G., Petty, K., Quie, P. G. (1978). Modification of otitis media following vaccination with the capsular polysaccharide of *Streptococcus pneumoniae* in chinchillas. *J. Infect. Dis.*, **38**, 480–487.

Giebink, G. S., Berzins, I. K., Schiffman, G., Quie, P. G. (1979). Experimental otitis media in chinchillas following nasal colonisation with type 7F *Streptococcus pneumoniae*: prevention after vaccination with pneumococcal capsular polysaccharide. *J. Infect. Dis.*, **140**, 716–723.

Giebink, G. S., Berzins, I. K., Marker, S. C., Schiffman, G. (1980). Experimental otitis media after nasal inoculation of *Streptococcus pneumoniae* and influenza A virus in chinchillas. *Infect. Immun.*, **30**, 445–450.

Giebink, G. S., Koskela, M., Vella, P. P., Harris, M., Le, C. T. (1993). Pneumococcal capsular polysaccharide-meningococcal outer membrane protein complex conjugate vaccines: immunogenicity and efficacy in experimental otitis media. *J. Infect. Dis.*, **167**, 347–355.

Giebink, G. S., Meier, J. D., Quartey, M. K., Ellis, R. W. (1995). Monovalent and polyvalent pneumococcal conjugate vaccines: antibody response and middle ear protection in chinchillas. In: *Recent Advances in Otitis Media—Proceedings of the Sixth International Symposium* (eds Lim, D., Bluestone, C., Casselbrant, M., Klein, J., Ogra, P.), pp. 507–510. B. C. Decker, Hamilton, Ontario.

Giebink, G. S., Meier, J. D., Quartey, M. K., Liebeler, C. L., Le, C. T. (1996). Immunogenicity and efficacy of *Streptococcus pneumoniae* polysaccharide-protein conjugate vaccines against homologous and heterologous serotypes in the chinchilla otitis media model. *J. Infect. Dis.*, **173**, 1119–1127.

Girard, A. E., Girard, D., English, A. R. *et al.* (1987). Pharmacokinetic and in vivo studies with azithromycin (CP-62,993), a new macrolide with an extended half-life and excellent tissue distribution. *Antimicrob. Agents Chemother.*, **31**, 1948–1954.

Girard, A. E., Cimochowski, C. R., Faiella, J. A. (1993). The comparative activity of azithromycin, macrolides and amoxycillin against streptococci in experimental infections. *J. Antimicrob. Chemother.*, **31** (Suppl. E), 29–37.

Girard, A. E., Cimochowski, C. R., Faiella, J. A. (1996). Correlation of increased azithromycin concentrations with phagocyte infiltration into sites of localized infection. *J. Antimicrob. Chemother.*, **37** (Suppl. C), 9–19.

Grote, J. J., van Blitterswijk, C. A. (1984). Acute otitis media. An experimental study. *Acta Oto-Laryngol*, **98**, 239–249.

Hardy, D. J., Swanson, R. N., Rode, R. A., Marsh, K., Shipkowitz, N. L., Clement, J. J. (1990). Enhancement of the *in vitro* and *in vivo* activities of clarithromycin against *Haemophilus influenzae* by 14-hydroxy-clarithromycin, its major metabolite in humans. *Antimicrob. Agents Chemother.*, **34**, 1407–1413.

Hardy, D. J., Swanson, R. N., Shipkowitz, N. L., Freiberg, L. A., Lartey, P. A., Clement, J. J. (1991). *In vitro* activity and *in vivo* efficacy of a new series of 9-deoxo-12-deoxy-9,12-epoxy-erythromycin A derivatives. *Antimicrob. Agents Chemother*, **35**, 922–928.

Herman, P., Friedlander, G., Tran Ba Huy, P., Amiel, C. (1992). Ion transport by primary cultures of Mongolian gerbil middle ear epithelium. *Am. J. Physiol.*, **262** (3), F373–F380.

Herman, P., Tu, T. Y., Loiseau, A. *et al.* (1995). Oxygen metabolites modulate sodium transport in gerbil middle ear epithelium: involvement of PG2. *Am. J. Physiol.*, **268** (3), L390–L398.

Herman, P., Tan, C. T., van den Abbeele, T., Escoubet, B., Friedlander, G., Tran Ba Huy, P. (1997). Glucocorticosteroids increase sodium transport in middle ear epithelium. *Am. J. Physiol.*, **272** (1), C184–C190.

Hermansson, A., Emgard, P., Prellner, K., Hellstöm, S. (1988). A rat model for pneumococcal otitis media. *Am. J. Otolaryngol.*, **9**, 97–101.

Hotaling, A. J., Doyle, W. J., Cantekin, E. L. (1987). Efficacy of a new cephalosporin for acute otitis media. *Arch. Otolaryngol. Head Neck Surg.*, **113**, 370–373.

Howie, V. M. (1992). Eradication of bacterial pathogens from middle ear infections. *Clin. Infect. Dis.*, **24** (Suppl. 2), 209–210.

Jacobs, M. R., Burch, D., Poupard, J., Moonsammy, G., Debanne, S., Appelbaum, P. (1996). Microbiology of acute otitis media—results of an international study of 917 patients. Abstract K26. In: *Program and Abstracts of the 36th Interscience Conference on Antimicrobial Agents and Chemotherapy*, p. 254. American Society for Microbiology, Washington, DC.

Jauris-Heipke, S., Leake, E. R., Billy, J. M., DeMaria, T. F. (1997). The effect of antibiotic treatment on the release of endotoxin during nontypable *Haemophilus influenzae*-induced otitis media in the chinchilla. *Acta Oto-Laryngol.*, **117**, 109–112.

Jossart, G. H., Canafax, D., Erdmann, G. R., Lovdhal, M. J., Russlie, H. Q., Juhn, S. K., Giebink, G. S. (1994). Effect of *Streptococcus pneumoniae* and influenza A virus on middle ear antimicrobial pharmacokinetics in experimental otitis media. *Pharm. Res.*, **11**, 860–864.

Juhn, S. K., Tolan, C. J., Antonelli, P. J., Giebink, G. S., Goycoolea, M. V. (1991). The significance of experimental animal studies in otitis media. *Otolaryngol. Clin. North Am.*, **24**, 813–827.

Kawana, M. (1995). Early inflammatory changes of the *Haemophilus influenzae* induced experimental otitis. *Auris, Nasus, Larynx*, **22**, 80–85.

Kawana, M., Kawana, C., Yokoo, T., Quie, P. G., Giebink, G. S. (1991). Oxidative metabolic products released from polymorphonuclear leukocytes in middle ear fluid during experimental pneumococcal otitis media. *Infect. Immun.*, **59**, 4084–4088.

Kawana, M., Kawana, C., Giebink, G. S. (1992). Penicillin treatment accelerates middle ear inflammation in experimental pneumococcal otitis media. *Infect. Immun.*, **60**, 1908–1912.

Klein, J. O. (1995). Antimicrobial therapy issues facing pediatricians. *Pediatr. Infect. Dis. J.*, **14**, 415–419.

Krekorian, T. D., Keithley, E. M., Fierer, D., Harris, J. P. (1991). Type B *Haemophilus influenzae*-induced otitis media in the mouse. *Laryngoscope*, **101**, 648–656.

Leibovitz, E., Piglansky, L., Raiz, S. *et al.* (1997). Bacteriologic efficacy of 3-day intramuscular ceftriaxone in non-responsive acute otitis media. Abstract K105. In: *Program and Abstracts of the 37th Interscience Conference on Antimicrobial Agents and Chemotherapy*, p. 346. American Society for Microbiology, Washington, DC.

Leizorovicz, A. (1995). Diagnosis and management of acute otitis media in children: results from a national survey in France. Abstract 2097. In: *Final Program and Abstracts of the 19th International Congress of Chemotherapy*, p. 359C. *Can. J. Infect. Dis.*, **6** (Suppl. C).

Lewis, D. M., Schram, J. L., Meadema S. J., Lime, D. J. (1980). Experimental otitis media in chinchillas. *Ann. Otol. Rhinol. Laryngol.*, **89**, 344–350.

Magit, A. E., Dolitsky, J. N., Doyle, W. J., Swarts, J. D., Seroky, J. T., Rosenfeld, R. M. (1994). An experimental study of cefixime in the treatment of *Streptococcus pneumoniae* otitis media. *Int. J. Pediatr. Otorhinolaryngol.*, **29**, 1–9.

Manninen, R., Huovinen, P., Nissinen, A. (1997). Increasing antimicrobial resistance in *Streptococcus pneumoniae, Haemophilus influenzae* and *Moraxella catarrhalis* in Finland. *J. Antimicrob. Chemother.*, **40**, 387–392.

Melhus, A., Janson, H., Westman, E., Hermansson, A., Forsgren, A., Prellner, K. (1997). Amoxicillin treatment of experimental acute otitis media caused by *Haemophilus influenzae* with non-β-lactamase-mediated resistance to β-lactams: aspects of virulence and treatment. *Antimicrob. Agents Chemother.*, **41**, 1979–1984.

Muffat-Joly, M., Barry, B., Hénin, D., Fay, M., Géhanno, P., Pocidalo, J. J. (1994). Otogenic meningoencephalitis induced by *Streptococcus pneumoniae* in gerbils. *Arch. Otolaryngol. Head Neck Surg.*, **120**, 925–930.

Ohashi, Y., Nakai, Y., Esaki, Y., Ohno, Y., Sugiura, Y., Okamoto, H. (1991). Influenza A-virus-induced otitis media and mucociliary dysfunction in the guinea pig. *Acta Oto-Laryngol. Suppl.*, **486**, 135–148.

Parks, R. R., Huang, C. C., Haddad, J. J. (1994). Evidence of oxygen radical injury in experimental otitis media. *Laryngoscope*, **104**, 1389–1392.

Parks, P. R., Huang, C. C., Haddad, J. J. (1995). Superoxide dismutase in an animal model of otitis media. *Eur. Arch. Otorhinolaryngol.*, **252**, 153–158.

Parks, R. R., Huang, C. C., Haddad, J. J. (1996). Middle ear catalase distribution in an animal model of otitis media. *Eur. Arch. Otorhinolaryngol.*, **253**, 445–449.

Pelton, S. I. (1996). New concepts in the pathophysiology and management of middle ear disease in childhood. *Drugs*, **52** (Suppl. 2), 62–67.

Post, J. C., Aul, J. J., White, G. J. *et al.* (1996). PCR-based detection of bacterial DNA after antimicrobial treatment is indicative of persistent, viable bacteria in the chinchilla model of otitis media. *Am. J. Otolaryngol.*, **17**, 106–111.

Reilly, J. S., Doyle, W. J., Cantekin, E. I. *et al.* (1983). Treatment of ampicillin-resistant acute otitis media in the chinchilla. *Arch. Otolaryngol.*, **109**, 533–535.

Retsema, J. A., Girard, A. E., Girard, D., Milisen, W. B. (1990). Relationship of high tissue concentrations of azithromycin to bactericidal activity and efficacy *in vivo*. *J. Antimicrob. Chemother.*, **25** (Suppl. A), 83–89.

Rosenfeld, R. M., Doyle, W. J., Swarts, J. D., Seroky, J., Pinero, B. P. (1992). Third-generation cephalosporins in the treatment of acute pneumococcal otitis media. An animal study. *Arch. Otolaryngol. Head Neck Surg.*, **118**, 49–52.

Saito, K., Suyama, K., Nishida, K., Sei, Y., Basile, A. S. (1996). Early increases in TNF-alpha, IL-6 and IL-1 beta levels following transient cerebral ischemia in gerbil brain. *Neurosci. Lett.*, **206**, 149–152.

Sato, K., Quartey, M. K., Liebeler, C. L., Giebink, G. S. (1995). Timing of penicillin treatment influences the course of *Streptococcus pneumoniae*-induced middle ear inflammation. *Antimicrob. Agents Chemother.*, **39**, 1896–1898.

Sato, K., Quartey, M. K., Liebeler, C. L., Le, C. T., Giebink, G. S. (1996). Roles of autolysin and pneumolysin in middle ear inflammation caused by a type 3 *Streptococcus pneumoniae* strain in the chinchilla otitis model. *Infect. Immun.*, **64**, 1140–1145.

Supance, J. S., Marshak, G., Doyle, W. J., Cantekin, E. I. (1982). Longitudinal study of the efficacy of ampicillin in the treatment of pneumococcal otitis media in a chinchilla animal model. *Ann. Otol. Rhinol. Laryngol.*, **91**, 256–260.

Suzuki, K., Bakaletz, L. O. (1994). Synergistic effect of adenovirus type 1 and nontypeable *Haemophilus influenzae* in a chinchilla model of experimental otitis media. *Infect. Immun.*, **62**, 1710–1718.

Swanson, R. N., Hardy, D. J., Chu, D. T. W., Shipkowitz, N. L., Clement, J. J. (1991). Activity of temafloxacin against respiratory pathogens. *Antimicrob. Agents Chemother.*, **35**, 423–429.

Thompson, T. A., Gardner, D., Fulghum, R. S. *et al.* (1981). Indigenous nasopharyngeal, auditory canal, and middle ear bacteria flora of gerbils: animal model for otitis media. *Infect. Immun.*, **32**, 1113–1118.

Thore, M., Burman, L. G., Holm, S. E. (1982). *Streptococcus pneumoniae* and three species of anaerobic bacteria in experimental otitis media in guinea pigs. *J. Infect. Dis.*, **145**, 822–828.

Thore, M., Sjoberg, W., Burman, L. G., Holm, S. E. (1985). Efficacy of Metronidazole in experimental *Bacteroides fragilis* otitis media. *Acta Oto-Laryngol*, **99**, 60–66.

Uhari, M., Mântysaari, K., Niemelä, M. (1996). A meta-analytic review of the risk factors for acute otitis media. *Clin. Infect. Dis.*, **22**, 1079–1083.

Van Buchem, F. L., Dunk, J. H. M., Van't Hof, M. A. (1981). Therapy of acute otitis media: myringotomy, antibiotics or neither? A double blind study in children. *Lancet*, **ii**, 883–887.

Von Unge, M., Decraemer, W. F., Bagger-Sjoback, D., Van den Berghe, D. (1997). Tympanic membrane changes in experimental purulent otitis media. *Hear. Res.*, **106**, 123–136.

Wang, W., Keller, K., Chadee, K. (1992). Modulation of tumor necrosis factor production by macrophages in *Entamoeba histolytica* infection. *Infect. Immun*, **60**, 3169–3174.

Zeevy, A. (1995). Early inflammatory events in a rat model of otitis media caused by infection with *Streptococcus pneumoniae*. In: *Recent Advances in Otitis Media — Proceedings of the Sixth International Symposium* (eds Lim, D., Bluestone, C., Casselbrant, M., Klein, J., Ogra, P.), pp. 258–261. B. C. Decker, Hamilton, Ontario.

Chapter 44

Bacterial Otitis Externa in the Guinea-pig Model

S. A. Estrem

Background of human infection

Otitis externa, known also as "swimmer's ear", is a common problem treated by primary-care practitioners and otolaryngologists. Acute bacterial otitis externa is usually caused by *Pseudomonas*, but sometimes *Proteus, Klebsiella*, or *Escherichia coli* is involved (Fairbanks, 1980). Topical gentamicin- or neomycin-containing eardrops have been used clinically because of their spectrum of activity against the above-mentioned bacteria. A variety of antibiotic ear drops are currently available. Applying ear drops to an ear canal partially or totally occluded with purulent debris dilutes the concentration of the drops and reduces penetration to the epithelium of the canal skin. Though widely believed, the theory that topical medications penetrate through edematous, inflamed, exuding membranes and reach the limits of infection at therapeutic concentration is unproven (Fairbanks, 1980).

Iontophoresis employs electrical current to drive ionically charged solutions into various tissues. Since Leduc first described iontophoresis in 1908, it has been used for a variety of medical purposes. It has been used to drive antibiotics into the cornea and treat intraocular infections (Von Sallman, 1943; Von Sallman and Mayes, 1944). Iontophoresis of histamine has been shown to be efficacious in local vasodilation and transplant viability of large composite grafts (DeHann and Stark, 1961). Dexamethasone may be concentrated into arthritic joints using iontophoresis, yielding local tissue concentrations of steroids second only to direct injection (Glass *et al.*, 1980). Otolaryngologists apply local anesthesia to the tympanic membrane and ear canal by lidocaine iontophoresis as described by Comeau *et al.* (1973). Iontophoresis of penicillin and gentamicin has been used with great success in treating auricular chondritis caused by *Staphylococcus aureus* and *Pseudomonas* (Rappaport *et al.*, 1965; LaForest and Cofrancesco, 1978; Greminger *et al.*, 1980).

Background of model

The rat and guinea-pig have both been used as animal models for external otitis. Emgard and Hellstrom (1997) developed otitis externa in the rat by mechanical stimulation of the ear canal. A 4 mm cut tip of a glass micropipette was rotated in the ear canal to abrade the skin. Hutchison and Wright (1975) created external otitis in the guinea-pig by introducing saline or water into the ear canal. The pinnas of the animals were taped down to occlude the ear canals on consecutive days until infection developed. The infections would develop in approximately 5 days. We elected to use an isolate of human *Pseudomonas aeruginosa* to insure external ear canal infection with *Pseudomonas aeruginosa*.

Animal spp.

Hartley guinea-pigs with an average weight of 300 g were used. Any strain of guinea-pig could be used, but the size should be large enough to allow examination of the external ear canal, which is quite tortuous in the guinea-pig. It is important that the guinea-pigs be examined and be free of middle and external ear infections prior to the experiment.

Preparation of animals

No specialized housing or care or special pretreatment is necessary.

Details of surgery

Overview

The goal was to transfer infection of *Pseudomonas aeruginosa* from a human with known otitis externa into a guinea-pig external ear canal to develop an external otitis in the guinea-pig. This would allow experimental manipulations in accordance with different treatment regimens.

Handbook of Animal Models of Infection
ISBN 0-12-775390-7

Materials required

Anesthetic, small ear speculum to allow visualization of the guinea-pig ear canal (1 mm diameter), small ear curette, small suction tips (no. 3 or smaller; 20–24 G), Lathbury probe used to make cotton swab, cotton, suture (2-0 silk suture or comparable strength), microscope to allow visualization of the ear.

Anesthesia

The guinea-pigs were anesthetized with an intramuscular injection of ketamine hydrochloride (35 mg/kg) and xylazine hydrochloride (7 mg/kg). Ketamine and xylazine hydrochloride in combination provide a short period of anesthesia, which is enough time and depth of anesthesia to allow cleaning of the ear canals and inoculation of the ears with bacteria. This procedure does not involve any significant discomfort, which makes this drug a good choice. It is safe as it does not result in respiratory depression when used in the recommended dosage. Inhalation agents such as ether, isofluorane or other agents can also be useful as they have a rapid onset of action and allow rapid recovery.

Storage, preparation of inocula

Pseudomonas aeruginosa is the most common organism isolated from the ear canal of patients with external otitis. This disease is prevalent in the summer months, with the source of bacteria emanating from any clinic of otolaryngology. The *Pseudomonas aeruginosa* is cultured by using a sterile cotton swab to cleanse a patient's infected ear. This swab is transferred to an aerobic culture medium and given to personnel in microbiology for culture and isolation of *Pseudomonas aeruginosa* from other organisms that are also present. A washed suspension of this isolate is used to infect the guinea-pig external ear canal.

Infection process

The ear canal of each guinea-pig was mechanically cleaned with the use of suction and cotton wrapped about the Lathbury probe, and inoculated with a washed suspension of *Pseudomonas aeruginosa*. The meatus was occluded with cotton plugs moistened with sterile saline. The pinnas were folded over the meatus with 2-0 silk suture used to maintain that position. Tape was also placed over the ear to assist in occlusion of the meatus. The cotton plugs were remoistened using sterile saline twice daily. Ears were re-examined and the procedure repeated on days 3, 5, and 7. All animals were reanesthetized with the same ketamine/xylazine combination during re-examination to minimize trauma to the ears. By the end of this 7-day period, significant external otitis was well established.

Key parameters to monitor infection

Otitis externa infection is evidenced by the presence of purulent debris in the ear canal, erythema (redness) and edema (swelling) of the ear canal skin. As the infection worsens, all the above parameters increase, with the edema of the ear canal progressing to the point of ear canal occlusion, such that the tympanic membrane (eardrum) cannot be visualized. Maceration of the meatal region may occur secondary to the infection. Otitis externa, when treated, will show gradual resolution of the above signs of infection, with edema being the last parameter to resolve.

All ears were examined daily. Inflammation, edema, and debris accumulation were each evaluated and graded as absent, mild, or marked. Infection was considered resolved when all three parameters had returned to normal; at that time, treatment was discontinued. The length of time required for each parameter of infection (inflammation, edema, and debris accumulation) to disappear was recorded.

Antimicrobial therapy

On day 7, ears have fully organized external otitis to allow experimentation with different treatment modalities. The author has compared mechanical cleaning alone, cleaning with antibiotic ear drops, and no treatment. Cleaning consisted of holding the pinna so as to straighten out the tortuous external ear canal. Suction and cotton swabs were employed using microscopic guidance to rid the canal of any debris. Non-sterile, clean technique was used for examination and cleaning of all ears throughout the experiment, simulating a clinic patient setting. A single mechanical cleaning, followed by gentamicin ophthalmic drops placed in the ear canal four times daily was compared with a single mechanical cleaning followed by a single treatment of gentamicin iontophoresis. This was performed under anesthesia by filling the ear canal with gentamicin sulfate in sterile water at a concentration of 5 mg/ml. A plastic speculum was placed in the ear canal, and a positive platinum electrode was attached to the animal's upper leg. The current was slowly increased over 1–2 minutes to 1.5 mA and was continued at this level for 20 minutes. The apparatus was removed and the ear canal was suctioned clean.

Pitfalls (advantages/disadvantages) of the model

Several laboratory animals have been evaluated as possible models for bacterial otitis externa, and the guinea-pig has appeared quite promising. This animal model provides a reliable, rapid simulation of the human condition.

Infections produced in the guinea-pig are remarkably similar in appearance to human ears. However, the duration of the disease in guinea-pigs, when left untreated, appears to be shorter than in humans who are untreated. The animals were observed to clear their ear canals by the centrifugal force generated by a violent rotational head shake. This may be at least partially responsible for the difference.

Contributions of the model to infectious disease therapy

This animal model allows varying types of treatment to be compared against the standard treatment of ear drops and mechanical cleaning. This model is a realistic comparison with the external otitis infection in humans with similar parameters of infection. The research with the guinea-pig animal model, as performed in our laboratory, supports the known clinical information in the management of ear canal infections.

References

Comeau, M. *et al.* (1973). Local anesthesia of the ears by iontophoresis. *Arch. Otolaryngol.*, **98**, 114–120.

DeHann, C. R., Stark, R. B. (1961). Changes in efferent circulation of tubed pedicles and in the transplantability of large composite grafts produced by histamine iontophoresis. *Plast. Reconstr. Surg.*, **28**, 577.

Emgard, P., Hellstrom, S. (1997). An animal model for external otitis. *Eur. Arch. Otorhinolaryngol.*, **254**, 115–119.

Fairbanks, D. N. F. (1980). Otic topical agents. *Otolaryngol. Head Neck Surg.*, **88**, 327–331.

Glass, J. M., Stephen, R. L., Jacobson, S. C. (1980). The quantity and distribution of radiolabeled dexamethasone delivered to tissue by iontophoresis. *Int. J. Dermatol.*, **19**, 519–525.

Greminger, M. D. *et al.* (1980). Antibiotic iontophoresis for the management of burned-ear chondritis. *Plast. Reconstr. Surg.*, **66**, 356–359.

Hutchison, J. L., Wright, D. N. (1975). Prophylaxis of predisposed otitis externa. *Ann. Otol.*, **84**, 16–21.

LaForest, N. T., Cofrancesco, C. (1978). Antibiotic iontophoresis in the treatment of ear chondritis. *Phys. Ther.*, **58**, 32–34.

Leduc, S. (1908). *Electronic Ions and Their Use in Medicine.* Rebman, London.

Rappaport, A. S. *et al.* (1965). Iontophoresis: a method of antibiotic administration in the burn patient. *Plast. Reconstr. Surg.*, **36**, 547.

Von Sallman, L. (1943). Penicillin and sulfadiazine in treatment of experimental intraocular infection with pneumococcus. *Arch. Ophthalmol.*, **30**, 246.

Von Sallman, L., Mayes, K. (1944). Penetration of penicillin into the eye. *Arch. Ophthalmol.*, **31**, 1.

Chapter 45

Otitis Media: The Chinchilla Model

D. M. Hajek, Z. Yuan, M. K. Quartey and G. S. Giebink

Introduction

Research using the chinchilla otitis media (OM) model has enhanced understanding of every aspect of this common childhood disease, including pathogenesis, immunoprophylaxis, and therapy. Otitis media represents a spectrum of disease with multifaceted etiology; consequently, there is not a single chinchilla OM model. Rather, a variety of methods have been used to induce middle ear (ME) inflammation under conditions of acute bacterial and viral infection, chronic inflammation, and eustachian tube dysfunction. Here we describe techniques for establishing and monitoring experimental OM in chinchillas, and discuss the adaptation of these experimental models to the study of antimicrobial therapy. In particular, we focus on methods for the study of ME pharmacokinetics and dynamics, issues that are central to rational antimicrobial therapy of OM.

Otitis media in humans

Otitis media is the most common pediatric diagnosis, with an estimated 24 million OM cases annually in the USA (Klein, 1994). The disease affects at least seven of every 10 children; one-third have repeated episodes; and 5–10% develop chronic otitis media with effusion (OME). Acute OM (AOM) initiates the continuum of OM, leading in many children to chronic OME and in some to chronic tissue sequelae such as mucosal granulation, mastoiditis, ossicular erosion and fixation, and cholesteatoma.

Streptococcus pneumoniae is the most frequent bacterium causing OM in infants and children (Luotonen *et al.*, 1981; Giebink *et al.*, 1982b; Bluestone *et al.*, 1992). There has been surprisingly little change in ME bacteriology reported during the past 18 years. Thirteen publications have enumerated the ME bacteriology of 4157 cases of AOM. Pneumococci were cultured from about one-third (20–37%) of acute middle-ear effusions (MEE); about 20% (6–31%) yielded *Haemophilus influenzae*, most of which were non-typable strains (NTHi), and smaller percentages yielded *Moraxella catarrhalis* (6%), *Streptococcus pyogenes* (2%), *Staphylococcus aureus* (2%), and others (6%). A study using fastidious bacteriologic techniques recovered pneumococci from 48% of AOM MEE samples (Del Beccaro *et al.*, 1992). Antigen and DNA detection methods have revealed pneumococci in 30–60% of cases with sterile MEE (Luotonen *et al.*, 1981; Post *et al.*, 1995).

Background of model

We developed the chinchilla OM model in 1975 to understand the pathophysiology of OM and to explore the efficacy of immunoprophylactic and chemotherapeutic interventions (Giebink *et al.*, 1976). First employed in the study of pneumococcal AOM, the chinchilla model has been adapted to study OM caused by NTHi (Doyle *et al.*, 1982; Giebink 1982), *Pseudomonas aeruginosa* (Antonelli *et al.*, 1992; Cotter *et al.*, 1996), and *Chlamydia trachomatis* (Weber and Koltai, 1991). The role of the respiratory viruses, influenza A and adenovirus, in AOM pathogenesis has also been investigated (Giebink *et al.*, 1980a; Bakaletz *et al.*, 1993). To advance the study of OME, we also developed a chinchilla model of chronic, non-suppurative OM (Canafax *et al.*, 1989).

Chinchilla langer (Figure 45.1) is ideally suited as an animal species for studying OM. Chinchillas are small, have auditory capabilities quite similar to those of humans, have a cochlea with a membranous architecture similar to the human cochlea, do not manifest presbycusis in long-term studies, and lack susceptibility to naturally occurring ME infections, which are common to the guinea-pig and rabbit. Chinchillas with fur defects can be obtained at reasonable cost, and are relatively inexpensive to maintain. Most studies have used 1–2-year-old chinchillas weighing 400–600 g. Figure 45.2 illustrates the radiographic anatomy of the chinchilla skull. The gross and microscopic anatomy of the chinchilla ME has been described (Daniel *et al.*, 1982; Hanamure and Lim, 1987; Vrettakos *et al.*, 1988). The ME volume is approximately 2 ml (Jossart *et al.*, 1990).

Handbook of Animal Models of Infection
ISBN 0-12-775390-7

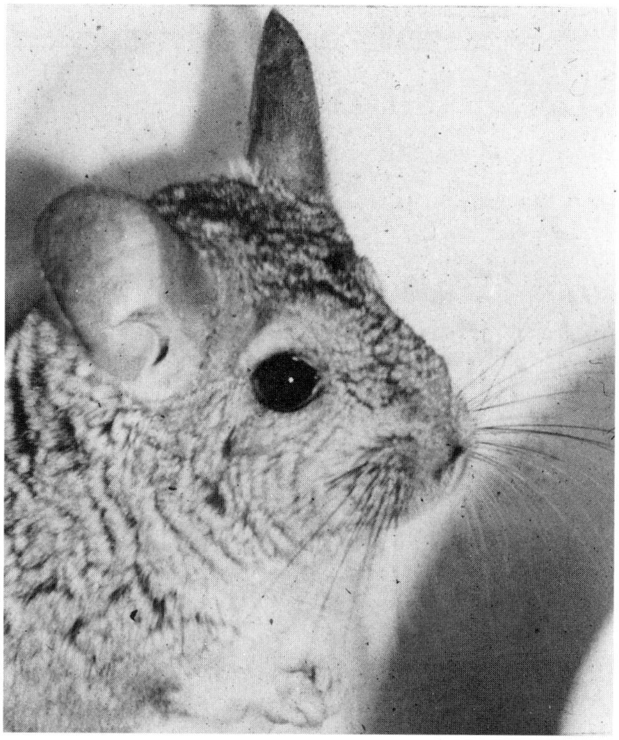

Figure 45.1 The chinchilla, *Chinchilla langer*.

Chinchilla model selection

A variety of chinchilla OM modeling techniques exist, and the selection of specific experimental conditions is guided by the purpose of the investigation. Most studies of OM antimicrobial therapy employ models involving acute bacterial infection. In general, the simplest of these models (direct ME inoculation) should be used when the outcome under study (ability of an antibiotic to sterilize the ME, for example) is thought to be independent of the means by which the experimental infection was established. AOM modeling techniques involving nasal colonization followed by ME deflation or respiratory virus attempt to simulate more closely the natural course of the disease in humans. They are useful in studying outcomes related to OM pathogenesis and prophylaxis. Acute OM sometimes leads to a chronic, suppurative OM that we have modeled successfully in chinchillas. The chinchilla OME model is applicable to the study of ME antimicrobial pharmacokinetics under conditions of chronic rather than acute inflammation.

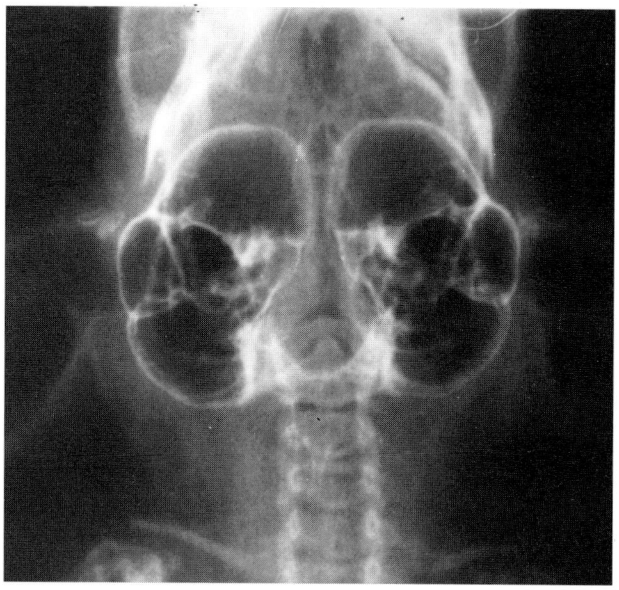

Figure 45.2 X-ray study demonstrating the structure of the chinchilla temporal bone.

General chinchilla use considerations

Preparation of chinchillas

Chinchillas are housed in stainless-steel wire cages (Hoeltge, Cincinnati, OH) except during experiments involving inoculation of respiratory virus, when filter-bonneted microisolator cages (Nalge Nunc, Rochester, NY) should be used during the period of viral shedding. Prior to the start of an experiment, ears should be examined by tympanometry and otoscopy to verify the absence of MEE or tympanic membrane pathology. We have reported the tympanometric configuration of healthy, uninfected chinchillas (Giebink *et al.*, 1982a). To prevent ME contamination during invasive procedures, hair should be shaved or plucked from the head over the area shown in Figure 45.3A.

Anesthesia

Chinchillas are effectively sedated with 10 mg ketamine hydrochloride injected into the anterior thigh. Complete anesthesia ensues in about 5 minutes and lasts for about 25 minutes. During longer procedures, a second dose may be necessary. Chinchillas should be observed carefully while anesthetized because ketamine occasionally causes respiratory depression.

Middle-ear access

Access to the chinchilla ME cavity is through the cephalad bulla underlying the dorsal aspect of the skull. Figure 45.3A illustrates the surface anatomy of the bulla with hair removed, and Figure 45.3B is the same view with the skin removed to reveal the underlying bony structure. The bone is very thin at the apex of the bulla, allowing the experimenter to insert a needle into the ME cleft in a relatively atraumatic fashion. We have successfully used needles as large as 15 G. Just prior to insertion of the needle, the skin overlying the bulla should be

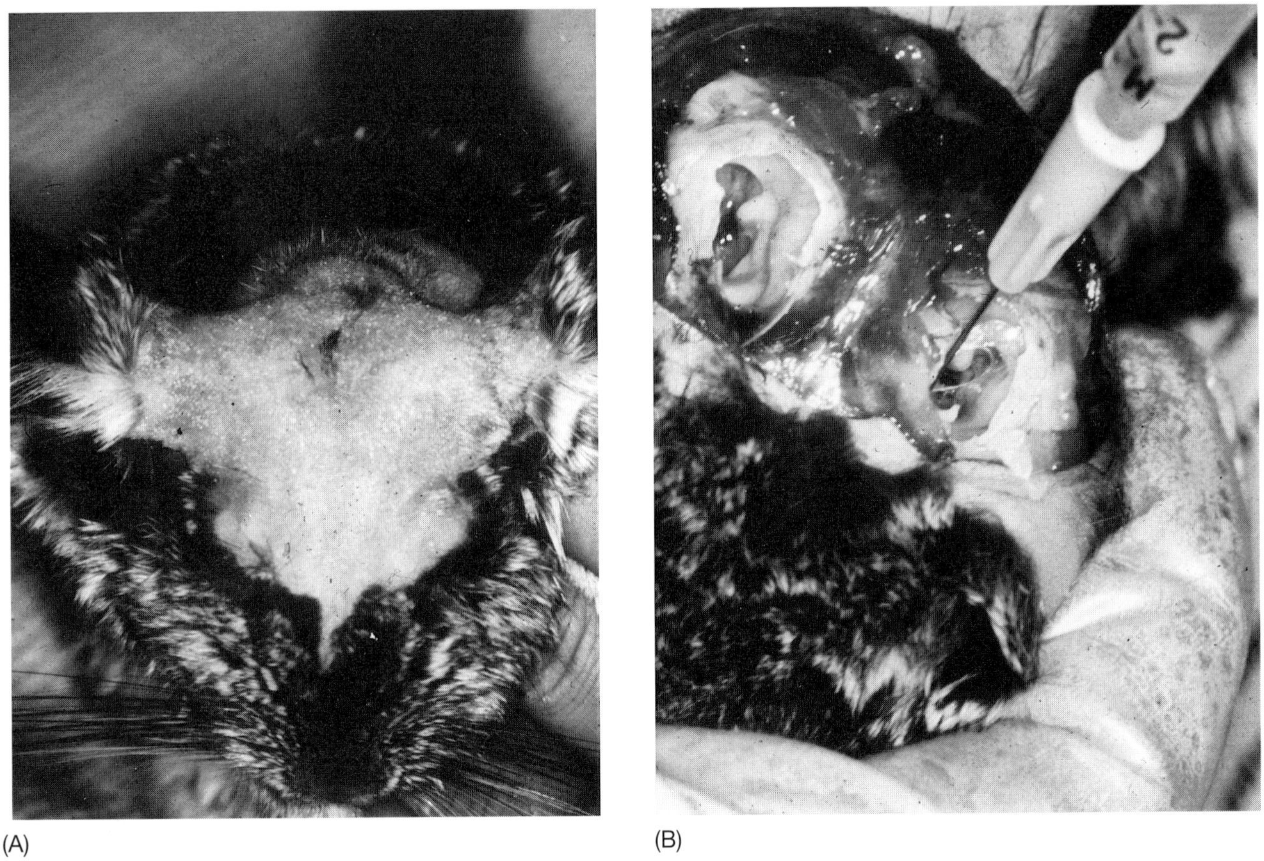

Figure 45.3 A Chinchilla head (with hair shaved) showing the surface anatomy of the cephalad bullae. B. The same view as (A), but with skin and bullar apices removed to demonstrate bony structure of the temporal bone surface and middle ear cleft.

displaced approximately 0.5 cm so that intact skin will overlie the defect in the bulla after needle withdrawal. The needle should be inserted with gentle pressure and a twisting motion to avoid fracturing the thin bone. In this way, the needle enters the ME cleft without penetrating so deeply that it encounters the middle ear ossicles. A fragment of bone may clog the needle bore. When this occurs, insertion of a sterile wire into the needle bore clears the blockage.

Postoperative care

Chinchillas should be kept warm and observed closely until they have completely recovered from anesthesia. Following recovery, chinchillas return to normal behavior and are given food and water *ad lib.*

Chinchilla acute otitis media models

Experimental AOM has been induced in chinchillas by at least three methods using live bacteria: direct ME inoculation of bacteria, nasal inoculation followed by ME deflation, and co-infection with bacteria and a respiratory virus.

Direct middle ear inoculation

Overview

Transbullar inoculation of live bacteria is the simplest and most direct means of inducing experimental AOM. In contrast to the other two AOM models described, transbullar inoculation allows for precise control over the number of live bacteria reaching the ME cavity and produces the most consistent rates of infection. We have employed this model to study AOM caused by pneumococcal types 3, 6A, 6B, 7F, 14, 18C, 19A, 19F, 23B, and 23F (Giebink *et al.*, 1980b, 1993, 1996) and by NTHi (Giebink, 1982).

Materials

Ketamine hydrochloride (50 mg/ml), 15 G needle, 23 G needles (two), tuberculin syringe containing bacterial inoculum, sterile polyethylene tubing (internal diameter, 0.58 mm; external diameter, 0.965 mm; 427420, Becton Dickinson, Lincoln Park, NJ), 70% ethanol, betadine solution.

Procedure

Following complete induction of anesthesia, the chinchilla is placed in a prone position. Hair is plucked from the skin covering the cephalad bulla, and the skin is scrubbed with betadine solution then rinsed with 70% ethanol. If the inoculum volume is more than 0.1 ml, a 23 G needle is inserted slightly off-center into the bulla to vent the ME during injection; this prevents membrane rupture during bullar injection. A 15 G needle is then inserted into the center of the bulla. A 23 G needle on a tuberculin syringe is inserted into one end of a 15 cm length of polyethylene tubing; the other end of the tubing is inserted through the bore of the 15 G needle and passed in a caudal direction to the hypotympanic bulla; and the inoculum is slowly infused into the ME. Both needles are withdrawn, and bleeding is controlled by direct pressure.

Nasal bacterial inoculation with middle-ear deflation

Overview

In this AOM model, intranasal inoculation of a ME pathogen establishes nasopharyngeal colonization. Subsequent application of negative pressure to the ME (deflation) aspirates a small number of bacteria into the ME cleft. In this respect, deflation may resemble the natural pathogenesis of AOM more closely than direct transbullar bacterial inoculation, and may be especially useful when the contribution of mucosal defense is of interest, as in vaccine evaluation. The model has been employed to study experimental OM due to pneumococcal types 6A, 7F, 18C, and 19F (Giebink *et al.*, 1979; Giebink, 1981).

Materials

Nasal inoculation. Ketamine hydrochloride (50 mg/ml); tuberculin syringe with 17 G blunt needle; bacterial inoculum.

Middle-ear deflation. Ketamine hydrochloride 50 mg/ml; tympanometer; otoscope; 19 G butterfly needle (P216119, Terumo, Elkton, MO) connected by firm plastic tubing and a three-way T-connector to a 30 ml syringe and an electronic and/or a liquid-column manometer; 70% ethanol; betadine solution; pressure valve grease; tuberculin syringe with 17 G blunt needle; 0.5 ml phosphate-buffered saline (PBS; pH 7.45, $[Na^+]$=150 mM, $[HPO_4^{-2}]$ = 15 mM).

Procedure

Nasal inoculation. Following induction of anesthesia, the chinchilla is placed in a supine position. The bacterial inoculum is drawn into the tuberculin syringe and, with the animal's muzzle slightly elevated, is instilled dropwise into the nares.

Middle-ear deflation. Following induction of anesthesia, the chinchilla is placed in a prone position. Hair is plucked from the skin covering the superior cephalad bulla, and the skin is scrubbed with betadine solution then rinsed with 70% ethanol. A 19 G butterfly needle attached by firm plastic tubing to a 30 ml syringe and a manometer is inserted into the center of the bulla. A small amount of vacuum grease should be applied to the skin and protruding needle shaft to ensure a tight seal. Just prior to the application of negative pressure, 0.5 ml PBS solution is instilled into the chinchilla's nares by means of a tuberculin syringe fitted with a blunt needle. This procedure transiently closes the eustachian tubes so the ME pressure does not equilibrate during deflation (Giebink *et al.*, 1979). The plunger of the deflating syringe is then withdrawn until a pressure of –30 mmHg is attained. More highly negative pressures cause barotrauma and acute histopathologic changes that may obscure experimental infection-related histopathology (Meyerhoff *et al.*, 1981). Great care must be taken to withdraw the syringe plunger slowly, as even brief periods of pressure below –30 mmHg may rupture the tympanic membrane. The inclusion of a liquid-column manometer results in a more compliant system and facilitates the maintenance of constant pressure throughout the 5 minute deflation period. The ear should be examined by tympanometry both during and after the procedure to ensure that negative ME pressure has been achieved. At the end of the deflation period, the needle is withdrawn and any bleeding is controlled by direct pressure. Otoscopic examination should follow to verify the integrity of the tympanic membrane.

Nasal inoculation of bacteria and respiratory virus

Overview

The association between respiratory virus infection and OM in children has been demonstrated (Henderson *et al.*, 1982; Klein *et al.*, 1982), and models have been developed in the chinchilla to explore these relationships. During respiratory viral infection, as in the middle-ear deflation model, bacteria enter the ME from the nasopharynx through the eustachian tube. In this case, however, virus-induced mucosal changes and negative ME pressure, rather than aspiration, promote bacterial ascent to the ME. Influenza A-virus-infected animals are significantly more susceptible to experimental pneumococcal OM than are non-virus-infected animals (Giebink *et al.*, 1980a, 1987). We studied the virulence of several influenza strains in chinchillas and found that an influenza A/Alaska/6/77 (H3N2) strain significantly increased susceptibility to pneumococcal OM (Giebink *et al.*, 1983). A similar interaction between adenovirus and NTHi has also been used to create experimental AOM (Bakaletz *et al.*, 1993; Suzuki and Bakaletz, 1994; Miyamoto and Bakaletz, 1997). However, adenovirus did not enhance the ability of nasally inoculated *M. catarrhalis* to induce experimental AOM in chinchillas (Bakaletz *et al.*, 1995). Moreover, adenovirus is only effective in very young chinchillas.

Materials

Ketamine hydrochloride (50 mg/ml), tuberculine syringe with 17 G blunt needle, bacterial inoculum, viral inoculum.

Procedure

We studied the effect of varying the time interval between inoculation with influenza virus and *S. pneumoniae* on the incidence of pneumococcal disease, and observed the highest incidence of pneumococcal AOM when intranasal pneumococcal inoculation occurred 4 days after intranasal influenza inoculation (Abramson *et al.*, 1982). Procedures for both viral and bacterial inoculation are essentially the same as those described under "Nasal bacterial inoculation with middle-ear deflation" above.

Chronic suppurative OM model

Tubal dysfunction or recent AOM may lead to the development of chronic suppurative OM by increasing tubotympanic susceptibility to opportunistic pathogens. These interactions have been explored and confirmed in chinchillas by intranasal *P. aeruginosa* inoculation of animals with previous experimental eustachian tube obstruction (Antonelli *et al.*, 1992) or concurrent pneumococcal AOM (Antonelli *et al.*, 1994). The surgical technique for eustachian tube obstruction is described in the following section.

OME model

Overview

Obstruction of the chinchilla eustachian tube creates a useful model for the study of OME (Juhn *et al.*, 1977; Canafax *et al.*, 1989). Because the penetration of antibiotics through chronically and acutely inflamed ME epithelium may differ, this model is important to the complete understanding of ME antimicrobial pharmacokinetics. Following surgery, fluid accumulates in the ME over a period of several weeks. The procedure is not uniformly effective: bilateral eustachian tube obstruction in 19 chinchillas resulted in a profound fall in ME pressure in only 53% of ears (Canafax *et al.*, 1989). We have experienced the most success obstructing the eustachian tube with silastic sponges (Dow Corning, Midland, MI). Obstruction with Coeflex paste (Coe Lab, Chicago, IL) caused an undesirable local inflammatory response (Canafax *et al.*, 1989).

Materials

Ketamine hydrochloride (50 mg/ml), atropine (0.5 mg/ml), gentamicin (50 mg/ml), spear-tip 45° offset myringotomy blade (Becton Dickinson, Lincoln Park, NJ), scalpel holder with no. 1 scalpel blade, two Barbara picks (N1705 79, Stortz, St Louis, MO), Hartman alligator ear forceps (380–150, Jarit, Hawthome, NY), Vienna nasal speculum (400–101, Jarit, Hawthorne, NY), column stand to hold speculum, silastic sponge (Dow Corning, Midland, MI), sterile cotton swabs, operating microscope (10–20 ×), suction pump, tissue glue (Nexaband, Tri-Point Medical, Raleigh, NC).

Procedure

The column stand is arranged to hold the nasal speculum at the mouth level of a supine chinchilla. Atropine i.m. (0.05 ml, 0.5 mg/ml) is used to control salivation during the procedure. Following complete induction of anesthesia, the chinchilla is placed ventral side up with its head towards the technician. The speculum is inserted into the animal's mouth and opened far enough for the palate to be exposed. After the soft palate has been visualized through the operating microscope, it is cleaned with an ethanol-soaked cotton swab. At all times during the procedure, fluid accumulations should be immediately aspirated from the posterior nasopharynx. Using a myringotomy blade, a longitudinal incision is made along the midline of the soft palate, extending 3 mm from the junction of the hard and soft palates. A no. 1 scalpel blade may be used to penetrate the fibrous connective tissue layer of the soft palate. When the incision through the soft palate is complete, the eustachian tube orifices may be visualized. Slivers of silastic sponge (approximately 4 mm × 1 mm × 0.5 mm) are grasped with Hartman alligator ear forceps and gently inserted into

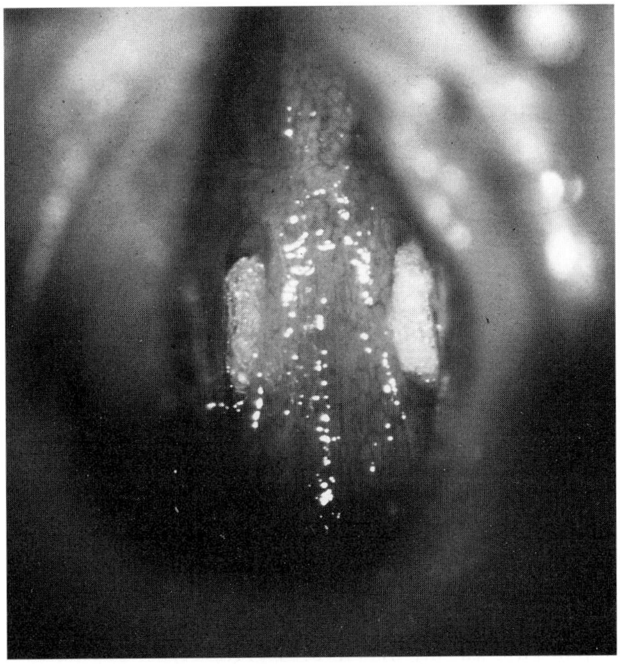

Figure 45.4 A view of the nasopharyngeal eustachian tube orifices after obstruction surgery. The obstructing masses of silastic sponge protrude from the orifices. The midline soft palate incision has not yet been closed.

the eustachian tube. A Barbara pick may be used along with the forceps to guide the sponge into the eustachian tube. Additional slivers are added until the mass of sponge is secure in the eustachian tube and no more sponge can be inserted gently. The procedure is repeated on the contralateral eustachian tube. Figure 45.4 illustrates the eustachian tube orifices and the protruding mass of silastic sponge. Once all bleeding has stopped, the palatal incision is closed with tissue glue and the chinchilla is removed from the speculum assembly.

Postoperative care

Gentamicin i.m. (0.02 ml, 50 mg/ml) is administered once after surgery to prevent postoperative infection. The chinchilla should be kept warm until it has recovered fully from anesthesia, and it should be observed closely on the day of the procedure. Frequent otoscopic and tympanometric examinations should be performed to monitor the accumulation of MEE, which peaks 3–4 weeks after eustachian tube obstruction (Canafax *et al.*, 1989).

Preparation of inocula

Chinchillas have been used to model OM caused by a variety of bacterial and viral pathogens. We have been most interested in OM caused by *S. pneumoniae*, and here we describe the growth of pneumococci and discuss appropriate intranasal and intrabullar inocula for the pneumococcal types we have used.

Prior to use in experimental infection, the virulence of a pneumococcal strain should be enhanced in mice by intraperitoneal inoculation (Briles *et al.*, 1992; Aaberge *et al.*, 1995). Pneumococci in mid-log-phase growth are injected into the peritoneal cavity of a mouse. After the mouse dies (usually 16–24 hours), an incision is made in the abdominal wall and a small quantity of peritoneal fluid is recovered on a sterile cotton swab. The swab is streaked on to a 5% sheep blood agar (SBA) plate, which is incubated at 37°C. If the resulting growth is pure, pneumococcal type is verified by the Quellung reaction using type-specific antisera (Statens Seruminstitut, Copenhagen, Denmark).

To grow pneumococci for inoculation, Todd–Hewitt broth (Difco Laboratories, Detroit, MI) supplemented with bovine serum albumin (Sigma, St Louis, MO) is inoculated with isolated colonies from an overnight subculture of the mouse-passed strain and incubated in a water bath at 37°C. Bacteria are grown to mid-log phase, and bacterial density is estimated using an optical density versus colony-forming unit (cfu) regression formula, which was generated using quantitative culture and spectrophotometric data taken at multiple timepoints from previous pneumococcal broth cultures. The culture is diluted appropriately in sterile 0.01 M PBS solution, and placed in an ice bath until inoculation occurs. Quantitative culture on 2% SBA confirms the cfu estimate. Frozen inocula should not be used in studies of ME inflammation involving direct inoculation because of possible inflammatory effects of glycerine on the ME mucosa.

We have reported direct ME inoculation of pneumococcal types 3, 6A, 6B, 7F, 14, 18C, 19F, 19A, 23B, and 23F (Giebink *et al.*, 1980b, 1993, 1996). For all types except 23F, inocula containing 20–50 cfu in 0.1–0.3 ml PBS produced 90–100% incidence of AOM. The type 23F strain used in these experiments was relatively avirulent in chinchillas: inocula as large as 10^8 resulted in very low AOM incidence (Giebink *et al.*, 1993).

Otitis media models involving nasopharyngeal colonization require much larger inocula. Colonization incidences after nasal inoculation of chinchillas with suspensions containing 10^3, 10^4, 10^5, and 10^6 cfu of type 7F pneumococci in 0.5 ml PBS were, respectively, 0%, 92%, 100%, and 100% (Giebink *et al.*, 1980a); 100% colonization has also been achieved with 2×10^7 cfu of types 3 and 6A pneumococci, and with 10^{15} cfu of types 6B and 19F pneumococci.

Key parameters to monitor experimental otitis and response to treatment

Dissemination of infection

Pneumococcal infection may not remain confined to the chinchilla ME. This is particularly true in experiments using the more virulent pneumococcal serotypes, such as 19F, when mortality due to invasive bacterial disease may be as high as 100% with ME inocula as small as 20 cfu (Giebink *et al.*, 1996). In such cases, high mortality may hinder attempts to monitor the ME infection over time. Extension of infection to the bony labyrinth presents as ataxia. We have observed that fatal infection invariably follows the onset of labyrinthitis. Therefore, we recommend that the mortality endpoint be defined as death or ataxia, and that chinchillas developing ataxia be euthanized to relieve suffering. Postmortem culture of blood, lung, and cerebrospinal fluid should be performed on any chinchilla that dies.

Quantifying nasopharyngeal colonization

Bacterial or viral nasopharyngeal colonization is verified by rinsing the nasal passages and culturing the resulting effluent on pathogen-specific media. Quantitative culture methods may be used to study the kinetics of bacterial nasopharyngeal colonization. Materials required for nasal rinse are: 3 ml syringe with sterile 17 G blunt needle, 1.5 ml sterile isotonic saline solution, sterile 15 ml tubes (containing appropriate transport media in the case of viral culture), and culture media. Following complete anesthesia induction with ketamine hydrochloride, the technician grasps the chinchilla around the thorax and forelimbs with the left hand. Holding the chinchilla with the left naris in the most dependent position, the technician inserts the blunt needle 2–3 mm into the right naris and slowly instills the sterile saline solution into the nasal passage. Approximately 50–75% of the solution will drip out of the left naris and into the sterile collection tube.

Middle-ear outcome

Non-invasive methods

Identification and monitoring of experimental otitis media in chinchillas is accomplished by non-invasive or invasive methods. The possibility of employing invasive methods is an inherent advantage of the animal model, since such procedures cannot be used when the disease process is studied in human subjects. The two primary non-invasive methods for identifying MEE are otoscopic evaluation of the tympanic membrane and objective assessment of the ME using an acoustic impedance measurement. This latter method is commonly known as tympanometry in clinical settings.

In several studies, the findings of these non-invasive techniques were compared with middle-ear findings after tympanocentesis or myringotomy, and the probability of MEE with various tympanic membrane characteristics and tympanometric patterns was determined. The diagnostic value of each measure was ascertained in the chinchilla model to determine the sensitivity and specificity of these non-invasive methods (Marshak *et al.*, 1981; Giebink *et al.*, 1982a). Recent experience in our laboratory suggests that the reported sensitivity and specificity of these methods apply only to adult chinchillas, and that tympanometry and otomicroscopic examination are unreliable methods for assessing experimental otitis in infant chinchillas (Giebink, unpublished data). Monitoring of experimental OM by magnetic resonance imaging of the ME cleft has been described in monkeys (Alper *et al.*, 1997) and chinchillas (Chan *et al.*, 1992, 1994).

Invasive methods

Temporal bone histopathology. Invasive methods are more comprehensive in assessing the total ME condition. Temporal bone harvest is the most definitive method for histopathological examination of the ME and related structures. The importance of histopathologic assessment was illustrated by Meyerhoff and Giebink (1982), and Giebink *et al.* (1985), who reported extensive residual ME disease in chinchillas with normal tympanic membranes and among chinchillas with disease-free ears according to otoscopic and tympanometric assessment. A disadvantage of histopathologic assessment is that it provides only one data point in time, and progression of the disease in individual chinchillas cannot be documented. This disadvantage may be overcome by serial sacrifice of chinchillas within an experimental group cohort at several timepoints during an experiment. However, this approach necessitates very large numbers of chinchillas.

MEE aspiration. Transbullar aspiration of the contents of the chinchilla ME cavity is used to follow the ME condition in a single chinchilla longitudinally. The procedure is useful for determining the presence or absence of MEE and to perform bacteriologic and biochemical tests on the MEE.

Materials required for MEE aspiration include: ketamine hydrochloride (50 mg/ml), 15 G needle, sterile polyethylene tubing (internal diameter, 0.58 mm; external diameter, 0.965 mm; 427420, Becton Dickinson, Lincoln Park, NJ), 70% ethanol, betadine solution, tuberculin syringe with 23 G needle, and sterile tubes in which to collect middle-ear effusion. After the induction of anesthesia, the chinchilla is placed in a prone position, the hair is plucked from the skin overlying the cephalad bulla, and the skin is scrubbed with betadine then rinsed with 70% ethanol. A 15 G needle is inserted into the center of the bulla until it just penetrates the bone. A 23 G needle on a tuberculin syringe is inserted into one end of a 15 cm length of polyethylene tubing, and the other end of the tubing is inserted through the bore of the 15 G needle to the base of the hypotympanic bulla. The plunger on the tuberculin syringe is slowly withdrawn until the desired volume of fluid has been aspirated, and the fluid is transferred to a sterile container. Volumes as large as 1.5 ml can be obtained from chinchillas with AOM.

AOM severity score

We have described a severity scoring scale for chinchilla models of pneumococcal AOM (Giebink *et al.*, 1993). This scale combines findings from tympanometry, ME aspiration and MEE culture, and is useful in assessing the efficacy of prophylactic or therapeutic interventions.

Middle-ear antimicrobial pharmacokinetics and pharmacodynamics

Background

The study of antibiotic dose, and blood and tissue concentration versus time profile is central to improving treatment of bacterial infections at extravascular sites in general and, specifically, in the ME (Barza and Cuchural, 1985; Nix *et al.*, 1991). Current AOM antibiotic treatment, however, is largely empirical and is not based on antibiotic ME pharmacokinetic (PK) and pharmacodynamic (PD) considerations. Treatment failure occurs in 5–10% of AOM cases, and a probable cause is inadequate

ME antibiotic concentration with failure to eradicate all the ME bacteria.

The transport of antibiotics through various tissue membrane barriers to reach the site of infection is a critical factor for their efficacy. Pharmacokinetic characteristics that describe antibiotic transport into the ME have begun to be evaluated in the chinchilla OM model (Juhn *et al.*, 1986; Canafax *et al.*, 1988, 1989, 1994; Jossart *et al.*, 1990, 1993).

Pharmacokinetic studies in the chinchilla OM model characterize antibiotic distribution into and elimination from the MEE. Pharmacodynamic analysis relates the PK behavior of a drug to its efficacy. Middle-ear antibiotic penetration and elimination are determined by the permeability of ME tissue membranes and drug elimination from the systemic circulation, among other relevant physiological factors such as plasma protein binding (Jossart *et al.*, 1993). Because drug permeability of a tissue membrane varies with the inflammatory state, PK studies have been conducted in chinchillas with both experimental AOM and experimental OME. Two approaches of evaluating antibiotic PK in the MEE, namely, "forward" penetration and "reverse" or local ME penetration models, have been employed in chinchilla experimental OM (Canafax *et al.*, 1994).

"Forward" penetration model

Overview

In the "forward" penetration model, antibiotics are administered systemically and antibiotic ME concentration versus time profile is determined. Initially, plasma and MEE dose-finding studies for each drug are conducted to guide dose and sampling time selection in subsequent PK and PD studies.

Animal preparation, antibiotic dosing, and sample collection

Dose-finding plasma and MEE pharmacokinetic protocols. These are performed by administering to chinchillas sequentially lower intramuscular doses of a drug and measuring the concentrations produced at the postdose times that are ideal for each drug, such as 0, 1, 2, 3, and 4 hours for amoxicillin. The administration of consistent drug volumes, typically 0.1 ml, eliminates the potential effects of dilution on absorption rate and extent. Blood samples are collected by cardiac puncture and placed in EDTA-containing glass tubes. The samples are immediately centrifuged at 1000 *g* for 10 minutes, and the plasma is decanted and frozen until analysis.

Pharmacokinetics in AOM. To study the behavior of antibiotic concentrations in MEE and plasma under steady-state conditions, studies of multiple antibiotic dosing are performed. Experimental AOM is induced 3–5 days prior to the PK study. The day before antibiotic administration, both ME of each chinchilla are aspirated to verify the presence of MEE, and 25 μl of MEE is collected for bacterial culture. Only culture-positive ears are used in the subsequent procedures. Each chinchilla receives an antibiotic every 8–12 hours, depending on the drug, for a total of four doses. After the fourth dose, serial plasma and MEE samples are collected to determine the ME antibiotic penetration at the steady state and to assess antibacterial efficacy. A sampling duration of 2–4 days is optimal; beyond that period AOM will resolve in some animals. At the conclusion of the experiment, chinchillas are euthanized and their temporal bones may be collected for histopathologic examination.

Pharmacokinetics in OME. To study penetration of the various antibiotics into the MEE in OME, eustachian tube obstruction is performed 3–4 weeks prior to the PK experiment. Just prior to antibiotic administration all ears are aspirated and the MEE is cultured to ensure that only ears with sterile effusions are included in the study. Following antibiotic administration, serial blood and MEE samples are obtained for analysis.

Pharmacokinetics in normal ME. To study PK in ME with normal mucosa, eustachian tube obstruction is performed within 24 hours of the PK experiment. A volume of 1.0 ml of phosphate-buffered saline, pH 7.4, is instilled into each ME. Antibiotic is administered i.m., and serial blood and MEE samples are obtained for analysis.

Quantitative analysis of antibiotics in chinchilla plasma and MEE

Quantitative analysis of antibiotics in chinchilla plasma and MEE is carried out by high-performance liquid chromatography (HPLC), because of its high sensitivity and specificity, which permits analysis of microliter MEE volumes. We have reported HPLC methods for measuring amoxicillin (Yuan *et al.*, 1997), trimethoprim and sulfamethoxazole (Erdmann *et al.*, 1988), ciprofloxacin (Lovdahl *et al.*, 1993) and cefpodoxime (Lovdahl *et al.*, 1994). These methods generally involve antibiotic extraction from chinchilla plasma or MEE by protein precipitation or solid-phase extraction, followed by separation and measurement on a C_8 or C_{18} reversed-phase column with UV or fluorescence detection.

Data analysis

Plasma and MEE antibiotic concentration versus time data are analyzed in the context of a PK model and associated parameters to gain insight into the behavior of the drug. MEE and plasma concentration–time data collected are analyzed by appropriate compartmental or non-compartmental methods (Rowland and Tozer, 1989). Pharmacokinetic parameters to be estimated are distribution volume (V), clearance (CL), and intercompartmental clearance (Q) in the case of multicompartmental analysis. Biological half-lives are estimated as secondary parameters of the model as $T_{1/2} = 0.693 \times V/CL$. If sufficient plasma and MEE concentration data are obtained in individual chinchillas, data are evaluated individually using software such as ADAPT II (D'Argenzio and Schumitzky, 1996). Where data are sparse, it may be necessary to estimate model parameters using population analysis, e.g., with NONMEM (Beal and Sheiner, 1980).

Because of small MEE volume, the plasma antibiotic profile may be used as a forcing function to model a MEE antibiotic profile (Wang and Welty, 1996). Relevant parameters such as the rate constants of penetration into and elimination out of the ME may be derived from the modeled profile. The rate constant elimination from the ME, k_e, can be estimated reliably from the logC_e-t plot. Because of sparse ME concentration versus time data, however, it has not been feasible to estimate reliably the kinetic parameters of antibiotic distribution from blood to the ME in the "forward" penetration model.

"Reverse" or local ME penetration model

Overview

The reverse or local penetration model is used to determine the elimination of antibiotics from the ME. This model employs chinchillas with MEE induced under conditions of acute or chronic inflammation by one of the methods described above.

Procedure

After complete induction of anesthesia, a 16 G needle is inserted through the center of the cephalad bulla and a 23 G needle is inserted adjacent to this needle as a pressure vent to prevent rupture of the tympanic membrane during injection of the antibiotic solution. All MEE is aspirated through a thin plastic catheter (internal diameter, 0.58 mm; external diameter, 0.965 mm; 427420, Becton Dickinson, Lincoln Park, NJ) inserted through the bore of the 16 G needle. A second catheter is passed through the 16 G needle to the base of the labyrinthine bulla. The antibiotic solution is slowly introduced at 1 ml/minute into the ME through the catheter, after which the catheter and both needles are removed. MEE samples are aspirated at selected time points, using the same needle–catheter technique, and antibiotic concentrations in these samples are measured by HPLC.

Data analysis

To obtain PK parameters from experiments with the "reverse" penetration model, we derived the relevant equations, assuming that antibiotics diffuse passively across the tissue space into the MEE. Fick's first law of diffusion states that the net rate of diffusion across a membrane is proportional to the surface area of the membrane and the concentration gradient across the membrane at steady state if the solutions are stirred. This relationship is expressed as: $J = -PS(C_b - C_e)$ where J is the net rate of diffusion in units of amount per time (i.e., flux); P is the membrane permeability coefficient in units of length per time and is proportional to the lipid solubility of the solute; S is the surface area of the ME in contact with the solute; C_b and C_e are the concentration of solute in the blood and in the MEE respectively. Because the antibiotic concentration in the blood is negligible, J can be expressed as dA_e/dt, A_e being the amount of a drug in the MEE. Substitution and rearrangement of the Fick equation yields:

$$dA_e/A_e = dC_e/C_e = -(PS/V)dt$$

where V is the MEE volume. The S/V ratio of an extravascular space is an important determinant of drug concentration behavior in that space.

Using quantitative histomorphometry, we measured the ME space in 10 chinchillas and found a V of 2.09 ± 0.08 ml (mean ± SD) and S of $14.41 \pm 1.48\,cm^2$ (Jossart *et al.*, 1990). The chinchilla ME S/V ratio of 6.93 is comparable with other species and humans (Van Etta *et al.*, 1982). A significant change in elimination rate constant (k_e) with small changes in the MEE volume is not expected because as V increases there is a proportional increase in S in contact with antibiotic solution and, thus, the S/V ratio remains constant if homogeneity of the MEE is assumed (Jossart *et al.*, 1990). Hence,

$$dC_e/C_e = -(PS/V)dt = (k_e)dt$$

where PS/V is equal to k_e, the elimination rate constant. Integration of this relationship yields:

$$k_e = -[ln\,(C_t/C_0)]/t$$

where C_0 and C_t are the concentrations of solute in the ME initially and at some later time (t), respectively. This equation is used to calculate the apparent k_e for studies that collect only two concentration–time points. For studies where multiple, serial concentration–time data points are collected, the Wagner–Nelson regression analysis is used to calculate apparent k_e. Elimination rate constants so derived agree well with those derived from experiments in the "forward" penetration model (Canafax *et al.*, 1994).

Pharmacodynamic analysis

Once ME antibiotic PK are characterized, the PD of single or multiple doses can be evaluated. The major determinant of bacterial killing efficacy with β-lactam and macrolide antimicrobials is the time during which antibiotic levels at the site of infection exceed the bacterial pathogen's MIC (Rotshafer *et al.*, 1994; Craig and Andes, 1996; Vogelman *et al.*, 1996). Treatment failure will result if the time over MIC is too short, and bacterial killing is incomplete. The MIC is also highly predictive of drug doses needed for bacterial efficacy based on a highly significant correlation between log MIC and log ED_{50} of penicillin in a mouse model of peritonitis (Knudsen *et al.*, 1995). Therefore, bacterial MIC and time over MIC in the ME will form the basis of experimental studies to optimize dosing regimens, i.e., to maximize the time over MIC of β-lactams and macrolides. Determination of antibiotic ME concentrations at one or two time points after a single dose is inadequate in assessing antimicrobial efficacy in OM. Contributions from host defense to bacterial killing are not separated from the effects of antibiotics in our current OM animal model.

Future PK investigation in the chinchilla OM model

In-vivo microdialysis

One major impediment to characterizing ME antibiotic pharmacokinetics is that the small and variable amount of MEE produced in the chinchilla OM model precludes a detailed analysis of antibiotic transport into the ME by conventional discrete sampling techniques. It is even more difficult to determine unbound (active) MEE antibiotic concentration, which is the most important parameter in assessing therapeutic efficacy (Nix *et al*, 1991).

During the last decade, *in-vivo* microdialysis as a unique tissue fluid space sampling technique has found increasing applications in the field of PK, particularly in the area of drug distribution and metabolism (Elmquist and Sawchuck, 1997). Microdialysis continuously monitors, for hours or even days, a specific fluid concentration in anesthetized, restrained, or freely moving animals. Since antibiotics and other small molecules diffuse through a polymeric membrane with a molecular weight cutoff of generally 10 kDa or greater, the concentration of analytes collected in the dialysate, which is protein-free, reflect their free (unbound) levels in extracellular fluid (ECF). This concentration can be determined by injecting the dialysate directly on to a chromatographic column. Most importantly, the free antibiotic concentration in the ME is a better measure of the antimicrobial efficacy.

With adequate analytical sensitivity, the time resolution of the measured concentration profile can be extremely good, e.g., of the order of 10 minutes with in-line HPLC coupled with UV or fluorescent detection. Such resolution is desirable where rapidly changing drug concentrations occur in plasma and the MEE, as in the case of rapidly eliminated antibiotics such as amoxicillin, which rapidly penetrate into MEE following single-dose administration. Microdialysis also reduces the number of study animals required and avoids problems associated with interanimal variability, which often complicate PK interpretation on a data set from a population of animals. *In-vivo* microdialysis is being developed in our laboratory to investigate antibiotic ME pharmacokinetics in the chinchilla OM model.

Allometric scaling

The clinical relevance and application of PK parameters obtained in the chinchilla OM model need to be evaluated with caution because of species-related PK differences. Allometric scaling is now an important technique to extrapolate animal data to humans in PK studies of new therapeutic agents (Lin, 1995; Kawakami *et al.*, 1994; Mordenti *et al.*, 1991). This technique is based on the observation that many anatomical and physiological parameters can be described as exponential functions of body weight: $Y = aW^b$ where Y is an anatomical and physiological parameter, W is body weight, a is the allometric coefficient, and b is the allometric exponent. A linear regression of a log–log plot of the kinetic parameters versus the body weight of the animal species produces an allometric relationship from which a particular human kinetic parameter can be calculated. Animal data in at least two animal species will be needed to perform meaningful extrapolations to human ME drug kinetics. Generally, a good allometric relationship is expected for kinetic parameters reflecting anatomical and physiologic function (Lin, 1995).

The PK parameters of interest in the chinchilla OM model are the antibiotic influx and efflux clearances into and out of the MEE. These clearances depend on permeability of the ME mucosa and the surface area available for antibiotic transport. They are related to the anatomical and physiological features of each species and, therefore, are excellent candidates for allometric scaling. The influx clearance also depends on systemic clearance because the driving force for distribution into the MEE is plasma concentration. This technique will be useful in making clinically relevant PK and PD evaluation of antibiotic dosing regimens based on available animal data.

Limitations of the model

Besides general precautions concerning cross-species generalization, there are several specific limitations inherent in the chinchilla OM model. Chinchillas are useful models for infection caused by *S. pneumoniae* and NTHi, but not by *M. catarrhalis* (Chung *et al.*, 1994; Bakaletz *et al.*, 1995). The virulence of several pneumococcal types is dissimilar in humans and chinchillas. For example, type 3 pneumococcus often causes severe disease in humans, but results in only local ME infection in chinchillas. Conversely, type 18C and 19F pneumococci are extremely virulent in chinchillas but less so in humans. Chinchillas are susceptible to the dissemination of initially localized AOM, an event that is uncommon in humans. High mortality rates among experimentally infected chinchillas may hinder between-group comparisons of middle-ear outcome over time. In such cases, histopathological assessment of temporal bones often provides a more sensitive delineation of middle-ear outcome. Finally, unimmunized chinchillas are immunologically naïve to the important middle-ear pathogens, while humans typically possess some humoral immunity to these endemic organisms, due to either persistent maternal antibody in the case of infants or prior exposure in the case of older children.

Conclusions

The emergence of antibiotic resistance among ME bacterial pathogens has been widely documented. To reduce the selective pressures that favor the development of drug resistance, immunoprophylaxis of AOM is a major health goal. Isotype-specific antibodies to chinchilla immunoglobulins, developed in our laboratory (Konietzko *et al.*, 1992), have made the chinchilla a useful model for studying immunogenicity and clinical efficacy of candidate vaccines against the pneumococcus and NTHi. Research in chinchillas will be an important component of the effort to prevent childhood AOM. Similarly, therapy of AOM must be improved both with respect to selection and dosing of antimicrobial agents. The PK methods described here will help to establish a rational approach to these questions. Research in the chinchilla model will continue to improve our understanding of ME biochemistry and the basic mechanisms of OM pathogenesis.

References

Aaberge, S. A., Eng, J., Lemark, G., Lovik, M. (1995). Virulence of *Streptococcus pneumoniae* in mice: a standardized method of preparation and frozen storage of the experimental bacterial inoculum. *Microb. Pathog.* **18**, 141–152.

Abramson, J. S., Giebink, G. S., Quie, P. G. (1982). Influenza A virus-induced polymorphonuclear leukocyte dysfunction in the pathogenesis of experimental pneumococcal otitis media. *Infect. Immun.*, **36**, 289–296.

Alper, C. M., Tabari, R., Seroky, J. T., Doyle, W. J. (1997). Magnetic resonance imaging of the development of otitis media with effusion caused by functional obstruction of the eustachian tube. *Ann. Otol. Rhinol. Laryngol.*, **106**, 422–431.

Antonelli, P. J., Juhn, S. K., Goycoolea, M. V., Giebink, G. S. (1992). *Pseudomonas* otitis media after eustachian tube obstruction. *Otolaryngol. Head Neck Surg.*, **107**, 511–515.

Antonelli, P. J., Juhn, S. K., Le, C. T., Giebink, G. S. (1994). Acute otitis media increases middle ear susceptibility to nasal injection of *Pseudomonas aeruginosa. Otolaryngol. Head Neck Surg.*, **110**, 115–121.

Bakaletz, L. O., Daniels, R. L., Lim, D. J. (1993). Modeling adenovirus type 1-induced otitis media in the chinchilla: effect on ciliary activity and fluid transport function of eustachian tube mucosal epithelium [published erratum appears in *J. Infect. Dis.* 1993; **168(6)**: 1605]. *J. Infect. Dis.*, **168**, 865–872.

Bakaletz, L. O., Murwin, D. M., Billy, J. M. (1995). Adenovirus serotype 1 does not act synergistically with *Moraxella (Branhamella) catarrhalis* to induce otitis media in the chinchilla. *Infect. Immun.*, **63**, 4188–4190.

Barza, M., Cuchural, G. (1985). General principles of antibiotic tissue penetration. *Antimicrob. Agents Chemother.*, **15**(Suppl.), 59–75.

Beal, S. L., Sheiner, L. B. (1980). The NONMEM system. *Am. Stat.*, **34**, 118–119.

Bluestone, C. D., Stephenson, J. S., Martin, L. M. (1992). Ten-year review of otitis media pathogens. *Pediatr. Infect. Dis. J.*, **11**(Suppl. 8), S7–S11.

Briles, D. E., Crain, M. J., Gray, B. M., Formann, C., Yother, J. (1992). Strong association between capsular type and virulence for mice among human isolates of *Streptococcus pneumoniae. Infect. Immun.*, **60**, 111–116.

Canafax, D. M., Giebink, G. S., Erdmann, G. R., Cipolle, R. J., Juhn, S. K. (1988). Penetration of trimethoprim and sulfamethoxazole into the middle ear in experimental otitis media. In: *Recent Advances in Otitis Media* (ed. Lim, D. J.). B. C. Decker, Philadelphia, PA.

Canafax, D. M., Nonomura, N., Erdmann, G. R., Le, C. T., Juhn, S. K., Giebink, G. S. (1989). Experimental animal models for studying antimicrobial pharmacokinetics in otitis media. *Pharm. Res.*, **6**, 279–285.

Canafax, D. M., Russlie, H., Lovdahl, M. J., Erdmann, G. R., Le, C. T., Giebink, G. S. (1994). Comparison of two otitis media models for the study of middle ear antimicrobial pharmacokinetics. *Pharm. Res.*, **11**, 855–859.

Chan, K. H., Swarts, J. D., Hashida, Y., Doyle, W. J., Kardatzke, D., Wolf, G. L. (1992). Experimental otitis media evaluated by magnetic resonance imaging: an *in vivo* model. *Ann. Otol. Rhinol. Laryngol.*, **101**, 248–254.

Chan, K. H., Swarts, J. D., Tan, L. (1994). Middle ear mucosal inflammation: an *in vivo* model [review]. *Laryngoscope*, **104**, t-80.

Chung, M. H., Enrique, R., Lim, D. J., De Maria, T. F. (1994). *Moraxella (Branhamella) catarrhalis*-induced experimental otitis media in the chinchilla. *Acta Otolaryngol.*, **114**, 415–422.

Cotter, C. S., Avidano, M. A., Stringer, S. P., Schultz, G. S. (1996). Inhibition of proteases in *Pseudomonas* otitis media in chinchillas. *Otolaryngol. Head Neck Surg.*, **115**, 342–351.

Craig, W. A., Andes, D. (1996). Pharmacokinetics and pharmacodynamics of antibiotics in otitis media. *Pediatr. Infect. Dis. J.*, **15**, 944–948.

Daniel, H. J., Fulghum, R. S., Brinn, J. E., Barrett, K. A. (1982). Comparative anatomy of eustachian tube and middle ear cavity in animal models of otitis media. *Ann. Otol. Rhinol. Laryngol.*, **91**, 82–89.

D'Argenzio, D. Z., Schumitzky, A. (1996). *ADAPT II User's Guide: Pharmacokinetic/Pharmacodynamic Systems Analysis Software*. Biomedical Simulations Resource, Los Angeles, CA.

Del Beccaro, M. A., Mendelmann, P. M., Inglis, A. F. *et al.* (1992). Bacteriology of acute otitis media: a new perspective. *J. Pediatr.*, **120**, 81–84.

Doyle, W. J., Supance, J. S., Marshak, G., Cantekin, E. I., Bluestone, C. D. (1982). An animal model of acute otitis media consequent to β-lactamase producing non-typable *Haemophilus influenzae. Otolaryngol. Head Neck Surg.*, **90**, 831–836.

Elmquist, W. F., Sawchuck, R. J. (1997). Application of microdialysis in pharmacokinetics. *Pharm. Res.*, **14**, 267–300.

Erdmann, G. R., Canafax, D. M., Giebink, G. S. (1988). High-performance liquid chromatographic analysis of trimethoprim and sulfamethoxazole in microliter volumes of chinchilla middle ear effusion and serum. *J. Chromatogr.*, **433**, 187–195.

Giebink, G. S. (1981). The pathogenesis of pneumococcal otitis media in chinchillas and the efficacy of vaccination in prophylaxis. *Rev. Infect. Dis.*, **3**, 342–353.

Giebink, G. S. (1982). Experimental otitis media due to *Haemophilus influenzae* in the chinchilla. In: *Haemophilus influenzae* (eds Sell, S. H., Wright, P. F.), pp. 73–80. Elsevier Science, Amsterdam.

Giebink, G. S., Wright, P. F. (1983). Different virulence of influenza A virus strains and susceptibility to pneumococcal otitis media in chinchillas. *Infect. Immun.* **41**, 913–920.

Giebink, G. S., Payne, E. E., Millis, E. L., Juhn, S. K., Quie, P. G. (1976). Experimental otitis media due to *Streptococcus pneumoniae*: immunopathogenic response in the chinchilla. *J. Infect. Dis.*, **134**, 595–604.

Giebink, G. S., Berzins, I. K., Schiffman, G., Quie, P. G. (1979). Experimental otitis media in chinchillas following nasal colonization with type 7F *Streptococcus pneumoniae*: prevention after vaccination with pneumococcal capsular polysaccharide. *J. Infect. Dis.*, **140**, 716–723.

Giebink, G. S., Berzins, I. K., Menken, S. C., Schiffman, G. (1980a). Experimental otitis media after nasal inoculation of *Streptococcus pneumoniae* and influenza A virus in chinchillas. *Infect. Immun.* **30**, 445–450.

Giebink, G. S., Berzins, I. K., Quie, P. G. (1980b). Animal models for studying pneumococcal otitis media and pneumococcal vaccine efficacy. *Ann. Otol. Rhinol. Laryngol. Suppl.*, **89**, t-43.

Giebink, G. S., Heller, K. A., Harford, E. R. (1982a). Tympanometric configurations and middle ear findings in experimental otitis media. *Ann. Otol. Rhinol. Laryngol.*, **91**, t-4.

Giebink, G. S., Juhn, S. K., Weber, M. L., Le, C. T. (1982b). The bacteriology and cytology of chronic otitis media with effusion. *Pediatr. Infect. Dis.*, **1**, 98–103.

Giebink, G. S., Ripley, M. L., Shea, D. A., Wright, P. F., Paparella, M. M. (1985). Clinical-histopathological correlations in experimental otitis media: implications for silent otitis media in humans. *Pediatr. Res.*, **19**, 389–396.

Giebink, G. S., Ripley, M. L., Wright, P. F. (1987). Eustachian tube histopathology during experimental influenza A virus infection in the chinchilla. *Ann. Otol. Rhinol. Laryngol.*, **96**, t-206.

Giebink, G. S., Koskela, M., Vella, P. P., Harris, M., Le, C. T. (1993). Pneumococcal capsular polysaccharide-meningococcal outer membrane protein complex conjugate vaccines: immunogenicity and efficacy in experimental pneumococcal otitis media. *J. Infect. Dis.*, **167**, 347–355.

Giebink, G. S., Meier, J. D., Quartey, M. K., Liebeler, C. L., Le, C. T. (1996). Immunogenicity and efficacy of *Streptococcus pneumoniae* polysaccharide-protein conjugate vaccines against homologous and heterologous serotypes in the chinchilla otitis media model. *J. Infect. Dis.*, **173**, 119–127.

Hanamure, Y., Lim, D. J. (1987). Anatomy of the chinchilla bulla and eustachian tube: I. Gross and microscopic study. *Am. J. Otolaryngol.*, **8**, 127–143.

Henderson, F. W., Collier, A. M., Sanyal, M. A. *et al.* (1982). A longitudinal study of respiratory viruses and bacteria in the etiology of acute otitis media with effusion. *N. Engl. J. Med.*, **306**, 2037–2045.

Jossart, G. H., Erdmann, G. R., Levitt, D. G. *et al.* (1990). An experimental model for measuring middle ear antimicrobial drug penetration in otitis media. *Pharm. Res.*, **7**, 1242–1247.

Jossart, G. H., Erdmann, G. R., Steury, J. C., Le, C. T., Giebink, G. S. (1993). Effects of pH and protein binding on middle ear antimicrobial drug penetration may explain acute otitis media treatment failures. *Recent Advances in Otitis Media* (ed. Lim, D. J.). B. C. Decker, Philadelphia, PA.

Juhn, S. K., Paparella, M. M., Kim, C. S., Goycoolea, M., Giebink, G. S. (1977). Pathogenesis of otitis media. *Ann. Otol. Rhinol. Laryngol.*, **86**, 481–492.

Juhn, S. K., Edlin, J., Jung, T. T., Giebink, G. S. (1986). The kinetics of penicillin diffusion in serum and middle ear effusions in experimentally induced otitis media. *Arch. Otorhinolaryngol.*, **243**, 183–185.

Kawakami, J. K., Yamamoto, K., Sawada, Y., Iga, T. (1994). Prediction of brain delivery of ofloxacin, a new quinolone, in the human from animal data. *J. Pharmacokinet. Biopharm.*, **22**, 207–227.

Klein, J. O. (1994). Lessons from recent studies on the epidemiology of otitis media. *Pediatr. Infect. Dis. J.*, **13**, 1031–1034.

Klein, B. S., Dollete, F. R., Yolken, R. H. (1982). The role of respiratory syncytial virus and other viral pathogens in acute otitis media. *J. Pediatr.*, **101**, 16–20.

Knudsen, J. D., Frimodt-Møller, N., Espersen, F. (1995). Experimental *Streptococcus pneumoniae* infection in mice for studying correlation of *in vitro* and *in vivo* activities of penicillin against pneumococci with various susceptibilities to penicillin. *Antimicrob. Agents Chemother.*, **39**, 1253–1258.

Konietzko, S., Koskela, M., Erdmann, G., Giebink, G. S. (1992). Isotype-specific rabbit antibodies against chinchilla immunoglobulins G, M, and A. *Lab. Anim. Sci.*, **42**, 302–306.

Lin, J. H. (1995). Species similarities and differences in pharmacokinetics. *Drug Metab. Dispos.*, **23**, 1008–1021.

Lovdahl, M. J., Steury, R. C., Russlie, H., Canafax, D. M. (1993). Determination of ciprofloxacin levels in chinchilla middle ear effusion and plasma by high-performance liquid chromatography with fluorescence detection. *J. Chromatogr.*, **617**, 329–333.

Lovdahl, M. J., Reher, K. E., Russlie, H. Q., Canafax, D. M. (1994). Determination of cefpodoxime levels in chinchilla middle ear fluid and plasma by high-performance liquid chromatography. *J. Chromatog. B Biomed. Sci. Appl.*, **653**, 227–232.

Luotonen, J., Herva, E., Karma, P. *et al.* (1981). The bacteriology of acute otitis media in children with special reference to *Streptococcus pneumoniae* as studied by bacterial and antigen detection methods. *Scand. J. Infect. Dis.*, **13**, 177–183.

Marshak, G., Cantekin, E. I., Doyle, W. J., Schiffman, G., Rohn, D. D., Boyler, R. (1981). Recurrent pneumococcal otitis media in the chinchilla. A longitudinal study. *Arch. Otorhinolaryngol.*, **107**, 532–539.

Meyerhoff, W. L., Giebink, G. S., Shea, D. (1981). Pneumococcal otitis media following middle ear deflation. *Ann. Otol. Rhinol. Laryngol.*, **90**, t-6.

Meyerhoff, W. L., Giebink, G. S. (1982). Panel discussion: pathogenesis of otitis media. Pathology and microbiology of otitis media. *Laryngoscope*, **92**, 273–277.

Miyamoto, N., Bakaletz, L. O. (1997). Kinetics of the ascension of NTHi from the nasopharynx to the middle ear coincident with adenovirus-induced compromise in the chinchilla. *Microb. Pathog.*, **23**, 119–126.

Mordenti, J., Chen, S. A., Moore, J. A. *et al.* (1991). Interspecies scaling of clearance and volume of distribution data from five therapeutic proteins. *Pharm. Res.*, **8**, 1351–1359.

Nix, D. E., Goodwin, S. D., Peloquin, C. A., Rotella, D. L., Schentag, J. J. (1991). Antibiotic tissue penetration and its relevance: Impact of tissue penetration on infection response. *Antimicrob. Agents Chemother.*, **35**, 1953–1959.

Post, J. C., Preston, R. A., Aul, J. J. *et al.* (1995). Molecular analysis of bacterial pathogens in otitis media with effusion. *J. A. M. A.*, **273**, 1598–1604.

Rotshafer, J. C., Walker, K. J., Maradas, Kelly, K. J., Sullivan, C. J. (1994). Antibiotic pharmacodynamics. In: *Pharmacodynamics and Drug Development: Perspectives in Clinical Pharmacology*, pp. 315–343. John Wiley, New York.

Rowland, M., Tozer, T. N. (1989). *Clinical Pharmacokinetics: Concepts and Applications*, 2nd edn. Lea & Febiger, Philadelphia.

Suzuki, K., Bakaletz, L. O. (1994). Synergistic effect of adenovirus type 1 and nontypeable *Haemophilus influenzae* in a chinchilla model of experimental otitis media. *Infect. Immun.*, **62**, 1710–1718.

Van Etta, L. L., Peterson, L. R., Fasching, C. E., Gerding, D. N. (1982). Effect of the ratio of surface area to volume on the penetration of antibiotics in to extravascular spaces in an *in vitro* model. *J. Infect. Dis.*, **146**, 423–428.

Vogelman, B., Gudmundsson, S., Leggett, J., Turnidle, Ebers, S., Craig, W. A. (1996). Correlation of antimicrobial pharmacokinetic parameters with therapeutic efficacy in animal models. *J. Infect. Dis.*, **158**, 831–847.

Vrettakos, P. A., Dear, S. P., Saunders, J. P. (1988). Middle ear structure of the chinchilla: a quantitative study. *Am. J. Otolaryngol.*, **9**, 58–67.

Wang, Y., Welty, D. F. (1996). The simultaneous estimation of the influx and efflux blood–brain barrier permeabilities of gabapentin using microdialysis-pharmacokinetic approach. *Pharm. Res.*, **7**, 398–403.

Weber, P. C., Koltai, P. J. (1991). *Chlamydia trachomatis* in the etiology of acute otitis media. *Ann. Otol. Rhinol. Laryngol.*, **100**, 616–619.

Yuan, Z., Russlie, H. Q., Canafax, D. M. (1997). High-performance liquid chromatographic analysis of amoxicillin in human and chinchilla plasma, middle ear fluid and whole blood. *J. Chromatogr. B Biomed. Sci. Appl.*, **692**, 361–366.

Chapter 46

A Guinea Pig Model of Acute Otitis Media

T. G. Takoudes and J. Haddad Jr

Background of model

Otitis media (OM) is a common pediatric infection that is the most frequent diagnosis for children younger than 15 years old in office practices in the United States (Schappert, 1992). Epidemiology, pathogenesis, classification, treatment and complications of OM have been reviewed elsewhere (Haddad, 1994; Bluestone and Klein, 1995). Several models of otitis media have been described using different animal species. These include the guinea-pig (Parks *et al.*, 1994), the chinchilla (Giebink *et al.*, 1976), the rat (Hermansson *et al.*, 1988), the gerbil (Fulghum and Marrow, 1996), the cat (Goycoolea *et al.*, 1980) and the monkey (Doyle, 1984). The guinea pig model of OM is commonly used because of the ease of access to the middle-ear cavity and because of its low cost. Most investigators use the Hartley strain. The anatomy of the guinea pig temporal bone is reviewed elsewhere (Asarch *et al.*, 1975).

There are several methods of inducing otitis media in the guinea pig. This chapter will describe a direct injection of bacteria into the middle-ear cavity through the tympanic membrane. Although the organism described in this chapter is *Streptococcus pneumoniae*, the section on the preparation of the inocula describes other organisms that may be used in this model. Alternative mechanisms of causing OM involve obstruction of the eustachian tube (Kim and Kim, 1995), transtympanic inoculation of lipopolysaccharide (Ohashi *et al.*, 1991a), and inoculation of an antigen into the middle ear cavity via a retroauricular approach (Gloddek *et al.*, 1992). It should be noted that the authors attempted to induce OM with *Streptococcus* inoculation into the guinea pig equivalent of the mastoid cavity via a retroauricular approach. Our experience with this model was disappointing, though, as the rate of otitis media created was low compared to the transtympanic method and the rate of wound infection was unacceptably elevated.

Preparation of animals

Upon delivery to the facility and prior to surgery, the guinea-pigs should be assessed otoscopically for ear disease. Otherwise the animals do not require specialized housing or treatment prior to surgery.

Details of surgery

Overview

Briefly, the animals are anesthetized with intraperitoneal ketamine and xylazine. Once sedated, an operating microscope and a pediatric speculum are used to examine the external ear canal, which is then cleaned with alcohol and dried with cotton. A blunt-tipped needle is then passed through the tympanic membrane and into the middle-ear cavity in order to inject the appropriate medium. The animals are observed for a brief period of time until they recover from anesthesia. They are then transported to specialized housing facilities, which include negative-pressure cubicles and isolation from other animals. The middle ears are intermittently assessed otoscopically before the animals are sacrificed as per the protocol.

Materials required

Anesthesia (preferably ketamine and xylazine from Henry Schein, Port Washington, NY), Lacri-Lube ophthalmic ointment (Allergan, Irvine, CA), operating microscope, pediatric ear speculum, alligator forceps, 70% vol./vol. ethanol, sterile cotton swabs, sterile syringes (1 ml and 5 ml) and needles (18 G and 22 G), 22 G blunt-tipped needle (preferably a lacrimal probe), injecting medium (i.e. bacterial suspension, sterile saline for control), pentobarbital (Henry Schein, Port Washington, NY), heavy scissors.

Anesthesia

The animals are sedated with an intraperitoneal injection of ketamine hydrochloride (87 mg/kg) and xylazine hydrochloride (13 mg/kg). Complete anesthesia occurs within 10–20 minutes and lasts for at least 30 minutes. The

Handbook of Animal Models of Infection
ISBN 0-12-775390-7

animals should have Lacri-Tube applied to their eyes to prevent corneal ulcerations. Respiratory depression is usually not seen during the effects of anesthesia. Post-operative recovery at this dose is typically uneventful. The animals should be kept under heat lamps or otherwise warmed during the recovery as the anesthesia wears off.

Surgical procedure

Following induction of anesthesia the animal is ready for surgery. If the operator is right-handed, then the left hand is used to hold the animal's head and position the speculum while the right hand is used to perform the procedure. When operating on the left ear, the animal is placed ventral side down with the head facing to the experimenter's left. The speculum is held between the thumb and first finger and placed into the external ear canal (Figure 46.1). The auricle is held between the index finger and the middle finger of the left hand. After some practice, the experimenter will be able to retract the auricle posteriorly and superiorly while maintaining the speculum in a good position in the canal. This maneuver allows for optimal visualization of the external ear canal and the tympanic membrane (TM). When operating on the animal's right ear, the left hand is again used to position the speculum and the right hand is used to operate. The animal and microscope are adjusted to bring the ear canal and the TM into view.

It is important to perform both the cleaning of the external ear canal and the injection of inocula under direct visualization using the microscope and speculum in order to prevent inadvertent injury to the TM. Debris within the canal is removed with the forceps and the TM is examined. If there is no obvious ear disease, then the canal is gently washed with 70% ethanol using a 5 ml syringe through either a 22 G needle or, preferably, a lacrimal probe. After approximately 1 minute the ear is tilted towards the ground to drain the ethanol. The remaining ethanol may be gently removed using a wisp of sterile cotton grasped by a forceps. Allow several minutes to elapse prior to proceeding in order to allow all the ethanol within the ear canal to evaporate. Otoscopy should confirm a clean canal with no residual ethanol, as well as an intact TM.

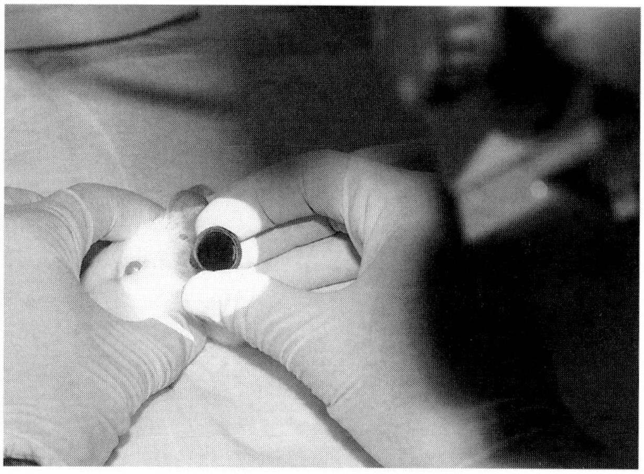

Figure 46.1 Positioning the speculum. With the operating microscope in place, the anesthetized guinea pig's head faces the experimenter's left. The left hand stabilizes the animal's head while the right hand places a pediatric ear speculum into the left ear.

Inoculation

Once the external ear canal is cleaned attention may be turned to the injection of either the bacteria or the sterile saline control into the middle-ear cavity. The amount of bacteria that we use is typically 100 μl of 10^8 cells of *Streptococcus pneumoniae* per ml. Please see the section on preparation of inocula for further details. The control fluid we use is 100 μl of sterilized phosphate-buffered saline (Sigma, St Louis, MD). 200–300 μl of the appropriate fluid is drawn into a 1 ml syringe, which is then capped with a 22 G lacrimal probe. Fluid should be ejected from the syringe through the probe until 100 μl remains in the syringe. This will eliminate any dead space in the long lacrimal probe.

Attention is then turned to the TM, which is visualized through the speculum. The anterior superior quadrant is identified and the lacrimal probe is gently guided across this portion of the TM and into the middle-ear cavity (Figure 46.2). The 100 μl suspension is injected into the middle-ear cavity under direct visual guidance. This amount of fluid can typically be injected into the middle ear cavity of a 300–400 g guinea-pig without causing overflow into the external ear canal. It should take 2–4 seconds to slowly inject the fluid. There may be a small hole remaining in the TM through which the fluid will not escape.

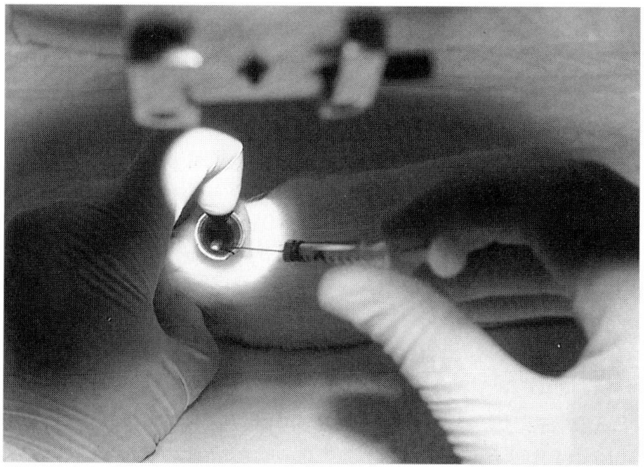

Figure 46.2 Inoculation into the middle ear cavity. With the tympanic membrane in view through the operating microscope, the left hand is used to stabilize the animal's head and the speculum. The right hand is free to introduce a 22 G lacrimal probe attached to a 1 ml syringe that contains the inoculum. Once the probe crosses the tympanic membrane into the middle-ear cavity, the right index finger is used to depress the plunger and inoculate the cavity.

Postoperative care

The animals should be kept warm immediately after surgery until they recover from anesthesia. They may then be housed in standard guinea pig cages with two to three animals per cage. The cages should be in a specially equipped negative-pressure room that houses only the infected animals. Animals typically return to normal activity within several hours. The animal's postoperative diet should consist of *ad lib* standard guinea pig chow and water.

A small percentage of animals will develop serious intratemporal, intracranial or systemic complications as a result of the local infection; the rate varies depending on the virulence of the bacterial strain but in our experience is typically less than 4%. The animals with infectious complications will demonstrate gait disturbances, lethargy and tachypnea. If the animal does not improve and an intracranial abscess or sepsis is suspected the animal may be euthanized with 100 mg/kg of intraperitoneal pentobarbitol. In addition, some of the saline-injected control ears will develop otitis media. Although this may be minimized by employing good antiseptic technique, 5–8% of control ears become infected in our experience.

Animal sacrifice and temporal bone recovery

The animals may be sacrificed with an excess dose of 100 mg/kg of intraperitoneal pentobarbital. Once death has occurred the research protocol often calls for removal of the temporal bones, since they house the middle-ear cavities. This may be achieved by using heavy scissors to decapitate the animal and then removing the skin overlying the skull. The scissors are then positioned with one edge under the caudal aspect of the midline portion of the skull and the other edge above the skull. The skull is divided in the midline to a point anterior to the temporal bones. Continuing to use the heavy scissors, lateral incisions are carried down to the edge of the skull bilaterally. The skull, brain, and any additional soft tissue surrounding the temporal bones are removed. The temporal bones may then be bluntly removed by prying with the edges of a pair of scissors or the operator's fingers.

Preparation and storage of inocula

Several species of bacteria may be used with this model. Most studies use the common human pathogens that cause acute otitis media (AOM), including *Streptococcus pneumoniae* (Yoshimura *et al.*, 1991; Naguib *et al.*, 1994; Parks *et al.*, 1994; Egusa *et al.*, 1995; Takoudes and Haddad, 1997), *Haemophilus influenzae* (Patel *et al.*, 1993), *Staphylococcus* (Goksu *et al.*, 1996; Sutbeyaz *et al.*, 1996) and *Pseudomonas* (Kim *et al.*, 1990) species. Other materials used to induce otitis media include lipopolysaccharide (Kim and Kim, 1995; Ohashi *et al.*, 1991a) and immunocomplex (Mogi *et al.*, 1990). We have used *S. pneumomiae* Type 3 (American Type Culture Collection, Rockville, MD). *Streptococcus* is the most common isolate from human AOM cultures (Bluestone *et al.*, 1992) and it consistently produces a purulent otitis media when used in the guinea pig model.

Our strain is reconstituted as per instructions from American Type Culture Collection in Todd–Hewitt broth. The organism is then cultured on to three plates: Columbia agar with 5% sheep's blood, mannitol and MacConkey's. Bacterial growth on the sheep's blood plate combined with no growth on the latter two plates confirms the isolation of *Streptococcus* without contaminants. The organism is then suspended in Todd–Hewitt broth and placed in an incubator at 37°C. After approximately 12 hours of growth, sterile glycerol is added to the suspension (20% glycerol and 80% suspension vol./vol.), which is then aliquoted and stored frozen at –70°C.

For use of the inoculum, the individual aliquot is thawed and cultured in Todd–Hewitt broth for 12–20 hours. At this point the concentration of bacterial cells per milliliter of broth is determined as follows: 80 μl of the suspension is pipetted into a test tube with 40 μl of methylene blue and 680 μl of formalin, creating a 10 : 1 dilution of the bacterial suspension. A small amount of this fluid is transferred to a hemocytometer, where 1000× magnification is used to count the number of methylene blue-stained bacterial cells in multiple fields. The number of cells per field is averaged and, using the known volume of the hemocytometer field, the concentration of cells in the original solution is calculated. The original bacterial suspension is then diluted in Todd–Hewitt broth to a concentration of 10^8 cells/ml, which is the concentration oi inocula injected into the middle ear.

A second method of titrating *Streptococcus pneumoniae* for inoculation is to adjust the concentration of the number of colony-forming units (cfu) per milliliter. This is achieved by performing an overnight culture of the bacteria on a Columbia agar 5% sheep's blood plate. Some 12–20 hours later bacterial cells from the plate are added to a Todd–Hewitt broth solution. This is placed in a spectrophotometer with the wavelength adjusted to 625 nm. *Streptococcus* is added until a suspension density of 0.5 McFarland is achieved. This corresponds to 10^7 cfu/ml, which may be diluted to the desired concentration for inoculation. Typically, a concentration of 10^5–10^8 cfu/ml is used as the inoculum (Yoshimura *et al.*, 1991; Naguib *et al.*, 1994).

Key parameters to monitor infection and response to treatment

After inoculation, the infection is confined to the middle-ear cavity and should be assessed by otoscopy for signs of acute otitis media. These signs include an erythematous and bulging TM and purulence either behind the TM or in the external ear canal. A small amount of anesthesia needs to be administered in order to examine the ears. Typically one-half to two-thirds of the dose of ketamine and xylazine given during surgery is adequate to sedate the animal for evaluation.

Although it would be beneficial to perform tympanometry on the guinea pigs in order to further assess the disease state, this has proved to be a difficult task in our experience. We have found that a tympanometer outfitted with a small speculum is ineffective because of the small volume of the external ear. However, a well performed otoscopic examination is typically adequate to assess the presence or absence of acute otitis media. Pneumatic otoscopy and the determination of the mobility of the tympanic membrane may also be helpful.

Some investigators may wish to confirm and quantify the growth of *Streptococcus* within the middle-ear cavity. This can be accomplished by employing the techniques already discussed for handling the bacteria. To confirm *Streptococcus* growth in the infected middle ear, the canal should be cleaned with 70% ethanol prior to culture with a sterile cotton swab. The swab can then be spread on to three plates: Columbia agar with 5% sheep's blood, mannitol and MacConkey's. If bacteria grow on the sheep's blood plate and not on the latter two plates, then it is likely that the *Streptococcus* infection is not contaminated with other bacterial species.

Additionally, it is possible to determine the cfu concentration within the middle-ear fluid by recovering the fluid and performing serial 1:10 or 1:100 dilutions. A quantity of 100 μl of each dilution can then be transferred on to a 5% sheep's blood plate. After overnight growth, the cfu concentration is determined by counting the individual colonies on the plates. For each plate, the cfu concentration of the original middle-ear fluid can be calculated based on the dilution and volume of fluid that was spread on to each plate. For an accurate cfu estimate, take the average of the calculated cfus from two or three plates at different dilutions.

Advantages and disadvantages of the model

The guinea pig model has several advantages. The animals are relatively inexpensive to purchase and house. In addition, although the inoculation and subsequent treatment of the animals is technically more difficult compared to larger animals such as the chinchilla and monkey, this is a straightforward procedure that may consistently and competently be performed in experienced hands.

As with other animal models for human disease, the guinea pig model for acute otitis media has disadvantages. These include problems related to all animal models that attempt to describe human disease, such as differences in anatomy, immune response and pharmacokinetic profiles of medications. This specific model has additional limitations. These include a sudden onset of infection after inoculation, which is not physiologic with respect to the mechanism of infection nor the bacterial load. Second, the eustachian tube dysfunction that is commonly seen in pediatric acute otitis media (Bluestone and Klein, 1995) is probably not a factor in this model, since the guinea pigs are noted to have normal TMs preoperatively. The effect of presumably normal eustachian tube function in this model of otitis media is unknown. Third, the middle-ear cavity has a smaller volume than that of the commonly used chinchilla model and is therefore technically more difficult to work with.

Contributions of the model to research

Evaluation of middle ear disease

The guinea pig is often used as a model to evaluate the pathogenesis of otitis media. One area of focus includes the role of oxygen-free radicals and their mediators in otitis media. Parks *et al.* (1994) demonstrated that the mucosa of *Streptococcus*-infected middle ears is characterized by elevated lipid hydroperoxide and malondialdehyde; both are byproducts of free radical damage. Further evaluation noted scant immunohistochemical staining of the anti-oxidant enzymes superoxide dismutase (Parks *et al.*, 1995) and catalase (Parks *et al.*, 1996) in the mucosa of infected middle ears compared to saline-injected control middle ears. A study by Haddad (1998) showed that middle-ear mucosal lipid hydroperoxide levels were elevated compared to controls up to 30 days after *Streptococcus* inoculation. In addition, evaluation of the middle-ear fluid in the first 24 hours following *Streptococcus* inoculation demonstrated elevated levels of hydrogen peroxide, a reactive oxygen intermediate with free-radical-like activity, compared to controls (Takoudes and Haddad, 1997).

This model has also been used by investigators to evaluate inflammatory mediators and growth factors in otitis media. Egusa *et al.* (1995) characterized the response of heat shock proteins in the middle-ear mucosa to acute *Streptococcus* infection. Others have studied the roles of lymphocytes (Patel *et al.*, 1993), growth factors (Ryan and Baird, 1993), and immunoglobins (Keithley *et al.*, 1990) in a guinea pig model of otitis media.

Investigators have also used this model to evaluate etiologic factors in the development of otitis media. Ohashi *et al.* (1993) showed that simultaneous inoculation of influenza A with endotoxin increased the chance of developing otitis media with effusion compared to endotoxin alone. The roles of allergy (Mogi *et al.*, 1992), vitamin A deficiency (Manning and Wright, 1992), and pneumococcal vaccine (Yoshimura *et al.*, 1991) in the pathogenesis of otitis media have also been studied. Finally, the guinea pig has been used to examine mucociliary function in the eustachian tube in otitis media (Ohashi *et al.*, 1991b, 1995).

Evaluation of inner ear disease

This model has been used to study the consequences of otitis media in the inner ear. Kim and Kim (1995) used auditory brain-stem responses to compare the effects of endotoxin on cochlear function in guinea pigs with and without otitis media. Naguib *et al.* (1994) demonstrated elevated levels of polyamines as markers of cochlear damage in OM. Gloddek *et al.* (1992) showed that interleukin-2 is present in perilymph in otitis media while Kim *et al.* (1990) characterized the permeability of the round window membrane at various stages of otitis media.

Potential future use of the model

While the guinea pig model of AOM is inexpensive and convenient, to our knowledge it has never been used for antibiotic sensitivity testing. However, the guinea pig has been used for microbial sensitivity assays in other models of infection. These include the testing of trovafloxacin in *Legionella pneumophila* pneumonia (Edelstein *et al.*, 1996) and the use of several fluoroquinolones against methicillin-resistant *Staphylococcus aureus* foreign body infections (Cagni *et al.*, 1995). Additionally, Barry *et al.* (1996) used a gerbil model of *Streptococcus*-induced acute otitis media to assess the efficacy of ceftriaxone. These studies suggest that the guinea pig model of AOM may be useful for *in vivo* studies of antibiotic efficacy.

Conclusion

Otitis media is a common pediatric infection for which several animal models exist. This chapter describes a guinea pig model of OM in which *Streptococcus pneumoniae* is injected into the middle-ear cavity through the tympanic membrane. Sterile saline may be similarly injected into the opposite middle-ear cavity to serve as a control. The model consistently produces a purulent otitis media with a low complication rate. The advantages of this model include its reliability as well as the low costs of purchasing and housing animals. Disadvantages include those that are typical to animal models, such as an unknown relevance to humans, as well as the fact that this procedure may be more technically difficult to perform on guinea pigs than on chinchillas and other animals. The contributions of the guinea pig model of otitis media to research are reviewed.

References

Asarch, R., Abramson, M., Litton, W. B. (1975). Surgical anatomy of the guinea pig ear. *Ann. Otol.*, **84**, 250–255.

Barry, B., Muffat-Joly, M., Bauchet, J. *et al.* (1996). Efficacy of single-dose ceftriaxone in experimental otitis media induced by penicillin- and cephalosporin-resistant *Streptococcus pneumoniae*. *Antimicrob. Agents Chemother*, **40**(9), 1977–1982.

Bluestone, C. D., Klein, J. O. (1995). Otitis media in infants and children, pp 1–240. W. B. Saunders, Philadelphia.

Bluestone, C. D., Stephenson, J. S., Martin, L. M. (1992). Ten-year review of otitis media pathogens. *Pediatr. Infect. Dis. J.*, **11**, S7–S11.

Cagni, A., Chuard, C., Vauduax, P. E., Schrenzel, J., Lew, D. P. (1995). Comparison of sparfloxacin, temafloxacin, and ciprofloxacin for prophylaxis and treatment of experimental foreign-body infection by methicillin-resistant *Staphylococcus aureus*. *Antimicrob. Agents Chemother.*, **39**(8), 1655–1660.

Doyle, W. J. (1984). Functional eustachian tube obstruction and otitis media in a primate model. A review. *Acta Otolaryngol. Suppl. (Stockh.)*, **414**, 52–57.

Edelstein, P. H., Edelstein, M. A., Ren, J., Polzer, R., Gladue, R. P. (1996). Activity of trovafloxacin (CP-99,219) against *Legionella* isolates: *in vitro* activity, intracellular accumulation and killing in macrophages, and pharmacokinetics and treatment of guinea pigs with *L. pneumophila* pneumonia. *Antimicrob. Agents Chemother.*, **40**(2), 314–319.

Egusa, K., Huang, C. C., Haddad, J. Jr (1995). Heat shock proteins in otitis media. *Laryngoscope*, **105**(7), 708–713.

Fulghum, R. S., Marrow, H. G. (1996). Experimental otitis media with *Moraxella (Branhamella) catarrhalis*. *Ann. Otol. Rhinol. Laryngol.*, **105**(3), 234–241.

Giebink, G. S., Payne, E. E., Mills, E. L., Juhn, S. K., Quie, P. G. (1976). Experimental otitis media due to *Streptococcus pneumoniae*: immunopathogenic response in the chinchilla. *J. Infect. Dis.*, **134**(6), 595–604.

Gloddek, B., Lamm, K., Haslov, K. (1992). Influence of middle ear immune response on the immunological state and function of the inner ear. *Laryngoscope*, **102**(2), 177–181.

Goksu, N., Ataoglu, H., Kemaloglu, Y. K., Ataoglu, O., Ozsokmen, D., Akyildiz, N. (1996). Experimental otitis media induced by coagulase negative staphylococcus and its L-forms. *Int. J. Pediatr. Otorhinolaryngol.*, **37**(3), 201–216.

Goycoolea, M. V., Paparella, M. M., Juhn, S. K., Carpenter, A. M. (1980). Otitis media with perforation of the tympanic membrane: a longitudinal experimental study. *Laryngoscope*, **90**(12), 2037–2045.

Haddad, J. Jr (1994). Treatment of acute otitis media and its complications. *Otolaryngol. Clin. North Am.*, **27**, 431–441.

Haddad, J. Jr (1998). Lipoperoxidation as a measure of free radical injury in otitis media. *Laryngoscope*, **108**, 44–48.

Hermansson, A., Emgard, P., Prellner, K., Hellstrom, S. (1988). A rat model for pneumococcal otitis media. *Am. J. Otolaryngol.*, **9**, 97–101.

Keithley, E. M., Krekorian, T. D., Sharp, P. A., Harris, J. P., Ryan, A. F. (1990). Comparison of immune-mediated models of acute and chronic otitis media. *Eur. Arch. Otorhinolaryngol.*, **247**(4), 247–251.

Kim, C. S., Kim, H. J. (1995). Auditory brain stem response changes after application of endotoxin to the round window membrane in experimental otitis media. *Otolaryngol. Head Neck Surg.*, **112**, 557–565.

Kim, C. S., Cho, T. K., Jinn, T. H. (1990). Permeability of the round window membrane to horseradish peroxidase in experimental otitis media. *Otolaryngol. Head Neck Surg.*, **103**(6), 918–925.

Manning, S. C., Wright, C. G. (1992). Incidence of otitis media in vitamin A-deficient guinea pigs. *Otolaryngol. Head Neck Surg.*, **107**(5), 701–706.

Mogi, G., Chaen, T., Tomonaga, K. (1990). Influence of nasal allergic reactions on the clearance of middle ear effusion. *Arch. Otolaryngol. Head Neck Surg.*, **116**(3), 331–334.

Mogi, G., Tomonaga, K., Watanabe, T., Chaen, T. (1992). The role of type I allergy in secretory otitis media and mast cells in the middle ear mucosa. *Acta Otolaryngol. (Stockh.) Suppl.*, **493**, 155–163.

Naguib, M. B., Hunter, R. E., Henley, C. M. (1994). Cochlear polyamines: markers of otitis media-induced cochlear damage. *Laryngoscope*, **104**(8), 1003–1007.

Ohashi, Y., Nakai, Y., Esaki, Y., Ohno, Y., Sugiura, Y., Okamoto, H. (1991a). Experimental otitis media with effusion induced by lipopolysaccharide from *Klebsiella pneumoniae*. *Acta Otolaryngol. (Stockh.) Suppl.*, **486**, 105–115.

Ohashi, Y., Nakai, Y., Esaki, Y., Ohno, Y., Sugiura, Y., Okamoto, H. (1991b). Influenza A virus-induced otitis media and mucociliary dysfunction in the guinea pig. *Acta Otolaryngol. (Stockh.) Suppl.*, **486**, 135–148.

Ohashi, Y., Nakai, Y., Ohno, Y. *et al.* (1993). Influenza A modification of endotoxin-induced otitis media with effusion in the guinea pig. *Eur. Arch. Otorhinolaryngol.*, **250**(1), 27–32.

Ohashi, Y., Nakai, Y., Ohno, Y., Sugiura, Y., Okamoto, H. (1995). Effects of human middle ear effusions on the mucociliary system of the tubotympanum in the guinea pig. *Eur. Arch. Otorhinolaryngol*, **252**(1), 35–41.

Parks, R. R., Huang, C. C., Haddad, J. Jr (1994). Evidence of oxygen radical injury in experimental otitis media. *Laryngoscope*, **104**, 1389–1392.

Parks, R. R., Huang, C. C., Haddad, J. Jr (1995). Superoxide dismutase in an animal model of otitis media. *Eur. Arch. Otorhinolaryngol.*, **252**(3), 153–158.

Parks, R. R., Huang, C. C., Haddad, J. Jr (1996). Middle ear catalase distribution in an animal model of otitis media. *Eur. Arch. Otorhinolaryngol*, **253**(8), 445–449.

Patel, J., Chonmaitree, T., Schmalstieg, F. (1993). Effect of modulation of polymorphonuclear leukocyte migration with anti-CD-18 antibody on pathogenesis of experimental otitis media in guinea pigs. *Infect. Immun.*, **61**(3), 1132–1135.

Ryan, A. F., Baird, A. (1993). Growth factors during proliferation of the middle ear mucosa. *Acta Otolaryngol. (Stockh.)*, **113**(1), 68–74.

Schappert, S. M. (1992). Office visits for otitis media: United States, 1975–90. *Vital Health Stat. Center Dis. Control/Natl. Center Health Stat.*, **214**, 1–18.

Sutbeyaz, Y., Yakan, B., Ozdemir, H., Karasen, M., Doner, F., Kufrevioglu, I. (1996). Effect of SC-41930, a potent selective leukotriene B4 receptor antagonist, in the guinea pig model of middle ear inflammation. *Ann. Otol. Rhinol. Laryngol.*, **105**(6), 476–480.

Takoudes, T. G., Haddad, J. Jr (1997). Hydrogen peroxide in acute otitis media in guinea pigs. *Laryngoscope*, **107**(2), 206–210.

Yoshimura, H., Watanabe, N., Bundo, I., Shinoda, M., Mogi, G. (1991). Oral vaccine therapy for pneumococcal otitis media in an animal model. *Arch. Otolaryngol. Head Neck Surg*, **117**(8), 889–894.

Chapter 47

Tissue Cage Infection Model

W. Zimmerli

Background of model

Since the 1940s, foreign materials such as metals or polymers have been increasingly used for multiple types of device (Küntscher and Maatz, 1945; Pudenz *et al.*, 1957; Starr and Edwards, 1961; Charnley, 1972). A feared complication associated with implanted devices is infection (Dougherty, 1988). Even minimal bacterial inocula can jeopardize implants (Elek and Conen, 1957). Once established, the infection always persists: there is no spontaneous healing. In addition, even with prolonged antimicrobial therapy, implant-associated infection is difficult to cure (Widmer *et al.*, 1990a).

Implants can become colonized by microorganisms by two possible routes, either by direct local contamination during surgery or by the hematogenous route (Zimmerli *et al.*, 1985). Infection due to contamination of prosthetic devices are characterized by (1) a low initial infecting inoculum of microorganisms (Elek and Conen, 1957); (2) a protracted clinical evolution; (3) the absence of spread of infection beyond the immediate vicinity of the implanted device; and (4) the persistence of the infectious process until removal of the prosthesis (Dougherty, 1988). During the last 40 years, different experimental models have been used to eludicate the multiple problems encountered in device-related infections (James and MacLeod, 1961; Garrison and Freedman, 1970; Busuttil *et al.*, 1979; Blomgren *et al.*, 1981; Zimmerli *et al.*, 1982, 1984; Glauser *et al.*, 1983; Vaudaux *et al.*, 1992; Zimmerli, 1993).

Whereas in the 1980s the main research effort was directed to the study of pathogenesis (Zimmerli *et al.*, 1982, 1984; Vaudaux *et al.*, 1992), within the last 10 years much has been learned about drug therapy and prophylaxis of device-related infection (Tshefu *et al.*, 1983; Widmer *et al.*, 1990b, 1991; Zimmerli, 1991, 1997; Zimmerli *et al.*, 1994; Schaad *et al.*, 1994a, b; Cagni *et al.*, 1995).

Depending on the problem under study, different models of foreign-body infection can be used (Elek and Conen, 1957; Blomgren and Lindgren, 1980; Grogan *et al.*, 1980; Blomgren *et al.*, 1981; Serota *et al.*, 1981; Zimmerli *et al.*, 1982, 1984; Tshefu *et al.*, 1983). Most of these models are suitable for the determination of the minimal infecting dose, for morphological studies of the offending microorganism (Grogan *et al.*, 1980; Christensen *et al.*, 1983), or for the analysis of the local bacterial clearance in the absence or presence of antibiotics (Katz *et al.*, 1981). Major disadvantages of most of these models include the high costs of the experimental animal and the sophisticated surgical technique required (Grogan *et al.*, 1980). Furthermore, the majority of these experimental systems do not allow the study of local host-defense mechanisms.

To be clinically relevant, an animal model for the study of the pathogenesis, the prevention and the management of foreign-body infections has to reproduce the common characteristics of human disease, i.e. the implant must be easy to infect and spontaneous healing should not occur. The animal model which best reproduces the clinical characteristics of human device-associated infection is the tissue cage infection model (Zimmerli *et al.*, 1982, 1984, 1994). In this model, perforated cylinders are implanted in the subcutaneous tissue. This model has been used in different animal species. However, guinea-pigs are the most commonly used species, especially for the study of pathogenesis, antimicrobial prophylaxis and short-term therapy. For the study of long-term therapy, the rat model is preferred, since guinea-pigs do not support prolonged antimicrobial therapy.

Animal species

Several types of animal have been used in the tissue cage model, such as guinea-pigs (Widmer *et al.*, 1991), rats (Schaad *et al.*, 1994a, b), mice (Dayer *et al.*, 1987), rabbits (Bamberger *et al.*, 1995), dogs (Gruet *et al.*, 1997) and calves (Bengtsson *et al.*, 1992). For the study of pathogenesis of foreign-body infection and host-defense mechanisms, tricolor guinea-pigs (500–800 g) have been used (Zimmerli *et al.*, 1982, 1984). For antimicrobial prophylaxis and short-term treatment, albino guinea pigs (600–800 g) have mainly been used (Widmer *et al.*, 1990b, 1991; Zimmerli, 1991, 1997; Zimmerli *et al.*, 1994). Guinea-pigs are not suitable for long-term treatment with antimicrobial agents because they die from diarrhea and weight loss. It should be considered that guinea-pigs do not tolerate β-lactam antibiotics

Handbook of Animal Models of Infection
ISBN 0-12-775390-7

at all. Other drugs, such as glycopeptides, quinolones and rifampin, can be given as prophylaxis or as short-term therapy up to 4 days. The main advantage of using guinea-pigs is the close similarity of the tissue-cage infection with human device-associated infection. The minimal infecting dose is extremely low (e.g. 100 cfu for most strains of *Staphylococcus aureus*), and spontaneous healing never occurs with *S. aureus* or coagulase-negative staphylococci.

For long-term studies, rats should be used as test animals, since they well tolerate antimicrobial agents. However, the disadvantages are the very low susceptibility to infection (high minimal infecting dose) and spontaneous healing of established infection.

Preparation of animal

No specialized housing or care nor specific pretreatment is required, as long as animals are not infected. They should get free access to food and water throughout the experiments. After implantation of the tissue cages, testing of antimicrobial agents should not be started before the cages are filled with clear interstitial fluid. It takes about 3 weeks until tissue-cage fluid is no longer hemorrhagic. After surgery or after infection, single housing is preferable because infected wounds are occasionally traumatized by cage-mates.

Details of surgery

Overview

Using a strictly aseptic technique, a 4 cm median dorsal incision is made in anesthetized animals (Figure 47.1A), and the subcutaneous, epifascial space is dissected bluntly (Figure 47.1B). In guinea pigs, two gas-sterilized tissue cages are inserted into each flank (47.1C), and the skin is closed with metal clips (Figure 47.1D). In rats only one cage is inserted into each flank. Tissue-cage infection can be induced either directly by injecting bacteria into the cages (Figure 47.1E), or hematogenously by injecting them intracardiacly (Figure 47.1F).

Materials required

Anesthetic (preferably intramuscular neuroleptanalgesia), opiate antidote, hair clippers or razor, skin disinfectant, a pair of curved scissors, surgical tweezers, scalpel handle plus blades, forceps for inserting and removing clips, three to five clips, four tissue cages, sterile towel with a hole, sterile surgical gloves, aerosol-based antiseptic plastic film, and three sterile gauze pads are required. Tissue cages are Teflon tubes with an internal and external diameter of 8 mm and 10 mm respectively and 32 mm long. These tubes have 130 regularly spaced holes 1 mm in diameter and are sealed at each end with a perforated (diameter 2 mm) cap of identical material. Such tissue cages are commercially available (Dr K. Vosbeck, c/o Novartis, Basel, Switzerland). In order to determine the number of surface-adherent bacteria, six to eight sinter glass beads containing pores of 60–300 μm (Sikuf 0.23/300/A; Schott Schleifer, Muttenz, Switzerland) were placed into the tissue cages before implantation (Schwank *et al.*, 1998).

Anesthesia

For surgery, guinea-pigs should have a weight of at least 500 g. In guinea-pigs, intramuscular neuroleptanalgesia (e.g. Innovar Vet®, Pitman-Moore, USA or Hypnorm®, Janssen Pharmaceuticals) is the preferable technique. Innovar Vet® contains droperidol, methylparaben and propylparaben. Guinea-pigs are anesthetized with an i.m. injection of 1.6–2 ml/kg of 1 : 4 diluted Innovar Vet®. One part Innovar Vet® is diluted with four parts of water for i.m. injection. After a minimum of 10 minutes, the animals are ready for surgery for a maximum time of 30 minutes. Respiration remains correct, but there is no deep muscle relaxation. However, for the implantation of tissue cages no deep relaxation is required. Pentobarbital should not be used in guinea-pigs, because the death rate due to respiratory failure is too high. Immediately after surgery, opiates should be antagonized with naloxone hydrochloride (Narcan®, DuPont, Bad Homburg, Germany), and the animals should be kept in a warm environment.

Surgical procedure

In guinea-pigs, skin shaving with electric clippers can be performed without anesthesia. Shaving should be performed immediately before surgery, in order to minimize the risk of wound infection. After shaving, the skin is disinfected with 70% vol./vol. ethanol or iodine-based solution (e.g. betadine). Following complete induction of anesthesia, the animal is placed dorsal side up. The shaved skin is again disinfected. In contrast to other surgical procedures in animals, a strictly aseptic technique is required in order to avoid spontaneous infection of tissue cages. A sterile towel with a hole is placed over the surgical field. A 4 cm median dorsal incision is made, and the subcutaneous space is dissected bluntly either with two fingers or with the closed scissors (Figure 47.1A, B). Two tissue cages are implanted into each flank (Figure 47.1C). The cages on one side of the flank should be about 6 cm from each other. The skin is closed with metal skin clips (9 mm, e.g. Clay Adams, Becton Dickinson; Figure 47.1D), which are removed 5 days after surgery. The wound is covered with an aerosol-based antiseptic plastic film (e.g. Nobecutan®, Astra).

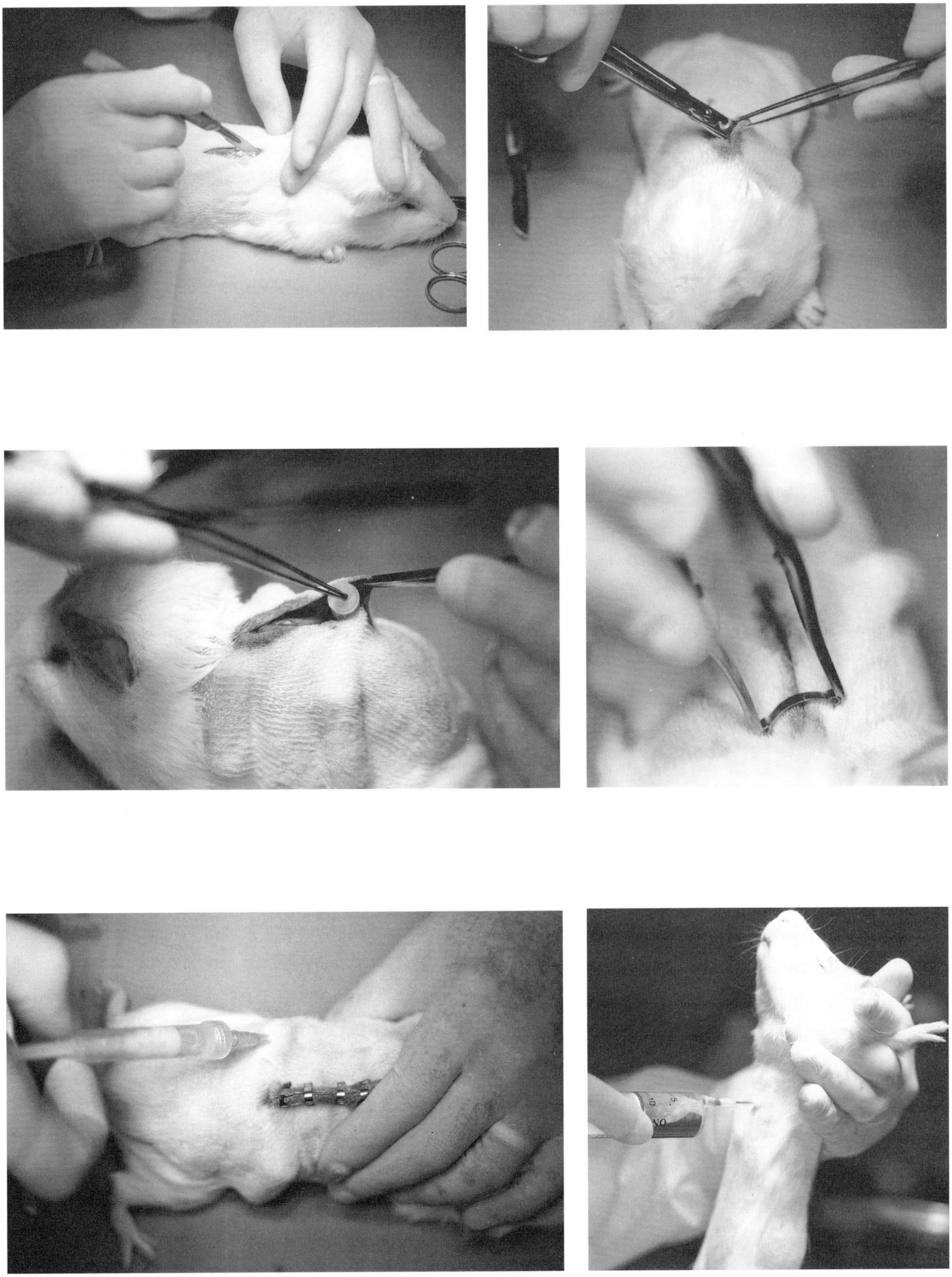

Figure 47.1 **A.** Primary surgical incision. **B.** Blunt dissection of the subcutaneous tissue with the closed scissors. **C.** Introduction of one of four tissue cages. **D.** Closure of the wound with metal skin clips. **E.** Percutaneous aspiration of tissue-cage fluid. **F.** Heart puncture in upright position.

Postoperative care

Animals should be kept warm after the surgery by placing them under an infrared lamp until they begin to recover from anesthesia. Animals should be housed individually until wound healing, in order to avoid the risk of lesions by cage-mates. Thereafter, they are housed in groups of at least three to facilitate a return to normal social activity. Animals usually return to normal behavior within 1 day. Postoperative analgesics are not required. The interstitial fluid accumulates progressively in the subcutaneous tissue cages. Generally, the cages are only inoculated after complete healing and accumulation of non-hemorrhagic interstitial fluid. In case of special questions, the cages can be infected at any time.

Tissue-cage puncture technique

The interstitial fluid accumulating progressively in the center of the subcutaneous tissue cages can be repeatedly aspirated by percutaneous aspiration (Figure 47.1E). To avoid bacterial contamination, a strictly aseptic technique is required. Prior to puncture, the skin over one extremity of the implanted tissue cage is shaved and epilated with Pelprep (Chemway, Wayne, NJ). Aspirations are done with a 0.5 × 16 mm subcutaneous needle in a 2 ml syringe. Care must be taken to avoid creating a markedly negative pressure, in order to reduce the risk of intracavitary hemorrhage due to capillary lesions. To this end, the syringe is separated from the needle after the withdrawal of 0.5 ml fluid, to equilibrate the negative tissue-cage pressure.

Characteristics of sterile tissue-cage fluid

During the first week after surgery, interstitial fluid accumulates inside the cages and becomes accessible to percutaneous aspiration. Table 47.1 summarizes the biochemical and biophysical characteristics of sterile tissue-cage fluid. In our experiment, several of these values differed significantly from serum values, especially the low pH and the high P_{O_2} and P_{CO_2}, when compared with mixed heart blood. Total protein level in tissue cages decreased after 10 weeks of implantation (Table 47.1). Cellular characteristics of tissue-cage fluid are summarized in Table 47.2. Granulocytes accumulating inside the tissue cages were tested to determine their functional properties. These tissue-cage granulocytes show a diminished overall phagocytic–bactericidal activity against *S. aureus* Wood 46 (Zimmerli *et al.*, 1982, 1984), a slightly decreased ingestion rate, granular enzymatic activity, and a markedly decreased oxidative metabolism (Zimmerli, 1984). These functional defects may be due to continuous stimulation of the tissue-cage granulocytes at the non-phagocytosible surface of the implanted foreign body. Histological sections performed 4 weeks after tissue-cage implantation, stained with hematoxylin–eosin, showed a richly vascularized granulation tissue containing lymphocytes, fibroblasts, and collagen fibers in close contact with the foreign body.

Table 47.1 Composition of normal sterile tissue-cage fluid, serum and heart blood; data are reported as means ± SEM (*n*);* Biorad method. NS, not significant ($p > 0.05$)

Component	*Tissue-cage fluid*	p *value*	*Serum or heart blood*	
Na (mmol/l)	136±0.4 (8)	NS	137±1.0	(11)
K (mmol/l)	5.0±0.009 (8)	<0.001	6.2±0.14	(8)
Ca (mmol/l)	2.4±0.03 (8)	<0.001	2.9±0.06	(5)
Mg (mmol/l)	1.1±0.08 (7)	NS	1.2±0.1	(5)
P (mmol/l)	1.2±0.07 (8)	<0.01	1.6±0.06	(10)
Zinc (μmol/l)	8.1±0.6 (6)	<0.001	13.9±0.7	(6)
pH	7.26±0.02 (10)	<0.001	7.43±0.03	(5)
P_{O_2} (kPa/l)	14.1±0.6 (10)	<0.001	8.8±0.5	(5)
P_{CO_2} (kPa/l)	8.2±0.3 (10)	<0.001	5.0±0.2	(5)
Hb (g/dl)	0.06±0.03 (5)	<0.001	13.6±0.7	(4)
Total protein* (g/l):				
Postoperative interval				
15 days	29.1±1.1 (10)	<0.001	52.3±0.9	(13)
20–25 days	33.0±1.4 (21)	<0.001		
26–34 days	33.0±1.3 (31)	<0.001		
75 days	22.8±1.1 (10)	<0.001		

Table 47.2 Number of erythrocytes and leukocytes in sterile tissue-cage fluid at 15, 20–25, and 26–34 days after implantation of tissue cage

	Postoperative interval		
Component	*15 days (n = 10)*	*20–25 days (n = 33)*	*26–34 days (n = 42)*
Erythrocytes/mm^3			
Mean	1.8×10^5	1.1×10^5	8.3×10^4
Median	1.5×10^5	1.0×10^5	5.0×10^4
Range	$0.3–3.6\times10^5$	$0.4–6.0\times10^5$	$0.1–3.9\times10^5$
Leukocytes/mm^3			
Mean	1.3×10^3	3.2×10^3	1.6×10^3
Median	1.3×10^3	2.4×10^3	1.0×10^3
Range	$0.4–2.2\times10^3$	$0.3–8.4\times10^3$	$0.2–7.6\times10^3$

Cardiac puncture technique

For hematogenous inoculation, pharmacokinetic studies, and microbiologic analyses, repeated cardiac blood punctures are required. Heart punctures are performed under CO_2-anesthesia. By holding the animal with one hand in an upright position, the heart action is easily palpable in the fourth–fifth left parasternal intercostal space (Figure 47.1F). By using a 22 G needle, atraumatic puncture can be performed up to more than five times a day. In guinea-pigs of 600 g body weight, the aspiration of more than 8 ml blood in one or several punctures should be avoided.

Storage and preparation of inocula, and infection process

In the tissue-cage infection model, different species of bacteria have been used. However, most experiments were performed with multiple strains of methicillin-sensitive and -resistant *Staphylococcus aureus* (Cagni *et al.*, 1995; Schaad *et al.*, 1994a, b; Tshefu *et al.*, 1983; Zimmerli., 1997; Zimmerli *et al.*, 1994), coagulase-negative staphylococci (Widmer *et al.*, 1990b), *Escherichia coli* (Widmer *et al.*, 1991), and *Salmonella dublin* (Widmer *et al.*, 1990a). With all these strains, an appropriate inoculum results in persistent tissue-cage infection in guinea-pigs. In contrast, with other species of bacteria, such as pyogenic streptococci, spontaneous healing occurred (unpublished observation). Therefore, before performing treatment or prophylaxis studies with a new strain, the spontaneous course of infection must be studied. The infecting strains (defined ATCC strains or clinical isolates cultured from patients with implant-associated infection) are stored in aliquots of the same initial culture. Some 15–24 hours before inoculation, the test strain is grown in Tryptic Soy Broth (Difco, Detroit, MI) at 37°C, washed and resuspended in saline at the desired spectrophotometrically determined density. A volume of 0.2 ml of this suspension is injected either directly into the tissue cages or intracardially. The minimal infecting dose for persisting infection after direct inoculation in the tissue cage is typically 10^2 cfu for *Staphylococcus aureus, Escherichia coli* and *Salmonella dublin*, and 10^3 cfu for coagulase-negative staphylococci. Most prophylaxis and treatment experiments were done with inocula of 10^4 cfu. Hematogenous infections have been studied only with one single test strain, namely *Staphylococcus aureus* Wood 46. With an intracardiac inoculation of 5×10^7 cfu 42% of the subcutaneous tissue cages could be permanently infected (Zimmerli *et al.*, 1985).

If other animal species such as rats are used for the experiments, higher inocula of *Staphylococcus aureus* (2×10^5–2×10^6 cfu) are needed (Schaad *et al.*, 1994b). Infection cannot be induced by any strain of *Staphylococcus aureus* and not at all by coagulase-negative staphylococci. The rat model does not really imitate human implant-associated infection, and should therefore be used only for special questions that cannot be studied in the guinea-pig model. Such a question is the role of long-term antimicrobial treatment of chronic device-related infection (Schaad *et al.*, 1994a, b).

Key parameters to monitor infection and response to treatment

Tissue-cage-associated infections are limited to the fluid in the cage and the tissue in the close vicinity of the implant. Even after hematogenous infection with an appropriate inoculum, blood cultures clear spontaneously, and no organs (kidney, lung, liver, spleen) are infected (Zimmerli *et al.*, 1985; Figure 47.2). This indicates that the implant causes a local host defense defect (Zimmerli *et al.*, 1982, 1984). Death due to systemic infection does not occur, as long as the above-mentioned local or systemic inocula are used. Guinea-pigs do not lose weight as a result of infection itself but during antimicrobial therapy. Therefore, death rate, bacteremia and body weight are not useful indices of infection or treatment success. To determine the efficacy of an antimicrobial regimen, the following tests can be performed: (1) bacterial counts in tissue-cage fluid before, during and after antimicrobial treatment, (2) counts of surface-adherent bacteria before and after therapy, and (3) growth of any bacteria on explanted tissue cages after therapy.

Bacterial counts in tissue-cage fluid

Implanted tissue cages can be used as *in-vivo* test tubes. The repetitive fluid sampling after infection, and during therapy allows to determine *in-vivo* growth and kill curves. During a 4-day course of antimicrobial therapy, the same tissue cage

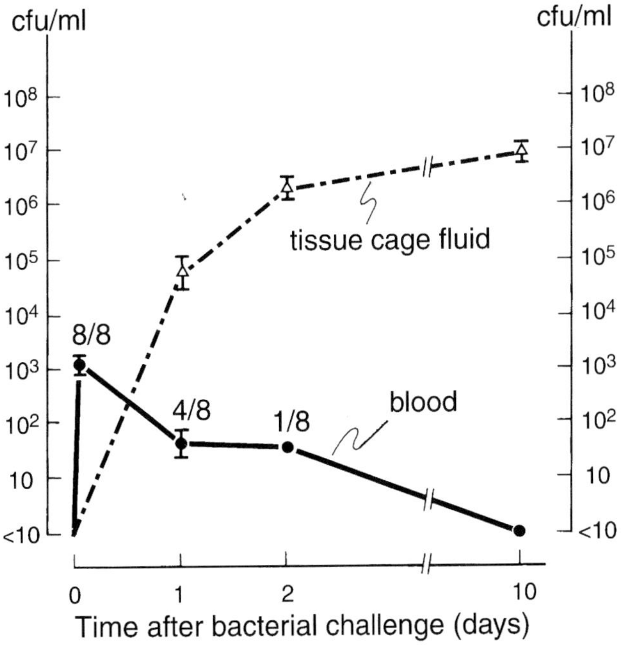

Figure 47.2 Quantitative culture of blood (●—●) and tissue-cage fluid (△– – –△) after intracardiac inoculation of 5×10^7 cfu *S. aureus* Wood 46 at zero time. Five minutes after bacterial inoculation 8/8 guinea-pigs had a positive blood culture. Numbers over symbols indicate positive blood cultures per total number of animals. Mean quantitative bacterial counts are calculated for the 11 tissue cage fluids that were infected. Values are reported as means ± SEM of the bacterial counts (cfu/ml) at the indicated times after experimental infection. Reproduced from Zimmerli *et al.*, 1985, with permission of the *Scandinvian Journal of Infectious Diseases*.

can be sampled every day, if the sampling is limited to about 0.2–0.4 ml. With this fluid serial quantitative cultures can be performed. If undiluted fluid should be cultured, a defined volume (0.1 ml) is passed through a filter which is washed with saline before culturing, in order to avoid carry-over of antibiotics resulting in false-negative cultures.

Semiquantitative and quantitative determination of surface-adherent microorganisms

The main problem in the treatment of implant-associated infection is the persistence of surface-adhering microorganisms. Therefore, the determination of bacterial counts in tissue-cage fluid may be misleading. Indeed, we could show that, despite sterile tissue-cage fluid, staphylococci may still persist on the tissue-cage surface (Zimmerli *et al.*, 1985). In the original technique, the tissue cages were excised under sterile conditions and semiquantitative bacterial counts were obtained by the agar roll technique (Zimmerli *et al.*, 1985). The colony count of the infecting strain can be determined after a 48 hour incubation at 37°C.

Since this semiquantitative technique does not take into account bacteria adhering to the inner surface of the tissue cage, we developed a new method using sinter glass beads (Sikug 023/300 A, Schott Schleifer AG, Muttenz, Switzerland) as implants (Schwank *et al.*, 1998). For this purpose, each tissue cage was filled with six sinter glass beads before implantation. In order to start infection, approximately 10^7 cfu of the test strain is inoculated in the implanted tissue cage. The number of bacteria adhering to the glass beads is determined as previously described (Zimmerli *et al.*, 1994). In brief, six beads are removed from each cage at each sampling point. Each bead is placed on a filter with sterile forceps and washed twice with saline. The number of bacteria adhering to the beads is determined by placing the washed glass beads in 2 ml of physiological saline containing EDTA (0.15%) and Triton-X (0.1%). The beads are vigorously vortexed within the tubes three times for 15 seconds to remove the adhering bacteria. Thereafter, the tubes are placed in an ultrasonic bath and sonicated for 3 minutes at 120 W (Labsonic 2000, Bender & Hobein, Zürich, Switzerland). After an additional mixing, a 100 μl aliquot is diluted for quantitative bacterial culture on tryptic soy agar or Mueller–Hinton plates. The detection limit is 20 cfu per bead. Using this procedure, 97 ± 2% of the adherent bacteria can be removed, as verified with [^{3}H]thymidine labelled *S. aureus* (unpublished own results).

Cure rate of tissue-cage infection

The ultimate test to prove the complete efficacy of an antimicrobial agent, is the determination of the cure rate. This test has been performed in most experimental treatment studies (Tshefu *et al.*, 1983; Widmer *et al.*, 1990b, 1991; Zimmerli, 1991, 1997; Zimmerli *et al.*, 1994). Several days, in general 1 week, after antimicrobial treatment, the animals are euthanized and the tissue cages are removed under strictly aseptic conditions. After semiquantitative culture on Mueller–Hinton agar (see above), tissue cages are incubated in tryptic soy broth for 48 hours. Any positive culture identified as the initially inoculated strain is defined as a treatment failure (Widmer *et al.*, 1990b).

Antimicrobial therapy

A variety of antimicrobial agents have been evaluated in the guinea-pig and rat models of tissue-cage infection. The most important studies are summarized in Table 47.3. Guinea pigs can be administered animicrobial agents by several routes (s.c., i.m., i.p., or p.o.). We never used the oral route because of the problem of dosing and the increased risk of diarrhea. In most studies, the i.p. or the i.m. route was used. In prophylaxis studies, one to several doses of the antimicrobial agent were applied at different time points in relation to the bacterial inoculation (Widmer *et al.*, 1991). All treatment studies in guinea-pigs were short-term regimens of up to 4 days, since these animals do not support prolonged therapy. During therapy, guinea pigs suffer from anorexia, often from diarrhea, and from important weight loss (about 10%). β-lactam antibiotics, clindamycin and possibly other drugs with a broad spectrum, including anaerobes, cannot be tested in guinea-pigs because of intolerance. Before each treatment with a novel antimicrobial substance, the dosing schedule and choice of dose should be tested in preliminary experiments considering the differing pharmacokinetics of antibiotics between humans and guinea-pigs. For practical reasons, in most experiments a twice-daily regimen was chosen, with doses resulting in serum peak levels in the range of human serum levels (Widmer *et al.*, 1990b).

Advantage and disadvantages of the model

The main advantage of the tissue-cage infection guinea-pig model is the fact that is closely resembles human disease. The model is readily available and inexpensive. The biological characteristics of the tissue-cage infection can be followed up very easily, since the fluid surrounding the foreign body can be aspirated repeatedly. As a paradigm of device-related infection, it allows:

1. documentation of the increase in pathogenicity of pyogenic microorganisms resulting from the presence of a foreign body (Zimmerli *et al.*, 1982; Vaudaux *et al.*, 1985);
2. study of the natural history of a foreign-body infection in subcutaneous tissue;

Table 47.3 Examples of antimicrobial therapy of tissue-cage infections in the guinea-pig

Reference	*Bacteria*	*Antibiotic*	*Dose mg/kg (b.i.d.)*	*No. of doses*	*Start of treatment**	*Cure rate†* (%)
Tshefu *et al.*, 1983	*S. aureus* (Wood 46)	Rifampin	7.5	4	−3 h	7/7 (100)
		Rifampin	7.5	4	+12 h	8/8 (100)
		Rifampin	7.5	4	+24 h	3/7 (43)
		Rifampin	7.5	4	+48 h	0/4 (0)
Widmer *et al.*, 1990b	*S. epidermidis* (B3972)	Rifampin	7.5	4	+24 h	18/24 (75)
		Rifampin	25	4	+24 h	12/12 (100)
		Ciprofloxacin	10	4	+24 h	0/12 (0)
		Vancomycin	15	4	+24 h	2/12 (17)
		Teicoplanin	6.6	4	+24 h	0/22 (0)
Widmer *et al.*, 1991	*E. coli* (ATCC25922)	Co-trimoxazole	10/50	4	+4 h	0/8 (0)
		Aztreonam	200	4	+4 h	20/24 (84)
		Aztreonam	200	4	+12 h	1/12 (8)
		Ciprofloxacin	10	4	+4 h	10/11 (91)
		Ciprofloxacin	10	4	+12 h	9/24 (37)
Zimmerli *et al.*, 1994	*S. aureus* (ATCC 29213)	Vancomycin	15	4	+24 h	0/12 (0)
		Teicoplanin	6.6	4	+24 h	0/12 (0)
		Ciprofloxacin	10	4	+24 h	2/12 (17)
		Fleroxacin	10	4	+24 h	0/8 (0)
		Rifampin	25	4	+24 h	10/20 (50)
		Rifampin plus	25	4		
		Ciprofloxacin	10	4	+24 h	11/12 (92)
Cagni *et al.*, 1995	*S. aureus* (MRSA)	None	NA	NA	NA	1/12 (8)
		Sparfloxacin	50	1	−3 h	8/8 (100)
		Ciprofloxacin	50	1	−3 h	0/8 (0)
		Temafloxacin	50	1	−3 h	8/8 (100)
		Vancomycin	30	1	−3 h	3/8 (38)

* Start with first dose in relation to bacterial inoculation (mostly inocula of 10^4 cfu into the tissue cage).
† Either negative tissue-cage fluid culture > 2 days after therapy or negative culture of the excised tissue cage.

3. study of the mechanisms and role of bacterial adherence *in vivo* (Vaudaux *et al.*, 1994);
4. testing of different device-materials;
5. analysis of the local host-defense mechanisms in the vicinity of a sterile or infected foreign body (Zimmerli *et al.*, 1982, 1984; Vaudaux *et al.*, 1992);
6. analysis of the risk of hematogenous seeding on extravascular devices (Zimmerli *et al.*, 1985);
7. determination of the efficacy of antimicrobial prophylaxis and therapy (Tshefu *et al.*, 1983; Zimmerli *et al.*, 1985, 1994; Zimmerli, 1997; Widmer *et al.*, 1990a,b, 1991; Cagni *et al.*, 1995);
8. correlation between drug efficacy *in vivo* and *in vitro* (Blaser *et al.*, 1995; Schwank *et al.*, 1998).

Despite the obvious advantages, the tissue-cage model has its limitations and pitfalls. It is somewhat artificial, since in contrast to most devices (sutures, valves, orthopedic implants) it does not fulfill mechanical functions. Furthermore, many devices are not in the subcutaneous tissue but in the bloodstream, in the vicinity of bone, or in brain tissue. Therefore, the observed local impairment of host-defence mechanisms may be relevant for extravascular devices such as subcutaneous catheters or breast implants but not for intravascular devices such as vascular prostheses or heart valves. A major pitfall is the fact that, during implantation of tissue cages into guinea-pigs, strict sterility is required, otherwise spontaneous infection is the rule. In studies of anti-microbial agents, it has to be considered that guinea-pigs do not support each antimicrobial agent, and that treatment periods have to be limited to a few days. In contrast, rats well support long-term treatment with any antibiotic; however, the experimental infection does not closely mimic human disease. Very high inocula are required for tissue-cage infection; furthermore, spontaneous healing occurs (unpublished observations).

Contributions of the model to infectious disease therapy

Elucidation of the pathophysiology of implant associated infection

The tissue-cage infection guinea-pig model has been widely used to analyze the role of local host factors in device-related infection (Zimmerli *et al.*, 1982, 1984; Vaudaux *et al.*, 1992). Various microbiogical, immunological, and cellular events preceding, or associated with implants and implant-associated infections have been studied. These studies have been recently reviewed by Vaudaux *et al.* (1994). In brief, a local granulocyte defect, a low local opsonic activity, and the adherence-promoting activity of foreign-body-bound fibronectin have been shown.

Profiling of antimicrobial treatment

Traditionally, it has been a dogma that infected devices have to be removed, because they resist even prolonged antimicrobial therapy (Dougherty, 1988). As a result of the tissue-cage infection model, the requirements of an efficacious antimicrobial therapy of device-associated infections could be defined (Widmer *et al.*, 1990a, b, 1991; Zimmerli *et al.*, 1994, 1997). In brief, efficacy on stationary-phase and adherent bacteria is required for complete elimination of device-associated microorganisms. As the result of these observations, rifampin is now routinely added in the treatment regimens of staphylococcal implant-associated infections (Widmer *et al.*, 1992; Drancourt *et al.*, 1993; Morscher *et al.*, 1995; Zimmerli *et al.*, 1995, 1998).

References

Bamberger, D. M., Herndon, B. L., Suvarna, P. R. (1995). Azithromycin in an experimental *Staphylococcus aureus* model. *J. Antimicrob. Chemother*, **35**, 623–629.

Bengtsson, B., Bredberg, U., Luthman, J. (1992). Mathematical description of the concentration of oxytetracycline and penicillin-G in tissue cages in calves as related to the serum concentration. *J. Vet. Pharmacol. Ther.*, **15**, 202–216.

Blaser, J., Vergères, P., Widmer A. F., Zimmerli, W. (1995). *In vivo* verification of *in vitro* model of antibiotic treatment of device-related infection. *Antimicrob. Agents Chemother.*, **39**, 1134–1139.

Blomgren, G., Lindgren, U. (1980). The susceptibility of total joint replacement to hematogenous infection in the early post-operative period. *Clin. Orthop.*, **151**, 308–312.

Blomgren, G., Lundquist, H., Nord, C. E., Lindgren, U. (1981). Late anaerobic haematogenous infection of experimental total joint replacement. A study in the rabbit using *Propionibacterium acnes*. *J. Bone Joint Surg*, **63-B**, 614–618.

Busuttil, R., Rees, W., Baker, D., Wilson, S. E. (1979). Pathogenesis of aortoduodenal fistula: experimental and clinical correlates. *Surgery*, **85**, 1–13.

Cagni, A., Chuard, C., Vaudaux, P. E., Schrenzel, J., Lew, D. P. (1995). Comparison of sparfloxacin, temafloxacin, and ciprofloxacin for prophylaxis and treatment of experimental foreign-body infection by methicillin resistant *Staphylococcus aureus*. *Antimicrob. Agents Chemother.*, **39**, 1655–1660.

Charnley, J. (1972). Postoperative infection after total hip replacement with special reference to air contamination in the operating room. *Clin. Orthop.*, **87**, 167–187.

Christensen, G. D., Simpson, W. A., Bisno, A. I., Beachey, E. H. (1983). Experimental foreign body infections in mice challenged with slime-producing *Staphylococcus epidermidis*. *Infect. Immun.*, **40**, 407–410.

Dayer, E., Yoshida, H., Izui, S., Lambert, P. H. (1987). Quantitation of retroviral gp70 antigen, autoantibodies, and immune complexes in extravascular space in arthritic MRL-lpr/lpr mice. Use of a subcutaneously implanted tissue cage model. *Arthritis Rheum.*, **30**, 1274–1282.

Dougherty, S. H. (1988). Pathobiology of infection in prosthetic devices. *Rev. Infect. Dis.*, **10**, 1102–1117.

Drancourt, M., Stein, A., Argenson, J. N., Zannier, A., Curvale, G., Raoult, D. (1993). Oral rifampin plus ofloxacin for treatment of *Staphylococcus*-infected orthopedic implants. *Antimicrob. Agents Chemother.*, **37**, 1214–1218.

Elek, S. D., Conen, P. E. (1957). The virulence of *Staphylococcus pyogenes* for man; a study of the problem of wound infection. *Br. J. Exp. Pathol.*, **38**, 573–586.

Garrison, P. K., Freedman, L. R. (1970). Experimental endocarditis. I. Staphylococcal endocarditis in rabbits resulting from placement of polyethylene catheter in the right side of the heart. *Yale J. Biol. Med.*, **42**, 394–410.

Glauser, M. P., Bernard, J. P., Morreillon, P., Francioli, P. (1983). Successful single-dose amoxicillin prophylaxis against experimental streptococcal endocarditis; evidence of two mechanisms of protection. *J. Infect. Dis.*, **146**, 568–575.

Grogan, E. L., Sande, M. A., Clark, R. E., Nolan, S. P. (1980). Experimental endocarditis in calf after tricuspid valve replacement. *Ann. Thorac. Surg.*, **30**, 64–69.

Gruet, P., Richard, P., Thomas, E., Autefage, A. (1997). Prevention of surgical infections in dogs with a single intravenous injection of marbofloxacin: an experimental model. *Vet. Rec.*, **140**, 199–202.

James, R. C., MacLeod, C. J. (1961). Induction of staphylococcal infections in mice with small inocula introduced on sutures. *Br. J. Exp. Pathol.*, **42**, 266–277.

Katz, S., Mordechai, I., Mirelman, D. (1981). Bacterial adherence to surgical sutures. *Ann. Surg.*, **194**, 35–41.

Küntscher, G., Maatz, R. (1945). *Technik der Marknagelung*. Georg Thieme, Leipzig.

Morscher, E., Herzog, R., Bapst, R., Zimmerli, W. (1995). Management of infected hip arthroplasty. *Orthopaed. Int.*, **3**, 343–351.

Pudenz, R. H., Russel, F. E., Hurd, A. H., Shelden, C. H. (1957). Ventriculo-auriculostomy. A technique for shunting cerebrospinal fluid into the right auricle. *J. Neurosurg.*, **14**, 171–179.

Schaad, H. J., Chuard, C., Vaudaux, P. *et al.* (1994a). Comparative efficacies of imipenem, oxacillin and vancomycin for therapy of chronic foreign body infection due to methicillin-susceptible and -resistant *Staphylococcus aureus*. *J. Antimicrob. Chemother.*, **33**, 1191–1200.

Schaad, H. J., Chuard, C., Vaudaux, P., Waldvogel, F. A., Lew, D. P. (1994b). Teicoplanin alone or combined with rifampin compared with vancomycin for prophylaxis and treatment of experimental foreign body infection in methicillin-resistant *Staphylococcus aureus*. *Antimicrob. Agents Chemother.*, **38**, 1703–1710.

Schwank, S., Rajacic, Z., Zimmerli, W., Blaser, J. (1998). Impact of the bacterial biofilm formation on *in-vitro* and *in-vivo* activity of antibiotics. *Antimicrobs. Agents Chemother.*, **42**, 895–898.

Serota, A. I., Williams, R. A., Rose, J. G., Wilson, S. E. (1981). Uptake in radiolabeled leukocytes in prosthetic graft infection. *Surgery*, **90**, 35–40.

Starr, A., Edwards, M. L. (1961). Mitral replacement; clinical experience with a ball-valve prosthesis. *Ann. Surg.*, **154**, 726–740.

Tshefu, K., Zimmerli, W., Waldvogel, F. A. (1983). Short-term administration of rifampin in the prevention or eradication of infection due to foreign bodies. *Rev. Infect. Dis.*, **5** (Suppl. 3), S474–S479.

Vaudaux, P. E., Zulian, G., Huggler, E., Waldvogel, F. A. (1985). Attachment of *Staphylococcus aureus* to polymethylmethacrylate increases its resistance to phagocytosis in foreign body infection. *Infect. Immun.*, **50**, 472–477.

Vaudaux, P., Grau, G. E., Huggler, E. (1992). Contribution of tumor necrosis factor to host defense against staphylococci in a guinea pig model of foreign body infections. *J. Infect. Dis.*, **166**, 58–64.

Vaudaux, P. E., Lew, D. P., Waldvogel, F. A. (1994). Host factors predisposing to and influencing therapy of foreign body infections. In: *Infections Associated with Indwelling Medical Devices*, 2nd edn. (eds Bisno, A. L., Waldvogel, F. A. pp. 1–29. American Society for Microbiology, Washington, DC.

Widmer, A. F., Colombo, V. E., Gächter, A., Thiel, G., Zimmerli, W. (1990a). *Salmonella* infection in total hip replacement: tests to predict the outcome of antimicrobial therapy. *Scand. J. Infect. Dis.*, **22**, 611–618.

Widmer, A. F., Frei, R., Rajacic, Z., Zimmerli, W. (1990b). Correlation between *in vivo* and *in vitro* efficacy of antimicrobial agents against foreign body infections. *J. Infect. Dis.*, **162**, 96–102.

Widmer, A. F., Wiestner, A., Frei, R., Zimmerli, W. (1991). Killing of nongrowing and adherent *Escherichia coli* determines drug efficacy in device-related infections. *Antimicrob. Agents Chemother.*, **35**, 741–746.

Widmer, A. F., Gächter, A., Ochsner, P. E., Zimmerli, W. (1992). Antimicrobial treatment of orthopedic implant-related infections with rifampin combinations. *Clin. Infect. Dis.*, **14**, 1251–1253.

Zimmerli, W. (1991). Rôle des antibiotiques dans les infections sur matériel étranger: modèle animal, tests microbiologiques et expériences cliniques. *Méd. Mal. Infect.*, **21**, 529–534.

Zimmerli, W. (1993). Animal models in the investigation of device-related infections. *J. Antimicrob. Chemother.*, **31** (Suppl. D), 97–102.

Zimmerli, W. (1995). Die Rolle der Antibiotika in der Behandlung der infizierten Gelenkprothesen. *Orthopäde*, **24**, 308–313.

Zimmerli, W. (1997). Experimental foreign-body associated infections. *Méd. Mal. Infect.*, **27**, 188–194.

Zimmerli, W., Waldvogel, F. A., Vaudaux, P., Nydegger, U. E. (1982). Pathogenesis of foreign body infection: description and characteristics of an animal model. *J. Infect. Dis.*, **146**, 487–497.

Zimmerli, W., Lew, P. D., Waldvogel, F. A. (1984). Pathogenesis of foreign body infection. Evidence for a local granulocyte defect. *J. Clin. Invest.*, **73**, 1191–1200.

Zimmerli, W., Zak, O., Vosbeck, K. (1985). Experimental haematogenous infection of subcutaneously implanted foreign bodies. *Scand. J. Infect. Dis.*, **17**, 303–310.

Zimmerli, W., Frei, R., Widmer, A. F., Rajacic, Z. (1994). Microbiological tests to predict treatment outcome in experimental device-related infections due to *Staphylococcus aureus*. *J. Antimicrob. Chemother.*, **33**, 959–967.

Zimmerli, W., Widmer, A. F., Blatter, M., Frei, R., Ochsner, P. E., for the FBI-study group (1998). Role of rifampicin for treatment of orthopedic implant-related staphylococcal infections: randomized controlled clinical trial. *J.A.M.A.*, **279**, 1537–1541.

Chapter 48

Rat Model of Bacterial Epididymitis

H. G. Schiefer and C. Jantos

Acute epididymitis is a common urological disease accounting for over 600 000 visits to physicians in the USA every year and affecting men of all ages. The disease is characterized by a unilateral painful scrotal swelling, often associated with dysuria and/or urethral discharge. Most cases of acute epididymitis normally result from retrograde, ascending bacterial infection from the urethra via the vas deferens. A correlation has been demonstrated between the age of patients and the causative organisms. *Eschericha coli* and other common urinary pathogens are most frequently associated with epididymitis in prepubertal children with predisposing structural or urological abnormalities, and in men over 35 years of age with underlying urinary tract pathology, e.g., benign prostatic hyperplasia. In contrast, in men up to 35 years of age, epididymitis is commonly caused by sexually transmitted pathogens, e.g. *Chlamydia trachomatis* or *Neisseria gonorrhoeae*. Occasionally, epididymitis may occur as complication of systemic infection with various pathogens, e.g., *Mycobacterium tuberculosis*, *Brucella* spp, *Streptococcus pneumoniae*, *Neisseria meningitidis*, *Treponema pallidum*, and various fungi. Complications of epididymitis include impairment of fertility, abscess formation, testicular infarction, orchitis, and chronic pain. Because of the grave complications, antibiotic treatment must be initiated immediately, even before the etiology is known, and should be effective against possible etiologic agents. In general, broad-spectrum parenteral antimicrobial therapy is recommended (Berger, 1990). However, epididymal ductal epithelium represents a blood–tissue barrier providing effective protection of sperm antigens against detection by the host immune system. On the other hand, it hinders antibiotic diffusion into epididymal stroma.

Several models of epididymitis using different animal species have been developed in order to allow more detailed studies into the pathogenesis, pathomorphology, kinetics of bacterial replication, and immune response of the disease, and to assess the effectiveness of both established and recently developed antimicrobial agents (Graves and Engel, 1950; Møller and Mårdh, 1980; Lucchetta *et al.*, 1983; Dathe and Vogt, 1985; Nielsen, 1987; Hackett *et al.*, 1988; Weidner *et al.*, 1989; See *et al.*, 1990).

Previously, we have described an experimental model for acute and chronic obstructive bacterial epididymitis in Wistar rats (Weidner *et al.*, 1989) using a uropathogenic *E. coli* strain. After surgical exposure, one epididymis was infected by intraepididymal injection of 1 ml of the *E. coli* suspension containing 10^8 cfu/ml into the epididymal stroma. During the follow-up of up to 6 months, epididymal tissue was analysed both bacteriologically and morphologically. In all infected animals acute purulent epididymitis was induced. Inflammatory response was related to reisolated *E. coli* only in the first few days after infection. Inflammation then progressed even in cases without positive cultural findings. Epididymitis ultimately resulted in chronic inflammation with granulation and fibrosis and obliteration of the ductus epididymidis. In some cases sperm granulomata were obvious.

However, infection by injecting the bacterial suspension into the epididymal stroma must be considered rather artificial. Therefore, the model described below was developed (Jantos *et al.*, 1989; Vieler *et al.*, 1993), which imitates the natural course of retrograde, ascending infection.

Animal species

Both Wistar and Sprague–Dawley strains of male rats, weighing 350–450 g, were used. Animals were kept separately in isolation cages in a natural day–night system and had free access to food and water.

Preparation of animals

No specialized care nor specific pre- and post-treatment were required.

Details of surgery

Overview

Following a scrotal incision, the right testis and epididymis were exteriorized, and the bacterial inoculum was instilled

Handbook of Animal Models of Infection
ISBN 0-12-775390-7

retrogradely into the lumen of the vas deferens by using a 27 G needle. After withdrawal of the needle, the wound was disinfected, the testis placed back into the scrotum, and the wound sewed up in a double layer with 5/0 chromatic catgut.

Materials required

Anaesthetic, skin disinfectant, scalpel handle plus blades, tweezers, tuberculin syringe, needle (27 G), suture material (5/0 chromatic catgut).

Anaesthesia

Animals were anaesthetized by intraperitoneal injection of a mixture of ketamine (0.8 ml/kg body weight) and xylazine (0.3 ml/kg body weight).

Surgical procedure

Following complete induction of anaesthesia (within about 15 minutes), the animal was placed and fixed on a cork board in a supine position. After skin disinfection (70% ethanol) and shave, the abdominal cavity was opened by a 1 cm inguinal–scrotal incision, and the right epididymis and testis were exteriorized. The testicular fatty appendix was perforated with a 20 G needle to apply a slight tension on the right testis and vas deferens.

Infection procedure

The vas deferens was punctured in the proximal third with a 27 G needle, and 0.1 ml of the *E. coli* suspension containing 10^6 colony forming units (cfu)/ml was carefully injected retrogradely into the lumen of the vas deferens. Subsequently the needle was withdrawn.

Wound closure

After disinfection, the testis was placed back into the scrotum, and the wound was sewed up in a double layer with 5/0 chromatic catgut, and finally sealed by spraying an antiseptic plastic film (Nobecutan, Astra, Hamburg, Germany).

Postoperative care

After wound closure, the animals were further housed in wire cages and required no special care. Reduced activity was observed in all animals on days 1 and 2 postinfection (p.i.)., but after day 3 normalized completely. One animal died at 2 days p.i. after a phase of apathy, shaggy coat, and haematuria. Weight loss (minus 5.6–7.2%) p.i. was obvious in all animals. Leukocyte numbers increased in all animals by 25–66%.

Storage and preparation of inocula

E. coli, serovar O:6, was a clinical isolate from a patient suffering from chronic bacterial prostatitis, and found to be a haemolytic, haemagglutinating (mannose-sensitive), serum-resistant, encapsulated strain. Stock cultures were made by incubating the organism in trypticase soy broth (TSB; Merck, Darmstadt, Germany) at 37°C for 24 hours. This suspension was then diluted 100-fold, and again incubated at 37°C for 3 hours. 1 ml samples were stored at –70°C. Fresh inocula were prepared by diluting a stock culture in 3 ml of TSB and incubating it overnight at 37°C. A 100-fold dilution of this culture was grown to a final concentration of approximately 10^7 cfu/ml, as determined by turbidimetry, serial 1:10 dilutions in TSB and plating aliquots on cystine-lactose-electrolyte-deficient (Sandys) agar plates. Minimal inhibitory (MIC) and minimal bactericidal (MBC) concentrations were determined by a broth macrodilution method. The MIC for the *E. coli* used was as follows: ofloxacin, 0.049 μg/ml; cefotaxime, 0.049 μg/ml; and doxycycline, 1.1 μg/ml. The MBC was 0.098 μg/ml for both ofloxacin and cefotaxime.

Key parameters to monitor infection

Evaluation of infection

To determine the numbers of organisms in the epididymal tissue before the initiation of therapy, ten randomly selected rats were sacrificed 24 hours p.i., at the time when treatment was started in the test rats. Epididymides were removed aseptically, weighed, and homogenized in sterile saline using a pestle and mortar. The number of bacteria per gram of epididymal tissue was determined by plating 0.1 ml of serial 10-fold dilutions of the homogenate in triplicate on cystine-lactose-electrolyte-deficient (Sandys) agar plates, incubating at 37°C for 24 hours, and counting the cfu. Results were expressed as $\log_{10}$ cfu of *E. coli* per gram of tissue. The method permitted the detection of 50 cfu/g of tissue.

At the start of therapy (24 hours p.i.), the epididymides of all 10 animals sacrificed were found to be infected with a median bacterial count of 4.4 $\log_{10}$ cfu/g of tissue (range, 2.8–6.8 $\log_{10}$ cfu/g).

Antimicrobial therapy

Pharmacokinetic studies

Antibiotic concentrations in blood, epididymides, and testes were determined for uninfected rats at various time points after a single administration of each drug: ofloxacin and cefotaxime were administered subcutaneously, and doxycycline was given orally by stomach tube. The concentrations in blood and homogenized organs were determined by an agar well diffusion method with the test organisms *Klebsiella aerogenes* for ofloxacin, *Streptococcus pyogenes* for cefotaxime, and *Bacillus cereus* for doxycycline, using standard curves with known concentrations of the respective antibiotics (Table 48.1).

Table 48.1 Pharmacokinetic data for ofloxacin, cefotaxime and doxycycline (from Vieler *et al.*, 1993); T_{max}, time to maximum concentration of drug in serum

Drug (dose [mg/kg])	*Site*	*Peak level (µg/ml [blood] or µg/g [tissue]; mean ± SD)**	T_{max} *(h)*
Ofloxacin (20)	Blood	3.63 ± 0.60	1.0
	Epididymis	4.72 ± 0.60	1.0
	Testis	2.94 ± 0.47	2.0
Cefotaxime (50)	Blood	49.34 ± 7.34	0.5
	Epididymis	14.40 ± 1.60	0.5
	Testis	7.72 ± 1.99	1.0
Doxycycline (100)	Blood	5.05 ± 3.26	6.0
	Epididymis	3.06 ± 0.79	2.0
	Testis	2.71 ± 0.57	4.0

* Blood levels were determined with groups of 10 animals. Tissue levels were determined with groups of four animals for each drug.

Therapeutic trials

Antimicrobial treatment was started 24 hours p.i., and continued for 7 days. Dosage regimens were designed to achieve peak blood drug levels similar to those observed in humans. Ofloxacin (20 mg/kg body weight), and cefotaxime (50 mg/kg) were given subcutaneously every 12 hours, and doxycycline (100 mg/kg) was administered orally by stomach tube every 24 hours. Control animals received sterile saline solution subcutaneously at 12 hour intervals.

Key parameters to monitor response to treatment

Animals were sacrificed 3 days after the termination of therapy. Epididymides were removed aseptically and cut into halves. One half was fixed in 10% buffered formalin, and the remaining half was weighed and homogenized in sterile saline. The number of organisms per gram of tissue was determined as described above. Bacteria recovered from the infected epididymides of treated rats were retested for MIC.

The results of treatment of experimental *E. coli* epididymitis with ofloxacin, cefotaxime and doxycycline are shown in Table 48.2.

Table 48.2 Results of treatment of experimental *E. coli* epididymitis (from Vieler *et al.*, 1993)

Therapeutic regimen	*No. of rats evaluated*	*No. of rats with sterile epididymides*	*Median* log_{10} *cfu/g, of epididymis (range)*
Ofloxacin	30	20	< 1.0 (< 1–5.0)*
Cefotaxime	28	3	4.4 (< 1–8.8)
Doxycycline	29	3	3.9 (< 1–7.9)†
Saline	30	0	5.5 (3–8.5)

* Significantly lower than the cefotaxime ($p < 0.0001$), doxycycline ($p < 0.01$) and control ($p < 0.0001$) groups.
† Significantly lower than the control group ($p < 0.01$).

Ofloxacin was the most effective drug. The numbers of *E. coli* recovered from the epididymides of ofloxacin-treated rats were significantly lower than those for cefotaxime-treated ($p<0.0001$), doxycycline-treated ($p<0.01$), and control ($p<0.0001$) animals. Cure of the infection was observed in 20 out of 30 ofloxacin-treated, animals, in contrast to three out of 28 cefotaxime-treated and three out of 29 doxycycline-treated rats. Doxycycline was less effective than ofloxacin; however, it reduced the bacterial counts significantly compared with the controls. In contrast, despite good *in vitro* activity, cefotaxime was not effective in reducing the magnitude of infection significantly. In addition, clearance of *E. coli* from the epididymides of any control animal was not observed. The *in vitro* susceptibilities of *E. coli* cultured from epididymides after treatment had not changed from MICs observed in pretreatment studies for all antibiotics.

Histopathology

Formalin-fixed tissues were dehydrated in graded alcohols, embedded in paraffin, and cut into 4 µm-thick sections. Mounted sections were stained for light microscopy with haematoxylin and eosin. Sections from the midline of both epididymides were examined in a blinded fashion for histopathological changes. Histopathological parameters were graded from 0 to 5 (0, not present; 5, most severe). All specimens were examined at least four times by the same investigator. Repeated examinations revealed minimal intraindividual variability in scoring.

Abscess formation, inflammation, fibrosis and sperm granulomata (Figures 48.1–48.5) were the most frequent histopathological findings in the epididymides of both treated and untreated rats (Table 48.3).

These changes were most prominent in the epididymides of the untreated control animals. In contrast, mean severity scores for abscess formation, inflammation, and sperm granulomata of the ofloxacin-treated animals were significantly lower than those of the control animals. Doxycycline had a significant effect only on the severity score of abscess formation. Pathology scores of the cefotaxime-treated rats were not different from those of the control animals. No treatment regimen had a significant effect on the occurrence of sperm granulomata. Histopathological changes were not observed in any of the left (uninfected) epididymides.

The numbers of bacteria recovered from the epididymides were significantly associated with the severity scores for abscess formation ($r = 0.352$; $p < 0.001$), inflammation ($r = 0.453$; $p < 0.001$), fibrosis ($r = 0.356$; $p < 0.001$), and sperm granulomata formation ($r = 0.280$; $p < 0.001$).

Table 48.3 Association between therapeutic regimen and histopathological findings (from Vieler *et al.*, 1993); severity scores graded from 0 (absent) to 5 (most severe).

Therapeutic regimen	*Histopathological findings (mean severity score/ no. of organs positive [%])*			
	Abscess	*Inflammation*	*Fibrosis*	*Granuloma*
Ofloxacin	0.38/13*	1.13/79*	1.00/63*	1.30/67
Cefotaxime	0.52/24	1.84/88	1.68/88	1.72/76
Doxycycline	0.28/12*	1.48/76	1.52/84	1.76/72
Saline	1.06/48	2.08/92	1.96/92	1.60/64

* Significantly lower than the control group ($P < 0.05$).

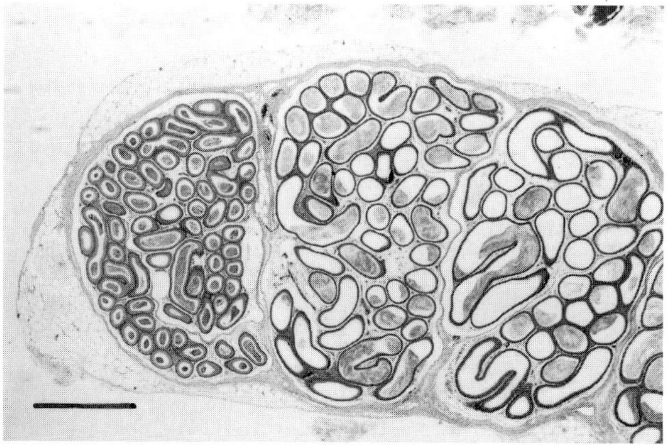

Figure 48.1 Epididymis of an uninfected rat. Normal tissue morphology. Bar, 1000 µm.

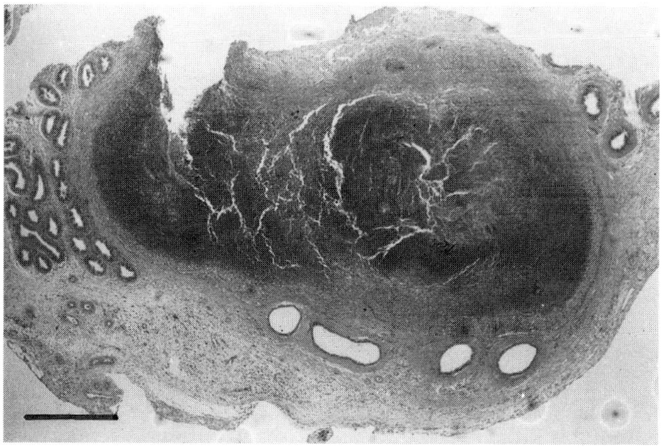

Figure 48.2 Epididymitis due to *E. coli* infection. Large abscess with complete destruction of epididymal tissue. Bar, 1000 µm.

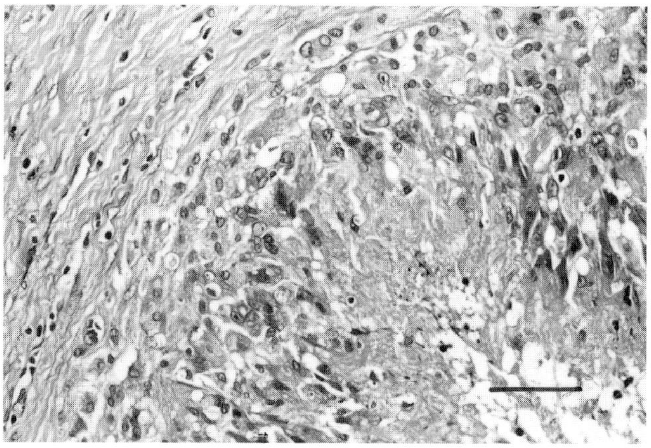

Figure 48.3 Epididymitis due to *E. coli* infection. Granulomatous inflammatory reaction with macrophages resembling epithelioid cells. Bar, 50 µm.

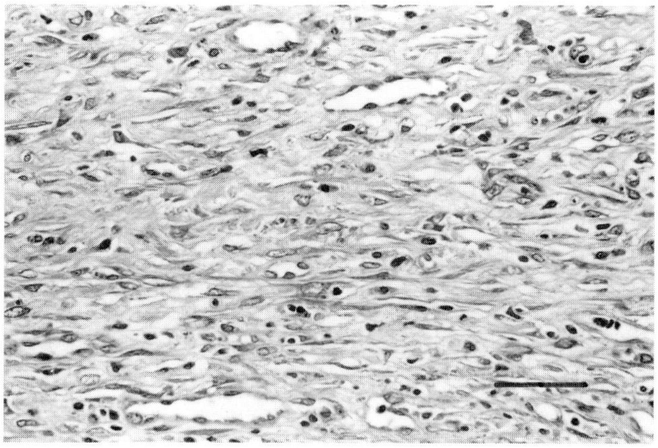

Figure 48.4 Epididymitis due to *E. coli* infection. Increase of fibrous tissue, numerous fibroblasts and proliferating capillaries. Bar, 50 µm.

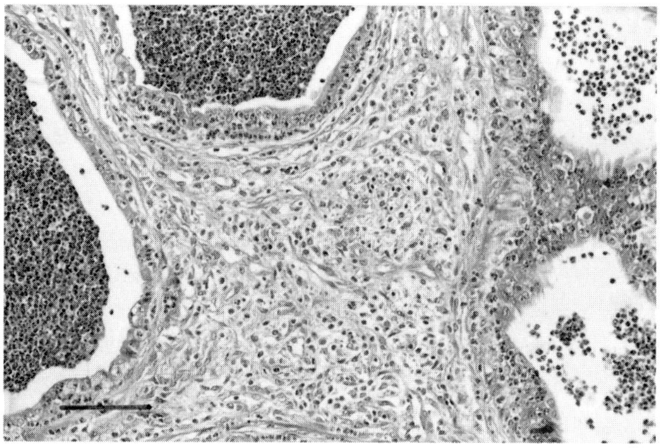

Figure 48.5 Epididymitis due to *E. coli* infection. Chronic inflammatory reaction in transition to fibrosis. Lumina of ductus epididymidis filled with polymorphonuclear granulocytes. Bar, 100 µm.

Critical review of the model

In our model of epididymitis, the natural way of infection by ascension from the urethra was simulated by retrograde injection of the bacterial inoculum into the vas deferens. Epididymitis could be induced in all experimental animals and, according to clinical, bacteriological and histopathological criteria, strongly resembled human disease.

Our model can readily be learned by a skillful experimentor who is able to puncture a tiny vessel with a small-gauge needle.

The results of the experimental therapeutic studies are in agreement with those of previous clinical trials with ofloxacin. In an open study, ofloxacin therapy of epididymitis due to common uropathogens or *Chlamydia trachomatis* cured 17 of 18 patients (Melekos and Asbach, 1987). In a similar trial, bacteriological cure, as determined by culture of midstream urinary tract specimens or urethral swabs, was achieved with ofloxacin in 16 of 20 patients. However, in six patients clinical symptoms were still present at 10 weeks after completion of therapy (Weidner *et al.*, 1987).

An unexpected finding in our study was the therapeutic failure of cefotaxime, which, despite good *in vitro* activity and tissue penetration, did not significantly reduce bacterial counts. This observation cannot be explained by selection of resistant mutants, since MICs of *E. coli* before and after treatment were identical. We assume that our dosing scheme, which was designed to achieve peak blood levels in experimental animals that mimic those in humans, might have been inappropriate in experimental animals, where clearance of antibiotics is more rapid than in humans. The bacteriological efficacy *in vivo* of β-lactam antibiotics is mainly influenced by the time during which serum levels exceed MIC. Therefore, shorter dosing intervals or continuously administered doses of cefotaxime could have been more appropriate and effective. In addition, poor penetration of the drug into epididymal abscesses might explain the failure of cefotaxime, which, in a similar experimental study, was effective only when administered before or immediately after the infection of rats (Nielsen, 1987).

Doxycycline was less effective than ofloxacin with respect to cure rates but treatment reduced *E. coli* numbers in the epididymides significantly as compared to controls.

In a similar rat model of *E. coli* epididymitis, tissue penetration of a single oral dose of sparfloxacin has been studied (Ludwig *et al.*, 1997).

Experimental chlamydial epididymitis could be induced in Wistar rats by injecting *Chlamydia trachomatis*, biovar mouse pneumonitis (Jantos *et al.*, 1992), into the vas deferens, and *Chlamydia psittaci*, guinea-pig inclusion conjunctivitis agent (Jantos *et al.*, 1989). Clinical, microbiological, immunological, and histopathological findings (Figures 48.6–48.8) in infected rats were found to closely resemble those seen in human disease.

Experimental *E. coli* epididymitis could also be induced in mice and allowed the detailed analysis of immunocompetent cells involved, i.e. the appearance of macrophages, MHC class II positive cells, T lymphocytes, and plasma cells, in epididymitis (Nashan *et al.*, 1993).

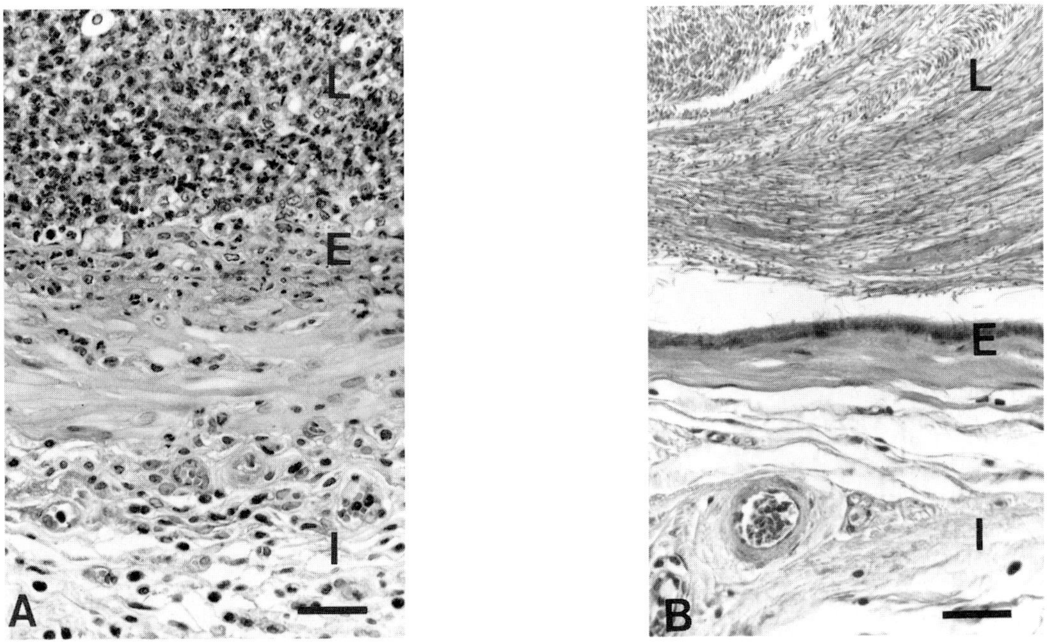

Figure 48.6 Epididymitis due to *C. trachomatis* infection. (**A**) Intratubular abscessing and mild interstitial mononuclear cellular infiltration at 7 days p.i. (**B**) Epididymis of a control rat. Normal morphology. L, lumen of the epididymal duct; E, epididymal epithelium; I, interstitium. Bars, 40 μm. (Reproduced from Jantos *et al.*, 1992, with permission from the American Society for Microbiology.

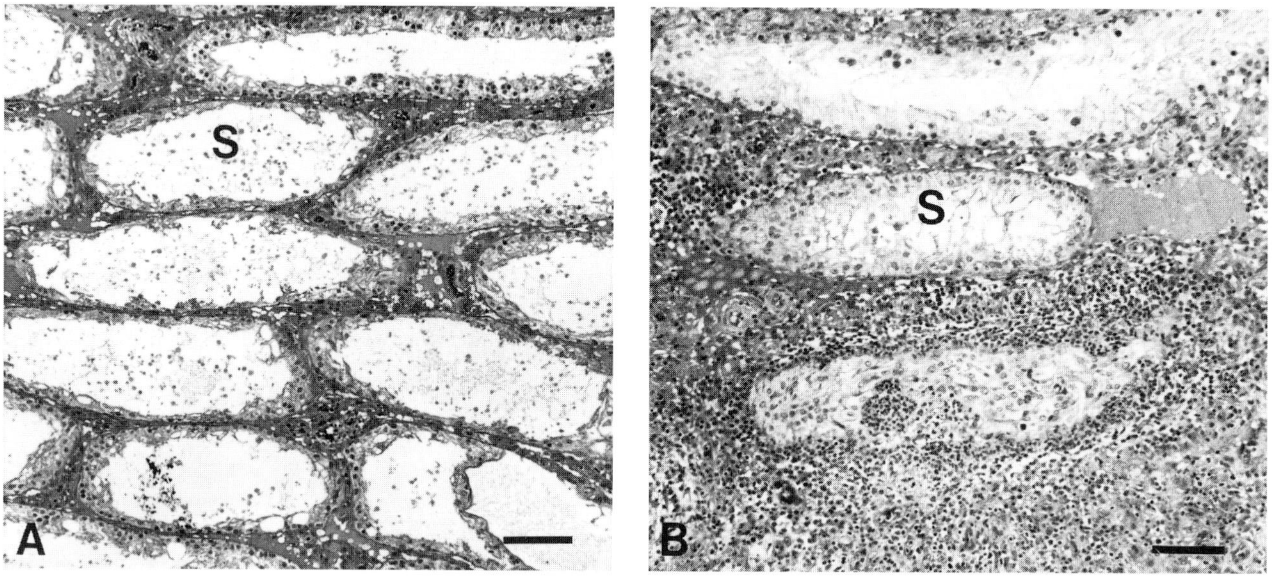

Figure 48.7 Epididymitis due to *C. trachomatis* infection. (**A**) Testis of a rat with ipsilateral *C. trachomatis* epididymitis at 30 days p.i. Severe loss of spermatogenic epithelium. (**B**) Testis of a control rat. No reduction of spermatogenesis. S, seminiferous tubule. Bars, 300 μm. (Reproduced from Jantos *et al.*, 1992, with permission from the American Society for Microbiology.)

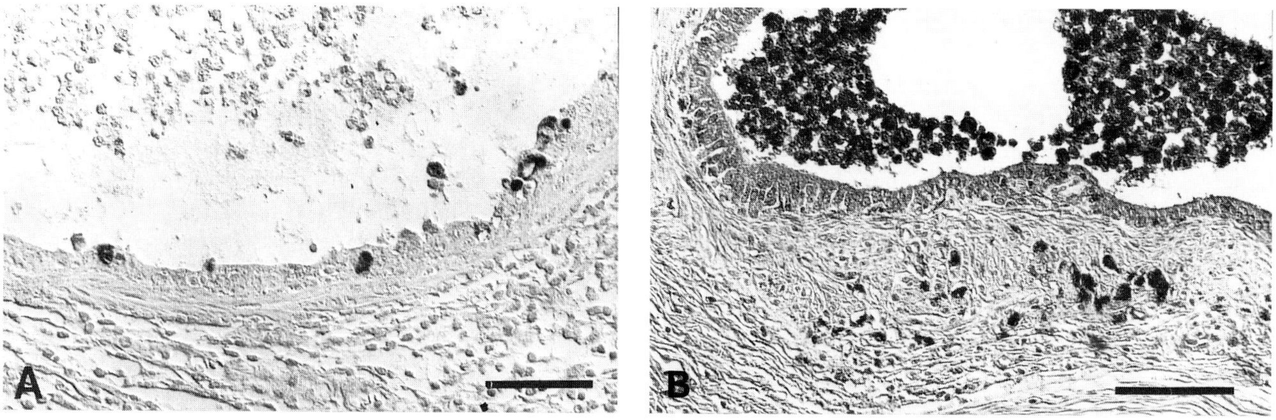

Figure 48.8 Epididymitis due to *C. trachomatis* infection. (**A**) Immunoperoxidase reaction at 7 days p.i. Darkly staining chlamydial antigen in inclusion bodies within epididymal epithelium and within single intratubular macrophages. (**B**) Immunoperoxidase reaction at 30 days p.i. Darkly staining chlamydial antigen in numerous intratubular macrophages and in macrophages within the surrounding interstitium. Bars, 50 μm. (Reproduced from Jantos *et al.*, 1992, with permission from the American Society for Microbiology.)

References

Berger, R. E. (1999). Acute epididymitis. In: *Sexually Transmitted Diseases* (eds Holmes, K. K., Mårdh, P.-A., Sparling, P. F., *et al*), 3rd edition, ch. 61, pp 847–858. McGraw-Hill, New York.

Dathe, G., Vogt, P. (1985). Effekt einer Antibiotikatherapie bei experimenteller, bakterieller Epididymitis. In: *Experimentelle Urologie* (eds Harzmann, R., Jacobi, H. G., Weißbach, L.), pp. 247–250. Springer Verlag, Berlin.

Graves, R. S., Engel, W. J. (1950). Experimental production of epididymitis with sterile urine; clinical implications. *J. Urol.*, **64**, 601–613.

Hackett, R. A., Huang, T. W., Berger, R. E. (1988). Experimental *Escherichia coli* epididymitis in rabbits. *Urology*, **32**, 236–240.

Jantos, C., Krauss, H., Altmannsberger, M., Thiele, D., Weidner, W., Schiefer, H. G. (1989). Experimental chlamydial epididymitis. *Urol. Int.*, **44**, 279–283.

Jantos, C., Baumgärtner, W., Durchfeld, B., Schiefer, H. G. (1992). Experimental epididymitis due to *Chlamydia trachomatis* in rats. *Infect. Immun.*, **60**, 2324–2328.

Lucchetta, R., Clavert, A., Meyer, J. M., Bollack, C. (1983). Acute experimental *Escherichia coli* epididymitis in the rat and its consequences on spermatogenesis. *Urol. Res.*, **11**, 117–120.

Ludwig, M., Jantos, C. A., Wolf, S. *et al.* (1997). Tissue penetration of sparfloxacin in a rat model of experimental *Escherichia coli* epididymitis. *Infection*, **25**, 178–184.

Melekos, M. D., Asbach, H. W. (1987). Epididymitis: aspects concerning etiology and treatment. *J. Urol.*, **138**, 83–86.

Møller, B. R., Mårdh, P. A. (1980). Experimental epididymitis and urethritis in Grivet monkeys provoked by *Chlamydia trachomatis*. *Fertil. Steril.*, **34**, 275–279.

Nashan, D., Jantos, C., Ahlers, D. *et al.* (1993). Immuno-competent cells in the murine epididymis following infection with *Escherichia coli*. *Int. J. Androl.*, **16**, 47–52.

Nielsen, O. S. (1987). An experimental study of the treatment of bacterial epididymitis. *Scand. J. Urol. Nephrol.*, **104** (Suppl.), 115–117.

See, W. A., Taylor, T. O., Mack, L. A., Tartaglione, T. A., Opheim, K. E., Berger, R. E. (1990). Bacterial epididymitis in the rat: a model for assessing the impact of acute inflammation on epididymal antibiotic penetration. *J. Urol*, **144**, 780–784.

Vieler, E., Jantos, C., Schmidts, H. L., Weidner, W., Schiefer, H. G. (1993). Comparative efficacies of ofloxacin, cefotaxime, and doxycycline for treatment of experimental epididymitis due to *Escherichia coli* in rats. *Antimicrob. Agents Chemother*, **37**, 846–850.

Weidner, W., Schiefer, H. G., Garbe, C. (1987). Acute non-gonococcal epididymitis. Aetiological and therapeutic aspects. *Drugs*, **34** (Suppl. 1), 111–117.

Weidner, W., Prudlo, J., Schiefer, H. G., Jantos, C., Altmannsberger, M., Aumüller, G. (1989). *Escherichia coli*-Epididymitis der Ratte: ein tierexperimentelles Modell der akut-eitrigen und chronisch-obstruktiven Entzündung. *Fertilität*, **5**, 151–155.

Chapter 49

Mouse Model of *Mycoplasma* Genital Infections

D. Taylor-Robinson and P. M. Furr

Introduction

Mycoplasmas are the smallest organisms capable of self-replication. Their small size (about 300 nm in diameter) is attributable to the presence of only a very small number of copies of even the most important genes. They differ from conventional bacteria in this respect and also in not having a rigid cell wall, being contained only by the plasma membrane. As a consequence, they are insensitive to antibiotics that act on bacterial cell walls, such as penicillins, but are sensitive to broad-spectrum antibiotics.

Although mycoplasmas tend to be species-specific, the use of sex hormones is known to predispose mice to genital-tract colonization with mycoplasmas of human origin as well as those of murine origin. The procedures required to induce such susceptibility and those required for the experimental inoculation of the genital tract are outlined. Since naturally occurring mycoplasmas should be excluded prior to the use of mice for experimental infection, consideration will be given to the techniques for the isolation of such organisms as well as those introduced experimentally. Finally, the methods involved in the administration of antibiotics and the effect they have on genital tract mycoplasmal infection in immunocompetent and immunosuppressed mice are presented.

Background of human infection

The first isolation of a mycoplasma from humans, reported about 60 years ago (Dienes and Edsall, 1937), was from the genital tract. Subsequently, 15 other genomically distinct mycoplasmas have been isolated from humans (Taylor-Robinson, 1996). However, about half of them appear to be commensals in that a pathogenic potential has not been demonstrated for them. Those mycoplasmas for which an association with disease has been shown consistently are presented in Table 49.1, together with their preferred site of colonization and metabolic activity.

Table 49.1 Primary sites of colonization and metabolism of mycoplasmas associated with disease in humans

	Primary site of colonization		*Metabolism of*		
Mycoplasma	*Genitourinary tract*	*Oropharynx*	*Glucose*	*Arginine*	*Association with disease*
Mycoplasma fermentans	+/–*	+	+	+	Detected in joints in arthritides and in lungs in HIV infection
Mycoplasma genitalium	+	+/–	+	–	A cause of acute and chronic non-gonococcal urethritis (NGU)
Mycoplasma hominis	+	+/–	–	+	A possible cause of pelvic inflammatory disease; causes infections in immunodeficiencies
Mycoplasma penetrans	+	–	+	+	Associated serologically with HIV infection
Mycoplasma pneumoniae	+/–	+	+	–	A cause of atypical pneumonia and sequelae
Mycoplasma salivarium	–	+	–	+	Usually non-pathogenic but causes arthritis in hypogammaglobulinaemia
Ureaplasma urealyticum†	+	+	–	–	A probable cause of acute NGU; causes chronic NGU, and arthritis in hypogammaglobulinaemia; detected in joints in inflammatory arthritides

* +/–, primary site occasionally.
† Metabolizes urea.

Handbook of Animal Models of Infection
ISBN 0-12-775390-7

The model described here has been developed in female mice. It is pertinent to point out, therefore, that, in women, there is some serological evidence to suggest that *Mycoplasma genitalium* may be a cause of some cases of acute pelvic inflammatory disease (Møller *et al.*, 1984). Furthermore, *M. hominis* has also been linked to pelvic inflammatory disease, but it is most strongly associated with bacterial vaginosis (Rosenstein *et al.*, 1996). *Ureaplasma urealyticum* organisms (ureaplasmas) are associated with bacterial vaginosis but less strongly. The proposal that the development of bacterial vaginosis is influenced by changing concentrations of female sex hormones (Taylor-Robinson and Hay, 1997) is relevant to the hormone-dependent mouse model of mycoplasmal infection.

Background of the model

Mycoplasma pulmonis is a respiratory pathogen of mice and rats (Cassell and Hill, 1979), but some strains are also arthritogenic and others have been recovered from the genital tract (Casillo and Blackmore, 1972). Experimentally, *M. pulmonis* will infect the genital tract of a proportion of female mice that have not had prior hormone treatment, but treatment with progesterone was found to greatly enhance their susceptibility (Furr and Taylor-Robinson, 1984, 1993a). This is a phenomenon similar to that of enhanced susceptibility of female mice to *Chlamydia trachomatis* induced by prior treatment with progesterone (Tuffrey and Taylor-Robinson, 1981). Later, it was discovered that the susceptibility of the genital tract of female mice to some other mycoplasmas, including *M. hominis* and *U. urealyticum*, was dependent on pretreatment with oestradiol (Furr and Taylor-Robinson, 1989a,b). The contrasting effects of the two hormones have been summarized previously (Furr and Taylor-Robinson, 1993b) and are presented in Table 49.2.

Mouse strains used

Various strains of female mice, 6–8 weeks old, have been used, most often BALB/c, TO and CBA strains. It is noteworthy that the susceptibility of mice of these three strains, following hormone treatment, to genital-tract colonization with *U. urealyticum* was different: mice of the BALB/c strain were more susceptible than those of the TO or CBA strains (Furr and Taylor-Robinson, 1989b). This implies that predisposition to colonization with ureaplasmas is, at least partially, under genetic control. Following intravaginal inoculation of *M. hominis*, larger numbers of organisms were recovered from TO mice than from BALB/c mice, but otherwise the pattern of colonization in the two strains of mouse was similar, the organisms persisting for 7–8 months (Furr and Taylor-Robinson, 1989a). The susceptibility of the mouse strains to *M. pulmonis* was similar (Furr and Taylor-Robinson, 1993a). The susceptibility of the mice to other mycoplasmas may vary according to the strains of mouse used, but has not been evaluated.

Preparation of the animals

Housing conditions

It is wise to maintain the mice in a suite designed for the containment of infectious agents. Otherwise, no special care is required, mice being boxed usually in groups of five and receiving a continuous supply of rodent diet and fresh water.

Screening for indigenous mycoplasmas

Screening prior to experimental infection to ensure that the mice are free of mycoplasmal organisms, especially

Table 49.2 The contrasting effects of hormones on colonization of the genital tract of female BALB/c mice by various mycoplasmal species

Mycoplasmal species	*Natural host*	*Substrate metabolized*	*Terminal structure*	*Colonization when treated with*	
				Progesterone	*Oestradiol*
M. hominis	Human	Arginine	–	–	+
M. salivarium	Human	Arginine	–	–	+
U. urealyticum	Human	Urea	–	–	+
M. fermentans	Human	Gluc/Arg	–	–	+
M. pirum	?Human	Gluc/Arg	+	–	+
M. penetrans	Human	Glucose	+	–	–
M. pneumoniae	Human	Glucose	+	+	–
M. genitalium	Human	Glucose	+	+	–
M. pulmonis	Murine	Glucose	+	+	–
M. neurolyticum	Murine	Glucose	–	–	–
M. gallisepticum	Avian	Glucose	+	–	–
Acholeplasma laidlawii	Ubiquitous	Glucose	–	–	–

Table 49.3 Naturally occurring mycoplasmal species isolated from laboratory rodents

	Indicated mycoplasma recovered from				
Mycoplasma	*Mouse*	*Rat*	*Guinea-pig*	*Hamster*	*Rabbit*
M. pulmonis	+	+	+	+	+
M. neurolyticum	+	+			
M. collis	+	+			
M. muris	+				
M. alvi	+				
M. arthritidis	+	+			
M. caviae		+	+		
Acholeplasma laidlawii		+	+		
A. cavigenitalium			+		
"Bovine group 7"			+		
M. oxoniensis				+	
A. multilocale					+

those to be used in the experimental procedure, is of obvious importance. In doing so, it is essential to examine not only the genital tract but also other anatomical sites. An existing mycoplasmal infection at an extragenital site might result in the spread of organisms to the genital tract or could render the mice refractory to genital colonization with the same mycoplasma (Furr and Taylor-Robinson, 1993c). *M. pulmonis* is the indigenous mycoplasma most likely to be met in mice, but vigil should be kept for others shown in Table 49.3. Most mycoplasmas are animal-species-specific, but occasionally they cross the species barrier and are recovered from closely related animals. This might occur especially in an animal house where there is close proximity, so it is helpful to appreciate the various mycoplasmas that have been isolated from rodents and the range of animals from which each has been recovered (Table 49.3), as an indication of what might be expected when screening.

Hormone treatment

As mentioned previously (Table 49.2), hormone treatment has a key role in predisposing female mice to colonization of the genital tract by various mycoplasmas. Progesterone or oestrogen is used and these hormones are given in the following regimes: 150 mg/ml of progesterone (Depo-Provera, Pharmacia & Upjohn, Milton Keynes, UK) is diluted 1 : 12 in isotonic saline and 0.2 ml (2.5 mg) is administered subcutaneously on four occasions at weekly intervals using a 1 ml syringe and 25 G × 5″ needle; or 5 mg/ml of oestradiol benzoate (Intervet UK, Cambridge, UK) is injected in a similar regime to that for progesterone, with each mouse receiving 0.1 ml (0.5 mg) subcutaneously on four occasions at weekly intervals.

Assessment of stage of reproductive cycle

This requires a vaginal cytological examination and is undertaken as follows. A nasopharyngeal swab (MW 142; Medical Wire and Equipment Co., Corsham, UK) is inserted in the vagina and rotated to abrade the epithelial mucosa. The swab is rolled along a 7.5 × 2.5 cm glass microscope slide to transfer the material for cytological examination, and the slide is left to dry at room temperature. Alternatively, a vaginal washing (see below) may be used. The cellular material is fixed by immersing the slide in methanol for 30 minutes. After fixation, slides are transferred to freshly prepared Giemsa stain for a further 30 minutes and then they are washed briefly in 30% methanol, followed by buffered distilled water, pH 6.8, and finally deionized water. The smears are dried at room temperature and are observed microscopically using a 10 × or 40 × objective.

The cellular contents of the slides are recorded and the stage of the murine reproductive cycle is determined by assessing whether the criteria of Rugh (1968) are fulfilled, as follows:

1. dioestrus: almost exclusively polymorphonuclear leukocytes (PMNL);
2. pro-oestrus: PMNL and nucleated epithelial cells in approximately equal proportions;
3. early oestrus: clearly defined squamous epithelial cells, some nucleated;
4. oestrus: almost exclusively enucleated squamous epithelial cells;
5. postoestrus: large epithelial cells with translucent nuclei and PMNL in approximately equal proportions.

The effect of the progesterone is to arrest the reproductive cycle in dioestrus, dominated by PMNL, while oestradiol maintains the oestrus stage characterized by enucleated squamous cells.

Preparation and storage of inocula

Mycoplasma strains

The mycoplasmas that have been shown to colonize the genital tract of female mice and their natural hosts of origin are shown in Table 49.2. The various strains are grown in the media described below and the number of organisms within each culture is determined, also as described below, before aliquots of a culture are stored at –70°C for use.

Media

The ability to culture mycoplasmas depends largely on the quality of the medium used and it will probably never be possible to produce a single medium in which it is possible to culture every mycoplasmal species. SP4 medium (Tully *et al.*, 1979) is more sensitive than other media for the isolation of some mycoplasmas but "Edward" broth medium (Freundt, 1983) is a useful medium for routinely culturing most mycoplasmas for experimental inoculation of animals and for isolating them thereafter. A litre of this medium is composed of PPLO broth base (Difco Labs, West Molesey, UK; 2.1% wt/vol.), 700 ml; benzylpenicillin (Glaxo Wellcome, Uxbridge, UK; 1000 IU/ml), 10 ml; thallous acetate (British Drug Houses (BDH), Dagenham, UK; 2.5% wt/vol.), 20 ml (often omitted or reduced in concentration, especially when culturing ureaplasmas); yeast extract (Distillers' Co., Burgess Hill, UK) (25% wt/vol.), 100 ml; horse serum (Imperial Labs, Andover, UK), 200 ml; and phenol red (BDH; 1%), 20 ml. This is regarded as a standard liquid medium (SLM), which is suitable for transportation and storage of specimens and to which may be added one of the following three substrates for culture of mycoplasmas; glucose (BDH; 10% wt/vol.), 10 ml; L-arginine hydrochloride (Sigma Chemical Co., Poole, UK; 10% wt/vol.), 20 ml; or urea (BDH; 10% wt/vol.), 10 ml. By the addition of normal sodium hydroxide or normal hydrochloric acid, the pH of the medium is adjusted to 7.8 for glucose-containing, 7.0 for arginine-containing, or 6.5–6.8 for urea-containing media, respectively.

Estimation of number of mycoplasmal organisms

In order to determine the number of organisms present in a culture, or in a specimen taken from a mouse, a series of 10-fold dilutions is prepared. This may be done in any volume, but 0.2 ml diluted in 1.8 ml of the required medium up to a dilution of 10^{-8} is convenient. The dilutions are contained in screw-cap vials of 2.5 ml capacity. Air-tight caps are required to prevent diffusion of gases from the medium causing non-specific colour changes. The vials are incubated aerobically at 37°C and observed for an alteration in the pH of the medium as denoted by a colour change from red to yellow in glucose-containing medium or from yellow to red in arginine- or urea-containing media. This denotes mycoplasmal growth and the last vial in which it occurs is deemed to contain 1 colour-changing unit. If, in examining murine specimens, quantitation is not required and it is necessary only to determine that mycoplasmas are present, it is still advisable to dilute the sample to at least 10^{-3}. The reasons for this are several, and are outlined elsewhere (Taylor-Robinson, 1989), the dilution of inhibitors being the principal one.

After experimental inoculation of animals with a known mycoplasma, it may not seem reasonable to confirm that all colour changes are due to mycoplasmal growth. The situation is different, however, in a screening programme, where a colour change cannot be assumed to be due to a mycoplasma and its existence is usually confirmed by seeking colonies on agar medium. For this purpose, mycoplasmal agar (Oxoid Unipath Ltd, Basingstoke, UK; 1% wt/vol.) is prepared in the PPLO broth base and the remaining components, as described earlier, are added aseptically. Agar plates are incubated under microaerophilic conditions (5% CO_2 + 95% N_2), usually at 36.5–37.5°C. All cultures should be inspected regularly and not discarded as negative for at least 3 weeks. Some acholeplasmas, such as *Acholeplasma multilocale* from guinea-pigs, do not metabolize glucose or arginine and their presence should be confirmed by subculture from broth to agar medium at 2- to 3-day intervals. Strict anaerobic conditions are required for the isolation of the arginine-metabolizing murine species *M. muris* (McGarrity *et al.*, 1983), and similar conditions are needed for a glucose-fermenting mycoplasma, resembling *M. alvi*, isolated from the gastrointestinal tract of wild mice (Gourlay and Wyld, 1976). Isolated mycoplasmas are identified specifically according to standard methods (Taylor-Robinson, 1989).

Infection process

Mice are used that, as a result of the screening process, are considered to be free of the mycoplasma to be inoculated. The stage of the reproductive cycle is checked, as described previously, and the organisms are given at a time that coincides with the second injection of hormone. The aliquot of organisms, stored at –70°C, is thawed rapidly and kept on wet ice during the inoculation procedure. Each mouse is held on its back in the palm of one hand of the operator, and 50 μl of the inoculum is introduced into the vagina using a Finnpipette (Labsystems; Jencons Scientific, Leighton Buzzard, UK). The mouse is held in this position for a further few seconds to reduce seepage of the inoculum from the vagina, and then returned to its box.

Monitoring infection

Infection does not cause death. The ability of the mycoplasma to colonize the genital tract, the stage of the reproductive cycle and the effect of antimicrobial therapy are assessed at intervals, perhaps weekly, by taking swab specimens from the vagina for isolation of organisms and for vaginal cytology. Mice may be sacrificed at intervals throughout the experiment and/or at its termination to determine whether mycoplasmas have spread to the upper genital tract. In addition, urine may be aspirated directly from the bladder to determine if urinary tract involvement has occurred (Taylor-Robinson and Furr, 1986).

Swab specimens

Plain cotton swabs are preferred to calcium-alginate- or serum-coated swabs, because they are cheaper to purchase and, more important, contain fewer inhibitors (Furr and Taylor-Robinson, unpublished observations). For collection of vaginal swab specimens or, indeed, throat swab specimens from mice, a nasopharyngeal swab, as described previously, is used. The swab is inserted and rotated to abrade the epithelial mucosa and the cellular contents of the swab are expressed in the appropriate broth (1.8 ml, for example) by squeezing the swab against the side of the vial. The swab is then discarded as the stick or cotton wool may contain resins that are toxic for the mycoplasmas. This sample, now regarded as a 10^{-1} dilution, should be transported rapidly, preferably on wet ice, to the laboratory and cultured or stored at –70°C until required.

Vaginal washings

These specimens may be collected for several reasons, which govern the choice of fluid with which the vagina is washed. A washing may be collected for mycoplasmal culture, in which case SLM or a complete mycoplasmal medium is used. If the washing is to be used for assessing the stage of the murine reproductive cycle, or for measuring local mycoplasmal antibody, then isotonic saline or phosphate-buffered saline is required. Vaginal washing is achieved by introducing 50 µl of the appropriate fluid into the vagina using a Finnpipette, withdrawing the fluid and reintroducing it twice more before collection. The volume of fluid used is determined by the size of the animal and may be increased to 100 µl for a fully adult mouse. Repeat samples may be obtained at regular intervals from the same mouse and there is evidence, from introducing methylene blue dye (P. M. Furr, personal observation), that the fluid does not pass the cervix.

Upper genital tract washings

Washings may be taken from the uterus after the mouse has been sacrificed (see below) using a technique similar to that described for the perfusion of mouse lungs (Taylor-Robinson *et al.*, 1972). By introducing a small cannula through the cervix and clamping it in place, the uterine horns may be aspirated with fluid, as described for vaginal washing, which is then used for the isolation of mycoplasmas or for antibody measurement. Also, histological fixative may be introduced.

Tissue samples

Genital samples for cultural or histological examination are taken at post-mortem. The animal is killed by administration of a lethal dose of barbiturate or by cervical dislocation. Exsanguination is undertaken, preferably by cardiac puncture, although, in the mouse, cutting blood vessels in the axilla may be preferred. The fur is swabbed with 70% ethanol and allowed to dry and an incision is made into the body cavity using sterile instruments, with aseptic technique being observed throughout. The tissue samples, such as uterine horns and ovaries, are removed and may be divided, one piece being placed in a preweighed sterile container and the other in an aliquot of formol saline or other fixative for histology. Samples for culture are weighed and a 10% (wt/vol.) suspension, regarded as a 10^{-1} dilution, is prepared by mascerating the tissue in a Ten Broeck tissue grinder. If electrically operated blenders are to be used, care should be taken to keep the vessel containing the tissue in wet ice, as the heat generated may have a deleterious effect on the mycoplasmas.

Antimicrobial therapy

In vitro susceptibility

As mentioned initially, mycoplasmas are susceptible to a variety of broad-spectrum antibiotics, most of which only inhibit their multiplication and do not kill them. The tetracyclines have a long history of usage, particularly for genital-tract infections, but erythromycin and the newer macrolides (except in the case of *M. hominis*), the ketolides and the newer quinolones have equal or sometimes greater activity. The two latter antibiotic groups also have some cidal activity. Detailed antibiotic susceptibility profiles of five mycoplasmas, four of which are found in the human genital tract, have been presented elsewhere (Taylor-Robinson and Bébéar, 1997). Some mycoplasmas may develop resistance, either by gene mutation or by acquisition of a resistance gene, to antibiotics to which they are usually sensitive. Resistance to tetracyclines is common and due to acquisition of the *tetM* gene. The antibiotic susceptibilities of mycoplasmas may be determined *in vitro* by two basic methods, the agar dilution method and the broth dilution method, usually in the form of the metabolism-inhibition test (Taylor-Robinson and Bébéar, 1997).

Table 49.4 The effect of treating female mice with various antibiotics on intravaginal colonization with *M. pulmonis* or *M. hominis*

Mouse strain	Mycoplasma	Antibiotic	Dose*	Frequency	Duration (days)	Days since end of treatment before test	No. of mice with vaginal organisms: Before treatment	No. of mice with vaginal organisms: After treatment	Titre of vaginal organisms: Before treatment	Titre of vaginal organisms: After treatment
TO	*M. pulmonis*	Oxytetracycline†	6 mg	Daily	5	7	10	1	10^4–10^8 (GM 1.5×10^5)	5×10^3
BALB/c	*M. pulmonis*	Oxytetracycline	6 mg	Daily	4	10	9	1	10^2–10^7 (GM 3.1×10^4)	1.2×10^0
BALB/c nude	*M. pulmonis*	Oxytetracycline	6 mg	Daily	4	10	8	8	10^5–10^7 (GM 3.9×10^6)	$> 10^3$
BALB/c	*M. hominis*	Oxytetracycline	6 mg	Daily	4	10	8	0	10^6–10^8 (GM 1.9×10^5)	
TO	*M. pulmonis*	Lymecycline	6 mg	Daily	10	10	18	1	10^5–10^7 (GM 1.4×10^6)	10^2
TO	*M. pulmonis*	Terramycin	0.3 mg	Daily	10	14	6	1	10^1–10^6 (GM 2.5×10^2)	5×10^3
TO	*M. pulmonis*	Terramycin	3 mg	Daily	10	14	17	3	10^3–10^7 (GM 1.2×10^6)	10^2–10^5 (GM 3.8×10^0)
BALB/c	*M. pulmonis*	Enrofloxacin‡	25 μg	Twice daily	5	12	10	10	10^6–10^8 (GM 2.5×10^6)	10^7–10^8 (GM 3×10^7)

* All antibiotics given subcutaneously.
† *In vitro* MIC 0.3–0.6 μg/ml.
‡ *In vitro* MIC 0.01 μg/ml.

Antimicrobials

The antibiotics that have been tested in TO and BALB/c strains of mice for their efficacy in clearing mycoplasmas from the genital tract are shown in Table 49.4.

Preparation and administration

Antibiotics were prepared as a solution in distilled water and were given subcutaneously in a volume of 0.1 ml. The dose, frequency and duration of administration are shown in Table 49.4.

Monitoring response to therapy

Since the model is one of mycoplasmal colonization, rather than disease production, the effect of antibiotic therapy that is sought is elimination of organisms from the genital tract. This is determined by taking intravaginal swab specimens at appropriate times after the conclusion of a course of therapy and undertaking quantitative estimations of the number of mycoplasmal organisms present, as described previously. In addition, the upper genital tract may be examined by taking specimens at autopsy.

Pitfalls and advantages of the model

The major pitfall is the one usually ascribed to small animal models, namely the likelihood that it does not fully reflect the human situation. However, from the point of view of assessing the effect of antibiotics, the results obtained so far are in keeping with expectations, namely that tetracyclines are effective in eliminating mycoplasmas from the genital tract.

The model has the obvious advantage that, if necessary, intravaginal swab specimens may be taken at frequent intervals and drugs in combination may be tested, as well as new drugs prior to clinical trials. In addition, the model has provided the opportunity to underline the importance of a competent immune system in eradicating mycoplasmas from the genital tract (Furr *et al.*, 1994; Franz *et al.*, 1997). This is sometimes a major problem in subjects who are hypogammaglobulinaemic. As shown in Table 49.4, oxytetracycline that was instrumental in eradicating mycoplasmas from the genital tract of immunocompetent mice failed to do so when it was administered to their nude counterparts. This model may be exploited further because it lends itself to an assessment of the exact cellular immunological defect that is important in the failure to eradicate and also allows examination of the value of administering high-titre specific mycoplasmal antibody to aid eradication, an approach that seems to have been of value in the human situation (Furr *et al.*, 1994; Franz *et al.*, 1997).

References

Casillo, S., Blackmore, D. K. (1972). Uterine infections caused by bacteria and mycoplasma in mice and rats. *J. Comp. Pathol.*, **82**, 477–482.

Cassell, G. H., Hill, A. (1979). Murine and other small animal mycoplasmas. In: *The Mycoplasmas* (eds Tully, J. G., Whitcomb, R. F.), vol. 2, pp. 235–273. Academic Press, New York.

Dienes, L., Edsall, G. (1937). Observations on the L-organism of Klieneberger. *Proc. Soc. Exp. Biol. Med.*, **36**, 740–744.

Franz, A., Webster, A. D. B., Furr, P. M., Taylor-Robinson, D. (1997). Mycoplasmal arthritis in patients with primary immunoglobulin deficiency: clinical features and outcome in 18 patients. *Br. J. Rheumatol.*, **36**, 661–668.

Freundt, E. A. (1983). Culture media for classic mycoplasmas. In: *Methods in Mycoplasmology* (eds Razin, S., Tully, J. G.) vol. 1, pp. 127–135. Academic Press, New York.

Furr, P. M., Taylor-Robinson, D. (1984). Enhancement of experimental *Mycoplasma pulmonis* infection of the mouse genital tract with progesterone. *J. Hyg.*, **92**, 139–144.

Furr, P. M., Taylor-Robinson, D. (1989a). Oestradiol-induced infection of the genital tract of female mice by *Mycoplasma hominis*. *J. Gen. Microbiol.*, **135**, 2743–2749.

Furr, P. M., Taylor-Robinson, D. (1989b). The establishment and persistence of *Ureaplasma urealyticum* in oestradiol-treated female mice. *J. Med. Microbiol.*, **29**, 111–114.

Furr, P. M., Taylor-Robinson, D. (1993a). The contrasting effects of progesterone and oestrogen on the susceptibility of mice to genital infection with *Mycoplasma pulmonis*. *J. Med. Microbiol.*, **38**, 160–165.

Furr, P. M., Taylor-Robinson, D. (1993b). Urogenital infections in rodents. In: *Molecular and Diagnostic Procedures in Mycoplasmology* (eds Tully, J. G., Razin, S.), vol. 2, pp. 337–347. Academic Press, New York.

Furr, P. M., Taylor-Robinson, D. (1993c). *Mycoplasma pulmonis* infection of the murine oropharynx protects against subsequent vaginal colonization. *Epidemiol. Infect.*, **111**, 307–313.

Furr, P. M., Taylor-Robinson, D., Webster, A. D. B. (1994). Mycoplasmas and ureaplasmas in patients with hypogammaglobulinaemia and their role in arthritis: microbiological observations over twenty years. *Ann. Rheum. Dis.*, **53**, 183–187.

Gourlay, R. N., Wyld, S. G. (1976). Ilsley-type and other mycoplasmas from the alimentary tracts of cattle, pigs and rodents. *Proc. Soc. Gen. Microbiol.*, **3**, 142.

McGarrity, G. J., Rose, D. L., Kwiatkowski, V., Dion, A. S., Phillips, D. M., Tully, J. G. (1983). *Mycoplasma muris*, a new species from laboratory mice. *Int. J. Syst. Bacteriol.*, **33**, 350–355.

Møller, B. R., Taylor-Robinson, D., Furr, P. M. (1984). Serological evidence implicating *Mycoplasma genitalium* in pelvic inflammatory disease. *Lancet*, **1**, 1102–1103.

Rosenstein, I. J., Morgan, D. J., Sheehan, M., Lamont, R. F., Taylor-Robinson, D. (1996). Bacterial vaginosis in pregnancy: distribution of bacterial species in different gram-stain categories of the vaginal flora. *J. Med. Microbiol.*, **44**, 1–7.

Rugh, R. (ed.) (1968). *The Mouse. Its Reproduction and Development*, pp. 38–39. Burgess, Minneapolis, MN.

Taylor-Robinson, D. (1989). Genital mycoplasma infections. *Clin. Lab. Med.*, **9**, 501–523.

Taylor-Robinson, D. (1996). Infections due to species of *Mycoplasma* and *Ureaplasma*: an update. *Clin. Infect. Dis.*, **23**, 671–684.

Taylor-Robinson, D., Bébéar, C. (1997). Antibiotic susceptibilities of mycoplasmas and treatment of mycoplasmal infections. *J. Antimicrob. Chemother.*, **40**, 622–630.

Taylor-Robinson, D., Furr, P. M. (1986). Urinary-tract infection by *Mycoplasma pulmonis* in mice and its wider implications. *J. Hyg.*, **96**, 439–446.

Taylor-Robinson, D., Hay P. E. (1997). The pathogenesis of the clinical signs of bacterial vaginosis and possible reasons for its occurrence. *Int. J. STD AIDS*, **8**, (Suppl. 1), 13–16.

Taylor-Robinson, D., Denny, F. W., Thompson, G. W., Allison, A. C., Mårdh, P-A. (1972). Isolation of mycoplasmas from lungs by a perfusion technique. *Med. Microbiol. Immunol.*, **158**, 9–15.

Tuffrey, M., Taylor-Robinson, D. (1981) Progesterone as a key factor in the development of a mouse model for genital-tract infection with *Chlamydia trachomatis*. *FEMS Microbiol. Lett.*, **12**, 111–115.

Tully, J. G., Rose, D. L., Whitcomb, R. F., Wenzel, R. P. (1979). Enhanced isolation of *Mycoplasma pneumoniae* from throat washings with a newly modified culture medium. *J. Infect. Dis.*, **139**, 478–482.

Chapter 50

Mouse Model of Ascending Urinary Tract Infection

W. J. Hopkins

Urinary tract infections in humans

Recurrent urinary tract infections (UTI) caused predominantly by *Escherichia coli* are a significant urologic disease for adult women. Our current understanding of the etiology of UTIs is that uropathogenic *E. coli* from the intestinal flora sequentially colonize the perineum, vagina, urethra, and bladder (Stamm *et al.*, 1991). Even in women without underlying anatomic or functional abnormalities, bacteria from the infected bladder may ascend to the kidneys, causing infection and renal scarring (Lomberg *et al.*, 1983). To most accurately study host or bacterial factors that influence the induction of bladder or kidney infections, animal models must accurately simulate these events.

Background of model

A variety of animal models have been used to define virulence factors of uropathogenic *E. coli* and mechanisms of host resistance to UTI (Kaijser and Larsson, 1982). In particular, intravesical inoculation of mice with uropathogenic bacteria has been used extensively as a model of unobstructed, ascending infection (Hagberg *et al.*, 1983). One of the most important considerations in insuring that the mouse model reflects events occurring during the infection process in humans is to insure that renal infection develops as the result of bacteria ascending from the infected bladder rather than from vesicoureteral reflux accompanying inoculation (Johnson and Manivel, 1991; Hopkins *et al.*, 1995; Johnson and Brown, 1996). Because inoculum volumes of 50 μl and greater have been found to introduce bacteria directly into the kidney, inoculation procedures have been refined to circumvent this problem (Hopkins *et al.*, 1995; Johnson and Brown, 1996). We have also instituted steps in the preparation of the inoculum and animals to maintain consistency between experiments.

Animal species

Female mice have typically been used to study ascending UTIs because the clinical problem of recurrent infections occurs predominantly in women. Male mice could also be used if a specific question relevant to UTIs in males was being investigated. The age of the mice generally ranges from 8–12 weeks. Many strains of inbred mice have been successfully used in this model, but susceptibility can vary in strains of different genetic backgrounds (Hopkins *et al.*, 1996). The strains chosen for a particular study would thus be dictated by the objectives of the investigations. Mice with differing genetic make-ups could be used to define host characteristics that influence resistance to infection, whereas a single mouse strain might be utilized to test the importance of different bacterial virulence factors.

Storage and preparation of the inoculum

Stocks of bacterial strains should be stored frozen at –70°C or lyophilized to provide a consistent source of bacteria for different experiments. A small aliquot of stock is used to inoculate a bacteriological medium that will promote development of virulence factors present in the bacteria being studied. For example, uropathogenic *E. coli* may carry genes for Type 1 or P fimbriae, and the choice of growth medium can influence which fimbrial type is expressed. Type 1 fimbriae are expressed well during growth in tryptose broth, with typical conditions being an initial 48 hour growth from stock, followed by overnight culture (Hopkins *et al.*, 1987). The P fimbriae are expressed better as a result of growth on colony-factor antigen agar (Evans *et al.*, 1977). It is also possible to use bacteria from continuous cultures, but this is not recommended since there may be a lack of consistency between experiments done over several weeks or months, and, furthermore, it is possible that bacteria could lose virulence factors when repeatedly subcultured over long periods of time.

The inoculum is prepared by harvesting cells either from liquid culture or from a suspension prepared by washing bacteria from solid medium. In either case, bacteria are collected by centrifugation at 9000 *g* for 8 minutes and then resuspended in approximately 20 ml of phosphate-buffered saline (PBS) as a means of removing residual media. This

Handbook of Animal Models of Infection
ISBN 0-12-775390-7

suspension is centrifuged as above to pellet the bacteria, after which they are resuspended in a small volume of PBS to obtain a concentrated suspension for preparing the inoculum. As a guideline, bacteria from 50–60 ml of overnight liquid culture can be resuspended in 2–3 ml to give a final concentration of greater than 10^{10}/ml. The exact concentration of bacteria can be determined spectrophotometrically using a standard curve that relates the OD_{420} of the bacterial suspension in PBS to the number of bacteria per milliliter. The concentrated suspension can now be diluted to prepare an inoculum that will contain sufficient bacteria in 10 μl to induce a UTI with the bacterial strain and animals being used. A minimal dose that will infect all the animals inoculated will need to be determined empirically by testing doses that cover a range less than and greater than 1×10^8 per mouse.

Preparation of animals

One potential problem with this model is that the inoculum may be voided and lost shortly after administration so that an infection does not have time to become established. This situation can be overcome by including two steps in the preparation of all animals. First, mice are taken off water for at least an hour before inoculation as a way to decrease urine output. Second, immediately prior to inoculation any residual urine is expressed from the bladder by applying gentle pressure with the thumb to the abdomen over the bladder (Figure 50.1).

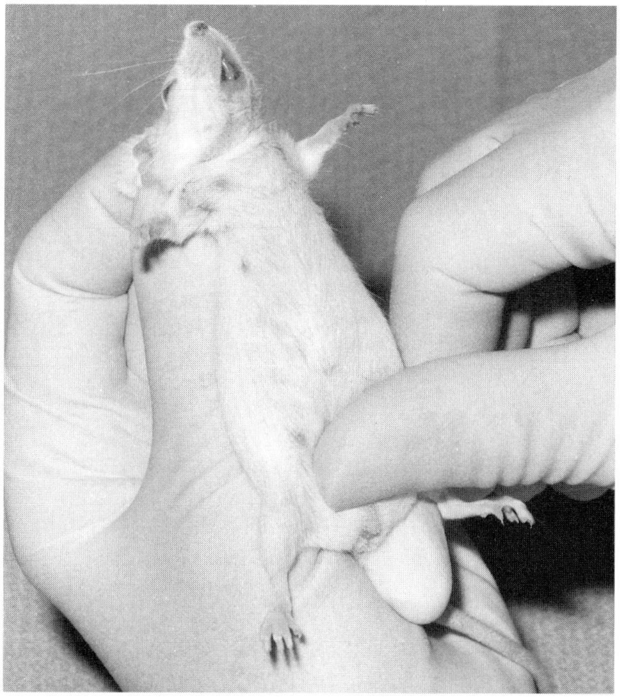

Figure 50.1 Expressing urine from the mouse bladder prior to inoculation. With the unanesthetized mouse held in one hand, urine is expressed using the thumb of the opposite hand. The thumb is pressed onto the abdomen over the bladder and drawn posteriorly towards the urethral opening.

Infection procedure

Urinary tract infections can be induced in mice by an intraurethral or intravesical inoculation of viable bacteria and, when performed correctly, either method will avoid the problem of direct infection of the kidneys. The equipment used is the same for both types of inoculation and consists of a 50 μl Hamilton® syringe with a 30 G needle that has been cut to approximately 2 cm and fitted with a short length of polyethylene tubing (Intramedic® PE-10, Becton-Dickinson, Parsippany, NJ). The tubing is cut 0.5 cm or 1.0 cm beyond the end of the needle for intraurethral or intravesical inoculations, respectively. To begin the inoculation, a mouse that has been anesthetized with ether or pentobarbital is placed onto its back, and its hind legs are held in place on the work surface with a scissors or hemostat handle, as shown in Figure 50.2. Grasp the papilla on one side with tissue forceps and pull it upwards slightly from the abdomen. The tubing is next inserted into the urethral opening while maintaining traction on the papilla, and advanced into the urethra (Figure 50.3). For urethral inoculation, only the 0.5 cm section of tubing extending past the end of the needle is placed into the urethra. When bacteria are instilled into the bladder, the 1.0 cm length of tubing is gently passed through the urethra. The tubing will slide easily through the urethra unless it encounters an obstruction, such as where the urethra passes adjacent to the pubic symphysis. In this case, withdraw the catheter 2–3 mm and advance it again while directing the tip slightly downward. Once the tubing is in place within the urethra or bladder, deliver the inoculum at a rate of less than 1 μl/s and then remove the catheter. The mice are allowed to recover from the anesthesia but are not given water for 1 hour after inoculation to decrease urine output and thereby loss of the inoculum through voiding.

Parameters to monitor infection

The intensity of a bladder or kidney infection can be quantified by determining the number of bacteria per milligram of bladder or kidney tissue. Peak infection levels generally occur by 1 day after inoculation and progressively decrease over a 2–3-week period, depending upon the bacterial strain and inbred mouse used. The organs are removed at predetermined time points and weighed to obtain a wet weight for the bladder and pool of both kidneys. The bladder can be placed into a 35 mm Petri dish and the kidneys into a sterile plastic bag (e.g., Whirlpak®)

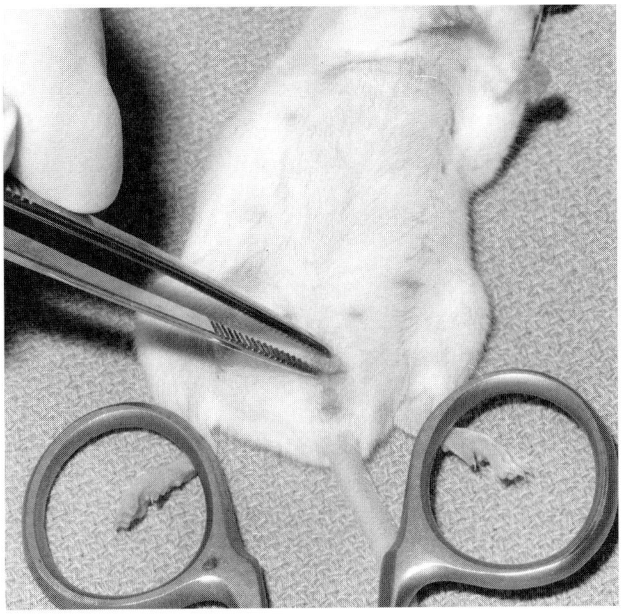

Figure 50.2 Position of the mouse for inoculation. The anesthetized mouse is placed onto its back, and the handle of a hemostat is used as shown to restrict movement of the hind legs. This photograph also shows the papilla and urethral opening.

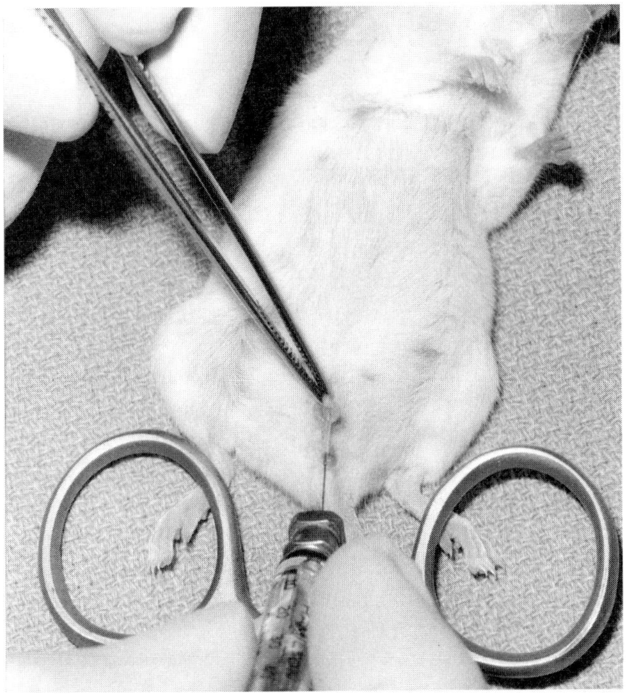

Figure 50.3 Inserting catheter tubing into the urethra. The papilla is grasped with light pressure and lifted away from the abdomen. While the papilla is maintained in this position, tubing attached to the inoculating syringe is advanced into the urethral opening and the urethra.

and kept on ice until homogenized. Tissues are homogenized in a small volume (2–4 ml) of PBS using a sterile Potter–Elvehjem tissue homogenizer, crushing in a Whirlpak® bag, or a Stomacher® homogenizer (Seward). Bladders are more effectively homogenized if they are first minced with a small scissors. Once the tissues are broken up, large pieces of debris can be removed by very low speed centrifugation and the supernatant transferred to a new sterile tube and kept cold. Aliquots of the supernatant are then serially diluted and plated on to Levine's eosin–methylene-blue (EMB) agar plates to obtain isolated colonies for counting. Plate count data are used to calculate the number of bacteria in the entire bladder and the combined kidneys, which, in turn, are used to compute the $\log_{10}$ (cfu/mg tissue) values needed to determine the mean and standard deviation for each group of animals in the experiment. As a point of reference for infection intensities, experiments using our uropathogenic *E. coli* strain at an inoculum dose of 1×10^8 in BALB/c mice will result in bladder mean $\log_{10}$ cfu values of approximately 3.5–4.0 one day after inoculation. The infection will gradually resolve over 14–21 days, and the mean $\log_{10}$ cfu will decline to less than 1.0, with the majority of animals having cleared their infections. These values are for infections induced with a minimal number of bacteria in one mouse strain, and the infection intensities will probably be different with other uropathogens, inbred mice, and the methods used in a particular laboratory. Whether the infection will ascend to the kidneys in the absence of inoculation-associated vesicoureteral reflux is in large part dependent on the bacterial strain. It is recommended that each team of investigators develops consistent inoculation and assay procedures for their particular studies.

While it is not a direct measure of infection, the intensity of inflammatory responses in the bladder and kidneys of infected animals correlates with the number of bacteria in these organs. A grading scale such as that shown in Table 50.1 can be used to assign a numerical value to each degree of inflammation. The scores for animals in experimental and control groups can then be statistically analyzed to determine the significance of a particular bacterial virulence factor or host characteristic in the pathophysiology of the infection.

Advantages/disadvantages of the model

The mouse model of UTI has several advantages for investigators interested in defining host or bacterial factors that affect infectivity, pathophysiology, or host defense mechanisms. Among these are the availability of inbred mouse strains with distinct genetic backgrounds. Mice with known genetic traits can be used to correlate individual genes with susceptibility to bladder or kidney colonization, pathology associated with host responses to

Table 50.1 Histopathological grading scale for degree of inflammation

Grade	*Bladder*	*Kidney*
0	Normal	Normal
1	Subepithelial cell inflammatory infiltration (focal and multifocal)	Focal inflammation (cellular infiltration and/or edema in the pelvis)
2	Edema and subepithelial inflammatory cell infiltration (diffuse)	Focal inflammation (more severe) in the pelvis to medulla with and without moderate edema
3	Marked subepithelial inflammatory cells with necrosis and neutrophils in and on bladder mucosal epithelium	Multifocal inflammation and cells from pelvis to medulla
4	Inflammatory cell infiltrate extends into muscle, in addition to criteria for grade 3	Extensive segmental inflammation and necrosis evident from pelvis to cortex
5	Loss of surface epithelium (necrosis with full-thickness inflammatory cell infiltration)	Diffuse tissue necrosis and inflammatory cell infiltration extending from pelvis to cortex

infection, and innate or adaptive defense mechanisms. Conversely, appropriate genetic crosses can be used to identify and map unknown genes associated with induction of bladder and kidney infections or with protective immune responses.

Another benefit of using mice is that animals are usually available in sufficient numbers to obtain experimental group sizes that will be large enough to provide meaningful statistical analysis. The relatively low acquisition and housing costs of mice are an economic advantage of this model.

Some disadvantages of the mouse model can be attributed to the small size of the animals. A considerable amount of expertise needs to be developed to perform bladder or urethral inoculations reproducibly. The need to use thin catheters and low volumes can pose technical problems when inoculating mice in which the external and internal anatomy can vary slightly between animals even within the same strain. There is also some limitation in the amount of tissue that is available for histologic examination, *in situ* immunologic or nucleic acid studies, or investigations requiring large numbers of inflammatory or immunologic cells recovered from the site of infection. The latter problem can sometimes be overcome by pooling tissue from several animals.

As with any animal model, there can always be the question of how relevant the studies are to the same disease in humans. In general, the many studies conducted thus far in mice infected with human uropathogens have contributed to validating this model on the conceptual level. Over the past several years our laboratory and other investigators have recognized that earlier models of ascending UTI in the mouse had the shortcoming of artificially inducing kidney infections by inoculation-associated vesicoureteral reflux. Efforts focused at alleviating this problem have resulted in a model with better-defined inoculation conditions that will increase consistency between experiments.

Contributions of the model to infectious disease therapy

Urinary tract infections in women are treated with short-term courses of antibiotics for acute episodes and prophylactic antibiotics to decrease the incidence of recurrent infections in susceptible patients. Antibiotic therapy does, however, have significant disadvantages in that bacteria acquire resistance to specific antibiotics and patients may develop allergic reactions to an effective drug. In view of such problems, there has been an effort to develop alternative approaches to treatment or prevention. The mouse model of ascending UTI has been used to explore the feasibility of prevention strategies based on blockade of bacterial proteins required for adherence to epithelial cells lining the vagina and bladder. Experiments in which α-D-mannopyranoside (Aronson *et al.*, 1979) or globoseries analogs (Svanborg Edén *et al.*, 1982) mixed with the *E. coli* inoculum decreased the intensities of bladder and kidney infections demonstrated that inhibition of bacterial adherence mediated by Type-1 and p fimbriae, respectively, might provide an alternative method of treatment or prophylaxis. More recently, we have used this model to develop a vaginal mucosal vaccine for UTIs. A multistrain, whole-cell vaccine containing several *E. coli* strains was instilled into the mouse vagina as a means of increasing urogenital mucosal immunity to *E. coli* (Uehling *et al.*, 1991). Immunized mice later challenged with an *E. coli* UTI had lower numbers of bacteria in the bladder than controls. Vaccine efficacy demonstrated in these animal studies formed the basis for clinical trials in patients with recurrent UTIs (Uehling *et al.*, 1994, 1997).

References

Aronson, M., Medalia, O., Schori, L., Mirelman, D., Sharon, N., Ofek, I. (1979). Prevention of colonization of the urinary tract of mice with *Escherichia coli* by blocking of bacterial adherence with methyl D-mannopyranoside. *J. Infect. Dis.*, **139**, 329–332.

Evans, D. G., Evans, D. J., Tjoa, W. (1977). Hemagglutination of human group A erythrocytes by enterotoxigenic *Escherichia coli* isolated from adults with diarrhea: correlation with colonization factor. *Infect. Immun.*, **18**, 330–337.

Hagberg, L., Engberg, I., Freter, R., Lam, J., Olling, S., Svanborg-Edén, C. (1983). Ascending, unobstructed urinary tract infection in mice caused by pyelonephritogenic *Escherichia coli* of human origin. *Infect. Immun.*, **40**, 273–283.

Hopkins, W. J., Uehling, D. T., Balish, E. (1987). Local and systemic antibody responses accompany spontaneous resolution of experimental cystitis in *Cynomolgus* monkeys. *Infect. Immun.*, **55**, 1951–1956.

Hopkins, W. J., Hall, J. A., Conway, B. P., Uehling, D. T. (1995). Induction of urinary tract infection by intraurethral inoculation with *Escherichia coli*: refining the murine model. *J. Infect. Dis.*, **171**, 462–465.

Hopkins, W. J., Gendron-Fitzpatrick, A., McCarthy, D. O., Haine, J. E., Uehling, D. T. (1996). Lipopolysaccharide-responder and nonresponder C3H mouse strains are equally susceptible to an induced *Escherichia coli* urinary tract infection. *Infect. Immun.*, **64**, 1369–1372.

Johnson, J. R., Brown, J. J. (1996). Defining inoculation conditions for the mouse model of ascending urinary tract infection that avoid immediate vesicoureteral reflux yet produce renal and bladder infection. *J. Infect. Dis.*, **173**, 746–749.

Johnson, J. R., Manivel, J. C. (1991). Vesicoureteral reflux induces renal trauma in a mouse model of ascending, unobstructed pyelonephritis. *J. Urol.*, **145**, 1306–1311.

Kaijser, B., Larsson, P. (1982). Experimental acute pyelonephritis caused by enterobacteria in animals. A review. *J. Urol.*, **127**, 786–790.

Lomberg, H., Hanson, L., Jacobsson, B., Jodal, U., Leffler, H., Svanborg Edén, C. (1983). Correlation of P blood group, vesicoureteral reflux, and bacterial attachment in patients with recurrent pyelonephritis. *N. Engl. J. Med.*, **308**, 1189–1192.

Stamm, W. E., McKevitt, M., Roberts, P. L., White, N. J. (1991). Natural history of recurrent urinary tract infections in women. *Rev. Infect. Dis.*, **13**, 77–84.

Svanborg Edén, C., Freter, R., Hagberg, L. *et al.* (1982). Inhibition of experimental ascending urinary tract infection by an epithelial cell-surface receptor analogue. *Nature*, **298**, 560–562.

Uehling, D. T., James, L. J., Hopkins, W. J., Balish, E. (1991). Immunization against urinary tract infection with a multivalent vaginal vaccine. *J. Urol.*, **146**, 223–226.

Uehling, D. T., Hopkins, W. J., Dahmer, L. A., Balish, E. (1994). Phase I clinical trial of vaginal mucosal immunization for recurrent urinary tract infection. *J. Urol.*, **152**, 2308–2311.

Uehling, D. T., Hopkins, W. J., Balish, E., Xing, Y. N., Heisey, D. M. (1997). Vaginal mucosal immunization for recurrent urinary tract infection: phase II clinical trial. *J. Urol.*, **157**, 2049–2052.

Chapter 51

Mouse Model of Ascending UTI Involving Short- and Long-term Indwelling Catheters

D. E. Johnson and C. V. Lockatell

Background of human infection

Every year millions of urinary catheters, most indwelling in the urethra, are used in hospitals and nursing homes to manage urinary retention or incontinence (Garibaldi, 1981). However, bacteria can enter the bladder and cause catheter-associated bacteriuria, the most common nosocomial infection in both types of institution (Garibaldi *et al.*, 1981; Haley *et al.*, 1985).

In hospitals, urethral catheters are usually in place for 2–4 days (Garibaldi *et al.*, 1981; Warren *et al.*, 1978). Bacteriuria develops in 10–30% of catheterized patients and its complications include fever, acute pyelonephritis, bacteremia, and death (Haley *et al.*, 1981). The prevalence of urinary incontinence among nursing home patients is estimated to be between 35% and 50% (Hing, 1981) and urinary catheters may be in place for months or years. Virtually all patients with long-term indwelling catheters (30 days or more) have polymicrobic bacteriuria, which appears to be the most common nosocomial infection in nursing home patients (Warren *et al.*, 1982). *Proteus mirabilis*, one of the most common uropathogens, is frequently isolated from the urine of elderly long-term-catheterized patients (Warren *et al.*, 1982; Mobley and Hausinger, 1989). Infection with this organism can result in serious complications, such as stone formation in the kidneys and bladder (Mobley and Hausinger, 1989) encrustation and obstruction of the catheter (Mobley and Warren, 1987), acute pyelonephritis (File *et al.*, 1985; Eriksson *et al.*, 1986; Rubin *et al.*, 1986), fever and bacteremia (Setia *et al.*, 1984). The portion of the urethral catheter that resides in the bladder lumen is a foreign body that provides a niche for bacterial colonization, elicits changes in bacterial attachment to uroepithelial cells (Daifuku and Stamm, 1986), reduces antibacterial leukocyte function (Zimmerli *et al.*, 1982), and becomes a nidus for infection-induced stone formation (Mobley and Warren, 1987).

Background of model

Many animal models have been used to study urinary tract infections (Hagberg *et al.*, 1983; Johnson *et al.*, 1987) including some with foreign bodies in the bladder. Vivaldi *et al.* (1959) and Cotran *et al.* (1963) demonstrated enhanced persistence of *P. mirabilis* or *Escherichia coli* in the urinary tracts of rats in the bladders of which glass beads had been surgically implanted. Musher *et al.* (1975), Satoh *et al.* (1984), and Nickel *et al.* (1987) surgically implanted zinc discs in rat bladders to study uropathogenicity of *P. mirabilis*. In all of those models the procedures for instillation of the foreign body in the animal bladders entailed incision of the bladder wall, resulting in trauma and disruption of the bladder mucosa. That disruption of the bladder mucosa in itself could result in altered bacterial adherence to the urinary tract, leading to enhanced bacteriuria, thus masking the contribution of the foreign body to infection of the urinary tract.

The mouse model is in many ways analogous to patients with urinary catheters in place. The foreign body is polyethylene tubing with a lumenal and external surface similar in construction to urinary catheters. As with human urinary catheters, organisms colonizing the luminal surface of the polyethylene tubing may evade host bacterial defense mechanisms. The tube is secured in such a way as to avoid urethral or ureteral obstruction. Finally, and importantly, the only trauma to the bladder is two punctures by the suture needle as it passes into and from the bladder lumen.

Animal species

This model uses the mouse CBA/J inbred (F200) substrain originating (Jax Notes, 1988) in a cross of a Bagg albino female and a DBA male. Female mice 6–8 weeks of age, 20–22 g, are used. Since the anatomy of the male urethra is prohibitive to catheterization, only females are used.

Preparation of animals

No special housing or care is required. Prior to experimental use a urine sample is obtained for culture to assure absence of bacteriuria. During handling of animals spontaneously

Handbook of Animal Models of Infection
ISBN 0-12-775390-7

voided urine can be obtained. However, background contamination from the periurethral area (approximately 10^2–10^3 cfu may be cultured, usually consisting of *Proteus, Enterobacter* and/or Gram-positive organisms) usually yields false positive results. A more reliable method is to anesthetize the mouse with minimal handling in an anesthetic jar containing methoxyflurane. Once anesthetized, remove the mouse from the jar and continue anesthesia as needed using a nose cone. Swab the periurethral area with povidone iodine (1.0% titratable iodine) solution and gently insert a catheter (25 mm long Clay Adams polyethylene PE-10 intramedic tubing i.d. 0.28 mm × o.d. 0.61 mm) into the urethra to approximately half the catheter length. Approximately 10–20 μl of urine may be expressed from the bladder and collected aseptically by gentle compression of the abdominal wall. While the mouse is anesthetized, shave fur from the mouse abdomen (No. 40 clipper blade) in preparation for the surgical procedure used in the long-term catheter model.

Details of surgery

Overview

Catheter segments are placed in the lumen of the mouse urinary bladder. In the short-term model the segments are not secured in the bladder and are expelled over time by micturition (Table 51.1). In the long-term model a single segment, secured in the bladder lumen with a suture, is retained for up to 1 year.

Materials required

Anesthetic, sterile catheter segments, stylet, hair clippers, skin disinfectant, scalpel handle and blades, forceps, hemostats, scissors, syringes and needles, suture material and wound clips, sterile surgical drapes, 2× binocular magnifier, heat lamp or mat, ampicillin.

Short-term indwelling catheter model

Mice are anesthetized with either methoxyflurane inhalation using an open-drop nose cone or with 50 mg/kg pentobarbital intraperitoneally (0.04–0.05 ml of 25 mg/ml pentobarbital solution; Wixson, 1990). A scavenging fume hood is used when anesthetizing mice with methoxyflurane. Animals respond to anesthetic within 5 minutes and remain anesthetized for 20–30 minutes. Prevention of hypothermia is particularly important with pentobarbital use. Ketamine (200 mg/kg i.m.) + xylazine (10 mg/kg i.p.) can be substituted for pentobarbital if use of a controlled substance is not desired.

The anesthetized mouse is placed in a supine position and the periurethral area is swabbed with povidone iodine solution, which is rinsed off with sterile water. Several 4-mm long (containing a radiopaque strip or other black marking) and a single 25-mm-long segment of polyethylene catheter (0.28 mm i. d., 0.61 mm o.d., Clay Adams, Parsippany, NJ; alternative catheters: polyurethane, Braintree Scientific, Braintree, MA and Argyle umbilical vessel catheter, Sherwood Medical, Allentown, PA) are fitted on to a metal sytlet (50 mm long, 0.25 mm o.d.; Bel-Art SMORC replacement wires, Pequannock, NJ; Figure 51.1A). Catheter segments are sterilized by using ethylene oxide before insertion into the bladder and sterility is maintained by using aseptic technique while handling the catheter segments. Care is taken to ensure that the 4-mm-long catheter segment protrudes slightly past the tip of the stylet and that the 25-mm-long segment (which is kept in place on the stylet with forceps) is snug against bottom of the 4-mm-long segment.

In preparation for catheterization the urethral meatus is gently grasped with forceps and pulled slightly forward parallel to the body. Use of a 2× binocular magnifier (OptiVision, Donegan Optical Co., Lenexa, KA) aids in visualization. The end of the stylet inserted through the 4-mm-long catheter segment is gently and carefully inserted into the urethra at a slight angle then moved parallel to animal's body below the pelvic arch. The entire apparatus is advanced until the leading end is in the bladder (Figure 51.1B). No resistance should be felt during this procedure. If resistance is felt, withdraw the apparatus and recheck the positioning. The 25-mm-long catheter is then advanced along the stylet to push the 4-mm-long segments off the stylet and into the bladder (Figure 51.1C). The stylet with the 25-mm-long catheter segment is removed, leaving the 4-mm-segments free in the bladder (Figure 51.1D).

Long-term model

In the long-term model a single 4-mm-long catheter is secured by a single suture within the bladder lumen. While the mice are still anesthetized, the mouse abdomen is

Table 51.1 Retention of unsecured catheter segments in bladder lumen of CBA mice

Days between catheter insertion and examination of mice	*No. mice with catheter segments in bladder/no. mice examined (%)*
2	9/10 (90)
3	7/10 (70)
7	4/9 (44)
14	2/7 (29)
21	0/10 (0)

disinfected with povidone iodine and alcohol scrub and the bladder is exposed by a midline abdominal incision. Visualization of the 4-mm-long catheter segment through the translucent bladder wall is assisted by black catheter markings. A sterile 5-0 polypropylene suture (Prolene, Ethicon, Somerville, NJ) is threaded through the bladder wall, through the lumen of the 4-mm-long bladder catheter segment (Figure 51.1E), back through the bladder wall, and through a second 4-mm-long catheter segment placed on the serosal side of the bladder, and is tied loosely (Figure 51.1F). The second catheter segment is used to prevent erosion of the suture into the serosal side of the bladder wall.

To accomplish this procedure, the bladder is gently grasped with straight forceps, holding the catheter segment stable within the lumen. Exercise extreme care when passing the 5-0 prolene suture through bladder wall and catheter segment, as the catheter can easily puncture the bladder wall. A second pair of forceps is used against the serosal surface of the bladder as the prolene suture is pulled slowly through the opposite side of the bladder wall; a hemostat is attached to the free end of the suture material to guard against the suture pulling out prior to being tied in place. The urinary bladder is returned to the abdomen, muscle and fascia are closed with 5-0 gut suture (Chromic Gut, Ethicon) by using a mattress suture pattern, and the skin is closed with 9 mm stainless steel staples (Autoclip, Clay Adams, Parsippany, NJ).

After surgery mice are placed in a clean cage with shredded nesting material (Kimwipes) under a heat lamp to maintain a temperature of 37.4°C until they have completely recovered from anesthesia. Ampicillin (0.5 mg/ml) is added to their drinking water daily for 7 days to eliminate any microorganisms introduced during the procedure. Staples are removed 7 days after surgery. Two weeks post-surgery bladders are histologically normal and mice are ready for experimental study.

Storage and preparation of inocula

Clinical isolates of *E. coli, P. mirabilis, Klebsiella pneumoniae* and *Candida albicans* are preserved by two methods; grown on trypticase soy agar (TSA, BBL, Hunt Valley, MD) and frozen at –70°C and lyophilized in trypticase soy broth (TSB) with 12% sucrose on sterile inert boiling stones (R. P. Cargille Laboratories, Cedar Grove, NJ). The study organism is revived from a lyophilized chip in 2 ml TSB grown static overnight at 37°C. Four large (20×150 mm) TSA culture slants are inoculated from the static culture and grown overnight at 37°C, the growth is harvested with 3 ml phosphate buffered saline for an inoculum of about $2–4\times10^{10}$ cfu/ml. Inoculum is enumerated quantitatively using TSA spread plates inoculated with serial 10-fold dilutions.

Infection process

Mice are anesthetized with 50 mg/kg pentobarbital i.p. and placed in a supine position. The periurethral area is swabbed with povidone iodine and a 25-mm-long polyethylene catheter is inserted into the urethra as previously described. A 30 G ½″ hypodermic needle is inserted into the catheter opening. The needle is attached to a 2.1 ml 20″

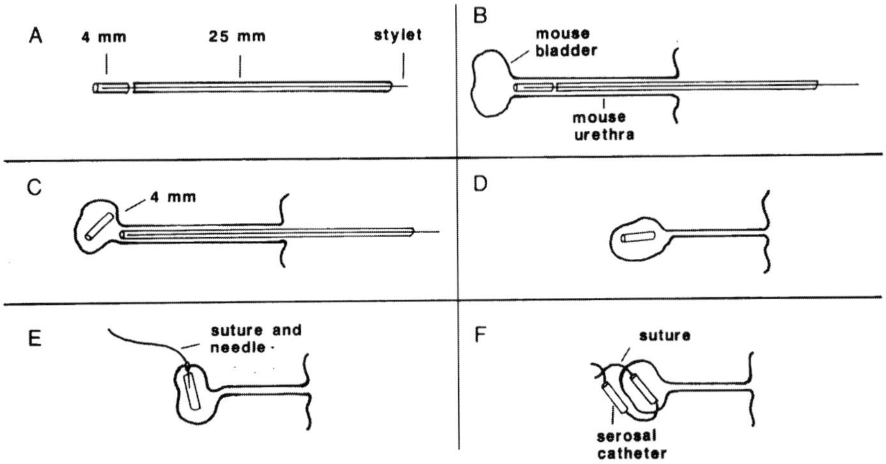

Figure 51.1 Procedure for placement of catheter segment into mouse urinary bladder. **A** Catheter segment fitted on to stylet. **B** Apparatus inserted into urethra and distal end advanced to bladder. **C** 25-mm-long catheter segment advanced along stylet, pushing 4-mm-long segment into bladder. **D** Stylet and 25-mm-long catheter segment removed, leaving 4-mm-long catheter segment in bladder lumen. **E, F** After bladder is exposed by midline incision, mucosal and serosal catheter segments are secured by suture. Reproduced with permission from Johnson, D. E., Lockatell, C. V., Hall-Craggs, M. and Warren, J. W. Mouse models of short- and long-term foreign body in the urinary bladder: analogies to the bladder segment of urinary catheters. *Lab. Anim. Sci.*, **41**, 452.

sterile Pharmaseal extension tube (Baxter Healthcare Corp., Valencia, CA) which is connected to a 1 ml syringe within an infusion pump (Harvard Apparatus, South Natick, MA) for delivery of a 50 μl inoculum over 30 seconds. The polyethylene tubing is withdrawn from the ureter immediately after challenge, and mice are returned to their cages and cared for by the normal routine.

Key parameters to monitor infection

Urinary tract infection is documented using quantitative cultures of bladder and kidney homogenates and light-microscopic examination of sections of bladder and kidney to assess histologic changes. Non-uropathogens are eliminated from the urinary tract within 2 days of challenge and no histologic changes are observed. Uropathogens persist in the urinary tract for 2 weeks or more and histologic changes are observed in bladder and kidney. The presence of catheter segments in the bladder prolongs persistence of the uropathogens and intensifies histologic changes. Results from urine cultures are not useful for assessing the progression of urinary tract infection. Dissemination of infection outside the urinary tract may be documented by quantitative cultures of blood and/or homogenates of liver and/or spleen. With the exception of studies with *P. mirabilis*, mortality is not a usual endpoint and mice with kidney counts exceeding 10^5 cfu/g appear normal.

Antimicrobial therapy

The model provides an excellent opportunity to assess the effect of antimicrobial therapy on catheter-associated urinary tract infection. Since pharmacokinetics of antimicrobials in rodents and man differ, results from preliminary kinetic studies may be used to guide dosing schedules in mice to conform with human pharmacokinetics. Oral compounds may be added to mouse drinking water or delivered by intragastric infusion. Subcutaneous, intramuscular, and intravenous dosing routes may also be used.

Key parameters to monitor response to treatment

Effective antimicrobial therapy will reduce or prevent (1) microbiological colonization and histologic changes of bladder and kidney, (2) dissemination of the challenge organism out of the urinary tract and (3) mortality (following *P. mirabilis* challenge).

Pitfalls (advantages/disadvantages) of the model

As with all animal models, the goal is to mimic human infection as closely as possible. The challenge dose used to induce urinary tract infection in mice (10^7–10^9 cfu/mouse) is probably higher than the dose required to induce human infection. However, urinary tract infection in mice is dose-responsive and high challenge doses are used to induce infection in a sufficient number of animals to obtain statistically significant results with a minimum number of animals. Lower doses would result in a lower percentage of animals infected, which may more closely mimic natural human infection but considerably more animals would have to be studied to answer the research question. Constant attention to the challenge volume (50 μl or less), the challenge dose delivery time (30 seconds or more) and consistency (use of an infusion pump) will be helpful in avoiding induction of reflux of the challenge dose into the ureters and kidneys during the challenge process.

The presence of the catheter segment in the bladder will reduce the dose required to induce urinary tract infection in mice, prolong persistence of the infecting organism in the urinary tract, increase histologic changes, and reduce antimicrobial efficacy. The short-term model is less time-consuming to establish but has the disadvantage of elimination of catheter segments over time by normal micturition. Securing the catheter segment in the bladder lumen in the long-term model is time-consuming (15 animals per day by an experienced technician), but the segment remains in place for up to 1 year.

Contribution of the model to infectious disease therapy

The model is useful in assessing the effect of an indwelling bladder catheter on:

1. microorganism uropathogenicity;
2. efficacy of antimicrobial therapy;
3. attachment patterns of uropathogens to uroepithelial cells;
4. host response to infection including changes in antibacterial leukocyte function, host inflammatory response and specific antibody responses;
5. *in vivo* microbial ecology of catheter surfaces and host urinary mucosa;
6. *in vivo* biocompatable and antimicrobial efficacy of novel catheter formulations.

References

Cotran, R. S., Thrupp, L. D., Hajj, S. N., Zangwell, D. P., Vivaldi, E., Kass, E. H. (1963). Retrograde *E. coli* pyelonephritis in the rat: a bacteriologic, pathologic, and fluorescent antibody study. *J. Lab. Clin. Med.*, **61**, 987–1004.

Daifuku, R., Stamm, W. (1986). Bacterial adherence to bladder uroepithelial cells in catheter-associated urinary tract infection. *N. Engl. J. Med.*, **314**, 1208–1213.

Eriksson, S., Zbornik, J., Dahnsio, H. *et al.* (1986). The combination of pivamoxicillin and pivmecillinam verses pivampicillin alone in the treatment of acute pyelonephritis. *Scand. J. Infect. Dis.*, **18**, 431–438.

File, T., Tan, J., Salstrom, S. J., Johnson, J. (1985). Timentin verses piperacillin in the therapy of serious urinary tract infections. *Am. J. Med.*, **79**, 91–95.

Garibaldi, R. A. (1981). Hospital acquired urinary tract infection. In: *CRC Handbook of Hospital Acquired Infections* (ed. Wenzel R. P.), pp. 513–537. CRC Press, Boca Raton, FL.

Garibaldi, R. A., Brodine, S., Matsumiya, S. (1981). Infections among patients in nursing homes. Policies, prevalence and problems. *N. Engl. J. Med.*, **305**, 731–735.

Hagberg, L., Engberg, I., Freter, R., Lam, J., Olling, S., Svanborg Edén, C. (1983). Ascending unobstructed urinary tract infection in mice caused by pyelonephritogenic *Escherichia coli* of human origin. *Infect. Immun.*, **40**, 273–283.

Haley, R., Culver, D., White, J. W., Morgan, W. M., Emori, T. G. (1985). The nationwide nosocomial infection rate: a need for vital statistics. *Am. J. Epidemiol*, **121**, 159–167.

Haley, R. W., Hooton, T. M., Culver, D. H. *et al.* (1981). Nosocomial infections in US hospitals, 1975–1976: estimated frequency by selected characteristics of patients. *Am. J. Med.*, **70**, 947–959.

Hing, E. (1981). Data from the National Survey; No. 51. In: *Vital Health Statistics*, Series 13. Pub. No. PHS 81-1712. Department of Health and Human Services, Washington, DC.

Jax Notes (1988). *CBA Substrains Maintained at the Jackson Laboratory*, No. 432. Jackson Laboratory, Bar Harbor, ME.

Johnson, D. E., Lockatell, C. V., Hall-Craggs, M., Mobley, H. L. T., Warren, J. W. (1987). Uropathogenicity in rats and mice of *Providencia stuartii* from long-term catheterized patients. *J. Urol.*, **138**, 632–635.

Mobley, H. L. T., Hausinger, R. P. (1989). Microbial ureases: significance, regulation, and molecular characterization. *Microbiol. Rev.*, **53**, 85–108.

Mobley, H. L. T., Warren, J. W. (1987). Urease-positive bacteriuria and obstruction of long-term urinary catheters. *J. Clin. Microbiol.*, **25**, 2216–2217.

Musher, D. M., Griffin, D. P., Yawn, D., Rossen, R. D. (1975). Role of urease in pyelonephritis resulting from urinary tract infection with *Proteus*. *J. Infect. Dis.*, **131**, 177–181.

Nickel, J. C., Olson, M., McLean, R. J. C., Grant, S. K., Costerton, J. W. (1987). An ecological study of infected urinary stone genesis in an animal model. *Br. J. Urol.*, **59**, 21–30.

Rubin, R. H., Tolkoff-Rubin, N. E., Cortan, R. S. (1986). Urinary tract infection, pyelonephritis, and reflux nephropathy. In: *The Kidney* (eds Brenner, B. H. and Rector, F. C.), pp. 1085–1141. W. B. Saunders, Philadelphia, PA.

Satoh, M., Munakata, K., Kitoh, K., Takeuchi, H., Yoshida, O. (1984). A newly designed model for infection-induced bladder stone formation in the rat. *J. Urol*, **132**, 1247–1249.

Setia, U., Serventi, I., Lorenz, P. (1984). Bacteremia in a long-term care facility: spectrum and mortality. *Arch. Intern. Med.*, **144**, 1633–1635.

Staats, J. (1985). Standardized nomenclature for inbred strains of mice, 8th listing. *Cancer Res.*, **45**, 945–977.

Vivaldi, E., Cotran, R., Zangwill, P., Kass, E. H. (1959). Ascending infection as a mechanism in pathogenesis of experimental non-obstructive pyelonephritis. *Proc. Soc. Exp. Biol. Med.*, **102**, 242–244.

Warren, J. W., Platt, R., Thomas, R. J., Rosner, B., Kass, E. H. (1978). Antibiotic irrigation and catheter-associated urinary-tract infections. *N. Engl. J. Med.*, **299**, 570–573.

Warren, J. W., Tenney, J. H., Hoopes, J. M., Muncie, H. L., Anthony, W. C. (1982). A prospective microbiologic study of bacteriuria in patients with chronic indwelling catheters. *J. Infect. Dis.*, **146**, 719–723.

Wixson, S. K. (1990). Current trends in rodent anesthesia and analgesia. Surgical Symposium NCAB/AALAS, Uniform Services, University of the Human Sciences, Bethesda, MD, June, 1990.

Zimmerli, W., Waldvogel, F. A., Vaudaux, P., Nydegger, U. E. (1982). Pathogenesis of foreign body infection: description and characteristics of an animal model. *J. Infect. Dis.*, **146**, 487–497.

Chapter 52

Rat Bladder Infection Model

Tetsuro Matsumoto

Background

Urinary tract infection (UTI) is one of the most common infectious diseases in humans (Kunin, 1994). UTI due to Gram-negative or -positive bacteria is thought to result from bacteria ascending the urethra after colonizing the meatal region or vaginal mucosa. Once in the urinary bladder, these bacteria may pass up the ureters to infect the kidney. Animal models of UTI have been devised using mice, rats, rabbits, pigs, dogs and monkeys (Johnson and Russell, 1997). The majority of animal studies of UTI have been conducted in rats for economic and technical reasons. The effects of vaccination and antibiotic therapy, and bacterial and host factors on the development of UTI have been studied in rat models. For the bladder infection model, routes of challenge have included transurethral and direct inoculation into the bladder. Although Vivaldi *et al.* (1959) were able to establish ascending infection in normal rats, most investigators have had to employ manipulations to make the rat urinary bladder susceptible to infection.

In this chapter, infection models of rat urinary bladder are reviewed and discussed.

Animal species

Sprague–Dawley (175–350 g), Wistar (130–250 g) or Dark Agouti (DA) (180–250 g) rats or hybrid rats such as DA × AS2 F1, DA × HO F1 aged 6–16 weeks are usually used (Miller, 1983). Sometimes, nude rats are used for a T-cell-deficient animal model (Miller and Findon, 1988; Cornish *et al.* 1987). Although both male and female rats are usually used in direct infection models, female rats are preferable in transurethral inoculation model for technical reasons.

Preparation of animals

Although no specialized housing or care is required, specific-pathogen-free conditions are preferable when the immunocompromised rats such as T-cell-deficient strains or those administered immunosuppressants are used. Food and water are supplied *ad libitum*.

Details of surgery

Overview

For bladder infection, rats are inoculated with bacteria through the urethra, using polyethylene tubes or catheters, or by direct inoculation via a suprapubic incision (Johnson and Russell, 1997). Many investigators have created complicated or chronic bladder infection models by inserting artificial materials in the bladder (Miller *et al.*, 1987).

Materials required

Urethral catheter

A polyethylene tube or urethral catheter is necessary for inoculation through the urethral meatus (Fader and Davis, 1980; Balish *et al.*, 1982; Issa *et al.*, 1992).

Direct inoculation

The bladder is exposed through a suprapubic incision. A bacterial inoculum is injected into the bladder using a fine-gauge needle.

Anesthesia

Inhalation anesthesia using ether or parenteral anesthesia using pentobarbital or ketamine hydrochloride by intraperitoneal or intramuscular injection have been used.

Surgical and infection procedures

Direct inoculation into the bladder

After the anesthesia, the lower abdomen is shaved and cleaned with povidone-iodine or alchohol. A midline

Handbook of Animal Models of Infection
ISBN 0-12-775390-7

incision is made to expose the bladder. A fine-gauge hypodermic needle is used to transmurally drain and collapse the bladder and to inoculate a bacterial suspension with or without artificial materials.

Transurethral infection

A polyethylene tube or a ureteral catheter (French no. 3) is inserted into the bladder through the urethral meatus. The bacterial suspension is injected into the bladder via the tube or the catheter.

Manipulation susceptible to infection

Urethral occlusion

The urethral meatus is frequently occluded by clamping to cause bacteriuria that persists for 2–4 hours after the bacterial inoculation. Investigators have obstructed the urethra by using warm wax, paraffin, a metal clip or a hemostat (Fader and Davis, 1980; Balish *et al.*, 1982).

Bladder massage

Heptinstall and Ramsden (1972) discovered that bladder massage increased vesicoureteral reflux and renal infection in rats. After transurethral challenge with bacteria, the bladder was palpated through the abdominal wall, and was gently squeezed between the fingers while the urethral meatus was compressed to prevent escape of the inoculum. Taylor and Koutsaimanis (1975) also massaged the bladder for 30 seconds after bacterial inoculation into the surgically exposed bladder of male rats and clamping of the penis.

Agar beads

Reid *et al.* (1985) challenged rats with organisms incorporated within agar beads. Urethral obstruction was accomplished by application of a wax seal to the urethral meatus or by clamping the meatus with a hemostat. Bacteria were incorporated into agar beads (20–150 μm in diameter); a combination of bacterial suspension, molten 2% agar, and peanut oil was vortexed and the mixture was placed in an ice bath. Reid *et al.* (1985) reported that there were approximately 5×10^3 agar beads in a 0.05 ml inoculum sample containing 5×10^9 bacteria.

Polyacrylamide sphere

Riedasch *et al.* (1981) used polyacrylamide spheres for a chronic cystitis model. Heparin-coated polyacrylamide spheres were inserted under sterile conditions into the bladder by cystostomy. The bladder urine was reported to be infected by transurethral injection of bacteria.

Glass beads

Cotran *et al.* (1963) described how the insertion of a single glass bead into the bladder both enhanced and prolonged renal infection in rats following retrograde challenge with bacteria. They theorized that the role of the glass bead was to maintain residual urine in the bladder after micturition. Vivaldi *et al.* (1959) also maintained that insertion of glass beads in rat bladders enhanced renal infection following both retrograde and hematogenous infection. Glass beads may injure the bladder mucosa, making it susceptible to bladder infection, and biofilm may form around the beads following bacterial challenge. Haraoka *et al.* (1995) used biofilm bacteria adherent to the glass beads to create a model of persistent infection of the bladder. Glass beads, 2 mm diameter, were cultivated with bacteria for 48 hours at 37°C. Bacteria multiplied and formed biofilm arround the beads *in vitro*. The rats were anesthetized and the bladder was exposed through a midline abdominal incision. One of the glass beads coated with biofilm was introduced into the bladder to create a persistent bladder infection, and refluxing pyelonephritis was induced by clamping the urethra.

Sponge core

Miller *et al.* (1988) and Holmang *et al.* (1993) made a chronic cystitis model using polyurethane sponge. A 4 mm core of sponge was punched from polyurethane foam and sectioned into 2 mm discs. The sponge disc was placed in the lumen of an 18 G needle, which was then pushed through the surgically exposed bladder wall and angled towards the dome. The sponge was expelled from the needle by a blunt trochar and deposited in the dome of the bladder. The needle was withdrawn and the puncture wound was closed with a single suture. The sponge remained in the dome of the bladder and became partially embedded in the bladder wall over the following 2 weeks. Chronic cystitis was then established by the direct inoculation of 100 μl of saline containing bacteria into the surgically exposed bladder.

Wire nidus

Davis *et al.* (1984, 1985, 1987) implanted wire nidus in the bladder dome. Female rats were incised in the fundus of the bladder. A 1.5 mm segment of surgical-steel wire coiled on itself was sutured into the bladder lumen with silk suture. The bladder incision was closed by a purse-string silk suture, which also anchored the stainless-steel wire inside the bladder lumen. Bacteria were inoculated through the external meatus.

Bladder diverticulum

Holmang *et al.* (1993) made a small diverticulum in the bladder to be complicated cystitis model. The bladder was exposed through a lower midline abdominal incision, punctured with a thin needle (external diameter 0.9 mm) and emptied of its urine. The bacterial suspension was injected into the bladder lumen. The puncture canal was thereafter closed with a cutgut ligature around the needle, which resulted in the development of a small diverticulum.

Xylene injection

Hohlbrugger *et al.* (1985) injected 0.5 ml of xylene into the bladder through the urethra to fix the mucosal surface. The xylene was removed and the bladder was washed out with normal saline 10 minutes after the injection. The bacterial suspension was then inoculated.

Water diuresis

The effect of water diuresis on susceptibility to UTI has been controversial. Whereas Mahoney and Persky (1963) demonstrated that, after injection of *Proteus vulgaris* into the bladder lumen, water diuresis reduced pyelonephritis by 47%, Freedman (1967) reported that, following bladder challenge, water diuresis increased susceptibility to *Escherichia coli* bacteriuria 1-million-fold.

Diabetic rats

Patients with diabetes mellitus are known to be susceptible to infectious diseases. To produce diabetes in rats, investigators dissolved alloxan monohydrate in sterile distilled water and injected 40 mg/kg body weght in a volume of 0.2 ml into each rat via the lateral tail vein. For glucose determination, 0.5 ml of serum samples were collected by venous puncture or by amputating the tips of the animals' tails 10 days after injecting alloxan (Levison and Pitsakis, 1987).

Postoperative care

Specific postoperative care is unnecessary.

Storage and preparation of inocula

Various bacterial species have been inoculated into the rat bladder. *E. coli* is the most common bacterial species used in UTI models. Other species of bacteria, such as *Klebsiella pneumoniae, Proteus mirabilis, P. vulgaris, Pseudomonas aeruginosa, Serratia marcescens, Salmonella typhimurium, Staphylococcus aureus* and *Enterococcus faecalis*, have been also used. Bacteria were stored deep-frozen and were then cultured on blood agar or other medium for 24 hours before being prepared at a concentration of 10^9–10^{10} organisms per ml.

Occasionally, *Candida albicans* was also used, for a fungal infection model (Levison and Pitsakis, 1987).

Infection process

Mechanisms of adherence to the urinary mucosa by means of pili or fimbriae have been investigated. Michaels *et al.* (1983) observed that *E. coli* bacteriuria was reduced when the challenge dose was suspended in D-mannose, possibly because of inhibition of adherence of *E. coli* to bladder epithelium by mannose-sensitive pili. Keith *et al.* (1986), using isogenic mutants that either expressed type 1 fimbriae or did not, reported that an inhibition of bacteriuria was observed when the organisms lacked type 1 fimbriae.

Fukushi *et al.* (1979), using electron microscopy to study the interaction between bacteria and the host urinary bladder, noted that the initial attachment of *E. coli* to the luminal membrane was followed by infolding of the membrane to envelop the organism, destruction of the membrane, apparent killing of the bacteria, desquamation of infected epithelial cells, and prompt regeneration of the desquamated epithelial layers. Bladder mucin appeared to play a protective role against bacterial infection by enmeshing and subsequently removing the infecting organisms from the urinary tract. Establishing of cystitis was associated with disruption of the mucin layer and bacterial adherence to the urothelial folds.

Davis *et al.* (1984) observed that hyperplastic alterations, dysplasia and early lesions consistent with neoplasia occurred in rats with bladder stainless steel wire implants and multiple bacterial injections. Uchida *et al.* (1989) also reported that the bladder epithelial hyperplasia was induced by instillation of live *E. coli* into the bladder.

Pathology

Urinary bladders are removed at each time interval after bacterial challenge, and are immediately fixed and dried. The tissues are stained with hematoxylin and eosin, Masson-Goldner, periodic-acid–Schiff or Giemsa stain by using standard techniques for light microscopy.

Tissue is also fixed for electron microscopy by immersion into glutaraldehyde 2.5% cacodylate buffer at 4°C.

Bacterial density

To determine the severity or persistency of infection and to assess antimicrobial treatment, a urine sample for bacterial count should be taken. Quantitative culture of urine samples is carried out. Urine samples are serially diluted, small volumes of diluted samples are placed on agar plates and the bacterial colonies are counted after 24 hours.

Antimicrobial therapy and determination of antibiotic penetration into the bladder or urine

The antibiotic concentration is usually measured in urine after the parenteral or oral administration of antibiotics by using bioassay or the high-performance liquid column method. Although numerous animal studies have been conducted assessing the *in-vivo* efficacy of antibiotic therapy for UTI, a few examples will suffice to demonstrate the utility of these models.

Pitfalls of the model

Either intentionally or unintentionally, retrograde challenge of rats results in reflux of the inoculum from the bladder into the ureter or the kidney. Vesicoureteral reflux, which occurs naturally in rats, can be increased by numerous manipulations of the rat urinary tract. Cotran (1963) reported that reflux routinely occurred with challenge volumes of 0.2 ml and occasionally at lower volumes. Andersen and Jackson (1961), using methylene blue as a tracer, reported that vesicoureteral reflux usually occurred in the range of 0.25–0.3 ml and that a volume of 0.6 ml or more caused reflux in all animals. Fierer *et al.* (1971) reported that 81% of rats were bacteremic 3 minutes after challenge with 1.0 ml of *E. coli* suspension. Larsson *et al.* (1980) did not detect bacteremia in 20 rats at 15–30 minutes after intravesical *E. coli* challenge at volumes of 0.5 ml but did detect bacteremia with increased challenge volumes.

Therefore, rat bladder infection is thought to be accompanied by renal infection when 0.5 ml or more volume is inoculated.

Contribution of the model to infectious disease therapy

Profiling of antimicrobial treatment

Parenteral or oral antibiotics are usually administered to treat UTI in humans. Bladder infection in rats has been treated experimentally with various types of antibiotic. As penicillins, cephems, aminoglycosides and fluoroquinolone antimicrobials are known to penetrate the urinary tract well, especially in the urine, these agents are suitable for treatment of UTI.

Vaccination

Elucidation of the pathophysiology of bacterial bladder infection

Uehling and Wolf (1969) demonstrated that the protective effect of parenteral immunization against *E. coli* cystitis did not correlate with levels of circulating antibody. Vesical immunization was not protective and did not induce circulating antibody.

Competitive inhibition of colonization in the bladder

Reid *et al.* (1985) reported that *Lactobacillus casei* isolated from the urethra of a healthy woman prevented onset of UTI in 84% of the animals tested, that the lactobacilli appeared to exclude the uropathogens from colonizing the urinary tract within the first 48 hours after challenge, and that the net effect was a complete eradication of bacteria from the uroepithelium.

Bladder irrigation

Wan *et al.* (1994) described a rat model of antibiotic instillation. The rat's abdomen was exposed after 1 week of bacterial inoculation and urine was aspirated from the bladder through a needle. A 1 ml of gentamicin sulfate solution (480 mg gentamicin sulfate in 10.9% NaCl) was injected back down the needle into the intact bladder. Venous blood was drawn from the inferior vena cava 30 minutes after intravesical instillation and gentamicin levels were determined using standard assay. Wan *et al.* reported increased absorption in 43% of the serum samples from infected rat bladders.

Conclusion

Rats have usually been used for bladder infection models for economical and technical reasons. Normal rat bladder is not very susceptible to bacterial infection. Various manipulations have been used to make rats more susceptible to infection. These persistent or complicated infection models may be useful for assessing antimicrobials or other preventive treatments in recurrent or severe cystitis in humans. However, ascending pyelonephritis or bacteremia are easily caused by bladder injection of bacterial suspension.

References

Andersen, B. R., Jackson, G. G. (1961) Pyelitis, an important factor in the pathogenesis of retrograde pyelonephritis. *J. Exp. Med.*, **114**, 375–384.

Balish, M. J., Jensen, J., Uehling, D. T. (1982) Bladder mucin: a scanning electron microscopy study in experimental cystitis. *J. Urol.*, **128**, 1060–1063.

Cornish, J., Vanderwee, M., Miller, T. (1987) Mucus stabilization in the urinary bladder. *Br. J. Exp. Pathol.*, **68**, 369–375.

Cotran, R. S. (1963) Retrograde *Proteus* pyelonephritis in rats. *J. Exp. Med.*, **117**, 813–822.

Cotran, R. S., Thrupp, L. D., Hajj, S. N., Zangwill, D. P., Vivaldi, E., Kass, E. H. (1963) Retrograde *E. coli* pyelonephritis in the rat; a bacteriologic, pathological, and fluorescent antibody study. *J. Lab. Clin. Med.*, **61**, 987–1004.

Davis, C. P., Cohen, M. S., Anderson, M. D., Gruber, M. B., Warren, M. M. (1985) Urothelial hyperplasia and neoplasia. II Detection of nitrosamines and interferon in chronic urinary tract infections in rats. *J. Urol.*, **134**, 1002–1006.

Davis, C. P., Cohen, M. S., Anderson, M. D., Reinarz, J. A., Warren, M. M. (1987) Total and specific immunoglobulin response to acute and chronic urinary tract infections in a rat model. *J. Urol.*, **138**, 1308–1317.

Davis, C. P., Cohen, M. S., Gruber, M. B., Anderson, M. D., Warren, M. M. (1984) Urothelial hyperplasia and neoplasia: A response to chronic urinary tract infections in rats. *J. Urol.*, **132**, 1025–1031.

Fader, R. C., Davis, C. P. (1980) Effect of piliation on *Klebsiella pneumoniae* infection in rat bladders. *Infect. Immun.*, **30**, 554–561.

Fierer, J., Talner, L., Braude, A. I. (1971) Bacteriuria in the pathogenesis of retrograde *E. coli* pyelonephritis in the rat. *Am. J. Pathol.*, **64**, 443–454.

Freedman, L. R. (1967) Experimental pyelonephritis. XIII On the ability of water diuresis to induce susceptibility to *E. coli* bacteriuria in the normal rat. *Yale J. Biol. Med.*, **39**, 255–266.

Fukushi, Y., Orikasa, S., Kagayama, M. (1979) An electron microscopic study of the interaction between vesical epithelium and *E. coli*. *Invest. Urol.*, **17**, 61–68.

Haraoka, M., Matsumoto, T., Takahashi, K., Kubo, S., Tanaka, M., Kumazawa, J. (1995) Effect of prednisolone on ascending renal infection due to biofilm disease and lower urinary obstruction in rats. *Urol. Res.*, **22**, 383–387.

Heptinstall, R. H., Ramsden, P. W. (1972) Antibody production in urinary tract infections in the rat. *Invest. Urol.*, **9**, 426–430.

Hohlbrugger, G., Lentsch, P., Pfaller, K., Maderbacher, H. (1985) Permeability characteristics of the rat urinary bladder in experimental cystitis and after overdistension. *Urol. Int.*, **40**, 211–216.

Holmang, S., Grenabo, L., Hedelin, H., Hugosson, J., Pettersson, S. (1993) Crystal adherence to rat bladder epithelium after long-term *E. coli* infection. *Scand. J. Urol. Nephrol.*, **27**, 71–74.

Issa, M. M., Shortliffe, L. M. D., Constantinen, C. E. (1992) The effect of bacteriuria on bladder and renal pelvic pressures in the rat. *J. Urol.*, **148**, 559–563.

Johnson, D. E., Russell, R. G. (1997) Animal models of urinary tract infection. In: *Urinary Tract Infections* (ed. Mobley, H. L. T., Warren, T. W.), pp. 377–403. American Society for Microbiology Press, Washington, DC.

Keith, B. R., Maurer, L., Spears, P. A., Orndorff, P. A. (1986) Receptor-binding function of type 1 pili effects bladder colonization by a clinical isolate of *Escherichia coli*. *Infect. Immun.*, **53**, 693–696.

Kunin, C. M. (1994) Urinary tract infections in females. *Clin. Infect. Dis.*, **18**, 1–12.

Larsson, P., Kaijser, B., Baltzer, I. M., Olbing, S. (1980) An experimental model for ascending acute pyelonephritis caused by *Escherichia coli* or *Proteus* in rats. *J. Clin. Pathol.*, **33**, 408–412.

Levison, M. E., Pitsakis, P. G. (1987) Susceptibility to experimental *Candida albicans* urinary tract infection in the rat. *J. Infect. Dis.*, **155**, 841–846.

Mahoney, S. A., Persky, L. (1963) Observations on experimental ascending pyelonephritis in the rat. *J. Urol.*, **89**, 779–783.

Michaels, E. K., Chimel, J. S., Plotkin, B. J., Schaffer, A. J. (1983) Effect of D-mannose and D-glucose on *Escherichia coli* bacteriuria in rats. *Urol. Res.*, **11**, 97–102.

Miller, T. (1983) Genetic factors and host resistance in experimental pyelonephritis. *J. Infect. Dis.*, **148**, 336.

Miller, T. E., Findon, G. (1988) Exacerbation of experimental pyelonephritis by cyclosporin A. *J. Med. Microbiol.*, **26**, 245–250.

Miller, T. E., Lecamwasam, T. P., Ormrod, D. J., Findon, G., Cornish, J. (1987) An animal model for chronic infection of the unobstructed urinary tract. *Br. J. Exp. Pathol.*, **68**, 575–583.

Reid, G., Chan, R. C. Y., Bruce, A. W., Costerton, J. W. (1985) Prevention of urinary tract infection in rats with an indigenous *Lactobacillus casei* strain. *Infect. Immun.*, **49**, 320–324.

Riedasch, G., Ritz, S. E., Bersch, W., Mohring, K. (1981) Antibody coating of bacteria in experimental infection of the urinary bladder. *Invest. Urol.*, **18**, 247–250.

Taylor, P. W., Koutsaimanis, K. G. (1975) Experimental *Escherichia coli* urinary tract infection in rat. *Kidney Int.*, **8**, 233–238.

Uchida, K., Samma, S., Rinsho, K., Warren, J. R., Oyasu, R. (1989) Stimulation of epithelial hyperplasia in rat urinary bladder by *Escherichia coli* cystitis. *J. Urol.*, **142**, 1122–1126.

Uehling, D. T., Wolf, L. (1969) Enhancement of the bladder defense mechanism by immunization. *Invest. Urol.*, **6**, 520–526.

Vivaldi, E., Cotran, R., Zangwill, D. P., Kass, E. H. (1959) Ascending infection as a mechanism in pathogenesis of experimental non-obstructive pyelonephritis. *Proc. Soc. Exp. Biol. Med.*, **102**, 242–244.

Wan, J., McGuire, E. J., Kozminski, M. *et al.* (1994) Intravesical instillation of gentamicin sulfate: *in vitro*, rat, canine, and human studies. *Urology*, **43**, 531–536.

Chapter 53

The Rabbit Model of Catheter-associated Urinary Tract Infection

D. W. Morck, M. E. Olson, R. R. Read, A. G. Buret and H. Ceri

Background of the model

Catheterization of the urinary bladder is a common procedure in a wide variety of patients (Stamm, 1975), including severely compromised individuals under intensive care. Bacteriuria and catheter-associated urinary tract infections (UTI) are inevitable sequelae in chronically catheterized patients, where the implications in costs and patient well-being are enormous (Warren *et al.*, 1981). Microorganisms gain entry into the urinary tract (UT) during the catheterization procedure, through the lumen of the catheter following insertion of the catheter, or periurethrally (Schaeffer, 1986) following catheterization. A variety of approaches have been employed to delay or prevent entry into the UT, including daily care and topical antibiotics at the urethral meatus (Burke *et al.*, 1981), closed drainage systems (Thornton and Andriole, 1970), antibiotic irrigation of these systems (Warren *et al.*, 1978), antibacterial agents in the urine bag (Maizels and Schaeffer, 1980), retrograde contamination guards (Khoury *et al.*, 1989), antimicrobial agents impregnated in catheters (Takeuchi *et al.*, 1993) and systemic antibiotics (Warren, 1990). Evaluation of the usefulness of these methods has been challenging. Lack of clear recommendations regarding the use of many of these strategies, including systemic antibiotics, is possibly the result of unequivocal data and conflicting data from clinical studies in which the full extent of the infection can not be assessed. Knowledge of the nature of the adherent bacteria (biofilm bacteria) on the urinary catheter and associated with adjacent tissues (Hessen and Kaye, 1989) and the potentially unique characteristics of these bacteria (Brown *et al.*, 1988; Costerton, 1988; Costerton and Lappin-Scott, 1989) is crucial in dealing with this type of infection. Several early descriptions of catheter-associated urinary tract models are available (Goodpasture *et al.*, 1982; Johnston *et al.*, 1983; Nickel *et al.*, 1985a; Murphy *et al.*, 1986; Virtanen and Andersson, 1986; Glahn *et al.*, 1988; Barsanti *et al.*, 1992); however, in our hands these models are less workable than the rabbit model of catheter-associated urinary tract infection described here. This model of catheter-associated bacteriuria and urinary tract infection has allowed a more complete clinical, gross, microscopic and bacteriological assessment of the use of antibiotic therapy in a system modelling chronic indwelling urinary catheter situations (Olson *et al.*, 1989; Morck *et al.*, 1993, 1994). Variations of the model have been used for evaluating other aspects of catheter care (Khoury *et al.*, 1989; Nickel *et al.*, 1991), but few groups use the infection model, primarily because of its complexity and the firm commitment required to make it function properly.

Animal species

Although a variety of rabbit strains could be employed within this model, we have used exclusively healthy, young, male, New Zealand White rabbits. Male rabbits (4.0 kg) are used because of their relatively large urethral size (8–10 French catheters can be used) and the relative ease of catheterizing the male animal. Large rabbits are essential for success; however, they should be young, healthy and free of pasteurellosis, bordetellosis, coccidiosis and other relatively common lapin infections. Animals must be free of other disease, as complicating problems such as latent pasteurellosis are intolerable in such a costly, labour-intensive model. Correct placement of a central venous catheter (described below) is essential for success of the model and delivery of intravenous antibiotics, if required in the study.

Preparation of animals

Animals should be purchased from certified suppliers that guarantee size and health status of the male rabbits. Typical quarantine should involve individual or group isolation for 14 days prior to the study and daily examination by a qualified laboratory animal veterinarian. Specialized housing is required following surgical procedures. Following surgery the rabbits are housed in Plexiglass cages, which allow maintenance of the urinary catheter and the intravenous catheter and feeding, but restrict the animal from turning and forcefully exteriorizing the catheters. Of course, all animal handling and experimental procedures must be in accordance with local guidelines and be approved by the

Handbook of Animal Models of Infection
ISBN 0-12-775390-7

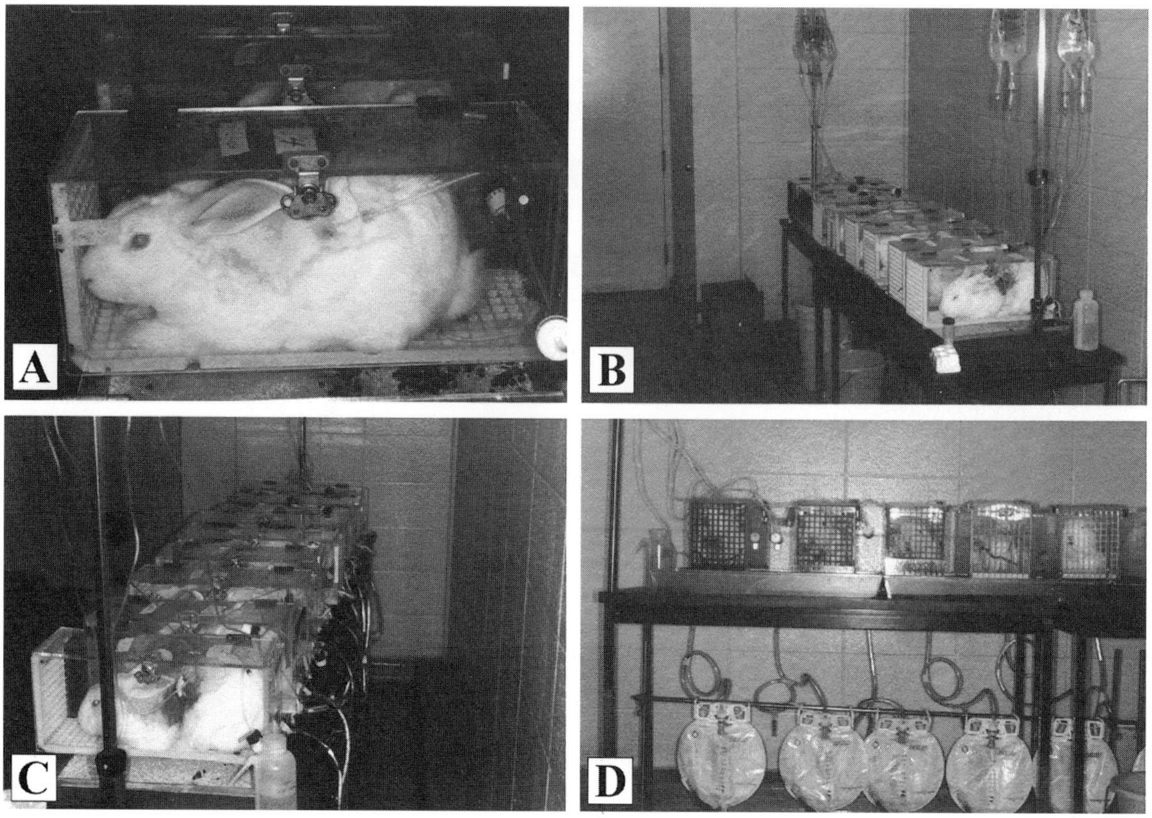

Figure 53.1 The specialized housing conditions required for the rabbit model of catheter-associated urinary tract infection. (**A**) A rabbit housed within the specially designed Plexiglass cage. (**B**) Eight cages set on a table, as they would be during an experiment involving the rabbit model. (**C**) The rear of the cages, with catheters and drainage collection bags. (**D**) Method of hanging collection bags beneath the table.

institutional Animal Ethics Committee. The specialized caging, housing system and general setup of the model are illustrated in Figure 53.1.

Details of surgery

Overview

Aseptic placement of the Foley urinary catheter and aseptic placement of a central venous catheter in the right jugular vein are essential for the success of this model, particularly if one is interested in a controlled bacteriological challenge. General surgical principles such as skill, speed, correct tissue handling technique and strict asepsis are required for the model to proceed smoothly.

Materials required

- Halothane gas anaesthesia and surgical supplies
- 8 French Foley catheters (Bard Urological: Bardex All Silicone Foley Catheter™–165808)
- Sterile lubricant
- Sterile saline
- Sterilized 5 French infant feeding tubes (Medicraft Limited: Cat. 100-515)
- High quality iris scissors
- Intravenous drip sets
- Extension sets (Baxter: Control-A-Flo Regulator with Injection Site 2C7591)
- 5% glucose and electrolytes (Travenol)
- Urinary drainage sets (Kendall: Mono-flo 30511)
- Specialized Plexiglass cages (Figures 53.1 and 53.3)
- Intravenous poles
- Acepromazine maleate
- *Escherichia coli* WE6933 (lactose-negative, streptomycin-resistant)
- Microbiological supplies and laboratory
- Histology supplies and laboratory
- Electron microscopy supplies and laboratory

Anaesthesia and preparation for surgery

Rabbits are placed in a surgical plane of anaesthesia with 4% halothane and oxygen gas (2 l/min) according to standard procedures. This involves placement of the animal in a "cat bag" and using an anaesthetic face mask to deliver the halothane gas. Premedications are not used as the surgery is very quick and delayed recovery is undesirable. Following successful anaesthesia, the hair on the right neck, dorsal cervical region and dorsal thoracic region is clipped using a No. 40 blade. A conventional surgical scrub is performed on the area with povidone-iodine soap (3 ×), 70% ethanol/water (vol./vol.; 1 ×), and it is painted with povidone-iodine (1 ×) in preparation for later placement of the intravenous catheter. The perineal region is draped with a sterile paediatric drape and the penis is manually externalized and painted with povidone-iodine.

Surgical procedure

Following preparation of the penis, sterile gloved hands are used to insert an 8–10 French urinary Foley catheter (tip lubricated) through the urethral meatus and urethra into the urinary bladder. Male rabbits have a urethral diverticulum and care must be taken to avoid placing the catheter tip in this diverticulum or causing urethral trauma with multiple attempts at the procedure. Fresh rabbit cadavers should be used to practise the procedure prior to attempting a

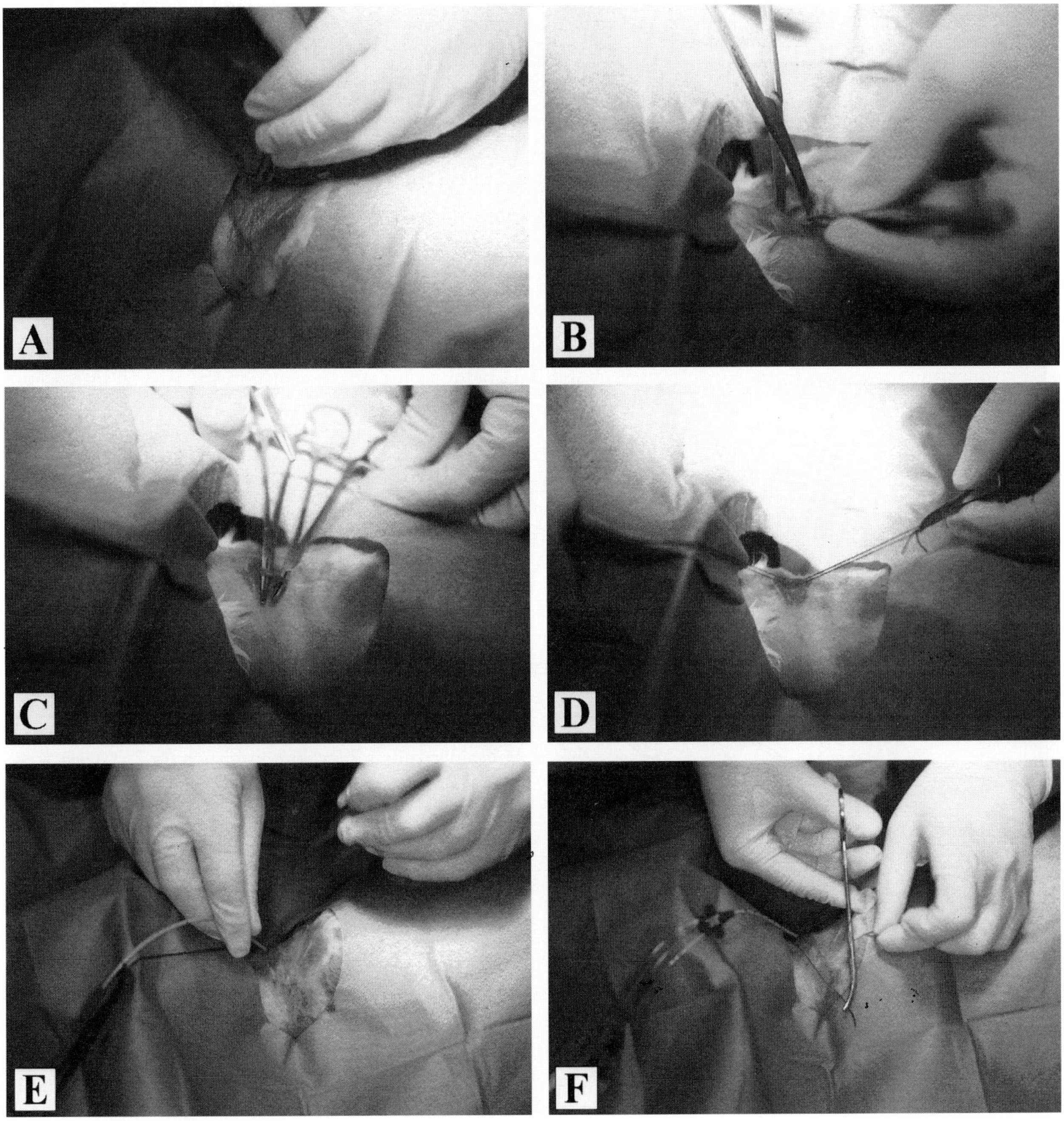

Figure 53.2 The correct surgical placement of a jugular intravenous catheter in a rabbit. (**A**) Making the skin incision in a properly prepared and draped animal. (**B**) Bluntly dissecting the jugular vein from tissues of the ventral lateral neck. (**C**) Isolation of the right jugular vein with two mosquito haemostats. (**D**) Placement of two silk ligatures around the right jugular vein. (**E**) Following incision of the vein with iris scissors, insertion of the catheter toward the heart. (**F**) Securing the catheter in place with silk ligatures (one cranial to the incision and the second caudal to the incision).

catheterization. Performing a urethral dissection on a specimen can aid the surgeon in appreciating the anatomy of the rabbit urethra. Once successfully placed in the bladder the catheter balloon is inflated with 10 ml of sterile saline solution. It is imperative to fill the balloon to this extent to prevent accidental expulsion by the animal following surgery. We have found that Bard catheters provide the best results; less expensive substitute catheters often can not survive 10 ml of balloon inflation and if deflated the catheter is very easily expelled by the conscious animal. A properly placed and properly inflated catheter can be readily palpated within the urinary bladder through the abdominal wall.

For the placement of the intravenous catheter a fenestrated paediatric sterile drape is placed over the cervical operative area, and a small (1 cm) incision made in the right lateral cervical region (jugular groove) directly over the right jugular vein. The jugular vein is isolated by blunt dissection and 2 silk sutures (3–0) are placed around the vein. The vein is incised (partially) with iris scissors, and the 5 French infant feeding tube (prefilled with sterile i.v. fluids) is advanced into the vein approximately 2–4 cm. Alternatively, commercially available 5–7 French central venous catheters can be used. It is crucial that the tip of the catheter is not located within the heart but is placed anterior to the chambers of the heart. The catheter is then secured in place with the two silk ligatures, one caudal to the venous incision and one cranial to the venous incision. Correct placement of an i.v. catheter is illustrated in Figure 53.2. The remainder of the catheter is tunnelled through the subcutaneous space dorsally and exited through the skin on the dorsal thoracic area of the animal. The catheter is secured to the skin using butterfly wings and 2–0 nylon monofilament suture material and a cutting needle. Excess space on this dorsal exit site can be abrogated using 4–0 polygalactin sutures and a tapered needle. The jugular surgical site is closed with a combination of 4–0 polygalactin suture material and a cutting needle, using a subcuticular pattern and/or tissue adhesive. The catheter is connected to a flow regulator and a standard intravenous drip set and bag. Intravenous glucose (5%) and electrolytes are used at a drip rate of 50–60 ml. The rate is adjusted to provide a urinary output of 800–1200 ml per 24 hours. The animal is placed in the specialized cage and the intravenous and urinary catheters are fixed in their respective locations at the top side edge and rear floor edge of the cage (Figure 53.3).

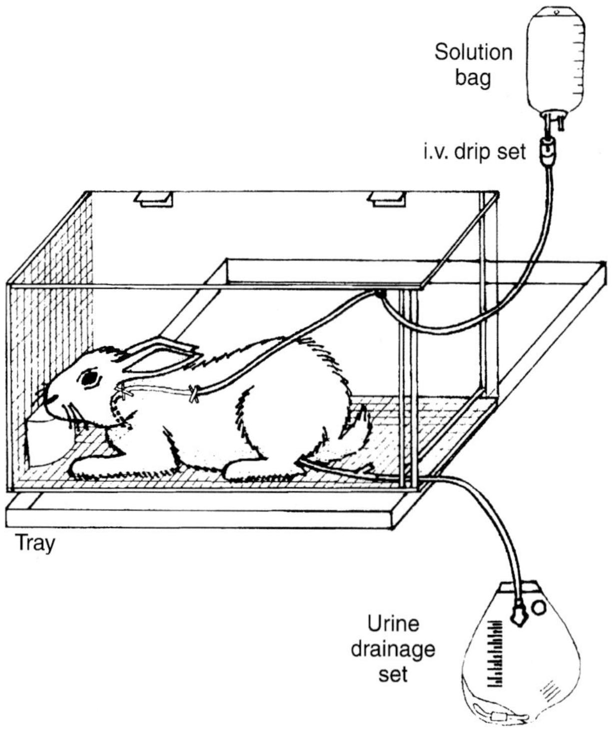

Figure 53.3 Diagram of the rabbit model of catheter-associated urinary tract infection, outlining the attachments of the intravenous and urinary drainage lines (arrows) and the positioning of the animal within the specialized Plexiglass cage.

Infection procedure

The infection procedure can be conducted in any of several fashions, including allowing normal bacterial species found in and on the rabbit to induce infection in a "natural" fashion, or the more controlled inoculation of "marked" bacteria such as *Escherichia coli* WE6933. This microorganism is a lactose-negative and streptomycin-resistant bacterium that can be monitored through the daily quantitative culture of collection bag urine and bladder urine on McConkey agar containing streptomycin. The animals can be infected by any of several methods including: (1) a single direct placement of bacteria (10^8 viable cells) in the catheter drainage tube followed by restricted urinary flow for 15 minutes; (2) direct contamination of the interior of the drainage bag with similar quantities of bacteria; or (3) painting the bacterial inoculum on the urethral meatus (this typically requires daily administration for 2–3 days). Viable bacterial counts (total bacteria and lact–/strep R *E. coli*) can be monitored during the approximate 72-hour period required for the infection to establish.

Postoperative care

The quality of the postoperative care in this animal model, as with most models, can be the difference between success and failure. The rabbit model of catheter-associated UTI is technically very demanding. Constant veterinary support of these animals is essential. Technical staff must monitor the animals 24 hours a day, which makes this model trying and expensive; particularly if the course of the study is chronic.

The rabbits require intravenous fluids with glucose, but will feed occasionally and quality certified rabbit chow must be offered. Fluid requirements can be easily met through the intravenous administrations. When animals become agitated due to infection and/or blockage, tranquilization can be used (acepromazine maleate 0.1–1.0 mg/kg q.8 hour); this can be added to the intravenous fluid to administer the agent slowly and uniformly. The flow rates can be revised and the Foley catheter can be gently aspirated to dislodge debris.

Storage and preparation of inocula

Escherichia coli WE6933 is stored frozen (–85°C) on brain–heart infusion slants. New slants are used for each inoculum preparation to avoid excessive subculture. The frozen culture is plated on tryptic soy agar and McConkey–streptomycin (100 µg/ml) agar, to confirm the markers, and is then grown for 14 hours in tryptic soy broth (TSB) at 37°C, washed three times in sterile buffered saline (pH 7.2) and suspended in TSB at 1.0×10^9 viable cells per millilitre. All inocula are estimated by McFarland nephelometry and confirmed by quantitative spread plating.

Key parameters to monitor infection and response to treatment

Close monitoring of the infection process is central to the success of the model. No single parameter is indicative of the state of the animal; however, the collective impression from all of the following assessments by an individual appreciative of clinical signs in animals can predict impending difficulties with animals and allow appropriate intervention. Maintaining adequate urine flow within the contaminated system is crucial for the animals to survive for the 5–8 days required to measure or establish a response to an antibiotic. Appreciation of normal rabbit behaviour is therefore imperative to determine early urinary blockage in these animals. The central parameters in evaluating response to therapy are those obtained at post-mortem examination of the animals.

Clinical signs

Clinical evaluation of animals in this model is a constant procedure. Demeanour and feeding behaviour are monitored continually. Animals are checked every 1–4 hours by technical staff and daily by the laboratory animal veterinarian.

Urine parameters

Urine is monitored daily for appearance (cloudiness, clots, clumps, debris), pH, specific gravity, protein content, glucose, ketones and occult blood. These parameters, along with behavioural changes in the animals, can be the initial indicators of blockage.

Urine microbiology

The daily evaluation of bacterial numbers in drainage bag urine and port urine (bladder urine) samples provide critical information on the state of infection. As illustrated in Figure 53.4, the numbers of viable bacteria rise rapidly (within 24 hours of inoculation) and any antibacterial response can be followed over time quite easily. Both total viable bacteria (cultured on a non-selective medium) and the marked *Escherichia coli* (McConkey–streptomycin agar) should be quantified daily to evaluate the success of the inoculation and the involvement of autochthonous bacteria (primarily Gram-positives from the prepuce and skin) in the infection. Effect of antibiotics on the planktonic (freely floating) bacteria in the urine can be assessed in this manner within the model, but it gives no appreciation of the extent of bacterial colonization of the catheter or adjacent tissues.

Gross pathology

At the termination of the experiment it is crucial to conduct a necropsy examination that is amenable to microbiological sampling. Since deflating the Foley balloon and pulling the catheter would clearly disturb biofilm structure, we remove the UT from the animal with the catheter *in situ* to avoid this disruption of biofilm adherent to the catheter and bacteria adherent to associated tissue adjacent to the catheter. This also allows reliable microbiological sampling from several regions within the UT. Following euthanasia with an intravenous overdose of sodium pentabarbital, the animal's ventral abdomen and perineum are clipped, washed, scrubbed, washed with ethanol and painted with povidone-iodine. A ventral midline incision is made and Allis forceps are used to manipulate the opening in the abdominal wall. A urine sample from the bladder is collected aseptically with a syringe and fine needle for quantitative bacteriology, quantification of antibiotic, etc. The kidneys and ureters are bluntly removed and the bladder and urethra are dissected to the pelvis. The pelvis is grasped with gloved hands and broken using a firm twist. The pelvic and perineal urethra, still containing the catheter, is dissected free of the carcass and the UT is placed on a sterile drape for dissection and photography. Sterile instruments (tissue forceps and scissors) are used to longitudinally split the urethra and open the urinary bladder with the catheter still in place. Gross pathology photographs can be taken at this time (Figure 53.5). Samples of urethra are taken for microbiology, histology, and electron microscopy if required.

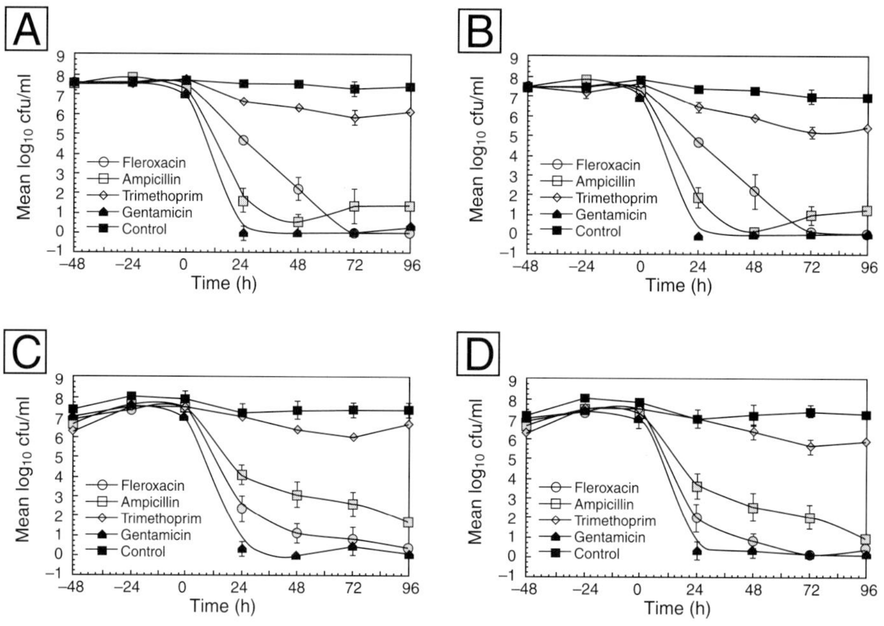

Figure 53.4 Graphical representation of the planktonic bacteria in urine within the rabbit model of catheter-associated urinary tract infection. (**A**) Total viable bacteria in urine obtained from the drainage collection bag from animals on various antibiotic treatments or controls. (**B**) Viable *E. coli* WE6933 from urine obtained from the drainage collection bag from animals on various antibiotic treatments or controls. (**C**) Total viable bacteria in urine obtained from the drainage collection port (i.e., bladder urine) from animals on various antibiotic treatments or controls. (**D**) Viable *E. coli* WE6933 in urine obtained from the drainage collection port (i.e., bladder urine) from animals on various antibiotic treatments or controls. Modified from Morck *et al.*, 1994.

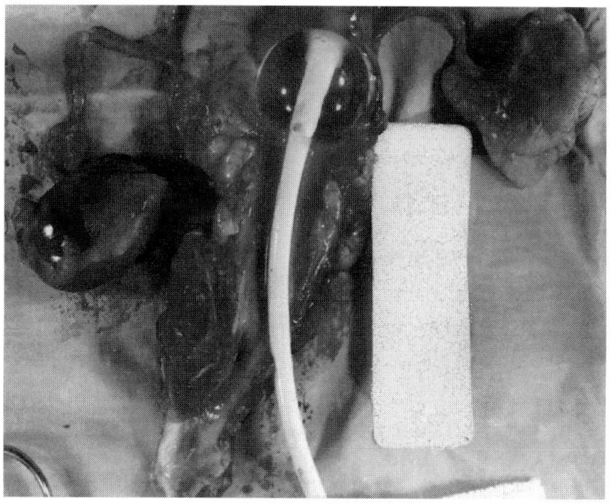

Figure 53.5 A photograph of the urinary tract of a rabbit experimentally infected with *E. coli* WE6933 in the rabbit model of catheter-associated UTI. This urinary tract has been removed from the euthanized animal aseptically and with the Foley catheter still in position within the bladder.

Tissue and catheter microbiology

At necropsy examination specimens of the urethral mucosa can be aseptically collected for microbiology. Samples are collected into pre-weighed sterile test tubes containing a precisely known volume of sterile buffer, homogenized and serially diluted. A spread plate method is used to evaluate total viable bacteria and numbers of viable *Escherichia coli* WE6933. Plates can be read and counted at 24 hours and the number of bacteria expressed in $\log_{10}$ viable bacteria per weight of tissue. For catheter specimens, the Foley balloon is deflated, a 1.0 cm length of the catheter is collected from the catheter tip (within the bladder) and an identical length of catheter is collected from the urethral region. Example data for tissue and catheters recovered from rabbits infected with *Escherichia coli* WE6933 and treated with one of several antibiotics are shown in Table 53.1. These data suggest that, despite antibiotic treatment and subsequent elimination of bacteria in urine and on the catheter (as assessed at necropsy), some bacteria can remain viable associated with tissues and act as a focus to reinfect the urinary drainage system.

Microscopic pathology

Samples for microscopic assessment for pathological abnormalities can be fixed in a variety of fashions, depending on the need. For standard histopathology we use conventional 10% neutral buffered formalin. However, if visualizing bacteria is essential and microscopic pathology of cells is not to be assessed, Carnoys fixative and methylene blue basic fuchsin staining of JB4 methacrylate (Polysciences) sections

Table 53.1 Tissue microbiology in the rabbit model of catheter-associated urinary tract infection (mean $\log_{10}$ viable *Escherichia coli* WE6933 per cm^2 or per gram*); from Morck *et al.*, 1994

	Fleroxacin (20 mg/kg)	*Ampicillin (10 mg/kg)*	*Trmp–Sulfa (10 mg/kg)*	*Gentamicin (2 mg/kg)*	*Untreated control*
Rabbits (*n* value)	10	8	8	8	9
Microbiology sample					
Urethral catheter exterior	0.00	0.00	3.81	0.00	4.57
Bladder catheter exterior	0.00	0.00	4.67	0.00	4.61
Urethral catheter lumen	0.00	0.00	4.10	0.00	4.24
Bladder catheter lumen	0.00	0.00	4.09	0.00	2.73
Bladder mucosa*	0.00	1.42	4.39	1.42	4.21
Urethra*	0.00	1.49	4.55	1.46	3.89

* From tissues obtained at post mortem.

is ideal. Fresh paraformaldehyde (4%) should be used if tissues are required for immunological labelling of components in the system (i.e. *Escherichia coli* WE6933). Pathological assessment of the urethra, bladder (several locations), ureters and kidneys can be performed. Histopathology scoring systems can be employed, whereby a histopathologist blinded to the study evaluates the urethra, bladder, ureter and kidney specimens for several pathological abnormalities (Morck *et al.*, 1993, 1994). With studies employing high *n* values this can be a useful exercise.

Table 53.2 Urine concentrations of antibiotic in the rabbit model of catheter-associated urinary tract infection – values are means (μg/ml); from Morck *et al.*, 1994

Antibiotic	*Peak*	*Trough*
Fleroxacin	131	15
Ampicillin	629	103
Trimethoprim–Sulfamethoxazole	361	0
Gentamicin	22	0.4

Electron microscopy

Scanning electron microscopy (SEM) can be routinely used to evaluate catheters. Catheter pieces (5 mm) obtained at post-mortem are placed into 5.0% glutaraldehyde in 0.1 M sodium cacodylate buffer and fixed for 48 hours at 4°C. The specimens are very gently washed in three changes of buffer, and very gently dehydrated in graded series of ethanol and freon. Specimens are air-dried from freon and sputter-coated with palladium gold. Photomicrographs generated by SEM of urinary catheter surfaces from this model are shown in Figure 53.6. Blinded scoring systems for biofilm presence and bacterial counts per field of view can be employed for less subjective assessment, if necessary.

Urine antibiotic concentrations

The measurement of peak and trough urine antibiotic concentrations is important to establish the efficacy of an antibiotic or to explain the lack of efficacy of an antibiotic. Depending on the antibiotic in question, bioassays, radiometric methods or chromatographic quantification (as discussed in Section I of this textbook) can be used. Table 53.2 illustrates the peak and trough concentrations of several antibiotics evaluated using the rabbit model of catheter-associated UTI. These data, together with the minimum inhibitory concentration (MIC), minimum bactericidal concentration (MBC) and minimum biofilm eradicating concentration (MBEC; Ceri *et al.*, 1996a,b,c), provide rationale for efficacy of particular antibiotics within the model. It is clear that if peak and trough concentrations of the antibiotic are not above the MBEC values one could suggest that the antibiotic may not be effective against the bacteria. MBEC may have predictive value in this model and in treatment of biofilm-associated infections (Ceri *et al.*, 1996b).

Minimum biofilm eradicating concentration of antibiotics

The MBEC value is a new concept to many infectious disease specialists. It is a measurement of the antibiotic concentrations required to significantly alter an established biofilm composed of bacteria. Classically, MIC and MBC have been used as indicators and predictors of effective antibiotics for specific microorganisms. MBEC provides similar information, but it pertains to bacteria existing adherent to surfaces, not bacteria floating freely in nutrient-rich medium, as in the MIC and MBC assays. MBEC has been determined using modified Robbins devices (MRD; Nickel *et al.*, 1985b,c; Morck *et al.*, 1994) but this device and method is cumbersome and very slow. We have developed the MBEC assay using the Calgary Biofilm Device (Ceri *et al.*,

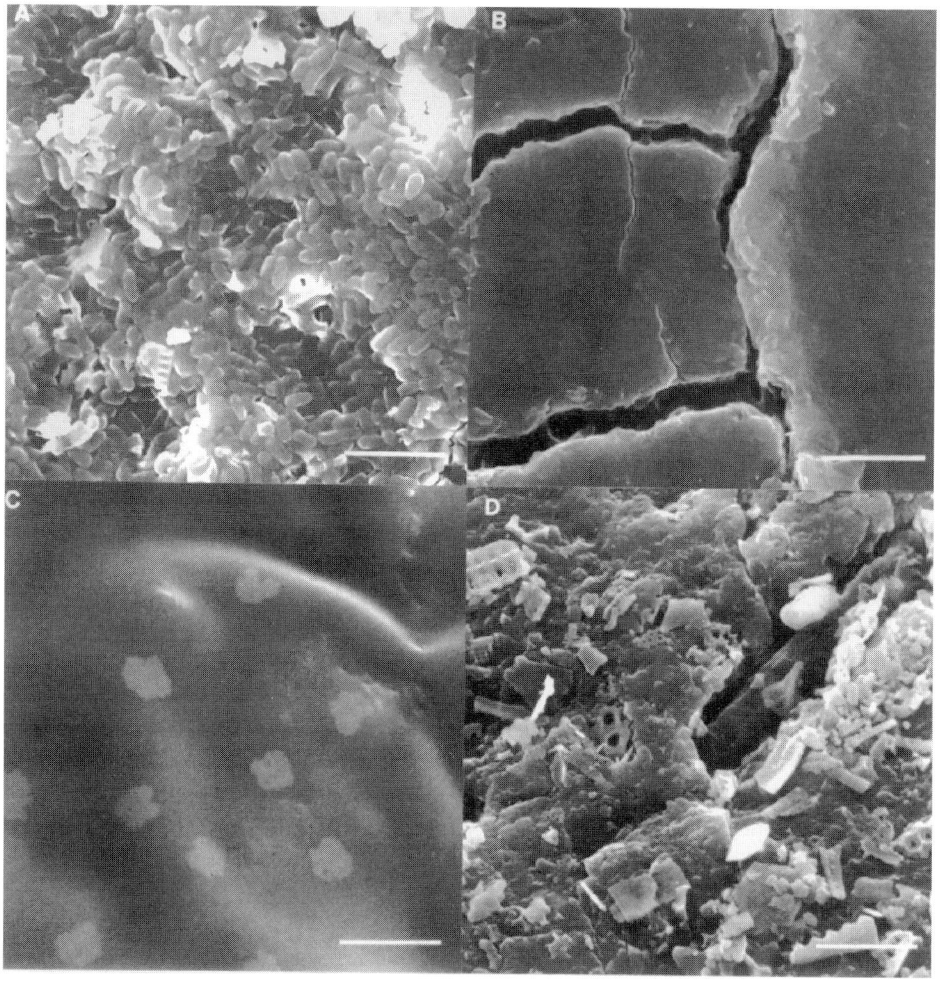

Figure 53.6 Scanning electron micrographs of urinary catheter surfaces (urethral region) from the rabbit model of catheter-associated urinary tract infection. Catheter surfaces were retrieved post mortem from a control rabbit (**A**), a rabbit treated with 30 mg/kg fleroxacin (**B**), a 20 mg/kg rabbit (**C**), and a rabbit treated with fleroxacin 10 mg/kg (**D**). Note the thick bacterial biofilm present in **A** with numerous rod-shaped bacterial cells enveloped in extracellular glycocalyx material. In **B** and **C** bare regions of catheter surface predominated. Catheters from rabbits treated with the lowest dosage of fleroxacin (**D**) had surfaces covered with debris-like material, but few recognizable bacterial cells were evident. Bars equal 5.0 μm. Modified from Morck *et al.*, 1993.

1996a,b,c; Olson *et al.*, 1996), which facilitates quick determination of the MBEC. Data generated on *Escherichia coli* WE6933 for MIC, MBC, MRD-derived MBEC and Calgary-Biofilm-Device-derived MBEC are illustrated in Table 53.3. On the basis of these data one could suggest that the MBEC assay quickly and reliably predicts the success of the fluorinated quinolone fleroxacin in this model of catheter-associated UTI.

Antimicrobial therapy

A variety of antimicrobial agents have been evaluated in the model (Olson *et al.*, 1989; Morck *et al.*, 1993; Morck *et al.*, 1994) and several strategies to prevent infection have also been examined (Nickel *et al.*, 1991; Khoury *et al.*, 1989).

Pitfalls (advantages/disadvantages) of the model

The rabbit model of catheter-associated urinary tract infection is an excellent model for examining antibiotic chemotherapeutics in an acute or chronic indwelling catheterized patient. It provides information regarding the effects of antibiotic therapy on planktonic bacteria floating freely in the urine, but also allows microbiological assessment of the bacteria adherent to the catheter and bacteria adherent to the adjacent tissues at post-mortem examination. The model provides crucial pathological information, as well as pharmacokinetic data and pharmacodynamic properties of an antibiotic in rabbits, if used for such purposes. The model can be difficult to operate, as the animals must be dealt with as if they were intensive care patients. Adequate technical training is essential for success of the model. Careful evaluation of the animals by these animal care technicians and correct decisions as regards seeking advice from the veterinarian can make this model proceed in a smooth fashion. A firm commitment must be made by the investigator, the veterinarian and the technical staff in order to successfully conduct a study using this model. Those wary of extended work days should not embark on such studies.

Contributions of the model to infectious disease therapy

Profiling of antibiotic treatment

The rabbit model of catheter-associated UTI has been useful in establishing efficacy of several antibiotics (Olson *et al.*, 1989; Morck *et al.*, 1994), conducting dose-titration studies (Morck *et al.*, 1993), and evaluating prototypes of infection prevention methods in closed drainage systems (Khoury *et al.*, 1989; Nickel *et al.*, 1991). The data from the model was used as a theoretical basis for human treatments of uncomplicated (Iravani, 1993; Whitby *et al.*, 1993; Pummer, 1993) and complicated UTI (Pittman *et al.*, 1993; Childs, 1993; Pummer, 1993; Naber and Sigl, 1993; Cox, 1993; Gelfand *et al.*, 1993). The model has generated specific data suggesting that fleroxacin may be useful for device-related infections of the urinary tract by susceptible bacteria (Morck *et al.*, 1994) and can be used to assess efficacy of antimicrobial compounds impregnated within urethral catheter materials. The potential for use in evaluating local therapy of the urinary tract, topical treatment of the urethral meatus, mechanisms of inflammation and immunity in catheter-associated urinary tract infection are enormous, but little work has been conducted on the model in these areas.

Confirmation of the importance of MBEC

One of the major contributions of the rabbit model of catheter-associated UTI has been the use of the model in confirmation of the inherent resistance of biofilm bacteria and tissue-adherent bacteria to certain antibiotics. MIC data in Table 53.3 suggest that any of fleroxacin, ampicillin, trimethoprim–sulfamethoxazole or gentamicin should be useful in treating an infection in rabbits with this *E. coli*. However, MBEC data, coupled with *in vitro* studies (Korber *et al.*, 1993) and pharmacokinetic and pharmacodynamic information, predict that fleroxacin is a superior antibiotic within this model. These data clearly support the difference in antimicrobial susceptibility between planktonic and adherent bacterial populations *in vivo* and demonstrate the more precise predictability of chemotherapeutic efficacy based on MBEC values determined *in vitro*. The MBEC concept has led to the more extensive evaluation of antibiotics against adherent populations of bacteria (see Section I), has contributed to the development of assays to evaluate susceptibility of adherent bacteria (Ceri *et al.*, 1996a,b) and has contributed to the confirmation of the complexity of interactions of bacterial communities (Shapiro, 1991) in a mammalian system.

Table 53.3 Antibiotic sensitivities of *Escherichia coli* WE6933 – values are μg/ml; modified from Morck *et al.*, 1994. MIC, minimum inhibitory concentration on planktonic bacteria; MBC, minimum bactericidal concentration on planktonic bacteria; MRD, minimum biofilm eradication concentration from the modified Robbins device (MRD); MBEC, minimum biofilm eradication concentration from the Calgary Biofilm Device)

Antibiotic	*MIC*	*MBC*	*MRD*	*MBEC*
Fleroxacin	0.16	0.32	2.0	2.0
Ampicillin	2.0	4.0	>16.0*	>1024*
Trimethoprim–sulfamethoxazole	1.0	>128	>256*	>256[a]
Gentamicin	0.5	1.0	>32.0*	512
Amdinocillin	2.0	32.0	>256*	256

* Experiment not continued therefore exact value not determined.

References

Barsanti, J. A., Shotts, E. B., Crowell, W. A., Finco, D. R., Brown, J. (1992) Effect of therapy on susceptibility to urinary tract infection in male cats with indwelling urethral catheters. *J. Vet. Int. Med.*, **6**, 64–70.

Brown, M. R. W., Allison, D. G., Gilbert., P. (1988) Resistance of bacterial biofilms to antibiotics: a growth-rate related effect? *J. Antimicrob. Chemother.*, **22**, 777–780.

Burke, J. P., Garibaldi, R. A., Britt, M. R., Jacobson, J. A., Conti, M., Alling, D. W. (1981) Prevention of catheter-associated urinary tract infections: efficacy of daily meatal care regimens. *Am. J. Med.* **70**, 655–658.

Ceri, H., Stremick, C., Olson, M. E., Read, R. R., Morck, D. W., Buret, A. G. (1996a) Antibiotic sensitivity of biofilm and planktonic human pathogens utilizing the Calgary Biofilm Device. Poster C19 Session C, *Abstracts of the Proceedings of the American Society for Microbiology Conference on Microbial Biofilms*, p. 36.

Ceri, H., Stremick, C., Olson, M. E., Read, R. R., Morck, D. W., Buret, A. G. (1996b) Antibiotic sensitivity of important veterinary pathogens: comparing planktonic and biofilm grown bacteria with the Calgary Biofilm Device. Poster C20 Session C, *Abstracts of the Proceedings of the American Society for Microbiology Conference on Microbial Biofilms*, p. 37.

Ceri, H., Kao, J., Stremick, C. *et al.* (1996c) Biocide sensitivity of important food and water contaminants: comparing planktonic and biofilm grown bacteria with the Calgary Biofilm Device. Poster C21 Session C, *Abstracts of the Proceedings of the American Society for Microbiology Conference on Microbial Biofilms*, p. 37.

Childs, S. (1993) Fleroxacin versus norfloxacin for oral treatment of serious urinary tract infections. *Am. J. Med.*, **94**, (Suppl. 3A), 105S–107S.

Costerton, J. W. (1988) Structure and plasticity at various organizational levels in the bacterial cell. *Can. J. Microbiol.*, **34**, 513–521.

Costerton, J. W., Lappin-Scott, H. M. (1989) Behaviour of bacteria in biofilms. *Am. Soc. Microbiol. News*, **55**, 650–654.

Cox, C. E. (1993) Comparison of intravenous fleroxacin with ceftazidime for the treatment of complicated urinary tract infections. *Am. J. Med.*, **94** (Suppl. 3A), 118S–125S.

Gelfand, M. S., Simmons, B. P., Craft, R. B., Grogan, J., Amarshi, N. (1993) A sequential study of intravenous and oral fleroxacin in the treatment of complicated urinary tract infection. *Am. J. Med.*, **94** (Suppl. 3A), 126S–130S.

Glahn, B. E., Braendstrup, O., Olseen, H. P. (1988) Influence of drainage conditions on mucosal bladder damage by indwelling catheters. II. Histological study. *Scand. J. Urol. Nephrol.*, **22**, 93–99.

Goodpasture, J. C., Cianci, J., Zaneveld, L. J. (1982) Long-term evaluation of the effect of catheter materials on urethral tissue in dogs. *Lab. Anim. Sci.*, **32**, 180–182.

Hessen, M. T., Kaye, D. (1989) Infections associated with foreign bodies in the urinary tract. In: *Infections Associated with Indwelling Medical Devices* (eds Bisno, A. L. and Waldvogel, F. A.), pp. 199–213. American Society for Microbiology, Washington, DC.

Iravani, A. (1993) Multicenter study of single-dose and multiple-dose fleroxacin versus ciprofloxacin in the treatment of uncomplicated urinary tract infections. *Am. J. Med.*, **94** (Suppl. 3A), 89S–96S.

Johnston, G. R., Stevens, J. B., Jessen, C. R., Osborne, C. A. (1983) Effects of prolonged distention of retention catheters on the urethra of dogs and cats. *Am. J. Vet. Res.*, **44**, 223–228.

Korber, D. R., James, G. A., Costerton, J. W. (1993) Early detection of fleroxacin incorporation and efficacy within *Pseudomonas fluorescens* biofilms using confocal scanning laser microscopy. Abstract P108, Canadian Society of Microbiologists Annual Meeting.

Khoury, A. E., Olson, M. E., Lam, K., Nickel, J. C., Costerton, J. W. (1989) Evaluation of the retrograde contamination guard in a bacteriologically challenged rabbit model. *Br. J. Urol.*, **63**, 384–388.

Maizels, M., Schaeffer, A. J. (1980) Decreased incidence of bacteriuria associated with periodic instillations of hydrogen peroxide into the urethral catheter drainage bag. *J. Urol.*, **123**, 841–845.

Morck, D. W., Olson, M. E., McKay, S. G. *et al.* (1993) Therapeutic efficacy of fleroxacin for eliminating catheter-associated urinary tract infection in a rabbit model. *Am. J. Med.*, **94**(Suppl 3A), 235–305.

Morck, D. W., Lam, K., McKay, S. G. *et al.* (1994) Comparative evaluation of fleroxacin, ampicillin, trimethoprim–sulfamethoxazole, and gentamicin as treatments of catheter-associated urinary tract infection in a rabbit model. *In. J. Antimicrob. Agents*, **4**(Suppl. 2), S21–S27.

Murphy, W. M., Blatnik, A. F., Shelton, T. B., Soloway, M. S. (1986) Carcinogenesis in mammalian urothelium: changes induced by non-carcinogenic substances and chronic indwelling catheters. *J. Urol.*, **135**, 840–844.

Naber, K. G., Sigl, G. (1993) Fleroxacin versus ofloxacin in patients with complicated urinary tract infection: a controlled clinical study. *Am. J. Med.*, **94** (Suppl. 3A), 114S–117S.

Nickel, J. C., Grant, S. K., Costerton, J. W. (1985a) Catheter-associated bacteriuria: an experimental study. *Invest. Urol.*, **26**, 369–375.

Nickel, J. C., Wright, J. B., Ruseska, I., Marrie, T. J., Costerton, J. W. (1985b) Antibiotic resistance of *Pseudomonas aeruginosa* colonizing a urinary catheter *in vitro*. *Eur. J. Clin. Microbiol*, **4**, 213–218.

Nickel, J. C., Ruseska, I., Wright, J. B., Costerton, J. W. (1985c) Tobramycin resistance of *Pseudomonas aeruginosa* cells growing as a biofilm on urinary catheter material. *Antimicrob. Agents Chemother.*, **27**, 619–624.

Nickel, J. C., Grant, S. K., Lam, K., Olson, M. E., Costerton, J. W. (1991) Bacteriologically stressed animal model of new closed catheter drainage system with microbicidal outlet tube. *Invest. Urol.*, **38**, 280–289.

Olson, M. E., Nickel, J. C., Khoury, A. E., Morck, D. W., Cleeland, R., Costerton, J. W. (1989) Amdinocillin treatment of catheter-associated bacteriuria in rabbits. *J. Infect. Dis.*, **159**, 1065–1072.

Olson, M. E., Ceri, H., Morck, D. W., Read, R. R., Buret, A. G. (1996) The Calgary Biofilm Device: a device for the determination of biofilm sensitivity to antibiotics and biocides. Poster C74 Session C, *Abstracts of the Proceedings of the American Society for Microbiology Conference on Microbial Biofilms*, p. 39.

Pittman, W., Moon, J. O., Hamrick, L. C., Jr *et al.* (1993) Randomized double-blind trial of high- and low-dose fleroxacin versus norfloxacin for complicated urinary tract infection. *Am. J. Med.*, **94** (Suppl. 3A), 101S–104S.

Pummer, K. (1993) Fleroxacin versus norfloxacin in the treatment of urinary tract infections: a multicenter, double-blind, prospective, randomized, comparative study. *Am. J. Med.*, **94** (Suppl. 3A), 108S–113S.

Schaeffer, A. J. (1986) Catheter associated bacteriuria. *Urol. Clin. North Am.*, **13**, 735–747.

Shapiro, J. A. (1991) Multicellular behavior of bacteria. *Am. Soc. Microbiol. News*, **57**, 247–253.

Stamm, W. E. (1975) Guidelines for the prevention of catheter-associated urinary tract infections. *Ann. Intern. Med.*, **82**, 386–390.

Takeuchi, H., Hida, S., Yoshida, O., Ueda, T. (1993) Clinical study on the efficacy of a Foley catheter coated with silver-protein in prevention of urinary tract infections. *Hinyokika Acta Urol. Jpn.*, **39**, 293–298.

Thornton, G. F., Andriole, V. T. (1970) Bacteriuria during indwelling catheter drainage: II. Effect of a closed sterile drainage system. *J.A.M.A.*, **214**, 339–342.

Virtanen, T. M., Andersson, L. C. (1986) Toxic catheters and diminished urethral blood circulation in the induction of urethral strictures. *Eur. Urol.*, **12**, 340–345.

Warren, J. W., Platt, R., Thomas, R. J., Rosner, B., Kass, E. H. (1978) Antibiotic irrigation and catheter-associated urinary-tract infections. *N. Engl. J. Med.*, **299**, 570–573.

Warren, J. W., Muncie, H. L. Jr, Bergquist, E. J., Hoopes, J. M. (1981) Sequelae and management of urinary infection in the patient requiring chronic catheterization. *J. Urol.*, **125**, 1–8.

Warren, J. W. (1990) Nosocomial urinary tract infections. In: *Principles and Practice of Infectious Diseases* (eds Mandell, G. L., Douglass, R. G., Bennett, J. E.), pp. 2205–2215. Churchill Livingstone, Edinburgh.

Whitby, M., Brown, P., Silagy, C., Champak, R. (1993) Comparison of fleroxacin and amoxicillin in the treatment of uncomplicated urinary tract infections in women. *Am. J. Med.*, **94** (Suppl. 3A), 97S–100S.

Chapter 54

Subclinical Pyelonephritis in the Rat

G. Findon

Background of model

Studies of patients with significant bacteriuria, who have undergone ureteric catheterization or bladder washout tests, have shown that in 30–60% of cases the infection originates in the upper urinary tract (Fairley *et al.*, 1971; Boutros *et al.*, 1972; Eykyn *et al.*, 1972; Harding *et al.*, 1978; Busch and Huland, 1984). The same studies have demonstrated a poor correlation between upper urinary tract localization and appropriate clinical symptoms. One explanation is that microorganisms may be present in the kidney but not initiate an inflammatory response with associated symptoms. We have developed a rat model of pyelonephritis which distinguishes between lesion-inducing and non-lesion-inducing infection and mimics the above situation. In the model, infection is induced in the left kidney and the non-manipulated contralateral kidney acquires a persistent microbial flora within 48 hours of the initial challenge. The bacterial invasion of the contralateral kidney is not associated with gross or histological changes within the renal parenchyma, but minor foci of inflammation are present beneath the epithelium lining the calyces. This model supports the concept of subclinical pyelonephritis and may explain the absence of symptoms in the clinically equivalent situation in humans.

Animal species

Adult male or female Hooded Oxford (HO) rats, 3–4 months old and weighing 200–250 g, were used to develop this model. In our experience, the HO strain readily acquires a subclinical infection in the contralateral kidney. Other strains, such as Dark Agouti, do not. Before using a specific rat strain, it would be advisable to carry out pilot experiments to assess the ability of the strain to develop subclinical pyelonephritis.

Preparation of animals

Animals are housed in polycarbonate cages on untreated wood shavings and fed Diet 86 and water *ad libitum*. No special care or treatment is required either pre- or post-surgery.

Details of surgery

Overview

Briefly, acute pyelonephritis is induced in the left kidney of HO rats. In this strain of rats, subclinical pyelonephritis develops in the contralateral non-manipulated kidney. The animals are left for various periods after infection before autopsy and analysis of both kidneys.

Materials required

Anaesthetic, hair clippers, skin disinfectant (95% ethanol), scalpel blade, scissors, forceps, sutures, glass micro-capillary pipette and syringe capable of delivering 5 μl samples.

Anaesthesia

Nembutal (pentobarbitone sodium, Techvet Laboratories, Auckland, NZ) was used as the anaesthetic for inducing the infection, but any other short-acting anaesthetic would be suitable.

Surgical procedure and induction of infection

After flooding the shaved skin with alcohol, a flank incision is made on the left side of the rat, just below the rib cage, through both the skin and muscle layer. The kidney is externalized by gently pulling on the renal fat pad and 5 μl of the inoculum is injected into each pole of the left kidney using an SMI Micro/Pettor and a glass microcapillary, pulled to a fine point. The total inoculum contains 10^6 viable *Escherichia coli* O75.

Handbook of Animal Models of Infection
ISBN 0-12-775390-7

Wound closure

The wound in the muscle layer and the skin is closed with a running suture.

Postoperative care

No special postoperative care is required. Animals recover from the surgery and resume normal activities within a few hours. The animals are housed in groups of four to six animals per cage, depending on size.

Storage and preparation of inocula

The strain of *E. coli* O75 used to characterize the model was a human isolate, maintained on Dorset Egg medium at 4°C. Weekly subcultures were made on to blood agar and, for use, a colony subcultured into nutrient broth for overnight incubation. After washing the culture three times in normal saline, the inoculum is made up to 10^6 cfu in 10 μl saline.

Key parameters to monitor infection and response to treatment

Bacteriological analysis

Autopsies were carried out at regular intervals over 60 days. The microbial numbers were quantified in the directly challenged left kidneys with classic pyelonephritis, and in the non-manipulated contralateral kidneys. The kidneys were removed aseptically, decapsulated and weighed, then homogenized in 9 ml of normal saline using a rotating PTFE pestle in a heavy-walled glass tube. Serial 10-fold dilutions of the homogenate were made in normal saline and Columbia agar pour plates were prepared to determine the bacterial numbers per gram of renal tissue.

Bacterial numbers in the directly challenged (left) kidneys reached a peak at 24 hours then declined slowly over the following week to reach a stable population of 10^4–10^5 bacteria per gram of renal tissue. The contralateral kidneys with subclinical pyelonephritis were sometimes colonized within 24 hours and, by 3 days, microbial invasion had occurred in all of them. Bacterial numbers were maintained at a constant level, but in a few kidneys spontaneous clearance occurred (Figure 54.1).

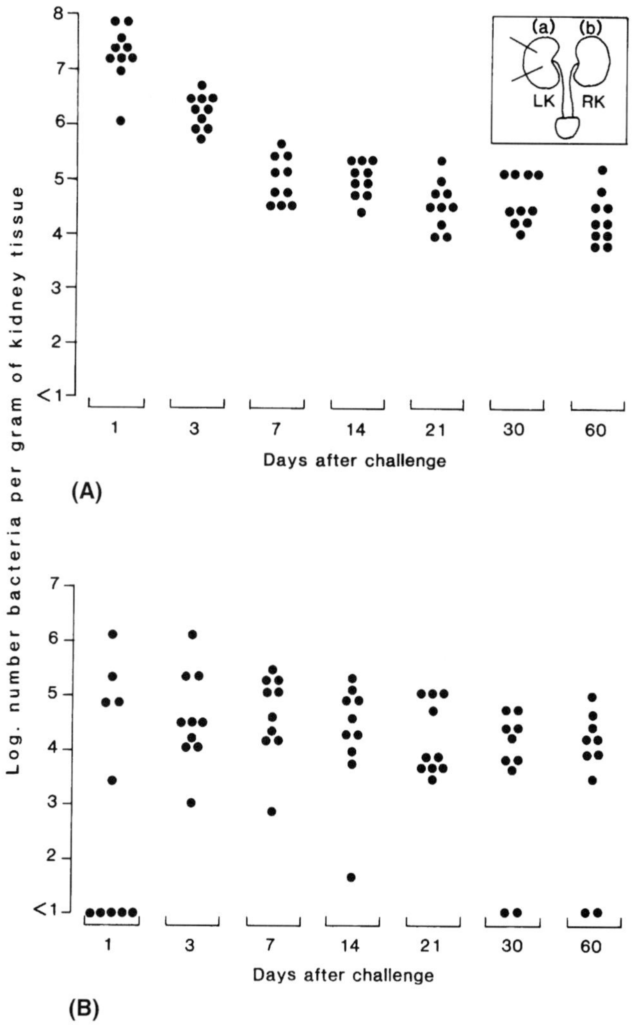

Figure 54.1 Bacteriological status, over a 60-day period, of the left and right kidneys in experimentally induced pyelonephritis. (**A**) Infection was established in the left kidney using a direct challenge with *E. coli* O75. (**B**) Right kidneys were not manipulated at any stage. (Reproduced with permission from Miller, T. E., Findon, G., Rainer, S. P. and Gavin, J. B. (1992). The pathology of subclinical pyelonephritis — an experimental evaluation. *Kidney International*, **41**, 1356–1365).

Microbial distribution

Both left and right kidneys were removed from the animals at autopsy 7, 30 or 60 days after the initial challenge and placed in a Petri dish on a thin bed of solidified wax. Molten wax was then poured over the entire kidney and allowed to solidify. A single-edged razor blade was used to make five cuts along the length of the kidney to form six slices. Each slice was then removed from the wax bed and laid on the flat surface of the wax. The slices of tissue were further dissected, each cut from top to bottom into portions weighing between 10 and 40 mg, so that, in total, each kidney was divided into 30–40 pieces (Figure 54.2). Segment numbers were then recorded on a paper profile and individual portions were weighed and homogenized in 1 ml normal saline using a flat-bottomed glass rod in a flat-bottomed glass vial. The homogenate was then poured on to a Columbia agar pour plate, mixed with molten agar and the total number of bacteria determined in each portion. This

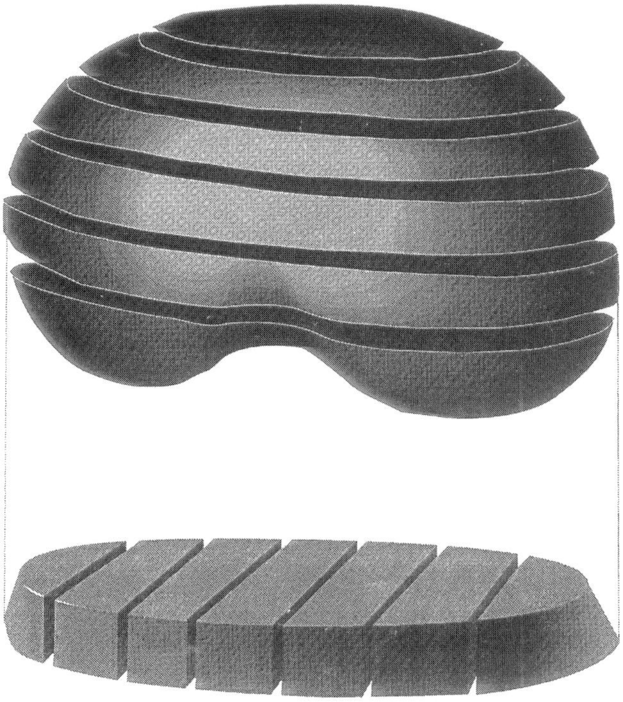

Figure 54.2 Diagramatic representation of the procedure used for kidney dissection. Each kidney yielded approximately 40 pieces, each weighing on average, 25 mg. (Reproduced with permission from Miller, T. E., Findon, G., Rainer, S. P. and Gavin, J. B. (1992). The pathology of subclinical pyelonephritis – an experimental evaluation. *Kidney International*, **41**, 1356–1365).

was expressed as a percentage of the total number of bacteria in the whole kidney and marked on the paper profile. For graphic representation, 2% of the total bacteria was represented as a single dot.

Microorganisms were usually found in focal aggregates in the directly challenged kidneys with classic pyelonephritis. The foci were mainly in the area of the gross lesion, up to 30 days after challenge, but thereafter they were more evenly distributed. A representation of this is seen in Figure 54.3A, showing the bacterial distribution after 7 days. In the unmanipulated kidneys, microorganisms were commonly spread evenly throughout the kidney, but focal aggregates were occasionally observed, especially in the early stages. This is shown in Figure 54.3B, 7 days after challenge of the contralateral kidney.

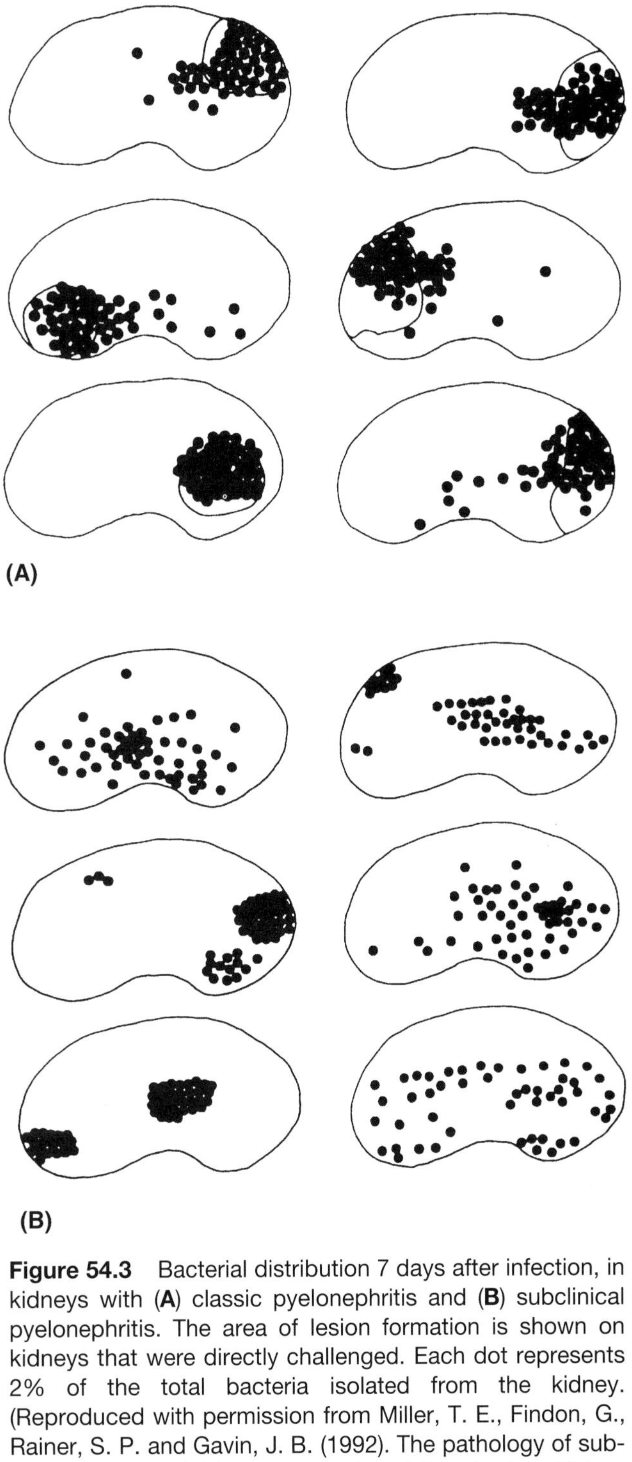

Figure 54.3 Bacterial distribution 7 days after infection, in kidneys with (**A**) classic pyelonephritis and (**B**) subclinical pyelonephritis. The area of lesion formation is shown on kidneys that were directly challenged. Each dot represents 2% of the total bacteria isolated from the kidney. (Reproduced with permission from Miller, T. E., Findon, G., Rainer, S. P. and Gavin, J. B. (1992). The pathology of subclinical pyelonephritis – an experimental evaluation. *Kidney International*, **41**, 1356–1365).

Gross pathology

At autopsy, both directly challenged and non-manipulated kidneys were examined for gross and histopathological changes. The gross pathology of each kidney was quantified using a computer program, written in BASIC and installed on a PC. The program produces two mirror image kidney profiles on the screen, representing both sides of one kidney. The dimensions of the profiles were scaled to those of the normal rat kidney and a grid was superimposed on to each kidney to aid orientation. Using a computer mouse, the scarred areas on the excised kidney were shaded in on the screen. The software then calculated the size of the shaded (scarred) areas on both screen profiles and expressed this as a percentage of the total kidney surface area. Histological analysis was done by haematoxylin-and-eosin-stained sections and electron microscopy.

Kidneys that had been directly challenged (classic pyelonephritis) all showed characteristic cortical lesions on gross examination (Figure 54.4). Histologically, these were seen to be wedge-shaped areas which traversed the cortex, medulla and renal pelvis. No surface lesions were apparent on the contralateral unmanipulated kidneys and the renal parenchyma was histologically normal in most cases, although minor changes were commonly observed in the mucosa lining the calyx. Localized inflammatory foci containing neutrophils and lymphocytes were present within the transitional cell epithelium and also in the subepithelial tissues.

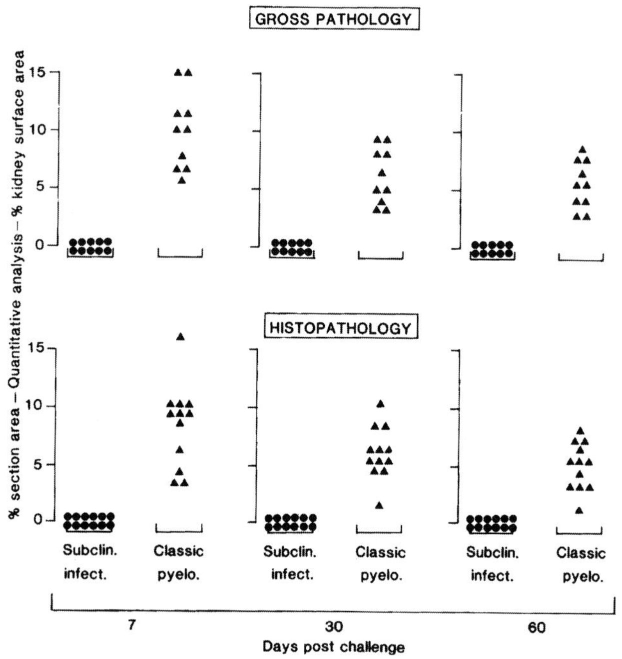

Figure 54.4 Quantitative analysis of gross and histopathological changes in directly challenged kidneys with classic pyelonephritis and unmanipulated kidneys with subclinical infection. (Reproduced with permission from Miller, T. E., Findon, G., Rainer, S. P. and Gavia, J. B. (1992). The pathology of subclinical pyelonephritis – an experimental evaluation. *Kidney International*, **41**, 1356–1365).

Antimicrobial therapy

Some 30 days after infection was established, groups of animals were treated with a variety of antibiotics for 10 days. The minimum inhibitory concentrations of each antimicrobial to the strain of *E. coli* O75 used to establish infection were determined before use. A dose was then chosen for each antibiotic that gave blood, urine and tissue concentrations of antibiotic in experimental animals, comparable to those found clinically (Table 54.1). The unmanipulated right kidneys were removed at autopsy for bacteriological evaluation using serial dilutions of a homogenate incorporated into Columbia agar pour plates.

Bacterial numbers in the right kidneys of animals with subclinical pyelonephritis treated with ampicillin, co-trimoxazole, ciprofloxacin, ceftriaxone and gentamicin were significantly reduced compared to untreated animals. However only ceftriaxone and ampicillin were able to sterilize 50% or more of the kidneys. Trimethoprim and aztreonam had no effect (Figure 54.5).

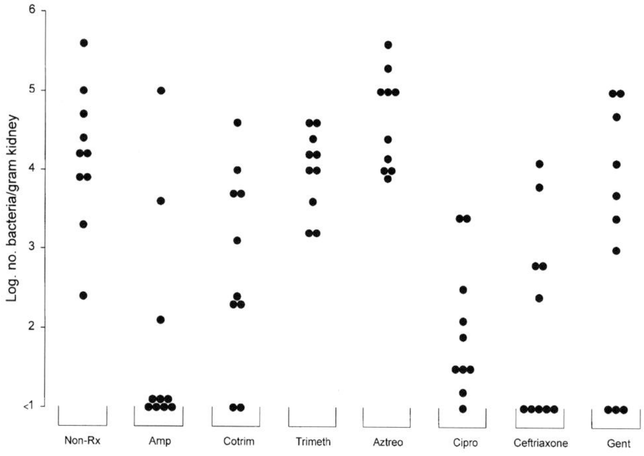

Figure 54.5 Bacterial numbers in renal tissue from animals with subclinical pyelonephritis treated for 10 days with various chemotherapeutic agents. Treatment commenced 30 days after the establishment of infection. (Reproduced with permission from Miller, T. E., Findon, G., Rainer, S. P. and Gavin, J. B. (1992). The pathology of subclinical pyelonephritis – an experimental evaluation. *Kidney International*, **41**, 1356–1365).

Pitfalls (advantages/disadvantages) of the model

There are no problems associated with the development of subclinical infection, provided that the Hooded Oxford rat is used. Other strains of rat may not be suitable for this model and should be assessed before use. The exact reason for these differences has not been established.

Use of model

The availability of the model of subclinical pyelonephritis characterized here should prove useful in developing more effective means for its diagnosis and treatment. Recognition that the kidney can be subclinically infected has important implications for antimicrobial therapy. A review of the management of acute bacterial cystitis, with and without evidence of upper urinary tract involvement, indicated that both ampicillin and co-trimoxazole can provide effective therapy when infection is confined to the lower urinary tract. In contrast, treatment is ineffective when the upper urinary tract is involved (Stamm *et al.*, 1989). The importance of successful

Table 54.1 Dosages and pharmacological parameters of antimicrobial agents; N/A, not available (reprinted with permission from Lecamwasan, J. P. and Miller, T. (1989). Antimicrobial agents in the management of urinary tract infection: an experimental evaluation. *J. Lab. Clin. Med.*, **114**, 510–519)

					Peak concentration		
Antibiotic	*MIC (μg/ml)*	*Dose (mg/kg)*	*Route*	*Frequency*	*Serum (μg/ml)*	*Urine (μg/ml)*	*Tissue (μg/g)*
Ampicillin (Ampicyn)	1.6	50	i.m.	2×daily	25	400	40
Gentamicin (Garamycin)	1.6	7.5	i.m.	2×daily	8*	40	45
Co-trimoxazole (Bactrim)	0.25	50	p.o.	2×daily	Trim. 3 Sulpha.† 4	Trim. 300 Sulpha. 20	Trim. 22 Sulpha. 2.4
Ceftriaxone (Rocephin)	0.1	200	i.m.	Daily	160‡	N/A	N/A
Ciprofloxacin (Ciproxin)	0.02	20	p.o.	2×daily	2.5**	345	4
Aztreonam (Azactam)	0.1	20	i.m.	2×daily	38*	632	190
Trimethoprim (Triprim)	N/A	10	p.o.	2×daily	3†	300	22

* Authors' unpublished data.
† *J. Clin. Pathol.* 1971; **24**: 430–437.
‡ *Chemotherapy* 1982; **28**: 410–416.
** *Drugs* 1988; **35**: 373–447.

management of subclinical pyelonephritis has been emphasized by Clark *et al.* (1969), who showed that infection-associated abnormalities of renal function are reversible if the microorganisms are eradicated. Failure to recognize persistent subclinical infection could explain why the management of cystitis has a 10–15% failure rate and a high incidence of recurrent infection (Stamey *et al.*, 1965; Fang *et al.*, 1978).

References

Boutros, P., Mourtada, H., Ronald, A. R. (1972). Urinary infection localization. *Am. J. Obstet. Gynecol.*, **112**, 379–381.

Busch, R., Huland, H. (1984). Correlation of symptoms and results of direct bacterial localization in patients with urinary tract infections. *J. Urol.*, **132**, 282–285.

Clark, H., Ronald, A. R., Cutler, R. E., Turck, M. (1969). The correlation between site of infection and maximal concentrating ability in bacteriuria. *J. Infect. Dis.*, **120**, 47–53.

Eykyn, S., Lloyd-Davies, R. W., Shuttleworth, K. E. D., Vinnicombe, J. (1972). The localization of urinary tract infection by ureteric catheterization. *Invest. Urol.*, **9**, 271–275.

Fairley, K. F., Grounds, A. D., Carson, N. E. *et al.* (1971). Site of infection in acute urinary-tract infection in general practice. *Lancet*, **ii**, 615–618.

Fang, L. S. T., Tolkoff-Rubin, N. E., Rubin, R. H. (1978). Efficacy of single-dose and conventional amoxicillin therapy in urinary-tract infection localized by the antibody-coated bacteria technic. *N. Engl. J. Med.*, **298**, 413–416.

Harding, G. K. M., Marrie, T. J., Ronald, A. R., Hoban, S., Muir, P. (1978). Urinary tract infection localization in women. *J.A.M.A.*, **240**, 1147–1150.

Stamey, T. A., Govan, D. E., Palmer, J. M. (1965). The localization and treatment of urinary tract infections: The role of bactericidal urine levels as opposed to serum levels. *Medicine*, **44**, 1–36.

Stamm, W. E., Hooten, T. M., Johnson, J. R. *et al.* (1989). Urinary tract infections: from pathogenesis to treatment. *J. Infect. Dis.*, **159**, 400–406.

Chapter 55

Models of Acute and Chronic Pyelonephritis in the Rat

G. Findon

Background of model

One problem in establishing a model of pyelonephritis is that the normal non-manipulated urinary tract of the rat is resistant to a challenge with microorganisms isolated from cases of human urinary tract infection. As a result, it has been necessary to obstruct the urinary tract or manipulate the kidney prior to challenge, before renal infection can be established. The present model of pyelonephritis developed from experiments that used micropuncture equipment, from the study of renal physiology, to deliver viable microorganisms into the kidney. This action led to the development of an acute wedge-shaped inflammatory lesion and formed the basis of the method described here, which uses a glass micropipette to deliver the challenging inoculum directly into the kidney. The gross appearances and histopathological changes are similar to those seen during the course of pyelonephritis in man.

Animal species

Female Wistar randomly bred rats weighing between 200 and 225 g were used initially to characterize the model, but other strains, notably the Dark Agouti, have given similar results.

Preparation of animals

No specified housing, care or specific pretreatment is required.

Details of surgery

Overview

Pyelonephritis is induced by the direct inoculation of *Escherichia coli* into the surgically exposed kidney using a glass microcapillary.

Materials required

Short-acting anaesthetic, hair clippers, skin disinfectant, scalpel handle and blades, forceps, scissors, suture materials, SMI Micro/Pettor, fine glass capillary formed from a Micro/Pettor capillary tube with an outer diameter of 1 mm, pulled out to give a fine tip with an external diameter of 0.25 mm, bacterial inoculum.

Anaesthesia

Various anaesthetics have been used successfully, including chloral hydrate, sodium barbiturate and ether. The procedure usually takes about 5 minutes to complete and causes minimal distress to the subject.

Surgical procedure and induction of infection

The kidney is exposed through an incision in the flank, just below the rib cage and externalized by gently pulling on the renal fat pad. A volume of 5 μl of the diluted bacterial inoculum containing approximately 3000 viable microorganisms is then introduced into the renal parenchyma at each of three sites (Figure 55.1). The capillary is inserted

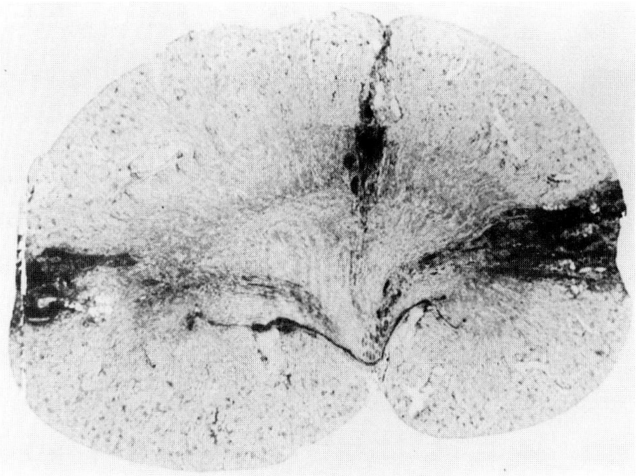

Figure 55.1 Haematoxylin-and-eosin-stained section of the rat kidney. The pyelonephritic lesions are associated with the three sites of direct challenge with *E. coli*.

Handbook of Animal Models of Infection
ISBN 0-12-775390-7

into the kidney to a depth of several millimetres and the inoculum is delivered along the needle track as the capillary is withdrawn from the kidney.

Wound closure

The wound is closed using a running suture, first through the muscle layer and then through the skin.

Postoperative care

Animals should be kept warm after the surgery until they have recovered from the anaesthetic. The animals return to full activity within a few hours and wound healing is quick and uneventful. They are usually housed in groups of two to five to facilitate a return to full social activity. Postoperative analgesics have not been routinely used. Weight gain, and food and water consumption are unaffected by the infection.

Storage and preparation of inocula

E. coli from a human case of acute pyelonephritis should be isolated using standard bacteriological procedures and serotyped to confirm that it is one of the serotypes characteristically associated with acute pyelonephritis. The strain selected can either be lyophilized or stored at 4 °C on Dorset Egg slopes. For use, a subculture is prepared on blood agar and a single colony is used to inoculate 10 ml of nutrient broth. After overnight incubation, the broth culture is washed twice in normal saline, resuspended with 10 ml of saline and stored at 4 °C. At this stage a bacterial count is carried out, using serial dilutions and Columbia agar pour plates to determine the number of viable microorganisms per mililitre of suspension. Then 24 hours later, the culture is diluted in saline to the point that 5 μl of inoculum contains approximately 3000 microorganisms.

Key parameters to monitor infection and response to treatment

Gross changes

Bacterial challenge to the kidney in the manner described consistently leads to infection of the renal parenchyma associated with histopathological changes characteristic of acute and chronic pyelonephritis. The histopathological changes are confined to the line of the inoculation and are characteristically wedge-shaped lesions originating in the medullary region and extending to the cortex. As the infection progresses, pitted and scarred areas characteristic of chronic pyelonephritis develop on the surface of the kidney (Figure 55.2).

Histopathological changes

The earliest changes in the infected kidneys are found 3 days after challenge. Areas of acute inflammation are confined to the line of inoculation and extend down into the papilla (Figure 55.3). As the infection progresses, the inflammatory infiltrate consists largely of small lymphocytes and plasma cells. Casts of degenerate pus cells can still be found in the distal and collecting tubules 30–60 days after challenge. Between 80 and 100 days after infection, infected kidneys show capsular scars and interstitial fibrosis with tubular atrophy. Plasma cells and lymphocytes remain the predominant inflammatory cells. The glomeruli remain

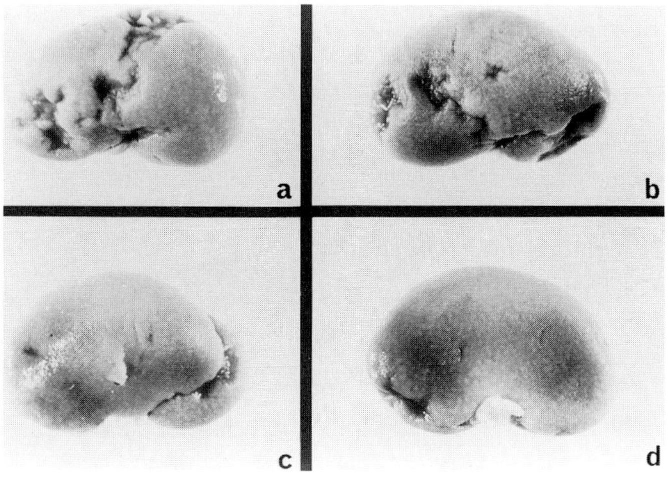

Figure 55.2 Pitted and scarred areas characteristic of chronic pyelonephritis. **a**, **b**, **c** and **d** represent the varying levels of severity commonly observed following a bacterial challenge.

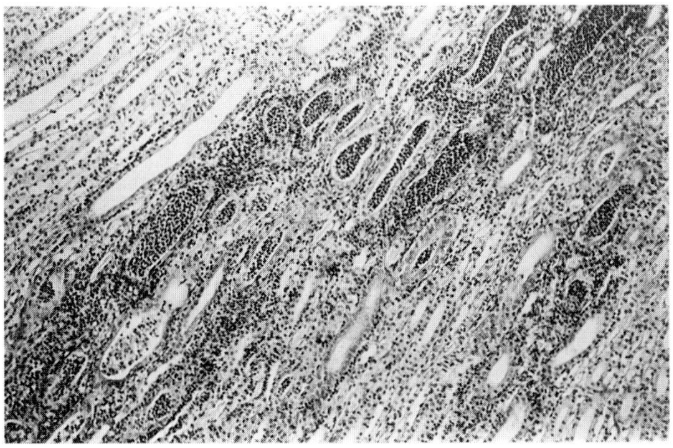

Figure 55.3 Areas of acute inflammation in the medulla of a rat kidney 5 days after challenge with *E. coli*.

normal and blood vessels show only medial proliferation of their walls.

Bacteriology

Bacterial numbers in the whole kidney are determined after autopsy by grinding the kidney with 9 ml of sterile saline in a motor-driven glass/Teflon mortar and pestle. Tenfold dilutions of the homogenate are then incorporated into nutrient agar pour plates. If more detail on the distribution of microorganisms is required, the kidney can be divided into cortical, medullary and papillary tissue using a hollow cylindrical blade to first remove a core of tissue, as illustrated in Figure 55.4. Cortical tissue can readily be distinguished from medullary in the core and can be removed with a sharp blade. The papilla is then removed from the medullary segment. The crescent of renal cortical tissue remaining after removal of the medullary core is then trimmed to remove residual medullary tissue. The remaining cortical tissue is pooled with the two samples of cortex previously removed from the medullary core. The results of the bacteriological examination of these three tissues over a 100-day period postchallenge can be seen in Figure 55.5.

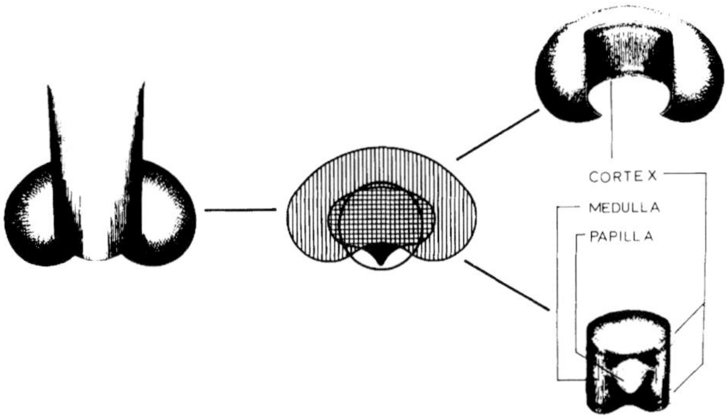

Figure 55.4 Illustration of the procedure for dissection of the rat kidney into the cortex, medulla and papilla. A hollow drill is used to remove a corticomedullary plug containing the papilla.

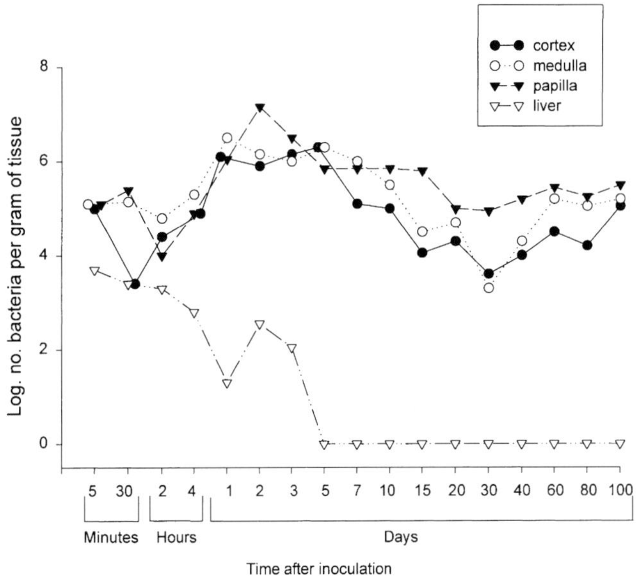

Figure 55.5 Bacterial numbers in the cortex, medulla and papilla of kidneys challenged with *E. coli*. The outcome is compared to the course of infection in liver subjected to a similar challenge.

Antimicrobial therapy

A variety of antimicrobial agents have been evaluated in this model for their effect on both acute and chronic infections of the kidney. The outcome of a sample experiment is shown in Figures 55.6 and 55.7. *In vitro* determination of the antimicrobial sensitivity pattern of the challenging strain, using the MIC of the antibiotic as an index, showed that the challenging strain of *E. coli* O75 employed was sensitive to all the antibiotics used in the study (Table 55.1). Antibacterial sensitivity of the microorganism determined by disc diffusion gave similar results. In the case of acute pyelonephritis, renal infection was established in 42 rats, which were then divided into seven groups of six. Antibiotic treatment commenced 7 days after the initiation of infection and continued for a further 7 days. Autopsies were carried out 48 hours after the final dose. The results (Figure 55.6) show that acute renal infection was eradicated by ceftriaxone and gentamicin in 12/12 kidneys, ampicillin in 6/12 and co-trimoxazole in 3/12. Aztreonam and norfloxacillin treatment reduced the bacterial numbers in the kidneys of treated animals but was unable to completely eradicate the infection. In the case of chronic pyelonephritis, the challenged animals were left for 30 days without any further manipulation. Antimicrobial agents were then administered to groups of seven or eight animals for 7 days and autopsies were carried out 2 days later. Bacteriological evaluation once again showed that ceftriaxone and gentamicin were the most effective agents and achieved eradication rates of 75% and 80% respectively. The remaining antibiotics had no statistically significant effect on bacterial numbers in the kidney, although infection was eradicated in individual cases (Figure 55.7).

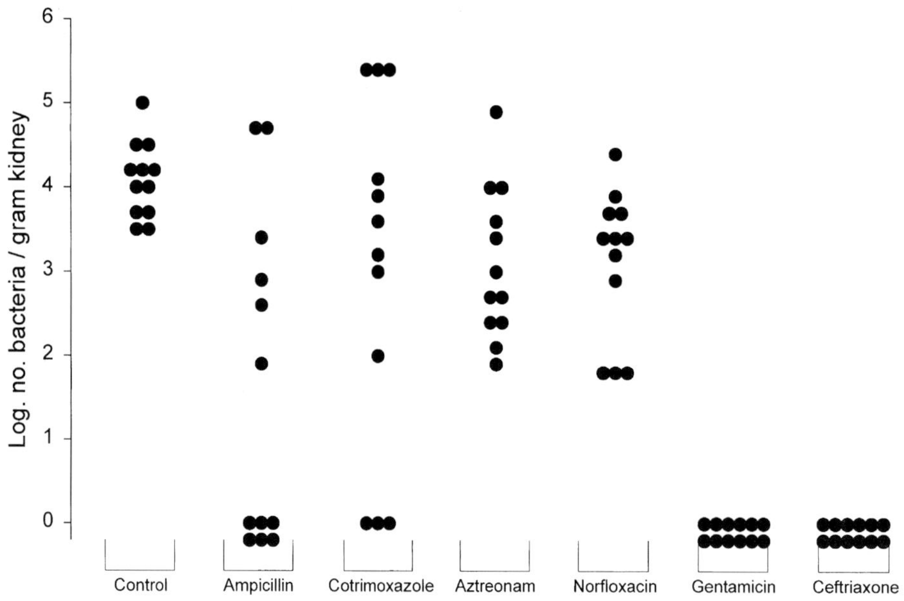

Figure 55.6 Effect of antimicrobial therapy on bacterial numbers in the kidneys of rats with acute pyelonephritis. See Table 55.1 for dosages.

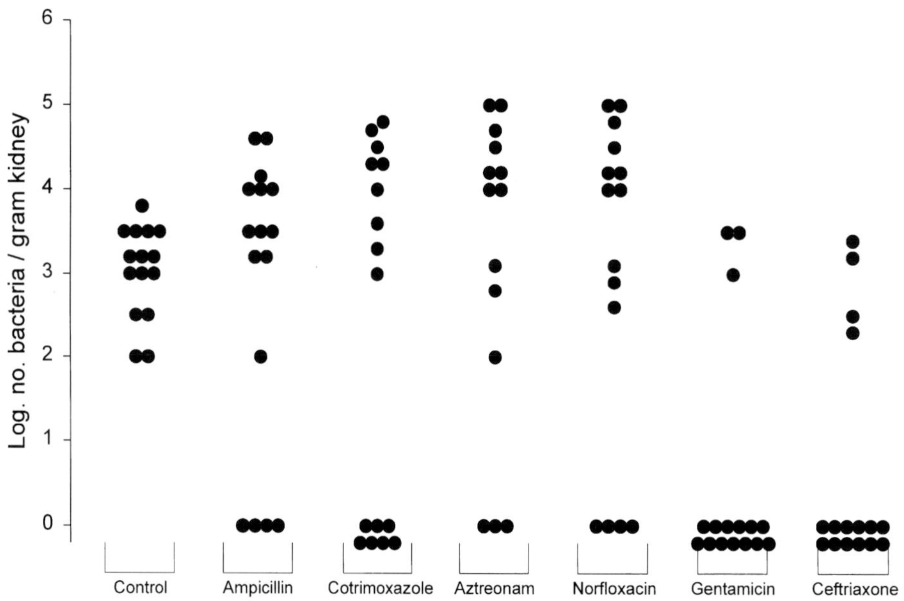

Figure 55.7 Effect of antimicrobial therapy on bacterial numbers in the kidneys of rats with chronic pyelonephritis. Treatment commenced 30 days after challenge. Details of the dosages are given in Table 55.1.

Table 55.1 Dosages and pharmacological parameters of antimicrobial agents; NA, not available. (Reprinted with permission from Lecamwasan, J. P. and Miller, T. (1989). Antimicrobial agents in the management of urinary tract infection: an experimental evaluation. *J. Lab. Clin. Med.*, **114**, 510–519)

					Peak concentration		
Antibiotic	*MIC (μg/ml)*	*Dose (mg/kg)*	*Route*	*Frequency*	*Serum (μg/ml)*	*Urine (μg/ml)*	*Tissue (μg/g)*
Ampicillin (Ampicyn)	1.6	50	i.m.	2 × daily	25	400	40
Gentamicin (Garamycin)	1.6	7.5	i.m.	2 × daily	8*	40	45
Co-trimoxazole (Bactrim)	0.25	50	oral	2 × daily	Trim. 3 Sulpha. 4†	Trim. 300 Sulpha. 20	Trim. 22 Sulpha. 2.4
Ceftriaxone (Rocephin)	0.2	200	i.m.	Daily	160‡	N/A	N/A
Norfloxacin (Noroxin)	0.24	50	oral	2 × daily	0.5–0.9**	20	10.5
Aztreonam (Azactam)	0.1	20	i.m.	2 × daily	38*	632	190

* Author's unpublished data.
† *J. Clin. Pathol.* 1971; 24: 430–437.
‡ *Chemotherapy* 1982; 28: 410–416.
** Manufacturer's information.

Pitfalls (advantages/disadvantages) of the model

The advantage of the model is that it predictably reproduces many features of the bacteriological, gross and histopathological features of the clinical disease. It is therefore a particularly useful model for the study of the pathobiology of renal infection. It can also be used to determine the ability of individual antibiotics to eradicate infection. The disadvantage is that it cannot be used to study early stages of the natural history of pyelonephritis, particularly the way in which microorganisms gain access to the kidney and establish infection.

Chapter 56

Rat Model of Chronic Cystitis

D. J. Ormrod and T. E. Miller

Background of model

Chronic cystitis due to *Escherichia coli* is frequently associated with anatomical or functional abnormalities of the lower urinary tract, but there is no satisfactory animal model to help resolve biological and management problems. To induce chronic infection of the unobstructed urinary tract in the rat, we implanted a small polyurethane sponge into the dome of the bladder and, 14 days later, introduced bacteria directly into the bladder. This manipulation provides a focus of infected urine and leads to the establishment of chronic cystitis, with pathological findings similar to those in man.

Animal species

Female animals from an inbred colony of Dark Agouti rats, aged 3–4 months, were used in the development of this model. Other strains may be suitable but have not been evaluated.

Preparation of animals

The animals are housed in polycarbonate cages on untreated wood shavings and fed Diet 86 and water *ad libitum*. No specific care or pre-treatment is required.

Details of surgery

Overview

A small polyurethane sponge is placed in the dome of the surgically exposed bladder and allowed to become embedded in the bladder wall. After 2 weeks, the bladder is infected to establish chronic cystitis.

Materials required

Anaesthetic, hair clippers, skin disinfectant (95% ethanol), scalpel blade, scissors, forceps, sutures, 16 G needle, blunt trochar to fit inside 16 G needle, Hamilton syringe and 26 G needle, bacterial inoculum. To make the sponges for implantation, a 4 mm core of sponge is punched from open-cell polyurethane foam, sectioned into 2 mm discs and sterilized by autoclaving.

Anaesthesia

Animals are anaesthetized with Nembutal (pentobarbitone sodium, Techvet Laboratories, Auckland, NZ) at the recommended dose, but any short-acting anaesthetic could be used.

Surgical procedure

An incision is made in the suprapubic region, through the skin and muscle layer to expose the bladder. The previously prepared sterile sponge disc is placed in the lumen of a 16 G needle, which is pushed through the bladder wall and angled towards the dome. The sponge is expelled from the needle with a blunt trochar and deposited in the dome of the bladder. The needle is withdrawn and the puncture wound is closed with a single suture (Figure 56.1). The sponge remains in the dome of the bladder and becomes partially embedded in the bladder wall over the following 2 weeks. The bladder is replaced in the abdominal cavity and the wound is closed.

Infection procedure

The bladder is re-exposed surgically 2 weeks after sponge implantation. Cystitis is established by the inoculation of 100 μl of saline containing 10^6 *Escherichia coli* O75 directly into the bladder using a Hamilton syringe fitted with a 26 G needle. Three groups of animals are required as controls. The first group receives the bacterial inoculum of 10^6 *E. coli*

Handbook of Animal Models of Infection
ISBN 0-12-775390-7

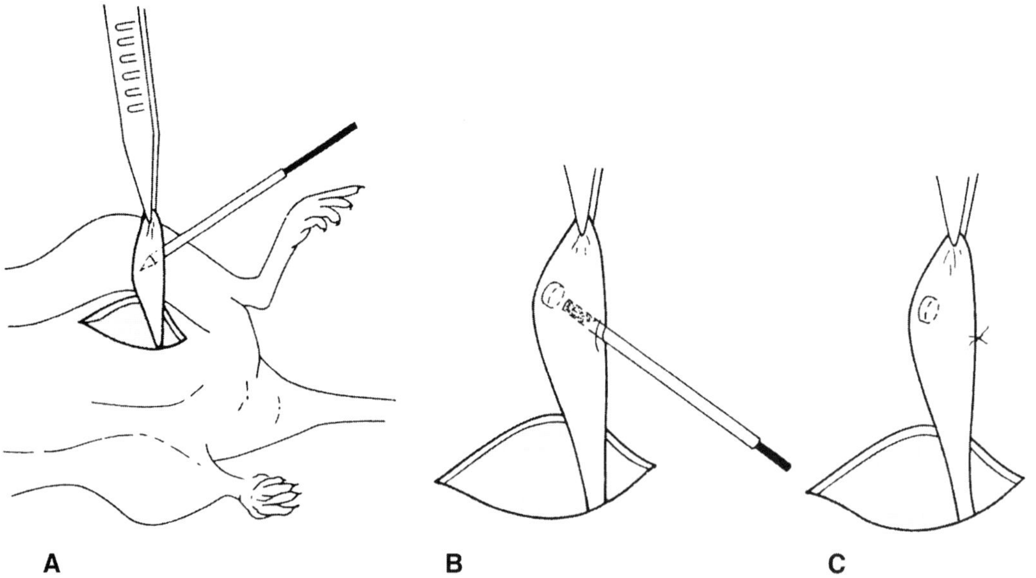

Figure 56.1 Procedure for the implantation of the polyurethane sponge into the bladder. (Reproduced with permission from Miller, T. E., Lecamwasam, J. P., Ormod, D. J., Findon, G. and Cornish, J. (1987). An animal model for chronic infection of the unobstructed urinary tract. *Br. J. Exp. Pathol*, **68**, 575–583.)

O75 directly into the normal, unmanipulated bladder lumen. A second group of animals is subjected to the surgical procedure (i.e. bladder punctured with a 16 G needle but no sponge is implanted), and the bacterial inoculum is introduced 2 weeks later. The third control group has the sponge implanted but the bacterial inoculum is replaced by sterile saline.

Wound closure

The skin and muscle layers are closed with a running suture after both procedures.

Postoperative care

No special postoperative care is required. Animals recover from the surgery and resume normal activities within a few hours. The animals are housed in groups of four to six animals per cage, depending on size. Sponge implantation and chronic urinary tract infection had no demonstrable effect on activity, appetite, weight gains or coat condition in the experimental animals. Infected animals remained well over the 12-week period of the study.

Storage and preparation of inocula

The strain of *E. coli* O75 chosen for this model fails to ferment arabinose and can therefore be differentiated from random lactose-fermenting Enterobacteriaceae found in normal rat urine (available on request). Examination of 1000 random rat urines confirmed that this strain of *E. coli* O75 was unique in this respect. The culture was stored on Dorset Egg medium and subcultured regularly on to blood agar. A single colony was placed in nutrient broth and incubated overnight, before being washed three times in saline. The final inoculum was diluted in saline to give an inoculum of 10^6 organisms in 100 μl.

Collection and handling of urine samples

Voided urine was collected at regular weekly intervals from all animals over a 12-week period. Urine was obtained by holding the animal over a Petri dish and, if necessary, applying gentle pressure to the lower abdomen. Urine samples were examined immediately after collection. Quantitative analyses were carried out on all urine samples. Dilutions of 1 : 10 and 1 : 100 were made in sterile saline and one microlitre of the diluted sample was spread on MacConkey agar containing crystal violet and 1% lactose (Difco Laboratories, Detroit, MI). Plates were incubated at 37 °C overnight before carrying out colony counts.

The ability of varying bacteria to ferment carbohydrates in differential media was used to distinguish the infecting from contaminating microorganisms. The identity of the isolates was then established as either *E. coli* O75, which fermented lactose, but not arabinose (*lac*$^+$ *ard*$^-$) or a contaminant, which ferments both lactose and arabinose (*lac*$^+$ *ara*$^+$). To carry out this test, 50 colonies fermenting lactose from the MacConkey agar plate were transferred on to a template of Columbia agar. After 16 hours incubation at 37 °C, replicate subcultures were made on to plates of MacConkey agar base containing 1% lactose or 1% arabinose (Sigma Chemical Co., St Louis, MO). These replica-plated Petri dishes were then incubated at 37°C for a further 16 hours and examined for lactose and arabinose fermentation.

Total white cell numbers in the urine samples were determined in wet preparations with a haemocytometer. Absolute numbers of leukocytes and epithelial cells were calculated after cell differentiation. Urine samples were first diluted in an equal volume of Hank's Basic Salt Solution containing 10% fetal calf serum and preparations were made from 0.5 ml aliquots delivered into chambers of a cytospin centrifuge (Shandon Southern, Runcorn, UK). The slides were dried at room temperature, fixed in cold methanol, stained with Leishman's stain and the percentage of each cell type was determined.

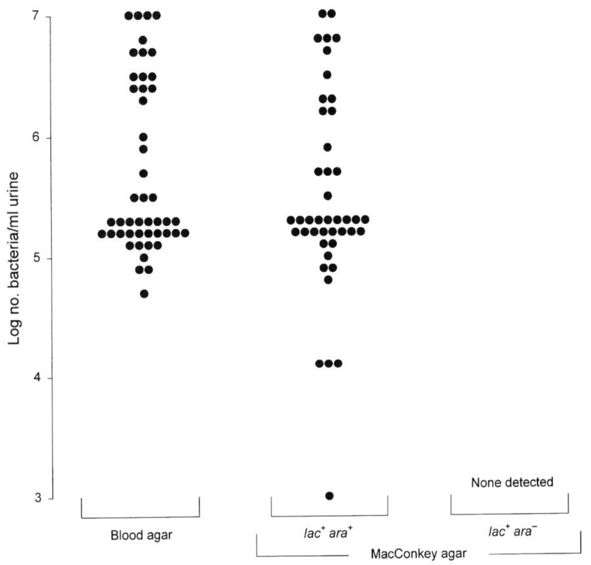

Figure 56.2 Bacteriuria in urine samples from normal animals. Lactose-fermenting colonies were differentiated into arabinose-fermenting and non-fermenting-microorganisms. (Reproduced with permission from Miller, T. E., Lecamwasam, J. P., Ormrod, D. J., Findon, G. and Cornish, J. (1987). An animal model for chronic infection of the unobstructed urinary tract. *Br. J. Exp. Pathol.*, **68**, 575–583.)

Key parameters to monitor infection and response to treatment

Gross observations of cystitis model

At autopsy, 12 weeks after challenge, the principal gross finding was a thickening of the dome of the bladder around the implanted sponge. There was no suggestion of obstruction of the urinary tract nor any evidence of ureteric or vesical dilation or hydronephrosis. Stone formation occurred in the bladders of a few animals. These animals were excluded from the study.

Urine bacteriology

Urine samples collected from normal animals and cultured on to blood agar and MacConkey agar containing lactose had bacterial counts between 10^5 and 10^7 ml of urine. The lactose-fermenting colonies obtained from MacConkey agar were confirmed as *lac⁺ ara⁺* (Figure 56.2). In animals containing a bladder sponge and infected with *E. coli* O75, infection was established in 79 of 90 animals (88%). The level of infection in the urine suggested a steady rise that peaked at week 8 (Figure 56.3). The isolates tested were almost always *lac⁺ ara⁻*, as few contaminants remained at the dilutions required to provide countable pour plates when the urines from infected animals were cultured. In control animals with unmanipulated bladders, the *E. coli* O75 challenge was cleared immediately. Animals that underwent the surgical manipulation (insertion of 16 G needle into the bladder) but without the sponge implant, also cleared the *E. coli* O75 inoculum in a similar way. Chronicity of infection can therefore be ascribed to the presence of the sponge implant and not to the surgical procedures. In the animals with sponge implants but no bacterial challenge, no *lac⁺ ara⁻* organisms were isolated.

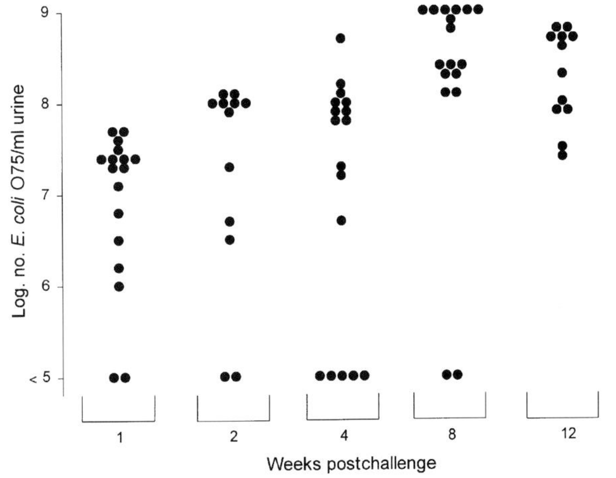

Figure 56.3 Bacterial numbers (*E. coli O75, lac⁺ ara⁻*) in the urine of animals containing sponge implants, up to 12 weeks after challenge with 10^6 *E. coli* O75. Each point on the graph represents bacterial counts in urine from a single animal sampled at different weeks. (Reproduced with permission from Miller, T. E., Lecamwasam, J. P., Ormrod, D. J., Findon, G. and Cornish, J. (1987). An animal model for chronic infection of the unobstructed urinary tract. *Br. J. Exp. Pathol.*, **68**, 575–583.)

Urine cytology

Prechallenge evaluation of urine samples showed less than 10 leukocytes/mm^3. A sharp increase in cell excretion occurred with challenge, and most urine samples contained between 10^2 and 10^3 cells/mm^3 during the first 8 weeks. The number of cells in the infected urine then decreased over the next 4 weeks (Figure 56.4). In control groups, the number of cells excreted was similar to normal animals.

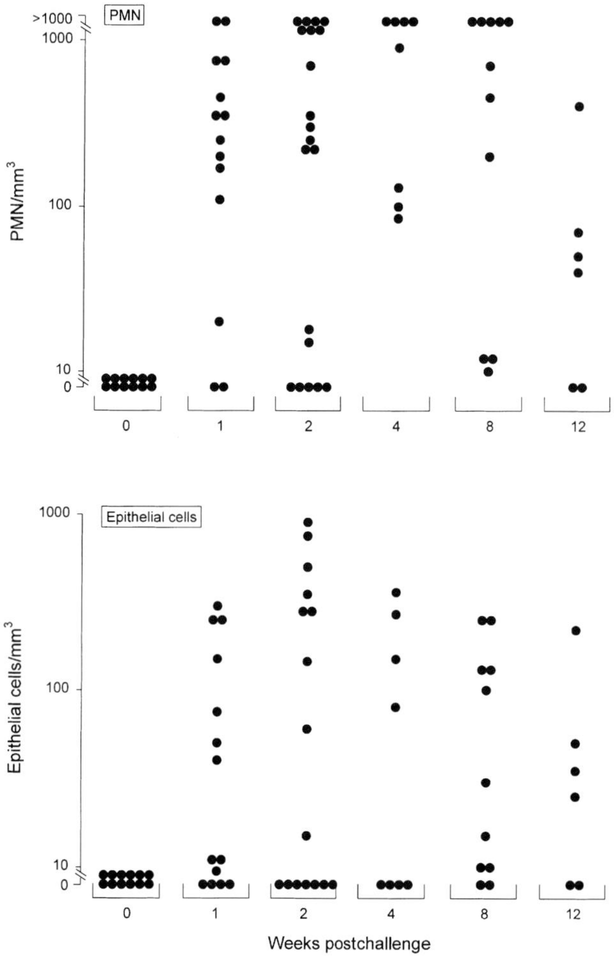

Figure 56.4 Quantitative counts of polymorphonuclear cells (PMN) and epithelial cells in the urine of animals with chronic cystitis. (Reproduced with permission from Miller, T. E., Lecamwasam, J. P., Ormrod, D. J., Findon, G. and Cornish, J. (1987). An animal model for chronic infection of the unobstructed urinary tract. *Br. J. Exp. Pathol.*, **68**, 575–583.)

Bladder histology

The bladders of challenged animals showed both focal and diffuse infiltrates in the epithelium and lamina propria. Neutrophils were the dominant cells in the early stages of infection, but lymphocyte nodules gradually became more common, and by 12 weeks, mononuclear cells were the main inflammatory cell present. No histopathological changes were seen in the control animals.

Antimicrobial evaluation in the model

Antibiotics commonly used to treat urinary tract infections were administered to animals with chronic cystitis at the doses shown in Table 56.1. Doses were chosen that gave blood, urine and tissue levels of the antibiotic similar to those found in humans. Treatment was initiated 14 days after infection and continued for 7 days. Using the MIC and disk diffusion as an index of sensitivity, *E. coli* O75 was shown to be susceptible to all the antibiotics tested. After the 7-day treatment, the bladders were removed and homogenized. Bacterial numbers were estimated from 10-fold dilutions made in Columbia agar pour plates. Norfloxacin and ceftriaxone were the most effective antimicrobials, eradicating bladder infection in 77% and 70%, respectively, of the animals treated. Aztreonam (53%), gentamicin (47%) and ampicillin (37%) were less effective, and co-trimoxazole cleared the infection in only one of the 13 animals treated (Figure 56.5).

Pitfalls (advantages and disadvantages) of model

Various animal models have been used to induce infection of the lower urinary tract. These include osmotic diuresis

Table 56.1 Dosages and pharmacological parameters of antimicrobial agents; NA, not available. (Reprinted with permission from Lecamwasam, J.P. and Miller, T. (1989). Antimicrobial agents in the management of urinary tract infection: an experimental evaluation. *J. Lab. Clin. Med.*, **114**, 510–519.)

					Peak concentration		
Antibiotic	*MIC (μg/ml)*	*Dose (mg/kg)*	*Route*	*Frequency*	*Serum (μg/ml)*	*Urine (μg/ml)*	*Tissue (μg/g)*
Ampicillin (Ampicyn)	1.6	50	i.m.	2 × daily	25	400	40
Gentamicin (Garamycin)	1.6	7.5	i.m.	2 × daily	8*	40	45
Co-trimoxazole (Bactrim)	0.25	50	Oral	2 × daily	Trim. 3 Sulpha. 4†	Trim. 300 Sulpha. 20	Trim. 22 Sulpha. 2.4
Ceftriaxone (Rocephin)	0.2	200	i.m.	Daily	160‡	N/A	N/A
Norfloxacin (Noroxin)	0.24	50	Oral	2 × daily	0.5–0.9**	20	10.5
Aztreonam (Azactam)	0.1	20	i.m.	2 × daily	38*	632	190

*Authors' unpublished data; † *J. Clin. Path.*, 1971; 24: 430–437; ‡ *Chemotherapy* 1982; 28: 410–416; **Manufacturer's information.

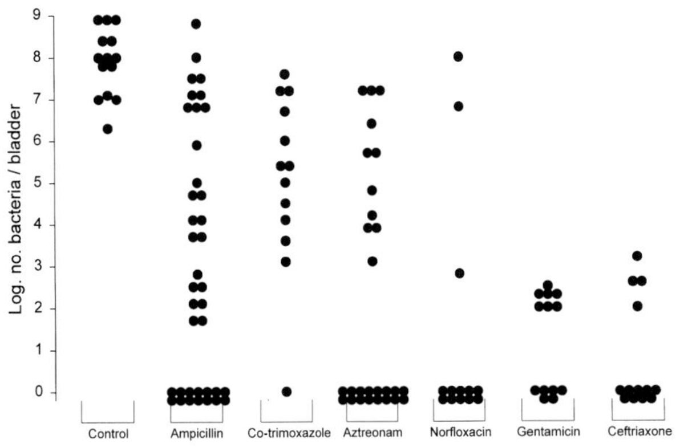

Figure 56.5 Effect of antimicrobial therapy on bacterial numbers in the bladder of animals with experimentally induced chronic cystitis. Treatment was carried out for 7 days. (Reproduced with permission from Lecamwasam, J. P. and Miller, T. (1989). Antimicrobial agents in the management of urinary tract infection: an experimental evaluation. *J. Lab. Clin. Med.*, **114**, 510–519).

(Freedman, 1967), placement of sutures in the bladder (Nicholson and Glynn, 1975), high challenging doses (Hagberg *et al.*, 1983) and infection with stone-inducing microorganisms (Lewis and Roberts, 1986), but the procedures have few clinical parallels and some obvious deficiencies. The model described here fulfils the need for a reproducible, non-obstructive manipulation of the lower urinary tract leading to chronic cystitis. The animals tolerate the surgical techniques well and show no ill effects from the implantation of the sponge or the infection. There was no evidence of obstruction to the urinary tract as long as the sponge was placed into the dome of the bladder, where it adheres to the mucosal lining.

Use of the model

This model will enable the evaluation of host defence mechanisms, both cellular and non-cellular, that are involved in the initiation and eradication of bladder infections. Existing antimicrobials showed a wide variation in bacterial eradication from the infected bladders in this model, which was similar to clinical efficacy. The model could therefore be useful in the evaluation of new antibiotics for the treatment of chronic cystitis.

References

Freedman, L. R. (1967). Experimental pyelonephritis. XIII. On the ability of water diuresis to induce susceptibility to *E. coli* bacteriuria in the normal rat. *Yale J. Biol. Med.*, **39**, 255–266.

Hagberg, L., Enberg, I., Freter, R., Lam, J., Olling, S., Svanborg Edén, C. (1983). Ascending, unobstructed urinary tract infection in mice caused by pyelonephritogenic *Escherichia coli* of human origin. *Infect. Immun.*, **40**, 273–283.

Lewis, R. W., Roberts, J. A. (1986). The effects of chronic cystitis on vesicoureteral reflux in an animal model. *J. Urol.*, **135**, 182–184.

Nicholson, A. M., Glynn, A. A. (1975). Investigation of the effect of K antigen in *Escherichia coli* urinary tract infections by use of a mouse model. *Br. J. Exp. Pathol.*, **56**, 549–553.

Chapter 57

Mouse Pneumococcal Pneumonia Models

E. Azoulay-Dupuis and P. Moine

Background of model

Streptococcus pneumoniae remains the most frequently isolated organism in community-acquired pneumonia and continues to be a significant cause of mortality in humans (Moine *et al.*, 1994a; Pallarés *et al.*, 1995; Lieberman *et al.*, 1996). Moreover, the worldwide incidence of infections caused by pneumococci resistant to penicillin, macrolides and other antimicrobials has increased at an alarming rate during the past two decades (Appelbaum, 1992, 1996). In some countries, the incidence of penicillin-resistant bacteria isolated from clinical specimens has risen to extremely high levels (50–70% in Spain and Hungary; Marton *et al.*, 1991). The emergence of multidrug-resistant *S. pneumoniae* strains has made these common infections more difficult to treat. There was thus an urgent need to identify compounds active against antibiotic-resistant pneumococci and usable for the treatment of pneumonia.

In the evaluation of new antibiotics, animal models of experimental infection are recognized as essential before clinical trials (Zak and Sande, 1986). Antibiotic treatment of bacterial infections creates a dynamic interplay between the drug, the host and the pathogen. Various factors may influence the activity of an antibiotic *in vivo*, some of which might not have been anticipated in *in-vitro* studies. Furthermore, respiratory tract infections require special consideration because of the complex pharmacokinetics of antibiotic penetration into tissues and the lung host defenses. The pulmonary defense system, the type of local disease, the course of the disease, the clearance rate of various bacterial inocula, and the penetration and pharmacokinetics of drugs in lung tissues are specific to this organ, and determine the local efficacy of antibiotics (Pennington, 1985).

Several animal models of pneumococcal pneumonia have been developed (Cleeland and Grunberg, 1986), although it has been difficult to create a reproducible model of fatal pneumococcal pneumonia resembling the pathophysiology of human disease. Fatal pneumonia is induced by deposition of virulent bacteria into the upper airways. Experimental therapeutics are assessed for their activity *in vivo* (bacterial counts in lung homogenates) and their efficacy (survival rate relative to untreated controls), but it must be borne in mind that laboratory animals have pharmacokinetic patterns different from those of humans (higher C_{max} and faster elimination half-life, in particular). Nevertheless, these models have proved to be useful for testing alternative treatments for penicillin-resistant and multidrug-resistant strains of *S. pneumoniae*. The emergence of resistant mutants following treatment of infected animals can also be studied.

Models of pneumococcal infection are devised according to the type of animal used, the inoculation technique and the virulence of the strain. When strains of different virulence and hosts of different lineages are combined, several experimental models with different natural histories are obtained. Thus, models of pneumococcal pneumonia have been described with different animal species.

The first method of lung inoculation used by many investigators was aerosol challenge (Green and Kass, 1964; Laurenzi *et al.*, 1964; Iizawa *et al.*, 1996). While this is a relatively easy method and offers the possibility of infecting a large number of animals simultaneously, delivery to the lungs of a large enough number of bacteria to ensure fatal pneumonia is very difficult in mice (Pennington, 1985), unless airways obstruction is produced (Iizawa *et al.*, 1996), and it is virtually impossible in larger animals (Berendt, 1977).

Acute pneumococcal pneumonia has been also induced by intranasal inoculation by several authors, including Muraoka *et al.* (1988), Tsuboi *et al.* (1988), Gisby *et al.* (1991), Girard *et al.* (1995), Tateda *et al.* (1996), and Van der Poll *et al.* (1996). Rolin and Bouanchaud (1987) developed a similar model of acute fatal infection induced by both intranasal and intrabuccal challenge with prior chemical injury of the pharynx. Methods using direct instillation of small-volume suspensions of bacteria into the lower airways were required to obtain fatal pneumonia. These latter methods of lung challenge require skill in animal intubation and/or surgery. Moreover, use of anesthesia during experimental inoculation makes the model non-physiological.

Intratracheal challenge after incision of the trachea has been used by Esposito and Pennington (1983), Girard *et al.* (1996), Ponte *et al.* (1996), and Candiani *et al.* (1997), in

Handbook of Animal Models of Infection
ISBN 0-12-775390-7

guinea pigs, rats, and mice. The intratracheal route allows the deposition of a standardized inoculum in 100% of the animals. Anesthesia inhibits coughing. This technique induces fatal pneumococcal pneumonia resembling the pathophysiology of human disease.

In our laboratory, direct *per oral* intratracheal challenge without incision has been used. This procedure avoids animal injury and facilitates the infection of a large number of animals (100 per hour).

Two models have been developed to study the antipneumococcal activity of antibiotics. With the same virulent challenge organism (serotype 1 and 3; inoculum of 10^4–10^5 cfu per mouse), the use of four genetically different mouse strains results in different disease patterns. Swiss mice, BALB/C mice and CBA mice develop acute pneumonia and all die within 2–3 days; C57BL/6 mice develop subacute pneumonia and die only after 8–10 days. Mice die when bacterial counts reach 10^8 cfu/lung (Azoulay-Dupuis *et al.*, 1991a; Moine *et al.*, 1994b). The comparative efficacy of antibiotics could thus be assessed at various stages of the disease in terms of the survival rate and bacterial clearance (Table 57.1).

A problem arises with the infection of immunocompetent mice by penicillin-resistant *S. pneumoniae* strains. We failed to induce reproducible 100% lethal pneumonia with an inoculum of 10^4–10^5 cfu of any penicillin-resistant *S. pneumoniae* strain in immunocompetent mice. Early reports by Briles *et al.* (1992) indicated that some strains of *S. pneumoniae* were relatively avirulent in mice. These authors failed to detect mouse-virulent isolates of capsular types 14, 19, and 23. They also demonstrated a very strong association between the virulence and capsular type of the infecting pneumococci. Furthermore, epidemiologic data showed that strains of *S. pneumoniae* with resistance to penicillin are found among serotypes 6, 9, 14, 19, and 23, and that all the penicillin-resistant strains belonging to these serotypes are naturally avirulent in immunocompetent mice, independently of their site of isolation in humans (Bédos *et al.*, 1991; Briles *et al.*, 1992).

We tested the virulence of 150 clinical strains of *S. pneumoniae* isolated from different clinical sites (in Hôpital Bichat-Claude Bernard, Paris, from April 1992 to September 1993), in a murine peritonitis model (mouse model of experimental septicemia) and were unable to find any virulent penicillin-resistant strains (virulence was defined by a dilution of 10^{-4} or less inducing 100% mortality; personal data). In Swiss mice, with an inoculum of 10^5 cfu, these penicillin-resistant strains are naturally cleared within

Table 57.1 Models of pneumococcal infection

Reference	*Strains*	*Bacteria*	*Mode of infection*	*Advantages/disadvantages*
Laurenzi *et al.*, 1964	Mice		Aerosol	Large number of animals simultaneously infected. Deposition of few bacteria
Iizawa *et al.*, 1996	CBA/J mice	*S. pneumoniae*	Aerosol after airway obstruction induced by 2% formalin	Organisms appeared in the lungs in greater numbers
Fridmodt-Møller *et al.*, 1987	Mice	*S. pneumoniae* type 3	Intraperitoneal inoculation	Septicemia model more than pneumonia
Girard *et al.*, 1995	CF1 mice	*S. pneumoniae* type 3		
Tateda *et al.*, 1996	CBA/J mice	*S. pneumoniae*	Intranasal challenge	
Van der Poll *et al.*, 1996	C57BL/6 mice	*S. pneumoniae* type 3		
Esposito and Pennington, 1983	Guinea-pigs			Deposition of a standardized inoculum reproducible in 100% of animals
Ponte *et al.*, 1996	Guinea-pigs	*S. pneumoniae* type 9	Intratracheal challenge after incision of trachea	Anesthesia allows no coughing by animals
Candiani *et al.*, 1996	CD rats			Induced initial pneumonia with secondary septicemia
Girard *et al.*, 1996	C57BL/6 mice	*S. pneumoniae* type 3		
Azoulay *et al.*, 1991a	C57BL/6 mice	*S. pneumoniae* types 1, 3, 6, 14, 19, 23	Intratracheal challenge without incision of trachea	No animal injury, large number of animals infected (100 per hour)
Moine *et al.*, 1994b	Swiss mice			

3–4 days and are not invasive. Thus, a 10^5cfu inoculum induces spontaneously resolving non-bacteremic pneumonia in immunocompetent Swiss mice and BALB/C mice. An inoculum higher than 10^8cfu of penicillin-resistant *S. pneumoniae* strains results in non-reproducible lethal pneumonia.

In order to obtain reproducible lethal pneumonia with the same inoculum size of any penicillin-resistant *S. pneumoniae* strain, pneumonia was induced in leukopenic Swiss mice or decomplemented CBA mice. With an inoculum of 10^7 cfu per mouse of any non-virulent penicillin-susceptible or -resistant *S. pneumoniae* strain, immunocompromised Swiss mice and CBA mice develop acute bacteremic pneumonia and all die within 3–4 days.

Tateda *et al.* (1996) described a non-compromised CBA/J mouse model of penicillin-resistant *S. pneumoniae* pneumonia. After intranasal instillation of 2×10^7 cfu of a penicillin-resistant *S. pneumoniae* strain serotype 19, 70% of mice died within 7 days. These results indicated that CBA/J mice were susceptible to penicillin-resistant strains. Nevertheless, this is not a satisfactory reproducible model of community-acquired penicillin-resistant *S. pneumoniae* pneumonia in immunocompetent animals. Indeed, a model has to be reproducible whatever the *S. pneumoniae* strains used, and results obtained by these authors may not apply to a wide range of *S. pneumoniae* strains, as the disease course is unpredictable. Among the 10 *S. pneumoniae* strains used in their CBA/J mice model, one penicillin-susceptible strain did not cause lethal pneumonia and one penicillin-resistant strain caused only a 40% death rate. Moreover, death was generally more rapid following infection with penicillin-susceptible strains (4–8 days following infection) than with penicillin-resistant strains (8–16 days following infection). Their findings and ours underline the difficulty in obtaining reproducible infection in healthy mice with non-virulent penicillin-susceptible and -resistant *S. pneumoniae* strains.

Animal species

Models of pneumococcal pneumonia have been described with different animals—guinea-pigs (Ponte *et al.*, 1996), rats (Schmidt and Walley, 1951; Bakker-Woundenerg *et al.*, 1984; Smith and Abbott, 1994; Gavalda *et al.*, 1997)—but the most common species used is the mouse. Female Swiss mice (Vogelman *et al.*, 1988; Azoulay-Dupuis *et al.*, 1992; Moine *et al.*, 1994b), CBA/J mice (Tateda *et al.*, 1996) and C57Bl/6J mice (Azoulay-Dupuis *et al.*, 1991a; Rubins and Pomeroy, 1997) have been used. Not all mouse strains can be used, and preliminary experiments are required to determine the optimal bacterial inocula. We describe in this chapter the Swiss and C57BL/6J mouse pneumonia models we have developed. We used female C57BL/6J mice (body weight 18–22 g) and Swiss mice (body weight 20–24 g).

Preparation of animals

After arrival, animals rest for a week before infection and treatment. No specialized housing, care or specific pretreatment is required.

Details of surgery

Overview

Briefly, after anesthesia, mice are infected by the intratracheal per-oral route, as described by Esposito and Pennington, 1983.

Anesthesia

Mice are anesthetized intraperitoneally with 0.2 ml of 0.65% sodium pentobarbital (Dolethal®, Vetoquinol Laboratories, France). Complete anesthesia ensues in about 1–2 minutes and lasts for about 30 minutes.

Infection procedure

Following complete induction of anesthesia, the animal is suspended vertically by placing the upper incisor teeth on a wire hook, with the ventral face turned towards the experimenter (Figure 57.1A). The trachea is cannulated via the mouth with a blunt metal needle (22–23 G). The feel of the needle tip against the tracheal cartilage confirms the location of the cannula (Figure 57.1B). A volume of 40–50 μl of bacterial suspension is delivered to each mouse by means of a microliter syringe (Hamilton Co., Reno, NV; Figure 57.1C). The animal is maintained in the vertical position for 5 minutes to facilitate distal alveolar migration of the bacteria by gravity (Figure 57.1D).

Leukocyte and complement depletion in mice

In Swiss mice sustained leukopenia is obtained by giving three daily intraperitoneal injections (150 mg/kg) of cyclophosphamide, starting 5 days before bacterial challenge. At this time counts of circulating leukocytes in blood are reduced from about 7000–1000/mm^3. CBA mice are decomplemented by five injections of cobra venon factor at 3.3 μg/injection, starting 54 hours before infection (T0, 5, 9, 24 and 48 hours). Infection is performed 6 hours after the last injection. Decreased C3 component (80% of the initial level) is observed from the first injection and is maintained as low during treatment; it increases to 50% only 130 hours after the first injection of cobra venom factor.

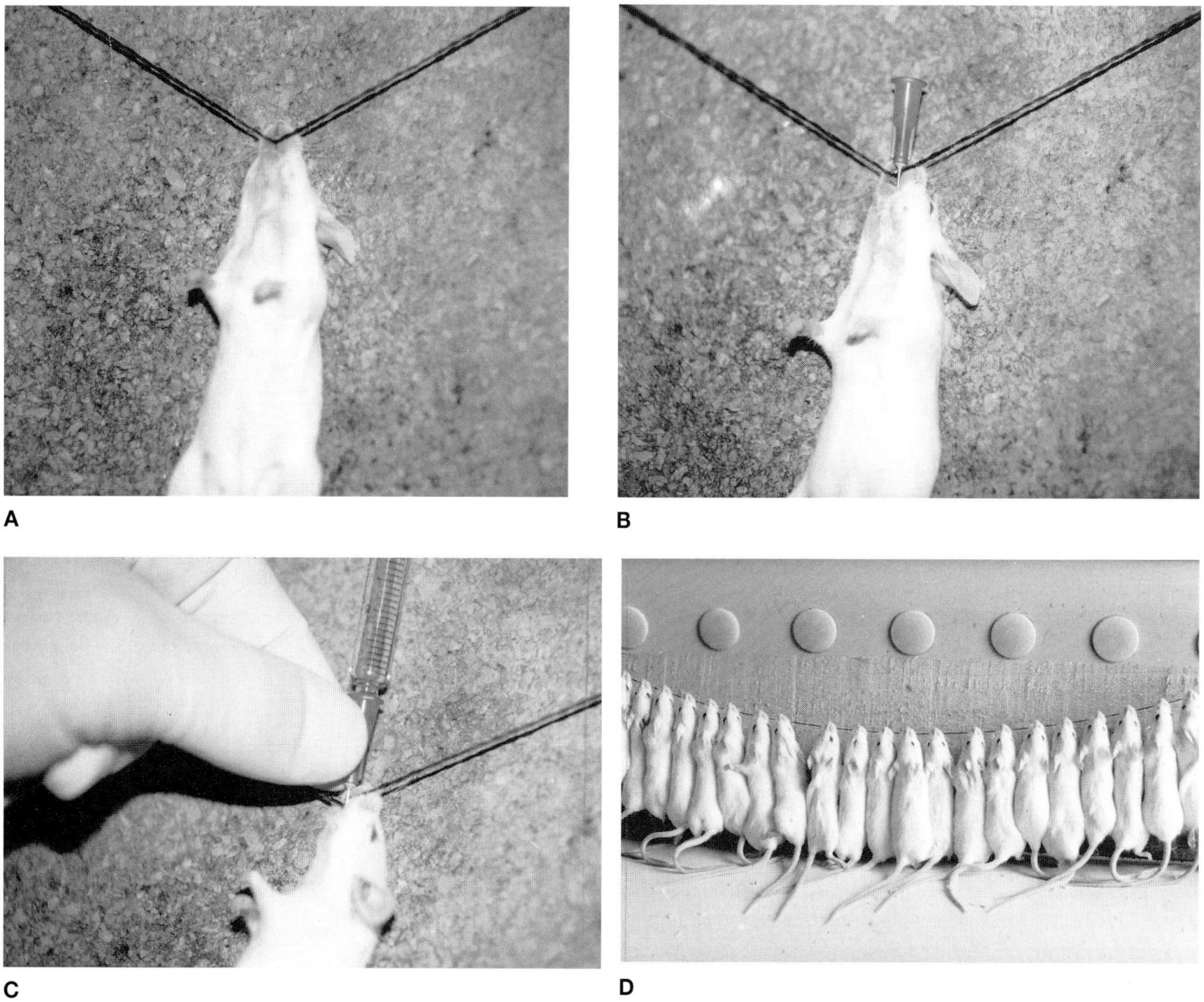

Figure 57.1 Intratracheal infection of mice. **A.** Vertical suspension of a mouse. **B.** Introduction of a needle in the trachea by the oral route. **C.** Delivery of a bacterial suspension into the lungs with a Hamilton microliter syringe. **D.** Mice maintained in vertical suspension to assist alveolar distal migration of bacteria.

Bacterial strains, storage and preparation of inocula

The *S. pneumoniae* strains we have used so far were clinical specimens, originally isolated from blood, sinus, tracheal aspiration, cerebrospinal fluid, and middle-ear fluid. The organisms are preserved at –70°C in brain–heart infusion broth supplemented with 5% filtered horse serum and nutrient broth (10% glycerol, 4% sorbitol). Virulence is maintained by monthly passage in mice. On the day of infection the bacteria are grown in brain–heart infusion broth with 5% filtered horse serum (incubation for 5–6 hours at 37°C until the culture is turbid to the naked eye). This exponential-growth culture is suspended in 0.9% saline to the desired concentration. Inoculum size is confirmed by plating serial 10-fold dilutions on to Columbia agar with 5% sheep blood and 18 hours incubation at 37°C.

Key parameters to monitor infection and response to treatment

During pneumonia, the clinical severity of infection can be measured by changes in individual animal weights and by scoring clinical symptoms as (1) no symptoms, (2) mild piloerection, (3) marked piloerection and listlessness, (4) minimal response to stimuli, and (5) death (Rubins and Pomeroy, 1997). In our immunocompetent and leukopenic Swiss mice models, body weight falls gradually to about 40% at the time of death. The clinical symptoms are rapid and marked piloerection and listlessness.

Survival studies

In immunocompetent mice, therapy is initiated 6 or 18 hours after bacterial challenge in the acute model (Swiss mice) and

18, 48, 72 or 96 hours in the subacute model (C57BL/6 mice). In leukopenic Swiss mice, the severity of the acute model requires earlier treatment, about 3 hours after bacterial challenge. The experiments must be repeated at least twice, with at least 15 animals in each treatment group; animals in the same experiment must be infected simultaneously. Antibiotics are given subcutaneously at various doses in a maximum volume of 0.5 ml. The observation period is 15 days, after which no further deaths occur. Cumulative survival rates are recorded daily and compared. Controls receive identical numbers of injections with isotonic saline.

Pulmonary clearance studies

Animals are killed with sodium pentobarbital. The thorax is opened and heart blood is removed for culture. The lungs are dissected free from the trachea and other structures and homogenized in 1 ml of sterile isotonic saline. Viable bacteria in the homogenates are quantified by plating 0.1 ml of serial 10-fold dilutions of the homogenates on Columbia agar with 5% sheep blood. Results are usually expressed as $\log_{10}$ cfu/ml of lung homogenate.

Pharmacokinetic studies

These models allow kinetic studies of antibiotics both in lungs and serum, in control mice as well as in infected mice. The lungs are removed from exsanguinated mice, weighed and homogenized in 1 ml of phosphate-buffered saline.

Histological studies

The lungs are fixed by *in-situ* intratracheal instillation of Bouin solution, under controlled pressure (15 cm) H_2O. The whole lungs are then removed and immersed in fixative for 2 hours. Slices are taken from different levels of each lobe, cut into small cubes, and processed routinely (Figure 57.2).

Antimicrobial therapy

β-lactams

Few animal studies have examined the relationship between *in-vitro* susceptibility testing and *in-vivo* efficacy against *S. pneumoniae* (Frimodt-Møller *et al.*, 1986; Frimodt-Møller and Thomsen, 1987a; Frimodt-Møller *et al.*, 1987). These studies did not account for differences between *S. pneumoniae* strains with regard to penicillin resistance. On the basis of these considerations, we asked whether the MIC results would be predictive of clinical outcome or therapeutic efficacy (Moine *et al.*, 1997a).

We used an E_{max} model to investigate the impact of penicillin resistance on the relative *in-vivo* efficacy of amoxicillin against strains of *S. pneumoniae* with penicillin MICs of < 0.01–16 μg/ml. The E_{max} model is described by the following equation (Gibaldi and Perrier, 1982):

$$E = (E_{max} \times D)/(P_{50} + D),$$

where E is the observed effect (maximal reduction in $\log_{10}$ cfu per lung compared with control mice just before therapy), D is the administered dose, E_{max} is a measure of relative efficacy indicated by the maximum antimicrobial effect attributable to the drug, and P_{50} is a measure of potency indicated by the dose producing 50% of E_{max}.

We found a highly significant correlation between P_{50} and MIC or MBC for amoxicillin against strains of *S. pneumoniae* covering a wide range of penicillin and amoxicillin MICs (Moine *et al.*, 1997a). Thus, the standard *in-vitro* MIC is an excellent predictor of the relative *in-vivo* potency of amoxicillin against pneumococci, including highly penicillin-resistant strains. Moreover, based on this model, the estimated MIC breakpoints for amoxicillin against *S. pneumoniae* should be 2 μg/ml for intermediate-resistant and 4 μg/ml for resistant strains. Pallarés *et al.* (1987) have suggested that only patients with bacteremic pneumococcal pneumonia due to penicillin-resistant strains (MICs ≥ 4 μg/ml) would not respond to therapy with a penicillin agent. Moreover, working from the simulation of human pharmacokinetics in a neutropenic murine thigh infection model, Andes *et al.* (1995) suggested that the estimated breakpoint for amoxicillin against *S. pneumoniae* would be 4 μg/ml for resistant strains.

Even if large doses of penicillin or amoxicillin can be given for strains more resistant to penicillin, alternative treatments must be considered. These include third-generation cephalosporins such as ceftriaxone and cefotaxime. In our leukopenic-mouse pneumonia model, we found that ceftriaxone had greater efficacy than amoxicillin against highly penicillin- and cephalosporin-resistant *S. pneumoniae* strains (penicillin MIC = 8 μg/ml and ceftriaxone MIC = 8 μg/ml) (Moine *et al.*, 1994b; Sauve *et al.*, 1996). The more favorable pharmacokinetics of ceftriaxone relative to amoxicillin was the determining factor involved in efficacy against highly resistant pneumococci *in vivo*. The time during which serum levels exceed the MIC (ΔtMIC) appears to be a major pharmacodynamic parameter correlating with ceftriaxone efficacy in this model. Results of the same studies showed that the high rate of protein binding of ceftriaxone, relative to that usually reported with amoxicillin, did not reduce its activity *in vivo*. On the contrary, the prolonged ΔtMIC suggested that the protein-bound fraction of the drug may serve as a reservoir and contribute to therapeutic efficacy (Moine *et al.*, 1994b).

Despite their similar MICs, cefotaxime was less active than ceftriaxone in our model (Sauve *et al.*, 1996; Moine *et al.*, 1997b). Furthermore, there was no major difference in the *in-vivo* efficacy of cefotaxime and amoxicillin against penicillin-resistant and cephalosporin-resistant *S. pneumoniae*.

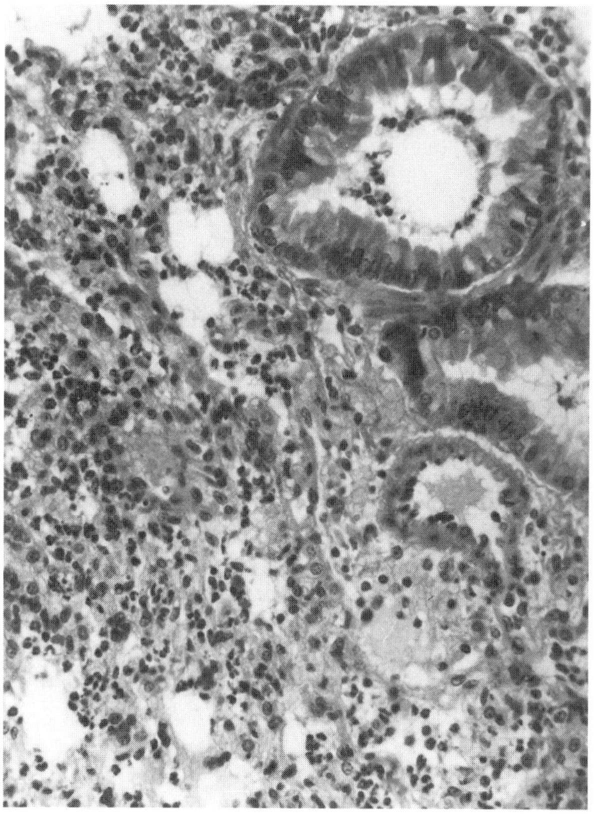
A

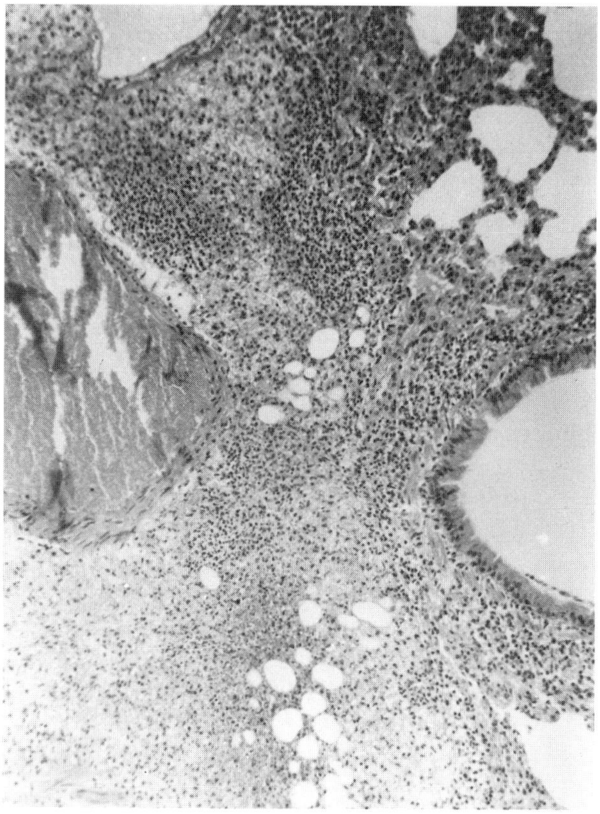
B

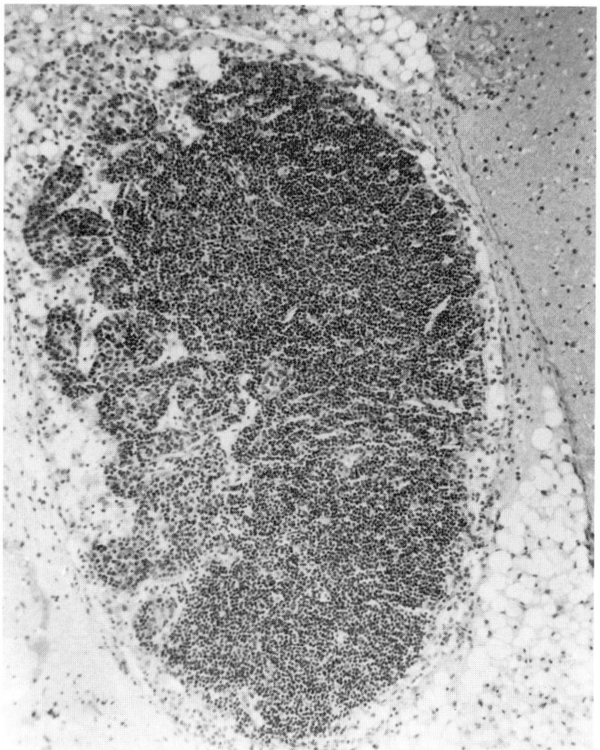
C

The pharmacokinetic properties of cefotaxime (shorter half-lives than amoxicillin in serum and lungs, and smaller AUCs) were probably a factor in its relative efficacy *in vivo* against highly penicillin-resistant pneumococci, regarding its low MIC for the test organisms (Tables 57.2 and 57.3).

Antibiotic combinations: β-lactams and aminoglycosides

Few studies of potential antipneumococcal synergy between β-lactam antibiotics and aminoglycosides have been reported (Haynes *et al.*, 1986; Fridmodt-Møller and Thomsen, 1987b; Gross *et al.*, 1995). Fridmodt-Møller *et al.* (1987) conducted an *in-vivo* study of interactions between β-lactam antibiotics and gentamicin against *S. pneumoniae* in mice infected intraperitoneally by a penicillin-susceptible strain. They found synergy between various β-lactam antibiotics and gentamicin. In our leukopenic-mouse model with acute bacteremic pneumonia induced by a highly penicillin-resistant *S. pneumoniae* strain (penicillin MIC = 4 μg/ml), synergy between amoxicillin and gentamicin was also demonstrated (Darras-Joly *et al.*, 1996).

The efficacy of various dosing regimens of aminoglycosides in combination with β-lactams has varied across models and bacteria. In experimental infections caused by aerobic

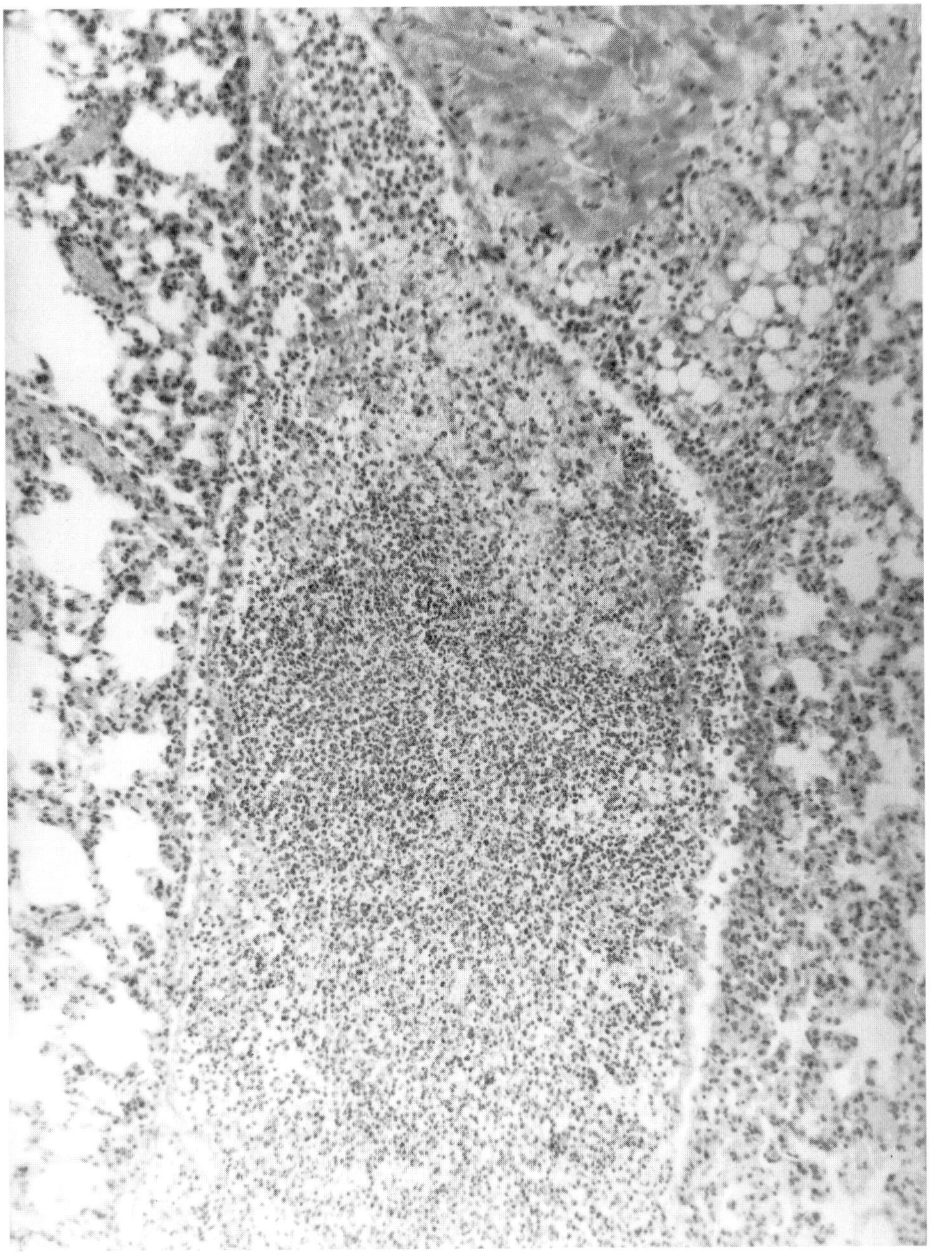

D

Figure 57.2 Histology of lungs in the subacute model of C57B1/6 mice infected with a virulent strain. **A.** By 24 hours after infection, there is invasion of the lungs by lymphocytes and polymorphonuclear cells, which have thickened the alveolar walls. **B.** By 72 hours after infection, peribronchovascular areas are distended and invaded by lymphoid patches. **C.** By 96 hours after infection, the lymphoid patches are clustered into follicles. **D.** Six days after infection, complete disorganization of lung structure is observed.

Gram-negative bacilli, a single daily dose of an aminoglycoside was at least as effective as conventional regimens (Fantin and Carbon, 1992). In our acute *S. pneumoniae* pneumonia model, the same daily dose of gentamicin had the same efficacy when it was given once daily as when it was given in two or three injections in combination (Darras-Joly *et al.*, 1996). This may be ascribable to the concentration-dependent bactericidal activity of aminoglycosides and their prolonged postantibiotic effect (Leggett *et al.*, 1989; Gilbert, 1991), which was confirmed *in vitro* and *in vivo* after a single injection of gentamicin (Darras-Joly *et al.*, 1996).

Quinolones

On the basis of their activity against *S. pneumoniae*, fluoroquinolones can be classified into three groups (Piddock, 1994):

- compounds whose *in-vitro* MICs are far above achievable concentrations in humans (8–16 μg/ml) and are therefore inconsistent with therapeutic efficacy, e.g. pefloxacin, fleroxacin, and lomefloxacin;
- compounds with lower MICs (1–4 μg/ml) that remain,

Table 57.2 Pharmacokinetic parameters of amoxicillin, ceftriaxone and cefotaxime in Swiss mice infected with a penicillin-resistant *Streptococcus pneumoniae* strain (P15986). Values are calculated from mean concentrations in serum and lung tissue samples taken at 0.5, 1, 2, 4, 6, 8, 10 and 24 hour post-dosing. Antibotics are determined by the agar well diffusion method. MICs for the challenge strain are 2 µg/ml for amoxicillin, 0.5 µg/ml for ceftriaxone and cefotaxime): C_{max}, maximum concentration observed, $T_{1/2}$, elimination half-life; Δ*t*MIC, time above the MIC; AUC, area under the curve. (Data are from Moine *et al.*, 1994b and 1997b.)

Drug (mg/kg)	*Site*	*Cmax (µg/ml or g)*	$T_{1/2}$ *(h)*	*ΔtMIC (h)*	AUC_{0-24} *(µg.h/ml or /g)*
Amoxicillin (50)	Serum	47±18	0.41	1.5	61
	Lung	15±9	0.79	1.5	15
Ceftriaxone (50)	Serum	47±1	1.41	9.0	97
	Lung	37±8	1.12	7.0	80
Amoxicillin (200)	Serum	118±12	0.41	3.3	92
	Lung	22±6	1.40	3.7	31
Cefotaxime (200)	Serum	101±10	0.35	3.8	80
	Lung	13±6	0.40	2.1	10

however, too close to achievable serum peak levels at standard regimens and carry a risk of clinical failure (Perrez-Tallero *et al.*, 1990) and selection of resistant mutants (Lafredo *et al.*, 1993; Bernard *et al.*, 1994) e.g. ciprofloxacin and ofloxacin;

- newer compounds, for instance sparfloxacin and trovafloxacin, whose *in-vitro* activity (MICs 0.25–0.5 µg/ml) and prolonged elimination half-life have resulted in greater efficacy in experimental mouse pneumonia models than ciprofloxacin (Azoulay-Dupuis *et al.*, 1992; Bédos *et al.*, 1998). The efficacy of sparfloxacin in these models was similar to previously reported data for temafloxacin compared with ciprofloxacin and ofloxacin (Azoulay-Dupuis *et al.*, 1991a). Against macrolide-resistant, penicillin-resistant and multiresistant strains, sparfloxacin was also highly effective (Table 57.4).

Few studies have been conducted in animals or humans to evaluate the significance of AUC/MIC, peak/MIC and Δ*t*MIC for predicting the bactericidal effect of fluoroquinolones (Leggett *et al.*, 1991; Drusano *et al.*, 1993). Recently, Forrest *et al.* (1993) and Hyatt *et al.* (1994) reported that the AUC/MIC ratio (AUIC) was the parameter most closely associated with bacterial eradication and clinical cure in nosocomial pneumonia. These authours determined that the threshold AUIC associated with clinical cure of patients with nosocomial pneumonia due to Gram-negative bacilli was equal to 125. Regardless of the species, bacterial killing was extremely rapid when AUIC exceeded 250. Using a dose fractionation approach in our models, we found that the AUIC was the best predictor of survival, and that an AUIC of 160 or more was associated with 100% clinical cure independently of the dosage schedule.

Macrolides

The comparative efficacy of five macrolides (erythromycin, roxithromycin, spiramycin, clarithromycin, and azithromycin) in our acute and subacute pneumonia models

Table 57.3 Outcome of betalactam and cephalosporin treatments in Swiss mice infected with resistant *Streptococcus pneumoniae* strains: bacterial clearance in lungs and survival rate after s.c. treatments initiated 3 hours postinfection. (Data are from Moine *et al.*, 1994b and 1997b, and Sauve *et al.*, 1996.)

S. pneumoniae *strains*	*Antibiotic (MIC/MBC µg/ml)*	*Dose (mg/kg)*	*Bacterial counts in lungs 8 h or *24 h post single dose (log_{10} cfu/ml)*	*Survival rate (%) 10 days after 3-day treatment: *q.12 h **q.8 h*	*Comments (see Table 57.2 for pharmacokinetic data)*
	None		7.42±0.40		Ceftriaxone versus amoxicillin
	Amoxicillin (2/4)	100	7.35±0.08		*Lower MICs against highly peni-R
		300		*67	strain
P15986	Ceftriaxone (0.5/1)	50	5.41±0.46	*75	*Greater efficacy
MIC/MBC		100	5.26±0.44		*Longer $T_{1/2}$ and greater AUC in lungs
peni: 4/8	None		7.42±0.40		Cefotaxime versus amoxicillin
µg/ml	Amoxicillin (2/4)	200	6.96±0.38		*Not as effective as expected from
		300	5.86±0.32	*67	its lower MIC
	Cefotaxime (0.5/1)	100	6.00±0.51		*Shorter $T_{1/2}$ in serum and lungs
		200	4.99±0.04	*67	
	None		*7.4±0.56		Ceftriaxone versus cefotaxime
P40984	Amoxicillin (4/8)	400	*5.32±0.74	**66	*Better efficacy against high-level
MIC/MBC	Cefotaxime (8/16)	400	*5.38±1.00	**75	cephalo-R strain despite similiar MICs
peni: 8/8	Ceftriaxone (8/16)	100	*4.08±0.80		*More favorable pharmacokinetics
µg/ml		200	*3.80±0.55	*83	

Table 57.4 Pharmacokinetic-pharmacodynamic parameters after a single s.c. dose of ciprofloxacin, sparfloxacin and temafloxacin in the serum and lungs of immunocompetent mice infected with a virulent penicillin-susceptible strain (P-4241). Values are calculated from the concentrations of six pooled samples of serum and lung taken at 0.5, 1, 3, 6, 8, and 24 hours post-dosing. $T_{1/2}$, elimination half-life; AUC, area under the curve. MICs/MBCs for the challenge strain: ciprofloxacin, 1/2 µg/ml, sparfloxacin: 0.25/0.5 µg/ml, temafloxacin: 1/1 µg/ml. (Data are from Azoulay-Dupuis *et al.*, 1991a, 1992.)

	C_{max} *(µg/ml or /g)*	$T_{1/2}$ *(h)*	AUC_{0-24} *(µg.h/ml or /g)*	*ΔtMIC/MBC (h)*	*AUC/MIC ratio*
Ciprofloxacin 100 mg/kg					
Serum	12.6	1.9	15	8/8	15
Lung	62.0	1.1	47	8/8	47
Sparfloxacin 50 mg/kg					
Serum	9.4	4.2	25	24/24	100
Lung	27.0	5.7	64	24/24	256
Temafloxacin 100 mg/kg					
Serum	14.7	3.8	142	8/8	142
Lung	124	2.8	490	24/24	490

induced by a virulent penicillin-susceptible strain has been reviewed with regard to the corresponding pharmacokinetic profiles (Veber *et al.*, 1993). Our data established the following hierarchy of *in-vivo* efficacy: azithromycin > spiramycin = clarithromycin > roxithromycin = erythromycin. In healthy mice the lung AUC of both azithromycin and spiramycin exceeded that of erythromycin (Table 57.5). This resulted not only from good penetration into the lung but also from the extended lung elimination half-life, which reflects their potential uptake, sequestration and durable delivery at the site of infection. These pharmacokinetic properties may explain the greater efficacy of both spiramycin and azithromycin relative to roxithromycin. Furthermore, in infected mice, the inflammatory process did not alter the pharmacokinetic profiles of roxithromycin, clarithromycin or erythromycin, while the pulmonary distribution of spiramycin and azithromycin increased markedly. Inflammatory cells served as a reservoir for cellular uptake and transport of macrolides. The role of leukocytes in the transport and release of azithromycin at the site of infection was confirmed (Veber and Pocidalo, 1995). Therefore, azithromycin was more effective than erythromycin whether given (*per os* or subcutaneously) prophylactically (as much as 24 hours before infection) or therapeutically (up to 48 hours postinfection; Azoulay-Dupuis *et al.*, 1991b). The pharmacokinetics of azithromycin in the lung (better tissue penetration, longer half-life than erythromycin) accounted for its superior efficacy. This could be explained by trapping of azithromycin in locally recruited phagocytic cells, whereas erythromycin is rapidly released (Gladue *et al.*, 1989). The better *in-vitro* activity of clarithromycin is probably the main determinant of its efficacy. In our models, clarithromycin lactobionate administered subcutaneously was more effective than erythromycin (Bédos *et al.*, 1992). The ^{14}OH metabolite of clarithromycin was as active as the parent compound and had an additive effect with clarithromycin.

By inducing pneumonia with a macrolide-resistant strain, we evaluated the efficacy of RP 59500 (Rhône-Poulenc-Rorer Laboratories), a semi-synthetic streptogramin (Azoulay-Dupuis *et al.*, 1991 c). We found that its two components, dalfopristine and quinupristine, were synergistic *in vivo*, despite a lack of efficacy when given individually. Low MIC and MBC values (0.125/0.25 µg/ml), rapid killing *in vitro* (<4 hours) and avid uptake by phagocytes (C/E ratio > 50) may account for the efficacy of dalfopristine + quinupristine in experimental pneumococcal pneumonia, despite a short half-life in the serum of infected and uninfected mice (30 minutes).

Advantages/disadvantages of the model

Besides the usual shortcomings of all animal models (differing pharmacokinetic profiles of antibiotics), the major shortcoming of this mouse model of *S. pneumoniae* pneumonia is the use of large inocula, directly delivered into the lung, which are required to establish the infection. This is unlikely to reflect the clinical situation. Moreover, the use of anesthesia during experimental inoculation makes the model non-physiological. Another limitation of the model is that artificial interventions are required to lower animal resistance. Because of these limitations, it is recommended that care be taken in the experimental design and subsequent interpretation of data on antimicrobial effectiveness.

Table 57.5 Pharmacokinetic parameters after a single s.c. dose (50 mg/kg) of erythromycin, azithromycin and azithromycin plus cyclophosphamide to healthy and infected Swiss mice. Mice were infected with a virulent penicillin-susceptible strain (P-4241) and treated 18 hours postinfection. Values are calculated from the concentrations of four pooled samples of serum and lung tissue taken at 0.5, 1, 2, 4, 6, 8, 10, 24, 48, 72 and 96 hours postdosing. Antibiotics are determined by microbiological assay. C_{max}, maximum concentration observed; $T_{1/2}$, elimination half-life; AUC; area under the curve. (Data are from Veber *et al.*, 1993.)

Drug	*Site*	C_{max} (*µg/ml or /g*)	$T_{1/2}$ (*h*)	AUC_{0-24} (*µg.h/ml or /g*)	*AUC Lung/serum*
Healthy Swiss mice					
Erythromycin	Serum	6	0.8	12	
	Lung	16	0.8	36	3.0
Azithromycin	Serum	5	7.3	32	
	Lung	36	17.1	676	21.1
Azithromycin + cyclophosphamide	Serum	1	2.6	4	
	Lung	18	18.1	289	72.2
Infected Swiss mice					
Erythromycin	Serum	7	0.8	12	
	Lung	17	0.9	34	2.8
Azithromycin	Serum	5	8.7	30	
	Lung	58	18.6	1041	34.7
Azithromycin + cyclophosphamide	Serum	1	2.0	5	
	Lung	16	16.5	214	42.8

Contributions of the model to infectious disease therapy

Profiling of antimicrobial treatment

Animal models of pneumococcal pneumonia have provided valuable information concerning human therapy. Apart from the establishment of the basic conditions of therapy, they have helped to define the use of new compounds. They have been used to verify the *in-vivo* effects and limits of combination regimens, to delineate the optimal conditions of administration, to evaluate pharmacokinetic differences in infected and non-infected animals, and to define the pharmacodynamic parameters predictive of efficacy. These models have also provided useful information concerning the pathophysiology of pneumococcal pneumonia.

Elucidation of the pathophysiology of pneumococcal pneumonia

Host response

Using an immunocompetent pneumococcal pneumonia mouse model induced by intranasal inoculation, it has been shown that TNFα was produced in the serum and lungs of CBA/J mice in parallel with bacterial growth in the lungs (Takashima *et al.*, 1997). Treatment with anti-TNFα antibody increased the number of infecting bacteria in the blood and accelerated the death of infected mice. These results suggested that TNFα plays a protective role in experimental pneumococcal pneumonia by preventing bacteremia. In the same model, Van der Poll *et al.* (1996) reported sustained and compartmentalized production of IL-10 within the lung during pneumococcal infection. Inhibition of IL-10 led to enhanced production of TNFα and interferon-γ, and reduced bacterial growth, while intrapulmonary administration of recombinant IL-10 resulted in reduced bacterial clearance and enhanced lethality.

Virulence factors

A strong association between capsular type and virulence for mice among human isolates of *S. pneumoniae* has been demonstrated in a mouse model (Briles *et al.*, 1992). A prospective study on 150 human isolates showed a relationship between serotype, virulence and susceptibility to penicillin. A negative correlation between penicillin resistance and virulence was also found (personal data). Apart from the capsule, other virulence factors such as pneumolysin and autolysin were investigated by the effect of insertional inactivation of the genes encoding these factors on the virulence of *S. pneumoniae* type 3. Both pneumolysin-negative and autolysin-negative strains had significantly reduced virulence in mice, as judged by survival time after intraperitoneal challenge (Berry *et al.*, 1992).

References

Andes, D., Urban, A., Craig, W. A. (1995). *In vivo* activity of amoxicillin and amoxicillin/clavulanate against penicillin-resistant pneumococci. In *Program and Abstracts of the 35th Interscience Conference on Antimicrobial Agents and Chemotherapy*, A82, p. 16, Annual Meeting of the American Society for Microbiology, San Francisco, CA.

Appelbaum, P. C. (1992). Antimicrobial resistance in *Streptococcus pneumoniae*: an overview. *Clin. Infect. Dis.*, **15**, 77–83.

Appelbaum, P. C. (1996). Emergence of resistance to antimicrobial agent in gram-positive bacteria-pneumococci. *Drugs*, **51**(Suppl. 1), 1–5.

Azoulay-Dupuis, E., Bédos, J. P., Vallée, E., Hardy, D. J., Swanson, R. N., Pocidalo, J. J. (1991a). Antipneumococcal activity of ciprofloxacin, ofloxacin, and temafloxacin in an experimental mouse pneumonia model at various stages of the disease. *J. Infect. Dis.*, **163**, 319–324.

Azoulay-Dupuis, E., Vallée, E., Bédos, J. P., Muffat-Joly, M., Pocidalo, J. J. (1991b). Prophylactic and therapeutic activities of azithromycin in a mouse model of pneumococcal pneumonia. *Antimicrob. Agents Chemother.*, **35**, 1024–1028.

Azoulay-Dupuis, E., Vallée, E., Veber, B., Pocidalo, J. J. (1991c). Activity of RP-59500, a new semi-synthetic streptogramin in mouse pneumonia induced by a macrolide-resistant strain of *Streptococcus pneumoniae*. In: *Program and Abstracts of the 31st Interscience Conference on Antimicrobial Agents and Chemotherapy*, A902. Annual Meeting of the American Society for Microbiology, Chicago, IL.

Azoulay-Dupuis, E., Vallée, E., Veber, B., Bédos, J. P., Bauchet, J., Pocidalo, J. J. (1992). *In vivo* efficacy of a new fluoroquinolone, sparfloxacin, against penicillin-susceptible and -resistant and multiresistant strains of *Streptococcus pneumoniae* in a mouse model of pneumonia. *Antimicrob. Agents Chemother.*, **36**, 2698–2703.

Bakker-Woundenberg, I. A. J. M., Van den Berg, J. C., Fontijne, P., Michel, M. F. (1984). Efficacy of continuous versus intermittent administration of penicillin G in *Streptococcus pneumoniae* in normal and immunodeficient rats. *Eur. J. Clin. Microbiol.*, **3**, 131–135.

Bédos, J. P., Rolin, O., Bouanchaud, D. H., Pocidalo J. J. (1991). Relation entre virulence et résistance aux antibiotiques des pneumocoques: apport des donnèes expérimentales sur un modèle animal. *Pathol. Biol.* **39**, 984–990.

Bédos, J. P., Azoulay-Dupuis, E., Vallée, E., Veber, B., Pocidalo, J. J. (1992). Individual efficacy of clarithromycin (A-56268) and its major human metabolite 14-hydroxy clarithromycin (A-62671) in experimental pneumococcal pneumonía in the mouse. *J. Antimicrob. Chemother.*, **29**, 677–685.

Bédos, J. P., Rieux, V., Bauchet, J., Muffat-Joly, M., Carbon, C., and Azoulay-Dupuis, E. (1998). Efficacy of trovafloxacin against penicillin-susceptible and multiresistant strains of *Streptococcus pneumoniae* in a mouse pneumonia model. *Antimicrob. Agents Chemother.*, **42**, 862–867.

Berendt, R. F. (1977). Relationship of method of administration to respiratory virulence of *Klebsiella pneumoniae* for mice and squirrel monkeys. *Infect. Immun.*, **20**, 581–583.

Bernard, L., Van Nguyen, J. C., Mainardi, J. L. (1994). *In vivo* selection of *Streptococcus pneumoniae* resistant to quinolones, including sparfloxacin. *Clin. Microbiol. Infect.*, **1**, 60–61.

Berry, A. M., Paton, J. C., Hansman, D. (1992). Effect of insertional inactivation of the genes encoding pneumolysin and autolysin on the virulence of *Streptococcus pneumoniae* type 3. *Microb. Pathogen.*, **12**, 87–93.

Briles, D. E., Crain, M. J., Gray, B. M., Forman, C., Yother. J. (1992). Strong association between capsular type and virulence for mice among human isolates of *Streptococcus pneumoniae*. *Infect. Immun.*, **60**, 111–116.

Candiani, G., Abbondi, M., Borgonovi, M., Williams, R. (1997). Experimental lobar pneumonia due to penicillin-susceptible and penicillin-resistant *Streptococcus pneumoniae* in immunocompetent and neutropenic rats: efficacy of penicillin and teicoplanin treatment. *J. Antimicrob. Chemother.*, **39**, 199–207.

Cleeland, R., Grunberg, E. (1986). Laboratory evaluation of new antibiotics *in vitro* and in experimental animal infections. In *Antibiotics in Laboratory Medicine*, 2nd edn (ed. Lorian V.) pp 825–876. Wiliams & Wilkins, Baltimore, MD.

Darras-Joly, C., Bédos, J. P., Sauve, C. *et al.* (1996). Synergy between amoxicillin and gentamicin in combination against a highly penicillin-resistant and -tolerant strain of *Streptococcus pneumoniae* in a mouse pneumonia model. *Antimicrob. Agents Chemother.*, **40**(9), 2147–2151.

Drusano, G. L., Johnson, D. E., Rosen, M., Standiford, H. C. (1993). Pharmacodynamics of a fluoroquinolone antimicrobial agent in a neutropenic rat model of *Pseudomonas sepsis*. *Antimicrob. Agents Chemother.*, **37**, 483–490.

Esposito, A. L., Pennington, J. E. (1983). Effects of aging on antibacterial mechanisms in experimental pneumonia. *Am. Rev. Resp. Dis. Suppl.*, **128**, 662–667.

Fantin, B., Carbon, C. (1992). *In vivo* antibiotic synergism: contribution of animal models. *Antimicrob. Agents. Chemother.*, **36**, 907–912.

Forrest, A., Nix, D. E., Ballow, C. H. *et al.* (1993). Pharmacodynamics of intravenous ciprofloxacin in seriously ill patients. *Antimicrob. Agents Chemother.*, **37**, 1073–1081.

Frimodt-Møller, N., Thomsen, V. F. (1987a). The pneumococcus and the mouse-protection test: correlation of *in vitro* and *in vivo* activity for beta-lactam antibiotics, vancomycin, erythromycin and gentamicin. *Acta Pathol. Microbiol. Immunol. Scand.*, **95**(5), 159–165.

Frimodt-Møller, N., Thomsen, V. F. (1987b). Interaction between betalactam antibiotics and gentamicin against *Streptococcus pneumoniae in vitro* and *in vivo*. *Acta Pathol. Microbiol. Immunol. Scand.*, **95**(5), 269–275.

Frimodt-Møller, N., Bentzon, M. W., Thomsen, V. F. (1986). Experimental infection with *Streptococcus pneumoniae* in mice: correlation of *in vitro* activity and pharmacokinetic parameters with *in vivo* effect for 14 cephalosporins. *J. Infect. Dis.*, **154**, 511–517.

Frimodt-Møller, N., Bentzon, M. W., Thomsen V. F. (1987) Experimental pneumococcus infection in mice: comparative *in vitro* and *in vivo* effect of cefuroxime, cefotaxime and ceftriaxone. *Acta Pathol. Microbiol. Immunol. Scand.*, **95**(5), 261–267.

Gavalda, J., Capdevila, J. A., Almirante, B. *et al.* (1997). Treatment of experimental pneumonia due to penicillin-resistant *Streptococcus pneumoniae* in immunocompetent rats. *Antimicrob. Agents Chemother.*, **41**, 795–801.

Gibaldi, M., Perrier, D. (1982). *Pharmacokinetics*, 2nd edn., pp. 221–232. Marcel Dekker, New York.

Gilbert, D. N. (1991). Once-daily aminoglycoside therapy. *Antimicrob. Agents Chemother.*, **35**, 399–405.

Girard, A. E., Girard, D., Gootz, T. D., Faiella, J. A., Cimochowski, C. R. (1995). *In vivo* efficacy of trovafloxacin (CP 99,219), a new quinolone with extended activities against gram-positive pathogens, *Streptococcus pneumoniae*, and *Bacteroides fragilis*. *J. Antimicrob. Chemother.*, **39**(10), 2210–2216.

Girard, A. E., Cimochowski, C. R., Faiella, J. A. (1996). Correlation of increased azithromycin concentrations with phagocyte infiltration into sites of localized infection. *J. Antimicrob. Chemother.*, **37**(Suppl. C), 9–19.

Gisby, J., Wightman, B. J., Beale, A. S. (1991). Comparative efficacies of ciprofloxacin, amoxicillin, amoxicillin–clavulanic acid, and cefaclor against experimental *Streptococcus pneumoniae* respiratory infections in mice. *Antimicrob. Agents Chemother.*, **35**(5), 831–836.

Gladue, R. P., Bright G. M., Isaacson, R. E., Newborg, M. F. (1989). *In vitro* and *in vivo* uptake of azithromycin (CP 62,993) by phagocytic cells: possible mechanism of delivery and release at sites of infection. *Antimicrob. Agents Chemother.*, **33**, 277–282.

Green, G. M., Kass, E. H. (1964). Factors influencing the clearance of bacteria by the lung. *J. Clin. Invest.*, **43**, 769–776.

Gross, M. E., Giron, K. P., Septimus, J. D., Mason, E. O., Musher, D. M. (1995). Antimicrobial activities of betalactam antibiotics and gentamicin against penicillin-susceptible and penicillin-resistant pneumococci. *Antimicrob. Agents Chemother.*, **39**, 1166–1168.

Haynes, J., Hawkey, P. M., Williams, E. W. (1986). The *in vitro* activity of combinations of penicillin and gentamicin against penicillin-resistant and penicillin-sensitive *Streptococcus pneumoniae*. *J. Antimicrob. Chemother.*, **18**, 426–428.

Hyatt, J. M., Nix, D. E., Schentag, J. J. (1994). Pharmacokinetics and pharmacodynamics of ciprofloxacin against similar MIC strains of *S. pneumoniae, S. aureus*, and of *P. aeruginosa* using kill curve methodology. *Antimicrob. Agents Chemother.*, **38**, 2730–2737.

Iizawa, Y., Kitamoto, N., Hiroe, K., Nakao, M. (1996). *Streptococcus pneumoniae* in the nasal cavity of mice causes lower respiratory tract infection after airway obstruction. *J. Med. Microbiol.*, **44**(6), 490–495.

Lafredo, S. C., Foleno, B. D., Fu, K. P. (1993). Induction of resistance of *Streptococcus pneumoniae* to quinolones *in vitro*. *Chemotherapy*, **39**, 36–39.

Laurenzi, G. A., Berman, L., First, M., Kass, E. H. (1964). A quantitative study of the deposition and clearance of bacteria in the murine lung. *J. Clin. Invest.*, **43**, 759–768.

Leggett, J. E., Fantin B., Ebert S. *et al.* (1989). Comparative antibiotic dose–effect relations at several dosing intervals in murine pneumonitis and thigh-infection models. *J. Infect. Dis.*, **159**, 281–292.

Leggett, J. E., Ebert S., Fantin B., Craig W. A. (1991). Comparative dose-effect relations at several dosing intervals for β-lactam, aminoglycoside and quinolone antibiotics against Gram-negative bacilli in murine thigh-infection and pneumonitis models. *Scand. J. Infect. Dis.*, **74**(Suppl.), 179–184.

Lieberman, D., Schlaeffer F., Boldur, H. *et al.* (1996). Multiple pathogens in adult patients admitted with community-acquired pneumonia: a one year prospective study of 346 consecutive patients. *Thorax*, **51**, 179–184.

Marton, A., Gulyas, M., Munos, R., Tomasz, A. (1991). Extremely high incidence of antibiotic resistance in clinical isolates of *Streptococcus pneumoniae* in Hungary. *J. Infect. Dis.*, **163**, 542–548.

Moine, P., Vercken, J. B., Chevret, S., Chastang, C., Gajdos, P., the French Study Group for Community-Acquired Pneumonia in the Intensive Care Unit (1994a). Severe community-acquired pneumonia. Etiology, epidemiology and prognosis factors. *Chest*, **105**, 1487–1495.

Moine, P., Vallée, E., Azoulay-Dupuis, E. *et al.* (1994b). *In vivo* efficacy of a broad-spectrum cephalosporin, ceftriaxone against a penicillin-susceptible and -resistant strains of *Streptococcus pneumoniae* in a mouse pneumonia model. *Antimicrob. Agents Chemother.*, **380**, 1953–1958.

Moine, P., Mazoit, J. X., Bédos, J. P., Vallée. E., Azoulay-Dupuis, E. (1997a). Correlation between *in vitro* and *in vivo* activity of amoxicillin against *Streptococcus pneumoniae* in a murine pneumonia model. *J. Pharmacol. Exp. Ther.*, **280**, 310–315.

Moine, P., Sauve, C., Vallée, E., Bédos, J. P., Pocidalo, J. J., Azoulay-Dupuis, E. (1997b). *In vivo* efficacy of cefotaxime and amoxicillin against penicillin-susceptible, penicillin-resistant and penicillin-cephalosporin-resistant strains of *Streptococcus pneumonia* in a mouse pneumonia model. *Clin. Microbiol. Infect.*, **3**, 608–615.

Muraoka, H., Tsuji, A., Ogawa, M., Goto, S. (1988). *In vitro* and *in vivo* antibacterial activity of roxithromycin against gram positive pathogens. *Br. J. Clin. Pract.*, **42** (Suppl. 55), 7–9.

Pallarés, R., Gudiol, F., Linarés, J. *et al.* (1987). Risk factors and response to antibiotic therapy in adults with bacteremic pneumonia caused by penicillin-resistant pneumococci. *N. Engl. J. Med.*, **317**, 18–22.

Pallarés, R., Linarés, J., Vadillo, M. *et al.* (1995). Resistance to penicillin and cephalosporin and mortality from severe pneumococcal pneumonia in Barcelona, Spain. *N. Engl. J. Med*, **333**, 474–480.

Pennington, J. E. (1985). Animal models of pneumonia for evaluation of antimicrobial therapy. *J. Antimicrob. Chemother.*, **16**, 1–6.

Perrez-Tallero, E., Garcia-Arenzana, J. M., Jiminez, J. A., Peris, A. (1990). Therapeutic failure and selection of resistance to quinolones in a case of pneumococcal pneumonia treated with ciprofloxacin. *Eur. J. Clin. Microbiol. Clin. Dis.*, **9**, 905–906.

Piddock, L. J. V. (1994). New quinolones and Gram-positive bacteria. *Antimicrob. Agents Chemother.*, **38**(2), 163–169.

Ponte, C., Parra A., Nieto E., Soriano F. (1996). Development of experimental pneumonia by infection with penicillin-insensitive *Streptococcus pneumoniae* in guinea pigs and their treatment with amoxicillin, cefotaxime, and meropenem. *Antimicrob. Agents Chemother.*, **40**, 2698–2702.

Rolin, O., Bouanchaud D. H. (1987). Antipneumococcal activity of erythromycin and spiramycin in 2 experimental models in mice. *Pathol. Biol.*, **35**(5), 742–745.

Rubins, J. B., Pomeroy, C. (1997). Role of gamma interferon in the pathogenesis of bacteremic pneumococcal pneumonia. *Infect. Immun.*, **65**, 2975–2977.

Sauve, C., Azoulay-Dupuis, E., Moine, P., Darras-Joly, C., Rieux, V., Bédos, J. P. (1996). Efficacy of cefotaxime and ceftriaxone in a mouse model of pneumonia induced by two penicillin- and cephalosporin-resistant strains of *Streptococcus pneumoniae*. *Antimicrob. Agents Chemother.*, **40**, 2829–2834.

Schmidt, L. H., Walley A. (1951). The influence of the dosage regimen on the therapeutic effectiveness of penicillin G in experimental lobar pneumonia. *J. Pharmacol. Exp. Ther.*, **103**, 479–488.

Smith, G. M., Abbott, K. H. (1994). Development of experimental respiratory infections in neutropenic rats with either penicillin-resistant *Streptococcus pneumoniae* or β-lactamase-producing *Haemophilus influenzae*. *Antimicrob. Agents Chemother.*, **38**, 608–610.

Tateda, K., Takashima, K., Miyazaki, H., Matsumoto, T., Hatori, T., Yamaguchi, K. (1996). Noncompromised penicillin-resistant pneumococcal pneumonia CBA/J mouse model and comparative efficacies of antibiotics in this model. *Antimicrob, Agents Chemother.*, **40**(6), 1520–1525.

Takashima, K., Tateda, K., Matsumoto, T., Iizawa, Y., Nakao, M., Yamaguchi, K. (1997). Role of tumor necrosis factor alpha in pathogenesis of pneumococcal pneumonia in mice. *Infect. Immun.*, **65**, 257–260.

Tsuboi, Y., Sakoda T., Mitsuhashi, S. (1988). *In vitro* and *in vivo* antibacterial activity of roxithromycin. *Br. J. Clin. Pract.*, **42**, (Suppl. 55), 30–32.

Van der Poll, T., Marchant, A., Keogh, C. V., Goldman, M., Lowry, S. F. (1996). Interleukin-10 impairs host defense in murine pneumococcal pneumonia. *J. Infect. Dis.*, **174**(5), 994–1000.

Veber, B., Vallée, E., Desmonts, J. M., Pocidalo, J. J., Azoulay-Dupuis, E. (1993). Correlation between macrolide lung pharmacokinetics and therapeutic efficacy in a mouse model of pneumococcal pneumonia. *J. Antimicrob. Chemother.*, **32**, 473–482.

Veber, B., Pocidalo, J. J. (1995). Le cas particulier des azalides: l'antibiodiapédèse. Données expérimentales à partir d'un modèle murin de pneumonie à pneumocoque. *Pathol. Biol.*, **43**(6), 524–528.

Vogelman, B., Gudmundsson, S., Leggett, J., Turnidge, J., Ebert, S., Craig, W. A. (1988). Correlation of antimicrobial pharmacokinetic parameters with therapeutic efficacy in an animal model. *J. Infect. Dis.*, **158**, 831–847.

Zak, O., Sande, M. A. (1986). Introduction: the role of animal models in the evaluation of new antibiotics. In: *Experimental Models in Antimicrobial Chemotherapy*, vol. 1 (eds Zac O., Sande, M. A.), pp. 1–5. Academic Press, London.

Chapter 58

Animal Models of Gram-negative Bacillary Experimental Pneumonia

M. Rouse and J. M. Steckelberg

Background of model

Respiratory tract infections remain among the most common clinical problems encountered by physicians. Pneumonitis or infection of the lung parenchyma itself is the most common infection-related cause of death and the sixth most common cause of death overall in the USA. A large variety of pathogens have been implicated in pneumonitis, depending on host factors, immunologic status, and nosocomial versus community acquisition. The challenges of pneumonitis management have been made more formidable by widespread distribution of antibiotic-resistant pathogens, such as penicillin- and macrolide-resistant pneumococci or multiply-resistant Gram-negative bacilli, and the proliferation of new antimicrobial agents. While suitably designed clinical trials remain the gold standard for assessing human efficacy, experimental models of pneumonitis have played a vital role in providing detailed data about the efficacy of specific antimicrobial–microorganism combinations, pharmacodynamic relationships, experimental pathogenesis, and data supporting clinical trial design. A variety of animal model study designs have been used to study pneumonia. Sauve *et al.* (1996) used an animal model of pneumococcal infection to study penicillin resistant pneumococci, *P. carinii* was studied by Garvy and Harmsen (1996), myocoplasma pneumonia was studied by Rodriguez *et al.* (1996) and a variety of Gram-negative bacilli by Kim *et al.* (1996), Mimoz *et al.* (1997), and Taba *et al.* (1996).

Animal species

In our laboratory, male C57bl/6 mice (25 g) and guinea-pigs have been used in experimental models of *H. influenzae* or *P. aeruginosa*. Joly-Guillou *et al.* (1997) used C3H/HeN mice to model *Acinetobacter* pneumonia and Tsai *et al.* (1997) used CBA/J mice to model *Klebsiella* experimental pneumonia. Any one of several inbred strains of mice can probably be used. We use same-sex (male) mice exclusively to eliminate problems keeping female and male mice separate when assigning animal subjects to treatment regimens. Other animal species have been used to model bacterial pneumonia, including rats, goats (Ole-Mapenay *et al.*, 1997), rabbits (Fox-Dewhurst *et al.*, 1997), dogs (Light, 1996), and guinea-pigs (Edelstein *et al.*, 1996). Mice are inexpensive, convenient to handle, and generally tolerate antimicrobial therapy well.

Study design

A basic approach to using these models to compare antimicrobial treatment regimens would be to perform *in-vitro* susceptibility studies and select representative strains of the microorganism for *in-vivo* use. Determine the pharmacokinetics of the compounds to be studied in the experimental animal chosen, perform *in-vivo* studies, and then analyze the data. An isolate to be studied *in-vivo* should be selected from a large collection based on *in-vitro* studies that include at least standardized *in-vitro* susceptibility testing (National Committee for Clinical Laboratory Standards, 1997). If a particular species or a specific antimicrobial susceptibility pattern is the primary focus of the study, this will provide information about the representation of the selected strain compared to the susceptibility patterns of the particular species. Appropriate antimicrobial dosages and treatment schedules can be chosen by determination of concentrations of antibiotics in serum at timed intervals after administration to healthy animals. The treatment regimens selected depend on the goals of the study (e.g., pharmacodynamic relationships versus efficacy comparisons among antimicrobials).

Preparation of animals

Mice are placed in our housing facility at least a week prior to use. This gives the animals time to acclimate to new surroundings and to recover from the stress associated with shipping. If the mice can be housed in a 'shoebox' type of cage with woodchip bedding, this offers the advantage of helping to maintain each animal's body temperature as they

Handbook of Animal Models of Infection
ISBN 0-12-775390-7

recover from anesthesia, and isolation lids can be used if needed.

Details of surgery

Overview

A suspension of viable bacteria is introduced into the alveoli via the trachea, infecting the lower respiratory tract. For assessing the rate of bacterial clearance as the primary endpoint, the animals are sacrificed and the number of viable bacteria surviving *in situ* is determined. The results of treatment are expressed as colony forming units per gram of lung and compared statistically using rank sum analysis. Alternative endpoints, for infections with appreciable mortality, are survival rates, either at the end of therapy or at some later point in time, or the proportion of animals microbiologically clear of the pathogen.

Materials

A list of materials needed would include: pentobarbital (26%, Fort Dodge Laboratories), 100 μl syringe (Hamilton Syringes no. 80600), feeding needle, 22 G × 2.5 cm with 1.25 mm ball (Popper BioMedical Needles No 7901), disposable ligature 20 cm stretched taut 20 cm above bench, sterile tissue forceps (6 cm) and scissors (6 cm) (× 2), sterile Petri plates, balance, sterile gauze, stomacher tissue homogenizer (Stomacher 80®, Tekmar Company), nutrient broth, agar plates, freezer vials (Microbank PL 160, Pro-Lab Diagnostics), bent glass rods, and capillary tubes.

Storage and preparation of inocula

When a specific isolate for study has been chosen, the isolate is passed through a mouse model of peritonitis to promote its virulence. This is done by intraperitoneally injecting 100 μl of overnight broth culture (containing 5×10^8 cfu) of the study isolate. The animals are returned to their cages and 24 hours later a blood sample is collected from the retro-orbital plexus. The isolate for study is recovered from the blood culture. This is repeated for a total of three passages through the mouse model; the bacteria recovered from the third blood culture are then stored in freezer vials at –70 °C. Bacteria saved by serial passage on agar plates or subjected to repeated freeze–thaw cycles may lose their virulence in time. Sufficient bacteria in aliquots should be frozen at the beginning of the study to complete all anticipated *in-vivo* experiments. Bacteria for intratracheal challenge are prepared from this frozen sample. A bead is removed from the freezer vial 24 hours prior to intratracheal challenge and rolled around on a third of an agar plate. The inoculated area is streaked for isolation and the plate is incubated overnight at 35 °C in 5% CO_2. The inocula is 15 colonies picked with a cotton tipped swab and emulsified in 1 ml of sterile saline. This preparation should be used within 2 hours of preparation. The purity and bacterial density should be confirmed before use of the inocula by plating 100 μl aliquots of serial dilutions on the surface of chocolate agar plates. Cash *et al.* (1979) used bacteria impregnated agar beads, prepared by chilling a solution of viable bacteria and 2% molten agar, to aid in establishing this infection.

Anesthesia

Pentobarbital 50 mg/kg administered intraperitoneally induces adequate anesthesia for intratracheal inoculation. Other anesthetics are easier to use but do not offer the advantage of the suppression of pulmonary reflexes secondary to barbiturate anesthesia, and this is desirable to promote establishing this infection.

Infection process

Intratracheal instillation is done using a 100 μl syringe with feeding needle. The feeding needle is bent 5 mm up from the ball to a 45° angle (Figure 58.1). The anesthetized mice are held in the left hand horizontally, ventral surface up, with the neck slightly hyperextended. The ball of the feeding needle is passed over the tongue, down the esophagus to the trachea (Figure 58.2). The trachea can be felt as a small bump. The trachea is occluded with the ball of the needle and 60 μl of viable bacteria is injected into the trachea. Tachypnea should be immediately obvious. The mice are then vertically suspended by their incisors on a taunt ligature for 2 minutes. This promotes diffusion of the bacteria throughout the lower respiratory tract. An alternate method is to simply drop the inocula on to the nostrils of the mice, allowing bacteria to be aspirated into the lungs. This method is technically simpler but allows less precise control of the pulmonary inocula.

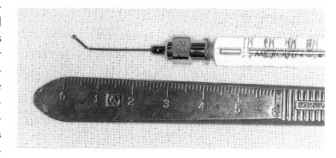

Figure 58.1 Feeding needle used for intratracheal instillations.

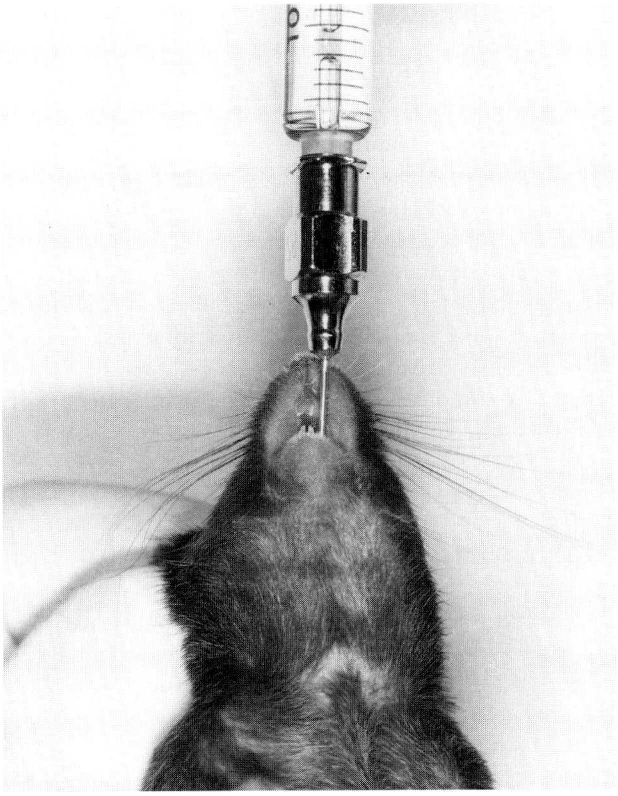

Figure 58.2 Intratracheal inoculation.

Postoperative care

The mice are then returned to their cages to recover from anesthesia.

Key parameters to monitor infection and response to drugs

Key parameters of the experimental model include the size and density of the bacterial inoculum, the time period between inoculation and the start of antimicrobial therapy, the antimicrobial dosage regimen and duration, and the time after antibiotic therapy until sacrifice or the assessment of an alternative endpoint such as survival. Optimization of these parameters depends on the study endpoints; for example, if quantitative culture results are the primary endpoint, survival up to the time of sacrifice should be relatively high in all treatment groups in order to avoid selection bias from only assessing survivors, which may have had less severe infection. On the other hand, high survival rates at the end of the study in all treatment groups would be non-discriminatory in a study with mortality as the primary endpoint.

In general, higher inocula will result in more fulminant infection and lower survival rates; an inoculum too low will result in failure to establish infection in all animals or in mild spontaneously resolving infection. Likewise, beginning antimicrobial therapy too early, before infection is established, results in over-optimistic estimates of efficacy. An excessive delay between inoculation and treatment results in either excessive experimental mortality or spontaneous self-cure. A time interval of 6–8 hours typically works well for many organisms.

H&E-stained sections of tissues from each treatment group should be microscopically examined for histopathological evidence of bacterial pneumonia to insure that the animals are actually infected rather than merely colonized with the inocula.

Lung pathology

The histopathology of mice infected with *H. influenzae* or *P. aeruginosa* is similar to that seen in human lung sections with bacterial pneumonia. An acute inflammatory response, intra-alveolar hemorrhage, alveolar septal edema, and consolidation of alveolar air spaces should be evident on examination of H&E-stained sections. Bacteria should be present in sufficient numbers in the untreated animals 24 hours after inoculation to be visible on Gram-stained touch preps or processed sections.

Bacterial density in lungs

The animals are sacrificed with a lethal dose of pentabarbital (100 mg/kg administered intraperitoneally). Sterile tissue forceps and scissors are used to open and remove the skin from the chest area, and the ventral chest wall (ribs and musculature) is removed with the same instruments. The contents of the chest cavity are readily apparent. Both lungs are removed lobe by lobe at the bronchi with a sterile set of instruments and placed together in a sterile Petri dish. The tissues are immediately weighed and immersed in 2 ml of nutrient broth in a tissue emulsifier bag and stored in the refrigerator (<2 hours) until processing for bacterial culture. The contaminated instrumentation is soaked in betadine, rinsed, sonicated in detergent, rinsed, wrapped and autoclaved.

Tissues from each animal are homogenized, and 100 μl of homogenate is diluted in 900 μl of nutrient broth in a serial dilution scheme of 10^{-0}–10^{-6}. A volume of 100 μl of the aliquots of each dilution is spread on the surface of a chocolate agar plate with a bent glass rod. The glass rods are sterilized by dipping in ethanol and flaming until the alcohol burns off. They are then cooled for a few minutes before use. The remaining broth in dilution tubes is poured back into the tissue homogenizer bags to be incubated along with the agar plates at 35 °C in 5% CO_2 for 48 hours.

After incubation, the agar plates are inspected for bacterial growth. Less than 15 colonies of respiratory flora on the undiluted plate is probably normal respiratory tract flora. More than 15 colonies, especially of a single species, may represent the contaminated tissues. If bacterial growth resembling the isolate inoculated into the lungs is present, inspect the plates to see that the bacterial distribution follows the dilution scheme. If this is not the case, then the plates may be mislabeled or an error may have been made in the dilution process. If the bacterial distribution follows the dilution scheme, find the dilution plate that has between 10 and 100 colonies present and count the actual number of colonies. The actual number of organisms per gram of tissue can be algebraically calculated as described by Washington (1985).

$$\text{cfu/g of lung} = [(N/V) \times D \times F]/\text{g of tissue},$$

where N = number of colonies counted, V = ml of each dilution plated (0.1), D = reciprocal of dilution inoculated on to agar plate and F = (ml of broth in homogenizer bag + tissue volume (1 g = 1 ml))/g of tissue

If no growth is present on any of the plates, then a swab from the incubated homogenizer bag should be subcultured on to the surface of a chocolate agar plate. If colonies resembling *H. influenzae* grow, the culture is positive from broth only; if no colonies resembling *H. influenzae* grow then the culture is negative for *H. influenzae*. Colonial morphology may change after *in-vivo* exposure to antibiotics. If there is doubt about the species seen growing on any of the plates, Gram's stain will reveal cell morphology to aid in deciding if the colonies are *H. influenzae* or not. The actual colony-forming units of bacteria per gram of lung can be a cumbersome number to work with; converting this number to $\log_{10}$ cfu per gram will make statistical analysis easier.

Determination of antimicrobial penetration into the lungs

Zelinka *et al.* (1997) recovered pulmonary fluid for antimicrobial assay, Fox-Dewhurst *et al.* (1997) homogenized the entire lungs and recovered pulmonary fluid for assay. Concentrations of antimicrobials in serum often correlate well with response to treatment.

Determination of additional markers of infection and response to treatment

Additional markers of infection have been studied. Survival and histopathology were monitored by Song *et al.* (1997); Rodriguez *et al.* (1996) studied immunohistochemistry associated with *Mycoplasma* experimental pneumonia. Other considerations may include blood gas analysis, radiographic examination or immunoglobulin concentrations.

Antimicrobial therapy

This model has been used by many investigators to compare results of treatment with various antimicrobials against many different species of bacteria. Antimicrobial therapy is usually initiated 4–24 hours after bacterial challenge and continued for a short time (one dose to eight doses). Mice can be administered antibiotics via a variety of routes and they tolerate many antimicrobials well. Gastric lavage for oral administration is not difficult. Repeated intravenous access is not practical; however, intraperitoneal administration may be an acceptable substitute for intravenous access. Mice metabolize many antibiotics very quickly and short dosing intervals can be used to provide adequate antimicrobial serum concentrations to compare antibiotic activity *in vivo*. Bergogne-Berezin (1995) has an informative discussion of the pulmonary disposition of antibiotics and how it relates to animal models of pneumonia.

Analysis of results

Comparisons of survival or complete killing of bacteria *in situ* (sterile cultures) can be done with Fishers' exact test. If the study is designed to compare the results of quantitative cultures, the numbers of study subjects may not provide a sensitive test of differences in survival or sterile cultures; this should be noted in any discussion. Significant differences may exist in this data and it should be analyzed. The results of the quantitative studies, expressed as $\log_{10}$ cfu/g of lung, can be analyzed by Student's T test or rank-sum tests, as appropriate (Ilstrup, 1990). When rank-sum analysis is preformed, sterile cultures are all assigned an equal value that is less than the lowest positive culture. If the investigator feels that cultures that are positive from broth only are different from other positive cultures, they can be assigned a value greater than the value assigned the sterile cultures and less than the lowest positive culture.

Advantages and disadvantages of the model

This model requires technical expertise to reliably occlude the trachea with a bent feeding needle, but once mastered, the procedure can be performed quickly. This can be learned by challenging a mouse with a dye solution and then sacrificing the mouse to see where the dye has been injected. After some practice, the dye should be exclusively in the lungs. The short dosing interval needed to provide adequate antimicrobial therapy requires that the technical staff be in the laboratory at unusual hours. If one uses immunosuppressed mice, caution must be used interpreting the data because of the role that an intact immune system plays in reducing number of viable bacteria *in situ*. If the mice are infected via the nostrils, rather than the trachea, upper respiratory infection and/or otitis media may also develop in some animals, possibly complicating the measured response to treatment. Using barbiturate as anesthesia carries with it an increased risk of accidental overdose; naloxone can be used as an antidote if needed. To what extent injecting large numbers of viable bacteria intratracheally imitates bacterial pneumonia in man is not completely known; certainly, aspiration pneumonia in man is not uncommon.

This model in mice offers several advantages over other species. Mice are inexpensive, they are easy to handle, and the histopathology of *H. influenzae* or *P. aeruginosa* experimental pneumonia is similar to that in man.

Contributions of the model to infectious disease therapy

This model has been used less than the frequency and diversity of serious bacterial pneumonia in man suggests. Several authors (Paranavitana *et al.*, 1996; Taba *et al.*, 1996; Mimoz *et al.*, 1997) have used this model to determine the relative efficacy of different antimicrobial therapies. Sauve *et al.* (1996) studied treatment of pathogens with novel antibiotic resistance patterns and Joly-Guillou *et al.* (1997) compared treatment regimens of emerging pathogens. It has also been used to compare intratracheal antimicrobial administration with intravenous antimicrobial administration (Zelinka *et al.*, 1997).

Elucidation of the pathophysiology of bacterial pneumonia

This model has been used by several investigators (Huffnagle, 1996; Paranavitana *et al.*, 1996; Pryhuber *et al.*, 1996) to study many different aspects of the expression and regulation of cytokines and host inflammatory response in bacterial pneumonia. Pulmonary vasoconstriction and blood flow in an infected host has been studied in this model by Light (1996).

References

Bergogne-Berezin, E. (1995). New concepts in the pulmonary disposition of antibiotics. *Pulm. Pharmacol.*, **8**, 65–81.

Cash, H. A., Woods, D. E., McCullough, B., Johanson, W. G., Bass, J. A. (1979). A rat model of chronic respiratory infection with *P. aeruginosa*. *Am. Rev. Resp. Dis.*, **119**, 453–459.

Edelstein, P. H., Edelstein, M. A. C., Lehr, K. H., Ren, J. (1996). *In-vitro* activity of levofloxacin against clinical isolates of *Legionella* spp, its pharmacokinetics in guinea pigs, and use in experimental *Legionella pneumophila* pneumonia. *J. Antimicrob. Chemother.*, **37**(1), 117–126.

Fox-Dewhurst, R., Alberts, M. K., Kajikawa, O. *et al.* (1997). Pulmonary and systemic inflammatory responses in rabbits with gram-negative pneumonia. *Am. J. Resp. Crit. Care Med.*, **155**(6), 2030–2040.

Garvy, B. A., Harmsen, A. G. (1996). Susceptibility to *Pneumocystis carinii* infection: host responses of neonatal mice from immune or naive mothers and of immune or naive adults. *Infect. Immun.*, **64**(10), 3987–3992.

Huffnagle, G. B. (1996). Role of cytokines in T cell immunity to a pulmonary *Cryptococcus neoformans* infection. *Biol. Signals*, **5**(4), 215–222.

Ilstrup, D. M. (1990). Statistical methods in microbiology. *Clin. Microbiol. Rev.*, **3**(3), 219–226.

Joly-Guillou, M.-L., Wolff, M., Pocidalo, J.-J., Walker, F., Carbon, C. (1997). Use of a new mouse model of *Acinetobacter baumannii* pneumonia to evaluate the postantibiotic effect of imipenem. *Antimicrob. Agents Chemother.*, **41**(2), 345–351.

Kim, H. J., Kim, S. K., Lee, W. Y. (1996). Monoclonal antibody against leucocyte CD11b (MAb 1B6) increase the early mortality rate in Sprague Dawley with *E. coli* pneumonia. *Tuberculosis Resp. Dis.*, **43**(4), 579–589.

Light, R. B. (1996). TI — effect of sodium nitroprusside and diethylcarbamazine on hypoxic pulmonary vasoconstriction and regional distribution of pulmonary blood flow in experimental pneumonia. *Am. J. Resp. Crit. Care Med.*, **153**(1), 325–330.

Mimoz, O., Jacolot, A., Padoin, C. *et al.* (1997). Cefepime and amikacin synergy against a cefotaxime-susceptible strain of *Enterobacter cloacae in vitro* and *in vivo*. *J. Antimicrob. Chemother.*, **39**(3), 363–369.

National Committee for Clinical Laboratory Standards (1997). *Methods for Dilution Antimicrobial Tests for Bacteria that Grow Aerobically. Approved Standard Publication,* 4th edn, M7-A4. NCCLS, Washington, DC.

Ole-Mapenay, I. M., Mitema, E. S. (1997). Efficacy of doxycycline in a goat model of *Pasteurella* pneumonia. *J. South Afr. Vet. Assoc.*, **68**(2), 55–58.

Paranavitana, C. M., Boyce, J. R., Burnstein, T. (1996). Protection by intratracheal immunoglobulins against *Pseudomanas aeruginosa* pneumonia in mice. *Serodiagn. Immunother. Infect. Dis.*, **8**(1–2), 25–32.

Pryhuber, G. S., Bachurski, C., Hirsch Racon, A., Whitsett, J. A. (1996). Tumor necrosis factor-alpha decreases surfactant protein B mRNA in murine lung. *Am. J. Physiol. Lung Cell. Molec. Physiol.*, **270**, L714–L721.

Rodriguez, F., Bryson, D. G., Ball, H. I., Forster, F. (1996). Pathological and immunohistochemical studies of natural and experimental *Mycoplasma bovis* pneumonia in calves. *J. Comp. Pathol.*, **115**(2), 151–162.

Sauve, C., Azoulay-Dupuis, E., Moine, P. *et al.* (1996). Efficacies of ceftoaxime and ceftriaxone in a mouse model of pneumonia induced by two penicillin- and cephalosporin-resistant strains of *Streptococcus pneumoniae*. *Antimicrob. Agents Chemother.*, **40**(12), 2829–2834.

Song, Z., Johansen, H. K., Faber, V. *et al.* (1997). Ginseng treatment reduces bacterial load and lung pathology in chronic *Pseudomonas aeruginosa* pneumonia in rats. *Antimicrob. Agents Chemother.*, **41**(5), 961–964.

Taba, H., Tohyama, M., Toyoda, K. *et al.* (1996). *In vitro* and *in vivo* antimicrobial activities of new fluoroquinolones against *Pseudomonas aeruginosa*. *Jpn. J. Chemother.*, **44**(7), 493–498.

Tsai, W. C., Strieter, R. M., Zisman, D. A. *et al.* (1997) Nitric oxide is required for effective innate immunity against *Klebsiella pneumoniae*. *Infect. Immun.*, **65**(5), 1870–1875.

Washington, J. A. (1985). Initial processing for cultures of specimens. In: *Laboratory Procedures in Clinical Microbiology,* 2nd

edn (ed. Washington, J. A.), pp. 101–102. Springer-Verlag, New York.
Zelinka, M. A., Wolfson, M. R., Calligaro, I., Rubenstein, S. D., Greenspan, J. S. (1997). A comparison of intratracheal and intravenous administration of gentamicin during liquid ventilation. *Eur. J. Pediatr.*, **156**(5), 401–404.

Chapter 59

Models of Pneumonia in Ethanol-treated Rats

M. J. Gentry and L. C. Preheim

Background of human infection

Bacterial infections are a major cause of morbidity and mortality in alcoholics, with alcoholism being one of the most potent predisposing factors to lobar pneumonia (Smith and Palmer, 1976; Adams and Jordan, 1984; Preheim and Gentry, 1996). The most common causative agents in these patients are *Streptococcus pneumoniae*, aerobic Gram-negative bacteria (particularly *Klebsiella pneumoniae*), anaerobes (often polymicrobial), and *Haemophilus influenzae*. Aside from an increased incidence of pneumonia, alcoholics also have a higher frequency of complications, including prolonged fever, abscess formation, empyema, and development of bacteremia. *Streptococcus pneumoniae*, the pneumococcus, is the leading cause of adult community-acquired pneumonia, and the most frequent cause of pneumonia in alcoholics (Austrian and Gold, 1964; Mufson *et al.*, 1974). Mortality rates from uncomplicated pneumococcal pneumonia in immunocompetent patients is 5%, but this increases to approximately 25% in patients who subsequently develop bacteremia (Mufson *et al.*, 1974; Hook *et al.*, 1983). Alcoholism predisposes humans to increased severity of pneumococcal pneumonia and subsequent development of bacteremia, with fatality rates in alcoholic patients reaching more than 50% in some studies (Gransden *et al.*, 1985; Perlino and Rimland, 1985).

Background of the model

Studies of the effects of ethanol on host defense against infections are difficult for several reasons. Human studies have been hindered by an inability to control for such variables as the amount of ethanol consumed, the duration of abuse, and the degree of accompanying malnutrition and liver disease. The mechanisms of ethanol-induced impaired resistance to bacterial pneumonia and the value of potential therapeutic agents therefore may be more appropriately studied in animal models. The metabolism of ethanol and its pathological effects in rats are similar in many respects to those in humans. The rat, therefore, is one of the most widely used animals for studies of human alcoholism (Lieber and DeCarli, 1994).

Two basic experimental methods of ethanol administration have been used in the study of ethanol's effects on susceptibility to and treatment of pneumonia. To test the effects of acute alcohol intoxication, ethanol is administered 30 minutes before infection, either by intraperitoneal injection of 20% ethanol at 5.5 g/kg (Nelson *et al.*, 1991) or by gastric lavage of 6 g/kg ethanol (Tuma *et al.*, 1986). Studies of the effects of chronic alcoholism are accomplished primarily by removing the rats' normal supply of food and water and replacing it with a totally liquid diet in which ethanol is incorporated. To control for nutritional deficiencies introduced by the rats' natural aversion to ethanol, a pair-feeding system is used, in which control animals are fed an equivalent amount of liquid diet with the ethanol calories replaced by carbohydrates. This feeding method is typically administered for 1–16 weeks before pneumonia is established by transtracheal infection of the animals.

Animal species and preparation

Although many strains of rats have been used in ethanol-feeding models, the majority of infectious disease studies have used 5–9-week-old male Sprague–Dawley rats. The diets used in our laboratory are the standard Lieber–DeCaril liquid diets containing 35% of their calories as fat (Dyets, Bethlehem, PA). The ethanol diet contains 5% wt/vol. of ethanol (36% of its total calories). In the control diet, the ethanol calories are replaced by dextrin-maltose. The rats are housed in group cages and fed standard rat chow and tap water until they reach an average weight of 150 g. They are then placed in individual cages, their water bottles are removed, and they are acclimatized to the liquid control diet *ad libitum* for 3 days. On the fourth day, the rats are matched by weight and paired so that one rat of each pair receives the ethanol diet *ad libitum* while the second rat is fed the amount of control diet equal to that consumed by his ethanol-fed mate on the previous day. A third group of

Handbook of Animal Models of Infection
ISBN 0-12-775390-7

nutritionally replete rats are maintained on rat chow and water *ad libitum* for the entire experimental period. In our laboratory, the rats are fed the ethanol diet for 7 days before being used in experiments, as this time period is sufficient to cause a significant increase in susceptibility to infection with *S. pneumoniae*. The pair-feeding regimen is then maintained for the remainder of each experiment, including the 10-day observation period after infection in mortality trials. The mean ± SD blood ethanol level after 7 days of ethanol treatment in our initial experiments was 24 ± 12 mmol/l (111 ± 55 mg/dl).

Storage and preparation of inocula

The rat model of lobar pneumonia after transtracheal infection has been used to study several different species of bacteria, including *S. pneumoniae* (Bakker-Woudenberg *et al.*, 1979a,b; Davis *et al.*, 1991; Lister *et al.*, 1993a,b; Gavalda *et al.*, 1997), *Klebsiella pneumoniae* (Bakker-Woudenberg *et al.*, 1982; Domenico *et al.*, 1982; Nelson *et al.*, 1991); *Haemophilus influenzae* (Slater, 1990), and *Pseudomonas aeruginosa* (Cripps *et al.*, 1994). Because the rat is inherently less susceptible to many of these human pathogens, the organisms have sometimes been injected in medium containing 0.7–1.0% agar (Slater, 1990; Gavaida *et al.*, 1997).

All studies in our laboratory have been performed with *S. pneumoniae*. The most virulent pneumococcal strains for rats, as well as most other laboratory animals, belong to the heavily encapsulated type 3 serotype. In our laboratory, the type 3 *S. pneumoniae* strain used is ATCC 6303 (American Type Culture Collection, Rockville, MD). The organism is passed periodically in white mice to maintain virulence, and stock cultures from the mice are stored at –70°C in Todd–Hewitt broth (THB; Difco Laboratories, Detroit, MI) containing 10% glycerol. For each experiment, the organisms are grown overnight in a 5% CO_2 atmosphere at 37°C in THB supplemented with 5% heat-inactivated rabbit serum (GIBCO, Grand Island, NY). The following morning the cultures are diluted with four volumes of fresh serum-supplemented THB and incubated for an additional 5 hours to logarithmic phase. The organisms are collected by centrifugation at 13 500*g*, washed once in phosphate-buffered saline (PBS), and resuspended in PBS to the appropriate concentration as determined by spectrophotometric estimation at 540 nm. Actual numbers of bacterial colony-forming units (cfu) are determined by a standard plate count technique using 5% sheep blood agar plates.

Infection process

Materials needed

Diethyl ether, 70% ethanol, scalpel handle plus size 10 blades, forceps, 20 G Cathlen IV™ i.v. catheters, 1 cc syringes and 9 mm metal wound clips.

Induction of pneumonia

Pneumonia is induced by transtracheal instillation of bacteria. The rats are anesthetized in an ether jar until unconscious. They are then placed in a supine position on a small animal surgery board with their forelegs secured with rubber bands and their necks mildly hyperextended by placing their upper teeth under a stretched rubber band. Anesthesia then is maintained with a nose cone containing a piece of ether-soaked cotton. The neck area is disinfected with 70% ethanol, and a small incision is made in the skin overlying the trachea. The trachea is exposed by blunt dissection with forceps, and a 20 G catheter is inserted into the mainstem bronchus (Figure 59.1A). The rat then is placed in an upright position and *S. pneumoniae* organisms, suspended in 0.3 ml of PBS, are injected into the lungs with a 1 cc syringe (Figure 59.1B). This is followed by injection of 0.3 ml of air to simulate aspiration. After removal of the catheter and the nose cone, the wound is closed with two wound clips and the rat is maintained in an inclined position until fully conscious. The rats tolerate this procedure well and do not require analgesics, as they resume normal activity, including eating, within a short time after infection.

Key parameters to monitor infection

Mortality

Mortality is determined for 10 days postinfection, with the majority of deaths occurring between days 3 and 7. In our laboratory, ingestion of ethanol for 7 days before challenge and during the postchallenge period decreased the LD_{50} of type 3 *S. pneumoniae* by 10 to 50-fold over that for pair-fed and chow-fed rats (Davis *et al.*, 1991; Lister *et al.*, 1993b).

Blood cell responses

Blood is obtained for determination of leukocyte counts and differentials by aseptic puncture of the foot veins under ether anesthesia (Snitily *et al.*, 1991). It should be noted that 6–7 days of pair-feeding causes a significant decrease in the percentage of circulating white blood cells that are polymorphonuclear lymphocytes (PMNLs), as well as the absolute number of PMNL in the peripheral blood of both ethanol-fed and pair-fed rats in comparison to chow-fed controls (Lister *et al.*, 1993a).

Pulmonary clearance of organisms

The efficacy of pulmonary host defense mechanisms and/or antibiotic treatment in clearance of pneumococci from the lungs is determined by quantitative lung cultures at selected

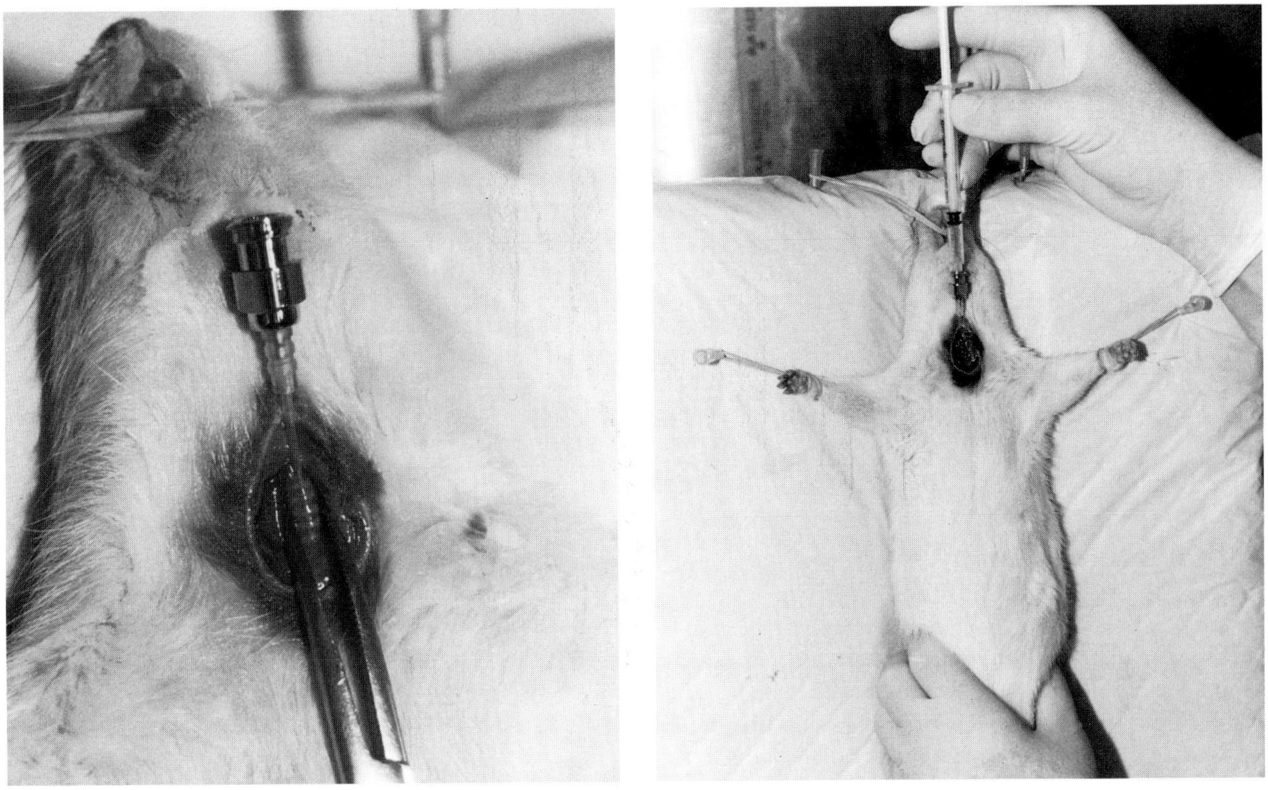
A B

Figure 59.1 **A.** Insertion of catheter into trachea exposed by blunt dissection. **B.** Injection of organism suspension into lungs via transtracheal catheter.

times post-infection (Jareo *et al.*, 1995). The rats are sacrificed by exsanguination via cardiac puncture under ether anesthesia. Their lungs are removed aseptically and homogenized in a total volume of 5 ml of PBS using a Ten Broeck tissue homogenizer (Fisher Scientific, Pittsburgh, PA). Serial 10-fold dilutions of the homogenate are cultured on 5% sheep blood agar plates for quantitication and expressed as cfu/total lungs. Log clearance is calculated as (log cfu in inoculum – log cfu remaining in the lungs at selected timepoints). In our laboratory, ingestion of ethanol for 7 days before challenge and during the postchallenge period significantly decreased the clearance of 1×10^6 type 3 *S. pneumoniae* from the lungs of ethanol-fed as compared to pair-fed rats at 48 and 72 hours postinfection (Davis *et al.*, 1991). In addition, *S. pneumoniae* types 10A, 14, and 19F were cleared well in 7 days from the lungs of pair-fed, but not ethanol-fed rats (Jareo *et al.*, 1995).

Histopathologic examination of lung tissue

Severity of pneumonia is determined histologically at selected time points postinfection. The lungs are removed as above and fixed in 10% formalin. Both lungs are sagittally sectioned and the mid-sagittal slice is embedded in paraffin, sectioned, and stained with hematoxylin and eosin. The slides are coded to prevent bias and examined by a pathologist. Each slide is estimated for the percentage of lung involved in the inflammatory process. The involved areas then are evaluated and scored for the presence and severity of inflammation, edema, hemorrhage, and necrosis according to the criteria listed in Table 59.1 (Davis *et al.*, 1991). The complete histologic presentation of pneumonia is determined by calculating a pneumonia severity score using the following formula: % lung involvement × (acute inflammation + hemorrhage + edema + necrosis). In our initial studies in this model, the lung severity score was significantly increased in ethanol-fed as compared to pair-fed rats on days 2 and 3 postinfection (Davis *et al.*, 1991).

Bacteremia

Failure to contain bacteria within the lungs leads to development of bacteremia, which considerably increases the mortality rate of pneumococcal pneumonia in rats and humans. Circulating pneumococci are quantified by plating 10-fold serial dilutions of blood collected by cardiac puncture or by the footstick method (Snitily *et al.*, 1991) onto 5% blood agar plates. In our laboratory, bacteremia occurred more frequently in ethanol-fed than in pair-fed rats, and intoxicated animals furthermore failed to clear the organisms from their bloodstream as effectively as their pair-fed controls (Davis *et al.*, 1991).

Table 59.1 Scoring system for histopathologic examination of lung tissue; PMNL, polymorphonuclear leukocyte. (Reprinted with permission from Davis *et al.*, 1991.)

Parameter	*Score*	*Interpretation*
Acute inflammation	0	No inflammatory cells in alveoli
	1	2–5 PMNLs in alveolar spaces
	2	> 5 PMNLs in alveolar spaces
Hemorrhage	0	< 10 erythrocytes in alveoli
	1	> 10 erythrocytes in alveoli
Edema	0	Absent
	1	Minimal swelling of alveolar walls and fibrinous exudate
	2	Widespread swelling of walls and fibrinous exudate
Necrosis	0	Absent
	1	Minimal; nuclei of alveolar cells missing
	2	Widespread; no remaining alveolar architecture

Antimicrobial therapy

To date, rat models of pneumonia have not been used extensively to determine the efficacy of antibiotic treatment. The majority of the published studies have been performed in the Netherlands, where the model was first used to test the efficacy of penicillin against the pneumococcus in rats with impaired phagocytosis (Bakker-Woudenberg *et al.*, 1979a,b). The rats were treated by intramuscular injection into the thigh muscle of the rear legs with 2–50 mg/kg of penicillin at 8–12 hour intervals, beginning 24–36 hours after infection. It was determined that the efficacy of penicillin therapy in rats pretreated with cobra venom factor to reduce their complement-mediated opsonization of the organisms was greatly reduced. The same group used the model to test the therapeutic activities of various antibiotics against *K. pneumoniae* in both normal and leukopenic rats (Bakker-Woudenberg *et al.*, 1982; Roosendaal *et al.*, 1986, 1987, 1989, 1991). These studies were performed using either intramuscular or intravenous injections into the tail vein.

With the worldwide emergence of antibiotic resistance in the pneumococcus (Lonks and Medeiros, 1995; Butler *et al.*, 1996), there has been increased interest in development of new antibiotics effective against this organism. One such study in a rat pneumonia model has been reported by Gavalda *et al.* (1997). In that study, male Wistar rats were treated with various antibiotic regimens that incorporated subcutaneous or intramuscular injection of the compounds beginning 4 hours after infection. To our knowledge there have been no published studies to date on antibiotic treatment of pneumonia in ethanol-ingesting rats. We are currently conducting such studies in our laboratory, using subcutaneous injection of antibiotics into the nape of the neck. We and others also have published studies on the efficacy of treatment with granulocyte-colony-stimulating factor (G-CSF) in ethanol-treated rats infected with *S. pneumoniae* (Lister *et al.*, 1993b) or with *K. pneumoniae* (Nelson *et al.*, 1991). Rats in those studies were treated subcutaneously twice daily with 50 μg/kg of G-CSF for 2 days before bacterial challenge. G-CSF was shown to mitigate the adverse effects of an acute dose of ethanol on susceptibility to *K. pneumoniae* (Nelson *et al.*, 1991). It did not, however, significantly protect rats ingesting ethanol for 7 days from a fatal infection with *S. pneumoniae* (Lister *et al.*, 1993b).

Advantages/disadvantages of the model

Experimental feeding of rats with alcohol in liquid diets was first described in 1963 by Lieber *et al.* (1963). This method has been modified a number of times since then, but is still widely used to study the pathological effects of human alcoholism (reviewed by Lieber and DeCarli, 1982). One advantage of this model is the compositional flexibility of liquid diets. This allows for manipulation and control of other specific dietary components, such as iron or fat content, which may be important in host defense against infections. Using this dietary model, the animals' blood alcohol concentrations are commonly 100–150 mg/dl (Lieber *et al.*, 1989). The concentrations fluctuate as a result of the animals' dietary patterns, but this also occurs in their human counterparts. Although the diets are nutritionally complete, the rats also decrease their overall food intake when ethanol is introduced into the diet. This causes them to become nutritionally depleted, again simulating the dietary habits of many chronic alcoholics. The pair-feeding method is used to show that the effects seen in the ethanol-fed animals are due to ingestion of ethanol *per se* rather than to their overall dietary restriction. However, the pair-fed control and ethanol-fed animals have different feeding patterns, with controls eating all or a major portion of their allotted food at the time of initial feeding and then effectively starving for the remainder of the 24 hour period. This can be avoided by administration of the liquid diet as a continuous intragastric infusion (Tsukamoto *et al.*, 1985), but this method is more costly and requires surgical implantation of gastric cannulae.

Rats fed alcohol in liquid diets develop a fatty liver, hyperlipidemia, and various other metabolic disturbances, but they do not generally progress to the more severe forms of liver injury, such as fibrosis and cirrhosis (Lieber and DeCarli, 1994). These animals consequently do not become hypocomplementemic like many humans with advanced alcohol-induced liver disease (Davis *et al.*, 1991). Because complement is such an important antibacterial host defense mechanism, this difference must be taken into account in interpretation of results as they relate to infections in long-term human alcoholics. Despite these limitations, however, it clearly has been shown that either acute alcohol administration (Nelson *et al.*, 1989, 1991) or 7 days of ethanol ingestion (Davis *et al.*, 1991) has a significant detrimental effect on host defenses against pneumonia.

The pathogenesis of bacterial pneumonia in humans generally begins with colonization of the nasopharynx. Pneumonia then develops after aspiration of the organisms into the lower respiratory tract if the host's pulmonary antibacterial defenses are unable to effectively kill and remove the bacteria. Experimental induction of pneumonia in rats by transtracheal instillation of large numbers of bacteria in a single bolus may not, therefore, represent the natural course of infection in humans. This may pose a significant problem in certain studies, such as those looking at protection by antibodies raised against bacterial components important in establishment of colonization (Henriksen *et al.*, 1997). However, it should not interfere with studies of antibiotic efficacy, which are meant to simulate acute disease in humans with fulminant pulmonary infections. Progression of human pneumonias to more serious complications such as development of bacteremia do occur in at least some rat models of transtracheal infection, including those with *K. pneumoniae* (Bakker-Woudenberg *et al.*, 1982) and *S. pneumoniae* (Davis *et al.*, 1991). These models, therefore, may be quite appropriate for determining efficacy of therapeutic interventions in prevention of extrapulmonary complications of pneumonia.

Contributions of the model to understanding ethanol-induced impairment of host defense against pneumonia

Studies of experimental pneumonia and the inflammatory response to bacterial lipopolysaccharide (LPS) in ethanol-treated rats has contributed to our understanding of ethanol's detrimental effects on host defense against pneumonia. Although macrophages comprise the initial cellular defense against bacterial pathogens reaching the lung, PMNLs recruited from the peripheral blood are very important in successful eradication of the organisms. Human PMNLs are relatively easy to obtain in large numbers from alcoholic patients. However, studies of PMNL bactericidal mechanisms from such patients have been conflicting, perhaps because of the inability to control for differences in drinking habits and amounts, nutritional factors, and underlying liver disease in these patients. Ethanol-treated rats, therefore, have been used to study the effect of alcohol ingestion on PMNL recruitment and antibacterial function.

Pulmonary neutrophil recruitment

Recruitment of PMNLs to the rats' lungs following transtracheal infection or instillation of bacterial products is measured by bronchoalveolar lavage (BAL). In our laboratory, the rats are euthanized by an intraperitoneal injection of pentobarbital (Nembutal, Abbott Laboratories, North Chicago, IL), and then exsanguinated by cardiac puncture. The thoracic cavity is opened and the lungs are flushed by gentle perfusion of 60 ml of PBS through the heart. The lungs and trachea are removed, the trachea is occluded at the severed end, and a 20 G catheter is inserted into the trachea. Next, 60–75 ml of cold (4 °C) PBS is injected into the lungs in 5–7 ml aliquots and recovered by dependent drainage until the total of returned volume of BAL fluid is 50 ml. The recovered cells are collected by centrifugation and resuspended in 5 ml of saline for counting in a hemocytometer and differential counts on stained cytospin preparations.

Seven days of ethanol ingestion did not reduce pulmonary recruitment of PMNLs in response to *S. pneumoniae* infection (Lister *et al.*, 1993a). However, acute ethanol treatment by intraperitoneal injection did significantly decrease PMNL recruitment to rats' lungs 4 hours postinfection with *K. pneumoniae* (Nelson *et al.*, 1991). Similarly, acute, but not chronic ethanol treatment significantly reduced pulmonary PMNL recruitment in response to an intratracheal instillation of LPS from *Escherichia coli* (Nelson *et al.*, 1989). This was related to complete suppression by ethanol of PMNL adhesion molecule CD11b/c upregulation in response to LPS treatment (Zhang *et al.*, 1997).

Neutrophil isolation

It is difficult to obtain sufficient numbers of PMNLs from the peripheral blood of young rats, especially those with leukopenia induced by the pair-feeding regimen. To increase the yield of PMNLs for functional studies from rats in ethanol-feeding trials, our laboratory has developed a magnetic cell-sorting isolation technique (Jareo *et al.*, 1997). The rats are exsanguinated by cardiac puncture using a hetastarch exchange transfusion method (Lister *et al.*, 1993a). Leukocytes are pelleted by centrifugation, washed, and labeled with a biotinylated anti-rat granulocyte monoclonal antibody (Pharmingen, San Diego, CA) followed by streptavidin-conjugated superparamagnetic microbeads

(Miltenyi Biotec, Auburn, CA). Labeled PMNLs are separated from unlabeled leukocytes in the suspension by passing them through a steel-wool column suspended within a powerful magnetic field. Unlabeled cells pass through the column, and the PMNLs are then collected by gently flushing buffer through the column after removing it from the magnetic field. In our laboratory, the magnetic cell sorting method yields two to five times more PMNLs per rat, with increased purity compared to density gradient centrifugation (Jareo *et al.*, 1997). PMNLs isolated by this method also produce almost no oxidative burst until they are appropriately stimulated, indicating negligible activation by the isolation procedure.

Neutrophil function

Most studies of PMNL function in human alcoholics have been performed with *Staphylococcus aureus*, which is easily taken up and killed by PMNLs without serum opsonization (Hof *et al.*, 1980). Encapsulated organisms, such as those most commonly causing pneumonias in alcoholic patients, are more refractory to PMNL uptake and killing. Using the magnetic cell separation technique, we have shown that PMNLs isolated from ethanol-fed rats have impaired ability *in vitro* to kill several strains of *S. pneumoniae*, but not *S. aureus* (Jareo *et al.*, 1995). This ethanol-induced defect was not due to a reduction in phagocytic uptake of the organisms (Jareo *et al.*, 1996). Spitzer and Zhang (1996) similarly have shown no decrease in phagocytic uptake of latex spheres by peripheral blood PMNLs from acutely intoxicated male Sprague–Dawley rats. In fact, PMNLs from female rats in that same study showed upregulated phagocytic activity in comparison to untreated controls.

The decreased antipneumococcal activity of PMNLs from chronically ethanol-fed rats was determined in our laboratory to be due to diminished production of oxygen radicals and decreased degranulation by PMNLs after stimulation with *S. pneumoniae* (Jareo *et al.*, 1996). Acute ethanol intoxication also has been shown to inhibit hydrogen peroxide generation by PMNLs and to inhibit induced nitric-oxide generation by alveolar macrophages of LPS-challenged male rats (Zhang *et al.*, 1997). It furthermore modulated PMNL F-actin polymerization in female rats (Zhang and Spitzer, 1997). These alterations in PMNL and macrophage bactericidal mechanisms may help to explain why alcoholics are especially prone to develop severe pneumonia with increased mortality rates.

Cytokine production

An acute dose of ethanol was shown to downregulate production of the inflammatory cytokine tumor necrosis factor (TNF) in response to either an intravenous or intratracheal challenge with bacterial LPS (Nelson *et al.*, 1989). In that study, intravenous injection of LPS caused a substantial increase in serum TNF in both normal and chronic alcoholic rats. However, this response was significantly suppressed in either type of rat if they were given an injection of 20% ethanol at a dose of 5.5 g/kg 30 minutes before LPS challenge. High levels of TNF also appeared in the BAL fluid of both normal and alcoholic rats challenged intratracheally with LPS, and these levels were significantly reduced by pretreatment with an acute dose of ethanol. The conclusion from these studies was that acute alcohol intoxication significantly inhibits the host's ability to produce and release TNF. This ethanol-induced disorder of inflammatory signaling may markedly impair normal host defenses against invasive bacteria in the alcoholic.

References

Adams, H. G., Jordan, C. (1984). Infections in the alcoholic. *Med. Clin. North Am.*, **68**, 179–200.

Austrian, R., Gold, J. (1964). Pneumococcal bacteremia with especial reference to bacteremic pneumococcal pneumonia. *Ann. Intern. Med.*, **60**, 759–776.

Bakker-Woudenberg, I. A. J. M., deJong-Hoenderop, J. Y. T., Michel, M. F. (1979a). Efficacy of antimicrobial therapy in experimental rat pneumonia: effects of impaired phagocytosis. *Infect. Immun*, **25**, 366–375.

Bakker-Woudenberg, I. A. J. M., vanGerwen, A. L. E. M., Michel, M. F. (1979b). Efficacy of antimicrobial therapy in experimental rat pneumonia: antibiotic treatment schedules in rats with impaired phagocytosis. *Infect. Immun.*, **25**, 376–387.

Bakker-Woudenberg, I. A. J. M., van den Berg, J. C., Michel, M. F. (1982). Therapeutic activities of cefazolin, cefotaxime, and ceftazidime against experimentally induced *Klebsiella pneumoniae* pneumonia in rats. *Antimicrob. Agents Chemother.*, **22**, 1042–1050.

Butler, J. C., Hofmann, J., Cetron, M. S., Elliott, J. A., Facklam, R. R., Breiman, R. F. (1996). The continued emergence of drug-resistant *Streptococcus pneumoniae* in the United States: an update from the Centers for Disease Control and Prevention's *Pneumococcal* Sentinel Surveillance System. *J. Infect. Dis.*, **174**, 986–993.

Cripps, A. W., Dunkley, M. L., Clancy, R. L. (1994). Mucosal and systemic immunizations with killed *Pseudomonas aeruginosa* protect against acute respiratory infection in rats. *Infect. Immun.*, **62**, 1427–1436.

Davis, C. C., Mellencamp, M. A., Preheim, L. C. (1991). A model of pneumococcal pneumonia in chronically intoxicated rats. *J. Infect. Dis.*, **163**, 799–805.

Domenico, P., Johanson W. G., Jr, Straus, D. (1982). Lobar pneumonia in rats produced by clinical isolates of *Kiebsiella pneumoniae*. *Infect. Immun.*, **37**, 327–335.

Gavalda, J., Capdevila, J. A., Almirante, B. *et al.* (1997). Treatment of experimental pneumonia due to penicillin-resistant *Streptococcus pneumoniae* in immunocompetent rats. *Antimicrob. Agents Chemother.*, **41**, 795–801.

Gransden, W. R., Eykyn, S. J., Phillips, I. (1985). Pneumococcal bacteremia: 325 episodes diagnosed at St Thomas' Hospital. *Br. Med. J.*, **290**, 505–508.

Henriksen, J. L., Preheim, L. P., Gentry, M. J. (1997). Vaccination with protein-conjugated and native type 3 capsular polysaccharide in an ethanol-fed rat model of pneumococcal pneumonia. *Alcoholism Clin. Exp. Res.*, **21**, 1630–1637.

Hof, D. G., Repine, J. E., Peterson, P. K., Holdal, J. R. (1980). Phagocytosis by human alveolar macrophages and neutrophils: qualitative differences in the opsonic requirements for uptake of *Staphylococcus aureus* and *Streptococcus pneumoniae in vitro*. *Am. Rev. Respir. Dis.*, **121**, 65–71.

Hook, E. W. III, Horton, C. A., Schaberg, D. R. (1983). Failure of intensive care unit support to influence mortality from pneumococcal bacteremia. *J.A.M.A.*, **249**, 1055–1057.

Jareo, P. W., Preheim, L. C., Lister, P. D., Gentry, M. J. (1995). The effect of ethanol ingestion on killing of *Streptococcus pneumoniae, Staphylococcus aureus,* and *Staphylococcus epidermidis* by rat neutrophils. *Alcohol Alcoholism*, **20**, 311–318.

Jareo, P. W., Preheim, L. C., Gentry, M. J. (1996). Ethanol ingestion impairs neutrophil bactericidal mechanisms against *Streptococcus pneumoniae*. *Alcoholism Clin. Exp. Res.*, **20**, 1646–1652.

Jareo, P. W., Preheim, L. C., Snitily, M. U., Gentry, M. J. (1997). Isolation of peripheral blood neutrophils from rats using magnetic cell sorting. *Lab. Anim. Sci.*, **47**, 414–418.

Lieber, C. S., DeCarli, L. M. (1982). The feeding of alcohol in liquid diets: two decades of applications and 1982 update. *Alcoholism Clin. Exp. Res.*, **6**, 523–531.

Lieber, C. S., DeCarli, L. M. (1994). Animal models of chronic ethanol toxicity. In: *Methods in Enzymology 233* (ed. Packet L.), pp. 585–595. Academic Press, New York.

Lieber, C. S., Jones, D. P., Mendelson, J., DeCarli, L. M. (1963). Fatty liver, hyperlipemia and hyperuricemia produced by prolonged alcohol consumption despite adequate dietary intake. *Trans. Assoc. Am. Physicians*, **76**, 289–300.

Lieber, C. S., DeCarli, L. M., Sorrell, M. F. (1989). Experimental methods of ethanol administration. *Hepatology*, **10**, 501–510.

Lister, P. D., Gentry, M. J., Preheim, L. C. (1993a). Ethanol impairs neutrophil chemotaxis *in vitro* but not adherence or recruitment to lungs in rats with experimental pneumococcal pneumonia. *J. Infect. Dis.*, **167**, 1131–1137.

Lister, P. D., Gentry, M. J., Preheim, L. C. (1993b). Granulocyte colony-stimulating factor protects control rats but not ethanol-fed rats from fatal pneumococcal pneumonia. *J. Infect. Dis.*, **168**, 922–926.

Lonks, J. R., Medeiros, A. A. (1995). The growing threat of antibiotic-resistant *Streptococcus pneumoniae*. *Med. Clin. North Am.*, **79**, 523–535.

Mufson, M. A., Kruss, D. M., Wasil, R. E., Metzger, W. I. (1974). Capsular types and outcome of bacteremic pneumococcal disease in the antibiotic era. *Arch. Intern. Med.*, **134**, 505–510.

Nelson, S., Bagby, G. J., Bainton, B. G., Summer, W. R. (1989). The effects of acute and chronic alcoholism on tumor necrosis factor and the inflammatory response. *J. Infect. Dis.*, **160**, 422–429.

Nelson, S., Summer, W., Bagby, G. *et al.* (1991). Granulocyte colony-stimulating factor enhances pulmonary host defenses in normal and ethanol-treated rats. *J. Infect. Dis.*, **164**, 901–906.

Perlino, C. A., Rimland, D. (1985). Alcoholism, leukopenia, and pneumococcal sepsis. *Am. Rev. Resp. Dis.*, **132**, 757–780.

Preheim, L. C., Gentry, M. J. (1996). Infection in the alcoholic. In: *Current Therapy of Infectious Disease* (ed. Schlossberg D.), pp. 308–311. C. V. Mosby, St Louis, MO.

Roosendaal, R., Bakker-Woudenberg, I. A. J. M., van den Berghe-van Raffe, M., Michel, M. F. (1986). Continuous versus intermittent administration of ceftazidime in experimental pneumonia in normal and leukopenic rats. *Antimicrob. Agents Chemother.*, **30**, 403–408.

Roosendaal, R., Bakker-Woudenberg, I. A. J. M., van den Berghe-van Raffe, M., Vink-van den Berg, J. C., Michel, M. F. (1987). Comparative activities of cirpfloxacin and ceftazidime against *Klebsiella pneumoniae in vitro* and in experimental pneumonia in leukopenic rats. *Antimicrob. Agents Chemother.*, **31**, 1809–1815.

Roosendaal, R., Bakker-Woudenberg, I. A. J. M., van den Berghe-van Raffe, M., Vink-van den Berg, J. C., Michel, M. F. (1989). Impact of the dosage schedule on the efficacy of ceftazidime, gentamicin and ciprofloxacin in *Klebsiella pneumoniae* pneumonia and septicemia in leukopenic rats. *Eur. J. Clin. Microbiol. Infect. Dis.*, **8**, 878–887.

Roosendaal, R., Bakker-Woudenberg, I. A. J. M., van den Berghe-van Raffe, M., Vink-van den Berg, J. C., Michel, M. F. (1991). Impact of the duration of infection on the activity of ceftazidime, gentamicin and ciprofloxacin in *Klebsielia pneumoniae* pneumonia and septicemia in leukopenic rats. *Eur. J. Clin. Microbiol. Infect. Dis.*, **10**, 1019–1025.

Slater, L. N. (1990). A rat model of prolonged pulmonary infection due to nontypable *Haemophilus influenzae*. *Am. Rev. Resp. Dis.*, **142**, 1429–1435.

Smith, F. E., Palmer, D. (1976). Alcoholism, infection and altered host defenses: a review of clinical and experimental observations. *J. Chron. Dis.*, **29**, 35–49.

Snitily, M. U., Gentry, M. J., Mellencamp, M. A., Preheim, L. C. (1991). A simple method for collection of blood from the rat foot. *Lab. Anim. Sci.*, **41**, 285–287.

Spitzer, J. A., Zhang, P. (1996). Gender differences in phagocytic responses in the blood and liver, and the generation of cytokine-induced neutrophil chemoattractant in the liver of acutely ethanol-intoxicated rats. *Alcoholism Clin. Exp. Res.*, **20**, 914–920.

Tsukamoto, H., French, S. W., Benson, N. *et al.* (1985). Severe and progressive steatosis and focal necrosis in rat liver induced by continuous intragastric infusion of ethanol and low fat diet. *Hepatology*, **5**, 224–232.

Tuma, D. J., Mailliard, M. E., Casey, C. A., Volentine, G. D., Sorrell, M. F. (1986). Ethanol-induced alterations of plasma membrane assembly in the liver. *Biochim. Biophys. Acta*, **856**, 571–577.

Zhang, P., Spitzer, J. A. (1997). Acute ethanol administration modulates leukocyte actin polymerization in endotoxic rats. *Alcoholism Clin. Exp. Res.*, **21**, 779–783.

Zhang, P., Nelson, S., Summer, W. R., Spitzer, J. A. (1997). Acute ethanol intoxication suppresses the pulmonary inflammatory response in rats challenged with intrapulmonary endotoxin. *Alcoholism Clin. Exp. Res.*, **21**, 773–778.

Chapter 60

Pneumococcal Pneumonia and Bacteremia in a Cirrhotic Rat Model

L. C. Preheim, G. L. Gorby and M. J. Gentry

Background of human infection

Recent emergence and global spread of antibiotic-resistant *Streptococcus pneumoniae* (the pneumococcus) emphasizes the need for greater understanding of the organism's pathogenic mechanisms and development of alternative therapies for pneumococcal disease (Lonks and Medieros, 1995). Numerous studies have demonstrated a relationship between alcoholic cirrhosis and an increased risk of bacterial infections, including pneumococcal pneumonia (Smith and Palmer, 1976; Adams and Jordan, 1984). Cirrhosis, like advanced age, predisposes patients to increased severity of pneumococcal disease, and raises the likelihood for subsequent development of bacteremia and death (Austrian and Gold, 1964). Fatality rates may exceed 50% in such patients despite hospitalization and appropriate therapy (Gransden *et al.*, 1985).

Background of model

Although the clinical relationship is well described, little is known about the intrinsic defect(s) that predispose cirrhotic patients to severe pneumococcal infections. To facilitate studies on cirrhosis, Procter and Chatamra (1982) successfully developed a rat model of carbon tetrachloride (CCl_4)-induced chronic liver disease that histologically is indistinguishable from Laënnec's cirrhosis in humans. Our laboratory was the first to adapt this model to study experimental pneumococcal pneumonia in a cirrhotic host (Mellencamp and Preheim, 1991). This model system can be used to study the host–pathogen interactions that facilitate pneumococcal pneumonia and bacteremia. It also can be used to study new approaches to the prevention and treatment of these infections. Additional information regarding rat models of pneumococcal pneumonia can be found in Chapter 59.

Animal species

The model originally was developed with Sprague–Dawley rats. If other species are used, preliminary experiments should be performed to establish that cirrhosis can be induced with CCl_4 gavage, and to determine the optimal inoculum of the organism to be used for experimental infection.

Preparation of animals

Cirrhosis is induced in 200 g male Sprague–Dawley rats by weekly gavage with the hepatotoxin CCl_4 (Mellencamp and Preheim, 1991). The rats are group-housed and fed standard rat chow *ad libitum* and water containing 1.5 m mol/l phenobarbital. The phenobarbital increases the hepatic toxicity of CCl_4. After 10–14 days of phenobarbital treatment, when the rats have reached a weight of at least 190 g, they are weighed and given an initial dose of 0.04 ml of CCl_4. The dose is administered by intragastric gavage to rats lightly anesthetized by ether inhalation. The doses are regulated in 0.02 ml increments using a Hamilton syringe with attached polypropylene tubing (Figure 60.1). Subsequent doses of CCl_4 are administered weekly and are adjusted

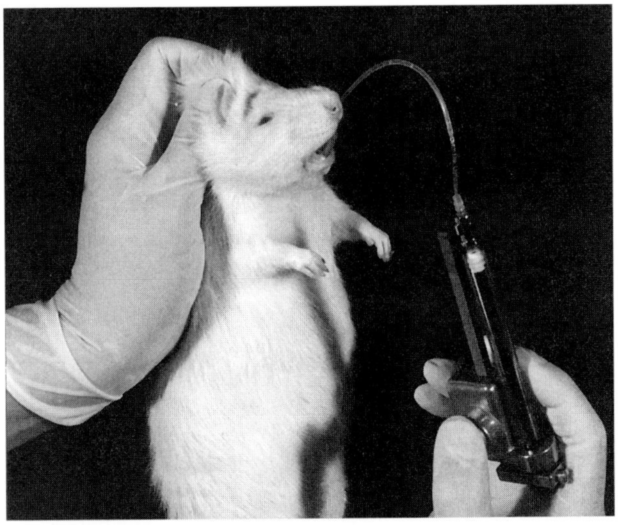

Figure 60.1 Intragastric installation of carbon tetrachloride using a syringe and attached tubing.

Handbook of Animal Models of Infection
ISBN 0-12-775390-7

based on the change in body weight occurring 48 hours after the previous dose (Table 60.1). All doses of CCl_4 are administered in a fume hood to prevent human inhalation of the hepatotoxic vapors. Age-matched control rats, maintained on rat chow and phenobarbital-treated water, are gavaged weekly with phosphate-buffered saline (PBS).

Table 60.1 Dosing schedule for induction of cirrhosis in rats by exposure to carbon tetrachloride (CCl_4). NC, no change in dose. (Reproduced with permission from Mellencamp, M. A. and Preheim, L. C. (1991). Pneumococcal pneumonia in a rat model of cirrhosis: effects of cirrhosis on pulmonary defense mechanisms against *Streptococcus pneumoniae*. *J. Infect. Dis.* **163**, 102–108)

Weight change 48 hours after last dose	*Change in CCl_4 dose (ml) by weeks of treatment*	
	<6	*≥6*
Weight gain	0.06	0.08
No change	0.06	0.08
Weight loss (%)		
2–6	0.04	0.06
7–10	0.02	0.04
11–15	NC	NC
>16	−0.04	−0.04

NB. Initial dose of intragastric CCl_4 is 0.04 ml. Subsequent doses are adjusted according to the above schedule and administered at weekly intervals. The indicated amount of CCl_4 is added to or subtracted from the previous dose. CCl_4 is discontinued when stable ascites is present.

In our laboratory, irreversible micronodular cirrhosis develops in 65–70% of rats after 8–12 weeks of CCl_4 treatment. They develop marked abdominal distention, due to the accumulation of intraperitoneal ascitic fluid, that is easily detected by visual inspection and palpation. Mortality due to CCl_4 toxicity during this treatment period is 10–15%. Cirrhotic rats with ascites have normal or elevated numbers of peripheral polymorphonuclear leukocytes, abnormal liver function tests, and decreased serum albumin levels (Table 60.2). The rats are rested for at least 1 week after their last treatment. A small minority of rats may lose their visible ascites during this rest period. Only those with stable abdominal distention due to ascites are used in experiments.

Storage and preparation of inocula

The type 3 *S. pneumoniae* used for the majority of the published studies was obtained from the American Type Culture Collection (ATCC; Rockville, MD). This strain, ATCC 6303, is a clinical isolate that is encapsulated and highly virulent for rats. A type 1 strain (ATCC 6301) has also been used successfully (Mellencamp and Preheim, 1991).

After serial intraperitoneal passage in white mice (BALB/c) to increase their virulence, the bacteria are stored at −70°C in Todd–Hewitt Broth (THB) containing 10% glycerol. For each experiment, the bacteria are grown to logarithmic phase in THB containing 5% heat-inactivated rabbit serum. The organisms are then collected by centrifugation, washed twice, and resuspended in PBS for infection of the animals. Actual bacterial inocula numbers are determined by plating serial 10-fold dilutions of the culture in duplicate on to 5% sheep blood agar plates.

Infection process

Pneumonia

Pneumonia is induced by transtracheal inoculation of *S. pneumoniae* under ether anesthesia (Mellencamp and Preheim, 1991). Materials needed include diethyl ether, 70% ethanol, scalpel handle plus size 10 blades, forceps, 20G Cathlon IV™ i.v. catheters, 1 ml syringes with 25G needles, and metal clips. Briefly, an incision is made over the trachea, which is then exposed by blunt dissection. A 20G catheter is inserted into the mainstem bronchus. Using a 1 ml syringe, bacteria suspended in 0.3 ml of PBS are introduced into the lungs, followed by 0.3 ml of air to simulate aspiration. After removal of the catheter, the wound is closed with two sterile metal clips and the rats are maintained in an inclined position until fully conscious. For illustrations and more details about the infection process see Figure 59.1 in Chapter 59.

Bacteremia

Rats are anesthetized with diethyl ether and placed in a lateral position on a small animal surgery board. The tail is disinfected with 70% ethanol and the bacteria are inoculated into a lateral tail vein in a total of 0.2 ml of PBS using a 1 ml syringe with a 25G ⅝″ needle. To ensure that the needle is in the vein, a small amount of blood is first withdrawn into the syringe before the inoculum is injected (Figure 60.2). Immediately after infection the rat is turned over and the contralateral tail vein is used for withdrawal of 200–300 µl of blood for determination of the initial number of colony forming units (cfu) per milliliter of blood.

Key parameters to monitor response to infection and treatment

Blood cell responses, clearance of bacteremia, and mortality

Peripheral white blood cell counts typically rise in animals during experimental pneumonia, and increases in the per-

Table 60.2 Laboratory test values for control and cirrhotic rats. Values are means ± SE for 10–15 rats in control group and 10–19 rats in the cirrhosis plus ascites group. (Reproduced with permission from Mellencamp, M.A. and Preheim, L.C. (1991). Pneumococcal pneumonia in a rat model of cirrhosis: effects of cirrhosis on pulmonary defense mechanisms against *Streptococcus pneumoniae*. *J. Infect. Dis.* **163**, 102–108)

Characteristic	*Control*	*Cirrhosis + ascites*
White blood cells × 10^3/mm^3)	11.3 ± 1.1	24.9 ± 2.3*
Red blood cells × 10^6/mm^3)	8.5 ± 0.1	7.6 ± 0.4
Hemaglobin (g/dl)	15.2 ± 0.2	14.6 ± 0.9
Hematocrit (%)	43.2 ± 1.0	41.7 ± 2.0
Platelets × 10^3/mm^3	1099 ± 54	1053 ± 116
Differential (%)		
Polymorphonuclear leukocytes	13.0 ± 1.1	20.8 ± 1.8*
Lymphocytes	86.4 ± 1.1	76.6 ± 1.8*
Monocytes	1.0 ± 0.2	1.1 ± 0.2
Eosinophils	0.4 ± 0.2	1.6 ± 1.0
Basophils	0	0.1 ± 0.1
Glucose (mg/dl)	194.5 ± 10.0	108.1 ± 10.8*
Sodium (meq/l)	148.5 ± 0.9	147.6 ± 0.7
Potassium (meq/l)	4.9 ± 0.2	5.0 ± 0.2
Chloride (meq/l)	106.7 ± 1.5	108.6 ± 1.1
Bicarbonate (meq/l)	22.6 ± 0.8	24.6 ± 0.9
Blood urea nitrogen (mg/dl)	18.8 ± 1.1	27.3 ± 2.5*
Creatinine (mg/dl)	0.52 ± 0.03	0.39 ± 0.04*
Calcium (mg/dl)	10.3 ± 0.2	9.8 ± 0.2*
Phosphorus (mg/dl)	5.9 ± 0.2	6.2 ± 0.5
Total protein (g/dl)	5.8 ± 0.3	4.5 ± 0.2*
Albumin (mg/dl)	4.2 ± 0.1	2.8 ± 0.1*
Total bilirubin (mg/dl)	0.21 ± 0.1	1.11 ± 0.32†
Lactate dehydrogenase (IU/l)	907 ± 159	1407 ± 371
Creatine kinase (IU/l)	621 ± 75	843 ± 181
Aspartate aminotransferase (IU/l)	172 ± 31	1519 ± 826
Alanine aminotransferase (IU/l)	48 ± 15	1036 ± 496†
Alkaline phosphatase (IU/l)	200 ± 22	1239 ± 681
γ-glutamyltransferase (IU/l)	7.0 ± 3.4	27.1 ± 6.4*

*Significantly different from control rats ($p < 0.05$) by analysis of variance [20].
†Difference approached significance ($p < 0.1$) when compared with controls.

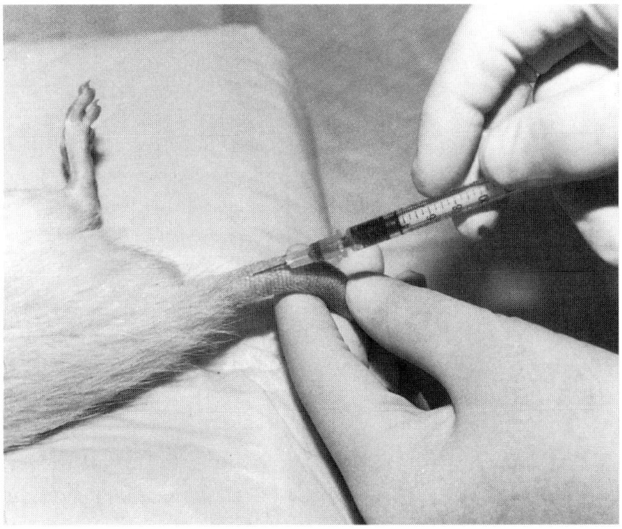

Figure 60.2 Method of injecting bacteria and obtaining blood samples from rat tail vein.

centage of circulating polymorphonuclear leukocytes are common. A variety of methods have been developed to obtain peripheral blood samples from rats (Cocchetto and Bjornsson, 1983; Kraus, 1980), but most of them have drawbacks. Cardiac puncture allows collection of large volumes of blood, but this technique may endanger the life of the rat when used for repeated sampling. Tail-tip or toe-tip amputation are potentially disfiguring and are distasteful to many technicians, as is puncture of the retro-orbital sinus. Phlebotomy via the jugular, penile, or lateral tail veins requires considerable technical skill.

By modifying a previously described technique (Nobunaga *et al.*, 1966), we devised a simple method for collection of blood from the rat foot utilizing the dorsal metatarsal vein (Snitily et al., 1991). With the rat under light ether anesthesia, one foot is washed with 70% alcohol and then rubbed vigorously with gauze moistened with methyl salicylate (oil of wintergreen; Humco Laboratory, Texarkana, TX). This stimulates blood flow to the foot so that the dorsal metatarsal vein and its tributaries become readily visible. A 20G needle is inserted into one of the veins between the digits. On withdrawal of the needle blood flows freely and can be collected in heparinized capillary tubes to allow quantitation of circulating white blood cells with a hemacytometer. If serum is desired, blood can be collected in an unheparinized capillary tube or a Pasteur pipette and expelled into an Eppendorf tube for clotting. Volumes of 0.2–0.5 ml of blood are easily recovered by this method. Gentle pressure is applied on the puncture site to stop the flow of blood. For repeated sampling, alternate sites on both feet can be used (Snitily *et al.*, 1991).

For experimental bacteremia studies, venous samples are taken from the tail vein to determine blood cell responses and clearance of the organism from the bloodstream. For clearance studies, at each sampling time 50 µl of blood is plated directly on to a blood agar plate, with an additional 50 µl sample diluted in 450 µl of PBS. This is further diluted in a series of six 10-fold dilutions with 100 µl of each dilution plated on to blood agar plates. Clearance is calculated as ((cfu at time 0–cfu at time X) ÷ (cfu at time 0)) × 100.

For determination of LD_{50}, mortality is recorded for at least 10 days postinfection. Similarly, animals are observed for 10 days if the model is used to determine the effectiveness of antimicrobial therapy for pneumococcal infection. In our laboratory the LD_{50} of type 3 *S. pneumoniae* (ATCC 6303) administered into the lung via transtracheal injection is 0.5–1.0 x 10^7 cfu in control rats and 0.5–1.0 × 10^6 cfu in cirrhotic rats with ascites.

Quantification of host defense factors in the lung

To monitor intra-alveolar responses to experimental pneumonia, bronchoalveolar lavage fluid (BALF) is collected from the rats' lungs by one of two methods. For studies of host factors such as pulmonary iron-binding capacity and cytokine levels, the lungs are lavaged *in situ* (Coonrod and Yoneda, 1981). The rats are exsanguinated after a lethal injection of sodium pentobarbital. An incision is made in the skin overlying the trachea, which is then exposed by blunt dissection. A catheter is inserted into the main stem bronchus and held in place with a string tied securely around the trachea. A volume of 6 ml of cold (4°C) PBS is injected, withdrawn, and reinjected into the lungs for one to five cycles, depending on the factor to be quantified. The return volume for such samples in our laboratory is routinely 4.5–5.0 ml. Any sample with a return volume of less than 4.5 ml is rejected. Cells are removed by centrifugation at 150 *g* for 7 minutes, and the supernatant is then assayed for host defense factors.

Studies of polymorphonuclear leukocyte (PMNL) recruitment and function are performed immediately after the rats are given a lethal injection of sodium pentobarbital and then exsanguinated by cardiac puncture. The lungs and trachea are removed *en bloc*, and cells are flushed from the lungs by repeated inflation with 8–10 ml volumes of cold PBS, which are collected by dependent drainage (Preheim *et al.*, 1991). Lavage is continued until a total volume of 50 ml is collected. The recovered cells are collected by centrifugation (150 *g* for 7 minutes), washed twice and suspended in 5 ml of buffer. The total number of cells recovered is determined by hemacytometer, and a differential cell count is done on stained slides prepared with a cytocentrifuge. Absolute numbers of PMNLs recovered are determined by multiplying the total number of cells collected per rat by the percentage of cells that were PMNLs.

Confocal microscopy and computerized image analysis

A method to quantify binding and internalization of pneumococci by pulmonary epithelial and endothelial cells was adapted from studies on gonococcal invasion of fallopian tubes (Gorby, 1994). This method uses computerized image analysis with digital confocal microscopy. Rats infected transtracheally with 1 × 10^8 cfu of type 3 *S. pneumoniae* are sacrificed at selected times postinfection (0, 1, 2, or 6 hours). The rats are exsanguinated and their lungs are fixed *in situ* by perfusion with 2% paraformaldehyde.

Grossly pneumonic areas of the lung are selected and embedded in paraffin for sectioning and microscopic analysis. Pneumococci in the sections are stained yellow-green with a fluoresceinated mouse monoclonal antibody to the pneumococcal cell-wall polysaccharide backbone (HASP8; Statens Seruminstitut, Copenhagen, Denmark). The cytoplasm of host cells is stained red with rhodamine-labeled phalloidin, which stains F-actin, and the host cell nuclei are stained blue with the DNA stain Hoechst 33342. Representative areas of lung parenchyma are selected using a transmitted polarized light technique (Nomarski DIC) without fluorescence to prevent bias in area selection. Images of the selected areas are collected at 0.5 µm intervals over 5 µm for each fluorochrome in a given tissue region.

At each plane of focus a corresponding fluorescein, rhodamine, and Hoechst image is collected by varying the excitation wavelength of light via a computer-controlled filter-switching mechanism. Ten different focal planes are evaluated in each region and five different regions are evaluated in each animal.

Digital confocal microscopy is used to remove out-of-focus haze from the images. In the rhodamine-stained actin images, the cytoplasmic boundaries of the cells are defined using a mouse-controlled cursor, and these boundaries are superimposed on the fluorescein-stained bacterial images. By automatically measuring the bacteria that are inside the epithelial or endothelial borders, differences between tissues from control and cirrhotic rats in the number of host-cell associated bacteria can be quantified.

To study pneumococcal penetration of the interstitium, sections of lung tissue are stained with an affinity-purified polyclonal rabbit antilaminin antibody followed by rhodamine-conjugated goat anti-rabbit IgG, which stains basement membrane instead of the cellular cytoplasm. In those sections, pneumococcal penetration of the interstitium is quantified by measuring the area occupied by organisms within each area of interstitium as defined by the designated border of the basement membrane.

This methodology can be used to follow the movement of pneumococci from alveoli through the pulmonary interstitium to blood vessels. We found that as early as 2 hours after experimental infection numerous pneumococci could already be seen within cells in the lung parenchyma and in the interstitium. Smaller numbers of bacteria remained in the alveoli or attached to alveolar walls. By 6 hours numerous PMNLs had infiltrated the parenchyma in both cirrhotic and normal rats. In normal rats pneumococci were seen predominantly inside PMNLs and macrophages. By contrast, in the cirrhotic lung an abundance of pneumococci were seen in macrophages and fibroblasts but not in PMNLs (unpublished observations). Thus confocal microscopy and computerized image analysis allows quantification of pneumococcal attachment to and invasion of lung cells. This method can be used to study the pathogenesis of pneumococcal pneumonia and to measure the effects of a variety of therapeutic interventions.

In vitro phagocytosis assay

PMNLs play an important role in host defense against pneumococci within the lung. In our laboratory, PMNL phagocytosis and bactericidal capacity were compared for cirrhotic and control rats both *in vitro* and *in vivo*. The phagocytic capacity of PMNLs from cirrhotic and control rats was compared *in vitro* using peripheral blood PMNLs isolated by density gradient centrifugation over Ficoll-Hypaque (Gentry *et al.*, 1995).

A washed, logarithmic-phase culture of the bacterium to be tested is labeled by incubation at 4°C for 1 hour in PBS containing 1 mg/ml fluorescein isothiocyanate (FITC). After being washed, the organisms are preopsonized for 30 minutes at 37°C with 20% pooled normal rat serum in Hank's balanced salt solution containing 0.1% gelatin (GHBSS). The organisms are then washed again and resuspended in GHBSS to a final concentration of 10^8 cfu/ml.

The phagocytosis mixture (200 μl of PMNL suspension (10^6 cells) and 200 μl of bacterial suspension (2×10^7 cfu)) is incubated for 15 minutes at 37°C with end-over-end rotation. The reaction is stopped by the addition of 2.5 ml of cold (4°C) PBS. Unassociated organisms are removed by differential centrifugation (three centrifugations at $180 \times g$). The final PMNL pellets are suspended in fixative (1% formaldehyde in 0.85% saline) for analysis of fluorescence on a FACScan flow cytometer.

In-vitro bactericidal assay

The comparison of the ability of cirrhotic and control rat PMNLs to kill bacteria *in vitro* was determined by a modification of the *in vitro* phagocytosis assay (Gentry *et al.*, 1995).

For this assay, bacteria are preopsonized as above but not FITC-labeled. After the organisms are incubated with isolated PMNLs for 15 minutes, unassociated bacteria are removed by differential centrifugation with cold PBS. The final PMNL pellet is resuspended in GHBSS (prewarmed to 37°C and supplemented with 20% normal rat serum). Aliquots of the suspension are removed immediately (time 0) and after 60 minutes of incubation at 37°C. Bacterial viability is determined by a plate-counting procedure after disruption of the PMNLs in distilled H_2O. Percentage of killing is calculated as (cfu at time 0–cfu at 60 minutes)/(cfu at time 0) × 100.

In-vivo phagocytosis assay

Like their human counterparts, cirrhotic rats are highly susceptible to experimental pneumococcal pneumonia, and commonly develop secondary bacteremia. Because PMNLs are the principal phagocytes that combat pulmonary pneumococcal infection, an *in vivo* assay was developed in order to study the effects of cirrhosis on phagocytosis within the lung (Gentry *et al.*, 1995).

Bacteria from a logarithmic-phase culture are heat-killed at 56°C for 1 hour. A portion of the suspension is labeled by overnight incubation at 4°C in $NaHCO_3$ buffer (pH 9.5) containing 1 mg/ml lucifer yellow VS. The remaining bacterial cells are incubated in buffer alone. Bacteria from each suspension are then washed and resuspended in PBS to an optical density of 2.5 at 540 nm ($\approx 3 \times 10^9$ organisms/ml). Rats are inoculated with 0.3 ml of labeled or unlabeled organism suspension by transtracheal injection as described above. Four hours later the rats are sacrificed by an overdose of pentobarbitol. Pulmonary cells are collected by bronchoalveolar lavage as described above, washed by differential centrifugation, and fixed for FACScan analysis as in the *in-vitro* phagocytosis assay.

In order to differentiate attached but uningested bacteria from those that have been truly phagocytosed, unfixed cell suspensions can be examined by dual-color FACScan analysis using a modification of a previously described method (Drevets and Campbell, 1991). The pulmonary cells are analyzed before and 5 minutes after the addition of 50 μg/ml ethidium bromide (EB). EB quenches the green fluorescence (FL1 detected at A_{530}) of external lucifer-yellow-labeled organisms and causes them to emit red fluorescence (FL2 detected at A_{585}). Internalized bacteria remain FL1-positive. After cytocentrifuge, slides of unfixed cells before and after addition of EB also are prepared for visual examination of red versus green organisms by fluorescent microscopy to confirm the intracellular location of the bacteria (Gentry et al, 1995).

Immunization studies

The effectiveness of pneumococcal vaccines is difficult to establish in many groups of patients who are susceptible to severe pneumococcal infection, including those who have severe liver disease. The rat cirrhosis model allows the study of the effects of cirrhosis on host antibody response to pneumococcal capsular polysaccharide and other antigens. The model also allows the study of protective efficacy of immunization in a cirrhotic host that is vaccinated and subsequently challenged by experimental pneumococcal infection. For example, in our laboratory studies of vaccination against type 3 *S. pneumoniae* were performed by subcutaneous injection of 25 μg of purified type 3 capsular polysaccharide (Preheim *et al.*, 1992).

Antibody levels postvaccination are determined by an indirect enzyme-linked immunosorbent assay (ELISA) after preadsorption of the sera for 30 minutes at room temperature with 100 μg of pneumococcal cell wall polysaccharide (Statens Seruminstitut, Copenhagen, Denmark) added to 1 ml of a 1:5 serum dilution (Henriksen *et al.*, 1998). Pneumococcal capsular polysaccharides often contain small amounts of contaminating cell-wall material. It has been shown previously that this preadsorption step effectively removes non-specific antibody to the cell-wall polysaccharide (Rodriguez-Barradas *et al.*, 1992).

Immulon I 96-well plates are coated overnight at 4°C with 1 μg/ml of type 3 pneumococcal capsular polysaccharide in 0.1 M HEPES buffer, pH 3.5 (Elkins *et al.*, 1990). The plates are blocked for 30 minutes at 37°C with 10% goat serum in distilled water (dH_2O). Serial dilutions of preadsorbed sera prepared in 0.15 N NaCl supplemented with 10% goat serum are added to the wells and the plates are incubated for 90 minutes at 37°C. The plates are rinsed and then incubated for an additional 90 minutes with horseradish peroxidase-conjugated immunoglobulin fraction of goat anti-rat immunoglobulin (IgG + IgA + IgM). After washing, antibody is detected with ABTS peroxidase substrate solution (2.2′-azino-di[3-ethyl-benzthiazoline sulfonate]) at a test wavelength of 410 and a reference wavelength of 490. Titers for individual rats are defined as the highest serum dilution providing an OD ≥ 0.1 (more than twice the standard deviation above control wells) after 20 minutes of incubation. Group geometric mean titers are calculated as the anti-log of the mean $\log_{10}$ titers for rats in that group.

For separate quantification of IgG and IgM antibody, peroxidase-conjugated goat anti-rat lgG or lgM is used as the detecting antibody. In addition, antibody levels are determined as ng/ml by comparison to a standard reference curve of rat lgG or lgM diluted in bicarbonate buffer (pH 9.6) and coated directly onto the plates.

Antibody titers to whole *S. pneumoniae* organisms also can be determined. For this assay, Immulon I plates are coated for 30 minutes at 37°C with 100 μl/well of Poly-L-lysine (0.5 mg/ml in PBS). After flicking the plates to remove the excess liquid, 50 μl/well of live *S. pneumoniae* suspended in PBS at an OD_{540} of 0.25 is added. The plates are centrifuged for 10 minutes at 1500 *g*, and 50 μl/well of 2.5% glutaraldehyde in dH_2O is added to fix the bacteria in the plate. After incubating the plates at 37°C for 30 minutes, they are washed with PBS and blocked overnight at 4°C with 10% goat serum in 0.15 N NaCl with 0.05% Tween 20 in dH_2O. After washing, the remainder of the ELISA procedure is performed as described above for the detection of antibodies to the capsular polysaccharide.

Antimicrobial and immunomodulator therapy

This model can be used to study the therapeutic efficacy of currently available or investigational antimicrobials or immunomodulators for the treatment of pneumococcal infections in a cirrhotic host. Studies in our laboratory with recombinant granulocyte colony-stimulating factor (50 μg/kg administered subcutaneously twice daily) demonstrated improved survival in treated control but not in treated cirrhotic rats with experimental pneumococcal pneumonia (Preheim *et al.*, 1996).

For studies involving antibiotic therapy, antimicrobials can be administered intravenously (via tail vein), intramuscularly (thigh), subcutaneously (nape of neck), intraperitoneally, or via intragastric installation. In addition to therapeutic studies, the effects of cirrhosis on the pharmacokinetics of antimicrobial agents can be evaluated in this model. Serial samples of blood for determinations of serum antibiotic concentrations are obtained via foot or tail veins if small volumes are adequate. Cardiac puncture is preferred for single, larger samples. To study intrapulmonary pharmacokinetics, groups of rats are sacrificed at separate timed intervals after administration of the antibiotic. Bronchoalveolar lavage is performed and the alveolar fluid and cells are separated as described above. Antibiotic concentrations in serum as well as in alveolar fluid and cells can be determined using high pressure liquid chromatography (Olsen *et al.*, 1996).

Advantages/disadvantages of the model

Advantages of this rat model of cirrhosis include its similarities to end-stage liver disease in humans. Little is known about how cirrhosis alters the host immune system to predispose patients to severe infections. This model allows, at relatively low cost, studies of how cirrhosis affects the pathogenesis, therapy, and prevention of pneumococcal disease. It could be adapted to investigate the other bacterial pathogens that cause infections in patients who are "immunocompromised" by chronic liver disease.

Disadvantages of the model would include the use of CCl_4, a hepatotoxin capable of inducing cirrhosis in humans as well as in rats if handled improperly. In addition, only limited numbers of pneumococcal serotypes are virulent in rats, and the majority of strains that are developing resistance to penicillin do not cause lethal infections in this model system (Briles *et al.*, 1992). Some investigators have successfully infected rats with less virulent strains by increasing the inoculum and/or administering the infective dose of organisms suspended in agar (Candiani *et al.*, 1997). Pneumococcal pneumonia typically results from aspiration of organisms into the lung from the oropharynx, and the usual initial inoculum in human infection is proportionately lower than that required to induce experimental pneumonia in this model.

References

Adams, H. G., Jordan, C. (1984). Infections in the alcoholic. *Med. Clin. North Am.*, **68**, 179–200.

Austrian, R., Gold, J. (1964). Pneumococcal bacteremia with especial reference to bacteremic pneumococcal pneumonia. *Ann. Intern. Med.*, **60**, 759–776.

Briles, D. E., Crain, M. J., Gray, B. M., Forman, C., Yother, J. (1992). Strong association between capsular type and virulence for mice among human isolates of *Streptococcus pneumoniae*. *Infect. Immun.*, **60**, 111–116.

Candiani, G., Abbondi, M., Borgonovi, M., Williams, R. (1997). Experimental lobar pneumonia due to penicillin-susceptible and penicillin-resistant *Streptococcus pneumoniae* in immunocompetent and neutropenic rats: efficacy of penicillin and teicoplanin treatment. *J. Antimicrob. Chemother.*, **39**, 199–207.

Cocchetto, D. M., Bjornsson, T. D. (1983). Methods for vascular access and collection of body fluids from the laboratory rat. *J. Pharm. Sci.*, **72**, 465–492.

Coonrod, J. D., Yoneda, K. (1981). Complement and opsonins in alveolar secretions and serum of rats with pneumonia due to *Streptococcus pneumoniae*. *Rev. Infect. Dis.*, **3**, 310–322.

Drevets, D. A., Campbell, P. A. (1991). Macrophage phagocytosis: use of fluorescence microscopy to distinguish between extracellular and intracellular bacteria. *J. Immunol. Methods*, **142**, 31–38.

Elkins, K. L., Stashak, P. W., Baker, P. J. (1990). Analysis of the optimal conditions for the adsorption of type III pneumococcal polysaccharide to plastic for use in solid-phase ELISA. *J. Immunol. Methods*, **130**, 123–131.

Gentry, M. J., Snitily, M. U., Preheim, L. C. (1995). Phagocytosis of *Streptococcus pneumoniae* measured *in vitro* and *in vivo* in a rat model of carbon tetrachloride-induced liver cirrhosis. *J. Infect. Dis.*, **171**, 350–355.

Gorby, G. (1994). Digital confocal microscopy allows measurement and three-dimensional multiple spectral reconstruction of *Neisseria gonorrheae* epithelial cell interactions in the human fallopian tube organ culture model. *J. Histochem. Cytochem.*, **42**, 297–306.

Gransden, W. R., Eykyn, S. J., Phillips, I. (1985). Pneumococcal bacteremia: 325 episodes diagnosed at St Thomas' Hospital. *Br. Med. J.*, **290**, 505–508.

Henriksen, J. L., Preheim, L. C., Gentry, M. J. (1998). Vaccination with protein-conjugated and native type 3 capsular polysaccharide in an ethanol-fed rat model of pneumococcal pneumonia. *Alcoholism Clin. Exp. Res.*, In press.

Kraus, A. L. (1980). Research methodology, p. 1–10. In *The Laboratory Rat*, Vol. II (eds Baker, H. J., Lindsey, J. R., Weisbroth, S. H.). Academic Press, New York.

Lonks, J. R., Medieros, A. A. (1995). The growing threat of antibiotic-resistant *Streptococcus pneumoniae*. *Med. Clin. North Am.*, **79**, 523–535.

Mellencamp, M. A., Preheim, L. C. (1991). Pneumococcal pneumonia in a rat model of cirrhosis: effects of cirrhosis on pulmonary defense mechanisms against *Streptococcus pneumoniae*. *J. Infect. Dis.*, **163**, 102–108.

Nobunaga, T., Nakamura, K., Imamichi, T. (1966). A method for intravenous injection and collection of blood from rats and mice without restraint and anesthesia. *Lab. Anim. Care*, **16**, 40–49.

Olsen, K. M., San Pedro, G. S., Gann, L. P., Gubbins, P. O., Halinski, D. M., Campbell, G. D. Jr (1996). Intrapulmonary pharmacokinetics of azithromycin in healthy volunteers given five oral doses. *Antimicrob. Agents. Chemother.*, **40**, 2582–2585.

Preheim, L. C., Gentry, M. J., Snitily M. U. (1991). Pulmonary recruitment, adherence, and chemotaxis of neutrophils in a rat model of cirrhosis and pneumococcal pneumonia. *J. Infect. Dis.*, **164**, 1203–1206.

Preheim, L. C., Mellencamp, M. A., Snitily, M. U., Gentry, M. J. (1992). Effect of cirrhosis on the production and efficacy of pneumococcal capsular antibody in a rat model. *Am. Rev. Resp. Dis.*, **146**, 1054–1058.

Preheim, L. C., Snitily, M.U., Gentry, M. J. (1996). Effects of granulocyte colony-stimulating factor in cirrhotic rats with pneumococcal pneumonia. *J. Infect. Dis.*, **174**, 225–228.

Proctor, E., Chatamra, K. (1982). High yield micronodular cirrhosis in the rat. *Gastroenterology*, **83**, 1183–1190.

Rodriguez-Barradas, M. C., Musher, D. M., Lahart, C. *et al.* (1992). Antibody to capsular polysaccharides of *Streptococcus pneumoniae* after vaccination of human immunodeficiency virus-infected subjects with 23-valent pneumococcal vaccine. *J. Infect. Dis.*, **165**, 553–556.

Smith, F. E., Palmer, D. L. (1976). Alcoholism, infection and altered host defenses: a review of clinical and experimental observations. *J. Chronic Dis.*, **29**, 35–49.

Snitily, M. U., Gentry, M. J., Mellencamp, M. A., Preheim, L. C. (1991). A simple method for collection of blood from the rat foot. *Lab. Anim. Sci.*, **41**, 285–287.

Chapter 61

Rat Model of Chronic *Pseudomonas aeruginosa* Lung Infection

H. K. Johansen and N. Høiby

Background of human infection

Cystic fibrosis (CF) is the most common lethal autosomal recessive genetic disorder affecting Caucasian populations. The incidence is approximately one in 2500–4700 live births; there are at least 50 000 CF patients in Europe, the USA and Latin America (Lewis, 1995). CF is caused by mutations in the cystic fibrosis transmembrane conductance regulator protein (CFTR), which principally functions as a chloride channel (Buchwald, 1996). This leads to a generalized dysfunction of exocrine glands primarily affecting the lungs as well as the gastrointestinal, hepatobiliary and reproductive tracts (Davis *et al.*, 1996). The CFTR mutation predisposes to viral and bacterial lung infections, predominantly with *Staphylococcus aureus, Streptococcus pneumoniae* and *Haemophilus influenzae* during childhood and later on with *Pseudomonas aeruginosa* (Koch and Høiby, 1993).

Once chronic *P. aeruginosa* lung infection is established, it is not possible to eliminate it with immune therapy or antibiotics (Pedersen, 1992). The main reasons are that *P. aeruginosa* produces mucoid exopolysaccharide (alginate) and forms microcolonies. The biofilm mode of growth protects the bacteria from antibiotics and from the host defence mechanisms such as antibody responses and polymorphonuclear leukocytes (PMNs). Together with the bacteria the recruited PMNs produce a large amount of elastase which, in concert with the excessive amounts of antibodies, maintains a chronic inflammation. This leads to progressive lung tissue damage and increased morbidity and mortality of the patients (Johansen, 1996a).

However, the course and prognosis have improved tremendously in recent years by intensive maintenance treatment of *P. aeruginosa*, cohort isolation and treatment of unstable patients with colistin inhalation and *ad hoc* oral ciprofloxacin between intravenous antibiotic therapy periods in chronically infected patients. In Denmark this treatment strategy has increased survival; the probability that newborn CF children reach their 45th birthday is now 80% (Frederiksen *et al.*, 1996).

Background of model

It has been difficult to establish an experimental model of the chronic *P. aeruginosa* lung infection, but the agar bead model developed by Cash *et al.* (1979) solved the problems. Cash enmeshed the bacteria in agar beads and installed the inoculum intratracheally into the lungs of normal rats. The serological and histopathological changes closely resembled the later stages of the lung infection in CF patients. Pathologically, the lungs were acutely inflamed, with numerous PMNs surrounding the bacteria-containing agar beads (microcolonies). For many years this model, or modified ones, were the only alternatives to the lung infections in the CF patients.

In our laboratory we have developed a similar rat lung model but we used seaweed alginate for embedding the bacteria instead of agar since it more closely resembles the alginate produced *in vivo* by mucoid *P. aeruginosa* (Pedersen *et al.*, 1990). The infection was established in normal as well as in athymic rats to elucidate the involvement of both the humoral and the cellular immune response subsequent to infection (Johansen *et al.*, 1993). We also used the model for evaluating the ability of different *P. aeruginosa* vaccines and Chinese medical herbs to prevent or modify the inflammatory response subsequent to infection (Johansen *et al.*, 1995; Song *et al.*, 1997). By using an alginate-containing vaccine we succeeded in improving survival and in changing the inflammatory response from an acute-type inflammation dominated by PMNs, as in CF patients and infected non-vaccinated rats, to a chronic-type inflammation dominated by mononuclear leukocytes (MNs) (Johansen, 1996a). In addition, we have established a similar model in mice, which we use for following the production of different cytokines during the chronic *P. aeruginosa* infection (Moser *et al.*, 1997).

Since the CF gene was found in 1989 several groups of experimenters have described different variants of transgenic CF mouse. Disease severity and tissue involvement are different but the major problem is that the mice fail to thrive and lose weight, and few of them survive for more than a couple of months (Snouwaert *et al.*, 1992; Ratcliff *et al.*, 1993). One model was very robust but it also had a low level of residual activity of the CF gene (Dorin *et al.*, 1992).

Handbook of Animal Models of Infection
ISBN 0-12-775390-7

Thus, until these mice become commercially available to all experimenters the lung infection established in normal rats and mice might be the best experimental alternative to the CF patients themselves.

Animal species

Since the first chronic *P. aeruginosa* lung infection model was established in rats (Sprague–Dawley: 140–160 g male, Cash *et al.*, 1979; 160–185 g female, Johansen *et al.*, 1993; Lewis: 160–180 g female, Johansen *et al.*, 1995; 200–220 g male, Woods *et al.*, 1982) several other species have been used to establish the infection such as mice (C57BL/6 and Balb/c, Stevenson *et al.*, 1995; C3H/HeN, Pier *et al.*, 1990), guinea pigs (Hartley strain 400 g, Pennington *et al.*, 1981), cats (3–5.4 kg either sex, Winnie *et al.*, 1982) and rhesus monkeys (*Macaca mulatta*, 6–12 kg, Cheung *et al.*, 1992). The most commonly used rat strain is Sprague–Dawley male rats; however, these rats are outbred and the response to the infection is more heterogeneous compared to inbred rats such as the Lewis rat. The rats we have used are 7 weeks old. We have chosen that age since we were interested in the immune response to different *P. aeruginosa* antigens before the animals' own immune apparatus is completely mature (Gruber, 1975). We established the model in female rats since they are less aggressive than males. Gender, however, does not seem to influence the outcome of the infection (Woods *et al.*, 1982; Johansen, 1996a).

Preparation of animals

No specific pretreatment is required, except that the animals should acclimatize at least 5–6 days before surgery and be kept under good conventional conditions with no more than three individuals in each cage.

Details of surgery

Overview

Briefly, anaesthetized rats are surgically tracheotomized, and the intratracheal challenge with 0.1 ml of *P. aeruginosa* embedded in alginate beads is installed in the lower part of the left lung with a curved, bead-tipped needle (1.0 G; Figure 61.1). The incision is sutured with silk and heals without complications.

Materials required

Anaesthetic, 10^8 cfu/ml *P. aeruginosa* alginate beads, operation table with a rubber band attached to the upper part of the table to keep the teeth and head of the rat fastened and the whole animal in a straight horizontal position, skin disinfectant (70% vol./vol. ethanol) and cotton, two pairs of scissors — small sharp ones (10 cm) and larger rounded ones (15 cm), two pairs of tweezers (a medical and a surgical), needles (18 G × 1.5″), a needle-holder, syringes, a curved bead-tipped needle (1.0 G) (Figure 61.2), and suture material (preferably surgical absorbable suture).

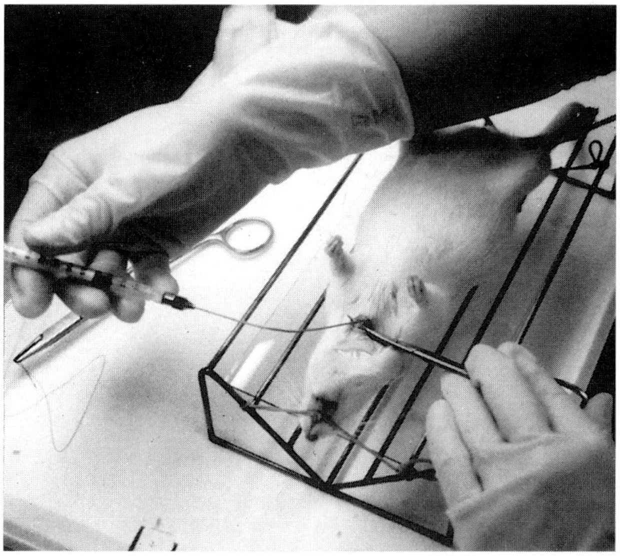

Figure 61.1 The surgical procedure. The rat is placed on the surgical table and the curved bead-tipped needle is installed in the lower part of the left lung.

Anaesthesia

Animals are sedated with an i.p. injection of 1.5 ml/kg body weight of a 1 : 1 mixture of etomidate (Janssen, Birkerød, Denmark) and midazolam (Roche, Hvidovre, Denmark). This is prepared by mixing one part of etomidate with one part of sterile water in one vial and one part of midazolam with one part of sterile water in another vial; finally, the two preparations are mixed together. Complete anaesthesia ensues in about 10 minutes and lasts for about 1 hour. No antidotes are required.

Surgical procedure

Following complete induction of anaesthesia, the animal is placed ventral side up on the operating table. The operating table is made by one of the sides of a wire basket on which we have welded a leg in each corner; on top of the table is placed a glass plate size 10 × 20 cm. The front teeth of the rat should be fastened under the rubber band that is attached to the upper end of the table (Figure 61.1). If the experimenter is right-handed, the rat should be placed straight and horizontal with its head towards the experi-

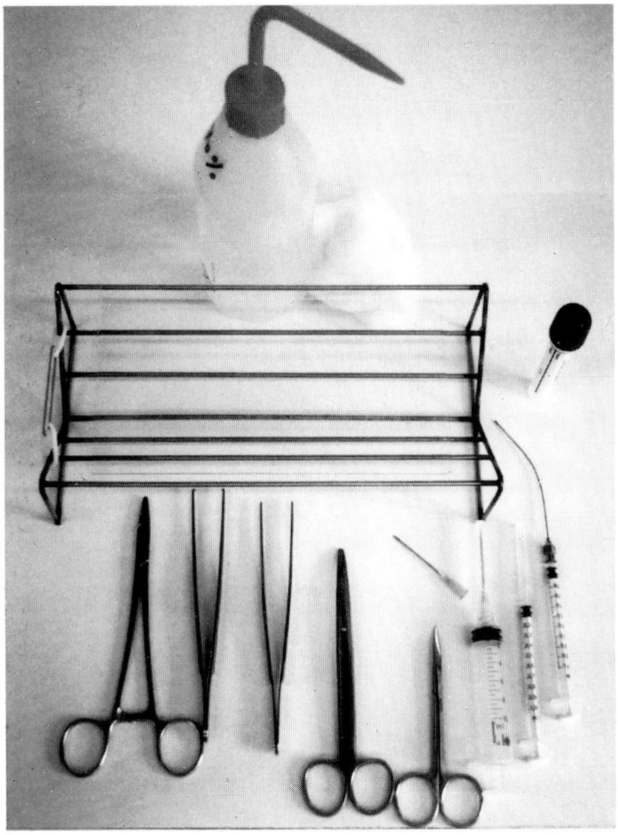

Figure 61.2 Materials required for the surgical procedure.

menter's left hand. The skin over the trachea is disinfected with 70% vol./vol. ethanol. Thereafter the skin is lifted with the surgical tweezers in the midline where it forms a triangle with the two claviculae, and a small incision is made through the skin with the large rounded scissors. The underlying muscles should be cut and dissected very carefully in order to minimize damage; we therefore use the small sharp scissors and the medical tweezers. There are major blood vessels on both sides of the paratracheal muscles so the experimenter should pay attention and always work in the midline right on top of the trachea.

When the trachea is exposed the small sharp scissors are placed under the trachea at right angles to keep the trachea fastened during the installation procedure. This is done by dissection using the sharp end of the scissors carefully against one of the legs of the tweezers on the opposite side of the trachea. While holding the scissors, a small incision (no more than 1 mm) is made with a needle in one of the cartilaginous rings of the trachea. Through this incision the curved bead-tipped needle with the syringe is entered. The tip of the needle is led through the left principal bronchus to the lower left bronchus (the curved bead-tipped needle is advanced approximately 2.5 cm into the trachea) and 0.1 ml of the bacterial inoculum is installed (Figure 61.1). The needle and the scissors are removed immediately after the procedure is completed, since the rat's respiration becomes somewhat impaired during the operation. The wound is closed with a suture.

It will take an experienced experimenter approximately 1–3 minutes to operate one rat, which amounts to 50 to 70 rats per day. The most time consuming procedure is to perform the anaesthesia and to prepare the inoculum in the syringe for each new operation.

Infection procedure

If necessary, the area can be cleaned of blood with a cotton swab soaked in adrenalin following the tracheal exposure. After the incision has been made in the cartilaginous rings of the trachea, the curved, bead-tipped needle is first placed into the trachea at right angles to the posterior wall; thereafter it is turned into position and advanced until resistance is felt. Then 0.1 ml of *P. aeruginosa* embedded in seaweed alginate is slowly installed in the lower part of the left lung. Following this procedure, the head of the rat should be in a slightly elevated position to ease respiration. Experiments with stained beads have revealed that the inoculum does not diffuse into other areas of the lung, although the rat will be moving around soon after the operation.

Wound closure

The wound is closed using synthetic, green-braided, polyglycolic acid, sterile, absorbable surgical suture, 3–0 Dexon 'S' (Davis & Beck, Mansti, USA) CE-6, 24 mm cutting. Normally, two sutures placed evenly over the wound are sufficient. The sutures fall out or are removed by the animal after 3–4 days. We have never seen any wound infections.

Postoperative care

The rats should be kept warm after surgery. This is done by placing them under paper towels until they begin to recover from the anaesthetic. Their heads are placed a little higher than the rest of their bodies to facilitate respiration. The animals can be housed together after surgery. Animals usually return to normal behaviour within 6–12 hours. Postoperative analgetics are not needed. Lubricant eye drops can be used to prevent the eyes from drying during the postoperative phase. Weight gain, food and water consumption appear to be unaffected by the infection.

Storage and preparation of the inocula

Although most studies have used *P. aeruginosa* (Johansen, 1996a; Table 61.1), the model is amenable to other species of bacteria such as *Bordetella pertussis* (Woods *et al.*, 1989) and *Haemophilus influenzae* (Slater, 1990; Maciver *et al.*, 1991). In our experiments we have used an alginate-producing *P. aeruginosa* strain PAO 579, International Antigenic Typing

Table 61.1 Examples of animal models of chronic *Pseudomonas aeruginosa* lung infection (amended from Johansen, 1996a) with permission from the editor of *Acta Pathologica, Microbiologica et Immunologica Scandinavica*); LPS, lipopolysaccharide; OMP F, outer membrane protein F; MEP, mucoid exopolysaccharide; BALT, bronchus-associated lymphoid tissue; D-ALG toxin A, depolymerized alginate toxin A conjugate; IFN-γ, recombinant rat interferon gamma; Ag, agar; As, agarose; HMPA, hexamethylphosphoramide; Sa, seaweed alginate; Rp, reserpine; El, elastase; Bi, brush injury metaperiodate or trypsin

Study (immobilization agent)	*Animals*	*Purpose of the study*
Cash *et al.* (1979) *Am. Rev. Resp. Dis.*, **119**, 453–459 (Ag)	Rats	First establishment of the chronic infection
Pennington *et al.* (1981) *J. Clin. Invest.*, **68**, 1140–1148 (Ag)	Guinea-pigs	Effect of immunization with LPS on chronic infection
Winnie *et al.* (1982) *Infect. Immun.*, **38**, 1088–1093 (As)	Cats	Characterization of the *P.-aeruginosa*-specific phagocytosis-inhibitor activity
Woods *et al.* (1982) *Infect. Immun.*, **36**, 1223–1228 (Ag)	Rats	Effect of elastase and toxin A on chronic infection
Boyd *et al.* (1983) *Infect. Immun.*, **39**, 1403–1410 (HMPA)	Rats	To establish the chronic infection after epithelial injury
Cash *et al.* (1983) *Can. J. Microbiol.*, **29**, 448–456 (Ag/As)	Rats	Demonstration of exoproducts produced during chronic infection
Klinger *et al.* (1983) *Infect. Immun.*, **39**, 1377–1384 (As)	Rats	Effect of a polyvalent PEV-01 vaccine on chronic infection
McArthur *et al.* (1983) *Infect. Immun.*, **42**, 574–578 (Ag)	Rats	Interaction of rat lung lectin with exopolysaccharides of *P. aeruginosa*
Nacucchio *et al.* (1984) *Pediatr. Res.*, **18**, 295–296 (Ag)	Rats	The role of agar beads in the pathogenicity of chronic infection
Thomassen *et al.* (1984) *Infect. Immun.*, **45**, 741–747 (As)	Cats	The role of macrophage activation during chronic infection
Woods *et al.* (1985) *J. Infect. Dis.*, **151**, 581–588 (Ag)	Rats	The ability of alginate to be protective in chronic infection
Starke *et al.* (1987) *Pediatr. Res.*, **22**, 698–702 (As)	Mice	Characterization of chronic *P. aeruginosa* and *P. cepacia* lung infection
Cochrane *et al.* (1988) *J. Med. Microbiol.*, **27**, 255–261 (Ag)	Rats	Antibody response to *P. aeruginosa* surface protein antigens
Döring and Dauner (1988) *Am. Rev. Respir. Dis.*, **138**, 1249–1253 (Rp/El)	Rats	Clearance of bacteria from structurally altered lung mucosa
Gilleland *et al.* (1988) *Infect. Immun.*, **56**, 1017–1022 (Ag)	Rats	Protective effect of OMP F in chronic infection
Freihorst *et al.* (1989) *Infect. Immun.*, **57**, 235–238 (Ag)	Rats	Effect of oral administration of *P. aeruginosa* before chronic infection
Grimwood *et al.* (1989) *J. Antimicrob. Chemother.*, **24**, 937–945 (Ag)	Rats	Effect of ciprofloxacin, tobramycin and ceftazidime on chronic infection
Graham *et al.* (1990) *J. Allergy Clin. Immunol.*, **93**, 100–107 (Ag)	Rats	Effect of chronic *P. aeruginosa* infection on pulmonary haemodynamics
Konstan *et al.* (1990) *Am. Rev. Resp. Dis.*, **141**, 186–192 (As)	Rats	Effect of ibuprofen on chronic infection
Pedersen *et al.* (1990) *Acta Pathol. Microbiol. Immunol. Scand.*, **98**, 203–211 (Sa/Ag)	Rats	Establishment of chronic infection and comparison of agar and alginate beads
Pier *et al.* (1990) *Science*, **249**, 537–540 (Ag)	Rats and mice	Protective effect of MEP against chronic infection
Iwata *et al.* (1991) *Infect. Immun.*, **59**, 1514–1520 (Ag)	Rats	The role of BALT in chronic infection
Woods *et al.* (1991) *J. Infect. Dis.*, **163**, 143–149 (Ag)	Rats	Induction of mucoid phenotype in the challenge strain
Yamaguchi and Yamada (1991) *Am. Rev. Resp. Dis.*, **144**, 1147–1152 (Bi)	Rats	Role of mechanical injury in the airways on chronic lung infection
Cheung *et al.* (1992) *J. Med. Primatol.*, **21**, 357–362 (Ag)	Rhesus monkeys	Effect of pentoxifylline on neutrophil influx
Cheung *et al.* (1993) *J. Med. Primatol.*, **22**, 257–262 (Ag)	Rhesus monkeys	Histopathological studies of chronic endobronchiolitis
Gilleland *et al.* (1993) *J. Med. Microbiol.*, **38**, 79–86 (Ag)	Rats	Vaccination efficacies of elastase, exotoxin A and OMP F
Hart *et al.* (1993) *Can. J. Microbiol.*, **39**, 1127–1134 (Ag)	Rats	Effect of r-human urokinase in chronic infection
Johansen *et al.* (1993) *Acta Pathol. Microbiol. Immunol. Scand.*, **101**, 207–225 (Sa)	Athymic and normal rats	Role of B and T cells in chronic infection
Fox *et al.* (1994) *Chest*, **105**, 1545–1550 (Ag)	Rats	Prevention of chronic lung infection by vaccination with OMP F
Johansen *et al.* (1994) *Infect. Immun.*, **62**, 3146–3155 (Sa)	Rats	Modulation of the inflammatory response in chronic infection

Table 61.1 *Continued*

Study (immobilization agent)	*Animals*	*Purpose of the study*
Tanaka *et al.* (1994) *Exp. Lung Res.*, **20**, 351–366 (As)	Mice	Role of neutrophil elastase on the pathogenesis of chronic infection
Lange *et al.* (1995) *Acta Pathol. Microbiol. Immunol. Scand.*, **103**, 367–374 (Sa)	Rats	Non-specific stimulation vs. vaccination in chronic infection
Morisette *et al.* (1995) *Infect. Immun.*, **63**, 1718–1724 (Ag)	Mice	Determination of genetic resistance to lung infection with *P. aeruginosa*
Stevenson *et al.* (1995) *Clin. Exp. Immunol.*, **99**, 98–105	Mice	*In-vivo* and *in-vitro* T cell responses in pulmonary infection
Johansen *et al.* (1995) *Am. J. Respir. Crit. Care Med.*, **152**, 1337–1346 (Sa)	Rats	Effect of D-ALG toxin A on the inflammatory response
Gosselin *et al.* (1995) *Infect. Immun.*, **63**, 3272–3278 (Ag)	Mice	Study of the underlying resistance to chronic *P. aeruginosa* lung infection
Beaulac *et al.* (1996) *Antimicrob. Agents Chemother.*, **40**, 665–669 (Ag)	Rats	Efficacy of free and liposomal encapsulated tobramycin on lung infection
Johansen *et al.* (1996) *Clin. Exp. Immunol.*, **103**, 212–218 (Sa)	Rats	Effect of IFN-γ on the severity of the inflammatory response
Song *et al.* (1996) *Acta Pathol. Microbiol. Immunol. Scand.*, **104**, 350–354 (Sa)	Rats	Effects of Chinese medical herbs on chronic lung infection
Song *et al.* (1997) *Acta Pathol. Microbiol. Immunol. Scand.*, **105**, 438–444 (Sa)	Rats	Effect of ginseng on chronic lung infection
Song *et al.* (1997) *Antimicrob. Agents Chemother.*, **41**, 961–964 (Sa)	Athymic and normal rats	Effect of ginseng on chronic lung infection

System O:2/5, which is a clinical isolate from sputum of a CF patient. It stably maintained the mucoid phenotype and the same antigenic type whether it was studied before challenge, after immobilization in seaweed alginate or in lung homogenate (Johansen, 1996a). This model has also been used with other mucoid *P. aeruginosa* strains isolated from sputum from children with CF (Cash *et al.*, 1979; Boyd *et al.*, 1983; Stevenson *et al.*, 1995).

Following overnight incubation at 35°C on Truche medium containing peptone (40 g/l; Orthana, Copenhagen, Denmark), NaCl (0.086 M) and glucose (0.01 M; pH 7.5), one colony is grown in 100 ml filtered ox broth for 18 hours on a gyratory shaker at 35°C. The culture is centrifuged by 4000–5000 *g* for 20 minutes, and the pellet is resuspended in 5 ml sterile serum bouillon. The inoculum is enumerated by double plate dilutions on modified Conradi–Drigalski's substrate (Johansen *et al.*, 1993).

The immobilization of *P. aeruginosa* is done in spherical alginate beads. Seaweed alginate with approximately 60% guluronic acid content (Protanal 10/60, Protan A/S, Drammen, Norway) is dissolved in 0.9% NaCl to a concentration of 11 mg/ml and autoclaved. A volume of 1 ml of the 18 hours *P. aeruginosa* culture is mixed with 9 ml sterile alginate solution, yielding approximately 10^9 cfu/ml, determined by double plate dilutions. The viscous suspension is placed in a cylindrical reservoir (made of Plexiglass) and forced through an 18 G cannula by compressed air (0.3 atm from the top of the reservoir), while another jet of air (1.0 atm, from the side of the reservoir) is forced coaxially along the needle to blow off the alginate droplets. The droplets are directed into a cross-linking solution of 0.1 M $CaCl_2$ in TRIS-HCl buffer (0.1 M, pH 7.0, (250 ml sterile water + 3.03 g trizmabase + 3.68 g $CaCl_2$); Figure 61.3). By adjusting the airflow the droplets can be varied to make spherical beads of different sizes. The beads cure for 1 hour in the calcium bath with continuous stirring, after which they are centrifuged at 180 *g* for 10 minutes and washed carefully twice in sterile NaCl 0.9%. The final number of cfu is determined by double plate dilutions. The beads are stored at 4°C until use (Pedersen *et al.*, 1990; Johansen *et al.*, 1993).

The size of 100 randomly selected alginate beads and the number of bacteria in them have been determined by phase-contrast light microscopy (Figure 61.4). The median and the 10th–90th percentile range for the size of the beads is 60 μm (30–110 μm). Beads of the median size contain approximately 100 bacteria. After 1 months storage at 4°C more than 80% of the initial inoculum remains (Pedersen *et al.*, 1990). Rats are infected with 0.1 ml of an inoculum containing 10^8 cfu/ml. The size of the challenge inoculum is determined every time it is used for a new procedure.

Infection process

A stable, reproductive chronic lung infection requires the presence of alginate, since the bacteria are rapidly cleared from the airways when they are suspended in saline without alginate (Johansen, 1996a). The role of the alginate is prob-

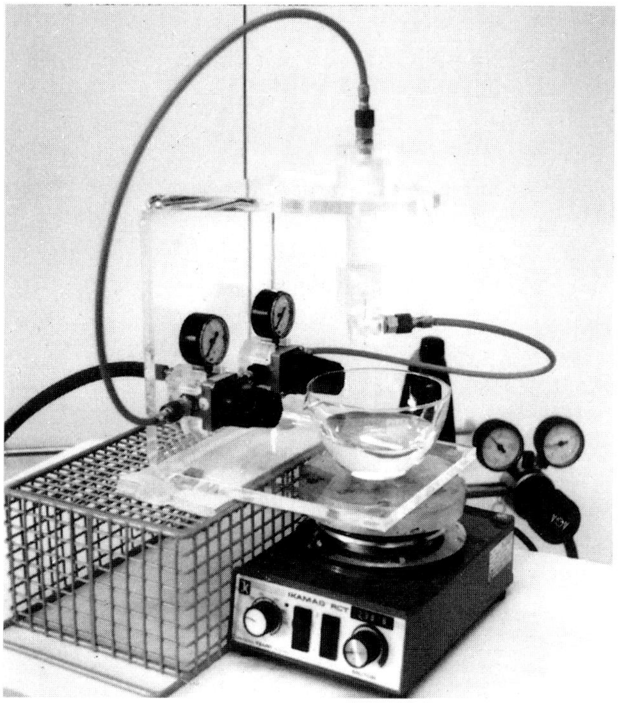

Figure 61.3 The set-up for alginate bead preparation, showing the cylindrical reservoir where the bacterial suspension is placed as well as the mounting of the tubes for air. The $CaCl_2$–TRIS–HCl buffer is placed on a magnet-stirrer.

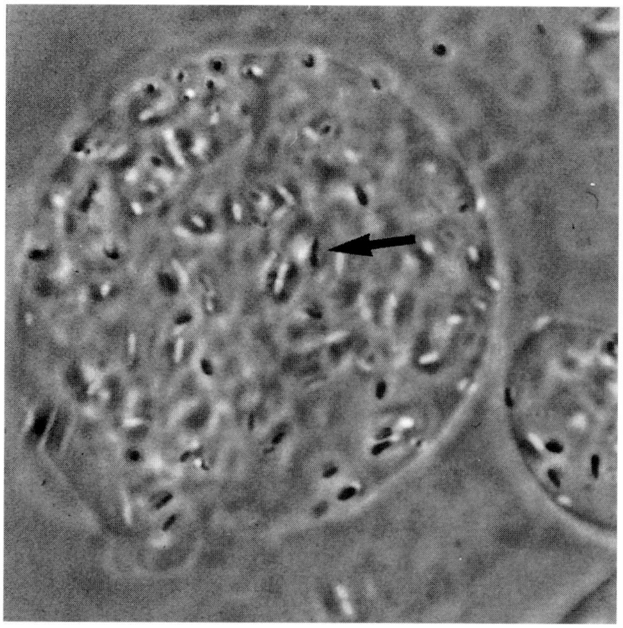

Figure 61.4 Phase contrast microphotograph of an alginate bead containing numerous *Pseudomonas aeruginosa* bacteria (arrow). Magnification × 1000. (From Johansen *et al.*, 1993, reproduced with permission from the editor of APMIS).

ably the same as that proposed for the bacterial biofilm, which provides protection from the host's defence mechanisms (Jensen *et al.*, 1990). Once chronic *P. aeruginosa* infection is established in CF patients it is not possible to eliminate it with immune therapy or antibiotics. The main reasons are that *P. aeruginosa* produces alginate (heteropolysaccharide composed of β-D-mannuronic acid and α-L-guluronic acid) and forms microcolonies (biofilm). Alginate may inhibit clearance of *P. aeruginosa* from the airways by interfering with phagocytosis by alveolar macrophages, inhibiting neutrophil chemotaxis and scavenging free radicals released from the PMNs (Pedersen, 1992). Although the rats do not suffer from CF the microorganisms immobilized in alginate gels are metabolically active and chemoattractants for PMNs must be liberated from the infected beads since numerous PMNs surround the alginate beads in the rat lungs. This corresponds to the paradoxical finding that histologically the inflammatory response in the lungs of CF patients who have been chronically infected for several years looks like an acute infection (PMNs) and not a chronic infection (MNs; Johansen, 1996a). The most frequently used materials for immobilization of the bacteria are agar or agarose (Johansen, 1996a). We prefer seaweed alginate, since it more closely resembles the alginate produced *in vivo* by mucoid *P. aeruginosa*.

Key parameters to monitor infection and response to treatment

The infection is confined to the lung. Most of the rats experience an increase of approximately 0.5°C in rectal temperature during the first day after operation. Deaths due to sepsis are rare, however, despite the fact that some animals have temporary bacteraemia. Body weight gain, food and water intake are similar to normal animals. We have used the following outcomes for monitoring the infection and judging the results of various vaccines and vaccination schedules: deaths, macroscopic lung pathology and histopathology, lung bacteriology and antibody responses. If the model is used for vaccine experiments or treatment purposes, several additional outcomes can be used to evaluate their effects such as: lung and body weight, cytokine production, cellular subsets and protein contents in bronchoalveolar lavage (Johansen, 1996a).

Deaths

The mortality rate in chronically infected rats is approximately 10–16% (Johansen *et al*, 1995).

Lung bacteriology

Determination of bacterial cfu within the lung tissue requires homogenization of the lung. After the rat is killed with 20% pentobarbital (DAK, Copenhagen, Denmark) at 3 ml/kg body weight the lung is removed aseptically from the rat and put into a sterile sputum container, which is stored at 4°C until homogenization. The lung should not be frozen or stored for more than 1–2 hours but homogenized immediately after it has been removed from the rat.

For quantitative bacteriology, 3 ml of sterile phosphate-buffered saline (PBS) is added to each lung. The homogenization is performed in a glass vial kept on ice. For homogenization, we use a blender (Heidolph Diax 600 type F 18 ml, Struers, Denmark) 30 seconds, 9000 rpm.

The total volume of each lung homogenate is 4.5 ml; 0.1 ml of appropriately diluted samples of the lung homogenate is plated to determine the number of cfu per lung (Johansen *et al.*, 1995). The lower limit of detection is 135 cfu/ml of lung homogenate (Song *et al.*, 1997).

Rats that are acutely infected with free *P. aeruginosa* either die or clear the inoculum within 48 hours (Johansen *et al.*, 1994a). Accordingly we have defined—rather arbitrarily—our rat lung model infection to be chronic if it persists for 2 weeks or more.

Although the infection is established in rats with a normal immune system and intact mucosal surfaces it is quite persistent. In one experiment we found that *P. aeruginosa* was still detectable in 45% of the lungs 13 weeks after challenge (Johansen *et al.*, 1995; Johansen, 1996a). Most authors who have used this model or similar ones have completed their experiments within 4 weeks after the infection was established (Boyd *et al.*, 1983; Pedersen *et al.*, 1990; Pier *et al.*, 1990; Morisette *et al.*, 1995).

Lung pathology (macroscopically)

Macroscopic description of the lung should be done both when the lung is *in situ* and after removal from the rat. We have defined four scoring groups according to the severity of the inflammation: 1, normal; 2, swollen lungs, hyperaemia, small atelectases (1 × 1 mm); 3, pleural adhesions, small haemorrhages, small abscesses (1 × 2 mm), atelectases (2 × 3 mm); 4, pleural adhesions, haemorrhages, abscesses (>2 mm) and atelectases (>3 mm). All scorings have been done "blinded" with two observers to avoid bias (Johansen *et al.*, 1995).

Lung histopathology

We cut a small piece approximately 1.5 × 1.5 cm that represents the macroscopic focus from the lower part of the left lung and put it into a plastic vial with formalin puffer (formaldehyde 4%, pH 7.0, Bie & Berntsen, Copenhagen, Denmark). After a couple of days the lung sample is embedded in paraffin wax, cut into 5–10 μm sections, and stained with haematoxylin and eosin (Johansen *et al.*, 1994b).

Microscopically the sections are assigned to one of four groups according to the severity of the inflammation: 1, normal; 2, mild focal inflammation; 3, moderate to severe focal inflammation with areas of normal lung tissue; 4, severe inflammation to necrosis and severe inflammation throughout the lung (Johansen *et al.*, 1994b).

The cellular changes are assigned to four groups according to acute and chronic inflammation by using a scoring system based on the percentage of PMNs and MNs present in the inflammatory foci in the sections. PMN response: ≥90% PMNs and ≤10% MNs; PMN/MN response: 50–90% PMNs and 10–50% MNs; MN/PMN response: 10–50% PMNs and 50–90% MNs; and MN response: ≤10% PMNs and ≥90% MNs. *Acute* inflammation is defined by inflammatory infiltrates dominated by neutrophils (PMN response: ≥90% PMNs). *Chronic* inflammation is defined as predominance of mononuclear cells (MN response: ≥90% MNs; lymphocytes and plasma cells) and presence of granulomas (Johansen *et al.*, 1994b).

Antibody responses

Production of antibodies against *P. aeruginosa* contained in the alginate beads is used as a marker of infection. We measure IgM, IgG and IgA in serum against *P. aeruginosa* sonicate, which contains lipopolysaccharide (LPS) and proteins from the bacteria, and against bacterial alginate by using an enzyme-linked immunosorbent assay (ELISA) (Johansen *et al.*, 1994b). During the first week after the infection has been established a pronounced IgM antibody response can be measured; in the following weeks it declines and is almost zero 4 weeks after challenge. IgG and IgA develop more slowly over the first 3 weeks of infection with peak values 7–8 weeks after challenge.

In brief, the ELISAs are performed in flat-bottomed 96-well microdilution plates (NUNC, Roskilde, Denmark), which are coated with antigen diluted in PBS (pH 7.2). Non-specific binding is eliminated by overnight blocking with Tween-20 (Sigma) and goat serum. Serum diluted in optimal dilutions is added to the wells. For determination of specific antibodies we use peroxidase conjugated goat-anti-rat immunoconjugate (IgG: Medac, Hamburg, Germany; IgM and IgA: Nordic Immunology, Tilburg, Netherlands). Colour reaction is performed with phenylenediamide (3.7 mM, Dakopatts A/S, Glostrup, Denmark) in sodium citrate buffer (0.1 M, pH 5.0) with H_2O_2 (6.5 mM). The colour reaction is stopped by adding H_2SO_4 to the wells; the optical density is measured at 492 nm on an ELISA reader (Johansen *et al.*, 1993).

Determination of additional markers of infection

Commercially available cytokines and kits for measuring cytokines in rats are still very expensive if on the market at all. This is probably because the rat is not very commonly used for experiments. For experimenters who are working with mice, many more products are available.

We found that spleen cells stimulated with concanavalin A from Balb/c mice (T-helper type 2 responders (TH2)) chronically infected with *P. aeruginosa* responded with high levels of interleukin-4 (IL-4) compared to C3H/HeN mice (TH1 responders), who responded with high IFNγ production (Moser *et al.*, 1997). Spleen cells from IL-12-treated Balb/c mice responded with increased levels of TNFα and decreased levels of IL-4 whereas IFNγ showed no significant change; in addition a large number of PMNs were found when the lungs were examined microscopically compared to non-treated Balb/c mice (Moser, 1997, personal communication).

We treated chronically infected rats with IFNγ either before or after challenge with *P. aeruginosa* in alginate beads and found a significantly reduced severity of macroscopic lung inflammation and a complete shift to a chronic-type inflammation dominated by MNs compared to control rats, who persistently showed an acute-type inflammation with PMNs, like that in CF (Johansen *et al.*, 1996b).

In a mouse model of chronic *P. aeruginosa* lung infection it was found that TNF αmRNA expression was increased to a greater extent in Balb/c mice than in C57BL/6 mice (Gosselin *et al.*, 1995). Treatment with anti-TNFα mRNA monoclonal antibodies significantly increased the number of *P. aeruginosa* bacteria after infection, indicating that TNFα might exert a protective role in response to pulmonary infection with *P. aeruginosa*.

Use of the model for other purposes

Since 1979 when the chronic *P. aeruginosa* lung infection model in rats was first described, more than 40 papers have been published where the model has been used for a huge variety of purposes (Johansen, 1996a; Table 61.1). In brief, the potentials of several bacterial vaccines have been investigated, e.g. LPS, outer membrane protein F, alginate, elastase, exotoxin A and sonicate. The effect of bacterial exoproducts has also been investigated, as well as the role of B cells and T cells during the chronic lung infection. The model is simple to perform and the materials needed to perform the surgical procedure, as well as the preparation of the alginate beads, are easily acquired. The model can easily be adopted for other purposes such as novel antimicrobials, effects of antimicrobial therapy — for which it has only rarely been used — and synergistic/antagonistic effects of antibiotics.

Pitfalls (advantages/disadvantages) of the model

Besides the usual shortcomings of all animal models (the rats we use as a model for CF do not have CF), the major shortcoming of the rat model that we describe here is the large inocula of alginate-embedded bacteria, directly implanted into the lungs, that are required to establish the infection. This is very different from the situation in most CF patients, since the majority of them (82%) have a period of approximately 1 year (range 0–5.5 years) of intermittent *P. aeruginosa* colonization preceding the onset of chronic infection (Johansen and Høiby, 1992). In only a minority of patients (18%) does the initial episode of colonization develop into chronic infection. When CF patients get chronically infected with *P. aeruginosa* their lung mucosa is often structurally altered as a result of local inflammation with other respiratory pathogens (*H. influenzae, S. pneumoniae, S. aureus*) or viral infections (Johansen *et al.*, 1994a). Therefore, for the model to be more similar to the situation in patients when they become colonized, some kind of respiratory damage should be added to the intact mucosal surface in the rat before the *P. aeruginosa* infection is established. Different methods have been investigated to mimic the CF epithelial injury, such as pretreatment of the rat lungs with reserpine, PMN elastase (Döring and Dauner, 1988), trypsin or brush injury (Yamaguchi and Yamada, 1991) or by adding hexamethylphosphoramide to the drinking water (Boyd *et al.*, 1983).

The alginate immobilization is somewhat artificial, since the initial colonizing strain of *P. aeruginosa* is usually non-mucoid. However, during the infection a shift to the mucoid phenotype usually occurs. It has been shown that even non-mucoid strains have some alginate present on the cell surface (Pier *et al.*, 1986), so this model is probably still the best alternative to the CF patients themselves until the CF transgenic mouse becomes commercially available to all experimenters.

The very aggressive infection we induce with the large bacterial inoculum may blur the effect of different treatment regimens that might be of benefit, but in order to get an infection we need to install a rather large inoculum into the lungs of the rats. The inoculum could therefore be too large to allow new regimens to exert an impact on the infection. Unfortunately, the model is very susceptible to inoculum size. If the inoculum contains too few bacteria the infection will not persist and if the inoculum is too high all rats will die. Careful inoculum titration is therefore very important for the outcome of the infection (Johansen, 1996a).

A very important parameter is the total time the experimenter uses from the time the skin is disinfected until the inoculum is delivered into the lung. This procedure should not take much longer than 3 minutes. If it does, it will adversely affect the mortality rate. Bleeding should also be avoided, since it can also reduce survival.

Concluding remarks

Although evaluations in animal models are not considered a prerequisite for clinical trials of novel treatment strategies, the rat model of chronic lung infection has often used for evaluation of possible new vaccine formulations. An LPS–toxin-A-containing vaccine that we have used in our rat model has also been used in CF patients (Cryz *et al.*, 1994). Many experiments have also dealt with the elucidation of the influence of bacterial exoproducts such as elastase, LPS, toxin A and alginate on the pathophysiology of the chronic *P. aeruginosa* lung infection. Although none of the different species used for this model have CF, the model has proved to be an invaluable supplement to the patients themselves by giving quick answers to questions that would have been nearly impossible to deal with without the model (Johansen, 1996a).

References

Boyd, R. L., Ramphal, R., Rice, R., Mangos, J. A. (1983). Chronic colonization of rat airways with *Pseudomonas aeruginosa*. *Infect. Immun.*, **39**, 1403–1410.

Buchwald, M. (1996). Cystic fibrosis: from the gene to the dream. *Clin. Invest. Med.*, **19**, 304–310.

Cash, H. A., Woods, D. E., McCullough, B., Johanson, W. G. Jr, Bass, J. A. (1979). A rat model of chronic respiratory infection with *Pseudomonas aeruginosa*. *Am. Rev. Respir. Dis.*, **119**, 453–459.

Cheung, A. T., Moss, R. B., Leong, A. B., Novick, W. J. Jr (1992). Chronic *Pseudomonas aeruginosa* endobronchitis in rhesus monkeys: I. Effects of pentoxifylline on neutrophil influx. *J. Med. Primatol.*, **21**, 357–362.

Cryz, S. J. Jr, Wedgwood, J., Lang, A. B. *et al.* (1994). Immunization of noncolonized cystic fibrosis patients against *Pseudomonas aeruginosa*. *J. Infect. Dis.*, **169**, 1159–1162.

Davis, P. B., Drumm, M., Konstan, M. W. (1996). Cystic fibrosis. *Am. J. Respir. Crit. Care Med.*, **154**, 1229–1256.

Dorin, J. R., Dickinson, P., Alton, E. W. *et al.* (1992). Cystic fibrosis in the mouse by targeted insertional mutagenesis. *Nature*, **359**, 211–215.

Döring, G., Dauner, H. M. (1988). Clearance of *Pseudomonas aeruginosa* in different rat lung models. *Am. Rev. Respir. Dis.*, **138**, 1249–1253.

Frederiksen, B., Lanng, S., Koch, C., Høiby, N. (1996). Improved survival in the Danish center-treated cystic fibrosis patients: results of aggressive treatment. *Pediatr. Pulmonol.*, **21**, 153–158.

Gosselin, D., DeSanctis, J., Boule, M., Skamene, E., Matouk, C., Radzioch, D. (1995). Role of tumor necrosis factor alpha in innate resistance to mouse pulmonary infection with *Pseudomonas aeruginosa*. *Infect. Immun.*, **63**, 3272–3278.

Gruber, F. (1975). Immunologie der Versuchstiere. Ein Beitrag zur Charakterisierung Immunologischer Modelle. In: *Schriftenreihe Versuchstierkunde*. (eds Merkenschlager, M., Gãrtner, K), pp 62–64. Paul Parey, Berlin.

Jensen, E. T., Kharazmi, A., Lam, K., Costerton, J. W., Høiby, N. (1990). Human polymorphonuclear leukocyte response to *Pseudomonas aeruginosa* grown in biofilm. *Infect. Immun.*, **58**, 2383–2385.

Johansen, H. K., Høiby, N. (1992). Seasonal onset of initial colonisation and chronic infection with *Pseudomonas aeruginosa* in patients with cystic fibrosis in Denmark. *Thorax*, **47**, 109–111.

Johansen, H. K., Espersen, F., Pedersen, S. S., Hougen, H. P., Rygaard, J., Høiby, N. (1993). Chronic *Pseudomonas aeruginosa* lung infection in normal and athymic rats. *Acta Pathol. Microbiol. Immunol. Scand.*, **101**, 207–225.

Johansen, H. K., Cryz, S. J. Jr, Høiby, N. (1994a). Clearance of *Pseudomonas aeruginosa* from normal rat lungs after immunization with somatic antigens or toxin A. *Acta. Pathol. Microbiol. Immunol. Scand.*, **102**, 545–553.

Johansen, H. K., Espersen, F., Cryz, S. J. Jr *et al.* (1994b). Immunization with *Pseudomonas aeruginosa* vaccines and adjuvant can modulate the type of inflammatory response subsequent to infection. *Infect. Immun.*, **62**, 3146–3155.

Johansen, H. K., Hougen, H. P., Cryz, S. J. Jr, Rygaard, J., Høiby, N. (1995). Vaccination promotes TH1-like inflammation and survival in chronic *Pseudomonas aeruginosa* pneumonia in rats. *Am. J. Respir. Crit. Care. Med.*, **152**, 1337–1346.

Johansen, H. K. (1996a). Potential of preventing *Pseudomonas aeruginosa* lung infections in cystic fibrosis patients: experimental studies in animals. *Acta Pathol. Microbiol. Immunol. Scand.*, **63**, 5–42 (Thesis).

Johansen, H. K., Hougen, H. P., Rygaard, J., Høiby N. (1996b). Interferon-gamma (IFN-γ) treatment decreases the inflammatory response in chronic *Pseudomonas aeruginosa* pneumonia in rats. *Clin. Exp. Immunol.*, **103**, 212–218.

Koch, C., Høiby, N. (1993). Pathogenesis of cystic fibrosis. *Lancet*, **341**, 1065–1069.

Lewis, P. A. (1995). The epidemiology of cystic fibrosis. In: *Cystic Fibrosis*. (eds Hodson, M. E., Geddes, D. M.), pp. 1–13. Chapman and Hall, London.

Maciver, I., Silverman, S. H., Brown, M. R., O'Reilly, T. (1991). Rat model of chronic lung infections caused by non-typable *Haemophilus influenzae*. *J. Med. Microbiol.*, **35**, 139–147.

Morissette, C., Skamene, E., Gervais, F. (1995). Endobronchial inflammation following *Pseudomonas aeruginosa* infection in resistant and susceptible strains of mice. *Infect. Immun.*, **63**, 1718–1724.

Moser, C., Johansen, H. K., Song, Z., Hougen, H. P., Rygaard, J., Høiby, N. (1997). Chronic *Pseudomonas aeruginosa* lung infection is more severe in TH2 responding Balb/c mice compared to TH1 responding C3H/HeN mice. *Acta Pathol. Microbiol. Immunol. Scand.*, **105**, 838–842.

Pedersen, S. S. (1992). Lung infection with alginate-producing, mucoid *Pseudomonas aeruginosa* in cystic fibrosis. *Acta Pathol. Microbiol. Immunol. Scand.*, **28**, 1–79 (thesis).

Pedersen, S. S., Shand, G. H., Hansen, B. L., Hansen, G. N. (1990). Induction of experimental chronic *Pseudomonas aeruginosa* lung infection with *P. aeruginosa* entrapped in alginate microspheres. *Acta Pathol. Microbiol. Immunol. Scand.*, **98**, 203–211.

Pennington, J. E., Hickey, W. F., Blackwood, L. L., Arnaut, M. A. (1981). Active immunization with lipopolysaccharide *Pseudomonas* antigen for chronic *Pseudomonas* bronchopneumonia in guinea pigs. *J. Clin. Invest.*, **68**, 1140–1148.

Pier, G. B., Small, G. J., Warren, H. B. (1990). Protection against mucoid *Pseudomonas aeruginosa* in rodent models of endobronchial infections. *Science*, **249**, 537–540.

Pier, G. B., DesJardins, D., Aguilar, T., Barnard, M., Speert, D. P. (1986). Polysaccharide surface antigens expressed by nonmucoid isolates of *Pseudomonas aeruginosa* form cystic fibrosis patients. *J. Clin. Microbiol.*, **24**, 189–196.

Ratcliff, R., Evans, M. J., Cuthbert, A. W. *et al.* (1993). Production of a severe cystic fibrosis mutation in mice by gene targeting. *Nat. Genet.*, **4**, 35–41.

Slater, L. N. (1990). A rat model of prolonged pulmonary infection due to nontypable *Haemophilus influenzae*. *Am. Rev. Respir. Dis.*, **142**, 1429–1435.

Snouwaert, J. N., Brigman, K. K., Latour, A. M. *et al.* (1992). An animal model for cystic fibrosis made by gene targeting. *Science*, **257**, 1083–1088.

Song, Z., Johansen, H. K., Faber, V. *et al.* (1997). Ginseng treatment reduces bacterial load and lung pathology in chronic *Pseudomonas aeruginosa* pneumonia in rats. *Antimicrob. Agents Chemother.*, **41**, 961–964.

Stevenson, M. M., Kondratieva, T. K., Apt, A. S., Tam, M. F., Skamene, E. (1995). *In vitro* and *in vivo* T cell responses in mice during bronchopulmonary infection with mucoid *Pseudomonas aeruginosa*. *Clin. Exp. Immunol.*, **99**, 98–105.

Winnie, G. B., Klinger, J. D., Sherman, J. M., Thomassen, M. J. (1982). Induction of phagocytic inhibitory activity in cats with chronic *Pseudomonas aeruginosa* pulmonary infection. *Infect. Immun.*, **38**, 1088–1093.

Woods, D. E., Cryz, S. J. Jr, Friedman, R. L., Iglewski, B. H. (1982). Contribution of toxin A and elastase to virulence of *Pseudomonas aeruginosa* in chronic lung infections of rats. *Infect. Immun.*, **36**, 1223–1228.

Woods, D. E., Franklin, R., Cryz, S. J. Jr, Ganss, M., Peppler, M., Ewanowich, C. (1989). Development of a rat model for respiratory infection with *Bordetella pertussis*. *Infect. Immun.*, **57**, 1018–1024.

Yamaguchi, T., Yamada, H. (1991). Role of mechanical injury on airway surface in the pathogenesis of *Pseudomonas aeruginosa*. *Am. Rev. Respir. Dis.*, **144**, 1147–1152.

Chapter 62

Hamster Model of *Mycoplasma* Pulmonary Infections

K. Ishida, M. Kaku and J. Shimada

Background of model

Mycoplasma pneumoniae is a major cause of pneumonia and accounts for as many as 20% of the total number of cases of pneumonia (Foy *et al.*, 1979). The recommended therapy is erythromycin, which is efficacious in reducing the duration of symptoms (Shames *et al.*, 1970). For studying the *in-vivo* effect of antimicrobial agents against *Mycoplasma pneumoniae*, there are some experimental models of mycoplasmal pneumonia. The most common species used for *Mycoplasma* pulmonary infection model is the hamster (Barile *et al.*, 1988). Infection by inhalation, pernasal infection, and intratracheal intubation are the methods used for infecting hamsters with *Mycoplasma pneumoniae*. We have compared these methods and the results show that the intratracheal intubation method is the most efficacious. Intratracheal intubation was described by Barile *et al.* (1988). Modifying this method, we devised the intrabronchial intubation method (Ishida *et al.,* 1994a,b), which is introduced in this chapter. This method can produce localized right lower lobe mycoplasmal pneumonia. By using this intrabronchial intubation method, both a precise histopathological evaluation of the effects of drugs against mycoplasmal pneumonia and evaluation of the mycoplasmal cell count can be obtained.

Animal species

The hamster is the most common species used for the mycoplasmal pulmonary infection model. Any hamster strain may be used for this infection, but the Syrian golden hamster is usually used.

Preparation of animals

No specialized housing or care nor specific pretreatment is required.

Details of infection

Overview

A flexible tube is introduced *per oral* into the right main bronchus and a suspension of *M. pneumoniae* is inoculated directly into the right lower lobe. This procedure produces a localized right lower lobe mycoplasmal pneumonia (Ishida *et al.*, 1994a,b; Figure 62.1).

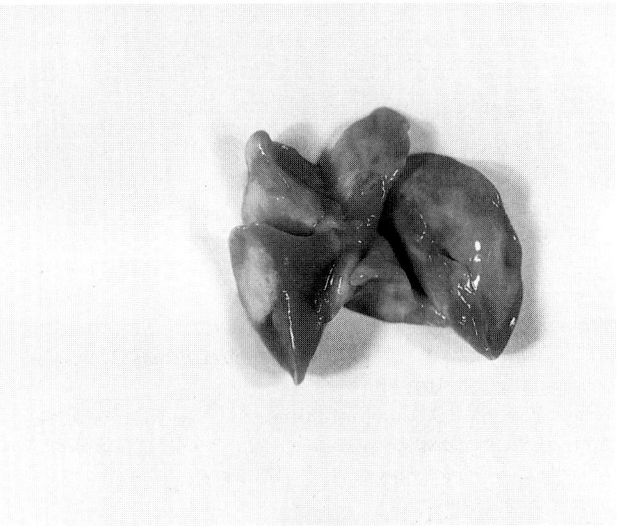

Figure 62.1 Macroscopic appearance of the lung after intrabronchial infection with *M. pneumoniae*. (Consolidation occupies about 75% of the right lower lobe.)

Materials required

- Anesthetic: ketamine hydrochloride.
- 2-inch 20 G Angiocath (Becton Dickinson Vascular Access) or similar.
- Ear-probe.
- Chanock broth medium (Chanock *et al.*, 1962) consisting of seven parts PPLO broth without crystal violet, two parts uninactivated horse serum, one part 25% fresh

Handbook of Animal Models of Infection
ISBN 0-12-775390-7

yeast extract, 1% glucose, and 0.002% phenol red adjusted to pH 7.8 with 1 N sodium hydroxide.

- Chanock agar plate consisting of seven parts PPLO agar, two parts uninactivated horse serum, one part 25% fresh yeast extract, 1000 U penicillin G per milliliter and 0.025% thallium acetate.
- Chanock broth medium containing about 10^7 cfu/ml of *M. pneumoniae* (strain M129).

Intrabronchial intubation method

Following anesthesia by intraperitoneal injection of 50–100 mg/kg ketamine, the animal is placed ventral side up at an angle of 45° to the vertical. When the shaft of an ear-probe is placed into the mouth and moved upward while pulling the tongue to the outside, the glottis and the orifice of the trachea will be seen. Having confirmed the tracheal orifice, a bent, blunted 2-inch 20 G Angiocath is inserted under the glottis into the trachea and the inner needle is withdrawn (Figures 62.2 and 62.3). The outer sheath, which is in the trachea, is advanced as far as possible. After wedging the sheath into the right lower bronchus, a 1 ml tuberculin syringe is attached to the outer catheter and a 0.15 ml inoculum of the suspension of *M. pneumoniae* M 129 strain at passage level 12 is inoculated directly into the right lower lobe. Control intrabronchial inoculations of Indian ink verified the limited distribution of the inoculum into the right lung, especially the lower righ lobe. For a negative control, broth not containing *M. pneumoniae* was also inoculated; no lung lesions were observed microscopically or macroscopically after that inoculation.

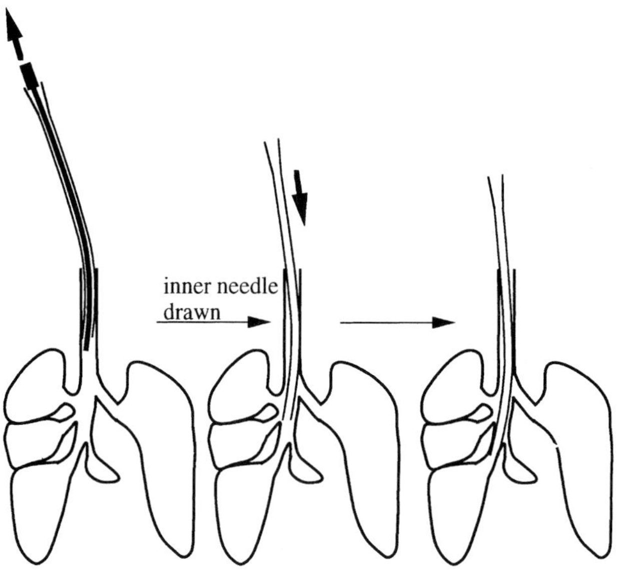

Figure 62.3 The right lung of the hamster is composed of four lobes. The outer sheath of the Angiocath is spontaneously wedged into the right lower lobe after this intrabronchial intubation, because of the angle of the carina.

Postprocedure care

There is no specific care for hamsters after this peroral intrabronchial infection procedure.

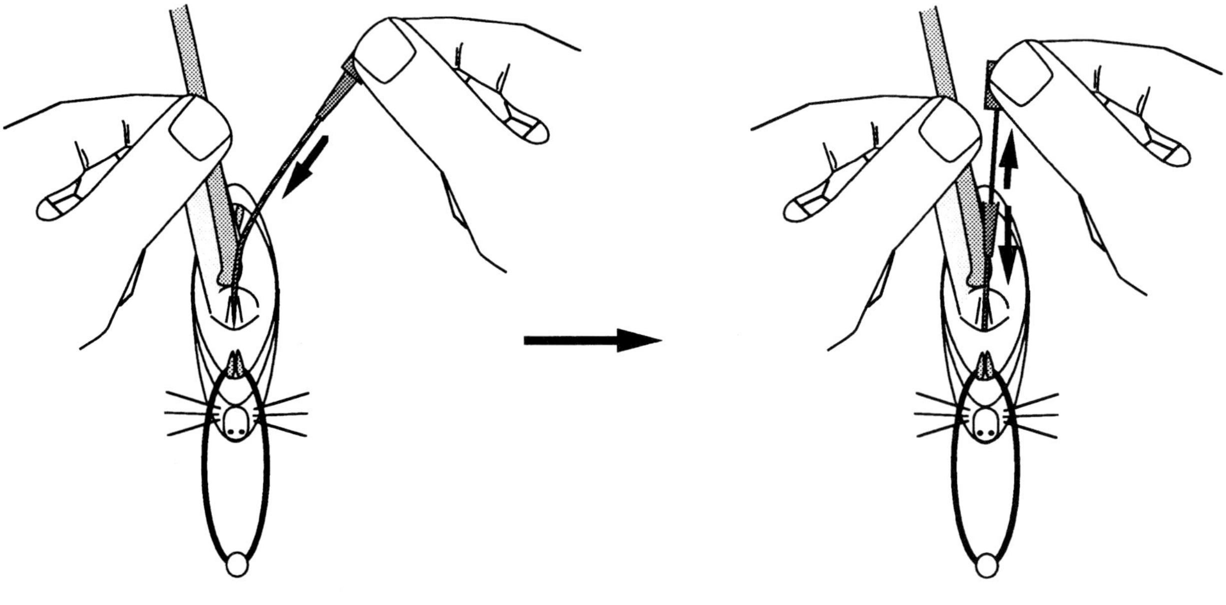

Figure 62.2 The procedure of intrabronchial intubation.

The course of pulmonary infection in the hamster

The lung lesion of the hamster is maximized about 1 week after infection. The lung lesion persists for about a month but it is self-limiting, so that no hamsters die of mycoplasmal pneumonia. During this time the hamster is less active because of the pneumonia.

Key parameters to monitor infection and response to treatment

The infection is self-limiting. There are no hamster deaths due to mycoplasmal pneumonia. Compared to healthy hamsters, infected hamsters are less active, but it is difficult to identify clear parameters for monitoring the infection without assessing lung pathologies and mycoplasmal cell count.

Mycoplasmal density in lung

Determination of the number of colony-forming units in the lungs requires homogenizing of the lung using a glass homogenizer.

Lung pathology

Counting the viable mycoplasmal cells after therapy was not sufficient to estimate the therapeutic effects of drugs, so it is necessary to estimate the severity of pneumonia by histological methods.

Therapeutic plan

The *in-vivo* activities of antimycoplasmal drugs against experimentally induced *M. pneumoniae* pneumonia in hamsters were examined by tube feeding. Oral administration of drugs is started on day 5 after inoculation with *M. pneumoniae*, taking the course of the pulmonary infection into consideration.

The efficacy of macrolides and quinolones have already been examined (Ishida *et al.*, 1994a,b; Kaku *et al.*, 1994). In this chapter *in-vivo* effects of macrolides against mycoplasmal pneumonia will be discussed as an example. Dosages, durations and results are shown in Figure 62.4.

Pulmonary clearance study

The lungs were removed aseptically, weighed, and homogenized in 5 ml of modified Chanock broth using a glass homogenizer. After homogenization, each batch was made up to 10 ml by addition of modified Chanock broth medium to the homogenized suspension. Serial 10-fold dilutions of the homogenized suspension with modified Chanock broth medium were made, from 10^0 to 10^{-8}. Twenty 0.01 ml portions (total, 0.2 ml) of each dilution were spotted on five Chanock agar plates (four spots on each plate). Plates were incubated aerobically at 37°C for 14 days in a tightly closed vinyl bag to maintain humidity, and the number of cfu per gram of lung was calculated from the number of colonies that developed on the five plates used for each dilution (Sasaki *et al.*, 1991). Some isolated colonies were identified by inhibition of growth in the presence of anti-*M-pneumoniae* rabbit serum and by their hemadsorption properties (Clyde *et al.*, 1964). For the untreated controls, oral administration of 5% gum arabic solution was started on day 5 after inoculation of *M. pneumoniae* and administered once only for 1 day and twice daily for 2.5 and 5 days. After 2 days, five hamsters in each control group were sacrificed and their lungs were quantitatively cultured. Before treatment (day 5 after infection), measurement of the number of viable mycoplasma cells in the lungs was also made.

Assessment of lung lesions

Counting the viable mycoplasmal cells after therapy was not sufficient to estimate the therapeutic effects of drugs, we attempted to estimate the severity of pneumonia microscopically and macroscopically.

Microscopic lung lesions

Five hamsters in each group were sacrificed and their lungs were fixed by *in-situ* intratracheal instillation of 10% phosphate-buffered formalin under controlled pressure. Thin sections of lung (in paraffin) were stained with hematoxylin and eosin. The severity of the lung lesions was scored microscopically by estimating the degree of peribronchial and peribronchiolar infiltrates and luminal exudates and by estimating the number of sites affected in three separate coronal sections that included the main bronchus of the right lower lobe, to which the lung lesions were limited, using the criteria in Table 62.1.

Macroscopic lung lesions

Lesions in the lung were limited to the right lower lobe, so the severity of pneumonia could be estimated macroscopically by measuring the extent of macroscopic lung lesions in the right lower lobe. The severity of pneumonia was classified into six groups according to the extent of macroscopic lesions, and a ratio representing each group was determined, as follows: lesions less than one-eighth of the lobe, 0; more than one-eighth and less than one-quarter, 1/8; more than one-quarter and less than one-half, 1/4; more than one-half and less than three-quarters, 1/2; more

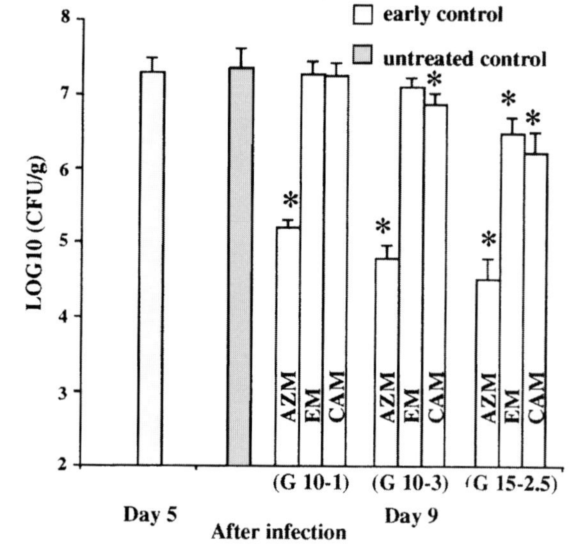

(A)

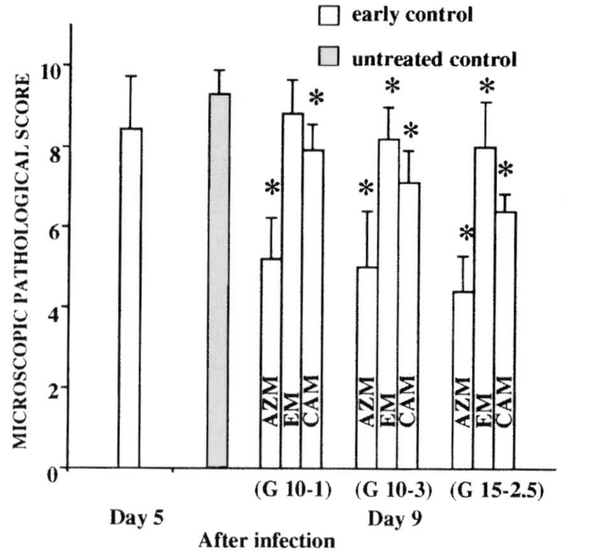

(C)

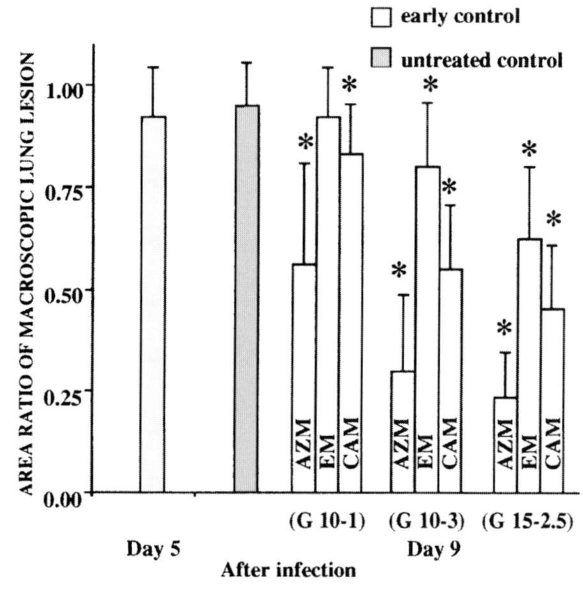

(B)

Figure 62.4 The effect of macrolides against *Mycoplasma pneumoniae* pulmonary infections. **A.** Number of viable *M. pneumoniae* cells ($n = 5$). **B.** Course of macroscopically assessed severity ($n = 10$). **C.** Course of microscopically assessed severity ($n = 5$). G 10–1, 10 mg/kg once daily for 1 day; G 10–3, 10 mg/kg once daily for 3 days; G 15–2.5, 15 mg/kg twice daily for 2.5 days; AZM, azithromycin; EM, erythromycin; CAM, clarithromycin; *, statistically significant.

Table 62.1 Microscopic parameters for evaluating the severity of mycoplasmal pneumonia in hamster lungs

*Score**	*No. of sites affected†*	*Peribronchial and peribronchiolar infiltrates*	*Luminal exudate*
0	None	None	None
1	Few	Mild	Mild
2	Many	Moderate	Moderate
3	Most	Severe	Severe
4	All		

*The microscopic pathological score was defined as the sum of the scores for number of sites affected, peribronchial and peribronchiolar infiltrates, and luminal exudate.
† Refers to the presence of any peribronchial and peribronchiolar infiltrate.

than three-quarters and less than one, 3/4; and more than one lobe, 1.

Antimicrobial therapy

The *in-vivo* activities of macrolides against *M. pneumoniae* pneumonia in hamsters is shown in Figure 62.4.

A variety of antimicrobial agents have been evaluated in this hamster pulmonary infection model. Basically the drugs are administered by tube feeding because mycoplasmal pneumonia in man is usually treated with oral antimicrobial agents. In this hamster model of *Mycoplasma* pulmonary infections the treatment window is about 5 days postinfection because the lung lesion has reached its peak by about 1 week postinfection.

Pitfalls (advantages/disadvantages) of the model

Infection by inhalation, pernasal infection, and intratracheal intubation have been used to infect hamsters with *Mycoplasma pneumoniae*. We tried all these methods when devising the hamster model of mycoplasmal pulmonary infection but it was very difficult to achieve uniform severity of pneumonia in each individual hamster with any of them. In order to localize the lung lesion, the peroral intrabronchial intubation method has been designed. This method can achieve uniform severity of pneumonia in individual hamsters and we could estimate the *in-vivo* effect of antimicrobial agents against *M. pneumoniae* precisely.

Histopathological features of humans and hamsters are similar, so it is possible to use histopathological techniques to estimate drug efficacy, and from this point of view the peroral intrabronchial intubation method has an advantage compared to other methods. A disadvantage of the model is that it takes a long time to infect each hamster compared to other methods, but this disadvantage is negligible when compared to its advantages.

Contribution of the model to infectious disease therapy

Some kinds of antimicrobial agents—macrolides, tetracyclines, and quinolones—are efficatious in the treatment of mycoplasmal pneumonia, but precise histopathological evaluation was difficult because of the difficulty of inducing the same severity of pneumonia in each animal. The peroral intrabronchial intubation method has made it possible to study histopathological features. It has revealed that some antimicrobial agents have the potential to reduce the histopathological severity of infection rather than numbers of viable mycoplasmal cells (Ishida *et al.*, 1994a,b). This result may reflect the mechanism of action of particular antimicrobial agents. Macrolides and minocycline in some concentrations only inhibit the virulence process of *M. pneumoniae*, which is complex, without killing it.

References

Barile, M. F., Chandler, D. K. F., Yoshida, H., Grabowski, M. W., Hayasawa, R., Razin, S. (1988). Parameters of *Mycoplasma pneumoniae* infection in Syrian hamsters. *Infect. Immun.*, **56**, 2443–2449.

Chanock, R. M., Hayflick, L., Barile, M. F. (1962). Growth on artificial medium of an agent associated with atypical pneumonia and its identification as PPLO. *Proc. Natl Acad. Sci. USA*, **48**, 41–49.

Clyde, M. A. Jr (1964). Mycoplasma species identification based upon growth inhibition by specific antisera. *J. Immunol.*, **92**, 958–965.

Foy, H. M., Kenny, G. E., Cooney, M. K., Allen, I. D. (1979). Long-term epidemiology of infections with *Mycoplasma pneumoniae*. *J. Infect. Dis.*, **139**, 681–687.

Ishida, K., Kaku, M., Irifune, K. *et al.* (1994). *In vitro* and *in vivo* activities of macrolides against *Mycoplasma pneumoniae*. *Antimicrob. Agents Chemother.*, **38**, 790–798.

Ishida, K., Kaku, M., Irifune, K. (1994). *In vitro* and *in vivo* activity of a new quinolone AM-1155 against *Mycoplasma pneumoniae*. *J. Antimicrob. Chemother.*, **34**, 875–883.

Kaku, M., Ishida, K., Irifune, K. (1994). *In vitro* and *in vivo* activities of Sparfloxacin against *Mycoplasma pneumoniae*. *Antimicrob. Agents Chemother.*, **38**, 738–741.

Sasaki, Y. A., Ogura, K., Nakayama, Y., Noguchi, K., Matuno, P., Saito, M. (1991). Susceptibility of newly established mouse strain MPS to *Mycoplasma pneumoniae* infection. *Microbiol. Immunol.*, **35**, 247–252.

Shames, J. M., George, R. B., Holliday, W. B., Rasch, J. R., Mogabgab, W. J. (1970). Comparison of antibiotics in the treatment of mycoplasmal pneumonia. *Arch. Intern. Med.*, **125**, 680–684.

Slotkin, R. I., Clyde, W. A. Jr, Denny, F. W. (1967). The effect of antibiotics on *Mycoplasma pneumoniae in vitro* and *in vivo*. *Am. J. Epidemiol.*, **86**, 225–237.

Chapter 63

Murine Models of Pneumonia Using Aerosol Inoculation

J. Leggett

Background of human infection

Pneumonia is currently the sixth most common cause of death in the USA and the leading infectious cause of death (Garibaldi, 1985; National Center for Health Statistics, 1993). Acute pneumonitis develops as a result of defective host defenses, invasion by a particularly virulent pathogen, or the presence of an overwhelming inoculum. A wide variety of microbial pathogens may cause pneumonia, and no single antimicrobial agent can be employed for all possibilities. A recent review of community-acquired pneumonia characterized the many disputes that still remain about diagnostic evaluation and therapeutic decisions, including justifications for antibiotic selection (Bartlett and Mundy, 1995).

Background of model

Most animal experiments aimed at improving the efficacy of antimicrobial therapy against Gram-negative bacillary infections have focused on new drugs. Studies that compare the efficacy of new and established antibiotics have usually employed a limited number of doses administered at intervals similar to those used in humans, without adjusting for the rapid elimination of these drugs in animals. The use of suboptimal, infrequent dosing intervals may lead to inaccurate comparisons among antibiotics (McColm *et al.*, 1986; Trautmann *et al.*, 1986; Johnson *et al.*, 1987).

Until recently, relatively few studies have evaluated the influences of a drug's pharmacokinetic profile or dosage regimen on its antimicrobial efficacy (Nishi and Tsuchiya, 1980; Roosendaal *et al.*, 1986). The use of multiple dosing intervals for each of several different cumulative drug doses has allowed these influences to be more explicitly examined (Vogelman *et al.*, 1988). By reducing the interdependence among pharmacokinetic parameters with such dosing strategies, these studies allow the determination of which parameter is of major importance in predicting efficacy. Moreover, by defining the maximum attainable bactericidal effect and the dose required to achieve a bacteriostatic effect or 50% of maximum at several dosing intervals, these studies allow more standardized comparisons among different antibiotics (Leggett *et al.*, 1989a; Craig, 1995).

Most pneumonia models have used intranasal or instillation of intratracheal inocula (Pennington, 1985; Roosendaal *et al.*, 1986). This methodology is mandatory for pathogens such as *S. pneumoniae*, which do not survive aerosolization (personal unpublished observations). Aerosol delivery has been employed for both infection and treatment purposes for the past three decades (Laurenzi *et al.*, 1964; Hashimoto *et al.*, 1996; Conley *et al.*, 1997). Most aerosolization studies have involved mycobacterial or viral infections (Johnson *et al.*, 1995; Kelly *et al.*, 1996; Phillpotts *et al.*, 1997), although some have targeted other atypical pneumonia pathogens (Vanrompay *et al.*, 1995). The majority of these studies have investigated vaccine strategies against airborne pathogens.

Animal species

A wide range of animal species has been employed in various aerosol infection models. We have used 6-week-old specific-pathogen-free female Swiss mice weighing 23–27 g (Leggett *et al.*, 1989a), as well as Sprague–Dawley rats (Leggett *et al.*, 1990). Presumably any species or strain could be used; preliminary experiments should be performed in order to determine optimal bacterial inocula and treatment delay.

Preparation of animals

No specialized housing or care is required. Immunocompetent mice may be used after appropriate acclimatization. In order to render mice neutropenic (<100 neutrophils/mm^3) for 48 hours after infection, two doses of cyclophosphamide are injected intraperitoneally 4 days (150 mg/kg) and 1 day (100 mg/kg) before the experiment. The duration of neutropenia may be prolonged by an additional 100 mg/kg intraperitoneal injection of cyclophosphamide every 48 hours as necessary. In order to simulate

Handbook of Animal Models of Infection
ISBN 0-12-775390-7

human pharmacokinetics for many renally cleared antimicrobial agents, uranyl nitrate may be administered 10 mg/kg i.p. 3 or 4 days prior to infection (Craig *et al.*, 1991).

Storage and preparation of inocula

Bacteria to be used can be kept frozen according to standard protocols (O'Reilly *et al.*, 1996). Stationary-phase broth cultures to be used for lung infection may be obtained by overnight incubation of approximately 1–2 liters of Mueller–Hinton or other nutrient broth. Cultures are centrifuged (10 000 *g* for 20 minutes) and washed twice in physiological saline before being resuspended at approximately 10^9 cfu/ml in sterile 0.85% saline. Many of our experiments have been performed with *Klebsiella pneumoniae* ATCC 43816 (capsular serotype 2).

Infection process

A diffuse pneumonia may be induced by exposing mice in a closed chamber for 45 minutes to a 4–5 l/minute rate of aerosol generated by a Collison nebulizer (Laurenzi *et al.*, 1964). Collison nebulizers are often used commercially for aerosolized drug delivery. Bacteria in the lung begin log phase growth approximately 6–8 hours following infection. Mice become bacteremic approximately 14–16 hours later, when there is histologic evidence of bronchopneumonia. The infection is lethal in control mice within 36–48 hours.

Key parameters to monitor infection

This pneumonia model may be used in survival studies looking at 50% protective doses (O'Reilly *et al.*, 1996) or in studies quantifying bacterial inocula at various time points (e.g. 24 hours following initiation of therapy). To evaluate antimicrobial activity at the end of 24 hours of therapy we generally sacrifice and quantify lung inocula in four mice for each specific antimicrobial regimen. Organism quantification is usually done in control mice immediately after aerosolization ($n = 2$) just before drug treatment ($n = 8$–10) and 24 hours after the onset of therapy ($n = 2$). Lungs are excised and homogenized in 5 ml of iced saline, serially diluted, and quantified using a spiral system plater and laser colony counter (Spiral System Instruments, Bethesda, MD). Manual dilution methods produce equivalent results. In experiments involving the largest doses of antibiotics, a cephalosporinase or 0.8% SDS may be added to iced saline to insure that drugs are inactivated (Klassen and Edberg, 1996). The extent of antimicrobial activity may be calculated by subtracting the $\log_{10}$ cfu per lung of each treated mouse at the end of therapy from the mean $\log_{10}$ cfu per lung of control mice either just before therapy (0 hours) or at the end of therapy (24 hours).

Antimicrobial therapy

Several antimicrobial agents have been evaluated in this murine aerosolization pneumonia model. Mice have generally been administered antibiotics subcutaneously in 0.2 ml volumes. Subcutaneous injection of antibiotics may be made anywhere on the dorsal aspect of the animal, without the need for anesthesia. The animal may be injected on the back or sides simply by grasping the tail as the animal stands on a grid (such as the cage top). For injections on the nape of the neck, the mouse may be placed on a flat surface and pressure applied to the back, as the skin of the neck is retracted with the thumb and forefinger. A 26 or 27G needle should be inserted at a shallow angle and wiggled to ensure that it is subcutaneous rather than intradermal. Intravenous administration via the tail vein is technically more difficult. When using this model to examine the variable impact of dosing interval on efficacy for different classes of antibiotics, a wide range of total doses and dosing intervals is necessary in order to minimize interdependent pharmacokinetic parameters. Dosing regimens may be selected by dividing 24 hour total doses into individual doses to be administered at 1- to 24-hourly intervals. We generally use 20–40 dosing regimens for each organism–drug combination, achieving a range of serum drug concentrations extending from less than to greater than that seen in humans, in order to describe the complete dose–response relationship from "no effect" to "maximum effect."

Key parameters to monitor response to treatment

Drug kinetics

Single-dose serum pharmacokinetic studies may be performed on the day of experimentation for the range of individual doses used. For each of three to nine typical individual doses examined, one or two groups of three or four mice may be sampled four or five times, each at 10–30 minute intervals over 4–5 half-lives. Blood may be obtained by retro-orbital puncture, collected into heparinized capillary tubes, and centrifuged for 10 minutes. Retro-orbital sinus puncture following induction of anesthesia by metofane or other agents may be performed by placing the heparinized pipette rostrally and gently rotating the pipette as it is inserted between the eye and the surrounding tissue. Blood flow stops upon removal of the pipette, and the area around the eye may be cleaned with saline or water. Plasma levels may be determined by agar-well microbiological or HPLC assay using standard media and methods (Klassen and Edberg, 1996). Standard samples are typically prepared in both buffer and pooled normal mouse serum. Pharmacokinetic constants (elimination rate constant, half-life and maximum concentration, apparent volume of distribution, area under the curve, and peak level) and steady-state time that serum levels remain above the MIC and the AUC above the MIC may be calculated (Vogelman *et al.*, 1988). Pharmacokinetic constants may be interpolated for values obtained in actual studies for doses that have no drug kinetics determined.

To determine the antibiotic tissue concentrations in lung, mice may be anesthetized (with metofane or other anesthetic agent) and perfused with 5–10 ml of phosphate-buffered saline until tissues are white in order to remove blood from the tissues. Exsanguinated tissue is then excised, homogenized, and centrifuged at 10 000 *g* for 5 minutes. The supernatant may then be collected and assayed as above. Standard samples may be prepared by adding known concentrations of drug to exsanguinated homogenates of uninfected tissues, incubating the samples for 1 hour or more at 37°C, centrifuging them at 10 000 *g* for 5 minutes, and collecting the supernatant. Tissue penetration may then be estimated from the ratio of area under the tissue concentration curve versus time to area under the serum concentration curve versus time, expressed as a percentage.

Statistical analysis

A sigmoid dose–effect model (E_{max}) may be used to evaluate the impact of dosing interval on antimicrobial activity. The model is described by the following equation:

$$E = (E_{max} \times D^n)/(P_{50}^n + D^n),$$

where E is the observed effect ($\Delta\log_{10}$ cfu/lung compared to controls at 24 hours), D is the cumulative 24 hour dose, E_{max} is a measure of relative activity indicated by the maximum antimicrobial effect attributable to the drug, P_{50} is a measure of potency indicated by the dose producing 50% of E_{max}, and n is a function describing a slope (Holford and Sheiner, 1981; Unadkat *et al.*, 1986).

A mean prediction error can be used to estimate bias and a root-mean-squared prediction error may be used to estimate the precision of the prediction (Sheiner and Beal, 1981). In order to allow meaningful comparison of efficacy among antimicrobial agents, the dose of each antibiotic required to achieve a fixed reduction (1 log cfu/lung at 24 hours or bacteriostatic dose, for example) may then be calculated. Linear regression analysis may be used to determine pharmacokinetic parameter–effect relations in the range of 20–80% of maximum efficacy (Holford and Sheiner, 1981; Vogelman *et al.*, 1988).

Pitfalls (advantages/disadvantages) of the model

The murine pneumonia model using aerosol inoculation may be used as a basic screening model, since up to 150 or more mice may be infected at the same time, and treatment begun either the same day or the following morning. Much like the murine peritonitis model (Knudsen *et al.*, 1995), it has the advantage of more closely resembling a common human correlate than the murine thigh model (Fantin *et al.*, 1991). Moreover, by incorporating measurement of drug pharmacokinetics and statistical analysis of maximal antimicrobial activity, the model may serve a more discriminative function in comparing antimicrobial agents or predicting optimal dosing strategies for new drugs (Zak and O'Reilly, 1990; Norby *et al.*, 1993; Wilson and Rayner, 1994). Host defenses (cyclophosphamide) or drug elimination (uranyl nitrate) are relatively easily manipulated.

Relative material disadvantages of this murine model include the need to obtain a Collison nebulizer and other equipment. In addition, an important bacterial pathogen, *S. pneumoniae*, is not suitable for aerosol inoculation because of its lability. Other perceived disadvantages include pharmacokinetic differences in drug elimination compared to humans (Mordenti, 1985; Mizen and Woodnutt, 1988), and the use of microbiological 24 hour endpoints rather than mortality. However, preliminary evidence suggests that these endpoints are correlated with survival (Leggett *et al.*, 1989b).

Contributions of the model to infectious disease therapy

This murine pneumonia model using aerosol inoculation has contributed to our understanding of the pharmacokinetic/pharmacodynamic principles leading to optimal antimicrobial dosing regimens in pneumonia. Studies using this model have shown that the use of infrequent dosing intervals that may be suboptimal for some of the drugs being investigated help to explain why some previous pneumonia model studies have found aminoglycosides to be more effective than β-lactams (McColm *et al.*, 1986; Trautmann *et al.*, 1986; Johnson *et al.*, 1987), while the opposite has been observed in a clinical trial (Smith *et al.*, 1984). Similarly, results using once-daily aminoglycoside dosing regimens in this model have been replicated in human clinical trials (Hatala *et al.*, 1996), and quinolone activity has to date shown remarkable similarity in human trials of pneumonia (Forrest *et al.*, 1993).

The FDA still requires preclinical animal testing (Beam *et al.*, 1992). Use of this type of model allows a more streamlined, focused approach to predict optimal dosing regimens of new drugs to later be tested in clinical trials, thereby reducing the time and expense involved in bringing new drugs to the market.

References

Bartlett, J. G., Mundy, L. M. (1995). Community-acquired pneumonia. *N. Engl. J. Med.*, **333**, 1618–1624.

Beam, T. R., Gilbert, D. N., Kunin, C. M. (1992). General guidelines for the clinical evaluation of anti-infective drug products. *Clin. Infect. Dis.*, **15** (Suppl. 1), S5–S32.

Conley, J., Yang, H., Wilson, T. *et al.* (1997). Aerosol delivery of liposome-encapsulated ciprofloxacin: aerosol characterization and efficacy against *Francisella tularensis* infection in mice. *Antimicrob. Agents Chemother.*, **41**, 1288–1292.

Craig, W. A. (1995). Interrelationship between pharmacokinetics and pharmacodynamics in determining dosage regimens for broad-spectrum cephalosporins. *Diagn. Microbiol. Infect. Dis.*, **22**, 89–96.

Craig, W. A., Redington, J., Ebert, S. C. (1991). Pharmacodynamics of amikacin in vitro and in mouse thigh and lung infections. *J. Antimicrob. Chemother.*, **27** (Suppl. C), 29–40.

Fantin, B., Leggett, J., Ebert, S., Craig, W. A. (1991). Correlation between in vitro and in vivo activity of antimicrobial agents against gram-negative bacilli in a murine infection model. *Antimicrob. Agents Chemother.*, **35**, 1413–1422.

Forrest, A., Nix, D. E., Ballow, C. H., Goss, T. F., Birmingham, M. C., Schentag, J. J. (1993). Pharmacodynamics of intravenous ciprofloxacin in seriously ill patients. *Antimicrob. Agents Chemother.*, **37**, 1073–1081.

Garibaldi, R. A. (1985). Epidemiology of community-acquired respiratory tract infections in adults: incidence, etiology, and impact. *Am. J. Med.*, **76**, 32S–37S.

Hashimoto, S., Wolfe, E., Guglielone, B. *et al.* (1996). Aerosolization of imipenem/cilastatin prevents *Pseudomonas*-induced acute lung injury. *J. Antimicrob. Chemother.*, **38**, 809–818.

Hatala, R., Dinh, T., Cook, D. J. (1996). Once-daily aminoglycoside dosing in immunocompetent adults: a meta-analysis. *Ann. Intern. Med.*, **124**, 717–725.

Holford, N. H. G., Sheiner, L. B. (1981). Understanding the dose–effect relationship: clinical application of pharmacokinetic-pharmacodynamic models. *Clin. Pharmacokinet.*, **6**, 429–453.

Johnson, E., Jaax, N., White, J., Jahrling, P. (1995). Lethal experimental infections of rhesus monkeys by aerosolized *Ebola* virus. *Int. J. Exp. Pathol.*, **76**, 227–236.

Johnson, M., Miniter, P., Andriole, V. T. (1987). Comparative efficacy of ciprofloxacin, azlocillin, and tobramycin alone and in combination in experimental *Pseudomonas* sepsis. *J. Infect. Dis.*, **155**, 783–788.

Kelly, B. P., Furney, S. K., Jessen, M. T., Orme, I. M. (1996). Low-dose aerosol infection model for testing drugs for efficacy against *Mycobacterium tuberculosis*. *Antimicrob. Agents Chemother.*, **40**, 2809–2812.

Klassen, M., Edberg, S. C. (1996). Measurement of antibiotics in human body fluids: techniques and significance. In: *Antibiotics in Laboratory Medicine* (ed. Lorian, V.), pp. 230–295. Williams & Wilkins, Baltimore, MD.

Knudsen, J. D., Frimodt-Møller, N., Espersen, F. (1995). Experimental *Streptococcus pneumoniae* infection in mice for studying correlation of *in vitro* and *in vivo* activities of penicillin against pneumococci with various susceptibilities to penicillin. *Antimicrob. Agents Chemother.*, **39**, 1253–1258.

Laurenzi, G. A., Berman, L., First, M., Kass, E. G. (1964). A quantitative study of the deposition and clearance of bacteria in the murine lung. *J. Clin. Invest.*, **43**, 759–768.

Leggett, J. E., Fantin, B., Ebert, S. *et al.*, (1989a). Comparative antibiotic dose–effect relations at several dosing intervals in murine pneumonitis and thigh-infection models. *J. Infect. Dis.*, **159**, 281–292.

Leggett, J. E., Ebert, S., Fantin, B., Craig, W. A. (1989b). A sigmoid dose–response model using bacterial counts predicts dose–survival results for *Klebsiella* pneumonia in neutropenic mice. In: *Program and Abstracts of the 29th Interscience Conference on Antimicrobial Agents and Chemotherapy*, Abstract 313, p. 153. American Society for Microbiology, Washington, DC.

Leggett, J., Kohlhepp, S., Swan, S., Ebert, S., Gilbert, D. (1990). Influence of azotemia on efficacy of infrequent gentamicin dosing regimens in experimental Klebsiella pneumonia. In: *Program and Abstracts of the 30th Interscience Conference on Antimicrobial Agents and Chemotherapy*, Abstract 153, p. 112. American Society for Microbiology, Washington, DC.

McColm, A. A., Shelley, E., Ryan, D. M., Acred, P. (1986). Evaluation of ceftazidime in experimental *Klebsiella pneumoniae* pneumonia: comparison with other antibiotics and measurement of its penetration into respiratory tissues and secretions. *J. Antimicrob. Chemother.*, **18**, 599–608.

Mizen, L., Woodnutt, G. (1988). A critique of animal pharmacokinetics. *J. Antimicrob. Chemother.*, **21**, 273–280.

Mordenti, J. (1985). Forecasting cephalosporin and monobactam antibiotic half-lives in humans from data collected in laboratory animals. *Antimicrob. Agents Chemother.*, **27**, 885–891.

National Center for Health Statistics (1993). *Vital Statistics of the United States, 1989, Vol. II: Mortality, Part A*. US Dept of Health and Human Services publication PHS 93–1101. Public Health Service, Washington, DC.

Nishi, T., Tsuchiya, K. (1980). Experimental respiratory tract infection with *Klebsiella pneumoniae* DT-S in mice: chemotherapy with kanamycin. *Antimicrob. Agents Chemother.*, **17**, 494–505.

Norby, S. R., O'Reilly, T., Zak, O. (1993). Efficacy of antimicrobial agent treatment in relation to treatment regimen: experimental models and clinical evaluation. *J. Antimicrob. Chemother.*, **31** (Suppl. D), 41–54.

O'Reilly, T., Cleeland, R., Squires, E. L. (1996). Evaluation of antimicrobials in experimental animal models. In: *Antibiotics in Laboratory Medicine* (ed. Lorian, V.), pp. 604–765. Williams & Wilkins, Baltimore, MD.

Pennington, J. E. (1985). Animal models of pneumonia for evaluation of antimicrobial therapy. *J. Antimicrob. Chemother.*, **16**, 1–4.

Phillpots, R. J., Brooks, T. J., Cox, C. S. (1997). A simple device for the exposure of animals to infectious microorganisms by the airborne route. *Epidemiol. Infect.*, **118**, 71–75.

Roosendaal, R., Bakker-Woudenberg, I. A. J. M., van den Berghe-van Raffe, J. C., Michel, M. F. (1986). Continuous versus intermittent administration of ceftazidime in experimental *Klebsiella pneumoniae* pneumonia in normal and leukopenic rats. *Antimicrob. Agents Chemother.*, **30**, 403–408.

Sheiner, L. B., Beal, S. L. (1981). Some suggestions for measuring predictive performance. *J. Pharmacokinet. Biopharm.*, **9**, 503–512.

Smith, C. R., Ambinder, R., Lipsky, J. J. *et al.*, (1984). Cefotaxime compared with nafcillin plus tobramycin for serious bacterial infections. *Ann. Intern. Med.*, **101**, 469–477.

Trautmann, M., Bruckner, O., Marre, R., Hahn, H. (1986). Comparative efficacy of different β-lactam antibiotics and gentamicin in *Klebsiella pneumoniae* septicaemia in neutropenic mice. *J. Microb. Chemother.*, **18**, 287–291.

Unadkat, J. D., Bartha, F., Sheiner, L. B. (1986). Simultaneous modeling of pharmacokinetics and pharmacodynamics with nonparametric kinetic and dynamic models. *Clin. Pharmacol. Ther.*, **40**, 86–93.

Vanrompay, D., Mast, J., Ducatelle, R., Haesebrouck, F., Goddeeris, B. (1995). *Chlamydia psittaci* in turkeys: pathogenesis of infections in avian servovars A, B and D. *Vet. Med.*, **47**, 245–256.

Vogelman, B., Gudmundsson, S., Leggett, J., Turnidge, J., Ebert, S., Craig, W. A. (1988). Correlation of antimicrobial pharmacokinetic parameters with therapeutic efficacy in an animal model. *J. Infect. Dis.*, **158**, 831–847.

Wilson, R., Rayner, C. (1994). Animal models of respiratory infection. *J. Antimicrob. Chemother.*, **33**, 381–386.

Zak, O., O'Reilly, T. (1990). Animal models as predictors of the safety and efficacy of antibiotics. *Eur. J. Clin. Microbiol. Infect. Dis.*, **91**, 472–478.

Chapter 64

Experimental Models of Infectious Arthritis

T. Bremell

Human bacterial arthritis

Bacterial arthritis denotes arthritis in which live bacteria are found within the joint structures. It is typically a rapidly progressive and highly destructive joint disease in humans. Symptoms of infection include a swollen, warm, often red joint in a patient with fever and general malaise. In the pre-antibiotic era, bacterial arthritis was not infrequently fatal (Heberling, 1941). Even today, in patients with rheumatoid arthritis, a fatal outcome is noted in 17–25% of the patients with concomitant septic arthritis (Goldenberg, 1989b; Ho, 1989; Kaandorp *et al.*, 1997). Early diagnosis, within a few days after the onset of symptoms, is clinically important for prognosis (Goldenberg and Reed, 1985; Ward *et al.*, 1960). In long-standing bacterial arthritis, joint destruction often continues even when the live bacterium is eradicated, indicating that the host inflammatory and immune responses participate in the process of joint erosion.

Risk factors for bacterial arthritis include: chronic systemic disorders, especially neoplasms, chronic liver diseases and diabetes mellitus; intravenous drug use; indwelling intravenous catheters or chronic haemodialysis; destructive joint diseases; joint implants; and immunosuppressive treatment (Mathews *et al.*, 1980; Goldenberg and Reed, 1985; Goldenberg 1989a; Brancos *et al.*, 1991).

Staphylococcus aureus is the dominating pathogen in human bacterial arthritis, except for children below the age of 2, where *Haemophilus influenzae* prevails (Goldenberg, 1989a).

Animal models of infectious arthritis

Well-defined human studies on infectious arthritis are generally not feasible because the exact onset of infection is often uncertain and adequate tissue samples are not readily available. Studies of infectious arthritis in experimental models overcome these limitations. In the majority of animal studies rabbits have been used to study joint infection because of their susceptibility to infection (Mahowald, 1986). However, dogs (Curtiss and Klein, 1965; Drutz, 1978), pigs (Pedersen *et al.*, 1981), rats (Nakamura *et al.*, 1991), mice (Friedlander *et al.*, 1951; Tissi *et al.*, 1990) and golden hamsters (Lindberg, 1969) have also been used. In general, arthritis has been induced by intra-articular bacterial inoculation. Intradermal, subcutaneous and intranasal administration of bacteria (Pedersen *et al.*, 1981), as well as intra-arterial inoculations and bone marrow injections, have occasionally been used (Wang *et al.*, 1983).

S. aureus is by far the most common bacterium used in animal studies but *S. albus, Escherichia coli, Neisseria gonorrhoeae*, group A streptococcus and type I *Streptococcus pneumoniae* have also been employed (Mahowald, 1986). Arthritis has also been induced by group B streptococci (Tissi *et al.*, 1990), RS streptococci (Pedersen *et al.*, 1981) and *Candida albicans* (Nakamura *et al.*, 1991).

Despite that fact the overwhelming majority of cases of human infectious arthritis have a haematogenic spread, relatively few studies have been performed on bacteraemia-induced septic arthritis (Table 64.1).

Intravenous inoculation of alpha-type streptococci, group B streptococci and *C. albicans* has previously been performed in mice and rats (Friedlander *et al.*, 1951; Tissi *et al.*, 1990; Nakamura *et al.*, 1992).

After a chance observation by a laboratory technician of an outbreak of *S. aureus* arthritis in different mouse strains (Bremell *et al.*, 1990) the animal model of septic *S. aureus* arthritis was elaborated in mice and rats (Bremell *et al.*, 1991, 1994a; Figure 64.1). The *S. aureus* models of septicaemia and arthritis in mice and rats were described with regard to clinical, serological, histopathological and immunohistochemical aspects of the disease (Bremell, 1991, 1992, 1994a). The i.v. administration of *S. aureus* permits the study of important virulence mechanisms of *S. aureus* such as factors influencing bacterial survival in the blood stream before the bacteria reach the joint and factors influencing bacterial ability to penetrate endothelium, synovium and bone and to interact with these tissues. One advantage of using rodents, apart from easy care, housing and accessibility, is the tremendous number of inbred and manipulated strains available, making the model suitable for studies of the interplay between host and bacterium. For a more detailed background review on host–bacterium relationship see Sheagren, 1988; Höök *et al.*, 1990; Sáez-Llorens *et al.*, 1990; Waldvogel, 1990; Dinarello, 1991; Sheagren and Schaberg, 1992; Gotschlich, 1993; Janeway, 1993; Paul, 1993; Schlievert, 1993.

Handbook of Animal Models of Infection
ISBN 0-12-775390-7

Table 64.1 Experimental models of haematogenously induced infectious arthritis

Infectious agent	*Animal species*	*Outcome*	*Reference*
Staphylococcus aureus	Rabbit	Arthritis in 100% of surviving rabbits	Kistler, 1936
	Rabbit	18/22 osteomyelitis 1/22 arthritis	Thompson and Dubos, 1938
	Rabbit	All died within 1–60 days, some with arthritis	Rigdon, 1942
	Rabbit	Arthritis; % not reported	Lewis and Cluff, 1965
	Rat	Arthritis in 90–100% Death in 0–50%	Bremell *et al.*, 1994a
	Mouse	Arthritis in 80–100% Death in 3%	Bremell *et al.*, 1991
Streptococci	Rabbit	Arthritis; % not reported	Lewis and Cluff, 1965
	Rat	Arthritis in 13–100% Death in 36%	Friedlander *et al.*, 1951
	Mouse	Arthritis in 41–78% Death in 35%	Friedlander *et al.*, 1951
	Mouse	Arthritis in 80%	Tissi *et al.*, 1990
Pneumococci	Rabbit	Arthritis; % not reported	Lewis and Cluff, 1965
Neisseria gonorrhoeae	Dog	Arthritis in 20%	Drutz, 1978
Candida albicans	Rat	Arthritis in >90%	Nakamura *et al.*, 1991

Animal species

Several strains of mouse and rat have been used in this model.

Mice: MRL 1pr/1pr, NZW, NZB, NZB/W, NZB/W/B, DBA/1, BALB/c, C3H/Hej, C3H/HeN, C57BL/6 and Swiss (NMRI) mice. Different inbred strains can also be used (Zhao *et al.*, 1995a,b; Zhao and Tarkowski, 1995; Abdelnour *et al.*, 1997).

Since Swiss mice displayed the highest occurrence of arthritis, when compared to the other mouse strains (Bremell *et al.*, 1991) this mouse strain has been often used.

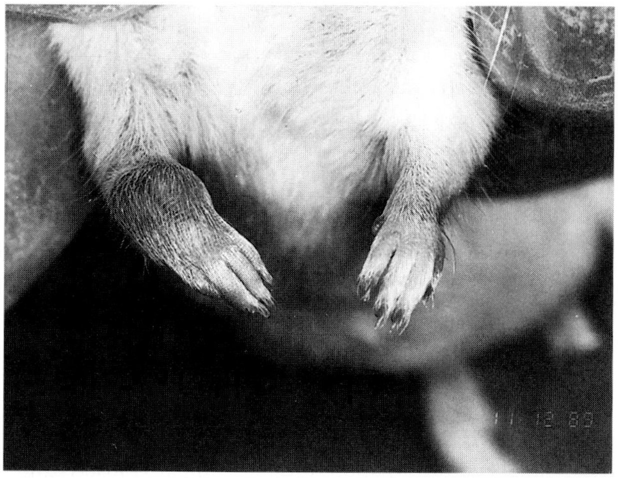

Figure 64.1 Photograph of a Swiss (NMRI) mouse showing arthritis of the right wrist 28 days after i.v. injection with 1×10^7 *S. aureus* LS-1. Note the non-arthritic left wrist! Reproduced with permission from Bremell *et al.* (1991). Experimental *Staphylococcus aureus* arthritis in mice. *Infect. Immun.*, **59**, Figure 3.

Rats: Sprague–Dawley rats have mostly been used (Bremell *et al.*, 1994a; Bremell and Tarkowski, 1995).

Presumably, any mouse or rat strain could be used but preliminary experiments should be performed to determine the optimal bacterial inoculum to ensure an adequate rate of infection. Sex and sex hormones did not affect the incidence or clinical course of *S. aureus*-induced septic arthritis (Abdelnour, 1994).

The age of the animals did not seem to affect the prevalence and course of arthritis. However, in general, at the start of experiments we used 4–6-week-old mice and 6–9-week-old rats.

Preparation of animals

No specialized animal housing or specific pretreatment are required when testing different bacterial virulence factors. However, when examining host factors of importance for the infection process, specialized housing and care (Wiedermann *et al.*, 1996) or specific pretreatment (Bremell *et al.*, 1994, and others) is required.

Details of bacterial inoculation

Briefly, the mice are inoculated with bacteria in the tail vein without any need for anaesthesia. During this procedure they are held in a mini-cage with the tail outside the cage. Blood samples during the experiments are obtained from the tail vein or orbital vein under light ether inhalation or barbitural anaesthesia, or at sacrifice from the aortic branch.

Rats are injected in the tail vein under anaesthesia (light ether inhalation or mebumal 60 mg/kg i.p.). Blood samples are obtained under anaesthesia from the tail vein or by cutting the distal end of the tail.

Bacterial samples from joints are obtained at sacrifice after skin disinfection and surgical incision with a pair of scissors or a scalpel. The samples are obtained using charcoal sticks and isolated on horse blood agar plates for 48 hours. The bacterial samples are easily obtained from the talocrural and ankle joints on one side, while the limbs from the contralateral side may be used for histopathology or immunohistochemistry (see below).

Storage and preparation of inocula

The model has been used for studying host and bacterial virulence factors in *S. aureus* arthritis, but it is amenable to using different species of bacteria. We have normally used *S. aureus* strains LS-1 and AB-1 (Bremell *et al.*, 1991, 1994a). Other strains may also be used (Patti *et al.*, 1994; Nilsson *et al.*, 1996). Using this model, we have not found an *S. aureus* strain unable to induce arthritis. However, the frequency and severity of arthritis depends on different properties of *S. aureus* strains as well as host factors.

S. aureus strains may be characterized with regard to reactivity to catalase and coagulase, type of capsular polysaccharide (Karakawa *et al.*, 1985), phage type (Williams and Rippon, 1952; Parker, 1983), binding ability to extracellular matrix proteins (Patti *et al.*, 1994; Abdelnour *et al.*, 1993), toxin production, and pattern of antibiotic resistance (Bremell *et al.*, 1991, 1994a).

The bacterial dose normally used is $1–2 \times 10^7$ cfu per mouse and 1×10^9 cfu per rat. The bacterium is stored in deep agar tubes, cultured on blood agar for 24 hours, reincubated on blood agar for another 24 hours and then kept frozen at –20°C in PBS containing 5% bovine serum albumin and 10% dimethylsulphoxide (DMSO) until use (Bremell *et al.*, 1994a). The storage of *S. aureus* at –20°C does not affect the viability or virulence of the bacteria (unpublished results). Culturing on solid agar is superior to culturing in broth with regard to expression of bacterial cell wall proteins (Cheung and Fischetti, 1988) and capsular polysaccharides (Lee *et al.*, 1993) — important bacterial virulence factors.

The frozen bacterial solution is thawed and washed in PBS. Viable counts are used to check the number of bacteria in each administered bacterial solution.

Infection process

Infection is initiated by an intravenous (i.v.) injection of 0.2 ml of an *S. aureus* suspension in PBS (phosphate-buffered saline) in the tail vein of the mouse (Bremell *et al.*, 1991) or 1 ml of *S. aureus* suspension in the tail vein of the rat (Bremell *et al.*, 1994a).

Intraperitoneal (i.p.) administration does not cause arthritis or osteitis at all, whereas subcutaneous (s.c.) injections give rise to arthritis in only a small proportion of the mice. Chequer-board titration showed that the optimal i.v. dose of *S. aureus* LS-1 was found to be 1×10^7 cfu/mouse. Lower doses failed to induce arthritis, whereas higher doses increased death rates. Other *S. aureus* strains may display different virulence and accordingly require different doses (Bremell *et al.*, 1991).

In rats, i.v. doses exceeding 1×10^9 cfu/rat of *S. aureus* AB-1 induce high mortality rates. Doses below 1×10^8 cfu/rat do not readily induce clinical arthritis (Bremell *et al.*, 1994a,b).

Key parameters to monitor response to treatment

Clinical evaluation

Mice and rats are labelled and monitored individually. Limbs are inspected visually at regular intervals, preferably by two blinded observers. Arthritis is defined as visible joint swelling or erythema of at least one joint. The overall condition was evaluated by assessment of weight, general appearance, alertness, and skin abnormalities. A clinical evaluation of arthritis was carried out using a system in which macroscopic evaluation of arthritis yields a score of 0–3 for each limb (0, normal appearance; 1, mild swelling and/or erythema; 2 moderate swelling and erythema; 3, marked swelling and erythema). An arthritic index was constructed for each animal by adding the scores from all four limbs for each animal. The arthritic index makes it possible to judge not only the prevalence but also the intensity of arthritis. The animals are examined daily during the first 3 days after inoculation and thereafter one to three times per week (Figure 64.2).

Histopathological examination

The histopathological processing includes routine fixation, decalcification, paraffin embedding, cutting and staining with haematoxylin and eosin. Tissue sections from upper extremities (elbow, wrist, carpal bones, fingers and occasionally shoulder) and lower extremities (knee, ankle, tarsal bones, and toes) are prepared. A segment of the tail may also permit the study of osteomyelitic lesions (Bremell *et al.*, 1992). The joints are evaluated by a blinded observer with regard to synovial hypertrophy (defined as a synovial membrane thickness of more than two cell layers (Goldenberg and Cohen, 1978)), pannus formation (synovial tissue overlying joint cartilage), and cartilage and subchondral bone destruction (Figure 64.4). In addition,

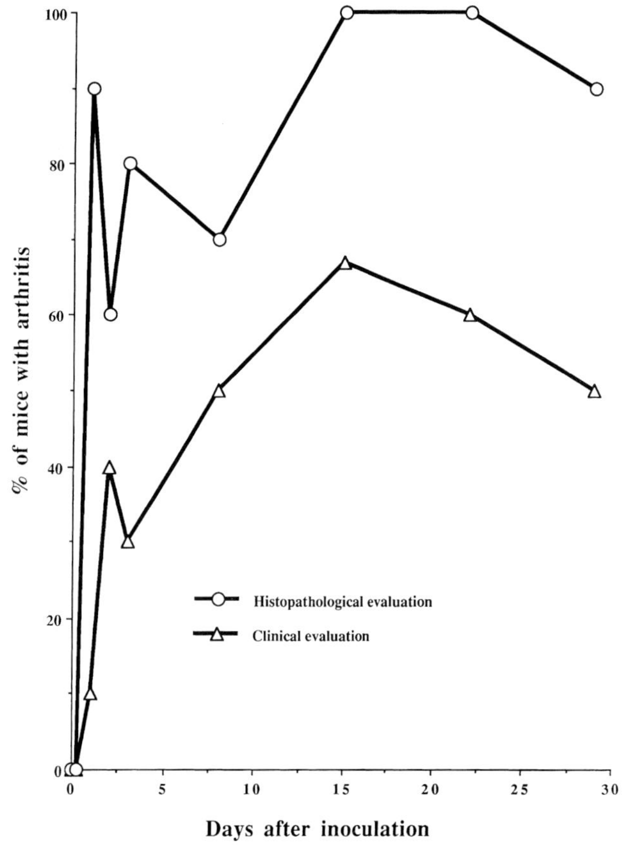

Figure 64.2 Prevalence of arthritis at sacrifice after a single i.v. injection of 1×10^7 *S. aureus* LS-1. Histopathological and clinical observations parallel each other, the former being more sensitive. Peak of arthritis is seen within 15–22 days. Reproduced with permission from Bremell *et al.* (1992). Histopathological and serological progression of experimental *Staphylococcus aureus* arthritis. *Infect. Immun.*, **60**, Figure 1.

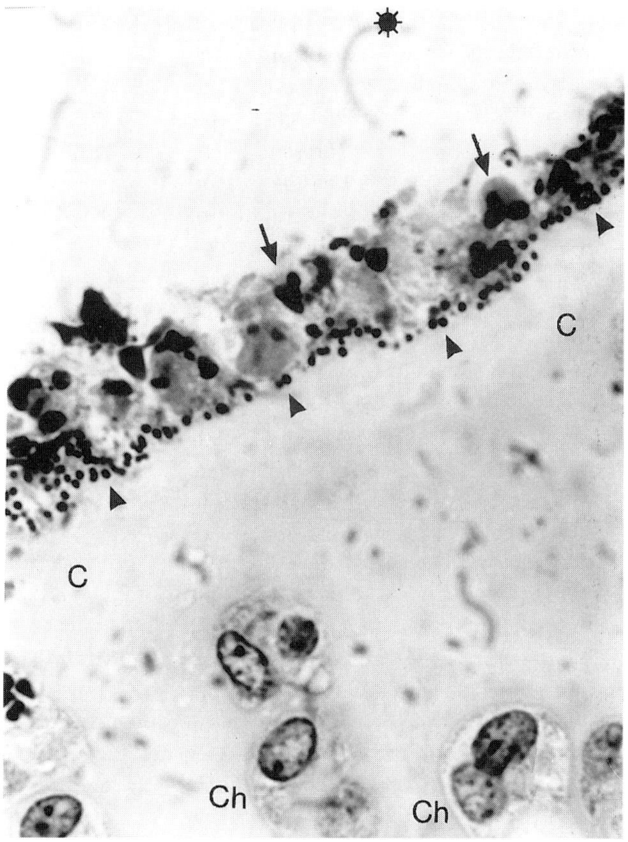

Figure 64.3 This picture gives an image of the local interactions between the host and the bacterium. Micrograph of an arthritic ankle in a male, 5-month-old Sprague–Dawley rat spontaneously infected with *S. aureus* AB-I. The bacterium displays strong binding to collagen type II. The bacterial particles (arrowheads) adhere to the joint cartilage (C). Inflammatory cells, mainly polymorphonuclear cells (arrows) engulf the bacteria. ✷ Represents the joint cavity. Magnification ×320. Reproduced with permission from Bremell *et al.* (1994a). Immunopathological features of rat *Staphylococcus aureus* arthritis. *Infect. Immun.*, **62**, Figure 1.

infiltration of inflammatory cells to the extra-articular space and types of invading cell can be evaluated (Figures 64.3 and 64.4).

Immunohistochemistry

After sacrifice, limbs are removed and demineralized by EDTA-treatment (Jonsson *et al.*, 1986). The demineralized specimens are mounted on cryostat chucks, frozen in isopentane prechilled by liquid nitrogen, and kept at –70°C until cryosectioned. Sections 6 μm thick are cut sagittally, fixed in cold acetone for 5 minutes and washed in PBS. The sections are incubated overnight in a humid atmosphere at 4°C with 50 μl portions of unlabelled mouse monoclonal antibodies or rat monoclonal antibodies diluted in PBS-BSA (1%). Biotin-labelled rabbit anti-Ig diluted in PBS-BSA is used as a secondary antibody. Depletion of endogenous peroxidase is performed with 0.3% H_2O_2 for 5 minutes. Binding of secondary antibodies is detected by stepwise incubation with avidin–biotin–peroxidase complexes (ABC; Hsu *et al.*, 1981) and a buffer containing 3-amino-9-ethyl-carbazole and H_2O_2 (Kaplow, 1975). All sections are counterstained with Mayer's haematoxylin. The proportion of a given cell type may be expressed as the percentage of stained cells in relation to all nucleated cells within a defined area of tissue.

Bacteriologic evaluation

After sacrifice the talocrural and radiocarpal joints are aseptically dissected. In the rat the knee and elbow joints may also be used for bacterial examination. Bacterial samples from other organs such as the liver, spleen, kidneys, heart, and lungs as well as blood samples from the heart cavity are also easily accessible. Bacterial samples from the joints are obtained using charcoal sticks. The samples are transferred to agar containing 5% human blood and incubated for

48 hours at 37°C. The colony appearance is monitored and bacterial colonies are tested for catalase and coagulase activity. To avoid false-positive results due to contamination, a joint or blood sample was considered positive if more than 20 colonies of *S.aureus* were present (Bremell *et al.*, 1991).

The bacteriologic examination performed as described above may underestimate bacterial growth because the method employed lacks a high degree of sensitivity and precision. However, empirical experience demonstrated that there was a good correlation ($p < 0.001$) between clinical signs of arthritis and bacterial growth in affected joints in a given animal population. In order to obtain more precise results, cultures of whole homogenized organs would have been preferred. This has been performed with parenchymatous organs like heart, lung, kidney and liver. These organs are aseptically removed and homogenized for 20 minutes in a Colworth Stomacher 80 homogenizer (A.J. Seward, London, UK) in a tissue homgenizer with 10 ml of PBS. Appropriate dilutions were made and 0.1 ml of tissue suspension was plated on agar plates containing 5% horse blood. After incubation for 48 hours, colonies were counted and the results were expressed as the number of cfu per organ.

Bacterial binding to radiolabelled proteins, such as bone sialoprotein, collagen type II, fibronectin, vitronectin, fibrinogen, trombospondin and others, may be performed (Holderbaum *et al.*, 1985; Ryden *et al.*, 1989; Switalski *et al.*, 1993; Figure 64.4).

The production of staphylococcal enterotoxin A, B, C, D, and TSST-1 can be tested using an RPLA diagnostic kit (Oxoid, Basingstoke, UK).

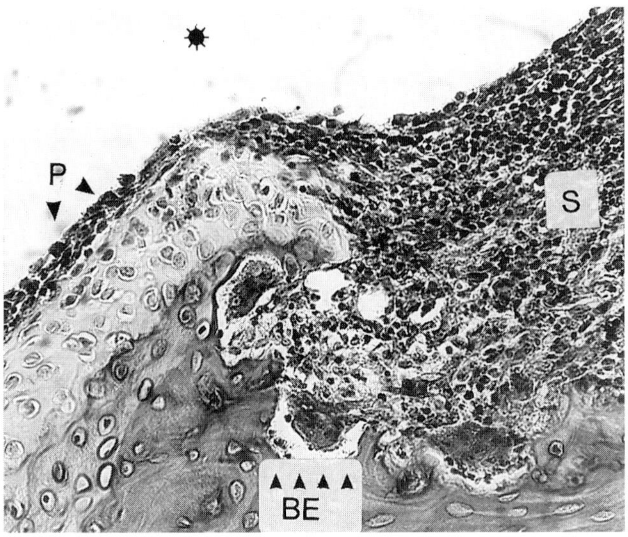

Figure 64.4 Micrograph showing an arthritic ankle joint in a male, 6–9-week-old Sprague–Dawley rat, sacrificed 11 days after bacterial inoculation. There is synovial hypertrophy (S), pannus formation (P), bone erosion (BE) at the cartilage–synovial junction and infiltration of PMNCs and MNCs. Magnification ×64. Reproduced with permission from Bremell *et al.* (1994a). Immunopathological features of rat *Staphylococcus aureus* arthritis. *Infect. Immun.*, **62**, Figure 5A.

Evaluation of host response

In order to assess causative as well as protective mechanisms in response to the bacterial infection there is a need for careful analysis of the immunological and inflammatory pattern in the diseased subjects.

A good marker of the degree of inflammation is measurement of serum IL-6 levels using the B9 bioassay (Lansdorp *et al.*, 1986; Aarden *et al.*, 1987; Brakenhoff *et al.*, 1987; Helle *et al.*, 1988). Also, serum immunoglobulin levels, measured by, for instance, the radial immunodiffusion technique (Mancini *et al.*, 1965), will significantly increase during the *S. aureus* infection (Bremell *et al.*, 1992).

The serum humoral immune response to the septicaemia and *S. aureus* arthritis can be monitored by measuring rheumatoid factors by diffusion-in-gel ELISA (DIG-ELISA) (Elwing and Nygren, 1979) as previously described (Tarkowski *et al.*, 1984) and anti-ssDNA antibodies by ELISA-technique (Carlsten *et al.*, 1990).

In order to test the specific humoral immune response to the whole bacterium we have used an ELISA procedure in which wells were precoated with poly-L-lysine followed by incubation with 100 μl of 10^8/ml of whole, formalin-treated (4%, 20 minutes) *S. aureus* LS-1 cells as an antigen. Serum levels of IgG antibodies to TSST-1 were estimated by ELISA procedure using 0.5 μg/ml of highly purified TSST-1 (Toxin Technology, Sarasota, FL) as a solid-phase coating.

Analyses of white blood cells, platelets and differential counts are performed in a Sysmex analyser (K-1000, Toa Medical Electronics, Japan). Differential counts are also performed on Giemsa-stained smears.

In-vivo cell mediated inflammatory responses may be measured.

(a) Induction and registration of delayed-type hypersensitivity (DTH)

Rats are sensitized by epicutaneous application of 450 μl of a mixture of absolute ethanol and acetone (3 : 1) containing 15% 4-ethoxymethylene-2-phenyloxazolone (OXA; BDH Chemicals, Poole, UK) on the shaved abdomen skin. Eleven days after sensitization all rats are challenged by topical application of 40 μl 10% OXA dissolved in olive oil on both sides of the left ear. The right ear is exposed to olive oil only (Carlsten *et al.*, 1986).

(b) Induction and registration of T-cell-independent inflammation

T-cell-independent inflammation is induced by intradermal injection of 100 μl olive oil (Apoteksbolaget AB, Gothenburg, Sweden) in the left hind footpad as previously described (Josefsson *et al.*, 1993). The same volume of PBS is administered to the right hind footpad.

Measurement: The thickness of ears and footpads is measured before and 24 hours after challenge using an Oditest spring caliper (Kröplin, Hessen, Germany) as previously described (Carlsten *et al.*, 1986). All challenges and measurements were performed under light-ether anaesthesia. The intensity of the reactions is expressed as increase of thickness times 10^{-3} cm.

In DTH responses, antigen-sensitized $CD4^+$ T cells extravasate at the site of antigen challenge, where they recruit a variety of other cell types, notably macrophages, with the subsequent release of inflammatory mediators and the development of oedema.

The T-cell-independent inflammation is primarily a test of PMNC migration and extravasation as shown by histopathological examination of footpad swelling 24 hours after s.c. administration of olive-oil-induced inflammation. *In-vivo* depletion of T cells does not affect the magnitude of the inflammatory response to olive oil (Josefsson *et al.*, 1992).

Therapy and immunization procedures

Experimental treatment of *S. aureus* arthritis with cloxacillin, cloxacillin+glucocorticoids and glucocorticoids alone favours the combined use of systemic corticosteroid administration along with antibiotic therapy with regard to the course and outcome of *S. aureus* arthritis (Sakiniene *et al.*, 1996). The results imply that the inflammatory response actively contributes to the joint deterioration in *S. aureus* arthritis. In this respect, the current use of corticosteroids to diminish sequelae in the treatment of bacterial meningitis should be mentioned (Sáez-Llorens *et al.*, 1990).

Different forms of anti-T-cell therapy, i.e. treatment with monoclonal antibodies to CD4 and the TCR, seem to reduce arthritis and mortality in severe *S. aureus* infection (Abdelnour *et al.*, 1994a,c; Bremell *et al.*, 1994a).

Immunization with genetically modified enterotoxins, devoid of superantigen effects (Nilsson, personal communication), or with a recombinant fragment of collagen adhesin (Nilsson *et al.*, 1998), may have protective properties and it is possible to test this using the animal model.

Pitfalls (advantages/disadvantages) of the model

The i.v. administration of *S. aureus* is in accordance with the most common transmission of human *S. aureus* arthritis and permits the study of important virulence mechanisms of *S. aureus* such as factors influencing bacterial survival in the blood stream before the bacteria reach the joint and factors influencing bacterial ability to penetrate endothelium, synovium and bone and to interact with these tissues. Therapeutic approaches, e.g. antimicrobial medication, vaccination and treatment with factors influencing the immune system, are easily performed. We have used treatment with monoclonal antibodies to CD4, CD8 and different Vβ regions of the TCR (Abdelnour *et al.*, 1994a,c), to the TCR (Bremell *et al.*, 1994a) and to CD43 (Bremell *et al.*, 1994b). Moreover, therapy with monoclonal antibodies to ICAM-1 (Verdrengh *et al.*, 1996) and cytokines (Zhao and Tarkowski, 1995) has been used. Also, inbred animals, e.g. B-cell-deficient xid-mice (Zhao *et al.*, 1995a) and MHC-class-II-deficient mice have been used (Abdelnour *et al.*, 1997). All these studies increase our knowledge of host responses to *S. aureus* infection.

As possible drawbacks of this model, as with all animal models, one may mention the small but obvious immunological discrepancies between mouse and man and some uncertainty as to whether the doses of bacteria used in rodents are equivalent to the bacterial load in human *S. aureus* arthritis.

Contributions of the model to infectious disease therapy

Antimicrobial treatment

Experimental treatment of *S. aureus* arthritis with antibiotics alone or in combination with different immune-modulating agents is readily performed using this model. The model offers a human-like disease where it is possible to follow the infection and the subsequent therapeutic measures from the start. Also, alternative antimicrobial therapies aiming at the neutralization of bacterial virulence factors such as the capsular polysaccharide and different adhesion molecules of the bacterial cell wall, as well as antitoxin therapies, may be tried.

Immune therapy and elucidation of the pathophysiology of *S. aureus* arthritis

This model has been used extensively for studies of host responses to the invading pathogen. It has been shown that vitamin A deficiency predisposes to *S. aureus* arthritis (Wiedermann *et al.*, 1996). The participation of T lymphocytes in the arthritic process is indicated by their appearance in the diseased synovium and by the milder course of arthritis following T-cell-targeted interventions (Abdelnour *et al.*, 1994a; Bremell *et al.*, 1994a). B cells also contribute to the pathogenesis of septic arthritis (Zhao *et al.*, 1995a), probably by efficient superantigen presentation.

Sialophorin (CD43)-expressing cells (Bremell *et al.*, 1994b), complement (Sakiniene *et al.*, 1999) and neutrophils (Verdrengh and Tarkowski, 1997) play a protective role in early *S. aureus* infection. MHC class II expression seems to be necessary for the development of *S. aureus* arthritis.

Moreover, different MHC class II haplotypes confer varying susceptibility to joint inflammation induced by staphylococci (Abdelnour *et al.*, 1997).

Bacterial virulence determinants include exotoxins, enzymes and different cell-wall-associated proteins. Genes that code for the expression of exoproteins and cell-wall proteins are controlled by different regulatory loci such as agr (accessory gene regulator) and sar (staphylococcus accessory regulator). These loci are important virulence determinants in the induction and progression of septic arthritis (Abdelnour *et al.*, 1993; Nilsson *et al.*, 1996).

There is a preferential induction of septic arthritis and mortality by superantigen-producing staphylococci (Bremell *et al.*, 1995). Toxic shock syndrome toxin 1 (TSST-1) contributes to the arthritogenicity of *S. aureus* (Abdelnour *et al.*, 1994b). Importantly, genetically modified enterotoxins, devoid of superantigen effects, protect against *S. aureus* arthritis (Nilsson *et al.*, unpublished results).

Although non-specific forces, such as hydrophobicity, seem to play some role, adhesion of bacteria is predominantly dependent upon specific interactions between bacterial surface receptors (adhesins) and ligands of the host tissue. In this regard, the collagen-adhesin-positive *S. aureus* strain Phillips more frequently induced arthritis, especially erosive arthritis, than its collagen-adhesin-negative mutant PH 100 (Patti *et al.*, 1994). Additionally, vaccination with a recombinant fragment of collagen adhesin provides protection against *S.-aureus*-mediated septic death (Nilsson *et al.*, 1998). For a detailed review concerning the possibilities of vaccination against *S. aureus* infection see Foster, 1991.

Another cell-wall factor of importance is the staphylococcal microcapsule expression (Fattom *et al.*, 1990). Expression of capsular polysaccharide (CP) 5 is a virulence determinant in *S. aureus* arthritis and septicaemia, probably by downregulatory properties of CP on the ingestion and intracellular killing capacity of phagocytes (Nilsson *et al.*, 1997).

In conclusion, the mouse and rat animal models of *S. aureus* arthritis have an obvious resemblance to human disease. Using these models we are able to study the pathogenetic features of septic arthritis as well as therapeutic and prophylactic measures.

References

Aarden, L. A., de Groot, E. R., Shaap, O. L., Lansdorp, P. M. (1987). Production of hybridoma growth factor by human monocytes. *Eur. J. Immunol.*, **17**, 1411–1416.

Abdelnour, A. (1994). Immunopathogenesis of experimental *Staphylococcus aureus* arthritis. Thesis, University of Göteborg, Sweden.

Abdelnour, A., Arvidsson, S., Bremell, T., Rydén, C., Tarkowski, A. (1993). The accessory gene regulator (agr) controls *Staphylococcus aureus* arthritis virulence in a murine arthritis model. *Infect. Immun.*, **61**, 3879–3885.

Abdelnour, A., Bremell, T., Holmdahl, R., Tarkowski, A. (1994a). Clonal expansion of T lymphocytes causes development of arthritis and mortality in mice infected with TSST-1-producing staphylococci. *Eur. J. Immunol.*, **24**, 1161–1166.

Abdelnour, A., Bremell, T., Tarkowski, A. (1994b). Toxic shock syndrome toxin-1 contributes to the arthritogenicity of *Staphylococcus aureus*. *J. Infect. Dis.*, **174**, 94–99.

Abdelnour, A., Bremell, T., Holmdahl, R., Tarkowski, A. (1994c). Role of T-lymphocytes in experimental *Staphylococcus aureus* arthritis. *Scand. J. Immunol.*, **39**, 403–408.

Abdelnour, A., Zhao, Y.-X., Holmdahl, R., Tarkowski, A. (1997). Major histocompatability complex class II region confers susceptibility to *Staphylococcus aureus* arthritis. *Scand. J. Immunol.*, **46**, 301–307.

Brakenhoff, J. P. J., de Groot, E. R., Evers, R. F., Pannekoek, H., Aarden, L. A. (1987). Molecular cloning and expression of hybridoma growth factor in *Escherichia coli*. *J. Immunol.*, **139**, 4116–4121.

Brancos, M. R., Peris, P., Miro, J. M. *et al.* (1991). Septic arthritis in heroin addicts. *Semin. Arthitis Rheum.*, **21**, 81–87.

Bremell, T., Tarkowski, A. (1995). Preferential induction of septic arthritis and mortality by superantigen-producing staphylococci. *Infect. Immun.*, **63**, 4185–4187.

Bremell, T., Lange, S., Svensson, L. *et al.* (1990). Outbreak of spontaneous staphylococcal arthritis and osteitis in mice. *Arthritis Rheum.*, **33**, 1739–1744.

Bremell, T., Lange, S., Yacoub, A., Rydén, C., Tarkowski, A. (1991). Experimental *Staphylococcus aureus* arthritis in mice. *Infect. Immun.*, **59**, 2615–2623.

Bremell, T., Abdelnour, A., Tarkowski, A. (1992). Histopathological and serological progression of experimental *Staphylococcus aureus* arthritis. *Infect. Immun.*, **60**, 2976–2985.

Bremell, T., Lange, S., Holmdahl, R., Rydén, C., Hansson, G. K., Tarkowski, A. (1994a). Immunopathological features of rat *Staphylococcus aureus* arthritis. *Infect. Immun.*, **62**, 2334–2344.

Bremell, T., Holmdahl, R., Tarkowski, A. (1994b). Protective role of sialophorin (CD43) expressing cells in experimental *Staphylococcus aureus* infection. *Infect. Immun.*, **62**, 4637–4640.

Carlsten, H., Nilsson, L.-Å., Tarkowski, A. (1986). Impaired cutaneous delayed-type hypersensitivity in autoimmune MRL lpr/lpr mice. *Int. Arch. Allergy Appl. Immun.*, **81**, 322–325.

Carlsten, H., Tarkowski, A., Jonsson, R., Nilsson. L.-Å. (1990). Expression of heterozygous *lpr* gene in MRL mice. II. Acceleration of glomerulonephritis, sialoadenitis and autoantibody production. *Scand. J. Immunol.*, **32**, 21–28.

Cheung, A. L., Fischetti, V. (1988). Variation in the expression of cell wall proteins of *Staphylococcus aureus* grown on solid and liquid media. *Infect. Immun.*, **56**, 1061–1065.

Curtiss, P. H., Klein. L. (1965). Destruction of articular cartilage in septic arthritis. II. In vivo studies. *J. Bone Joint Surg.*, **47A**, 1595–1604.

Dinarello, C. (1991). The proinflammatory cytokines interleukin-1 and tumor necrosis factor and treatment of the septic shock syndrome. *J. Infect. Dis.*, **163**, 1177–1183.

Drutz, D. (1978). Hematogenous gonococcal infections in rabbits and dogs. In: *Immunobiology of* Neisseria Gonorrhoeae (ed. Brooks, B. F.), pp. 307–313. American Society for Microbiology, Washington, DC.

Elwing, H., Nygren, H. (1979). Diffusion-in-gel enzyme-linked immunosorbent assay (DIG-ELISA): a simple method for quantification of class-specific antibodies. *J. Immunol. Meth.*, **31**, 101–107.

Fattom, A., Schneerson, R., Szu, S. C. *et al.* (1990). Synthesis and immunologic properties in mice of vaccines composed of *Staphylococcus aureus* type 5 and type 8 capsular polysaccharides conjugated to *Pseudomonas aeruginosa* exotoxin A. *Infect. Immun.*, **58**, 2367–2374.

Foster, T. J. (1991). Potential for vaccination against infections caused by *Staphylococcus aureus*. *Vaccine*, **9**, 221–227.

Friedlander, H., Habermann, R. T., Parr, L. W. (1951). Experimental arthritis in albino rats and mice produced by alpha type streptococci. *J. Infect. Dis.*, **88**, 298–304.

Goldenberg, D. L. (1989a). Bacterial arthritis. In: *Textbook of Rheumatology*, 3rd edn (eds Kelly, W. O., Harris, E. D., Ruddy, S., Sledge, C. B.), pp. 1567–1585. W. B. Saunders, Philadelphia, PA.

Goldenberg, D. L. (1989b). Infectious arthritis complicating rheumatoid arthritis and other chronic rheumatic disorders. *Arthritis Rheum.*, **32**, 496–502.

Goldenberg, D. L., Cohen, A. S. (1978). Synovial membrane histopathology in the differential diagnosis of rheumatoid arthritis, gout, pseudogout, systemic lupus erythematosus, infectious arthritis and degenerative joint disease. *Medicine*, **57**, 239–252.

Goldenberg, D. L., Reed, J. I. (1985). Bacterial arthritis. *N. Engl. J. Med.*, **312**, 764–771.

Goldenberg, D. L., Chisholm, P. L., Rice, P. A. (1983). Experimental models of bacterial arthritis. *J. Rheumatol.*, **10**, 5–11.

Gotschlich, E. C. (1993). Immunity to extracellular bacteria. In: *Fundamental Immunology*, 3rd edn (ed. Paul, W. E.), pp. 1287–1308. Raven Press, New York.

Heberling, J. A. (1941). A review of two hundred and one cases of suppurative arthritis. *J. Bone Joint Surg.*, **23**, 917–921.

Helle, M., Boeije, L., Aarden, L. A. (1988). Functional discrimination between interleukin-6 and interleukin-1. *Eur. J. Immunol.*, **18**, 1535–1540.

Ho, G. (1989). Bacterial arthritis. In: *Arthritis and Allied Conditions — A Textbook of Rheumatology*, 11th edn (ed. McCarty, D.J.), pp. 1892–1914. Lea & Febiger, Philadelphia, PA.

Holderbaum, D., Spech, R. A., Ehrhart, L. A. (1985). Specific binding of collagen to *Staphylococcus aureus*. *Collagen Relat. Res.*, **5**, 261–276.

Höök, M., McGavin, M. J., Switaiski, L. M. *et al.* (1990). Interactions of bacteria with extracellular matrix proteins. *Cell Different. Devel.*, **32**, 433–438.

Hsu, S., Raine, L., Fanger, H. (1981). Use of avidin–biotin–peroxidase complex (ABC) in immunoperoxidase techniques: a comparison between ABC and unlabelled antibody (PAP) procedures. *J. Histochem. Cytochem.*, **29**, 577–580.

Janeway, C. A. Jr (1993). How the immune system recognizes invaders. *Sci. Am.*, **269(3)**, 41–47.

Jonsson, R., Tarkowski, A., Klareskog, L. (1986). A demineralization procedure for immunohistopathological use: EDTA treatment preserves lymphoid cell surface antigens. *J. Immunol. Meth.*, **88**, 109–114.

Josefsson, E., Tarkowski, A., Carlsten, H. (1992). Anti-inflammatory properties of estrogen. I. *In vivo* suppression of leukocyte production in bone marrow and redistribution of peripheral blood neutrophils. *Cell Immunol.*, **142**, 67–78.

Josefsson, E., Carlsten, H., Tarkowski, A. (1993). Neutrophil mediated inflammatory response in murine lupus. *Autoimmunity*, **14**, 251–257.

Kaandorp, C. J. E., Krijnen, P., Moens, H. J. B., Habbema, J. D. K., van Schaardenburg, D. (1997). The outcome of bacterial arthritis — a prospective community-based study. *Arthritis Rheum.*, **40**, 884–892.

Kaplow, L. (1975). Substitute for benzidine in myeloperoxidase stains (letter). *Am. J. Clin. Pathol.*, **63**, 451.

Karakawa, W. W., Fournier, J. M., Vann, W. F., Arbeit, R., Schneerson, R. S., Robbins, J. B. (1985). Method for the serological typing of the capsular polysaccharides of *Staphylococcus aureus*. *J. Clin. Microbiol.*, **22**, 445–447.

Kistler, G. H. (1936). Embolic *Staphylococcus* arthritis and endocarditis in rabbits. *Proc. Soc. Exper. Biol. Med.*, **34**, 829–831.

Lansdorp, P. M., Aarden, L. A., Calafat, J., Zeijlemaker, W. P. (1986). A growth factor dependent B cell Hybridoma. *Curr. Topics Microbiol. Immunol.*, **132**, 105–113.

Lee, J., Takeda, S., Livolsi, P. J., Paoletti, L. C. (1993). Effects of in vitro and in vivo growth conditions on expression of type 8 capsular polysaccharide by *Staphylococcus aureus*. *Infect. Immun.*, **61**, 1853–1858.

Lewis, G. W., Cluff, L. E. (1965). Synovitis in rabbits during bacteremia and vaccination. *Bull. Johns Hopkins Hosp.*, **116**, 175–190.

Lindberg, L. (1969). Experimental staphylococcal arthritis in golden hamsters (*Mesocricetuers auratus*). *Acta Pathol. Microbiol. Immunol. Scand.*, **76**, 117–125.

Mahowald, M. L. (1986). Animal models of infectious arthritis. *Clin. Rheum. Dis.*, **12**, 403–421.

Mancini, G., Carbonara, A. O., Heremans, J. F. (1965). Immunochemical quantitation of antigens by single radial immunodiffusion. *Immunochemistry*, **2**, 235–254.

Mathews, M., Shen, F. H., Linder, A., Sherrard, D. J (1980). Septic arthritis in hemodialyzed patients. *Nephron.*, **25**, 87.

Nakamura, Y., Masuhara, T., Ito-Kuwa, S., Aoki, S. (1991). Induction of experimental *Candida* arthritis in rats. *J. Med. Vet. Mycol.*, **29**, 179–192.

Nilsson, I.-M., Bremell, T., Rydén, C., Cheung, A. L., Tarkowski, A. (1996). The staphylococcal accessory gene regulator (sar) controls arthritogenicity. *Infect. Immun.*, **64**, 4438–4443.

Nilsson, I.-M., Lee, J. C., Bremell, T., Rydén, C., Tarkowski, A. (1997). The role of staphylococcal polysaccharide microcapsule expression in septicemia and septic arthritis. *Infect. Immun.*, **65**, 4216–4221.

Nilsson, I.-M., Patti, J. M., Bremell, T., Höök, M., Tarkowski, A. (1998). Vaccination with a recombinant fragment of collagen adhesin provides protection against *Staphylococcus aureus* mediated septic death. *J. Clin. Invest.*, **101**, 2640–2649.

Parker, M. T. (1983). The significance of phage typing patterns in *Staphylococcus aureus*. In: *Staphylococci and Staphylococcal Infections* (eds Easmon, C. S. F., Adlam, C.), vol 1, pp. 33–62. Academic Press, London.

Patti, J. M., Bremell, T., Krajewska-Pietrasik, D. *et al.* (1994). The *Staphylococcus aureus* collagen adhesin is a virulence determinant in experimental septic arthritis. *Infect. Immun.*, **62**, 152–161.

Paul, W. E. (1993). Infectious diseases and the immune system. *Sci. Am.*, **269**, 3:57–63.

Pedersen, K. B., Kjems, E., Perch, B., Slot, P. (1981). Infection with RS streptococci in pigs. *Acta Pathol. Microbiol. Scand. Sect. B*, **89**, 161–165.

Rigdon, R. H. (1942). Pathogenesis of arthritis following the intravenous injection of Staphylococci in the adult rabbit. *Am. J. Surg.*, **55**, 553–561.

Rydén, C., Yacoub, A., Maxe, I. *et al.* (1989). Specific binding of bone sialoprotein to *Staphylococcus aureus* isolated from patients with osteomyelitis. *Eur. J. Biochem.*, **184**, 331–336.

Sáez-Llorens, X., Ramilo, O., Mustafa, M. M., Mertsola, J., McCracken, G. H. Jr (1990). Molecular pathophysiology of bacterial meningitis: current concepts and therapeutic implications. *J. Pediatr.*, **5**, 671–684.

Sakiniene, E., Bremell, T., Tarkowski, A. (1996). Addition of corticosteroids to antibiotic treatment ameliorates the course of experimental *Staphylococcus aureus* arthritis. *Arthritis Rheum.*, **39**, 1596–1605.

Sakiniene, E., Bremell, T., Tarkowski, A. (1999) Complement depletion aggravates *Staphylococcus aureus* septicemia and arthritis. *Clin. Exp. Immunol.*, **115**, 95–102.

Schlievert, P. M. (1993). Role of superantigens in human disease. *J. Infect. Dis.*, **167**, 997–1002.

Sheagren, J. N. (1988). Inflammation induced by *Staphylococcus aureus*. In: *Inflammation: Basic Principles and Clinical Correlates* (eds Gallin, J. I., Goldstein, I. M., Snyderman, R.), pp. 829–840. Raven Press, New York.

Sheagren, J. N., Schaberg, D. R. (1992). Staphylococci. In: *Infectious Diseases* (eds. Gorbach, S. L., Bartlett, J. G., Blacklow, N. R.), pp. 1395–1400. W. B. Saunders, Philadelphia, PA.

Switalski, L. M., Patti, J. M., Butcher, W. G., Gristina, A. G., Speziale, P., Höök, M. (1993). A collagen receptor on *Staphylococcus aureus* isolated from patients with septic arthritis mediates adhesion to cartilage. *Molec. Microbiol.*, **7**, 99–107.

Tarkowski, A., Czerkinsky, C., Nilsson, L.-Å. (1984). Detection of IgG rheumatoid factor secreting cells in autoimmune MRL/1 mice: a kinetic study. *Clin. Exp. Immunol.*, **58**, 7–12.

Thompson, R. H. S., Dubos, R. J. (1938). Production of experimental osteomyelitis in rabbits by intravenous injection of *Staphylococcus aureus*. *J. Exp. Med.*, **68**, 191–209.

Tissi, L., Marconi, P., Mosci, P. *et al.* (1990). Experimental model of Type IV *Streptococcus agalactiae* (Group B Streptococcus) infection in mice with early development of septic arthritis. *Infect. Immun.*, **58**, 3093–3100.

Verdrengh, M., Tarkowski, A. (1997). Role of neutrophils in experimental septicemia and septic arthritis induced by *Staphylococcus aureus*. *Infect. Immun.*, **65**, 2517–2521.

Verdrengh, M., Springer, T. A., Gutierrez-Ramos, J.-C., Tarkowski, A. (1996). Role of ICAM-1 in the pathogenesis of staphylococcal arthritis and in host defence against staphylococcal bacteriemia. *Infect. Immun.*, **64**, 2804–2807.

Waldvogel, F. A. (1990). *Staphylococcus aureus* (including toxic shock syndrome). In: *Principles and Practice of Infectious Diseases*, 3rd edn (eds Mandell, G. L., Douglas, R. G. Jr, Bennett, J. E.), pp. 1489–1510. Churchill Livingstone, New York.

Wang, Y., Xu, J., Xue, D. (1983). Experimental study of acute suppurative bone and joint infection. II. Suppurative arthritis. *Chinese Med. J.*, **96**, 907–912.

Ward, J., Cohen, A. S., Bauer, W. (1960). The diagnosis and therapy of acute suppurative arthritis. *Arthritis Rheum.*, **3**, 522–535.

Wiedermann, U., Tarkowski, A., Bremell, T., Kahu, H., Hanson, L.-Å., Dahlgren, U. (1996). Increased susceptibility to *Staphylococcus aureus* arthritis in Vitamin A deficient rats. *Infect. Immun.*, **64**, 209–214.

Williams, R. E. O., Rippon, J. E. (1952). Bacteriophage typing of *Staphylococcus aureus*. *J. Hyg.*, **50**, 320–353.

Zhao, Y.-X., Tarkowski, A. (1995). Impact of interferon-γ receptor deficiency on experimental *Staphylococcus aureus* septicemia and arthritis. *J. Immunol.*, **155**, 5736–5742.

Zhao, Y.-X., Abdelnour, A., Holmdahl, R., Tarkowski, A. (1995a). Mice with the xid B cell defect are less susceptible to develop *Staphylococcus aureus*-induced arthritis. *J. Immunol.*, **155**, 2067–2076.

Zhao, Y. -X., Abdelnour, A., Kalland, T., Tarkowski, A. (1995b). Overexpression of the T cell receptor Vβ3 in transgenic mice increases mortality during infection by enterotoxin A-producing *Staphylococcus aureus*. *Infect. Immun.*, **63**, 4463–4469.

Chapter 65

Experimental Group B *Streptococcus* Arthritis in Mice

L. Tissi

Background of human infection

Group B streptococci (GBS) are the leading cause of life-threatening infection in neonates and young infants (Baker and Barrett, 1974; Baker and Edwards, 1995). Invasive neonatal GBS infection has either an early (usually during the first 24 hours after birth) or late (7 days after birth) onset. Early-onset disease is acquired from the mother by the ascending route in the uterus during labor or by direct contact at delivery (Baker, 1978; Gerards *et al.*, 1985; Dillon *et al.*, 1987; Weisman *et al.*, 1992). Late-onset disease may be acquired at birth from the mother or later in life from other individuals (Anthony *et al.*, 1979; Gardner *et al.*, 1980). The attack rates reported in the last two decades for the early-onset infection vary from 0.7 to 3.7 per 1000 live births (Baker and Edwards, 1995). The attack rates for late-onset infection range from 0.5 to 1.8 per 1000 (Baker and Edwards, 1995). Common manifestations of GBS disease in newborns include pneumonia, septicemia, meningitis, bacteremia, and bone or joint infections (Ancona *et al.*, 1979; Memon *et al.*, 1979; Lai *et al.*, 1980; Baker and Edwards, 1995). Septic arthritis is described as a clinical manifestation of late-onset GBS disease in newborns (Ancona *et al.*, 1979; Memon *et al.*, 1979; Gardner *et al.*, 1980; Dan, 1983; Yagupsky *et al.*, 1991; Baker and Edwards, 1995) and requires prolonged antibiotic treatment to insure an uncomplicated outcome. Drainage of the suppurative focus is an adjunct to antibiotic therapy. Recently, septic arthritis due to GBS has been demonstrated in adults (Small *et al.*, 1984; Schwartz *et al.*, 1991; Farley *et al.*, 1993) and is often associated with age and risk factors such as diabetes mellitus, cancer, cardiovascular diseases, chronic renal insufficiency, and other underlying severe illnesses (Jackson *et al.*, 1995). In adults also, extended antibiotic therapy together with aspiration or open drainage are necessary for successful treatment (Goldenberg and Cohen, 1976; Small *et al.*, 1984).

Background of model

Relevant results for a better understanding of human GBS disease have been obtained through the study of GBS serotype (Ia/c, Ib/c, II, III, IV, V, VI) infections in adult mice (Furtado, 1976; Wennerstrom and Schutt, 1978; Baltimore *et al.*, 1979; Fleming, 1980; Edwards and Fuselier, 1983; Poultrel and Dore, 1985; Molinari *et al.*, 1987; von Hunolstein *et al.*, 1993), as well as in infant and adult rats (Ferrieri *et al.*, 1980; Schuit and Debiaso, 1980; Rubens *et al.*, 1987; Martin *et al.*, 1988, 1992; Wessels *et al.*, 1989, 1992).

Experimental arthritis was previously obtained with streptococci of group A and D or with GBS in rats. In such cases, articular lesions were also reproduced by inoculation of inactivated cells or bacterial sonic extracts containing high doses of rhamnose (Cromartie *et al.*, 1977; Spitznagel *et al.*, 1983). Polyarthritis was produced in mice by intravenous (i.v.) injection of the cell-wall or peptidoglycan fractions of group A streptococci (Koga *et al.*, 1984). Our model of GBS arthritis was first performed in a study on virulence of type IV GBS in mice; articular lesions were obtained only with live micro-organisms (Tissi *et al.*, 1990, 1991). The mouse model of GBS septic arthritis was used to investigate arthritogenicity of different GBS serotypes (Cornacchione *et al.*, 1992) to compare the activity of different antibiotics (azithromycin, erythromycin, and penicillin G; Tissi *et al.*, 1994, 1995) and to study the protective activity of a murine monoclonal antibody in type IV GBS infection (Ricci *et al.*, 1996).

Animal species

Outbred CD-1 mice, 8–10 weeks old, weighing 20–25 g, obtained from Charles River Breeding Laboratories (Calco, Milan, Italy) were used (Tissi *et al.*, 1990, 1991, 1994, 1995; Cornacchione *et al.*, 1992; Ricci *et al.*, 1996). No differences in results were observed between males and females. All strains of mice can be employed, but, because of the differences in susceptibility to GBS infection between mouse strains (Cornacchione *et al.*, 1992), preliminary experiments should be performed to determine the optimal bacterial inoculum to insure a high frequency of arthritis and a low mortality rate.

Handbook of Animal Models of Infection
ISBN 0-12-775390-7

Preparation of animals

No specialized housing, care, or specific treatment is required.

Storage and preparation of inocula

This model can be employed with all strains of different GBS serotypes. We used both reference strains and GBS clinical isolates. Reference strains included NCTC 11079 (type II), NCTC 11080 (type III), strain 1/82 (type IV), 10/84 (type V), 118754 (type VI), and 7271 (type VII) obtained from the Czech National Type Culture Collection (Prague, Czech Republic). Clinical isolates of different GBS serotypes were obtained from the Istituto Superiore di Sanità (ISS) Culture Collection (Rome, Italy). All GBS strains were grown in Todd–Hewitt broth (THB, Oxoid, Basingstoke, UK) and on GBS agar base Islam (Oxoid), additioned with 5% heat inactivated horse serum or on Columbia colistin-nalidixic acid (CNA) agar (Oxoid) additioned with 5% defibrinated sheep blood. GBS and CNA agar were incubated at 37°C in anaerobic conditions. Growth of GBS on GBS agar was characterized by colonies with a typical orange carotenoid pigment. This peculiarity made cfu enumeration easy (particularly from organ homogenates).

All GBS strains used in our studies were grown at 37°C in THB and samples were stored at –70°C until use. For experimental infections the organisms were cultured overnight in THB and then washed and diluted in serum-free RPMI 1640 medium (Flow Laboratories, McLean, VA). The inoculum size was estimated turbidimetrically at 540 nm in a Beckman DV-68 spectrophotometer (Beckman instruments, Fullerton, CA). The number of live bacterial cells was confirmed by enumeration of cfu on GBS or CNA agar. The desired number of bacteria was diluted in RPMI medium and injected i.v. via the tail vein in a volume of 0.5 ml per mouse. Four doses were used to establish the optimal inoculum size: 10^9, 10^8, 10^7, and 10^6 cfu/mouse. The dose of 10^7 cfu/mouse gave a high frequency of arthritis and a relatively low mortality rate. All mice died within a few days (1–7 days) with 10^9 or 10^8 cfu/mouse, while no mice died with 10^6 cfu/mouse; however, arthritis was not evident. Thus, an inoculum of 10^7 cfu/mouse was used throughout our studies on GBS arthritis. GBS could also be diluted in THB and then injected. We used RPMI 1640 as the suspension medium because we diluted in the same medium the samples of blood or organs and joints that had been recovered from mice at selected times during infection.

Infection process

As mentioned above, the infection was initiated by direct i.v. injection of 1×10^7 GBS/mouse via the tail vein in a volume of 0.5 ml. At the beginning of our studies, we also injected the bacteria intraperitoneally (i.p.) but no strain of the GBS serotypes was able to induce articular lesions by the i.p. route with the exception of type VII (only at the dose of 10^8 cfu/mouse; unpublished results).

Key parameters to monitor infection

The infection was not confined to the joints. Systemic invasion occurred. The parameters for monitoring infection were death, general appearance, cfu enumeration in the blood and organs, blood cell responses, and macroscopic evaluation of articular lesions and histology.

Deaths

Groups of mice (usually 20 mice) were observed at 24-hour intervals after GBS challenge and mortality was recorded. With type IV GBS, death occurred in 30–40% of mice (Tissi *et al.*, 1990).

Appearance

Arthritis was not the only clinical feature of GBS infection. After type IV GBS infection, mice manifested neurological disorders (7.7%), panophthalmitis (9.2%), and flaccid paresis of the hind legs (4.6%; Tissi *et al.*, 1991). Loss of body weight was also observed and was recorded.

Blood and organ CFU

Blood and organ infections were determined by cfu evaluation at different times after bacterial inoculation. Spleen, liver, kidneys, lungs, and brain were aseptically removed and placed in a tissue homogenizer with 3 ml of RPMI medium. Blood samples were obtained by retro-orbital bleeding before sacrifice. Appropriate 10-fold dilutions of blood and organs were plated on GBS agar plates and incubated under anaerobic conditions. Results were expressed as the number of cfu per milliliter of blood or per whole organ. Joints were prepared by removing the skin and separating the limb below the joint. Then the joints were removed, weighed, and homogenized in 1 ml of sterile RPMI medium. Homogenization was performed by hand using a cold mortar and pestle. The joint samples were plated on GBS agar and the results were expressed as the number of cfu per milliliter of joint homogenate. Growth of type IV GBS in the organs and joints is shown in Table 65.1 and Figure 65.1.

Table 65.1 Growth of type IV GBS in the joints of CD-1 mice

	*cfu recovered on postinfection days**
Days	*Joints*
2	$1.5 \times 10^6 \pm 0.1 \times 10^6$
5	$1.6 \times 10^9 \pm 0.1 \times 10^9$
10	$5.6 \times 10^{10} \pm 0.5 \times 10^{10}$
20	$7.1 \times 10^{12} \pm 0.5 \times 10^{12}$

* Mice were inoculated i.v. with 1×10^7 micro-organisms on day 0. Values represent the means ± standard errors of three separate experiments. Ten mice per group were sacrificed at each time-point. Number of cfu per milliliter of joint homogenate is reported.

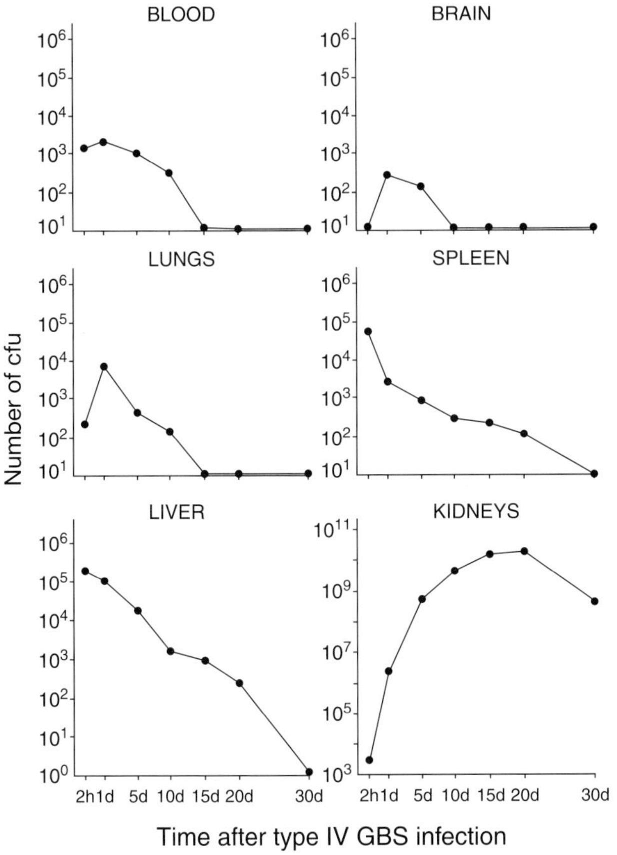

Figure 65.1 Growth kinetics of type IV GBS 1/82 in various organs of CD-1 mice. Mice were inoculated i.v. with 10^7 micro-organisms on day 0. Values are the mean cfu recovered per milliliter of blood of whole organs of mice. Standard errors, always <10%, have been omitted. Reproduced with permission from Tissi *et al.* (1990) *Infect. Immun.*, **58**, Figure 1.

Blood cell responses

During infection with type IV GBS, high numbers of monocytes and polymorphonuclear leukocytes were observed between 10 and 15 days after challenge. At this time, the number of GBS in the spleen, liver, lungs, and brain started to decrease while there was a progressive increase in the kidneys and joints. During infection, high neutrophilia was observed but was not sufficient to eliminate the micro-organisms from the joints.

Clinical evaluation of arthritis

Mice challenged with GBS were examined numerous times during day 1 after infection and then daily for 2 months to evaluate the presence of joint inflammation. Time of onset, number of joints involved, incidence, duration of arthritis, and occurrence of ankylosis were recorded. Arthritis was defined as visible erythema or swelling of at least one joint. To evaluate the intensity of arthritis, a clinical score (arthritic index) was used for each limb as determined by macroscopic inspection: 1 point, mild swelling and erythema; 2 points, moderate swelling and erythema; 3 points, marked swelling, erythema, and occasionally ankylosis. Thus, a mouse could have a maximum score of 12. The arthritis index was constructed by dividing the total score by the number of animals used in each experimental group. The incidence and severity of arthritis induced by type IV GBS are shown in Figure 65.2. This type of clinical score has been used by many authors in different mouse models of arthritis (Wolley *et al.*, 1993; Abdelnour *et al.*, 1994; Kasama *et al.*, 1995).

Joint pathology

Joint histology is used to monitor progress of arthritis. Pathology scores have been employed by several investigators. The joints were examined with regard to synovial hypertrophy, pannus formation, and cartilage and bone destruction. In addition, infiltration of inflammatory cells in joints and periarticular tissues were evaluated (Bremell *et al.*, 1991; Williams *et al.*, 1992; Abdelnour *et al.*, 1993). In our studies, joint specimens were prepared for histology by fixing in 10% formalin for 24 hours and decalcifying in 5% trichloroacetic acid for 7 days. All specimens were dehydrated, embedded in paraffin, sectioned at 3–5 μm and stained with hematoxylin and eosin. The major histopathological changes in the joints of mice injected with type IV GBS were the presence of an acute exudative synovitis, which started as early as 48 hours after infection, and a polymorphonuclear-leukocyte–monocyte infiltrate of the subsynovium and periarticular connective tissues. One week later, articular cavities of involved joints were filled with purulent exudate. In mice with persistent articular inflammation, joint destruction progressed rapidly with loss of cartilage and proliferation of granulation tissue. Fibrous ankylosis was observed on day 60.

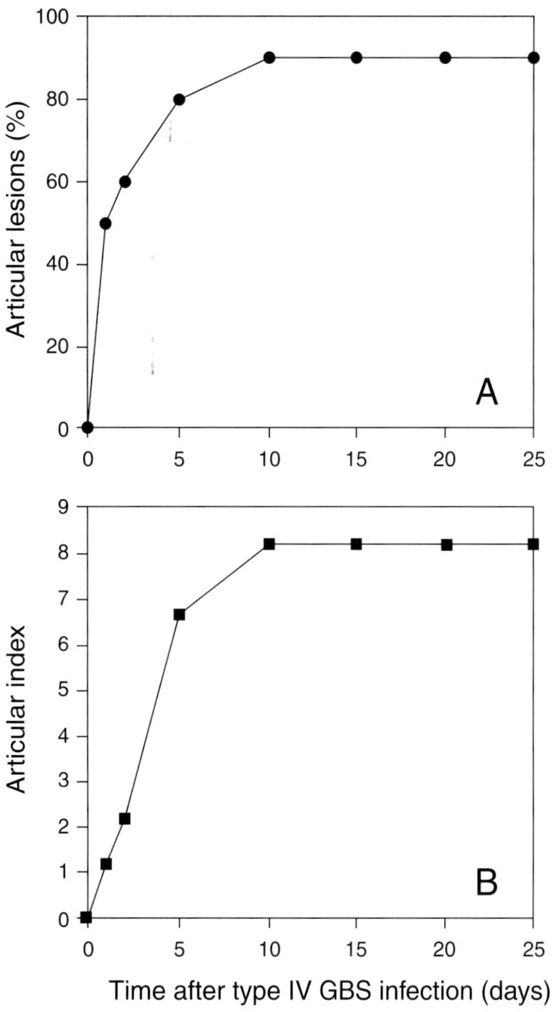

Figure 65.2 Incidence (**A**) and severity (**B**) of septic arthritis in CD-1 mice after i.v. inoculation of 1 × 10^7 type IV GBS. Values are the means of three separate experiments. In each experiment 40 mice were used. Standard errors, always <10%, have been omitted.

Cytokines

The contribution of cytokines to the pathogenesis of arthritis in mice has been demonstrated by several authors (Sáez-Llorens *et al.*, 1991; Thorbecke *et al.*, 1992; Kasama *et al.*, 1995). Cytokine appearance was evaluated during the whole course of arthritis. Quantification of cytokine levels (TNFα, IL-1β, and IL-6) in the joints and serum of mice injected with type IV GBS was performed by enzyme-linked immunosorbent assay (ELISA; unpublished data). Joint tissues were prepared by first removing the skin and separating the limb below the joints. Joint tissues were then homogenized in 3 ml of lysis medium (RPMI containing 2 mM PMSF and 1 μg/ml each of aprotinin, leupeptin, and pepstatin A, final concentration). Then the homogenized tissues were centrifuged at 2000*g* for 10 minutes and supernatants were sterilized using a Millipore filter (0.45 μm) and stored at –80°C until analysis. Blood samples were taken at selected intervals and the sera were stored at –80°C until analysis. A strong local production of all cytokines tested in the joints was detected between 5 and 15 days after type IV GBS inoculation. It is important to point out that during this period the incidence and the severity of arthritis reached the maximum. We hypothesize that these cytokines may contribute to the pathogenesis of septic arthritis.

Antimicrobial therapy

Although other penicillins are equally active against group B streptococci *in vitro*, penicillin G remains the drug of choice for the treatment of GBS infection in infants and adults (Baker and Edwards, 1995). To ensure rapid bactericidal effects, particularly in the cerebrospinal fluid, relatively high doses of intravenous penicillin G (500 000 U/kg/day) or ampicillin (300–400 mg/kg/day) are recommended for both early and late onset (Baker and Edwards, 1995). However, in the typical clinical setting, antimicrobial therapy is started before definite identification of the micro-organism. Therefore, initial therapy includes ampicillin and an aminoglycoside appropriate for the treatment of neonatal pathogens in addition to group B streptococci. This combination has been shown to be more effective than penicillin or ampicillin alone in the *in-vitro* and *in-vivo* killing of group B streptococci (Cooper *et al.*, 1979; Baker *et al.*, 1981). Once the bacteriologic diagnosis of group B streptococci is known, treatment can be changed to penicillin G or ampicillin alone. The antimicrobial regimen recommended for treatment of GBS septic arthritis in infants is intravenous penicillin G (200 000 U/kg/day) for at least 2–3 weeks. Infants with osteomyelitis should be given intravenous penicillin G at the dose of 200 000 U/kg/day for 3–4 weeks (Baker and Edwards, 1995). Examples of chemotherapy in GBS arthritis in adults are shown in Table 65.2. In this case also, penicillin G or other penicillins were the drugs most frequently used. In our experimental model of septic arthritis induced by type IV GBS, we used penicillin G as the standard effective drug, azithromycin (CP-62, 993) as the new antibiotic to be studied, and erythromycin as the semisynthetic derivative of azithromycin from erythromycin. The antibiotics were first tested *in vitro* for their efficacy against type IV GBS by standard techniques and the MICs were determined in the late exponential phase or stationary phase of growth.

For *in-vivo* treatment, sodic penicillin G (Squibb) was reconstituted in sterile distilled water and diluted to the desired concentration in saline solution. Azithromycin (Pfizer) and erythromycin (Sigma) were used in a free-base form and prepared by homogenizing the powder in a standard diluent containing methocell 15 (0.5 g), polysorbate 80 (Tween 80, 1.0 g), carboxymethyl cellulose (low viscosity: 10.0 g) sodium chloride (9.0 g) and water. All drugs were administered i.p. Different treatment schedules were employed. In a first set of experiments, each antibiotic

Table 65.2 Examples of chemotherapy of GBS arthritis in human adults

Reference	*Therapy*	*Outcome*	*Comments*
Tabatabai *et al.*, 1986	Penicillin G i.v. 800 000 U/day × 4 weeks	Favorable	Underlying illness: alcoholic liver disease
Small *et al.*, 1984	Penicillin i.v. × 2, 4 or 6 weeks Penicillin i.v. and i.m. × 44 days + streptomycin i.m. × 20 days Ampicillin i.v. × 6 weeks + gentamycin i.v. × 4 weeks Others	Favorable 58%	Joint aspiration 21%, surgical drainage 12%, Girdlestone procedure needed 8%. Underlying illness: diabetes mellitus 33%
Galloway *et al.*, 1993	Benzylpenicillin i.v. 2.4 g every 6 hours × 2 weeks followed by oral amoxicillin 500 mg every 6 hours × 2 weeks	Favorable	Bilateral endophthalmitis No underlying illness
Mateo Soria *et al.*, 1992	Penicillin IV i.v.	Favorable	Only 1 patient with GBS arthritis Underlying illness: rheumatoid arthritis
Carrascosa *et al.*, 1996	Ciprofloxacin i.v. 200 mg 2 × daily × 2 weeks followed by penicillin G i.v. 400 000 U 4 × daily × 2 weeks and then phenoxymethyl penicillin 4 × daily × 7 days	Favorable	Underlying illness: diabetes, cancer
Steinbrecher, 1981	Parenteral penicillin G i.v. × 2 weeks or penicillin + gentamycin × 2 weeks	Favorable	Organisms tolerant to penicillin Antimicrobial synergy was evident
Hartmann and Nauseef, 1990	Penicillin G i.v. 12 000 000 U daily × 10 days Tobramycin + penicillin G i.v. 24 000 000 U daily ×4 weeks	Favorable	Underlying illness: hemophilia B, alcohol abuse
Stark, 1987	Parenteral penicillin G i.v. × 4 weeks	Favorable	Underlying illness: diabetes

was administered to mice in a single dose at 30 minutes or 6, 12, and 24 hours after i.v. infection with 1×10^7 type IV GBS. Penicillin G was administered at the doses of 400 000 and 600 000 U/kg; azithromycin and erythromycin were administered at 50 or 100 mg/kg. Because of the poor efficacy of a single dose of antibiotics on the prevention of septic arthritis, in other experiments multiple doses of antibiotics were given at 12-hour intervals starting 30 minutes or 6 hours after infection. Other studies were performed to establish the effect of the antibiotics on recovery from septic arthritis. In this case, all animals positive for articular lesions were treated with different drugs on days 7, 8, and 9 after infection.

Key parameters to monitor response to treatment

The parameters to monitor response to treatment were the same used to monitor infection. In particular, reduction in mortality rates, GBS clearance from blood and joints, reduction in the incidence or prevention of arthritis, and changes in histopathologic features were evaluated both in protection studies and therapeutic regimens.

Death

Differences in mortality rates after treatment with antibiotics were observed. An example of the effect of multiple treatment with penicillin G, erythromycin, and azithromycin on survival of CD-1 mice injected with type IV GBS is shown in Table 65.3. These results point out the importance of the dose and the time of drug administration on mouse mortality. No significant differences in survival times were observed when high doses of antibiotics were given immediately after infection. On the contrary, differences in antibiotic activities were more evident when lower doses were employed and drugs were injected later in infection.

Table 65.3 Effect of multiple treatments with penicillin G, erythromycin, and azithromycin on survival of CD-1 mice inoculated i.v. with 10^7 type IV GBS (S/T, number of surviving mice at 60 days/total number of animals tested)

Drug	*Time administered**	*Dose*	*S/T†*
Penicillin G	30 min, 12 h, 24 h	200 000 U/kg	40/40
		400 000 U/kg	40/40
	6 h, 18 h, 30 h	200 000 U/kg	37/40
		400 000 U/kg	40/40
Erythromycin	30 min, 12 h, 24 h	25 mg/kg	36/40
		50 mg/kg	40/40
	6 h, 18 h, 30 h	25 mg/kg	35/40
		50 mg/kg	40/40
Azithromycin	30 min, 12 h, 24 h	25 mg/kg	40/40
		50 mg/kg	40/40
	6 h, 18 h, 30 h	25 mg/kg	40/40
		50 mg/kg	40/40
None (control)			26/40

* Times after GBS infection.
† Penicillin G treatment versus erythromycin treatment, $p < 0.001$; azithromycin treatment versus erythromycin treatment, $p < 0.001$; azithromycin treatment versus penicillin G treatment, $p > 0.1$.

Table 65.4 Effect of penicillin G, erythromycin, and azithromycin on recovery from arthritis of CD-1 mice inoculated i.v. with 1×10^7 type IV GBS

*Drug**	*Dose*	*No. (%) of cured mice†*
Penicillin G	400 000 U/kg	5 (16.6)
	600 000 U/kg	8 (26.6)
Azithromycin	50 mg/kg	16 (53.3)‡
	100 mg/kg	21 (70.0)‡
Erythromycin	50 mg/kg	5 (16.6)
	100 mg/kg	5 (16.6)
None	–	0 (0.0)

* Antibiotics were administered daily, 7, 8, and 9 days after GBS infection. All treated animals had articular lesions.
† Number (%) of cured animals at the end of the observation period (60 days). In each experiment 30 mice were used. Values are the means of 3 separate experiments. Standard errors, always <10%, have been omitted.
‡ $p < 0.001$ (azithromycin versus erythromycin and penicillin G treatment).

Prevention of and recovery from arthritis

Prevention of articular lesions, reduction in the incidence of arthritis, or recovery from arthritis were direct evidence of the efficacy of the antibiotics. Macroscopic evaluation of treated and untreated mice were performed as described above. The effect of multiple administrations of different antibiotics on arthritis induced by type IV GBS is reported in Figure 65.3. Azithromycin showed the best activity in prevention of arthritis, and in this case also its greater efficacy was demonstrated when the lower dose was used and when the drug was administered 6 hours after infection. The study on the therapeutic effect of antibiotics involved the use of animals with severe articular lesions. In our experimental model the incidence and the severity of arthritis reached a maximum 8–10 days after infection. Drug administration started at this time. Recovery from arthritis was evaluated by macroscopic evaluation and histology. Only azithromycin significantly reduced the percentage of articular lesions (Table 65.4).

Bacterial clearance from blood and joints

Measurement of cfu reduction after treatment with antibiotics was another criterion for efficacy. For this purpose, samples from blood and joints were collected at different intervals after infection from treated and untreated mice. Samples were processed as described for monitoring infection and the results were expressed as number of cfu per milliliter of blood and per milliliter of joint homogenate. In our experimental model the administration of penicillin G, erythromycin, and azithromycin after infection with type IV GBS resulted in complete sterilization of blood (Table 65.5). Only azithromycin was able to clear GBS from the joints at any time-point examined and dose employed (Table 65.6).

Drug concentration in sera and joints

The antibiotic concentration can be determined by various methods. The agar well diffusion bioassay with *Micrococcus luteus* (ATCC 9341) as the bioassay organism is frequently used (Simon and Yin, 1970; Girard *et al.*, 1987; Azoulay-Dupuis *et al.*, 1991; O'Reilly *et al.*, 1992). Serum and joint samples were collected from groups of 8 to 10 mice at different intervals after multiple treatments at 30 minutes and 1, 2, 4, 8, 24, 48, 96, and 120 hours after the last inoculation. The antibiotic concentration was examined simultaneously in healthy controls and infected mice. Sera were collected as described and joints were removed, weighed, ground in a mortar, and then resuspended in 1 ml of RPMI medium. The homogenate was centrifuged and supernatants were used for the assays.

Standard curves for each antibiotic were made with sera and supernatants of joint homogenate from control mice. Pharmacokinetic parameters were estimated by standard methods (Greenblatt and Koch-Weser, 1975). In our model, azithromycin pharmacokinetic data showed that the drug maintained continuously high levels in the joints of both infected and uninfected mice with respect to those observed in erythromycin- and penicillin G-treated mice (Tissi *et al.*, 1995). Joint concentrations of azithromycin were significantly greater than serum concentrations, and the former persisted while the latter declined. This pharmacokinetic behavior of azithromycin was in agreement with that

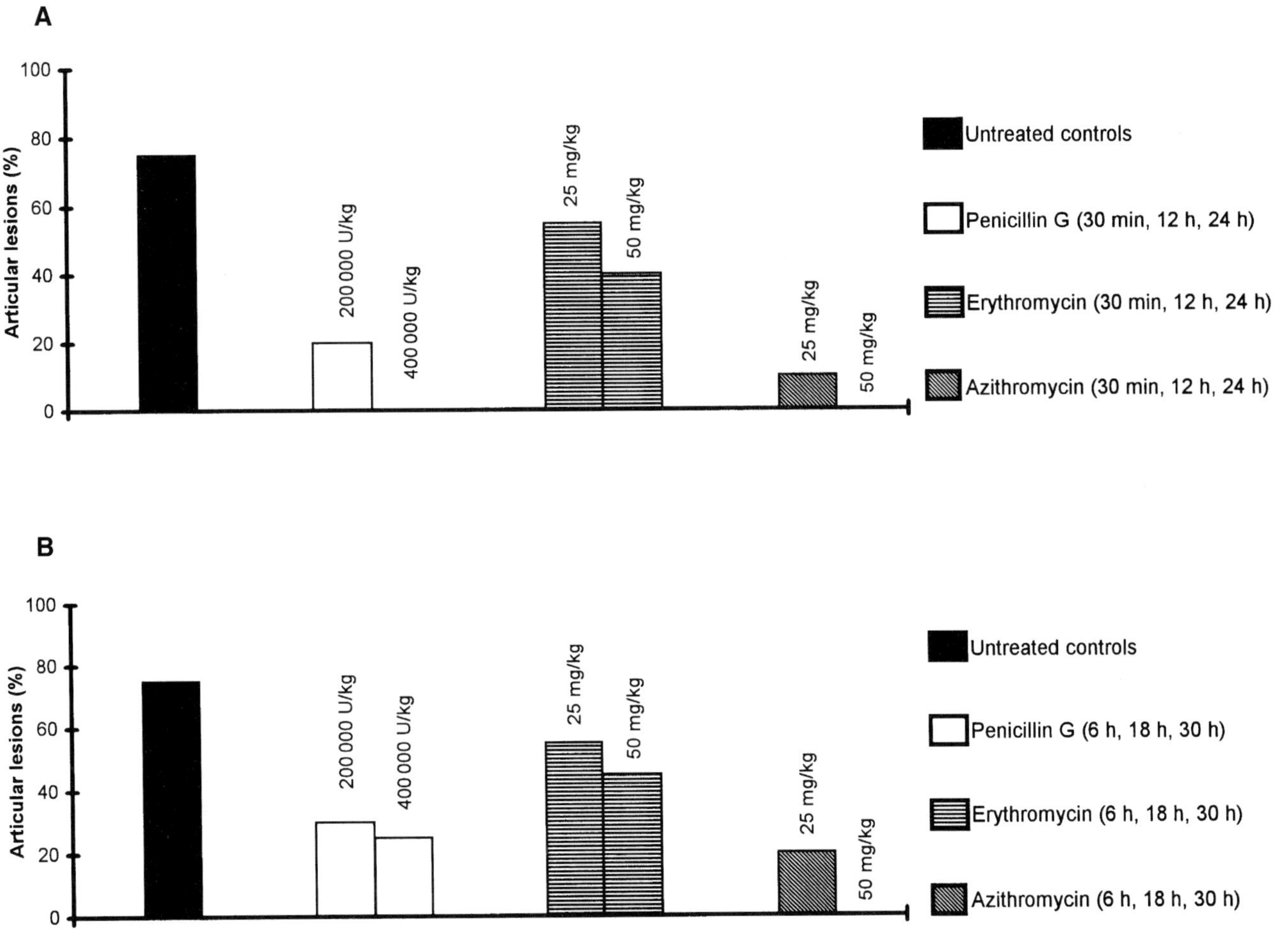

Figure 65.3 Effect of multiple administrations of penicillin G, erythromycin, or azithromycin on the incidence of articular lesions induced by i.v. inoculation of 10^7 type IV GBS. Mice received three i.p. doses of antibiotics at 12-hour intervals starting 30 minutes (**A**) or 6 hours (**B**) after GBS infection. In each experiment 40 mice were used. Values are the means of three experiments. Standard errors, always $<10\%$, have been omitted. Penicillin G treatment versus erythromycin treatment, $p < 0.05$; azithromycin treatment versus erythromycin treatment, $p < 0.001$; azithromycin treatment versus penicillin G treatment, $p < 0.05$. Reproduced with permission from Tissi *et al.* (1995). *Antimicrob. Agents Chemother.*, **39**, Figure 2.

Table 65.5 Effect of multiple treatments with penicillin G, erythromycin, and azithromycin on recovery of GBS from blood of CD-1 mice inoculated i.v. with 10^7 type IV GBS

			cfu recovered on postinfection day†			
			2	*6*	*10*	*20*
Drug	*Time administered**	*Dose*	*Blood*	*Blood*	*Blood*	*Blood*
Penicillin G	30 min, 12 h, 24 h	400 000 U/kg	0	0	0	0
	6 h, 18 h, 30 h	400 000 U/kg	0	0	0	0
Erythromycin	30 min, 12 h, 24 h	50 mg/kg	0	0	0	0
	6 h, 18 h, 30 h	50 mg/kg	0	0	0	0
Azithromycin	30 min, 12 h, 24 h	50 mg/kg	0	0	0	0
	6 h, 18 h, 30 h	50 mg/kg	0	0	0	0
None (control)			2.1×10^3 $\pm0.2\times10^3$	7.2×10^2 $\pm0.6\times10^2$	2.4×10^2 $\pm0.2\times10^2$	0

* Time of antibiotic treatment after GBS infection.
† Values represent the means ± standard errors of three separate experiments. Eight mice per group were sacrificed at each time point. Numbers of cfu per milliliter of blood are reported.

Table 65.6 Effect of multiple treatments with penicillin G, erythromycin, and azithromycin on recovery of GBS from joints of CD-1 mice inoculated i.v. with 10^7 type IV GBS

		cfu recovered on postinfection day†			
Drug/dose	*Time administered**	*2*	*6*	*10*	*20*
Penicillin G 400 000 U/kg	30 min, 12 h, 24 h	0	0	0	0
	6 h, 18 h, 30 h	8.1×10^3 $\pm0\times10^3$‡	3.1×10^3 $\pm0.3\times10^3$‡	2.3×10^2 $\pm0.1\times10^2$‡	0
Erythromycin 50 mg/kg	30 min, 12 h, 24 h	3.4×10^5 $\pm0.3\times10^5$‡	2.5×10^6 $\pm0.2\times10^6$‡	6.3×10^6 $\pm0.4\times10^6$‡	9.3×10^6 $\pm0.9\times10^6$‡
	6 h, 18 h, 30 h	8.5×10^5 $\pm0.5\times10^5$‡	5.3×10^6 $\pm0.5\times10^6$‡	8.5×10^6 $\pm0.5\times10^6$‡	9.9×10^6 $\pm0.8\times10^6$‡
Azithromycin 50 mg/kg	30 min, 12 h, 24 h	0	0	0	0
	6 h, 18 h, 30 h	0	0	0	0
None (control)		1.5×10^6 $\pm0.1\times10^6$	1.6×10^9 $\pm0.1\times10^9$	5.6×10^{10} $\pm0.5\times10^{10}$	7.1×10^{12} $\pm0.5\times10^{12}$

* Time of antibiotic treatment after GBS infection.
† Values represent the means ± standard errors of three separate experiments. Ten mice per group were sacrificed at each time point. Number of cfu per milliliter of joint homogenate are reported.
‡ $p < 0.001$ (treated mice versus controls).

observed in human tissues (Foulds *et al.*, 1990). Penicillin G reached very high levels in the sera and joints of mice but was eliminated rapidly ($T_{1/2}$ ranging between 0.3 and 0.4 hours). However, when administered at high doses (400 000 U or 600 000 U/kg) immediately after infection (30 minutes), penicillin G confirmed its efficacy as in human therapy of septic arthritis. Erythromycin was never detected in the joints of healthy or infected mice (Figure 65.4).

Histology

Joint histology was used to investigate the effect of antibiotics both on the prevention and therapeutic treatments. The absence or presence and features of lesions were evaluated. In particular, in the therapeutic treatments, recovery from arthritis was observed. In our study (Tissi *et al.*, 1995), complete recovery was observed in the joints of azithromycin-treated mice 10 days after the conclusion of therapy, while articular lesions similar to those of untreated infected mice were observed in penicillin G- and erythromycin-treated mice.

Pitfalls (advantages and disadvantages)

The use of CD-1 mice in our experimental model of GBS arthritis included several advantages with respect to other models proposed in different animal species (rats and rabbits). CD-1 mice are relatively inexpensive and a large number of animals can be employed. This possibility allows appropriate controls at any time during infection and a reliable statistical evaluation. The use of the outbred strain of mouse provides a heterogeneous population, which allows for immunologic and other host factor variations (O'Reilly *et al.*, 1996), thus mimicking human situations. CD-1 mice are tolerant to prolonged antibiotic administration and, because of the size of the animals, small quantities of antimicrobial agents can be used. Furthermore, reduced housing space and no particular care are required. Clinical evaluation of arthritis is easy to perform and has been standardized by many authors (Bremell *et al.*, 1991; Abdelnour *et al.*, 1993; Wolley *et al.*, 1993; Kasama *et al.*, 1995). The joints are small and easy to handle. This is important because of the large number of joints involved in the experimentation. In addition to these technical aspects, the major advantage of this model is the hematogenous spread of bacteria that is typical in human septic arthritis (Goldenberg and Reed, 1985). Bacteremia without a detectable focus of infection and bone or joint foci, or both, is the second most important clinical expression of late-onset group B streptococcal disease in infants (Yagupsky *et al.*, 1991; Baker and Edwards, 1995).

In adults, GBS suppurative arthritis may occur as a localized infection or as a manifestation of generalized septicemia (Lerner *et al.*, 1977). However, 50% of patients ended up bacteremic (Small *et al.*, 1984) and GBS-positive blood cultures have also been observed in some case reports (Galloway *et al.*, 1993; Carrascosa *et al.*, 1996). Localization of arthritis seems to mimic human situations. In mice, the most involved joints are the ankle, wrist, and coxafemoral joints and, less frequently, the tarsophalangeal, carpophalangeal, and interphalangeal joints (Tissi *et al.*, 1990, 1995). In both infants and adults, the hip, ankle, and wrist are frequently affected joints (Small *et al.*, 1984; Stark, 1987; Hartmann and Nauseef, 1990; Galloway *et al.*, 1993; Baker and Edwards, 1995).

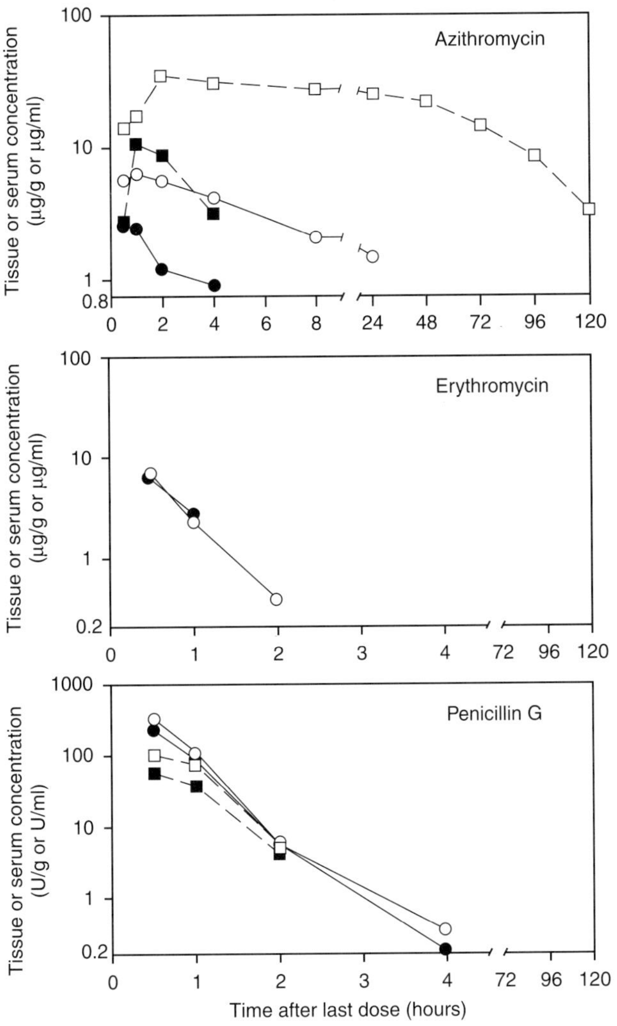

Figure 65.4 Mean antibiotic concentrations in sera and joints of CD-1 mice infected or not with 10^7 type IV GBS. Mice received three i.p. doses of azithromycin (50 mg/kg), erythromycin (50 mg/kg), or penicillin G (400 000 U/kg) at 12-hour intervals starting 30 minutes after infection or saline injection. Samples were collected at 30 minutes and 1, 2, 4, 8, 24, 48, 72, 96, and 120 hours after the last drug administration. Values are the means of three separate experiments. Standard errors, always <10%, have been omitted. ●, sera of non-infected mice; ○, sera of infected mice; ■, joints of non-infected mice; □, joints of infected mice. The missing time-points on the curves correspond to undetectable levels of antibiotics. Erythromycin was never detected in the joints of non-infected mice.

The mouse model of GBS arthritis does have some drawbacks. The major limitation of this model is the use of a relatively high inoculum, which is required for a sufficient number of bacteria to reach the joints and establish arthritis. The injection of 10^7 cfu/mouse induces 80–85% of articular lesions; mortality occurs in 30–40% of the animals. This is not representative of the clinical situation, even if in neonates the risk of penetration of the mucous membrane barriers, with resultant bacteremia, correlates directly with the intensity of maternal genital infection (inoculum size; Baker and Edwards, 1995). Another relevant fact is that not all strains of the different GBS serotypes are able to induce arthritis at the same inoculum size. Thus, the study of a new strain requires cautious preliminary experiments and comparisons with reference strains.

The onset of GBS arthritis in mice is rapid (within 48 hours) and this again may not fully represent the human situation, where GBS bone and joint infections often have an indolent onset. However, the mean age at diagnosis of GBS septic arthritis in infants is 20 days (Baker and Edwards, 1995). The course of treatment used in our study is considerably shorter than that clinically employed (2–3 days versus 3–4 weeks). The short treatment was effective in prevention studies but not in therapeutic regimens, where not all animals were cured. In this case, prolonged treatment should have been employed to mimic human therapy. Despite the limitations mentioned above, the clinical features of GBS infection in mice are very similar to those observed in humans. Moreover, this model could represent a valid test for antimicrobial therapy.

Contribution of the model to infectious disease therapy

In spite of the differences in pharmacokinetic parameters of antibiotics in humans and animals (O'Reilly *et al.*, 1996) and considering that animal models of septic arthritis are not sufficiently standardized to use as a prerequisite for clinical trials (Nelson *et al.*, 1992), the mouse model of GBS infection could be useful for the evaluation of new antimicrobial agents as compared to clinically employed agents. In our study, azithromycin proved to be a very promising drug for the therapy of arthritis as compared to penicillin G (Tissi *et al.*, 1995). Because of the increase of GBS infection in adults, and in particular in patients with severe chronic diseases (diabetes, cancer, cirrhosis, etc.), it would be interesting to study the induction and course of GBS arthritis in mice with altered immune defenses (as in diabetic mice). This would result not only in a better understanding of the pathogenesis of arthritis in this particular host, but also in the development of novel therapeutic approaches for immunocompromised patients. This model could also be applied to studies on immunotherapy, as demonstrated by the efficacy of a murine monoclonal antibody against the type-specific antigen of type IV GBS in preventing septic arthritis (Ricci *et al.*, 1996).

Acknowledgments

I am grateful to Eileen Mahoney Zannetti for dedicated editorial and secretarial assistance.

References

Abdelnour, A., Arvidson, S., Bremell, T., Rydén, C., Tarkowski, A. (1993). The accessory gene regulator (agr) controls *Staphylococcus aureus* virulence in a murine arthritis model. *Infect. Immun.*, **61**, 3879–3885.

Abdelnour, A., Bremell, T., Tarkowski, A. (1994). Toxic shock syndrome toxin 1 contributes to the arthritogenicity of *Staphylococcus aureus*. *J. Infect. Dis.*, **170**, 94–99.

Ancona, R. J., McAuliffe, J., Thompson, T. R., Speert, D. P. Ferrieri, P. (1979). Group B streptococcal sepsis with osteomyelitis and septic arthritis. Its occurrence with acute heart failure. *Am. J. Dis. Child.*, **133**, 919–920.

Anthony, B. F., Okada, D. M., Hobel, C. J. (1979). Epidemiology of the group B streptococcus: maternal and nosocomial sources for infant acquisitions. *J. Pediatr.*, **95**, 431–436.

Azoulay-Dupuis, E., Vallée, E., Bedos, J. P., Muffat-Joly, M. Pocilado, J. J. (1991). Prophylactic and therapeutic activities of azithromycin in a mouse model of pneumococcal pneumonia. *Antimicrob. Agents Chemother.*, **35**, 1024–1028.

Baker, C. J. (1978). Early onset group B streptococcal disease. *J. Pediatr.*, **93**, 124–125.

Baker, C. J., Barrett, F. F. (1974). Group B streptococcal infection in infants, the importance of various serotypes. *J.A.M.A.*, **230**, 1158–1160.

Baker, C. J., Edwards, M. S. (1995). Group B streptococcal infection. In: *Infectious Diseases of the Fetus and Newborn Infant* (eds. Remington, J. S., Klein, J. O.), pp. 980–1054. W. B. Saunders, Philadelphia, PA.

Baker, C. J., Thornsberry, C., Facklam, R. R. (1981). Synergism killing kinetics and antimicrobial susceptibility of group A and B streptococci. *Antimicrob. Agents Chemother.*, **19**, 716–725.

Baltimore, R. S., Kasper, D. L., Vechitto, J. (1979). Mouse protection test for group B streptococcus Type III. *J. Infect. Dis.*, **140**, 81–88.

Bremell, T., Lange, S., Yacoub, A. Rydén, C., Tarkowski, A. (1991). Experimental *Staphylococcus aureus* arthritis in mice. *Infect. Immun.*, **59**, 2615–2623.

Carrascosa, M., Pascual, F., Corrales, A., Martinez, J., Valle, R., Perez-Castrillon, J. L. (1996). Septic sternoclavicular arthritis caused by group B *Streptococcus*: case report and review. *Clin. Infect. Dis.*, **22**, 579–580.

Cooper, M. D., Keeney, R. E., Lyons, S. F., Cheatle, E. L. (1979). Synergistic effects of ampicillin-aminoglycoside combinations on group B streptococci. *Antimicrob. Agents Chemother.*, **15**, 484–486.

Cornacchione, P., Bistoni, F., Tissi, L. *et al.* (1992). Septic arthritis induced in mice by systemic infection with different serotypes of group B streptococci (GBS). In: *New Perspectives on Streptococci and Streptococcal Infection* (ed. Orefici, G.), pp. 446–447. G. Fischer, New York.

Cromartie, W. J., Craddock, J. G., Schwab, J. H., Anderle, S. K., Yang, C. H. (1977). Arthritis in rats after systemic injection of streptococcal cells or cell walls. *J. Exp. Med.*, **146**, 1585–1602.

Dan, M. (1983). Neonatal septic arthritis. *Isr. J. Med. Sci.*, **19**, 967–971.

Dillon H. C. Jr, Khare, S., Gray, B. M. (1987). Group B streptococcal carriage and disease: a 6-year prospective study. *J. Pediatr.*, **110**, 31–36.

Edwards, M. S., Fuselier, P. A. (1983). Enhanced susceptibility of mice with streptozotocin-induced diabetes to type II group B streptococcal infection. *Infect. Immun.*, **39**, 580–585.

Farley, M. M., Harvey, R. C., Stull, T. *et al.* (1993). A population-based assessment of invasive disease due to group B *Streptococcus* in nonpregnant adults. *N. Engl. J. Med.*, **328**, 1807–1811.

Ferrieri, P., Burke, B., Nelson, J. (1980). Production of bacteremia and meningitis in infant rats with group B streptococcal serotypes. *Infect. Immun.*, **27**, 1023–1032.

Fleming, D. O. (1980). Mucin model for group B type III streptococcal infection in mice. *Infect. Immun.*, **27**, 449–454.

Foulds, G., Shepard, R. M., Johnson, R. B. (1990). The pharmacokinetics of azithromycin in human serum and tissues. *J. Antimicrob. Chemother.*, **25** (Suppl. A), 73–82.

Furtado, D. (1976). Experimental group B streptococcal infection in mice: hematogenous virulence and mucosal colonization. *Infect. Immun.*, **13**, 1315–1320.

Galloway, A., Deighton, C. M., Deady, J., Marticorena, I. F., Efstratiou, A. (1993). Type V group B streptococcal septicaemia with bilateral endophthalmitis and septic arthritis. *Lancet*, **341**, 960–961.

Gardner, S. E., Mason E. O. Jr, Yow, M. D. (1980). Community acquisition of group streptococcus infants of colonized mothers. *Pediatrics*, **66**, 873–875.

Gerards, L. J., Cats, B. P., Hoogkamp-Korstanje, J. A. A. (1985). Early neonatal group B streptococcal disease: degree of colonization as an important determinant. *J. Infect.*, **11**, 119–124.

Girard, A. E., Girard, D., English, A. R. *et al.* (1987). Pharmacokinetic and *in vivo* studies with azithromycin (CP-62, 993), a new macrolide with an extended half-life and excellent tissue distribution. *Antimicrob. Agents Chemother.*, **33**, 277–282.

Goldenberg, D. L., Cohen, A. S. (1976). Acute infectious arthritis: a review of patients with nongonococcal joint infections. *Am. J. Med.*, **60**, 369–377.

Goldenberg, D. L., Reed, J. I. (1985). Bacterial arthritis. *N. Engl. J. Med.*, **312**, 764–771.

Greenblatt, D. J., Koch-Weser, J. (1975). Clinical pharmacokinetics. *N. Engl. J. Med.*, **297**, 702–705.

Hartmann, L., Nauseef, C. V. (1990). Group B streptococcal polyarthritis complicating hemophilia B. *Acta Haematol.*, **84**, 95–97.

Jackson, L. A., Hilsdon, R., Farley, M. M. *et al.* (1995). Risk factor for group B streptococcal diseases in adults. *Ann. Intern. Med.*, **123**, 415–420.

Kasama, T., Strieter, R. M., Likacs, N. W., Lincoln, P. M., Burdick, M. D., Kunkel, S. L. (1995). Interleukin-10 expression and chemokine regulation during the evolution of murine type II collagen-induced arthritis. *J. Clin. Invest.*, **95**, 2868–2876.

Koga, T., Kakimoto, K., Hirofuji, T. *et al.* (1984). Polyarthritis in mice following the systemic injection of the cell wall or its peptidoglycan fraction from group A streptococci and other bacteria. In: *Recent Advances in Streptococci and Streptococcal Diseases* (eds Kimura, Y., Kotami, S., Shiokawa, Y.), pp. 350–351. Reedbooks, Berkshire, UK.

Lai, T. K., Hingston, J., Scheifele, D. (1980). Streptococcal neonatal osteomyelitis. *J. Dis. Child.*, **134**, 711.

Lerner, P., Gopalakrisna, K., Wolinsky, E., McHenry, M., Tan, J., Rosenthal, M. (1977). Group B *Streptococcus* (*S. agalactiae*) bacteremia in adults: analysis in 32 cases and reviews of the literature. *Medicine*, **56**, 457–473.

Martin, T. R., Rubens, C. E., Wilson, C. B. (1988). Lung antibacterial defense mechanisms in infants and adult rats: implication for the pathogenesis of group B streptococcal infection in the neonatal lung. *J. Infect. Dis.*, **157**, 91–100.

Martin, T. R., Ruzinski, J. T., Rubens, C. E., Chi, E. Y., Wilson, C. B. (1992). The effect of type-specific polysaccharide capsule on the clearance of group B streptococci from the lungs of infants and adult rats. *J. Infect. Dis.*, **165**, 306–314.

Mateo Soria, L., Nolla Solé, J. M., Rodazilla Sacanell, A., Valverde García, J., Roig Escofet, D. (1992). Infectious arthritis in patients with rheumatoid arthritis. *Ann. Rheum. Dis.*, **51**, 402–403.

Memon, I. A., Jacobs, N. M., Yeh, T. F., Lilien, L. D. (1979). Group B streptococcal osteomyelitis and septic arthritis. Its occurrence in infants less than 2 months old. *Am. J. Dis. Child.*, **133**, 921–923.

Molinari, A., von Hunolstein, C., Donnelli, G., Paradisi, S., Arancia, G., Orefici, G. (1987). Effect of some capsular components on pathogenicity of type IV and provisional type V group B streptococci. *FEMS Microbiol. Lett.*, **41**, 69–72.

Nelson, J. D., Norden, C., Mader, J. T., Calandra, B. (1992). Evaluation of new anti-infective drugs for the treatment of acute suppurative arthritis in children. *Clin. Infect. Dis.*, **15**, (Suppl. 1), S172–S176.

O'Reilly, T., Kunz, S., Sande, E., Zak, O., Sande, M. A., Tauber, M. G. (1992). Relationship between antibiotic concentration in bone and efficacy of treatment of staphylococcal osteomyelitis in rats: azithromycin compared with clindamycin and rifampin. *Antimicrob. Agents Chemother.*, **36**, 2693–2697.

O'Reilly, T., Cleeland, R., Squires, E. L. (1996). Evaluation of antimicrobials in experimental animal infections. In: *Antibiotics in Laboratory Medicine*, 4th edn (ed. Lorian, V.), pp. 599–759. Williams & Wilkins, Baltimore, MD.

Poultrel, B., Dore, J. (1985). Virulence of human and bovine isolates of group B streptococci (type la and III) in experimental pregnant mouse models. *Infect. Immun.*, **47**, 94–97.

Ricci, M. L., von Hunolstein, C., Gomez, M. J., Parisi, L., Tissi, L., Orefici, G. (1996). Protective activity of a murine monoclonal antibody against acute and chronic experimental infection with type IV group B *Streptococcus*. *J. Med. Microbiol.*, **44**, 475–481.

Rubens, C. E., Wessels, M. R., Heggen, L. M., Kasper, D. L. (1987). Transposon mutagenesis of type III group B *Streptococcus*: correlation of capsule expression with virulence. *Proc. Natl. Acad. Sci. USA*, **84**, 7208–7212.

Sáez-Llorens, X., Jafari, H. S., Olsen, K. D., Nariuchi, H., Hansen, E. J., McCracken, G. H. Jr. (1991). Induction of suppurative arthritis in rabbits by *Haemophilus* endotoxin, tumor necrosis factor-α, and interleukin-1β. *J. Infect. Dis.*, **163**, 1267–1272.

Schuit, K. E., Debiaso, R. (1980). Kinetics of phagocyte response to group B streptococcal infection in newborn rats. *Infect. Immun.*, **28**, 319–324.

Schwartz, B., Schuchat, A., Oxtoby, M. J., Cochi, S. L., Hightower, A., Broome, C. V. (1991). Invasive group B streptococcal disease in adults. *J.A.M.A.*, **266**, 1112–1114.

Simon, H. J., Yin, E. J. (1970). Microbioassay of antimicrobial agents. *Appl. Microbiol.*, **19**, 573–579.

Small, C. B., Slater, L. N., Lowy, F. D., Small, R. D., Salvati, E. A., Casey, J. I. (1984). Group B streptococcal arthritis in adults. *Am. J. Med.*, **76**, 367–375.

Spitznagel, J. K., Goodrum, K. J., Warejcka, D. J. (1983). Rat arthritis due to whole group B streptococci. Clinical and histopathologic features compared with groups A and D. *Am. J. Pathol.*, **112**, 37–47.

Stark, R. H. (1987). Group B β-hemolytic streptococcal arthritis and osteomyelitis of the wrist. *J. Hand Surg.*, **12A**, 296–299.

Steinbrecher, V. P. (1981). Serious infection in an adult due to penicillin tolerant group B *Streptococcus*. *Arch. Intern. Med.*, **141**, 1714–1715.

Tabatabai, M. F., Sapico, F. L., Canawati, H. N., Harley, H. A. (1986). Sternoclavicular joint infection with group B *Streptococcus*. *J. Rheumatol.*, **13**, 466.

Thorbecke, G. J., Shah, R., Kuruvilla, A. P., Hardison, A. M., Palladino, M. A. (1992). Involvement of endogenous tumor necrosis factor α and transforming growth factor β during induction of collagen type II arthritis in mice. *Proc. Natl. Acad. Sci. USA*, **89**, 7375–7379.

Tissi, L., Marconi, P., Mosci, P. *et al.* (1990). Experimental model of type IV *Streptococcus agalactiae* (group B *Streptococcus*) infection in mice with early development of septic arthritis. *Infect. Immun.*, **58**, 3093–3100.

Tissi, L., Marconi, P., Mosci, P. *et al.* (1991). Experimental septic arthritis induced by type IV group B *Streptococcus* in mice. *J. Chemother. Suppl.*, **4**, 195–196.

Tissi, L., Campanelli, C., Mosci, P. *et al.* (1994). Effect of penicillin G, erythromycin and azithromycin on septic arthritis induced by type IV group B *Streptococcus*. In: *Pathogenic Streptococci: Present and Future* (ed. Totolian, A.), pp. 81–84. Lancer Publications, St Petersburg, Russia.

Tissi, L., von Hunolstein, C., Mosci, P., Campanelli, C., Bistoni, F., Orefici, G. (1995). *In vivo* efficacy of azithromycin in treatment of systemic infection and septic arthritis induced by type IV group B *Streptococcus* strains in mice: comparative study with erythromycin and penicillin G. *Antimicrob. Agents Chemother.*, **39**, 1938–1947.

Von Hunolstein, C., D'Ascenzi, S., Wagner, B. *et al.* (1993). Immunochemistry of capsular type polysaccharide and virulence properties of type VI *Streptococcus agalactiae* (group B streptococci). *Infect. Immun.*, **61**, 1271–1280.

Weisman, E., Stoll, B. J., Cruess, D. F. *et al.* (1992). Early-onset group B streptococcal sepsis: a current assessment. *J. Pediatr.*, **121**, 428–433.

Wennerstrom, D. E., Schutt, R. W. (1978). Adult mice as a model for early onset group B streptococcal disease. *Infect. Immun.*, **19**, 741–744.

Wessels, M. R., Rubens, C. E., Benedì V. J., Kasper, D. L. (1989). Definition of a bacterial factor: sialylation of the group B streptococcal capsule. *Proc. Natl. Acad. Sci. USA*, **86**, 8983–8987.

Wessels, M. R., Haft, R. F., Heggen, L. M., Rubens, C. E. (1992). Identification of a genetic locus essential for capsule sialylation in type III group B streptococci. *Infect. Immun.*, **60**, 392–400.

Williams, R. O., Feldmann, M., Maini, R. N. (1992). Anti-tumor necrosis factor ameliorates joint disease in murine collagen induced arthritis. *Proc. Natl. Acad. Sci. USA.*, **89**, 9784–9788.

Wolley, P. H., Dutcher, J., Widmer, M. B., Gillis, S. (1993). Influence of a recombinant human soluble tumor necrosis factor receptor Fc fusion protein on type II collagen-induced arthritis in mice. *J. Immunol.*, **151**, 6602–6607.

Yagupsky, P., Menegus, A., Powell, K. R. (1991). The changing spectrum of group B streptococcal disease in infants: an eleven-year experience in a tertiary care hospital. *Pediatr. Infect. Dis.*, **10**, 801–808.

Chapter 66

Rat Model of Bacterial Osteomyelitis of the Tibia

T. O'Reilly and J. T. Mader

Background of model

Osteomyelitis is a difficult infection to treat requiring prolonged treatment times and, in the case of chronic infection, surgical intervention. Classification of the infection, diagnosis and antimicrobial and adjunct therapy have been reviewed previously (Norden, 1989, 1996; Laughlin *et al.*, 1994; Haas and McAndrew, 1996; Mader and Calhoun, 1996; Lew and Waldvogel, 1997; Mader *et al.*, 1997; Wall, 1998). Several models of osteomyelitis using different animal species have been described (for reviews see Norden, 1988; Mader, 1985; Mader and Wilson, 1986; Mader and Adams, 1988; Rissing 1990; O'Reilly *et al.*, 1996), but the most common species used are the rabbit (Chapter 68), mice (Matsushita *et al.*, 1997), and rats. Establishment of experimental osteomyelitis in rats by the haematogenous route is described in Chapter 67. Production of experimental osteomyelitis by direct infection of the bone is described in the present chapter. This rat model is based upon the rabbit model described by Scheman *et al.* (1941) and refined by Norden (1970), and was originally described by Zak *et al.* (1981) and more fully described by Rissing *et al.* (1985b). A thorough study of bone pathology, demonstration of the microcolony mode of growth of *Staphylococcus aureus* within the bone, and partial characterization of the humoral immune response has been provided by Power *et al.* (1990) and a review of the pathogenesis of *S. aureus* osteomyelitis (including a list of the disadvantages of existing animal models) by Cunningham *et al.* (1996). An excellent review of the more commonly used osteomyelitis models and their relevance for evaluation of pathology and therapeutic intervention is available (Crémieux and Carbon, 1997).

Animal species

Several strains of rat have been used in this model: Sprague–Dawley (300–400 g, male; Rissing *et al.*, 1985b; Nelson, 1990a), Wistar (≈ 200 g, male; Zak *et al.*, 1981; Henry *et al.*, 1987), Wistar (≈ 275 g, female; Gomis *et al.*, 1990b), Madorin (a Wistar subline, ≈ 165–200 g, male; Dworkin *et al.*, 1990; O'Reilly *et al.*, 1992) and CD (200–250 g, male; Gisby *et al.*, 1994). Presumably, any rat strain could be used, but preliminary experiments should be performed to determine the optimal bacterial inocula to ensure infection (Nelson *et al.*, 1990a).

Preparation of animals

No specialized housing or care nor specific pretreatment is required.

Details of surgery

Overview

Briefly, the tibial bones of anaesthetized animals are surgically exposed, a hole approximately 1 mm in diameter is drilled in the medullary cavity, into which sodium morrhuate and bacterial suspension are injected and the hole is subsequently sealed. Sodium morrhuate is a sclerotizing agent to induce a mild inflammatory response and resultant bone damage; not all variants of this model use sodium morrhuate (see below).

Materials required

Anaesthetic (preferably injectable, incorporating a pain killer), anaesthetic antidote, hair clippers or razor, skin disinfectant, scalpel handle plus blades, forceps, high-speed dental drill with 1 mm (or smaller) drill bit, sodium morrhuate, syringes and needles (23G), suture material (preferably metal clips).

Anaesthesia

Animals are sedated with an intraperitoneal (i.p.) or subcutaneous (s.c.) injection of 0.5 mg fluanisone plus 0.01 mg

Handbook of Animal Models of Infection
ISBN 0-12-775390-7

fentanyl citrate (Hypnorm®, Janssen Pharmaceuticals). This is prepared by diluting one part Hypnorm with three parts water for injection. After a minimum of 10 minutes and a maximum of 30 minutes, the animals are administered a mixture of 0.3 mg midozalam (Dormicum®, Roche), 0.67 mg fluanisone plus 0.013 mg fentanyl citrate by intramuscular (i.m.) injection. This solution is prepared by mixing one part Hypnorm with one part Dormicum and one part water for injection. Complete anaesthesia ensues in about 10 minutes and lasts for about 30 minutes. Respiration should be strong, but this anaesthetic does not produce the deep muscle relaxation that other anaesthetics do (e.g. pentobarbital). A suitable antidote for respiratory decline is naloxone hydrochloride (Narcan®, Du Pont Pharma) and normally 0.05–0.1 ml s.c. is sufficient. Higher doses will cause reversal of anaesthesia and preclude the animal from immediate application of more anaesthetic. Alternative anaesthetics used by other reports include pentobarbital, ketamine (≈28 mg/kg) plus xylazine (≈5.7 mg/kg; Nelson *et al.*, 1990a). An excellent review on the use of anaesthetics for experimental animals is provided by Bertens *et al.* (1993).

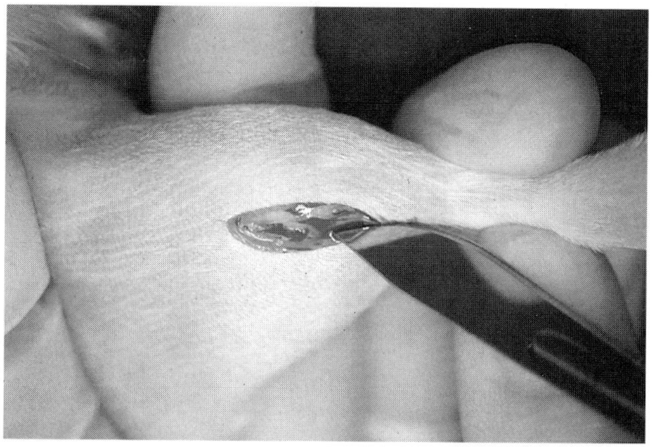

Figure 66.1 Primary surgical incision.

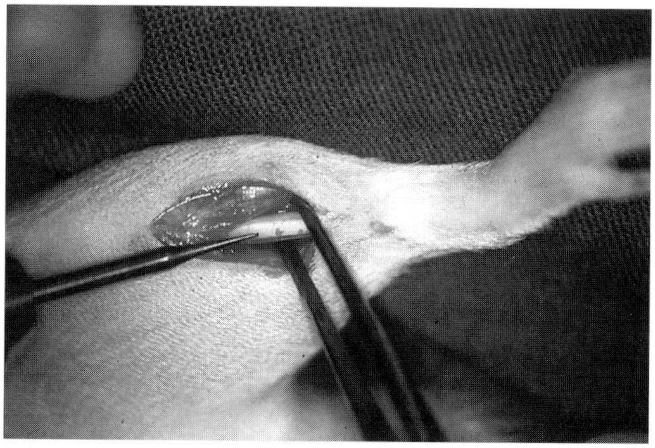

Figure 66.2 Exposure of the tibia, and drilling into the medullary cavity.

Surgical procedure

Following complete induction of anaesthesia, the animal is placed ventral side up. If the experimenter is right-handed (holding the drill with the favoured hand), performing the entire procedure with the left leg of the animal is easier. The skin over the leg and over the opposing peritoneum is shaved with electric clippers and disinfected. We normally use 70% vol./vol. ethanol, but other disinfectants (e.g. isopropanol- or iodine-based solutions) are also applicable. The leg is pushed away from the investigator and the longitudinal incisions below the knee joint are made in the sagittal plane. Exposure of the tibiofibular bone can be made with a minimum of two incisions, first through the skin and second through the muscle (Figure 66.1). A number curved edge surgical blade is used, and we normally use a large scapel handle fitted with a No. 10 or No. 22 curved scapel blade (Aesculap AG, Tuttlingen, Germany). The incision is made over the bone below the 'knee' joint (tibia head or condyle) but not completely extending to the ankle (malleolus–tarsal area). Surgical exposure of the bone requires cutting through the superficial muscles covering the bone (biceps femoris overlying the tibialis cranialis), but care should be taken to damage the muscle as little as possible by clean incisions. There are no major blood vessels in the region, but the greater saphenous and popliteal veins branching from the femoral vein lie on the inner side of the leg, towards the peritoneum and genitalia. A high-speed dental drill (e.g. Technobox, Bien Air, Switzerland) fitted with a 1 mm bulb bit is used to drill a hole into the medullar cavity of the tibia (Figure 66.2). Others have used a smaller (0.42 mm) bit (Nelson *et al.*, 1990a) and a Hamilton syringe is used to inject 5 μl of bacterial suspension. The bone drilling is the most difficult part of the procedure. The bone is held firmly with forceps and the drill should be held at a low angle relative to the tibia in order to avoid drilling completely through the bone. The drill should be in operation prior to contacting the bone and must be held steady, particularly at first contact with the bone. Resistance to the drill dissipates when the compact bone is completely pierced. The hole need only be a few millimeters deep, depending upon the size of the animal.

Infection procedure

Following the drilling procedure, the area is cleaned using a sterile cellulose swab soaked in sterile physiological saline. A volume of 0.05 ml of 5% wt/vol. sodium morrhuate (e.g. QUAD Pharmaceuticals, Indianapolis, IN) is injected into the medullar cavity using a tuberculin syringe fitted with a 25G 5/8″ needle (0.5 × 15 mm; (Figure 66.3). The needle is first placed into the hole perpendicular to the bone, and then lowered to be as parallel to the bone as possible while

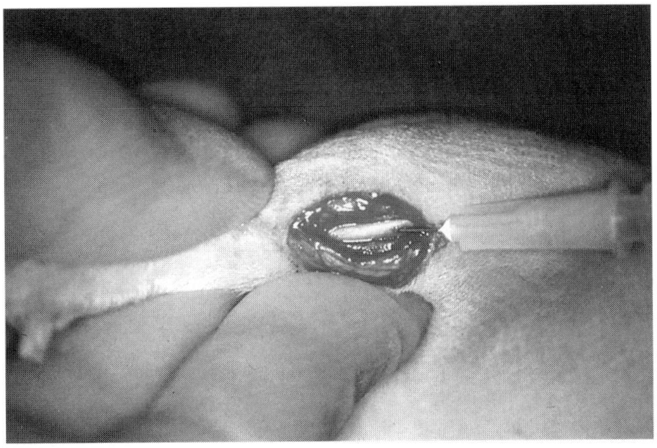

Figure 66.3 Injection of sodium morrhuate or bacterial suspension.

being pushed into the medullar cavity. The solution is slowly injected, and a small portion of the liquid will always be lost to the bone surface. Immediately following this, the bacterial inoculum is similarly injected in a 0.05 ml volume. The hole is sealed by applying a small amount of dry dental gypsum (e.g. Contura, Zürich, Switzerland) over the hole (Figure 66.4), which immediately absorbs fluids and adheres to the site. Other investigators have used bone wax (e.g. Ethicon; Nelson *et al.*, 1990b) or fibrin glue (reconstituted lyophilized human fibrin, e.g. Immuno, Vienna, Austria; Spagnolo *et al.*, 1993) to seal the hole.

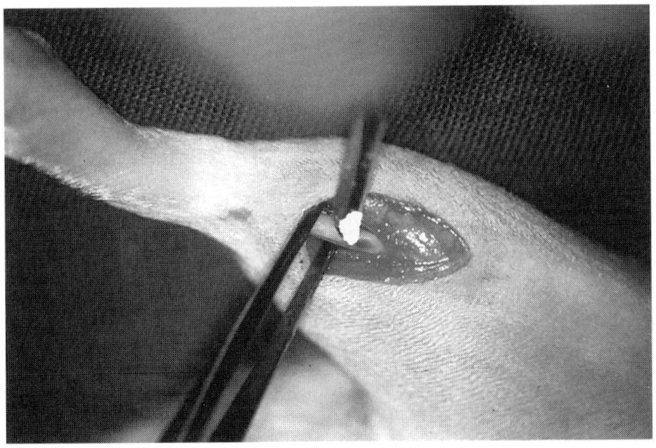

Figure 66.4 Sealing the hole with dental gypsum.

Wound closure

The wound is closed using metal skin clips (9 mm, e.g. Becton Dickinson & Co., Franklin Lakes, NJ Figure 66.5). Normally two clips placed evenly over the wound are sufficient. The wound can also be covered with an aerosol-based antiseptic plastic film (Nobecutan®, Astra). The clips often fall out (or are removed by the animal) after 1 or 2 weeks, but they can remain in place during the course of infection.

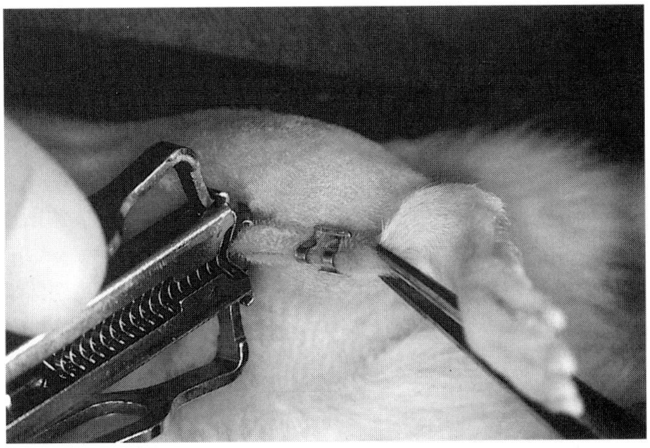

Figure 66.5 Wound closure with metal clips.

Sutures can also be used but, in our experience, this lengthens the procedure and sutures are easier for the rats to bite through, risking removal before the wound is fully healed.

Postoperative care

Animals should be kept warm after the surgery by placing them under an infrared lamp until they begin to recover from the anaesthetic. Animals should be housed individually on wire supports for 1–2 days following the surgery in order to facilitate wound healing and prevent infection of the wound. Animals return to normal behaviour usually within 1 day. Following surgery, animals are housed individually for 2–3 days to allow incisional wound healing. Subsequently the animals are housed in groups of two to five to facilitate a return to normal social activity. Postoperative analgesics have not been reported to be routinely used. This is worthy of further study, since the inclusion of such agents during the course of treatment would probably reflect the clinical situation because osteomyelitis is a painful infection. Buprenorphine (0.01–0.05 mg/kg, s.c., q. 8–12 hours) is recommended as a general postoperative analgesic (Bertens *et al.*, 1993). Owing to the involvement of prostaglandins in the pathology of bacterial-induced bone destructive diseases (Nair *et al.*, 1996), routine use of nonsteroidal anti-inflammatory agents (NSAIDs) is not recommended. Weight gain, and food and water consumption appear to be unaffected by this infection.

Storage and preparation of inocula

The model is amenable to using several different species of bacteria, although most studies have used *S. aureus*, *Escherichia coli* or *Pseudomonas aeruginosa*. We have normally used the antibiotic-susceptible *S. aureus* strain 1098 in our experiments, and methicillin-resistant *S. aureus* has also been used (Henry *et al.*, 1987). Our strain has been adapted to growth in bone by passing the organism in bone, followed by a single culture in brain–heart infusion (BHI) broth (Difco). Following cessation of growth, the culture was centrifuged, washed once with 0.9% wt/vol. saline and resuspended in BHI broth supplemented with 10% (vol./vol.) fetal bovine serum at a density of approximately 10^{10} colony forming units (cfu)/ml (based upon turbidimetry). The suspension was aliquoted, and a portion was used to check that a monoculture was obtained by subculturing on BHI agar plates. The culture was stored frozen (–125°C) and was used without subculture. For use as an inoculum, the culture is thawed, diluted in saline and rats are inoculated with approximately $2–5 \times 10^7$ cfu/tibia and the inoculum is enumerated by plating dilutions on BHI agar. The culture has proved to be stable for at least 6 years, and consistently results in more than 95% of the rats developing osteomyelitis. Freshly prepared inocula could also be used, but the culture should be of proven high viability, as dead bacteria and their products will probably induce a sterile osteomyelitis.

Infection process

The infection is initiated by direct injection of bacterial culture into the medullary cavity of the bone as described above. A comparison of the effectiveness of different infection procedures (Rissing *et al.*, 1985b) indicated that the combination of drilling, sodium morrhuate and sealing the drill hole produced the most reliable infection. Bone wax appears to promote infection by acting as a foreign body (Nelson *et al.*, 1990b). Sodium morrhuate is a complex of fatty acids (Mader and Wilson, 1986) and is rich in arachdonic acid (Rissing *et al.*, 1985a); arachdonic acid and metabolites (prostaglandins) are known to induce bone damage (Nair *et al.*, 1996) and participate in bone destruction during bacterial osteomyelitis in the rat (Rissing *et al.*, 1985a). The low-level, but reversible damage to the bone probably facilitates infection. Not all investigators use sodium morrhuate. Hienz *et al.* (1995 and Chapter 54) have used a modified procedure where the mandible or tibial bones of rats are exposed, drilled and injected only with sodium morrhuate and left unsealed, the incision being closed with skin staples. Infection was initiated by the intravenous (i.v.) injection of between 5×10^6 and 5×10^8 cfu of *S. aureus* via the femoral vein. Another variant of this model was described by Elson *et al.* (1977), where contaminated bone cement plugs were implanted into the tibia of rats.

Key parameters to monitor infection and response to treatment

The infection is confined to the bone. Deaths due to sepsis are rare, as is bacteriaemia, and therefore these are not suitable indices of infection. Appearance, body weight gain, food and water intake are similar to normal animals, and therefore changes in these parameters should indicate an abnormal infection process or treatment-associated toxicity. We have not thoroughly investigated haematological changes, but slight neutrophilia is common.

Bone pathology

Bone histology is normally used to monitor progress of infection and response to antimicrobial therapy (Spagnolo *et al.*, 1993; Hienz *et al.*, 1995). Typically, bones are fixed in 5% neutral-buffered formalin for at least 48 hours at room temperature, then decalcified by immersion in 4 N formic acid (with or without 10% wt/vol. sodium citrate). Bones are then embedded in paraffin wax and sectioned at 4–6 µm. Staining is normally with haematoxylin and eosin, but other specialized stains can be used. Pathology scores have been used by several investigators. Rissing *et al.* (1985b) and Nelson *et al.* (1990a) scored gross pathology according to: 0, no apparent pathology; 1 minimal erythema without abscess or new bone formation; 2, erythema and widening of tibial shaft (new bone formation); 3, purulent exudate or abscess; and 4, severe bone destruction involving more than 50% of diaphysis. Gisby *et al.* (1994) have used a histopathology scoring system as follows: 0, no apparent pathology; 1, small microabcesses containing bacteria-free neutrophils; 2, small microabcesses containing bacteria and areas of necrotic bone and bone resorption; 3, moderate microabcesses with a central area containing bacteria and large area of necrotic bone and extensive bone resorption; and 4, large microabcesses with purulent exudate. *In situ* hybridization techniques have been used to monitor the appearance of TNFα mRNA during the course of osteomyelitis in the rat and associate its upregulation with areas of bone turnover (Littlewood-Evans *et al.*, 1997a,b).

Bone weights increase during infection, disproportionally to increase in body weight, and can serve as a crude measure of infection progress but apparently do not rapidly respond to chemotherapy (Table 66.1).

Use of imaging techniques

Among others, X-rays have been used by Rissing *et al.* (1985b) to assess infection. A composite score devised upon radiographic evidence of periosteum structure, architecture of bone, shaft dimension and new bone formation was used to assess progress of infection and outcome of antibacterial therapy. X-rays have also been used to profile variations of the basic model (Spagnolo *et al.*, 1993; Hienz *et al.*, 1995). Denistometric analysis of X-ray films is recommended to add a quantitative element to the analysis. Modern imaging techniques are extensively used in the diagnosis of osteomyelitis (Crim and Seeger, 1994; Abernethy and Carty, 1997; Ma *et al.*, 1997), but experimental use is limited, probably because of restricted access to the required equipment. Ultrasonic imaging has been applied in the rabbit model of osteomyelitis (Abiri *et al.*, 1992). Magnetic resonance imaging has been used to monitor progress of infection in the rabbit model of osteomyelitis (Volk *et al.*, 1994) but apparently has not been used in the rat model. Determination of bone mineral density by dual-energy X-ray absorptiometry (DXA) is commonly used for evaluation of rat models of bone diseases (e.g. Järvinen *et al.*, 1998), but has not been used in osteomyelitis. Computed tomography (CT) is also used in experimental models of bone disease in rats (Ferretti, 1995; Gasser, 1995; Breen *et al.*, 1996; Ferretti *et al.*, 1996), but not yet in osteomyelitis models. Recently, fluoro-2-deoxy-D-glucose positron emission tomography (FDG PET) has been used to image clinical osteomyelitis (Guhlmann *et al.*, 1998) and presumably could also be used for experimental evaluations.

Bacterial density in bone

Determination of bacterial cfu within bone normally requires homogenizing the bone. This can be easily accomplished by deep-freezing the bones (liquid nitrogen) and then crushing them in a metal ball mill (e.g. Mixer Mill, Retsch, Haan, Germany, see Dworkin *et al.*, 1990) or pulverized by hand using a frozen (–70°C) mortar and pestle (Rissing *et al.*, 1985b; Gomis *et al.*, 1990b). The bone powder should remain cold, or preferably frozen, during the procedure and is then resuspended in a small volume of liquid. Standard dilution and bacterial enumeration procedures can then be used. When establishing the model, a control of spiking uninfected bones with a known inoculum of bacteria should be performed in order to ensure that the bacteria do indeed survive the freezing and extraction process. As the method cannot 'dissolve' bone, a fine suspension is produced and care must be taken to use a representative sample for dilutions, and interpret data carefully, as bacteria are likely to be adsorbed to particulate bone material. Typical results from a treatment experiment are shown in Table 66.1.

Determination of antibiotic penetration into bone

Bone can be pulverized as described above. Antibiotic concentrations can be determined by a variety of methods, although bioassay appears to be the most common method used. The bone antibiotic levels determined should be carefully interpreted owing to the nature of the material used for determinations (suspension rather than solution) and the degree of adsorption of antibiotics to bone being generally either not determined or not reported. Presumably, autoradiographic methods to determine radiolabelled antibiotic penetration into cardiac vegetations during endocarditis (Crémieux *et al.*, 1989) or joint prosthesis infection (Crémieux *et al.*, 1996) could be used in this model.

It should be noted that some antibiotics are sensitive to freeze–thaw cycles. Therefore, preliminary experiments where bone powder is spiked with a known concentration of antibiotic should be tested for a reduction in antibiotic concentration and/or activity following freezing and thawing.

The concentrations achieved in rat bone (Table 66.2) appear to correlate well with those found in humans. Ciprofloxacin levels in rat bone (≈ 2.1 mg/kg, Henry *et al.*, 1987; ≈ 1.7 mg/kg, Dworkin *et al.*, 1990) are similar to those found in osteomyelitic patients (≈ 1.4 mg/kg, Fong *et al.*, 1986). High-dose pefloxacin achieves bone levels of approximately 5–8 mg/kg in humans (Dan *et al.*, 1990; Green, 1993) and approximately 8.3 mg/kg was found in rat

Table 66.1 Activity of azithromycin and rifampicin treatment on bone weights and bacterial densities in the bone during chronic *S. aureus* osteomyelitis in rats. Rats were infected by direct inoculation with $\log_{10}$ 7.7 cfu *S. aureus* 1098 and treatment with saline or azithromycin (Az; 50 mg/kg, p.o.) plus rifampicin (Rif; 20 mg/kg s.c.) for 21 days commencing either 14 or 56 days postinfection. All values are means ± SEM. Data are from Littlewood-Evans *et al.* (1997a,b and unpublished)

	Grams of bone			*Log (cfu/g bone)*		
Day	*Uninfected*	*Saline*	*Az/Rif*	*Uninfected*	*Saline*	*Az/Rif*
0	0.34 ± 0.01			< 1.8		
Starting treatment 14 days postinfection						
14	0.52 ± 0.04	0.9 ± 0.1		< 1.8	6.4 ± 0.2	
39	0.65 ± 0.05	1.0 ± 0.2	1.0 ± 0.2	< 1.8	5.9 ± 0.3	2.4 ± 0.6
49	0.63 ± 0.06	0.8 ± 0.2	1.0 ± 0.2	< 1.8	5.9 ± 0.3	4.4 ± 0.5
Starting treatment 56 days postinfection						
56	0.64 ± 0.09	1.23 ± 2.3		< 1.8	6.2 ± 0.4	
81	0.71 ± 0.05	0.96 ± 0.11	1.09 ± 0.2	< 1.8	6.0 ± 0.1	< 1.8
84	0.70 ± 0.04	0.9 ± 0.09	1.06 ± 0.25	< 1.8	5.8 ± 0.4	4.6 ± 1
98	0.73 ± 0.06	1.01 ± 0.27	0.92 ± 0.05	< 1.8	6.0 ± 0.3	5.5 ± 0.3

Table 66.2 Examples of chemotherapy of experimental bacterial osteomyelitis infections in the rat (NR, not reported)

			Peak antibiotic concentration			
Reference	*Bacteria*	*Antibiotics (dose, route)*	*Serum (mg/l)*	*Bone (mg/kg)*	*Bone cfu (log cfu/g bone)*	*Comments*
Zak *et al.*, 1981	*S. aureus*	Control			6.5 ± 0.4	Bone cfu determined 48 hr following cessation of treatment
		Cloxacillin (100 mg/kg, q. 12 hr, s.c. for 5 days starting 21 days postinfection)	14.3 ± 2.1	4.3 ± 3.2	5.8 ± 0.3	Prolongation of treatment to 21 days improved outcome only of combination
		Rifampicin (20 mg/kg, q. 24 hr, s.c. for 5 days starting 21 days postinfection)	15.8 ± 1.4	12.6 ± 2.2	4.2 ± 0.5	
		Cloxacillin + rifampicin			3.7 ± 1.2	
Rissing *et al.*, 1985b	*S. aureus*	Control			6.28	Bone cfu determined 14 days following cessation of treatment
		Oxacillin (120 mg/kg, q. 12 hr, s.c. for 14 days starting 21 days postinfection)	NR	NR	4.89	Prolonged treatment improves outcome only for cloxacillin
		Ceftriaxone (50 mg/kg, q. 24 hr, s.c. for 14 days starting 21 days postinfection)	NR	NR	4.07	
Henry *et al.*, 1987	*S. aureus* (MRSA)	Control			4.4 ± 0.7	Cfu determined 12 hr after completion of therapy
		Vancomycin (50 mg/kg, q. 12 hr, i.m. for 21 days starting 28 days postinfection)	38.0 (peak at 30 min)	5.0 (peak at 30 min)	4.4 ± 0.7	Regrowth determined 28 days after completion of therapy Significant only in ciprofloxacin + rifampicin group (to log 2.9 ± 2.0 cfu/g bone)
		Ciprofloxacin (50 mg/kg, q. 12 hr, s.c. for 21 days starting 28 days postinfection)	4.9 (peak at 30 min)	2.1 (peak at 30 min)	3.9 ± 1.2	
		Rifampicin (5 mg/kg, q. 12 hr, s.c. for 21 days starting 28 days postinfection)	2.9 (peak at 60 min)	0.65 (peak at 60 min)	2.1 ± 1.5	
		Vancomycin + rifampicin			2.0 ± 1.3	
		Ciprofloxacin + rifampicin			0.5 ± 0.5	

Table 66.2 *Continued*

Reference	*Bacteria*	*Antibiotics (dose, route)*	*Peak antibiotic concentration* Serum (mg/l)	Bone (mg/kg)	*Bone cfu (log cfu/g bone)*	*Comments*
Luu *et al.*, 1989	*S. aureus*	Control			5.13 ± 1.58	1/16 culture-negative
		Daptomycin (10 mg/kg, q. 12 hr, s.c. for 24 days starting 21 days postinfection)	43.6 μg/ml (peak) 2.3 μg/ml (trough)	NR	5.36 ± 0.43	0/16 culture-negative, treatment reduced histopathology score
		Vancomycin (80 mg/kg, q. 12 hr, s.c. for 24 days starting 21 days postinfection)	17 μg/ml (peak) 1.8 μg/ml (trough)	NR	4.33 ± 1.72	2/17 culture-negative, treatment reduced histopathology score
Gomis *et al.*, 1990b	*E. coli*	Control			6.72 ± 0.3 (0% sterile)	Cfu determined 28 days after completion of therapy
		Cefotaxime (100 mg q. 12 hr, s.c.?, for 14 or 28 days starting 14 days postinfection)	NR	NR	3.66 ± 0.95 (28.5% sterile)	Prolongation of treatment improved outcome (to 85.6% sterility of bones); Pathology analyses included
Nelson, *et al.*, 1990a	*P. aeruginosa*	Control			5.08 ± 0.37	Bone cfu determined 14 days post therapy
		Ceftazidime (50 mg/kg, q. 8 hr, s.c. for 14 days starting 35 days postinfection)	ceftazidime: 17.4 ± 4.3 (2 hr)	NR	4.91 ± 0.8	Antibiotics ineffective in reducing cfu or bone pathology
		Ceftazidime + tobramycin (40 mg/kg, q. 12 hr, s.c.)	tobramycin: 30.7 ± 2 (2 hr)	NR	4.57 ± 0.66	
Dworkin, *et al.*, 1990	*S. aureus*	Control			6.1 ± 0.43	Cfu determined 4 days after completion of therapy
		Ciprofloxacin (30 mg/kg, q. 12 hr, s.c. for 24 days starting 25 days postinfection)	3.4 ($T_{1/2}$ 2.4 hr)	1.7 ($T_{1/2}$ 2.5 hr)	6.06 ± 0.26	
		Pefloxacin (60 mg/kg, q. 12 hr, s.c. for 24 days starting 25 days postinfection)	12.1 ($T_{1/2}$ 2.8 hr)	8.3 ($T_{1/2}$ 3.5 hr)	6.04 ± 0.38	
		Vancomycin (60 mg/kg, q. 12 hr, s.c. for 24 days starting 25 days postinfection)	42.2 ($T_{1/2}$ 1.4 hr)	9.0 ($T_{1/2}$ 2.2 hr)	5.43 ± 1.05	
		Rifampicin (20 mg/kg, q. 24 hr, s.c. for 24 days starting 25 days postinfection)	12.6 ($T_{1/2}$ 5.5 hr)	19.1 ($T_{1/2}$ 5.0 hr)	4.74 ± 0.60	
		Ciprofloxacin + rifampicin			2.88 ± 1.25	
		Pefloxacin + rifampicin			3.07 ± 1.15	
		Vancomycin + rifampicin			2.76 ± 1.49	

Table 66.2 *Continued*

Reference	*Bacteria*	*Antibiotics (dose, route)*	*Peak antibiotic concentration* Serum (mg/l)	*Peak antibiotic concentration* Bone (mg/kg)	*Bone cfu (log cfu/g bone)*	*Comments*
O'Reilly *et al.*, 1992	*S. aureus*	Control			6.54 ± 0.20 (0% sterile)	Cfu determined 4 days after completion of therapy
		Azithromycin (50 mg/kg, q. 24 hr, p.o. starting 10 days postinfection)	0.63 ± 0.24 ($T_{1/2}$ 8.5 hr)	20.7 ± 15.6 ($T_{1/2}$ > 24 hr)	5.31 ± 0.46 (0% sterile)	
		Clindamycin (90 mg/kg, q. 8 hr, p.o. starting 10 days postinfection)	2.98 ± 1.1 ($T_{1/2}$ 4.2 hr)	3.22 ± 0.5 ($T_{1/2}$ 7.2 hr)	3.26 ± 2.14 (20% sterile)	
		Rifampicin (20 mg/kg, q. 24 hr, s.c., starting 10 days postinfection)	12.6 ($T_{1/2}$ 5.5 hr)	19.1 ($T_{1/2}$ 5.0)	1.50 ± 1.92 (53% sterile)	
		Clindamycin + rifampicin			0.87 ± 1.34 (66% sterile)	
		Azithromycin + rifampicin			0.51 ± 1.08 (80% sterile)	
Gisby *et al.*, 1994	*S. aureus*	Control			4.89 ± 0.55 (0% sterile)	Cfu determined 6 days after completion of therapy
		Amoxycillin/clavulanic acid (200/50 mg/kg, q. 8 hr, s.c., for 28 days starting 14 days postinfection)	121.8 ± 17.9 ($T_{1/2}$ 0.76 hr) (amoxycillin)	22.5 ± 4.0 ($T_{1/2}$ 1.34 hr) (amoxycillin)	2.95 ± 1.88 (50% sterile)	Re-growth at 28 days after completion of therapy
		Flucloxacillin (200 mg/kg, q. 8 hr, s.c., for 28 days starting 14 days postinfection)	51.4 ± 17.5 ($T_{1/2}$ 1.46 hr)	9.5 ± 6.9 ($T_{1/2}$ 0.68)	3.5 ± 1.5 (17% sterile)	Pathology analyses included
		Clindamycin (50 mg/kg, q. 8 hr, s.c., for 28 days starting 14 days postinfection)	7.6 ± 1.1 ($T_{1/2}$ 1.68 hr)	10.7 ± 2.3 ($T_{1/2}$ 2.41 hr)	2.92 ± 1.0 (25% sterile)	
Littlewood-Evans *et al.*, 1997b	*S. aureus*	Controls			6.0 ± 0.1	Cfu determined 4 days after completion of therapy
		Azithromycin (50 mg/kg, q. 24 hr, p.o.) and rifampicin (20 mg/kg, q. 24 hr, s.c.) (for 21 days starting 56 days postinfection)	NR	NR	< 1.8	Regrowth at 8 days after completion of therapy

bone (Dworkin *et al.*, 1990). However, levels of azithromycin were considerably higher in infected rat bone (≈18.7 mg/kg, O'Reilly *et al.*, 1992) than those observed in non-osteomyelitic humans given a single dose (≈1 mg/kg, Foulds *et al.*, 1990), but this may reflect greater azithromycin penetration within diseased bones. The penetration of antibiotics into bone has been reviewed (Hughes and Anderson, 1985).

Determination of additional markers of infection and response to treatment

Littlewood-Evans *et al.* (1997a,b) measured TNF concentrations in aqueous extracts of bone powder using the WEHI 164 bioassay method (Espevik and Nissen-Meyer, 1986). TNF was persistently elevated during infection and responded little to antimicrobial therapy, at least during the observation period. Increased TNFα mRNA (extracted by acid-guanidinium-thiocyanate–phenol/chloroform extraction (Chomczynski and Sacchi, 1987) and analysed by northern blot procedure) was correlated to TNF activity (Littlewood-Evans *et al.*, 1997a,b). Elevated TNF activity is also found in clinical samples of bone (Klosterhalfen *et al.*, 1996), but not in serum (Klosterhalfen *et al.*, 1996; Evans *et al.*, 1998); TNF activity has not been detected in the serum of rats with experimental osteomyelitis (Littlewood-Evans and O'Reilly, unpublished). For the RNA extraction procedure to be reliable, the bone should remain frozen throughout the pulverizing procedure, owing to the labile nature of RNA.

Antimicrobial therapy

A variety of antimicrobial agents have been evaluated in the rat model of osteomyelitis and have been summarized in Table 66.2. Rats can be administered antibiotics by a variety of routes (s.c., i.p., p.o. or gavage) but i.v. administration (via the tail vein) is technically difficult, especially on older animals used in this model. Care should be taken, especially with oral application routes, to formulate the antibiotic appropriately to ensure maximal exposure of the drug. Antimicrobial therapy is usually begun 14 or 21 days postinfection (Table 66.2), but a variety of pretreatment periods have been used, including up to 56 days postinfection (Tables 66.1 and 66.2). Determination of the bone cfu is usually done several days after the cessation of treatment in order to allow for elimination of the antibiotic from the bone (thus reducing any effects of antibiotic carry-over during the determination of cfu) and to determine if the bones have indeed been sterilized (Table 66.2); alternatively, all the bone powder used for the dilution series can be cultured in broth to determine sterility (Dworkin *et al.*, 1990; O'Reilly *et al.*, 1992). Inactivation of antibiotics during dilution of the bone suspension is recommended (e.g. use of β-lactamase for β-lactam antibiotics. Duration of treatment can dramatically affect the outcome of therapy (Table 66.2, see also Norrby *et al.*, 1993), and experimental design should consider varying the length of treatment. The dosing schedule and choice of dose should be given careful consideration, given the differing pharmacokinetics of antibiotics between humans and rats (O'Reilly *et al.*, 1996). Although fractional dosing of antibiotics to mimic human pharmacokinetics is probably best, consideration of the intensity of labour and the long duration of treatment may require a compromise. Interestingly, measurement of serum bactericidal titres is performed by many clinicians as a means of monitoring the effectiveness of antimicrobial chemotherapy of osteomyelitis (Haas and McAndrew, 1996), but this is apparently rarely done during experimental studies (Table 66.2).

Local antibiotic therapy of osteomyelitis using antibiotic-impregnated beads or bone cements is employed clinically (Henry *et al.*, 1991; Popham *et al.*, 1991; Henry and Galloway, 1995; Wininger and Fass, 1996). Laurencin *et al.*, (1993) have used the rat model to evaluate the activity of gentamicin in locally implanted slow-release formulation, demonstrating the superiority of a polyanhydride/gentamicin over a polymethylmethacrylate (PMMA)/gentamicin formulation. Similar studies have been performed evaluating gentamicin plus vancomycin in PMMA as prophylaxis against *S. aureus* osteomyelitis (Gerhart *et al.*, 1988, 1993). The rat osteomyelitis model has been used to show that hydroxyapatite provides a suitable slow-release substrate for arbekacin (Itokazu *et al.*, 1997a) as does a freeze-dried fibrin-based adhesive agent (Itokazu *et al.*, 1997b). The clinical use of gentamicin beads has been associated with the selection of persistent small colony variants and recurrence of osteomyelitis (von Eiff *et al.*, 1997); the rat model should be useful for elucidating this process.

Pitfalls (advantages/disadvantages) of the model

Besides the usual shortcomings of all animal models (e.g. differing pharmacokinetic profiles of antibiotics; Craig, 1998), the major shortcoming of the rat model of osteomyelitis is the use of the large inocula, directly implanted into bone, that are required to establish the infection. This most probably does not represent the clinical situation, nor the natural progression of infection in humans, except perhaps for osteomyelitis arising from grossly contaminated fractures. This model may not always represent the normal location of osteomyelitis: infection of the tibia appears most commonly in children, where haematogenous origin of disease is likely, whereas in adults the appendicular skeleton is usually involved (Haas and McAndrew, 1996). Another limitation of the model is that the drilling procedure is likely to contaminate the infection site with

Table 66.3 Correlation of antibiotic efficacy in the experimental rat model of osteomyelitis with clinical experience

Experimental experience				*Clinical experience*				
Reference	*Infection*	*Agent(s)*	*Outcome*	*Reference*	*Infection*	*Agent(s)*	*Outcome*	*Comments*
Gomis *et al.*, 1990b	*E. coli*	Cefotaxime	Favourable, long-term treatment sterilizes bones	Gomis *et al.*, 1990a	Gram-negative and mixed Gram-negative and Gram-positive	Cefotaxime	Favourable. Outcome influenced by surgical intervention	Positive correlation
				LeFrock and Carr, 1982	Gram-positive and Gram-negative	Cefotaxime	76% clinical response rate	
				Mader *et al.*, 1982		Cefotaxime	Clinical, responses in 93%, of acute, and 89% of chronic cases	
				Loffler *et al.*, 1986	*S. aureus*	Cefotaxime vs ampicillin/ sulbactam	100% clinical response 85% bacteriological cure; agents considered equivalent	
				Maucer, 1994	Gram positive and Gram-negative	Cefotaxime with ambulatory delivery system	83% clinical response; 78% bacteriological response	
Gisby *et al.*, 1994	*S. aureus*	Amoxicillin plus clavulanic acid	Favourable, but failed to eradicate infection	Bassey, 1992	*S. aureus, E. coli* and *Proteus* spp.	Amoxycillin plus clavulanic acid	82.9% clinical cure rate and 94.4% bacteriological cure rate in *S. aureus* infections	Positive correlation

Table 66.3 *Continued*

Experimental experience				Clinical experience				
Reference	*Infection*	*Agent(s)*	*Outcome*	*Reference*	*Infection*	*Agent(s)*	*Outcome*	*Comments*
Gisby *et al.*, 1994; O'Reilly *et al.*, 1992; Dworkin *et al.*, 1990; Henry *et al.*, 1987	*S. aureus*	Clindamycin	Favourable, improved activity in combination with rifampicin	Pontifex and McNaught, 1973; Hugo *et al.*, 1977		Clindamycin (as sole agent)	84% clinical response rate, some side effects	Positive correlation
Dworkin *et al.*, 1990; Henry *et al.*, 1987	*S. aureus*	Quinolones	Poor as single agents, improved activity in combination with rifampicin or vancomycin	Lew and Waldvogel, 1995; Rissing, 1997		Quinolones	Reviews of clinical data; good clinical responses	Negative or weak correlation with single agents results, positive correlation with combination results
Laurencin *et al.*, 1993	*S. aureus*	Locally implanted gentamicin in biodegradable implants	Favourable	Wininger and Fass, 1996		Several	Review of clinical data; good clinical responses	Positive correlation
Nelson *et al.*, 1990a; Rouse *et al.*, 1992	*P. aeruginosa*	β-lactam ± gentamicin	β-lactam alone poor, but combination improves outcome	Lucht *et al.*, 1994	*P. aeruginosa*	β-lactam quinolone combinations for prolonged treatment	Combinations result in 73% clinical cure and 93% bacteriological cure	Model predicted that monotherapy is inadequate

dead bone, which may become a "foreign" body that promotes infection; as infections of foreign bodies are extremely resistant to infection this model may represent a severe test of antimicrobial efficacy. Furthermore, the large inocula will most probably initiate a rapid induction of pathology associated with bacteria or excreted products (Nair *et al.*, 1996) that may not fully reflect the clinical situation. This aggressive infection may also contribute to the lack of bacteriological cures in the rat model (Tables 66.1 and 66.2) despite the clinical success of the antibiotics used (Table 66.3); a compounding factor is that treatment courses used experimentally (~3–4 weeks) are considerably shorter that those clinically (up to several months in adult osteomyelitis).

Because of these limitations, it is recommended that care be taken in the experimental design and subsequent interpretation of data on antimicrobial effectiveness when using this rat model of osteomyelitis. At a minimum, the evaluation of new agents or novel combinations of agents should be taken in comparison with clinically used reference agents. Furthermore, the dose levels of the antibiotics should be "reasonable", at least in so far as the serum levels of antibiotics are those attainable in humans. In the case of compounds with unknown pharmacokinetics in humans, the dose and schedule used should be comparable to reference compounds or the same or similar chemical class.

The rat model is apparently less used than the rabbit model of osteomyelitis (Chapter 68). Some advantages of the rat model over the rabbit model include the size of the animal, thus necessitating smaller quantities of antimicrobial agents and reduced housing space, and the fact that rats are generally more tolerant to prolonged antibiotic administration (Morris, 1995). Although repeated bleeding for determination of plasma pharmacokinetics is easier in rabbits, use of the sublingual vein (under the tongue) allows repeated bleeding of the rat. Rats are mildly anaesthetized by inhalation (isoflurane), the tongue is held back and the vein is punctured with a 23G needle; blood is collected in a tube containing anti-coagulant. Volumes of 1 ml can be routinely collected, and the bleeding can be stopped by wiping the underside of the tongue with a cotton swab soaked in 10% (wt/vol.) $FeCl_3$.

Contributions of the model to infectious disease therapy

Profiling of antimicrobial treatment

Although animal model evaluations are not considered a prerequisite to clinical trials of novel antimicrobial agents in osteomyelitis (Mader *et al.*, 1992; but see Beam *et al.*, 1992) the rat model of osteomyelitis is often used in the profiling of new drugs or evaluation of novel combinations. A comparison of some of the efficacy data from the rat model with efficacy data from clinical studies is presented in Table 66.3; the data presented was selected to match the experimental and clinical studies and therefore should be considered illustrative rather than exhaustive. The outcome of antimicrobial therapy appears to correlate well with the clinical situation (Table 66.3), although bacteriological cures are generally not achieved in the rat model, probably because of the lack of prolonged duration of treatment, which has been shown in a number of infection models to affect outcome (Norrby *et al.*, 1993).

Elucidation of the pathophysiology of bacterial osteomyelitis

This model has also been used to identify the involvement of prostaglandins (PGE_2) in pathogenesis and has demonstrated the potential value of an adjunct anti-inflammatory agent (ibuprofen) in osteomyelitis (Rissing and Buxton, 1986). Elevated bone TNF levels in osteomyelitis occurs clinically (Klosterhalfen *et al.*, 1996) and in infected rats (Littlewood-Evans *et al.*, 1997a), but is not rapidly reduced by antibiotic treatment (Littlewood-Evans *et al.*, 1997b); TNF is implicated in bacteria-associated bone destruction (Nair *et al.*, 1996) and therefore may be a suitable target for novel adjunct treatment of osteomyelitis.

References

Abernethy, L. J., Carty, H. (1997). Modern approach to the diagnosis of osteomyelitis in children. *Br. J. Hosp. Med.*, **58**, 464–468.

Abiri, M. M., DeAngelis, G. A., Kirpekar, M., Abou, A. N., Ablow, R. C. (1992). Ultrasonic detection of osteomyelitis. Pathologic correlation in an animal model *Invest. Radiol.*, **27**, 111–113.

Bassey, L. (1992). Oral and parenteral amoxycillin/clavulanic acid in conjunction with surgery for the management of chronic osteomyelitis and severe bone infection. *Curr. Ther. Res.*, **52**, 922–928.

Beam, T. R., Gilbert, D. N., Kunin, C. M. (1992). General guidelines for the clinical evaluation of anti-infective drugs. *Clin. Infect. Dis.*, **15** (Suppl. 1), S5–S32.

Bertens, A. P. M. G., Booij, L. H. D. J., Flecknell, P. A., Lagerweij, E. (1993). Anaesthesia, analgesia and euthanasia. In: *Principles of Laboratory Animal Science* (eds van Zutphen, L. F. M., Baumans, V., Beyen, A. C.), pp. 267–298. Elsevier, Amsterdam.

Breen, S. A., Millest, A. J., Loveday, B. E., Johnstone, D., Waterton, J. C. (1996). Regional analysis of bone mineral density in the distal femur and proximal tibia using peripheral quantitative computed tomography in the rat *in vivo*. *Calcif. Tissue Int.*, **58**, 449–453.

Chomzynski, P., Sacchi, N. (1987). Single-step method of RNA isolation by acid guanidinium thiocyanate-phenol-chloroform extraction. *Anal. Biochem.*, **162**, 156–159.

Craig, W. A. (1998). Pharmacokinetic/pharmacodynamic parameters: rationale for the dosing of mice and men. *Clin. Infect. Dis.*, **26**, 1–12.

Crémieux, A. C., Carbon, C. (1997) Experimental models of bone and prosthetic joint infections. *Clin. Infect. Dis.*, **25**, 1295–1302.

Crémieux, A. C., Mazière, B., Vallois, J. M. *et al.* (1989). Evaluation of antibiotic diffusion into cardiac vegetations by quantitative autoradiography. *J. Infect. Dis.*, **159**, 938–944.

Crémieux, A. C., Saleh-Mghir, A., Bleton, R. *et al.* (1996). Efficacy of sparfloxacin and autoradiographic diffusion pattern of [^{14}C] sparfloxacin in experimental *Staphylococcus aureus* joint prosthesis infection. *Antimicrob. Agents Chemother.*, **40**, 2111–2116.

Crim, J. R., Seeger, L. L. (1994). Imaging evaluation of osteomyelitis. *Crit. Rev. Diagn. Imaging*, **35**, 201–256.

Cunningham, R., Cockaayne, A., Humphreys, H. (1996). Clinical and molecular aspects of the pathogenesis of *Staphylococcus aureus* bone and joint infections. *J. Med. Microbiol.*, **44**, 157–164.

Dan, M., Siegman, I. Y., Pitlik, S. *et al.* (1990). Oral ciprofloxacin treatment of *Pseudomonas aeruginosa* osteomyelitis. *Antimicrob. Agents Chemother.*, **34**, 849–852.

Dworkin, R., Modin, G., Kunz, S., Rich, R., Zak, O., Sande, M. (1990). Comparative efficacies of ciprofloxacin, pefloxacin, and vancomycin in combination with rifampin in a rat model of methicillin-resistant *Staphylococcus aureus* chronic osteomyelitis. *Antimicrob. Agents Chemother.*, **34**, 1014–1016.

Elson, R. A., Jephcott, A. E., McGechie, D. B., Verettas, D. (1977). Bacterial infection and acrylic cement in the rat. *J. Bone Joint Surg.*, **59B**, 452–457.

Espevik, T., Nissen-Meyer, J. (1986). A highly sensitive cell line, WEHI 164 subclone 13, for measuring cytotoxic factor/tumor necrosis factor from human monocytes. *J. Immunol.*, **95**, 99–105.

Evans, C. A. W., Jellis, J., Hughes, S. P. F., Remick, D. G., Friedland, J. S. (1998). Tumor necrosis factor-α, interleukin-6 and interleukin-8 secretion and the acute phase response in patients with bacterial and tuberculous osteomyelitis. *J. Infect. Dis.*, **177**, 1582–1587.

Ferretti, J. L. (1995). Perspectives of pQCT technology associated to biomechanical studies in skeletal research employing rat models. *Bone*, **17** (4 Suppl.), 353S–364S.

Ferretti, J. L., Capozza, R. F., Zanchetta, J. R. (1996). Mechanical validation of a tomographic (pQCT) index for noninvasive estimation of rat femur bending strength. *Bone*, **18**, 97–102.

Fong, I. W., Ledbetter, W. H., Vandenbroucke, A. C., Simbul, M., Rahm, V. (1986). Ciprofloxacin concentration in bone and muscle after oral dosing. *Antimicrob. Agents Chemother.*, **29**, 405–408.

Foulds, G., Shepard, R. M., Johnson, R. B. (1990). The pharmacokinetics of azithromycin in human serum and tissues. *J. Antimicrob. Chemother.*, **25** (Suppl. A), 473–482.

Gasser, J. A. (1995). Assessing bone quantity by pQCT. *Bone*, **17** (4 Suppl.), 145S–154S.

Gerhart, T. N., Roux, R. D., Horowitz, G., Miller, R. L., Hanff, P., Hayes, W. C. (1988). Antibiotic release from an experimental biodegradable bone cement. *J. Orthoped. Res.* **6**, 585–592.

Gerhart, T. N., Roux, R. D., Hanff, P. A., Horowitz, G. L., Renshaw, A. A., Hayes, C. (1993). Antibiotic-loaded biodegradable bone cement for prophylaxis and treatment of experimental osteomyelitis in rats. *J. Orthoped. Res.*, **11**, 250–255.

Gisby, J., Beale, A. S., Bryant, J. E., Toseland, C. D. N. (1994). Staphylococcal osteomyelitis—a comparison of co-amoxiclav with clindamycin and flucloxacillin in an experimental model. *J. Antimicrob. Chemother.*, **34**, 755–764.

Gomis, M., Herranz, A., Aparicio, P., Alonso, M. J., Filloy, L., Pastor, J. (1990a). Cefotaxime in the treatment of chronic osteomyelitis caused by Gram-negative bacilli. *J. Antimicrob. Chemother.*, **26** (Suppl. A), 45–52.

Gomis, M., Herranz, A., Aparicio, P., Alonso, M. J., Prieto, J., Martinez, T. (1990b). An experimental model of chronic osteomyelitis caused by *Escherichia coli* treated with cefotaxime. *J. Antimicrob. Chemother.*, **26** (Suppl. A), 15–21.

Green, S. L. (1993). Efficacy of oral fleroxacin in bone and joint infections. *Am. J. Med.*, **94** (Suppl. A), S174–S176.

Guhlmann, A., Brecht-Krauss, D., Suger, G. *et al.* (1998) Chronic osteomyelitis: detection with FDG PET and correlation with histopathologic findings. *Radiology*, **206**, 749–754.

Haas, D. W., McAndrew P. (1996). Bacterial osteomyelitis in adults: evolving considerations in diagnosis and treatment. *Am. J. Med.*, **101**, 550–561.

Henry, S. L., Galloway, K. P. (1995). Local antibacterial therapy for the management of orthopaedic infections. Pharmacokinetic considerations. *Clin. Pharmacokinet.*, **29**, 36–45.

Henry, N. K., Rouse, M. S., Whitesell, A. L., McConnell, M. E., Wilson, W. R. (1987). Treatment of methicillin-resistant staphylococcal osteomyelitis with ciprofloxacin or vancomycin alone or in combination with rifampin. *Am. J. Med.*, **82** (Suppl. 4A), 73–75.

Henry, S. L., Seligson, D., Mangino, P., Popham, G. L. (1991). Antibiotic-impregnated beads. Part I: Bead implantation versus systemic therapy. *Orthoped., Rev.*, **20**, 272–247.

Hienz, S. A., Sakamoto, H., Flock, J.-I. *et al.* (1995). Development and characterization of a new model of hematogenous osteomyelitis in the rat. *J. Infect. Dis.*, **171**, 1230–1236.

Hughes, S. P., Anderson, F. M. (1985). Penetration of antibiotics into bone. *J. Antimicrob. Chemother.*, **15**, 517–519.

Hugo, H., Dornvusch, K., Sterner, G. (1977). Studies on the clinical efficacy, serum levels and side effects of clindamycin phosphate administered intravenously. *Scand. J. Infect. Dis.*, **9**, 221–226.

Itokazu, M., Ohno, T., Tanemori, T., Wada, E., Kato, N., Watanabe, K. (1997a). Antibiotic-loaded hydroxyapatite blocks in the treatment of experimental osteomyelitis in rats. *J. Med. Microbiol.*, **46**, 779–783.

Itokazu, M., Yamamoto, K., Yang, W. Y., Aoki, T., Kato, N., Watanabe, K. (1997b). The sustained release of antibiotic from freeze-dried fibrin-antibiotic compound and efficacies in a rat model of osteomyelitis. *Infection*, **25**, 359–363.

Järvinen, T. L. N., Sievänen, H., Kannus, P., Järvinen, M. (1998). Dual-energy X-ray absorptiometry in predicting mechanical characteristics of rat femur. *Bone*, **22**, 551–558.

Klosterhalfen, B., Peters, K. M., Töns, C., Hauptmann, S., Klein, C. L., Kirkpatrick, C. J. (1996). Local and systemic inflammatory mediator release in patients with acute and chronic post-traumatic osteomyelitis. *J. Trauma*, **40**, 372–378.

Laughlin, R. T., Sinha, A., Calhoun, J. H., Mader, J. T. (1994). Osteomyelitis. *Curr. Opin. Rheumatol.*, **6**, 401–407.

Laurencin, C. T., Gerhart, T., Witschger, P. *et al.* (1993). Bioerodible polyanhydrides for antibiotic drug delivery: *In vivo* osteomyelitis treatment in the rat model system. *J. Orthoped. Res.*, **11**, 256–262.

LeFrock, J. L., Carr, B. B. (1982). Clinical experience with cefotaxime in the treatment of serious bone and joint infections. *Rev. Infect. Dis.*, **4** (Suppl.), S465–S471.

Lew, D. P., Waldvogel, F. A. (1995). Quinolones and osteomyelitis: state-of-the-art. *Drugs*, **49** (Suppl. 2), 100–111.

Lew, D. P., Waldvogel, F. A. (1997). Osteomyelitis. *N. Engl. J. Med.*, **336**, 999–1007.

Littlewood-Evans, A. J., Hattenberger, M., Lüscher, C., Pataki, A., Zak, O., O'Reilly, T. (1997a). Local expression of TNFα during an experimental rat model of osteomyelitis. *Infect. Immun.*, **65**, 3438–3443.

Littlewood-Evans, A. J., Hattenberger, M., Zak, O., O'Reilly, T. (1997b). Effect of combination therapy of rifampicin and azithromycin on TNF levels during a subacute rat model of osteomyelitis. *J. Antimicrob. Chemother.*, **39**, 493–498.

Loffler, L., Bauernfeind, A., Keyl, W., Hoffstedt, B., Peirfies, A., Lenz, W. (1986). An open, comparative study of sulbactam plus ampicillin vs. cefotaxime as initial therapy for serious soft tissue and bone and joint infections. *Rev. Infect. Dis.*, **8** (Suppl. 5), S593–S598.

Lucht, R. F., Fresard, A., Berthelot, P. *et al.* (1994). Prolonged treatment of chronic *Pseudomonas aeruginosa* osteomyelitis with a combination of two effective antibiotics. *Infection*, **22**, 276–280.

Luu, Q. N., Buxton, T. B., Nelson, D. R., Rissing, J. P. (1989) Treatment of chronic experimental *Staphylococcus aureus* osteomyelitis with LY146032 and vancomycin. *Eur. J. Clin. Microbiol. Infect. Dis.*, **8**, 562–563.

Ma, L. D., Frassica, F. J., Bluemke, D. A., Fishman, E. K. (1997). CT and MRI evaluation of musculoskeletal infection. *Crit. Rev. Diagn. Imaging.*, **38**, 535–568.

Mader, J. T. (1985). Animal models of osteomyelitis. *Am. J. Med.*, **78**(6B), 213–217.

Mader, J. T., Wilson, K. J. (1986). Models of osteomyelitis. In: *Antimicrobial Chemotherapy, Experimental Models* (eds Zak, O. and Sande, M. A.) vol. 2, pp. 155–173. Academic Press, London.

Mader, J. T., Adams, K. R. (1988). Experimental osteomyelitis. In: *Orthopedic Infection* (ed. Schlossberg, D.), pp. 39–48. Springer Verlag, New York.

Mader, J. T., Calhoun, J. (1996). Osteomyelitis. In: *Principles and Practice of Infectious Disease*, 4th edn (eds Mandell, G. L., Douglas, R. G., Bennett, J. E.), pp. 1039–1051. Churchill Livingstone, New York.

Mader, J. T., LeFrock, J. L., Hyams, K. C., Molavi, A., Reinarz, J. A. (1982). Cefotaxime therapy for patients with osteomyelitis and septic arthritis. *Rev. Infect. Dis.*, **4** (Suppl.), S472–S480.

Mader, J. T., Norden, C., Nelson, J. D., Calandra, G. B. (1992). Evaluation of new anti-infective drugs for the treatment of osteomyelitis in adults. *Clin. Infect. Dis.*, **15** (Suppl. 1), S155–S161.

Mader, J. T., Mohan, D., Calhoun, J. (1997). A practical guide to the diagnosis and management of bone and joint infections. *Drugs*, **54**, 253–264.

Matsushita, K., Hamabe, M., Matsuoka, M. *et al.* (1997). Experimental hematogenous osteomyelitis by *Staphylococcus aureus*. *Clin. Orthopaed. Rel. Res.*, **334**, 291–297.

Maucer, A. A. (1994). Treatment of bone and joint infections utilizing a third-generation cephalosporin with an outpatient drug delivery device. *Am. J. Med.*, **97** (Suppl. 2A), 14–22.

Morris, T. H. (1995). Antibiotic therapeutics in laboratory animals. *Lab. Anim.*, **29**, 16–36.

Nair, S. P., Medghi, S., Wilson, M., Reddi, K., White, P., Henderson, B. (1996). Bacterially induced bone destruction: mechanisms and misconceptions. *Infect. Immun.*, **64**, 2371–2380.

Nelson, D. R., Buxton, T. B., Luu, Q. N., Rissing, J. P. (1990a). An antibiotic resistant experimental model of *Pseudomonas* osteomyelitis. *Infection*, **18**, 246–248.

Nelson, D. R., Buxton, T. B., Luu, Q. N., Rissing, J. P. (1990b). The promotional effect of bone wax on experimental *Staphylococcus aureus* osteomyelitis. *J. Thorac. Cardiovasc. Surg.*, **99**, 977–980.

Norden, C. W. (1970). Experimental osteomyelitis. I. A description of the model. *J. Infect. Dis.*, **122**, 410–418.

Norden, C. W. (1988). Lessons learned from animal models of osteomyelitis. *Rev. Infect. Dis.*, **10**, 103–110.

Norden, C. W. (1989). Bone and joint infection. *Curr. Opin. Infect. Dis.*, **2**, 213–217.

Norden, C. W. (1990). Osteomyelitis. In: *Principles and Practice of Infectious Disease*, 3rd edn (eds Mandell, G. L., Douglas, R. G., Bennett, J. E.), pp. 921–930. Churchill Livingstone, New York.

Norden, C. W. (1996). Bone and joint infection. *Curr. Opin. Infect. Dis.*, **9**, 109–114.

Norrby, S. R., O'Reilly, T., Zak, O. (1993). Efficacy of antimicrobial agent treatment in relation to treatment regimen: experimental models and clinical evaluation. *J. Antimicrob. Chemother.*, **31** (Suppl. D), 41–54.

O'Reilly, T., Kunz, S., Sande, E., Zak, O., Sande, M. A., Tauber, M. G. (1992). Relationship between antibiotic concentration in bone and efficacy of treatment of staphylococcal osteomyelitis in rats: azithromycin compared with clindamycin and rifampin. *Antimicrob. Agents Chemother.*, **36**, 2693–2697.

O'Reilly, T., Cleeland, R., Squires, E. (1996). Evaluation of antimicrobials in experimental animal infections. In: *Antibiotics in Laboratory Medicine* 4th edn. (ed. Lorian, V.), pp. 599–759. Williams & Wilkins, Baltimore, MD.

Pontifex, A. H., McNaught, D. R. (1973). The treatment of chronic osteomyelitis with clindamycin. *Can. Med. Assoc. J.*, **109**, 105–107.

Popham, G. L., Mangino, P., Seligson, D., Henry, S. L. (1991). Antibiotic-impregnated beads. Part II: Factors in antibiotic selection. *Orthoped. Rev.*, **20**, 331–337.

Power, M. E., Olson, M. E., Domingue, P. A. G., Costerton, J. W. (1990). A rat model of *Staphylococcus aureus* chronic osteomyelitis that provides a suitable system for studying the human infection. *J. Med. Microbiol.*, **33**, 189–198.

Rissing, J. P. (1990). Animal models of osteomyelitis. Knowledge, hypothesis, and speculation. *Infect. Dis. Clin. North. Am.*, **4**, 377–390.

Rissing, J. P. (1997). Antimicrobial therapy for chronic osteomyelitis in adults: role of the quinolones. *Clin. Infect. Dis.*, **25**, 1327–1333.

Rissing, J. P., Buxton, T. B. (1986). Effect of ibuprofen on gross pathology, bacterial count, and levels of prostaglandin E_2 in experimental staphylococcal osteomyelitis. *J. Infect. Dis.*, **154**, 627–630.

Rissing, J. P., Buxton, T. B., Fisher, J., Harris, R., Shockley, R. K. (1985a). Arachidonic acid facilitates experimental chronic osteomyelitis in rats. *Infect. Immun.*, **49**, 141–144.

Rissing, J. P., Buxton, T. B., Weinstein, R. S., Schockley, R. K. (1985b). Model of experimental chronic osteomyelitis in rats. *Infect. Immun.*, **47**, 518–586.

Rouse, M. S., Tallan, B. M., Henry, N. K., Steckelberg, J. M., Wilson, W. R. (1992). Animal models as predictors of outcome of therapy with broad spectrum cephalosporins. *J. Antimicrob. Agents*, **29** (Suppl. A), 39–45.

Scheman, L., Janota, M., Lewin, P. (1941). The production of experimental osteomyelitis. Preliminary report. *J.A.M.A.* **117**, 1525–1529.

Spagnolo, N., Greco, F., Rossi, A., Ciolli, L., Teti, A., Posteraro, P. (1993). Chronic staphylococcal osteomyelitis. A new experimental model. *Infect. Immun.*, **61**, 5225–5230.

Volk, A., Crémieux, A. C., Belmatoug, N., Vallois, J. M., Pocidalo, J. J., Carbon, C. (1994). Evaluation of a rabbit model for osteomyelitis by high field, high resolution imaging using the chemical-shift-specific-slice-selection technique. *Magn. Reson. Imaging*, **12**, 1039–1046.

Von Eiff, C., Bettin, D., Proctor, R. A. *et al.* (1997). Recovery of small colony variants of *Staphylococcus aureus* following gentamicin bead placement for osteomyelitis. *Clin. Infect. Dis.*, **25**, 1250–1251.

Wall, E. J. (1998). Childhood osteomyelitis and septic arthritis. *Curr. Opin. Pediatr.*, **10**, 73–76.

Wininger, D. A., Fass, R. J. (1996). Antibiotic-impregnated cement and beads for orthopedic infections. *Antimicrob. Agents Chemother.*, **40**, 2675–2679.

Zak, O., Zak, F., Rich, R., Tosch, W., Kradolfer, F., Scheld, W. M. (1981). Experimental staphylococcal osteomyelitis in rats: therapy with rifampin and cloxacillin, alone or in combination. In: *Current Chemotheraphy and Immunotherapy. Proceedings of the 12th International Congress of Chemotherapy, Florence, Italy, 19–24 July*.

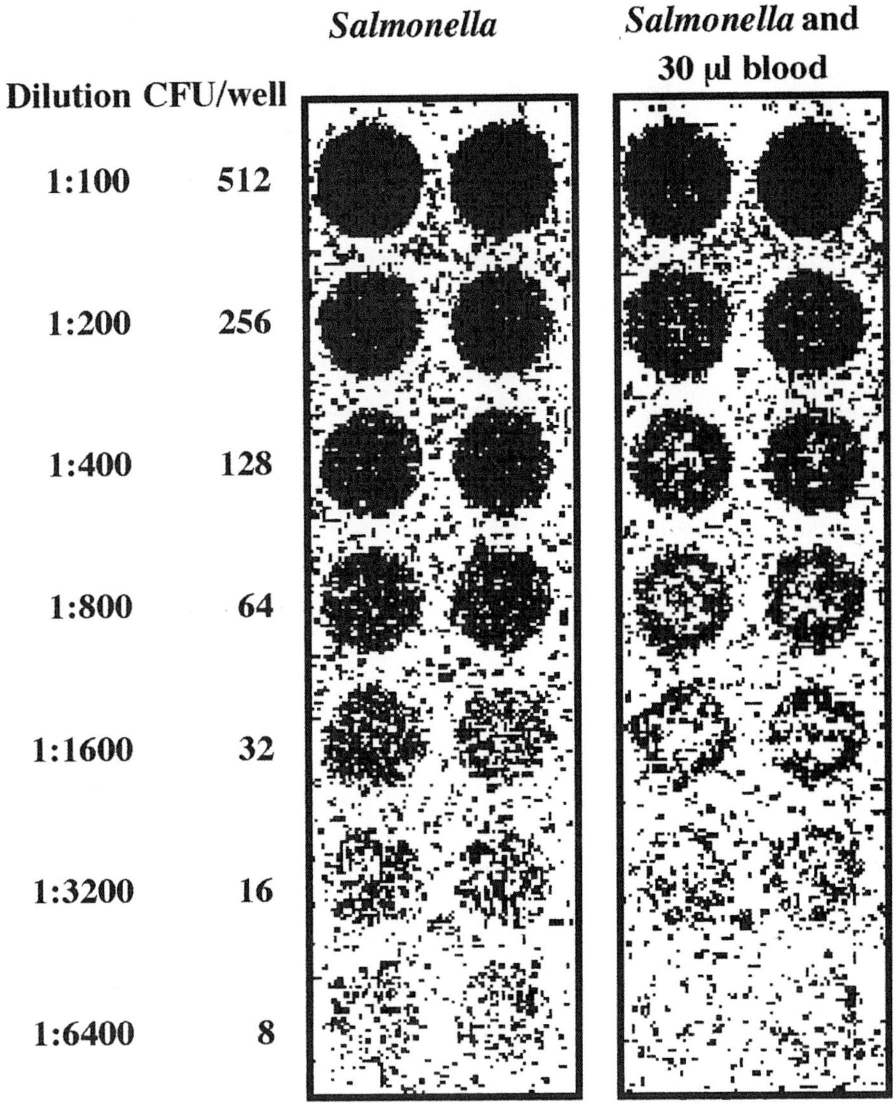

ure 7.1 Sensitivity of bacterial tection *in vitro*. Short-term cultures of luminescent *Salmonella* were diluted ofold in 96-well plates and imaged to termine the least number of bacterial ls that could be detected without the sorbing and scattering effects of blood ft) and then in the presence of blood an example of mammalian tissue ght). Cells were imaged at 37°C with integration time of 30 min. The ution factor and the predicted number colony-forming units (CFU: based on terminations at lower dilutions of the me series) are shown for the duplicate lls.

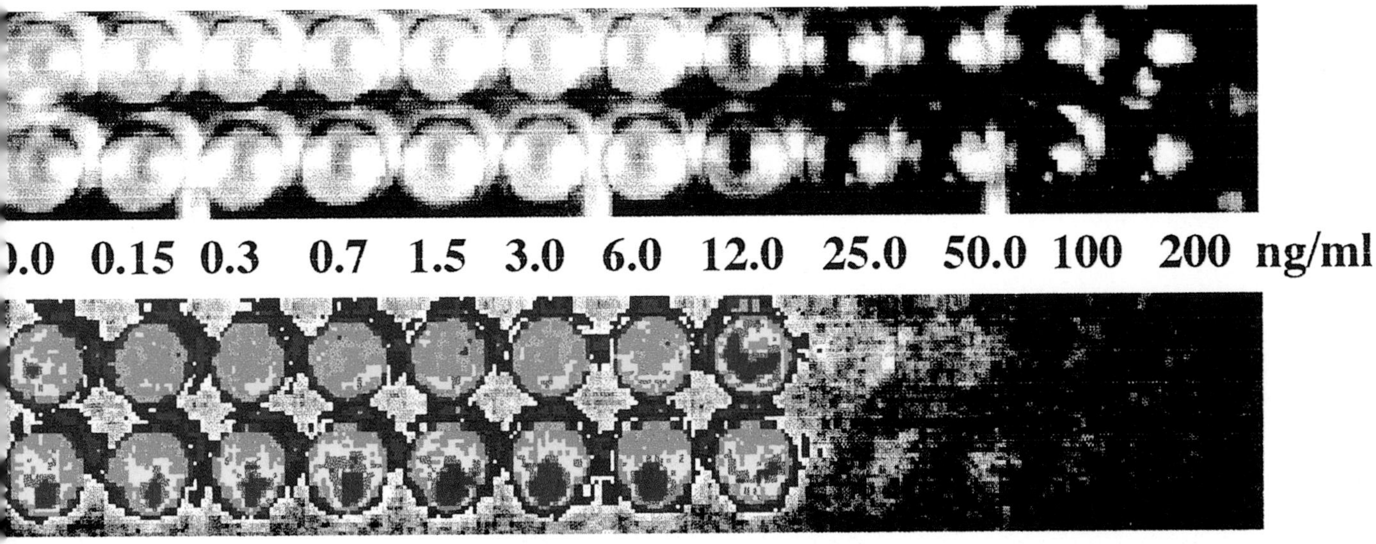

gure 7.2 *In vitro* assays: bioluminescent minimum inhibitory concentration (MIC). Growth of *Salmonella typhimurium* ;L1344) in the presence of various concentrations of ciprofloxicin (0.002-0.1 mg/ml) was assessed. Turbidity of cultures, the pical read-out in MIC, was apparent in the gray-scale reference image (top). In this reference image, bacterial growth at 12 g/ml is apparent but less than that observed at lower drug concentrations. In the pseudocolor image of this plate (bottom), rowth and metabolic activity are apparent at this concentration where turbidity reading is ambiguous.

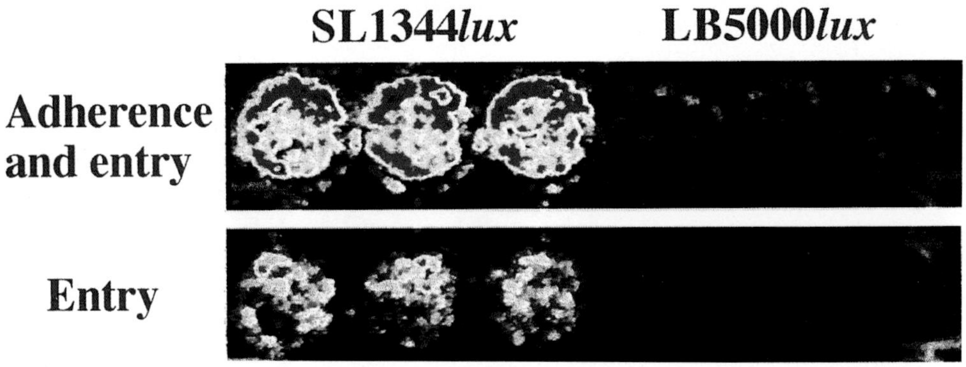

Figure 7.3 Bioluminescence in cell culture correlates for bacterial virulence. Adherence and entry of *Salmonella typhimurium* strains SL1344*lux* and LB5000*lux* were assayed via bacterial bioluminescence in cultures of HEp-2 cells, and colony-forming units for adherence and entry were determined for the luminescent strains (data not shown). Images of a representative well of internalized (gentamicin-treated) and adherent and internalized *Salmonella* (untreated co-cultures) are shown for both *Salmonella* strains.

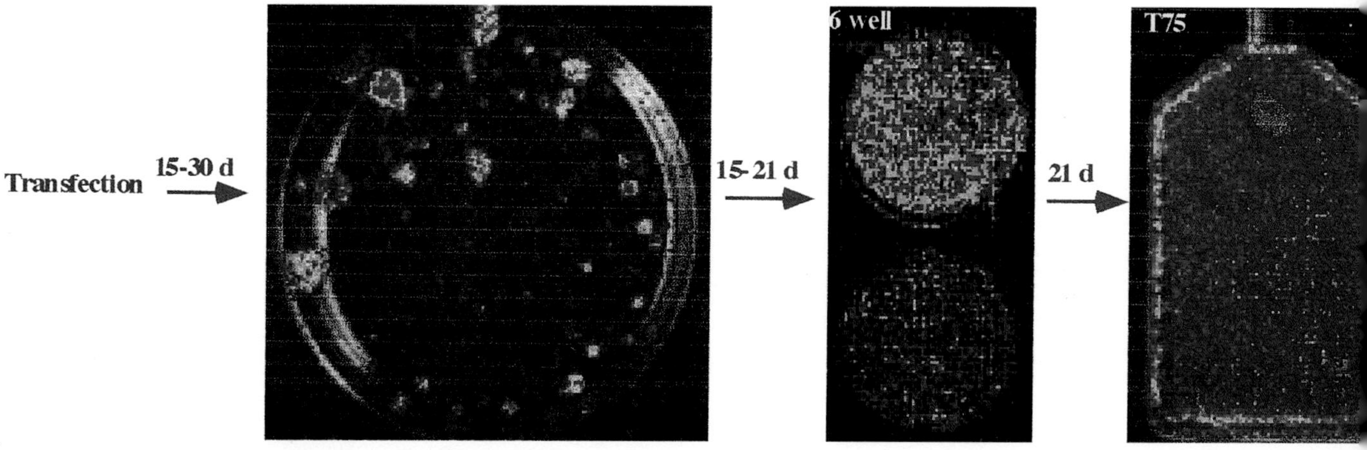

Figure 7.4 Monitoring transfection, selection and propagation of mammalian cells in culture. Using bioluminescent marker each step in the preparation of stable cell lines can be monitored. Two plasmids, one encoding a modified firefly luciferase (pGL-3, Promega), and the other resistance to the antibiotic Zeocin (pZEO, Invitrogen) were co-transfected into NIH 3T3 cell using cationic liposomes (Lipofectamine, Gibco-BRL). Stable transfectants were selected in 500 mmol/l Zeocin (Invitrogen). various steps the substrate, luciferin (0.5 mmol/l final, Molecular Probes), was added to the cultures just prior to imaging. Cultures were imaged with 2 min. integration time.

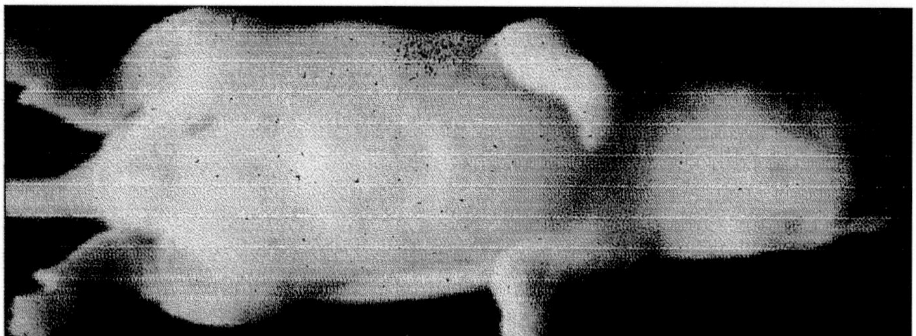

Figure 7.5 Monitoring *in vivo* gene deliver The luciferase expression vector, pGL3 control, was delivered to cells in the lung of neonatal rats via cationic liposome delivery, and sustrate was injected i.p. at 24 and 48 h post DNA delivery. A representative image at the 48 h time point is shown. This image indicates that luciferase expression from transient *in vivo* transfections of the lung, represente as a pseudocolor image, can be monitore externally.

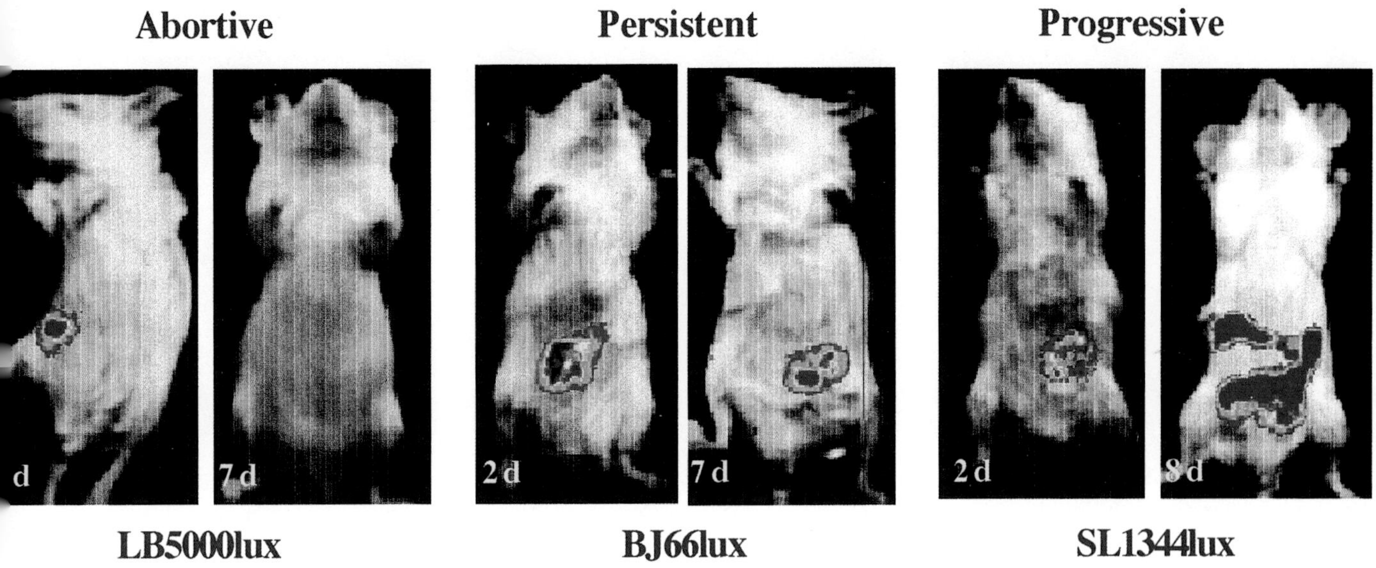

gure 7.6 Patterns of disease. The disease course following oral infection with three different strains of *Salmonella* was onitored over a period of about 1 week; early and late time points are shown (Contag *et al.*,1995). Infections with each rain caused a focal infection in the cecum early and then were either cleared (LB5000*lux*), persisted (BJ66*lux*) or progressed L1344*lux*). This type of analysis can be used to monitor potential virulence factors or the efficacy of therapies aimed at pecific molecular targets.

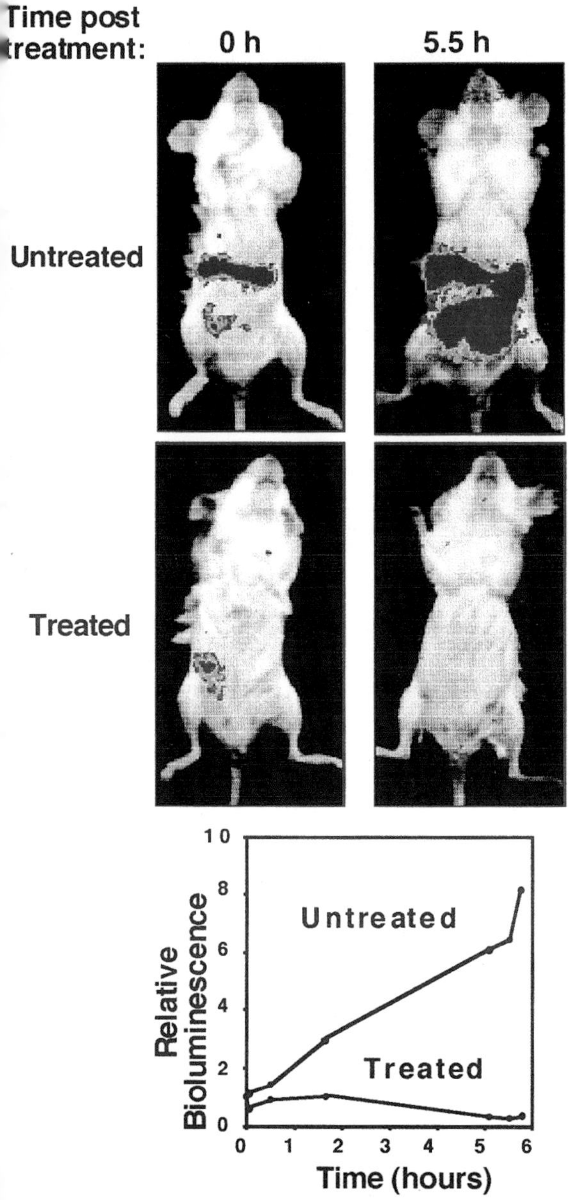

Figure 7.7 (*left*) Monitoring antibiotic therapy. The effects of antibiotic therapy on *Salmonella* infection of the gastro-intestinal track were determined at 7 days post-infection when the infection, following oral inoculation, was at the threshold of systemic disease. Ciprofloxacin was administered (10 mg/kg) via intraperitoneal injection (A). Companion mice were left untreated (B). Animals were imaged at times post-treatment (2, 5, 30, 60, 120, 180, 240, 330 min) and total bioluminescent signals over the abdominal surface of treated and untreated animals were determined for each time point, normalized to that of time 0 and plotted with respect to time (C). Reduction in bioluminescent signals was immediately apparent in treated animals and was reduced to background levels over the 5.5-h period of time. In contrast, signals in untreated animals increased 7.5-fold over the same time period.

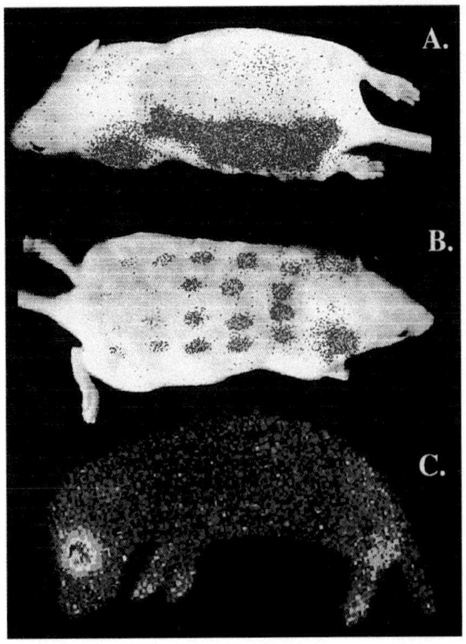

Figure 7.8 (*right*) Use of promoter-luciferase gene fusions as transgenes in mice to monitor gene expression. The HIV-LTR luciferase gene fusion was used to demonstrate multifactor assays (A), control of gene expression (B) and development (C).

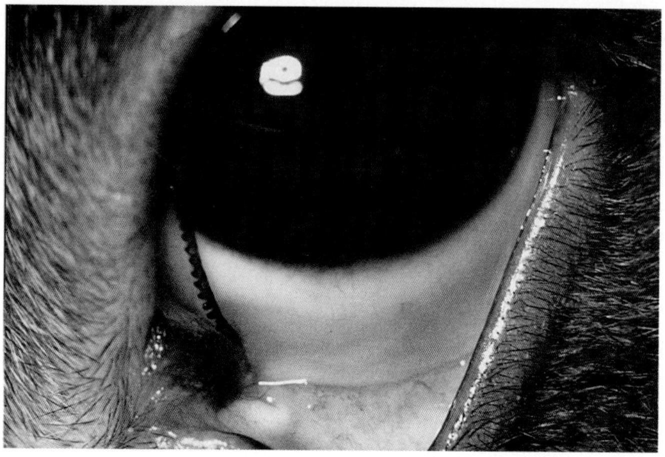

Figure 40.1 (A) Clinical score for conjunctival hyperemia. No conjunctival hyperemia (grade 0).

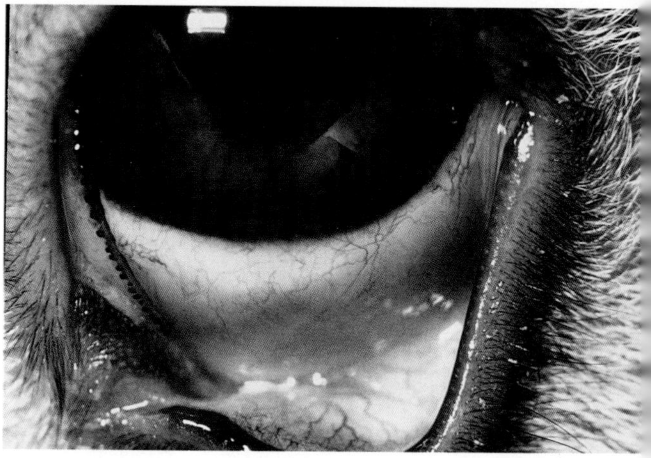

Figure 40.1 (B) Clinical score for conjunctival hyperemia. Slight conjunctival hyperemia (grade 1).

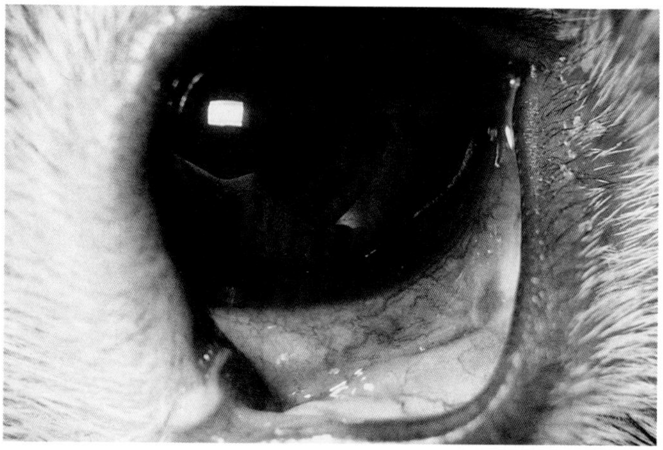

Figure 40.1 (C) Clinical score for conjunctival hyperemia. Moderate conjunctival hyperemia (grade 2).

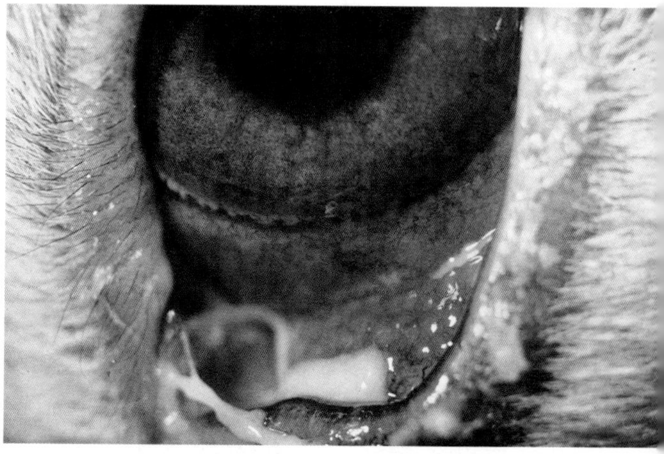

Figure 40.1 (D) Clinical score for conjunctival hyperemia. Severe conjunctival hyperemia (grade 3) with purulent discharge.

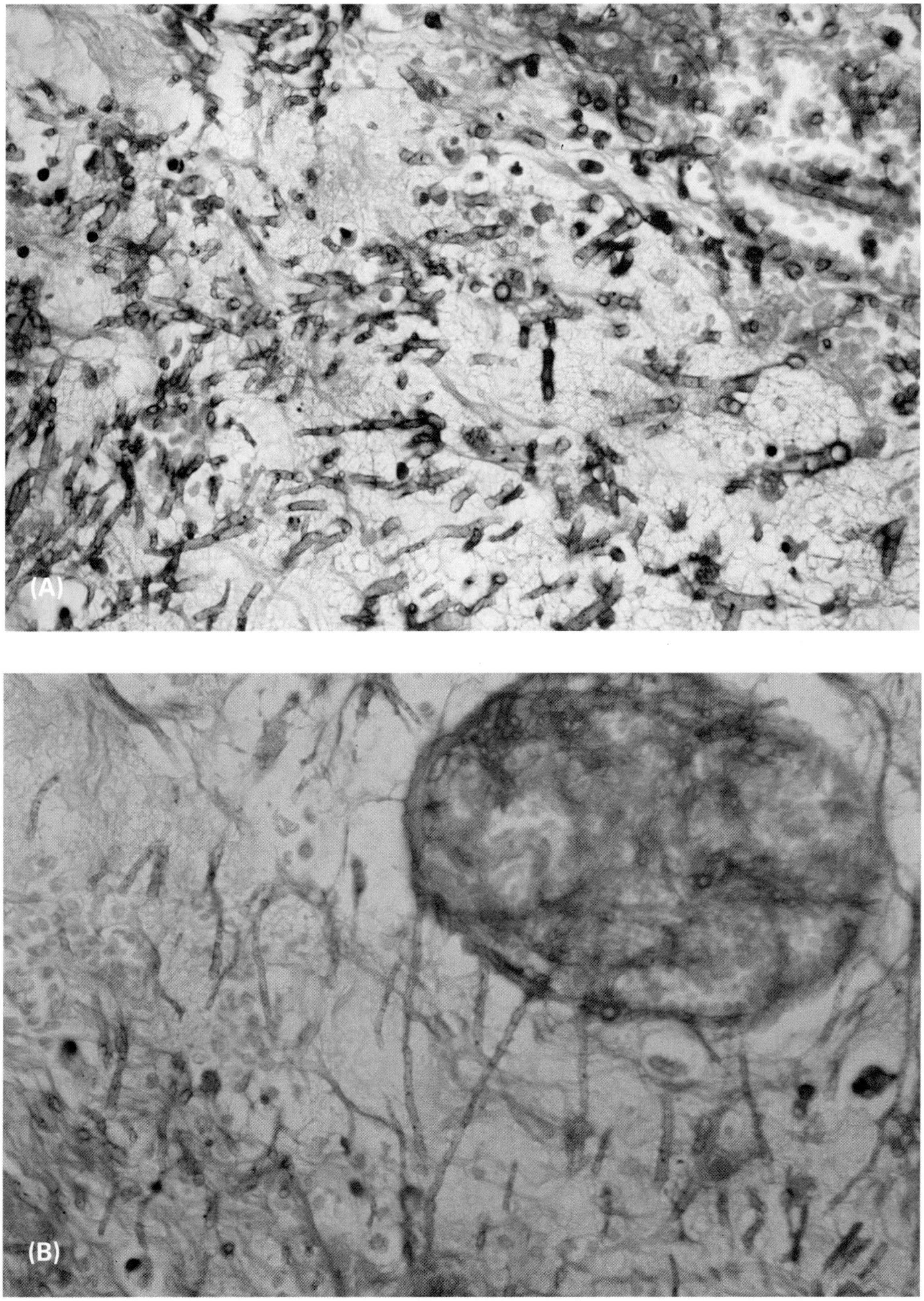

Figure 84.3 (A) and (B) Histopathologic features of invasive pulmonary aspergillosis in persistently granulocytopenic rats at 40 h after inoculation with 1×10^4 *Aspergillus fumigatus* conidia. (A) Invasion of lung tissue by long, slender, dichotomously branching septate hyphae of *A. fumigatus* with paucity of surrounding cellular infiltrate (periodic acid-schiff stain, original magnification: ×400). (B) Invasion of a pulmonary blood vessel by hyphae of *A. fumigatus* resulting in intravascular thrombosis (periodic acid-schiff stain, original magnification: ×400).

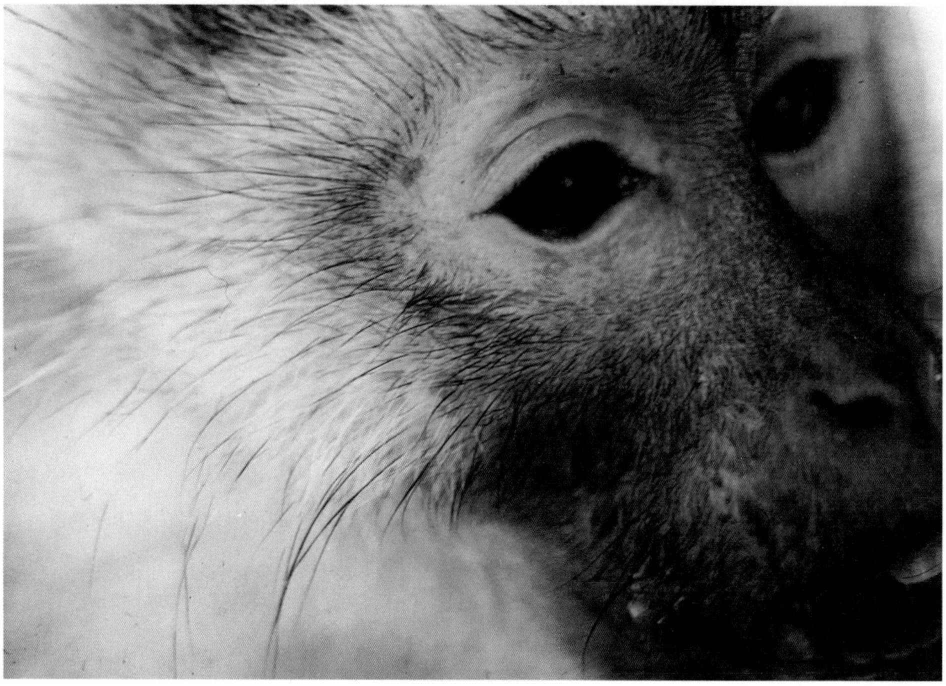

Figure 116.1 (A) Vesicular lesions of simian varicella virus on the face of an African green monkey.

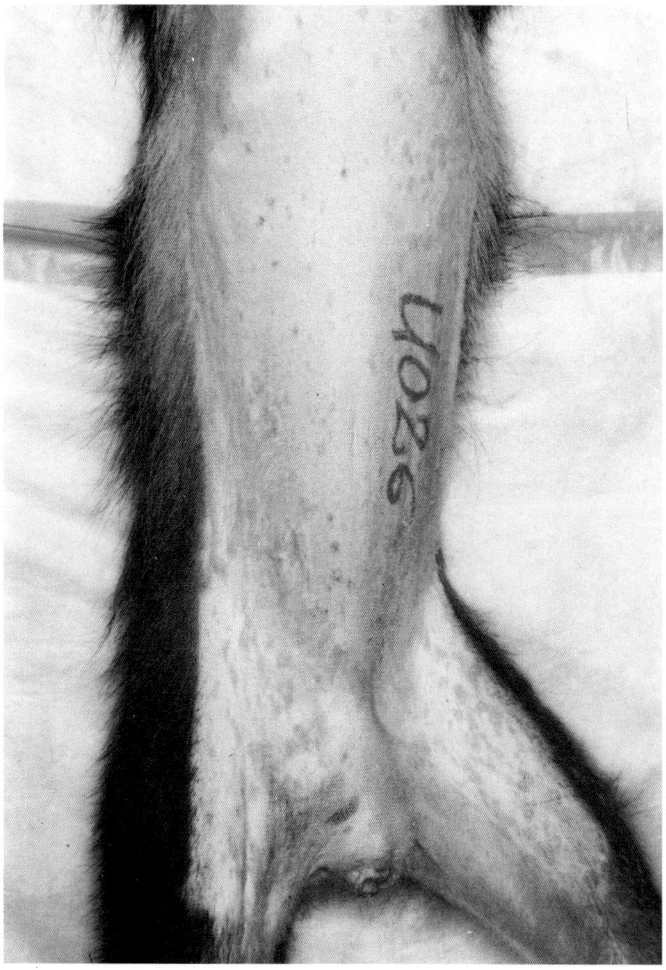

Figure 116.1 (B) Simian varicella rash on the body of an African green monkey at 9 days after virus inoculation.

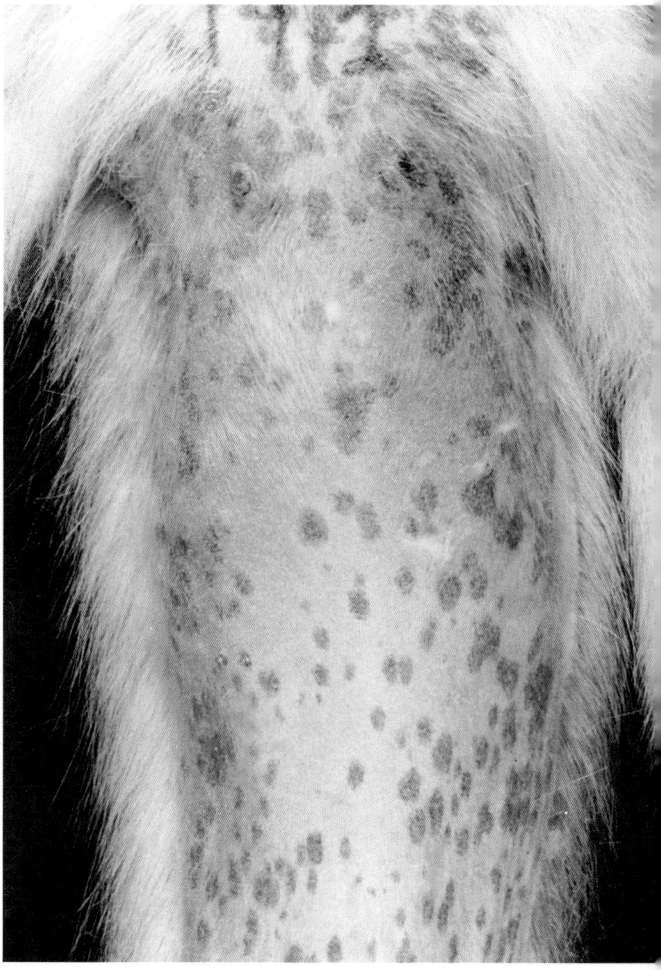

Figure 116.1 (C) Hemorrhagic rash seen 10 days post-infection.

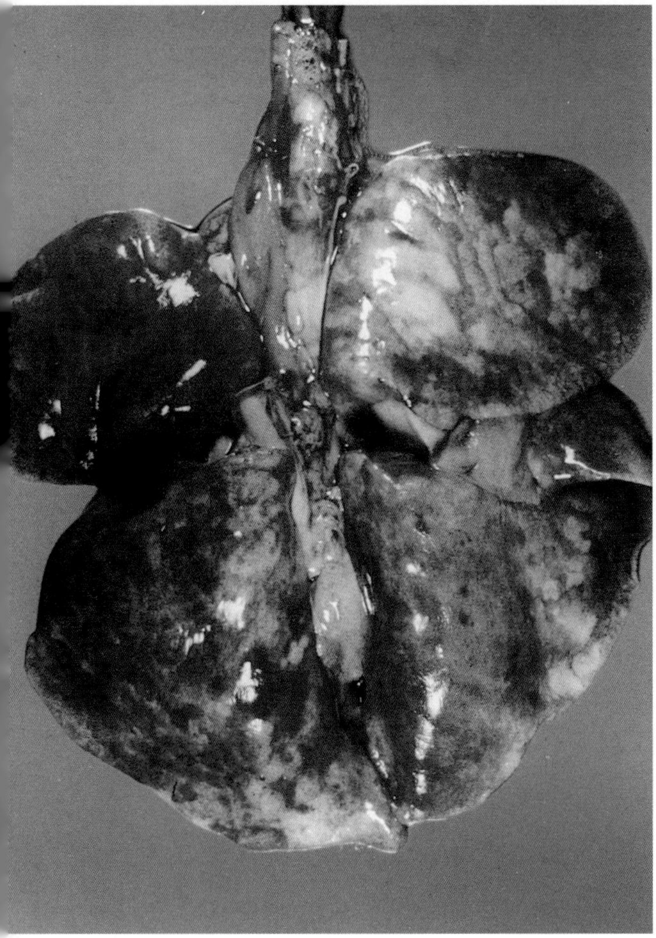

igure 116.2 (A) Lung of an African green monkey which ed with simian varicella infection. The lung is wet and ottled with areas of collapse, congestion and hemorrhage.

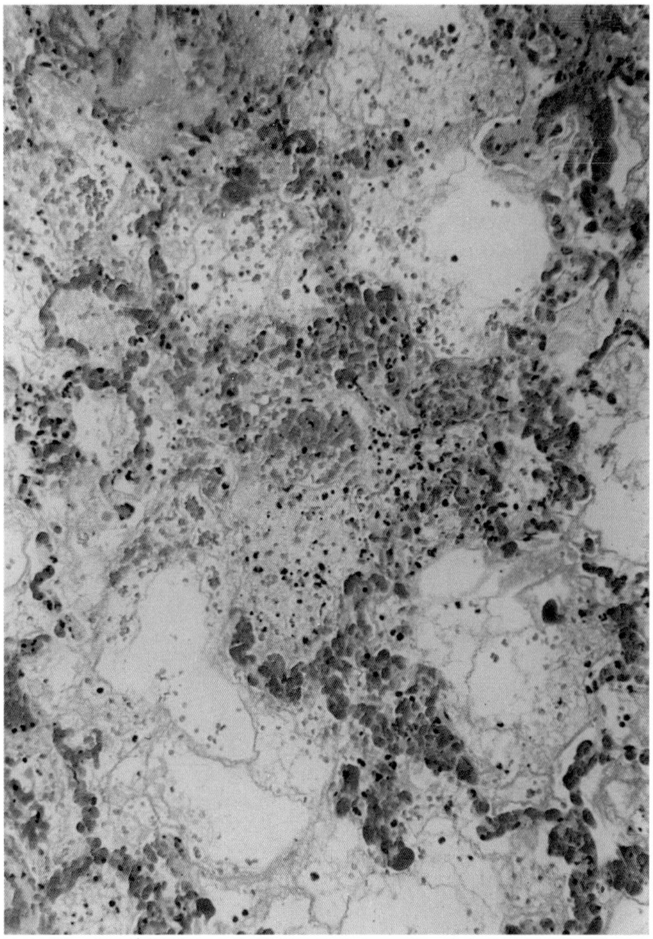

Figure 116.2 (B) Sections from affected lung are severely congested with patchy areas of alveolar septal necrosis and fibrinohemorrhagic exudate in alveolar spaces.

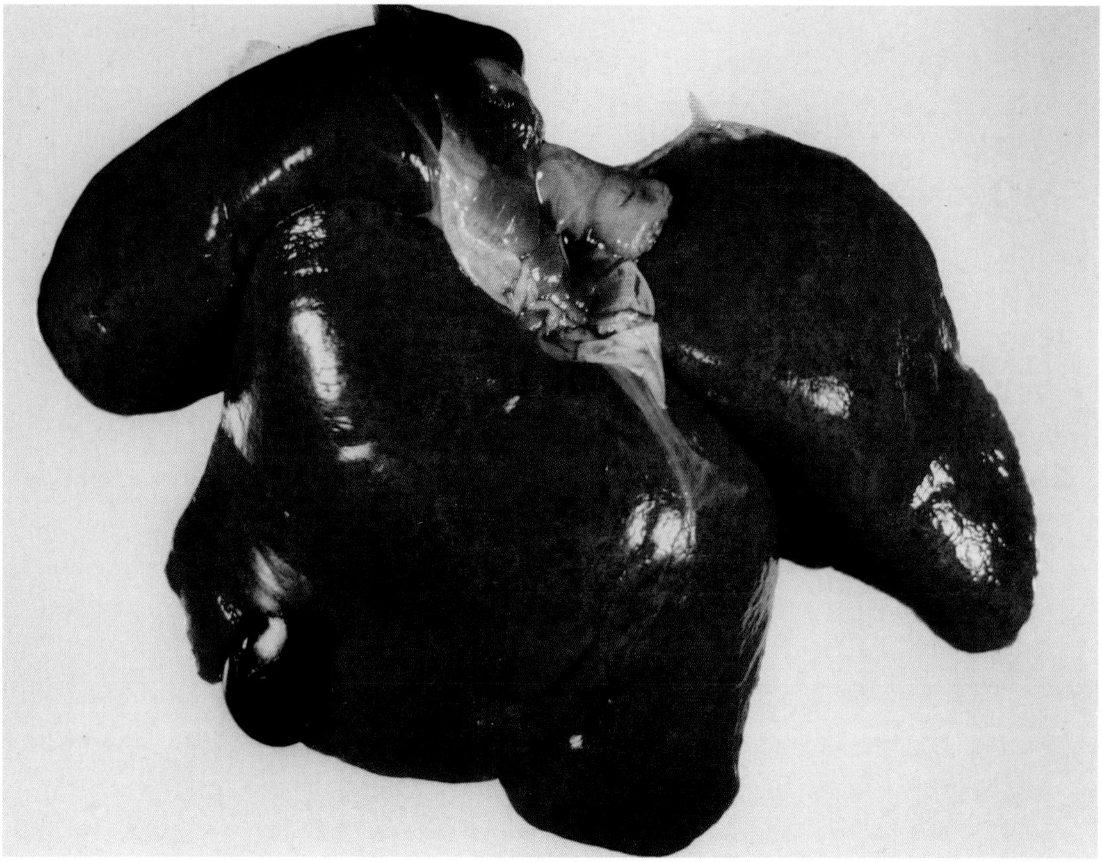

Figure 116.3 (A) Liver shows numerous petichial hemorrhages on surface.

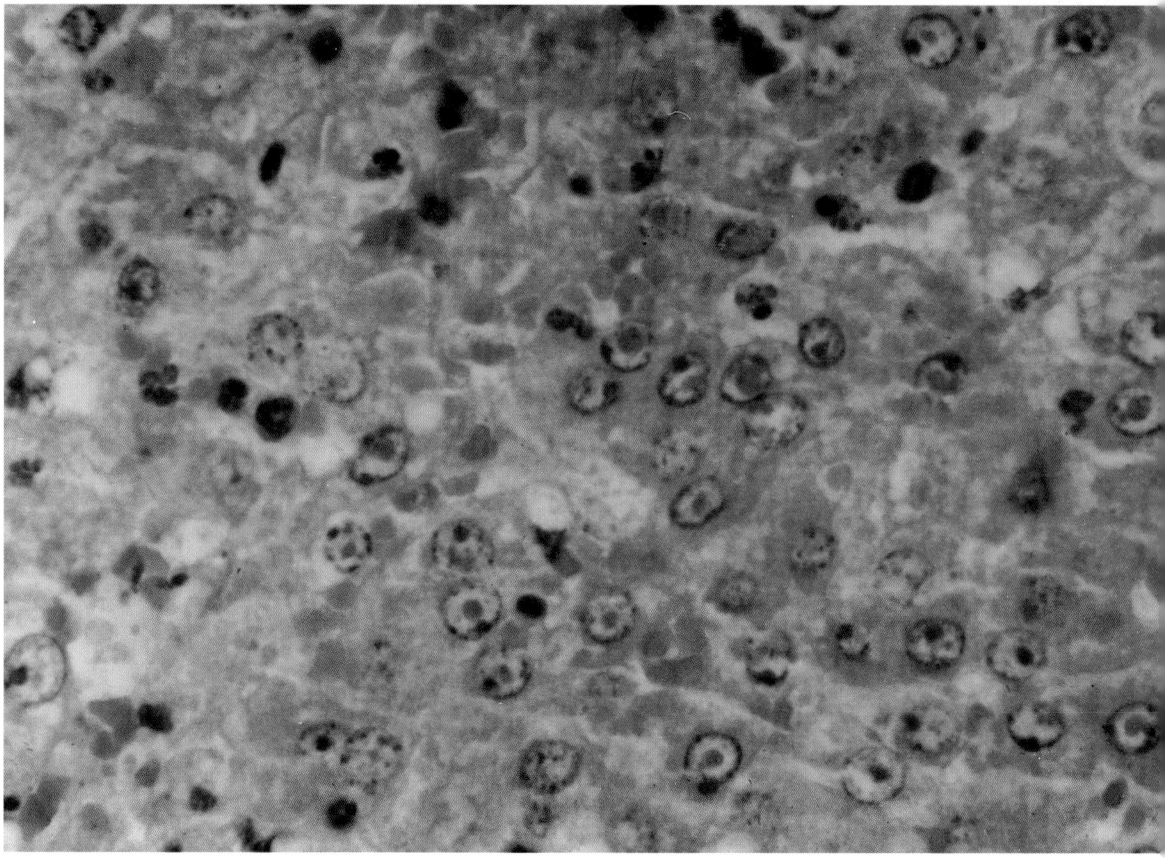

Figure 116.3 (B) Section of liver stained with hematoxylin and eosin shows numerous intranuclear inclusions.

Chapter 67

Hematogenous Osteomyelitis in the Rat

S. A. Hienz, C. Erik Nord, A. Heimdahl, F. P. Reinholt and J.-I. Flock

Background

Osteomyelitis, a progressive infectious process involving various compartments of bone such as the medullary cavity, cortex and periosteum, continues to be a common disease (Mader *et al.*, 1992; Dirschl and Almekinders, 1993). To initiate osteomyelitis, an inoculum of microorganisms and a susceptible focus are required. Delivery of the microorganisms might take place during direct contamination, as in trauma or surgery, or during hematogenous or contiguous spread (Gristina *et al.*, 1985). As noted by others, bacteremia alone is not sufficient by itself to cause osteomyelitis, and trauma is often associated with the onset of the disease (Morrissy and Haynes, 1989; Mader *et al.*, 1992; Nelson *et al.*, 1992; Dirschl and Almekinders, 1993). About one-third of the patients have a known history of trauma to the area of infection, and in about 75% of the patients with clinical diagnosis of osteomyelitis, a bacterial etiology can be established (Mader *et al.*, 1992; Nelson *et al.*, 1992; Meadows *et al.*, 1993). In addition, it was recently noted that tissue status in the environment of the injury, such as the presence of dead bone, influences the incidence of experimental osteomyelitis (Evans *et al.*, 1993).

The aim of this chapter is to present an improved model of hematogenous osteomyelitis in the rat, without the use of any foreign body or sclerosing agent.

Bacteria and preparation of inoculum

Staphylococcus aureus strain Phillips, supplied by Joe Patti (Texas A&M University, Houston, TX) has been used in the majority of experiments. In addition, *S. aureus* L9, a clinical osteomyelitis isolate, provided by B. Christensson (Department of Infectious Diseases, University Hospital, Lund, Sweden) has also been used. Bacteria are grown in brain–heart infusion, diluted in phosphate-buffered saline to a concentration of 10^9 colony forming units (cfu) per milliliter and stored at −70°C. Bacteria are then diluted in PBS to desired inoculum sizes and 1 ml is injected i.v.

Animals and anesthesia

Female Wistar (BK Universal, Sollentuna, Sweden), weighing approximately 185 g, are anesthetized (pentobarbital or chloral hydrate i.p.).

Induction of osteomyelitis

In short, animals are anesthetized, surgically manipulated at the tibia or mandible, and subsequently challenged with *S. aureus* (Figure 67.1). Two weeks postinoculation, rats are sacrificed and assessed.

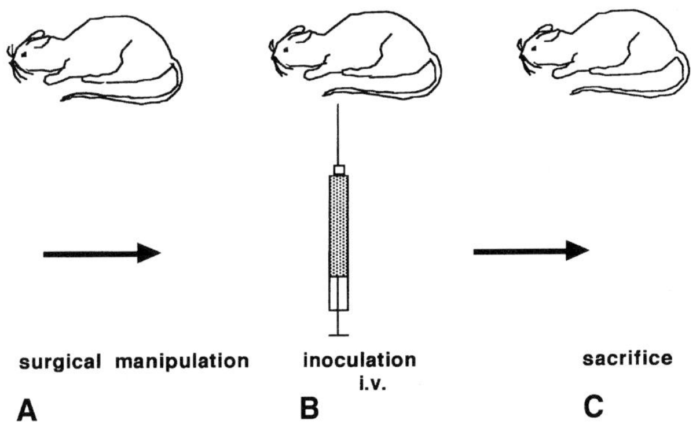

Figure 67.1 Basic stages of animal model. **A.** Rats receive damage to osseous tissue. **B.** Rats are inoculated intravenously with *S. aureus*. **C.** Rats are sacrificed 14 days postinoculation.

Model A

The mandibular ramus or the proximal medial surface of the tibia is shaved and disinfected with 70% EtOH. The area is surgically exposed and the cortex is carefully penetrated with a 1.2 mm bit using a hand drill. A volume of 50 µl of sodium morrhuate (Palisades Pharmaceuticals, Tenafly, NJ), a sclerosing agent, are slowly injected into the

Handbook of Animal Models of Infection
ISBN 0-12-775390-7

medullary space without causing the agent to overflow. The skin incisions are then closed with skin staples to eliminate time-consuming suturing and the animals are then inoculated intravenously as described below.

Model B

The proximal medial surface of the tibia is exposed as described in model A. A cylindrical piece of cortical bone is removed from the proximal medial part of the tibia (Figure 67.2) using a hollow drill (2 mm in diameter). To prevent the cortical piece from getting stuck inside the drill, it is important to irrigate intermittently with sterile saline. The cortical piece is removed from the drill and devitalized by submerging it in boiling PBS for 20 seconds. It is then carefully reinserted into the cortical window. The incisions are tightly closed with resorbable suture (Ethicon, Norderstedt, Germany) to ensure that the cortical piece remains in its position. Rats are intravenously challenged as outlined below.

Inoculation of bacteria

The medial aspect of the thigh of the opposite leg is shaved and disinfected. The skin is stretched and, using a short cut, the great saphenous vein is exposed at the medial side of the thigh before it passes through the saphenous hiatus. If the injection fails, blunt dissection helps to expose the femoral vein, which allows for an easy i.v. access.

Sampling methods

Blood samples are best obtained before the animals are challenged i.v. Collection of blood from the orbit can also be performed but requires a high level of skill and sedation of the animal. Direct heart punctures can be easily performed during sacrifice and usually yield a large volume of blood. In order to collect bone samples, the site needs to be shaved, disinfected, and the specimen removed under sterile conditions, especially when it will be used to determine the actual bacterial count (cfu) of the infective foci. Specimens collected for histopathological examination do not need to be harvested under strictly sterile conditions.

Assessment of osteomyelitis

Radiographic examination

Between inoculation and sacrifice, animals may be monitored radiographically for signs of infection. Since radiographic evaluation requires anesthesia of the animals and is

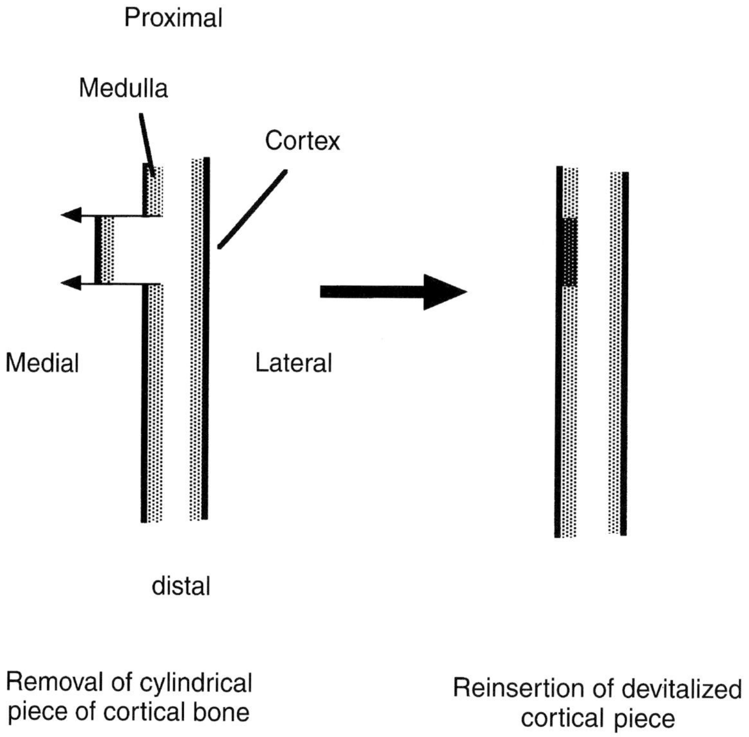

Figure 67.2 Schematic of surgical procedure to induce osteomyelitis. After exposure of the proximal medial surface of the tibia, a cylindrical piece of the cortical bone is removed. The piece is devitalized in boiling PBS and reinserted into the cortical window.

very time-consuming, we recommend using a direct digital method that will cut down on working time. Radiographic parameters of osteomyelitis include general loss of bone architecture, bone destruction, sequestrum formation, and presence of periosteal bone. Radiographic evaluation has only been employed in model A to monitor bone changes. Most dramatic changes can be observed after 2 weeks and include diffuse loss of bone from the medulla extending to the cortex. Also, elevation and disruption of the normally smooth periosteum and new bone deposition become apparent.

Microbiological examination

Bone samples are suspended in PBS and homogenized with a mortar and pestle. This procedure has to be done carefully to minimize false-negative results. Homogenates are serially diluted, plated on blood agar and incubated for 24 hours at 37°C. Microbiological assessment has only been performed in model A and it can be expected that the infection rate is inoculum-dose-dependent (Figure 67.3).

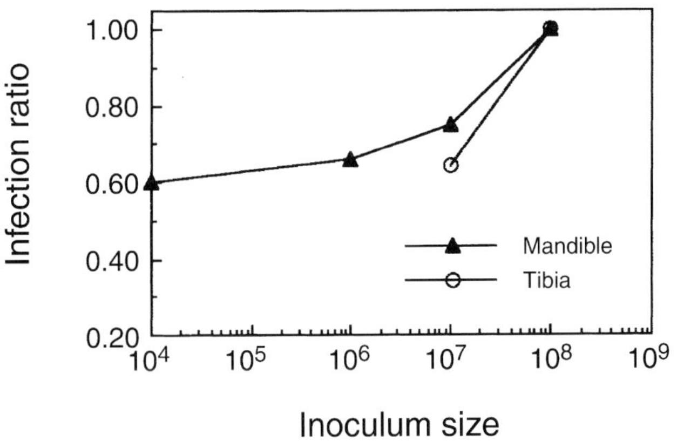

Figure 67.3 Infection rates versus challenging dose in the mandibles or tibiae of rats sacrificed 14 days after inoculation.

Histopathologic examination

The freshly isolated tibiae or mandibles from experimental animals are prepared for histopathologic examination according to conventional techniques. Sections cut at several levels are stained with hematoxylin and eosin and examined for signs of osteomyelitis (i.e. suppurative inflammation, bone necrosis, delayed repair, periosteal new bone formation). Since this method allows the presence of bacteria to be identified and aid in differentiation between different levels of infection, the authors believe that this is the most valuable tool to assess the infectious process in the model.

Animals injected with normal saline solution instead of bacteria do not present with signs of infection. Instead, they display normal healing of the tissue with osseous callus formation. Likewise, rats challenged with bacteria at low concentrations proceed through the early stages of the healing process, exemplified by fibrocartilaginous callus formation. However, with an increase in bacterial challenge dose, histopathologic examination reveals suppurative reaction with necrosis, involucrum formation, and presence of bacterial microcolonies.

Using these criteria, rats have been classified as having osteomyelitis and the infection rates of groups challenged with two strains at different concentrations have been compared. The results are depicted in Figure 67.4. Challenging rats with *S. aureus* strain L9 resulted in higher infection rates than strain Phillips at an inoculum size of 1×10^7 cfu ($p < 0.05$; Fisher exact test).

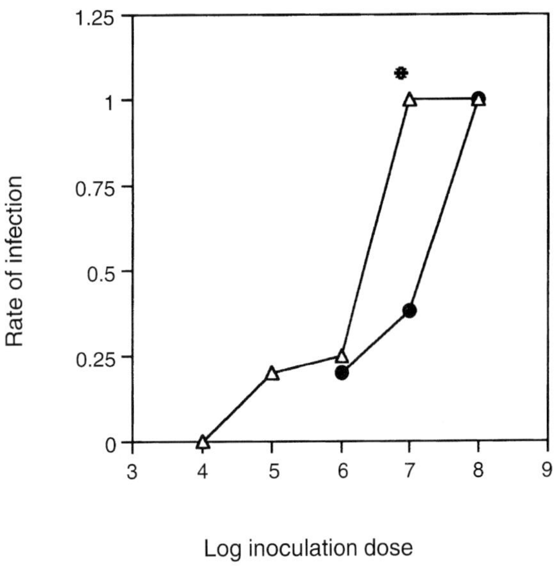

Figure 67.4 Infection rates of L9 (triangles) and Phillips (filled circles), at different challenge doses. $^*p < 0.05$.

Discussion

Both models produce osteomyelitis in the rat. They mimic human infection with respect to its natural course and bone changes, and display a dose-dependent infection rate. Both models avoid a common problem that arises when directly inoculating into the medulla of the bone. This latter procedure entails the probable spillage of bacteria from the injection site into the surrounding tissues, which represents an undesirable contamination with subsequent infection of soft tissues.

The modification of model A to model B allowed us to completely omit the use of artificial components, such as bone wax, fibrin glue, foreign body implants, or sclerosing agents. Models requiring the presence of such agents or materials do not naturally simulate the injured tissue environment in which osteomyelitis commonly arises and may alter the immune response of the host to the bacteria inoculated. Also, they may alter factors that are of importance during bacterial adhesion. The second modification of model A entailed the reinsertion of devitalized cortical bone into the tibia of the animals. The dead bone resembles a susceptible focus, which is necessary for infection. The combination of dead bone and pathogenic bacteria within a surgical wound has been proposed as the most important factor in the production of postsurgical osteomyelitis (Evans *et al.*, 1993). In the majority of the remaining non-postsurgical cases, a disturbance in normal bone architecture is suspected to favor the initiation of osteomyelitis (Mader *et al.*, 1992) and it has been shown that the presence of a hematoma-filled dead space resulted in a reduced incidence of osteomyelitis compared to the presence of necrotic bone (Evans *et al.*, 1993). In agreement with the findings of Evans *et al.* (1993), we observed that a focus of necrotic bone, as is used in model B, favors the induction of osteomyelitis. Histopathologic examination revealed that lesions in the tibiae of rats from the modified model were similar to those obtained by the original model. Earlier experience also showed that sampling of bone in conjunction with microbiological examination remained difficult and represented a source of large variations among estimated cfu values. Since we assume that radiographic and microbiologic examinations would have matched those of the described model earlier and since it is not possible to do a microbiologic and pathohistologic examination on the same sample, we omitted radiographic and microbiologic examinations in the present study.

The modified model, model B, also enabled the investigators to detect a difference in infection rates between *S. aureus* strain L9 and strain Phillips. Strain L9 yielded a significantly higher incidence of osteomyelitis than the other strain when 1×10^7 cfu/ml bacteria were injected. This finding demonstrates that the model displays a sensitivity that is suitable for virulence comparisons of different bacterial strains.

Both models, especially model B, enable the exploration of unanswered questions concerning the initiating events of osteomyelitis, such as bacterial adherence, and point the way to new antimicrobial treatment regimens. For example, mutant strains that differ only in the expression of one binding function and have already been successfully used in other infection models can be employed (Patti *et al.*, 1994; Hienz *et al.*, 1996).

In summary, the model can induce hematogenous osteomyelitis without the use of artificial factors and is sensitive enough to detect a difference in virulence between two unrelated bacterial strains.

References

Dirschl, D. R., Almekinders, L. C. (1993). Osteomyelitis. Common causes and treatment recommendations. *Drugs*, **45**, 29–43.

Evans, R. P., Nelson, C. L., Harrison, B. H. (1993). The effect of wound environment on the incidence of acute osteomyelitis. *Clin. Orthop.*, **286**, 289–297.

Gristina, A. G., Oga, M., Webb, L. X., Hobgood, C. D. (1985). Adherent bacterial colonization in the pathogenesis of osteomyelitis. *Science*, **228**, 990–993.

Hienz, S. A., Schennings, T., Heimdahl, A., Flock, J.-I. (1996). Collagen binding of *Staphylococcus aureus* is a virulence factor in experimental endocarditis. *J. Infect. Dis.*, **174**, 83–88.

Mader, J. T., Norden, C., Nelson, J. D., Calandra, G. B. (1992). Evaluation of new anti-infective drugs for the treatment of osteomyelitis in adults. Infectious Diseases Society of America and the Food and Drug Administration. *Clin. Infect. Dis.*, **15**, S155–S161.

Meadows, S. E., Zuckerman, J. D., Koval, K. J. (1993). Posttraumatic tibial osteomyelitis: diagnosis, classification, and treatment. *Bull. Hosp. Joint. Dis.*, **52**, 11–16.

Morrissy, R. T., Haynes, D. W. (1989). Acute hematogenous osteomyelitis: a model with trauma as etiology. *J. Pediatr. Orthop.*, **9**, 447–456.

Nelson, J. D., Norden, C., Mader, J. T., Calandra, G. B. (1992). Evaluation of new anti-infective drugs for the treatment of acute hematogenous osteomyelitis in children. Infectious Diseases Society of America and the Food and Drug Administration. *Clin. Infect. Dis.*, **15**, S162–S166.

Patti, J. M., Bremell, T., Krajewska-Pietrasik, D. *et al.* (1994). The *Staphylococcus aureus* collagen adhesin is a virulence determinant in experimental septic arthritis. *Infect. Immun.*, **62**, 152–161.

Chapter 68

The Rabbit Model of Bacterial Osteomyelitis of the Tibia

J. T. Mader and M. E. Shirtliff

Background of model

The treatment of acute and chronic orthopedic infections is difficult, time-consuming and expensive (Waldvogel *et al.*, 1970; Patzakis, 1982; Cierny and Mader, 1983; Cierny *et al.*, 1985; Kelly, 1986). Therefore, a variety of animal models have been used in order to test the latest antimicrobial therapies, surgical procedures, and implants for osteomyelitis. The rabbit model of osteomyelitis has been preferred by many because of the low cost, effectiveness and ease of infection initiation, low mortality rate, and close approximation to human histological and pathological patterns of this disease (Mader, 1985).

There are two current rabbit models used for the study of osteomyelitis. The Andriole model of progressive osteomyelitis was developed in 1974 and is extremely useful in the assessment of therapies associated with a complication of internal fixation (Andriole *et al.*, 1973; 1974). In this model, *Staphylococcus aureus* is injected into the intramedullary cavity through a hole drilled through the proximal tibia. The progressive bone infection is produced through local bone trauma. This local trauma can be developed by disrupting local blood supply with the insertion of an intramedullary stainless steel nail or through a fracture of the middle third of the tibia followed by nail insertion. Both techniques were efficient in inducing progressive osteomyelitis at levels of 100% and 88% in the nail-alone rabbit group and the fracture-plus-nail rabbit group, respectively, but required a very high inoculum of *S. aureus* (10^8 cfu); Andriole *et al.*, 1973, 1974). The model produces an osteomyelitis that is classed as a Stage 1A osteomyelitis (according to the Cierny–Mader classification system) and a chronic contiguous focus osteomyelitis (according to the Waldvogel system; Table 68.1; Mader, 1985; Waldvogel *et al.*, 1970). Recent studies have demonstrated that a more efficient infection model uses slotted intramedullary nails instead of the solid nail that was originally used by Andriole (Melcher *et al.*, 1994). An interesting variation of the Andriole model was developed by Morrissy and Haynes (1989). This model combines tibial trauma and bacteremia for the production of acute hematogenous osteomyelitis in rabbits. In this variation to the Andriole model, the rabbit receives both a fracture injury to the proximal tibial physeal plate and a standardized bacteremia. Rabbits developed significant osteomyelitis in almost all cases. The second (and most widely used) rabbit osteomyelitis model was developed by Scheman *et al.* (1941), and was refined by Norden and Kennedy (1970). This model initially produces a contiguous-focus osteomyelitis that closely mimics acute–subacute osteomyelitis in humans. However, as the infection progresses, the development of macronecrosis is observed, consistent with chronic osteomyelitis in humans. Therefore, between 1 and 3 weeks postinfection, this model produces a diffuse 3A or 4A osteomyelitis, according to the Cierny–Mader staging system (Mader, 1985). Since the Andriole model is rarely used by investigators, we will limit our discussion to the contiguous-focus rabbit osteomyelitis model that is based upon the Norden–Kennedy rabbit model or the rabbit modification of the Fitzgerald dog model (Fitzgerald, 1983; Mader, 1985).

Table 68.1 Osteomyelitis classification systems

Waldvogel system
Hematogenous osteomyelitis
Contiguous-focus osteomyelitis
Osteomyelitis associated with vascular disease
Chronic osteomyelitis
Cierny–Mader system
Anatomic stage
Stage 1: Medullary osteomyelitis
Stage 2: Superficial osteomyelitis
Stage 3: Localized osteomyelitis
Stage 4: Diffuse osteomyelitis
Physiologic stage
A host: Normal host
B host: Systemic compromise (BS)
Local compromise (BL)
C host: Treatment worse than the disease

Animal species

New Zealand White rabbits are the species of choice and females weighing 2–3.5 kg (8–12 weeks in age) are often used because of their low level of aggressive behavior.

Handbook of Animal Models of Infection
ISBN 0-12-775390-7

Preparation of animals

Once animals have arrived at your animal care facility, it is important to wait 7–10 days between arrival and initiation of infection in order to screen for those animals with infections (mainly *Pasteurella multocida*, but also *Clostridium pilaformi*, or antibiotic-induced overgrowth of intestinal Gram-negatives) obtained from the commercial rabbitry source. Also, this waiting period is required for the animals to adjust to their new housing environment. In our experience, this adjustment period makes the animals more relaxed during handling.

Storage, preparation of inocula

The strain of *S. aureus* that we currently use was obtained from a patient with osteomyelitis undergoing treatment at the University of Texas Medical Branch, Galveston, TX. The strain is a methicillin-sensitive *S. aureus* (MSSA) and is stored at –70°C in defibrinated sheep blood until needed. *Pseudomonas aeruginosa*, an increasingly important pathogen in human osteomyelitis, has also been used with this rabbit model (Van Wingerden *et al.*, 1974; Dominguez *et al.*, 1975; Norden and Keleti, 1980; Norden *et al.*, 1980; Mader, 1985; Tomczak *et al.*, 1989). The infection produced by this organism is less destructive, more indolent, typified by significant bony and fibrous tissue repair, and rarely leads to extraosseous expression (Mader, 1985; Norden and Keleti, 1980). These histologic elements are somewhat different from the suppurative destructive necrosis found in the *S. aureus* osteomyelitis model. Other organisms used with this model include *Staphylococcus epidermidis*, methicillin-resistant *Staphylococcus aureus* and *Bacteroides* spp. (including *thetaiotamicro, melanogenicus* and *fragilis*; Mader and Adams, 1989; Mayberry-Carson *et al.*, 1990; 1992; Johansson *et al.*, 1991; Lambe *et al.*, 1991; Petri, 1991). However, in our experience at the University of Texas Medical Branch, *Staphylococcus aureus* and coagulase-negative staphylococci are the most common organisms isolated from patients with osteomyelitis (Mader *et al.*, 1997a). We will focus the description of this rabbit model using *S. aureus* as the representative infective species of bacterium.

The minimum inhibitory concentration of *S. aureus* to the test antibiotic or the antibiotic impregnated into the implant is determined using an antibiotic-tube–dilution method in cation-supplemented Mueller–Hinton broth (CSMHB; Difco Laboratories, Detroit, MI; Ericcson and Sherris, 1971). The antibiotic of interest is serially diluted twofold in tubes containing 5.0 ml of CSMHB. The *S. aureus* inocula for a series of tubes is 0.1 ml of a 1.0×10^6 cfu/ml dilution in CSMHB of an overnight CSMHB culture. The minimum inhibitory concentration is the lowest concentration of antibiotic that prevents turbidity after 20 hours of incubation at 37°C. After the minimum inhibitory concentration is determined, 0.01 ml of each clear tube is streaked on to the surface of a TSA II 5% defibrinated sheep's blood agar plate (Fisher Scientific, Pittsburgh, PA). The minimum bactericidal concentration is the lowest concentration of antibiotic that results in 10 or fewer cfu in a plated 0.010 ml sample after 24 hours of 37°C incubation.

The infective media is prepared by incubating 1 cfu of *S. aureus* overnight in CSMHB at 37°C. The bacterial concentration of the culture is adjusted to 0.5 McFarlands (10^2 cfu/ml) using a turbidimeter (Abbott Laboratories, Chicago, IL). The culture is be further diluted in 0.85% saline to a final concentration of 10^7 cfu/ml.

Infection process

A contiguous focus *S. aureus* osteomyelitis is surgically induced in the left lateral tibial metaphysis of all rabbits within all study groups.

Rabbit infection

New Zealand White rabbits (Ray Nicholl's Rabbitry, Lumberton, TX), 8–12 weeks old and weighing 2.0–3.5 kg, are used for the study. After the mandatory 7-day wait following delivery to the on-site animal resources center, rabbits are anesthetized using an intramuscular injection of 45 mg/kg ketamine (Fort Dodge Laboratories, Fort Dodge, IA) and 5.0–8.0 mg/kg xylazine (Rugby Laboratories, Rockville Center, NY). An 18 G needle is then inserted percutaneously through the lateral aspect of the left tibial metaphysis into the intramedullary cavity. Then 0.15 ml of 5% sodium morrhuate (Eli Lilly, Indianapolis, IN), 0.1 ml of *S. aureus* (10^7 cfu/ml) and 0.2 ml of sterile 0.85% saline are injected sequentially (Mader *et al.*, 1987; Norden, 1970; Mader and Wilson, 1983). However, it is important when working with different strains of *S. aureus* to perform preliminary experiments in order to determine the optimal bacterial inoculum. The optimum bacterial inoculum is one that induces a significant localized osteomyelitis within 2 weeks of infection without a significant increase in mortality for the 6–8 weeks of postinfection survival required for study completion.

After saline injection, the needle is removed and the animal is returned to its cage. The animal must be carefully monitored for pain for the next 24 hours. If tibial tenderness is noted (demonstrated by a resistance to bear weight on the infected leg), analgesic relief may be provided to animals by a subcutaneous dose of buprenorphine 0.01–0.05 mg/kg rabbit body weight. This dose is effective for a period of 6–12 hours. The infection is allowed to progress for 2 weeks, at which time the severity of osteomyelitis is determined radiographically (Table 68.2) and the rabbits that demonstrate a 3+ to 4+ radiographic severity are divided into treatment groups.

Table 68.2 Criteria for grading the radiographic severity of *Staphylococcus aureus* osteomyelitis in rabbits; percentages of disrupted bone are visually estimated from the radiographs

Severity rating	*Radiographic characteristics*
0	Normal
1+	Elevation or disruption of periosteum, or both; soft tissue swelling
2+	< 10% disruption of normal bone architecture
3+	10–40% disruption of normal bone architecture
4+	> 40% disruption of normal bone architecture

Antimicrobial therapy

Treatment groups when testing new antibiotics

At the end of 2 weeks postinfection, the rabbits with *S. aureus* contiguous-focus proximal tibial osteomyelitis are separated into study groups with each group being sacrificed at 8 weeks postinfection. Group One; this group of rabbits is infected but left untreated for the duration of the study. Group Two; rabbits in this group are treated for 4 weeks with a standard antibiotic used in the treatment of human *S. aureus* osteomyelitis (such as parenteral vancomycin or nafcillin). Group Three; rabbits are treated for 4 weeks with the experimental antibiotic. The dose levels of the standard and experimental antibiotics should be chosen so they will achieve serum levels in the rabbits that are comparable to optimum serum levels suggested for humans. If desired, one may include other groups in which rabbits are treated for 4 weeks at a different dose level. Each group should consist of 15–20 rabbits in order to attain statistical significance of results and compensate for the estimated 5–10% of rabbits that die before study completion.

At the end of 28 days (a total of 42 days postinfection) of therapy, radiographs are repeated and compared to the pretreatment radiographs to evaluate the extent of osteomyelitis, the effect of therapy, and the bone growth. In order to accurately assess the effectiveness of the treatment modalities, the rabbits remain untreated for 2 more weeks following the end of antibiotic treatment (a total of 42 days postinfection). Animals are radiographed and sacrificed at this time and all tibias are harvested for bone *S. aureus* concentration determination. The rabbits are monitored daily for stool character, weight loss, caloric intake, tibial tenderness, and overall general health.

Treatment regimen when testing new implant material

In-vitro *antibiotic elution from implant material*

If the *in-vitro* elution rate of the study antibiotic from the implant material has not previously been done, an *in-vitro* determination of elution rates should be performed (Mader *et al.*, 1997b). An antibiotic-loaded bead (3–6 mm diameter) constructed of the implant material of interest is submerged in 1 ml of phosphate-buffered saline (pH 7.4) for 24 hours at 37°C. At 24-hour intervals, the phosphate buffer is drawn off and the antibiotic-loaded beads are then resubmerged in fresh buffer. This procedure is repeated until the bead is fully dissolved or the eluted antibiotic levels are undetectable. All samples should be kept frozen at –70°C until antibiotic concentrations are determined by the disk diffusion method (see below). Obviously, if implants are to be constructed into shapes or sizes different than beads, appropriately adjust the amount of PBS added.

Treatment groups when testing novel antibiotic-impregnated implants

At the end of 2 weeks postinfection, the rabbits with contiguous-focus proximal tibial osteomyelitis are separated into study groups and each group is sacrificed at 8 weeks postinfection. We list here some suggested groups that researchers may wish to include in a study that evaluates novel antibiotic-impregnated implants.

- Group One; this group of rabbits is infected but left untreated for the 8-week duration of the study.
- Group Two; rabbits in this group undergo debridement surgery 2 weeks postinfection.
- Group Three; rabbits undergo debridement surgery (2 weeks after infection) and 4 weeks of appropriate antibiotic administration (subcutaneous, intramuscular, intravenous, or oral).
- Group Four; rabbits have only 4 weeks of appropriate antibiotic administration starting 2 weeks postinfection.
- Group Five; rabbits have the experimental implant material (without antibiotic impregnation) implanted into the dead space created by the debridement surgery (performed 2 weeks after infection).
- Group Six; rabbits in this group have the experimental implant material (without antibiotic impregnation) implanted into the debrided region and appropriate antibiotic administration at the correct dosage for 28 days following debridement surgery.
- Group Seven; rabbits in this group have the experimental material (impregnated with tobramycin, vancomycin, or other standard implanted antibiotic) implanted into the dead space left after the debridement surgery.
- Group Eight; rabbits in this group have polymethylmethacrylate (PMMA) beads (impregnated with the same standard implanted antibiotic as above) implanted into the dead space after debridement surgery.

The choice of appropriate subcutaneous, intramuscular, intravenous, or oral antibiotic will depend upon the type of infecting organism that will be tested (see Table 68.3 for some examples) and dosages of these antibiotics are chosen to produce serum levels that are consistent with those

Table 68.3 Type of bacterial species used for disk diffusion seed organism when testing specific antibiotics

Test antibiotic	*Bacterial species for disk diffusion*
Cefamanadole	*Bacillus subtilus* (Difco)
Cephazolin	*Bacillus subtilus* (Difco)
Cephalothin	*Bacillus subtilus* (Difco)
Ciprofloxacin	*Bacillus subtilus* (Difco)
Ofloxacin	*Bacillus subtilus* (Difco)
Rifampicin	*Bacillus subtilus* (Difco)
Ticarcillin	*Bacillus subtilus* (Difco)
Tobramycin	*Bacillus subtilus* (Difco)
Vancomycin	*Bacillus subtilus* (Difco)
Azithromycin	*Micrococcus luteus* (ATCC 9341)
Clarithromycin	*Micrococcus luteus* (ATCC 9341)
Clindamycin	*Micrococcus luteus* (ATCC 9341)
Daptomycin	*Micrococcus luteus* (ATCC 9341)
Nafcillin	*Micrococcus luteus* (ATCC 9341)
Tosufloxacin	*Escherichia coli*
Cefdinir	*Providencia stuartii* (ATCC 43665)

suggested for human. However, for all implants, we generally use vancomycin as the impregnating antibiotic, but other implanted antibiotics, including tobramycin, clindamycin, and gentamicin (in Europe), have been successfully used. As was true with group sizes when testing new antibiotics, each group should consist of 15–20 rabbits in order to attain statistical significance of results and compensate for the estimated 5–10% of rabbits that die before study completion.

The implant material should not be packed within the dead space to fill the entire debrided area, but should be placed within the defect as beads of less than 6 mm diameter, for a number of reasons. First, by packing the entire dead space with the implant material, the immune system and blood supply become locally compromised, leading to hindrance of infection resolution. Second, by having a single, solid implant, the amount of surface area is much lower than that derived from beads (Gristina, 1994). This lower surface area prevents effective resorption, dissolution, or degradation of a degradable implant. Also, in antibiotic impregnated material, a lower surface area reduces the effectiveness of antibiotic elution from the material, leaving a largely unused residual amount of antibiotic in the center of the single, solid implant (Calhoun and Mader, 1997; Mader *et al.*, 1997b). Lastly, by implanting a large amount of material that has a slow dissolving rate and a low antibiotic elution profile, one provides a long-term substrate for the attachment and eventual colonization of pathogenic organisms (Gristina, 1994).

A 4-week therapy period is required for each group (except Group One). At the end of 4 weeks of therapy, radiography is repeated and compared to the pretreatment radiographs to evaluate the extent of osteomyelitis, the effect of therapy and the bone growth. The rabbits are sacrificed after the radiographic study. All groups should have histologic studies performed on the tibias (three per group) at the end of the 8-week study. The remaining tibias from each group are used for quantitative *S. aureus* concentration determination.

Treatment regimen when testing a new osteomyelitis detection method

Rabbits should initially be divided into two groups before the infection procedure. Group one rabbits are infected as above in the left leg. However, when testing a new detection method, it is prudent to also insert an 18 G needle, percutaneously through the lateral aspect of the left tibial metaphysis into the intramedullary cavity. This is followed by sequential injections of 0.15 ml of 5% sodium morrhuate (Eli Lilly, Indianapolis, IN) and 0.3 ml of sterile 0.85% saline. This internal control is necessary since radiographic, magnetic resonance imaging, and some nuclear studies (unpublished data) demonstrate that the sclerosing agent of sodium morrhuate alone can yield images that are similar to those produced by infection at early stages after inoculation (Volk *et al.*, 1994). These Group One rabbits should be imaged before infection and at 3, 7, 14, 21, and 28 days following infection, using a gold-standard method of osteomyelitis detection (such as radiographs or three-phase bone scans). Group Two rabbits should have identical procedures performed on them, except that detection uses the new osteomyelitis detection method. All rabbits upon sacrifice or death should have histological and culture results that prove the existence of *S. aureus* osteomyelitis.

Details of surgery

Required personnel

Surgery requires three participants. A surgeon and an assistant are required to perform the actual surgery and both are sterile from the waist to the neck, including arms and hands. A 'third out' is also required; s/he is responsible for animal preparation, animal and material transport, preparing the surgical area, casting, and all other actions that would compromise the sterility of the surgical area and attendees.

Materials for surgery

The materials that must be sterilized and remain in the sterile surgical field are placed on a surgical tray, double-wrapped with cotton sheets, and sterilized (121°C, 15 Atm, 30 minutes). This allows the sheet to be opened without compromising the sterility of any of the instruments on the tray. These surgical tray instruments are used to perform the surgery and include: two disposable aluminum cups, gauze pads, two scalpel handles, two surgical blades (no. 10), two short-straight needle holders, two short-curved needle holders, two surgical straight scissors, four towel clips, two spatulas, one forceps, one tissue forceps, one bone saw blade, and four surgical towels. Other tools needed for the surgery but not placed on the surgical tray include: two ready-made syringes of the ketamine/xylazine cocktail (see above), scissors for leg preparation, gauze pads (non-sterile), 70% isopropyl alcohol swabs, surgical gowns, surgical gloves, surgical hats, 4–0 chromic gut sutures, 4–0 prolene sutures, implant material, powdered antibiotic, extra 1 ml and 3 ml syringes, and a weighing scale (0.1–20.0 g). All surgical materials can be obtained from Fisher Scientific (Pittsburgh, PA) except sutures, which may be obtained from Ethicon (Somerville, NJ).

Surgical preparation

It is highly recommended that personnel performing the surgeries be well-trained by individuals proficient in sterile technique surgery. The entire surgical area is cleaned and sterilized using a 10% bleach solution. Immediately before the animal is brought into the surgical area, the surgical tray is placed on one end of the table. It is then opened so that none of the sterile field inside the sheet is broken, and the inside of the sheet is facing up. This provides a sterile area for the rabbit to be placed upon. The bone saw is also wiped, cleaned and sterilized with a 10% bleach solution by the third out. A final pass with bleach is made starting at the cutting end of the saw and working down towards the cable, never moving back toward the tip of the saw. The saw is then placed in its designated area within the sterile field.

Animal preparation

The animal is first anesthetized using an intramuscular injection consisting of ketamine (45 mg/kg) and xylazine (6.0–8.0 mg/kg). After the animal is properly anesthetized, it is brought into the preparation area. All the hair from the infected leg should be removed, using scissors to cut the hair, getting as close to the skin as possible. The entire leg is then washed with liberal amounts of alcohol, followed by a thorough washing and wiping of the area with clinidine solution. This should remove any of the normal skin flora that could enter the surgical site during the procedure. The cleaned leg should now be considered a sterile site, and should not come into contact with other objects, including the rabbit itself. After the rabbit is prepared for surgery, it is brought in and laid down on to the sterile sheet. During the transit, care is taken that the prepared leg does not come into contact with any contaminants.

Surgeon preparation

The surgeon opens the package containing the surgical gown and the package containing the surgical gloves in a designated area of the surgical suite. First, the surgeon puts on a surgical hat and mask, then scrubs in using an iodine scrub pack, following standard surgical procedure. This consists of making sure that the hands remain at an elevation above the rest of the forearm at all times. The hands receive the most attention, scrubbing each surface area of the hand at least ten times. The forearm is also scrubbed vigorously to remove any microorganisms. When rinsing the hands, the hands still remain above the elbows so that the water rinses from the hands down, to keep any microbes from less scrubbed areas from landing on the hands. Maintaining the sterility of the hands, the surgeon enters the surgical suite and dons the surgical gown and gloves. The sterile field on the surgeon is considered to be from the waist up to the neck including the lengths of both arms.

Surgical procedure

All sterile materials for surgery are derived from the surgical tray. Once the rabbit is placed on the sterile sheet, the surgical towels are draped over and attached to the animal (using towel clips) to expose the infected leg. Then, using the number 10 scalpel, an incision is made just through the dermal layer along the medial aspect of the tibia, approximately 2 cm below the tibial metaphysis, and about 1 cm in length. After a small incision has been made, surgical scissors are used to cut the skin through the cutaneous layer, but not through the underlying fascia. The total length of the incision is about 4 cm. The plane of dissection can be seen and felt grossly by locating the lateral edge of the tibia and the incision should follow this edge. The exact line of dissection is a white line separating the two muscles, musculus tibialis cranialis and musculus extensor digiti II (Figure 68.1). The next incision is made through the muscle fascia. A very slight pressure is applied to the scalpel so that the blade does not injure the underlying muscle. The next incision is through the aforementioned white line in order to separate the two muscles (Figure 68.2). If at any time, bleeding ensues, then sterile gauze is placed on the site, and pressure is applied until the bleeding stops.

A gauze dissection is then performed to separate the muscles and expose the tibia. A piece of sterile gauze from the surgery tray is placed on each muscle at the zone of incision, and the two muscle groups are pulled away from each other (Figure 68.3 and 68.4). This method allows the muscles to be moved while sustaining little or no damage and leaving none of the underlying fascia attached to the bone. At this point, the area of bone window to be removed is determined and this is usually done by examining the radiograph of the animal. The area that contains the

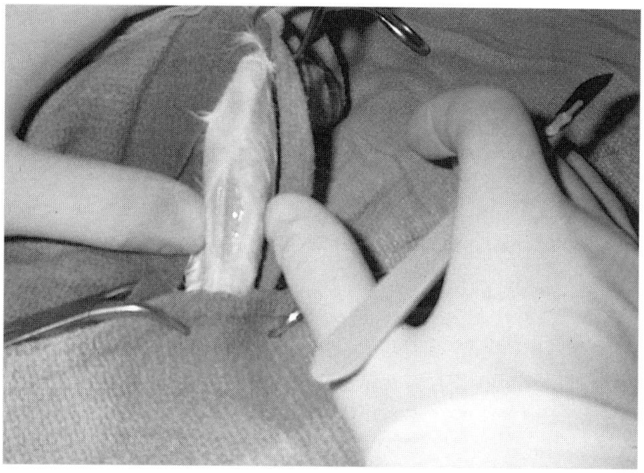

Figure 68.1 After a small incision has been made, surgical scissors are used to cut the skin through the cutaneous layer, but not through the underlying fascia. The total length of the incision is about 4 cm. The plane of dissection can be seen and felt grossly by locating the lateral edge of the tibia; the incision should follow this edge. The exact line of dissection is a white line separating the two muscles, musculus tibialis cranialis and musculus extensor digiti II.

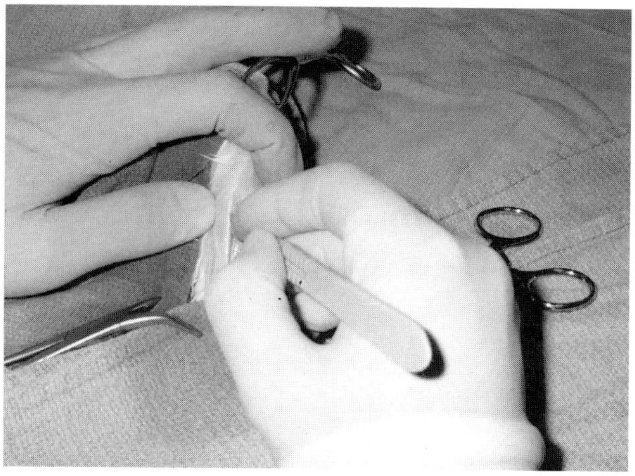

Figure 68.2 The next incision is made through the muscle fascia. A very slight pressure is applied to the scalpel so that the blade does not injure the underlying muscle. The next incision is through the aforementioned white line in order to separate the two muscles.

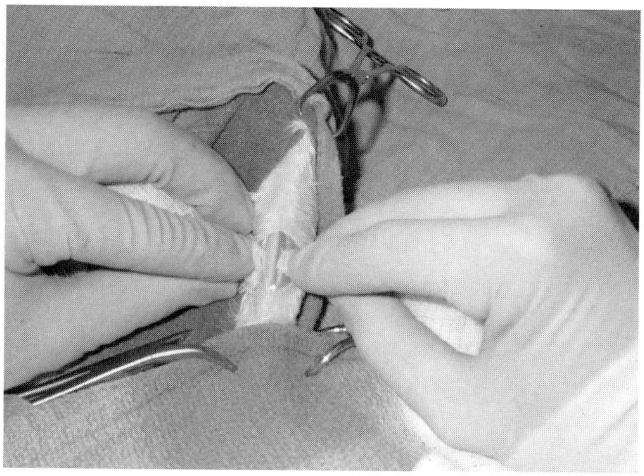

Figure 68.3 A gauze dissection is then performed to separate the musculus tibialis cranialis and musculus extensor digiti II and expose the tibia. A piece of sterile gauze from the surgery tray is placed on each muscle at the zone of incision, and the two muscle groups are pulled away from each other.

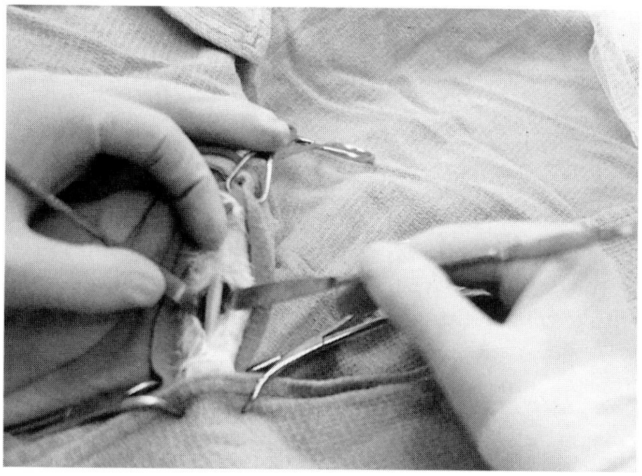

Figure 68.4 Using two spatulas, the surgeon displays the infected left tibia following gauze dissection.

greatest concentration of infection should be the initial location of debridement.

Debridement starts with the removal of a bone flap approximately 0.5 cm wide by 2.0 cm long using a Stryker 40 k command sag saw, with the angle saw blade (Styker Corporation, Kalamazoo, MI). The bone flap is then removed. The infected area is easily detected as a yellow purulent substance, as opposed to the deep red color of normal marrow. Using the narrow end of the spatula, the visible infection is removed. However, there is usually more infection spread throughout the bone. The amount of debrided bone should be kept to a minimum, as large areas of removal will decrease bony stability. However, it is imperative to remove all sequestra and discolored bone marrow (Figure 68.5).

Surgical excision of the bone is carried down to uniform haversian or cancellous bleeding, termed the paprika sign. While this technique will not completely sterilize the bone, it will remove the bone sequestra and the majority of the infecting pathogen, allowing antibiotics, implants, and/or the rabbit immune system to suppress the infection.

The implant should now be placed into the dead space created by the debridement. The implant material is prepared in the sterile field, so that it will not pose any threat by becoming contaminated and infecting the host. The experimental implant material is prepared by placing a known amount of the material into one of the aluminum cups on the surgery tray. Then, saline (and antibiotic if

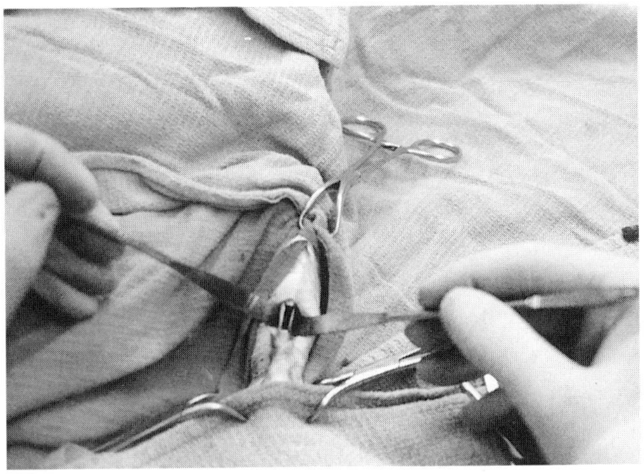

Figure 68.5 The area that contains the greatest concentration of infection should be the initial location of debridement. Debridement starts with the removal of a bone flap approximately 0.5 cm wide by 2.0 cm long using a Stryker 40k command sag saw. The bone flap is then removed. The amount of debrided bone should be kept to a minimum, as large areas of removal will decrease bony stability. However, it is imperative to remove all sequestra and discolored bone marrow.

necessary) are added to the cup, and the substances are mixed together until the desired consistency is obtained. It will often require preliminary experiments to determine the amount of antibiotic and saline to add to the experimental implant in order to obtain a correct consistency. PMMA beads are prepared in a similar manner (see manufacturer's directions, Howmedica, Rutherford, NJ). As soon as a homogenous PMMA consistency is reached, small beads are rolled out by hand. This must be done quickly, as the PMMA reacts quickly, and becomes solid within a few minutes. While it is up to the discretion of the researcher as to what shape the implant will take, we have had numerous successes with multiple beads (3–6 mm diameter) implanted into the dead space left by debridement surgery.

After the implant is in place, the surgical site is closed. It is suggested that the bone flap is left out and not secured with bone wax since this material seems to act as an attachment material for *S. aureus* to colonize and form a biofilm. The muscles that were separated at the beginning of the surgery are reattached with Ethicon chronic gut sutures, size 4–0 (tapered needle). The suturing line should be along the same line as the original ligament that was dissected. The sutures are tied so that the muscles are touching, but are not pulled together so tight that there is a bunched group of muscle at the dissection line. This procedure is repeated down the dissection line every 0.5 cm. The end result should be muscles that are attached along the same line as the original dissection. The outer cutaneous skin is sutured in the same manner, except that Ethicon Prolene 4–0 sutures with a cutting needle are used. If, at any time during the procedure, bleeding starts to occur, the procedure is halted and sterile gauze is placed over the bleeding site with adequate pressure until the bleeding stops. After the bleeding stops, the procedure resumes.

Postoperative procedure

After the surgical incision has been closed, and no bleeding is visible, the animal is removed from the surgical suite and taken back into the preparation area. The debridement of the tibia usually causes some bone instability so a fiberglass cast can be used to provide external fixation. Once inside the preparation room, the leg is thoroughly washed with saline. Four-inch gauze is then wrapped around the leg, starting at the proximal portion of the leg and working down towards the foot. Next, a cotton layer (2 in Sof-Rol cast padding, Fisher Scientific, Pittsburgh, PA) is wrapped around the leg in the same fashion as the gauze. Then 3M Scotchcast Plus fiberglass casting, size 3 in (3M Healthcare, St Paul, MN) is applied, starting from the proximal end of the leg and progressing down towards the distal end of the foot. Care must be taken to ensure proper cast fit. A cast too loose will not serve its purpose, and a cast too tight can cause a number of problems, including tissue ischemia and inflammation. Also, at the proximal end of the cast, there needs to be ample amounts of Sof-Rol protruding from the top, so that it can provide a barrier against the cast rubbing the leg and side of the animal to a point of infection. After the casting has been applied, it is shaped to the curvature of the animal's leg. The shape of the casting should correspond to the position of the rabbit sitting at rest, allowing the rabbit to rest comfortably once it recovers from surgery (Figure 68.6). Before the rabbit is returned to its cage, it is given a postsurgical analgesic, Buprenex, at 0.1 mg/kg. administered subcutaneously. Another 0.1 mg dose is given 24 hours postoperatively.

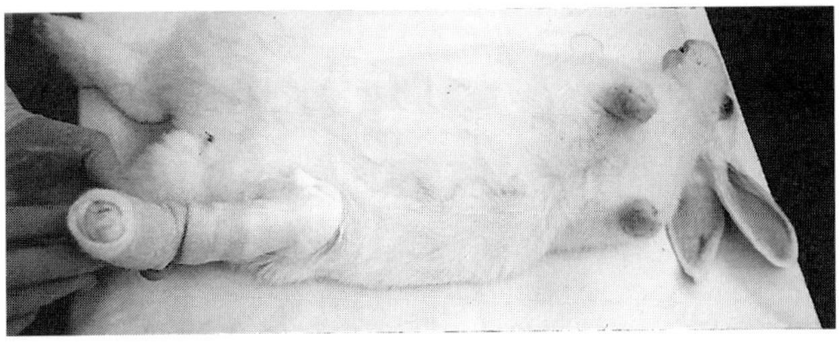

Figure 68.6 Following Soft-Rol application, 3M Scotchcast Plus fiberglass casting (size 3″) is applied, starting from the proximal end of the leg, and progressing down towards the distal end of the foot. Care must be taken to ensure proper cast fit. A cast too loose will not serve its purpose, and a cast too tight can cause a number of problems, including tissue ischemia and inflammation. Also, at the proximal end of the cast, there must be an ample amount of Sof-Rol protruding from the top, so that it can provide a barrier against the cast rubbing the leg and side of the animal to a point of infection. After the casting has been applied, it is shaped to the curvature of the animal's leg. The shape of the casting should correspond to the position of the rabbit sitting at rest, allowing the rabbit to rest comfortably once it recovers from surgery.

Key parameters to monitor response to treatment

Simultaneous level measurements in serum and bone

A group of uninfected animals and a group of 3-week-infected animals are administered a single dose of a standard antibiotic used in the treatment of osteomyelitis. To another group of uninfected animals and a group of 3-week-infected animals, the experimental antibiotic is administered by a single dose. The antibiotic dose amounts of each antibiotic are given at a level that provides serum concentrations that correspond to desired human serum levels in treating osteomyelitis. At 1, 3, 6, 12, and 24 hours following administration of either antibiotic animals for each group at each timepoint have serum drawn, and the right and left tibias are harvested (following sacrifice) for antibiotic concentration determination (see below). In order to obtain serum samples, the animal is first anesthetized with the normal dose of 45 mg/kg ketamine plus 5.0–8.0 mg/kg xylazine. The animal is then laid in a lateral position with the ear in an inferior position to the rabbit body, until the major ear vein becomes prominent. Blood is slowly drawn (0.5 ml) from the major ear vein with a 25 g needle and placed in a sterile microcentrifuge tube in order to allow for coagulation at room temperature. It is very important to note if the test antibiotic is light-sensitive because these require storage and testing in minimal ambient light.

However, in studies testing the efficacy of antibiotic-impregnated implants, a systemic and intramedullary serum samples must be obtained. In this case, a group of uninfected animals and a group of 2-week-infected animals undergo debridement surgery followed by standard antibiotic-impregnated experimental material implantation on the same day. Another group of uninfected animals and a group of 2-week-infected animals undergo debridement surgery followed by standard antibiotic-impregnated polymethylmethacylate (PMMA; Howmedica, Rutherford, NJ) bead implantation on the same day. Serum and bone biopsy samples are obtained at this time then again at 1, 2, 3 and 4 weeks following debridement surgery and material implantation. Systemic serum samples are obtained from anesthetized animals as described above. However, the procedure for obtaining the intramedullary serum sample is somewhat more complex. A small window (2.0 × 2.0 cm) is cut into the cast about 3–4 cm distally from the surgical site. The underlying cotton wrap is then removed to provide clear access to the leg. An 18 G needle is inserted into the bone allowing for approximately 0.5 ml of blood to be drawn from the intramedullary space. The inserted needle may need to be removed and a fresh one inserted through the newly created channel, as the first needle may have a bone blockage, preventing the blood from being drawn. This procedure may be repeated on the non-casted, uninfected right tibia. All blood samples are allowed to coagulate at room temperature, then the serum is drawn off and stored at –70°C until the antibiotic concentrations in these samples can be determined (see below).

Drug assays in serum and bone

An agar-disc-diffusion bioassay is used to measure antibiotic concentrations in both serum and bone eluates. Seeded agar test organisms are usually chosen in conjunction with the pharmaceutical company. Standards of antibiotic are created by serially diluting a 1000 μg/ml stock solution in 100% normal rabbit serum (Sigma Chemical Co., St Louis, MO) and storing these dilutions at –70°C. Volumes of 20 μl of serum standards or serum samples are placed on blank concentration discs (¼″; Difco Laboratories, Detroit, MI), placed on the seeded agar plates, and incubated at 37°C overnight. The diameters of the zones of inhibition of test bacteria growth are measured for both samples and standards. Unknown concentrations are determined from semi-log standard curves plotted from standards. Also, blood is drawn from each rabbit prior to the administration of antibiotic. If inhibitors are detected the animal should not be used in the study.

Infected and uninfected bone is prepared for assay by dissecting them free of all soft tissue, breaking them into small pieces, and removing the marrow. The small bone pieces and the marrow are then separately weighed and suspended in 50% 0.1 M sodium phosphate buffer, pH 7.5, and 50% normal rabbit serum at a rate of 1 ml of the buffer–serum solution per 0.5 g bone/marrow. The bone and marrow samples are then agitated in an Erlenmeyer flask for 1 hour at 4°C with a magnetic stirring bar. Antibiotic levels in the supernatant fluid are assayed by the previously described disc-diffusion methods. Standard solutions of the drugs are prepared by adding normal uninfected bone to the buffer–serum solution containing known amounts of antibiotic.

Radiographs

Radiographs are taken at infection surgery, at initiation of therapy (2 weeks postinfection), and at therapy termination (6 weeks postinfection). Radiographs may also be taken just before sacrifice (eight weeks post-infection). Radiographs should be comparatively scored (see Table 68.2) by double blind evaluation.

Determination of bacterial concentration per gram of bone

Quantitative counts of *S. aureus* cfu per gram of tibial bone should be determined for all study groups. After animals are sacrificed, the right and left tibias are stripped free of all soft tissue, broken into small fragments, and all adhering bone marrow is removed and separated. Bone and marrow samples are separately weighed, physiologic (0.85%) saline is added in a 3 : 1 ratio (3 ml saline/g of bone or marrow) and the suspension is vortexed for 5 minutes. Seven 10-fold dilutions of each of the saline–bone or saline–marrow suspension are prepared with sterile 0.85% NaCl solution; 20 μl samples of each of the seven dilutions are spotted in triplicate on to 5% TSA II defibrinated sheep's blood agar plates (Fischer Scientific, Pittsburgh, PA) and incubated at 37°C for 24 hours. Be sure to give the 20 μl samples enough time at room temperature to absorb into the agar before moving them to the incubator. Cfus are then counted for each tibia sample. The mean log of the cfu for the dilutions and the mean *S. aureus* concentration for each treatment group can then be calculated. It is always important to check isolated colonies with a Becton Dickson Staphyloside test (Becton Dickson & Co., Cockeysville, MA) or other related coagulase test to accurately identify *S. aureus* colonies. If contamination of the plate occurs, streak the most concentrated dilution on to mannitol salt agar plates (Fischer Scientific, Pittsburgh, PA), a media that is selective for *S. aureus* colonies.

Pitfalls (advantages/disadvantages) of the model

While this rabbit model has proved to be successful in the study of osteomyelitis, it does have a few limitations. First, the model requires a very high inoculum of bacteria (10^5–10^6cfu for *S. aureus* and up to 10^8cfu for *P. aeruginosa*) to produce an adequate experimental osteomyelitis. This is unlike the human condition, in which even a very small bacterial inoculum can produce osteomyelitis. Second, in previous studies while testing a new antibiotic in rabbits with experimental osteomyelitis, rabbits were treated with the antimicrobial without any surgical intervention. In the clinical setting, however, osteomyelitis is best resolved by the combination or surgical procedures and antibiotic management. Infected patients usually undergo debridement surgery in order to remove bony sequestra and areas of aseptic necrosis. Therefore, this procedure should be considered when testing the *in-vivo* efficacy of a new antimicrobial agent in rabbits. Third, sodium morrhuate produces a small focus of fibrosis at the inoculation site, often mimicking the histological and imaging characteristics of the osteomyelitis infection pattern. Therefore, when testing a new imaging procedure, it is extremely important to employ negative control rabbits that have sodium morrhuate but not bacteria injected into their tibias. Fourth, the infection has been shown to clear with time, especially when osteomyelitis is induced by *P. aeruginosa*. Fifth, because of the rabbit's small size, many surgical procedures are either impossible or require extreme precision. Some of these procedures include saucerization, local muscle flaps, and free flaps. Lastly, the use of long-term, high-dose antibiotic regimens in rabbits results in significant morbidity and mortality due to the heightened sensitivity of rabbits to some of the toxic effects of certain antibiotics. In particular, nafcillin administration often leads to excessive diarrhea, weight loss, and dehydration in rabbits treated for more than 7 days. If left untreated, mortality rates can reach as high as 50–60% (unpublished data). However, treatment with a daily subcutaneous injection of 50 ml of lactate Ringer's solution and a special diet including cabbage and alfalfa reduces the mortality rates to acceptable levels (5–10%).

While it is obvious that this model is imperfect, the rabbit model of osteomyelitis has been used for a number of studies because of its significant advantages. First, it has lent support to many clinical observations and has successfully been used to assess pathophysiologic and therapeutic variables. Next, rabbit osteomyelitis has the advantage of demonstrating very similar pathologic characteristics and histologic properties to human osteomyelitis. This model is also reliable, inexpensive, and the animals are large enough to allow some surgical manipulation. Lastly, this model also allows for the specific and controlled study of osteomyelitis without the interference of multiple variables. Although a perfect model has yet to be invented, the rabbit model of osteomyelitis has definite advantages when human studies are inappropriate or unavailable.

Contributions of the model to infectious disease therapy

Besides being used for a variety of molecular investigations of infectious agents (Mayberry-Carson *et al.*, 1986; Costerton *et al.*, 1987; Gillaspy *et al.*, 1987), this rabbit model of osteomyelitis has been used to determine the effectiveness of experimental antibiotics and the levels attained by these antibiotics in the tissue and serum of infected and non-infected animals (Van Wingerden *et al.*, 1974; Norden, 1975, 1978; Mader *et al.*, 1978; Mader and Wilson, 1983; Norden and Shaffer, 1983; Norden *et al.*, 1986a,b; Mader *et al.*, 1987, 1989; Mader and Adams, 1989; Mayberry-Carson *et al.*, 1990; Norden and Budinsky, 1990). Other studies that have utilized this model include histologic studies (Crane *et al.*, 1977), hyperbaric oxygen evaluation (Mader *et al.*, 1978, 1980; Esterhai *et al.*, 1986), determination of tissue oxygen tensions (Mader *et al.*, 1978; Aro *et al.*, 1984), blood flow studies (Mader *et al.*, 1980), new bone implants (either alone or impregnated with antibiotics) (Laky *et al.*, 1983; Tomczak *et al.*, 1989; Thomas *et al.*, 1989; Lambe *et al.*, 1991; Wei *et al.*, 1991; Mayberry-Carson *et al.*, 1992; Melcher *et al.*, 1994; Tsourvakas *et al.*, 1995; Heard *et al.*, 1997), and new osteomyelitis detection methods (Risky *et al.*, 1977; Alazraki *et al.*, 1984; Hartshorne *et al.*, 1985; Abiri *et al.*, 1992; Volk *et al.*, 1994).

References

Abiri, M. M., DeAngelis, G. A., Kirpekar, M., Abou, A. N., Ablow, R. C. (1992). Ultrasonic detection of osteomyelitis. Pathologic correlation in an animal model. *Invest. Radiol.*, **27**, 111–113.

Alazraki, N., Moitoza, J., Heaphy, J., Taylor, A. Jr (1984). The effect of iatrogenic trauma on the bone scintigram: an animal study: concise communication. *J. Nucl. Med.*, **25**, 978–981.

Andriole, V. T., Nagel, D. A., Southwick, W. O. (1973). A paradigm for human chronic osteomyelitis. *J. Bone Joint Surg.*, **55A**, 1511–1515.

Andriole, V. T., Nagel, D. A., Southwick, W. O. (1974). Chronic staphylococcal osteomyelitis: an experimental model. *Yale J. Biol. Med.*, **47**(1), 33–39.

Aro, H., Eerola, E., Aho, A. J., Niinikoski, J. (1984). Tissue oxygen tension in externally stabilized tibial fractures in rabbits during normal healing and infection. *J. Surg. Res.*, **37**, 202–207.

Calhoun, J. H., Mader, J. T. (1997). Treatment of osteomyelitis with a biodegradable antibiotic implant. *Clin. Orthop.*, **341**, 206–214.

Cierny, G. III, Mader, J. T. (1983). Management of adult esteomyelitis. In: *Surgery of the Musculoskeletal System* (ed. Evarts, C. M.) vol. 10, pp. 15–35. Churchill Livingstone, New York.

Cierny, G. III, Mader, J. T., Penninck, J. J. (1985). A clinical staging system for adult osteomyelitis. *Contemp. Orthop.*, **10**(5), 17–37.

Costerton, J. W., Lambe, D. W. Jr, Mayberry-Carson, K. J., Tober-Meyer, B. (1987). Cell wall alterations in staphylococci growing in situ in experimental osteomyelitis. *Can. J. Microbiol.*, **33**, 142–150.

Crane, L. R., Kapdi, C. C., Wolfe, J. N., Silberberg, B. K., Lerner, A. M. (1977). Xeroradiographic, bacteriologic, and pathologic studies in experimental staphylococcal osteomyelitis. *Proc. Soc. Exp. Biol. Med.*, **156**, 303–314.

Dominguez, J., Crase, D., Soave, O. (1975). A case of *Pseudomonas* osteomyelitis in a rabbit. *Lab. Anim. Sci.*, **25**, 506.

Ericcson, H. M., Sherris, J. C. (1971). Antibiotic sensitivity testing. Report of an international collaborative study. *Acta Pathol. Microbiol. Scand. Sect. B. Suppl*, **217**, 1–90.

Esterhai, J. L. Jr, Clark, J. M., Morton, H. E., Smith, D. W., Steinbach, A., Richter, S. D. (1986). Effect of hyperbaric oxygen exposure on oxygen tension within the medullary canal in the rabbit tibial osteomyelitis model. *J. Orthop. Res.*, **4**, 330–336.

Fitzgerald, R. H. Jr (1983). Experimental osteomyelitis: description of a canine model and the role of depot administration of antibiotics in the prevention and treatment of sepsis. *J. Bone Joint Surg. (Am.)*, **65**(3), 371–380.

Gillaspy, A. F., Hickmon, S. G., Skinner, R. A., Thomas, J. R., Nelson, C. L., Smeltzer, M. S. (1995). Role of the accessory gene regulator (agr) in pathogenesis of staphylococcal osteomyelitis. *Infect. Immun.*, **63**, 3373–3380.

Gristina, A. G. (1994). Implant failure and the immunoincompetent fibro-inflammatory zone. *Clin. Orthop.*, **298**, 106–118.

Hartshorne, M. F., Graham, G., Lancaster, J., Berger, D. (1985). Gallium-67/technetium-99 m methylene diphosphonate ratio imaging: early rabbit osteomyelitis and fracture. *J. Nucl. Med.*, **26**, 272–277.

Heard, G. S., Oloff, L. M., Wolfe, D. A., Little, M. D., Prins, D. D. (1997). PMMA bead versus parenteral treatment of *Staphylococcus aureus* osteomyelitis. *J. Am. Pod. Med. Assoc.*, **87**, 153–164.

Johansson, A., Svensson, O., Blomgren, G., Eliasson, G., Nord, C. E. (1991). Anaerobic osteomyelitis. A new experimental rabbit model. *Clin Orthop. Rel. Res.*, 297–301.

Kelly, P. J. (1986). Chronic osteomyelitis. In: *Musculoskeletal Infections* (eds Hughes, S. P. F., Fitzgerald, R. H. Jr), pp. 136–150. Year Book Medical Publishers, Chicago, IL.

Laky, R., Kocsis, B., Kubatov, M., Szittyai, B. (1983). [Model studies of the antibiotic release of gentamycin, tobramycin and cephalothin polymethylmethacrylate beads *in vivo*]. [German]. *Zschr. Exp. Chirurg. Transpl. Kunstl. Org.*, **16**, 306–314.

Lambe, D. W. Jr, Ferguson, K. P., Mayberry-Carson, K. J., Tober-Meyer, B., Costerton, J. W. (1991). Foreign-body-associated experimental osteomyelitis induced with *Bacteroides fragilis* and *Staphylococcus epidermidis* in rabbits. *Clin. Orthop. Rel. Res.*, 285–294.

Mader, J. T. (1985). Animal models of osteomyelitis. *Am. J. Med.*, **78**, 213–217.

Mader, J. T., Adams, K. (1989). Comparative evaluation of daptomycin (LY146032) and vancomycin in the treatment of experimental methicillin-resistant *Staphylococcus aureus* osteomyelitis in rabbits. *Antimicrob. Agents Chemother.*, **33**, 689–692.

Mader, J. T., Wilson, K. J. (1983). Comparative evaluation of cefamandole and cephalothin in the treatment of experimental *Staphylococcus aureus* osteomyelitis in rabbits. *J. Bone J. Surg.*, **65**, 507–513.

Mader, J. T., Guckian, J. C., Glass, D. L., Reinarz, J. A. (1978). Therapy with hyperbaric oxygen for experimental osteomyelitis in rabbits. *J. Bone Joint Surg.*, **65A**, 507–513.

Mader, J. T., Brown, G. L., Guckian, J. C., Wells, C. H., Reinarz, J. A. (1980). A mechanism for the amelioration by hyperbaric oxygen of experimental staphylococcal osteomyelitis in rabbits. *J. Infect. Dis.*, **142**, 915–921.

Mader, J. T., Morrison, L. T., Adams, K. R. (1987). Comparative evaluation of A-56619, A-56620, and nafcillin in the treatment of experimental *Staphylococcus aureus* osteomyelitis. *Antimicrob. Agents Chemother.*, **31**, 259–263.

Mader, J. T., Adams, K., Morrison, L. (1989). Comparative evaluation of cefazolin and clindamycin in the treatment of experimental *Staphylococcus aureus* osteomyelitis in rabbits. *Antimicrob. Agents Chemother.*, **33**, 1760–1764.

Mader, J. T., Mohan, D., Calhoun, J. (1997a). A practical guide to the diagnosis and management of bone and joint infections. *Drugs*, **54**(2), 253–264.

Mader, J. T., Calhoun, J., Cobos, J. (1997b). *In vitro* evaluation of antibiotic diffusion from antibiotic-impregnated biodegradable beads and polymethylmethacrylate beads. *Antimicrob. Agents Chemother.*, **41**(2), 415–418.

Mayberry-Carson, K. J., Tober-Meyer, B., Lambe, D. W. Jr, Costerton, J. W. (1986). An electron microscopic study of the effect of clindamycin therapy on bacterial adherence and glycocalyx formation in experimental *Staphylococcus aureus* osteomyelitis. *Microbios*, **48**, 189–206.

Mayberry-Carson, K. J., Tober-Meyer, B., Gill, L. R., Lambe, D. W. Jr, Hossler, F. E. (1990). Effect of ciprofloxacin on experimental osteomyelitis in the rabbit tibia, induced with a mixed infection of *Staphylococcus epidermidis* and *Bacteroides thetaiotaomicron Microbios*, **64**, 49–66.

Mayberry-Carson, K. J., Tober-Meyer, B., Lambe, D. W. Jr,

Costerton, J. W. (1992). Osteomyelitis experimentally induced with *Bacteroides thetaiotaomicron* and *Staphylococcus epidermidis*. Influence of a foreign-body implant. *Clin. Ortho. Rel. Res.*, 289–299.

Melcher, G. A., Claudi, B., Schlegel, U., Perren, S. M., Printzen, G., Munzinger, J. (1994). Influence of type of medullary nail on the development of local infection. An experimental study of solid and slotted nails in rabbits. *J. Bone Joint Surg.*, **76B**, 955–959.

Morrissy, R. T., Haynes, D. W. (1989). Acute hematogenous osteomyelitis: a model with trauma as an etiology. *J. Pediatr. Orthop.*, **9**, 447–456.

Norden, C. W. (1975). Experimental osteomyelitis. VI. Therapeutic trials with rifampin alone and in combination with gentamicin, sisomicin, and cephalothin. *J. Infect. Dis.*, **132**, 493–499.

Norden, C. W. (1978). Experimental osteomyelitis. V. Therapeutic trials with oxacillin and sisomicin alone and in combination. *J. Infect. Dis.*, **137**, 155–160.

Norden, C. W., Budinsky, A. (1990). Treatment of experimental chronic osteomyelitis due to *Staphylococcus aureus* with ampicillin/sulbactam. *J. Infect. Dis.*, **161**, 52–53.

Norden, C. W., Keleti, E. (1980). Experimental osteomyelitis caused by *Pseudomonas aeruginosa*. *J. Infect. Dis.*, **141**, 71–75.

Norden, C. W., Kennedy, E. (1970). Experimental osteomyelitis. A description of the model. *J. Infect. Dis.*, **122**, 410–418.

Norden, C. W., Shaffer, M. (1983). Treatment of experimental chronic osteomyelitis due to *Staphylococcus aureus* with vancomycin and rifampin. *J. Infect. Dis.*, **147**, 352–357.

Norden, C. W., Myerowitz, R. L., Keleti, E. (1980). Experimental osteomyelitis due to *Staphylococcus aureus* or *Pseudomonas aeruginosa:* a radiographic-pathological correlative analysis. *Br. J. Exp. Pathol.*, **61**, 451–460.

Norden, C. W., Niederreiter, K., Shinners, E. M. (1986a). Treatment of experimental chronic osteomyelitis due to *Staphylococcus aureus* with teicoplanin. *Infection*, **14**, 136–138.

Norden, C. W., Shinners, E., Niederreiter, K. (1986b). Clindamycin treatment of experimental chronic osteomyelitis due to *Staphylococcus aureus*. *J. Infect. Dis.*, **153**, 956–959.

Patzakis, M. J. (1982) Management of open fracture wounds. In: *Instructional Course Lectures* (ed. Griffin, P. P.), Vol. XXXVI, pp. 367–369. American Academy of Orthopaedic Surgeons, Rosemont, IL.

Petri, W. H. III (1991). Evaluation of antibiotic-supplemented bone allograft in a rabbit model. *J. Oral Maxillofacial Surg.*, **49**, 392–396.

Risky, L., Goris, M. L., Schurman, D. J., Nagel, D. A. (1977). ^{99}Technetium bone scanning in experimental osteomyelitis. *Clin. Orthop. Rel. Res.*, 361–366.

Scheman, L., Janota, M., Lewin, P. (1941). The production of experimental osteomyelitis. *J.A.M.A.* **117**, 1525–1529.

Thomas, V. L., Sanford, B. A., Keogh, B. S., Triplett, R. G. (1989). Antibody response to *Staphylococcus aureus* surface proteins in rabbits with persistent osteomyelitis after treatment with demineralized bone implants. *Infect. Immun.*, **57**, 404–412.

Tomczak, R. L., Dowdy, N., Storm, T., Lane, J., Caldarella, D. (1989). Use of ceftazidime-impregnated polymethyl methacrylate beads in the treatment of *Pseudomonas* osteomyelitis. *J. Foot Surg.*, **28**, 542–546.

Tsourvakas, S., Hatzigrigoris, P., Tsibinos, A., Kanellakopoulou, K., Giamarellou, H., Dounis, E. (1995). Pharmacokinetic study of fibrin clot-ciprofloxacin complex: an *in vitro* and *in vivo* experimental investigation. *Arch. Orthop. Trauma Surg.*, **114**, 295–297.

Van Wingerden, G. I., Lolans, V., Jackson, G. G. (1974). Experimental *Pseudomonas* osteomyelitis. Treatment with sisomicin and carbenicillin. *J. Bone Joint Surg.*, **56A**, 1452–1458.

Volk, A., Cremieux, A. C., Belmatoug. N., Vallois, J. M., Pocidalo, J. J., Carbon, C. (1994). Evaluation of a rabbit model for osteomyelitis by high field, high resolution imaging using the chemical-shift-specific-slice-selection technique. *Mag. Res. Imag.*, **12**, 1039–1046.

Waldvogel, F. A., Medoff, G., Swartz, M. N. (1970). Osteomyelitis: a review of clinical features, therapeutic considerations and unusual aspects (3 parts). *N. Engl. J. Med.*, **282**(4), 198–206; **282**(5), 260–266; **282**(6), 316–322.

Wei, G., Kotoura, Y., Oka, M. *et al.* (1991). A bioabsorbable delivery system for antibiotic treatment of osteomyelitis. The use of lactic acid oligomer as a carrier. *J. Bone Joint Surg.*, **73**B, 246–252.

Chapter 69

The Arthroplasty Model in Rats

J. Renneberg, K. Piper and M. Rouse

Background of the model

Deep infection after total knee arthroplasty is one of the most costly and devastating complications of arthroplasty. The incidence of infection after knee arthroplasty ranges from 1.1% to 5% (Walker and Schurman, 1984; Rand, 1993). Classification of the infections is generally divided in relation to the time of development after primary implantation: acute infections, less than 3 months after implantation of the prosthesis; delayed infection, between 3 months and 1 year after implantation; and late infection, more than 1 year after implantation (Antti-Poika *et al.*, 1990). The above classification, as well as antimicrobial chemotherapy and surgical intervention have all been based upon experience from retrospective evaluation of cases of human infection (Fitzgerald, 1992; Buccholz *et al.*, 1989).

The infections, as mentioned, are of low frequency but each infection is extremely expensive and devastating, so intensive research in this field is necessary, preferably in the form of prospective studies. In order to carry out prospective studies, animal models have been developed. The rat prosthesis arthritis model is one of these models; it makes it possible to get experience of infection development and treatment using a functioning arthroplastic joint.

Animal species

Several strains of rat have been used as models for osteomyelitis: Sprague–Dawley (Rissing *et al.*, 1985), Wistar (Zak *et al.*, 1981; Gomis *et al.*, 1990) and others. We used Wistar rats, both males and females, weighing 200–270 g.

Preparation of animals

It is recommended that, as far as possible, rats should be housed so as to control extremes of environment, and that all cages and feed equipment be of metal construction to facilitate sanitation; that water systems should be of the dewdrop type; and that floors and walls be impervious to liquid and moisture, and vermin-proof. The animals should be obtained 2 weeks prior to receiving anaesthesia (Loew *et al.*, 1987).

Details of surgery

Overview

Briefly, the right joint capsule of anaesthetized rats is exposed, the patella is dislocated laterally and the articulating cartilage is osteotomized from the distal femur and the proximal tibia. A 4 mm hole is bored into the femur and a 2 mm hole into the tibia to fit the joint components. The joint capsule and skin are closed after placement of the knee prosthesis.

Materials required

A rat-sized, non-constrained condylar knee prosthesis, the femoral component milled out of aluminium bar stock and the tibial component milled from high-density polyethylene stock (Figure 69.1; Rouse *et al.*, 1997). Anaesthetic (preferably injectable, and including an analgesic), hair clippers or razor, skin disinfectant, scalpel handle plus blades, forceps, high-speed dental drill with 4 mm and 2 mm drill bits, polymethylmethacrylate bone cement, suture material (preferably prolene sutures).

Anaesthesia

Rats are sedated with an i.m. (or i.p.) injection of 30 mg/kg ketamine hydrochloride (Ketaset; US Pharmacopeia) and 4 mg/kg xylazine (Rompun; Bayer Corporation). This is prepared by adding 1 ml of xylazine, 100 mg/ml, to a 10 ml bottle of ketamine, 100 mg/ml. Complete anaesthesia ensues in about 1–8 minutes and lasts for about 30–45 minutes. The analgesic, xylazine, provides muscle relaxation; respiration should be strong in rodents under the dissociative anaesthetic ketamine (Loew *et al.*, 1987).

Handbook of Animal Models of Infection
ISBN 0-12-775390-7

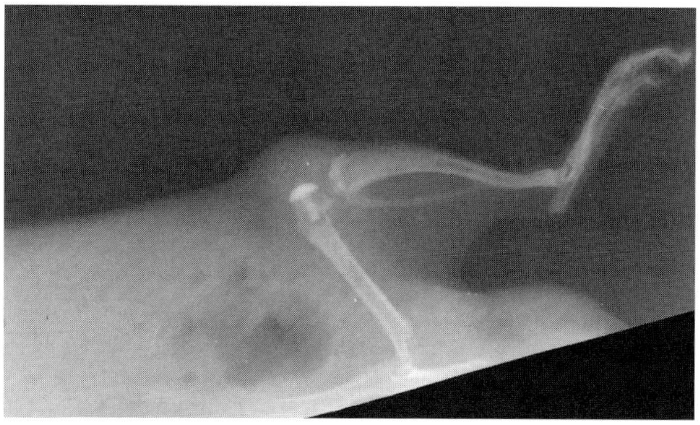

Figure 69.1 X-ray view of the prosthesis *in situ*.

Surgical procedure

Following complete induction of anaesthesia, the rat is placed ventral side up. The skin over the knee is shaved with electric clippers fitted with a no. 10 blade; the skin is disinfected with a topical antiseptic. We normally use a povidone–iodine 10% solution (Betadine, equivalent to 1% iodine). The right joint capsule is opened with a medial parapatellar incision over the flexed knee. We use a no. 3 scalpel handle fitted with a no. 10 curved scalpel blade (Bard-Parker; Becton Dickinson). The patella is dislocated laterally using forceps. The articulating cartilage is osteotomized from the distal femur and proximal tibia by slicing off the end of the bone with the scalpel blade until a flat surface the size of the prosthesis is obtained. A high-speed dental drill fitted with a 4 mm bulb bit is used to drill a hole into the medullar cavity of the femur. The femoral component is then cemented in place with polymethylacrylate (Surgical Simplex P; Howmedica). While the cement is drying, a 2 mm bulb bit is used to drill a hole into the medullar cavity of the tibia. The tibial component is then press-fitted (Rouse *et al.*, 1997).

Infection procedures

Following the preparing of the medullary canals, if needed, the area is cleaned using sterile physiological saline, and 0.1 ml of bacterial inoculum is injected into the medullar cavity of the femur and tibia using a 1 ml ONCE syringe and a 26 G ½″ needle. The solution is slowly injected; a small portion of the liquid will always be lost to the joint surface. Immediately following this, the cement is similarly injected with a 2 ml ONCE syringe, without a needle, into the medullary canals and the prosthesis is inserted and held for about 10 minutes until the cement hardens.

In some experiments the rats are inoculated intravenously. The inoculum is then injected after or during the operative procedure—nothing is given locally in these experiments. If both knees are used, the inoculation should be given after both arthroplastic operations are finished.

Wound closure

The joint capsule and the skin are both closed with Prolene sutures (3–0; Ethicon). Normally, two sutures over the joint capsule and three spaced evenly over the skin are sufficient. The wound can also be covered with antiseptic liquid bandage (New Skin; Medtech Laboratories). Prolene sutures are preferred because they minimize tissue reaction and will not absorb the bacterial inoculum (Davis *et al.*, 1987).

Postoperative care

Rats should be returned to their individual cages, laid on their sides and covered to keep them warm. The rats usually return to normal behaviour within 1 day. Subsequently, the rats may be housed in groups. Postoperative analgesics have not been routinely used. The knee arthroplasties are radiographed and examined at 1 week and 1 month postoperatively to assess component alignment, presence or absence of periprosthetic fracture or postoperative complications of component failure or migration (Figure 69.2).

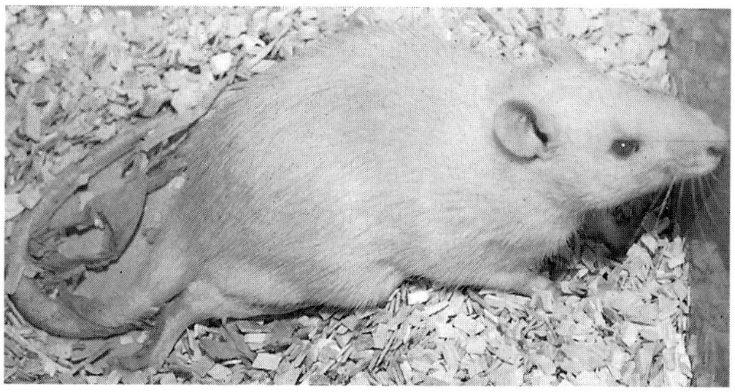

Figure 69.2 The prosthesis *in situ*.

Storage and preparation of inocula

The model is amenable to using several different species of bacteria, although most studies have used *Staphylococcus aureus, S. epidermidis, E. coli* or *Pseudomonas aeruginosa*. We used the antibiotic-susceptible *Staphylococcus aureus* strain 1369, phage type 85 in our experiments. This strain has been adapted to the rat environment by passaging in rats (peritoneum) two or three times. The inoculum is prepared by growing overnight in serum broth, washing three times in 0.9% NaCl and finally resuspending to 10 times the required inoculum per millilitre. The inoculum was checked before each experiment by plating. The inocula for the whole experimental series (30–40 animals) was produced at the same time and frozen at –80°C in small amounts for each experiment day. This procedure must be used to ensure that the bacteria keep their virulence against the rats. The inocula were used within 4 months and thereby reduction of the quantity of living bacteria was avoided.

Infection process

The infection is initiated by direct injection of bacterial suspension into the medullary canals of femur and tibia. Using strain 1369, we achieved an osteomyelitis with an inoculum of 10^4 and 10^5cfu but with 10^7cfu the rats develop sepsis. Inoculum of 10^8 resulted in death in up to 56% of cases. A comparison of the effectiveness of different infection procedures showed that *S. aureus* strain 1369 used locally at a rate of 10^5 was as effective as *S. aureus* strain TB (Bremell *et al.*, 1990) and *S. aureus* 1369 given i.v. to inoculate arthroplastic joints (infection rate 95%).

Histological signs of disease were observed within 1 week and remained essentially unchanged throughout 3–4 weeks.

Key parameters to monitor infection and response to treatment

With an inoculum of 10^{4-5}cfu of *S. aureus* the infection is confined to the bone. Death due to sepsis is extremely rare; inocolum as large as 10^{8-9}cfu of *S. aureus* will be needed. Sedimentation rates cannot be used as an infection parameter in rats: HGB and WBC most often both fall within the first week of infection.

Appearance, body-weight gain, food and water intake are affected by the infection process or treatment-associated toxicity. We found that the operative procedure itself caused weight loss in the animals and this, together with local infection, gave a weight loss of about 15%. The animals never regained their initial body weight (Batchelor and Giddins, 1995).

Bone pathology

Bone histology is normally used to monitor progress of infection and response to antimicrobial therapy (Spagnolo *et al.*, 1993; Cordero *et al.*, 1994; Heinz *et al.*, 1995).

Pathology scores have been used by several investigators. We used the suggestions of Rissing *et al.* (1985), who scored gross pathology as: 0, no apparent pathology; 1, minimal erythema without abscess or new bone formation; 2, erythema and widening of tibial shaft (new bone formation); 3, purulent exudate or abscess; and 4, severe bone destruction involving more than 50% of the diaphysis in rats. Blomgren (1981; rabbits) found that the bone cement was more or less surrounded by a thin membrane consisting of fibroblasts and inflammatory cells. In most cases the innermost layer of the cortical bone, close to the cement or membrane, seemed dead, with empty osteocyte lacunae.

Bacterial density in bone

Determination of bacterial cfu within bone normally requires homogenizing of the bone. This can be easily accomplished by deep-freezing the bones and then crushing the bones by hand using a mortar (Rissing *et al.*, 1985; Calhoun and Mader, 1997). The bone powder should remain cold or preferably frozen during the procedure and is then resuspended in a small volume of PBS. Standard dilution and bacterial enumeration procedures can then be used.

Table 69.1 Inoculum, inoculation site and refinding of bacteria in rats; when an inoculum below 10^7 was used none of the animals died

Inoculum bacteria	*Local joint injection*	*Intravenous injection*	*Joint + prosthesis (%)*	*Joint – prosthesis (%)*	*Arthritis*	*Osteitis*	*Other locations*
10^2	×		–	–	–	–	–
10^2		×	–	–	–	–	–
10^3	×		–	–	–	–	–
10^3		×	–	–	–	–	–
10^4	×		80	–	–	–	–
10^4		×	50	–	–	–	–
10^5	×		90	–	–	–	–
10^5		×	95	10	–	–	–
10^6	×		96	39	+	–	Sepsis, liver, kidney and lungs
10^6		×	95	28	+	+	Sepsis, brain, heart, kidney and liver

Determination of antibiotic penetration into bone

The bone specimens should be cleaned free of adherent tissue, rinsed in PBS (pH 7.2) and crushed into small pieces, which are weighed and immersed in PBS. The total volume is measured and the mixture shaken at 4°C for 5 hours to equilibrate the antibiotic concentration in bone with that of the supernatant (Leigh, 1986), which was used for the assays. Joint fluids should be sampled and used for the assays as well.

Two assays can be used: a standard microbiological assay or high-performance liquid chromatography (HPLC) method (Renneberg *et al.*, 1993).

Antimicrobial therapy

Rats have been used in osteomyelitis models in several models (Zak *et al.*, 1981; Henry *et al.*, 1987; Dworkin *et al.*, 1990; Gisby *et al.*, 1994). There has been good experience using antibiotics and achievable antibiotic concentrations in bone and joint fluid from these models. We have used dicloxacillin and aminoglycosides locally (in the cement) and found that dicloxacillin concentrations reached levels of around 5 mg/kg bone and 7.4 mg/l joint fluid and that gentamicin peak levels in bone are 4.0 mg/kg and 10 g/l in joint fluid.

Pitfalls (advantages/disadvantages) of the model

Besides the usual shortcomings of all animal models, the major shortcoming of the rat model of orthopaedic prosthesis infections has been that it was impossible to get a functioning prosthesis. This problem has now been solved. We produce our own prosthesis parts both at the Mayo Clinic and in Copenhagen. All are made of the same materials as human prostheses. The model's advantages are as follows.

- It is easy to handle.
- It is reproducible.
- The rat can survive several operative procedures without dying.
- It is cheap.
- It is easy to handle large numbers of animals and thus obtain information that is unachievable in the clinical situation.
- The model tests a prosthesis in its intended functioning place.

The disadvantage of the model is that the antibiotic therapy regimens demand frequent administration.

Contributions of the model to infectious disease therapy

The model is rather new (1997), so any results concerning therapy regimens cannot be presented, but it has potential for profiling antibiotic treatment regimens, both for prophylaxis and for real infection treatment. Furthermore, it can be used to look at the pathophysiology of bacterial infection in a prosthetic joint. We have so far used the model to show that, after intravenous inoculation with 10^5cfu of *S. aureus*, infection occurs more frequently in a joint with a prosthesis fitted (100%) than in a joint with no prosthesis, but after the same surgical procedure (12%).

References

Antti-Poika, I., Josefsson, G., Konttinen, Y., Lidgren, L., Santavirta S., Sanzén, L. (1990). Hip arthroplasty infection. *Acta Orthop. Scand.*, **61**(2), 163–169.

Batchelor, G. R., Giddins, G. E. B. (1995). Bodyweight changes in laboratory rabbits following operative procedures. *Anim. Technol.*, **46**(3), 153–161.

Blomgren, G. (1981). Hematogeneous infection of total joint replacement. *Acta Orthop. Scand.*, **52** (Suppl. 187).

Brause, B. D. (1989). Infected orthopedic prostheses. In: *Infections Associated with Indwelling Medical Devices* (eds Bisno, A. L.; Waldvogel, F. A.), pp. 111–127. American Society of Microbiology, Washington, DC.

Bremell, T., Lange, S., Svensson, L. *et al.* (1990). Outbreak of spontaneous staphylococcal arthritis and osteitis in mice. *Arthritis Rheum.*, **33**(11), 1739–1744.

Buccholz, H. W., Elson, R. A., Engebracht, B. (1989). Management of deep infection of total hip replacement. *J. Bone Joint Surg.*, **63(B)**, 342–347.

Calhoun, J. H., Mader, J. T. (1997). Treatment of osteomyelitis with a biodegradable antibiotic implant. *Clin. Orthop.*, **341**, 206–214.

Cordero, J., Munuera L., Folgueira M. (1994). Influence of metal implants on infection: an experimental study in rabbits. *J. Bone Joint Surg.*, **76**B.

Croft, P. G. (1964). Chapter 5 in *Small Animal Anesthesia* (ed. Graham-Jones, O.), pp. 99–102. Oxford University Press, Oxford.

Davis, C. V. *et al.* (1987). *Clinical Surgery*. C. V. Mosby, St Louis, MO.

Dworkin, R., Modin, G., Kunz, S., Rich, R., Zak, O., Sande, M. (1990). Comparative efficacies of ciprofloxacin, perfloxacin and vancomycin in combination with rifampicin in a rat model of methicillin resistant *Staphylococcus aureus* chronic osteomyelitis. *Antimicrob. Agents Chemother.*, **34**, 1014–1016.

Fitzgerald, R. H. (1992). Total hip arthroplasty sepsis. *Orthop. Clin. North Am.*, **23**(2), 259–264.

Gisby, J., Beale, A. S., Bryant, J. E., Toseland, C. D. N. (1994). Staphylococcal osteomyelitis — a comparison of co-amoxiclav with clindamycin and flucloxacillin in an experimental model. *J. Antimicrob. Chemother.*, **34**, 755–764.

Gomis, M., Herranz, A., Aparicio, P., Alonso, M. J., Prieto, J., Martinez, T. (1990). An experimental model of chronic osteomyelitis caused by *Escherichia coli* treated with cefotaxime. *J. Antimicrob. Chemother.*, **26** (Suppl. A), 15–21.

Heinz, S. A., Sakamoto, H., Flock, J.-I. *et al.* (1995). Development and characterization of a new model of haematogeneous osteomyelitis in the rat. *J. Infect. Dis.*, **171**, 1230–1236.

Henry, N. K., Rouse, M. S., Whitesell, A. L., McConnell, M. E., Wilson, W. R. (1987). Treatment of methicillin-resistant staphylococcal osteomyelitis with ciprofloxacillin or vancomycin alone or in combination with rifampin. *Am. J. Med.*, **82** (Suppl. 4A), 73–75.

Leigh, D. A. (1986). Serum and bone concentrations of cefuroxime in patients undergoing knee arthroplasty. *J. Antimicrob. Chemother.,* **18**, 609–611.

Loew, F. M., Bennett, B. T., Blake, D. A. *et al.* (1987). Surgery: protecting your animals and your study. In: *The Biomedical Investigator's Handbook for Researchers Using Animal Models*, pp. 19–27. Foundation for Biomedical Research, Washington.

Rand, J. A. (1993). Sepsis following total knee arthroplasty. In: *Total Knee Arthroplasty* (ed. Rand, J. A.), pp. 349–375. Raven Press, New York.

Renneberg, J., Christensen, O. M., Thomsen, N. O. B., Tørholm, C. (1993). Cefuroxime concentrations in serum, joint and bone in elderly patients undergoing arthroplasty after administration of cefuroxime axetil. *J. Animicrob. Chemother.*, **32**, 751–755.

Rissing, J. P., Buxton, T. B., Weinstein, R. S., Schockley, R. K. (1985). Model of experimental chronic osteomyelitis in rats. *Infect. Immun.*, **47**, 518–586.

Rouse, M., Huddleston, P., Hanssen, A., Wilson, W., Patel. R., Steckelberg, J. (1997). *A Novel Model of Total Knee Arthroplasty in Rats.* Poster P809 at the 8th European Congress of Clinical Microbiology and Infectious Diseases, Vienna.

Spagnolo, L., Greco, F., Rossi, A., Ciolli, L., Teti, A., Posteraro, P. (1993). Chronic staphylococcal osteomyelitis. A new experimental model. *Infect. Immun.*, **61**, 5225–5230.

Walker, L. A., Schurmann, D. J. (1984). Management of infected total knee arthroplasties. *Clin. Orthop.*, **186**, 81–89.

Zak, O., Zak, F., Rich, R., Tosch, W., Kradolfer, F., Scheld, W. M. (1981). Experimental staphylococcal osteomyelitis in rats: therapy with rifampin and cloxacillin, alone or in combination. In: *Current Chemotherapy and Immunotherapy. Proceedings of the 12th International Congress of Chemotherapy, Florence, Italy, 19–24 July*.

Chapter 70

The Arthroplasty Model in Rabbits

N. Søe-Nielsen and J. Renneberg

Background of model

Deep infection after total knee arthroplasty is one of the most costly and devastating complications of arthroplasty. The incidence of infection after knee arthroplasty ranges from 1.1% to 5% (Walker and Schurman, 1984; Rand, 1993).

These infections are mostly caused by staphylococci (Brause, 1989), particularly *Staphylococcus epidermidis*, which has emerged during the last decade as the leading cause of bioimplant-related infections (Needham and Stempsey, 1984). The infections are generally grouped according to the time of development after primary implantation. Acute infection is defined as an infection that develops less than 3 months after implantation of the prosthesis. Delayed infection is defined as infection that develops between 3 months and 1 year after implantation of the prosthesis. Late infection is defined as infection that develops more than 1 year after implantation of the prosthesis (Antti-Poika *et al.*, 1990). Because of the relatively low incidence of the infections it is nearly impossible to include enough patients in prospective clinical trials, if treatment regimens or procedures are to be evaluated. Therefore, several animal models have been developed to test the various infection/treatment situations. The model described in this chapter deals with acute infection, in which staphylococci, both *Staphylococcus aureus* and *S. epidermidis*, account for 85–90% of infections.

Animal species

Rabbits were used as experimental animals because they have proved to be suitable for experiments concerning osteomyelitis and septic arthritis. Thompson and Dubos (1938), Blomgren (1981) and Schurman *et al.* (1978) all used New Zealand White rabbits 15–20 weeks old weighing 2.5–4.0 kg.

Previous studies have demonstrated that indolent orthopaedic implant-related infection could be reproducibly caused by inoculating adult New Zealand rabbits with a bacterial inoculum of 10^5–10^7 cfu (Isiklar *et al.*, 1993, 1996; Belmatoug *et al.*, 1996; Cremieux *et al.*, 1996; Nielsen *et al.*, 1996).

Preparation of animals

It is recommended that, as far as possible, rabbits should be housed so as to control extremes of environment. All cages and feed equipment should be of metal construction to facilitate sanitation; water systems should be of dewdrop type; and floors and walls should be impervious to liquid and moisture and be vermin-proof.

Table 70.1 is presented as a guide to the allowable floor space for holding rabbits.

Animal room temperatures should be maintained between 15.5 °C and 21 °C. Relative humidity levels should be maintained between 40% and 60% (Weisbroth *et al.*, 1974).

Table 70.1 Allowable floor space for holding rabbits

Weight	*Square centimetres per animal*	*Maximum population/cage*
1.35–2.25 kg	1160	6
Over 2.25 kg	2320	3

Males over 2.25 kg and females over 2.7 kg should be caged separately.

Details of surgery

Overview

Preoperative washing and disinfecting procedure results in a cutaneous surface free of bacteria (Smith, 1979). In the deeper cutaneous layers and in the sebaceous glands bacteria remain and cannot be removed without damage to the tissue. The operative incision allows the escape of these bacteria, and Smith (1979) has theoretically calculated the number as being between 10^2 and 10^3 organisms in an

Handbook of Animal Models of Infection
ISBN 0-12-775390-7

incision of 15 cm length. In our study (Nielsen *et al.*, 1996) we found about 10^4 organisms from contamination during surgery on the rabbit knee.

The knee of anaesthetized animals is surgically exposed. An inoculum of 10^5 *S. aureus* strain 1369 is injected into the medullar canal to induce a moderate inflammatory response locally without harming the animal.

Materials required

Anaesthetic, anaesthetic antidote, hair clippers or razors, skin disinfectant, scalpel handle plus blades (Sabre D/15), forceps, bone spoon, bone forceps, bacteria inoculum, syringes and needles (21 G × ½″ and 26 G × ½″), suture materials (Ticron 4-0 and Dermalon 5-0). Commercial orthopaedic bone cement (Palacos, Schering Corp.), St Georg prosthesis for the PIP joint (Valdemar Link). Ice to prevent the cement from hardening.

Anaesthesia

Preanaesthetics are effective in the rabbit and should be used — atropine sulphate 1–3 mg/kg and diazepam 5–10 mg/kg (Sedgwick, 1986). Short-acting anaesthetics — propanidid (Sombrevin®) 10–20 mg/kg — can be give intravenously into the lateral ear veins. Venous catheterization is recommended, duration 1–2 minutes. Thiomebumal (1.2%) 30 mg/kg should be administered, half of the dose as a bolus and the rest slowly until fully effective, duration 10–15 minutes. Long-acting anaesthetics that can be given are Fentanyl-fluanisone (Hypnorm vet®) 0.3 ml/kg intramuscularly and diazepam 2 mg/kg intravenously, duration 20–40 minutes. Repeat intramuscular injections of Hypnorm (0.1 ml/kg) can be given every half-hour. The antidote is naloxone (0.1 mg/kg) or buprenorphine (0.01 mg/kg) intravenously. Side-effects: respiratory depression.

Pentobarbital sodium is perhaps the most widely used barbiturate (Marston *et al.*, 1965). The dosage is generally 20–40 mg/kg body weight and should be administered as follows.

1. Gently place the rabbit in a restraint box (the animal is forced to sit still).
2. Slowly inject one-half to three-quarters of the calculated dose of a 2% pentobarbital solution intravenously over a 2 minute period.
3. Keep the needle in the vein as you remove the rabbit from the restraint box and lay it on its side.
4. Slowly continue the injection until the desired plane of anaesthesia is reached.
5. Try to avoid giving a second dose of pentobarbital to prolong anaesthesia.

The normal duration of anaesthesia is 30–45 minutes. Even though this method has been successfully used, many investigators have been troubled by excessive deaths due to the anaesthetic procedure when using this agent (Field, 1957; Ling, 1957; Dolowy and Hesse, 1959; Croft, 1964; Gardner, 1964).

Inhalation anaesthesia is not required but an oxygen mask is recommended and endotracheal intubation should be performed when anaesthesia has to be prolonged (Sedgwick, 1986).

Surgical procedure

Following complete induction of anaesthesia, the animal is placed ventral side up and kept on the operating table with a vacuum cushion. The skin over the knee and most of the leg is shaved with electric clippers; the skin is disinfected with 70% vol./vol. ethanol. Other disinfectants (e.g., isopropanol- or iodine-based solutions) are also applicable. The leg is pushed away from the surgeon and a longitudinal parapatellar medial incision is made at the knee. The patella is dislocated laterally and the knee joint is exposed. The joint surfaces are cleaned and the medullar canals are opened. Commercial orthopaedic bone cement should be used for fixation. Using a syringe, the cement is injected into the prepared medullar canals and the prosthesis is inserted (Blomgren, 1981; Nielsen *et al.*, 1996; Figure 70.1).

Surgical exposure of the knee joint should be done without damaging the inferior patellar tendon and the collateral ligament. The cruciate ligament should be excised, together with the joint surfaces. The medullar canals are prepared with a high-speed dental drill (Micro-Aire) or a small bone spoon.

Infection procedures

Following the preparing of the medullar canals, if needed, the area is cleaned using sterile physiological saline. A volume of 0.1 ml of bacterial suspension is injected into the medullar cavity of the femur and tibia using a 1 ml syringe fitted with a 260 × ½″ needle. The solution is slowly injected; a small portion of the liquid will always end upon the joint surface. Immediately following this, the cement is similarly injected with a 2 ml syringe, without a needle, into the medullar canals and the prosthesis is inserted and maintained in place for about 10 minutes until the cement hardens.

Wound closure

The capsule is closed using Ticron 4-0 with square knots. The skin is closed using Novafil 5-0 with square knots. We do not use metal skin clips. The wound can also be covered with an aerosol-based antiseptic plastic film (Nobecutan, Astra).

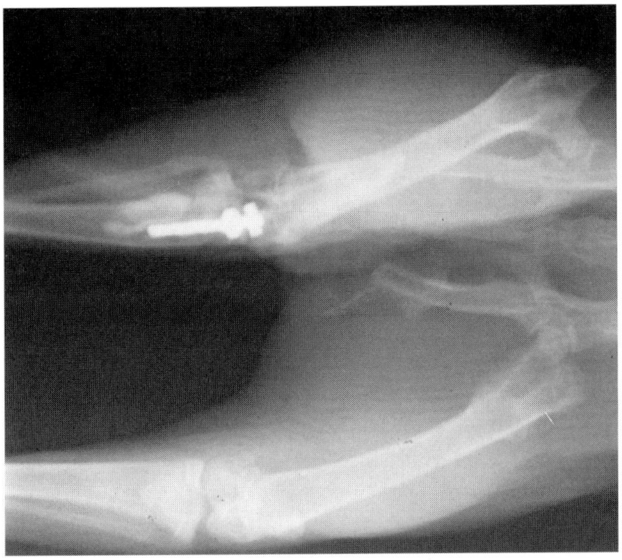

A

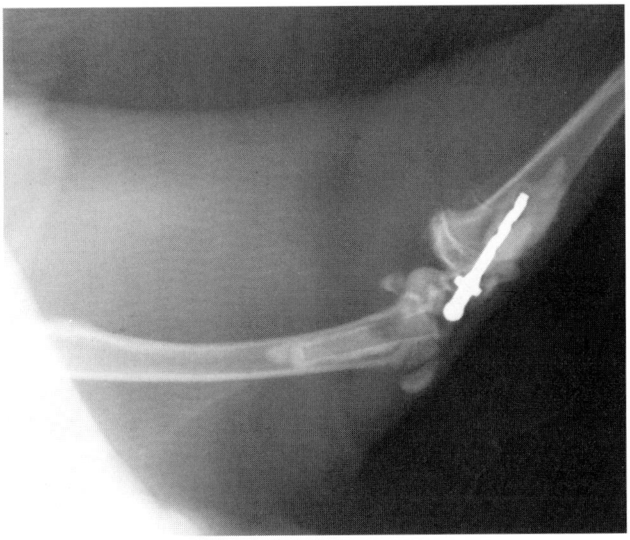

B

Figure 70.1 **(A & B)** X-ray view of the prosthesis *in situ*.

Postoperative care

The rabbit should be kept warm after surgery; therefore it should be placed under an infrared lamp until it begins to recover from the anaesthetic. Animals should be housed individually on wire supports for 1–2 days following surgery in order to facilitate wound healing and to prevent infection of the wound. After 2 days the animals are housed in groups. Postoperative analgesics should be used routinely, especially after painful infection and osteomyelitis. We prefer to use buprenorphine (0.01–0.05 mg/kg s.c.) or fentanyl (Haldid) 10 μg/kg i.v. The antidote is nalorphine 0.1 mg/kg (Sedgwick, 1986). Weight and temperature should be measured. The normal temperature is 38.3°C (37–39.4°C). The animals normally lose about 10% in weight over the first week after the surgery.

Storage and preparation of inoculate

Several bacteria species have been used in the model. Most studies have used *Staphylococcus aureus, Streptococcus pneumoniae* or *Mycobacterium* sp. A few have used β-haemolytic streptococci (Piepmeier *et al.,* 1995) or slime-producing *S. epidermidis* (Isiklar *et al.*, 1996).

We used antibiotic-susceptible *S. aureus* strain 1369, phage type 85 in our experiments. The inoculum was prepared by growing the bacteria overnight in serum broth, washing three times in 0.9% NaCl and finally resuspending to 10 times the required inoculum per millilitre. The inoculum was checked before each experiment by plating. The inocula for the whole experimental series (30–40 animals) was produced at the same time and frozen at –80°C in small amounts for each experiment day. The inocula were used within 4 months and thereby reduction of the quantity of living bacteria was avoided. Before use, each inoculum was tested to ensure stability of the viability of the bacteria over time.

Infection process

The infection is initiated by direct injection of bacterial culture into the medullar canals of femur and tibia. Using *S. aureus* strain 1369, we achieved an osteomyelitis with an inoculum of 10^4 and 10^5 cfu, but with an inoculum of 10^6 cfu the rabbits develop sepsis (Nielsen *et al.*, 1996). Using another *S. aureus* strain, UAMS-1, Smeltzer *et al.* (1997) achieved an osteomyelitis rate of 75% with an inoculum as small as 2×10^3 cfu and significant radiographic and histological signs of disease with an inoculum of at least 2×10^4 cfu. In their silicone–elastomer model, Belmatoug *et al.* (1996) used 5×10^6 cfu to achieve infection. They confirmed development of the infection using magnetic resonance, and evaluated the infection process both macroscopically and microscopically during an observation period of 8 weeks. We found histological signs of disease within 1 week and this remained essentially unchanged throughout 3–4 weeks (Figure 70.2).

After daily intravenous injections of a suspension of the experimental strain (*S. aureus*, Wood 46), Blomgren (1981) found obvious infection with pus in the prosthetic joint after 1 week. After a 4-week observation period 8/10 endoprostheses had positive cultures. Although macroscopic signs of infection were lacking in the kidneys, *S. aureus* was isolated in four cases out of 10.

Key parameters to monitor infection and response to treatment

With an inoculum of 10^4 cfu of *S. aureus* the infection is confined to the bone. Death due to sepsis (bacteriaemia) is seen with inoculum of more than 10^6 cfu, given locally or

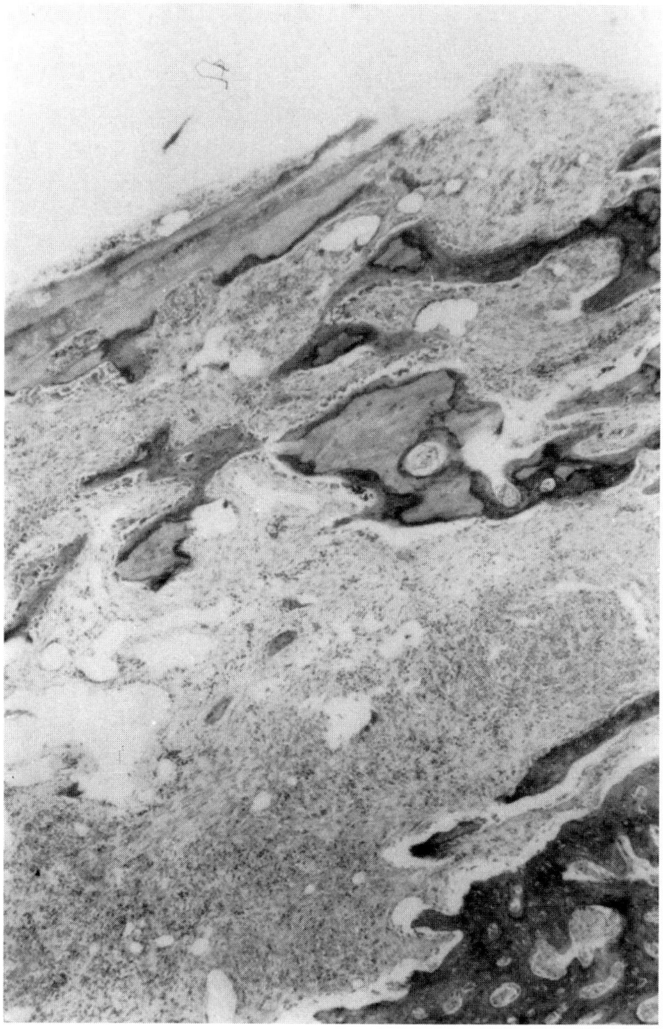

Figure 70.2 Histological presentation of the acute infection 2 weeks after implantation of the prosthesis.

intravenously. Erythrocyte sedimentation rates are increased after a few days with an inoculum of 10^4 without supplemental antibiotics. Appearance, body-weight gain, food and water intake are affected by the infection process or treatment-associated toxicity. We found that the surgical procedure itself caused weight loss in the animals and this, together with a local infection, gave a weight loss of about 15%. The animals never regained their initial body weight (Batchelor and Giddins, 1995).

A rise in antibody titre against the infecting micro-organism, lowering of HGB and rise of WBC numbers after 1 week are parameters that can be used to decide if the animal is infected or not.

Bone pathology

Bone histology is normally used to monitor progress of infection and response to antimicrobial therapy (Thompson *et al.*, 1938; Mader and Adams, 1989; Lambe *et al.*, 1991; Cordero *et al.*, 1994).

Several investigators have used pathology scores. Rissing *et al.* (1985) scored gross pathology thus: 0, no apparent pathology; 1, minimal erythema without abscess or new bone formation; 2, erythema and widening of tibial shaft (new bone formation); 3, purulent exudate or abscess; and 4, severe bone destruction involving more than 50% of the diaphysis in rats. Blomgren (1981) found that the bone cement was more or less surrounded by a thin membrane consisting of fibroblasts and inflammatory cells. In most cases the innermost layer of the cortical bone, close to the cement or membrane, seemed dead, with empty osteocyte lacunae.

Among others, sonographic findings seemed to correlate well with histological findings in a rabbit model of osteomyelitis (Abiri *et al.*, 1992).

Bacterial density in bone

Determination of bacterial cfu within bone normally requires homogenizing of the bone. This can be easily accomplished by deep-freezing the bones and then crushing the bone by hand using a mortar (Rissing *et al.*, 1985; Calhoun and Mader, 1997). The bone powder should remain cold or preferably frozen during the procedure and is then resuspended in a small volume of PBS. Standard serial dilution technique and bacterial enumeration procedures can then be used.

Determination of antibiotic penetration into bone

The bone specimens should be cleaned and freed from adhering tissue, rinsed in PBS (pH 7.2) and crushed into small pieces, which are weighed and immersed in PBS. The total volume is then measured and the mixture is shaken at 4°C for 5 hours to equilibrate the antibiotic concentration in bone with that of the supernatant (Leigh, 1986) that was used for the assays. Joint fluids should be sampled and used for the assays as well.

Two assays can be used, a standard microbiological assay or high-performance liquid chromatography (HPLC; Renneberg *et al.*, 1993).

Antimicrobial therapy

The model has been validated using several antibiotic combinations, primarily β-lactam antibiotics and aminoglycosides (Blomgren, 1981; Nielsen *et al.*, 1996); but drugs such as sparfloxacin (Cremieux *et al.*, 1996) have also been tested. Dicloxacillin concentrations reach levels of around 5 mg/kg in bone and 7.4 mg/l in joint fluid, and gentamicin reaches peak levels of 4.0 mg/kg in bone and 10 mg/l in joint fluid. These concentrations are in the same levels as reached in the same site in humans.

Pitfalls (advantages/disadvantages) of the model

Besides the usual shortcomings of all animal models, the major shortcoming of the rabbit model of osteomyelitis is the need for large inoculae (10^7–10^8 cfu) directly implanted into bone, which are required to establish the infection. Furthermore, the rabbit's resistance to infection is considerable (Wadstrøm, 1981). The rabbits are mildly anaesthetized by the intravenous technique but are somewhat more sensitive than rats. A considerable problem is the high cost compared to mice or rat experiments.

Contributions of the model to infectious disease therapy

The rabbit prosthesis arthritis models have contributed well to our knowledge of the pathophysiology of bacterial infection around orthopaedic prostheses and the use of antibiotics as prophylactics by implantation (Blomgren, 1981; Nielsen *et al.*, 1996). Further, diagnostic methods — serological methods, reactive proteins, WBC and other SR as well as direct specimens — have been evaluated. Results from these studies are part of the basis of diagnostic strategy used today (Blomgren, 1981).

References

Abiri, M. M., DeAngelis, G. A., Kirpekar, M., Abou, A. N., Ablow, R. C. (1992). Ultrasonic detection of osteomyelitis. Pathologic correlation in an animal model. *Invest. Radiol.*, **27**(2), 111–113.

Antti-Poika, I., Josefsson, G., Konttinen, Y., Lidgren, L., Santavirta, S, Sanzén, L. (1990). Hip arthroplasty infection. *Acta Orthop. Scand.*, **61**(2), 163–169.

Batchelor, G. R., Giddins, G. E. B. (1995). Bodyweight changes in laboratory rabbits following operative procedures. *Anim. Technol.*, **46**(3), 153–161.

Belmatoug, N., Cremieux, A. C., Bleton, R. *et al.* (1996) A new model of experimental prosthetic joint infection due to methicillin-resistant *Staphylococcus aureus*: a microbiologic, histopathologic and magnetic resonance imaging characterization. *J. Infect. Dis.*, **174**(2), 414–417.

Blomgren, G. (1981). Hematogenous infection of total joint replacement. *Acta Orthop. Scand.*, **52** (Suppl. 187).

Brause, B. D. (1989). Infected orthopedic prostheses. In: *Infections Associated with Indwelling Medical Devices* (eds Bisno, A. L., Waldvogel, F. A.), pp. 111–127. American Society of Microbiology, Washington, DC.

Calhoun, J. H., Mader, J. T. (1997) Treatment of osteomyelitis with a biodegradable antibiotic implant. *Clin. Orthop.*, **34**(1), 206–214.

Cordero, J., Munuera, L., Folgueira, M. (1994). Influence of metal implants on infection: an experimental study in rabbits. *J. Bone Joint Surg.*, **76B**.

Cremieux, A. C., Mghir, A. S., Bleton, R. *et al.* (1996). Efficacy of sparfloxacin and autoradiographic diffusion pattern of [^{14}C]sparfloxacin in experimental *Staphylococcus aureus* joint prosthesis infection. *Antimicrob. Agents Chemother.*, **21**, 11–16.

Croft, P. G. (1964). In *Small Animal Anesthesia* (ed Graham-Jones, O.), pp. 99–102. Oxford University Press, Oxford.

Dolowy, W. C., Hesse, A. L. (1959). Chlorpromazine premedication with pentobarbital anesthesia in the rabbit. *J. Am. Vet. Med. Ass.*, **134**, 183.

Field, E. J. (1957). Anesthesia in rabbits. *J. Anim. Tech. Ass.*, **8**, 47.

Gardner, A. F. (1964). The development of general anesthesia in the albino rabbit for surgical procedures. *Lab. Anim. Care*, **14**, 214–225.

Isiklar, Z. U., Darouiche, R. O., Landon, G. C., Beck, T. (1996) Efficacy of antibiotics alone for orthopaedic device related infections. *Clin. Orthop. Rel. Res.*, **322**, 184–189.

Isiklar, Z. U., Landon, G. C., Saleh, G., Musher, D. M., Femau, R. (1993) Increasing bacteriologic accuracy in orthopedic implant infection: *in vivo* study. *Orthop. Trans.*, **18**, 453.

Lambe, D. V., Ferguson, K. P., Mayberry-Carson, K. J., Tober-Meyer, B., Costerton, J. W. (1991) Foreign-body-associated osteomyelitis experimentally induced with *Bacteroides fragilis* and *Staphylococcus epidermidis* in rabbits. *Clin. Orthop.*, **266**, 285–294.

Leigh, D. A. (1986). Serum and bone concentrations of cefuroxime in patients undergoing knee arthroplasty. *J. Antimicrob. Chemother.*, **18**, 609–611.

Ling, H. W. (1957). Anesthesia in rabbits. *J. Anim. Tech. Ass.*, **8**, 58.

Mader, J. T., Adams, K. (1989) Comparative evaluation of daptomycin (Lyl46032) and vancomycin in the treatment of experimental methicillin-resistant *Staphylococcus aureus* osteomyelitis in rabbits. *Antimicrob. Agents Chemother.*, **33**, 689–692.

Marston, J. H., Rand, G., Chang, M. C. (1995). The care, handling and anesthesia of the snowshoe hare (*Lepus americanus*). *Lab. Anim. Care*, **15k**, 325–327.

Needham, C. A., Stempsey, W. (1984). Incidence of adherence and antibiotic resistance of coagulase negative staphylococcus species causing human disease. *Diagn. Microbiol. Infect. Dis.*, **2**, 293–299.

Nielsen, N. H. S., Renneberg, J., Nürnberg, B. M., Tørholm, C. (1996) Experimental implant-related osteomyelitis induced with *Staphylococcus aureus. Eur. J. Orthop. Surg. Traumatol.*, **6**, 97–100.

Piepmeier, E., Hammett-Stabler, C., Price, M. *et al.* (1995) Myositis and faciitis associated with group A beta hemolytic streptococcal infections: development of a rabbit model. *J. Lab. Clin. Med.*, **126**(2), 137–143.

Rand, J. A. (1993). Sepsis following total knee arthroplasty. In: *Total Knee Arthroplasty* (ed. Rand, J. A.), 349–375. Raven Press, New York.

Renneberg, J., Christensen, O. M., Thomsen, N. O. B., Tørholm, C. (1993). Cefuroxime concentrations in serum joint fluid and bone in elderly patients undergoing arthroplasty after administration of cefuroxime axetil. *J. Antimicrob. Chemother.*, **32**, 751–755.

Rissing, J. P., Buxton, T. B., Weinstein, R. S., Schockley, R. K. (1985). Model of experimental chronic osteomyelitis in rats. *Infect. Immun.*, **47**, 518–586.

Schurmann, D. J., Trindade, C., Hirshman, H. P., Moser, K., Kajiyama, G., Stevens, P. (1978) Antibiotic–acrylic bone cement composites. Studies of gentamicin and Palacos. *J. Bone Joint Surg.*, **60A**, 978–984.

Sedgwick, C. J. (1986). Anesthesia for rabbits. *Vet. Clin. North Am. Food Anim. Pract.*, **2**(3), 731–736.

Smeltzer, M. S., Beck, T. Jr, Hickmon, S. G. *et al.* (1997) Characterization of a rabbit model of staphylococcal osteomyelitis. *J. Ortop. Res.*, **15**(3), 414–421.

Smith, G. (1979) Primary postoperative wound infection due to *Staphylococcus pyogenes. Curr. Prob. Surg.*, **16**(7).

Thompson, R. H. S., Dubos, R. I. (1938) Production of experimental osteomyelitis in rabbits by intravenous injection of *Staphylococcus aureus. J. Exp. Med.*, **68**, 191–209.

Wadström, T. (1981) Aktuelt om stalylokokinfektionens patogenes. *Läkertidningen*, **78**, 690–693.

Walker, L. A., Schurmann, D. J. (1984). Management of infected total knee arthroplasties. *Clin. Orthop.*, **186**, 81–89.

Weisbroth, S. H., Flatt, R. E., Kraus, A. L. (1974) *The Biology of the Laboratory Rabbit*. Academic Press, New York.

Chapter 71

Mouse Model of Streptococcal Fasciitis

S. Sriskandan and J. Cohen

Background of human infection

Reports from around the world suggest that the incidence and severity of fasciitis and toxic shock due to *Streptococcus pyogenes* have increased (Cone *et al.*, 1987). Patients present with severe pain due to localized soft tissue inflammation and necrosis, in association with multiorgan failure. A history of surgical or traumatic wound infection suggests local entry of bacteria; other cases appear to arise as a result of bacteraemic seeding within soft tissues recently subjected to blunt trauma (Stevens, 1992). Recent clinical descriptions of streptococcal disease show that myonecrosis can accompany severe cases of fasciitis. Most cases are associated with bacteraemia; rash and hypotension may additionally be present, fulfilling the criteria for streptococcal toxic shock syndrome (STSS; Stevens *et al.*, 1989; Working Group on Severe Streptococcal Infections, 1993).

Although invariably associated with multiorgan failure, histopathological changes in tissues other than affected fascia and muscle are poorly documented from cases of severe streptococcal fasciitis. In staphylococcal TSS, sloughing of gastrointestinal and vaginal epithelia is seen. Hepatic fatty change with central or single cell necrosis, varying degrees of renal acute tubular necrosis, and splenic lymphoid depletion with haemophagocytosis are also documented. Some patients show evidence of myocarditis or acute respiratory distress syndrome (Larkin *et al.*, 1982). In both streptococcal and staphylococcal TSS thrombocytopenia and lymphopenia are common; coagulopathy is variably present (Larkin *et al.*, 1982; Hoge *et al.*, 1993). Increases and decreases in superantigen-reactive T-cell subsets have been described (Michie *et al.*, 1994; Watanabe-Ohnishi *et al.*, 1995).

Mortality from severe streptococcal fasciitis remains high (30–60%), even in cases treated with combined surgery and penicillin. It is of note that several streptococcal virulence factors are implicated in disease pathogenesis, in particular superantigenic exotoxins and protease production. A number of adjuvant treatments that target these virulence factors have been proposed, such as protein synthesis inhibitors, intravenous immunoglobulin, antitoxins and prevention by vaccination (Stevens *et al.*, 1994), although, to date, none has been fully analysed in experimental or clinical settings.

Background to model

The mouse model was developed to allow investigation of streptococcal disease pathogenesis, in particular the role of bacterial toxins in multi-organ failure, and also to evaluate novel therapeutic approaches to invasive streptococcal disease. This model of lethal streptococcal fasciitis/myositis leads to bacteraemia, multiorgan failure and evidence of a systemic inflammatory response.

Streptococcal fasciitis has previously been successfully reproduced in the rabbit by subcutaneous or intradermal inoculation of toxin-producing bacterial strains or co-administration of staphylococcal α-lysin with live streptococci (Seal and Kingston, 1988; Piepmeier *et al.*, 1995). Intramuscular inoculation of *S. pyogenes* in mice, using the Selbie thigh lesion approach, results in spreading fasciitis associated with varying degrees of cellulitis and muscle degeneration (Selbię and Simon, 1952; Stevens *et al.*, 1988). Lethality depends very much on the bacterial strain used and size of mouse. Other murine models of severe invasive group A streptococcal infection have been described involving peritoneal, pulmonary and intravenous infection (Martin *et al.*, 1992; Kapur *et al.*, 1994; Husmann *et al.*, 1996) Animal models of TSS are reviewed elsewhere; baboon and rabbit models appear the most successful (de Azavedo, 1989).

Animal species and preparation

Outbred male 6–8-week-old CD1 mice (Charles River, Margate, UK) weighing 22–25 g have been used to characterize this model. Female BALB/c mice (20 g) have also been successfully used. No specialized housing or preparation is required.

Storage and preparation of inocula

Reference strain 8198 *S. pyogenes* (serotype M1T1, streptococcal pyrogenic exotoxin A [SPEA] producer:

Handbook of Animal Models of Infection
ISBN 0-12-775390-7

Streptococcal Reference Laboratory, Central Public Health Laboratory, Colindale, UK) has been used in this model. Other strains can be used, provided dose–response analysis is performed. For intramuscular (i.m.) challenge, a fresh colony from a blood agar plate is inoculated into Todd–Hewitt broth (Oxoid, Basingstoke, UK) and incubated for 8 hours at 37°C. Bacteria are harvested by centrifugation, washed in pyrogen-free saline, re-centrifuged and concentrated to yield a final density of 2×10^{10} cfu/ml in saline. Standard growth curves are used to relate measured optical density (OD_{325}) to bacterial concentration. Culture purity is ascertained by inoculation of 100 μl aliquots on to blood agar plates and overnight incubation at 37°C. Dilutions (10^6–10^9-fold) of the suspension are similarly plated on to blood agar to precisely enumerate the inoculum dose. The inoculum should then be used promptly, at room temperature; if a delay is necessary, the inoculum should be stored at 4°C.

N.B. serial passage of group A streptococci in the laboratory is not recommended, as phenotypic changes may occur over 6-month periods, in particular the expression of bacterial capsule, M protein and SPEA production. It is suggested that frozen aliquots of the original experimental strain are thawed for use approximately every 3 months.

Infection process

Bacteria are inoculated into the right thigh by i.m. injection of 0.05 ml of suspension ($LD_{90} = 1 \times 10^9$ cfu/mouse), using the mouse thigh lesion approach. The suspension should be well mixed prior to use; 0.5 ml can be drawn into a single syringe and used to inoculate 10 mice at a time using a 25 G needle. *Caution*: inadvertent autoinoculation of live *S. pyogenes* can lead to severe infection; medical attention and antibiotic treatment with penicillin are recommended if this occurs.

Key parameters to monitor infection

Eight hours following injection, mice develop visible swelling and erythema in the infected thigh, causing altered gait. After 15 hours there is evidence of systemic illness (lethargy, ruffled fur, conjunctivitis) with extension of erythema from the thigh to the foot, groin and tail base in some animals. Deaths occur between 24 and 72 hours; 90% mortality should be expected (Figure 71.1). Smaller animals may be more rapidly sensitive to infection; older, larger animals may demonstrate 50% mortality. Bacteraemia is present in 80–100% of animals at 15 hours and 25 hours (range 10^3–10^6 cfu/ml blood). Though mortality is the major parameter to follow in this model, systemic markers of sepsis and well-defined histopathological changes can also be monitored.

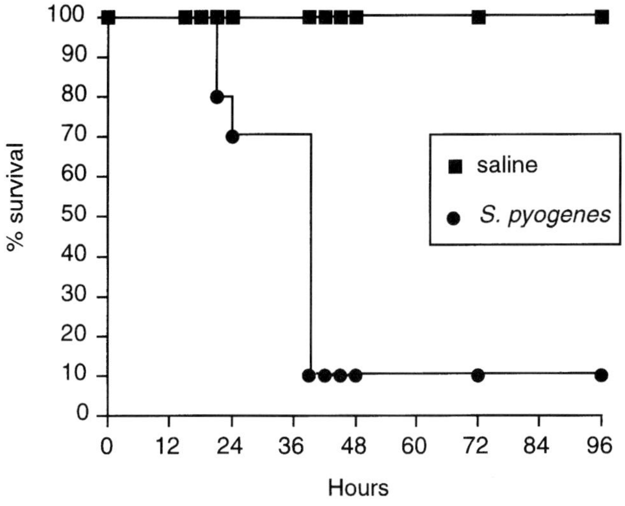

Figure 71.1 Kaplan Meier plot showing survival pattern following intramuscular inoculation of 2×10^9 cfu *Streptococcus pyogenes*.

Cardiovascular abnormalities

These have been recorded in this model using invasive monitoring techniques. Systemic hypotension occurs during sepsis, coupled with a decline in heart rate which is associated with lethal outcome (D. Rees, personal communication).

Haematological anomalies

Haematological anomalies at 50 hours include thrombocytopenia (mean levels: controls 1228×10^9/l SD 275, infected 898×10^9/l SD 153) and leukopenia due mainly to profound lymphopenia (mean levels: controls 9.68×10^9/l; infected 0.73×10^9/l). Prolongation of the partial thromboplastin time, reflecting activation of the contact system, also occurs (mean time: controls 25.1 seconds SD 3.9; infected 84.6 seconds SD 32).

Serum cytokine responses

These have been measured by ELISA from tail bleeds taken at predetermined time points during streptococcal sepsis. In early experiments, it was shown that raised interleukin-6 levels were associated with shortened survival in this model (Sriskandan *et al.*, 1996). Serum TNFα and interferon-γ have been detected only in mice with the highest IL-6 levels, during later stages of sepsis (Figure 71.2). Serum nitrite plus nitrate, a surrogate marker of nitric oxide production, rises 15 hours after infection (mean 170 μmol/l) compared with uninfected controls (mean 75 μmol/l); levels measured by modified Griess reaction

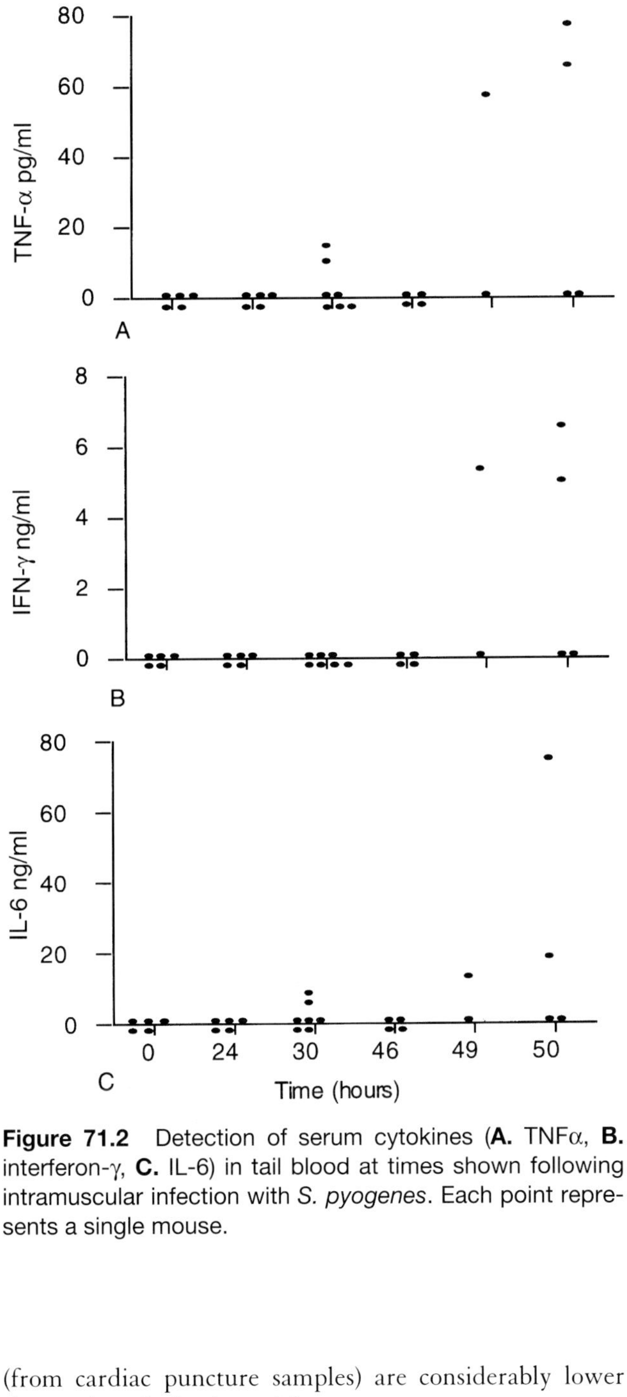

Figure 71.2 Detection of serum cytokines (**A.** TNFα, **B.** interferon-γ, **C.** IL-6) in tail blood at times shown following intramuscular infection with *S. pyogenes*. Each point represents a single mouse.

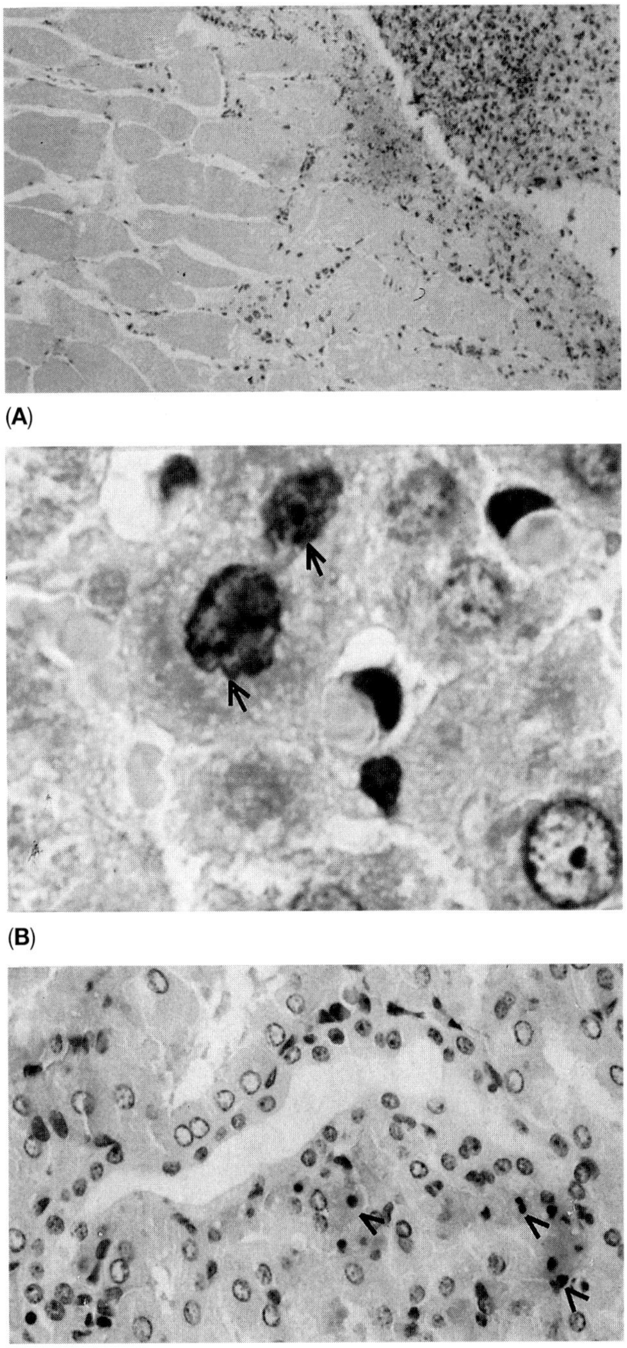

(A)

(B)

(C)

Figure 71.3 Histopathology following infection with *S. pyogenes*. **A.** Thigh muscle and fascla. × 340. **B.** Liver tissue from infected mouse. Apoptotic hepatocytes arrowed. × 680. **C.** Renal tissue from infected mouse. Dead tubular cells arrowed. × 680.

(from cardiac puncture samples) are considerably lower than endotoxic shock models.

Histopathology

Thigh lesion

An infiltrate of degenerate neutrophils can be seen throughout the connective tissues of the thigh, with fibrin deposition, occasional colonies of Gram-positive cocci and some haemorrhage. Muscle fibres adjacent to inflamed fascia will be necrotic, although fibres distant from inflamed connective tissue are normal (Figure 71.3A).

Liver

Extensive fatty change can be seen throughout the liver with increased numbers of apoptotic hepatocytes displaying clumping of chromatin and dense pink cytoplasm. Bacterial colonies are not seen within the hepatic parenchyma (Figure 71.3B).

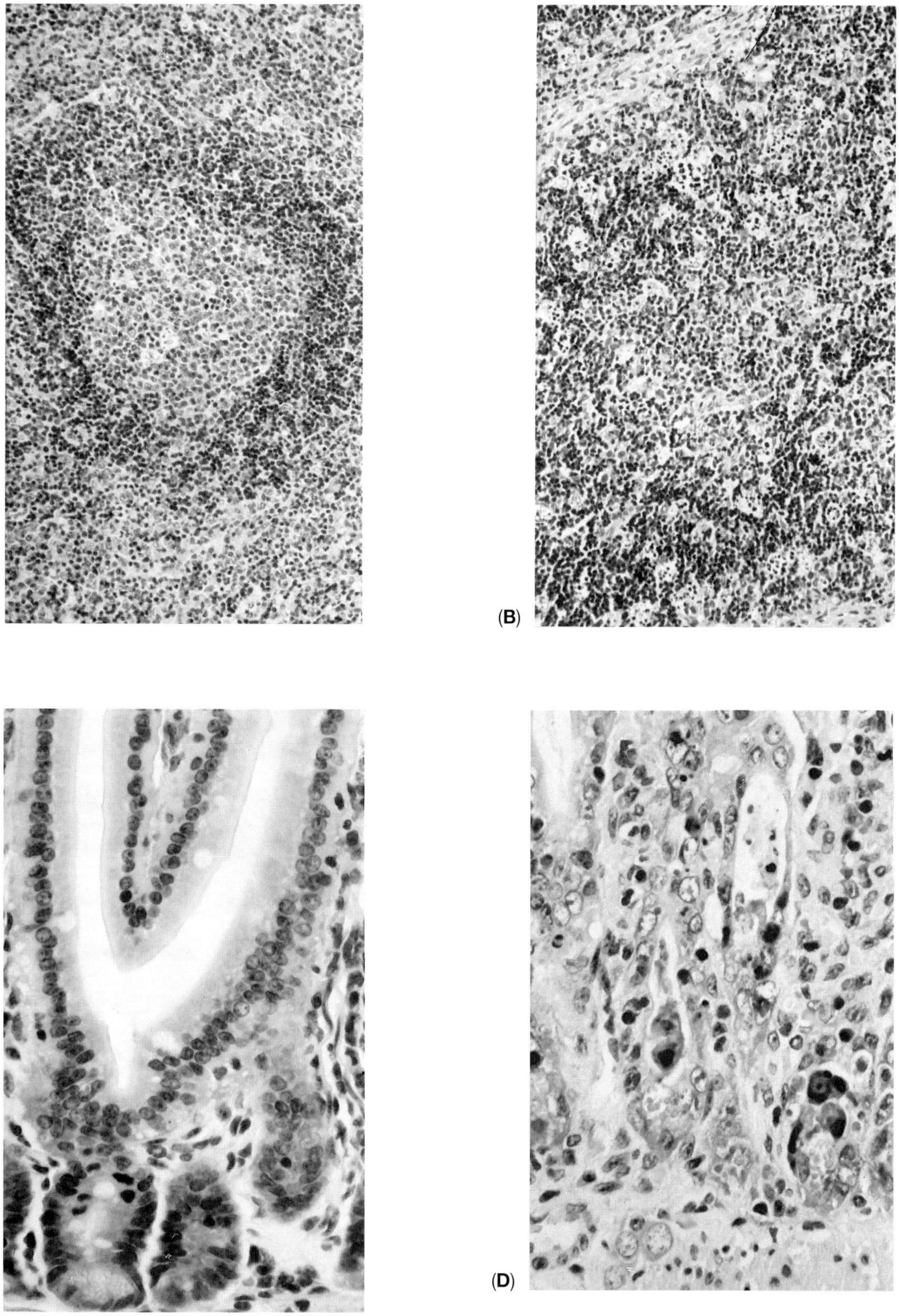

(A) (B) (C) (D)

Figure 71.4 **A.** Spleen tissue from normal mouse, with normal follicular architecture. **B.** Spleen from infected mouse showing loss of follicular margins and lymphoid depletion. **C.** Normal intestinal epithelium from normal mouse. **D.** Intestinal epithelium from infected mouse showing sloughing of mucosal cells into the lumen. × 400.

Kidney

Changes of acute tubular necrosis are seen in infected animals (Figure 71.3C).

Spleen

Marked structural disorganization of the splenic architecture occurs; splenic tissue appears pale because of large gaps in follicles of the white pulp. Loss of follicular margins and macrophages containing chromatin-dense bodies are strongly suggestive of lymphoid depletion due to apoptosis (Figure 71.4A, B).

Small bowel

Cells from intestinal crypts are shed into the small bowel lumen of infected animals. Cells at crypt bases are necrotic, without evidence of normal mitotic activity (Figure 71.4C, D).

Lung, heart

No consistent abnormality has been found in these organs.

Pitfalls (advantages/disadvantages) of the model

There are a number of reasons for choosing the mouse as a model of severe streptococcal infection. Firstly, it is undesirable to induce such destructive infection in larger animals. Secondly, use of mice permits mortality studies, which are of considerable importance when assessing novel treatment modalities. Lastly, a wide range of reagents are commercially available to study immunological responses in mice; the immunology of larger animals has not been characterized to the same degree. Use of inbred murine strains such as BALB/c can facilitate study of T-cell responses to the streptococcal superantigen SPEA.

The model described adequately reproduces clinical disease, in that soft-tissue infection is progressive and is associated with bacteraemia, multiorgan failure, systemic inflammatory response and death. We know that the bacterial toxin SPEA is produced systemically in this model and have been able to measure levels of SPEA in blood and localize SPEA immunohistochemically (Sriskandan *et al.*, 1996a). The model can therefore be used for study of other, potentially important bacterial products.

A major drawback of this model is that cardiovascular function is not easily monitored without intricate invasive monitoring devices. In addition, mice are relatively insensitive to bacterial toxins; the potency of streptococcal superantigens in human and rabbit cell culture is around two logs greater than in murine cell culture. This is because of differences between human and murine HLA class II molecules and, possibly, T-cell receptor subsets. The clinical manifestations of superantigen exposure may therefore be absent in a murine streptococcal sepsis model.

Contributions of the model to infectious disease therapy

This model has not been used in extensive studies of antimicrobial therapy. Intravenous administration of penicillin at onset of infection results in complete cure; attenuation of infection severity can be achieved by delayed dosing with antibiotics. In a similar model of streptococcal myositis, Stevens *et al.* (1988) have shown that clindamycin administered by the intraperitoneal (i.p.) route demonstrated significantly superior efficiency to penicillin therapy, even when treatment was delayed. Novel therapeutic agents (anti-TNF monoclonal antibodies, soluble TNF receptors, IVIG, anti-SPEA and inhibitors of nitric oxide synthase) have been tested in this model but have not demonstrated any positive benefit, even when administered at the time of infection (Sriskandan *et al.*, 1996a, 1997). This is in spite of the fact that some of these agents are of proven benefit in endotoxic shock. This may relate to rapid clearance of agents administered as a bolus i.v., or to the overwhelming nature of the sepsis syndrome in this model. Systematic testing of potential therapies in this model is in progress, using antibiotic administration to attenuate severity.

This model has been important in increasing our understanding of the pathogenesis of streptococcal soft-tissue infection and the associated sepsis syndrome. The production and localization of bacterial superantigenic toxin (SPEA) during sepsis was characterized for the first time using this model; this has been followed by detection of superantigen in serum obtained from patients using the same methodology (Sriskandan *et al.*, 1996b). Immunological responses to streptococcal soft-tissue infection contrast with those seen following i.v. endotoxin injection and Gram-negative infection. This raises important therapeutic issues when considering uniform approaches to management of septic shock.

References

Cone, L. A., Woodard, D. R., Schlievert, P. M., Tomory, G. S. L. (1987). Clinical and bacteriologic observations of a toxic shock-like syndrome due to *Streptococcus pyogenes*. *New Engl. J. Med.*, **317**, 146–149.

De Azavedo, J. C. S. (1989). Animal models for toxic shock syndrome: overview. *Rev. Infect. Dis.*, **11** (suppl. 1), S205–S209.

Hoge, C. W., Schwartz, B., Talkington, D. F. *et al.* (1993). The changing epidemiology of invasive group A streptococcal infections and the emergence of Streptococcal Toxic Shock-like syndrome. *J. A. M. A.*, **269**, 384–389.

Husmann, L. K., Dillehay, D. L., Jennings, V. M., Scott, J. R. (1996). *Streptococcus pyogenes* infection in mice. *Microb. Pathog.*, **20**, 213–224.

Kapur, V., Maffei, J. T., Greer, R. S. *et al.* (1994). Vaccination with streptococcal extracellular cysteine protease (interleukin-1β convertase) protects mice against challenge with heterologous group A streptococci. *Microb. Pathols.*, **16**, 443–450.

Larkin, S. M., Williams, D. N., Ostreholm, M. T., Tofte, R. W., Posalaky, Z. (1982). Toxic shock syndrome: clinical, laboratory, and pathologic findings in fatal cases. *Ann. Intern. Med.*, **96**, 858–864.

Martin, R. A., Silva, A. T., Cohen, J. (1992). Effect of anti-TNF-α treatment in an antibiotic treated murine model of infection due to *Streptococcus pyogenes*. *FEMS Microbiol. Lett.*, **79**, 313–322.

Michie, C., Scott, A., Cheesbrough, J., Beverley, P., Pasvol, G. (1994). Streptococcal toxic shock-like syndrome: evidence of superantigen activity and its effects on T lymphocyte subsets *in vivo*. *Clin. Exp. Immunol.*, **98**, 140–144.

Piepmeler, E., Hammet-Stabler, C., Price, M. *et al.* (1995). Myositis and fasciitis associated with group A beta-hemolytic streptococcal infections; development of a rabbit model. *J. Lab. Clin. Med.*, **126**, 137–143.

Seal, D. V., Kingston, D. (1988). Streptococcal necrotizing fasciitis: development of an animal model to study its pathogenesis. *Br. J. Exp. Pathol.*, **69**, 813–831.

Selbie, F. R., Simon, R. D. (1952). Virulence to mice of *Staphylococcus pyogenes*: its measurement and its relation to certain *in vitro* properties. *Br. J. Exp. Pathol.*, **33**, 315–326.

Sriskandan, S., Moyes, D., Buttery, L. K. *et al.* (1996a). Streptococcal pyrogenic exotoxin A (SPEA) release, distribution and role in a murine model of fasciitis and multi-organ failure due to *Streptococcus pyogenes*. *J. Infect. Dis.*, **173**, 1399–1407.

Sriskandan, S., Moyes, D., Cohen, J. (1996b). Detection of circulating bacterial superantigen and lymphotoxin-α in patients with streptococcal toxic-shock syndrome. *Lancet*, **348**, 1315–1316.

Sriskandan, S., Moyes, D., Buttery, L. K. *et al.* (1997). The role of nitric oxide in experimental murine septic shock due to SPEA-producing *Streptococcus pyogenes*. *Infect. Immun.*, **65**, 1767–1772.

Stevens, D. L. (1992). Invasive group A streptococcal infections. *Clin. Infect. Dis.*, **14**, 2–13.

Stevens, D. L., Bryant, A. E., Yan, S. (1994). Invasive group A streptococcal infection: new concepts in antibiotic treatment. *Int. J. Antimicrob. Agents*, **4**, 297–301.

Stevens, D. L., Gibbons, A. E., Bergstrom, R., Winn, V. (1988). The Eagle Effect revisited: efficacy of clindamycin, erythromycin, and penicillin in the treatment of streptococcal myositis. *J. Infect. Dis.*, **158**, 23–28.

Stevens, D. L., Tanner, M. H., Winship, J. *et al.* (1989). Severe group A streptococcal infections associated with a toxic shock-like syndrome and scarlet fever toxin A. *N. Engl. J. Med.*, **321**, 1–7.

Watanabe-Ohnishi, R., Low, D. E., McGreer, A. *et al.* (1995). Selective depletion of Vβ-bearing T cells in patients with severe invasive group A streptococcal infections and streptococcal toxic shock syndrome. *J. Infect. Dis.*, **17**, 74–84.

Working Group on Severe Streptococcal Infections (1993). Defining the group A streptococcal toxic shock syndrome. Rationale and consensus definition. *J. A. M. A.*, **269**, 390–391.

Chapter 72

Rabbit Model of Bacterial Endocarditis

A. Lefort and B. Fantin

Background of human infection

Bacterial endocarditis remains a life-threatening infection despite the recent development of potent anti-infective agents. Diagnosis of this infection has become easier since the use of precise and objective criteria (Durack *et al.*, 1984). Antibiotics at high doses either alone or in combination and maintained for a prolonged period are sometimes insufficient to sterilize cardiac vegetations, preventing therapeutic failures and emergence of bacterial populations resistant to the treatment. Mortality remains important, for hemodynamic and infectious reasons. Clinical trials of endocarditis are limited by the relative infrequency of the disease, the difficulties to evaluate precise parameters of therapeutic efficacy and the heterogeneity of the cases.

Background of model

Since the first report in 1885 (Ribbert, 1885), several models of experimental infective endocarditis have been attempted in a variety of animals (dogs, horses, pigs, opossum, rats and rabbits; Tunkel and Scheld, 1992). Early researchers recognized the ability of valve trauma or other manipulations to facilitate colonization of cardiac valves by injected bacteria. The rabbit model, as currently used, is based upon the simple and reproducible model described by Garrison and Freedman (1970). This model, using a polyethylene catheter to induce valvular damage, was modified by Durack and Beeson (1972), Sande and Irwin (1974) and others (Gutschik and Christensen, 1978a,b).

Animal species

White New Zealand female rabbits weighting 2–2.5 kg are most commonly used.

Preparation of animals

No specialized housing or care nor specific pretreatment is required.

Details of surgery

Overview

To induce non-bacterial thrombotic endocarditis (NBTE), a polyethylene catheter is inserted into the right carotid artery and advanced towards the heart until the catheter crosses the aortic valve, where it is left in place.

Material required

Anesthetic, syringes and needles, scalpel handle plus blades, electric clippers, skin disinfectant (ethyl alcohol), suture material, pliers, dissection pliers, clips, silk thread, catheter guide, polyethylene catheter, bounds, compresses. Hollow polyethylene catheters are usually used (external diameter 0.96 mm, internal diameter 0.58 mm), but solid cannulae are also appropriate (Tanphaichitra *et al.*, 1974).

Anesthesia

Animals are most commonly anesthetized by intramuscular injection with ketamine hydrochloride (15 mg/kg). This protocol produces transient anesthesia lasting 15–20 minutes, which corresponds to the duration necessary for surgery.

Production of sterile aortic valve vegetations

The anesthetized rabbit is maintained in a dorsal position with the help of cords fixing the animal's paws to the surgical table (Figure 72.1). The skin over the neck is shaved with electric clippers, disinfected with ethyl alcohol and a 3–5 cm longitudinal incision is made along the right side of the trachea. The right carotid artery is then exposed and ligated with a silk suture. A clamp is placed on the artery 1–2 cm upstream of the ligature, demarcating a portion of the artery where the polyethylene catheter is cautiously introduced via a guide-wire. The catheter is then inserted up to the clamp and the guide can be removed (Figure 72.2).

Handbook of Animal Models of Infection
ISBN 0-12-775390-7

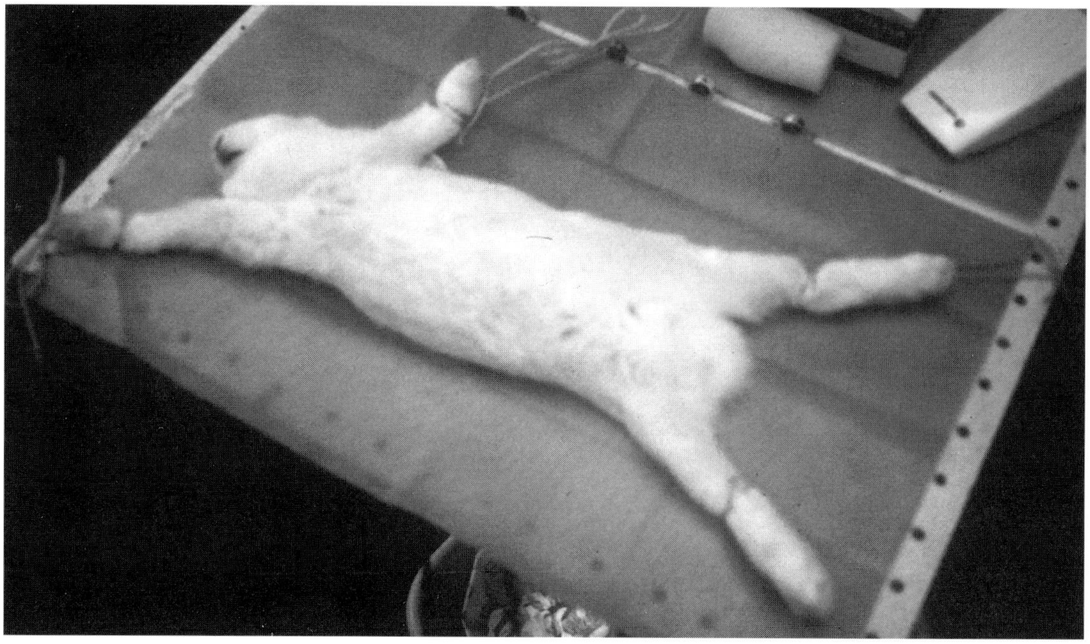

Figure 72.1 Anesthetized rabbit in a supine position prior to surgery.

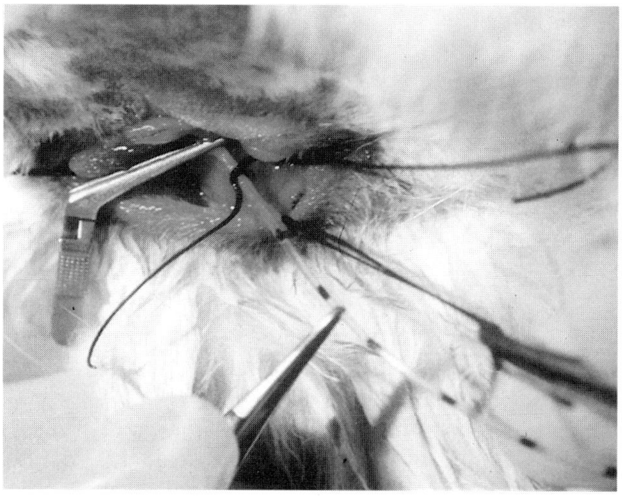

Figure 72.2 Polyethylene catheter inserted into the right carotid artery.

Before removing the clamp and inserting the catheter further, a silk thread should be placed around the catheter-containing artery and slightly tightened. This will help prevent ebbing of blood after clamp removal. After removing the clamp, the catheter is inserted a distance of about 8 cm until resistance is met. Characteristic pulsations of the catheter are detected when the catheter is in the heart. It is then withdrawn slightly, just a few millimeters, and secured with a loop of silk suture. The tip thus remains in a position just above the semilunar cusps of the aortic valves (Perlman and Freedman, 1971a). The catheter is then tied upon itself and the skin incision is closed with silk over the free end of the catheter.

The catheter is left *in situ* until the death or sacrifice of the animal, as several studies have shown that withdrawal of the catheter before treatment reduces the incidence of infective endocarditis (Durack *et al.*, 1973), slows down bacterial multiplication within the vegetations (Heraief *et al.*, 1982) and facilitates eradication of organisms with therapy (Pelletier and Petersdorf, 1976). Similar catheterization can be applied to the right side of the heart, via the femoral vein, as described originally by Garrison and Freedman (1970), or via the jugular route (Archer and Fekety, 1976). However, infections of the tricuspid valve have been shown to heal more rapidly, even without treatment, and to be less virulent than left-sided infections (Perlman and Freedman, 1971b). The presence of the catheter in the right or left side of the heart provokes valvular lesions predisposing to infective endocarditis within few minutes (Durack *et al.*, 1973). The deposition of fibrin and platelets on these lesions leads to the formation, in 24 hours, of macroscopic aseptic vegetations characteristic of NBTE. These lesions are highly susceptible to colonization by circulating microorganisms. This surgical technique is an easy, reproducible and reliable method of producing NBTE, and does not require rigorous sterile care because rabbit serum bactericidal activity prevents local or general surgery-related infections.

Storage and preparation of inocula

The ability of a microorganism to produce bacterial endocarditis depends on its adhesion properties, its virulence and its susceptibility to serum bactericidal systems. However, increasing the inoculum size of an organism can improve its ability to produce the disease. Thus, the size of the required inoculum varies considerably according to the various species and strains. Freedman and Valone (1979) determined the inoculum size necessary to cause infection in 50% of rabbits (ID_{50}). For aortic valve vegetations with the catheter in place for 2 days, the ID_{50} for *Staphylococcus aureus* was $10^{3.75 \pm 0.060}$, *Streptococcus sanguis* $10^{4.67 \pm 0.52}$ and *Escherichia coli* $10^{6.29 \pm 0.58}$. For enterococci, the ID_{50} in rabbits was estimated at 5.9×10^6 cfu (Durack *et al.*, 1977). Rabbits with right or left heart catheters in place for 2–7 days are generally 100% susceptible to inocula of at least 10^6 streptococci or staphylococci. They are less susceptible to infection with *Escherichia coli* and cell-wall-deficient bacterial forms (Freedman and Valone, 1979). The inoculum is usually prepared from an overnight culture in broth, centrifuged and resuspended in saline. Appropriate dilutions of the inoculum are administered 24 hours after catheter insertion.

Infection process

Two main routes of administration have been used. In the model of Garrison and Freedman (1970), the catheter itself was filled with a suspension of *Staphylococcus aureus* before insertion and the vegetations became infected. However, these experiments were not reproducing the clinical situation, in which cardiac valves are colonized by circulating microorganisms. Durack and Beeson (1972) first described the production of bacterial endocarditis after a single intravenous injection of bacteria through a peripheral ear vein. This technique is currently the most frequently used and is very easy to perform. After shaving the external part of the ear and producing vasodilatation by applying ethyl alcohol, the inoculum is injected into the vein with a fine needle.

Key parameters to monitor infection

The infected rabbits develop clinical and biological manifestations that are very close to those observed in humans with infective endocarditis, such as fever, weight loss, anemia, positive blood cultures, peripheral emboli and secondary septic localizations. Spontaneous mortality may rapidly occur, particularly when infective bacteria are highly virulent. However, the most reliable parameter to monitor infection is the count of colony forming units (cfu) of bacteria per gram of vegetations (for technical description see below).

Antimicrobial therapy

The rabbit model of experimental endocarditis has proved to be highly valuable in evaluating the *in-vivo* effectiveness of antibiotics for treatment as well as for prophylaxis. Many drugs have been tested, either alone or in combination. The intramuscular and intravenous routes of administration are the most commonly used. Dosage, treatment interval and duration of therapy depend on the question being studied and must be in agreement with the pharmacodynamic and pharmacokinetic properties of the antibiotic being used.

If the results are to have clinical relevance, the rate of administration must produce serum concentrations close to those achieved in humans. For most antibiotics, determination of the optimal schedule of administration should be facilitated by evaluating serum levels and the half-life of the drug, which is usually shorter in rabbits than in humans. The time between bacterial seeding of the valve and initiation of antimicrobial therapy is a major factor that influences the therapeutic efficacy of various antimicrobial regimens. Early treatment of infective endocarditis enhances the apparent efficacy of antimicrobial therapy, even with drugs that are clinically ineffective (Carrisoza and Kaye, 1976). Most investigators initiate therapy when bacterial density in vegetations is maximal 24–48 hours after infection. For prophylaxis studies, the antimicrobial drug is usually administered at least 1 day after insertion of the catheter and 30 minutes prior to injection of the microorganism (Durack and Petersdorf, 1973; Pelletier *et al.*, 1975).

Key parameters to monitor response to treatment

The rabbit model of aortic endocarditis provides precise endpoints, which can be relevant in the evaluation of response to treatment: number of cfu of bacteria per unit weight of tissue (vegetation or other organ being secondarily infected), detection of resistant bacterial populations, presence or absence of positive blood cultures, incidence of death, percentage of relapse after therapy has been stopped.

The most reliable parameter to evaluate therapeutic efficacy is the reduction of cfu per gram of vegetation in comparison with control animals. Determination of bacterial cfu requires the analysis of the aortic vegetations. The animals are sacrificed (by intravenous injection of 100 mg of sodium thiopental). The heart is removed and opened (Figure 72.3), being careful to preserve sterility, the vegetations are excised, weighed, homogenized, serially diluted and cultured on appropriate agar plates. The best homogenization is obtained by using an electric tissue grinder. Animals are usually sacrificed 12–48 hours after the last antibiotic injection according to serum elimination half-life, in order to avoid carry-over. However, the absence of bacteriologic relapse following therapeutic discontinuation represents the most relevant parameter for predicting therapeutic cure (Fantin *et al.*, 1989). In studies of prophylaxis, rabbits that had been treated prior to bacterial challenge are sacrificed 24–72 hours after injection of the microorganism.

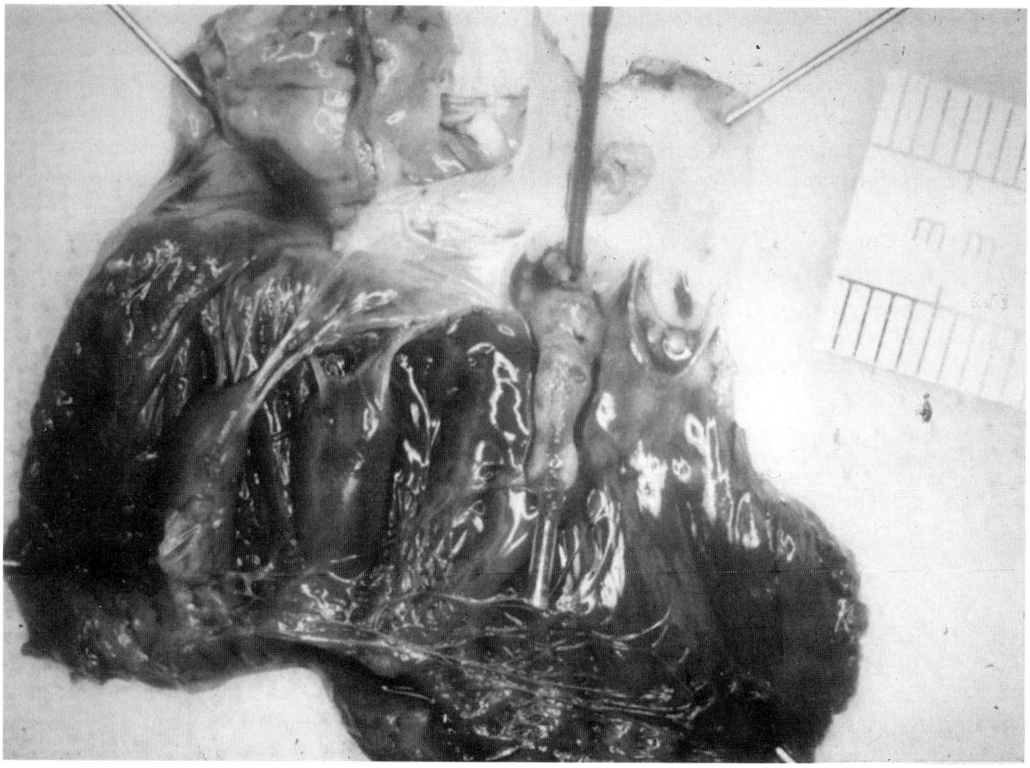

Figure 72.3 Cardiac vegetations surrounding a transaortic catheter and located on the aortic valves.

Determination of the cfu per gram of vegetations allows the evaluation of the percentage of infected animals for each prophylactic regimen.

Pitfalls (advantages/disadvantages) of the model

The rabbit model has been shown to produce infection in animals that closely mimicks the characteristics of infective endocarditis in humans. However, several differences exist between the animal and the human disease and one must be particularly careful in extrapolating the animal results to human therapy. First, the valvular trauma induced by the polyethylene catheter is pathophysiologically very different from the underlying conditions that predispose humans to infective endocarditis. Also, in most studies the catheter has been left in the heart until the animal is killed. Presence of this foreign body mimicks the clinical situation of patients with prosthetic valve endocarditis rather than infection on native valves. In addition, the size of the inoculum required to produce infection in animals and the subsequent bacteremia are much greater than the number of circulating bacteria responsible for human endocarditis. The intravenous portal of entry is relatively rare in humans. Moreover, bacteremia induced by invasive procedures in humans are often polymicrobial, whereas animals are usually infected with a single microorganism. In the experimental model, no more than one or two strains of a species of bacteria are tested on animals and extrapolation to the other strains of the species must be done very cautiously. For this reason, the choice of the infective bacteria must be representative of the clinical question being elucidated. Finally, the infection induced in animals is probably more severe and more acute than the human disease, as shown by the rapid mortality of infected animals left untreated.

As compared with the rat model, the rabbit model of endocarditis presents two major advantages. First, the size of the vegetations is much larger in rabbits; this is of great importance because diffusion of antibiotics through the vegetations is clearly influenced by their size. Second, the higher bacterial inoculum per vegetation in rabbits allows the selection of resistant mutant during therapy (Fantin *et al.*, 1993). On the other hand, rabbits are more expensive and more difficult to handle than rats. For these reasons, the rabbit model is more appropriate for therapeutic studies and the rat model for prophylactic studies.

Contributions of the model to infectious disease therapy

Efficacy of antimicrobial agents

Evaluation of the effectiveness of various antimicrobials in the rabbit model of endocarditis is an essential prerequisite that greatly influences therapeutic options in humans. Several general principles concerning endocarditis therapy have been established by using the rabbit model of endocarditis, in accordance with the very specific pathophysiological characteristics of the disease, which may influence the pharmacokinetic and pharmacodynamic properties of antimicrobial agents (Carbon, 1993). First, the treatment should be bactericidal. Bacteriostatic therapy would produce an apparent clinical cure, but bacterial growth leading to relapse of the disease would occur as soon as therapy was discontinued. Secondly, treatment should be administered for a prolonged period in order to eradicate infection. Third, therapy should be given at high doses in order to achieve a bactericidal effect. Fourth, synergistic combinations should be preferred to monotherapies. As results of time-kill methods *in vitro* are predictive of *in-vivo* effect (Scheld, 1987), optimal therapeutic regimens should be screened *in vitro* before evaluation in animals and subsequent extrapolation to human therapy.

Routes of administration and schedule

The rabbit experimental model of endocarditis has provided essential informations about the optimal antibiotic schedules being administered according to the pharmacodynamic characteristics of antiinfective drugs (Carbon, 1991; Potel *et al.*, 1991). As many variables can be carefully controlled, it becomes relatively easy to determine the predictive values of *in-vitro* tests on the *in-vivo* outcome. For antibiotics with a slow bactericidal rate (i.e. with a time-dependent effect) and no or negligible postantibiotic effect (PAE), such as β-lactams, intervals between administrations must be short in order to maintain local inhibitory levels throughout the entire dose interval. In contrast, antibiotics with a rapid killing rate (i.e., with a concentration effect) and a PAE, such as aminoglycosides and fluoroquinolones on some organisms, can be administered at longer dosing intervals and at high doses (Carbon, 1993). Recently, simulation of human pharmacokinetics in rabbits has improved the clinical relevance of the rabbit model to optimization of schedule and dosing regimens in humans (Bugnon *et al.*, 1996).

Indication of synergistic/antagonist effects of antibiotic combinations

The rabbit model of endocarditis has been extensively used for the evaluation of the potential consequences of antibiotic combinations (Fantin and Carbon, 1992). The benefit of selected antibiotic combinations has been demonstrated in terms of increased bactericidal activity and reduction of the emergence of resistant mutants as compared with single drug therapy (Aslangul *et al.*, 1997).

Pharmacokinetics, pharmacopharmacodynamics, tissue penetration

The rabbit model represents an excellent way of studying the pharmacokinetic properties of antibiotics in vegetations (Carbon *et al.*, 1995). To reach microorganisms, antibiotics must diffuse through layers of fibrin and platelets that separate bacteria from the bloodstream. Although equilibration between antibiotic concentrations in the serum and in the vegetations is rapid and complete with a similar elimination half-life in both compartments (Contrepois *et al.*, 1986), significant differences in the diffusion patterns of antibiotics through vegetations have been observed by autoradiography (Crémieux *et al.*, 1989). To date, three different types of diffusion have been characterized. The most frequent pattern is homogeneous diffusion through the vegetation, as observed with tobramycin, spiramycin, quinupristin, perfloxacin, sparfloxacin, amoxicillin and clavulanic acid. Diffusion may describe a gradient of decreasing concentrations between the periphery and the core of the vegetation, as observed for ceftriaxone, penicillin and dalfopristin. The third pattern of diffusion is observed with teicoplanin, which remains at the periphery of the vegetation and fails to diffuse into the core.

The model of rabbit endocarditis has also generated information concerning the *in-vivo* pharmacodynamics of antimicrobials. This included the measure of the duration of *in-vivo* PAE (Hessen *et al.*, 1989) and adaptative resistance into the vegetation (Xiong *et al.*, 1997), the determination of the pharmacokinetic/pharmacodynamic parameter that best predicts the *in-vivo* activity of a given antibiotic (Fantin *et al.*, 1995), and the value of the rate of killing and serum elimination half-life to predict the feasibility of a single daily dose of antibiotics (Potel *et al.*, 1991).

Development of novel therapeutic approaches

The model of rabbit endocarditis has provided information in the development of novel therapeutic approaches in two ways; first, in the investigation of new compounds, since the pharmacokinetic, pharmacodynamic and microbiologic parameters may be investigated for a given antimicrobial, creating understanding of the factor(s) that limit the *in vivo* efficacy (Belmatoug and Fantin, 1997). Second, the model has been used for the investigation of other therapeutic approaches than antibiotics, based on knowledge of the physiopathology of the disease, such as the investigation of the potential benefit of the use of human recombinant granulocyte colony-stimulating factor in rabbits with aortic endocarditis due to *Pseudomonas aeruginosa*, in addition to antimicrobial therapy (Vignes *et al.*, 1995).

Understanding of pathology and pathogenesis

The rabbit model of endocarditis has yielded much information about the adhesive properties of different microorganisms. It has been demonstrated that the production of dextran, an exopolysaccharide produced by various strains of streptococci and enterococci, correlates both to adherence *in vitro* and infectivity *in vivo* (Scheld *et al.*, 1978), facilitates bacterial persistence in the vegetations, limits access of antibiotics to their bacterial targets and protects microorganisms against host phagocytic defenses (Dall *et al.*, 1987).

The role of platelets in the pathogenesis of streptococcal endocarditis has also been well studied. Platelets act as a major promoter of the disease. On the one hand, platelets contribute to the formation of NBTE and increase the adhesion of *Streptococcus sanguis* to fibrin. On the other hand, activation of platelets leads to the production of microbicidal peptides, which play a protective role against colonization of NBTE (Dankert *et al.*,1995; Yeaman, 1997).

Several characteristics of endocardial vegetations have also been well established by the use of the rabbit endocarditis model. The heterogeneous distribution of bacteria in the vegetations, with a platelet–fibrin matrix surrounding the bacterial colonies, is an important morphological characteristic of the lesions, which can explain therapeutic failures. Histological studies have demonstrated the absence, at the early stage of the disease, of phagocytic cells in the vegetation, whereas bacterial multiplication is rapid (Durack and Beeson, 1972). Titers of bacteria in vegetations are particularly high and often exceed 10^8 cfu per gram (Walker and Hamburger, 1959). This could account for a reduced antibiotic activity related to that 'inoculum effect'. The metabolic activity of bacterial populations located in the core of vegetations is reduced compared with that of peripheral bacteria, as shown by autoradiography of vegetations after [^{3}H]-L-alanine incorporation (Durack and Beeson, 1972). This may explain a reduced effectiveness of antibiotics active on rapidly dividing microorganisms, because most of the deep-seated bacteria are in the resting state (Cozens *et al.*, 1986). In addition, important morphological modifications of bacteria have been observed in the vegetations from rabbits with streptococcal endocarditis (Frehel *et al.*, 1988), similar to the reversible abnormalities observed *in vitro* when growth conditions are unfavorable. Altogether, these physiopathologic characteristics of the vegetation may account for the phenotypic-tolerance phenomenon, which is characterized by the lack of bactericidal activity of antibiotics *in vivo* despite good sensitivity *in vitro*.

The kinetics of the production of tumor necrosis factor have been monitored during the course of rabbit endocarditis, in untreated controls and in animals receiving antibiotics with different mechanisms of action (Mohler *et al.*, 1994). The model provided an opportunity to study the inflammatory response during the course of a subacute infectious process.

In conclusion, it is quite clear that the model of experimental endocarditis has greatly contributed to our understanding of the physiopathologic process of bacterial endocarditis and of the *in-vivo* activity and limitations of antibiotics.

References

Archer, G., Fekety, F. R. (1976). Experimental endocarditis due to *Pseudomonas aeruginosa*. Description of a model. *J. Infect. Dis.*, **134**, 1–7.

Aslangul, E., Baptista, M., Fantin, B. *et al.* (1997). Selection of glycopeptide-resistant mutants of VanB-type *Enterococcus faecalis* BM4281 *in vitro* and in experimental endocarditis. *J. Infect. Dis.*, **175**, 598–605.

Belmatoug, N., Fantin, B. (1997). Contribution of animal models of infection for the evaluation of the activity of antimicrobial agents. *Int. J. Antimicrob. Agents*, **9**, 73–82.

Bugnon, D., Potel, G., Xiong Y. Q. *et al.* (1996). *In vivo* antibacterial effects of simulated human serum profiles of once-daily versus thrice-daily dosing of amikacin in a *Serratia marcescens* endocarditis experimental model. *Antimicrob. Agents Chemother.*, **40**, 1164–1169.

Carbon, C. (1991). Impact of the antibiotic dosage schedule on efficacy in experimental endocarditis. *Scand. J. Infect. Dis.*, **74**, 163–172.

Carbon, C. (1993). Experimental endocarditis: a review of its relevance to human endocarditis. *J. Antimicrob. Chemother.*, **31**, 71–85.

Carbon, C., Crémieux A. C., Fantin B. (1995). Pharmacokinetic and pharmacodynamic aspects of therapy of experimental endocarditis. In: *Infectious Disease Clinics of North America, Infective Endocarditis* (eds Wilson, W. R., Steckelberg, J. M.), pp. 37–51. W. B. Saunders, Philadelphia, PA.

Carrisoza, J., Kaye, D. (1976). Antibiotic synergism in enterococcal endocarditis. *J. Lab. Clin. Med.*, **88**, 132–141.

Contrepois, A., Vallois, J. M., Garaud, J. J. *et al.* (1986). Kinetics and bactericidal effect of gentamicin and latamoxef (moxalactam) in experimental *Escherichia coli* endocarditis. *J. Antimicrob. Chemother.*, **17**, 227–237.

Cozens, R. M., Tuomanen, E., Tosch, W., Zak, O., Suter, J., Tomasz, A. (1986). Evaluation of the bactericidal activity of β-lactam antibiotics on slowly growing bacteria cultured in the chemostat. *Antimicrob. Agents Chemother.*, **29**, 797–802.

Crémieux, A. C., Maziére, B., Vallois, J. M. *et al.* (1989). Evaluation of antibiotic diffusion into cardiac vegetations by quantitative autoradiography. *J. Infect. Dis.*, **159**, 938–944.

Dall, L., Barnes, W. G., Lane J. W., Mills, J. (1987). Enzymatic modification of glycocalyx in the treatment of experimental endocarditis due to viridans streptococci. *J. Infect. Dis.*, **156**, 736–740.

Dankert, J., Van Der Werff, J., Zaat, S. A., Joldersma, W., Klein, D., Hess, J. (1995). Involvement of bactericidal factors from thrombin-stimulated platelets in clearance of adherent *viridans* streptococci in experimental infective endocarditis. *Infect. Immun.*, **63**, 663–671.

Durack, D. T., Beeson, P. B. (1972). Experimental bacterial endocarditis. I: Colonization of a sterile vegetation. *Br. J. Exp. Pathol.*, **53**, 44–49.

Durack, D. T., Petersdorf, R. G. (1973). Chemotherapy of experimental infective endocarditis. I: Comparison of commonly recommended prophylaxis regimens. *J. Clin. Invest.*, **52**, 592–598.

Durack, D. T., Beeson, P. B., Petersdorf, R. G. (1973). Experimental bacterial endocarditis. III: Production and progress of the disease in rabbits. *Br. J. Exp. Pathol.*, **54**, 142–151.

Durack, D. T., Starkebaum, M. K., Petersdorf, R. G. (1977). Chemotherapy of experimental streptococcal endocarditis. VI: Prevention of enterococcal endocarditis. *J. Lab. Clin. Med.*, **90**, 171–179.

Durack, D. T., Lukes, A. S., Bright, D. K., the Duke Endocarditis Service (1994). New criteria for diagnosis of infective endocarditis: utilization of specific echocardiographic findings. *Am. J. Med.*, **96**, 200–209.

Fantin, B., Carbon, C. (1992). *In vivo* antibiotic synergism: contribution of animal models. *Antimicrob. Agents Chemother.*, **36**, 907–912.

Fantin, B., Pangon, B., Potel, G. *et al.* (1989). Ceftriaxone–netilmicin combination in single-daily-dose treatment of experimental *Escherichia coli* endocarditis. *Antimicrob. Agents Chemother.*, **33**, 767–770.

Fantin, B., Leclerc, R., Duval, J., Carbon, C. (1993). Fusidic acid alone or in combination with vancomycin for therapy of experimental endocarditis due to methicillin-resistant *Staphylococcus aureus*. *Antimicrob. Agents Chemother.*, **37**, 2466–2469.

Fantin B., Leclercq, R., Merle, Y. *et al.* (1995). Critical influence of resistance to streptogramin B-type antibiotics on activity of RP 59500 (quinupristin/dalfopristin) in experimental endocarditis due to *Staphylococcus aureus*. *Antimicrob. Agents Chemother.*, **39**, 400–405.

Freedman, L. R., Valone, J. (1979). Experimental infective endocarditis. *Prog. Cardiovasc. Dis.*, **22**, 169–180.

Frehel, C., Hellio, R., Crémieux, A. C., Contrepois, A., Bouvet, A. (1988). Nutritionally variant streptococci develop ultrastructural abnormalities during experimental endocarditis. *Microb. Pathog.*, **4**, 247–255.

Garrison, P. K., Freedman, L. R. (1970). Experimental endocarditis. I: Staphylococcal endocarditis in rabbits resulting from placement of a polyethylene catheter in the right side of the heart. *Yale J. Biol. Med.*, **42**, 394–410.

Gutschik, E., Christensen, N. (1978a). Experimental endocarditis in rabbits. I: Techniques and spontaneous course of non-bacterial thrombotic endocarditis. *Acta Pathol. Microbiol. Immunol. Scand. Sect. B Microbiol.*, **86**, 215–221.

Gutschik, E., Christensen, N. (1978b). Experimental endocarditis in rabbits. II: Course of untreated *Streptococcus faecalis* infection. *Acta Pathol. Microbiol. Immunol. Scand. Sect. B Microbiol.*, **86**, 223–228.

Heraief, E., Glauser, M. P., Freedman, L. R. (1982). Natural history of aortic valve endocarditis in rats. *Infect. Immun.*, **37**, 127–131.

Hessen, M. T., Pitsakis, P. G., Levison, M. E. (1989). Postantibiotic effect of penicillin plus gentamicin versus *Enterococcus faecalis in vitro* and *in vivo*. *Antimicrob. Agents Chemother.*, **33**, 608–611.

Mohler, J., Fantin, B., Mainardi, J. L., Carbon, C. 1994. Influence of antimicrobial therapy on kinetics of tumor necrosis factor levels in experimental endocarditis caused by *Klebsiella pneumoniae*. *Antimicrob. Agents Chemother.*, **38**, 1017–1022.

Pelletier, L. L., Petersdorf, R. G. (1976). Chemotherapy of experimental streptococcal endocarditis. V: Effect of duration of infections and retained intracardiac catheter on response to treatment. *J. Lab. Clin. Med.*, **87**, 692–702.

Pelletier, L. L., Durack, D. T., Petersdorf, R. G. (1975). Chemotherapy of experimental streptococcal endocarditis. IV: Further observations on prophylaxis. *J. Clin. Invest.*, **56**, 319–330.

Perlman, B. B., Freedman, L. R. (1971a). Experimental endocarditis. II: Staphylococcal infection of the aortic valve following placement of a polyethylene catheter in the left side of the heart. *Yale J. Biol. Med.*, **44**, 206–213.

Perlman, B. B., Freedman, I. R. (1971b). Experimental endocarditis. III: Natural history of catheter induced staphylococcal endocarditis following catheter removal. *Yale J. Biol. Med.*, **44**, 214–224.

Potel, G., Chau, N. P., Pangon, B. *et al.* (1991). Single daily dosing of antibiotics: importance of *in vivo* killing rate, serum half life, and protein binding. *Antimicrob. Agents Chemother.*, **35**, 2085–2090.

Ribbert, M. W. H. (1885). Beitrage zur Localisation der Infectionskrankheiten. *Dtsch. Med. Wochenschr.*, **11**, 717–719.

Sande, M. A., Irwin, R. G. (1974). Penicillin–aminoglycoside synergy in experimental *Streptococcus viridons* endocarditis. *J. Infect. Dis.*, **129**, 572–576.

Scheld, W. M. (1987). Therapy of streptococcal endocarditis: correlation of animal models and clinical studies. *J. Antimicrob. Chemother.*, **20** (Suppl. A), 71–85.

Scheld, W. M., Valone, J. A., Sande, M. A. (1978). Bacterial adherence in the pathogenesis of infective endocarditis. *J. Clin. Invest.*, **61**, 1394–1404.

Tanphaichitra, D., Ries, K., Levison, M. E. (1974). Susceptibility to *Streptococcus viridans* endocarditis in rabbits with intracardiac pacemaker electrodes or polyethylene tubing. *J. Lab. Clin. Med.*, **84**, 726–730.

Tunkel, A. R., Scheld, W. M. (1992). Experimental models of endocarditis. In: *Infective Endocarditis*, 2nd edn (ed Kaye D.), pp. 37–56. Raven Press, New York.

Vignes, S., Fantin, B., Elbim, C., Walker, F., Gougerot-Pocidalo, M. A., Carbon, C. (1995). Critical influence of timing of administration of granulocyte colony-stimulating factor on antibacterial effect in experimental endocarditis due to *Pseudomonas aeruginosa*. *Antimicrob. Agents Chemother.*, **39**, 2702–2707.

Xiong, Y. Q., Caillon J., Kergueris, M. F. *et al.* (1997). Adaptative resistance of *Pseudomonas aeruginosa* induced by aminoglycosides and killing kinetics in a rabbit endocarditis model. *Antimicrob. Agents Chemother.*, **41**, 823–826.

Yeaman, M. R. (1997). The role of platelets in antimicrobial host defense. *Clin. Infect. Dis.*, **25**, 951–970.

Chapter 73

Infant Rat Model of Acute Meningitis

U. Vogel and M. Frosch

Background of human infection

The 1997 epidemic of meningococcal meningitis in Africa again demonstrated that bacterial meningitis continues to be a major world-wide health problem. Despite the availability of potent antimicrobial agents, adjunctive glucocorticoid therapy and vaccines (Quagliarello and Scheld, 1997), the case fatality rates are considered to be 10% and more, even in the northern hemisphere, and persistent neurological sequelae in survivors are frequent (Unhanand *et al.*, 1993). Causative agents of bacterial meningitis are *Escherichia coli* K1, group B streptococci, *Neisseria meningitidis, Streptococcus pneumoniae* and *Haemophilus influenzae*, with an age-related distribution of bacterial meningitis due to a specific pathogen (Wenger *et al.*, 1990). There have been tremendous research efforts on the pathophysiology of disease and the molecular biology of pathogenicity; however, questions of outstanding interest, e.g., how the bacteria pass the blood–brain barrier, could not be answered until now. Antibiotic therapy of the disease remains to be a matter of discussion, because resistance to standard antibiotics has become prevalent in certain parts of the world (Marton *et al.*, 1991), and controversy exists with regard to the implementation of adjunctive corticosteroid therapy (Quagliarello and Scheld, 1997). The dramatic decline of *Haemophilus influenzae* meningitis, due to widespread vaccination (Adams *et al.*, 1993), again illustrates the need for potent vaccines directed against other major causes of bacterial meningitis, e.g., against serogroup B meningococci.

Background of model

In order to analyse the pathophysiology of bacterial meningitis, and to evaluate the therapeutic effects of antibiotics and immunization, efforts were made to establish suitable animal models, one of which was the infant rat model of bacterial meningitis presented in the 1970s (Smith *et al.*, 1973; Glode *et al.*, 1977; Moxon *et al.*, 1977; Ferrieri *et al.*, 1980; Rodriguez *et al.*, 1991). Because of the narrow host range of the causative agents of bacterial meningitis and the restricted susceptibility of hosts depending on their age (Smith *et al.*, 1973; Glode *et al.*, 1977), the infant rat model proved to be an approach followed until now. The animals used in the model exhibit a disease with features that closely resemble human disease.

1. The development of bacteraemia and meningitis is frequently inversely related to age (Smith *et al.*, 1973; Glode *et al.*, 1977; Bortolussi *et al.*, 1978; Salit *et al.*, 1984).
2. Infection may be achieved by the natural route, e.g., following intranasal colonization (Moxon *et al.*, 1977; Salit *et al.*, 1984) or oral ingestion (Glode *et al.*, 1977; Moxon *et al.*, 1977).
3. Meningitis occurs as a result of bacteraemia (Smith *et al.*, 1973; Moxon *et al.*, 1977; Ferrieri *et al.*, 1980; Rodriguez *et al.*, 1991).
4. Histopathological findings close to those demonstrated in infected humans have been reported (Smith *et al.*, 1973; Ferrieri *et al.*, 1980; Wiedermann *et al.*, 1986; Kaplan *et al.*, 1989; Rodriguez *et al.*, 1991; Kim *et al.*, 1995).

There are now several prototype studies and reviews (Smith *et al.*, 1973; Moxon *et al.*, 1974, 1977; Glode *et al.*, 1977; Bortolussi *et al.*, 1978; Ferrieri *et al.*, 1980; Salit *et al.*, 1984; Martin *et al.*, 1986; Tauber and Zwahlen, 1994; Rodriguez *et al.*, 1991; Kim *et al.*, 1995), which demonstrate an extensive knowledge of the handling of the animals. Natural infection routes have been reported (Glode *et al.*, 1977; Moxon *et al.*, 1977), as well as repeated sampling of blood and cerebrospinal fluid (CSF) from an individual animal (Ferrieri *et al.*, 1980; Kim *et al.*, 1984; Kim, 1985). The model has been used for analysing the efficacy of antibiotic treatment (Kim *et al.*, 1984, 1995; Kim, 1985; Kaplan *et al.*, 1989) and passive immunization (Greenberg *et al.*, 1989). Furthermore, valuable information could be drawn from studies investigating the virulence of genetically engineered mutants (Vogel *et al.*, 1996a).

Animal species and preparation

Most workers purchase specific-pathogen-free, pregnant rats from commercial suppliers. Exact prediction of the

Handbook of Animal Models of Infection
ISBN 0-12-775390-7

date of birth can be offered. The rats should give birth to the pups not earlier than 3 days after arrival. Wistar and Sprague–Dawley rats are most frequently used. If several litters are included in the experiments, the pups can be randomly distributed between the mothers (Rodriguez *et al.*, 1991). To our knowledge, the influence of sex on the course of bacterial meningitis in the infant rat has not been studied. The pups are used for infection at the age of approximately 5 days, when they weigh 10–20 g. Significant delay in the time of infection results in reduced rates of infected animals (Smith *et al.*, 1973; Glode *et al.*, 1977; Ferrieri *et al.*, 1980; Salit *et al.*, 1984).

Preparation of inocula

Bacterial species used in the infant rat model of bacterial meningitis should be stored at –70°C in broth (e.g. tryptic soy broth or brain–heart infusion broth) with 10% glycerol. From the permanent stocks, bacteria are streaked on to appropriate agar plates for overnight growth, or an overnight broth culture is performed. Five to 10 small agar-plate-grown colonies are either directly resuspended in saline and diluted further to achieve the desired infectious dose (Vogel *et al.*, 1996a), or they are grown to mid-logarithmic growth phase for another couple of hours in broth (Kim, 1985). Overnight broth cultures should be diluted in broth the next morning, in order to achieve a mid-logarithmic growth phase by continuous incubation (Saukkonen *et al.*, 1988). Thereafter, the bacteria should be washed several times with 0.9% saline or phosphate-buffered saline (PBS). Inoculum sizes reported in the literature vary tremendously and depend on bacterial species, strain, age of animals, challenge route and aim of the study.

Infection process

Five modes of challenge of infant rats with bacteria have been described:

- the intraperitoneal route, which is easy to perform and which yields very reproducible results (Smith *et al.*, 1973; Ferrieri *et al.*, 1980; Saukkonen, 1988).
- the intranasal route (Moxon *et al.*, 1974, 1977), which mimics natural infection by *Haemophilus influenzae* and *Neisseria meningitidis*, both of which spread to the bloodstream from the nasopharynx;
- oral infection, which mimics natural infection by *E. coli* K1 (Glode *et al.*, 1977; Moxon *et al.*, 1977),
- subcutanous infection (Kim, 1985; Greenberg *et al.*, 1989);
- intracisternal infection (Kim *et al.*, 1995).

Intraperitoneal infection

A volume of 100 μl of a suspension of bacteria in sterile saline is injected intraperitoneally (i.p.) using a 26 G or 30 G needle attached to a tuberculin syringe. The head and hindlegs should be carefully immobilized with the index and little fingers (Figure 73.1).

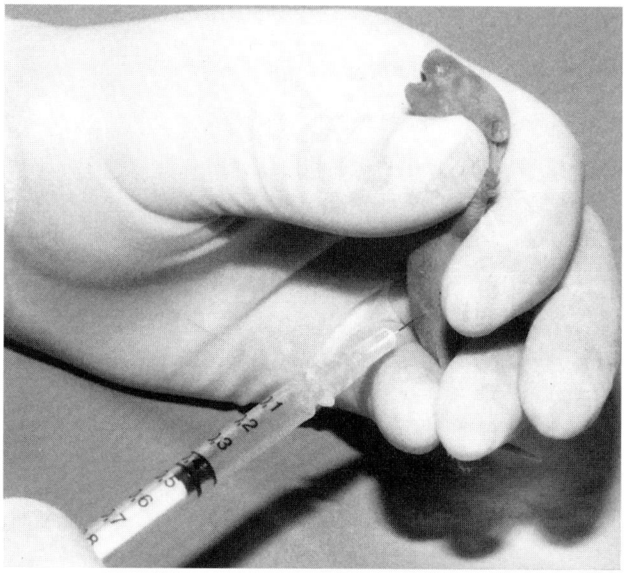

Figure 73.1 Intraperitoneal challenge of an infant rat with a 30 G needle connected to a tuberculin syringe.

Intranasal infection

Intranasal infection has been described in detail by Moxon *et al.* (1974, 1977). Briefly, 10–20 μl is slowly injected with a 23 G or 25 G needle into the anterior nares (Moxon *et al.*, 1974).

Oral infection

Oral infection has been described as an effective and natural challenge route for the *E. coli* K1 meningitis in infant rats (Moxon *et al.*, 1977). Application of 10^3–10^4 cfu via an oro-gastric tube resulted in bacteraemia in half of the animals (Moxon *et al.*, 1977). A comparison reported by Moxon *et al.* of the intranasal and the oral route of challenge revealed that the natural route was most effective for both *H. influenzae* (intranasal) and *E. coli* (oral). A detailed description of the technique of oral infection was provided in a separate paper by the same group. An 0.061 cm gastric tube was used, which was placed directly in the stomach (Glode *et al.*, 1977). There have been attempts to orally infect the pups by letting them suck on pipette tips (Ferrieri *et al.*, 1980); however, group B streptococcal bacteraemia at least was not achieved by this method.

Subcutaneous infection

Subcutaneous infection was used in studies on the group B streptococcal (GBS) and the staphylococcal infant rat model (Kim, 1985; Greenberg *et al.*, 1989). The variability of the blood cfu counts seemed to be unacceptably high in the GBS study (Kim, 1985). Nevertheless, the method appears to be technically simple and mimics possible natural routes of infection.

Intracisternal infection

For technical reasons, intracisternal infection is more frequently used in the rabbit model than in the infant rat model of bacterial meningitis (see Chapter 75). Nevertheless, intracisternal challenge of infant rats has been reported for GBS infection (Kim *et al.*, 1995). The animals were challenged by an intracisternal injection of 10 μl of a bacterial suspension using a 32 G needle. Bypass of the primary steps of pathogenesis, i.e., entering the blood stream, systemic spread and crossing the blood–brain barrier is achieved. Using intracisternal infection of GBS, meningitis with inflammation of the subarachnoid space and of the ventricles, as well as vasculopathy and neuritis, was efficiently induced (Kim *et al.*, 1995).

Repeated infection

Because of interspecies and interstrain variability in bacterial virulence for infant rats, repeated infection should be considered if there are difficulties in achieving appropriate levels of bacteraemia or meningitis. A schedule of three infections performed at 24 hour intervals gave rise to pneumococcal meningitis with one out of six pneumococcal strains tested (Rodriguez *et al.*, 1991).

Enhancers of infection

Iron dextran has been used to supplement meningococci with iron during systemic spread in the infant rat. Iron dextran is injected i.p. in a volume of 100 μl at a dosage of 250 μg/g body weight (Salit *et al.*, 1984). Iron dextran is given simultaneously with the application of the bacteria. In our hands, untreated animals developed a moderate meningococcal bacteraemia of 10^3 cfu/ml 14 hours after an i.p. challenge with 10^5 cfu. In contrast, two of three iron-dextran-treated animals died; the third was severely ill, with 10^6 cfu/ml blood (unpublished data). Since bacteraemia is a prerequisite of the development of meningitis, iron supplementation has the potential to increase the rates of meningitis achieved in experimental infection. An intriguing finding by Novicki *et al.* (1995) points to an alternative enhancer: gonococci, which do not cause systemic infection in infant rats, were protected from the bactericidal effects of infant rat serum by the presence of human complement factor Clq. Furthermore, human Clq acted as an enhancer of the systemic infection when given i.p. prior to infection (Nowicki *et al.*, 1995). However, the mechanism of Clq-promoted gonococcal virulence remains obscure.

Keeping animals after infection

Keeping the pups with their mothers during the course of the disease is desirable. If separation is necessary, the animals should be kept warm and given glucose-water by tube to prevent dehydration (Glode *et al.*, 1977). The oral route of infection is especially prone to interlitter transmission (Glode *et al.*, 1977), and separation should therefore be taken into consideration. Separation furthermore results in colostrum deprivation, which has been described to increase the number of bacteraemic pups (Glode *et al.*, 1977), and thus can be considered as a potential enhancer of infection.

Key parameters to monitor infection

Monitoring of disease includes (1) sampling and analysis of body fluids, (2) post-mortem analysis of organs and tissues and (3) additional tests such as peripheral blood cell monitoring or behavioural examination of reconvalescent animals.

Sampling

Sampling of blood

Several reports on the infant rat model of bacterial meningitis demonstrated that bacteraemia is indispensable for the development of meningitis (Moxon *et al.*, 1977; Bortolussi *et al.*, 1978; Ferrieri *et al.*, 1980; Saukkonen, 1988). Thus, careful analysis of bacterial counts in the blood is absolutely necessary for the interpretation of meningitis in the infant rat. Several methods have been described of blood sampling, which is performed either *intra vitam* for repeated sampling or while animals are sacrificed. In one study, *intra-vitam* monitoring was performed by external jugular venipuncture (Kim *et al.*, 1984), and was repeated daily for 5 days (Kim, 1985). Other veins accessable for puncture are the tail veins and the lateral marginal vein of the hind leg (Saukkonen *et al.*, 1988). Ferrieri *et al.* (1980) reported repeated amputation of the pups tails to consecutively sample blood in 10–50 μl amounts. If the pups are sacrificed by decapitation, sampling of blood is very easy and volumes of 500 μl can be obtained (Vogel *et al.*, 1996a). Following sacrifice of animals with a lethal dose of ether (Ferrieri *et al.*, 1980) or intraperitoneal pentobarbital

(Rodriguez *et al.*, 1991), blood is obtained by cardiac puncture (Ferrieri *et al.*, 1980).

Sampling of cerebrospinal fluid (CSF)

CSF is obtained either *intra vitam*, which was reported to be repeatedly possible to monitor the course of disease in an individual animal (Kim, 1985), or immediately post-mortem, e.g. following sacrifice with ether (Ferrieri *et al.*, 1980). The unanaesthetized rat is held in an immobile position with the head flexed forward. A 30 G needle is inserted into the cisterna magna through the open fontanelle, which is identified by palpation (Figure 73.2). CSF appears in the hub of the needle. The needle is then removed, and the CSF collected in the hub can be extracted using a tuberculin syringe or allowed to fill a small glass capillary tube (Saukkonen *et al.*, 1988; Tauber and Zwahlen, 1994). Approximately 5–20 μl of CSF will be obtained by this method.

Monitoring of blood contamination of CSF. Blood contamination of sampled CSF occurs during the insertion of the needle into the cisterna magna. Therefore, most workers use 30 G needles in order to minimize the trauma. The problem of blood contamination of CSF is mentioned in some reports (Glode *et al.*, 1977); however, the definition of meningitis in the infant rat model as the detection of bacteria in the CSF occurs frequently. One should be aware that contamination of a pup's sample of 10 μl CSF with 0.01 μl blood is not visible macroscopically. However, in an animal with a level of bacteraemia of 10^6 cfu/ml, this contamination would give rise to 10 cfu/10 μl CSF or 10^3 cfu/ml CSF. This is well within levels reported as proving bacterial meningitis in some papers. It is crucial, therefore, to monitor the blood contamination of the CSF. This can be achieved by the use of Neubauer's chambers for counting of the red blood cells in serial dilutions of the CSF and the blood. Nassif *et al.* used haemocytometers for more rapid processing of the samples (Nassif *et al.*, 1992). We recently established the use of test strips that are commercially available for the quantitative determination of haematuria (Vogel *et al.*, 1996b). Serial dilutions of both CSF and blood are tested for intact erythrocytes, which give rise to coloured spots due to an enzymatic reaction (Figure 73.3). We proposed a meningitis factor, which is defined as the ratio between the total number of CFU/ml CSF and that due to blood contamination. Meningitis was assumed, when the meningitis factor was at least five (Vogel *et al.*, 1996b). The procedure proved to be rapid, and only minimal amounts of CSF were required for determination of the level of blood contamination.

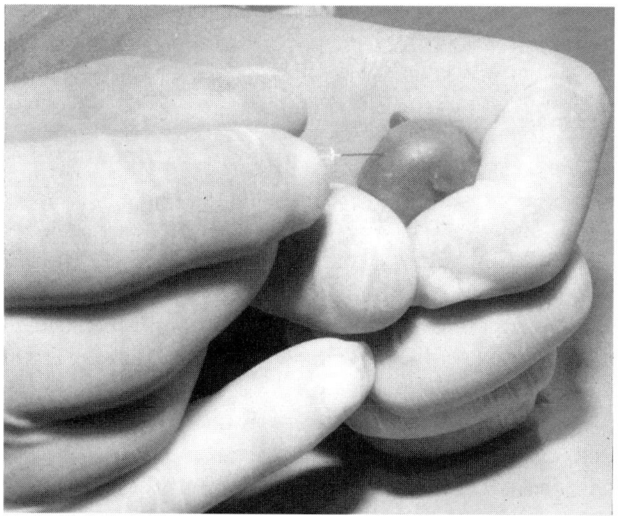

Figure 73.2 Sampling of cerebrospinal fluid (CSF): a 30 G needle is inserted into the cisterna magna through the fontanelle.

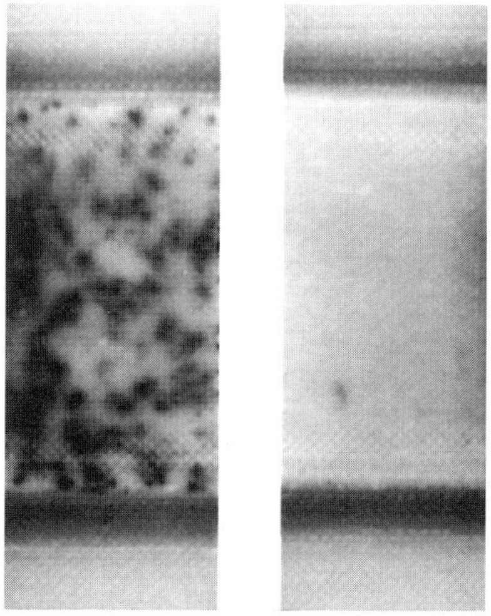

Figure 73.3 Use of Sangur™ test strips for the detection of blood contamination of the CSF. CSF was diluted 1:300 and 1:30 000, respectively, and 10 μl of the diluted CSF were dropped on to the test strip fields. Coloured spots, which theoretically represent one erythrocyte, became visible after approximately 1 minute. In this example, one spot appeared in the case of a dilution of 1:30 000. Thus, approximately $10^{6.5}$ red blood cells (RBC) were present per millilitre of CSF. Assuming a number of $10^{9.4}$ RBC/ml, which was determined by Sangur test strips in infant rat blood (Vogel *et al.*, 1996b), the example is indicative of blood contamination of approximately 1:800.

Intraperitoneal lavage

The peritoneal cavity is the site of the primary proliferation of bacteria if intraperitioneal challenge is applied (Saukkonen *et al.*, 1988). The number of intraperitoneal bacteria is determined by intraperitoneal lavages with 0.5–1.0 ml saline injected into the peritoneal cavity (Saukkonen *et al.*, 1988; Vogel *et al.*, 1996a). Following careful massage of the abdomen, approximately 100–300 μl of the fluid can be recovered and processed for plating on agar plates for the quantitative determination of bacterial counts. Since peritoneal phagocytes are the first line of cellular defence against infection, it is informative to analyse the phagocytosis of bacteria by peritoneal phagocytes (Vogel *et al.*, 1996a). This can be achieved by centrifugation of 200–300 μl lavage fluid on to glass slides, e.g. with a Shandon cytospin 3 centrifuge (Life Sciences International, UK), and staining of the cells with appropriate dyes (Vogel *et al.*, 1996a).

Brain pathology

Histology of the brain and the meninges is used as a definite proof of bacterial invasion of the infant rat's central nervous system. Furthermore, the cellular response to a bacterial pathogen is analysed by histology. Although studies on *H.-influenzae*-b-induced meningitis (Smith *et al.*, 1973; Moxon *et al.*, 1974) and on pneumococcal meningitis demonstrated the recruitment of polymorphonuclear leukocytes and monocytes in 5-day-old infant rats, conflicting results were obtained using group B streptococci as the infective pathogen (Ferrieri *et al.*, 1980), in which case cellular infiltration of the meninges was demonstrated exclusively in animals older than 5 days. The authors suggested that chemotaxis is impaired at the age of 5 days. Furthermore, despite the presence of bacteria in the meninges of intraperitoneally infected Wistar rats, we were not able to demonstrate cellular infiltration in the meningococcal system (unpublished data, Figure 73.4).

Histology is performed on complete heads of the infant rats, which are fixed in 10% buffered formaldehyde and embedded in paraffin. Appropriate dyes are, for instance, haematoxylin and eosin (H & E) or Giemsa (Smith *et al.*, 1973). Alternatively, cryosections can be tried, which are very well suited for immunohistology. In our hands, however, it was difficult to maintain the structure of the brain tissue and the meninges in cryosections, resulting in large gaps between the brain tissue and the pachymeninx. Smith *et al.* proposed the use of a peroxidase assay of sonically treated tissue preparations to quantify the brain polymorphonuclear infiltration by a spectrophotometric assay (Smith *et al.*, 1974). The assay takes advantage of the fact that myeloperoxidase is present in polymorphonuclear cells but not in *H. influenzae* and in brain tissue.

Another way to demonstrate leukocyte infiltration is scanning or transmission electron microscopy of the subarachnoid space, which has been used to study the meningeal inflammatory response in the *H. influenzae* model (Merchant *et al.*, 1981, 1982). Vasculopathy and neuronal injury was studied extensively in one study of group B streptococcal meningitis (Kim *et al.*, 1995).

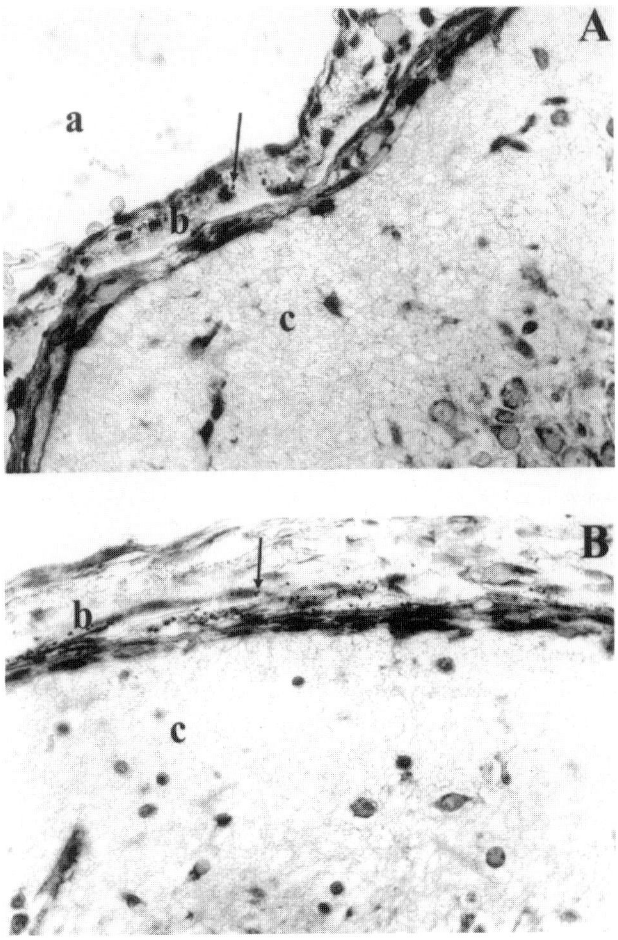

Figure 73.4 Haematoxylin-and-eosin-stained sections of the brain of a 6-day-old infant rat 30 hours after experimental infection with serogroup B meningococci. (**A**) Detail of a section through the sagittal sinus (a), with adjacent subarachnoidal space (b) and cortex (c). The arrow indicates the presence of meningococci in the subarachnoidal space. (**B**) Section through the superior frontal subarachnoidal space (b), which is infiltrated by meningococci (arrow), and the adjacent cortex (c). The sections demonstrate that, in the serogroup B meningococcal meningitis of the infant rat, cellular infiltration of the subarachnoidal space is virtually absent.

Additional markers of infection

To complete the understanding of bacterial meningitis in the infant rat, parameters other than brain pathology and brain or blood bacterial counts should be included. Some studies analysed bacterial counts in organs of secondary interest, such as spleen and liver (Smith *et al.*, 1973). The same study also presented the use of behavioural tests to analyse cerebral dysfunction in survivors of the infection, as well as total DNA content analysis to study the severity of brain damage following infection. Clinical presentation (seizures, activity, righting) and body weight should be determined during the course of infection (Kim *et al.*, 1995). Using blood smears, we were able to demonstrate that the infant rat's peripheral blood cell counts rapidly reacted to meningococcal infection. Within 7 hours of a lethal infection with serogroup B meningococci, peripheral leukocyte counts declined to a third of control values, which was accompanied by an increase of the number of normoblasts (unpublished data). Taking into account that we were not able to demonstrate meningeal inflammation in infant rats infected with meningococci, the rapid decline of peripheral leukocytes probably explains the lack of cerebral recruitment of leukocytes.

Antimicrobial therapy

There are some reports in the literature demonstrating the effectiveness of antimicrobial therapy in the infant rat model of bacterial meningitis. Intraperitoneal injection of cefotaxime and latamoxef was proved to be bactericidal for *E. coli* Kl in both CSF and blood (Kim *et al.*, 1984). Imipenem was equally effective against group B streptococcal sepsis and meningitis (Kim, 1985). Adjunctive corticosteroid therapy prevented CSF inflammation induced by *H. influenzae* or group B streptococci (Kaplan *et al.*, 1989; Kim *et al.*, 1995). However, the effectiveness of ampicillin was reduced by steroid administration (Kaplan *et al.*, 1989). Using the infant rat model, a beneficial effect of subinhibitory concentrations of polymyxin B, which binds to the lipid A fraction of the lipopolysaccharide molecule, has been postulated, if polymyxin B was used alone or in combination with ampicillin for the treatment of *Haemophilus* meningitis (Walterspiel *et al.*, 1986). Accordingly, inactivation of *H. influenzae* LOS by polymyxin B also resulted in reduced meningeal inflammation in rabbits (Syrogiannopoulos *et al.*, 1988) and in blocking of the detrimental effects of *H. influenzae* LOS on brain endothelial cells *in vitro* (Patrick *et al.*, 1992). Nevertheless, because of the limited number of studies on the effects of antibiotics in the infant rat model of bacterial meningitis, and the varying causative agents and therapeutical approaches used, general statements about combination therapy, pharmacodynamics, tissue penetration, etc. cannot be made in this chapter.

Pitfalls of the model and its contribution to infectious disease therapy

As highlighted in the above paragraph, there have been some studies demonstrating the efficiency of antimicrobial therapy and adjuvant corticosteroid therapy in the infant rat, indicating that the model has the potential to be a test for antimicrobial therapy. However, there are several pitfalls that have to be addressed if the model is to be used for the evaluation of therapeutical approaches.

1. Challenge routes that lead to severe bacteraemia prior to meningitis often result in multiorgan failure and death, whereas intracisternal injection of bacteria simply bypasses major steps of the pathogenesis (Kim *et al.*, 1995).

2. To our knowledge, neither is the interaction of bacteria with the infant rat's blood–brain barrier understood nor is there a large body of evidence supporting the suggestion that the pathophysiological and pharmacological features of the infected human neonate's blood–brain barrier resemble those of the infant rat. The same holds true for the immunology of the infant rat, which has been discussed above concerning the variable recruitment of PMN to the CNS. Another example is the tremendous effect of human complement factor Clq in the infant rat model of gonococcal infection (Nowicki *et al.*, 1995), which demonstrates that there is a divergence of central effectors of the different hosts' immune systems.

3. The small size of the animals provides the investigator with problems in several aspects of the model, including the route of drug application and the monitoring of essential parameters such as drug concentrations in CSF and blood, bacterial load and intracranial pressure. Novel routes of drug application cannot be tested in the infant rat model of bacterial meningitis. Sophisticated methods of monitoring CSF dynamics during infection and therapy will be studied in the rabbit model (Lauritsen and Oberg, 1995) or in the adult rat (Koedel *et al.*, 1996), rather than in the infant rat model of bacterial meningitis.

The strength of the infant rat model is its contribution to our understanding of bacterial virulence and the pathophysiology of menigitis. It is not surprising that the model's impact on antimicrobial therapy has been low, and will be restricted to the primary testing of completely new antimicrobial substances or, more generally, new concepts of treatment. Most causative agents of bacterial meningitis, like meningococci, are restricted to the human host. Therefore, recent developments in vaccine design for serogroup B meningococci were tested in primates, in order to mimic the exclusive natural host of meningococci as closely as possible (Devi *et al.*, 1997; Fusco *et al.*, 1997; Zollinger *et al.*, 1997). Finally, for the study of bacterial meningitis, the implementation of advances in animal models, like the use of genetically deficient or transgenic animals or the use of severely immunocompromised animals grafted with human phagocytes, is still restricted to mice (Tan *et al.*, 1995; Westerink *et al.*, 1997).

References

Adams, W. G., Deaver, K. A., Cochi, S. L. *et al.* (1993). Decline of childhood *Haemophilus influenzae* type b (Hib) disease in the Hib vaccine era. *J. A. M. A.*, **269**, 221–226.

Bortolussi, R., Ferrieri, P., Wannamaker, L. W. (1978). Dynamics of *Escherichia coli* infection and meningitis in infant rats. *Infect. Immun.*, **22**, 480–485.

Devi, S. J., Zollinger, W. D., Snoy, P. J. *et al.* (1997). Preclinical evaluation of group B *Neisseria meningitidis* and *Escherichia coli* K92 capsular polysaccharide-protein conjugate vaccines in juvenile rhesus monkeys. *Infect. Immun.*, **65**, 1045–1052.

Ferrieri, P., Burke, B., Nelson, J. (1980). Production of bacteremia and meningitis in infant rats with group B streptococcal serotypes. *Infect. Immun.*, **27**, 1023–1032.

Fusco, P. C., Michon, F., Tai, J. Y., Blake, M. S. (1997). Preclinical evaluation of a novel group B meningococcal conjugate vaccine that elicits bactericidal activity in both mice and non-human primates. *J. Infect. Dis.*, **175**, 364–372.

Glode, M. P., Sutton, A., Moxon, E. R., Robbins, J. B. (1977). Pathogenesis of neonatal *Escherichia coli* meningitis: induction of bacteremia and meningitis in infant rats fed *E. coli* K1. *Infect. Immun.*, **16**, 75–80.

Greenberg, D. P., Bayer, A. S., Cheung, A. L., Ward, J. I. (1989). Protective efficacy of protein A-specific antibody against bacteremic infection due to *Staphylococcus aureus* in an infant rat model. *Infect. Immun.*, **57**, 1113–1118.

Kaplan, S. L., Hawkins, E. P., Kline, M. W., Patrick, G. S., Mason, E. O. Jr (1989). Invasion of the inner ear by *Haemophilus influenzae* type b in experimental meningitis. *J. Infect. Dis.*, **159**, 923–930.

Kim, K. S. (1985). Efficacy of imipenem in experimental group B streptococcal bacteremia and meningitis. *Chemotherapy*, **31**, 304–309.

Kim, K. S., Manocchio, M., Bayer, A. S. (1984). Efficacy of cefotaxime and latamoxef for *Escherichia coli* bacteremia and meningitis in newborn rats. *Chemotherapy*, **30**, 262–269.

Kim, Y. S., Sheldon, R. A., Elliott, B. R., Liu, Q., Ferriero, D. M., Tauber, M. G. (1995). Brain injury in experimental neonatal meningitis due to group B streptococci. *J. Neuropathol. Exp. Neurol.*, **54**, 531–539.

Koedel, U., Bernatowicz, A., Frei, K., Fontana, A., Pfister, H. W. (1996). Systemically (but not intrathecally) administered IL-10 attenuates pathophysiologic alterations in experimental pneumococcal meningitis. *J. Immunol.*, **157**, 5185–5191.

Lauritsen, A., Oberg, B. (1995). Adjunctive corticosteroid therapy in bacterial meningitis. *Scand. J. Infect. Dis.*, **27**, 431–434.

Martin, P. V., Laviotola, A., Ohayon, H., Riou, J. Y. (1986). Presence of a capsule in *Neisseria lactamica*, antigenically similar to the capsule of *N. meningitidis*. *Ann. Inst. Pasteur Microbiol.*, **137A**, 279–285.

Marton, A., Gulyas, M., Munoz, R., Tomasz, A. (1991). Extremely high incidence of antibiotic resistance in clinical isolates of *Streptococcus pneumoniae* in Hungary. *J. Infect. Dis.*, **163**, 542–548.

Merchant, R. E., Willard, J. E., Daum, R. S. (1981). Ultrastructural histopathology of experimental *Haemophilus influenzae* type b. Meningitis in the infant rat. I. Leukocytes of the spinal leptomeninges. *J. Submicrosc. Cytol.*, **13**, 501–514.

Merchant, R. E., Daum, R. S., Willard, J. E. (1982). Ultrastructural histopathology of experimental *Haemophilus influenzae* type b meningitis in the infant rat. II. Phagocytosis and lysis of microorganisms by leptomeningeal leukocytes. *J. Submicrosc. Cytol.*, **14**, 215–225.

Moxon, E. R., Smith, A. L., Averill, D. R., Smith, D. H. (1974). *Haemophilus influenzae* meningitis in infant rats after intranasal inoculation. *J. Infect. Dis.*, **129**, 154–162.

Moxon, E. R., Glode, M. P., Sutton, A., Robbins, J. B. (1977). The infant rat as a model of bacterial meningitis. *J. Infect. Dis.*, **136** (Suppl.), S186–S190.

Nassif, X., Mathison, J. C., Wolfson, E., Koziol, J. A., Ulevitch, R. J., So, M. (1992). Tumour necrosis factor alpha antibody protects against lethal meningococcaemia. *Mol. Microbiol*, **6**, 591–597.

Nowicki, S., Martens, M. G., Nowicki, B. J. (1995). Gonococcal infection in a nonhuman host is determined by human complement Clq. *Infect. Immun.*, **63**, 4790–4794.

Patrick, D., Betts, J., Frey, E. A., Prameya, R., Dorovini Zis, K., Finlay, B. B. (1992). *Haemophilus influenzae* lipopolysaccharide disrupts confluent monolayers of bovine brain endothelial cells via a serum-dependent cytotoxic pathway. *J. Infect. Dis.*, **165**, 865–872.

Quagliarello, V. J., Scheld, W. M. (1997). Treatment of bacterial meningitis. *N. Engl. J. Med.*, **336**, 708–716.

Rodriguez, A. F., Kaplan, S. L., Hawkins, E. P., Mason, E. O. Jr (1991). Hematogenous pneumococcal meningitis in the infant rat: description of a model. *J. Infect. Dis.*, **164**, 1207–1209.

Salit, I. E., Van Melle, E., Tomalty, L. (1984). Experimental meningococcal infection in neonatal animals: models for mucosal invasiveness. *Can. J. Microbiol.*, **30**, 1022–1029.

Saukkonen, K. (1988). Experimental meningococcal meningitis in the infant rat. *Microb. Pathog*, **4**, 203–211.

Saukkonen, K. M., Nowicki, B., Leinonen, M. (1988). Role of type 1 and S fimbriae in the pathogenesis of *Escherichia coli* O18 : K1 bacteremia and meningitis in the infant rat. *Infect. Immun.*, **56**, 892–897.

Smith, A. L., Smith, D. H., Averill, D. R. Jr, Marino, J., Moxon, E. R. (1973). Production of *Haemophilus influenzae* b meningitis in infant rats by intraperitoneal inoculation. *Infect. Immun.*, **8**, 278–290.

Smith, A. L., Rosenberg, I., Averill, D. R., Moxon, E. R., Stossel, T., Smith, D. H. (1974). Brain polymorphonuclear leukocyte quantitation by peroxidase assay. *Infect. Immun.*, **10**, 356–360.

Syrogiannopoulos, G. A., Hansen, E. J., Erwin, A. L. *et al.* (1988). *Haemophilus influenzae* type b lipooligosaccharide induces meningeal inflammation. *J. Infect. Dis.*, **157**, 237–244.

Tan, T. Q., Smith, C. W., Hawkins, E. P., Mason, E. O. Jr, Kaplan, S. L. (1995). Hematogenous bacterial meningitis in an intercellular adhesion molecule-l-deficient infant mouse model. *J. Infect. Dis.*, **171**, 342–349.

Tauber, M. G., Zwahlen, A. (1994). Animal models for meningitis. *Methods Enzymol.*, **235**, 93–106.

Unhanand, M., Mustafa, M. M., McCracken, G. H. Jr, Nelson, J. D. (1993). Gram-negative enteric bacillary meningitis: a twenty-one-year experience. *J. Pediatr.*, **122**, 15–21.

Vogel, U., Hammerschmidt, S., Frosch, M. (1996a). Sialic acids of both the capsule and the sialylated lipooligosaccharide of *Neisseria meningitidis* serogroup B are prerequisites for virulence of meningococci in the infant rat. *Med. Microbiol. Immunol. Berl.*, **185**, 81–87.

Vogel, U., Steinmetz, I., Frosch, M. (1996b). Avoiding artifacts in the infant rat model for bacterial meningitis: use of Sangur test strips for the rapid quantification of blood contamination in cerebrospinal fluid. *Med. Microbiol. Immunol. Berl.*, **185**, 27–30.

Walterspiel, J. N., Kaplan, S. L., Mason, E. O. Jr, Walterspiel, J. W. (1986). Protective effect of subinhibitory polymyxin B alone and in combination with ampicillin for overwhelming *Haemophilus influenzae* type B infection in the infant rat: evidence for *in vivo* and *in vitro* release of free endotoxin after ampicillin. *Pediatr. Res.*, **20**, 237–241.

Wenger, J. D., Hightower, A. W., Facklam, R. R., Gaventa, S., Broome, C. V. (1990). Bacterial meningitis in the United States, 1986: report of a multistate surveillance study. The Bacterial Meningitis Study Group. *J. Infect. Dis.*, **162**, 1316–1323.

Westerink, M. A., Metzger, D. W., Hutchins, W. A. *et al.* (1997). Primary human immune response to *Neisseria meningitidis* serogroup C in interleukin-12-treated severe combined immunodeficient mice engrafted with human peripheral blood lymphocytes. *J. Infect. Dis.*, **175**, 84–90.
Wiedermann, B. L., Hawkins, E. P., Johnson, G. S., Lamberth, L. B., Mason, E. O., Kaplan, S. L. (1986). Pathogenesis of labyrinthitis associated with *Haemophilus influenzae* type b meningitis in infant rats. *J. Infect. Dis.*, **153**, 27–32.
Zollinger, W. D., Moran, E. E., Devi, S. J., Frasch, C. E. (1997). Bactericidal antibody responses of juvenile rhesus monkeys immunized with group B *Neisseria meningitidis* capsular polysaccharide-protein conjugate vaccines. *Infect. Immun.*, **65**, 1053–1060.

Chapter 74

Adult Rat Model of Meningitis

G. C. Townsend and W. M. Scheld

Background of human infection

Bacterial meningitis is a disease characterized by the invasion of the subarachnoid space (SAS), normally a sterile space, by pathogenic bacteria. Although its incidence has declined in many industrialized countries, it remains a relatively common disease in many parts of the developing world. Morbidity and mortality remain high despite advances in diagnosis and antimicrobial therapy.

Animal models have been instrumental in the study of bacterial meningitis. Research conducted using these models has enabled investigators to identify the pathogenic and pathophysiologic mechanisms responsible for the neurologic damage observed in patients with bacterial meningitis. Several animal models have been employed in the study of bacterial meningitis, including the infant rat model and the rabbit model. This chapter will review the adult rat model.

Background of model

The infant rat model and the most commonly used rabbit model of bacterial meningitis were both originally described in 1974 (Dacey and Sande, 1974; Moxon *et al.*, 1974). Both have provided vital information on the pathophysiology of bacterial meningitis, and the rabbit model especially has been important in examining antimicrobial therapy in bacterial meningitis. As with any animal model of a human disease, however, there are some limitations to each model.

The adult rat model most commonly used was first described by Quagliarello *et al.* (1986). In its original description, 150 g Wistar rats were used. Subsequent models have primarily employed Wistar or Sprague–Dawley rats, generally from 175–330 g (Lorenzl *et al.*, 1993; Boje, 1995b; Wispelwey *et al.*, 1988).

Preparation of animals

No specialized housing or care nor specific pretreatment is generally required.

Details of surgery

The procedures described below are primarily those described in the model defined and refined in the Scheld laboratory (Quagliarello *et al.*, 1986; Wispelwey *et al.*, 1988; Tunkel *et al.*, 1990).

Materials required

Anesthetic, hair clippers, skin disinfectant, 1.0 ml tuberulin syringe, catheter with 25 G needle, forceps, scissors.

Anesthesia

Rats are anesthetized with intramuscular injections of ketamine (75 mg/kg) and xylazine (5 mg/kg). This is prepared by mixing three parts of ketamine with one part of xylazine. Complete anesthesia occurs in approximately 10 minutes and lasts for about 30–60 minutes. The achievement of anesthesia is determined by the loss of corneal reflexes. Other investigators have used alternative anesthetics such as intraperitoneal thiobutabarbiturate (Lorenzl *et al.*, 1993).

Surgical procedure

After induction of anesthesia, the dorsal surface of the animal's neck is shaved with electric clippers and cleaned with a skin disinfectant (e.g., ethanol). The animal is then placed dorsal side up on a stable platform with its neck flexed over a small roll of cotton. The hub of a flexible catheter with a 25 G butterfly needle is connected to a 1.0 ml tuberculin syringe, and the needle is fitted to a "micromanipulator" directly above the animal's neck. Using this device, the operator uses a rotary knob to move the needle in the vertical plane. The cisterna magna is located by gentle probing with blunt forceps. The needle is slowly lowered through the skin at the cisterna magna until there is an abrupt drop in resistance, indicating penetration through the posterior longitudinal ligament into the SAS. At this point, the plunger of the syringe is withdrawn a short distance to decrease pressure within the syringe and the catheter tubing. If no

Handbook of Animal Models of Infection
ISBN 0-12-775390-7

cerebrospinal fluid (CSF) enters the tubing, the needle is slowly withdrawn until there is the return of CSF within the tubing. CSF is allowed to flow until approximately 75 μl of CSF is evident within the tubing; it may be necessary to withdraw the plunger of the syringe to facilitate this process. When the desired amount of CSF has been withdrawn, the tubing is clamped in two places and the tubing is cut between the forceps. The bacterial inoculum is then administered into the SAS via the tubing, after which the needle is completely withdrawn and CSF remaining within the tubing is collected. If any blood is noted in the CSF at any time, the experiment is terminated.

In another model, investigators have placed the rats in a stereotaxic frame and implanted a closed cranial window (Morii *et al.*, 1986; Lorenzl *et al.*, 1993). A catheter is then placed in the cisterna magna for sampling.

Postoperative care

Animals are returned to their cages and supplied with food and water until sacrificed.

Storage and preparation of inocula

The bacteria most commonly used in these studies have been *Streptococcus pneumoniae* and *Haemophilus influenzae*; in addition to whole live bacteria, killed bacteria, bacterial components such as lipopolysaccharide and cell-wall fragments, and proinflammatory mediators such as cytokines have been used.

In experiments in our laboratory involving *H. influenzae*, type b strain Rd–b$^+$/O2 has been used. Strain Rd–b$^+$/O2 is a transformant of the Rd strain created by using donor DNA from type b strain Eagan (Hoiseth *et al.*, 1985) and has been used extensively in this model (Lesse *et al.*, 1988). The bacteria are stored at –70°C in skim milk. The bacteria are plated on to chocolate agar and incubated overnight at 37°C in 5% CO_2. Suspensions of the bacteria are made in phosphate-buffered saline at a concentration of 10^8 cfu/ml.

Infection process

The bacterial inoculum is drawn up in a 1.0 ml tuberculin syringe. When the SAS has been successfully entered and the desired amount of CSF has been withdrawn, approximately 50 μl (5×10^6 cfu) of this suspension is inoculated by injection into the catheter tubing.

Key parameters to monitor infection and response to treatment

CSF may be obtained for analysis at various times after inoculation, depending on the inoculum and the parameter under study. The successful induction of meningitis is confirmed by the appearance of leukocytes in the CSF; this process generally begins approximately 2 hours after inoculation. The concentration of leukocytes within the CSF (cells/μl) is used as the primary indicator of the degree of inflammation. Blood leukocyte concentrations have also been measured.

A number of other parameters have been used to assess the response to infection. The integrity of the BBB has been assessed by a variety of methods, including measurement of the percentage of systemically administered radiolabeled iodine present in the CSF (Lesse *et al.*, 1988), extravasation of Na^+-fluorescein (Lorenzl *et al.*, 1993), and transfer of radiolabeled sucrose (Boje, 1995a). In one model, intracranial pressure and brain edema are measured (Koedel *et al.*, 1995a; Lesse *et al.*, 1988).

In addition to these parameters, response to treatment may be monitored by determination of CSF and blood bacterial concentrations. Antibiotic concentrations may be determined by high-performance liquid chromatography (Meulemans *et al.*, 1989).

Antimicrobial therapy

The adult rat model of bacterial meningitis has been used almost exclusively to examine the pathophysiology of bacterial meningitis, with very little published data on its use in trials of antimicrobial therapy. In a model of *Pseudomonas aeruginosa* meningitis (Meulemans *et al.*, 1989), a single arterial dose of cefsulodin was administered after intraventricular inoculation of bacteria. CSF was sampled every 15 minutes through an indwelling ventricular cannula. Peak CSF concentrations and area under the curve (AUC) of cefsulodin were approximately three to four times greater in infected animals than in uninfected animals.

In a model of pneumococcal meningitis (Strake *et al.*, 1996), rats were administered one of three regimens of cefotaxime, one of two regimens of ampicillin and gentamicin, or no antibiotics, subcutaneously every 12 hours after intracisternal inoculation of *S. pneumoniae* in 5% hog gastric mucin. The mucin was used to facilitate the induction of meningitis, as in this model meningitis was not induced with *S. pneumoniae* alone; a control group of animals was administered mucin without bacteria or antibiotics. The animals were sacrificed at postinoculation day 4, then CSF was collected by intracisternal puncture. Brains were also collected, and brain homogenates were cultured. Brain tissue from animals receiving cefotaxime 25 mg/kg, no antibiotics, or mucin only was examined for histopathologic changes. Higher doses of cefotaxime (25 or 6.25 mg/kg) resulted in reduced CSF and brain concentrations of bacteria compared with infected animals that did not receive antibiotics ($p < 0.05$); there was complete bacterial killing in CSF and brain in animals that received cefotaxime 25 mg/kg. Lower doses of cefotaxime and both regimens of ampicillin and gentamicin used were ineffective in reducing CSF and brain bacterial concentrations. Leptomeningeal inflammation was observed in brain tissue from rats receiving no antibiotics, but not in rats receiving cefotaxime or mucin only.

Pitfalls (advantages/disadvantages) of the model

The adult rat model suffers from some significant drawbacks as a discriminative model of human infection. Bacteremia and subsequent invasion of the central nervous system cannot be achieved in this model by intranasal inoculation of bacteria, necessitating direct inoculation into the SAS. This obviously does not represent the hypothesized clinical scenario, in which colonization of the upper respiratory tract is followed by bacteremia and ultimately meningitis. Although the amount of CSF that may be withdrawn (approximately 50–75 μl) is greater than is possible with the infant rat model, it is still far less than may be withdrawn in the rabbit model.

Despite these shortcomings, the model has some advantages over other models. The animals are less expensive to purchase and to maintain than are rabbits, and require less space for housing. As noted above, the amount of CSF that may be withdrawn is greater than is possible with the infant rat model. The procedure used is also less complex than that used in the rabbit model.

Contributions of the model to infectious disease therapy

Indication of efficacy of antimicrobial agents

Although great care must be taken in applying the results of animal trials to potential clinical applications, the results of the study by Strake *et al.* (1986) indicate that this model may be used to preliminarily assess the relative efficacies of different antimicrobial regimens and doses.

Increase in understanding of pathology and pathogenesis, host inflammatory (cytokine, cellular), and immune response

The adult rat model has been used in numerous studies to examine the pathogenetic and pathophysiologic changes that occur in bacterial meningitis. Evidence from these studies supports a role for cytokines (such as interleukin-1; Quagliarello *et al.*, 1991), platelet-activating factor (Townsend and Scheld, 1994), and nitric oxide (Boje 1995b; Buster *et al.*, 1995; Koedel *et al.*, 1995b).

References

Boje, K. M. (1995a). Cerebrovascular permeability changes during experimental meningitis in the rat. *J. Pharmacol. Exp. Ther.*, **274**, 1199–1203.

Boje, K. M. (1995b). Inhibition of nitric oxide synthase partially attenuates alterations in the blood–cerebrospinal fluid barrier during experimental meningitis in the rat. *Eur. J. Pharmacol*, **272**, 297–300.

Buster, B. L., Weintrob, A. C., Townsend, G. C., Scheld, W. M. (1995). Potential role of nitric oxide in the pathophysiology of experimental bacterial meningitis in rats. *Infect. Immun.*, **63**, 3835–3839.

Dacey, R. G. Jr, Sande, M. A. (1974). Effect of probenecid on cerebrospinal fluid concentrations of penicillin and cephalosporin derivatives. *Antimicrob. Agents Chemother.*, **6**, 437–441.

Hoiseth, S. K., Connelly, C. J., Moxon, E. R. (1985). Genetics of spontaneous, high-frequency loss of b capsule expression in *Haemophilus influenzae*. *Infect. Immun.*, **49**, 389–395.

Koedel, U., Bernatowicz, A., Paul, R., Frei, K., Fontana, A., Pfister, H. W. (1995a). Experimental pneumococcal meningitis: cerebrovascular alterations, brain edema, and meningeal inflammation are linked to the production of nitric oxide. *Ann. Neurol.*, **37**, 313–323.

Koedel, U., Bernatowicz, A., Paul, R., Frei, K., Fontana, A., Pfister, H. W. (1995b). Experimental pneumococcal meningitis: cerebrovascular alterations, brain edema, and meningeal inflammation are linked to the production of nitric oxide. *Ann. Neurol.*, **37**, 313–323.

Lesse, A. J., Moxon, E. R., Zwahlen, A., Scheld, W. M. (1988). Role of cerebrospinal fluid pleocytosis and *Haemophilus influenzae* type b capsule on blood brain barrier permeability during experimental meningitis in the rat. *J. Clin. Invest.*, **82**, 102–109.

Lorenzl, S., Koedel, U., Dirnagl, U., Ruckdeschel, G., Pfister, H. W. (1993). Imaging of leukocyte–endothelium interaction using in vivo confocal laser scanning microscopy during the early phase of experimental pneumococcal meningitis. *J. Infect. Dis.*, **168**, 927–933.

Meulemans, A., Vicart, P., Pangon, B., Mohler, J., Bocquet, L., Vulpillat, M. (1989). Pharmacokinetics of cefsulodin in rat cerebrospinal fluid during experimental *Pseudomonas aeruginosa* meningitis. *Chemotherapy*, **35**, 237–241.

Morii, S., Ngai, C., Winn, H. R. (1986). Reactivity of rat pial arterioles and venules to adenosine and carbon dioxide: with detailed description of the closed cranial window technique in rats. *J. Cereb. Blood Flow Metab.*, **6**, 34–41.

Moxon, E. R., Smith, A. L., Averill, D. R., Smith, D. H. (1974). *Haemophilus influenzae* meningitis in infant rats after intranasal inoculation. *J. Infect. Dis.*, **129**, 154–162.

Quagliarello, V. J., Long, W. J., Scheld, W. M. (1986). Morphologic alterations of the blood–brain barrier with experimental meningitis in the rat: temporal sequence and role of encapsulation. *J. Clin. Invest.*, **77**, 1084–1095.

Quagliarello, V. J., Wispelwey, B., Long, W. J., Scheld, W. M. (1991). Recombinant human interleukin-1 induces meningitis and blood–brain injury in the rat: characterization and comparison with tumor necrosis factor. *J. Clin. Invest.*, **87**, 1360–1366.

Strake, J. G., Mitten, M. J., Ewing, P. J., Alder, J. D. (1996). Model of *Streptococcus pneumoniae* meningitis in adult rats. *Lab. Anim. Sci.*, **46**, 524–529.

Townsend, G. C., Scheld, W. M. (1994). Platelet-activating factor augments meningeal inflammation elicited by *Haemophilus influenzae* lipooligosaccharide in an animal model of meningitis. *Infect. Immun.*, **62**, 3739–3744.

Wispelwey, B., Lesse, A. J., Hansen, E. J., Scheld, W. M. (1988). *Haemophilus influenzae* lipopolysaccharide-induced blood brain barrier permeability during experimental meningitis in the rat. *J. Clin. Invest.*, **82**, 1339–1346.

Chapter 75

Rabbit Model of Bacterial Meningitis

J. Tureen and E. Tuomanen

Background of human infection

Bacterial meningitis is a medical emergency with high morbidity and mortality. For example, the mortality rate from pneumococcal infection is 20–30% despite the availability of highly bactericidal antibiotics, and over half of the survivors have severe sequelae (Schuchat *et al.*, 1997). The three common bacterial causes of meningitis in childhood are *Streptococcus pneumoniae, Haemophilus influenzae* and *Neisseria meningitidis*. Neonatal meningitis is more frequently caused by *Escherichia coli* or Group B streptococcus. With the advent of the *Haemophilus* type b vaccine, the incidence of disease has shifted more heavily to the pneumococcus. The increasing spread of antibiotic resistance, particularly among pneumococci to penicillin and cephalosporins, has complicated predicting effective therapy (Quagliarello and Scheld, 1997).

Breakdown of the blood–brain barrier is the central pathological element of meningitis. Morphologically the barrier consists of the arachnoid membrane and tight junctions between choroid plexus epithelial cells and between endothelial cells of cerebral capillaries. During meningitis, gaps are opened between endothelial cells, pinocytotic activity increases, leukocytes are recruited to the CSF, cytochemical parameters in CSF change, and autoregulation of cerebral vascular perfusion pressure is lost. All these changes are demonstrable in the rabbit meningitis model.

Background of model

The central method of the rabbit meningitis model was developed by Drs Dacey and Sande over 20 years ago (Dacey and Sande, 1974). By using an acrylic helmet affixed to the calvarium, animals can be stabilized in a stereotaxic frame. This allows precise placement of a spinal needle into the cisterna magna, where it can remain for several hours of sequential sampling. Upon inoculation of inflammatory agents into the cisterna magna, repeated CSF sampling is undertaken to determine multiple parameters of injury: cytochemical abnormalities, intracranial pressure, cerebral edema, blood–brain barrier permeability, and cerebral perfusion pressure. The ability to obtain multiple samples over time from one animal is a major advantage of the model as it allows the investigator to follow the course of disease comparing values in the same animal. This results in the use of fewer animals to document an experimental finding when compared to rodent models.

The focus of the experimental effort using the rabbit model for the past 20 years has been to establish the pathophysiology of the infection, determine optimal use of antibiotics for this sequestered site and control the vigorous host inflammatory response in the closed subarachnoid space so as to minimize neuronal damage.

Animal species

Several species of rabbit have been used, most frequently New Zealand White rabbits. Either sex is appropriate. For optimal handling, particularly when using the stereotaxic frames, animals should weigh between 1.8 and 2.2 kg.

Preparation of animals

No specialized housing or care is required.

Placement of acrylic helmet to calvarium

Surgical time: 10 minutes

A dental acrylic helmet is affixed to the skull so as to provide a stable mechanism to hold the animal in the stereotaxic frame. The helmet is placed 24–48 hours prior to the experiment. The screws that hold the helmet to the skull penetrate the calvarium but not the dura and therefore the helmet can be left in place for up to 5 days without the appearance of a reactive inflammatory response in the underlying CSF.

The animals are anesthetized with 35 mg/kg ketamine i.m., 0.6 mg/kg acepromazine i.m. and 2.5 mg/kg xylazine i.m. The sites of the helmet and future cisternal punctures

Handbook of Animal Models of Infection
ISBN 0-12-775390-7

are prepared by shaving the backs of the ears (for later insertion of an intravenous line), the superior aspect of the skull between the ears, and the nape of the neck. Disinfect the area between the ears with an iodine scrub and alcohol.

Identify the coronal and sagittal sutures by palpation (Figure 75.1). Make a 3–4 cm longitudinal incision in the skin along the sagittal suture. The incision should start behind the level of the eyes and stop in front of the ears. Spread the skin and make two incisions into the periosteum in a cross pattern along both sutures. Push the periosteum to the sides using a scalpel and a gauze pad.

Drill four holes into the skull, one in each quadrant of the cross formed by the sutures (Figure 75.1). The holes should be approximately 0.7–1 cm apart, forming a rectangle. With a scalpel, clean the holes of small bone pieces and smooth the edges. Place a screw into each hole and screw down about two full turns until it is tight. Place a turnbuckle so that one arm is between the screws and the other is posterior to the rectangle (Figure 75.2). The screw hole for the bolt should be on the left side of the rabbit.

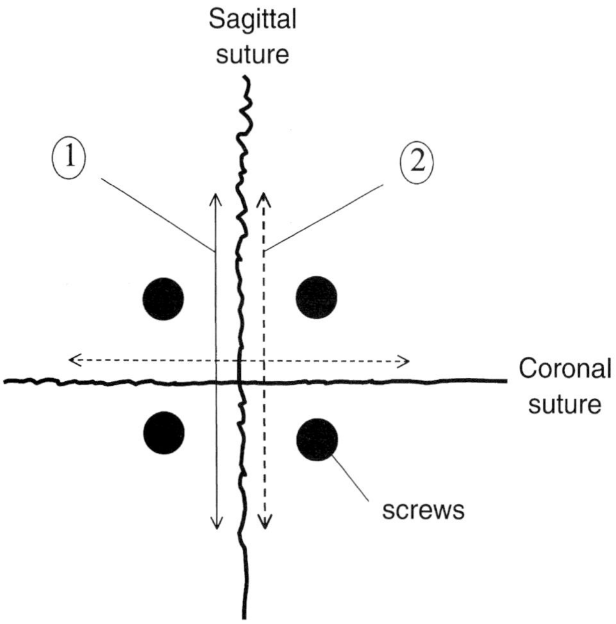

Figure 75.1 Placement of incisions and screws with regards to suture lines. The incision in the skin (1) is made parallel to the sagittal suture, extending from the level of the eyes to behind the ears. Two incisions are made in a cross pattern in the periosteum (2). The four screws are situated in each quadrant created by the intersection of the sagittal and coronal sutures.

Mix Fastray powder plus Fastray liquid and form a plum-sized aliquot on top of the turnbuckle. Work the material to cover all screws and the incision, the final shape being a rectangle on the top of the rabbit's skull. Make sure not to cover the opening for the bolt on the turnbuckle that

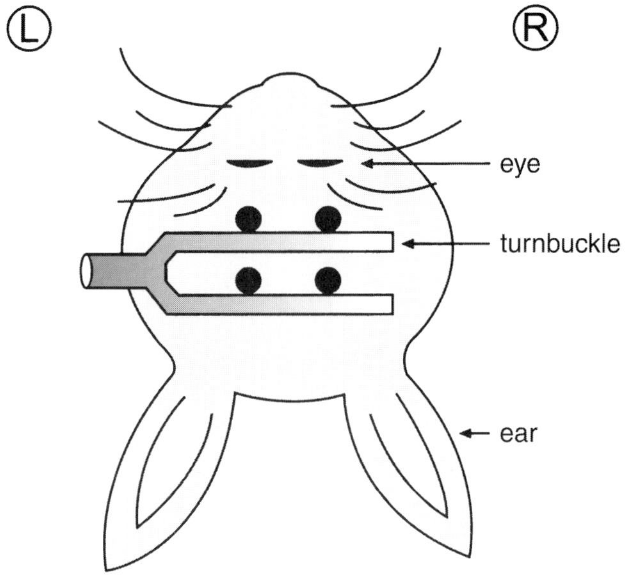

Figure 75.2 Orientation of the turnbuckle and screws in relation to the head of the rabbit. Note the turnbuckle should be placed with the anterior arm between the screws and the shaft to the left of the rabbit so as to fit the frame properly.

is going to attach to the frame later on (see Figure 75.3C). Several minutes after mixing, the cap gets hot as the acrylic sets. Cool it by allowing tap water to run over the top of the rabbit's head for at least 30 seconds or until the cap remains cool to touch. Allow the animal to awaken under supervision then return to cage with food/water *ad libitum*.

Storage and preparation of inocula

Bacteria are stored frozen in bacteriological medium plus 10% vol./vol. glycerol. To prepare an inoculum of living bacteria, 100 μl of stock culture is inoculated into liquid bacteriological medium and incubated at 37°C until the mid-logarithmic phase of growth. The culture is centrifuged at 20 000 *g* and bacteria are resuspended in pyrogen-free saline. The inoculum is adjusted to a desired concentration in 100 μl. A brisk inflammatory response will occur when the concentration in CSF exceeds 10^5 bacteria/ml. The total volume of CSF in a rabbit is about 5 ml. Thus, as an initial recommendation, inoculation of 10^6 bacteria in 100 μl will initiate inflammation within several hours.

Bacterial components have also been used extensively to induce a meningeal inflammatory response. The most common is lipopolysaccharide with a dose of 50 ng invoking a reproducible inflammatory response. Heat-killed pneumococci, when not encapsulated (i.e., strain R6), have also been used. The inoculum is heated to 50°C for 20 minutes, cooled and adjusted to 10^6 bacteria per 100 μl.

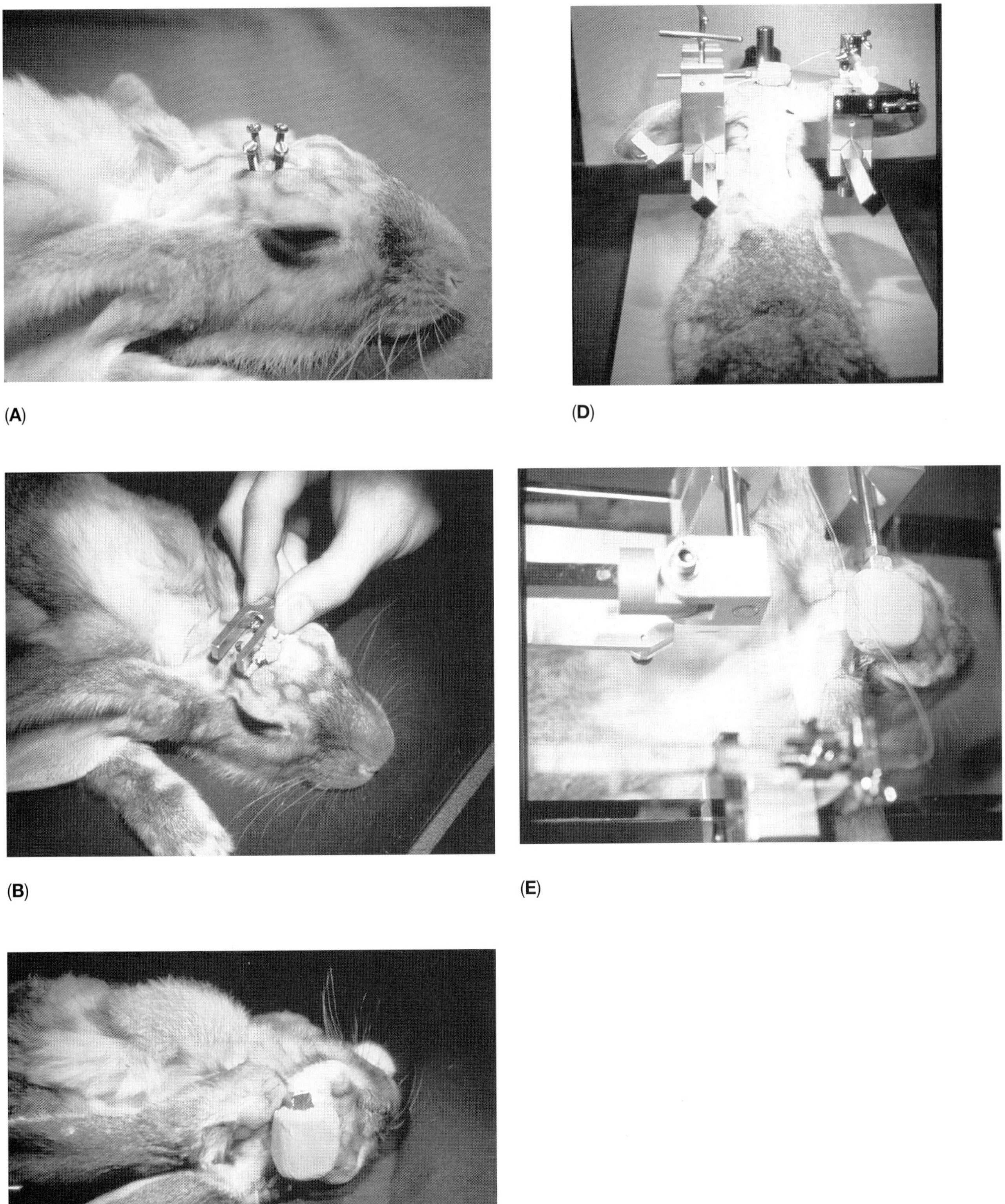

(A) (D) (B) (E) (C)

Figure 75.3 Preparation and placement of animal in sterotaxic frame. **A.** Placement of four screws as per Figure 75.1. **B.** Placement of turnbuckle as per Figure 75.2. **C.** Securing the turnbuckle to the skull using dental acrylic. **D.** Placement of rabbit in frame. **E.** Insertion of spinal needle into the cisterna magna.

Details of the surgery and infection process

Anesthetize with nembutal 0.5 ml/kg i.v. and 25% urethane 3.5 ml/kg s.c. Place a catheter in an ear vein and flush with heparinized saline. The animal is ready for placement in the stereotaxic frame approximately 30 minutes after the urethane. For manipulation, the animal may require additional anesthesia, in which case nembutal (0.1–0.4 ml i.v.; 15 mg/kg) can be administered by the ear vein catheter.

The specifications for making the stereotaxic frame can be obtained from the Instrument Shop, Rockefeller University, 1230 York Ave, New York, NY 10021, USA.

Place the animal in the frame, stabilizing the head via the turnbuckle in the helmet (Figure 75.3D, E). Insert a 25 G ½″ spinal needle into the needle holder of the frame. Stretch the skin of the nape of the neck down with one finger and palpate the spinous process of the second cervical vertebra (C2). Advance the needle at an angle approximately 0.5 cm above the spinous process of C2 so that it passes over C1. Advance the needle to a depth of 1.5 cm. Pull on a tuberculin syringe attached to the needle hub to determine when CSF flow is initiated. A reasonable initial sample size of CSF is 0.3 ml. Instill bacteria intracisternally in a volume of 100 μl.

Placement of intravascular catheters

Catheters can be used for repeated blood sampling, physiologic measurements, or measurement of cerebral blood flow.

Supplies: 2% lidocaine, povidone-iodine, scalpel, iris scissors, hemostats, curved forceps, syringe with heparinized saline, attached to PE-90 or PE-100 tubing, suture material.

Femoral artery catheter

Animals should be anesthetized with urethane or another long-acting drug. With animal in supine position, shave groin bilaterally. At the midpoint of the junction of thigh and body, locate the femoral artery by palpation. Sterilize skin with povidone-iodine and infiltrate with 2% lidocaine. Make a 1–1.5 cm transverse incision over the artery and dissect with scissors to expose. The femoral artery, vein and nerve will be visible together. Free the artery and insert two 30 cm lengths of suture beneath. Place a single loose knot in the proximal suture and tie off the artery distally with the second tie. Insert curved forceps under the artery to elevate it and occlude the blood supply with the proximal tie. (This can be done either with the help of an assistant or by clamping the free ends of the suture with a hemostat and positioning to apply gentle tension.) Make an arteriotomy with the iris scissors approximately one-third the diameter of the vessel. Insert the PE tubing, remove the forceps from beneath the artery and advance gently. When the catheter enters the vessel, remove tension on the proximal tie and advance 10 cm. Tie the proximal suture securely over catheter and vessel and inspect for bleeding. Blood should withdraw freely.

Carotid artery catheter

This is performed in a similar manner to the above procedure. Shave the neck in the midline and right side ventrally. Working from above the head, infiltrate with lidocaine and make a 2 cm midline incision above the thyroid cartilage. Dissect on the right side until the carotid artery is isolated. Using two sutures as above, tie off distally and occlude proximally. Perform an arteriotomy (taking care to insure that the vessel is occluded) and advance the catheter 6–8 cm until a 'pop' is felt. This occurs when the catheter crosses the aortic valve. Tie securely in place. The right carotid artery can be sacrificed because the rabbit has an intact circle of Willis.

Key parameters to monitor infection

CSF measurements

- Remove 0.3 ml of CSF at each sampling time, e.g., 2, 4, and 6 hours.
- Leukocyte density.
- Protein, glucose, lactate concentration.
- Bacterial density.
- Intracranial pressure.

Brain edema

- Sacrifice with pentobarbital (150 mg/kg i.v.).
- Rapidly remove top of cranium with hemostat and heavy scissors.
- Remove brain, cut into sections and place into tared Pyrex dish, and reweigh (W_1).
- Vacuum dry (–50 kPa) at 110–120°C for 7 days or until stable weight is reached (W_2).
- Brain water content (g water/100 g dry weight) = $[(W_1 - W_2)/(W_2 - \text{tare})] \times 100$.

Blood–brain barrier permeability

- Blood–brain barrier injury is measured by increased accumulation of a blood-borne marker in the brain.
- Animals receive [^{3}H]penicillin or [^{125}I]albumin 30 minutes before the end of the experiment.
- Collect 2 ml blood from ear catheter for cpm determination in blood.
- Sacrifice with nembutal i.v. 2.5 ml/kg.
- Remove brain and homogenize a preweighed sample for counting of cpm.

Cerebral perfusion pressure

Cerebral perfusion pressure (CPP) is the calculated difference between mean blood pressure (MBP) and intracranial pressure (ICP).

- MBP is calculated from blood pressure measured with a pressure transducer from the femoral artery. MBP = [diastolic blood pressure + 1/3 (systolic blood pressure–diastolic blood pressure)].
- ICP is measured by placing a spinal needle in the cisterna magna. When free flow of CSF is established, connect directly to a pressure transducer. A valid measurement will show variations of 1–2 mmHg with respiration.
- CPP = MBP – ICP.

Cerebral blood flow

Performed by method of Heymann *et al.* (1977). May also be measured by [C^{14}]-antipyrine.

- Prepare animal with left ventricle and femoral artery catheter.
- Connect tared, heparinized syringe to Harvard pump®; set withdrawal rate at ≈2.5 ml/min; connect to femoral catheter.
- Begin timed withdrawal of 1.5 minutes; after 0.25 minutes, inject 10^5–10^6 labeled, 15 μm microspheres into LV catheter over 0.25 minutes. When withdrawal complete, disconnect syringe (reference sample) and reweigh.
- This may be repeated multiple times in the same animal with different fluorescent colors or different isotopes.
- Sacrifice, remove brain and cut into sections to be counted.
- Blood flow is calculated as a ratio between counts in tissue and counts in the reference sample.

Microspheres are available from Molecular Probes, Eugene, OR (fluorescent), or Du Pont/NEN Research Products, Boston, MA (radioactive).

Brain histopathology

- Sacrifice with nembutal i.v. 2.5 ml/kg.
- Remove brain for histopathologic analysis.

Antimicrobial and adjunctive therapy and key parameters to monitor response to treatment

Response to treatment is monitored over the course of hours to days. For replacement on the frame, animals are re-anesthetized each day and sampled as needed. Changes in parameters of infection as monitored above are sufficient to determine response to therapy.

The model is suitable for assessment of the pharmcokinetics of a drug in CSF as well as potency of antimicrobial activity. Determination of the concentration of drug and the decrease in viable bacteria over time describe this potency. This is a straightforward and common use of the model.

In addition to monitoring the bacteriologic response to treatment, the model has offered the opportunity to correlate bacterial killing with the state of injury in the subarachnoid space and brain. The most important application of this possibility is the testing of adjunctive therapies directed at reducing host response to bacterial products, some of which are released transiently in large amounts as bacteria lyse and die. By comparing the improvement in CSF cytochemistry and other measures of injury, the model has tested the potential advantage of using many anti-inflammatory modalities, together with antibiotics including steroids, non-steroidal anti-inflammatory agents, cytokine inhibitors, inhibitors of leukocyte transmigration, inhibitors of nitric oxide generation, etc.

Advantage and disadvantages of the model

These experiments cannot be conducted using non-animal alternatives. *In-vitro* cultivation of the blood–brain barrier has been achieved in several laboratories but simultaneous measurement of migration of leukocytes and leakage of protein across such a monolayer has not been demonstrated. Thus, *in-vitro* approximation of the blood–brain barrier can provide a measure of the potential of a bacterial component to increase vascular permeability in the brain but it can not approximate the situation in the presence of leukocytes, which are important mediators of brain injury.

There exist four models of meningitis in animals: rat, rabbit, piglet and monkey. The rat model has a phase of bacteremia and is therefore more useful in studies of events leading to seeding of the CSF space that can not be studied in rabbits or monkeys. On the other hand, rabbits can provide multiple CSF samples per animal, eliminating the need for many animals to construct a time course of inflammation as is the case for rats. One rabbit can give up to eight sequential samples without a change in CSF pressure, thus sparing the need to study many animals. Monkeys, piglets and rabbits are predictive of the course of disease in man; choice of the animal lower on the phylogenetic order, the rabbit, is generally recommended. General use of the model has built an extensive data base for this infection, with strong indicators that what happens in the rabbit is predictive of the outcome in children. The rabbit and the human share antigenically identical CD18 adhesion molecules on leukocytes and the blood–brain barrier, a useful cross-reactivity not shared by any other laboratory animal.

Contributions of the model to infectious diseases therapy

Improvements in therapeutic regimens

Antibiotics

The general principles guiding antibiotic therapy of meningitis have been established in the animal model. Initially, detailed study of the pharmcokinetics and pharmacodynamics of antibiotics in the CSF space were undertaken (Dacey and Sande, 1974; Tauber *et al.*, 1984). The need for bactericidal versus bacteriostatic antibiotics in this environment was clearly documented using the model (Scheld and Sande, 1983). As new antibiotics were developed, they were tested in the model for potential use in meningitis (i.e., imipenem, meropenem, aztreonam). Similarly, when clinical failures were encountered, for example that of cefuroxime, an explanation was sought by investigation of treatment outcome in the model. Antibiotic regimens were also developed for particularly difficult infections, such as *Proteus* or *Listeria* meningitis and ventriculitis in neonatal meningitis. More recently, the model has provided a mechanism to optimize therapy for emerging antibiotic-resistant bacteria, particularly the penicillin-resistant pneumococcus (Friedland and Istre, 1992; Friedland *et al.*, 1993b).

Fluid management

Since the 1970s, it has been recognized that the syndrome of inappropriate secretion of antidiuretic hormone (SIADH) occurs in some children with meningitis. This led to an almost universal practice of fluid restriction in children with meningitis, sometimes to an excessive degree. Work with the rabbit model of meningitis demonstrated first that cerebral vascular reactivity (autoregulation) was abnormal in bacterial meningitis (Tureen *et al.*, 1990) and could result in cerebral ischemia if blood pressure was not maintained. This was further elucidated in subsequent studies, which compared fluid restriction to maintenance hydration in the rabbit and which demonstrated significantly reduced brain blood flow and increased cerebrospinal fluid lactate (a marker for cerebral anaerobic metabolism) in fluid-restricted rabbits (Tureen *et al.*, 1992). These findings were reinforced by a clinical study that compared these two types of fluid regimen in children with meningitis and demonstrated improved outcome in the children that were not fluid-restricted (Singhi *et al.*, 1995).

Novel adjunctive antiinflammatory therapy

Work using the rabbit meningitis model demonstrated that the inflammatory burst generated by antibiotic-induced bacterial lysis was associated with tissue injury (Tuomanen *et al.*, 1985, 1987; Tauber *et al.*, 1987; Friedland *et al.*, 1993a). This finding was documented in pneumococcal, *Haemophilus* and *E. coli* meningitis in the animal. This period of enhanced inflammation upon initiation of antibiotic therapy was then documented in children (Arditi *et al.*, 1989). These findings suggested that a rational approach to improved outcome might involve attenuation of inflammation at the time of onset of antibiotic therapy. This was tested in the model using an extensive array of anti-inflammatory modalities. Administration of steroids, non-steroidal agents, antioxidants, cytokine antagonists, and inhibitors of leukocyte trafficking dramatically enhanced outcome in the experimental meningitis model (Tauber *et al.*, 1985; Tuomanen *et al.*, 1987, 1989; Tureen *et al.*, 1991; Pfister *et al.*, 1992; Granert *et al.*, 1994; Paris *et al.*, 1994). Several studies were then conducted introducing this concept into clinical care of children by the use of steroids together with antibiotics (Lebel *et al.*, 1988; Girgis *et al.*, 1989; Odio *et al.*, 1991; Syrogiannopoulos *et al.*, 1994). Days of fever and the incidence of hearing loss decreased for *Haemophilus* meningitis, indicating that the therapy was of tangible benefit, but the results were not uniformly dramatic. For instance, the mortality rate in most clinical studies using steroids has decreased but the degree of impact varies from mortality rates below 5% up to as high as 20%. A general recommendation for the use of steroids in clinical meningitis has been instituted by the American Academy of Pediatrics.

Increased understanding of the pathogenesis of inflammation and CNS damage

The meningitis model has revolutionized understanding of the pathogenesis of meningitis. Virtually every step in the infectious process after the bacteria have entered the subarachnoid space has been defined in this model (Pfister *et al.*, 1994; Tunkel and Scheld, 1995). The rates of multiplication of bacteria in CSF versus blood have been defined, as well as the rates of killing and clearance in the presence of antibiotics. The influence of host physiology on disease, such as fever, neutropenia, complement deficiency, etc., have been documented. The bacterial components that induce inflammation have been identified as lipopolysaccharide and peptidoglycan/teichoic acid. The effects of these agents on generation of cytokines, leukocytosis, brain edema and blood–brain barrier abnormalities have been quantitated and dissected mechanistically. Inflammatory cytokines (interleukin-1, interleukin-6, tumor necrosis factor) have been shown to have a role in generating the inflammation of meningitis, as evidenced by a recapitulation of disease when the agents are instilled directly into the CSF (Ramilo *et al.*, 1990; Saukkonen *et al.*, 1990) and by the amelioration of inflammation when inhibitors are administered.

The final common mediators of neuronal toxicity have been more difficult to define, but the model has implicated both direct effects of bacterial products (e.g., pneumolysin or teichoic acids, Tauber *et al.*, 1992; Friedland *et al.*, 1995) and side-effects of host defense systems, such as excitatory amino acids (Guerra-Romero *et al.*, 1992, 1993), radicals of molecular oxygen (Lieb *et al.*, 1996) and nitric oxide (Koedel *et al.*, 1995; Tureen, 1995). The development of sequelae such as hearing loss has been studied by sequentially measuring auditory evoked responses during meningitis, with the conclusion that CSF inflammation directly relates to events in the cochlea (Bhatt *et al.*, 1991, 1993, 1994).

The pathophysiology of cerebrovascular abnormalities

The study of cerebral hemodynamics in humans is technically difficult and potentially dangerous in patients with a life-threatening intracranial disorder. Therefore, animal models have been an extremely useful means of identifying intracranial circulatory abnormalities, including loss of cerebrovascular autoregulation (Tureen *et al.*, 1990), cerebral ischemia (Tureen *et al.*, 1990), and cerebral infarction (Lorenzl *et al.*, 1993). These abnormalities arise from multiple mechanisms, including physiologic changes (i.e. intracranial hypertension), release of cytokines (TNFα, IL-1β, PAF), the action of reactive species of oxygen and nitrogen (oxygen radical, nitric oxide, peroxynitrite), and possibly upregulation of leukocyte adhesion molecules on vascular endothelium. All these abnormalities identified in animal models have been described in humans with meningitis, and elucidation of the underlying mechanisms may permit adjunctive forms of therapy to ameliorate these adverse consequences.

References

Arditi, M., Ables, L., Yogev, R. (1989). Cerebrospinal fluid endotoxin levels in children with *H. influenzae* meningitis before and after administration of intravenous ceftriaxone. *J. Infect. Dis.*, **160**, 1005–1011.

Bhatt, S., Halpin, C., Hsu, W. *et al.* (1991). Hearing loss and pneumococcal meningitis: an animal model. *Laryngoscope*, **101**, 1285–1292.

Bhatt, S., Lauretano, A., Cabellos, C. *et al.* (1993). The progression of hearing loss in experimental pneumococcal meningitis: correlation with cerebrospinal fluid cytochemistry. *J. Infect. Dis.*, **167**, 675–683.

Bhatt, S., Cabellos, C., Nadol, J. *et al.* (1994). The impact of dexamethasone on hearing loss in experimental pneumococcal meningitis. *Pediatr. Infect. Dis. J.*, **14**, 93–96.

Dacey, R., Sande, M. A. (1974). Effect of probenecid on cerebrospinal fluid concentrations of penicillin and cephalosporin derivatives. *Antimicrob. Agents Chemother.*, **6**, 437–441.

Friedland, I., Istre, G. (1992). Management of penicillin-resistant pneumococcal infections. *Pediatr. Infect. Dis. J.*, **11**, 433–435.

Friedland, I., Jafari, H., Ehrett, S., McCracken, G. (1993a). Comparison of endotoxin release by different antimicrobial agents and the effect on inflammation in experimental *Escherichia coli* meningitis. *J. Infect. Dis.*, **168**, 1342–1348.

Friedland, I. R., Shelton, S., Paris, M. *et al.* (1993b). Dilemmas in diagnosis and management of cephalosporin-resistant *Streptococcus pneumoniae* meningitis. *Pediatr. Infect. Dis. J.*, **12**, 196–200.

Friedland, I., Paris, M., Hickey, S. *et al.* (1995). Limited role of pneumolysin in the pathogenesis of pneumococcal meningitis. *J. Infect. Dis.*, **172**, 805–809.

Girgis, N., Farid, Z., Mikhail, I., Farrag, I., Sultan, Y., Kilpatrick, M. (1989). Dexamethasone treatment for bacterial meningitis in children and adults. *Pediatr. Infect. Dis.*, **8**, 848–851.

Granert, C., Raud, J., Xie, X., Lindquist, L., Lindbom, L. (1994). Inhibition of leukocyte rolling with polysaccharide fucoidin prevents pleocytosis in experimental meningitis in the rabbit. *J. Clin. Invest.*, **93**, 929–936.

Guerra-Romero, L., Tauber, M. G., Fournier, M. A., Tureen, J. H. (1992). Lactate and glucose concentrations in brain interstitial fluid, cerebrospinal fluid, and serum during experimental pneumococcal meningitis. *J. Infect. Dis.*, **166**, 546–550.

Guerra-Romero, L., Tureen, J., Fournier, M., Makrides, V., Tauber, M. (1993). Amino acids in cerebrospinal and brain interstitial fluid in experimental pneumococcal meningitis. *Pediatr. Res.*, **33**, 510–513.

Heymann, M. A., Payne, B. D., Hoffman, J. I. E., Rudolph, A. M. (1977). Blood flow measurements with radionuclide-labeled particles. *Prog. Cardiovasc. Dis.*, **20**, 55–70.

Koedel, U., Bernatowicz, A., Paul, R., Frei, K., Fontana, A., Pfister, H. W. (1995). Experimental pneumococcal meningitis: cerebrovascular alterations, brain edema, and meningeal inflammation are linked to the production of nitric oxide. *Ann. Neurol.*, **37**, 313–323.

Lebel, M. H., Freij, B. J., Syrogiannopoulos, G. A., McCracken, G. J. (1988). Dexamethasone therapy for bacterial meningitis. *N. Engl. J. Med.*, **15**, 964–971.

Leib, S. L., Kim, Y. S., Chow, L. L., Sheldon, R. A., Tauber, M. G. (1996). Reactive oxygen intermediates contribute to necrotic and apoptotic neuronal injury in an infant rat model of bacterial meningitis due to group B streptococci. *J. Clin. Invest.*, **98**, 2632–2639.

Lorenzl, S., Koedel, U., Dirnagl, U., Ruckdeschel, G., Pfister, H. W. (1993). Imaging of leukocyte–endothelium interaction using *in vivo* confocal laser scanning microscopy during the early phase of experimental pneumococcal meningitis. *J. Infect. Dis.*, **168**, 927–933.

Odio, C. M., Faingezicht, I., Paris, M. *et al.* (1991). The beneficial effects of early dexamethasone administration in infants and children with bacterial meningitis [see comments]. *N. Engl. J. Med.*, **324**, 1525–1531.

Paris, M., Friedland, I., Ehrett, S. *et al.* (1994). Effect of Interleukin-1 receptor antagonist and soluble tumor necrosis factor receptor is animal models of infection. *J. Infect. Dis.*, **171**, 161–169.

Pfister, H. W., Koedel, U., Lorenzl, S., Tomasz, A. (1992). Antioxidants attenuate microvascular changes in the early phase of experimental pneumococcal meningitis in rats. *Stroke*, **23**, 1798–1804.

Pfister, H., Fontana, A., Tauber, M., Tomasz, A., Scheld, W. (1994). Mechanisms of brain injury in bacterial meningitis: workshop summary. *Clin. Infect. Dis.*, **19**, 463–479.

Quagliarello, V., Scheld, W. (1997). Treatment of bacterial meningitis. *N. Engl. J. Med.*, **336**, 708–716.

Ramilo, O., Saezllorens, X., McCracken, G. J. (1990). Tumor necrosis factor-alpha cachectin and Interleukin-1-β initiate meningeal inflammation. *J. Exp. Med.*, **172**, 497–507.

Saukkonen, K., Cioffe, C., Wolpe, S., Sherry, B., Cerami, A., Tuomanen, E. (1990). The role of cytokines in the generation of inflammation and tissue damage in experimental Gram-positive meningitis. *J. Exp. Med.*, **171**, 439–448.

Scheld, W., Sande, M. (1983). Bactericidal versus bacteriostatic antibiotic therapy of experimental pneumococcal meningitis in rabbits. *J. Clin. Invest.*, **71**, 411–419.

Schuchat, A., Robinson, K., Wenger, J. (1997). Bacterial meningitis in the United States. *N. Engl. J. Med.*, **337**, 970–976.

Singhi, S. C., Singhi, P. D., Srinivas, B. *et al.* (1995). Fluid restriction does not improve the outcome of acute meningitis. *Pediatr. Infect. Dis. J.*, **14**, 495–503.

Syrogiannopoulos, G., Lourida, A., Theodoridou, M. *et al.* (1994). Dexamethasone therapy for bacterial meningitis in children: 2 versus 4 day regimen. *J. Infect. Dis.*, **169**, 853–858.

Tauber, M., Doroshow, C., Hackbarth, C., Rusnak, M., Ta, D., Sande, M. (1984). Antibacterial activity of beta-lactam antibiotics in experimental meningitis due to *Streptococcus pneumoniae*. *J. Infect. Dis.*, **149**, 568–574.

Tauber, M. G., Khayam-Bashi, H., Sande, M. A. (1985). Effects of ampicillin and corticosteroids on brain water content, cerebrospinal fluid pressure, and cerebrospinal fluid lactate levels in experimental pneumococcal meningitis. *J. Infect. Dis.*, **151**, 528–534.

Tauber, M. G., Shibl, A. M., Hackbarth, C. J., Larrick, J. W., Sande, M. A. (1987). Antibiotic therapy, endotoxin concentration in cerebrospinal fluid, and brain edema in experimental *Escherichia coli* meningitis in rabbits. *J. Infect. Dis.*, **156**, 456–462.

Tauber, M. G., Sachdeva, M., Kennedy, S. L., Loetscher, H., Lesslauer, W. (1992). Toxicity in neuronal cells caused by cerebrospinal fluid from pneumococcal and gram-negative meningitis. *J. Infect. Dis.*, **166**, 1045–1050.

Tunkel, A., Scheld, W. (1995). Acute bacterial meningitis. *Lancet*, **346**, 1675–1680.

Tuomanen, E., Liu, H., Hengstler, B., Zak, O., Tomasz, A. (1985). The induction of meningeal inflammation by components of the pneumococcal cell wall. *J. Infect. Dis.*, **151**, 859–868.

Tuomanen, E., Hengstler, B., Rich, R., Bray, M., Zak, O., Tomasz, A. (1987). Nonsteroidal anti-inflammatory agents in the therapy of experimental pneumococcal meningitis. *J. Infect. Dis.*, **155**, 985–990.

Tuomanen, E., Saukkonen, K., Sande, S., Cioffe, C., Wright, S. D. (1989). Reduction of inflammation, tissue damage, and mortality in bacterial meningitis in rabbits treated with monoclonal antibodies against adhesion-promoting receptors of leukocytes. *J. Exp. Med.*, **170**, 959–969.

Tureen, J. (1995) Effect of recombinant human tumor necrosis factor-alpha on cerebral oxygen uptake, cerebrospinal fluid lactate, and cerebral blood flow in the rabbit: role of nitric oxide. *J. Clin. Invest.*, **95**, 1086–1091.

Tureen, J., Dworkin, R., Kennedy, S., Sachdeva, M., Sande, M. (1990). Loss of cerebrovascular autoregulation in experimental meningitis in rabbits. *J. Clin. Invest.*, **85**, 577–581.

Tureen, J. H., Tauber, M. G., Sande, M. A. (1991). Effect of indomethacin on the pathophysiology of experimental meningitis in rabbits. *J. Infect. Dis.*, **163**, 647–649.

Tureen, J., Tauber, M., Sande, M. (1992). Effect of hydration status on cerebral blood flow and cerebrospinal fluid lactic acidosis in rabbits with experimental meningitis. *J. Clin. Invest.*, **89**, 947–953.

Chapter 76

Escherichia coli Brain Abscess Method in Rat

J. M. Nazzaro and E. A. Neuwelt

Background of model

The most common method of establishing brain abscesses in the laboratory is the direct inoculation of brain with bacteria using stereotactic techniques. Other techniques, such as the intracarotid artery injection of bacterial emboli to brain in dogs (Molinari, 1972; Molinari *et al.*, 1973) or primates (Long *et al.*, 1978, 1979), and brain abscess formation following systemic yeast infection in mice (Ashman and Papadimitriou, 1994), have also been described. Winn *et al.* (1978, 1979) reported characterization of an *Escherichia coli (E. coli)* brain abscess model in rat achieved by inoculation of bacteria suspended in saline into brain. Our laboratory used this model, although in early work both moderate acute morbidity and mortality among inoculated animals were encountered (Neuwelt *et al.*, 1984a,b). Approximately 25% of the animals demonstrated acute weight loss and systemic sepsis. Histologic examination of these animals demonstrated meningitis, ventriculitis and, in several cases, evidence of sepsis, which included septic emboli to brain and lung. In addition, 25% of the remaining inoculated animals failed to demonstrate abscess lesions (Neuwelt *et al.*, 1984a,b). Schroeder *et al.* (1987) also reported similar acute mortality rates attributable to complications such as meningitis and sepsis in rats inoculated with *Staphylococcus aureus (S. aureus)*.

In subsequent work, the *E. coli* brain abscess model in rat has been further refined (Nazzaro *et al.*, 1990, 1991). Several parameters used in this model, such as strain of bacterium, inoculation preparation and placement techniques, and lesion location, differ from previous studies (Winn *et al.*, 1978, 1979; Mendes *et al.*, 1980; Costello *et al.*, 1983; Neuwelt *et al.*, 1984a,b). Using the present system, histologic, quantitative culture, and permeability drug studies suggest consistent formation of viable abscesses without evidence of meningitis or systemic sepsis (Nazzaro *et al.*, 1991, 1992).

Laboratory animals

Laboratory animals used in brain abscess research include cat, dog, monkey, mouse, rabbit, and rat (Table 76.1). Concerning the rat, the most common strain used in this work is the Sprague–Dawley, although others, including the Wistar and the Lewis, have been investigated (Table 76.2). Either male or female animals are used and they generally weigh between 250 and 400 g (Table 76.2). In our laboratory, female Sprague–Dawley rats weighing 250–800 g have been used in all work. Investigators have not addressed whether rat strain or sex affects the characteristics of the experimental brain lesions produced. However, it has been demonstrated, for example, that patterns of resistance to *Candida albicans* infections differs among inbred mouse strains (Ashman *et al.*, 1993). Also, important differences in intracerebral abscess formation, characteristics, and response to pharmacologic agents between species have been suggested (Long and Meacham, 1968; Quartey *et al.*, 1976; Britt *et al.*, 1981; Neuwelt *et al.*, 1984a,b).

Bacteria used to create brain abscess

Several bacterial strains as well as fungal and yeast species have been used for laboratory intracerebral abscess study (Table 76.1). Strains of *Staphylococcus, Streptococcus*, and *E. coli* have been most frequently studied. Less commonly, other infectious agents such as *Pseudomonas aeruginosa, Bacteroides fragilis*, and *Candida albicans* have been reported (Table 76.1). In the rat, *S. aureus* and *E. coli* have been most commonly investigated (Table 76.1). In our laboratory, *E. coli* has been used exclusively. It is important to note that significant differences regarding the ability to establish an intracerebral abscess as well as the specific characteristics of the lesions produced often exist within a bacterial strain and thus it is imperative that the type and possibly subtype of infectious agent employed be specified. In pilot work in our laboratory, different strains of *E. coli* isolated from human blood cultures were investigated for brain abscess formation before consistent lesions were established (Nazzaro, unpublished data). Lo and colleagues (1991, 1993, 1994) selected a strain of *S. aureus* characterized by increased vascular permeability and inflammatory response in comparison to other strains of *S. aureus*. Thus, the characteristics of an individual brain abscess model pertain only to the specific bacterial isolate used.

Handbook of Animal Models of Infection
ISBN 0-12-775390-7

Table 76.1 Brain abscess animal studies

Study	*Species*	*Infection*
Ashman *et al.*, 1993; Ashman and Papadimitriou, 1994	Mouse	*Candida albicans*
Bohl *et al.*, 1981	Cat	*Staphylococcus aureus*
Bothe *et al.*, 1984; Bothe and Paschen, 1986	Cat	*Staphylococcus aureus*
Britt *et al.*, 1981	Dog	α-streptococcus
Britt *et al.*, 1984	Dog	*Bacteroides fragilis*
Costello *et al.*, 1983	Rat	*Streptococcus pyogenes* *Staphylococcus aureus* *Escherichia coli* *Pseudomonas aeruginosa* *Candida albicans* *Streptococcus intermedius* *Streptococcus-MG-intermedius* *Bacteroides fragilis* *Peptostreptococcus anaerobius* *Peptococcus asacchasolyticus* *Peptococcus prevotil*
Enzmann *et al.*, 1979, 1982a,b	Dog	α-streptococcus
Enzmann *et al.*, 1986	Dog	*Staphylococcus aureus*
Essick, 1919	Cat	Not known
Falconer *et al.*, 1943	Rabbit	*Staphylococcus aureus* *Staphylococcus albus* *Streptococcus pyogenes* Pneumococcus *Proteus vulgaris* *Bacteroides friedlanderi* Anaerobic streptococci Fusiform bacilli
Flaris and Hickey, 1992	Rat	*Staphylococcus aureus*
Groff, 1934	Cat	*Staphylococcus aureus* *Streptococcus haemolyticus* *Staphylococcus* and *Streptococcus* Colon bacilli Pneumococcus
Hassler and Forsgren, 1964	Rabbit	*Staphylococcus aureus* *Proteus vulgaris* *Neisseria flavescens* *Streptococcus zymogenes* *Streptococcus faecalis*
Housmann *et al.*, 1983	Cat	*Staphylococcus aureus*
Kretzschmar *et al.*, 1981	Cat	*Staphylococcus aureus*
Kurzydlowski *et al.*, 1987	Dog	*Streptococcus-MG-intermedius*
Lo *et al.*, 1991, 1993, 1994	Rat	*Staphylococcus aureus*
Long and Meacham, 1968	Dog	*Staphylococcus aureus*
Mendes *et al.*, 1980	Rat	*Staphylococcus aureus* *Escherichia coli*
Molinari, 1972; Molinari *et al.*, 1973	Dog	*Streptococcus faecalis* *Streptococcus viridans* *Staphylococcus aureus* *Escherichia coli*
Nakagawa *et al.*, 1990	Rat	*Staphylococcus aureus*
Nazzaro *et al.*, 1991, 1992	Rat	*Escherichia coli*
Neuwelt *et al.*, 1984a,b	Rat	*Escherichia coli*
Obana *et al.*, 1986	Dog	α-streptococcus
Quartey *et al.*, 1976	Rabbit	*Streptococcus pyogenes* *Staphylococcus aureus*
Runge *et al.*, 1985	Dog	α-streptococcus
Schroeder *et al.*, 1987	Rat	*Staphylococcus aureus*
Thomas, 1942	Cat	*Staphylococcus aureus* *Streptococcus haemolyticus* Pneumococcus
Wallenfang *et al.*, 1979	Rabbit	*Staphylococcus aureus*
Wallenfang *et al.*, 1980	Cat	*Staphylococcus aureus*
Winn *et al.*, 1978, 1979	Rat	*Staphylococcus aureus* *Escherichia coli*
Wood *et al.*, 1978, 1979	Monkey	*Staphylococcus aureus*

Table 76.2 Brain abscess studies in rat; strain/sex/weight of experimental animals (SD, Sprague–Dawley; W, Wistar; L, Lewis; F, female; M, male; nr, not reported)

Study	*Strain*	*Sex*	*Weight (g)*
Costello *et al.*, 1983	SD	F	250–300
Lo *et al.*, 1991, 1993, 1994	SD	M	250–390
Mendes *et al.*, 1980	SD	F	300
Nazzaro *et al.*, 1991, 1992	SD	F	250–300
Neuwelt *et al.*, 1984a,b	SD	F	250–300
Schroeder *et al.*, 1987	SD	nr	250–400
Winn *et al.*, 1978, 1979	SD	F	300
Nakagawa *et al.*, 1990	W	M/F	300–400
Flaris and Hickey, 1992	W, L	nr	nr

Bacteria preparation and characterization

E. coli (specimen no. 6702) used in our experiments was isolated from a human blood culture in the bacteriology laboratory of the Providence Medical Center, Portland, OR, USA. Bacteria are preserved in aliquots of a 10% (vol./vol.) glycerol in normal saline (NS) solution and maintained at –70°C. The day prior to surgery, 10 ml of an inoculated trypticase soy (T. soy) broth culture is grown statically for 21 hours at 37°C and then centrifuged at 50 000 *g* for 20 minutes. The broth is discarded and the bacterial pellet is resuspended in 2 ml of sterile NS and recentrifuged for 20 minutes at 50 000 *g*. The supernatant is discarded and the bacterial pellet is resuspended in 0.5 ml of sterile NS. This suspension contained approximately 1.8×10^7 colony-forming units (cfu)/μl.

Bacteria characterization for surface antigen

Bacteria used in the present abscess model are K1 surface antigen negative as tested by the latex slide agglutination test using *Neisseria meningitidis* group B antisera (Difco Laboratories, Detroit, MI) or by the K1 antibody identification technique (Burroughs Wellcome Diagnostic Division, Temple Hill, Dartford, UK) with *E. coli* serotype 07:K1 (American Type Culture Collection, Rockville, MD) serving as the positive control for K1 identification.

Abscess production

Female Sprague–Dawley rats weighing 250–275 g are anesthetized with i.p. ketamine (11.6 mg/kg) and xylazine (1.16 mg/kg). Subsequently, the head is shaved with electric shears, and cleansed with antiseptic solution, and the animals are placed in a stereotactic frame (David Kopf Instruments, Tajunga, CA). The right caudate putamen (coordinates from dura at bregma = 0 and interaural = 9.2 mm: –5.5 mm vertical, –3.0 mm lateral, +0.02 mm AP; Paxinos *et al.*, 1982) is inoculated with 3 μl (5.4×10^7 cfu) of the bacteria/saline preparation through a 2 mm burr hole using a 10 μl Hamilton syringe with a 30 G needle mounted on a micromanipulator (David Kopf Instruments, Tajunga, CA). Craniectomy is accomplished with a hand held power drill using a 1 mm cutting burr. Remaining remnants of bone overlying the dura are removed with a microhook. A small opening is made in the dura with the tip of a No. 11 surgical blade. Particular care is taken during bone drilling and dural opening not to disturb the underlying brain.

Immediately prior to stereotactic inoculation needle placement, care is taken to insure that there is no dripping or collection of the inoculation medium at the needle tip. Inoculations are performed using a Burleigh Inchworm injection apparatus (Burleigh Instruments, Fishers, NY) with a 90 minute total infusion time. In control animals, sterile saline is inoculated at an identical volume and rate. Five minutes after completion of the inoculation, the needle is withdrawn 0.5 mm every 5 minutes. Bone wax is applied to close the burr hole. Following saline irrigation, the skin is closed in a single layer. Animals are housed individually and have *ad libitum* access to food and water. Experiments are conducted 6 days after surgery. Animals should appear healthy, with a normal food and water intake, as well as normal activity and general response to the environment. Evidence of weight loss, shaking, decreased activity, and lethargy are signs of meningitis and/or systemic sepsis.

Characterization of abscess

Histopathologic examination

Following pentobarbital anesthesia (50 mg/kg, i.p.) *E. coli*-inoculated animals are perfused with 10% buffered, neutralized formaldehyde (EM Science, Cherry Hill, NJ) for histopathologic characterization. Brains are fixed in the formalin solution for at least 1 week and subsequently embedded in paraffin. Blocks of brain are sectioned at 7 μm on a vibratome, mounted on glass slides and stained with hematoxylin and eosin, Mallory's aniline blue collagen, or Gram's stain. Animals inoculated only with sterile saline also have histopathologic examination.

Quantitative abscess culture analysis

For quantitative culture analysis, a group of *E.-coli*-inoculated animals are killed with i.p. pentobarbital (70 mg/kg). The brains are sectioned sagittally along the midline, homogenized in sterile NS (1 : 10; wt/vol.), serially diluted and cultured on T. soy culture plates. Bacterial counts are recorded following 24 and 48 hours of incubation at 37°C. To detect the presence of contaminating species, the colonies are identified as being *E. coli* by routine methods. As an additional control, a separate group of animals are inoculated with sterile saline, killed 6 days following surgery and analyzed for bacterial growth as described.

Results

Morbidity and mortality

In our laboratory, systemic infection or meningitis is not associated with the described *E. coli* brain abscess model.

Histopathology

The histopathology of the lesions produced has been described (Nazzaro *et al.*, 1992). Lesions are characterized by a central area of purulent material, surrounded by an area of neovascularity, necrosis, brain macrophages, and microcystic changes. More peripherally, lesions characteristically demonstrate inflammatory cells, macrophages, and early collagen fibers. In brains studied, there has been no evidence of ventriculitis, meningitis, or septic emboli.

Quantitative cultures

Six days following brain inoculation with *E. coli*, quantitative bacterial culture analysis demonstrated 1×10^6 to 4.6×10^7 cfu of *E. coli* (Nazzaro *et al.*, 1992). There was no bacterial growth from control animals inoculated with sterile saline.

Permeability characteristics of abscess and brain adjacent to abscess

Study of the permeability characteristics of experimental brain abscesses and brain around abscess to drugs and other substances, is another method used to investigate the lesions produced. Low- and high-molecular-weight substances such as fluorescein, Evans blue albumin, radiolabeled drugs such as gentamicin and antigentamicin antibody have been used (Nazzaro *et al.*, 1991 and see below, Discussion). This work has demonstrated that consistent brain lesions are produced using the present model, and are characterized by a differential permeability to low- and high-molecular-weight substances (Nazzaro *et al.*, 1991).

Discussion

The reduction in acute morbidity and mortality rates reported here might be due to several experimental factors. These include the method of bacteria preparation, specific *E. coli* bacterial strain used, brain area inoculated, as well as parameters concerning the inoculation technique. Each of these factors was adjusted based on pilot work to optimize bacteria localization.

Brain area inoculated

In the present model, the deep basal ganglion area is targeted for bacteria inoculation. This may be contrasted with previous work by this laboratory as well as others (in which direct inoculation technique was used) when the rodent abscess model of Winn *et al.* (1979) was used, though brain areas only 2–3 mm beneath the cortical surface were inoculated (Mendes *et al.*, 1980; Costello *et al.*, 1983; Neuwelt *et al.*, 1984a,b; Schroeder *et al.*, 1987). Such studies reported a relatively high acute mortality rate, viewed as secondary to meningitis, ventriculitis, or overwhelming sepsis. For example, Schroeder *et al.* (1987) reported that, in rats who received *S. aureus* brain inoculations 3 mm below the dura, five of 17 (29%) rats died within 4 days from such complications. Earlier work in our laboratory, in which relatively superficial brain implant coordinates were implanted with *E. coli* suspension, also encountered relatively high morbidity and mortality rates (Neuwelt *et al.*, 1984a,b). Our histologic data suggest that the transverse white matter of the corpus callosum aided in preventing backtracking of bacteria along the needle tract, thereby minimizing contamination of the subarachnoid space.

Inoculum injection time and volume

Our total time for the brain inoculation injection is longer than in previous reports (Winn *et al.*, 1979; Mendes *et al.*, 1980; Costello *et al.*, 1983; Neuwelt *et al.*, 1984a,b; Schroeder *et al.*, 1987). This parameter was established through pilot work, which suggested that quicker injection times contributed to backtracking of the bacteria along the needle track during inoculation. However, according to the

Table 76.3 Brain abscess rat studies — intracerebral inoculation parameters (nr, not reported)

Study	*Needle gauge*	*Injection depth* (mm)*	*Inoculum concentration (cfu/μl)*	*Inoculum volume (μl)*	*Injection time (min)*	*Needle withdrawal time (min)*
Costello *et al.*, 1983	27	2.0	10^2–10^7	1	50	nr
Flaris and Hickey, 1992	21	5–6†	nr	20–40‡	0.25	nr
Lo *et al.*, 1991, 1993, 1994	27	2.5	10^7	1	30	10
Mendes *et al.*, 1980	26	2.0	10^1–10^6	1	60	nr
Nakagawa *et al.*, 1990	27	3.0	nr	nr	30	15
Nazzaro *et al.*, 1991, 1992	30	5.5	1.8×10^7	3	90	30
Neuwelt *et al.*, 1984a,b	30	2.5	10^5	1	30	10
Schroeder *et al.*, 1987	27	3.0	2.4–5×10^7	1	30	15
Winn *et al.*, 1978	27	2.0	10^4–10^6	0.2	nr	nr
Winn *et al.*, 1979	27	2.0	10^6	1	60	nr

* From outer cortical brain surface unless otherwise indicated.
† From skull surface.
‡ Agarose beads.

present method, total inoculum volume is also greater than that used by most other investigators using the rat. Most laboratories using direct intracerebral injection techniques in rat have reported an inoculation volume of 1 μl. In the present *E. coli* brain abscess model discussed in this chapter, in order to increase number of bacteria/inoculum (see also below), 3 μl is injected (Table 76.3).

Time taken to withdrawal needle from brain after bacterial inoculation

After bacterial inoculation into brain, the inoculating needle is left in place for 5 minutes, following which the needle is withdrawn 5 mm every 5 minutes (Table 76.2). On the basis of pilot work, the deliberate withdrawal of the injecting needle over an extended time course is employed to help insure localization of the bacterial inoculum. While several investigators do not report this variable, among those that do, 5 minutes of withdrawal time per approximately 5 mm of needle depth is followed (Table 76.2).

Specific strain of *E. coli* used

Acute sepsis in earlier work (Winn *et al.*, 1979; Neuwelt *et al.*, 1984a,b) may also be related, in part, to the presence of the K1-positive surface antigen on the *E. coli* used in those experiments. This antigen is immunochemically related to the meningococcal group-B capsular polysaccharide (Kasper *et al.*, 1973; Grados *et al.*, 1989) which is associated with increased virulence in rodent models (Wolberg and DeWitt, 1969; Howard *et al.*, 1981; Raff *et al.*, 1988) and *E. coli* meningitis in the pediatric population (Robbins *et al.*, 1974; McCracken *et al.*, 1976; Raff *et al.*, 1988). In the present model, the *E. coli* used were K1-surface-antigen-negative.

Suspension media for bacteria and host alterations at time of inoculation

In previous studies using *E. coli* and other bacteria, inoculated animals often failed to demonstrate viable bacteria in lesions when examined histopathologically or upon quantitative culture analysis within 1 week of surgery (Neuwelt *et al.*, 1984b; Kurzydlowski *et al.*, 1987; Schroeder *et al.*, 1987). In order to circumvent such difficulties, several investigators have inoculated brain with bacteria suspended in growth medium (Quartey *et al.*, 1976; Wood *et al.*, 1978; Bohl *et al.*, 1981; Kretzschmar *et al.*, 1981; Runge *et al.*, 1985; Lo *et al.*, 1991, 1993, 1994) or nutrient agarose (Wood *et al.*, 1978; Enzmann *et al.*, 1979, 1986; Wallenfang *et al.*, 1980; Britt *et al.*, 1981, 1984; Kurzydlowski *et al.*, 1987; Flaris and Hickey, 1992). In other brain abscess models, investigators have traumatized tissue prior to inoculation or have altered host defenses (Falconer *et al.*, 1943; Long and Meacham, 1968; Kurzydlowski *et al.*, 1987). Such inoculation techniques, however, may alter bacterial growth as well as the histologic and permeability characteristics of the lesions produced (Falconer *et al.*, 1943; Neuwelt *et al.*, 1989). In our laboratory, bacteria are suspended in sterile saline, and host defenses are not altered or traumatized prior to inoculation.

Bacteria number and inoculum volume

In the rat, *E. coli, S. aureus*, and streptococcal strains have most commonly been investigated (Table 76.1). Available evidence suggests that experimental brain abscess formation in rat requires an inoculum of 10^5–10^7 bacteria (Table 76.3). Whether other pathogen strains require this number at time of inoculation is not clear. Costello *et al.* (1983) investigated multiple infectious strains (Table 76.1), and reported that, for most strains studied, intracerebral injection of 10^2–10^7 pathogens produced infection. However, it is unclear in that study if true abscess lesions were created, as the dependent variable measured was number of bacteria on culture analysis; histopathologic examinations of brains were not performed. Culture analyses may be contaminated by and reflect, in part, bacterial spread to other areas such as the subarachnoid space. It is noted that Costello *et al.* (1983) reported that the experimental animals studied were often cachectic and generally sickly in appearance, which, in our experience and that of others, is most commonly seen in settings of meningitis and sepsis.

Applicability of model

The brain abscess model has been used to study multiple processes associated with these lesions. These include: the stages of abscess formation; the brain edema associated with abscess lesions; lesion response to medical therapies; the influence of steroids and host-immune-related factors on lesion characteristics and response to medical interventions; the effect of osmotic blood–brain barrier (BBB) disruption on antibiotic delivery; and the radiographic characteristics of the lesions produced. Many infectious pathogens have been studied (Table 76.1).

The present model has been employed by our laboratory to study the possibility of a therapeutic drug-rescue method based on the differential permeability of the lesions produced to low- and high-molecular-weight substances (Nazzaro *et al.*, 1991). In other words, substantial evidence suggests that CNS lesions such as brain abscesses and tumors have increased permeability to substances and drugs of low molecular weight compared to those of high molecular weight (Neuwelt, 1989). These observations suggest that advantage might be taken of this differential permeability in the delivery of therapeutic agents to brain lesions. In the case of cerebral abscess, one could administer a large intravenous dose of a low-molecular-weight antibiotic. Subsequently, antibody that is directed against the drug is administered. The antibody would inactivate (i.e., "rescue") a circulating potentially toxic drug, but not the drug that is active in the brain abscess since the antibody, because of its relatively high molecular weight, is excluded from the brain abscess. The idea of abscess differential permeability, then, might allow for a new method of elevated dosage schedules in the medical therapy of brain abscess, which have hitherto been limited by the toxicity associated with high dosages. Prevention of systemic toxicity otherwise associated with high drug dosages by intravenously administered drug-specific antibodies has been demonstrated with the cardiac glycoside digoxin (Smith *et al.*, 1976, 1982). Indeed, the binding of systemic drug, but not drug within brain abscess, by antidrug immunoglobulin (IgG) was demonstrated in the *E. coli* brain abscess model (Nazzaro *et al.*, 1991). In the *E. coli* abscess model we were able to demonstrate that, while 95% of the circulating systemic serum gentamicin was bound by antibody directed against gentamicin, there was no significant binding of the drug by antibody in brain abscess. Further, and importantly, this method with brain abscess has been applied to tumors, the results of which support the hypothesis of blood–brain barrier tumor differential permeability to low- and high-molecular-weight substances (Kroll *et al.*, 1994; Kroll and Neuwelt, 1998).

Acknowledgements

The authors wish to thank Dr Susan Kohlhepp for providing the bacteria used in these experiments, Dr Gleb Budzilovich for the preparation and review of histopathologic specimens, and Dr Joel Oppenheim for further characterization of the *E. coli* used in this work. This work was supported by the Preuss Foundation, Veterans Administration Merit Review Grant, and the National Institutes of Health, Grant No. CA3 1770.

References

Ashman, R. B., Papadimitriou, J. M. (1994). Endothelial cell proliferation associated with lesions of murine systemic candidiasis. *Infect. Immun.*, **62**, 5151–5153.

Ashman, R. B., Bolitho, E. M., Papadimitriou, J. M. (1993). Patterns of resistance to *Candida albicans* in inbred mouse strains. *Immunol. Cell Biol.*, **71**, 221–225.

Bohl, I., Wallenfang, T. H., Bothe, H., Schurmann, K. (1981). The effect of glucocorticoids in the combined treatment of experimental brain abscess in cats. *Adv. Neurosurg.*, **9**, 125–133.

Bothe, H. W., Paschen, W. (1986). Regional morphology and biochemistry in experimental brain abscesses. *Acta Neuropathol. (Berl.)*, **69**, 17–22.

Bothe, H. W., Bodsch, W., Hossman, K. A. (1984). Relationship between specific gravity, water content, and serum protein extravasation in various types of vasogenic brain edema. *Acta Neuropathol. (Berl.)*, **64**, 37–42.

Britt, R. H., Enzmann, D. R., Yaeger, A. S. (1981). Neuropathological and computerized tomographic findings in experimental brain abscess. *J. Neurosurg.*, **55**, 590–603.

Britt, R. H., Enzmann, D. H., Placone, R. C., Obana, W. G., Yaeger, A. S. (1984). Experimental anaerobic brain abscess. *J. Neurosurg.*, **60**, 1148–1159.

Costello, G. T., Heppe, R., Winn, H. R., Scheld, W. M., Rodeheaver, G. T. (1983) Susceptibility of brain to aerobic, anaerobic, and fungal organisms. *Infect. Immun.*, **41**, 535–539.

Enzmann, D. R., Britt, R. H., Yeager, A. S. (1979). Experimental brain abscess evolution: computed tomographic and neuropathologic correlation. *Radiology*, **133**, 113–122.

Enzmann, D. R., Britt, R. H., Lyons B., Carroll, B., Wilson, D. A., Buxton, J. (1982a). High-resolution ultrasound evaluation of experimental brain abscess evolution: comparison with computed tomography and neuropathology. *Radiology*, **142**, 95–102.

Enzmann, D. R., Britt, R. H., Placone, R. C., Lyons, B., Yeager, A. S. (1982b). The effect of short-term corticosteroid treatment on the CT appearance of experimental brain abscesses. *Radiology*, **145**, 79–84.

Enzmann, D. R., Britt, R. H., Obana, W. G., Stuart, J., Murphy-Owen, K. (1986). Experimental *Staphlococcus aureus* brain abscess. *A. J. N. R.*, **7**, 395–402.

Essick, C. R. (1919). Pathology of experimental traumatic abscess of the brain. *Arch. Neurol. Psychiatr.*, **1**, 673–707.

Falconer, M. A., McFatlan, A. M., Russell, D. S. (1943). Experimental brain abscesses in the rabbit. *Br. J. Surg.*, **30**, 245–260.

Flaris, N. A., Hickey, W. F. (1992). Development and characterization of an experimental model of brain abscess in the rat. *Am. J. Pathol.*, **141**, 1299–1307.

Grados, O., Ewing, W. H. (1970). Antigenic relationship between *Escherichia coli* and *Neisseria menigitidis. J. Infect. Dis.*, **122**, 100–103.

Groff, R. A. (1934). Experimental production of abscess of the brain in cats. *Arch. Neurol. Psychiatr. (Chic.)*, **31**, 199–204.

Hassler, O., Forsgren, A. (1964). Experimental abscesses in brain and subcutis. A microangiographic study in the rabbit. *Acta Pathol. Microbiol. Scand.*, **62**, 59–67.

Housmann, K. A., Bothe, H. W., Bodsch, W., Paschen, W. (1983). Pathophysiological aspects of blood–brain barrier disturbances in experimental brain tumors and brain abscesses. *Acta Neuropathol (Berl.) Suppl.*, **8**, 89–102.

Howard, C. J., Given, A. A. (1971). The virulence for mice of strains of *Escherichia coli* related to the effects of antigen on their resistance to phagocytosis and killing by complement. *Immunology*, **20**, 767–777.

Kasper, D. L., Winkelhake, J. L., Zollinger, W. D., Brandt, B. L., Artenstein, M. S. (1973). Immunochemical similarity between polysaccharide antigens of *Escherichia coli* 07, K1(L) and Group B *Neisseria meningitidis. J. Immunol.*, **110**, 262–268.

Kretzschmar, K., Wallenfang, T., Bohl, J. (1981). CT studies of brain abscesses in cats. *Neuroradiology*, **22**, 93–98.

Kroll, R. A., Neuwelt, E. A. (1998). Outwitting the blood–brain for therapeutic purposes: osmotic opening and other means. *Neurosurgery*, **42**, 1083–1100.

Kroll, R. A., Pagel, M. A., Langone, J. J., Neuwelt, E. A. (1994). Differential permeability of the blood–tumor barrier in intracerebral tumor-bearing rats: anti-drug antibody to achieve systemic drug rescue. *Therap. Immunol.*, **1**, 333–341.

Kurzydlowski, H., Wollenschlager, C., Venezio, F. R., Ghobrial, M., Soriano, M. M., Reichman, O. H. (1987). Reevaluation of an experimental streptococcal canine brain abscess model. *J. Neurosurg.*, **67**, 717–720.

Lo, W. D., McNeely, D. L., Boesel, C. W. (1991). Blood–brain barrier permeability in an experimental model of bacterial cerebritis. *Neurosurgery*, **29**, 888–892.

Lo, W. D., Wolny, A. C., Timan, C., Shin, D., Hinkle, G. H. (1993). Blood–brain barrier permeability and the brain extracellular space in acute inflammation. *J. Neurol. Sci.*, **118**, 188–193.

Lo, W. D., Wolny, A., Boesel, C. (1994). Blood–brain barrier permeability in staphylococcal cerebritis and early brain abscess. *J. Neurosurg.*, **80**, 897–905.

Long, W. D., Meacham, W. F. (1968). Experimental model for producing brain abscess in dogs with evaluation of the effect of dexamethasone and antibiotic therapy on pathogenesis of intracerebral abscess. *Surg. Forum*, **19**, 437–438.

McCracken, G. H. Jr, Mize, S. G. (1976). A controlled study of intrathecal antibiotic therapy in Gram-negative enteric meningitis of infancy. *J. Pediatr.*, **89**, 66–72.

Mendes, M., Moore, P., Wheeler, C. B., Winn, H. R., Rodeheaver, G. (1980). Susceptibility of brain and skin to bacterial challenge. *J. Neurosurg.*, **52**, 772–775.

Molinari, G. F. (1972). Septic cerebral embolism. *Stroke*, **3**, 117–122.

Molinari, G. F., Smith, L., Goldstein, M. N., Satran, R. (1973). Brain abscess from septic cerebral embolism: an experimental model. *Neurology*, **23**, 1205–1210.

Nakagawa, Y., Shinno, K., Okajima, K., Matsumoto, K. (1990). Perifocal brain oedema in experimental brain abscess in rats. *Acta Neurochir. Suppl.*, **51**, 381–382.

Nazzaro, J. M., Rosenbaum, L. C., Pagel, M. A., Neuwelt, E. A. (1991). A new model of systemic drug rescue based on permeability characteristics of the blood–brain barrier in intracerebral abscess-bearing rats. *J. Neurosurg.*, **74**, 467–474.

Nazzaro, J. M., Pagel, M. A., Neuwelt, E. A. (1992). Further refinement of the *Escherichia coli* brain abscess model in rat. *J. Neurosci. Methods*, **44**, 85–90.

Neuwelt, E. A. (ed.) (1989). *Implications of the Blood–Brain Barrier and its Manipulation*. Plenum Press, New York.

Neuwelt, E. A., Barnett, P. A. (1989). Blood–brain barrier disruption in the treatment of brain tumors: animal studies. In: *Implications of the Blood–Brain Barrier and its Manipulation* (ed. Neuwelt, E. A.), vol. 2, *Clinical Aspects*, pp. 107–194. Plenum Press, New York.

Neuwelt, E. A., Dahlborg, S. A. (1989). Blood-brain barrier disruption in the treatment of brain tumors: clinical implications. In: *Implications of the Blood–Brain Barrier and its Manipulation* (ed. Neuwelt, E. A.), vol. 2, *Clinical Aspects*, pp. 195–262. Plenum Press, New York.

Neuwelt, E. A., Baker, D. E., Pagel, M. A., Blank, N. K. (1984a). The cerebrovascular permeability and delivery of gentamicin to normal brain and experimental brain abscess in rats. *J. Neurosurg.*, **61**, 430–439.

Neuwelt, E. A., Lawrence, M. S., Blank, N. K. (1984b). The effect of gentamicin and dexamethasone on the natural history of the rat *Escherichia coli* brain abscess model with histopathological correlation. *Neurosurgery*, **15**, 475–483.

Neuwelt, E. A., Enzmann, D. R., Pagel, M., Miller, G. (1989). Bacterial and fungal brain abscesses and the blood–brain barrier. In: *Implications of the Blood–Brain Barrier and its Manipulation* (ed. Neuwelt, E. A.), vol. 2, *Clinical Aspects*, pp. 263–306. Plenum Press, New York.

Obana, W. G., Britt, R. H., Placone, R. C., Stuart, J. S., Enzmann, D. R. (1986) Experimental brain abscess development in the chronically immunosuppressed host. Computerized tomographic and neuropathological correlations. *J. Neurosurg.*, **65**, 382–391.

Paxinos, G., Watson, C., Emson, P. (1982). *The Rat Brain in Stereotaxic Coordinates*, pp. 129–149. Academic Press, New York.

Quartey, G. R. C., Johnston, J. A., Rozdilsky, B. (1976). Decadron in the treatment of cerebral abscess: an experimental study. *J. Neurosurg.*, **45**, 301–310.

Raff, H. V., Devereux, D., Shuford, W., Abbott-Brown, D., Maloney, G. (1988) Human monoclonal antibody with protective activity for *Escherichia coli* K1 and *Neisseria meningitides* group B infections. *J. Infect. Dis.*, **157**, 118–126.

Robbins, J. B., McCracken, G. H. Jr, Gotschlich, E. C., Orskov, F., Orskov, L., Hanson, L. A. (1974). *Escherichia coli* K1 capsular polysaccharide associated with neonatal meningitis. *N. Engl. J. Med.*, **290**, 1216–1220.

Runge, V. M., Clanton, J. A., Price, A. C. *et al.* (1985). Dyke Award. Evaluation of contrast-enhanced MR imaging in a brain-abscess model. *Am. J. Neuroradiol.*, **6**, 139–147.

Schmidt, D. H., Butler, V. P. Jr (1971). Reversal of digoxin toxicity with specific antibodies. *J. Clin. Invest.*, **50**, 1738–1744.

Schroeder, K. A., McKeever, P. E., Schaberg, D. R., Hoff, J. T. (1987). Effect of dexamethasone on experimental brain abscess. *J. Neurosurg.*, **66**, 264–269.

Smith, T. W., Haber, E., Yeatman, *et al.* (1976). Reversal of advanced digoxin intoxication with Fab fragments of digoxin-specific antibodies. *N. Engl. J. Med.*, **294**, 797–800.

Smith, T. W., Butler, V., Haber, E., *et al.* (1982). Treatment of life threatening digitalis intoxication with digoxin-specific Fab antibody fragments: experience in 26 cases. *N. Engl. J. Med.*, **307**, 1357–1362.

Thomas, L. (1942). A single stage to produce brain abscess in cats. *Arch. Pathol.*, **33**, 472–476.

Wallenfang, T., Bohl, J., Schreiner, G. (1979). Experimental brain edema in acute and chronic brain abscess in rabbits and its morphological alterations. In: *Advances in Neurosurgery* 7 (eds Marguth, F., Brock, M., Kazner, E., Klinger, M., Schmiedek, P.), pp. 304–310. Springer-Verlag, Berlin.

Wallenfang, T., Bohl, J., Kretzschmar, K. (1980). Evolution of brain abscess in cats: formation of capsule and resolution of brain edema. *Neurosurg. Rev.*, **3**, 101–111.

Winn, H. R., Rodeheaver, G., Moore, P., Wheeler, C. (1978). A new model for experimental brain abscess in rats and mice. *Surg. Forum*, **29**, 500–502.

Winn, H. R., Mendes, M., Moore, P., Wheeler, C., Rodeheaver, G. (1979). Production of experimental brain abscess in the rat. *J. Neurosurg.*, **51**, 685–690.

Wolberg, G., DeWitt, C. W. (1969). Mouse virulence of K(L) antigen-containing strains by *Escherichia coli*. *J. Bacteriol.*, **100**, 730–737.

Wood, J. H., Doppman, J. L., Lightfoote, W. E., Girton, M., Ommaya, A. K. (1978). Role of vascular proliferation on angiographic appearance and encapsulation of experimental traumatic and metastatic brain abscesses. *J. Neurosurg.*, **48**, 264–273.

Wood, J. H., Lightfoote, W. E. II and Ommaya, A. K. (1979). Cerebral abscesses produced by bacterial implantation and septic embolisation in primates. *J. Neurol. Neurosurg. Psychiatr.*, **42**, 62–69.

Section III

Mycotic Infection Models

Chapter 77

Rodent Models of *Candida* Sepsis

V. Joly and P. Yeni

Background of human infection

Rates of candidemia have increased substantially, and *Candida* has become the fourth most common isolate recovered from blood cultures (Jarvis, 1995). Candidal infections occur on both medical and surgical services. The mortality rate is high. A noticeable shift in the species of *Candida* causing infection toward non-*albicans* species has occurred (Wingard *et al.*, 1991; Rex *et al.*, 1994; Wingard, 1995); some of these non-*albicans* emerging species are relatively resistant to the available antifungal agents, whereas there are few new antifungal agents under development.

Numerous risk factors for candidemia have been identified, including use of antibiotics, neutropenia, indwelling catheters, chemotherapy, parenteral alimentation, immunosuppressive therapy after organ transplantation and colonization with *Candida* species. Knowledge of the most appropriate therapeutic strategies remains limited because few studies have compared the usefulness of recent antifungal agents (fluconazole, itraconazole, lipid-based amphotericin B formulations) to that of the traditional agent, deoxycholate amphotericin B. The study of serious candidal infection remains difficult due to the complex disease profile of these patients.

Background of model

In 1885, Klemperer produced disseminated candidosis in guinea-pigs by intravenous inoculation of *C. albicans*. This experiment was soon repeated and expanded by others. Of the many types of animal used to study disseminated candidosis, the mouse is the most popular species, followed by the rabbit, the rat and the guinea-pig.

Animal species

Several strains of mice have been used in this model. The use of outbred mice (male, 25–30 g) is generally appropriate — ICR mice (Barchiesi *et al.*, 1996), Swiss mice (Van't Wout *et al.*, 1989), CF1 mice (Karyotakis *et al.*, 1993), CD-1 mice (Hanson *et al.*, 1991). Inbred mice from different strains (C3H/He, BALB/c, C57BL etc.) including congenitally immunodeficient mice (nude mice, beige mice) have also been used to study the strain-dependent difference susceptibility of mice to candidosis (Marquis *et al.*, 1986; Cantorna and Balish, 1990).

Disseminated candidosis can also be obtained in guinea-pig (Hurley and Fauci, 1975) and outbred strains of rats (male, 200–250 g; Rogers and Balish, 1978; Fisher *et al.*, 1989; Vitt *et al.*, 1994).

Preparation of animals

No specialized housing or care is required. Rodents may be immunocompromised, with the aim of studying the efficacy of antifungal drugs in the context of severe underlying immunosuppression, or to obtain infection with non-*albicans Candida* species. Neutropenia can be induced with cyclophosphamide (Van't Wout *et al.*, 1989; Anaissie *et al.*, 1993), 5-fluorouracil or gold sodium thionalate (Atkinson *et al.*, 1994). Hydrocortisone acetate is used to decrease phagocytic activity (Anaissie *et al.*, 1993; Kullberg *et al.*, 1993).

Storage and preparation of inocula

Various *albicans* and non-*albicans Candida* strains have been used, usually clinical isolates from blood cultures. Isolates are maintained on Sabouraud dextrose agar slants or in water stocks. Organisms are freshly subcultured over 18–24 hours on Sabouraud agar or in appropriate broth, then suspended in 0.9% NaCl or phosphate buffered saline, washed and adjusted to the appropriate inoculum spectrophotometrically or by enumeration with a hemocytometer. Counts of colony-forming units (cfu) per ml are verified by plating serial 10-fold dilutions on to Sabouraud dextrose agar plates. It is possible to prepare aliquots of a stock solution of the inoculum, kept frozen at –80°C and thawed before the experiment.

Handbook of Animal Models of Infection
ISBN 0-12-775390-7

A large range of inoculum has been tested. In normal, unpretreated mice, the intravenous LD_{50} is of the order of 10^4–10^6 yeasts; the LD_{50} depends on the virulence of the *Candida* strain and on the susceptibility of the mouse strain to disseminated candidosis.

Infection process

Intravenous injection of a suitable *Candida* inoculum is sufficient to establish a disseminated visceral infection that resembles disseminated candidosis in humans. This route of experimental infection is used the most widely. Intraperitoneal inoculation also commonly leads to disseminated infection, but the inoculum needs to be 10 times higher than an intravenous inoculum to induce similar extensive pathology. Finally, experimental gastrointestinal candidosis in immunosuppressed mice may be used as a model for dissemination of *Candida* infection from the gut to the blood stream (Wingard *et al.*, 1980).

Key parameters to monitor infection and response to treatment

Main key parameters to monitor infection are death and organ cfu. The rate of progression of experimental disseminated candidosis varies with the inoculum size: in mice, most investigators have found that an intravenous inoculum greater than 10^6 *C. albicans* yeasts is rapidly lethal whereas an inoculum less than 10^4 gives a low-grade, chronic infection with spontaneous resolution. With *Candida* species of low virulence, the severity of the disease can be modulated by the use of immunosuppressed animals. Mortality data can be expressed in terms of LD_{50} value, that is, the dose inducing death in 50% of animals until the end of the observation, as mean survival time or as a Kaplan–Meier survival curve.

The course of experimental disseminated candidosis has been described in many papers. In mice, enumeration of viable *C. albicans* in body organs shows a peak in cfus recoverable from liver, lungs and spleen shortly after intravenous infection (Evans and Mardon, 1977). This peak is followed by a gradual decline in the number of organisms in all organs except the kidney. Viable counts in the kidney steadily increase and this organ becomes the most severely infected. The high predilection of *Candida* for the kidney remains unexplained. Maximum involvement of the kidney after intravenous inoculation is also observed in the rat (Rogers and Balish, 1978) and the guinea-pig (Hurley and Fauci, 1975). The kidney is thus the site of choice to assess cfu counts. Organs are sampled aseptically, weighed and homogenized in phosphate buffered saline. Ten-fold dilutions of the homogenate are plated on Sabouraud agar and the colonies are counted after 24–48 hours incubation at 37°C. Results are expressed in log cfu per gram of tissue. Histopathology can be performed to assess host inflammatory reaction and morphologic form of the organism (yeasts, pseudohyphae, hyphae).

An increase in peripheral blood granulocytes and monocytes is noted after infection of mice (Kullberg *et al.*, 1992, 1993) and rats (Vitt *et al.*, 1994). Tumor necrosis factor (TNF) and interleukin-6 are produced systemically during *C. albicans* infection (Riipi and Carlson, 1990; Steinshamn and Waage, 1992). Complement plays a major role in innate resistance (Hector *et al.*, 1982).

Antimicrobial therapy

In vitro tests for the assessment of antifungal agents activity have some limitations, particularly in the case of azole compounds, for which it is difficult to define interpretative breakpoints predicting *in vivo* efficacy. Animal models of candidosis are largely used to study simultaneously the *in vivo* efficacy and toxicity of antifungal agents.

Several laboratories have developed the mouse/intravenous model to the point at which reproducible inoculation of intermediate doses of *Candida* is routinely achieved, so that the model can be used to evaluate antifungal action with manageable mortality rates and times. Most systemic antifungal agents have been evaluated in the rodent model of disseminated candidosis. Table 77.1 lists some examples of a variety of rodent models of disseminated candidosis and the antifungals tested in them. Treatment can be administered by a variety of routes: subcutaneously, by oral gavage, intravenously, intraperitoneally. The intravenous route allows one to know precisely the amount of administered drug, but is less convenient than intraperitoneal route for repeated administrations. The intraperitoneal route may be considered to be representative of a slow-rate infusion. The choice of the route also depends on the drug: some lipophilic triazoles are not available for the parenteral route, and the intravenous route is the only valuable route for liposomal-amphotericin B to respect the conditions of administration in humans.

Drug therapy is begun within the hours following infection, ranging from 1 hour to several days after challenge. By contrast, drug therapy begins before infection in some studies. The duration of therapy varies from a single administration to several days. The dosing schedule and choice of dose should be given careful consideration, as stated below. Determination of cfus in kidney is usually done one to several days after the end of therapy.

Pitfalls of the model

The models of systemic candidosis in rodents have several limitations. After intravenous challenge of a high inocu-

Table 77.1 Examples of rodent models of candidosis used for the evaluation of antifungal agents *in vivo*

Model	*AmB*	*L-AmB*	*Flucytosine*	*Ketoconazole*	*Triazoles*	*Echinocandins*	*Drug combination*
Mouse	Rabinovich *et al.* (1974)	Lopez-Berestein *et al.* (1983)	Pitillo and Ray (1969)	Heel *et al.* (1982)	Richardson *et al.* (1985)	Abruzzo *et al.* (1997)	Rabinovich *et al.* (1974)
	Lopez-Berestein *et al.* (1983)	Lopez-Berestein *et al.* (1984a)	Rabinovich *et al.* (1974)	Hare and Loebenberg, 1988	Van't Wout *et al.* (1989)	Graybill *et al.* (1997a)	Polak *et al.* (1982)
	Lopez-Berestein *et al.* (1984a)	Tremblay *et al.* (1984)	Anaissie *et al.* (1994)	Richardson *et al.* (1985)	Kullberg *et al.* (1992)	Graybill *et al.* (1997b)	Atkinson *et al.* (1994)
	Van't Wout *et al.* (1989)	Gondal *et al.* (1989)			Karyotakis *et al.* (1993)		Sugar *et al.* (1994a)
	Tremblay *et al.* (1984)	Clark *et al.* (1991)			Anaissie *et al.* (1994)		Sugar *et al.* (1994b)
	Gondal *et al.* (1989)	Pahls and Schaffner (1994)			Sugar *et al.* (1994a)		Hanson *et al.* (1991)
	Hanson *et al.* (1991)				Sugar *et al.* (1994b)		Sanati *et al.* (1997)
	Clark *et al.* (1991)				Barchiesi *et al.* (1996)		
	Karyotakis *et al.* (1993)						
	Sugar *et al.* (1994a)						
	Sugar *et al.* (1994b)						
	Anaissie *et al.* (1994)						
Rat	Galgiani and Van Wyck (1984)			Rogers and Galgiani (1986)	Richardson *et al.* (1985)		
	Richardson *et al.* (1985)			Richardson *et al.* (1985)	Rogers and Galgiani (1986)		
	Fisher *et al.* (1989)				Fisher *et al.* (1989)		

lum, infection is so rapidly fatal that drugs cannot be expected to exert a significant effect. On the other hand, infection with a low inoculum leads to spontaneous high survival rates and, therefore, drug effects are likely to be inapparent. Even in the case of a reproducible inoculation of intermediate doses of *C. albicans*, leading to an acute infection, many studies involve a prophylactic treatment with the drug given before or at the same time as the *Candida* inoculum, rarely beyond 24 hours after infection — far from the usual situation in humans. Furthermore, the model remains difficult to adapt to infections caused by species other than *C. albicans*.

The results of therapeutic trials vary substantially with the timing and route of administration of the antifungal drug. The route of administration of antifungals in animal models is not always the same as that used in humans: non-traumatic oral administration of drugs is a frequent problem with laboratory animals and repeated intravenous injections may be difficult in rodents. The intraperitoneal route is therefore often preferred, but is not suitable for some antifungal agents such as lipidic formulations of amphotericin B.

Pharmacokinetics of drugs may be different in rodents and humans. The serum half-life of drugs is often decreased in rodents compared to humans; this can lead to therapeutic failure if the interval between doses is too long. Animals with systemic candidosis typically develop renal infection and drugs with good renal concentrations and low nephrotoxicity may have a special advantage, which does not necessarily extrapolate to human illness.

In experimental candidosis in rodents, evaluation of toxicity is basic, mainly founded on survival. Histologic studies and biochemical monitoring of hepato- and nephrotoxicity are infrequently performed and do not bring as much information as in studies performed in more sophisticated models. For example, the first studies performed with liposomal-amphotericin B reported a dramatically favorable degree of protection against toxicity of the drug (Lopez-Berestein *et al.*, 1983), but further studies performed in rabbits (Lee *et al.*, 1994) or rats (Longuet *et al.*, 1991) and clinical data showed that nephrotoxicity could be observed for daily dosages which were clearly lower than the higher dose tolerated in mice as assessed on the basis of survival.

Contributions of the model to infectious disease therapy

Indication of efficacy of antimicrobial agents

Disseminated candidosis in rodents is the first step in the evaluation of the *in vivo* activity of antifungal compounds. This model allowed the demonstration that fluconazole was superior to ketoconazole in the treatment of disseminated candidosis; this result has been further confirmed by clinical data. In mice infected with 10^6 cfu and treated daily for 10 days, the percentage of mice alive on day 11 was used to determine the effective dose 50 (ED_{50}) of fluconazole, i.e., the dose inducing survival of 50% of the animals Fluconazole had an ED_{50} of 0.15 mg/kg, and ketoconazole was far less active, with an ED_{50} of 20 mg/kg. Although mice treated with azoles appeared to be cured, a more quantitative analysis did not support this conclusion. The average log cfu per kidney was stabilized at about 10^4 cfu per kidney, the level found 6 hours after inoculation, showing that azoles acted as fungistatic agents in this model. By contrast, 3 subcutaneous doses of amphotericin B reduced pathogen levels to below the level of detection of 100 CFU per kidney. The model could therefore distinguish between fungicidal and fungistatic compounds (Hare and Loebenberg, 1988). In rats infected with *C. albicans*, ketoconazole (10 mg/kg daily for 3 days) was more efficient than fluconazole (0.5 mg/kg daily for 3 days) to increase survival, although ketoconazole was 16-fold more active *in vitro* on the *Candida* strain (Rogers and Galgiani, 1986).

Disseminated candidosis has also been used extensively to study the comparative efficacy of deoxycholate-amphotericin B (Fungizone) and lipidic formulations of amphotericin B (Lopez-Berestein *et al.*, 1983; Gondal *et al.*, 1989; Clark *et al.*, 1991). Most studies have shown that lipidic-amphotericin B was as effective as free amphotericin B when used at similar dosage, but the reduced toxicity of lipidic-amphotericin B allowed the dose to be increased. Thus, lipidic-amphotericin B had an improved therapeutic index. Clinical data have confirmed that lipidic-amphotericin B could be used at higher dosages (3–5 mg/kg daily) than Fungizone, and allowed to obtain good clinical and mycological response in patients in whom Fungizone would have been toxic.

Experimental disseminated candidosis has been used to study the role of resistance to antifungal agents in therapeutic failure. Some *Candida* species exhibit intrinsic resistance to antifungal agents: *C. krusei*, for example, is resistant *in vitro* to fluconazole, and fluconazole is not active in the model of disseminated candidosis due to *C. krusei* (Anaissie *et al.*, 1993; Karyotakis *et al.*, 1993). This species has been shown to emerge under prophylaxis in patients (Wingard *et al.*, 1991). In mice infected with clinical strains isolated from oral cavities of patients with acquired immunodeficiency syndrome (AIDS), the therapeutic efficacy of fluconazole was clearly decreased when mice were challenged with strains isolated from patients in the post-therapy period (Barchiesi *et al.*, 1996). In addition, correlations between *in vitro* activity and *in vivo* efficacy of a given antifungal agent have been established in the model of systemic candidosis (Anaissie *et al.*, 1993): it has been shown that the breakpoint level of *in vitro* activity necessary to obtain *in vivo* antifungal effect had to been defined for each drug. However, for a given drug, *in vitro* activity predicted *in vivo* therapeutic efficacy, providing that appropriate tests were used (Rogers and Galgiani, 1986; Anaissie *et al.*, 1993).

Disseminated candidosis in rodents is a simple way of evaluating the *in vivo* anticandidal activity of new compounds. MK-991 is a new echinocandin candidate for clinical development. It acts through inhibition of the 1,3-β-D-glucan synthase, which synthetizes a critical structural cell wall component in certain pathogenic fungi. It has shown efficacy in the treatment of murine disseminated candidiasis, including fluconazole-resistant strain (Graybill *et al.*, 1997a) and non-*albicans* strains (Abruzzo *et al.*, 1997; Graybill *et al.*, 1997b).

Indication of probably synergistic/antagonistic effects of antifungal agent combinations

The availability of different antifungal drugs makes the administration of combinations of drugs desirable in an attempt to achieve synergy, to broaden their combined spectrum of activity or to lower the dose, because of the toxicity of amphotericin B. Studies in experimental candidosis have shown that amphotericin B and flucytosine are usually additive or indifferent (Polak *et al.*, 1982). There were theoretical concerns that combinations of polyenes, whose affinity for fungal cell membrane is increased by the presence of ergosterol, and azoles, which inhibit the synthesis of ergosterol, would be antagonistic. The study performed by Schaffner and Frick (1985) in experimental aspergillosis suggested that the combination of ketoconazole and amphotericin B might be antagonistic. In murine disseminated candidosis, however, it was shown that the combination of fluconazole and amphotericin B was not antagonistic (Sugar *et al.*, 1994b; Sanati *et al.*, 1997). The same type of results were found with the combination of amphotericin B and saperconazole (Sugar *et al.*, 1994a) or DO870 (Atkinson *et al.*, 1994). The interactions of azoles and polyenes may differ according to fungus strain, test model and the drug used, and a general rule concerning the concomitant or sequential use of these drugs cannot be drawn from these experimental data. Additional studies are necessary, including clinical studies.

Differential pathogenicity of *Candida*

Candida species involved in various human infections differ in their inherent capacity to cause disease. *C. albicans* isolates are consistently the most virulent, causing morbidity or mortality. Experimentally, non-*albicans* strains are usually less pathogenic (Bistoni *et al.*, 1984; Anaissie *et al.*, 1993), although some studies suggest that *C. tropicalis* could be more pathogenic than *C. albicans*, particularly in disseminated infection resulting from gastrointestinal colonization (Wingard *et al.*, 1980; De Repentigny *et al.*, 1992).

Species and strain differences in yeast pathogenicity can be exploited as a basis for determining molecular virulence attributes in the pathogenesis of candidosis. Some experimental studies suggest that enzymes produced by *C. albicans* (hyaluronidase, chondroitin sulphatase, proteinase and phospholipase) might be virulence factors (Macdonald and Odds, 1983; Kwon-Chung *et al.*, 1985; Shimizu *et al.*, 1996). The relationship between virulence and drug resistance has been studied in mice: the multidrug-resistant strain colonized kidneys to high levels, but had a markedly reduced virulence compared to parental strain on the view of mice mortality (Becker *et al.*, 1995).

Indication of pharmacokinetics, pharmacodynamics and tissue penetration

Plasma and tissue distribution of antifungal drugs can be studied in uninfected or infected mice. This has been particularly important for new formulations of amphotericin B in an attempt to evaluate the alteration of drug pharmacokinetics in the presence of different lipidic vehicles. Lopez-Berestein *et al.* (1984a) showed that liposomal encapsulation of amphotericin B improved the delivery of the drug to the liver, spleen, lung and kidney in both normal and infected mice. Furthermore, amphotericin B was measurable in brain tissue of infected animals after injection of the encapsulated drug, but not in the brain of normal animals, nor infected animals treated with free amphotericin B (Lopez-Berestein *et al.*, 1984b); these results suggested that vectorization, capillary endothelial damage and phagocytic uptake during infection may alter tissue amphotericin B delivery.

Increase in understanding host inflammatory response

The role of phagocytosis as a major defense mechanism against disseminated candidosis is reinforced by the susceptibility to systemic *Candida* challenge of animals pretreated with agents that reduce the number of phagocytic leukocytes (Moser and Domer, 1980; van't Wout *et al.*, 1988). Conversely, modifications leading to enhanced neutrophils or phagocytic killing, such as treatment with granulocyte or macrophage colony-stimulating factor may improve the evolution of disseminated experimental candidosis (Polak-Wiss, 1991; Vitt *et al.*, 1994). The role of complement in the elimination of *C. albicans* has been shown in mice — higher LD_{50} for C5-sufficient than for C5-deficient mice (Morelli and Rosenberg, 1971).

Evidence has been obtained for a major role of T helper type I response in controlling invasive forms of candidosis in mice with demonstrable immunity (Cenci *et al.*, 1990). Cytokine production by CD4 T cells, particularly interferon-γ, contribute to resistance to infection (Romani *et al.*, 1992). Administration of recombinant interferon-γ enhances host resistance against acute disseminated *C. albicans* infection in mice through activation of polymorphonuclear cells (Kullberg *et al.*, 1993). Interleukin-1 administered therapeutically with fluconazole in neutropenic mice can lead to a significant clearance of *C. albicans* from the kidney and the spleen (Kullberg *et al.*, 1992).

Tumor necrosis factor (TNF) and interleukin-6 are systemically produced during *Candida albicans* infection (Riipi and Carlson, 1990; Steinshamn and Waage, 1992). Neutralization of TNF with an anti-TNF antibody results in an increased proliferation of *C. albicans* in normal mice, but not in granulocytopenic mice, suggesting the role of TNF for granulocyte antifungal activity *in vivo* (Steinshamn and Waage, 1992).

Conclusion

Disseminated candidosis in rodents is a simple and reproducible model that can be used to study the mechanisms involved in *Candida* species pathogenicity and to predict clinical efficacy of antifungal drugs, providing that different parameters (inoculum size, *Candida* species, therapeutic regimens) are controlled and that large numbers of animals are used. To enhance the predictibility of these models, it is necessary to use as many *Candida* strains as possible, including clinical isolates due to new resistance patterns. Results obtained in rodents could be confirmed in more sophisticated models of experimental infections that allow a more accurate evaluation of drug pharmacokinetics and toxicity.

References

Abruzzo, G. K., Flattery A. M., Gill, C. J. *et al.* (1997). Evaluation of the echinocandin antifungal MK-991 (L-743,872): efficacies in mouse models of disseminated aspergillosis, candidiasis, and cryptococcosis. *Antimicrob. Agents Chemother.*, **41**; 2333–2338.

Anaissie, E., Hachem, R., K-Tin-U, C., Stephens, L. C., Bodey, G. P. (1993). Experimental hematogenous candidiasis caused by *Candida krusei* and *Candida albicans*: species differences in pathogenicity. *Infect. Immun.*, **61**, 1268–1271.

Anaissie, E. J., Karyotakis, N. C., Hachem, R., Dignani, M. C., Rex, J. H., Paetznick, V. (1994). Correlation between *in vitro* and *in vivo* activity of antifungal agents against *Candida* species. *J. Infect. Dis.*, **170**, 384–389.

Atkinson, B. A., Bocanegra, R., Colombo, A. L., Graybill, J. R. (1994). Treatment of disseminated *Torulopsis glabrata* infection with DO870 and amphotericin B. *Antimicrob. Agents Chemother.*, **38**, 1604–1607.

Barchiesi, F., Najvar, L. K., Luther, M. F., Scalise, G., Rinaldi, M. G., Graybill, J. R. (1996). Variation in fluconazole efficacy for *Candida albicans* strains sequentially isolated from oral cavities of patients with AIDS in an experimental murine candidiasis model. *Antimicrob. Agents Chemother.*, **40**, 1317–1320.

Becker, J. M., Henry, L. K., Jiang, W., Koltin, Y. (1995). Reduced virulence of *Candida albicans* mutants affected in multidrug resistance. *Infect. Immun.*, **63**, 4515–4518.

Bistoni, F., Vecchiarelli, A., Cenci, E., Puccetti, P., Marconi, P., Cassone, A. (1984). A comparison of experimental pathogenicity of *Candida* species in cyclophosphamide-immunodepressed mice. *Sabouradia J. Med. Vet. Mycol.*, **22**, 409–418.

Cantorna, M. T., Balish, E. (1990). Mucosal and systemic candidiasis in congenitally immunodeficient mice. *Infect. Immun.*, **58**, 1093–1100.

Cenci, E., Romani, I., Vechiarelli, A., Puccetti, P., Bistoni, F. (1990). T cell subsets and IFN-γ production in resistance to systemic candidiasis in immunized mice. *J. Immunol.*, **144**, 4333–4339.

Clark, J. M., Whitney, R. R., Olsen, S. J. *et al.* (1991). Amphotericin B lipid complex therapy of experimental fungal infections in mice. *Antimicrob. Agents Chemother.*, **35**, 615–621.

De Repentigny, L., Phaneuf, M., Mathieu, L. G. (1992). Gastrointestinal colonization and systemic dissemination by *Candida albicans* and *Candida tropicalis* in intact and immunocompromised mice. *Infect. Immun.*, **60**, 4907–4914.

Evans, Z. A., Mardon, D. N. (1977). Organ localization in mice challenged with a typical *Candida albicans* strain and a pseudohyphal variant. *Proc. Soc. Exp. Biol. Med.*, **155**, 234–238.

Fisher, M. A., Shen, S., Haddad, J., Tarry, W. F. (1989). Comparison of *in vivo* activity of fluconazole with that of amphotericin B against *Candida tropicalis, Candida glabrata* and *Candida krusei*. *Antimicrob. Agents Chemother.*, **33**, 1443–1446.

Galgiani, J. N., Van Wyck, D. B. (1984). Ornithyl amphotericin B methyl ester treatment of experimental candidiasis in rats. *Antimicrob. Agents Chemother.*, **26**, 108–109.

Gondal, J. A., Swartz, R. P., Rahman, A. (1989). Therapeutic evaluation of free and liposome-encapsulated amphotericin B in the treatment of systemic candidiasis in mice. *Antimicrob. Agents Chemother.*, **33**, 1544–1548.

Graybill, J. R., Najvar, L. K., Luther, M. F., Fothergill, A. W. (1997a). Treatment of murine disseminated candidiasis with L-743,872. *Antimicrob. Agents Chemother.*, **41**, 1775–1777.

Graybill, J. R., Bocanegra, R., Luther, M., Tothergill, A., Ribaldi, M. J. (1997b). Treatment of murine *Candida krusei* or *Candida glabrata* infection with L-743,872. *Antimicrob. Agents Chemother.*, **41**, 1937–1939.

Hanson, L. H., Perlman, A. M., Clemons, K. V., Stevens, D. A. (1991). Synergy between cilofungin and amphotericin B in a murine model of candidiasis. *Antimicrob. Agents Chemother.*, **35**, 1334–1337.

Hare R. S., Loebenberg, D. (1988). Animal models in the search for antifungal agents. *ASM News*, **54**, 235–239.

Hector, R. F., Domer, J. E., Carrow, E. W. (1982). Immune responses to *Candida albicans* in genetically distinct mice. *Infect. Immun.*, **38**, 1020–1028.

Heel, R. C., Brogden, R. N., Carmine, A., Morley, P., Speight, A., Avery, G. S. (1982). Ketoconazole: a review of its therapeutic efficacy in superficial and systemic fungal infections. *Drugs*, **23**, 1–36.

Hurley, D. L., Fauci, A. S. (1975). Disseminated candidiasis. I. An experimental model in the guinea-pig. *J. Infect. Dis.*, **131**, 516–521.

Jarvis, W. R. (1995). Epidemiology of nosocomial fungal infections, with emphasis on *Candida* species. *Clin. Infect. Dis.*, **20**, 1526–1530.

Karyotakis, N. C., Anaissie, E. J., Hachem, R., Dignani, M. C., Samonis, G. (1993). Comparison of the efficacy of polyenes and triazoles against hematogenous *Candida krusei* infection in neutropenic mice. *J. Infect. Dis.*, **168**, 1311–1313.

Klemperer, G. (1885). Ueber die natur des soorpilzes. *Zbl. Klin. Med.*, **6**, 849–851.

Kullberg, B-J., van't Wout, J. W., Poell, R. J. M., van Furth, R. (1992). Combined effect of fluconazole and recombinant human interleukin-1 on systemic candidiasis in neutropenic mice. *Antimicrob. Agents Chemother.*, **36**, 1225–1229.

Kullberg, B-J., van't Wout, J. W., Hoogstraten, C., van Furth, R. (1993). Recombinant interferon-gamma enhances resistance to acute disseminated *Candida albicans* infection in mice. *J. Infect. Dis.*, **168**, 436–443.

Kwon-Chung, K. J., Lehman, D., Good, C., Magee, P. T. (1985). Genetic evidence for role of extracellular proteinase in virulence of *Candida albicans*. *Infect. Immun.*, **49**, 571–575.

Lee, J. W., Amantea, M. A., Francis, P. A. *et al.* (1994). Pharmacokinetics and safety of a unilamellar liposomal formulation of amphotericin B (AmBisome) in rabbits. *Antimicrob. Agents Chemother.*, **38**, 713–718.

Longuet, P., Joly, V., Amirault, P., Seta, N., Carbon, C., Yeni, P. (1991). Limited protection by small unilamellar liposomes against the renal tubular toxicity induced by repeated amphotericin B infusions in rats. *Antimicrob. Agents Chemother.*, **35**, 1303–1308.

Lopez-Berestein, G., Mehta, R., Hopfer, R. L. *et al.* (1983). Treatment and prophylaxis of disseminated infection due to *Candida albicans* in mice with liposome-encapsulated amphotericin B. *J. Infect. Dis.*, **147**, 939–945.

Lopez-Berestein, G., Hopfer, R. L., Mehta, R., Mehta, K., Hersh, E. M., Juliano, R. (1984a). Liposome-encapsulated amphotericin B for treatment of disseminated candidiasis in neutropenic mice. *J. Infect. Dis.*, **150**, 278–283.

Lopez-Berestein, G., Rosenblum, M. G., Mehta, R. (1984b). Altered tissue distribution of amphotericin B by liposomal encapsulation: comparison of normal mice to mice infected with *Candida albicans*. *Cancer Drug Deliv.*, **1**, 199–205.

Macdonald, F., Odds F. C. (1983). Virulence for mice of a proteinase-secreting strain of *Candida albicans* and a proteinase-deficient mutant. *J. Gen. Microbiol.*, **129**, 431–438.

Marquis, G., Montplaisir, S., Pelletier, M., Moisseau, S., Auger, P. (1986). Strain-dependent differences in susceptibility to experimental candidosis. *J. Infect. Dis.*, **154**, 906–909.

Morelli, R., Rosenberg, L. T. (1971). Role of complement during experimental *Candida* infection in mice. *Infect. Immun.*, **3**, 521–523.

Moser, S. A., Domer, J. E. (1980). Effects of cyclophosphamide on murine candidiasis. *Infect. Immun.*, **27**, 376–386.

Pahls, S., Schaffner, A. (1994). Comparison of the activity of free and liposomal amphotericin B *in vitro* and in a model of systemic and localized murine candidiasis. *J. Infect. Dis.*, **169**, 1057–1061.

Pitillo, R. F., Ray, B. J. (1969). Chemotherapeutic activity of 5-fluorocytosine against a lethal *Candida albicans* infection in mice. *Appl. Microbiol.*, **17**, 773–774.

Polak, A., Scholer, H. J., Wall, M. (1982). Combination therapy of experimental candidiasis, cryptococcosis and aspergillosis in mice. *Chemotherapy*, **28**, 461–479.

Polak-Wiss, A. (1991). Protective effect of human granulocyte colony stimulating factor (hG-CSF) on *Candida* infections in normal and immunosuppressed mice. *Mycoses*, **34**, 109–118.

Rabinovich, S., Shaw, B. D., Bryant, T., Donta, S. T. (1974). Effect of 5-fluorocytosine and amphotericin B on *Candida albicans* infection in mice. *J. Infect. Dis.*, **130**, 28–31.

Rex, J. H., Bennett, J. E., Sugar, A. M. *et al.* (1994). A randomized trial comparing fluconazole with amphotericin B for the treatment of candidemia in patients without neutropenia. *N. Engl. J. Med.*, **331**, 1325–1330.

Richardson, K., Brammer, K. W., Marriott, M. S., Troke, P. F. (1985). Activity of UK-49,858, a bis-triazole derivative, against experimental infections with *Candida albicans* and *Trichophyton mentagrophytes*. *Antimicrob. Agents Chemother.*, **27**, 832–835.

Riipi, L., Carlson, E. (1990). Tumor necrosis factor (TNF) is induced in mice by *Candida albicans*: role of TNF in fibrinogen increase. *Infect. Immun.*, **58**, 2750–2754.

Rogers, T. J., Balish, E. (1978). Immunity to experimental renal candidiasis in rats. *Infect. Immun.*, **19**, 737–740.

Rogers, T. E., Galgiani, J. N. (1986). Activity of fluconazole and ketoconazole against *C. albicans in vitro* and *in vivo*. *Antimicrob. Agents Chemother.*, **30**, 418–422.

Romani, L., Mencacci, A., Cenci, E. *et al.* (1992). Course of primary candidiasis in T cell-depleted mice infected with attenuated variant cells. *J. Infect. Dis.*, **166**, 1384–1392.

Sanati, H., Ramos, C. F., Bayer, A. S., Ghannoum, M. A. (1997). Combination therapy with amphotericin B and fluconazole against invasive candidiasis in neutropenic-mouse and infective-endocarditis rabbit models. *Antimicrob. Agents Chemother.*, **41**, 1345–1348.

Schaffner, A., Frick, P. G. (1985). The effect of ketoconazole on amphotericin B in a model of disseminated aspergillosis. *J. Infect. Dis.*, **151**, 902–910.

Shimizu, M. T., Almeida, N. Q., Fantinato, V., Unterkircher, C. S. (1996). Studies on hyaluronidase, chondroitin sulfatase, proteinase and phospholipase secreted by *Candida* species. *Mycoses*, **39**, 161–167.

Steinshamn, S., Waage, A. (1992). Tumor necrosis factor and interleukin-6 in *Candida albicans* infection in normal and granulocytopenic mice. *Infect. Immun.*, **60**, 4003–4008.

Sugar, A. M., Salibian, M., Goldani, L. Z. (1994a). Saperconazole therapy of murine disseminated candidiasis: efficacy and interactions with amphotericin B. *Antimicrob. Agents Chemother.*, **38**, 371–373.

Sugar, A. M., Hitchcock, C. A., Troke, P. F., Picard, M. (1994b). Combination therapy of murine invasive candidiasis with fluconazole and amphotericin B. *Antimicrob. Agents Chemother.*, **39**, 598–601.

Tremblay, C., Barza, M., Fiore, C., Szoka, F. (1984). Efficacy of liposome-intercalated amphotericin B in the treatment of systemic candidiasis in mice. *Antimicrob. Agents Chemother.*, **26**, 170–173.

Van't Wout, J. W., Linde, I., Leijhh, P. C., van Furth, R. (1988). Contribution of granulocytes and monocytes to resistance against experimental disseminated *Candida albicans*. *Eur. J. Clin. Microbiol. Infect. Dis.*, **7**, 736–741.

Van't Wout, J. W., Mattie, H., Van Furth, R. (1989). Comparison of the efficacies of amphotericin B, fluconazole, and itraconazole against a systemic *Candida albicans* infections in normal and neutropenic mice. *Antimicrob. Agents Chemother.*, **33**, 147–151.

Vitt, C. R., Fidler, J. M., Ando, D., Zimmerman, J., Aukerman, S. L. (1994). Antifungal activity of recombinant human macrophage colony-stimulating factor in models of acute and chronic candidiasis in the rat. *J. Infect. Dis.*, **169**, 369–374.

Wingard, J. R. (1995). Importance of *Candida* species other than *C. albicans* as pathogens in oncology patients. *Clin. Infect. Dis.*, **20**, 115–125.

Wingard, J. R., Dick, J. D., Merz, W. G., Sandford, G. R., Saral, R., Burns, W. H. (1980). Pathogenicity of *Candida tropicalis* and *Candida albicans* after gastrointestinal inoculation in mice. *Infect. Immun.*, **29**, 808–813.

Wingard, J. R., Merz, W. G., Rinaldi, M. G., Johnson, T. R., Karp, J. E., Saral, R. (1991). Increase in *Candida krusei* infection among patients with bone marrow transplantation and neutropenia treated prophylactically with fluconazole. *N. Engl. J. Med.*, **325**, 1274–1277.

Chapter 78

A Generalized *Candida albicans* Infection Model in the Rat

A. Schmidt

Introduction

A reproducible intravenous infection model of a generalized *Candida albicans* infection was established in rats. In contrast to the intravenous *C. albicans* model in mice which has a high affinity of the infective organism to the kidney, the infectious process in rats is much more severely spread into organs other than the kidneys, such as brain, heart, liver, lung, eye, and spleen. Apart from a severe granulomatous nephritis beginning 1 day after infection, a severe pneumonitis was observed 3 days after infection with a mass of extravasal erythrocytes in the interstitium and the alveolar space. Furthermore, multiple nodular lesions could be observed in the brain, heart, liver, retina and spleen on the first day after infection. Total mortality occurred within 1 week; most deaths occurred from the fifth to the seventh day. Antifungal therapy with amphotericin B (1 mg/kg body weight twice daily over 5 days given by oral gavage) or fluconazole (2.5 mg/kg respectively) led to long-term survival over 4 months. Such long-term survival could not be achieved in mice, even after sufficient therapy with high doses of the two substances, as was shown in previous tests performed in our laboratory.

Background of human fungal infections

Life-threatening systemic fungal infections are rapidly increasing in incidence (Wegmann, 1994). The increased incidence of these opportunistic infections is mainly due to factors such as more aggressive cancer and post-transplantation chemotherapy, the emergence of acquired immunodeficiency syndrome (AIDS) as well as predisposing factors resulting from improved intensive care treatment, and the widespread use of antibiotics, corticosteroids, and immunosuppressives (Ellis, 1994). All these factors result in a lowered resistance of the host towards a wide range of infectious agents, including fungal pathogens (Kwon-Chung and Bennett, 1992).

Most of these fungal infections are caused by yeasts, mainly due to *Candida* species, though hyphomycetes such as different *Aspergillus* species — and in particular *A. fumigatus* (Schmidt, 1995, 1997) — also are outstandingly important. Further, a mass of emerging rare fungal pathogens (e.g., *Fusarium* species and other miscellaneous moulds, dermatiaceous fungi, Zygomycetes, etc.) are an increasing cause of opportunistic mycoses (Anaissie *et al.*, 1989; Kwon-Chung and Bennett, 1992; de Hoog and Guarro, 1995). Yeasts account for up to 10% of nosocomial blood stream infections and are a serious cause of mortality among different patient populations (Komshian *et al.*, 1989; Banerjee *et al.*, 1991; Wenzel and Pfaller, 1991). Most of the systemic fungal infections are due to yeasts of the genus *Candida* and *C. albicans* is extremely important (Pfaller, 1989; Pfaller and Wenzel, 1992). To a lesser extent, although increasing in importance, yeast infections are also caused by non-*albicans Candida* species such as *C. tropicalis*, *C. krusei*, *C. parapsilosis*, and *Candida* (formerly also *Torulopsis*) *glabrata* (Solomon *et al.*, 1984; Gomez-Mateos *et al.*, 1988; McQuillen *et al.*, 1992). The increase of non-*albicans Candida* species in systemic yeast infections can partly be explained by a selection process due to antifungal prophylaxis regimen (Komshian *et al.*, 1989; Wingard *et al.*, 1991; Hitchcock *et al.*, 1993).

For the treatment of systemic mycotic infections, even nowadays, only a limited spectrum of antimycotic agents is available. Apart from amphotericin B and 5-fluorocytosine, broad-spectrum azoles such as fluconazole and itraconazole are the mainstay of therapy (Denning and Stevens, 1989; Simon and Stille, 1997). Especially for 5-fluorocytosine and azole antimycotics, the development of resistance of the fungi is a major problem (Fan-Havard *et al.*, 1991; Horn *et al.*, 1995; Rex *et al.*, 1995). The increased medical need as a result of increased incidence of systemic fungal infections as well as development of resistance in the causative fungal pathogens has stimulated research on new antifungal agents, for which animal testing is an indispensable necessity (Zak and Sande, 1986; Miyaji, 1987).

Background of the model

Animal experiments are essential for the study of infectious diseases, including mycotic infections (Miyaji, 1987). They

Handbook of Animal Models of Infection
ISBN 0-12-775390-7

offer the possibility of studying the parasitic forms of fungi (Miyaji and Nishimura, 1980), the host defence mechanisms in mycotic infections (Nishimura and Miyaji, 1985), and the effects of antimycotic agents *in vivo* (Zak and Sande, 1986).

In view of the increased systemic fungal infections, standardized *in vivo* models for these infections with human pathogenetic relevance (Yamaguchi, 1987), which are easy to handle, are of outstanding importance.

Where therapy has failed after antimycotic therapy together with good *in vitro* sensitivity of the strain to the corresponding antimycotic agent, animal models are useful for the assessment *in vivo* of experimental antifungal susceptibility of these selected patient strains. Furthermore, the evaluation of the correlation of different antifungal susceptibility-testing methods (Rinaldi and Troke, 1991; NCCLS, 1997) with the *in vivo* outcome of therapy requires reproducible and standardized animal tests and is a good complement to clinical studies (Ghannoum *et al.*, 1996; Denning *et al.*, 1997).

In vivo models for systemic *Candida* infections caused by the intravenous route, with or without induced immunosuppression, which have been described previously include generalized infections in mice (Louria *et al.*, 1960, 1963; Rogers and Balish, 1976; Schmidt, 1996a; Schmidt and Geschke, 1996; Schmidt *et al.* 1996), rats (Rogers and Balish, 1976; Balk *et al.*, 1978; Nakamura *et al.*, 1991; Schmidt, 1996b; Schmidt and Geschke, 1996), guinea-pigs (Winner, 1960; Hurley and Fauci, 1975), and rabbits (Walsh *et al.*, 1992).

In this chapter, a reproducible infection model of a generalized systemic *C. albicans* infection, caused by the strain *C. albicans* ATCC (American Type Culture Collection) 200498, which is identical to *C. albicans* strain BSMY 212 (Bayer Strain collection Mycology Yeasts; Schmidt, 1996b), in the Sprague-Dawley rat will be described, evaluated histopathologically and discussed. This model has been established without immunosuppression, which guarantees that no undesired interference of immunosuppressive drugs with tested antimycotic agents can occur.

Animals used

Eight-week-old specific pathogen-free male Sprague-Dawley rats (HsdOla: Sprague-Dawley SD, Winkelmann, Paderborn, Germany) weighing 200 g were housed in groups of five in type four Macrolon cages (Ehret, Emmendingen, Germany) on wood chips (Rettenmaier, Ellwangen, Germany). Photo periods were adjusted to 12 hours of light and 12 hours of darkness daily. The environmental temperature was constantly 23 °C. Altromin standard diet (Altromin, Lange, Germany) and water were provided *ad libitum*. Acclimatization time before the experiment was 1 week.

Storage of the organism and preparation of inoculum

The strain *C. albicans* ATCC 200498 (= BSMY 212) used in this model is a clinical isolate obtained from a patient with gastrointestinal candidiasis and was deposited at the American Type Culture Collection by the author. The strain was maintained by cryoconservation in liquid nitrogen over the whole time period.

For the preparation of the infection inoculum, the strain was cultivated on malt extract agar slopes (Difco, Detroit, MI) inoculated the day before harvesting and incubated at 28°C in the dark. Yeast cells were scraped off and washed twice with phosphate-buffered saline (PBS) by resuspending them in PBS on a Vortex-Mixer (Heidolph, Kelheim, Germany) with a consecutive centrifugation step at 3000 *g* for 5 minutes each time. This suspension was then further diluted with PBS to an optical density which corresponds to an inoculum of 10^7 colony-forming units (cfu) of *C. albicans* per millilitre as determined by quantitative plating in 1 : 10 dilution steps on Sabouraud 2% dextrose agar (Merck, Darmstadt, Germany), and 0.5 ml of this suspension, which corresponds to an inoculum of 5×10^6 cfu of *C. albicans*, was administered to each rat by the intravenous route.

In earlier tests performed in our laboratory, this inoculum (5×10^6 cfu per rat) was shown to be the lowest infectious dose to achieve a standardized 100% mortality within 1 week. Half of this infectious dose (2.5×10^6 cfu per rat) still leads to almost 100% mortality within 2 weeks, but the mortality pattern cannot be well-standardized with this infectious dose: there is a high degree of variation in mortality between different experiments, which was shown in previous tests in our laboratory.

Infection process

Animals were infected by injecting 5×10^6 cfu of *C. albicans* per animal in 0.5 ml PBS intravenously through the lateral tail vein using a tuberculin syringe of 1 ml with a 0.6 × 30 mm needle with a translucent hub. For the infection process, the animals were suitably restrained in a plastic rat restrainer (our own construction). Dilatation of the tail vein was achieved by a tourniquet around the base of the tail which was released after puncture, just before the injection was performed, as described by Waynforth and Flecknell (1992).

Animals were observed twice daily for mortality and behavioural abnormalities until the end of the experiment.

Histopathological technique

For the histopathological examination, animals were sacrificed by CO_2 inhalation. Organs were removed immediately after mortification and fixed in 5% aqueous formaldehyde solution. After paraffin embedding and sectioning, the material was simultaneously stained by haematoxylin and eosin (H&E), and periodic acid–Schiff (PAS; all reagents: Sigma, Deisenhofen, Germany) as a special staining method for fungal elements (for further information on the staining methods, see Bancroft *et al.*, 1990).

Parameters to monitor infection

Course of infection

Early reactions within 24 hours of infection could not be observed. On the second day after infection, encrustations of blood could be seen in the medial eye angle as well as on the tip of the nose. The first animals died on the fifth day after infection; mortality was 100% on the seventh day after infection (Figure 78.1).

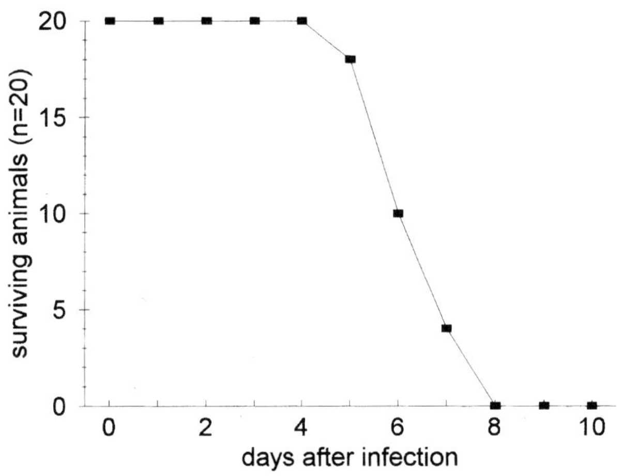

Figure 78.1 Mortality of Sprague-Dawley rats. Infection established by intravenous injection of *Candida albicans* ATCC 200498 (= BSMY 212); infectious dose: 5×10^6 colony-forming units per animal. From Schmidt (1996b), with permission.

Necropsy status

In the necropsy of rats that died of the infection, white foci were found on the surface of both kidneys. On other organs, no foci could be observed macroscopically. On the surface of the lung lobes, petechial bleedings and, in some cases also, free blood in the pleural cavity were evident.

Histopathological evaluation of the model

Kidney

One day after infection, nodular lesions with an average diameter of 0.15 mm could be found in the cortical region of the parenchyma. Centrally, a necrotic tissue reaction could be observed with polymorphonuclear leukocytes surrounding the foci. Most lesions were located interstially between the tubuli; no penetration into the glomeruli was found. Mycelial structures were predominant towards blastospores. PAS-positive red spots, which were seen in the peripheral areas of the lesions, can be interpreted as degeneration products of fungal elements. Only a few lesions were located in the medulla, and these were mostly in subcortical areas.

On the second day, granulomata had an average diameter of 0.3 mm. In the whole medulla, granulomata could be observed, penetrating from the papilla into the kidney pelvis. All granulomata contained mycelia, blastospores and chlamydospore-like structures. Tubuli were filled with protein cylinders.

On the third day, the diameter of granulomata increased significantly with severe necrotic tissue reactions. Glomeruli were mostly unaffected. Fungal elements were significantly reduced in the granulomata, combined with a massive proliferation of fungal elements in the kidney pelvis, containing mycelia, blastospores and chlamydospore-like structures. There was only a sparse infiltration of the fungal mass in the kidney pelvis with polymorphonuclear leukocytes.

On the fifth day, the whole kidney was covered with granulomata with a diameter up to 0.5 mm. Granulomata were predominantly free of fungal elements; sometimes a few chlamydospore-like structures of PAS-positive red spots could be found. The kidney pelvis was still filled with a mass of fungal elements containing blastospores and hyphae.

Spleen

On the first day after infection, only a few germinating blastospores without any inflammatous tissue reaction were found in the spleen. There was no special affinity to the white or red pulpa; foci in the spleen were self-limiting and were not seen after the second day after infection.

Liver

On the first day after infection, a few nodular lesions with a diameter of up to 0.1 mm could be seen in the liver with a quite homogeneous distribution. There was no specific affinity to central or periportal areas of the hepatic lobe. These lesions contained only a few fungal elements, mostly only a few PAS-positive red spots. A diffuse leukocyte infiltration could be observed in the foci, with a relatively high PAS density of the hepatocytes surrounding the foci. Lesions in the liver subsequently regressed during the following days.

Heart

On the first day after infection, necrotic tissue reactions with a diameter of up to 0.2 mm could be observed in the heart with quite homogeneous distribution. Only a sparse leukocyte infiltration could be seen; the foci contained mycelia, blastospores and PAS-positive red spots. The surrounding myocardial parenchyma showed a red tinge in the PAS stain. Fungal elements became rarer during the following days, with remaining focal myocardial necrotizations.

Brain

On the first day after infection, cortical and subcortical small necrotic lesions with a diameter of up to 0.05 mm were seen in the cerebrum as well as in the cerebellum and the brain stem. Some foci contained mycelia and blastospores without any inflammatory tissue response, although highly reactive processes with an extended polymorphonuclear leukocytic infiltration and PAS-positive red spots could be observed. Fungal elements in the foci decreased from the second day with remaining necrotic tissue reactions.

Eye

On the third day after infection, lesions with predominantly blastospores could be observed in the retina.

Lung

The lung showed only discrete pathological changes with a few erythrocytes and lymphocyte-like cells in an apparently widened interstitium up to the second day after infection. On the third day after infection, an extensive pneumonitis with a mass of extravasal erythrocytes in the interstitium as well as the alveolar space could be seen. The lung was histopathologically free of fungal elements.

Antimycotic therapy and response to treatment

For therapy, five rats in each group were treated with fluconazole (CAS 86386-73-4; Pfizer, Karlsruhe, Germany) or amphotericin B (CAS 1397-89-3 [Na-desoxycholate complex]; Squibb-Heyden, Munich, Germany) at doses of 0.5, 1, 2.5 and 5 mg of free substance per kg body weight. Substances were administered in 0.5 ml glucose (5%) agar (0.2%) solution (chemicals: Sigma) by oral gavage, twice daily with an interval of 8 hours in between over 5 days starting 1 hour after infection. Animals were examined twice daily up to 4 months after infection.

The minimal oral dose to achieve a 100% survival rate over 4 months was 2.5 mg/kg body weight twice daily over 5 days for fluconazole. With the dose of 1 mg/kg under the same administration regimen, an 80% survival rate was achieved within 1 month (4 of 5 animals survived); the dose of 0.5 mg/kg led to a 20% survival rate (1/5 animals) respectively. The minimal dose to achieve a 100% survival rate over 4 months was 1 mg/kg body weight twice daily over 5 days for amphotericin B. With the dose of 0.5 mg/kg under the same administration regimen, a 60% survival rate was achieved within 1 month (3 of 5 animals survived).

Validation of the model

In this rat model of a generalized *C. albicans* infection, a severe progressive granulomatous nephritis was found starting 1 day after infection. A severe progressive kidney dysfunction seems to be a main reason for the mortality. Apart from that, a severe bleeding disorder started around the second day after infection, with an extended pneumonitis with a mass of extravasal erythrocytes in the interstitium and the alveolar space. In addition, petechial bleedings on the surface of the lung lobes as well as free blood in the pleural cavity, in the eye angles, and on the tip of the nose could be observed. The mechanisms of these bleeding phenomena still need further clarification and may possibly help to understand bleeding disorders as they occur in fungal diseases such as aspergillomata in humans (Raper and Fennell, 1977; Wegmann, 1994). The focal myocarditis could have haemodynamic consequences and could be another mortality factor. Although infectious foci were found in the central nervous system, no severe behavioural disorders of the animals could be observed apart from a slowly progressive sedation.

The course of infection in liver and spleen was rather limited. On the fifth day after infection, almost no fungi could be microscopically found in the kidney parenchyma, though the kidney pelvis was massively filled with fungal elements. This is the reason why the worth of quantitative plating of mashed organs, such as kidneys, in order to assess the fungal organ burden seems to have severe limitations and should be combined with histopathological methods. Quantitative plating does not allow the differentiation between parenchymal infection and cavital proliferation/persistence of the organisms which can be achieved by histopathological evaluations.

In contrast to the model in Wistar rats, as described by Balk *et al.* (1978), who observed a continuous distribution of deaths from the first day up to the 10th day after infection, this model shows a later but much more rapid pattern of mortality with nearly all deaths occurring from the fifth to the seventh day after infection. Further, cases of *C. albicans* arthritis as described by Nakamura *et al.* (1991) were not observed in this model.

Contributions to antimycotic therapy

This *C. albicans* infection in rats offers the possibility to induce a more generalized, systemic experimental infection with dissemination into several other organs apart from the kidneys. This model mimics a generalized invasive candidosis much more than the intravenous *C. albicans* infection model in mice, which is much more restricted to the kidneys as almost the only target organ.

The rat model described shows a highly standardized mortality pattern which occurs within 1 week. With only five animals per group, even weak signals of efficacy of substances can be detected in a highly reproducible way in the search for new antimycotic lead structures, as is known from the routine use of this model in our laboratory. Further, the 1-week mortality allows one to perform substance evaluations within a suitable period of time. Apart from the oral substance application described above, subcutaneous, intravenous, and intraperitoneal applications can easily be performed and show efficacy with the two antimycotics tested, as we know from our experience in our laboratory.

The long-term survival over 4 months we observed after antimycotic therapy with amphotericin B (1 mg/kg body weight twice daily over 5 days) and fluconazole (2.5 mg/kg respectively) could not be achieved after therapy of the systemic *C. albicans* infection in mice with the same two antimycotics. Also, after sufficient therapy with high doses of these two substances, almost all mice showed late cases of deaths, even occurring 1 month or later after infection. From our histopathological and blood chemical observations (unpublished data), these late deaths seem not to be due to the primary mycotic infection but rather can be attributed to aseptic necrotic residues of nephritis of still unknown origin causing a progressive renal failure. This phenomenon was not observed in the rat model described.

This rat model of a systemic *C. albicans* infection is a good complement to the familiar systemic candidiasis model in mice and is especially appropriate in second-line *in vivo* testings in the search for new antimycotic lead structures as well as for the development and evaluation of the potency of antimycotic agents in a species other than mice.

A relative disadvantage of this model over the *Candida*-infection model in mice is that on average 10 times more substance is required to treat rats with an average body weight of 200 g per animal, in contrast to young mice which have a body weight of 15–20 g per animal on average.

Acknowledgement

All animal tests reported here were performed in strict accordance with federal regulations. Results of this work have already partly been published by the author (Schmidt, 1996b).

References

Anaissie, E. J., Bodey, G. P., Rinaldi, M. G. (1989). Emerging fungal pathogens. *Eur. J. Clin. Microbiol. Infect. Dis.,* **8**, 323–330.

Balk, M. W., Crumrine, M. H., Fischer, G. W. (1978). Evaluation of miconazole therapy in experimental disseminated candidiasis in laboratory rats. *Antimicrob. Agents Chemother.*, **13**, 321–325.

Bancroft, J. D., Stevens, A., Turner, D. R. (1990). *Theory and Practice of Histological Techniques*. Churchill Livingstone, Edinburgh, UK.

Banerjee, S. N., Emori, T. G., Culver, D. H. *et al.* (1991). Secular trends in nosocomial primary blood stream infections in the United States, 1980–1989. *Am. J. Med.*, **91** (suppl. 3B), 86–89.

de Hoog, G. S., Guarro, J. (1995). *Atlas of Clinical Fungi*. Centraalbureau voor Schimmelcultures, Baarn/Delft, Netherlands.

Denning, D. W., Stevens, D. A. (1989). New drugs for systemic fungal infections. Greater choice and more difficult clinical decisions. *Br. Med. J.*, **299**, 407–408.

Denning, D. W., Baily, G. G., Hood, S. V. (1997). Azole resistance in *Candida*. *Eur. J. Microbiol. Infect. Dis.*, **16**, 261–280.

Ellis, D. H. (1994). *Clinical Mycology. The Human Opportunistic Mycoses*. Gillingham Prints, Underdale, Australia.

Fan-Havard, P., Capano, D., Smith, S. M., Mangia, A., Eng, R. H. K. (1991). Development of resistance in *Candida* isolates from patients receiving prolonged antifungal therapy. *Antimicrob. Agents Chemother.*, **35**, 2302–2305.

Ghannoum, M. A., Rex, J. H., Galgiani, J. N. (1996). Susceptibility testing of fungi: current status of correlation of *in vitro* data with clinical outcome. *J. Clin. Microbiol.*, **34**, 489–495.

Gomez-Mateos, J. M., Porto, A. S. *et al.* (1988). Disseminated candidiasis and gangrenous cholecystitis due to *Candida* spp. *J. Infect. Dis.*, **158**, 653–655.

Hitchcock, C. A., Pye, G. W., Troke, P. F., Warnock, D. W. (1993). Fluconazole resistance in *Candida glabrata*. *Antimicrob. Agents Chemother.*, **37**, 1962–1965.

Horn, C. A., Washburn, R. G., Givner, L. B., Peacock, J. E. Jr, Pegram, P. S. (1995). Azole-resistant oropharyngeal and esophageal candidiasis in patients with AIDS. *AIDS*, **9**, 533–535.

Hurley, D. L., Fauci, A. S. (1975). Disseminated candidiasis. An experimental model in guinea-pigs. *J. Infect. Dis.*, **131**, 516–521.

Komshian, S. V., Uwaydak, A. K., Sobel, J. D., Crane, L. R. (1989). Fungemia caused by *Candida* species and *Torulopsis glabrata* in the hospitalized patient: frequency, characteristics, and evaluation of factors influencing outcome. *Rev. Infect. Dis.*, **145**, 45–56.

Kwon-Chung, K. J., Bennett, J. E. (1992). *Medical Mycology*. Lea & Febiger, Philadelphia, USA.

Louria, D. B., Fallon, N., Browne, H. G. (1960). The influence of cortisone on experimental fungus infections in mice. *J. Clin. Invest.*, **39**, 1435–1449.

Louria, D. B., Brayton, R. G., Finkel, G. (1963). Studies on the pathogenesis of experimental *Candida albicans* infection in mice. *Sabouraudia*, **2**, 271–283.

McQuillen, D. P. *et al.* (1992). Invasive infections due to *Candida krusei*. Report of ten cases of fungemia that include three cases of endophthalmitis. *Clin. Infect. Dis.*, **14**, 472–478.

Miyaji, M. (1987). *Animal Models in Medical Mycology*. CRC Press, Boca Raton, USA.

Miyaji, M., Nishimura, K. (1980). Parasitic forms of pathogenic fungi. *Jpn. J. Clin. Med.*, **38**, 14–19.

Nakamura, Y., Masuhara, T., Ito-Kuwa, S., Aoki, S. (1991). Induction of experimental *Candida* arthritis in rats. *J. Med. Vet. Mycol.*, **29**, 179–192.

NCCLS (National Committee for Clinical Laboratory Standards; 1997). *Reference Method for Broth Dilution Antifungal Susceptibility Testing of Yeast; Approved Standard M 27-A*, vol. 17 (no. 9). NCCLS, Wayne, USA.

Nishimura, K., Miyaji, M. (1985). Tissue responses against *Cladosporium trichoides* and its parasitic forms in congenitally athymic nude mice and their heterozygous littermates. *Mycopathologia*, **90**, 21–27.

Pfaller, M. A. (1989). Infection control: opportunistic fungal infection — the increasing importance of *Candida* species. *Infect. Control Hosp. Epidemiol.*, **9**, 408–416.

Pfaller, M. A., Wenzel, R. (1992). Impact of the changing epidemiology of fungal infections in the 1990s. *Eur. J. Clin. Microbiol. Infect. Dis.*, **11**, 287–291.

Raper, K. B., Fennell, D. I. (1977). *The Genus Aspergillus*. Krieger, Malabar, FL, USA.

Rex, J. H., Rinaldi, M. G., Pfaller, M. A. (1995). Resistance of *Candida* species to fluconazole. *Antimicrob. Agents. Chemother.*, **39**, 1–8.

Rinaldi, M. G., Troke, P. F. (1991). *Antifungal Susceptibility Testing; A Manual of Methods in Development*. Pfizer, New York, USA.

Rogers, T., Balish, E. (1976). Experimental *Candida albicans* infection in conventional mice and germfree rats. *Infect. Immun.*, **14**, 33–38.

Schmidt, A. (1995). Diagnostics in aspergillosis [Diagnostik der Aspergillose; in German]. *Münch. Med. Wochenschr.*, **137**, 607–610.

Schmidt, A. (1996a). Pathogenicity of *Candida tropicalis*, *Candida krusei*, and *Torulopsis (Candida) glabrata* for outbred CFW1 mice. *J. Mycol. Med.*, **6**, 133–135.

Schmidt, A. (1996b). Systemic candidiasis in Sprague-Dawley rats. *J. Med. Vet. Mycol.*, **34**, 99–104.

Schmidt, A. (1997). *Pathogenetic Relevance of Molds of the Genus* Aspergillus *with Studies Concerning the Morphological Variability and Chemosensitivity of* Aspergillus fumigatus [Pathogenetische Bedeutung von Schimmelpilzen der Gattung *Aspergillus* mit Untersuchungen zur morphologischen Variabilität und *in vitro* Sensibilität von *Aspergillus fumigatus*; in German]. Deutsche Hochschulschriften 2449. Hänsel-Hohenhausen, Egelsbach, Germany.

Schmidt, A., Geschke, F. U. (1996). Comparative virulence of *Candida albicans* strains in CFW1 mice and Sprague-Dawley rats. *Mycoses*, **39**, 157–160.

Schmidt, A., Geschke, F. U., Jaroch, M., Osterkamp, M., Rühl-Hörster, B. (1996). Virulence of *Candida albicans* in outbred and C5-deficient inbred mice [Virulenz von *Candida albicans* in Auszucht- und C5-defekten Inzucht-Mäusestämmen; in German]. Abstract V15, MYK 96–30. Sc. Symp. DMyKG.

Simon, C., Stille, W. (1997). *Antibiotics in Clinic and Practice* [Antibiotikatherapie in Klinik und Praxis; in German]. Schattauer, Stuttgart, Germany.

Solomon, S. L. *et al.* (1984). An outbreak of *Candida parapsilosis* bloodstream infections in patients receiving parenteral nutrition. *J. Infect. Dis.*, **149**, 98–102.

Walsh, T. J., Lee, J. W., Roilides, E. *et al.* (1992). Experimental antifungal chemotherapy in granulocytopenic animal models of disseminated candidiasis. Approaches to understanding investigational antifungal compounds for patients with neoplastic diseases. *Clin. Infect. Dis.* **14** (suppl. 1), 139–147.

Waynforth, H. B., Flecknell, P. A. (1992). *Experimental and Surgical Technique in the Rat*. Academic Press, London, UK.

Wegmann, T. (1994). *Medical Mycology, A Practical Guideline* [Medizinische Mykologie — ein praktischer Leitfaden; in German]. Editiones Roche, Basel, Switzerland.

Wenzel, R. P., Pfaller, M. A. (1991). *Candida* species: emerging hospital blood stream pathogens. *Infect. Control Hosp. Epidemiol.*, **12**, 523–524.

Wingard, J. R., Merz, W. G., Rinaldi, M. G., *et al.* (1991). Increase in *Candida krusei* infections among patients with bone marrow transplantation and neutropenia treated prophylactically with fluconazole. *N. Engl. J. Med.*, **325**, 1274–1277.

Winner, H. J. (1960). Experimental moniliasis in the guinea-pig. *J. Pathol. Bacteriol.*, **79**, 420–423.

Yamaguchi, H. (1987). Opportunistic fungal infections. In *Animal Models in Medical Mycology* (ed. Miyaji, M.), pp. 101–158. CRC Press, Boca Raton, USA.

Zak, O., Sande, M. A. (1986). Introduction: the role of animal models in the evaluation of new antibiotics. In: *Experimental Models in Antimicrobial Chemotherapy* (eds Zak, O., Sande, M. A.), vol. 1. Academic Press, London, UK.

Chapter 79

Experimental Oropharyngeal and Gastrointestinal *Candida* Infection in Mice

A. M. Flattery, G. K. Abruzzo, C. J. Gill, J. G. Smith and K. Bartizal

Background of model

Maintenance of *Candida albicans* in the gastrointestinal tract of non-immune-compromised mice is short-lived, with the host immune system rapidly clearing the yeast. Evidence suggests that cell-mediated immunity, and more specifically, CD4$^+$ T lymphocytes, play an important role in resistance to mucosal candidiasis (Rogers and Balish, 1980; Balish *et al.*, 1990; Cantorna and Balish, 1991). In particular, the decrease in CD4$^+$ T-lymphocyte counts associated with human immunodeficiency virus (HIV) infection and acquired immunodeficiency syndrome (AIDS) has been correlated with the rise in cases of alimentary tract candidiasis (Epstein *et al.*, 1984; Klein *et al.*, 1984; Glatt *et al.*, 1988; Pankhurst and Peakman, 1989). Other mouse models of mucosal candidiasis have been described which utilize combinations of chemically induced immune suppression, elimination or alteration of host microflora by administration of antibiotics, high inocula, trauma, infant animals or animals with congenital, functional, physiological, immunological or metabolic defects to facilitate colonization of the gastrointestinal tract by *C. albicans* (Nolting, 1975; Jorizzo *et al.*, 1980; Field *et al.*, 1981; Guentzel and Herrera, 1982; Ekenna and Fader, 1989; Cole *et al.*, 1990; Narayanan *et al.*, 1991). This model of oropharyngeal and gastrointestinal candidiasis was designed to represent the immune status of the patient population in which mucosal candidiasis is prevalent. It employs a combination of selective CD4$^+$ T-cell depletion to initiate a specific immune deficiency and antibiotic reduction of the normal gastrointestinal microflora to allow colonization of the alimentary tract by *Candida*. Efficacy of therapy with a variety of known and novel antifungals may then easily be assessed in this model.

Animal species

Mice deficient in CD4$^+$ T-lymphocyte function facilitate prolonged colonization of the alimentary tract by *C. albicans*. This immune deficiency may be created in a variety of ways in both immune-competent and immune-deficient mice as well as in outbred and inbred strains. Mice may be selectively depleted of CD4$^+$ T-lymphocytes by treatment with a rat immunoglobulin G_2 monoclonal antibody (MAb) secreted by GK1.5 hybridoma cells (American Type Culture Collection, Rockville, MD, Culture #TIB 207), which is specific for mouse CD4$^+$ T cells. Alternately, mice may be injected directly with the hybridoma cells subcutaneously, leading to *in vivo* secretion of MAb and depletion of the CD4$^+$ population. We commonly use complement component 5-deficient female DBA/2 mice (Taconic, Germantown, NY) since these mice show acceptable duration of depletion of CD4$^+$ T cells and are easily colonized with *C. albicans* both orally and in a disseminated model of candidiasis (Bartizal *et al.*, 1992). Similar methods of CD4$^+$ T-cell depletion have also been used in C3Heb/FeJ mice with comparable results and in BALB/c mice with a shorter duration of CD4$^+$ T-cell depletion (McFadden *et al.*, 1994). Presumably other strains of mice may be used; however the efficacy of the immune suppression would need to be verified by fluorescence-activated cell sorter (FACS) analysis and perhaps the level of MAb administered adjusted accordingly. Recently, transgenic mice have been developed which lack certain genes necessary for CD4$^+$ T-lymphocyte production and thus which are CD4$^+$ T-lymphocyte deficient. These mice are available from multiple sources, including Jackson Laboratories (Bar Harbor, ME) and eliminate the need for antibody treatment altogether. Fecal *C. albicans* colonization levels after oral inoculation in these transgenic animals are similar to those in MAb-induced CD4$^+$ T-lymphocytopenic mice.

Immune suppression

In DBA/2 mice three intraperitoneal (i.p.) injections of 300 μg purified MAb per mouse in sterile saline administered 3 days prior to, the day of and 1 week after challenge maintain depletion of CD4$^+$ T cells through 14 days after challenge. A single subcutaneous (s.c.) injection of 9×10^6 GK1.5 hybridoma cells 4 days prior to challenge maintains CD4$^+$ T-cell depletion through 21 days after challenge.

ISBN 0-12-775390-7

Cell culture

GK1.5 hybridoma cells are cultured in high glucose Dulbecco's Modified Eagle's Medium (D-MEM, Sigma, St Louis, MO) supplemented with 10% fetal bovine serum (Sigma), 1% L-glutamine (Sigma), 100 units/ml penicillin (Sigma) and 100 µg/ml streptomycin (Sigma) at 37°C under 5% CO_2. Cells for injection should be passaged and incubated for log phase growth. Cultures are then harvested, centrifuged for 8 minutes at 400*g* to pellet cells and washed twice with the above medium. Resuspended cells should be counted on a hemacytometer, with viability confirmed by trypan blue dye exclusion, and adjusted to the appropriate concentration for $CD4^+$ T-cell depletion by s.c. injection or for generation of MAb by ascites production.

Ascites production

Athymic *nu/nu* mice are used for ascites production. In our experience outbred athymic *nu/nu* Swiss Webster mice (Taconic) work well. Mice are primed by i.p. injection with 0.5 ml of pristane (2,6,10,14-tetramethylpentadecane, Sigma). Ten days after pristane priming, mice are injected i.p. with 5×10^6 GK1.5 hybridoma cells prepared as stated above. Ascites is then collected from the mice, centrifuged for 10 minutes at 400*g* to remove cellular debris, and then stored frozen at –20°C until use for antibody purification. MAb is purified from ascites by passage over a protein G column (MabTrap G, Pharmacia LKB, Piscataway, NJ) and quantitated by Bio-Rad Protein Assay (Bio-Rad, Rockville Center, NY) using a bovine gamma globulin standard. Purified MAb is then stored at –20°C until used for immune suppression.

FACS analysis

The depletion of $CD4^+$ T lymphocytes is monitored using FACS analysis. Splenic T lymphocytes may be stained with fluoresceinated MAbs specific for mouse $CD4^+$, $CD8^+$ or $CD3^+$ T cells. Spleens are removed aseptically and using frosted glass microscope slides splenic tissue is teased apart to create a cell suspension. Cells are suspended in 3 ml phosphate-buffered saline (PBS, GIBCO, Grand Island, NY) and centrifuged for 5 minutes at 400*g*. The supernatant is decanted and the cell pellet resuspended in 1 ml ACK lysing buffer (GIBCO), vortexed for 1 minute to lyse red blood cells, then diluted in 3 ml PBS and centrifuged for 5 minutes at 400*g*. The supernatant is again decanted and cells are resuspended in 4 ml PBS. Then, 100 µl of the spleen cell suspension is incubated with rat anti-mouse MAb at a concentration of 5 µg/ml for 30 minutes at room temperature. The cells are stained either with fluorescein isothiocyanate (FITC) conjugated L3T4 (CD4) (PharMingen, San Diego, CA) and R-phycoerythrin (PE) conjugated Ly-2 (CD8a) (PharMingen), or with FITC conjugated Thy-1.2 (PharMingen), which reacts with 100% of T cells in mice expressing the Thy-1.2 allele. Cells are washed with 3 ml PBS and centrifuged for 5 minutes at 400*g*. The supernatant is then decanted and the cell pellet resuspended in 200 µl propidium iodide (Sigma) at 1 µg/ml. Samples are run on the FACScan analyzer (Becton Dickinson, San Jose, CA) to determine percentage of total lymphocytes which are $CD4^+$, $CD8^+$ or $CD3^+$ T cells.

Preparation of animals

Housing

Animals should be housed in sterile microisolator cages and given sterile feed and water *ad libitum*. Mice are pretreated with gentamicin (Garamycin injectable, Schering, Kenilworth, NJ), a non-absorbable broad-spectrum antibacterial, at 0.1 mg/ml in the drinking water from 4 days prior to *C. albicans* challenge through 3 days post-challenge to reduce the normal gastrointestinal microflora, allowing less competition for colonization of the gastrointestinal tract by *C. albicans*.

Preparation of inocula and infection process

Stock cultures of *C. albicans* MY1055 (Merck Culture Collection) are maintained by monthly transfers on Sabouraud dextrose agar (SDA, BBL, Cockeysville, MD). Growth from an 18–24-hour SDA culture of *C. albicans* is suspended in sterile saline to a concentration of 10^8 cells/ml determined by hemacytometer count and verified by plate counts. Mice are challenged by gavage with 0.2 ml of the yeast suspension (approximately 2×10^7 cells/mouse) and additionally by swabbing their oral cavities with the yeast suspension, while gently abrading the buccal mucosa by rotation of the swab. This model should be amenable to use of other strains of *C. albicans* as well as other species of *Candida*; however, colonization levels may vary between strains and species and challenge amounts may need to be adjusted accordingly.

Key parameters to monitor infection and response to treatment

Infection in this model is confined to the oropharynx and gastrointestinal tract and translocation of the yeast out of the gastrointestinal tract is rare, and in fact difficult to induce. Despite high levels of colonization mice show no outward signs of illness. The extent of colonization of the alimentary tract after challenge may then easily be monitored throughout the study by culturing fecal samples and by swabbing the oral mucosa. This model also allows the investigator to follow individual mice throughout the course of treatment since mice need not be euthanized for sampling.

Fresh fecal pellets are collected from each mouse, weighed, homogenized in sterile saline, serially diluted and plated on Sabouraud's dextrose agar containing 50 µg/ml chloramphenicol (SDAC) for inhibition of bacterial growth. Oral swabs may also be taken from each mouse and plated on SDAC for a qualitative estimate of oral *Candida* load. At termination of the study segments of the alimentary tract may also be cultured as above to determine levels of colonization. Typically DBA/2 mice carry a *Candida* load of between 4 and 5 log colony-forming units (cfu) per gram feces through at least 14 days. Oral *Candida* cfu may vary markedly depending on the technique of the individual swabbing the oral cavity and the method of applying the swabbed material to the culture plate.

Antifungal therapy

Many different routes of administration of antifungal therapy may be used in this model, including parenteral injection, administration by gavage (p.o.) and administration in the drinking water. Administration i.p., s.c. and p.o. allows exactly measured amounts of drug to be delivered, while administration in the drinking water gives an approximate dose level, given that mice drink approximately 5 ml of water per day (Ralston Purina Company, 1961). However, the drinking water route of administration allows for therapy similar to the "swish-and-swallow" routes now used clinically with both Diflucan (fluconazole for oral suspension) and Fungizone (amphotericin B oral suspension). The route chosen may be affected by the properties of the compound given. For example, administration in the drinking water requires that the compound be relatively soluble in a formulation suitable for oral administration to mice, since insoluble compounds tend to settle at the bottom of the water bottle. These compounds must also be relatively stable at room temperature. An experimental chemotherapeutic not having these characteristics may need to be administered by an alternate route.

Typically, antifungal therapy in this model is begun 3 days after challenge so that the yeast has sufficient time to colonize the alimentary tract and also so that a pretreatment colonization level may be determined. Therapy is then begun by any of the above routes and continued for varying lengths of time based on the characteristics of the antifungal under study. A dosing regimen of 10 days (days 3–13 after challenge) allows maximal time for the antifungal to reduce the *Candida* load, while also maintaining the test to within the 14 days in which antibody-treated mice are CD4-depleted and colonization levels in control mice are still high. Since mice are monitored throughout the course of therapy the rapidity and extent of efficacy may easily be determined. Also, rebound in growth of any uncleared yeasts may be seen if fecal cultures are taken 24 hours after the termination of therapy at day 14 after challenge.

We have used this model to assess the efficacy of novel antifungal agents as compared to the efficacy of agents currently marketed for oropharyngeal candidiasis. The activity of currently marketed agents in this model correlated well with their efficacy in humans. Fluconazole administered for 10 days in the drinking water at 100–400 µg/ml (approximately 25–100 mg/kg per day) rapidly decreases the fecal and oral *Candida* load in this model. The polyene nystatin administered in the drinking water was somewhat less efficacious, with fecal and oral colonization levels decreasing more slowly and to a lesser extent than the azole (Flattery *et al*., 1996).

Advantages and disadvantages of the model

There is no difference in gastrointestinal colonization by *Candida* in mice depleted of $CD4^+$ T cells by either the hybridoma or antibody method; however, there are certain side-effects associated with each method. In long-term experiments, multiple injections of rat GK 1.5 MAb may cause some mortality in mice. A single injection of hybridoma cells requires no antibody purification and less animal handling but it also has some disadvantages. Approximately 50% of DBA/2 mice injected with hybridoma cells develop visible tumors beginning 3 weeks following hybridoma injection. Many of these tumors become so large that they are lethal, while approximately one in eight tumors regresses.

All mice with visible tumors are depleted of $CD4^+$ T cells and depletion may continue for up to 5 weeks after tumor regression. In mice with no visible tumors $CD4^+$ T cells begin to reappear at 4 weeks after injection of hybridoma cells. For experimental procedures requiring less than 3 weeks to complete, these tumors pose no problem. However, in long-term experiments, significant mortality and the reappearance of $CD4^+$ T cells after 4 weeks must be anticipated. These tumors are observed using DBA/2 mice; however, in other strains of mice such as BALB/c and C3Heb/FeJ mice, visible tumors have not formed and the length of $CD4^+$ T-cell depletion varied (McFadden *et al*., 1994). For these reasons use of transgenically $CD4^+$ T-lymphocyte deficient mice, although possibly more costly, might be preferable to either antibody or hybridoma treatment, especially for long-term experiments.

Although this model mimicks the immune status of the patient population in which the infection commonly occurs, it may not be truly representative of the disease state seen in humans. Histology of the oropharynx and gastrointestinal tract of DBA/2 mice shows little adhesion of *C. albicans* to the mucosal surface and little, if any, inflammatory response, indicating a colonization rather than true infection. In 85% of DBA/2 mice intragastrically inoculated with *C. albicans* and administered antibiotics over a long term, small self-limiting foci of mucosal involvement in the stomach were seen, but *C. albicans* was eventually cleared without dissemination (Bistoni *et al.*, 1993). In fact it has been shown that in mice a combination of both cell-mediated immunity and phagocytic cell defects is necessary for extensive infection or invasion of the gastrointestinal tract (Cantorna and Balish, 1990). The CD4+ T-lymphocyte-deficient DBA/2 mice offer a good model of chronic *Candida* without dissemination, in which *in vivo* antifungal activities correlate well with *in vitro* susceptibility of the isolate to the antifungal compounds.

Contributions of the model to infectious disease therapy

With the increase in numbers of immunocompromised patients, there has been a significant rise in the incidence of fungal infections, including alimentary tract candidiasis. Unlike antibacterials, few antifungal agents exist which are both safe and efficacious. Also, with the incidence of azole-resistant *Candida* strains on the rise, the need for novel antifungal agents is increasing. This model of oropharyngeal and gastrointestinal candidiasis may give a good indication of the efficacy of novel agents as compared to currently marketed antifungals. These agents may then also be useful in more severe models of disseminated fungal disease.

References

Balish, E., Filutowicz, H., Oberly, T. D. (1990). Correlates of cell-mediated immunity in *Candida albicans*-colonized gnotobiotic mice. *Infect. Immun.*, **58**, 107–113.

Bartizal, K., Abruzzo, G., Trainor, C. *et al.* (1992). *In vitro* antifungal activities and *in vivo* efficacies of 1,3-β-D glucan synthesis inhibitors L-671,329, L-646991, tetrahydroechinocandin B, and L-687,781 a papulacandin. *Antimicrob. Agents Chemother.*, **36**, 1648–1657.

Bistoni, F., Cenci, E., Mencacci, A. *et al.* (1993). Mucosal and systemic T helper cell function after intragastric colonization of adult mice with *Candida albicans*. *J. Infect. Dis.*, **168**, 1449–1457.

Cantorna, M. T., Balish, E. (1990). Mucosal and systemic candidiasis in congenitally immunodeficient mice. *Infect. Immun.*, **58**, 1093–1100.

Cantorna, M. T., Balish, E. (1991). Role of CD4+ lymphocytes in resistance to mucosal candidiasis. *Infect. Immun.*, **59**, 2447–2455.

Cole, G. T., Lynn, K. T., Seshan, K. R. (1990). An animal model for oropharyngeal, esophageal, and gastric candidosis. *Mycoses*, **33**, 7–19.

Ekenna, O., Fader, R. C. (1989). Effect of thermal injury and immunosuppression on the dissemination of *Candida albicans* from the mouse gastrointestinal tract. *J. Burn Care Rehabil.*, **10**, 138–145.

Epstein, J. B., Truelove, E., Izutzu, K. T. (1984). Oral candidiasis: pathogenesis and host defense. *Rev. Infect. Dis.*, **6**, 96–106.

Field, L. H., Pope, L. M., Cole, G. T., Guentzel, M. N., Berry, L. J. (1981). Persistence and spread of *Candida albicans* after intragastric inoculation of infant mice. *Infect. Immun.*, **31**, 783–791.

Flattery, A. M., Abruzzo, G. K., Gill, C. J., Smith, J. G., Bartizal, K. (1996). New model of oropharyngeal and gastrointestinal colonization by *Candida albicans* in CD4+ T-cell-deficient mice for evaluation of antifungal agents. *Antimicrob. Agents Chemother.*, **40**, 1604–1609.

Glatt, A. E., Chirgwin, K., Landesman, S. H. (1988). Treatment of infections associated with human immunodeficiency virus. *N. Engl. J. Med.*, **3**, 1439–1448.

Guentzel, M. N., Herrera, C. (1982). Effects of compromising agents on candidosis in mice with persistent infections initiated in infancy. *Infect. Immun.*, **35**, 222–228.

Jorizzo, J. L., Sams, W. M., Jr, Jegasothy, B. V., Olansky, A. J. (1980). Cimetedine as an immunomodulator: chronic mucocutaneous candidiasis as a model. *Ann. Intern. Med.*, **92**, 192–195.

Klein, R. S., Harris, C. A., Small, C. B., Moll, B., Lesser, M., Friedland, G. H. (1984). Oral candidiasis in high-risk patients as the initial manifestation of the acquired immunodeficiency syndrome. *N. Engl. J. Med.*, **311**, 354–358.

McFadden, D. C., Powles, M. A., Smith, J. G., Flattery, A. M., Bartizal, K., Schmatz, D. M. (1994). Use of anti-CD4+ hybridoma cells to induce *Pneumocystis carinii* in mice. *Infect. Immun.*, **62**, 4887–4892.

Narayanan, R., Joyce, W. A., Greenfield, R. A. (1991). Gastrointestinal candidiasis in a murine model of severe combined immunodeficiency syndrome. *Infect. Immun.*, **59**, 2116–2119.

Nolting, S. (1975). Effect of antibiotics and cytostatic drugs on experimental candidiasis in mice. *Mykosen*, **18**, 309–313.

Pankhurst, C., Peakman, M. (1989). Reduced CD4+ T cells and severe oral candidiasis in absence of HIV infection. *Lancet*, **i**, 672.

Ralston Purina Company (1961). *Laboratory Animal Care: Management of Laboratory Animals, the Mouse*, section 2, p. 5. Ralston Purina Co., St. Louis, MO.

Rogers, T. J., Balish, E. (1980). Immunity to *Candida albicans*. *Microbiol. Rev.*, **44**, 660–682.

Chapter 80

Paw Oedema as a Model of Localized Candidiasis

G. Findon

Background of model

Existing models of localized *Candida albicans* infections are semiquantitative and do not allow continuous observations to be made in individual animals. These models include the mouse thigh model (Pearsall and Lagunoff, 1974), cutaneous lesions in rodents (Ray and Wuepper, 1976) and guinea-pigs (Sohnle *et al.*, 1976), subcutaneous lesions (Giger *et al.*, 1978) and diffusion chambers (Poor and Cutler, 1981). This chapter describes a procedure for the quantitation of experimentally induced candidiasis which uses the inflammatory response to infection as a surrogate measure of living yeast cell numbers in the footpad. The model allows consecutive readings in the same animal and correlates well with the number of yeast cells in the local lesion. The animals tolerate the procedure well and show few signs of discomfort.

Animal species

Male and female animals from an inbred colony of Dark Agouti rats, weighing between 200 and 250 g were used to characterize this model. Other strains may be suitable but have not been evaluated. The animals are housed in polycarbonate cages on litter of untreated wood shavings and fed pelleted food (Diet 86) and water *ad libitum*.

Preparation of animals

No specific care or pretreatment is required.

Details of surgery

Overview

Live *Candida albicans* is injected intradermally directly into the footpad and the resulting inflammatory response, measured as swelling, is used as a correlate of yeast cell numbers.

Materials required

Anaesthetic, skin disinfectant (95% alcohol), Hamilton syringe and 26 G needle and engineer's thickness gauge are required.

Anaesthesia

Animals are anaesthetized with Nembutal (pentobarbitone sodium, Techvet Laboratories, Auckland, NZ) or other suitable short-acting anaesthetic. No pain relief is necessary.

Infection procedure

After anaesthesia, the hind footpads are sterilized with 95% alcohol. The inoculum of yeast cells is suspended in 20 µl of sterile normal saline and injected intradermally into the plantar region of each hind footpad, using a 26 G needle and a 0.1 ml Hamilton syringe. Great care must be taken to ensure that the inoculum is injected intradermally, not subcutaneously (Gray and Jennings, 1955). This is confirmed by the appearance of a skin blister at the site of injection.

Postoperative care

No special postoperative care is required and the animals resume normal activities immediately after recovery from the anaesthetic.

Storage and preparation of inocula

The inoculum consists of a culture of *Candida albicans* from a clinical specimen, maintained in Sabouraud dextrose broth at –20°C and subcultured monthly on to Sabouraud

Handbook of Animal Models of Infection
ISBN 0-12-775390-7

dextrose agar for storage at 4°C. The cultures are grown overnight in nutrient broth at 37°C and washed three times in saline. Cell concentration is adjusted after counting the yeast cells in an improved Neubauer counting chamber. The number of viable organisms is confirmed by colony counts in serial 10-fold dilutions in pour plates of Sabouraud dextrose agar.

Measurement of footpad oedema

Footpad thickness is measured centrally between the dorsal and plantar surfaces using an engineer's pocket thickness gauge (model 7309, range 0.01–9.00 mm ± 0.01 mm, Mitutoyo, Tokyo, Japan), previously modified to avoid pressure on the swollen tissue by reducing the spring tension (Winter *et al.*, 1962). Before injecting the yeast suspensions, a baseline thickness for each footpad is measured. The footpads are measured daily and the increase in footpad thickness expressed as a percentage of the baseline value. To establish the error of the procedure, we measured the footpad thickness of 10 normal untreated rats over a period of 10 days. The readings were within 1.8% of the mean baseline thickness.

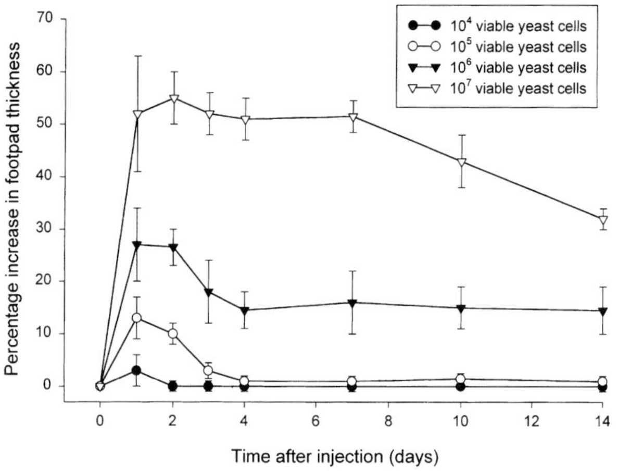

Figure 80.1 Footpad responses to a local infection induced by increasing doses of *Candida albicans* injected directly into the plantar tissue. $n = 20$, bars represent 2 SD. From Miller and Findon (1985), with permission.

Key parameters to monitor infection and response to treatment

Inflammatory response to local infection

When rats were injected in both hind feet with 10^4, 10^5, 10^6 or 10^7 viable yeast cells and the footpad thickness measured daily for 14 days, the peak inflammatory response for all inocula occurred in the first 2 days and then gradually subsided. Host responses to inocula of 10^4 and 10^5 yeast cells were minimal, but 10^6 and 10^7 yeast cells induced maximum increases in footpad thicknesses of 27% and 55% respectively (Figure 80.1). Apart from the footpad swelling and some redness, no other signs of infection were visible and no surface lesions developed. Control animals were challenged with heat-killed (100°C for 1 hour) yeast cells at various dilutions (10^5, 10^6 and 10^7). The peak swelling was only one-third of that observed with the equivalent inoculum of viable cells. Footpad thickness returned to normal within 4 days in the control groups, whereas footpads injected with viable yeast cells remained swollen for more than 10 days.

Determination of yeast cell numbers in the footpad

At post-mortem, animals were placed in a prone position with the hind legs extended toward the operator. Each foot was flooded with 95% alcohol and left to dry. The plantar tissue was removed from the ventral surface of the foot with a sterile scalpel blade and forceps; the tissue was weighed then chopped coarsely with a scalpel blade. After homogenizing the tissue in 9 ml sterile saline with a rotating Teflon pestle in a heavy-walled glass tube (Tri-R Instruments, Long Island, New York, USA), serial 10-fold dilutions were made in pour plates of Sabouraud dextrose agar containing 0.1 mg/ml chloramphenicol (Chloromycetin, Parke Davis). Numbers of yeast cells per gram of footpad were determined by colony counts.

Correlation of footpad oedema with yeast cell numbers

Before footpad swelling could be accepted as a substitute for yeast cell numbers, it was necessary to demonstrate a close correlation between the inflammatory response and the course of infection. This was achieved through the manipulation of the inflammatory response by treating animals with cyclophosphamide to depress immune capability, or amphotericin to limit yeast cell replication. The effect of these manipulations on yeast cell numbers in the footpads and on footpad oedema was then compared.

Cyclophosphamide treatment

Animals were made neutropenic by the subcutaneous implantation of 3×75 mg of cyclophosphamide bone cement discs, 4 days before footpad challenge. The procedure has been fully described by Ormrod *et al.* (1984). In animals treated in such a way, the peripheral leukocyte count reduces to $1.7 \pm 0.6 \times 10^9$/l and remains at this level for at least 10 days. After a challenge of 10^6 viable yeast cells, large increases in the inflammatory response were observed in the immunocompromised group compared with an untreated control group. The responses in the two groups were similar 24 hours after challenge but thereafter inflammation subsided in the controls. In contrast, inflammation continued to increase in the cyclophosphamide-treated animals and was maintained at a significantly higher level for at least 10 days (Figure 80.2A). In a further group of animals, numbers of viable yeast cells in the footpads of cyclophosphamide-treated and control animals (Figure 80.2B) closely followed the inflammatory oedema profiles. Yeast cell numbers up to 100 times that of the control group were found in the cyclophosphamide treated animals. By the end of the study, infection had resolved in the control group and in 50% of the cyclophosphamide-treated animals.

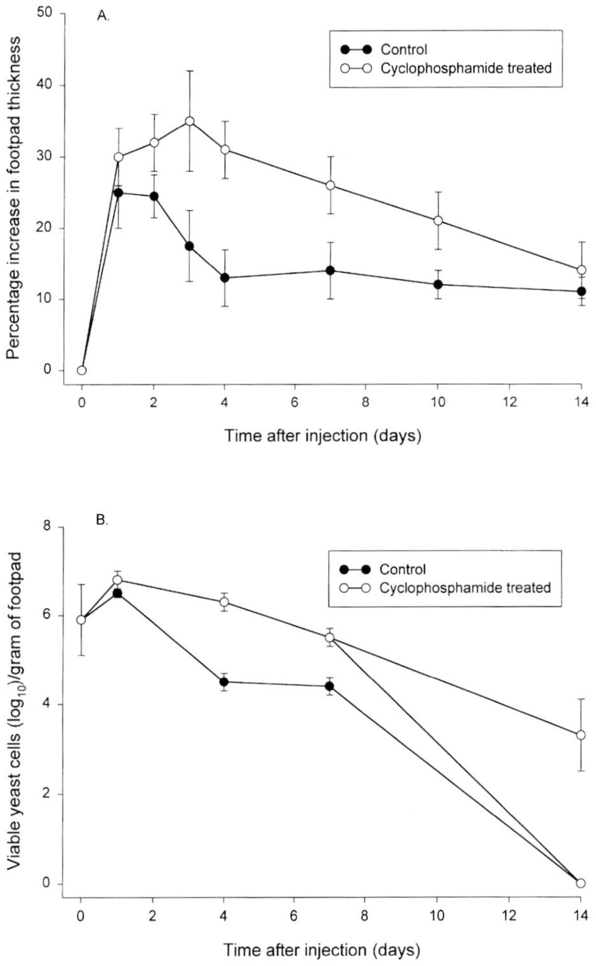

Figure 80.2 Effect on (**A**) footpad swelling and (**B**) number of viable yeast cells in untreated controls and in animals treated with 225 mg cyclophosphamide after challenge with 10^6 colony-forming units *C. albicans*. **A:** n = 20; **B:** n = 12; bars represent 2 SD. From Miller and Findon (1985), with permission.

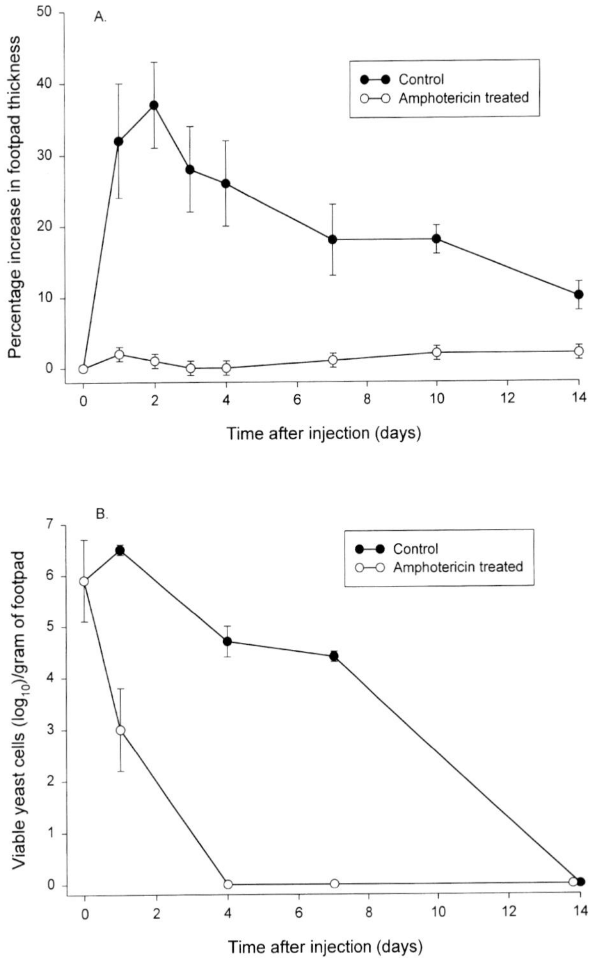

Figure 80.3 Effect of (**A**) footpad swelling and (**B**) number of viable yeast cells in untreated controls and in animals treated with 5 mg/kg amphotericin after challenge with 10^6 *C. albicans*. **A:** n = 20; **B:** n = 12; bars represent 2 SD. From Miller and Findon (1985), with permission.

Amphotericin treatment

Animals were given 5 mg/kg amphotericin (Fungizone, ER Squibb, Hounslow, Middlesex) daily i.p. for 2 days before footpad infection and throughout the course of the study. Control animals were given 1 ml saline instead of amphotericin and all animals were challenged with 10^6 viable yeast cells/footpad. Challenge of amphotericin-treated animals with yeast cells failed to induce footpad swelling (Figure 80.3A). The result corresponded with the quantitative mycological examination, which showed that the infection had failed to become established and that yeast cells had been progressively eliminated during a 4-day period (Figure 80.3B). A pilot experiment using a carrageenan-induced footpad swelling technique (Miller and Ormrod, 1980), showed that the drug had a marginal anti-inflammatory effect at the dose used.

Table 80.1 Comparison of the features of animal models of *Candida albicans* infection

Model	*Inoculum size (cfu)*	*Lesion size*	*Animal discomfort*	*Peak response*	*No. animals for experiment*	*Environmental contamination*	*Systemic involvement*	*Ease of measurement*	*Accuracy*
Thigh lesion*	5×10^8 in 0.2 ml	Large	High	6 days	Few	Unlikely	Possible	Difficult	Not reported Range up to ± 9%
Flank cutaneous lesion†	10^6	Small	Low	6 days	Few	Unlikely	Unlikely	Difficult	Not reported Range up to ± 24%
Cutaneous lesion: rats‡	10^4 in 10 µl	Small	Low	21 hours	Many	High	Unlikely	Not quantitative	Not measurable
Cutaneous lesions: guinea-pigs§	10^7 and 10^9 in 50 µl	Moderate	Low	24 hours	Few	High	Unlikely	Not quantitative	Not measurable
Diffusion chambers**	4×10^7 in 0.15 ml	Nil	Medium	24 hours	Many	Possible	Unlikely	Time consuming	Satisfactory ± 3%
Footpad inflammation (this chapter)	10^7 in 20 µl	Small	Low	24 hours	Few	Unlikely	Unlikely	Simple	Satisfactory ± 1.8%

* Pearsall and Lagunoff (1974).
† Giger *et al.* (1978).
‡ Ray and Wuepper (1976).
§ Sohnle *et al.* (1976).
** Poor and Cutler (1981).
From Miller and Findon (1985), with permission.

Pitfalls (advantages/disadvantages) of the model

This model has several advantages over other models of localized yeast infection. These are summarized in Table 80.1. The infection is localized to the footpad and allows continuous assessment in the same animal. Because both hind feet are used, the number of observations for a given number of animals is doubled. Laboratory contamination is minimized by eliminating the need for tissue homogenization and pour plate counting. Environmental contamination from open lesions is also avoided and the animals suffer minimal discomfort, so it is more ethically acceptable.

Measurement of the swelling is simple and accurate, with a margin of error of 1.8%. The only care needed is to ensure that the spring tension on the gauge is reduced to minimize pressure on the swollen footpad, which may reduce the accuracy of the reading by forcing the fluid away from the site. Repeated measurements at the same timepoint should be avoided for similar reasons.

Use of the model

The measurement of the footpad swelling as described in this model does not require special equipment or skills to achieve reproducible results. The simple measurement of the footpad thickness accurately correlates with the number of viable yeast cells in the footpad. This model is therefore useful in the study of *C. albicans* and perhaps other fungal infections and in the assessment of new antifungal agents.

References

Giger, D. K., Domer, J. E., McQuitty, J. T. (1978). Experimental murine candidiasis: pathological and immune responses to cutaneous inoculation with *Candida albicans*. *Infect. Immun.*, **19**, 499–509.

Gray, D. F., Jennings, P. A. (1955). Allergy in experimental mouse tuberculosis. *Am. Rev. Tuberculosis*, **72**, 171–195.

Miller, T. E., Findon, G. (1985). Experimental candidosis: paw oedema in the analysis of local infection. *J. Med. Microbiol.*, **20**, 283–290.

Miller, T. E., Ormrod, D. (1980). The anti-inflammatory activity of *Perna canaliculus* (NZ green lipped mussel). *N.Z. Med. J.*, **92**, 187–193.

Ormrod, D. J., Cawley, S., Miller, T. E. (1984). Extended immunosuppression with cyclophosphamide using controlled-release polymeric implants (bone cement). *Int. J. Immunopharmacol.*, **7**, 443–448.

Pearsall, N. N., Lagunoff, D. (1974). Immunological responses to *Candida albicans*. I. Mouse-thigh lesion as a model for experimental candidiasis. *Infect. Immun.*, **9**, 999–1002.

Poor, A. H., Cutler, J. E. (1981). Analysis of an *in vivo* model to study the interaction of host factors with *Candida albicans*. *Infect. Immun.*, **31**, 1104–1109.

Ray, T. L., Wuepper, K. D. (1976). Experimental cutaneous candidiasis in rodents. *J. Invest. Dermatol.*, **66**, 29–33.

Sohnle, P. G., Frank, M. M., Kirkpatrick, C. H. (1976). Mechanisms involved in elimination of organisms from experimental cutaneous *Candida albicans* infections in guinea pigs. *J. Immunol.*, **117**, 523–530.

Winter, C. A., Risley, E. A., Nuss, G. W. (1962). Carrageenin-induced edema in hind paw of the rat as an assay for anti-inflammatory drugs. *Proc. Soc. Exp. Biol. Med.*, **111**, 544–547.

Chapter 81

Murine Model of Allergic Bronchopulmonary Aspergillosis

P. Dussault, M. Laviolette and G. M. Tremblay

Background of the model

Aspergillus species are ubiquitous fungi that cause in humans a number of lung-associated diseases. They are largely distributed and found in construction sites, compost piles, potting soil, mulches, sewage facilities, water and bird excreta (Fink and Kurup, 1997). The lungs constitute a major potential site of growth by *Aspergillus*, particularly *Aspergillus fumigatus* (*Af*). The spectrum of pulmonary aspergillosis includes first, saprophytic diseases such as airway colonization and aspergilloma; second, hypersensitivity diseases such as asthma, hypersensitivity pneumonitis and allergic bronchopulmonary aspergillosis (ABPA); and third, acute and chronic forms of invasive pulmonary aspergillosis (IPA). Morphologic and growth characteristics for *Aspergillus* as well as the determinants of pulmonary aspergillosis have been reviewed recently (MacLean, 1996; Fink and Kurup, 1997).

ABPA is usually associated with mild to moderate asthma with productive cough (Patterson *et al.*, 1995). Lung histological changes concentrate essentially around the bronchi and membranous bronchioles of the upper respiratory tree. These changes are characterized by mucoid impaction of the bronchi, bronchocentric granulomatosis, central bronchiectasis, eosinophilic and exudative bronchitis and bronchiolitis, and eosinophilic pneumonia (Myers, 1995). To document the diagnosis of ABPA, several serological tests are used, including total serum immunoglobulin E (IgE), precipitating antibodies to *Af* antigen and serum specific IgE and IgG (Roberts and Greenberger, 1995).

IPA usually affects patients whose immune system is suppressed by either disease or medication (Saral, 1991). IPA is mostly characterized by the presence of fungal hyphae invading lung tissues (Fraser, 1993). The invasiveness of IPA distinguishes this form of aspergillosis from the saprophytic and hypersensitivity forms where tissue invasion is absent (MacLean, 1996).

To our knowledge, only ABPA and IPA possess reliable animal models. Our laboratory and the one of Dr Kurup in Wisconsin developed a murine model of ABPA which mimicks two major immunological features of human ABPA, namely pulmonary eosinophilia (Wang *et al.*, 1994) and elevated total and specific serum IgE and IgG (Kurup *et al.*, 1990; Wang *et al.*, 1994). This ABPA model is the focus of the present chapter. Readers interested in an IPA model are referred to the following articles: Eisenstein *et al.* (1990); Mondon *et al.* (1996); Nawada *et al.* (1996).

Animal species

Two different animal species have been used as an ABPA model. While primates have been used in the past (Slavin *et al.*, 1978), mice are preferred today (Kurup *et al.*, 1992; Wang *et al.*, 1994). Strains of mice used include C3H/HeN (Kurup *et al.*, 1990), Balb/c (Kurup *et al.*, 1992) and C57Bl/6 (Wang *et al.*, 1994). C57B1/6 has been used extensively in our laboratory (Wang *et al.*, 1993, 1994, 1996; Chu *et al.*, 1995, 1996a,b).

Animal model establishment

Preparation and storage of inoculum

Strains and forms of Aspergillus

Several strains and forms of *Af* (spores, soluble antigen or particulate antigen) can be used to induce ABPA in mice. We isolated a strain of *Af* (local *Af*) from an air sample of swine confinement buildings (Cormier *et al.*, 1990). Once instilled, the local *Af* was comparable to a commercial strain of *Af* (ATCC 13073) to induce a strong pulmonary inflammatory response (Wang *et al.*, 1994). This response was not due to contamination by bacteria or endotoxins, as shown by microbiological culture and smear examination and by a limulus amebocyte lysate assay (Wang *et al.*, 1994). Recently, Kurup *et al.* (1997) demonstrated the antigenic potential of different forms of *Af* antigen. In contrast to soluble antigen, particulate antigen (soluble antigen coupled to inert particles) produced a superior peripheral blood eosinophilia and pulmonary inflammation.

Antigen preparation from mycelium

The method used for antigen preparation is modified from Sang and Chaparas (1978). The inoculum comes from a

Handbook of Animal Models of Infection
ISBN 0-12-775390-7

3-day-old culture of *Af* on Sabouraud malt extract agar slant tubes. The conidia are recuperated by scraping the surface with a sterile solution of sodium dodecyl sulfate (0.5%). Then, 200 ml of Czapek medium is seeded with conidia in a cotton-closed bottle of 500 ml. After 4 days of incubation with agitation at 37°C, the mycelium is recuperated, separated from the culture filtrate by passage through a Whatman no. 42 paper, washed three times with sterile distilled water, resuspended in sterile distilled water and stored at 4°C. The day after, the mycelium is harvested by filtrating on Buchner. Finally, the mycelium is resuspended with two parts of sterile distilled water and frozen at –70°C to facilitate the later rupture of cells. After defrosting, the suspension is pre-broken with an Omni mixer homogenizer (Omni International, Gainesville, Florida). Then, 25 ml of the suspension is transferred to a 50-ml bottle with 20 ml of 0.45-mm glass beads. The mixture is homogenized on a Braun mechanical cell homogenizer for 5 minutes with intermittent chilling with CO_2. The crude antigen preparation is finally freeze-dried and resuspended in sterile physiological saline to a final concentration of 2 mg/ml. The suspension is aliquoted and stored at –70°C until used.

Administration of inoculum

C57Bl/6 mice are lightly anesthetized with isoflurane (AErrane, Ohmeda Pharmaceuticals Products, Mississauga, Ontario), then 50 µl of antigen (100 µg dry weight) is administered at the tip of the nose with a micropipette (e.g. Pipetman). The mice are held upright during instillation. The *Af* antigen is instilled on 3 successive days per week for 3–12 weeks. Control mice are instilled with 50 µl of sterile physiological saline (NaCl 0.9%). Other routes of inoculation, such as intraperitoneal, do not induce blood and lung eosinophilia despite increased IgE and IgG levels (Kurup *et al.*, 1992). Repeated intranasal instillations should therefore be preferred because this is closer to the normal way of sensitization in humans. Interestingly, a similar instillation protocol in C57Bl/6 mice with *Saccharopolyspora rectivirgula*, an antigen responsible for farmer's lung disease, induces an alveolar mononuclear cell and neutrophil infiltration instead of an eosinophilia (Tremblay *et al.*, 1993a,b). This highlights the specificity of the pulmonary response depending on the antigen used.

Bronchoalveolar lavage (BAL) and serum collection

BAL is used to assess the alveolar cell population and to measure the different immunoglobulins, cytokines and mediators present in the epithelial lining fluid. After euthanasia by cervical dislocation, the trachea is exposed, incised and cannulated with a 21-gauge catheter (Wang *et al.*, 1994). Hank's balanced salt solution (1 ml), at room temperature, is instilled with a syringe and recuperated. This procedure is repeated twice and at least 85% of the 3-ml solution administered is usually recovered. All recuperated BAL fluids are kept on ice until further processing. Cells are separated from BAL fluid by centrifugation at 400 *g* for 10 minutes. Total cell count and viability are determined on a hemacytometer by crystal violet and exclusion of trypan blue respectively (Wang *et al.*, 1994). The excessive presence of erythrocytes in the BAL fluid indicates a contamination by peripheral blood and such samples should be discarded. Cell differential counts are done by the glass cover slip method (Laviolette *et al.*, 1988), stained with Diff-Quik (Baxter Healthcare Corporation, McGaw Park, IL) and examined for cellular morphology. Serum is collected from blood by cardiac puncture. Aliquots of BAL supernatants and serum are made and stored at –70°C.

Histological examination

Lung index and histology

Unlavaged lungs are removed, trimmed of extraneous tissues and weighed. Lung index (LI) is calculated according to the following formula:

$$LI = (LW/BW)_{test}/(LW/BW)_{control}$$

where LW = lung weight and BW = body weight (Wilson *et al.*, 1982). The same lung can be processed to lung histology in order to evaluate the nature and progression of the pathology.

Several methods can be used to prepare lungs for histology. After fixation with Bouin liquid, lung tissues are embedded in paraffin, sectioned and then stained with hematoxylin and eosin for routine coloration (Wang *et al.*, 1993, 1994). Sirius red is used to show the presence of eosinophils (Bogomoletz, 1980), while Silver stain or periodic acid–Schiff (PAS) reveals the presence of *Aspergillus* (Luna, 1968; Hould, 1984). Instead of Bouin liquid, ice-cooled acetone containing phenylmethylsulfonyl fluoride (PMSF) and iodoacetamide as protease inhibitors may be used to fix tissues overnight at 4°C (Chu *et al.*, 1995, 1996a,b). Then, lung tissues are transferred into acetone and methylbenzoyl at room temperature followed by immersion in glycol methacrylate (GMA) monomer at 4°C for 7 hours. Finally, tissues are embedded in GMA resin prepared by mixing GMA monomer and benzoyl peroxide and polymerized overnight at 4°C. Britten *et al.* (1993) demonstrated that GMA embedding preserves antigenic sites for immunodetection.

The blocks are stored in airtight containers at –20°C for later analysis. The same routine staining as carried out by Wang *et al.* (1993, 1994) could be performed on these lung preparations except for the specific coloration of eosinophils, which is 1% chromotrope 2R staining (Broide *et al.*, 1992). Other investigators prefer to fix lung tissues in 10% buffered formalin followed by paraffin embedding (Kurup *et al.*, 1990, 1992, 1994, 1997). In addition to standard staining for histological examination, Wright–Giemsa and Gomori methenamine silver are used to stain eosinophils and fungi respectively.

Histopathology scores are utilized by several investigators to demonstrate the severity and progression of lung pathology. Wang *et al.* (1994) scored the severity of each histopathological event as follows: 0+ (no lesion), 1+ (focal), 2+ (mild), 3+ (moderate), and 4+ (massive). The events observed include focal alveolar lesion, focal alveolar lesion with epithelioid cells, granuloma with giant cells, peribronchovascular infiltration with lymphoid cells, and eosinophil infiltration. Kurup *et al.* (1990) used two histopathological scoring systems: first, alveolar macrophage infiltration: + (minimal reaction and inflammatory cells), ++ (mild increase), +++ (moderate increase), and ++++ (severe increase); second, peribronchiolar and perivascular inflammatory cells: + (minimal), ++ (mild), +++ (moderate), and ++++ (severe). Recently, Kurup *et al.* (1997) have used a different scoring system for eosinophil count: 0 (no), 1–3 (occasional), 3–10 (few), and 11–30 (numerous eosinophils).

Serum

Serological assays, such as biotin-streptavidin-linked immunosorbent assay (BSELISA), are used to detect total IgE, total IgG_1 and specific IgG_1 antigen in BAL fluid and sera (Wang *et al.*, 1993, 1994). Kurup *et al.* (1990, 1992, 1994, 1997) also used different variants of ELISA techniques.

Monitoring the inflammation

The induced inflammation is mostly restricted to the lung. Any modifications in physical appearance, body weight loss and food and water intake should indicate the occurrence of another inflammatory or infectious process. Although *Af* could be grown from our *Af* antigen preparation, lung histology examination of *Af*-treated mice did not show any evidence of *Af* lung invasion (Wang *et al.*, 1994). Moreover, culture of lung samples and BAL fluid obtained from *Af*-treated mice did not allow any *Af* colony to grow (Wang *et al.*, 1994).

Basic parameters

Normal control mice demonstrate a low number of cells in which macrophages constitute the majority of BAL cells (>97%). Lymphocytes and neutrophils count for less then 1.5% each of total BAL cells respectively. After repeated exposures to *Af* antigen, there is a 10-fold increase of total BAL cells (1×10^6 versus 1×10^5 cells), predominantly constituted of eosinophils (65–70%; Wang *et al.*, 1994). On a 3-week experiment basis, Wang *et al.* (1993) observed a kinetic in the different cell populations recovered by BAL. The first week, the population was constituted mainly of

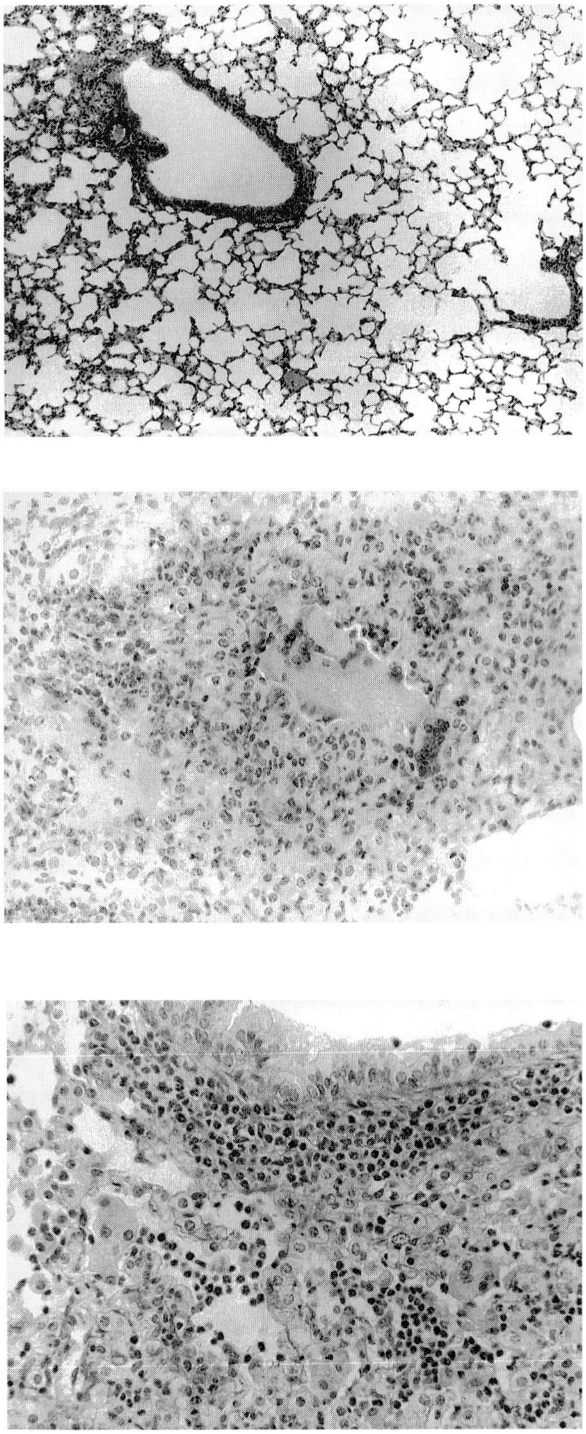

(A) (B) (C)

Figure 81.1 Histological findings. (**A**) Section of lung of a mouse treated with saline instillations showing normal bronchi and parenchyma. Haematoxylin & eosin stain; original magnification ×150. (**B**) Section of lung of a mouse, treated with *Aspergillus* antigen at week 2, showing a diffuse cellular infiltration and a few epithelial cells forming granuloma with giant cells. Hematoxylin & eosin stain; original magnification ×320. (**C**) Section of lung of a mouse treated with *Aspergillus* antigen showing a diffuse infiltration of lymphocytes, epithelioid cells and mainly numerous eosinophils. Sirius red stain; original magnification ×320.

macrophages, lymphocytes and neutrophils. The second and third week showed a different population pattern with a predominance of eosinophils followed by macrophages and lymphocytes, while neutrophil numbers diminished markedly (Figure 81.1). The same authors observed an elevation in total IgE and IgG in both BAL and serum. Moreover, they detected the presence of *Af*-specific IgG in the *Aspergillus*-treated mice group. Histopathological findings presented an inflammatory pulmonary response characterized by focal alveolar lesions with peribronchial and perivascular infiltration of lymphoid cells, numerous eosinophils and epithelioid cells, and a few granulomas with giant cells (Figure 81.2). These findings are similar to human ABPA, but without tissue necrosis (Greenberger, 1984; Slavin *et al.*, 1988).

A 12-week experiment was done to evaluate more chronic histopathological features (Wang *et al.*, 1994). The first weeks demonstrated the same results as mentioned above. The peak of eosinophilia and pulmonary infiltration was observed at 4 weeks. Thereafter, the repeated antigen exposure did not exacerbate the eosinophilia in the BAL fluid, nor did it worsen histological damage. No bronchocentric granulomatosis nor bronchiectasis was observed. The results obtained with this chronic model show that the inflammatory process is self-limited, in terms of lung necrosis and fibrosis, even if antigen is given for 12 weeks, which represents a significant period in the life of a mouse.

Immunohistochemistry showed that different cell markers and cytokine-positive cells are increased in murine ABPA. Such information gives clues to facilitate a better understanding of the different mechanisms leading to the development and establishment of ABPA.

More specific parameters

In *Af*-challenged mice, the expression of intercellular adhesion molecule-1 (ICAM-1) is up-regulated on endothelial cells of arteries, arterioles, veins and venules as on bronchial and alveolar epithelial cells and alveolar macrophages (Chu *et al.*, 1995). The elevated expression of ICAM-1 seems more important in the alveolar walls than in the venules.

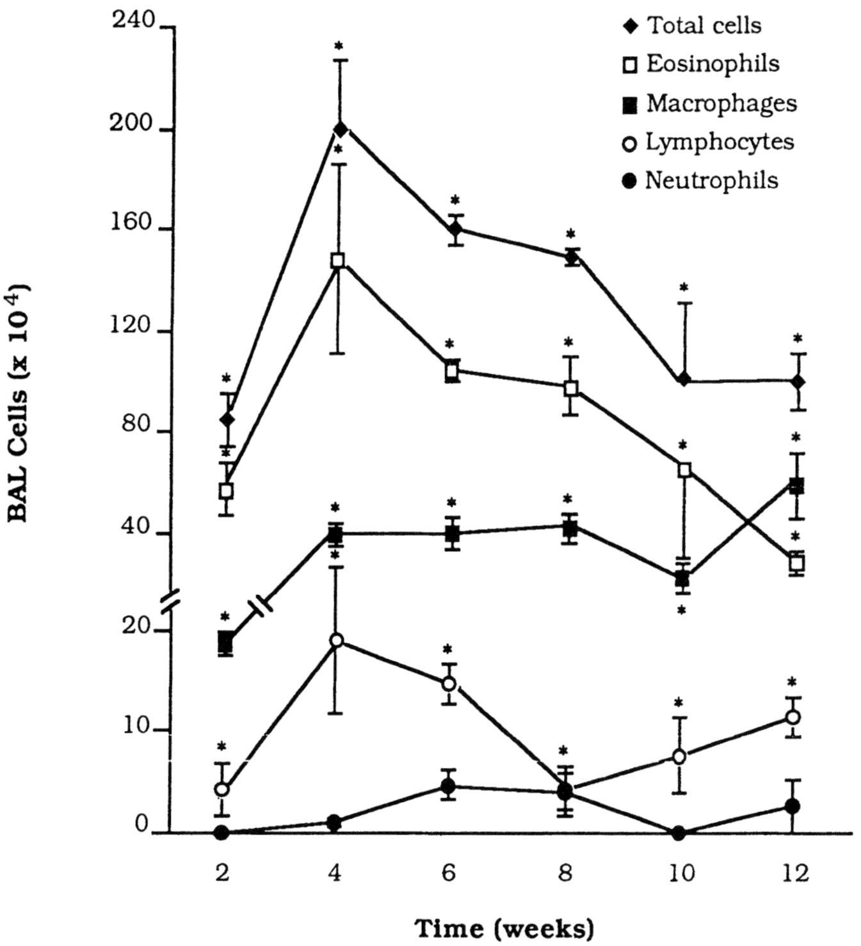

Figure 81.2 Bronchoalveolar lavage (BAL) total cell and the differential cell counts of *Aspergillus*-antigen-treated C57Bl/6 mice for the 12-week study. BAL total cell, eosinophil, macrophage, lymphocyte and neutrophil counts of the saline-treated group remained low during this period: 8.5–17.6 × 10^4, 0.0–0.2 × 10^4, 8.0–14.8 × 10^4, 0.2–2.3 × 10^4 and 0.1–0.7 × 10^4 cells. Results are expressed as means ± SEM (n = 3). *P < 0.05 compared to control mice.

This observation suggests that in inflamed lung, inflammatory cells transmigrate, probably from the alveolar capillary sites to the interstitium and air spaces. In *Af*-challenged lung, clusters of inflammatory cells are seen where ICAM-1 is increased. No eosinophil positive for ICAM-1 is observed. Other cell-membrane markers, such as CD3 (T lymphocytes) and LFA-1 (hematopoietic cells, lymphocytes, monocytes, macrophages, neutrophils and eosinophils) are also increased in *Af*-challenged mice and are differentially expressed throughout the experiment. On day 14, levels of CD3, CD4 (thymocytes, helper T cells, monocytes and bone marrow cells) and CD8 (thymocytes, cytotoxic T lymphocytes and some dendritic cells) are increased. At the end of the treatment (day 21), only CD3 and CD4 are increased. The Th_1 profile, which is represented principally by CD8 T lymphocytes, seems to be suppressed.

Proinflammatory cytokines, including tumor necrosis factor-α (TNF-α) and interleukin-1α (IL-1α), have been investigated by us (Chu *et al.*, 1996b). We found an increased expression of both TNF-α and IL-1α in the same cellular population expressing ICAM-1. Alveolar macrophages represented more than 90% of cells expressing these cytokines. In contrast, bronchial epithelial cells in *Af*-treated mice only expressed TNF-α. The increased expression of TNF-α and IL-1α was observed on days 4, 14 and 21. There was no eosinophil positive for TNF-α and IL-1α in the *Af*-treated mouse lung. This suggests that eosinophils do not significantly produce these cytokines.

We also examined other cytokines that may play a significant role in tissue eosinophilia in the context of an allergic inflammatory response, namely granulocyte–macrophage colony-stimulating factor (GM-CSF), IL-4 and IL-5 (Chu *et al.*, 1996a). On a 3-week basis, the expression of these cytokines is increased differently. Two weeks after the beginning of the challenge, GM-CSF, IL-4 and IL-5 are up-regulated. In contrast, on day 21, only GM-CSF expression is up-regulated. Most of the positive-expressing cells are T lymphocytes. Few eosinophils are positive for GM-CSF and IL-5. Moreover, we observed a strong correlation between *in situ* GM-CSF and IL-5 expression and pulmonary eosinophilia.

ABPA therapy

To date, there have been few records of treatments administered to stop the inflammatory reaction generated by *Af* in an animal model. We know that IL-5 is a potent eosinophil chemoattractant and activator. Use of an anti-IL-5 antibody proved to abrogate the peripheral blood, lung and bone marrow eosinophilia generated by an *Af* challenge (Murali *et al.*, 1993). Also, the utilization of anti-IL-4 antibody diminished the level of IgE (Kurup *et al.*, 1994). We used dexamethasone to inhibit the Af-induced eosinophil pulmonary infiltration (Figure 81.3) and the serum IgE production (Wang *et al.*, 1996). Treatment with this glucocorticoid also reduced the number of T cells and IL-4, IL-5 and GM-CSF-positive cells.

Pitfalls

Not all the human histopathological findings are found in the murine model and the reasons are still unknown. Anti-IL-5, as described above, only diminish the number of eosinophils. There is still a moderate interstitial chronic

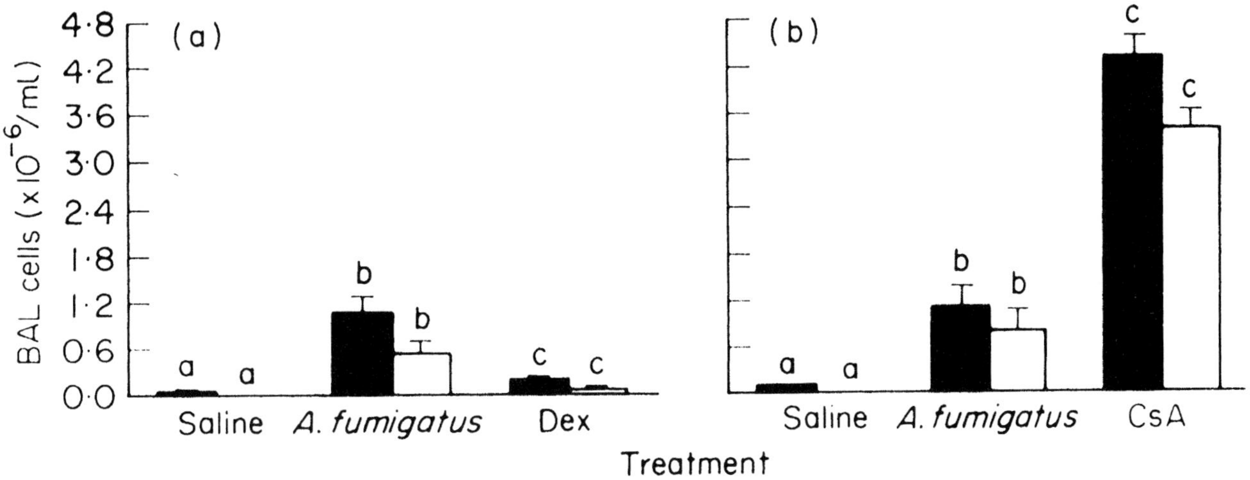

Figure 81.3 (**a**) Effect of dexamethasone (Dex) and of (**b**) cyclosporin A (CsA) on the bronchoalveolar lavage (BAL) total cell counts and BAL eosinophil numbers in C57Bl/6 mice. Dex, Mice treated with dexamethasone before instillation with *Aspergillus fumigatus*; CsA, mice treated with CsA before instillation with *A. fumigatus*. The results are expressed as mean ± SEM (n = 3 or 4). The statistical differences between individual groups are as follows: (**a**) a–b, $P < 0.01$; a–c, $P < 0.05$; b–c, $P < 0.01$; (**b**) a–b, $P < 0.01$; a–c, $P < 0.01$; b–c, $P < 0.01$. ■, Total cells; □, eosinophils.

inflammatory infiltrate consisting of lymphocytes, plasma cells and histiocytes (Kurup *et al.*, 1994). Cyclosporin A (CsA) is an immunosuppressive agent that acts primarily on T cells (Bierer, 1994). This product interacts specifically with differentiation of T lymphocytes into helper/inducer and cytotoxic subsets. CsA is proven to be useful in suppressing the cytotoxic subset implicated in organ transplantation.

In murine hypersensitivity pneumonitis, CsA modulates positively the progression of the disease (Takizawa *et al.*, 1988; Denis *et al.*, 1992). In contrast, CsA induces a paradoxical reaction in the murine ABPA model. All the inflammatory parameters mentioned above, such as IL-4-, GM-CSF- and IL-5-positive cells, are increased. Moreover, in *Af*-treated mice, CsA increases the total number of BAL cells (Figure 81.3). Therefore, we hypothesize that CsA, by inhibiting Th_1-derived cytokine expression, may favor the Th_2 pathway despite the fact that we never found any IL-2 and interferon-γ. The exact mechanism by which CsA acts on a Th_2-like response is still unknown (Wang *et al.*, 1993, 1996).

Contributions of the model to disease therapy

The ABPA murine model presents a significant limitation. No lung tissue necrosis and remodeling is induced by the *Af* antigen. Various explanations could be proposed. The *Af* antigen used in the model does not allow *Af* growth whereas, in humans, a bronchial colonization happens. This colonization might promote the presence of different *Af* antigens and the combination of various immunological stimulations which could then induce the tissue remodeling seen in humans. Growth of *Af* could also liberate toxins (MacLean, 1996) not present in the *Af* antigen preparation used in our murine model. These toxins could, with the concomitant antigenic stimulation, induce tissue lesions. It is also possible that mice develop tolerance against *Af* and won't present any lung necrosis and remodeling.

Despite the fact that the murine ABPA model cannot be used to evaluate the mechanisms underlying tissue necrosis and remodeling, it is useful to understand the immunological mechanisms involved in ABPA. The role of different cell populations and of various mediators and cytokines in the inflammatory process could be evaluated. Moreover, the *Af* antigens involved in these phenomena could be identified. The model is therefore useful to define ABPA pathogenesis better (MacLean, 1996) and eventually to define therapeutic immunological strategies (Murali *et al.*, 1993).

In humans, corticosteroids constitute the main therapeutic avenue to treat ABPA (Greenberger, 1995). These drugs are also very effective in the mouse model (Wang *et al.*, 1996). However, CsA proved to increase the inflammatory reaction in the *Af*-challenged mouse. We are investigating this paradoxical action of CsA. Since the inflammatory response observed is mostly of the Th_2 type in this model, whereas CsA is mostly used to damper Th_1 host response to graft, this suggests that CsA should be used with caution in any Th_2-driven inflammatory condition.

References

Bierer, B. A. (1994). Biology of cyclosporin A and FK506. *Prog. Clin. Biol. Res.*, **390**, 203–223.

Bogomoletz, W. (1980). Avantages de la coloration par le rouge Sirius de l'amyloïde et des éosinophiles. *Arch. Anat. Cytol. Pathol.*, **4**, 250–253.

Britten, K. M., Howarth, P. H., Roche, W. R. (1993). Immunohistochemistry on resin sections: a comparison of resin embedding techniques for small mucosal biopsies. *Biotech. Histochem.*, **68**, 271–280.

Broide, D. H., Paine, M. M., Firestein, G. S. (1992). Eosinophils express interleukin 5 and granulocyte macrophage-colony-stimulating factor mRNA at sites of allergic inflammation in asthmatics. *J. Clin. Invest.*, **90**, 1414–1424.

Chu, H. W., Wang, J. M., Boutet, M., Boulet, L.-P., Laviolette, M. (1995). Increased expression of intercellular adhesion molecule-1 (ICAM-1) in a murine model of pulmonary eosinophilia and high IgE level. *Clin. Exp. Immunol.*, **100**, 319–324.

Chu, H. W., Wang, J. M., Boutet, M., Boulet, L.-P., Laviolette, M. (1996a). Immunohistochemical detection of GM-CSF, IL-4 and IL-5 in a murine model of allergic bronchopulmonary aspergillosis. *Clin. Exp. Allergy*, **26**, 461–468.

Chu, H. W., Wang, J. M., Boutet, M., Boulet, L.-P., Laviolette, M. (1996b). Tumor necrosis factor-α and interleukin-1α expression in a murine model of allergic bronchopulmonary aspergillosis. *Lab. Anim. Sci.*, **46**, 42–47.

Cormier, Y., Tremblay, G., Meriaux, A., Brochu, G., Lavoie, J. (1990). Airborne microbial contents in two types of swine confinement buildings in Quebec. *Am. Ind. Hyg. Assoc. J.*, **51**, 304–309.

Denis, M., Cormier, Y., Laviolette, M. (1992). Murine hypersensitivity pneumonitis: a study of cellular infiltrates and cytokine production and its modulation by cyclosporin A. *Am. J. Respir. Cell Mol. Biol.*, **6**, 68–74.

Eisenstein, D. J., Biddinger, P. W., Rhodes, J. C. (1990). Experimental murine invasive pulmonary aspergillosis. *Am. J. Clin. Pathol.*, **93**, 510–515.

Fink, J. N., Kurup, V. P. (1997). Allergic bronchopulmonary aspergillosis. In *Asthma* (eds Barnes, P. J., Grunstein, M. M., Leff, A. R., Woolcock, A. J.), pp. 2077–2087. Lippincott-Raven, Philadelphia.

Fraser, R. S. (1993). Pulmonary aspergillosis: pathologic and pathogenic features. *Pathol. Annu.*, **28**, 231–277.

Greenberger, P. A. (1984). Allergic bronchopulmonary aspergillosis. *J. Allergy Clin. Immunol.*, **74**, 645–653.

Greenberger, P. A. (1995). Management of allergic bronchopulmonary aspergillosis. In: *Allergic Bronchopulmonary Aspergillosis* (eds Patterson, R., Greenberger, P. A., Roberts, M. L.), pp. 25–33. OceanSide Publications, Providence.

Hould, R. (1984). Méthode de digestion du glycogène sur coupes par l'amylase. In: *Techniques d'Histopathologie et de Cytopathologie* (ed. Hould, R.), pp. 190–191. Décarie, Montréal.

Kurup, V. P., Choi, H., Resnick, A., Kalbfleisch, J., Fink, J. N. (1990). Immunopathological response of C57Bl/6 and C3H/HeN mice to *Aspergillus fumigatus* antigens. *Int. Arch. Allergy Appl. Immunol.*, **91**, 145–154.

Kurup, V. P., Mauze, S., Choi, H., Seymour, B. W. P., Coffman, R. L. (1992). A murine model of allergic bronchopulmonary aspergillosis with elevated eosinophils and IgE. *J. Immunol.*, **148**, 3783–3788.

Kurup, V. P., Choi, H., Murali, P. S., Coffman, R. L. (1994). IgE and eosinophil regulation in a murine model of allergic aspergillosis. *J. Leukoc. Biol.*, **56**, 593–598.

Kurup, V. P., Choi, H., Murali, P. S., Resnick, A., Fink, J. N., Coffman, R. L. (1997). Role of particulate antigens of *Aspergillus* in murine eosinophilia. *Int. Arch. Allergy Immunol.*, **112**, 270–278.

Laviolette, M., Carreau, M., Coulombe, R. (1988). Bronchoalveolar lavage cell differential on microscope glass cover: a simple and accurate technique. *Am. Rev. Respir. Dis.*, **138**, 451–457.

Luna, L. G. (1968). Grocott's method for fungi (GMF). In *Manual of Histologic Staining Methods of the Armed Forces Institute of Pathology* (ed. Luna, L. G.), pp. 230–232. McGraw-Hill Book Company, New York.

MacLean, J. A. (1996). The spectrum of pulmonary responses to *Aspergillus*. In *Immunopathology of Lung Disease* (eds Kradin, R. L., Robinson, B. W. S.), pp. 281–299. Butterworth-Heinemann, Newton.

Mondon, P., De Champs, C., Donadille, A., Ambroise-Thomas, P., Grillot, R. (1996). Variation in virulence of *Aspergillus fumigatus* strains in a murine model of invasive pulmonary aspergillosis. *J. Med. Microbiol.*, **45**, 186–191.

Murali, P. S., Kumar, A., Choi, H., Bansal, N. K., Fink, J. N., Kurup, V. P. (1993). *Aspergillus fumigatus* antigen induced eosinophilia in mice is abrogated by anti-IL-5 antibody. *J. Leukoc. Biol.*, **53**, 264–267.

Myers, J. L. (1995). Pathology of allergic bronchopulmonary aspergillosis. In *Allergic Bronchopulmonary Aspergillosis* (eds Patterson, R., Greenberger, P. A., Roberts, M. L.), pp. 39–46. OceanSide Publications, Providence.

Nawada, R., Amitani, R., Tanaka, E. *et al.* (1996). Murine model of invasive pulmonary aspergillosis following an earlier stage, noninvasive *Aspergillus* infection. *J. Clin. Microbiol.*, **34**, 1433–1439.

Patterson, R., Greenberger, P. A., Roberts, M. L. (1995). The diagnosis of allergic bronchopulmonary aspergillosis. In *Allergic Bronchopulmonary Aspergillosis* (eds Patterson, R., Greenberger, P. A., Roberts, M. L.), pp. 1–3. OceanSide Publications, Providence.

Roberts, M., Greenberger, P. A. (1995). Serologic analysis of allergic bronchopulmonary aspergillosis. In *Allergic Bronchopulmonary Aspergillosis* (eds Patterson, R., Greenberger, P. A., Roberts, M. L.), pp. 11–15. OceanSide Publications, Providence.

Sang, J. K., Chaparas, S. D. (1978). Characterization of antigens from *Aspergillus fumigatus*: preparation of antigen from organisms grown in completely synthetic medium. *Am. Rev. Respir. Dis.*, **118**, 547–551.

Saral, R. (1991). *Candida* and *Aspergillus* infections in immunocompromised patients: an overview. *Rev. Infect. Dis.*, **13**, 487–492.

Slavin, R. G., Fischer, V. W., Levin, E. A., Tsai, C. C., Winzenburger, P. A. (1978). A primate model of allergic bronchopulmonary aspergillosis. *Int. Arch. Allergy Appl. Immunol.*, **56**, 325–333.

Slavin, R. G., Bedrossian, C. W., Hutcheson, P. S. *et al.* (1988). A pathologic study of allergic bronchopulmonary aspergillosis. *J. Allergy Clin. Immunol.*, **81**, 718–725.

Takizawa, H., Suko, M., Kobayashi, N. *et al.* (1988). Experimental hypersensitivity pneumonitis in the mouse: histologic and immunologic features and their modulation with cyclosporin A. *J. Allergy Clin. Immunol.*, **81**, 391–400.

Tremblay, G. M., Israel-Assayag, E., Sirois, P., Cormier, Y. (1993a). Murine hypersensitivity pneumonitis: evidences for the role of eicosanoids and platelet activating factor. *Immunol. Invest.*, **22**, 341–352.

Tremblay, G. M., Thérien, H.-M., Rocheleau, H., Cormier, Y. (1993b). Liposomal dexamethasone effectiveness in the treatment of hypersensitivity pneumonitis in mice. *Eur. J. Clin. Invest.*, **23**, 656–661.

Wang, J. M., Denis, M., Fournier, M., Laviolette, M. (1993). Cyclosporin A increases the pulmonary eosinophilia induced by inhaled *Aspergillus* antigen in mice. *Clin. Exp. Immunol.*, **93**, 323–330.

Wang, J. M., Denis, M., Fournier, M., Laviolette, M. (1994). Experimental allergic bronchopulmonary aspergillosis in the mouse: immunological and histological features. *Scand. J. Immunol.*, **39**, 19–26.

Wang, J. M., Chu, H. W., Bossé, M., St-Pierre, J., Boutet, M., Laviolette, M. (1996). Dexamethasone and cyclosporin A modulation of cytokine expression and specific antibody synthesis in an allergic bronchopulmonary aspergillosis murine model. *Eur. J. Clin. Invest.*, **26**, 951–959.

Wilson, B. D., Sternick, J. L., Yoshizawa, Y., Katzenstein, A.-L., Moore, V. L. (1982). Experimental murine hypersensitivity pneumonitis: multigenic control and influence by genes within the *I-B* subregion of the *H-2* complex. *J. Immunol.*, **129**, 2160–2163.

Chapter 82

Experimental Pulmonary Cryptococcal Infection in Mice

M. F. Lipscomb, C. R. Lyons, A. A. Izzo, J. Lovchik and J. A. Wilder

Background of model

Cryptococcus neoformans is a fungus which infects the host as an encapsulated yeast acquired from the environment, probably in a desiccated form. Infection is generally thought to be initiated in the human host as a respiratory tract infection, but causes only minimal respiratory tract symptoms, if any. The organism then disseminates to extrapulmonary sites in the predisposed. The most common clinical presentation is as a chronic meningococcal infection with headache and fever, and is particularly likely to occur in T lymphocyte-deficient hosts. Nevertheless, infections in the lung, bone, and other tissues are not uncommon and the organism often produces a widespread systemic infection in the severely immunocompromised host.

The pathogenic mechanisms of *C. neoformans* are still incompletely understood, but three major ones have been identified (Mitchell and Perfect, 1995). First is the ability to produce a capsule which allows the organism to escape phagocytosis and exert immunosuppressive activity. The second is the ability to grow at 37°C. A third important mechanism is melanin production. Thus, the organism expresses a unique phenol oxidase which oxidizes several substrates, including norepinephrine and dopamine, which are constituents of the central nervous system. Melanin has antioxidant activity and may function to protect the yeast from destruction by oxidative phagocyte killing mechanisms (Jacobson and Tinnell, 1993).

C. neoformans is classified for purposes of human disease into two varieties, var. *neoformans* and var. *gattii*, as well as by serologically defined capsular epitopes A, B, C, D, and AD. A and D are closely related and occur in var. *neoformans* (Mitchell and Perfect, 1995). Similarly, C and D are also closely related and occur in var. *gattii*. Serotype A, var. *neoformans* produces the overwhelming majority of all infections in humans. Although for many years *C. neoformans* was not known to exist in a perfect state, less than 25 years ago Kwon-Chung demonstrated that *C. neoformans* represented the anamorphic (asexual) form of the telomorphic (sexual) *Filobasidiella neoformans* (Kwon-Chung, 1976).

The increased incidence of cryptococcal infections with the acquired immunodeficiency syndrome (AIDS) epidemic has heightened interest in this ubiquitous organism, which is generally considered an opportunist. The difficulty of treating severely immunocompromised patients with cryptococcosis has renewed interest in new vaccine strategies to prevent the infection. In addition, new antibiotics and innovative new adjunctive therapies such as monoclonal antibodies and recombinant cytokines could be extremely useful.

While *in vitro* studies with the organism are useful in identifying promising chemotherapeutic agents, animal models are finally necessary to demonstrate *in vivo* efficacy. To that end, central nervous system, systemic and pulmonary infection models of cryptococcosis are very important.

This chapter will discuss methods to use the respiratory route of infection to produce a cryptococcal pneumonia in mice. In our opinion, this route of infection more closely mimicks the route of infection in humans. However, unlike the situation in humans, the lung route of infection always produces a prominent pneumonia in mice, unless an acapsular strain is used (Lipscomb, unpublished observations). Thus, this initial lung infection, which either becomes chronic or clears very slowly depending on the virulence of the yeast, provides an excellent model to study the efficacy of drug delivery via the aerosol route for treatment of chronic persisting lung infections. Furthermore, dissemination of the organism to extrapulmonary organs, including the brain, can occur in those mice whose pneumonia is persistent. Therefore, systemic agents can be tested in these mice to treat the later stages of disease.

Animal species

Both rats and mice have been used to study cryptococcal infections initiated by the respiratory route. However, the murine model offers the greatest advantage and is the model described here. Advantages include lower cost and the availability of many genetically characterized mouse strains as compared to rats. Either outbred or inbred mouse strains may be used and are infected between 8 and 12 weeks of age. However, it is very important to appreciate that responses to infection differ among inbred strains.

Handbook of Animal Models of Infection
ISBN 0-12-775390-7

Indeed, these defined genetic differences are among the advantages of using mice, since genetic factors in resistance may then be examined. Following intratracheal inoculation of a low-virulence *C. neoformans* strain 52D (24067 from the American Type Culture Collection, Rockville, MD, USA), BALB/c, C.B-17, and CBA mice exhibit resistance in the lung with gradual clearance of the yeast. In contrast, C57BL/6, C3H, and B10.D2 mice exhibit much less resistance to this organism and are considered susceptible (Huffnagle *et al.*, 1991b; Hoag *et al.*, 1995; Lipscomb *et al.*, personal observations). Severe combined immunodeficient C.B-17 scid/scid (SCID) mice (congenic with C.B-17) and athymic nude mice on a BALB/c background develop chronic persisting and widely disseminated infections which are often lethal (Huffnagle *et al.*, 1991b). These observations have been used to argue that acquired T cell immunity is necessary for optimal lung clearance in resistant mouse strains.

Care and housing

Mice must be housed under pathogen-free conditions in order to avoid developing common rodent infections with agents such as mouse hepatitis virus, Sendai virus, and *Mycoplasma pulmonis*. Furthermore, if mice are ordered from a vendor, it is recommended that they be shipped in special filtered boxes that prevent contamination from any other animals that might be shipped or received at the same time. Sentinel animals should be housed with the mice and tested regularly (we test every 3 months) for serologic conversion to the above micro-organisms. Thus, concurrent infections with these microorganisms both alter the course of the cryptococcal infection, as measured by survival rates and colony-forming units (cfu) in the various organs, and also change the tempo and character of the inflammatory response. In order to guarantee protection against inadvertently infecting experimental animals with these pathogens, all handlers must gown, glove and don masks before entering the immediate animal housing quarters. All food, water, bedding and cages should be autoclaved before use. Otherwise animals are allowed to run freely on bedding and are given standard chow and water *ad libitum*.

Organism

Any cryptococcal strain may be used. However, the course of the infection in the lung must be carefully analyzed for each new cryptococcal strain and inbred mouse strain before using the model for testing a new therapeutic agent, because there are clear differences depending on each of these parameters. The yeast strain used in the majority of our experiments, i.e., strain 52D, has a relatively low virulence for mice and requires T cell depletion in resistant mice to circumvent the lung clearance. Thus, murine infection with 52D may more closely resemble *C. neoformans* infections in humans, i.e., T cell immunodeficiency makes the infection both much more frequent and much more serious. The course of a lung infection with strain 52D in C.B-17, BALB/c and C57BL/6 mice is shown in Figure 82.1: this illustrates the typical gradual clearance of

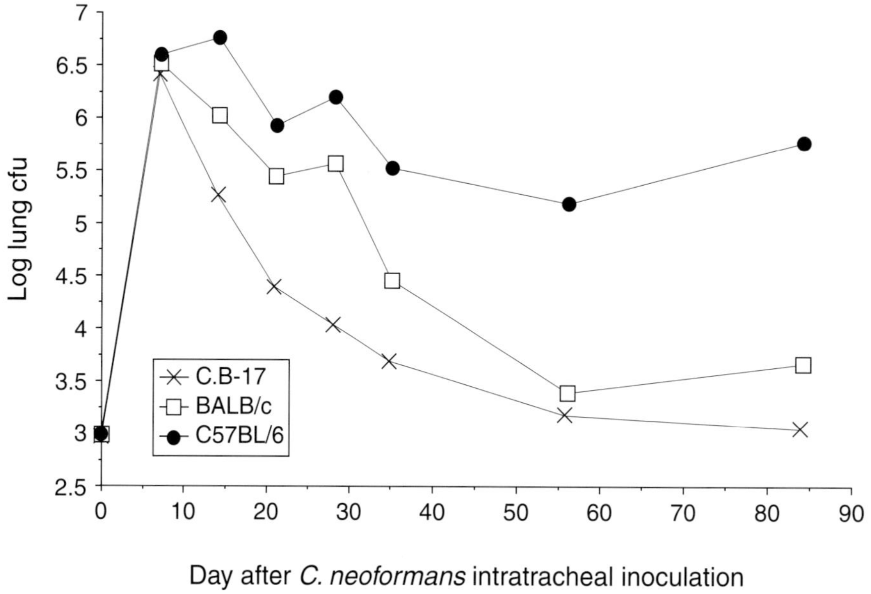

Figure 82.1 Pulmonary clearance pattern of *Cryptococcus neoformans*, strain 52D, after intratracheal infection of three mouse strains. At the days indicated, lungs were removed and colony-forming units (cfu) per lung were assessed. Data are expressed as mean ± standard error. After day 7, C.B-17 clearance is different from C57BL/6 at all points, and from BALB/c on days 14 through 35. BALB/c clearance is different from C57BL/6 at all time points except days 7 and 21. n = 11–61 mice per strain per time point.

the lung infection in both C.B-17 and BALB/c mice and the persistence in C57BL/6.

Frozen stock cultures are prepared from ATCC stock in Sabouraud's dextrose broth (SDB, 1% neopeptone, 2% dextrose; Difco, Detroit, MI, USA), sedimented, resuspended in 15% glycerol in saline, and stored at −70°C. Resting stock cultures are grown from frozen stock cultures on SD agar slants and stored refrigerated for up to 6 months. New resting stock cultures are prepared from frozen stock every 6 months. Fresh working cultures are prepared monthly by streaking from the refrigerated resting stock culture on SD agar (SDA) slants, growing at room temperature, and storing in the refrigerator. The return to frozen stock on a regular basis for preparing resting stock cultures is essential in order to guard against progressive selection of yeast with a changed virulence when inoculated into mice.

To prepare organisms for infections and for growth inhibition studies, the organism is grown in SDB for 36–48 hours at room temperature on a shaker. For the standard 5×10^3 inoculum, prior to inoculation the organism is washed in sterile saline, counted on a hemocytometer, and diluted to approximately 10^5 cfu/ml in sterile nonpyrogenic saline (Baxter Healthcare Corporation, Deerfield, IL, USA).

Surgical procedures and initiation of infection

Overview

Mice infected via the respiratory tract always develop a pulmonary infection. The pulmonary infection can be initiated by several methods. First, mice have been allowed to run over bedding which was admixed with the organisms (Smith *et al.*, 1964). Aerosol exposures have also been used (Smith *et al.*, 1964). Intranasal infection also successfully infects mice (Ritter and Larsh, 1963) and is the method of choice of some investigators because of ease of performance, although animals must be lightly anesthetized. However, the intranasal route may result in a more variable lung deposition. Furthermore, significant ingestion of the organism may require that attention be paid to the role of the gastrointestinal tract in regulating the immunologic response to the infection.

Intratracheal inoculation may be administered via two methods: directly into the trachea or orally via an endotracheal catheter. We have chosen the former because of ease of performance. This method will be described.

It is essential when initiating an infection for studying treatment effects on groups of animals that all animals be approximately the same age. We attempt to mix sexes equally unless all are of the same sex. The inoculum of *C. neoformans* should always be prepared in a standard way, and both the inoculum and the amount deposited initially should be accurately quantitated.

Anesthesia

All mice receive anesthesia prior to undergoing intratracheal inoculation. They are given Avertin (44 µmol/l tribromoethanol and 71 µmol/l tert-amyl alcohol in distilled water) via the intraperitoneal route in the amount of 0.02 ml/g. Mice are allowed to lose consciousness to the extent that there is no visible reaction to gentle pressure exerted on the extremities before undergoing surgery.

For euthanizing mice for harvest of organs and/or body fluids, they are generally given 100% CO_2 by inhalation, although an overdose of Avertin may also be used.

Intratracheal inoculation

Some investigators have used an oral method of intratracheal inoculation by placing a rigid needle into the trachea and threading a small catheter through the needle. This method is quite satisfactory, but placement of the needle is sometimes uncertain and delivery of the inoculum can be into the stomach. Proper placement can be assured when utilizing this method by cutting down over the trachea and examining the trachea as the needle is inserted.

We prefer to use a direct intratracheal inoculation through the tracheal wall just below the pharynx. To achieve this, anesthetized mice are gently restrained in the supine position on a small clean styrofoam board. A good exposure of the trachea can be achieved by attaching the upper incisors to a rubber band which encircles the board, thus arching the neck slightly upwards. A small incision is made over the trachea, and the trachea is exposed by blunt dissection. A slightly bent 30 G needle attached to a tuberculin syringe via PE-10 tubing (Intramedic; Clay Adams, Parsippany, NJ, USA) is inserted into the trachea, and 50 ml of fluid containing the yeast is delivered. After injection, the skin is closed with cyanoacrylate adhesive.

Following surgery, the mice should be kept warm with a heating pad and allowed to recover from the anesthesia under direct observation prior to being put back into cages.

Analysis of infection

Overview

In resistant C.B-17 and BALB/c mice administered strain 52D (5×10^3 cfu), the organism grows rapidly in the lungs for the first 7 days, peaking between 10^6 and 10^7 cfu. In C.B-17 mice, clearance begins at that time (Figure 82.1). The clearance in the BALB/c is initially more gradual than the C.B-17 and numbers of lung cfu vary more. By 3 months both C.B-17 and BALB/c have either completely cleared the infection or retain less than 10^4 cfu. Inflammation is not detectable until about 5 days in either strain. It is first prominently neutrophilic, but admixed with lymphocytes and macrophages (Lovchik *et al.*, 1995). Inflammation peaks at day 14 when macrophages and lymphocytes predominate. At this time the inflammation is clearly granulomatous with cells surrounding the organism in alveolar spaces but with prominent perivascular lymphocytic cuffing (Huffnagle *et al.*, 1991a). The inflammation gradually resolves over time as the yeast cfu decrease.

In the more sensitive C57BL/6 strain, the number of cfu peaks at days 7 and 14, decreases slightly and then persists over the ensuing 3 months (Figure 82.1); in some mice of this strain the infection is lethal. A distinguishing feature of the C57BL/6 mice is that in addition to granulomatous inflammation, a persisting and prominent eosinophilia is present (Huffnagle *et al.*, 1998).

Survival studies

These are useful when a new inoculum dose, cryptococcal strain or mouse strain is used. Time of death and even whether death occurs or not is variable, depending on each of these parameters. Mice given a lethal inoculum of cryptococci intravenously usually succumb to an overwhelming meningeal infection. In contrast, death following an intratracheal infection may occur with high lung cfu and lower brain cfu than occurs in animals dying following the parenteral route. Thus, pneumonia appears to play a major role in causing death. A high dose of a very virulent organism may inflict death on half the animals by 35 days, which contrasts with the intravenous route, where death in a mouse strain deficient in the fifth complement component may occur as early as 5 days, or in complement-sufficient mice between 12 and 25 days (Rhodes *et al.*, 1980). In both the intravenous and pulmonary infection model, depletion of T cells may hasten death. Animals must not be allowed to suffer any excessive morbidity due to the infection. When infected mice demonstrate behavior consistent with near death (significant respiratory distress, ruffled fur, hunching and decreased mobility), the animals should be immediately euthanized.

Following the progress of infection: assessing colony-forming units

Mice are euthanized. Their lungs, brain, spleen, or other organs of interest are removed into 2–10 ml of sterile water and homogenized. Several techniques for homogenization are acceptable, but we prefer to use a tissue homogenizer with cleansing between samples in 70% alcohol and sterile water to remove traces of alcohol. We have found that carry-over from one sample to the next is not a problem if care in cleansing the blades is taken. Serial 10-fold dilutions are made and 50 μl is placed on to SDA plates containing chloramphenicol. Cfu are assessed following 48–72 hours of growth at room temperature.

Following the progress of infection: assessing inflammation

The progress of the infection may be monitored by histological assessment of inflammation and location of the organisms in various tissues using standard techniques. Inflation with buffered formalin via the trachea will enhance the histologic preparation of the lung. Routine sections are stained with hematoxylin and eosin (H&E) or with mucicarmine to visualize the organisms optimally. If immunohistochemical assessment is desired, preparing frozen sections will insure preservation of antigenic structure. In this circumstance, inflation with a solution of embedding medium for frozen sections (Tissue Tek; Miles, Eckhart, IA, USA) diluted 1:3 in phosphate-buffered saline (PBS), pH 7.4, prior to excision is desirable. The lungs should then be immediately placed into 2-methylbutane (Aldrich Chemicals, Milwaukee, WI, USA) that has been cooled in a methanol/dry ice bath. The lungs can then be stored at –70°C.

We have developed a semiquantitative assessment of the total inflammatory response for each lung using H&E sections as follows: a low-power objective is used to scan the entire H&E-stained lung and the lung is scored depending on the percent of the lung that contains inflammatory cells. The scoring system is as follows: 1 = 6–20% involvement, 2 = 21–40%, 3 = 41–60%, 4 = 61–80%, 5 = 81–100%. Lungs are also scanned to determine the relative density of infiltrating cells in the inflamed areas using a graded coding system similar to that used to assess the percentage of inflamed lung, i.e., 1 = 6–20% of the inflamed area is occupied by cells, etc. The two values for each lung can then be multiplied to obtain a possible total score of 25. For example, if 50% of the lung were involved with inflammation, but in each of the inflamed areas the cells occupied only 15% of the area on average, the total score would be $3 \times 1 = 3$. Sections are best read blinded and by two individuals. A score for each mouse lung is the average of both readers' scores, and the data is expressed as the mean ± SD for each mouse lung per group per time point.

Another good way to characterize the inflammatory response in the infected mouse lung is to isolate, count and determine the cell types by either Wright–Giemsa smears or, more accurately, by flow cytometry using monoclonal antibodies specific for a given cell type. To obtain cells from the bronchoalveolar space, bronchoalveolar lavage can be performed using PBS containing 0.6 mm ethylenediaminetetraacetic acid in a 1 ml syringe connected to a small piece of PE-50 tubing. Five 1 ml aliquots of fluid are injected into the lungs via the trachea and drawn out through the same syringe, with 0.5–0.8 ml aliquots being withdrawn each time. To enumerate and characterize all lung cells, the lungs are perfused from the right heart ventricle with saline, minced and treated for 90 minutes at 37°C with Collagenase A (0.7 mg/ml, Boehringer Mannheim Biochemicals) and DNase I (30 μg/ml, Sigma) in 15 ml RPMI with 5% fetal calf serum per lung. Digested tissue is tapped through a wire screen (60 mesh), and cells are collected in Hanks Balanced Salt Solution (HBSS) or complete medium (RPMI 1640, supplemented with 100 U/ml penicillin, 100 μg/ml streptomycin, 2 μg/ml amphotericin B, 2 mmol/l L-glutamine non-essential amino acids, and 1 mmol/l sodium pyruvate) and washed. Coarse clumps may be removed by passing the cell suspension through a small, loose, nylon wool plug. Red cells may be lysed using 0.383% ammonium chloride lysing buffer. Cells may then be counted with live cells assessed using trypan blue. Cell types are determined by differential counting after staining cytospin preparations with either a Leukostat Stain Kit (Fisher Diagnostic, Pittsburgh, PA, USA) or any acceptable Wright–Giemsa staining method.

Contribution of the model to infectious disease therapy

Use of model to study pulmonary immunoregulation

Our primary use of the pulmonary cryptococcal infection model has been to gain a better understanding of how the host mounts an effective immune response in the lung against a chronic, opportunistic infectious agent. It was our hypothesis that the less virulent 52D cryptococcal strain would initiate a primary immune response on inoculation into the murine lung (Huffnagle *et al.*, 1991b). Thus, the strain grows progressively in the lungs and disseminates, but after day 7, the cfu in the lung steadily decreases. The decrease in lung cfu correlates in time with the development of a delayed-type hypersensitivity (DTH) reaction as measured by a positive foot pad swelling assay in resistant BALB/c mice, although little DTH is detected in susceptible C57BL/6 (Huffnagle *et al.*, 1991b). As discussed, by comparing immunocompetent BALB/c mice and congenic C.B-17 mice with athymic nude and SCID mice, T cells were demonstrated to be required for optimal lung clearance. Furthermore, adoptive transfer experiments demonstrated that immune lung T cells or splenocytes were effective in protecting SCIDs from pulmonary growth of *Cryptococcus neoformans* following an intratracheal inoculation of the yeast.

We have also demonstrated a role for CD4 and CD8 T cells in pulmonary clearance by depleting BALB/c mice of these cells with the relevant monoclonal antibodies (clones GKI.5 and YTSI69.4 respectively; Huffnagle *et al.*, 1991a). Both CD4 and CD8 cells are required for optimal clearance, and examination of the lungs histologically demonstrated that there was a marked decrease in inflammation in lungs of both CD4- and CD8-depleted mice. In mice depleted of both CD4 and CD8 T cells, virtually no inflammation occurs at 14 days, the time when normal mice demonstrate a peak inflammatory response. Additional studies have shown that an effective T helper 1 (Th1) response is essential for clearance and that the effector mechanism is related to the ability of lung macrophages to make nitric oxide (Hoag *et al.*, 1995; Lovchik *et al.* 1995).

Use of model in studying new therapeutic agents

We have used the infection with 52D strain in C.B-17 mice to study the role of the cytokines interleukin (IL)-12 and interferon-γ (IFN-γ) in regulating the development of a protective Th1 response which is required to resolve the infection (Hoag *et al.*, 1997). We have found that sensitive C57BL/6 mice develop a non-protective Th2 response (Hoag *et al.*, 1997; Huffnagle *et al.*, 1998) and studied the effect of antibodies to IL-4 in shifting the Th2 to a Th1 response—so far without success (Izzo and Lipscomb, unpublished data). Both of these models, i.e., the resistant and susceptible variants, could be used to assess the role of systemic and/or aerosol delivery of various adjuvant cytokine therapies in altering the course of infection.

We have also examined the effect of cyclosporin A (CsA) in altering the course of a pulmonary infection using a highly virulent, lethal, serotype A cryptococcal strain, 145A (Mody *et al.*, 1988, 1989). CsA was originally developed as a possible antifungal agent, but was almost immediately recognized as a remarkable T-cell-suppressive drug and gained wide use in prevention of allograph rejection. Early literature suggested that Cne replication was not directly affected by this drug (Perfect and Durack, 1985), and our initial studies were designed to determine the effect of T-cell suppression by CsA on infection with Cne 145A. Mice were pretreated with the drug prior to being infected by either the intravenous or intratracheal routes. The surprising result was that survival of mice was indefinitely prolonged in both infection models. *In vitro* studies established that CsA was directly cytotoxic to the organism at doses achieved in mouse blood (Mody *et al.*, 1988). Additional studies established that the doses of CsA used were sufficient to immunosuppress the mice (20, 50, and 75 mg/kg per day; Mody *et al.*, 1988), so the results were doubly surprising because they implied that a drug that is immunosuppressive might still protect against an infection if it had a greater toxic effect on the micro-organism.

In a second series of experiments, we assessed whether mice could be protected if CsA were administered after the infection was already established or if they were profoundly immunosuppressed, i.e., had no T cells (Mody *et al.*, 1989). Both infected immunocompetent BALB/c mice and congenic infected athymic nude mice were studied after an intratracheal infection. In *normal* mice after 28 days of infection (just prior to the time of the usual demise of the animals), initiation of CsA treatment effectively prolonged survival. CFU in organs just prior to death were significantly less than they were at day 28 when treatment was initiated; that is, with the exception of in the brain. Apparently, once the organism reached the brain, CsA was relatively ineffective, and treated animals still eventually died of meningitis. To verify this interpretation, normal mice were inoculated with cryptococci intravenously to initiate a rapidly developing brain infection. In these mice, CsA was unable to reduce the numbers of brain CFU, verifying the poor access of CsA into cerebrospinal fluid. In athymic nude mice infected via the respiratory tract, treatment also prolonged survival, but did not prevent death.

Our conclusions were that CsA could kill Cne *in vivo*, but could not sterilize the host if there were a complete absence of T cells. Nor could CsA penetrate the central nervous system to afford protection once the organism was implanted at this site. These experiments illustrate the utility of the lung infection model in assessing the efficacy of a therapeutic modality, but also suggest the need to explore various routes of initiating infection in order to understand completely the full effect of the intervention.

Summary

Using a low-virulence cryptococcal strain, a cryptococcal lung infection model in mice has been extensively studied, describing survival and lung clearance data. Inflammatory and effector mechanisms in the lungs of resistant and sensitive immunocompetent mice have also been characterized. Using a highly virulent cryptococcal strain, an example of the effect of a therapeutic agent on the evolution of a cryptococcal lung infection was also discussed. Either model or a variation of these models would be useful in examining new therapeutics for this difficult-to-eradicate infection in the immunocompromised host.

Acknowledgment

The authors gratefully acknowledge the contributions of Drs Chris Mody, Gary Huffnagle, Nancy Street, and Kathy Hoag to the development of the described model. All studies were supported by a National Institutes of Health grant, #RO1A121951.

References

Hoag, K. A., Street, N. E., Huffnagle, G. B., Lipscomb, M. F. (1995). Early cytokine production in pulmonary *Cryptococcus neoformans* infections distinguishes susceptible and resistant mice. *Am. J. Respir. Cell Mol. Biol.*, **13**, 487–495.

Hoag, K. A., Lipscomb, M. F., Izzo, A. A., Street, N. E. (1997). IL-12 and IFN-γ are required for initiating the protective Th1 response to pulmonary cryptococcosis in resistant C.B-17 mice. *Am. J. Respir. Cell Mol. Biol.*, **17**, 733–739.

Huffnagle, G. B., Yates, J. L., Lipscomb, M. F. (1991a). Immunity to a pulmonary *Cryptococcus neoformans* infection requires both CD4+ and CD8+ T cells. *J. Exp. Med.*, **173**, 793–800.

Huffnagle, G. B., Yates, J. L., Lipscomb, M. F. (1991b). T cell-mediated immunity in the lung: a *Cryptococcus neoformans* pulmonary infection model using SCID and athymic nude mice. *Infect. Immun.*, **59**, 1423–1433.

Huffnagle, G. B., Boyd, M. B., Street, N. E., Lipscomb, M. F. (1998). IL-5 is required for eosinophil recruitment, crystal deposition, and mononuclear cell recruitment during a pulmonary *Cryptococcus neoformans* infection in genetically "susceptible" mice (C57BL/6). *J. Immunol.*, **160**, 2393–2400.

Jacobson, E. S., Tinnell, S. B. (1993). Antioxidant function of fungal melanin. *J. Bacteriol.*, **175**, 7102–7104.

Kwon-Chung, K. J. (1976). Morphogenesis of *Filobasidiella neoformans*, the sexual state of *Cryptococcus neoformans*. *Mycologia*, **68**, 821–833.

Lovchik, J. A., Lyons, C. R., Lipscomb, M. F. (1995). A role for gamma interferon-induced nitric oxide in pulmonary clearance of *Cryptococcus neoformans*. *Am. J. Respir. Cell Mol. Biol.*, **13**, 116–124.

Mitchell, T. G., Perfect, J. R. (1995). Cryptococcosis in the era of AIDS – 100 years after the discovery of *Cryptococcus neoformans*. *Clin. Microbiol. Rev.*, **8**, 515–548.

Mody, C. H., Toews, G. B., Lipscomb, M. F. (1988). Cyclosporin A inhibits the growth of *Cryptococcus neoformans* in a murine model. *Infect. Immun.*, **56**, 7–12.

Mody, C. H., Toews, G. B., Lipscomb, M. F. (1989). Treatment of murine cryptococcosis with cyclosporin-A in normal and athymic mice. *Am. Rev. Respir. Dis.*, **139**, 8–13.

Perfect, J. R., Durack, D. T. (1985). Effects of cyclosporin in experimental cryptococcal meningitis. *Infect. Immun.*, **50**, 22–26.

Rhodes, J. D., Wicker, L. S., Urba, W. J. (1980). Genetic control of susceptibility to *Cryptococcus neoformans* in mice. *Infect. Immunol.*, **29**, 494–499.

Ritter, R. C., Larsh, H. W. (1963). The infection of white mice following an intranasal instillation of *Cryptococcus neoformans*. *Am. J. Hygiene*, **78**, 241–246.

Smith, C. D., Ritter, R., Larsh, H. W., Furcolow, M. L. (1964). Infection of white Swiss mice with airborne *Cryptococcus neoformans*. *J. Bacteriol.*, **87**, 1364–1368.

Chapter 83

Experimental Pulmonary *Cryptococcus neoformans* Infection in Rats

D. L. Goldman and A. Casadevall

Background of human infection

Cryptococcus neoformans is a fungus which can cause life-threatening infections in humans. Most cases occur in immunosuppressed individuals, but occasional cases occur in persons with no obvious immune deficiency. Cryptococcosis occurs in 6–8% of patients with acquired immunodeficiency syndrome (AIDS) and characteristically involves the meninges and brain parenchyma (Dismukes, 1988; Lee *et al.*, 1996). Despite antifungal therapy, the acute mortality associated with cryptococcal infection in AIDS patients is approximately 15% (Saag *et al.*, 1992). Complete cure is rare in patients with AIDS and those who survive the initial presentation require lifetime suppressive therapy to prevent recurrence of disease (Chuck and Sande, 1989).

Cryptococcal infection is believed to be acquired by inhalation of desiccated yeast cells or basidiospores. Pulmonary infection in immunocompetent individuals characteristically produces no or little symptoms. The histopathologic findings of pulmonary cryptococcal infection in the immune-competent host consist primarily of well-defined granulomas and suggest the importance of cellular immunity in controlling infection. It has been hypothesized that the impaired host response of certain immunodeficiency states allows extrapulmonary dissemination, which can in turn result in cryptococcal meningoencephalitis. *C. neoformans* infections are often chronic and the organism can cause latent infections that have the potential to reactivate, especially in the setting of acquired immune deficiency.

Much of what we know about the pathogenesis of cryptococcal infection was learned by animal experimentation. Animal models of pulmonary cryptococcosis are important for elucidating the immune mechanisms responsible for confining infection to the lungs and for the development of therapies that will enhance local host responses.

Background of model

Rats have been used as models of *C. neoformans* infection for over a century. Early investigators focused on systemic models of disease. In 1896, Curtis observed that rats inoculated subcutaneously with *C. neoformans* developed large tumor-like subcutaneous lesions in association with lung, spleen and kidney involvement. In 1935, Benham described systemic cryptococcal infection in rats following intraperitoneal inoculation of *C. neoformans*.

The first description of a pulmonary model of *C. neoformans* infection in rats was made by Gadebusch and Gikas in 1964. These investigators induced pulmonary infection in Holtzman rats by aerosolization of a clinical strain of *C. neoformans*. Graybill *et al.* (1983) induced pulmonary infection in Sprague–Dawley and nude rats by direct injection of a clinical isolate of *C. neoformans* into the right main stem bronchus following tracheostomy. In the method described here, rats are infected under direct visualization of the larynx using an otoscope without the need for surgery (Goldman *et al.*, 1994).

In addition to *in-vivo* models, rat cells and tissues have been used to study various aspects of *C. neoformans* pulmonary infection. Alveolar macrophages obtained from bronchioalveolar lavage of Lewis, Fischer and Sprague–Dawley rats have been used in *in-vitro* phagocytosis and killing assays (Bolanos and Mitchell, 1989; Chen *et al.*, 1994). Furthermore, primary rat lung cell cultures prepared from the lungs of Sprague–Dawley rats have been used to study *C. neoformans* adhesion properties (Merkel and Cunningham, 1992).

Animal species

Male Fischer 344 rats obtained from the National Cancer Institute (Frederick, MD) and ranging in weight from 150 to 250 grams were used in the model described here. Other rat strains used to study pulmonary infection, include Sprague–Dawley, nu/nu rats of the Rowett strain (Graybill *et al.*, 1983) and Holtzman rats (Gadebusch and Gikas, 1964).

Preparation of animals

No special feed or housing preparations are needed. Rats should be free of *Mycoplasma pulmonis* and cilia-associated

Handbook of Animal Models of Infection
ISBN 0-12-775390-7

respiratory bacillus, both of which can induce spontaneous pulmonary infection in rats and could interfere with the interpretation of pulmonary findings.

Overview

The basic protocol involves the inoculation of *C. neoformans* cells directly into the lungs through the intubated trachea. This procedure is accomplished with an otoscope and provides a mechanism for direct pulmonary inoculation without the need for surgery to expose the trachea. Briefly, anesthetized rats are placed on an incline board. An otoscope is placed in the mouth of the rat and used to visualize the trachea and vocal cords (Figure 83.1). Rats are then infected by placing a blunted spinal needle or an angiocath (Becton-Dickinson, Franklin Lakes, NJ) through the mouth into the trachea with the tip resting at a site distal to the vocal cords and injecting a suspension of yeast cells into the lungs.

Materials required

The materials required include 23 G spinal needles with blunted tips or 14 G angiocaths, otoscope, ear speculums, incline board, methoxyflurane (Pitman-Moore, Mundelein, IL), tuberculin syringes and anesthetic chamber. The speculum is modified by cutting a groove in it to accommodate the angiocath or spinal needle. Angiocaths are softer than the spinal needle and less likely to induce local trauma. They are therefore preferred by some investigators to intubate the trachea.

Organism strain and preparation

C. neoformans American Type Culture Collection (ATCC, Rockville, MD) strain 24067 (serotype D) was used in this model. ATCC 24067 has also been used extensively in murine models of cryptococcal disease, including pulmonary infection models (Curtis *et al.*, 1994; Feldmesser and Casadevall, 1997). While serotype D constitutes approximately 7% of clinical isolates in the USA, this serotype is the most common clinical isolate in some European countries (Kwon-Chung and Bennett, 1984). Various *C. neoformans* isolates have been used to induce pulmonary infection in mice. Since *C. neoformans* strains can differ significantly in their virulence, experimentation would be required to determine the optimal inoculum needed to induce infection, if other rat or *C. neoformans* strains are to be used.

Cells are grown in Sabouraud's media in a rotary shaker at 30°C until they reach stationary phase (2–3 days). They are then collected by centrifugation, washed three times in sterile phosphate-buffered saline (PBS) and counted in a hemocytometer. The number of cells is then adjusted to $0.5–1 \times 10^8$/ml. A 0.2 ml volume of PBS solution containing *C. neoformans* (inoculum dose $1–2 \times 10^7$ cells) is drawn up into a tuberculin syringe with a small volume (0.1–0.2 ml) of air behind the PBS solution.

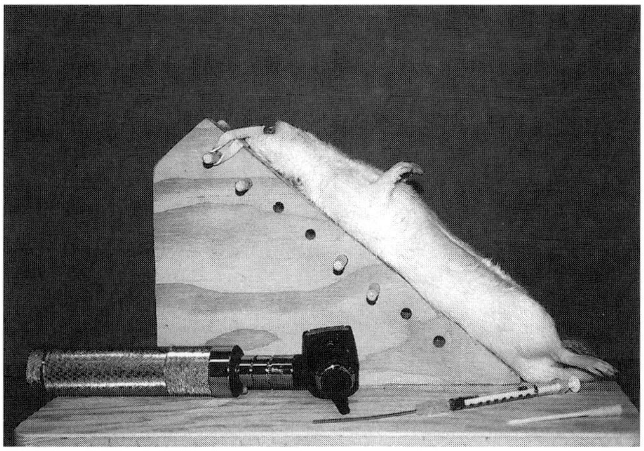

Figure 83.1 Rat on incline board with otoscope, spinal needle and angiocath.

Anesthesia

Rats are placed in an anesthetic chamber in which a gauze, lightly saturated with methoxyflurane, has been placed. Prior to the onset of respiratory depression, rats are removed from the chamber. Adequate anesthesia can be insured by squeezing a footpad and observing for lack of withdrawal. The time needed to induce appropriate anesthesia is of the order of several minutes but there is significant variation between individual animals and each rat should be observed carefully.

Procedure

The anesthetized rat is placed supine on the incline board with the head of the rat at the top of the incline. A rubber band is placed underneath the upper, central incisors to prevent the rat from slipping down the incline (Figure 83.1). The tongue of the rat is pulled out and to the side by hand and an otoscope with the modified speculum is inserted into the pharynx. Using the edge of the otoscope to exert light pressure on the tongue, the visualized field is kept clear and the vocal cords are visualized. An angiocath or blunted spinal needle attached to a tuberculin syringe containing *C. neoformans* inoculum is inserted into the groove of the speculum and placed between the vocal cords under direct visualization. Occasionally mucus secretions may obstruct visualization of the vocal cords. These secretions can be removed by use of a cotton-tip applicator. The procedure usually takes 2–3 minutes.

Infection process

With the tip of the catheter in place immediately below the vocal cords, the inoculum of *C. neoformans* in a volume of 0.2 ml of PBS is instilled. The entire air contents of the syringe are injected, including the air bubble, to insure that the complete inocula is uniformly delivered to the lungs.

Clinical course

Rats infected in this manner recover quickly from the actual infection procedure. Despite successful inoculation, there are no acute clinical signs of pulmonary or extrapulmonary disease. Death secondary to crytpococcal infection using the experimental parameters described here (i.e strain 24067 at infecting dose of 1×10^7 in Fischer rats) is rare. In the context of immunosuppression or following the administration of very high inoculum (i.e., 10^9) of *C. neoformans*, the mortality rate may be greatly increased (Gadebusch and Gikas, 1964; Graybill *et al.*, 1983).

Key parameters to monitor infection and response to treatment

It is difficult to ascertain whether infection was successful by clinical observation alone. Rats will appear well and the investigator may wonder whether infection was successful. In our hands, the protocol described here results in 100% acute infection using an inocula of 1×10^7 yeast cells per rat. Because immunocompetent rats are capable of containing infection in the lungs, there are few parameters to monitor infection in live rats. In fact, the only definitive means of establishing that the rat was infected is to kill the animal and harvest the lung tissue for colony count determination.

Organ fungal burden

In the immunocompetent rat, inoculation of *C. neoformans* into the respiratory tract produces pulmonary infection with low levels of extrapulmonary dissemination (Graybill *et al.*, 1983; Goldman *et al.*, 1994). Occasional colonies can sometimes be recovered from the brain, liver, spleen and kidneys. In contrast, infection of nude rats or administration of glucocorticoids to immunocompetent rats significantly increases levels of extrapulmonary dissemination. Organ fungal burdens are determined by culturing organ homogenates on Sabouraud's agar plates and incubating plates at 30°C for 3–4 days. Briefly, this is done by homogenizing tissue in PBS and spreading an aliquot (100 μl) on Sabouraud agar or a comparable fungal media. In heavily infected lungs, serial dilutions of tissue homogenates are necessary for quantitative measurements of organ fungal burden.

Serum polysaccharide levels

Cryptococcal polysaccharide (CNPS) antigen is seldom detectable in the serum of rats given pulmonary infection unless dissemination has occurred. Hence serum CNPS levels can be used to monitor extrapulmonary dissemination, but this assay is not likely to be useful for determining whether pulmonary infection was successful. The presence of CNPS in the serum is believed to be the result of shedding capsular polysaccharide by yeast cells. The antigen is not degradable by host enzymes and tends to accumulate in tissues. Since the level of serum antigen reflects the infection burden, CNPS levels can be used to monitor the response of disseminated disease to therapy.

Serum CNPS antigen levels may be determined by latex agglutination or enzyme-linked immunosorbent assays (ELISAs). Commercial assays are available for measuring CNPS levels in biological fluids. A double-sandwich ELISA assay which utilizes monoclonal antibodies to capture and immobilize CNPS has been developed by Casadevall *et al.* (1992). This system provides quantitative measurement of CNPS levels. Prior to CNPS determinations, serum specimens should be treated with pronase to remove endogenous antibodies that bind CNPS and could potentially interfere with antigen detection.

Antibody titers

Antibody responses to CNPS are dependent upon the immune status of the host. Significant interspecies variation in antibody responses are also likely; however this has not been thoroughly investigated. In patients with cryptococcal meningitis, the development of anti-CNPS response has been reported to be a positive prognostic indicator (Diamond and Bennett, 1974).

Antibody responses to CNPS in rats have been detected after intravenous and subcutaneous administration of CNPS (Gadebusch and Gikas, 1964; Goldman *et al.*, 1995). Pulmonary infection also induces an antibody response. Using a hemagglutination inhibition assay, Gadebusch and Gikas (1964) demonstrated that administration of cortisone reduces anti-CNPS titers in rats immunized with CNPS. Using ELISA, Goldman *et al.* (1994) demonstrated immunoglobulin M (IgM) and IgG responses to CNPS during the course of pulmonary infection in the rat. The

IgG response was maximum 25 days after infection and correlated with an increase in opsonizing activity of rat sera in an *in-vitro* assay. The same authors have demonstrated local antibody production in the lungs of rats infected with *C. neoformans* using immunohistochemical techniques (Goldman *et al.*, 1994).

Pathology

The pathology of cryptococcal pulmonary infection in the rat model is remarkable for extensive granuloma formation. Areas of inflammation may be apparent on gross examination of lungs and appear as consolidated, whitish discolored regions. With the experimental conditions described here, maximum granuloma formation is present at day 14 of infection with some resolution of the inflammatory response by day 25. Granulomas tend to localize to peribronchial regions and protrusion of the inflammatory response into the bronchiolar lumen may be evident. Some granulomas may be necrotic. There is extensive lymphocytic infiltration with large amounts of perivascular cuffing as early as day 7 of infection.

Immunohistochemistry

Immunohistochemical techniques have also been used to characterize various aspects of the pulmonary model of *C. neoformans* infection. Some of these techniques are suitable for paraffin sections, while others require frozen sections. A list of some of the antibodies used in this model, along with technical information and important findings, is given in Table 83.1 (Goldman *et al.*, 1994, 1996).

Antimicrobial therapy

No studies have been done to examine the effect of antimicrobial therapy in the rat pulmonary model of *C. neoformans* infection. However, Finquelievich *et al.* (1988) and

Table 83.1 Summary of immunohistochemical reagents used in rat model of cryptococcosis.

Antibody directed against	*Tissue Processing*	*Source*	*Primary antibody dilution*	*Important findings*
Cryptococcal polysaccharide (CNPS) (Mukherjee *et al.*, 1994)	Frozen or paraffin	Casadevall Laboratory (Rabbit polyclonal antibody commercially available)	1:1000	• Demonstration of *C. neoformans* cells • Release of CNPS into lung parenchyma • Localization of CNPS to macrophages and epithelioid cells
Inducible nitric oxide synthase (*NOS*$_2$)	Frozen or paraffin	Transduction Laboratories	1:200	• *NOS*$_2$ expression in granulomas, which varies between granulomas and with the stage of infection
Proliferating cell nuclear antigen	Paraffin	Sigma (St Louis, MO)	1:50	• Proliferation of macrophages and lymphocytes within the inflammatory response. Decreased proliferative activity during the later stages of infection
Granulocyte macrophage colony-stimulating factor (GM-CSF)	Frozen	Sigma	1:500	• GM-CSF immunoreactive cells adjacent to inducible NOS$_2$-positive granulomas
Interferon-γ	Frozen	Biosource, (Camarillo, CA)	1:200	• Cell immunoreactive for interferon-γ adjacent to NOS$_2$-positive granulomas
Transforming growth factor-β	Frozen	Genzyme (Cambridge, MA)	1:200	• TGF-β immunoreactive cells within NOS$_2$ positive granulomas

Negroni *et al.* (1991) have evaluated the efficacy of various antimicrobial agents, including amphotericin B, itracanazole, fluconazole and SCH 39304 (an experimental azole) in a rat model of systemic *C. neoformans* infection and demonstrated that these antimicrobials are effective. This systemic model involves intracardiac injection of *C. neoformans* in Wistar rats which produces a subacute, progressive and disseminated infection involving brain, lungs, and thymus. The efficacy of antimicrobial therapy was evaluated by examination of histopathology and fungal burden. Hence, antifungal agents can be effective in rats even though these drugs have not been specifically evaluated in the pulmonary infection model described here. The rat model may be particularly useful in evaluating the efficacy of drug regimens to prevent central nervous system infection in patients at risk for cryptococcosis.

Advantages and disadvantages

The rat is one of several animal species that can be used to study *C. neoformans* infection. In the past, rabbits and mice have provided the primary animal models to study the pathogenesis of cryptococcal infection and the efficacy of drug regimens against cryptococcosis. While mice have been used primarily to study pulmonary and systemic infection, rabbits have been used to study central nervous infection. In general, the rat as a species offers several advantages compared to rabbits and mice in the study of cryptococcal pathogenesis. Unlike rabbits, rats are naturally susceptible to cryptococcal infection and do not require immunosuppressive therapy to establish infection. Furthermore, the body temperature of the rat approximates that of humans, while the body temperature of rabbits is significantly higher (38.5–40°C) and may alter cryptococcal pathogenesis (Perfect *et al.*, 1980). In comparison to mice, rats are larger and certain procedures including endotracheal intubation and intracisternal puncture can be performed more easily.

With respect to a pulmonary model of infection, the rat model offers several advantages when compared to the murine model. As noted above, rats can be infected without the need for surgery. In addition, the pathogenesis of pulmonary infection in the rat seems to resemble more closely infection in immunocompetent humans than does infection in mice. Mice are very susceptible to *C. neoformans* infection and pulmonary inoculation characteristically produces disseminated disease. In contrast, rats like immunocompetent humans are able to contain *C. neoformans* infection to the respiratory tract. The rat model therefore provides an opportunity to study the host defense mechanisms that are important in containing infection and preventing dissemination.

The disadvantages of the rat pulmonary model are related to the lack of clinical signs and non-invasive assays available to ascertain the course of pulmonary infection. Since rats are asymptomatic and seldom die from pulmonary cryptococcal infection, survival cannot be used as an endpoint for antimicrobial studies, unless massive inocula are used. A definitive assessment of pulmonary infection usually requires killing the animal and studying the fungal burden in lung tissue and/or pathology. In addition, rats are more expensive than mice and gene-deficient (knockout) rats are currently unavailable. There are also fewer reagents available to study the immunological aspects of rat infection compared to murine infection.

Contributions of the model to infectious disease therapy

The rat model of pulmonary *C. neoformans* infection has provided important insights into the pathogenesis of cryptococcal infection. This model has been used to demonstrate the shedding of CNPS antigen by cryptococcal organisms into the lung parenchyma and to demonstrate that serum CNPS antigen is absent, without significant extrapulmonary dissemination (Goldman *et al.*, 1994). This model has also been used to demonstrate and to characterize inducible nitric oxide synthase (NOS_2) expression by granulomas during the course of pulmonary infection. Local expression of cytokines which regulate NOS_2 expression (interferon-γ, transforming growth factor granulocyte-β and macrophage colony-stimulating factor), within granulomas has also been shown (Goldman *et al.*, 1996). Other important findings include demonstration of inhibition of alveolar macrophage migration following pulmonary infection with *C. neoformans* (Graybill *et al.*, 1983) and a characterization of the effects of glucocorticoids on antibody responses to CNPS (Gadebusch and Gikas, 1964).

Key References

Benham, R. W. (1935). Cryptococci – their identification by morphology and by serology. *J. Infect. Dis.*, **57**, 255–274.

Bolanos, B., Mitchell, T. G. (1989). Phagocytosis and killing of *Cryptococcus neoformans* by rat alveolar macrophages in the absence of serum. *J. Leukoc. Biol.*, **48**, 521–528.

Casadevall, A., Mukherjee, J., Scharff, M. D. (1992). Monoclonal antibody ELISAs for cryptococcal polysaccharide. *J. Immunol. Meth.*, **154**, 27–35.

Chen, G.-H., Curtis, J. L., Mody, C. H., Christensen, P. J., Armstrong, L. R., Toews, G. B. (1994). Effect of granulocyte-macrophage colony stimulating factor on rat alveolar macrophage anticryptococcal activity *in vitro*. *J. Immunol.*, **152**, 724–734.

Chuck, S. L., Sande, M. A. (1989). Infections with *Cryptococcus neoformans* in the acquired immunodeficiency syndrome. *N. Engl. J. Med.*, **321**, 794–799.

Curtis F. (1896) Contribution à l'ètude de la saccharomycose humaine. *Ann. Inst. Pasteur.*, **10**, 449–468. (cited in) Drouhet, E. (1997) Milestones in the history of *Cryptococcus* and cryptococcosis. *J. Mycol. Med.*, **7**, 10–27.

Curtis, J. L., Huffnagle, G. B., Chen, G. H. *et al.* (1994). Experimental murine pulmonary cryptococcosis. *Lab. Invest.*, **71**, 113–126.

Diamond, R. D., Bennett, J. E. (1974). Prognostic factors in cryptococcal meningitis. *Ann. Intern. Med.*, **80**, 176–181.

Dismukes, W. E. (1988). Cryptococcal meningitis in patients with AIDS. *J. Infect. Dis.*, **157**, 624–628.

Feldmesser, M., Casadevall, A. (1997). Effect of serum IgG1 to *Cryptococcus neoformans* glucuronoxylomannan on murine pulmonary infection. *J. Immunol.*, **158**, 790–799.

Finquelievich, J. L., Negroni, R., Bava, A. J., Iovannitti, C. (1988). Treatment of experimental sub-acute cryptococcosis in the Wistar rat. *Medicina*, **48**, 506–510.

Gadebusch, H. H., Gikas, P. W. (1964). The effect of cortisone upon experimental pulmonary cryptococcosis. *Am. Rev. Resp. Dis.*, **92**, 64–74.

Goldman, D., Lee, S. C., Casadevall, A. (1994). Pathogenesis of pulmonary *Cryptococcus neoformans* infection in the rat. *Infect. Immun.*, **62**, 4755–4761.

Goldman, D. L., Lee, S. C., Casadevall, A. (1995). Tissue localization of *Cryptococcus neoformans* glucuronoxylomannan in the presence and absence of specific antibody. *Infect. Immun.*, **63**, 3448–3453.

Goldman, D., Cho, Y., Zhao, M.-L., Casadevall, A., Lee, S. C. (1996). Expression of inducible nitric oxide synthase in rat pulmonary *Cryptococcus neoformans* granulomas. *Am. J. Pathol.*, **48**, 1275–1282.

Graybill, J. R., Ahrens, J., Nealon, T., Paque, R. (1983). Pulmonary cryptococcosis in the rat. *Am. Rev. Respir. Dis.*, **127**, 636–640.

Kwon-Chung, K. J., Bennett, J. E. (1984). Epidemiologic differences between the two varieties of *Cryptococcus neoformans*. *Am. J. Epidemiol.*, **120**, 123–130.

Lee, S. C., Dickson, D. W., Casadevall, A. (1996). Pathology of cryptococcal meningoencephalitis: analysis of 27 patients with pathogenic implications. *Hum. Pathol.*, **27**, 839–847.

Merkel, G. J., Cunningham, R. K. (1992). The interaction of *Cryptococcus neoformans* with primary rat lung cell cultures. *J. Med. Vet. Mycol.*, **30**, 115–121.

Mukherjee, S., Lee, S., Mukherjee, J., Scharff, M. D., Casadevall, A. (1994). Monoclonal antibodies to *Cryptococcus neoformans* capsular polysaccharide modify the course of intravenous infection in mice. *Infect. Immun.*, **62**, 1079–1088.

Negroni, R., De Elias Costa, M. R. I., Finquelievich, J. L. *et al.* (1991). Treatment of experimental cryptococcosis with SCH 39304 and fluconazole. *Antimicrob. Agents Chemother.*, **35**, 1460–1463.

Perfect, J. R., Lang, S. D. R., Durack, D. T. (1980). Chronic cryptococcal meningitis: a new experimental model in rabbits. *Am. J. Pathol.*, **101**, 177–193.

Saag, M. S., Powderly, W. G., Cloud, G. *et al.* (1992). Comparison of amphotericin B with fluconazole in the treatment of acute AIDS-associated cryptococcal infections. *N. Engl. J. Med.*, **326**, 83–89.

Chapter 84

Rat Model of Invasive Pulmonary Aspergillosis

A. C. A. P. Leenders, E. W. M. van Etten and I. A. J. M. Bakker-Woudenberg

Background of the model

Aspergillus species are opportunistic fungi, giving rise to pulmonary and other invasive infections in immunocompromised patients (Rinaldi 1983; Denning and Stevens, 1990; McWhinney *et al.*, 1993). The number of invasive *Aspergillus* infections is steadily increasing (Denning *et al.*, 1991; McWhinney *et al.*, 1993; Khoo and Denning, 1994). This is due to growing numbers of susceptible patients—patients with prolonged granulocytopenia after aggressive chemotherapy and acquired immunodeficiency syndrome (AIDS) patients. A major problem with these infections is the lack of sufficient effective therapy. Even when treated promptly with the gold standard therapy (amphotericin B deoxycholate), these infections have a high morbidity and mortality (Denning and Stevens, 1990). Several animal models have been used to study these infections and their treatment, mostly mice models of systemic aspergillosis (Leenders and de Marie, 1997). Pulmonary aspergillosis has been studied in models using rabbits and rats (Leenders and de Marie, 1997).

The rat model here described is based on the rat model of *Klebsiella pneumoniae* pneumonia, described by Bakker-Woudenberg *et al.* (1982).

The main features of *A. fumigatus* pneumonia in this model are one-sided onset of infection of the lung, persistent granulocytopenia and start of treatment at a time when hyphal growth is firmly established.

General aspects of the model

Animal species and care

The rats used were female R-strain albino rats (specified pathogen-free; 18–26 weeks old; weight 185–225 g; bred at Harlan CPB, Austerlitz, The Netherlands). Animals receive a normal, pathogen-free diet and water *ad libitum* and, during experiments, are housed separately under filter caps. The microbiological status of the rats is of importance when choosing an antibacterial prophylactic regimen for prevention of opportunistic infections (see below). Because animals are rendered severely granulocytopenic, without antibacterial prophylaxis superinfections with Enterobacteriaceae and even enterococcal species occur.

To prevent bacterial superinfections strict hygienic care was applied and animals received ciprofloxacin (660 mg/l) and polymyxin B (100 mg/l) in their drinking water throughout the experiment. Starting 1 day before inoculation daily amoxycillin (40 mg/kg i.m.) was added to this regime for the remainder of the experiment. Finally gentamicin (40 mg/kg i.m.) was administered shortly before and after inoculation.

Aspergillus strain and inoculum preparation

A clinical isolate of *A. fumigatus* from an immunocompromised patient with invasive pulmonary aspergillosis is used (strain number 1040403005). This strain is kept stored in oil on short agars during the study. The minimum inhibitory concentration (MIC) and minimum fungicidal concentration (MFC) for amphotericin B, determined by the method described by Schmitt *et al.* (1988), are 0.4 and 0.8 μg/ml respectively. The strain is subcultured on Sabouraud dextrose agar (SDA) for 96 hours until a dense mycelium with abundant conidia has formed. These conidia are harvested by adding 5 ml of phosphate-buffered saline (PBS) and rubbing the surface gently. This suspension is washed in PBS. After this, conidia are resuspended in PBS and counted with a hemocytometer. The number of viable *A. fumigatus* used to inoculate the left lung in each experiment is confirmed by plate counts on SDA plates. Before each experiment the strain is passed through a rat to keep virulence as constant as possible.

Although we did not study this, when using other strains of *A. fumigatus*, different infection characteristics can be expected (time to mortality, dissemination, etc.).

Induction of leukopenia

Leukopenia is induced and maintained by multiple doses of cyclophosphamide (Sigma Chemical, St Louis, MO, USA) intraperitoneally. A first dose of 90 mg/kg is administered

Handbook of Animal Models of Infection
ISBN 0-12-775390-7

5 days before fungal inoculation and a second of 60 mg/kg 1 day before fungal inoculation. This scheme is sufficient to keep animals persistently granulocytopenic (granulocytes $< 0.1\ 10^9/l$) for at least 5 days. To lengthen the granulocytopenic period the second dose of 60 mg/kg was followed by similar doses every fourth day. Quantitation of blood leukocytes was done in blood samples obtained by orbital puncture under light CO_2 anesthesia. Total leukocyte counts were determined from diluting these samples in Türk solution in a hemocytometer. Total numbers of granulocytes were calculated after differential counts of leukocytes in cytospin preparations of buffy coats. Differential counts showed that less than 10% of the leukocytes consisted of granulocytes. Experiments using other dosages of cyclophosphamide are currently being performed.

Infection with *A. fumigatus*

Anesthesia

Animals first receive a dose of 0.1 ml Hypnorm (Janssen Pharmaceutical, Oxford, UK) i.m. Afterwards animals are sedated via an i.p. injection of 0.3 ml Nembutal (Sanofi Sante BV, Maassluis, The Netherlands).

Because laryngeal muscle relaxation develops, the animal's tongue is gently pulled outwards to prevent occlusion of the respiratory tract. To prevent hypothermia, animals are covered with a piece of cloth.

After surgery nalorfine hydrochloride 0.1 ml i.m. is administered as an antidote to shorten the duration of further anesthesia. To prevent aspiration and backflow of inoculum (see later) animals are positioned with their heads high on the bedding material during their recovery.

Surgical/inoculation procedure

Following control of the depth of anesthesia via corneal reflexes, animals are placed in a laminar flow cabinet with their ventral sides upwards. Without shaving, the skin of the neck is disinfected using 70% ethanol. A longitudinal incision of approximately 1.5 cm is made in the midline of the neck just above the larynx. Subcutis and muscles overlying the larynx are separated longitudinally using two pincets. In this way the larynx is exposed without blood loss. A metal tube is passed gently through the mouth of the animal into the larynx. The cricoid is passed by rotating the tube and applying mild pressure. The tube is then passed on a few centimeters and left in place while the wound is closed, using three stitches. When all animals are intubated, the inoculation procedure is started.

Animals are put in the vertical position by hanging them at their teeth using a small piece of string. In this way the laryngeal/broncheal tree is stretched maximally. The tube is now passed further downwards into the left main bronchus. A cannula is placed on to a syringe containing the inoculation suspension. This cannula is passed through the tube until 1–2 mm is outside the tube end. A small volume of 20 μl containing the inoculum is slowly injected directly into the left lung. Tube and cannula are withdrawn at the same time. Animals are left hanging in place until at least three strong inhalations have occurred.

After this procedure animals are given naloxone hydroxychloride and are laid into their filtercapped housing. Normally animals start to move again within 10 minutes, and are drinking within a few hours. Within a day animals have returned to their normal behavior. Wound healing takes longer, so the stitches are left in place throughout the experiment. Wound infections have not been recorded, even when wound dehiscence developed.

The efficacy of inoculation of the left lung with a volume of 20 μl after intubation of the left main stem brochus was verified with Dionosil, a contrast medium for bronchography. As shown in Figure 84.1, the volume inoculated is rapidly distributed exclusively in the left lung. A control experiment revealed that the technique of intubation and inoculation in itself did not influence total body weight or body temperature.

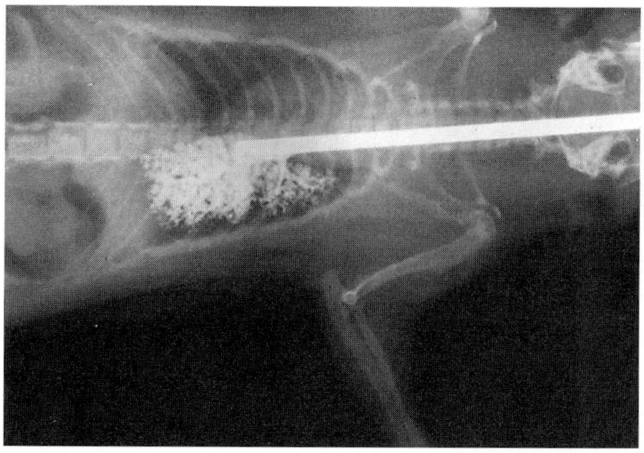

(a)

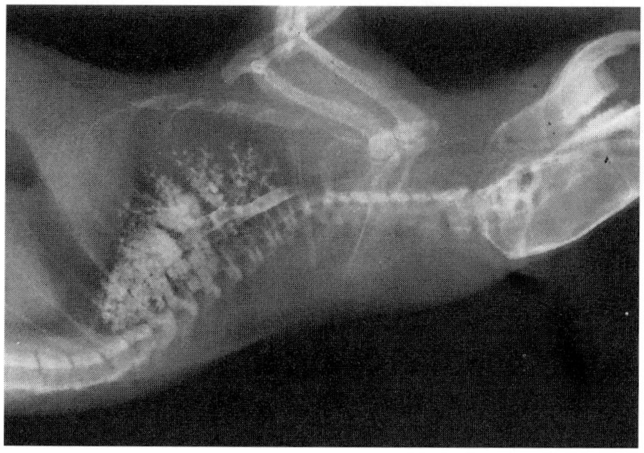

(b)

Figure 84.1 Chest roentgenograms of rats after inoculation of the left lung with a volume of 20 μl of Dionosil, a contrast medium for bronchography. (a), Ventral. (b), lateral.

Untreated infectious process

If left untreated, inoculation initially results in a left-sided invasive pulmonary infection. After several days the infection disseminates and *A. fumigatus* can be detected in the right lung, the liver and the spleen. Untreated animals inoculated with a large enough number of conidia eventually die after a short period of respiratory distress. Unfortunately, the exact cause of mortality is not clear at this moment.

Influence of inoculum size on infectious process

Inoculation of relatively low numbers of conidia (10^2 or 10^3) results in survival of 100% or 80% of the animals respectively, for at least 12 days after fungal inoculation. Inoculation of higher numbers of conidia (10^4 or 5×10^4) resulted in 100% mortality of animals; the first animals died on day 4 after fungal inoculation (Figure 84.2).

The size of the inoculum also influences the time at which histologically proven mycelial disease can be observed. Inocula of 10^2, 10^4 and 5×10^4 result in mycelial disease at ≥48, 40 and 30 hours after inoculation, respectively. Histological examination reveals abundant septate branching hyphae invading lung tissue and blood vessels (Figure 84.3—see colour plate). Macroscopically, often gross hemorrhagic infarctions can be seen, even at a time when infectious lesions are still small. All these data were obtained in the model using the cyclophosphamide scheme described above.

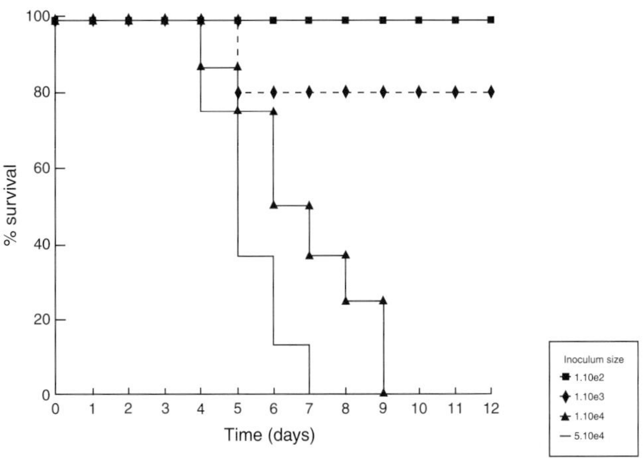

Figure 84.2 Survival of granulocytopenic rats after inoculation of the left lung with various numbers of *Aspergillus fumigatus* conidia: 10^2 or 10^3 ($n=5$), 10^4 or 5×10^4 ($n=8$). Modified from Leenders *et al.* (1996).

Parameters to monitor efficacy of antifungal therapy

The parameters used in this model are mortality rate and time to death. With the proper antibacterial prophylactic regimens, death from bacterial superinfection does not occur.

During the infectious process, the infection tends to spread from the inoculated left lung to the right lung and to other organs. Therefore the dissemination of infection is also used as a parameter of efficacy of antifungal therapy. Furthermore, the number of viable *A. fumigatus*, determined by the number of colony-forming units (CFU) after culture of the infected organs, is a parameter.

Antifungal therapy

In studies carried out so far, the efficacy of amphotericin B deoxycholate (AmB; Fungizone: Bristol Myers-Squibb, Woerden, The Netherlands) and AmBisome (NeXstar, San Dimas, California, USA) have been compared to each other and to untreated control animals (Leenders *et al.*, 1996). This was done as follows: therapeutic preparations were administered by i.v. infusion in a volume of 1 ml via the tail vein. This way of administering therapy is technically limited to a maximum of 10 dosages. After this number of intravenous injections, some animals do not allow more injections because of thrombosis of the two tail veins. Control animals are given 5% glucose in water. Mortality is recorded every day. Animals that die are autopsied. Animals that survive are sacrificed 1 day after the last dosage of antifungal therapy. Lungs, liver and spleen are removed to study the number and nature of the pathologic lesions, to determine the number of viable fungi, and to exclude bacterial superinfection. At present, the efficacy of a new type of liposomal amphotericin B is compared with that of Fungizone in this model (Van Etten *et al.*, 1997). Besides intravenous therapy, within this model it is possible to administer antifungal intraperitoneally and, via aerosol inhalation, directly intrapulmonary.

Semiquantative analysis of tissue burden of *A. fumigatus*

This method was described earlier by Patterson *et al.* (1989), and has been slightly modified. To determine the number of viable fungi in the lungs, liver, spleen and kidneys, these organs are removed and homogenized in 20 ml phosphate-buffered saline (PBS) for 30 seconds (10 000 rpm) with an electric tissue homogenizer. Serial 10-fold dilutions of the homogenates are prepared. Volumes of 0.2 ml of each dilution and of the undiluted homogenate are spread on SDA plates. To prove the sterility of an organ, a volume of 2 ml and the remainder of the homogenate are cultured using the pour-plate method. All plates are incubated for 24 hours at 37°C, and for an extra 24 hours at room temperature, and colonies are counted.

Advantages and disadvantages of this particular model

The major advantage of this rat model is that it is characterized by a one-sided pneumonia. In humans pulmonary aspergillosis is often initially detected in one of both lungs. By intubating the rats' left main bronchus, we are able to inoculate only the left lung, which allows the study of progression of the infection to the right lung—a phenomenon which is often seen in clinical situations. Furthermore, this way of inoculating allows one to study drug pharmacokinetics in infected lung tissue versus uninfected lung tissue in the same animal. Another advantage of inoculating the lung in this way is that the progression of the infection in the left lung is highly reproducible, with only minor variations between individual rats. Also at other points this model very closely mimicks the clinical situation in humans; for example, the observation that, despite treatment with amphotericin B, survival percentages are less than 10–20% in patients with persistent granulocytopenia.

Although not very well-documented, corticosteroids (which are not used in this model) have been reported to cause combined toxicity with amphotericin B (Graybill *et al.*, 1983; Graybill and Kaster, 1984).

The major disadvantage of this model is that the exact cause of death is not yet clear. We do not know if animals die from respiratory insufficiency, acute septicemia, toxins produced by the fungus or some other reason. Investigations into this matter are being carried out at this moment.

Contribution of this model to infectious disease therapy

At this moment, antifungal agents with sufficient efficacy are not yet available for the treament of invasive fungal infections in immunocompromised patients. The model described here is, and can be, used to study the efficacy of newly developed antifungal agents and new formulations of existing agents. At the moment, prophylaxis of fungal infections is of major importance. In a study we performed (Leenders *et al.*, 1996) it was shown that relatively low dosages of liposomal AmB reduced dissemination of infection from the inoculated lung. We speculate that these dosages could have a place in a prophylactic setting.

References

Bakker-Woudenberg, I. A. J. M., Berg van den, J. C., Michel, M. F. (1982). Therapeutic activities of cefazolin, cefotaxime, and ceftazidime against experimentally induced *Klebsiella pneumoniae* pneumonia in rats. *Antimicrob. Agents Chemother.*, **22**, 1042–1050.

Denning, D. W., Stevens, D. A. (1990). Antifungal and surgical treatment of invasive aspergillosis: review of 2121 published cases. *Rev. Infect. Dis.*, **12**, 1147–1201.

Denning, D. W., Follansbee, S. E., Scolaro, M., Norris, S., Edelstein, H., Stevens, D. A. (1991). Pulmonary aspergillosis in the acquired immunodeficiency syndrome. *N. Engl. J. Med.*, **324**, 654–662.

Graybill, J. R., Kaster, S. R. (1984). Experimental murine aspergillosis: comparison of amphotericin B and a new polyene antifungal drug, SCH 28191. *Am. Rev. Respir. Dis.*, **129**, 292–295.

Graybill, J. R., Kaster, S. R., Drutz, D. J. (1983). Treatment of experimental murine aspergillosis with BAY n7133. *J. Infect. Dis.*, **148**, 898–906.

Khoo, S. H., Denning, D. W. (1994). Invasive aspergillosis in patients with AIDS. *Clin. Infect. Dis.*, **19** (suppl. 1), 41–48.

Leenders, A. C. A. P., de Marie, S. (1997). The use of lipid formulations of amphotericin B for systemic fungal infections. *Leukemia*, **10**, 1570–1575.

Leenders, A. C. A. P., de Marie, S., Kate ten, M., Bakker-Woudenberg I. A. J. M., Verbrugh, H. A. (1996). AmBisome reduces dissemination of infection as compared to Fungizone in a model of one-sided pulmonary aspergillosis. *J. Antimicrob. Chemother.*, **38**, 215–225.

McWhinney, P. H. M., Kibbler, C. C., Hamon, M. D. *et al.* (1993). Progress in the diagnosis and management of aspergillosis in bone marrow transplantation: 13 years' experience. *Clin. Infect. Dis.*, **17**, 397–404.

Patterson, T. F., Miniter, P., Dijkstra, J., Szoka, F. C., Ryan, J. L., Andriole, V. T. (1989). Treatment of experimental invasive aspergillosis with novel amphotericin B/cholesterol-sulphate complexes. *J. Infect. Dis.*, **159**, 717–724.

Rinaldi, M. G. (1983). Invasive aspergillosis. *Rev. Infect. Dis.*, **5**, 1061–1077.

Schmitt, H. J., Bernard, E. M., Hauser, M., Armstrong, D. (1988). Aerosol amphotericin B is effective for prophylaxis and therapy in a rat model of pulmonary aspergillosis. *Antimicrob. Agents Chemother.*, **32**, 1676–1679.

Van Etten, E. W. M., ten Kate M. T., Bakker-Woudenberg, I. A. J. M. (1998). Efficacy of a new type of liposomal amphotericin B in severe invasive pulmonary aspergillosis in leukopenic rats. Abstract B-15, 37th Interscience conference on antimicrobial agents and chemotherapy, Toronto, Canada.

Chapter 85

Rabbit Model of *Candida* Keratomycosis

M. Motschmann and W. Behrens-Baumann

Background of human infection

Mycotic infection of the cornea is one of the most dangerous diseases of the eye. An ensuing liquefaction of the corneal tissue can result in perforation and ultimately lead to blindness of the affected eye. Up to the middle of this century, keratomycosis was a rare disease: until 1955 only 120 cases had been reported in the literature (Hoffmann and Schmitz, 1963a). Since then, there has been a considerable increase in the incidence of fungal infections of the cornea (Thygeson and Okumoto, 1974; Ishibashi, 1982). In most cases the infection is induced by an injury from objects contaminated by fungi (Leber, 1879; Naumann *et al*., 1967; Newmark *et al*., 1971a; Polack *et al*., 1971). The development of keratomycosis is favoured by both topical diseases of the eye, such as herpes corneae or trachoma, and general resistance-diminishing factors (Naumann *et al*., 1967; Thygeson and Okumoto, 1974; Verin *et al*., 1984). The increased incidence during the past few decades is mainly ascribed to fungal contamination of soft contact lenses (Yamamoto *et al*., 1979; Alfonso *et al*., 1986; Wilson and Ahearn, 1986) and also to the widespread indiscriminate use of corticosteroids (Mitsui and Hanabusa, 1955; Ley and Sanders, 1956; Zimmermann, 1963). Pharmacotherapy of keratomycosis is in many cases extremely protracted and the results are often unsatisfactory, despite the availability of numerous antimycotics. The longer the infectious process in the cornea persists, the greater is the danger of secondary intraocular changes such as iritis, the formation of synechiae, secondary glaucoma, or corneal vascularizations. Corneal revascularizations in particular reduce the prospects of success of corneal transplantation (Behrens-Baumann, 1984), which in many cases of keratomycosis must be performed at an early stage, not only for visual rehabilitation, but above all to save the eye (De Voe, 1971). Although the prognosis for keratoplasty *à chaud* in cases of mycotic infection of the cornea is favourable (Doden, 1973; Böke and Thiel, 1974; Hallermann, 1975; Thiel, 1978; Behrens-Baumann, 1984), efforts to optimize conservative drug therapy appear worthwhile with a view to avoiding the need for surgical intervention, including the intensive and lengthy after-treatment of the affected patients.

Background of the model

Even during the last century, Leber (1879) and Stoewer (1899) conducted studies of experimental keratomycosis in the rabbit eye. Later, experiments were also performed in other species such as rats (Burda and Fischer, 1959), mice (Mühlhäuser *et al*., 1975) and monkeys (Forster and Rebell, 1975a). The rabbit, however, appears to be the most suitable animal model for studies of experimental keratomycosis. The morphology of the rabbit eye is very similar to that of the human eye. Apart from irrelevant differences in the dimensions (Table 85.1), there are only minor anatomical

Table 85.1 Corneal dimensions in the fully grown rabbit (Davis, 1929; Prince, 1964; Watsky *et al*., 1988) and in adult humans (Duke-Elder and Wybar, 1961)

	Rabbit	*Humans*
Thickness (mm) central/peripheral	0.37/0.45	0.5/0.74
Diameter (mm) horizontal/vertical	15.6/13.8	11.7/10.6
Radius of curvature (mm)	7.3	7.8 (anterior surface) 6.6 (posterior surface)
Surface of cornea (cm^2)	1.55–2.03	1.04
Fresh weight (mg)	approx. 160	approx. 180
Epithelial height (µm)	30–40	50–100
Thickness of Descemet's membrane (µm) central/peripheral	7-22/11-45	5-7/8-10
Temperature (°C) surface/bulb	32/38	31-34/approx. 36.5

Handbook of Animal Models of Infection
ISBN 0-12-775390-7

disparities (Davis, 1929; Duke-Elder and Wybar, 1961; Prince, 1964; Watsky *et al.*, 1988).

As in humans, the natural immune status of the rabbit generally prevents the development of keratomycosis if the uninjured eye becomes contaminated by fungi. The simple instillation of a fungal suspension into the conjunctival sac and on the cornea therefore does not normally give rise to an infection (Stoewer, 1899; Hervouet and Lenoir, 1953; Graf, 1963a; Chowchuvech *et al.*, 1973). Even if the corneal epithelium is previously abraded or the upper layers of the stroma scarified, a mycotic infection does not develop (Stoewer, 1899). At the most, slight superficial keratitis can ensue (Oji, 1982; O'Day *et al.*, 1983).

The prospects of inducing a keratomycosis are greatest if the fungal suspension is injected directly into the rabbit cornea (Montana and Sery, 1958; Agarwal *et al.*, 1963; Ellison and Newmark, 1970, 1973; Forster and Rebell, 1975b; Kunze, 1979; Thiel *et al.*, 1982). Even with this procedure, however, in many cases a manifest fungal infection of the cornea fails to develop.

Many authors have therefore had recourse to simultaneous immunosuppression of the experimental animals. Procedures adopted for this purpose were fractionated cobalt whole-body irradiation (Oggel and de Decker, 1975a,b), the application of antilymphocyte serum (Smolin and Okumoto, 1969) and the induction of alloxan diabetes (Hirose *et al.*, 1957). In many cases, the immune status of the rabbits was also suppressed by topical or systemic administration of corticosteroids (Burda and Fischer, 1959; Agarwal *et al.*, 1963; Berson *et al.*, 1965, 1967; Ellison *et al.*, 1969; Smolin and Okumoto, 1969; Stern *et al.*, 1979).

Impairment of the experimental animal's defence mechanisms by any form of artificially induced immunosuppression, however, misrepresents the natural conditions. Valid conclusions concerning the human situation therefore cannot be drawn from such experiments (Head *et al.*, 1990).

Besides the inoculation technique and the preparation of the experimental animal, a further factor decisive for the reproducibility of an experimental keratomycosis is the nature of the fungal suspension used. Up to now, *Candida albicans* and *Aspergillus* have been used in most cases. Table 85.2 gives an overview of the fungal strains employed thus far, with or without simultaneous immunosuppression by corticosteroids.

In many of the studies carried out so far, however, no precise information is given as to the origin of the fungal strains used (Ley, 1956; Graf, 1963b; Ellison *et al.*, 1969; Hasany *et al.*, 1973; Ivandic, 1973; Stern *et al.*, 1979; Ohno *et al.*, 1983). In other cases, smears of human pathogenic isolates were used (Montana and Sery, 1958; Agarwal *et al.*, 1963; Rheins *et al.*, 1965, 1966; Smolin and Okumoto, 1969; Mühlhäuser *et al.*, 1975; Oggel and de Decker, 1975b; Kunze, 1979; Oji, 1981; Ishibashi and Matsumoto, 1984; Ishibashi and Kaufman, 1985, 1986; Serdarevic *et al.*, 1985). Pathogenicity for the human eye, however, does not necessarily warrant conclusions as to the same effect in the rabbit. Moreover, it is not enough only to indicate the species of the fungal strain applied without any further characterization, since the experiment is then not reproducible. Yet in the models employed so far, only in a few cases was the strain used characterized by the designation of the microbiological laboratory (Hoffmann and Schmitz, 1963a,b; Berson *et al.*, 1965, 1967; O'Day *et al.*, 1983; Garcia de Lomas *et al.*, 1985; Behrens-Baumann *et al.*, 1987b). This has a bearing on the reproducibility in as much as fungi of the same species can show differing degrees of virulence. This important fact has long been known (Ley, 1956; Burda and Fischer, 1959; François and Rijsselaere, 1974), yet in the hitherto reported model experiments it has been accorded scant consideration or none at all. Only O'Day *et al.* (1983) and Behrens-Baumann *et al.* (1987b) pointed out that when various subgroups of the same fungal species are used, totally different results can be obtained.

A further problem is the avoidance of a mixed infection in the course of the mycotic infection. A bacterial superinfection can falsify the results obtained with antimycotics tested in the experimental keratomycosis model. In 2 of 8 fungally infected eyes, Uter (1987) found histological evidence of coccal colonization. In certain circumstances, clinical progression of the bacterial infection can falsely indicate inefficacy of the tested antimycotic, although a fungicidal or fungistatic activity may in fact be present. Burda and Fischer (1959) and Kunze (1979) drew attention to a possibly bacterial origin of corneal ulcers observed in their keratomycosis experiments. Both in the rabbit (Uter, 1987) and in humans (Behrens-Baumann *et al.* 1984), bacteria can be detected in the conjunctival sac even in the absence of any manifest ocular irritation. Berson *et al.* (1967) excluded all rabbits showing bacteria in conjunctival smears from their experiment. Although partially synergistic or antagonistic effects can occur upon simultaneous administration of antibiotics and antimycotics (Meyer-Rohn and Lange-Brock, 1957; Van Winckle *et al.*, 1964; Hoffmann, 1965; Rheins *et al.*, 1965; Stern, 1978), it would appear that antibacterial prophylaxis is necessary to avoid bacterial superinfection.

In sum, the criteria that must be satisfied by a reproducible model of experimental keratomycosis are as follows:

1. A high incidence of synchronously progressive and adequately severe infection
2. A sufficiently long florid state of the infection before onset of reparative partial recovery
3. Development of infection without simultaneous immunosuppression of the experimental animals
4. Exact characterization of the fungal strain used with the designation of the microbial laboratory
5. Prevention of mixed infection or superinfection by bacteria

These criteria are satisfied by the model described below (Behrens-Baumann *et al.*, 1987b).

Table 85.2 Overview of fungal species used to date with various inoculation techniques

Superficial inoculation with corticosteroids	*Superficial inoculation without corticosteroids*	*Intracorneal inoculation with corticosteroids*	*Intracorneal inoculation without corticosteroids*
Candida albicans Berson *et al.*, 1965 Berson *et al.*, 1967 Hasany *et al.*, 1973 Hoffmann and Schmitz, 1963a Ley, 1956 Mühlhäuser *et al.*, 1975 Ohno *et al.*, 1983 Rheins *et al.*, 1965 Serdarevic *et al.*, 1985 Smolin and Okumoto, 1969 Stern *et al.*, 1979	*Candida albicans* Berson *et al.*, 1967 Hasany *et al.*, 1973 Hoffmann and Schmitz, 1963a Kunze, 1979 Mühlhäuser *et al.*, 1975 O'Day *et al.*, 1983 Rheins *et al.*, 1966 Richards *et al.*, 1969 Smolin and Okumoto, 1969 Stoewer, 1899	*Aspergillus* Agarwal *et al.*, 1963 Ellison *et al.*, 1969 Forster *et al.*, 1975a François and Rijsselaere, 1974 Hasany *et al.*, 1973 Ley 1956 O'Day, 1991	*Aspergillus* Agarwal *et al.*, 1963 Ellison and Newmark, 1970 Ellison and Newmark, 1973 Forster *et al.*, 1975a François and Rijsselaere, 1974 Garcia de Lomas *et al.*, 1985 Hasany *et al.*, 1973 Ivandic, 1973 Ley, 1956 Newmark *et al.*, 1971b Singh *et al.*, 1974 O'Day, 1992
C. tropicalis and *C. pseudotropicalis* Graf, 1963a	*C. tropicalis* and *C. pseudotropicalis* Graf, 1963a	*Candida albicans* Agarwal *et al.*, 1963 Ellison *et al.*, 1969 Hasany *et al.*, 1973 Ishibashi and Matsumoto, 1984 Ishibashi and Kaufman, 1985 Montana and Sery, 1958 O'Day *et al.*, 1984	*Candida albicans* Agarwal *et al.*, 1963 Graf ,1963b Hasany *et al.*, 1973 Ivandic, 1973 Montana and Sery, 1958 Oggel and De Decker, 1975b
Aspergillus Hasany *et al.*, 1973 Ley, 1956	*Aspergillus* Hasany *et al.*, 1973 Ley, 1956 Oji, 1982 Rheins *et al.*, 1966		
Cephalosporium Burda and Fisher, 1959 Ley, 1956	*Cephalosporium* Ley, 1956		
		C. krusei Tandon, 1984	
Allescheria boydii Ley, 1956	*Allescheria boydii*: Ley, 1956 Rheins *et al.*, 1966	*C. tropicalis* and *C. pseudotropicalis* Graf, 1963a O'Day, 1991	*C. tropicalis* and *C. pseudotropicalis* Graf, 1963a
Geotrichum Ley, 1956	*Geotrichum* Ley, 1956		
	Sporotrichum schenkii and *Scopulariopsis brevicaulis* Rheins *et al.*, 1966	*Fusarium solani* Forster *et al.*, 1975a François and Rijsselaere, 1974 Ishibashi, 1979 O'Day, 1991	*Fusarium solani* Forster *et al.*, 1975a François and Rijsselaere, 1974 Fiscella, 1997
	Pityrosporium ovale Chowchuvech *et al.*, 1973		
		Cephalosporium Burda and Fischer, 1959	*Cephalosporium* Burda and Fischer, 1959 Burda and Fischer, 1960
		Lasiodiplodia Forster *et al.*, 1975	*Lasiodiplodia* Forster *et al.*, 1975a
			Fusarium moniliforme Dudley *et al.*, 1964

Animal species/preparation of animals

The experimental animals used were male rabbits of 3.5–4.5 kg body weight. To exclude bacterial superinfection, gentamicin sulphate 0.5% eye drops or chloramphenicol 0.2% eye drops, both without preservatives, were instilled twice daily. The efficacy of this measure was verified by weekly bacterial examinations including an antibiogram (DIN 59840). The antibody titre of the rabbits before the beginning of the experiment was 1:20 in the haemagglutination test (Ansorg, 1981). No measures were taken to influence the immune status of the animals.

Infection process/details of surgery

The animals were anaesthetized with 5 mg/kg body weight each of Rompun and Ketavet. The fungal suspension was injected directly into the cornea through a 27 G cannula. The injection volume in each case was 10 μl (2.5 × 10 cells). No significant loss from the puncture channel occurred during the injection.

Storage and preparation of inocula

In 17 eyes, the infective strain used was *Candida albicans* DSM 70010 (DSM = Deutsche Sammlung für Mikroorganismen), which *in vitro* displayed a strong germ-tube formation after 24-hour incubation at 37°C in glucose bouillon with aqueous humour (Uter *et al.*, 1987). Parallel experiments *in vitro* and *in vivo* were performed on six eyes using *Candida albicans* CBS 2730 (CBS = Centraal Bureau voor Schimmelcultures). This is a less virulent variant of *C. albicans* DSM 70010 likewise belonging to seroclass A, but displaying less marked filament formation. Under the same conditions, this strain showed no germ-tube formation. In preliminary experiments, the individual components of the yeast suspension were likewise each injected intracorneally into 2 eyes to differentiate between their effects and that of the full suspension.

Key parameters to monitor infection

To check the progress of the infection, the findings were examined by slit-lamp microscopy and documented photographically daily during the first week and thereafter every other day from the second week until the end of the experiment.

The evaluation criteria adopted were the size of infiltrates and ulcers, the presence of a hypopyon, the extent of reactive corneal vascularization, and the occurrence of a descemetocele/corneal perforation. After the end of the observation period, the animals were killed with an overdose of Rompun and the eyes enucleated.

In 6 eyes the corneal ulcer was excised and incubated *in toto* at room temperature on gentamicin-containing beerwort agar to demonstrate fungal growth. The remaining eyes were fixed in 3% glutaraldehyde for 3 days, rinsed, and stained with haematoxylin and eosin, periodic acid–Schiff and methenamine silver.

Results of clinical findings

An overview of the clinical course of keratomycosis in the 23 eyes investigated is given in Table 85.3.

- In all 17 eyes infected with the DSM strain, a corneal infiltrate appeared on Day 2 and thereafter continued to increase in size (Figure 85.1).
- In all eyes, the infiltrate developed into a corneal ulcer between Day 3 and Day 10.
- A hypopyon appeared between Day 3 and Day 14, likewise in all eyes.
- A descemetocele was observed in 3 eyes between Day 7 and Day 10.
- A corneal perforation occurred in 9 eyes between Day 10 and Day 21.

By contrast, the infection in the 6 eyes infected with the CBS strain followed a much blander course:

- Although a corneal infiltrate developed in all eyes between Day 2 and Day 5, the size of the infiltrate was distinctly smaller than with the DSM strain (Figure 85.2).
- In contrast to the findings with the DSM strain, in none of the eyes did the infiltrate lead to corneal ulceration ($P < 0.001$, χ-square test).

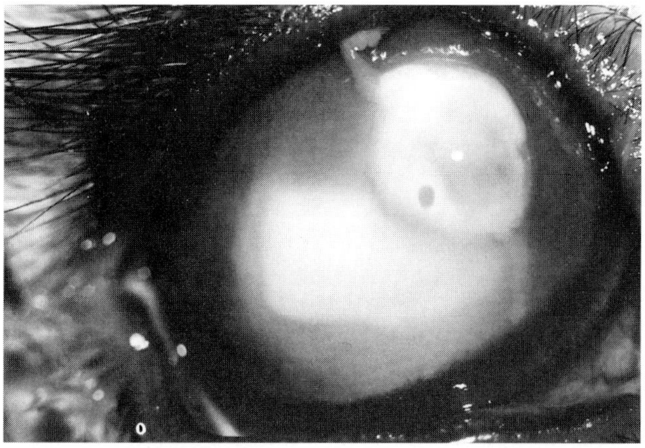

Figure 85.1 *Candida albicans* (DSM 70010): infection on Day 7.

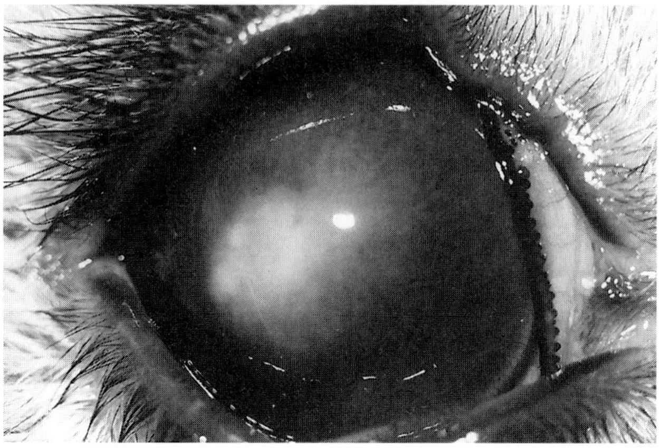

Figure 85.2 *Candida albicans* (CBS 2730): infection on Day 7.

Table 85.3 Overview of clinical course after inoculation of two strains of *Candida albicans*

Candida albicans strain	*Eye*	*Infliltrate beginning (day)*	*Max. growth (mm)*	*Reached from (day)*	*Ulcer beginning (day)*	*Max. growth (mm)*	*Hypopyon beginning (day)*	*Complication (on day)*	*Duration of observation (days)*
DSM 70010	1	2	3×4	5	4	2×3	4	Perforation (14)	14
	2	2	1×3	5	5	1×2	3		8
	3	2	3×4	10	7	2×3	4	Descemetocele (14)	42
	4	2	4×4	7	7	3×4	7	Perforation (10)	42
	5	2	4×5	10	10	2×3	7	Perforation (11)	11
	6	2	4×4	14	6	2×2	14		14
	7	2	5×5	10	3	3×4	3	Perforation (14)	14
	8	2	2×2	7	5	1×1	3	Descemetocele (7)	7
	9	2	2×3	10	7	1×2	7		11
	10	2	2×2	7	5	1×1	5	Perforation (17)	24
	11	2	3.5×4	17	10	2×3	7		17
	12	2	4×4	23	5	3×3	3	Perforation (17)	23
	13	2	2.5×3	17	5	2×2	7	Perforation (17)	23
	14	2	3.5×5	14	3	1×1	7		23
	15	2	3×4	14	10	2×2	3	Perforation (17)	17
	16	2	3×4	17	3	1×1.5	5	Perforation (21)	21
	17	2	3×3	10	7	1.5×2	5	Descemetocele (10)	21
CBS 273019	18	3	2×3	14			17		42
	19	2	3×4	14			7		42
	20	3	3×3	7					42
	21	5	3×3	14					42
	22	3	2.5×3	7			7		42
	23	5	1.5×4	10					42

- A hypopyon only developed in 3 eyes between Day 7 and Day 17.
- No instances of descemetocele or corneal perforation were observed.

In 0.9% NaCl or glucose bouillon, *C. albicans* DSM 70010 purified of its metabolites likewise produced a severe keratomycosis in all cases, although the course of its development was slower than with the full suspension. The non-vital components of the full suspension on the other hand led to no significant inflammatory reaction (Table 85.4).

Results of microbiological examination

In 4 of 6 eyes infected with the DSM strain, it was possible to reculture *C. albicans* from the corneal infiltrate. The sprouting of the hyphae through the cornea into the anterior chamber is exemplified in Figures 85.3 and 85.4.

Antimicrobial therapy/key parameters to monitor response to treatment

In the described model, the pharmacokinetic and pharmacodynamic properties of various antimycotics have been

Table 85.4 Clinical course after intracorneal injection of the full suspension and its various components

Inoculate	*Number of eyes*	*Course*
(a) Full suspension (vital yeasts in 24 hour culture medium glucose bouillon)	17	Infiltrate regularly progressive from Day 2 for about 1 week. Ulcer on average about 5–6 days. Regularly hypopyon. In 12 of 17 eyes perforation or descemetocele,
(b) Washed vital yeasts in fresh glucose bouillon	2	similarly to full suspension, but 2–3 days later and with less inflammatory reaction
(c) Washed vital yeasts in 0.9% NaCl	2	Same as (b)
(d) Washed killed yeasts in 0.9% NaCl	2	Deposition of yeasts, no inflammatory reaction
(e) Culture supernatant with metabolites, cell-free	2	Central opacity for 3–5 days; no inflammatory reaction
(f) Glucose bouillon	1	Opacity for 2–3 hours; no infiltrate
(g) 0.9% NaCl	1	Same as (f)

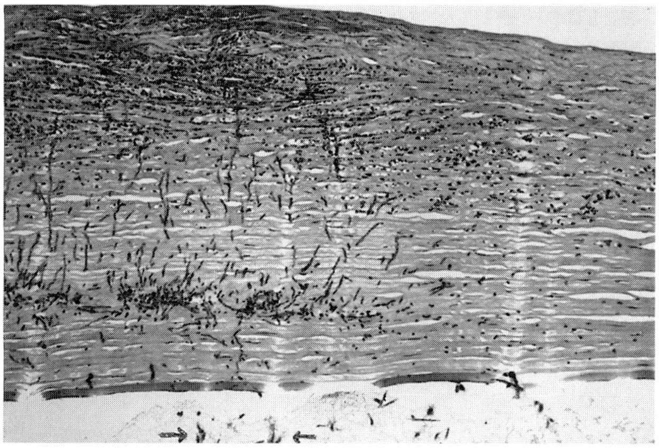

Figure 85.3 Infiltration of the cornea and invasion of the anterior chamber by *Candida albicans* DSM 70010 (arrow). Periodic acid Schiff staining. Magnification approx. ×110.

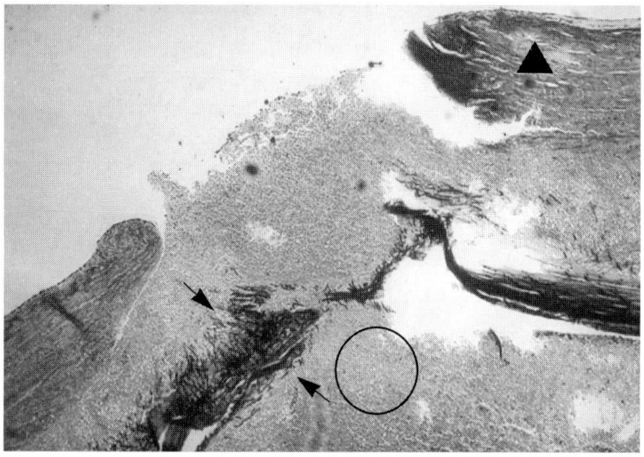

Figure 85.4 Central perforation of the cornea after infection with *Candida albicans* DSM 70010. Fibrinocellular anterior chamber exudate (circled), intracorneal abscess (triangle) and hypha-shaped fungal elements (arrows). Methenamine silver staining. Magnification ×50.

investigated. These are exemplified below by the studies conducted with amphotericin B (Behrens-Baumann *et al.*, 1987a). Also examined in this model was the influence of corticosteroids on the treatment of keratomycosis (Behrens-Baumann and Küster, 1987). In this context, only the more virulent DSM 70010 strain was used.

Amphotericin B

In the literature, amphotericin B is unanimously regarded as being active against *C. albicans* infections.

Bioavailability

Methods. Primarily, it was of interest to determine whether and to what extent Amphotericin B penetrates the cornea. For this purpose Amphotericin B was instilled 10 times at intervals of 1 hour into 16 eyes showing no signs of irritation. Under general anaesthesia (5 mg/kg body weight each of Rompun and Ketavet), the corneal epithelium of the right eye had been totally abraded. The corneal epithelium of the left eye was left intact. Fifteen minutes after the last instillation, again under general anaesthesia, 0.2 ml of aqueous humour was withdrawn from each eye with a 24 G cannula.

In the serial dilution test using microtitre plates, dilutions of aqueous humour in increments of 50 μl were diluted with 100 μl of blastospore suspension in glucose bouillon. As growth controls, 50 μl 0.9% NaCl and aqueous humour from untreated eyes were used. The commercially available preparation amphotericin B Squibb was used as reference standard in a parallel dilution series. After 24 hours' incubation at 37°C fungal growth and/or growth inhibition was assessed macroscopically. The lower limit of detectability for amphotericin B in aqueous humour is 0.06 μg/ml. In the aqueous humour controls microscopic examination revealed increased formation of mycelia. At the end of the experiment pairs of corneas (with and without abrasion) were excised, homogenized in a Potter homogenizer and centrifuged, and the supernatant was then examined for antimycotic activity in the diffusion test. For this purpose 50 μl samples of this solution were each incubated with 100 μl of blastspore suspension and the inhibition of growth assessed after 24 hours.

Results. After instillation of 10 drops of amphotericin B at hourly intervals, concentrations of active substance were not detectable in either the cornea or the aqueous humour of eyes with intact epithelia. When the epithelium had previously been abraded, concentrations near the limit of detectability (0.06 μg/ml) were found in 4 out of 8 eyes. All 4 corneal samples studied revealed a qualitatively inhibitory effect on fungal growth.

Pharmacodynamics

Methods. Forty-eight hours after bilateral intracorneal induction of infection, the animals were assigned to the groups indicated in Table 85.5. Thereafter the drops were instilled (10 times daily at hourly intervals). In the control group, 6 eyes were treated with deoxycholate, the solubi-

Table 85.5 Group assignments in main experiment. The respective numbers of eyes are shown in brackets

Type of treatment	*With abrasion*	*Without abrasion*
1 drop amphotericin B Squibb 10 × daily	Group I (18 eyes)	Group II (18 eyes)
1 drop deoxycholate 10 × at hourly intervals	Group IIIa (3 eyes)	Group IIIb (3 eyes)
No treatment	Group IIIc (9 eyes)	Group IIId (6 eyes)

lizer used with the commercial preparation. Additionally, gentamicin sulphate 0.5% was applied twice daily and atropine borate 1.0% once daily to all eyes. As soon as more than three-quarters of the corneal surface had become re-epithelialized in at least 1 eye of one of the abrasion group (I, IIIa and IIIc in Table 85.5), abrasion was repeated under general anaesthesia in all eyes of that group. This procedure was necessary every 2–3 days. At the end of the experiment after 3 weeks, microscopic and histological examinations were performed.

Results. The incidence of complications (descemetocele or perforation) is shown in Table 85.6. It is evident that in Group I (amphotericin B + abrasion) the integrity of the eyeball was unimpaired, whereas in the other groups the majority of the eyes showed perforations. Between Groups I and II and I and III, the differences were respectively significant and highly significant ($P < 0.05$ and $P < 0.01$, χ-square test).

The same significant differences between Group I and the other groups were also found in regard to the incidences of hypopyon (Table 85.7) and positive reculture from the cornea or anterior chamber (Table 85.8). A significant difference was likewise found in the size of the infiltrate ($P<0.05$, Wilcoxon test) between Group I (amphotericin B + abrasion) and Group III (controls).

Corticosteroids

Corticosteroids possess on the one hand anti-inflammatory and on the other hand immunosuppressive effects. It was therefore of interest to determine whether the unspecific complications (descemetocele and perforations) of an infectious corneal process in the case of keratomycosis would be retarded or reinforced upon simultaneous administration of amphotericin B and corticosteroids.

Table 85.6 Frequency of complications (descemetocele and/or perforation)

Treatment		Descemetocele and/or perforation		Number of eyes
		Yes	No	
With abrasion	Group I (Am B)	0	18	18
	Group IIIa (Deoxy)	3	0	3
	Group IIIc (no drops)	6	3	9
Without abrasion	Group II (Am B)	9	9	18
	Group IIIb (Deoxy)	2	1	3
	Group IIId (no drops)	5	1	6

Am B, amphotericin B; Deoxy, deoxycholate.

Table 85.7 Incidence of hypopyon

Treatment		Hypopyon		Number of eyes
		positive	negative	
With abrasion	Am B	3	15	18
	Deoxy	2	1	3
	None	6	3	9
Without abrasion	Am B	14	4	18
	Deoxy	3	0	3
	None	6	0	6

Am B, amphotericin B; Deoxy, deoxycholate.

Table 85.8 Frequency of positive and negative reculture tests with *Candida* from cornea and anterior chamber after 3 weeks' treatment with amphotericin B

Treatment		Fungal Culture		Number of eyes
		positive	negative	
With abrasion	Am B	1	17	18
	Deoxy	2	0	3
	None	2	2	9
Without abrasion	Am B	7	4	18
	Deoxy	2	1	3
	None	2	1	6

Am B, amphotericin B; Deoxy, deoxycholate.

Methods

Seventeen animals were prepared with amphotericin B according to the procedure described above. In addition to this basic therapy, one eye of each animal was injected subconjunctivally every 2 days with 4 mg (0.4 ml) of aqueous dexamethasone phosphate. In preliminary experiments this dose had proved favourable. A lower dose (2 mg every other day) had led to no change in the ocular findings. A higher dose (4 mg daily) resulted in an unfavourable course of infection.

Twenty-one infected eyes without antimycotic therapy, of which 12 had been abraded parallel to the therapy groups, served as controls. The course of the infection was documented photographically and the area of infiltrate/ulceration calculated. After 3 weeks, microbiological and histological examinations were performed and an immunofluorescence test carried out to detect *C. albicans*.

Results

In eyes given supplementary dexamethasone therapy the corneal infiltrate/ulceration was less pronounced on Day 21.

The area of infiltrate/ulceration in the eyes treated only with the antimycotic averaged 5.69 mm^2, as against 5.23 mm^2 in those also treated with dexamethasone.

The results concerning corneal revascularization are presented in Table 85.9. The eyes given the combined therapy showed significantly ($P < 0.05$) less vascularization of the cornea than those treated only with amphotericin B. Complications such as hypopyon, descemetocele or perforation did not occur.

In both groups, both histological examination and the indirect immunofluorescence technique revealed fungal elements after termination of the therapy. These appear to be non-vital yeasts, since an attempt to reculture them failed. Positive reculture was nevertheless obtained in the group described above (amphotericin B without abrasion, Table 85.8).

Table 85.9 Frequency of neovascularization. The figures indicate the number of eyes

	Neovascularization		*Total*
	Yes	*No*	
Amphotericin B with abrasion	11	6	17
Amphotericin B with abrasion + dexamethasone	2	17	17

Pitfalls of the model: advantages and disadvantages

A certain disadvantage of the model consists in the need for intracorneal application of the fungal suspension. That is, however, only a theoretical disadvantage.

In the natural environment, superficial injuries to the cornea caused by plants or branches colonized by fungi are not uncommon. Nevertheless, in both humans and animals a keratomycosis seldom develops after such injuries. Plants frequently display colonizations of moulds (Emmons, 1975), and fungal infections could therefore be expected to occur frequently. However, given an unimpaired immune status, the mechanical rinsing effect of lacrimal fluid is evidently sufficient to prevent the development of keratomycosis in most cases. Apparently keratomycosis can only develop if the corneal injury is deep enough, the eye is infected by a virulent strain, and incipient vascularization over the defect hinders rinsing with lacrimal fluid. This is, however, exactly the situation simulated by the described model, which is therefore altogether realistic.

One technical difficulty lies in ensuring the complete injection of the extremely small volume of the suspension (10 μl). Theoretically, a loss of volume through the puncture channel cannot be ruled out, so that the infective dose injected may not be uniform in all eyes (Kunze, 1979; Thiel *et al.*, 1982). The conceivable loss of injection volume might account for the slight variations in the size of infiltrates and ulcerations, but would have no influence on the induction of the infection in the described model.

Contribution of the model to infectious disease therapy

While there are numerous antimycotics available for the treatment of keratomycosis, assessments of the efficacy of the individual agents in the literature vary. This is largely because the findings were obtained from differing and unreproducible animal models. In contrast to earlier models, the present model fulfils all criteria of reproducibility with an exactly characterized, highly virulent fungal strain, without immunosuppression of the animals. The course of the infection is virtually uniform, intensive, and sufficiently long to allow reliable testing of the efficacy of drugs.

Amphotericin B is unanimously considered in the literature to be effective against infections with *C. albicans* (Foster *et al.*, 1958; Anderson *et al.*, 1959; Kaufman and Wood, 1965; De Voe, 1971; Jones *et al.*, 1972; Böke and Thiel, 1973; Wood and Williford, 1976; Shukla *et al.*, 1984). For that reason, it was chosen as the first agent to be tested in this model. The results obtained demonstrate that "*debridement*" (Polack, 1973) of the cornea augments the efficacy of amphotericin B. In terms of human medicine, this means that during topical therapy of a keratomycosis with amphotericin B, an open corneal epithelium should be present, or corneal abrasion performed to ensure an adequate concentration of the active substance at the site of infection. Under certain circumstances, abrasion may have to be repeated several times, if the epithelium closes again. This fact runs counter to the usual therapeutic concepts regarding "open" corneal ulcers, which aim at achieving re-epithelialization as rapidly as possible. According to the findings made in the rabbit model, however, it appears advisable to adopt a different therapeutic approach in the case of keratomycosis. The same applies to the further antimycotics, natamycin, clotrimazole, and bifonazole, likewise tested in this model and found to be less effective than amphotericin B (Behrens-Baumann and Klinge, 1990; Behrens-Baumann *et al.*, 1990b). Only fluconazole shows efficacy despite the presence of a closed epithelium, although even then corneal abrasion proved favourable (Behrens-Baumann *et al.*, 1990a). The reason for this lies in the good penetrative properties of fluconazole due to its low molecular weight of 381.68 Da (amphotericin B = 924.10 Da). Substances with molecular weights above 500 Da either cannot penetrate or are only minimally capable of penetrating the cornea, since frictional forces increasingly hinder diffusion (Benson, 1974).

A full overview of the therapeutic efficacy ratings of all

Table 85.10 Efficacy rankings of various antimycotics tested in the keratomycosis model

Drugs of first choice
Fluconazole
Amphotericin B
Drugs of second choice
Clotrimazole/bifonazole
Natamycin

the substances tested in this model is presented in Table 85.10. The antimycotics of first choice in *Candida* infections of the cornea are accordingly fluconazole and amphotericin B, whereas the remaining substances can only be considered second-choice drugs.

In cases of keratomycosis, many authors recommend that antimycotic therapy should be supplemented by topical administration of corticosteroids to reduce unspecific inflammatory processes (e.g. corneal oedema, formation of fibrin and synechiae; Maylath and Leopold, 1955; Newmark *et al.*, 1971b; Mühlhäuser *et al.*, 1975; Ishibashi, 1982).

On the other hand, in many experimental studies in animals, corticosteroids have led to progression of the fungal infection (Louria *et al.*, 1960; Graf, 1963a; Hoffmann and Schmitz, 1963a,b; Hasany *et al.*, 1973; François and Rijsselaere, 1974).

According to the results obtained in our model, a slight dose of a corticosteroid in low concentration exerts a favourable effect on the course of keratomycosis, in particular with regard to the suppression of neovascularization.

Besides the concentration and frequency of administration of corticosteroids, however, it is especially the time of onset of the treatment that seems to be decisive for its successful outcome. If corticosteroid therapy is begun 2 days before infection, its effect is maximally negative. It is distinctly less so if begun 1 day after infection. If the corticosteroid is not given until Day 7, there is no further impairment, as Louria *et al.* (1960) found in the mouse model of systemic *Candida* infection.

So far, however, no studies have been undertaken to determine the significance of the time factor in regard to topical *Candida* keratomycosis.

We are now carrying out relevant experiments in our animal model. Initial tendencies confirm the assumed importance of the time factor for the further course of keratomycosis (Behrens-Baumann *et al.*, 1998).

References

Agarwal, L.P., Malik, S.R.K., Mohan, M., Koshla, P.K. (1963). Mycotic corneal ulcers. *Br. J. Ophthalmol.*, **47**, 109–115.

Alfonso, E., Mandelbaum, S., Fox, M.J., Forster, R.K. (1986). Ulcerative keratitis associated with contact lens wear. *Am. J. Ophthalmol.*, **101**, 429–433.

Anderson, B., Roberts, S.S., Gonzalez, C., Chick, E.W. (1959). Mycotic ulcerative keratitis. *Arch. Ophthalmol.*, **62**, 169–179.

Ansorg, R. (1981). Mikromethode zur Bestimmung der Candida-Antikörper im indirekien Hämagglutinationstest. *Arzd. Lab.*, **27**, 106–110.

Behrens-Baumann, W. (1984). Ergebnisse der Keratoplastik a chaud. *Klin. Monatsbl. Augenheilkd.*, **185**, 25–27.

Behrens-Baumann, W., Dobrinski, B., Zimmermann, O. (1984). Bakterienflora der Lider nach präoperativer Desinfektion. *Klin. Monatsbl. Augenheilkd.*, **192**, 40–43.

Behrens-Baumann, W., Klinge, B. (1990). Natamycin (Pimaricin) in der Behandlung der experimentellen Keratomykose. *Fortschr. Ophthalmol.*, **87**, 237–240.

Behrens-Baumann, W., Klinge, B., Rüchel, R. (1990a). Topical fluoconazole for experimental candida keratitis in rabbits. *Br. J. Ophthalmol.*, **74**, 40–42.

Behrens-Baumann, W., Klinge, B., Uter, W. (1990b). Topical clotrimazole and bifonazole in the treatment of Candida keratitis in rabbits. *Mycoses*, **33**, 567–573.

Behrens-Baumann, W., Küster, M. (1987). Der Einfluß von Kortikosteroiden bei der antimykotischen Therapie der Candida-Keratitis. *Klin. Monatsbl. Augenheilkd*, **191**, 222–225.

Behrens-Baumann, W., Uter, W., Ansorg, R. (1987a). Experimentelle Untersuchungen zur lokalen Therapie der Candida-Keratomykose mit Amphotericin B. *Klin. Monatsbl. Augenheilkd.*, **191**, 125–128.

Behrens-Baumann, W., Uter, W., Vogel, M., Ansorg, R. (1987b). Tierexperimentelles Modell einer Keratomykose. *Klin. Monatsbl. Augenheilkd.*, **190**, 496–500.

Benson, H. (1974). Permeability of the cornea to topically applied drugs. *Arch. Ophthalmol.* **91**, 313–327.

Berson, E.L., Kobayashi, G.S., Oglesby, R.B. (1965). Treatment of experimental fungal keratitis. *Arch. Ophthalmol.*, **74**, 403–411.

Berson, E.L., Kobayashi, G.S., Becker, B., Rosenbaum, L. (1967). Topical corticosteroids and fungal keratitis. *Invest. Ophthalmol.*, **6**, 512–517.

Böke, W., Thiel, H.-J. (1973). Zur Therapie der Hypopyonkeratitis und des Hornhautabszesses. *Klin. Monatsbl. Augenheilkd.*, **163**, 125–131.

Böke, W., Thiel, H.-J. (1974). Tektonische und kurative Keratoplastik nach perforierender herpetischer Keratitis. *Klin. Monatsbl. Augenheilkd.*, **165**, 153–159.

Burda, C.D., Fisher, E. (1959). The use of cortisone in establishing experimental fungal keratitis in rats. *Am. J. Ophthalmol.*, **48**, 330–335.

Burda, C.D., Fisher, E. (1960). Corneal destruction by extracts of cephalosporium mycelium. *Am. J. Ophthalmol.*, **50**, 926–934.

Chowehüvech, E., Sawicki, L., Tenenbaum, S., Galin, M.A. (1973). Effect of various microorganisms found in cosmetics on the normal and injured eye of the rabbit. *Am. J. Ophthalmol.*, **75**, 1004–1009.

Davis, F.A. (1929). The anatomy and histology of eye and orbit of the rabbit. *Trans. Am. Ophthalmol. Soc.*, **27**, 401–441.

De Voe, A.G. (1971). Keratomycosis. *Am. J. Ophthalmol.*, **71**, 406–414.

Doden, W. (1973). Dringliche Keratoplastiken bei entzündlichen Hornhaustprozessen. *Ophthalmologica*, **167**, 402–407.

Dudley, M.A., Chick, E.W. (1964). Corneal lesions produced in rabbits by an extract of fusarium moniliforme. *Lab. Sci.*, **72**, 346–350.

Duke-Elder, S., Wybar, K.C. (1961). The anatomy of the visual system. In: Duke-Elder, S. (ed.) *System of Ophthalmology*, **2**, 92–94.

Ellison, A.C., Newmark, E. (1970). Potassium iodine in mycotic keratitis. *Am. J. Ophthalmol.,* **69**, 126–129, Supplement 70, 152.

Ellison, A.C., Newmark, E. (1973). Effecets of subconjunctival pimaricin in experimental keratomycosis. *Am. J. Ophthalmol.,* **75**, 760–764.

Ellison, A.C., Newmark, M.S., Kaufman, H.E. (1969). Chemotherapy of experimental keratomycosis. *Am. J. Ophthalmol.,* **68**, 812–819.

Emmons, C.W. (1975). The natural occurrence of pathogenic fungi. In: *Opportunistic fungal infections*. Proceedings of the 2nd International Conference (eds. Chick, E.W., Balows, A. *et al.*), Thornas, Springfield, Ill, pp 22–30.

Fiscella, R.G., Moshifar, M., *et al.* (1997). *Cornea* **16**, 447–449.

Foster, J.T.B., Almeda, E., Littmann, M.L., Wilson, M.E. (1958). Some intraocular and conjunctival effects of Amphotericin B in man and in the rabbit. *Arch. Ophthalmol.*, **60**, 555–564.

Forster, R.K., Rebell, G. (1975a). Animal model of fusarium solani keratitis. *Am. J. Ophthalmol.*, **79**, 510–516.

Forster, R.K., Rebell, G. (1975b). Therapeutic surgery in failures of medical treatment for fungal keratitis. *Br. J. Ophthalmol.*, **59**, 366–371.

Francois, J., Rijsselacre, M. (1974). Corticosteroids and ocular mycoses: experimental study. *Ann. Ophthalmol.*, **6**, 207–217.

Garcia de Lomas, J., Fons, M.A., Nogueira, J.M., *et al.* (1985). Chemotherapy of *aspergillus fumigatus* keratitis: an experimental study. *Mycopathologia,* **89**, 135–138.

Graf, K. (1963a). Über den Einfluß von Kortison auf das Entstehen von Keratomykosen am Kaninehenauge durch saprophyfare Pilze des menschlichen Bindehautsackes. *Klin. Monatsbl. Augenheilkd.,* **143**, 356–361.

Graf, K. (1963b). Zur Frage der Hornhautimmunität des Kaninehenauges gegenüber Candida albicans. *Graefe Arch. Ophthalmol.*, **166**, 331–334.

Hallermann, W. (1975) Keratoplastik aus akuter Indikation. *Klin. Monatsbl. Augenheilkd.*, **167**, 345–352.

Hasany, S.M., Basu, P.K., Kazdan, J.J. (1973). Production of corneal ulcer by opportunistic and saprophytic fungi 1. The effect of pretreatment of fungi with steroids. *Can. J. Ophthalmol.*, **8**, 119–131.

Head, W.S., O'Day, D.M., Robinson, R.D., Williams, T.E. (1990). Effect of corticosteroid administration on modeling keratomycosis in the rabbit. *Invest. Ophthalmol. Vis. Sci.*, **31**, 447.

Hervouet, F., Lenoir, A. (1953). Aspergillose corneenne. *Bull. Soc. Franc. Ophthalmol.*, **66**, 287–292.

Hirose, K., Yoshioku, H., Abe, S., Kanemitsu, J., Kiya, K. (1957). Effect of cortisone on experimental keratomycosis. *Acta Soc. Ophthalmol. Jap.*, **61**, 1106–1133.

Hoffmann, D.H. (1965). Pilzinfektionen des Auges. *Fortschr. Augenheilkkd.*, **16**, 63–217.

Hoffmann, D.H., Schmitz, R. (1963a). Untersuchtungen zum Einfluß des Cortisons auf die experimentelle Candida-Mykose der Kaninchenhornhaut Graefe. *Arch. Ophthalmol.*, **166**, 260–276.

Hoffmann, D.H., Schmitz, R. (1963b). Die experimentelle Keratomykose als Beitrag zur Frage des Cortisonschadens am Auge. Ein vorläufiger Bericht. *Mycosen*, **6**, 12–20.

Ishabashi, Y. (1979). Experimental fungal keratitis due to Fusarium: Studies on animal model and inoculation technique. In: *Proc. 23rd Congr. Ophthalmol. Kyoto 1978*, pp. 1705–1707.

Ishibashi, Y. (1982). The difference of the effects on experimental keratomycosis due to the length of medicational period of corticosteroid. *Nippon Ganka Gakkai Zasshi 86* (English abstract).

Ishibashi, Y., Kaufman, H.E. (1985). The effects of subconjunctival miconszole in the treatment of experimental Candida keratitis in rabbits. *Arch. Ophthalmol.*, **103**, 1570–1573.

Ishibashi, Y., Kaufman, H.E. (1986). Topical ketoconazole for experimental Candida keratitis in rabbits. *Am. J. Ophthalmol.*, **102**, 522–526.

Ishibashi, Y., Matsumoto, Y. (1984). Oral ketoconazole therapy for experimental Candida albicans keratitis in rabbits. *Sabouraudia*, **22**, 323–330.

Ivandic, T. (1973). Zur Behandlung von Augenmykosen. *Klin. Monatsbl. Augenheilkd.*, **162**, 634–637.

Jones, D.B., Forster, R.K., Rebell, G. (1972). Fusarium solani keratitis treated with Pimaricin: 18 cases. *Arch. Ophthalmol.*, **88**, 147–154.

Kaufman, H.E., Wood, R.M. (1965). Mycotic keratitis. *Am. J. Ophthalmol.*, **59**, 993–1000.

Kunze, M. (1979). Experimentelle Keratomykose beim Kaninchen durch *Candida albicans*. Inaugural-Dissertation, Kiel.

Leber, T. (1879). Keratomycosis aspergillina als Ursache von Hypopyonkeratitis. *Graefe Arch. Ophthalmol.*, **25**, 285–301.

Ley, A.P. (1956). Experimental fungus infections of the cornea. A preliminary report. *Am. J. Ophthalmol.*, **42**, 59–71.

Ley, A.P., Sanders, T.E. (1956). Fungus keratitis – a report of three cases. *Arch. Ophthalmol.*, **56**, 257–264.

Louria, D.B., Fallon, N., Browne, H.G. (1960). The influence of cortisone on experimental fungus infections in mice. *J. Clin. Invest.*, **39**, 1435–1499.

Maurice, D.M. (1960). The permeability of the cornea. In: *The transparency of the cornea*, (eds Duke-Elder, S., Perkins, E.S.), pp. 67–71.

Maylath, F.R., Leopold, I.H. (1955). Study of experimental intraocular infection. I. The recoverability of organisms inoculated into ocular tissues and fluids. II. The influence of antibiotics and cortisone, alone and combined, on intraocular growth of these organisms. *Am. J. Ophthalmol.*, **40**, 86–101.

Meyer-Rohn, J., Lange-Brock, T. (1957). Untersuchungen zur Frage der Wachstumssimulierung von *Candida albicans* durch Antibiotica. *Arch. Klin. Exp. Dermatol.*, **204**, 58–69.

Mitsui, Y., Hanabusa, J. (1955). Corneal infections after cortisone therapy. *Br. J. Ophthalmol.*, **39**, 244–250.

Montana, J.A., Sery, T.W. (1958). Effect of fungistatic agents on corneal infections with *Candida albicans*. *Arch. Ophthalmol.*, **60**, 1–6.

Mühlhäuser, J., Wildfeuer, A., Meister, H. *et al.* (1975). Die Candida-Keratitis im Tierexperiment. *Graefe Arch. Ophthalmol.*, **195**, 251–262.

Naumann, G., Green, W.R., Zimmermann, L.E. (1967). Mycotic keratitis. *Am. J. Ophthalmol.*, **64**, 668–682.

Newmark, E., Kaufman, H.E., Polack, F.M., Ellison, A.C. (1971a). Clinical experience with Pimaricin-therapy in fungal keratitis. *Sth. Med. J.*, **64**, 935–941.

Newmark, E., Ellison, A.C., Kaufman, H.E. (1971b). Combined pimaricin and dexamethason therapy of keratomycosis. *Am. J. Ophthalmol.*, **71**, 718–722.

O'Day, D.M., Robinson, R., Head, W.S. (1983). Efficacy of antifungal agents in the cornea. *Invest. Ophthalmol. Vis. Sci.*, **24**, 1098–1102.

O'Day, D.M., Ray, W.A., Robinson, R., Head, W.S. (1984). Efficacy of antifungal agents in the cornea. II. Influence of corticosteroids. *Invest. Ophthalmol. Vis. Sci.*, **25**, 331–335.

O'Day, D.M., Ray, W.A., Robinson, R.D., Head, W.S., Williams,

T.E. (1991a). Differences in response in vivo to Amphotericin B among *Candida albica* strains. *Invest. Ophthalmol. Vis. Sci.*, **32**, 1569–1572.

O'Day, D.M., Ray, W.A., Head, W.S. *et al.* (1991b). Influence of corticosteroid on experimentally induced keratomycosis. *Arch. Ophthalmol.*, **109**, 1601–1604.

O'Day, D.M., Head, W.S., Robinson, R.D. *et al.* (1991c). The evaluation of therapeutic responses in experimental keratomycosis. *Eye Research*, **11**, 35–44.

O'Day, D.M., Head, W.S., Robinson, R.D. *et al.* (1992). Ocular pharmacokinetics of saperconazole in rabbits. A potential agent against keratomycoses. *Arch. Ophthalmol.*, **110**, 550–554.

Oggel, K., de Decker, W. (1975a). Standard-Candida-Mykose am Kaninschenange. *Graefe Arch. Ophthalmol.*, **193**, 193–195.

Oggel, K., de Decker, W. (1975b). Untersuchungen zur Frage der Wirksamkeit eines neuen Antimykotikums bei experimenteller Keratomykose an Kaninehen. *Graefe Arch. Ophthalmol.*, **193**, 189–192.

Ohno, S., Fuerst, D.J., Okumoto, M. *et al.* (1983). The effect of K-582, a new antifungal agent, on experimental candida keratitis. *Invest. Ophthal. Vis. Sci.*, **24**, 1626–1629.

Oji, F.O. (1981). Development of quantitative methods of measuring antifungal drug effects in the rabbit cornea. *Br. J. Ophthalmol.*, **65**, 89–96.

Oji, E.O. (1982). Study of ketoconazole toxicity in rabbit cornea and conjunctiva. *Int. Ophthal.*, **5**, 169–174.

Polack, F.M., Kaufman, H.E., Newmark, E. (1971). Keratomycosis. *Arch. Ophthalmol.*, **85**, 410–416.

Polack, F.M. (1973). Diagnosis and management of keratomycoses. *International Ophthalmology Clinics,* **13**, 75–91.

Prince, J.H. (1964). Cornea, trabecular region, and sclera. In: *The rabbit in eye research*, (ed Prince, J.H.), pp. 86–139.

Rheins, M.S., Suie, T., van Winkle, M.G. (1965). Further investigation of the effects of antibiotics. *Am. J. Ophthalmol.*, **59**, 221–225.

Rheins, M.S., Pixley, P.A., Suie, T., Keates, R.H. (1966). Diagnosis of experimental fungal corneal ulcers by fluorescent antibody techniques. *Am. J. Ophthalmol.*, **62**, 892–900.

Richards, A.B., Clayton, Y.M., Jones, B.R. (1969). Antifungal drugs for oculomycosis: IV. The evaluation of antifungal drugs in the living animal cornea. *Trans. Ophthalmol. Soc. U.K.*, **89**, 847–861.

Serdarevic, O., Darell, R.W., Krueger, R.R., Trokel, S.L. (1985). Excimer laser therapy for experimental candida keratitis. *Am. J. Ophthalmol.*, **99**, 534–538.

Shukla, P.K., Khan, Z.A., Lal, P. *et al.* (1984). A study on the association of fungi in human corneal ulcers and their therapy. *Mykosen,* **27**, 385–390.

Singh, G., Malik, S.R.K., Bhatnagar, P.K. (1974). Therapeutic value of keratoplasty in keratomycosis – an experimental study. *Arch. Ophthalmol.*, **92**, 48–50.

Smolin, G., Okumoto, M. (1969). Potentiation of *Candida albicans* keratitis by antilymphocyte serum and corticosteroids. *Am. J. Ophthalmol.*, **68**, 675–682.

Stern, G.A. (1978). In vitro antibiotic synergism against ocular fungal isolates. *Am. J. Ophthalmol.*, **86**, 359–367.

Stern, G.A., Okumoto, M., Smolin, G. (1979). Combined amphotericin B and rifampin treatment of experimental *Candida albicans* keratitis. *Arch. Ophthalmol.*, **97**, 721–722.

Stoewer (1899). Über die Wirkung pathogener Hefen am Kaninchenauge. *Graefe Arch. Opthalmol.*, **68**, 178–191.

Tandon, R.N., Wahab, S., Srivastava, O.P. (1984). Experimental infection by candida krusei (east). Berkhout isolated from a case of corneal ulcer and its sensitivity to antimycotics. *Mykosen*, **27**, 355–360.

Thiel, H.J. (1978). Keratoplastik bei akuten infektiösen Hornhautprozessen. *Klin. Monatsbl. Augenheilkd.*, **173**, 171–181.

Thiel, H.J., Kunze, M., Pülhörn, G. (1982). Experimentelle Keratomykose beim Kaninchen durch Candida albicans. *Zentralbl. Ophthalmol.*, **123**, 141.

Thygeson, P., Okumoto, M. (1974). Keratomycosis – a preventable disease. *Trans. Amer. Acad. Ophthalmol. Otolaryng.*, **78**, 433–439.

Urer, W. (1987). Untersuchungen zur Frage der Wirksamkeit einer lokalen Behandlung mit Amphotericin B bei experimenteller Candida-Albicans-Keratitis am Kaninchenauge. Inaugural Dissertation, Göttingen.

Van Winckle, M.G., Rheins, M.S., Suie, T. (1964). Effects of antibiotics on experimental candida corneal infections. *Am. J. Ophthalmol.*, **51**, 84–87.

Verin, P., Nguyen, D.H., Comte, P., Nguyen, D.T. (1984). Durch Reisähren verursachte Keratitis – eine wichtige Ursache der Erblindung in den Entwicklungsländern. *Fortschr. Augenheilkd.*, **81**, 418–420.

Watsky, M.A., Jablonski, M.M., Eldershausen, H.F. (1988). Comparison of conjunctival and corneal surface areas in rabbit and human. *Curr. Eye Res.*, **7**, 483–486.

Wilson, L.A., Ahearn, D.G. (1986). Association of fungi with extended wear soft contact lenses. *Am. J. Ophthalmol.*, **101**, 434–436.

Wood, T.O., Williford, W. (1976). Treatment of keratomycosis with Amphotericin B 0.15%. *Am. J. Ophthalmol.*, **81**, 847–849.

Yamamoto, G.K., Pavan-Langston, D., Stowe, G.C., Albert, D.M. (1979). Fungal invasion of a therapeutic soft contact lens and cornea. *Am. Ophthalmol.*, **11**, 1731–1735.

Zimmerman, L.E. (1963). Keratomycosis. *Surv. Ophthalmol.*, **8**, 1–25.

Chapter 86

Experimental *Candida* Endocarditis

M. R. Yeaman, J. Lee and A. S. Bayer

Background: *Candida* endocarditis in humans

Fungal infective endocarditis (IE) is a life-threatening condition which has been observed with increasing incidence (Pelletier and Petersdorf, 1977; Bayliss *et al.*, 1983; Griffin *et al.*, 1985). Diverse organisms have now been recognized as potential pathogens in fungal IE; however, *Candida* species continue to be among the most common etiologic agents (Torack, 1957; Seelig *et al.*, 1974; Rubenstein *et al.*, 1975). Although *Candida* IE occurs in a wide array of clinical settings, specific factors have been correlated with predisposition of individuals to this and other fungal IE. These include use of long-term indwelling intravenous cathetherization, implantation of prosthetic cardiac valves, immunocompromising conditions such as neutropenia or diabetes mellitus, long-term and/or broad-spectrum antibiotic therapy, parenteral hyperalimentation, and intravenous drug abuse (Andriole *et al.*, 1962; Kammer and Utz, 1974; Seelig *et al.*, 1979; Vo *et al.*, 1981). A review of *Candida* IE in humans is available elsewhere (Meyer and Edwards, 1992).

The epidemiology of fungal IE, including that due to *Candida* species, reveals that male patients outnumber females by a ratio of 3:1, and overall, patients tend to be relatively young in comparison to those with other life-threatening endovascular infections (approximately 30–40 years; Kammer and Utz, 1974; Woods *et al.*, 1989). Nearly two-thirds of all patients with *Candida* IE have a history of underlying cardiac valve disease (Rubenstein *et al.*, 1975; Pelletier and Petersdorf, 1977; Griffin *et al.*, 1985; Moyer and Edwards, 1992). Individuals suffering from *Candida* IE frequently develop serious embolic sequelae due to the propensity of fungi to produce massive, yet friable, valvular vegetations (Rubenstein *et al.*, 1975). It is noteworthy that the majority of cases of *Candida* IE in intravenous drug abusers have been due to non-*albicans* species such as *C. tropicalis* or *C. parapsilosis* (Rubenstein *et al.*, 1975; Garvey and Nell, 1978).

The diagnosis of *Candida* IE is difficult, principally due to the fact that patients with this condition often have negative blood cultures (Seelig *et al.*, 1974; Edwards, 1991). As a result, extension of the infection from the cardiac valve to periannular tissue is common, as are metastatic sequelae. These events present additional problems regarding placement of prosthetic valves, and prevention of their subsequent infection (Andriole *et al.*, 1962; Robboy and Kaiser, 1975; Edwards, 1991).

Even when the diagnosis is made relatively early in the course of infection, *Candida* IE is difficult to treat with antifungal therapy alone. Management of *Candida* IE usually involves high-dose initial therapy with potentially toxic agents (e.g. amphotericin B), and frequently necessitates long-term or even lifelong suppressive antifungal therapy (Seelig *et al.*, 1979; Edwards, 1991). Furthermore, surgical intervention is usually required to resolve extensive valvular or perivalvular destruction or intracardiac fistulae, compounded as a result of delayed diagnosis. Despite advances in antifungal therapy and surgical technologies, the morbidity and mortality associated with *Candida* IE remain unacceptably high. Several studies have demonstrated an overall mortality associated with *Candida* and other fungal IE approaching 75% (Kammer and Utz, 1974; McLeod and Remington, 1978; Vo *et al.*, 1981; Moyer and Edwards, 1992). Furthermore, the emergence of *Candida* resistance to antifungal agents magnifies these concerns. In this regard, there is an urgent need to understand better the pathogenesis and therapy of *Candida* IE, with the goal of improving its diagnosis, prevention and treatment (Walsh and Pizzo, 1988a,b). Thus, until improvements in serodiagnosis of *Candida* IE occur, and reliable methods for evaluating antifungal therapies *in vitro* are established, relevant animal models will continue to play an invaluable role in optimizing therapeutic strategies for *Candida* IE.

Background of *Candida* endocarditis experimental models

Pathogenesis

Calderone *et al.* (1978, 1985, 1988; Calderone and Scheld, 1987), Klotz *et al.* (1983, 1985, 1989; Klotz, 1992; Klotz and Maca, 1988), Maisch and Calderone (1980, 1981), and Rotrosen *et al.* (1985, 1986), have provided much of the current knowledge of pathogenesis in *Candida* IE through investigations using experimental animal models and tissue

Handbook of Animal Models of Infection
ISBN 0-12-775390-7

culture systems. Filler *et al.* (1991, 1995, 1996) and Edwards *et al.* (1975, 1987) have contributed the majority of recent information regarding endothelial cell interactions with *Candida* that may influence disease progression or antifungal host defense mechanisms. As determined through these studies, the pathogenesis of *Candida* IE represents a multifactorial process, initiated by fungemia, with subsequent colonization of cardiac valve endothelium. The above studies, as well as data from bacterial IE models (Archer and Fekety, 1972; Durack and Beeson, 1972a,b, 1973; Scheld *et al.*, 1978, 1985; Baddour *et al.*, 1989), have led to the traditional view of the pathogenesis of *Candida* IE in humans, believed to involve the following sequence of events:

1. Predisposed cardiac valve endothelium (e.g. rheumatic fever, prosthetic heart valve)
2. Presence of a sterile thrombus (vegetation) on the damaged or abnormal cardiac valve endothelium
3. Fungemia facilitating hematogenous access to the endothelium/sterile vegetation
4. Direct or indirect adhesion of the pathogen to the valve or sterile vegetation
5. *Candida* survival at the vegetation/endothelium interface
6. Proliferation of the organism within the vegetative lesion
7. Evolution of the infected vegetation, with extracardiac hematogenous dissemination.

As with experimental bacterial IE, in experimental *Candida* IE, catheter-induced trauma to cardiac valves greatly facilitates initiation of the infection (Tunkel and Scheld, 1992). Injury to vascular endothelial cells or exposure of subendothelial stroma leads to the generation of cytokines (Filler *et al.*, 1991, 1995, 1996), along with procoagulant molecules such as platelet-activating factor and tissue factor (Drake and Pang, 1988, 1989; Drake *et al.*, 1984; Bansci *et al.*, 1994). In the presence of other coagulation pathway constituents (e.g. factors VII and X, and calcium), tissue factor catalyzes the conversion of prothrombin to thrombin (Weksler, 1992). Thrombin, among the most potent of all platelet agonists, activates scenescent, discoid platelets to become spheroid cells with multiple pseudopodia (Colman, 1991; Weksler, 1992). Such activation is associated with platelet expression of numerous surface ligands, adhesion of the platelet to damaged endothelium, and ultimate platelet degranulation. Platelet degranulation, in turn, releases an array of bioactive substances, including secondary platelet agonists, such as adenosine diphosphate, and thromboxane A_2 (Colman, 1991; Weksler, 1992). Therefore, the cardiac valvular vegetation evolves as repetitive waves of activated platelets are deposited on to the damaged endothelium. It should be emphasized that it is highly likely that platelets function in a key antimicrobial host defense role to limit microbial colonization of vascular endothelium (Yeaman *et al.*, 1992, 1996, 1997; Christin *et al.*, 1998; reviewed in Yeaman, 1997). Thus, *Candida* must presumably first circumvent relevant antifungal host defense mechanisms to proliferate within the vegetation.

The platelet–fibrin matrix constitutes the majority of the mass of the cardiac vegetation. Previous studies have suggested that *C. albicans* expresses receptors which promote adhesion to platelets or adhesive proteins such as fibrinogen and fibrin (Edwards *et al.*, 1986; Robert *et al.*, 1992). Recent evidence indicates that *C. albicans* possesses specific adhesion molecules which may also mediate its attachment to vascular endothelial cells (Fu *et al.*, 1998). Thus, *Candida* organisms which are capable of circumventing the antimicrobial host defense properties of platelets (Yeaman *et al.*, 1996) may conceivably utilize the vegetation as an adhesive surface. Given the minimal contribution of neutrophils to limiting microbial proliferation within left-sided IE vegetations (Bayer *et al.*, 1988, 1989), and the observation that left-sided IE is more common and generally more severe than right-sided infection, it is highly likely that an important interaction exists between platelets and neutrophils in preventing and limiting infections in these settings. Furthermore, the absence of neutrophils does not significantly diminish the efficacy of prophylaxis of IE (Berney and Francioli, 1990), thus implicating a key role for platelets in antimicrobial host defense against infections involving the vascular endothelium (Yeaman, 1997; Yeaman *et al.*, 1997).

Evaluation of antifungal efficacy in the *Candida* IE model

Experimental IE due to *Candida* represents a significant challenge regarding efficacy of antifungal agents. This relates to the typically high fungal densities within cardiac vegetations, the propensity of infections involving catheters or other intravascular devices to be refractory to sterilization, and the frequent and often serious metastatic sequelae that accompany *Candida* IE. As a result, antifungal agents (other than amphotericin B) tested in this model generally require relatively high serum concentrations (above the minimal inhibitory concentration (MIC) of the infecting strain) and/or prolonged therapeutic duration to achieve even modest antifungal efficacy. The complex pharmacokinetics of amphotericin B preclude the ability to define a relationship between serum levels and therapeutic efficacy. None the less, complete sterilization of vegetations infected with *Candida*, even in the presence of the most potent fungicidal agents, is rarely observed in this model.

Considerations for appropriate animal for model

Investigations examining the pathogenesis and therapy of experimental disseminated *Candida* infection have been performed using a variety of animals models, including the mouse, rat, and rabbit (Witt and Bayer, 1991; Witt *et al.*, 1993; Cole *et al.*, 1995; De Bernardis *et al.*, 1995; Futenma *et al.*, 1995; Rozalska *et al.*, 1995; Sawyer *et al.*, 1995a,b; Flattery *et al.*, 1996; Hata *et al.*, 1996; Kretschmar *et al.*, 1996; Plotkin *et al.*, 1996; Schmidt, 1996; Fallon *et al.*, 1997;

Han and Cutler, 1997; Herzyk *et al.*, 1997; Matuschak and Lechner, 1997; Ohtsuka *et al.*, 1997). In addition, murine (Futenma *et al.*, 1995; Flattery *et al.*, 1996) or Sprague–Dawley or Zucker rat models (Plotkin *et al.*, 1996; Schmidt, 1996) have now been used to investigate disseminated candidiasis in the context of immunodeficiency. The comparative use of various animal models of IE has been previously reviewed (Tunkel and Scheld, 1992). The mouse and rat models require less antifungal agent overall, and the costs of obtaining and caring for small rodents are less than those of rabbits. In addition, more numerous and varied models of gene knockout-mediated immunosuppression exist in mice and rats than in rabbits. None the less, the rabbit model is generally viewed as the principal tool for studying *Candida* IE *in vivo* (Filler *et al.*, 1991; Witt and Bayer, 1991; Witt *et al.*, 1993; Yeaman *et al.*, 1996). Given their similarities to humans in cardiovascular function, as well as their ability to survive unilateral carotid artery ligation, the ease of surgical manipulation and serial blood sampling, and overall maintenance and hardiness, the rabbit is an ideal model system. Although many rabbits species may be suitable for use in the model of *Candida* IE, the New Zealand White rabbit has been most extensively used (Filler *et al.*, 1991; Witt and Bayer, 1991; Witt *et al.*, 1993; Yeaman *et al.*, 1996).

Preparation of animals for use in the experimental IE model

Investigators should consult appropriate references and guidelines (Fraser, 1991; Brown *et al.*, 1993; Harkness and Wagner, 1995; National Center for Research Resources, 1996) for animal surgery in research prior to utilizing this or any animal model. There are no unusual preparations necessary for rabbit care. Animals should arrive and be housed at least 24 hours prior to catheterization surgery to allow acclimatization to new ambient conditions. The length of time required for stabilization may be longer, depending on the type and duration of animal transportation (National Center for Research Resources, 1996). This reduces animal stress, and appears to improve survival post-surgery. Food (rabbit chow) and water are generally provided *ad libitum* before and after surgery, as well as during the course of infection and treatment studies. A time-controlled lighting system should be used to insure a uniform diurnal lighting cycle. Although rabbits generally exhibit a high degree of resistance to most environmental pathogens, they are susceptible to *Pasteurella multocida*, *Bordetella bronchisepticum*, or related pathogens when housed in crowded or damp conditions (Harkness and Wagner, 1995). Symptoms of these upper respiratory infections include catarrhalic discharges from eyes and nose, nystagmus, torticollis, diarrhea, and/or dysphagia (Fraser, 1991; Harkness and Wagner, 1995). Animals contracting such infections cannot be used in experimental models, and frequently expire. Additionally, animals developing signs of illness should be quarantined immediately. Thus, it is imperative that veterinary staff conduct daily animal health evaluation rounds, and a reliable communication network exist between the veterinarian and study co-ordinators or principal investigators.

Catheterization surgery in the rabbit model

Overview

Under aseptic conditions, surgery is performed to insert a sterile polyethylene catheter into either the right carotid artery or right jugular vein, after which it is passed across the aortic (for left-sided IE) or tricuspid (for right-sided IE) cardiac valve, respectively. In either case, the catheter is sutured in place for at least 24 hours to insure development of sterile vegetations. The catheter may either be removed prior to infection, or remain indwelling for the duration of the study (the latter case provides a particularly rigorous test of antifungal efficacy). Animals are subsequently infected intravenously with a test *Candida* strain, for which an inoculum capable of producing infection in $\geq 90\%$ of all animals (infective dose, 90%: ID_{90}) has been determined (Filler *et al.*, 1991; Witt and Bayer, 1991; Witt *et al.*, 1993; Yeaman *et al.*, 1996). Antifungal therapy may be initiated prior to infection to examine prophylactic efficacy, or may be initiated at any time point following establishment of *Candida* IE to define therapeutic efficacy. Specific parameters used in evaluating disease progression or therapeutic efficacy include gross organ histopathology and/or quantitive culture of cardiac vegetations and distant target organs to monitor fungal density and hematogenous seeding.

Required materials

Materials must be prepared prior to catheterization surgery. As described in detail below, these include general and local anesthetics, anesthetic antidotes (if any), skin antiseptics, sterilized surgical supplies (surgical scalpel, curved hemostat, and forceps, catheter introducer, polyethylene catheter, sutures, gauze, surgical stapler), and high intensity lighting. In addition, postoperative use of a heating pad or low-intensity heat lamp enhances animal recovery from catheterization surgery and anesthesia by facilitating maintenance of adequate core body temperature (National Center for Research Resources, 1996). It should be noted that the guidelines and technical procedures outlined below are also relevant to experimental models of bacterial IE.

Anesthesia and animal preparation

New Zealand White rabbits (females are generally more docile; 2.5 kg in size) should be ear-tagged to facilitate identification. Rabbits are anesthetized for catheterization surgery with concomitant intramuscular injections of 40 mg/kg ketamine hydrochloride and 5 mg/kg xylazine.

When animals are kept warm, surgical-level anesthesia usually follows within 5–7 minutes. More detailed information related to animal anesthesia is provided in the review by Bertens *et al.* (1993). The pericarotid areas of the rabbit neck should be shaved prior to catheterization. This maneuver facilitates surgical location of the carotid artery or jugular vein for investigators utilizing this model for the first time, and promotes skin disinfection and wound healing. The hair can be removed with an electric clipper equipped with a number 40 blade, usually after the animal has been anesthetized.

Catheterization surgery

Once anesthetized, the rabbit is placed in a supine position, ventral side facing upward. The rabbit vertebral column is particularly vulnerable when the animal is anesthetized; care should be taken to minimize torsion or flexion. The cervical area is set into a immobilizing device which provides maximal exposure of the neck. Catheterization surgery is efficiently performed by two-person teams, where individuals performing and assisting with the surgery are seated opposite one another across the procedure table. A third person is desirable if available, to assist in administration of anesthetics, observing animals during the immediate postoperative period, and co-ordinating surgical activities. The surgeon is seated at the head of the rabbit, such that visual access of the neck is optimized; the primary assistant is seated across, in position to immobilize the rabbit limbs and manipulate vessel ligatures (Figure 86.1). The neck is then disinfected with 70% isopropyl alcohol or ionophore such as betadine, and the incision site (see below) locally infused with approximately 1.0 ml of a 1.0% lidocaine solution (using a tuberculin syringe with 24 G needle) to reduce pain and bleeding.

A 5 cm incision is made approximately 1 cm dextrolateral to the trachea for either the right- or left-sided IE models. The incision should only be sufficiently deep to expose the superficial fascia, particularly since the right jugular vein lies close to the fascial surface in the rabbit. A surgical hemostat is used in conjunction with forceps to reflect adipose tissue, fascia, and muscle by blunt dissection (Figure 86.1). The carotid artery must be localized to establish aortic valve (left-sided) *Candida* IE. This vessel is situated within the carotid sheath, which is beneath the sternocleido-mastoideus muscle, and medial to the vagus nerve (Figure 86.2). For tricuspid valve (right-sided) IE, the jugular vein will be found situated superior to the carotid artery and vagus nerve bundle, just beneath the superficial fascia. A curved hemostat in the closed position is used to isolate and elevate the internal carotid artery or jugular vein (at its bifurcation), respectively (Figure 86.3). The hemostat is then opened sufficiently to grasp a sterile suture inserted by the surgical assistant, and withdrawn such that the suture is drawn beneath the vessel of interest to isolate it from other tissues (Figure 86.4).

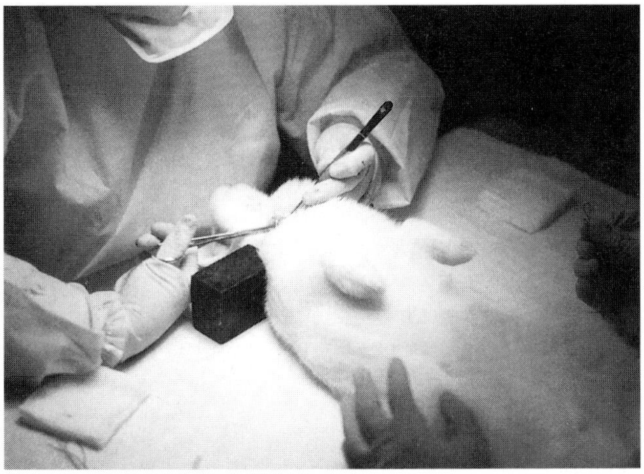

Figure 86.1 Positioning of the surgical team members for the rabbit model of infective endocarditis. Note the surgical assistant in position to provide necessary sutures and surgical instruments, and to control the limbs of the animal during surgery.

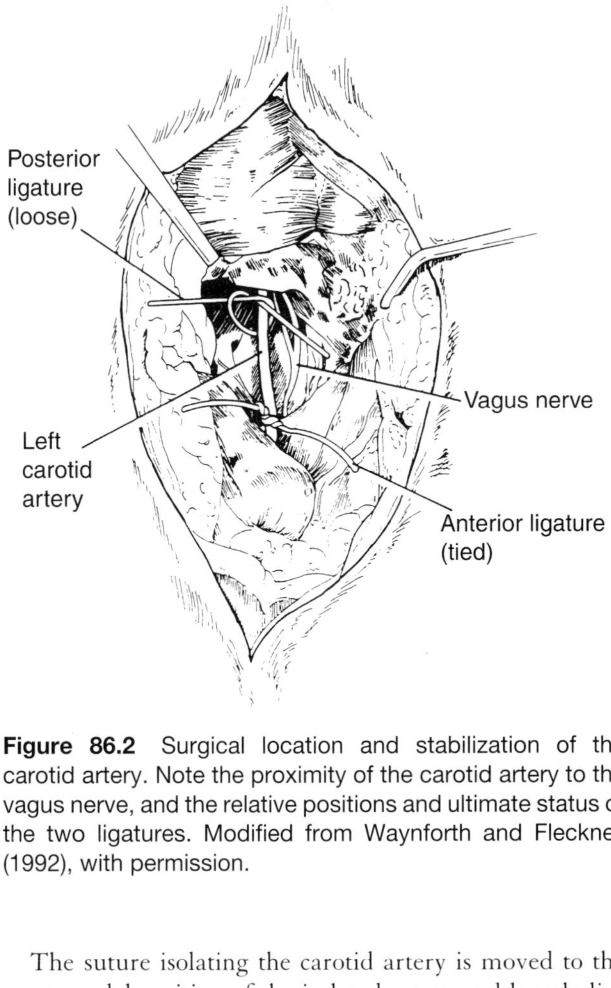

Figure 86.2 Surgical location and stabilization of the carotid artery. Note the proximity of the carotid artery to the vagus nerve, and the relative positions and ultimate status of the two ligatures. Modified from Waynforth and Flecknell (1992), with permission.

The suture isolating the carotid artery is moved to the most caudal position of the isolated artery and loosely ligated; this will facilitate catheter insertion. A second suture is similarly placed, but moved cranially to the limit of the exposed artery, and tightly ligated to prevent excessive bleeding during catheter placement (Figure 86.5). This latter suture terminates blood flow to the brain via the right

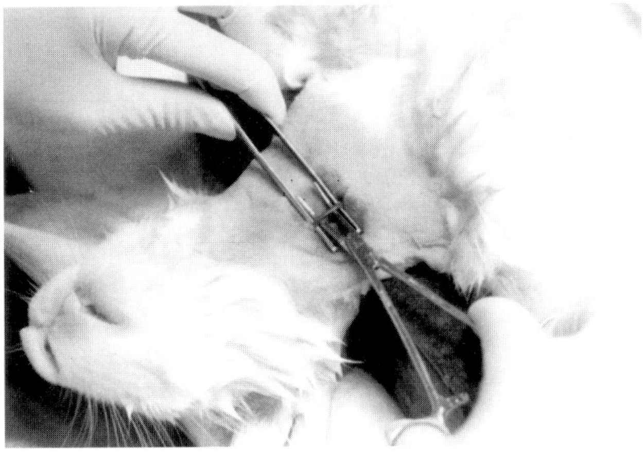

Figure 86.3 Isolation of the carotid artery. Note how the surgeon uses the curved hemostat to isolate the vessel, then opens the hemostat to receive the ligature from the assistant.

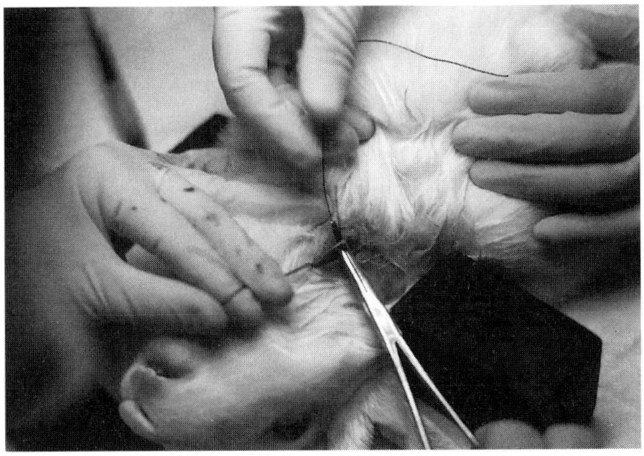

Figure 86.4 Placement of the anterior and posterior ligatures. After the anterior suture is positioned and ligated, the posterior suture is drawn beneath the carotid artery.

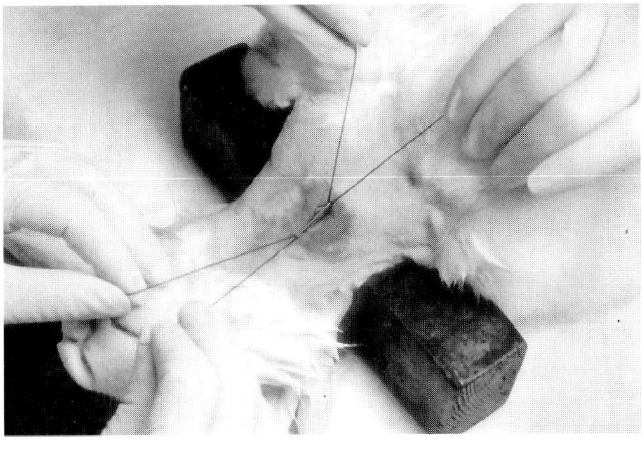

Figure 86.5 Stabilization of the carotid artery for catheter introduction. The surgeon and the assistant manipulate the anterior and posterior sutures, respectively. The anterior suture is ligated, while the posterior suture remains loose. The assistant applies tension to the posterior suture during catheter introduction to minimize blood loss from the vessel.

carotid artery. In rabbits, this is not a lethal event, as the complementary carotid artery provides compensatory blood supply to both brain hemispheres via the circle of Willis. This feature makes the rabbit somewhat unique in its use in this model of cardiac catheterization (for example, this often leads to stroke and/or death in dogs and pigs). The opposing sutures (usually about 2 cm apart) are then used to lift and stabilize the carotid artery for catheterization (Figure 86.5). While the surgical assistant uses the loose distal suture to provide tension to stabilize both the artery or vein, and prevent retrograde blood ejection, the surgeon lifts the proximal (anterior) ligated suture, and punctures the vessel with a sterile introducer, point facing caudally (Figure 86.6). The introducer is transferred from the hand of the surgeon to that of the assistant (while remaining in the vessel), allowing the surgeon to insert the catheter into the carotid, below the introducer. This same procedure would be used for cannulating the right jugular vein (for the tricuspid valve IE model), with the exception that the proximal suture ligation occurs just below the bifurcation point of this vessel.

The polyethylene catheter (internal diameter, 0.86 mm; external diameter, 1.27 mm; Becton Dickinson, Sparks, MD; sterilized by 12–24 hours immersion in Cidex dialdehyde solution; Johnson & Johnson Medical, Arlington, TX) is routinely 30 cm in length, and attached via a 21 G needle to a 5-ml syringe containing 1 ml of sterile normal saline. Once the catheter is within the vessel, the introducer is removed. The catheter is passed retrograde down the artery approximately 15–20 cm, at which point there is resistance, indicating that the tip of the catheter has crossed the aortic valve and reached the left ventricle. Proper placement of the catheter is crucial to achieve adequate valvular trauma associated with high rates of IE induction following microbial challenge. If placed properly, there is strong resistance to withdrawing blood into the syringe when the catheter tip is lodged against the ventricle. Strong pulsation of the catheter serves as a secondary indicator that proper placement into the ventricle has been achieved. Additionally, by withdrawing the catheter slightly, blood return up the catheter is indicative of its transvalvular and intraventricular position. Some investigators connect the proximal end of the catheter to a pressure transducer, where recording of pressure wave-form profiles consistent with either the left or right ventricle confirms proper intraventricular position of the catheter.

Once the catheter is verified for successful placement, the surgeon then uses the hemostat to grasp and fold the proximal catheter above the loose suture, and the assistant then tightly ligates the suture over the folded catheter to secure it in place. The portion of the catheter having been folded over is then cut to a minimal length, buried within the pericarotid fascia, and the neck incision is closed by surgical stapling (Figure 86.7). The complete surgical process, from initial anesthesia to wound closure, usually requires 15–20 minutes per surgical team per animal. The same procedure would be used to induce tricuspid valve trauma (right-sided

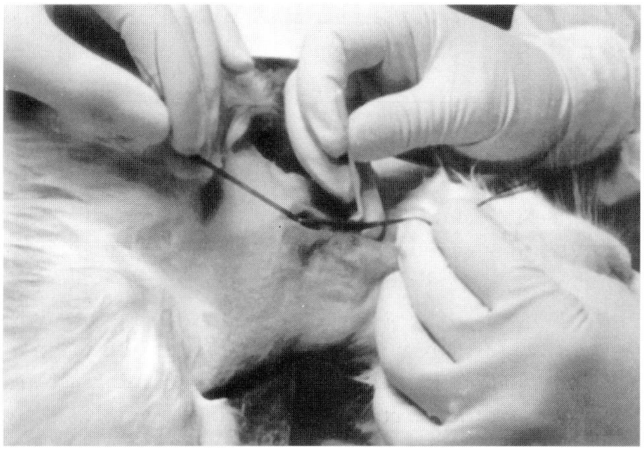

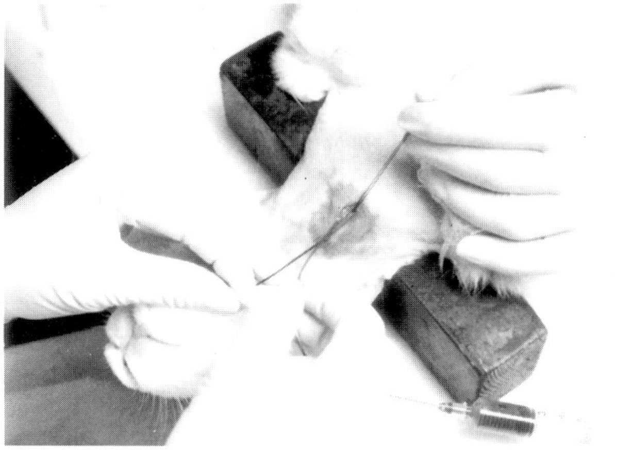

A B

Figure 86.6 Catheter introduction into the carotid artery. While stabilized with the opposing sutures, the carotid artery is accessed with the introducer (held by the surgeon) (**A**). The catheter is then inserted (**B**), and the assistant relieves tension on the posterior suture, allowing the catheter to be passed retrograde down the vessel and into the ventricle.

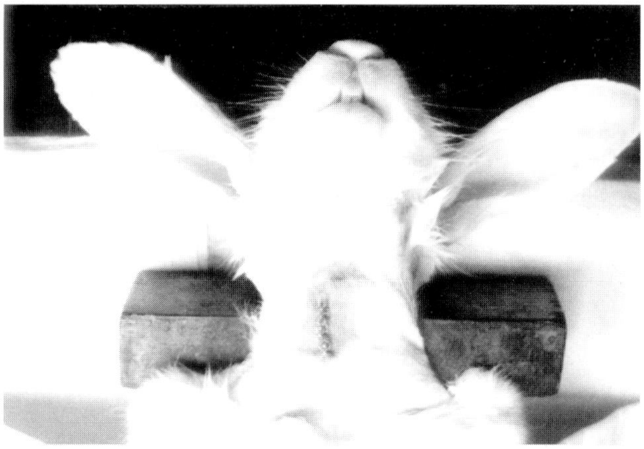

Figure 86.7 Closure of the surgical incision. Once proper catheter placement is assessed and the catheter has been trimmed, folded, and ligated in place, and buried in the peri-carotid tissue, the surgical incision is closed by surgical stapling.

IE model) via retrograde catheterization of the right ventricle (via the jugular vein) across the tricuspid valve. It should be noted that the degree of catheter pulsation associated with the right ventricle, even with proper placement of the catheter, is less than that observed with the left ventricle via aortic catheterization.

Postsurgical recovery

Animals should be placed in the lateral position on warming pads or under low-intensity heating lamps immediately following catheterization and closure of the neck incision. Normally, rabbits begin to recover from anesthesia within 10–15 minutes, and have completely awakened by 45–60 minutes post-anesthesia (Bertens *et al.*, 1993). Therefore, animals can generally be returned to their housing approximately 20 minutes after surgery, where they should be provided immediate access to water (water and food *ad libitum*). Animals should be monitored regularly until they have fully recovered from anesthesia, and then daily for any post-surgical complications. Animals routinely show no signs of pain or discomfort following surgery; thus, analgesics are not normally administered. If analgesia is given, anti-inflammatory agents are discouraged, since they may exert effects on platelets and endothelium that could confound the interpretation of subsequent data related to IE pathogenesis or antifungal efficacy.

Euthanasia

At selected timepoints post-induction of IE (depending on the specific study), animals are euthanized using a rapid, intravenous sodium pentobarbitol injection (100 mg/kg; Abbott Laboratories, Chicago, IL). This results in a rapid cessation of breathing, and a pain-free death (Bertens *et al.*, 1993; National Center for Research Resources, 1996).

Establishment of infection

Few studies have examined the pathogenesis of *Candida* IE or antifungal efficacy after catheter removal following valve traumatization (generally for 24–48 hours). This is due to the fact that the presence of the catheter prevents self-sterilization of the valvular infection by antimicrobial host defense mechanisms. In contrast, most contemporary studies are carried out using the indwelling catheter model (Filler *et al.*, 1991; Witt and Bayer, 1991; Witt *et al.*, 1993; Yeaman *et al.*, 1996). The following discussion will focus on this latter model, representative of a more severe infection, and a rigorous test of antifungal efficacy.

Preparation of the organism

It is essential to conduct pilot studies to determine the growth kinetics, infectivity (ID_{90}), and *in vitro* antifungal susceptibility of each *Candida* strain to be studied in the animal model. Growth kinetics are determined by dilution quantitative culture of a known inoculum in a selected medium over a time-range encompassing lag through stationary growth phases. For many *Candida* isolates cultured in enriched medium, the lag phase is observed 2–4 hours post-inoculation, the logarithmic phase from 4–20 hours, and the stationary phase ensuing beyond 20 hours of culture. These periods will vary depending on the specific strain, growth medium, and incubation temperature utilized. Determination of the specific growth kinetics in this manner will facilitate identification of logarithmic-phase cells, which should be used for all experimental animal model challenges. The ID_{90} of each study organism must be determined in pilot studies of no less than 5 animals. Induction of IE is confirmed for a given inoculum by positive fungal cultures of cardiac vegetations performed at necropsy 24 hours post-challenge (Filler *et al.*, 1991; Witt and Bayer, 1991; Witt *et al.*, 1993; see below). Antifungal susceptibility profiles for each *Candida* strain should be determined as per National Committee on Clinical Laboratory Standards (NCCLS) guidelines.

Organisms are stored in stock suspensions at −70°C on yeast extract agar (YEA) or Sabouraud dextrose agar (SDA; Difco Laboratories, Detroit, MI). These preserved cultures are thawed and sufficiently resuscitated in advance of use in animal challenge to permit their reliable purification, and phenotypic or genotypic confirmation. Routinely, organisms should be passaged several times in fresh medium. Organisms are cultured to logarithmic phase (as determined by growth kinetic studies), and blastospores harvested by centrifugation, washed in phosphate-buffered saline (PBS; pH 7.2) to remove exhausted medium and metabolic waste products, sonicated to insure singlet cells, and enumerated. Enumeration is achieved either by hemocytometer count of a dilution of the washed blastospores, or by spectrophotometric assessment (optical density 420 or 600 nm), for which a standard curve has been constructed comparing turbidity data with quantitative culture data for each particular strain.

Induction of IE by intravenous challenge

In most instances, animals will be challenged with the study organism within 48 hours of catheterization surgery. At that time, animals appearing healthy are randomly assigned into selected study groups. To prevent undue stress or injury during intravenous challenge, animals should be placed into specialized restraining cages as appropriate for their size. The untagged ear should be disinfected with 70% isopropyl alcohol prior to challenge. Washed *Candida* blastospores, diluted to the desired concentration (colony-forming units; cfu), are injected intravenously via the marginal ear vein. Previous studies have shown that inocula ranges from 10^4 to 10^7 cfu are sufficient to establish *Candida* IE (ID_{90}), depending on the specific strain studied (Filler *et al.*, 1991; Witt and Bayer, 1991; Witt *et al.*, 1993; Yeaman *et al.*, 1996). However, the ID_{90} of each candidate study isolate should be determined in pilot experiments prior to initiation of larger-scale studies. The intravenous injection should proceed slowly to prevent pain and minimize inoculum extravasation, with no more than 1 ml of organism suspension injected. Slight direct pressure is then applied to the injection site with sterile gauze to stop bleeding should it occur. Animals are placed back into their housing immediately following challenge injection, which is recorded as "time zero" of the infection model.

Monitoring evolution of experimental *Candida* IE

There are numerous parameters used to monitor the evolution of experimental *Candida* IE. The rabbit model allows evaluation of the relative contribution of each step in sequential pathogenesis of this infection:

1. Organism access to endothelium (fungemia)
2. Adherence to damaged endothelial surfaces (early adhesion to cardiac vegetations)
3. Induction and evolution of IE (vegetation fungal densities)
4. Hematogenous dissemination and metastatic sequelae (seeding and proliferation of *Candida* within distant target organs)
5. Overall outcome of infected animals (morbidity and mortality).

Additionally, several investigators have used animal models of *Candida* IE to evaluate the influence of neutropenia or other immunodeficient states on the evolution of infection (Berney and Francioli, 1990; Futenma *et al.*, 1995; Flattery *et al.*, 1996; Plotkin *et al.*, 1996; Schmidt, 1996). More recent studies have also begun to demonstrate the utility of this rabbit model in examining the role of various host defense mechanisms in limiting *Candida* IE or its metastatic sequelae (Yeaman *et al.*, 1996).

Morbidity and mortality

Experimental *Candida* IE is a serious infection which often leads to animal death if untreated. Therefore, regular (e.g. twice-daily) inspection of infected animals for overt morbidity and impending mortality is essential (National Center for Research Resources, 1996). Common symptoms of fulminant *Candida* IE in rabbits may include lethargy, dysphagia, tachypnea, and diarrhea. Untreated animals challenged with an ID_{90} inocula often begin demonstrating such symptoms within 72 hours post-infection. Careful consideration and consultation with veterinary staff should be used to guide endpoints for animals exhibiting severe morbidity or imminent mortality. Investigators must be prepared to perform euthanasia and conduct necropsy, tissue excision, and quantitative organ culture of profoundly morbid animals on a daily basis to minimize animal discomfort and maximize data recovery. Delays in animal necropsy should be minimized, and animals should not be used for necropsy if 6 hours or more have elapsed after death without refrigeration (4°C). If delays in necropsy are unavoidable, euthanized animals should be refrigerated to reduce tissue destruction and post-mortem microbial proliferation, which could impact upon relevant data interpretation.

Fungemia

It is important to determine the influence of fungemia and its rate of clearance on the evolution of *Candida* IE. For example, differences observed among strain ability to induce IE or produce distant organ metastases via hematogenous dissemination may be due to differential levels and durations of fungemia (Filler *et al.*, 1991; Witt and Bayer, 1991; Witt *et al.*, 1993). Therefore, blood samples are normally drawn from the central ear artery (ear opposite from that used for infection) at 1, 30, and/or 60 or 120 minutes post-challenge. These are quantitatively cultured on to SDA or 6.6% sheep blood agar (BA) for determination of fungemia extent and duration. Beyond 120 minutes post-challenge, the intermittent nature of positive blood cultures, even in well-established *Candida* IE, precludes use of this parameter as a marker of therapeutic efficacy.

Adherence to valvular vegetations

Adherence of *Candida* blastospores to the damaged cardiac endothelium or adjacent thrombus is essential for the establishment and evolution of infection (Calderone *et al.*, 1978; Mausch and Calderone, 1980; Klotz *et al.*, 1983, 1989; Rotrosen *et al.*, 1985; Klotz and Maca, 1988; Edwards and Mayer, 1990; Klotz, 1992). Different study organisms may vary in their capacity to initially seed the sterile cardiac vegetation. In this regard, differences in fungal densities within vegetations (see below) could result from differential adherence (seeding), survival, and/or proliferation in the setting of the cardiac vegetation. Thus, it is important to evaluate the ability of study organisms to adhere at this site initially. Routinely, 4–6 animals are used in this determination for each study strain and each challenge inoculum. Rabbits are challenged with a defined inoculum (as above), and euthanized 30 minutes post-challenge (timepoint considered to represent maximal extent of valvular adhesion; Dhawan *et al.*, 1997). Necropsy is performed immediately to excise and quantitatively culture all cardiac vegetations (as described below).

Gross pathology

It is important to thoroughly inspect experimental and control animals immediately following euthanasia for gross pathology associated with *Candida* IE at pre-selected endpoints of the experiment. Upon opening the chest cavity, the heart is inspected for evidence of pericarditis or other adjacent pathology. The heart is then bisected by vertical incision, and proper catheter position confirmed in the ventricle of interest. This is particularly important, since data from animals with misplaced or dislodged catheters should not be included in statistical analyses. Vegetation site (valvular, mural), quantity, and relative size should be noted prior to their excision and culture (see below). In addition, perianular extension or sinus tract formation may also be observed at this time. The spleen, liver, kidneys and lungs (in tricuspid IE) should also be carefully inspected for external evidence of infarct or fungal abscess prior to their removal for quantitative culture (see below). Additionally, *Candida* may cause endophthalmitis secondary to IE (Edwards *et al.*, 1975). Thus, ophthalmoscopic inspection and/or quantitative culture of the retinas may be performed as a complementary parameter in evaluation of metastatic sequelae associated with *Candida* IE. Other tissues may be of interest for inspection depending on the focus of the particular study.

Vegetation fungal density

Fungal densities within specific target organs are the standard parameters for evaluating induction, proliferation, and hematogenous dissemination resulting from *Candida* IE (Filler *et al.*, 1991; Witt and Bayer, 1991; Witt *et al.*, 1993; Yeaman *et al.*, 1996). At selected study endpoints, cardiac vegetations excised from individual infected animals are weighed, pooled, homogenized in a motorized Teflon-pestel/glass-tube apparatus (or equivalent), and quantitatively cultured on to SDA or BA. Following incubation for up to 48 hours at 37°C, colonies are enumerated, and results expressed as logarithm_{10} cfu/gram vegetation ± SD.

Hematogenous dissemination to distant target organs

Candida density within the spleen (a relatively friable organ) is likewise determined by *en bloc* homogenization and quantitative culture of this organ. For larger or more solid organs such as kidneys, liver, or lungs, samples consistent in size and location are excised from the liver, and from each of the paired organs. The uniform distribution of abscesses within these organs as a result of hematogenous seeding allows for sampling of small sections of the organ (sample size of at least 1 gram of tissue), which serve as representative of the whole organ fungal density. These tissues are homogenized as for the vegetations and spleen (pooled if desired in the case of paired organs), and quantitatively cultured. As with vegetations, *Candida* densities in each of these tissues is determined from colony counting adjusted for dilution, and expressed as logarithm_{10} cfu/gram ± SD.

It is important to use a sterile homogenizer for each different tissue of each animal. To maximize sensitivity and accuracy, the homogenate, as well as sterile 1 ml PBS washings of the homogenizer and homogenization tube, should be pooled and cultured for each tissue sampled. Tissue homogenization which produces excessive heat should be conducted on ice.

Parameters in monitoring response to therapy

Any or all of the parameters listed above may be evaluated in the presence or absence of antifungal prophylaxis or therapy. Studies focusing on the efficacy of antifungal prophylaxis or therapy have principally utilized incidence of IE induction and proliferation of *Candida* within infected cardiac vegetations as sentinel markers of antifungal efficacy (Filler *et al.*, 1981; Witt and Bayer, 1991; Witt *et al.*, 1993). Others have included hematogenous dissemination (metastatic sequelae) as indicators of therapeutic efficacy of *Candida* IE (Filler *et al.*, 1991; Witt and Bayer, 1991; Witt *et al.*, 1993; Yeaman *et al.*, 1996). Few studies have reported mortality as a principal indicator of antifungal efficacy.

Considerations for evaluation of antifungal agents

The *in vitro* antifungal susceptibility of a specific organism to be studied is a principal consideration in studies intended to evaluate antifungal prophylactic or therapeutic efficacy in the animal model. In addition, the pharmacokinetics, pharmacodynamics, and route of administration of the antifungal agent to be evaluated are important considerations in the strategic design of experiments using the experimental model of *Candida* IE. For example, the pharmacokinetics of the particular drug or drugs to be studied are important to consider as it relates to determining frequency of administration of the antifungal agent needed to achieve desired serum or tissue levels (Bayer *et al.*, 1988; Filler *et al.*, 1991; Witt and Bayer, 1991; Witt *et al.*, 1993). Route of administration is pivotal, since some methods are technically difficult (e.g. intraperitoneal injection), and require additional staff training and time. In addition, there may be substantial time required for preparation of antifungal agents prior to administration. For example, fluconazole is usually dissolved in an amphipathic carrier (Cremophore; Sigma Chemical, St Louis; Witt and Bayer, 1991; Witt *et al.*, 1993) prior to dispensing for intravenous, subcutaneous, or intraperitoneal injection.

The majority of studies examining antifungal efficacy in experimental *Candida* IE have focused on evaluation or comparison of amphotericin B with 5-fluorocytosine, or triazole agents such as fluconazole (Filler *et al.*, 1991; Witt and Bayer, 1991; Witt *et al.*, 1993). In contrast to bacterial models of IE, where reductions in organism densities in tissues are often demonstrable within 48 hours of therapy, antifungal agents frequently require ≥ 7–10 days to exhibit substantial efficacy as compared with their counterparts in untreated control groups. Thus, studies to determine the absolute or comparative effects of various antifungal regimens in experimental models must be conducted over a prolonged time period. This necessitates a considerable expense in housing and animal care costs, amounts of antifungal agent used, as well as technical and veterinary staff time, as compared with studies of shorter duration. This is particularly true when antifungal therapies require unusual routes of administration (e.g. intraperitoneal), or multiple administrations per day.

Potential pitfalls of the model

The experimental model of *Candida* IE is subject to standard limitations inherent to any animal model of infection. These include the fact that some differences exist in physiology (e.g. tissue response to anesthesia and trauma, innate and acquired immune functions), as well as pharmacokinetic profiles and efficacy between experimental animals and humans. In addition, investigators must guide their interpretations of data derived from such studies by considerations of potential pitfalls unique to these models. For example, experimental IE is an infection which is induced by first traumatizing the cardiac valve to be studied. Thus, while the establishment of such an infection may be highly relevant to human infections in patients with rheumatic heart disease or prosthetic valves, the relevance of the model to community-acquired IE in patients lacking underlying valve disease is less clear. Furthermore, the catheter model creates a setting in which fungal pathogenesis may be substantially influenced by the presence of an indwelling foreign body.

In addition to the above stipulations, large inocula are required to initiate IE, even following valvular trauma. These high inocula probably initiate artificially aggressive infections which do not realistically simulate the clinical situation. None the less, these factors make the rabbit and rat models of IE even more rigorous tests of the prophylactic and therapeutic efficacy of antifungal agents. Alternatively, such large inocula may activate host defenses which would otherwise remain quiescent in less dramatic challenges, giving rise to misintrepretation of fungal virulence. This possibility makes the potential use of *ex vivo* models (Rybak *et al.*, 1997) meritorious, given that they can minimize potential confounding variables related to immune system function in limiting infection. For example, individual components integral to IE pathogenesis, such as platelets, organisms, or plasma, may be tested in such models for their relative influence on organism survival and proliferation.

Given the above limitations, appropriate experimental design is essential to address a specific study hypothesis.

Thus, studies involving untreated control groups, as well as animal groups treated with standard agents of known efficacy, are crucial for assessing the antifungal activities of novel agents. Antifungal regimens in animals should be designed to best approximate parameters of dosage, route and frequency of administration, and physiologically relevant and achievable serum or tissue concentrations in humans (if known) in the context of the *in vitro* antifungal susceptibility profiles of the infecting strain.

Contributions of the model to pathogenesis and therapy

Pathogenesis of *Candida* IE

Use of the animal model of experimental *Candida* IE has revealed remarkable insights into the pathogenesis of infection due to this organism, as well as related fungal pathogens. These observations have focused on the interrelationship between the organism, endothelium, neutrophils, and platelets. The fact that *Candida* must traverse the endothelial barrier to gain access to the deeper tissue parenchmya makes this organism an invaluable probe for determining microbial mechanisms of immunoavoidance, endothelial cell invasion and penetration, and the role of platelet and neutrophil interactions in antimicrobial host defense of the vascular endothelium. Likewise, immune responsiveness (e.g. cytokine expression, neutrophil chemotaxis, endogenous antimicrobial peptides) to *Candida* may provide increased awareness of surveillance mechanisms for detecting and responding to fungemia. Such information will be integral to the development of new approaches for prevention (e.g. vaccines, inhibitors of adhesion) and therapy of systemic fungal infections, including *Candida* IE, and perhaps infections caused by unrelated organisms.

Antifungal efficacy

There is no certainty that antifungal agents will act in humans as they do in animal models. However, the overall outcome of antifungal therapy in humans appears to reflect that of the rabbit and rat models of *Candida* IE (Filler *et al.*, 1991; Witt and Bayer, 1991; Witt *et al.*, 1993). Thus, these animal models of experimental *Candida* IE have proven extremely useful in profiling the relative prophylactic and therapeutic effects of many novel antifungal agents, particularly in comparison to agents with known degrees of efficacy. Moreover, newer antifungal agents identified as being efficacious in these models (e.g. fluconazole) have proven themselves to be significantly less toxic than previously existing agents (e.g. amphotericin B). In this regard, use of animal models such as these has directly led to a significant improvement in the prevention and therapy of life-threatening disseminated *Candida* infections in humans.

References

Andriole, V. T., Kravetz, H. M., Roberts, W. C., Utz, J. P. (1962). *Candida* endocarditis: clinical and pathologic studies. *Am. J. Med.*, **23**, 251–285.

Archer, G., Fekety, F. R. (1972). Experimental endocarditis due to *Pseudomonas aeruginosa*. I. Description of a model. *J. Infect. Dis.*, **134**, 1–7.

Baddour, L. M., Christensen, G. D., Lowrance, L. H., Simpson, W. A. (1989). Pathogenesis of experimental endocarditis. *Rev. Infect. Dis.*, **11**, 452–463.

Bansci, M. F., Thompson, J., Bertina, R. M. (1994). Stimulation of monocyte tissue factor expression in an *in vitro* model of bacterial endocarditis. *Infect. Immun.*, **62**, 5669–5672.

Bayer, A. S., Crowell, D. J., Yih, J., Bradley, D. W., Norman, D. C. (1988). Comparative pharmacokinetics and pharmacodynamics of amikacin and ceftazidime in tricuspid and aortic vegetations in experimental *Pseudomonas* endocarditis. *J. Infect. Dis.*, **158**, 355–359.

Bayer, A. S., O'Brien, T., Norman, D. C., Nast, C. C. (1989). Oxygen-dependent differences in exopolysaccharide production and aminoglycoside-inhibitory or bactericidal interactions with *Pseudomonas aeruginosa* – implications for endocarditis. *J. Antimicrob. Chemother.*, **23**, 21–35.

Bayliss, R., Clarke, C., Oakley, C. M., Somerville, W., Whitfield, A. G. W., Young, S. E. J. (1983). The microbiology and pathogenesis of infective endocarditis. *Br. Heart J.*, **50**, 513–519.

Berney, P., Francioli, P. (1990). Successful prophylaxis of experimental streptococcal endocarditis with single-dose amoxicillin administered after bacterial challenge. *J. Infect. Dis.*, **161**, 281–285.

Bertens, A. P. M. G., Booij, L. H. D. J., Fleckness, P. A., Lagerweij, E. (1993). Anesthesia, analgesia, and euthanasia. In *Principles of Laboratory Animal Science* (eds van Zutphen, L. F. M., Baumans, V. *et al.*), pp. 267–298. Elsevier, New York, NY.

Brown, M. J., Pearson, P. T., Tomson, F. N. (1993). Guidelines for animal surgery in research and teaching. *J. Am. Vet. Assn.*, **54**, 1544–1559.

Calderone, R. A., Rotondo, M. F., Sande, M. A. (1978). *Candida albicans* endocarditis – ultrastructural studies of vegetation formation. *Infect. Immun.*, **20**, 279–289.

Calderone, R. A., Cihlar, R. L., Lee, D. D. S., Hoberg, K., Scheld, W. M. (1985). Yeast adhesion in the pathogenesis of endocarditis due to *Candida* albicans: studies with adherence-negative mutants. *J. Infect. Dis.*, **152**, 710–715.

Calderone, R. A., Scheld, W. M. (1987). Role of fibronectin in pathogenesis of candidal infections. *Rev. Infect. Dis.*, **9**, S400–S403.

Calderone, R. A., Linehan, L., Wadsworth, E., Sandberg, A. L. (1988). Identification of C3d receptors on *Candida albicans*. *Infect. Immun.*, **56**, 252–258.

Christin, L., Wysong, D. R., Meshulam, T., Hastey, R., Simons, E. R., Diamond, R. D. (1998). Human platelets damage *Aspergillus fumigatus* hyphae and may supplement killing by neutrophils. *Infect. Immun.*, **66**, 1181–1189.

Cole, M. F., Bowen, W. H., Zhao, X. J., Cihlar, R. L. (1995). Avirulence of *Candida albicans* auxotrophic mutants in a rat model of oropharyngeal candidiasis. *FEMS Microbiol. Lett.*, **126**, 177–180.

Colman, R. W. (1991). Receptors that activate platelets. *Proc. Soc. Exp. Biol. Med.*, **197**, 242–248.

De Bernardis, F., Cassone, A., Sturtevant, J., Calderone, R. (1995).

Expression of *Candida albicans* SAP-1 and SAP-2 in experimental vaginitis. *Infect. Immun.*, **63**, 1887–1892.

Dhawan, V. K., Yeaman, M. R., Cheung, A. L., Kim, E., Sullam, P. M., Bayer, A. S. (1997). Phenotypic resistance to thrombin-induced platelet microbicidal protein *in vitro* is correlated with enhanced virulence in experimental endocarditis due to *Staphylococcus aureus*. *Infect. Immun.*, **65**, 3293–3299.

Drake, T. A., Pang, M. (1988). *Staphylococcus aureus* induces tissue factor expression in cultured human cardiac valve endothelium. *J. Infect. Dis.*, **157**, 749–756.

Drake, T. A., Pang, M. (1989). Effects of interleukin-1, lipopolysaccharide, and streptococci on procoagulant activity of cultured human cardiac valve endothelial and stromal cells. *Infect. Immun.*, **57**, 507–512.

Drake, T. A., Rodgers, G. M., Sande, M. A. (1984). Tissue factor is a major stimulus for vegetation formation in enterococcal endocarditis in rabbits. *J. Clin. Invest.*, **73**, 1750–1753.

Durack, D. T., Beeson, P. B. (1972a). Experimental bacterial endocarditis. I. Colonization of a sterile vegetation. *Br. J. Exp. Pathol.*, **53**, 44–49.

Durack, D. T., Beeson, P. B. (1972b). Experimental bacterial endocarditis. II. Survival of bacteria in endocardial vegetations. *Br. J. Exp. Pathol.*, **53**, 50–53.

Durack, D. T., Beeson, P. B. (1973). Experimental bacterial endocarditis. III. Production and progress of the disease in rabbits. *Br. J. Exp. Pathol.*, **54**, 142–151.

Edwards, J. E., Jr. (1991). Invasive *Candida* infections: evolution of a fungal pathogen. *N. Engl. J. Med.*, **324**, 1060–1062.

Edwards, J. E., Jr., Mayer, C. L. (1990). Adherence of *Candida albicans* to mammalian cells. In *Microbial Determinants of Virulence and Host Response* (eds Ayoub, E. M., Cassel, G. H., Branche, W. C. Jr., Henry, T. J.), pp. 179–194. American Society for Microbiology, Washington, DC.

Edwards, J. E., Jr., Montgomerie, J. Z., Foos, R. Y., Shaw, V. K., Guze, L. B. (1975). Experimental hematogenous endophthalmitis caused by *Candida albicans*. *J. Infect. Dis.*, **131**, 649–657.

Edwards, J. E., Jr., Gaither, T. A., O'Shea, J. J. *et al.* (1986). Expression of specific binding sites on *Candida* with functional and antigenic characteristics of human complement receptors. *J. Immunol.*, **137**, 3577–3583.

Edwards, J. E., Jr., Rotrosen, D., Fontaine, J. W., Haudenschild, C. C., Diamond, R. D. (1987). Neutrophil-mediated protection of cultured human vascular endothelial cells from damage by growing *Candida albicans* hyphae. *Blood*, **69**, 1450–1457.

Fallon, K., Bausch, K., Noonan, J., Huguenel, E., Tamburini, P. (1997). Role of aspartic proteases in disseminated *Candida albicans* infection in mice. *Infect. Immun.*, **65**, 551–556.

Filler, S. G., Crislip, M. A., Mayer, C. L., Edwards, J. E. Jr. (1991). Comparison of fluconazole and amphotericin B for treatment of disseminated candidiasis and endophthalmitis in rabbits. *Antimicrob. Agents Chemother.*, **35**, 288–292.

Filler, S. G., Ibe, B. O., Luckett, P. M., Usha Raj, J., Edwards, J. E. Jr. (1991). *Candida albicans* stimulates endothelial cell eicosanoid production. *J. Infect. Dis.*, **164**, 928–935.

Filler, S. G., Swerdloff, J. N., Hobbs, C., Luckett, P. M. (1995). Penetration and damage of endothelial cells by *Candida albicans*. *Infect. Immun.*, **63**, 976–983.

Filler, S. G., Pfunder, A. S., Spellberg, B. J., Spellberg, J. P., Edwards, J. E. Jr. (1996). *Candida albicans* stimulates cytokine production and leukocyte adhesion molecule expression by endothelial cells. *Infect. Immun.*, **64**, 2609–2617.

Flattery, A. M., Abruzzo, G. K., Gill, C. J., Smith, J. G., Bartizal, K. (1996). New model of oropharyngeal and gastrointestinal colonization by *Candida albicans* in CD4+ T-cell-deficient mice for evaluation of antifungal agents. *Antimicrob. Agents Chemother.*, **40**, 1604–1609.

Fraser, C. M. (1991). Management, husbandry, and diseases of rabbits. In *The Merck Veterinary Manual: A handbook of diagnosis, therapy, and disease prevention and control for the veterinarian* (ed. Fraser, C. M.), pp. 1060–1061. Merck & Co., Rahway, NJ.

Fu, Y., Filler, S. G., Spellberg, B. J. *et al.* (1998). Cloning and characterization of CAD1/AAF1, a gene from *Candida albicans* that induces adherence to endothelial cells after expression in *Saccharomyces cerevisiae*. *Infect. Immun.*, **66**, 2078–2084.

Futenma, M., Kawakami, K., Saito, A. (1995). Production of tumor necrosis factor-alpha in granulocytopenic mice with pulmonary candidiasis and its modification with granulocyte colony-stimulating factor. *Microbiol. Immunol.*, **39**, 411–417.

Garvey, G. J., Neu, H. C. (1978). Infective endocarditis – an evolving disease. *Medicine*, **57**, 105–127.

Griffin, M. R., Wilson, W. R., Edwards, W. D., O'Fallon, W. M., Kurland, L. T. (1985). Infective endocarditis. Olmstead County, Minnesota, 1950–1981. *JAMA*, **254**, 1199–1202.

Han, Y., Cutler, J. E. (1997). Assessment of a mouse model of neutropenia and the effect of an anti-candidiasis monoclonal antibody in these animals. *J. Infect. Dis.*, **175**, 1169–1175.

Harkness, J. E., Wagner, J. E. (1995). Specific diseases and conditions: *Pasteurella multocida* infections in rabbits. In *The Biology and Medicine of Rabbits and Rodents* (eds Harkness, J. E., Wagner, J. E.), pp. 265–271. Williams & Wilkins, Baltimore, MD.

Hata, K., Kimura, J., Miki, H., Toyosawa, T., Moriyama, M., Katsu, K. (1996). Efficacy of ER-30346, a novel oral triazole antifungal agent, in experimental models of aspergillosis, candidiasis, and cryptococcosis. *Antimicrob. Agents Chemother.*, **40**, 2243–2247.

Herzyk, D. J., Ruggieri, E. V., Cunningham, L. *et al.* (1997). Single-organism model of host defense against infection: a novel immunotoxicologic approach to evalulate immunomodulatory drugs. *Toxicol. Pathol.*, **25**, 351–362.

Kammer, R. B., Utz, J. P. (1974). *Aspergillus* species endocarditis: the new face of a not so rare disease. *Am. J. Med.*, **56**, 506–521.

Klotz, S. A. (1992). Fungal adherence to the vascular compartment: a critical step in the pathogenesis of disseminated candidiasis. *Clin. Infect. Dis.*, **14**, 340–347.

Klotz, S. A., Drutz, D. J., Harrison, J. L., Huppert, M. (1983). Adherence and penetration of vascular endothelium by *Candida* yeasts. *Infect. Immun.* **42**, 374–384.

Klotz, S. A., Drutz, D. J., Zajic, J. E. (1985) Factors governing adherence of *Candida* species to plastic surfaces. *Infect. Immun.*, **50**, 97–101.

Klotz, S. A., Maca, R. D. (1988). Endothelial cell contraction causes *Candida* adherence to exposed extracellular matrix. *Infect. Immun.*, **56**, 2495–2498.

Klotz, S. A., Harrison, J. L., Misra, R. P. (1989). Aggregated platelets enhance adherence of *Candida* yeasts to endothelium. *J. Infect. Dis.*, **160**, 669–677.

Kretschmar, M., Nichterlein, T., Hannak, D., Hof, H. (1996). Effects of amphotericin B incorporated into liposomes and in lipid suspensions in the treatment of murine candidiasis. *Arzneimittelforschung*, **46**, 711–715.

Maisch, P. A., Calderone, R. A. (1980). Adherence of *Candida albicans* to a fibrin-platelet matrix formed *in vitro*. *Infect. Immun.*, **27**, 650–656.

Maisch, P. A., Calderone, R. A. (1981). Role of surface mannans in the adherence of *Candida albicans* to fibrin-platelet clots formed *in vitro*. *Infect. Immun.*, **32**, 92–97.

Matuschak, G. M., Lechner, A. J. (1997). The yeast to hyphal transition following hematogenous candidiasis induces shock and organ injury independent of circulating tumor necrosis factor-alpha. *Crit. Care Med.*, **25**, 111–120.

McLeod, R., Remington, J. S. (1978). Fungal endocarditis. In *Infective Endocarditis* (ed. Rahimtoola, S. H.), pp. 211–290. Grune & Stratton, New York.

Moyer, D. V., Edwards, J. E. Jr. (1992). Fungal endocarditis. In *Infective Endocarditis* (ed. Kaye, D.), Raven Press, New York.

National Center for Research Resources (1996). *National Institutes of Health (NIH; USA) Guidelines for the Care and Use of Laboratory Animals*. National Academy Press, Washington, DC.

Ohtsuka, K., Watanabe, M., Orikasa, Y. *et al.* (1997). The *in vivo* activity of an antifungal antibiotic, benanomicin A, in comparison with amphotericin B and fluconazole. *J. Antimicrob. Chemother.*, **39**, 71–77.

Pelletier, L. L., Petersdorf, R. G. (1977). Infective endocarditis: a review of 125 cases from the University of Washington Hospitals. *Medicine*, **56**, 287–313.

Plotkin, B. J., Paulson, D., Chelich, A. *et al.* (1996). Immune responsiveness in a rat model for type II diabetes (Zucker rat, fa/fa): susceptibility to *Candida albicans* infection and leucocyte function. *J. Med. Microbiol.*, **44**, 277–283.

Robboy, S. J., Kaiser, J. (1975). Pathogenesis of fungal infection on heart valve prostheses. *Hum. Pathol.*, **6**, 711–715.

Robert, R., Senet, J. M., Mahaze, C. *et al.* (1992). Molecular basis of the interactions between *Candida albicans*, fibrinogen, and platelets. *J. Mycol. Med.*, **2**, 19–25.

Rotrosen, D., Edwards, J. E., Jr., Gibson, T. R., Moore, J. C., Cohen, A. H. (1985). Adherence of *Candida* to cultured vascular endothelial cells: mechanisms of attachment and endothelial cell penetration. *J. Infect. Dis.*, **152**, 1264–1274.

Rotrosen, D., Calderone, R. A., Edwards, J. E. Jr. (1986). Adherence of *Candida* species to host tissues and plastic surfaces. *Rev. Infect. Dis.*, **8**, 73–85.

Rozalska, B., Ljungh, A., Burow, A., Rudnicka, W. (1995). Biomaterial-associated infection with *Candida albicans* in mice. *Microbiol. Immunol.*, **39**, 443–450.

Rubenstein, E., Noriega, E. R., Simberkoff, M. S., Holzman, R., Rahal, Jr. J. J. (1975). Fungal endocarditis: analysis of 24 cases and review of the literature. *Medicine*, **54**, 331–344.

Rybak, M. J., Houlihan, H. H., Mercier, R. C., Kaatz, G. W. (1997). Pharmacodynamics of RP 59500 (quinupristin-dalfopristin) administered by intermittent versus continuous infusion against *Staphylococcus aureus*-infected fibrin-platelet clots in an *in vitro* infection model. *Antimicrob. Agents Chemother.*, **41**, 1359–1363.

Sawyer, R. G., Adams, R. B., Rosenlof, L. K., May, A. K., Pruett, T. L. (1995a). Effectiveness of fluconazole in murine *Candida albicans* and bacterial *C. albicans* peritonitis and abscess formation. *J. Med. Vet. Mycol.*, **33**, 131–136.

Sawyer, R. G., Adams, R. B., May, A. K., Rosenlof, L. K., Pruett, T. L. (1995b). Development of *Candida albicans* and *C. albicans/Escherichia coli/Bacteroides fragilis* intraperitoneal abscess models with demonstration of fungus-induced bacterial translocation. *J. Med. Vet. Mycol.*, **33**, 49–52.

Scheld, W. M., Valone, J. A., Sande, M. A. (1978). Bacterial adherence in the pathogenesis of endocarditis. Interaction of bacterial dextran, platelets, and fibrin. *J. Clin. Invest.*, **61**, 1394–1404.

Scheld, W. M., Strunk, R. W., Balian, G., Calderone, R. A. (1985). Microbial adhesion to fibronectin *in vitro* correlates with production of endocarditis in rabbits. *Proc. Soc. Exp. Biol. Med.*, **180**, 474–482.

Schmidt, A. (1996). Systemic candidiasis in Sprague-Dawley rats. *J. Med. Vet. Mycol.*, **34**, 99–104.

Seelig, M. S., Speth, C. P., Kozinn, P. J., Taschdjian, C. L., Toni, E. F., Goldberg, P. (1974). Patterns of *Candida* endocarditis following cardiac surgery: importance of early diagnosis and therapy (an analysis of 91 cases). *Prog. Cardiovasc. Dis.*, **XVII**, 125–160.

Seelig, M. S., Kozinn, P. J., Goldberg, P., Berger, A. R. (1979). Fungal endocarditis: patients at risk and their treatment. *Postgrad. Med. J.*, **55**, 632–641.

Torack, R. M. (1957). Fungus infections associated with antibiotic and steroid therapy. *Am. J. Med.*, **22**, 872–882.

Tunkel, A. R., Scheld, W. M. (1992). Experimental models of endocarditis. In *Infective Endocarditis* (ed. Kaye, D.), pp. 37–56. Raven Press, New York.

Vo, N. M., Russell, J. C., Becker, D. R. (1981). Mycotic emboli of the peripheral vessels: analysis of 44 cases. *Surgery*, **90**, 541–545.

Walsh, T. J., Pizzo, A. (1988). Nosocomial fungal infections: a classification for hospital-acquired infections and mycoses arising from endogenous flora or reactivation. *Annu. Rev. Microbiol.*, **42**, 517–545.

Walsh, T. J., Pizzo, A. (1988). Treatment of systemic fungal infections: recent progress and current problems. *Eur. J. Clin. Microbiol. Infect. Dis.*, **7**, 460–475.

Waynforth, H. B., Flecknell, P. A. (1992). Specific surgical operations. In *Experimental and Surgical Technique in the Rat*, pp. 203–312. Academic Press, New York, NY.

Weksler, B. B. (1992). Platelets. In *Inflammation: Basic Principles and Clinical Correlates* (eds Gallin, J. I., Goldstein, I. M., Snyderman, R.). Raven Press, New York.

Witt, M. D., Bayer, A. S. (1991). Comparison of fluconazole and amphotericin B for prevention and treatment of experimental *Candida* endocarditis. *Antimicrob. Agents Chemother.*, **35**, 2481–2485.

Witt, M. D., Imhoff, T., Li, C., Bayer, A. S. (1993). Comparison of fluconazole and amphotericin B for treatment of experimental *Candida* endocarditis caused by non-*C. albicans* strains. *Antimicrob. Agents Chemother.*, **37**, 2030–2032.

Woods, G. L., Wood, R. P., Shaw, B. W. Jr. (1989). *Aspergillus* endocarditis in patients without prior cardiovascular surgery: report of a case in a liver transplant recipient and review. *Rev. Infect. Dis.*, **11**, 263–272.

Yeaman, M. R. (1997). The role of platelets in antimicrobial host defense. *Clin. Infect. Dis.*, **25**, 951–970.

Yeaman, M. R., Puentes, S. M., Norman, D. C., Bayer, A. S. (1992). Partial characterization and staphylocidal activity of thrombin-induced platelet microbicidal protein. *Infect. Immun.*, **60**, 1202–1209.

Yeaman, M. R., Soldan, S. S., Ghannoum, M. A., Edwards, J. E., Jr., Filler, S. G., Bayer, A. S. (1996). Resistance to platelet microbicidal protein results in increased severity of experimental *Candida albicans* endocarditis. *Infect. Immun.*, **64**, 1379–1384.

Yeaman, M. R., Tang, Y. Q., Shen, A. J., Bayer, A. S., Selsted, M. E. (1997). Purification and *in vitro* activities of rabbit platelet microbicidal proteins. *Infect. Immun.*, **65**, 1023–1031.

Chapter 87

Rabbit Model of Cryptococcal Meningitis

J. R. Perfect

Background of model

Cryptococcal meningitis has become the most common cause of fungal meningitis worldwide. This clinical infection has become more frequent because of several factors. First, the yeast has been found in a variety of environments and therefore exposure to the infectious propagule is common. Second, there is an enlarging worldwide immunocompromised population with the pandemic of human immunodeficiency virus infections. Third, treatment of some modern maladies has frequently included immunosuppressive drugs such as corticosteroids and the immunophilins cyclosporine and tacrolimus. These agents can make the host particularly susceptible to infection with *Cryptococcus neoformans*. It was also this clinical insight which helped develop the rabbit model of cryptococcal meningitis.

Rabbits are considered a corticosteroid-sensitive species similar to humans. Large doses of exogenous corticosteroids rapidly cause lymphopenia in rabbits by lyses of lymphocyte populations in blood and tissue. Reduction in cell populations can be rapid and can occur within 5–8 hours of a corticosteroid dose. This ability to immunosuppress rabbits is crucial to the establishment of a prolonged and fatal infection in this species with *C. neoformans*. However, normal rabbits and those in which corticosteroids are discontinued can generally recover from a high inoculum-induced infection.

The rabbit model focuses on the active proliferation stage of cryptococcal infection as the yeast primarily produces infection within the central nervous system (CNS) in this model (Perfect *et al.*, 1980). Similar to the standard rabbit model of bacterial meningitis, this model also introduces yeasts directly into the subarachnoid space of anesthetized rabbits through a cisternal inoculation. Although infection is primarily expressed in the subarachnoid space and adjacent brain parenchyma, yeast can disseminate from CNS and grow in other organs such as liver, lung, spleen, and kidney.

Although the CNS is the focus of this particular model, the rabbit has been used for other cryptococcosis models, including endophthalmitis (Weiss *et al.*, 1948; Kligman and Weidman, 1949; Fujita, 1983), lung infection (Felton *et al.*, 1966), and intratesticular infection (Bergman, 1996). Corneal and intraocular infections can be easily produced but the lung is difficult to cause infection without immunosuppression. Direct inoculation into the lung can result in short-term pneumonia, fungemia and evidence of meningitis but the CNS features are not prominent (Felton *et al.*, 1966). However, more frequent involvement of the CNS can probably be improved with severe immunosuppression treatment regimens. The use of testicular tissue reflects the general observation that *C. neoformans* strains generally do not tolerate high environmental temperatures (i.e. greater than 39°C). The rabbit has an average body temperature of 39.3–39.5°C. However, despite the high body temperature in rabbits, *C. neoformans* strains can grow in this host under corticosteroid immunosuppression.

Animal species

Pathogen-free 2–3 kg male New Zealand White rabbits are used and fed standard diets of Purina rabbit chow and water *ad lib*. Male rabbits have been used since it has been known that male species tend to be more susceptible to cryptococcosis. However, it is likely that under corticosteroid immune suppression female animals will also be susceptible to infection.

Preparation of animals

There is no specialized housing of animals. Although yeasts will disseminate to multiple body tissue sites in this model, except for cerebrospinal fluid (CSF), *C. neoformans* has not been routinely recovered from other body fluids. Also, when bedding and fecal pellets from rabbits were examined, *C. neoformans* could not be recovered. However, rabbit cages catch waste in removable paper pads which are sprayed with 10% Clorox before discarding. When handling infected rabbits, disposable latex gloves and surgical facial masks are worn.

Rabbits are treated with cortisone acetate 1 day prior to inoculation of *C. neoformans*. The corticosteroid is given intramuscularly in the hind quarters and injections are repeated daily throughout the infection. Cortisone doses can significantly affect the progression of infection and are

Handbook of Animal Models of Infection
ISBN 0-12-775390-7

the major factor in the temporal control of CSF yeast counts within a single strain. Doses ranging from 0.25 to 25 mg/kg per day of cortisone acetate have been used. A standard dose of 2.5 mg/kg per day has been most frequently administered to produce a consistent infection with persistence of high concentrations of yeasts in CSF over a 2-week observation period and without animals experiencing neurological deficits or death. Alternatives for cortisone acetate have not been commonly used but would include triamcinolone at 4 mg/kg per day or a hydrocortisone suspension. All corticosteroid preparations except cortisone acetate will likely need to be adjusted in dose and duration to allow for consistency of infection. It should be noted that, because of corticosteroid administration, rabbits will lose 10–20% of their initial body mass despite excellent oral intake. Despite this weight loss from the steroids, daily corticosteroid dosing is based on initial weight at the beginning of the experiment.

Details of inoculation

Overview

Inocula are injected percutaneously into the cisternal magna of anesthetized rabbits and CSF is withdrawn through a similar percutaneous needle-stick. Inoculation is performed with rabbits in a prone position and with flexion of the head. The needle is inserted behind the occipital notch in the midline. The syringe is aspirated until CSF flows into the barrel and inocula of 0.3 ml are injected. CSF can be used for quantitating yeasts present in subarachnoid space and isolating RNA from these cells; CSF host cells and contents of CSF can also be studied. CSF can be withdrawn daily from the same animals during infection. Some animals (<10%) on serial CSF withdrawals will eventually produce "dry taps" (i.e. CSF cannot be removed). This feature may occur secondary to hematoma around the epidural space or other factors. If accidental human inoculation with a needle containing yeast occurs during the procedure, it is recommended that the individual immediately receives a short course of fluconazole therapy (Casadevall *et al.*, 1994).

Materials required

Tuberculin syringe is used to administer daily intramuscular doses of steroids. Inocula and CSF withdrawals are administered through 3-ml 25 G 5/8-inch syringes (Becton-Dickinson). For intravenous injections or blood withdrawal, 25 G scalp-vein sets are used.

Anesthesia

Rabbits are sedated for all inoculations, blood or CSF withdrawals, or any intravenous drug infusions. A mixture of 65 mg/kg of ketamine and 3 mg/kg of xylazine is given intramuscularly into the hind limb. Sedation is complete in 5–7 minutes and rabbits will remain sedated for 30–45 minutes. With repeated doses, rabbits may require slightly higher doses for similar anesthesia levels. After the procedure, rabbits are placed into cages and observed until they reach sternal recumbency. Mortality due to anesthesia is less than 1%.

Storage, preparation of inocula

The model has been best characterized with a single strain (H99). H99 is a serotype a, MAT-alpha *Cryptococcus neoformans* strain which was isolated from a patient with cryptococcal meningitis and Hodgkin's disease in 1978. This strain is presently stored at –70°C in 15% glycerol and yeast extract peptone dextrose agar (YEPD). However, at times over the years it has been passed from a series of agar slants and stored at 4°C. It has been shown that some strains can change karyotype through passages and it is known that a strain such as H99 can contain a heterogeneous population of cells. Therefore, all investigators should carefully store their pathogenic *C. neoformans* strains. H99 over years of passage (approximately 20) has slightly attenuated its virulence, but its karyotype has remained stable for over 10 years. For example, it requires higher inocula to achieve a similar state of infection compared to initial reported studies (Perfect *et al.*, 1980). Similar to some bacteria, passage of H99 through rabbits can increase its virulence.

Inocula are important. In early studies, inoculum sizes of H99 from 10^2–10^7 cfu were similarly successful in producing infection. Presently, inoculum sizes for producing consistent infection over 2 weeks ranges from 5×10^7–1×10^8 cfu. Yeasts are grown on Sabouraud or YEPD agar plates for 4 days. The yeasts are swabbed off plates on cotton applicators placed in 0.15 mol/l phosphate-buffered saline at pH7.4, resuspended, counted by hemocytometer, and quantitative plate counts are performed to insure correct viable yeast concentrations. The inoculum can be used successfully for 2–3 hours after preparation and probably longer, but past this time range there are no data. The H99 strain can be obtained from the author's laboratory.

The vast majority of data has been accumulated with strain H99 and previous work does show that, even in this immunocompromised model, strains vary in their ability to produce disease (Perfect *et al.*, 1980). Each *C. neoformans* strain to be used in this model will need to be specifically examined for its state of virulence.

Infection process

The rapidity of infection depends on inocula and corticosteroid dose. In early studies, it was found that a 10^7 inoculum of H99 with 2.5–25 mg/kg per day of cortisone acetate administered would produce approximately 100% mortality within 6 weeks with a range of 11–80 days (Perfect *et al.*, 1980). With improved animal health (i.e. pathogen-free animals) and some virulence attenuation of the H99 yeast strain, this survival period may actually be longer. However, it has not recently been examined since present infections are not carried past 14–18 days at this time and survival is no longer used as an endpoint for the model.

Generally, rabbits have few observable symptoms during the first 2 weeks of infection. Neurologically, they are intact and only occasionally will an animal die. Food and water intake is similar to normal rabbits. As infection progresses in the third through fifth weeks, rabbits will develop neurological symptoms such as cerebellar signs and occasional lateralizing neurological signs with limb paralysis or neck tilts. Death can be sudden or associated with a 12–24-hour coma.

Key parameters to monitor infection and response to therapy

Although infection does disseminate to other organs from the subarachnoid space, the primary advantage of this model is serial examination of the subarachnoid space for both host and yeast factors. The host CNS immune response can be followed serially in both normal and immunosuppressed animals. Immunosuppression is a key feature of this model and can be produced by corticosteroids (Perfect *et al.*, 1980), cyclosporine (Perfect and Durack, 1985), and tacrolimus (Odom *et al.*, 1997). Corticosteroid treatment generally produces a severe CSF leukopenia in rabbits with cryptococcal meningitis compared to control rabbits. This finding is similar to histopathological observations noted in patients with acquired immunodeficiency syndrome (AIDS) and cryptococcal meningitis. Several CSF measurements for viable yeast counts during various treatment regimens have been made. CSF host cells have been analyzed and the CSF itself examined for both host immune and treatment factors. Finally, the yeasts have been recovered from CSF to capture gene expressions and certain yeast phenotypes under genetic control can be examined for their effect on infection. The ability to follow serially with a variety of parameters this infection at the actual infectious site without sacrificing the rabbit is an extremely attractive feature of this model. Table 87.1 lists the factors which have been followed in this model. It is clear that the potential exists for further host and yeast studies.

Table 87.1 Cerebrospinal fluid parameters in rabbit model examined

Host factors	Lymphocyte cell populations
	Macrophage cytotoxicity
	Antibodies
	Chemotactic factors
Yeast factors	Antigens (polysaccharide)
	Metabolites (mannitol)
	Identification of regulated genes
	Site-directed mutants for virulence
Treatment	Drug treatment effects on quantitative counts
	Drug levels in cerebrospinal fluid
	Effects of immunosuppressives

Histopathology

The infection produces a subacute meningitis with evidence of all host cell lines (polymorphonuclear cells, lymphocytes, and macrophages) participating in the successful elimination of the yeasts from the subarachnoid space. In rabbits without steroid treatment an impressive granulomatous inflammatory response is produced within the subarachnoid space and infection is limited to this space. On the other hand, there is a paucity of inflammatory cells within the subarachnoid space of cortisone-treated animals and there is evidence of yeast extension within the brain parenchyma as cystoid collections of yeasts (Perfect *et al.*, 1980).

Determination of fungal density

The burden of viable organisms ranges from 10^3 to 10^5 cfu/ml of CSF during the first 2 weeks of infection under corticosteroid treatment. In contrast, normal rabbits will clear the CSF of viable yeasts during that time period. CSF counts can rise above 10^5 cfu and reach as high as 10^7 cfu/ml of CSF if higher doses of cortisone are administered.

Determination of antifungal penetration into CSF compartment

CSF and blood drug levels can be measured simultaneously at multiple time periods over several days or at several time periods during a day. It is recommended not to perform more than three cisternal taps in 24 hours because with each tap it raises the level of "dry taps". CSF drug levels can be measured by bioassay or high-performance liquid chromatography and compared to readily available and simultaneous plasma or serum levels (Perfect *et al.*, 1986, 1989).

Determination of markers of infection and response to therapy

CSF fluid is readily available for biological fluid analysis such as antibody, cytokine, chemokine and metabolite analysis (Perfect *et al.*, 1981; Perfect and Durack, 1985;

Wong *et al.*, 1990). CSF taps commonly yield 1–2 ml of fluid and polysaccharide antigen titers can be measured (Perfect *et al.*, 1980). Large numbers of yeast cells are required for isolation of yeast RNA determination for differential gene expressions (Perfect and Rude, 1993) and this may require multiple animals. However, reverse transcriptase-polymerase chain reaction (RT-PCR) can be used to identify single gene expression from cells during infection in this model (Salas *et al.*, 1996). CSF host cells can be used for both phenotyping studies and/or specific functional assays such as macrophage cytotoxicity assays (Perfect *et al.*, 1988). Following response to therapy it has been convenient to use a 2-week observation period and follow drug effects on quantitative yeast counts. Successful regimens generally require at least 4 days of treatment before a significant drop of CSF yeast counts can be detected. This has been a good surrogate marker for success of treatment both in the rabbit model and in its comparison to human infection. Serial quantitative CSF measurements are made and individual rabbits can be followed for their drop in CSF yeast counts.

Antimicrobial therapy

A variety of antifungal agents have been evaluated in this model and many of the results with these agents have been published (Perfect and Durack, 1982, 1985a,b; Perfect *et al.*, 1986, 1989, 1996; Perfect, 1990; Wright *et al.*, 1990; Perfect and Wright, 1994; Kartalija *et al.*, 1996). The breadth of experience and consistency of this model allows for lower numbers of animals to be used for each dose to be studied. In most cases between 5 and 8 animals can determine efficacy of an antifungal agent. Control rabbits which do not receive treatment are always performed concurrently with each inoculum to insure quality control. Treatment regimens can be given daily or multiple times each day. The route of administration has been either oral or intravenous and both routes are easy to perform. Intravenous antifungal drugs are given under xylazine/ketamine sedation. Also, intracisternal medication can be successfully administered in this model. The pharmacokinetics and pharmacodynamics of the agents are routinely studied and can be performed easily for several body fluids such as CSF and blood. Attention to these factors is important in interpreting treatment results. It should be emphasized that the endpoint of treatment in this model is a positive or negative effect on quantitative CSF yeast counts and not animal survival. At the end of this experiment rabbits are sacrificed with a lethal injection of pentobarbital. Any animal during infection which develops neurological disease or difficulties with maintaining daily activities is sacrificed before completion of the experiment.

Determination of strain virulence

This model has been extremely successful in comparing virulence between strains and among isogenic mutant pairs. Strains can be inoculated at the same time and compared for their virulence potential by the use of molecular probes such as karyotyping to identify each strain (Elias *et al.*, 1995). Isogenic mutants for adenine (Perfect *et al.*, 1993), calcineurin A (Odom *et al.*, 1997), n-myristoyl transferase (Lodge *et al.*, 1994), urease (Cox *et al.*, 1996), and g-protein (Alspaugh *et al.*, 1997) have been studied in this model and identification of the contribution of these genetic loci for virulence has been documented for *C. neoformans*. Avirulent mutants are generally eliminated from CSF within 1–2 weeks of inoculation.

Pitfalls (advantages/disadvantages) of the model

Besides the usual shortcomings of all animal models which are countered by the controlled nature of the model, the advantages and disadvantages of this particular model are described in Table 87.2 and Table 87.3.

Table 87.2 Advantages of the rabbit model of cryptococcal meningitis

Reproducible infection which mimicks the human infection in histopathology and response to treatments
Size of animal allows easy drug administration and withdrawal of body fluids
Pharmacokinetics/pharmacodynamics can be studied concurrently
Immunosuppression is an important feature of the model, which is similar to humans (corticosteroids/immunophilins)
Serial measurements from the same animal
Low number of animals needed to make evaluations
Twenty years of experience with treatment, host immunology, and virulence studies in this model
Examines the most important site of infection with *Cryptococcus neoformans*
Endpoint is quantitative counts and not survival; animals are not apparently sick by observation during study period
Access to host and yeast cells at the site of infection

Table 87.3 Disadvantages of the rabbit model of cryptococcal meningitis

Expensive to purchase and care for animals
Immunological markers not as well-developed as mice
Requires high inocula for infection and inoculation site is not normally a pathological entry point
Generally examines only one site of infection
Only one yeast strain has been well-characterized
Relatively large amounts of drug or compound must be used in treatment experiments
Limited numbers of investigators have used the model
Specific immune modulators are generally not available

Contribution of the model to infectious disease therapy

Profiling of antimicrobial treatment

Because of size and cost this animal model is not an attractive model for the early screening of large numbers of antifungal compounds. However, as a drug approaches clinical trials in humans, it has become an excellent marker for the success of treatment in humans. For example, similar to humans, amphotericin B is the most powerful single antifungal agent available in this model despite undetectable levels of drug in the CSF. This model predicted that amphotericin B would sterilize CSF faster than azoles such as fluconazole, and this was also shown in humans. It is also suggested that itraconazole, despite low CSF penetration, would have activity against *C. neoformans* in the CNS and this was later confirmed in humans. It was initially shown that polyene and azoles would not be antagonistic in cryptococcal meningitis and although in humans these agents are not routinely given concurrently, it has become common practice to use them sequentially. The model has also screened compounds which did not have antifungal activity in this model and these compounds were not taken further in clinical development. Finally, this model demonstrated the importance of dosing for the lipid-based amphotericin B products. This model suggests that they are not as potent as amphotericin B on a mg/kg basis, but the ability to give high doses can eventually allow a possible therapeutic advantage for these preparations.

The model at present continues to have room to identify new and better antifungal agents. It is clear that most available agents are not yet rapidly fungicidal in this model and, since most available agents have been screened in the model, a new agent which produces more rapid sterilization of the CSF than amphotericin B would be extremely attractive to study in humans. Also, antifungal combinations and their pharmacodynamics could be easily evaluated in this model.

Elucidation of the pathobiology of cryptococcal meningitis

This model is ideal for the study of dynamic host factors at the site of infection. There is a direct window into the host–fungal interaction which can be repeatedly observed in the same animal. The only limitation is the number of rabbit-specific immunological reagents. The factor of inbred versus outbred genetic background in this model has both positive and negative features.

The model has been extremely successful in defining certain features of the yeast. It can be used to look at yeast gene expressions in the host, which will allow understanding of how the yeast produces diseases. It also allows for screening of strains with specific gene disruptions to help understand the composite virulence phenotype of *C. neoformans*.

References

Alspaugh, J. A., Perfect, J. R., Heitman, J. (1997). *Cryptococcus neoformans* mating and virulence are regulated by the G-protein gamma subunit GPA1 and cAMP. *Genes Dev.*, **11**, 3206–3217.

Bergman, F. (1996). Effect of temperature on intratesticular cryptococcal infection in rabbits. *Sabouraudia*, **5**, 54–58.

Casadevall, A. J., Mukherjee, J., Ruong, R., Perfect, J. R. (1994). Management of *Cryptococcus neoformans* contaminated needle injuries. *Clin. Infect. Dis.*, **19**, 951–953.

Cox, G. M., Cole, G. T., Perfect, J. R. (1996). Identification and disruption of the *Cryptococcus neoformans* urease gene. Proceedings of 3rd International Conference on *Cryptococcus* and Cryptococcosis, Paris **1**, 6. 150 (abstract).

Elias, S., Cox, G. M., Meyer, W., Mitchell, T. G., Perfect, J. R. (1995). Use of molecular techniques to evaluate fungal virulence and antifungal resistance in experimental cryptococcal meningitis. Proceedings of 35th Interscience Conference on Antimicrobial Agents and Chemotherapy, Anaheim, CA. (abstract 1206) B68.

Felton, F. G., Maldonado, W. E., Muchmore, H. G., Rhoades, E. R. (1966). Experimental cryptococcal infection in rabbits. *Am. Rev. Respir. Dis.*, **94**, 589–594.

Fujita, N. (1983). Experimental hematogenous endophthalmitis due to *Cryptococcus neoformans*. *Invest. Ophthalmol. Vis. Sci.*, **24**, 368–375.

Kartalija, M., Kaye, K., Tureen, J. H., Liu, Q. X., Tauber, M. G., Elliott, B. R., Sande, M. A. (1996). Treatment of experimental Cryptococcal meningitis with fluconazole: impact of dose and addition of flucytosine on mycologic and pathophysiologic outcome. *J. Infect. Dis.*, **173**, 1216–1221.

Kligman, A. M., Weidman, F. D. (1949). Experimental studies on treatment of human torulosis. *Arch. Dermatol. Syph.*, **60**, 726–741.

Lodge, J. K., Jackson-Machelski, E., Toffaletti, D. L., Perfect, J. R., Gordon, J. I. (1994). Targeted gene replacement demonstrates that myristoyl CoA: protein N-myristoyl-transferase is essential for the viability of *Cryptococcus neoformans*. *Proc. Natl. Acad. Sci. USA*, **91**, 12008–12012.

Odom, A., Del Poeta, M., Perfect, J., Heitman, J. (1997). The immunosuppressant FK506 and its non-immunosuppressive analog L-685,818 are toxic to *Cryptococcus neoformans* by inhibition of a common target protein. *Antimicrob. Agents Chemother.*, **41**, 156–161.

Odom, A., Muir, S., Lim, E., Toffaletti, D. L., Perfect, J. R., Heitman, J. (1997). Calcineurin is required for virulence of *Cryptococcus neoformans*. *EMBO J.*, **16**, 2576–2589.

Perfect, J. R. (1990). Fluconazole therapy for experimental cryptococcosis and candidiasis in the rabbit. *Rev. Infect. Dis.*, **12** (suppl. 3), 299–302.

Perfect, J. R., Durack, D. T. (1982). Treatment of experimental cryptococcal meningitis with amphotericin B, 5-fluorocytosine and ketoconazole. *J. Infect. Dis.*, **146**, 429–435.

Perfect, J. R., Durack, D. T. (1985). Effects of cyclosporine in experimental cryptococcal meningitis. *Infect. Immun.*, **50**, 22–26.

Perfect, J. R., Durack, D. T. (1985a). Chemotactic activity of cerebrospinal fluid in experimental cryptococcal meningitis. *J. Med. Vet. Mycol.*, **23**, 37–45.

Perfect, J. R., Durack, D. T. (1985b). Comparison of amphotericin B and N-D-ornithyl amphotericin B methyl ester in experimental cryptococcal meningitis and *Candida albicans* endocarditis with pyelonephritis. *Antimicrob. Agents Chemother.*, **28**, 751–755.

Perfect, J. R., Rude, T. H. (1993). Identifying *Cryptococcus neoformans* by differential expression at the site of infection. Proceedings of 2nd International Conference on *Cryptococcus* and Cryptococcosis L3, p. 24 (abstract).

Perfect, J. R., Wright, K. A. (1994). Amphotericin B lipid complex in the treatment of experimental cryptococcal meningitis and disseminated candidiasis. *J. Antimicrob. Chemother.*, **33**, 73–81.

Perfect, J. R., Lang, S. D. R., Durack, D. T. (1980). Chronic cryptococcal meningitis: a new experimental model in rabbits. *Am. J. Pathol.*, **101**, 177–194.

Perfect, J. R., Lang, D. R., Durack, D. T. (1981). Influence of agglutinating antibody in experimental cryptococcal meningitis. *Br. J. Exp. Pathol.*, **62**, 595–599.

Perfect, J. R., Savani, D. V., Durack, D. T. (1986). Comparison of itraconazole and fluconazole in treatment of cryptococcal meningitis and *Candida* pyelonephritis in rabbits. *Antimicrob. Agents Chemother.*, **29**, 579–583.

Perfect, J. R., Hobbs, M. M., Granger, D. L., Durack, D. T. (1988). Cerebrospinal fluid macrophage cytotoxicity: *in vitro* and *in vivo* correlation. *Infect. Immun.*, **56**, 849–854.

Perfect, J. R., Wright, K. A., Hobbs, M. M., Durack, D. T. (1989). Treatment of experimental cryptococcal meningitis and disseminated candidiasis with SCH 39304. *Antimicrob. Agents Chemother.*, **33**, 1735–1740.

Perfect, J. R., Toffaletti, D. L., Rude, T. H. (1993). The gene encoding for phosphoribosylaminoimidazole carboxylase (ADE2) is essential for growth of *Cryptococcus neoformans* in cerebrospinal fluid. *Infect. Immun.*, **61**, 4446–4451.

Perfect, J. R., Cox, G. M., Dodge, R. K., Schell, W. A. (1996). *In vitro* and *in vivo* efficacies of the azole SCH56592 against *Cryptococcus neoformans*. *Antimicrob. Agents Chemother.*, **40**, 1910–1913.

Salas, S. D., Bennett, J. E., Kwon-Chung, K. J., Perfect, J. R., Williamson, P. R. (1996). Effect of the laccase gene, CNLAC1, on virulence of *Cryptococcus neoformans*. *J. Exp. Med.*, **184**, 377–386.

Weiss, C., Perry, I. H., Sheury, M. C. (1948). Infections of the human eye with *Cryptococcus neoformans*. *Arch. Ophthalmol.*, **39**, 739.

Wong, B., Perfect, J. R., Beggs, S., Wright, K. A. (1990). Production of the hexitol D-mannitol by *Cryptococcus neoformans in vitro* and in rabbits with experimental meningitis. *Infect. Immun.*, **58**, 1664–1670.

Wright, K. A., Perfect, J. R., Ritter, W. (1990). The pharmacokinetics of BAY R3783 and its efficacy in the treatment of experimental cryptococcal meningitis. *J. Antimicrob. Chemother.*, **26**, 387–397.

Chapter 88

Rat Models of Ascending Pyelonephritis due to *Candida albicans*

M. Ohkawa, M. Takashima, T. Nishikawa and S. Tokunaga

Background of model

Although fungal infections of the urinary tract are far less common than bacterial infections, we have seen a marked increase in their incidence in the last three or four decades (Schönebeck and Ånséhn, 1972; Michigan, 1976). These are often attributable to the widespread use of antimicrobial agents, immunosuppressive drugs and anticancer agents, prolonging the life of patients who have serious underlying conditions, and the development of more aggressive surgical procedures. *Candida albicans* is the most prevalent and pathogenic of the fungi affecting the urinary tract (Roy *et al.*, 1984). *Candida* infection of the lower urinary tract is believed to be caused by ascent through the urethra (Schönebeck, 1972). In contrast upper urinary tract candidiasis has generally been accepted as a secondary phenomenon in *Candida* septicemia since experimental studies have clearly demonstrated that hematogenous spread has the potential to cause renal candidiasis (Hurley and Winner, 1963; Louria *et al.*, 1963).

Tennant *et al.* (1968) and Schönebeck and Winbland (1971) reported cases of primary renal candidiasis, which is defined by the presence of renal candidal involvement without concomitant invasion of the blood stream or any other internal organs. Tennant *et al.* (1968) collected 11 cases of this condition, including their 2 cases, and suggested the existence of an ascending infection route, since 9 patients were female and 6 had renal involvement localized to the renal pelvis and calyces.

However, little convincing experimental evidence demonstrating that *Candida* cells instilled into the bladder can produce ascending *Candida* pyelonephritis is available. Against this background, we have investigated and established two experimental rat models of ascending *Candida* pyelonephritis with combination pretreatments of leukopenia and vesicourethral reflux (VUR; model 1; Tokunaga *et al.*, 1993a), and of unilaterally incomplete ureteral obstruction and VUR (model 2; Takashima, 1992).

Animal species and preparation of animals

Female Wistar rats weighing 200–250 g each have been used in these models. The animals are allowed free access to standard laboratory chow (Oriental Yeast, Tokyo, Japan) and tap water.

Preparation of organism

Candida albicans ATCC 10259 strain (serotype A) is used. *Candida* cells grown for 48 hours at 37°C in Sabouraud's broth are harvested by centrifugation at 4°C for 10 minutes, washed twice with distilled water, and resuspended in sterile physiological saline to reach a final concentration of 2.0×10^7 colony-forming units (cfu)/ml.

Model 1

Infection procedures

Cyclophosphamide injection

The rats are rendered leukopenic by a single transperitoneal injection of 200 mg/kg body weight of cyclophosphamide (Shionogi, Osaka, Japan) 3 days before inoculation of the *Candida* cells. By this procedure, leukopenia less than $1000/mm^3$ was induced in the following 3–7 days in a preliminary study (Figure 88.1).

Chemically induced cystitis

One day before *Candida* inoculation, using an 18G tip needle, 0.75% acetic acid solution (0.5 ml) is instilled transurethrally into the bladder to produce chemically induced cystitis. Two of our colleagues, Sugata (1980) and Orito (1981), demonstrated that VUR frequently occurs after a single intravesical instillation of 1.0 ml of contrast medium in normal rats; in addition, VUR occurred in approximately 80% of rats whose bladders were treated with 0.5 ml of a 0.75% acetic acid solution, resulting in destruction of the antireflux mechanism.

Ascending pyelonephritis

An inoculation of the suspension containing 10^7 *Candida* cells (0.5 ml) is instilled transurethrally into the bladder

Handbook of Animal Models of Infection
ISBN 0-12-775390-7

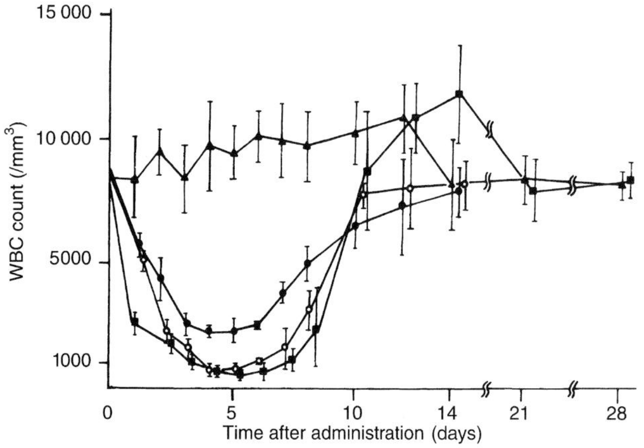

Figure 88.1 Effect of cyclophosphamide administration on peripheral white blood cell (WBC) counts of normal Wistar rats. Cyclophosphamide administration: filled circles, 100 mg/kg; open circles, 150 mg/kg; filled squares, 200 mg/kg; filled triangles, non-administered. Data represent (n = 5) the mean ± SD.

using the needle tip and then the external urethral meatus is clipped with a meatal clip for 1 hour. Gentamicin (8 mg/kg) is administered once a day for 1 week following the administration of cyclophosphamide to prevent contaminating bacterial infections.

Key parameters to monitor infection

Kidney pathology

Kidney histology is normally used to monitor the progress of infection. Pyelonephritis was classified into four grades as follows: grade 0 (no inflammation); grade 1 (inflammation localized to the renal pelvis; pyelitis); grade 2 (inflammation invading the renal medulla or slightly into the cortex, but not reaching the renal capsule); and grade 3 (microscopic or macroscopic renal abscess). Histological studies of half a kidney are performed on hematoxylin and eosin, and periodic acid–Schiff-stained slides.

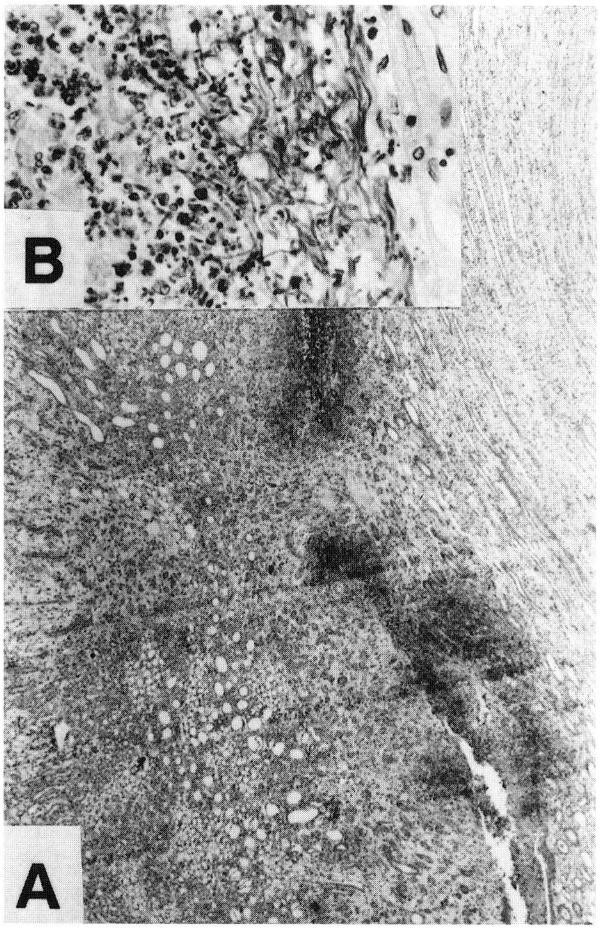

Figure 88.2 Photomicrographs showing acute pyelitis (grade 1) of a rat 3 days after candidal inoculation. Inflammation is observed mainly at the fornix (**A**; periodic acid–Schiff, ×10), where numerous mycelia are seen with polymorphonuclear cell infiltration and edema (**B**: periodic acid–Schiff, ×200).

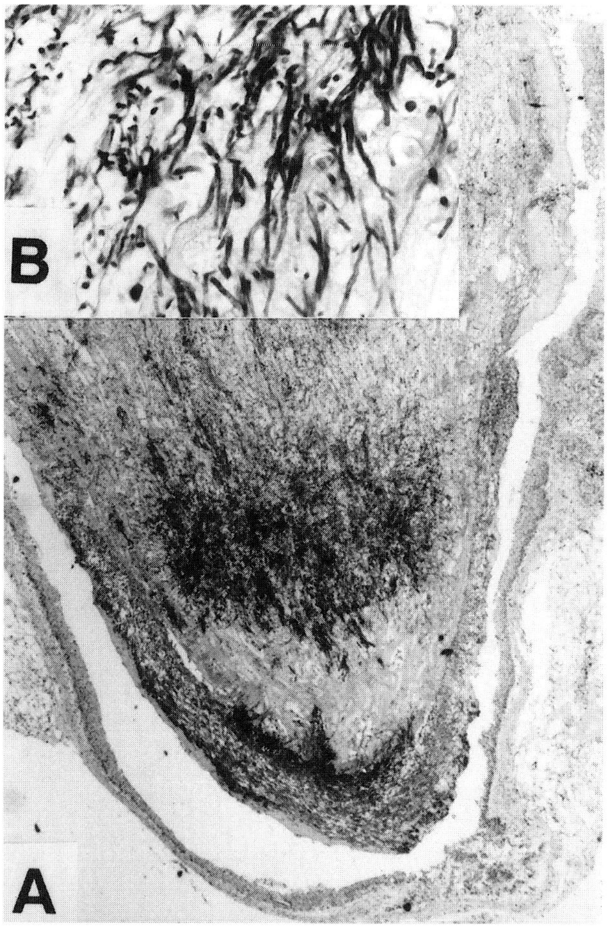

Figure 88.3 Photomicrographs showing acute pyelonephritis (grade 2) of a rat 3 days after candidal inoculation. Marked inflammation is observed at the papilla and medulla, showing necrotizing papillitis (**A**: periodic acid–Schiff, ×10), where numerous mycelia are also seen (**B**: periodic acid–Schiff, ×400).

In our experiment (Tokunaga *et al.*, 1993a), a total of 128 kidneys were investigated. One, 3, 7, 14 and 28 days after inoculation, pyelonephritis was noted in 27, 42, 79, 71 and 80% of the kidneys tested. Within 3 days of inoculation, microcopy revealed acute pyelonephritis showing polymorphonuclear leukocyte infiltration, mucosal edema, and bleeding, together with *Candida* cells invading from the renal fornix in about half of the infected kidneys (pyelitis: grade 1; Figure 88.2). At the papilla, acute inflammatory changes and partial necrosis which extended into the renal medulla (grade 2; Figure 88.3) were observed in the other half. In the period from 7 to 28 days, chronic inflammatory changes associated with small cell infiltration and fibrosis were observed up to the cortex of 19% of the kidneys tested (grade 3; Figure 88.4). *Candida* bezoars containing necrotic materials and *Candida* cells were observed in the renal pelvis of 24% of the infected kidneys (Figure 88.4). However, grade 1 pyelonephritis was observed most frequently (67%) in this period.

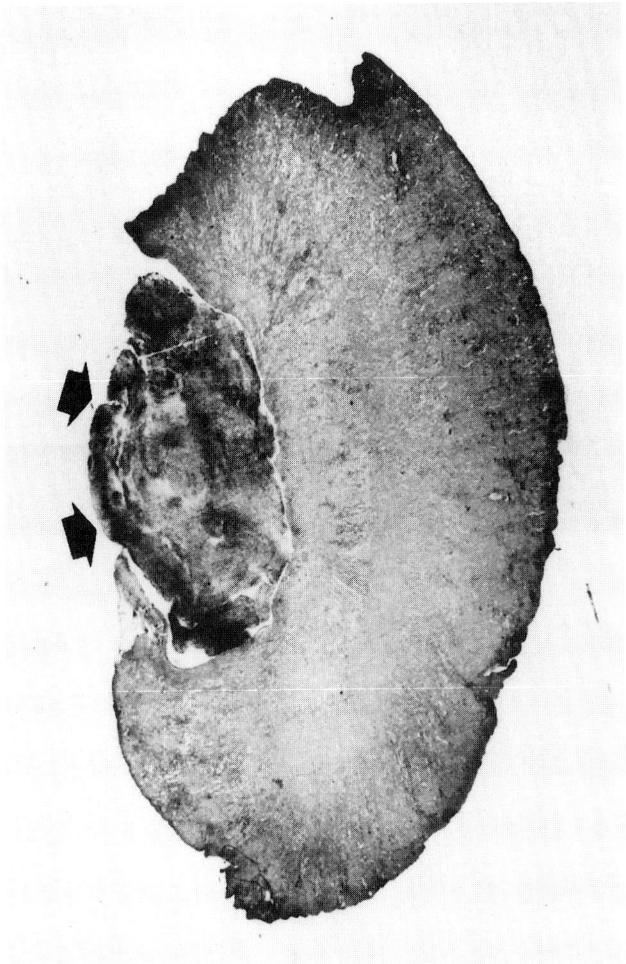

Figure 88.4 Photomicrograph showing chronic pyelonephritis (grade 3) of a rat 14 days after candidal inoculation. The renal pelvis is filled with *Candida* bezoar (arrows), and diffuse inflammation extends deeply into the cortex (periodic acid–Schiff, ×1).

Candidal density in kidney and urine

A 100-mg portion of the other half of the kidney is homogenized in 1 ml of sterile physiological saline. Aliquots of undiluted homogenate (0.1 ml) and of 200-fold dilutions of the homogenate are then cultured in duplicate on *Candida* GE agar plate (Nissui, Tokyo, Japan) and are incubated for 72 hours at 37°C. Aliquots of urine (0.1 ml), which is directly obtained by bladder puncture 1 hour after ligation of the external urethral meatus at autopsy, are also cultured, and *Candida* cells are counted by the dilution method.

In our experiment (Tokunaga *et al.*, 1993a), *Candida* colony counts of bladder urine were calculated according to the severity of pyelonephritis in 47 rats in which bladder puncture was successfully performed. The renal candidal population was $\geq 10^4$ cfu/g in more than 90% of the infected kidney, with all of the kidneys with grade 2 and 3 pyelonephritis showing $\geq 10^6$ cfu/g. There was a significant relationship between renal and urinary *Candida* populations ($P<0.01$); the urinary counts of grades 2 and 3 were significantly higher than those of grades 0 and 1 ($P < 0.01$).

Determination of serum D-arabinitol

Small quantities of D-arabinitol, a major candidal metabolite, have been found in healthy human sera, with high concentrations of serum D-arbinitol reported in patients with invasive candidiasis (Kiehn *et al.*, 1979; Wong *et al.*, 1982a, b; Gold *et al.*, 1983), including *Candida* pyelonephritis (Eng *et al.*, 1981; Tokunaga *et al.*, 1992). Based on these results, serum D-arabinitol has been recognized as a quantitative and specific marker substance which is useful in distinguishing invasive candidiasis from benign colonization.

D-Arabinitol is measured by gas–liquid chromatography according to the method of Kiehn *et al.* (1979). We employ a JGC-1100 gas–liquid chromatograph with a flame ionization detector (JOEL, Tokyo, Japan). Methyl-α-D-mannopyranoside (P-L Biochemicals, Milwaukee, USA) is used as an internal standard. Trimethylsilyl ether derivatives are prepared according to the method of Sweeley *et al.* (1963). For analysis of trimethylsilyl ethers, the instrument is fitted with 2-m coiled columns (6 mm in diameter) packed with 3% SE-30 on 80/100 mesh GAS-Chrom Q (Nihon Chromato Works, Tokyo, Japan). The temperatures of the injector and detector are maintained at 210 and 220°C, respectively, and the column oven temperature is set at 165°C. In Tokunaga's report (1983), elevated serum D-arbinitol levels were found in the model 1 with acute-phase pyelonephritis: serum D-arbinitol determined by gas–liquid chromatography was considered to be a useful marker for the diagnosis of *Candida* pyelonephritis.

D-Arabinitol concentrations have been measured using gas–liquid chromatography by many investigators. However, the routine clinical use of this method is complicated and time-consuming, and only a single sample can be assayed. To resolve these problems, an enzymatic fluorometric assay has been developed (Soyama and Ono, 1985), with which good results have been obtained as regards quantity, specificity and sensitivity for D-arbinitol in patients with invasive candidiasis (Soyama and Ono, 1985; Fujita *et al.*, 1989). Recently we have determined serum D-arbinitol by the enzymatic assay, which is commercially available in kits (Nacalai Tesque, Kyoto, Japan), using a fluorescence spectrophotometer (Model 850; Hitachi, Tokyo, Japan) with a circulating thermobath at 37°C. The exiting and emission wavelengths are 340 and 460 nm, respectively. In our study (Tokunaga *et al.*, 1992), a significant correlation was demonstrated between the results of the two methods in 25 human serum samples: the equation for the line for the best fit was values by gas–liquid chromatography = 0.956 × values by the enzymatic fluorometric assay: the correlation coefficient was 0.943 ($P<0.01$).

Model 2

Infection procedure

Surgical procedure for incomplete ureteral obstruction

Under satisfactory ether anesthesia rats are placed supine. A midline incision is made at the lower abdomen after the skin is shaved with an electric clipper and then disinfected with an iodine-based solution. In order to produce open hydronephrosis, incomplete occlusion of the left ureter is induced by exposing and detaching the left ureter and applying a drop of an instant adhesive (α-cyanoacrylate monomer: Aronalpha A, Sankyo, Tokyo, Japan) to the inferior end of the ureter (Ikeda, 1981). Using this method open hydronephrosis could be produced in a high proportion of rats. An intravenous pyelogram shows moderate hydronephrosis (Figure 88.5) 7 days after operation in a rat.

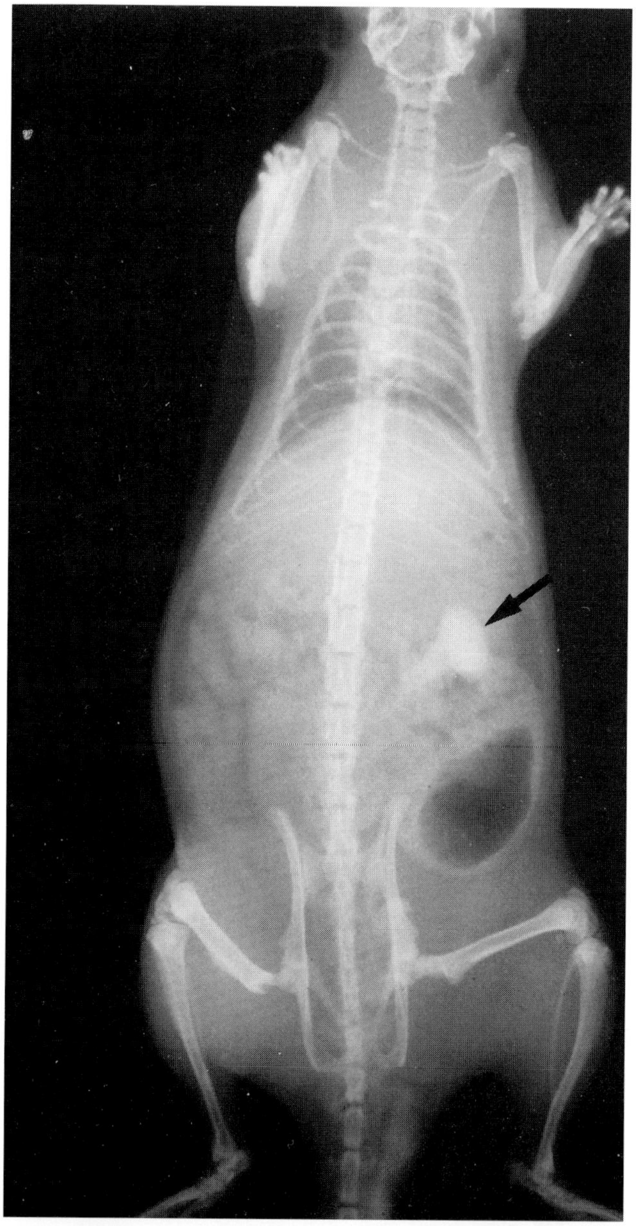

Figure 88.5 An intravenous pyelogram showing moderate hydronephrosis (arrow) 7 days after placing a drop of α-cyanoacrylate monomer on the left lower ureter.

Ascending pyelonephritis

Acetic acid solution and then *Candida* cell solution are instilled into the bladder, and additionally, the external meatus is clipped in the same procedure as in model 1. Furthermore, incomplete occlusion of the left ureter is created as described above. Gentamicin (8 mg/kg) is administered once a day for 1 week following the instillation of acetic acid.

Key parameters to monitor infection

Kidney pathology

Histological examinations were performed as described in model 1. In Takashima's study (1992), a total of 84 hydronephrotic kidneys were examined. The incidence of ascending pyelonephritis in hydronephrotic kidneys increased depending on the period after inoculation; 1, 3, 7, 14, 21 and 28 days after inoculation, pyelonephritis was noted in 53, 71, 87, 87, 92 and 100% of the kidneys examined. Ascending pyelonephritis first occurred at the renal fornix and papilla and then gradually spread into the cortex; only in the period from 7 days to 28 days, *Candida* bezoars were observed in the distended pelvis (Figure 88.6).

Candidal density in kidney and urine

Candida colony counts are calculated as described in model 1. In Takashima's experiment (1992), the number of

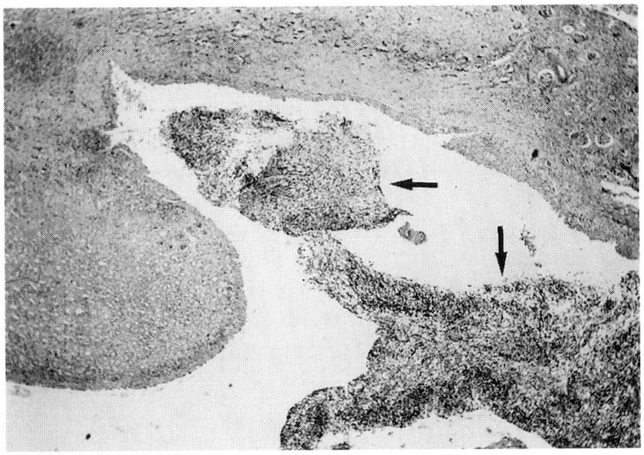

Figure 88.6 *Candida* bezoars (arrows) filled the renal pelvis distended by incomplete ureteral obstruction (periodic acid–Schiff, ×400).

Candida cells increased significantly depending on the histological grade in the infected kidneys (P<0.05), revealing a significant relationship between renal and urinary *Candida* populations (P<0.05).

Determination of serum D*-arabinitol*

D-Arabinitol is reported to be eliminated by urinary excretion and is cleared at virtually the same rate as creatinine (Wong *et al.*, 1982a). Thus, elevated concentrations have also been found in sera in non-candidiasis patients with renal failure (Eng *et al.*, 1981; Wong *et al.*, 1982b; Gold *et al.*, 1983). Therefore, to correct for the effect of renal function, the use of the serum D-arabinitol/creatinine ratio has been recommended.

Serum D-arabinitol was measured in model 2 with pyelonephritis by the enzymatic fluorometric method (Takashima, 1992). The serum D-arabinitol/creatinine ratio in rats with *Candida* pyelonephritis was significantly greater than that in the non-pyelonephritis rats, both within 3 days after inoculation (P<0.01) and also during the period from 7 to 28 days after inoculation (P<0.05). The results suggest that the serum D-arabinitol/creatinine ratio in *Candida* pyelonephritis rats may be useful to monitor infection, not only in the acute phase but also in the chronic phase.

Determination of serum mannan

The antigen mannan is a polysaccharide which is the main component of the *Candida* cell wall. Mannan can be detected in the sera of patients with invasive candidiasis by several methods with high diagnostic sensitivity and specificity (Bailey *et al.*, 1985; de Repentigny *et al.*, 1985; Fujita *et al.*, 1986; Bougnoux *et al.*, 1990; Lemieux *et al.*, 1990). We reported that serum mannan in candiduria patients might be a useful parameter in diagnosing *Candida* pyelonephritis and deciding on effective treatment (Tokunaga *et al.*, 1993b).

We measure serum mannan by an enzyme-linked immunosorbent assay (ELISA) using a biotin-streptoavidin procedure. The ELISA technique for serum mannan is similar in principle to that described by Guedson *et al.*, (1979) and Fujita *et al.* (1986). Antisera and *Candida* mannan are prepared as follows: Japan White rabbits are immunized intravenously with heat-treated cells of a *C. albicans* serotype A strain, and antisera are collected. Immunoglobulin G (IgG), prepared by fractionation of the sera with Na_2SO_4 followed by passage through a diethylaminoethyl (DEAE) cellulose column, is dialyzed overnight at 4°C. Biotin-N-hydroxysuccinimide (Pierce, Rockford, USA) is dissolved in dimethylformamide; this biotin preparation is added to IgG and allowed to react. Polystyrene microtiter plates (Immulon 2: Dynatech, Chantilly, USA) are coated with the IgG fraction of antiserum, and then supernatant of the heat-treated sample is added to the wells and incubated. Diluted biotin-linked IgG, dilution of streptoavidin-horseradish peroxidase conjugate (BRL, Geithersburg, USA) and then a substrate is added to the wells. The enzyme substrate fraction is allowed to proceed at room temperature, and the absorbance at 450 nm is measured. From Takashima's report (1992), increased serum mannan concentrations were found in model 2 within 3 days of inoculation. Accordingly, serum mannan is considered to be useful for diagnosing *Candida* pyelonephritis, especially in the acute phase.

Pitfalls of these models

Although several efforts to develop an experimental model of invasive candidiasis have been made by compromising the animals by a variety of means prior to challenge with the micro-organisms, there has not yet been an experimental study to demonstrate clearly ascending *Candida* pyelonephritis, as mentioned in Michigan's review (1976). Leukocytes are well-known to play an important role in the host's defense against *Candida* species. VUR is accepted as a major condition that perpetuates and aggravates ascending bacterial pyelonephritis. In our preliminary study, more than the inoculation size (10^7 *Candida* cells) and volume (0.5 ml) applied in model 1 could cause higher-grade pyelonephritis at a higher incidence, resulting in a high incidence of candidal involvement of organs other than the kidneys, including candidemia. The normal capacity of rat bladder is reported to be less than 0.5 ml (Andersen and Jackson, 1961). Therefore, Takashima (1992) also instilled the same size and volume as in model 1 into the bladder of rat with hydronephrotic kidney, and established model 2.

The natural history of hematogenous *Candida* pyelonephritis produced experimentally has been reported in detail (Hurley and Winner, 1963; Louria *et al.*, 1963). The infection first localizes in the renal cortex, in which multiple abscesses develop by 48 hours. After 9–11 days the cortical area eventually heals, and instead the tubules and collecting tubules become filled with fungi, and papillary necrosis develops. Fungi collect in the renal pelvis, and form *Candida* bezoars. In contrast, the ascending *Candida* pyelonephritis observed in our studies starts from the renal fornix where pyelitis exists, and then extends to the parenchyma. These findings are almost identical to those of ascending *Candida* pyelonephritis in the rabbit model reported previously (Parkash *et al.*, 1970).

Contribution of the models to infectious disease study

Profiling of antimicrobial treatment

As yet there has been no study of antimicrobial treatment using our models. We think that these models could be used in profiling of new drugs or evaluating combinations, using parameters including kidney pathology and candidal density in the kidney and urine sample. In addition, serum levels of D-arabinitol/creatinine ratio and mannan could be useful for monitoring *Candida* pyelonephritis condition.

Elucidation of the pathophysiology of *Candida* pyelonephritis

Using model 2, Nishikawa *et al.* (1997) investigated the pathogenicity of the hyphal form of *C. albicans* in pyelonephritis due to *C. albicans* KD 4900 strain (showing only the yeast form but not the hyphal form when induced by ultraviolet mutagenesis) and *C. albicans* KD 4907 (a revertant strain from *C. albicans* KD 4900 showing bimorphism), which were provided by Shimokawa and Nakayama (1984). The revertant strain showed a significantly higher frequency of pyelonephritis as compared with the mutant strain ($P<0.01$), and the grade of inflammation was also higher in the revertant strain group. In addition, higher renal tissue and serum levels of both lipid peroxide and superoxide dismutase, which are related to marked renal oxidant injury, tended to be correlated with the degree of neutrophil infiltration. From this study, it is suggested that the hyphal form plays an important role in the development of *C. albicans* pyelonephritis and also that the oxygen radicals from the neutrophils appearing at the sites of inflammation play a major part in the further extension of inflammatory lesions.

References

Andersen, B. R., Jackson, G. G. (1961). Pyelitis, an important factor in the pathogenesis of retrograde pyelonephritis. *J. Exp. Med.*, **114**, 375–384.

Bailey, J. W., Sada, E., Brass, C., Bennet, J. E. (1985). Diagnosis of systemic candidiasis by latex agglutination for serum antigen. *J. Clin. Microbiol.*, **21**, 749–752.

Bougnoux, M. E., Hill, C., Moissenet, D. *et al.* (1990). Comparison of antibody, antigen, and metabolic assays for hospitalized patients with disseminated or peripheral candidiasis. *J. Clin. Microbiol.*, **28**, 905–909.

de Repentigny, L., Marr, L. D., Keller, J. W. *et al.* (1985). Comparison of enzyme immunoassay and gas–liquid chromatography for the rapid diagnosis of invasive candidiasis in cancer patients. *J. Clin. Microbiol.*, **21**, 972–979.

Eng, R. H. K., Chmel, H., Buse, M. (1981). Serum levels of arabinitol in the detection of invasive candidiasis in animals and humans. *J. Infect. Dis.*, **143**, 677–683.

Fujita, S., Matsubara, F., Matsuda, T. (1986). Enzyme-linked immunosorbent assay measurement of fluctations in antibody titer and antigenemia in cancer patients with and without candidiasis. *J. Clin. Microbiol.*, **23**, 568–575.

Fujita, S., Maeno, K., Soyama, K. (1989). Mannan and D-arabinitol concentrations in serum from a patient with *Candida albicans* endocarditis. *Mycopathologia*, **105**, 87–92.

Gold, J. W. M., Wong, B., Bernard, E. M., Kiehn, T. E., Armstrong, D. (1983). Serum arabinitol concentrations and arabinitol/creatinine ratios in invasive candidiasis. *J. Infect. Dis.*, **147**, 504–513.

Guedson, J-L., Ternynck, T., Avrameas, S. (1979). The use of avidin-biotin interaction in immunoenzymatic techniques. *J. Histochem. Cytochem.*, **27**, 1131–1139.

Hurley, R., Winner, H. I. (1963). Experimental renal moniliasis in the mouse. *J. Pathol. Bacteriol.*, **86**, 75–82.

Ikeda, A. (1981). A method of producing open hydronephrosis in the rat. *Jpn. J. Urol.*, **72**, 1056–1063.

Kiehn, T. E., Bernard, E. M., Gold, J. W. M., Armstrong, D. (1979). Candidiasis: detection by gas–liquid chromatography of D-arabinitol, a fungal metabolite, in human serum. *Science*, **206**, 577–580.

Lemieux, C., St-Germain, G., Vincelette, J., Kaufman, L., de Repentigny, L. (1990). Collaborative evaluation of antigen detection by a commercial latex agglutination test and enzyme immunoassay in the diagnosis of invasive candidiasis. *J. Clin. Microbiol.*, **28**, 249–253.

Louria, D., Brayton, R., Finkel, G. (1963). Studies on the pathogenesis of experimental *Candida albicans* infections in mice. *Sabouraudia*, **2**, 271–283.

Michigan, S. (1976). Genitourinary fungal infections. *J. Urol.*, **116**, 390–397.

Nishikawa, T., Tokunaga, S., Fuse, H. *et al.* (1997). Experimental study of ascending *Candida albicans* pyelonephritis focusing on the hyphal form and oxidant injury. *Urol. Int.*, **58**, 131–136.

Orito, M. (1981). Experimental studies on the chemotherapeutic effect on retrograde pyelonephritis in rats. *Jpn. J. Urol.*, **72**, 680–693.

Parkash, C., Chu, T. D., Gupta, S. P., Thanic, K. D. (1970). *Candida* infection of urinary tract: an experimental study. *J. Ass. Phys. India*, **18**, 492–502.

Roy, J. B., Geyer, J. R., Mohr, J. A. (1984). Urinary tract candidiasis: an update. *Urology*, **23**, 533–537.

Schönebeck, J. (1972). Studies on *Candida* infection of the urinary tract and on the antimycotic drug 5-fluorocytosine. *Scand. J. Urol.* (suppl. 11), 7–48.

Schönebeck, J., Ånséhn, S. (1972). The occurrence of yeast-like fungi in the urine under normal conditions and in various types of urinary tract pathology. *Scand. J. Urol. Nephrol.*, **6**, 123–128.

Schönebeck, J., Winbland, B. (1971). Primary renal *Candida* infection. *Scand. J. Urol. Nephrol.*, **5**, 281–284.

Shimokawa, O., Nakayama, H. (1984). Isolation of *Candida albicans* mutant with reduced content of cell wall mannan and deficient mannan phosphorylation. *Sabouraudia*, **22**, 315–321.

Soyama, K., Ono, E. (1985). Enzymatic fluorometric method for determination of D-arabinitol in serum by initial rate analysis. *Clin. Chim. Acta.*, **149**, 149–154.

Sugata, T. (1980). Experimental studies in pyelonephritis in special reference to stone formation. *Jpn. J. Urol.*, **71**, 113–131.

Sweeley, C. C., Bently, R., Makita, M., Wells, W. W. (1963). Gas–liquid chromatography of trimethylsilyl derivatives of sugars and related substances. *J. Am. Chem. Soc.*, **85**, 2497–2507.

Takashima, M. (1992). An experimental study of ascending *Candida albicans* pyelonephritis in rats with open hydronephrosis. *Juzen Med. Soc.*, **101**, 395–404.

Tennant, F. S. Jr., Remmers, A. R. Jr., Perry, J. E. (1968). Primary renal candidiasis. *Arch. Intern. Med.*, **122**, 435–440.

Tokunaga, S. (1983). Experimental studies of ascending *Candida* pyelonephritis in rats. *Jpn. J. Urol.*, **74**, 683–697.

Tokunaga, S., Ohkawa, M., Takashima, M., Hisazumi, H. (1992). Clinical significance of measurement of serum D-arabinitol levels in candiduria patients. *Urol. Int.*, **48**, 195–199.

Tokunaga, S., Ohkawa, M., Nakamura, S. (1993a). An experimental model of ascending pyelonephritis due to *Candida albicans* in rats. *Mycopathologia*, **123**, 149–154.

Tokunaga, S., Ohkawa, M., Takashima, M. (1993b). Diagnostic value of determination of serum mannan concentrations in patients with candiduria. *Eur. J. Clin. Microbiol. Dis.*, **12**, 542–545.

Wong, B., Bernard, E. M., Gold, J. W. M., Fong, D., Silber, A., Armstrong, D. (1982a). Increased arabinitol levels in experimental candidiasis in rats: arabinitol appearance rate, arabinitol/creatinine ratios, and severity of infection. *J. Infect. Dis.*, **146**, 346–352.

Wong, B., Bernard, E. M., Gold, J. W. M., Fong, D., Armstrong, D. (1982b). The arabinitol appearance rate in laboratory animals and humans: estimation from the arabinitol/creatinine ratio and relevance to the diagnosis of candidiasis. *J. Infect. Dis.*, **146**, 353–359.

Chapter 89

Rat Model of *Candida* Vaginal Infection

F. De Bernardis, R. Lorenzini and A. Cassone

Background of the model

Vulvovaginal candidiasis is a widespread, common disease affecting a large proportion of otherwise healthy women. Some subjects suffer from recurrent, severe and often intractable forms of the disease (Sobel, 1989).

Symptomatic candidal vaginitis presents with characteristic vaginal discharge with cottage-cheese appearance and itching, burning, vaginal and vulvar erythema and edema. Variations — even remarkable ones — in this clinical picture are, however, present (Sobel, 1985). *Candida albicans,* a dimorphic commensal organism of the genital and gastrointestinal tract, is the causative agent of candidal vaginitis in approximately 80–90% of patients. Other prevalent *Candida* isolates are *C. glabrata*, *C. krusei*, *C. tropicalis* and *C. parapsilosis.* It is notable that only for *C. glabrata* and, mostly, *C. parapsilosis* is there clear clinical and experimental evidence for these species being true, causative agents of the disease rather than commensals.

The classification of infection, etiology, diagnosis and therapy have been reviewed previously. The reader may consult one of these excellent reviews, in particular those by Sobel (1989) and Fidel and Sobel (1996).

The clinical importance of *Candida* infections has stimulated advances in understanding host–fungus interaction. In fact, the pathogenesis of this disease remains unknown. The transition from asymptomatic colonization to symptomatic candidiasis occurs in the presence of factors that enhance *Candida* virulence and/or as a result of loss of defense mechanisms.

In order to gain insight into the mechanisms of candidiasis and to understand the fungal and host factors involved in the pathogenesis of *Candida* vaginitis, mouse and rat models of vaginal infections have been employed. The aim of using animal models is to study various aspects of the pathogenesis, immune response and treatment of *Candida* vaginitis under conditions possibly mimicking human infection. In these models *Candida* infections may be established under controlled and reproducible conditions.

The mouse model of candidal vaginitis has been used by Taschdjian *et al.* (1960), Ryley and McGregor (1986), Fidel *et al.* (1993, 1994, 1995a,b). A correspondent rat model was generated to study the efficacy of antifungal therapy (Scholer, 1960; McRipley *et al.,* 1979; Thienpont *et al.,* 1980; Ryley *et al.*, 1981; Sobel and Muller, 1983) and has recently been adapted to study virulence factors of *Candida* and their role in the pathogenesis of vulvovaginal candidiasis (De Bernardis *et al.*, 1993, 1995). Studies of anti-*Candida* immunity have also been initiated in this model to understand better the immunological mechanisms underlying protection from or facilitation of this infection (Cassone *et al.*, 1995; De Bernardis *et al.*, 1997). The same is being approached in mouse models of *Candida* vaginitis (Sobel *et al.*, 1984, 1985; Kinsman *et al.*, 1986; Ryley and Ryley, 1990). It is expected that, from a more incisive adoption of the above models, better insight into the various aspects of host–*Candida* relationship in vaginitis will soon be obtained.

Animal species and maintenance

Several strains of rats have been used in this model: Sprague–Dawley (175–200 g; Sobel and Muller, 1983), CD (150–200 g; Kinsman *et al.*, 1986), Wistar (100 g; Ichise *et al.*, 1986; Van Cutsen *et al.,* 1987), Alderley Park (100–125 g; Ryley and McGregor, 1986). No evidence for any specific advantage of some rat strain can be inferred by the published data, but no specific studies have been carried out. In our hands, Wistar and Sprague–Dawley rats perform comparably well in the vaginitis model.

No specialized housing or feeding is required, and caring is under the ordinary standards for outbred animal care.

Animal predisposition and estrogen treatment

Rat vaginal infection by *Candida* is under strict hormonal control. Namely, constant estrus is essential to establish infection and, for this to occur, the induction and maintenance of a "pseudoestrus", i.e., an artificial manipulation consisting of oophorectomy and estrogen administration, is

Handbook of Animal Models of Infection
ISBN 0-12-775390-7

required (Kinsman and Collard, 1986). Within 48 hours of estrogen administration the vaginal mucosa is composed of stratified, squamous epithelium overlaid by keratin. *C. albicans* only colonizes the vagina when the epithelium is fully keratinized. Induction of germ-tube (hyphal) formation, strong adherence and proteinase activity (see below) are made possible or greatly facilitated by the production of the keratin layer. In addition, at the estrous phase, there is no or low-grade leukocyte infiltration—a fact that greatly helps fungal growth and the establishment of infection. Finally, estrogen treatment reduces the consistency and possibly the antifungal activity of the columnar epithelial layers of the vagina (Fidel and Sobel, 1996).

The hormonal dependence of candidal vaginitis in rats has a precise correlate in the role of estrogen in human vaginitis. Although not exactly defined, this role can easily be inferred by the rarity of *Candida* vaginitis in premenarchal and post-menopausal women, while its incidence greatly increases and is exacerbated in the premenstrual week (Sobel and Muller, 1984).

In our model, oophorectomized Wistar rats weighing 80–100 g (Charles River Breeding Laboratories, Calco, VA, Italy) are maintained in pseudoestrus by injection of estradiol benzoate (0.5 mg subcutaneously; Benzatron; Samil, Rome, Italy) every 2 days (De Bernardis *et al.,* 1989).

Preparation of inocula

For the infectious challenge, cells of each *Candida* strain tested are grown in Winge broth (2% glucose; 0.3% yeast extract) for 48 hours at 28°C on a shaker at 200 rpm (New Brunswick Scientific Co, Edison, NJ). There is no evidence that medium selection plays a critical role in the establishment of infection. However, prolonged storage as stock cultures in artificial media may influence the vaginopathic potential of the strain (De Bernardis *et al.*, unpublished observations). Growth is measured by hemocytometer counts and the yeast suspension is appropriately diluted in physiological saline before animal inoculation. At least for *C. parapsilosis*, the source of the isolate (human vagina, blood, skin or environment) may be important for the extent of rat vaginal infection (De Bernardis *et al.*, manuscript in preparation 1999).

Infection procedure

Six days after the first estradiol dose, the animals are inoculated intravaginally with 10^7 *Candida* cells in 0.1 ml of saline solution, which is administered to each animal through a syringe equipped with a multipurpose calibrated tip (Combitip; PBI, Milan Italy). To favor intravaginal contact and adsorption of fungal cells, the rat is held head-down for 1 minute.

Key parameters to monitor infection

To monitor the infection vaginal fluids are taken from each animal as desired. However, since repeated tapering and lavages, mostly in the early days, may disturb the turnover of the keratinized epithelia to which *Candida* adheres, and because of the kinetics of the infection itself, we prefer to take fluids at a maximum of once every 2 days. Vaginal scrapings are taken with a calibrated (1 μl) plastic loop (Dispoinoc; PBI, Milan, Italy) inserted and removed from the vagina and streaked on a slide, as reported elsewhere (De Bernardis *et al.*, 1989). Samples are then air-dried and stained by the periodic acid–Schiff (PAS) method for microscopic examination, which reveals *Candida* multiplication and adherence, as well as hyphal growth and penetration of the keratinized layers (Figure 89.1). These cytohistological features are, together with assessment of fungal burden, critical criteria for evaluating infection rate and extent of tissue involvement.

Other fluids (one per rat) are used for measurement of colony-forming units (cfu). For this purpose, vaginal samples recovered with the calibrated loop are plated on

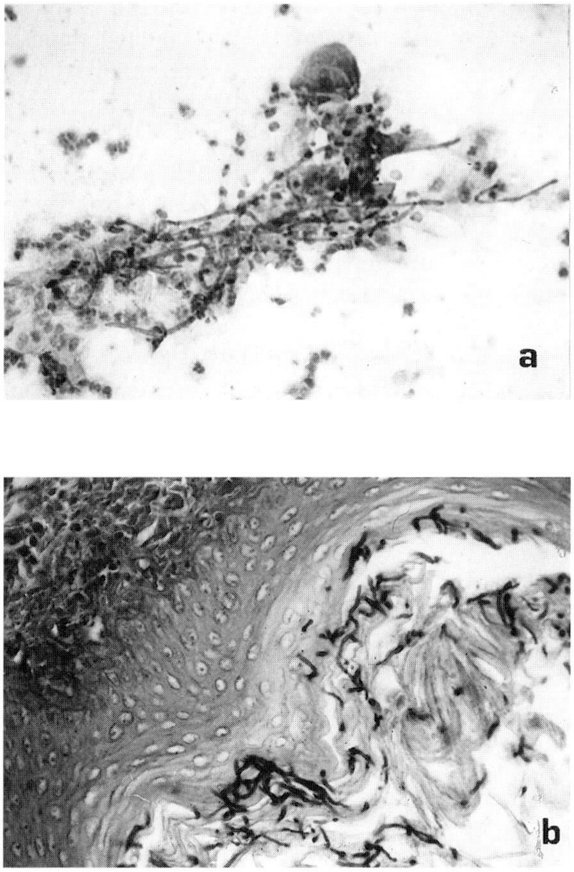

Figure 89.1 Periodic acid–Schiff-stained vaginal smear (**a**) and section of vagina (**b**) from rat infected with a vaginopathic strain of *Candida albicans*. Magnification (**a**) ×250 and (**b**) ×400.

Sabouraud dextrose agar containing the antibacterial antibiotic chloramphenicol (20 μg/ml) and plates are incubated at 30°C for 48 hours. This procedure offers a reproducible and consistent determination of fungal burden in vagina and avoids misreading due to bacterial contamination. Another useful procedure is washing the vaginal cavity with a given amount of sterile saline, then counting the cfu. In our initial approaches to this model, we compared the number of *Candida* colonies recovered by plating vaginal samples with a calibrated loop on Sabouraud agar with yeasts recovered using a vaginal-washing technique. Vaginal lavage generated lower counts than those obtained by using loop sampling, which eventually resulted in a more sensitive method of monitoring infection. By using the 1 μl loop sampling, vaginal fluids containing $<10^3$ cfu/ml are obviously negative and the corresponding animals are conventionally defined as "non-infected" or "cleared" from infection as opposed to the "infected" animals (from 10^3 to $>10^5$/ml of vaginal fluid).

For immunological studies, i.e., detection of antibodies, samples of vaginal fluids are taken at regular intervals from each animal by gently washing the vaginal cavity with 0.5 ml of sterile saline solution, as described elsewhere (Cassone *et al.*, 1995). The collected fluid is centrifuged at 3500*g* for 15 minutes in a refrigerated Biofuge and the supernatant is conveniently assayed.

For the histological examination of vaginal tissue, vaginas from euthanized *Candida*-infected rats and control are aseptically removed and fixed in 10% (vol/vol) formalin. Paraffin sections are examined after treatment with PAS and van Giesen stains. Cfu enumeration of *Candida* cells present in the whole vagina can also be used, but this obviously negates the possibility of continued use of the same animal, is expensive and is probably non-ethical.

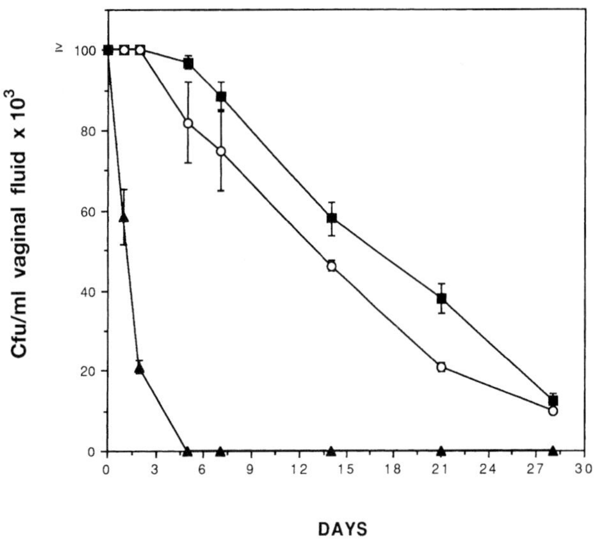

Figure 89.2 Experimentally induced rat vaginitis with clinical vaginal isolates of *Candida albicans* (squares), *Candida parapsilosis* (circles) and *Saccharomyces cerevisiae* (triangles). Kinetics of vaginal infections are expressed as mean colony-forming units (cfu) ± SE (bar)/ml vaginal fluid of infected (> 10^3 cfu/ml) rats.

Experimental vaginal infection

Figure 89.2 shows the kinetics of infection, i.e. the viable counts of representative isolates of *C. albicans*, *C. parapsilosis* and *Saccharomyces cerevisiae*. Both *C. albicans* and *C. parapsilosis* produce vaginal infection, with a sustained and prolonged high number of viable cells for around 1 week, followed by a declining period. In contrast, *S. cerevisiae* is eliminated from the vagina of the infected rats within 3–5 days after the intravaginal challenge. In our experience, this behavior is quite reproducible in distinguishing between vaginopathic and non-vaginopathic species and strains of *Candida* and other genera.

Advantages and disadvantages of the model

The rat model of *Candida* vaginitis is less used than the mouse model.

Some of its disadvantages include size, cost and maintenance of the animal and more difficult handling compared to the mouse. Moreover, the mouse is, at present, more useful for immunological studies because of the better basic knowledge of mouse immunology and availability of immunological tools together with genetically modified strains. However, in our hands, mouse infection is more severely dependent on continued estrogen treatment and *Candida* counts are more variable if the loop-sampling method is used. In fact, several authors prefer to use the whole vagina *Candida* counting, which necessitates animal sacrifice. The rat model avoids this substantial inconvenience and is more suitable for monitoring infection and reinfection.

Another great advantage of the rat model is the possibility of obtaining a relatively large amount of vaginal fluid and other materials (blood, for instance) for the studies of pathogenesis and immunity in candidiasis. It is expected that current research will soon cover the gap between the rat and mouse in the availability of gene knockout animals and immunological tools.

Contributions of the model to evaluate the efficacy of antifungal therapy

The studies of the efficacy of antifungal drugs in rat vaginitis offer more accurate data of the *in vivo* outcome than tests *in vitro* or in other models involving mice or other animals. Thus, the model has been largely used to study the therapeutic effect of antifungal agents in eradicating or controlling acute vaginitis (Mc Ripley *et al.*, 1979; Thienpont *et al.*, 1980; Ryley *et al.*, 1981; Sobel *et al.*, 1981; Sobel and Muller, 1983, 1984; Kinsman *et al.*, 1986; Van Cutsen *et al.*, 1987; Jansen *et al.*, 1991). In particular, several problems can be encountered in determining the sensitivity of *C. albicans* isolates to imidazole and triazole antifungals *in vitro*, as no standard methods exist. Moreover, the rat model uses animals without inducing systemic morbidity and lethality, so that the animals can be reused for other *in vivo* experiments.

Sometimes *C. albicans* isolates, which cause relapse, are no less sensitive to antimycotics than any other isolate *in vitro*. The sensitivity depends upon the method employed. The evaluation of drug sensitivity in the animal model of vaginitis gave additional confirmation and a more accurate correlation with the clinical observation. In fact, studies with animals offer one solution to the problem of *in vitro* compensation for the effect of host factors.

As described above, experimental *Candida* vaginitis in rats is a relatively short-lived (1–2 months), self-limiting and spontaneously healing infection, nonetheless, the model is reasonably good for the assessment of antifungal activity *in vivo* and has become a model correlate of human therapy of vaginal candidiasis.

Most studies involve prophylactic treatment with an antifungal given before or at the same time as the *Candida* inoculum. In therapeutic models it is rare for treatment to be delayed much beyond 24 hours after infection—quite the opposite of the usual situation in humans.

The route of administration of antifungal in animal tests is not always the same as that intended for humans: non-traumatic oral administration of drugs is a frequent problem with laboratory animals.

The data from animals may not be readily extrapolated to humans, for the pharmacokinetics of these antifungal agents in animals are usually different from those in humans.

Despite the limitations in the use of this and other animal models, they offer the most reliable information in the preclinical evaluation of antifungal drugs.

Contributions of the model to studying aspects of the pathogenesis and immunity in *Candida* vaginitis

The rat vaginitis model has been particularly useful for studying fungal virulence factors. For *C. albicans* (the most virulent and most frequently isolated species of *Candida*) the putative virulence attributes are dimorphism, adherence, enzyme secretion, phenotypic switching, antigen variation and possession of complement-binding receptors (Sobel, 1989; Cutler, 1991). However, the actual contribution of each of these factors to the pathogenesis and severity of the disease awaits elucidation. Recently, the use of mutant strains and their parental counterparts has been important to investigate the role of virulence factors of *Candida* in the pathogenesis of candidal vaginitis.

We studied the vaginopathic potential and the intravaginal morphology of a non-germinative variant of *C. albicans*, strain CA-2. This mutant expressed low virulence in systemic infections while, in contrast, being capable of causing a vaginal infection of the same duration and extent as that obtained in rats challenged with the parental germ-tube-forming strain *C. albicans*, 3153. During the experimental infection, the CA-2 cells did not maintain their yeast morphology but developed hyphal filaments (De Bernardis *et al.* 1993). The elevated vaginopathic potential of strain CA-2, in contrast to its low virulence in systemic infection, suggested that different *Candida* virulence factors come into play in local and disseminated candidal infections (De Bernardis *et al.*, 1993).

Secreted aspartyl proteinases (Sap) appear to be another virulence-associated attribute of *Candida* species (Cutler, 1991). Most members of this family of isoenzymes are expressed at acidic vaginal pH and can cleave several proteins which are important in host defense, such as antibodies of both immunoglobulin G (IgG) and IgA isotype. They may also promote the colonization, penetration and invasion by *C. albicans* (Hube, 1996).

Since, because of the pH constraint, the expression of these enzymes in mucosal surface may be more important than in systemic candidiasis, we have been studying Sap secretion and activity in experimental vaginitis. In the rat model, a clear correlation between the ability of strains to secrete aspartyl proteinase *in vivo* and to cause disease has been observed. In fact we found that a stable non-proteolytic mutant of *C. albicans* was less vaginopathic than the proteolytic parent strain (Ross *et al.*, 1990).

Experimental vaginitis with associated Sap secretion in vaginal fluids was produced in rats challenged with proteolytic species (*C. albicans* and *C. parapsilosis*) but not on challenge by non-proteolytic species (*C. glabrata*, *C. krusei*, *S. cerevisiae*; Agatensi *et al.*, 1991).

In this model we also studied the expression of two major Sap genes (*SAP1* and *SAP2*). Northern blot analysis with RNA extracted from the vaginal fluid of rats infected with highly vaginopathic strains demonstrated the expression of both *SAP1* and *SAP2* during the first week of infection. In contrast, neither gene was expressed during infection by a non-vaginopathic strain (which did not adhere to and infect vaginal cells to the same extent as other strains tested). The results demonstrated that both *SAP1* and *SAP2* were expressed *in vivo* during vaginal infections. It is important that these genes are expressed *in vivo* only by strains which are more virulent in this animal model (De Bernardis *et al.*, 1995).

The *in vivo* expression of SAP genes was also monitored by a polymerase chain reaction (PCR) detection with primers selected for each of the *SAP* genes. RNA extracted from vaginal fluids taken from rats at different days post-infection with *C. albicans* was converted in complementary

DNA and amplified by PCR. The identity of PCR products generated by amplification was confirmed by hybridization with the corresponding probes. Thus, *SAP1*, *SAP2* and *SAP3* were expressed by *C. albicans* during vaginal infection while no *SAP4*, *SAP5* and *SAP6* vaginal transcripts were found. Thus, the rat vaginitis model proved to be the first animal model of candidiasis where the expression of virulence genes could be rapidly and conveniently monitored during infection (De Bernardis *et al.*, 1997; Figure 89.3).

More refined and definitive evidence of the role of the virulence factors of *Candida* in the pathogenesis of infection may be obtained using this animal model to evaluate the virulence of gene knockout mutants and revertant fungal strains.

Animal models of *C. albicans* vaginitis have also been adopted to study host defense mechanisms associated with vaginal infections. In fact, some authors have suggested that women with recurrent vulvovaginal candidiasis have local — rather than systemic — specific immune deficiency leading to a high susceptibility to the disease (Sobel, 1989; Fidel and Sobel, 1996).

In particular, using the mouse model, Fidel and colleagues (1994, 1995a,b) showed that systemic cell-mediated immunity (CMI) was not protective against vaginitis, and suggested that local host defense mechanisms could be operative at the vaginal mucosa, as distinct from more general anti-*Candida* mechanisms.

However, the nature of the immunoprotective factors remains undefined (Steele *et al.*, 1997).

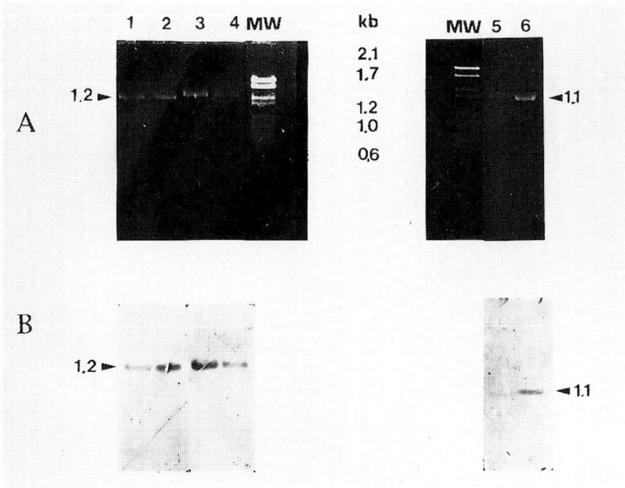

Figure 89.3 (A) RNA extracted from *Candida albicans* grown *in vitro* (lanes 1, 3 and 5) and from rat vaginal fluids 2 days post-infection with *C. albicans* (lanes 2, 4 and 6) was converted into DNA and amplified by polymerase chain reaction (PCR) with specific primers for *SAP1* (lanes 1 and 2), *SAP2* (lanes 3 and 4) and *SAP3* (lanes 5 and 6) genes. (B) The identity of PCR products generated by amplification was confirmed by hybridization with the corresponding probes.

In the rat model of vaginitis we showed that, after clearing the primary *C. albicans* infection, the animals were highly resistant to a second vaginal challenge with the fungus. The vaginal fluid of *Candida*-resistant rats contained antibodies directed against mannan constituents and Sap enzymes of *C. albicans* and was capable of transferring a degree of anti-*Candida* protection to naive, non-immunized rats (Cassone *et al.*, 1995).

Pre-absorption of the antibody-containing fluids with either — or both — mannan and proteinase sequentially reduced or abolished the level of protection. A degree of protection against vaginitis was also conferred by post-infection intravaginal administration of anti-proteinase and anti-mannan monoclonal antibodies and by intravaginal immunization with a mannan extract or with a highly purified *SAP2* preparation. No anti-*Candida* antibodies were elicited during *C. albicans* vaginal infection of congenitally athymic nude rats and these animals did not show increased resistance to a rechallenge, demonstrating that induction of anticandidal protection in normal rats was a thymus-dependent antibody response. Overall, the results obtained using the rat vaginitis model demonstrated that antibodies against some defined *Candida* antigens are relevant in the mechanism of acquired anti-candidal protection in vaginitis (De Bernardis *et al.*, 1997).

In conclusion, our data show that the rat vaginitis model is perfectly suitable to study the vaginal immunity to *Candida*. It is hoped that genetic manipulation in the rat to a level similar to the one currently achievable with the mouse will soon bring about the discovery of mechanisms of immune response to *Candida* in the vagina, in order to facilitate evidence-based and rational approach to immunotherapy or immunoprophylaxis of *Candida* vaginitis in humans.

Acknowledgments

The authors wish to thank Drs M. Boccanera, L. Morelli and D. Adriani for their excellent technical support and participation in the referenced work. Special thanks are due to Mrs G. Mandarino, for help in the preparation of the manuscript. The authors' work quoted in this paper has been supported by a grant from the National AIDS Project, contract no. 10/A/U.

References

Agatensi, L., Franchi, F., Mondello, F. *et al.* (1991). Vaginopathic and proteolytic *Candida* species in outpatients attending a gynaecology clinic. *J. Clin. Pathol.*, **44**, 826–830.

Cassone, A., Boccanera, M., Adriani, D., Santoni, G., De Bernardis, F. (1995). Rats clearing a vaginal infection by *Candida albicans* acquire specific antibody-mediated resistance to vaginal reinfection. *Infect. Immun.*, **63**, 2619–2624.

Cutler, J. E. (1991). Putative virulence factors of *Candida albicans*. *Annu. Rev. Microbiol.*, **45**, 187–218.

De Bernardis, F., Lorenzini, R., Morelli, L., Cassone, A. (1989). Experimental rat vaginal infection with *Candida parapsilosis* from outpatients with vaginitis. *J. Clin. Microbiol.*, **27**, 2598–2603.

De Bernardis, F., Adriani, D., Lorenzini, R., Pontieri, E., Carruba, G., Cassone A. (1993). Filamentous growth and elevated vaginopathic potential of a non-germinative variant of *Candida albicans* expressing low virulence in systemic infection. *Infect. Immun.*, **61**, 1500–1508.

De Bernardis, F., Cassone, A., Startevant, J., Calderone, R., (1995). Expression of *Candida albicans* SAP1 and SAP2 in experimental vaginitis. *Infect. Immun.*, **63**, 1887–1892.

De Bernardis, F., Boccanera, M., Adriani, D., Spreghini, E., Santoni, G., Cassone, A. (1997). Protective role of antimannan and anti-asparyl proteinase antibodies in an experimental model of *Candida albicans* vaginitis in rats. *Infect. Immun.*, **65**, 3399–3405.

Fidel, P. L., Sobel, J. D. (1996). Immunopathogenesis of recurrent vulvovaginal candidiasis. *Clin. Microbiol. Rev.*, **9**, 335–348.

Fidel, P. L., Lynch, M. E., Sobel, J. D. (1993). *Candida*-specific cell-mediated immunity is demonstrable in mice with experimental vaginal candidiasis. *Infect. Immun.*, **61**, 1032–1038.

Fidel, P. L., Lynch, M. E., Sobel, J. D. (1994). Effect of preinduced *Candida*-specific systemic cell-mediated immunity on experimental vaginal candidiasis. *Infect. Immun.*, **62**, 1032–1038.

Fidel, P. L., Cutright, J. L., Sobel, J. D. (1995a). Effects of systemic cell-mediated immunity on vaginal candidiasis in mice resistant and susceptible to *Candida albicans* infections. *Infect. Immun.*, **63**, 4191–4194.

Fidel, P. L., Lynch, M. E., Sobel, J. D. (1995b). Circulating CD4 and CD8 T-cells have little impact on host defence against experimental vaginal candidiasis. *Infect. Immun.*, **63**, 2403–2408.

Hube, B. (1996). *Candida albicans* secreted aspartyl proteinases. *Curr. Top. Med. Mycol.*, **7**, 55–69.

Ichise, K., Tanio, T., Saji, I., Okuda, T. (1986). Activity of SM-4470, a new imidazole derivative, against experimental fungal infections. *Antimicrob. Agents Chemother.*, **30**, 366–369.

Jansen, T. M., Van De Ven, M. A., Borgers, M. J., Odds, F. C., Van Cutsen, J. M. P. (1991). Fungal morphology after treatment with itraconazole as a single oral dose in experimental vaginal candidosis in rats. *Am. J. Obstet. Gynecol.*, **165**, 1552–1557.

Kinsman, O. S., Collard, A. E. (1986). Hormonal factors in vaginal candidiasis in rats. *Infect. Immun.*, **53**, 498–504.

Kinsman, O. S., Collard A. E., Savage, T. J. (1986). Ketoconazole in experimental vaginal candidosis in rats. *Antimicrob. Agents Chemother.*, **30**, 771–773.

McRipley, R. J., Schwind, R. A., Erhard, P. J., Whitney, R. R. (1979). Evaluation of vaginal antifungus formulations *in vivo*. *Postgrad. Med. J.*, **55**, 648–652.

Ross, J. K., De Bernardis, F, Emerson, G. W., Cassone, A., Sullivan, P. A. (1990) The secreted aspartate proteinase of *Candida albicans*: physiology of secretion and virulence of a proteinase-deficient mutant. *J. Gentile. Microbiol.*, **136**, 687–694.

Ryley, J. F., McGregor, S. (1986). Quantification of vaginal *Candida albicans* infections in rodents. *J. Med. Vet. Mycol.*, **24**, 455–460.

Ryley, J. F., Ryley, N. G. (1990). *Candida albicans* – do mycelia matter? *J. Med. Vet. Mycol.*, **28**, 225–239.

Ryley, J. F., Wilson, R. G., Gravestock, M. B., Poyser, J. P. (1981). Experimental approaches to antifungal chemotherapy. *Adv. Pharm. Chemother.*, **18**, 49–176.

Ryley, J. F., Wilson, R. G., Barrett-Bee, K. S. (1984). Azole resistance in *Candida albicans*. *Sabouraudia*, **22**, 53–63.

Scholer, H. J. (1960). Experimentalle vaginal candidiasis der ratte. *Pathol. Microbiol.*, **23**, 62–68.

Sobel, J. D., (1985). Epidemiology and pathogenesis of recurrent vulvovaginal candidiasis. *Am. J. Obstet. Gynecol.*, **152**, 924–935.

Sobel, J. D., (1989). Pathogenesis of *Candida* vulvovaginitis. In *Current Topics in Medical Mycology* (eds McGinnis, M. R., Borges, M.), pp. 86–108. Springer-Verlag, Stuttgart, Germany.

Sobel, J. D., Muller, G. (1983). Comparison of ketoconazole, BAY N7133 and BAY L9139 in the treatment of experimental vaginal candidiasis. *Antimicrob. Agents Chemother.*, **24**, 434–436.

Sobel, J. D., Muller, G. (1984). Comparison of itraconazole and ketoconazole in the treatment of experimental candidal vaginitis. *Antimicrob. Agents Chemother.*, **26**, 266–267.

Sobel, J. D., Meyers, P. G., Kaye, D., Levison, M. E. (1981). Adherence of *Candida albicans* to human vaginal and buccal epithelial cells. *J. Infect. Dis.*, **143**, 76–83.

Sobel, J. D., Muller, G., Buckley, H. R. (1984). Critical role of germ tube formation in the pathogenesis of candidal vaginitis. *Infect. Immun.*, **44**, 576–580.

Sobel, J. D., Muller, G., McCormick, J. F. (1985). Experimental chronic vaginal candidosis in rats. *Sabouraudia, J. Med. Vet. Mycol.*, **23**, 199–206.

Steele, C., Ozenci, H., Luo, W., Scott, M., Fidell, P. L. (1997). Anti-*Candida* host resistance by vaginal epithelial cells. Abstract P 15. Fourth NIAID Workshop in Medical Mycology, Lake Tahoe, CA, August, 1997.

Taschdjian, C. L., Reiss, F., Kozinn, P. J. (1960). Experimental vaginal candidiasis in mice; its implications for superficial candidiasis in humans. *J. Invest. Dermatol.*, **34**, 89–94.

Thienpont, D., Van Cutsen J., Borger, S. M. (1980). Ketoconazole in experimental candidosis. *Rev. Infect. Dis.*, **2**, 570–577.

Van Cutsen, J., Van Gerven, F., Janssen, A. J. (1987) Activity of orally, topically and parenterally administered itraconazole in the treatment of superficial and deep mycoses animal models. *Rev. Infect. Dis.*, **9** (Suppl. 1), S15–S32.

Chapter 90

Murine Models of *Candida* Vaginal Infections

P. L. Fidel Jr. and J. D. Sobel

Background on the human infection

Vulvovaginitis (VVC) is a significant problem in women of child-bearing age. It can be expected that 75% of all women will experience at least one episode of VVC during their lifetime, with between 5 and 10% of women experiencing recurrent vulvovaginal candidiasis (RVVC; Sobel, 1988). Greater than 85% of vaginal fungal infections are caused by the dimorphic fungal organism, *Candida albicans*, with *C. glabrata*, *C. tropicalis*, *C. parapsilosis*, and *C. krusei* identified in the remaining diagnosed symptomatic infections (Sobel, 1988). *C. glabrata*, the lone non-dimorphic *Candida* species (formerly known as *Torulopsis glabrata*), is the most common non-*albicans Candida* species identified and often occurs under conditions of diabetes mellitus (Sobel, 1988; Redondo-Lopez *et al.*, 1990). Signs and symptoms of vaginitis include vaginal and vulvar erythema and/or edema, vaginal itching, burning, soreness, an abnormal cottage-cheese-like discharge, and dyspareunia (Sobel, 1988).

C. albicans is considered a vaginal commensal organism with detectable asymptomatic colonization rates in approximately 15–25% of normal healthy women (Sobel, 1988). Although there are several predisposing factors for acute episodes of VVC (i.e., antibiotic usage, oral contraceptives, pregnancy, uncontrolled diabetes mellitus, and tight underclothing), chronic or recurrent VVC is often unexplained or idiopathic (Sobel, 1988). It is proposed that women with idiopathic RVVC share some immunological deficiency or dysfunction that enhances susceptibility to the local infection (Fidel and Sobel, 1996), but no specific systemic or local deficiency has been identified to date. Nevertheless, most cases of symptomatic acute or chronic vaginitis require antimycotic drug intervention. There is no evidence, though, that antifungal drug resistance contributes to the susceptibility to repeated episodes of RVVC caused by *C. albicans* (Lynch *et al.*, 1996).

Background of the model

Animal models have been widely used to test experimental topical or systemic antifungal therapies for a variety of fungal infections, both systemic and mucosal. Included are the rodent models of experimental *C. albicans* and *C. glabrata* vaginitis (Sobel *et al.*, 1985; Ryley and McGregor, 1986; Fidel *et al.*, 1996a). In fact, the original rodent vaginitis models were developed for the sole purpose of testing therapeutic agents (McRipley *et al.*, 1979; Sobel *et al.*, 1985; Ryley and McGregor, 1986). A critical property of the model is the dependence upon estrogen (pseudoestrus) for the infection to be acquired and persist long enough to evaluate therapeutic regimens (Taschdjian *et al.*, 1960; Sobel *et al.*, 1985; Ryley and McGregor, 1986). Despite this somewhat artificial requirement, the experimental animal models have been extremely useful and predictive of clinical efficacy. Recently, experimental vaginitis models have been expanded to examine immunological issues in the pathogenesis of *Candida* vaginitis (Fidel *et al.*, 1993a,b, 1994, 1995b; Cassone *et al.*, 1995; De Bernardis *et al.*, 1997). Specifically, the models have been employed to study systemic- and mucosal-associated innate and acquired host defense mechanisms, including effector function by phagocytes, antibodies, and T cells. This chapter will review the technical aspects of the experimental models of *C. albicans* and *C. glabrata* vaginal infections in mice.

Animal species

Female mice have been employed in the experimental model of *C. albicans* vaginal infection. The age of the animals can vary, but should be >6 weeks of age. In general, most mice used for vaginitis are >20 g. The infection in mice is relatively easy to induce and numerous strains (CBA/J, DBA/2, C57BL/6, C3H/HEN, BALB/c) covering several haplotypes ($H\text{-}2^{k,b,d}$) can be employed (Fidel *et al.*, 1993a, 1995a). In contrast to experimental models of *C. albicans* systemic infection (Hector *et al.*, 1982), there does not appear to be any difference in susceptibility of various strains of mice to experimental *C. albicans* vaginitis (Fidel *et al.*, 1995b). In contrast to the relative ease of inducing experimental *C. albicans* vaginal infections, it is extremely difficult to induce experimental *C. glabrata* infections. To date, the *C. glabrata* vaginal infections have only been

Handbook of Animal Models of Infection
ISBN 0-12-775390-7

induced in mice genetically susceptible to a hyperglycemic condition (non-obese diabetic mice, NOD/Lt; Fidel *et al.*, 1996a), or CBA/J mice given streptozocin to generate an artificial state of hyperglycemia (Fidel, unpublished observations).

Preparation of the animals

An important prerequisite for a *C. albicans* or *C. glabrata* vaginal infection is a persistent hormonal state of estrus. Since rodents have a short 4-day estrus cycle with a state of estrus lasting for not more than 2 days at a time, the mice must be placed in a state of constant estrus (pseudoestrus) with or without ovariectomy. The estrogen converts the columnar epithelium to thick stratified squamous epithelium (Figure 90.1), and increases the glycogen content, pH, and growth substrates (CHO, N_2), all of which increases the avidity of the yeast for the tissue and enhances its growth (Powell and Drutz, 1983; Kinsman and Collard, 1986). Additionally, estrogen is known to inhibit innate and/or acquired host defenses (Carlsten *et al.*, 1991; Styrt and Sugarman, 1991). Together, the organism grows in population numbers under the pseudoestrus condition, overwhelms any remaining host defense mechanisms, and causes a superficial infection of the vaginal squamous epithelium. This critical role of estrogen in the animals is analogous to its role in humans. VVC is rare in premenarchal females and occurs considerably less frequently in postmenopausal women unless they are receiving estrogen hormone replacement therapy (Sobel, 1988).

To induce the pseudoestrus condition, estradiol valerate (Sigma) is dissolved in sesame oil (this requires heating at 37°C with intermittent vortexing) and injected subcutaneously into the abdomen of each mouse in a volume of 0.1 ml (Ryley and McGregor, 1986; Fidel *et al.*, 1993a; Valentin *et al.*, 1993). The first injection is usually given 48–72 hours prior to organism inoculation and should continue weekly for the duration of the experiment. In addition to the pseudoestrus-mediated histological changes within the tissue (Figure 90.1), one can usually visualize the pseudoestrus condition as the anterior vaginal tissue shows visual signs of swelling and redness. A wide range of estrogen concentrations (0.01–0.5 mg/ms) can be used depending upon the condition desired. For *C. albicans* infections, 0.2–0.5 mg/ms should be used if a long-lived persistent infection is desired (i.e., efficacy of therapeutic agents). On the other hand, 0.01–0.02 mg/ms should be employed if a less persistent infection is warranted (3–4 weeks; i.e., immunological analyses). Figure 90.2 illustrates the relative

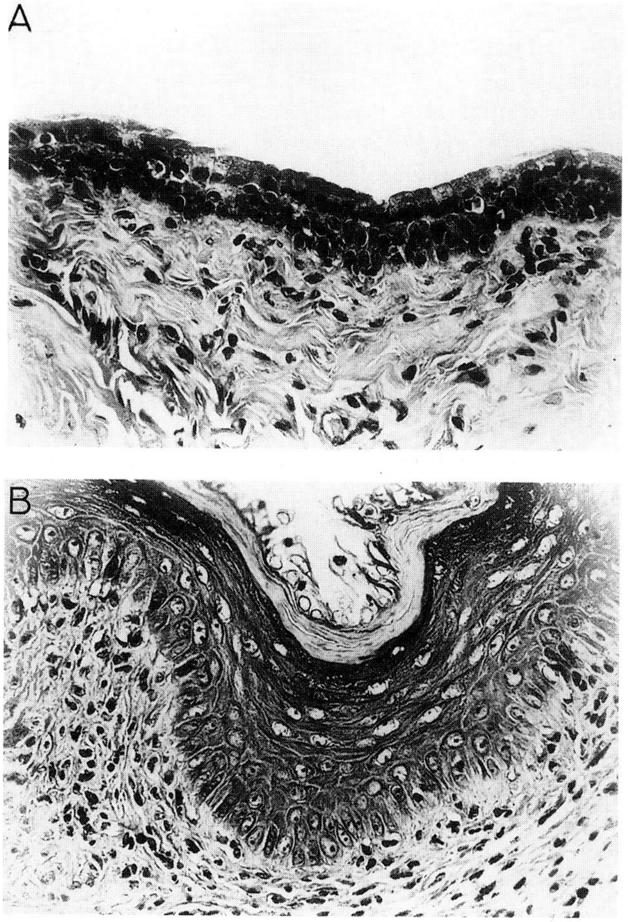

Figure 90.1 Histological illustration of the mouse vaginal mucosa and submucosa under estrogen and non-estrogen conditions. (A) Vaginal section under non-estrogenized conditions revealing columnar epithelium. (B) Vaginal section under estrogenized conditions (0.5 mg/week) revealing keratinized squamous epithelium.

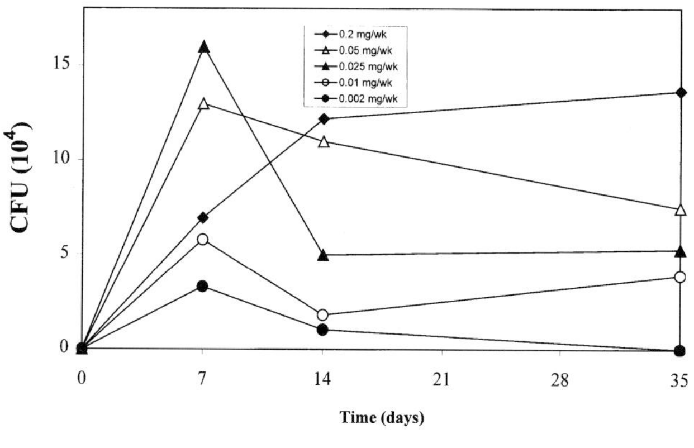

Figure 90.2 Vaginal *Candida albicans* titers in mice under varying estrogen concentrations. Mice were induced into a state of pseudoestrus with varying concentrations of estrogen (0.002–0.2 mg/ms per week) injected weekly beginning 48 hours prior to vaginal inoculation. Animals were inoculated with 5×10^4 *C. albicans* blastospores from a stationary-phase culture. At specified times, the animals were sacrificed, vaginal lavages were performed, and quantitative lavage cultures were used to assess vaginal fungal burden. This figure shows the mean colony-forming units (cfu) per group of animals ($n = 5$–10).

vaginal fungal burden during a *C. albicans* infection in mice under various estrogen concentrations. The levels of estrogen can also be decreased or terminated indefinitely during the course of an infection. In such cases, the vaginal fungal titers will decline or become undetectable over time depending on the experimental design. It should be recognized, though, that depending on the weekly concentration employed, the effects of estrogen can persist for approximately 2–3 weeks following its termination. Since *C. glabrata* vaginal infections are difficult to achieve in rodents (Sobel, 1988; Redondo-Lopez *et al*., 1990), subcutaneous injections of 0.2–0.5 mg/ms per week estradiol into NOD mice is suggested (Fidel *et al*., 1996a). The estrogen is usually made fresh for each injection, but is stable in sesame oil for several months at room temperature.

In the absence of pseudoestrus, a short-lived (2–3 weeks) experimental *C. albicans* and *C. glabrata* vaginal infection will ensue with fungal titers in most animals declining to undetectable levels within the second week post-inoculation (Fidel *et al*., 1993a, 1996a). Moreover, the rates of infection in a group of mice are usually erratic under such conditions due to the varied levels of endogenous estrogen.

Infection procedure

Storage and preparation of inocula

Laboratory, commercial, or clinical isolates can be used to induce the experimental vaginal infection providing they are virulent strains. The sources of these isolates (American Type Culture Collection, clinical infections, etc.) usually insure virulence. *C. albicans* or *C. glabrata* isolates can be frozen indefinitely in 50% glycerol or litmus milk freezing medium (Becton Dickinson). The organisms can also be subcultured for long periods without any detectable loss of pathogenicity. In preparation of the infecting inocula, the organisms should be freshly subcultured for 48 hours. Several colonies are then inoculated into 1% phytone-peptone broth–0.1% dextrose and incubated for 18–24 hours at 25°C in a shaking water bath to establish a stationary-phase culture. Finally, the organisms are collected, washed twice in phosphate-buffered saline (PBS), counted using a hemocytometer, and resuspended in PBS to the desired concentration.

Anesthesia and inoculation

Mice are usually anesthetized with ether for a short period of time (20–30 seconds), during which 20 μl of the inocula is administered from a pipetman into the vaginal lumen. The *C. albicans* inocula for mice is usually 5×10^4–5×10^5 stationary-phase blastospores (2.5×10^6–2.5×10^7 cells/ml; Fidel *et al*., 1993a), while the *C. glabrata* inocula is usually 1×10^7 blastospores (5×10^8 cells/ml; Fidel *et al*., 1996a).

All inocula should be routinely serially diluted and plated on Sabouraud-dextrose agar to verify the actual inocula in the animals. Our experience shows that the hemocytometer count and actual inocula can vary slightly.

Monitoring of infection

There are several methods to monitor the *C. albicans* or *C. glabrata* vaginal infection. The most effective method is a combination of quantitative and qualitative measures. The quantitative method of choice is the culture of vaginal lavage fluid. To achieve this, the animals are sacrificed or anesthetized and subjected to a vaginal lavage using 100 μl sterile PBS. The lavage is performed with gentle to intermediate agitation and frequent aspiration. The lavage fluid is serially diluted (usually 1:10) and plated on to Sabouraud dextrose agar using 50–100 μl per plate. The plates are incubated at 30–35°C for 48 hours and enumerated. The mean colony-forming units (cfu) are tabulated per 100 or 300 μl, respectively, for each animal. The mean cfus are then tabulated for each group ± SEM.

An additional quantitative measure of vaginal fungal burden can be achieved by the homogenization of excised vaginal tissue, followed by a similar plating of serially diluted tissue homogenates. The cfu can be expressed as a function of volume of homogenate or, more appropriately, by the gram tissue weight.

A qualitative or semiquantitative culture can also be achieved by the swab technique. This is somewhat difficult in mice due to the small size of the vaginal lumen relative to that of a cotton swab. Nevertheless, if performed, the vaginal lumen is sampled with a dampened sterile cotton swab and either plated directly onto Sabouraud-dextrose agar (qualitative) or placed in 0.3 ml of PBS for serial dilution and plating (semiquantitative). The cfus are tabulated as described above. It should be stressed that, in any of these culture techniques, a certain level of bacterial contamination from the vagina is inevitable. However, the bacterial organisms will be limited on the Sabouraud's agar and, although they may be detectable, will not affect the growth of the yeast. Alternatively, antibiotics (chloramphenicol, gentamicin, etc.) can be added to the agar to inhibit any bacterial growth.

A quick qualitative measure of infection by *C. albicans* can be achieved with a wet-mount slide preparation of the lavage fluid (10 μl) and scoring the relative degree of hyphae. Since *C. albicans* is a dimorphic organism whose mycelial phase occurs at 37°C, the majority of organisms causing an infection in the animals will be in the form of hyphae rather than blastospores. Hyphal formation is easily detectable either free in the fluid or more often associated with sheets of epithelial cells. The hyphae can be qualitatively scored (0–4+) for the relative intensity of the infection. Figure 90.3 shows *C. albicans* hyphae amidst a squamous epithelial cell in vaginal lavage fluid.

The most common qualitative measure of infection,

although more labor intensive, is histological analysis. For this the animal is sacrificed and the vaginal tissue is excised, fixed in formalin, and serially sectioned using a cryostat. Sections can be silver-stained (PAS) using a commercial kit (Sigma) or stained with hematoxylin and eosin (H&E). Due to the superficial nature of the vaginal infection, the *C. albicans* hyphae can be very focal in the vaginal lumen. Thus, it is critical that several serial sections be examined before interpretations are made. Histological analyses are also useful for detection of *C. glabrata*. Since *C. glabrata* do not form hyphae, one must identify the blastospores. Figure 90.4A shows a representative PAS stain of *C. albicans* hyphae superficially associated with the vaginal mucosa of mice, while Figure 90.4B illustrates a similar superficial association by *C. glabrata*.

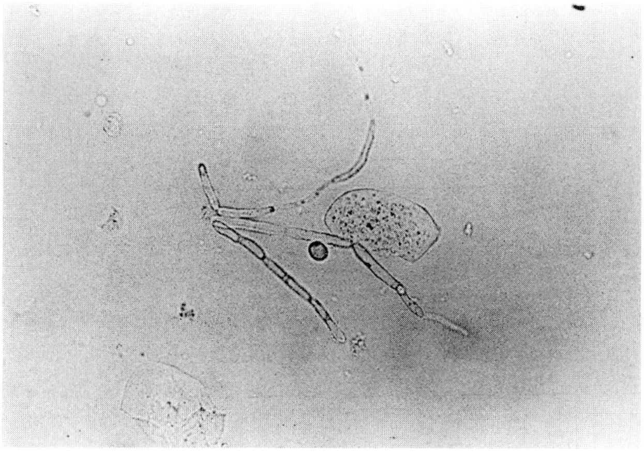

Figure 90.3 *Candida albicans* hyphae in vaginal lavage fluid from an infected mouse. A sample of lavage fluid from an infected mouse shows hyphae amidst epithelial and leukocytic cells. Magnification 400×.

Antimicrobial therapy

Several topical and systemic antifungal regimens have been employed for *C. albicans* and *C. glabrata* vaginal infections in mice. Usually a 10–30-day infection experiment is performed, with treatments beginning on day 4–5 post-infection and continuing once or twice daily for a predetermined period depending on the drug and pharmacokinetics. In most cases a 10–12-day infection is used with the drug terminated at a predetermined time prior to sacrifice of the animals. Longitudinal analyses can also be performed in order to examine the long-term effects of a particular drug.

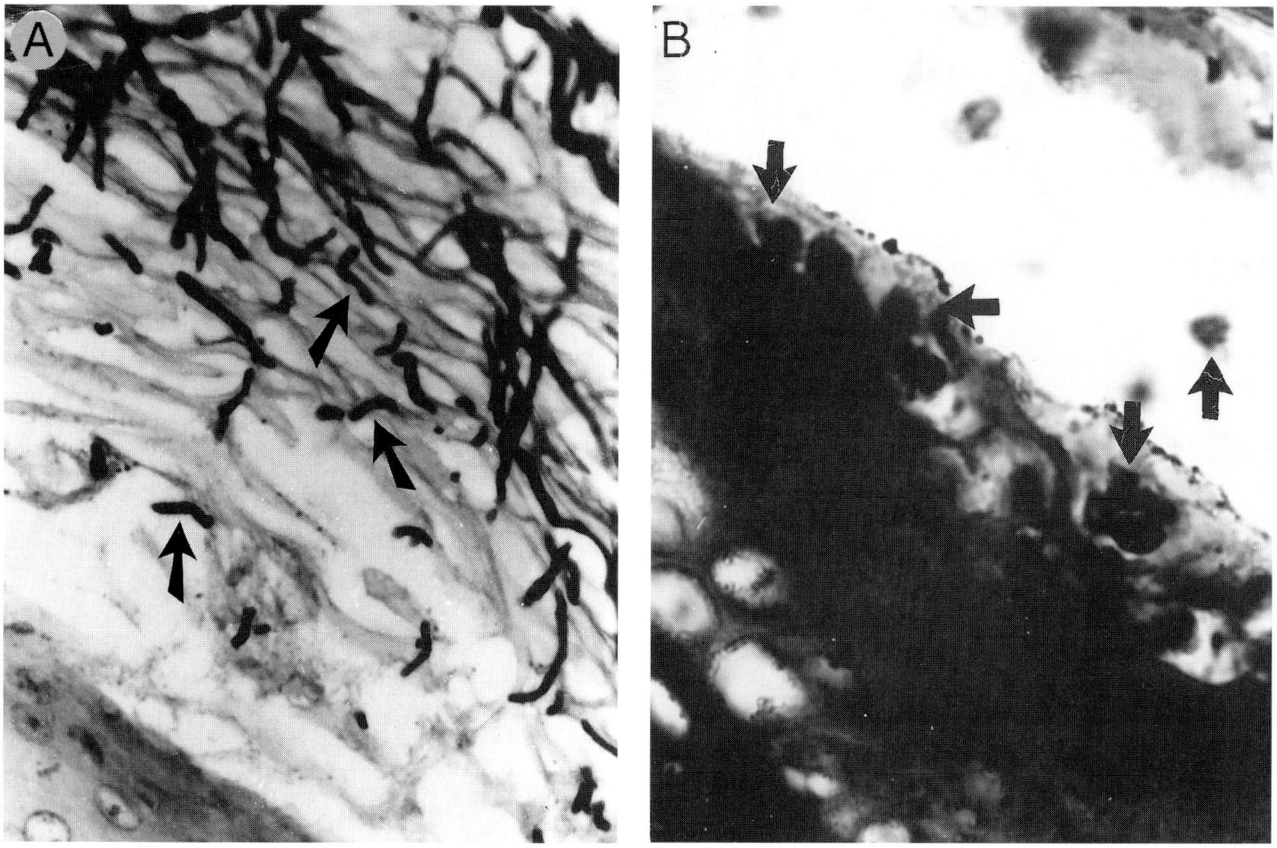

Figure 90.4 Silver (periodic acid–Schiff) stain of *Candida albicans* and *C. glabrata* in vaginal sections of infected mice under pseudoestrus conditions. (A) *C. albicans*-infected mice. Hyphae are shown superficially associated with the vaginal epithelium. (B) *C. glabrata*-infected mice. *C. glabrata* blastospores are shown superficially associated with the vaginal epithelium. Arrows point to the representative organisms.

However, it is not recommended that vaginal lavages (method of choice for determinating fungal burden) be performed longitudinally or more than once in the same animal, although economics and feasibility might dictate the contrary. Despite evidence that considerable numbers of organisms remain in the vaginal lumen after multiple lavages, the effects the lavage may have on the vaginal environment make longitudinal results showing a decrease in fungal titers in the same animal difficult to interpret. Thus, separate groups of animals for each time point are recommended for comparison at each time point when quantitative lavage cultures are being employed. On the other hand, longitudinal swabbing can be used and may be the method of choice when separate groups of animals are not feasible. Regardless of the method of monitoring the infection, a minimum of 5–10 animals should be included in a group for each therapeutic variable.

Topical treatments for mice usually involve the administration of the drug in an oil-based vehicle (i.e., polyethylene glycol) given in a small volume (20 μl) using a pipetman or small gavage needle. Topical treatments with creams can also be used and are administered through a syringe and gavage needle. Oral treatment is administered to restrained animals by gavage needle in a volume of 100 μl. Diluents for oral treatment are usually water-based, supplemented with HCl or Tween 80 (1%), but can be oil-based as well. It is important to include the appropriate positive and negative controls in each experiment. These include a positive control for infection (no therapeutic treatment regimen) and a negative control for the treatment regimen, in which animals are given the diluent under the identical treatment regimen.

Most antifungal drugs used in mice are usually tested in several doses and formulations for preclinical evaluation as the first measure of *in vivo* efficacy. Regimens are selected somewhat arbitrarily with regard to dose (dry concentration), frequency of administration, and duration of experimental therapy. Such preliminary studies are termed dose–range analyses. Those drugs and formulations demonstrating the greatest efficacy in the experimental model are potentially used in clinical trials providing that they exhibit no or minimal levels of animal toxicity. Marketed drugs are often then employed again in the experimental models for comparison with new antifungal preparations/formulations.

Systemic treatments for experimental *C. albicans* vaginal infections in mice have included fluconazole, ketoconazole, itraconazole, and most recently the triazole D0870 (Zeneca). The daily or single oral treatments of ketoconazole, itraconazole, and fluconazole (10–100 mg/kg) have been effective, with itraconazole showing superior clearance rates in mice (Fidel *et al.*, 1997; Richardson *et al.*, 1985; Fidel, unpublished observations). This may be reflective of the fact that fluconazole has a very short half-life compared to itraconazole and ketoconazole in mice. As a result, higher doses of fluconazole compared to itraconazole and ketoconazole are required in mice to achieve equivalent therapeutic efficacy. In contrast, the longer half-life of fluconazole in humans indicates equivalent or even greater effectiveness at similar doses of each drug. One of the most effective oral treatments to date has been with D0870 (Zeneca; Fidel *et al.*, 1997). In a study comparing its effects to fluconazole, D0870 showed clear superiority to fluconazole using a 10-fold lower concentration (Fidel *et al.*, 1997). Sterile lavage cultures were achieved at doses as low as 2.5 mg/kg given daily, but statistical efficacy was achieved at doses as low as 0.5 mg/kg (Fidel *et al.*, 1997). D0870 was also effective against fluconazole-resistant clinical isolates, albeit at higher concentrations (25 mg/kg; Fidel, *et al.*, 1997).

Topical treatments have included miconazole, butoconazole, and itraconazole. Results showed that daily treatment with each (10–25 mg/kg) has considerable efficacy against experimental *C. albicans* vaginitis (Valentin *et al.*, 1993; Fidel, unpublished observations).

In contrast to the effects of these drugs on *C. albicans* experimental vaginitis, no drug to date has been effective for experimental *C. glabrata* vaginitis, including D0870 (Fidel *et al.*, 1997) and boric acid which is often used clinically (Redondo-Lopez *et al.*, 1990; Sobel and Chaim, 1997).

The results of azole antifungal therapy in the treatment of experimental *C. albicans* vaginitis in rodents are generally consistent with results predicted on the basis of *in vitro* minimal inhibitory concentration (MIC) for each (Fidel *et al.*, 1997). This appears to include antifungal resistance as well, since positive correlations have been made between the lack of *in vitro* effectiveness for fluconazole and the lack of its efficacy against the *C. albicans* experimental vaginal infection induced by isolates defined as fluconazole-resistant by clinical outcome (Fidel *et al.*, 1997). Together, the *in vitro* or *in vivo* experimental results are generally a good indication of its clinical potential, provided that toxicity studies and human pharmacokinetics are favorable. The only exception to date in the animal models has been with therapy of *C. glabrata* experimental vaginal infection. Although *in vitro* MICs showed that *C. glabrata* was susceptible to D0870, D0870 was ineffective in the murine model (Fidel *et al.*, 1997). More studies are required to understand this general lack of antifungal efficacy against the experimental *C. glabrata* infection.

Key parameters to monitor response to treatment

The most important parameter in monitoring a response to treatment is to utilize a reproducible technique to quantitate vaginal fungal burden. To date, the most reliable technique is quantitative vaginal lavage cultures. Quantitative lavage cultures are relatively simple to perform, are more sensitive than swabbing, and can reliably identify significant differences in vaginal fungal titers. Although the ultimate response to treatment is undetectable vaginal fungal burden, it should be recognized that this is infrequently observed with fungistatic agents such as the azoles. Therefore, one must be prepared to measure efficacy through statistically significant reductions in vaginal fungal titers. With respect to data analysis, since the vaginal fungal titers can be quite variable within a group of animals, a non-parametric test such as the Mann–Whitney U rank sum test is usually employed to evaluate significance, although parametric analysis of variance (ANOVA) and Student's *t*-test may also be acceptable. A log or greater difference in fungal load can be significant depending on the variability between animals. It is suggested that a minimum of 5 mice per group are included, although 10 mice per group will increase confidence in the results. It is also suggested that experiments be repeated at least once under identical conditions using the appropriate positive and negative controls.

Pitfalls of the model

The major pitfall of the mouse model of vaginitis is the requirement for pseudoestrus. Although estrogen facilitates and guarantees a predictable and persistent infection to collect therapeutic or immunological data, its presence creates an artificial setting of infection in as much as the promoting influence of estrogen may overwhelm the natural immune response or mask weaker ameliorating effects of the antifungal therapy. Pretreatment with reduced estrogen concentrations may remedy these problems, but may also result in lower rates of infection and a less predictable infection. A second disadvantage of the model is the relatively small size of the animal. Although the use of mice is feasible for inclusion of larger numbers of animals/group, sampling can be difficult from the small vaginal lumen. Third, the size of the vaginal lumen in mice poses difficulties for testing solid topical agents (i.e., creams, etc.). Lastly, there are several distinctions between humans and rodents that may preclude predicting efficacy for humans based on the results in animals. In contrast to humans, no detectable yeast colonization has been detected in rodents (Fidel *et al.*, 1993a). Thus, *C. albicans* is not considered a commensal organism of the rodent mucosa. This is further supported by the lack of demonstrable immunity in uninfected mice (antibodies, and lymphocyte proliferation or cutaneous skin test reactivity in response to *Candida* antigen) that is detectable in 80–90% of normal healthy uninfected humans. Other differences include the high pH of the vagina in mice (7–9) compared to the acid pH of humans (4–4.5) and the pH-dependent differences in bacterial flora, and the potential differences in pharmocokinetics of specific drugs in each species. With respect to pharmacokinetics, short half-lives of certain drugs in animals may result in poor to intermediate efficacy despite exquisite *in vitro* efficacy. For example, fluconazole shows excellent *in vitro* efficacy and has the best record in humans (Rex *et al.*, 1997), yet is unimpressive in mice, where its half-life is very short (Fidel *et al.*, 1997). Also of importance is the keratinization of the vaginal epithelium that occurs in the mouse vagina but that is completely absent in humans. Thus, although the actual infection is similar to humans (a prerequisite of an animal model for testing preclinical treatment regimen), difficulties may arise in interpreting the results in mice as a predictive indicator of the drug's efficacy in humans. However, to date, this potential problem has not stood the test of time as most antifungal drugs showing efficacy in the rodent models have also been successful in clinical trials.

Contributions of the model to infectious disease

Contributions of the model to infectious disease can currently only be realized for *C. albicans* vaginitis since the model for *C. glabrata* vaginitis has not yet been studied to any significant depth. For *C. albicans* vaginitis though, the experimental model has made significant contributions both to drug therapy and to identification of important host defense mechanisms.

Drug therapy

If the preclinical drug tested is efficacious in the animal model and reflects the *in vitro* sensitivity, there is a high predictability of comparable efficacy in human clinical trials. Similarly, the mouse model has also been useful in confirming *in vitro* observations of antifungal drug resistance (Fidel *et al.*, 1997). Additionally, the routes employed for administration of the drugs to the animals (topical or oral) for treatment of experimental infection mirror those desired for human use. Virtually all the drugs available on the market have been initially screened and tested in animal models. Taken together, although the rodent vaginitis models pose some limitations to the human condition, they have been instrumental in bringing new antifungal drugs to clinical trials, establishing the most efficient treatment regimens, and assessing the clinical predictive value of each drug tested.

Immunological host defense evaluation

As stated earlier, the mouse model has recently been employed for immunological evaluation to identify specific host defenses important for protection against infection. Immunological evaluation can include innate resistance or acquired immunity. Acquired immunity can include both humoral (antibodies) and cell-mediated (T-cell) immunity. Either type of immune reactivity can be measured at both systemic and local levels. However, since the infection is local and data from our laboratory suggest that systemic immunity plays little role in host defense against experimental vaginal candidiasis (Fidel *et al.*, 1994, 1995b), only technical aspects regarding local immune response evaluation will be discussed here. The analysis of antibodies is relatively straightforward, involving the identification of antibodies (total or *Candida*-specific) in lavage fluid. Enzyme-linked immunosorbent assays (ELISA) have been developed for these determinations or can be designed with commercial antibodies/reagents. The limiting factor is the small volume of lavage fluid that can be obtained from mice compared to that required to complete a panel of antibody analyses. Generally, lavages from multiple mice are required to achieve an appropriate volume. Alternatively, tissue homogenates (discussed below) can be used to obtain larger volumes of fluid, but analyses cannot be performed longitudinally.

In contrast to humoral immunity, innate and cell-mediated immunity requires the isolation of cells from the vaginal mucosa by a mechanical or enzymatic digestion (Fidel *et al.*, 1996b). Since the condition of the tissue is not critical, a relatively easy method for excising the vaginal tissue in mice is by involution. For this, the peritoneum of a sacrificed animal is opened and the cervix is cut from the fallopian tubes. Forceps are then inserted in the vaginal lumen and the vaginal tissue is involuted outward by the cervix and excised. The cervix is removed and discarded, and the tissue is finely minced and mechanically disrupted or enzymatically digested for the enrichment of lymphoid as well as epithelial and endothelial cells. Lymphoid cells are extracted mechanically by ethylene diaminetetraacetic acid (EDTA; Nandi and Allison, 1991) or enzymatically by collagenase (Type IV: 0.25%; Fidel *et al.*, 1996b) for 30 minutes at 37°C in a shaking water bath with intermittent Stomacher homogenization (paddle homogenization of tissue in sterile bags). To recover the lymphoid-like cellular fraction, the digested tissue is clarified through a gauze mesh and the cellular fraction is subjected to low-speed (200*g*) short (1 minute) centrifugations to pellet the larger epithelial cells. The supernatants containing the lymphoid-enriched cell fraction are then clarified at 800*g* and the cells layered over Ficoll-Hypaque and subjected to density gradient centrifugation. The cells collected from the interface are then enumerated by trypan blue dye exclusion. One can expect approximately 1×10^4–1×10^5 lymphoid-like cells per mouse. It is recommended that not less than 8 mice be used to obtain a reasonable cell recovery.

The larger epithelial or endothelial fraction can also be recovered from the collagenase digestion. For this, the pelleted cells are recovered from the slow-speed short centrifugations and layered over the Ficoll. The pelleted cells are then recovered from the density gradient centrifugation and enumerated by trypan blue dye exclusion. An alternative means of obtaining a highly enriched population of epithelial cells is to treat the tissue with Dispase II or EDTA for 30 minutes at 37°C which mechanically disrupts the epithelium from the lamina propria. The epithelium can then be minced and subjected to a collagenase (type I or IV) or trypsin digestion with the cells recovered following clarification by the gauze mesh and pelleting by density gradient centrifugation (Ficoll). The lymphoid or epithelial cells can be used for a variety of purposes, including flow cytometry, *in vitro* tissue culture assays cultured with antigen, or assays to assess the effects of the cells on *C. albicans* growth. It should be cautioned, though, that any of the enzymatic digestions may affect the cell surface markers or functional activity of the cells and one must determine which method and reagent is most appropriate for the specific application.

Immunohistochemical analyses can be performed on intact vaginal tissue sections. For this, the tissue is excised without involution, snap-frozen in liquid nitrogen and/or frozen immediately in OCT medium at −70°C. Sections can then be cut (10 μm) and stained with antibodies to cell surface markers or connective tissue by standard protocols.

Vaginal tissue (whether involuted or not) can also be excised and processed for molecular analyses such as polymerase chain reaction (PCR) and Southern or Northern blot analyses. For this the RNA or DNA is extracted from the tissue by standard protocols. From the RNA, complementary DNA can be easily synthesized and used in PCR to examine host defense parameters, including molecular evidence for cytokines, chemokines and cellular infiltration. RNA or DNA can also be hybridized with specific radiolabeled probes in Northern or Southern blot analyses, respectively.

Finally, vaginal tissue can be excised and homogenized in a lysis buffer containing a detergent (NP-40 or triton x-100) supplemented with protease inhibitors. The clarified supernatant can then be used to identify soluble immunomodulators such as antibodies, chemokines, or cytokines.

References

Carlsten, H., Holmdahl, R., Tarkowski, A. (1991). Analysis of the genetic encoding of oestradiol suppression of delayed-type hypersensitivity in (NZB × NZW) F1 mice. *Immunology*, **73**, 186–190.

Cassone, A., Boccanera, M., Adriani, D. A., Santoni, G., De Bernardis, F. (1995). Rats clearing a vaginal infection by *Candida albicans* acquire specific, antibody-mediated resistance to vaginal infection. *Infect. Immun.*, **63**, 2619–2624.

De Bernardis, F., Boccanera, M., Adriani, D., Spreghini, E., Santoni, G., Cassone, A. (1997). Protective role of antimannan and anti-aspartyl proteinase antibodies in an experimental model of *Candida albicans* vaginitis in rats. *Infect. Immun.*, **65**, 3399–3405.

Fidel, P. L. Jr, Sobel, J. D. (1996). Immunopathogenesis of recurrent vulvovaginal candidiasis. *Clin. Microbiol. Rev.*, **9**, 335–348.

Fidel, P. L. Jr, Lynch, M. E., Sobel, J. D. (1993a). *Candida*-specific cell-mediated immunity is demonstrable in mice with experimental vaginal candidiasis. *Infect. Immun.*, **61**, 1990–1995.

Fidel, P. L. Jr, Lynch, M. E., Sobel, J. D. (1993b). *Candida*-specific Th1-type responsiveness in mice with experimental vaginal candidiasis. *Infect. Immun.*, **61**, 4202–4207.

Fidel, P. L. Jr, Lynch, M. E., Sobel, J. D. (1994). Effects of preinduced *Candida*-specific systemic cell-mediated immunity on experimental vaginal candidiasis. *Infect. Immun.*, **62**, 1032–1038.

Fidel, P. L. Jr, Cutright, J. L., Sobel, J. D. (1995a). Effects of systemic cell-mediated immunity on vaginal candidiasis in mice resistant and susceptible to *Candida albicans* infections. *Infect. Immun.*, **63**, 4191–4194.

Fidel, P. L. Jr, Lynch, M. E., Conaway, D. H., Tait, L., Sobel, J. D. (1995b). Mice immunized by primary vaginal *C. albicans* infection develop acquired vaginal mucosal immunity. *Infect. Immun.*, **63**, 547–553.

Fidel, P. L. Jr, Cutright, J. L., Tait, L., Sobel, J. D. (1996a). A murine model of *Candida glabrata* vaginitis. *J. Infect. Dis.*, **173**, 425–431.

Fidel, P. L. Jr, Wolf, N. A., KuKuruga, M. A. (1996b). T lymphocytes in the murine vaginal mucosa are phenotypically distinct from those in the periphery. *Infect. Immun.*, **64**, 3793–3799.

Fidel, P. L. Jr, Cutright, J. L., Sobel, J. D. (1997). Efficacy of D0870 treatment of experimental *Candida* vaginitis. *Antimicrob. Agents Chemother.*, **41**, 1455–1459.

Hector, R. F., Domer, J. E., Carrow, E. W. (1982). Immune responses to *Candida albicans* in genetically distinct mice. *Infect. Immun.*, **38**, 1020–1028.

Kinsman, O. S., Collard, A. E. (1986). Hormonal factors in vaginal candidiasis in rats. *Infect. Immun.*, **53**, 498–504.

Lynch, M. E., Sobel, J. D., Fidel, P. L. Jr (1996). Role of antifungal drug resistance in the pathogenesis of recurrent vulvovaginal candidiasis. *J. Med. Vet. Mycol.*, **34**, 337–339.

McRipley, R. J., Erhard, P. J., Scwind, R. A., Whitney, R. R. (1979). Evaluation of vaginal antifungal formulations *in vivo*. *Postgrad. Med. J.*, **55**, 648–652.

Nandi, D., Allison, J. P. (1991). Phenotypic analysis and gamma/delta-T cell receptor repertoire of murine T cells associated with the vaginal epithelium. *J. Immunol.*, **147**, 1773–1778.

Powell, B. L., Drutz, D. I. (1983). Estrogen receptor in *Candida albicans*: a possible explanation for hormonal influences in vaginal candidiasis. In: *Abstracts from the 23rd Interscience Conference on Antimicrobial Agents and Chemotherapy*, p. 222.

Redondo-Lopez, V., Lynch, M., Schmitt, C., Cook, R., Sobel, J. D. (1990). *Torulopsis glabrata* vaginitis: clinical aspects and susceptibility to antifungal agents. *Obstet. Gynecol.*, **76**, 651–655.

Rex, J. H., Pfaller, M. A., Galgiani, J. N. *et al.* (1997). Development of interpretive breakpoints for antifungal susceptibility testing: conceptual framework and analysis of *in vitro–in vivo* correlation data for fluconazole, itraconazole, and *Candida* infections. *Clin. Infect. Dis.*, **24**, 235–247.

Richardson, K., Brammer, K. W., Marriott, M. S., Troke, P. F. (1985). Activity of UK-49,858, a bis-triazole derivative, against experimental infections with *Candida albicans* and trichophyton mentagrophytes. *Antimicrob. Agents Chemother.*, **27**, 832–835.

Ryley, J. F., McGregor, S. (1986). Quantitation of vaginal *Candida albicans* infections in rodents. *J. Med. Vet. Mycol.*, **24**, 455–460.

Sobel, J. D. (1988). Pathogenesis and epidemiology of vulvovaginal candidiasis. *Ann. N.Y. Acad. Sci.*, **544**, 547–557.

Sobel, J. D., Chaim, W. (1997). Treatment of *Torulopsis glabrata* vaginitis: retrospective review of boric acid therapy. *Clin. Infect. Dis.*, **24**, 649–652.

Sobel, J. D., Muller, G., McCormick, J. F. (1985). Experimental chronic vaginal candidosis in rats. *Sabouraudia*, **23**, 199–206.

Styrt, B., Sugarman, B. (1991). Estrogens and infection. *Rev. Infect. Dis.*, **13**, 1139–1150.

Taschdjian, C. L., Reiss, F., Kozinn, P. J. (1960). Experimental vaginal candidiasis in mice; its implications for superficial candidiasis in humans. *J. Invest. Dermatol.*, **34**, 89–94.

Valentin, A., Bernard, C., Mallie, M., Huerre, M., Bastide, J. M. (1993). Control of *Candida albicans* vaginitis in mice by short-duration butoconazole treatment *in situ*. *Mycoses*, **36**, 379–384.

Chapter 91

Sporotrichosis

A. Polak-Wyss

Background of the model

Sporotrichosis is a chronic mycotic disease of humans and animals. *Sporotrix schenckii,* the etiological agent of sporotrichosis, is a dimorphic fungus (mold and yeast phase) which occurs naturally on vegetation or as a soil saprophyte. This mycosis develops after the traumatic introduction of *S. schenckii* into the dermis. Clinically, classical sporotrichosis is usually seen as an ascending lymphangitis of the extremities following development of the primary lesion (shunt) at the site of inoculation. Various other clinical types of sporotrichosis are also diagnosed—self-limiting pulmonary infection, subacute classical form of cutaneolymphatic complex and severe hematogenous dissemination. The diagnosis is usually made on the basis of clinical symptoms rather than histological findings or cultivation of the fungus. The epidemiology, classification of the disease, diagnosis, histopathology of the lesions and taxonomy of the agent have been summarized by Lavalle and Mariat in 1983.

Several animal models are described using different animal species, including mice, guinea-pigs and hamsters. Additionally, various methods of inoculation—intraperitoneal, intracutaneous, intratestis, intragastric and intravenous—are practiced. Characteristic disease is produced within 2–3 weeks, with involvement of the spleen, liver and testes as target organs. Sections of these organs stained by periodic acid–Schiff (PAS) or Gomorri-Grocott methods reveal abundant intracellular yeast-like forms. Asteroid bodies are usually frequently seen in histological sections taken from the testes. It is interesting to note that the yeast-like tissue forms of the fungus are significantly easier to demonstrate in experimental animal infections than in naturally acquired disease.

Animal species

Various animal species and animal strains have been used over the last few years.

Mice

Mice strains include: BALB/c, 4.5 months of age, weighing 20 g (Kan and Bennett, 1988); male nu/nu back-crossed BALB/c weighing 16–19 g and nu/+ mice weighing 19–23 g (Shiraishi *et al.*, 1979; Miyaji and Nishimura, 1982); male nude mice of BALB/c background (Hachisuka and Sasai, 1981); male NYLAR mice weighing 18–20 g (Dixon *et al.*, 1992); Swiss mice weighing 18–25 g (Carlos *et al.*, 1992); female NMRI weighing 22–25 g (Schaude and Meingassner, 1986; Schaude *et al.*, 1990); female homozygous type 1 hairless (hr/hr) mice back-crossed to ICR strains, 10 weeks old (Lei *et al.*, 1993); female and male ICR Swiss albino mice (Kennedy *et al.*, 1982); ICR Swiss albino Webster mice weighing 20–30 g (Kazanas, 1986); ddy male mice weighing 34–40 g (Miyaji and Nishimura, 1986); DD male mice (Hachisuka and Sasai, 1981).

Hamsters

Goldhamster *Mesocricetos auratus*, weighing 50–60 g (Gonzalez *et al.*, 1990) has been used.

Guinea-pigs

Male Hartley strain weighing 500–700 g (Kwong-Chung, 1979; Hachisuka and Sasai, 1980) and male albino guinea-pigs (Van Cutsem *et al.*, 1987) have been used.

Preparation of animals

In most of the experiments the animals are normally housed at room temperature with water and food *ad libitum* without special precautions. In the experiment using gastrointestinal inoculation (Kennedy *et al.*, 1982) mice were housed in sterile plastic cages and supplied with Purina mouse chow and sterile distilled water containing vancomycin (500 mg/ml), ampicillin (1 mg/ml) and gentamicin (100 mg/ml) *ad libitum* 2 days before the inoculation and

Handbook of Animal Models of Infection
ISBN 0-12-775390-7

throughout the whole experiment. Hachisuka and Sasai (1981) immunosuppressed some of their animals with X-ray irradiation (699 R) or with a dose of 200 mg/kg cyclophosphamide given 2 days before the fungal challenge.

Storage, preparation of the inoculum

S. schenckii is maintained on Sabouraud dextrose agar at 25°C.

Two different forms of inoculum are used: for virulence studies conidia are inoculated, whereas the yeast form of *S. schenckii* is mainly used in chemotherapeutic studies.

For the conidia preparation, *S. schenckii* is grown on potato dextrose agar for 7 days at 30°C (Dixon *et al.*, 1992). Colonies are flooded with sterile saline while gently rubbing the surface growth with a pipet tip. The resulting suspension is filtered through a sterile paper wipe contained in a funnel. This procedure results in a suspension of conidia devoid of hyphal fragments. The conidia are sedimented at 1400*g* for 10 minutes and are resuspended in saline to the desired hemocytometer counts. In other experiments conidia are harvested from a Sabouraud culture, filtered through a gauze, washed three times and centrifuged at 400*g* for 30 minutes and then resuspended in saline to the desired concentration (Hachisuka and Sasai, 1980).

For the yeast-phase preparation *S. schenckii* is either grown on blood–glucose–cystein agar at 37°C for 14 days (Kan and Bennett, 1988) or is cultured at 37°C for 4–8 days in brain–heart infusion broth with constant shaking at 150 cycles/min (Kennedy *et al.*, 1982; Carlos *et al.* 1992). Schaude and Meingassner (1986) cultivated the fungus in Sabouraud glucose broth for 7 days at 37°C in a shaker at 150 rev/min. Yeast cells are harvested, washed three times in phosphate-buffered saline (PBS; pH 7.4) containing 6.7 mmol/l phosphate and resuspended in PBS to yield the final concentration, which is controlled by hemocytometer count.

The inoculum is freshly used in most experiments, however, a stock solution of yeast culture stored in liquid nitrogen can be prepared by adding 5% dimethyl sulfoxide (DMSO; vol/vol) as a cryptoprotective agent (Schaude and Meingassner, 1986).

Infection process

Intravenous

For chemotherapeutic experiments mice receive 1×10^6, 4×10^6 or 1×10^7 yeast cells of *S. schenckii* in 0.1 ml of PBS via the lateral tail vein (Block *et al.*, 1973; Shiraisi *et al.*, 1979; Miyaji and Nishimura 1982; Dickerson *et al.*, 1983; Kan and Bennett, 1988; Carlos *et al.*, 1992). For virulence studies 6×10^5 to 2×10^8 conidia are injected in 0.2 ml saline in the lateral vein of the tail (Dixon *et al.*, 1992).

Intraperitoneal

Mice are infected intraperitoneally with a range of inocula, namely, 5×10^4 to 9×10^9 yeast. The results of this method varied considerably and therefore the intravenous route of inoculation was selected for further experiments (Dickerson *et al.*, 1983). In virulence studies, mice have been inoculated with 10^6 conidia (Kwong-Chung, 1979). Half of the mice used in the experiment of Kazanas (1986) were inoculated intraperitoneally with 6×10^6 or 2×10^7 conidia in 1 ml of saline.

Subcutaneous

Hamsters were inoculated subcutaneously with 50 µl of PBS containing 2500 yeast cells of *S. schenckii* (Gonzalez *et al.*, 1990).

Intracutaneously or intradermally

Mice or guinea-pigs are inoculated intracutaneously with 0.1–0.2 ml of conidia suspension (5×10^6 conidia per ml) into the dorsal portion of the right foot (Hachisuka and Sasai, 1980, 1981). Guinea-pigs receive an intradermal injection of 10^4 or 10^5 conidia in 0.1 ml saline on the shaved dorsal area of the animals (Kwong-Chung, 1979).

Intragastric

Newborn mice are intubated with 0.5 ml tuberculin syringe equipped with a blunt 27.5 G needle tipped with flexible polyethylene tubing. The tuberculin syringe, calibrated in 0.01 ml divisions, is used with a holder in a calibrated syringe microburet. The proper length of tubing for a newborn mouse is equal to the distance between its nose and the xiphoid of its externum. This distance is marked with tape on the tubing before introduction. Adult mice are intubated with the rigid cannula technique: 50 µl of a conidia suspension is used as inoculum (Kazanas, 1986). Kennedy *et al.* (1982) inoculated lightly anesthetized mice with 0.5 ml saline containing 1×10^7 colony-forming units intragastrically using 5-cm 18 G plastic catheters.

Intratestis

Guinea-pigs are infected in the left testicle with 9.8×10^4 yeast cells of *S. schenckii* in a volume of 0.25 ml saline (Van Cutsem *et al.*, 1987).

Diffusion chamber technique

Chambers are constructed from plastic rings 2 mm thick with an outer diameter of 14 mm and an internal diameter of 10 mm. Sartorius filters with pore size of 0.45 µm are glued to both sides of the ring with MF cement. The larger pore size allows host cells actively to immigrate into the chamber, but *S. schenckii* cells are not able to escape. The total volume of the chamber is 0.15 ml. The chambers are sterilized with ethylene oxide for 6 hours at 37°C. By means of a syringe and a 27 G needle, 0.1 ml of yeast suspension is added through a hole in the ring which is then sealed with a small amount of melted paraffin wax. Each chamber is tested to avoid leaks. The chambers are then implanted subcutaneously into anesthetized mice through a dorsal near-midline incision, which is closed after the procedure with surgical clips (Schaude and Meingassner, 1986).

Key parameters to monitor infection and response to therapy

Disseminated sporotrichosis

S. schenckii affects bone, joints and most organs like kidney, liver, spleen, lung and testes. The *in vivo* growth rate is the highest in the testes (Kwong-Chung, 1979; Kazanas, 1986; Dixon *et al.*, 1992). The course of disseminated infection is monitored by mortality data, macroscopical and radiological examination of the bone and joint involvement, and by histological and cultural examination of inner organs, mainly spleen and liver.

Death occurs in mice infected with *circa* 10^7 conidia or yeasts of a virulent strain within 20 days of inoculation (Kan and Bennett, 1988; Dixon *et al.*, 1992). Mice infected intravenously with *S. schenckii* manifest the full spectrum of the disease, with culturally positive organs, cachexia, swelling of the footpads and distal extremities, and multiple nodular tail swellings which progress to ulcerated, draining, necrotic lesions. Whole-body radiographs demonstrate osteoporosis, articular involvement, and extensive destruction of bones most prominent in the vertebral column, tail and small bones of the distal extremities. Histopathological examinations reveal microfoci in the liver and spleen 5 days after inoculation. The liver is generally more affected than the spleen; higher numbers of foci are counted and higher numbers of fungal cells are recovered from the liver than from the spleen. The chief histological features of these foci are purulent inflammatory reactions. Later on, mononuclear cells accumulate at the periphery of the foci and histopathological features change from pyogenic to granulomatous lesions. The number of foci and yeasts decreases abruptly after day 14 (Miyaji and Nishimura, 1982, 1986).

Microscopic examinations of liver foci or of Gomori-Grotcott methenamine silver-stained histological sections of portions of various organs reveal numerous oval and elongated (cigar-shaped) budding yeasts, the characteristic tissue form of *S. schenckii* in animal models. These cigar-shaped yeast cells change their form roughly after the 10th day; they swell and become large round or ovoid cells (Block *et al.*, 1973; Dixon *et al.*, 1992).

Cutaneous infections

The infection is monitored macroscopically for evidence of inflammation, crusting and scaling. The number of recognized nodules and the size of each nodule are recorded weekly over a period of 10 weeks after inoculation. The skin of each inoculated site is biopsied weekly for histology and culturing. Tissue sections are stained with hematoxylin and eosin (H&E) or PAS (Lei *et al.*1993).

Gastrointestinal infection

Kennedy *et al.* (1982) tried to clarify whether the gastrointestinal tract can become colonized with *S. schenckii*. The survival rate of the fungus in the gastrointestinal tract was measured during the first hours after intragastrical inoculation. In this study it was proven that *S. schenckii* is not able to colonize the gastrointestinal tract, nor to cause a dissemination, thus the gastrointestinal tract is not a normal portal of entry for this fungus.

Kazanas (1986) worked with a food-borne strain of *S. schenckii*. Cultural recovery of the fungus from visceral organs, intestines, stomach and cecum–anal canal was done up to 31 days post-inoculation. Additionally, histological examinations of the intestine were performed. Kazanas found that neonatal mice became infected after oral inoculation, whereas adult mice were susceptible to *S. schenckii* only by intraperitoneal inoculation, but not by gavage.

Diffusion chamber

The growth rate of *S. schenckii* in the implanted chambers is studied as follows. Chambers are removed from the animal 24, 48, 96, 144 and 192 hours after subcutaneous implantation and put into glass tubes with 0.9 ml of sterile saline. After mechanical destruction of the filter membranes and addition of approximately 1 g sterile glass beads (diameter 200 μm) the tubes are sonified for 30 seconds. This procedure does not affect the recovery of the fungus and produces uniform suspensions of fungal cells. Dilution of this suspension is plated on Sabouraud glucose agar containing streptomycin sulfate (0.1 mg/ml). Colony-forming units of *S. schenckii* are counted after 4 days incubation at 30°C (Schaude and Meingassner, 1986).

Antifungal therapy

Treatment schedules

A variety of antifungals have been evaluated in systemic sporotrichosis. Drugs have been given once daily by the intraperitoneal or oral route (Van Cutsem *et al.*, 1987; Kan and Bennett, 1988). For systemic sporotrichosis, the antimycotic therapy was started on day 2 (Block *et al.*, 1973), day 3 (Kan and Bennett, 1988) or day 7 (Van Cutsem, 1989) after fungal challenge and always lasted for 28 days.

The effect of proteinase inhibitors on the cutaneous lesions of *S. schenckii* has been studied by Lei *et al.* (1993). The inhibitors solubilized in ointments were applied daily directly on to the inoculation sites; the lesions were observed over a period of 9 weeks after the inoculation.

In experiments measuring the prophylactic effect, the antimycotic treatment was initiated at the day of infection (Van Cutsem *et al.*, 1987).

The treatment schedule for the experiments with the implanted diffusion chambers was as follows: 2 hours before and 2, 4, 24, 48 and 72 hours after implantation. Chambers are removed 24 hours after the last dose and populations of surviving fungi are determined by culture.

Standard solutions of the antifungal drugs

Amphotericin B (Fungizone) is dissolved in distilled water at a concentration of 0.9 mg/ml, and divided into aliquots in vials, and frozen at −70°C. Flucytosine is dissolved in water and diluted to the appropriate dose in water (oral application) or in a solution containing 5% dextrose (intraperitoneal application). Itraconazole is dissolved in polyethylene glycol 200 by heating to yield final concentrations of 2 or 8 mg/ml. Terbinafine is suspended in 1% Tween 20–5% dimethyl sulfoxide (DMSO) in distilled water to yield 10 and 40 mg/ml. The triazole SDZ89-485 is suspended in 0.2% Tween 20–10% DMSO in distilled water at 5 mg/ml (Kan and Bennett, 1988). Chymostatin (0.1%) and/or pepstatin ointment was prepared as a mixture of 9 g of hydrophilic ointment and 1 ml of inhibitors solubilized in 0.01 mol/l acetic acid (chymostatin) and/or 0.01 mol/l NaOH (pepstatin).

Pitfalls (advantages/disadvantages) of the model

The animal model of sporotrichosis has not become a worldwide adjusted screening model. The number of publications dealing with this *in vivo* model are scarce. The model is mainly used to study the defense mechanisms against *S. schenckii* and to investigate the virulence and pathogenicity of various isolates. For this purpose, both models — cutaneous and disseminated — proved to be very useful.

The formation of granuloma plays a major defensive role in humans as well as in animals (Lavalle and Mariat, 1983; Miyaji and Nishimura, 1982; Mohri, 1987). The foci in the disseminated human disease are rich in elongated yeast cells: the same phenomenon is observed in animals. The role of proteinases as virulence factor was confirmed by Lei *et al.* (1993) in the cutaneous model; these authors additionally proved the efficacy of proteinase inhibitors (applied topically) to inhibit the fungal growth in the tissue.

The disseminated model of sporotrichosis provided useful data to identify virulent strains isolated during the largest outbreak of sporotrichosis in the USA (Dixon *et al.*, 1991), and to distinguish virulent clinical and environmental *S. schenckii* isolates from other dematiaceous, apathogenic fungi like *Ophiostoma stenoceras* (Dixon *et al.*, 1992).

Thus, the animal model for sporotrichosis is highly predictive of the study of immunological questions and helps to understand the epidemiology of *S. schenckii*.

The disseminated animal model would also produce predictive data for human chemotherapy, but the experience in evaluating new antifungal agents is rather limited; the model has not yet found its way into the worldwide screening routine. Only a few researchers in universities have employed the model to study antifungal chemotherapy and only one pharmaceutical company, Janssen, has established sporotrichosis as a screening model. The lack of enthusiasm may lie in the fact that the prognosis of sporotrichosis in humans has always been excellent. Potassium iodide has been successfully used as unique therapy for some time. This agent was inactive *in vitro*, and there was never a need to investigate its mode of action or an urge to find new and more powerful regimens of therapy. The scientific world was more interested in finding new drugs for the emerging opportunistic fungi. Nevertheless, some investigators studied the potency of old and new antifungal drugs in this model.

The disseminated model, which is the easiest and most predictive to investigate the power of antifungal therapy, has a few shortcomings. The inoculation route is unnatural (intravenous versus cutaneous) and the animal disease is mimicking the most severe form of sporotrichosis — the disseminated form — and not the most common one — the cutaneolymphatic form. Thus, an evaluation of an antifungal drug in this model is the most rigorous test one can perform to characterize the efficacy of a new antifungal drug for this indication.

Contributions of the model to infectious disease therapy

Block *et al.* (1973) studied the efficacy of flucytosine in cladosporiosis and sporotrichosis. Flucytosine is efficacious in infections caused by dematiaceous fungi (Dixon and Polak, 1987), but it is inactive in sporotrichosis. Itraconazole was the first antifungal drug to be thoroughly evaluated in the laboratories of a pharmaceutical company (Van Cutsem *et al.*, 1987). Itraconazole proved to be highly active in the guinea-pig model in preventing the dissemination of the disease from the testes (Van Cutsem *et al.*, 1987; Van Cutsem, 1989). Later on, this drug has revolutionized the therapy of sporotrichosis in humans; it is now used as first-line therapy for this indication (Lyman and Walsh, 1992;

Table 91.1 Characteristic lesions and culture results in surviving mice sacrificed 38 days after fungal challenge with *Sporotrix schenckii*

Treatment	*Dose* (mg/kg)	*Survivors* (%)	*Positive culture* (%)	*Tail and paw nodules* (%)	*Ataxia* (%)
None		10	100	100	0
Amphotericin B	4.5	100	40	30	0
Itraconazole	20	80	100	100	37.5
	80	90	88	100	22.2
Terbinafine	50	10	100	100	0
	200	10	100	100	
SDZ 89-485	25	100	90	100	40
	50	100	90	90	0

From Kan and Bennett (1988), with permission.

Como and Dismuskes, 1994; Karyotakis and Anaissie, 1994; Kauffman, 1996). Potassium iodide has lost its unique position.

Kan and Bennett (1988) studied the efficacies of four antifungal drugs in murine sporotrichosis (Table 91.1). Amphotericin B, itraconazole and the preclinical triazole SDZ 89-485 significantly prolonged the survival of infected animals, but these agents did not wholly protect the infected mice, as tail and paw lesions, whole-body radiographs and positive culture from survivors showed evidence of dissemination. Terbinafine had no effect on survival despite documented drug absorption.

These animal data show that the murine model of disseminated sporotrichosis is a rigorous test to detect high efficacy of new antifungal agents and therefore its predictive value for human chemotherapy is surprisingly high. Amphotericin B, which is still the drug of choice for severe disseminated sporotrichosis (Karyotakis and Anaissie, 1994), induced 100% survival of all mice tested at a relatively high dose of 4.5 mg/kg i.p., but in 30% of the animals, tail lesions were observed and 40% of the survivors showed positive culture data. Itraconazole, which is nowadays the first-line therapy of cutaneous and disseminated sporotrichosis (Como and Dismuskes, 1994; Kauffman, 1996), had a significant effect on the survival time but all animals remained culturally positive and had tail abnormalities, and ataxia was seen in approximately 30% of the animals. Thus, it seems that all antifungals tested are more active against human disease than in murine sporotrichosis.

Ketoconazole, itraconazole and the preclinical triazole SDZ 89-485 were tested in the artificial model of implanted chambers. All three drugs caused a significant reduction of colony counts of *S. schenckii* with ED_{50} values of 19.0, 7.97 and 2.98 mg/kg respectively.

In conclusion, the disseminated murine sporotrichosis is a helpful animal model to predict the therapeutic value of a new antifungal drug for human chemotherapy.

References

Block, E. R., Jennings, A. E., Bennett, J. E. (1973). Experimental therapy of cladiosporiosis and sporotrichosis with 5-fluorocytosine. *Antimicrob. Agents Chemother.*, **3**, 95–98.

Borelli, D. (1984). Micopathologia experimental: models multiples. *Rev. Fundacion José Maria Vargas*, **8**, 188–192.

Carlos I. Z., Da Graca Sgarbi, D. B., Angluster, J. Alviano, C. S., Lopes-Silva, C (1992). Detection of cellular immunity with soluble antigen of the fungus *Sporothrix schenckii* in the systemic form of the disease. *Mycopathologia*, **117**, 139–144.

Como, J. A., Dismuskes, W. E. (1994). Oral azole drugs as systemic antifungal therapy. *N. Engl. J. Med.*, **330**, 263–272.

Dickerson, C. L., Taylor, R. L., Drutz, D. J. (1983). Susceptibility of congenitally athymic (nude) mice to sporotrichosis. *Infect. Immun.*, **40**, 417–420.

Dixon, D. M., Polak, A. (1987). *In vitro* and *in vivo* drug studies with three agents of central nervous system phaeohyphomycosis. *Chemotherapy*, **33**, 129–140.

Dixon, D. M., Salkin, I. F., Duncan, R. A. *et al.* (1991). Isolation and characterisation of *Sporotrix schenckii* from clinical and environmental sources associated with the largest U. S. epidemic of sporotrichosis. *J. Clin. Microbiol.*, **29**, 1106–1113.

Dixon, D. M., Duncan, R. A., Hurd, N. J. (1992). Use of a mouse model to evaluate clinical and environmental isolates of *Sporothrix* spp. from the largest U. S. epidemic of sporotrichosis. *J. Clin. Microbiol.*, **30**, 951–952.

Gonzalez, L. A., Alzate, A., Saravia, N. (1990). Comportamiento experimental del *Sporothrix schenckii* y la *Leishmania mexicana* en el hamster. *Rev. Inst. Med. Trop. Sao Paulo*, **32**, 319–324.

Hachisuka, H., Sasai, Y. (1980). Subpopulation of lymphocytes in the filtrate of experimental sporotrichosis. *Mycopathologia*, **71**, 167–169.

Hachisuka, H., Sasai, Y. (1981) Development of experimental sporotrichosis in normal and modified animals. *Mycopathologia*, **76**, 79–82.

Kan, V. L., Bennett, J. E. (1988). Efficacies of four antifungal agents in experimental murine sporotrichosis. *Antimicrob. Agents Chemother.*, **32**, 1619–1623.

Karyotakis, N. C., Anaissie, E. J. (1994). The new antifungal azoles: fluconazole and itraconazole. *Curr. Op. Infect. Dis.*, **7**, 658–666.

Kauffman, C. A. (1996). Role of the azoles in antifungal therapy. *Clin. Infect. Dis.*, **22** (suppl. 2), S148–S153.

Kazanas, N. (1986). Foodborne *Sporothrix schenckii*: infectivity for mice by intraperitoneal and intragastric inoculation with conidia. *Mycopathologia*, **95**, 3–16.

Kennedy, A. J., Bajwa, P. S., Volz, P. A. (1982). Gastointestinal inoculation of *Sporothrix schenckii* in mice. *Mycopathologia*, **78**, 141–143.

Kwong-Chung, K. J. (1979). Comparison of isolates of *Sporotrix schenckii* obtained from fixed cutaneous lesions with isolates from other types of lesions. *J. Infect. Dis.*, **139**, 424–451.

Lavalle, P., Mariat, F. (1983). Sporotrichosis. *Bull. Inst. Pasteur*, **81**, 295–322.

Lei, P-Ch. P., Yoshikii, T., Ogawa, H. (1993). Effects of proteinase inhibitors on the cutaneous lesion of *Sporothrix schenckii* inoculated haireless mice. *Mycopathologia*, **123**, 81–85.

Lyman, C. A., Walsh, T. J. (1992). Systemically administered antifungal agents, a review of their clinical pharmacology and therapeutic applications. *Drugs*, **44**, 9–35.

Miyaji, M., Nishimura, K. (1982). Defensive role of granuloma against *Sporotrix schenckii* infection. *Mycopathologia*, **80**, 117–124.

Miyaji, M., Nishimura, K. (1986). Increased resistance of splenectomised mice to *Sporotrix scheckii* infection. *Mycopathologia*, **96**, 143–151.

Mohri, S. (1987). Study in sporotrichosis III. Histological and immunochemical study in experimental cutaneous sporotrichosis in man. *Yokohama Med. Bull.*, **38**, 37–48.

Schaude, M., Meingassner, J. (1986). A diffusion chamber technique for testing antifungal drugs against *Sporothrix schenckii in vivo*. *J. Med. Vet. Mycol.*, **24**, 297–304.

Schaude, M., Petranyi, G., Ackerbauer, H., Meingassner, J. G., Mieth, H. (1990). A diffusion chamber technique for testing of antifungal drugs against *Sporothrix schenckii in vivo*. *J. Med. Vet. Mycol.*, **28**, 445–454.

Shiraishi, A., Nakagaki, K., Arao, T. (1979). Experimental sporotrichosis in congenitally athymic (nude) mice. *J. Retic. Soc.*, **26**, 333–336.

Van Cutsem, J. (1989). Oral, topical and parenteral antifungal treatment with itraconazole in normal and immunocompromised animals. *Mycoses*, **32** (suppl. 1), 14–24.

Van Cutsem, J., Van Gerven, P., Janssen, P. A. (1987). Activity of orally, topically and parenterally administered itraconazole in the treatment of superficial and deep mycoses: animal models. *Rev. Infect. Dis.*, **9** (suppl. 1), S15–S32.

Section IV

Parasitic Infection Models

Chapter 92

Malaria

W. Peters and B. L. Robinson

Malaria in humans

The genus *Plasmodium* in humans

Seasonal, intermittent, paroxysmal fever associated with a swampy environment has been recorded in medical history since time immemorial. Commonly known as the "ague" or "mal'aria", the disease which is now called malaria was first shown to be caused by protozoa living in human erythrocytes by Laveran in 1880. In 1897 Ross, in India, observed cysts on the stomachs of mosquitoes of the genus *Anopheles* that had fed on infected patients. Subsequently he described similar cysts of avian malaria parasites in culicine mosquitoes and described their cyclical transmission back to birds in 1898. In the same and following years, the Italian scientists, Grassi and Bignami, demonstrated unequivocally that anopheline mosquitoes transmit human malaria. (See Harrison, 1978, for an interesting historical account of human malaria and attempts to control it.)

Four species of the Apicomplexan protozoan parasite in the genus *Plasmodium* are responsible for the syndrome of fever, splenomegaly, hepatomegaly and anaemia that is known as malaria. The commonest species is *Plasmodium falciparum* which causes "malignant tertian malaria" and has the highest fatality rate. Two species produce "benign tertian malaria", namely *P. vivax* and *P. ovale*. The fourth species, *P. malariae*, is associated with "quartan malaria". Rare cases of malaria caused by simian parasites such as *P. cynomolgi bastianellii* have also been recorded.

Female anopheline mosquitoes carrying infective sporozoites of *Plasmodium* bite humans in order to acquire blood that they use as a source of nutrients with which to mature their eggs (Figure 92.1). Sporozoites (A) are released into the blood stream from where they migrate rapidly into the liver. There they enter hepatocytes (B), within which they undergo an asexual reproduction phase that leads, within 5–15 days (depending upon the parasite species), to the formation of a pre-erythrocytic (also called primary exoerythrocytic) schizont (C). On maturation this disgorges its contents, several thousand microscopic merozoites, into the circulation, where each young parasite attaches to and penetrates into an erythrocyte. Within the erythrocyte the parasite evolves by a further asexual reproductive process from a simple ring-like organism to a schizont containing from eight to 24 daughter cells, also called merozoites. These erupt on maturation (in 48–72 hours depending upon the species) and each daughter merozoite then invades a fresh erythrocyte, thus continuing the asexual, intraerythrocytic cycle (D). The haemolysis associated with this periodic cycle leads to anaemia and certain parasite products released along with the merozoites serve as pyrogens and mediate the intermittent fever, as well as other pathophysiological processes. In malaria caused by *P. falciparum*, severe changes in the cerebral microcirculation and other changes associated with the generation of cyclokines may bring about the syndrome known as cerebral malaria which, inadequately treated,

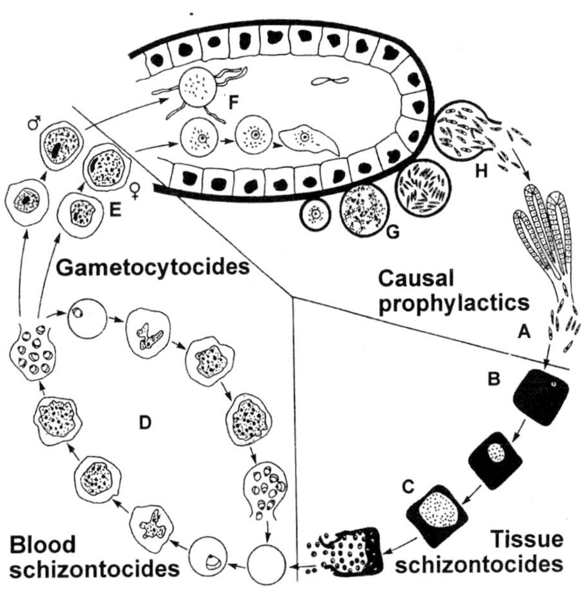

Figure 92.1 Schematic diagram of the life cycle of rodent *Plasmodium*, indicating the classification in relation to the sites of action of antimalarial compounds. A, Infective sporozoites; B, earliest stage of infection in a hepatocyte; C, maturing pre-erythrocytic (exoerythrocytic) schizont in a hepatocyte; D, the asexual intraerythrocytic schizogonic cycle; E, formation of male and female gametocytes; F, sexual conjugation in the stomach of the mosquito vector; G, maturation of the oocysts; H, sporozoites migrating from the oocyst into the salivary gland of the vector.

Handbook of Animal Models of Infection
ISBN 0-12-775390-7

may be fatal. *P. malariae* is the slowest developer, requiring 15 days for the hepatic cycle and 72 hours for that in the erythrocyte.

A so-far unidentified trigger diverts the development of certain intraerythrocytic merozoites from asexual to sexual stages known as gametocytes (E). If these are picked up by another mosquito feeding on the host they undergo a complex cycle, first of sexual conjugation (F), then asexual reproduction within cysts (oocysts) attached to the stomach wall (G). Infective sporozoites developed inside the oocysts erupt when mature to migrate into the salivary glands (H) from where they are poised to infect a new human host. This sporogonic cycle occupies from about 10 days to 1 month depending upon the species and microclimate.

In two of the malaria parasites of humans, *P. vivax* and *P. ovale*, some of the sporozoites that invade hepatocytes remain latent in a single-celled form known as the hypnozoite. These commence development into secondary exoerythrocytic schizonts after an interval that may last from weeks to months, when the maturing merozoites invade the circulation to cause the relapses that characterize these two types of "benign tertian" malaria.

Prevalence and incidence

It is estimated that 40% of the world's population in 91 countries occupy areas endemic for malaria which is believed to kill annually between 1.5 and 2.7 million people, the majority infants and young children. In tropical Africa, where *P. falciparum* is the dominant pathogen, malaria is responsible for 10–30% of all hospital admissions and between 15 and 25% of all deaths in infants under 5 years of age. International endeavours to eliminate malaria, first initiated in the 1950s, failed for reasons that included resistance of mosquito vectors to insecticides, resistance of the parasites to drugs, socioeconomic problems and the lack of any immunological weapons, i.e., vaccines. Indeed, it is now recognized that an impasse has been reached in the fight against malaria (World Health Organization (WHO), 1993).

Mechanisms of pathogenesis

In addition to the direct consequences of erythrocyte destruction associated with malaria, the induction of cytokines and immune mediators such as nitric oxide (NO) can trigger a cascade of pathophysiological changes, especially in infection with *P. falciparum* (Pasvol *et al.*, 1995). While this topic is beyond the scope of this chapter, it should be noted that, in addition to exploring immunological methods, efforts are currently being made to investigate the use of drugs that can counteract some of these changes. Because of the importance of cerebral malaria caused by *P. falciparum*, animal models have been sought for this condition. In spite of claims by a number of investigators (e.g. Grau *et al.*, (1987) described cerebral pathology in mice infected with *P. berghei* ANKA as a relevant model), it is our contention that the rodent malarias described below do not provide adequate models for all the severe pathological changes seen in falciparum malaria of humans.

Management and control

Brief mention was made above to the international efforts to control malaria on a global basis — the period of "malaria eradication" that commenced in the 1950s. This campaign, based on the extensive dispersion of the insecticide DDT and the mass distribution of drugs, notably chloroquine and pyrimethamine, reduced transmission in certain fringe areas but failed to remove malaria from the most highly endemic parts of the world. Indeed, recent years have seen a serious resurgence of transmission, even in countries such as India where the initial campaigns brought about a massive reduction in the incidence of the disease and, consequently, a greatly diminished acquired immunity in the survivors. This increase in transmission is manifested in many areas now as severe epidemics in communities, the younger individuals of which are no longer protected by partial immunity.

In 1992 a new global strategy was adopted by WHO giving priority to the prevention of mortality and a reduction in morbidity, as well as socioeconomic loss (WHO, 1993). By this time, however, a spiralling rise in the resistance of malaria parasites to safe, affordable drugs, resistance of anopheline vectors to insecticides and sociopolitical disturbances in some of the worst malaria problem areas had culminated in a serious fading of interest in malaria by national governments, international donor agencies and by industry. On the other hand, a surge of scientific interest in immunology and vaccine development for malaria did receive substantial financial support. However, in spite of several decades of imaginative and innovative research that has culminated in the production of several promising leads (see review by Gilbert and Hill, 1997), to date no vaccine that has reached the stage of clinical trial has provided a reproducibly high level of protection against challenge. Moreover, almost all these efforts have been directed, quite justifiably, at the prevention of infection with *P. falciparum*.

The progressive abandonment of antimalarial drug research and development by the greater part of the pharmaceutical industry, for a variety of reasons, has resulted in a serious shortage in the antimalarial armamentarium. A major contribution to new drug development resulted from the war in Vietnam which exposed large numbers of non-immune soldiers to chloroquine-resistant falciparum malaria in the 1960s. In the monumental programme set up by the United States Army Research and Development Command and run by the Walter Reed Army Institute of Research (WRAIR), valuable old leads were re-explored and new leads discovered, resulting in two of today's most valuable drugs, mefloquine and halofantrine (Peters, 1987). Fortunately that programme remains operative, albeit at a reduced level. However, apart from the WRAIR programme in which over 300 000 chemical entities were studied, relatively little has been achieved. Were it not for the exploitation by modern techniques of some of their traditional herbal remedies against malaria that resulted, in 1979, in the development of artemisinin by Chinese investigators, many people currently infected with multidrug-resistant strains of *P. falciparum* would die. A new global initiative has now been launched to counter the major threat that malaria continues to pose to health and life, especially of the people of the African continent (Butler *et al.*, 1997).

Background of animal models for chemotherapy

The priority target for both drug and vaccine development is, as stated above, *P. falciparum*. Fortunately, invaluable techniques for the continuous culture of *P. falciparum in vitro*, first described in 1976, and more recent advances in molecular biology have greatly facilitated the study of many aspects of malaria, including drug discovery, but animal models remain indispensable for the in-depth evaluation of both antimalarials and vaccines. From the turn of the century until World War II, a number of pioneering studies of the mode of action of the then only useful antimalarial drug, quinine, as well as the search for new drugs were carried out in birds infected with various species of *Plasmodium* or the related protozoan genus, *Haemoproteus*. A few studies were conducted in rhesus monkeys infected with *P. knowlesi*. In the second largest antimalarial screening and development programme ever conducted, namely the World War II programme co-ordinated by the US Public Health Service in which over 4000 compounds were examined, the majority of studies were made in chicks infected with *P. gallinaceum* or ducklings with *P. lophurae* (Wiselogle, 1946).

The greatest contribution to malaria chemotherapy research, however, was the discovery in 1948 by the Belgian parasitologist, Ignace Vincke, of a malaria parasite in thicket rats of the Katanga region of the then Belgian Congo (Zaire; see reviews in Killick-Kendrick and Peters, 1978). This parasite, *P. berghei*, which proved to be infective to laboratory mice, rats and hamsters, has been widely exploited for the past 50 years, not only for chemotherapy research, but also for research on malaria immunology. *P. berghei*, unlike the avian malarias, has a life cycle that is essentially the same as that of the malaria parasites of humans, but with differences in the duration of the different stages of the life cycle. The pre-erythrocytic cycle, for example, requires between 42 and 72 hours, while the asexual, intraerythrocytic cycle occupies between 22 and 25 hours (Landau and Boulard, 1978). This parasite, however, differs from both *P. vivax* and *P. ovale* in not possessing a hypnozoite stage in the liver. Since 1948, other murine malaria parasites have been discovered and a number of these have been exploited for chemotherapy research (Table 92.1). The durations of the different stages of the life cycles of other murine species are in the same order of magnitude as those of *P. berghei*.

The rodent malarias provide invaluable models for the investigation of drugs against any of the stages of the life cycle (except hypnozoites) but have been most widely exploited for the study of blood schizontocides. A relevant and major review on the biology of *Plasmodium* in rodents has recently been published by Landau and Gautret (1998). In this chapter attention is focused on the use of rodent models for the *in vivo* investigation of antimalarial drugs. Research using liver stages in monolayer cultures or intraerythrocytic stages *in vitro* will be mentioned mainly to offer key references to the literature.

Reference has already been made to the absence of a hypnozoite stage in the rodent malarias. For studies on compounds with hypnozoitocidal activity, the only valid model is *P. cynomolgi* in the rhesus monkey (*Macaca mulatta*). Drug action against the blood stages of both *P. falciparum* and *P. vivax* can be studied in the South America monkeys, *Saimiri sciureus* or various subspecies of *Aotus*. Outlines of the methods of animal maintenance and handling as well as the procedures used for drug testing will be found in the report by WHO (1973) and in Peters (1987). It is obvious that simian models are quite unsuitable for general screening because of the specialized skills required, as well as the cost in terms of animals and drugs. Moreover, ethical considerations demand that the use of such experimental animals should be strictly limited.

In vitro techniques

The seminal work of Trager and Jensen (1976) in developing a relatively simple system for the continuous *in vitro* cultivation of *P. falciparum* provided the basis upon which several extensively used *in vitro* models have been produced subsequently for the evaluation of compounds against the asexual intraerythrocytic stages and gametocytes of this parasite. It is clearly of immense value to have a model of the prime target pathogen that is adaptable to semi-automated, computerized drug screening and for which a very wide range of isolates and clones are available with a broad spectrum of responses to standard blood schizontocides.

Unfortunately, the rodent plasmodia have proved remarkably refractory to attempts to develop similar continuous cultivation techniques and the few *in vitro* systems that have been described (e.g., Janse *et al.*, 1984) have been little used in experimental chemotherapy to date. Procedures to cultivate the exoerythrocytic stages in

Table 92.1 Murine *Plasmodium* species used as models for chemotherapy research

Species	*Subspecies*	*Isolate or strain*	*Place of origin*	*Extent of use*
Plasmodium berghei (Vincke and Lips, 1948)		Keyberg 173	Zaire (Katanga)*	+++++
		NK65	Zaire (Katanga)*	++
		ANKA	Zaire (Katanga)*	++
		RLL	Zaire (Katanga)*	–
		SP11	Zaire (Katanga)*	+
P. yoelii	*P. y. yoelii* (Landau and Killick-Kendrick, 1966)	17X	Central African Republic	++
	P. y. killicki (Landau *et al.*,1968)		Brazzaville	–
	P. y. nigeriensis (Killick-Kendrick, 1973)	N67	S. W. Nigeria	+++
	P. yoelli ssp. NS †		Zaire (Katanga)*	++++
P. vinckei	*P. v. vinckei* (Rodhain, 1952)		Zaire (Katanga)*	+
	P. v. brucechwatti *(Killick-Kendrick, 1975)*		S. W. Nigeria	–
	P. v. lentum (Landau *et al.*, 1970)		Brazzaville	–
	P. v. petteri *(Carter and Walliker, 1977*		Central African Republic	+
	P. vinckei ssp.		E. Cameroon	–
P. chabaudi	*P. c. chabaudi* (Landau, 1965)	AS	Central African Republic	++
	P. c. adami (Carter and Walliker, 1977)		Brazzaville	–

* Zaire has been renamed the Democratic Republic of Congo.
† See Peters *et al.* (1978) for a history of the isolation of *P. yoelii* ssp. NS-type parasites.

hepatocytes or hepatoma cells *in vitro* have met with some success (Millet *et al.*, 1985; Suhrbier *et al.*, 1987) but they are highly labour-intensive and expensive, hence of limited value other than for more advanced stages of drug development such as studies on the mode of drug action.

Rodent models: host species and strains

For the great majority of experimental therapy using murine malaria, one or other strain of albino laboratory mouse has been employed. Whereas in the early days of such research it was believed that it was essential to use specific, inbred lines of mouse such as NMRI or Balb-C, long experience has shown us that this is not essential, provided that the mice are appropriately housed, nourished and free of concomitant pathogens. Random-bred albino Swiss mice are perfectly adequate for most purposes and we currently use male animals weighing between 18 and 20 g of the TFW strain supplied by A. Tuck and Son, Rayleigh, Essex. Since it is rarely, if ever, necessary to use rats or hamsters for studies on chemotherapy, this chapter will focus on the mouse as the mammalian host.

For routine chemotherapy research there is no need to use immunologically deprived animals. However there have been several reports of the deployment of severe combined immunodeficiency (SCID) mice to obtain infections with liver or blood stages of *P. falciparum*. Williman *et al.*, (1995), for example, attempted to develop a model of cerebral malaria in SCID mice while Butcher *et al.* (1993) unsuccessfully tried to implant human hepatocytes in SCID mouse kidneys in the hope of inoculating them with *P. falciparum* sporozoites.

The practice of maintaining strains of rodent *Plasmodium in vivo* by the passage of infected blood from one host to another has resulted in the loss from many such strains of the ability to produce gametocytes. Thus they can no longer be employed to infect mosquitoes and ensure cyclical transmission. On the contrary, cyclical transmission carried out frequently, even if irregularly, usually maintains or even restores gametocytogenesis (Mons *et al.*, 1985). Fortunately, all the gametocyte-producing strains currently available can be passaged through laboratory-bred *Anopheles stephensi*, which is one of the most widely maintained vectors employed in experimental malaria. It is easily reared in the insectary but infected insects must be held at different environmental temperatures depending upon the parasite species. The sporogonic stages of gametocyte-producing strains of *P. berghei* such as NK65 and ANKA mature at a temperature of 19–21°C, whereas *P. yoelii* 17X and *P. chabaudi* require about 24°C.

Parasite species and strains

Table 92.2 indicates the characteristics of the most widely used drug-sensitive and drug-resistant lines derived from various species and subspecies of rodent *Plasmodium* in our laboratory.

For the majority of experiments on known or putative blood schizontocides, *P. berghei* provides a simple and satisfactory model. Whereas we have long believed that the fact that the asexual, intraerythrocytic cycle is asynchronous is not an obstacle to its use, recent experience has taught us that a synchronous infection such as that yielded by *P. vinckei* or *P. chabaudi* can give a better response to a few types of compounds. Studies with certain iron chelating agents such as desferrioxamine or with antagonists of phospholipid metabolism have revealed more activity than did experiments using either *P. berghei* or *P. yoelii* ssp. NS. *P. v. petteri* has been valuable in pinpointing the precise stages upon which a drug exerts its action (Caillard *et al.*, 1992).

It is relatively easy, using the appropriate technique, to select lines of rodent malaria that are resistant to most antimalarial drugs (Peters, 1987). *P. berghei*, in fact, was the first mammalian malaria species from which parasites were developed with a high level of resistance to chloroquine. Subsequently it was shown that *P. yoelii* 17X possesses a low-level, inherent resistance to this drug and another parasite currently labelled *P. yoelii* ssp. NS is now widely deployed in our laboratory as a model for chloroquine-resistant *P. falciparum*. Moreover, from this has been derived a battery of lines resistant to most of the standard drugs and to a number of novel antimalarial compounds (see below).

Storage and preparation of parasites

Intraerythrocytic stages of plasmodia are easily conserved in liquid nitrogen provided that the blood containing them

Table 92.2 Characteristics of drug-resistant lines

Parent species/ subspecies	*Strain or line*	*Primary resistance*	*ED_{90} mg/kg per day s.c. ×4*	*I_{90}**	*Asynchronous/ synchronous*
Plasmodium berghei	K173	Sensitive	3.1	1.0	A
	RC	Chloroquine	230	75	A
	NAM	Amodiaquine	350	135	A
	NPN	Pyronaridine	13.5	19.3	A
	Q	Quinine	>600	>5.0	A
	N/1100	Mefloquine	540	117	A
	NH	Halofantrine	3.6	3.3	A
	P	Primaquine	74.0	15.4	A
	B	Cycloguanil	>100	>30	A
	PYR	Pyrimethamine	3.4	28.3	A
	ORA	Sulphaphenazole	196	70	A
	MEN	Menoctone	3700	2643	A
P. y. yoelii	17X	Chloroquine	16		A
P. y. nigeriensis	N67	Chloroquine	6.7		A
P. yoellii ssp.	NS	Chloroquine	56	1.0	A
	SAM	Amodiaquine	112	6.2	A
	MEF	Mefloquine	1000	139	A
	SH	Halofantrine	375	375	A
	SPN	Pyronaridine	33.5	27.9	A
	ART	Artemisinin	165	16.5	A
	SAT	Atovaquone	5.6	112	A
P. c. chabaudi	AS	Sensitive		1.0	S
	ASCQ	Chloroquine		>1.0	S

* Index of cross-resistance $(I_{90}) = \dfrac{ED_{90}\text{ Resistant line}}{ED_{90}\text{ Parent strain}}$

For the purposes of determining I_{90} values, lines are compared with the parent strain from which they are derived, e.g. lines derived from chloroquine-resistant *P. yoelii* ssp. NS are compared with this and not with chloroquine-sensitive *P. berghei*. When *P. yoelii* ssp. NS is compared directly with *P. berghei* K173 it has an I_{90} of 18. If I_{90} values for NS-derived strains were also calculated in this way, the degree of cross-resistance would be grossly distorted.

is protected with a suitable cryopreservant. Blood is collected from donor animals as described below in a heparinized syringe. It is diluted with Alsever-glycerol solution that is prepared as follows. Alsever solution contains 4.66 g glucose, 1.05 g NaCl and 2.0 g Na citrate in 200 ml distilled water. To this is added glycerol to 10% final concentration. One part of blood is gently mixed with three parts of the Alsever-glycerol, then inserted into 1 ml plastic cryotubes or capillary tubes which are then sealed. These are appropriately labelled, then snap-frozen in liquid nitrogen. After freezing they are stored immersed in the liquid phase of a liquid nitrogen container. For use the containers are removed, rapidly thawed in warm water, then injected undiluted into mice by the intraperitoneal (i.p.) route. When parasitaemia reaches an appropriate level in the first recipients, their blood can be used, after appropriate dilution, to infect further animals. It is likely that only the young ring stages survive cryopreservation. In principle, malaria blood-stage parasites can be cryopreserved indefinitely. In our hands strains have been recovered from liquid nitrogen after more than 20 years. However, there is some evidence that cryopreservation may select for certain characteristics, e.g. chloroquine resistance, presumably because these clones within an uncloned, preserved parasite population are more durable than the others. As a precaution it is advisable, every year or so, to unfreeze a sample of such material, passage it through clean mice, then refreeze.

Rodent parasites can also be maintained *in vivo*, either by simple passage of blood containing asexual, intraerythrocytic stages (for the technique, see below) or, for gametocyte-producing strains, by intermittent cyclical passage through *A. stephensi* (see following sections).

Although isolates of *P. falciparum* are routinely maintained *in vitro* in continuous passage, it has proved very difficult to conserve any rodent species in this way. As the procedure described by Janse *et al.* (1984) is labour-intensive and gives inconsistent results, it is better to employ either cryopreservation or *in vivo* blood passage for routine maintenance of murine malaria.

Infection techniques

Blood stages

The two routes for infecting mice with donor blood are i.p. and intravenous (i.v.). The latter is preferable since the resulting infection in any group of recipient animals shows little variation whereas, on the contrary, a wide spread of parasitaemia levels may be found in animals inoculated i.p. With practice the i.v. route is as simple and as rapid to use as the i.p. route, especially if the mice are gently warmed with a heat source such as an infrared lamp and they are immobilized with a simple device such as that shown in Figure 92.2.

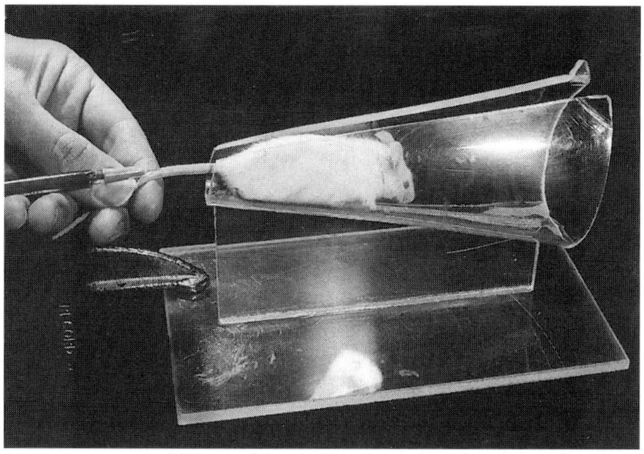

Figure 92.2 A simple device made of Perspex to immobilize mice while performing an intravenous inoculation via the dorsal tail vein. From Peters (1987), with permission.

Blood is collected from a donor animal that has a parasitaemia of approximately 20% in a 10 ml heparinized syringe. The donor is first anaesthetized with intravenous pentobarbitone sodium (60 mg/kg) after which the axilla is opened in a sterile manner, the subclavian artery severed, and the emerging blood drawn up into the syringe. Alternatively, blood may be obtained from the anaesthetized animal by cardiac puncture. The blood is gently mixed, then diluted with the appropriate volume of sterile physiological saline such that each 0.2 ml contains approximately 2×10^6 parasitized erythrocytes. The recipient mice are held in a large holding cage and infected in random order, care being taken gently to mix the contents of the syringe at frequent intervals to avoid the red cells depositing. After infection the animals are caged in groups of five in the test cages.

Sporozoite infections

The adult *Anopheles stephensi* stock colony is maintained at 24.0±1.0°C and 70.0%±5.0% relative humidity in autoclavable, wire-framed cages 60×60×60 cm in size, covered with black mosquito netting. The females are offered a blood feed from anaesthetized, uninfected TFW mice each day and sugar solution is provided at all times. Water bowls are provided for egg-laying and eggs are collected each day and set up in polypropylene water trays. The larvae are fed Biodin PL, which is a proprietary brand of fish food made from freeze-dried plankton. Pupae are harvested each day and transferred in netting-covered bowls, either to the stock cages or into experimental cages. These are similar in construction to the stock cages but are only 30 cm^3. The netting covering is used to deny access to gravid females seeking to lay eggs and the newly emerged mosquitoes trapped with the bowls are released each morning and afternoon.

When sporozoites are required, the female mosquitoes are allowed to feed on mice that have been infected with the required parasite species, at a time when gametocytes are just becoming visible in thin blood films. This is usually about 3 days after the mice are infected from donors. The mosquitoes are then held in plastic cups, about 20 per cup. The cups are closed with mosquito netting on top of which is placed a wad of cotton wool soaked in 5% glucose. The cups are held in the insectary at an environmental temperature of 19–20°C for *P. berghei* strains and 24°C for *P. yoelii* or *P. chabaudi* subspecies or strains for a minimum of 11 days. One or two insects are removed from the 11th day onwards and dissected to confirm the presence of sporozoites in the salivary glands.

The sporozoite inoculum is prepared from *A. stephensi* fed 12–14 days earlier on infected mice. The mosquitoes are stunned by concussion and homogenized by hand in a Teflon grinder with Locke's solution with added 3% w/v bovine serum albumin (fraction V). The suspension is lightly centrifuged, decanted and 0.2 ml inocula are given intravenously to mice. Approximately 500 mosquitoes are used to infect 50 mice.

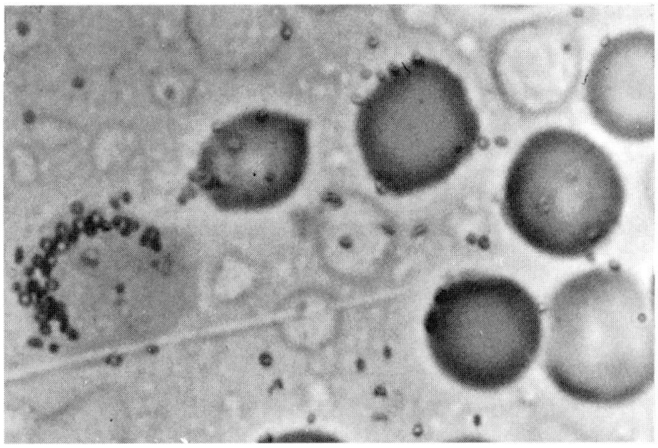

Figure 92.3 *Eperythrozoon coccoides* lining the surface of a mouse erythrocyte. Other isolated organisms appear to be free in the plasma or attached to other red cells. From Peters (1987), with permission.

Maintenance and treatment of animals

Animal accommodation and diet

Malaria infection in mice rapidly results in a fall in their body temperature which must be countered by maintaining them at a sufficiently high environmental temperature and humidity. Animals are normally caged in groups of five per cage which should be held in a dedicated unit. In our institute the unit is held at a temperature of 20±2°C and 55% relative humidity (±10%) with a dedicated air-conditioning system monitored by a computerized control system. The rooms are kept under positive pressure with an exchange flow of 320 cubic feet/minute, providing 20 air changes each hour. The animals are housed in North Kent Plastic RM2 cages (32×20×20 cm) and are maintained on a diet of SDS RM3 expanded diet and water *ad libitum*. The nature of the diet is very important since it must contain adequate *p*-aminobenzoic acid to permit the parasites to develop, but not enough to antagonize the action of such compounds as sulphonamides and sulphones.

Contaminating mouse pathogens

Although it is not necessary to use specific pathogen-free (SPF) animals, it is essential that the mice should come from well-controlled breeding colonies that are uncontaminated by such organisms as *Eperythrozoon coccoides*, *Haemobartonella muris* or the commoner pathogenic murine viral and bacterial infections. *E. coccoides*, which is transmitted by mouse ectoparasites, is a particularly insidious pathogen since it may be present undetected in a mouse colony and only come to light when animals are infected with malaria. The organisms are frequently mistaken by the microscopist for granules of Romanowsky stain (Figure 92.3).

If malaria-infected blood from a mouse contaminated with *E. coccoides* is used to infect new animals or is cryopreserved, the contaminants are carried with the plasmodia and thus perpetuated in new hosts or in the cryobank. The most obvious effect they exert on the host is to haemolyse erythrocytes to which they are attached. They also serve as stimulants of the immune system. Since the course of the intraerythrocytic malaria infection is influenced by the proportions of normocytes and reticulocytes in the circulation, animals contaminated with *E. coccoides* and *H. muris* show great diversity in the levels of parasitaemia from mouse to mouse. Critical review of many early publications on chemotherapy and immunity in rodent malaria suggests that many bizarre results were probably attributable to the authors' unawareness of the presence and/or significance of such contaminants.

Unfortunately, only two types of compound have been identified that will clear these pathogens, namely organic arsenicals such as neoarsphenamine and the tetracyclines. The latter have the disadvantage that they will also remove the malaria parasites, whereas compounds such as neoarsphenamine and oxophenarsine are no longer available. If these contaminating pathogens are detected, therefore, it is best to discard both parasites and hosts and start from clean material. If the status of the mice is not known, e.g. when a new supplier is used, *Eperythrozoon* infection may be readily revealed by splenectomizing a few animals and preparing Giemsa-stained blood films daily. If present, substantial numbers of *Eperythrozoon* will be apparent in the blood within 2–4 days.

Specific techniques for compounds acting on different stages of the life cycle

Causal prophylactics

General. For the routine examination of compounds for causal prophylactic activity we use the modified form of the technique described by Gregory and Peters (1970) and later elaborated by Peters *et al.* (1975a). This has an inherent drawback in its inability to differentiate between direct action on the hepatic schizont and an effect on the emerging blood in the presence of both causal prophylactic and marked residual activity. Nevertheless we feel that, for the majority of compounds, this has proved to be the most satisfactory of the tests available other than the use of *P. cynomolgi* in the rhesus monkey with all its intrinsic problems. In this context it should also be stressed that, unlike the rhesus monkey/*P. cynomolgi* system, the rodent model is designed to demonstrate causal prophylactic action and not antirelapse activity.

Host and parasite species. The most commonly used vector for rodent malaria is *A. stephensi*. The methods used for producing and harvesting the infective sporozoites are described above. The most suitable parasites for studies on causal prophylaxis are:

1. *P. yoelii nigeriensis* (N67): transmissible, slightly chloroquine-resistant strain
2. *P. yoelii yoelii* (17X): transmissible, slightly chloroquine-resistant strain
3. *P. berghei* (ANKA): transmissible, chloroquine-sensitive strain of *P. berghei*.

All three are maintained by cyclical transmission through *A. stephensi* without drug pressure.

Preliminary screening test. This test is designed to indicate the presence of any form of prophylactic activity in mice infected with *P. y. nigeriensis*. Three groups of TFW mice (3 mice/group) are used in this test for each compound:

1. Group 1: sporozoite inoculum at D + 0
2. Group 2: sporozoite inoculum at D + 0; 30 mg/kg test s.c. compound 2 hours post-infection
3. Group 3: sporozoite inoculum at D + 0; 100 mg/kg test s.c. compound 2 hours post-infection.

Mice are infected with sporozoites as described above. Test compounds are prepared for use as described below for the "4-day test". Unless an alternative route is indicated, the compounds routinely are administered subcutaneously. The dose levels used have been arbitrarily selected to give an indication of the doses to be used in the full causal prophylactic test, and may be varied when necessary, e.g. where the test compound is toxic at the proposed dose. Stained blood films from each animal are examined at D + 7 and D + 14. The results are expressed only as positive or negative and four categories of activity are recognized:

1. Grade 1: ? fully active — 0/3 positive
2. Grade 2: ? active — 1/3 positive
3. Grade 3: ? slightly active — 2/3 positive
4. Grade 4: inactive — 3/3 positive.

This preliminary screen affords a simple method of determining the presence or absence of activity and also, by extension of the dose range, it enables a large number of compounds to be screened rapidly to determine the probable effective dose prior to examination in the more time-consuming, full causal prophylactic test.

Definitive test protocol. This test is designed to differentiate between prophylactic activity and residual suppressive effect of test compounds in rodent malaria infections. It does not indicate the presence of antirelapse activity since there is no comparable stage in this infection to the hypnozoite of *P. cynomolgi* or *P. vivax*. The test is based on the inverse linear relationships between the logarithm of the sporozoite inoculum and the mean time taken for the resulting erythrocytic infection rate in groups of mice to reach 2%. This relationship is valid only in an established range and breaks down if the sporozoite inoculum is insufficient to give 100% patency, or if the sporozoite inoculum is extremely large. Furthermore, the test depends on the finding that the minimum prepatent period is between 47 and 50 hours (48 hours has been assigned for purposes of calculation) and the growth rate and drug sensitivity of the erythrocytic stage of the parasite are independent of the source, i.e., whether derived from injected sporozoites or parasitized red blood cells (RBC). The original test contained a control group which received infected RBC only, as well as the control which was infected with both sporozoites and RBC, but a statistical analysis of several hundred control animals has shown that there is no significant difference in the mean prepatent period between these groups. Therefore the control receiving only RBC has been dropped from the test and, as a result, the calculation of residual and prophylactic activity has become much simpler. Four groups of mice are used in the test:

1. Group 1: sporozoite inoculum on D0; saline at D0 + 2 hours
2. Group 2: sporozoite inoculum on D0; test compound at D0 + 2 hours
3. Group 3: as Group 1 + RBC at D0 + 48 hours
4. Group 4: as Group 2 + RBC at D0 + 48 hours.

The sporozoite inoculum is prepared and administered as described for the preliminary screening technique. The blood inoculum, from infected TFW strain mice, is given intravenously in a volume of 0.2 ml and contains 2×10^6 infected RBC in isotonic saline. The parasite we normally use is *P. y. nigeriensis* N67. Test compounds are prepared and administered as previously described for the blood schizontocidal test. Primaquine phosphate is included in all tests as a positive control. Daily blood films are made and examined from D + 3 until the parasitaemia reaches 2%. Any animals which do not show patent infection by D + 14 are considered to be negative.

Differences in the pre-2% patency period (P2PP) between control and treated, sporozoite-inoculated animals can reflect a drug action on exoerythrocytic stages, erythrocytic forms or both. Cross-inoculation in parallel series of groups with infected RBC allows the residual drug action on erythrocytic forms to be assessed, leaving a value proportional to the effect on the EE stages only. Hence if P2PP Group 1 = a, P2PP Group 2 = b, P2PP Group 3 = c, P2PP Group 4 = d, then:

(b − a) = Total activity
(d − c) = Residual activity
(b − a) − (d − c) = Causal prophylactic activity.

Where negative values or positive values less than 0.5 are obtained from these calculations they should be regarded as 0 (zero). Table 92.3 summarizes data on causal prophylaxis

Table 92.3 The activity of primaquine and a primaquine analogue (tafenoquine) in the definitive test for causal prophylaxis. Parasite *P. yoelii nigeriensis* N67

Compound name: Primaguine diphosphate
Route: s.c.
Formulation: Distilled water
Maximum tolerated dose: > 100 mg/kg × 1

Dose	*Control*	*30.0*	*60.0*	*100.0*
Patency — infected with:				
Sporozoites	5/5	5/5	2/5	0/5
Blood stages	5/5	5/5	5/5	5/5
Mean 2% time	(a)	(b)	(b)	(b)
Sporozoite-infected	4.03	4.86	>11.49	>14.00
	(c)	(d)	(d)	(d)
Blood-stage infected	3.59	3.28	3.32	3.29
Total activity				
(b − a)		0.83	>7.46	>9.97
Residual activity				
(d − c)		0	0	0
Prophylactic activity				
(b − a) − (d − c)		0.83	>7.46	>9.97
Comment		Slightly active	Active	Fully active

Minimum fully active dose: 60–100 mg/kg × 1
Residual activity: Nil at 100 mg/kg × 1

Compound name: Tafenoquine
Route: s.c.
Formulation: Distilled water
Maximum tolerated dose: > 100 mg/kg × 1

Dose	*Control*	*10.0*	*30.0*	*100.0*
Patency — infected with:				
Sporozoites	5/5	4/5	1/5	0/5
Blood stages	5/5	5/5	5/5	5/5
Mean 2% time	(a)	(b)	(b)	(b)
Sporozoite-infected	5.55	>7.43	>12.58	>14.00
	(c)	(d)	(d)	(d)
Blood-stage infected	3.30	2.96	5.71	9.84
Total activity				
(b − a)		>1.88	>7.03	>8.45
Residual activity				
(d − c)		0	2.41	6.54
Prophylactic activity				
(b − a) − (d − c)		>3.60	>4.62	>1.91
Comment		Active	Active. Some residual activity	Fully active. Much residual activity

Minimum fully active dose: 30–100 mg/kg × 1
Residual activity present at: 30 mg/kg × 1
Marked at: 100 mg/kg × 1

obtained in the definitive test with two active 8-aminoquinolines.

The amount of compound normally required is 500 mg for preliminary and full test.

Blood schizontocides

The "4-day test". Of all the variations of the test that have been published since the early 1950s, that described here (Peters *et al.*, 1975b) has been the most widely used. Reference to similar procedures will be found in Thompson and Werbel (1972) and Peters (1987).

Preliminary "4-day test" for detection of suppressive activity. For each compound under test, two batches of five male, random-bred Swiss albino mice weighing 18–22 g are inoculated intravenously with 2×10^6 RBC parasitized with *P. berghei* N and a further two groups with *P. chabaudi* AS strain. Animals are then given a fixed dose of 30 mg/kg once daily for 4 consecutive days beginning on the day of infection (D0). Compounds are dissolved or suspended, using ultrasonication to achieve an even suspension, in sterile distilled water with Tween 80 and administered subcutaneously to one group of each infection and orally to a second group. Where exceptional difficulty is encountered in preparing an aqueous preparation, the test compound is first dissolved in dimethyl sulphoxide and subsequently an aqueous dilution is prepared for use. Some highly lipophilic compounds may be dissolved in sesame oil and dilutions made from this for s.c. or oral administration. In this case sesame oil must also be used as a control. The parasitaemia is determined on the day following the last treatment (D + 4) to determine qualitatively the presence and degree of activity at the screening dose.

The amount of compound required is 100 mg.

"4-day test" for quantitative assessment of blood schizontocidal activity. Male, random-bred Swiss albino mice weighing 18–22 g are inoculated intravenously with 2×10^6 RBC parasitized with one of the above strains. Animals are then dosed once daily for 4 consecutive days beginning on the day of infection (i.e. D0 to D + 3), using a dose range determined by the results obtained in the preliminary test. Compounds dissolved or suspended as described above are administered s.c., i.p., p.o. or by such other route as may be required. The parasitaemia is determined on the day following the last treatment (D + 4) and the ED_{50} and ED_{90} values, i.e. 50% and 90% suppression of parasitaemia when compared with untreated controls, are estimated from a plot of log-dose against probit-activity. Standard deviations are determined using Microsoft Excel. The degree of cross-resistance (I_{50} or I_{90}) is determined by comparing activity in the sensitive and resistant strains using the following formula:

$$\text{Index of cross-resistance} = \frac{ED_{50}/ED_{90}\ \text{Resistant line}}{ED_{50}/ED_{90}\ \text{Sensitive parent strain}}$$

The amount of compound required is 250–1500 mg depending on active dose level found in the preliminary screen.

"Extended test" for blood schizontocidal activity. Male, random-bred, Swiss albino mice weighing 18–22 g are inoculated intravenously with 2×10^6 *P. chabaudi*-infected RBC. Animals are then dosed daily for 14 consecutive days beginning on the day of infection. Compounds prepared as above are administered s.c., i.p., p.o. or by such other route as may be required. The parasitaemia is monitored daily from D+1 to D+35 using Giemsa-stained blood films and a graph illustrating the course of infection is plotted. *P. chabaudi* AS strain infections can be expected to prove lethal by D+7 to D+8 in untreated controls and activity can be demonstrated by an extension of life. Radical cure can be proved by subinoculation of blood from animals which are aparasitaemic at D+35 into clean mice. In addition, the parasitaemias on D+4 may be used to determine the ED_{50} and ED_{90} of active compounds as described above.

A further application of extended observation using this parasite or *P. berghei* N is to observe the effect of administering a single large drug dose on the level of parasitaemia, rate of recrudescence and morphology of the treated parasites. In this case the drug is given at a time when the parasitaemia has reached between 10 and 20% and daily parasite counts are made in thin blood films. This is similar to the initial stage of the "2% relapse technique" described below.

The amount of compound required is 1500 mg.

Suppressive curative activity. Compounds shown to possess suppressive activity can be further tested at a dose range determined from the "4-day test" results and including one dose step beyond the minimum fully suppressive dose, for suppressive curative activity and influence on survival times. TFW mice are infected with *P. berghei* (N strain) and treated in the same manner as described above for the "4-day test". The parasitaemia is monitored on D + 4 and D + 7 and a record of survival times is kept. Any mice surviving beyond D + 7 have their parasitaemia monitored by examination of blood films every 7 days until D + 35. On D + 35, pooled blood from each blood film-negative group is subinoculated into clean mice and the absence of patent infection in these mice within 21 days is regarded as evidence of curative activity. This technique is very similar to the so-called "Thompson test" (Thompson and Werbel, 1972).

The advantage of infecting mice i.v. can be seen from Figure 92.4 which is made from part of the data obtained in a test of the suppressive curative activity of chloroquine phosphate. The figure shows individual parasitaemias in mice infected with *P. berghei* N strain on D0 treated with the subcurative dose of 3 mg/kg per day ×4 s.c. of chloroquine, compared with those in untreated control animals. The points to be noted are the relative closeness of the counts in individual control mice; the marked action of chloroquine which, at this dose, represents approximately

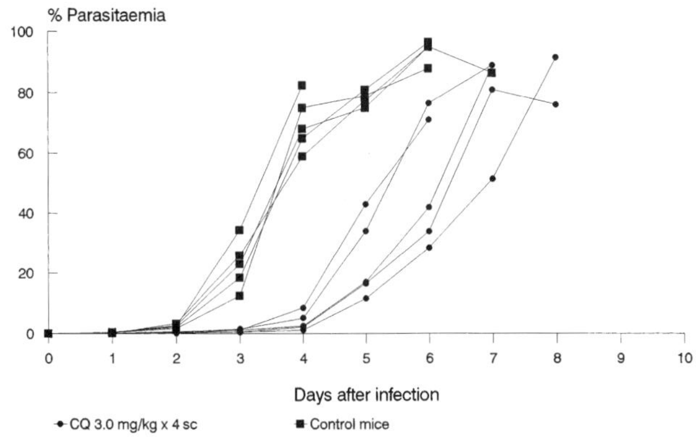

Figure 92.4 The action of 3 mg/kg per day×4 s.c. of chloroquine phosphate in the suppressive curative test, on the parasitaemia and survival of mice infected with *Plasmodium berghei* N strain (see text for comments).

the ED_{95} as it would be assessed if this were a "4-day test", and the rapid rise of parasitaemia of this virulent parasite after D + 4. All control animals died from days +4 to +7 and treated animals died between days +6 and +8.

"Rane test". This is a simple test that was used to screen the bulk of the 300 000 compounds that entered the WRAIR antimalarial programme. It had the advantage of requiring minimal labour since it was based essentially on the observation of the levels of mortality of *P. berghei*-infected mice and did not, in its early form at least, require the preparation and microscopic examination of blood films. Although it could be criticized as being a crude procedure, the Rane test did, in fact, yield an unprecedented volume of data from which promising new lead compounds were identified for more detailed evaluation in other test systems (Peters, 1987). Basically the test was conducted as follows.

Mice are inoculated i.p. with 10^6 infected donor RBC on D0. Test drugs prepared as solutions or ultrasonicated in arachis oil are administered on a single occasion at an initial dose range of 80, 160, 320 and 640 mg/kg s.c. on D + 3. The animals are observed daily for survival or death. Survival time to more than twice that of the controls is taken as evidence of activity and survival beyond 60 days is considered as a cure. The minimum effective dose (MED) is compared with the maximum tolerated dose, i.e. the dose that produces no more than 1 in 5 toxic deaths. Dose levels are titrated downwards to find the MED and to give an indication of the therapeutic index. Promising compounds revealed by the Rane test can then be submitted to the "4-day test" or similar procedure to obtain more accurate information on their activity.

Test for repository action. The duration of action of a compound depends, firstly, on its rate of absorption from the site of administration (i.e. bioavailability) and then on other factors such as tissue binding, rate of metabolism and rate of excretion. Poorly soluble compounds, those that are highly bound to body tissues and others that are only slowly metabolized may have a long duration of action. The rodent model can be adapted to give an indication of this "repository" effect as follows.

Seven groups of 5 TFW mice are treated with the maximum tolerated dose of the compound under test. An untreated control group is set up in parallel with each treated group. Pairs of groups are challenged with infective inocula of 2×10^6 parasitized RBC from *P. berghei* N strain-infected donor mice at intervals of 1, 2, 4, 7, 14, 21 and 28 days after treatment. The development of infection after challenge is monitored to the 2% level by examination of Giemsa-stained, thin blood films and the results compared with that of the control group to determine the days' delay to 2% parasitaemia produced by the compound. In the event of a compound still showing a repository effect at the 28-day challenge, a further test is performed using an extended dosing regime, i.e. 35, 42, 49, 56, 63, etc. days. Any compound possessing a significant repository effect at the dosage employed in this test is re-examined through a range of doses using the same technique.

The amount of compound required is 500–1500 mg, depending on the maximum tolerated dose.

Induction of drug resistance—the "2% relapse technique" (Peters et al., *1970).* The rapid spread of resistance to antimalarials over the past few years has made it imperative to have means of protecting not only the present generation of drugs, but also those new compounds that are currently being developed. One of the important features of a novel antimalarial, therefore, is the rate at which resistance to it can be selected. We employ the following procedure. This may be used not only on single compounds (e.g. Ridley *et al.*, 1996) but also on effective combinations, identified in drug interaction tests. It is our experience that this technique is a far more realistic model for resistance acquisition than that employing steadily increasing drug doses. The latter can produce artificially high levels of resistance that have no counterpart in the resistant strains of human malaria. The best example of this is our *P. berghei* RC strain (Peters, 1965) which has a very high level of resistance to chloroquine but many abnormal morphological and physiological characteristics. Studies of the influence of drug mixtures on the rate of development of resistance to mefloquine using the "2% relapse" technique have been reported by Peters *et al.* (1977) and more recently the combination of mefloquine and a mixture of pyrimethamine and sulfadoxine were investigated by the same means (Peters and Robinson, 1984). It is essential that further studies should be performed to investigate this important means of prolonging the useful life of those drugs presently in use, as well as those which are due to be introduced and which appear already to be vulnerable to the development of resistance.

As a prelude to the "2% technique", a range of single doses of the test drug is given an hour following the infection of groups of mice and the development of parasitaemia is followed in daily blood films. That dose that prolongs the attainment of 2% for a period of 7–10 days is then selected for administration at the time of each passage. Daily blood films are made and the development of the parasitaemia is monitored. Comparison of the time taken to reach a 2% parasitaemia in this treated group with that in an untreated control group reveals the "2% delay time". This value is expressed as a percentage of the delay time of the parent strain as found at passage 1 of the series and may be plotted against time in a graph which shows the course of development of resistance. The strain is passaged from the treated group as soon as possible after 2% patency is reached and each successive passage is treated with the same, single, fixed dose as was used in the first passage of the series. Resistant lines are developed in this technique from both drug-sensitive *P. berghei* N strain and chloroquine-resistant *P. yoelii* ssp. NS strain. It has been our experience that, generally, resistance is far more readily acquired by the NS strain than by a drug-sensitive line (Peters and Robinson, 1992). This reflects the picture seen in the development and spread of multidrug-resistant *P. falciparum* and, indeed, this technique has a proven record of predicting the ease with which resistance to new antimalarials will develop. Figure 92.5 illustrates the rapid development of resistance to atovaquone using this technique and the stability of resistance when the drug selection pressure is withdrawn.

Once maximum resistance has been acquired using the relapse technique, the strain is passaged under drug pressure until stability of the resistance is apparent from consistent 2% delay times. At this point, the process is reversed and a subline is started in which the infection is serially passaged using the untreated control mice as donors. This subline reveals whether the resistant parasite is capable of maintaining resistance in the absence of drug selection pressure. The line is cryopreserved at the first passage and at intervals during its development to allow a comparison of actual levels of resistance in the developing strain to be made with the parent strain by the use of the "4-day test". The maximally resistant strain is added to the reference cryobank for possible future use in drug testing.

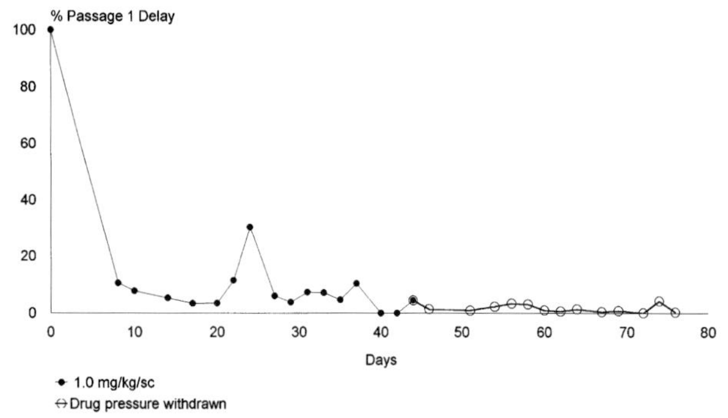

Figure 92.5 Resistance developing in a single passage to atovaquone in a line of *Plasmodium yoelii* ssp. NS in the "2% relapse technique". The acquisition of resistance is indicated by a decrease in the "% Passage 1 delay time". In this example the level of resistance remained constant even after the release of drug selection pressure.

The amount of compound required depends on initial doses and outcome.

Combination studies. There are a number of reasons for deploying drug combinations. These include the exploitation of combinations that display synergism, thus enabling lower doses of the individual compounds to be administered and of drug pairs that are able to reverse drug resistance (e.g. verapamil with chloroquine)—so-called resistance modulators, some of which may possess no inherent antimalarial activity. Since some drug combinations, however, may also prove to be antagonistic, it is essential to be able to detect these at an early stage.

The procedures we follow are as follows. The "4-day test" technique has proved itself to be a sensitive system for detecting interactions between drugs. When two compounds are simultaneously administered in an appropriate series of dilutions, it is possible to determine the influence of one compound upon the ED_{90} of the other in a series of combination ratios. The ED_{90} values obtained with combinations in a test of this type may be compared with those of the individual compounds to obtain an isobolar equivalent. These are plotted for each compound in an isobologram in order to demonstrate the presence of synergism, antagonism or a simple additive action.

It is, however, frequently the case that compounds which themselves have little or no inherent antimalarial activity may influence the action of, for example, chloroquine on chloroquine-resistant parasites (e.g. verapamil), providing combinations where only one compound has a direct action on the parasite. This precludes the use of the conventional isobologram for the analysis of the interaction between the two compounds, since it is necessary to know the ED_{90} of both compounds for this technique. To overcome this difficulty we have devised a method to illustrate graphically and to quantify the influence of an antimalarially inactive compound on the activity of a known antimalarial in the rodent model (Peters and Robinson, 1991).

Graphically, this is done by plotting the ED_{90} of the active partner against varying dosage of the antimalarially inert drug on a simple graph which has a line indicating the ED_{90} of the active drug drawn on it. Figure 92.6 illustrates this principle and shows how the different types of interaction present themselves by this technique. The single compound ED_{90} line is bounded by the extreme limits of confidence and points falling within this zone indicate a failure to influence the activity of the antimalarial partner compound. Synergism is shown by a graph which drops away from the ED_{90} line toward the bottom axis whilst, if antagonism is present between the two compounds, the curve will rise progressively as the dose increases. A quantitative assessment of the degree of influence of the inactive compound on its antimalarial partner is made by calculating the activity enhancement index (AEI). This involves a simple calculation in which the ED_{90} of the active compound alone is divided by the ED_{90} of the paired drugs at each dose of the "inert" compound. For example, in order to quantify the influence of a

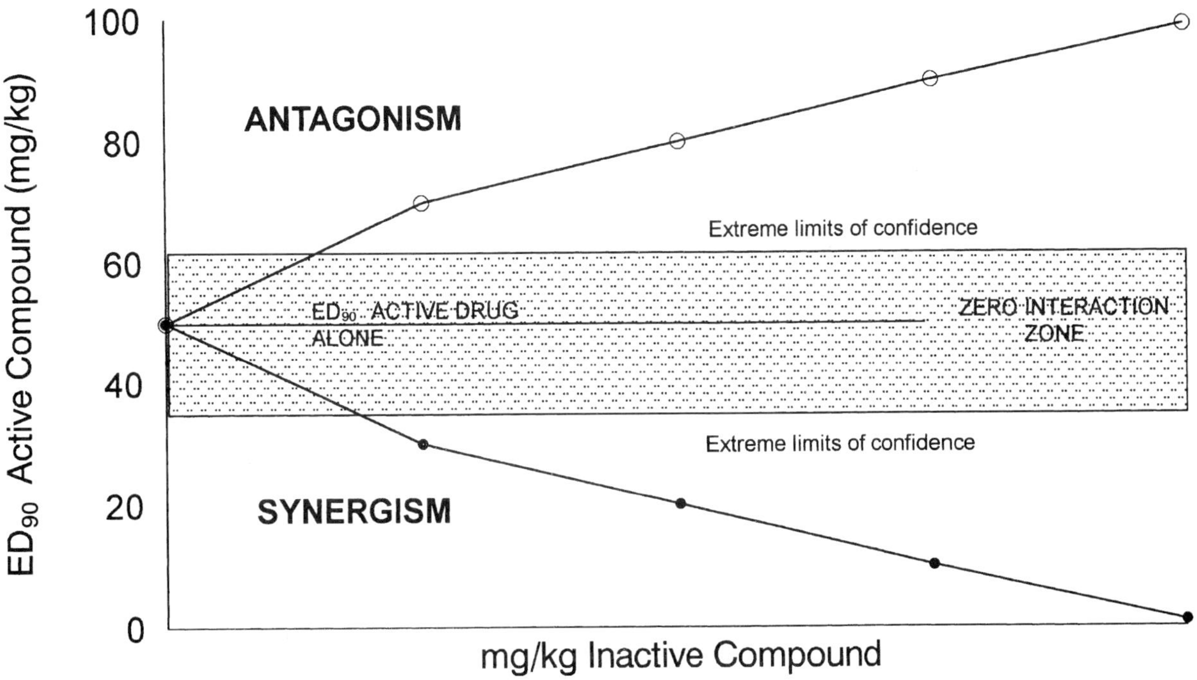

Figure 92.6 Graphic illustration of the possible drug interactions when only one of a pair of drugs administered concurrently possesses innate blood schizontocidal activity.

second compound on the activity of chloroquine, we compare the ED_{90} of chloroquine alone (CQ) with that of chloroquine combined with the test substance (CQ + Drug X):

$$AEI = \frac{ED_{90}\,CQ}{ED_{90}(CQ + \text{Drug X})}.$$

In effect, this regards the activity of chloroquine alone as representing an AEI of 1.0 and enhanced activity resulting from the combination of chloroquine with a second compound produces an AEI value greater than 1.0. Similarly, an antagonistic interaction would be indicated by the AEI being reduced to a value lower than 1.0. This will apply whether the companion compound possesses antimalarial activity in its own right or not and regardless of the cause of enhancement, e.g. synergism, or reversal of resistance. The use of AEI analysis not only permits a direct comparison to be made between a series of compounds in separate experiments, but may also be used to indicate the dose of an individual compound which produces optimal enhancement of activity in a second compound (Peters *et al.*, 1990, 1993).

The amount of compound required is 1 g.

Gametocytocides

Comparative studies of gametocytocidal and sporontocidal action can give a valuable indication of the need for a compound to be metabolized to an active form, e.g. primaquine is highly active as a gametocytocide, but it is inactive in the sporontocide test (Peters and Robinson, 1987). As described previously, mice are used in this test. It is essential that both the donor mice and the vectors should receive adequate *p*-aminobenzoic acid in their diets in order to obtain optimal infection rates in the mosquitoes (Peters and Ramkaran, 1982). On D + 0 mice are intravenously infected with 2×10^6 infected red cells from a *P. y. nigeriensis*-infected donor. On the third day after infection the animals are given a single s.c. or i.p. dose of test compound. For screening purposes we use the arbitrarily selected dose of 30 mg/kg in the first test and subsequently extend the range as indicated in a further test. Groups of female *A. stephensi* are permitted a blood meal of 30 minutes' duration from the treated mice at intervals of 1, 3, 6 and 24 hours after treatment. Approximately 25 mosquitoes are dissected from batches fed at each of these intervals and the oocysts are counted on the individual midguts after staining with Mayer's haemalum and mounting in Gurr's Aquamount. Mean oocyst counts from mosquitoes fed on treated mice are compared with counts from those fed on control animals and the percentage suppression of infection is calculated. Primaquine phosphate is included in each test as a positive drug control and chloroquine as a negative. Ramkaran and Peters (1969a) observed that the latter actually enhanced the infectivity of a chloroquine-resistant parasite.

The amount of compound required is 50–100 mg.

Sporontocides

The same procedure as described for the gametocytocidal test is followed, except that the mosquitoes are fed on untreated rather than treated mice (Ramkaran and Peters, 1969b). After feeding on the gametocyte carriers the mosquitoes are held in waxed cardboard cartons at 19–21 or

24°C (depending on the parasite species) at 75% relative humidity and are fed 0.05% solutions of the test drugs in 4% sucrose solution supplied in cotton wool pads placed on top of the gauze covers of the containers. The pads are replaced with fresh ones every day. Mosquitoes are dissected on the seventh day after the blood meal and the oocysts are counted as in the gametocytocidal test.

The amount of compound required is 50–100 mg.

Key parameters for evaluating response to drugs

General parameters

Acute toxicity of a test compound in, for example, the "4-day test", may become evident at any time from D0 when the first dose is administered up to D + 4 when these experiments are terminated. Toxicity may be manifested in many ways, for example by a decreased intake of food, physical signs such as rough hair, agitation or decreased movement or actual death. Careful observation and experience are needed to distinguish signs of toxicity from the signs that accompany an increasing malaria parasitaemia. The latter include listlessness, shivering (due to a decrease in body temperature), pallor due to the malaria-induced haemolysis. The most virulent strains such as *P. berghei* N usually kill untreated control mice that have received an inoculum of 2×10^6 parasites in 6–8 days.

Assessment of parasitaemia

Thin blood films are made on cleaned slides from a drop of blood collected by snipping a few millimetres from the tip of the tail with a very sharp, well-cleaned pair of scissors. The films are rapidly air-dried, fixed with methanol, then stained with a good, well-filtered Romanowsky stain. We employ Giemsa stain that is prepared from commercially available powder by dissolving 0.8 g in 100 ml of a 1:1 Analar methanol/glycerol mixture with the aid of ultrasonication. The concentrated solution is diluted to 5% in distilled water that has been buffered to pH 7.2 by the addition of 3.0 g Na_2HPO_4 and 0.6 g KH_2PO_4 per litre. Thick blood films are not suitable as many murine parasites reside not in mature normocytes but in reticulocytes, the nuclear residues of which can be mistaken for parasites.

The numbers of asexual parasites are counted and recorded in relation to the numbers of erythrocytes, using a ×100 oil immersion lens. In heavy infections (e.g. 20% or more of the RBC infected) it may only be necessary to count parasites in up to five microscope fields. For lighter infections, parasites will have to be counted in relatively more fields. In the lowest levels of infection it may be necessary to examine up to 100 or 200 fields before declaring a film to be negative. Care must be taken not to mistake for parasites, granular deposits of stain from poorly prepared stock solutions or from badly washed films. Attention is drawn also to the possible presence of contaminants such as *E. coccoides* (see above) that can also resemble stain deposits.

In the "4-day test", parasite counts are routinely made only on D + 4 but, when required for other tests, they may be made daily. The mean values of the D + 4 counts for mice treated at each dose level are expressed as a percentage of control values. An appropriate computer program such as Microsoft Excel can be used to calculate these values and their standard deviations. The percentages, with their confidence limits, are then plotted against dose with the aid of log-dose/probit-activity graph paper. From the graph which yields a straight line relating dose and activity, it is simple to read off the dose levels that produce 50 or 90% reduction in parasitaemia, i.e, the ED_{50} and ED_{90} values, as well as the extreme ranges. Furthermore, an index of resistance (ID_{90}) can be calculated as indicated above.

Careful examination of the morphology of parasites from treated mice may also give a clue to the general mode of drug action. Chloroquine and other 4-aminoquinolines, for example, cause clumping of the haemozoin granules within trophozoites (Figure 92.7), although this is more likely to be detected if animals in which parasitaemia is already advanced are then given an effective drug dose in order to observe the drug's action on the numbers and morphology of the parasites.

Assessment of activity on liver stages

The procedures outlined above for the definitive causal prophylactic test permit an evaluation of a test drug by following parasitaemia in sporozoite-infected animals. Known positive controls (e.g. primaquine) and negative controls (e.g. chloroquine) should be included when assessing a causal prophylactic compound.

It is especially important to distinguish between genuine activity against the liver schizonts and action on first-generation merozoites emerging from them; this is especially likely with a compound that possesses a significant repository action. However, pre-erythrocytic schizonts can also be sought directly in Giemsa-stained microscope sections of the livers of treated and control animals, but only if they receive exceptionally heavy sporozite inocula (Boulard *et al.*, 1986). This procedure, like that referred to above for studying drug action on parasites in hepatic cell or hepatoma monolayers, is very labour-intensive and only of practical value in a specialized laboratory.

Advantages and limitations of rodent models

The main advantages of the rodent malarias relate both to the parallels of their life cycles with those of the *Plasmodium*

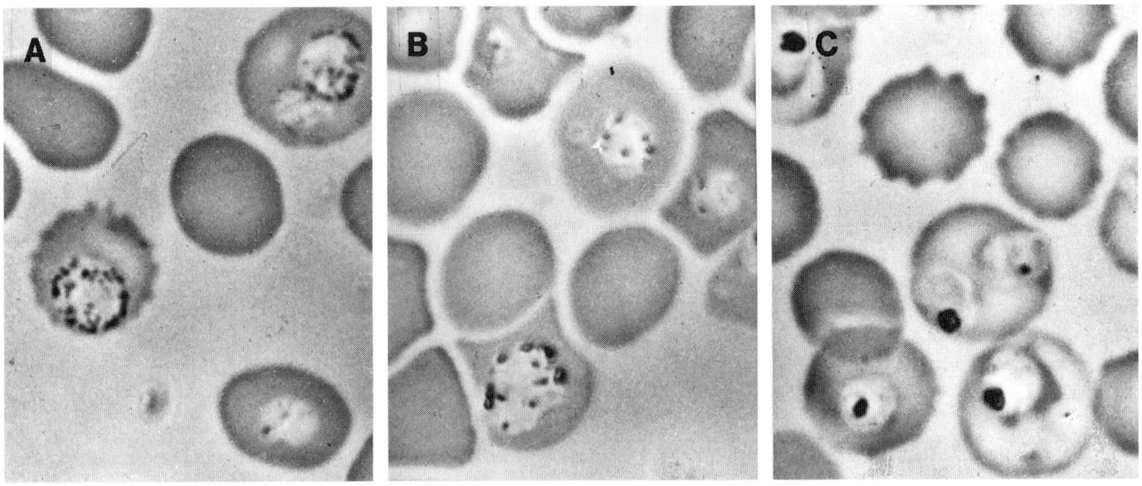

Figure 92.7 Morphological changes in the haemozoin granules of *Plasmodium berghei* in a thin film of unstained fresh blood seen by phase contrast microscopy (**A**) before and (**B**, **C**) after exposure to chloroquine *in vitro*. Note the clumping of the haemozoin which is already apparent after 10 minutes (**B**) and marked clumping seen after 40 minutes (**C**). Magnification ×3000. Courtesy of Dr D. C. Warhurst.

species of humans, and to practical considerations of cost and ease of manipulation. As described above, the life cycle of all murine species is essentially the same, other than in the duration of the different stages, as those of *P. falciparum* and *P. malariae* but differs from *P. vivax* and *P. ovale* in the absence of a hypnozoite stage. The rodent infections can serve, therefore, as models for causal prophylactic, blood schizontocidal, gametocytocidal and sporontocidal drugs.

Since the identification of novel antimalarials that are effective against multidrug-resistant falciparum malaria is one of the current prime objectives, the ready availability of different drug-resistant lines derived from *P. berghei*, *P. yoelii* and *P. chabaudi* subspecies and strains is a marked advantage for these models. Moreover, the ease or difficulty of developing lines resistant to novel compounds has proved invaluable in forecasting their resistance-developing potential once deployed in humans in the field.

The ability to use random-bred mice which are easy to handle and cheap to maintain is a distinct advantage over alternative vertebrate hosts. Avian malaria parasites which have a different life cycle have now been abandoned both for this reason and because baby chicks and ducks are more difficult to handle and maintain than mice.

The different species of monkeys that are hosts to simian or human *Plasmodium* parasites have advantages in that they provide data that may be more directly relevant to the human malarias but also have obvious disadvantages in terms of availability, cost, relative difficulty of handling and maintenance and ethical considerations.

Rodent models have several disadvantages also, particularly as regards drug metabolism. One classical example is proguanil, which is rapidly cyclicized in the liver to an active triazine, cycloguanil. Unlike the situation in humans, the biological serum half-life of cycloguanil in the mouse is far greater than that of proguanil, whereas in humans cycloguanil is rapidly excreted (unpublished data). A second example that shows the contrary situation is mefloquine, which has a prolonged biological half-life in humans, but has only a short duration of action in the mouse. A further disadvantage is that the drug responses of the various murine species to certain types of antimalarial may differ. The examples mentioned earlier are iron chelators and phospholipid synthesis inhibitors. This may necessitate employing more than one rodent *Plasmodium* species for screening.

Contributions of rodent models to prevention and treatment

Since 1948 when *P. berghei* was first discovered, extensive use has been made of the rodent malaria models for new drug discovery and development. Their contribution is reflected by comparing the 4000 compounds screened against avian malaria during World War II and the more than 300 000 compounds screened against rodent models between about 1963 and 1990. Table 92.4 shows some of the compounds that have reached clinical trial or that are currently in clinical use, and the potential of which was first made clear by studies in mice.

By using a battery of drug-resistant lines derived from *P. yoelii* ssp. NS, we have been able to identify and forecast the potential value of all the compounds listed in Table 92.4. On the basis of studies on the development of resistance in mice we have also been able to forecast the ease with which resistance to compounds such as mefloquine and pyronaridine can arise when these compounds are deployed singly against multidrug-resistant *P. falciparum* or when antimetabolites such as sulphonamides, cycloguanil, pyrimethamine or atovaquone are deployed singly against chloroquine-sensitive parasites. Moreover, the value of

Table 92.4 Antimalarials evaluated in rodent models that entered trials or are currently used in humans

Compound	Current status	
	In trial	*In clinical use*
Single compounds		
Mefloquine		+
Halofantrine		+
Benflumetol	+	
Artemisinin		+
Dihydroartemisinin	+	
Artemether		+
Arteether	+	
Artesunate		+
Sulfadoxine		+
WR 99,210	+	
PS-15	+	
Menoctone	+	
Atovaquone		+
Tafenoquine	+	
Combinations		
Sulfadoxine–pyrimethamine		+
Sulfalene–pyrimethamine		+
Sulfadoxine–pyrimethamine–mefloquine		+
Atovaquone–proguanil		+
Chlorproguanil–dapsone	+	
Mefloquine–artemether		+
Benflumetol–artemether	+	

selected drug combinations in delaying the onset of resistance was clearly demonstrated with, for example, pyrimethamine–sulfadoxine in the mouse model and fully confirmed when this combination came to be used extensively in the clinic.

Finally, rodent malaria models have proved invaluable in studies on the molecular biology and genetics of drug resistance (e.g. Walliker, 1983; Van Dijk *et al.*, 1996). The relative ease of crossing different resistant lines in the anopheline host has yielded much information on the nature of the transfer and dissemination of resistance which is relevant to the situation of multidrug-resistant *P. falciparum* in nature.

References

Boulard, Y., Landau, I., Miltgen, F., Peters, W., Ellis, D. S. (1986). The chemotherapy of rodent malaria XLI. Causal prophylaxis part V. Effect of mefloquine on exoerythrocytic schizogony in *Plasmodium yoelli yoelli*. *Ann. Trop. Med. Parasitol.*, **80**, 577–580.

Butcher, G. A., Couchman, A. F., Van Pelt, J. F., Fleck, S. L., Sinden, R. E. (1993). The SCID mouse as a laboratory model for development of the exoerythrocytic stages of human and rodent malaria. *Exp. Parasitol.*, **77**, 57–260.

Butler, D., Maurice, J., O'Brien, C. (1997). Briefing malaria. Time to put malaria control on the global agenda. *Nature*, **386**, 535–540.

Caillard, V., Beauté-Lafitte, A., Chabaud, A., Landau, I. (1992). *Plasmodium vinckei petteri*: identification of the stages sensitive to arteether. *Exp. Parasitol.*, **75**, 449–456.

Carter, R., Walliker, D. (1975). New observations on the malaria parasite of rodents of the Central African Republic—*Plasmodium vinckei petteri* subsp. nov. and *Plasmodium chabaudi*, Landau 1965. *Ann. Trop. Med. Parasitol.*, **69**, 187–196.

Carter, R., Walliker, D. (1977). Malaria parasites of rodents of the Congo (Brazzaville): *Plasmodium chabaudi adami* subsp. nov. and *P. vinckei lentum*. Landau, Michel, Adam, Boulard (1970). *Ann. Parasitol. Hum. Comp.*, **51**, 637–646.

Gilbert, S. C., Hill, A. V. S. (1997). Protein particle vaccines against malaria. *Parasitol. Today*, **13**, 302–306.

Grau, G. E., Fajardo, L. F., Piguest, P.-F., Allet, B., Lambert, P.-H. Vassalli, P. (1987). Tumor necrosis factor (cachectin) as an essential mediator in murine cerebral malaria. *Science*, **237**, 1210–1212.

Gregory, K. G., Peters, W. (1970). The chemotherapy of rodent malaria, IX. Causal prophylaxis, Part 1. A method for demonstrating drug action on erythrocytic stages. *Ann. Trop. Med. Parasitol.*, **64**, 15–24.

Harrison, G. (1978). *Mosquitoes, Malaria and Man: A History of the Hostilities since 1880*. John Murray, London.

Janse, C. J., Mons, B., Croon, J. J. A. B., Van der Kaay, J. J. (1984). Long-term *in vitro* cultures of *Plasmodium berghei* and preliminary observations on gametocytogenesis. *Int. J. Parasitol.*, **14**, 317–320.

Killick-Kendrick, R. (1973). Parasitic protozoa of the blood of rodents. 1. The life-cycle and zoogeography of *Plasmodium berghei nigeriensis* subsp. nov. *Ann. Trop. Med. Parasitol.*, **67**, 261–277.

Killick-Kendrick, R. (1975). Parasitic protozoa of the blood of rodents. V. *Plasmodium vinckei bruchechwatti* subsp. nov. A malaria parasite of the thicket rat *Thamnomys rutilans* in Nigeria. *Ann. Parasitol. Hum. Comp.*, **50**, 251–264.

Killick-Kendrick, R., Peters, W. (ed.) (1978). *Rodent Malaria*. Academic Press, London.

Landau, I. (1965). Description de *Plasmodium chabaudi* n. sp., parasite de rongeurs africains. *C. R. Hebd. Seance Acad. Sci. Paris*, **260**, 3758–3761.

Landau, I., Boulard, Y. (1978). Life cycles and morphology. In *Rodent Malaria* (eds Killick-Kendrick, R. Peters, W.), pp. 53–84. Academic Press, London.

Landau, I., Gautret, P. (1998). Animal models; rodents. In *Malaria Parasite Biology, Pathogenesis and Protection* (ed Sherman, I. W.), pp. 401–417. ASM Press, Washington.

Landau, I., Killick-Kendrick, R. (1966). Rodent plasmodia of the République Centrafricaine: the sporogony and tissue stages of *Plasmodium chabaudi* and *P. berghei yoelii*. *Trans. R. Soc. Trop. Med. Hyg.*, **60**, 633–649.

Landau, I., Michel, J. C., Adam, J. P. (1968). Cycle biologique au laboratoire de *P. berghei killicki* n. subsp. *Ann. Parasitol. Hum. Comp.*, **43**, 545–550.

Landau, I., Michel, J. C., Adam, J. P., Boulard, Y. (1970). The life cycle of *Plasmodium vinckei lentum* subsp. nov. in the laboratory: comments on the nomenclature of the murine malaria parasites. *Ann. Trop. Med. Parasitol.*, **64**, 315–323.

Millet, F., Landau, I., Baccam, D., Miltgen, F., Mazier, D., Peters, W. (1985). Mise au point d'un modèle expérimental "rongeur" pour l'étude *in vitro* des schizonticides exo-érythrocytaires. *Ann. Parasitol. Hum. Comp.*, **60**, 211–212.

Mons, B., Boorsma, E. G., Van der Kaay, H. J. (1985). Studies on the loss of sexual capacity in the ANKA isolate of *Plasmodium berghei* and the reversibility of the process. *Ann. Soc. Belge Méd. Trop.*, **65** (suppl. 2), 1–6.

Pasvol, G., Clough, B., Carlsson, J., Snounou, G. (1995). The pathogenesis of severe falciparum malaria. In *Clinical Infectious Diseases*, vol. 2, no. 2. *Malaria* (ed Pasvol, G.), pp. 249–270. Ballière Tindall, London.

Peters, W. (1965). Drug resistance in *Plasmodium berghei*, Vincke and Lips, 1948. I. Chloroquine resistance. *Exp. Parasitol.*, **17**, 89–90.

Peters, W. (1987). *Chemotherapy and Drug Resistance in Malaria*, 2 vols, 2nd edn. Academic Press, London.

Peters, W., Ramkaran, A. E. (1982). The chemotherapy of rodent malaria, XXXII. The influence of *p*-aminobenzoic acid on the transmission of *Plasmodium yoelii* and *P. berghei* by *Anopheles stephensi*. *Ann. Trop. Med. Parasitol.*, **74**, 275–282.

Peters, W., Robinson, B. L. (1984). The chemotherapy of rodent malaria, XXXV. Further studies on the retardation of drug resistance by the use of a triple combination of mefloquine, pyrimethamine and sulfadoxine in mice infected with *P. berghei* and "*P. berghei* NS". *Ann. Trop. Med. Parasitol.*, **78**, 459–466.

Peters, W., Robinson, B. L. (1987). The activity of primaquine and its possible metabolites against rodent malaria. In: *Primaquine: Pharmocokinetics, Metabolism, Toxicity and Activity* (eds Wernsdorfer, W. H., Trigg, P. I.), pp. 93–101. John Wiley, New York.

Peters, W., Robinson, B. L. (1991). The chemotherapy of rodent malaria. XLVI. Reversal of mefloquine resistance in rodent *Plasmodium*. *Ann. Trop. Med. Parasitol.*, **85**, 5–10.

Peters, W., Robinson, B. L. (1992). The chemotherapy of rodent malaria. XLVII. Studies on pyronaridine and other Mannich base antimalarials. *Ann. Trop. Med. Parasitol.*, **86**, 455–465.

Peters, W., Bafort, J., Ramkaran, A. E. (1970). The chemotherapy of rodent malaria, XI. Cyclically transmitted, chloroquine-resistant variants of the Keyberg 173 strain of *Plasmodium berghei*. *Ann. Trop. Med. Parasitol.*, **64**, 41–51.

Peters, W., Davies, E. E., Robinson, B. L. (1975a). The chemotherapy of rodent malaria, XXIII. Causal prophylaxis, Part 2. Practical experience with *Plasmodium yoelii nigeriensis* in drug screening. *Ann. Trop. Med. Parasitol.*, **69**, 311–328.

Peters, W., Portus, J. H., Robinson, B. L. (1975b). The chemotherapy of rodent malaria, XXII. The value of drug resistant strains of *P. berghei* in screening for blood schizontocidal activity. *Ann. Trop. Med. Parasitol.*, **69**, 155–171.

Peters, W., Portus, J., Robinson, B. L. (1977). The chemotherapy of rodent malaria, XXVIII. The development of resistance to mefloquine (WR142,490). *Ann. Trop. Med. Parasitol.*, **71**, 419–427.

Peters, W., Chance, M., Lissner, R., Momen, H., Warhurst, D. C. (1978). The chemotherapy of rodent malaria. XXX. The enigmas of the "NS" lines of *P. berghei*. *Ann. Trop. Med. Parasitol.*, **72**, 23–36.

Peters, W., Ekong, R., Robinson, B. L., Warhurst, D. C., Pan X-Q. (1990). The chemotherapy of rodent malaria. XLV. Reversal of chloroquine resistance in rodent and human *Plasmodium* by antihistaminic agents. *Ann. Trop. Med. Parasitol.*, **84**, 541–555.

Peters, W., Robinson, B. L., Milhous, W. K. (1993). The chemotherapy of rodent malaria. LI. Studies on a new 8-aminoquinoline, WR 238,605. *Ann. Trop. Med. Parasitol.*, **87**, 547–552.

Ramkaran, A. E., Peters, W. (1969a). Infectivity of chloroquine resistant *Plasmodium berghei* to *Anopheles stephensi* enhanced by chloroquine. *Nature*, **223**, 635–636.

Ramkaran, A. E., Peters, W. (1969b). The chemotherapy of rodent malaria, VIII. The action of some sulphonamides alone or with folic reductase inhibitors against malaria vectors and parasites. Part 3: The action of sulphormethoxine and pyrimethamine on the sporogonic stages. *Ann. Trop. Med. Parasitol.*, **63**, 449–454.

Ridley, R. G., Hofheinz, W., Matile, H. *et al.* (1996). 4-Aminoquinoline analogs of chloroquine with shortened side chains retain activity against chloroquine-resistant *Plasmodium falciparum*. *Antimicrob. Agents Chemother.*, **40**, 1846–1854.

Rodhain, J. (1952). *Plasmodium vinckei* n. sp. Un deuxième *Plasmodium* parasite de rongeurs sauvage au Katanga. *Ann. Soc. Belg. Méd. Trop.*, **32**, 275–279.

Suhrbier, A., Janse, C., Mons, B. *et al.* (1987). The complete development *in vitro* of the vertebrate phase of the mammalian malarial parasite *Plasmodium berghei*. *Trans. R. Soc. Trop. Med. Hyg.*, **81**, 907–909.

Thompson, P. E., Werbel, L. M. (1972). *Antimalarial Agents. Chemistry and Pharmacology*. Academic Press, New York.

Trager, W., Jensen, J. B. (1976). Human malaria parasites in continuous culture. *Science*, **193**, 673–675.

Van Dijk, M. R., Janse, C. J., Waters, A. P. (1996). Expression of a *Plasmodium* gene introduced into subtelomeric regions of *P. berghei* chromosomes. *Science*, **271**, 662–665.

Vincke, I. H., Lips, M. (1948). Un nouveau *Plasmodium* d'un rongeur sauvage du Congo *Plasmodium berghei* n. sp. *Ann. Soc. Belge Méd. Trop.*, **28**, 97–104.

Walliker, D. (1983). *The Contributions of Genetics to the Study of Parasitic Protozoa*. Research Studies Press, Letchworth/John Wiley, New York.

WHO (1973). *Chemotherapy of malaria and Resistance to Antimalarials*. Report of a WHO Scientific Group. WHO Technical Report Series no. 529. WHO, Geneva.

WHO (1993). *Implementation of the Global Malaria Control Strategy*. Report of a WHO study group on the implementation of the Global Plan of Action for Malaria Control 1993–2000. WHO Technical Report Series no. 839. WHO. Geneva.

Williman, K., Matile, H., Weiss, N. A., Imhof, B. A. (1995). *In vivo* sequestration of *Plasmodium falciparum*-infected human erythrocytes: a severe combined immunodeficiency mouse model for cerebral malaria. *J. Exp. Med.*, **182**, 643–653.

Wiselogle, F. Y. (ed) (1946). *A Survey of Antimalarial Drugs, 1941–1945*, vol. 1. Edwards, Ann Arbor, Michigan.

Chapter 93

Animal Models of Cutaneous Leishmaniasis

V. Yardley and S. L. Croft

Background of cutaneous leishmaniasis

Leishmaniasis is a disease complex caused by the obligate, intracellular, protozoan parasites of the genus *Leishmania*. Depending upon the species, infection can manifest itself in humans viscerally, mucocutaneously or cutaneously. Recent reviews of the clinical spectrum of the disease cover the geographic distribution of the syndromes and the species of *Leishmania* involved in cutaneous leishmaniasis (World Health Organization (WHO), 1990; Pearson and de Queiroz Sousa, 1996; Table 93.1) as well as current chemotherapy regimes (Bryceson, 1996; Berman, 1997; Croft *et al.*, 1997; Van Gompel and Vervoort, 1997). Sensitivity to the drugs used for treatment varies between species. In general, New World (NW) species tend to cause more severe and longer-lasting infections than Old World (OW) species. Cutaneous leishmaniasis (CL) usually takes the form of a self-limiting cutaneous lesion, the appearance and severity of which can depend upon the species involved as well as the immune status of the host.

The standard drugs, pentavalent antimonials (Sb^{V}), Pentostam (sodium stibogluconate) and Glucantime (meglumine antimonate) have been used for over 50 years. *Leishmania mexicana* and *L. major* have low response rates to these antimonials. Intralesional treatment of some infections can hasten healing. Nodules can be removed by surgery, cryotherapy or heat therapy. *L. tropica* infections may require higher concentrations of antimonials to effect a clinical cure. *L. aethiopica* infections do not respond to conventional antimonial treatment. *L. aethiopica* lesions can be treated with pentamidine or aminosidine. Infections caused by *L. braziliensis* and *L. panamensis* have the potential to develop into mucocutaneous leishmaniasis (MCL), which occurs in about 6% of infected individuals (WHO, 1990) and so require vigorous systemic treatment with high doses of antimonials (Ballou *et al.*, 1987). If MCL does develop, i.v. applied amphotericin B is the second line of treatment. These regimes also apply to *L. amazonensis* infections, in order to prevent diffuse CL from developing (Pearson and de Queiroz Sousa, 1996; Berman, 1997).

Background of model

Two main animal models, the mouse and the hamster, are used in the study of leishmaniasis. Different strains of mouse display a range of susceptibilities which can be employed to reflect the diverse manifestations of the disease in humans. The hamster model can also be used for both visceral and cutaneous studies (see Chapter 94 for the visceral leishmaniasis (VL) model). Several strains of inbred mice have been used in studies of cutaneous leishmaniasis. A frequently used strain which self-cures when infected with *L. major* is C57B1/6; for others see Table 93.1 (Leclerc *et al.*, 1981; Beil *et al.*, 1992; Sunderkotter *et al.*, 1993; Scott and Scharton, 1994; Stenger *et al.*, 1994; Dabes-Guimaraes *et al.*, 1996; Shankar *et al.*, 1996). However, when this strain of mouse is infected with *L. amazonensis*, the parasite that can cause diffuse CL in humans, there is no self-cure (Alfonso and Scott, 1993). Other strains of mice, CBA (Leclerc *et al.*, 1981; Goto *et al.*, 1995), C3H (Leclerc *et al.*, 1981; Fortier *et al.*, 1990; Scharton-Kersten and Scott, 1995; Evans *et al.*, 1996) and DBA/2 (Leclerc *et al.*, 1981; Roberts *et al.*, 1990) will self-cure when infected with *L. amazonensis*. BALB/c mice are highly susceptible to cutaneous infections of *L. major* and develop chronic, fulminating lesions which will not resolve without intervention. This strain of mouse mounts a Th2 response to *L. major* infections, resulting in anergy and progression of the disease (reviewed in McElrath *et al.*, 1987; Müller *et al.*, 1989; Locksley and Louis, 1992; Scott and Scharton, 1994; Reiner and Locksley, 1995; Solbach and Laskay, 1995).

Another variable to be considered is the sex of mice; differences in the susceptibility and course of infection have been described for *L. mexicana* and *L. major* in BALB/c, DBA/2 and C57B1 strains (Roberts *et al.*, 1990). Severe combined immune-deficient (SCID) mice have also been used to study the immunology of cutaneous leishmaniasis (Guy and Belosevic, 1995; Doherty and Coffman, 1996; Seydell and Stanley, 1996). SCID mice can be maintained in the same way as the other model mice but their caging requirements are different: they must be maintained in a sterile environment. The golden hamster (*Mesocricetus auratus*) model has been used for strains of *Leishmania* with a low infectivity to mice, for example, *L. braziliensis* and

Handbook of Animal Models of Infection
ISBN 0-12-775390-7

Table 93.1 Treatments for the different forms of cutaneous leishmaniases and mouse models

Leishmania species	*Primary pathology in humans*	*Secondary pathology in humans*	*Treatment in humans*	*Mouse model* (susceptible^S/resistant^R)	*References for model*
L. (v) braziliensis NW	CL	5% of infections develop MCL ("Espundia")	Sb^V i.m. × 28 days, amphotericin B i.v.	BALB/c^S, C57BL/6^R, DBA^S, C3H^R, SWR^S C57BL/6^R, CBA^R,	Perez *et al.* (1979); Neal and Hale, (1983) Childs *et al.* (1984) MCL Barral *et al.* (1983)
L. (v) panamensis NW	CL ("ulcera de bejuco")	5% of infections develop MCL	Sb^V i.m. × 28 days, amphotericin B i.v. allopurinol, ketoconazole	CBA^S, BALB/c^S,	Neal and Hale (1983)
L. (v) guyanensis NW	CL ("pain bois")		Sb^V		
L. peruviana NW	CL ("uta")		Sb^V		
L. mexicana complex NW					
L. mexicana mexicana NW	CL ("chiclero's ear")		Sb^V	BALB/c^S, C57BL/6^R, CBA^S, C3H^R, SWR^S, DBA^S,	Perez *et al.* (1979); Neal and Hale (1983); Childs *et al.* (1984)
L. mex. amazonensis NW	CL	Diffuse CL	Systemic Sb^V and aminosidine, pentamidine i.m.	BALB/c^S, C57B1/6^S	Alfonso and Scott (1993)
L. tropica OW	CL	Recidivans	Sb^V, local or systemic	BALB/c^S, DBA^S, C3H^R C57BL/6^R	Bjorvatn and Neva (1979); Leclerc *et al.* (1981)
L. aethiopica OW	CL	Diffuse CL	Systemic Sb^V and aminosidine, pentamidine i.m.	BALB/c^S, SWR^S, DBA^S, C3H^R, C57BL/6^R, CBA^R,	Childs *et al.* (1984)
L. major OW	CL		Localized Sb^V, aminosidine ointment	BALB/c^S, CBA^R, C3H^R, C57B1/6^R,	Neal (1984)

CL, Cutaneous leishmaniasis; MCL, mucocutaneous leishmaniasis; OW, Old World; NW, New World; Sb^V, pentavalent antimony.
Adapted from Bryceson (1996) and Pearson and de Queiroz Sousa (1996).
Other more obscure strains which have been utilized are: wild type 129/Sv/Ev (Mattner *et al.*, 1996) and SWR (Nabors and Farrell, 1994).

L. panamensis (Herrer *et al.*, 1979; Wilson *et al.*, 1979; Neal and Hale, 1983). Other rodent models that have been used in the past for chemotherapy studies include the guinea-pig for the non-human pathogen *L. enriettii* (Kadivar and Soulsby, 1975).

Animal species

Inbred mice strains are most commonly used for experimental CL infections. Different strains of mouse vary in their susceptibility to strains and species of *Leishmania* (Neal and Hale, 1983). The *L. major*–BALB/c non-healing model provides a rigorous test for any drug, especially against established lesions. In this model even the standard antimonials have little effect. Drugging at the time of parasite infection is a more sensitive test of chemotherapeutic activity. Self-cure models should be used in further studies to evaluate lead compounds.

The Syrian or golden hamster (*Mesocricetus auratus*) is the recommended species of hamster for infections of *L. mexicana*, *L. panamensis* and *L. braziliensis*, although other types of hamster have been used previously (Beveridge, 1963).

Depending upon the model being used, the most common inoculation sites are subcutaneously (s.c.) on the rump above the tail or on the dorsal surface of the footpad in the mouse model. In the hamster model the footpad or nose has been used as the site of inoculation.

Preparation of animal

Most species of *Leishmania* that cause CL are regarded as category 2 pathogens, with the exception of *L. braziliensis*, which is classed as a category 3 pathogen (UK Control of Substances Hazardous to Health Regulations 1994). Basic protection such as a high-necked laboratory coat with elastic cuffs, gloves and face visor should be worn when inoculating with any strain of *Leishmania* (UK Health and Safety at Work Act 1994).

Details of surgery

Overview

Although not classed as highly infectious pathogens, work with *L. major*, *L. tropica* and *L. aethiopica* and *L. braziliensis* should be carried out wearing basic protective clothing (high-necked laboratory coat, gloves and a full face visor) when inoculating animals with parasites. For drug activity assays groups of 5 mice (same sex) per evaluation point should be used. The average weight of the mouse should be 20 g (6–8 weeks old). To prepare the site for s.c. inoculation an area of fur, about 2 cm in diameter on the rump, should be shaved. Electric animal clippers are recommended for this. Care must be taken to restrain the mouse so that the skin is not cut following any sudden movement during the operation. On parasite inoculation, a rounded blister should form; this will diminish in about 5 minutes. The mice should be checked twice weekly for the development of a lesion that initially appears as a raised, circular area. Lesions form approximately 2–4 weeks after infection depending upon the site of inoculum and the strain of *Leishmania* used. The ideal time to harvest amastigotes or to start drug treatment is at this time, *before* the lesion ulcerates and complicating factors such as secondary bacterial infections occur.

Materials required

Materials required include skin disinfectant, dissecting kit, to include scissors, forceps and scalpel and pins, needles (23G 1½) and syringes (1 ml).

Anaesthesia

Mice do not require anaesthesia for infection or drug administration by the s.c. route. A restrainer that gives full access to the tail vein and the base of the tail is recommended for ease of handling and for i.v. access. Hamsters may be sedated or anaesthetized with halothane or injectable pentobarbitone BP (Sagatal, Rhône-Poulenc Rorer, UK) prior to inoculation or drug administration. The recovery time for halothane is 2–5 minutes in 100% oxygen, whereas the recovery time following pentobarbitone anaesthesia is over 1 hour. Hamsters should be monitored throughout the recovery period.

Storage and preparation of inocula

This protocol is recommended for all CL models and requires forward planning. Cultured promastigotes may be used to infect animals. However, if the promastigotes have been maintained in culture for >6–8 passages, their infectivity to animals decreases (Neal, 1984). It is advisable to maintain infective parasites in animals to ensure a supply of fresh isolates. From these lesions amastigotes can be isolated, transformed to promastigotes in culture medium (for example Schneider's Drosophila medium plus 10% heat in activated fetal calf serum; hiFCS) at 24°C. From this, a bulk culture can be grown and low-passage (P1–P3) promastigotes can be used to reinfect the experimental animals. If parasites are to be stored long-term, it is recommended that they are kept in liquid nitrogen fridges and frozen in complete medium with 10% glycerol or 7.5% dimethyl sulphoxide as a cryoprotectant.

To isolate parasites from a lesion, place a sacrificed mouse in aseptic conditions, i.e. a class II hood. The mouse fur should be thoroughly wetted with 70% alcohol. Care must be taken to ensure that mouse hair is not allowed to contaminate the infected tissue during isolation. The lesion should be excised with scissors and inverted. The dermal tissue can then be removed by scraping the underside of the skin with a sharp scalpel, taking care to remove the material at the edge of the lesion where the majority of parasites reside. The pieces of tissue should then be placed in 2 ml of tissue culture medium (RPMI 1640 with gentamicin 100 μg/ml and penicillin/streptomycin (aq.) 100 U/ml) and kept on ice (4°C). To extract the amastigotes the medium and tissue are put into a 10 ml, safe-seal, glass homogenizer with excess tissue culture medium and macerated vigorously for about 5 minutes. The homogenate is transferred into a 50 ml centrifuge tube, topped up with fresh medium and centrifuged at 700 rpm for 10 minutes. The supernatant is removed to another 50 ml tube and centrifuged at 3000 rpm for 15 minutes. This will produce a pellet of amastigotes. The amastigotes may then be resuspended in fresh medium.

If the amastigotes are to be used to produce a promastigote culture they should be resuspended in Schneider's medium with 10% hiFCS, and placed in a 24°C cooled incubator. If the amastigotes are to be used to infect macrophages for an *in vitro* assay, they should be resuspended in a small amount of RPMI 1640, counted using either a Neubauer or a Thoma haemocytometer, and diluted to the required concentration with medium +10% hiFCS. Other methods, which are more time-consuming, can be used, to harvest amastigotes without host material (Glaser *et al.*, 1990). The parasites are now ready for use, either for infection or for culture.

The promastigote cultures should be monitored daily. The stationary phase of the cycle, where infective metacylic forms are present, is the point at which the promastigotes may be counted and used to infect mice. Promastigotes grown in culture are prepared by washing in serum-free, cold, medium, resuspended in a low volume of fresh, serum-free medium, and counted using a haemocytometer. The parasites should be kept on ice until inoculation.

Infection procedure

The mouse should be placed in a restrainer which enables access to the rump and base of the tail. Using a small-gauge needle (23G 1½), inject $1-2\times10^7$ promastigotes s.c. The inocula volume should be no more than 0.2 ml. An alternative method of infection is to inject the parasites into the dorsal surface of the footpad of the rear leg. This method is not recommended as it can cause unnecessary distress to the animal and the site is not as clean as the base of the tail. On s.c. inoculation an initial blister appears and is rapidly absorbed.

The promastigotes evade the potentially lethal serum components, most significantly the complement system, and attach and enter host cells, namely tissue macrophages, Langerhans cells and polymorphonuclear granulocytes. The ensuing inflammatory response results in the recruitment of macrophages to the site of infection. This sequence of events has been extensively reviewed by Solbach and Laskay (1996).

After inoculation the animals should be randomly sorted into groups of 5 or 4 mice or hamsters respectively. After 1 week the inoculation site should be checked for appearance of a lesion. With the more virulent strains of CL, the lesion will appear 10–14 days post-infection, so the animals should be monitored twice weekly. The lesion should appear as a discrete raised area of unbroken skin, or nodule. It is advisable to infect more animals than required to allow for instances when lesions may not develop.

Key parameters to monitor infection

Biopsy of the skin is not recommended for monitoring the course of infection as external indicators of infection are sufficient. The main parameter for monitoring infection is the external measurement of the lesion. The lesion is measured in two dimensions with Vernier callipers and the mean diameter is calculated. The percentage increase or decrease of the lesion can be evaluated and, if monitoring drug therapy, can be compared to untreated controls. A possible confounding factor of lesion assessment is the presence of secondary infection when the nodule ulcerates, and this can make the lesion appear larger. However, measurement should be made from the base of the outside edge of the lesion. The parasites reside around the periphery of the lesion and not in the central part.

Antimicrobial therapy

In general, if treatment is to be given, high doses of pentavalent antimony are recommended for Old and New World forms of cutaneous leishmaniasis. However high doses of pentavalent antimonials have poor activity in the *L. major*–BALB/c model. Topical treatment with paromomycin ointment has shown activity in this model (El-On *et al.*, 1984). In other models antimony treatment can accelerate cure (Neal, 1987; Travi *et al.*, 1993). In general, pentavalent antimony treatment tends to be less effective against CL than VL. CL infections vary extensively in their susceptibility to the drug. For example, *L. amazonensis* tends to be more susceptible than *L. major*, whereas *L. braziliensis* may be the least sensitive to treatment in experimental infections (Beveridge *et al.*, 1980). This situation is reversed for clinical infections.

Treatment of infection may begin at the same time as inoculation of parasites or when the infection is fully established, i.e. when a lesion has formed — it depends upon the purpose of the study. Dosing at the time of infection gives a compound a better chance of being effective. In the mouse model, drugs may be administered s.c., intraperitoneally (i.p.). and orally (p.o.) with ease. Intravenous formulations should be given via the lateral tail veins, which require warming for vasodilation and increased visibility. The cutaneous infection with *L. major* in the BALB/c mouse is a non-curing one, and is consequently a stringent test for any treatment. Any amelioration of the lesion should be regarded as a significant result. In a self-cure model, such as C57BL/6, the rapidity of healing should be compared to that of the untreated control group.

Key parameters to monitor response to treatment

At 2–4 weeks post-infection, depending upon the model being used, the lesions should be measured with callipers in two dimensions. The mean diameter can then be calculated. If drug therapy is to start at this point the ideal lesion diameter is approximately 5 mm. A quantitative method of measuring the parasite load of a lesion can only be made at necropsy. Tissue is measured, weighed and the parasites isolated as previously described. A microdilution assay is employed to evaluate parasite numbers. This method is detailed elsewhere (Hill *et al.*, 1983; Titus *et al.*, 1985; Lima *et al.*, 1997).

Pitfalls (advantages/disadvantages) of the model

Although there are obvious pharmacokinetic differences to humans, the mouse and hamster models offer a reproducible method of evaluating drug activity using all the standard routes of administration. Therapy can be monitored in real time: necropsy is not required to evaluate activity, but assessment of activity is not easily or accurately quantifiable. Measurement of the dimensions of a lesion gives an indication of activity at best and is not a quantitative parameter. The measurements can be misleading due to the inflammatory response of some strains of mouse to infection, e.g. *L. major*–BALB/c, and secondary infections. Reproducibility is best achieved through a standard inoculum given subcutaneously using an inbred strain of mouse. Mice are susceptible/resistant to a wide variety of laboratory-adapted strains and recent isolates of *Leishmania* and these models can be tailored to mimic various infections in humans. Immunodeficient mice with the same genetic background can be used in further studies on drug efficacy as well as knockout mice. Mice are not suitable for all strains of CL. Hamsters require larger quantities of drug for tests (most hamsters are 50–70 g, in contrast to 18–20 g mice) and are more difficult to handle, to infect and to dose. The availability of inbred strains of hamsters is limited and use of outbred animals can result in a greater variation in levels of infection.

Contribution of the model to infectious disease therapy

The mouse model of CL has mainly been used as a tool to elucidate the immunological profile of the disease and to clarify cellular events in genetically different mice, rather than as a model for chemotherapy studies. The majority of CL cases in humans will self-cure eventually, but there is still a need for safe, fast and effective therapies of CL and this model is used for such important studies. For example, three topical formulations of paromomycin have been produced for use in humans following tests in the mouse model, and intravenous versus subcutaneous liposomal amphotericin B has been assessed experimentally against *L. major* in BALB/c mice — the former route of administration achieves cure, the latter has no effect (Yardley and Croft, 1997). The non-cure models of CL provide a rigorous test for novel therapies. If a compound is found to be effective in such a model it should be tested against different strains of CL and also in a self-cure model.

The mouse/CL model is a valuable tool for immunology and chemotherapy as the genetic and cellular backgrounds of the model are now well-characterized and understood. The immunology of CL in mice has been extensively reported (McElrath *et al.*, 1987; Müller *et al.*, 1989; Locksley and Louis, 1992; Elhassan *et al.*, 1994; Reiner and Locksley, 1995; Solbach and Laskay, 1995) and to a lesser extent in hamsters (Travi *et al.*, 1996). The increased use of knockout and immunodeficient mice as well as the established models now provide the basis for extensive studies on the interactions between drugs and the immune response, as well as the detection of immunomodulators and assessment of vaccines.

References

Alfonso, L. C., Scott, P. (1993). Immune responses associated with susceptibility of C57BL/10 mice to *Leishmania amazonensis*. *Infect. Immun.*, **61**, 2952–2959.

Ballou, W. R., McClain, J. B., Gordon, D. M. *et al.*, (1987). Safety and efficacy of high dose sodium stibogluconate therapy of American cutaneous leishmaniasis. *Lancet*, **ii**, 13–16.

Barral, A., Petersen, E. A., Sacks, D. L., Neva, F. A. (1983). Late metastatic leishmaniasis in the mouse. *Am. J. Trop. Med. Hyg.*, **32**, 277–285.

Beil, W. J., Meinardus-Hager, G., Neugebauer, D. C., Sorg C. (1992). Differences in the onset of the inflammatory response to cutaneous leishmaniasis in resistant and susceptible mice. *J. Leukoc. Biol.*, **52**, 135–142.

Berman, J. D. (1997). Human leishmaniasis: clinical, diagnostic and chemotherapeutic developments in the last 10 years. *Clin. Infect. Dis.*, **24**, 684–703.

Beveridge, E. (1963). Chemotherapy of leishmaniasis. In *Experimental Chemotherapy*, vol. 1 (eds Schnitzer, R. J., Hawking, F.), pp. 257–287. Academic Press, London.

Beveridge, E., Caldwell, I. C., Latter, V. S., Neal, R. A., Udall, V., Waldron, M. M. (1980). The activity against *Trypanosoma cruzi* and cutaneous leishmaniasis, and toxicity, of moxipraquine (349 C59). *Trans. R. Soc. Trop. Med. Hyg.*, **74**, 43–51.

Bjorvatn, B., Neva, F. A. (1979). A model in mice for experimental leishmaniasis with a West African strain of *Leishmania tropica*. *Am. J. Trop. Med. Hyg.*, **28**, 472–479.

Bryceson, A. D. M. (1996). Leishmaniasis. In *Manson's Tropical Diseases*, 20th edn (ed Cook, G. C.) pp. 1214–1245. Saunders, London.

Childs, G. E., Lightner, L. K., McKinney, L., Groves, M. G., Price, E. E. Hendricks, L. D. (1984). Inbred mice as model hosts for cutaneous leishmaniasis. I. Resistance and susceptibility to infection with *Leishmania braziliensis, L. mexicana* and *L. aethiopica*. *Ann. Trop. Med. Parasitol.*, **78**, 25–34.

Croft, S. L., Urbina, J. A., Brun, R. (1997). Chemotherapy of human leishmaniasis and trypanosomiasis. In *Trypanosomiasis and Leishmaniasis* (eds Hide, G., Mottram, J. C., Coombs, G. H., Holmes, P. H.) pp. 245–257. CAB International, Wallingford, UK.

Dabes-Guimaraes, T. M., de Toledo, V. D., da Costa, C. A. *et al.* (1996). Assessment of immunity induced in mice by glycoproteins derived from different strains and species of *Leishmania*. *Mem. Inst. Oswaldo Cruz.* **91**, 63–70.

Doherty, T. M., Coffman, R. L. (1996). *Leishmania major*: effect of infectious dose on T cell subset development in BALB/c mice. *Exp. Parasitol.*, **84**, 124–135.

Elhassan, A. M., Gaafar, A., Theander, T. G. (1994). Antigen-presenting cells in human cutaneous leishmaniasis due to *Leishmania major*. *Clin. Exp. Immunol.*, **99**, 445–453.

El-On, J., Jacobs, G. P., Wiztum, E., Greenblatt, C. L. (1984). Development of topical treatment for cutaneous leishmaniasis caused by *Leishmania major* in experimental animals. *Antimicrob. Agents Chemother.*, **26**, 745–751.

Evans T. G., Reed S. S., Hibbs, J. B. Jr. (1996). Nitric oxide production in murine leishmaniasis: correlation of progressive infection with increasing systemic synthesis of nitric oxide. *Am. J. Trop. Med. Hyg.*, **54**, 486–489.

Fortier, A. H., Tong, A., Nacy, C. A., (1990). Susceptibility of inbred mice to *Leishmania major* infection: genetic analysis of macrophage activation and innate resistance to disease in individual progeny of P/J (susceptible) and C3H/HeN (resistant) mice. *Infect. Immun.*, **58**, 4149–4152.

Glaser, T. A., Wells, S. J., Spithill, T. W., Pettott, J. M., Humphries, D. C., Mukkada, A. J. (1990). *Leishmania major* and *L. donovani*: a method for rapid purification of amastigotes. *Exp. Parasitol.*, **71**, 343–345.

Goto, H., Rojas, J. I., Sporrong, L., de Carreira, P., Sancez, C., Orn, A. (1995). *Leishmania (viannia) panamensis*-induced cutaneous leishmaniasis in susceptible and resistant mouse strains. *Rev. Inst. Med. Trop. Sao Paulo*, **37**, 475–481.

Guy, R. A., Belosevic, M. (1995). Response of *scid* mice to establishment of *Leishmania major* infection. *Clin. Exp. Immunol.*, **100**, 347–353.

Herrer, A., Telford, S. R. Jr., Christensen, H. A. (1979). *Leishmania braziliensis*: dissemination of Panamanian strains in golden hamsters. *Exp. Parasitol.*, **48**, 356–359.

Hill, J. O., North, R. J., Collins, F. M. (1983). Advantages of measuring changes in the number of viable parasites in murine models of experimental cutaneous leishmaniasis. *Infect. Immun.*, **39**, 1087–1094.

Kadivar, D. M. H., Soulsby, E. J. L. (1975). Model for disseminated cutaneous leishmaniasis. *Science*, **190**, 1198–1200.

Leclerc, C., Modabber F., Deriaud, E., Cheddid L. (1981). Systemic infection of *Leishmania tropica (major)* in various strains of mice. *Trans. R. Soc. Trop. Med. Hyg.*, **75**, 851–854.

Lima, H. C., Bleyenberg, J. A., Titus, R. G. (1997). A simple method for quantifying *Leishmania* in tissues of infected animals. *Parasitol. Today* **13**, 80–81.

Locksley, R. M., Louis, J. A. (1992). Immunology of leishmaniasis. *Curr. Op. Immunol.*, **4**, 413–418.

Mattner, F., Magram, J., Ferrante, J. *et al.* (1996). Genetically resistant mice lacking interleukin-12 are susceptible to infection with *Leishmania major* and mount a polarized Th2 response. *Eur. J. Immunol.*, **26**, 1553–1559.

McElrath, M. J., Kaplan, G., Nusrat, A., Cohn, Z. A. (1987). Cutaneous leishmaniasis: the defect in T cell influx in BALB/c mice. *J. Exp. Med.*, **165**, 546–559.

Müller, I., Pedrazzini, T., Farrell, J. P., Louis J. (1989). T-cell responses and immunity to experimental infection with *Leishmania major*. *Annu. Rev. Immunol.*, **7**, 561–578.

Nabors, G. S., Farrell, J. P. (1994). Site-specific immunity to *Leishmania major* in SWR mice: the site of infection influences susceptibility and expression of the antileishmanial immune response. *Infect. Immun.*, **62**, 3655–3662.

Neal, R. A. (1984). *Leishmania major*: culture media, mouse strains and promastigote virulence and infectivity. *Exp. Parasitol.*, **57**, 269–273.

Neal, R. A. (1987). Experimental chemotherapy. In *The Leishmaniases*, vol. II (eds Peters, W., Killick-Kendrick, R.), pp. 793–845. Academic Press, London.

Neal, R. A., Hale, C. (1983). A comparative study of inbred and outbred mouse strains compared with hamsters to infection with New World cutaneous leishmaniasis. *Parasitology*, **87**, 7–13.

Pearson, R. D., de Queiroz Sousa A. (1996). Clinical spectrum of leishmaniasis. *Clin. Infect. Dis.*, **22**, 1–13.

Perez, H., Labrador, F. Torrealba, J. W. (1979). Variations in the response of five strains of mice to *Leishmania mexicana*. *Int. J. Parasitol.*, **9**, 27–32.

Reiner S. L., Locksley R. M. (1995). The regulation of immunity to *Leishmania major*. *Annu. Rev. Immunol.*, **13**, 151–177.

Roberts M., Alexander, J., Blackwell, J. M. (1990). Genetic analysis of *Leishmania mexicana* infection in mice: single gene (Scl-2) controlled predisposition to cutaneous lesion development. *J. Immunogenet.*, **17**, 89–100.

Scharton-Kersten, T., Scott, P. (1995). The role of the innate immune response in Th1 cell development following *Leishmania major* infection. *J. Leukoc. Biol.*, **57**, 515–522.

Scott, P., Scharton, T. (1994). Interaction between the innate and the acquired immune system following infection of different mouse strains with *Leishmania major*. *Ann. NY Acad. Sci.*, **15**, 84–92.

Seydell, K. B., Stanley, S. L. Jr. (1996). SCID mice and the study of parasitic disease. *Clin. Microbiol. Rev.*, **9**, 126–134.

Shankar A. H., Morin P., Titus, R. G. (1996). *Leishmania major*: differential resistance to infection in C57BL/6 (high interferon-alpha/beta) and congenic B6.C-H-28c (low interferon-alpha/beta) mice. *Exp. Parasitol.*, **84**, 136–143.

Solbach, W., Laskay, T. (1995). *Leishmania major* infection: the overture. *Parasitol. Today*, **11**, 394–397.

Solbach, W., Laskay, T. (1996). Evasion strategies of *Leishmania* parasites. In *Molecular and Immune Mechanisms in the Pathogenesis of Cutaneous Leishmaniasis* (eds Tapia, F. J., Cáceres-Dittmar, G., Sánchez, M. A.), pp. 25–47. R. G. Landes, New York, USA.

Stenger, S., Thuring, H., Rollinghoff, M., Bogdan C. (1994). Tissue expression of nitric oxide synthase is closely associated with resistance to *Leishmania major*. *J. Exp. Med.*, **180**, 783–793.

Sunderkotter, C., Kunz, M., Steinbrink, K. *et al.* (1993). Resistance in mice to experimental leishmaniasis is associated with more rapid appearance of mature macrophages *in vitro* and *in vivo*. *J. Immunol.*, **151**, 4891–4901.

Titus, R. G., Marchand, M., Boon, T., Louis, J. A. (1985). A limiting dilution assay for quantifying *Leishmania major* in tissues of infected mice. *Parasit. Imm.*, **7**, 545–555.

Travi, B. L., Martinez, J. E., Zea, A. (1993). Antimonial treatment of hamsters infected with *Leishmania (viannia) panamensis*: assessment of parasitological cure with different therapeutic schedules. *Trans. R. Soc. Trop. Med. Hyg.*, **87**, 567–569.

Travi, B. L., Osorio, Y., Saravia, N. G. (1996). The inflammatory response promotes cutaneous metastasis in hamsters infected with *Leishmania (viannia) panamensis*. *J. Parasitol.*, **82**, 454–457.

Van Gompel, A., Vervoort, T. (1997). Chemotherapy of leishmaniasis and trypanosomiasis. *Curr. Op. Infect. Dis.*, **10**, 469–474.

Wilson, H. R., Dieckmann, B. S., Childs, G. E. (1979). *Leishmania braziliensis* and *Leishmania mexicana*: experimental cutaneous infections in golden hamsters. *Exp. Parasitol.*, **47**, 270–283.

WHO (1990). *Report of a WHO Expert Committee. Control of the Leishmaniases*. Technical Report Series 793. World Health Organization, Geneva.

Yardley, V., Croft, S. L. (1997). Activity of liposomal amphotericin B against experimental cutaneous leishmaniasis. *Antimicrob. Agents Chemother.*, **41**, 752–756.

Chapter 94

Animal Models of Visceral Leishmaniasis

S. L. Croft and V. Yardley

Background of visceral leishmaniasis

Visceral leishmaniasis (VL) is caused by haemoflagellate protozoan parasites belonging to the genus *Leishmania*, in particular the species *Leishmania donovani, L. infantum* and *L. chagasi*. VL occurs in tropical and subtropical regions throughout the world with an estimated incidence of 500 000 cases/year. In the mammalian host *Leishmania* are obligate intracellular parasites of macrophages, where they survive and multiply as 1–2 μm amastigote forms in the phagolysosome. Parasites are transmitted between mammalian hosts by female phlebotomine sandflies and live in the gut as extracellular flagellated promastigotes. The aetiology, incidence, clinical symptoms, pathology and immunology of VL have been extensively reviewed (Grimaldi and Tesh, 1993; Bryceson, 1996; Pearson and de Queiroz Sousa, 1996; Alvar *et al.*, 1997; Berman, 1997).

Control of leishmaniasis is mainly by vector control and chemotherapy. There are no vaccines currently available for clinical leishmaniasis (Modabber, 1995). The recommended drugs for the treatment of VL are the pentavalent antimonials, sodium stibogluconate (Pentostam, Glaxo Wellcome) and meglumine antimoniate (Glucantime, Rhône Poulenc Rorer), with the diamidine pentamidine and the polyene antibiotic amphotericin B as second-line drugs (World Health Organization (WHO), 1990). Parenteral formulations of the aminoglycoside antibiotic aminosidine (paromomycin) and lipid formulations of amphotericin B are important new alternative therapies (Olliaro and Bryceson, 1993; Gradoni *et al.*, 1995; Berman, 1997). Treatment with the standard drugs is inadequate due to variable efficacy, increasing levels of resistance to antimonials, toxicity, the requirement for long courses of parenteral administration and cost.

Background of model

Hamster and mouse models of infection have been used in studies on the immunology, pathology and chemotherapy of VL. The hamster was the accepted model for laboratory studies on *L. donovani* (Beveridge, 1963) until the 1980s (Hanson *et al.*, 1977). Since then inbred strains of mice have been the most extensively used model. Initially, outbred mice were used in chemotherapy studies (for example, Peters *et al.*, 1980; Trotter *et al.*, 1980) but the work of Bradley and colleagues (see Blackwell, 1988) defined the susceptibility and resistance of inbred strains of mice to *L. donovani*. Immunodeficient nude and severe combined immunodeficient (SCID) mice have been used in studies to investigate the immune response (Kaye and Bancroft, 1992) and the immune dependence of drugs (Murray *et al.*, 1989). Rats have been used in a limited number of studies but are not preferred. Primate models have been developed for drug and vaccine trials (Matindou *et al.*, 1985; reviewed by Neal, 1987) but have been used rarely. Few products have required primate tests, as lead compounds have frequently been used in humans for other indications.

Animal species

Several varieties of hamster have been used (Beveridge, 1963), but the most common model is the Syrian or golden hamster, *Mesocricetus auratus*. Both male and female animals, weighing 50–100 g, are suitable.

Inbred mice strains show a wide variation in susceptibility to *L. donovani* and *L. infantum* (Blackwell, 1988). BALB/c mice are the most commonly used strain for chemotherapy studies, although drugs must be tested during the first 3 weeks of infection as after this time the liver infection becomes chronic and gradually resolves. A difference in the course of infection of *Leishmania* in male and female mice is observed in some strains (Roberts *et al.*, 1996). Immunodeficient Nu/nu and SCID mice can be used to answer specific questions on immune dependence of chemotherapy (Murray *et al.*, 1989; Kaye and Bancroft, 1992).

Preparation of animals

Restricted-access housing is required for rodent hosts, as *L. donovani* is considered to be a category 3 pathogen (UK

Handbook of Animal Models of Infection
ISBN 0-12-775390-7

Control of Substances Hazardous to Health Regulations 1994).

Details of surgery

Overview

As *L. donovani* is a category 3 pathogen, all operating procedures should take place in a laboratory with limited access, negative pressure and wash facilities. All experimenters should wear protective clothing, including a high-necked washable laboratory coat, gloves and a visor (UK Health and Safety at Work Act 1994). There are many sources of health and safety information available on the Internet World-Wide Web.

Materials required

Materials required include anaesthetic, skin disinfectant, forceps, scalpels, scissors, syringes and needles (23G 1½).

Anaesthesia

Mice do not require anaesthesia for normal handling, infection or administration of drugs or antigens. A restrainer designed to enable access to the tail vein is sufficient. If required, mice can be sedated with halothane. Hamsters require anaesthesia by halothane or pentobarbarbitone sodium BP (Sagatal, Rhône-Poulenc Rorer, UK) for intracardiac inoculation. The recovery time for halothane is 2–5 minutes in oxygen, whereas the recovery time following pentobarbitone anaesthesia is over 1 hour. Hamsters should be monitored throughout the recovery period.

Storage and preparation of inocula

The hamster and mouse model can be used for studies on *L. donovani, L. infantum* and *L. chagasi*. The hamster is the best donor for parasite material as 10^9–10^{10} amastigotes can be obtained from the spleen of this rodent with relatively little contaminatory tissue. The liver is also a source of amastigotes but the preparation of a pure inoculum of amastigotes is more difficult.

The spleen is removed from a freshly killed hamster under asceptic conditions and placed in tissue culture medium containing 50 µg/ml gentamicin. In the laboratory a small section of spleen is removed and the remainder of the spleen is weighed. Smears of the spleen section are made on a glass microscope slide, fixed for 1 minute in methanol and stained with 10% Giemsa stain for 45 minutes. The remainder of the spleen is transferred to a 15 ml safe-seal tissue homogenizer containing 10 ml of medium and macerated. The tissue suspension is placed in a 50 ml centrifuge tube and centrifuged at 1400 *g* for 15 minutes at 4 °C. The cell suspension is removed to a fresh, 50 ml centrifuge tube and recentrifuged at 24–25 000 *g* for 15 minutes at 4 °C. This procedure will produce a pellet of amastigotes which is then resuspended in 10 ml of medium or phosphate-buffered saline (PBS) and stored at 37 °C for a maximum of 24 hours. The number of amastigotes in the spleen is calculated from the formula: weight of spleen (mg) × ratio of number of amastigotes to number of spleen cells (determined from examination of at least 500 host cells on the prepared microscope slide) × the constant, 200 000 (Stauber *et al.*, 1958). The pellet of amastigotes is resuspended in medium or PBS to the appropriate concentration for the inoculum.

An alternative method of determining the number of amastigotes derived from the spleen is to take 10 µl of the final suspension, spread evenly in 4 cm^2, marked on a microscope slide, allow to dry, fix in methanol for 1 minute and stain with 10% Giemsa stain for 10 minutes. Stained amastigotes can easily be counted using a 100 × oil immersion lens. Knowing the area of the field of the 100 × lens (measured by a micrometer slide) it is possible to estimate the number of amastigotes spread over the 4 cm^{-2} area (i.e. in 10 µl) and calculate the number of amastigotes per millilitre.

For longer periods of storage, the suspension of amastigotes should be mixed with an equal volume of 15% dimethyl sulphoxide and kept in either a –70 °C freezer (for maximum 2 weeks) or stored in a liquid nitrogen freezer.

The flagellated promastigote form of the parasite, which is found naturally in the female sandfly vector, can also be grown in culture medium and can be used to infect animals. The promastigote stage of the life cycle can be grown in either RPMI 1640 medium or 199 medium supplemented with 50 mmol/l glutamine and 10% heat-inactivated calf serum at 24 °C. The growth of the culture is monitored microscopically. After several days the culture changes from the logarithmic phase of growth, where the parasites have low infectivity, to the stationary phase of growth where infective metacyclic forms are present. The time for this change to be reached (3–7 days) and the concentration of promastigote forms (10^7–10^8/ml) depends upon the strain of parasite and the culture conditions used. Promastigotes can be grown readily and short-term storage should not be required. Long-term storage at –70 °C or in liquid nitrogen follows the same procedures used for amastigotes. After several weeks of passage in culture the parasites can begin to lose infectivity and regular passage of strains through a rodent model to maintain infectivity is recommended.

Infection procedure

In the mouse model the intravenous route of infection gives the quickest and most reproducible infection in the liver, spleen and bone marrow. Prior to infection mice are warmed in a cage by a light or warm water to raise the tail veins. The inoculum of parasites, either amastigote or promastigotes, is loaded into a 1 ml syringe fitted with a 23G 1½ needle. Mice are placed in a restrainer with the tail vein emerging towards the experimenter. The tail is held at full stretch and, with the bevel of the needle facing outwards, the needle is inserted at a low angle into the tail vein and a volume of 0.1 or 0.2 ml of inoculum is slowly injected. In the hamster model intraperitoneal infection is frequently used and intrasplenic infection was also originally recommended. However, the quickest and most reproducible infections are achieved by inoculating by the intracardiac route, a technique initially demonstrated by Stauber *et al.* (1958). This is the recommended route of infection. The anaesthetized hamster is placed on the bench, dorsal side upwards, and the dorsal surface is swabbed with mild disinfectant, for example 70% ethanol. The inoculum of 0.1 or 0.2 ml parasites is loaded into a 1 ml syringe fitted with a 23 G 1½ needle. The needle is inserted vertically into the chest at the midpoint of the sternum on the leftside. The presence of the heart is detectable when the needle is correctly inserted in intracardiac position and a backflow of blood will enter the syringe. The inoculum is then injected slowly.

The number of parasites injected in an inoculum will depend upon the purpose of the study, the model used and the strain of parasite. Typically an inoculum to maintain a laboratory-adapted strain for passage in a hamster by the intracardiac route is 5×10^7 amastigotes or 10^8 promastigotes to produce a progressive infection in 8 weeks. For experimental infection an inoculum of 10^7 amastigotes, intravenously in mice and by the intracardiac route in hamsters, will produce a microscopically detectable infection in the liver of mice and liver and spleen of hamster after 1 week of infection. This level is suitable for drug tests.

Key parameters to monitor infection

In the early stages of infection VL in mice presents no obvious external symptoms. In any study it is advisable to infect extra mice or hamsters to be sacrificed prior to drugging to check that the infection is established. Microscopical examination of stained slides prepared from the liver and/or spleen of rodents will indicate whether the inoculum was satisfactory and that infection has been established. The appearance of hamsters does change in the later stages of infection. The most noticeable features are loss in weight and dulling of the hair. Occasionally hamsters may develop ascites.

Antimicrobial therapy

The standard drugs for the treatment of visceral leishmaniasis are the pentavalent antimonials sodium stibogluconate and meglumine antimoniate. The activities of these and other recommended drugs in mouse and hamster models are given in Table 94.1. Antiprotozoal drugs can be administered to mice by a variety of routes (s.c., i.p. and p.o.) and for some formulations i.v. administration by the tail vein is required. In the mouse model treatment is best evaluated against an established infection on days 7–11 post-infection. If a lower infection inoculum is used, tests can be carried out on days 14–18 post-infection. The patterns of infection by *L. donovani* strains have been well-characterized in inbred strains of mice (Blackwell, 1988) but should be reassessed as the characteristics of parasite strains change as they adapt to hosts. In the commonly used BALB/c mouse the infection in the liver increases linearly until days 21–28 following infection by 10^7 amastigotes or promastigotes;

Table 94.1 Activities of standard anti-leishmanial drugs in animal models

Leishmania species	*Experimental animal models*	*Drug*	*Regime/dose*
L. donovani	Mouse	Sodium stibogluconate (Pentostam)	ED_{50} 65 mg Sb^V/kg per day × 1
			ED_{50} 11.2 mg Sb^V/kg per day × 5
			High activity at 40 mg Sb^V/kg per day × 6;
	Hamster		50–125 mg Sb^V/kg per day × 5
	Mouse	Meglumine antimoniate (Glucantime)	ED_{50} 11.6 mg Sb^V/kg per day × 5
			High activity at 104 mg Sb^V/kg per day × 4
	Hamster		Low activity at 50 mg Sb^V/kg per day × 6
	Dog		High activity at 104 mg Sb^V/kg per day × 5
	Mouse	Amphotericin B	ED_{50} 11.0 mg/kg per day × 5
	Hamster		
	Mouse	Pentamidine	Inactive at maximum tolerated dose of 50 mg/kg per day × 5
	Hamster		Active at 25 mg/kg per day × 6; active at 52 mg/kg per day × 4

Adapted from Neal (1987).

after this point the liver infection will become chronic and eventually cure. The spleen infection, although microscopically detectable from week 1, will only become fully established after the 4th week of infection. The hamster model requires 2 weeks for an infection to be established in the liver and spleen for antimicrobial studies. The same procedure for treatment used in the mouse model can be followed, although intracardiac dosing of the hamster is much more difficult than intravenous tail vein dosing in mice. At necropsy, more importance is given to spleen infection in the hamster than in the mouse, where infection is often very low during the period of drug testing. If the infection is left for several weeks prior to treatment, then chronic granulomatous infection is established in both models, and this has been shown to be less sensitive to standard drugs.

In an alternative approach using the mouse and hamster models, drug treatment is started immediately after infection. This protocol is used by some groups and is more likely to detect antileishmanial activity in the compound under test.

Key parameters to monitor response to infection

At the end of treatment the mice are weighed to give an estimation of drug toxicity. The livers and spleens are removed from freshly sacrificed animals and weighed. Smears are prepared from the livers and spleens on microscope slides, fixed in methanol for 1 minute and stained with Giemsa stain for 45 minutes. The number of parasites/500 liver and/or spleen cells is determined microscopically for each experimental animal. This figure is multiplied by total organ weight (mg) and this figure, the Leishman–Donovan unit (LDU) is used as the basis for calculating the difference in parasite load between treated and untreated animals. Drug activity is evaluated in the same way in the hamster model.

In some studies the activity of novel compounds is compared with that of the standard antimonial drugs and expressed as a therapeutic ratio.

Pitfalls (advantages/disadvantages) of the model

In addition to the normal pharmacokinetic limitations both models rely upon necropsy and microscopical evaluation of drug activity. The mouse model offers a reproducible method of determining drug activity with the possibility of testing drugs by all the standard routes of administration. Reproducibility is best achieved through a standard inoculum given intravenously using a susceptible inbred strain of mice. Mice are susceptible to a wide variety of laboratory-adapted strains and recent isolates of *Leishmania*. A 5-day course of drug treatment is sufficient to determine relative drug potencies. Immunodeficient mice with the same genetic background can be used in further studies on drug efficacy; this is important as VL is now regarded as an opportunistic infection in human immunodeficiency virus (HIV) patients. The major limitations of the model are first, the differences in the immunology and pathology of VL in mice and humans, in particular the low infection found in murine spleen until late on in the infection; second, the difficulty in monitoring levels of infection without biopsy and only one time point to evaluate the parasite burden at necropsy, and third, the time-consuming effort in the microscopic determination of the number of parasites in the liver and spleen of treated and untreated animals. The number of parasites in the bone marrow is even more difficult to evaluate as this tissue is not as accessible as the liver and spleen.

The course of infection and the pathology of VL in hamsters is closer to human disease. However, hamsters require a greater quantity of drug for tests (most hamsters weigh 50–70 g, in contrast to 18–20 g mice), and are more difficult to handle, to infect (via the intracardiac route: there is no i.v. route in female hamsters) and to dose. The availability of inbred strains of hamsters is limited and use of outbred animals can result in a greater variation in levels of infection.

Contributions of the model to infectious disease therapy

The hamster and the mouse models were used in the experimental studies on most of the current therapies for VL. These models have also been used to test a variety of novel formulations of antileishmanial drugs, including the liposomal formulations of pentavalent antimonial drugs and amphotericin B, as well as for testing drug combinations. Both models offer opportunities to examine different routes of drug administration. The mouse model has also been used in studies on the variation of the sensitivities to drugs of different strains and species of *Leishmania* causing VL. The mouse model is used widely in the identification and evaluation of novel compounds. The hamster model, used from the 1940s to the 1980s, still has an important role in antileishmanial chemotherapy. All lead compounds should be tested in the hamster model to confirm results obtained in the primary mouse models. It is also possible that compounds have low activity in the mouse model, and due to pharmacokinetics have a higher activity in the hamster model. The two models in tandem offer possibilities to indicate lead compounds clearly.

Leishmaniasis has offered immunologists a model for examining the Th1 and Th2 arms of cell-mediated immunity. This has been described in depth in murine models of VL (Kaye *et al.*, 1995) and a few studies in hamsters (Rodrigues *et al.*, 1992). These studies have established a basis for research on the interactions between drugs and the immune response, as well as the detection of immunomodulators. Immunodeficient mice have been used in studies related to examining the potential of drugs in HIV leishmaniasis cases (Murray *et al.*, 1993).

References

Alvar, J., Canavate, C., Gutierrez-Solar, B. *et al.* (1997). Leishmania and human immunodeficiency virus coinfection: the first 10 years. *Clin. Microbiol. Rev.*, **10**, 298–319.

Berman, J. D. (1997). Human leishmaniasis: clinical, diagnostic and chemotherapeutic developments in the last 10 years. *Clin. Infect. Dis.*, **24**, 684–703.

Beveridge, E. (1963). Chemotherapy of leishmaniasis. In *Experimental Chemotherapy*, vol. 1 (eds Schnitzer, R. J., Hawking, F.), pp. 257–287. Academic Press, London.

Blackwell, J. M. (1988) Protozoan infections. In *Genetics of Resistance to Bacterial and Parasitic Infection* (eds Wakelin, D., Blackwell, J. M.), pp. 103–151. Taylor & Francis, London.

Bryceson, A. D. M. (1996). Leishmaniasis. In *Manson's Tropical Diseases*, 20th edn, (ed Cook, G. C.) pp. 1214–1245. Saunders, London.

Gradoni, L., Bryceson, A., Desjeux, P. (1995). Treatment of Mediterranean visceral leishmaniasis. *Bull. WHO*, **73**, 191–197.

Grimaldi, G. Jr., Tesh, R. B. (1993). Leishmaniasis in the New World: current concepts and implications for future research. *Clin. Microbiol. Rev.*, **6**, 230–250.

Hanson, W. L., Chapman, W. L., Kinnamon, K. E. (1977). Testing of drugs for antileishmanial activity in golden hamsters infected with *Leishmania donovani*. *Int. J. Parasitol.*, **7**, 443–447.

HMSO (1994). *Essentials of Health and Safety at Work*. HMSO, London.

HMSO (1995). *Categorisation of Biological Agents According to Hazard and Categories of Containment*, 4th edn. HMSO, London.

Kaye, P. M., Bancroft, G. (1992). *Leishmania donovani* infection in *scid* mice: lack of tissue response and *in vivo* macrophage activation correlates with failure to trigger natural killer cell-derived gamma interferon production *in vitro*. *Infect. Immun.*, **60**, 4335–4342.

Kaye, P. M., Gorak, P., Murphy, M., Ross, S. (1995). Strategies for immune intervention in visceral leishmaniasis. *Ann. Trop. Med. Parasitol.*, **89** (suppl. 1), 75–81.

Matindou, T. J., Hanson, W. L., Chapman, W. L. Jr (1985). Chemotherapy of visceral leishmaniasis (*Leishmania donovani*) in the squirrel monkey (*Saimiri sciureus*). *Ann. Trop. Med. Parasitol.*, **79**, 13–19.

Modabber, F. (1995). Vaccines against leishmaniasis. *Ann. Trop. Med. Parasitol.*, **89**, 83–88.

Murray, H. W., Oca, M. J., Granger, A. M., Schreiber, R. D. (1989). Requirement for T cells and effect of lymphokines in successful chemotherapy for an intracellular infection. *J. Clin. Invest.*, **83**, 1253–1257.

Murray, H. W., Hariprashad, J., Fichtl, R. E. (1993). Treatment of experimental visceral leishmaniasis in a T-cell-deficient host: response to amphotericin B and pentamidine. *Antimicrob. Agents Chemother.*, **37**, 1504–1505.

Neal, R. A. (1987). Experimental chemotherapy. In *The Leishmaniases*, vol. II (eds Peters, W., Killick-Kendrick, R.), pp. 793–845. Academic Press, London.

Olliaro, P., Bryceson, A. D. M. (1993). Practical progress and new drugs for changing patterns of leishmaniasis. *Parasitol. Today*, **9**, 323–328.

Pearson, R. D., de Queiroz Sousa, A. (1996). Clinical spectrum of leishmaniasis. *Clin. Infect. Dis.*, **22**, 1–13.

Peters, W., Trotter, E. R., Robinson, B. L. (1980). The experimental chemotherapy of leishmaniasis, V. The activity of potential leishmanicides against "*L. infantum* LV9" in NMRI mice. *Ann. Trop. Med. Parasitol.*, **74**, 289–298.

Roberts, C. W., Satoskar, A., Alexander, J. (1996). Sex steroids, pregnancy-associated hormones and immunity to parasitic infection. *Parasitol. Today*, **12**, 382–388.

Rodrigues, V., Jr, da Silva, J. S., Campos-Neto, A. (1992). Selective inability of spleen antigen presenting cells from *Leishmania donovani* infected hamsters to mediate specific T cell proliferation to parasite antigens. *Parasite Immunol.*, **14**, 49–58.

Stauber, L. A., Franchino, E. M., Grun, J. (1958). An eight-day method for screening compounds against *Leishmania donovani* in the golden hamster. *J. Protozool.*, **5**, 269–273.

Trotter, E. R., Peters, W., Robinson, B. L. (1980). The experimental chemotherapy of leishmaniasis IV. The development of a rodent model for visceral infection. *Ann. Trop. Med. Parasitol.*, **74**, 127–138.

WHO (1990). *Report of a WHO Expert Committee. Control of the Leishmaniases*. Technical Report Series 793. World Health Organization, Geneva.

Chapter 95

Animal Models of Acute (first-stage) Sleeping Sickness

R. Brun and R. Kaminsky

Background of human infection

Human African trypanosomosis (HAT) or African sleeping sickness is caused by two closely related protozoan parasites belonging to the genus *Trypanosoma*. Its distribution is restricted to sub-Saharan African countries. *Trypanosoma brucei rhodesiense* causes the more acute form of HAT in East Africa: this can lead to death within a few weeks or months, while *T. brucei gambiense* causes a chronic form of the disease in Central and West Africa which may last for several years. While HAT due to *T. b. rhodesiense* currently shows a prevalence of a few thousand patients per year, *T. b. gambiense* occurs in more than 30 countries and causes epidemics in Congo (formerly Zaire), Angola and Sudan, with a prevalence of 200 000–300 000 patients. Different species of tsetse flies (*Glossina* spp.) transmit the disease, and wild and domestic animals can act as reservoir hosts for the trypanosomes. After an infective bite, parasites multiply at the site of injection and a chancre may be formed. Thereafter the trypanosomes invade the haematolymphatic system (first stage). Initial symptoms are fever, headache, adenopathy, joint pains, endocrinological problems and weight loss. Once the parasites cross the blood–brain barrier (second stage), neurological symptoms appear (reversal of sleeping pattern, loss of appetite), followed by meningoencephalitis, mental deterioration, wasting and coma. The disease is fatal if untreated.

Since no vaccination for HAT is available today, control of the disease is mainly based on vector control and chemotherapy, which depends on a few old drugs developed decades ago. The only new drug which came on to the market over the last 10 years is Eflornithine (DL-α-difluoromethylornithine, DFMO; McCann *et al.*, 1981) but its use is restricted to *T. b. gambiense* infections due to an innate tolerance of *T. b. rhodesiense* (Iten *et al.*, 1995). The only drug effective against the second stage (late stage) of both forms of HAT is the melaminylphenyl arsenical melarsoprol (Arsobal, Specia, France; Friedheim, 1949). However, melarsoprol shows severe side-effects (the worst being a reactive encephalopathy) and a relapse rate of 1–10% (Kazyumba *et al.*, 1988; Wellde *et al.*, 1989). Recently, relapse rates of over 30% have been observed in *T. b. rhodesiense* patients in North-west Uganda (Maiso, personal communication). New, safe and affordable drugs are urgently needed for second-stage sleeping sickness. Appropriate animal models for testing the therapeutic efficacy as well as the pharmacodynamics of new drugs are therefore a crucial part in the improvement of trypanosomosis management. Such models can also be used to study aspects of the disease in the mammalian hosts that may lead to anti-disease treatment and vaccine development.

Background of model

African trypanosomes can be kept as living cells in the laboratory in three different ways: first, in liquid nitrogen as cryopreserved stabilates; second, in continuous culture as blood stream forms and third, in rodents by continuous passage. Before techniques of cryopreservation became available and *in vitro* cultivation of the blood stream stage was possible, passaging the parasite in rodents was the only way to maintain trypanosome populations in the laboratory. But continuous syringe passages in rodents represent a constant selection which may change the characteristics of the original population, in that slow-growing variants are being lost. During this process virulence for the mammalian host usually increases, while the ability to express pleomorphism in the blood stream and to complete the cycle in the vector, the tsetse fly, gradually decreases. After years of rodent passages *T. brucei* spp. is likely to become monomorphic, highly virulent and to lose the potential to complete the cycle in the tsetse fly vector.

While *T. b. rhodesiense* (and the closely related *T. b. brucei*) isolates can easily be propagated in various mice and rats exhibiting fair to fulminant parasitaemia, *T. b. gambiense* isolates do not grow well or even at all in laboratory rodents. Selected rodent species or immunocompromised animals are required to obtain a detectable infection; however, the number of trypanosomes is very modest and approximately 100 times lower than for *T. b. rhodesiense*. There is a great variety in virulence among different trypanosome isolates. But it is not only the parasite that determines the virulence; the host also does. It was observed that different mouse breeds have different susceptibilities for a given *T. b. rhodesiense* population (Black *et al.*, 1985).

Handbook of Animal Models of Infection
ISBN 0-12-775390-7

Models for acute sleeping sickness due to *T. b. rhodesiense*

Animal species

A large number of different breeds of mice and rats expressing different degrees of susceptibility have been used to propagate *T. b. rhodesiense*, e.g. Swiss, CD1, Balb/c and other breeds of mice. Generally, all breeds of mice and rats are susceptible and can be used for the propagation of *T. b. rhodesiense*. Young adults are preferred over older animals. Mice 8 weeks or older and weighing 20–35 g are ideal. Both sexes can be equally infected but females are preferred over males because they are easier to keep in small groups without fighting. The mice or rats are kept in standard cages under standard conditions, and they are given commercial pellets and water *ad libitum*.

Storage, preparation of inocula and infection process

Bloodstream form trypanosomes are stored as cryopreserved stabilates in liquid nitrogen. Stabilates are thawed and diluted with phosphate saline glucose (PSG). Mice are injected either intraperitoneally or intravenously with 0.1–0.5 ml. The number of parasites injected determines the prepatent period and the course of parasitaemia. Rats are infected in the same way; the volume for injection may be larger.

Key parameters to monitor infection

After infection the parasites can be found in the blood stream. Parasitaemia is usually monitored three times a week by examination of tail blood. For very virulent strains a daily examination may be more appropriate. Packed cell volume and animal weight can also be used as a parameter to assess pathogenesis. In an acute infection with a virulent strain the animals may not look sick at all, despite high parasitaemia, and may be found dead the next morning.

Antiparasitic therapy

Standard drugs are suramin (Germanin, Bayer, Germany), diminazene aceturate (Berenil, Hoechst, Germany) and isometamidium chloride (Trypamidium or Samorin, Rhône Mérieux, France). The dose, duration and route are shown in Table 95.1.

Table 95.1 Standard drugs for acute sleeping sickness model

Drug	*Dose (mg/kg)*	*Duration*	*Route*
Suramin	10 or 20	Single dose	Intraperitoneally
Diminazene aceturate	7 or 15*	Single dose	Intraperitoneally
Isometamidium chloride	1	Single dose	Intraperitoneally

*Calculated for active principle.

Pitfall of the model

While *in vitro* assays measure the activity of a more or less constant concentration of a drug for the parasite, the mouse model has the advantage of changing drug concentrations based on the distribution, metabolism and excretion of the active molecule and its metabolites. However, the pharmacokinetic properties of a drug are different in the mouse model as compared to the human patient: the generally shorter terminal half-life of a compound in the mouse is in part compensated for by using a higher initial dose. Experience with standard drugs used in humans shows that the acute mouse model for *T. b. rhodesiense* can be cured by such drugs. Experimental compounds with distinct *in vitro* activity are often successful in curing infected mice. However, the rodent model cannot always predict toxicity of a compound for the host. An example is sinefungin, a natural nucleoside, which eliminated trypanosomes *in vitro* (Kaminsky and Zweygarth, 1989) and in mice (Dube *et al.*, 1983). In mice, a dose of 80 mg/kg sinefungin was administered five times without detectable adverse effects (Zweygarth and Röttcher, 1989). Neither the *in vitro* results, nor the experiments in mice predicted the pronounced nephrotoxicity of sinefungin in goats (Zweygarth *et al.*, 1986), which rendered the compound unsuitable for further evaluation.

For pharmacokinetic investigations the model is by no means ideal, because the mouse has a metabolism divergent from that of humans, and the size of the animal limits the sampling procedure. Pathogenesis in the mouse model differs from that in the human patient by a much more fulminant course of infection, a much higher parasitaemia and a short survival time of the infected animal. These features render the mouse not an ideal model for all aspects of pathogenesis. Indications of the ability of a drug to pass the blood–brain barrier cannot be obtained from the acute mouse model either. However, in our experience this model is a suitable test for assessment of antitrypanosomal activity of experimental compounds.

Contributions of the model to antiparasitic therapy

The acute mouse model is a logical step subsequent to observed activity in *in vitro* screening assays. The mouse is

Table 95.2 New experimental drugs which were evaluated in the rodent model

Compound	*In vitro activity* (IC_{50})	*Curative dose in mice against* T. b. rhodesiense	*Curative dose in* Mastomys *against* T. b. gambiense	*Reference*
CGP 40215	2.31 ng/ml	4×5 mg/kg i.p.	4×5 mg/kg i.p.	Brun *et al.* (1996)
(Novartis)		2×10 mg/kg i.p.	2×10 mg/kg i.p.	Bacchi *et al.* (1996)
S-HPMPA	28 ng/ml	2×10 mg/kg i.p.	No cure	Kaminsky *et al.* (1996)
Trybizine HCl	0.04 ng/ml	4×0.25 mg/kg i.p.	4×1 mg/kg i.p.	Kaminsky and Brun (1998)
(SIPI 1029)*		4×20 mg/kg oral		

* Shanghai Institute of Pharmaceutical Industry, Shanghai, China.

not simply an organic culture vessel but an organism with a defence (immune) system. For a cytostatic drug such as DFMO the *in vitro* assay reveals a poor activity, whereas in the mouse model the immune system will eliminate the non-proliferating parasites (Bitonti *et al.*, 1986). In the mouse model not only antitrypanosomal activity is assessed but also the pharmacodynamic properties of a drug, i.e. the model shows whether the drug concentration can be maintained for a required time over therapeutic level. Taking all characteristics into consideration the acute mouse model has a good predictive ability for activity in larger mammals or humans.

Table 95.2 demonstrates the usefulness of rodent models. All compounds were highly active *in vitro* and against the reference *T. b. rhodesiense*. However, the purine analogue (S)-HPMPA could not cure *T. b. gambiense*-infected *Mastomys*, while trybizine-HCl could not cure mice infected with a multidrug-resistant *T. b. brucei*.

Models for acute sleeping sickness due to *T. b. gambiense*

Animal species

Most of the commonly used laboratory rodents are more or less refractory to *Trypanosoma b. gambiense*. Exceptions are nursling rats, *Cricetomys gambianus* (Lariviere, 1957), *Microtus montanus* and *Mastomys natalensis*. Seed and Negus (1970) described the mountain vole *Microtus montanus* as a suitable laboratory model. Drawbacks are that infections only reach moderate parasitaemias after a few subpassages, the animals are hard to bleed and the amount of blood obtained from one animal is minute. Several *T. b. gambiense* isolates from the Ivory Coast could be propagated in *M. montanus* (Brun and Jenni, 1984). The multimammate rat *Mastomys natalensis* has been used by Mehlitz (1978) to isolate stocks from patients in West Africa. This rodent species is easy to breed and handle and gives slightly higher parasitaemias as compared to *Microtus*. However, it also requires several subpassages to reach a trypanosome density higher than 10^6/ml. Immunosuppression with cyclophosphamide (200 mg/kg i.p. 1 day before infection; Diffley and Scott, 1984) or dexamethasone (5 mg/l in drinking water, starting 3 days before infection and throughout the experiment) usually helps to obtain higher parasitaemias (Brun, unpublished observation). Both sexes of animals can become infected but females are easier to maintain in small groups. Young adults 8 weeks of age are ideal.

Severe combined immunodeficient (SCID) mice appear to have a much better susceptibility to *T. b. gambiense* than all the rodents used so far (Inoue *et al.*, 1998). These authors found that parasitaemia in SCID mice increased continuously until it reached approximately 10^9/ml, at which level it remained for a few days until the death of the animal. Mice were housed under standard conditions and used for infection when 5–6 weeks old.

Storage, preparation of inocula and infection process

T. b. gambiense stocks have initially to be isolated from human patients. For that purpose a blood or cerebrospinal fluid sample is inoculated intravenously or intraperitoneally into the rodents. *Microtus montanus* and *Mastomys natalensis* should be immunosuppressed by cyclophosphamide or dexamethasone. Once cryopreserved blood samples from infected animals are available, mice are infected i.p. with trypanosomes from such a stabilate suspended in PSG. In order to obtain a reproducible infection a fixed number of trypanosomes (e.g. 5×10^4/mouse) has to be recommended for injection.

Key parameters to monitor infection

The parasites can first be detected in the blood stream after a few days up to 3 weeks depending on the degree of adaptation of the trypanosome population. Parasitaemia is usually monitored three times a week by examination of tail blood using the haematocrit centrifugation technique (Woo, 1970). Once the number of trypanosomes observed in the buffy coat of the haematocrit capillary exceeds a few hundred, a wet blood film should be prepared to score the parasitaemia. In *Microtus* and *Mastomys* it may take

5–8 days before parasites can be found. This prepatent period is not much shorter in SCID mice but once the parasitaemia becomes detectable the number of trypanosomes increases far beyond that of other rodent species. Other criteria used to monitor infection in *T. b. rhodesiense* can be applied to *T. b. gambiense* as well.

Antiparasitic therapy

Standard drugs are suramin (Germanin, Bayer, Germany), pentamidine isethionate (Pentacarinat, Rhône-Poulenc Pharma, France), diminazene aceturate (Berenil, Hoechst, Germany) and melarsoprol (Arsobal, Specia, France; 1–10 mg/kg). The dose, duration and route are as described for *T. b. rhodesiense*.

Pitfall of the model

Mastomys natalensis give a reliable parasitaemia of *T. b. gambiense* isolates once they are subpassaged a few times. It has to be kept in mind that during this adaptation phase a selection process for rapidly proliferating subpopulations is taking place. For drug screening this is less of a problem since cloned populations are used which are well-adapted.

Immunosuppression and innate immunodeficient (SCID) mice could mask the outcome of an experiment in that trypanostatic drugs may not be able to cure due to the missing immune response (e.g. DFMO). On the other hand, the extremely low parasitaemia in normal (immunocompetent) rodents makes it almost impossible to perceive cure as such, since control animals may also have a parasitaemia below detection level. In our experience cure with standard drugs and experimental compounds can reliably be observed in immunosuppressed *M. natalensis*.

Contributions of the model to antiparasitic therapy

M. natalensis as well as SCID mice produce a moderate (10^6–10^7/ml for *Mastomys*) to high (10^9/ml for SCID) parasitaemia which is a prerequisite for monitoring the success of therapy. For drug testing both models appear suitable as long as we keep in mind that rodents immunosuppressed by chemicals or by knockout techniques (SCID mice lack a B- and T-cell system) do not have the supportive action of an immune defence. This characteristic of the SCID mice should not lead to false-positive results. It would rather contribute to missed cure; however, active compounds should at least show up on a longer survival time or by lowering the parasitaemia. While *Mastomys* have been used for drug tests in our laboratory, the SCID mice may have great potential which has yet to be demonstrated.

References

Bacchi, C. J., Brun, R., Croft, S. L., Alicea, K., Bühler, Y. (1996). *In vivo* trypanocidal activities of new S-adenosylmethionine decarboxylase inhibitors. *Antimicrob. Agents Chemother.*, **40**, 1448–1453.

Bitonti, A. J., McCann, P. P., Sjoerdsma, A. (1986). Necessity of antibody response in the treatment of African trypanosomiasis with alpha-difluoromethylornithine. *Biochem. Pharmacol.*, **35**, 331–334.

Black, S. J., Sendashonga, C. N., O'Brien, C. O., *et al.* (1985). Regulation of parasitaemia in mice infected with *Trypanosoma brucei*. *Curr. Top. Microbiol. Immunol.*, **117**, 93–118.

Brun, R., Jenni, L. (1984). *In vivo* and *in vitro* production of *Trypanosoma b. gambiense*. Proceedings of the Symposium On the diagnosis of African sleeping sickness due to *T. gambiense* (ed Crooy, P. J.), pp. 33–35. Smithkline-RIT, Belgium.

Brun, R., Bühler, Y., Sandmeier, U. *et al.* (1996). *In vitro* trypanocidal activities of new S-adenosylmethionine decarboxylase inhibitors. *Antimicrob. Agents Chemother.*, **40**, 1442–1447.

Diffley, P., Scott, J. O. (1984). Immunological control of chronic *Trypanosoma brucei gambiense* in outbred rodents. *Acta Trop.*, **41**, 335–342.

Dube, D. K., Mpimbaza, G., Allison, A. C., Lederer, E., Rovis, L. (1983). Antitrypanosomal activity of sinefungin. *Am. J. Trop. Med. Hyg.*, **32**, 31–33.

Friedheim, E. A. H. (1949). Mel B in the treatment of human trypanosomiasis. *Am. J. Trop. Med. Hyg.*, **20**, 173–180.

Inoue, N., Narumi, D., Mbati, P. A., Hirumi, K., Situakibanza, N-T. H., Hirumi, H. (1998). Susceptibility of severe combined immuno-deficient (SCID) mice to *Trypanosoma brucei gambiense* and *T. b. rhodesiense*. *Trop. Med. Intern. Hlth*, **3**: 408–412.

Iten, M., Matovu, E., Brun, R., Kaminsky, R. (1995). Innate lack of susceptibility of Ugandan *Trypanosoma brucei rhodesiense* to DL-α-difluoromethylornithine (DFMO). *Trop. Med. Parasitol.*, **46**, 190–194.

Kaminsky, R., Zweygarth, E. (1989). Feeder layer-free *in vitro* assay for screening antitrypanosomal compounds against *Trypanosoma brucei brucei* and *T. evansi*. *Antimicrob. Agents Chemother.*, **33**, 881–885.

Kaminsky, R., Brun, R. (1998). In vitro and in vivo activities of trybizine hydrochloride against various pathogenic Trypanosome species. *Antimicrob. Agents Chemother.*, **42**, 2858–2862.

Kaminsky, R., Schmid, C., Bühler, Y. *et al.* (1996). (S)-9-(3-hydroxy-2-phosphonylmethoxypropyl)adenine[(S)-HPMPA]: a purine analogue with trypanocidal activity *in vitro* and *in vivo*. *Trop. Med. Intern. Hlth.*, **1**, 255–263.

Kazyumba, G. L., Ruppol, J. F., Tshefu, A. K., Nkanga, N. (1988). Arsenoresistance et difluoromethylornithine dans le traitment de la trypanosomiase humaine africaine. *Bull. Soc. Pathol. Exot.*, **81**, 591–594.

Lariviere, M. (1957). Etude de l'infection expérimentale à *T. gambiense* du *Cricetomys gambianus*. Note préliminaire. *Bull. Méd. Afr. Occid. Francaise*, **2**, 122–125.

McCann, P. P., Bacchi, C. J., Clarkson, A. B. *et al.*, (1981). Further studies on difluoromethylornithine in African trypanosomes. *Med. Biol.*, **59**, 434–440.

Mehlitz, D. (1978). Untersuchungen zur Empfänglichkeit von *Mastomys natalensis* fur *Trypanozoon brucei gambiense*. *Tropenmed. Parasitol.*, **29**, 101–107.

Seed, J. R., Negus, N. C. (1970). Susceptibility of *Microtus montanus* to infection by *Trypanosoma gambiense*. *Lab. Anim. Care*, **20**, 657–661.

Wellde, B. T., Chumo, D. A., Reardon, M. J. *et al.* (1989). Treatment of Rhodesian sleeping sickness in Kenya. *Ann. Trop. Med. Parasitol.*, **83** (suppl. 1), 99–109.

Woo, P. T. K. (1970). The haematocrit centrifuge technique for the diagnosis of African trypanosomiasis. *Acta Trop.*, **27**, 384–386.

Zweygarth, E., Röttcher, D. (1989). Efficacy of experimental trypanocidal compounds against a multiple drug-resistant *Trypanosoma brucei brucei* stock in mice. *Parasitol. Res.*, **75**, 178–182.

Zweygarth, E., Schillinger, D., Kauffmann, W., Röttcher, D. (1986). Evaluation of sinefungin from the treatment of *Trypanosoma (Nannomonas) congolense* infections in goats. *Trop. Med. Parasitol.*, **37**, 255–257.

Chapter 96

Animal Models of CNS (second-stage) Sleeping Sickness

C. Gichuki and R. Brun

The mouse model of African sleeping sickness

Background of model

Jennings *et al.* (1977) used two diminazene aceturate-(Berenil, Hoechst) susceptible stabilates of *Trypanosoma brucei brucei* to investigate the source of relapse of parasitaemia in mice. They found that relapses consistently occurred if treatment was delayed by 21 or more days following infection, independent of dosage of drug used or the size of the parasite inoculum. This also occurred when other trypanocides (isometamidium, ethidium and prothidium) that do not cross the blood–brain barrier were used instead of diminazene. The brain was demonstrated to be the source of recrudescent infections in such mice (Jennings et al., 1979; Jennings and Gray, 1983) as a result of the parasites reinvading the blood stream once the blood drug levels decreased below protective concentrations. This hypothesis was supported by the finding that parasites isolated from such infections were not drug-resistant (Brun, personal communication). This mouse model is now widely used to evaluate the efficacy of new chemotherapeutic agents for efficacy against second-stage human African trypanosomiasis (HAT).

Although *T. b. brucei* is not infective to humans, it belongs to the same species (*T. brucei*) as the two human infective parasite subspecies. The similarities include tissue invasiveness and causing chronic infection in both domestic and laboratory animals with similar characteristics to the human disease, such as involvement of the central nervous system (CNS; Goodwin, 1974; Poltera *et al.*, 1980; Jennings *et al.*, 1989).

Animal species

Several strains of white inbred mice have been used: NIH, CD-1 and Swiss white (Jennings and Gray, 1983; Hunter *et al.*, 1991; Gichuki *et al.*, 1997). Adult mice weighing between 25 and 35 g are used. The sex is not important but females are more often used because they are more docile and easier to handle and exhibit less cannibalism when kept in groups. The mice are maintained in standard accommodation and are given commercial pellet feed and water *ad libitum*.

Storage, preparation of inocula and the infection process

The mouse model as described by Jennings *et al.* (1977) is based on infection of mice with *T. b. brucei* strain GVR 35. Mice are infected intraperitoneally with 4×10^4 trypanosomes from a cryopreserved stabilate suspended in phosphate saline glucose (PSG). The stabilate may be expanded in immunosuppressed mice, after which infected mouse blood may be used.

Key parameters to monitor infection

After infection, parasitaemia is monitored three times a week by examination of tail blood. After 21 days of infection the mice are treated with an early-stage trypanocide, such as diminazene aceturate, to clear the parasites in the systemic circulation and in tissues other than the CNS. The parasites that remain within the brain parenchyma continue to replicate and give rise to recrudescent parasitaemia (Jennings *et al.*, 1977, 1989; Jennings and Gray, 1983). In addition, the clinical status of the mice is closely observed. Raised hair coats, dullness, excessive sweating and reduced appetite have been observed to coincide with peak parasitaemia while hindlimb paralysis has been observed to occur after the infection is established in the CNS (Jennings, personal communication). To measure the response to treatment the same parameters are used, i.e. monitoring of parasitaemia. A 60-day post-treatment follow-up on parasitaemia is recommended before cure can be assumed.

Antiparasitic therapy

A combination of the arsenical drug, Mel Cy (Cymelarsen, Rhône Mérieux, France) with the nitroimidazole com-

Handbook of Animal Models of Infection
ISBN 0-12-775390-7

pound, MK 436 (Merck Institute for Research, New Jersey) is used as a curative treatment in this mouse model. The drugs are administered in two daily doses of 5 mg/kg Mel Cy + 15 mg/kg MK 436 intraperitoneally (Jennings, personal communication). Late-stage disease in the mouse model can also be curatively treated using melarsoprol at an intraperitoneal dose of 14.4 mg/kg on 4 consecutive days (Brun, personal communication).

Pitfall of the model

The major problem of the CNS mouse model is based on the different metabolic rates and pharmacology of the drug in the small mammal as compared to humans. There is the possibility that a compound which cannot cure the CNS mouse model due to such pharmacological problems would still be able to build up therapeutic levels in the CNS of a larger mammalian host.

Contributions of the model to antiparasitic therapy

The CNS mouse model is a useful step beyond the acute mouse model, which usually follows observed activity in *in vitro* screening assays. The mouse as a small mammal has metabolic rates which may differ significantly from those of humans, and thus it must be expected that the pharmacokinetics of drugs in the mouse model would be very different from that of humans. The size of this animal limits the amounts of blood that can be taken for analysis of drug levels. Taken together, this limits the ability of this animal model to have a predictive value for pharmacokinetics in humans.

The CNS mouse model gives an indirect indication of blood–brain barrier permeability. Cures indicate that the drug penetrated the blood–brain barrier, indicating that such a compound has the potential to cure late-stage sleeping sickness in humans. However, if a compound needs longer exposure time to kill the trypanosomes, the time–dose relationship becomes a crucial parameter, which might be rather different in the mouse as compared to humans. The predictive ability of this mouse model remains uncertain, especially with compounds that do not cure late-stage infections in mice.

The vervet monkey model of African sleeping sickness

Background of model

The formation of the KETRI monkey model for sleeping sickness started in December 1979, when vervet monkeys (African green monkey, *Cercopithecus aethiops*) were introduced into a quarantine division, which had been put into place at the Veterinary Research Department of Kenya Agricultural Research Institute (Sayer *et al.*, 1980). Considerable assistance in planning and designing the unit together with training in animal handling techniques was obtained from the director and staff of the Institute of Primate Research in Kenya. The aim of the formation of such a model was to investigate the therapeutic efficacy and pharmacodynamics of drugs and experimental compounds for the treatment of human trypanosomosis.

As a first step, it was intended to develop an artificially induced disease syndrome using trypanosomes of the *brucei* group which mimics as precisely as possible the disease as it is seen in humans. Expectations were such that the study of the disease syndrome had to take second place and drug testing had to begin immediately. Professor Helmut Schmidt, a neuropathologist from West Germany, was already in Kenya waiting to test two diamidines in subhuman primates for toxicity and efficacy against trypanosome-induced encephalitis. As a result of that study, it was found that the strains of trypanosomes used only rarely produced nervous symptoms in 30 infected monkeys and that only 1 monkey showed true pathological evidence of encephalitis on post-mortem examination. Mild meningitis occurred in monkeys that survived the infection for more than 45 days. Encephalitis was consistently found in animals that showed relapse of infection following treatment with diamidines. It was also found that the monkeys conditioned for 12 weeks had an average survival time of 50 days while those conditioned for a longer duration had a survival time of 80 days on average. This was of great importance because the longer the survival time, the higher were the chances of encephalitis occurring.

After the initial diamidine experiments had been carried out, it was clear that a more suitable strain of trypanosome had to be found in order to produce the late-stage encephalitis syndrome, while the *T. b. rhodesiense* EATRO 1989 was to be used in the interim. Later it was found that using a derivative of the same *T. b. rhodesiense* stock, the animals died early due to cardiac failure and that this could be avoided by treating the infected monkeys with the diamidine diminazene aceturate (Berenil, Hoechst, Germany), on or after day 28 of infection. This resulted in relapse of parasitaemia 50–70 days later (80–100 days of infection) accompanied by clinical signs of late-stage encephalitis, which closely simulated signs seen in humans (Schmidt and Sayer, 1982 a,b).

The model has been improved over the years and currently it is the only one of its kind in the world. It has been the subject of various studies, including drug evaluation and pharmacokinetic investigations. Currently, infection of the vervet monkeys with *T. b. rhodesiense* KETRI 2537 produces three disease models:

1. The early-stage disease model, which is formed by the infection between days 0 and 35
2. The terminal disease model, in which the infection is allowed to develop up to day 42 of infection

3. The advanced late-stage model in which the infection is treated on days 33–35 with diminazene aceturate and allowed to relapse 50–70 days later (days 80–100 of infection).

Animal species

The African green (vervet) monkey (*C. aethiops*) was chosen because it is readily available and a suitable subhuman primate, present in large numbers in the Kenyan wild, where it is frequently an agricultural and hotel pest. In addition, there has been an increase in captive breeding. Improvements in housing and handling of these monkeys have made them an acceptable laboratory species. Dangerous zoonotic diseases continue to be a major hazard, for which extreme caution must be exercised if the lives of animal handlers and research workers are not to be endangered.

Acquisition of monkeys

Monkeys that are available commercially can be grouped into three categories:

1. Raw or unconditioned monkeys
2. Conditioned monkeys: these are usually captured and held in the facilities through a quarantine period, during which they get acclimatized to laboratory conditions and human presence
3. Captive-bred monkeys: these are defined and disease-free monkeys.

Receiving and initial examination

Animals are received by experienced personnel and placed in an isolated (quarantine) facility some distance from the regular holding facility. All animals are isolated for diagnosis and treatment. Moribund animals are euthanized. A full clinical examination is carried out by an experienced veterinarian and all the monkeys are properly labelled with tag marks or tattoos.

Housing and food

Holding rooms are constructed in such a way that there are two screening doors between the wards and the corridor to the other rooms in the building. In between these doors, there is a foot bath in which antiviral disinfectant should be placed at all times. In addition, the holding rooms should be maintained at a temperature between 20 and 26°C, with a relative humidity of between 40 and 60%. Relative humidity of below 30% predisposes to respiratory disorders. Acoustic ceilings are preferred to reduce the amount of noise generated by the animals. Protective screens should be placed on the ceiling and the wall fixtures. The entrance to the animal facility is designated hazardous and entry is restricted to authorized staff only. The monkeys are housed in individual squeeze-back cages to minimize the stress of handling (Figure 96.1). They are provided with fresh water *ad libitum* and are fed on commercial monkey pellets, fresh vegetables and fruit twice everyday. The vegetables consist of fresh carrots, kale, sweet potato, maize and tomatoes while the fruit mainly consists of bananas. These are purchased every other day to ensure freshness.

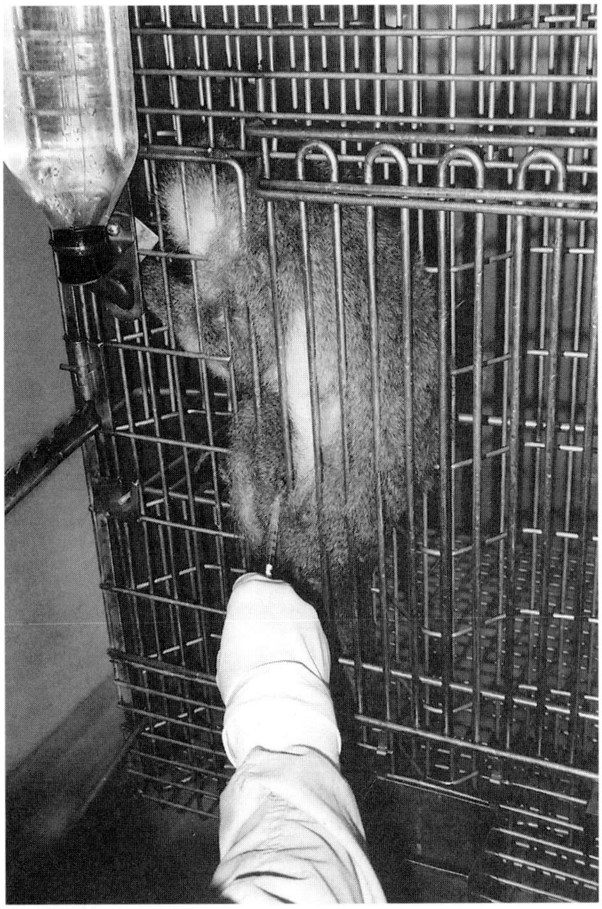

Figure 96.1 Squeeze-back cages ease the handling of monkeys for anaesthesia and intramuscular drug-administration. They are also useful for *per os* administration of drugs without anaesthesia.

Health records

Health records are kept for individual monkeys throughout their life from the first day in quarantine and throughout their life in the facility. The records should indicate the findings of each clinical examination, any treatments administered, vaccinations and results of any laboratory tests, and should indicate the date, dosage and recommendations put into place.

Quarantine

The purpose of the quarantine is to isolate new arrivals from the present, assumed healthy monkeys, until their health status is established. This is particularly important for the monkeys that are trapped in the wild and transported in cages to the quarantine facility.

Many simian viral or bacterial diseases found in the vervets are hazardous to humans. The quarantine period is designed to safeguard both the human personnel and the monkeys already present and presumed healthy. Apart from zoonotic diseases, there is the major problem of conditioning these wild monkeys to cage life, to a constant human presence and the various daily routine procedures. The quarantine takes a minimum of 12 weeks (approximately 90 days) during which time, the monkeys are screened for Marburg virus antibody, subjected to repeated and detailed examinations: clinical, haematological, serological, faecal and tuberculin-tested (Table 96.1). During this time, the monkeys are also screened for simian immunodeficiency virus (SIV) antibodies and treated for intestinal parasites, helminths and protozoa if detected by faecal examination, and also for ectoparasites such as lice (Table 96.1).

Storage, preparation of inocula and infection

Today, the monkey model is based on infections of the vervet monkeys with *T. b. rhodesiense*, KETRI 2537, a derivative of EATRO 1989 which was isolated by direct injection of a patient's blood into a monkey and originated from Busoga, Uganda. Cryopreserved trypanosomes are gently thawed and passaged once through sublethally irradiated mice to raise the numbers of surviving trypanosomes. When peak parasitaemia is attained, the mice are exsanguinated. Trypanosomes are separated from blood cells through DEAE cellulose (DE 52) in PSG. Trypanosomes are diluted and injected intravenously into the monkeys at a concentration of 10^4 trypanosomes per monkey. At this concentration, monkeys infected with *T. b. rhodesiense*, KETRI 2537, usually develop parasitaemia on day 4 of infection.

Key parameters to monitor infection

Every day following infection, the monkeys are seen by a veterinarian who records their demeanour, state of sleepiness, their appetite and stool consistency. Parasitaemia is monitored daily by examination of wet blood film collected by pricking the ear pinnae. Every fortnight, the animals are anaesthetized using ketamine hydrochloride at 10 mg/kg, a full clinical examination is carried out, 5 ml of blood is collected for haematology, serology and biochemistry (Figure 96.2) and cerebrospinal fluid is collected via lumbar puncture for determination of parasitosis, protein concentration, white cell counts and blood contamination (Figure 96.3). In our experience, the normal parameters in vervet monkeys are as shown in Table 96.2.

These parameters are monitored throughout experimentation. Deviations from the normal ranges are used as indicators of sickness. After treatment, these parameters revert to the normal range, indicating chemotherapeutic success. Animals that maintain normal parameters for at least 600 days after treatment are considered cured.

Standard drugs for antiparasitic therapy

Early-stage treatment is instituted using three daily doses of diminazene aceturate at 5 mg/kg intramuscularly. This drug is no longer curative after day 28 of infection with *T. b. rhodesiense* KETRI 2537. An alternative treatment can be instituted by treating the animals intravenously with suramin at 20 mg/kg five times. The suramin injections are spread over a 10-day period such that the animals receive suramin on days 1, 3, 5, 7 and 10 of treatment.

So far, the only curative drug for late-stage infections with *T. b. rhodesiense* KETRI 2537 is melarsoprol, which is injected daily at 3.6 mg/kg intravenously on 4 consecutive days.

Table 96.1 Procedures and tests carried out during quarantine of wild-caught vervet monkeys

Procedure/test	*Times procedure/test carried out*	*Weeks in quarantine*	*Procedure if positive*
Tuberculin test	3×, 2 weeks apart	2, 4 and 6	Euthanized
Haemorrhagic fevers (Marburg, Ebola, rotaviruses) tests	2×	Before introduction into facility and week 2	Euthanized
Simian immunodeficiency virus test	1×	1	Euthanized
Dewormed	3× once a month	2, 6 and 10	Anthelminthics
Faecal bacteriology	In case of diarrhoea		Antibiotic treatment or antiprotozoal
Ectoparasitic washes	3× once a month	2, 6 and 10	Insecticide washes
Cage training	3× every week	From week 2	

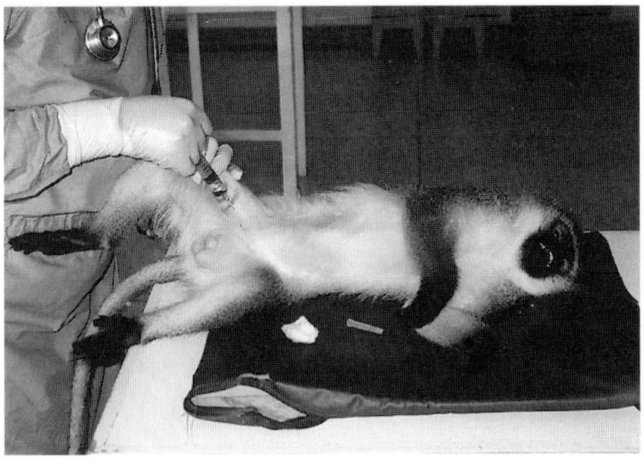

Figure 96.2 A blood sample is being drawn from the inguinal vein of an anaesthetized monkey. Personnel need to wear protective clothing and rubber gloves.

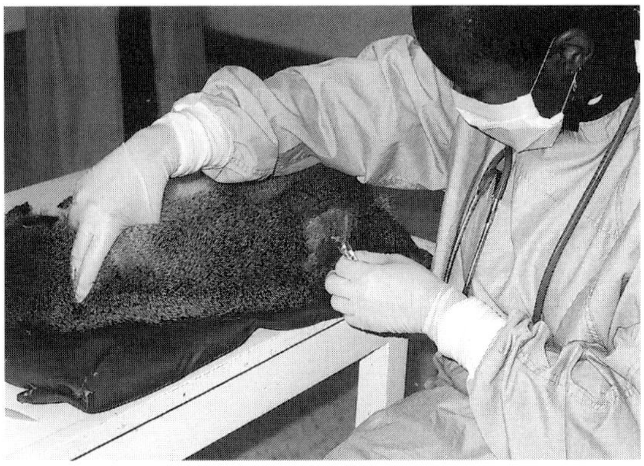

Figure 96.3 A sample of cerebrospinal fluid is being tapped from a lumbar space. Personnel need to wear protective clothing, rubber gloves and a face mask.

Pitfalls of the model

The follow-up after treatment before a test compound can be declared curative is lengthy. The accepted duration by World Health Organization (WHO) is 600 days. In addition, the model is costly with daily maintenance costs of US$5/day per monkey during quarantine and US$3/day per monkey after quarantine and purchasing costs of US$200/monkey. The lengthy after-treatment follow-up can be shortened by use of serological and molecular biological techniques. It has been shown that the use of antigen detection enzyme-linked immunosorbent assay (ag ELISA) can shorten the post-treatment follow-up by more than half (Gichuki *et al.*, 1994) and that it can predict relapse of parasitaemia 11 weeks before demonstration of parasites in the blood or cerebrospinal fluid (Njue *et al.*, 1997). However, the parasite antigens persist for several months in the serum of monkeys following curative trypanocidal therapy. The post-treatment follow-up may be further shortened by introduction of trypanosome RNA detection tests, since RNA has a short half-life compared to antigens.

Contributions of the model to antiparasitic therapy

The fact that standard drugs for human sleeping sickness therapy (e.g. melarsoprol) are also able to cure infected vervet monkeys suggests that this model is a good indicator of the efficacy of new antiparasitic compounds. The closeness of the disease presentation in the model compared to that of humans clinically, pathologically and immunologically indicates that this model also offers an excellent opportunity for studying the pathogenesis of the disease. The vervet monkey is an excellent model for human sleeping sickness (Schmidt and Sayer, 1982a,b), showing symptoms similar to those found in human patients, and an ideal animal model for pharmacokinetic investigations due to its close relationship to humans. For the trypanocidal drug melarsoprol it could be shown that the pharmacokinetics is very similar in the vervet monkey as compared to humans (Burri *et al.*, 1993, 1994). The ability repeatedly to draw cerebrospinal fluid from the monkeys' lumbar spaces enables one to follow the progression of CNS invasion by the parasites and the pathology induced by such invasion to be studied. Determination of drug levels in the cerebrospinal fluid indicates the ability of such drugs to cross the blood–brain barrier and provides pharmacokinetic data in both the systemic circulation and the CNS.

In summary, it appears that the predictive ability of the vervet monkey model for the chemotherapy of sleeping sickness is the best that is currently available.

Table 96.2 Parameters used to monitor infection in the vervet monkey model

Parameter	*Blood*	*Cerebrospinal fluid*
Total protein	5.36–6.05 g/dl	28 mg/dl
Albumin	3.06–3.73 g/dl	
Globulins		
Total white cells	$3.0–8.0 \times 10^3$ cells/μl	0–10 cells/μl
Red cells	$4.0–6.0 \times 10^6$ cells/μl	0
Packed cell volume	45–56%	Not applicable
Parasites	0	0

References

Burri, C., Baltz, T., Giroud, C., Doua, F., Welker, H. A., Brun, R. (1993). Pharmacokinetic properties of the trypanocidal drug melarsoprol. *Chemotherapy*, **39**, 225–234.

Burri, C., Onyango, J. D., Auma, J. E., Burudi, E. M. E., Brun, R. (1994). Pharmacokinetics of melarsoprol in uninfected vervet monkeys. *Acta Trop.*, **58**, 35–49.

Gichuki, C. W., Nantulya, V. M., Sayer, P. D. (1994). *Trypanosoma brucei rhodesiense*: use of an antigen detection enzyme immunoassay for evaluation of response to chemotherapy in infected vervet monkeys (*Cercopithecus aethiops*). *Trop. Med. Parasitol.*, **45**, 237–242.

Gichuki, C. W., Jennings, F. W., Kennedy, P. G. E. *et al.* (1997). The effect of azathiprine on the neuropathology associated with experimental murine African trypanosomiasis. *Neurol. Infect. Epidemiol.*, **2**, 53–61.

Goodwin, L. G. (1974). The African scene: mechanisms of pathogenesis. In *Trypanosomiasis and Leishmaniasis*, pp. 107–124. Ciba Foundation Symposium 20 (new series). Elsevier Excerpta Medica, North Holland.

Hunter, C. A., Gow, J. W., Kennedy, P. G. E., Jennings, F. W., Murray, M. (1991). Immunopathology of experimental African sleeping sickness: detection of cytokine mRNA in the brains of *Trypanosoma brucei brucei*-infected mice. *Infect. Immun.*, **59**, 4636–4640.

Jennings, F. W., Gray, G. D. (1983). Relapsed parasitaemia following chemotherapy of chronic *T. brucei* infections in mice and its relation to cerebral trypanosomes. *Contrib. Microbiol. Immunol.*, **7**, 147–154.

Jennings, F. W., Whitelaw, D. D., Urquhart, G. M. (1977). The relationship between duration of infection with *Trypanosoma brucei* in mice and the efficacy of chemotherapy. *Parasitology*, **75**, 143–153.

Jennings, F. W., Whitelaw, D. D., Holmes, P. H., Chizyuak, H. G. B., Urquhart, G. M. (1979). The brain as a source of relapsing *Trypanosoma brucei* infection after chemotherapy. *Int. J. Parasitol.*, **9**, 381–384.

Jennings, F. W., McNeil, P. E., Ndung'u, J. M., Murray, M. (1989). Trypanosomiasis and encephalitis: possible aetiology and treatment. *Trans R. Soc. Trop. Med. Hyg.*, **83**, 518–519.

Njue, A. I., Olaho-Mukani, W., Ndung'u, J. M. (1997). Use of antigen ELISA to evaluate response to treatment in *T. rhodesiense* infected vervet monkeys. pp. 227. Proceedings of the 18th African Health Sciences Congress, Capetown, South Africa.

Poltera, A. A., Hochmann, A., Rudin, W., Lambert, P. H. (1980). *Trypanosoma brucei brucei*: a model for cerebral trypanosomiasis in mice. An immunological, histopathological and electron microscopic study. *Clin. Exp. Immunol.*, **40**, 496–507.

Sayer, P. D., Njugo, A. R., Losos, G. J. (1980). The development of a primate unit for the study of chemotherapy in African trypanosomiasis. pp. 47–52. KETRI Annual Report 1980.

Schmidt, H., Sayer, P. (1982a). *Trypanosoma brucei rhodesiense* infection in vervet monkeys. I. Parasitologic, hematologic, immunologic and histologic results. *Tropenmed. Parasitol.*, **33**, 249–254.

Schmidt, H., Sayer, P. (1982b). *Trypanosoma brucei rhodesiense* infection in vervet monkeys. II. Provocation of the encephalitic late phase by treatment of infected monkeys. *Tropenmed. Parasitol.*, **33**, 255–259.

Chapter 97

Animal Models of *Trypanosoma cruzi* Infection

M. M. do Canto Cavalheiro and L. L. Leon

Background of human infection

Incidence and etiology

The infection caused by *Trypanosoma cruzi* is normally transmitted by a hematophagous insect of the subfamily Triatominae. Infection can also occur through blood transfusions: in endemic countries, not all blood samples are controlled as well as by congenital transmission (Brener, 1984). Chagas' disease affects about 16–18 million people in Latin America (Dias, 1997).

The life cycle of *T. cruzi* involves obligatory passage through vertebrate and invertebrate hosts, in a series of different developmental stages. The trypomastigote ingested by the insect differentiates into the proliferative epimastigote form which, on reaching the lower intestine, differentiates into the metacyclic trypomastigote form. The latter, following invasion of vertebrate host cells, undergoes differentiation into the amastigote form which, after several reproductive cycles, transforms into the trypomastigote form which is responsible for the dissemination of infection.

Pathology

Three phases in the natural history of human infection with *T. cruzi* are recognized clinically.

Initial phase

The initial phase of acute infection is characterized by myocarditis with foci of invasion of the cardial myocytes by the parasites. The acute form of human Chagas' disease is rare in adults and usually non-fatal; in addition to patent blood parasitemia, symptoms may include a chagoma (tissue reaction at the site of parasite entry), fever, muscle pain, vomiting, diarrhea, enlarged liver, spleen and lymph nodes, and occasionally, myocarditis.

Indeterminate phase

In most cases acute Chagas' disease subsides within 2–4 months and the disease enters its latent phase; serology remains positive but patients are otherwise asymptomatic. Persisting low levels of parasitemia are only evident from occasional positive results using sensitive diagnostic techniques.

Chronic phase

The duration of the latent phase in those infected people (around 10%) who go on to develop the chronic form of Chagas' disease has not been accurately determined but may take 10–30 years (Santos-Buch, 1979). The chronic phase involves progressive myocarditis and congestive heart failure with myocardial hypertrophy, myocyte degeneration, severe intestinal fibrosis and thickening of the basement membranes of the cardiac myocytes, endothelial cells, and vascular smooth muscle cells.

The exact relationship between *T. cruzi* infection and Chagas' disease is not known (Tarleton *et al.*, 1994); however, both direct (parasite-mediated tissue destruction) and indirect (mostly immunological or autoimmune) etiologies have been proposed. Support for an immunological basis of pathology in chronic *T. cruzi* infection comes mainly from the difficulty in detecting intracellular and blood forms of *T. cruzi* in patients with chronic infection.

A number of mechanisms have been proposed to account for chronic pathology, including direct tissue destruction, loss of nervous tissue function, intravascular platelet aggregation and generation of autoimmune reactivity. The earliest proposed and still the simplest explanation for chronic-phase pathology is the cumulative damage due to parasite invasion of muscle cells, lysis of those cells, release of additional parasites, and a consequent inflammatory reaction which terminates in fibrosis. Sterilizing immunity and total clearance of *T. cruzi* are rarely reported, so it is presumed that most infected individuals harbor the parasite for life and may experience significant cumulative organ damage due to cycles of parasitosis.

Immune response

Specific antibodies against *T. cruzi* (detected by complement fixation, immunofluorescence and hemagglutination tech-

Handbook of Animal Models of Infection
ISBN 0-12-775390-7

niques) are found in human infection: they develop from the acute phase and persist throughout the chronic phase. During the initial acute phase immunoglobulin M (IgM) predominates and later IgG and IgA are major antibody classes.

Cell-mediated immunity to *T. cruzi* is evident from both positive delayed hypersensitivity skin reactions and *in vitro* leukocyte transformation in contrast with the consistent presence of antibodies after infection, the expression of cell-mediated immunity seems to be more variable; for that reason, there is a limit to the potential of tests using this parameter (Scott and Snary, 1982).

Diagnosis

Diagnosis of *T. cruzi* infection relies on the identification of either the organism in the blood stream or specific antibodies produced during the infection.

Identification of the parasite

During the acute phase of infection, when large numbers of parasites are present in the blood, diagnosis is easily performed by the identification of trypanosomes in blood smears. During the chronic phase, when low numbers of parasites are found, the usual tests are xenodiagnosis and the hemoculture. Xenodiagnosis is effected by feeding Triatomids (the natural vector) on infected patients and subsequently examining the bugs for the presence of *T. cruzi* in their hind gut; this method has the disadvantage that only a small percentage of known infections are identified (20–40%) and repeated testing is often necessary to achieve a positive result. In the hemoculture technique, which is equally as sensitive as xenodiagnosis, blood samples are incubated in a medium which is capable of supporting the growth of *T. cruzi* in the epimastigote form. However, to confirm chronic *T. cruzi* infections in humans, there is a need for a modern technique, such as polymerase chain reaction, that is shown to be highly sensitive and specific (Junqueira *et al.*, 1996).

Serodiagnosis

Since infection with *T. cruzi* results in a humoral immune response, detection of specific antibodies is the best test for the identification of a chronic phase of the disease. The correlation between serological diagnosis and infection is so high that some authors have used the disappearance of sero-positive reactions as an indication of an effective chemotherapeutic cure. The complement fixation test was the first method for serodiagnosis of Chagas' disease described by Guerreiro-Machado, showing a high degree of sensitivity (>90%). Direct agglutination of specific antibody by the agluttination of epimastigote forms is complicated by the ability of epimastigotes to agglutinate spontaneously in the presence of normal sera. Indirect agglutination is a sensitive method of diagnosis and normally becomes positive later after infection. However, the IgM produced during the acute phase can be detected using a polysaccharide fraction from *T. cruzi* to coat erythrocytes. Indirect immunofluorescence is an extremely sensitive and rapid assay, which becomes positive before most others when acute infections are studied; this method has the added advantage that fluorescent classes of specific reagents can be used to identify IgM and IgG antibody, to differentiate between recent and chronic infections. The enzyme-linked immunosorbent assay (ELISA) is simple, sensitive, inexpensive and readily adaptable to field conditions; and ELISA adaptation of the complement fixation assay (CELISA) has already been reported.

Autoantibody is present in many patients with Chagas' disease and can be used diagnostically. The antibody gives a characteristic staining pattern for mammalian heart, binding to endocardium, vascular tissue and interstitium of striated muscle (Scott & Snary, 1982).

Animal models

Chagas infection has been observed in different mammal species (Chagas, 1909). Several animal models have been used experimentally (WHO, 1984; Brener and Ramirez, 1985), such as mice (Pereira da Silva and Nussenzweig, 1953; Brener, 1962; Federici *et al.*, 1964), hamsters (Cariola *et al.*, 1950; Osimani and Gurri, 1954; Ramirez *et al.*, 1991, 1993, 1994), dogs (Vichi, 1961; Castro and Brener, 1985; Lana *et al.*, 1992), rats (Ciconelli, 1963; Chapadeiro *et al.*, 1988), rabbits (Chiari *et al.*, 1980; Teixeira *et al.*, 1983; Figueiredo *et al.*, 1985; Ramirez and Brener, 1987) and monkeys (Seah *et al.*, 1974; Marsden *et al.*, 1976; Miles *et al.*, 1979; Pung *et al.*, 1988; Rosner *et al.*, 1988, 1989; Bonecini-Almeida *et al.*, 1990). The course of the *T. cruzi* infection varies widely between those laboratory animals, depending upon the host and parasite strains used, the route of inoculation and the size of the inoculum.

Parasite strain

It has been shown that different strains can behave quite differently in experimental Chagas' disease with regard to characteristics such as the course of infection, the degree of parasitemia, tissue tropisms, histopathological changes and mortality (Boneccini-Almeida *et al.*, 1990).

Several strains of *T. cruzi* have been used in different animal models: Y (isolated from a human acute case; Pereira da Silva and Nussenzweig, 1953); CL (from *Triatoma infestans*, Brener *et al.*, 1976); Ernane (from a patient with cardiac form and megaesophagus); Benedito (isolated through xenodiagnosis from a patient with the indeterminate form) and Vicentina (isolated through xenodiagnosis from a patient with the chronic cardiac form; Ramirez *et al.*, 1994).

Strains of *T. cruzi*, from different geographical areas, had previously been characterized into various types according to their infectivity rate and tropism in mice (Andrade, 1976; Andrade *et al.*, 1985).

The classification includes the following:

1. Type I, characterized by a rapid course of infection in mice, high levels of parasitemia and mortality around the 9th and 10th day of infection, with predominance of slender forms and macrophage tropism during the acute phase of the infection
2. Type II shows increasing parasitemia from the 12th to the 20th day of infection, low mortality rate, predominance of broad forms of the parasite and myocardial tropism
3. Type III shows a slow development of parasitemia that reaches a high level 20–30 days after inoculation, low mortality and predominance of parasitism in skeletal muscles.

Inoculation

The inoculum range is usually from 1×10^3 to 1×10^7 trypomastigotes (obtained from infected animals) or $2–4\times10^3$ metacyclic trypomastigotes (obtained from triatomid bugs). Several inoculation routes are used; the most common are intraperitoneal and conjunctival.

Mouse model

The use of mice as the main experimental model for *T. cruzi* infection is particularly valuable in view of the large amount of information available about their immunological system (Torrico *et al.*, 1991; Ribeiro *et al.*, 1992) and their resistance to different parasite strains.

However, various mice strains differ markedly in their resistance to *T. cruzi*. More resistant strains might provide a good model for the chronic disease (WHO, 1974). At this stage, the murine model of Chagas' disease has usually been used in experimental chemotherapy (Filardi, 1988).

Several strains of mice have been used in this model: Swiss, weight 18–20 g, female (McCabe *et al.*, 1983); Balb/c, 8–10 weeks, female (Queiroz da Cruz *et al.*, 1991); albino, weight 18–20 g, male (Filardi, 1988). In each case preliminary experiments were performed to determine the optimal parasite inocula to insure infection.

The parasites were maintained by serial passage through female C3H/He mice, which is a resistant strain. Mice with parasitemia were bled into heparinized (1000 U/l) phosphate-buffered saline (50 : 50) and cryopreserved (Scott and Matthews, 1987).

Hamster model

Hamsters (non-isogenic Syrian hamsters, *Mesocricetus auratus*, male/female) can be infected with *T. cruzi*. During the acute phase an inflammatory reaction can be observed characterized by mononuclear and polymorphous leukocyte infiltration of variable degree in the majority of tissues and organs. In the chronic phase the same kind of lesions can be observed, but the inflammatory process is less severe and characterized by mononuclear infiltration in the myocardium (Ramirez *et al.*, 1994). The authors noted high levels of parasitemia in the beginning of the infection, which varied with the strain used.

Canine model

Chagas' disease induced experimentally in dogs is said to resemble human disease closely in all aspects (WHO, 1974). It is the only experimental animal model in which the indeterminate phase progresses to the late phase of severe chronic myocarditis (Andrade *et al.*, 1997).

Mongrel dogs (3–4 months old, of both sexes) were used as an experimental model. In these dogs, the acute infection subsided and was followed by a prolonged, clinically silent interval, after which a late phase of chronic myocarditis and heart failure developed. However, an enhancement of chronic myocarditis was obtained after they were treated with low doses of cyclophosphamide (Andrade *et al.*, 1987).

Rat model

Rivera-Vanderpas *et al.* (1983) described the susceptibility of different inbred strains of rats to *T. cruzi* infection. Eight strains of male rats (AUG, BN, LEW, LIS, WAG, F344, LOU/M, DA) and three strains of female rats (LEW, F344, KGH) were challenged with 15×10^4 trypomastigotes of the Tchuantepec strain of *T. cruzi* (isolated from Mexico by Brumpt). Both male and female, showed a high degree of parasitemia, but all the males died between the 34th and 42nd day of the infection, whereas all the females survived the infection with total disappearance of the parasitemia.

These results confirm observations by other investigators (Scorza and Scorza, 1972; Rodriguez *et al.*, 1981) with regard to the resistance of rats to *T. cruzi* infection. However, contradictory results have been cited in the literature (Bonecini-Almeida *et al.*, 1990).

Rabbit model

Rabbits (outbred male, 2–4 months old) infected with *T. cruzi* show lesions resembling the pathology found in chronic human Chagas' disease, such as diffuse myocarditis, signs of heart failure and megacolon (Teixeira and

Santos-Buch, 1975). In contrast, Ramirez and Brener (1987) described a parasitological study of the rabbit as a model for Chagas' disease and found that the animals showed a high level of parasitemia in the early stage of the infection. The course of the parasitemia was highly influenced by the parasite strain and the route of inoculation. Furthermore, through xenodiagnosis and/or hemoculture the parasites have been recovered from 40% of the animals during the chronic phase of the infection.

Monkey model

The first study with a non-human primate as an animal model for Chagas' disease was carried out by Chagas (1909) using *Callitrix penicillata*, which developed the infection 20–30 days after contact with infected triatomids. The behavior of several *T. cruzi* strains was evaluated in *Cebus apella* using metacyclic trypomastigotes (introduced by conjunctival route) and blood trypomastigotes (introduced by intraperitoneal route), where patent parasitemia but no acute clinical manifestations were observed (Fallasca *et al.*, 1986). Follow-up of the infection with Y. strain (subcutaneous route) in *Cebus apella* (Rosner *et al.*, 1988) showed a very characteristic acute phase with parasitemia, weight loss, fever, radiological and electrocardiographic alterations which were similar to those found in human infection. In an experiment with *Saimiris aciureus*, 66% of infected animals also developed electrocardiographic alterations similar to those found in human infection: high titers of specific antibodies were also present.

Rhesus monkeys (*Macaca mulatta*) were studied initially by Muniz *et al.* (1946) in a vaccination trial using dead epimastigotes in a challenge with triatomids feces (conjunctival route): no difference was observed compared with controls. Marsden *et al.* (1976) described periorbital edema, high levels of IgM and circulating parasites in rhesus monkeys infected subcutaneously; histopathological and electrocardiographic alterations were observed in rhesus monkeys after 6–8 years of infection with *T. cruzi* by Miles *et al.* (1979). The inoculum used by different routes, in all experiments of rhesus monkeys infection ranged from 1×10^4 to 1×10^5. Circulating specific antibodies (IgG/IgM) have been described as early as the second week after inoculation (Bonecini-Almeida *et al.*, 1990), and IgG levels persisted until the end of the experiment (3 years), but IgM antibodies were detectable 9 months after inoculation. Hematological alterations comprised leukocytosis and lymphocytosis. A chagoma in the beginning of the infection (between the 3rd and 13th days) and patent parasitemia (between the 13th and 59th days) was related; after this period the parasitemia could be demonstrated only by hemoculture and/or xenodiagnosis. These authors also found minor and transient electrocardiographic alterations, such as those detected in non-lethal human acute chagasic myocarditis and suggest that the rhesus monkey model reproduces the acute and indeterminate phases of human Chagas' disease.

Infection process

Previous experiments have demonstrated that mice weighing 18–20 g and inoculated by the intraperitoneal route with 5×10^4–1×10^5 trypomastigotes, present quite homogeneous infection. Daily trypanosome counts provided the following pattern for the parasitemia: parasites appear from the 4th or 5th day after inoculation, their number decreases markedly on the 6th day, increases until the 7th or 8th day, and finally decreases again around the 9th day. From the 10th day onwards the pattern of parasitemia is quite irregular. These data are very helpful not only in the choice of the best time for the strain passage but also in the assessment of therapeutic activity.

Most infected animals die in the period from the 5th to the 20th day after inoculation; the highest mortality rates are observed around the 15th day. General mortality rates are about the same with regard to both sexes and only a small number of infected animals will outlive 40 days. These characteristics have remained relatively stable for more than 4 years of successive transfers in mice. Temporary exacerbation of virulence was controlled by adequate modification of the number of trypanosomes and weight of the animals (Brener, 1962).

Key parameters to monitor infection

The infection of all these animal models is monitored by parasitological methodology, since the parasites can be found in the blood in the acute phase. The parasitemia is usually high (with little variation, depending on the strain used) and follow-up is done by daily fresh blood examination. The blood is collected and the parasites counted in a Neubauer's chamber.

In all animal models for *T. cruzi* infection, the parasitemia is strongly influenced by the parasite strain used. In general, animals inoculated with the CL strain have higher or more prolonged parasitemia than those inoculated with the Y strain. Also, inoculation with blood trypomastigotes resulted in higher levels of parasitemia than inoculation with metacyclic trypomastigotes. The route of inoculation is also an important parameter, since the intraperitoneal route yields higher levels of parasite than the conjunctival route.

In the chronic phase the parasitemia is low and the procedures to follow the infection are first, xenodiagnosis: at different periods after inoculation using first- and third-nymphs of several triatomine bugs, such as *Dipetalogaster maximus, Triatoma infestans, Rhodnius neglectus* or *Panstrongylus megistus*. The bugs are maintained at 26–28°C and examined after 30–40 days. The second procedure is hemoculture; during which the animals are bled and 10 ml of heparinized blood is collected aseptically. Blood is centrifuged at 2000 rpm/30 minutes and the cell pellet is resuspended in liver infusion tryptose (LIT) medium. The cultures are maintained at 28°C and examined at 30 and 60 days for living flagellates.

The level and nature of cytokine production may be required to monitor acute and chronic infection with *T. cruzi*, since they are an indication of a heightened production of both interferon-γ and tumor necrosis factor-α (Rivera *et al.*, 1995). The production of both of these cytokines, although induced by infection, is regulated before the peak parasitemia is reached.

Antimicrobial therapy

The chemotherapy of Chagas' disease is still inadequate. Both drugs in clinical use — nifurtimox {4-[(5-nitrofurfurylidene)-amino]-3-methylthiomorpholine-1,1-dioxide]} and benznidazole (*N*-benzyl-1-2-nitro-1-imidazole-acetamide), — are associated with side-effects.

A large variety of antimicrobial agents have been evaluated in the mouse model of *T. cruzi* (DeCastro, 1993; Urbina, 1997). However, few compounds have reached the stage of a clinical trial for Chagas' disease and none has been shown to be suitable as an effective chemotherapeutic agent which could be used extensively in humans (Filardi and Brener, 1982). A review of chemotherapy for Chagas disease in animal experiments is given in Table 97.1.

Anti-protozoal activity has been characterized for a wide range of nitroheterocyclic compounds, including nitrothiazoles, nitroimidazoles (McCabe, 1988) and nitrofurans (Brener, 1984). As mentioned above, only nifurtimox and benznidazol are being used in patients with Chagas' disease. Both compounds have significant side-effects and should be reserved for the management of acute cases or the treatment of limited numbers of chronic patients (Brener, 1984).

Clinic trials have been performed with allopurinol (Avila *et al.*, 1987), which interferes with nucleic acid synthesis, and the antifungals ketoconazole and itraconazole, which are inhibitors of sterol biosynthesis. These drugs were able to suppress but not eliminate the infection. In contrast the compound bis-triazole D0870, an inhibitor of sterol biosynthesis, promises to be a more recent chemotherapeutic agent for the treatment of Chagas disease (Urbina, 1997).

The administration of drugs begins on the day after inoculation and doses corresponding to about one-fifth of the LD_{50} are given for 10 consecutive days. On the 5th day after inoculation the number of parasites in 5 mm^3 of blood is determined. On the 8th day, when the number of parasites in the inoculated animals is generally higher, a new count is performed. Comparison of the data thus obtained with those from the controls is generally quite sufficient for a good evaluation of the drug activity. Daily records of the mortality rates must be kept so that a clear picture may be provided (Brener, 1962).

Key parameters to monitor response to treatment

The best initial criteria for therapeutic activity in the experimental Chagas' disease should be based on mortality and parasitemia (Brener, 1962). However, since the acute phase of the infection may be followed by a chronic stage in which parasites are reduced to submicroscopic level, then indirect laboratory methods should be used, such as subinoculation, xenodiagnosis and serological techniques. Recently, the polymerase chain reaction has been used as a complementary criterion for therapeutic activity in the chronic stage of experimental Chagas disease (Junqueira *et al.*, 1996; Urbina *et al.*, 1996; Urbina, 1997).

The following techniques were used to establish reliable criteria for cure in the mouse model of Chagas' disease:

1. Fresh blood examination: a drop of blood from the mouse's tail was carefully examined in a Neubauer's chamber daily or every other day
2. Blood subinoculation: mice were killed about 1 or 2 months after treatment and 0.4–0.6 ml of citrated blood, collected from the severed axillary artery, was inoculated intraperitoneally into susceptible mice. From the 5th day of inoculation, fresh blood examinations were performed daily or every other day for a period of at least 6 weeks
3. Blood culture: blood from treated animals was inoculated into Noeller's culture medium and culture was frequently examined for at least 30 days after inoculation
4. Xenodiagnosis: 1 or 2 months after treatment, mice were anesthetized and 4 triatomine nymphae were allowed to feed on them. After 45–50 days, the bugs were carefully examined for trypanosomes
5. Histological examination: histological sections of the heart of treated animals were stained with hematoxylin and eosin and carefully examined
6. Re-inoculation: some of the treated animals were re-inoculated at different periods after treatment with about 4000 blood parasitic forms per gram of weight; daily counts of trypanosomes were performed so that a new acute phase of the disease might be detected.

In spite of differences in the biological behavior of *T. cruzi* populations kept in the laboratory, a method was described which permits one to determine *in vivo* and in a short period of time (4–6 hours) the sensitivity of *T. cruzi* strains to active chemotherapeutic agents (Filardi and Brener, 1984; Filardi, 1988). The results obtained by these authors show good correlation with those obtained by

Table 97.1 Examples of chemotherapy in mice infection with *Trypanosoma cruzi*

Reference	*Strain of mice*	T. cruzi *strain*	*Antimicrobial agent*	*Dose (mg/kg)*	*Median survival (days)*	*Cure (%)*	*Comments*
McCabe *et al.* (1983)	Swiss Webster female	Y (10^5 b.t.)	Control				Dying 14 days p.i.
			Ketoconazole	30 mg/kg per day twice daily for 7 days, t.s. 24 hours p.i.	100% (150 days)	Rare parasites in the blood	Fully protected from death by first 19 days
Urbina *et al.* (1993)	NMRI albino female	Y (10^5 b.t.)	Control		0% (24 days p.i.)		
			Mevinolin + ketoconazole	20 mg/kg per day once daily for 7 days t.s. 24 hours p.i.	100% (40 days p.i.)	Almost complete parasitemia suppression	Protected from death for 40 days
Maldonado *et al.* (1993)	NMRI albino female	Y (10^5 b.t.)	Control				Died 20–24 days p.i.
			Terbinfine + ketoconazole	100 mg/kg per day + 15 mg/kg per day, once daily, for 7 days t.s., 24 hours p.i.	100% (35 days p.i.)	No circulating parasites were found 55 days p.i.	
			Bistriazole (ICI 195,739)	1 mg/kg per day once daily for 5 days	90% (30 days p.i.)	Almost complete parasitemia suppression	
McCabe *et al.* (1986)	Swiss Webster female	Y,CL,Tu (10^5 b.t.)	Control		0% (10–15 days p.i.)		
			Itraconazole	120 mg/kg per day two daily dose, for 7–9 weeks, t.s. 24 hours p.i.	100% (35 days p.i.)	100% Parsitologic cure	Completely protected against death
Scott and Matthews (1987)	C_3H/He female	Brazil (10^3 and 5×10^4 b.t.)	Control		0% (23 days p.i.)		
			Imidazole (RS-49676)	100 mg/kg per day, twice daily for 5 days, t.s. 24 hours p.i.	60% (100 days p.i.)	20–25%	Also reduced the number of amastigotes present within the heart and spleen of infected mice
Urbina *et al.* (1996)	NMRI albino female	Y (10^5 b.t.)	Control		0% (23 days p.i.)		
			Triazol (D0870)	20 mg/kg per day, e.o.d., t.s. 24 hours p.i., t.d. 28	100% (105 days p.i.)	60%	Almost completely (–) blood PCR (+) serological tests
			Nifurtimox	50 mg/kg per day daily, t.d. 43	60% (105 days p.i.)	NR	
		Bertoldo (10^4b.t.)	Control		50% (145 days p.i.)		
			Triazol (D0870)	20 mg/kg per day, e.o.d., t.s. 40–50 days p.i.	100% (145 days p.i.)	80–90%	A parasitological cure of experimental long-term Chagas' disease
			Nifurtimox	50 mg/kg per day, daily, t.d. 43	80–90% (145 days p.i.)	11%	

Table 97.1 — Continued

Reference	*Strain of mice*	T. cruzi *strain*	*Antimicrobial agent*	*Dose (mg/kg)*	*Median survival (days)*	*Cure (%)*	*Comments*
Filardi and Brener (1982)	Albino male	Y (5×10^4 b.t.)	Control				
			Nitroimidazol (CL,64,855)	100 mg/kg per day, daily for 20 days		100%	
Andrade *et al.* (1989)	Swiss	Type II; Type III ($1–5 \times 10^4$ b.t.)	Control				
			2-substituted 5-nitroimidazole (MK 436)	250 mg/kg, two daily doses for 30 days or 60 days		90% Type II, 95.7% Type III of parasitological cure	Action of MK 436 against chronic infection; for acute infection, see Andrade *et al.* (1987)
Avila and Avila (1981)	C57B1/6J (inbred) IVIC–NMRI	Y (10^6 b.t.)	Control				Dying 16 days p.i.
			Allopurinol	32–64 mg/kg per day, for 10 days t.s. 2 days p.i.	90% (310 days)	NR	Parasitemia (direct counts became negative for more than 300 days a.t.i.)
Avila *et al.* (1987)	NMRI	Ma and FL isolates (5×10^6 b.t.)	Control		81.73 (FL) 92.24 (Ma) days (mst)		
			Allopurinol (FOB)	0.1–32 mg/kg per day for 10 days, t.s. 4 days p.i.	371.4 days (mst)	NR	FOB showed significant modification of mst and in parasitemia levels
Kinnamon *et al.* (1996)	Albino CF1, female	Brazilian strain (5×10^4 b.t.)	Control				
			Primaquine analogs	0.8125–104 mg/kg per day, twice daily for 4 days, t.s. 11 days p.i.	NR	Suppressing parasitemia for 52 members of 78 analogs tested	Broad range of activity; one member was 14 times more active than the standard furtimox

b.t., Blood trypomastigotes; t.s., treatment started; p.i., post-inoculation; NR, not reported; Tu, Tulahuen; a.t.i., after treatment initiation; e.o.d., every other day; PCR, polymerase chain reaction; t.d., total number doses; mst, median survival time.

prolonged treatment schedules used to assess the action of drugs in experimental Chagas' disease and may be used to study the sensitivity of *T. cruzi* strains to active drugs.

Pitfalls (advantages/disadvantages) of the model

One is faced with several difficulties when attempting to find a reliable animal model that allows one to follow the time course of human Chagas' disease. The main problem is the lack of knowledge of the immunopathological evolution of the disease.

The ideal model is defined as one that would do the following:

1. Support a long-lasting subclinical parasitemia which can be detected by xenodiagnosis and/or hemoculture, as well as by conventional serology
2. Present cellular and/or humoral immune reactions
3. Develop the cardiac and digestive forms of the disease with typical histopathologic lesions
4. Survive the acute phase of the infection
5. Display lesions in a relatively short period of time
6. Develop the disease in a manner which is more or less independent of the age and sex of the infected animal involved
7. Utilize animals native to the endemic area and which are easy to obtain
8. Be available at a reasonable cost.

Based on the data in the literature, we suggest that the best animal model for *T. cruzi* infection is the canine model. Chagas' disease induced experimentally in dogs is said to resemble the human disease in all its phases (WHO, 1974). In the dog the indeterminate phase progresses to the late phase of severe chronic myocarditis (Andrade *et al.*, 1997). The rhesus monkey model also develops an indeterminate phase; however, the disadvantage is the fact that it is very expensive to work with this kind of animal. Most studies used the mouse model because it is cheaper, easy to work with and it can produce both the acute and chronic phase of the disease.

Contribution of the model to infectious disease therapy

Traditionally, most efforts to identify novel compounds for Chagas' disease have relied on cure of acutely infected animals. Parasitemic mice can be considered to be the indicator of success. However, the lack of reliable criteria of cure in the human disease, the difficulties in carrying out prolonged follow-up of treated cases and the possible participation of autoimmune reactions in the pathogenesis of Chagas' disease has prevented an assessment of the influence of specific treatment in the outcome of the disease. However, some experimental data and a few clinical investigations are worth mentioning (Brener, 1984).

In reality, among the drugs that have reached the stage of clinical trials, none has been considered safe and completely effective, with the possible exception of allopurinol (Gallerano, 1985). Therefore, the identification of alternative compounds, and even of substitutes for nifurtimox and benznidazole, remains a vital and open-ended field of research (De Castro, 1993).

In contrast, comparative studies on the sensitivity of *T. cruzi* strains with active compounds have been extensively performed in murine models (Andrade *et al.*, 1985; Filardi, 1988).

In addition, the contribution of a T-cell subpopulation to immunopathology in murine *T. cruzi* infection was studied by Tarleton *et al.* (1994). Data from this study provide solid evidence for the role of CD4+ and CD8+ T cells in immune control of *T. cruzi* in the acute and chronic stages of the infection and additional support for the hypothesis that the parasite load is the determining factor in the severity of chronic Chagas' disease.

Finally, the results of the combination of azoles with other ergosterol biosynthesis inhibitors (Maldonado *et al.*, 1993; Urbina, 1997) acting at different points of the biosynthetic pathway may be useful in the treatment of human Chagas' disease because they may allow lower levels of those compounds to be used. The special activity of this class of compounds against *T. cruzi* is due to a dual mechanism of action which is not restricted to sterol biosynthesis inhibition; in such a case, potentiation by other sterol biosynthesis inhibitors that act at different points of the pathway is not necessarily expected.

Philosophy of model: is the model a true test of antimicrobial therapy?

Based on all these data from different animal models, it is very hard to establish which is the best for antimicrobial therapy. Before 1960, the main approach to Chagas' disease drug development was completely empirical, based on clinical observation. No rigid criteria related to treatment schedule or to evaluation of results were applied in clinical trials. In contrast, in animal experiments, most treatments were carried out for only short periods and the results only expressed decreasing parasitemia, mortality and acute-phase extension. Since some animal models can be useful in the development of one specific phase of the disease and not good enough for another phase, the choice of model to work with trypanocidal drugs will rely on two main criteria — effectiveness and absence of toxicity. It is important to consider that no matter which model, anti- *T. cruzi* chemotherapy should be done during the acute phase, to avoid all the sequelae which are observed in the chronic phase.

References

Andrade, S. G. (1976). Tentative for grouping different *T. cruzi* strains in some types. *Rev. Inst. Med. Trop. S. Paulo*, **18**, 114–141.

Andrade, S. A., Magalhães, J. B., Pontes, A. L. (1985). Evaluation of chemotherapy with benznidazole and nifurtimox in mice infected with *T. cruzi* strains of different types. *Bull. WHO*, **63**, 721–726.

Andrade, Z. A., Andrade S. G., Sadigursky, M. (1987). Enhancement of chronic *Trypanosoma cruzi* myocarditis in dogs treated with low doses of cyclophosphamide. *Am. J. Pathol.*, **127**, 467–473.

Andrade, S. G., Silva, R. C., Santiago, C. M. G. (1989). Treatment of chronic experimental *Trypanosoma cruzi* infections in mice with MK-436, a 2-substituted 5-nitroimidazole. *Bull. WHO*, **67**, 509–514.

Andrade, Z. A., Andrade, S. G., Sadigurski, M., Wenthold, R. J., Hilbert, S. L., Ferrans, V. J. (1997). The indeterminate phase of Chagas' disease: ultrastructural characterization of cardiac changes in the canine model. *Am. J. Trop. Med. Hyg.*, **57**, 328–336.

Avila, J. L., Avila, A. (1981). *Trypanosoma cruzi*: allopurinol in the treatment of mice with experimental acute Chagas' disease. *Exp. Parasitol.*, **51**, 204–208.

Avila, J. L., Polegre, M. A., Robins, R. K. (1987). Biological action of pyrazolopyrimidime derivatives against *Trypanosoma cruzi* studies *in vitro* and *in vivo*. *Comp. Biochem. Physiol.*, **86C**, 49–54.

Bonecini-Almedia, M. G. (1991). Doença de Chagas em macacos rhesus (*Macaca mullata*): avaliação de um modelo experimental, (M.Sci. thesis). Instituto Oswaldo Cruz, Brazil.

Bonecini-Almeida, M. G., Galvão-Castro, B., Pessoa, M. H. R., Pirmez, C., Laranja, F. (1990). Experimental Chagas' disease in Rhesus monkeys. I. Clinical, parasitological, hematological and anatomo-pathological studies in the acute and indeterminated phase of the disease. *Mem. Inst. Oswaldo Cruz.*, **85**, 163–171.

Brener, Z. (1962). Therapeutic activity and criterion of cure on mice experimentally infected with *Trypanosoma cruzi*. *Rev. Inst. Med. Trop. S. Paulo*, **4**, 389–396.

Brener, Z. (1984). Recent advances in the chemotherapy of Chagas' disease. *Mem. Inst. Oswaldo Cruz*, **79**, 149–155.

Brener, Z., Ramirez, L. E. (1985). Modelos crônicos da Doença de Chagas experimental. In: *Cardiopatia Chagasica* (eds Cançado, J. R., Chuster, M.), pp. 29–32. Fundação Carlos Chagas, Belo Horizonte, Brasil.

Brener, Z., Costa, C. A. G., Chiari, E. (1976). Differences in the susceptibility of *Trypanosoma cruzi* strains to active chemotherapeutic agents. *Rev. Inst. Med. Trop. S. Paulo*, **18**, 450–455.

Cariola, J., Prado, R., Agosin, M., Christen, R. (1950). Susceptibilidad del hamster (*Cricetus auratus*) y la infección experimental por *Trypanosoma cruzi*, cepa Tulahen. *Bol. Inform. Parasit. Chilenas*, **V**, 44–45.

Castro, M. A., Brener, Z. (1985). Estudo parasitológico e anatomopatológico da fase aguda da Doença de Chagas em cães inoculados com duas differentes cepas de *Trypanosoma cruzi*. *Rev. Soc. Bras. Med. Trop.*, **18**, 223–229.

Chagas, C. (1909). Nova Tripanosomiase humana. Estudo sobre a morfologia e o ciclo do *Schizotrypanum cruzi*, n.g., n.sp. agente etiológico de uma nova entidade mórbida do homem. *Mem. Inst. Oswaldo Cruz*, **1**, 158–218.

Chapadeiro, E., Beraldo, P. S. S., Jesus, P. C., Oliveira W. P., Jr., Junqueira L. F. Jr. (1988). Lesões cardiacas em ratos Wistar infectados com diferentes cepas do *Trypanosoma cruzi*. *Rev. Soc. Bras. Med. Trop.*, **21**, 95–103.

Chiari, E., Tafuri, W. L., Bambirra, E. A. *et al.* (1980). The rabbit as a laboratory model for studies on Chagas' disease. *Rev. Inst. Med Trop. S. Paulo*, **22**, 207–208.

Ciconelli, A. J. (1963). Estudo quantitativo dos neurônios do plexo hipogástrico inferior em ratos normais e em infectados experimentalmente pelo *Trypanosoma cruzi* (D.Sci thesis). Universidade de S. Paulo, Ribeirão Preto, SP.

De Castro, S. L. (1993). The challenge of Chagas' disease chemotherapy: an update of drugs assayed against *Trypanosoma cruzi*. *Acta Trop.*, **53**, 83–98.

Dias, J. C. P. (1997). Present situation and future of human Chagas' disease in Brazil. *Mem. Inst. Oswaldo Cruz*, **92**, 13–15.

Fallasca, A., Grana, J., Buccolo, M. *et al.* (1986). Susceptibility of the *Cebus apella* monkey to different strains of *T. cruzi* after single or repeated inoculations. *Panam. Health Org. Bull.*, **20**, 117–137.

Federici, E. E., Abelman, W. H., Neva, F. A. (1964). Chronic and progressive myocarditis and myositis in C3H mice infected with *Trypanosoma cruzi*. *Am. J. Trop. Med. Hyg.*, **13**, 272–280.

Figueiredo, F., Rossi, M. A., Ribeiro dos Santos, R. (1985). Evolução da cardiopatia experimentalmente induzida em coelhos infectados com *Trypanosoma cruzi*. *Rev. Soc. Bras. Med. Trop. S. Paulo*, **18**, 133–141.

Filardi, L. S. (1988). Quimioterapia da Doença de Chagas murina experimental (D. Sci. thesis). Universidade Federal de Minas Gerais, Brazil.

Filardi, L. S., Brener, Z. (1982). A nitroimidazole-thiadiazole derivative with curative action in experimental *Trypanosoma cruzi* infections. *Ann. Trop. Med. Parasitol.*, **76**, 293–297.

Filardi, L. S. Brener, Z. (1984). A rapid method for testing "*in vivo*" the susceptibility of different strains of *Trypanosoma cruzi* to active chemotherapeutic agents. *Mem. Inst. Oswaldo Cruz*, **79**, 221–225.

Gallerano, R. H. (1985). Estúdio epidemiológico de la enfermedad de Chagas en estudiantes de la Universidad Nacional de Córdoba. Manifestaciones iniciales. *Ver. Fed Arg Cardiol.*, **14**, 37.

Junqueira, A. C. V., Chiari, E., Wincker, P. (1996). Comparison of the polymerase chain reaction with two classical parasitological methods for the diagnosis of Chagas' disease in an endemic region of north-eastern Brazil. *Trans. R. Soc. Trop. Med. Hyg.*, **90**, 129–132.

Kinnamon, K. E., Poon, B. T., Hanson, W. L., Waits, V. B. (1996). Primaquine analogues that are potent anti-*Trypanosoma cruzi* agents in a mouse model. *Ann. Trop. Med. Parasitol.*, **90**, 467–474.

Lana, M., Chiari, E., Tafuri, W. L. (1992). Experimental Chagas' disease in dogs. *Mem. Inst. Oswaldo Cruz*, **87**, 59–71.

Maldonado, R. A., Molina, J., Payares, G., Urbina, J. A. (1993). Experimental chemotherapy with combinations of ergosterol biosynthesis inhibitors in murine models of Chagas' disease. *Antimicrob. Agents Chemother.*, **37**, 1353–1359.

Marsden, P. D., Seah, S. K. K., Draper, C. C., Pettitt, L. E., Miles, M. A., Volter, A. (1976). Experimental *Trypanosoma cruzi* infection in Rhesus monkey. II. The early chronic phase. *Trans. R. Soc. Trop. Med. Hyg.*, **70**, 247–251.

McCabe, R. E. (1988). Failure of ketoconazole to cure murine Chagas' disease. *J. Infect. Dis.*, **158**, 1408–1409.

McCabe, R. E., Araujo, F. G., Remington, J. S. (1983). Ketoconazole protects against infection with *T. cruzi* in a murine model. *Am. J. Trop. Med. Hyg.*, **32**, 960–962.

McCabe, R. E., Remington, J. S., Araujo, F. G. (1986). *In vitro* and *in vivo* effects of itraconazole against *Trypanosoma cruzi*. *Am. J. Trop. Hyg.*, **35**, 280–284.

Miles, M. A., Marsden, P. D., Pettitt, L. E. *et al.* (1979). Experimental *Trypanosoma cruzi* infection in Rhesus monkeys. *Trans. R. Soc. Trop. Med. Hyg.*, **73**, 355–410.

Muniz, J., Nobrega, G., Cunha, M. (1946). Ensaios de vacinação preventivos e curativos na infecção pelo *Schizotrypanum cruzi*. *Mem. Inst. Oswaldo Cruz.*, **44**, 529–541.

Osimani, J. J., Gurri, J. (1954). Infectión experimental del hamster dorado (*Mesocrycetus auratus*) con algunas cepas uruguayas de *Trypanosoma cruzi*. *Arch. Soc. Biol. Montevideo*, **XVIII**, 73–78.

Pereira da Silva, L. H., Nussenzweig, V. (1953). Sobre uma cepa de *Trypanosoma cruzi* altamente virulenta para o camundongo branco. *Folia Cli. Biol.*, **20**, 191–208.

Pung, O. J., Hulsebos, L. H., Kuhn, R. E. (1988). Experimental Chagas' disease (*Trypanosoma cruzi*) in the Brazilian squirrel (*Saimiri sciureus*): hematology, cardiology, cellular and humoral immune responses: *Int. J. Parasitol.*, **18**, 115–120.

Queiroz da Cruz, M., Brascher, H. M., Vargens, J. R., Oliveira-Lima, A. (1991). Effect of actinomycin D on *T. cruzi*. *Experientia*, **47**, 89–92.

Ramirez, L. E., Brener, Z. (1987) Evaluation of the rabbits as a model for Chagas' disease. I. Parasitological studies. *Mem. Inst. Oswaldo Cruz,* **82**, 531–536.

Ramirez, L. E., Lages-Silva, E., Chapadeiro, E. (1991). Infecção do hamster pelo *Trypanosoma cruzi*. *Rev. Soc. Bras. Med Trop.*, **24**, 119–120.

Ramirez, L. E., Lages-Silva, E., Soares J. M. Jr., Chapadeiro, E. (1993). Infecção experimental do hamster pelo *Trypanosoma cruzi:* fase crônica. *Rev. Soc. Bras. Med Trop.*, **26**, 253–254.

Ramirez, L. E., Lages-Silva, E., Soares J. M., Jr. Chapadeiro, E. (1994). The hamster (*Mesocricetus auratus*) as experimental model in Chagas' disease: parasitological and histopathological studies in acute and chronic phases of *Trypanosoma cruzi* infection. *Rev. Soc. Bras. Med Trop.*, **27**, 163–169.

Ribeiro, R. S., Rossi, M. A., Laos, J. L., Silva, J. S., Savino, W., Mengel, J. (1992). Anti-CD4 abrogates rejection and re-establishes long-term tolerance to singeneic newborn hearts grafted in mice chronically infected with *T. cruzi*. *J. Exp. Med.*, **175**, 28–39.

Rivera, M. T., Araujo, S. M., Lucas, R. *et al.* (1995). High tumor necrosis factor alpha (TNF-α) production in *Trypanosoma cruzi* infected pregnant mice and increased TNF-γ gene transcription in their offspring. *Infect. Immun.*, **63**, 591–595.

Rivera-Vanderpas, M. T., Rodriguez, A. M., Afchan, D., Bazin, H., Capron, A. (1983). *Trypanosoma cruzi*: variation in susceptibility of inbred strain of rats. *Acta Trop.*, **40**, 5–10.

Rodriguez, A. M., Santoro, F., Afchain, D., Bazin, H., Capron, A. (1981). *T. cruzi* infection in B-cell-deficient rats. *Infect. Immun.*, **31**, 524–529.

Rosner, J. M., Shinini, A., Rovira, T. *et al.* (1988). Acute Chagas' disease in non-human primates. I. Chronology of parasitemia and immunological parameters in the *Cebus apella* monkeys. *Trop. Med. Parasitol.*, **39**, 51–55.

Rosner, J. M., Belassai, J., Shinini, A. *et al.* (1989). Cardiomyopathy in *Cebus apella* monkeys experimentally infected with *Trypanosoma cruzi*. *Trop. Med. Parasitol.*, **40**, 24–31.

Santos-Buch, C. A. (1979). American trypanosomiasis: Chagas' disease. *Int. Rev. Exp. Pathol.*, **19**, 63.

Scorza, C., Scorza, J. V. (1972). Acute myocarditis in rats innoculated with *T. cruzi:* study of animals sacrified between the fourth and twenty-ninth day after infection. *Ver. Inst. Med. Trop. São Paulo*, **14**, 171–177.

Scott, V. R., Matthews, T. R. (1987). The efficacy of an *N*-substituted imidazole, RS-49676, against a *Trypanosoma cruzi* infection in mice. *Am. J. Trop. Med. Hyg.,* **37**, 308–313.

Scott, M. T., Snary, D. (1982). American trypanosomiasis (Chagas' disease). In *Immunology of Parasitic Infections* (eds Cohen, S., Warren, K. S.), pp. 261–298. Blackwell Scientific Publications, Oxford.

Seah, S. K. K., Marsden, P. D., Voller, A., Pettitt, L. E. (1974). Experimental *Trypanosoma cruzi* infection in Rhesus monkeys: the acute phase. *Trans. R. Soc. Trop. Med. Hyg.*, **68**, 63–69.

Tarleton, R. L., Sun, J., Zhang, L., Postan, N. (1994). Depletion of T-cell subpopulations results in exacerbation of myocarditis and parasitism in experimental Chagas' disease. *Infect. Immun.*, **62**, 1820–1829.

Teixeira, A. R. L., Santos-Buch, C. A. (1975). The immunology of Chagas' disease. IV. Production of lesions in rabbits similar to those of chronic Chagas' disease in man. *Am. J. Pathol.*, **80**, 163–180.

Teixeira, A. R. L., Figueiredo, F., Rezende Filho, J., Macedo, V. (1983). Chagas' disease: a clinical, parasitological, immunological and pathological study in rabbits. *Am. J. Trop. Med. Hyg.*, **3**, 37–41.

Torrico, F., Heremans, H., Rivera, M. T., Marck, E. V., Billiau, A., Carlier, Y. (1991). Endogenous IFN-γ is required for resistance to acute *T. cruzi* infection in mice. *J. Immunol.*, **146**, 3626–3632.

Urbina, J. (1997). Lipid biosynthesis pathways as chemotherapeutic targets in kinetoplastid parasites. *Parasitology*, **114**, S91–S99.

Urbina, J. A., Lazardi, K., Marchan, E. *et al.* (1993). Mevinolin (Lovastatin) potentiates the antiproliferative effects of ketoconazole and terbinafine against *Trypanosoma (Schizotrypanum) cruzi: in vitro* and *in vivo* studies. *Antimicrob. Agents Chemother.*, **37**, 580–591.

Urbina, J., Payares, G., Molina J. *et al.* (1996). Cure of short- and long-term experimental Chagas' disease using D0870. *Science*, **273**, 969–971.

Vichi, F. I. (1961). Estudo do parasitismo na medula espinal de ratos na fase aguda da molestia de Chagas. *Rev. Inst. Med. Trop. S. Paulo*, **3**, 37–42.

WHO (1974). Immunology of Chagas' disease. *Bull. WHO*, **50**, 459–472.

WHO (1984). *Report of the Scientific Working Group on the Development and Evaluation of Animal Models for Chagas' disease*. WHO, Geneva, Switzerland.

Chapter 98

Animal Models of *Toxoplasma* Infection

K. Janitschke

Background of human infection

Toxoplasma infection is caused by the protozoon parasite *Toxoplasma gondii*. *Toxoplasma* infections are widespread in warm-blooded animals and humans. In particular, there is a high prevalence in animals that will enter the food chain — up to 90% (Dubey and Beattie, 1988). The prevalence in humans varies in different parts of the world. In parts of Germany the rate in pregnant women reaches 55% (Janitschke *et al.*, 1988), whereas in the USA it is around 20%. Clinically, the frequently inapparent (inactive, latent, chronic) *Toxoplasma* infection must be distinguished from the relatively rare toxoplasmosis (active, acute).

The causative organism can occur in humans in the following two forms:

1. Tachyzoite: an ameboid, bent form approximately 6 μm long, which primarily presents in cells of the reticuloendothelial system
2. Cyst: a round permanent stage, about 300 μm in diameter, containing thousands of individual parasites (bradyzoites); the cysts are largely non-reactive and are found in tissue.

A third form can be seen in cats: oocysts. Cats excrete another form, oocysts, in their feces. This is round, approximately 10×12 μm, and able to survive in the soil for more than 1 year.

Depending on the host, the causative organism can reproduce itself sexually and/or asexually. In all warm-blooded animals and in humans, proliferation of the organism on oral ingestion of cysts or oocysts or prenatal transmission of tachyzoites occurs asexually.

The rate of proliferation decreases with the host's increasing defense reaction and cysts will develop in tissue, primarily in the musculature or in the brain. After initial oral infection of cats, additional sexual reproduction takes place in the intestines. The resulting oocysts are excreted with the feces for 1–2 weeks. These oocysts only become infectious after a maturing phase (sporulation) of about 3 days.

The most frequent path of postnatal transmission to humans is the intake of *Toxoplasma* cysts by ingestion of infected raw or inadequately heated meat, especially pork.

Another important source of infection is *Toxoplasma* oocysts, which get into garden or field soil via cat feces. Infection can then occur by finger-to-mouth transmission. Direct contact with cats does not have any consequences.

The disease (toxoplasmosis) after primary infection manifests mostly as lymphadenopathy in immunocompetent patients. In immunocompromised patients an inactive infection can be reactivated. These are patients with Hodgkin's disease, cytostatic therapy, bone marrow transplantation and patients with acquired immunodeficiency syndrome (AIDS; Pohle, 1994). *Toxoplasma* infections are also highly significant for pregnant women. If women get a primary infection during pregnancy, in about 50% the parasite can be transmitted to the fetus. If the infection does not induce abortion or stillbirth, manifestation in the prenatal neonate can range from (the very rare) severe lesions to a subclinical disease, which is at first only detectable serologically. In non-clinical infections, however, lesions involving the central nervous system and the eyes may become obvious after months or many years.

Because *Toxoplasma* infection is medically significant, sufficient chemotherapy is essential. There are many reports on chemotherapeutical studies in humans and also in artificially infected animals.

Background of the models

The first chemotherapeutical experiments in animals were done by Sabin and Warren (1941) and Biocca and Pasqualin (1942). They tested sulfonamides in experimentally infected mice. Eyles and Coleman (1952) reported on pyrimidin. After the establishment of the Sabin–Feldman dye test in 1948 the medical community became aware of toxoplasmosis and many chemotherapeutical studies have been done since then. Jira and Kozojed (1970, 1983) list publications documented from 1908 to 1967 and 1968 to 1975. The publications are documented in the following chapters: general, sulfonamides and sulfones, pyrimethamine antibiotics, spiramycin and toxic side-effects, other drugs. The reader who is interested in these subjects should use this excellent collection of literature. For publications after the year 1983

Handbook of Animal Models of Infection
ISBN 0-12-775390-7

electronic programs will give newer information. Therefore we will not report here all the published techniques of chemotherapeutical studies, but we will select the principal models:

1. Tachyzoite infections in rodents
2. Bradyzoite infections in rodents
3. Connatal infections
4. Infections in immunodeficient mice
5. Oocyst studies in cats
6. Studies on other animal species.

Tachyzoite infections in rodents

Animal species

Some species and strains have been used for this model. Mice examples are: NMRI (Werner and Dannemann, 1972), Swiss-Webster (Tabbara *et al.*, 1982), HSD:ICR (Lindsay *et al.*, 1995), NIH (Harper, 1985). In the selection of mice strains their differences in susceptibility to *Toxoplasma* in general are important (Araujo et al., 1976). Mice are ideal experimental hosts for this parasite, whereas white rats are less suitable, but some therapeutic experiments with this rodent have been performed. Wistar rats were used by Brun-Pascaud *et al.* (1994), and other rats have been used by Eyles and Jones (1955). The weight, age and sex seem to have no influence on susceptibility for *Toxoplasma* infection and drug testing, but female mice are easier to keep in cages.

Preparation of animals

No specialized housing or care or specific pretreatment is required.

Selection, storage and preparation of inocula

Many *Toxoplasma* strains isolated from humans or animals can be used for experimental infections, depending on the objective of a study. The BK and RH strains are mostly used for tachyzoite infections. These strains are in general subcultured in mice two or three times a week by intraperitoneal injections; if not, they will kill mice within a few days. Also the intraperitoneal exudates are not useful, because dead toxoplasms and bacterial superinfection are useless for further inoculations. In Appendix 1 we report on our system for the BK strain in NMRI mice, which we use for the production of tachyzoites for the Sabin–Feldman dye test. This technique is standardized (for Germany; Bundesgesundheitsamt, 1989) and with it is supplied an exudate containing a relatively fixed number of living parasites. 1–3 hours after harvesting, about 10% of the tachyzoites are not alive, as determined by dye test.

The peritoneal exudate must be kept in a refrigerator until 4 hours after collection, because longer storage reduces the number of killed toxoplasms. For chemotherapeutical studies a fixed or different numbers of tachyzoites can be used. For this purpose the parasites should be counted in a hemocytometer and diluted in buffered saline (Appendix 1). The addition of antibiotics to the inoculum to depress bacterial infections is not necessary and would influence the results of chemotherapeutic studies. Lindsay *et al.* (1995) used toxoplasms cloned in tissue culture. The authors scraped off the monolayer, mixed the tissue fluid with a needle (0.26 mm inner diameter), filtered it through a 3 μm polycarbonate filter and adjusted the number of tachyzoites at 1×10^4 in 0.5 ml per mice. Eyles and Jones (1955) adjusted the inoculum at 1 million tachyzoites per rat.

Infection process

Immediately before injection the inoculum should be mixed three times with a syringe. A plastic syringe of 1 or 2 ml with a needle of 0.7×32 mm 22 G×1¼ in (number 12, DIN/Germany 1997) or smaller is recommended. Foaming should be avoided. The abdomen of mice should be disinfected with isopropylic alcohol and the inoculum can be injected intraperitoneally. The size of inoculum should be between 0.2 and 1.0 ml. Animals can be marked on their fur with a water solution of picrinic acid, in order to differentiate between them. This marking will be readable for about 6 weeks. For a special lung model Garin *et al.* (1963) infected mice intranasally.

Key parameters to monitor response to treatment

The BK or RH *Toxoplasma* strains only form tachyzoites under normal conditions, not cysts, and infected animals will be killed within a few days. Most authors (Werner *et al.*, 1972) used the survival rate to monitor the effect of drugs on *Toxoplasma* tachyzoites. The ratio of surviving and killed animals can be calculated using tests for significance. All the animals used for an experiment should be examined for toxoplasms. In moribund or dead animals the peritoneal exudate and/or internal organs can be checked microscopically for parasites. For this the material can be examined natively or stained with Giemsa or fluorescein-labeled *Toxoplasma* antibodies. Brun-Pascaud *et al.* (1994) used rats and counted the number of tachyzoites after serial dilutions in tissue culture. They calculated the parasite burden as a reciprocal titer per ml. The burden per organ was mean log value ±1 SD. Also Piketty *et al.* (1990) used the tissue culture technique (Appendix 2).

All animals alive at the end of the trial must also be examined. They can be tested for *Toxoplasma* antibodies by dye test. If there is a titer of ≥1:4096 then an infection is confirmed. In cases of 1:1024 a subinoculation in clean mice is recommended. Lower titers in the original mice may exclude an infection.

The next method of examining mice at the end of a trial is to kill living mice at least 4 weeks after infection. The brain must be checked for *Toxoplasma* cysts. For this a small piece of brain can be put on a microscopic slide and pressed with a cover slip. The preparation should be analysed by magnification of 100× or 400× for cysts. Cysts are about 300 μm and contain thousands of bradyzoites. If the microscope examiner is experienced he or she may see very young cysts from the third day after infection (Werner and Dannemann, 1972).

As well as this quality test for cysts, quantitative estimation is possible — the so-called index of cysts (Werner and Egger, 1968). The technique is as follows: a mouse brain will be minced with 1 ml physiological saline in a mortar. After this, 4 ml of the suspension will be put on a slide, covered with a slip of 18×9 mm and examined microscopically. At least three such preparations must be counted. The mean number of cysts is the index of cysts. Nguyen and Stadtsbaeder (1983) used as a criterion the resistance to lethal challenge with the virulent RH *Toxoplasma* strain.

Pitfalls (advantages/disadvantages) of the model

The advantages of the model are that a counted number of tachyzoites can be injected and that white mice are excellent hosts for this parasite. A certain problem is that the BK or RH strain of *Toxoplasma* in particular is very virulent and will kill mice within a few days. Other strains can be used, but they are cyst-forming parasites, which are more useful for the cyst model (bradyzoites infection). Huskinson *et al.* (1991) used relatively avirulent parasite strains and pretreated mice to make them more susceptible to injection of 0.2 ml rabbit–antimouse interferon-γ 1 day before infection.

Bradyzoite infections in rodents

Animal species

Several strains of mice have been used for this model (see tachyzoite model, above). *Mastomys natalensis* has also been used (Werner and Egger, 1974).

Preparation of animals

No special housing or care or specific pretreatment is required.

Selection, storage and preparation of inocula

Many cyst-forming strains isolated from humans or animals can be used for experimental infections. Nguyen and Stadtsbaeder (1983) used the Beverly, Espinas *et al.* (1982) S 273, Werner and Dannemann (1972) DX and ALT strains. These strains are subcultured in mice periodically between every 2 months and 2 years. Mice are killed and the brains minced in a mortar adding 0.5 ml physiological saline or shaken in a bottle with glass beads.

Infection process

The inoculum should contain about 20–30 cysts per mouse. They can be given to the animals intraperitoneally or orally using a bent button needle of 1 mm diameter. This technique needs some experience, but then is very easy to perform. There are two possibilities for chemotherapeutic studies on bradyzoites. According to Werner and Egger (1975), one can study the efficacy of drugs on very young cysts; this is possible from the third day of infection. The other possibility is to do examinations on older cysts, i.e., from 4 weeks after infection. In this phase cysts are fully developed and contain thousands of bradyzoites within the cystic wall.

Key parameters to monitor response to treatment

During and at the end of studies on very young cysts the key parameter to monitor is mainly examination for tachyzoites and cysts. In studies on well-developed cysts (bradyzoites) cyst examination is carried out. All the monitoring techniques are described in detail above. Additionally Werner and Matuschka (1979) and Matuschka (1977) monitored the efficacy by electron microscopic studies on cyst walls and bradyzoites.

Pitfalls (advantages/disadvantages) of the model

The differentiation between both models — tachyzoites and bradyzoites — cannot be very strict, because there is fluent development from tachyzoites to bradyzoites. However there are two reasons why the models should be differentiated. First, one takes cyst-forming or tachyzoite-forming *Toxoplasma* strains. Werner and Dannemann (1972) were able to show that very young cysts can be seen from a few days after infection. These authors showed that drug efficacy depends on the intensity of bradyzoite metabolism within cysts and on the structure of the nature of cyst membranes (Werner and Egger, 1974). Studies on well-developed cysts are more significant than on very young cysts. The reason for this is that the reactivation of a latent infection in immunocompromised patients is highly clinically significant. Until now we do not have any potent drug against cysts and therefore we need chemotherapeutic studies on bradyzoite (cyst) models. Mice are very useful, but

Mastomys develop many more and larger cysts in their brains. A disadvantage is the difficult handling of this rodent.

Connatal infections

Animal species

Experiments were done on mice and on sheep.

Preparation of animals

No special housing, care or specific pretreatment is required.

Selection, storage and preparation of inocula

Beverly *et al.* (1973) used a cyst suspension for experiments on mice, whereas Dubreuil (1972) used tachyzoites. The preparation of inocula is as described above.

Infection process

Mice were infected by subcutaneous injection of 20 cysts per animal and mated 6–8 weeks later. Each sheep got 1500 tachyzoites.

Key parameters to monitor response to treatment

Litter mice were killed at 56 days, and their brains were emulsified and searched for *Toxoplasma* cysts. If these were present a count was made; if not, the remainder of the emulsion was passaged into clean mice, which were examined 8 weeks later for toxoplasms. Beside this, histological sections of mice brain, lymph nodes, thymus, heart, liver, lung, spleen and skeletal muscle were prepared, stained and checked for the parasite.

The studies on ewes were not successful, with the exception of antibody detection by dye test in lambs.

Pitfalls (advantages/disadvantages) of the model

Results of a mouse model cannot be transferred to humans, because of a different type of placenta and the fact that, in contrast to humans, infected female mice can transmit toxoplasms connatally in more than one generation to their offspring. A better model is sheep, but there are no other publications on drug studies.

Infections in immunodeficient mice

This model has some significance as regards immunosuppression (AIDS and transplants patients) in humans.

Animal species

Murray *et al.* (1993) showed in athymic nude mice, infected with *Leishmania donovani*, that the chemotherapeutic efficacy is T-cell-dependent. These authors also did such studies on athymic nude mice, from a BALB background, infected with *Toxoplasma*.

Preparation of animals

Specialized housing and care were necessary for these immunodeficient mice, but no specific pretreatment is required.

Selection, storage and preparation of inocula

The authors used the RH *Toxoplasma* strain and diluted it to 100 tachyzoites in 0.5 ml saline as an inoculum.

Infection process

The animals were infected intraperitoneally. Some mice were reconstituted immunologically: 1 day before infection with toxoplasms, mice were injected in the tail vein with 0.2 ml saline, containing 1×10^7 non-adherent nu/+ spleen cells, which were unfractionated or first depleted of either $L3T4^+$ or $Lyt\text{-}2^+$ cells. Depletion of more than 88% was accomplished by treatment with hybridoma culture supernatants, containing anti L3T4 monoclonal antibody or anti-Lyt-2 monoclonal antibody, followed by incubation with complement.

Key parameters to monitor response to treatment

The authors used only the survival rate to monitor infection in both groups of athymic mice.

Pitfalls (advantages/disadvantages) of the model

The advantages are that one can investigate chemotherapeutic efficacy as a model for T-cell-deficient patients, but there is no rodent model for the same pathologic reactions as in humans.

Oocyst studies in cats

Animal species

This model has some significance with relation to protecting humans from oocyst infection by the oral route.

Felidae, the final host for toxoplasms, can be used for oocyst studies. House cats are sufficient for drug-testing.

Preparation of animals

Cats should be serologically tested for *Toxoplasma* antibodies by indirect immunofluorescence, dye test or others. Cats free of *Toxoplasma* antibodies should also be monitored at least once by microscopical examination of their feces for *Toxoplasma* oocysts (see below). The daily cleaning of cage grills, food and water containers by boiling water or autoclaving is necessary. Cats should be fed with canned or dry food only and feces should be autoclaved to prevent staff infection. Cats serologically and coproscopically free of *Toxoplasma* infection can then be used for experiments.

Selection, storage and preparation of inocula

Cats can be infected with all the three stages of the parasite. The following strains of bradyzoites (cysts) were used: C-56 (Sheffield and Melton, 1976), M-7741 (Dubey and Yeary, 1977; Frenkel and Smith, 1982), Gail (Frenkel and Smith, 1982), and a strain from a pig (Rommel *et al.*, 1987).

Infection process

Cats can be infected orally by feeding food mixed with mice brains or carcasses or by administration of a brain suspension in the posterior pharynx by a syringe.

Key parameter to monitor response to treatment

The parameters are described in detail by Dubey and Yeary (1977). Efficacy of drugs was judged by comparing the duration of oocyst shedding and the number of oocysts. For this feces from each cat were collected daily in litter and this was changed daily. Feces were moistened with water and floated in a sucrose solution of 1.15 specific gravity in 50 ml centrifuge tubes at 2000 rpm. After removing a drop of the supernatant from the top for microscopic examination, 5 ml of the supernatant was mixed with 45 ml of water and centrifuged at 2000 rpm for 10 minutes. After discarding the supernatant, the sediment was suspended in 10 ml of 2% H_2SO_4 and the number of oocysts was counted in 0.4 µl of oocyst suspension in a hemocytometer. The total number of oocysts present in the entire daily fecal sample was obtained by multiplying the number of oocysts in 0.4 µl of fecal suspension by 25 000. Thus, the lower threshold of countable oocysts was 25 000 in the daily sample. A cat was presumed to be shedding 10 000 oocysts daily if several oocysts were detected microscopically but were absent in 0.4 µl of fecal suspension; it would be shedding 1000 oocysts if only one to five oocysts were found in the microscopic examination and none in the hemocytometer chamber. All fecal samples were inoculated in mice and the mice were examined for *Toxoplasma* infectivity.

Beside oocyst excretion the cats were monitored for infectivity of their organs for mice. For that, cats were killed between 10 and 30 days post-infection and pooled suspensions of their brains, hearts, mesenteric lymph nodes, lungs, livers, spleens and skeletal muscles (hereafter referred to as tissues) were inoculated intraperitoneally into 6 mice. Mice were examined for *Toxoplasma* infectivity. Briefly, smears of internal organs of the mice that died were examined for evidence of *Toxoplasma* infection. Survivors were bled and killed 21 days after inoculation and their brains were examined for *Toxoplasma* cysts. Mice were considered not infected if cysts and antibody to *T. gondii* were not demonstrable 21 days after inoculation with feline tissues. Cats were also serologically tested for *Toxoplasma* antibodies by dye test (also immunofluorescence antibody test) and above all, cat organs were microscopically examined after hematoxylin and eosin staining for toxoplasms.

Pitfalls (advantages/disadvantages) of the model

The house cat model is the only one used for drug-testing for oocyst excretion. The disadvantages are that one must use phylogenetically higher developed animals, and caging and care-taking are expensive. There is also a certain danger of infection for other animals or staff. Immunosuppressed people or pregnant women should not take care of such cats.

Studies of other animal species

Animal species

Rabbits and guinea-pigs (Jacobs *et al.*, 1964), pigs (Tsunoda *et al.*, 1966, Shimizu and Shirahata, 1968), sheep (Dubreuil, 1972), and monkeys (Harper, 1985) have been studied.

Preparation of animals

No specialized housing or care or specific pretreatment is required.

Selection, storage and preparation of inocula

Several *Toxoplasma* strains were used and inocula were prepared from tachyzoites or cysts (see rodent models, above).

Infection process

The *Toxoplasma* inocula were given subcutaneously, intraperitoneally, intramuscularly or orally (pigs, sheep). A special infection process was done by Jacobs *et al.* (1964). They injected 5000–10 000 tachyzoites directly into the anterior chamber of rabbit eyes (Appendix 3) and of guinea-pigs.

Squirrel monkeys were sedated by intramuscular injection of 10 mg/kg ketamine hydrochloride. A 20% suspension of *Toxoplasma*-infected mice brain was given by stomach catheter and after this an additional 5 ml phosphate-buffered saline (Harper, 1985) was given.

Key parameters to monitor response to treatment

To monitor infection, the same parameters as described before were used — survival rate, examination of organs for toxoplasms and antibody testing. In the eye model the degree of uveitis was measured and graded (Tabbara *et al.*, 1974; Appendix 3).

Pitfalls (advantages and disadvantages) of the models

There seem to be no advantages in using animals other than rodents as models. The only reason is if one would like to transfer the results from an animal species which is experimentally used directly to the same species.

Antimicrobial therapy

A variety of antimicrobial agents have been evaluated in animal models. This is documented by Jira and Kozojed (1970, 1983) for the period 1908–1967 and 1968–1975. Newer drugs have been reviewed and discussed by Ruf (1994). Studies on the combination of drugs and immune-serum have been published by Werner *et al.* (1977) and on interferon-γ by Israelski and Remington (1990). These drugs are: sulfonamides, sulfones, pyrimethamine, antibiotics and others. Depending on solubility and animal species, the drugs can be given orally, intraperitoneally, intravenously or subcutaneously (Werner *et al.*, 1972).

Drugs can be given some days before (in prophylactic studies) or after infection (in therapeutic studies). In experimental studies on well-developed cysts (bradyzoites) one should begin with treatment at least 4 weeks after infection. The duration of treatment depends on the study design, but in general should be daily between 5 days (Werner and Egger, 1975) and 8 weeks (Araujo *et al.*, 1991). Examples of studies on tachyzoites, bradyzoites and oocysts are given in Tables 98.1–98.3.

Contributions of the models to infection disease therapy

Studies on animals with *Toxoplasma* infections are essential for preliminary evaluations of candidate drugs, whereas the transfer of results from animals (mostly mice) to humans is limited. The treatment of chronic infection (bradyzoites, cysts), which would be highly significant for transplant and AIDS patients is still in question. It may be that the combination of immunoreactive components with drugs could open new avenues for the treatment of chronic infection in the future.

Table 98.1 Efficacy of a drug combination to an acute toxoplasmosis (proliferative phase) of mice

Treatment	*Dose*	*11th day after end of therapy*			*22nd day after end of therapy*			*Parasite-*	*Chronic*
	(mg/kg)	*Dead*	*Survived*	*CD_{50} (mg/kg)*	*Dead*	*Survived*	*DC_{50} (mg/kg)*	*free*	*infection*
Sulfamethoxazole	100 + 10	3/20	17/20	57.75 + 5.75	11/20	9/20	114.2 + 11.4	0/20	9/20
+ pyrimethamine	50 + 5	14/20	6/20	(49.8 + 80.8)	18/20	2/20	(79.3 – 198.8)	0/20	2/20
10 + 1	25 + 2.5	17/20	3/20		20/20	0/20		0/20	0/20
Sulfadoxine	100 + 10	1/20	19/20	29.1 + 2.9	8/20	12/20	65.9 + 6.5	3/20	9/20
+ pyrimethamine	50 + 5	6/20	14/20	(22.0 – 46.3)	9/20	11/20	(53.1 – 99.1)	2/20	9/20
10 + 1	25 + 2.5	11/20	9/20		19/20	1/20		1/20	0/20
Sulfamethoxypyrazine	100 + 10	1/20	19/20	6.13 + 0.61	7/20	13/20	53.1 + 5.3	0/20	13/20
+ pyrimethamine	50 + 5	1/20	19/20	(0.5 – 85.7)	8/20	12/20	(39.7 – 85.7)	0/20	12/20
10 + 1	25 + 2.5	4/20	16/20		16/20	4/20		0/20	4/20

Onset of therapy: 3 days post-infection; current: 5 days.
From Werner and Egger (1975), with permission.

Table 98.2 Results of chemoimmunotherapy against *Toxoplasma gondii* cysts in latent infected mice

Treatment	*Toxoplasma strain*										
	Witting			*Alt*			*Gail*			*Heart passage*	
	*Cyst index**	*Reduction (%)†*	P	*Cyst index**	*Reduction (%)†*	P	*Cyst index**	*Reduction (%)†*	P	+	–
SDDS	4.80± 1.62	47.83	<0.001	6.70± 1.89	33	0.05	7.50± 1.90	5.07	NS	5	5
SDDS + pyrimethamine	3.60± 2.32	60.87	<0.001	5.90± 1.97	41	0.02	4.50± 1.84	43.04	<0.001	5	5
SDDS + hyperimmune serum	4.10± 2.64	55.44	<0.001	4.20± 2.15	58	0.002	5.13± 0.99	35.07	<0.001	6	4
SDDS + pyrimethamine + hyperimmune serum	3.40± 2.22	63.05	<0.001	3.50± 1.58	65	<0.001	4.00± 2.16	49.37	<0.001	6	4
Hyperimmune serum	3.70± 1.77	59.79	<0.001	4.90± 2.23	51	0.01	6.60± 1.96	16.46	NS	6	4
None (control)	9.20± 2.04			10.00 ±4.55			7.90± 1.45			7	3

SDDS, 2-Sulfamoyl-4,4-diaminodiphenylsulfone.
* Mean; † compared with untreated controls; ± standard deviation obtained from 10 animals; NS, not significant.
From Werner *et al.* (1977), with permission.

Table 98.3 Efficacy of 2-sulfamoyl-4,4-diaminodiphenylsulfone (SDDS), sulfadiazine (SD) and pyrimethamine (PR) therapy in *Toxoplasma*-infected cats

	Cat no.					
	7	*8*	*9*	*10*	*11*	*12*
Drug and level (mg/kg per day)*	SDDS 160	SDDS 320	SDDS, PR 160, 1	SD 120	SD, PR 120, 1	None
Oocyst shedding days Shed	5–6	None shed	5–6, 16–20	4–7	4–6, 17–21	4–10
Total shed	20 000		70 000, 546 000	40 000	30 000 195 000	4 000 000
Antibody titer 30 DPIT	64	64	<2	16	16	4096
Isolation in mice	Positive	Negative	Positive	Positive	Negative	Positive
Hemograms DPIT						
0	Normal	Normal	Normal	Normal	Normal	Normal
10	Normal	Normal	Leukopenic	Normal	Leukopenic	Normal
Weight gain or loss (%) *DPIT*						
10	2	18	12	15	18	7
18	3	19	5	22	4†	22
28	15	47	15	40	18	40

* Drugs administered twice daily fourth through ninth day post-infection with toxoplasms (DPIT). Cats were 86–89 days old on day 0.
† Weight loss. Other values are weight gain.
From Dubey and Yeary (1977), with permission.

Appendix 1: *Toxoplasma* tachyzoite (BK strain) model in mice (Bundesgesundheitsamt, 1989)

Reagents	*Phosphate-buffered saline (PBS)*		
Solution A	$Na_2HPO_4 \times 2\ H_2O$		23.76 g
	Aqua bidestillata		2000 ml
Solution B	KH_2PO_4		9.08 g
	Aqua bidestillata		1000 ml
Solution C	NaCl		18 g
Aqua bidestillata			2000 ml
Final solution	Solution A	143 ml	
	Solution B	57 ml	
	Solution C	2000 ml	
Adjust pH by adding 1 mol/l HCl or 1 mol/l NaOH at 7.2			

Animal inoculation

Female or male NMRI mice, 18–20 g, can be used. Depending on the time between the subinoculations two possibilities exist.

Subinoculation of brain

Mice infected between 3 and 5 days before and killed by ether can be used. The head of a mouse should be disinfected by alcohol. After 2 minutes the skin over the skull can be opened using two pinsettes and then the skull is opened with scissors. The brain should be taken out and put into a glass-stopped bottle containing glass beads (diameter 4 mm) and 1.5 ml PBS. Shake the bottle to get a suspension. A syringe (1 or 2 ml, needle of 226×1¼in) can be filled with it and 0.5 ml can be injected intraperitoneally into a clean mouse. After 5 days the peritoneal exudate contains millions of tachyzoites.

Subinoculation of peritoneal exudate

Mice infected 3 days before can be used. The abdomen of killed mice must be disinfected with alcohol and then the skin can be opened using two pincettes, but the abdomen must be unopened. Fill the abdomen with 2 ml PBS using a syringe. Aspirate the contents slowly. One will get about 1.5–2 ml peritoneal exudate containing millions of tachyzoites.

If there is bacterial infection mice can be treated with antibiotics. Add about 0.4 mg penicillin and 0.6 mg streptomycin per mouse to the peritoneal exudate, which will be subinoculated.

Appendix 2: Quantification of parasites in the blood, lungs and brains of infected mice treated with antimicrobial agents against *Toxoplasma* (Piketty *et al.*, 1990)

Parasitic loads were determined by subculturing serial dilutions of blood, brain, and lung homogenates on fibroblast tissue cultures. Cultures of MRC5 fibroblasts were prepared by using minimum essential medium supplemented with glutamine (290 μg/ml), penicillin (50 IU/ml), streptomycin (50 μg/ml), and 10% heat-inactivated fetal bovine serum. A total of 20 000 cells in 200 μl of culture medium were seeded into each well of 96-well tissue culture plates and grown to confluence at 37°C in a moist 5% CO_2–95% air atmosphere. Culture plates were used within 10 days of preparation.

Infected mice were killed by cervical dislocation. Blood was collected immediately from the retro-orbital sinus, and then lungs and brain were removed and washed with phosphate-buffered saline (PBS) solution–0.15 mol/l NaCl (pH 7.2). Each organ was gently wiped, weighed, and then homogenized in a tissue grinder with 4 ml of PBS supplemented with penicillin and streptomycin. From each blood and organ suspension, serial fourfold dilutions were prepared in the culture medium, and then 40 μl of each dilution was inoculated into duplicate wells of the tissue culture plates. After 72 hours of incubation at 37°C, the plates were emptied, washed by immersion in PBS, fixed with cold methanol for 10 minutes, and air-dried.

Demonstration of *Toxoplasma gondii* in the culture was performed by indirect immunofluorescence assay. Briefly, 50 μl of a dilution of a rabbit anti-*T. gondii* antibody was added to each well and incubated for 30 minutes at 37°C. After two washings with PBS, 50 μl of a dilution of a fluorescein-labeled anti-rabbit immunoglobulin G conjugate was added and incubated for an additional 30 minutes. After two washings, the plates were emptied and examined under an inverted fluorescence microscope at a magnification of ×100. The presence of parasitic foci was recorded in each well. The final titer was the last dilution for which the tissue culture contained at least one parasitic focus. The number of parasites per gram or milliliter (parasitic load) was calculated as follows: parasitic load = reciprocal titer in tissue culture/volume (ml) or weight (mg) ×1000.

Appendix 3: Ocular model (Jacobs *et al.*, 1964; Tabbara *et al.*, 1974)

For the purposes of the blood tests the procedure was to inoculate doses of 5000–10 000 parasites directly into the anterior chamber of the eye. The inocula were prepared from the chorioallantoic membranes of 14-day-old chick embryos which had received intrayolk sac inoculations of parasites 7 days earlier. After the membranes were ground in a mortar with saline, the heavy portions of the tissue suspensions were allowed to settle, the parasites in the supernatant fluid were removed and counted in a Neubauer-Levy counting chamber, and appropriately diluted to give the desired number of organisms per 0.1 ml.

Inoculations were made into the right eye. The rabbit was previously anesthetized with pentobarbital (Nembutal) and/or ether and the eye further anesthetized with 1% tetracaine. A needle (0.21 mm inner diameter) was inserted through the cornea close to the limbus and passed across the anterior chamber so that the injected fluid would have little likelihood of seeping out when the needle was withdrawn. In some tests the left eye was left undisturbed, to serve as a reference point for grading the lesions in the right eye; in others, it received 1/10th the number of parasites in order to discern drug effectiveness, if any, against a low inoculum, even if it was not apparent against a larger dosage of organisms.

Methods for study of treatment effects

One-half of each piece of excised tissue was fixed in a solution containing 19 parts of absolute ethanol and one part of glacial acetic acid. This portion was subjected to evaluation and immunofluorescent staining with fluorescein-tagged antibodies. The other half was prepared for isolation attempts, as described below.

Histopathology

The histopathologic changes in both the treated and untreated eyes were recorded according to a histopathologic grading system devised for this experimentally produced toxoplasmic retinochoroiditis.

Depending on the degree of mononuclear cell infiltration, the architecture of the affected choroidal and retinal layers, and the presence or absence of necrosis, inflammation was graded from 0 (normal) to 4+ (severe) as follows: 0 (normal), normal retina or choroid; 1+ (mild), mild infiltration, normal architecture; 2+ (moderate), moderate infiltration, normal architecture; 3+ (marked), marked infiltration, loss of normal architecture (with or without areas of focal necrosis); and 4+ (severe), severe infiltration, massive destruction, loss of architecture, necrosis. The retinas and choroids of both the treated and control groups were graded separately.

Eye tissue was examined by fluorescent antibody technique as well as mouse inoculation.

References

Araujo, F. G., Williams, D. M., Grumet, F. C., Remington, J. S. (1976). Strain-dependent differences in murine susceptibility to *Toxoplasma*. *Infect. Immun.*, **13**, 1528–1530.

Araujo, F. G., Huskinson, J., Remington, J. S. (1991). Remarkable *in vitro* and *in vivo* activities of the hydroxynaphthoquinone 566C80 against tachyzoites and tissue cysts of *Toxoplasma gondii*. *Antimicrob. Agents Chemother.*, **35**, 293–299.

Beverly, J. K. A., Freeman, A. P., Henry, L. (1973). Prevention of pathological changes in experimental congenital *Toxoplasma* infections. *Lyon Med.*, **230**, 491–498.

Biocca, E., Pasqualin, R. (1942). Ricerche preliminari sull'-azione terapeutica di alcuni composti sulfanilamidici nell'infezioni sperimentale da *Toxoplasma*. *Ann. Igiene*, **52**, 12–14.

Brun-Pascaud, M., Chau, F., Simonpoli, A. M. *et al.* (1994). Experimental evaluation of combined prophylaxis against murine pneumocystosis and toxoplasmosis. *J. Infect. Dis.*, **170**, 653–658.

Bundesgesundheitsamt (1989). Sabin-Feldman-Test (SFT) zum Nachweis von Antikörpern gegen *Toxoplasma gondii* (Routinemethode). *Bundesgesundhbl.*, **32**, 553–555.

Dubey, J. P., Beattie, C. P. (1988). *Toxoplasmosis of Animals and Man*. CRC Press, Boca Raton, Florida.

Dubey, J. P., Yeary, R. A. (1977). Anticoccidial activity of 2-sulfamoyl-4,4-diaminodiphenylsulfone, sulfadiazine, pyrimethamine and clindamycin in cats infected with *Toxoplasma gondii*. *Can. Vet. J.*, **18**, 51–57.

Dubreuil, G. (1973). Toxoplasmosis in the pregnant ewe. Trial of spiramycin in the prevention of experimental toxoplasmosis. Thèse Ecole Nationale Veterinaire, Lyon. *Vet. Bull.*, **43**, 560.

Espinas, F. M., Takei, Y., Sakurai, H. *et al.* (1982). Studies on the therapeutics of experimental toxoplasmosis. II. Effect of acetylspiramycin alone or in combination with an immunopotentiator (CSP-II) or sulfamethopyrazine on *Toxoplasma* multiplication in the heart of mice acutely and chronically infected with *Toxoplasma gondii*. *Jpn. J. Ant.*, **35**, 362–368.

Eyles, D. E., Coleman, N. (1952). Tests of 2,4-diaminopyrimidines on toxoplasmosis. *Publ. Hlth. Rep.*, **67**, 249–252.

Eyles, D. E., Jones, F. E. (1955). The chemotherapeutic effect of pyrimethamine and sulfadiazine on toxoplasmosis of the Norway rat. *Antibiotics Chemother.*, **12**, 731–734.

Frenkel, I. K., Smith, D. D. (1982). Inhibitory effects of monensin on shedding of *Toxoplasma* oocysts by cats. *J. Parasitol.*, **68**, 851–855.

Garin, J. P., Perrin-Fayolle, M., Paliard, P. (1963). Toxoplasmose pulmonaire expérimentale. *Lyon Méd.*, **209**, 1269–1273.

Harper, J. S. (1985). Five drug regimens for treatment of acute toxoplasmosis in squirrel monkeys. *Am. J. Trop. Med. Hyg.*, **34**, 50–57.

Huskinson, J., Araujo, F. G., Remington, J. S. (1991). Evaluation of the effect of drugs on the cyst form of *Toxoplasma gondii*. *J. Infect. Dis.*, **164**, 170–177.

Israelski, D. M., Remington, J. S. (1990). Activity of gamma interferon in combination with pyrimethamine of clindamycin in treatment of murine toxoplasmosis. *Eur. J. Clin. Microb. Infect. Dis.*, **9**, 358–360.

Jacobs, L., Melton, M. L., Kaufman, H. E. (1964). Treatment of experimental ocular toxoplasmosis. *Arch. Ophthalmol.*, **71**, 111–118.

Janitschke, K., Busch, W., Kellershofen, C. (1988). Untersuchungen zur Anwendbarkeit der direkten Agglutination zur

Toxoplasmose-Überwachung im Rahmen der Schwangerenvorsorge. *Immun. Infect.*, **16**, 189–191.

Jira, J., Kozojed, V. (1970). *Toxoplasmosis 1908–1967. Documentation of Literature*, Ser. 3, Part 1. Gustav Fischer Verlag, Stuttgart.

Jira, J., Kozojed, V. (1983). *Toxoplasmosis 1968–1975. Documentation of Literature*, Ser. 3, Part 2. Gustav Fischer Verlag, Stuttgart.

Lindsay, D. S., Rippey, N. S., Blagburn, B. L. (1995). Treatment of acute *Toxoplasma gondii* infections in mice with diclazuril or a combination of diclazuril and pyrimethamine. *J. Parasitol.*, **81**, 315–318.

Matuschka, F. R. (1977). The effect of sulfomethoxypyrazine-pyrimethamine therapy on the fine structure of *Toxoplasma gondii*-cysts in the brain of *Mastomys natalensis. Zbl. Bakt. I. Orig.*, **238**, 419–429.

Murray, H. W., Teitelbaum, R., Hariprashad, J. (1993). Response to treatment for an intracellular infection in a T cell-deficient host: toxoplasmosis in nude mice. *J. Infect. Dis.*, **167**, 1173–1177.

Nguyen, B. T., Stadtsbaeder, S. (1983). Comparative effects of cotrimoxazole (trimethoprimsulphamethoxazole), pyrimethamine-sulphadiazine and spiramycin during avirulent infection with *Toxoplasma gondii* (Beverly strain) in mice. *Br. J. Pharm.*, **79**, 923–928.

Piketty, Ch., Derouin, F., Rouveix, B., Pocidalo, J.-J. (1990). *In vivo* assessment of antimicrobial agents against *Toxoplasma gondii* by quantification of parasites in the blood, lungs, and brain of infected mice. *Antimicrob. Agents Chemother.*, **34**, 1467–1472.

Pohle, D. (1994). Toxoplasmose bei immunsuppression. In *Toxoplasmose – Erreger und Krankheit* (eds Pohle, H. D., Remington, J. S.) pp. 141–164. SMV-Verlag, Gräfelfing, Germany.

Rommel, M., Schneider, T., Krause, H. D., Westerhoff, J. (1987). Versuche zur Unterdrückung der Bildung von Oozysten und Zysten von *Toxoplasma gondii* in Katzen durch Medikation des Futters mit Toltrazuril. *Vet. med. Nachr.*, **2**, 141–153.

Ruf, B. (1994). Therapie und Prophylaxe der Toxoplasmose. In *Toxoplasmose – Erreger und Krankheit* (eds Pohle, H. D., Remington, J. S.) pp. 195–234. SMV-Verlag, Gräfelfing, Germany.

Sabin, A. B., Warren, J. (1941). Therapeutic effect of the sulfonamides on infection by an intracellular protozoan (*Toxoplasma*). *J. Bacteriol.*, **41**, 80.

Sheffield, H. G., Melton, M. L. (1976). Effects of pyrimethamine and sulfadiazine on the intestinal development of *Toxoplasma gondii* in cats. *Ame. J. Trop. Med. Hyg.*, **25**, 379–383.

Shimizu, K., Shirahata, T. (1968). Therapeutic effect of SDDS (2-sulfamoyl-4,4'-diaminodiphenyl sulfone) on experimental toxoplasmosis. *Jpn. J. Vet. Sci.*, **30**, 183–195.

Tabbara, K. F., Nozik, R. A., O'Connor, G. R. (1982). Clindamycin effects on experimental ocular toxoplasmosis in rabbits. *Arch. Ophthalmol.*, **92**, 244–247.

Tsunoda, K., Suzuki, K., Ito, S., Tsutsumi, Y. (1966). Isolation of *Toxoplasma* from experimental pigs medicated with sulfamonomethoxine. *Nat. Inst. Anim.*, **6**, 83–88.

Werner, H., Dannemann, R. (1972). Experimentelle Untersuchungen über die Wirksamkeit von Therapeutika auf Zoite and Zystenentwicklung von *Toxoplasma gondii* in NMRI-Mäusen. *Z. Tropenmed. Parasitol.*, **23**, 63–77.

Werner, H., Egger, I. (1968). Die latente Toxoplasma-Infektion der Uterus und ihre Bedeutung für die Schwangerschaft. II. Mitteilung. Experimentelle Untersuchungen über den Einfluβ der latenten Toxoplasma-Infektion auf den Verlauf der Trächtigkeit bei Mäusen. *Zbl. Bakt. I. Orig.*, **208**, 122–135.

Werner, H., Egger, L. (1974). Experimentelle Untersuchungen über die Wirkung von Sulfamethoxypyrazine (Longum = Kelfizina) mit Pyrimethamin (Daraprim) auf die Zystenbildungsphase von *Toxoplasma gondii* in *Mastomys natalensis. Zbl. Bakt. I. Orig.*, **226**, 554–560.

Werner, H., Egger, I. (1975). Vergleichende chemotherapeutische Untersuchungen über die Wirksamkeit von BACTRIM und anderen Kombinationspräparaten auf die proliferative und Zystenbildungsphase von *Toxoplasma gondii* in NMRI-Mäusen. Ein weiterer Beitrag zur experimentellen Chemotherapie der Toxoplasmose. *Zbl. Bakt, I. Orig.*, **231**, 349–364.

Werner, H., Matuschka, F. R. (1979). Structural changes of *Toxoplasma gondii* bradyzoites and cysts following therapy with sulfamethoxypyrazine-pyrimethamine: studies by light and electron microscopy. Consequence for chemotherapy. *Zbl. Bakt. I. Orig.*, **245**, 240–253.

Werner, H., Dannemann, R., Egger, I. (1972). Über die Wirkung des Ultralangzeit-Sulfonamids Sulfamethoxypyrazine (Longum® Kelfizina®) auf die proliferative Vermehrungsphase von *Toxoplasma gondii* in NMRI-Mäusen. Ein Beitrag zur experimentellen Chemotherapie der Toxoplasmose. *Zbl. Bakt. I. Orig.*, **220**, 126–141.

Werner, H., Masihi, K. N., Tischer, I., Aduso, E. (1977). The effect of chemo-immunotherapy with SDDS, Pyrimethamine and Anti-Toxoplasma Serum on *Toxoplasma gondii* cysts in latent infected NMRI mice. *Tropenmed. Parasitol.*, **28**, 528–532.

Chapter 99

Animal Models of Coccidia Infection

A. Haberkorn and G. Greif

Background information on infection

Coccidiosis, an infection of the small intestine, is relatively rarely diagnosed in humans, where it is caused by *Isospora belli*. However, humans are also the final host of at least two other cyst-forming coccidial species (*Sarcocystis suihominis* and *S. bovihominis*). Consumption of raw or inadequately cooked pork or beef containing such cysts can lead to severe diarrhoea (Rommel and Heydorn, 1972; Lindsay, 1990), the cause of which is probably seldom diagnosed correctly. Coccidia (phylum Apicomplexa, suborder Eimeriina) are one of the most successful groups of parasitic protozoans, having conquered virtually every class of Metazoa. The ones that are of particular importance for humans are the 60–100 species which parasitize domestic animals, and which in some instances can cause very severe losses—especially in poultry, though also in lambs, calves, piglets, rabbits, and other animals (Table 99.1).

Most of the pathogenic species are strictly host-specific. They have a complex life cycle (Figure 99.1), with two asexual reproduction phases (schizogony or merogony, and sporogony) and a sexual development phase (gametogony). In view of the major importance of coccidiosis numerous reviews are available, for example, by Davies *et al.* (1963), Hammond and Long (1973), Long (1982, 1990), and Pellerdy (1974). The economically important species sometimes differ considerably in their sensitivity to medicinal active ingredients. The sensitivity of the different developmental stages to medicinal agents also varies enormously (McDougald, 1982; Maes *et al.*, 1989).

As far as the use of drugs is concerned, prophylaxis is the main approach in poultry, when symptoms do not appear

Table 99.1 Causes of intestinal coccidiosis in domestic animals

Animal	*Number of Eimeria and/or Isospora species**	*Most pathogenic and/or very common species*
Chicken *(Gallus gallus)*	7	*Eimeria tenella, E. necatrix, E. maxima E. acervulina*
Turkey (Meleagris gallopavo)	7	*E. meleagrimitis, E. adenoides*
Goose (*Anser anser*)	6	*E. anseris, E. truncata, E. nocens, E. kotlani*
Duck (*Anas platyrhynchas*)	3	*Tyzzeria perniciosa, E. anatis*
Pigeon (*Columba livia*)	2??	*E. columbarum, E. labbeanea*
Rabbit (*Oryctolagus cuniculus*)	11 (12)	*E. intestinalis, E. flavescens, E. stiedai, E. magna, E. perforans*
Sheep (*Ovis arius*)	11 (16)	*E. ovinoidalis, E. ashata, E. ovina*
Goat (*Capra hircus*)	12 (15)	*E. ninakohlyakimovae, E. arloingi*
Cattle (*Bos taurus*)	12 (15)	*E. zuernii E. bovis, E. auburnensis*
Pig (*Sus scofra*)	7 (14)	*Isospora suis, E. debliecki, E. scabra*
Dog (*Canis familiaris*)	5	*I. canis, I. (Cystisospora) burrowsi*
Cat (*Felis catus*)	2 + 6	*I. felis, I. rivolta,* as final host: *Sarcocystis bovifelis, S. ovifelis, S. fusiformis, S. muris, S. cuniculi, Toxoplasma gondii*

* According to Pellerdy (1974), Eckert *et al.* (1995b), Levine and Ivens (1970) and Mehlhorn (1988).

Handbook of Animal Models of Infection
ISBN 0-12-775390-7

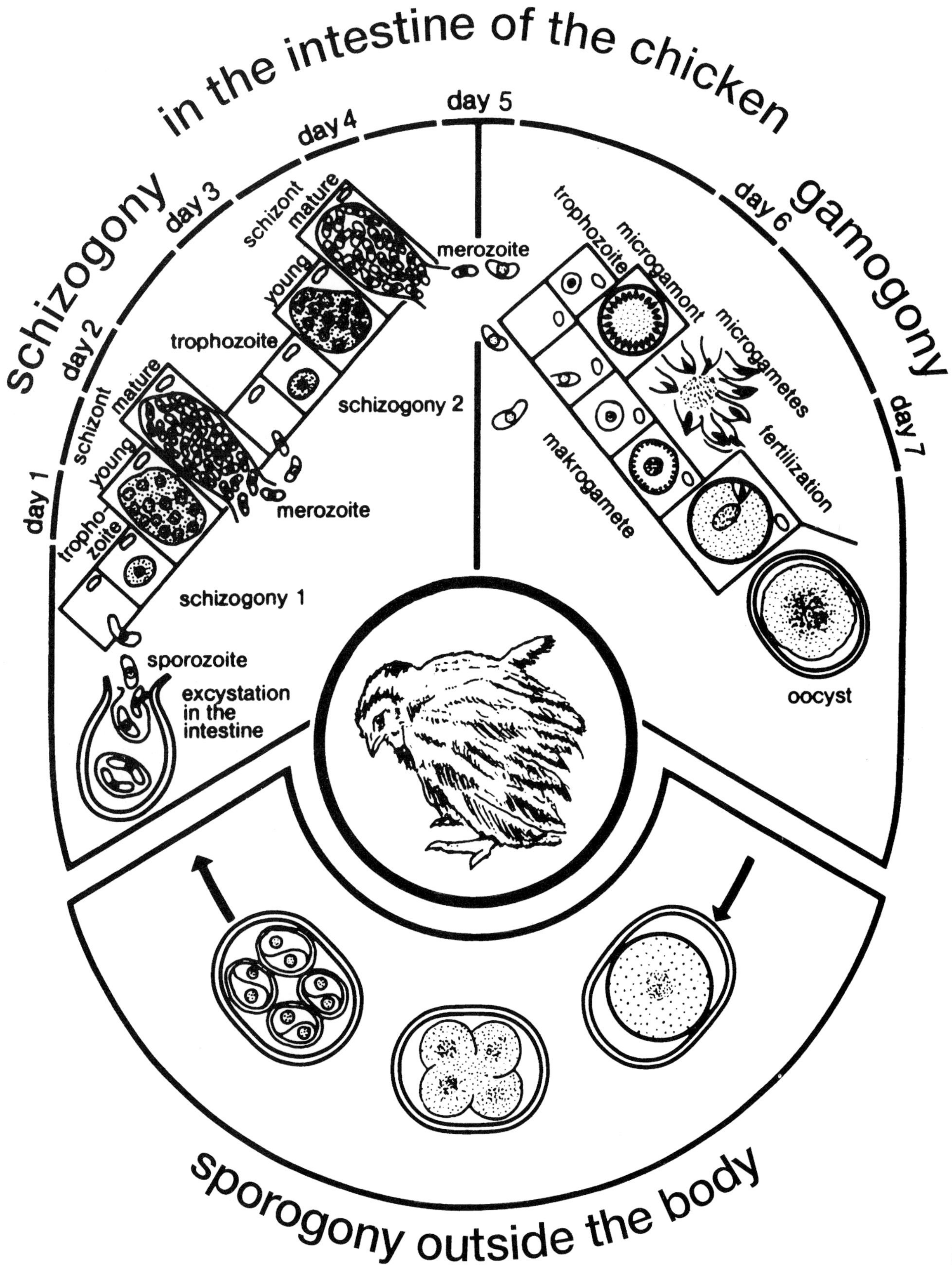

Figure 99.1 Lifecycle of Coccidia (*Eimeria tenella*).

until the phase of increased morbidity, and therapy is the principal strategy in mammals (McDougald, 1990). Given the multiplicity of pathogens and hosts, there is no ideal model for identifying and testing anticoccidial agents. For example, most of the many substances used for preventing coccidiosis in poultry are insufficiently effective or even completely ineffective against mammalian coccidia (Haberkorn and Mundt, 1989; Haberkorn, 1996). Numerous works and sets of instructions have been published on testing of active ingredients in animals for anticoccidial efficacy and for immunization. One particularly important and comprehensive example is the survey of current methods published by Eckert *et al.* (1995a,b). Two models which have proved particularly successful in work to identify and evaluate new anticoccidials are described below.

Coccidiosis of poultry

Background of the model

Initially a reduction in the mortality rate from the infection was considered to be the most important (and occasionally the only) criterion, and *Eimeria tenella* was used as the model. As time went by, with increasing success reduction in clinical symptoms and in oocyst excretion were also incorporated as criteria (Lynch, 1961; Clarke, 1962; Ryley, 1980). Cage studies with small groups of animals (3–10) and a duration of 5–7 days (or 14 days at the very outside) are normally used for the identification and initial testing of anticoccidials (Raether, 1980; Ryley, 1980; McDougald, 1982). Two groups of untreated infected and untreated non-infected controls (four groups in all) should be used, particularly when the number of birds in each group is small. Floor-housing studies involving groups of 20 to over 100 animals and usually lasting 4–8 weeks are then used to simulate conditions close to those found in practice. These studies—and also cage studies with various *Eimeria* species and field strains and different treatment regimens—are used for more detailed testing, e.g., investigations to ascertain against which developmental stages a product is effective, and investigations of experimental resistance selection. The registration of anticoccidials is usually subject to very stringent conditions. Some countries therefore have guidelines that must be followed in the more detailed testing of potential commercial products (Raines, 1978; Cameron, 1985).

Animal species

Any defined breed of domestic fowl (*Gallus gallus*) is suitable in principle. There is no need to use birds from specific pathogen-free (SPF) flocks. However, a guaranteed supply of birds of the same breed (inbred lines as far as possible) from a well-run breeding operation is essential. When the groups used in the studies are small, the chicks should be of the same sex. Egg-laying breeds are good ones to use, for two reasons: since the birds are sorted by sex and the males removed on the first day of life, these are available ready-sexed and they can be purchased cheaply; they also consume nearly a third less food than broilers. This means a corresponding reduction in the amount of active ingredient that is required, which is particularly helpful in the initial phase of testing. Seven- to 14- or, at the very most, 21-day-old chicks are used as a rule (Ryley, 1980; McDougald, 1982), though 1–3-day-old (broiler) chicks are sometimes used as well (Raether, 1980). Different strains can show different degrees of susceptibility to the infection material, so it is essential to carry out pilot studies to establish the best dose for infection (see below).

Preparation of the animals

It is essential that the animals are kept separate from others before the start of the study in order to prevent spontaneous infection. For this reason, and to avoid any prior intake of feeds containing anticoccidial agents, the birds should be purchased as day-old chicks. Day-old chicks require a high ambient temperature, 30–33°C, which can be reduced to 28–30°C after a week. To meet this requirement, one should use standard commercial radiator heaters or suitable heating plates over the cages or, in the case of floor-housed birds, over the pens. The light phase should last at least 18 hours. Other than this, the conditions for housing chicks apply (European Convention for the Protection of Vertebrate Animals used for Experimental and other Scientific Purposes, 1986; see Ausems, 1986). Feed and water are available *ad libitum*. It is advisable to keep the birds in fairly large groups (e.g., 30–50 animals) before the study begins. For behavioural reasons, the minimum group size during the study is 3. The birds must be at least 3 weeks old before they can be kept on their own.

At the start of the study the birds are weighed and, as far as possible, placed in groups of the same weight. In the case of small groups (up to 10), it is recommended to mark the birds individually, e.g. with dye markings as follows: no "A", head "B", back "C", tail "D", head-back "E", head-tail "F", or something along these lines. The group number is provided by the cage number. Methylene blue and/or acridine red have proved particularly useful. Standard commercial wing tags can also be used.

Details of surgery

Surgical interventions before and during the study are not necessary.

Infection procedure

As far as possible, the birds are infected on a Monday morning, e.g. on the 7th to 12th day of life. The day of the week is important if the amount of work to be done at weekends is to be kept to a minimum. A known number of sporulated oocysts is introduced into the chick's crop via a flexible rubber oesophageal tube or plastic tubing (approximately 4 mm diameter). In many instances studies are carried out with monoinfections, i.e. with only one species of *Eimeria* (usually *E. tenella*; Lynch, 1961; Clarke, 1962; Ryley, 1980). We prefer to use mixed infections initially. Infections with *E. tenella*, *E. acervulina*, and *E. maxima* have proved particularly satisfactory. These three species, which parasitize different parts of the intestine, are particularly important from an economic point of view and, with a little practice, their oocysts are readily distinguishable from one another under a microscope.

Storage and preparation of inocula

To obtain the infection material 2–3-week-old chicks raised under coccidia-free conditions are infected with oocysts of one species. In the case of pathogenic strains or species the amount of inoculum used is so small that deaths are unlikely. In general 10^3–10^4 sporulated oocysts per bird is sufficient, depending on the chick's age. If the infection material is relatively old, or from resistant strains, the inoculum should be increased by a factor of 2–10. To prevent contamination with other strains, the stock should be kept in isolation units as far as possible, or at least in separate freshly disinfected rooms (Shirley, 1995a).

The faeces, which contain the oocysts, are collected 4½–7 days after infection in the case of *E. acervulina*, *E. mitis*, and *E. praecox*, 6–9 days after infection with *E. maxima* and *E. necatrix*, and 6–8 days after infection with *E. tenella*. With *E. tenella* it is possible—and also easier as far as working-up is concerned—to sacrifice the birds 7 days after infection and to remove the caecum. The intestine is cut open and the contents scraped out and suspended in 2% potassium dichromate ($K_2Cr_2O_7$). With this suspension oocyst enrichment by flotation is usually unnecessary.

It is best to carry out oocyst enrichment before sporulation. This is certainly a sensible step when working up large quantities of faeces, but it is not absolutely necessary when working up small quantities of faeces or the caecum contents of birds infected with *E. tenella*. The faeces, suspended in water with the aid of an electric stirrer, are passed through a sieve to separate off the coarser constituents. We have found that a sieve arrangement such as the Vibrator sieve shaker (Fritsch) performs well here. One can also use single sieves with mesh sizes of 710 μm and then around 150 and 50 μm when smaller volumes are involved. The suspension is allowed to settle for about 30 minutes in a glass cylinder of suitable size, or is centrifuged for 5 minutes at 1500–3000 rpm. The supernatant is discarded. The sediment is suspended in approximately 3 volumes of saturated sodium chloride (NaCl; specific gravity 1.2). A saturated $MgSO_4$, $ZnCl_2$, or sugar solution (specific gravity 1.5) may be used instead. There are various methods of harvesting the oocysts—which rise to the top because of their lower specific gravity—depending on the volumes to be worked up:

1. Large volumes: suspension in a suitably sized glass cylinder; after 30–60 minutes the topmost layer, which contains the oocysts, is siphoned off. The NaCl is then washed out with at least 3 portions of tap water; this is done with a centrifuge as far as possible. The oocysts are in the sediment.
2. The suspension in the saturated NaCl, $MgSO_4$, $ZnCl_2$, or sugar is introduced into centrifuge tubes, carefully overlayered with a few millilitres of water, and centrifuged twice at 1500 rpm (100*g*), each time for 5 minutes. In each case the supernatant containing the oocysts is collected and washed out with tap water (see above).
3. Petri dishes are filled to the rim with the sediment suspended in the saturated solution (see above) and after 10 minutes covered with a Petri dish lid (the lid must be absolutely grease-free). Care must be taken that no air bubbles are left between the liquid and the lid. After about 30 minutes the lid is carefully lifted, and the oocyst-containing solution on the underside of the lid is rinsed into a beaker with water, using a wash bottle. The process is repeated two or three times, until the oocysts have been collected almost quantitatively (checks under a microscope). This somewhat laborious method yields very clean oocyst suspensions. This is followed by centrifuging and further washing (see above).

The purified suspension is treated with the same volume of 4% potassium dichromate solution.

Sporulation

For the oocysts to sporulate the suspension is kept in a water bath (30–35°C), or simply in a warm room (temperature up to 30°C). When small volumes are involved the suspension is introduced into large (covered) Petri dishes. The sediment should not be more than about a third of the filling depth. For larger volumes it is advisable to use covered glass cylinders with a capacity of 1–10 litres; here too the sediment layer should not account for more than a third of the volume. The contents are kept in constant motion with stirrers (mechanical or magnetic) or small pumps. The time required for sporulation depends on the species and on temperature. At temperatures up to 30°C 3 days should be allowed, particularly in the case of *E. maxima*. It is important that sporulation begins without any appreciable delay and can proceed without interruption, i.e. unsporulated oocysts must not be stored in a refrigerator! If immediate collection and enrichment of the oocysts (see above) is not possible, for reasons of time, for example, the sieved faecal suspension can if necessary be mixed with $K_2Cr_2O_7$ solution and left to sporulate with stirring. In this case the sediment fraction should not exceed 20% by volume.

Cleaning of the oocysts

Sodium hypochlorite is added to a thoroughly sieved and enriched suspension of sporulated oocysts until the NaOCl fraction represents 10–20% by volume. After thorough stirring, the tube is left to stand in an ice bath for 10–15 minutes. The sodium hypochlorite must then be quickly and quantitatively washed out with water (small check). It is important to notice that the cleaning has to be done *after* sporulation! *E. tenella* oocysts from the caecum can also be purified by treatment with trypsin (Difco 1 : 250 powder, final concentration approximately 1.5% wt/vol.). After 30 minutes at a temperature of 41°C the suspension is filtered through gauze and centrifuged for 10 minutes (1500 rpm). The supernatant is discarded and the centrifugate suspended.

Preparation and storage of the inoculation material

The number of sporulated oocysts is determined (see section on excretion of oocysts, below). Purified infection material of the strains used is mixed to produce the inoculum that was found to work well in pilot studies. The suspension volume given to each bird should not be much more than 1 ml in the case of 7–14-day-old chicks. An oocyst suspension of this kind can be kept in a refrigerator without any problems for 3–6 months, and usually for up to 9 months to a year. The potassium dichromate fraction must be drastically reduced before infection, by centrifugation and a single washing operation. The infection material can be kept much longer by deep-freezing (e.g. in liquid nitrogen). However, there is no method for freezing the oocysts intact. The oocyst walls must first be ruptured and the sporocysts released and separated on a glass bead column. For details of the deep-freezing method, see Raether *et al.* (1995) and Shirley (1995b). Although we regularly store the infection material used for cell culture studies (sporozoites) deep-frozen, the relatively high cost of this method is not justified in routine *in vivo* studies; however, it is justified for the storage of coccidial strains that are not in constant demand and for the maintenance of a reserve.

Infection process

Each chick is given the appropriate volume of a defined oocyst suspension orally, through an oesophageal tube. The infection dose depends on the strain's pathogenicity, on the age of the chicks, and to a much smaller degree on the susceptibility of the chicks used. The infection dose chosen should produce readily discernible clinical symptoms such as diarrhoea, pathological changes in the intestine, impaired growth, and an infection-related mortality of some 30–70% (mortality due mostly to *E. tenella*) in the infected and untreated controls. Because of the substantial differences in strain pathogenicity, the quantity of inoculum used varies considerably, from 30 000 sporulated oocysts per bird in Haberkorn and Stoltefuss (1987), for example, to 500 000 per bird in Ryley (1980).

Key parameters for monitoring infection and response to treatment

Tables 99.2 and 99.3 list the parameters which can be taken into account with the various *Eimeria* species. In evaluation of a study these are considered in relation to the findings for the infected untreated and non-infected untreated control groups. The individual parameters are weighted according to their importance. There are various rating systems for this purpose, and they can also depend *inter alia* on the question under investigation. Table 99.4 gives one example which, with slight adjustments, has proved its value in our own work over more than 25 years. It was originally developed on the basis of a method described by Jones (1946) and by Thompson *et al.* (1950) for animal studies with amoebas.

It is important, particularly when evaluating macroscopic changes (diarrhoea and gross pathological autopsy findings), that the evaluation should always be done by the same person as far as possible. Unpublished investigations of our own demonstrate that findings of this kind, recorded in this way, show less variation—even over many years—than the number of excreted oocysts per gram of faeces, for example.

Parameters

Mortality

Deaths from infection occur 4½–6½ days after infection in the case of *E. tenella*, *E. necatrix*, and *E. brunetti*. Earlier deaths are attributed to other factors, such as intolerance of the test substance, and are not taken into account in the evaluation of efficacy.

Diarrhoea/excretion of blood in the faeces

Bloody diarrhoea is very conspicuous in infections with *E. tenella* (and *E. necatrix*). In mixed infections it masks diarrhoea caused by other species. About half a day before the expected onset of diarrhoea, the dishes for collecting faeces are replaced with fresh ones, or the faeces are removed and the dishes lined with white plastic-coated absorbent paper. The evaluation is carried out macroscopically (Table 99.3) 4½–5 (or 6–7) days after infection; in groups in which there is no detectable bloody diarrhoea at this time, the faeces should be inspected again 7 days after infection. Delayed excretion of blood occurs, for example, if a test substance is not completely effective against the schizonts, particularly those of the first generation of *E. tenella*. The evaluation is carried out with the 0–6 rating

Table 99.2 Chicken coccidiosis model: important parameters and data*

Eimeria sp.	*Morbidity†*	*Mortality†*	*Diarrhoea* Days p.i.	*Diarrhoea* Intensity†	*Oocyst shedding‡ (days p.i.)*	*Oocysts length (L) × width (W) (average) mini.–max. form§*	*Oocyst production** Maximum reproduction/oocyst Total output/bird*	*Evidence of macrosopic lesions‡ (cf. Table 99.3)*
E. acervulina	++	Rare +	4–5	++	**4–7 (8)**	(18 × 15) L = 17.7–20.2 W = 13.7–16.3 ovoid 1.25	72 000 432 000 000	+++
E. brunetti	++	**Moderate ++**	4–6	++	**6–8**	(25 × 19) L = 20.7–30.3 W = 18.1–24.2 ovoid 1.31	400 000 53 000 000	+++
E. maxima	+++	rare +	(5)	(+)	**6–8**	(31 × 21) L = 21.5–42.5 W = 16.5–29.8 ovoid 1.47	12 000 36 000 000	++
E. mitis	+	seldom (+)	(4–5)	(+)	**4–7**	(16 × 14) L = 11.7–18.7 W = 11.0–18.0 suspherical 1.09	? ?	+
E. necatrix	+++	**High +++**	**7–8**	**++ Brown, bloody**	**6–9**	(20 × 17) L = 13.2–22.7 W = 11.3–18.3 oblong ovoid 1.19	58 000 12 000 000	++
E. praecox	+	none –	(4)	(+)	**4–7**	(21 × 17) L = 19.8–24.7 W = 15.7–19.8 ovoidal 1.24	? ?	–
E. tenella	**+++**	**High +++**	**4.5–6**	**+++ Bloody**	**7–10**	(22 × 19) L = 19.5–26.0 W = 16.5–22.8 ovoid 1.16	400 000 65 000 000	+++

* cf. Davies *et al.* (1963), Hilbrich (1967), Janssen Pharmaceutica (1990), Long and Reid (1982) and Tyzzer (1929).
† Depending largely on the properties of the strain as well as on the dose of inoculum and age and susceptibility of the chicken strain.
‡ There is some variation between strains but also in relation to intensity of infection: cf. Haberkorn (1970) and unpublished data.
§ Long and Reid (1982).
** according to Janssen Pharmaceutica (1990); also cf. † and ‡.
p.i., Post-infection; bold text, evident criterion.

Table 99.3 Dropping and lesion scores

		Score 1	*Score 2*	*Score 3*	*Score 4*	*Score 5*	*Score 6*
Eimeria acervulina	Localization	Upper duodenum	Upper duodenum	Upper duodenum	Beyond pancreas slope	Up to yolk sac	Up to yolk sac
	Aspect	Small white patches	Long white patches ladder-like, traversing	Long white patches ladder-like, traversing	Long white patches ladder-like, traversing, but partly coalescent	Coated aspect	Greyish to bright red
	Frequency	Singular to very few	Few	Numerous	Very numerous	Coalescent	Completely coalescent
	Intestinal wall	Normal	Normal	Normal	Thickened	Thickened	Greatly thickened
	Intestinal content	Normal	Normal	Normal	Normal	Watery, slimy	Creamy exudate
	Faeces	A little soft	25% soft	50% soft	All soft	Soft, greyish, partly fluid	A lot of fluid
Eimeria maxima	Localization	Central small intestine	Central small intestine	Central small intestine	Central small intestine	Central to lower small intestine	Central to lower small intestine
	Aspect	A few small, pink spots Bleeding	Small, pink spots bleeding on serosal side	Small petechiae on serosal side	Red petechiae on both serosal and mucosa side	Mucosa side rough and bloody red	Mucosa side rough and bloody red
	Frequency		Scattered	Rather numerous	Numerous	Coalescent	Coalescent
	Intestinal wall	Normal	Normal	Slightly thickened	Thickened	Thickened and swollen, rough mucosa	Greatly thickened and swollen, mucosa rough
	Intestinal content	Normal	Sometimes orange mucus	Sometimes orange slime	Orange slime	Slime, pinpoint clots	Many blood clots
Eimeria tenella	Localization	One caecum	Caeca	Caeca	Caeca	Caeca	Caeca
	Aspect	Petechiae	Petechiae	Petechiae	Petechiae	Bloody	Bloody
	Frequency	A few, only one caecum affected	Few, scattered: both caeca affected	Numerous	More numerous	Coalescent	Coalescent
	Intestinal wall	Normal	Normal	A little thickened	Somewhat thickened	Greatly thickened	Greatly swollen and thickened mucosa, rough and hard
	Intestinal content (day 7 post-infection)	Normal	Normal (a few caseous granules)	A little blood in caeca (caseous granules)	Blood in the caeca (caseous)	Much blood and many fibrin clots, few or no remains of faeces (caseous)	Blood No faecal remains (caseous, hard)
	Day 5 post-infection consistency of faeces	Partly soft	Mostly soft	Soft	Soft	Fluid still a little soft; intestinal content	Fluid
	Blood in faeces	1 to 5 small tips	6–10 tips or drops	>10 tips or drops	>50% with blood	All bloody	Granular, bloody and pure blood

Table 99.3 — *(Continued)*

		Score 1	*Score 2*	*Score 3*	*Score 4*	*Score 5*	*Score 6*
Eimeria brunetti	Localization	Lower small intestine	Lower small intestine	Lower small intestine	Lower small intestine	Lower small intestine, caeca, rectum	Small intestine, caeca, rectum
	Aspect	Normal	Bloody, transverse red streaks in lower intestine	Bloody, transverse red streaks in rectum	Bloody and necrotic lesions	Bloody transverse red streaks in rectum	Bloody and necrotic lesions
	Frequency	A few salmon-coloured spots	Some salmon-coloured spots	Scattered	Scattered	Coalescent	Coalescent
	Intestinal wall	Normal	Normal	Nearly normal	Thickened up to lower part	Thickened and swollen	Thickened, caseous cores may plug the caeca
	Intestinal content	Normal	Normal	Normal	A little watery	Bloody and watery, sometimes soft slime balls	Sometimes caseous clots obstruct caeca
Eimeria mitis	Localization	Duodenum	Duodenum	Duodenum	Upper half of small intestine	Up to lower small intestine	Entire small intestine
	Aspect	Round white patches; red-coloured	Round white patches; red-coloured	Round white patches; red-coloured	Round white patches; red-coloured but more coalescent	Round white patches; red-coloured but more coalescent	Round white patches; red-coloured but more coalescent
	Frequency	Singular	A few	Numerous	Many	No coalescence	No coalescence
	Intestinal wall	Normal	Normal	Normal	A little thickened	Thickened	Greatly thickened
	Intestinal content	Normal	Normal	Normal	Normal	Watery, slimy	Creamy exudate
Eimeria necatrix	Localization	Small intestine	Small intestine	Small intestine	Small intestine	Small intestine	Whole intestine
	Aspect	A few scattered petechiae	Petechiae; white patches seen from serosal side; no damage on mucosal surface	Petechiae; white patches on serosa	Petechiae on serosa; violent red	Red petechiae and white plaques on serosa	Dark colour, extensive haemorrhages
	Frequency	A few	Scattered	Scattered	Numerous	Very numerous	Coalescent
	Intestinal wall	Normal	Normal	Central small intestine slightly swollen	Central small intestine ballooned	Rough, thickened and swollen up to lower small intestine	Massively swollen
	Intestinal content	Normal	Normal	Normal	Normal	Extensive haemorrhages	Red or brown mucus

Based on Janssen Pharmaceutica (1990), Johnson and Reid (1970), Shirley (1995a) and own data.
Score 2 identical with 1 after Johnson and Reid (1970); score 4,5,6 identical with 2,3,4 respectively after Johnson and Reid (1970).

system given in Table 99.3. Other authors (Johnson and Reid, 1970; Shirley, 1995a) use a point-scale with scores of 0–4.

Excretion of oocysts

Before the excretion of oocysts begins, the excrement in the faeces collection dishes is removed and the dishes are lined with plastic-coated absorbent paper. To determine the oocyst excretion, the pooled faeces for the day to be investigated are weighed and thoroughly mixed and a sample of about 10 g generally worked up directly for counting. We prefer to get a rough idea of the excreted oocysts' sporulation capacity as well (see below). For initial routine testing of substances for efficacy we have restricted ourselves to examining faeces from day 7 after infection. If the substance is found to be effective the pooled faeces for days 4–6 and 7–8 after infection are examined for oocysts in repeat studies. The total number of excreted oocysts can then be determined as well.

The number of oocysts is ascertained with the aid of counting chambers. We use the McMaster chamber (Davis, 1973; Janssen Pharmaceutica, 1990; Taylor *et al.*, 1995): 2 g of well-mixed faeces is suspended in 60 ml of saturated saline and passed through a fine sieve to eliminate the coarser constituents. The sieved suspension is mixed thoroughly so that the oocysts are uniformly dispersed. The suspension, thoroughly mixed, is introduced into the chamber with a pipette. A second and, if necessary, a third chamber are filled with fresh samples of the suspension. The count can be made after only a few minutes. A microscopic magnification of about 100× is needed for this purpose. The objective and eyepiece chosen must be such that the field of view covers the space between two lines. The counting chamber is 1.5 mm deep, and an area of 10 × 10 mm on its cover glass is subdivided by lines running from top to bottom. One field thus corresponds to 0.15 ml of suspension. The number of oocysts found in such a field, multiplied by 200, gives the number of oocysts per g of (Opg) faeces. When the level of infection is low, 2–3 fields must be counted for an accurate determination of the Opg. In the determination of Opg values in samples containing abundant oocysts in the screening test we restrict ourselves to counting only a few rows (except in the case of infected controls). The reduction in accuracy is justified, as this simplified counting saves time and as oocyst excretion in excess of 50% relative to infected untreated controls is no longer included in the rating calculation (cf. Table 99.4, below).

Sporulation rate

A faeces sample (approximately 200 g) is mixed with the same weight of a 2% solution of potassium dichromate and stirred with an electric stirrer; the container, a 500 ml disposable plastic beaker marked with the group number, is covered with aluminium foil and left to stand in the test room (at about 28°C) until the following day, by which time some of the oocysts are already sporulating (except in the case of *E. maxima*). In microscopic determination of the Opg the fraction of sporulated oocysts is ascertained at the same time, at least for one species. Only products with which the sporulation rate is virtually 0% are of interest here. Precise counting is therefore only necessary in a few instances, i.e. when there is an obvious drastic reduction.

Growth

The birds are weighed at the start and end of the study, and the weight gain of the various groups is compared with that

Table 99.4 Rating system for chemotherapeuticals with chicken coccidiosis

		Dropping scores					
Score	*Mortality*	*Day 5*	*Retarded day 5–7*	*Lesion scores*	*Oocysts (% of infected controls)*	*Weight (% of non-infected controls)*	*Sporulating rate (%)*
0	0	0	0	0	< 1	> 90	< 1
0.5		1	0–3	1–2	1–5	75–89	
1.0		2	1–3 2–3		6–10	60–74	1–10
1.5	1/3 (< 50%)	3	1–4 0–6	3	11–20	< 60	
2.0	2–3/3 (> 50%)	4–6	4–6	4–6	21–50		> 10
3.0					> 50		

Calculation

The quotient: $\dfrac{\text{total score for test group}}{\text{total score for infected control group}}$

expressed as a decimal is the rating:
0.00–0.29 = effect (100–71% effect)
0.30–0.79 = slight effect (70–21% effect)
0.80–1.00 = ineffective (20–0% effect)

of the non-infected untreated controls. In the case of mixed infections the impairment of growth cannot be attributed to a particular species. When small groups are used, conclusions can only be drawn from the results in the presence of fairly major differences. For example, if, despite good parasitological efficacy, it is found that growth is distinctly poorer than in untreated non-infected controls, this indicates that the dose in question is poorly tolerated. Very good data for growth, with suppression of deaths but otherwise with minimal effect on the parasitological parameters (diarrhoea, excretion of oocysts, autopsy findings), can also be a significant indication of efficacy, even if inadequate efficacy. The efficiency of feed conversion (quotient of feed consumption and weight gain in grams) can be calculated in supplementary studies and using broiler chicks.

Autopsy findings (lesion scoring)

At the end of the study (8 days after infection in the case of mixed infections) the birds are sacrificed with CO_2 in a closed container or by intraperitoneal injection of T61 (Hoechst AG, 1 ml for chicks weighing around 100 g). At autopsy the part of the intestine typically parasitized by the species in question (Figure 99.2) is removed, subjected to an internal and external macroscopic inspection for lesions, and evaluated according to the specified rating system (Table 99.3). In routine tests to identify new active ingredients or possible bases for these (compound screening), many active ingredients do not show any efficacy whatsoever. It is sufficient in these cases — provided that the birds have been infected with *E. tenella* — to evaluate only the caecum, which does not necessitate removal of the intestine *in toto*. In any event, earlier inspection — 6–7 days after infection — is better for the evaluation of lesions in the small intestine. As a rule, therefore, we have restricted this inspection to more detailed supplementary studies. Additional microscopic examinations of mucosal samples (stopped-down bright field or interference phase contrast) are a possibility, though we largely restrict their use to more detailed studies.

Calculation of findings

The various criteria are considered in relation to one another and compared with the control groups. A rating system was developed on the basis of a method proposed by Jones (1946) and Thompson *et al.* (1950) for studies with amoebas (Table 99.4; Haberkorn and Schulz, 1981; Haberkorn, 1986).

Anticoccidial medication

The active ingredients to be tested are mixed into ground, anticoccidial-free, complete chick feed with an electric feed mixer (e.g., from Lödige, Paderborn). It is advisable to prepare a premix first, by mixing the weighed-out active ingredient with about 30 g in an electric coffee mill. Treatment of the chicks is begun on the day of infection at the latest. We prefer to start giving the medication on Friday, i.e., 3 days before infection on Monday. In this way any effect on sporozoites is also detected. The treatment goes on until the end of the study. For three 12-day-old chicks of laying breeds and a study duration of 11 days, 1 kg of medicated feed is sufficient. Feed consumption is determined by weighing the feed remaining at the end of the study, and can thus be included in the evaluation. The dose of the standard commercial anticoccidials ranges from 125 ppm (e.g., Zoalene) down to 1 ppm (diclazuril). For special purposes an oral treatment using a stomach tube is more convenient than medication via the feed, for example, to determine on exactly which developmental stages a drug is acting.

Pitfalls (advantages and disadvantages) of the model

The model permits reliable conclusions on the efficacy of an active ingredient in the target animal. It also yields rough initial information on tolerability. Consideration of the specified criteria enables the investigator to see, in the very first study, whether all three of the *Eimeria* species are controlled. It is often already possible at this stage to deduce the phase of parasite development affected by the active ingredient, particularly in the case of *E. tenella*. Consideration of several criteria also has the advantage that studies in which, for example, all the infected untreated controls survive (for whatever reason) can still be fully evaluated. The disadvantage of using small numbers of birds can easily be eliminated by carrying out repeat studies and increasing the numbers if necessary. Seasonal extremes of temperature together with long travelling distances can occasionally make it difficult to obtain day-old chicks. Cage studies with chicks are much more informative and realistic than *in vitro* studies, e.g., with *E. tenella*-infected cell cultures (monolayer tissue cultures). In cell culture studies one can at most form conjectures on *in vivo* efficacy within a few narrowly delimited groups of chemical substances. A particular disadvantage is the high frequency of false-negative findings: ineffectiveness *in vitro* does not exclude efficacy in the chick (Ryley and Wilson, 1972, 1976; Ricketts, 1992). The advantage is that little active ingredient is required for cell culture tests. *In vitro* studies can be valuable for more detailed supplementary investigations. This is also true for experimental *Eimeria* infections in embryonating eggs (Shirley, 1995b).

Contributions of the model to infectious disease prophylaxis and therapy

All commercially available anticoccidials were discovered in cage studies of this kind and a similar — though some-

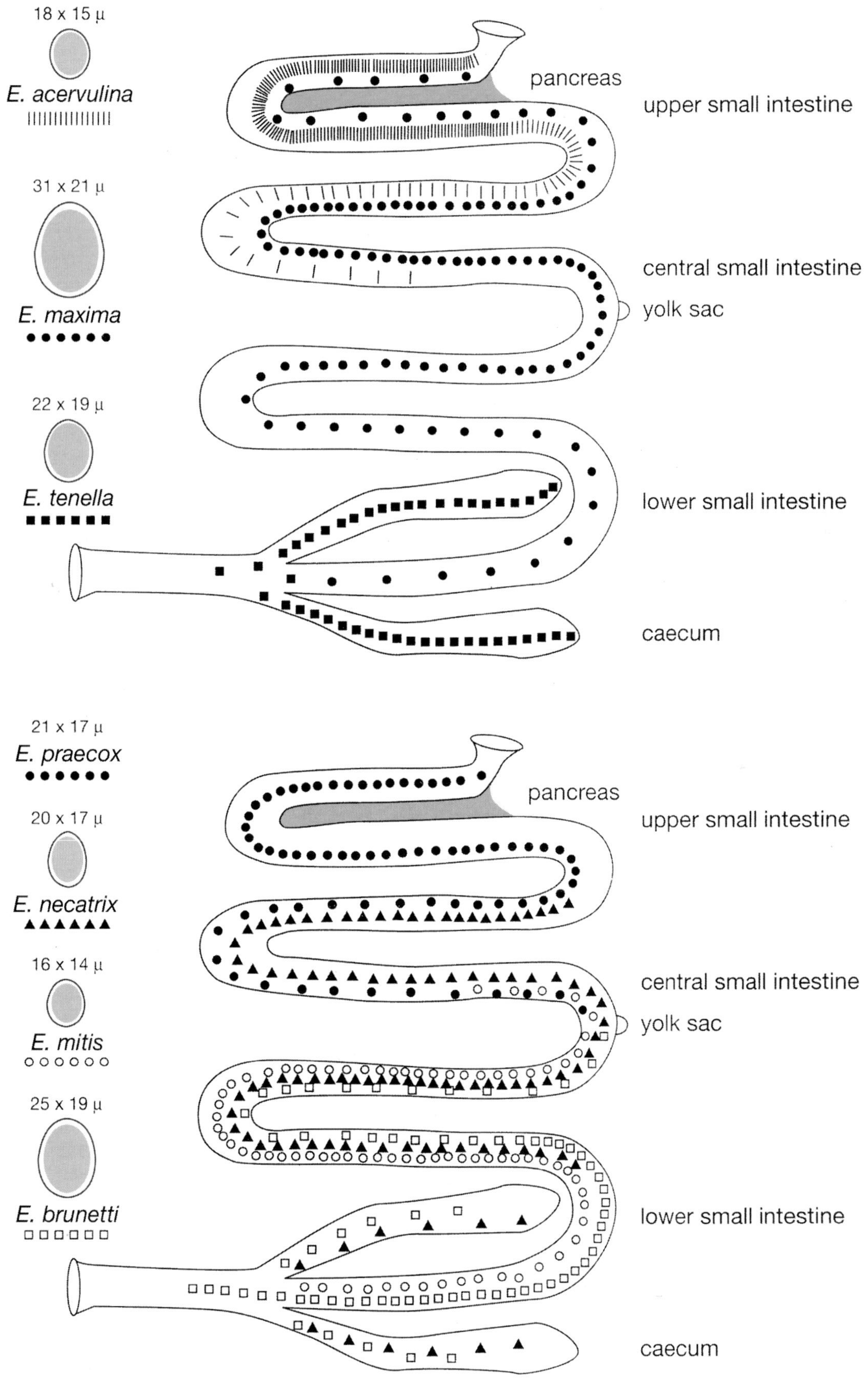

Figure 99.2 Comparative localization of chicken coccidia in the intestine.

times less sophisticated — experimental design (Ryley and Betts, 1973; Ryley, 1980). For more detailed investigations, floor-housing studies are then needed as well.

Elucidation of the pathophysiology of coccidiosis

Cage studies of this kind with *Coccidia*-infected chicks also allow further investigations of pathophysiology, development of immunity, and development of resistance. The study conditions described here can be adapted for these without any problem (Hammond and Long, 1973; Rose, 1995; Stephan *et al.*, 1997).

Mammalian coccidiosis

Very up-to-date descriptions of target-animal models are also available for the principal mammalian coccidioses of our domestic animals, e.g., for rabbits (Coudert *et al.*, 1995), lambs (Taylor *et al.*, 1995), calves (Bürger *et al.*, 1995), piglets (Hendriksen, 1995), and dogs and cats (Rommel *et al.*, 1986), though these are not screening models but experimental designs for more detailed clinical investigations in the target animal species.

Background of the model

Given the importance of mammalian coccidiosis and the sometimes considerable variation in sensitivity to drugs, we developed a mouse model which we have been using successfully since 1963 (Haberkorn, 1986). We use *E. falciformis* as the parasite. *E. falciformis* develops in the caecum and especially in the upper colon of the laboratory mouse (*Mus musculus*) (Haberkorn, 1970). Other *Eimeria* species of the mouse, such as *E. vermiformis* and *E. f. pragensis*, are also reasonably suitable (Haberkorn, 1993). Occasionally investigations are successfully carried out with rat coccidia (e.g., *E. nieschulzii*, *E. separata*, *E. contorta*; Haberkorn and Schulz, 1981). For chemotherapeutic studies, e.g., for the identification of new active ingredients, these models do not in our experience offer any advantage over the mouse model, particularly as the higher body weight means that more active ingredient is required (Haberkorn, unpublished work).

Animal species

SPF mice of one sex, weighing 15–18 g, e.g., Bor: CFW are used. Some mouse strains (e.g., C57BL/10ScCr Bom) show very high mortality when exposed to coccidial infection (Haberkorn, 1983).

Preparation of the animals

For an initial screening programme a group of 2–3 animals per dose is sufficient. As in every other study, the mice should be delivered at least 1–2 days earlier, so that the course of the infection is not affected by stress due to transport and adaptation to a new environment. Standard commercial pelletized feed and drinking water should be provided *ad libitum*. Other than this, the conditions for the housing of laboratory mice apply (European Convention for the Protection of Vertebrate Animals used for Experimental and other Scientific Purposes, 1986; see Ausems, 1986). The animals are individually marked (see above).

Details of surgery

No surgical interventions are necessary, either before or during the study.

Infection procedure

Oral infection via oesophageal tube is best carried out on Tuesday morning, with medication with the feed on Monday.

Storage and preparation of inocula

Suspensions of *E. falciformis* oocysts can be kept for 12 months without any problems in a refrigerator at 4–6°C. To obtain the infection material SPF mice (e.g., CFW) weighing 22–30 g are each infected with approximately 5000–10 000 sporulated oocysts and sacrificed on day 9 after infection. The caecum and colon are removed and opened, and the contents are scraped out with a spatula and suspended in 2% potassium dichromate. The suspension, in Petri dishes or beakers, is left to sporulate for 2 days, e.g., on a magnetic stirrer hotplate. The oocysts can of course also be obtained from the faeces (day 8–10 after infection). For cleaning of the oocysts, see the relevant section above.

Infection process

The infection dose is established in a pilot study. The rate of mortality from the infection should be around 50%. With our *E. falciformis* strain the dose of inoculum used in mice has for many years been 18 000 sporulated oocysts per 15–18 g body weight. The suspension used for the infection must be adjusted so that no more than 0.5 ml per mouse has to be administered.

Table 99.5 *Eimeria falciformis* mouse scoring of diarrhoea, lesions and oocysts

		0.1–0.5	*1*	*2*	*3*	*4*	*5*	*6*
Lesions	Localization		Upper third of colon	Upper third of colon	Caecum and upper third of colon	Caecum and more than ½ colon	Caecum and more than ½ colon	Caecum and more than ½ colon
	Intestinal wall		A little swollen	A little swollen	Swollen	Markedly swollen	Swollen,thickened wall	Swollen, thickened wall
	Content		Normal	Softened	Soft	Soft	Slime	Slime, bloody
Diarrhoea	Faeces		Some soft pellets	⅓ of pellets soft	⅔ of pellets soft	All pellets soft	A few soft pellets still detectable	No pellets were detectable
	Anus		Clean	Diarrhoea just detectable	A little dirty	Smeared	Very dirty	Extremely dirty, including some blood
Oocysts >20/MF	Oocysts per microscopic fields (100× magnification)	1–5/25 MF		6–24/25 MF	1–2/MF	3–5/MF	6–10/MF	11–20/MF

Key parameters for monitoring infection and response to treatment

The main parameters are mortality due to the infection, growth, diarrhoea, and macroscopic and microscopic autopsy findings. The evaluation is carried out as in the chick studies.

Parameters

Mortality

Deaths occurring before day 6 after infection are interpreted as toxicity-related or intercurrent. Animals dying on days 6–8 after infection are autopsied.

Diarrhoea

Diarrhoea occurs from day 7 after infection. It is readily discernible around the anus. Since in any event the mice have to be picked up by hand for the oral treatment, the individual inspection can be made without any additional work. The evaluation is carried out as shown in Table 99.5.

Autopsy findings (lesion scoring)

On day 9 after infection the animals are sacrificed, the caecum and colon are removed, and the lesions assessed in accordance with Table 99.5.

Oocyst excretion

The contents of the upper colon are scraped out with a spatula and suspended in saturated sodium chloride in a tube (65×25 mm) or small dish (separately or pooled groupwise). The tube or dish is filled to the rim with saturated NaCl. After approximately 30 minutes a drop is lifted from the surface with a cover glass of 18×18 mm, transferred to a microscope slide, and examined for oocysts under a microscope (magnification ×100 in a stopped-down bright field or with interference phase contrast). For semiquantitative evaluation we use the system given in Table 99.5. In more detailed studies the pooled faeces from days 6–9 after infection are examined and the number of oocysts is ascertained in a counting chamber. The collected faeces are obtained by keeping the mice in a wire cage suspended in a suitable plastic cage. It is a good idea to put moist absorbent paper in the plastic cage to prevent the droppings from drying out.

Growth

Weighing of the animals, which is carried out anyway for the purposes of establishing the exact dose, takes place on days 1, 3, 6, and 8 and (at the end of the study) on day 9 after infection. The difference between the initial and final weights compared with the corresponding difference for non-infected and untreated controls is used in the evaluation (Table 99.5). If the initial weights are the same, one can also compare the final weights.

Calculation of findings

This is carried out in the same way as the evaluation of coccidiosis studies in chicks (Table 99.6).

Anticoccidial medication

The test substances are dissolved or suspended in water so that the mice receive 1.0 ml per 20 g body weight. The treatment is generally given orally by oesophageal tube at 24-hour intervals on days 1, 2, 3, 6, 7 and 8 after infection. On

Table 99.6 Rating system for chemotherapeuticals with mouse coccidiosis

Score	*Mortality*	*Lesion scores*	*Oocyst scores*	*% Opg**	*Weight % of non-infected controls*
0	0	0	0	<1	>90
1.0		1–3		1–5	75–89
1.5				6–10	60–74
2.0	1–3/3	4	0.1–3	11–20	<60
2.5		5–6		21	
3.0			4	22–50	
4.5			5–6	>50	

* % of infected control if oocysts per gram (Opg) were determined.
This system was also used for rating experiments with *Eimeria vermiformis* and *E. pragensis* (mouse) as well as for rat coccidiosis (*E. nieschulzi*, *E. contorta*, *E. separata*).

Calculation

The quotient: $\dfrac{\text{total score for test groups}}{\text{total score for infected control group}}$

expressed as a decimal is the rating:
0.00–0.29 = effect (100–71% effect)
0.30–0.79 = slight effect (70–21% effect)
0.80–1.00 = ineffective (20–0% effect)

Table 99.7 Efficacy of anticoccidials in the chicken and in the mouse model

Drug	Test model	Effective	Unit	Maximum tolerated dose (no mortality)	Dose range: Fully effective	Dose range: Slightly effective	Highest not effective dose
Aklomide	1	yes	ppm	>250		250–150*	100
	2	yes	mg/kg	500	500	250–125	100
	3	yes	ppm	>1000	1000–500	250	125
Amprolium	1	yes	ppm	>1000	1000–100†	50†	25
	2	yes	mg/kg	250	250–125	100–50	25
	3	yes	ppm	500		500–250	125
Amprolium	1	yes	ppm	>1000	1000–250	100–50	25
+	2	yes	mg/kg	250		250–125	10
Ethopabate	3	yes	ppm	250	1000	500–250	125
Aprinocid	1	yes	ppm	250	500–100	50–25	10
	2	yes	mg/kg	25	10	5	2.5
	3	yes	ppm	250	250–100	75–50	25
Buquinolate	1	yes	ppm	>1000	1000–250	100	50
	2	no	mg/kg	>1000			1000
	3	yes	ppm	250	250	250–100	50
Clopidol	1	yes	ppm	>250	250	100	50
	2	yes	mg/kg	500	500–250	100	50
	3	yes	ppm	250	250	100	50
Clopidol	1	yes	ppm	>250	250–125	100–50	25
+	2	yes	mg/kg	100	100–50	25–10	5
Methylbenzoquate	3	yes	ppm	500	500–50		25
Decoquinate	1	yes	ppm	>500	500–25	10	5
	2	no	mg/kg	>1000			1000
	3	yes	ppm	>1000	1000–25		10
Diclazuril	1	yes	ppm	>250	250–0.5	0.25–0.05	0.025
	2	yes	mg/kg	>100	100–5	2.5–1	0.5
Glycamide	1	yes	ppm	>1000	1000–100 (50†)	50–25†	10
	2	no	mg/kg	>1000			1000
	3	yes	ppm	250	250–125	100–75	50
Halofuginone	1	yes	ppm	10	10–3	2.5–2.0	1
	2	yes	mg/kg	1	2.5–1.5	1–0.5	0.25
	3	yes	ppm	15	15	10–1.5	1
Lasalocid	1	yes	ppm	>150	150–100	75	50
	2	yes	mg/kg	100		100–30	25
	3	yes	ppm	500	500	250–100	50
Maduramicin	1	yes	ppm	10	10–5	3–2	1.0
	2	yes	mg/kg	10	10–3	2.5–1.5	1.0
	3	yes	ppm	15	15–5	3–2.5	1.0
Monensin	1	yes	ppm	>500	500–250 (125*)	125–75	50
	2	yes	mg/kg	500	500	250	100
	3	yes	ppm	250	500–125	100–75	50
Narasin	1	yes	ppm	>250	250–(125*)		100
	3	yes	ppm	250	250–75	50–25	10
Nicarbazin	1	yes	ppm	>500	500–250	100	50
	2	yes	mg/kg	500		250–50	25
	3	yes	ppm	50		25	10

Table 99.7 — *(Continued)*

Drug	*Test model*	*Effective*	*Unit*	*Maximum tolerated dose (no mortality)*	*Dose range*		*Highest not effective dose*
					Fully effective	*Slightly effective*	
Nitrofurazone	1	yes	ppm	>500			500
	2	yes	mg/kg	100		100–50	25
	3	yes	ppm	500	500	250–100	50
Pyrimethamine	1	yes	ppm	>1000		1000–250	100
	2	yes	mg/kg	250		250–10	5
	3	yes	ppm	250		250–100	50
Robenidine	1	yes	ppm	500	500–25	10	5
	2	yes	mg/kg	250	250–50	25–10	5
	3	yes	ppm	100	100–50	25–15	10
Salinomycin	1	yes	ppm	>150	150–100 (60*)		30
	2	no	mg/kg	50			50
	3	yes	ppm	500	500–125	100–50	30
Sulfachlorpyrazine	1	yes	ppm	>1000		1000–250	100
	2	yes	mg/kg	500	500		250
Sulfaquinoxaline	1	yes	ppm	500	500–250	100	50
	2	yes	mg/kg	250		1000–500	250
	3	yes	ppm	500		250	100
Toltrazuril	1	yes	ppm	1000	15	10–7.5	5
	2	yes	mg/kg	250	1000–5	2.5–1	0.5
	3	yes	ppm	250	1000–5	2.5–1	0.5
Zoalen	1	yes	ppm	500		250–150	100
	2	yes	mg/kg	100	100	50–25	10
	3	yes	ppm	500	250–125	100–50	25

1, Chicken model (*E. acervulina* + *E. maxima* + *E. tenella*) treatment via feed from day –3 to day 8 post-infection.
2, Mouse model (*E. falciformis*): oral treatment on day 1, 2, 3, 6, 7, 8 after infection; maximum dosage 1000 mg/kg body weight per day.
3, Mouse model (*E. falciformis*): treatment via feed from 3 days before until 9 days after infection; maximum dosage 1000 ppm.
* Not *Eimeria tenella*; † *Eimeria tenella* only.

days 1, 3, 6 and 8 after infection the animals are weighed immediately before treatment so that the dose can be brought into line with the current body weight through adjustment of the administration volume (0.05 ml more or less for each 1 g above or below 20 g).

If the test substances are administered continuously with the feed, they are mixed in with ground mouse feed. In this case it is advisable to start the treatment 2–3 days before the infection.

Pitfalls (advantages and disadvantages) of the model

The model can be standardized to a very high degree: there is a much better guarantee of animal material of uniform quality, all year round, than is the case with chicks. A further major advantage of the model is that the amount of active ingredient required is much smaller (about one-third); this also means that work can begin with the highest doses possible (e.g., 1000 mg/kg; Table 99.7). The animals are also easier to keep, both in terms of space and room temperature requirements.

With oral treatment according to the described system, only 4 out of 22 known products currently or formerly used to combat coccidiosis in chicks did not show any effect against *E. falciformis* at the maximum tolerated dose: glycamide (1000 mg/kg), salinomycin (50 mg/kg), and buquinolate and decoquinate (1000 mg/kg in each case; Haberkorn, 1986). This is not surprising in the case of the last two quinolone products, given their purely static effect against sporozoites. When they were administered in the feed from 3 days before the infection, they proved effective, like all polyethers and like glycamide. The risk that an active ingredient of potential relevance to coccidial infections in poultry may be missed in a preliminary test with *E. falciformis* is thus small, and practically non-existent with medication via the feed. However, since there are often considerable differences in efficacy against coccidia of poultry and mammalian coccidia such as *E. falciformis*, products which show even the slightest sign of efficacy in the mouse test should be tested in chicks. In our experience a high degree of efficacy against *E. falciformis* is a very reliable indication of a broad spectrum of action, and in particular of high efficacy against other mammalian coccidia (Haberkorn and Schulz, 1981).

Contributions of the model to infectious disease prophylaxis and therapy

In 1971 we were the first to discover the anticoccidial symmetrical triazinetriones in the mouse. This work ultimately led to the discovery and development of toltrazuril, an anticoccidial with an exceptionally broad spectrum of action (Aichinger *et al.*, 1978; Haberkorn and Stoltefuss, 1987).

Elucidation of the pathophysiology of coccidiosis

The model described is also highly suitable for the investigation of other questions relating to the biology and immunology of coccidia (Rose, 1995).

References

Aichinger, G., Haberkorn, A., Kölling, H., Kranz, E., Reisdorf, J., Stoltefuß, J. (1978). Anticoccidial activity of Bay G 7183. *4th ICOPA 19–26 Aug. 1978*. Short communications section D p. 96. Warsaw, Poland.

Ausems E. J. (1986). The European Convention for the protection of vertebrate animals used for experimental and other scientific purposes. *Z. Versuchstierk.*, **28,** 219.

Bürger, H.-J., Fiege, N., Gahr, A., Heise, A., Roloff, H. (1995). *Eimeria* species of cattle. In *COST 89/820 Biotechnology, Guidelines on Technics in Coccidiosis Research* (eds Eckert J., Braun, R., Shirley, M. W. *et al.*), pp. 40–51. Office for Official Publications of the European Communities, Luxembourg.

Cameron, R. S. (1985). Procedures for drug approval by the EEC. In *Research in Avian Coccidiosis. Proceedings of the Georgia Coccidiosis Conference*, (eds McDougald, L. R., Joyner, L. P., Long, P. L.) pp. 385–393. University of Georgia Dept. Poultry Science, Athens.

Clarke, M. L. (1962). A mixture of diaveridine and sulfaquinoxaline as a coccidiostat for poultry. *Vet. Rec.*, **74,** 845–847.

Coudert, P., Licois, D., Drouet-Viard, F. (1995). *Eimeria* species and strains of rabbits. In *COST 89/820 Biotechnology, Guidelines on Technics in Coccidiosis Research* (eds Eckert, J., Braun, R., Shirley, M. W. *et al.*) pp. 52–73. Office for Official Publications of the European Communities, Luxembourg.

Davies, S. F. M., Joyner, L. P., Kendall, S. B. (1963). *Coccidosis*. Oliver and Boyd, Edinburgh.

Davis, L. R. (1973). Techniques. In *The Coccidia Eimeria, Isospora, Toxoplasma and Related Genera* (eds Hammond, D. M., Long, P. L.), pp. 411–458. University Park Press, Butterworths, London.

Eckert, J., Braun, R., Shirley M. W., Coudert, P. (eds) (1995a). *COST 89/820 Biotechnology, Guidelines on Technics in Coccidiosis Research*. Office for Official Publications of the European Communities, Luxembourg.

Eckert, J., Taylor, M., Catchpole, J., Licois, D., Coudert, P., Bucklar, H. (1995b). Morphological characteristics of oocysts. In *COST 89/820 Biotechnology, Guidelines on Technics in Coccidiosis Research* (eds Eckert, J., Braun, R., Shirley, M. W. *et al.*), pp. 103–119. Office for Official Publications of the European Communities, Luxembourg.

Haberkorn, A. (1970). Die Entwicklung von *Eimeria falciformis* (EIMER 1870) in der weissen Maus. *Z. Parasitenkd.*, **34,** 49–67.

Haberkorn, A. (1983). Infektionsverlauf einer *Eimeria*—Mischinfektion in verschiedenen Mäusestämmen. *Mitt. Oesterr. Ges. Tropenmed. Parasitol.*, **5,** 39–44.

Haberkorn, A. (1986). Use of a mouse coccidiosis model for predicting anticoccidial efficacy in poultry. In *Research in Avian Coccidiosis* (eds McDougald, L. R., Long, P. L., Joyner, L. P.) pp. 263–270. Georgia Coccidiosis Conference, Nov. 19–21, 1985, Athens, GA, USA. Department of Poultry Science, University of Georgia, AL.

Haberkorn, A. (1993). Differences in sensitivity of the mouse-coccidia *Eimeria f. falciformis, E. f. pragensis* and *E. vermiformis* to toltrazuril and its sulfon. IX. Intern. Congr. Protozoology, July 24–31. 1993, Berlin Abstracts p. 49 (ed. Moltmann, U). German Society of Parasitology, Stuttgart, Germany.

Haberkorn, A. (1996). Chemotherapy of human and animal coccidioses: state and perspectives. *Parasitol. Res.*, **82,** 193–199.

Haberkorn, A., Mundt, H.-C. (1989). Progress in the control of mammalian coccidiosis. In *Coccidia and Intestinal Coccidiomorphs*, (ed. Yvore, P). pp. 435–440. Les Colloques de l' INRA, INRA, Paris.

Haberkorn, A., Schulz, H. P. (1981). Experimental chemotherapy of mammalian coccidiosis with Bay G 7183. *Zbl. Bakt. Hyg. I. Abt. Orig.*, **A250,** 260–267.

Haberkorn, A., Stoltefuss, J. (1987). Studies on the activity spectrum of toltrazuril, a new anti-coccidial agent. *Vet. Med. Rev.*, **1/87,** 22–32.

Hammond, D. M., Long P. L. (eds) (1973). *The Coccidia Eimeria, Isospora, Toxoplasma and Related Genera*. University Park Press, Butterworths, London.

Hendriksen, S. A. (1995). *Isospora suis* of swine. In *COST 89/820 Biotechnology, Guidelines on Technics in Coccidiosis Research* (eds Eckert, J., Braun, R., Shirley, M. W. *et al.*) pp. 74–78. Office for Official Publications of the European Communities, Luxembourg.

Hilbrich, P. (1967). *Krankheiten des Geflügels*. Verlag Hermann Kuhn KG, Schwenningen Neckar.

Janssen Pharmaceutica (1990). *Diagnosis of Coccidiosis in Chickens*. Janssen Pharmaceutica Animal Health Department B-2340 Beerse, Belgium.

Johnson, J., Reid, W. M. (1970). Anticoccidial drugs: lesion scoring techniques in battery and floor-pen experiments with chickens. *Exp. Parasitol.*, **28,** 30–36.

Jones, W. R. (1946). The experimental infection of rats with *Entamoeba histolytica* with a method for evaluation the anti-amebic properties of new compounds. *Ann. Trop. Med. Parasitol.*, **40,** 130–140.

Levine, N. D., Ivens, V. (1970). *The Coccidian Parasites (Protozoa, Sporozoa) of Ruminants*. Illinois biological monographs no. 44. University of Illinois Press, Illinois.

Lindsay, D. S. (1990). *Isospora*: infections of intestine: biology. In *Coccidiosis of Man and Domestic Animals* (ed Long, P. L.), pp. 77–89. CRC Press, Boca Raton.

Long, P. L. (ed) (1982). *The Biology of the Coccidia*. University Park Press, Baltimore.

Long, P. L. (ed) (1990). *Coccidiosis of Man and Domestic Animals*. CRC Press, Boca Raton.

Long, P. L., Reid, W. M. (1982). *A Guide for the Diagnosis of Coccidiosis in Chickens* (ed. Sparer, D). Research Report 404. University of Georgia College of Agriculture, Athens.

Lynch, J. E. (1961). A new method for the primary evaluation of anticoccial activity. *Am. J. Vet. Res.*, **22**, 324–326.

Maes, L., Coussement, W., Vanparijs, O., Verheyen, F. (1989). Species specificity action of diclazuril (Clinacox®) against different *Eimeria* species in the chicken. Coccidia and intestinal coccidiomorphs (ed Yuore, P.) Vth Intern. Cocc. Conf. Tours (France) 17–20. Octob. 1989. INRA, Paris.

McDougald, L. R. (1982). Chemotherapy of coccidiosis. In *The Biology of the Coccidia* (ed Long, P. L.), pp. 373–427. University Park Press, Baltimore.

McDougald, L. R. (1990). Control of coccidiosis: chemotherapy. In *Coccidiosis of Man and Domestic Animals* (ed Long, P. L.), pp. 307–320. CRC Press, Boca Raton.

Mehlhorn H (ed.) (1988). *Parasitology in Focus*, pp. 1–50. Springer, Berlin.

Pellerdy, L. P. (1974). *Coccidia and Coccidiosis*. Paul Parey, Berlin.

Raether, W. (1980). Salinomycin, a polyether antibiotic with marked activity against experimental coccidiosis in chickens and turkeys. 6th European Poultry Conference, Hamburg, 8–12 September 1980, Vol. II, pp. 507–514. World's Poultry Science Association, Braunschweig, Germany.

Raether, W., Hofmann, J., Uphoff, M. (1995). *In vitro* cultivation of avian *Eimeria* species: *Eimeria tenella*. In *COST 89/820 Biotechnology, Guidelines on Technics in Coccidiosis Research* (eds Eckert, J., Braun, R., Shirley, M. W. *et al.*) pp. 79–92. Office for Official Publications of the European Communities, Luxembourg.

Raines, T. V. (1978). Guidelines for the evaluation of anticoccidial drugs. In: *Avian Coccidiosis* (eds Long P. L., Boorman, K. N., Freeman, B. M.) pp. 339–346. British Poultry Science, Edinburgh.

Ricketts, A. P. (1992). *Eimeria tenella:* growth and drug sensitivity in tissue culture under reduced oxygen. *Exp. Parasitol.*, **74**, 463–469.

Rommel, M., Heydorn, A. O. (1972). Beiträge zum Lebenszyklus der Sarcosporidien III. *Isospora hominis* (Raillet and Lucet, 1891; Wenyon, 1923) eine Dauerform der Sarkosporidien des Rindes und des Schweins. *Berl. Müchn. Tierärztl. Wchschr.*, **85**, 143–145.

Rommel, M., Schnieder, T., Westerhoff, J., Krause, H. D., Stoye, M. (1986). The use of toltrazuril-medicated food to prevent the development of *Isospora* and *Toxoplasma* oocysts in dogs and cats. *Symp. Biol Hung.*, **33**, 445–449.

Rose, M. E. (1995). Immunity to *Eimeria* infections: characterisation of cells involved in protection. Proceedings of 1995 Annual Workshop COST 820, Vaccines against animal coccidioses (ed. Bedrnik, P.) Biopharm Research Institute of Biopharmacy and Veterinary Drugs, Jilove, near Prague, Czech Republic.

Ryley, J. F. (1980). Screening for and evaluation of anticoccidial activity. *Adv. Pharmacol. Chemother.*, **17**, 1–23.

Ryley, J. F., Betts, M. J. (1973). Chemotherapy of chicken coccidiosis. *Adv. Pharmacol. Chemother.*, **11**, 221–293.

Ryley, J. F., Wilson, R. G. (1972). Comparative studies with anticoccidials and three species of chicken coccidia *in vivo* and *in vitro*. *J. Parasitol.*, **58**, 664–668.

Ryley, J. F., Wilson, R. G. (1976). Drug screening in cell culture for the detection of anticoccidial activity. *Parasitology*, **73**, 137–148.

Shirley, M. W. (1995a). *Eimeria* species and strains of chickens. In *COST 89/820 Biotechnology, Guidelines on Technics in Coccidiosis Research* (eds Eckert, J., Braun, R., Shirley, M. W. *et al.*) pp. 1–24. Office for Official Publications of the European Communities, Luxembourg.

Shirley, M. W. (1995b). Cultivation of avian *Eimeria* species in embryonating eggs. In *COST 89/820 Biotechnology, Guidelines on Technics in Coccidiosis Research* (eds Eckert, J., Braun, R., Shirley, M. W. *et al.*) pp. 93–94. Office for Official Publications of the European Communities, Luxembourg.

Stephan, B., Rommel, M., Daugschies, A., Haberkorn, A. (1997). Studies of resistance to anticoccidials in *Eimeria* field isolates and pure *Eimeria* strains. *Vet. Parasitol.*, **69**, 19–29.

Taylor, M., Catchpole, J., Marshall, R., Norton, C. C., Green, J. (1995). *Eimeria* species of sheep. In *COST 89/820 Biotechnology, Guidelines on Technics in Coccidiosis Research* (eds Eckert, J., Braun, R., Shirley, M. W. *et al.*) pp. 25–39. Office for Official Publications of the European Communities, Luxembourg.

Thompson, P. E., Dunn, M. C., Bayeles, A., Reinertson, W. (1950). Action of chloramphenicol (chloromycetin) and other drugs against *Entamoeba histolytica in vitro* and in experimental animals. *Am. J. Trop. Med.*, **30**, 203–215.

Tyzzer, E. E. (1929). Coccidiosis in gallinaceous birds. *Am. J. Hyg.*, **10**, 269–383.

Chapter 100

Animal Models of *Trichomonas vaginalis* Infection with Special Emphasis on the Intravaginal Mouse Model

S. F. Hayward-McClelland, K. L. Delgaty and G. E. Garber

Background of human infection

Trichomoniasis is a sexually transmitted disease (STD) of worldwide importance, which is caused by the parasitic protozoan *Trichomonas vaginalis*. Trichomoniasis was once considered a "nuisance disease" of women, but has now been accepted to have important medical, social, and economic implications. It is the most common non-viral STD, with more than 170 million cases occurring annually throughout the world (World Health Organization, 1995). In North America alone, more than 8 million new cases are reported each year (World Health Organization, 1995). These estimates may be low, since up to 50% of infections are asymptomatic (Fouts and Kraus, 1980). This is of particular public health importance, since asymptomatic individuals are carriers who may unknowingly transmit the disease to others (Nicoletti, 1961). *T. vaginalis* is associated with many perinatal complications, genitourinary disorders, and the transmission of other STDs such as human immunodeficiency virus (HIV; Cameron and Padian, 1990; Laga *et al.*, 1993, 1994).

T. vaginalis is an obligate parasite which lacks a cystic stage. Although the trichomonad can survive outside the host for several hours, it is transmitted almost exclusively through coitus (Heine and McGregor, 1993). Trichomoniasis has a cosmopolitan distribution and has been found in all continents, racial groups and socioeconomic strata. Nonetheless, it is associated with exposure to multiple sexual partners, low education and income, and it appears more prevalent in certain populations, although it is not known whether this is related to biologic or socioeconomic status (Heine and McGregor, 1993). The true prevalence of *T. vaginalis* infection is unknown, since it is not a reportable disease. *T. vaginalis* strains are quite variable with respect to virulence, antigenic properties, and geographic distribution (Lossick, 1990), but only eight serotypes have been identified (Ackers, 1990).

This protozoan primarily infects the squamous epithelium of the genital tract. The incubation period is usually 3–28 days in women and infection itself may persist for an extended amount of time. It is chiefly a disease of the reproductive years, and is rarely observed before the menarche or after menopause. Clinical presentation in women (for recent reviews see Wolner-Hanssen *et al.*, 1989; Rein, 1990; Heine and McGregor, 1993) ranges from asymptomatic to acute or chronic, although one-third of asymptomatic women develop clinical disease within 6 months. Acute trichomoniasis can include the following: severe vaginitis, pruritus, dysuria, dyspareunia, and diffuse vulvitis associated with copious leukorrhea. The vaginal discharge is often malodorous, frothy, mucopurulent, and yellowish-green in colour. The classic "strawberry cervix", which is characterized by punctate hemorrhages, is observed in only 2% of cases by the naked eye (Fouts and Kraus, 1980), but in 90% of cases by colposcopy (Wolner-Hanssen *et al.*, 1989). Symptoms and parasitic load are cyclical and worsen during menstruation; the pH of the vagina is usually elevated from the normal 4.5 to greater than 5.0. Chronic infection is mild and presents with pruritus and dyspareunia, with little or scant vaginal secretion mixed with mucus. Complications of *T. vaginalis* infection include premature rupture of the placental membranes, premature labour, low-birth-weight infants, adnexitis, pyosalpinx, endometritis, atypical pelvic inflammatory disease (Heine and McGregor, 1993), cervical cancer (Wolner-Hanssen *et al.*, 1989), cervical erosion (Rein and Chapel, 1975), and infertility (Rein, 1990).

T. vaginalis infection in males is poorly understood and there exists many conflicting reports with regard to its clinical picture. Male trichomoniasis is generally thought to be asymptomatic, although non-gonococcal urethritis and prostatitis are not uncommon (Holmes *et al.*, 1975; Mardh and Colleen, 1975; Krieger, 1984; Krieger *et al.*, 1993). It has also been associated with a number of other inflammatory conditions of the male urogenital tract (Krieger *et al.*, 1993). It is thought that in men the infection lasts about 10 days, although the prevalence of spontaneous resolution and the consequence of chronic infection remain undefined (Krieger, 1990).

Diagnosis is often difficult because the symptoms of trichomoniasis mimick those of candidiasis and bacterial vaginosis (Rein and Holmes, 1983). Diagnosis is therefore made on the basis of both clinical picture and the identification of trichomonads in the urogenital tract. Detection methods include the gold-standard broth culture, various microscopic staining techniques and, more recently, antibody,

Handbook of Animal Models of Infection
ISBN 0-12-775390-7

antigen, and DNA detection (Lossick and Kent, 1991; Bhatt *et al.*, 1996; Petrin *et al.*, 1998). Once diagnosed, both the patient and sexual partners are treated orally with metronidazole (single 2 g dose, or 250 mg three times a day for 7 days) to prevent recurrence (Lossick, 1982). Metronidazole is the only licensed treatment in North America for trichomoniasis, but other nitroimidazoles are available around the world.

Antibiotic resistance and refractory cases are becoming increasingly problematic, however. New drug and vaccine development may not only provide better treatment, but may offer prophylaxis and protective immunity. Unfortunately, this research has been limited by the lack of understanding of the pathogenesis of the disease, and the role of the host immune response to *T. vaginalis* infection (for a recent review of these factors, see Petrin *et al.*, 1998). A good animal model is required for standardized controlled research into the pathogenesis, host immune response, and treatment of *T. vaginalis* infection.

Background of animal models

Many animal models for *T. vaginalis* infection have been proposed (Jirovec and Petru, 1968; Honigberg, 1978; Kulda, 1990). The intraperitoneal models of Teras and Roigas (1966) and Cavier *et al.* (1972) and the subcutaneous model of Honigberg *et al.* (1966) have been used extensively for pathogenesis and drug studies. Intraperitoneal injection of trichomonads produces visceral lesions which can result in death, while subcutaneous injection of *T. vaginalis* results in the development of a localized abscess. The extent and severity of these indices can be used to estimate the level of virulence, but their correlation with clinical disease is poor (Garber and Lemchuk-Favel, 1990). Vaginal infections are preferable to those of ectopic nature since they more closely parallel the human disease. However, many of the intravaginal infections in laboratory animals have not met the expectations of a suitable model.

Vaginal infection of the squirrel monkey produces symptomatic disease of up to 3 months, and horizontal transmission between animals has been achieved (Street *et al.*, 1983; Gardner *et al.*, 1987). Immune response is weak, however, and the presence of indigenous trichomonads which require eradication is problematic (Hollander and Gonder, 1985). Recent studies have been too small to assess adequately its appropriateness as a model, and more research is required. Unfortunately, associated cost and upkeep are prohibitive.

Intravaginal infection of the guinea-pig is relatively simple to accomplish, requiring no supportive estrogen treatment, and results in a symptomatic infection similar to that of women. From this model, a series of reports focusing on the pathology of trichomoniasis was published (Soska *et al.*, 1962; Kazanowska *et al.*, 1983; Honigberg, 1978). The model never really progressed beyond the original research, although guinea-pigs intravaginally infected with *Tritrichomonas foetus* (which is the bovine urogenital pathogen) have been used in some drug studies (Maestrone and Semar, 1967; Michaels, 1968). This probably reflects the decline in interest for finding novel human treatments, and a surge of research by the veterinary community into bovine trichomoniasis.

The intravaginal infection of the hamster (also achieved without supportive hormonal therapy) persists up to several months but is asymptomatic. It was used mainly for drug assays for *T. foetus* infection (Ryley and Stacey, 1963), and for a few drug assays for *T. vaginalis* infection (Michaels, 1968; Wildfeuer, 1974; Kulda, 1990). Unfortunately, contamination of the vagina with indigenous trichomonads of the gut render the model and results obtained from it potentially unreliable (Kulda, 1990). Intravaginal infection of the rat is only accomplished after surgical removal of the ovaries, followed by estrogenization (Cavier and Mossion, 1956; Meingassner *et al.*, 1981). Infection lasts approximately 2–10 weeks, and is generally asymptomatic. The rat model is somewhat applicable to drug studies (Michaels, 1968; Meingassner *et al.*, 1981; Kulda, 1990), but researchers have sought a more convenient model. With the resurgence of interest in finding novel antitrichomonal therapies, the mouse has gained favor as an appropriate animal for use in models of *T. vaginalis* infection. This is due, in part, to their availability and simple upkeep. Furthermore, surgical preparation of the animal is not required, and mice seem to be free of indigenous trichomonads. Clearly, the animal models which achieve a consistent, long-lasting, symptomatic genital infection would be ideal.

The intravaginal mouse model

Mouse intraperitoneal and subcutaneous *T. vaginalis* infections have provided a wealth of useful information on drug pharmacology and parasite pathogenesis. However, the intravaginal infection model has the most relevance to human vaginal infections, and it is this model which will be discussed most extensively in this chapter.

The intravaginal route of inoculation in the mouse has met with limited success due to technical difficulties encountered by early investigators. Research has been impeded by the fact that the animal models employed did not closely replicate the environment found in the human vagina, where the epithelium is exposed to estrogen throughout most of the menstrual cycle (Larsen, 1993).

It has been established that susceptibility of laboratory animals to intravaginal infection is related to their hormonal status and varies with changing phases of the sexual cycle (Kulda, 1990). The estrous cycle of mice (and other laboratory animals) includes four separate phases — they are proestrus, estrus, metestrus and diestrus. Optimal conditions for intravaginal growth of *T. vaginalis* are found during proestrus and estrus, when an increased level of estrogen is present, and neutrophils and bacteria are at a minimum (Corbeil *et al.*, 1985). In fact, Jirovec and Petru (1968) suggested that *T. vaginalis* can grow only in an environment that includes estrogen. To keep the animals in protracted estrus, researchers have employed a supportive treatment with estrogen (thus preventing the animals from cycling into metestrus and diestrus; Kulda, 1990).

The first attempt at murine intravaginal inoculation was made by Patten *et al.* (1963) when they repeatedly introduced *T. vaginalis* into the mouse vagina over a number of weeks. They examined pathology in response to continuous introduction of the organism, but they did not verify the duration of infection or presence of trichomonads between inoculations. Subsequent researchers, however, were more interested in the establishment of an infection after a primary introduction of the parasite. Cappuccinelli's group (1974) was the first to establish successfully an intravaginal *T. vaginalis* infection in the mouse with an initial inoculation of the organism, and exogenously administered estrogen. They suggested that estrogen is necessary for induction of infection, but not for maintenance. A number of subsequent studies have supported the need for estrogenization (Coombs *et al.*, 1986; McGrory and Garber, 1992; Meysick and Garber, 1992; Abraham *et al.*, 1996). There have been inconsistencies, however, in the duration of infection in estrogenized mice. The reasons for failure of long-term murine genital *T. vaginalis* infection are probably multifactorial.

Factors directly responsible for the enhancement of growth of trichomonads in estrogenized animals are unknown and the exact role of this hormone is still under debate (Van Andel *et al.*, 1996b). It has been suggested that there is a possible association between increased susceptibility to infection and accumulation of a substrate (glycogen) that could be used as a nutrient source by *T. vaginalis* (Kulda, 1990). The relationship is probably more complex and may include factors such as the modification of vaginal epithelial cells, changes in the normal vaginal flora, and the number of neutrophils present in the vagina (Corbeil *et al.*, 1985).

While it is widely accepted that estrogen levels and stage of estrous are important in the establishment of an intravaginal *T. vaginalis* infection, it has also been shown that sustained estrogenization alone is insufficient to support long-term genital *T. vaginalis* infection in mice (Van Andel *et al.*, 1996b). Investigators have therefore made an attempt to alter the mouse vaginal environment so that it more closely reflects the human vaginal milieu. Meingassner (1977) co-inoculated mice with *Candida albicans* in order to sustain a more reliable model for use in drug testing. Vaginal flora and pH are quite dissimilar in mice and humans; high numbers of lactobacilli and low pH are hallmarks of the human vagina (Paavonen, 1983; Hanna *et al.*, 1985; Larsen, 1993), while only a small percentage of mice harbor lactobacilli and mouse vaginal pH is neutral (Meysick and Garber, 1992). McGrory and Garber (1992) introduced *Lactobacillus acidophilus* into the mouse vagina prior to inoculation with *T. vaginalis*, and observed a significantly more consistent and sustained infection. Of mice preinoculated with *L. acidophilus* prior to introduction of *T. vaginalis*, 69% were still infected (with *T. vaginalis*) after 24 days; only 11% of the control group (no *L. acidophilus*) maintained the infection. They also noted that the addition of *L. acidophilus* did not otherwise significantly alter the resident mouse vaginal flora.

The exact nature of the relationship between *L. acidophilus* preinfection and greater duration of *T. vaginalis* infection is not clearly understood. It may be that the lactobacilli become a nutrient source for the *T. vaginalis*, when the invading trichomonads phagocytose the bacteria. The prolonged infection seen in the presence of the *L. acidophilus* may also be attributed to direct or indirect actions of the lactobacilli on the vaginal environment, providing conditions which help facilitate the establishment of a long-term *T. vaginalis* infection (McGrory and Garber, 1992).

Animal strains

Female BALB/c mice (Charles River Co., Montreal, Canada) are used in experiments with the vaginal mouse model described here. Experiments done by Landolfo *et al.* (1981) support the use of the BALB/c strain. While studying the genetic control of resistance or susceptibility to *T. vaginalis* infection in various strains of mice, they found that female BALB/c and DBA/2 mice are highly susceptible to intraperitoneal, subcutaneous, and intravaginal infection with the parasite. Common lab strains C57BL/6 and C3H female mice display intermediate levels of resistance following intraperitoneal or subcutaneous inoculation, whereas they display high levels of resistance to intravaginal infection and would thus be inappropriate for use in the intravaginal infection model.

The animals used in the infection protocol are approximately 6–8 weeks old (22–24 g) at the beginning of the experiment. It has been shown that the age of the mice plays an important role in establishment of an infection. Landolfo *et al.* (1981) found that susceptibility of BALB/c females to *T. vaginalis* decreases with age; susceptibility is maximal at 3–4 weeks and minimal at 40–42 weeks.

Preparation of animals

Mice used in the protocol are housed in plastic cages and are supplied with food and tap water *ad libitum*. They are provided with 12 hours of artificial illumination each day. All aspects of housing and husbandry are performed in accordance with the regulations set forth by the University of Ottawa Animal Care Committee and the Canadian Council on Animal Care (Olfert *et al.*, 1993). In addition to nutritional provisions, mice are given nesting material and polyvinyl chloride (PVC) tubes which are believed to provide environmental enrichment.

In this infection protocol, the mice are treated with estrogen, then inoculated with *L. acidophilus* prior to inoculation with *T. vaginalis* (see Table 100.1 for an outline of the protocol). On Day –9 of the infection protocol (2 days prior to vaginal inoculation with *L. acidophilus* and 9 days before inoculation with *T. vaginalis*), each mouse is given a subcutaneous injection of 0.05 ml of Delestrogen (estradiol valerate 10 mg/ml; Squibb Canada, Montreal, Canada, Cappuccinelli *et al.*, 1974; McGrory and Garber, 1992; Meysick and Garber, 1992; Abraham *et al.*, 1996). The Delestrogen is suspended in sesame oil and is, therefore, viscous and somewhat difficult to inject. Possible dosage problems can, in part, be overcome by use of a 26 G needle, which is large enough to allow relatively easy injection, while minimizing leakage of estrogen. In addition, leakage can be avoided by massaging the site of injection, allowing immediate subcutaneous distribution of the material. The estrogen treatment is repeated on Day –2 (2 days before vaginal inoculation with *T. vaginalis*). Other investigators have developed alternate protocols for estrogen treatment (Table 100.2).

Before the mice are vaginally inoculated with *L. acidophilus*, the stage of estrous of each mouse is determined by vaginal smear. The swabs are made by wrapping a small amount of cotton around the end of a round toothpick (which is made dull by snipping the point with scissors). The swab is then moistened with saline, making insertion into the mouse vagina easier and less traumatic for the animal. Once inserted gently, the swab can be carefully rolled and turned in such a way that cells in the vagina are collected on the cotton tip. The swab is removed and lightly rolled on to a glass slide, which is then fixed with CytoPrep (Fisher Scientific, Nepean, Canada). The smear is stained with the Diff-Quick Stain Set (Dade Diagnostics, Aguada, PR), and examined to ascertain stage of estrous.

Stage of estrous can easily be determined through comparison with established criteria (Fox and Laird, 1970; Jacoby and Fox, 1984). Characterization of the cycle stage by means of vaginal smear is based on the detection of changes in the vaginal epithelium. During the estrous cycle, the epithelium becomes cornified, and is eventually shed; this is followed by an influx of leukocytes. The exfoliative cytology at estrus is characterized by the presence of anucleated cornified epithelial cells and a lack of leukocytes. In metestrus, numerous leukocytes are found, together with scattered cornified cells and nucleated squamous cells. Diestrus is characterized by a predominance of leukocytes, while in proestrus, the leukocytes are sparse, and round nucleated cells predominate (Kulda, 1990).

Microbiological organisms

T. vaginalis

Organisms are grown as previously described (Garber *et al.*, 1987) in 10 ml Diamond's TYI-S-33 medium (pH 6.2; Diamond *et al.*, 1978) supplemented with 10% heat-inactivated fetal bovine serum (FBS; Gibco BRL, Life Technologies Inc. Grand Island, NY), 100 U/ml penicillin, 100 μg/ml streptomycin (penicillin–streptomycin solution, Gibco BRL) and 2.5 μg/ml amphotericin B (Fungizone, Gibco BRL). Growth media containing the *T. vaginalis* is incubated at 37 °C with 5% CO_2 in glass screw-capped tubes held at a 45° angle. Cultures are passaged every 2–3 days.

Viable organisms are counted on a hemacytometer, using trypan blue exclusion (Takahashi *et al.*, 1970). To insure that the culture is healthy and growing as expected, it is advisable to passage the *T. vaginalis* two to three times (from frozen stock), before use in the infection protocol. It should be noted, however, that a marked decrease in infectivity and pathogenicity was observed by Coombs *et al.* (1986) and Bremner *et al.* (1986b) respectively, after prolonged *in vitro* axenic cultivation (several months) of a strain. For this reason, it is not desirable to passage an isolate over an extended period. For long-term storage, axenic cultures are frozen at

Table 100.1 Experimental timeline for mouse intravaginal infection model

Day	Procedure
Day –9	Subcutaneous injection of Delestrogen (estradiol valerate 10 mg/ml; 0.05 cc)
Day –7/Day –6	Intravaginal inoculation with *Lactobacillus acidophilus* (10^{10} cfu/ml; 20 μl) Vaginal smear to confirm estrous stage
Day –2	Subcutaneous injection of Delestrogen (estradiol valerate 10 mg/ml; 0.05 cc)
Day –1	Vaginal washes to confirm presence of *L. acidophilus* infection (50 μl MRS)
Day 0/Day 1	Intravaginal inoculation with *Trichomonas vaginalis* (9×10^6 cells/ml; 20 μl)
Days 7, 14, 21, 28	Vaginal washes to confirm presence of *T. vaginalis* infection (50 μL TYI + 10% FBS)

Cfu, Colony-forming units; FBS, fetal bovine serum.

Table 100.2 Procedures resulting in the establishment of lasting *Trichomonas vaginalis* infections in laboratory mice

Estrogen treatment				*Infection protocol*	*Results*			
Form of hormone and method of administration	*Dosage*	*Mouse weight or age*	*Period after estrogen treatment*	*No. of trichomonads and volume of inoculation*	*Total no. of mice*	*Initial percentage of animals infected*†	*Maximum period of infection (days)*†	*Reference*
Estradiol valerate (s.c. inoculation in sesame oil)	0.5 mg/animal every 10 days	30 g	3 days	One inoculation of 5×10^3 or 2×10^6 in 0.05 ml	48	52 (37–75)	>40	Cappuccinelli *et al.* (1974)
Estradiol valerate (s.c. inoculation in sesame oil)	0.5 mg/animal in a single dose	30 g	3 days	Two inoculations of 5×10^5 at 24-hour intervals in 0.05 ml	76	99	50	Cappuccinelli *et al.* (1974)
Estradiol undecylate (Progynon-Depot; s.c. and intraperitoneally in sesame oil)	40 mg/kg in two equal doses	25–30 g	3 days	One inoculation of 10^5 trichomonads + *Candida albicans** in 0.05 ml	27	94	NA	Meingassner† (1977)
Estradiol benzoate microcrystalline suspension (s.c. inoculation)	0.5 mg/animal weekly	18–20 g	7 days	One inoculation of 1.5×10^5 (4 different strains) in 0.05 ml	77	26 (15–40)	14–21	Wildfeuer (1974)
Estradiol benzoate microcrystalline suspension (s.c. inoculation)	0.5 mg/animal weekly	18–20 g	7 days	One inoculation of 5×10^5 or two inoculations of 5×10^5 and 3×10^5 at 24-hour or 48-hour intervals all in 0.05 ml	48	62 (50–65)	42	Wildfeuer (1974)
Estradiol cypionate (Sigma; s.c. inoculation in corn oil)	40 mg/kg in a single dose	2–3 months	2 days	One inoculation of 10^5 in 0.02 ml	180	52	92	Coombs *et al.* (1986)
Estradiol valerate (Squibb; s.c. inoculation in sesame oil)	0.5 mg/animal weekly	22–24 g	2 days	Two inoculations of 1×10^4 trichomonads in 0.02 ml	24	83	4	McGrory and Garber (1992)
					18	11	24	
Estradiol valerate (Squibb; s.c. inoculation in sesame oil)	0.5 mg/animal weekly	22–24 g	2 days	Two inoculations of 1×10^9 *Lactobacillus acidophilus* 1 week prior to the	24	79	4	McGrory and Garber (1992)
				two inoculations of 1×10^4 trichomonads in 0.02 ml	16	69	24	
Estradiol valerate (Squibb; s.c. inoculation in sesame oil)	0.5 mg/animal weekly	22–24 g	2 days	Two inoculations of 2×10^8 *Lactobacillus acidophilus* 1 week prior to the	33	91	7	Abraham *et al.* (1996)
				two inoculations of 5×10^5 trichomonads in 0.02 ml	29	52	28	
17β-Estradiol (s.c. implants)	0.015 mg/animal	6 weeks	2 days	Two inoculations of 8×10^5 trichomonads	20	NA	42	Van Andel *et al.* (1996b)

*Trichomonads and *C. albicans* administered in a mixed inoculum.
†Percentage of animals infected at initial sampling point does not necessarily reflect percentage infected at maximum day of infection — see individual references for more details.
s.c., subcutaneous; NA, data not available in original report.
Adapted from Kulda (1990), with permission.

–70 °C (cryopreserved with 20% heat-inactivated FBS and 10% dimethyl sulfoxide (BDH-Associate of E. Merck, Darmstadt, Germany)).

L. acidophilus

L. acidophilus (American Type Culture Collection (ATCC), Rockville, MD; strain #4356) is used in the protocol (McGrory and Garber, 1992; Abraham *et al.*, 1996). A start-up culture is streaked on to Bacto Lactobacilli MRS agar plates (Difco Laboratories, Detroit, MI), and is then subcultured into 10 ml Bacto Lactobacilli MRS broth (Difco Laboratories) in glass screw-capped tubes. Cultures are incubated in 5% CO_2 at 37 °C, and are passaged every 2–3 days. At least two to three passages are performed before use in the protocol.

Preparation of inocula

L. acidophilus

On the day before *L. acidophilus* inoculation into the mouse, a 1 liter volume of MRS broth (at 37 °C) is seeded with 1 ml of pure *L. acidophilus* subculture (grown to log phase). The culture is then left to grow overnight at 37 °C in 5% CO_2.

On the days of inoculation (Days –7 and –6 of the protocol), the culture is quantified by measuring absorbance at 650 nm in a spectrophotometer (Beckman DU-88). Absorbance can be correlated with colony-forming units (cfu) per ml with a previously plotted titration curve). The culture is quantified before and after harvest of the *L. acidophilus* in order to achieve a more accurate estimate of bacterial concentration.

Organisms are harvested by centrifugation for 10 minutes at 5000*g* at 4 °C (Beckman centrifuge, Model J2-21M). The cells are then washed with cold, sterile phosphate-buffered saline (PBS; pH 7.2) three times, and the resulting pellet is resuspended in MRS broth to achieve a final concentration of 10^{10} *L. acidophilus*/ml. Before inoculation, a Gram stain is done to insure that the culture is not contaminated.

To verify the concentration of *L. acidophilus* inoculated into the mouse, serial 10-fold dilutions in PBS are made from the inoculum. The actual number of *L. acidophilus* cfu is determined by spreading 100 µl of each dilution on to MRS plates, which are then incubated at 37 °C (5% CO_2); bacterial cfu are counted at 48 hours. All mice are inoculated twice (Days –7 and –6; McGrory and Garber, 1992; Abraham *et al.*, 1996).

T. vaginalis

T. vaginalis culture (100 ml) is grown to log phase in preparation for inoculation. On the days of inoculation (Days 0 and 1), the *T. vaginalis* culture (at log phase) is counted on the hemacytometer and the total number of organisms is calculated. The cells are then harvested by centrifugation (Omnifuge RT) for 10 minutes at 900 *g* (at 4 °C), and washed three times in cold, sterile PBS (pH 7.2). The final pellet is resuspended in media to a final concentration of 2.5×10^7 cells per ml. The prewarmed (37 °C) resuspension media is composed of TYI-S-33 supplemented with 10% FBS and 0.32% agar (QueBact, Quelab Laboratories, Montreal, Canada; McGrory and Garber, 1992; Meysick and Garber, 1992; Abraham *et al.*, 1996). Once the pellet is resuspended, the cells are counted again in order to confirm the concentration of cells in the sample. Necessary adjustments to the concentration are made at this point. Once the concentration of cells is established, it is advisable to passage a small amount of the sample into fresh media to ensure the viability of cells.

Inoculation process

All mice are given two intravaginal inoculations (Days –7 and –6) of 20 µl of *L. acidophilus* (10^{10} cfu/ml). Similarly, they are inoculated on Days 0 and 1 with the same volume of *T. vaginalis* (2.5×10^7 cells/ml; McGrory and Garber, 1992; Abraham *et al.*, 1996).

Intravaginal inoculation of the mice is a delicate process, but can be relatively stress-free for the mouse when it is done quickly in the following manner. It is important to be prepared before starting any manipulations with the mice. The inocula are introduced into the mouse vagina with an Eppendorf pipette (Brinkmann Instruments, Westbury, NY), which is prepared before removing the mouse from its cage. Once the pipette is ready, the mouse can be picked up by the base of the tail and set on the top of the cage in such a way that it is able to grasp the lid with its front feet. While holding the base of the tail, the mouse is lifted slightly so that the vagina can be viewed easily. The inoculum can then be delivered by carefully inserting the tip of the pipette into the vagina and dispelling the contents. Inoculation protocols developed by other researchers are presented in Table 100.2.

Key parameters to monitor infection

Presence of intravaginal infection with *L. acidophilus* and *T. vaginalis* is confirmed by vaginal wash (McGrory and Garber, 1992; Meysick and Garber, 1992; Abraham *et al.*, 1996). The wash technique is carried out in the same fashion as the inoculation procedure. Once the mouse is restrained, the pipette tip is inserted into the vagina and 50 µl of wash media is injected and drawn back into the pipette tip several times until it is turbid.

To confirm the presence of *L. acidophilus* infection, a

control group of mice is washed with pre-warmed (37 °C) MRS broth 1–2 days before *T. vaginalis* inoculation. The wash material is inoculated into a glass screw-capped tube containing MRS supplemented with 5 μg/ml ciprofloxacin (Cipro I.V., Bayer, Etobicoke, Canada) and 20 μg/ml metronidazole, which inhibit the growth of other vaginal bacteria. The tubes are incubated at 37 °C in 5% CO_2 for 24–48 hours.

Duration of infection with *T. vaginalis* is determined by washing with 50 μl pre-warmed TYI medium supplemented with 10% FBS. The washes are inoculated into glass screw-capped tubes containing pre-warmed TYI supplemented with 10% FBS, 300 U/ml penicillin, 300 μg/ml streptomycin, 2.5 μg/ml amphotericin B and 10 μg/ml gentamicin (Gentamicin Reagent Solution, Gibco BRL); the antibiotics are included to reduce bacterial contamination. The tubes are incubated at 37 °C (5% CO_2) at a 45 ° angle, and examined daily for the presence of motile *T. vaginalis* using an inverted microscope (McGrory and Garber, 1992; Meysick and Garber, 1992; Abraham *et al.*, 1996). Because false negatives can be misleading, lack of infection in an individual mouse should be concluded only after two consecutive negative washes (Coombs *et al.*, 1986; Abraham *et al.*, 1996).

Antimicrobial therapy

Most preclinical trials involving antitrichomonal chemotherapy agents were conducted during the 1960s and 1970s. During this time, the mouse subcutaneous and intraperitoneal models of *T. vaginalis* infection were most popular, followed by the intravaginal rat and hamster models (for reviews, including references to these trials, see Ryley and Stacey, 1963; Michaels, 1968; Kulda, 1990). Metronidazole, the current therapy for trichomoniasis, was originally tested using the subcutaneous mouse model (Cosar and Julou, 1959). Because a satisfactory cure had been found, therapy has not expanded beyond the nitroimidazoles. By the time an intravaginal mouse model became available, interest in preclinical trials for antitrichomonal drugs had faded, and thus few new studies were conducted.

In an effort to test the validity of the model itself, some researchers have used metronidazole and other commercial anti-trichomonal agents (Wildfeuer, 1974; Meingassner, 1977) to cure mouse intravaginal infections. Others have employed the model to study the efficacy of new drugs (Bremner *et al.*, 1986a, 1987; Wang, 1993; Table 100.3). Because of occasional intolerance to metronidazole, and the emergence of metronidazole-resistant strains of *T. vaginalis*, the need for novel treatment and protection strategies is becoming more clear. This may renew interest in the use of the intravaginal mouse model for preclinical drug trials.

The intravaginal mouse model can also be used in the search for a suitable *T. vaginalis* vaccine. Such a strategy may prove to be an important alternative to current therapies. While the research has been relatively limited, promising data have been published (Abraham *et al.*, 1996) that demonstrate the efficacy of a whole-cell vaccine preparation against *T. vaginalis*.

The vaccine protocol works around the timeline of the infection protocol, and begins 56 days prior to the intravaginal inoculation with *T. vaginalis*. On Day –56, log-phase *T. vaginalis* is harvested (as per isolation procedure above), with the final pellet resuspended in PBS so that the desired concentration of cells is achieved. A volume of 100 μl of the cell preparation is suspended in an equal volume of Freund's complete adjuvant (FCA; Gibco BRL), and the total volume (200 μl) is used for subcutaneous immunization in the side of the abdominal/groin area. The same amount of the cell preparation suspended in an equal volume of Freund's incomplete adjuvant (FIA; Gibco BRL) is used for booster immunization (in the side opposite the first immunization) at Day –28. Once the *T. vaginalis* is suspended in the adjuvant, the cells are rendered immobile and will cease to grow. Experiments have shown that a concentration of 9×10^6 *T. vaginalis*/ml (before addition of adjuvant) is optimal for immunization. The same study illustrated the importance of the booster immunization as well as the need for an adjuvant in eliciting an immune response in the mice (Abraham *et al.*, 1996).

Abraham *et al.* (1996) showed that mice that were vaginally infected, treated with metronidazole, and reinfected vaginally, did not develop protective immunity or an increased ability to clear infection. This mirrors the human condition in which women suffer numerous reinfections without developing protective immunity after vaginal exposure. However, mice which had received subcutaneous immunization and boosting with the *T. vaginalis* vaccine prior to intravaginal challenge were significantly less likely to develop infection. Only 25% of immunized mice were infected after 7 days, while 91% of sham-immunized and naive mice were infected at this timepoint. The mice also demonstrated a significantly increased rate of trichomonad clearance when compared with animals which had received no immunization. While only 4% of mice receiving the vaccine and booster still harbored the organism after 28 days, 52% of those not immunized with *T. vaginalis* vaccine were still infected. In addition, a measurable serum and vaginal antibody response was observed in mice immunized with *T. vaginalis* previous to intravaginal challenge (Abraham *et al.*, 1996).

Thus, the method of antigen presentation to the immune system (systemic versus vaginal) may be important in inducing an immune response. This study supports the notion that a vaccine could be an effective anti-trichomonal therapy.

Table 100.3 Drug efficacy in the intravaginal and subcutaneous *Trichomonas vaginalis* infections in the mouse model

Reference	*Drug*	*Dosage of drug*	*Effect on vaginal infection*	*Dosage of drug*	*Effect on subcutaneous infection**
Meingassner (1977)	Metronidazole	3.71 mg/kg p.o. 2, 18, 24 p.i.	Dosages required to clear infection in 50% of animals (ED_{50})	10.9 mg/kg p.o. 2, 18, 24 hours p.i.	Dosages required to clear infection in 50% of animals (ED_{50})
	Tinidazole	1.41 mg/kg		7.5 mg/kg	
	Nimorazole	5.62 mg/kg		33.9 mg/kg	
	Ornidazole	4.57 mg/kg		10.6 mg/kg	
Bremner *et al.* (1986a)†	Metronidazole	50 mg/kg p.o. o.d. × 5 days‡	100% curative	15 mg/kg i.v. 2, 18, 24 hours p.i.	No abscess
	Leupeptin	50 mg/kg i.v. 2, 18, 24 hours p.i.	Ineffective	50 mg/kg i.v. 2, 18, 24 hours p.i.	Abscess delayed to 50 hours p.i.
Bremner *et al.* (1987)†	Metronidazole	50 mg/kg p.o. o.d. × 5 days‡	100% curative	15 mg/kg p.o. t.i.d. × 5 days p.i.	100% curative
	Diflouromethylornithine	750 mg/kg p.o. t.i.d. × 5 days p.i.§	Ineffective	750 mg/kg p.o. t.i.d. × 5 days p.i.	Abscess delayed to 75 hours p.i.
Wang (1993)†	Metronidazole	50 mg/kg p.o. o.d. × 7 days‡	100% curative	75 mg/kg p.o. b.i.d. × 7 days‡	100% curative
	Emodin	500 mg/kg p.o. o.d. × 5 days‡	8/14 mice infected after 18 days, but 11/14 mice infected after 42 days	100 mg/kg s.c. b.i.d. × 2 days p.i.	Abscess delayed to 48 hours p.i.
				500 mg/kg p.o. or 200 mg/kg s.c. b.i.d. × 7 days‡	Decreased abscess volume
				200 mg/kg s.c. b.i.d. × 3 days**	Decreased no. of trichomonads retrieved from abscess

* Method of Honigberg *et al.* (1966).
† See methods of Coombs *et al.* (1986).
‡ Administration of drug began 3 days post-infection (p.i.).
§ Administration of drug at −2, 2, 4, 6, 21, 26, 30, 46, 50, 54, 70, 74, 77, 94, 98, 102 hours post-infection.
** Administration of drug began 2 days post-infection.

Key parameters to monitor response to treatment

The vaccine protocol includes vaginal washes with TYI to monitor the presence and persistence of vaginal *T. vaginalis* infection (described above). Vaginal washes with PBS are done 3 days after the TYI washes and are used to detect the level of antibody present in the vagina. Both TYI and saline vaginal washes are performed on a weekly basis following intravaginal inoculation with *T. vaginalis* (Abraham *et al.*, 1996).

Mice are also tail-bled for serological observation throughout the vaccine experiment. A volume of 15–20 μl of blood is taken from each mouse and is centrifuged to separate the serum (Abraham *et al.*, 1996). The enzyme-linked immunosorbent assay (ELISA) is used to assess antibody reactivity in serum and vaginal washes (Abraham *et al.*, 1996).

Aside from the specific methods used to monitor response to vaccine treatment, there are a number of other parameters used to observe the overall well-being and health of the mice involved in the experiment. Some of these are rather subjective (activity level and general appearance), while other criteria can be assessed more objectively (weight, for example).

Advantages and disadvantages of the intravaginal model

The ideal animal model of *T. vaginalis* infection should mimick human vaginal trichomoniasis, producing in the animal the symptoms, pathology and immune response which are similar to the human disease. The infection should be persistent, consistent, and reproducible. In this way, the variables can be controlled so that only one aspect of the disease can be studied at a given time. Many of the requirements of a good model are met in the intravaginal model.

The intravaginal model has advantages over the subcutaneous and intraperitoneal models, since this route is more comparable to the human infection. The vagina is a more appropriate site for the *T. vaginalis* infection, since it is thought that virulence factors may be tissue-specific or only expressed within the vaginal milieu (Krieger *et al.*, 1990; Corbeil, 1995). Another advantage of this model is the fact that the infection produces a reliable immunoglobulin G response, which correlates with the presence or absence of vaginal infection (Abraham *et al.*, 1996).

Some intravaginal mouse models have used *T. foetus*, the causative agent of bovine trichomoniasis (St Claire *et al.*, 1994; Van Andel *et al.*, 1996a). However, those which employ the human pathogen (*T. vaginalis*) are more relevant to the human disease. *T. vaginalis* does not occur naturally in mice, however, and this can present a disadvantage. There may be host specificity with respect to virulence factors and immune response (Corbeil, 1995). Certainly, it would be ideal to study *T. vaginalis* infection in humans, but this is usually not feasible, particularly for pilot intervention studies.

The intravaginal mouse model which employs lactobacilli (and thus more closely mimicks the human milieu) enjoys an added advantage over some previous intravaginal mouse models. This model results in increased consistency and duration of infection. McGrory and Garber (1992) found that, when they preinfected mice with lactobacilli, 69% were still infected with *T. vaginalis* after 28 days. Previously, the highest reported frequency of infection at this timepoint was 53% (Coombs *et al.*, 1986).

One disadvantage of the model is that the pathology of the infection has not been adequately described in the literature, and there is some debate as to the extent of clinical symptoms seen in the animals. Van Andel *et al.* (1996b) found that all mice remained clinically normal throughout their study, but they did observe histologic changes, which they attributed to estrogen therapy. Patten *et al.* (1963) documented cytopathology in 22% of non-estrogenized mice, as well as delayed maturation of the vaginal epithelium. A more recent intravaginal mouse model of *T. foetus* infection (St Claire *et al.*, 1994; Hook *et al.*, 1995; Van Andel *et al.*, 1996a), as well as the guinea-pig model of *T. vaginalis* infection (Soska *et al.*, 1962; Kazanowska *et al.*, 1983), produces severe signs and symptoms of vaginal trichomoniasis, and for this reason may be more useful in studies of pathogenesis.

Contributions of the model to infectious disease therapy

Pathology, pathogenesis and transmission

Intravaginal models of trichomoniasis, particularly the guinea-pig model, have facilitated studies of pathogenesis, and have provided researchers with valuable information regarding virulence, initial events of infection, gross pathology, cytopathology, histopathology, and the role of environmental factors on *T. vaginalis* infection. The intravaginal mouse model has the potential for studies of transmission, since sexual transmission between male and female mice has been accomplished (Cappucinelli *et al.*, 1974). A clearer understanding of the pathogenesis of *T. vaginalis* will lead to the development of potential intervention strategies.

Chemotherapeutic agents

Animals have been vital in preclinical drug trials to assess efficacy, routes of administration, mechanism of action, and various other aspects of drug pharmacology. Often, these studies follow the preliminary *in vitro* work and provide information as to whether further research is warranted. Selected drug studies which compared drug efficacy in intravaginal versus subcutaneous and intraperitoneal infections have shown that the latter two infections are, by themselves, inadequate for the assessment of antitrichomonal agents. Variation in drug efficacy is seen when comparing subcutaneous and intravaginal infections, and this may be related to a combination of differences in drug delivery, drug pharmacology, or survival of *T. vaginalis* between the tissue sites (Bremner *et al.*, 1986a, 1987).

Immunology and vaccine research

The intravaginal mouse model has great potential for use in continued research concerning the dynamics of the host immune response and parasite immune evasion. Cappuccinelli's group (1974) was one of the first to look at the connection between intravaginal infection with *T. vaginalis* and the immune system. Application of this model to vaccine research is the most recent and perhaps the most exciting step in the model's evolution. Employment of the intravaginal mouse model in vaccine research enables novel investigation into vaccine preparation, protocols, routes of administration, adjuvants, immune response, and protection (Abraham *et al.*, 1996).

With continued research, a human clinical vaccine trial may well be on the horizon. The eventual refinement of a human vaccine for *T. vaginalis* will be the culmination of many years of research, with animal models playing a key role in the process.

References

Abraham, M. C., Desjardins, M., Filion, L. G., Garber, G. E. (1996). Inducible immunity to *Trichomonas vaginalis* in a mouse model of vaginal infection. *Infect. Immun.*, **64**, 3571–3575.

Ackers, J. P. (1990). Immunologic aspects of human trichomoniasis. In *Trichomonads Parasitic in Humans* (ed Honigberg B. M.), pp. 36–52. Springer-Verlag, New York.

Bhatt, R., Abraham, M., Petrin, D., Garber, G. E. (1996). New concepts in the diagnosis and pathogenesis of *Trichomonas vaginalis*. *Can. J. Infect. Dis.*, **7**, 321–325.

Bremner, A. F., Coombs, G. H., North, M. J. (1986a). Antitrichomonal activity of the protease inhibitor leupeptin. *IRS Med. Sci.*, **14**, 555–556.

Bremner, A. F., Coombs, G. H., North, M. J. (1986b). Studies on Trichomonad pathogenicity: The effects of *Trichomonas vaginalis* on mammalian cells and mice. *Acta Univ. Carolinae (Prague) Biol.*, **30**, 381–386.

Bremner, A. F., Coombs, G. H., North, M. J. (1987). Antitrichomonal activity of α-diflouromethylornithine. *J. Antimicrob. Chemother.*, **20**, 405–411.

Cameron, D. W., Padian, N. S. (1990). Sexual transmission of HIV and the epidemiology of other sexually transmitted diseases. *AIDS*, **4** (suppl. 1), S99–S103.

Cappuccinelli, P., Lattes, C., Gagliani, I., Ponzi, A. N. (1974). Features of intravaginal *Trichomonas vaginalis* infection in the mouse and the effect of oestrogen treatment and immunodepression. *G. Batteriol. Virol. Immunol. Ann. Osp. Maria Vittoria Torino (Parte I: Ser Microbiol.)*, **67**, 31–40.

Cavier, R., Mossion, X. (1956). Essais d'infestation experimentale de la ratte par *Trichomonas vaginalis* (Donne, 1837). *C. R. Acad. Sci. (Paris)*, **242**, 2412–2414.

Cavier, R. E., Gobort, J. G., Savel, J. (1972). Application of a method of intraperitoneal infection of the mouse with *Trichomonas vaginalis* for the pharmacological study of trichomonacides (in French). *Ann. Pharm. Fr.*, **30**, 637–642.

Coombs, G. H., Bremner, A. F., Markham, D. J., Latter, V. S., Walters, M. A., North, M. J. (1986). Intravaginal growth of *Trichomonas vaginalis* in mice. *Acta Univ. Carolinae (Prague) Biol.*, **30**, 387–392.

Corbeil, L. B. (1995). Use of an animal model of trichomoniasis as a basis for understanding this disease in women. *Clin. Infect. Dis.*, **21** (suppl. 2), S158–S161.

Corbeil, L. B., Chatterjee, A., Foresman, L., Westfall, J. A. (1985). Ultrastructure of cyclic changes in the murine uterus, cervix and vagina. *Tissue Cell*, **17**, 53–68.

Cosar, C., Julou, L. (1959). Activité de 1(2-hydroxyethyl)-2-methyl-5-mitroimidazole (8823 R.P.) vis-à-vis des infections expérimentales à *Trichomonas vaginalis*. *Ann. Inst. Pasteur (Paris)*, **96**, 238–241.

Diamond, L. S., Harlow, D. R., Cunnick, C. C. (1978). A new medium for the axenic cultivation of *Entamoeba histolytica* and other *Entamoeba*. *Trans. R. Soc. Trop. Med. Hyg.*, **72**, 431–432.

Fouts, A. C., Kraus, S. J. (1980). *Trichomonas vaginalis*: reevaluation of its clinical presentation and laboratory diagnosis. *J. Infect. Dis.*, **141**, 137–143.

Fox, R. R., Laird, C. W. (1970). Sexual cycles. In *Reproduction and Breeding Techniques for Laboratory Animals* (ed Hafez, E.S.E.), pp. 107–122. Lea & Febiger, Philadelphia.

Garber, G. E., Lemchuk-Favel, L. T. (1990). Association of production of cell-detaching factor with the clinical presentation of *Trichomonas vaginalis*. *J. Clin. Microbiol.*, **28**, 2415–2417.

Garber, G. E., Sibau, L., Ma, R., Proctor, E. M., Shaw, C. E., Bowie, W. R. (1987). Cell culture compared with broth for detection of *Trichomonas vaginalis*. *J. Clin. Microbiol.*, **25**, 1275–1279.

Gardner, W. A. Jr, Culberson, D. E., Scimeca, J. M., Brady, A. G., Pindak, F. F., Abee, C. R. (1987). Experimental genital trichomoniasis in the squirrel monkey (*Saimiri sciureus*). *Genitourin. Med.*, **63**, 188–191.

Hanna, N. F., Taylor-Robinson, D., Kalodiki-Karamanoli, M., Harris, J. R. W., McFadyen, I. R. (1985). The relation between vaginal pH and the microbiological status in vaginitis. *Br. J. Obstet. Gynaecol.*, **92**, 1267–1271.

Heine, P., McGregor, J. A. (1993). *Trichomonas vaginalis*: a reemerging pathogen. *Clin. Obstet. Gynecol.*, **36**, 137–144.

Hollander, D. H., Gonder, J. D. (1985). Indigenous intravaginal pentatrichomonads vitiate the usefulness of squirrel monkeys (*Saimiri sciurius*) as models for human trichomoniasis in men. *Genitourin. Med.*, **61**, 212.

Holmes, K. K., Handsfield, H. H., Wang, S. P., Wentworth, B. B., Turck, M., Anderson, J. B., Alexander, E. R. (1975). Etiology of nongonococcal urethritis. *N. Engl. J. Med.*, **292**, 1199–1205.

Honigberg, B. M. (1978). Trichomonads of importance in human medicine. In *Parasitic Protozoa* (ed. Kreier, J. P.), pp. 275–454. Academic Press, New York.

Honigberg, B. M., Livingstone, M. C., Frost, J. K. (1966). Pathogenicity of fresh isolates of *Trichomonas vaginalis*. The "mouse assay" versus clinical and pathologic findings. *Acta Cytol.* **10**, 353–361.

Hook, R. R. Jr, St Claire, M. C., Riley, L. K., Franklin, C. L., Besch-Williford, C. L. (1995). *Tritrichomonas foetus*: comparison of isolate virulence in an estrogenized mouse model. *Exp. Parasitol.*, **81**, 202–207.

Jacoby, R. O., Fox, J. G. (1984). Biology and diseases of mice. In *Laboratory Animal Medicine* (ed. Fox, J., Cohen, B., Loew, F.), pp. 32–89. American College of Laboratory Animal Medicine Series. Academic Press, New York.

Jirovec, O., Petru, M. (1968). *Trichomonas vaginalis* and trichomoniasis. *Adv. Parasitol.*, **6**, 117–188.

Kazanowska, W., Kuczynska, K., Skrrzypiec, R. (1983). Pathology of *T. vaginalis* infection in experimental animals. *Wiad. Parazytol.*, **29**, 63–66.

Krieger, J. N. (1984). Prostatitis syndromes: pathophysiology, differential diagnosis, and treatment. *Sex. Transm. Dis.*, **11**, 100–112.

Krieger, J. N. (1990). Epidemiology and clinical manifestations of urogenital trichomoniasis in men. In *Trichomonads Parasitic in Humans* (ed Honigberg, B. M.), pp. 235–245. Springer-Verlag, New York.

Krieger, J. N., Wolner-Hanssen, P., Stevens, C., Holmes, K. K. (1990). Characteristics of *Trichomonas vaginalis* isolates from women with and without colpitis macularis. *J. Infect. Dis.*, **161**, 307–311.

Krieger, J. N., Jenny, C., Verdon, M. *et al.* (1993). Clinical manifestations of trichomoniasis in men. *Ann. Intern. Med.*, **118**, 844–849.

Kulda, J. (1990). Experimental animals in studies of *T. vaginalis* infection. In *Trichomonads Parasitic in Humans* (ed. Honigberg, B. M.), pp. 112–153. Springer-Verlag, New York.

Laga, M., Manoka, A., Kivuvu, M. *et al.* (1993). Non-ulcerative sexually transmitted diseases as risk factors for HIV-1 transmission in women: results from a cohort study. *AIDS*, **7**, 95–102.

Laga, M., Alary, M., Nzila, N. *et al.*, (1994). Condom promotion, sexually transmitted diseases treatment, and declining incidence of HIV-1 infection in female Zairian sex workers. *Lancet*, **344**, 246–248.

Landolfo, S., Martinotti, M. G., Martinetto, P., Forni, G., Rabagliata, A. M. (1981). *Trichomonas vaginalis*: dependence of resistance among differential mouse strains upon the non-H-2 gene haplotype, sex, and age of recipient host. *Exp. Parasitol.*, **52**, 312–318.

Larsen, B. (1993). Vaginal flora in health and disease. *Clin. Obstet. Gynecol.*, **36**, 107–121.

Lossick, J. G. (1982). Treatment of *Trichomonas vaginalis* infections. *Rev. Infect. Dis.*, **4** (suppl.), S801–S818.

Lossick, J. G. (1990). Epidemiology of urogenital trichomoniasis. In *Trichomonads Parasitic in Humans* (ed Honigberg, B. M.), pp. 311–323. Springer-Verlag, New York.

Lossick, J. G., Kent, H. L. (1991). Trichomoniasis: trends in diagnosis and management. *Am. J. Obstet. Gynecol.*, **165**, 1217–1222.

Maestrone, G., Semar, R. (1967). Experimental infection with *Trichomonas foetus* in guinea pigs. *Chemotherapia*, **12**, 137–145.

Mardh, P. A., Colleen, S. (1975). Search for uro-genital tract infections in patients with symptoms of prostatitis. Studies on aerobic and strictly anaerobic bacteria, mycoplasmas, fungi, trichomonads and viruses. *Scand. J. Urol. Nephrol.*, **9**, 8–16.

McGrory, T., Garber, G. E. (1992). Mouse intravaginal infection with *Trichomonas vaginalis* and role of *Lactobacillus acidophilus* in sustaining infection. *Infect. Immun.*, **60**, 2375–2379.

Meingassner, J. G. (1977). Comparative studies on the trichomonacidal activity of 5-nitroimidazole-derivatives in mice infected s.c. or intravaginally with *T. vaginalis*. *Experientia*, **33**, 1160–1161.

Meingassner, J. G., Nesvedba, H., Meith, H. (1981). New chemotherapeutic nitroheterocycles active against 5-nitroimidazole-resistant strains of trichomonads. *Arzneim.-Forsch.*, **31**, 6–8.

Meysick, K. C., Garber, G. E. (1992). Interactions between *Trichomonas vaginalis* and vaginal flora in a mouse model. *J. Parasitol.*, **78**, 157–160.

Michaels, R. M. (1968). Chemotherapy of trichomoniasis. *Adv. Chemother.*, **3**, 39–108.

Nicoletti, N. (1961). The problem of trichomoniasis of the lower genital tract in the female. *Br. J. Vener. Dis.*, **37**, 222–228.

Olfert, E. D., Cross, B. M., McWilliam, A. A. (eds) (1993). *Guide to the Care and Use of Experimental Animals*, vol. 1, 2nd edn. Canadian Council on Animal Care, Bradda Printing Services, Ottawa, Ontario.

Paavonen, J. (1983). Physiology and ecology of the vagina. *Scand. J. Infect. Dis.*, **40** (suppl.), 31–35.

Patten, S. F. Jr., Hughes, C. P., Reagan, J. W. (1963). An experimental study of the relationship between *Trichomonas vaginalis* and dysplasia in the uterine cervix. *Acta Cytol.*, **7**, 187–190.

Petrin, D., Delgaty, K., Bhatt, R., Garber, G. E. (1998). Clinical and microbiological aspects of *Trichomonas vaginalis*. *Clin. Microbiol. Rev.*, **11**, 300–317.

Rein, M. F. (1990). Clinical manifestations of urogenital trichomoniasis in women. In *Trichomonads Parasitic in Humans* (ed Honigberg, B. M.), pp. 225–234. Springer-Verlag, New York.

Rein, M. F., Chapel, T. A. (1975). Trichomoniasis, candidiasis, and the minor venereal diseases. *Clin. Obstet. Gynecol.*, **18**, 73–88.

Rein, M. F., Holmes, K. K. (1983). Non-specific vaginitis, vulvovaginal candidiasis and trichomoniasis: clinical features, diagnosis and management. In *Current Clinical Topics in Infectious Diseases* (eds Remington, J., Swartz, M. N.), pp. 281–315. Blackwell Scientific Publications, New York.

Ryley, J. F., Stacey, G. J. (1963). Experimental approaches to the chemotherapy of trichomoniasis. *Parasitology*, **53**, 303–320.

Soska, S., Kazanowska, W., Kuczynska, K. (1962). Injury of the epithelium of the vagina caused by *Trichomonas vaginalis* in experimental animals. *Wiad. Parazytol.*, **8**, 209–215.

St Claire, M. C., Riley, L. K., Franklin, C. L., Besch-Williford, C. L., Hook, R. R., Jr. (1994). Experimentally induced intravaginal *Tritrichomonas foetus* infection in the estrogenized mouse. *Lab. Animal Sci.*, **44**, 430–435.

Street, D. A., Taylor-Robinson, D., Hetherington, C. M. (1983). Infection of female squirrel monkeys (*Saimiri sciureus*) with *Trichomonas vaginalis* as a model of trichomoniasis in women. *Br. J. Vener. Dis.*, **59**, 249–254.

Takahashi, T., Old, L. J., Boyse, E. A. (1970). Surface alloantigens of plasma cells. *J. Exp. Med.*, **131**, 1325–1341.

Teras, J., Roigas, E. (1966). Characteristics of the pathomorphological reaction in cases of experimental infection with *Trichomonas vaginalis*. *Wiad. Parazytol.*, **12**, 161–172.

Van Andel, R. A., Franklin, C. L., St Claire, M. C., Riley, L. K., Besch-Williford, C. L., Hook, R. R., Jr. (1996a). Lesions of experimental genital *Tritrichomonas foetus* infections in estrogenized BALB/c mice. *Vet. Pathol.*, **33**, 407–411.

Van Andel, R. A., Kendall, L. V., Franklin, C. L., Riley, L. K., Besch-Williford, C. L., Hook, R. R., Jr. (1996b). Sustained estrogenization is insufficient to support long-term experimentally induced genital *Trichomonas vaginalis* infection on BALB/c mice. *Lab. Animal Sci.*, **46**, 689–690.

Wang, H-W. (1993). Antitrichomonal action of emodin in mice. *J. Ethnopharmacol.*, **40**, 111–116.

Wildfeuer, V. A. (1974). Die chemotherapie der vaginalen Trichomoniasis und Candidosis der Maus. *Arzneim.-Forsch.*, **24**, 937–943.

Wolner-Hanssen, P., Krieger, J. N., Stevens, C. E. *et al.* (1989). Clinical manifestations of vaginal trichomoniasis. *JAMA*, **261**, 571–576.

World Health Organization. (1995). An overview of selected curable sexually transmitted diseases, pp. 2–27. In *Global Program on AIDS*. World Health Organization, Geneva.

Chapter 101

Animal Models of *Cryptosporidium* Gastrointestinal Infection

D. S. Lindsay, B. L. Blagburn and S. J. Upton

Introduction

Human cryptosporidiosis is caused by an Apicomplexan parasite, *Cryptosporidium parvum*. Infection can cause life-threatening diarrheal disease in immunocompromised humans. The infection is persistent and many acquired immunodeficiency syndrome (AIDS) and other severely immunocompromised patients remain infected for several years or for their life. Immunocompetent individuals also suffer clinical disease but develop immunity and usually clear the parasite. Laboratory models, especially the neonatal mouse, immunocompromised rat, and severe combined immunodeficiency (SCID) mouse, have helped identify potential effective agents for treatment and to elucidate many of the mechanisms of immunity to this intracellular parasite. The present chapter summarizes these three models.

Background of human infection

Interest in human cryptosporidiosis was initiated with the advent of the AIDS epidemic and the large numbers of cases seen in these immunocompromised hosts (Current *et al.*, 1983; Fayer and Ungar, 1986; O'Donoghue, 1995). Profuse, watery diarrhea is the main clinical manifestation of human cryptosporidiosis. Associated symptoms are cramps, abdominal pain, low-grade fever, nausea, and vomiting; malaise, fatigue, headache, myalgia, and anorexia are less common symptoms (O'Donoghue, 1995). The length of time the diarrhea remains is dependent on the immune status of the individual. The duration is generally a few days to a few weeks in immunocompetent individuals. The diarrhea persists for months to years in immunocompromised patients and can be life-threatening. Many antimicrobial compounds have been examined and some are highly effective in animal models (Blagburn and Soave, 1997).

Background of animal models

The genus *Cryptosporidium* was first described from laboratory mice by Tyzzer in 1907 (Tyzzer, 1907, 1910). The type species is *C. muris* which was observed in the gastric glands of the stomach. This species has received much less attention than *C. parvum* which was described from the small intestines of mice and rabbits by Tyzzer in 1912 (Tyzzer, 1912). Soon after *C. parvum* was recognized as a serious pathogen of humans, scientists began work on developing mouse models for the parasite and it was determined that most outbred strains of weaned or adult mice were resistant to experimental infection but nursing animals could be readily infected (Tzipori *et al.*, 1980; Sherwood *et al.*, 1982; Current and Reese, 1986).

Animal species

Mice and rats are the animal species most often used. Hamsters, guinea-pigs, cats, dogs, calves, pigs, opossums, primates, and chickens have been used as animal models of cryptosporidiosis (Lindsay, 1997). However, mice and rats have been the most frequently used in antimicrobial chemotherapy studies and are the only species that will be discussed here.

Nursing mice are more susceptible to infection than are weaned or adult mice. Congenitally immunosuppressed nursing mice develop persistent or prolonged infections (Sherwood *et al.*, 1982; Heine *et al.*, 1984; Harp and Moon, 1991; Kuhls *et al.*, 1992; Aguirre *et al.*, 1994). Immunocompetent nursing mice develop patent infections if inoculated when less than 12–14 days of age (Sherwood *et al.*, 1982; Novak and Sterling, 1991; Upton and Gillock, 1996). Nursing rats (Tzipori *et al.*, 1980; Reese *et al.*, 1982) are more susceptible to *C. parvum* than are adult rats (Regh *et al.*, 1987; Meulbroek *et al.*, 1991).

Immunocompetent adult mice and rats are resistant to infection (Rehg *et al.*, 1987; Enriquez and Sterling, 1991; Johansen and Sterling, 1991; Rasmussen and Healey, 1992). Chemically immunosuppressed (Rehg *et al.*, 1987, 1988a; Brasseur *et al.*, 1988; Meulbroek *et al.*, 1991; Yang and Healey, 1993) and genetically immunosuppressed mice and rats (Heine *et al.*, 1984; Ungar *et al.*, 1990, 1991; Gardner *et al.*, 1991; Harp and Moon, 1991; Mead *et al.*, 1991a; Kuhls *et al.*, 1992; Aguirre *et al.*, 1994) are susceptible to infection.

Handbook of Animal Models of Infection
ISBN 0-12-775390-7

Storage, preparation of inocula

Large numbers of *C. parvum* oocysts can be generated in calves. Calves are obtained when 2–3 days old. It is best if the calves have nursed or been fed colostrum. Calves are orally inoculated when 4–6 days of age with 5–20 million oocysts in water or any physiological saline. Calves are fed commercial milk replacer 3 times daily until midday of the fourth day postinfection. The calves are then placed on a commercial oral electrolyte solution 4 times daily. The use of oral electrolyte solution diminishes the amount of fecal fat that is present and aids in eventual concentration and clean-up of oocysts.

The calves are placed in stainless steel cages that are 48 in long × 36 in wide × 42 in high with vinyl-coated mesh gratings and collecting pans underneath (LGL Animal Products, Byran, Texas, USA). Feces are collected daily for 6 days beginning on day 5 after inoculation. Feces collected from the pans is brought back to the laboratory and strained through a graded series of wire mesh sieves, with the final exclusion being 100 μm. Approximately 15 liters of material is produced by a calf in a day. The suspension is concentrated by centrifugation at 800–2000*g* for 10–15 minutes and stored as a sludge in 2.5% (w/v) potassium dichromate solution (approximately 500 ml from a total of 15 liters) at 4°C until used for further purification by sugar flotation or cesium chloride gradient centrifugation (Current, 1990). Oocysts will remain viable for 3–6 months. From 10 to 60 billion oocysts can be collected from a single calf using these methods.

Prior to inoculation, the potassium dichromate solution is removed from oocysts by repeated centrifugations (2000 *g* for 10–15 minutes) in water or saline solution. Oocysts can be sterilized at 4°C in 10% (v/v) bleach solution for 5 minutes. The bleach solution is removed by washing by centrifugation in sterile saline or water solution. Oocysts are counted in a hemocytometer (Figure 101.1) prior to being used for experimental inoculations.

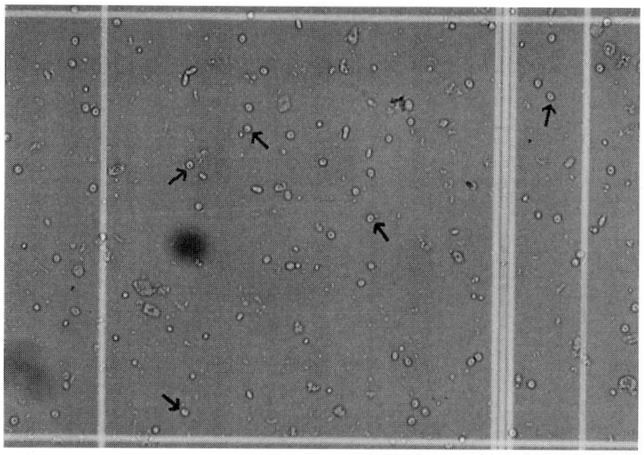

Figure 101.1 Oocysts of *Cryptosporidium parvum* (several are marked by arrows) in a hemocytometer. Bar = 50 μm.

Infection process

The oral 50% infectious dose (ID_{50}) for 5-day-old Swiss-Webster mice is between 100 and 500 oocysts (Ernest *et al.*, 1986), for suckling BALB/c mice 60 oocysts (Korich *et al.*, 1990) or 1000 oocysts (Riggs and Perryman, 1987), 600 oocysts for 3–5-day-old C57BL/6J mice (Aguirre *et al.*, 1994), and 79 oocysts for 4-day-old CD-1 mice (Finch *et al.*, 1993). The numbers of oocyst orally inoculated can influence such parameters as prepatent period, site of intestinal colonization, and the numbers of oocyst excreted by infected hosts (Mead *et al.*, 1994). Little difference in clinical signs is usually noted with different doses of inoculum.

Patent intestinal tract infections can be established in immunosuppressed C57BL/6N mice by intraperitoneal (i.p.) or intravenous (i.v.) injection but not subcutaneous (s.c.) or intramuscular (i.m.) injection of 1×10^6 oocysts (Yang and Healy, 1994a,b). The patent infections are delayed by several days in some i.p. and s.c. inoculated mice. The uterus is also colonized in some i.v. inoculated mice. Direct inoculation of 2×10^5 oocysts into the uterus will also produce intrauterine infections in BALB/c mice (Liebler *et al.*, 1986). Patent infections can be produced in nursing mice by direct inoculation of oocysts in the colon (Riggs and Perryman, 1987). Direct intratracheal inoculation of 1×10^6 oocysts will produce respiratory cryptosporidiosis in immunosuppressed but not immunocompetent rats (Meulbroek *et al.*, 1991).

Key parameters to monitor infection and response to treatment

Oocyst excretion is the most reliable parameter to monitor in nursing mice and immunosuppressed rats because no clinical signs are present. If no oocysts are excreted at the appropriate time postinoculation by non-medicated non-treated animals, then no infection or a substandard infection has occurred. Clinical signs of dehydration, icterus and lethargy are helpful in SCID mice but take about 6 weeks to develop. Reduction in oocyst numbers between treated and non-medicated controls are the best indicator of a positive response to treatment.

Antimicrobial therapy

Antimicrobial agents must be given orally by gavage when using the nursing mouse model. Appropriate controls must be given the vehicle used to solubilize the test agent. Antimicrobial agents can be given in the drinking water or in food if the immunosuppressed rat or SCID mouse models are being used. Appropriate controls for vehicle must be used if in-water or in-feed methods of treatment are employed.

Immunocompetent outbred nursing mouse model

Outbred nursing mice should be between 5 and 9 days old when used because older mice do not develop reliable infections. Mice that are 7–8 days old appear to produce the most consistent results (Upton and Gillock, 1996). Additionally, litters should be cross-fostered based on weight to help negate any maternal influences (Blagburn *et al.*, 1991; Upton and Gillock, 1996). If either of these parameters are not met, then the results of chemotherapy studies can be misleading.

Blagburn *et al.* (1991) developed an outbred ICR strain nursing mouse model for chemotherapy studies. In this model, mouse pups are cross-fostered when 2 days old based on weight prior to allocation of litters in to treatment groups. This produces litters with pups of approximately equal weight prior to initiation of the study. Pups are weighed daily and treated orally with a microdoser (Traco Atlas Microdoser Model 1003, Micro-jector Model 1003) and syringe pump equipped with a tuberculin syringe fitted with a 27 G needle with microbore tubing (1 mm outer diameter; 0.5 mm inner diameter). A mark is placed on the tubing corresponding to the approximate length from its tip to the stomach of the mouse pup. The technician gently holds the pup's head in one hand and inserts the tube into the pup's mouth with the other hand. Gentle pressure is used and the tube slides down the esophagus. If the tube is forced with too much pressure there is a danger of killing the pup. The tube is inserted in the esophagus until the black line is in the pup's mouth (Figure 101.2).

A foot pedal is pressed and the appropriate dose is administered by the microdoser. Because the pups are weighed daily, this allows for the volume of treatment compound to be adjusted daily for the increase or decrease in weight of each mouse pup. This keeps the dose agent constant throughout the study. Treatments are initiated when the mice in each group are 5 days old (usually 3 g in body weight). They are orally inoculated with 2×10^5 oocysts 1 day later (6 days old). Pups are treated 2 hours after inoculation, and then daily for 6 days. Pups are killed 7 days post-infection and the intestinal tract from the pylorus to the rectum is removed and homogenized in 10 ml of 2.5% (w/v) potassium dichromate solution for 30 seconds using a Tekmar tissumizer (Model SDT 1810S1) homogenizer. The numbers of oocysts present are determined by counting, using a hemocytometer (Figure 101.1). Comparisons of oocyst production in treated mice and controls are used to determine the efficacy of anticryptosporidial agents.

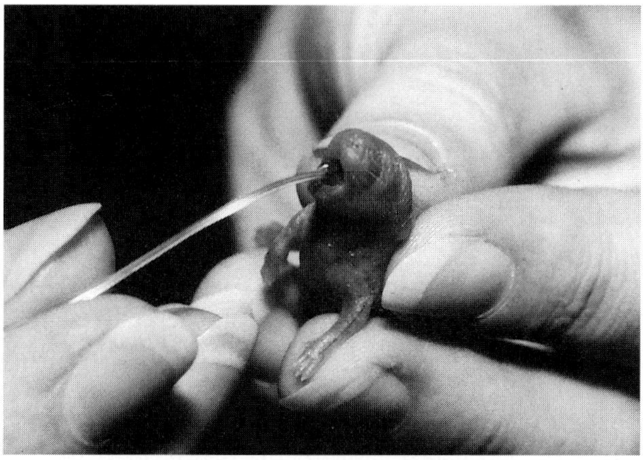

Figure 101.2 Insertion of microbore tubing in to the esophagus of a neonatal mouse.

Modifications to the model described above can be made to suit individual laboratory needs. Small amounts of inoculum or treatment volumes can be administered with a 10–20 μl automatic pipetter. Intestinal tissues can be homogenized in a motor-driven Teflon-coated tissue grinder or similar apparatus. Oocyst numbers can be counted by flow cytometery instead of manually on a hemocytometer (Mead *et al.*, 1995).

Advantages and disadvantages of nursing mouse model

The primary advantage of this model is that large numbers of animals can be used. The infection dynamics have been well-characterized. No immunosuppressive treatment is needed. Outbred mice are readily available and require no special housing needs. They are more hardy than inbred mice. Because of the small size (body weight), little test compound is needed. In general, results obtained in the neonatal mouse model have correlated with those observed in other animals.

Disadvantages to this model are that neonatal mice are small and some practice is needed in handling and inoculating mice. The mice grow fast and adjustments for weight gain must be made. Occasionally some dams will stop producing milk or cannibalize their young, leading to their removal from the study.

Contributions of neonatal mouse model to anticryptosporidial therapy

Several anticryptosporidial agents have been examined in the nursing mouse model (Table 101.1; Tzipori *et al.*, 1982; Angus *et al.*, 1984; Blagburn *et al.*, 1991; Fayer and Ellis, 1993, 1994; Rohlman *et al.*, 1993a; Cama *et al.*, 1994; Fayer and Fetterer, 1995; Waters *et al.*, 1997). The neonatal mouse model has been used to examine the developmental biology (Current and Reese, 1986; Tilly *et al.*, 1990) and maternal and other immune mechanisms (Moon *et al.*, 1988; Harp and Whitmire, 1991; Uhl *et al.*, 1992; Waters and Harp, 1996) or pathogen interactions (Vitovec *et al.*, 1991) of *C. parvum* infections. Additionally, the mouse model is used as a bioassay for the effects various treatments on *C. parvum* oocyst viability (Upton *et al.*, 1988; Korich *et al.*, 1990; Villacorta-Martinez *et al.*, 1992; Fayer *et al.*, 1991, 1997; Fayer, 1994, 1995).

Table 101.1 Antimicrobial agents examined against *Cryptosporidium parvum* in neonatal mice

Mouse strain	*Agent*	*Dose (mg/kg)**	*Results*
BALB/c	Clarithromycin	200	±
BALB/c	Azithromycin	100 or 200	+
BALB/c	Erythromycin	100 or 200	±
BALB/c	Paromomycin	100 or 200	+
Inbred/outred	Clopidol	0.25 or 2.5	–
SCID	Atovaquone	≥100	–
Inbred/outbred	Decoquinate	2.0 or 4.0	–
Inbred/outbred	Methylbenzoquate	0.25 or 2.5	–
C57	Ethopabate	28	–
Inbred/outbred	Dinitolmide	0.125	±
Inbred/outbred	Robenidine	0.003 or 0.3	–
Inbred/outbred	Furaltedone	0.2 or 2.0	–
C57	Furaltadone	0.1	–
Inbred/outbred	Furazolidone	0.4 or 2.0	–
Inbred/outbred	Amprolium	0.25	±
C57	Amprolium	0.02 or 8	–
C57	Arprinocid	8	–
Inbred/outbred	Arprinocid	0.06	+
Inbred/outbred	Sulfaquinoxaline	2.8	±
C57	Sulfaquinoxaline	30	–
C57	Sulfamethazine	5	–
Outbred	Alborixin	1.5 or 2.5	+
Outbred	Lasalocid	20 or 30	+
Outbred	Maduramicin	1.0 or 2.5	+
C57	Monensin	0.04	–
Inbred/outbred	Salinomycin	0.06	+
C57	Salinomycin	0.02	–
C57	Nicarbazin	0.1	–
Inbred/outbred	Nicarbazin	0.125	–
C57	Enterolyte-N	0.02	–
C57	Trinamide	3	–
C57	Phenamidine	0.01	–
C57	Zoaquin	0.5	–
C57	Halafuginone	6	–
Outbred	Enrofloxacin	1.0 or 3.0	–
Outbred	Aromatic amidines	2.8 or 11.3	±
BALB/c	Artemisinin	200 or 400	–
C57	Emtryl	4	–
BALB/c	Thiabendazole	100 or 200	–
BALB/c	Parabendazole	30	–
BALB/c	Oxibendazole	5 or 10	–
BALB/c	Mebendazole	8 or 15	–
BALB/c	Albendazole	7.5, 10 or 15	–

Adapted from Blagburn and Soave (1997), with permission.
*Some studies were prophylactic and some were therapeutic.
– = No activity; ± = partial activity; + = demonstrable activity.

Artificially immunosuppressed rat model

The dexamethasone-treated rat model was developed by Rehg *et al.* (1988a). In this model female Sprague–Dawley rats, 200–250 g, are immunosuppressed by giving 0.25 mg/kg dexamethasone daily in the drinking water for 10 days prior to oral inoculation with 1–6×10^5 *C. parvum* oocysts (Rehg, 1991a,b). Dexamethasone must be continuously administered or the rats will self-cure the infection. Anticryptosporidial agents are given in the water, in medicated feed or by gavage. The weight of the rats, the amount of water and of medicated feed consumed is monitored every 1 or 2 weeks to monitor potential changes in treatment/immunosuppression parameters. Clinically, some rats will develop loose stools but diarrhea is not present. Cryptosporidial infections peak at 7 days post-infection and most parasites are present in the terminal 2 cm of the ileum. Extraintestinal involvement is not common in this model. Histological scoring of ileal tissue sections taken at 11 days is used to judge the efficacy of prophylactic treatments and 21 days post-infection to judge therapeutic treatments which are initiated on day 10 post-infection.

Advantages and disadvantages of immunosuppressed rat model

Dexamethasone is readily obtainable and in-water immunosuppressive treatment is easy to accomplish. Outbred rats are also easy to obtain and work with. Young adult animals are used and rapid increases in weight, observed in nursing mice, is not a potential problem. Numerous studies have been done with this model and a large database is available for comparative purposes.

Because of their size, more test agents must be available for study in rats than in mice. More room is also needed to house and maintain rats than mice. Rats will self-cure if dexamethasone is not continuously administered. Agents are more likely to have activity in this model than in the neonatal mouse or SCID mouse model and may reflect this tendency to self-cure. The currently used method of histological scoring should be replaced by quantitative oocyst counts.

Contributions of immunosuppressed rat model to anticryptosporidial therapy

Several anticryptosporidial agents have been examined in the immunosuppressed rat model (Table 101.2; Rehg *et al.*, 1988b; Rehg, 1991a,b,c, 1993, 1994, 1995; Verdon *et al.*, 1994, 1995). It has also been used to examine the effects of immunopotentiators on *C. parvum* infections (Rasmussen *et al.*, 1991, 1992, 1993).

SCID mouse model

SCID mice have been used as models of cryptosporidiosis in AIDS patients (Mead *et al.*, 1991a,b, 1994, 1995; Kuhls *et al.*, 1992; Rohlman *et al.*, 1993a). Mead *et al.* (1994) critically examined the infection dynamics and clinical features of cryptosporidiosis in 6–8-week-old SCID mice. The mice were orally inoculated with 10^3, 10^4, 10^5, 10^6, or 10^7 oocysts. Mice inoculated with the lower numbers of oocysts have longer prepatent periods. All mice developed chronic

Table 101.2 Antimicrobial agents examined against *Cryptosporidium parvum* in dexamethasone-treated immunosuppressed rats

Agent	*Dose (mg/kg)**	*Results*
Halfuginone	37.5 or 75	–
Halfuginone	150, 300, 600 or 900	+
Paromomycin	50	–
Paromomycin	100, 200 or 400	+
Gentamicin	200	–
Neomycin	200	–
Kanamycin A	200	–
Polymyxin B	200	–
Streptomycin	200	–
Azithromycin	50, 100 or 200	±
Azithromycin	400	+
Clarithromycin	400	±
Oleandomycin	400	±
Spiramycin	200	±
Arprinocid	12.5	–
Arprinocid	25 or 50	+
Lasalocid	1.12	–
Lasalocid	2.25, 4.5, 9 or 18	+
Monensin	9	–
Salinomycin	9	–
Diethyldithiocarbamate	37.5	–
Diethyldithiocarbamate	75, 150, 300, 600, or 900	+
Amphotericin B	20	–
Eflornithine	3000	–
Ivermectin	0.4	–
Levamisole	4	–
Thiabendazole	50	–
Thalidomide	150	–
Succinylsulfathiazole	360	–
Sulfacetamide	120	–
Sulfabenzamide	240	–
Sulfachloropyridazine	360	–
Sulfadiazine	250	–
Sulfadimethoxine	10	–
Sulfadimethoxine	20,40,80,120	+
Sulfadoxine	160	–
Sulfaguanidine	120	–
Sulfamerazine	200	+
Sulfameter	120	+
Sulfamethazine	175	+
Sulfamethizole	480	–
Sulfamethoxazole	320	–
Sulfamethoxypyridazine	160	–
Sulfanilamide	120	–
Sulfanilic acid	120	–
Sulfanitran	200	–
Sulfapyridine	240	–
Sulfasalazine	400	–
Sulfathiazole	120	–
Sulfisomidine	240	–
Sulfisoxizole	120	+

Adapted from Blagburn and Soave (1997), with permission.
*Some studies were prophylactic and some were therapeutic.
– = No activity; ± = partial activity; + = demonstrable activity.

infections. Hepatic involvement was present in all inoculation groups. Thirty to 40% of mice given 10^3–10^5 oocysts had hepatic involvement, while 80–90% of the mice inoculated with 10^7 oocysts had hepatic cryptosporidiosis. Some of these mice died or were euthanized due to hepatic cryptosporidiosis.

The SCID mouse model presented by Mead *et al.* (1995) uses 6–8-week-old female SCID mice. Mice are maintained under pathogen-free conditions and housed in microisolator cages (Nalgene Labware, Rochester, NY) in high-efficiency laminar flow units (Labline, Maywood, NJ). All cages, water, and bedding are sterilized before use and sterilized surgical clothing is worn when mice are handled (Mead *et al.*, 1995). All animal manipulations are done on a high-efficiency particulate air-filtered bench.

Because the SCID mouse is both an infection and disease model it can be used to examine both the prophylactic and therapeutic activities of an antimicrobial agent. Prophylactic studies are done by beginning treatment prior to oral inoculation of 10^6 oocysts. The SCID mice are treated daily for 28 days. Fecal samples are examined for oocysts at weekly intervals. Portions of liver, gallbladder, ileum and large intestine are taken for histologic examination and microscopic lesion scoring. Therapeutic studies are done by orally inoculating mice with 10^6 oocysts and letting the infection become established for 4 weeks before treatments are begun. Mice are treated with test agents for 3 weeks. Fecal samples are examined for oocysts at weekly intervals. Portions of liver, gallbladder, ileum and large intestine are taken for histologic examination and microscopic lesion scoring.

Advantages and disadvantages of the SCID mouse model

The primary advantage of the SCID mouse model is that it is both an infection and a disease model. However, diarrhea is not a common clinical sign. Extraintestinal infections are common and mimick what is observed in AIDS patients. Adult mice are used and are easier to handle than are neonatal mice. Additionally, test agents can be administered in the water or food. The main disadvantage is that this model is expensive and requires specialized animal-handling facilities. Because of their body size, more test agent is needed than in neonatal mice but less than in immunosuppressed rats.

Contributions of the SCID mouse model to anticryptosporidial therapy

Few antimicrobial agents have been examined in the SCID mouse model. Rohlman *et al.* (1993a) tested atovaquone against *C. parvum* infections in neonatal SCID mice and found little activity at doses of ≥100 mg/kg. Mead *et al.* (1994) examined two ionophores in adult SCID mice and determined that maduramicin and alborixin were both effective when given at 3 mg/kg. Cama *et al.* (1994) examined clarithromycin and its metabolite 14-OH clarithromycin in SCID mice and observed moderate activity for both agents.

Several groups of researchers have used *C. parvum* in SCID mice to examine immune mechanisms (Mead *et al.*, 1991a,b; Kuhls *et al.*, 1992, 1994; McDonald *et al.*, 1992; Chen *et al.*, 1993a,b; Rohlman *et al.*, 1993b) or importance of intestinal microflora (Harp *et al.*, 1992).

References

Aguirre, S. A., Masson, P. H., Perryman, L. E. (1994). Susceptibility of major histocompatibility complex (MHC) class I- and class II-deficient mice to *Cryptosporidium parvum* infection. *Infect. Immun.*, **62**, 697–699.

Angus, K. W., Hutchison, G., Campbell, I., Snodgrass, D. R. (1984). Prophylactic effects of anticoccidial drugs in experimental murine cryptosporidiosis. *Vet. Rec.*, **114**, 166–168.

Blagburn, B. L., Soave, R. (1997). Prophylaxis and chemotherapy: human and animal. In Cryptosporidium *and cryptosporidiosis* (ed. Fayer, R.), pp. 111–128. CRC Press, Boca Raton, Florida.

Blagburn, B. L., Sundermann, C. A., Lindsay, D. S., Hall, J. E., Tidwell, R. R. (1991). Inhibition of *Cryptosporidium parvum* in neonatal Hsd:(ICR)BR Swiss mice by polyether ionophores and aromatic amidines. *Antimicrob. Agents Chemother.*, **35**, 1520–1523.

Brasseur, P., Lemeteil, D., Ballet, J. J. (1988). Rat model for human cryptosporidiosis. *J. Clin. Microbiol.*, **26**, 1037–1039.

Cama, V. A., Marshall, M. M., Shubitz, L. F., Ortega, Y. R., Sterling, C. R. (1994). Treatment of acute and chronic *Cryptosporidium parvum* infections in mice using clarithromycin and 14-OH clarithromycin. *J. Euk. Microbiol.*, **41**, 25S.

Chen, W., Harp, J. A., Harmsen, A. G. (1993a). Requirements for CD4+ cells and gamma interferon in resolution of established *Cryptosporidium parvum* infections. *Infect. Immun.*, **61**, 3928–3932.

Chen, W., Harp, J. A., Harmsen, A. G., Havell, E. A. (1993b). Gamma interferon functions in resistance to *Cryptosporidium parvum* infection in severe combined immunodeficient mice. *Infect. Immun.*, **61**, 3548–3551.

Current, W. L. (1990). Techniques and laboratory maintenance of *Cryptosporidium*. In *Cryptosporidiosis of Man and Animals* (eds Dubey, J. P., Speer, C. A., Fayer, R.), pp. 31–49. CRC Press, Boca Raton, Florida.

Current, W. L., Reese, N. C. (1986). A comparison of endogenous development of three isolates of *Cryptosporidium* in suckling mice. *J. Protozool.*, **33**, 98–108.

Current, W. L., Reese, N. C., Ernst, J. V., Bailey, W. S., Heyman, M. B., Weinstein, W. M. (1983). Human cryptosporidiosis in immunocompetent and immunodeficient persons. Studies of an outbreak and experimental transmission. *N. Engl. J. Med.*, **308**, 1252–1257.

Enriquez, F., Sterling, C. R. (1991). *Cryptosporidium* infections in inbred strains of mice. *J. Protozool.*, **38**, 100S–102S.

Ernest, J. A., Blagburn, B. L., Lindsay, D. S., Current, W. L. (1986). Infection dynamics of *Cryptosporidium parvum* in neonatal mice (*Mus musculus*). *J. Parasitol.*, **72**, 796–798.

Fayer, R. (1994). Effect of high temperature on infectivity of *Cryptosporidium parvum* oocysts in water. *Appl. Environ. Microbiol.*, **60**, 2732–2735.

Fayer, R. (1995). Effect of sodium hypochlorite exposure on infectivity of *Cryptosporidium parvum* oocysts for neonatal BALB/c mice. *Appl. Environ. Microbiol.*, **61**, 844–846.

Fayer, R., Ellis, W. (1993). Glycoside antibiotics alone and combined with tetracyclines for prophylaxis of experimental cryptosporidiosis in neonatal BALB/c mice. *J. Parasitol.*, **79**, 533–558.

Fayer, R., Ellis, W. (1994). Qinghaosu (artemisinin) and derivatives fail to protect neonatal BALB/c mice against *Cryptosporidium parvum* (CP) infection. *J. Euk. Microbiol.*, **41**, 41S.

Fayer, R., Fetterer, R. (1995). Activity of benzimidazoles against cryptosporidiosis in neonatal BALB/c mice. *J. Parasitol.*, **81**, 794–795.

Fayer, R., Ungar, B. P. L. (1986). *Cryptosporidium* species and cryptosporidiosis. *Microbiol. Rev.*, **50**, 458–483.

Fayer, R., Nerad, T., Rall, W., Lindsay, D. S., Blagburn, B. L. (1991). Studies on the cyropreservation of *Cryptosporidium parvum*. *J. Parasitol.*, **77**, 357–361.

Fayer, R., Farley, C. A., Lewis, E. J., Trout, J. M., Graczyk, T. K. (1997). Potential role of the eastern oyster, *Crassostrea virginica*, in the epidemiology of *Cryptosporidium parvum*. *Appl. Environ. Microbiol.*, **63**, 2086–2088.

Finch, G. R., Daniels, C. W., Black, E. K., Schaefer, F. W., Belosevic, M. (1993). Dose response of *Cryptosporidium parvum* in outbred neonatal CD-1 mice. *Infect. Immun.*, **59**, 3661–3665.

Gardner, A. L., Roche, J. K., Weikel, C. S., Guerrant, R. L. (1991). Intestinal cryptosporidiosis: pathophysiologic alterations and specific cellular and humoral immune responses in RNU/+ and RNU/RNU (athymic) rats. *Am. J. Trop. Med. Hyg.*, **44**, 49–62.

Harp, J. A., Moon, H. W. (1991). Susceptibility of mast cell-deficient W/W^{v} mice to *Cryptosporidium parvum*. *Infect. Immun.*, **59**, 718–720.

Harp, J. A., Whitmire, W. M. (1991). *Cryptosporidium parvum* infection in mice: inability of lymphoid cells or culture supernatants to transfer protection from resistant adults to susceptible infants. *J. Parasitol.*, **77**, 170–172.

Harp, J. A., Chen, W., Harmsen, A. G. (1992). Resistance to severe combined immunodeficient mice to infection with *Cryptosporidium parvum*: the importance of intestinal microflora. *Infect. Immun.*, **60**, 3509–3512.

Heine, J., Moon, H. W., Woodmansee, D. B. (1984). Persistent *Cryptosporidium* infection in congenitally athymic (nude) mice. *Infect. Immun.*, **43**, 856–859.

Johansen, G. A., Sterling, S. R. (1994). Detection of a prolonged *C. parvum* infection in immunocompetent adult C57BL/6 mice. *J. Euk. Microbiol.*, **41**, 45S.

Korich, D. G., Mead, J. R., Madore, M. S., Sinclar, N. A., Sterling, C. R. (1990). Effects of ozone, chlorine dioxide, chlorine, and monochloramine on *Cryptosporidium parvum* oocyst viability. *Appl. Environ. Microbiol.*, **56**, 1423–1428.

Kuhls, T. L., Greenfield, R. A., Mosier, D. A., Crawford, D. L., Joyce, W. A. (1992). Cryptosporidiosis in adult and neonatal mice with severe combined immunodeficiency. *J. Comp. Pathol.*, **106**, 399–410.

Kuhls, T. L., Mosier, D. A., Abrams, V. L., Crawford, D. L., Greenfield, R. A. (1994). Inability of interferon-gamma and aminoguanidine to alter *Cryptosporidium parvum* infection in mice with severe combined immunodeficiency. *J. Parasitol.*, **80**, 480–485.

Liebler, E. M., Pohlenz, J. F., Woodmansee, D. B. (1986). Experimental intrauterine infection in adult BALB/c mice with *Cryptosporidium* sp. *Infect. Immun.*, **54**, 255–259.

Lindsay, D. S. (1997). Laboratory models of cryptosporidiosis. In Cryptosporidium *and cryptosporidiosis* (ed Fayer, R.) pp. 209–224. CRC Press, Boca Raton, Florida.

McDonald, V., Deer, R., Uni, S., Iseki, M., Bancroft, G. J. (1992). Immune response to *Cryptosporidium muris* and *Cryptosporidium parvum* in adult immunocompetent or immunocompromised (nude and SCID) mice. *Infect. Immun.*, **60**, 3325–3331.

Mead, J. R., Arrowood, M. J., Healey, M. C., Sidwell, R. W. (1991a). Chronic *Cryptosporidium parvum* infections in congenitally immunodeficient SCID and nude mice. *J. Infect. Dis.*, **163**, 1297–1304.

Mead, J. R., Arrowood, M. J., Healey, M. C., Sidwell, R. W. (1991b). Cryptosporidial infections in SCID mice reconstituted with human or murine lymphocytes. *J. Protozool.*, **38**, 59S–61S.

Mead, J. R., Ilksoy, N., You, X. *et al.* (1994). Infection dynamics and clinical features of cryptosporidiosis in SCID mice. *Infect. Immun.*, **62**, 1691–1695.

Mead, J. R., You, X., Pharr, J. E. *et al.* (1995). Evaluation of maduramicin and alborixin in a SCID mouse model of chronic cryptosporidiosis. *Antimicrob. Agents Chemother.*, **39**, 854–858.

Meulbroek, J. A., Novilla, M. N., Current, W. L. (1991). An immunosuppressed rat model of respiratory cryptosporidiosis. *J. Protozool.*, **38**, 113S–115S.

Moon, H. W., Woodmansee, D. B., Harp, J. A., Abel, S., Ungar, B. L. P. (1988). Lacteal immunity to enteric cryptosporidiosis in mice: immune dams do not protect their suckling pups. *Infect. Immun.*, **56**, 649–653.

Novak, S. M., Sterling, C. R. (1991). Susceptibility dynamics in neonatal BALB/c mice infected with *Cryptosporidium parvum*. *J. Protozool.*, **38**, 102S–104S.

O'Donoghue, P. J. (1995). *Cryptosporidium* and cryptosporidiosis in man and animals. *Int. J. Parasitol.*, **25**, 139–195.

Rasmussen, K. R., Healey, M. C. (1992). Experimental *Cryptosporidium parvum* infections in immunosuppressed adult mice. *Infect. Immun.*, **60**, 1648–1652.

Rasmussen, K. R., Martin, E. G., Arrowood, M. J., Healey, M. C. (1991). Effects of dexamethasone and dehydroepiandrosterone in immunosuppressed rats infected with *Cryptosporidium parvum*. *J. Protozool.*, **38**, 157S–159S.

Rasmussen, K. R., Arrowood, M. J., Healey, M. C. (1992). Effectiveness of dehydroepiandrosterone in reduction of cryptosporidial activity in immunosuppressed rats. *Antimicrob. Agents Chemother.*, **36**, 220–222.

Rasmussen, K. R., Martin, E. G., Healey, M. C. (1993). Effects of dehydroepiandrosterone in immunosuppressed rats infected with *Cryptosporidium parvum*. *J. Parasitol.*, **79**, 364–370.

Reese, N. C., Current, W. L., Ernst, J. V., Bailey, W. S. (1982). Cryptosporidiosis of man and calf: a case report and results of experimental infections in mice and rats. *Am. J. Trop. Med. Hyg.*, **31**, 226–229.

Rehg, J. E. (1991a). Anticryptosporidial activity is associated with specific sulfonamides in immunosuppressed rats. *J. Parasitol.*, **77**, 238–240.

Rehg, J. E. (1991b). Anti-cryptosporidial activity of macrolides in immunosuppressed rats. *J. Protozool.*, **38**, 228S–230S.

Rehg, J. E. (1991c). Activity of azithromycin against cryptosporidia in immunosuppressed rats. *J. Infect. Dis.*, **163**, 1293–1296.

Rehg, J. E. (1993). Anticryptosporidial activity of lasalocid and other ionophorous antibiotics in immunosuppressed rats. *J. Infect. Dis.*, **168**, 1293–1296.

Rehg, J. E. (1994). A comparison of anticryptosporidial activity of paromomycin with that of other aminoglycosides and azirthromycin in immunosuppressed rats. *J. Infect. Dis.*, **170**, 934–938.

Rehg, J. E. (1995). The activity of halofuginone in immunosuppressed rats infected with *Cryptosporidium parvum*. *J. Antimicrob. Chem.*, **35**, 391–397.

Rehg, J. E., Hancock, M. L., Woodmansee, D. B. (1987). Characterization of cyclophosphamide-rat model of cryptosporidiosis. *Infect. Immun.*, **55**, 2669–2774.

Rehg, J. E., Hancock, M. L., Woodmansee, D. B. (1988a). Characterization of dexamethasone-treated rat model of cryptosporidial infection. *J. Infect. Dis.*, **158**, 1406–1407.

Rehg, J. E., Hancock, M. L., Woodmansee, D. B. (1988b). Anticryptosporidial activity of sulfadimethoxine. *Antimicrob. Agents Chemother.*, **32**, 1907–1908.

Riggs, M. W., Perryman, L. E. (1987). Infectivity and neutralization of *Cryptosporidium parvum* sporozoites. *Infect. Immun.*, **55**, 2081–2087.

Rohlman, V. C., Kuhls, T. L., Mosier, D. A. *et al.* (1993a). Therapy with atovaquone for *Cryptosporidium parvum* infection in neonatal severe combined immunodeficiency mice. *J. Infect. Dis.*, **168**, 258–260.

Rohlman, V. C., Kuhls, T. L., Mosier, D. A., Crawford, D. L., Greenfield, R. A. (1993b). *Cryptosporidium parvum* infection after abrogation of natural killer cell activity in normal and severe combined immunodeficiency mice. *J. Parasitol.*, **79**, 295–297.

Sherwood, D., Angus, K. W., Snodgrass, D. R., Tzipori, S. (1982). Experimental cryptosporidiosis in laboratory mice. *Infect. Immun.*, **38**, 471–475.

Tilly, M., Upton, S. J., Freed, P. S. (1990). Comparative study on the biology *Cryptosporidium serpentis* and *Cryptosporidium parvum* (Apicomplexa: Cryptosporidiidae). *J. Zoo. Wildl. Med.*, **21**, 463–467.

Tyzzer, E. E. (1907). A sporozoan found in the peptic glands of the common mouse. *Proc. Soc. Exp. Med.*, **5**, 12–13.

Tyzzer, E. E. (1910). The extracellular coccidium, *Cryptosporidium muris* (gen. et sp. nov.), of the gastric glands of the common mouse. *J. Med. Res.*, **18**, 487–509.

Tyzzer, E. E. (1912). *Cryptosporidium parvum* (sp. nov.) a coccidium found in the small intestine of the common mouse. *Arch. Protistenkd.*, **26**, 394–412.

Tzipori, S., Angus, K. W., Campbell, I., Gray, E. W. (1980). *Cryptosporidium*: evidence for a single-species genus. *Infect. Immun.*, **30**, 884–886.

Tzipori, S., Campbell, I., Angus, K. (1982). The therapeutic effects of 16 antimicrobial agents on *Cryptosporidium* infection in mice. *Aust. J. Exp. Biol. Med. Sci.*, **60**, 187–190.

Uhl, E. W., O'Connor, R. M., Perryman, L. C., Riggs, M. W. (1992). Neutralization-sensitive epitopes are conserved among geographically diverse isolates of *Cryptosporidium parvum*. *Infect. Immun.*, **60**, 1703–1706.

Ungar, B. L. P., Burris, J.A., Quinn, C. A., Finkelman, F. D. (1990). New mouse models for chronic *Cryptosporidium* infection in immuno deficient hosts. *Infect. Immun.*, **58**, 961–969.

Ungar, B. L. P., Kao, T. C., Burris, J. A., Finkelman, F. D. (1991).

Cryptosporidium infection in an adult mouse model: Independent roles of INF-gamma and CD4+ lymphocytes in protective immunity. *J. Immunol.*, **147**, 1014–1022.

Upton, S. J., Gillock, H. H. (1996). Infection dynamics of *Cryptosporidium parvum* in ICR outbred suckling mice. *Folia Parasitol.*, **43**, 101–106.

Upton, S. J., Tilley, M. E., Marchin, G. L., Fine, L. R. (1988). Efficacy of a pentaiodide resin disinfectant on *Cryptosporidium parvum* (Apicomplexa: Cryptosporidiidae) oocysts *in vitro*. *J. Parasitol.*, **74**, 719–721.

Verdon, R., Polianski, J., Gaudebout, C., Marche, C., Garry, L., Pocidalo, J. J. (1994). Evaluation of curative anticryptosporidial activity of paromomycin in a dexamethasone-treated rat model. *Antimicrob. Agents Chemother.*, **38**, 1681–1682.

Verdon, R., Polianski, J., Gaudebout, C. *et al.* (1995). Evaluation of high-dose regimen of paromomycin against cryptosporidiosis in the dexamethasone-treated rat model. *Antimicrob. Agents Chemother.*, **39**, 2155–2157.

Villacorta-Martinez, I., Ares-Mazas, M., Duran-Oreiro, D., Lorenzo-Lorenzo, M. J. (1992). Efficacy of activated sludge in removing *Cryptosporidium parvum* oocysts from sewage. *Appl. Environ. Microbiol.*, **58**, 3514–3516.

Vitovec, J., Koudela, B., Vladik, P., Hausner, O. (1991). Interaction of *Cryptosporidium parvum* and *Campylobacter jejuni* in experimentally infected neonatal mice. *Zbl. Bakt.*, **274**, 548–559.

Waters, W. R., Harp, J. A. (1996). *Cryptosporidium parvum* infection in T-cell receptor (TCR)-α- and TCR-δ-deficient mice. *Infect. Immun.*, **64**, 1854–1857.

Waters, W. R., Reinhardt, T. A., Harp, J. A. (1997). Oral administration of putrescine inhibits *Cryptosporidium parvum* infection of neonatal C57BL-6 mice and is independent of nitric oxide synthesis. *J. Parasitol.*, **83**, 746–750.

Yang, S., Healey, M. C. (1993). The immunosuppressive effects of dexamethasone administered in drinking water to C57BL/6N mice infected with *Cryptosporidium parvum*. *J. Parasitol.*, **79**, 626–630.

Yang, S., Healey, M. C. (1994a). Patent gut infections in immunosuppressed adult C57BL/6N mice following intraperitoneal injection of *Cryptosporidium parvum* oocysts. *J. Parasitol.*, **80**, 338–342.

Yang, S., Healey, M. C. (1994b). Development of patent gut infections in immunosuppressed adult C57BL/6N mice following intravenous inoculations of *Cryptosporidium parvum* oocysts. *J. Euk. Microbiol.*, **41**, 67S.

Chapter 102

Animal Models of *Entamoeba histolytica* Infection

S. L. Stanley Jr, T. Zhang and K. B. Seydel

Background of human infection

Amebic dysentery and amebic liver abscess are major causes of morbidity and mortality worldwide, with approximately 50 000 000 cases of diarrhea and 50 000 deaths each year from these diseases. (For reviews, see Ravdin, 1995; Li and Stanley, 1996). The causative agent of amebiasis, *Entamoeba histolytica*, is a protozoan parasite whose only natural host appears to be humans and some non-human primates. *E. dispar* is a morphologically identical species but is a non-pathogenic gut commensal which can be differentiated from *E. histolytica* by molecular techniques (Jackson, 1998).

The life cycle of *E. histolytica* is simple. Disease begins with the ingestion of the cyst form of the parasite, which is found in fecally contaminated water and food. The hardy cyst survives the gastric acidity of the stomach, and in the distal small intestine or proximal colon excysts to form the motile trophozoite. Under some conditions, *E. histolytica* trophozoites in the distal colon encyst, and these excreted cysts continue the life cycle. *E. histolytica* trophozoites can adhere to colonic epithelial cells, and are capable of invading into the colonic mucosa, causing colitis, with characteristic flask-shaped ulcers. In as many as 5–10% of individuals with intestinal infection, *E. histolytica* trophozoites reach the submucosal layers of the gut, and from there enter the portal circulation, which transports them to the liver. There they can cause extensive tissue damage, resulting in an amebic liver abscess.

Background of models

Animal models for amebiasis have been limited by the fact that *E. histolytica* only naturally infects humans, and it has been impossible to establish infections in small animals by the administration of *E. histolytica* cysts. Over the years there have been scattered reports of the successful establishment of intestinal disease in animals by direct intracecal inoculation of amebic trophozoites, but none of these methods have yet achieved widespread success (Vinayak *et al.*, 1979; Chadee and Meerovitch, 1985; Ghosh *et al.*, 1994; Shibayama *et al.*, 1997). In most cases these techniques employed *E. histolytica* trophozoites that were co-cultured with bacteria (xenic cultures), further complicating the interpretation of these studies. Small animal models of amebic liver abscess have been more reliable, and have been successfully established in hamsters, gerbils, and severe combined immunodeficient (SCID) mice (Thompson *et al.*, 1954; Sepúlveda *et al.*, 1974; Chadee and Meerovitch, 1984; Cieslak *et al.*, 1992). Infection is not established by the natural route (hematogenous spread to the liver by a foci of intestinal disease), but rather by the direct intrahepatic inoculation of axenically cultured (without bacteria) amebic trophozoites into the liver. In some cases, inoculation of trophozoites into the portal vein has been used as a more "physiologic" approach to establishing amebic liver abscess, but this is technically more difficult, and is probably not necessary for most studies of disease (Tsutsumi *et al.*, 1984).

Recently, a new model for intestinal amebiasis which employs *E. histolytica* infection of human intestinal xenografts in SCID mice (SCID-HU-INT mice; Seydel *et al.*, 1997a) has been developed. The intestinal xenografts produce mucin, and differentiate into anatomically normal-appearing small intestine or colon, depending on the tissue of origin. This model provides a reproducible system for establishing intestinal amebiasis, and for directly studying the interactions of *E. histolytica* trophozoites with human intestinal tissue.

In this chapter we will describe the methods for producing SCID-HU-INT mice and the infection of their human intestinal xenografts with axenically cultured *E. histolytica* trophozoites. We will also outline the approach to establishing amebic liver abscesses in gerbils, hamsters, and SCID mice by direct hepatic inoculation of axenically cultured *E. histolytica* trophozoites. The reader is referred to other references for details on alternative approaches to establishing amebiasis in animals (Miller, 1952; Tsutsumi *et al.*, 1984; Anaya-Velazquez *et al.*, 1985; Meerovitch and Chadee, 1988; Shibayama-Salas *et al.*, 1992; Ghosh *et al.*, 1994; Bhopale *et al.*, 1995; Sohni and Bhatt, 1996).

Handbook of Animal Models of Infection
ISBN 0-12-775390-7

SCID-HU-INT model of intestinal amebiasis

Animal species

Male or female CB-17 SCID mice, aged 6–10 weeks old, are used as xenograft recipients. Since maturation of the intestinal xenograft requires at least 10 weeks post-surgery, use of younger mice is recommended.

Preparation of animals

All SCID mice are screened for the "leaky" phenotype at the age of 5 weeks by measuring serum antibodies using an enzyme-linked immunosorbent (ELISA) assay (Seydel *et al.*, 1997a). Any mouse with serum immunoglobulin G (IgG) levels greater than 4 μg/ml is considered "leaky" and is not used. SCID mice are housed in a sterile barrier environment and animals are fed with autoclaved food and water.

Details of surgery

Preparation of intestine

Fetal human intestinal tissue of 12–15 weeks developmental age is stored at 4°C in Dulbecco's Modified Eagle's media supplemented with penicillin/streptomycin until the time of surgery. All surgical procedures should be carried out in a laminar flow hood, and all instruments should be autoclaved between uses. At the time of surgery, the fetal intestine is decanted into a small Petri dish. All mesentery is removed from the intestine. Care should be taken during this cleaning procedure not to injure the tissue by grasping or pulling on it with forceps. The best technique involves grasping the mesentery with blunt forceps, lifting it off gently to visualize its connection to the intestine and then severing the connection with scissors, thus completely eliminating the manipulation of the intestine itself. This is the time to differentiate small intestine from the large intestine based on proximity to the stomach, and the location of the appendix. In subsequent steps, sections of the intestine will be prepared, and there are no gross differences in appearance between small intestine and large intestine at this time point in development.

Once the intestine is cleaned of adjoining organs and mesentery it should be placed into a clean Petri dish with fresh media, and divided into lengths of approximately 4 cm. This dish should be kept on ice until the time of implantation.

Anesthesia

Mice are prepared for surgery by anesthesia with a mixture of ketamine and xylazine. A stock solution of concentration 5.8 mg/ml ketamine and 0.87 mg/ml xylazine is used. Mice are weighed and 12 μl per gram of body weight of the stock solution is administered intraperitoneally via a 29 G needle. This should provide approximately 15 minutes of anesthesia.

Surgical procedure

A small strip of fur from the hip to the scapular area is shaved parallel to the spine on both sides. Animals are immobilized in the prone position, and the shaved area is scrubbed with a povidone-iodine scrub solution. With iris scissors, a 0.5 cm incision is made in the flank and scapular region of each side. The incision should run perpendicular to the axis of the spine, and penetrate just the skin and not any of the underlying muscle layers. A closed pair of 6-in dressing forceps is then inserted into the incision in the flank region (Figure 102.1), and by blunt dissection is tunneled up to the opening at the scapular region.

The forceps are forced out of the scapular incision and opened. Often times there will be fascia still covering the forceps at this point. This should be removed, allowing for a patent opening of the forceps' tip. While maintaining the forceps in this subcutaneous position, a section of intestine is chosen with another forceps. The section should be grasped at one extreme end. It should then be dangled above the subcutaneous forceps, allowing the opposite end to be grasped by the subcutaneous forceps (Figure 102.2). The subcutaneous forceps should then be slowly backed out of the subcutaneous tunnel, drawing the intestine through the tunnel. The trailing edge of the intestine should be maintained in the other forceps until the point where both forceps are out of the tunnel and the ends of the intestine are equally spaced extending from the two incisions. At this point one should move to the contralateral side and thread an intestine through that tunnel. This gives the inserted

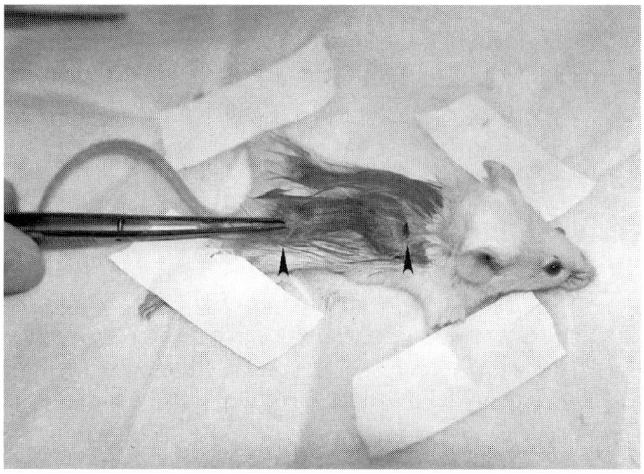

Figure 102.1 SCID mouse under anesthesia showing the paired skin incisions on both the right and left sides, and the forceps introduced into the posterior incision on the right flank.

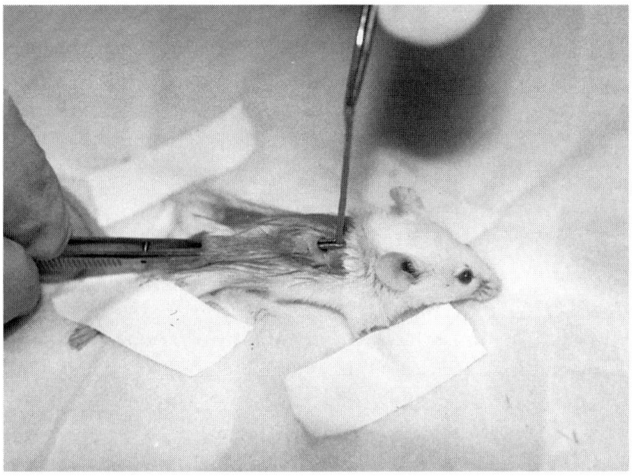

Figure 102.2 SCID mouse under anesthesia showing the forceps introduced through both incisions, and grasping the end of a 5 cm length of intestine before pulling it gently into the subcutaneous tunnel.

intestine time to recover from the stretching effect of pulling it through the tunnel. After this respite, the ends of the intestine should be trimmed to where they are even with the incision.

Wound closure/postoperative care

Each incision is closed with a 7.5 mm Michel clip. The clip should incorporate both skin flaps of the incision as well as the end of the intestine, thus anchoring the end of the intestine. Failure to incorporate the intestine in this clip will result in the shrinkage of the length of the intestine as it matures. The animals should then be returned to their cages and allowed to recover. The success of the engraftment can be assessed after 10 weeks. At this point the grafts can be palpated through the skin. SCID-HU-INT mice should be used for experiments between 10 and 14 weeks post-engraftment; grafts allowed to develop for longer periods of time grow too large, endangering the health and well-being of the mouse.

Storage, preparation of inocula

E. histolytica strain HM1:IMSS trophozoites are grown in B1-S33 media. This strain has been passaged bi-monthly through mouse livers to maintain virulence. Culture tubes or flasks containing *E. histolytica* are chilled on ice for 5 minutes, then the media containing *E. histolytica* is gently centrifuged (10 minutes × 500*g*), and the pellet of amebic trophozoites resuspended in B1-S-33 media at a concentration of 10^6 *E. histolytica* trophozoites per 100 μl. This suspension is kept on ice until used for inoculation.

Infection of intestinal xenografts

SCID-HU-INT animals are anesthetized using ketamine/xylazine as described above. They are immobilized on a surgical board in the prone position. A 3–4 cm incision is made with a surgical blade along the spine of the mouse, and the skin on one side is retracted and pinned to the board with a 31 G needle to expose the subcutaneous graft (Figure 102.3). Care should be taken to insure that the vascular supply to the graft is not compromised during the retraction of the skin.

With the graft visualized, the inocula of *E. histolytica* trophozoites is resuspended and taken up into a 1 ml syringe with a 26 G needle. The needle is advanced directly through the graft wall at an angle to avoid the possibility of advancing the needle all the way through the lumen and into the opposite wall. There should be very little resistance to injection, as the inocula should be going into the lumen of the xenograft. High resistance indicates the needle is in either of the graft walls, and needs to be relocated. Upon successful inoculation, the needle is removed, and a small square of Gelfoam (Upjohn, Kalamazoo, MI) is placed on the injection site to prevent the leakage of the *E. histolytica* suspension out of the xenograft. The contralateral xenograft is inoculated using the same protocol. The two skin flaps are then conjoined and closed by stapling with 7.5 mm Michel clips.

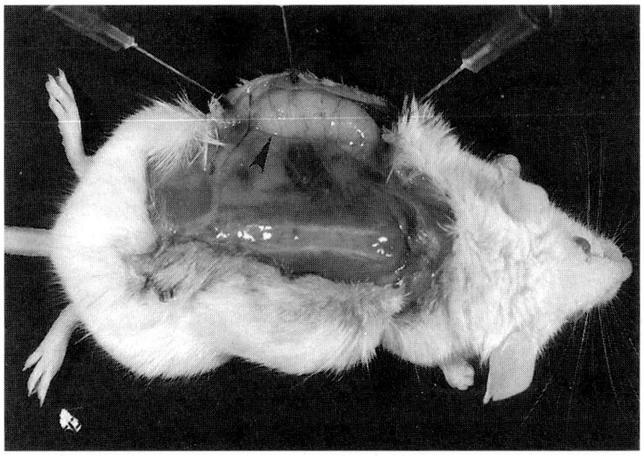

Figure 102.3 SCID mouse under anesthesia shown 10 weeks after engraftment, with the cutaneous tissue peeled back and one of the human intestinal xenografts exposed. Note the extensive vascular supply to the graft, and its increase in width since implantation.

Key parameters to monitor infection

Our studies indicate that infection is established within 4–8 hours of inoculation, and significant tissue damage can be seen histologically by 24 hours after infection. Infection is generally monitored by histologic examination of hematoxylin and eosin-stained sections of the intestinal xenograft. For studies of the inflammatory response to infection we have also monitored cytokine production by ELISA and reverse transcriptase polymerase chain reaction (RT-PCR; Seydel *et al.*, 1997a).

Antimicrobial therapy

To date, this model has been used to test the efficacy of anti-inflammatory therapy in amebic infection, and to study the role of the inflammatory process in amebiasis.

Pitfalls (advantages/disadvantages) of the model

The major advantage of this model is that it allows one to study *E. histolytica* intestinal disease *in vivo*. It is ideal for studies of the interactions of *E. histolytica* with human intestine, as one can experimentally observe and manipulate amebic adherence to intestinal epithelial cells, invasion through mucosal surfaces, and the host inflammatory response to infection. The disadvantages are the need for human fetal intestinal tissue, the fact that the intestinal xenografts differ from normal bowel in the lack of bacterial flora, peristalsis, and excretion of luminal contents, and the inability to study lymphocyte-based immunity in this model.

Contributions of the model to infectious disease therapy

This model could be used to evaluate new anti-amebic agents, and would be especially useful for the assessment of agents designed to work in the gut lumen (e.g. compounds designed to inhibit amebic adherence to target cells). Because the intestinal xenografts are extensively vascularized, this model could also be used to evaluate the efficacy of parenterally and orally administered anti-amebic agents. The major use of this model to date has been in improving our understanding of the pathogenesis of amebic infection, and especially the role of the host inflammatory response in the tissue damage seen with *E. histolytica* infection (Seydel *et al.*, 1997a).

Gerbil and hamster models of amebic liver abscess

Animal species

Male or female mongolian gerbils (*Meriones unguiculatus*) or golden hamsters (*Mesocricetus auratus*) aged 6–12 weeks are used in these studies.

Preparation of animals

Gerbils and hamsters are housed in pathogen-free barrier environment and are fed with autoclaved food and water.

Details of surgery and infection

Anesthesia

Gerbils and hamsters are prepared for surgery by anesthesia with sodium pentobarbital. Hamsters receive 0.8 μl of a 65 mg/ml solution of sodium pentobarbital per gram body weight, while gerbils receive 2 μl of a 22 mg/ml solution of sodium pentobarbital per gram/body weight, administered intraperitoneally via a 29 G needle.

Surgical procedure

Following the induction of anesthesia, the fur on the anterior abdominal wall and peritoneum is shaved. The gerbil or hamster is then placed ventral side up on a surgical board, and the shaved area is scrubbed with a povidone-iodine scrub solution. Under sterile conditions, using sterile drapes, gloves and instruments, a 1.5 cm vertical incision is made in the anterior abdominal wall and peritoneum with a surgical blade and scissors, and the liver is visualized (Figure 102.4). Using the index finger, the liver is gently pushed out for full exposure. Next, 100 μl of amebic inoculum containing 10^5–10^6 *E. histolytica* HM1:IMSS trophozoites is loaded into a 1-ml disposable syringe fitted with a 26 G needle that is kept in the vertical, needle-down position. This suspension is then injected slowly into the left lobe of the liver such that a visible subcapsular bleb is raised at the inoculum site of the liver surface (Figure 102.5). The needle is gently withdrawn and the injection site is blotted

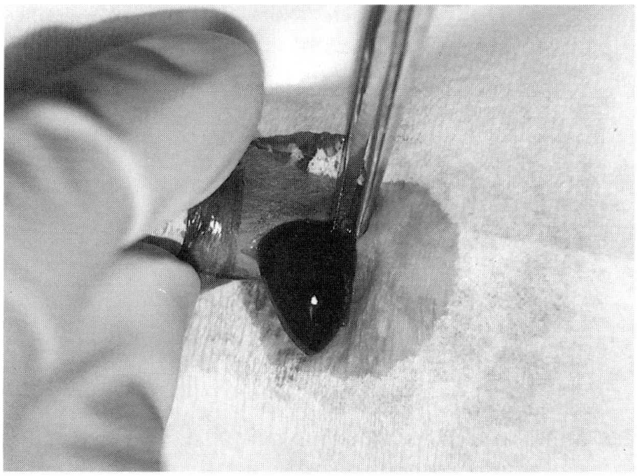

Figure 102.4 Gerbil under anesthesia showing the draped operative field and exposure of the liver immediately prior to inoculation.

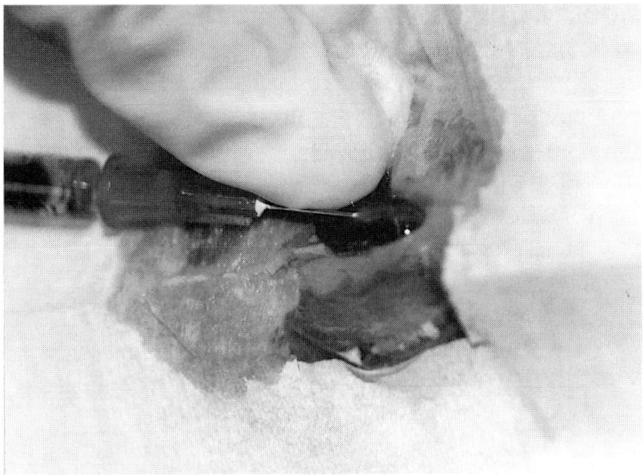

Figure 102.5 Gerbil under anesthesia showing inoculation of the gerbil's liver with *Entamoeba histolytica* trophozoites using a 26 G needle, which is guided just under the liver capsule such that a bleb forms with injection.

with a small square of sterile Gelfoam (Upjohn) to preclude leakage of the inoculum and stop hemorrhage. The liver is gently placed back into the peritoneal cavity, and the peritoneum is closed using 4–0 chromic gut suture. The two skin flaps are closed by stapling with 7.5 mm Michel clips.

Storage, preparation of inocula

Entamoeba histolytica strain HM1:IMSS trophozoites are cultured in BI-S33 medium. This strain has been passaged bi-monthly through gerbil and hamster livers to maintain virulence. For inoculation of gerbils or hamsters, culture tubes or flasks containing 60-hour *E. histolytica* cultures (mid-log phase) are chilled for 5 minutes on ice and centrifuged at 500 g for 10 minutes, and the amebic trophozoites in the pellet are resuspended in BI-S-33 medium and counted with a hemocytometer. The concentration of the amebic inoculum is adjusted to 10^5–10^6 *E. histolytica* trophozoites per 0.1 ml of fresh medium. The dosage of *E. histolytica* needed to achieve successful infection appears to vary from batch to batch of *E. histolytica*, hence we routinely inoculate 1 to 2 animals 1 week before a large experiment to determine the minimum dosage needed to produce an abscess of at least 10% of the liver at the 7-day point. This is the dosage then used for the large-scale study. The *E. histolytica* suspension is kept on ice until used for inoculation; ideally it should be used within 30 minutes of preparation.

Key parameters to monitor infection

Generally, animals are sacrificed 7 days after infection. The entire liver is removed and weighed, and any abscess seen grossly is resected and weighed. The percent liver abscessed is calculated as the weight of the abscess divided by the liver weight before abscess removal. A small specimen from each abscess is cultured in BI-S33 medium to look for the growth of *E. histolytica* trophozoites. This is done to confirm the etiology as amebic liver abscess, and may be performed as part of our continuous bi-monthly passage of *E. histolytica* trophozoites through hamster and gerbil livers to maintain virulence.

Liver abscesses, as well as normal appearing regions of livers, are fixed in formalin, sectioned, and stained with hematoxylin and eosin for histological examination. Thus, liver abscesses in hamsters and gerbils are monitored on the basis of the gross appearance (Figure 102.6) and quantified by the percentage of the liver that is occupied by the abscess. The diagnosis of amebic liver abscess is confirmed by a positive culture for *E. histolytica* and compatible histologic findings. Disease appears to be more severe in hamsters, and mortality has been used as an endpoint in studies of amebic liver abscess in that species.

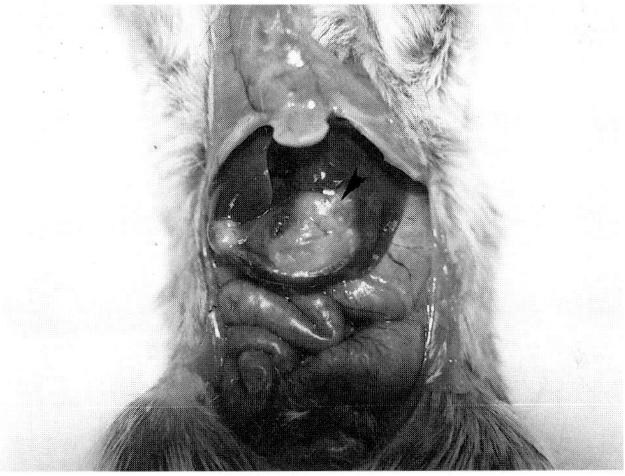

Figure 102.6 Post-mortem picture of gerbil showing the gross appearance of an amebic liver abscess 7 days following initial inoculation. This was a large abscess and constituted 33% of the liver by weight.

Antimicrobial therapy

The hamster model of amebic liver abscess has been used to test the anti-amebic efficacy of a number of new compounds (Pargal *et al.*, 1993a,b; Sohni and Bhatt, 1996; Gonzalez-Garza *et al.*, 1997). Compounds have been administered by the oral, intravenous, and intraperitoneal route.

Pitfalls (advantages/disadvantages) of the model

Neither model represents the physiologic way in which amebic liver abscess normally develops, i.e. introduction of trophozoites into the liver through the portal circulation. In addition, neither animal is a natural host for amebiasis, hence any conclusions about pathogenesis must be qualified. Finally, compared to the mouse or rat, few reagents are available for characterizing or manipulating the immune responses of gerbils and hamsters, limiting studies of the immunopathogenesis of infection.

Contributions of this model to infectious disease therapy

The gerbil and hamster models of infection have been used to understand better the pathogenesis of amebic liver abscess, and represent the standard models for studying vaccines to prevent amebic liver abscess (Stanley, 1997). They are also standard models for assessing the efficacy of anti-amebic agents *in vivo*.

SCID mouse model of amebic liver abscess

Animal species

Male or female CB-17 SCID mice, aged 6–10 weeks old are used as the experimental host for *E. histolytica*.

Preparation of animals

All SCID mice are screened for the "leaky" phenotype at the age of 5 weeks, as described above.

Details of surgery and infection

Anesthesia

Mice are prepared for surgery by anesthesia with a mixture of ketamine and xylazine using doses identical to those described above for the SCID-HU-INT procedure.

Surgical procedure

The procedure for creating amebic liver abscesses in SCID mice is identical to that described for gerbils and hamsters, except the initial incision is smaller (1.0 cm), and the dosage of ameba inoculated is always 100 μl of amebic inoculum containing 1×10^6 *E. histolytica* HM1:IMSS trophozoites.

Storage, preparation of inocula

E. histolytica strain HM1:IMSS trophozoites are cultured in BI-S33 medium. This strain has been passaged bi-monthly through SCID mouse livers to maintain virulence. For inoculation of SCID mice, culture tubes or flasks containing 60-hour *E. histolytica* cultures (mid-log phase) are chilled for 5 minutes on ice and centrifuged at 500*g* for 10 minutes, and the amebic trophozoites in the pellet are resuspended in B1-S33 medium and counted with a hemocytometer. The concentration of the amebic inoculum was adjusted to 1×10^6 *E. histolytica* per 0.1 ml of fresh medium. The suspension is kept on ice until used for inoculation.

Key parameters to monitor infection/interventions

SCID mice are sacrificed at varying times after infection, depending on the nature of the study. Liver abscesses appear to be maximum in size at 4–7 days, but significant abscesses are visible at earlier time points (48 hour; Seydel *et al.*, 1997b). Abscess size decreases after 10 days, and abscesses are no longer visible in most animals by 28 days following infection. Upon sacrifice, the liver is removed and weighed, and any abscess seen grossly is resected and weighed. The percent of liver abscess is calculated as the weight of the abscess divided by the liver weight before abscess removal. A small specimen from each abscess is cultured in BI-S33 medium. Liver abscesses, as well as visually normal livers, are fixed in formalin, sectioned, and stained with hematoxylin and eosin for histological examination. Liver abscesses in the SCID mice are generally monitored on the basis of the gross appearance and disease is quantified by measuring the percentage of the liver (by weight) occupied by an abscess. The diagnosis is confirmed by the finding of a positive culture for ameba and compatible histologic findings.

Antimicrobial therapy

The SCID model of amebic liver abscess has been used to test the efficacy of proteinase inhibitors in blocking the damage seen with amebic liver abscess (Li *et al.*, 1995; Stanley *et al.*, 1995). Compounds have been administered to mice by the oral and intraperitoneal routes.

Pitfalls (advantages/disadvantages) of the model

The SCID mouse model offers the ability to use the many reagents available to characterize and manipulate the murine immune response. This is an advantage for the study of amebic pathogenesis, especially the role of innate immunity in disease. As with the gerbil and hamster models of disease, it offers the disadvantage of a "non-physiologic" mode of infection. In addition, mice appear to be less susceptible to amebic liver abscess than hamsters or gerbils, and spontaneous cure of the lesions is common in SCID mice, and more unusual in hamsters and gerbils.

Contributions of the model to infectious disease therapy

The SCID mouse model of amebic liver abscess has been most useful in studies of the host inflammatory and innate immune responses to amebic liver abscess. The model has been used to study the role of antibody in mediating protection against amebic liver abscess (Cieslak *et al.*, 1992; Zhang *et al.*, 1994; Seydel *et al.*, 1996; Lotter *et al.*, 1997; Marinets *et al.*, 1997), the role of cysteine proteinases and laminin in the damage seen in amebic liver abscess (Li *et al.*, 1995; Stanley, *et al.*, 1995), and the role of neutrophils in host defense against amebic liver abscess (Seydel *et al.*, 1997b).

References

Anaya-Velazquez, F., Martinez-Palomo, A., Tsutsumi, V., Gonzalez-Robles, A. (1985). Intestinal invasive amebiasis: an experimental model in rodents using axenic or monoxenic strains of *Entamoeba histolytica*. *Am. J. Trop. Med. Hyg.*, **34**, 723–730.

Bhopale, K. K., Pradhan, K. S., Masani, K. B., Kaul, C. L. (1995). Additive effect of diloxanide furoate and metronidazole (Entamizole) in experimental mouse caecal amoebiasis. *Ind. J. Exp. Biol.*, **33**, 73–74.

Chadee, K., Meerovitch, E. (1984). The pathogenesis of experimentally induced amebic liver abscess in the gerbil (*Meriones unguiculatus*). *Am. J. Pathol.*, **117**, 71–80.

Chadee, K., Meerovitch, E. (1985). The pathology of experimentally induced cecal amebiasis in gerbils (*Meriones unguiculatus*). *Am. J. Pathol.*, **119**, 485–494.

Cieslak, P. R., Virgin, H. W., IV, Stanley, S. L., Jr. (1992). A severe combined immunodeficient (SCID) mouse model for infection with *Entamoeba histolytica*. *J. Exp. Med.*, **176**, 1605–1609.

Ghosh, P. K., Mancilla, R., Ortiz-Ortiz, L. (1994). Intestinal amebiasis: histopathologic features in experimentally infected mice. *Arch. Med. Res.*, **23**, 297–302.

Gonzalez-Garza, M. T., Castro-Garza, J., Anaya-Velazquez, F. *et al.* (1997). Gossypol anti-amebic effect *in vivo*. *Arch. Med. Res.*, **28**, 298–299.

Jackson, T. F. (1998). *Entamoeba histolytica* and *Entamoeba dispar* are distinct species; clinical, epidemiological and serological evidence. *Int. J. Parasitol.*, **28**, 181–186.

Li, E., Stanley, S. L., Jr. (1996). Protozoa: amebiasis. *Gastro. Clin. North Am.*, **25**, 471–492.

Li, E., Yang, W. G., Zhang, T. H., Stanley, S. L., Jr. (1995). Interaction of laminin with *Entamoeba histolytica* cysteine proteinases and its effect on amebic pathogenesis. *Infect. Immun.*, **63**, 4150–4153.

Lotter, H., Zhang, T., Seydel, K. B., Stanley, S. L., Jr., Tannich, E. (1997). Identification of an epitope on the *Entamoeba histolytica* 170-kDa lectin conferring antibody mediated protection against invasive amebiasis. *J. Exp. Med.*, **185**, 1793–1801.

Marinets, A., Zhang, T., Guillen, N. *et al.* (1997). Protection against invasive amebiasis by a single monoclonal antibody directed against a lipophosphoglycan antigen localized on the surface of *Entamoeba histolytica*. *J. Exp. Med.*, **186**, 1557–1565.

Meerovitch, E., Chadee, K. (1988). *In vivo* models of pathogenicity in amebiasis. In *Amebiasis. Human Infection by Entamoeba histolytica* (ed Ravdin, J. I.), pp. 425–437. John Wiley, New York.

Miller, M. J. (1952). The experimental infection of *Macaca mulatta* with human strains of *Entamoeba histolytica*. *Am. J. Trop. Med. Hyg.*, **1**, 417–428.

Pargal, A., Kelkar, M. G., Bhopale, K. K., Phaltankar, P. G., Kaul, C. L. (1993a). Pharmacokinetics and amoebicidal activity of (±)-(E)-3-(4-methylsulphinylstyryl)-1,2,4-oxadiazole (BTI 2286[E]) and its sulphone metabolite (BTI 2571[E]) in the golden hamster, *Mesocricetus auratus*. *J. Antimicrob. Chemother.*, **32**, 109–115.

Pargal, A., Rao, C., Bhopale, K. K., Pradhan, K. S., Mansani, K. B., Kaul, C. L. (1993b). Comparative pharmacokinetics and amoebicidal activity of metronidazole and satrianidazole in the golden hamster *Mesocricetus auratus*. *J. Antimicrob. Chemother.*, **32**, 483–489.

Ravdin, J. I. (1995). Amebiasis. *Clin. Infect. Dis.*, **20**, 1453–1466.

Sepúlveda, B., Tanimoto-Weki, M., Calderón, P. (1974). Induccion de inmunidad pasiva antiamibiana en el hamster por la injeccion de suero immune. *Arch. Invest. Méd. (Méx.)*, **5** (suppl 2), 451–456.

Seydel, K. B., Braun, K., Zhang, T., Jackson, T. F. H. G., Stanley, S. L., Jr. (1996). Human anti-amebic antibodies provide protection against amebic liver abscess formation in the SCID mouse. *Am. J. Trop. Med. Hyg.*, **55**, 330–332.

Seydel, K. B., Li, E., Swanson, P. E., Stanley, S. L., Jr. (1997a). Human intestinal epithelial cells produce pro-inflammatory cytokines in response to infection in a SCID mouse-human intestinal xenograft model of amebiasis. *Infect. Immun.*, **65**, 1631–1639.

Seydel, K. B., Zhang, T., Stanley, S. L., Jr. (1997b). Neutrophils play a critical role in early resistance to amebic liver abscesses in SCID mice. *Infect. Immun.*, **65**, 3951–3953.

Shibayama, M., Navarro-Garcia, F., Lopez-Revilla, R., Martinez-Palomo, A., Tsutsumi, V. (1997). *In vivo* and *in vitro* experimental intestinal amebiasis in Mongolian gerbils (*Meriones unguiculatus*). *Parasitol Res.*, **83**, 170–176.

Shibayama-Salas, M., Tsutsumi, V., Martinez-Palomo, A. (1992). Early invasive intestinal amebiasis in mongolian gerbils. *Arch. Med. Res.*, **23**, 187–190.

Sohni, Y. R., Bhatt, R. M. (1996). Activity of a crude extract formulation in experimental hepatic amoebiasis and in immunomodulation studies. *J. Ethnopharmacol.*, **54**, 119–124.

Stanley, S. L., Jr. (1997). Progress towards development of a vaccine for amebiasis. *Clin. Microbiol. Rev.*, **10**, 637–649.

Stanley, S. L., Jr., Zhang, T., Rubin, D., Li, E. (1995). Role of the *Entamoeba histolytica* cysteine proteinase in amebic liver abscess formation in severe combined immunodeficient (SCID) mice. *Infect. Immun.*, **63**, 1587–1590.

Thompson, P. E., McCarthy, D., Reinertson, J. W. (1954). Observations on the virulence of *Entamoeba histolytica* during prolonged subcultivation. *Am. J. Hyg.*, **59**, 249–261.

Tsutsumi, V., Mena-Lopez, R., Anaya-Velazquez, F., Martinez-Palomo, A. (1984). Cellular bases of experimental amebic liver abscess formation. *Am. J. Pathol.*, **117**, 81–91.

Vinayak, V. K., Chitkara, N. L., Chhuttani, P. N. (1979). Effect of corticosteroid and irradiation on caecal amoebic infection in rats. *Trans. R. Soc. Trop. Med. Hyg.*, **73**, 266–268.

Zhang, T., Cieslak, P. R., Foster, L., Kunz-Jenkins, C., Stanley, S. L., Jr. (1994). Antibodies to the serine rich *Entamoeba histolytica* protein (SREHP) prevent amebic liver abscess in severe combined immunodeficient (SCID) mice. *Parasite Immunol.*, **16**, 225–230.

Chapter 103

Animal Models of Giardiasis

R. C. A. Thompson

Background of human infection

Giardia is a flagellated protozoan parasite in the order Diplomonadida, and is considered to be one of the most primitive eukaryotes (Thompson *et al.*, 1993). Species of *Giardia* inhabit the intestinal tracts of virtually all classes of vertebrates, including humans. *G. duodenalis* (sometimes referred to erroneously as *G. lamblia* or *G. intestinalis*) is the only species found in humans and is also common in other mammals. At present, the only other species recognized from mammalian hosts is *G. muris*, a form which appears to be restricted to rodents.

G. duodenalis has a global distribution. About 200 million people in Asia, Africa and Latin America have symptomatic giardiasis and there are some 500 000 new cases a year (WHO, 1996). *Giardia* infection may cause acute and persistent diarrhoea, abdominal pain and rapid weight loss, but risk factors for severe giardiasis are poorly understood and symptoms are highly variable. Children in developing countries and among disadvantaged groups such as Australian Aborigines are most frequently infected, with rates of infection of up to 50% in children under the age of 5 (Meloni *et al.*, 1993; Hopkins *et al.*, 1997). A large proportion of these children have clinical histories of poor weight gain and failure to thrive (Roberts *et al.*, 1988; Gracey, 1994) and it is evident that *Giardia* infections, especially in young children between the ages of 6 months and 4 years, can interfere with growth and development (Islam, 1990; Rabbani and Islam, 1994). In the developed world, the disease has become a serious problem in day-care centres where it is one of the most commonly recognized enteropathogens (Thompson, 1994; CDC, 1995; WHO, 1996).

G. duodenalis has also been isolated from numerous domestic animals and it is an economically significant pathogen in ruminants (Xiao and Herd, 1994a; Xiao *et al.*, 1994; Olson *et al.*, 1995). Mixed infections of *Giardia* and *Cryptosporidium* have been reported in pigs (Xiao *et al.*, 1994b), horses (Xiao and Herd, 1994b) and sheep (Xiao *et al.*, 1994b).

Until recently, there has been a relatively limited range of drugs available for the treatment of *Giardia* infection. These drugs comprise the nitroimidazoles, quinacrine (rarely used) and furazolidone, all of which have problems associated with their use (Reynoldson *et al.*, 1992; Thompson *et al.*, 1993). In particular, in endemic areas where the frequency of *Giardia* transmission is high, there are some drawbacks associated with long dose regimes, poor palatability, patient compliance and potential toxicity, especially in children who may be given frequent treatments because of reinfection. The situation has been alleviated to some extent with the demonstration of the efficacy of benzimidazoles against *Giardia* (Reynoldson *et al.*, 1992, 1997, 1998; Thompson *et al.*, 1993). However, their mode of action, optimum dosages and treatment regimes remain to be fully determined. Furthermore, apart from benzimidazoles, such as albendazole, there are no alternative, or new, antigiardial agents available. Appropriate animal models for screening potential chemotherapeutic agents against *Giardia* are therefore essential.

Background of model

Over the last 40 years, experimental infections of *Giardia* have been studied in a variety of animals. Mice, rats and gerbils have been most commonly used in the development of laboratory models, although dogs, cats, rabbits and sheep have also been utilized (reviewed by Faubert and Belosevic, 1990; Stevens, 1990; Thompson *et al.*, 1990, 1993; Olson *et al.*, 1995).

The reasons for developing a model include the following:

1. Establishment and amplification of isolates for further studies
2. Studying mechanisms of pathogenesis
3. Drug efficacy studies
4. Analysing characteristics of *Giardia* infections
5. Determining viability of field isolates
6. Determining the zoonotic potential and infectivity of particular isolates
7. Studying immune responses
8. Continuous supply of cysts.

In relation to the development of laboratory animal models a number of criteria have been proposed on which

Handbook of Animal Models of Infection
ISBN 0-12-775390-7

the selection of the model should be based (Meyer *et al.*, 1984; Faubert and Belosevic, 1990):

1. Sensitivity to infection
2. Successful infection, including colonization and multiplication of trophozoites in the small intestine
3. Formation and release of cysts in the faeces
4. Infection transferable from humans or other hosts by oral inoculation
5. Cysts excreted infective to other animals
6. *In vivo* susceptibility complemented by ability of *Giardia* isolate to grow *in vitro*
7. Disease resulting from experimental infection with a given strain mimics giardiasis in humans
8. Animal model should be small, inexpensive, readily available, genetically reproducible, able to breed under laboratory conditions and be easy to maintain in the laboratory
9. Background knowledge of the biological properties of the animal.

Not all of the above criteria are essential in a model system to study the antigiardial activity of different drugs. However, it is important that the animal chosen should reflect, as much as possible, the sensitivity to the infection as it occurs in nature and the pathology that will develop later, without any prior manipulation of the animal before the experimental infection. The course of the infection and associated pathology in the animal host should mimic the changes observed in human giardiasis. In addition, it is important that *in vitro* cultivation of the isolate of *Giardia* chosen for animal infection is possible. The ability to study the same organism *in vivo* and *in vitro* clearly offers powerful research advantages in the determination of drug efficacy. This is because initial drug sensitivity testing is usually undertaken *in vitro* and not all isolates of *G. duodenalis* that can be grown *in vitro* are necessarily infective to rodents.

There has been considerable variability in the results of experimental infections between different laboratories. Some of the reasons for this are summarized in Table 103.1. Procedural factors have undoubtedly contributed to this variability, although the contribution of parasite and host factors has often not been taken into account sufficiently. *G. duodenalis* exhibits a considerable degree of genetic heterogeneity which is reflected in a diversity of phenotypic differences between genotypes/strains (Thompson and Meloni, 1993; Thompson *et al.*, 1993; Meloni *et al.*, 1995). This is manifested in characters such as host specificity, growth rate, infectivity, virulence, drug sensitivity, antigenic characteristics and metabolism. Unfortunately, the majority of studies involving experimental infections of non-human hosts have examined only a single isolate of *Giardia*. When different isolates have been compared in the same animal species, variation in the nature and course of infection has usually been reported (Sharma and Mayrhofer, 1988; Visvesvara *et al.*, 1988; Chochillon *et al.*, 1990; Thompson *et al.*, 1990; Udezulu *et al.*, 1992; Cevallos *et al.*, 1995; Thompson and Lymbery, 1996). Consequently, it is essential, that for any animal model system, the characteristics of the isolate(s) of *Giardia* used and their behaviour in a particular laboratory animal are clearly determined and defined.

Table 103.1 Reasons for variability in the results of experimental infections with *Giardia duodenalis*

Experimental animals not proven to be free of infection before inoculation
Experimental animals immune to *Giardia* infection due to prior exposure
Experimental animals exposed to environmental contamination with *Giardia* before and/or during experimental infection
Previous chemotherapeutic treatment not completely effective
Previous chemotherapeutic treatment may have residual effect on experimental infection
Viability of inoculating cysts not assessed
Variability in origin (i.e. strain/genotype) of isolate of *Giardia duodenalis*
Variability in number and viability of cysts used in different experiments
Variability in diet of experimental animals by different laboratories
Inconsistency between laboratories in the number, age, sex and strain of experimental host used
Inadequate controls used
Assessment of the results of experimental inoculation may be in doubt due to the use of unreliable indicators of infection

Unfortunately, the ideal non-human model with which to study *G. duodenalis* has yet to be described, and more research in this area is urgently required. At the present time, the most useful animal models for drug studies are rodents, primarily because of cost, availability and standardization. However, most genotypes of *G. duodenalis* are not natural parasites of rodents and there are problems obtaining reproducible infections, particularly in mature animals (see below). The rodent parasite *G. muris* may appear to be a logical alternative. However, *G. muris* is unlikely to reflect human giardiasis. It does not infect humans and differs demonstrably from the *G. duodenalis* organisms that do. In addition, it will not develop *in vitro*, which suggests that major physiological and biochemical differences exist between *G. muris* and *G. duodenalis*.

Animal species

Mouse

There have been numerous attempts to infect different strains of mice with isolates of *G. duodenalis*, but successful, reproducible infections can only be obtained in young animals, particularly neonates, suckling or weanling mice, although infections are of short duration (Thompson *et al.*, 1993). Alternatively, immunodeficient or immunocompromised animals can be used to prolong infections (Gottstein and Nash, 1991; Watson, 1993).

These limitations, particularly the age-dependent susceptibility, reduce the usefulness of the murine model. Results obtained using immature hosts are often difficult to interpret and many authorities question the validity of the mouse as a model for human giardiasis. Of particular significance is the observation that mice have a rapid gut transit time which is thought to limit their usefulness in studies on drug efficacy (Reynoldson *et al.*, 1991a,b).

Rat

As with mice, rats exhibit variable susceptibility to infection with different isolates of *G. duodenalis*, although success has been achieved with mature as well as immature animals (Craft *et al.*, 1987; Sharma and Mayrhofer, 1988; Cevallos *et al.*, 1995). However, the gut transit time of rats is much slower than that of mice and more closely reflects the situation in humans (Reynoldson *et al.*, 1991a,b; Dow *et al.*, 1998). Consequently, the rat has been advocated as a more useful model host in studies on anti-giardial agents (Thompson *et al.*, 1993).

Gerbil

The Mongolian gerbil (*Meriones unguiculatus*) is widely regarded as being the best experimental host of *G. duodenalis* infections to date and offers a much better alternative to mice as a laboratory model. Many studies on *G. duodenalis* have been greatly advanced with the development of the gerbil model (Belosevic and Faubert, 1983; Belosevic *et al.*, 1983; Faubert *et al.*, 1983; Wu *et al.*, 1989; Faubert and Belosevic, 1990; Udezulu *et al.*, 1992). Gerbils make an excellent model because:

1. They can be infected with isolates from a variety of hosts and are highly susceptible to infection with *G. duodenalis*-type organisms (Wallis and Wallis, 1986; Roach and Wallis, 1988; Swabby *et al.*, 1988)
2. Adult animals can be infected with either cysts or trophozoites cultured *in vitro*
3. Their prepatent time and pathogenesis are similar to those of the original hosts (Swabby *et al.*, 1988; Buret *et al.*, 1991) and so permit pathological studies
4. They are capable of maintaining strains of *G. duodenalis* in the laboratory by serial passage
5. They are relatively inexpensive and easy to care for
6. They are prolific and easy to breed
7. They can be infected with a low infectious dose.

The limitations of the gerbil model are that specific pathogen-free (SPF) animals are not yet available and therefore they may harbour infections with other species of protozoans. Visvesvara *et al.* (1988) reported that not all strains of *G. duodenalis* from humans would reliably infect gerbils. Availability may also be a problem for workers in countries such as Australia where quarantine restrictions do not at present allow the importation of gerbils.

Preparation of animals

If laboratory rodents are not SPF, then it is essential to exclude previous contacts with *Giardia*. This will require collection of several faecal samples over a period of 7 days and microscopic and/or coproantigen analysis to determine *Giardia* status. If naturally infected, it may be necessary to treat animals at least 10 days before experimental infection, orally with metronidazole (20 mg/day). However, if animals are found to be infected before experimentation, it is preferable to exclude such animals because of the possibility of resistance developing to subsequent infections as well as the effects that previous metronidazole treatment may have on subsequent drug efficacy studies.

Upon arrival and following experimental infection, animals should be kept in isolation from non-infected animals and preferably in filter-top cages.

Details of surgery

No surgical procedures or anaesthetics are required since parasites and drugs are administered orally and no invasive procedures are involved in establishing experimental infections in this model.

Storage, preparation of inocula

For the production of reproducible infections, axenically *in vitro* grown trophozoite stages of *G. duodenalis* should be used to establish experimental infections in rodents. *In vitro* cultivation procedures for the establishment and maintenance of isolates of *G. duodenalis* in axenic culture have been described in detail (Meyer, 1976; Meloni and Thompson, 1987; Radulescu and Meyer, 1990; Thompson *et al.*, 1993). Not all isolates of the parasite are amenable to *in vitro* cultivation and for the production of trophozoites to establish infections in laboratory animals it is important to use isolates of *G. duodenalis* which are known to grow reproducibly well *in vitro*, with the rapid and consistent production of monolayer cultures.

In order to prepare inocula, monolayer cultures of *Giardia* trophozoites are placed on ice for 20 minutes and rolled between the palms in order to dislodge trophozoites which normally adhere to the wall of the culture vessel. The trophozoites are then collected by centrifugation at 2500 rpm for 10 minutes at 4°C, counted and resuspended in culture medium for inoculation (Reynoldson *et al.*, 1991b).

Infection process

Animals are infected orally (*per os*) through a stomach tube (intragastric gavage) with 0.1–0.5 ml culture medium containing trophozoites of *G. duodenalis*. This is accomplished most effectively using a plastic tube (diameter 0.7–0.8 mm) attached to the needle of a 1 ml syringe (Reynoldson *et al.*, 1991b). The volume of liquid will vary depending upon the age and size of the animals. We have used between 10^5 and 10^7 trophozoites of strain BAC1 but the ideal numbers of trophozoites used will also vary depending on the species/strain of host and parasite isolate used.

Key parameters to monitor infection

Infection is confined to the intestine. Trophozoites multiply in the small intestine where they attach to the surface of the mucosa. The parasite is not invasive. Infective stages (cysts) are voided in the faeces and this commences from as early as 5 days post-infection. Deaths due to giardiasis are rare and the only signs of infection are diarrhoea and body weight loss. However, these are non-specific symptoms and may not be apparent even in heavily infected animals. The only accurate way to monitor infection in the living animal is the detection of parasite cysts in the faeces. The possibility of using coproantigen detection (Hopkins *et al.*, 1993) appears to be the most accurate method to monitor the status of infection, but this needs to be fully evaluated in the rodent model.

Antimicrobial therapy

It has only been over the last 10 years that any real attempts have been made to develop model systems to determine the efficacy of drugs against *Giardia*, and most of these have been *in vitro* systems (Boreham, 1994). The first *in vivo* system utilized neonatal mice and it is only recently that rats have been used. The mouse model has been used successfully to examine the activity of known antigiardial agents as well as to determine the activity of potentially active compounds (Boreham *et al.*, 1986; Boreham and Upcroft, 1991; Reynoldson *et al.*, 1991b). The gerbil model has still to be evaluated for studying the antigiardial activity of drugs, but, as emphasized above, it offers the most promise.

Drugs are administered once experimental infections are established; this is confirmed by the presence of cysts in the faeces. This is usually between 6 and 10 days after oral administration of trophozoites. Drugs are administered orally by stomach tube in an appropriate vehicle depending on solubility (e.g. phosphate-buffered saline (PBS) or dimethyl sulphoxide). Animals should be split into two groups with one half receiving drug and the other vehicle alone to act as control.

Key parameters to monitor response to treatment

The day before treatment and every day during treatment, all animals are screened for cyst excretion by placing individual animals into a container until faeces are passed. Direct smears prepared on a microscope slide can be screened rapidly for cysts and a semiquantitative assessment of cyst excretion can be made by grading smears from 0 to 4+. However, although a decrease/cessation of cyst excretion in the faeces may give an indication of the anti-giardial activity of a drug, the only definitive and quantitative method of drug efficacy evaluation is to count the numbers of parasites in the intestine.

Depending on the drug being evaluated, animals are usually sacrificed between 4 and 48 hours after the last dose of drug. The proximal two-thirds of the small intestine are removed from each animal, cut longitudinally and placed in screw-capped vials containing a known volume of cold (4°C) PBS. Immersion in cold PBS causes the parasites to detach from the mucosal surface. Vials are shaken vigorously for 30 seconds before a small aliquot is loaded on to a haemocytometer and trophozoite numbers are counted in five squares ($0.1 \times 1 \times 1$ mm) of an improved Neubauer chamber.

Pitfalls (advantages/disadvantages) of the model

The main advantage of an animal model is that potential anti-giardial drugs can be screened in an *in vivo* system before proceeding to a clinical trial in humans. This is important since *in vitro* activity is not necessarily an indicator that a compound may show similar anti-giardial activity *in vivo*, and this has been demonstrated previously (Reynoldson, 1994). Consequently, the provision of an animal model will obviate the unnecessary trialling of drugs in humans. Drugs with anti-giardial activity will as a result become available more quickly for clinical use.

The limitations of an animal model for screening potential anti-giardial agents primarily relate to differences in pharmacodynamics between rodents and humans. As discussed above, this is particularly so for mice, whereas rats offer a more acceptable alternative. Although the gerbil is the best model in terms of susceptibility to, and the reproducibility of, experimental infections, little is known of how the pharmacodynamics of drugs in gerbils relate to the situation in humans. Another disadvantage is that we still do not fully understand how the host–parasite relationship in rodents compares with that in humans, particularly with respect to immune responses and pathogenesis. This is important since the anti-giardial efficacy and mode of action of some drugs may depend, in part, on synergistic relationships with host disease and/or defence processes.

Contributions of the model to infectious disease therapy

In the case of *Giardia*, animal models have only recently been used for evaluating drug efficacy. However, they have already made a significant contribution in terms of the availability of new therapeutics for giardiasis. The discovery a few years ago that benzimidazoles, particularly albendazole, had pronounced anti-giardial activity *in vitro* (Edlind *et al.*, 1990; Meloni *et al.*, 1990) was quickly followed by *in vivo* screening in mice and rats (Reynoldson *et al.*, 1991a,b) which confirmed the anti-giardial action of albendazole. These findings enhanced the possibilities of undertaking clinical trials in humans in a number of endemic regions in the world (Hall and Nahar, 1993; Reynoldson *et al.*, 1997, 1998). As a consequence, from 1997, albendazole is being marketed throughout the world for the treatment of giardiasis. Therefore, in the case of albendazole, the use of animal models allowed this safe, broad-spectrum antiparasitic agent to be available more quickly for the therapy of giardiasis in humans than if only *in vitro* screening had been available.

In addition to enhancing the discovery and availability of new drugs to treat infections with *Giardia*, animal models have played and will continue to play a fundamental role in understanding the pharmacodynamics and mode of action of potential anti-giardial agents.

References

Belosevic, M., Faubert, G. M. (1983). *Giardia muris*: correlation between oral dosage, course of infection, and trophozoite distribution in the mouse small intestine. *Exp. Parasitol.*, **56**, 352–359.

Belosevic, M., Faubert, G. M., MacLean, J. D., Law, C., Croll, N. A., (1983). *Giardia lamblia* infections in Mongolian gerbils: an animal model. *J. Infect. Dis.*, **147**, 222–226.

Boreham, P. F. L. (1994). The current status of chemotherapy for Giardiasis. In Giardia: *From Molecules to Disease* (eds Thompson, R. C. A., Reynoldson, J. A., Lymbery, A. J.), pp. 317–328. CAB International, Wallingford, UK.

Boreham, P. F. L., Upcroft, J. A. (1991). The activity of azithromycin against stocks of *Giardia intestinalis in vitro* and *in vivo*. *Trans. R. Soc. Trop. Med. Hyg.*, **85**, 620–621.

Boreham, P. F. L., Phillips, R. E., Shepherd, R. W. (1986). The activity of drugs against *Giardia intestinalis* in neonatal mice. *J. Antimicrob. Chemother.*, **18**, 393–398.

Buret, A., Gall, D. G., Olson, M. E. (1991). Growth activities of enzymes in the small intestine, and ultrastructure of microvillous border in gerbils infected with *Giardia duodenalis*. *Parasitol. Res.*, **77**, 109–114.

CDC (1995). *Division of Parasitic Diseases Program Review*. National Center for Infectious Diseases, Centers for Disease Control, Atlanta, Georgia, USA.

Cevallos, A., Carnaby, S., James, M., Farthing, J. G. (1995). Small intestinal injury in a neonatal rat model of giardiasis is strain dependent. *Gastroenterology*, **109**, 766–773.

Chochillon, C., Favennec, L., Gobert, J. G., Savel, J. (1990). *Giardia intestinalis*: study of infestation procedures of young mice before weaning. *C. R. Soc. Biol. Filiales*, **184**, 150–157.

Craft, J. C., Holt, E. A., Tan, S. H. (1987). Malabsorption of oral antibiotics in humans and rats with giardiasis. *Paed. Infect. Dis. J.*, **6**, 832–836.

Dow, G. S., Reynoldson, J. A., Thompson, R. C. A. (1998). *Plasmodium berghei: in vivo* efficacy of albendazole in different rodent models. *Exp. Parasitol.*, **88**, 154–156.

Edlind, T. D., Hang, T. L., Chakraborty, P. R. (1990). Activity of anthelmintic benzimidazoles against *Giardia lamblia in vitro*. *J. Infect. Dis.*, **162**, 1408–1411.

Faubert, G. M., Belosevic, M. (1990). Animal models for *Giardia duodenalis* type organisms. In *Giardiasis* (ed Meyer, E. A.), pp. 77–90. Elsevier, Amsterdam.

Faubert, G. M., Belosevic, M., Walker, T. S., MacLean, J. D., Meerovitch, E. (1983). Comparative studies on the pattern of infection with *Giardia* spp. in Mongolian gerbils. *J. Parasitol.*, **69**, 802–805.

Gottstein, B., Nash, T. E. (1991). Antigenic variation in *Giardia lamblia* infection of congenitally athymic nude and SCID mice. *Parasite Immunol.*, **13**, 649–660.

Gracey, M. (1994). The clinical significance of giardiasis in Australian Aboriginal children. In Giardia: *From Molecules to Disease* (eds Thompson, R. C. A., Reynoldson, J. A., Lymbery, A. J.) pp. 281–293. CAB International, Wallingford, UK.

Hall, A., Nahar, Q. (1993). Albendazole as a treatment for infections with *Giardia duodenalis* in children in Bangladesh. *Trans. R. Soc. Trop. Med. Hyg.*, **87**, 84–86.

Hopkins, R. M., Deplazes, P., Meloni, B. P., Reynoldson, J. A., Thompson, R. C. A. (1993). A field and laboratory evaluation of a commercial ELISA for the detection of *Giardia* coproantigens in humans and dogs. *Trans. R. Soc. Trop. Med. Hyg.*, **87**, 39–41.

Hopkins, R. M., Gracey, M., Hobbs, R. P., Spargo, R. M., Yates, M., Thompson, R. C. A. (1997). The prevalence of hookworm (*Ancylostoma duodenale*) infection, iron deficiency and anaemia in an Aboriginal community in north-west Australia. *Med. J. Aust.*, **166**, 241–244.

Islam, A. (1990). Giardiasis in developing countries. In *Giardiasis* (ed Meyer, E. A.), pp. 235–266. Elsevier, Amsterdam.

Meloni, B. P., Thompson, R. C. A. (1987). Comparative studies on the axenic *in vitro* cultivation of *Giardia* of human and canine origin: evidence for intraspecific variation. *Trans. R. Soc. Trop. Med. Hyg.*, **81**, 637–640.

Meloni, B. P., Thompson, R. C. A., Reynoldson, J. A., Seville, P. (1990). Albendazole: a more effective antigiardial agent *in vitro* than metronidazole or tiniazole. *Trans. R. Soc. Trop. Med. Hyg.*, **84**, 375–379.

Meloni, B. P., Thompson, R. C. A., Hopkins, R. M., Reynoldson,

J. A., Gracey, M. (1993). The prevalence of *Giardia* and other intestinal parasites in children, dogs and cats from Aboriginal communities in the west Kimberley region of Western Australia. *Med. J. Aust.*, **158**, 157–159.

Meloni, B. P., Lymbery, A. J., Thompson, R. C. A. (1995). Genetic characterisation of isolates of *Giardia duodenalis* by enzyme electrophoresis: implications for reproductive biology, population structure, taxonomy, and epidemiology. *J. Parasitol.*, **81**, 368–383.

Meyer, E. A. (1976). *Giardia lamblia*: isolation and axenic cultivation. *Exp. Parasitol.*, **39**, 101–105.

Meyer, E. A., Erlandsen, S. L., Radulescu, S. A. (1984). Animal models for giardiasis. In Giardia *and Giardiasis* (eds Erlandsen, S. L., Meyer, E. A.), pp. 233–240. Plenum Press, New York.

Olson, M. E., McAllister, T. A., Deselliers *et al.* (1995). Effects of giardiasis on production in a domestic ruminant (lamb) model. *Am. J. Vet. Res.*, **56**, 1470–1474.

Rabbini, G. H., Islam, A. (1994). Giardiasis in humans: populations most at risk and prospects for control. In Giardia: *From Molecules to Disease* (eds Thompson, R. C. A., Reynoldson, J. A., Lymbery, A. J.), pp. 217–249. CAB International, Wallingford, UK.

Radulescu, S., Meyer, E. A. (1990). *In vitro* cultivation of *Giardia* trophozoites. In *Giardiasis* (ed. Meyer, E. A.), pp. 99–110. Elsevier, Amsterdam.

Reynoldson, J. A. (1994). New approaches in chemotherapy. In Giardia: *From Molecules to Disease* (eds Thompson, R. C. A., Reynoldson, J. A., Lymbery, A. J.), pp. 339–355. CAB International, Wallingford, UK.

Reynoldson, J. A., Thompson, R. C. A., Meloni, B. P. (1991a). The mode of action of benzimidazoles against *Giardia*, and their chemotherapeutic potential against *Giardia* and other protozoa. In *Biochemical Protozoology* (eds Coombs, G. H., North, M. J.), pp. 587–593. Taylor and Francis, London.

Reynoldson, J. A., Thompson, R. C. A., Meloni, B. P. (1991b). *In vivo* efficacy of albendazole against *Giardia duodenalis* in mice. *Parasitol. Res.*, **77**, 325–328.

Reynoldson, J. A., Thompson, R. C. A., Meloni, B. P. (1992). The potential and possible mode of action of the benzimidazoles against *Giardia* and other protozoa. *J. Pharm. Med.*, **2**, 35–50.

Reynoldson, J. A., Behnke, J. M., Pallant, L. J., Macnish, M. G., Gilbert, F., Thompson, R. C. A. (1997). Failure of pyrantel in treatment of human hookworm infections (*Ancylostoma duodenale*) in the Kimberley region of north west Australia. *Acta Trop.*, **68**, 301–312.

Reynoldson, J. A., Behnke, J. M., Gracey, M. *et al.* (1998). Efficacy of albendazole against *Giardia* and hookworm in a remote Aboriginal community in the north of Western Australia. *Acta Trop.*, **71**, 27–44.

Roach, P. D., Wallis, P. M. (1988). Transmission of *Giardia duodenalis* from human and animal sources in wild mice. In *Advances in* Giardia *Research* (eds Wallis, P. M., Hammond, B. R.), pp. 79–82. University of Calgary Press, Calgary.

Roberts, D., Gracey, M., Spargo, R. M. (1988). Growth and morbidity in children in a remote Aboriginal community in North-West Australia. *Med. J. Aust.*, **148**, 68–71.

Sharma, A. W., Mayrhofer, G. (1988). A comparative study of infections with rodent isolates of *Giardia duodenalis* in inbred strains of rats and mice and in hypothymic nude rats. *Parasite Immunol.*, **10**, 169–179.

Stevens, D. P. (1990). Animals model of *Giardia muris* in the mouse. In *Giardiasis* (ed Meyer, E. A.) pp. 91–97. Elsevier, Amsterdam.

Swabby, K. D., Hibler, C. P., Wegrzyn, J. G. (1988). Infection of Mongolian gerbils (*Meriones unguiculatus*) with *Giardia* from human and animal sources. In *Advances in* Giardia *Research* (eds Wallis, P. M., Hannon, B. R.), pp. 75–77. University of Calgary Press, Calgary.

Thompson, S. C. (1994). *Giardia lamblia* in children and the child care setting: a review of the literature. *J. Paed. Child Hlth.*, **30**, 202–209.

Thompson, R. C. A., Lymbery, A. J. (1996). Genetic variability in parasites and host–parasite interactions. *Parasitology*, **112**, S7–S22.

Thompson, R. C. A., Meloni, B. P. (1993). Molecular variation in *Giardia* and its implications. *Acta Trop.*, **53**, 167–184.

Thompson, R. C. A., Lymbery, A. J., Meloni, B. P. (1990). Genetic variation in *Giardia* Kunstler, 1882: taxonomic and epidemiological significance. *Protozool. Abs.*, **14**, 1–28.

Thompson, R. C. A., Reynoldson, J. A., Mendis, A. H. W. (1993). *Giardia* and giardiasis. *Adv. Parasitol.*, **32**, 71–160.

Udezulu, I. A., Visvesvara, G. S., Moss, D. M., Leitch, G. J. (1992). Isolation of two *Giardia lamblia* (WB strain) clones with distinct surface protein and antigenic profiles and differing infectivity and virulence. *Infect. Immun.*, **60**, 2274–2280.

Visvesvara, G. S., Dickerson, J. W., Healy, G. R. (1988). Variable infectivity of human-derived *Giardia lamblia* cysts for Mongolian gerbils (*Meriones unguiculatus*). *J. Clin. Micro.*, **26**, 837–841.

Wallis, P. M., Wallis, H. M. (1986). Excystation and culturing of human and animal *Giardia* spp. by using gerbils and TYI-S-33 medium. *Appl. Environ. Microbiol.*, **51**, 647–651.

Watson, R. R. (1993). Resistance to intestinal parasites during murine AIDS: role of alcohol and nutrition in immune dysfunction. *Parasitology*, **107**, S69–S74.

Wu, Y. H., Wang, Z. Y., Lu, S. Q., Ji, A. P., Zhang, C. H. (1989). Further studies on the Mongolian jird model of *Giardia lamblia*. *Chin. J. Parasitol. Parasitic Dis.*, **7**, 187–190.

WHO (1996). *The World Health Report 1996. Fighting Disease Fostering Development*. World Health Organization, Geneva.

Xiao, L., Herd, R. P. (1994a). Infection pattern of *Cryptosporidium* and *Giardia* in calves. *Vet. Parasitol.*, **55**, 257–262.

Xiao, L., Herd, R. P. (1994b). Epidemiology of equine *Cryptosporidium* and *Giardia* infections. *Equine Vet. J.*, **26**, 14–17.

Xiao, L., Herd, R. P., McClure, K. E. (1994a). Periparturient rise in the excretion of *Giardia* sp. cysts and *Cryptosporidium parvum* oocysts as a source of infection for lambs. *J. Parasitol.*, **80**, 55–59.

Xiao, L., Herd, R. P., Bowman, G. L. (1994b). Prevalence of *Cryptosporidium* and *Giardia* infections on two Ohio pig farms with different management systems. *Vet. Parasitol.*, **52**, 331–336.

Chapter 104

Schistosomosis

G .C. Coles

Introduction

The human blood fluke is estimated to infect about 200 million people worldwide, with the most important species of worms being *Schistosoma mansoni* in Africa and parts of South America, *S. haematobium* in Africa and parts of the Middle East and *S. japonicum* primarily in the Philippines and China. The major damage caused by the worms is due to the host response to the eggs (granuloma formation) in the liver and intestines, or in the bladder with *S. haematobium*. Infection can be controlled by:

1. Killing the snail intermediate host with molluscicides
2. Altering the environment so it is unsuitable for snails, e.g. with concrete irrigation channels
3. Improving hygiene through provision of toilets and clean water for washing.

However because of its ease of application and relatively low cost the major emphasis on control has been on chemotherapy of infected people.

The history of development of anti-schistosomal compounds has been reviewed several times, for example, by Sturrock *et al.*, (1986). The first widely used drugs were the antimonials, followed by hycanthone and niridazole. More recently therapy has relied on oxamniquine (*S. mansoni* only), metrifonate (*S. haematobium* only) and praziquantel (all species). The excellent activity of praziquantel against all three species has resulted in a loss of interest in the production of new drugs. As a result basic research on *Schistosoma* has been largely directed to either gene sequencing or the discovery of vaccines. Oxamniquine and metrifonate have almost been forgotten as practical anti-schistosomal drugs and their continued production could be placed in jeopardy as a result.

Resistance

The major problem with reliance on chemotherapy for any infective organism is the development of drug resistance, and schistosomes are no exception to this rule. A definition of resistance in *Schistosoma* has been given by Coles and Kinoti (1997) and drug resistance in *Schistosoma* has been reviewed by Coles and Bruce (1990), Cioli *et al.* (1993; primarily oxamniquine) and Fallon *et al.* (1996a; praziquantel). The reason that resistance has been slow to develop is that the contribution of eggs laid by schistosomes surviving therapy to the overall contamination of the environment, and thus infection of snails, will be very small with sporadic treatment of infected people. However, where mass chemotherapy is practised as, for example, in Egypt with praziquantel, or where infection is introduced to a new area of irrigation from people who have been treated with schistosomicides, then resistance can be expected to develop as a practical problem. Work in Kenya demonstrated that low numbers of worms were resistant to oxamniquine despite its lack of widespread use (Coles *et al.*, 1987). Thus resistance to oxamniquine could rapidly become a practical problem if mass therapy was used with this drug in East Africa. In Egypt not all patients were cured despite multiple treatments with praziquantel (Ismail *et al.*, 1996). This is of concern if mass therapy with praziquantel continues. Lower than expected cure rates were found after a divided dose of praziquantel in Senegal (Guisse *et al.*, 1997). Laboratory studies showed a lower cure rate in mice than with three other isolates (Fallon *et al.*, 1996b, 1997), although these studies and clinical trials in Senegal confirmed susceptibility to oxamniquine (Stelma *et al.*, 1997).

There is a need, therefore, for the development and discovery of novel schistosomicides. The poor market size for schistosomicides, due to the poverty of those infected, combined with the excellent activity and safety of praziquantel, has discouraged pharmaceutical companies from searching for new products. Further screening of chemicals and evaluation of new actives will have to be sponsored by the World Health Organization if additional schistosomicides with novel mechanisms of action are to be produced. The chemicals to be tested will come from three sources: plant extracts, microbial fermentations and synthetic chemistry.

Primary screens

Enzyme screens

In the pharmaceutical industry the major emphasis on drug discovery has returned from attempting to design drugs to

Handbook of Animal Models of Infection
ISBN 0-12-775390-7

serendipity. Very large numbers of chemicals are made in small amounts (combinatorial chemistry) and are tested on robotically run screens employing receptors or enzymes as the targets. There is no reason why this cannot be used for testing schistosome enzymes or receptors, but the application of this type of technology to metazoan parasites has problems. Most robotic screens are designed for finding synthetic chemicals or natural products that would affect a particular physiological function in a mammal/human and thus correct a disorder. However the physiological/biochemical functions which will make a really good target for killing *Schistosoma* are not known. For example, the exact mode of action of praziquantel has not yet been determined (Redman *et al.*, 1996). Therefore praziquantel could not have been discovered by robotic screening. Hopefully the sequencing of the schistosome genome combined with a logical investigation of schistosome biochemistry (Thompson *et al.*, 1996) will produce meaningful targets for robotic screening. Until a wide range of schistosome enzymes and receptors are available for screening, tests based on use of the whole organism will be required.

In vitro tests

The advantage of using the whole organism in screening, rather than using enzymes or receptors, is that chemicals can be identified which act on unknown targets. Three problems emerge when using whole schistosomes. The first is the difficulty of obtaining sufficiently large numbers of chemicals given the way the pharmaceutical industry has changed recently. The second is the activation of chemicals by the host or the loss of activity of compounds, that are active *in vitro*, when given to the mammalian host (for discussion, see, for example, Pellegrino and Katz, 1968). The third is the immunological role of the host in the killing of schistosomes (Brindley and Sher, 1987, 1990; Doenhoff *et al.*, 1987, 1991; Fallon *et al.*, 1995, 1996b).

In vitro tests were examined by Coles (1975). The tests were run before praziquantel was generally available. It was concluded that the screen was not sufficiently selective for use with random compounds either using schistosomula (the easiest stage to prepare in large numbers) or 3-week-old or adult worms; the latter two were collected by perfusion from mice. It would be worth investigating the response of schistosomula to drugs in the improved medium of Basch (1981), but whether a more selective *in vitro* screen would result remains to be determined. An alternative method to looking for death of worms might be to look at tegumental damage in the presence of anti-schistosomal serum, as revealed by binding of host cells to cultured schistosomes. However, the complexity of this approach as well as the cost of production of antiserum and host cells may make the test impractical.

The mouse model

Adult mice have been and remain the best small mammal model for evaluating chemicals against schistosomes. They have been used in all recent research on potential or actual schistosomicides that dose not involve the use of primates (Foster *et al.,* 1970, 1971; Andrews, 1981; Sabah *et al.,* 1986; Munro and McLaren, 1990; el Kouni, 1991; Fallon *et al.*, 1995; Pendino *et al.*, 1995). The most comprehensive review of the maintenance of schistosomes in the laboratory and their use for testing of compounds is given by Pellegrino and Katz (1968), but other descriptions and reviews include those by Lee and Lewert (1956) and Fransden (1981).

Preparation of cercariae

S. mansoni is usually used in screens as it is much easier to maintain the snail intermediate host (*Biomphalaria* species, and particularly *B. glabrata*) than the intermediate hosts for *S. japonicum* (*Oncomelania* sp.) or *S. haematobium* (*Bulinus* sp). *B. glabrata* can readily be maintained in shallow dishes or aquaria containing hard (calcium-rich) water maintained at 25–27°C and fed a diet of green lettuce supplemented with fish food. Water is changed weekly. Small (c. 5 mm diameter) snails are infected with miracidia either individually in multi-well plates or *en masse*, in either case using about 10 miracidia per snail. After about 4 weeks (the exact time depends on the temperature), snails are transferred to the dark to prevent cercarial shedding. When these snails are put in a beaker in strong light for 1–2 hours, cercariae are shed. Numbers are counted in a gently stirred suspension by taking aliquots and staining with Lugol's iodine. Cercariae should be used as soon as they have been shed as infectivity decreases with time. Since the cercariae are highly infective, active steps must be taken to prevent any contact with the skin of the personnel involved and safety precautions for handling infected snails and cercariae should be prepared and enforced.

Mice used can be either outbred or inbred, male or female. Mice can be infected in a number of ways.

Infection of mice

The numbers of cercariae used for infection will depend on the age and breed of mice and the route of infection and will probably be in the region of 100–130, but will not usually exceed 200 per mouse.

1. Cercariae are injected intraperitoneally using a reasonably broad-gauge needle to prevent destruction of the cercariae during injection. Use of injection to infect mice with *S. japonicum* has been described by Moloney and Webbe (1982).
2. Mice are injected with barbiturate as an anaesthetic. The abdomen of the mice is shaved and a stainless steel ring is placed on the abdomen. The cercariae are pipetted into the ring and allowed to remain there for about 20 minutes (Smithers and Terry, 1965).
3. Mice are confined in a suitable container and allowed to paddle in shallow water. This causes defecation and

urination which may damage cercariae. Mice are then transferred individually to suitably vented bottles containing cercariae and allowed to paddle for a further 20 minutes (Standen, 1949).

4. Mice are infected by dangling the tail in a testtube containing cercariae for more than 30 minutes (Olivier and Stirewalt, 1952). Mice can either be anaesthetized (Pellegrino and Katz, 1968) or held in a restrainer (Berrios Duran, 1955; Stirewalt and Bronson, 1955).

Where percutaneous infection is being used it is important that, prior to infection, mice are not kept on soft-wood shavings, as oils in the shavings may adversely affect cercarial penetration through the skin.

Therapy of mice

Mice are kept for 6 weeks to allow worms to mature and are then treated with drug, usually by oral gavage on 5 consecutive days, although drug injection may be appropriate in certain circumstances. The effectiveness of the drug treatment can be checked by determining whether eggs have stopped appearing in the faeces (if no effect has been observed on egg production animals can be re-used). Alternatively, the effectiveness of drug treatment can be evaluated using the oogram technique (Pellegrino and Faria, 1965), in which changes in miracidial development are investigated within eggs in the tissues. In small mammals this will usually be on tissue taken at post-mortem. In larger mammals oograms can be determined from biopsy samples taken from the large intestine. The oogram technique will detect all compounds affecting egg laying by worms but some of these chemicals may not be killing adult worms. The third method of assessing treatment effectiveness requires mice to be killed with barbiturate and the worms removed by perfusion and counted. Methods for the easy recovery of worms have been described by Radke *et al.* (1961) and Smithers and Terry (1965).

Other models

Compounds showing activity in mice will then be evaluated in hamsters and possibly gerbils. Infection of these species will usually be by paddling in cercarial suspension, although infection via the cheek pouch has been used in hamsters (Pellegrino *et al.*, 1965). Compounds showing activity in two species of rodents against *S. mansoni* can then be evaluated in primates. However, on economic grounds it is probably essential to have activity against all three major species of schistosomes if further evaluation is to be justified. Both hamsters (James *et al.*, 1972) and gerbils have the advantage that they can be used with all three major species of schistosomes. The rabbit can also be used for infections of *S. japonicum* (Xiao *et al.*, 1992).

Use of primates

Only compounds showing activity and acceptable levels of toxicity in rodents will be tested in primates. A wide range of monkeys have been investigated as models for infection with *S. mansoni* (reviewed by Pellegrino and Katz, 1968). The preferred species may depend on availability of primates but can include *Cebus apella macrocephalus* (capuchin monkey), *Macaca mulatta* (rhesus monkey), *Cercopithecus aethiops* (vervet monkey) and *Papio anubis* (baboon; Foster *et al.*, 1971; Webbe *et al.*, 1981; Crawford *et al.* 1983; Sturrock *et al.*, 1985). After treatment faecal egg counts, rectal snips and oograms will be performed on live animals, and worm counts and tissue egg burdens determined at post-mortem.

Contributions of the model to infectious disease therapy

Apart from antimony and metrifonate, which were directly tested on people, all recent experimental (e.g. oltipraz and amoscanate) and successful schistosomicides were discovered using the mouse model and then further evaluated in primates.

References

Andrews, P. (1981). A summary of the efficacy of praziquantel against schistosomes in animal experiments and notes on its mode of action. *Drug Res.*, **31**, 538–541.

Basch, P. F. (1981). Cultivation of *Schistosoma mansoni in vitro*. I. Establishment of cultures from cercariae and development until pairing. *J. Parasitol.*, **67**, 179–185.

Berrios Duran, L. A. (1955). An efficient device for exposing mice to schistosome cercariae and holding of small animals for post mortem examination. *J. Parasitol.*, **41**, 641–642.

Brindley, P. J., Sher, A. (1987). The chemotherapeutic effect of praziquantel against *Schistosoma mansoni* is dependent on host antibody response. *J. Immunol.*, **139**, 215–220.

Brindley, P. J., Sher, A. (1990). Immunological involvement in the efficacy of praziquantel. *Exp. Parasitol.*, **71**, 245–248.

Cioli, D., Pica-Mattoccia, L., Archer, S. (1993). Drug resistance in schistosomes. *Parasitol. Today*, **9**, 162–166.

Coles, G. C. (1975). The response of *Schistosoma mansoni* maintained *in vitro* to schistosomicidal compounds. *J. Helminthol.*, **49**, 205–209.

Coles, G. C., Bruce, J. I. (1990). Drug resistance in *Schistosoma*. In *Resistance of Parasites to Antiparasitic Drugs* (eds Boray, J. C., Martin, P. J., Roush, R. T.), pp. 51–60. MSD AGVET, Rahway, New Jersey.

Coles, G. C. Kinoti, G. K. (1997). Defining resistance in *Schistosoma*. *Parasitol. Today*, **13**, 157–158.

Coles, G. C., Mutahi, W. T., Kinoti, G. K., Bruce, J. I., Katz, N. (1987). Tolerance of Kenyan *Schistosoma mansoni* to oxamniquine. *Trans. R. Soc. Trop. Med. Hyg.*, **81**, 782–785.

Crawford, K. A., Asch, H. L., Bruce, J. I., Bueding, E., Smith, E. R. (1983). Efficacy of amoscanate against experimental schistosomal infections in monkeys. *Am. J. Trop. Med. Hyg.*, **32**, 1055–1064.

Doenhoff, M. J., Sabah, A. A., Fletcher, C., Webbe, G., Bain, J. (1987). Evidence for an immune-dependent action of praziquantel on *Schistosoma mansoni* in mice. *Trans. R. Soc. Trop. Med. Hyg.*, **81**, 947–951.

Doenhoff, M. J., Modha, J., Lambertucci, J. R., Mclaren, D. J. (1991). The immune dependence of chemotherapy. *Parasitol. Today*, **7**, 16–18.

el Kouni, M. H. (1991). Efficacy of combination therapy with tubericidin and nitrobenzylthioinosine 5'-monophosphate against chronic and advanced stages of schistosomiasis. *Biochem. Pharmacol.*, **41**, 815–820.

Fallon, P. G., Hamilton, J. V., Doenhoff, M. J. (1995). Efficacy of treatment of murine *Schistosoma mansoni* infections with praziquantel and oxaminiquine correlates with infection intensity — role of antibody. *Parasitology*, **111**, 59–66.

Fallon, P. G., Tao, L. F., Ismail, M. M., Bennett, J. L. (1996a). Schistosome resistance to praziquantel — fact or artifact. *Parasitol. Today*, **12**, 316–320.

Fallon, P. G., Fookes, R. E., Wharton, G. A. (1996b). Temporal differences in praziquantel induced and oxamniquine induced tegumental damage to adult *Schistosoma mansoni* — implications for drug antibody synergy. *Parasitology*, **112**, 47–58.

Fallon, P. G., Mubarak, J. S., Fookes, R. E. *et al.* (1997). *Schistosoma mansoni*: maturation rate and drug susceptibility of different geographic isolates. *Exp. Parasitol.*, **86**, 29–36.

Foster, R., Cheetham, B. L., Mesmer, E. T., Ming, D. F. (1970). Comparative studies of the action of mirasan, lucanthone, hycanthone and niridazole against *Schistosoma mansoni* in mice. *Ann. Trop. Med. Parasitol.*, **65**, 45–58.

Foster, R., Cheetham, B. L., King, D. F., Mesmer, E. T. (1971). The action of UK 3883, a novel 2-aminomethyl tetrahydroquinoline derivative, against mature schistosomes in rodents and primates. *Ann. Trop. Med. Parasitol.*, **65**, 59–70.

Fransden, F. (1981). Cultivation of schistosomes for chemotherapeutic studies. *Acta Pharmacol. Toxicol.*, **49** (suppl. 5), 118–122.

Guisse, F., Polman, K., Stelma, F. F. *et al.* (1997). Therapeutic evaluation of two different dose regimens of praziquantel in a recent *Schistosoma mansoni* focus in northern Senegal. *Am. J. Trop. Med. Hyg.*, **56**, 511–514.

Ismail, M., Metwally, A., Farghaly, A., Bruce, J., Tao, L. F., Bennett, J. L. (1996). Characterisation of isolates of *Schistosoma mansoni* from Egyptian villagers that tolerate high doses of praziquantel. *Am. J. Trop. Med. Hyg.*, **55**, 214–218.

James, C., Webbe, G., Preston, J. M. (1972). A comparison of the susceptibility to metrifonate of *Schistosoma haematobium*, *S. mattheei* and *S. mansoni* in hamsters. *Ann. Trop. Med. Parasitol.*, **66**, 467–474.

Lee, C. L., Lewert, R. (1956). The maintenance of *Schistosoma mansoni* in the laboratory. *J. Infect. Dis.*, **99**, 15–20.

Moloney, N. A., Webbe, G. (1982). A rapid method of infecting mice with *Schistosoma japonicum*. *Trans. R. Soc. Trop. Med. Hyg.*, **76**, 200–203.

Munro, G. H., McLaren, D. J. (1990). Toxicity of cyclosporin A (CsA) against developmental stages of *Schistosoma mansoni* in mice. *Parasitology*, **100**, 29–34.

Olivier, L., Stirewalt, M. A. (1952). An efficient method for exposure of mice to cercariae of *Schistosoma mansoni*. *J. Parasitol.*, **38**, 19–23.

Pellegrino, J., Faria, J. (1965). The oogram method for the screening of drugs in schistosomiasis mansoni. *Am. J. Trop. Med. Hyg.*, **14**, 363–369.

Pellegrino, J., Katz, N. (1968). Experimental chemotherapy of schistosomiasis mansoni. *Adv. Parasitol.*, **6**, 233–290.

Pellegrino, J., De Maria, M., Fariae, J. (1965). Infection of the golden hamster with *Schistosoma mansoni* through the cheek pouch. *J. Parasitol.*, **51**, 1015.

Pendino, M. L. O., Nelson, D. L., Vieira, L. Q., Watson, D. G., Kusel, J. R. (1995). Metabolism by *Schistosoma mansoni* of a new schistosomicide 2-[(1-methylpropyl)amino]-1-octanethiosulphuric acid. *Parasitology*, **111**, 177–185.

Radke, M. G., Berrios-Durran, L. A., Moran, K. (1961). A perfusion procedure (perf-o-suct-ion) for recovery of schistosome worms. *J. Parasitol.*, **47**, 366–368.

Redman, C. A., Robertson, A., Fallon, P. G. *et al.* (1996). Praziquantel — an urgent and exciting challenge. *Parasitol. Today*, **12**, 14–20.

Sabah, A. A., Fletcher, C., Webbe, G., Doenhoff, M. J. (1986). *Schistosoma mansoni*: chemotherapy of infections of different ages. *Exp. Parasitol.*, **61**, 294–303.

Smithers, S. R., Terry, R. J. (1965). The infection of laboratory hosts with cercariae of *Schistosoma mansoni* and the recovery of adult worms. *Parasitology*, **55**, 695–700.

Standen, O. D. (1949). Experimental schistosomiasis. II. Maintenance of *Schistosoma mansoni* in the laboratory, with some notes on experimental infection with *S. haematobium*. *Ann. Trop. Med. Parasitol.*, **43**, 268–283.

Stelma, F. F., Sall, S., Daff, B., Sow, S., Niang, M., Gryseels, B. (1997). Oxamniquine cures *Schistosoma mansoni* infection in a focus in which cure rates with praziquantel are unusually low. *J. Infect. Dis.*, **176**, 304–307.

Stirewalt, M. A., Bronson, J. F. (1955). Description of a plastic mouse restraining case. *J. Parasitol.*, **41**, 328.

Sturrock, R. F., Otieno, M., James, E. R., Webbe, G. (1985). A note on the efficacy of a new class of compounds 9-acridanone-hydrazones, against *Schistosoma mansoni* in a primate — the baboon. *Trans. R. Soc. Trop. Med. Hyg.*, **79**, 129–131.

Sturrock, R. F., Doenhoff, M. J., Webbe, G. (1986). Schistosomiasis. In *Experimental Models in Antimicrobial Chemotherapy*, vol 3, (eds Zak, O., Sande, M. A.), pp. 241–279. Academic Press, London.

Thompson, D. P., Klein, R. D., Gery, T. G., (1996). Prospects for rational approaches to anthelminthic discovery. *Parasitology*, **113**, S217–238.

Webbe, G., James, C., Nelson, G. S., Sturrock, R. F. (1981). The effect of praziquantel on *Schistosoma haematobium, S. japonicum* and *S. mansoni* in primates. *Arzneimittelforschung*, **31**, 542–544.

Xiao, S. H., You, J. Q., Guo, H. F. (1992). Plasma pharmacokinetics and therapeutic efficacy of praziquantel and 4-hydroxypraziquantel in *Schistosoma japonicum* infected rabbits after oral, rectal, and intramuscular administration. *Am. J. Trop. Med. Hyg.*, **46**, 582–588.

Chapter 105

Animal Models for Echinococcosis

T. Romig and B. Bilger

Background of model

Overview

The taeniid cestodes of the genus *Echinococcus* are characterized by their small body size as adult tapeworms and their massive growing capacity and asexual propagation as metacestodes. Like most taeniids, they require two different host species (both mammals) in order to complete their life cycle: the adult cestodes are intestinal parasites of carnivores (final hosts), the larval stages (metacestodes) are tissue parasites of a large variety of intermediate host species which — depending on the species of parasite and host — may occur in almost every location of the body.

Intermediate hosts acquire the infection by oral uptake of cestode eggs from the final host's faeces. After hatching from the eggs, the oncosphere larvae penetrate the intestinal wall and are transported by the circulatory system to various organs where they settle and grow to the metacestode stage. Eventually, large numbers of protoscolices are produced within the metacestode which, after ingestion, grow into adult cestodes in the final host's intestine (predator–prey systems). *Echinococcus* metacestodes in suitable intermediate hosts are characterized by rapid growth and profuse production of protoscolices (fertile metacestodes), whereas in less suitable species growth is delayed and protoscolex production is impaired or absent (sterile metacestodes).

Four species of *Echinococcus* are currently recognized; the metacestodes of all species are known to be pathogenic to humans. *E. oligarthrus* and *E. vogeli* are exclusively distributed in South and Central America. Their life cycle, range and medical importance have recently been reviewed (D'Alessandro, 1997). *E. multilocularis* is a wildlife parasite occurring throughout the temperate and arctic regions of the northern hemisphere. Its public health importance is generally low, although in some regions it constitutes a serious problem (for review see Schantz *et al.*, 1995; Craig *et al.*, 1996). *E. granulosus* is of worldwide distribution. Its life cycle typically involves domestic animals and, consequently, its medical impact on a worldwide scale is by far the most serious of all *Echinococcus* species (Schantz *et al.*, 1995; Craig *et al.*, 1996). It has to be noted, however, that *E. granulosus* is most likely a paraphyletic assemblage of some seven species showing distinct genetic, morphological and ecological differences. Since a formal recognition of these species is still pending (Thompson *et al.*, 1994; Lymbery, 1995), the term *E. granulosus* is used here in its conventional sense.

Clinical appearance

The terminology of clinical disease caused by the four *Echinococcus* species is shown in Table 105.1; for further details see Pawlowski (1997). Human cystic echinococcosis usually presents as large cysts that may contain up to several litres of hydatid fluid; fertility is high and the growth pattern resembles that in animal intermediate hosts (not all strains of *E. granulosus*, however, are supposed to be infectious to humans). Alveolar echinococcosis mostly appears as a solid parasitic mass which is composed of small fluid-filled vesicles; in humans, the growth rate is usually very slow and fertility is low or absent. Polycystic echinococcosis is intermediate in appearance; it is not further considered

Table105.1 Overview of *Echinococcus* spp. and nomenclature of the disease

Species	*Name of disease*	*Natural intermediate hosts*
E. granulosus	Cystic echinococcosis, hydatidosis, hydatid disease	Domestic and wild ruminants and other herbivores
E. multilocularis	Alveolar echinococcosis	Voles and other rodents
E. oligarthrus	Polycystic echinococcosis	Rodents
E. vogeli	Polycystic echinococcosis	Rodents

Handbook of Animal Models of Infection
ISBN 0-12-775390-7

here since clinical cases are rare and not well-known. Due to its growth characteristics and capacity for metastases formation, alveolar echinococcosis is considered the most serious disease and, likewise, the most difficult to diagnose and treat.

All types of echinococcosis may present as inoperable conditions. The currently available drugs (albendazole, mebendazole) in most cases do improve the patient's condition and arrest the progression of the disease, but do not reliably destroy the parasite and may have to be administered in long-term courses (with alveolar echinococcosis, life-long chemotherapy is the rule). For a review of clinical disease and treatment, see Ammann and Eckert (1995).

The present state of clinical management of echinococcosis is generally unsatisfactory and requires further efforts both in drug development and in immunological studies with a view to vaccination or immunotherapy.

Existing animal models

Due to the low host specificity of *Echinococcus* spp. in the metacestode stage, a wide range of mammal species are known to be natural intermediate hosts (Rausch, 1995) and, consequently, a long list of species has been used for experimental infections. Due to the partially atypical growth in humans, experimental infections of rodents cannot be considered satisfactory models for human echinococcosis. Further, the two-host transmission cycle would ideally require oral infections of rodents using *Echinococcus* eggs in order to model the infection mode and early development of the disease. To avoid the safety precautions necessary when handling eggs, the oral route of infection is usually replaced by intraperitoneal transplantation of metacestode material (secondary echinococcosis). Various species of animals and various ways of infection have been described and will be briefly reviewed in the following (for examples and references, see Tables 105.2 and 105.3). Experimental infections of final hosts (dogs, cats and other carnivores) are not considered here (for the development of adult worms in laboratory rodents, see Kamiya and Sato, 1990).

E. granulosus

Oral infection

This mode of infection has been routinely performed with sheep, although sporadically other mammals have been infected for various purposes. Studies with monkeys (vervets and baboons) as model hosts were not entirely conclusive due to large differences in infection rates and intensity. Little work has been done to establish a rodent model: various strains of mice (*Mus musculus*) showed unsatisfactory infection rates which improved when eggs were not administered orally but were artificially hatched, activated and injected intraperitoneally or intravenously. In all

Table 105.2 Selected studies with metacestodes of *Echinococcus granulosus*

Reference	*Animal spp. (strain)*	*Route of application*	*Inoculum per animal*	*Subject of study*
Heath *et al.* (1992)	Sheep	Oral	500–2000 eggs	Immunology
Dempster *et al.* (1995)	Sheep	Oral	200–5000 eggs	Vaccination
Rogan *et al.* (1993)	Baboons, vervet monkeys	Oral	2000–9000 eggs	Model development
Jenkins and Thompson (1995)	Rabbits	Oral	500–6000 eggs	Host specificity
Dempster *et al.* (1992)	Mice (DBA/2, CBA, BALB/c, C57B16, CF-1)	Oral, intraperitoneal, intravenous	50–1000 eggs, 60–900 activated oncospheres	Model development
Hernández and Nieto (1994)	Mice (CD1)	Intraperitoneal	2000 protoscolices	Immunology, vaccination
Liu *et al.* (1992)	Mice (BALB/c)	Intraperitoneal	2000 protoscolices	Immunlogy
Rogan and Richards (1989)	Mice (BALB/c)	Intraperitoneal	4000 protoscolices	Parasite development
Denegri *et al.* (1995)	Mice (NMRI)	Intraperitoneal	1500 protoscolices	Serodiagnosis
Pennoit-De Cooman and De Rycke (1978)	Mice (NMRI)	Intraperitoneal	17–100 cysts (<1 mm diameter)	Model development
Pérez-Serrano *et al.* (1997)	Mice (NMRI)	Intraperitoneal	1500 protoscolices	Chemotherapy
Casado *et al.* (1992)	Mice (NMRI)	Intraperitoneal	15 cysts (1–2.5 mm diameter)	Viability test
Taylor *et al.* (1989a)	Gerbils	Intraperitoneal	5000 protoscolices	Chemotherapy
Taylor *et al.* (1989b)	Gerbils	Intraperitoneal	4–15 cysts (0.3–6.6 mm diameter)	Viability test
Riley *et al.* (1985)	Mice (BALB/c)	Subcutaneous	200 protoscolices	Immunology

Table 105.3 Selected studies with metacestodes of *Echinococcus multilocularis*

Reference	*Animal spp. (strain)*	*Route of application*	*Inoculum per animal*	*Subject of study*
Rausch (1954)	Field vole (*Microtus pennsylvanicus*)	Oral	400–2000 eggs	Parasite development
Ohbayashi (1960)	Voles (*M. montebelli*), cotton rats, mice (DBA, CF-1, C57BL/6)	Oral	120–1000 eggs	Parasite development
Veit *et al.* (1995)	Voles (*M. arvalis*)	Oral	2000 eggs	Viability test
Bauder *et al.* (1998)	Mice (C57BL/6, C57BL/10)	Oral	1700 eggs	Immunology
Taylor *et al.* (1988)	Cotton rats	Intraperitoneal	0.5 ml metacestode homogenate	Chemotherapy
Liance *et al.* (1992)	Mice (AKR)	Intraperitoneal	0.5 ml metacestode homogenate	Immunology
Schantz *et al.* (1990)	Gerbils	Intraperitoneal	0.1 ml metacestode homogenate	Chemotherapy
Kroeze and Tanner (1987)	Mice (C57L, SJ/L, C57BL10Sn, C57BL/6, BALB/c)	Intraperitoneal	3 vesicles (< 1 mm diameter)	Model development
Ali-Khan *et al.* (1988)	Mice (C57BL/6, CBA, C3H/He, C3H/HeSn)	Intraperitoneal	50–60 vesicles	Immunology
Playford *et al.* (1992)	Mice (SCID)	Intraperitoneal	0.5 ml metacestode homogenate	Immunology
Novak *et al.* (1991)	Gerbils	Intraperitoneal	0.5 ml metacestode homogenate	Imaging research
Modha *et al.* (1997)	Gerbils	Intraperitoneal	0.5 ml metacestode homogenate	Biochemistry
Sarciron *et al.* (1995)	Gerbils	Intraperitoneal	50 mg piece of metacestode (transplant)	Chemotherapy
Liance *et al.* (1984)	Mice (AKR, CBA, C57BL/6, C57BL/10)	Intrahepatic	0.1 ml metacestode homogenate (1 : 4)	Model development
Nakaya *et al.* (1997)	Mice (A, AKR, BALB/c, C3H, C57BL/6, C57BL/10, CBA, DBA/1, DBA/2, B10.D2)	Intrahepatic (transportal)	0.1 ml metacestode homogenate (1 : 20)	Model development
Ohnishi (1984)	Rats (Wistar)	Transportal	5000 protoscolices	Model development
Eckert *et al.* (1983)	Gerbils	Subcutaneous	150–200 mg piece of metacestode (transplant)	Metastases formation
Ali-Khan and Siboo (1980)	C57/L	Subcutaneous	10–12 vesicles	Immunology

studies cited, egg material was obtained from experimentally infected dogs. Infection doses range from 500 to 9000 eggs per animal.

Intraperitoneal infection

Injection of protoscolices into the peritoneum is the most widely used technique to establish cystic echinococcosis in rodents. Various strains of mice (*M. musculus*) and Mongolian gerbils (*Meriones unguiculatus*) have been used as model hosts. Hydatid cysts usually develop free in the peritoneal cavity, but may be attached to or grow into neighbouring organs. Growth is slow; in the BALB/c mouse, after 15 months only a few cysts reach >20 mm in diameter. Even then, most cysts remain sterile. A more rapid development has been reported by intraperitoneal implantation of cysts cultured *in vitro* (Casado *et al.*, 1992); the latter technique has also been used to test the viability of cysts from other sources and for the serial passage of sterile cysts in mice. Generally, there is very little hard information on the influence of mouse strain, parasite strain and condition of the material used for infection on the subsequent development. Since protoscolex formation is poor or absent in laboratory animals, protoscolices for infection are usually obtained from cysts of ovine, equine or bovine origin. The infection doses used by most authors range from 2000 to 5000 protoscolices per animal. Injection of small sterile cysts was found to be equally successful in mice as injection of protoscolices (Pennoit-De Cooman and De Rycke, 1978). For discussion on the role of protoscolices to establish secondary echinococcosis, see also Gottstein *et al.* (1992).

Subcutaneous infection

For immunological studies, secondary cystic echinococcosis has also been established by subcutaneous injection of protoscolices.

E. multilocularis

Oral infection

Several species of voles (*Microtus* spp.), cotton rats (*Sigmodon hispidus*) and various strains of mice (*Mus musculus*) have been used for oral infection. Since rodents are natural intermediate hosts for this cestode species, development of the metacestode (the primary location is invariably the liver) is usually rapid and leads to numerous protoscolices. However, due to difficulties in obtaining infective eggs (mostly from shot foxes) and the safety measures required for handling material infective to humans, few laboratories use oral infection as a routine method. Few comparative studies on the metacestode development in different host species or strains are available.

In *Microtus*, the first metacestode lesions are already visible 2 days post-infection (p.i.). Fully developed protoscolices start appearing from day 50 p.i., and at 5 months p.i. fertility has reached its maximum. By then, the liver is enormously enlarged and completely infiltrated by the parasite; death of the host may occur. In cotton rats, the metacestode grows even more rapidly with infectious protoscolices appearing from day 20 p.i.; in some mouse strains (e.g. C57BL/6) they appear not earlier than 5–7 months p.i. Infection intensity in any rodent species depends on the viability of egg material used for infection, varies considerably and has to be established in preliminary experiments before embarking on a large series of infections. Inocula reported range from 400 to 2000 eggs per animal.

Intraperitoneal infection

Injection of homogenized metacestode material into the peritoneal cavity is the most widely used method to establish secondary alveolar echinococcosis in rodents. The most rapid growth occurs in Mongolian gerbils (*Meriones unguiculatus*), cotton rats (*S. hispidus*) and common voles (*Microtus arvalis*); in gerbils, a 0.5 ml inoculum leads to 12 g of metacestode material within 1 month; at 4 months the fertile parasitic mass causes gross enlargement of the abdomen and constitutes some 50% of the host's weight. Gerbils have been proposed as good models for drug screening due to the absence of effective regulatory immune responses (Schantz *et al.*, 1990). In cotton rats development may be even faster with first mature protoscolices present 4 weeks p.i.; the host animals usually die from echinococcosis within 10 weeks.

In mice, various strains show considerable differences in metacestode development. Growth is slow and protoscolex formation is delayed, poor or absent. Various strains have been proposed as susceptible, e.g. C57L, C57BL/6, AKR and immunodeficient strains (e.g. severe combined immunodeficiency (SCID) mice); in AKR, an inoculum of 0.5 ml leads to 3 g of metacestode mass containing some protoscolices after 4 months; in SCID mice protoscolices are present at 10 weeks p.i. C57BL/6 mice, in which the parasite develops few protoscolices and grows in a vesicular form, have been proposed as suitable models for human infections, although considerable differences in the growth pattern are obvious (for discussion, see Gottstein and Felleisen, 1995). However, published results from various host species are in some cases contradictory, which may be explained by differences concerning the source and isolate of the parasite material and mode of infection.

It has been demonstrated that after several i.p. passages in rodents *E. multilocularis* isolates tend to grow more rapidly (Lubinsky, 1960) and after years of i.p. passage may lose their ability to produce protoscolices even in host species that support fertile growth of the parasite. Likewise, it is not yet clear which component of the injected metacestode suspension is responsible for the establishment of secondary echinococcosis. Conventionally, this has been attributed to protoscolices, whose ability for vesicular growth has been demonstrated both *in vitro* (Howell and Smyth, 1995) and *in vivo* by implanting protoscolices contained in micropore chambers into the peritoneal cavity of gerbils (Al Nahhas *et al.*, 1991). However, sterile metacestodes can also be passaged without difficulty and as a rule, increased purification of protoscolices from metacestode suspensions leads to decreasing success of infection. Gottstein *et al.* (1992) failed to infect C57BL/6 mice with a carefully purified protoscolex suspension and discussed the possibility that small vesicular cysts present in less purified suspensions may be responsible for subsequent development (which may explain contradictory results: Yamashita *et al.*, 1962). The usual sources of infection material are gerbils, cotton rats or voles in which isolates can be maintained by serial intraperitoneal passage (for a method of cryopreservation see Bretagne *et al.*, 1990). Infection dose is usually 0.2–0.5 ml of metacestode suspension.

An additional method to establish secondary alveolar echinococcosis is the transplantation of solid, non-homogenized metacestodes into the peritoneal cavity. In gerbils (*Meriones unguiculatus*), a 50 mg piece of metacestode grows to some 10 g after 3.5 months and some 20 g after 5 months.

Subcutaneous infection

Subcutaneous infections have been used to study host–parasite interactions. Injection of 10 metacestode vesicles under the skin of C57L mice grew to a mass of 37 mm diameter after 22 weeks (Ali-Khan and Siboo, 1980).

Eckert *et al.* (1983) described a method to implant subcutaneously pieces of metacestode tissue in gerbils; 200 mg of tissue had grown to some 4 g at 12 weeks p.i.; metastases were present in lymph nodes and lungs by that time.

Intrahepatic infection

To model the intrahepatic growth of *E. multilocularis* metacestodes in the naturally infected host without suffering the hazard of handling eggs, several authors report methods to introduce metacestode material into the liver. Following the method described by Liance *et al.* (1984), homogenized metacestodes are injected directly into the superior liver lobe after midline laparotomy; AKR and C57BL/6 mice were found to be susceptible to this method with 100% infection success, while only 31% of C57BL/10 mice could be so infected. Four months p.i. in susceptible strains the hepatic lesions are fertile and account for >50% of the liver mass; metastases are present in the peritoneal cavity. In resistant mice, growth is delayed, protoscolices are not produced and metastases are often absent. Recent data of Bauder *et al.* (1998) using orally infected C57BL/6 and C57BL/10 mice, however, do not show such marked strain differences in infectivity, which emphasizes the importance of the infection route for the establishment of the parasite (see also Gottstein *et al.*, 1994).

Infection dose for intrahepatic injections is 0.1 ml of a metacestode homogenate diluted 1:4 in saline.

Alternatively, metacestode suspensions containing protoscolices injected into the superior mesenteric vein induced echinococcosis of the liver in Wistar rats and various strains of mice (method described by Ohnishi, 1984; Nakao *et al.*, 1990). Differences between the metacestode's development in different strains of mice were mostly confined to the development of protoscolices: they were present at week 13 p.i. in some strains (e.g. AKR, BALB/c, C57BL/6) but failed to develop in others (e.g. C57BL/10, C3H, B10.D2). The intravenous inoculum for mice was 0.1 ml of a 5% suspension of homogenized metacestode in phosphate-buffered saline (PBS).

Secondary alveolar echinococcosis: description of model

From the large variety of models summarized above, one will be described in detail: secondary alveolar echinococcosis in gerbils, cotton rats or voles by intraperitoneal injection of homogenized metacestode material. This model was selected because it is the simplest to maintain in the non-specialized laboratory and has frequently been used for drug screening and evaluation of chemotherapy schedules (Table 105.4).

Animal species

Mongolian gerbils (*Meriones unguiculatus*), cotton rats (*Sigmodon hispidus*) and voles (*Microtus* spp., e.g. *M. arvalis*) are suitable species for the method described. Laboratory strains (in- and outbred) exist in numerous parasitological laboratories, but are not always available from commercial breeders.

Table 105.4 Examples of chemotherapy trials using the secondary alveolar echinococcosis model

Reference	*Agent and application*	*Outcome*	*Clinical application of the agent**
Taylor *et al.* (1988)	Albendazole (20 and 50 mg/kg per day) and praziquantel (500 mg/kg per day) for 1 month, oral	Favourable, but viable metacestodes still present; combination not more effective than drug given singly	Yes
Taylor *et al.* (1989c)	Albendazole (50 mg/kg per day), mebendazole (50 mg/kg per day), praziquantel (500 mg/kg per day) for 25 weeks, oral	Albendazole most effective in inhibiting metacestode growth	Yes
Novak (1990)	Mitomycin C (0.04–0.3 mg total dose) for 3 weeks, intraperitoneal	Reduced parasite growth, but no cure	No
Schantz *et al.* (1990)	Albendazole (0.05–0.1% in medicated food) for 3.5 months	Favourable with reduced growth and partial tissue degeneration; still viable metacestodes in 60% of animals	Yes
Duriez *et al.* (1991)	Oltipraz (50 mg/kg per day) for 43 days, oral	Favourable, but less efficient than mebendazole	No
Liance *et al.* (1992)	Cyclosporin A (40 mg/kg per day) for 80 days, intraperitoneal	No antiparasitic effect	No
Miyaji *et al.* (1993)	Alpha-difluoromethylornithine (2% in drinking water) for 76 days	No effect on metacestode development	No
Marchiondo *et al.* (1994)	Praziquantel (300 mg/kg per day) for 40 days, oral	Unfavourable (enhanced metastases formation)	No
Ochieng'-Mitula *et al.* (1994)	Ivermectin (30 mg/kg every 4th day) for 1 month, intraperitoneal	Reduced parasite growth, but no cure	No

* For discussion, see Ammann and Eckert (1995).

Preparation of animals

No preparation is required.

Source and preparation of inocula

Most frequently, i.p.-passaged metacestode material from the same or other host species will be used for inoculation. Rodents with mature peritoneal echinococcosis are sacrificed using CO_2 chambers or vertebral dislocation in ether narcosis, according to the prevailing official regulations. All following steps should be performed under sterile conditions.

After opening the peritoneal cavity using scissors or scalpel blades, the metacestode material is removed. Those parts of the metacestode which have grown into host tissue should be discarded. Metacestodes are frequently covered by vascularized connective tissue of the host which has to be carefully removed. The resulting material is minced into small pieces using scissors and pressed through a metal or plastic sieve (≤0.5 mm mesh) into a beaker containing sterile PBS (pH 7.4). Addition of penicillin (100 U/ml) and streptomycin (100 µg/ml) is required if a partial necrosis and bacterial infection of the metacestode is suspected.

After allowing the material to settle, the supernatant is discarded and an aliquot of the sediment is examined microscopically. It will contain a mixture of small vesicles, brood capsules, free protoscolices and tissue fragments. The viability of the material can be examined by checking for fully developed structures and flame cell activity of the protoscolices (see also Marchiondo and Anderson, 1984); if in doubt, various stain exclusion tests (e.g. trypan blue exclusion: Clegg and Smithers, 1968) can be used with protoscolices. Per animal, 0.2 ml (up to 0.5 ml) of the sediment, diluted 1:1 in PBS, is administered into the abdominal cavity using a 1 ml tuberculin syringe with a needle of 1.1 mm diameter (caudal third of the abdomen, needle pointing forward). Immediately before injection, the animal has to be briefly anaesthesized in ether or chloroform vapour. For variations and further details of this technique, see Lubinsky (1960) and Hinz (1972).

Safety measures

Metacestode material of *E. multilocularis* is not infective to humans via the oral route. However, contact with open wounds must be prevented and, since orbital infection cannot be ruled out if material reaches the eye, protective glasses should be worn when handling the material. As a matter of course, potential final hosts (dogs or cats) must never have access to metacestodes, so all material (including vessels and plastic ware) has to be autoclaved before it is disposed of.

Post-inoculation care

No specific care is required.

Evaluation of response to treatment

In most studies, metacestode weight was used as the main parameter to measure drug efficiency. In addition, depending on the study purpose, various histological methods, viability tests by flame cell activity or stain exclusion tests, and further inoculation into laboratory animals may be employed. Recently, an *in vitro* viability assay using enzymatic activity was proposed (Emery *et al.*, 1995). Whatever the study, it has to be taken into account that the rodent species named have a limited life expectancy after inoculation. To ensure survival of the untreated controls until the end of the study, the entire period p.i. should not exceed 2 months (cotton rats), 3–4 months (gerbils) and 6–9 months (voles), respectively.

Pitfalls of the model

As explained above, secondary echinococcosis in rodents is not a perfect model of the human disease, since the route of infection, the growth rate and the development characteristics of the parasite within the host exhibit considerable differences. Its main value lies in the maintenance and propagation of parasite isolates for further study purposes, and the primary screening of drugs and treatment schedules. Comparison of the results with published data has to be done with care, since different isolates of the parasite show different growth patterns.

References

Ali-Khan, Z., Siboo, R. (1980). Pathogenesis and host response in subcutaneous alveolar hydatidosis I. Histogenesis of alveolar cyst and a qualitative analysis of the inflammatory infiltrates. *Z. Parasitenkd.*, **62**, 241–254.

Ali-Khan, Z., Sipe, J. D., Du, T., Riml, A. (1988). *Echinococcus multilocularis*: relationship between persistent inflammation, serum amyloid, protein response and amyloidosis in four mouse strains. *Exp. Parasitol.*, **67**, 334–345.

Al Nahhas, S., Gabrion, C., Walbaum, S., Petavy, A. F. (1991). *In vivo* cultivation of *Echinococcus multilocularis* protoscoleces in micropore chambers. *Int. J. Parasitol.*, **21**, 383–386.

Ammann, R., Eckert, J. (1995). Clinical diagnosis and treatment of echinococcosis in humans. In Echinococcus *and Hydatid Disease* (eds Thompson, R. C. A., Lymbery, A. J.), pp. 411–463. CAB International, Wallingford.

Bauder, B., Auer, H., Schilcher, F., Gabler, C., Aspöck, H. (1998). Experimental investigations on the B and T cell immune response in alveolar echinococcosis. *Alpe Adria Microbiol. J.*, **7**, 66–67.

Bretagne, S., Beaujean, F., Uner, A., Bresson-Hadni, S., Liance, M., Houin, R. (1990). Rapid technique for cryopreservation of *Echinococcus multilocularis* metacestodes. *Int. J. Parasitol.*, **20**, 265–267.

Casado, N., Criado, A., Jimenez, A. *et al.* (1992). Viability of *Echinococcus granulosus* cysts in mice following cultivation *in vitro*. *Int. J. Parasitol.*, **22**, 335–339.

Clegg, J. A., Smithers, S. R. (1968). Death of schistosome cercariae during penetration of the skin. II. Penetration of mammalian skin by *Schistosoma mansoni*. *Parasitology*, **58**, 111.

Craig, P. S., Rogan, M. T., Allan, J. C. (1996). Detection, screening and community epidemiology of taeniid cestode zoonoses: cystic echinococcosis, alveolar echinococcosis and neurocysticercosis. *Adv. Parasitol.*, **38**, 170–250.

D'Alessandro, A. (1997). Polycystic echinococcosis in tropical America: *Echinococcus vogeli* and *E. oligarthrus*. *Acta Trop.*, **67**, 43–65.

Dempster, R. P., Harrison, G. B. L., Berridge, M. V., Heath, D. D. (1992). *Echinococcus granulosus*: use of an intermediate host mouse model to evaluate sources of protective antigens and a role for antibody in the immune response. *Int. J. Parasitol.*, **22**, 435–441.

Dempster, R. P., Harrison, G. B. L., Berridge, M. V. (1995). Maternal transfer of protection from *Echinococcus granulosus* infection in sheep. *Res. Vet. Sci.*, **58**, 197–202.

Denegri, G., Perez-Serrano, J., Casado, N., Rodriguez-Caabeiro, F. (1995). ^{13}C-Nuclear magnetic resonance spectral profiles of serum from normal and *Echinococcus granulosus*-infected mice: a kinetic study. *Parasitol. Res.*, **81**, 170–172.

Duriez, T., Afchain, D., Thuillier, P., Vaccher, C., Berthelot, P., Debaert, M. (1991). Activite de l'Oltipraz et de six derives sur l'hydatide de *Echinococcus multilocularis*. *Bull. Soc. Franc. Parasitol.*, **9**, 211–217.

Eckert, J., Thompson, R. C. A., Mehlhorn, H. (1983). Proliferation and metastases formation of larval *Echinococcus multilocularis*. *Z. Parasitenkd.*, **69**, 737–748.

Emery, I., Bories, C., Liance, M., Houin, R. (1995). *In vitro* quantitative assessment of *Echinococcus multilocularis* metacestode viability after *in vivo* and *in vitro* maintenance. *Int. J. Parasitol.*, **25**, 275–278.

Gottstein, B., Felleisen, R. (1995). Protective immune mechanisms against the metacestode of *Echinococcus multilocularis*. *Parasitol. Today*, **11**, 320–326.

Gottstein, B., Deplazes, P., Aubert, M. (1992). *Echinococcus multilocularis*: immunological study on the Em2-positive laminated layer during *in vitro* and *in vivo* post-oncospheral and larval development. *Parasitol. Res.*, **78**, 291–297.

Gottstein, B., Wunderlin, E., Tanner, I. (1994). *Echinococcus multilocularis*: parasite-specific humoral and cellular immune response subsets in mouse strains susceptible (AKR, C57BI/6J) or "resistant" (C57/BI10) to secondary alveolar echinococcosis. *Clin. Exp. Immunol.*, **86**, 245–252.

Heath, D. D., Lawrence, S. B., Yong, W. K. (1992). *Echinococcus granulosus* in sheep: transfer from ewe to lamb of ARC 5 antibodies and oncosphere-killing activity, but no protection. *Int. J. Parasitol.*, **22**, 1017–1021.

Hernández, A., Nieto, A. (1994). Induction of protective immunity against murine secondary hydatidosis. *Parasite Immunol.*, **16**, 537–544.

Hinz, E. (1972). Die Aufbereitung des Infektionsmaterials für die intraperitoneale Infektion der Maus mit *Echinococcus multilocularis*. *Z. Tropenmed. Parasit.*, **23**, 387–390.

Howell, M. J., Smyth, J. D. (1995). Maintenance and cultivation of *Echinococcus* species *in vivo* and *in vitro*. In Echinococcus *and Hydatid Disease*. (eds Thompson, R. C. A., Lymbery, A. J.) pp. 201–232. CAB International, Wallingford.

Jenkins, D. J., Thompson, R. C. A. (1995). Hydatid cyst development in an experimentally infected wild rabbit. *Vet. Rec.*, **137**, 148–149.

Kamiya, M., Sato, H. (1990). Survival, strobilation and sexual maturation of *Echinococcus multilocularis* in the small intestine of golden hamsters. *Parasitology*, **100**, 125–130.

Kroeze, W. K., Tanner, C. E. (1987). *Echinococcus multilocularis*: susceptibility and responses to infection in inbred mice. *Int. J. Parasitol.*, **17**, 873–883.

Liance, M., Vuitton, D. A., Guerret-Stocker, S., Carbillet, J. P., Grimaud, J. A., Houin, R. (1984). Experimental alveolar echinococcosis. Suitability of a murine model of intrahepatic infection by *Echinococcus multilocularis* for immunological studies. *Experientia*, **40**, 1436–1439.

Liance, M., Bresson-Hadni, S., Vuitton, D. A., Lenys, D., Carbillet, J. P., Houin, R. (1992). Effects of cyclosporin A on the course of murine alveolar echinococcosis and on specific cellular and humoral immune responses against *Echinococcus multilocularis*. *Int. J. Parasitol.*, **22**, 23–28.

Liu, D., Lightowlers, M. W., Rickard, M. D. (1992). Examination of murine antibody response to secondary hydatidosis using ELISA and immunoelectrophoresis. *Parasite Immunol.*, **14**, 239–248.

Lubinsky, G. (1960). The maintenance of *Echinococcus multilocularis sibiricensis* without the definitive host. *Can. J. Zool.*, **38**, 149–151.

Lymbery, A. J. (1995). Genetic diversity, genetic differentiation and speciation in the genus *Echinococcus* Rudolphi 1801. In Echinococcus *and Hydatid Disease* (eds Thompson, R. C. A., Lymbery, A. J.), pp. 51–87. CAB International, Wallingford.

Marchiondo, A. A., Anderson, F. L. (1984). Light microscopy and scanning electron microscopy of the *in vitro* evagination process of *Echinococcus multilocularis* protoscoleces. *Int. J. Parasitol.*, **14**, 151–157.

Marchiondo, A. A., Ming, R., Anderson, F. L., Slusser, J. H., Conder, G. A. (1994). Enhanced larval cyst growth of *Echinococcus multilocularis* in praziquantel-treated jirds (*Meriones unguiculatus*). *Am. J. Trop. Med. Hyg.*, **50**, 120–127.

Miyaji, S., Katakura, K., Matsufuji, S. *et al.* (1993). Failure of treatment with alpha-difluoromethylornithine against secondary multilocular echinococcosis in mice. *Parasitol. Res.*, **79**, 75–76.

Modha, A., Novak, M., Blackburn, B. J. (1997). Alteration in brain metabolites of jirds infected with alveolar *Echinococcus*. *J. Parasitol.*, **83**, 764–766.

Nakao, M., Nakaya, K., Kutsumi, H. (1990). Murine model for hepatic alveolar hydatid disease without biohazard. *Jpn J. Parasitol.*, **39**, 296–298.

Nakaya, K., Nakao, M., Ito, A. (1997). *Echinococcus multilocularis*: mouse strain difference in hydatid development. *J. Helminthol.*, **71**, 53–56.

Novak, M. (1990). Efficacy of mitomycin C against alveolar *Echinococcus*. *Int. J. Parasitol.*, **20**, 119–120.

Novak, M., Kornovski, B., Shimizu, K. Y., Buist, R. J. (1991). *Echinococcus multilocularis* — a model for imaging research. *J. Parasitol.*, **77**, 803–805.

Ochieng'-Mitula, P. J., Burt, M. D. B., Tanner, C. E. *et al.* (1994). The effect of ivermectin on the hydatid cyst of *Echinococcus multilocularis* in gerbils (*Meriones unguiculatus*). *Can. J. Zool.*, **72**, 812–818.

Ohbayashi, M. (1960). Studies on echinococcosis X. Histological observations on experimental cases of multilocular echinococcosis. *Jpn J. Vet. Res.*, **8**, 134–160.

Ohnishi, K. (1984). Transportal, secondary hepatic alveolar echinococcosis of rats. *J. Parasitol.*, **70**, 987–988.

Pawlowski, Z. S. (1997). Terminology related to *Echinococcus* and echinococcosis. *Acta Trop.*, **67**, 1–5.

Pennoit-De Cooman, E., De Rycke, P. H. (1978). Serial passages of larval *Echinococcus granulosus* from equine origin in mice. II. Infections with sterile cysts. *Z. Parasitenkd.*, **56**, 39–45.

Perez-Serrano, J., Denegri, G., Casado, N., Rodriguez-Caabeiro, F. (1997). *In vivo* effect of albendazole and albendazole sulphoxide on development of secondary echinococcosis in mice. *Int. J. Parasitol.*, **27**, 1341–1345.

Playford, M. C., Ooi, H.-K., Oku, Y., Kamiya, M. (1992). Secondary *Echinococcus multilocularis* infection in severe combined immunodeficient (scid) mice: biphasic growth of the larval cyst mass. *Int. J. Parasitol.*, **22**, 975–982.

Rausch, R. L. (1954). Studies on the helminth fauna of Alaska XX. The histogenesis of the alveolar larva of *Echinococcus* species. *J. Infect. Dis.*, **94**, 178–186.

Rausch, R. L. (1995). Life cycle patterns and geographic distribution of *Echinococcus* species. In Echinococcus *and Hydatid Disease* (eds Thompson, R. C. A., Lymbery, A. J.), pp. 89–134. CAB International, Wallingford.

Riley, E. M., Dixon, J. B., Kelly, D. F., Cox, D. A. (1985). The immune response to *Echinococcus granulosus*: sequential histological observations of the lymphoreticular and connective tissues during early murine infection. *J. Comp. Path.*, **95**, 93–104.

Rogan, M. T., Richards, K. S. (1989). Development of the tegument of *Echinococcus granulosus* (Cestoda) protoscoleces during cystic differentiation *in vivo*. *Parasitol. Res.*, **75**, 299–306.

Rogan, M. T., Marshall, I., Reid, G. D. F., Macpherson, C. N. L. (1993). The potential of vervet monkeys (*Cercopithecus aethiops*) and baboons (*Papio anubis*) as model for the study of the immunology of *Echinococcus granulosus* infections. *Parasitology*, **106**, 511–517.

Sarciron, M. E., Walbaum, S., Petavy, A. F. (1995). Effects of Isoprinosine on *Echinococcus multilocularis* and *E. granulosus* metacestodes. *Parasitol. Res.*, **81**, 329–333.

Schantz, P. M., Brandt, F. H., Dickinson, C. M., Allen, C. R., Roberts, J. M., Eberhard M. L. (1990). Effects of albendazole on *Echinococcus multilocularis* infection in the mongolian jird. *J. Infect. Dis.*, **162**, 1403–1407.

Schantz, P. M., Chai, J., Craig, P. S., Jenkins, D. J., Macpherson, C. N. L., Thakur, A. (1995). Epidemiology and control of hydatid disease. In Echinococcus *and Hydatid Disease* (eds Thompson, R. C. A., Lymbery, A. J.), pp. 233–331. CAB International, Wallingford.

Taylor, D. H., Morris, D. L., Richards, K. S., Reffin, D. (1988). *Echinococcus multilocularis: in vivo* results of therapy with albendazole and praziquantel. *Trans. R. Soc. Trop. Med. Hyg.*, **82**, 611–615.

Taylor, D. H., Morris, D. L., Richards, K. S. (1989a). Albendazole is effective against established *Echinococcus granulosus* in gerbils: comparison of serum concentrations achieved by gavage and feed administration. *Ann. Trop. Med. Parasitol.*, **83**, 483–488.

Taylor, D. H., Morris, D. L., Richards, K. S. (1989b). *Echinococcus granulosus: in vitro* maintenance of whole cysts and the assessment of the effects of albendazole sulphoxide and praziquantel on the germinal layer. *Trans. R. Soc. Trop. Med. Hyg.*, **83**, 535–538.

Taylor, D. H., Morris, D. L., Reffin, D., Richards, K. S. (1989c). Comparison of albendazole, mebendazole and praziquantel chemotherapy of *Echinococcus multilocularis* in a gerbil model. *Gut*, **30**, 1401–1405.

Thompson, R. C. A., Lymbery, A. J., Constantine, C. C. (1994). Variation in *Echinococcus*: towards a taxonomic revision of the genus. *Adv. Parasitol.*, **35**, 145–176.

Veit, P., Bilger, B., Schad, V., Schäfer, J., Frank, W., Lucius, R. (1995). Influence of environmental factors on the infectivity of *Echinococcus multilocularis* eggs. *Parasitology*, **110**, 79–86.

Yamashita, J., Ohbayashi, M., Sakamoto, T., Orihara, M. (1962). Studies on echinococcosis. XIII. Observation on the vesicular development of the scolex of *E. multilocularis in vitro*. *Jpn J. Parasitol.*, **10**, 85–96.

Chapter 106

Intestinal Worm Infections

S. S. Johnson, E. M. Thomas and T. G. Geary

GI helminthiases

"Worms" in the context of parasites is a definition that includes members of several different phyla, with a corresponding array of host niche habitats, feeding and reproductive strategies, and pathological consequences. Broadly speaking, these organisms may be divided into flatworms (the Platyhelminthes) and round worms (the Nematoda). By far the most important metazoan parasites of the gastrointestinal (GI) tract are nematodes, and they will be the focus of most of this chapter. Among the flatworms, tissue-dwelling trematodes (or flukes; e.g., *Schistosoma* spp. and *Fasciola* spp.) are of enormous significance in infectious disease, but no representatives of this group are of major consequence in the GI tract. Similarly, cestodes (tapeworms) are much more pathologic as larval stages in tissues than as adults in the gut. However, since adult tapeworms are legitimate targets for chemotherapy in strategic control therapy, they will also be addressed here.

GI nematodes of humans

It is sobering to acknowledge that upwards of 25% of humans harbor GI nematodes. Although estimates of prevalence are imprecise, there may be 1.5×10^9 humans infected with *Ascaris lumbricoides*, 1×10^9 with the whipworm, *Trichuris trichiura*, approximately the same with hookworms (*Necator americanus* and *Ancylostoma duodenale*), 5×10^8 with the pinworm, *Enterobius vermicularis*, and another 10^8 with *Strongyloides stercoralis* (e.g., Chan *et al.*, 1994). Many people harbor more than one species of parasite. Most, but not all, infections occur in developing nations. Consequences of infection for human health depend upon many variables, including parasite species and intensity of infection. None the less, hookworm infection is a major cause of anemia and reasonably convincing data suggest that infection with hookworms *A. lumbricoides* and *T. trichiura* contribute to poor growth and poor school performance in infected children (Nokes and Bundy, 1994).

Except for *S. stercoralis*, which has an autoinfective cycle, GI nematodiases are typically initiated by ingestion of infective larvae from an environment contaminated with parasite eggs shed in the feces of an already infected host (hookworm larvae penetrate the skin directly, as can larvae of *Strongyloides* spp.). Improved sanitation could have profound effects on the disease burden. In the absence of investment in the infrastructure of affected regions that would provide better sanitation, chemotherapy continues to be the mainstay strategy for control of GI nematodiases.

GI nematodes of production animals

Animals are raised for meat and dairy production throughout the world. Keeping animals in herds with limited grazing area, as in modern production systems, is almost inevitably associated with infection by one or more species of parasitic nematodes, most of which reside as adults in the gut. In much of the developed world, GI nematodiases are of major economic concern mainly in ruminant animals (cattle, sheep and goats). However, in developing regions, GI parasitism is a significant cause of economic losses in virtually all production animals.

GI nematodes of ruminant animals

By far the most significant parasitism of ruminant animals is due to trichostrongylid nematodes, which are a global problem. Though phylogenetically closely related, different species occupy niches ranging from the abomasum (true stomach) to the small intestine and employ a variety of feeding strategies, including direct blood-feeding. Pathological consequences of infection are correlated with the intensity of infection and range from poor performance to death. Predominant species for control include *Haemonchus* spp., *Trichostrongylus* spp., *Ostertagia* spp., *Cooperia* spp. and *Nematodirus* spp. Other GI nematodes of importance in ruminants include hookworms (*Bunostomum* spp.) and the strongylid nodular worms of the genus *Oesophagostomum*, which reside as adults in the large intestine.

Swine

Although modern swine-rearing facilities are designed to abrogate the transmission of GI nematodes, parasitism remains an important health problem in much of the world

Handbook of Animal Models of Infection
ISBN 0-12-775390-7

and, indeed, in any place where pigs are raised outdoors. Of most importance are the ascarid, *Ascaris suum* (closely related to the human parasite *A. lumbricoides*), the whipworm, *Trichuris suis*, *Strongyloides ransomi*, *Oesophagostomum* spp. and the trichostrongylid, *Hyostrongylus rubidus*.

GI nematodes of companion animals

Dogs and cats

GI nematodes are nearly ubiquitous parasites of dogs and cats in the absence of chemotherapy. Of particular significance for human health are ascarids (*Toxocara canis*, *T. cati* and *Toxascaris leonina*). Ingestion of *Toxocara* spp. eggs by humans leads to hatching in the intestinal tract. Second stage larvae (L_2) migrate through the tissues of the "wrong" host, causing considerable pathology. Of more importance for the health of the pet is infection with the hookworms *Ancylostoma caninum* (the larvae of which are transmitted from the bitch to the pup when nursing), other *Ancylostoma* spp. and *Uncinaria stenocephala*. The canine whipworm *Trichuris vulpis* is also a target for chemotherapy.

Horses

GI nematodes cause serious pathology in horses, especially strongyles in the genus *Strongylus* ("large strongyles"), particularly *S. vulgaris*. Related parasites ("small strongyles" or cyathostomes) in genera such as *Cyathostomum*, *Cylicostephanus* and *Cylicocyclus*, among others, are also problematic. Both groups cause pathology from the small intestine to the cecum; tissue reactions to migrating larval stages result in the most severe damage. Horses are also parasitized by the ascarid, *Parascaris equorum*, the trichostrongylid parasite of the stomach, *Trichostrongylus axei*, and the threadworm, *Strongyloides westeri*.

Cestodiases

Intestinal stages of tapeworms (i.e., adults) do not usually have serious pathological consequences. However, the larval stages of the pork tapeworm, *Taenia solium*, cause cysticercosis in humans, and larvae of canid tapeworms in the genus *Echinococcus* cause devastating pathology in humans following oral ingestion of proglottids. In any case, the appearance of tapeworm proglottids in pet feces often triggers treatment. The most important tapeworms in companion animals include *Dipylidium caninum* (acquired by dogs and cats from a flea intermediate host) and various species in the genus *Taenia*.

Current anthelmintic chemotherapy

Unlike many areas of chemotherapy, treatment of GI nematodiases is strongly constrained by price. Human infections occur primarily in less developed countries in which per capita health budgets are very small; GI nematodes compete for these limited funds with far more serious infectious diseases, such as acquired immunodeficiency syndrome (AIDS), tuberculosis, malaria and schistosomiasis. Programs for treatment of GI helminthiases therefore require broad-spectrum, inexpensive drugs. Similarly, production animal agriculture is constrained by the need to make a profit on the product; use of anthelmintics is strictly dependent on the cost–benefit profile. Drugs that control a very broad spectrum of parasites are thus preferred. Only for companion animals in developed nations is cost less of an immediate concern, but the spectrum of action is still a crucial feature.

Since GI nematodes from many different families, with distinct niches and feeding habits, must be controlled by a single compound, it is perhaps not surprising that only three classes of broad-spectrum anthelmintics are currently available. The limited number of chemical templates and the spectre of the inevitable spread of drug-resistant strains of parasites are factors that drive the search for new compounds. The major classes of broad-spectrum anthelmintic drugs are discussed in the following sections, after which the animal models preferred for anthelmintic discovery and evaluation are presented.

Benzimidazoles

Thiabendazole, the prototype benzimidazole reported in 1961, was the first broad-spectrum anthelmintic and the first to be discovered by screening. Scientists at Merck found the compound by screening for larvicidal activity against gastrointestinal nematodes of ruminants in culture (see below); active compounds were tested in sheep parasitized with several trichostrongyle species (Brown *et al.*, 1961; Egerton, 1961). Since then, a number of derivatives with improved potency and spectrum have been introduced into clinical use, which now predominate (McKellar and Scott, 1990). The benzimidazoles have found wide use in both human and veterinary medicine, being generally inexpensive, safe and efficacious for a wide variety of nematodes (and other helminths, including cestodes). There is solid evidence that benzimidazoles selectively destabilize helminth microtubules as a mechanism of action (Lacey, 1988). Restricted delivery options in veterinary applications now limit the market for these drugs. In addition, the appearance of benzimidazole-resistant strains of parasites, particularly trichostrongyles of ruminants (Prichard, 1994) and cyathostomes of horses (Wescott, 1986), has become a serious problem in many areas (Conder and Campbell, 1995). A recent report indicating that human hookworms have developed mebendazole resistance in Mali (DeClercq *et al.*, 1997) suggests that the problems encountered in livestock may be spreading to human populations.

Nicotinic cholinergic agonists

Scientists at Janssen Pharmaceutica reported the anthelmintic activity of tetramisole (a racemic compound) in a screen against *Ascaridia galli, Heterakis gallinae* and a *Capillaria* spp. in chickens (Janssen, 1976); the activity was later found to be mostly associated with the L-isomer, levamisole, which is now preferred. This compound is a selective agonist at acetylcholine receptors in nematode muscle (Martin *et al.*, 1997). Subsequent screening at Pfizer in mice infected with *Nematospiroides dubius* (*Heligmosomoides polygyrus*) and rats infected with *Nippostrongylus brasiliensis* (*N. muris*) identified pyrantel (Austin *et al.*, 1966) and morantel as compounds that, while structurally distinct from levamisole, act via a similar mechanism (Martin, 1997). These compounds have broad efficacy against nematodes (especially levamisole). Levamisole is used primarily in ruminants, while pyrantel (as the pamoate salt) finds relatively wide use in swine and companion animals. Levamisole resistance occurs in sheep trichostrongyles (Prichard, 1994; Conder and Campbell, 1995; Sangster, 1996), and pyrantel resistance has been reported in horse cyathostomes and in *Oesophagostomum dentatum* in swine (Conder and Campbell, 1995).

Macrocyclic lactones

Screening of fermentation extracts in mice infected with the blood-feeding trichostrongylid *H. polygyrus* led to the discovery at Merck of the avermectins, prototypes of the macrocyclic lactone class of anthelmintics (Campbell *et al.*, 1983). Targeted screening programs subsequently identified the milbemycins as additional members of this family, and several analogs of both templates are now available (Shoop *et al.*, 1995). The macrocyclic lactones act, at least in part, by opening glutamate-gated chloride channels in the nematode pharynx (Arena *et al.*, 1995; Martin *et al.*, 1997). Such receptors are not present in vertebrates, but these drugs also affect γ-aminobutyric acid (GABA)-gated chloride channels in the mammalian central nervous system (Fisher and Mrozik, 1992). These compounds are exceptionally potent and have an extremely broad spectrum of action that includes many species of ectoparasites. However, they are not active against trematodes or cestodes. Resistance to macrocyclic lactones has appeared in trichostrongylid parasites of sheep (Sangster, 1996) and, more recently, in cattle *Cooperia* spp. (Vermunt *et al.*, 1995; Coles *et al.*, 1998).

In vitro techniques

Although thiabendazole was discovered in an *in vitro* screen against larval stages of a parasitic nematode, subsequent anthelmintics, including novel templates such as paraherquamide (Ostlind *et al.*, 1990; immature *T. colubriformis* in the jird) and PF1022A (Sasaki *et al.*, 1992; *A. galli* in chickens) have all been discovered in animal models. None the less, the pharmaceutical industry has almost completely abandoned *in vivo* screening as a discovery strategy and there are nearly irresistible pressures to convert anthelmintic discovery to an *in vitro*, mechanism-based process (Thompson *et al.*, 1996). Primary among them are the high cost of random screening in animal models, the requirement for relatively large amounts of compound and the incompatibility of *in vivo* screening with combinatorial chemistry. Most companies began using an *in vitro* screen with the free-living nematode, *Caenorhabditis elegans* (see below) at least 15 years ago in response to these concerns. Considering that macrocyclic lactones, the most recently commercialized anthelmintic template, were discovered some 20 years ago, the utility of this particular *in vitro* screen is still in question. The development of mechanism-based anthelmintic assays (Thompson *et al.*, 1996; Klein and Geary, 1997) may supercede the development of new high-throughput whole-animal *in vitro* screens.

Adult stages of parasites

It is not yet possible routinely to grow parasitic nematodes in continuous culture. In order to test drugs against adult parasites, the target stage for chemotherapy, infected animal hosts must be kept and sacrificed to supply the worms. It is thus expensive and labor-intensive to work with adult nematodes in culture. Furthermore, adult GI worms typically show slow, continuous declines in viability in culture. None the less, many systems have been developed that employ adult stages of GI nematodes for compound screening and evaluation. These systems have been thoroughly reviewed (Jenkins, 1982; Coles, 1986, 1990). One major problem with the use of adult worms is the difficulty of devising a reliable quantitative measure of drug effects on their viability or behavior. A motility meter which provides such a measure has proven to be of value for drug evaluation (Bennett and Pax, 1986; Geary *et al.*, 1993), but unfortunately is no longer in manufacture.

Larval development assays

Since the discovery of thiabendazole, it has been apparent that anthelmintic drugs inhibit the hatching of nematode eggs and/or subsequent larval development. Larval development assays using a number of GI parasites have been described (Jenkins, 1982; Coles, 1986, 1990). It has also become evident that motility and behavior of the free-living L_3 stage can be used as a discriminator of drug action (Folz *et al.*, 1987; Geary *et al.*, 1993). An important application of these test systems is in the characterization of anthelmintic-resistant populations of GI nematodes (Coles, 1990; Lacey *et al.*, 1990; Gill *et al.*, 1995, 1998; Hoekstra *et al.*, 1997).

C. elegans

Since Simpkin and Coles (1981) described the use of *C. elegans* in anthelmintic screening, most pharmaceutical companies with an interest in such drugs incorporated this free-living nematode into the discovery paradigm.

Although whole-organism screening assays, such as *C. elegans*, are becoming increasingly incompatible with modern drug discovery strategies (Klein and Geary, 1997; Geary *et al.*, 1998) and screening programs that rely on it have not produced a new anthelmintic class, this organism remains a powerful tool in anthelmintic discovery and evaluation.

There are myriad undeniable continuities between this free-living nematode and GI parasites (Bürglin *et al.*, 1998). The complete sequence of the *C. elegans* genome is expected to be available by the end of 1998 and will provide an immensely valuable resource for identifying and characterizing anthelmintic drug targets in parasitic species (Geary *et al.*, 1998). The availability of *C. elegans* mutants that are resistant to known anthelmintics provides a background against which new leads can be evaluated. Techniques for genetic manipulation (knockout and overexpression) allow the modeling of agonist and antagonist effects on putative drug targets as a way to validate them prior to the launch of a specific discovery effort (Geary *et al.*, 1998). The value of *C. elegans* as a research model for parasitic species will greatly outstrip its value in whole-organism screening.

Background of animal models

Whether or not used as primary screens, animal models for GI nematodes will continue to play an extremely important role in lead evaluation. As should be clear from previous sections, there is a continuity among parasites of various hosts: ascarids, hookworms, whipworms, and trichostrongyles (+ horse strongyles). There are close evolutionary connections between parasites of model animal and target animal species, and for many, the target parasite species can be induced to live in a model host. If broad-spectrum activity against all important types of GI nematodes is desired, any model can be used as the starting point for lead evaluation. An advantage of anthelmintic evaluation is that compounds active in model systems can quickly be evaluated against a wide array of target parasites in target hosts. The considerable history of animal models in anthelmintic discovery and evaluation has been thoroughly covered in previous reports (Cavier, 1973; Düwel, 1975; Theodorides, 1976; Campbell, 1986; Coles, 1986; Katiyar *et al.*, 1989) and will not be repeated here. Instead, we provide experimental details of animal models we judge most useful for drug discovery or evaluation.

Trichostrongyles (*Haemonchus contortus*, *Trichostrongylus colubriformis* and *Ostertagia ostertagi*) in the jird, *Meriones unguiculatus*

Background of model

On strictly economic considerations, ruminant trichostrongyles are the most important target for new anthelmintics and have therefore received highest priority in drug discovery programs. This is also an area in which most recent progress has been made. The ability to test compounds against target parasites maintained in a laboratory animal model is a powerful advantage of this model.

Kates and Thompson (1967) first showed that jirds infected with *T. colubriformis* could be used in anthelmintic screening. Subsequent studies confirmed these observations (Ostlind and Cifelli, 1981; Panitz and Shum, 1981; Court and Lees, 1985). The utility of the model for evaluation of breadth of spectrum was enhanced by the work of Conder *et al.*, who showed that *H. contortus* could be maintained in jirds (1990) and could be combined with concurrent infections with *T. colubriformis* (1991a). *H. contortus* infections in jirds closely mimick those in sheep (Conder *et al.*, 1992); this parasite is found in the abomasum of ruminants, and resides in the glandular portion of the jird stomach. Importantly, *T. colubriformis* is found in the small intestine of both animals. That both species occupy similar niches in the GI tract of the model and target animal boosts confidence in the predictive ability of the model. Extension of the jird as a model came from studies with *Ostertagia circumcincta* (Court *et al.*, 1988) and *O. ostertagi*, which is found in the abomasum of ruminants (primarily cattle). This species can also be established in the jird stomach and offers an additional indication of spectrum for potential anthelmintics (Conder and Johnson, 1996). Because necropsy schedules are different for *O. ostertagi* and *H. contortus*/*T. colubriformis* (see below), the three species are not combined in concurrent infections.

Animal species

Outbred female jirds weighing 30–35 g (Charles River Laboratories, Stone Ridge, NY) are used.

Preparation of host

Three animals each are randomly assigned upon arrival to translucent polypropylene cages (26.7 × 21.6 × 14.0 cm) containing wood shavings. Commercial rodent chow (Purina #5002) and water are given *ad libitum*. After a 2-day acclimatization period, jirds are switched to a modification of the chow containing 0.02% hydrocortisone for the duration of the study.

Storage, preparation of inocula

Inocula used are *H. contortus* originally obtained in 1982 from Dr K.S. Todd, Jr. (University of Illionis, Urbana, Illinois), *T. colubriformis* originally obtained in 1987 from Dr L.W. Bone (Regional Parasite Research Lab, USDA, Auburn, Alabama), both of which have since been maintained continuously in sheep at Pharmacia & Upjohn, and *O. ostertagi* supplied by Dr L.R. Ballweber (Mississippi

State College of Veterinary Medicine, Mississippi). The *O. ostertagi* culture is maintained at Mississippi State University. Infective larvae of each parasite (L_3) cultured ≈1–2 months prior to inoculation are used. The larvae are kept separately in culture flasks containing 50 ml tap water and stored at 4 °C until used. For exsheathment, aliquots of 10^5–10^6 L_3 are centrifuged. The supernatant is removed and replaced with 10–15 ml of Earle's balanced salt solution (EBSS) and prepared as follows. For *H. contortus* and *O. ostertagi*, CO_2 is bubbled into a Parafilm-covered centrifuge tube containing L_3 in EBSS for 10 minutes. The Parafilm is replaced with screw-on caps and the larvae are incubated at 37 °C in a shaking water bath for 1 hour for *H. contortus* and 18 hours for *O. ostertagi*. Following incubation, percentage exsheathment is determined by averaging the counts of exsheathed larvae in the first 100 worms observed in two separate samples. For good infectivity, ~70% of *H. contortus* and ~50% of *O. ostertagi* should be exsheathed. Each larval preparation is then centrifuged for 10 minutes and the supernatant removed and replaced with sufficient EBSS to resuspend the larvae, which are then transferred to a 1000 ml beaker. Sufficient EBSS is added to the beaker so that 0.2 ml contains ~1000 L_3 of *H. contortus* (exsheathed) or *O. ostertagi* (exsheathed + ensheathed).

For *T. colubriformis*, the larvae suspended in 10–15 ml of EBSS are transferred to a 1000 ml beaker and sufficient EBSS is added so that 0.2 ml contains ~1000 L_3 (unexsheathed). Exsheathed *T. colubriformis* larvae infect jirds poorly (they may not survive well in the acid pH of the stomach), but ensheathed larvae are infective for the jird without pretreatment.

Infection process

Jirds are inoculated with ~1000 L_3 5 days after being placed on hydrocortisone chow. Inoculation is *per os* using an 18 G dosing needle fitted to a 1 ml syringe. Animals infected with either *H. contortus* or *T. colubriformis* are treated on day 10 post-inoculation (p.i.) and necropsied on day 13 p.i. Animals infected with *O. ostertagi* are treated on day 6 p.i. and necropsied on day 8 p.i. By the time of necropsy, each of the nematode species will have developed to at least the L_4 stage. However, *O. ostertagi* infections begin to diminish after about day 10, requiring animals infected with this parasite to be treated and necropsied earlier than those infected with the other parasites. Numbers of *H. contortus* and *T. colubriformis* are reasonably stable through day 14 p.i. but fall off after day 21, with only an occasional worm found by day 28 p.i. Remaining *H. contortus* do not develop past immature adults, while *T. colubriformis* reach sexual maturity and gravid females can be seen at necropsy on day 13 p.i.

Monitor infection

One day prior to treatment, animals in three randomly selected cages are killed and examined for worms in the stomach and/or small intestine (see below). If all are infected and the mean number of worms is >40 for each species (generally many more worms will be present, with an upper limit of ~400), the animals are assigned randomly by cage to treatment groups (3–9 jirds/treatment group, consistent within each experiment).

Anthelmintic therapy and response to treatment

Following allocation of animals to treatment groups, compounds to be tested are suspended or dissolved in an appropriate vehicle using a vortex mixer. Each jird is given 0.2 ml *per os* containing the desired dose, although other routes of administration (intramuscular, intraperitoneal, subcutaneous, topical) can also be used. Drugs are routinely administered once. However, compounds can be given in multiple doses on the same day or daily over a period of days.

Jirds are routinely necropsied 3 days post-treatment for *H. contortus* and/or *T. colubriformis*, or 2 days post-treatment for *O. ostertagi* infections, but this can be modified to answer specific questions about drug activity (time to clear worms, etc.). Jirds are killed with CO_2 and the stomach and/or small intestine removed. Stomach contents are removed by quickly rinsing the stomach in water; the parasites will not be detached from the mucosa. However, small intestine samples cannot be rinsed without loss of *T. colubriformis* from the lumen. Each tissue is put in a scintillation vial containing 14 ml distilled water and the vials are vortexed and placed in a 37 °C water bath for 5 hours.

Following incubation, 1 ml of formaldehyde solution is added to each vial and the vials are vortexed and stored for subsequent examination.

For examination, the vials are vortexed and the tissues removed. The entire contents of the vial are examined for worms in 1–3 aliquots (depending on clarity) using a dissecting microscope (15–45 ×). Percentage clearance for each drug is determined based on the following formula:

[(mean number of worms recovered from vehicle control group)–(mean number of worms recovered from treated group)] ÷ (mean number of worms recovered from vehicle control group) × 100.

Most modern broad-spectrum anthelmintics are detected in this model and are effective at doses similar to those efficacious in sheep and cattle. The minimum doses of standard anthelmintics necessary to clear 95% of *H. contortus* and *T. colubriformis* and from jirds, sheep and cattle are shown in Table 106.1 and for *O. ostertagi* in jirds and cattle in Table 106.2. Closantel, which is active only against *H. contortus* in ruminants, is also active only against *H. contor-*

Table 106.1 Minimum doses (mg/kg) of standard anthelmintics effective in clearing 95% of adult *Haemonchus contortus/Trichostrongylus colubriformis* from jirds, sheep and cattle

Anthelmintic	*Jirds**	*Sheep†*	*Cattle†*
Albendazole	1.875/3.125	5.0/5.0	7.5/7.5
Fenbendazole	1.875/3.125	5.0/5.0	7.5/7.5
Oxfendazole	1.875/3.125	5.0/5.0	2.5/2.5
Thiabendazole	187.5/187.5	75.0/75.0	100.0/100.0
Ivermectin	0.125/0.0625	0.2/0.2	0.2/0.2
Levamisole hydrochloride	10.0/1.875	7.5/7.5	7.5/7.5

* Values for jirds are from Conder *et al.* (1991a).
† Values for sheep and cattle are from Boersema (1985).

Table 106.2 Minimum doses (mg/kg) of standard anthelmintics effective in clearing 95% of *Ostertagia ostertagi* from jirds and cattle

Anthelmintic	*Jirds*	*Cattle**
Albendazole	62.5	10.0
Fenbendazole	75.0	5.0
Oxfendazole	75.0	2.5
Thiabendazole	187.5	66.0
Ivermectin	0.5	0.2
Levamisole hydrochloride	10.0	8.0†

* Values for cattle are from Lynn (1995).
† Adults only, oral dose.

tus in jirds. Bithionol, diethylcarbamazine and piperazine, which are not active in ruminants, are also inactive in the jird. Anthelmintic resistance is readily detectable in this model (Conder *et al.*, 1991b, 1993).

Alternative trichostrongyle models

Background

Anthelmintic models using the rodent trichostrongyles *H. polygyrus* and *N. brasiliensis* have been frequently used in drug discovery and lead evaluation programs. Detailed descriptions of these models are available (Cavier, 1973; Theodorides, 1976) and will not be repeated here. Since target species can now be used in convenient laboratory models, the value of incorporating a model parasite and a model host into anthelmintic experiments can be questioned. However, an alternative target parasite/model host system has been reported and deserves consideration.

Trichostronglyus colubriformis in the rat

Although attempts have been made to develop infections of *T. colubriformis* in several small mammal hosts other than jirds, the best documented alternative is the rat (Gration *et al.*, 1992). Young male rats (3–4 weeks old; 40–50 g) are maintained on a standard rat chow diet supplemented with 60 ppm hydrocortisone beginning 1 week before infection. Each animal receives 1500 L_3-stage *T. colubriformis* in a small volume of water by oral gavage. These parasites will produce eggs in the feces of infected rats; similar results are obtainable in the jird (Johnson *et al.*, unpublished results). Rats are treated with anthelmintics at day 14 p.i. by oral gavage or subcutaneous injection. Animals are necropsied 4 days later and the number of worms present in the small intestine determined by counting under a dissecting microscope. Drugs tested include several benzimidazoles, levamisole, morantel, dihydroavermectin B1I and I-milbemycin. In each case, potency in the rat was very similar to that observed in ruminants and quite comparable to those found in the jird model.

Hookworms (*Necator americanus* and *Ancylostoma ceylonicum*) in the hamster, *Mesocricetus auratus*

Background of model

Laboratory animal models of hookworm infections have been reviewed (Behnke, 1990). In our opinion, the hamster is the preferred host for experimental chemotherapy studies. Target parasite species can be assayed conveniently in the hamster and can be maintained in a continuous cycle in the laboratory, unlike the situation with other rodent hosts.

Host

Either outbred or inbred hamsters can serve as hosts. For infections with *N. americanus*, very young animals are required and synchronous breeding may be required to provide sufficient numbers of hosts for experiments. Adult animals can be infected with *A. ceylonicum*. Male and female hamsters can be used for both species of parasites. Hamsters are kept under standard laboratory conditions with food (standard hamster chow) and water available *ad libitum*.

Parasites

An important aspect of the hookworm–hamster model is that hamster-adapted strains of *N. americanus* and *A. ceylonicum* are available (Behnke, 1990). While neither isolate is as fecund as the corresponding wild-type strain in the natural host, both show complete development in the hamster to egg-laying adults. Eggs begin to appear in the feces of hamsters infected with *N. americanus* at about 6 weeks p.i.; eggs from *A. ceylonicum* first appear at about 2 weeks p.i. The eggs are fertile and can be cultured from the feces of infected hamsters to the infective larval stage (L_3) to permit the infection to be maintained in the laboratory. Adult parasites persist in the small intestine for more than a year, although burdens typically diminish over time.

Anthelmintic studies

Detailed studies are available on hamster infections with *N. americanus* (Sen, 1972) and *A. ceylonicum* (Garside and Behnke, 1989).

Necator americanus

Baby hamsters (1–3 days old) are infected percutaneously with ~100–500 L_3 larvae and returned to the mother until 21 days of age. Drug effects on immature worms are monitored in animals dosed at 5 weeks p.i., whereas a period of 50 days p.i. is sufficient to test drugs against adult parasites. A few infected animals may be sacrificed prior to treatment to verify infection and provide an estimate of expected worm burden in untreated control animals. Drugs can be administered orally in single or multiple doses (up to 3 days). Hamsters are necropsied 2–3 days post-treatment; the small intestine is removed and opened and hookworms scraped into a Petri dish for counting under a dissecting microscope.

Immature worms are about as sensitive to chemotherapy (Rajasekariah *et al.* 1986) as adult parasites (Rajasekariah *et al.*, 1989). Immature stages require single doses of 20 mg/kg pyrantel and 15 mg/kg mebendazole to achieve 90–95% clearance. Pyrantel exhibited similar potency against adult worms, but only 5 mg/kg mebendazole was required for 90% clearance. Single doses of 7.5 mg/kg ivermectin were necessary for good efficacy against *N. americanus*. This relative lack of potency reflects the relatively weak activity of ivermectin against this species in other hosts. Thiabendazole required very large single (700 mg/kg) or repeated (2 × 500 mg/kg) doses for efficacy against immature stages in this model. Bephenium, known to have activity against hookworms in target hosts (with very poor potency) was only active in the hamster model when given in 2 or 3 consecutive daily doses (400 or 200 mg/kg per day, respectively).

Ancylostoma caninum

Adult hamsters 8–12 weeks old are infected by oral gavage, using a 21G blunted needle. Animals typically receive 50–150 L_3 larvae to initiate the infection. Dosing can be initiated aganist adult stages beginning about 18 days p.i. Subcutaneous and oral administration of compounds are acceptable routes. Necropsy is scheduled 3–7 days post-treatment. Drug effects are measured by counting worms remaining in the intestine, as described above.

Ivermectin was shown to be about 300 times more potent against *A. ceylonicum* than *N. americanus* in this model (Behnke *et al.*, 1993). Subsequent experiments showed that a similar difference in susceptibility is noted when these parasites are exposed to the drug *in vitro* (Richards *et al.*, 1995). In contrast, both hookworm species were equally susceptible to pyrantel in hamsters.

Ascarids: chicken/*Ascaridia galli* model

Random screening against chickens infected with *A. galli* has led to the discovery of two anthelmintic classes. However, ascarids *per se* are not as commercially important for anthelmintic spectrum as trichostrongyles and the ascarid models are not as convenient as the jird model described above. Compounds with anthelmintic activity can be efficiently evaluated for ascarididical effects in a target parasite–target host secondary model, such as the dog infected with *Toxocara canis*.

Nonetheless, a model ascarid system can play a role in anthelmintic screening and lead evaluation. The preferred ascarid model is the chick infected with *A. galli*. Descriptions of several variations of it have been published (Nilsson and Alderin, 1988; Sharma *et al.*, 1990; Sasaki *et al.* 1992). Although adult chickens can be used, lower amounts of drugs are needed if 1-week-old chicks are chosen. After acclimatization and routine vaccinations (about 1 week), infections are initiated by oral administration of 200–1500 embryonated eggs of *A. galli* obtained from gravid female worms removed from infected chickens or from feces dropped by infected birds. Drugs can be administered by oral gavage, subcutaneous injection or in the drinking water beginning 10 days p.i. (for immature stages) or 35 days p.i. (mature adults). Drug effects are monitored by measuring egg excretion from infected birds and/or by direct count of worms remaining in the intestinal tract following necropsy. The major disadvantages of the system include a host animal that is atypical in most laboratory settings and the approximately 8-week duration of the trial.

Whipworms

Laboratory models for whipworm infections are not used in primary anthelmintic screens. Secondary models for direct evaluation of compounds with activity in other anthelmintic assays against target parasites in target hosts are available, such as the dog infected with *Trichuris vulpis*. A *Trichuris muris*/mouse model has been developed for anthelmintic evaluation (Keeling, 1961; Worley *et al.*, 1962), but few further studies are available.

Strongyloides

Species in the genus *Strongyloides* are serious pathogens of humans and horses, and sometimes in dogs and swine. As such, they are legitimate targets for anthelmintic drugs and deserve early consideration in efforts to identify the spectrum of candidate compounds. However, animal models of infection with *Strongyloides* spp. are not preferred for primary anthelmintic evaluation. A model employing *Strongyloides ratti*, and its natural host, the rat, has been used in anthelmintic testing (Cavier, 1973; Katiyar *et al.*, 1989).

An interesting development in this area is the demonstration that the jird can serve as a host for *Strongyloides* spp.; perhaps most important is a model of the jird infected with the human pathogen, *S. stercoralis* (Nolan *et al.*, 1993). The infection closely mimicks the human disease in many aspects, including the ability to generate a model of hyperinfection (Nolan *et al.*, 1993; Kerlin *et al.*, 1995). While trials of anthelmintics have not apparently been reported in this model, the inclusion of *S. stercoralis* (or another *Strongyloides* spp.) in the group of target parasites that infect the jird could provide an additional indication of spectrum.

Pinworms

A pinworm-only anthelmintic would have little commercial value and so pinworm models are not preferred over models that incorporate target parasites. None the less, a mouse model employing the murine pinworm, *Syphacia obvelata*, has proven useful for both screening and compound evaluation (Cavier, 1973; Theodorides, 1976; Coles, 1986). The pinworm was often used in combination with other parasites, such as *Heligmosomoides polygyrus* and the cestode *Hymenolepis nana*, so that early indications of anthelmintic spectrum could be obtained in the mouse (Theodorides, 1976). Details of the model are available (Cavier, 1973; Theodorides, 1976) for the interested reader.

An interesting option for further development of a pinworm model is the observation that *S. obvelata* can be maintained in the jird (Wightman *et al.*, 1978). Complete development occurs in the jird and the infection can be transmitted from jird to jird both naturally and experimentally. Chemotherapeutic trials on pinworms in jirds have not apparently been reported, but would provide an important addition to the model.

Cestode model: *Hymenolepis diminuta* in the jird

Background of model

Gravid adult cestodes have been recovered from jirds infected with *Taenia crassiceps* (Sato and Kamiya, 1989, 1990), *Echinococcus multilocularis* (Kamiya and Sato, 1990) and *Hymenolepis diminuta* (Johnson and Conder, 1996), which suggested that this animal could serve as a useful model for anthelmintic studies. *H. diminuta* was found to establish, grow and develop to the gravid adult stage 2 weeks post-infection in immunosuppressed jirds. The *H. diminuta*/jird model has the advantage of cost-effectiveness, since the infected intermediate host beetle, *Tenebrio molitor*, is commercially available at low cost and the cysticercoids are easily recovered from the intermediate host.

Animal species

Outbred female jirds weighing 30–35 g (Charles River Laboratories, Stone Ridge, NY) are used. Cysticercoids of *H. diminuta* are obtained from infected *T. molitor* beetles (Carolina Biological Supply Company, Burlington, NC).

Preparation of animals

Upon arrival, 3 animals each are randomly assigned to translucent polypropylene cages (26.7 × 21.6 × 14.0 cm) containing wood shavings and fed commercial rodent chow and water *ad libitum*.

Storage, preparation of inocula

Infected *T. molitor* beetles should be acquired within 1 week of use and maintained at room temperature in the carton in which they were shipped.

Infection process

Cysticercoids are harvested from live *T. molitor* beetles. The beetles are placed dorsal side up in a dish containing 0.65% physiological saline (insect physiological saline solution; IPSS). The head, wings and dorsal integument of the beetle are removed and the body contents scraped out. Cysticercoids, which are usually found near the gut, will be released into the saline. For each jird, 5 cysticercoids are drawn into an 18 G dosing needle fitted to a 1 ml syringe preloaded with 0.1 ml IPSS. The cysticercoids are drawn into the needle only, not into the syringe barrel. Each jird is inoculated orally and after inoculation the needle is flushed and the wash examined to be certain all the cysticercoids are delivered. Immediately after inoculation, the jirds are fed medicated rodent chow containing 0.02% hydrocortisone and this feed is continued for the remainder of the experiment.

Anthelmintic therapy and response to treatment

Compounds to be tested are suspended or dissolved in an appropriate vehicle using a vortex mixer. The jirds are treated on day 4 or days 4, 5, and 6 p.i and necropsied on day 7 p.i. At necropsy, the jirds are killed by CO_2 inhalation. The small intestine is removed and placed in 0.85% physiological saline solution, opened longitudinally and examined grossly for tapeworms. If no worms, or fewer than 5 worms are observed, the intestine should be stripped between thumb and forefinger and the intestine and stripped material re-examined for the presence of scolices using a stereoscopic microscope.

Praziquantel, albendazole and niclosamide — compounds with known cestocidal activity — are active in the *H. diminuta*/jird model. Praziquantel (10 mg/kg) clears 100% of the worms (>95% clearance at 5 mg/kg). Albendazole given orally at 90 mg/kg per day for 3 consecutive days has 95% efficacy and niclosamide at 360 mg/kg given for 3 consecutive days clears 80% of *H. diminuta* from the jird.

Advantages/disadvantages

The *H. diminuta*/jird model has the advantage of low cost of labor and animals, ease of handling and less space required for housing than the *H. diminuta*/rat model. A comparison of *H. diminuta* in the jird and the rat shows that the worms can live for the life span of the rat (several months) while in the immunosuppressed jird they survive at least 21 days. In both species, the worms grow the length of the intestine and produce gravid adults (≤7 days in jirds; ≥16 days in rats; Roberts, 1961).

Based on results for praziquantel, albendazole and niclosamide, it appears that the *H. diminuta*/jird model generally requires higher doses than those needed for efficacy against *H. diminuta* in the rat. In rats, praziquantel completely cleared *H. diminuta* at 0.5 mg/kg (Fan and Ito, 1995), and at 2.5–5 mg/kg (Thomas and Andrews, 1977; Thomas and Gönnert, 1977), although 30 mg/kg was required by Dixon and Arai (1991). Therefore, it appears that the strain of rat and/or *H. diminuta* may contribute to the efficacy of praziquantel.

McCracken and Lipowitz (1990) found that a single 50 mg/kg dose of albendazole cleared *H. diminuta* from rats, whereas this drug is active in the *H. diminuta*/jird model only when given in consecutive, multiple doses (90 mg/kg per day p.o. for 3 consecutive days). Dixon and Arai (1991) achieved 100% clearance of *H. diminuta* in rats treated with niclosamide at 1500 mg/kg, although 95% clearance was observed at 300 mg/kg. Lower doses (≥50 mg/kg) resulted in destrobilated worms. In the *H. diminuta*/jird model, a single oral dose of 360 mg/kg produced < 50% clearance, and 360 mg/kg per day for 3 consecutive days resulted in a percentage clearance of 80.5 ± 11.6 (three experiments). One of the worms recovered from the animals receiving multiple doses was destrobilated.

Conclusions

There remains an acute need for new anthelmintic drugs. Useful anthelmintics must demonstrate an almost unreasonable spectrum of action (considering the diverse group of target organisms), coupled with a high margin of safety and low cost. Laboratory animal models are essential for the discovery of such drugs, since they permit testing of candidate compounds against a wide array of target species in convenient and compound-sparing formats. That so many target species can now be maintained in laboratory model hosts is evidence that experiments in this area of parasitology have borne remarkable fruit.

The most promising area for future development is the inclusion of new target species in the jird. This small mammal has proven to be of exceptional value for anthelmintic testing. The possibility that pinworms and *Strongyloides* spp. could be added to the model requires only straightforward drug trials in already established models. It would be of benefit to determine if the hamster-adapted hookworm strains could infect jirds as well. We predict that the jird will become the standard laboratory host for GI nematodes.

Finally, we must note that the large and small strongyles of horses, which are a significant source of equine disease, cannot currently be modeled in laboratory animals. Although there is reason to believe that compounds active in models of ruminant trichostrongyles and hookworms in the genus *Ancylostoma* (both of which are classified in the same order as the horse strongyles) will also be active against the horse parasites (Bowman, 1995), it will still be a useful addition to the parasitologists' tool kit to be able to grow these parasites in a more convenient host.

References

Arena, J. P., Liu, K. K., Paress, P. S. *et al.* (1995). The mechanism of action of avermectins in *Caenorhabditis elegans*: correlation between activation of glutamate-sensitive chloride current, membrane binding, and biological activity. *J. Parasitol.*, **81**, 286–294.

Austin, W. C., Courtney, W., Danilewicz, J. C. *et al.* (1966). Pyrantel tartrate, a new anthelmintic effective against infections of domestic animals. *Nature*, **212**, 1273–1274.

Behnke, J. M. (1990). Laboratory animal models. In *Hookworm Disease: Current Status and New Directions* (eds Schad, G. A., Warren, K. S.), pp. 105–128. Taylor and Francis, London.

Behnke, J. M., Rose, R., Garside, P. (1993). Sensitivity to ivermectin and pyrantel of *Ancylostoma ceylanicum* and *Necator americanus*. *Int. J. Parasitol.*, **23**, 945–952.

Bennett, J. L., Pax, R. A. (1986). Micromotility meter: an instrument designed to evaluate the action of drugs on motility of larval and adult nematodes. *Parasitology*, **93**, 341–346.

Boersema, J. H. (1985). Chemotherapy of gastrointestinal nematodiasis in ruminants. In *Handbook of Experimental Pharmacology*, vol. 77, *Chemotherapy of Gastrointestinal Helminths* (eds Vanden Bossche, H., Thienpont, D., Janssens, P.G.), pp. 407–442. Springer-Verlag, New York.

Bowman, D. D. (1995). Helminths. In *Georgi's Parasitology for Veterinarians* (ed. Bowman, D. D.), pp 113–246. W. B. Saunders, Philadelphia, PA.

Brown, H. D., Matzuk, A. R., Ilves, I. R. *et al.* (1961). Antiparasitic drugs. IV. 2-(4′-thiazolyl)-benzimidazole, a new anthelmintic. *J. Am. Chem. Soc.*, **83**, 1764–1765.

Bürglin, T. R., Lobos, E., Blaxter, M. L. (1998). *Caenorhabditis elegans* as a model for parasitic nematodes. *Int. J. Parasitol.*, **28**, 395–411.

Campbell, W. C. (1986). Historical introduction. In *Chemotherapy of Parasitic Diseases* (eds Campbell, W. C., Rew, R. E.), pp. 3–21. Plenum Press, New York.

Campbell, W. C., Fisher, M. H., Stapley, E. O., Alberts-Schonberg, G., Jacob, T. A. (1983). Ivermectin: a potent new antiparasitic agent. *Science*, **221**, 823–828.

Cavier, R. (1973). Chemotherapy of intestinal nematodes. In *Chemotherapy of Helminthiasis*, vol. 1 (eds Cavier, R., Hawking, F.), pp. 215–436. Pergamon Press, Oxford.

Chan, M. S., Medley, G. F., Jamison, D., Bundy, D. A. P. (1994). The evaluation of potential global morbidity attributable to intestinal nematode infections. *Parasitology*, **109**, 373–387.

Coles, G. C. (1986). Models of infection for intestinal worms. *Exp. Models Antimicrob. Chemother.*, **3**, 333–351.

Coles, G. C. (1990). Recent advances in laboratory models for evaluation of helminth chemotherapy. *Br. Vet. J.*, **146**, 113–119.

Coles, G. C., Stafford, K. A., MacKay, P. H. (1998). Ivermectin-resistant *Cooperia* species from calves on a farm in Somerset. *Vet. Rec.*, **142**, 255–256.

Conder, G. A., Campbell, W. C. (1995). Chemotherapy of nematode infections of veterinary importance, with special reference to drug resistance. *Adv. Parasitol.*, **35**, 1–84.

Conder, G. A., Johnson, S. S. (1996). Viability of infective larvae of *Haemonchus contortus, Ostertagia ostertagi* and *Trichostrongylus colubriformis* following exsheathment by various techniques. *J. Parasitol.*, **82**, 100–102.

Conder, G. A., Jen, L-W., Marbury, K. S., *et al.* (1990). A novel anthelmintic model utilizing jirds, *Meriones unguiculatus*, infected with *Haemonchus contortus*. *J. Parasitol.*, **76**, 168–170.

Conder, G. A., Johnson, S. S., Guimond, P. M., Cox, D. L., Lee, B. L. (1991a). Concurrent infections with the ruminant nematodes *Haemonchus contortus* and *Trichostrongylus colubriformis* in jirds, *Meriones unguiculatus*, and use of this model for anthelmintic studies. *J. Parasitol.*, **77**, 621–623.

Conder, G. A., Johnson, S. S., Guimond, P. M. *et al.* (1991b). Utility of a *Haemonchus contortus*/jird (*Meriones unguiculatus*) model for studying resistance to levamisole. *J. Parasitol.*, **77**, 83–86.

Conder, G. A., Johnson, S. S., Hall, A. D., Fleming, M. W., Mills, M. D., Guimond, P. M. (1992). Growth and development of *Haemonchus contortus* in jirds, *Meriones unguiculatus*. *J. Parasitol.*, **78**, 492–497.

Conder, G. A., Thompson, D. P., Johnson, S. S. (1993). Demonstration of co-resistance of *Haemonchus contortus* to ivermectin and moxidectin. *Vet Rec.*, **132**, 651–652.

Court, J. P., Lees, G. M. (1985). The efficacy of benzimidazole anthelmintics against late fourth stage larvae of *Trichostrongylus colubriformis* in gerbils and *Nippostrongylus brasiliensis* in rats. *Vet. Parasitol.*, **18**, 359–365.

Court, J. P., Lees, G. M., Coop, R. L., Angus, K. W., Beesley, J. E. (1988). An attempt to produce *Ostertagia circumcincta* infection in Mongolian gerbils. *Vet. Parasitol.*, **28**, 79–91.

deClercq, D., Sacko, M., Behnke, J., Gilbert, F., Dorny, P., Vercruysse, J. (1997). Failure of mebendazole in treatment of human hookworm infections in the southern region of Mali. *Am. J. Trop. Med. Hyg.*, **57**, 25–30.

Dixon, B. R., Arai, H. P. (1991). Anthelmintic-induced destrobilation and its influence on calculated drug efficacy in *Hymenolepis diminuta* infections in rats. *J. Parasitol.*, **77**, 769–774.

Düwel, D. (1975). Laboratory methods in the screening of anthelmintics. *Prog. Drug Res.*, **19**, 48–63.

Egerton, J. R. (1961). The effect of thiabendazole upon *Ascaris* and *Stephanurus* infections. *J. Parasitol.*, **47**, (suppl.), 37.

Fan, P. C., Ito, A. (1995). The minimum effective dose of praziquantel in treatment of *Hymenolepis diminuta* in rats. *J. Helminthol.*, **69**, 91–92.

Fisher, M. H., Mrozik, H. (1992). The chemistry and pharmacology of avermectins. *Annu. Rev. Pharmacol. Toxicol.*, **32**, 537–553.

Folz, S. D., Pax, R. A., Thomas, E. M., Bennett, J. L., Lee, B. L., Conder, G. A. (1987). Detecting *in vitro* anthelmintic effects with a micromotility meter. *Vet Parasitol.*, **24**, 241–250.

Garside, P., Behnke, J. M. (1989). *Ancylostoma ceylanicum* in the hamster: observations on the host–parasite relationship during primary infection. *Parasitology*, **98**, 283–289.

Geary, T. G., Sims, S. M., Thomas, E. M. *et al.* (1993). *Haemonchus contortus*: ivermectin-induced paralysis of the pharynx. *Exp. Parasitol.*, **77**, 88–96.

Geary, T. G., Thompson, D. P., Klein, R. D. (1999). Mechanism-based screening: discovery of the next generation of anthelmintics depends upon more basic research. *Int. J. Parasitol.*, *Int. J. Parasitol.*, **29**, 105–112.

Gill, J. H., Redwin, J. M., Van Wyk, J. A., Lacey, E. (1995).

Avermectin inhibition of larval development in *Haemonchus contortus* — effects of ivermectin resistance. *Int. J. Parasitol.*, **25**, 463–470.

Gill, J. H., Kerr, C. A., Shoop, W. L., Lacey, E. (1998). Evidence of multiple mechanisms of avermectin resistance in *Haemonchus contortus* — comparison of selection protocols. *Int. J. Parasitol.*, **28**, 783–789.

Gration, K. A. F., Bishop, B. F., Martin-Short, M. R., Herbert, A. (1992). A new anthelmintic assay using rats infected with *Trichostrongylus colubriformis*. *Vet Parasitol.*, **42**, 273–279.

Hoekstra, R., Borgsteede, F. H. M., Boersema, J. H., Roos, M. H. (1997). Selection for high levamisole resistance in *Haemonchus contortus* monitored with an egg-hatch assay. *Int. J. Parasitol.*, **27**, 1395–1400.

Janssen, P. A. J. (1976). The levamisole story. *Prog. Drug Res.*, **20**, 347–383.

Jenkins, D. C. (1982). *In vitro* screening tests for anthelmintics. In *Animal Models in Parasitology* (ed. Owen, D. G.), pp. 173–186. MacMillan Press, London.

Johnson, S. S., Conder, G. A. (1996). Infectivity of *Hymenolepis diminuta* for the jird, *Meriones unguiculatus*, and utility of this model for anthelmintic studies. *J. Parasitol.*, **82**, 492–495.

Kamiya, M., Sato, H. (1990). Complete life cycle of the canid tapeworm, *Echinococcus multilocularis*, in laboratory rodents. *FASEB J.*, **4**, 3334–3339.

Kates, K. C., Thompson, D. E. (1967). Activity of three anthelmintics against mixed infections of two *Trichostrongylus* species in gerbils, sheep and goats. *Proc. Helm. Soc. Wash.*, **34**, 228–236.

Katiyar, J. C., Gupta, S., Sharma, S. (1989). Experimental models in drug development for helminthic diseases. *Rev. Infect. Dis.*, **11**, 638–654.

Keeling, J. E. D. (1961). Experimental trichuriasis. II. Anthelmintic screening against *Trichuris muris* in the albino mouse. *J. Parasitol.*, **47**, 647–651.

Kerlin, R. L., Nolan, T. J., Schad, G. A. (1995). *Strongyloides stercoralis*: histopathology of uncomplicated and hyperinfective strongyloidiasis in the Mongolian gerbil, a rodent model for human strongyloidiasis. *Int. J. Parasitol.*, **25**, 411–420.

Klein, R. D., Geary, T. G. (1997). Recombinant microorganisms as tools for high throughput screening for nonantibiotic compounds. *J. Biomol. Screen.*, **2**, 41–49.

Lacey, E. (1988). The role of the cytoskeletal protein tubulin in the mode of action and mechanism of drug resistance to benzimidazoles. *Int. J. Parasitol.*, **18**, 885–936.

Lacey, E., Redwin, J. M., Gill, H. J., Demargheriti, V. M., Waller, P. J. (1990). A larval development assay for the simultaneous detection of broad spectrum anthelmintic resistance. In *Resistance of Parasites to Antiparasitic Drugs* (eds Foray, J. C., Martin, P. J., Roush, R. T.), pp. 117–184. MSD Agvet, Rahway, NJ.

Lynn, R. C. (1995). Antiparasitic drugs. In *Georgi's Parasitology for Veterinarians* (ed Bowman, D. D.), pp. 264–272. W. B. Saunders, Philadelphia, PA.

Martin, R. J. (1997). Modes of action of anthelminitic drugs. *Vet. J.*, **154**, 11–34.

Martin, R. J., Robertson, A. P., Bjorn, H. (1997). Target sites of anthelmintics. *Parasitology*, **114**, S111–S124.

McCracken, R. O., Lipowitz, K. B. (1990). Experimental and theoretical studies of albendazole, oxibendazole and tioxidazole. *J. Parasitol.*, **76**, 180–185.

McKellar, Q. A., Scott, E. W. (1990). The benzimidazole anthelmintic agents — a review. *J. Vet. Pharmacol. Ther.*, **13**, 223–247.

Nilsson, O., Alderin, A. (1988). Efficacy of piperazine dihydrochloride against. *Ascaridia galli* in the domestic fowl. *Avian Pathol.*, **17**, 495–500.

Nokes, C., Bundy, D. A. P. (1994). Does helminth infection affect mental processing and educational achievement? *Parasitol. Today.*, **10**, 14–18.

Nolan, T. J., Megyeri, Z., Bhopale, V. M., Schad, G. A. (1993). *Strongyloides stercoralis:* The first rodent model for uncomplicated and hyperinfective strongyloidiasis, the Mongolian gerbil (*Meriones unguiculatus*). *J. Infect. Dis.*, **168**, 1479–1484.

Ostlind, D. A., Cifelli, S. (1981). Efficacy of thiabendazole, levamisole hydrochloride and the major natural avermectins against *Trichostrongylus colubriformis* in the gerbil (*Meriones unguiculatus*). *Res. Vet. Sci.*, **31**, 255–256.

Ostlind, D. A., Mickle, W. G., Ewanciw, D. V. *et al.* (1990). Efficacy of paraherquamide against immature *Trichostrongylus colubriformis* in the gerbil (*Meriones unguiculatus*). *Res. Vet. Sci.*, **48**, 260–261.

Panitz, E., Shum, K. L. (1981). Efficacy of four anthelmintics in *Trichostrongylus axei* or *T. colubriformis* infections in the gerbil, *Meriones unguiculatus*. *J. Parasitol.*, **67**, 135–136.

Prichard, R. (1994). Anthelmintic resistance. *Vet. Parasitol.*, **54**, 259–268.

Rajasekariah, G. R., Deb, B. N., Dhage, K. R., Bose, S. (1986). Response of preadult *Necator americanus* to some known anthelmintics in hamsters. *Chemotherapy*, **32**, 75–82.

Rajasekariah, G. R., Deb, B. N., Dhage, K. R., Bose, S. (1989). Response of adult *Necator americanus* to some known anthelmintics in hamsters. *Ann. Trop. Med. Parasitol.*, **83**, 279–285.

Richards, J. C., Behnke, J. M., Duce, I. R. (1995). *In vitro* studies on the relative sensitivity to ivermectin of *Necator americanus* and *Ancylostoma ceylanicum*. *Int. J. Parasitol.*, **25**, 1185–1191.

Roberts, L. S. (1961). The influence of population density on patterns and physiology of growth in *Hymenolepis diminuta* (Cestoda: Cyclophyllidea) in the definitive host. *Exp. Parasitol.*, **11**, 332–371.

Sangster, N. C. (1996). Pharmacology of anthelmintic resistance. *Parasitology*, **113**, S201–S216.

Sasaki, T., Takagi, M., Yaguchi, T., Miyadoh, S., Okada, T., Koyama, M. (1992). A new anthelmintic cyclodepsipeptide, PF1022A. *J. Antibiot.*, **45**, 692–697.

Sato, H., Kamiya, M. (1989). Deleterious effect of prednisolone on the attachment of *Taenia crassiceps* cysticerci to the intestine of gerbils. *Jpn J. Vet. Sci.*, **51**, 1099–1101.

Sato, H., Kamiya, M. (1990). Establishment, development and fecundity of *Taenia crassiceps* in the intestine of prednisolone-treated Mongolian gerbils and inbred mice. *J. Helminth.*, **64**, 217–222.

Sen, H. G. (1972). *Necator americanus*: behaviour in hamsters. *Exp. Parasitol.*, **32**, 26–32.

Sharma, R. L., Bhat, T. K., Hemaprasanth (1990). Anthelmintic activity of ivermectin against experimental *Ascaridia galli* infection in chickens. *Vet. Parasitol.*, **37**, 307–314.

Shoop, W. L., Mrozik, H., Fisher, M. H. (1995). Structure and activity of avermectins and milbemycins in animal health. *Vet. Parasitol.*, **59**, 139–156.

Simpkin, K. G., Coles, G. C. (1981). The use of *Caenorhabditis elegans* for anthelmintic screening. *J. Chem. Tech. Biotech.*, **31**, 66–69.

Theodorides, V. J. (1976). Anthelmintics: from laboratory animals to the target species. In *Chemotherapy of Infectious Disease* (ed Gadebusch, H. H.), pp. 71–96. CRC Press, Cleveland, OH.

Thomas, H., Andrews, P. (1977). Praziquantel—a new cestocide. *Pest. Sci.*, **8**, 556–560.

Thomas, H., Gönnert, R. (1977). The efficacy of praziquantel against cestodes in animals. *Z. Parasitenk.*, **52**, 117–127.

Thompson, D. P., Klein, R. D., Geary, T. G. (1996). Prospects for rational approaches to anthelmintic discovery. *Parasitology*, **113**, S217–S238.

Vermunt, J. J., West, D. M., Pomroy, W. E. (1995). Multiple resistance to ivermectin and oxfendazole in *Cooperia* species of cattle in New Zealand. *Vet. Rec.*, **137**, 43–45.

Wescott, R. B. (1986). Anthelmintics and drug resistance. *Vet. Clin. N. Am. Equine Pract.*, **2**, 367–380.

Wightman, S. R., Wagner, J. E., Corwin, R. M. (1978). *Syphacia obvelata* in the Mongolian gerbil (*Meriones unguiculatus*): natural occurrence and experimental transmission. *Lab. Anim. Sci.*, **28**, 51–54.

Worley, D. E., Meisenhelder, J. E., Sheffield, H. G., Thompson, P. E. (1962). Experimental studies on *Trichuris muris* in mice with an appraisal of its use for evaluating anthelmintics. *J. Parasitol.*, **48**, 433–437.

Section V

Viral Infection Models

Chapter 107

Animal Models for Central Nervous System and Disseminated Infections with Herpes Simplex Virus

E. R. Kern

Background of human infections

The herpesviruses are the etiologic agents of a wide variety of diseases in humans, many of which have high mortality or morbidity rates in normal or immunocompromised patients. Herpes simplex virus type 1 (HSV-1) is the most common cause of sporadic, life-threatening encephalitis in the United States. In untreated patients, mortality rates are about 70% and severe neurologic sequelae occur in most survivors (Whitley *et al.*, 1977). In patients with herpes encephalitis, treatment with idoxuridine, the first antiviral agent tested, did not influence the disease and was associated with toxic side-effects (Boston Interhospital Virus Study Group, 1975). In subsequent studies, adenine arabinoside (ara-A) therapy was shown to reduce significantly mortality rates and the frequency of neurologic sequelae (Whitley *et al.*, 1977, 1981). In the next series of clinical trials, acyclovir (ACV) was shown to be more effective than ara-A for treatment of this disease (Whitley *et al.*, 1986). HSV-2 infection of the newborn infant, which is usually acquired on passage through the mother's infected birth canal, is also associated with a high mortality rate and severe neurological sequelae in a significant number of the survivors (Whitley *et al.*, 1980a). Treatment of neonatal herpes with ara-A also significantly reduced mortality and neurologic sequelae (Whitley *et al.*, 1983), and therapy with ACV has been reported to be at least as effective as ara-A (Whitley *et al.*, 1991).

Although there have been significant advances in the treatment of severe HSV infections with ACV, there is still a need to develop more effective modes of therapy for these infections. In particular, for herpes encephalitis, neonatal herpes, and mucocutaneous infections in acquired immunodeficiency syndrome (AIDS) patients caused by ACV-resistant viruses, optimal therapy is currently not available. Hopefully in the next few years, new more active and/or less toxic agents will be developed and the use of better delivery methods or combinations of existing drugs will improve the outcome of these serious diseases.

Background of animal model infections

The use of mice inoculated intranasally (i.n.) with HSV-1 as a model for herpes encephalitis was initially described by De Clercq and Luczak (1976) and has been used extensively in our laboratory for evaluation of antiviral agents directed against this disease. After HSV-1 inoculation the virus travels from the nasopharynx via olfactory and trigeminal nerve tracts to the brain, resulting in death from an acute encephalitis. In this model, treatment with ara-A initiated within 24–48 hours of infection resulted in significant protection (Kern, 1988). The protection against mortality observed with ara-A treatment also correlated with alteration of HSV-1 replication in the central nervous system (CNS). Therapy with 250 mg/kg of ara-A twice daily for 7 days initiated 48 hours after viral inoculation reduced considerably virus titers in olfactory lobes, cerebral cortex, cerebrum, diencephalon, and the pons medulla when compared with placebo-treated mice (Kern, 1990). Treatment also reduced viral replication in spleen but had only a marginal effect in lung. In clinical trials of HSV-1 encephalitis, treatment with ara-A also resulted in significant reduction in mortality (Whitley *et al.*, 1977, 1981).

In the animal model infection, treatment with ACV was considerably more effective than ara-A in preventing mortality (Kern, 1988). Concentrations as low as 15 mg/kg significantly reduced mortality when therapy was begun 72 hours after infection, a time when virus is replicating in the CNS. In a comparison between ara-A and ACV treatment on the pathogenesis of HSV-1 encephalitis in mice, virus titers in ACV-treated animals were lower in all target organs than in placebo-treated animals (Kern, 1990). In the cerebellum, diencephalon and pons medulla, ACV treatment was superior to ara-A in reducing the magnitude of viral replication and promoted a more rapid clearance of HSV-1 in these tissues. ACV therapy was also more effective than ara-A in reducing viral replication in lung tissue (Kern, 1990). Based on these experimental data, it was predicted that ACV would be more effective than ara-A in treatment of HSV-1 encephalitis in humans. Subsequent clinical trials confirmed observations from the animal model studies as ACV was shown to be significantly more effective than ara-A (Whitley *et al.*, 1986).

Mice inoculated i.n. with HSV-2 appear to be a good model for disseminated neonatal herpes with CNS involvement. After viral inoculation, replication of virus is initially detectable in lung with subsequent dissemination to

Handbook of Animal Models of Infection
ISBN 0-12-775390-7

visceral organs. Virus is also transmitted concomitantly by neural routes from the nasopharynx to olfactory lobe and cerebellum — brain stem, cerebrum and spinal cord (Kern *et al.*, 1978). This animal model provides a severe test of antiviral efficacy and treatment with ara-A failed to provide any protection against mortality; however, there was some reduction of virus replication in brain, lung, liver, and spleen (Overall *et al.*, 1975; Kern, *et al.*, 1975, 1978; Kern, 1988).

This is a complex model of infection involving both CNS and non-CNS tissues and historically has not responded well to any of the antiviral agents tested. In placebo-controlled clinical trials, however, ara-A did reduce mortality (Whitley *et al.*, 1980b) but many survivors had permanent neurological sequelae (Whitley *et al.*, 1983). Parenteral treatment with ACV in the murine model was very effective in reducing mortality when therapy was initiated 48–96 hours after infection (Kern *et al.*, 1982; Kern, 1988), and also reduced significantly viral replication in target organs (Kern *et al.*, 1986). These results from experimental HSV-2 infection in mice predicted that ACV would be more effective than ara-A in the treatment of neonatal herpes; however, in clinical trials in which ara-A and ACV were compared for efficacy, the two agents were judged to be comparable (Whitley *et al.*, 1991). The experimental HSV-2 infection, therefore, accurately predicted efficacy for treatment of human disease with ara-A or ACV but apparently failed to determine which of the two drugs would be most active.

It should be stressed, however, that disseminated HSV-2 infections of mice and humans are complex and the animal model may not accurately represent the entire pathogenesis of human disease, which includes not only CNS and disseminated visceral components, but also may have skin, eye, or mouth involvement which are not seen in the mouse model.

Rationale for use of animal models

The importance of experimental viral infections in animal models for the development and evaluation of new antiviral agents prior to their use in humans should not be minimized. While tissue culture systems are of great value in determining if a new drug has activity against a particular virus, these systems should not be used as indicators or predictors of activity in humans. Only where suitable animal models are not available should a compound be taken from tissue culture directly into human trials. Although one can legitimately argue that most, if not all, animal model infections are not identical to the human disease, it can be demonstrated that a compound does in fact have activity in an *in vivo* system and early indications of its antiviral activity, tissue distribution, metabolic disposition, pharmacokinetics, and acute toxicity can be realized. Importantly, all of these parameters of drug pharmacodynamics can be correlated with inhibition of viral replication in target organs. Additionally, our understanding of the pathogenesis of many viral infections, the response of the host to infection and interaction between the viral infection, the host's immune response, and a therapeutic agent has been enhanced greatly through the use of animal model systems.

When selecting or developing an animal model for determining efficacy of an antiviral, one should utilize systems that best simulate the corresponding human disease. Some of the factors that need to be considered when selecting an animal model have been reviewed previously (Field, 1988; Kern, 1990). It is obvious that there is currently no experimental viral infection that meets all of the properties of an ideal animal model. The most common animal species used for experimental HSV infections are rodents and these differ from humans by most criteria. There are significant differences in drug metabolism and pharmacokinetics between humans and animals that must be considered when extrapolating results obtained in animal models to the design of initial clinical trials. Since the herpesviruses, and particularly HSV, are the etiologic agents of a wide variety of diseases in humans, it is important that new antiviral agents be tested in animal models that simulate each of those disease states in humans. In addition, since animal models for HSV infections have been the most widely used, the predictability for most of these infections has been established.

Methods

Animals and animal care

Three-week-old Balb/c or outbred mice such as CD-1 or Swiss-Webster weighing approximately 10–12 g are used for infection. While we have used females exclusively, males can also be used. In our laboratory, Balb/c mice are slightly more susceptible to infection with HSV than the outbred animals, and the results appear to be more consistent. Other animal strains such as C_3H or C57/BL6 are more resistant to infection with HSV and virus replicates less well in target organs such as lung, liver, spleen, and kidney than in Balb/c or outbred mice.

All mice in our facility are housed in standard filter top polycarbonate shoebox cages (155 square inches). The maximum number of mice housed per cage is 15. They are given standard rodent chow and water *ad libitum*. Contact bedding is pine chips, and cages and water are changed twice a week. They are maintained on computer-controlled 12-hour off/on light cycle at 35% relative humidity. Cages are checked daily for dead or sick animals.

Virus stocks, assays, and cell cultures

Stock pools of HSV are prepared routinely in primary cultures of rabbit kidney cells. Virus preparations are quantified using a standard plaque assay in primary rabbit kidney cells or low passaged human foreskin fibroblast cells under a liquid overlay containing a sufficient amount of antibody (gamma-globulin) to retard extracellular spread or alternatively an agarose or methylcellulose overlay media can be used. Primary rabbit kidney fibroblast cells are prepared from rabbits aged 3 weeks. Kidneys are removed aseptically and washed in saline; the medulla is removed and discarded and the cortex tissue is minced and washed repeatedly with saline to remove debris and red blood cells. Individual cell suspensions are prepared by multiple extractions using 0.25% trypsin. After the tissue is digested, the cells are centrifuged, re-suspended in tissue culture media with 10% fetal bovine serum (FBS) and placed in 175 cm^2 tissue culture flasks and incubated at 37°C with 95% CO_2 and 90% humidity. After the cells become confluent, they are trypsinized and added to multi-well tissue culture plates for use in the plaque assay.

Inoculation of animals

Intracerebral

Female Balb/c mice (21 days old) are inoculated intracerebrally with various dilutions of HSV by administering 0.03 ml volume with a ½ cc glass syringe and a 26 G 0.5 in needle into the right cerebral hemisphere. The syringe is held perpendicular to the mouse's head and inserted until the needle just pierces the skull. Care should be taken not to hit one of the major blood vessels. Animals are anesthetized prior to inoculation with 5 mg/ml of ketamine containing 15 mg/ml xylazine in a 0.1 ml volume.

Intraperitoneal

Female Balb/c mice (21 days old) are inoculated with 0.1 ml of 10^2–10^3 plaque-forming units (pfu) of HSV. A 1 cc tuberculin syringe and 25 G 5/8 in needle are used to deliver virus into the lower left quadrant of the abdomen. No anesthesia is used.

Intranasal

Female Balb/c mice (21 days old) are inoculated with 10^4–10^5 pfu/animal. Virus is administered in a 0.04 ml volume with a 20 μl Eppendorf pipette tip by allowing the mouse to inhale 20 μl in each nostril. Mice do not need to be anesthetized for this procedure. Male Fischer rats weighing 175–200 g are inoculated with HSV-1 at about 10^4 pfu/animal. As with mice, rats are given 0.02 ml of HSV in each nostril, with a 20 μl pipette. The rats are anesthetized for this procedure with 100 mg/ml of ketamine containing 15 mg/ml of xylazine at the rate of 0.1 ml/100 g of body weight.

Experimental infections and pathogenesis

Intracerebral inoculation

Intracerebral inoculation of young mice has been used for evaluation of antiviral therapy; however, it does not represent a natural route of infection and it is not a suitable model for this purpose. Although direct inoculation of HSV into the brain might simulate herpes encephalitis that is a consequence of viral reactivation within the CNS, there is concern that this method of inoculation may disrupt tissue architecture and the blood–brain barrier, making it more permeable to antiviral drugs. While the model is not optimal for evaluating antiviral therapy, it is very useful in determining the virulence of viruses and in particular HSV, as wild-type isolates will kill a young mouse with as little as 1 pfu inoculated directly into the brain. We have utilized the model extensively for characterization of HSV mutants, particularly those that are resistant to antiviral agents such as ACV and have a mutation in the thymidine kinase (TK) or polymerase gene. Additionally, the model is useful for determining the virulence and stability of attenuated HSV vaccines.

Virulence is determined according to the number of pfu of HSV required to kill 50% of inoculated animals. Wild-type isolates that are TK + will kill mice at 1–10 pfu, whereas those viruses that no longer produce a functional TK require many orders of magnitude more virus, if in fact any mortality at the highest possible dose inoculated can be demonstrated. We have evaluated the virulence of numerous isolates of HSV-1 or HSV-2 that are ACV-resistant and TK-deficient, compared to MS, a laboratory-passaged standard wild-type strain of HSV-2. The results of some of these isolates inoculated intracerebrally in mice are summarized in Table 107.1. The pfu median lethal dose (LD_{50}) ratios were calculated using combostat, version 3, (Belen'kii and Schinazi, Emory University School of Medicine). The pfu/LD_{50} ratio for the HSV-2, MS strain was 0.78. The HSV-2 TK-altered strain, 12247, proved to be as virulent as wild-type with a pfu/LD_{50} ratio of 3.2. Of particular interest is the isolate AG-3 which expresses less than 3% (low producer) of normal TK but is still highly virulent. Two isolates in which the TK has been deleted (Δ305, SR6008) and an ACV-resistant isolate (DM 21) required 10^4 or greater pfu to produce mortality in mice.

Intraperitoneal inoculation

Although the i.n. HSV infection models have been predictive of efficacy for human disease, they are severe tests of efficacy and i.n. inoculation of mice is a tedious and time-consuming process. As a result, it is difficult to use large numbers of animals for screening several antiviral agents and/or various dosage concentrations and treatment regimens. For these reasons, we first screen compounds for *in*

Table 107.1 Virulence of herpes simplex virus (HSV) isolates in mice inoculated intracerebrally

Strain	*TK*	*pfu/ mouse*	*Dead/ total*	*pfu/ LD_{50}*
HSV-2, MS	+	4.2×10^2	10/10	0.78
		4.2×10^1	10/10	
		4.2	9/10	
		0.42	3/10	
HSV-2, 12247	A	1.2×10^3	10/10	3.2
		1.2×10^2	10/10	
		1.2×10^1	7/10	
		1.2	5/10	
		0.12	0/10	
HSV-2, AG-3	LP	1.1×10^3	9/10	1.1×10^1
		1.1×10^2	9/10	
		1.1×10^1	3/10	
		1.1	2/10	
HSV-1, DM21	–	1.3×10^5	10/10	1.6×10^4
		1.3×10^4	2/10	
		1.3×10^3	0/10	
		1.3×10^2	0/10	
HSV-1, Δ305	–	6.0×10^4	2/10	>6.0×10^4
		6.0×10^3	0/10	
		6.0×10^2	0/10	
		6.0×10^1	0/10	
		6.0	0/10	
HSV-1, SR6008	–	4.5×10^4	2/10	>4.5×10^4
		4.5×10^3	0/10	
		4.5×10^2	0/10	
		4.5×10^1	0/10	
		4.5	0/10	

TK, Thymidine kinase phenotype; pfu, plaque-forming unit; LD_{50}, median lethal dose; +, positive, wild-type; A, altered; LP, low producer; – negative.

Table 107.2 Effect of treatment with acyclovir on the mortality of mice inoculated intraperitoneally with herpes simplex virus type 2 (HSV-2)

	Mortality				
*Treatment**	*No.*	*%*	*P Value*	*MDD*	*P value*
Control	14/15	93		9.0	
Placebo	15/15	100	NS	8.5	NS
Acyclovir					
80 mg/kg + 24 h	4/15	27	<0.001	11.8	NS
80 mg/kg + 48 h	2/15	13	<0.001	17.0	<0.05
80 mg/kg + 72 h	3/15	20	<0.001	12.3	0.05
80 mg/kg + 96 h	5/15	33	<0.001	9.6	0.05
80 mg/kg – Tox†	1/10	10		8.0	
40 mg/kg + 24 h	4/15	27	<0.001	17.8	<0.01
40 mg/kg + 48 h	3/15	20	<0.001	13.3	<0.05
40 mg/kg + 72 h	6/15	40	<0.001	13.5	0.01
40 mg/kg + 96 h	7/15	47	<0.01	7.9	NS
40 mg/kg – Tox	0/10	0			
20 mg/kg + 24 h	7/15	47	<0.01	10.9	0.01
20 mg/kg + 48 h	7/15	47	<0.01	10.9	0.08
20 mg/kg + 72 h	9/15	60	<0.05	10.6	<0.05
20 mg/kg + 96 h	12/15	80	NS	10.8	NS
20 mg/kg – Tox	0/10	0			
10 mg/kg + 24 h	11/15	73	NS	8.1	NS
10 mg/kg + 48 h	9/15	60	<0.05	9.0	NS
10 mg/kg + 72 h	11/15	73	NS	9.7	NS
10 mg/kg + 96 h	15/15	100	NS	9.5	NS

* Animals were treated intraperitoneally twice a day for 7 days with the above concentrations and initiated at the times indicated.
† Uninfected animals treated as stated above.
MDD, Mean day of death; NS, not statistically significant when compared to the appropriate placebo-treated group.

vivo HSV activity in outbred Swiss, CD-1, or Balb/c mice inoculated intraperitoneally (i.p.) with HSV-1 or HSV-2. After i.p. inoculation with HSV-1 or HSV-2, virus replicates in the gut, liver, and spleen and then spreads to the CNS by viremia and probably also by peripheral nerves (Kern *et al.*, 1978, 1982). Virus is first detected in the brain by about day 5, and animals die on about day 9, which allows ample time to demonstrate an antiviral effect.

This model system has consistently been the most sensitive system for determining efficacy of a new antiviral. Although it does not simulate a natural route of infection, it is an ideal system for screening new compounds and for determining optimal dosages and treatment regimens prior to testing in mice inoculated i.n. that more closely simulate human infections. The effects of ACV in mice inoculated i.p. with HSV-2 are shown in Table 107.2 as an example of the use of these models in screening new compounds for HSV activity. Treatment with 20 mg/kg twice daily was highly effective even when treatment was delayed until 72 hours post-infection. The effect of ACV treatment on virus replication in target organs is shown in Figure 107.1. When treatment was initiated 24 hours after infection, virus replication was completely inhibited in all of the critical target organs. Similar results were obtained with mice inoculated i.p. with HSV-1 (data not shown). If the experimental drug has activity in this model, it is next evaluated in mice inoculated by the i.n. route. If the compound is not active in one of these two i.p. models, no further evaluation is performed.

Intranasal inoculation

Intranasal inoculation of 3-week-old Swiss-Webster mice with HSV-1 provides a model for herpes encephalitis of humans which utilizes a natural route of infection. After inoculation of approximately 5 × 10^5 pfu of the E-377 strain of HSV-1, virus replicates in the nasopharynx and spreads to the CNS by way of olfactory and trigeminal nerves. Virus is detected in olfactory bulb, trigeminal nerves, and trigeminal ganglia prior to its detection in the cortex, cerebrum, or cerebellum (Kern *et al.*, 1978; Field *et al.*, 1984). Untreated animals generally die on days 6–10. A representative experiment illustrating the use of the model to determine the effi-

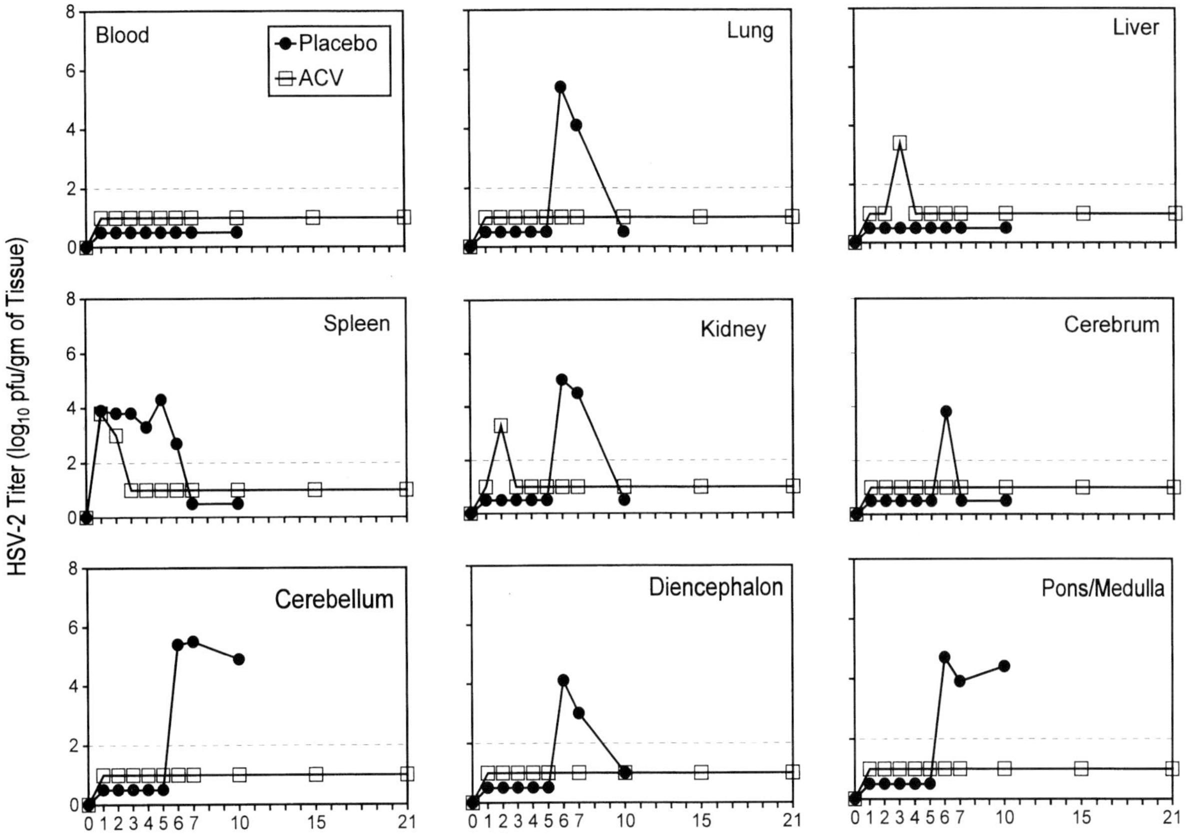

Figure 107.1 Pathogenesis of herpes simplex virus type 2 (HSV-2) infection in mice inoculated intraperitoneally and treated with acyclovir (ACV).

cacy of an antiviral agent, ACV, on mortality of mice inoculated i.n. with HSV-1 is show in Table 107.3. Treatment was highly effective when initiated 96 hours after infection, a time when virus is replicating to high titers in all parts of the brain. The effect of ACV on the replication of HSV-1 in CNS and other tissues is presented in Figure 107.2. In the animals treated with 20 mg/kg of ACV beginning 24 hours after infection, virus titers in all the organs were reduced compared with placebo-treated mice.

One of the major problems in the treatment of HSV encephalitis in humans is the inability of most antiviral drugs readily to penetrate the blood–brain barrier. In order to investigate the ability of blood–brain barrier penetration enhancers to facilitate the passage of antiviral drug such as ACV, a larger animal than mice is required so compounds can readily be administered by way of arteries or veins. Therefore, we have inoculated 170 g Fisher rats (Charles River Laboratories) intranasally with 10^2–10^5 pfu of HSV by the i.n. route. In these animals 5×10^3 pfu of HSV inoculated intranasally into male Fisher rats produces about 90% mortality. A study of the pathogenesis of HSV-1 replication in brains of rats indicated viral replication occurred in all areas of the brain with peak titers occurring on days 6–7 (Figure 107.3). Virus titers increased to maximum values on day 6 with a range of approximately log 4.0–6.0, and all areas of the brain were infected. Virus was not detected in any visceral organs until day 6 with titers of $10^{1.9}$ and $10^{3.3}$ in the kidney and lung, respectively. It is interesting that in rats inoculated i.n. with HSV-1, initial replication appears to occur in the pons medulla rather than in the olfactory tracts seen in the mouse and replication seems to progress more slowly in the rat than in mice. Although we have not utilized this model extensively for antiviral drug therapy, it certainly can be used for this purpose, and treatment with ACV does provide protection against mortality and reduces virus replication in CNS tissues (data not presented).

The pathogenesis of HSV encephalitis has also been studied in rabbits. If rabbits are inoculated in the olfactory bulb (similar to i.n., but avoids the trigeminal tract) with HSV-1, virus tended to localize in the olfactory bulb, olfactory cortex (analogous to human temporal lobe) and the frontal lobe (Schlitt *et al.*, 1986). In another rabbit model in which HSV-1 was inoculated i.n., reactivation of a latent infection in trigeminal ganglia by immunosuppression resulted in focal lesions in olfactory and temporal lobes of the brain (Stroup and Schaefer, 1986). Although the rabbit is not a practical model for evaluating drug treatment of HSV encephalitis, it has been a valuable model for the study of this disease.

There is a variety of evidence that supports the use of i.n. inoculation as a natural route of infection. Although the pathogenesis of acute HSV encephalitis in humans is not entirely understood, the two most popular ideas presently are first, that encephalitis is due to a newly acquired infection, most probably by way of the nasopharynx, and

Table 107.3 Effect of treatment with acyclovir on the mortality of mice inoculated intranasally with herpes simplex virus type 1 (HSV-1)

	Mortality				
*Treatment**	*No.*	*%*	*P Value*	*MDD*	*P value*
Control	14/15	93		5.2	
Placebo + 24 h	14/15	93	NS	5.7	NS
Acyclovir					
80 mg/kg + 24 h	5/15	33	<0.01	4.6	NS
80 mg/kg + 48 h	3/15	20	<0.001	4.0	NS
80 mg/kg + 72 h	5/15	33	<0.01	5.4	NS
80 mg/kg + 96 h	7/15	47	0.01	6.6	NS
80 mg/kg – Tox†	0/10	0			
40 mg/kg + 24 h	0/15	0	<0.001		
40 mg/kg + 48 h	3/15	20	<0.001	4.0	NS
40 mg/kg + 72 h	5/15	33	<0.01	5.2	NS
40 mg/kg + 96 h	7/15	47	0.01	5.1	NS
40 mg/kg – Tox	1/10	10		8.0	
20 mg/kg + 24 h	5/15	33	<0.01	5.8	NS
20 mg/kg + 48 h	7/15	47	0.01	5.1	NS
20 mg/kg + 72 h	8/15	53	<0.05	4.8	NS
20 mg/kg + 96 h	9/15	60	NS	5.9	NS
20 mg/kg – Tox	0/10	0			
10 mg/kg + 24 h	5/15	33	<0.01	9.0	<0.05
10 mg/kg + 48 h	7/15	47	0.01	4.7	NS
10 mg/kg + 72 h	12/15	80	NS	5.7	NS
10 mg/kg + 96 h	14/15	93	NS	5.4	NS

* Animals were treated intraperitoneally twice a day for 7 days with the above concentrations and initiated at the times indicated.
† Uninfected animals treated as stated above.
MDD, Mean day of death; NS, not statistically significant when compared to the appropriate untreated control or placebo.

second, that encephalitis is the result of reactivation of a latent infection, probably from the trigeminal ganglia or some other site within the CNS (e.g., temporal lobe). There are a number of new pieces of evidence that support the postulate that HSV spreads to the frontal or temporal lobes by way of olfactory and/or trigeminal nerve tracts. During HSV encephalitis in patients, virus has also been detected in olfactory and limbic areas (Eseri, 1982). In addition some patients have virus in the nasopharynx and in one series of patients using restriction endonuclease techniques, the same virus was isolated from the nose and the CNS (Whitley *et al.*, 1982). These data in animal models and in humans strongly support an olfactory and/or trigeminal route of spread to the temporal lobe and indicate that i.n. inoculation simulates the natural route of infection.

Evaluation of antiviral therapy—key parameters

These models lend themselves well to the use of i.p., subcutaneous, or oral therapy. When first evaluating a new antiviral agent in animal models one does not generally know the pharmacokinetic profile of the agent in that species. For all initial evaluations, we utilize a standard treatment regimen consisting of i.p. treatment twice daily for 7 days. A sufficient number of drug concentrations should be used so that the maximum tolerated dose and the minimal effective dose can be calculated. A minimum of 10–15 animals per group should be used. The initial and most common endpoint for efficacy evaluation is usually death. Examples of typical mortality experiments using ACV are presented in Tables 107.1 and 107.2. Before beginning efficacy studies in mice, an initial experiment should be conducted in which uninfected animals are treated i.p. twice daily for 7 days with various concentrations of test compound to establish the maximum tolerated dose. After completing the toxicity study, mice were allocated to placebo or ACV treatment groups, with each animal receiving drug i.p. twice daily for 7 days at doses below that which could be maximally tolerated. For each treatment group, therapy was initiated at 24, 48, 72, or 96 hours after viral challenge. Two control groups were included in the study (one untreated and one given placebo beginning 24 hours post-challenge). Most of the mice given ACV at the highest concentrations survived, even when therapy was delayed 24–72 hours post-infection, a time when virus was replicating in brain tissue. An excellent dose response was obtained as well as a correlation between dose and time of initiation of therapy.

Since mortality is not a very sensitive endpoint and may not accurately reflect antiviral efficacy in a particular target organ, the optimal measure of a drug's activity in an animal is to determine the effect of treatment on viral replication in the target organs such as brain, lung, etc. Examples of experiments of this type comparing ACV with placebo are shown in Figures 107.1 and 107.2. In both experiments, acyclovir was administered twice daily at 20 mg/kg for 5 days beginning 24 hours after infection. Virus titers were determined in target organ tissues at various intervals for 21 days post-inoculation. The results in either model indicate that ACV administered 24 hours after HSV infection and at a time when virus replication is occurring can alter replication in most target organs. This provides a sensitive measure for determining antiviral efficacy in tissues such as brain and lung, which are primary organs affected in severe HSV infections in humans.

Potential pitfalls (advantages/disadvantages) of the models

One of the major advantages to the models is that HSV isolated from humans will readily infect mice, rats, and most other small animals. It is still important, however, to determine the sensitivity of the virus strain that will be used in animals in tissue culture prior to the initiation of animal experiments. Murine models are of particular value since,

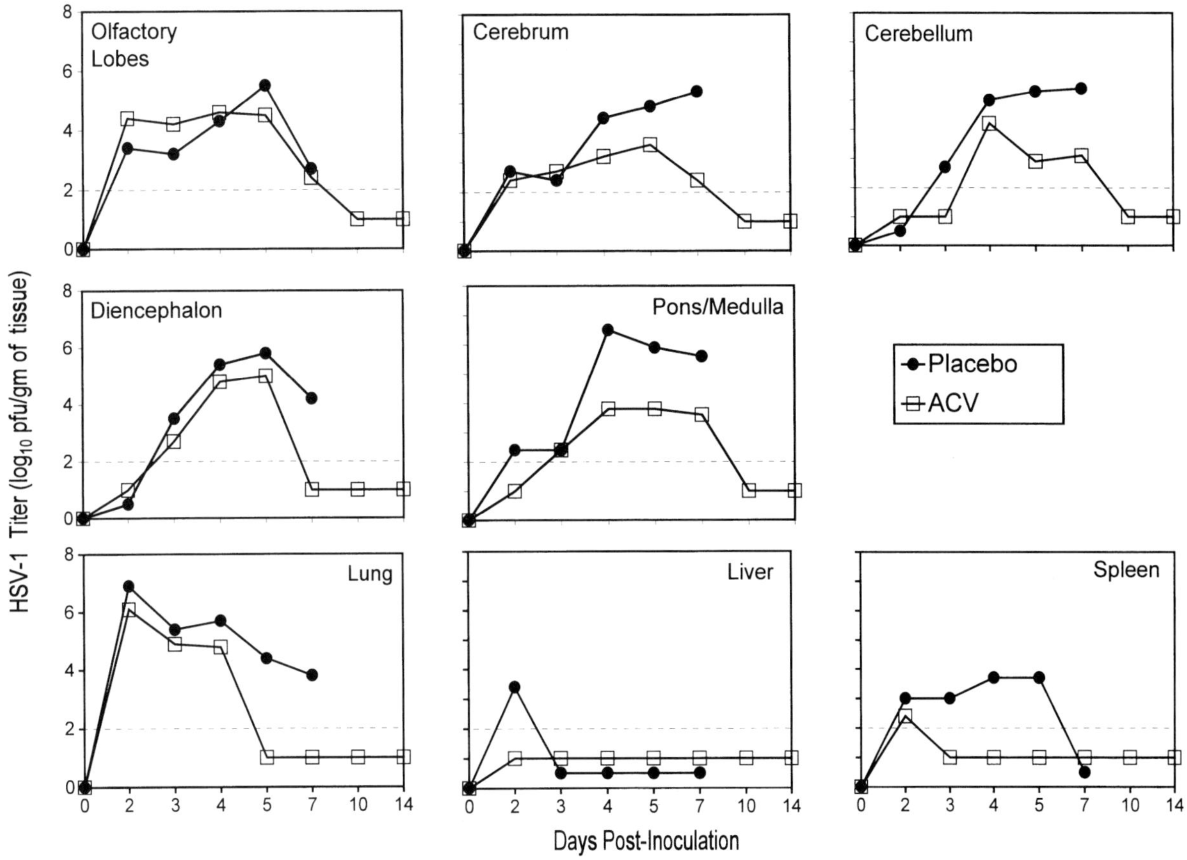

Figure 107.2 Pathogenesis of herpes simplex virus type 1 (HSV-1) infection in mice inoculated intranasally and treated with acyclovir (ACV).

because of their small size and relatively low cost, large numbers can be used to determine optimal dosage concentrations and treatment regimen accurately and they require small amounts of drug. The HSV model infections are particularly useful because they can be used to simulate a variety of different disease states in humans ranging from cutaneous infections to a fatal encephalitis. The advantage that the murine models can be readily infected with HSV can also be one of the disadvantages. Regardless of the route of inoculation, the virus has a propensity for invading into peripheral nerves and is transmitted to the CNS, resulting in a fatal encephalitis. The virus also has the ability to establish latent infections in regional ganglia of surviving animals; however, unfortunately, spontaneous reactivation does not generally occur, nor are they easily induced. The rodent models, therefore, do not generally lend themselves to the evaluation of the effect of therapy on the maintenance and reactivation of HSV infections.

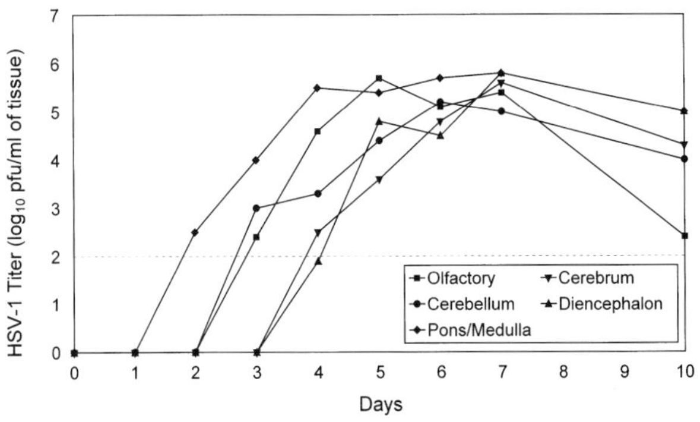

Figure 107.3 Pathogenesis of herpes simplex virus type 1 (HSV-1) infection in rats inoculated intranasally.

As with all murine models and other rodents, the adsorption, distribution, metabolism, excretion, and toxicity of antiviral agents will undoubtedly be different than observed in humans. These differences, however, have not detracted from the value of these models in the preclinical evaluation of new drugs prior to their use in humans.

Contributions of the models and their predictability

Much of our understanding of the pathogenesis and latency of HSV infections in humans has come from the study of experimental infections in the mouse. Since the animal model utilizes a human virus and simulates many aspects of

HSV infections in humans, a lot of valuable information obtained in the animal models can be extended to human disease. In general, results obtained from preclinical studies of antiviral agents in our animal models preceded and were predictive of efficacy or lack of efficacy in subsequent trials of the corresponding human disease. In the models of herpes encephalitis and neonatal herpes used in our laboratory, idoxuridine and cytosine arabinoside were not effective, while ara-A was moderately active and ACV was highly active. Similar results were obtained in human studies (Kern, 1997). Therefore, the mouse models for herpes encephalitis, and neonatal herpes, have a record of predictability for the outcome of treatment of these infections in humans.

References

Boston Interhospital Virus Study Group, and the NIAID-Sponsored Cooperative Antiviral Clinical Study (1975). Failure of high dose 5-iodo-2′deoxyuridine in the therapy of herpes simplex virus encephalitis. *N. Engl. J. Med.*, **292**, 599–603.

De Clercq, E., Luczak, M. (1976). Intranasal challenge of mice with herpes simplex virus: an experimental model for evaluation of the efficacy of antiviral drugs. *J. Infect. Dis.*, **133** (suppl), A226–A236.

Eseri, N. M. (1982). Herpes simplex encephalitis, an immunohistological study of the distribution of viral antigen within the brain. *J. Neural Sci.*, **54**, 209–226.

Field, H. J. (1988). Animal models in the evaluation of antiviral chemotherapy. In *Antiviral Agents: The Development and Assessment of Antiviral Chemotherapy* (ed Field, H. J.), pp. 67–84. CRC Press, Boca Raton.

Field, H. G., Anderson, J. R., Efstathiou, J. (1984). A quantitative study of the effects of several nucleoside analogues on established herpes encephalitis in mice. *J. Gen. Virol.*, **65**, 707–719.

Kern, E. R. (1988). Animal models as assay systems for the development of antiviurals. In *Antiviral Drug Development* (eds De Clercq, E., Walker, R. T.), pp. 149–172. Plenum Press, New York.

Kern, E. R. (1990). Preclinical evaluation of antiviral agents: *in vitro* and animal model testing. In *Antiviral Agents and Viral Diseases of Man* (eds Galasso, G. J., Whitley, R. J., Merigan, T. C.), pp. 87–123. Raven Press, New York.

Kern, E. R. (1997). Preclinical evaluation of antiviral agents. In *Antiviral Agents and Human Viral Diseases* (eds Galasso, G. J., Whitley, R. J., Merigan, T. C.), pp. 79–111. Lippincott-Raven, Philadelphia.

Kern, E. R., Overall, J. C. Jr, Glasgow, L. A. (1975). Herpesvirus hominis infection in newborn mice: comparison of the therapeutic efficacy of 1-β-D-arabinofuranosylcytosine and 9-β-D-arabinofuranosyladenine. *Antimicrob. Agents Chemother.*, **7**, 587–595.

Kern, E. R., Richards, J. T., Overall, J. C. Jr, Glasgow, L. A. (1978). Alteration of mortality and pathogenesis of three experimental herpesvirus hominis infections of mice with adenine arabinoside 5′monophosphate, adenine arabinoside, and phosphonoacetic acid. *Antimicrob. Agents Chemother.*, **13**, 53–60.

Kern, E. R., Richards, J. T., Glasgow, L. A., Overall, J. C. Jr, De Miranda, P. (1982). Optimal treatment of herpes simplex virus encephalitis in mice with oral acyclovir. Symposium on acyclovir. *Am. J. Med.*, **73**, 125–131.

Kern, E. R., Richards, J. T., Overall, J. C. Jr. (1986). Acyclovir treatment of disseminated herpes simplex virus type 2 infection in weanling mice: alteration of mortality and pathogenesis. *Antiviral Res.*, **6**, 189–195.

Overall, J. C. Jr., Kern, E. R., Glasgow, L. A. (1975). Herpesvirus hominis type 2 infections in mice with adenine arabinoside. In *Adenine Arabinoside: An Antiviral Agent* (eds Pavan-Langston, D., Buchanan, R. A., Alford, Jr C. A.), pp. 95–110. Raven Press, New York.

Schlitt, M., Lakeman, A. D., Wilson, E. R. *et al.* (1986). A rabbit model of focal herpes simplex encephalitis. *J. Infect. Dis.*, **153**, 732–735.

Stroup, W. G., Schaefer, D. C. (1986). Production of encephalitis restricted to the temporal lobes by experimental reactivation of herpes simplex virus. *J. Infect. Dis.*, **153**, 721–731.

Whitley, R. J., Soong, S., Dolin, R. *et al.* (1977). Adenine arabinoside therapy of biopsy-proved herpes simplex encephalitis: National Institute of Allergy and Infectious Diseases collaborative antiviral study. *N. Engl. J. Med.*, **297**, 289–294.

Whitley, R. J., Nahmias, A. J., Visintine, A. M., Fleming, C. L., Alford, C. A. (1980a). The natural history of herpes simplex virus infection of mother and newborn. *Pediatrics*, **66**, 489–494.

Whitley, R. J., Nahmias, A. J., Soong, S.-J., Galasso, G. G., Fleming, C. L., Alford, C. A. (1980b). Vidarabine therapy of neonatal herpes simplex virus infections. *Pediatrics*, **66**, 495–501.

Whitley, R. J., Soong, S.-J., Hirsch, M. S. *et al.* (1981). Herpes simplex encephalitis: vidarabine therapy and diagnostic problems. *N. Engl. J. Med.*, **304**, 313–318.

Whitley, R. J., Lakeman, A. D., Nahmias, A. J., Roizman, B. (1982). DNA restriction enzyme analysis of herpes simplex virus isolates obtained from patients with encephalitis. *N. Engl. J. Med.*, **307**, 1060–1062.

Whitley, R. J., Yeager, A., Kartus, P. *et al.* (1983). Neonatal herpes simplex virus infection: follow-up evaluation of vidarabine therapy. *Pediatrics*, **72**, 778–785.

Whitley, R. J., Alford, C. A., Hirsch, M. S. *et al.* (1986). Vidarabine versus acyclovir therapy in herpes simplex encephalitis. *N. Engl. J. Med.*, **314**, 144–149.

Whitley, R. J., Arvin, A., Prober, C. *et al.* (1991). A controlled trial comparing vidarabine with acyclovir in neonatal herpes simplex virus infection. *N. Engl. J. Med.*, **324**, 444–449.

Chapter 108

Animal Models of Herpesvirus Genital Infection: Guinea-Pig

N. Bourne and L. R. Stanberry

Background of human infection

Genital herpes is a common sexually transmitted disease caused by herpes simplex virus (HSV) either type 1 or 2. It occurs worldwide and is recognized as a major public health problem in many developed countries (Stanberry *et al.*, 1997). Recent seroprevalence data indicate that more than one in five adult Americans have been infected with HSV-2, the predominant cause of genital herpes in the United States (Fleming *et al.*, 1997).

The natural history of genital herpes includes the initial or primary infection, the establishment of a latent infection in paraspinal sensory ganglia, and the periodic reactivation of latent infection which causes recurrent infections (Wald and Corey, 1996). Initial and recurrent infections in humans may be either symptomatic or asymptomatic. Symptomatic infection is characterized by the development of vesicles and/or ulcers at mucocutaneous sites in the anogenital area. Initial infection is generally more severe than recurrent infections and may be associated with systemic findings such as fever, urinary retention or dysuria, and meningitis. In humans, HSV-1 and HSV-2 produce indistinguishable initial genital infections but the type 2 virus typically causes more frequent symptomatic recurrent infections than HSV-1. Anecdotal reports suggest that a variety of stimuli, including stress and onset of menses, can trigger recurrent infections. To date only experimental ultraviolet radiation exposure has been proven to trigger symptomatic recurrent infections. Asymptomatic initial or recurrent infections are common and probably play a major role in the transmission of virus since people experiencing asymptomatic infections shed virus from anogenital sites but are unaware they are contagious.

Intravenous acyclovir and oral acyclovir, valacyclovir, and famciclovir have been proven effective in the treatment of initial genital herpes and impact both the clinical and virological course of the infection (CDC, 1998). Oral acyclovir, valacyclovir, and famciclovir have a modest effect on recurrent infections when used episodically to treat recurrences but all are highly effective at preventing symptomatic and asymptomatic recurrent infections when taken daily as suppressive therapy (CDC, 1998). While there are animal data to suggest that treatment with anti-HSV nucleoside analogs early in the course of the initial infection can impact latent infection, the clinical trials conducted to date suggest that antiviral drugs like acyclovir cannot prevent the establishment of the latent infection that occurs during the initial infection, nor can they eliminate the latent infection once it is established.

Background of model

Originally described in the 1970s by Austrian scientists, the female guinea-pig model of genital herpes was extensively characterized in the 1980s and has found wide use in exploring the viral pathogenesis and in the evaluation of antiviral drugs and experimental HSV vaccines. The model mimicks many of the clinical, virologic and immunologic features of human infection (Stanberry *et al.*, 1982, 1985; Stanberry, 1991). Intravaginal inoculation of guinea-pigs with HSV-1 or 2 produces either symptomatic or asymptomatic infection depending on the inoculum titer. Infection results in high titer replication in the lower genital tract and intraneural spread to lumbosacral dorsal root ganglia and spinal cord. Symptomatic infection is characterized by vesiculoulcerative lesions on and about the perineum and some animals exhibit urologic (urinary retention) and neurologic (hind limb paralysis) findings. Initial infection results in the establishment of latent HSV infection in lumbosacral dorsal root ganglia.

After recovery from the initial infection, guinea-pigs experience spontaneous recurrent infections manifested as either shedding of virus in vaginal secretions in the absence of recognizable lesions (subclinical viral shedding) or discrete erythematous or vesicular lesions on the perineal skin. The frequency of recurrent infections declines over time and animals infected with HSV-2 exhibit significantly more recurrences than HSV-1-infected guinea-pigs. In addition to spontaneous recurrences, recurrent infections can be induced by exposure of the perineal skin to ultraviolet radiation (Stanberry, 1989). Local and systemic immune responses to the genital infection have been reported.

A male model of genital herpes involving intraurethral HSV inoculation has also been described (Stephanopoulos

Handbook of Animal Models of Infection
ISBN 0-12-775390-7

et al., 1989). The male model has been used in the evaluation of candidate antiviral drugs (Stanberry *et al.*, 1992), but compared to the female model has a higher mortality rate, is difficult to assess recurrent infections, and lacks the ease of repetitive sampling for virology specimens afforded by vaginal swabbing.

Animal species

Symptomatic genital infection can be produced in either inbred or outbred female animals but outbred female Hartley guinea-pigs (250–350 g) are typically used. Fowler *et al.* (1992) showed that the viral inoculum required to produce infection in 50% of inbred (strain 2) and outbred (Hartley) animals was similar but that the inoculum required to produce clinical disease in 50% of the animals (CD_{50}) was 10 times greater for the strain 2 animals. In addition, while the lethal dose for 50% of the animals was 100-fold greater than the CD_{50} in Hartley animals, the two values were equivalent in strain 2 animals.

Preparation of animals

No specialized housing, care nor pretreatment is required.

Storage, preparation of inocula

Viral inocula can be prepared on any susceptible cell monolayer, it is not necessary to grow the virus in guinea-pig cells. Virus stocks are titrated, aliquoted and stored frozen at –80°C. When ready for use the virus is quickly thawed, diluted to the appropriate titer and kept on ice until used. Both HSV-1 and HSV-2 strains have been used in genital herpes studies in guinea-pigs. The characteristics of primary and recurrent disease produced by a number of laboratory-adapted strains are shown in Table 108.1. Both virus types produce comparable primary infections. However, nearly all HSV-2-infected animals subsequently manifest recurrent disease with recurrences developing frequently, while many HSV-1-infected animals experience no recurrent disease and those that do have only infrequent recurrences. In general, HSV-2 strain MS and strain 333 both cause reproducible primary and recurrent infections and are widely used for the testing of antiviral drugs. Low passage clinical virus isolates can also be used; however, care should be taken since they are frequently more virulent than laboratory-adapted strains and can produce high mortality during primary infection.

Infection process

Prior to virus inoculation, the vaginal closure membrane is ruptured using a moistened calcium alginate-tipped swab. The vaginal vault is then swabbed gently with the moistened swab and subsequently with a dry swab. The viral inoculum (typically ~5.7 $\log_{10}$ pfu for HSV-2 strain MS or strain 333) is instilled into the vagina in a volume of 0.1–0.2 ml using a 1 cc tuberculin syringe and plastic catheter or a micropipettor. To optimize the number of animals that become infected, virus instillation can be performed twice within a short time interval. Some investigators anesthetize the animals prior to inoculation and some plug the vaginal opening with gelfoam after inoculation in an attempt to minimize leakage of the inoculum. Neither of these procedures is necessary.

Table 108.1 Genital herpes simplex virus (HSV) infection in female guinea-pigs

*Virus**	*Number infected†*	*Number with skin diseases‡*	*Severity§*	*Number with recurrences***	*Frequency of recurrences††*
HSV-1					
17 syn +	11/12	11/11	8.0 ± 0.4	7/10	1.4 ± 0.4
KOS(W)	10/12	8/10	1.4 ± 0.3	2/6	0.5 ± 0.3
MP	7/12	5/7	4.3 ± 1.6	2/5	0.2 ± 0.2
HSV-2					
HG52	11/12	10/11	2.6 ± 0.5	9/9	7.8 ± 1.7
333	11/12	11/11	15.7 ± 1.4	8/8	6.8 ± 1.2
MS	11/12	11/11	11.7 ± 1.5	9/9	11.6 ± 1.7
186	11/12	11/11	14.3 ± 1.7	2/2	17.5 ± 7.5

* Animals received 5×10^5 plaque-forming units (pfu) intravaginally.
† Defined as infected if virus was recovered from genital tract (vaginal specimens) 24 hours post-infection.
‡ Symptomatic animals/infected animals.
§ Area under the lesion score day curve for all animals infected (mean ± SE).
** Number of animals with recurrences/number of symptomatic animals that could be evaluated.
†† Lesion days for valuable animals — days 15–63 (mean ± SE).

Key parameters to monitor infection and evaluate antimicrobial therapy

Virus replication in the genital tract and neural tissue

The magnitude and duration of acute virus replication in the genital tract can be quantified by plaque titration of vaginal swab samples. Peak virus titers (usually 6.0–7.0 $\log_{10}$ pfu/ml) are typically seen on day 1 or 2 post-inoculation (p.i.) and then gradually decline until approximately day 10 p.i. (Stanberry *et al.*, 1982).

Following intravaginal inoculation the virus travels via the genitofemoral nerve, reaching the lumbosacral dorsal root ganglia by day 2, the spinal cord by day 3 and the brain stem and cerebral cortex by day 5. Virus replication in the lumbrosacral dorsal root ganglia and spinal cord continues until about day 10 p.i. and can be quantified by plaque titration of tissue homogenates.

Measurement of the effects of treatment on the incidence, magnitude and duration of virus replication in the genital tract and neural tissues can provide a quantitative assessment of the efficacy of antiviral treatment.

Primary disease

Following virus inoculation animals are examined daily during acute infection for evidence of primary disease. Genital skin disease typically develops by day 4 p.i. beginning as discrete lesions with an erythematous base and progressing to multiple vesiculoulcerative lesions by days 5–8 p.i. Hemorrhagic crusts cover the lesions by days 8–10 p.i. with complete healing of the external genitalia by days 13–15 p.i. The severity of the primary genital skin disease can be quantified using a lesion score-scale ranging from 0 for no disease to 4 for severe vesiculoulcerative disease of the perineum (Stanberry *et al.*, 1982). The time to appearance of lesions, severity of lesions, peak lesion score and time to healing are all assessed and can be used as indices of therapeutic efficacy. During the primary infection some animals may experience urinary retention and/or hind limb paralysis, both of which are generally transient. Urinary retention can be detected by palpation of the bladder and must be relieved by manual expression of the retained urine.

Recurrent disease

Following recovery from acute genital infection, guinea-pigs typically develop periodic spontaneous recurrent lesions on the perineum (Stanberry *et al.*, 1985). Individual lesions may be erythematous and/or vesicular and generally last 1–2 days. The frequency of spontaneous recurrent disease declines over time. By 100 days after HSV-2 inoculation spontaneous recurrences are rare, although they may be induced by exposure to ultraviolet radiation (Stanberry, 1989).

In addition to recurrent lesions animals experience cervicovaginal shedding of virus which can occur independently of recognizable recurrences (Stanberry *et al.*, 1985). During the initial week after recovery from primary infection, virus can be isolated by culture in the cervicovaginal secretions of 70–90% of animals with the frequency of isolation decreasing with time after virus inoculation.

The incidence and frequency of both recurrent lesions and asymptomatic shedding provide indices to measure the efficacy of antiviral treatment. Animals treated during primary infection can be evaluated to determine whether treatment results in any subsequent effect on recurrent disease. In addition, latently infected animals can be treated to evaluate the effects of therapy both during treatment and once treatment has ceased.

Antimicrobial therapy

Guinea-pigs can be administered antiviral compounds by a variety of routes including orally (in drinking water or by gavage), topically to the perineal skin, intravaginally, or by intraperitoneal injection.

Pitfalls (advantages/disadvantages) of the model

Advantages

Guinea-pigs are docile and widely available. The pathogenesis of genital herpes in the guinea-pig appears very similar to the infection in humans and involves virus replication in genital and neural tissues. Unlike with the mouse, viremia does not occur and infection is generally self-limiting. A unique feature of the model is the ability of guinea-pigs to experience spontaneous and induced recurrent infections. The model can be used to test experimental therapies for the treatment of either primary or recurrent genital herpes and there are easily quantifiable clinical and virological parameters that can be used to assess efficacy. The model offers an advantage over less sophisticated lethal challenge models in that the numerous efficacy parameters can be used to distinguish between different antiviral compounds, thus providing a basis for identifying a lead compound for further clinical evaluation.

The effects of acyclovir in the guinea-pig model of genital herpes parallels what has been observed in human clinical trials, proving strong validation for the model.

Disadvantages

Guinea-pigs are expensive and the number of immunological reagents available for the guinea-pig is limited. Occasionally infected animals can develop a secondary bacterial or fungal infection of the perineal skin, making evaluation of recurrent infections impossible. Metabolism of nucleoside analogs can be more rapid in guinea-pigs than in humans, making evaluation of systemically administered drugs more complex.

Contributions of the model to infectious disease therapy

Much of our understanding of the pathogenesis of genital herpes comes from studies using the guinea-pig model (Stanberry, 1996). The model has also proven valuable in the assessment of antiviral drugs, vaccines and immunomodulators. Studies in this model accurately predicted the effectiveness of acyclovir in treating HSV-1 and HSV-2 primary genital infection and the usefulness of suppressive acyclovir therapy in controlling recurrent disease (Kern, 1982). The model also demonstrated that treatment of the primary infection did not prevent the establishment of latent infection and hence did not prevent the subsequent development of recurrent infections (Bernstein *et al.*, 1986). Studies using the guinea-pig model of recurrent genital herpes established that therapeutic vaccines administered to latently infected animals could reduce the frequency of recurrent infections (Stanberry *et al.*, 1988). This important observation was subsequently confirmed in a clinical trial of patients with frequently recurring genital herpes (Straus *et al.*, 1994). The effectiveness of imiquimod, a potent immunomodulator, in the treatment of a viral sexually transmitted disease was first shown in the guinea-pig model of genital herpes (Bernstein and Harrison, 1989). Imiquimod was subsequently shown to be effective in the treatment of genital warts (human papillomavirus infection) for which it is currently licensed. The model has also shown the effectiveness of several novel prophylactic vaccine strategies, including genetically attenuated and replication-incompetent vaccine viruses and nucleic acid-based (DNA) vaccines (Meignier *et al.*, 1988;' Bourne *et al.*, 1996; Da Costa *et al.*, 1997).

References

Bernstein, D. I., Harrison, C. J. (1989). Effects of the immunomodality agent R837 on acute and latent herpes simplex virus type 2 infection. *Antimicrob. Agents Chemother.*, **33**, 1511–1515.

Bernstein, D. I., Stanberry, L. R., Harrison, C. J., Kappes, J., Myers, M. G. (1986). Effects of oral acyclovir treatment of initial genital HSV-2 infection upon antibody response, recurrence patterns and subsequent HSV-2 reinfection in guinea pigs. *J. Gen. Virol.*, **67**, 1601–1612.

Bourne, N., Stanberry, L. R., Bernstein, D. I., Lew, D. (1996). DNA immunization against experimental genital herpes simplex virus infection. *J. Infect. Dis.*, **173**, 800–807.

Centers for Disease Control and Prevention (1998). Guidelines for treatment of sexually transmitted diseases. *MMWR*, **47**, 20–26.

Da Costa, X. J., Bourne, N., Stanberry, L. R., Knipe, D. M. (1997). Construction and characterization of a replication-defective herpes simplex virus 2 ICP8 mutant strain and its use in immunization studies in a guinea pig model of genital disease. *Virology*, **232**, 1–12.

Fleming, D. T., McQuillan, G. M., Johnson, R. E. *et al.* (1997). Herpes simplex virus type 2 in the United States, 1976 to 1994. *N. Engl. J. Med.*, **337**; 1105–1111.

Fowler, S. L., Harrison, C. J., Myers, M. G., Stanberry, L. R. (1992). Outcome of herpes simplex virus type 2 infection in guinea pigs. *J. Med. Virol.*, **36**, 303–308.

Kern, E. R. (1982). Acyclovir treatment of experimental genital herpes simplex virus infections. *Am. J. Med.*, **73**, 100–107.

Meignier, B., Longnecker, R., Roizman, R. (1988). *In vivo* behavior of genetically engineered herpes simplex viruses R7017 and R7020: construction and evaluation in rodents. *J. Infect. Dis.*, **58**, 602–614.

Stanberry, L. R. (1989). An animal model of ultraviolet radiation-induced recurrent herpes simplex virus infection. *J. Med. Virol.*, **28**, 125–128.

Stanberry, L. R. (1991). Herpes simplex virus vaccine evaluation in animals: the guinea pig model. *Rev. Infect. Dis.*, **13** (suppl. 11), S920–S923.

Stanberry, L. R. (1996). The pathogenesis of herpes simplex virus infections. In *Genital and Neonatal Herpes* (ed Stanberry, L. R.), pp. 31–48. John Wiley, London.

Stanberry, L. R., Kern, E. R., Richards, J. T., Abbott, T. M., Overall, J. C. Jr. (1982). Genital herpes in guinea pigs: pathogenesis of the primary infection and description of recurrent disease. *J. Infect. Dis.*, **146**, 397–404.

Stanberry, L. R., Kern, E. R., Richards. J. T., Overall, J. C., Jr. (1985). Recurrent genital herpes simplex virus infection in guinea pigs. *Intervirol.*, **24**, 226–231.

Stanberry, L. R. Burke, R. L., Myers, M. G. (1988). Herpes simplex virus glycoprotein treatment of recurrent genital herpes. *J. Infect. Dis.*, **157**, 156–163.

Stanberry, L. R., Bourne, N., Bravo, F. J., Bernstein, D. I. (1992). Capsaicin sensitive peptidergic neurons are involved in the zosteriform spread of herpes simplex virus infection. *J. Med. Virol.*, **38**, 142–146.

Stanberry, L. R., Jorgensen, D. M., Nahmias, A. J. (1997). Herpes simplex viruses 1 and 2. In *Viral Infection of Humans: Epidemiology and Control*, 4th edn (eds Evans, A. S., Kaslow, R. A.), pp. 419–454. Plenum Press, New York.

Stephanopoulos, D. E., Myers, M. G., Bernstein, D. I. (1989). Genital infections due to herpes simplex virus type 2 in male guinea pigs. *J. Infect. Dis.*, **159**, 89–95.

Straus, S. E., Corey, L., Burke, R. L. *et al.* (1994). Placebo-controlled trial of vaccination with glycoprotein D of herpes simplex virus type 2 immunotherapy of genital herpes. *Lancet*, **343**, 1460–1463.

Wald, A., Corey, L. (1996). The clinical features and diagnostic evaluation of genital herpes. In *Genital and Neonatal Herpes* (ed Stanberry, L. R.), pp. 109–137. John Wiley, London.

Chapter 109

Animal Models of Herpes Skin Infection: Guinea-pig

M. B. McKeough and S. L. Spruance

Background of human infection

Recurrent mucocutaneous herpes simplex virus (HSV) infections (herpes labialis, herpes genitalis) afflict approximately one-third of the adult population and have proved to be difficult diseases to treat. Descriptions of the diseases, diagnosis and therapy have been reviewed previously (Ship *et al.*, 1960; Embil *et al.*, 1975; Grout and Barber, 1976; Young *et al.*, 1976; Corey *et al.*, 1983; Spruance, 1995). Despite 25 years of clinical research with established antiviral substances, only a few treatments are available and these have modest efficacy (Spruance *et al.*, 1990a,b, 1997; Corey *et al.*, 1983; Sacks *et al.*, 1996).

Background of the model

Progress in developing effective antivirals has been slow, in part because of the time-consuming nature of large, patient-initiated clinical trials. The dorsal cutaneous guinea-pig model is a rapid and efficient means of identifying topical antiviral formulations with clinical promise. A critical factor for efficacy of topically-administered antiviral substances is penetration of the intact epithelium of the involved area. Based on this and other considerations, the predictive power of an animal model testing topical treatments for recurrent human HSV infections depends in part on the application of drug to intact skin.

Teague and Goodpasture (1924) first showed that the skin of the guinea-pig dorsum could be infected with HSV. Tomlinson and MacCallum (1968) also did early work with the model. In 1974, Hubler and coworkers, concerned about rear leg paralysis and death from systemic viremia and central nervous system infection in a high percentage of the animals, improved the model in several important ways. The depilated dorsal skin of the guinea-pig was demarcated into 6–8 squares with a marking pen. Then 0.02 ml of high-titered HSV-1 stock (10^7 plaque-forming units (pfu)/ml) was applied to each square and introduced into the skin with a 6-pronged (0.75 mm depth), spring-loaded, vaccination instrument (Sterneedle). The procedure resulted in multiple vesicular lesions which crusted and progressed to complete healing in 10–12 days, clinically similar to human cutaneous HSV disease. All of the animals survived. Virus stock containing less than 10^7 pfu/ml failed to produce lesions consistently and inoculation of previously infected animals resulted in abortive, rapidly healing lesions. This basic procedure for producing a uniform, non-fatal infection of guinea-pig skin has been utilized by multiple investigators and has been referred to as the Hubler model.

Whereas Hubler preferred to concentrate the inoculation stabs in one area, we have found it advantageous to space the stab wounds carefully, in order to be able to enumerate the number of lesions developing and to minimize the trauma to the epithelium. To this end, a higher-titered virus inoculum was necessary ($5–10 \times 10^7$ pfu/ml), which may be conveniently prepared by virus concentration with polyethylene glycol (Lancz, 1973). Selection of an effective virus strain is also an important factor. The Sterneedle is no longer commercially available, but another vaccination instrument can be used (Bignell Surgical Instruments, Littlehampton, W. Sussex, UK).

Animal species

Animals used are female, Hartley outbred, albino guinea-pigs, 500 g each (Charles River, Baltimore, MA). Hairless guinea-pigs can also be used but their availability is limited, they are considerably more expensive, and based on a small experience in our laboratory, their susceptibility to our inoculation procedure may be different.

Preparation of animals

Anesthesia

Each 500 g animal is injected with about 0.24 ml of 50 mg ketamine/ml plus 10 mg xylazine/ml, subcutaneous (s.c.), high on the back between the shoulder blades using a tuberculin syringe with a 25 G 5/8 in needle. After 5–10 minutes the animals are sedated.

Handbook of Animal Models of Infection
ISBN 0-12-775390-7

Hair removal

The hair on the back from the shoulder to the "tail" and from flank to flank is clipped close with electric grooming clippers using a no. 80 clipper head (Oster Model). The hair stubble is then thoroughly wetted with warm water poured from a cupped hand and a depilatory (Nair) is liberally applied and rubbed into the stubble. After about 10 minutes the depilatory is removed with paper towels, the remaining stubble rinsed and rewetted and a second application of depilatory applied. About 5–10 minutes later the depilatory is removed with paper towels, the skin is rinsed thoroughly with warm water and dried with paper towels. The skin should be dry, smooth and stubble-free.

Housing

Animals are caged individually, before and after infection.

Details of infection

Overview

Briefly, the dorsal skin of anesthetized animals is infected with HSV-1 in four quadrants with 10 well-spaced activations of a vaccination instrument in each quadrant. Ideally, the infection in each quadrant should develop over the course of 4 days into 10 small, well-spaced circles of six lesions in each circle for a total of 60 lesions per area. The day of infection is Day 0. Treatment of each quadrant with either drug or matching vehicle begins on the morning of Day 1 and continues through Day 3. On the morning of Day 4 regrown hair on the dorsum is removed using the depilatory, lesions in each quadrant are counted, photographs are taken, the animals are sacrificed and full-thickness skin from each quadrant is then taken from the sacrificed animal. The pictures are used to measure the diameters of a representative sample of lesions to determine total lesion area for each drug and its placebo. Virus is harvested from the skin and the amount present determined in a plaque assay. The clinical parameters for each drug versus placebo pair are total lesion number, total lesion area and virus titer.

Materials required

The materials required for infection are gloves, anesthetic, syringes and needles, hair clippers, depilatory, paper towels, microliter pipetter with tips, six-pronged vaccination instrument with tips, virus. For taking photographs you need: camera (Polaroid CU5 Land Camera with 3 in distance lens which produces life-size images), film (Polaroid 669 color film). For harvesting full-thickness skin sections: scalpel handle and blades, forceps, 50 ml conical tubes containing 15 ml of media, ice bath, Stomacher Lab Blender (Tekmar Co., Cincinnati, OH), Stomacher bags. The materials required for virus assay include confluent vero cells in six-well culture plates, diluent, dilution tubes, micropipetter and tips, semisolid overlay, stain and plaque reader.

Dermal irritation evaluation

The active compound and matching placebo are evaluated for any dermal irritation they might cause in a small group of uninfected guinea-pigs to determine the appropriate dosing regimen for infected animals.

After hair removal the guinea-pig's back from the mid-back to the rump is divided into four approximately equal areas with a black marking pen. The animals are left uninfected. Day 0 is the day when hair is removed. The active compound and its placebo are applied once, twice or four times a day at the same rostral–caudal level starting on Day 1 and continuing through Day 3. Dermal irritation is scored on the morning of Day 4 after depilation using a scale of 0–4 (0, no irritation; 1, minor redness; 2, moderate redness; 3, severe redness; 4, severe redness with punctate bleeding). Three application sites are used to evaluate each formulation. Both drug and vehicle control formulations are tested since we have found several instances where the vehicle was more irritating than the active preparation. An average score >2.5 shows that the active compound and/or its placebo is too irritating to use in the model. Severe irritation and redness prevent visualization of lesions, making them very hard to count and measure.

Inoculation

Guinea-pigs, after hair removal, are infected in groups of 4–6 animals, with 8–12 animals being the number typically infected in one day. The animals are prepared as described above. The guinea-pig's back from the mid-back to the rump is divided into four approximately equal areas with a black marking pen. Each area is infected by placing 0.035 ml of high-titer (1×10^8 pfu/ml or greater) virus, HSV-1 E115, into the area. The Sterneedle tip is placed in the virus fluid and activated 10 times per area in a well-spaced, non-overlapping pattern. The day of infection — usually a Monday — is Day 0.

The guinea-pigs are placed in clean, individual cages and watched until they have recovered from anesthesia. Recovery occurs in 30–45 minutes after injection when they regain their feet and balance and can move around their cages. Losing animals to anesthetic overdose should be a rarity — a less than 1% occurrence.

Application of drugs and matching vehicles

Drugs and matching placebos are applied with gloved finger tips in the amount of about 250 mg per area if the drug is a cream or ointment. If the compounds are liquids one can use a soaked, cotton-tipped applicator to cover each area to deliver about 100 μl of liquid.

Storage, preparation of inocula

A low-passage, high-titer HSV-1 E115 pool is grown on mink lung cells (MV-1-Lu) modified after the method of Roizman and Spear (1968). Pools with a titer greater than 1×10^8 pfu/ml can be tested on guinea-pigs to see if they produce an acceptable infection. An acceptable infection is 50 or more individual lesions (out of a possible 60) per quadrant on the animal's back.

Infection process

Histology of a cutaneous HSV-1 lesion, 72 hours after inoculation of the skin, shows that infection primarily involves the epidermal layer in the skin with focal epidermal destruction, leukocytic infiltration of the epidermis and adjacent dermis, and cytologic changes in infected epidermal cells typical for HSV. While the "stab" inoculation procedure with 0.75 mm vaccination prongs potentially may introduce virus deep into the skin, the primary impact of inoculation was to induce epidermal cell infection, histologically similar to a recurrent lesion of herpes labialis (Huff *et al.*, 1981). The kinetics of clinical lesion development and virus titers in the skin show that virus replication rises rapidly to near maximal levels within 24 hours of inoculation, while the peak clinical lesion score occurs on Day 5 (Figure 109.1; Schaefer *et al.*, 1977). On Day 1 (Day 0 is the day of infection) one can see distinct macular areas that develop into papules on Day 2. By Day 3 individual vesicles have evolved. If left untreated the lesions would progress through the crust stage and heal by about Day 10.

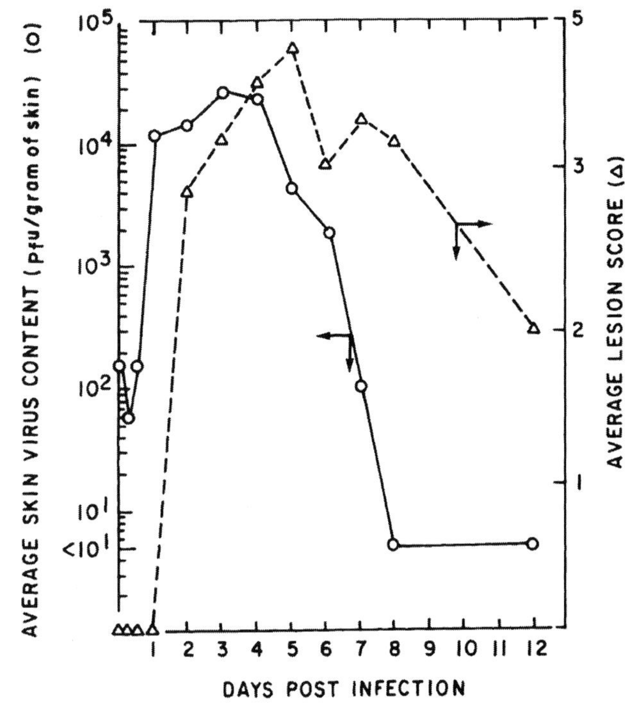

Figure 109.1 Herpes simplex virus type 1 (HSV-1) average growth curve and lesion development in guinea-pig skin (Schaefer *et al.*, 1977).

Key parameters to monitor infection

The infection is confined to the skin of the animal's back. Food and water intake are normal. General health is maintained. Very rarely does a systemic infection, typically causing rear leg paralysis, occur necessitating sacrificing the animal. Any lethargy or abnormal behavior on the part of a guinea-pig is usually associated with an overdose of or reaction to anesthesia rather than systemic infection. The guinea-pigs do not chew, lick or scratch at the infected areas.

Antimicrobial therapy

Topical treatments

A variety of novel antivirals and drug vehicles have been evaluated in this guinea-pig model and are summarized in Table 109.1. Therapy begins on the morning of Day 1 and continues for Days 2 and 3. (Day 0 is the day of infection.) The regimen is the maximum one tolerated during the dermal irritation phase of the study that allows for visualization of lesions on the morning of Day 4. Gel, lotion, ointment and liquid antivirals can be evaluated. Careful attention should be paid to compounds that dry and adhere to the skin because this occlusive effect can exacerbate any dermal irritation that might occur. Dried compounds can be washed off with wet paper towels. Hair does regrow during antiviral treatment, so attention must be paid to insure that all of the infected quadrant is being adequately covered by the compounds. Drug and matching vehicle are always placed opposite each other on the same rostral—caudal level. Two different drug and vehicle pairs can be tested on one animal as there is no evidence to date of systemic "contamination" of the animal by any compound used, even when a known penetration enhancer like dimethyl sulfoxide (DMSO) is the drug vehicle.

Enteral treatments

Because of the widespread use of acyclovir (ACV) capsules for the treatment of HSV infections, including two clinical trials in recurrent herpes labialis, it was of interest for us to study oral ACV therapy in the guinea-pig model and compare efficacy with the topical ACV treatments (Raborn *et al.*, 1987; Spruance *et al.*, 1990b). Attempts to administer

Table 109.1 Efficacy of topical antiviral formulations in the dorsal cutaneous guinea-pig model of herpes simplex virus type 1 (HSV-1) infection

Experiment no.	*Test treatment*[a]			*Percent reduction in lesion severity by test treatment compared to vehicle control*[b]			*Percent reduction in lesion severity by 5% ACV ointment compared to ointment base control*			*Source*
				Lesion						
	Antiviral	*Conc.*	*Vehicle*	*No.*	*Area*	*Virus titer*	*No.*	*Area*	*Virus titer*	
1	ACV	5%	DMSO	58*	73*	84	8	35	21	Spruance *et al.* (1984a)
2	Bromovinyl deoxyuridine (BVDU)	5%	Aqueous cream	20	40	90*	13	28	55	Freeman *et al.* (1985)
	BVDU	5%	DMSO	55*	69*	94*				
3	EDU	3%	Aqueous cream	29	44	68	15	32	60	Spruance *et al.* (1985)
	EDU	5%	DMSO	39	60*	90*				
4	ACV	5%	Aqueous cream	19*	31	75	–8	19	60	Spruance *et al.* (1986)
	Sodium phosphonoformate (PFA)	0.3%	Aqueous cream	36*	52*	80				
	PFA	3%	Aqueous cream	54*	73*	90*				
5	ARA-A	5%	DMSO	44*	53*	60	–13	8	37	
	Iododeoxyuridine (IDU)	5%	DMSO	75*	81*	99*				
6	Trifluorothymidine (TFT)	1%	Azone and propylene glycol	32	70*	97	18	46	90	Spruance *et al.* (1984b)
7	Triacontanol	5%	Branch chain ester base	9	17	41	7	22	64	
8	Ribavirin	10%	Site release base	17	36	50	16	32	56	
	Ribavirin	10%	Ointment	10	25	53				
	Ribavirin	10%	Aqueous cream	15	33	37				
	Ribavirin	5%	Propylene glycol lotion	10	34	38				
9	Interferon***		Gel	10	2**	21**	11	30	68	Freeman *et al.* (1987)
10	Heparin/zinc		Gel	1	8	2	2	16	26	
			Lipstick	2	6	15				
11	Tioconazol	1%	Gel	0	6	0**	4	21	59	
	Folic acid	1–2%	Gel	5	0**	0**				
	Tioconazol	1%	Ointment	6	0**	0**				
	Folic acid	1–2%	Ointment	7	0**	0**				
12	MMUdR		Topical formula	0	0	8	6	16	53	
	ARA-A		Topical formula	0	7	40				
	MMUdR + ARA-A		Topical formula	2	7	0				
13	Oligonucleotide antisense cmpds	1%	Cream	2	13	0	1	19	5	
	ACV	5%	Cream (European)	8	32*	75*				

14	Proanthocyanidin polymer (SP303)	15%	95%DMSO	3	13	19	5	29	59
15	Lidakol	10%	Stearic acid cream	0	8	0	5	22	72
		12%	Stearic acid cream	0	0	ND	2	23	ND
16	ACV	5%	Vehicle A	2	19	83*	2	24	63
	ACV	5%	Vehicle B	5	21	69			
17	ACV	5%	Vehicle F2	1	12**	6**	0	22	68
	ACV	5%	Vehicle G2	8	14	37**			
18	Experimental polyaxomer	2.5%	Ointment	0	0**	32**	0	12	68
	Experimental polyaxomer	1%	100% DMSO	0	0**	13**			
19	ACV Phospho-ester prodrug 1A		Cream	45*	59*	96*	2	17	68
	ACV Phospho-ester prodrug 1B		Cream	35*	57*	97*	0	25	61
20	ACV Ester prodrug (P4010)		Vehicle A	2	5**	48	7	33	72
	DHPG Ester prodrug (P4018)		Cream	0	8**	46	0	24	73
21	Penciclovir cream	1%	Cream	11	32	72	ND	ND	ND
22	ACV monophosphate		T1 cream	7	19	81	5	25	65
	ACV monophosphate		T2 cream	ND	ND	96*			

[a] Treatments were given 4 times a day for 3 days beginning 24 hours after inoculation, except for formulations which were irritating to the skin and administered 1–2 times a day (PFA, TFT). ACV, BVDU, PFA (foscarnet), EDU, TFT and recombinant interferon-alpha A were obtained through pharmaceutical industry sponsors as previously documented (see references documented under Source). Triacontanol, a mixture of long-chain hydrocarbons, was provided by Royal Pharmaceutical Products, Inc., Bountiful, UT; ribavirin (Virazole) in different vehicles by Viratek, Costa Mesa, CA; and the combination of heparin and zinc (Bonactin) by Ciba Consumer Pharmaceuticals, Edison, NJ. ARA-A was obtained from Sigma Chemical Co., St. Louis, MO; IDU and DMSO were provided by Research Industries, Salt Lake City, UT. MMUdR and ARA-A were supplied by Spencer-Jones, Inc., Pointe Claire, Quebec, Canada. Oligonucleotide antisense compound was provided by ISIS Pharmaceuticals, Carlsbad, CA. Proanthocyanidin polymer was provided by Shaman Pharmaceuticals, Inc., San Carlos, CA. Lidakol cream (*n*-docosanol) was supplied by Lidak Pharmaceuticals, La Jolla, CA. ACV in vehicles A, B, F2 and G2 supplied by TheraTech, Inc., Salt Lake City, UT. ACV phosphoester prodrugs 1A and 1B supplied by Drug Innovation and Design, Inc., Needham, MA. ACV ester prodrug (P4010; (9-(2′-(trans-9″-octadecenoyloxy)-ethoxymethyl)-guanine)) and DHPG ester prodrug (P4018;(9-(2′-(trans-9″-octadecenoyloxy)-1′-(hydroxymethyl)-ethoxymethyl)-guanine)) provided by Norsk Hydro ASA, Porsgrunn, Norway. Penciclovir cream (Denavir) and vehicle cream provided by SmithKline Beecham Pharmaceuticals, King of Prussia, PA. ACV monophosphate and T1 and T2 placebos provided by Triangle Pharmaceuticals, Inc., Durham, NC. Five percent ACV ointment (Zovirax Ointment) was generously supplied by Burroughs-Wellcome Co., Research Triangle Park, NC.

[b] Percent differences between mean lesion severity at drug-treated infection sites compared to the vehicle-treated sites are shown. Data are derived from comparison of 8–15 drug- and vehicle-treated sites 4 days after inoculation. A positive value indicates a reduction in lesion severity by the test compound. For statistical analysis, paired data were evaluated by the Wilcoxon signed-rank test, utilizing the percent differences between $\log_{10}$ derivatives of the drug and drug vehicle results. These percent differences were also used as a means to compare the efficacy of different antivirals by the Mann–Whitney rank sum procedure. All probability determinations were two-tailed, and $P \leq 0.05$ was considered to be significant.

ACV, Acyclovir, DMSO, dimethyl sulfoxide; BVDU, EDU, 5-ethyl-2′-deoxyuridine; PFA, ARA-A, arabenofuranosyl-adenine; IDU, TFT, MMUdR, 5-methoxymethyl-2-deoxyuridine, ND, not detected.

* Significantly better ($P \leq 0.05$) therapeutic effect compared to ACV ointment.

** Significantly less ($P \leq 0.05$) therapeutic effect compared to ACV ointment.

*** Recombinant interferon-alpha A.

ACV to guinea-pigs by intermittent gavage led to variable and unpredictable blood levels (Spruance, unpublished data). Therefore, ACV was administered in the animals' water bottles as a 5 mg/ml solution, a procedure which results in blood levels of ACV in the guinea-pig comparable to peak serum levels in human subjects following ingestion of a 200 mg capsule. Treatment was begun 24 hours after infection. The results of the experiment are shown in Table 109.1. Animals receiving ACV in drinking water had significantly lower total lesion area (31% reduction) and lesion virus titers (84% reduction) compared to control animals. The efficacy of ACV by the oral route of administration was slightly more than that which we have seen by topical therapy with 5% ACV ointment in the model (greater reduction in lesion virus titers); comparable to the efficacy of 5% ACV cream; and considerably less than the efficacy of topical 5% ACV in DMSO. The similarity between the efficacy of oral ACV and ACV cream in the model is of interest since these preparations have both had small positive effects on the severity of recurrent herpes labialis. Since enhanced delivery of ACV by formulation in DMSO is associated with better therapeutic efficacy in the model, increased delivery of ACV by the oral or topical route is a logical means to pursue a better treatment for herpes labialis.

Intramuscular treatments

The activity of systemic (intramuscular, in the hip, 1 cc tuberculin syringes with 25 G 5/8 in needles) interferon has been tested in the model. Results are shown in Table 109.1.

Key parameters to monitor response to treatment

For each experiment, three measures of lesion severity were used to assess drug effect: lesion number, lesion area and lesion virus titer. The measures were affected by antiviral drug therapy in the following order: virus titer > lesion area > number of lesions. Lesions are measured once at the point of maximum severity, 4 days after infection, by counting the number of lesions; measuring lesion diameter with the aid of Polaroid photographs and a calibrated magnifying lens; and calculating the total lesion area. The same lesions were then excised for virus quantitation by sacrificing the animal, removing full-thickness skin from the quadrant and putting it into a 50 ml conical tube on ice with 15 ml of media in the tube. The skin and media were then placed in a Stomacher bag and put into the Stomacher Lab Blender for about 1 minute. The action of the Stomacher essentially squeezes the virus out of the lesions into the media. The media plus virus is put back into the conical tube while the skin and bag are discarded. The debris in the tube is pelleted in a refrigerated centrifuge, a 5–7 ml sample of the media is collected into another tube and then frozen at −70°C for later virus assay.

For statistical analysis, paired data were evaluated by the Wilcoxon signed-rank test, utilizing the percent differences between $\log_{10}$ derivatives of the drug and drug vehicle results. These percent differences were also used as a means to compare the efficacy of different antivirals by the Mann–Whitney rank-sum procedure. All probability determinations were two-tailed and $P \leq 0.05$ was considered to be significant.

The finding that lesion virus titers were affected to the largest degree supports the contention that the antiviral activity of the test compounds was responsible for the clinical benefits. Using 48 data points from experiment no. 4 (Table 109.1), we found a significant correlation between reduction in total area and reduction in lesion virus titer ($r = 0.61$; $P < .001$).

In our experience with the model, we have observed a rostral–caudal gradient in lesion severity; more severe lesions develop on the rump than at higher locations on the back. However, drug efficacy (difference between active substance and contralateral vehicle control) appears to be independent of position on the rostral–caudal axis.

Advantages and disadvantages of the model

Evaluation of clinical lesion severity in the Hubler guinea-pig model has been most commonly done by a lesion scoring system, assigning an incremental numeric value as lesions evolved in size and pathologic maturity (Schaefer *et al.*, 1977; Alenius and Oberg, 1978). The total lesion experience was expressed by a plot of average daily subjective lesion severity scores and time or by cumulative scores. This system takes all phases of the experimental disease into account, but has the disadvantage of subjective lesion observations and arbitrary quantitative relationships between disease scores. Our procedure, measuring lesion severity at one time point, is quantitative and objective and permits a direct correlation of the clinical and virologic outcomes, which enables us to distinguish between antiviral and anti-inflammatory drugs. The disadvantage is a lack of information about the impact of treatment on time to lesion healing.

With regard to the skin as a barrier to topically administered drugs, there are several aspects of existing animal model systems that deserve comment. The process of virus inoculation into the skin of the experimental animal may alter the integrity of the stratum corneum and create an abnormal percutaneous route of drug absorption, such as by needle scratches, thermal injury with a surgical cautery, or injury produced by tape-stripping of mouse ears. Likewise, preparation of the dorsal skin of guinea-pigs has included shaving and chemical depilation. Examination of the effect of these various procedures on the barrier function of the skin is essential to determine if the model tests the ability of antivirals to diffuse into undisrupted skin. In the female guinea-pig genitalia model, topically applied drugs could be absorbed across the introital, vaginal or peritoneal mucosa. Topical therapy in this model might lead to significant systemic drug levels and demonstration of efficacy does not necessarily prove direct drug penetration of the stratum corneum of external cutaneous genital lesions. In our system, we have examined the effect of the inoculation process and found no changes in drug penetration (Freeman *et al.*, 1986). Compounds that are relatively insoluble in water, creating problems for *in vitro* assay of antiviral activity, can be screened for activity in the dorsal cutaneous guinea-pig model.

Because this model is a primary infection, efficacy is greater in the model than is anticipated in human recurrent disease. For this reason, we have always used a control compound, Zovirax Ointment, to gauge the relative efficacy of test treatments. Penetration of compounds through human skin may be less than through guinea-pig skin—another factor to take into consideration (Spruance *et al.*, 1984), as well as differences in the concentration of nucleosides which may compete with test nucleoside analog antivirals for viral DNA polymerase (Harmenberg, 1983; Larsson *et al.*, 1983; Ericson *et al.*, 1985).

Contributions of the model to infectious disease therapy

Studies in the dorsal cutaneous guinea-pig model have defined the importance of drug permeation through intact, keratinized skin to achieve efficacy against herpes simplex virus infection. A formulation without efficacy in immunocompetent patients with recurrent herpes, Zovirax Ointment, has minimal efficacy in the guinea-pig model, and other formulations of acyclovir and idoxuridine with more penetration (cream with propylene glycol, dimethyl sulfoxide) have had significantly greater activity in the guinea-pig and were efficacious in clinical trials against recurrent herpes labialis (Freeman and Spruance, 1986). The vast superiority of intramuscular versus topical interferon contributed to the abandonment of the topical route of administration of this compound (Freeman *et al.*, 1987). Compounds that failed in the guinea-pig model (less activity than Zovirax Ointment) have been failures in clinical trials (Safrin *et al.*, 1994; Anonymous, 1996), lending credibility to the model as a screening tool.

References

Alenius, S., Oberg, B. (1978). Comparison of the therapeutic effects of five antiviral agents on cutaneous herpesvirus infection in guinea pigs. *Arch. Virol.*, **58**, 277–288.

Anonymous (1996). Topical lidakol phase III trials for oral HSV fail to show efficacy vs placebo. *Antiviral Agents Bulletin.*, **9**, 104–105 (abstract).

Corey, L., Adams, H. G., Brown, Z. A., Holmes, K. K. (1983). Genital herpes simplex virus infections: clinical manifestations, course, and complications. *Ann. Intern. Med.*, **98**, 958–972.

Embil, J. A., Stephens, R. G., Manuel, F. R. (1975). Prevalence of recurrent herpes labialis and aphthous ulcers among young adults on six continents. *Can. Med. Assoc. J.*, **113**, 627–630.

Ericson, A. C., Larsson, A., Aoki, F. A. *et al.* (1985). Antiherpes effects and pharmacokinetic properties of 9-(4-hydroxybutyl) guanine and the (R) and (S) enantiomers of 9-(3,4-dihydroxybutyl)guanine. *Antimicrob. Agents Chemother.*, **27**, 753–759.

Freeman, D. J., Sacks, S. L., DeClercq E., Spruance, S. L. (1985). Preclinical assessment of topical treatments for herpes simplex virus infection: 5% (E)-5-(2-bromovinyl)-2′-deoxyuridine (BVDU) cream. *Antiviral Res.*, **5**, 169–177.

Freeman, D. J., Spruance, S. L. (1986). Efficacy of topical treatment for herpes simplex virus infections: predictions from an index of drug characteristics *in vitro*. *J. Infect. Dis.*, **153**, 64–70.

Freeman, D. J., Sheth, N. V., Spruance, S. L. (1986). Failure of topical acyclovir (ACV) in ointment to penetrate human skin. *Antimicrob. Agents Chemother.*, **29**, 730–732.

Freeman, D. J., McKeough, M. B., Spruance, S. L. (1987). Recombinant human interferon-alpha A treatment of an experimental cutaneous herpes simplex virus type 1 infection of guinea pigs. *J. Interferon Res.*, **7**, 213–222.

Grout, P., Barber, V. E. (1976). Cold sores – an epidemiologic survey. *J. R. Coll. Gen. Pract.*, **26**, 428–434.

Harmenberg, J. (1983). Intracellular pools of thymidine reduce the antiviral action of acyclovir. *Intervirology*, **20**, 48–51.

Hubler W. R. Jr., Felber, T. D., Troll, D., Jarratt, M. (1974). Guinea pig model for cutaneous herpes simplex infection. *J. Invest. Dermatol.*, **62**, 92–95.

Huff, J. C., Krueger, G. G., Overall J. C. Jr., Copel, J., Spruance, S. L. (1981). The histopathologic evolution of recurrent herpes simplex labialis. *J. Am. Acad. Dermatol.*, **5**, 550–557.

Lancz, G. J. (1973). Rapid method for the concentration and partial purification of herpes simplex virus types 1 and 2. *Arch. Ges. Virusforsch.*, **42**, 303–306.

Larsson, A., Brannstrom, G., Oberg, B. (1983). Kinetic analysis in cell culture of the reversal of antiherpes activity of nucleoside analogs by thymidine. *Antimicrob. Agents Chemother.*, **24**, 819–822.

Raborn, G. W., McGaw, W. T., Grace, M., Tyrrell, L. D., Samuels, S. M. (1987). Oral acyclovir and herpes labialis: a randomized, double-blind, placebo-controlled study. *I. Am. Dent. Assoc.*, **115**, 38–42.

Roizman, G., Spear, P. G. (1968). Preparation of herpes simplex virus of high titer. *Virology*, **2**, 83–84.

Sacks, S. L., Aoki, F. Y., Diaz Mitoma, F., Sellors, J., Shafran, S. D. (1996). Patient-initiated, twice-daily oral famciclovir for early recurrent genital herpes. A randomized, double-blind multicenter trial. Canadian Famciclovir Study Group. *J. A. M. A.*, **276**, 44–49.

Safrin, S., McKinley, G., McKeough, M., Robinson, D., Spruance, S. L. (1994). Treatment of acyclovir-unresponsive cutaneous herpes simplex virus infection with topically applied SP-303. *Antiviral Res.*, **25**, 185–192.

Schaefer, T. W., Lieberman, M., Everitt, J., Came, P. O. (1977). Cutaneous herpes simplex virus infection of guinea pigs as a model for antiviral chemotherapy. *Ann. N. Y. Acad. Sci.*, **284**, 624–631.

Ship, I. I., Morris, A. L., Durocher, R. T, Burket, L. W. (1960). Recurrent aphthous ulcerations and recurrent herpes labialis in a professional school student population. *Oral Surg. Oral Med. Oral Pathol.*, **13**, 1191–1202.

Spruance, S. L. (1995). Herpes simplex labialis. In *Clinical Management of Herpes Viruses* (eds Sacks, S. L., Straus, S. E., Whitley, R. J., Griffiths, P. D.) pp. 3–42. IOS Press, Amsterdam.

Spruance, S. L., McKeough, M. B., Cardinal, J. R. (1984). Penetration of guinea pig skin by acyclovir in different vehicles and correlation with the efficacy of topical therapy of experimental cutaneous herpes simplex virus infection. *Antimicrob. Agents Chemother.*, **25**, 10–15.

Spruance, S. L., McKeough, M. B., Sugibayashi, K., Robertson, F., Gaede, P., Clark, D. S. (1984b). Effect of azone and propylene glycol on penetration of trifluorothymidine through skin and the efficacy of different topical formulations against cutaneous herpes simplex virus infection in guinea pigs. *Antimicrob. Agents Chemother.*, **26**, 819–823.

Spruance, S. L., Freeman, D. J., Sheth N. V. (1986). Comparison of topically applied 5-ethyl-2′-deoxyuridine and acyclovir in the treatment of cutaneous herpes simplex virus infection in guinea pigs. *Antimicrob. Agents Chemother.*, **28**, 103–106.

Spruance, S. L., Freeman, D. J., Sheth N. V. (1986). Comparison of topical foscarnet, acyclovir (ACV) cream and ACV ointment in the treatment of experimental cutaneous herpes simplex virus (HSV) infection. *Antimicrob. Agents Chemother.*, **30**, 196–198.

Spruance, S. L., Stewart, J. C. B., Freeman, D. J. *et al.* (1990a). Early application of topical 15% idoxuridine in dimethyl sulfoxide shortens the course of herpes simplex labialis: a multicenter placebo-controlled trial. *J. Infect. Dis.*, **161**, 191–197.

Spruance, S. L., Stewart, J. C. B., Rowe, N. H., McKeough, M. B., Wenerstrom, G., Freeman, D. J. (1990b) Treatment of recurrent herpes simplex labialis with oral acyclovir. *J. Infect. Dis.*, **161**, 185–190.

Spruance, S. L., Rea, T. L., Thoming, C., Tucker, R., Saltzman, R., Boon, R. (1997). Penciclovir cream for the treatment of herpes simplex labialis: a randomized, multicenter, double-blind, placebo-controlled trial. *J. A. M. A.*, **277**, 1374–1379.

Teague, O., Goodpasture, E. W. (1924). Experimental herpes zoster. *J. Med. Res.*, **44**, 185–200.

Tomlinson, A. H., MacCallum, F. O. (1968). The effect of 5-iodo-2′-deoxyuridine on herpes simplex virus infections in guinea-pig skin. *Br. J. Exp. Pathol.*, **49**, 277–282.

Young, S. K., Rowe, N. H., Buchanan, R. A. (1976). A clinical study for the control of facial mucocutaneous herpes virus infections. I. Characterization of natural history in a professional school population. *Oral Surg. Oral Med. Oral Pathol.*, **41**, 498–507.

Chapter 110

Animal Models of Ocular Herpes Simplex Virus Infection (Rabbits, Primates, Mice)

B. M. Gebhardt, E. D. Varnell, J. M. Hill and H. E. Kaufman

Background of human infection

Approximately 500 000 people are affected by ocular herpesvirus infection in the United States (National Advisory Eye Council, 1993). Superficial infection of the corneal epithelium is known as herpetic keratitis and is characterized by a branching lesion in the epithelium referred to as a *dendrite*. Herpetic lesions of the cornea can also appear as large geographic epithelial ulcers, clusters of small epithelial defects, or a deeper corneal stromal infection characterized by an infiltrate of inflammatory and immune cells and corneal edema, which is usually accompanied by visual impairment (Liesegang, 1988; Kaufman *et al.*, 1998).

Epithelial infections are usually self-limited and the disease interval can be significantly shortened by treatment with topical antiviral drugs. Stromal disease is of a more chronic nature and does not respond as well to short-term application of a topical antiviral (Liesegang, 1988; Kaufman *et al.*, 1998). Treatment involves long-term antiviral therapy, usually given in combination with corticosteroids, to prevent the outbreak of epithelial disease and to palliate symptoms while the disease runs its course (Liesegang, 1988; Kaufman *et al.*, 1998).

The real threat of ocular herpetic infection is its recurrent nature. Patients who experience corneal infection caused by herpes simplex virus types 1 or 2 (HSV-1 or -2) may develop one or more episodes of recurrent disease in the months to years after the initial infection, as a result of the reactivation of the virus from the ganglionic site of latency (Nesburn and Green, 1976; Kaufman *et al.*, 1998). These recurrent infections may lead to scarring of the cornea and permanent loss of vision. One of the leading indicators for corneal transplantation is the restoration of vision in an eye with postherpetic scarring of the cornea (National Advisory Eye Council, 1993). This and other associated morbidities have provided the impetus for research into the development of effective therapies for the prevention and prophylaxis of ocular herpes infections.

Background of model

Publications dealing with ocular herpetic infection in experimental animals date back to 1923 (Goodpasture and Teague, 1923–1924; Laibson and Kibrick, 1966; Kaufman, 1976, 1993). The true dawning of the age of animal models of ocular herpes simplex infection, however, occurred in the early 1960s when two of us (Kaufman and Varnell) showed for the first time that a viral thymidine kinase inhibitor, idoxuridine, could, specifically and with a minimum of side-effects, shorten the disease interval in the rabbit eye (Kaufman, 1962, 1993; Kaufman and Maloney, 1962; Kaufman *et al.*, 1962a). Subsequently, we showed that the results of studies of epithelial disease treatment in the rabbit correlated very well with the results obtained in humans (Kaufman *et al.*, 1962b). Since then, we and others have used rabbits, primates, and mice to characterize the pathogenesis of many strains of herpesvirus and have studied acute and recurrent infection in the animal eye from the point of view of developing safe and effective antiviral therapies.

In this chapter we will summarize the animal models of anterior segment ocular herpes that we feel are the most useful and readily available to investigators. Studies of herpetic infection of the retina, including the von Szily model, will not be covered.

Animal species

Rabbits

We and other investigators have used the New Zealand white (NZW) rabbit for studies of ocular herpes (Kaufman, 1962; Kaufman and Maloney, 1962; Kaufman *et al.*, 1962a; Nesburn *et al.*, 1974; Smith *et al.*, 1984). The rabbit infects readily with most strains of HSV, the rabbit eye is only slightly smaller than the human eye, and when the inoculation is done under controlled circumstances, the infectious process is prototypical of the human disease. Rabbits develop classic corneal epithelial dendrites that respond to topical antiviral therapy, and the infection follows a well-

Handbook of Animal Models of Infection
ISBN 0-12-775390-7

defined and reproducible clinical course. Dutch Belted and other pigmented rabbits can also be used in ocular herpetic studies, but in general, the NZW rabbit is preferable because of its low cost and availability.

Disease-free, certified, and inspected NZW rabbits are widely bred for supermarket and restaurant supply and for medical research. We use rabbits of either sex weighing 2–5 kg. No detailed information has been reported regarding sex differences in rabbits relative to the severity of ocular herpes infection or to responsiveness to antiviral drug therapy. In the guinea-pig, young animals are reported to be easier to infect than older animals (Mani *et al.*, 1996).

Primates

Old World monkeys, such as Rhesus and Green monkeys, frequently have antibodies to herpesviruses and it is not possible to produce epithelial disease reliably with HSV-1 or -2 in these animals. Also, they are almost prohibitively expensive and their availability has become increasingly limited in recent years. Studies in these species have primarily been concerned with recurrent infections.

We have used New World monkeys, including the owl monkey (*Aotus trivirgatus*) and the squirrel monkey (*Saimiri sciureus*), in studies of acute and recurrent ocular herpetic infection (Varnell *et al.*, 1987, 1995; Kaufman *et al.*, 1991). Squirrel monkeys can still be obtained from a few suppliers. The University of South Alabama maintains a National Institutes of Health (NIH) Squirrel Monkey Breeding Research Resource in Mobile, AL, and investigators doing NIH-sponsored research may request animals from them. The Primate Supply Information Clearinghouse at the University of Washington in Seattle provides listings of available animals.

A detailed medical history of each monkey is important to insure that it has not been used in other virological studies. For the health and safety of the animals and the investigators, tuberculosis testing is a must. The squirrel monkey can readily be infected with any of the common laboratory strains of HSV-1 and, in general, the pathology and time course of the infection are similar to those seen in rabbits.

Mice

The laboratory mouse is available in two basic genetic versions: first, random-bred or non-inbred lines (often erroneously referred to as strains) and second, genetically homogeneous, inbred strains. Random-bred or non-inbred mice such as the ICR, Swiss Webster, and NIH lines are readily available and less expensive than the inbred strains. Some investigators feel that the genetic inhomogeneity in these lines insures that their response to ocular HSV infection is similar to that seen in genetically inhomogeneous human beings.

We have used inbred strains of mice, including C57BL/6, A, and BALB/c, because of their genetic purity, freedom from both horizontally and vertically transmitted pathogens, and ready availability from suppliers. The genetically determined susceptibility to herpesvirus infection has been extensively investigated in inbred strains of mice (Lopez, 1980a,b; Kastrukoff *et al.*, 1982). The C57BL/6 strain is relatively resistant to viral infection, whereas the A and BALB/c strains are more susceptible. Resistance and susceptibility differences are only relative and it is possible to vary the strain of inbred mouse, the strain of HSV, and the titer of the inoculum to produce a range of pathology from no perceptible infection to 100% mortality (Table 110.1).

Preparation of animals

In most university-based animal care facilities, animals infected with a human pathogen such as herpesvirus must be housed in a room separate from other animals. In general, infection from one animal to another does not occur when they are housed in separate cages, and infection of personnel working with infected animals does not seem to occur. Monkeys should be housed individually during treatment and observation; when treatments are not taking place, they may be housed in gang cages to allow social interaction. Otherwise, no specialized housing or preparation of animals for use in antiviral chemotherapy studies is required.

Table 110.1 The role of the laboratory mouse (*Mus musculus*) in ocular herpesvirus research

Mouse strain/line	*Resistance/ susceptibility**	*References*
Inbred†		
C57BL/6	Resistant	Lopez, 1980a,b
BALB/c	Moderately resistant	Lopez, 1980b
A	Susceptible	Lopez, 1980b
C3H	Moderately resistant	Kastrukoff *et al.*, 1982
NIH	Susceptible	Shimeld *et al.*, 1990; Blatt *et al.*, 1993
SJL	Susceptible	Knotts *et al.*, 1974
Outbred‡		
CD1	Susceptible	Mayo *et al.*, 1979
Swiss Webster	Susceptible	Cook *et al.*, 1991; Sawtell and Thompson, 1992

* Resistance and susceptibility are not absolute but relative, and depend on other factors such as age and sex of the animals and the titer and strain of the virus.
† Available from The Jackson Laboratory, Bar Harbor, ME, USA.
‡ Available from Charles River Laboratories, Wilmington, MA; Harlan Sprague Dawley, Inc., Indianapolis, IN, USA.

Virus strains, storage, preparation of inocula

Sources of strains of HSV-1 and HSV-2 abound. Most investigators freely exchange wild-type and genetically engineered variants of herpesviruses. In addition, it is possible to isolate a new strain of virus from clinical lesions in animals or humans, characterize the virus in terms of growth properties, and establish its genetic relationship to other known viruses that have been mapped. Such founder strains can be used in one's own studies. Herpesvirus can also be obtained from commercial sources. The American Type Culture Collection (ATCC, Rockville, MD) catalog lists several human isolates of HSV-1 and -2 and numerous isolates of other herpes simplex viruses from various animal species. Some of the commonly used laboratory isolates of HSV-1 and HSV-2, the animal species in which they have been used for ocular infection studies, and representative publications are given in Table 110.2.

The McKrae strain of HSV-1, which we isolated 35 years ago from a patient with ocular herpes, is often the strain of choice for studies in rabbits and mice. McKrae strain infects these species readily, establishes a self-limited corneal infection, and has variable mortality.

The Rodanus strain of HSV-1 is our choice for primate studies because this strain readily infects squirrel monkeys but does not cause serious systemic disease. The virus produces spontaneous corneal epithelial recurrences and can be reactivated from latency with appropriate stimulation. Many other strains have been tested in primates but do not produce spontaneous recurrences.

Propagation of most strains of HSV-1 in tissue culture is not difficult. We routinely use primary rabbit kidney cells isolated from newborn animals or the Vero cell variant of the African green monkey kidney line (ATCC, Rockville, MD). Monolayer cultures are infected at a multiplicity of infection of 0.01. When the cytopathic effect (CPE) becomes maximal, usually within 48–72 hours, the cells and tissue culture supernatant are collected separately, titered on CV-1 cells or other appropriate indicator cells, and stored frozen as concentrated stocks. We obtain viral titers in the culture supernatants of 5×10^6 to 5×10^8 plaque-forming units (pfu)/ml.

Many investigators perform plaque purification of virus, the details of which can be found in various virology methods manuals. The basic technique involves selecting infectious virus from a single plaque, propagating it in a permissive cell line, performing a plaque assay to establish that the selected virus yields a homogeneous plaque size, growing large quantities, and storing the plaque-purified virus as a stock suspension in liquid nitrogen and/or in an ultralow temperature freezer.

The viral titer used to inoculate rabbits, primates, and mice depends on the strain of the virus, the age of the animals, and the particular approach to be used in the actual infection (see below). To prepare an inoculum, the stored virus is thawed and used undiluted for rabbits and primates. For mice, the virus is diluted to a predetermined pfu/ml in culture medium and held on ice until it is used to infect the animals.

Table 110.2 Herpesvirus (HSV) strains used in ocular infection studies

Virus	*References*
HSV-1	
RE	Wander *et al.*, 1980; Hill *et al.*, 1987
F	Wander *et al.*, 1980; Hill *et al.*, 1987
17Syn+	Cook *et al.*, 1991; Sawtell and Thompson, 1992
Shealey	Wander *et al.*, 1980
CGA-3	Wander *et al.*, 1980; Hill *et al.*, 1987; Shimeld *et al.*, 1990; Blatt *et al.*, 1993
McKrae	Kaufman and Varnell, 1976; Wander *et al.*, 1980; Hill *et al.*, 1987
Rodanus	Knotts *et al.*, 1974; Hill *et al.*, 1987
MacIntyre	Hill *et al.*, 1987
HSV-2	
Hicks	Wander *et al.*, 1980
G	Wander *et al.*, 1980

Infection process

Inoculation of rabbits, primates, and mice by simply placing a suspension of viral particles on the ocular surface without scarification produces a highly variable frequency of clinical infection, sometimes as low as 10–15%. In some animals, an initial infection is visible 3 days after inoculation while in others, signs of infection may not appear for 5–7 days.

To regulate the initial appearance of clinical disease and to increase the frequency of clinical infection, the corneal scarification technique is used. With this approach, 90–100% of the eyes develop herpetic keratitis, usually within 3 days. In rabbits, the cornea is anesthetized with a drop of topical 0.5% proparacaine hydrochloride (Ophthetic, Allergan; AK-Taine, Akorn) or 0.5% tetracaine hydrochloride (AK-T-Taine, Akorn; Bausch & Lomb Pharmaceuticals; CIBA Vision Ophthalmics). After 3–5 minutes (which allows adequate anesthesia to develop and the preservative in the eye drop to exit the cul-de-sac), four superficial epithelial scratches are made in the form of a # using the tip of a sterile 27 or 30 G hypodermic needle. For rabbits, 25 µl of McKrae strain virus suspension at concentrations from 5×10^6 to 5×10^8 pfu/ml is dropped on the scarified corneal surface, the eyelids are closed and gently massaged over the corneal surface for approximately 10 seconds, and the animal is returned to its cage. Corneal scarification and infection with the McKrae strains in rabbits usually results in 100% of the corneas exhibiting ocular herpes by day 3.

The scarification procedure, inoculum of virus, and frequency and time-course of initial signs of infection are the same for monkeys as for rabbits, except that the Rodanus strain is used. Small monkeys such as the squirrel or owl monkey do not require anesthesia during handling, infection, or examination. They may be gently restrained by one person while another performs the necessary procedures.

Mice are fully anesthetized by an intraperitoneal injection of xylazine (10 mg/kg) and ketamine (200 mg/kg) and a 10 μl droplet of McKrae, 17Syn+, or one of various genetically engineered constructs at a concentration of 1×10^5 pfu/ml is placed on the ocular surface that has been scarified with a sterile 27 G needle. This process usually results in infection of 100% of the animals with a mortality of less than 10%. All of the animals surviving the infection become latent for HSV-1 and are thus useful in latency and reactivation studies.

Key parameters to monitor infection and response to treatment

The severity of herpetic infection is graded using a slit lamp biomicroscope. The only modification of the slit lamp that is necessary for use with animals is the removal of the chin rest used with humans and its replacement with a horizontal support that is large and sturdy enough to accommodate the body of a large rabbit (Figure 110.1).

The corneal ulcers are first delineated with fluorescein stain (Fluor-i-Strip, Wyeth-Ayerst, West Point, PA). A drop of sterile balanced salt solution is used to wet the strip, and a droplet is allowed to fall on to the cornea. Care is taken to avoid touching the cornea with the strip because this causes staining that is often difficult to differentiate from the stained ulcer. With the cobalt blue filter in the light path of the slit lamp, the epithelial dendrite stands out from the surrounding intact epithelium (Figure 110.2). As each cornea is stained, it is examined and graded immediately because the fluorescein stain dissipates into the tissue surrounding the ulcer with time. The grading system is based on the proportion of the cornea covered by the epithelial ulceration, ranging from 0 (no ulcer) to 4.0 (entire cornea ulcerated; Table 110.3).

It is essential that all evaluations of severity of keratitis be done in a masked fashion, because it is virtually impossible to keep bias out of the evaluations if treatment is known. The evaluator should not know the treatment, animals should not be presented together in a complete treatment group, nor should one eye be chosen for treatment with the other always chosen as control (right always control and left always treatment) because opinions are formed easily as the pattern emerges and this can produce statistically significant changes in the grading. It is important to use non-parametric statistical analyses to analyze differences in treatments.

Another parameter that may be evaluated, particularly in studies of drugs designed to prevent reactivation of latent

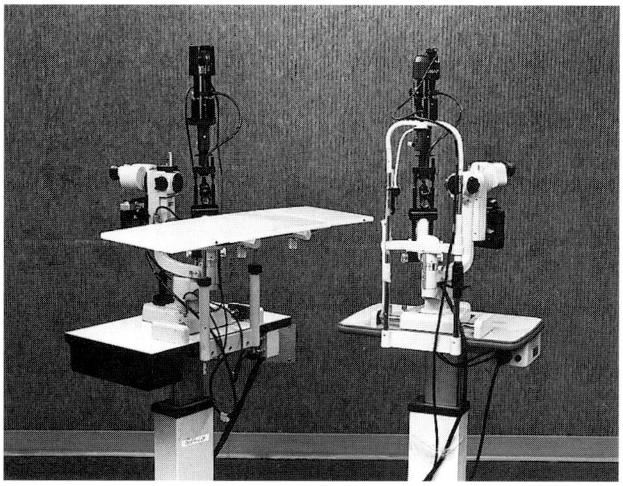

Figure 110.1 Slit lamp modified for examining animal eyes (left) and standard, unmodified slit lamp used for examining human eyes (right). For examination of animal eyes, the standard chin rest and fixation light are replaced with a rectangular platform on which the animal is placed.

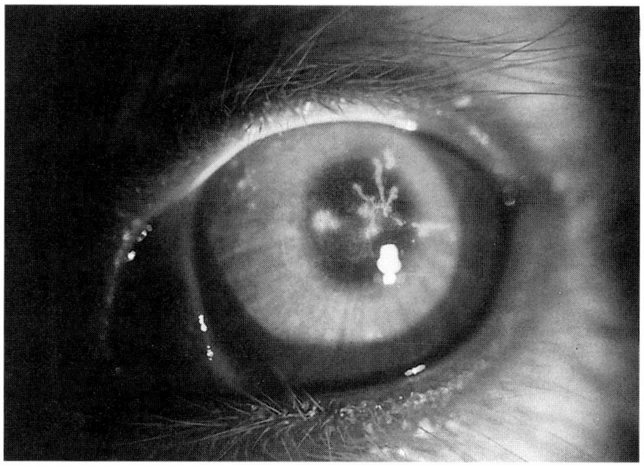

Figure 110.2 Herpetic corneal epithelial dendrite in a New Zealand white rabbit. Fluorescein dye on the cornea highlights the dendrite in this photograph taken with a blue filter.

Table 110.3 Grading system for corneal herpes simplex virus infection in rabbits and primates

Severity of keratitis	*Area of epithelium ulcerated*
0	None
0.25	One tiny dendrite involving less than 1/8 of epithelial area
0.5	1/8 of epithelial area
1.0	1/4 of epithelial area
1.5	3/8 of epithelial area
2.0	1/2 of epithelial area
2.5	5/8 of epithelial area
3.0	3/4 of epithelial area
3.5	7/8 of epithelial area
4.0	Entire epithelial area

virus, is virus shedding into the tear film. For this purpose, a sterile Dacron applicator (VWR Scientific Products, Houston, TX) is placed in the lower cul-de-sac and allowed to remain there for 30 seconds. The applicator is then placed into a tissue culture tube containing 2 ml of tissue culture medium and a monolayer of indicator cells such as CV-1 cells or primary rabbit kidney cells. The applicator is left in the tube overnight, then compressed against the side of the tube to remove the culture medium, and discarded. The tubes are incubated for a minimum of 14 days, during which time they are examined daily by a masked observer who scores for the presence or absence of CPE.

Death in rabbits and primates is not a useful measure of drug efficacy because most drugs applied to the cornea would not be expected to have a systemic effect sufficient to prevent death from generalized viremia.

Antiviral therapy

To some extent, the choice of animal species is dictated by the investigator's interests, i.e., the study of acute viral infection or recurrent viral infection.

Rabbits

Rabbits can be used for the study of both acute and recurrent ocular herpes.

Acute infection

For antiviral therapeutic efficacy studies, treatment is usually begun 3 days after virus inoculation, at which time dendritic keratitis that mimics human clinical disease is apparent. The severity of the keratitis is graded, and animals with disease of comparable severity are randomized into treatment and control groups. Conjunctivitis, iritis, and uveitis are not graded unless there is an indication that the drug will penetrate into the deeper layers of the eye. In some studies, virus shedding into the tear film is monitored by culture of tears obtained from the lower cul-de-sac every other day.

When commercially available eye drops are being evaluated, the concentration of the topical medication used is usually the same as that used in humans. New drugs are dissolved in balanced salt solution or phosphate-buffered saline solution at a pH of 5–7.5. One or two drops of the highest concentration of drug that is non-toxic to the external eye are administered to the cornea five to seven times a day. Treatment is continued for a minimum of 5 days. Corneas are examined daily and the severity of keratitis graded in a masked fashion. A concurrent control group is always included with test groups to monitor for vehicle toxicity and possible variations in animals or virus. It is important to remember that, with small animals, the absorption of drug from the mucous membranes may be sufficient to produce a systemic effect. For example, systemic absorption of drugs such as dexamethasone and cyclophosphamide has been sufficient to confuse the interpretation of topical therapy in rabbits.

If the test drug cannot be formulated for topical administration, it must be tested as a systemically administered drug. Theoretically, the dose of a systemic medication can be calculated roughly on the basis of surface area rather than weight, as in a very much larger animal. When the dosage of a new systemic drug is not known, toxicity testing is performed to determine the maximal tolerated dosage; a single injection of the drug is given, blood samples are compared with normal profiles, and adverse reactions, if any, are noted. The initial dose and dilutions thereof are then administered to groups of uninfected animals and serum samples are evaluated to determine the actual amount of drug in the serum. The serum concentration is generally judged to be adequate if it approximates 100 times the concentration that is effective against the virus in tissue culture or in other *in vitro* systems. In general, it is desirable to use a trough concentration (concentration of drug in serum or plasma immediately before the next dose of medication) for this determination; this also helps to determine the necessary frequency of administration. Systemic treatment can be administered intravenously, intraperitoneally, or by gavage (for orally active drugs). Usually, intraperitoneal injection is the easiest and gives a relatively good duration of absorption and effect.

Ultimately, the data derived from such experiments can be subjected to statistical analysis and relatively small differences between antiviral drug and placebo treatment can be reliably discerned.

Recurrent infection

The NZW rabbit undergoes a low but measurable ocular recurrence rate following infection with some strains of HSV-1. For example, the Rodanus, 17Syn+, and McKrae strains produce spontaneous recurrent infections of the cornea at a frequency of approximately 5% over a 30-day period (Hill *et al.*, 1987). The rate of recurrence decreases with time after infection. Other strains such as MacIntyre, CGA-3, F, and KOS do not produce spontaneous recurrences.

It is difficult to study antiviral drugs designed to prevent recurrences in models with low spontaneous reactivation frequencies. Therefore, a variety of stimuli are used in animals latent for HSV to increase the incidence of recurrences to a level at which it is possible to obtain statistically significant differences between treatments (Kaufman *et al.*, 1991; Varnell *et al.*, 1995). One such approach is the iontophoresis of epinephrine, which involves the placement of an eye cup containing 0.01% epinephrine on the rabbit's eye, followed by the application of a mild electric current (Hill *et al.*, 1983). Epinephrine iontophoresis performed once a day for 3 days induces viral shedding and results in the presence of infectious virus at the ocular surface in 70–80% of latent rabbits and in 90% of eyes (Hill *et al.*, 1987). Although the frequency of virus reactivation is excellent, the procedure does produce some local tissue damage in the cornea.

Other protocols that have been used in rabbits and mice to induce reactivation involve the intravenous injection of anti-inflammatory and immunosuppressive drugs, such as cyclophosphamide and dexamethasone (Sekizawa and Openshaw, 1984; Green *et al.*, 1987; Cook *et al.*, 1991). These models may be somewhat less effective than the iontophoresis procedure, but they do not cause any associated corneal tissue damage.

Reactivation of latent herpesvirus can result in shedding of virus in the tears and/or the development of actual corneal lesions. However, virus shedding is much more frequent than true recurrent disease; thus, the detection of virus in the tears is not necessarily a direct measure of recurrent disease, although the two are often equated. It must be remembered that the ultimate goal of treatment is the suppression of disease. When treating with agents such as interferon, antibodies, and very potent drugs, carry-over of therapeutic agents may reduce or eliminate detectable virus, and this needs to be considered in the experimental design.

Primates

Because of the scarcity of primates and the expense of using them in research, special precautions must be taken to insure that there are no, or at least minimal, losses of animals due to the experimental treatment.

Acute infection

Today, primates are rarely used to test antivirals in acute studies unless the compound to be tested is active only in primates.

Recurrent infection

Primates given primary ocular infections with HSV-1 characteristically become latent for the virus and, like humans, undergo random, sporadic, and infrequent spontaneous recurrent keratitis (Varnell *et al.*, 1987). Because the reactivation frequency is low, it is very expensive to study spontaneous recurrent infection in such animals. A few years ago, we found that the frequency of recurrent disease increased in Rodanus-infected squirrel monkeys when the ambient temperature in their quarters was lowered from the normal temperature of 24–27°C to 18°C overnight (Varnell *et al.*, 1995). Using this model, we were able to demonstrate the efficacy of 9-(4-hydroxybutyl)-N^2-phenylguanine (HBPG), a new thymidine kinase inhibitor (Kaufman *et al.*, 1996), for the prevention of recurrent disease. Thirty days after the primary infection, 150 mg/kg of HBPG dissolved in corn oil was injected intraperitoneally every 8 hours for 48 hours, beginning one morning at 8 a.m. and ending two mornings later at 8 a.m., for a total of seven treatments. Control animals received injections of the drug vehicle only. On the second night, the animal room was chilled to 18°C and then returned to normal by morning. The monkeys' eyes were examined daily for 8 days. The presence of corneal epithelial lesions was recorded and statistical analysis of the data performed.

Mice

Acute infection

Various strains of mice and outbred lines of mice have been used in herpesvirus research over the years. The laboratory mouse can readily be infected by virtually all of the available isolates of HSV-1 and -2 (Brandt *et al.*, 1992), although we consider the rabbit a superior model for assessing topical antiviral drugs because of the ease of evaluating corneal lesions in this species.

Acute infection studies using newly synthesized antiviral drugs have been reported (Brandt *et al.*, 1992; Gebhardt *et al.*, 1996). The essentials of establishing the primary infection are similar to those described above for rabbits and primates. Documentation of the success of infection is usually most accurately accomplished in the laboratory mouse by culturing the ocular tear film on days 3, 5, and 7 after the primary infection. Visual examination using a stereoscopic microscope can be performed, but because of the small size of the mouse eye, it is difficult to see the corneal lesions. In addition, because of the hyperactivity of mice grouped together in cages, we find that there is a relatively high frequency of spontaneously occurring, non-viral lesions in uninfected animals. Thus, it is most reliable to culture for infectious virus by swabbing the ocular surface.

The formulation, dosages, treatment schedules, and route of administration of newly acquired antiviral compounds must be empirically established in tissue culture experiments and ultimately in preliminary experiments in mice. For example, a newly synthesized antiviral compound that reduces the tissue culture infectious dose (TCID) of HSV-1 by several logs must nevertheless be titrated *in vivo* in animals as well. Typically, we find it necessary to perform acute and chronic toxicity studies, followed by dose–response titration studies in animals.

In general, for topical treatment with a new drug, the concentration applied to the surface of the eye must be at least 100 times the effective concentration in tissue culture and may need to be as much as 100 times the $TCID_{50}$. However, the small size of the ocular surface area suggests that this is a relatively poor delivery system in mice.

Systemic medications can be administered in the drinking water. In a series of studies with acyclovir, we dissolved the drug in the drinking water to a concentration of 3.5 mg/ml. In a typical experiment, the animals were given the drug for 3–5 days before HSV-1 infection. Each day the acyclovir solution was replaced by freshly prepared solution. Using this approach, we found that the serum concentrations of acyclovir reached a plateau by the time the animals were infected. Continuation of the acyclovir treatment for 10 days (the duration of the acute ocular infection) significantly reduced the incidence of corneal disease and greatly attenuated the disease interval in the animals in which the infection was successful. This type of protocol can be modified and adjusted according to the solubility of the antiviral agent and the ability of the compound to be transported across the gut wall.

Recurrent infection

It has been widely reported that mice latently infected with HSV-1 and HSV-2 do not undergo spontaneous reactivation or exhibit recurrent infection (Nesburn and Green, 1976; Mester and Rouse, 1991). The biological basis for this resistance to reactivation is not known. As a corollary it has long been noted that many of the stimuli used to induce reactivation in rabbits successfully do not work reliably and reproducibly in mice. However, in recent years several experimental protocols for inducing viral reactivation in mice have been reported. Various investigators, including Shimeld *et al.* (1990) and Blatt *et al.* (1993), reported that ultraviolet irradiation of certain strains of mice results in reactivation of infection in approximately 35–50% of the animals. More recently, Sawtell and Thompson (1992) reported that hyperthermic stress involving immersion of the mice in 43°C water for 10 minutes results in a relatively high frequency of viral reactivation.

In collaboration with Dr George Wright of the University of Massachusetts, we tested HBPG in this thermal stress model (Gebhardt *et al.*, 1996). Mice latently infected with HSV were given intraperitoneal injections of HBPG or vehicle control for a period of 2 days, stressed, and treated for 2 additional days, during which time the appearance of infectious virus at the ocular surface was measured and viral DNA synthesis in the trigeminal ganglia and corneas was quantitated. This model was found to be useful for testing the capacity of an antiviral compound to suppress viral reactivation both centrally in the ganglion and peripherally at the ocular surface.

Pitfalls (advantages/disadvantages) of the models

Problems with these models have been discussed as each model has been described. In general, the most important factors are the choice of appropriate species and virus strain. A major advantage of the rabbit and primate models of ocular herpes is that the infection can be graded visually and tear samples for virus culture can be obtained with relative ease, providing both qualitative and quantitative data. The remaining pitfall is the predictability factor, i.e., the degree of correspondence between the results of the therapeutic modality in the model and the results in humans.

Contributions of the model to ocular herpesvirus therapy

The rabbit model for ocular herpes infection was used to verify the efficacy of the first specific agent (idoxuridine) against any human viral infection (Kaufman, 1962). The efficacy of the next generation of effective antivirals (vidarabine, trifluridine, and acyclovir) was also proven in this rabbit ocular model (Kaufman and Heidelberger, 1964; Schaeffer *et al.*, 1978). In addition, the model was used to demonstrate the effectiveness of combination therapy of antivirals with corticosteroids for the treatment of deeper corneal stromal herpetic infections (Kaufman *et al.*, 1963). To this day, the rabbit model of ocular herpetic infections is widely used to screen potential agents because of the ease of determining efficacy in unanesthetized animals and the proven applicability of the results to the treatment of human disease.

Studies in primates have been especially important in determining the efficacy of newly synthesized antiviral agents in an animal species evolutionarily and genetically close to humans. Topical and systemic toxicity studies have also generally been done in primates before clinical studies are begun.

Increased use of the mouse model of ocular HSV infection in recent years has resulted from the availability of both acute and recurrent infection protocols in this species. In particular, studies that involve molecular biological analysis of ganglionic tissues for viral nucleic acids are facilitated in this model because of the ready availability and relative low cost of obtaining and maintaining the necessary numbers of animals to achieve statistically significant results.

Acknowledgment

This work was supported in part by US Public Health Service grants EY02672 (HEK), EY02377 (HEK), and EY06311 (JMH) from the National Eye Institute, National Institutes of Health, Bethesda, MD.

References

Blatt, A. N., Laycock, K. A., Brady, R. H., Traynor, P., Krogstad, D. J., Pepose, J. S. (1993). Prophylactic acyclovir effectively reduces herpes simplex virus type 1 reactivation after exposure of latently infected mice to ultraviolet B. *Invest. Ophthalmol. Vis. Sci.*, **34**, 3459–3465.

Brandt, C. R., Coakley, L. M., Grau, D. R. (1992). A murine model of herpes simplex virus-induced ocular disease for antiviral drug testing. *J. Virol. Method.*, **36**, 209–222.

Cook, S. D., Paveloff, M. J., Doucet, J. J., Cottingham, A. J., Sedarati, F., Hill, J. M. (1991). Ocular herpes simplex virus reactivation in mice latently infected with latency-associated transcript mutants. *Invest. Ophthalmol. Vis. Sci.*, **32**, 1558–1561.

Gebhardt, B. M., Wright, G. E., Xu, H., Focher, F., Spadari. S., Kaufman, H. E. (1996). 9-(4-Hydroxybutyl)-N^2-phenylguanine (HBPG), a thymidine kinase inhibitor, suppresses herpes virus reactivation in mice. *Antiviral Res.*, **30**, 87–94.

Goodpasture, E. W., Teague, O. (1923–1924). Transmission of the virus of herpes febrilis along nerves in experimentally infected rabbits. *J. Med. Res.*, **44**, 139–184.

Green, M. T., Dunkel, E. C., Pavan-Langston, D. (1987). Effect of immunization and immunosuppression on induced ocular shedding and recovery of herpes simplex virus in infected rabbits. *Exp. Eye Res.*, **45**, 375–383.

Hill, J. M., Kwon, B. S., Shimomura, Y., Colborn, G. L., Farivar Y., Gangarosa, L. P. (1983). Herpes simplex virus recovery in neural tissues after ocular HSV shedding induced by epinephrine iontophoresis to the rabbit cornea. *Invest. Ophthalmol. Vis. Sci.*, **24**, 243–247.

Hill, J. M., Rayfield, M. A., Haruta, Y. (1987). Strain specificity of spontaneous and adrenergically induced HSV-1 ocular reactivation in latently infected rabbits. *Curr. Eye Res.*, **6**, 91–97.

Kastrukoff, L., Hamada, T., Schumacher, U., Long, C., Doherty, P. C., Koprowski, H. (1982). Central nervous system infection and immune response in mice inoculated into the lip with herpes simplex virus type 1. *J. Neuroimmunol.*, **2**, 295–305.

Kaufman, H. E. (1962). Clinical cure of herpes simplex keratitis by 5-iodo-2′-deoxyuridine. *Proc. Soc. Exp. Biol. Med.*, **109**, 251–252.

Kaufman, H. E. (1976). Relation of animal herpes to man. *Trans. Am. Acad. Ophthalmol. Otolaryngol.*, **81**, OP629–OP631.

Kaufman, H. E. (1993). Introduction: the first effective antiviral. In *The Search for Antiviral Drugs* (eds Adams, J., Merluzzi V. J.), pp. 1–21. Birkhauser, Boston.

Kaufman, H. E., Heidelberger, C. (1964). Therapeutic antiviral action of 5-trifluoromethyl-2′-deoxyuridine. *Science*, **145**, 585–586.

Kaufman, H. E., Maloney, E. D. (1962). IDU and hydrocortisone in experimental herpes simplex keratitis. *Arch. Ophthalmol.*, **68**, 396–398.

Kaufman, H. E., Varnell, E. D. (1976). Effect of 9-β-D-arabinofuranosyladenine 5′-monophosphate and 9-β-D-arabinofuranosylhypoxanthine 5′-monophosphate on experimental herpes simplex keratitis. *Antimicrob. Agents Chemother.*, **10**, 885–888.

Kaufman, H. E., Nesburn, A. B., Maloney, E. D. (1962a). IDU therapy of herpes simplex. *Arch. Ophthalmol.*, **67**, 583–591.

Kaufman, H. E., Martola, E.-L., Dohlman, C. H. (1962b). The use of 5-iodo-2′deoxyuridine (IDU) in the treatment of herpes simplex keratitis. *Arch. Ophthalmol.*, **68**, 235–239.

Kaufman, H. E., Martola, E-L., Dohlman, C. H. (1963). Herpes simplex treatment with IDU and corticosteroids. *Arch. Ophthalmol.*, **69**, 468–472.

Kaufman, H. E., Varnell, E. D., Cheng, Y. C. *et al.* (1991). Suppression of ocular herpes recurrences by a thymidine kinase inhibitor in squirrel monkeys. *Antiviral Res.*, **16**, 227–232.

Kaufman, H. E., Varnell, E. D., Wright, G. E. *et al.* (1996). Effect of 9-(4-hydroxybutyl)-N^2-phenylguanine (HBPG), a thymidine kinase inhibitor, on clinical recurrences of ocular herpetic keratitis in squirrel monkeys. *Antiviral Res.*, **33**, 65–72.

Kaufman, H. E., Rayfield, M. A., Gebhardt, B. M. (1998). Herpes simplex viral infections. In *The Cornea*, 2nd edn (eds Kaufman, H. E., Barron, B. A., McDonald, M. B.), pp. 247–277. Butterworth-Heinemann, Boston.

Knotts, F. B., Cook, M. L., Stevens, J. G. (1974). Pathogenesis of herpetic encephalitis in mice after ophthalmic inoculation. *J. Infect. Dis.*, **130**, 16–27.

Laibson, P. R., Kibrick, S. (1966). Reactivation of herpetic keratitis by epinephrine in rabbit. *Arch. Ophthalmol.*, **75**, 254–260.

Liesegang, T. J. (1988). Ocular herpes simplex infection: pathogenesis and current therapy. *Mayo Clin. Proc.*, **63**, 1092–1105.

Lopez, C. (1980a). Genetics of natural resistance to herpesvirus infections in mice. *Nature*, **258**, 152–153.

Lopez, C. (1980b). Resistance to HSV-1 in the mouse is governed by two major, independently segregating, non-H-2 loci. *Immunogenetics*, **11**, 87–92.

Mani, C. S., Bravo, F. J., Stanberry, L. R., Myers, M. G., Bernstein, D. I. (1996). Effect of age and route of inoculation on outcome of neonatal herpes simplex virus infection in guinea pigs. *J. Med. Virol.*, **48**, 247–252.

Mayo, D. R., Richards, J. C., Rapp, F. (1979). Acycloguanosine treatment of herpesvirus infections in footpads and nervous tissue of normal and immunosuppressed mice. *Intervirology*, **12**, 345–348.

Mester, J. C., Rouse, B. T. (1991). The mouse model and understanding immunity to herpes simplex virus. *R.I.D.*, **13** (suppl. 11), S935–S945.

National Advisory Eye Council (1993). *Vision Research. A National Plan: 1994–1998. Report of the Corneal Diseases Panel*, p. 138. National Eye Institute, National Institutes of Health, U.S. Department of Health and Human Services, Washington, DC.

Nesburn, A. B., Green, M. T. (1976). Recurrence in ocular herpes simplex infection. *Invest. Ophthalmol.*, **15**, 515–518.

Nesburn, A. B., Robinson, C., Dickinson, R. (1974). Adenine arabinoside effect on experimental idoxuridine-resistant herpes simplex infection. *Invest. Ophthalmol.*, **13**, 302–304.

Sawtell, N. M., Thompson, R. L. (1992). Rapid *in vivo* reactivation of herpes simplex virus in latently infected murine ganglionic neurons after transient hyperthermia. *J. Virol.*, **66**, 2150–2156.

Schaeffer, H. J., Beauchamp, L., de Miranda, P., Elion, G. B., Bauer, D. J., Collins, P. (1978). 9-(2-Hydroxyethoxymethyl)guanine activity against viruses of the herpes group. *Nature*, **272**, 583–585.

Sekizawa, T., Openshaw, H. (1984). Encephalitis resulting from reactivation of latent herpes simplex virus in mice. *J. Virol.*, **50**, 263–266.

Shimeld, C., Hill, T. J., Blyth, W. A., Easty, D. L. (1990). Reactivation in latent infection and induction of recurrent herpetic eye disease in mice. *J. Gen. Virol.*, **71**, 397–404.

Smith, K. O., Hodges, S. L., Kennell, W. L. *et al.* (1984). Experimental ocular herpetic infections in rabbits. Treatment with 9-([2-hydroxy-1-(hydroxymethyl)ethoxy]methyl)guanine. *Arch. Ophthalmol.*, **102**, 778–781.

Varnell, E. D., Kaufman, H. E., Hill, J. M., Wolf, R. H. (1987). A primate model for acute and recurrent herpetic keratitis. *Curr. Eye Res.*, **6**, 277–279.

Varnell, E. D., Kaufman, H. E., Hill, J. M., Thompson, H. W. (1995). Cold stress-induced recurrences of herpetic keratitis in the squirrel monkey. *Invest. Ophthalmol. Vis. Sci.*, **36**, 1181–1183.

Wander, A. H., Centifanto, Y. M., Kaufman, H. E. (1980). Strain specificity of clinical isolates of herpes simplex virus. *Arch. Ophthalmol.*, **98**, 1458–1461.

Chapter 111

Animal Models for Cytomegalovirus Infection: Murine CMV

E. R. Kern

Background of human infection

Several investigators have reported that congenital cytomegalovirus (CMV) infections occur at a frequency of 1–2% of all live births (Alford, 1984). Initial studies indicated that only 10% of congenitally infected infants manifested disease in the neonatal period and suffered serious neurologic sequelae. Further investigation, however, demonstrated that minor degrees of central nervous system (CNS) damage, such as hearing loss or mental retardation, occur in a significant proportion of the congenitally infected, but otherwise normal-appearing, newborn infants (Reynolds *et al.*, 1974). Several studies in renal transplant recipients indicate that acute CMV infection, either as an acquired or as a reactivated infection, may occur in 80–90% of patients (Ho, 1984). Infection with CMV may result in significant clinical disease and death in organ transplant recipients (Glenn, 1981), and may also enhance susceptibility to bacterial and fungal infections (Rand *et al.*, 1976), which are often a common cause of death in organ transplant recipients and other immunocompromised patients. In patients with acquired immunodeficiency disease syndrome (AIDS), CMV is the most common cause of life-threatening disease (Jacobson, 1989). Fifty percent or greater of AIDS patients may have CMV viremia, or shed virus in semen, and 90% or more have histopathologic evidence of CMV on autopsy (Mintz *et al.*, 1983; Reichert *et al.*, 1983; Quinnan *et al.*, 1984).

For treatment of CMV infections, ganciclovir (GCV) has been shown to reduce the severity of retinitis, gastrointestinal disease and to a lesser extent, pneumonia, in AIDS and organ or bone marrow transplant patients (Buhles *et al.*, 1988; Crumpacker *et al.*, 1988; Dieterich *et al.*, 1988; Mills *et al.*, 1988; Winston *et al.*, 1988; Spector *et al.*, 1993). However, at least for retinitis, when therapy is discontinued the beneficial effect is soon lost, necessitating the use of long-term maintenance therapy. Additionally, there is still a need for more effective treatment of CMV infections in organ and bone marrow transplant recipients and the development of resistance to GCV has been a common problem with long-term therapy (Drew *et al.*, 1991). Foscarnet (phosphonoformic acid) (PFA) has also been approved for treatment of CMV infections (Walmsley *et al.*, 1988), and it has been shown to be effective against CMV isolates that are resistant to GCV (Jacobson and O'Donnell, 1991).

A new class of phosphonyl nucleotide analogs has been studied extensively in recent years and one of these, Cidofovir (CDV), has excellent activity against CMV infections in tissue culture and in animal studies (Snoeck *et al.*, 1988; Kern, 1991; Neyts *et al.*, 1992). This compound has been effective in the treatment of CMV infections; however, its nephrotoxicity has limited its use primarily to retinitis in AIDS patients (Polis *et al.*, 1995; Jacobson, 1997; Lalezari *et al.*, 1997; Safrin *et al.*, 1997). The development of resistance of CMV to all these agents (Chou *et al.*, 1997; Smith *et al.*, 1997) necessitates the continued development of new and less toxic antiviral agents for treatment of CMV infections.

Although there have been numerous significant advances in the treatment of CMV infections with GCV, PFA, and CDV, there is still a need to develop more effective modes of therapy for these infections. In particular, for CMV infections in the immunocompromised patient, optimal therapy is currently not available. Hopefully, in the next few years, new more active and/or less toxic agents will be developed and the use of better delivery methods or combinations of existing drugs will improve the outcome of these serious diseases.

Background of animal model

As pointed out previously, CMV is the causative agent of a variety of clinical syndromes in the fetus, neonate, and particularly in patients who are immunosuppressed for organ transplantation, during chemotherapy for malignancies, or as a result of another infection such as human immunodeficiency virus (HIV). Due to the strict species-specificity of human CMV, it has not generally been possible to test potential antiviral agents against this virus in experimental animals. There are a number of natural CMV infections in various animal species, however, that have been utilized for antiviral evaluation. These include murine CMV (MCMV; Glasgow *et al.*, 1982), rat CMV (Stals *et al.*, 1991), and guinea-pig CMV (Li *et al.*, 1990). The complete DNA

Handbook of Animal Models of Infection
ISBN 0-12-775390-7

sequence of MCMV is now available and, despite significant differences in the overall arrangement of the genomes of MCMV and human CMV (HCMV), they share much similarity at the genetic and nucleotide level. Counterparts of all the human enzyme homologs in HCMV are also present in the MCMV genome (Rawlinson *et al.*, 1996). Rat CMV and guinea-pig CMV are less well-characterized.

Inoculation of mice with MCMV provides a model infection that shares many characteristics with the human disease. Both acute lethal and chronic non-lethal MCMV infections have been used in our laboratory to determine efficacy of antivirals (Overall *et al.*, 1976; Glasgow *et al.*, 1982; Kern, 1997). After intraperitoneal (i.p.) inoculation of 3-week-old Swiss-Webster female mice with 1×10^6 plaque-forming units (pfu) of MCMV, 90–100% of animals die with a mean day of death of 5–6 days; however, if the inoculum is reduced to 1×10^5 pfu, all animals survive. With either inoculum, high titers of virus are present in lung, liver, spleen, kidney, and blood within 24 hours and in the salivary gland by 48–72 hours after viral inoculation. In the non-lethal infection, persistent viral replication occurs in lung, liver, kidney, and spleen for 45–60 days and in salivary glands for months. In Balb/c mice infected with MCMV, immunosuppression with cyclophosphamide results in a severe interstitial pneumonitis (Shanley and Pesanti, 1985). Immunosuppression may also result in the induction of CMV retinitis in a small percentage of infected animals (Bale *et al.*, 1984). The MCMV infection, therefore, involves many of the same target organs as HCMV and has been shown to be an excellent model for predicting the outcome of therapy in human CMV diseases (Kern, 1991, 1997).

Methods

Animals

All mice in our facility are housed in standard filter top polycarbonate shoebox cages (155 sq in). The maximum number of mice housed per cage is 15. They are given standard rodent chow and water *ad libitum*. Contact bedding is pine, and cages and water are changed twice a week. Complete cage change-outs are every 2 weeks. They are maintained on a computer-controlled 12-hour off/on light cycle at 35% relative humidity. Cages are checked daily for dead or ill animals.

Severe combined immunodeficient (SCID) mice and other immunocompromised mice are housed in controlled environments. All cages, food, and water are autoclaved prior to use. Animal handling and cage changing is done in sterile hoods. Personnel working with SCID mice must wear disposable gloves, mask, head covers, and shoe covers, that have not been worn in rooms with other mice. Gowns specifically for that room are kept separate from other animal rooms. All items are sprayed with disinfectant prior to their placement in the sterile hood and sterile technique is followed at all times. All apparel worn in animal rooms are not to be worn in other rooms and are to be removed and disposed of before exiting the area.

Three-week-old Balb/c or outbred mice such as CD-1 or Swiss-Webster weighing approximately 10–12 g are used for infection. While we have used females exclusively, males can also be used. In our laboratory, Balb/c mice are slightly more susceptible to infection with MCMV than outbred animals, and the results appear to be more consistent. Other animal strains such as C_3H or C57/BL6 are extremely resistant to infection with MCMV and virus replicates much less well in target organs such as lung, liver, spleen, and kidney than in Balb/c or outbred mice.

CB-17 SCID mice (4–6-week-old) which lack T cells and B cells are exquisitely sensitive to infection with MCMV. This experimental infection provides an excellent model for CMV infections of the immunocompromised host to evaluate the potential of an antiviral agent prior to use in humans (Neyts *et al.*, 1992; Smee *et al.*, 1992).

Virus stocks, assays and cell cultures

Stock pools of MCMV are prepared routinely as 10% wt/vol homogenates of salivary glands harvested 10–12 days after 3–4-week-old mice are inoculated i.p. with a non-lethal dose of MCMV. It is important to use tissue-derived virus as tissue culture-propagated virus has been shown to have reduced virulence. Virus preparations are quantified using a standard plaque assay in primary mouse embryo fibroblast cells using sequential agarose overlays. While established murine cell lines such as 3T3 and others can be used (Smee *et al.*, 1989), we find that the primary cells are considerably more sensitive for the detection and quantitation of MCMV. Primary mouse embryo fibroblast cells are prepared from embryos of 16–19 days gestational age. Embryos are removed aseptically, washed in saline, and the head and limbs removed. The tissue is minced and washed with saline to remove debris and red blood cells. Individual cell suspensions are prepared by multiple extractions using 0.25% trypsin. After the tissue is digested, the cells are centrifuged, re-suspended in tissue culture media with 10% fetal bovine serum (FBS) and placed in 175 cm^2 tissue culture flasks and incubated at 37°C with 95% CO_2 and 90% humidity. After the cells become confluent, they are trypsinized and added to multi-well tissue culture plates for use in the plaque assay for MCMV.

Experimental infection and pathogenesis

Intraperitoneal inoculation of 3-week-old Balb/c mice with 10^6 pfu of MCMV results in an acute, lethal infection with rapid virus replication in lung, liver, spleen, kidney, intestine, salivary gland, and other visceral and glandular tissue, and animals die in about 4–7 days. Since this is a lethal infection, the model can be used for rapid identification of potential antiviral compounds. Reduction of the virus inoculation to 10^5 pfu of MCMV results in a non-lethal, chronic, generalized infection which has many similarities to human CMV infections. Virus can be readily isolated from blood, lung, liver, spleen, kidney, urine, intestine, pancreas, adrenals, and salivary gland (Figure 111.1). Virus replication persists in some of these target organs for 45–60 days and in salivary glands for months.

The histopathology of the MCMV infection has been described in detail previously (Ho, 1991) and our own observations agree with these reports. Briefly, the most extensive lesions occur in the liver, spleen, adrenals, lymph nodes, and subperitoneal connective tissue and fat. Focal areas of necrosis, inflammation, and cells with intranuclear inclusions are observed in these tissues. Less extensive changes are found in the lungs, kidneys, intestines, and pancreas. Lesions are found in the salivary glands a few days later than in the other tissues. The persistent virus replication in tissues during chronic infection makes it an ideal model for evaluating long-term or maintenance therapy.

SCID mice inoculated i.p. with from 1.0 pfu to 10^5 pfu and left untreated all die in a dose-dependent manner. The mortality rates of SCID mice inoculated i.p. with 3 pfu – 3×10^4 pfu of MCMV are shown in Figure 111.2. Animals that received 3×10^4 had a mean day of death of about 14 days, whereas those inoculated with 3 pfu survived an average of about 25 days. Every $\log_{10}$ increase in virus inoculum decreased survival time by about 3 days. This result has been very predictable for a number of experiments and can be used to predict the mortality rate for a particular inoculum.

To determine the pathogenesis of this infection, mice were inoculated with 10 pfu and on various days post-infection, three mice per day were sacrificed, their tissues removed, homogenized, and assayed for MCMV. Virus was first detected in salivary glands by day 6 followed by lung, spleen, kidney, adrenals, and pancreas on days 9–12. Liver, which is one of the most permissive organs in normal mice, did not have detectable virus until day 18. Brain also was infected by day 18 (Figure 111.3). Mortality also occurs 7–14 days later than Balb/c mice given a similar inoculum size. The histopathology in these mice followed a distinct pattern. At post-infection day 5, and at all subsequent time points (days 8, 12, 15), the liver and spleen were the prime targets of the viral infection. Initially there is mild, multifocal necrosis in these organs, and very little disease in other

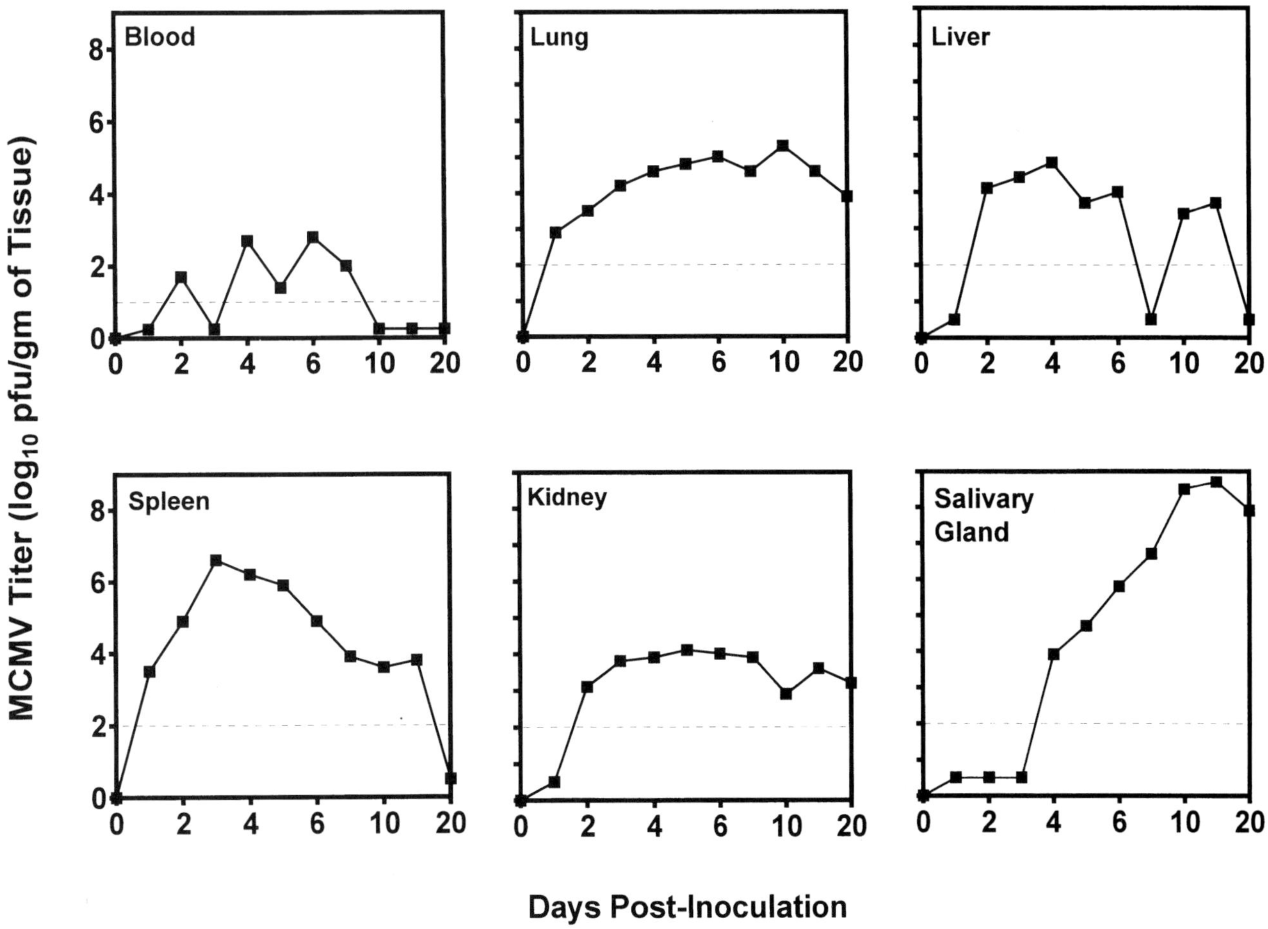

Figure 111.1 Pathogenesis of murine cytomegalovirus (MCMV) infection in Balb/c mice.

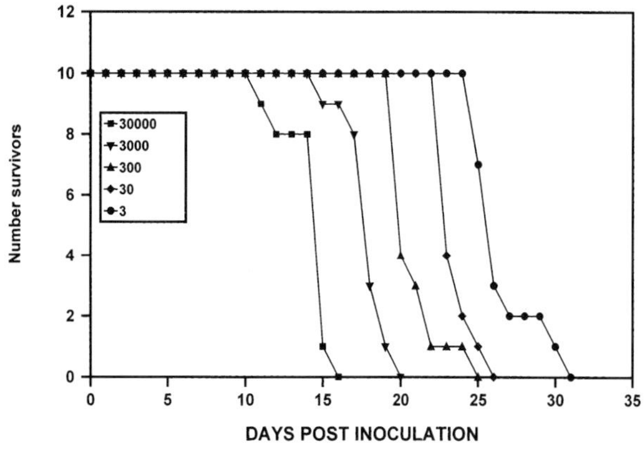

Figure 111.2 Mortality of severe combined immunodeficient (SCID) mice inoculated intraperitoneally with murine cytomegalovirus.

organs. The severity of disease in liver and spleen slowly increases to maximum at day 15. At this time, the lesions in liver are still multifocal but the number of necrotic foci is very great and the individual foci are large. At day 15, the spleen is diffusely affected and severe loss of total spleen parenchyma has occurred. Throughout the infection, there are numerous large intranuclear inclusions typical of CMV in all affected tissue.

At day 5, organs other than liver and spleen are unaffected histologically. By day 8, mild focal necrosis is present in salivary glands, stomach, small intestine, large intestine, lymph nodes, and kidney. The lungs have small foci of histiocytic pneumonitis in alveoli. Subsequently (at 8, 12, and 15 days), these foci enlarge but they never reach the severity of those in liver and spleen. Foci of necrosis also appear in pancreas and adrenal glands at day 12 and nasal passages and bone marrow at day 15. In some cases significant ulcers form in the stomach and large intestinal mucosa at day 12 and 15, and these gut lesions probably contribute to mortality through sepsis from the gut flora.

These results indicate that inoculation of SCID mice with low concentrations of MCMV result in a disseminated infection with viral replication in the same target organs as seen in immunodeficient patients. These animals have high levels of virus in their tissues for at least 2–3 weeks, which should allow adequate time to document an antiviral response in treated animals compared with placebo animals. It is interesting to note that, although SCID mice are more susceptible to infection and death than normal mice, the rate of virus replication in target organs is considerably reduced compared to normal mice and mortality is delayed by 1–2 weeks.

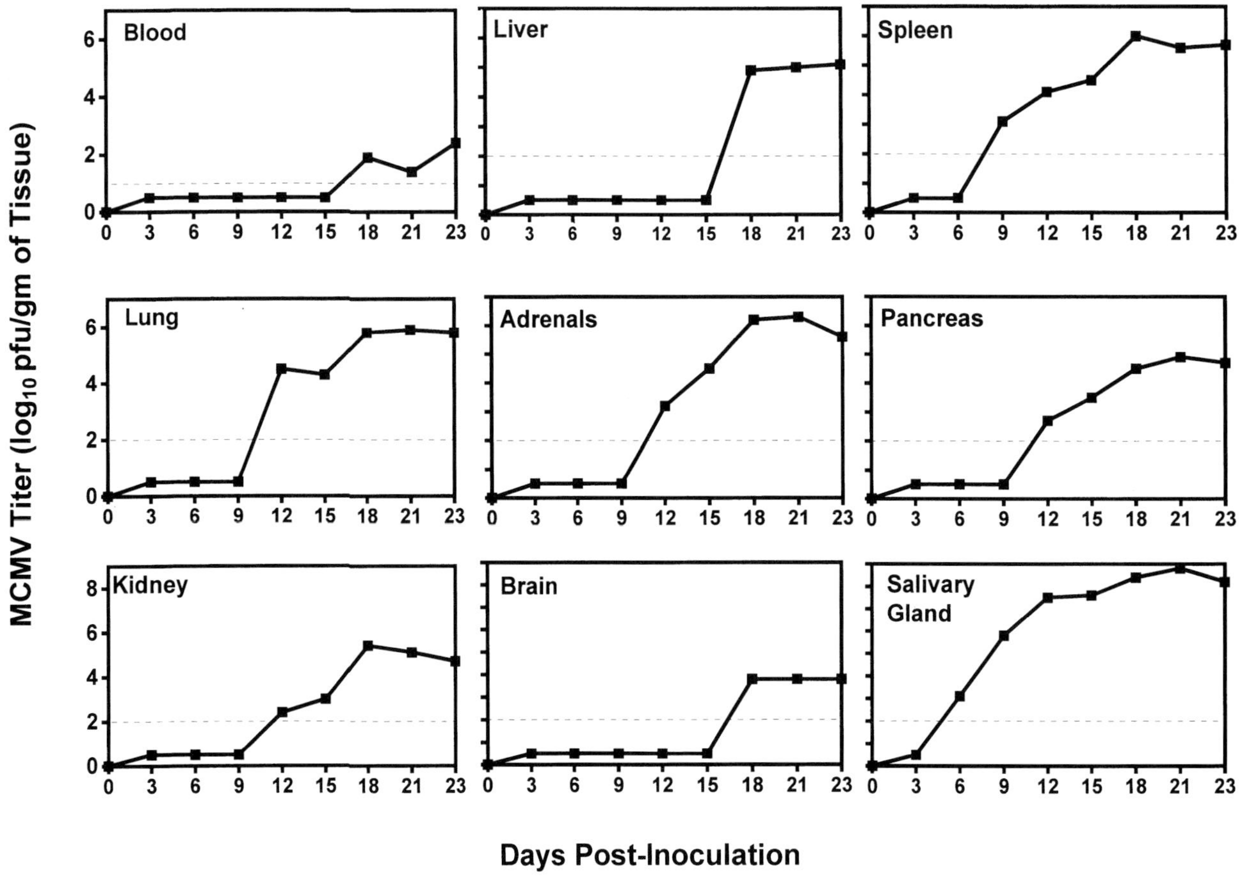

Figure 111.3 Pathogenesis of murine cytomegalovirus (MCMV) infection in severe combined immunodeficient (SCID) mice.

Evaluation of antiviral therapy — key parameters

These murine models lend themselves well to the use of i.p., subcutaneous, or oral therapy. When first evaluating a new antiviral agent in animal models, one does not generally know the pharmacokinetic profile of the agent in that species. For all initial evaluations, we utilize a standard treatment regimen consisting of twice-daily treatment for 5 days. A sufficient number of drug concentrations should be used so that the maximum tolerated dose and the minimal effective dose can be calculated. The initial and most common endpoint for efficacy evaluation is usually death. An example of a typical mortality experiment using GCV and CDV is presented in Table 111.1. Before beginning efficacy studies in mice, an initial experiment should be conducted in which uninfected animals are treated i.p. twice daily for 7 days, with various concentrations of test compound to establish the maximum tolerated dose.

Table 111.1 Effect of treatment with Cidofovir or ganciclovir on the mortality of mice inoculated with murine cytomegalovirus

	Mortality			
*Treatment**	*No. %*	P *Value*	*MDD*	P *value*
Control	11/15 (73)		5.5	
Placebo + 24 h	12/15 (80)	NS	5.1	NS
Cidofovir				
16.7 mg/kg + 6 h	0/15 (0)	<.001		
16.7 mg/kg + 24 h	2/15 (13)	<.001	5.5	NS
16.7 mg/kg + 48 h	10/15 (67)	NS	8.0	<.001
5.6 mg/kg + 6 h	0/15 (0)	<.001		
5.6 mg/kg + 24 h	1/15 (7)	<.001	5.0	NS
5.6 mg/kg + 48 h	5/15 (33)	<.05	5.4	NS
1.9 mg/kg + 6 h	0/15 (0)	<.001		
1.9 mg/kg + 24 h	7/15 (47)	NS	4.6	NS
1.9 mg/kg + 48 h	5/15 (33)	<.05	6.4	NS
0.6 mg/kg + 6 h	1/15 (7)	<.001	7.0	NS
0.6 mg/kg + 24 h	6/15 (40)	.06	4.7	NS
0.6 mg/kg + 48 h	6/15 (40)	.06	6.5	<.05
Ganciclovir				
16.7 mg/kg + 6 h	1/15 (7)	<.001	7.0	NS
16.7 mg/kg + 24 h	5/15 (33)	<.05	4.6	NS
16.7 mg/kg + 48 h	10/15 (67)	NS	6.0	NS
5.6 mg/kg + 6 h	5/15 (33)	<.05	4.0	NS
5.6 mg/kg + 24 h	3/15 (20)	<.01	4.3	NS
5.6 mg/kg + 48 h	4/15 (27)	<.01	4.3	NS
1.9 mg/kg + 6 h	7/15 (47)	NS	5.1	NS
1.9 mg/kg + 24 h	13/15 (87)	NS	4.1	<.05
1.9 mg/kg + 48 h	15/15 (100)	NS	5.4	NS
0.6 mg/kg + 6 h	14/15 (93)	NS	5.1	NS
0.6 mg/kg + 24 h	12/15 (80)	NS	5.7	NS
0.6 mg/kg + 48 h	12/15 (80)	NS	5.1	NS

MDD, Mean day of death; NS, not significant.
* Animals were treated intraperitoneally twice daily for 5 days with the above-stated concentrations and initiated at the times indicated.

After completing the toxicity study, mice were allocated to placebo, CDV or GCV treatment groups, with each animal receiving drug i.p. twice daily for 5 days at doses (0.6, 1.9, 5.6, or 16.7 mg/kg) considerably below that which could be maximally tolerated. For each treatment group, therapy was initiated at either 6, 24, or 48 hours after viral challenge. Two control groups were included in the study (one untreated and one given placebo at 24 hours post-challenge). Irrespective of dosage, almost all of the mice given CDV 6 hours after inoculation with MCMV survived, in contrast to the GCV-treated group, which exhibited increasing mortality as the drug dose was progressively reduced from 16.7 to 0.6 mg/kg. The better outcome for the animals receiving CDV, in terms of reduced mortality, was also evident when treatment was begun 24 hours post-challenge. If treatment was delayed for 48 hours, the mortality rate for the two groups was comparable at the highest doses tested, but substantially lower for mice given 0.6 and 1.9 mg/kg of CDV. These latter data were particularly exciting as they indicated that treatment initiated in mice as late as 48 hours following exposure to MCMV (especially with CDV) can still have a mortality-sparing effect on the course of disease — a remarkable finding considering that untreated MCMV-infected animals begin to die as early as 3–4 days post-infection.

Since mortality is not a very sensitive endpoint and may not accurately reflect antiviral efficacy in a particular target organ, the optimal measure of a drug's activity in an animal is to determine the effect of treatment with CDV on viral replication in the target organs such as lung, liver, spleen, kidney, glandular tissue, etc. An example of an experiment of this type comparing CDV and GCV in a chronic non-lethal infection is shown in Figure 111.4. Both compounds were administered twice daily at 10 mg/kg for 5 days beginning 4 days after infection. Virus titers were determined in target organ tissues at various intervals for 21 days post-inoculation. The results indicate that CDV administered beginning 4 days after MCMV infection and at a time when virus titers are at peak levels can drastically alter replication soon after therapy is initiated. In other experiments, the model has been used to investigate the efficacy of twice-a-week dosing for 60 days. Therefore, the chronic infection of MCMV is a very sensitive and useful model for determining the effect of maintenance therapy on viral replication in target organs.

As indicated previously, MCMV infection of SCID mice provides an ideal model for CMV infections in the immunocompromised host. When this model infection is utilized for evaluation of treatment such as the use of twice-weekly CDV for 3 weeks, animals survive on an effective regimen as long as drug is being administered (Figure 111.5). However, because the animals do not have a normal immune system, the animals cannot clear the virus and all die soon after drug is discontinued and do so in a dose-dependent fashion. This provides an excellent example of the need for an immune system to help clear viral infections, as effective antiviral agents alone cannot do the job.

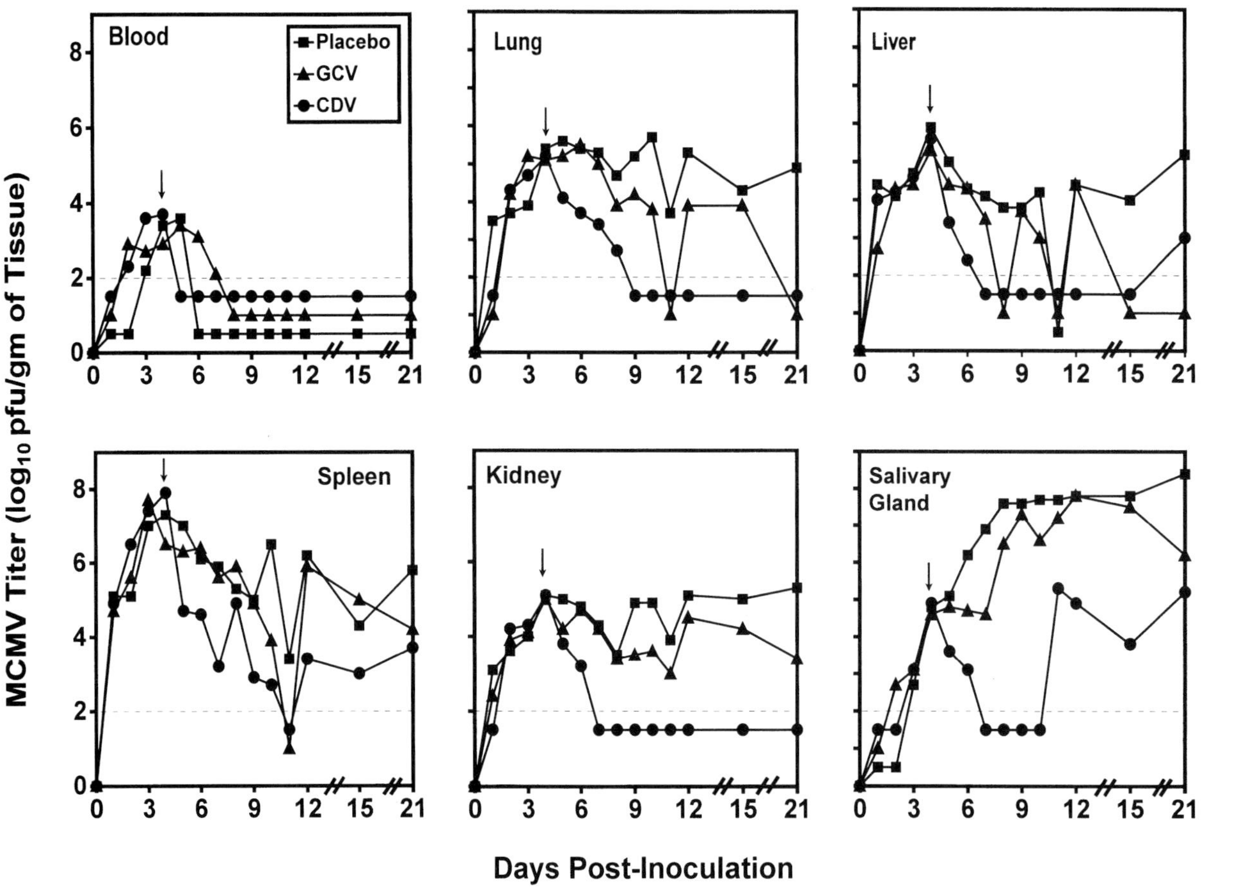

Figure 111.4 Effect of treatment with ganciclovir (GCV) or Cidofovir (CDV) on the replication of murine cytomegalovirus (MCMV) in target organs of Balb/c mice.

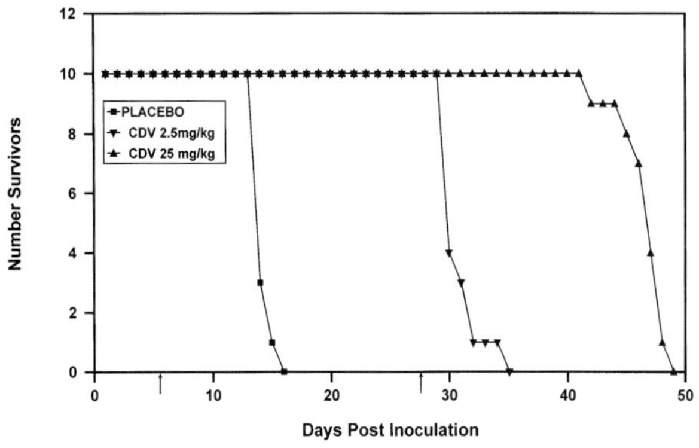

Figure 111.5 Effect of twice-weekly treatment with Cidofovir (CDV) on mortality of severe combined immunodeficient (SCID) mice inoculated with murine cytomegalovirus.

The effect of treatment with CDV on viral replication in target organs is summarized in Figure 111.6. Whereas placebo mice all have high levels of virus replicating in their tissues and subsequently die on about day 14, tissues from mice that received CDV had little or no MCMV in tissues other than their salivary glands.

Potential pitfalls (advantages/disadvantages) of the models

One of the major limitations to the model is that MCMV may not necessarily be susceptible to the same antiviral agents that human CMV is. It is very important, therefore, to determine the sensitivity of the virus in tissue culture prior to the initiation of animal experiments. Historically, MCMV has been at least as sensitive as, if not more so, than HCMV for GCV, PFA, and CDV, the three agents currently approved for use in treatment of CMV infection of humans. As with all murine models and other rodents, the adsorption, distribution, metabolism, excretion, and toxicity of antiviral agents will undoubtedly be different than observed in humans. These differences, however, have not detracted from the value of these models in the preclinical evaluation of new drugs prior to their use in humans.

Murine models are of particular value since, because of their small size and relatively low cost, large numbers can be used to determine optimal dosage concentrations and treatment regimen accurately and they require small amounts of drug. The MCMV model infections are particularly useful because they can be used not only as a lethal infection to determine the efficacy of a new drug rapidly,

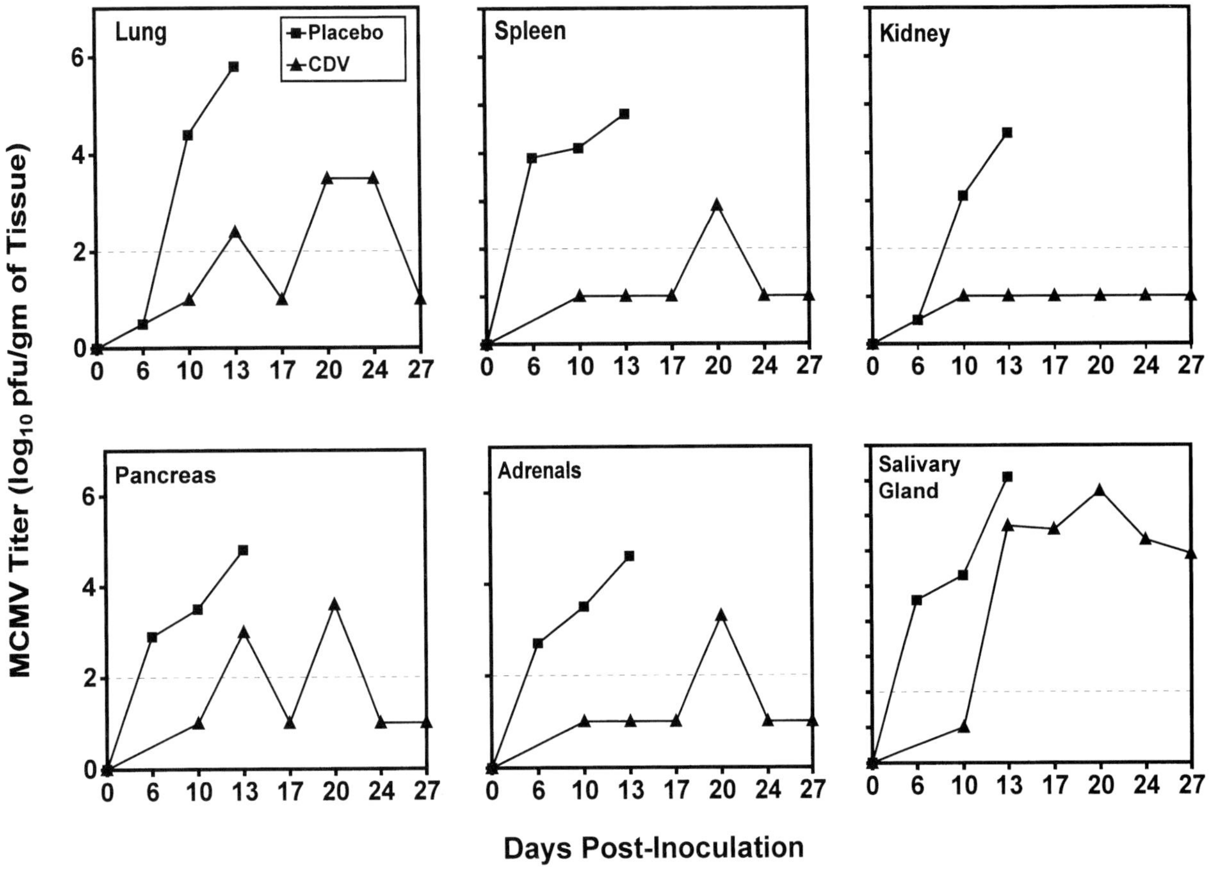

Figure 111.6 Effect of treatment with Cidofovir (CDV) on replication of murine cytomegalovirus in severe combined immunodeficient (SCID) mice.

but can also be utilized as a chronic infection to determine the effect of therapy on virus replication in the same tissues as are infected in humans. The model is particularly useful for determining the efficacy of long-term maintenance therapy. Although the quantitation of MCMV replication in target organs is quite labor-intensive, it provides a very sensitive system for evaluating efficacy in target tissues.

Although the SCID mice are relatively expensive and in short supply, they offer a model using a natural immunocompromised host that is not dependent on the use of chemical immunosuppressive agents. This insures that every animal is of uniform immune status and provides extremely consistent results. Overall, the MCMV infection of normal or SCID mice provides an excellent model for CMV infection of humans and its advantages as a model far outweigh the few disadvantages.

Contributions of the model and its predictability

Much of our understanding of the pathogenesis and latency of CMV infections in humans has come from the study of MCMV infections. The murine virus and the animal model closely resemble HCMV and many aspects of CMV infection of humans, so a lot of valuable information obtained in the animal can be extended to human disease.

Since there are only three antiviral drugs approved for treatment of CMV infection in humans, there is not a lot of information available to establish the predictability of the experimental infection. All three of these drugs, GCV, PFA, and CDV, were shown to be very effective in the MCMV infection and these models played a major role in the preclinical development of these therapies (Kern, 1997).

References

Alford, C. A. (1984). Chronic intrauterine and perinatal infections. In *Antiviral Agents and Viral Diseases of Man* (eds Galasso, G. J., Merigan T. C., Buchanan, R. A.), pp. 443–486. Raven Press, New York.

Bale, J. F. Jr., O'Neil, M. E., Hogan, R. N., Kern, E. R. (1984). Experimental murine cytomegalovirus infection of ocular structures. *Arch. Ophthalmol.*, **102**, 1214–1219.

Buhles, W. C., Mastre, B. J., Tinker, A. J. *et al.* (1988). Ganciclovir treatment of life- or sight threatening cytomegalovirus infection: experience in 314 immunocompromised patients. *Rev. Infect. Dis.*, **10** (suppl), S495–S506.

Chou, S., Marousek, G., Guentzel, S. *et al.* (1997). Evolution of mutations conferring multidrug resistance during prophylaxis and therapy for cytomegalovirus disease. *J. Infect Dis.*, **176**, 786–789.

Crumpacker, C., Marlowe, S., Zhang, J. L. *et al.* (1988). Treatment of cytomegalovirus pneumonia. *Rev. Infect. Dis.*, **10** (suppl), S538–S546.

Dieterich, D. T., Chachoua, A., Lafleur, F., Worrell, C. (1988). Ganciclovir treatment of gastrointestinal infections caused by cytomegalovirus in patients with AIDS. *Rev. Infect. Dis.*, **10** (suppl), S532–S537.

Drew, W. L., Miner, R. C., Busch, D. F. *et al.* (1991). Prevalence of resistance in patients receiving ganciclovir for serious cytomegalovirus infection. *J. Infect. Dis.*, **163**, 716–719.

Glasgow, L. A., Richards, J. T., Kern, E. R. (1982). Effect of acyclovir treatment on acute and chronic murine cytomegalovirus infection. Symposium on acyclovir. *Am. J. Med.*, **73**, 132–137.

Glenn, J. (1981). Cytomegalovirus infections following renal transplantation. *Rev. Infect. Dis.*, **3**, 1151–1178.

Ho, M. (1984). Systemic viral infections and viral infections in immunosuppressed patients. In *Antiviral Agents and Viral Diseases of Man* (eds Galasso, G. J., Merigan, T. C., Buchanan, R. A.), pp. 487–516. Raven Press, New York.

Ho, M. (1991). Murine cytomegalovirus. In *Cytomegalovirus: Biology and Infections*, pp. 327–353. Plenum, New York.

Jacobson, M. A. (1989). Cytomegalovirus disease in acquired immunodeficiency disease. In *Opportunistic Infections in Patients with the Acquired Immunodeficiency Syndrome* (eds Leoung, G., Mills, J.), pp. 195–213. Marcel Dekker, New York.

Jacobson, M. A. (1997). Treatment of cytomegalovirus retinitis in patients with the acquired immunodeficiency syndrome. *Drug Ther.*, **337**, 105–114.

Jacobson, M. A., O'Donnell, J. J. (1991). Approaches to the treatment of cytomegalovirus retinitis: ganciclovir and foscarnet. *J. Acquired Immune Defic.*, **4** (suppl), S11–S15.

Kern, E. R. (1991). The value of animal models to evaluate agents with potential activity against human cytomegalovirus. *Transplant Proc.*, **23** (suppl. 3), 152–155.

Kern, E. R. (1997). Preclinical evaluation of antiviral agents. In *Antiviral Agents and Human Viral Diseases* (eds Galasso, G. J., Whitley, R. J., Merigan, T. C.), pp. 79–111. Lippincott-Raven, Philadelphia.

Lalezari, J. P., Stagg, R. J., Kuppermann, B. D. *et al.* (1997). Intravenous cidofovir for peripheral cytomegalovirus retinitis in patients with AIDS. A randomized controlled trial. *Ann. Intern. Med.*, **126**, 257–263.

Li, S. B., Yang, Z. H., Feng, J. S., Fong, C. K. Y., Lucia, H. L., Hsiung, G. D. (1990). Activity of (S)-1-(3-hydroxy-2-phosphonylmethoxypropyl) cytosine (HPMPC) against guinea pig cytomegalovirus infection in cultured cells and in guinea pigs. *Antiviral Res.*, **13**, 237–252.

Mills, J., Jacobson, M. A., O'Donnell, J. J., Cederberg, D., Holland, G. N. (1988). Treatment of cytomegalovirus retinitis in patients with AIDS. *Rev. Infect. Dis.*, **10** (suppl), S522–S531.

Mintz, L., Drew, W. L., Miner, R. C., Braff, E. H. (1983). Cytomegalovirus infections in homosexual men: an epidemiologic study. *Ann. Intern. Med.*, **98**, 326–329.

Neyts J., Balzarini, J., Naesens, L., De Clercq, E. (1992). Efficacy of (S)-1-(3-hydroxy-2-phosphonylmethoxypropyl)cytosine and 9-(1,3-dihydroxy-2-propoxymethyl) guanine for the treatment of murine cytomegalovirus infection in severe combined immunodeficiency mice. *J. Med. Virol.*, **37**, 67–71.

Overall, J. C. Jr, Kern, E. R., Glasgow, L. A. (1976). Effective antiviral chemotherapy in cytomegalovirus infection of mice. *J. Infect. Dis.*, **133** (suppl), A237–A244.

Polis, M. A., Spooner, K. M., Baird, B. F. *et al.* (1995). Anticytomegaloviral activity and safety of cidofovir in patients with human immunodeficiency virus infection and cytomegalovirus viruria. *Antimicrob. Agents Chemother.*, **39**, 882–886.

Quinnan, G. V. Jr, Masur, H., Rook, A. H. *et al.* (1984). Herpesvirus infections in the acquired immune deficiency syndrome. *J.A.M.A.*, **252**, 72–77.

Rand, K. H., Rasmussen, L. E., Pollard, R. B., Arvin, A., Merigan T. C. (1976). Cellular immunity and herpesvirus infections in cardiac-transplant patients. *N. Engl. J. Med.*, **296**, 1372–1377.

Rawlinson, W. D., Farrell, H. E., Barrell B. G. (1996). Analysis of the complete DNA sequence of murine cytomegalovirus. *J. Virol.*, **70**, 8833–8849.

Reichert, C. M., O'Leary, T. J., Levens, D. L., Simrell, C. R., Macher, A. M. (1983). Autopsy pathology in the acquired immune deficiency syndrome. *Am. J. Pathol.*, **12**, 357–382.

Reynolds, D. W., Stagno, S., Stubbs, K. G. *et al.* (1974). Inapparent congenital cytomegalovirus infection with elevated cord IgM levels. *N. Engl. J. Med.*, **290**, 291–296.

Safrin, S., Cherrington, J., Jaffe, H. S. (1997). Clinical uses of cidofovir. *Rev. Med. Virol.*, **7**, 145–156.

Shanley, J. D., Pesanti, E. L. (1985). The relation of viral replication to interstitial pneumonitis in murine cytomegalovirus lung infection. *J. Infect. Dis.*, **151**, 454–458.

Smee, D. F., Colletti, A., Alaghamanda, H. A., Allen, L. B. (1989). Evaluation of continuous cell lines in antiviral studies with murine cytomegalovirus. *Arch. Virol.*, **107**, 253–260.

Smee, D. F., Morris, J. L. B., Leonhardt, J. A., Mead, J. R., Holy, A., Sidwell, R. W. (1992). Treatment of murine cytomegalovirus infections in severe combined immunodeficient mice with ganciclovir, (S)-1-[3-hydroxy-2-(phosphonylmethoxy) propyl] cytosine, interferon and bropiramine. *Antimicrob. Agents Chemother.*, **36**, 1837–1842.

Smith, I. L., Cherrington, J. M., Jiles, R. E., Fuller, M. D., Freeman, W. R., Spector, S. A. (1997). High-level resistance of cytomegalovirus to ganciclovir is associated with alterations in both UL97 and DNA polymerase genes. *J. Infect. Dis.*, **176**, 69–77.

Snoeck, R., Sakuma, T., De Clercq, E., Rosenberg, I., Holy A. (1988). (S)-1-(3-hydroxy-2-phosphonylmethoxypropyl) cytosine, a potent and selective inhibitor of human cytomegalovirus replication. *Antimicrob. Agents Chemother.*, **32**, 1839–1844.

Spector, S. A., Weingeist, T., Pollard, R. B. *et al.* (1993). A randomized controlled study of intravenous ganciclovir therapy for cytomegalovirus peripheral retinitis in patients with AIDS. *J. Infect. Dis.*, **168**, 557–563.

Stals, F. S., De Clercq, E., Bruggeman, C. A. (1991). Comparative activity of (S)-1-(3-hydroxy-2-phosphonylmethoxypropyl) cytosine and 9-(1,3-dihydroxy-2-propoxymethyl) guanine against rat cytomegalovirus infection *in vitro* and *in vivo*. *Antimicrob. Agents Chemother.*, **35**, 2262–2266.

Walmsley, S. L., Chew, E., Read, S. E. *et al.* (1988). Treatment of cytomegalovirus retinitis with trisodium phosphonoformate hexahydrate (foscarnet). *J. Infect. Dis.*, **157**, 569–572.

Winston, D. J., Ho, W. G., Bartoni, K. *et al.* (1988). Ganciclovir therapy for cytomegalovirus infections in recipients of bone marrow transplants and other immunosuppressed patients. *Rev. Infect. Dis.*, **10** (suppl), S547–S553.

Chapter 112

Animal Models for Cytomegalovirus Infection: Guinea-pig CMV

D. I. Bernstein and N. Bourne

Introduction

Cytomegalovirus (CMV) infections are common throughout the world but only rarely lead to serious disease in immunocompetent subjects (Eddleston *et al.*, 1997). Indeed, most CMV infections are subclinical. As with other herpesvirus infections, primary infection is followed by persistent and/or recurrent infections, as well as less frequent reinfections. CMV is species-specific so that replication of human CMV (hCMV) only occurs in human cells. Thus, most animal models utilize species-specific CMV that are similar but not identical to hCMV—a distinct drawback of any animal model, as discussed below.

CMV infections are of clinical importance for two reasons. Infection of pregnant women can lead to congenital infection of the unborn fetus which can have devastating results, especially if the pregnant woman is experiencing a primary infection (reviewed in Alford *et al.*, 1990; Demmler, 1996). Approximately 1% of all fetuses are infected *in utero*, making it the most common congenital infection; and 10–15% will develop significant sequelae including death, mental retardation, and most commonly, sensorineural hearing deficits. Second, infection of immunocompromised patients, including solid organ and bone marrow transplant recipients, as well as human immunodeficiency virus (HIV)-infected patients, can have severe consequences (reviewed in Smith and Brennessel, 1994; Hibberd and Snydman, 1995; Zaia, 1996). Infections are most severe when immunocompromised subjects experience a primary infection, but disease due to persistent or reinfections also occur. Interstitial pneumonia following transplantation and retinitis leading to blindness in HIV-infected patients are two of the most common serious infections in these populations.

Background

The development of guinea-pig models of CMV (gpCMV) have been guided by the clinical manifestations described above. Thus, two basic models have been developed to study the pathogenesis of CMV disease and to examine the effects of antivirals or vaccines. In one, pregnant dams are infected, resulting in infection and frequently death of the fetus. In the other, immunocompromised animals are infected with virulent gpCMV, leading to the death of the animal. Other models, such as infection of the cochlea, mimick the major outcome of congenital CMV infection, deafness, and have been used more rarely. In this chapter we will detail some of these models and describe their use in the evaluations of antivirals and vaccines.

gpCMV was identified over 75 years ago and the historical background is well-reviewed (Bia *et al.*, 1983). The original observations centered on the identification of intranuclear inclusions in the salivary glands of guinea-pigs (Jackson, 1920) and were followed shortly by the recognition that these were the result of a virus infection (Kuttner, 1927). In 1936, the susceptibility of the guinea-pig fetus to *in utero* infection was demonstrated (Markham and Hudson, 1936). By 1957 gpCMV had been isolated from infected salivary glands with the use of explants from guinea-pig embryos and this virus was provided to the American Type Tissue Culture Collection (ATCC 22122), where it serves as the source for most investigators working with gpCMV today (Hartley *et al.*, 1957).

The initial study of the molecular biology of gpCMV was published in 1984 (Isom *et al.*, 1984). The virus shares DNA homology (Isom *et al.*, 1984; Schleiss, 1994, 1995) with hCMV and antibody to one virus cross-reacts with proteins of the other (Kacica *et al.*, 1990). The major neutralization target for CMV antibody, gB, also shares homology at the amino acid level (Schleiss, 1994), but the viruses reportedly do not share cross-reactive neutralizing epitopes (Adler *et al.*, 1995).

Congenital model

The discoveries described above and the increased understanding of the role of CMV in congenital infections led, in 1978–1979, to the development by several investigators of the guinea-pig model of congenital CMV infection (Choi and Hsiung, 1978; Kumar and Nankervis, 1978; Johnson and Connor, 1979). Subsequent studies have increased our understanding of the pathogenesis and immunology of this

Handbook of Animal Models of Infection
ISBN 0-12-775390-7

common congenital infection. The later development of a model of sensorineural hearing loss in the guinea-pig further increased our knowledge of this disease (Woolf, 1991).

Immunocompromised model

At about the same time, other guinea-pig models of mononucleosis (Griffith *et al.*, 1981), pneumonia (Bia *et al.*, 1982), and disseminated infection in immunocompromised animals (Aquino-de Jesus and Griffith, 1989) were described. The most frequently used models are those which utilize immunosuppressive agents like cyclosporin A or cyclophosphamide to compromise the animals so that infection produces a disseminated infection and frequently death.

Methods

Preparation of animals

It is important to screen animals for CMV antibody before their use to insure that they have not previously been infected with gpCMV. CMV screening is included in some but not all commercial breeding facilities. The volume of blood (~0.5 ml) required to provide serum for screening can be easily obtained by toenail clipping, with the cut nail being coated with ferric subsulfate once bleeding is complete. Screening is performed using an enzyme-linked immunosorbent assay (ELISA) with solubilized gpCMV-infected tissue culture as the capture antigen, and uninfected tissue culture as the control (Bratcher *et al.*, 1995). When larger volumes of blood are required they are readily obtained by intracardiac puncture.

In our studies of immunocompromised animals, 1 day prior to virus challenge animals receive an initial immunosuppressive treatment of 100 mg/kg cyclophosphamide by intraperitoneal (i.p.) injection. A second dose of 50 mg/kg cyclophosphamide is administered 6 days after virus inoculation (see below for other immunosuppressive regimens). Animals are not anesthetized prior to administration of cyclophosphamide or virus.

Storage and preparation of inocula

gpCMV (strain 22122, American Type Culture Collection, Rockville, MD) requires sequential *in vivo* passage to attain the virulence required for studies of mortality in either immunocompromised or pregnant animals. Initial tissue culture stocks of the virus are prepared on either primary fetal guinea-pig cells, or guinea-pig lung cells (ATCC CCL 158). These stocks are then used undiluted as the inoculum to derive the initial *in vivo* virus stock which is prepared as a clarified salivary gland homogenate supernatant, aliquoted and stored frozen (−80°C) prior to use. In our experience, a number of factors can influence the outcome of infection with gpCMV, including the number of *in vivo* passages that the virus has undergone (virulence increases with increasing passage number), the virus titer of the homogenate, and the use of inbred (strain 2, strain 13, JY-9) or outbred (Hartley) guinea-pigs. It should be noted that stored frozen virus declines in titer with time, and retitration prior to the initiation of new studies is advisable.

To prepare *in vivo* passaged virus, small groups (4–6 animals) of inbred strain 2 guinea-pigs (~400 g) are inoculated by subcutaneous injection with undiluted gpCMV either prepared in tissue culture or from previous salivary gland passaged stocks. Between 21 and 28 days later animals are sacrificed, the salivary glands harvested aseptically and placed immediately on ice. While on ice, the tissues are pooled, finely minced and homogenized in a glass Dounce homogenizer. The final volume is adjusted to ~50 ml with cold sterile phosphate-buffered saline (PBS) and the resulting suspension clarified by low-speed centrifugation for 10 minutes at 4°C. The supernatant is removed, aliquoted and immediately frozen (−80°C). Dimethyl sulfoxide (DMSO) can be added to the virus pool before it is aliquoted and frozen.

Infection process

Animals are inoculated by either intraperitoneal or subcutaneous injection. Subcutaneous inoculation is recommended for pregnant animals to avoid fetal injection. If the subcutaneous route is used, the inoculum is administered in the ventral neck. For either route, the virus inoculum is given in a volume of 0.5–1.0 ml.

In our hands, an inoculum of 5.7 $\log_{10}$ plaque-forming units (pfu) of virus which has undergone ~7 passages *in vivo* is sufficient to produce nearly universal mortality in Hartley-strain guinea-pigs immunocompromised with the use of cyclophosphamide as described above.

Key parameters

Congenital CMV

The guinea-pig is the only small animal model of CMV in which virus crosses the placenta to infect and damage the fetus. The guinea-pig placenta resembles the human placenta with a single trophoblast layer (Enders, 1965). Further, guinea-pigs have a relatively long gestation of 65–70 days, allowing investigations in a trimester-like basis (Choi and Hsiung, 1978; Griffith and Hsiung, 1980; Kumar and Prokay, 1983; Harrison and Myers, 1990). Primary infection of pregnant animals results in a vertical transmission rate of 40–80%, similar to the infection rate in humans.

The highest vertical transmission rates and most severe outcomes (highest newborn mortality) occur with gpCMV inoculation during late gestation. Infection in early pregnancy or just prior to conception is associated with a high rate of fetal resorptions (Harrison and Myers, 1990). Congenital infection of pups has also been associated with significant auditory deficits and identification of CMV antigen in the inner ear (Woolf *et al.*, 1989; Woolf, 1991).

The outcome of maternal challenge depends on the timing of challenge (Choi and Hsiung, 1978; Griffith and Hsiung, 1980; Kumar and Prokay, 1983; Harrison and Myers, 1990), as well as on both the virulence and titer of the inocula (Bratcher *et al.*, 1995). As discussed above, gpCMV increases in virulence with each subsequent passage *in vivo*. If high-enough titers of a virulent salivary gland virus are used, most pregnant dams will not survive to delivery of their pups. The duration and titer of gpCMV in the blood and selected organs, including the lungs, spleen, kidney and salivary gland, can be followed and the impact of treatment evaluated. Virus is also easily isolated from the placentas of infected animals (Choi and Hsiung, 1978; Griffith *et al.*, 1980b, 1990; Goff *et al.*, 1987).

Pup mortality and infection are the most common endpoint for evaluations of interventional strategies. Virus can be isolated from pups at about a week following maternal inoculation and frequently persist through the newborn period. Virus is most often recovered from the spleen and salivary gland but can also be recovered from the blood, liver, lungs, kidneys and brain of infected newborns (Griffith and Hsiung, 1980; Griffith *et al.*, 1982; Harrison and Myers, 1990). More recently polymerase chain reaction (PCR) assays have been developed for identification of gpCMV in blood and solid organs (Bratcher *et al.*, 1995; Harrison *et al.*, 1995). Pup mortality manifests in three ways. Pups are most commonly stillborn following maternal challenge in the second or third trimester while fetal resorptions, determined by comparing ultrasound determination of the number of fetuses to those delivered, is common following early challenge (Harrison and Myers, 1990). Perinatal deaths, within 7 days of delivery, and intrauterine growth retardation and, thus, decreased birth weights, have also been reported (Harrison and Myers, 1990).

Immunocompromised model

Following gpCMV inoculation, adult immunocompetent outbred Hartley guinea-pigs experience transient splenomegaly, a self-limited viremia and peripheral blood mononucleosis similar to that observed in humans (Griffith *et al.*, 1981). Inbred strain 2 animals are more susceptible to gpCMV infection with mortality rates as high as 44% following infection with salivary gland-passaged virus (Bia *et al.*, 1982). Some investigators have relied on weight loss, spleen index measurement and viral titers in the blood and internal organs of immunocompetent animals to evaluate treatment strategies (Fong *et al.*, 1987; Yang *et al.*, 1989). Others have used immunosuppressive agents in attempts to increase the severity of CMV infection in guinea-pigs and more closely mimick the clinical situations in which CMV infections are important. Bia *et al.* found that a 14-day regimen of cyclosporin A treatment increased the rate of virus isolation from the internal organs of gpCMV-infected strain 2 and Hartley guinea-pigs compared to controls despite similar rates of viremia (Bia *et al.*, 1985). In 1989, Aquino-de Jesus and colleagues evaluated several cyclophosphamide (CY) treatment regimens in guinea-pigs to produce a model of acute CMV infection in the immunocompromised host (Aquino-de Jesus *et al.*, 1989). Using a single dose of 300 mg/kg, they were able to produce universal mortality in CY-treated, gpCMV-infected Hartley-strain guinea-pigs without producing increased mortality in animals that received CY alone. In our hands, a single dose of CY is insufficient to produce long-lasting immunosuppression in Hartley guinea-pigs. We have evaluated several other regimens and found that a dose of 100 mg/kg given 1 day prior to virus challenge followed by a dose of 50 mg/kg on day 6 post-challenge maintained prolonged immunosuppression. Using this regimen, we achieve a nearly universal mortality in Hartley guinea-pigs inoculated with salivary gland-derived gpCMV with minimal mortality in non-infected, immunocompromised animals (Bourne *et al.*, 1997a).

The principal clinical endpoint used in immunocompromised animals is mortality and mean number of days to death. Animals are also monitored daily for evidence of fever and weight loss. Blood samples can be collected to monitor for evidence of hepatitis using serum chemistries (serum glutamic oxaloacetic or pyruvic transaminase). Peak hepatitis normally occurs at about 7–10 days post-virus inoculation. As for CMV antibody screening, the volume of blood required is easily obtained by toenail clipping.

The model also provides direct virologic endpoints. Blood samples can be obtained at various times to measure the magnitude and duration of viremia. Viral quantification of visceral organ infection (liver, lung, spleen, and salivary gland) can also be used to monitor infection and evaluate the efficacy of therapy. If animals are to be sacrificed, the spleen weight index provides another useful endpoint.

We typically have found that quantification of organ virus titers on days 7 and 10 post-inoculation provides a good measure of infection and the effects of intervention. More recently, PCR assays have been developed for identification of gpCMV in blood and solid organs (Bratcher *et al.*, 1995; Harrison *et al.*, 1995).

Pitfalls (advantages/disadvantages of the model)

The chief drawback to the guinea-pig model of CMV infection is that the species-specific nature of CMV requires that the model utilizes gpCMV rather than the human virus. Therefore, in experiments to evaluate new antiviral compounds, it is important to conduct pre-screening *in vitro* to insure that agents with promising activity against hCMV have similar *in vitro* activity against gpCMV.

The need to use *in vivo* passaged virus as the inoculum in these models presents another drawback. Production of serially passaged virus is a time-consuming process that can take several months to prepare a suitably passaged stock. Once the stock virus is prepared it is important to establish its virulence and to retiter it periodically since titer declines with storage even at low temperature (−80°C).

Contributions of models

Congenital model

The model of congenital CMV has been used to investigate both active and passive immune protection by several investigators (Bia *et al.*, 1980, 1984; Bratcher *et al.*, 1995; Harrison *et al.*, 1995), but has not been used in antiviral investigations. We have shown that high-titered anti-CMV antibody given prior to or just after CMV challenge can protect the mother and fetus from intrauterine death but could not prevent infection of the dams or pups when challenged in the late second to early third trimester (Table 112.1). Administration of antibody prior to challenge protected against both high- and low-dose challenge while administration after challenge protected only against low-dose challenge. Previously, Bia *et al.* (1980) had also shown that antibody could protect against pup mortality, but they also found some protection against maternal viremia and disseminated infection. Recently, Chatterjee *et al.* reported that using an inbred JY-9 model and first-trimester challenge antibody given after maternal challenge decreased the duration of maternal viremia and *in utero* fetal deaths and also increased the weight of the pups *in utero* (Chatterjee *et al.*, 1997).

Both live attenuated viruses (Bia *et al.*, 1980, 1984) and subunit vaccines (Bia *et al.*, 1980; Harrison *et al.*, 1995; Bourne *et al.*, 1997b) have been shown to be effective in this model. In an early investigation, animals given a live, attenuated (serial passages in tissue culture) virus developed antibody and were protected from acute viremia, death and disseminated infection in the pups (Bia *et al.*, 1980, 1984). More recently, Harrison *et al.* (1995) and our group (Bourne *et al.*, 1997b) described the use of a subunit vaccine. Using their inbred JY-9 model and challenge in the first trimester, Harrison *et al.* showed that vaccination with a purified CMV glycoprotein B homolog induced neutralizing antibody and lymphocyte proliferation responses and reduced the number of days of maternal viremia (17.3 versus 3.3; Table 112.2). Further, immunization significantly improved pup survival (48 versus 95%), increased the mean birth

Table 112.1 Effect of passive antibody on maternal and congenital guinea-pig cytomegalovirus (gpCMV) infection

	Maternal mortality	*Pup mortality*	*Pup infection*
High dose (5.2 log_{10} pfu)			
Negative sera	3/5	19/22 (86%)	18/21 (86%)
Pre-inoculation antibody	1/7	20/42* (48%)	32/35 (91%)
Post-inoculation antibody	4/5	28/28	25/28 (89%)
Low dose (3.5 log_{10} pfu)			
Negative sera	0/8	22/45 (49%)	2/20 (10%)
Pre-inoculation antibody	0/8	10/43* (23%)	4/21 (19%)
Post-inoculation antibody	0/8	8/42* (19%)	8/20 (40%)

* Significantly different from control: *P* <0.05.
Pfu, Plaque-forming units.
Pregnant guinea-pigs were given three doses of 5 ml of guinea-pig anti-gpCMV antibody on either days −3, −1 and 7 or 1, 3 and 7 in relation to maternal CMV challenge. Animals were then challenged in the early second to late third trimester with high or low doses of virulent salivary gland-passaged (SG-18) virus.

Table 112.2 Effect of glycoprotein B (gB) immunization on congenital guinea-pig cytomegalovirus (gpCMV) infection

	Infected	*Stillborn*	*Survivor*	*Mean weight (g)*
Not immunized	12/23 (52%)*	10/23 (43%)*	11/23 (48%)*	74.3*
Immunized	4/22 (18%)	1/22 (5%)	21/22	99.1

* $P < 0.05$.
Female Jy-9-strain guinea-pigs were immunized with 25 µg of gp 60–90 complex plus CFA followed by a repeat dose with incomplete FA and then bred. On days 21–25 of pregnancy, animals were challenged with 6.2×10^4 plaque-forming units (pfu) of a virulent SG7 passage gpCMV.

weight (74.3 versus 99.1 g), and decreased the infection rate (52 versus 18%). In our studies using purified gpCMV glycoproteins (Bourne *et al.*, 1997b), we found high levels of antibody in all immunized animals and significant protection from pup mortality (52 versus 18%).

Immunocompromised model

Guinea-pig models of CMV infection have been used to evaluate a number of antiviral agents, immunomodulators and combination therapies (Lucia *et al.*, 1984; Fong *et al.*, 1987; Chen *et al.*, 1988; Yang *et al.*, 1989; Li *et al.*, 1990; Yamamoto *et al.*, 1990). The efficacy of 9-(1,3-dihydroxy-2-phosphonylmethyl) guanine (DHPG) treatment has been examined in both immune-competent and immunocompromised animals (Fong *et al.*, 1987), where it provided only a modest antiviral effect. However, *in vitro* data revealed that gpCMV is less susceptible to DHPG than human CMV, thus predicting a decreased efficacy. This emphasizes the fact that the model employs a virus other than the human one and, thus, care must be taken when interpreting results. DHPG also had some effect when evaluated in the guinea-pig model of CMV labyrinthitis (Woolf *et al.*, 1988).

The guinea-pig model has also been useful in predicting the toxicity of certain antiviral compounds. Thus, while (s)-1-(3-dihydroxy-2-phosphonyl methoxypropyl) cytosine (HPMPC) is active against hCMV and has shown similar activity against gpCMV *in vitro*, evaluation *in vivo* revealed significant renal toxicity in guinea-pigs (Li *et al.*, 1990; Bravo *et al.*, 1993). Renal toxicity due to HPMPC was later identified as a significant problem in patients. The cyclic derivative of HPMPC (CHPMPC), however, has reduced toxicity in guinea-pigs; and we have recently evaluated this compound against gpCMV infection in immunocompromised animals (Bourne *et al.*, 1997a). Guinea-pigs were treated either once daily for 7 days with 5 mg/kg per day or received two doses of 17.5 mg/kg administered on days 1 and 4 post-inoculation.

The results of two studies are shown in Table 112.3. In the first study, each of the two antiviral treatment regimens provided significant protection against mortality. Overall, only 1 of 32 treated animals died compared to 13 of 15 controls ($P < 0.001$). In the second study, only the daily treatment regimen was evaluated. In this experiment, mortality in the control group was unexpectedly low and we were unable to show significant protection. However, when the results of the two daily treatment experiments were combined, CHPMPC treatment provided significant protection against mortality. Among animals that died there was no difference in the mean day of death between treated animals (16.6 ± 4.0) and controls (16.3 ± 2.2).

Table 112.3 Effect of CHPMPC on mortality in cytomegalovirus (CMV)-infected immunocompromised guinea-pigs

	Group	*n*	*Mortality*
Experiment 1	Placebo	15	13 (87%)
	CHPMPC (q.d.)	16	1 (6%)*
	CHPMPC (2×/week)	16	0*
Experiment 2	Placebo	10	6 (60%)
	CHPMPC (q.d.)	11	4 (36%)
Combined	Placebo	25	19 (76%)
	CHPMPC (q.d.)	27	5 (19%)*

* $P < 0.001$ versus placebo. Immunocompromised Hartley-strain guinea-pigs were treated with CHPMPC (see below) or placebo daily or twice (day 1 and 4) following gpCMV challenge.

We also evaluated the effect of CHPMPC treatment on virus replication in three tissues in which CMV disseminates—liver, lung and spleen—in the two studies described above. Animals were sacrificed on days 7 (the last day of therapy) or day 10 (3 days after discontinuation of drug) post-virus challenge. There were no significant differences comparing the two treatments or the two studies and so results have been combined (Table 112.4). On day 7 post-inoculation, the incidence of virus recovery from the spleen was significantly reduced and from the liver marginally reduced in treated animals. In contrast, there was no reduction in the incidence of virus recovery from lungs. When all tissues were combined, CHPMPC treatment significantly reduced the incidence of virus recovery. In contrast, by day 10 post-inoculation treatment did not significantly reduce the incidence of virus isolation from any of the three tissues or the overall incidence of isolation

Table 112.4 Effect of CHPMPC viral recovery guinea-pig cytomegalovirus (gpCMV) infected in immunocompromised guinea-pigs

			*% Recovered (Titer)**		
	Group	*n*	*Liver*	*Lung*	*Spleen*
Day 7	Placebo	9	33% (0.33)	100% (2.4)	89% (2.0)
	CHPMPC	13	0	92% (2.2)	15%† (0.3†)
Day 10	Placebo	8	75% (1.9)	100% (3.9)	100% (3.3)
	CHPMPC¶	13	46% (0.8‡)	100% (3.2)	85% (2.0‡)

* Geometric mean titer $\log_{10}$.
† $P < 0.005$ versus placebo.
‡ $P < 0.05$ versus placebo.
¶ C is for cyclic compound see above.

when the results for all tissues were combined. Further, on both days 7 and 10, virus titers in the spleens of treated animals were lower than those in controls, while virus replication in the liver of treated animals was significantly reduced on day 10. Virus was not isolated from the liver of any treated animal. In lung tissue, CHPMPC failed to decrease virus replication.

References

Adler, S. P., Shaw, K. V., McVoy, M. *et al.* (1995). Guinea pig and human cytomegaloviruses do not share cross-reactive neutralizing epitopes. *J. Med. Virol.*, **47**, 48–51.

Alford, C. A., Stagno, S., Pass, R. F. *et al.* (1990). Congenital and perinatal cytomegalovirus infections. *Rev. Infect. Dis.*, **7**, S745–S753.

Aquino-de Jesus, M. J., Griffith, B. P. (1989). Cytomegalovirus infection in immunocompromised guinea pigs: a model for testing antiviral agents *in vivo*. *Antiviral Res.*, **12**, 181–193.

Bia, F. J., Griffith, B. P., Tarsio, M. *et al.* (1980). Vaccination for the prevention of maternal and fetal infection with guinea pig cytomegalovirus. *J. Infect. Dis.*, **142**, 732–738.

Bia, F. J., Lucia, H. L., Fong, C. K. *et al.* (1982). Effects of vaccination on cytomegalovirus-associated interstitial pneumonia in strain 2 guinea pigs. *J. Infect. Dis.*, **145**, 742–747.

Bia, F. J., Griffith, B. P., Fong, C. K. *et al.* (1983). Cytomegaloviral infections in the guinea pig: experimental models for human disease. *Rev. Infect. Dis.*, **5**, 177–195.

Bia, F. J., Miller, S. A., Lucia, H. L. *et al.* (1984). Vaccination against transplacental cytomegalovirus transmission: vaccine reactivation and efficacy in guinea pigs. *J. Infect. Dis.*, **149**, 355–362.

Bia, F. J., Lucia, H. L., Bia, M. J. *et al.* (1985). Effects of cyclosporine on the pathogenesis of primary cytomegalovirus infection in the guinea pig. *Intervirology*, **24**, 154–165.

Bourne, N., Hitchcock, M., Bernstein, D. I. (1997a). Cyclic HPMPC is safe and highly effective against guinea pig cytomegalovirus infection in immunocompromised animals. *Antiviral Res.*, **34**, A85 (abstract).

Bourne, N., Rosteck, R., Fox, D. *et al.* (1997b). Immunization with a cytomegalovirus (CMV) glycoprotein vaccine improves pregnancy outcome in an animal model of congenital CMV infection. *Abstracts of 36th ICAAC* (#H87), 179 (abstract).

Bratcher, D. F., Bourne, N., Bravo, F. J. *et al.* (1995). Effect of passive antibody on congenital cytomegalovirus infection in guinea pigs. *J. Infect. Dis.*, **172**, 944–950.

Bravo, F. J., Stanberry, L. R., Kier, A. B. *et al.* (1993). Evaluation of HPMPC therapy for primary and recurrent genital herpes in mice and guinea pigs. *Antiviral Res.*, **21**, 59–72.

Chatterjee, A., Harrison, C. J., Bewtra, C. *et al.* (1997). Effects of primary maternal CMV infection on early pregnancy and the immature placenta in guinea pigs. *Abstracts of 35th IDSA Annual Meeting* (#106), 91 (abstract).

Chen, M., Griffith, B. P., Lucia, H. L. *et al.* (1988). Efficacy of S26308 against guinea pig cytomegalovirus infection. *Antimicrob. Agents Chemother.*, **32**, 678–683.

Choi, Y. C., Hsiung, G. D. (1978). Cytomegalovirus infection in guinea pigs. II. Transplacental and horizontal transmission. *J. Infect. Dis.*, **138**, 197–202.

Demmler, G. J. (1996). Congenital cytomegalovirus infection and disease. *Adv. Pediatr. Infect. Dis.*, **11**, 135–162.

Eddleston, M., Peacock, S., Juniper, M. *et al.* (1997). Severe cytomegalovirus infection in immunocompetent patients. *Clin. Infect. Dis.*, **24**, 52–56.

Enders, A. C. (1965). A comparative study of the fine structure of the trophoblast in several hemochorial placentas. *Am. J. Anat.*, **116**, 29–67.

Fong, C. K., Cohen, S. D., McCormick, S. R. *et al.* (1987). Antiviral effect of 9-(1, 3-dihydroxy-2-propoxymethyl) guanine against cytomegalovirus infection in a guinea pig model. *Antiviral Res.*, **7**, 11–23.

Goff, E., Griffith, B. P., Booss, J. (1987). Delayed amplification of cytomegalovirus infection in the placenta and maternal tissues during late gestation. *Am. J. Obstet. Gynecol.*, **156**, 1265–1270.

Griffith, B. P., Hsiung, G. D. (1980). Cytomegalovirus infection in guinea pigs. IV. Maternal infection at different stages of gestation. *J. Infect. Dis.*, **141**, 787–793.

Griffith, B. P., Lucia, H. L., Bia, F. J. *et al.* (1981). Cytomegalovirus-induced mononucleosis in guinea pigs. *Infect. Immun.*, **32**, 857–863.

Griffith, B. P., Lucia, H. L., Hsiung, G. D. (1982). Brain and visceral involvement during congenital cytomegalovirus infection of guinea pigs. *Pediatr. Res.*, **16**, 455–459.

Griffith, B. P., Chen, M., Isom, H. C. (1990). Role of primary and secondary maternal viremia in transplacental guinea pig cytomegalovirus transfer. *J. Virol.*, **64**, 1991–1997.

Harrison, C. J., Myers, M. G. (1990). Relation of maternal CMV viremia and antibody response to the rate of congenital infection and intrauterine growth retardation. *J. Med. Virol.*, **31**, 222–228.

Harrison, C. J., Britt, W. J., Chapman, N. M. *et al.* (1995). Reduced congenital cytomegalovirus (CMV) infection after maternal immunization with a guinea pig CMV glycoprotein before gestational primary CMV infection in the guinea pig model. *J. Infect. Dis.*, **172**, 1212–1220.

Hartley, J. W., Rowe, W. P., Huebner, R. J. (1957). Serial propagation of the guinea pig salivary gland virus in tissue culture. *Proc. Soc. Exp. Biol. Med.*, **96**, 281–285.

Hibberd, P. L., Snydman, D. R. (1995). Cytomegalovirus infection in organ transplant recipients. *Infect. Dis. Clin. North Am.*, **9**, 863–877.

Isom, H. C., Gao, M., Wigdahl, B. (1984). Characterization of guinea pig cytomegalovirus DNA. *J. Virol.*, **49**, 426–436.

Jackson, L. (1920). An intracellular protozoan parasite of the ducts of the salivary glands of the guinea pig. *J. Infect. Dis.*, **26**, 347–350.

Johnson, K. P., Connor, W. S. (1979). Guinea pig cytomegalovirus: transplacental transmission. *J. Exp. Med.*, **59**, 263–267.

Kacica, M. A., Harrison, C. J., Myers, M. G. *et al.* (1990). Immune response to guinea pig cytomegalovirus polypeptides. *J. Med. Virol.*, **32**, 155–159.

Kumar, M. L., Nankervis, G. A. (1978). Experimental congenital infection with cytomegalovirus: a guinea pig model. *J. Infect. Dis.*, **138**, 650–654.

Kumar, M. L., Prokay, S. L. (1983). Experimental primary cytomegalovirus infection in pregnancy: timing and fetal outcome. *Am. J. Obstet. Gynecol.*, **145**, 56–60.

Kuttner, A. G. (1927). Further studies concerning the filterable virus present in the submaxillary glands of guinea pigs. *J. Exp. Med.*, **46**, 935–956.

Li, S. B., Yang, Z. H., Feng, J. S. *et al.* (1990). Activity of (S)-1-(3-

hydroxy-2-phosphonylmethoxypropyl) cytosine (HPMPC) against guinea pig cytomegalovirus infection in cultured cells and in guinea pigs. *Antiviral Res.*, **13**, 237–252.

Lucia, H. L., Griffith, B. P., Hsiung, G. D. (1984). Effect of acyclovir and phosphonoformate on cytomegalovirus infection in guinea pigs. *Intervirology*, **21**, 141–149.

Markham, F. S., Hudson, N. P. (1936). Susceptibility of the guinea pig fetus to the submaxillary gland virus of guinea pigs. *Am. J. Pathol.*, **12**, 175–182.

Schleiss, M. R. (1994). Cloning and characterization of the guinea pig cytomegalovirus glycoprotein B gene. *Virology*, **202**, 173–185.

Schleiss, M. R. (1995). Sequence and transcriptional analysis of the guinea pig cytomegalovirus DNA polymerase gene. *J. Gen. Virol.*, **76**, 1827–1833.

Smith, M. A., Brennessel, D. J. (1994). Cytomegalovirus (review). *Infect. Dis. Clin. North Am.*, **8**, 427–438.

Woolf, N. K. (1991). Guinea pig model of congenital CMV-induced hearing loss: a review. *Transplant. Proc.*, **23**, 32–34.

Woolf, N. K., Ochi, J. W., Silva, E. J. *et al.* (1988). Ganciclovir prophylaxis for cochlear pathophysiology during experimental guinea pig cytomegalovirus labyrinthitis. *Antimicrob. Agents Chemother.*, **32**, 865–872.

Woolf, N. K., Koehrn, F. J., Harris, J. P. *et al.* (1989). Congenital cytomegalovirus labyrinthitis and sensorineural hearing loss in guinea pigs. *J. Infect. Dis.*, **160**, 929–937.

Yamamoto, N., Yamada, Y., Daikoku, T. *et al.* (1990). Antiviral effect of oxetanocin G against guinea pig cytomegalovirus infection *in vitro* and *in vivo*. *J. Antibiotics*, **43**, 1573–1578.

Yang, Z. H., Lucia, H. L., Tolman, R. L. *et al.* (1989). Effect of 2′-nor-cyclic GMP against guinea pig cytomegalovirus infection. *Antimicrob. Agents Chemother.*, **33**, 1563–1568.

Zaia, J. A. (1996). Prophylaxis and treatment of CMV infections in transplantation. *Adv. Exp. Med. Biol.*, **394**, 117–134.

Chapter 113

Animal Models for Cytomegalovirus Infection: Rat CMV

F. S. Stals

Background of human infection

Cytomegalovirus (CMV) infection is ubiquitous in humans and it mostly runs an asymptomatic course in the immunocompetent host. In contrast, during immunocompromised conditions severe disease develops with high rates of morbidity and mortality. CMV occurs in about 50–60% of patients after organ and bone marrow transplantation (BMT; Applebaum *et al.*, 1982; Rasing *et al.*, 1990). After allogeneic BMT, CMV-induced interstitial pneumonitis (IP) is observed in 15–18% of cases and represents a major cause of death, with a mortality rate of up to 83% (Winston *et al.*, 1990; Ho, 1991).

Ganciclovir combined with immunoglobulin reduces mortality in human CMV pneumonitis in BMT recipients (Mills and Corey, 1986; Emanuel *et al.*, 1988; Reed *et al.*, 1988), and administration at the time that viral shedding is identified reduces the subsequent risk of CMV disease (Goodrich *et al.*, 1991; Ljungman *et al.*, 1992). However, the clinical use of this agent is limited by adverse reactions (Collaborative DHPG Treatment Study Group, 1986). Moreover, treatment of IP after allogeneic BMT remains a major clinical problem, partly because of the poor understanding of the pathogenesis of CMV-induced IP (Shanley *et al.*, 1982, 1987) and partly because of the unavailability of potent and selective inhibitors of human CMV replication *in vivo* (Balfour, 1990). The pathogenesis of disease is poorly understood and treatment of disease is limited. Therefore, we searched for an animal model in which the pathology of CMV infection resembles that in humans and which is suitable for studies on the treatment of CMV infections.

Background of model

Rat-specific cytomegalovirus strain Maastricht (further referred to as RCMV) was isolated from the salivary glands of wild brown rats (Bruggeman *et al.*, 1982, 1983) and characterized morphologically and biochemically (Bruggeman *et al.*, 1983; Meijer *et al.*, 1984, 1986). Infection of rats with RCMV results in asymptomatic infection in immunocompetent rats, whereas under immunocompromised conditions severe symptoms develop. The behavior and reactivation of the virus were previously studied after several allogeneic organ transplants, such as kidneys, hearts and lungs, were carried out from a latently infected donor into a non-infected recipient rat (Bos *et al.*, 1989; Bruning *et al.*, 1989a,b; Lemstrom *et al.*, 1993, 1994; Yagyu *et al.*, 1993; Li *et al.*, 1995) and bone marrow (Bos *et al.*, 1989; Bruggeman, 1991).

Two models for CMV infection which have been developed in rats will be described in this chapter: in the first model generalized CMV infection was induced in immunocompromised rats (Stals *et al.*, 1990b) and in the second model interstitial lung disease (ILD) was induced in allogeneic BMT recipient rats (Stals *et al.*, 1993a). In the later model additional single-lung transplantation was performed to include an internal syngeneic control (Stals *et al.*, 1996b). In both infection models the pathogenesis of disease was studied. Additionally, the effect of different treatment modalities was studied in the two models.

Animal species

Eight-week-old inbred specified pathogen-free (SPF) male Brown Norway (BN) rats with a total body weight between 140 and 180 g were used in all experiments of generalized infection. These animals were also used for donor tissue and organ transplant experiments. Lewis and BN rats were used as donors for syngeneic and allogeneic transplants. This combination was used in bone marrow as well as lung transplantations. Probably more species combinations are possible, but they should be examined for the development of pneumonia and the amount of virus in the inoculum required (e.g., inflammatory parameters and virus load). Because of total body irradiation (TBI) standardization of body weight is essential. Animals were free from RCMV at the start of each experiment, as determined by enzyme immunosorbent assay.

Handbook of Animal Models of Infection
ISBN 0-12-775390-7

Preparation of animals

To prevent bacterial infections all animals received enrofloxacin 2 mg/kg per day (Baytril, Bayer, Mijdrecht, the Netherlands) and acidified drinking water (Stals *et al.*, 1990a; Vancutsem *et al.*, 1990) throughout the experiment, starting 1 week before the onset of the experiment. In addition, all animals were kept in sterile cages in groups of 2–3 animals depending on the design of the experiments and they received sterile food. (After virus inoculation, infected animals were kept separated from the non-infected ones.)

Immunosuppression

For immunosuppression rats received TBI 1 day before infection. In generalized infection experiments a sublethal dose of TBI (5.0 Gy) was used to perform a moderate degree of immunosuppression. TBI was performed with a Philips roentgen X-ray irradiation apparatus, type MU 15 F-225 kV. Rats were placed in a Perspex irradiation container and received their irradiation dose. The container had a 20 cm diameter and a height of 6 cm. The container has little holes for fresh air. Four to 5 rats could be placed in each container. The irradiation dose was measured in an ionization chamber (Ionex).

In the second model, IP was induced in allogeneic BMT recipient rats. One day before RCMV inoculation and BMT, animals received an otherwise lethal dose of TBI (e.g., 9.6 Gy). On the day of infection RCMV was administered intraperitoneally (i.p.), followed by intravenous (i.v.) administration of freshly harvested viable bone marrow cells (Bos *et al.*, 1989).

Preparation of donor bone marrow cells

Bone marrow cells from donors were prepared from the femur of donor rats as described earlier (Bos *et al.*, 1989; Stals *et al.*, 1993a). One day after 9.6 Gy TBI and 6 hours post-infection (p.i.), each rat received 5×10^7 viable cells intravenously. Donor animals were killed by dislocation of the cervical spinal cord, while anesthetized. Femur and tibia were aseptically removed. Bone marrow cells were harvested. After recipient rats received ether anesthesia, 5×10^7 viable bone marrow cells in 1 ml of cell culture medium were injected into the dorsal vein of the penis.

Lung transplantation

Animals received Nembutal anesthesia and atrophine sulfate 50 μg intramuscularly (i.m.) to prevent airway secretion during transplantation.

Donor procedure

The animal is not intubated. Bilateral subcostal laparotomy is performed to achieve access to the inferior vena cava. Heparin 1000 IU is administered into the vena. Section of the diaphragm to both sides is performed. Under visualization of the thoracic organs careful section of the sternum up to the jugulum is performed. The lungs are exposed by retraction of the two parts of the thoracic wall. Cuts are made into the right and the left atrial appendages to eliminate systemic blood return and to allow free flow of the pulmonary perfusate, 10 ml Ringer's solution (4°C) administered at a pressure of 15 cmH_2O. The left lung is harvested by careful gross dissection of the left hilum. The organ is placed in 4°C Ringer's solution.

Recipient preparation

The animal is intubated transorally and positioned on its left side. Ventilation is presumed with oxygen. A lateral thoracotomy is undertaken by visualizing and dissecting the subcutaneous fat tissue, the lattisimus dorsi muscle, and several origins of the anterior serratus muscle. The thorax is entered via the fifth intercostal space. A rib retractor is positioned. The left native lung is trapped in gauze and the hilar structures (pulmonary vein, pulmonary artery and the bronchus) are separated. The bronchus is ligated by Mersilene size 5.0. A Castaneda vascular clamp is placed on the vessels and the native lung is eliminated.

Implantation

The donor lung is positioned in a cold wet swab opposite to the clamped recipient vessels. Then, pulmonary venous anastomoses are performed with a closed technique and continuous running suture with 10.0 Ethilone. The pulmonary artery anastomosis is performed by the same technique. Next, the vessels are unclamped to re-establish pulmonary perfusion (end of ischemic time). Finally, bronchial anastomosis is performed using the same technique. Transplant ventilation is presumed and bleeding sites are taken care of. Pulmonary drainage is performed by means of a 14 G Venflon through the sixth intercostal space. Closure of the thoracic wound is performed in three layers (ribcage, muscle, skin). Finally, the drain is removed after free thoracic air has been removed.

At the end of the procedure spontaneous ventilation will quickly start when oxygen supply is being withdrawn from the animal. Extubation is possible within a few minutes, as soon as pharyngeal reflexes have returned. The animal is placed in a provisional hyperbaric oxygen chamber until fully awake. Thereafter, the animal can return to the cage with free access to food and water *ad libitum*.

Anesthesia

For short-term anesthesia (for example, during intravenous injection of bone marrow cells) animals were placed in a closed glass container with diethylether (Merck, The Netherlands). For surgical procedures animals received pentobarbital (Narcovert, Apharma, Arnhem, The Netherlands), i.p. at 0.1 ml/100 g of total body weight. During lung transplantation animals received phenobarbital (Nembutal) at 50 mg/kg for anesthesia. The donor and recipient rats receive 100% and 80% of this dose, respectively. To minimize airway secretions atropine sulfate is administered to both animals at a dose of 50 μg i.m.

When the recipient starts to awake from anesthesia before the end of the procedure, isoflurane inhalation 0.4–0.8% is performed. In this way the timing of the duration of anesthesia can be optimized.

Infection process

In vivo RCMV infection is mostly established by i.p. injection of 10^5 plaque-forming units (pfu) of a virus suspension (Bruggeman *et al.*, 1983). Early after i.p. infection an influx of mononuclear cells into the peritoneal cavity is observed. The virus is rapidly cleared from the peritoneal cavity by an uptake into peritoneal macrophages (Bruggeman *et al.*, 1985a), which are altered in function at 3 days p.i. (Engels *et al.*, 1989). Intraperitoneal RCMV inoculation induces acute RCMV infection in a wide range of cells, mainly in cells from mesothelial origin (Bruggeman *et al.*, 1983, 1985b). When the virus was inoculated locally, for example in the hind paw, an infection was generated localized to the hind paw. For evaluation of meningoencephalitis symptoms 50 μl of virus at 5×10^4 was inoculated in the cerebrum (Stals *et al.*, 1996a; Kloover, *et al.*, 1997).

Storage, preparation of inocula

RCMV Maastricht strain was firstly isolated from the salivary gland of wild rats (Bruggeman *et al.*, 1982). After several *in vivo* passages virus stocks were harvested from the salivary glands of BN rats. The salivary gland was homogenized in minimal essential medium with 2% newborn calf's serum by a tissue grinder and centrifuged for 10 minutes at 1000*g*. Small portions of the supernatant were frozen and stored immediately at −70°C. Plaque assays were performed to determine the amount of virus. Generally, the number of virus particles yielded from the salivary glands varied between 10^5 and 5×10^6.

Key parameters to monitor infection

Survival

The effect of (combined) treatment on survival was recorded over a period of 21 days p.i. The *in vivo* activity of DHPG is expressed as the minimal effective dose or the daily dose at which at least 50% of the animals survived the lethal infection (ED_{50}). Combined interaction was calculated from the fractional effective dose (FED):

$$FED = (ED_A^{comb}/ED_A^{alone}) + (ED_B^{comb}/ED_B^{alone})$$

In addition, the mean day of death (MDD) was calculated.

Viral load

The effect of treatment on virus titers in organs was determined at 8 days p.i., as described earlier (Stals *et al.*, 1991a). For this purpose spleen, liver and lungs were removed aseptically, homogenized in a tissue grinder and suspended in basal medium Eagle's (BME) + 2% newborn calf serum (NCS) and plaque assays were performed. The amount of virus was expressed as the number of pfu per gram organ tissue.

Histological examination

Samples of the organs mentioned above were fixed in paraformaldehyde lysine periodate (PLP) and embedded in paraffin. Serial 4-μm-thick sections were prepared for hematoxylin and eosin (H&E) staining and immunoperoxidase staining using monoclonal antibodies to RCMV antigens, as described earlier (Stals *et al.*, 1990b, 1991a). Specificity of the monoclonal antibodies (numbers 8 and 35) has been described before (Bruning *et al.*, 1987). Monoclonal antibodies ED-1 and W3/13 were used for identification of rat monocytes/macrophages and rat T cells, respectively (Brown *et al.*, 1981; Dijkstra *et al.*, 1985).

For morphometric analyses of the alveolar septa a reticulin staining was performed and the size of the alveolar wall was measured, as described before (Stals *et al.*, 1991a). The relative amount of inflammatory infiltrate in alveolar septa was assessed semiquantitatively in each section and was expressed in score units: absence of cells was scored as 0, slight inflammatory infiltrate as 1, moderate dense infiltrate as 2, and severe dense inflammatory infiltrate as 3.

Antimicrobial therapy

The effect of different antivirals was investigated: treatment with drugs such as 9-(1,3-dihydroxy-2-propoxymethyl)guanine (ganciclovir, DHPG, Cymevene, Syntex Inc.; Stals *et al.*, 1991a, 1993a), trisodium phosphonoformatic acid (PFA, foscarnet, foscavir; Astra Pharmaceutical Products, Inc.; Stals *et al.*, 1991a), (*S*)-1-(3-hydroxy-2-phosphonylmethoxypropyl)cytosine (HPMPC; Stals *et al.*, 1991a,b, 1993a,b), aciclovir (ACV; Bruggeman *et al.*, 1987) was evaluated.

Other treatment modalities, which were (partly) based on an immunomodulatory effect were studied with hyperimmune serum (HIS), control immune serum (CIS; Stals *et al.*, 1990b, 1993a, 1996b) tumor necrosis factor-alpha (TNF-alpha; Haagmans *et al.*, 1994a), and interferon-gamma (IFN-gamma; Haagmans *et al.*, 1994b).

DHPG treatment (administered i.p.) was started at 6 hours p.i. and continued twice daily for 5 days. Treatment for 3 days and less was not effective at any dose.

HPMPC was administered in a bolus injection of different dosages i.p. at 6 hours post-infection. PME-A and 9-(2-phosphonyl-methoxyethyl)-2,6-diaminopurine (PME-DAP) were administered twice daily for 5 days (Neyts *et al.*, 1993). Foscarnet was administered three times a day for 5 days.

One ml of HIS, diluted twofold in BME + 2% NCS, was administered i.v. at 6 hours p.i. To determine the effect of combined treatment, DHPG was administered twice daily for 5 days starting at 6 hours p.i. and HIS was administered in a single dose at 6 hours p.i.

Pitfalls (advantages/disadvantages) of the model

The animal most frequently used to study CMV infections *in vivo* is the mouse model. The major advantages of the rat model over the mouse model when studying the treatment and pathology of CMV infections is the possibility of performing organ transplantations (Bruggeman *et al.*, 1995). The pathogenesis of RCMV infections in rats shows striking similarities with that of human CMV in humans, which underlines the fact that the rat model may be important to give greater insight into the *in vivo* behavior of CMV and its clinical consequences. The rat models developed in this study can also be helpful when designing appropriate treatment modalities for CMV disease.

The disadvantages of the rat model are the poor knowledge about the structure and behavior of the virus *in vitro* and the lack of molecular biologic tools (only a few groups are working with the virus). However, sequence analyses of the virus are in progress (Bruggeman, Maastricht, personal communication). Another disadvantage of the model is that a rat is relatively expensive compared to a mouse.

Contributions of the model to infectious disease therapy

Efficacy of antimicrobial agents

The following questions were the subject of our antiviral treatment studies in the rat model. What is the optimal therapeutic regimen and in what circumstances should combined treatment with ganciclovir and HIS be advised? Should the serum preparations used for treatment contain specific anti-CMV activity or can non-specific antibodies be used? Are there alternative antiviral agents which are effective against clinical CMV infections?

Studies were performed on HIS treatment in rats suffering from generalized RCMV infection (Stals *et al.*, 1990b). A few lessons were learned from these studies. The HIS that had high neutralizing antibody titers reduced mortality significantly, when administered early during infection (within 24–48 hours) and there was a dose–effect relationship. Virus titers were reduced in many organs, such as the lungs and liver, but not in the spleen. When HIS treatment was started late (3 days or later p.i.), it did not affect mortality, nor did it affect virus titers in the internal organs. Treatment with CIS had no effect on mortality, indicating that specific anti-CMV-directed antibodies formed the basis of the effect on generalized CMV infection. Interestingly, variation in the virus titer of the inoculum scarcely influenced the effect of HIS treatment, but an increased dose of TBI abolished the antiviral effect of HIS. This suggests that, although antibodies against CMV act as neutralizing antibodies *in vivo*, cell-dependent antibody actions such as antibody-dependent cell toxicity may be of great importance for the recovery of virus infection. After allogeneic BMT, treatment with HIS had a beneficial effect on the course of CMV-induced ILD and it reduced the CMV titers in the lung. CIS had no effect on virus titers and on the development of ILD (Stals *et al.*, 1994).

Studies on the effect of DHPG treatment revealed that the drug was effective against generalized RCMV infection in a dose-dependent manner (Stals *et al.*, 1991a). The effective dose was 20 mg/kg per day given in two daily doses for 5 days. Shortening the duration of treatment and delaying treatment for 3 days led to therapy failure and all animals died from infection. Toxicity was not observed at that dose. DHPG treatment in the allogeneic BMT model led to a significant decrease of virus titers in the lung of these animals, but had almost no effect on ILD (Stals *et al.*, 1993a). This effect is very similar to observations from the studies described in humans (Collaborative DHPG Treatment Study Group, 1986; Balfour, 1990).

Indications of probably synergistic effects of antimicrobial agents

Combined treatment with ganciclovir and HIS was studied in both the generalized infection model (Stals *et al.*, 1990b, 1994) and the ILD model (Stals *et al.*, 1994). In the generalized infection model, the combined treatment was synergistic, since the effective dose of both ganciclovir and HIS during combined treatment was reduced to 25% of the effective dose for treatment with a single antiviral agent. The FED is 0.25. In the BMT model, combined treatment with DHPG and HIS reduced the virus titers in lung and spleen to below detection levels and ILD was totally prevented. Interestingly, treatment with both DHPG and CIS reduced the virus titers in lung and spleen only slightly, but markedly reduced ILD. Administration of CIS as a single drug had no effect on virus titers and ILD. From these experiments it can be concluded that specific antibodies to CMV have an important antiviral effect during immunocompromised conditions. Since the effect of antiviral treatment is dependent on the degree of TBI, we conclude that the antiviral effect has to be supported by the cellular immune response.

Development of novel therapeutic approaches

A recently designed nucleotide analog, (*S*)-1-(3-hydroxy-2-phosphonylmethoxypropyl) cytosine (HPMPC) was evaluated for the treatment of CMV infections *in vivo* (Stals *et al.*, 1991a,b, 1993a,b, 1994; Neyts *et al.*, 1993). *In vitro*, HPMPC seems to be a candidate for the treatment of CMV infections, since the selectivity index (the ratio of cell toxic concentration and effective antiviral concentration) is 1400, which is 8 times higher than that of DHPG. The drug also has some other advantages over DHPG, such as the long intracellular half-life of the active metabolites and the low grade of resistance *in vitro*. The selectivity index of HPMPC for RCMV was about 500. Therefore, the drug turned out to be also a selective inhibitor of RCMV infection. The *in vivo* antiviral activity of the drug against RCMV was assessed in the generalized infection model as well as in the ILD model. In the generalized infection model, HPMPC was far more active than DHPG: the effective dose was 2 mg/kg given in a single dose as compared to DHPG, which had to be administered at a dose of 20 mg/kg per day twice daily for at least 5 days. One single administration of HPMPC at 20 mg/kg (which is 10 times the effective dose) completely inhibited viral replication in all organs. As was the case for DHPG, HPMPC administered as late as 3 days p.i. was not effective. Preventive administration with HPMPC, in contrast to DHPG, was effective in reducing mortality from generalized infection. Toxicity was not observed at this dose. These properties of HPMPC make the drug suitable for long-term preventive treatment of CMV infections. After allogeneic BMT in rats, HPMPC at 20 mg/kg completely inhibited viral replication in organs, such as the spleen and lungs, and the development of ILD. This study indicates that HPMPC is a potential suitable drug for the treatment of CMV infections. Although the first studies with this drug have been performed in acquired immunodeficiency syndrome (AIDS) patients, other applications in humans have to be studied.

Contribution of the model in understanding of the pathogenesis of disease

For induction of generalized infection animals were immunosuppressed by a sublethal challenge (5 Gy) of TBI and infected with RCMV. At 8 days p.i. a generalized RCMV infection developed with macroscopic and microscopic signs of splenitis, hepatitis and, to a lesser extent, pneumonia (Stals *et al.*, 1990b). Infection was also associated with hemostasis disorders. High virus titers were found in the spleen and liver and lower titers in the lung and kidney. The bone marrow was also heavily infected. Ultimately, most animals died from infection between 8 and 10 days p.i. The symptoms in these animals were similar to those in patients with severe immunological impairment, such as AIDS patients, in whom involvement of the bone marrow is described, and in perinatally infected newborns, which suffer from hepatitis and splenitis. Thrombocytopenia and hemostasis disorders are also frequently described in neonates. The multiple bleedings found in the internal organs point in the direction of diffuse vascular coagulation and coagulopathy. A typical example of vascular involvement during CMV infection is CMV-induced retinitis in AIDS patients, which basically affects the vascular layer of the retina—the chorion. There is increasing evidence that CMV-infected endothelium becomes activated and ultimately loosens from the subendothelial layer, resulting in vascular leakage and hemostatic disorders. This led to the hypothesis that changes in the vascular endothelium form the basis of CMV-induced disease (Bruggeman *et al.*, 1995).

Pneumonia is a serious complication of CMV infection in immunocompromised patients. In particular, ILD frequently occurs in BMT recipients. In these patients ILD is associated with CMV infection and graft-versus-host disease (GvHD), and is thought to be an immunopathologic process. To study the pathogenesis of CMV-induced ILD, we performed a BMT in rats (Stals *et al.*, 1993a). In this model, animals received potentially lethal TBI, followed by replacement of the bone marrow and CMV infection. ILD occurred in rats which had received allogeneic BMT and RCMV infection, but not in animals which had received syngeneic BMT and RCMV infection, nor in mock-infected animals after allogeneic BMT. Therefore, these studies clearly demonstrated that a combination of CMV infection and allogeneic BMT is required for the development of ILD.

However, these experiments did not allow a definitive answer to the question of whether the occurrence of ILD in allogeneic, but not syngeneic, BMT recipients is due to differences in the host, such as homing of the bone marrow cells, or that it is based on immunological factors, such as differences in major histocompatibility complex (MHC) antigens of the lung. Therefore, we performed a series of experiments in which, prior to allogeneic BMT, an allogeneic lung was transplanted in the host (Stals *et al.*, 1996b). In this model the allogeneic bone marrow recognized the donor lung as syngeneic tissue and the recipient lung as allogeneic. After RCMV infection, significantly more CMV antigen-expressing cells in the interstitial septa were found in the allogeneic recipient lung than in the syngeneic donor lung. The higher viral load in the septal wall of the lung was accompanied by severe diffuse ILD, characterized by congestion of the interstitial septa with mononuclear cells and erythrocytes, swelling of endothelial cells in the pulmonary capillaries and endothelial leakage. In contrast, in the donor lung that was syngeneic to the bone marrow, only minor changes were observed in the interstitial septa and in the capillary endothelium.

It was concluded that a high virus load in the interstitium, as is detected in the recipient lung, is associated with interstitial and endothelial changes in the capillaries and interstitial congestion. The splenic weight of these animals was not increased and an inflammatory response in the perivascular and peribronchial region of the recipient lung was absent. It was therefore concluded that these CMV-induced interstitial changes in the allogeneic lung are not associated with GvHD.

In order to evoke a strong immune response, additional lymphocytes that were syngeneic to both the bone marrow and the donor lung, were administered. After lymphocyte donation, the cellular inflammatory response in the alveolar septa of the allogeneic lung was less than in the syngeneic lung. After RCMV infection this response was not increased.

Since no signs of GvHD were observed in these experiments, it can be concluded that GvHD is not a prerequisite for the development of ILD in these acutely infected rats. In this study the first particular finding was the presence of vascular leakage and subsequent congestion of the interstitial septa in the lung, which occurs in the absence of an inflammatory response. These changes are related to high local virus titers after allogeneic BMT (Stals *et al.*, 1996b). Therefore, we hypothesize that vascular activation and subsequent interstitial congestion are the primary pathological findings during CMV-induced ILD. This vascular activation is followed by an inflammatory response, rather than an aggressive local immune response to CMV, as suggested by others (Grundy *et al.*, 1985, 1987). In our hands it seemed that GvHD was even inhibited by the CMV infection. An explanation for these different findings may be the relation between GvHD and reactivation of latent CMV infection, that confounds data in humans: the combination of allogeneic transplantation and immunosuppression evokes the reactivation of latent CMV infection. This leads to the hypothesis that GvHD and subsequent immunological cell proliferation permit the virus to reactivate from a latent state. GvHD, in turn, has to be treated with a high-level immunosuppressive regimen, making escape from immune surveillance possible. Ultimately, this results in a generalized CMV infection in which the lung is often involved. After allogeneic organ transplantation, CMV seems to accelerate rejection by enhanced class II antigen expression (You *et al.*, 1996; Lautenschlager *et al.*, 1997; Lemstrom *et al.*, 1997).

We concluded that CMV-induced ILD is a pathological process characterized by extensive microvascular damage, and that it occurs under conditions of impaired immunologic control, such as after allogeneic BMT. In addition, lymphocytes seem to be only secondarily involved in the pathogenesis of ILD, and GvHD is not a prerequisite for induction of ILD.

References

Applebaum, F. R., Meyers, J. D., Fefer, A. *et al.* (1982). Nonbacterial nonfungal pneumonia following marrow transplantation in 100 identical twins. *Transplantation*, **33**, 265–268.

Balfour, H. H., Jr (1990). Management of cytomegalovirus disease with antiviral drugs. *Rev. Infect. Dis.*, **12** (suppl. 7), S849–S860.

Bos, G. M., Majoor, G. D., Bruggeman, C. A., Grauls, G., van-de-Gaar, M. J., van-Breda-Vriesman, P. J. (1989). Rat cytomegalovirus can be transferred by bone marrow cells but does not affect the course of acute graft-versus-host disease. *Transplant. Proc.*, **21**, 3050–3052.

Brown, W. R. A., Barclay, A. N., Sunderland, C. A., Williams, A. F. (1981). Identification of a glycophorin-like molecule at the cell-surface of rat thymocytes. *Nature*, **289**, 456–460.

Bruggeman, C. A. (1991). Reactivation of latent CMV in the rat. *Transplant. Proc.*, **23** (suppl. 3), 22–24.

Bruggeman, C. A., Meijer, H., Dormans, P. H. J., Debie, W. M. H., Grauls, G. E. L. M., van-Boven, C. P. A. (1982). Isolation of a cytomegalovirus-like agent from wild rats. *Arch. Virol.*, **73**, 231–241.

Bruggeman, C. A., Debie, W. M., Grauls, G., Majoor, G., van-Boven, C. P. (1983). Infection of laboratory rats with a new cytomegalo-like virus. *Arch. Virol.*, **76**, 189–199.

Bruggeman, C. A., Grauls, G., van-Boven, C. P. A. (1985a). Susceptibility of peritoneal macrophages to rat cytomegalovirus infection. *FEMS*, **27**, 263–266.

Bruggeman, C. A., Meijer, H., Bosman, F., van-Boven, C. P. (1985b). Biology of rat cytomegalovirus infection. *Intervirology*, **24**, 1–9.

Bruggeman, C. A., Engels, W., Endert, J. (1987). Treatment of experimental cytomegalovirus infections with acyclovir. *Arch. Virol.*, **97**, 3527–3537.

Bruggeman, C. A., Li, F., Stals, F. S. (1995). Pathogenicity: animal models. *Scand. J. Infect. Dis.* (suppl. 99), 43–50.

Bruning, J. H., Debie, W. H., Dormans, P. H., Meijer, H., Bruggeman, C. A. (1987). The development and characterization of monoclonal antibodies against rat cytomegalovirus induced antigens. *Arch. Virol.*, **94**, 55–70.

Bruning, J. H., Bruggeman, C. A., van-Boven, C. P., van-Breda-Vriesman, P. J. (1989a). Reactivation of latent rat cytomegalovirus by a combination of immunosuppression and administration of allogeneic immunocompetent cells. *Transplantation*, **47**, 917–918.

Bruning, J. H., Bruggeman, C. A., van-Breda-Vriesman, P. J. (1989b). The effect of cytomegalovirus infection on renal allograft function and rejection in a rat model. *Transplantation*, **47**, 742–744.

Collaborative DHPG Treatment Study Group (1986). Treatment of serious cytomegalovirus infections with 9-(1,3-dihydroxy-2-propoxymethyl)guanine in patients with AIDS and other immunodeficiencies. *N. Engl. J. Med.*, **314**, 801–805.

Dijkstra, C. D., Döpp, E. A., Joling, P., Kraal, G. (1985). The heterogeneity of mononuclear phagocytes in lymphoid organs:

distinct macrophage subpopulations in the rat recognized by monoclonal antibodies ED1, ED2 and ED3. *Immunology*, **54**, 589–599.

Emanuel, D., Cunningham, I., Jules-Elysee, K. *et al.* (1988). Cytomegalovirus pneumonia after bone marrow transplantation successfully treated with the combination of ganciclovir and high-dose intravenous immune globulin. *Ann. Intern. Med.*, **109**, 777–782.

Engels, W., Grauls, G., Lemmens, P. J., Mullers, W. J., Bruggeman, C. A. (1989). Influence of a cytomegalovirus infection on functions and arachidonic acid metabolism of rat peritoneal macrophages. *J. Leukoc. Biol.*, **45**, 466–473.

Goodrich, J. M., Mori, M., Gleaves, C. A. *et al.* (1991). Early treatment with ganciclovir to prevent cytomegalovirus disease after allogeneic bone marrow transplantation. *N. Engl. J. Med.*, **325**, 1601–1607.

Grundy, J. E., Shanley, J. D., Shearer, G. M. (1985). Augmentation of graft-versus-host reaction by cytomegalovirus infection resulting in interstitial pneumonitis. *Transplantation*, **39**, 548–553.

Grundy, J. E., Shanley, J. D., Griffiths, P. D. (1987). Is cytomegalovirus interstitial pneumonitis in transplant recipients an immunopathological condition? *Lancet*, **2**, 996–999.

Haagmans, B. L., Stals, F. S., van-der-Meide, P. H., Bruggeman, C. A., Horzinek, M. C., Schijns, V. E. (1994a). Tumor necrosis factor alpha promotes replication and pathogenicity of rat cytomegalovirus. *J. Virol.*, **68**, 2297–2304.

Haagmans, B. L., van-der-Meide, P. H., Stals, F. S. *et al.* (1994b). Suppression of rat cytomegalovirus replication by antibodies against gamma interferon. *J. Virol.*, **68**, 2305–2312.

Ho, M. (1991). *Cytomegalovirus. Biology and Infection*. Plenum, New York.

Kloover, J. S., Vanagt, W. Y., Stals, F. S., Bruggeman, C. A. (1997). Effective treatment of experimental cytomegalovirus-induced encephalo-meningitis in immunocompromised rats with HPMPC. *Antiviral Res.*, **35**, 105–112.

Lautenschlager, I., Soots, A., Krogerus, L. *et al.* (1997). CMV increases inflammation and accelerates chronic rejection in rat kidney allografts. *Transplant. Proc.*, **29**, 802–803.

Lemström, K., Persoons, M., Bruggeman, C., Ustinov, J., Lautenschlager, I., Hayry, P. (1993). Cytomegalovirus infection enhances allograft arteriosclerosis in the rat. *Transplant. Proc.*, **25**, 1406–1407.

Lemström, K., Bruning, J., Koskinen, P., Bruggeman, C., Lautenschlager, I., Hayry, P. (1994). Triple-drug immunosuppression significantly reduces chronic rejection in noninfected and RCMV-infected rats. *Transplant. Proc.*, **26**, 1727–1728.

Lemström, K., Sihvola, R., Bruggeman, C., Hayry, P., Koskinen, P. (1997). Cytomegalovirus infection-enhanced cardiac allograft vasculopathy is abolished by DHPG prophylaxis in the rat. *Circulation*, **95**, 2614–2616.

Li, F., Grauls, G., Yin, M., Bruggeman, C. (1995). Initial endothelial injury and cytomegalovirus infection accelerate the development of allograft arteriosclerosis. *Transplant. Proc.*, **27**, 3552–3554.

Ljungman, P., Engelhard, D., Link, H. *et al.* (1992). Treatment of interstitial pneumonitis due to cytomegalovirus with ganciclovir and intravenous immune globulin: experience of European Bone Marrow Transplant Group. *Clin. Infect. Dis.*, **14**, 831–835.

Meijer, H., Dormans, P. H., Geelen, J. L., van-Boven, C. P. (1984). Rat cytomegalovirus: studies on the viral genome and the proteins of virions and nucleocapsids. *J. Gen. Virol.*, **65**, 681–695.

Meijer, H., Dreesen, J. C., van-Boven, C. P. (1986). Molecular cloning and restriction endonuclease mapping of the rat cytomegalovirus genome. *J. Gen. Virol.*, **67**, 1327–1342.

Mills, J., Corey, L. (1986). *Antiviral Therapy: New Directions for Clinical Application and Research*, pp. 195–203. Elsevier Science, New York.

Neyts, J., Stals, F. S., Atherton, S., Persoons, M., Bruggeman, C. A., de-Clerq, E. (1993). Efficacy of HPMPC in the treatment of CMV infections in various animal models. In *Multidisciplinary Approach to the Understanding of Cytomegalovirus Disease* (eds Michelson, S., Plotkin, S. A.) pp. 279–285. Elsevier Science, Amsterdam.

Rasing, L. A., de-Weger, R. A., Verdonck, L. F. *et al.* (1990). The value of immunohistochemistry and *in situ* hybridization in detecting cytomegalovirus in bone marrow transplant recipients. *Apmis.*, **98**, 479–488.

Reed, E. C., Bowden, R. A., Dandliker, P. S., Lilleby, K. E., Meyers, J. D. (1988). Treatment of cytomegalovirus pneumonia with ganciclovir and intravenous cytomegalovirus immunoglobulin in patients with bone marrow transplants. *Ann. Intern. Med.*, **109**, 783–788.

Shanley, J. D., Pesanti, E. L., Nugent, K. M. (1982). The pathogenesis of pneumonitis due to murine cytomegalovirus. *J. Infect. Dis.*, **146**, 388–396.

Shanley, J. D., Via, C. S., Sharrow, S. O., Shearer, G. M. (1987). Interstitial pneumonitis during murine cytomegalovirus infection and graft-versus-host reaction. Characterization of bronchoalveolar lavage cells. *Transplantation*, **44**, 658–662.

Stals, F. S., van-den-Bogaard, A. E. J. M., Bruggeman, C. A. (1990a). Selective bowel decontamination and the incidence of bacteraemia in irradiated and cytomegalovirus infected rats. *Microecol. Ther.*, **20**, 153–156.

Stals, F. S., Bosman, F., van-Boven, C. P., Bruggeman, C. A. (1990b). An animal model for therapeutic intervention studies of CMV infection in the immunocompromised host. *Arch. Virol.*, **114**, 91–107.

Stals, F. S., de-Clercq, E., Bruggeman, C. A. (1991a). Comparative activity of (S)-1-(3-hydroxy-2-phosphonylmethoxypropyl)cytosine and 9-(1,3-dihydroxy-2-propoxymethyl) guanine against rat cytomegalovirus infection *in vitro* and *in vivo*. *Antimicrob. Agents. Chemother.*, **35**, 2262–2266.

Stals, F. S., de-Clerq, E., Bruggeman, C. A. (1991b). Activity of (S)-1-(3-Hydroxy-2-phosphonylmethoxypropyl)-cytosine against rat cytomegalovirus (RCMV) in cell culture and in immunocompromised rats. In *Progress in Cytomegalovirus Research* (ed Landini, M. P.), pp. 349–352. Excerpta Medica, New York.

Stals, F. S., Zeytinoglu, A., Havenith, M., de-Clercq, E., Bruggeman, C. A. (1993a). Rat cytomegalovirus-induced pneumonitis after allogeneic bone marrow transplantation: effective treatment with (S)-1-(3-hydroxy-2-phosphonyl-methoxypropyl)cytosine. *Antimicrob. Agents Chemother.*, **37**, 218–223.

Stals, F. S., Zeytinoglu, A., Havenith, M., de-Clerq, E., Bruggeman, C. A. (1993b). Comparative activity of (S)-1-(3-hydroxy-2-phosphonylmethoxypropyl)cytosine (HPMPC) and 9-(1,3-dihydroxy-2-propoxymethyl) (DHPG) treatment on interstitial pneumonia in allogeneic bone marrow recipients. *Transplant. Proc.*, **25**, 1248–1249.

Stals, F. S., Wagenaar, Sj. Sc., Bruggeman, C. A. (1994). Generalized cytomegalovirus (CMV) infection and CMV-induced pneumonitis in the rat: combined effect of 9-(1,3-dihydroxy-2-propoxymethyl)guanine and specific antibody treatment. *Antiviral Res.*, **25**, 147–160.

Stals, F. S., Wagenaar, Sj. Sc., Kloover, J. S., Vanagt, W. Y., Bruggeman, C. A. (1996a). Combinations of ganciclovir and antibody for experimental CMV infections. *Antiviral Res.*, **29**, 61–64.

Stals, F. S., Steinhoff, G., Wagenaar, S. S. *et al.* (1996b). Cytomegalovirus induces interstitial lung disease in allogeneic bone marrow transplant recipient rats independent of acute graft-versus-host response. *Lab. Invest.*, **74**, 343–352.

Vancutsem, P. M., Babish, J. G., Schwark, W. S. (1990). The fluoroquinolone antimicrobials: structure, antimicrobial activity, pharmacokinetics, clinical use in domestic animals and toxicity. *Cornell. Vet.*, **80**, 173–186.

Winston, D. J., Ho, W. G., Champlin, R. E. (1990). Cytomegalovirus infections after allogeneic bone marrow transplantation. *Rev. Infect. Dis.*, **12** (suppl. 7), S776–S792.

Yagyu, K., van-Breda-Vriesman, P. J., Duijvestijn, A. M., Bruggeman, C. A., Steinhoff, G. (1993). Reactivation of cytomegalovirus with acute rejection and cytomegalovirus infection with obliterative bronchiolitis in rat lung. *Transplant. Proc.*, **25**, 1152–1154.

You, X. M., Steinmuller, C., Wagner, T. O., Bruggeman, C. A., Haverich, A., Steinhoff, G. (1996). Enhancement of cytomegalovirus infection and acute rejection after allogeneic lung transplantation in the rat: virus-induced expression of major histocompatibility complex class II antigens. *J. Heart Lung Transplant.*, **15**, 1108–1119.

Chapter 114

Human Cytomegalovirus Infection of the SCID-hu (thy/liv) Mouse

G. W. Kemble, G. M. Duke and E. S. Mocarski

Background on human infection

Infection with human cytomegalovirus (HCMV) reaches a high proportion of the population. Serological status indicates prior infection in 50 and 80% of the adult population by mid-life (Alford and Britt, 1995). Person-to-person transmission occurs by direct contact and rates are higher in developing, as opposed to developed, countries and urban, as opposed to suburban, settings in developed countries. Infection of otherwise healthy seronegative children or adults is generally inapparent; however, symptoms indistinguishable from Epstein–Barr virus mononucleosis, including fever, headache, and malaise, occur in a small proportion of post-pubescent individuals (Alford and Britt, 1995; Britt, 1996). Once infected, the virus remains in a life-long latent infection which is associated with frequent reactivation and shedding in urine and saliva without any clinical symptoms (Alford and Britt, 1995; Collier *et al.*, 1995; McVoy and Adler, 1989). Latency may be maintained in hematopoietic progenitors within the granulocyte/macrophage/dendritic lineage (Hahn *et al.*, 1998; Kondo *et al.*, 1994; Mendelson *et al.*, 1996; Soderberg-Naucler *et al.*, 1997).

In contrast to the generally benign infection of immunocompetent children and adults, HCMV is an important congenital pathogen causing significant fetal morbidity and mortality in expectant mothers who undergo primary infection during pregnancy. An estimated 4000–8000 births annually (0.1% of all live births) in the United States alone suffer the sequelae of congenital infection (Fowler *et al.*, 1992; Istas *et al.*, 1995). The principal damage from congenital infection affects the central nervous system, causing hearing loss and mental retardation. An estimated 10% of congenitally infected children with signs of CMV disease do not survive childhood (Stagno *et al.*, 1982; Fowler *et al.*, 1992; Alford and Britt, 1995; Istas *et al.*, 1995). Congenital disease could be largely controlled through universal childhood vaccination using the same strategy that succeeded in controlling rubella-associated congenital disease (Plotkin, 1994; Britt, 1996).

Starting in the 1960s, HCMV attained increased medical significance as an opportunistic pathogen in immunocompromised individuals where disease may follow primary infection or reactivation of latent infection (Meyers *et al.*, 1982; Winston *et al.*, 1990; Drew, 1992; Alford and Britt, 1995). HCMV produces a range of severe end-organ disease in different settings. Human immunodeficiency virus (HIV)-infected individuals with low CD4+ T-cell levels develop HCMV retinitis as well as ulcerative gastroenteritis, diseases that respond to antiviral therapy (Drew, 1992) which may be compromised by antiviral resistance (Drew, 1996). Bone marrow allograft recipients succumb to HCMV pneumonia (Meyers *et al.*, 1982; Winston *et al.*, 1990). Prior to the widespread use of ganciclovir prophylaxis and therapy, this population experienced a mortality rate as high as 80% (depending upon the transplant center), primarily from pneumonia (Alford and Britt, 1995; Emanuel *et al.*, 1988). Even in the face of antiviral therapy, CMV disease remains a problem in allograft recipients (Boeckh *et al.*, 1995). Ganciclovir, along with selection of appropriate immunosuppressive regimens, suppresses CMV disease in most solid organ transplant settings (Falagas and Snydman, 1995); however, transfer of a liver, heart–lung, pancreas or combinations of organs from a seropositive donor to a seronegative recipient still leads to complex disease patterns, including HCMV hepatitis and pneumonia, as well as other organ-specific diseases that are refractory to antiviral treatment (reviewed in Alford and Britt, 1995; Britt, 1996). Interestingly, vaccination has already been shown to be effective in preventing post-transplantation complications from CMV disease in a solid organ transplant population (Plotkin *et al.*, 1989).

Primary human diploid fibroblasts, generally prepared from neonatal foreskins or fetal lung tissue, have been the standard cell culture system for laboratory propagation of HCMV. In the normal host, however, HCMV infects a broad range of cell types, from leukocytes in the blood to epithelial cells in salivary glands and kidneys. The principal sites of infection responsible for the high level in saliva, urine, and breast milk (Britt, 1996), as well as for the recurrent shedding of HCMV throughout life, are epithelial cells in the salivary glands and kidneys. Studies on HCMV gene function and tropism have relied on the generation of recombinant CMVs carrying gene disruptions (Mocarski and Kemble, 1996), but progress in understanding the role of specific HCMV genes in the interaction with polarized

Handbook of Animal Models of Infection
ISBN 0-12-775390-7

epithelial cells is just beginning. HCMV US9 has been ascribed a role in viral maturation in polarized epithelial cells (Maidji *et al.*, 1996) and murine CMV *sgg*1 has been ascribed a role in controlling the level of replication in salivary gland epithelium during natural infections of mice (Manning *et al.*, 1992; Lagenaur *et al.*, 1994).

Background of the SCID-hu (thy/liv) mouse model

HCMV has a narrow host range both in cell culture and in the host. Because commonly used laboratory animals cannot be infected, murine, rat and guinea-pig CMV that naturally infect small animals have been employed for decades as surrogate systems to understand the parameters of replication, pathogenesis and latency (Staczek, 1990).

In 1988, McCune and colleagues described a method for surgical implantation of human fetal thymus and liver (thy/liv) under the kidney capsule of a severe combined immunodeficient CB-17 *scid/scid* mouse that resulted in the development and establishment of an implant resembling human fetal thymus (McCune *et al.*, 1988; Namikawa *et al.*, 1988). In addition to general histological similarity to human thymus, the implant carried out activities inherent to human thymus, including the release of human CD4+ and CD8+ T cells into the peripheral blood of the animal. Within a short time after being developed, SCID-hu (thy/liv) mice were shown to be able to support HIV replication (Namikawa *et al.*, 1988; McCune *et al.*, 1990).

In addition to thy/liv, a variety of human tissues, including thymus alone, bone, lung, lymph node, skin, brain and retinal tissue have been implanted into various body sites of immunodeficient rodents. Although HCMV will replicate in any of these implants (Mocarski *et al.*, 1993; Mocarski, Bonyhadi and Kaneshima, unpublished observations), two systems have been best studied. The most detailed studies have been carried out in thy/liv implants where high titer, sustained growth and viral strain differences have been noted (Mocarski *et al.*, 1993; Brown *et al.*, 1995). Virus replication localizes to a very small percentage of keratin-positive medullary epithelial cells within implants which may be sustained in a small percentage of cells for several months with no discernible cytopathology. Consistent with the restricted host range of HCMV, surrounding mouse tissues remained virus-free. Less detailed investigations of HCMV replication in human retinal tissue implants placed into the anterior chamber of an immunodeficient rodent's eye (DiLoreto *et al.*, 1994; Laycock *et al.*, 1997) have suggested viral replication and pathogenesis mimicking retinitis that occurs in acquired immunodeficiency syndrome (AIDS) patients.

The SCID-hu (thy/liv) mouse allows evaluation of replication and tropism of HCMV strains in epithelial cells within a tissue architecture. Interestingly, epithelial cells cultured directly from fetal thymus do not retain the level of permissiveness observed within the tissue implant (Brown and Mocarski, unpublished observations). One of the first observations made with SCID-hu (thy/liv) mice was the striking differences in titers among different HCMV strains (Mocarski *et al.*, 1993; Brown *et al.*, 1995). Strains such as Toledo (Plotkin *et al.*, 1989) that had undergone very limited propagation in cultured fibroblasts were shown to have a much greater capacity for replication in thy/liv implants. Laboratory-adapted strains Towne or AD169 were found to replicate poorly (Brown *et al.*, 1995), probably due to the significant loss of genome sequences during their extensive laboratory propagation in cultured cells (Cha *et al.*, 1996). Analysis of viral growth in SCID-hu (thy/liv) mice also revealed differences even among very closely related variants of the same virus strain. Depending upon the source, strain AD169 showed a different replication capability, undoubtedly due to different sets of mutations that have accumulated while being carried in different locations (Brown *et al.*, 1995; Mocarski *et al.*, 1997). Altogether, studies in SCID-hu (thy/liv) mice have revealed dramatic differences between and within strains of virus that had previously been shown to behave the same way in cultured human fibroblasts. These differences are likely to reflect functions important for replication in the naturally infected human host. Genetic loci contributing to epithelial cell targets may now be investigated.

Animal species

The CB-17 *scid/scid* strain was originally described as the recipient of human tissue implants; however, other immunodeficient strains (such as *rag* knockouts) would probably support the same range of implants. The inherent immune-privileged nature of the anterior chamber of the eye has allowed human retinal tissue implants to be made T-cell-deficient mice and rats (DiLoreto *et al.*, 1994; Laycock *et al.*, 1997). The CB-17 *scid/scid* strain has a leaky phenotype in that mature murine B and T cells can be detected by 10–14 months in non-implanted animals. This leakiness is likely to have little consequence on studies of HCMV, since implantation and infection can be carried out within an 8–10-week time frame.

Preparation of animals

All immunodeficient animals should be housed in microisolator units and only manipulated in laminar flow hoods using aseptic techniques. The bedding, food and water should be autoclave-sterilized or ultrapasteurized. These animals are particularly susceptible to infection with *Pneumocystis carinii* and helminths. *P. carinii* can be controlled by including prophylactic trimethoprim (0.25 mg/ml)/sulfamethoxazole (1.25 mg/ml; pediatric suspension) in drinking water (Namikawa *et al.*, 1988). Care must be taken to maintain both drugs in suspension. Utilizing the proper husbandry techniques described above, including quarantine of new animals and sentinal monitoring animal rooms, reduces the need for drug prophylaxis.

Details of surgery

Overview

Briefly, one kidney of the anesthetized animal is exposed through an incision on the back of the animal, a blister is made with phosphate-buffered saline (PBS) under the kidney, and human fetal tissue is inserted into the blister using forceps. The wound is closed and the animal is allowed at least 3 weeks to recover, during which time the implant grows and differentiates (McCune *et al.*, 1988; Namikawa *et al.*, 1988, 1990; Stanley *et al.*, 1993). We have found that successful implantation depends upon the batch of fetal tissue, with approximately 80% of animals harboring implants ranging from 20% to 150% of the size of the mouse kidney at 3 weeks following surgery. For virus inoculation, the animal is anesthetized a second time; the kidney is exposed as before; virus is inoculated in a small volume (20–50 μl) and the wound is closed (Mocarski *et al.*, 1993). Following 2 weeks of incubation, the animal is euthanized and the virus is harvested from the implanted tissue.

Materials used

Materials used include anesthetic, hair clippers, skin disinfectants, including iodine solutions and 70% alcohol, scalpel handle plus blades, blunt forceps, straight blunt standard scissors, 30 G needle, 1 ml syringe, microdissecting spring scissors, microforceps, 4–0 adsorbable suture, 9 mm wound clips, autoclip, and operating or dissecting binocular microscope. The human fetal thymus and liver is obtained from the same donor at 16–22 weeks of gestational age (Advanced Bioscience Resources, Alameda, CA). The tissue is shipped in RPMI medium on wet ice and used within 24 hours of harvest.

Anesthesia and preparation

Anesthesia is administered intramuscularly (i.m.) into the muscle of the left hind leg using a 1 ml syringe fitted with a 30 G needle from a solution of 10 mg ketamine and 10 mg xylazine solution per ml in PBS at a dose of 50 mg ketamine and 50 mg xylazine per kg body weight (a 25 g mouse requires 125 μl of this solution). Complete anesthesia ensues after approximately 5 minutes and lasts for approximately 30 minutes. Following complete induction of anesthesia, the animal is placed on its right side and the hair is removed from the left side with clippers. The skin is disinfected by wiping successively with iodine scrub (povidone scrub), 70% alcohol, and iodine solution (povidone solution). An ophthalmic ointment (Artificial Tears, OH) is placed on to the eyes to prevent drying under the microscope illumination. Animals are kept from cooling during procedures until signs of anesthesia wear off.

Surgical procedure

The animal is placed right-side-down on the surgical platform and skin is drawn taught over the disinfected left side. The underlying abdominal muscle wall is exposed by making a 2–3 cm incision through the skin with a scalpel fitted with a number 10 blade or a pair of surgical scissors. The operating field is then brought into focus under an operating binocular microscope. A pair of blunt forceps is used to hold the abdominal muscle away from the kidney while a small incision is then made laterally above the kidney through the wall of the abdomen with a pair of standard blunt surgical scissors. The incision is extended for a total of 1–2 cm and a pair of blunt forceps is used to grasp the connective tissue at the base of the kidney and expose the organ through the incision.

The kidney is held in place by the disinfected skin flap. A 30 G needle attached to a 1 ml syringe containing PBS is inserted, bevel side up, just under the kidney capsule and approximately 200 μl of PBS is injected to form a blister. To avoid piercing the kidney, the point of the needle entry should be along the flat side of the kidney. The needle is withdrawn and an incision is made at the point of needle entry to expel the PBS. A pocket is formed and is used to hold the inserted human tissue.

The human fetal thymus and liver tissue is minced in tissue culture media containing Dulbecco's modified minimum essential medium (DMEM, GIBCO/BRL) supplemented with 10% fetal calf serum, penicillin G (100 units/ml) and streptomycin (100 μg/ml). One piece (2–4 mm^3) of thymus, preferably corresponding to a single lobe, is placed in the pocket under the capsule with the aid of a pair of microforceps. One piece of liver, of approximately similar size, is then placed adjacent to the thymus in the pocket. Using the side of a forcep, the thymus and liver sandwich is pushed away from the pocket opening. The kidney is reinserted into the abdomen and the muscle wall is closed with 4–0 suture. The outer skin is then closed with 9 mm wound clips.

Postoperative care

The animal is placed in a sterile, empty cage on top of a water blanket warming pad. The animals generally recover activity within 15–30 minutes and are returned to fresh sterile bedding and housed together with 3–4 other animals operated on the same day. Animals generally have returned to normal activity by the next day. No drop in weight or food and water consumption should occur.

Storage, preparation of virus inocula

Several different strains of HCMV have been evaluated for growth in thy/liv implants in SCID-hu mice. High-passage, laboratory strains AD169 and Towne show dramatic differences when compared to low-passage Toledo strain, suggesting that the laboratory strains have lost important determinants for epithelial cell growth during extensive passage in fibroblasts (Brown *et al.*, 1995). Other low-passage isolates of HCMV also replicate well in thy/liv implants, consistent with the data initially generated with the Toledo strain. Biological differences have also been detected between stocks of the same strain (AD169) obtained from different sources. AD169 was an early isolate from adenoids of a child (Rowe *et al.*, 1956). One variant carried at the American Type Culture Collection does not replicate to any detectable extent in thy/liv implants even after an inoculum of 10^6 plaque-forming units (pfu). Another variant, carried for decades in the United Kingdom and used as the starting point for extensive genome sequence analysis (Chee *et al.*, 1990), retains a limited, but detectable growth capability in thy/liv implants (Brown *et al.*, 1995). Although not considered to be directly related to their propensity for growth in thy/liv implants, these two variants also differ in genetic complexity (Mocarski *et al.*, 1997). The ATCC variant carries an invariant 929 bp of DNA in the UL42–UL43 region of the viral genome not represented in a majority of genomic DNA molecules from the UK variant which was not reported in the HCMV (AD169) genome sequence (Chee *et al.*, 1990).

The inoculum used in these studies is prepared by a standard method used for harvesting infectious HCMV in our laboratories. In brief, cultures of infected human foreskin or lung fibroblasts are infected with HCMV at a multiplicity of infection (MOI) of 0.001–0.01. Extensive cytopathic effects (CPE) develop gradually over a 7–10-day period, during which time medium is changed every 3–4 days. Infected cells are harvested 3–4 days after 100% CPE is attained, combined with supernatant virus and sonicated with a probe-type sonifier on wet ice to release virus. Virus stocks are usually stabilized by the addition of an equal volume of sterile (triply autoclaved) 9% skim milk in water and maintained frozen at −80°C. Suspension of virus stocks in 10% fetal calf serum instead of sterile milk yields virus with similar stability to freezing but not as stable as in repeated freeze–thaw cycles.

In addition to the use of aseptic technique during culture and harvest, all plasticware is free of pyrogens. As few as 50 pfu of Toledo or low-passage isolates that replicate efficiently in this tissue can be used to initiate the infection. In contrast, at least 10^3 pfu of Towne and AD169-UK are required to recover progeny from the implant after 12–14 days. The nature of these replication differences has yet to be elucidated but probably reflects the tropism of these strains for human epithelial cells.

Infection procedure

Three or more weeks following implantation of the human tissue, the infection of the implant is performed in a manner similar to the implantation procedure. Anesthesia is administered in a regimen similar to that described above. In addition, we have found methoxyflurane a useful anesthetic for this procedure, since infection is not as lengthy a procedure as the initial surgery. Following complete induction of anesthesia, the animal is placed on its right side, the hair is removed, and the skin is disinfected. The left kidney is exposed again in a manner identical to that for implantation.

After exposing the kidney and locating the implant, 10–25 µl of inoculum is injected into the kidney. A syringe and 30 G needle are loaded with inoculum. The needle is inserted with the bevel facing upward directly into the implant, rotated 180°, and the material is injected smoothly. Following the injection, the needle is held in place for 3–5 seconds, rotated 180°, returning the bevel to its upward orientation and withdrawn along its path of entry. Little to no leakage should be evident.

The kidney is reinserted into the abdomen and the muscle wall is closed with 4–0 sutures. The outer skin is then closed with an autoclip and 9 mm wound clips. Following wound closure, the animal is recovered in a sterile, empty cage on top of a water blanket warming pad. The animals generally recover activity within 15–30 minutes and are returned to fresh sterile bedding and housed with 3–4 other animals operated on the same day. Animals generally have returned to normal activity by the next day. No drop in weight or food and water consumption should occur.

Key parameters to monitor infection

There should be no overt signs of infection in these mice. The narrow host range of HCMV restricts detectable replication to the human medullary epithelium in this implant (Mocarski *et al.*, 1993). HCMV has not been detected in any of the associated murine tissues. Because of this restriction, no weight loss or other indications of infection should be present in the infected mouse. At an appropriate time after infection, the mouse is euthanized and the implant is harvested. In general, we monitor replication 2 weeks after inoculation. Initially, HCMV enters an eclipse period when the virus is undetectable. Following this time, the titer of infectious virus reaches sustainable levels that can be maintained for several months (Mocarski *et al.*, 1993).

The human tissue can be treated in many ways to detect infectious virus. The most efficient method to detect viral replication is to homogenize the implant, sonicate the resulting homogenate on wet ice and determine the amount of infectious virus by plaque assay (pfu/ml) or tissue culture infectious dose 50 ($TCID_{50}$). Additionally, subsets of human cells can be acquired or the implants can be sectioned and used for immunohistochemistry. The morphology of the implant does not change following infection.

Antimicrobial therapy

One important parameter in the establishment of this model was documenting the ability of the viral replication to be inhibited by an HCMV-specific antiviral compound. Ganciclovir administered either by direct intraperitoneal injection or in the drinking water was shown to reduce the amount of infectious virus in a dose-dependent manner (Mocarski *et al.*, 1993). Indeed, these studies showed that the 2–4 log reduction in titer with administration of ganciclovir at 8 or 40 mg/kg per day is similar to the profile seen in humans (Mocarski *et al.*, 1993). These studies set the stage for further evaluation of antivirals in this model.

Pitfalls (advantages/disadvantages) of the model

Unlike tissue culture systems, the SCID-hu (thy/liv) mouse predicts the ability of HCMV to replicate in human epithelial cells. The implant, present in an intact organism, allows assessment of the impact of antivirals after they have been exposed to several of the modifications and alterations possibly encountered in a mammalian host. One of the main disadvantages of this model is the lack of histopathology associated with HCMV infection of the implant. Because there are no overt signs of viral infection on either the mouse or on the implant, quantification of virus or viral antigen or nucleic acid must be used to reveal effects of either antiviral agents or mutations present in the virus. Limitations to these parameters may allow important effects of an agent or a mutation to be missed.

Contributions of the model to infectious disease therapy

Currently, many agents proposed for treatment of HCMV infection and disease are studied either solely *in vitro* or tested on mice infected with murine CMV (MCMV). Although MCMV is useful for understanding the limited effects of these compounds *in vivo*, the extensive genetic and biological differences between MCMV and HCMV may result in excluding compounds that may have significant activity against HCMV. The SCID-hu (thy/liv) model allows direct analysis of the antiviral on HCMV and may prove to be a useful tool in evaluating the activity of a compound *in vivo*.

References

Alford, C. A., Britt, W. J. (1995). Cytomegalovirus. In *Fields Virology* (eds Fields, B. N., Knipe, D. M., Howley, P. M.), pp. 2493–2534. Lippincott-Raven, New York.

Boeckh, M., Gooley, T. A., Reusser, P., Buckner, C. D., Bowden, R. A. (1995). Failure of high-dose acyclovir to prevent cytomegalovirus disease after autologous marrow transplantation. *J. Infect. Dis.*, **172**, 939–943.

Britt, W. J. (1996). Vaccines against human cytomegalovirus: time to test. *Trends Microbiol.*, **4**, 34–38.

Brown, J. M., Kaneshima, H., Mocarski, E. S. (1995). Dramatic interstrain differences in the replication of human cytomegalovirus in SCID-hu mice. *J. Infect. Dis.*, **171**, 1599–1603.

Cha, T. A., Tom, E., Kemble, G. W., Duke, G. M., Mocarski, E. S., Spaete, R. R. (1996). Human cytomegalovirus clinical isolates carry at least 19 genes not found in laboratory strains. *J. Virol.*, **70**, 78–83.

Chee, M. S., Bankier, A. T., Beck, S. *et al.* (1990). Analysis of the protein-coding content of the sequence of human cytomegalovirus strain AD169. *Curr. Top. Microbiol. Immunol.*, **154**, 125–170.

Collier, A. C., Handsfield, H. H., Ashley, R. *et al.* (1995). Cervical but not urinary excretion of cytomegalovirus is related to sexual activity and contraceptive practices in sexually active women. *J. Infect. Dis.*, **171**, 33–38.

DiLoreto, D. Jr, Epstein, L. G., Lazar, E. S., Britt, W. J., del Cerro, M. (1994). Cytomegalovirus infection of human retinal tissue: an *in vivo* model. *Lab. Invest.*, **71**, 141–148.

Drew, W. L. (1992). Cytomegalovirus infection in patients with AIDS. *Clin. Infect. Dis.*, **14**, 608–615.

Drew, W. L. (1996). Cytomegalovirus resistance to antiviral therapies. *Am. J. Health Syst. Pharm.*, **53**, S17–S23.

Emanuel, D., Cunningham, I., Jules, E. K. *et al.* (1988). Cytomegalovirus pneumonia after bone marrow transplantation successfully treated with the combination of ganciclovir and high-dose intravenous immune globulin. *Ann. Intern. Med.*, **109**, 777–782.

Falagas, M. E., Snydman, D. R. (1995). Recurrent cytomegalovirus disease in solid-organ transplant recipients. *Transplant. Proc.*, **27**, 34–37.

Fowler, K. B., Stagno, S., Pass, R. F., Britt, W. J., Boll, T. J., Alford, C. A. (1992). The outcome of congenital cytomegalovirus infection in relation to maternal antibody status [see comments]. *N. Engl. J. Med.*, **326**, 663–667.

Hahn, G., Jores, R., Mocarski, E. S. (1998). Cytomegalovirus remains latent in a common progenitor of dendritic and myeloid cells. *Proc. Natl Acad. Sci. USA*, **95**, 3937–3942.

Istas, A. S., Demmler, G. J., Dobbins, J. G., Stewart, J. A. (1995). Surveillance for congenital cytomegalovirus disease: a report from the National Congenital Cytomegalovirus Disease Registry. *Clin. Infect. Dis.*, **20**, 665–670.

Kondo, K., Kaneshima, H., Mocarski, E. S. (1994). Human cytomegalovirus latent infection of granulocyte-macrophage progenitors. *Proc. Natl. Acad. Sci. USA*, **91**, 11879–11883.

Lagenaur, L. A., Manning, W. C., Vieira, J., Martens, C. L., Mocarski, E. S. (1994). Structure and function of the murine cytomegalovirus sgg1 gene: a determinant of viral growth in salivary gland acinar cells. *J. Virol.*, **68**, 7717–7727.

Laycock, K. A., Fenoglio, E. D., Hook, K. K., Pepose, J. S. (1997). An *in vivo* model of human cytomegalovirus retinal infection. *Am. J. Ophthalmol.*, **124**, 181–189.

Maidji, E., Tugizov, S., Jones, T., Zheng, Z., Pereira, L. (1996). Accessory human cytomegalovirus glycoprotein US9 in the unique short component of the viral genome promotes cell-to-cell transmission of virus in polarized epithelial cells. *J. Virol.*, **70**, 8402–8410.

Manning, W. C., Stoddart, C. A., Lagenaur, L. A., Abenes, G. B., Mocarski, E. S. (1992). Cytomegalovirus determinant of replication in salivary glands. *J. Virol.*, **66**, 3794–3802.

McCune, J. M., Namikawa, R., Kaneshima, H., Shultz, L. D., Lieberman, M., Weissman, I. L. (1988). The SCID-hu mouse: murine model for the analysis of human hematolymphoid differentiation and function. *Science*, **241**, 1632–1639.

McCune, J. M., Namikawa, R., Shih, C. C., Rabin, L., Kaneshima, H. (1990). Suppression of HIV infection in AZT-treated SCID-hu mice. *Science*, **247**, 564–566.

McVoy, M. A., Adler, S. P. (1989). Immunologic evidence for frequent age-related cytomegalovirus reactivation in seropositive immunocompetent individuals. *J. Infect. Dis.*, **160**, 1–10.

Mendelson, M., Monard, S., Sissons, P., Sinclair, J. (1996). Detection of endogenous human cytomegalovirus in CD34+ bone marrow progenitors. *J. Gen. Virol.*, **77**, 3099–3102.

Meyers, J. D., Flournoy, N., Thomas, E. D. (1982). Nonbacterial pneumonia after allogeneic marrow transplantation: a review of 10 years' experience. *Rev. Infect. Dis.*, **4**, 1119–1132.

Mocarski, E. S. J., Kemble, G. W. (1996). Recombinant cytomegaloviruses for study of replication and pathogenesis. *Intervirology*, **39**, 320–330.

Mocarski, E. S., Bonyhadi, M., Salimi, S., McCune, J. M., Kaneshima, H. (1993). Human cytomegalovirus in a SCID-hu mouse: thymic epithelial cells are prominent targets of viral replication. *Proc. Natl Acad. Sci. USA*, **90**, 104–108.

Mocarski, E. S., Prichard, M. N., Tan, C. S., Brown, J. M. (1997). Reassessing the organization of the UL42-UL43 region of the human cytomegalovirus strain AD169 genome. *Virology*, **239**, 169–175.

Namikawa, R., Kaneshima, H., Lieberman, M., Weissman, I. L., McCune, J. M. (1988). Infection of the SCID-hu mouse by HIV-1. *Science*, **242**, 1684–1686.

Namikawa, R., Weilbaecher, K. N., Kaneshima, H., Yee, E. J., McCune, J. M. (1990). Long-term human hematopoiesis in the SCID-hu mouse. *I. Exp. Med.*, **172**, 1055–1063.

Plotkin, S. A. (1994). Vaccines for varicella-zoster virus and cytomegalovirus: recent progress. *Science*, **265**, 1383–1385.

Plotkin, S. A., Starr, S. E., Friedman, H. M., Gonczol, E., Weibel, R. E. (1989). Protective effects of Towne cytomegalovirus vaccine against low-passage cytomegalovirus administered as a challenge. *J. Infect. Dis.*, **159**, 860–865.

Rowe, W. P., Hartley, J. W., Waterman, S., Turner, H. C., Huebner, R. J. (1956). Cytopathogenic agent resembling human salivary gland virus recovered from tissue cultures of human adenoids. *Proc. Soc. Exp. Biol. Med.*, **92**, 418–424.

Soderberg-Naucler, C., Fish, K. N., Nelson, J. A. (1997). Reactivation of latent cytomegalovirus by allogeneic stimulation of blood cells from healthy donors. *Cell*, **91**, 119–126.

Staczek, J. (1990). Animal cytomegaloviruses. *Microbiol. Rev.*, **54**, 247–265.

Stagno, S., Pass, R. F., Dworsky, M. E. *et al.* (1982). Congenital cytomegalovirus infection: the relative importance of primary and recurrent maternal infection. *N. Engl. J. Med.*, **306**, 945–949.

Stanley, S. K., McCune, J. M., Kaneshima, H. *et al.* (1993). Human immunodeficiency virus infection of the human thymus and disruption of the thymic microenvironment in the SCID-hu mouse. *J. Exp. Med.*, **178**, 1151–1163.

Winston, D. J., Ho, W. G., Champlin, R. E. (1990). Cytomegalovirus infections after allogeneic bone marrow transplantation. *Rev. Infect. Dis.*, **12** (suppl 7), 776–792.

Chapter 115

Animal Model for Ocular Human Cytomegalovirus Infections in SCID-hu Mice

D. J. Bidanset, M. del Cerro, E. S. Lazar, O. M. Faye-Petersen and E. R. Kern

Background of human infection

Human cytomegalovirus (HCMV) is a member of the Betaherpesvirus subfamily of Herpesviridae with a ubiquitous serological distribution. This seropositivity normally follows an unrecognized infection acquired during childhood. (For review see Sullivan and Hanshaw, 1982; Ho, 1991; Alford and Britt, 1993; Sinzger and Jahn, 1996; Gershon *et al.*, 1997). However, in immunodeficient or immunosuppressed individuals, infection with HCMV can cause severe clinical presentations (Ho, 1990). For example, congenital HCMV infections in neonates can manifest as petechiae accompanied by hepatomegaly or splenomegaly (Dobbins and Stewart, 1992). Additional manifestations include intracranial calcifications, microencephaly, thrombocytopenia and neurological sequelae which can lead to outcomes such as psychomotor retardation, neuromuscular disorders, hearing loss, and death (Fowler *et al.*, 1992). In immunodeficient or immunosuppressed adults, HCMV primary infection or reactivation can result in infectious mononucleosis-like syndrome (Klemola and Kaariainen, 1965), pneumonitis (Ljungman *et al.*, 1994), hepatitis (Barkholt *et al.*, 1994), gastrointestinal disorders (Sinzger *et al.*, 1995), retinitis (Jabs *et al.*, 1989; Heinemann, 1992), encephalopathies (Dorfman, 1973) and the various multi-system diseases characterized under the general heading of cytomegalic inclusion disease (Gershon *et al.*, 1997). Cytopathology includes both nuclear and cytoplasmic inclusions similar to those seen with other herpesviruses (Ho, 1990).

The HCMV particle ranges in size from 180 to 250 nm and has 162 capsomeres. The inclusion bodies present in the nucleus and later in the cytoplasm of HCMV-infected cells contain aggregated chromatin and virus particles. HCMV has a slow reproductive cycle leading to slowly forming foci in culture. It is primarily cell-associated and, like other herpesviruses, HCMV is able to establish a latent infection which can reactivate under appropriate stimuli. Unlike other herpesviruses, this reactivation does not occur sporadically but rather under conditions of suppressed immunity. (For review see Mathews, 1979; Sullivan and Hanshaw, 1982); Ho, 1990.)

The severity of HCMV infection in immunocompromised patients first became evident in 1982 when the first case of CMV retinitis in acquired immunodeficiency syndrome (AIDS) patients was reported (for review, see Jacobson, 1997). This disease is characterized by retinal necrosis and edema and, if left untreated, can result in severe visual impairment or blindness. By 1984, ganciclovir was being widely used as an antiviral drug therapy for CMV infections (Goodrich *et al.*, 1991; Cinque *et al.*, 1995) and since then only two other drugs, namely foscarnet (for review see Kuppermann, 1997) and cidofovir (for review, see Safrin *et al.*, 1997), have been approved and licenced by the Food and Drug Administration (FDA) for use in patients with HCMV infections.

Since the portions of the retina damaged by HCMV do not regain function, the goal of antiviral therapy is to prevent further retinal damage. However, this is not often the case with any of the approved antiviral therapies. In many patients, the progression of the disease is only slowed down, and in others a resistance of the HCMV to the antiviral drug develops. In addition, serious and irreversible toxic effects caused by the antiviral drug occur which warrant cessation of treatment. Taken together, these problems suggest the need for the development of new drugs and the development of *in vivo* models which can accurately predict the outcome of various antiviral therapies.

One of the major problems in HCMV research has been the fact that the virus exhibits a narrow host specificity. HCMV does not replicate in non-human tissues at a detectable level and investigations using respective animal CMVs may not be predictive of the immune responses seen with HCMV infections (Sinzger and Jahn, 1996). For these reasons, new approaches to developing and establishing HCMV models in animals have been investigated. One such model is presented here.

Background for animal model

Both cyclosporin A immunosuppressed rats (Cvetkovich *et al.*, 1992; Epstein *et al.*, 1992; DiLoreto *et al.*, 1996a) and athymic (nude) rats (Epstein *et al.*, 1994; Laycock *et al.*, 1997) have been used as host animals for human retinal

Handbook of Animal Models of Infection
ISBN 0-12-775390-7

tissue transplants. Recently the use of severe combined immunodeficient (SCID) mice as host animals for human retinal xenografts (SCID-hu) has been described and appears to be a preferable method (Diloreto *et al.*, 1994; Epstein *et al.*, 1994). This method will be described here.

Preparation of animals

SCID mice are immunodeficient, lacking B-cell (immunoglobulin variable chain rearrangement) and T-cell (T-cell receptor) immune functions. As such, they require sterilized housing, bedding, food and water to decrease chances of infection in their surroundings.

Cage changes in a sterile designated hood and housing in filter-covered cages in a facility with limited access are also recommended. As contamination can occur from both animals and humans, great care should be taken to keep SCID mice separate from other animals and to limit exposure to them by humans. Care-takers and researchers should always wear a designated gown, mask, goggles and gloves before opening the cages. Surgery should be performed in a designated clean room using aseptic techniques.

Tissue preparation and implantation

Human fetal eyes (16–20 weeks gestation) were temporarily stored in balanced salt solution until dissection to prevent drying out. For dissection, the eyes were transferred to a sterile 35×10 mm Petri dish and held in place with conjunctival forceps (Stortz Instrument Co., St Louis, MO), being careful not to damage the underlying retina. A small incision was made posterior to the ora serrata using a #11 surgical blade to relieve the global pressure. The incision was continued using small stitch removal scissors (Stortz Instrument Co., St. Louis, MO) all the way around posterior to the ora serrata, and the cornea and lens were removed. A second incision perpendicular to the first was made towards the optic nerve to allow the posterior globe to be opened up. The vitreous was removed and the retina was carefully detached from the underlying retinal pigment epithelium using the smaller end of a zaufal (Martin, Tuttlingen, Germany). The retina was clipped at the optic nerve and carefully removed using arruga curved forceps (Stortz Instrument Co., St. Louis, MO). The retina was then stored in Optisol (Chiron Ophthalamics, Irvine, CA) at 4°C for up to 7 days prior to implantation (DiLoreto *et al.*, 1996b).

In preparation for transplantation, the fetal human retinas were mechanically dissociated in Optisol using small stitch removal scissors. A 1 cc tuberculin syringe was filled with fresh Optisol using a 27 G × 3/8 inch winged infusion set and retinal tissue was slowly drawn into the needle and tubing.

SCID mice (4–8 weeks old) were anesthetized with an intraperitoneal injection of ketamine (125 mg/kg) and xylazine (20 mg/kg). Eyes were further anesthetized with a topical anesthetic, proparacaine HCl (0.05%; Allergan Inc., Hormigueros, Puerto Rico). If necessary, eyes were kept moist with balanced salt solution during the transplant procedure.

For transplantation, eyes were exposed by gently pushing on the skin above and below the eye. The butterfly needle, containing the minced retinal tissue, was then inserted through the nasal sclera with the bevel up. The needle was carefully manipulated above the lens and iris toward the temporal portion of the anterior chamber (Figure 115.1). At this point, approximately 5 μl of the retinal suspension was injected into the anterior chamber of the mouse eye. Using this procedure, approximately 20 animals (40 eyes) can be transplanted per human fetal retina. Over the week following transplantation, the retinal tissue attaches to the host iris and begins to grow. By 2 months after transplantation, the retinal graft can be seen to have increased considerably in size and become infiltrated with iridal vasculature (Figure 115.2).

Preparation of virus pools

The HCMV strains, AD169, Toledo and Towne were all grown in human foreskin fibroblast (HFF) cells. The HFF cells were passaged from primary cultures and maintained in minimal essential media (MEM) containing 10% fetal bovine serum (FBS), 2 mM/l L-glutamine, 100 U/ml penicillin, and 25 μg/ml gentamicin for up to 10 passages. Confluent cell monolayers were infected by removing the media and adding an appropriate inoculum of virus (approximately 10^4 plaque-forming units), which was allowed to adsorb for 1 hour, shaking every 15 minutes. Media was then replaced and cultures incubated at 37°C for approximately 1–2 weeks. When a cytopathic effect of 90% was reached, supernatant was removed and cells were washed and trypsinized. Cell-associated virus was then suspended in MEM containing 10% FBS and 10% dimethyl sulfoxide and stored in 0.5 ml aliquots at −70°C until further use.

Infection process

Approximately 1–2 weeks after transplantation of the human retinal tissue, animals were again anesthetized with the Ketamine/Xylazine mixture and the topical Proparacaine HCl. Stock solutions of HCMV were diluted to contain between 1×10^5 and 5×10^5 plaque-forming units (pfu)/ml in MEM containing 10% FBS and kept on ice. Once again, a 1 cc tuberculin syringe was loaded with virus and the winged infusion set was filled. The needle was inserted into the mouse eye as previously described, being

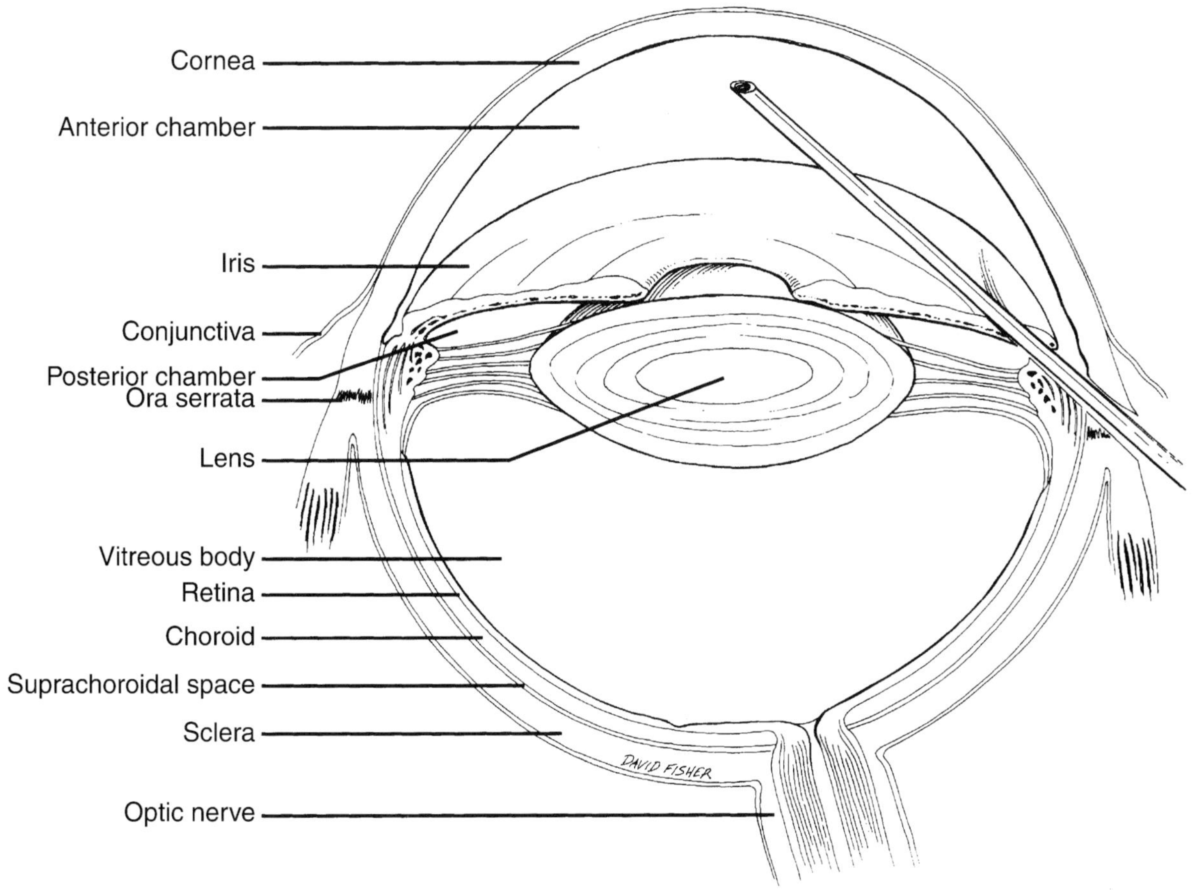

Figure 115.1 Transplantation of human fetal retinal tissue into the anterior chamber of the severe combined immunodeficiency (SCID) mouse eye.

careful not to disturb the implant. Once the needle was temporal to the implant, approximately 20 μl of HCMV was injected. Since about 50% of the inoculum flows back out of the eye, we estimate that only 10 μl of the virus remains in the anterior chamber of the eye, giving a viral input of between 1×10^3 and 5×10^3 pfu/eye.

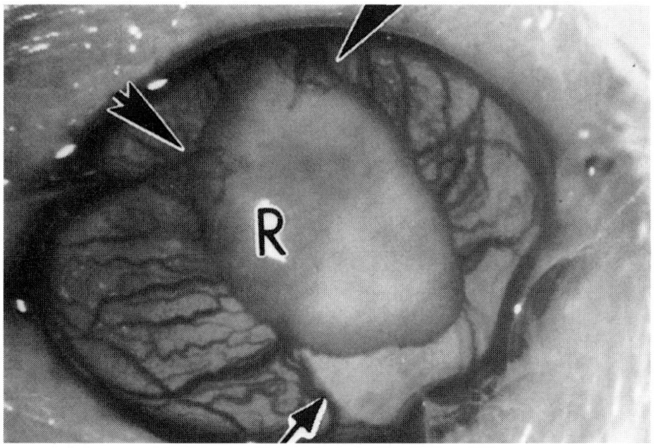

Figure 115.2 Photograph of human fetal retina (R) transplanted into the anterior chamber of a severe combined immunodeficiency (SCID) mouse eye. Arrowheads = iridal vessels; arrow = pupil. From DiLoreto *et al*. (1994), with permission.

Parameters for monitoring infection and response to treatment

Since HCMV does not infect murine tissue, monitoring of infection is accomplished by examination of the human retinal implant. At various times after infection, mice were sacrificed with a lethal injection of Ketamine/Xylazine and their eyes removed for further analysis. Briefly, enucleation was performed as follows. The whiskers were cut and the inferior conjunctiva was grasped with caliper-style forceps (Stortz Instrument Co., St Louis, MO) at either the nasal or temporal corner of the eye. Small, fine, curved iris scissors (Stortz Instrument Co., St Louis, MO) were inserted into the conjunctiva (curved away from the eye) to separate the globe from the surrounding tissue. Once the conjunctiva was cut, the tissue underneath the eye (i.e., muscle and optic nerve) was cut until the eye could be removed from the cavity.

Eyes were temporarily kept in balanced salt solution and then homogenized in 1.0 ml clinical MEM (MEM containing 10% FBS, 2 mM L-glutamine, 200 U/ml penicillin, 50 μg/ml gentamicin, and 3 μg/ml fungizone). The homogenate was centrifuged at 1500 rpm for 15 minutes at 4°C using a Beckman AH-4 rotor. The supernatant was removed and stored at −70°C.

For quantitation of HCMV replication, viral titers were determined by plaque assay in HFF cells. Homogenate supernatants were thawed quickly at 37°C and placed on ice. Each sample was diluted 1/3, 1/10, and 1/100 and was assayed in duplicate using confluent HFF cells in 6-well plates. The media was removed and cells were inoculated with 0.2 ml of the various dilutions of eye homogenate supernatant. Virus was allowed to adsorb for 1 hour at 37°C, shaking the plates every 15 minutes, after which 2.0 ml of MEM containing 2% FBS was added to each well. Cultures were incubated for 8 days and were fed every 4–5 days by addition of 1.0 ml of MEM containing 2% FBS. After 8 days, the monolayers were stained with 6% Neutral Red in phosphate-buffered saline for 6–8 hours, and plaques were counted using a stereomicroscope.

Determination of HCMV replication kinetics

To determine the replication rates of three well-characterized strains of HCMV, namely AD169, Toledo and Towne, the SCID-hu retinal implants were infected with 2000 pfu of each strain. At 14 and 28 days after infection, mice were sacrificed, eyes were enucleated and homogenized in MEM containing 10% FBS, 2 mM/l L-glutamine, 200 U/ml penicillin, 50 μg/ml gentamicin and 3 μg/ml fungizone, and homogenates were stored at −70°C for plaque assay analysis. The results (Table 115.1) indicate that all three strains of HCMV were able to replicate in the SCID-hu retinal implants with a similar infection rate at each time point. However, at the times assayed, neither AD169 nor Towne were able to replicate to the titers observed with the Toledo strain of HCMV.

The growth of the Toledo strain of HCMV was also determined at various times up to 70 days after the initial infection. As shown in Figure 115.3, the Toledo strain replicated gradually to a peak titer of about 300 pfu/ml by about 28 days after the initial infection. This titer then gradually

Table 115.1 Replication of various strains of human cytomegalovirus in human fetal retinal cell implants in severe combined immunodeficiency (SCID) mice

	Day 14		*Day 28*	
Strain	*% Positive*	*pfu/ml**	*% Positive*	*pfu/ml**
AD169	12.5 (1/8)	15 ± 0	50 (4/8)	36 ± 8
Toledo	25 (2/8)	225 ± 286	50 (2/8)	189 ± 160
Towne	37.5 (3/8)	60 ± 57	25 (2/8)	90 ± 85

* Values are expressed as plaque-forming unit (pfu)/ml ± SD.

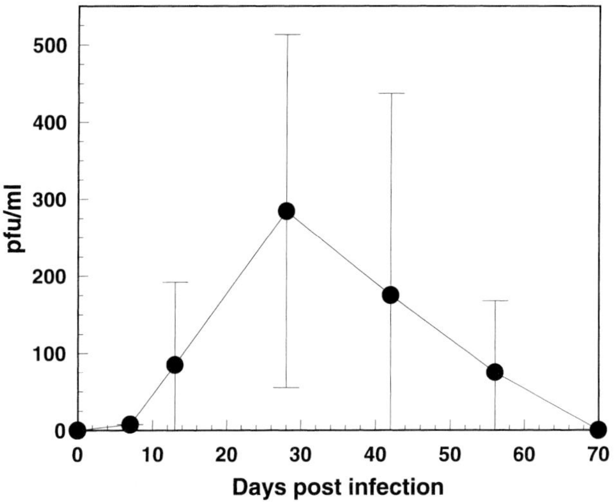

Figure 115.3 Replication of human cytomegalovirus (HCMV) in SCID-hu retinal implants.

decreased to undetectable levels by 70 days post-infection. In other experiments, peak HCMV levels of 3000 pfu/ml have been detected. It is important to note that 7 days after infection of the retinal implant there is no or only a small amount of virus in the implant, indicating that we are detecting newly replicating virus and not just recovering input virus.

Antiviral therapy

In order to validate this model for use in evaluating antiviral therapies, the effect of ganciclovir (GCV) on replication of the Toledo strain of HCMV in SCID-hu retinal implants was examined. Two weeks after implantation, SCID-hu retinal implants were infected with 2000 pfu of the Toledo strain of HCMV. Starting 1 day after infection, mice were injected intraperitoneally with either 15 or 45 mg/kg of GCV in sterile water or sterile water alone. This treatment was given twice daily for 14 days and then once daily for an additional 14 days. Eyes from placebo- and GCV-treated mice were removed on the indicated days after infection and HCMV titers were determined by plaque assay. The results are summarized in Table 115.2.

In the group that received placebo, 83% of the implants were positive for HCMV on day 28 with a mean plaque count of 2938 ± 4816. In animals that received 15 mg of GCV per kg, 88% of the implants were positive for HCMV with a mean plaque count of 461 ± 758 pfu per ml of homogenized sample. In the mice that were treated with 45 mg GCV per kg, only 25% of the implants had recoverable HCMV on day 28, with an average plaque count of 31.9 ± 63.8 pfu/ml. These results indicate that GCV treatment at 45 and even at 15 mg GCV per kg markedly reduced HCMV replication in the SCID-hu retinal implants when compared to titers obtained from implants in placebo treated mice.

Table 115.2 Effect of ganciclovir treatment on human cytomegalovirus (HCMV) replication in SCID-HU retinal implants

Day	*Treatment*	*Number positive/total*	*Mean ± SD (pfu/ml)*	*Median (pfu/ml)*	*Wilcoxon*
1	Placebo	1/6 (17%)	6.3 ± 15.3	0	
1	GCV-15	3/6 (50%)	12.5 ± 15.5	7.5	NS
1	GCV-45	4/6 (67%)	85 ± 153	18.8	NS
4	Placebo	0/6 (0%)	0 ± 0	0	
4	GCV-15	0/6 (0%)	0 ± 0	0	NS
4	GCV-45	0/6 (0%)	0 ± 0	0	NS
7	Placebo	1/6 (17%)	10.0 ± 24.5	0	
7	GCV-15	0/6 (0%)	0 ± 0	0	NS
7	GCV-45	1/6 (17%)	1.3 ± 3.1	0	NS
10	Placebo	0/6 (0%)	0 ± 0	0	
10	GCV-15	2/6 (33%)	9.6 ± 20.0	0	NS
10	GCV-45	0/6 (0%)	0 ± 0	0	NS
14	Placebo	2/6 (33%)	55.0 ± 103	0	
14	GCV-15	2/6 (33%)	36.3 ± 85.2	0	NS
14	GCV-45	0/6 (0%)	0 ± 0	0	NS
21	Placebo	3/6 (50%)	100 ± 123	52.5	
21	GCV-15	1/6 (17%)	4.2 ± 10.2	0	NS
21	GCV-45	1/6 (17%)	8.8 ± 21.4	0	NS
28	Placebo	5/6 (83%)	2938 ± 4816	1360	
28	GCV-15	7/8 (88%)	461 ± 758	139	NS
28	GCV-45	1/4 (25%)	31.9 ± 63.8	0	0.08

NS, Not significant.

Furthermore, these results suggest that this model of infection may be extremely valuable for *in vivo* evaluation of drugs for HCMV prior to their use in humans.

Pitfalls (advantages/disadvantages) of the model

There are several advantages of the SCID-hu retinal implant model over other *in vivo* models of HCMV. First, the mouse model compared to the athymic or immunosuppressed rat-hu model uses animals of smaller sizes, thus utilizing reduced housing space and smaller quantities of antiviral drugs. Second, this model utilizes an inbred strain of mice that are naturally immunocompromised, thus eliminating the need consistently to immunosuppress animals so that the human implant is not rejected. Third, using the SCID-hu retinal implant model, one is able specifically to elucidate HCMV infection in human ocular tissue, a recent problem in immunocompromised individuals.

In comparison, the SCID-hu Thy/Liv model in which human fetal thymus and liver are implanted under the kidney capsule provides information on HCMV visceral disease which is generally not a problem in these infections. Using the ocular model, drug can be delivered easily via intraperitoneal injection or via the oral route. As the implant becomes vascularized, drug administered by either route will be able to reach the retinal implant. As infection is limited to the human retinal implant, monitoring of the infection is easily performed by removal and plaque assay analysis of the implanted mouse eye. Additionally, although the experimental infection is not a model for HCMV retinitis, it provides an excellent model for HCMV replication in ocular tissues and can be utilized to determine the effectiveness of the transport of a drug from the circulation into the eye.

One of the major shortcomings of the SCID-hu retinal implant model is the variability of the results. The large standard deviations seen in a majority of the results can be attributed to several factors. One of these factors is the inability precisely to control how much tissue is transplanted into the anterior chamber of the SCID mouse eye. Another consideration is the inability to determine how much tissue will implant on to the host iris and how large the implant will grow. And lastly is the inability to determine whether an implant is infected with virus and to what extent prior to the sacrifice of the animal. It is impossible to determine the amount of virus which comes in contact with the implant.

Because of these factors and their effect on variability, interpretation of data obtained from this model should be carefully evaluated and compared to previously tested reference controls.

Contributions of the model to infectious disease therapy

The treatment of HCMV retinitis with GCV is well-documented (for review, see Jacobson, 1997). The median effective inhibitory dose of 1.5 μg/ml is achieved in human intravitreal space with twice-daily intravenous dosings of

5 mg/kg. This treatment regime and outcome appear to correlate well with that seen in the treatment of SCID-hu retinal implant.

Although animal models of CMV infection do not always predict precisely what will happen in the clinical situation, the use of the immunosuppressed or immunocompromised animal as a host for human tissue and HCMV infection appears to parallel closely the clinical presentation. Since other drugs such as cidofovir were also efficacious in the model and correlated with their use in the clinical situation, this model has the potential of providing predictable information on the efficacy of new antiviral therapies. Additionally, the SCID-hu retinal implant model can be used to elucidate further the biology of HCMV infection *in vivo* as well as to provide insight into the pathophysiology, virulence, pathogenesis and effect of antiviral treatment on drug-resistant HCMV strains.

Acknowledgments

We wish to thank Caroll B. Hartline and Rachel J. Rybak for technical assistance. This work was supported by the Public Health Service grant U19-Al-31718 and by the Public Health Service contract NO1-Al-65290 from NIAID, NIH.

References

Alford, C. A., Britt, W. J. (1993). Cytomegalovirus. In *The Human Herpesviruses* (eds Roizman, B., Whitley, R. J., Lopez, C.), pp. 227–255. Raven Press, New York.

Barkholt, L. M., Ehrnst, A., Varess, B. (1994). Clinical use of immunohistopathologic methods for the diagnosis of cytomegalovirus hepatitis in human liver allograft biopsy specimens. *Scand. J. Gastroenterol.*, **29**, 553–560.

Cinque, P., Baldanti, F., Vago, L. *et al.* (1995). Ganciclovir therapy for cytomegalovirus (CMV) infection of the central nervous system in AIDS patients: monitoring by CMV DNA detection in cerebrospinal fluid. *J. Infect. Dis.*, **171**, 1603–1606.

Cvetkovich, T. A., Lazar, E. S., Blumberg, B. M. *et al.* (1992). Human immunodeficiency virus type 1 (HIV-1) infection of neural xenografts. *Proc. Natl Acad. Sci. USA*, **11**, 5162–5166.

DiLoreto, D., Epstein, L. G., Lazar, E. S., Britt, W. J., del Cerro, M. (1994). Cytomegalovirus infection of human retinal tissue: an *in vivo* model. *Lab Invest.*, **71**, 141–148.

DiLoreto, D., del Cerro, C., del Cerro, M. (1996a). Cyclosporine treatment promotes survival of human fetal neural retina transplanted to the subretinal space of light-damaged Fischer 344 rat. *Exp. Neurol.*, **140**, 37–42.

DiLoreto, D. A., del Cerro, C., Lazar, E. S., Cox, C., del Cerro, M. (1996b). Storage of human fetal retina in optisol prior to subretinal transplantation. *Cell Transplant.*, **5**, 553–561.

Dobbins, J. G., Stewart, J. A. S. (1992). Surveillance of congenital cytomegalovirus disease 1990–1991. *Morbid. Mortal. Week. Rep.*, **41**, SS-2.

Dorfman, L. J. (1973). Cytomegalovirus encephalitis in adults. *Neurology*, **23**, 136–144.

Epstein, L. G., Cvetkovich, T. A., Lazar, E. *et al.* (1992). Successful xenografts of second trimester human fetal brain and retinal tissue in the anterior chamber of the eye of adult immunosuppressed rats. *J. Neural Transplant. Plasticity*, **3**, 151–158.

Epstein, L. G., Cvetkovich, T. A., Lazar, E. S. *et al.* (1994). Human neural xenografts: progress in developing an *in-vivo* model to study human immunodeficiency virus (HIV) and human cytomegalovirus (HCMV) infection. *Adv. Neuroimmunol.*, **4**, 257–260.

Fowler, K. B., Stagno, S., Pass, R. F., Britt, W. J., Boll, T. J., Alford, C. A. (1992). The outcome of congenital cytomegalovirus infection in relation to maternal antibody status. *N. Engl. J. Med.*, **326**, 663–667.

Gershon, A. A., Gold, E., Nankervis, G. A. (1997). Cytomegalovirus. In *Viral Infections of Humans* (eds Evans A. S., Kaslow, R. A.), pp. 229–251. Plenum, New York.

Goodrich, J. M., Mori, M., Gleaves, C. *et al.* (1991). Early treatment with ganciclovir to prevent cytomegalovirus disease after allogenic bone marrow transplantation. *N. Engl. J. Med.*, **325**, 1601–1607.

Heinemann, M. H. (1992). Characteristics of cytomegalovirus retinitis in patients with acquired immunodeficiency syndrome. *Am. J. Med.*, **92**, 12S–16S.

Ho, M. (1990). Epidemiology of cytomegalovirus infections. *Rev. Infect. Dis.*, **12** (suppl.), 701–710.

Ho, M. (ed) (1991). *Cytomegalovirus; Biology and Infection.*, 2nd edn. Plenum, New York.

Jabs, D. A., Engler, C., Bartlett, J. G. (1989). Cytomegalovirus retinitis and acquired immunodeficiency syndrome. *Arch. Opthalmol.*, **107**, 75–80.

Jacobson, M. A. (1997). Drug therapy: treatment of cytomegalovirus retinitis in patients with acquired immunodeficiency syndrome. *N. Engl. J. Med.*, **337**, 105–114.

Klemola, E., Kaariainen, L. (1965). Cytomegalovirus as a possible cause of a disease resembling infectious mononucleosis. *Br. Med. J.*, **2**, 1099–1102.

Kuppermann, B. D. (1997). Therapeutic options for resistant cytomegalovirus retinitis. *J. AIDS Human Retrovirol.*, **14** (suppl.), S13–S21.

Laycock, K. A., Feneglio, E. D., Hook, K. K., Pepose, J. S. (1997). An *in vivo* model of human cytomegalovirus retinal infection. *Am. J. Opthalmol.*, **124**, 181–189.

Ljungman, P., Biron, P., Bosi, A. *et al.* (1994). Cytomegalovirus interstitial pneumonia in autologous bone marrow transplant recipients. Infectious disease working party of the European group for bone marrow transplantation. *Bone Marrow Transplant.*, **13**, 209–212.

Mathews, R. E. F. (1979). Classification and nomenclature of viruses. Third report of the International Committee on Taxonomy of Viruses. *Intervirology*, **12**, 132–296.

Safrin, S., Cherrington, J., Jaffe, H. S. (1997). Clinical use of cidofovir. *Rev. Med. Virol.*, **7**, 145–156.

Sinzger, C., Jahn, G. (1996). Human cytomegalovirus cell tropism and pathogenesis. *Intervirology*, **39**, 302–319.

Sinzger, C., Grefte, A., Plachter, B., Gouw, A. S. H., The, T. H., Jahn, G. (1995). Fibroblasts, epithelial cells, endothelial cells and smooth muscle cells are the major targets of human cytomegalovirus infection in lung and gastrointestinal tissues. *J. Gen. Virol.*, **76**, 741–750.

Sullivan, J. L., Hanshaw, J. B. (1982). Human cytomegalovirus infections. In *Human herpesvirus infections: Clinical Aspects* (eds Glaser, R., Gotlieb-Stematsky, T.), pp. 57–83. Marcel Dekker, New York.

Chapter 116

Animal Models for Varicella-zoster Infections: Simian Varicella

K. F. Soike

Background of human infection

Varicella-zoster virus (VZV) infection is most commonly seen as a benign disease of children causing fever and vesicular skin lesions (Weller, 1983). In the adolescent or adult it may result in more severe disease. Complications, when they occur, include pneumonia, hepatitis, and encephalitis. Children with cancer (Feldman and Lott, 1987), the immunocompromised (Morgan, 1983; Perronne *et al.*, 1990), and patients with impaired cell-mediated immunity may display visceral or central nervous system involvement (Weller, 1983). The virus is a herpesvirus and is capable of producing a latent infection of ganglia (Gilden *et al.*, 1983, 1987; Hyman *et al.*, 1983; Vafai *et al.*, 1988). As a consequence, years after primary infection, the latent virus may be reactivated and appears again as a vesicular rash with accompanying pain, with its distribution limited to an area served by a specific dermatome. This condition, referred to as zoster or shingles, may be complicated by a persisting post-herpetic neuralgia (Weller, 1983).

The availability of an animal model for evaluation of potential antiviral therapies for varicella-zoster infection would be desirable. However, VZV is a highly species-specific virus producing clinical disease only in humans or in higher primate species. An exception is a report by Myers *et al.* (1980), who have shown infection of guinea-pigs with VZV by intranasal or subcutaneous injection of virus propagated in fetal guinea-pig tissue. Recovery of VZV from nasal swabs or from blood was possible in only a small percentage of guinea-pigs sampled and clinical disease was not observed. A subsequent study in hairless or depilated guinea-pigs (Myers *et al.*, 1991) inoculated with VZV by subcutaneous injection resulted in an erythematous or papular rash in some of the guinea-pigs which could be prevented by administration of VZV convalescent-phase guinea-pig serum. Rash, however, was not consistently reported in the inoculated animals. Lowry *et al.* (1993) using a polymerase chain (PCR) reaction, were able to show viremia in approximately 50% of three strains of guinea-pigs (strain 2, hairless, and Hartley guinea-pigs) using isolated peripheral blood mononuclear cells from the animals. PCR was also used to show the presence of latent virus in ganglia from 6 of 8 guinea-pigs tested.

An available alternative for testing efficacy of potential antivirals from VZV infection lies in the discovery and characterization of a closely related virus that does infect and produces clinical disease in several species of non-human primates and is referred to as simian varicella virus (SVV).

Background of model

Simian varicella was first recognized following an outbreak in a colony of African green monkeys housed at the Liverpool School of Tropical Medicine (Clarkson *et al.*, 1967). Of 17 African green monkeys (*Cercopithecus aethiops*) at risk, 9 deaths occurred. Clinically, a papular to vesicular rash was seen which on histological examination was marked by cells bearing herpes-type nuclear inclusions. Lung, liver and spleen from monkeys dying from the infection showed marked hemorrhagic necrosis and cells containing intranuclear inclusions. The authors described the virus as a possible VZV and named the virus the Liverpool vervet virus. A number of similar outbreaks have been reported around the world resulting in similar disease syndromes and the isolation of similar viruses (Table 116.1). The source of these outbreaks is unknown but evidence suggests that in some it may have been the consequence of exposure of naive animals to virus shed by a reactivated case. It is evident from the table that infection in the African green monkey (*C. aethiops*) and the patas monkey (*Erythrocebus patas*) is the most severe with the highest mortality. Infection in the macaque species such as the pigtail macaque (*Macaca nemestrina*), cynomolgus (*M. fascicularis*), stumptail (*M. arctoides*) and the rhesus (*M. mulatta*) is less severe.

The various SVV isolates, when compared to each other and to human VZV by cross-neutralization and complement fixation (CF) tests using sera prepared in rhesus monkeys to the Delta, Liverpool, Glaxo, and Medical Lake viruses, resulted in similar titers among the four simian viruses (Felsenfeld and Schmidt, 1977). The monkey sera were also effective in neutralization and CF tests with human VZV, although at lower titer than was seen with the

Handbook of Animal Models of Infection
ISBN 0-12-775390-7

Table 116.1 Occurrence of natural outbreaks of simian varicella virus infections in non-human primate colonies: species affected and mortalities

Date	*Location*	*Monkey species*	*Monkeys infected*	*Fatalities (%)*	*References*
1966	Liverpool School of Tropical Medicine	*Cercopithecus aethiops*	17	53	Clarkson *et al.*, 1967
1967	Glaxo Laboratories (UK)	*Erythrocebus patas*	47	36	McCarthy *et al.*, 1968
			48	40	
1968	Delta Primate Center	*E. patas*	39	50	Ayres, 1971
1969–71	Medical Lake, Washington	*Macaca nemistrina*	41	7	Blakey *et al.*, 1973
	Primate Center	*M. fascicularis*	9	11	
1973	Delta Primate Center	*E. patas*	24	58	Allen *et al.*, 1974
1981	Litton Bionetics	*E. patas*	101	44	Schmidt *et al.*, 1983
1983	Bowman Gray School	*C. aethiops*	9	56	Lehner *et al.*, 1984
		M. fascicularis	0	0	
		M. arctoides	0	0	
1984	Hazelton Laboratories	*C. aethiops*	35	23	
		M. fascicularis		5	
		M. mulatta		2	
1989	Japan National Institute of Health	*M. fascicularis*	111	14	Takasaka *et al.*, 1990

simian strains. Sera from human varicella and zoster patients resulted in higher neutralization and CF titers against VZV than against the simian viruses. Harbour and Caunt (1979), using neutralization, CF, immunofluorescence and gel precipitation tests, came to the conclusion that the Delta virus differed slightly from the Glaxo virus, the Liverpool vervet virus and the Medical Lake macaque virus. They also showed that they all share antigens in common with VZV. Gray and Gusick (1996) demonstrated that the genomic DNA restriction endonuclease profiles of epidemiologically distinct simian varicella isolates are similar, but not identical.

The pathology observed in African green monkeys infected with the Delta strain of SVV is similar to that seen in severe disseminated human disease (Roberts *et al.*, 1984; Dueland *et al.*, 1992). Lung, liver, spleen and lymph nodes all show inflammation, hemorrhagic necrosis and intranuclear inclusions. Lesions are also seen in the adrenal, the bone marrow, digestive tract and kidneys. The brain and spinal cord are not involved, although ganglia may be mildly inflamed and intranuclear inclusions are present in non-neuronal cells of some ganglia (Dueland *et al.*, 1992). The virus does establish latent infection in trigeminal and dorsal root ganglia demonstrated by PCR in African green monkeys following recovery from infection (Mahalingam *et al.*, 1991).

Molecularly, similarities in the genomes of VZV and SVV also show their relatedness. Gray and Oakes (1984) reported that the viruses have a conserved nucleotide sequence homology of at least 70% base pair matching. Restriction endonuclease analysis of the genomes of VZV and SVV show them to be similar in size and structure (Gray *et al.*, 1992; Pumphrey and Gray, 1992). The molecular size of SVV is 121.0 ± 0.8 kilobase pairs (kbp), approximately 76.8 megadaltons (MD), while VZV is 124.8 kbp or approximately 79.2 MD. The genome of both viruses contain an ~100 kbp long (L) component covalently linked to an invertible short (S) component. The S components contain a unique short sequence bracketed by inverted repeat sequences (Gray *et al.*, 1992; Fletcher and Gray, 1993). Immunoprecipitation analysis demonstrated cross-reactivity between the polypeptides and glycoproteins of VZV and SVV (Fletcher and Gray, 1992). These studies support the use of SVV as a model for VZV infection in humans.

Animal species (strains, sex, age)

Monkeys used in experimental infection with SVV have been patas monkeys and the African green monkey. Monkeys of either sex respond similarly to infection. The age of the monkey is not a major factor. Both the patas and African green monkeys are native to Africa and are present in Eastern, Western and South Africa. African green monkeys generally range in weight from 2 to 4 kg with an occasional adult male weighing more. Patas monkeys are larger, weighing between 3 and 7 kg. African green monkeys have been the species primarily used because of this smaller size and their easier availability. However, since the appearance of human infections with hemorrhagic fever viruses (Ebola, Marburg and Lessa fever) the importation of monkeys from Africa has been difficult.

Recently we have used African green monkeys from the Caribbean islands of St Kitts or Barbados. These monkeys, which were pets on slave ships, were released on the Caribbean islands in the late 1600s and early 1700s and thrived, with the numbers increasing to greater than 10 000 at present (McQuire, 1974). Essentially no additional monkeys have been added since the mid 1700s. They are readily available and respond well to SVV infection, developing rash, viremia and occasional mortality with pneumonia or hepatitis. Naturally occurring antibody to simian varicella virus has not been observed in either the African or Caribbean monkeys, suggesting that simian varicella is not endemic in the wild population. Although simian varicella virus has been observed to infect several macaque species, these monkeys have not been used in many experimental infections. Experience has shown the macaques to exhibit less rash and viremia with a lower mortality.

Monkeys purchased from whatever source require a period of quarantine for 60–90 days, during which time they are allowed to stabilize and are treated for concurrent bacterial and parasitic infections and undergo skin tests for tuberculosis. Importing of monkeys from outside the United States requires involvement of the United States Customs Department and personnel from the Centers for Disease Control and Prevention. The responsibility for clearing the monkeys for importation is best handled by the importer. The following is a partial list of commercial importers who can assist with the acquisition of monkeys.

- African or Caribbean monkeys: James Sears, Three Springs Scientific, 1730 West Rock Rd, Purkasee, PA; tel: (215) 757–6055
- Barbados African green monkeys: Jean Baulu, Barbados Primate Research Center, Farley Hill, St Peter, Barbados; tel: (809) 422–8826
- St Kitts African green monkeys: St Kitts Biomedical Research Foundation, Care of Maryanne Johnson, 100 Deepwood Drive, Hamden, CT 06517; tel: (203) 773–9300
- A source of African green monkeys (originally from St Kitts but bred and maintained in the United States) is: John Hardcastle, Head, Animal Resources, University of Southwestern Louisiana, New Iberia Research Center, 4401 West Admiral Doyle Drive, New Iberia, LA 70560; tel: (318) 482–0250; fax: (318) 373–0073.

Preparation of animals

Upon arrival from the importer the monkeys should be singly housed in an isolated area away from contact with other animals. A minimum cage size of 26.5 × 24 × 34 inches should be used and the monkeys should be observed daily for any symptoms of disease or any abnormal behavior. Following completion of the quarantine period the monkeys should continue to be housed singly to limit the spread of infection. Upon completion of an experiment with SVV the possibility of a persisting latent varicella infection exists.

Anesthesia used

To minimize stress to the monkeys and to limit the chance of biting or scratching of support personnel, the monkeys are sedated with ketamine prior to handling. Ketamine is administered by intramuscular injection at 10 mg/kg; this sedates the monkey within 5 minutes and persists for approximately 20–30 minutes. Manipulations such as clinical evaluation, inoculation, specimen collection, bleeding and treatment should be done with the monkey under anesthesia.

Storage, preparation of inocula

While a number of isolates of SVV have been recovered from patas, vervet and macaque species of monkeys, the strain of virus most thoroughly characterized is the Delta simian varicella virus. None of these viruses is available through the American Type Culture Collection. The Delta strain was isolated from a patas monkey infected at the Delta (Tulane) Regional Primate Research Center. It can be obtained by contacting the Center at the following address: Tulane Regional Primate Research Center, Department of Microbiology, 18703 Three Rivers Road, Covington, LA 70433; tel: (504) 892–2040.

The virus is a very strongly cell-associated virus and tends to attenuate upon passage in tissue culture. The original isolate has been passaged a limited number of times in patas and African green monkeys. The virus for preparation of a virus stock has been propagated from peripheral blood mononuclear cells (PBMC) separated on a ficol-hypaque gradient from heparinized blood from an infected African green monkey. The washed PBMC are suspended in RPMI-1640 medium with 15% fetal bovine serum and antibiotics and inoculated into a 75 cm^2 tissue culture flask seeded with Vero cells 24 hours earlier. The Vero cells are grown in minimum essential medium (MEM) with 10% fetal bovine serum and antibiotics. Following 5–7 days incubation in a CO_2 incubator at 37°C when numerous plaques are seen microscopically, the culture is expanded from one flask to five. This is accomplished by removal of the culture fluid, washing the monolayer with 10 ml of phosphate-buffered saline, pH 7.2, and trypsinization of the monolayer. The trypsinized cells are suspended in MEM and used for inoculation of five 75 cm^2 flasks containing Vero cells seeded 24 hours earlier.

Five to 7 days later, when adequate virus growth is seen, a second and final 1:5 expansion is made, inoculating 25 75 cm^2 flasks from the five infected flasks. After another 5–7 days' incubation, the infected cells are scraped from the surface of the flask with a policeman or by the addition of sterile 3 mm glass beads which are shaken to detach the cells. The culture fluids and cells are collected in 50 ml centrifuge tubes and centrifuged at 1500 rpm for 20 minutes. The supernate is discarded and the sedimented cells suspended in freezing medium using 3 ml of freezing medium for each flask pooled for the virus stock. Vortexing or pipetting may serve to break up any cell clumps or aggregates. The freezing medium is composed of MEM tissue culture medium made with 10% fetal bovine serum and 10% dimethyl sulfoxide (DMSO). The virus is ampouled in cryovials at 1 ml per vial and frozen at −70°C.

Following freezing, the virus stock is titrated by preparation of 10-fold dilutions from 10^{-1} to 10^{-4}. One ml volumes of each dilution are inoculated into duplicate or triplicate 25 cm^2 flasks seeded with Vero cells 24 hours earlier. An additional 4 ml of growth medium is added to provide adequate fluid volume to each flask. After 5–7 days' incubation, the culture fluids are discarded, the monolayers washed with 5 ml phosphate-buffered saline and fixed with 3 ml methanol per flask for 10–15 minutes. The methanol is discarded and the cell monolayers are stained with methylene blue–basic fuchsin stain. Stock solutions of methylene blue and basic fuchsin are prepared by dissolving 5 grams of methylene blue in 500 ml of methanol and 5 grams of basic fuchsin in 500 ml of methanol. For staining of methanol-fixed monolayers, 60 ml of methylene blue is mixed with 20 ml basic fuchsin and 20 ml of methanol. Two to 3 ml of the mixed stain is added to the culture flask for 30 minutes and then washed in running tap water. An adequate stock for inoculation of monkeys contains 10^4 plaque-forming units (pfu) or better per ml of the stock virus.

Infection process

Prior to infection of the monkeys with SVV it is necessary to determine that the monkeys are seronegative. This is accomplished by a serum neutralization assay. A 3 ml volume of blood is drawn from the femoral vein and the serum separated and heat inactivated. A 1 : 10 dilution of serum is made in tissue culture medium (0.1 ml serum plus 0.9 ml medium). A 1 ml volume of a dilution of stock SVV containing approximately 100 pfu of virus is added to each serum dilution, mixed and held at room temperature for 1 hour. A control tube contains 1 ml of medium plus 1 ml of virus. At the end of the incubation period, the serum–virus mixtures are divided between two Vero cultures. The Vero cells may be grown in 25 cm^2 flasks or in 24, 12, or 6-well culture plates. Tissue culture medium should be added to provide sufficient culture medium for the culture. Following 5–7 days' incubation at 37°C the culture fluid is discarded, the cells washed with phosphate-buffered saline, fixed for 10 minutes with methanol and stained for 30 minutes with methylene blue–basic fuchsin stain. After drying, the numbers of plaques in each culture is counted. The numbers of plaques in serum–virus cultures should be approximately the number counted in the cultures incubated with the virus control. Cultures in which the numbers of plaques is appreciably reduced suggest the presence of prior antibody and this monkey is eliminated from the study.

Infection of seronegative monkeys is by intratracheal inoculation of SVV. The virus stock is thawed and diluted to contain approximately 10^4 pfu/ml. The monkey is anesthetized with ketamine (10 mg/kg) and laid on its back with the shoulders slightly elevated. A sterile French 8 feeding tube is cut to the length of a stylet, and this provides rigidity to the feeding tube. With a laryngoscope, the feeding tube is introduced into the trachea. The stylet is removed and a 3 ml syringe containing 1.5 ml of the viral inoculum is used to introduce the fluid slowly. The feeding tube is flushed with 3 ml of air and the tube is removed.

Key parameters to monitor infection

With intratracheal inoculation of SVV a lymphocyte associated viremia occurs about day 3 post-infection (p.i.). The viremia peaks around day 7 and essentially clears by day 11 p.i. For assessment of viremia, 3 ml of blood is drawn from the femoral vein in heparin on days 3, 5, 7, 9 and 11 p.i. The blood is mixed with 3 ml sterile PBS and layered over 4 ml of ficoll-hypaque lymphocyte separation medium in a 15 ml centrifuge tube. The tubes are centrifuged at 1400 rpm for 30 minutes and the lymphocyte band is carefully collected and resuspended in 5 ml RPMI-1640 medium with antibiotics (100 units penicillin, 100 μg streptomycin and 100 μg gentamicin per ml). The suspended lymphocytes are centrifuged at 1400 rpm for 15 minutes and the supernate discarded. The sedimented cells are resuspended in 11 ml of RPMI-1640 medium with 15% fetal bovine serum and antibiotics by mixing with a pipette and vortexing to dispense aggregated cells. One ml of the cell suspension is diluted 1 : 10 in 9 ml of the culture medium. Each of the two cell suspensions is inoculated in 5 ml volumes into duplicate 25 cm^2 tissue culture flasks containing Vero cells and incubated at 37°C in a CO_2 incubator for 7 days. No agar overlay is required due to the firm cell-associated nature of the virus. Midway in the 7-day incubation, the culture fluid is changed, replacing the RPMI-1640 medium with MEM and 2% fetal bovine serum with antibiotics. This helps to preserve the cell monolayer and removes the floating lymphocytes. After 7 days' incubation, culture fluids are discarded and the monolayers washed with phosphate-buffered saline, fixed with methanol and stained with methylene blue–basic fuchsin. The number of viral plaques per flask are counted and the mean number of plaques in the paired flasks is determined.

Rash is assessed by careful examination of the skin of each monkey. A papular to vesicular rash appears anywhere on the body around day 7 p.i. and increases to a peak around day 9 or 10 p.i. (Figure 116.1A,B—figures in this chapter are in the colour plate). The rash begins to dry by day 14 p.i. and resolves completely in time. Rash is scored subjectively on a scale of ± to 4+. A ± score is given to monkeys with 5 or fewer skin lesions and increases with severity to 4+ where the body is covered with rash. In some cases, the rash may become hemorrhagic, with the reddened lesions increasing in size and visibility (Figure 116.1C). This is a poor prognostic sign: the monkey frequently dies with severe systemic disease.

In a number of untreated or inadequately treated cases of SVV infection, the lungs and the liver become involved. Pneumonia is seen as labored and rapid respiration and hepatitis can be assessed by clinical chemistry tests. Aminotransferase values increase with liver involvement while all other chemistry values remain within normal limits. Hematology values are unremarkable.

Approximately 25–50% of infected and untreated African green monkeys die with severe simian varicella infection. Necropsy shows pulmonary involvement with edematous reddened and firm areas of the lung (Figure 116.2). Histologically, inflammation and necrosis are seen and many cells contain intranuclear inclusion bodies (Roberts *et al.*, 1984; Dueland *et al.*, 1992). The bronchioles, terminal bronchi and alveoli all show necrosis and infiltration of neutrophils and macrophages. All visceral tissues, including gastrointestinal tract, liver (Figure 116.3), spleen, kidney, adrenals, lymph nodes and bone marrow are involved and show gross and/or histologic signs of infection.

Effective antiviral drug treatment frequently results in decreased antibody titers in serum neutralization assays. Three ml of blood is taken from the femoral vein on days 14 and 21 p.i. After the sera are heat-inactivated at 56°C for 30 minutes, each serum is diluted in twofold dilutions from 1:10 to 1:640 in 1 ml volumes of MEM with 2% fetal bovine serum and antibiotics. The virus is diluted to contain approximately 200 pfu/ml and added in equal volumes to the 1 ml serum dilutions. Control cultures should be set up with the virus dilution mixed 1:1 with normal culture medium. After 1 hour incubation at room temperature, the virus–serum mixtures are inoculated into duplicate wells (1 ml/well) of 6-well tissue culture plates containing Vero cells. Three ml of growth medium is added to each well. After 7 days' incubation at 37°C in a CO_2 incubator, the culture fluid is discarded and the cell monolayers washed with phosphate-buffered saline, fixed in methanol and stained with methylene blue–basic fuchsin and the number of plaques counted. The antibody titer is the dilution of serum resulting in 20% or less of the mean number of plaques counted in control cultures.

Therapy

Potential antiviral compounds are generally provided by the supplier with information on solubility and a recommended dosing regimen. *In vitro* antiviral tests should be performed using SVV as the inoculum. *In vitro* assays are performed by preparation of the drug at a concentration of 100 μg/ml or greater. Less soluble drugs are prepared at their maximum soluble concentration. The drugs are dissolved in MEM with 10% fetal bovine serum and antibiotics. Serial twofold dilutions are prepared in the growth medium.

Culture fluids are decanted from 24 well plates seeded earlier with Vero cells. Each of the wells is inoculated with 0.5 ml of a dilution of SVV stock containing approximately 10^2 pfu per ml. After 4 hours incubation at 37°C, the viral inoculum is aspirated with a pipette and replaced with 1.0 ml volumes of the drug dilutions using three wells per drug concentration. One set of three wells receives growth medium without drug and serves as the virus control. After incubation at 37°C for 5–7 days, the plates are examined microscopically for the development of viral plaques. If the plaques are countable, the plates are fixed with methanol and stained with methylene blue–basic fuchsin, as previously described. Following the counting of numbers of plaques, the median effective dose (ED_{50}) can be calculated by determination of the concentration of drug inhibiting 50% of the number of plaques in the control wells without drug. Data from the *in vitro* assays correlate to some degree with *in vivo* results, with many of the compounds tested for antiviral activity against SVV (Table 116.2). These data are therefore useful to establish a range of doses to be evaluated for antiviral activity in monkeys.

Table 116.2 Correlation of *in vitro* and *in vivo* effective doses of selected antiviral drugs against simian varicella virus

	Effective dose		
Antiviral drug	In vitro *(mg/ml)*	In vivo *(mg/kg per day)*	*Route of administration*
1-β-D-Arabinofuranosyl-E-5-(2-bromovinyl)uracil (BV-ara-U)	0.001	0.2	p.o.
(E)-5-(2-Bromovinyl)-2′-deoxyuridine (BVDU)	0.01	5.0	p.o.
1-(2-Deoxy-2-fluoro-β-D-arabinofuranosyl)-5-ethyl uracil (FEAU)	0.33	10.0	p.o.
Adenine arabinoside-5′-monophosphate (Ara-AMP)	3.3	>18.4	i.v.
9-(2-Hydroxyethoxy methyl)guanine (acyclovir)	10.0	45–60	i.v.
9-(1,3-Dihydroxy-2-propoxymethyl)guanine (ganciclovir)	10.0	30–45	i.v.
(S)-2-(3-Hydroxy-2-phosphonylmethoxy)propyl)cytosine (HPMPC)	6.25	5.0	i.v.
Phosphonoacetic acid (PFA)	15	200	i.m.

p.o., By mouth; i.v., intravenous; i.m., intramuscular.

Drug treatment is administered to monkeys on a mg/kg per day basis. With this in mind, the number of treatments per day and the volume to be administered must be considered for preparation of the drug. The drug is dissolved in distilled water, saline, or other suitable vehicle on a predetermined mg/ml concentration. The pH should be adjusted to near neutrality if the solution is very acidic or alkaline. Prior to administration of the drug it is sterilized by filtration. Ideally the drug solution should be prepared daily.

Prior to infection of monkeys with SVV, they are moved to an isolated area where infection cannot be transmitted to susceptible animals within the area. They are best allowed to stabilize in their new surroundings for 5–7 days before virus inoculation. After virus administration, they are assigned to groups of three or more animals per treatment group, attempting to equalize weight and sex within the groups.

Treatment of SVV infection is most commonly begun on day 2, 48 hours after virus inoculation. Initiation of treatment can be varied depending on the drug under evaluation or the objective of the study. Treatment is routinely continued for 10 days through day 12 p.i. The drug may be given in divided doses up to three times daily. If more frequent treatment is given, the monkey eats and drinks poorly due to long periods of sedation. Administration of the drug may be by intramuscular injection into the thigh, by subcutaneous injection, by intravenous injection into the saphenous vein, or orally by gavage. Oral treatment by gavage employs a French 5 or 8 nasogastric tube which is passed via the mouth into the esophagus and then into the stomach. The drug is given by syringe and the nasogastric tube flushed with air or water before withdrawing.

Key parameters to monitor infection

Viremia, rash, death and antibody response are the major parameters used to determine antiviral activity of a prospective antiviral drug (Table 116.3). These responses in treated monkeys must be compared with responses in infected control monkeys treated with vehicle without drug. The mean number of plaques counted from blood samples taken on days 3, 5, 7, 9, and 11 p.i. are compared between treated and control monkeys. The variability in plaque number between monkeys in the respective groups and the number of monkeys per group limits the use of statistical methods in evaluation of efficacy. A highly effective drug, however, is capable of completely preventing the appearance of plaques in cultures inoculated with blood specimens.

Table 116.3 Parameters to evaluate antiviral drugs against simian varicella virus infection in African green monkeys

	Expected appearance
Clinical signs	
Viremia	3–11 days p.i.
Rash	7–14 days p.i.
Anorexia	7–14 days p.i.
Antibody response	14–21 days p.i.
Death	8–14 days p.i.
Clinical chemistry	
Aminotransferases	5–14 days p.i.

p.i., Post-infection.

The evaluation of rash is a subjective evaluation. It appears first as a few vesicles seen on the chest or groin, increasing in numbers during the next 2 or more days. Virus is present in the vesicles, disappearing as the lesion crusts. The severity of the rash varies between monkeys but does become well-dispersed over the body surface in the majority of untreated control monkeys. The appearance of hemorrhagic rash in some monkeys has no explanation but is usually associated with severe systemic disease. The development of rash can be completely prevented by effective drug treatment. Rash has been inhibited by treatment with suboptimal doses of an antiviral which have only reduced and not eliminated viremia.

Death, when it occurs in the monkey with SVV infection, results from severe pneumonia and liver involvement. A complete necropsy shows lesions in many organs, including the gastrointestinal tract, heart, adrenals and kidneys. Treatment with an effective antiviral can prevent death even at doses which allow some viremia and rash to appear.

Successful treatment with antiviral drugs may — but does not always — lead to lower titers of serum-neutralizing antibody in the treated monkeys. A rare monkey fails to seroconvert following successful prevention of clinical disease but the majority give rise to titers of serum-neutralizing antibody in the range of 1 : 20 to 1 : 80. Untreated control monkeys have titers of around 1 : 320 or greater. In dose–response studies higher titers are seen in monkeys receiving suboptimal doses of the antiviral.

Anorexia can be induced by SVV infection. Food intake generally declines markedly in monkeys at around day 5 p.i. However, considerable variability occurs between individual monkeys. Furthermore, some drugs also reduce food intake, particularly in monkeys receiving the higher doses. When treatment calls for frequent daily drug administration, three or four times per day, the repeated sedation with ketamine also affects the amount of food consumed with the monkeys choosing fruit rather than the more easily counted biscuits of monkey chow. The use of anorexia as a parameter of clinical disease or the beneficial effects of treatment is not highly recommended.

Pitfalls of the model

While it is obvious that SVV is not the virus causing disease in humans, the viruses are sufficiently closely related at the molecular level so that the use of SVV infection in monkeys can be predictive of useful therapy for varicella infection of humans. Although, in rare instances, a compound found to be active against human VZV *in vitro* has not been effective in inhibiting the replication of SVV *in vitro*, the majority of antiviral compounds developed for potential use in human VZV infections have been found to be active in the monkey model. The dose of antiviral drug required for treatment of SVV infection usually exceeds the concentration required to treat VZV in human patients.

The cost of doing experiments in monkeys is expensive when you consider the purchase of monkeys and the daily charges for care. However, the monkey as a non-human primate provides an animal model that is phylogenetically closer to the human. This offers advantages in the physiologic assessment of the drug with respect to antiviral efficacy, pharmacokinetics and to detection of potential toxicity. The size of the monkey requires a larger quantity of drug for treatment than is needed for the more commonly used laboratory animals. With mg/kg doses, gram quantitites of drug are required. Laboratory synthesis of promising new antiviral compounds may make it difficult to prepare adequate quantities of drug for studies in 2–4 kg monkeys. However, if adequate quantities of drug are available, the size of the monkey permits frequent and fairly good-size specimens to be collected and pharmacokinetic studies are easily performed in monkeys and show agreement with data from humans. Possible toxicity of new antivirals may be evaluated using blood specimens taken during treatment for hematology and clinical chemistry tests.

Of concern is that SVV infection of the monkey results in a latent infection. Reactivation of the virus is occasionally seen in a monkey under stress. Reactivated virus results in development of a mild rash which does not appear to be restricted to the distribution of a dermatome but may appear as a minimal number of vesicles anywhere on the body. They could be easily overlooked. Virus has been isolated from vesicle fluid, and the monkey with reactivated virus can conceivably serve as a source of infection of susceptible monkeys. Attempts to isolate virus from trigeminal and dorsal root ganglia by co-cultivation with Vero cells have been consistently negative. Virus has been demonstrated in ganglion cells by PCR techniques (Mahalingam *et al.*, 1991). Attempts to reactivate the virus with the production of clinical symptoms of reactivated disease has been unsuccessful using a variety of means to immunosuppress the monkey, specifically X-irradiation, cortisone, cyclosporin A, cyclophosphamide, and simian immunodeficiency virus infection.

SVV is very likely not infectious for humans. The virus has been used in the laboratory for many years in both *in vitro* and *in vivo* experiments with no one showing any sign of clinical disease. However, with the severity of infection in African green and patas monkeys, appropriate measures should be taken to handle the virus as a potential human pathogen.

Contributions of the model to infectious disease therapy

Early in the preclinical studies with acyclovir (virazole: 9-(2-hydroxyethoxymethyl)guanine) this compound was administered to patas monkeys or to African green monkeys by intravenous or intramuscular injection (Soike *et al.*, 1981a; Soike and Gerone, 1982) with moderate activity at doses of 20 or 45 mg/kg per day inhibiting development of rash but with minimal activity against viremia. Higher doses of 50 or 100 mg/kg per day given intramuscularly or intravenously were effective in suppressing both rash and viremia. Fyfe *et al.* (1982) showed that SVV thymidine kinase was able to activate acyclovir to the monophosphate similar to VZV. *In vitro* studies showed a greater resistance of SVV than VZV to acyclovir. Ten μg/ml was required for 80% inhibition of SVV when VZV was inhibited by 2.5 μg/ml using Vero cells for the simian virus and WI-38 cells for VZV.

Ganciclovir (9-(1,3-dihydroxy-2-propoxy-methyl)guanine) at 60 mg/kg per day given by intravenous injection in the saphenous vein twice daily was effective in preventing both rash and viremia (Soike *et al.*, 1987). When given at 20 mg/kg per day, activity was seen to be only minimally effective.

A number of nucleosides evaluated as potential antivirals for treatment of VZV infection have shown activity when tested against SVV infection in monkeys. These include (E)-5-(2-bromovinyl)-2′-deoxyuridine (BVDU); 1-β-D-arabinofuranosyl-E-5-(2-bromovinyl)uracil (BV-ara-U); 1-β-D-arabinofuranosyl-thymine (ara-T); 6-dimethylamino-9-(β-D-arabinofuranosyl)-9H-purine (ara-DMA-P); and [3S-{3α(E), 4β,5α}]-5-(2-bromoethunyl-1- [titrahydro-4,5-bis (hydroxymethyl)-3-furanyl]-2,4 (1H, 3H) pyrimidinedione (BMS-191165); Soike *et al.*, 1981b, 1984, 1991, 1993a, 1994). BVDU inhibited rash and viremia following oral, intramuscular and intravenous routes of administration at 10 mg/kg per day in divided doses twice daily. Treatment with BVDU was partially effective at 15 mg/kg per day when delayed as late as 6 days p.i. (Soike *et al.*, 1981b). BV-ara-U was the most active compound tested against SVV infection. A dose of 0.1 mg/kg given twice daily by gavage was capable of essentially preventing infection (Soike *et al.*, 1992). BV-ara-U given orally at 0.4 mg/kg twice daily prevented rash development and markedly reduced viremia even when treatment was delayed until 4 days following virus inoculation.

The fluorinated pyrimidines 1-(2-deoxy-2′-fluoro-β-D-arabinofuranosyl)-5-idouracil (FIAU) and the 5-methyl (FMAU) and 5-ethyl (FEAU) analogs were each highly effective in preventing the development of simian varicella following oral dosing (Soike *et al.*, 1986, 1990). No toxicity was evident in monkeys treated for 10 days with FIAU at 50 mg/kg per day given by gavage or by intravenous injection twice daily. Treatment with FIAU could be delayed as late as 5 days after virus inoculation and still inhibit the progression of infection.

(*S*)-1-[3-Hydroxy-2-(phosphonylmethoxy)propyl]-cytosine (S-HPMPC), an antiviral compound which does not require phosphorylation by the virus thymidine kinase and a compound with a long half-life, was shown to be effective against SVV. A single intravenous injection of S-HPMPC given as late as 4 days p.i. prevented the development of disease (Soike *et al.*, 1991).

The simian varicella disease model was used to evaluate the anti-varicella compound 9-[4-hydroxy-2-(hydroxymethyl)butyl]guanine (2HM-HBG) as the racemic mixture, the active (–) isomer, and as a pro-drug (Lake-Bakaar *et al.*, 1988; Soike *et al.*, 1993, 1997).

SVV infection responds to treatment with recombinant human lymphokines. Both recombinant human α interferon A and recombinant human interferon-β were able to prevent infection (Soike *et al.*, 1983; Chiang *et al.*, 1993). Interferon-β prepared in liposomes permitting sustained release of interferon-β was also able to reduce viremia and rash following inoculation of African green monkeys with virus (Eppstein *et al.*, 1989). For unknown reasons, recombinant human tumor necrosis factor-α was seen to enhance simian varicella infection resulting in increased mortality, viremia, and histopathology (Soike *et al.*, 1989).

The monkeys also have the advantage of allowing collection of pharmacokinetic data. Blood levels of many compounds approach concentrations close to the *in vitro* ED_{50} for these compounds and may provide data bearing on the frequency of treatment which will result in optimal effects (Soike *et al.*, 1983, 1992, 1994; Lake-Bakaar *et al.*, 1988; Chiang *et al.*, 1993).

Studies of combinations of drugs to demonstrate synergy or antagonism between two or more effective antivirals have been done in monkeys (Soike *et al.*, 1987, 1990). The numbers of monkeys required for such studies, however, make these studies expensive. Statistical evaluation of the results of drug combinations have employed the medium effect equation as developed by Chou and Talalay (1984) and Chou and Chou (1987).

The number of anti-VZV compounds that have gone to clinical trials is limited. Where pursued with acyclovir, ganciclovir, and BV-ara-U, studies performed with SVV have been predictive of results in humans. The SVV model for VZV has served to compare various antivirals with respect to dose, route of administration, and to evaluate therapeutic efficacy when treatment was delayed in the clinical course of infection. The close similarities of SVV with VZV provides an appropriate alternative viral disease model for evaluation of antiviral drug treatment of VZV as well as a potential model for investigation of pathogenesis and viral latency.

Acknowledgment

Appreciation is expressed to Dr Donald Gilden and Dr Ravi Mahalingam, University of Denver School of Medicine, and to Dr Wayne Gray, University of Arkansas for Medical Sciences, for their help in characterizing similarities between VZV and SVV. Thanks are given to Dr Wayne Gray for his critical review of the manuscript. This work has been supported by contracts with the Antiviral Substances program of the National Institute of Allergy and Infectious Disease and by grant support of the Tulane Regional Primate Research Center from Division of Research Resources; National Institute of Health, Grant RR-00164.

References

Allen, W. P., Felsenfeld, A. D., Wolf, R. H., Smetana, H. F. (1974). Recent studies on the isolation and characterization of delta herpesvirus. *Lab. Animal Sci.*, **24**, 222–228.

Ayres, J. P. (1971). Studies of the delta herpesvirus isolated from the patas monkey (*Erythrocebus patas*). *Lab. Animal Sci.*, **21**, 685–695.

Blakely, G. A., Lourie, B., Morton, W. G., Evans, H. H., Kaufmann, A. F. (1973). A varicella-like disease in macaque monkeys. *J. Infect. Dis.*, **127**, 617–626.

Chiang, J., Gloff, C. A., Soike, K. F., Williams, G. (1993). Pharmacokinetics and antiviral activity of recombinant human interferon-β_{ser17} in African green monkeys. *J. Interferon Res.*, **13**, 111–120.

Chou, J., Chou, T.-C. (1987). *Dose–effect Analysis with Microcomputors: Quantitation of ED_{50}, LD_{50}, Synergism and Antagonism, Low-dose Risk, Receptor Binding and Enzyme Kinetics*. A computer software for Apple II series or IBM PC and manual. Elsevier Science, Cambridge.

Chou, T. C., Talalay, P. (1984). Quantitative analysis of dose–effect relationships: the combined effects of multiple drugs or enzyme inhibitors. *Adv. Enzyme Regul.*, **22**, 27–55.

Clarkson, M. J., Thorpe, E., McCarthy, K. (1967). A virus disease of captive vervet monkeys (*Cercopithecus aethiops*) caused by a new herpesvirus. *Arch. Virusforsch.*, **22**, 219–234.

Dueland, A. N., Martin, J. R., Devlin, M. E. *et al.* (1992). Acute simian varicella infection. Clinical, laboratory, pathologic and virologic features. *Lab. Invest.*, **66**, 762–773.

Eppstein, D. A., van der Pas, M. A., Gloff, C. A., Soike, K. F. (1989). Liposomal interferon β: sustained release treatment of simian varicella virus infection in monkeys. *J. Infect. Dis.*, **159**, 616–620.

Feldman, S., Lott, L. L. (1987). Varicella in children with cancer: impact of antiviral therapy and prophylaxis. *Pediatrics*, **80**, 465–472.

Felsenfeld, A. D., Schmidt, N. J. (1977). Antigenic relationships among several simian varicella-like viruses and varicella-zoster virus. *Infect. Immun.*, **15**, 807–812.

Fletcher, T. M., Gray, W. L. (1992). Simian varicella virus: characterization of virion and infected cell polypeptides and the antigenic cross-reactivity with varicella-zoster virus. *J. Gen. Virol.*, **73**, 1209–1215.

Fletcher, T. M., Gray, W. L. (1993). DNA sequence and genetic organization of the unique short (Us) region of the simian varicella genome. *Virology*, **193**, 762–773.

Fyfe, J. A., Biron, K. K., McKae, S. A., Kelly, C. M., Elion, G. B., Soike, K. F. (1982). Activation and antiviral effect of acyclovir in cells infected with a varicella-like simian virus. *Am. J. Med.*, **73**, (suppl.), 58–61.

Gilden, D. H., Vafai, A., Shtram, Y., Becker, Y., Devlin, M., Wellish, M. (1983). Varicella-zoster virus DNA in human sensory ganglia. *Nature*, **306**, 478–480.

Gilden, D. H., Rozenman, Y., Murray, R., Devlin, M., Vafai, A. (1987). Detection of varicella-zoster virus nucleic acid in neurons of normal thoracic ganglia. *Ann. Neurol.*, **22**, 377–380.

Gray, W. L., Gusick, N. J. (1996). Viral isolates drived from simian varicella epizooties are genetically related but are distinct from other primate herpesviruses. *Virology*, **224**, 161–166.

Gray, W. L., Oakes, J. E. (1984). Simian varicella virus DNA shares homology with human varicella-zoster virus DNA. *Virology*, **136**, 241–246.

Gray, W. L., Pumphrey, C. Y., Ruyechan, W. T., Fletcher, T. M., (1992). The simian varicella virus and varicella zoster virus genomes are similar in size and structure. *Virology*, **186**, 562–572.

Harbour, D. A., Caunt, A. E. (1979). The serological relationship of varicella-zoster virus to other primate herpesviruses. *J. Gen. Virol.*, **45**, 469–477.

Hyman, R. W., Ecker, J. R., Tenser, R. B. (1983). Varicella-zoster virus RNA in human trigeminal ganglia. *Lancet*, (Oct 8), 814–816.

Lake-Bakaar, D. M., Abele, G., Lindborg, B., Soike, K. F., Datema, R. (1988). Pharmacokinetics and antiviral activity in simian varicella virus-infected monkeys by (R,S)-9-[4-hydroxy-2-(hydroxymethyl)butyl]guanine, an anti-varicella zoster drug. *Antimicrob. Agents Chemother.*, **32**, 1807–1812.

Lehner, N. D. M., Bullock, B. C., Jones, N. D. (1984). Simian varicella infection in the African green monkey (*Cercopithecus aethiops*). *Lab. Animal Sci.*, **34**, 281–285.

Lowry, P. W., Sabella, C., Koropchak, C. M. *et al.* (1993). Investigation of the pathogenesis of varicella-zoster virus infection in guinea pigs by using polymerase chain reaction. *J. Infect. Dis.*, **167**, 78–83.

Mahalingam, R., Smith, D., Wellish, M. *et al.* (1991). Simian varicella virus DNA in dorsal root ganglia. *PNAS*, **88**, 2750–2752.

McCarthy, K., Thorpe, E., Laursen, A. C., Geymann, C. S., Beale, A. J. (1968). Exanthematous disease in patas monkeys caused by a herpesvirus. *Lancet*, (Oct. 19), 856–857.

McQuire, M. T. (1974). The St Kitts vervet (*Cercopithecus aethiops*). *J. Med. Primatol.*, **3**, 285–297.

Morgan, E. R. (1983). Varicella in immunocompromised children. *Am. J. Dis. Child.*, **137**, 883–885.

Myers, M. G., Duer, H. L., Hausler, C. K. (1980). Experimental infection of guinea pigs with varicella-zoster virus. *J. Infect. Dis.*, **142**, 414–420.

Myers, M. G., Connelly, B. L., Stanberry, L. R. (1991). Varicella in hairless guinea pigs. *J. Infect. Dis.*, **163**, 746–751.

Perronne, C., Lazanas, M., Leport, C. *et al.* (1990). Varicella in patients infected with the human immunodeficiency virus. *Arch. Dermatol.*, **126**, 1033–1036.

Pumphrey, C. Y., Gray, W. L. (1992). The genomes of simian varicella virus and varicella zoster virus are colinear. *Virus Res.*, **26**, 255–266.

Roberts, E. D., Baskin, G. B., Soike, K., Gibson, S. V. (1984). Pathologic changes of experimental simian varicella (delta herpesvirus) infection in African green monkeys. *Am. J. Vet. Res.*, **45**, 523–530.

Schmidt, N. J., Arvin, A. M., Martin, D. P., Gard, E. A. (1983). Serological investigation of an outbreak of simian varicella in *Erythrocebus patas* monkeys. *J. Clin. Microbiol.*, **18**, 901–904.

Soike, K. F., Gerone, P. J. (1982). Acyclovir in the treatment of simian varicella virus infection of the African green monkey. *Am. J. Med.*, **73**, (suppl.), 112–117.

Soike, K. F., Felsenfeld, A. D., Gerone, P. J. (1981a). Acyclovir treatment of experimental simian varicella infection of monkeys. *Antimicrob. Agents Chemother.*, **20**, 291–297.

Soike, K. F., Gibson, S., Gerone, P. J. (1981b). Inhibition of simian varicella virus infection of African green monkeys by (E)-5-(2-bromovinyl)-2′-deoxyuridine (BVDU). *Antiviral Res.*, **1**, 325–337.

Soike, K. F., Kramer, M. J., Gerone, P. J. (1983). *In vivo* antiviral activity of recombinant type α interferon A in monkeys with infections due to simian varicella virus. *J. Infect. Dis.*, **147**, 933–938.

Soike, K. F., Baskin, G. B., Cantrell, C., Gerone, P. J. (1984). Investigation of antiviral activity of 1-β-D-arabinofuranosylthymine (ara T) and 1-β-D-arabinofuranosyl-E-5-(2-bromovinyl)uracil (BV-ara-U) in monkeys infected with simian varicella virus. *Antiviral Res.*, **4**, 245–257.

Soike, K. F., Cantrell, C., Gerone, P. J. (1986). Activity of 1-(2′-deoxy-2′-fluoro-β-D-arabinofuransoyl)-5-iodouracil against simian varicella virus infections in African green monkeys. *Antimicrob. Agents Chemother.*, **29**, 20–25.

Soike, K. F., Eppstein, D. A., Gloff, C. A., Cantrell, C., Chou, T.-C. Gerone, P. J. (1987). Effect of 9-(1,3-dihydroxy-2-propoxymethyl)guanine and recombinant human β interferon alone and in combination on simian varicella virus infection in monkeys. *J. Infect. Dis.*, **156**, 607–614.

Soike, K. F., Czarniecki, C. W., Baskin, G., Blanchard, J. B., Liggitt, D. (1989). Enhancement of simian varicella virus infection in African green monkeys by recombinant human tumor necrosis factor α. *J. Infect. Dis.*, **159**, 331–335.

Soike, K. F., Chou, T.-C., Fox, J. J., Watanabe, K. A., Gloff, C. A. (1990). Inhibition of simian varicella virus infection of monkeys by 1-(2-deoxy-2-fluoro-1-β-D-arabinofuranosyl)-5-ethyluracil (FEAU) and synergistic effects of combination with human recombinant interferon-β. *Antiviral Res.*, **13**, 165–174.

Soike, K. F., Huang, J.-L., Zhang, J.-Y., Bohm, R., Hitchcock, M. J., Martin, J. C. (1991). Evaluation of infrequent dosing regimens with (S)-1-[3-hydroxy-2-(phosphonyl-methoxy)propyl] cytosine (S-HPMPC) on simian varicella infection in monkeys. *Antiviral Res.*, **16**, 17–28.

Soike, K. F., Huang, J.-L., Stouffer, B. *et al.* (1992). Oral bioavailability and anti-simian varicella virus efficacy of 1-β-D-arabinofuranosyl-E-5-(2-bromovinyl)uracil (BV-araU) in monkeys. *J. Infect. Dis.*, **165**, 732–736.

Soike, K. F., Bohm, R., Huang, J.-L., Oberg, B. (1993). Efficacy of (–)-9-[4-hydroxy-2-(hydroxymethyl)butyl]guanine in African green monkeys infected with simian varicella virus. *Antimicrob. Agents Chemother.*, **37**, 1370–1372.

Soike, K. F., Huang, J. L., Russell, J. W. *et al.* (1994). Pharmacokinetics and antiviral activity of a novel isonucleoside, BMS-181165, against simian varicella virus infection in African green monkeys. *Antiviral Res.*, **23**, 219–224.

Soike, K. F., Alder, J., Shioskaki, K. (1997). Antiviral efficacy of orally administered ABT-606, a diester prodrug of H2G, in African green monkeys infected with simian varicella virus. *Interscience Conference on Antimicrobial Agents and Chemotherapy*, Sept. 28–Oct 1, p. 223 (abstract).

Takasaka, M., Sakakibara, I., Mukai, R., Suzuki, M. (1990). An outbreak of nonhuman primate varicella-like herpesvirus infection in the established breeding colony of cynomolgus monkeys. *TPC News*, **9**, 5–9.

Vafai, A., Murray, R. S., Wellish, M., Devlin, M., Gilden, D. H. (1988). Expression of varicella-zoster virus and herpes simplex virus in normal human trigeminal ganglia. *P.N.A.S.*, **85**, 2362–2366.

Weller, T. H. (1983). Varicella and herpes zoster. Changing concepts of the natural history, control and importance of a not-so-benign virus. *N. Engl. J. Med.*, **309**, 1362–1368, 1434–1440.

Chapter 117

Varicella-zoster Virus Infection of T cells and Skin in the SCID-hu Mouse Model

J. F. Moffat and A. M. Arvin

Background of human infection

Varicella-zoster virus (VZV) is a human α-herpesvirus that causes varicella, or chickenpox, in susceptible individuals (Arvin, 1996). The critical events in the pathogenesis of primary VZV infection include inoculation of respiratory mucosa, the occurrence of cell-associated viremia and the transfer of infectious virus to skin, resulting in the characteristic vesicular exanthem (Ozaki *et al.*, 1986; Koropchak *et al.*, 1989, 1991; Sawyer *et al.*, 1992). Like other α-herpesviruses, VZV establishes latency in sensory ganglia (Croen *et al.*, 1988; Mahalingam *et al.*, 1990). VZV reactivation from the latent state causes herpes zoster, manifesting as a localized rash in a unilateral, dermatomal distribution that is often associated with severe neuropathic pain (Whitley, 1990; Arvin, 1996).

The Oka strain, a Japanese clinical isolate of VZV, was used to develop a live attenuated varicella vaccine which is the first licensed herpesvirus vaccine (Krause and Klinman, 1995). The varicella vaccine was attenuated by passage in semi-permissive guinea-pig embryo fibroblasts (Takahashi *et al.*, 1974). While most healthy children and adults who are given the varicella vaccine develop immunity without experiencing any signs of disease, the virologic basis for this clinical attenuation of the vaccine Oka strain (V-Oka) is not known. The pattern of replication of V-Oka in tissue culture cells resembles other VZV strains. There are no known intrinsic strain differences among VZV isolates, and restriction endonuclease analysis of genomic DNA does not reveal obvious differences between V-Oka and other geographically related VZV isolates (LaRussa *et al.*, 1992; Gelb, 1993b). Naturally occurring VZV mutants that lack expression of glycoprotein C or thymidine kinase have been identified and genetically altered mutants of VZV can now be generated from cosmids (Cohen and Seidel, 1993; Mallory *et al.*, 1997). Animal models are needed to evaluate how naturally occurring mutants and VZV recombinants differ in pathogenicity. Whether particular genes are critical for virus–cell interactions must be determined in intact tissue *in vivo*.

The severe combined immunodeficiency-human (SCID-hu) mouse model has proved to be invaluable for analyzing genetic factors that alter VZV pathogenesis *in vivo* (Moffat *et al.*, 1995, 1998) and is the focus of this review. Although it is otherwise highly species-specific in its infectivity, VZV also replicates in inbred strain 2, outbred Hartley and euthymic, hairless guinea-pigs when it has been adapted by serial passage in guinea-pig embryo fibroblasts (Myers *et al.*, 1985; Pavan-Langston and Dunkel, 1989). Guinea-pigs infected with VZV have viremia; the virus can be detected in dorsal root ganglia and infection induces humoral and cellular immunity to viral proteins (Matsunaga *et al.*, 1982; Jenski and Myers, 1987; Hayward *et al.*, 1991; Lowry *et al.*, 1992; Sabella *et al.*, 1993; Sadzot *et al.*, 1995). Hairless guinea-pigs infected with VZV develop a transient maculopapular rash (Myers *et al.*, 1991). Infection of ganglia can also be demonstrated in rats after subcutaneous inoculation with VZV (Sadzot *et al.*, 1995).

Background of SCID-hu model

SCID-hu mice implanted with thymus/liver implants are effective models for human immunodeficiency virus (HIV)-1, human cytomegalovirus (HCMV), and measles virus infections in intact human tissues (Kaneshima *et al.*, 1991; Aldrovandi *et al.*, 1993; Mocarski *et al.*, 1993; Auwaerter *et al.*, 1996). The SCID-hu mouse model provides a unique opportunity to examine the attenuation of V-Oka and other aspects of VZV pathogenesis independently of the effects of the host immune response on viral replication (Moffat *et al.*, 1995). The inoculation of human skin implants in SCID-hu mice with a low-passage clinical isolate of VZV induces histopathologic changes typical of VZV skin lesions observed in patients with varicella or herpes zoster (Cheatham *et al.*, 1956; Moffat *et al.*, 1995). VZV replicates in CD4+ and CD8+ human T cells in thymus/liver (thy/liv) implants in SCID-hu mice, demonstrating that it must be classified as a lymphotropic as well as a neurotropic herpesvirus (Croen *et al.*, 1988; Mahalingam *et al.*, 1990; Croen and Straus, 1991; Moffat *et al.*, 1995). The SCID-hu model provides a new system for analyzing the relationships between viral gene expression and human cell tropism, using naturally arising VZV variants and recombinant VZV strains that carry directed mutations.

Handbook of Animal Models of Infection
ISBN 0-12-775390-7

Animal species

Male homozygous CB-17 *scid/scid* mice are bred, maintained and implanted with human fetal tissues at SyStemix, Inc, Palo Alto, CA. When the mice are 8 weeks old, co-implants of human fetal thymus and liver tissue from 18–23-week fetuses are introduced under the kidney capsule as a conjoint implant using an 18 G trocar (Namikawa *et al.*, 1990). The thymus/liver implants develop for 6–8 weeks before use. Human skin is introduced subcutaneously as full-thickness dermal grafts and is allowed to engraft for 3–5 weeks before use. Human fetal tissues are obtained with informed consent according to federal and state regulations and are screened for HIV.

Preparation of animals

Due to the impaired immunity of SCID-hu mice, the bedding, food and water are sterilized by autoclaving before use. In addition, SCID-hu mice receive trimethoprim sulfamethoxazole (40 mg trimethoprim, 200 mg sulfamethoxazole per 5 ml suspension with 0.125 ml suspension per 4 ml drinking water/mouse per day) in the drinking water for 3 days of each week to prevent opportunistic infection by *Pneumocystis carinii* and other parasites.

Details of surgery

Overview

Thymus/liver tissue under the kidney capsule or subcutaneous skin tissue implanted in SCID-hu mice is surgically exposed, then VZV-infected cells are injected into the implant to initiate virus replication in the human tissue. The mice remain free of disease during the course of the experiment.

Materials required

The materials required include anesthetic, 70% ethanol in a spray bottle to disinfect skin, scalpel, forceps, hemostat, syringes, needles, sutures, and metal clips.

Anesthesia

Mice are grasped firmly in one hand, then sedated with an intraperitoneal (i.p.) injection of 100 mg/kg ketamine and 20 mg/kg xylazine. This is prepared by diluting ketamine to 20 mg/ml and xylazine to 4 mg/ml in phosphate-buffered saline (PBS) and injecting 0.1–0.2 ml per mouse depending on weight (range 20–40 g). Complete anesthesia occurs in about 5 minutes and lasts for approximately 45 minutes. Younger mice occasionally experience a burst of agitation following administration of the anesthetic and must be kept in a tightly closed cage until they are unconscious to prevent escape.

Surgical procedure for inoculating thymus/liver implants

A sedated SCID-hu mouse (thy/liv) is placed on its right side so that the left side of the animal is facing up. The skin and fur of the left flank are disinfected by spraying with 70% v/v ethanol until moistened. Too much ethanol can cool the mouse excessively and cause hypothermia.

The approximate location of the implant is determined by palpating the abdomen and feeling for a small lump on the kidney. A 1–2 cm incision is made in the skin with small scissors (4.5 in, delicate tip) over the ventral side of the implant. Another 1 cm incision is made through the peritoneum over the ventral edge of the implant, being careful to avoid damaging the liver or spleen which lie anterior to the kidney. (The incision should not lie directly over the implant because the peritoneum can adhere to the implant during the healing process and hinder excision.) The entire kidney or a portion of it is brought through the peritoneal incision by gently squeezing the abdomen. The thy/liv implant appears as a pale pink hemisphere, about 0.5–1.0 cm in diameter, on the dark red kidney (Figure 117.1 top).

To infect the implant, the kidney is stabilized with one hand while the thy/liv implant is injected with the VZV inoculum. Inject the implant in two or three locations with a total volume of 20–50 μl of inoculum per implant, depending on size. The implant swells slightly during the injection and care should be taken not to inject so much that the implant bursts. After inoculation of the implant, the kidney is pushed back through the incision. Blood loss is usually minimal but can be heavy if the peritoneal incision is large.

Following injection of the inoculum, the wound is closed by placing two or three sutures in the peritoneum using non-dissolving silk with a tapered, ½ in RB-1 needle (Ethicon, Inc.). The skin is closed with two or three small metal clips (9 mm, Clay Adams). The mouse is kept warm until it begins to recover from the anesthetic, then it is placed in its cage with up to 4 littermates. No medication is given for postoperative pain and the animals do not appear to be in distress.

Procedure for harvesting infected thy/liv implants

At 7, 14, and 21 days post-inoculation, at least 3 mice are terminated and thy/liv implants are removed by cutting the implant away from the underlying kidney tissue. The

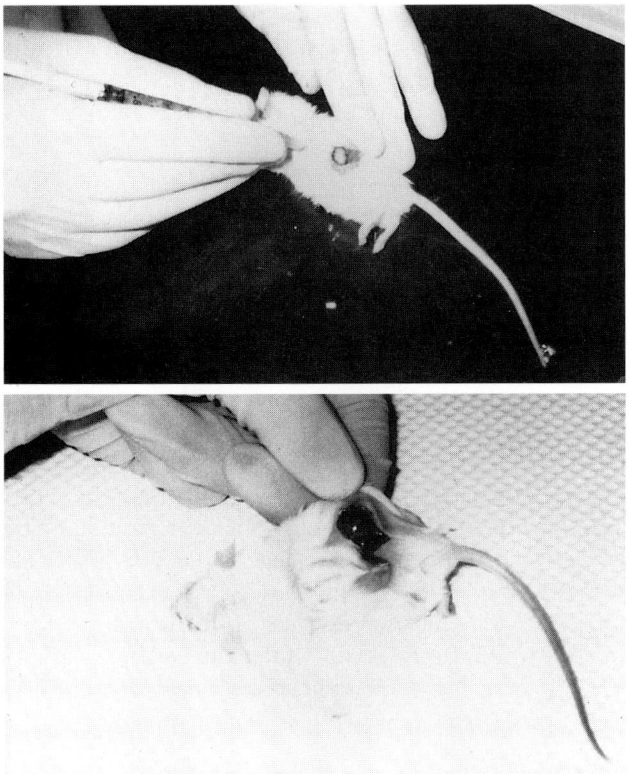

Figure 117.1 Surgical exposure of implants in SCID-hu mice. (Top) Access to thymus/liver implants is through incisions in the left flank and the peritoneum which reveal the implant growing under the kidney capsule. (Bottom) Bilateral, subcutaneous skin implants are exposed through a single incision in the back.

implants are divided for analysis: half of each tissue block is fixed in 10% formalin or 4% paraformaldehyde for histology; the other half is compressed and disrupted between ground-glass slides to release thymocytes into PBS with 2% fetal calf serum. The cell suspension is filtered through a sterile nylon mesh to remove large debris and the released cells are counted, washed, and used for FACS analysis and virus titration.

Surgical procedure for inoculating skin implants

A sedated mouse is placed face-down and the hair on its back is moistened with a small amount of 70% ethanol. Bilateral skin implants are exposed through a single 1 cm incision in the middle of the back. One implant, which appears as a thumbnail-sized block of tissue, is pushed through the incision to expose the epidermis (Figure 117.1 bottom). The epidermal side of the implant is gray to pink and may have visible hair growth while the dermal side is smooth, white adipose tissue. The inoculum (approximately 10 μl) is gently scratched and injected at multiple sites into the epidermis using a 27 G needle. Both implants are inoculated, and then the incision is closed with two small clips. The mouse is kept warm until it has recovered from the anesthesia, then it is placed in a cage with up to 5 littermates. No postoperative treatments for pain or distress are necessary.

Procedure for harvesting infected skin implants

At 14, 21 and 28 days after inoculation, at least 2 mice are terminated and the implants are removed and divided; one 2.0 mm central slice is fixed in 10% formalin for histology and *in situ* hybridization, approximately one-third is frozen in PBS at −20°C for later extraction of total protein, and the other third is placed in SPGA buffer (218 mmol/l sucrose, 3.8 mmol/l KH_2PO_4, 7.2 mmol/l K_2HPO_4, 4.9 mmol/l sodium glutamate, 1% bovine albumin, and 10% fetal calf serum) for virus isolation. SPGA buffer is optimized to stabilize VZV virions which are extremely labile in standard solutions. Each implant is minced and vortexed thoroughly in 1.0 ml SPGA buffer; an aliquot of the suspension is titered directly as described below and a second aliquot is filtered through a 0.45 μm membrane before titration to detect cell-free virus.

Storage, preparation of inocula

VZV strains are grown in MRC-5 cells (fetal lung fibroblasts) and stored at −80°C in 10% dimethyl sulfoxide. Before inoculation, VZV strains are passed three times in MRC-5 cells in minimal essential medium (MEM) supplemented with 50 IU penicillin, 50 μg streptomycin, and 0.5 μg fungizone with 10% fetal calf serum. When at least 90% of the MRC-5 cells appear infected, the monolayer from a 75 cm^2 flask is trypsinized; the cells are counted, centrifuged, resuspended in 2–3 ml growth medium, and briefly stored on ice before injection into the SCID-hu mouse implants; mock-infected implants are injected with an equal number of uninfected MRC-5 cells. The titer of each inoculum is determined by infectious focus assay and usually ranges from 10^4 to 10^5 plaque-forming units (pfu)/ml (method described below). To inoculate implants, a 1 ml tuberculin syringe fitted with a 27 or 30 G needle is loaded with the VZV-infected cell suspension immediately prior to use to avoid settling of the cells in the syringe. Implants from several mice can be inoculated with a single syringe if the cell suspension is mixed well before each use.

Infection process

Thy/liv and skin implants are infected by direct injection of a VZV-infected cell suspension into the graft as described above. The initial cell types infected in each implant are not precisely known and the sequence of infection remains to be elucidated. VZV-infected fibroblasts presumably initiate viral replication by fusing with T cells in thy/liv implants or with epithelial cells in skin implants. Alternatively, VZV-infected MRC-5 cells lyse soon after inoculation and release virions attached to cellular debris which bind to target cells and initiate infection. Histological analysis shows that by 7 days post-infection no evidence of the inoculum remains.

Key parameters to monitor infection

VZV infection is restricted to the human tissue implants, hence VZV does not infect mouse tissue nor does it cause disease in the animal. In thy/liv implants, T cells are the primary site of viral replication. The extent of VZV infection is determined by titration of infected T cells in an infectious focus assay and by FACS analysis using VZV-specific antibodies and conjugates to T-cell markers such as CD3, CD4, and CD8 (methods described below). A decrease in the number of viable T cells is another indication of viral replication. Thymocytes are lysed and the implants become shrunken and necrotic as VZV infection spreads (Figure 117.2). Consequently, infected thy/liv implants appear hard, red and much smaller than mock-infected controls. The depletion of target cells limits VZV growth to a 3-week period with maximal titers reached between 7 and 14 days post-inoculation.

Histological analysis of infected skin implants shows early lesions in the epidermis which develop into large vesicles that penetrate the basement membrane and involve dermal fibroblasts (Figure 117.3). By 3–4 weeks post-inoculation, the keratinized outer layer of the skin may slough off and expose a necrotic exudate that contains large amounts of cell-associated virus and some cell-free virions. Implants are finely minced to release virions and infected cells, then the number of pfu per implant is calculated by titering the cell suspension in an infectious focus assay. Total protein can also be extracted from the minced implants by sonication in a detergent buffer, then the viral proteins are quantitated on Western blots. Briefly, extracts are separated by sodium dodecylsulfate polyacrylamide gel electrophoresis (SDS-PAGE), transfered to nylon membranes, and VZV proteins detected using a high-titer polyclonal human serum and visualized by enhanced chemiluminescence (ECL). Phosphorimager analysis or densitometry can be used to quantitate VZV proteins.

FACS analysis of VZV-infected thymocytes

Aliquots of approximately 10^6 thymocytes obtained from infected and uninfected thy/liv implants are washed and resuspended in PBS with 2% fetal calf serum before staining with various antibodies and conjugates. VZV protein synthesis is detected with VZV-immune human polyclonal immunoglobulin G and a goat anti-human FITC conjugate for the second antibody. T-cell markers are detected with mouse monoclonal anti-CD4-phycoerythrin (PE) and anti-CD8-tricolor (TRI) conjugates. Cell suspensions are analyzed with a Becton/Dickinson FacScan apparatus.

Infectious focus assay

Virus titrations of infected MRC-5 cell inocula, infected thymocytes from thy/liv implants, skin implant suspensions, or cell-free filtrates of skin implant suspensions are done using specimens serially diluted 10-fold in growth medium. A 0.1 ml cell suspension, mixed with 1.5×10^5 Vero cells in 0.9 ml medium, is added to 24-well plates in triplicate. The plates are incubated for 6 days at 37°C in 5% CO_2; 1.0 ml fresh medium is added on Day 3. Following aspiration of the supernatant, the wells are flooded with crystal violet stain (5% ethanol, 5% formaldehyde, 0.13% crystal violet in PBS) for 2–5 minutes. The stain is aspirated, the wells are air-dried, and plaques are counted using an inverted light microscope (magnification ×40). The level of detection of the infectious focus assay is 10 pfu per specimen.

Histology, immunohistochemistry and *in situ* hybridization

Thy/liv and skin implants are fixed in 10% formalin or 4% paraformaldehyde overnight, embedded in paraffin, cut into 3 μm sections and stained with hematoxylin and eosin (H&E). Unstained sections are deparaffinated in xylene and rehydrated in graded ethanols before use. Immunohistochemistry and *in situ* hybridization are used to detect VZV protein and DNA respectively according to previously published methods (Moffat *et al.*, 1995).

Antimicrobial therapy

No antimicrobial agents have been studied in the SCID-hu model of VZV replication. However it should be possible to evaluate antiviral compounds by a variety of methods. Thy/liv and skin implants are fully vascularized, thus agents administered to the mouse reach the thy/liv or skin implant via the circulatory system. The per oral, subcutaneous and intraperitoneal routes are the simplest, yet intravenous administration is possible by the tail vein. Skin implants are amenable to application of topical agents since they can be repeatedly and easily accessed through the dorsal incision. Two nucleoside analogs, acyclovir and vidarabine, and the chemokine interferon-α have been used successfully to treat varicella or zoster infections in humans (Gelb, 1993a). Oral acyclovir is well-tolerated and is most effective when given early during the illness. Other antiviral agents have not been developed for VZV, partly due to the lack of small animal models.

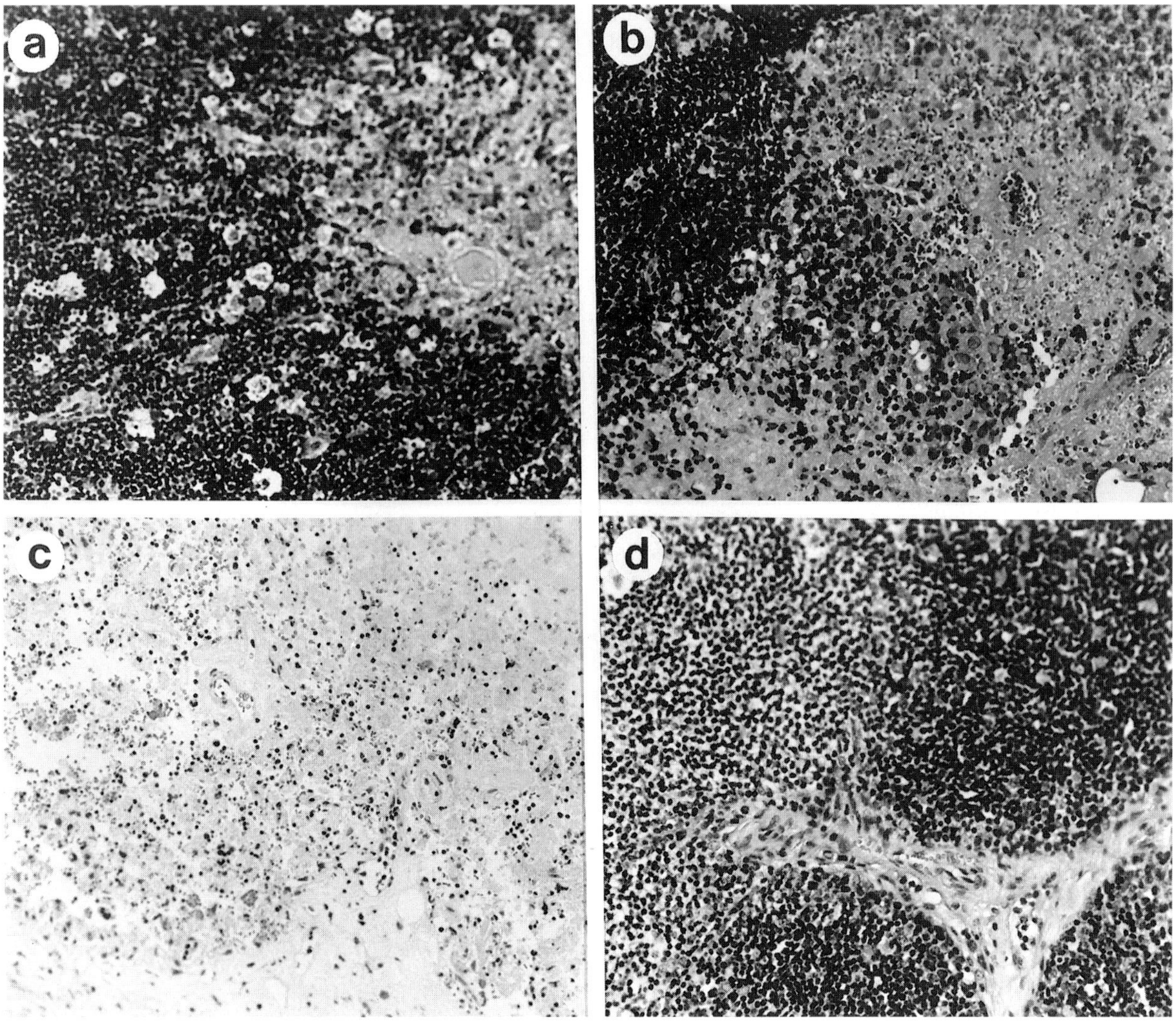

Figure 117.2 Histology of varicella-zoster virus (VZV)-infected thymus/liver implants. Implants were fixed in paraformaldehyde, paraffin-embedded, cut into 3 μm sections, and stained with hematoxylin and eosin (H&E). (**a**) Early evidence of VZV infection is shown by infiltration of macrophages that have engulfed infected T cells and give a "starry night" appearance to the thymus. (**b**) By Day 7 post-infection, lymphocyte depletion, fibrosis, and red blood cells indicate extensive VZV replication. (**c**) Complete destruction of the implant occurs by Day 21 post-infection and VZV replication ceases. (**d**) Mock-infected implants have a normal appearance. Magnification ×164.

A SCID-hu model was used to study the effects of ganciclovir on replication of HCMV, a related herpesvirus, in thy/liv implants (Mocarski *et al.*, 1993). Mice were given oral or intraperitoneal ganciclovir from 4–6 hours after inoculation and continuing daily for 12 days. The highest concentrations of ganciclovir inhibited CMV replication by three to four orders of magnitude, demonstrating the value of this system for evaluating antimicrobial therapies.

Pitfalls (advantages/disadvantages) of the model

The very feature of SCID mice that enables them to be implanted with human tissue gives rise to the greatest disadvantage of using this model to study VZV pathogenesis, that is, the lack of an immune response to the virus. This major drawback precludes testing components of VZV for immunogenicity and makes challenge studies impossible. The SCID-hu mouse implanted with thy/liv or skin tissue is thus solely a model of VZV replication and does not reproduce the entire spectrum of VZV disease. Aspects of varicella and zoster that cannot be studied in the SCID-hu model are the initial steps of inoculation by aerosol route, infection of regional lymph nodes, spread to internal organs, infection of neural tissue, and reactivation. Attempts to implant ganglia from cadavers and study VZV replication in neural tissue were unsuccessful due to failure of the grafts. When designing experiments using skin or thy/liv implants, adequate numbers of implants must be infected to allow for implant variability and the inability to

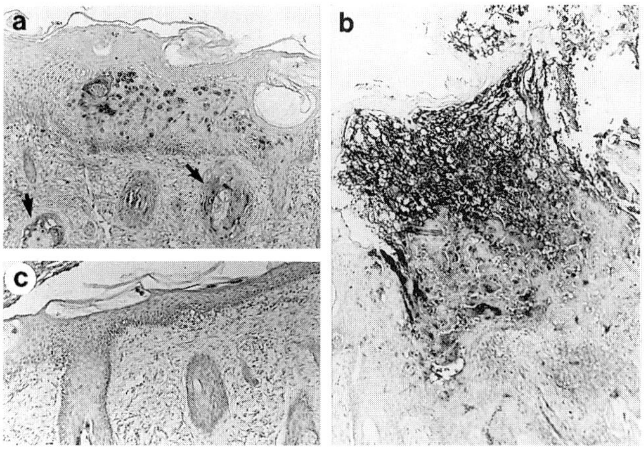

Figure 117.3 Histology of varicella-zoster virus (VZV)-infected skin implants. Subcutaneous skin implants infected with VZV were fixed in formalin, paraffin-embedded, and cut into 3 μm sections before *in situ* hybridization was performed. Darkly stained cells indicate VZV DNA and a light hematoxylin counterstain reveals tissue histology. (**a**) Infected cells are present in the epidermis and in the epithelial cells surrounding hair follicles (arrows) during the early stages of VZV infection. (**b**) By 21 days post-inoculation, large amounts of VZV DNA are detected in vesicular lesions that extend deep into the dermis. (**c**) Mock-infected implants have normal skin structure. Magnification ×82.

monitor the course of infection in individual implants. For these reasons, at least three implants, or preferably four, must be allocated for each strain tested at each time point.

Despite the limitations of SCID-hu mice, this model provides unique opportunities to study VZV in differentiated tissue *in vivo*. Skin tissue is notoriously difficult to grow in culture and studies of VZV replication in skin have been hindered by the difficulty of reproducing the multilayered structure of the skin *in vitro*. Skin can be separated into epidermal cells (keratinocytes) and dermal fibroblasts (human foreskin fibroblasts) (HFFs) and cultured by standard techniques; however, these systems do not reveal the differences in VZV virulence that can be seen in SCID-hu skin implants. Skin equivalents (raft cultures) can be made by embedding fibroblasts in a collagen matrix then allowing keratinocytes to adhere, but these lack hair follicles and glandular epidermal cells that VZV preferentially infects. Our group and others have shown that VZV infects peripheral blood mononuclear cells (PBMC) *in vitro*, yet propagation of the virus has not been demonstrated (Gilden *et al.*, 1987; Koropchak *et al.*, 1989). SCID-hu thy/liv implants are currently the only system available to study the interaction of VZV with T cells.

Contributions of the model to infectious disease therapy

The SCID-hu model of VZV replication has advanced our understanding of the cell tropism and virulence determinants of this virus. The capacity to investigate VZV interactions with T lymphocytes in thy/liv implants is essential for experiments to identify VZV gene functions that account for this important tropism of the virus. Recent work using this model has identified a viral protein kinase encoded by ORF 47 that is required for VZV replication in T cells (Moffat, unpublished data). The comparative analysis of V-Oka and the parent Oka virus along with other VZV strains in the SCID-hu mouse model established that V-Oka has a diminished capacity to replicate in human skin (Moffat *et al.*, 1998). The clinical attenuation of V-Oka can be explained by this decreased virulence for human skin, a characteristic which is associated with prolonged growth in tissue culture cells. Analysis of a variant of V-Oka that does not express glycoprotein C (gC) showed that gC was a specific virulence determinant in VZV infection of human skin cells. The findings that VZV gC is required for effective viral replication in skin and ORF 47 is required in T cells are the first evidence of essential roles for any VZV gene product in the pathogenesis of human infection. The SCID-hu model is clearly beneficial to those who wish to test potential vaccine strains and to test antiviral agents effective against VZV.

References

Aldrovandi, G. M., Feuer, G., Gao, L. *et al.* (1993). The SCID-hu mouse as a model for HIV-1 infection. *Nature*, **363**, 732–736.

Arvin, A. M. (1996). Varicella-zoster virus. In *Virology* (eds Fields, B. N., Knipe, D. N., Howley, P. M.), pp. 2547–2585. Lippincott-Raven, Philadelphia.

Auwaerter, P. G., Kaneshima, H., McCune, J. M., Wiegand, G., Griffin, D. E. (1996). Measles virus infection of thymic epithelium in the SCID-hu mouse leads to thymocyte apoptosis. *J. Virol.*, **70**, 3734–3740.

Cheatham, W. J., Weller, T. H., Dolan, T. F. J., Dower, J. C. (1956). Varicella: report of two fatal cases with necropsy, virus isolation, and serologic studies. *Am. J. Pathol.*, **32**, 1015–1035.

Cohen, J. I., Seidel, K. E. (1993). Generation of varicella-zoster virus (VZV) and viral mutants from cosmid DNAs: VZV thymidylate synthetase is not essential for replication *in vitro*. *Proc. Natl Acad. Sci. USA*, **90**, 7376–7380.

Croen, K., Straus, S. (1991). Varicella-zoster virus latency. *Annu. Rev. Microbiol.*, **4**, 2391–2399.

Croen, K., Ostrove, J., Dragovic, L., Straus, S. (1988). Patterns of gene expression and sites of latency in human nerve ganglia are different for varicella-zoster and herpes simplex viruses. *Proc. Natl Acad. Sci. USA*, **85**, 9773–9777.

Gelb, L. D. (1993a). Varicella-zoster virus: clinical aspects. In *The Human Herpesviruses* (eds Roizman, B., Whitley, R. J., Lopez, C.), pp. 281–308. Raven Press, New York.

Gelb, L. D. (1993b). Varicella-zoster virus: molecular biology. In *The Human Herpes viruses* (eds Roizman, B., Whitley, R. J., Lopez, C.), pp. 257–279. Raven Press, New York.

Gilden, D. H., Hayward, A. R., Krupp, J., Hunter-Laszlo, M., Huff, J. C., Vafai, A. (1987). Varicella-zoster virus infection of human mononuclear cells. *Virus Res.*, **7**, 117–129.

Hayward, A. R., Burger, R., Scheper, T., Arvin, A. M. (1991). Major histocompatibility complex restriction of T cell responses to varicella zoster virus in guinea pigs. *J. Virol.*, **65**, 1491–1495.

Jenski, L., Myers, M. G. (1987). Cell-mediated immunity to varicella-zoster virus infection in strain 2 guinea pigs. *J. Infect. Dis.*, **156**, 430–435.

Kaneshima, H., Shih, C.-C., Namikawa, R. *et al.* (1991). Human immunodeficiency virus infection of human lymph nodes in the SCID-hu mouse. *Proc. Natl Acad. Sci. USA*, **88**, 4523–4527.

Koropchak, C. M., Diaz, P. S., Arvin, A. M. (1989). Investigation of varicella-zoster virus infection of lymphocytes by *in situ* hybridization. *J. Virol.*, **63**, 2392–2395.

Koropchak, C. M., Graham, G., Palmer, J. *et al.* (1991). Investigation of varicella-zoster virus infection by polymerase chain reaction in the immunocompetent host with acute varicella. *J. Infect. Dis.*, **163**, 1016–1022.

Krause, P. R., Klinman, D. M. (1995). Efficacy, immunogenicity, safety, and use of live attenuated chickenpox vaccine. *J. Pediatr.*, **127**, 518–525.

LaRussa, P., Lungu, O., Hardy, I., Gershon, A., Steinberg, S. P., Silverstein, S. (1992). Restriction fragment length polymorphism of polymerase chain reaction products from vaccine and wild-type varicella-zoster virus isolates. *J. Virol.*, **66**, 1016–1020.

Lowry, P. W., Solem, S., Watson, B. N. *et al.* (1992). Immunity in strain 2 guinea pigs inoculated with vaccinia virus recombinants expressing varicella-zoster virus glycoproteins I, IV, V, or the protein product of the immediate early gene 62. *J. Gen. Virol.*, **73**, 811–819.

Mahalingam, R., Wellish, M., Wolf, W. *et al.* (1990). Latent VZV DNA in human trigeminal and thoracic ganglia. *N. Engl. J. Med.*, **323**, 627–631.

Mallory, S., Sommer, M., Arvin, A. M. (1997). Mutational analysis of the role of glycoprotein I in varicella-zoster virus replication and its effects on glycoprotein E conformation and trafficking. *J. Virol.*, **71**, 8279–8288.

Matsunaga, Y., Yamanishi, K., Takahashi, M. (1982). Experimental infection and immune response of guinea pigs with varicella-zoster virus. *Infect. Immun.*, **37**, 407–412.

Mocarski, E. S., Bonyhadi, M., Salimi, S., McCune, J. M., Kaneshima, H. (1993). Human cytomegalovirus in a SCID-hu mouse: thymic epithelial cells are prominent targets of viral replication. *Proc. Natl Acad. Sci. USA*, **90**, 104–108.

Moffat, J. F., Stein, M. D., Kaneshima, H., Arvin, A. M. (1995). Tropism of varicella-zoster virus for human CD4+ and CD8+ T lymphocytes and epidermal cells in SCID-hu mice. *J. Virol.*, **69**, 5236–5242.

Moffat, J. F., Zerboni, L., Kinchington, P. R., Grose, C., Kaneshima, H., Arvin, A. M. (1998). The attenuation of the vaccine Oka strain of varicella-zoster virus and the role of glycoprotein C in alphaherpesvirus virulence demonstrated in the SCID-hu mouse. *J. Virol.*, **72**, 965–974.

Myers, M. G., Stanbury, L. R., Edmond, B. J. (1985). Varicella-zoster virus infection of strain 2 guinea pigs. *J. Infect. Dis.*, **151**, 106–113.

Myers, M. G., Connelly, B. L., Stanberry, L. R. (1991). Varicella in hairless guinea pigs. *J. Infect. Dis.*, **163**, 746–751.

Namikawa, R., Weilbaecher, K. M., Kaneshima, H., Yee, E. J., McCune, J. M. (1990). Long term human hematopoiesis in the SCID-hu mouse. *J. Exp. Med.*, **172**, 1055–1063.

Ozaki, T., Ichikawa, T., Matsui, Y. *et al.* (1986). Lymphocyte-associated viremia in varicella. *J. Med. Virol.*, **19**, 249–253.

Pavan-Langston, D., Dunkel, E. (1989). Ocular varicella-zoster virus infection in the guinea pig: a new *in vivo* model. *Arch. Ophthalmol.*, **107**, 1068–1072.

Sabella, C., Lowry, P. W., Abbruzzi, G. M. *et al.* (1993). Immunization with the immediate-early tegument protein (open reading frame 62) of varicella-zoster virus protects guinea pigs against virus challenge. *J. Virol.*, **67**, 7673–7676.

Sadzot, D. C., Debrus, S., Nikkels, A., Piette, J., Rentier, B. (1995). Varicella-zoster virus latency in the adult rat is a useful model for human latent infection. *Neurology*, S18–S20.

Sawyer, M. H., Wu, Y. N., Chamberlin, C. J. *et al.* (1992). Detection of varicella-zoster virus DNA in the oropharynx and blood of patients with varicella. *J. Infect. Dis.*, **166**, 886–888.

Takahashi, M., Otsuka, T., Okuno, Y., Asano, Y., Yazaki, T., Isomura, S. (1974). Live vaccine used to prevent the spread of varicella in children in hospital. *Lancet*, **2**, 1288–1290.

Whitley, R. J. (1990). Varicella-zoster virus. In *Antiviral Agents and Viral Diseases of Man* (eds Galasso, G., Whitley, R., Merigan, T. C.), pp. 235–264. Raven Press, New York.

Chapter 118

The Mouse Model of Influenza Virus Infection

R. W. Sidwell

Background of human infection

Influenza, the highly contagious acute respiratory illness caused by influenza type A, B, or C viruses, has been a disease afflicting mankind since antiquity (Noble, 1982). In humans, the disease occurs periodically as epidemics, with occasional pandemics afflicting millions (Monto *et al.*, 1997). This repeated occurrence of major outbreaks of influenza has been due to antigenic shifts in the virus, resulting in the emergence of new virus subtypes containing a novel hemagglutinin (HA) or neuraminidase (NA) against which currently circulating antibodies have no effect (Lamb and Krug, 1996).

Background of mouse infection model

The public health importance of influenza has resulted in much effort being invested in developing therapies for the disease. Such antiviral efforts have required the use of suitable animal models which will be predictive of the effects such experimental therapies may have in the clinic. Since the initial observations by Andrewes *et al.* (1934) that anesthetized mice inoculated intranasally (i.n.) with nasal turbinate material from influenza virus-infected ferrets developed pneumonia, the mouse model of influenza has been used extensively in such experimental evaluations.

The Collaborative Antiviral Testing Group (CATG) of the National Institute of Allergy and Infectious Diseases, National Institutes of Health, has used the model for *in vivo* evaluation of potential influenza virus inhibitors since the inception of that program (Glasgow and Galasso, 1972). We have continued to use the mouse, with some modifications, for these antiviral evaluations (Sidwell *et al.*, 1994, 1995a,b, 1996, 1998).

Experimental infections of the mouse can be achieved with both influenza A and B viruses, although the viruses have to be adapted to the animal by multiple passages through their lungs. These mouse-adapted viruses have developed an ability to infect alveolar cells (Mulder and Hers, 1972) and are associated with changes in the immunological specificity of the surface HA (Gitelman *et al.*, 1983).

It is unclear how much animal adaptation of the influenza virus affects its sensitivity to antiviral drugs. Tisdale and Bauer (1976) have reported that decreases in antiviral activity in the animal occur after continued virus passage through mice, this effect possibly being related to altered growth characteristics of the virus in the animal. The recent clinically isolated influenza A/Beijing/32/92 (H3N2) virus that was not adapted to mice induced in the animals lung consolidation and a decline in arterial oxygen saturation that was inhibited by amantadine, but not by ribavirin therapy (Sidwell *et al.*, 1995a). Ribavirin was effective against this virus *in vitro*, and has exhibited inhibition of other influenza virus infections in mice (Sidwell *et al.*, 1972). It was concluded that use of a non-mouse-adapted virus in mice may produce false-negative results.

Infection of mice

Animals

Most strains of the laboratory mouse are susceptible to influenza virus infection. We have used ICR Swiss, Swiss-Webster, and BALB/c mice, which are all sensitive to the virus, in influenza antiviral experiments (Sidwell *et al.*, 1968, 1986, 1996). Specific pathogen-free animals are recommended. Female and male mice yield similar results, but females are usually employed because of the fighting tendencies inherent with the males. The age of the mouse to be used in antiviral studies will depend upon the influenza virus being studied; viruses that are poorly animal-adapted are more infectious in the weanling (8–10 g) animals, whereas established laboratory strains remain infective and capable of inducing a fatal pneumonitis in 5–8-week-old (18–21 g) mice.

Preparation of animals

It is recommended that all mice received from outside suppliers undergo a 24–48 hour quarantine to enable the animal to overcome the stress of transit and acclimatize to

Handbook of Animal Models of Infection
ISBN 0-12-775390-7

its new surroundings. We seek to avoid the potential of secondary bacterial infections, which may mask the antiviral effects sought, by maintaining the mice on drinking water containing 0.006% oxytetracycline (Pfizer, New York, NY) after exposure to the virus. The animals are fed standard mouse chow *ad libitum*.

Storage, preparation, and animal adaptation of inocula

All influenza viruses used in murine antiviral studies are stored as either cell culture or lung homogenates in cryovials maintained at –80°C. Prior to preparing a final virus pool for use in animals, the virus will need to be titrated in mice.

Many influenza viruses, particularly if they are clinical isolates not previously used in mice, will require multiple passages through mice. Care should be taken with new isolates not to expose the animal to too high a viral inoculum in the early passages, since high concentrations of such viruses will produce a toxic pneumonitis with little viral replication (Sugg, 1949).

We have had the greatest success by exposing 8–10 g mice intranasally (i.n.) to at least four dilutions of the virus (e.g., 1:4, 1:8, 1:16, 1:32). Lungs are harvested 2, 4 and 6 days later. These times are selected because not all virus strains achieve maximal titers at the same time in the mouse. The lungs are homogenized to a 10% suspension in minimum essential medium (MEM), and a sample of each assayed for infectious virus in Maden Darby canine kidney (MDCK) cells grown in 96-well flat-bottomed microplates. The remainder of each homogenate is frozen at –80°C. Development of discernible viral cytopathic effect 72 hours after incubation at 37°C is indicative of virus infection. This initial titration will reveal those individual lung homogenates having the highest titers; these are then thawed, pooled, and again passaged through mice until a virus titer is obtained that is at least 10^5–10^6 cell culture infectious doses ($CCID_{50}$)/ml. In our experience, it may require 6–10 passages of a clinical influenza isolate before sufficient adaptation occurs to induce fatal pneumonitis in mice.

Infection process

Mice can be infected by influenza virus either by i.n. instillation of the virus on the nares of the anesthetized animal or by exposure to a small-particle aerosol (s.p.a.). The latter procedure will provide a somewhat more uniform infection but may utilize larger quantities of virus; the method has a greater potential for infecting laboratory personnel, and is not practical if the animals require therapy prior to virus exposure.

The i.n. process is achieved by anesthetizing the animal by parenteral injection of a recognized anesthetic such as Avertin, ketamine HCl or sodium pentobarbital. Avertin (1.5 g of tribromethanol in 1.0 ml of isoamyl alcohol; Aldrich Chemical Co., Milwaukee, WI) is used intraperitoneally (i.p.) as a 1:80 dilution of 94 mg in sterile saline (Sidwell *et al.*, 1992). Ketamine HCl (Phoenix Scientific, Inc., St Joseph, MO) is recommended for use at a dose of 400 mg/kg administered intramuscularly (i.m.; Hughes, 1981). Sodium pentobarbital (Abbott Laboratories, North Chicago, IL) is injected i.p. in a dose of 60 mg/kg. When the mice are fully anesthetized, 60 µl of solution containing influenza virus is placed as droplets on the nares; this is aspirated into the lower respiratory tract. Occasionally a mouse will not inhale the virus properly, exhibiting bubbling or frothing; these animals are discarded.

The s.p.a. method utilizes an aerosol-producing apparatus which generates particles having a mass median diameter of < 5 µm. We have used a Puritan Bennett nebulizer (Puritan-Bennett Co., Lenexa, KS). Compressed air at a rate of 10 l/minute (50 psi) is used. A drying chamber between the generator and the aerosol chamber will reduce the variation in aerosol particle size (Knight *et al.*, 1986). In our studies, mice are housed in either shoebox-style polycarbonate cages which are closed with a plastic top or in two stainless-steel wire cages placed in a specially constructed rectangular stainless-steel box. Using either chamber, similar antiviral results have been obtained (Sidwell *et al.*, 1994, 1996).

When aerosols are produced, there will be a need to characterize them. A model 3510 split-laser sensing aerodynamic particle sizer (TSI, Inc., St Paul, MN) has been used in our laboratory (Sidwell *et al.*, 1994). The instrument was equipped with Distfit software to provide particle distribution analysis.

Key parameters to monitor the murine infection

General considerations

Many parameters have been used to monitor the infection in mice (Table 118.1); some are dependent upon the infectivity and challenge dosage of the virus. Infections that cause no visible, clinically manifested illness, can be studied using quantitation of recoverable virus from the lung. Often in such infections, slight to moderate lung consolidation accompanied by declines of less than 10% in arterial oxygen saturation may occur.

Ideally, an antiviral experiment should use multiple parameters to assess the efficacy of test materials; this may involve using two groups of infected mice treated with each dose of test compound or with placebo; one group is held for the duration of the experiment for monitoring of daily changes in some parameters, whereas, from the second group, mice are killed at selected times and their lungs removed for assay of disease manifestations and recoverable virus.

Table 118.1 Parameters used for antiviral experiments in mouse influenza virus infection

Parameter	*Reference**
Pneumonia-associated death	Davies *et al.* (1964)
Mean time to death	Davies *et al.* (1964)
Arterial oxygen saturation	Sidwell *et al.* (1992)
Pulmonary gas exchange	Tschorn *et al.* (1978)
Rales	Kaji and Tani (1967)
Lung score	Sidwell *et al.* (1968)
Lung weight	Sidwell *et al.* (1994)
Ratio of lung weight to body weight	Schulman (1968)
Lung water content	Arensman *et al.* (1977)
Change in host weight	Hoffmann (1973)
Change in water intake	Hoffmann (1973)
Hypothermia	Arensman *et al.* (1977)
Lung virus assayed by hemagglutinin	Davies *et al.* (1964)
Lung virus assayed by virus-induced plaques or cytopathic effect	Finter (1970)
Lung virus assayed by allantois-on-shell assay	Tisdale and Bauer (1976)
Histopathogical changes in lung	Arensman *et al.* (1977)

* Initial report of use in antiviral experiment.

Clinically manifested illness

Following exposure to a highly infective influenza virus, mice begin, within 48–96 hours, to show decreased activity and less food and water intake; they will fail to gain or begin to lose weight, show tendencies to huddle; their respiration increases, and ruffled fur is evident. Hypothermia develops and this can be measured rectally (Arensman *et al.*, 1977; Sidwell *et al.*, 1986). Death may occur in 5–12 days, depending on the viral inoculum (Grunert *et al.*, 1965; Sidwell *et al.*, 1992).

Several of these clinical signs of influenza disease can be monitored as quantitative assays of the efficacy of antiviral agents. Daily animal weights, water consumption, rectal temperature, mean day to death, and number of mice dying prior to the end of the experiment have been successfully used. Rectal temperatures are readily determined using a telethermometer (e.g., Yellow Springs Instrument Co., Inc., Yellow Springs, OH).

Lung consolidation

The lungs of infected mice gradually become consolidated due to a combination of viral damage to alveolar cells causing necrosis of the capillary walls leading to lung hemorrhage, as well as to late-occurring vascular phenomena resulting from immune response to the infection (for review, see Sidwell *et al.*, 1995a). This consolidation can be assigned a score according to the percentage of the lung displaying the plum coloration typical of viral pneumonia: scores range from 0 (normal) to 4 (100% of lung affected). The extent of pulmonary lesions expressed as a percentage of total lung surface as described by Schulman (1968) has also been used. The lung weight increases significantly as consolidation progresses, so comparisons of mean lung weights of infected drug- and placebo-treated controls is a quantitative measure of this disease manifestation.

As the pulmonary disease progresses in the animal, the lung function deteriorates and the arterial oxygen saturation (SaO_2) declines; this can be readily measured by pulse oximetry (Sidwell *et al.*, 1992) as illustrated in Figure 118.1. In our studies, a Biox 3740 pulse oximeter (Ohmeda, Louisville, OH) has been used, with the human ear probe applied to the inner-thigh muscle of the mouse (Figure 118.1). The slow instrument mode is selected, with recordings made after a 30 seconds stabilization on each animal. This parameter has proven of great value in monitoring the effects of therapy on disease progress in the intact animal. Actual blood-gas measurements can also be obtained from arterial blood of the mice (Tschorn *et al.*, 1978; Sidwell *et al.*, 1992).

Viral replication in the lung

Significant influenza virus-inhibitory compounds should reduce the replication of the virus in the lung, although in our experience this inhibition of virus titer does not have to be overwhelming for protection to the animal to occur. When varying dilutions of influenza virus are administered i.n. to mice, it can be shown that a dilution of one-half $\log_{10}$ of the challenge virus can markedly reduce the percentage of deaths occurring.

Viral assays can be performed on lung lavage fluids or on lung homogenates. The latter are prepared as 10% wt/vol.

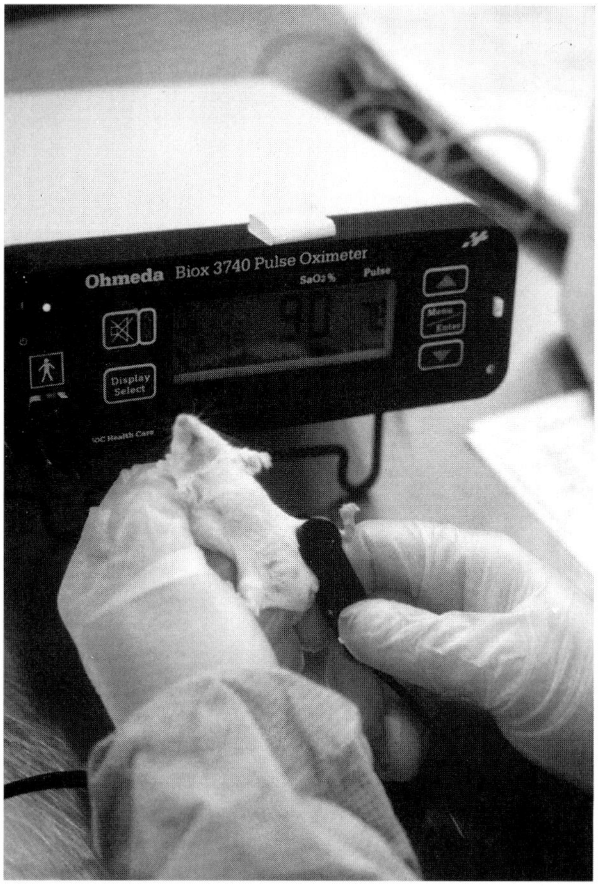

Figure 118.1 Pulse oximeter being used to determine arterial oxygen saturation levels in mice.

suspensions in MEM, which are then serially diluted and inoculated into embryonated hens' eggs (Wood, 1965), on to chorioallantoic membranes attached to eggshell (Tisdale and Bauer, 1976), or on to a monolayer of MDCK cells (Sidwell *et al.*, 1986). The viral titers are expressed as 50% egg infectious doses (EID_{50}), 50% allantois-on-shell infectious doses (AOS ID_{50}), $CCID_{50}$, or plaque-forming units (pfu) per volume of lavage or per gram of lung. Another method for assay of lung virus titer determines viral HA units in the lung lavages or supernatant fluids from homogenates (Squires, 1970); this method uses 0.5% suspensions of guinea-pig or chicken red blood cells. This HA method may not correlate with the titers of infectious virus in the lungs, however, so it may not be as acceptable as the infectious virus assays.

Influenza virus that has been mouse-adapted replicates rapidly in the lungs of mice. It is advantageous to assay for effects on virus titers at varying times after initiation of infection. We commonly use days 2, 4, 6, and 8 for evaluating lung virus titers, often finding a significant titer inhibition early, but not late, in the infection despite antiviral therapy of relatively long duration (Sidwell *et al.*, 1996).

Test drug administration

Potential antiviral agents may be administered to influenza virus-infected mice by parenteral injection (i.p., i.m., subcutaneous, intravenous), orally using gavage or incorporation into the drinking water, or by inhalation using i.n.-instilled droplets or by s.p.a. The selection of the appropriate treatment route will be dependent upon the agent's pharmacologic properties. If the i.n. route is used, the virus should first be titrated in mice treated i.n. with the drug diluent, using the same treatment schedule anticipated to be employed in the antiviral experiment. This is done because it has been shown that i.n. treatment will enhance viral replication and cause deaths to occur earlier than in infected animals which are not treated (Taylor, 1941; Takano *et al.*, 1963).

Advantages/disadvantages of the model

Particularly when pneumonia-associated lethal infections are used, the mouse model is somewhat unrelated to the more "typical" influenza virus infection seen in humans: the mouse infection is a greater challenge to the antiviral compound being evaluated. In addition, the mouse does not develop the fever and nasal discharge commonly seen in the human disease.

Despite these differences from the human disease, however, the mouse infection has been quite predictive for antiviral activity seen in the clinic with antiviral drugs. Amantadine and rimantadine have exerted striking influenza-inhibitory efficacy in the mouse model using a variety of disease parameters, and have shown similar efficacy against the disease induced by the same viral serotypes in the clinic (Hoffmann, 1973). Similarly, ribavirin has been considered to have significant efficacy against a spectrum of both influenza A and B virus infections in the mouse model, but only at high dosages, indicating a less-than-acceptable therapeutic index; human studies have yielded mixed results which appear dependent on the dosage of the drug (for review, see Sidwell, 1996). Zanamivir (GG167), a potent influenza NA inhibitor, has been shown to be highly inhibitory to the murine disease when administered i.n. (Ryan *et al.*, 1994), and significant disease-inhibitory efficacy has also been seen using the i.n.-administered drug in clinical challenge studies (Hayden *et al.*, 1996).

This predictability of the mouse model, coupled with its small size, ease of maintenance, reproducible response, multiple disease evaluation parameters, and relatively low cost allowing large numbers to be used, has led to the mouse being widely accepted for influenza antiviral studies.

Contributions of the mouse influenza model to infectious disease therapy

Transmission of infection

Schulman (1968) has shown that when normal mice are caged with animals 24 hours after the latter were infected by s.p.a. with influenza A (H2N2) virus, the infection can be sufficiently transmitted to produce viral titers in the lungs. A transmission rate of approximately 60% occurred. Treatment of the initially infected mice with rimantadine significantly prevented this viral transmission from occurring.

Demonstration of drug-resistant virus development

A problem with influenza virus-inhibitory drugs is the development of viruses which are resistant to them. Treatment of clinical cases of influenza with amantadine or rimantadine can result in over half of the patients shedding drug-resistant virus (for review, see Hayden and Couch, 1992). This rapid development of amantadine- and rimantadine-resistant virus has also been demonstrated in the mouse influenza virus model. Oxford *et al.* (1970) showed this in mice infected with influenza A (H2N2) virus, with viral resistance increasing with passage through amantadine-treated animals. Amantadine- and rimantadine-resistant viruses can be readily recovered from the lungs of mice challenged with influenza A/Port Chalmers/1/73 (H3N2) virus treated through a single course with high dosages of either compound, despite what appeared to be a strong antiviral effect induced in the animals (Sidwell *et al.*, 1995b). The virus recovered from amantadine-treated mice was 100-fold less sensitive to amantadine, and the virus taken from rimantadine-treated mice was 1000-fold less sensitive to rimantadine, than control viruses recovered from placebo-treated animals. These data indicate the potential of this infection model to predict the development of viral resistance in the clinic.

Influence of immunomodulating substances

Extensive studies have been reported on the primary and secondary immunological responses of the mouse to challenge with influenza virus (for review, see Murphy and Webster, 1996), and efforts have been made to identify immunomodulating substances as potential influenza inhibitors.

Interferon and interferon inducers have been reported to induce resistance to influenza infections in mice (Takano *et al.*, 1963; Came *et al.*, 1969; Gerone *et al.*, 1971). The hydrochloride salt of octadecyl D-alanyl L-glutamine (BCH-527), an analog of the recognized immunomodulator muramyl dipeptide (MDP), was found to be a moderate stimulator of natural killer (NK) cell, cytotoxic T lymphocyte (CTL), and macrophage activity in mice infected with influenza virus (Sidwell *et al.*, 1995c); all of these immune functions play a role in the primary protective response to influenza virus infection (for review, see Murphy and Webster, 1996). The therapy with BCH-527 only inhibited lung consolidation in mice lethally infected with influenza A (H1N1) virus (Sidwell *et al.*, 1995c). It is probable that the immune stimulation induced by BCH-527 was insufficient to protect the animals adequately from the infection. Another MDP analog, *N*-acetylmuramyl-L-alanyl-D-glutaminyl-*n*-butyl ester (murabutide), significantly protected mice from influenza virus challenge when administered from 4 days before to up to 2 days after challenge with influenza A (H3N2) virus (Chomel *et al.*, 1988). MDP, which is recognized as a strong macrophage activator (for review, see Fogler and Fidler, 1984), induced a protective effect in mice infected with influenza A (H1N1) virus when administered intravenously and was effective therapeutically when administered i. n. to the animals (Masihi *et al.*, 1983; Morin *et al.*, 1985).

Methionine enkephalin (Met-Enk), an endogenous opioid peptide, stimulates NK cell and CTL activity, increases interferon-γ and interleukin-1 production, activates macrophages, increases T-cell response to mitogens, and up-regulates expression of interleukin-2 receptors (for review, see Burger *et al.*, 1995). The effects of this material on NK and CTL activity were especially profound in influenza virus-infected mice, and the infection in the animals was significantly reduced by Met-Enk therapy (Burger *et al.*, 1995).

Pharmacokinetic utility

Although most pharmacokinetic studies on drugs are performed in rats and larger animals, the mouse can also be used for such investigations, which will provide data directly related to the performance of a drug against the experimentally induced influenzal disease in these animals.

A recent example of this correlation has been in the development of the potent influenza virus NA inhibitor, GS4104. High blood levels seen in mice following oral administration of this compound (Li *et al.*, 1998) correlated well with the material's influenza-inhibitory effects in mice when administered by the same route (Sidwell *et al.*, 1998).

References

Andrewes, C. H., Laidlaw, P. P., Smith, W. C. (1934). The susceptibility of mice to the viruses of human and swine influenza. *Lancet*, **2**, 859–862.

Arensman, J. B., Dominik, J. W., Hilmas, D. E. (1977). Effects of small-particle aerosols of rimantadine and ribavirin on arterial blood pH and gas tensions and lung water content of A2 influenza-infected mice. *Antimicrob. Agents Chemother.*, **12**, 40–46.

Burger, R. A., Warren, R. P., Huffman, J. H., Sidwell, R. W. (1995). Effect of methionine enkephalin on natural killer cell and cytotoxic T lymphocyte activity in mice infected with influenza A virus. *Immunopharmacol. Immunotoxicol.*, **17**, 323–334.

Came, P. E., Lieberman, M., Pascale, A., Shimonaski, G. (1969). Antiviral activity of an interferon-inducing synthetic polymer. *Proc. Soc. Exp. Biol. Med.*, **131**, 443–446.

Chomel, J. J., Simon-Lavoine, N., Thouvenot, D. *et al.* (1988). Prophylactic and therapeutic effects of murabutide in OF1 mice infected with influenza A/H3N2 (A/Texas/1/77) virus. *J. Biol. Response Mod.*, **7**, 581–586.

Davies, W. L., Grunert, R. R., Hoff, R. F. *et al.* (1964). Antiviral activity of 1-adamantanamine (amantadine). *Science*, **144**, 862–864.

Finter, N. B. (1970). Methods for screening *in vitro* and *in vivo* for agents active against myxoviruses. *Ann. N. Y. Acad. Sci.*, **173**, 131–138.

Fogler, W. E. and Fidler, I. J. (1984). Modulation of the immune response by muramyl dipeptide. In *Immune Modulation Agents and Their Mechanisms* (eds Fenichel, R. L., Chirigos, M. A.), pp. 499–512. Marcel Dekker, New York.

Gerone, P. J., Hill, D. A., Appell, L. H., Baron, S. (1971). Inhibition of respiratory virus infections of mice with aerosols of synthetic double-stranded ribonucleic acid. *Infect. Immun.*, **3**, 323–327.

Gitelman, A. K., Kavepin, N. V., Kharitonenkov, I. G., Rudneva, I. A., Zhdanov, U. M. (1983). Antigenic changes in mouse-adapted influenza virus strains. *Lancet*, **1**, 1229.

Glasgow, L. A. and Galasso, G. J. (1972). Isoprinosine: lack of antiviral activity in experimental model infections. *J. Infect. Dis.*, **126**, 162–169.

Grunert, R. R., McGahen, J. W., Davies, W. L. (1965). The *in vivo* antiviral activity of L-adamantanamine (amantadine). I. Prophylactic and therapeutic activity against influenza viruses. *Virology*, **26**, 262–269.

Hayden, F. G., Couch, R. B. (1992). Clinical and epidemiological importance of influenza A viruses resistant to amantadine and rimantadine. *Rev. Med. Virol.*, **2**, 89–96.

Hayden, F. G., Treanor, J. J., Betts, R. F., Lobo, M., Esinhart, J. D., Hussey, E. K. (1996) Safety and efficacy of the neuraminidase inhibitor GG167 in experimental human influenza. *J.A.M.A.*, **275**, 295–299.

Hoffmann, C. E. (1973). Amantadine HCl and related compounds. In *Selective Inhibitors of Viral Functions* (ed Carter, C. A.), pp. 199–212. CRC Press, Boca Raton.

Hughes, H. C. (1981). Anesthesia of laboratory animals. *Lab. Animal*, **10**, 40–56.

Kaji, M., Tani, H. (1967). New parameter in the screening test for antiinfluenza agent. *Proc. 5th Int. Congr. Chemother.*, **2**, 19–21.

Knight, V., Gilbert, B. E., Wilson, S. Z. (1986). Ribavirin small particle aerosol treatment of influenza and respiratory syncytial virus infections. In *Studies with a Broad-Spectrum Antiviral Agent* (ed. Stapleton, T.), pp. 37–56. Royal Society of Medicine Services, New York.

Lamb, R. A., Krug, R. M. (1996). Orthomyxoviridae: the viruses and their replication. In *Fields Virology*, 3rd edn (eds Fields, B. N., Knipe, D. M., Howley, P. M. *et al.*), pp. 1353–1395. Lippincott-Raven, Philadelphia.

Li, W., Escarpe, P. A., Eisenberg, E. J. *et al.* (1998). Identification of GS4104 as an orally bioavailable prodrug of the influenza neuraminidase inhibitor GS4071. *Antimicrob. Agents Chemother.*, **42**, 647–653.

Masihi, K. N., Brehmer, W., Lange, W., Ribi, E. (1983). Protective effect of muramyl dipeptide analogs in combination with trehalose dimycolate against aerogenic influenza virus and *Mycobacterium tuberculosis* infections in mice. *J. Biol. Response Mod.*, **3**, 663–671.

Monto, A. S., Iacuzio, D. A., LaMontague, J. R. (1997). Pandemic influenza – confronting a reemergent threat. *J. Infect. Dis.*, **176** (suppl. 1), 51–53.

Morin, A., Charley, B., Petit, A., Chedid, L. (1985). Protective effects of muramyl peptides against experimental viral infections. *Int. J. Immunopharmacol.*, **7**, 345.

Mulder, J., Hers, J. F. P. (1972). *Influenza*, pp. 214–238. Walters-Noordhoff, Groningen, The Netherlands.

Murphy, B. R., Webster, R. G. (1996). Orthomyxoviruses. In *Fields Virology*, 3rd edn (eds Fields, B. N., Knipe, D. M., Howley, P. M. *et al.*), pp. 1397–1445. Lippincott-Raven, Philadelphia.

Noble, G. R. (1982). Epidemiology and clinical aspects of influenza. In *Basic and Applied Influenza Research* (ed. Beare, A. S.), pp. 11–50. CRC Press, Boca Raton.

Oxford, J. S., Logan, I. S., Potter, C. W. (1970). *In vivo* selection of an influenza A2 strain resistant to amantadine. *Nature*, **226**, 82–83.

Ryan, D. M., Ticehurst, J., Dempsey, M. H., Penn, C. R. (1994). Inhibition of influenza virus replication in mice by GG167 (4-guanidino-2, 4-dideoxy-2, 3-dehydro-*N*-acetylneuraminic acid) is consistent with extracellular activity of viral neuraminidase (sialidase). *Antimicrob. Agents Chemother.*, **38**, 2270–2275.

Schulman, J. (1968). Effect of L-amantanamine hydrochloride (amantadine HCl) and methyl-L-adamantanethylamine hydrochloride (rimantadine HCl) on transmission of influenza virus infections in mice. *Proc. Soc. Exp. Biol. Med.*, **128**, 1173–1178.

Sidwell, R. W. (1996). Ribavirin: a review of antiviral efficacy. *Rec. Res. Dev. Antimicrob. Agents Chemother.*, **1**, 219–256.

Sidwell, R. W., Dixon, G. J., Sellers, S. M., Schabel, F. M. Jr (1968). *In vivo* antiviral properties of biologically active compounds. II. Studies with influenza and vaccinia viruses. *Appl. Microbiol.*, **16**, 370–392.

Sidwell, R. W., Huffman, J. H., Khare, G. P., Allen, L. B., Witkowski, J. T., Robins, R. K. (1972). Broad-spectrum antiviral activity of Virazole: 1-β-D-ribofuranosyl-1,2,4-triazole-3-carboxamide. *Science*, **177**, 705–706.

Sidwell, R. W., Huffman, J. H., Call, E. W., Alaghamandan, H., Cook, P. D., Robins, R. K. (1986). Effect of selenazofurin on influenza A and B virus infections in mice. *Antiviral Res.*, **6**, 343–353.

Sidwell, R. W., Huffman, J. H., Gilbert, J. *et al.* (1992). Utilization of pulse oximetry for the study of the inhibitory effects of antiviral agents on influenza virus in mice. *Antimicrob. Agents Chemother.*, **36**, 473–476.

Sidwell, R. W., Huffman, J. H., Moscon, B. J., Warren, R. P. (1994). Influenza virus-inhibitory effects of intraperitoneal- and aerosol-administered SP-303, a plant flavonoid. *Chemotherapy*, **40**, 42–50.

Sidwell, R. W., Bailey, K. W., Wong, M. H., Huffman, J. H. (1995a). *In vitro* and *in vivo* sensitivity of a non-mouse-adapted influenza A (Beijing) virus infection to amantadine and ribavirin. *Chemotherapy*, **41**, 455–461.

Sidwell, R. W., Bailey, K. W., Wong, M. H., Morrison, A. G., Huffman, J. H. (1995b). Rapid *in vivo* development of influenza A (H3N2) virus resistance to amantadine and rimantadine. *Antiviral Res.*, **26**, A328.

Sidwell, R. W., Smee, D. F., Huffman, J. H. *et al.* (1995c). Antiviral activity of an immunomodulatory lipophilic desmuramyl dipeptide analog. *Antiviral Res.*, **26**, 145–159.

Sidwell, R. W., Huffman, J. H., Bailey, K. W., Wong, M. H., Nimrod, A., Panet, A. (1996). Inhibitory effects of recombinant manganese superoxide dismutase in influenza virus infections in mice. *Antimicrob. Agents Chemother.*, **40**, 2626–2631.

Sidwell, R. W., Huffman, J. H., Barnard, D. L. *et al.* (1998). Inhibition of influenza virus infections in mice by GS4104, an orally effective influenza virus neuraminidase inhibitor. *Antiviral Res.*, **37**, 107–120.

Squires, S. L. (1970). The evaluation of compounds against influenza viruses. *Ann. N. Y. Acad. Sci.*, **173**, 239–248.

Sugg, J. Y. (1949). An influenza virus pneumonia of mice that is non-transferable by serial passage. *J. Bacteriol.*, **57**, 399–403.

Takano, K., Jensen, K. E., Warren, J. (1963). Passive interferon protection in mouse influenza. *Proc. Soc. Exp. Biol. Med.*, **114**, 472–475.

Taylor, R. M. (1941). Experimental infection with influenza A virus in mice. *J. Exp. Med.*, **73**, 43–55.

Tisdale, M., Bauer, D. J. (1976). A comparison of test methods in influenza chemotherapy. In *Chemotherapy and Control of Influenza* (eds Oxford, J. S., Williams, J. D.), pp. 55–62. Academic Press, London.

Tschorn, R. R., Arensman, J. B., Hilmas, D. E. (1978). Effect of small-particle aerosols of rimantadine and ribavirin on pathophysiological changes associated with A/NJ influenza in mice. *Proc. Soc. Exp. Biol. Med.*, **158**, 454–457.

Wood, T. R. (1965). Methods useful in evaluating 1-adamantanamine hydrochloride—a new orally active synthetic antiviral agent. *Ann. N. Y. Acad. Sci.*, **130**, 419–431.

Chapter 119

The Ferret as an Animal Model of Influenza Virus Infection

C. Sweet, R. J. Fenton and G. E. Price

Background of the model

Influenza, a pandemic viral respiratory disease, results in high morbidity and mortality in both humans and domestic animals. Human influenza is predominantly an upper respiratory tract (URT) infection, although influenza pneumonia is a rare complication (Douglas, 1975). Typical signs and symptoms are nasal discharge, cough, fever, neuralgia, myalgia, malaise, anorexia, depression and occasionally neurological and gastrointestinal disturbance (Nicholson, 1992). It is rarely lethal except in the very young and the elderly; deaths in the elderly are primarily due to bacterial pneumonia.

Influenza virus was first isolated from ferrets infected with throat washes from human patients (Smith *et al.*, 1933). The disease observed in ferrets closely resembles uncomplicated human influenza, and the ferret remains a valuable model system for influenza. Ferrets inoculated intranasally with virus remain healthy until about 24 hours post-infection (p.i.), when they develop observable symptoms including fever, listlessness, sneezing and nasal discharge. Animals appear ill, retreating to the back of their cage and refusing food once symptoms develop, leading to anorexia. Fever is associated with the polymorphonuclear leukocyte inflammatory influx into the URT, and results from the production of "endogenous pyrogen" (probably pro-inflammatory, pyrogenic cytokines) by nasal inflammatory cells (Sweet *et al.*, 1979). Fever typically persists for 24–36 hours, often peaking between 40 and 42°C (3–4°C above baseline temperatures). However the magnitude and duration of the fever depend on the virus strain used (Coates *et al.*, 1986). Virus is shed from the URT starting from 12 hours p.i., with peak viral titres seen around 24–48 hours p.i. and gradually decreasing from then onwards, although with some strains virus may still be detectable 3 days or more p.i. (Toms *et al.*, 1977). Lower respiratory tract infection also occurs with peak virus titres 3–4 days p.i. but replication and damage are essentially limited to the airways of the URT and bronchi (Cavanagh *et al.*, 1979; Sweet *et al.*, 1981; Husseini *et al.*, 1983b). The use of the ferret as a model for influenza virus pathogenesis has been comprehensively reviewed (Sweet and Smith, 1980; Smith and Sweet, 1988).

Animal species

The domestic ferret, *Mustela putorius furo*, is a carnivore of the family Mustelidae, believed to have been derived from the wild (European) polecat, *Mustela putorius putorius*. Introduced into the USA in the late 19th century, it is distinct from its North American counterpart, the black-footed ferret, *Mustela nigripes*. Ferrets are outbred, with no standardized breeds or strains. Two varieties are recognized based on coloration; the Fitch (common or polecat) with a brown coat and mask-like face and albino (English), which is cream to yellow with pink eyes.

Whilst ferrets are regarded as vicious and dangerous laboratory animals, this has not been our experience. Once acclimatized to their new environment, they can usually be handled without difficulty and without gloves. Young animals and frequently handled adults are quite friendly, but care should be taken with nursing mothers as they are defensive of their litters. Newborn animals, called kits, should be handled with disposable gloves as mothers can detect human scent on their offspring, and may kill them. The ferret is best restrained by gently but securely holding it behind its head with one hand, and supporting the body with the other hand.

Feeding and housing

We have found that ferrets thrive on a mixture of complete dried and tinned commercial cat food. However, this is insufficient for pregnant animals who frequently deliver abnormal offspring (with exencephaly, anencephaly, spina bifida and gastroschisis) on this diet. Supplementing diet with fresh fish or meat, and milk or lactol (a powdered milk for pregnant and lactating bitches) overcomes this problem. Commercial complete ferret diets are also available.

It is better to house ferrets in communal groups until required; in this way they can remain playful and active. However, cages are desirable for experimental ferrets or where breeding cycles are controlled artificially. The UK Home Office code of practice (Animals (Scientific Procedures) Act 1986) stipulates a minimum cage size with

Handbook of Animal Models of Infection
ISBN 0-12-775390-7

floor area 4500 cm^2 and height 50 cm for singly housed ferrets over 800 g. For ferrets less than 800 g the cage size can be 2250 cm^2/50 cm; in groups < 800 g it is 1500 cm^2/50 cm. Prior to experimentation, the wire screen base of the cage is removed so that the absorbent bedding (sawdust) keeps down the ferret odour and provides warmth. During experiments, however, animals are kept on the wire screens to prevent contamination of ferret airways with bedding material. Adult males (hobs) weigh between 1200 and 2500 g and females (jills) between 800 and 1200 g. Optimal room temperature is 16–20°C but unweaned animals should be kept at a minimum of 18–20°C. Optimal humidity is 40–65% and generally ferrets are kept on a 12-hour light cycle.

Breeding

Ferrets are sexually mature around 8–10 months of age and the breeding season extends from March to August for females and November to July for males. Jills are seasonally polyoestrous, and increasing day length induces the onset of oestrous. However, by manipulation of the photoperiod, the breeding season may be prolonged to allow mating throughout the year. Oestrous can be delayed or eliminated by maintaining the jills on a 6-hour light period. When required for mating they are placed on a 14-hour light cycle to induce onset of oestrous, which is apparent approximately 11 weeks later as a swelling of the vulva. Mating is carried out after full vulval swelling (1–2 cm diameter) has been obtained. An oestrous female is taken to the male's cage for breeding. The breeding pair should be observed for a few minutes as copulation is very involved and occasionally violent, with the male grabbing the female at the back of the neck and dragging her around the cage for up to 1 hour. It is important to differentiate this behaviour from fighting, and if the latter ensues on initial introduction, the female should be removed. Females are generally mated with 2–3 males over a 48-hour period to maximize successful mating.

Day 1 of pregnancy is taken as the day after introduction of the first male and pregnancy is indicated by a reduction in the size of the vulva, confirmed by abdominal palpation on days 17–21 post-coitus.

Male ferrets can receive a constant 6-hour photoperiod to maintain them sexually active. If testicular regression or failure to copulate is noticed, they should be rested for 6–12 weeks on a 14-hour photoperiod.

The gestation period is 42 ± 2 days. Pregnant jills are given litter boxes some days before term and generally the litters are nested in these boxes, out of view and disturbed as little as possible; this is important as upset or frightened nursing mothers may cannibalize their young. Litter size varies from 5 to 15 kits; newborn animals weigh approximately 7–10 g and are weaned in about 6 weeks. Significant mortality (12–30%) can occur among newborn ferrets (Collie *et al.*, 1980; Husseini *et al.*, 1983a). Many of these deaths are attributed to killing by the females as they are sometimes poor mothers, or failure of lactation (Willis and Barrow, 1971). However, some may arise from bacterial colonization, as antibiotic treatment has been shown to be beneficial (Husseini *et al.*, 1983a).

Infection procedure

An important advantage of ferrets is that they are susceptible to human influenza virus isolates. As a consequence they must be bled before use to ensure they have not been previously infected (examination of serum anti-influenza virus antibody levels by haemagglutination inhibition tests achieves this aim; Toms *et al.*, 1974). Anaesthesia of ferrets prior to inoculation is essential for restraint, and also to subdue the sneezing response, preventing ejection of the inoculum, and to aid inoculum penetration. In our experience recovery from diethyl ether is rapid (within minutes) but care should be taken as diethyl ether stimulates overproduction of mucus which may cause the tongue to block the airway. If necessary, this may be relieved by inserting a large (10 ml) syringe over the tongue to remove secretions. Alternatively, isoflurane (3%) administered from a mini-Boyle anaesthetic apparatus is very effective and does not stimulate mucus production. While other anaesthetics have been used during infection (Moody *et al.*, 1985), we have been reluctant to change from diethyl ether as other anaesthetics have been shown to alter influenza pathogenesis, at least in mice (Tait *et al.*, 1993).

The inoculum (in a volume of 0.25–0.5 ml) is administered to anaesthetized adult animals dropwise to each nostril from a 1 ml syringe fitted with a 25 G 5/8-in needle or a Gilson-type pipette. We use a relatively high inoculum (10^6 egg infectious doses (EID_{50})/0.5 ml) as this produces a consistent, reproducible infection with our strains. However, the inoculum required for reproducible infections varies with strain and passage history of the virus. Titrations of stock pools in ferrets are normally required prior to experimentation, although usually 10^4–10^6 EID_{50}/ml (or tissue culture infectious doses; $TCID_{50}$/ml) will produce a classical infection. Newborn ferrets (1 day old) may be inoculated without anaesthesia by intranasal instillation of 50 μl of Dulbecco A phosphate-buffered saline (PBSA) containing 10^1 egg-bit infectious doses ($EBID_{50}$) of virus (Collie *et al.*, 1980). Some isolates are readily transmitted between ferrets (Squires and Belyavin, 1975), but provided infected ferrets are isolated from stock ferrets by placing them in separate rooms and separate gowns and gloves (if needed) are used for different rooms, then this presents little problem. Indeed, we have held infected and uninfected animals in the same room and have encountered no problems with our viruses, provided uninfected animals are handled first.

Drug dosing procedures

Ferrets are frequently dosed by the intraperitoneal, subcutaneous, or intranasal routes. In each case, volumes between 0.25 and 0.5 ml/kg appear optimum. Other typical volumes are as follows: 0.6 ml/kg oral, 1 ml/kg intraperitoneal, 0.2 ml/kg intramuscular and 2.5 ml/kg intravenous. Aerosolized drug administration is also effective (Fenton *et al.*, 1977a), though while this method has the advantage of dosing most of the respiratory tree, the proportion of the dose actually inhaled is considerably less than that administered (<5%). The preferred method of aerosol administration is to dose ferrets whilst under isoflurane anaesthesia (1–3%), over 5–10 minutes. We use a mini-Boyle anaesthetic apparatus with a rigid face mask (International Market Supplies, Congleton, Cheshire, UK), modified to accept a nebulizer (sidestream nebulizer, Medic-Aid, Pagham, Sussex, UK). Solubilized drug (10 ml) can then be slowly administered via the face mask.

Parameters for monitoring infection and response to treatment

The ferret is an excellent model for studying influenza since the clinical manifestations resemble the normal adult human disease (see above). In addition, its unusual tracheal length allows easy compartmentalization of the upper and lower respiratory tracts and the animal is large enough to allow easy monitoring of parameters such as temperature, respiratory rate and nasal congestion (Chen *et al.*, 1995) as well as ferret mobility/activity (Reuman *et al.*, 1989). We routinely monitor virus and inflammatory cells in nasal washings and pyrexia. Proteins (Potter *et al.*, 1972; Toms *et al.*, 1976), kinins (Barnett *et al.*, 1990) and cytokines may also be monitored in nasal washings (Jakeman *et al.*, 1991) and in plasma (Price *et al.*, 1997). Antibody levels may be measured in ferret sera, normally by haemagglutination-inhibition (HAI) tests.

Nasal washings

This is accomplished swiftly with little distress by holding the unanaesthetized animal on its back at an angle of approximately 45°, with one hand holding its back legs and the other holding the animal's head with the thumb placed under its lower jaw. With its nose pointing down towards a plastic Petri dish, 5–10 ml of PBSA is dropped on to each nostril alternately so that the ferret breathes in the saline solution and expels it into the Petri dish either as a sneeze or a snuffle. The entire solution can be administered within a few minutes and the animal seems not to resent the procedure. Nasal washings can be used for titration of virus and inflammatory cell counts. Generally nasal washings are centrifuged (300 *g*, 10 minutes, room temperature) to pellet the cells which are resuspended in 1 ml of PBSA for counting. A sample of wash fluid is frozen at –70°C after addition of bovine serum albumin (1% wt/vol. final concentration) for later titration of virus. Total cell counts are obtained using a haemocytometer, while differential counts are obtained after methanol fixation and May–Grünwald–Giemsa staining (Toms *et al.*, 1977). Protein levels in nasal washings increase gradually between 5 and 15 days p.i.; peak concentrations found 7–9 days p.i. are three- to five-fold greater than pre-infection levels. The predominant protein found is albumin, but immunoglobulin G (IgG) and trace amounts of other proteins are also found. However, no proteins analogous to secretory IgA or IgM have been detected (Potter *et al.*, 1972).

Temperature monitoring

The parameter measured most often in influenza virus-infected ferrets is body temperature. As described above, fever is seen between days 1 and 3 following infection, and may peak at between 41 and 43°C depending on the viral strain used. Two main options exist for monitoring body temperature. The classical method is the use of rectal temperature probes (either a clinical thermometer or an electronic thermometer; (R.S. Components, Corby, Northants, UK) with a thin, flexible insulated probe (type K thermocouple)) which should be inserted (with Vaseline lubrication if required) to a depth of around 10–14 cm and gently moved back and forth until a constant temperature is recorded. This method has the drawback that infected animals must be frequently handled, and time points for temperature sampling must be chosen judiciously as the peak in fever may be easily missed (typical time points used are 12, 24, 30, 36, 42, 48, 54, 72 and 96 hours p.i.; Coates *et al.*, 1985). Temperature monitoring should be carried out on conscious ferrets wherever possible, as anaesthesia may significantly depress body temperature. For example, ferrets anaesthetized with 3% isoflurane for 10 minutes lose around 2°C and take over 2 hours to return to baseline body temperatures (Price and Fenton, unpublished observations).

The second method available is the use of surgically implanted temperature sensors (telemetry units) which transmit core temperature data to a receiver (usually in the cage) at predefined intervals (every 15 minutes is typical). This has the disadvantage of high initial expense and the requirement for surgical implantation. However, implanted sensors allow several animals to be monitored simultaneously, which saves time during long-term experiments (Fenton *et al.*, 1993b). The remote (and automatic) monitoring of infected animals also removes any temperature variations due to handling, and some sensors also allow measurement of animal activity (by recording movement) with data collected automatically on computer. The procedure for surgical implantation of sensors is outlined below. With both methods, it is critical to obtain reliable pre-infection mean temperatures for each animal (with rectal probes, readings should be taken morning and afternoon, to account for diurnal variation, for 1 week prior to infection; Toms *et al.*, 1977). Fever is considered significant if it exceeds 0.7°C above baseline (pre-infection mean) temperature, with a rise of 0.7°C being twice the standard deviation of the average pre-infection mean (Campbell *et al.*, 1979). Temperatures can be plotted against time p.i., and it may be valuable to calculate the area beneath the fever curve >0.7° (expressed as °C hours) which is known as the fever index (Coates *et al.*, 1986). The fever index provides a numerical measure of both magnitude and duration of fever, and is a convenient representation of virus-induced constitutional symptoms. With telemetry systems an equivalent to the fever index is sometimes known as the area under the curve (AUC) value.

Withdrawal of blood

The blood volume is assumed to be 5–7% of the body weight (ml/kg; Moody *et al.*, 1985) depending on species. In the ferret it is 7.5%. UK guidelines suggest that 1% of an animals circulating blood volume may be removed every 24 hours (Morton *et al.*, 1993). The superficial veins of the ferret are small, inaccessible and prone to spasm. Whilst bleeding may be accomplished easily by cardiac puncture, this also carries a risk to the animal, and some regulatory bodies may only allow cardiac puncture as a terminal procedure. The use of the caudal (tail) artery for bleeding conscious ferrets has been described (Bleakley, 1980), as has the retro-orbital technique for blood collection (Fox *et al.*, 1984), but the volume of blood easily obtained by these routes is limited, and the latter method is regarded as unacceptable. Although blood is obtainable by jugular puncture, location of the jugular vein is often difficult as it is not tethered and so can easily move as the needle is inserted. However, with practice this is the method of choice. For jugular puncture, the anaesthetized animal is laid on its back with the head slightly tilted away from the operator so as to stretch the neck, which is then shaved. The jugular vein generally follows an imaginary line between the base of the ear and the sternum. For the right-handed operator, the middle finger of the left hand is used to apply pressure to occlude the vein as it emerges from the thoracic cavity to the side of the sternum. Once the vein is dilated, the thumb of the left hand is used to reduce the mobility of the vein. A 23 G needle attached to a syringe is inserted slightly to the side of the vein at a 45° angle. A technique for jugular catheterization, which allows repeated blood sampling, is given below.

Assessment of signs of illness

Nasal symptoms and levels of activity are evaluated three times daily from 4 days before to 4 days after infection. Nasal symptoms may be scored according to the following criteria: 1, for nasal rattling or sneezing during transport from cage to evaluation area; 2, for evidence of nasal discharge on external nares; 3, for mouth breathing; 0, for none of the above. Similarly, activity levels are scored: 0, fully playful; 1, responded only to play overtures and did not initiate any play activity; 2, alert but not at all playful; 3, neither playful nor alert (Reuman *et al.*, 1989).

Assessment of lung infection

Lower respiratory tract infection is normally assessed by virus titration of whole-lung homogenates and is not routinely monitored since it requires sacrifice of the animal; homogenates of trachea, bronchi and subregions of the lung may also be examined (Sweet *et al.*, 1981). Whole lungs can be macerated safely in a Sorvall Omnimixer in 10 ml of tissue culture medium, and after centrifugation (1400 *g*, 10 minutes, 4°C), the supernatants are assayed for virus in embryonated hens' eggs. Because of the large size of the ferret lung it is also possible to divide the lower respiratory tract into trachea, external bronchi and subdivisions of the lung enriched for airway or alveolar tissue (Sweet *et al.*, 1981).

Assessment of nasal obstruction

The nasal passages become obstructed during infection due to increased protein secretion. The degree of nasal obstruction may be measured as an increase in the nasal airway resistance (Wardell *et al.*, 1968; Boyd and Beeson, 1975). In these studies, the amount of obstruction was measured by passing air through the nasal passages of anaesthetized ferrets and relating the transnasal pressure to the rate of airflow. This technique has been utilized to demonstrate the nasal decongestant activity of aspirin in infected ferrets (Haff and Pinto, 1972).

Specialist surgical procedures

Relevant procedures which have been used successfully include peritoneal implantation of telemetry units for temperature monitoring, intravenous catheterization and the surgical generation of anatomically distinct tissues (tracheal pouches) susceptible to influenza virus infection (this last method is very seldom used). These procedures will be described briefly below, but readers should consult the cited literature for more detailed information.

Surgical implantation of telemetry units

The ferret is anaesthetized under isoflurane, the abdomen is shaved, and the skin is swabbed with disinfectant. A 3 cm midline laparotomy is performed, and the peritoneal cavity opened. A sterile reusable telemetry unit (Dataquest IV; Data Sciences, St Pauls, MN) is inserted into the peritoneal cavity through the incision. The peritoneal cavity is closed with simple interrupted sutures and the dermis is closed with steel wound clips. Animals are monitored for visible signs of postoperative infection, and as a precautionary measure antibacterials (e.g., Zinacef) are administered intraperitoneally at 25 mg/kg twice daily for 1 week following the procedure. At least 1 week recovery time is allowed before further procedures are performed, although the telemetry unit may be activated 2 days after implantation by "swiping" a magnet across the abdomen of the animal.

Jugular catheterization

If blood must be drawn from ferrets on repeated occasions, the use of an indwelling jugular catheter may be more suitable than repeated cardiac or jugular puncture (Mesina *et al.*, 1988). Briefly, animals are anaesthetized by intramuscular injection (ketamine 30 mg/kg–xylazine 3 mg/kg), a midline incision is made on the neck and the right jugular is exposed and tied above the site of venesection. A 19 G needle is inserted below the ligature and a polyethylene catheter filled with heparin–saline is passed through the needle. The needle is then withdrawn and the catheter tied in place. A second incision is then made behind the right ear of the ferret so a 17 G cannula can be inserted such that it emerges adjacent to the jugular. The catheter is then passed up through the 17 G cannula, which is then withdrawn. The catheter is now coiled and secured in place on the head of the ferret by the suture used to close the ear incision and adhesive tape. The catheter is thus secured in a position where the animal cannot reopen the wound deliberately.

The catheter must be kept clear by daily flushing with heparin–saline and may be heat-crimped to ensure that it is sealed (which prevents infection). Blood is withdrawn by cutting the sealed end of the catheter and inserting a blunted 23 G needle attached to a three-way adapter. A 3 ml syringe half-filled with heparin–saline is attached to the horizontal arm of the adapter and is used to fill the catheter with blood (and remove air bubbles), whilst blood is withdrawn via a second syringe attached to the vertical arm. After sampling the catheter is filled with heparin–saline and the end is heat-sealed (Mesina *et al.*, 1988). Catheterization of the jugular artery can also be useful for repeated administration of substances into the blood stream. This technique can be applied at the same time as catheterization of the jugular vein and is particularly useful in experiments where a kinetic profile of the drug administered is required.

Tracheal pouches

The unusual length of the ferret respiratory tract allows the surgical separation of a section of trachea to provide an anatomically isolated portion of the respiratory tract (a tracheal pouch). This gives two separate sites (nose and pouch) for infection and sample collection (Barber and Small, 1977). Tracheal pouches are generated by transecting the trachea at two sites about 12 cartilaginous "rings" apart. The middle section of the trachea (which forms the pouch) is displaced laterally and the posterior end is closed with two sutures, whilst the anterior end is sutured to a length of tubing with a recessed multipuncture seal at the end. Once the pouch has been formed, the airway is re-established by joining the two ends of the trachea, and the skin is closed so that the pouch is subcutaneous. It is then possible to infect, or withdraw samples from the pouch by puncturing the seal with a 16 G needle (Barber and Small, 1977). Viral dissemination is not seen between the pouch and the URT or vice versa. Tracheal pouches have proven valuable in the demonstration that infection of the pouch does not induce immunity in the URT, suggesting that systemic immunity (via IgG) is not responsible for protection of the URT (Barber and Small, 1978).

Extraction of tissues for *in vitro* work

Organ cultures of ferret tissues have been extensively used to study virulence and pathogenicity of influenza virus (Toms *et al.*, 1974; Sweet *et al.*, 1977, 1978a,b, 1984; Husseini *et al.*, 1983b) but have seldom been used to examine the effects of antiviral compounds. Amantadine and its analogues suppress viral replication in ferret tracheal organ cultures (Arroyo and Reed, 1977, 1978) and these observations have been extended by Burlington and colleagues (1982) to examine the toxicity of rimantadine and amantadine on cilial activity. Ferret cell lines are now also available for such studies (Trowbridge *et al.*, 1982).

Immune response

The immune response has been poorly studied in ferrets, largely due to the lack of reagents and inbred strains. The cellular composition of ferret blood is similar to that of humans. Ferrets can mount a delayed-type hypersensitivity response to influenza (Potter *et al.*, 1974; Kauffman and Bergman, 1981) and their lymphocytes undergo *in vitro* proliferation in response to mitogens and antigens (McLaren and Butchko, 1978; Kauffman *et al.*, 1978; Kauffman and Bergman, 1981). Serum contains anti-influenza virus antibody as early as 6 days p.i. (McLaren and Butchko, 1978), comprising IgA and IgM with IgG predominating (Suffin *et al.*, 1979; Sweet *et al.*, 1987b). Antibody (some of it produced locally) can be detected in nasal washings but it is of low titre and short-lived (Shore *et al.*, 1972). Antibody (IgG) directed entirely to the viral haemagglutinin (Jakeman *et al.*, 1989) is passed from mother to offspring in colostrum and milk and selectively protects the lung rather than the nasal turbinates (Sweet *et al.*, 1987a,b) as protection of the latter in both adults and neonates is due to local antibody (Barber and Small, 1978; Sweet *et al.*, 1987b). The predominant antibody subtype found in ferret nasal washings 5–15 days p.i. is IgG, although IgM predominates in the serum at this time (Potter *et al.*, 1972) and as in humans, the ferret only later switches to serum IgG at around 17 days p.i. (Shore *et al.*, 1972). This further supports the hypothesis that nasal IgG is produced locally early in infection. In contrast to guinea-pigs, sera from infected ferrets contain relatively low levels of non-specific inhibitors.

The degree to which cell-mediated responses play a role in immunity to influenza in the ferret is not known. However, the passive transfer of specific influenza anti-sera failed to protect recipient ferrets from subsequent homologous virus challenge (Small *et al.*, 1976). Similarly, although equivalent titres of antibody were induced by both live and inactivated influenza vaccines, immunity (as determined by the lack of virus secretion on re-infection with homologous virus) was seen only in animals inoculated with the live vaccine (Potter *et al.*, 1973a,b). This suggests that, in ferrets, antibody alone does not protect from re-infection with homologous virus.

Vaccine studies

The ferret has been widely used to evaluate the performance of a variety of live and killed vaccine preparations (Potter *et al.*, 1972; Herlocher *et al.*, 1996), the effects of adjuvants (Potter *et al.*, 1972; Fenton *et al.*, 1981), the immunogenicity of whole and split vaccines (Fenton *et al.*, 1977b), and the effect of a "priming" infection with a heterologous virus on the immunity induced by a subsequent killed virus vaccine (McLaren and Potter, 1974; Sweet *et al.*, 1974). The results of these and other studies indicate firstly that killed influenza vaccines alone in unprimed ferrets do not induce immunity, although this could be improved if given with adjuvant. Furthermore, complete protection (as measured by a lack of recoverable virus and suppression of symptoms) was seen only in ferrets given live vaccines. Prior priming with a heterologous virus does allow some immunization of ferrets with unadjuvanted killed vaccines, although the protection was not as good as with live vaccines. Serum antibody is induced by killed vaccines in primed ferrets, and some decreased virus excretion and decreased nasal wash protein levels are seen.

Antiviral therapy studies in ferrets

Several studies have shown the usefulness of the ferret as a model for examining anti-influenza virus compounds (Table 119.1). Initial screening is made in tissue culture, then promising compounds are tested in the mouse, which has played an important role in evaluating potential antiviral agents (Grunnert *et al.*, 1965; Wilson *et al.*, 1980). Infection of mice with mouse-passaged influenza virus models the more serious pneumonia that occurs largely in the elderly (Sweet and Smith, 1980). Also infection of mice may be lethal, providing death as an unequivocal endpoint for drug evaluation. However, ferrets are used more frequently because of the similarity of the disease to humans and because effects of compounds on respiratory and constitutional symptoms can also be examined.

Several compounds have been evaluated in ferrets, including amantadine (Potter and Schofield, 1975; Fenton

Table 119.1 Anti-influenza virus compounds tested in ferrets: a summary

Generic name	*Systematic name*	*Mode of action*	*Reference*
Amantadine (Symmetrel)	1-Amino-adamantane hydrochloride	M2 Ion channel inhibitor (prevents viral uncoating)	Cochran *et al.*, 1965 Fenton *et al.*, 1977a Potter and Schofield, 1975 Potter *et al.*, 1976 Squires, 1970
Rimantadine	α-Methyl-1-adamantanemethylamine	Similar to amantadine	Burlington *et al.*, 1982
Ribavirin	1-β-D-ribofuranosyl-1,2,4-triazole-3-carboxamide	Nucleoside analogue	Fenton and Potter, 1977 Fenton *et al.*, 1993b
2′-flurodGuo	2′-Deoxy-2′-fluoroguanosine	Nucleoside analogue	Jakeman *et al.*, 1994
LY 217896	1,3,4-Thiadiazol-2-ylcyanamide	Nucleoside analogue	Colacino *et al.*, 1990
GG 167 (Zanamivir)	4-Guanidino-2,4-dideoxy-2,3-dehydro-*N*-acetylneuraminic acid	Neuraminidase inhibitor (prevents viral release)	von Itzstein *et al.*, 1993 Fenton *et al.*, 1993a Ryan *et al.*, 1995
GS 4104	ethyl (3R,4R,5S)-4-acetamide-5-amino-3-(1-ethylpropoxy)-1-cyclohexene-1-carboxylate	As GG167	Mendel *et al.*, 1997

et al., 1977a), ribavirin (Schofield *et al.*, 1975; Fenton and Potter, 1977), 1,3,4-thiadiazol-2-ylcyanamide (LY217896) (Colacino *et al.*, 1990), 2′-deoxy-2′-fluoroguanosine and its pro-drug 2,6-diamino-purine-2′-fluororiboside (Jakeman *et al.*, 1994) and the anti-neuraminidase compounds 4-guanidino-Neu5Ac2en (GG167 or Zanamivir; von Itzstein *et al.*, 1993) and GS 4071 and GS 4104 (Mendel *et al.*, 1997).

Amantadine, which together with rimantadine are the only antiviral agents approved for the prophylaxis and treatment of influenza A infections, has very limited activity against influenza in ferrets (Potter and Schofield, 1975). When given orally twice daily at a dose of 100 mg/kg amantadine is quite toxic: ferrets develop tremors and mild convulsive attacks within 1 hour of dosing (Fenton *et al.*, 1977a). This suggested initially that the ferret might be a poor model for testing antiviral drugs. However, when given twice daily as an aerosol at 6 mg/kg, no toxic effect was observed and the compound consistently reduced pyrexia and virus shedding without suppressing the immune response (Fenton *et al.*, 1977a). Nevertheless, it is now known that these drugs have significant side-effects in humans (Hayden, 1994).

Similarly, at a concentration of 100 mg/kg, ribavirin abrogated fever and reduced levels of virus excretion and nasal wash proteins (Schofield *et al.*, 1975). However, again ribavirin was found to be toxic in that the animals were immunosuppressed, failing to develop either local or serum antibody despite demonstrable virus replication (Potter *et al.*, 1976). Ribavirin too has been found to be toxic for human epithelial explants at a dose of 10 mg/ml (Rollins *et al.*, 1993) and clinical trials were disappointing (Smith *et al.*, 1980).

The importance of the route of antiviral treatment has recently been emphasized with the neuraminidase inhibitors GG167 and GS 4104. GG167 is a potent inhibitor of influenza virus replication *in vitro* and in the ferret reduces viral titres and suppresses pyrexia (Fenton *et al.*, 1993a). However, the compound must be given intranasally or by aerosol because of the poor bioavailability of the drug in both ferrets and humans. GS 4104 showed increased bioavailability following oral administration, and twice-daily oral administration of 25 mg/kg reduced virus excretion and prevented or reduced constitutional (fever and a decrease in animal activity) and respiratory (nasal inflammatory cell count and nasal wash proteins) signs and symptoms (Mendel *et al.*, 1997).

The suppression of viral symptoms, especially fever (Husseini *et al.*, 1982) and nasal congestion (Haff and Pinto, 1972) by aspirin, has also been investigated in ferrets. However, although aspirin was effective at reducing symptoms, treated animals had a prolonged period of viral shedding (Husseini *et al.*, 1982). Other studies have described the toxic effects of aspirin in influenza virus-infected ferrets in conjunction with either food withdrawal (Linnemann *et al.*, 1979) or arginine-deficient diets (Deshmukh and Thomas, 1985). In the latter case, infected ferrets on an arginine-deficient diet treated with aspirin showed hyperammonaemia, encephalitis and many of the clinical and histopathologic features of Reye's syndrome, making it a good animal model for this condition (Deshmukh and Thomas, 1985). Reye's syndrome is a post-viral encephalopathy, characterized by a mild prodromal illness, followed abruptly by vomiting, encephalopathy, mitochondrial abnormalities in hepatocytes and neurones, and fatty degeneration of the viscera. Studies in children have shown that taking salicylates seems to potentiate the occurrence of Reye's syndrome in a dose-dependent manner (Reynolds *et al.*, 1972; Starko *et al.*, 1980).

Advantages/disadvantages of the model

As with all experimental models for human infection, the ferret has its particular disadvantages. Probably the most significant of these is the lack of ferret-specific reagents, which has seriously hampered the use of ferrets in immunological studies and promoted the use of murine models for influenza. Other disadvantages are the expenses associated with ferrets (both the high cost of purchase and of upkeep). Ferrets have — rather unfairly in our experience — gained a reputation as difficult animals to work with, although we have not found this to be the case, provided they are familiar with their surroundings and frequently handled. Finally, the size of the cages (and hence the physical space) required may also be off-putting, although single caging of animals is only strictly necessary once procedures have begun: communal housing in a pen is preferable before experimentation. Indeed, the size of ferrets is a positive point as regards experimental procedures and histology. The ability of natural human isolates of influenza virus to cause symptomatic infection (which may be easily monitored) is of special benefit. Natural isolates cause only a mild subclinical infection of the upper and lower respiratory tract in mice, and mouse-passaged (or adapted) strains cause markedly more lung involvement; animals have died of severe pneumonia (making mice a more suitable model for human influenza pneumonia, but a poor model for uncomplicated infection; Renegar, 1992).

Additionally, mice do not develop a febrile response during influenza and instead lose body temperature (Conn *et al.*, 1995). Thus, despite the above disadvantages of the ferret, it remains the best available model for human influenza.

References

Arroyo, M., Reed, S. E. (1977). The use of ferret tracheal organ culture for therapeutic studies of anti-influenzal drugs. I. Evaluation of the model in comparison with infection in humans. *J. Antimicrob. Chemother.*, 3, 601–607.

Arroyo, M., Reed, S. E. (1978). The use of ferret tracheal organ culture for therapeutic studies of anti-influenzal drugs. II. Comparison of 4 compounds in tissue culture and organ culture. *J. Antimicrob. Chemother.*, **4**, 363–371.

Barber, W. H., Small, P. A. Jr (1977). Construction of an improved tracheal pouch in the ferret. *Am. Rev. Respir. Dis.*, **115**, 165–179.

Barber, W. H., Small, P. A. Jr (1978). Local and systemic immunity to influenza infections in ferrets. *Infect. Immun.*, **21**, 221–228.

Barnett, J. K. C., Cruse, L. W., Proud, D. (1990). Kinins are generated in nasal secretions during influenza A infection in ferrets. *Am. Rev. Respir. Dis.*, **142**, 162–166.

Bleakley, S. P. (1980). Simple technique for bleeding ferrets (*Mustela putorius furo*). *Lab. Anim.*, **14**, 59–60.

Boyd, M. R., Beeson, M. F. (1975). Animal models for the evaluation of compounds against influenza viruses. *J. Antimicrob. Chemother.*, **1** (suppl.), 43–47.

Burlington, D. B., Meiklejohn, G., Mostow, S. P. (1982). Anti-influenza A virus activity of amantadine hydrochloride and rimantadine hydrochoride in ferret tracheal ciliated epithelium. *Antimicrob. Agents Chemother.*, **21**, 794–799.

Campbell, D., Sweet, C., Smith, H. (1979). Comparisons of virulence of influenza virus recombinants in ferrets in relation to their behaviour in man and their genetic constitution. *J. Gen. Virol.*, **44**, 37–44.

Cavanagh, D., Mitkis, F., Sweet, C., Collie, M. H., Smith, H. (1979). The localization of influenza virus in the respiratory tract of ferrets: susceptible nasal mucosa cells produce and release more virus than susceptible lung cells. *J. Gen. Virol.*, **44**, 505–514.

Chen, K.-S., Bharaj, S. S., King, E. C. (1995). Induction and relief of nasal congestion in ferrets infected with influenza virus. *Int. J. Exp. Pathol.*, **76**, 55–64.

Coates, D. M., Sweet, C., Quarles, J. M., Overton, H. A., Smith, H. (1985). Severity of fever in influenza: studies on the relation between viral surface antigens, pyrexia, level of nasal virus and inflammatory response in the ferret. *J. Gen. Virol.*, **66**, 1627–1631.

Coates, D. M., Sweet, C., Smith, H. (1986). Severity of fever in influenza: differential pyrogenicity in ferrets exhibited by H1N1 and H3N2 strains of differing virulence. *J. Gen. Virol.*, **67**, 419–425.

Cochran, K. W., Maassab, H. F., Tsunoda, A., Berlin, B. S. (1965). Studies on the antiviral action of amantadine hydrochloride. *Ann. N.Y. Acad. Sci.*, **130**, 432–439.

Colacino, J. M., Delong, D. C., Nelson, J. R. *et al.* (1990). Evaluation of the anti-influenza virus activities of 1,3,4-thiadiazol-2-ylcyanamide (LY217896) and its sodium salt. *Antimicrob. Agents Chemother.*, **34**, 2156–2163.

Collie, M. H., Rushton, D. I., Sweet, C., Smith, H. (1980). Studies of influenza virus infection in newborn ferrets. *J. Med. Virol.*, **13**, 561–571.

Conn, C. A., Mcclellan, J. L., Maassab, H. F., Smitka, C. W., Majde, J. A., Kluger, M. J. (1995). Cytokines and the acute phase response to influenza virus in mice. *Am. J. Physiol.*, **37**, R78–R84.

Deshmukh, D. R., Thomas, P. B. (1985). Arginine deficiency, hyperammonemia and Reye's syndrome in ferrets. *Lab. Anim. Sci.*, **35**, 242–245.

Douglas, R. G. (1975). Influenza in man. In *The Influenza Viruses and Influenza* (ed Kilbourne, E. D.), pp. 395–481. Academic Press, London.

Fenton, R. J., Potter, C. W. (1977). Dose–response activity of ribavirin against influenza virus infection in ferrets. *J. Antimicrob. Chemother.*, **3**, 263–271.

Fenton, R. J., Bessell, C. J., Spilling, C. R., Potter, C. W. (1977a). The effects of peroral or local aerosol administration of L-aminoadamantane hydrochloride (amantadine hydrochloride) on influenza infections of the ferret. *J. Antimicrob. Chemother.*, **5**, 463–472.

Fenton, R. J., Jennings, R., Potter, C. (1977b). The serological response of experimental animals to inactivated whole and split influenza virus vaccines. *J. Biol. Stand.*, **5**, 263–271.

Fenton, R. J., Clark, A., Potter, C. W. (1981). Immunity to influenza in ferrets. XIV. Comparative immunity following infection or immunisation with live or inactivated vaccine. *Br. J. Exp. Pathol.*, **62**, 297–307.

Fenton, R. J., Ashton, C. J., Dempsey, M. H. *et al.* (1993a). Anti-influenza virus activities of two novel neuraminidase (sialidase) inhibitors in both mouse and ferret animal models. *Recent Advance in Chemotherapy* (Proceedings of the 18th International Congress of Chemotherapy), pp. 646–648. American Society for Microbiology, Bethesda.

Fenton, R. J., Dempsey, M. H., Ryan, D. M. (1993b). The use of telemetry to compare the pyrexic responses of ribavirin-treated and untreated influenza virus-infected ferrets. *Welfare and Science* (Proceedings of the Fifth FELASA Symposium), pp. 367–370. Royal Society of Medicine Press, London.

Fox, J. G., Hewes, K., Niemi, S. M. (1984). Retro-orbital technique for blood collection from the ferret (*Mustelo putorius furo*). *Lab. Anim. Sci.*, **34**, 198–199.

Grunnert, R. R., McGahen, J. W., Davies, W. L. (1965). The *in vivo* antiviral activity of L-adamantanamine (amantadine). 1. Prophylactic and therapeutic activity against influenza viruses. *Virology*, **26**, 262–269.

Haff, R. F., Pinto, C. A. (1972). The nasal decongestant action of aspirin in influenza infected ferrets. *Life Sci.*, **12**, 9–14.

Hayden, F. G. (1994). Amantadine and rimantadine resistance in influenza A viruses. *Curr. Op. Infect. Dis.*, **7**, 674–677.

Herlocher, M. L., Clavo, A. C., Treanor, J., Maassab, H. F. (1996). Phenotypic characteristics of A/AA/6/60 viruses and reassortants. In *Proceedings of the Third International Conference on Options for the Control of Influenza* (eds Brown, L. E., Hampson, A. W., Webster, R. G.), pp. 634–646. Elsevier, Amsterdam.

Home Office (1986). *Animals (Scientific Procedures) Act.* Her Majesty's Stationery Office, London.

Husseini, R. H., Sweet, C., Collie, M. H., Smith, H. (1982). Elevation of nasal virus levels by suppression of fever in ferrets infected with influenza viruses of differing virulence. *J. Infect. Dis.*, **145**, 520–524.

Husseini, R. H., Collie, M. H., Rushton, D. I., Sweet, C., Smith, H. (1983a). The role of naturally acquired bacterial infection in influenza-related death in neonatal ferrets. *Br. J. Exp. Pathol.*, **64**, 559–569.

Husseini, R. H., Sweet, C., Bird, R. A., Collie, M. H., Smith, H. (1983b). Distribution of viral antigen within the lower respiratory tract of ferrets infected with a virulent influenza virus: production and release of virus from corresponding organ cultures. *J. Gen. Virol.*, **64**, 589–598.

Jakeman, K. J., Smith, H., Sweet, C. (1989). Mechanism of immunity to influenza: maternal and passive neonatal protection following immunization of adult ferrets with live vaccinia-influenza virus haemagglutinin recombinant but not with recombinants containing other influenza virus proteins. *J. Gen. Virol.*, **70**, 1523–1531.

Jakeman, K. J., Bird, C. R., Thorpe, R., Smith, H., Sweet, C. (1991). Nature of the endogenous pyrogen (EP) induced by influenza viruses: lack of correlation between EP levels and content of the known pyrogenic cytokines, interleukin 1, interleukin 6 and tumour necrosis factor. *J. Gen. Virol.*, **72**, 705–709.

Jakeman, K. J., Tisdale, M., Russell, S., Leone, A., Sweet, C. (1994). Efficacy of 2′-deoxy-2′-fluororibosides against influenza A and B viruses in ferrets. *Antimicrob. Agents Chemother.*, **38**, 1864–1867.

Kauffman, C. A., Bergman, A. G. (1981). Lymphocyte subpopulations in the ferret. *Dev. Comp. Immunol.*, **5**, 671–678.

Kauffman, C. A., Schiff, G. M., Phair, J. P. (1978). Influenza in ferrets and guinea pigs: effect on cell mediated immunity. *Infect. Immun.*, **19**, 547–552.

Linnemann, C. C. Jr., Ueda, K., Hug, G., Schaeffer, A, Clark, A., Schiff, G. M. (1979). Salicylate intoxication and influenza in ferrets. *Pediatr. Res.*, **13**, 44–47.

McLaren, C., Butchko, G. M. (1978). Regional T- and B-cell responses in influenza-infected ferrets. *Infect. Immun.*, **22**, 189–194.

McLaren, C., Potter, C. W. (1974). Immunity to influenza in ferrets. VII. Effect of previous infection with heterotypic and heterologous influenza viruses on the response of ferrets to inactivated influenza virus vaccines. *J. Hygiene*, **72**, 91–100.

Mendel, D. B., Tai, C. Y., Escarpe, P. A. *et al.* (1998). Oral administration of a prodrug of the influenza neuraminidase inhibitor GS 4071 protects mice and ferrets against influenza infection. *Antimicrob. Agents Chemother.*, **42**, 640–646.

Mesina, J. E., Sylvina, T. J., Hotaling, L. C., Goad, J. M. P., Fox, J. G. (1988). A simple method for chronic jugular catheterization in ferrets. *Lab. Anim. Sci.*, **38**, 89–90.

Moody, K. D., Bowman, T. A., Lang, C. M. (1985). Laboratory management of ferret for biomedical research. *Lab. Anim. Sci.*, **35**, 272–279.

Morton, D. B., Abbot, D., Barclay, R. *et al.* (1993). Removal of blood from laboratory mammals and birds. First report of the BVA/FRAME/RSPCA/UFAW joint working group on refinement. *Lab. Anim.*, **27**, 1–22.

Nicholson, K. G. (1992). Clinical features of influenza. *Semin. Resp. Infect.*, **7**, 26–37.

Pinto, C. A., Haff, R. F. (1969). Antiviral activity of cyclooctylamine hydrochloride in influenza virus-infected ferrets. *Antimicrob. Agents Chemother.*, 201–206.

Potter, C. W., Schofield, K. P. (1975). The effect of amantadine on the response of ferrets to influenza virus infection. Ciba-Geigy Symposium. *Symmetrel in Virology*, (ed. Wink, C. A. S), pp. 24–31. Hunters Armlay Ltd, Leeds.

Potter, C. W., Oxford, J. S., Shore, S. L., McLaren, C., Stuart-Harris, C. H. (1972). Immunity to influenza in ferrets. I. Response to live and killed virus. *Br. J. Exp. Pathol.*, **53**, 153–167.

Potter, C. W., Jennings, R., Marine, W. M., McLaren C. (1973a). Potentiation of the antibody response to inactivated A2-Hong Kong vaccines by previous heterotypic influenza virus infection. *Microbios*, **8**, 101–110.

Potter, C. W., Jennings, R., McLaren, C. (1973b). Immunity to influenza in ferrets. VI. Immunization with adjuvanted vaccines. *Arch. Gesamte Virusforsch.*, **42**, 285–296.

Potter C. W., Rees, R. C., Shore, S. L., Jennings, R., McLaren, C. (1974). Immunity to influenza in ferrets. IX. Delayed hypersensitivity following infection or immunization with A2/Hong Kong. *Microbios*, **10A** (suppl.), 7–21.

Potter, C. W., Phair, J. P., Vodinelich, L., Fenton, R. J., Jennings, R. (1976). Antiviral, immunosuppressive and antitumour effect of ribavirin. *Nature*, **259**, 496–497.

Price, G. E., Fenton, R. J., Smith, H., Sweet, C. (1997). Are known pyrogenic cytokines responsible for fever in influenza? *J. Med. Virol.*, **72**, 336–340.

Renegar, K. B. (1992). Influenza virus infections and immunity: a review of human and animal models. *Lab. Anim. Sci.*, **42**, 222–232.

Reuman, P. D., Keely, S., Schiff, G. M. (1989). Assessment of signs of influenza illness in the ferret model. *J. Virol. Methods*, **24**, 27–34.

Reynolds, D. W., Riley, H. D. Jr, Lafont, D. S., Vorse, H., Stout, C., Carpenter, R. L. (1972). An outbreak of Reye's syndrome associated with influenza B. *J. Pediatr.*, **81**, 429–432.

Rollins, B. S., Elkhatieb, A. H. A., Hayden, F. G. (1993). Comparative antiinfluenza virus activity of 2′-deoxy-2′-fluororibosides *in vitro*/*Antiviral Res.*, **21**, 357–368.

Ryan, D. M., Ticehurst, J., Dempsey. M. H. (1995). GG167 (4-guanidino-2,4-dideoxy-2,3-dehydro-*N*-acetylneuraminic acid) is a potent inhibitor of influenza virus in ferrets. *Antimicrob. Agents Chemother.*, **39**, 2583–2584.

Schofield, K. P., Potter, C. W., Edey, D., Jennings R., Oxford, J. S. (1975). Antiviral activity of ribavirin in influenza infection in ferrets. *J. Antimicrob. Chemother.*, **1** (suppl.), 63–69.

Shore, S. L., Potter, C. W., Mclaren, C. (1972). Immunity to influenza in ferrets. IV. Antibody in nasal secretions. *J. Infect. Dis.*, **126**, 394–400.

Small, P. A. Jr, Waldman, R. H., Bruno, J. C., Gifford, G. E. (1976). Influenza infection in ferrets: role of serum antibody in protection and recovery. *Infect. Immun.*, **13**, 417–424.

Smith, H., Sweet, C. (1988). Lessons for human influenza from pathogenicity studies with ferrets. *Rev. Infect. Dis.*, **10**, 56–75.

Smith, W., Andrewes, C. H., Laidlaw, P. P. (1933). A virus obtained from influenza patients. *Lancet*, **2**, 66–68.

Smith, C. B., Charette, R. P., Fox, J. P., Cooney, M. K., Hall. C. B. (1980). Lack of effect of oral ribavirin in naturally occurring influenza A virus (H1N1) infection. *J. Infect. Dis.*, **141**, 548–554.

Squires, S. L. (1970). The evaluation of compounds against influenza viruses. *Ann. N.Y. Acad. Sci.*, **173**, 239–248.

Squires, S., Belyavin, G. (1975). Free contact infection in ferret groups. *J. Antimicrob. Chemother.*, **1** (suppl.), 35–42.

Starko, K. L., Ray, C. G., Dominguez, L. B., Stromberg, W. L., Woodall, D. F. (1980). Reye's syndrome and salicylate use. *Pediatrics*, **66**, 859–864.

Suffin, S. S., Prince, G. A., Muck, K. B., Porter, D. D. (1979). Ontogeny of the ferret humoral immune response. *J. Immunol.*, **123**, 6–9.

Sweet, C., Smith, H. (1980). Pathogenicity of influenza virus. *Microbiol. Rev.*, **44**, 303–330.

Sweet, C., Stephen, J., Smith, H. (1974). Immunization of ferrets against influenza: a comparison of killed ferret grown and egg grown virus. *Br. J. Exp. Pathol.*, **55**, 296–304.

Sweet, C., Toms, G. L., Smith, H. (1977). The pregnant ferret as a model for studying the congenital effects of influenza virus infection *in utero*: infection of foetal tissues in organ culture and *in vivo*. *Br. J. Exp. Pathol.*, **58**, 116–129.

Sweet, C., Cavanagh, D., Collie, M. H., Smith, H. (1978a). Sensitivity to pyrexial temperatures: a factor contributing to virulence differences between two clones of influenza virus. *Br. J. Exp. Pathol.*, **59**, 373–380.

Sweet, C., Cavanagh, D., Collie, M. H., Smith, H. (1978b). Yields of a virulent and an attenuated clone of influenza virus from organ cultures of ferret nasal tissue: divergence with time of incubation. *FEMS Microbiol. Lett.*, **4**, 191–194.

Sweet, C., Bird, R. A., Cavanagh, D., Toms, G. L., Collie, M. H., Smith, H. (1979). The local origin of the febrile response induced in ferrets during respiratory infection with a virulent influenza virus. *Br. J. Exp. Pathol.*, **60**, 300–308.

Sweet, C., Macartney, J. C., Bird, R. A. *et al.*, (1981). Differential distribution of virus and histological damage in the lower respiratory tract of ferrets infected with influenza viruses of differing virulence. *J. Gen. Virol.*, **54**, 103–114.

Sweet, C., Bird, R. A., Husseini, R. H., Smith, H. (1984). Differential replication of attenuated and virulent influenza viruses in organ cultures of ferret bronchial epithelium. *Arch. Virol.*, **80**, 219–224.

Sweet, C., Bird, R. A., Jakeman, K. J., Coates, D. M., Smith, H. (1987a). Production of passive immunity in neonatal ferrets following maternal vaccination with killed influenza A virus vaccines. *Immunology*, **60**, 83–89.

Sweet, C., Jakeman, K. J., Smith, H. (1987b). Role of milk-derived IgG in passive maternal protection of neonatal ferrets against influenza. *J. Gen. Virol.*, **68**, 2681–2686.

Tait, A. R., Davidson, B. A., Johnson, K. J., Remick, D. G., Knight, P. R. (1993). Halothane inhibits the intra-alveolar recruitment of neutrophils, lymphocytes and macrophages in response to influenza virus infection in mice. *Anesth. Analg.*, **76**, 1106–1113.

Toms, G. L., Rosztoczy, I., Smith, H. (1974). The localization of influenza virus: minimal infectious dose determinations and single cycle kinetic studies on organ cultures of respiratory and other ferret tissues. *Br. J. Exp. Pathol.*, **55**, 116–129.

Toms, G. L., Bird, R. A., Kingsman, S. M., Sweet, C., Smith, H. (1976). The behaviour in ferrets of two closely related clones of influenza virus of differing virulence for man. *Br. J. Exp. Pathol.*, **57**, 37–48.

Toms, G. L., Davies, J. A., Woodward, C. G., Sweet, C., Smith, H. (1977). The relation of pyrexia and nasal inflammatory response to virus levels in nasal washings of ferrets infected with influenza viruses of differing virulence. *Br. J. Exp. Pathol.*, **58**, 444–458.

Trowbridge, R. S., Lehmann, J., Brophy, P. (1982). Establishment and characterization of ferret cells in culture. *In Vitro*, **18**, 952–960.

von Itzstein, M., Wu, W.-Y., Kok, G. B. *et al.* (1993). Rational design of potent sialidase-based inhibitors of influenza virus replication. *Nature*, **363**, 418–423.

Wardell, J. R., Fimiliar, R. G., Haff, R. F. (1968). A technique for measuring nasal airway resistance in ferrets. *J. Allergy*, **40**, 100–106.

Willis, L. S., Barrow, M. V. (1971). The ferret (*Mustela putorius furo*) as a laboratory animal. *Lab. Anim. Sci.*, **21**, 712–716.

Wilson, S. Z., Knight, V., Wyde, P. R., Drake, S., Couch, R. B. (1980). Amantadine and ribavirin aerosol treatment of influenza A and B infection in mice. *Antimicrob. Agents Chemother.*, **17**, 642–648.

Chapter 120

The Cotton Rat as a Model of Respiratory Syncytial Virus Pathogenesis, Prophylaxis and Therapy

G. A. Prince

Background of human infection

Since respiratory syncytial virus (RSV) was first identified as a human pathogen in 1956, its importance as a cause of human disease has continued to increase (reviewed in Collins *et al.*, 1996). It causes serious disease primarily in the very young, the very old and the immunologically compromised. Two antigenic groups of RSV, designated A and B, co-circulate with neither having an apparent selective advantage over the other, nor being consistently more virulent.

The pathogenesis of RSV disease is not well-understood. Most cases are self-resolving, and except for patients in whom underlying conditions exacerbate the severity of disease, fatal cases are uncommon. In the immunocompetent host, infection is restricted to respiratory epithelium and the host inflammatory and immune responses account for a major portion of the disease process. By contrast, infection in immunodeficient patients may progress to tissues outside the respiratory tract, and direct viral damage of host tissues is a major factor in pathogenesis.

Efforts to prevent RSV infection by vaccination began shortly after the discovery of the virus. Clinical trials of a formalin-inactivated vaccine, designated "Lot 100", were initiated in 1965 (Kim *et al.*, 1969). The lack of a small animal model of RSV infection precluded safety and efficacy testing in animals. Although Lot 100 stimulated a moderate serum antibody response, vaccinees were not protected from natural infection the following winter, and those who became infected with RSV experienced dramatically more severe disease than RSV-infected control patients. The development in the cotton rat of a model of vaccine-enhanced RSV disease (Prince *et al.*, 1986) has laid the groundwork for understanding the mechanisms underlying the Lot 100 experience, and for developing a safety profile against which candidate vaccines may be compared.

Background of model

In 1971 Soviet scientists reported that RSV could infect and produce histopathologic changes in cotton rats (Dreizin *et al.*, 1971). Other investigators, both before and after the failed Lot 100 trials, had recognized the importance of developing a small animal model of RSV, but previous attempts using a variety of species had failed. Our later report (Prince *et al.*, 1978) confirmed and extended the findings of the Soviet scientists, who laid the foundation for all subsequent studies in cotton rats, yet never extended their own initial observations.

Cotton rat natural history

The cotton rat, genus *Sigmodon*, is a New World rodent represented by seven species in North and Central America (*S. hispidus*, *S. fulviventer*, *S. arizonae*, *S. leucotis*, *S. ochrognathis*, *S. alleni*, *S. mascotensis*), and an eighth (*S. alstoni*) in Venezuela. Though all are phenotypically similar, karyotypes are widely disparate, ranging from 22 chromosomes (*S. arizonae*) to 52 (*S. hispidus*). *S. hispidus* has been most widely used in biomedical research. The number and diversity of human pathogens which have been shown to infect cotton rats are impressive, yet the basis of its susceptibility remains unknown.

Maintenance and handling

A widespread misconception is that cotton rats are aggressive. They move quickly and can jump as much as 45 cm vertically from a standing start, and improper technique in handling cages can easily result in rapid escape of rats. They will attempt to bite the handler once they are picked up, and we highly recommend the wearing of protective gloves, but in over two decades of work with cotton rats we have yet to see aggressive behavior on their part toward the handlers.

Cotton rats are mid-range in size between laboratory mice (genus *Mus*) and laboratory rats (genus *Rattus*), reaching an adult weight of 200–250 grams. We house them in polycarbonate rat cages with an appropriate absorbent bedding material. For breeding colonies a material such as

Handbook of Animal Models of Infection
ISBN 0-12-775390-7

recycled paper products which can be shaped into a nest is preferred. We maintain our animals on a diet of standard rodent chow and water, *ad libitum*.

Cotton rats reach sexual maturity by 2 months of age and produce litters averaging 5 pups following gestation of 27 days. Females ovulate the same day a litter is born, and if the male is left with the female throughout the breeding cycle, a female may produce one litter per month for as many as 8 months. We have had best results when we wean and separate offspring by sex at 3 weeks of age, then reintroduce males and females into breeding set-ups (2 females per male) at 6 weeks of age. Establishment of breeding set-ups at an earlier age results in dramatically reduced fertility, while establishment at a later age results in substantial mortality (generally of males) due to fighting. Offspring may be weaned as early as 2 weeks of age if their diet is supplemented with apples. Females may be housed as many as 4 to a cage indefinitely with little fighting; males will generally begin to fight by about 6 weeks of age and must thereafter be housed separately.

Although several ingenious techniques have been developed for the changing of cages, we have found that the combination of quick reflexes, sliding the cage lid just enough to allow entry of the gloved hand, and grasping the animal about the torso is the quickest and easiest. Animals should not be grasped by the tail, as the skin easily strips.

Anesthesia, inoculation, surgery, euthanasia

We use methoxyflurane (Metofane, Mallinckrodt Veterinary) for most procedures requiring anesthesia. Ether and isoflurane are highly toxic to cotton rats, for reasons unknown. We employ a glass instrument jar (stainless-steel lid) as an anesthesia chamber, to which we add methoxyflurane-impregnated gauze. Induction time is rapid — generally less than 1 minute — and anesthesia is maintained for about 1 additional minute. If longer anesthesia is required (such as for surgical procedures), one may place a 50 ml plastic centrifuge tube, with methoxyflurane-soaked gauze in its bottom, over the nose of the animal, varying the distance from nose to tube to regulate the depth of anesthesia. Alternatively, systemic anesthetics (pentobarbital or ketamine) have been used.

Intranasal inoculation requires anesthesia, otherwise the inoculum will be swallowed instead of aspirated. Other forms of inoculation, including intramuscular, subcutaneous, intradermal and intraperitoneal, may be accomplished without anesthesia, though we recommend its use by the novice. Intravenous inoculation is problematic, as there is not an accessible tail vein, but may readily be accomplished (in an anesthetized animal) by cardiac puncture, with the needle (23 G/1 inch or smaller) being inserted at a low angle below and to the left side of the xyphoid process.

The only type of survival surgery which has been reported in cotton rats is parabiosis, a technique dating back more than a century in other species, in which two animals are surgically linked (side-to-side) in order to establish a common blood circulation (Piazza *et al.*, 1995). This has proven to be an extremely valuable tool in dissecting RSV immunity and was the basis of critical animal studies showing an effector role for serum antibody in preventing RSV disease. Daily postoperative attention by laboratory personnel results in a high rate of success (over 90%), even in protocols in which the parabiotic linkage is maintained for nearly 1 month. The cotton rat is remarkably resistant to infection by environmental bacteria, and even though we have never used antibiotics postoperatively, we have never seen a case of bacterial sepsis.

The quickest and simplest method of euthanasia, approved by the Panel on Euthanasia of the American Veterinary Medical Association, is carbon dioxide intoxication. This procedure takes less than 1 minute and maintains host tissues well for all purposes which we have examined.

Preparation and administration of viral inocula

RSV is a labile virus and care must be taken to maintain viability prior to inoculating animals, generally by placing virus suspensions on wet ice. Most laboratories prepare virus stocks in HEp-2 cells, with minimum essential medium. Titers of about 10^6 plaque-forming units (PFU) per milliliter are not difficult to achieve. We recommend a challenge volume of 1 μl per gram of body weight (e.g., 100 μl/100 g animal), which will reach both nasal and pulmonary tissues. We recently described an alternative method of infecting cotton rats, in which a small volume of virus suspension (2 μl) is introduced via a cannula into one side of the nose (Johnson *et al.*, 1996). Under these conditions an infection in the inoculated tissues is established which extends neither to the contralateral side of the nose nor to the lungs.

Harvesting of tissues

Within about 6 hours after inoculation into cotton rats, RSV enters an eclipse phase during which the process of viral replication proceeds but no infectious virus can be recovered from infected tissues. By about 16 hours post-inoculation and for up to 10 days thereafter in immunocompetent animals, infectious virus can be recovered from respiratory tissues (lungs, trachea and nose). We recommend homogenizing tissue in 10 volumes of diluent, generally balanced salt solution containing sucrose, which helps maintain the viability of the virus in suspension. Soft tissues (lung and trachea) may be homogenized in glass-on-glass (TenBroeck) grinders, while nasal tissues are best homogenized in sterile mortars and pestles, to which have been added diluent and sterile sand. Following homogenization lysates are clarified by low-speed centrifugation then stored at –70 °C until assay. Virus is more stable within infected tissues than in homogenates, and lungs maintained at 4 °C showed no reduction in virus titers for up to 6 hours.

Tissues for histopathologic studies are routinely fixed in 10% neutral buffered formalin prior to processing and embedding in paraffin. Lungs should be inflated intratracheally with formalin (with the trachea then being tied with suture to prevent escape of the formalin), and immersed in the same in order to preserve the normal expanded architecture of the lungs. When histochemical studies (including immunofluorescence or immunoenzymology) are required on lungs, cryoembedding compound, diluted with an equal part of phosphate-buffered saline to reduce viscosity, may be substituted for formalin (Prince and Porter, 1975). After the trachea is tied off, the lungs may be quick-frozen and stored at –20 °C or lower prior to cryostat sectioning. It is often practical to harvest tissues from a single animal for multiple purposes.

The nasal tissues may easily be bisected by passing a scalpel or razor blade along the mid-saggital suture of the skull. We have routinely trisected the lungs for simultaneous viral titration, paraffin-embedded histology and cryosectioning. This is best accomplished by first tying off the two left lobes of the lungs and separating them by cutting through the bronchi. One of these lobes may be inflated with cryoembedding compound and frozen, and the other homogenized for viral titration. The three lobes of the right lung may then be inflated with formalin for paraffin embedding.

Blood may be obtained at the time of euthanasia by cardiac puncture, or at any time during an experimental protocol by puncture of the retro-orbital venous plexus. We recommend methoxyflurane anesthesia for the latter procedure. Once anesthesia is induced, the skin above and below the eye is pulled back slightly, causing the eye to protrude. A sterile Pasteur pipet may then be inserted gently at the front angle of the eye and passed behind the eye (with care being taken to avoid traumatizing the eye) until it reaches the venous plexus. The pipet is then pushed gently and rotated against the plexus, causing rupture of the superficial veins and bleeding into the pipet. Simultaneous gentle pressure against the jugular vein, by the other hand which is stabilizing the animal, will cause sufficient venous backpressure to move blood into the pipet. As much as 1 ml of whole blood may thus be obtained at one bleeding from an adult animal.

Cotton rat lymphoid tissue has not been studied nor described in detail, due largely to the lack of specific reagents required for characterization of lymphocyte populations. However, an elegant study by a group studying cotton rat filariasis produced detailed line drawings of the gross anatomy of the cotton rat lymphatic system (Wenk, 1964).

Pathogenesis

The prototypical model of RSV infection in cotton rats (Prince *et al.*, 1978) consists of an intranasal inoculation with approximately 10^5 PFU/animal. After a brief eclipse phase viral titers rise rapidly, achieving about 10^5 PFU/g in the lungs and slightly higher in the nose by the fourth day post-infection. Lung titers then drop quickly to undetectable levels by day 7, while nasal virus is still detectable until day 10. Viral antigens may persist beyond those times.

Infection is not age-dependent. That is, the amount of virus recovered from lungs and noses, when normalized to the weight of the animal, does not vary from infants to adults. This is one of the main advantages of the cotton rat model, permitting long-term studies without concern for changes in tissue permissiveness. This is in sharp contrast to the ferret, the first small animal model of RSV, whose nasal tissues remain susceptible to infection throughout life, while pulmonary tissues become refractory to infection by 4 weeks of age (Prince and Porter, 1976).

The cotton rat is semi-permissive for RSV infection, which is to say that the amount of virus recovered from the tissues (either pulmonary or nasal) is directly proportional to the amount of input virus (Piazza *et al.*, 1993). Suppression of the immune system of cotton rats by thrice-weekly injections of cyclophosphamide (Johnson *et al.*, 1982) will increase peak viral titers as much as 100-fold higher as long as the immunosuppressive regimen is continued, and may result in spread of the virus to non-respiratory tissues. This protocol has gained new significance recently, as RSV has been shown to be a major cause of morbidity and mortality in patients undergoing bone marrow transplantation, nearly all of whom receive cyclophosphamide as the primary immunosuppressive agent (Garcia *et al.*, 1997).

Histopathologic changes in nasal tissues are mild, consisting primarily of destruction of epithelial cells in the pseudostratified columnar cells of the olfactory epithelium (Prince *et al.*, 1978). Immunofluorescence studies showed that at the height of infection the majority of these epithelial cells may contain viral antigen. By contrast, very few pulmonary cells (< 0.1%) display viral antigen, even at the time of peak viral titers. Damage to these cells, which are primarily in the bronchiolar epithelium, is mild. The primary histopathologic process in the lungs is the inflammatory response of the host to infection, rather than direct viral damage of pulmonary tissues. This finding, as will be detailed later in this chapter, carries major implications regarding therapy of RSV disease.

We have compared RSV infection in two species of cotton rats (Prince *et al.*, 1978; Piazza *et al.*, 1992). RSV causes primarily peribronchiolitis in *S. hispidus* and interstitial pneumonitis in *S. fulviventer*, the same pattern we reported with parainfluenza virus type 3 infection (Porter *et al.*, 1991). The primary parameters which have been used to study experimental RSV disease and to monitor the efficacy of prophylactic and therapeutic strategies have been viral titers (primarily from tissue homogenates), histopathologic changes in formalin-fixed tissues, and serologic markers. Of the latter, the most useful has been the neutralization assay. Other serologic assays include an enzyme-linked immunosorbent assay (ELISA), with single or multiple viral antigens as targets, and a syncytium-inhibition assay.

The immune response to RSV, whether in humans or cotton rats, is not well-understood. While early studies, including the Lot 100 trials, relied upon neutralizing serum antibody as the primary marker of immunity, it is now clear that other immunologic effectors are of importance in clearing infectious virus, in preventing reinfection, and in mediating vaccine-enhanced RSV disease. These effectors remain unidentified largely because of the lack of cotton rat-specific reagents. Indeed, state-of-the-art studies of cellular (T-cell, B-cell, natural killer cell, macrophage, dendritic cell) and humoral (antibody classes and subclasses, cytokines, chemokines) effectors will not be possible until such reagents can be developed. We have begun to address this need in our laboratory by cloning and expressing relevant cotton rat genes, and plan to develop an extensive library of such reagents.

Prevention and therapy

Efforts to prevent RSV disease were dealt a serious setback by the Lot 100 trials, which not only delayed by decades major efforts to develop new candidate vaccines but led to the erroneous conclusion that serum antibody to RSV caused, rather than prevented, severe disease (Chanock *et al.*, 1970). Conclusive demonstration of a protective role for serum immunoglobulin G (IgG) derived from studies in the cotton rat which demonstrated pulmonary protection and correctly predicted the titer of circulating antibody necessary to achieve it (Prince *et al.*, 1983, 1985a,b). Subsequent clinical trials of plasma-derived IgG, which were based upon data generated in cotton rats, resulted in the licensure of RespiGam as an RSV-preventive in high-risk infants (PREVENT Study Group, 1997). Recent clinical trials of a humanized monoclonal antibody against RSV (designated MEDI-493), also based on cotton rat studies, showed it to be even more efficacious than RespiGam as an RSV-preventive.

Candidate vaccines of several major companies, including Wyeth-Lederle Vaccines and Pediatrics (Piedra *et al.*, 1996), Smith-Kline Beecham Biologicals (Neuzil *et al.*, 1997), Pasteur Mérieux Connaught (Du *et al.*, 1994), and the Institut de Recherche Pierre Fabre (Power *et al.*, 1997) are now in or near clinical testing, and all have utilized the cotton rat as the primary model of efficacy and safety. However, the legacy of the Lot 100 trials continues to cast a shadow over non-replicating vaccines, and in the three decades since the cessation of the Lot 100 trials, no non-replicating RSV vaccine has been allowed to proceed to efficacy trials in immunologically naive infants. It is likely that approval to proceed to that stage of testing will rely heavily upon the model of vaccine safety currently being defined in the cotton rat.

In 1986 our laboratory demonstrated that cotton rats immunized either with Lot 100 vaccine (stored under refrigeration since the clinical trials two decades earlier) or freshly prepared formalin-inactivated RSV induced a hypersensitivity state manifested as enhanced disease upon subsequent viral challenge (Prince *et al.*, 1986). The hallmark of vaccine-enhanced disease, seen in the two Lot 100 fatalities and in the cotton rat and subsequently verified in models of enhancement in the African green monkey (Kakuk *et al.*, 1993) and the calf (Gershwin *et al.*, 1998), is the presence of neutrophils in alveolar spaces. We presume that the absence of such a condition in animals primed with candidate vaccines will be a crucial marker of safety.

Over a decade ago Ribavirin was licenced for use as an RSV therapeutic. The initial enthusiasm over its licensure has gradually been replaced by disappointment, as a succession of controlled clinical studies have cast doubts on its efficacy (Moler *et al.*, 1996). Recent studies in our laboratory, either with Ribavirin or IgG as an antiviral agent, showed that neither agent reversed the histopathologic process (Prince and Hemming, 1994). Only when a glucocorticosteroid was added to the antiviral compound was there a concomitant reduction in virus load and reversal of the inflammatory process. We reported a similar finding with studies of parainfluenza virus type 3 in the cotton rat (Prince and Porter, 1996), and hypothesize that viral pneumonias in which the host inflammatory response is the major component of disease will best be treated with a combined antiviral/anti-inflammatory formulation. Although reduction in virus titer upon experimental challenge is an accurate marker of prophylaxis, it appears that the best correlate of therapy in an animal model is the histopathologic status of the lungs.

Future directions

Several factors have limited the utilization of the cotton rat as a model of RSV infection, including inbred strains, commercial availability, specific reagents and database. We have addressed, or are addressing, each of these. In collaboration with Dr Carl Hansen of the National Institutes of Health, we developed an inbred strain of *S. hispidus*, and have now taken a strain of *S. fulviventer* into the 22nd generation of brother–sister mating. Since 1990 we have been a United States Department of Agriculture-licenced seller of cotton rats, providing animals on a limited basis to laboratories throughout the world. We are working with Harlan Sprague Dawley to transfer to them the responsibility for continued commercial distribution of cotton rats. We have developed polyclonal and/or monoclonal (MAb) antibodies against cotton rat IgG, IgM, IgA (whole molecule and/or heavy chain) and the C3 component of complement. We have cloned fragments of the CD8 gene; are working to clone the CD4 gene; have targeted over a dozen cytokine and chemokine genes for cloning, expression, and MAb development; will develop MAb-based immunologic assays for cytokines and chemokines and MAb-based reagents for the targeted T-cell antigens; and are developing an RNA-based assay system for quantitating messenger RNA coding for cytokines and chemokines. Finally, we have developed a database consisting of over 1300 scientific reports concerning cotton rats, nearly 100 of which are masters' theses and doctoral dissertations.

Although there have been notable recent successes in passive RSV prophylaxis, many important goals and questions remain to be resolved. RespiGam is currently licenced for two high-risk populations which represent only a small fraction of the population adversely affected by RSV. It is not yet clear how many other populations can be effectively and economically protected against disease by passive prophylactic regimens. Active immunization remains an elusive goal due to issues of safety and to the realization that no single vaccine formulation is likely to be effective for the many different populations at risk of severe RSV disease. Indeed, the fastest-growing of these groups, bone marrow transplant recipients, are at highest risk of fatal RSV disease when they have no functioning immune system with which to respond to a vaccine. Furthermore, even a vaccine which is shown to be safe and effective may be of limited utility because natural infection induces immunity of very limited duration, and it is not likely that an RSV vaccine will induce more durable immunity. Given these factors, it is doubtful that we will see vaccine-induced "herd immunity" to RSV, and thus there will continue to be a need for passive prophylaxis and for therapy. As awareness of the medical and economic burdens of RSV continues to grow, efforts to develop commercially viable strategies for prevention and treatment will increase. And as the utility of the cotton rat model increases, particularly with the increasing availability of immunologic and genetic reagents, it is likely that it will remain a key tool in the development of such strategies.

References

Chanock, R. M., Kapikian, A. Z., Mills, J., Kim, H. W., Parrott, R. H. (1970). Influence of immunological factors in respiratory syncytial virus disease of the lower respiratory tract. *Arch. Environ. Health*, **21**, 347–355.

Collins, P. L., McIntosh, K., Chanock, R. M. (1996). Respiratory syncytial virus. In *Fields Virology*, 3rd edn (eds Fields, B. N., Knipe, D. M., Howley, P. M. *et al.*), pp. 1313–1351. Lippincott-Raven, New York.

Dreizin, R. S., Vyshnevetskaya, L. O., Bagdamyan, E. E., Yankevich, O. D., Tarasova, L. B., Klenova, A. V. (1971). Study of experimental RS virus infection in cotton rats. Virologic and immunofluorescent studies. *Vop. Virusol.*, **16**, 670–676.

Du, R.-P., Jackson, G. E. D., Wyde, P. R. *et al.* (1994). A prototype recombinant vaccine against respiratory syncytial virus and parainfluenza virus type 3. *Bio/Technology*, **12**, 813–818.

Garcia, R., Raad, I., Abi-Said, D. *et al.*, (1997). Nosocomial respiratory syncytial virus infections: prevention and control in bone marrow transplant patients. *Infect. Control Hosp. Epidemiol.*, **18**, 412–416.

Gershwin, L. J., Schelegle, E. S., Gunther, R. A. *et al.* (1998). A bovine model of vaccine enhanced respiratory syncytial virus pathophysiology. *Vaccine*, **16**, 1225–1236.

Johnson, R. A., Prince, G. A., Suffin, S. C., Horswood, R. L., Chanock, R. M. (1982). Respiratory syncytial virus infection in cyclophosphamide-treated cotton rats. *Infect. Immun.*, **37**, 369–373.

Johnson, S. A., Ottolini, M. G., Darnell, M. E. R., Porter, D. D., Prince, G. A. (1996). Unilateral nasal infection of cotton rats with respiratory syncytial virus allows assessment of local and systemic immunity. *J. Gen. Virol.*, **77**, 101–108.

Kakuk, T. J., Soike, K., Brideau, R. J. *et al.* (1993). A human respiratory syncytial virus (RSV) primate model of enhanced pulmonary pathology induced with a formalin-inactivated RSV vaccine but not a recombinant FG subunit vaccine. *J. Infect. Dis.*, **167**, 553–561.

Kim, H. W., Canchola, J. G., Brandt, C. D. *et al.* (1969). Respiratory syncytial virus disease in infants despite prior administration of antigenic inactivated vaccine. *Am. J. Epidemiol.*, **89**, 422–434.

Moler, F. W., Steinhart, C. M., Ohmit, S. E., Stidham, G. L. (1996). Effectiveness of ribavirin in otherwise well infants with respiratory syncytial virus-associated respiratory failure. *J. Pediatr.*, **128**, 422–428.

Neuzil, K. M., Johnson, J. E., Tang, Y.-W. *et al.* (1997). Adjuvants influence the quantitative and qualitative immune response in BALB/c mice immunized with respiratory syncytial virus FG subunit vaccine. *Vaccine*, **15**, 525–532.

Piazza, F. M., Johnson, S. A., Ottolini, M. G. *et al.* (1992). Immunotherapy of respiratory syncytial virus infection in cotton rats (*Sigmodon fulviventer*) using IgG in a small-particle aerosol. *J. Infect. Dis.*, **166**, 1422–1424.

Piazza, F. M., Johnson, S. A., Darnell, M. E. R., Porter, D. D., Hemming, V. G., Prince, G. A. (1993). Bovine respiratory syncytial virus protects cotton rats against human respiratory syncytial virus infection. *J. Virol.*, **67**, 1503–1510.

Piazza, F. M., Schmidt, H. J., Johnson, S. A. *et al.* (1995). A cotton rat model of effectors of immunity to respiratory syncytial virus other than serum antibody. *Pediatr. Pulmonol.*, **19**, 355–359.

Piedra, P. A., Grace, S., Jewell, A. *et al.* (1996). Purified fusion protein vaccine protects against lower respiratory tract illness during respiratory syncytial virus season in children with cystic fibrosis. *Pediatr. Infect. Dis. J.*, **15**, 23–31.

Porter, D. D., Prince, G. A., Hemming, V. G., Porter, H. G. (1991). Pathogenesis of human parainfluenza virus type 3 infection in two species of cotton rats: *Sigmodon hispidus* develops bronchiolitis while *Sigmodon fulviventer* develops interstitial pneumonia. *J. Virol.*, **65**, 103–111.

Power, U. F., Plotnicky-Gilquin, H., Huss, T. *et al.* (1997). Induction of protective immunity in rodents by vaccination with a prokaryotically expressed recombinant fusion protein containing a respiratory syncytial virus G protein fragment. *Virology*, **230**, 155–166.

PREVENT Study Group. (1997). Reduction of respiratory syncytial virus hospitalization among premature infants and infants with bronchopulmonary dysplasia using respiratory syncytial virus immune globulin prophylaxis. *Pediatrics*, **99**, 93–99.

Prince, G. A., Hemming, V. G. (1994). Method for treating infectious respiratory diseases. United States Patent #5,290,540.

Prince, G. A., Porter, D. D. (1975). Cryostat microtomy of lung tissue in an expanded state. *Stain Technol.*, **50**, 43–45.

Prince, G. A., Porter, D. D. (1976). The pathogenesis of respiratory syncytial virus infection in infant ferrets. *Am. J. Pathol.*, **82**, 339–352.

Prince, G. A., Porter, D. D. (1996). Treatment of parainfluenza virus type 3 bronchiolitis and pneumonia in a cotton rat model using topical antibody and glucocorticosteroid. *J. Infect. Dis.*, **173**, 598–608.

Prince, G. A., Jenson, A. B., Horswood, R. L., Camargo, E., Chanock, R. M. (1978). The pathogenesis of respiratory syncytial virus infection in cotton rats. *Am. J. Pathol.*, **93**, 771–791.

Prince, G. A., Horswood, R. L., Camargo, E., Koenig, D., Chanock, R. M. (1983). Mechanisms of immunity to respiratory syncytial virus in cotton rats. *Infect. Immun.*, **42**, 81–87.

Prince, G. A., Hemming, V. G., Horswood, R. L., Chanock, R. M. (1985a). Immunoprophylaxis and immunotherapy of respiratory syncytial virus infection in the cotton rat. *Virus Res.*, **3**, 193–206.

Prince, G. A., Horswood, R. L., Chanock, R. M. (1985b). Quantitative aspects of passive immunity to respiratory syncytial virus infection in infant cotton rats. *J. Virol.*, **55**, 517–520.

Prince, G. A., Jenson, A. B., Hemming, V. G. *et al.* (1986). Enhancement of respiratory syncytial virus pulmonary pathology in cotton rats by intramuscular inoculation of formalin inactivated virus. *J. Virol.*, **57**, 721–728.

Wenk, P. (1964). The topographic anatomy of the lymphatic system of the cotton rat *Sigmodon hispidus* Say et Ord. *Z. Versuchstierkd.*, **4**, 80–97.

Chapter 121

Animal Models for Coxsackievirus Infections

C. J. Gauntt

Background

Newborn mice were used initially to isolate non-polioviruses from patients with paralytic diseases and these viruses were subsequently classified into the coxsackieviruses group A (CVA; Dalldorf and Sickles, 1948) or coxsackieviruses group B (CVB; Melnick *et al.*, 1949). Mice have played a major role in isolation of many of the classified 23 CVA and 6 CVB strains (Melnick, 1985) and have been found to be excellent animal models for several CVB-induced diseases of humans. Major variables that affect the outcome of a CVB inoculation of mice include (Gauntt, 1997b): first, the genetic background (strain), age and gender of the mouse, factors that in part probably reflect non-specific and immune-resistance capabilities; second, genotype of the challenge virus; and third general health of the host and extrinsic manipulations imposed on the mouse (deficient diets, exercise stresses, suppressive manipulations of the immune systems, etc.).

Murine models for the study of CVB-induced myocarditis and pancreatitis/diabetes mellitus will be discussed in this chapter. These models have been extensively characterized over several decades, as described in several reviews (Gauntt *et al.*, 1993a,b; Khatib *et al.*, 1986; Toniolo *et al.*, 1988; Leslie *et al.*, 1989; Yoon, 1990; Rose *et al.*, 1992). Less well-characterized murine models of CVB-induced brain cavity destruction (porencephaly and hydranencephaly) will be noted, as well as CVB models in hamsters, swine and subhuman primates.

Animal models

Murine models

Myocarditis

Establishment of a CVB-, particularly a coxsackievirus B3 (CVB3)-, murine model of myocarditis is a straightforward process if one carefully selects male mice of the right age and strain to match the selected cardiovirulent virus strain (Woodruff, 1980; Khatib *et al.*, 1987; McManus *et al.*, 1987; Gauntt *et al.*, 1993b). Virus challenge usually occurs by the intraperitoneal route, although per oral challenge with a CVB3 will induce myocarditis (Woodruff, 1980). A pediatric model of CVB3-induced myocarditis has been studied by Rose and colleagues (Rose *et al.*, 1992). In this model, 2-week-old mice are challenged with a CVB3 strain that is essentially non-cardiovirulent in adolescent (4–6 weeks old) mice (Chow *et al.*, 1991). Studies with this model established that the major histocompatibility locus was not associated with susceptibility to induction of myocarditis, but genetic background determined: first, severity of myocarditis induced, which generally correlated with capacity for rapid production of higher titers of protective antibodies, and second, whether the acute disease resolved or there was transit to chronic myocarditis (Beisel *et al.*, 1985; Herskowitz *et al.*, 1987; Beisel and Traystman, 1988; Rose *et al.*, 1992). These and subsequent studies with cardiovirulent CVB3 strains that were generally lethal for a significant proportion of mice younger than 4–5 weeks of age in many inbred strains also identified murine strains whose fate post-CVB3 inoculation would be complete recovery with resolved disease or development of chronic myocarditis (Gauntt *et al.*, 1984, 1993a; Lerner and Reyes, 1985; Khatib *et al.*, 1986; Leslie *et al.*, 1989; Chow *et al.*, 1991). These studies showed that challenge of adolescent male mice with an A, C3H or SWR background with a cardiovirulent CVB3 at a dose of $1–5 \times 10^5$ plaque-forming units (PFU) per mouse by intraperitoneal route would result in chronic myocarditis that could last for months. In contrast, inoculation of many inbred or outbred murine strains, e.g., C57B1/6 or CD-1 mice, with the same inoculum would induce acute inflammation of the myocardium that is resolved within 3 weeks. Except for high mortality in A background mice (≥50%), deaths in these models is generally < 10%.

Challenge of all BALB/c strains with cardiovirulent CVB3 results in 100% fatalities within 2 weeks (Leslie *et al.*, 1989; Gauntt *et al.*, 1993b). Studies with CVB3-inoculated transgenic knockout mice show that deletion of MIP-1α prevents disease (Cook, 1996), whereas deletion of CD4 or β-microglobin resulted in more severe myocarditis (Henke *et al.*, 1995). Generally, female mice are more resistant to induction of myocarditis by cardiovirulent CVB3 than males of the same strain (Woodruff, 1980; Leslie *et al.*, 1989;

Handbook of Animal Models of Infection
ISBN 0-12-775390-7

Gauntt *et al.*, 1993a), although there are exceptions (Gauntt *et al.*, 1984). This resistance has been shown to result from females responding to the virus infection with a Th_2 response rather than the Th_1 response of males (Huber and Pfaeffle, 1994).

Genotype of the CVB3 strain is of major importance in determining whether myocarditis is induced, and severity of disease induced (Gauntt *et al.*, 1984, 1993b; Gauntt, 1988, 1997b). Well-characterized CVB3 strains are known to induce myocarditis in every murine strain challenged (Gauntt, 1988; Leslie *et al.*, 1989); however, most CVB3 strains isolated from humans are non-cardiovirulent (Gauntt and Pallansch, 1996). Persistence of CVB3 genomes in heart tissues of some murine strains most likely contributes to the chronic disease state induced in these mice, perhaps via continued stimulation of production of pro-inflammatory cytokines and autoimmune responses to cryptic or shared antigens (molecular mimickry) between the virus and normal cells in heart tissues (Gauntt, 1997b).

Pancreatitis, insulitis and diabetes mellitus

The CVB, and particularly some coxsackievirus B4 (CVB4) strains, are known to infect cells of the exocrine and endocrine pancreas (Toniolo *et al.*, 1988; Gauntt, 1988, 1997b; Yoon, 1990). Murine models of acute pancreatitis with hyperglycemia have been established with specific murine strains and specific pancreatotropic CVB4 strains (Loria, 1988; Yoon, 1990). Diagnostic parameters of diabetes mellitus are induced in these models, i.e., hyperglycemia, insulitis, and autoantibodies against insulin, glutamic acid decarboxylase (GAD_{65}), and islet cell cytoplasmic autoantigens. A shared immunogenic sequence between GAD_{65} and a CBV-common sequence in the nonstructural P2-C protein induces both autoantibodies and sensitized $CD8^+$ T cells in diabetogenic murine strains, with the latter immune response thought to be associated with destruction of the islets (Chatterjee, 1994; Gauntt, 1997b). Recent studies show that both cardiovirulent and non-cardiovirulent CVB3 strains can induce a progressive unrelenting insulitis and destruction of exocrine pancreatic cells in both CD-1 and C3H/HeJ mice from 2 weeks through day 90 post-inoculation (Reyna, Woods and Gauntt, unpublished data). A reversible diabetes-like syndrome has been described for CVB4-inoculated mice (Khatib *et al.*, 1986).

Brain cavity diseases

Intracranial challenge of neonatal (<24 hours old) CD-1 or C57Bl/6J mice with one of two temperature-sensitive (ts) mutants of CVB3 induced porencephaly and/or hydranencephaly in approximately two-thirds of mice surviving (60–70%) to adolescence (Marlin *et al.*, 1985; Gauntt, 1988). Lack of control of body temperature in the infants was thought to contribute to the extensive virus replication. This was not the only factor, as intraperitoneal challenge of gravid females in the second trimester with one ts mutant resulted in a few (4%) adolescents presenting with porencephaly.

Influences of diet suppression of non-specific protective mechanisms on outcomes in murine models of coxsackievirus myocarditis

Adolescent mice maintained on severe nutrient limitation, hypercholesteremic-, vitamin E- or selenium-deficient diets develop more severe myocarditis when challenged with a cardiovirulent CVB strain than mice maintained on normal lab chow and challenged with that CVB (reviewed in Woodruff, 1980; Loria, 1988; Gauntt, 1997a). Mice maintained on two deficient diets and challenged with a non-cardiovirulent CVB3 also developed myocarditis as a result of rapid selection of cardiovirulent variants under these conditions. Immunosuppression of mice by injection of cyclophosphamide, cyclosporin, or corticosteroids (Woodruff, 1980; McManus *et al.*, 1987; Leslie *et al.*, 1989; Gauntt *et al.*, 1993b) generally exacerbated cardiac inflammation. Thymectomy significantly reduced the severity of myocarditis induced in some but not all murine strains, suggesting that different pathogenic mechanisms were operative (Woodruff, 1980; Leslie *et al.*, 1989; Gauntt *et al.*, 1993b). *In vivo* antibody-induced lysis of natural killer cells increased severity of CVB3-induced myocarditis and promoted extensive cardiac calcification (Gauntt *et al.*, 1993a,b). Antibody inactivation of pro-inflammatory cytokines *in vivo* can ameliorate CVB3-induce myocarditis (Gauntt *et al.*, 1993b; Huber and Pfaeffle, 1994; Gauntt, 1997b), whereas injection of pro-inflammatory cytokines can induced myocarditis in infant mice normally resistant to induction by a CVB3 (Rose *et al.*, 1992). Enforced exercise (swimming) in CVB3-inoculated adolescent mice significantly enhanced myocarditis and calcification in the facial lesions (Khatib *et al.*, 1986; Loria, 1988; Gauntt *et al.*, 1993b). In the majority of the above-cited manipulations that increased cardiopathology, virus titers were generally increased in heart tissues and the duration of infectious virus in the heart was increased.

Other animal models

Experimental models of CVB3- or CVB4-inoculated hamsters, cynomolgus monkeys, chimpanzees, baboons and swine have been reported (Woodruff, 1980; Khatib *et al.*, 1986; McManus *et al.*, 1987). Electrocardiographic and hemodynamic changes associated with progression from CBV3-induced myocarditis to dilated cardiomyopathy were reported for the hamster (McManus *et al.*, 1987). Efforts to establish CVB3 rat (several strains) models of myocarditis have gone unrewarded in many labs.

Technical problems

Passage of a cardiovirulent CVB3 repeatedly in tissue culture cells can also select for an attenuated stock, so each stock must be assayed in mice to ascertain cardiovirulence capabilities and stored at –70°C before use in extensive studies. Storage of diluted stock in the absence of proteins at –20°C can reduce the titer in 24 hours. Re-isolation of a CVB3 strain from heart tissue of a mouse early in infection (3–4 days post-inoculation) is a useful procedure for selection of a cardiovirulent pool of viruses (Woodruff, 1980). Attention must be given to the challenge virus, as these small RNA genome viruses such as the enteroviruses have high mutation rates (McManus *et al.*, 1987; Gauntt, 1988). The virus titer range over which acute myocarditis is induced in a selected murine strain is quite narrow, usually 1×10^5 to 5×10^6 PFU/mouse, and higher-titer inocula can result in a reduction in focal lesion numbers in the heart (Gauntt *et al.*, 1993b). Each CVB strain and selected murine strain should be examined for unwanted experimental deaths due to challenge of the male mice at a young age (up to 50% of mice in some strains succumb to infections if challenged between 3 and 5 weeks of age with a cardiovirulent CVB).

References

Beisel, K. W., Traystman, M. D. (1988). Viral myocarditis: immunogenetic and autoimmune aspects. In *New Concepts in Viral Heart Disease* (ed Schultheiss, H.-P.), pp. 148–159. Springer-Verlag, Berlin.

Beisel, K., Wolfgram, L., Herskowitz, A., Rose, N. (1985). Differences in severity of coxsackievirus B3-induced myocarditis among H-2 congenic mouse strains. *Prog. Leuko. Biol.*, **3**, 195–201.

Chatterjee, N. K. (1994). Antibodies to glutamic acid decarboxylase and P2-C peptides in sera from coxsackievirus B4-infected mice and IDDM patients. *Diabetes*, **43**, 1260–1266.

Chow, L. H., Gauntt, C. J., McManus, B. M. (1991). Differential effects of myocarditic variants of coxsackievirus B3 in inbred mice. A pathologic characterization of heart tissue damage. *Lab. Invest.*, **64**, 55–64.

Cook, D. N. (1996). The role of MIP-1α on inflammation and hematopoiesis. *J. Leukoc. Biol.*, **59**, 61–66.

Dalldorf, G. E., Sickles, G. M. (1948). An unidentified filterable agent isolated from the faeces of children with paralysis. *Science*, **108**, 61–62.

Gauntt, C. J. (1988). The possible role of viral variants in pathogenesis. In *Coxsackieviruses — A General Update*, vol. 1 (eds Friedman, H., Bendinelli, M.), pp. 159–179. Plenum Press, New York.

Gauntt, C. J. (1997a). Nutrients often influence viral diseases. *ASM News*, **63**, 133–135.

Gauntt, C. J. (1997b). Roles of the humoral response in coxsackievirus B-induced disease. In *Current Topics in Microbiology and Immunology. The Coxsackie B Viruses*, vol. 223 (eds Tracy, S., Chapman, N., Mahy, B.), pp. 259–282. Springer Verlag, Berlin.

Gauntt, C. J., Pallansch, M. A. (1996). Coxsackievirus B3 clinical isolates and murine myocarditis. *Virus Res.*, **41**, 89–99.

Gauntt, C. J., Gomez, P. T., Duffey, P. S. *et al.* (1984). Characterization and myocarditic capabilities of coxsackievirus B3 variants in selected mouse strains. *J. Virol.*, **52**, 598–605.

Gauntt, C. J., Higdon, A. L., Arizpe, H. M. *et al.* (1993a). Specific and nonspecific heart defenses in enteroviral infections. In *Idiopathic Dilated Cardiomyopathy* (eds Figulla, H. R., Kandolf, R., McManus, B.), pp. 310–321. Springer-Verlag, Berlin.

Gauntt, C., Higdon, A., Bowers, D., Maull, E., Wood, J., Crawley R. (1993b). What lessons can be learned from animal model studies in viral heart diseases? *Scand. J. Infect. Dis.*, **88** (suppl), 49–65.

Henke, A., Huber, S., Stelzner, A., Whitton, J. L. (1995). The role of $CD8^+T$ lymphocytes in coxsackievirus B3-induced myocarditis. *J. Virol.*, **69**, 6720–6728.

Herskowitz, A., Wolfgram, L. J., Rose, N. R., Beisel, K. W. (1987). Coxsackievirus B3 murine myocarditis-marked strain differences in histopathologic features of early and late phase myocarditis. *J. Am. Coll. Cardiol.*, **9**, 1311–1319.

Huber, S. A., Pfaeffle, B. (1994). Differential Th_1 and Th_2 cell responses in male and female BALB/c mice infected with coxsackievirus group B type 3. *J. Virol.*, **68**, 5126–5132.

Khatib, R., Reyes M. P., Lerner A. M. (1986). Animal models for coxsackievirus infections. In *Experimental Models in Antimicrobial Chemotherapy*, vol. 2 (eds Zak, O., Sande, M.), pp. 175–193. Academic Press, London.

Khatib, R., Probert, A., Reyes, M. P., Khatib, G., Chason, J. L. (1987). A review. Mouse strain-related variation as a factor in the pathogenesis of coxsackievirus B3 murine myocarditis. *J. Gen. Virol.*, **68**, 2981–2988.

Lerner, A. M., Reyes, M. P. (1985). Coxsackievirus myocarditis with special reference to acute and chronic effects. *Prog. Cardiovasc. Dis.*, **27**, 373–394.

Leslie, K., Blay, R., Haisch, C., Lodge, P., Weller, A., Huber, S. (1989). Clinical and experimental aspects of viral myocarditis. *Clin. Microbiol. Rev.*, **2**, 191–203.

Loria, R. M. (1988). Host conditions affecting the course of coxsackievirus infections. In *Coxsackieviruses. A General Update* (eds Bendinelli, M., Friedman, H.), pp. 135–157. Plenum, New York.

Marlin, A. E., Huntington, W. H., Arizpe, H. M., Gudvangen, R. J., Brans, Y. W., Gauntt, C. J. (1985). Coxsackieviruses group B and hydranencephaly. *Concepts Pediatr. Neurosurg.*, **6**, 147–160.

McManus, B. M., Gauntt, C. J., Cassling, R. S. (1987). Immunopathologic basis of myocardial injury. *Cardiovasc. Clin.*, **18**, 163–184.

Melnick, J. L. (1985). Enteroviruses: polioviruses, coxsackieviruses, echoviruses, and newer enteroviruses. In *Fundamental Virology* (ed. Fields, B.), pp. 739–794. Raven Press, New York.

Melnick, J. L., Shaw, E. W., Curnen, E. C. (1949). A virus from patients diagnosed as non-paralytic poliomyolitis or aseptic meningitis. *Proc. Soc. Exp. Biol. Med.*, **71**, 344–349.

Rose, N. R., Neumann, D. A., Herskowitz, A. (1992). Coxsackievirus myocarditis. *Adv. Intern. Med.*, **34**, 411–429.

Toniolo, A., Federico, G., Basolo, F., Onodera, T. (1988). Diabetes mellitus. In *Coxsackieviruses. A General Update* (eds. Bendinelli, M., Friedman, H.), pp. 351–382. Plenum, New York.

Woodruff, J. F. (1980). Viral myocarditis. A review. *Am. J. Pathol.*, **101**, 425–483.

Yoon, J. W. (1990). The role of viruses and environmental factors in the induction of diabetes. *Curr. Top. Microbiol. Immunol.*, **164**, 95–123.

Chapter 122

Animal Models for HBV Infections — Transgenic Mice

J. D. Morrey, R. W. Sidwell and B. A. Korba

Background of human infection

The background of the human hepatitis B virus (HBV) infection is amply described in previous chapters; however, there are some features of the human infection that are modeled by the transgenic mice. Some of the salient comparisons between infected human patients and transgenic mice are as follows. Different morphologies of viral particles are identified in the serum of infected human subjects which include infectious Dane particles. Transgenic mice generated with redundant sequences of the HBV genome complete sufficient portions of the HBV life cycle to result in the assembly of Dane particle-like virions at appreciable titers in the serum (Guidotti *et al.*, 1995). The HBV life cycle in the transgenic mice creates the opportunity to study viral replication and to evaluate the potential of antiviral agents on the human virus. A detailed comparison of these features is reviewed by Chisari (1995a,b, 1996).

Regulatory sequences and promoters of the HBV genome, used to generate the transgenic mice expressing the complete virion in their serum, contain some determinants of tissue specificity (Guidotti *et al.*, 1995). In human subjects, the liver is the main target tissue for virus replication. Likewise, virus expression in these transgenic mice is predominant in the liver, as well as in the kidney. Extrahepatic expression in the kidney might be biologically significant. Such organs with trace levels of infection might allow HBV to sequester in tissues that are not as bioavailable to drug therapy as the liver tissue (Guidotti *et al.*, 1995; Cavanaugh *et al.*, 1997). The HBV regulatory sequences within the transgenic mice also contain responsive elements, such as for sex steroids (Farza *et al.*, 1987; DeLoia *et al.*, 1989), that might contribute to differences of disease incidence and progression that occur in human patients. This provides the opportunity to investigate the efficacy of therapeutic treatments between the sexes.

Lastly, the HBV-induced pathology in the natural infection is mediated in part by immune response to the virus-infected liver cells (Hollinger, 1996). By using adoptive transfer of hepatitis B surface antigen (HBsAg)-specific cytotoxic T cells within transgenic mice, the dynamics between the pathology and antiviral effects of T cells and other effector cells have been more clearly elucidated (Ando *et al.*, 1994a,b; Guidotti *et al.*, 1996). Such experiments have also demonstrated that there is a non-cytolytic suppression of viral gene expression and replication that is mediated by interferon-γ, tumor necrosis factor-α and interleukin-2 (Guidotti *et al.*, 1994b, 1996; Cavanaugh *et al.*, 1997).

The immunological aspects of HBV in relation to transgenic mice have been reviewed (Chisari, 1995b, 1996) and will not be described in detail within this chapter except to mention that such an approach may provide opportunities to evaluate therapeutic immunomodulators in this animal model.

Background of model

The development of the transgenic mouse model carrying the infectious HBV genome has been motivated in part by the expense and minimal availability of the HBV-chimpanzee model, and the absence of more convenient, small non-primate animal models that can be infected with HBV. Much of the knowledge about HBV has been extrapolated from the infection of natural hosts with non-human hepatitis viruses that possess similar characteristics to HBV. The infections of ducks, woodchucks, and squirrels with their respective animal hepatitis viruses (Ganem, 1996b) have been enormously important in formulating much of our knowledge of HBV infection, but small animal models carrying the human virus have not been available previous to HBV-transgenic mice because of the limited host range of the virus.

The limitation of the host range of viruses can be partially overcome by genetically engineering animals to possess viral genomes such as has been demonstrated with human immunodeficiency virus (Leonard *et al.*, 1988; Morrey *et al.*, 1991a), papillomavirus (Lacey *et al.*, 1986), and human HBV (Farza *et al.*, 1987, 1988; DeLoia *et al.*, 1989; Nagahata *et al.*, 1992; Chisari, 1995a; Guidotti *et al.*, 1995). Many transgenic mice have been produced that express portions of the HBV genome coding for the envelope (Chisari *et al.*, 1986, 1989; Gilles *et al.*, 1992b; Ando *et al.*, 1993), core (Guidotti *et al.*, 1994c; Milich *et al.*, 1994), precore (Guidotti *et al.*, 1995), and X (Koike *et al.*, 1994; Perumo *et al.*, 1992; Lee *et al.*, 1990)

Handbook of Animal Models of Infection
ISBN 0-12-775390-7

proteins. The studies with these animals yielded good information on the viral induction, protein transport, assembly, secretion and HBV-specific immune responses. Such animals are generally not considered suitable for antiviral studies on viral replication, however, because they only express portions of the viral genome.

Initial attempts have been made to generate transgenic mice carrying the complete linear viral genome (Farza *et al.*, 1987; Araki *et al.*, 1989; Guidotti *et al.*, 1994a), with the anticipation that the effort would provide information on viral replication. These transgenic mice demonstrated that HBV can replicate in murine hepatocytes, but the virus was at low levels in the serum. Guidotti *et al.* (1995) have been successful in generating transgenic mice that replicate high levels of human HBV in clinically important target organs of the liver and kidney. Pooled sera from these mice contain high titers of viral DNA approaching that found in the natural chronic human infection (Guidotti *et al.*, 1995). Moreover, the profiles of HBV DNA replicative intermediate forms and RNA species were far more similar to those species identified in human and primate hosts as compared to the earlier HBV transgenic mice (Farza *et al.*, 1987; Araki *et al.*, 1989; Guidotti *et al.*, 1994a).

A number of chemotherapeutic studies have been performed in a variety of HBV-transgenic mice (Nagahata *et al.*, 1992; Guilhot *et al.*, 1993; Cavanaugh *et al.*, 1997). The results from these experiments suggest that such small animal models might be predictive indicators of therapeutic efficacy in HBV-infected people. This chapter will describe the features of the HBV-transgenic mice (Guidotti *et al.*, 1995) that produce appreciable levels of serum hepatitis B virions as determined by the presence of HBV DNA and electron microscopy.

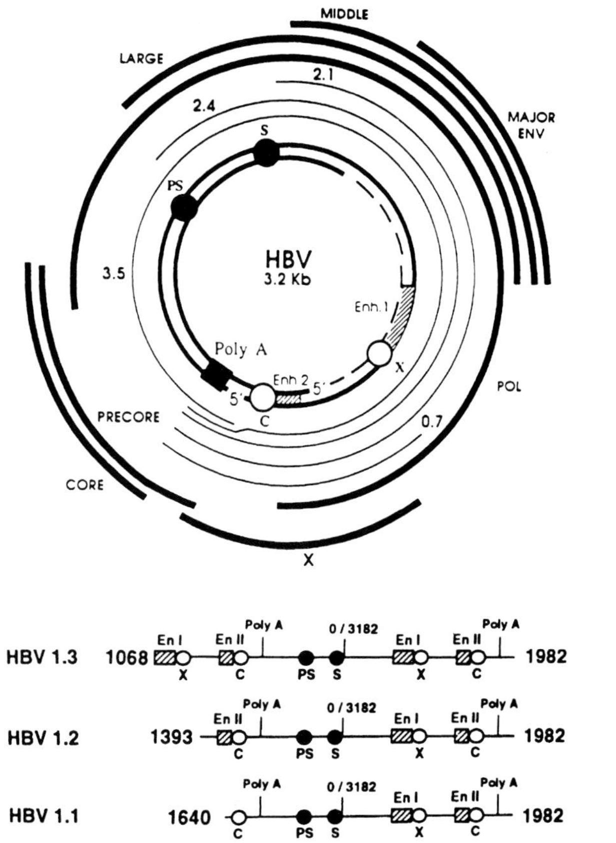

Figure 122.1 (Top) Schematic representation of the hepatitis B virus (HBV) genome (awy subtype) used to generate transgenic mice. (Bottom) Schematic representations of the HBV-derived constructs that contain terminally redundant, greater-than-length (1.1-, 1.2-, 1.3-eq) copies of the complete HBV genome used for the generation of transgenic mice. Enh. and En, Enhancer; Poly A, polyadenylation signal; X, C, PS, and S, X, core, pre-S, and S promoter, respectively. From Guidotti et *al.* (1995), with permission.

Animal species

The two lineages of high-expressing HBV-transgenic mice that have been characterized are 1.3.32 (official designation, Tg[HBV 1.3 genome]Chi32) and 1.3.46 (Tg[HBV 1.3 genome]Chi46) (Guidotti *et al.*, 1995). The transgene consisted of a terminally redundant, 1.3× greater-than-genome length construct (Figure 122.1). From the down-stream to up-stream direction, it consisted of the 3′ terminus at nucleotide position 1982 downstream of the polyadenylation signal, through the entire length of the genome, plus a terminally redundant section of the nucleocapsid promoter, enhancer II, X promoter and enhancer I. Transgenic mice generated with shorter repeating sequences containing only the nucleocapsid promoter and the enhancer II sequence did not generate high levels of serum virus. Consequently, they would not be as useful for antiviral experiments.

The 1.3.32 and 1.3.46 lineages of transgenic mice originated from (C57BL/6 X SJL)F_2 or inbred congenic B10.D2 embryos, respectively. The mice were expanded by repeated backcrossing against either C57BL/6 or B10.D2 strains depending on the genetic background of the founder transgenic mice (Guidotti *et al.*, 1995). Therefore, backcrossing with appropriate strains of mice should be done when expanding the numbers of these mice.

Preparation of animals

Other than standard care for mice (1996d), one needs to consider the biological containment of transgenic mice due to the presence of potentially infectious HBV, screening for transgenic mice, and breeding inefficiencies of C57BL/6 mice.

The existing transgenic mice derived from founder 1.3.32 (Guidotti *et al.*, 1995) have been backcrossed many generations with C57BL/6 mice so that they possess mostly this genetic background. Young sexually mature C57BL/6 mice do not readily breed at efficiencies seen with other common laboratory mouse strains so that obtaining an experienced older breeding colony substantially delays the researcher's effort to establish a productive colony. These delays should be planned for accordingly. HBV-transgenic mice have successfully been bred to homozygosity (personal communications, Francis Chisari, Scripps Institute, La Jolla, CA) which markedly increases the efficiency of the transgenic production colony, since all mice would be transgenic.

We screen for the identity of transgenic mice by assaying serum, obtained from tail bleeding, for HBsAg or HBeAg. Assays for both antigens are available as kits from Abbott Laboratories (Abbott Park, IL). IncStar Corporation (Stillwater, MN) also sells a kit for HBeAg. In our laboratory, we detect both antigens from Abbott kits because of convenience, and the same washing instrumentation and spectrophotometer is required for both kits. We also prefer to screen with HBeAg, as compared to HBsAg, since antibodies specific for HBsAg can mask the detection of HBsAg. We do not screen for the presence of the transgene by Southern blot hybridization or the polymerase chain reaction (PCR) from tail tissue, as is frequently done with other transgenic mice (Morrey *et al.*, 1991a), because the antigen assays are easy to perform and the presence of the transgene does not necessarily insure the production of viral products in which we are most interested in terms of antiviral studies.

The safety classification of the human hepatitis B virus is biosafety level 2 or 3 (BSL-2 or -3), depending on the potential for droplet or aerosol production (1993a, 1994b). However, when greater than two-thirds of infectious viral genomes are used to generate transgenic animals, the recommendation is for the biosafety level to increase the containment conditions (1994c). Dr Francis Chisari's group has determined that the transgenic viral particles are infectious by transferring hepatitis in the chimpanzee model. The biosafety level for the transgenic mice of this review, containing the infectious virus, is BSL3-N, where the N connotes additional precautions for animals. Additional precautions must be added to insure that the animals do not inadvertently escape from the designated animal facility. This is accomplished in our laboratory by sealing all drains, spaces around plumbing pipes, or otherwise slight holes exiting the animal rooms. A 45 cm Plexiglas barrier is placed at the entrance of the doors to the animal rooms to prevent mice from exiting under the door. The lab attendants simply step over the barrier to enter the room. Additionally, we identified each individual mouse by toe clipping (Murphy, 1993). Not only does this allow us to associate each data point with a particular mouse, but it allows us accurately to control the location or disposition of animals in the colony.

Since the transgenic animals contain infectious HBV, those people working with these animals must take appropriate safety precautions. In our laboratory, all personnel who handle the animals are required to receive the commercially available hepatitis B vaccination (Engerix-B, SmithKline Beecham Pharmaceuticals, Philadelphia). Vaccinated individuals are monitored for seroconversion by the local health department to verify that personnel are protected. The personnel receive special training on blood-borne pathogen-handling by our institution's Environmental Health and Safety Office on a continuous annual basis.

Details of surgery

No surgeries are required for the standard use of this transgenic model. However, options do exist for researchers to remove hormone-producing organs, such as ovaries, to evaluate their effects on viral production, to explant liver or hematopoietic tissues, or to perform passive immune transfer, but none of these procedures are required for standard use of the transgenic mouse model described in this chapter.

Storage, preparation of inocula

As mentioned in the background section, transgenic mice carry the infectious HBV genome integrated into the host genome. Consequently, the transgene is inherited to progeny animals by classical Mendelian genetics and the virus is not injected with exogenous inocula.

Infection process

Viral inoculum is not administered due to the presence of the HBV transgene. The life cycle of HBV and how it relates to the transgene, however, is important for understanding this transgenic model. In the natural host, infection is initiated by viral ligand–receptor interactions (Ganem, 1996b). In transgenic mice (Guidotti *et al.*, 1995), viral RNA and subsequent viral proteins and virions are generated from the HBV transgene integrated in the host genome. Therefore, an initial infectious event is not operative in transgenic mice as it is in natural infection.

The ability for transgenically produced virions to infect nuclei within the same cells in which they are produced (recycling; Ganem, 1996a) is a different type of cellular evident as compared to natural infection. The apparent absence of the replicative intermediate (covalently closed circular DNA or cccDNA; Ganem, 1996c) in these transgenic mice (Guidotti *et al.*, 1995) suggests that infection of the nuclei of transgenic mice does not occur. The infection or recycling processes in transgenic mice is speculative, however, and requires more experimentation.

Key parameters to monitor infection

Many animal models for infectious human disease use death or pathology as necessary and important indictors of therapeutic efficacy. No death or even pathology has been identified in the transgenic mice that efficiently produces the complete viral particle (Guidotti *et al.*, 1995). Consequently, viral parameters are used as indicators of drug efficacy for these mice. HBV RNA and protein products of replication are illustrated in Figure 122.1. Four species of HBV RNA — 3.5 kb, 2.4 kb, 2.1, kb, and 0.7 kb — are produced in normal infection. The protein products are envelope (large, middle, major), polymerase (pol), X, precore and core proteins. The parameters to measure these viral products in the transgenic mice are listed in Table 122.1.

Viral particles and their serological determinants are similar between the transgenic mice described in this chapter and in human patients. The infectious Dane-like particles are found in the transgenic mice; consequently the antigens and viral DNAs associated with the Dane particles are found in the sera (Guidotti *et al.*, 1995), namely the HBV DNA, pre-S protein, HBsAg, HBe/cAg (Table 122.1), as they are in human subjects. In previously generated transgenic mice, infectious viral Dane particles and its associated HBV DNA were only present in trace or undetectable amounts (Farza *et al.*, 1988; Araki *et al.*, 1989; Guidotti *et al.*, 1995).

The precore antigen, detected immunologically as the HBeAg, is rapidly secreted from the cells of these transgenic mice and is detected in the sera (Guidotti *et al.*, 1995; Table 122.1). It is not found in the cytoplasm or nuclei of liver cells, indicating that intracellular retention of the precore protein is very low in these mice, probably due to rapid secretion. The viral parameters typically measured in the serum of the transgenic mice are HBeAg, HBsAg and HBV DNA. We use HBV DNA as the marker for the presence of the Dane particle.

In natural infections, supercoiled or cccDNA is considered to be a replication intermediate (Ganem, 1996a) and is important for generating virions in regenerating daughter liver cells (Ganem, 1996c). There is no *a priori* reason not to find cccDNA in these transgenic mice; nevertheless, cccDNA is not readily detected in these mice. The integrated transgene provides the template for expression in the transgenic mice, so cccDNA is probably not an obligate replication intermediate in the mice. The absence of supercoiled cccDNA in transgenic mice suggests that species-specific differences at this level may play a role in determining the host range of HBV (Chisari, 1996). The ability to measure the effect of drug treatment on cccDNA is considered advantageous for determining the ability of the drug to eliminate infection; however, this parameter cannot be directly measured in this model.

The profile of DNA species, besides cccDNA, is very similar to that found in infected human patients and chimpanzees (Guidotti *et al.*, 1995). Figure 122.2 typifies such a profile with relaxed circular (RC), double-stranded circular (DS) and single-stranded circular (SS) DNAs from transgenic mice. The transgenic mouse liver HBV DNA (lane 4)

Table 122.1 Viral parameters of hepatitis B virus (HBV) transgenic mouse model

Tissue	*Parameter*	*Type of assay*	*Feature*	*Reference*
Serum	HBsAg (surface antigen)	Enzyme-linked antigen (Abbott Laboratories)	Easy to perform, commercial kits available, ~10 μl serum required	Guidotti *et al.*, 1995
Serum	HBeAg	Enzyme-linked antigen (Abbott Laboratories or IncStar)	Easy to perform, commercial kits available, ~10 μl serum required	Guidotti *et al.*, 1995
Serum	HBV DNA	Manifold dot blot hybridization	Larger volumes of sera, > 50 μl	Guidotti *et al.*, 1995; Will *et al.*, 1985
Serum	HBV DNA	Quantitative polymerase chain reaction	Small volumes of sera, ~1 μl	Kaneko *et al.*, 1993
Tissue (liver or kidney)	HBV DNA	Southern blot hybridization	Measures various DNA replicative intermediates to compare with normal HBV infection	Guidotti *et al.*, 1995
Tissue (liver or kidney)	HBV RNA	Northern blot hybridization	Measures various RNA transcripts to compare with normal HBV infection	Guidotti *et al.*, 1994a, 1995
Tissue (liver or kidney)	HBcAg, HBeAg, HBsAg	Western blot analysis	Quantitative, requires HBV-specific antibody reagent	Araki *et al.*, 1989; Guidotti *et al.*, 1994c, 1995
Tissue (liver or kidney)	HBcAg	Immunohistochemistry	Paraffin-embedded tissues, technically more difficult, not quantitative, requires HBV-specific antibody reagent	Guidotti *et al.*, 1994c

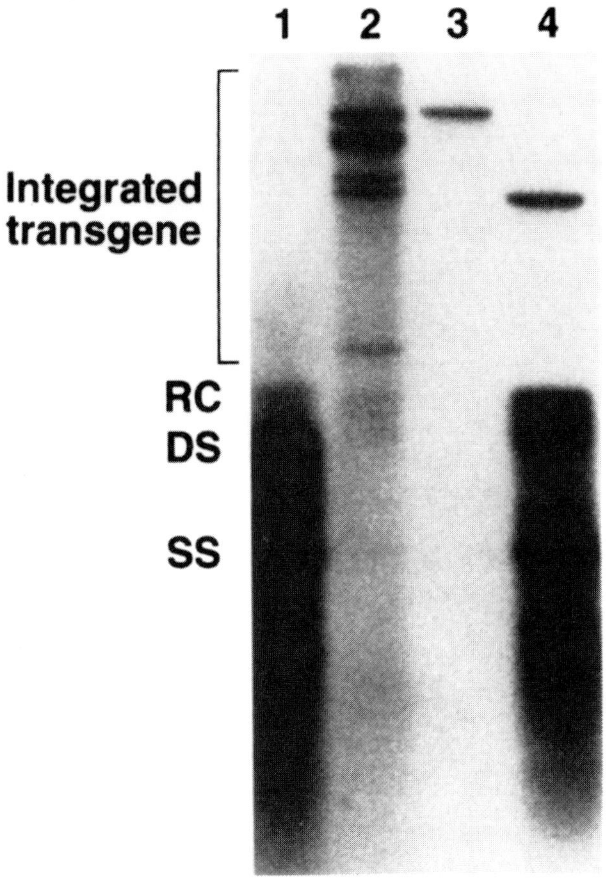

Figure 122.2 Southern blot analysis of 20 µg of total hepatic DNA extracted from chronically hepatitis B virus (HBV)-infected human liver (lane 1), previously generated transgenic-mouse lineage of PC21 (Farza *et al.*, 1988; lane 2), transgenic-mouse lineage 1.2-eq (Figure 122.1), transgenic-mouse lineage 1.2HB-BS (Araki *et al.*, 1989; lane 3), transgenic-mouse lineage 1.3.32 (lane 4). All DNA samples were RNase-treated before gel electrophoresis. The filters were hybridized with a ^{32}P-labeled HBV-specific DNA probe. Bands corresponding to the integrated transgenes were used to normalize the amount of DNA bound to the membrane. Bands corresponding to the expected size of the integrated transgene, relaxed circular (RC), double-stranded linear (DS), and single-stranded (SS) HBV DNAs are indicated. From Guidotti *et al.* (1995), with permission.

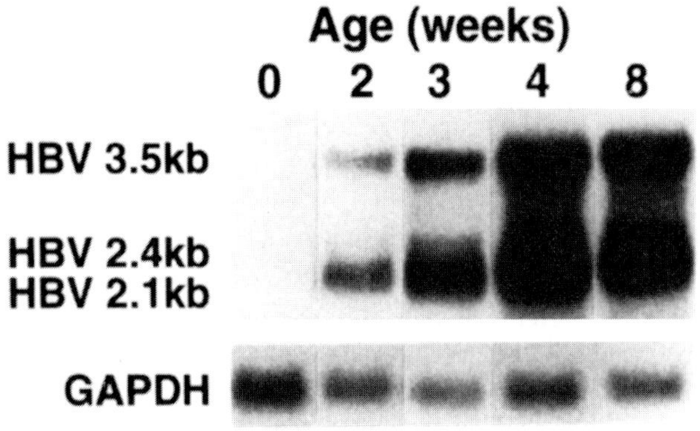

Figure 122.3 Developmental regulation of hepatic hepatitis B virus (HBV) gene expression. Northern blot analysis of 20 µg of total hepatic RNA extracted from lineage 1.3.32. The filters were hybridized with HBV- and glyceraldehyde-3-phosphate dehydrogenase (GAPDH)-specific probes. The results obtained from a representative transgenic mouse at each time point are indicated. The house-keeping GAPDH gene was used to normalize the amount of RNA bound to the membrane. From Guidotti *et al.* (1995), with permission.

is similar to the chronically infected human liver HBV DNA (lane 1), however, and it is quite dissimilar to the HBV DNA from the previously generated transgenic mice (lanes 2 and 3). The types, abundance, and ratios of HBV RNA transcripts, shown by Northern blot analysis (Figure 122.3; Guidotti *et al.*, 1995), are also remarkably similar to that found in normal HBV infection (Ganem, 1996b). The presence of 3.5 kb RNAs in the tissues of transgenic mice is important for the value of this model, because a 3.5 kb RNA viral genome is necessary for assembly of the Dane particle. The 3.5 kb RNA species are also the messenger RNA transcripts for the nucleocapsid proteins (HBeAg and HBcAg) and are reverse-transcribed to produce the first strand of viral DNA during viral replication (Summers *et al.*, 1982). The presence of 3.5 kb RNAs correlates with the presence of virions in the serum of these mice (Guidotti *et al.*, 1995). Conversely, the absence of this RNA species correlates with the absence of such virions in previously generated transgenic mice (Farza *et al.*, 1988; Araki *et al.*, 1989). In both transgenic mice and normal viral replication, the 3.5 and 2.1 kb transcripts are more abundant than the 2.4 or 0.7 kb transcripts. This is similar to normal infection in which the 2.4 kb transcript is present as a minor RNA species and the 0.7 kb transcript is not readily detectable by Northern blot analysis (Ganem, 1996b). As a consequence of these RNA data, it is not surprising to find viral particles in the serum of these transgenic mice that are indistinguishable from human Dane particles.

The core/precore antigens (HBc/eAg) are readily detectable in the livers or kidneys by Western blot analysis or immunocytochemistry (Guidotti *et al.*, 1995). Even though Western blot analysis can yield quantitative data, the qualitative immunocytochemistry might be more instructive, depending on the molecular target of the antiviral drug. For example, intraperitoneal interleukin-12 treatment eliminated HBcAg within the cytoplasm of centrilobular hepatocytes, but not within the nuclei of those same cells (Cavanaugh *et al.*, 1997). In the same study, Western blot results were not reported, but one might anticipate that homogenized tissues prepared for Western analysis may not yield the same stark results as the immunocytochemistry if the cytoplasm and nuclei extracts were not delineated.

Antiviral therapy

Cytokines and nucleoside analogs have been used in limited studies in the transgenic mice of this review (Cavanaugh *et al.*, 1997; Guidotti *et al.*, 1997) and in previously generated transgenic mice not expressing the complete infectious virus (Gilles *et al.*, 1992a,c; Nagahata *et al.*, 1992; Guilhot *et al.*, 1993; Guidotti *et al.*, 1994). The purpose of only one (Nagahata *et al.*, 1992) of these studies was directly to test the efficacy of antiviral agents in earlier generated transgenic mice. In the remainder of the studies, the agents were used to address basic virological events and not primarily to develop a well-defined chemotherapeutic model. Nevertheless, such studies have shown that chemotherapeutic agents can reduce or eliminate viral parameters in transgenic mice. In the earliest therapeutic study (Nagahata *et al.*, 1992), the nucleoside analog, oxetanocin G (30 mg/kg/day, i.p., bid for 7 days) and mouse interferon-α (400 000 units/kg/day, i.p., bid for 3 days) markedly reduced HBV DNA intermediates in the liver. Adenine arabinoside was not efficacious. These were preliminary studies since small number of animals (2–3 per group) and limited dosages and routes of administration were evaluated.

Chisari's group (Guilhot *et al.*, 1993; Guidotti *et al.*, 1994a, 1997; Cavanaugh *et al.*, 1997) have done elaborate mechanistic studies to determine the roles of cytotoxic T and effector cells in reducing viral load that involves T-cell killing or non-cytotoxic elimination of viral RNAs and prevention of nucleocapsid assembly. In the process of these experiments, they determined that interleukin-12 (Cavanaugh *et al.*, 1997), interleukin-2 (Guilhot *et al.*, 1993, 1994b), interferon-α/β (Guidotti *et al.*, 1994b), interferon-γ (Gilles *et al.*, 1992c), tumor necrosis factor α (Gilles *et al.*, 1992a) and lipopolysaccharide (Gilles *et al.*, 1992a,c) down-regulate viral load or gene expression.

HBV DNA detection by PCR

Since antiviral therapy can significantly lower HBV serum DNA levels, PCR-based analysis can be valuable to follow accurately the progress of viremia, as illustrated in a chemotherapeutic study described in a subsequent section. In that experiment, a 360 bp region was chosen within the core gene sequence of the HBV genome to be amplified using PCR. This region is highly conserved among published HBV isolates and is also highly conserved between HBV and the woodchuck hepatitis virus (WHV). Of several primer sets examined that targeted various parts of the viral genome, the primer set used for these experiments consistently gave the highest sensitivity in detecting the presence of virus using PCR in our experience (B. Korba, unpublished observations). The DNA primers utilized for these analyses were: CATTGTTCACCTCACCATACTGCAC ("forward primer", initiation site 2041) and GATTGAGACCTTCGTCTGCGAG ("backward primer", initiation site 2411).

Mouse serum samples were analyzed in duplicate. HBV DNA was prepared for amplification by extraction using GeneReleaser matrix (Bioventures, Murfreesboro, TN). For the extraction step, 5 μl of serum and 15 μl of GeneReleaser were added to 0.2 ml PCR reaction tubes and incubated in the thermocycler (Touchdown, Hybaid, Middlesex, UK) using the following program: 37°C, 30 seconds/ 8°C, 30 seconds/ 65°C, 90 seconds/ 97°C, 180 seconds/8°C, 60 seconds/ 65°C, 180 seconds/ 97°C, 60 seconds/ 65°C, 60 seconds/ 80°C, 30 minutes. Thus, only one glove change per serum sample was required for the lysis/DNA extraction step and the chance of loss due to pipetting errors during multiple organic extraction steps was eliminated. Following the lysis incubation procedure, the remaining components of the PCR reaction mixture were added (final volume of 40 μl; Kaneko *et al.*, 1989), and the reaction tubes were subjected to the following amplification program: 94°C, 2 minutes/ 94°C, 1 minute/ 55°C, 1 minute/ 72°C, 1 minute/ (repeat steps 2–4, 39 times)/ hold at 4°C. Since the thermocycler was equipped with a heated lid, the need for oil or "quick-start" style reaction tubes was eliminated.

Following PCR amplification, 60 μl distilled, sterile, filtered water was added to each tube. Following 1 minute of centrifugation (15 000 *g*), the contents were applied to nitrocellulose membranes using a 96-well dot blot manifold (GIBCO-BRL, Gaithersburg, MD) and hybridized to a ^{32}P-labeled, 3.2 kb cloned HBV DNA fragment as previously described (Korba *et al.*, 1986). HBV DNA content in the mouse sera were quantitated by comparison to a dilution series of chronic HBV carrier chimpanzee serum that contained a previously determined concentration of HBV DNA using an InstantImager β scanner (Packard Instrument, Downers Grove, IL). This standard series, as well as appropriate negative control samples, was PCR-amplified and dot-blotted, in duplicate, each time the mouse sera were analyzed. The sensitivity of this HBV DNA detection procedure was approximately 500 HBV genome equivalents/ml serum.

Key parameters to monitor response to treatment

Since the mice of this review do not display death or pathology, only viral parameters can be monitored in response to antiviral therapy. An experiment was devised to help validate these transgenic mice as models for therapeutic intervention. HBV and all retroviruses have reverse transcriptase activities so nucleoside analogs could conceivably be efficacious for both these viruses. However, the nucleoside analog zidovudine (AZT) does not reduce HBV infection in humans, but it is highly efficacious for reducing human immunodeficiency virus (HIV) titers in humans (Clumeck, 1993), and retrovirus titers in mice (Morrey *et al.*, 1990, 1991b; Ruprecht *et al.*, 1990) and in other animal species (Fazely *et al.*, 1991; Tavares *et al.*, 1997). The nucleoside analog, (-)2′-deoxy-3′thiacytidine (3TC) has previously been shown to be effective for both treatment of HBV (Ling *et al.*, 1996; Schnittman and Pierce, 1996), and HIV infections (Hart *et al.*, 1992; Clumeck, 1993). Therefore, we tested the validity of this transgenic model by evaluating both AZT and 3TC in these mice. Male and female transgenic mice were also monitored to determine any differences or disadvantages between the sexes. Fertilized embryos from transgenic mice (1.3.32) were obtained from Dr Frank Chisari (Scripps Institute, La Jolla, CA) to establish a colony of the animals at Utah State University.

The 3TC (Dr Bud Tennant, Cornell University, Ithaca, NY) was prepared in sterile physiological saline. The AZT (Glaxo-Wellcome, Research Triangle Park, NC) was prepared in sterile double-distilled water at a concentration of 0.2 mg/ml. Both compounds were stored at 4°C until used. The male or female transgenic mice were treated *per os* (p.o.) twice a day for 21 days with 3TC (100 mg/kg/day). The AZT was administered *ad libitum* via the drinking water, which was changed daily. Previous studies (Morrey *et al.*, 1991b) indicated that the mice would drink sufficient water to achieve a dosage of about 22 mg/kg per day. A group of 8–13 mice were included in each treatment, depending on the sex and treatment. Blood was collected from the tails weekly for 4 weeks. As a control, transgenic mice treated p.o. with saline were run in parallel. The viral parameters of individual mice were recorded so that the effect of drug could be monitored over time for each individual animal. Differences between mean values were analyzed using the *t*-test. Standard deviations were also determined.

Figure 122.4 graphically represents the anti-HBV effect of 3TC in both sexes of transgenic mice. The results and conclusions were essentially the same for each sex. At days 14 and 21 after initial treatment, 3TC treatment was efficacious in reducing HBV DNA titers. Clear reduction in titers occurred as the treatments progressed from days 7 through 21.

AZT was administered in parallel with the 3TC at a concentration shown to be highly efficacious in Friend virus-infected mice (Morrey *et al.*, 1990, 1991b). As might be predicted, however, AZT treatment did not affect HBV

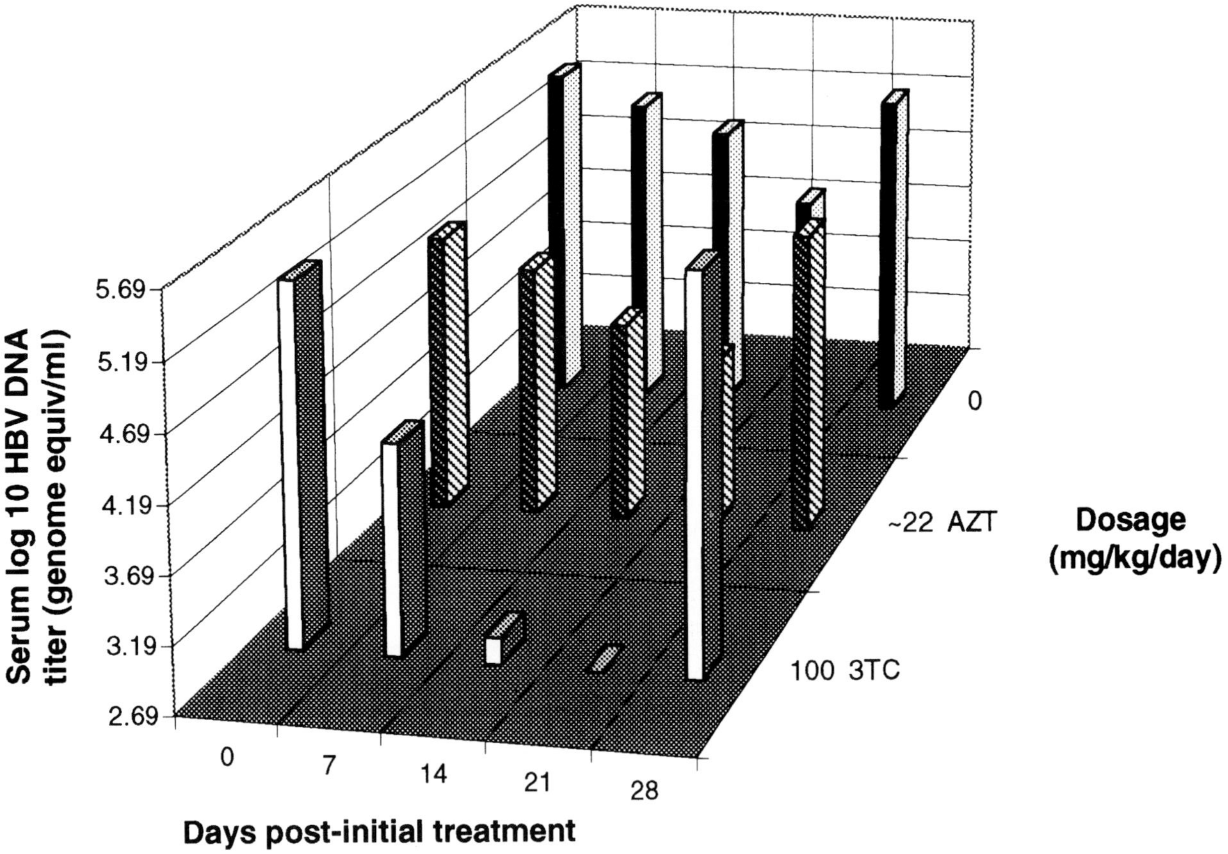

Figure 122.4 Effect of orally treated (-)2′-deoxy-3′thiacytidine (3TC) on serum hepatitis B virus (HBV) DNA in transgenic mice during and after last treatment at 21 days. 3TC (100 mg/kg per day) was administered p.o. b.i.d. for 21 days, at which time treatment ceased. Serum was assayed for HBV DNA by quantitative polymerase chain reaction every 7 days until day 28, 1 week after the last treatment. AZT (0.2 mg/ml or about 22 mg/kg per day) was administered in parallel with 3TC in the drinking water for 21 days.

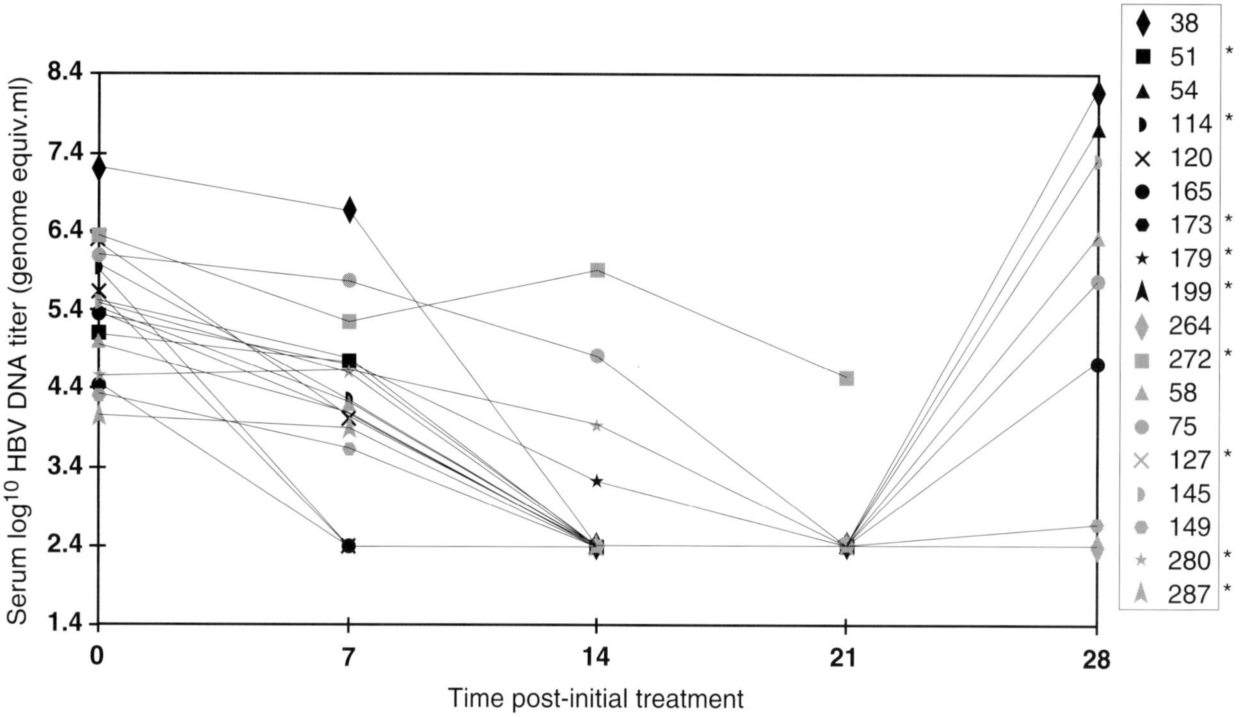

Figure 122.5 Effect of orally treated 3TC on individual hepatitis B virus (HBV) transgenic mice. Refer to legend in Figure 122.4 for experiment description. The serum HBV DNA titers were plotted for individual mice of both sexes. The identification numbers for the mice are shown in the legend. * Died or were sacrificed during the treatment.

DNA titers in transgenic mice when the titers were compared statistically to the values at day 0 of the same group. There was some diminution of titers as the experiment progressed, but this reduction was the same as seen with the saline-treated animals and was not nearly as profound as seen with 3TC-treated animals.

We found it particularly instructive to trace the antiviral effects of 3TC in each individual mouse, as shown in Figure 122.5. By doing this, it was found that the detection of positive mean HBV DNA titer at day 21 was actually due to a single positive mouse (#264). The HBV DNA titers of all other mice (n = 9) were below detectable limits. After treatments were terminated, the HBV DNA titers had rebounded 7 days later, which was the next assay period. This rebound effect is also seen after cessation of 3TC treatment in human subjects (Marinos *et al.*, 1996), although not necessarily with the same kinetics. In conclusion, the results of this transgenic mouse experiment were similar to what would be predicted from treatment of HBV-infected human patients with AZT and 3TC; such data help to validate this transgenic mouse model.

Variability of the sexes

HBsAg in transgenic mice was found to be regulated by sex steroids and glucocorticoids (Farza *et al.*, 1987). This raises the possibility that the female murine values might vary due to hormone fluctuations. We did observe high variability in both the serum antigen and HBV DNA titers in female mice, but we also observed that male transgenic mice had the same HBV DNA mean and variability (4.2 ± 1.0, mean serum titer in $\log_{10} \pm$ SD, $n=13$) as the female mice (4.3 ± 1.0, $n=13$). The variability of the serum HBsAg titers were also essentially the same between the males (0.90 OD ± 0.82, 1:30 dilution) and females (1.0 OD ± 0.85). The mean of the serum HBeAg titers in the females (0.36 OD ± 0.15, 1:30 dilution) were actually lower ($P<0.01$) than the mean of male animals (0.62 OD ± 0.22, 1:30 dilution). These data indicate that females are just as suitable for HBV chemotherapeutic studies as males in terms of variability. The fact that females are used is fortuitous because they do not need to be housed in separate cages, like the males. This is a great benefit when experiments can consist of greater than 100 mice.

Pitfalls (advantages/disadvantages) of the model

The HBV transgenic mouse model of this review has advantages and disadvantages as compared with woodchucks, ducks and squirrels infected with their respective animal viruses, and which are currently used as small animal models. The following are the advantages of the transgenic mouse model used in this review:

- The transgenic mouse model uses the human virus as compared to animal hepatitis viruses (Ganem, 1996b). The human virus may respond differently to certain antiviral agents than the animal hepatitis viruses currently used.
- Low cost for animal maintenance: in our animal facility at time of publication, *per diem* rate for BL3-N mice is \$0.15/day. The *per diem* rate for the woodchuck is \$2.00/day, or a 14-fold increase.
- Less drug is required for treatment of mice as compared with larger animals.
- Once an HBV transgenic mouse colony is established, large numbers of the animals could be used in an experiment, providing greater statistical value.
- Breeding of mice is not seasonal and gestation is short (3 weeks).
- Inbred mice have homogenous genetics for immunological and virological experimentation.
- HBV is expressed at appreciable levels in the serum of the transgenic mice of this review (Guidotti *et al.*, 1995), but it is not found at such levels in previously generated HBV-transgenic mice (Farza *et al.*, 1987; Araki *et al.*, 1989; Guidotti *et al.*, 1994).
- Chisari's group (Guilhot *et al.*, 1993; Guidotti *et al.*, 1994a, 1997; Cavanaugh *et al.*, 1997) have done elaborate mechanistic studies to determine that virus load can be reduced through T-cell killing, or non-cytotoxic elimination of viral RNAs and prevention of nucleocapsid assembly. Sophisticated immunological studies are possible by employing the knowledge of mouse immunogenetics.

Some disadvantages are listed below:

- No HBV-induced pathology or death is observed with these animals (Guidotti *et al.*, 1995).
- In strict terms, the transgenic mouse model is not an infectious model. The animals do not require inoculation with virus since the viral transgene is integrated into the mouse genome.
- Supercoiled DNA intermediate (cccDNA) is not present (Guidotti *et al.*, 1995).
- Greater biological containment is required for HBV transgenic mice (1994c).
- Immunological studies may be limited since the mice are relatively immunotolerant to HBV proteins (Milich *et al.*, 1994).

Contributions of the model to infectious disease therapy

This transgenic mouse model that expresses the Dane-like particle is relatively new: the first report was in 1995 (Guidotti *et al.*, 1995). Consequently, the mice are currently being more fully developed as an HBV-chemotherapeutic model. HBV transgenic mice have contributed to date in the following important areas:

- Determinants of HBV host range and tissue specificity
- Assembly, transport and secretion of HBV proteins
- Pathological potential of viral proteins
- Lack of transport of capsids in or out of nuclei
- Pathologic potential of cellular immune response to HBV
- Non-cytopathic antiviral potential of immune response
- Potential mechanisms of hepatocellular carcinoma involving X protein or surface antigens, which are reviewed elsewhere (Chisari, 1995a,b, 1996).

One anticipates that as the HBV transgenic mouse becomes more fully developed as a chemotherapeutic model, questions about efficacy of different agents, routes of administration, synergy of antiviral combinations, and novel drug therapies will be determined.

Acknowledgement

We thank Francis Chisari and Luca Guidotti at The Scripps Research Institute for providing embryos from which to generate transgenic mice. This work was supported by contract No1-A1-65291 from the Virology Branch, National Institute of Allergy and Infectious Diseases, National Institutes of Health.

References

(1993a). *Biosafety in Microbiological and Biomedical Laboratories*. US Government Printing Office, Washington, DC pp. 106–107.

(1994b) Appendix B-II-D. Class 2 viral, rickettsial, and chlamydial agents. *Guidelines for Research Involving Recombinant DNA Molecules* (NIH Guidelines). United States Government, Washington, DC.

(1994c). Section III-C-4. Experiments involving whole animals. *Guidelines for Research Involving Recombinant DNA Molecules* (NIH Guidelines). United States Government, Washington, DC.

(1996d). *Guide for the Care and Use of Laboratory Animals*. National Academy Press, Washington, DC.

Ando, K., Moriyama, T., Guidotti, L. G. *et al.* (1993). Mechanisms of class I restricted immunopathology: a transgenic mouse model of fulminent hepatitis. *J. Exp. Med.*, **178**, 1541–1554.

Ando, K., Guidotti, L. G., Cerny, A., Ishikawa, T., Chisari, F. V. (1994a). CTL access to tissue antigen is restricted *in vivo*. *J. Immunol.*, **153**, 482–488.

Ando, K., Guidotti, L. G., Wirth, S. *et al.* (1994b). Class I-restricted cytotoxic T lymphocytes are directly cytopathic for their target cells *in vivo*. *J. Immunol.*, **152**, 3245–3253.

Araki, K., Miyazaki, J.-I., Hino, O. *et al.* (1989). Expression and replication of hepatitis B virus genome in transgenic mice. *Proc. Natl Acad. Sci. USA*, **86**, 207–211.

Cavanaugh, V. J., Guidotti, L. G., Chisari, F. V. (1997). Interleukin-12 inhibits hepatitis B virus replication in transgenic mice. *J. Virol.*, **71**, 3236–3243.

Chisari, F. V. (1995a). Hepatitis B virus transgenic mice: insights into the virus and the disease. *Hepatology*, **22**, 1316–1325.

Chisari, F. V. (1995b). Hepatitis B virus immunopathogenesis. *Annu. Rev. Immunol.*, **13**, 29–60.

Chisari, F. V. (1996). Hepatitis B virus transgenic mice: models of virus immunobiology and pathogenesis. *Curr. Topics Microbiol. Immunol.*, **206**, 149–173.

Chisari, F. V., Filippi, P., McLachlan, A. *et al.* (1986). Expression of hepatitis B virus large envelope polypeptide inhibits hepatitis B surface antigen secretion in transgenic mice. *J. Virol.*, **60**, 880–887.

Chisari, F. V., Klopchin, K., Moriyama, T. *et al.* (1989). Molecular pathogenesis of hepatocellular carcinoma in hepatitis B virus transgenic mice. *Cell*, **59**, 1145–1156.

Clumeck, N. (1993). Current use of anti-HIV drugs in AIDS. *J. Antimicrob. Chemother.*, **32** (suppl. A), 133–138.

DeLoia, J. A., Burk, R. D., Gearhart, J. D. (1989). Developmental regulation of hepatitis B surface antigen expression in two lines of hepatitis B virus transgenic mice. *J. Virol.*, **63**, 4069–4073.

Farza, H., Salmon, A. M., Hadchouel, M. *et al.*, (1987). Hepatitis B surface antigen gene expression is regulated by sex steriods and glucocorticoids in transgenic mice. *Proc. Natl Acad. Sci. USA*, **84**, 1187–1191.

Farza, H., Hadchouel, M., Scotto, J., Tiollais, P., Babinet, C., Pourcel, C. (1988). Replication and gene expression of hepatitis B virus in a transgenic mouse that contains the complete viral genome. *J. Virol.*, **62**, 4144–4152.

Fazely, F., Haseltine, W. A., Rodger, R. F., Ruprecht, R. (1991). Postexposure chemoprophylaxis with ZDV combined with interferon-alpha: failure after inoculating rhesus monkeys with a high dose of SIV. *J. AIDS*, **4**, 1093–1097.

Ganem, D. (1996a). *Hepadnaviridae* and their replication. In *Field's Virology* (eds Fields, B. N., Knipe, D. M., Howeley, P. M.), p. 2709. Lippencott-Raven, Philadelphia.

Ganem, D. (1996b). *Hepadnaviridae* and their replication. In *Field's Virology* (eds Fields, B. N., Knipe, D. M., Howeley, P. M.), pp. 2703–2727. Lippencott-Raven, Philadelphia.

Ganem, D. (1996c). *Hepadnaviridae* and their replication. In *Field's Virology* (eds Fields, B. N., Knipe, D. M., Howeley, P. M.), pp. 2724. Lippencott-Raven, Philadelphia.

Gilles, P. N., Fey, G., Chisari, F. V. (1992a). Tumor necrosis factor alpha negatively regulates hepatitis B virus gene expression in transgenic mice. *J. Virol.*, **66**, 3955–3960.

Gilles, P. N., Guerrette, D. L., Ulevitch, I. J., Schreiber, R. D., Chisari, F. V. (1992b). Hepatitis B surface antigen retention sensitizes the hepatocyte to injury by physiologic concentrations of gamma interferon. *Hepatology*, **16**, 655–663.

Gilles, P. N., Guerrette, D. L., Ulevitch, R. J., Schreiber, R. D., Chisari, F. V. (1992c). Hepatitis B surface antigen retention sensitizes the hepatocyte to injury by physiologic concentrations of gamma interferon. *Hepatology*, **16**, 655–663.

Guidotti, L. G., Ando, K., Hobbs, M. V. *et al.* (1994a). Cytotoxic T lymphocytes inhibit hepatitis B virus gene expression by a noncytolytic mechanism in transgenic mice. *Proc. Natl Acad. Sci. USA*, **91**, 3764–3768.

Guidotti, L. G., Guilhot, S., Chisari, F. V. (1994b). Interleukin-2 and alpha/beta interferon down-regulate hepatitis B virus gene expression *in vivo* by tumor necrosis factor-dependent and -independent pathways. *J. Virol.*, **68**, 1265–1270.

Guidotti, L. G., Martinez, V., Loh, Y. -T., Rogler, C. E., Chisari, F. V. (1994c). Hepatitis B virus nucleocapsid particles do not cross the hepatocyte nuclear membrane in transgenic mice. *J. Virol.*, **68**, 5469–5475.

Guidotti, L. G., Matzke, B., Schaller, H., Chisari, F. V. (1995). High-level hepatitis B virus replication in transgenic mice. *J. Virol.*, **69**, 6158–6169.

Guidotti, L. G., Ishikawa, T., Hobbs, M. V., Matzke, B., Schreiber, R., Chisari, F. V. (1996). Intracellular inactivation of the hepatitis B virus by cytotoxic T lymphocytes. *Immunity*, **4**, 25–36.

Guidotti, L. G., Matzke, B., Chisari, F. V. (1997). Hepatitis B virus replication is cell cycle independent during liver regeneration in transgenic mice. *J. Virol.*, **71**, 4804–4808.

Guilhot, S., Guidotti, L. G., Chisari, F. V. (1993). Interleukin-2 downregulates hepatitis B virus gene expression in transgenic mice by a posttranscriptional mechanism. *J. Virol.*, **67**, 7444–7449.

Hart, G. J., Orr, D. C., Penn, C. R. *et al.* (1992). Effects of (-)-2′-deoxy-3′-thiacytidine (3TC) 5′-triphosphate on human immunodeficiency virus reverse transcriptase and mammalian DNA polymerases alpha, beta, and gamma. *Antimicrob. Agents Chemother.*, **36**, 1688–1694.

Hollinger, F. B. (1996). Hepatitis B virus. In *Field's Virology* (eds Fields, B. N., Knipe, D. M, Howeley, P. M.), pp. 2739–2807. Lippincott-Raven, Philadelphia.

Kaneko, S., Feinstone, S. M., Miller, R. H. (1989). Rapid and sensitive method for the detection of serum hepatitis B virus DNA using the polymerase chain reaction technique. *J. Clin. Invest.*, **27**, 1930–1933.

Koike, K., Moriya, K., Iino, S. *et al.* (1994). High level expression of hepatitis B virus HBx gene and hepatocarcinogenesis in transgenic mice. *Hepatology*, **19**, 810–819.

Korba, B. E., Wells, F., Tennant, B. C., Yoakum, G. H., Purcell, R. H., Gerin J. L. (1986). Hepadnavirus infection of peripheral blood lymphocytes *in vivo*: woodchuck and chimpanzee models of viral hepatitis. *J. Virol.*, **58**, 1–8.

Lacey, M., Alpert, S., Hanahan, D. (1986). Bovine papillomavirus genome elicits skin tumours in transgenic mice. *Nature*, **322**, 609–612.

Lee, T. H., Finegold, M. J., Shen, R. F., DeMayo, J. L., Woo, S. L. C., Butel, J. S. (1990). Hepatitis B virus transactivator X protein is not tumorigenic in transgenic mice. *J. Virol.*, **64**, 5939–5947.

Leonard, J. M., Abramczuk, J. W., Pezen, D. S. *et al.* (1988). Development of disease and virus recovery in transgenic mice containing HIV proviral DNA. *Science*, **242**, 1665–1670.

Ling, R., Mutimer, D., Ahmed, M. *et al.* (1996). Selection of mutations in the hepatitis B virus polymerase during therapy of transplant recipients with lamivudine. *Hepatology*, **24**, 711–713.

Marinos, G., Naoumov, N. V., Williams, R. (1996). Impact of complete inhibition of viral replication on the cellular immune response in chronic hepatitis B virus infection. *Hepatology*, **24**, 991–995.

Milich, D. R., Jones, J. E., Hughes, J. L. *et al.* (1994). Extrathymic expression of the intracellular hepatitis B core antigen results in T cell tolerance in transgenic mice. *J. Immunol.*, **152**, 455–466.

Morrey, J. D., Warren, R. P., Burger, R. A., Okleberry, K. M., Johnston, M. A., Sidwell, R. W. (1990). Effects of zidovudine on Friend virus complex infection in Rfv-3r/s genotype-containing mice used as a model for HIV infection. *J. AIDS*, **3**, 500–510.

Morrey, J. D., Bourn, S. M., Bunch, T. D. *et al.* (1991a). *In vivo* activation of human immunodeficiency virus type 1 long terminal repeat by UV light type-A (UV-A) light plus psoralen and UV-B light in skin of transgenic mice. *J. Virol.*, **65**, 5045–5051.

Morrey, J. D., Okleberry, K. M., Sidwell, R. W. (1991b). Early-initiated zidovudine therapy prevents disease but not low levels of presistent retrovirus in mice. *J. AIDS*, **4**, 506–512.

Murphy, D. (1993). Isolation of genomic DNA from tail tissue. In *Transgenesis Techniques: Principles and Protocols* (eds Murphy, D., Carter, D. A.), p. 311. Humana Press, Totowa.

Nagahata, T., Araki, K., Yamamura, K.-I., Matsubara, K. (1992). Inhibition of intrahepatic hepatitis B virus replication by antiviral drugs in a novel transgenic mouse model. *Antimicrob. Agents Chemother.*, **36**, 2042–2045.

Perumo, S., Amicone, C., Colloca, S., Giorgio, M., Pozzi, I., Tripodi, M. (1992). Recognition efficiency of the hepatitis B virus polyadenylation signals is tissue specific in transgenic mice. *J. Virol.*, **66**, 6819–6823.

Ruprecht, R. M., Chou, T.-C., Chipty, F. *et al.* (1990). Interferon-α and 3′-azido-3′-deoxythymidine are highly synergistic in mice and prevent viremia after acute retrovirus exposure. *J. AIDS*, **3**, 591–600.

Schnittman, S. M., Pierce, P. F. (1996). Potential role of lamivudine (3TC) in the clearance of chronic hepatitis B virus infection in a patient coinfected with human immunodeficient virus type. *Clin. Infect. Dis.,* **23**, 638–639.

Summers, J., Smith, P. M., Horwich, A. L. (1982). Replication of the genome of a hepatitis B-like virus by reverse transcription of an RNA intermediate. *Cell*, **29**, 403–415.

Tavares, L., Roneker, C., Johnson, K., Nusinoff-Lehrman, S., de Noronha, F. (1997). 3′-Azido-3′-deoxythymidine in feline leukemia and lyphocyte decline but not primary infection in feline immunodeficiency virus-infected cats: a model for therapy and prophylaxis of AIDS. *Cancer Res.*, **47**, 3190–3194.

Will, H., Cattaneo, R., Darai, G., Deinhardt, F., Shcellekens, H., Schaller, H. (1985). Infectious hepatitis B virus from cloned DNA of known nucleotide sequence. *Proc. Natl Acad. Sci. USA*, **82**, 891–895.

Chapter 123

Animal Models for Hepatitis B Infections — Duck Hepatitis

T. Shaw, C. A. Luscombe and S. A. Locarnini

Background of human infection

Infection with hepatitis B virus (HBV) is currently a major public health problem, being endemic in many different parts of the world. Although effective anti-HBV vaccines have been developed, they are of no benefit to individuals who are already persistently infected with HBV. Recent estimates place the global population of HBV carriers at close to 350 million, or about 5% of the world's total human population. These individuals constitute a large reservoir of potentially infectious virus. Although variable and determined partly by the host's genetic make-up, age of infection, and partly by environmental factors, sequelae of uncontrolled HBV infection may include chronic active hepatitis, cirrhosis, and primary hepatocellular carcinoma (see Hollinger, 1996, for a comprehensive review). Unfortunately, antiviral drugs which are able to eliminate completely chronic HBV infection have yet to be developed, and those available for its control were, until very recently, only partially effective (for review see Shaw and Locarnini, 1995). There is therefore a need for screening systems which are able rapidly to identify safer and more effective antihepadnaviral drugs.

Early studies of HBV were hampered not only by the lack of laboratory animal and cell culture systems able to support productive HBV infection *in vivo* and *in vitro* respectively, but also by potential hazards and ethical problems associated with attempts to develop such systems. Primary cultures of fetal or adult human hepatocytes can be productively infected with HBV *in vitro,* and some primate species, including chimpanzees (see Chapter 131) and tree shrews (Walter *et al.*, 1996), are susceptible to experimental *in vivo* infection with HBV, but cost, ethical and safety considerations preclude their routine use.

Background of model

Summers and colleagues first described a virus similar to HBV which they found associated with hepatitis and hepatoma in woodchucks (Summers *et al.*, 1978). Other workers subsequently identified HBV-like animal viruses in ground squirels (Marion *et al.*, 1980), ducks (Mason *et al.*, 1980), tree squirrels (Feitelson *et al.*, 1986), grey herons (Sprengel *et al.*, 1988) and arctic ground squirrels (Testut *et al.*, 1996). It seems likely that further members of this family, the Hepadnaviridae, await discovery in other species. All hepadnaviruses have circular genomes in the 3.0–3.2 kilobase pair range, making them the smallest double-stranded DNA viruses of vertebrates, and all share several common features, including hepatotropism, restricted host range and similar virion structure, genome organization and replication strategy (reviewed by Ganem, 1996). The unusual replication strategy of the hepadnaviruses was first described by Summers and Mason (1982) based on studies of the *in vivo* replication of duck HBV (DHBV). In contrast to most double-stranded DNA genomes, which replicate semi-conservatively, hepadnaviral genomes are reverse-transcribed from an intracellular genome-length RNA intermediate before being replicated by the multifunctional viral polymerase (for review see Ganem *et al.*, 1994). Some other features of hepadnaviral replication are also unusual and may present targets for specific inhibition by antiviral agents (see below; also see Nassal and Schaller, 1996). Partly for historical reasons, the two most popular animal models for HBV infection are infection by HBV viruses of woodchucks (see Chapter 124) and ducks (for earlier review see Sprengel and Will, 1988).

Ducks may be chronically or acutely infected with DHBV. Chronic infection is usually associated with mild hepatitis which rarely progresses to cirrhosis or hepatocellular carcinoma in the absence of environmental risk factors (Omata *et al.*, 1983; Marion *et al.*, 1984; Cullen *et al.*, 1990), a situation which mimicks the long-lasting healthy carrier state for chronic human HBV infection. Acute DHBV infection is usually self-limiting and also resembles the human HBV counterpart (Jilbert *et al.*, 1992; Vickery and Cossart, 1996).

Animal species

Many subspecies of duck derived from the mallard (*Anas platyrhnchos*) are known to be susceptible to DHBV

Handbook of Animal Models of Infection
ISBN 0-12-775390-7

infection, including domestic breeds such as the Pekin and Pekin-Aylesbury, which are readily obtainable from commercial breeders (Mason *et al.*, 1983; Marion *et al.*, 1984; Tsiquaye *et al.*, 1985a,b). Some members of the genus *Anser*, which includes domestic and wild geese, can also be experimentally infected with DHBV (Marion *et al.*, 1987; Sprengel and Will, 1988).

The incidence of natural DHBV infection of different domestic duck flocks appears to vary considerably with geographical location but may approach 60% (Sprengel *et al.*, 1984; Freiman and Cossart, 1986). Up to 12% of ducks in predominantly uninfected domestic and wild flocks have also been found to be harbor DHBV (Mason *et al.*, 1983; Cova *et al.*, 1986), but it is possible to establish true DHBV-free breeding colonies as a source of eggs and ducklings which are susceptible to acute infection (Mason *et al.*, 1983). Congenital DHBV infection is more efficient than acute infection, and congenitally DHBV-infected ducklings from a single supplier are preferred for antiviral studies, since they can be expected to have a consistent viral and genetic background. Overt morbidity associated with DHBV carriage in naturally infected ducks has been reported, but histological evidence of mild to moderately severe hepatitis has been found only in a minority of infected birds; furthermore, direct correlations between the severity of hepatitis and the severity of viremia are unusual (Omata *et al.*, 1983; Freiman and Cossart, 1986). Antiviral treatment of congenitally infected ducklings is usually started at about 5 or 6 weeks post-hatch, when growth rates, body weight, and viremia begin to stabilize. Younger ducklings (2–3 weeks post-hatch) are preferred as a source of primary hepatocytes (see below).

Housing and preparation of animals

Ducklings younger than 1 week post-hatch must be kept warm at about 35°C (90°F). Commercial or improvised cages warmed by heat lamps and carpeted with wood shavings are suitable for housing ducklings until they are old enough to be transferred to outdoor facilities. Provided that the climate is temperate, the transfer can be made at about 2 weeks post-hatch. Ducklings can be fed commercial dried poultry feed supplemented with fresh green vegetables. Uninfected and infected and treated and untreated ducklings should be housed separately and cages of treated ducks should be cleaned frequently to avoid drug re-ingestion. Water containers deep enough to allow head-dunking (total immersion of ducks' heads) should be provided, since ducks need to be able to head-dunk regularly to clear their eyes and respiratory passages. In geographical locations subject to climatic extremes, provision should also be made to regulate temperature and humidity. There are no specialized housing or handling requirements other than these.

Details of surgery

Surgical procedures are rarely required during routine short-term antiviral studies in live ducklings. However, during long-term studies (Mason *et al.*, 1994; Luscombe *et al.*, 1996) it may be desirable to obtain sequential liver biopsies from individual birds. Detailed procedures for duck liver biopsy have been described by Carp *et al.* (1991). For most studies, it is usually adequate to sample representative individuals from different treatment groups at various time points during each study. Ducklings are anesthetized and killed humanely by lethal intravenous injection with phenobarbital, after which the peritoneal cavity is opened and the condition of abdominal organs assessed by inspection before selection and removal of tissue samples for subsequent processing.

Bile duct ligation

Particular aspects of pathogenesis can be readily studied in the duck model, such as biliary hyperplasis produced by bile duct ligation (Nicoll *et al.*, 1997). The technique for general anesthesia in this situation is described by Nicoll and Koppinen (1996).

Preparation and storage of inocula

Ducklings congenitally infected with DHBV (usually of uncharacterized genotype) are frequently used for *in vivo* antiviral experiments, but high-efficiency persistent infection with virus of defined genotype can readily be achieved provided the ducklings are inoculated before they are a week old (Tsiquaye *et al.*, 1985a,b; Vickery and Cossart, 1996). Several different naturally occurring DHBV strains have been sequenced (Wildner *et al.*, 1991) and infectious clones produced (Horwich *et al.*, 1990; Lenhoff and Summers, 1994; Sprengel *et al.*, 1984). However, strain-related differences in the histological and virological progression of infection, if they occur, have not been documented. Although there are no established interlaboratory standards, intralaboratory strain use can be assumed to be standardized, since individual groups tend to be consistent in their choice of animal supply and viral inocula. Useful sources of DHBV inocula include sera obtained from infected ducklings and medium collected from cultures of DHBV-infected primary duck hepatocytes (Tuttleman *et al.*, 1986a) or DHBV-transfected hepatoma cells such as the avian LMH line (Condreay *et al.*, 1990; Gong *et al.*, 1996) or the human HepG2 and HuH-7 lines, which may also be productively transfected with DHBV (Galle *et al.*, 1988; Horwich *et al.*, 1990; von Weizsacker *et al.*, 1996).

Virus particles may be aggregated with 10% (w/v) polyethylene glycol and concentrated by ultracentrifugation through a 20% sucrose cushion. Viral titers are routinely estimated by slot-blot or spot-blot hybridization in parallel with a standard dilution series containing known amounts of DHBV DNA, assuming that each picogram of DHBV DNA contains approximately 3.3×10^6 viral genome equivalents (Jilbert *et al.*, 1988). Titration in primary duck hepatocytes using immunofluoresence or plaque assays (Tuttleman *et al.*, 1986a; Davis, 1987) to detect infectious foci is an alternative, more time-consuming method by which the number of infectious particles in viral stocks may be calculated. Virus stocks should be stored at –70°C and resuspended in sterile phosphate-buffered saline for injection.

Routes of inoculation for *in vivo* infection

Infection is most easily initiated by intravenous (i.v.) or intraperitoneal (i.p.) injection, although direct injection of DHBV or naked DHBV DNA into the liver has also been used (Mason *et al.*, 1983; Sprengel *et al.*, 1984; Tagawa *et al.*, 1996). The jugular vein or heart are preferred sites for intravenous inoculation of young ducklings, because, although leg veins can also be used, their small size makes injection difficult. The technically simplest method, intraperitoneal inoculation, is performed by pinching the ventral skin over the peritoneal cavity and lifting it to form a small intraperitoneal cavity into which the inoculum is injected. Care is required to avoid perforation or laceration of abdominal viscera. To ascertain that the viral inoculum is delivered into the intraperitoneal cavity, the syringe plunger should be withdrawn slightly before injection to check for absence of blood or intestinal contents. Much larger inoculum volumes (up to 5 ml) can be delivered more safely by i.p. injection than by i.v. injection.

Key parameters to monitor infection and response to treatment

The same markers are used to monitor infection and response to treatment. Ducklings congenitally infected with DHBV for use in antiviral studies should be selected on the basis of moderate, stable and persistent viremia which is routinely monitored by slot-blot or spot-blot DNA hybridization at regular intervals pre-treatment. Successful *de novo* DHBV infection is likewise indicated by the development of persistent viremia, the rapidity of onset of which is directly related to the concentration of infectious particles in the inoculum (Jilbert *et al.*, 1996; Vickery and Cossart, 1996). Uninfected and DHBV-infected ducklings' appearance, as well as food and water intake, should be similar, and it is useful to record ducklings' body weights at regular intervals before, during and after drug therapy. Differences in body weight between normal age-matched individuals of the same sex and strain reared under identical conditions are invariably small. Failure of treated ducklings to gain weight, or losses in body weight in treated groups relative to untreated matched controls is therefore indicative of abnormal development, underlying infection, drug-related toxicity or other problems.

Deaths due to treatment or infection are rare and can usually be avoided by careful monitoring for signs of ill health. In order to gauge the significance of therapy-related changes it is essential to establish baseline levels of viremia and other markers in all individual ducklings before experimental therapy begins, which requires that individual ducklings are readily distinguishable. Coded leg bands and/or indelible head markings are usually used to identify individual animals.

During antiviral studies, regular (weekly or more frequent) blood samples should be collected to monitor viremia (as above) and to check for possible drug-related toxicity. Before feather growth begins, blood samples can easily be obtained from the jugular vein of young ducklings without causing trauma; older, larger birds can be easily bled from wing or leg veins. Cardiac bleeds can be performed at any age. To prepare sera, blood should be collected into plain tubes and allowed to clot for several hours or overnight before centrifugation to remove residual cells. Sera can be kept for short periods at –20°C but should otherwise be stored at –70°C. At regular intervals (at least monthly), additional blood samples should also be taken into appropriate anticoagulant(s) to check for possible drug-related effects on renal, hepatic and hematopoietic function (Lin *et al.*, 1996). Suitable tests may already be routinely available in many laboratories but such routine diagnostic procedures and reference values are likely to differ between laboratories. Procedures which have been developed for analyses of human blood and plasma may need modification in order to produce meaningful results. Many published procedural modifications and reference values are available for guidance (Blackmore, 1988; Dieterlen-Lievre, 1988).

Significant and sustained dose- and time-dependent reductions in viremia in the absence of apparent or measurable toxicity is regarded as initial evidence of potentially useful *in vivo* antiviral activity. Similarly, dose-dependent reductions in intracellular DHBV DNA at drug concentrations which are not cytotoxic indicate *in vitro* activity. The more potent antiviral drugs identified by initial screening *in vitro* or *in vivo* may be selected for more rigorous and comprehensive testing (see below).

Antiviral therapy: routes of administration for *in vivo* tests

I.p. injection is the preferred route of administration for drug screening in the duck model because it is technically simple, it is the most assured method (other than direct intrahepatic injection) of drug delivery to the liver, and it minimizes unknown variables such as extrahepatic uptake and metabolism associated with other routes. Compounds known to be stable in the gut and to be absorbed efficiently may be given orally, and routes of administration (oral, subcutaneous, i.v. or intramuscular) may be used in bioavailability studies and must also be considered when it is necessary to test drugs which are unsuitable for oral or i.p. delivery. Ideally, antihepadnaviral drugs destined for human use must be orally available; unfortunately, radical and unpredictable species-dependent differences in pharmacokinetics, and especially in oral availability, have often been observed and this limits the usefulness of all animal models. Even in cases where theoretical predictions of pharmacokinetic parameters for the parent compound have been confirmed experimentally, this problem may re-emerge when a prodrug is used, as a result of species-dependent differences in activities and specificities of the enzyme systems required for uptake and metabolism. For example, qualitative and quantitative species-dependent differences in metabolism of famciclovir, the orally available form of penciclovir, have recently been identified (Rashidi *et al.*, 1997). On the other hand, essentially identical antiviral effects were produced in DHBV-infected ducklings by i.p. penciclovir (Lin *et al.*, 1996) and oral famciclovir (Tsiqueaye *et al.*, 1994). However, in the absence of detailed pharmacological data, each drug and prodrug should be regarded as a separate entity.

Additional parameters to monitor response to treatment

Drugs which show potentially useful antiviral activity in initial *in vitro* and/or *in vivo* screens may be subjected to more intensive studies to define better the mechanisms and sites of action. DNA extracted from small samples of liver and other tissues are analyzed for the presence of the various DHBV replicative species by agarose gel electrophoresis and Southern blot hybridization using standard techniques. Southern analysis may also be performed after extraction of DNA using modified procedure designed specifically to enrich for the minor covalently closed circular (ccc DNA) species (Wu *et al.*, 1990), which is known to be most resistant to conventional antiviral therapy (Yokusuka *et al.*, 1985; Omata *et al.*, 1986; Mason *et al.*, 1994; Wang *et al.*, 1995). Total RNA analysis by slot-blot or Northern blot hybridization as well as immunoblot analysis of DHBV specific proteins are useful techniques to aid identification of sites of action (Wang *et al.*, 1995; Colledge *et al.*, 1997). *In vitro* assays for reverse transcriptase and DNA polymerase activities are in fairly widespread use (Table 123.1) and more sophisticated techniques, although not routinely available or applicable, have been designed to identify and monitor many specific steps in the viral replication cycle. These include virus attachment and uncoating (Pugh *et al.*, 1995; Kock *et al.*, 1996), priming of reverse transcription (Staschke and Colacino, 1994; Dannaoui *et al.*, 1997), and RNAse H activity (Oberhaus and Newbold, 1995).

Other techniques currently under development are likely to be increasingly useful in the future; they include assays capable of detecting individual stages in the complex sequence of reactions which couple intracellular replication to viral protein synthesis and virus particle assembly (Chen *et al.*, 1994; Lenhoff and Summers, 1994; and see reviews by Ganem, 1996; Nassal and Schaller, 1996), as well as assays aimed at identifying key reactions associated with viral chromatin assembly and regulation of its topology and transcriptional activity (Newbold *et al.*, 1995). These latter methods, together with the standard molecular techniques listed above, can provide quantitative and qualitative (but not demographic) information about viral replication in specific tissues. The complementary demographic data, i.e., the nature and extent of tissue damage (if present) and the extent of viral replication in tissues of interest at regional, cellular and subcellular levels, can be obtained from histological examination together with *in situ* DNA and RNA hybridization and immunohistochemical or immunofluorescence techniques (Luscombe *et al.*, 1994, 1996; Mason *et al.*, 1994; Lin *et al.*, 1996; Nicoll *et al.*, 1997). In particular, immunofluorescence techniques are useful for identification and investigation of extrahepatic sites of HBV replication (Halpern *et al.*, 1983) which may be viral sanctuaries during antiviral therapy (Luscombe *et al.*, 1994; Lin *et al.*, 1996; Nicoll *et al.*, 1997). Although it is technically possible to obtain multiple tissue biopsies for these analyses from individual ducklings during the course of an *in vivo* drug trial, it is more usual to sacrifice matched groups of treated and control (untreated) animals at 3- or 4-week intervals (usually pre-treatment, end-of-treatment and follow-up, in short-term trials). For statistically interpretable results, the number of ducklings in each group at each time point must be at least 3.

In vitro assays for anti-DHBV activity in primary duck hepatocytes

Primary duck hepatocytes (PDH) are isolated using a modification (Civitico *et al.*, 1990) of the method originally described by Tuttleman *et al.* (1986a). The isolation procedure is identical regardless of whether ducklings are DBHV-infected or uninfected, but for use in antiviral assays, uninfected PDH must be acutely infected *in vitro* within 3 days of isolation. Briefly, 10–14-day-old ducklings are anesthetized with 30 mg/kg intramuscular ketamine, and the liver is surgically removed and perfused with 200 ml of pre-warmed (37°C) Hanks balanced salt solution (calcium- and magnesium-free) containing 0.5 mmol/l ethylene glycol-*bis* (β-aminoethylether)-N,N,N′,N′,-tetra

Table 123.1 Antihepadnaviral activities of various compounds in duck hepatitis B virus (DHBV)-based tests

Test compound	*Assay system; dose; duration*	*Outcome and comments*	*Reference*
Purine nucleoside or nucleotide analogs			
Acyclovir	CI ducklings; 10 or 30 mg/kg per day i.p. for 3 weeks	Reduction in serum DNA; not sustained post-treatment	Tsiquaye *et al.*, 1986
Various, including ganciclovir and acyclovir	CI PDH; ≤ 12 days	Reduction in all DHBV DNA replicative species except cccDNA	Civitico *et al.*, 1990
Ganciclovir	CI ducklings; 10 or 30 mg/kg per day i.p. for 3 weeks	Reduced serum and liver DNA; no effect on liver cccDNA or serum sAg	Wang *et al.*, 1991
Ganciclovir	CI ducklings; 10 mg/kg i.p. for 4 weeks in combination with ampligen and coumermycin A1	Reduction in serum and liver DNA; minimal effect on liver cccDNA, not sustained post-treatment	Niu *et al.*, 1993
Ganciclovir	CI ducklings; 10 mg/kg per day i.p. for 4 weeks	Study of serum and liver DNA rebound kinetics during first post-treatment week	Dean *et al.*, 1995
Ganciclovir	CI ducklings; 10 mg/kg per day i.p. for 3 weeks; included in *in situ* DNA hybridization and immunohistochemical studies	Reduced serum and liver DNA; no effect on liver cccDNA or serum sAg; extra- and intrahepatic DHBV sanctuaries identified	Luscombe *et al.*, 1994
Ganciclovir, penciclovir	CI PDH; ≤ 50 μmol/l ≤ 10 days	Penciclovir more effective than ganciclovir on a molar basis	Shaw *et al.*, 1994
Ganciclovir	CI ducklings; 10 mg/kg per day i.p. for 3 or 4 weeks; also with nalidixic acid	Reduction in serum and liver DNA and RNA; reduction in liver cccDNA by combination	Wang *et al.*, 1995
Ganciclovir	CI ducklings; 10 mg/kg per day i.p. for 24 weeks	Extension of earlier study; confirmed earlier results; found intrahepatic accumulation of pre-S Ag	Luscombe *et al.*, 1996
Ganciclovir (+ foscarnet)	CI PDH; alone and in combination	Activities are additive or synergistic at clinically relevant concentrations	Civitico *et al.*, 1996
Famciclovir (penciclovir)	CI ducklings infected *in ovo*; 25 mg/kg famciclovir given orally for ≤ 12 days	Treatment for > 2 days significantly reduced serum DNA and DNA replicative intermediates in liver as well as other tissues	Tsiquaye *et al.*, 1994
Penciclovir	CI ducklings; 10 mg/kg per day i.p. for 4 weeks	Reduction in serum DNA and liver RNA and DNA including cccDNA; pre-S and core Ag also reduced; not sustained post-treatment	Lin *et al.*, 1996
Penciclovir triphosphate (septate enantiomers); other dGTP analogs	Transcription/translation-coupled cell-free assays for reverse transcriptase (RT) and RT priming reactions	Penciclovir-triphosphate inhibited both RT and RT priming (R) > (S); related dGTP analogs also active to different extents	Dannaoui *et al.*, 1997
Penciclovir (+ lamivudine)	CI PDH	Activities are additive or synergistic at clinically relevant concentrations	Colledge *et al.*, 1997
2′,3′-Dideoxy-3′-fluoroguanosine	Ducklings infected *de novo*; 1 or 5 mg/kg per day i.v. for 9 days CI PDH, ≤10 μmol/L, ≤7 days	Reduction in serum DNA *in vivo*; IC_{90} = 0.1 μmol/l *in vitro*	Hafkemeyer *et al.*, 1996
2′ Carbodeoxyguanosine	CI ducklings infected *in ovo*; 10 or 100 μg on alternate days for 12 weeks	90% reduction in liver DNA at higher dose associated with hepatocellular toxicity	Mason *et al.*, 1994
Pyrimidine nucleoside analogs			
Thymidine analogs including 2′-azido-2′,3′-dideoxythymidine (AZT)	CI PDH	AZT inhibited replication at 20–50 μmol/l	Suzuki *et al.*, 1991
2′-azido-2′,3′-dideoxythymidine (AZT)	CI ducklings	No antiviral effect	Haritani *et al.*, 1989
Lamivudine triphosphate; lamivudine 2′,3′-dideoxy-β-L-5-fluorocytidine 3TC)	Cell-free assay for viral polymerase/RT activity using DHBV core particles	3TC-TP competes with dCTP; blocks DNA chain elongation; K_i for 3TC-TP ≈ 0.8 μmol/l in cell-free RT assay	Severini *et al.*, 1995
2′,3′-Dideoxy-β-L-5-fluorocytidine (FTC); FTC-TP related analog TPs	CI PDH, ≤ 100 μmol/l 4–13 days; cell-free assay for viral polymerase/RT activity using rabbit reticulocyte system; CI ducklings	Inhibited replication and virion release by 50% at ≈ 1 μmol/l in cell culture; 2 μm in cell-free assays; serum DHBV reduced to undetectable levels in < 1 week by 50 mg/kg twice daily	Zoulim *et al.*, 1996
Nucleotide analogs, oligonucleotides, cytokines and cytokine inducers			
Mismatched dsRNA	CI ducklings; 5 mg/kg single i.v. dose or 0.2 mg/kg per day i.v. for 7 days	Interferon induction; decrease in serum DNA not sustained post-treatment	Ijichi *et al.*, 1994
Recombinant human interferon-γ	Human hepatoma (HuH-7) cells transfected with DHBV	Interferon induction; inhibition of viral DNA, RNA and protein synthesis due to increased RNA degradation	Lavine and Ganem, 1993
Recombinant duck interferon	CI PDH *in vitro*	Inhibition of viral DNA, RNA and protein synthesis	Schultz *et al.*, 1995
Various, including (R)- and (S)-9-(3-Hydroxy-2-phosphonylmethoxypropyl) adenine [HPMPA]	CI PDH *in vitro*	Lowest IC_{50} = 0.5 μg/ml for (S)-[HPMPA]	Yokota *et al.*, 1990
9-(2-Phosphonylmethoxy-ethyl) adenine (PMEA)	CI PDH *in vitro*; ducklings infected *de novo*	Reduction in serum DNA *in vivo*; IC_{50} = 0.2 μmol/l *in vitro*	Heijtink *et al.*, 1993

Table 123.1 *continued*

Test compound	*Assay system; dose; duration*	*Outcome and comments*	*Reference*
Bacteriophage DNA; denatured dsDNA	Ducklings infected *de novo*; I2.45 mg/kg per day i.v. for 10 days	Interferon induction; decrease in serum DNA not sustained post-treatment	Iizucka *et al.*, 1994
Antisense oligonucleotides	CI PDH *in vitro*; ducklings infected *de novo*	Most inhibition by antisense oligo to 5′-end of pre-S gene; replication inhibited by ≈ 90% both *in vivo* and *in vitro*	Offensperger *et al.*, 1993a
Miscellaneous, including natural products			
Phyllanthus amarus extract	CI ducklings, oral or i.p. for 10–12 weeks	No effect on serum or liver DNA; minor decrease in DHBsAg after i.p. treatment	Niu *et al.*, 1990
Phyllanthus spp. extracts	Day-old ducklings infected *de novo*; 100-mg/kg i.p. for 4 weeks	No significant antiviral effects	Munshi *et al.*, 1993
DHBV cAg mutants	DHBV-transfected avian hepatoma (LMH) cells	COOH-, but not amino-terminal mutants inhibited replication by ≤ 90% by interfering with virion assembly	von Weizsacker *et al.*, 1996
Suramin	Ducklings infected *de novo*; per- or post-treated with suramin	Suramin protected against infection by blocking early steps in replication; not protective when given post-infection	Offen sperger *et al.*, 1993b
Hypericin	DHBV-transfected cells	Prevented secretion of infectious DHBV from infected cells; probably inhibits a late step in virion assembly	Moraleda *et al.*, 1993

The list of tested compounds is intended to be representative, not exhaustive. CI, Congenitally infected; *CI, Congenitally infected; i.p., intraperitonal; PDH, primary duck hepatocytes; cccDNA, covalently closed circular DNA; sAg, surface antigen; pre-S aG, pre-S antigen; dGTP, deoxyguanosine triphosphate; IC_{50}, IC_{90}, concentrations causing 50% or 90% inhibition, respectively; K_i, inhibition constant.

acetic acid (EGTA), followed by 200 ml of pre-warmed serum-free Eagle's minimum essential medium (MEM) supplemented with 100 mg collagenase type 1 (Boehringer Mannheim, West Germany) and 2.5 mmol/l $CaCl_2$. A single cell suspension is prepared in MEM by gently extruding the perfused liver through a fine wire-gauze mesh. Hepatocytes are isolated and purified from the cell suspension using Percoll density gradients (Pharmacia, Sweden) following a modification of the manufacturer's specifications. The gradient medium stock solution (stock isotonic Persoll–SIP) consists of nine parts Percoll mixed with one part 1.5 mol/l NaCl. Isotonic Percoll, having a density of 1.05 g/ml, is prepared by diluting six parts SIP with four parts MEM at a final pH of 7.4. Five ml of hepatocyte cell suspension is layered on to 30 ml of this colloidal solution before centrifugation at 20 000 rpm in a JA-20 fixed-angle rotor (Beckman, USA) for 20 minutes at 20°C. The bands of cells corresponding to the density of hepatocytes (1.07–1.09 g/cm^3) are collected and washed in Leibovitz L 15 medium supplemented with 5% fetal bovine serum and counted in a hemocytometer. Cell viability is established by trypan blue dye exclusion. Hepatocyte suspensions are diluted and subsequently seeded in L 15 complete medium (L 15) which consists of standard L 15 supplemented with 15 mmol/l Tris, insulin, glucose, hydrocortisone hemisuccinate, penicillin and streptomycin and 5% (v:v) fetal bovine serum. Hepatocytes are seeded at approximately 1.5–2.0 million cells per well into 6-well multiplates (Greiner, West Germany) and allowed to attach overnight before the first medium change (on day 1 post-plating). They are maintained with L 15 complete medium at 37°C in a 5% CO_2-humidified incubator. The culture medium in both control and treated cultures (see below) is changed every second day. We have no longer found it necessary to use feeder cell layers as originally specified by Civitico *et al.*, (1990).

For routine antiviral testing, cells are harvested on day 10 post plating, DNA extracted, transferred to a nylon membrane and assayed for DHBV DNA by spot-blot hybridization using standard methods. A variety of more specific assays (see above) may be used to investigate the mechanisms and sites of action for individual drugs (see above) and to define other important parameters such as the duration of persistence of antiviral activity after drug removal. More detailed descriptions of methodology and data analysis can be found in recent publications from this laboratory (Civitico *et al.*, 1990, 1996; Shaw *et al.*, 1994; Colledge *et al.*, 1997).

Advantages and disadvantages of the model

For reasons outlined above, animal models, in particular the duck and woodchuck models, have been extensively relied upon for testing anti-HBV agents. Although neither model is ideal, each has particular advantages and disadvantages.

Some advantages of the duck model include:

1. Ducklings are small, available globally, inexpensive and readily available commercially, easy to handle and economical to keep; unlike woodchucks, they neither bite nor hibernate.
2. It is not difficult to maintain separate populations of uninfected, congenitally infected and acutely infected ducklings for study.
3. Experimental DHBV infection of young, uninfected ducklings, which is very efficient, causes a rapid onset of

viremia followed by a high incidence of asymptomatic persistent infection, mimicking early and intermediate stages of chronic HBV infection in humans (Omata *et al.*, 1983).

4. Older ducklings can be acutely infected with DHBV, which mimicks acute HBV infection in humans (Jilbert *et al.*, 1992; Vickery and Cossart, 1996).
5. Infection with DHBV causes no apparent discomfort or adverse physiological effects in the host (Freiman and Cossart, 1986).
6. Integration of DHBV into the hosts' nuclear DNA is a relatively rare event which occurs only at a very late stage of infection and, in the absence of environmental and/or genetic stimuli, progression to hepatocellular carcinoma is also slow and infrequent; both characteristics are shared by human HBV infection (Marion *et al.*, 1984; Cullen *et al.*, 1990).
7. DHBV is not transmissible to humans and special precautions and biological containment facilities are not needed for its laboratory study and manipulation (Sprengel and Will, 1998).
8. Reliable *in vitro* cell culture systems procedures have been already developed for antiviral assays and studies of most stages of the viral replication cycle, and *in vitro* cell-free assays for investigation of specific stages of the DHBV replication strategy either already exist or are being developed (see previous sections).
9. The system can be used to investigate extrahepatic replication, which is believed to contribute to refractoriness of chronic HBV infections to antiviral therapy (Lin *et al.*, 1996; Luscombe *et al.*, 1996; Nicoll *et al.*, 1997).
10. Drug interactions can be studied *in vitro* and *in vivo* in this system (Table 123.1).
11. Finally, a large body of data and experience pertaining to both the molecular biology of DHBV and the relative activities and mechanisms of action of many antiviral drugs has already been accumulated during the past decade (Ganem, 1996; Nassal and Schaller, 1996).

Identifiable disadvantages of the duck model are all related to evolutionary distance. Of the commonly used animal models, DHBV and its host are the most evolutionarily remote from their human equivalents, having diverged from common ancestors an estimated 30 000 years and 300 million years ago respectively (Orito *et al.*, 1989). Birds, including ducks, have higher metabolic rates than mammals of corresponding size, gastrointestinal systems which are structured differently, and genomes which are less than half the size of mammalian genomes. In common with lower vertebrates, they have a renal portal system which is absent in mammals, and they are uricotelic, that is, ammonia excretion is coupled to purine turnover rather than to urea synthesis as in mammals (Campbell *et al.*, 1987). Mammalian and avian hepadnaviruses are classified into separate genera on the basis of biophysical, immunological and genetic differences. Mammalian HBV genomes have four open reading frames, whilst their avian counterparts, which lack a complete X coding region, have only three, (Mandart *et al.*, 1984). The duck model is therefore unsuitable for direct studies of the X gene product. Its usefulness is also limited where immune functions are involved: while avian immune systems are well-characterized down to the cellular, and in some cases molecular level (Dieterlen-Lievre, 1988), the systems' inherent complexity and evolutionary separation make cross-species comparisons difficult. The mammalian HBV core and envelope proteins are antigenically cross-reactive with each other, but not with their avian homologs (Cote *et al.*, 1982; Feitelson *et al.*, 1983). Virion ultrastructure and buoyant density of mammalian and avian hepadnaviruses also differ, and whereas the former generate non-infectious surface antigen filaments during infection, the latter do not. Moreover, most of the relevant biochemical, pharmacological and physiological data published to date serve mainly to emphasize differences between mammalian and avian systems.

Although factors listed above reduce the predictive power of the duck model in pharmacokinetic, immunological and toxicological studies, cumulative experience has confirmed, rather than diminished, its usefulness as a tool for investigating hepadnaviral replication and its inhibition by antiviral drugs.

Contributions of the duck model to infectious disease therapy

The duck model has contributed immeasurably to our current understanding of the hepadnaviral replication strategy. Besides the initial demonstration by Summers and Mason (1982) of the obligatory reverse transcription requirement, key features of HBV replication identified in the duck model include intracellular genome amplification of cccDNA (Tuttleman *et al.*, 1986b; Wu *et al.*, 1990); demonstration of domain structure and RNAse H activity of the viral polymerase (Radziwill *et al.*, 1990); identification of the unusual mechanism of reverse transcription (Wang and Seeger, 1992); characterization of the viral minichromosome (Newbold *et al.*, 1995) and, more recently, elucidation of complex mechanisms which co-ordinate viral replication, transcription and translation with virion assembly and release (reviewed by Nassal and Schaller, 1996). The extensive reliance on DHBV as a replication model has provided a solid foundation for antiviral studies and large numbers of antiviral agents have been tested for efficacy against DHBV both *in vitro* and *in vivo* (see Table 123.1 for summary). Experience gained during the past decade has amply vindicated the utility and reliability of the model since, in general, agents identified as being most efficacious show similar activity in other models and in the clinic and vice versa. Conversely, findings of poor antiviral activity in the duck model have usually pre-empted or confirmed similar findings in other systems.

The utility of the duck model is exemplified by the development of famciclovir as an orally available anti-HBV agent. Almost 10 years ago, Locarnini *et al.* (1989) made the fortuitous observation that ganciclovir chemotherapy caused a significant reduction in HBV viremia acquired immune deficiency syndrome (AIDS) patients who were co-infected with HIV and HBV as well as human cytomegalovirus (HCMV), and who were treated with ganciclovir to control the latter. This finding prompted *in vitro* and *in vivo* tests of ganciclovir in chronically DHBV-infected primary hepatocytes (Civitico *et al.*, 1990, 1996) and ducklings (Wang *et al.*, 1991; Luscombe *et al.*, 1994), which confirmed its antihepadnaviral activity. The results were rather unexpected, since intracellular activation of ganciclovir and related acyclic guanine nucleoside analogs was assumed to require a viral deoxypyrimidine kinase which is not encoded by HBV genomes (see Shaw and Locarnini, 1995, for discussion). Ganciclovir was later successfully used to control HBV in liver transplant patients (Gish *et al.*, 1996). Subsequently, penciclovir, a nucleoside analog which is structurally very similar to ganciclovir, was found to possess even more potent anti-DHBV activity when tested in primary duck hepatocytes, apparently due to the longer intracellular half-life of its active metabolites (Shaw *et al.*, 1994). Penciclovir was subsequently shown to be effective against DHBV *in vivo* (Lin *et al.*, 1996), as was its prodrug, famciclovir (Tsiquaye *et al.*, 1994).

Additional studies in human hepatoma cells (Korba and Boyd, 1996; Shaw *et al.*, 1996) and tests in cell-free assays (Shaw *et al.*, 1996; Dannaoui *et al.*, 1997) confirmed penciclovir's anti-(D)HBV activity and partly identified its mechanisms of action. Famciclovir, the orally available form of penciclovir, was successfully used to control HBV in clinical trials (Main *et al.*, 1996), which are continuing.

Antihepadnaviral activities of at least two other drugs, lamivudine and PMEA (adefovir), which have also entered phase II or III clinical trials against HBV, were first demonstrated or confirmed using the duck model. The model has also been used to investigate drug interactions both *in vivo* and *in vitro*, and to investigate activities of several novel agents (Table 123.1). However, for reasons outlined above, interspecies extrapolation of the results obtained should be made only with extreme caution, at least until confirmatory data from other systems become available.

In summary, whilst the duck model has recognized limitations, it has proven to be an exceptionally useful system for the elucidation of hepadnaviral replication strategy and antiviral drug screening.

References

Blackmore, D. J. (1988). *Animal Clinical Biochemistry*. Cambridge University Press, Cambridge, U.K.

Campbell, J. W., Vorhaben, J. E., Smith, D. D. (1987). Uricoteley: its nature and origin during the evolution of tetrapod vertebrates. *J. Exp. Zool.*, **243**, 349–363.

Carp, N. Z., Saputelli, J., Halbherr, T. C., Mason, W. S., Jilbert, A. R. (1991). A technique for liver biopsy performed in Pekin ducks using anesthesia with Telazol. *Lab. Anim. Sci.*, **41**, 474–475.

Chen, Y., Robinson, W. S., Marion, P. L. (1994). Selected mutations of the duck hepatitis B virus P gene RNase H domain affect both RNA packaging and priming of minus-strand DNA synthesis. *J. Virol.*, **68**, 5232–5238.

Civitico, G., Wang, Y. Y., Luscombe, C. *et al.* (1990). Antiviral strategies in chronic hepatitis B virus infection: II. Inhibition of duck hepatitis B virus *in vitro* using conventional antiviral agents and supercoiled-DNA active compounds. *J. Med. Virol.*, **31**, 90–97.

Civitico, G., Shaw, T., Locarnini, S. (1996). Interaction between ganciclovir and foscarnet as inhibitors of duck hepatitis B virus replication *in vitro. Antimicrob. Agents Chemother.*, **40**, 1180–1185.

Colledge, D., Locarnini, S., Shaw, T. (1997). Synergistic inhibition of hepadnaviral replication by lamivudine in combination with penciclovir *in vitro. Hepatology*, **26**, 216–225.

Condreay, L. D., Aldrich, C. E., Coates, L., Mason, W. S., Wu, T. T. (1990). Efficient duck hepatitis B virus production by an avian liver tumor cell line. *J. Virol.*, **64**, 3249–3258.

Cote, P. J., Dapolito, G. M., Shih, J. W., Gerin, J. L. (1982). Surface antigenic determinants of mammalian "hepadnaviruses" defined by group- and class-specific monoclonal antibodies. *J. Virol.*, **42**, 135–142.

Cova, L., Lambert, V., Chevallier, A. *et al.* (1986). Evidence for the presence of duck hepatitis B virus in wild migrating ducks. *J. Gen. Virol.*, **67**, 537–547.

Cullen, J. M., Marion, P. L., Sherman, G. J., Hong, X., Newbold, J. E. (1990). Hepatic neoplasms in aflatoxin B1-treated, congenital duck hepatitis B virus-infected, and virus-free pekin ducks. *Cancer Res.*, **50**, 4072–4080.

Dannaoui, E., Trepo, C., Zoulim, F. (1997). Inhibitory effect of penciclovir-triphosphate on duck hepatitis B virus reverse transcription. *Antiviral Chem. Chemother.*, **8**, 38–46.

Davis, D. (1987). Duck hepatitis virus: adaptation of a plaque assay to determine 50 per cent end points with duck sera. *Res. Vet. Sci.*, **42**, 167–169.

Dean, J., Bowden, S., Locarnini, S. (1995). Reversion of duck hepatitis B virus DNA replication *in vivo* following cessation of treatment with the nucleoside analogue ganciclovir. *Antiviral Res.*, **27**, 171–178.

Dieterlen-Lievre, F. (1988). Birds. In *Vertebrate Blood Cells* (eds Rowley, A. F, Ratcliffe, N. A.), pp. 257–334. Cambridge University Press, Cambridge, U. K.

Feitelson, M. A., Marion P. L., Robinson, W. S. (1983). The nature of polypeptides larger in size than the major surface antigen components of hepatitis B and like viruses in ground squirrels, woodchucks, and ducks. *Virology*, **130**, 76–90.

Feitelson, M. A., Millman, I., Halbherr, T., Simmons, H., Blumberg, B. S. (1986). A newly identified hepatitis B type virus in tree squirrels. *Proc. Natl Acad. Sci. USA*, **83**, 2233–2237.

Freiman, J. S., Cossart, Y. E. (1986). Natural duck hepatitis B virus infection in Australia. *Aust. J. Exp. Biol. Med. Sci.*, **64**, 477–484.

Galle, P. R., Schlicht, H. J., Fischer, M., Schaller, H. (1988). Production of infectious duck hepatitis B virus in a human hepatoma cell line. *J. Virol.*, **62**, 1736–1740.

Ganem, D. (1996) Hepadnaviridae and their replication. In *Fields*

Virology, 3rd edn (eds Fields, B. N., Knipe, D. M., Hawley, P. M. *et al.*) pp. 2739–2807. Lippincott-Raven, Philadelphia.

Ganem, D., Pollack, J. R., Tavis, J. (1994). Hepatitis B virus reverse transcriptase and its many roles in hepadnaviral genomic replication. *Infect. Agents. Dis.*, **3**, 85–93.

Gish R. G., Lau, J. Y., Brooks, L. *et al.* (1996). Ganciclovir treatment of hepatitis B virus infection in liver transplant recipients. *Hepatology*, **23**, 1–7.

Gong, S. S., Jensen, A. D., Rogler, C. E. (1996). Loss and acquisition of duck hepatitis B virus integrations in lineages of LMH-D2 chicken hepatoma cells. *J. Virol.*, **70**, 2000–2007.

Hafkemeyer, P., Keppler-Hafkemeyer, A., al Haya, M. A. *et al.* (1996). Inhibition of duck hepatitis B virus replication by 2′,3′-dideoxy-3′-fluoroguanosine *in vitro* and *in vivo*. *Antimicrob. Agents Chemother.*, **40**, 792–794.

Halpern, M. S., England, J. M., Deery, D. T. *et al.* (1983). Viral nucleic acid synthesis and antigen accumulation in pancreas and kidney of Pekin ducks infected with duck hepatitis B virus. *Proc. Natl Acad. Sci. USA*, **80**, 4865–4869.

Haritani, H., Uchida, T., Okuda, Y., Shikata, T. (1989). Effect of 3′-azido-3′ deoxythymidine on replication of duck hepatitis B virus *in vivo* and *in vitro*. *J. Med. Virol.*, **29**, 244–248.

Heijtink, R. A., De Wilde, G. A., Kruining, J. *et al.* (1993). Inhibitory effect of 9-(2-phosphonylmethoxyethyl)-adenine (PMEA) on human and duck hepatitis B virus infection. *Antiviral. Res.*, **21**, 141–153.

Heijtink, R. A., Kruining, J., de Wilde, G. A., Balzarini, J., de Clercq, E., Schalm, S. W. (1994). Inhibitory effects of acyclic nucleoside phosphonates on human hepatitis B virus and duck hepatitis B virus infections in tissue culture. *Antimicrob. Agents Chemother.*, **38**, 2180–2182.

Higgins, D. A. (1990). Duck lymphocytes III. Transformation responses to some common mitogens. *Comp. Immunol. Microbiol. Infect. Dis.*, **13**, 13–23.

Higgins, D. A., Cromie, R. L., Srivastava, G. *et al.* (1993). An examination of the immune system of the duck (*Anas platyrhynchos*) for factors resembling some defined mammalian cytokines. *Dev. Comp. Immunol.*, **17**, 341–355.

Hollinger, F. B. (1996). Hepatitis B virus. In *Fields Virology*, 3rd edn (eds Fields, B. N., Knipe, D. M., Hawley, P. M. *et al.*), pp. 2739–2807. Lippincott-Raven, Philadelphia.

Horwich, A. L., Furtak, K., Pugh, J., Summers, J. (1990). Synthesis of hepadnavirus particles that contain replication-defective duck hepatitis B virus genomes in cultured HuH7 cells. *J. Virol.*, **64**, 642–650.

Howe, A. Y., Robins, M. J., Wilson, J. S., Tyrrell, D L. (1996). Selective inhibition of the reverse transcription of duck hepatitis B virus by binding of 2′,3′-dideoxyguanosine 5′ triphosphate to the viral polymerase. *Hepatology*, **23**, 87–96.

Ijichi, K., Mitamura, K., Ida, S., Machida, H., Shimada (1994). *In vivo* antiviral effects of mismatched double-stranded RNA on duck hepatitis B virus. *J. Med. Virol.*, **43**, 161–165.

Jilbert, A. R., Freiman, J. S., Burrell, C. J. *et al.* (1988). Virus–liver cell interactions in duck hepatitis B virus infection. A study of virus dissemination within the liver. *Gastroenterology*, **95**, 1375–1382.

Jilbert, A. R., Wu, T. T., England, J. M. *et al.* (1992). Rapid resolution of duck hepatitis B virus infections occurs after massive hepatocellular involvement. *Virology*, **66**, 1377–1388.

Jilbert, A. R., Miller, D. S., Scougall, C. A., Turnbull, H., Burrell, C. J. (1996). Kinetics of duck hepatitis B virus infection following low dose virus inoculation: one virus DNA genome is infectious in neonatal ducks. *Virology*, **226**, 338–345.

Kock, J., Borst, E. M., Schlicht, H. J. (1996) Uptake of duck hepatitis B virus into hepatocytes occurs by endocytosis but does not require passage of the virus through an acidic intracellular compartment. *J. Virol.*, **70**, 5827–5831.

Korba, B. E., Boyd, M. R. (1996). Penciclovir is a selective inhibitor of hepatitis B virus replication in cultured human hepatoblastoma cells. *Antimicrob. Agents Chemother.*, **40**, 1282–1284.

Lavine, J. E., Ganem, D. (1993). Inhibition of duck hepatitis B virus replication by interferon-gamma. *J. Med. Virol.*, **40**, 59–64.

Lenhoff, R. J., Summers, J. (1994). Coordinate regulation of replication and virus assembly by the large envelope protein of an avian hepadnavirus. *J. Virol.*, **68**, 4565–4571.

Lin, E., Luscombe, C., Wang, Y. Y., Shaw, T., Locarnini, S. (1996). The guanine nucleoside analog penciclovir is active against chronic duck hepatitis B virus infection *in vivo*. *Antimicrob. Agents Chemother.*, **40**, 413–418.

Locarnini, S. A., Guo, K., Lucas, C. R., Gust, I. D. (1989). Inhibition of HBV replication by ganciclovir in patients with AIDS. *Lancet*, **2**, 1225–1226.

Luscombe, C., Pedersen, J., Bowden, S., Locarnini, S. (1994). Alterations in intrahepatic expression of duck hepatitis B viral markers with ganciclovir chemotherapy. *Liver*, **14**, 182–192.

Luscombe, C., Pedersen, J., Uren, E., Locarnini, S. (1996). Long-term ganciclovir chemotherapy for congenital duck hepatitis B virus infection *in vivo*: effect on intrahepatic-viral DNA, RNA, and protein expression. *Hepatology*, **24**, 766–773.

Main, J., Brown, J. L., Howells, C. *et al.* (1996) A double blind, placebo-controlled study to assess the effect of famciclovir on virus replication in patients with chronic hepatitis B virus infection. *J. Viral. Hepato.*, **3**, 211–215.

Mandart, E., Kay, A., Galibert, F. (1984). Nucleotide sequence of a cloned duck hepatitis B virus genome: comparison with woodchuck and human hepatitis B virus sequences. *J. Virol.*, **49**, 782–792.

Marion, P. L., Oshiro, L. S., Regnery, D. C., Scullard, G. H., Robinson, W. S. (1980). A virus in Beechey ground squirrels that is related to hepatitis B virus of humans. *Proc. Natl Acad. Sci. USA*, **77**, 2941–2945.

Marion, P. L., Knight, S. S., Ho, B. K., Guo, Y. Y., Robinson, W. S., Popper, H. (1984). Liver disease associated with duck hepatitis B virus infection of domestic ducks. *Proc. Natl Acad. Sci. USA*, **81**, 898–902.

Marion, P. L., Cullen, J. M., Azcarraga, R. R., Van Davelaar, M. J., Robinson, W. S. (1987). Experimental transmission of duck hepatitis B virus to Pekin ducks and to domestic geese. *Hepatology*, **7**, 724–731.

Mason, W. S., Seal, G., Summers, J. (1980). Virus of Pekin ducks with structural and biological relatedness to human hepatitis B virus. *J. Virol.*, **36**, 829–836.

Mason, W. S., Halpern, M. S., England, J. M. *et al.* (1983). Experimental transmission of duck hepatitis B virus. *Virology*, **131**, 375–384.

Mason, W. S., Cullen, J., Saputelli, J. *et al.* (1994). Characterization of the antiviral effects of 2′ carbodeoxyguanosine in ducks chronically infected with duck hepatitis B virus. *Hepatology*, **19**, 398–411.

Moraleda, G., Wu, T. T., Jilbert, A. R. *et al.* (1993). Inhibition of duck hepatitis B virus replication by hypericin. *Antiviral Res.*, **20**, 235–247.

Munshi, A., Mehrotra, R., Panda, S. K. (1993). Evaluation of *Phyllanthus amarus* and *Phyllanthus maderaspatensis* as agents for postexposure prophylaxis in neonatal duck hepatitis B virus infection. *J. Med. Virol.*, **40**, 53–58.

Nassal, M., Schaller, H. (1996). Hepatitis B virus replication – an update. *J. Viral Hepatitis*, **3**, 217–226.

Newbold, J. E., Xin, H., Tencza, M. *et al.* (1995). The covalently closed duplex form of the hepadnavirus genome exists *in situ* as a heterogeneous population of viral minichromosomes. *J. Virol.*, **69**, 3350–3357.

Nicoll, A. J., Koppinen, J. A. (1996). A technique for general anaesthesia and bile duct ligation in Pekin ducklings. *Scand. J. Animal Lab. Sci.*, **23** (suppl. 1), 459–462.

Nicoll, A. J., Angus, P. W., Chou, S. T., Luscombe, C. A., Smallwood, R. A., Locarnini, S. A. (1997). Demonstration of duck hepatitis B virus in bile duct epithelial cells: implications for pathogenesis and persistent infection. *Hepatology*, **25**, 463–469.

Niu, J. Z., Wang, Y. Y., Qiao M. *et al.* (1990). Effect of *Phyllanthus amarus* on duck hepatitis B virus replication *in vivo*. *J. Med. Virol.*, **32**, 212–218.

Niu, J., Wang, Y., Dixon, R. *et al.* (1993). The use of ampligen alone and in combination with ganciclovir and coumermycin A1 for the treatment of ducks congenitally-infected with duck hepatitis B virus. *Antiviral Res.*, **21**, 155–171.

Oberhaus, S. M., Newbold, J. E. (1995). Detection of an RNase H activity associated with hepadnaviruses. *J. Virol.*, **69**, 5697–5704.

Offensperger, W. B., Offensperger, S., Walter, E., Blum, H. E., Gerok, W. (1993a). Suramin prevents duck hepatitis B virus infection *in vivo*. *Antimicrob. Agents Chemother.*, **37**, 1539–1542.

Offensperger, W. B., Offensperger, S., Walter, E. *et al.* (1993b). *In vivo* inhibition of duck hepatitis B virus replication and gene expression by phosphorothioate modified antisense oligodeoxynucleotides. *EMBO J.*, **12**, 1257–1262.

Omata, M., Uchiumi, K., Ito, Y. *et al.* (1983). Duck hepatitis B virus and liver diseases. *Gastroenterology*, **85**, 260–267.

Omata, M., Hirota, K., Yokosuka, O. (1986). *In vivo* study of the mechanism of action of antiviral agents against hepadna virus replication in the liver. Resistance of supercoiled viral DNA. *J. Hepatol.*, **3** (suppl. 2), S49–S55.

Orito, E., Mizokami, M., Ina, Y. *et al.* (1989). Host-independent evolution and a genetic classification of the hepadnavirus family based on nucleotide sequences. *Proc. Natl Acad. Sci. USA*, **86**, 7059–7062.

Pugh, J. C., Di, Q., Mason, W. S., Simmons, H. (1995). Susceptibility to duck hepatitis B virus infection is associated with the presence of cell surface receptor sites that efficiently bind viral particles. *J. Virol.*, **69**, 4814–4822.

Radziwill, G., Tucker, W., Schaller, H. (1990). Mutational analysis of the hepatitis B virus P gene product: domain structure and RNase H activity. *J. Virol.*, **64**, 613–620.

Rashidi, M. R., Smith, J. A., Clarke, S. E., and Beedham, C. (1997). *In vitro* oxidation of famciclovir and 6-deoxypenciclovir by aldehyde oxidase from human, guinea pig, rabbit, and rat liver. *Drug Metab. Dispos.*, **25**, 805–813.

Schultz, U., Kock, J., Schlicht, H. J., Staeheli, P. (1995). Recombinant duck interferon: a new reagent for studying the mode of interferon action against hepatitis B virus. *Virology*, **212**, 641–649.

Severini, A., Liu, X. Y., Wilson, J. S., Tyrrell, D. L. (1995). Mechanism of inhibition of duck hepatitis B virus polymerase by (-)-beta-L-2′,3′-dideoxy-3′-thiacytidine. *Antimicrob. Agents Chemother.*, **39**, 1430–1435.

Shaw, T., Locarnini, S. A. (1995). Hepatic purine and pyrimidine metabolism: implications for antiviral chemotherapy of viral hepatitis. *Liver*, **15**, 169–184.

Shaw, T., Amor, P., Civitico, G., Boyd, M., Locarnini, S. (1994). *In vitro* antiviral activity of penciclovir, a novel purine nucleoside, against duck hepatitis B virus. *Antimicrob. Agents Chemother.*, **38**, 719–723.

Shaw, T., Mok, S. S., Locarnini, S. A. (1996). Inhibition of hepatitis B virus DNA polymerase by enantiomers of penciclovir triphosphate and metabolic basis for selective inhibition of HBV replication by penciclovir. *Hepatology*, **24**, 996–1002.

Sprengel, R., Will, H. (1988). Duck hepatitis virus. In: *Virus diseases in Laboratory and Captive Animals*, pp. 363–386. Martinus Nijhoff, Boston.

Sprengel, R., Kuhn, C., Manso, C., Will, H. (1984). Cloned duck hepatitis B virus DNA is infectious in Pekin ducks. *J. Virol.*, **52**, 932–937.

Sprengel, R., Kaleta, E. F., Will, H. (1988). Isolation and characterization of a hepatitis B virus endemic in herons. *J. Virol.*, **62**, 3832–3839.

Staschke, K. A., Colacino, J. M. (1994). Priming of duck hepatitis B virus reverse transcription *in vitro*: premature termination of primer DNA induced by the 5′-triphosphate of fialuridine. *J. Virol.*, **68**, 8265–8269.

Summers, J., Mason, W. S. (1982). Replication of the genome of a hepatitis B-like virus by reverse transcription of an RNA intermediate. *Cell*, **29**, 403–415.

Summers, J., Smolec, J. M., Snyder, R. (1978). A virus similar to human hepatitis B virus associated with hepatitis and hepatoma in woodchucks. *Proc. Natl Acad. Sci. USA*, **75**, 4533–4537.

Suzuki, S., Lee, B., Luo, W., Tovell, D., Robins, M. J., Tyrrell, D. L. (1988). Inhibition of duck hepatitis B virus replication by purine 2′,3′-dideoxynucleosides. *Biochem. Biophys. Res. Commun.*, **156**, 1144–1151.

Suzuki, S., Tyrrell D. L. J., Saneyoshi, M. (1991). Inhibition of duck hepatitis B virus replication *in vitro* by 2′,3′-dideoxy-3′-azidothymidine and related compounds. *Acta Virol.*, **35**, 430–437.

Tagawa, M., Yokosuka, O., Imazeki, F., Ohto, M., Omata, M. (1996). Gene expression and active virus replication in the liver after injection of duck hepatitis B virus DNA into the peripheral vein of ducklings. *J. Hepatol.*, **24**, 328–334.

Testut, P., Renard, C. A., Terradillos, O. *et al.* (1996). A new hepadnavirus endemic in arctic ground squirrels in Alaska. *J. Virol.*, **70**, 4210–4219.

Tsiquaye, K. N., McCaul, T. F., Zuckerman, A. J. (1985a). Maternal transmission of duck hepatitis B virus in pedigree Pekin ducks. *Hepatology*, **5**, 622–628.

Tsiquaye, K. N., Rapicetta, M., McCaul, T. F., Zuckerman, A. J. (1985b). Experimental *in ovo* transmission of duck hepatitis B virus. *J. Virol. Methods*, **11**, 49–57.

Tsiquaye, K. N., Collins, P., Zuckerman, A. J. (1986). Antiviral activity of the polybasic anion, suramin and acyclovir in Hepadna virus infection. *J. Antimicrob. Chemother.*, **18** (suppl. B), 223–228.

Tsiquaye, K. N., Slomka, M. J., Maung, M. (1994). Oral famciclovir against duck hepatitis B virus replication in hepatic and nonhepatic tissues of ducklings infected in ovo. *J. Med. Virol.*, **42**, 306–310.

Tuttleman, J. S., Pugh, J. C., Summers, J. W. (1986a). *In vitro* experimental infection of primary duck hepatocyte cultures with duck hepatitis B virus. *J. Virol.*, **58**, 17–25.

Tuttleman, J. S., Pourcel, C., Summers, J. (1986b). Formation of the pool of covalently closed circular viral DNA in hepadnavirus-infected cells. *Cell*, **47**, 451–460.

Vickery, K. Cossart, Y. (1996). DHBV manipulation and prediction of the outcome of infection. *J. Hepatol.*, **25**, 504–509.

von Weizsacker, F., Wieland, S., Blum, H. E. (1996). Inhibition of viral replication by genetically engineered mutants of the duck hepatitis B virus core protein. *Hepatology*, **24**, 294–299.

Walter, E., Keist, R., Niederost, B., Pult, I., Blum H. E. (1996). Hepatitis B virus infection of tupaia hepatocytes *in vitro* and *in vivo*. *Hepatology*, **24**, 1–5.

Wang, G. H., Seeger, C. (1992). The reverse transcriptase of hepatitis B virus acts as a protein primer for viral DNA synthesis. *Cell*, **71**, 663–670.

Wang, Y., Bowden, S., Shaw, T. *et al.* (1991). Inhibition of duck hepatitis B virus replication *in vivo* by the nucleoside analogue ganciclovir(9-[2-hydroxy-1-(hydroxymethyl)ethoxymethyl]-guanine). *Antiviral Chem. Chemother*., **2**, 107–114.

Wang, Y., Luscombe, C., Bowden, S., Shaw, T., Locarnini, S. (1995). Inhibition of duck hepatitis B virus DNA replication by antiviral chemotherapy with ganciclovir-nalidixic acid. *Antimicrob. Agents Chemother.*, **39**, 556–558.

Wildner, G., Fernholz, D., Sprengel, R., Schneider, R., Will, H. (1991). Characterization of infectious and defective cloned avian hepadnavirus genomes. *Virology*, **185**, 345–353.

Wu, T. T., Coates, L., Aldrich, C. E., Summers, J., Mason, W. S. (1990) In hepatocytes infected with duck hepatitis B virus, the template for viral RNA synthesis is amplified by an intracellular pathway. *Virology*, **175**, 255–261.

Yokosuka, O., Omata, M., Imazeki, F., Okuda, K., Summers, J. (1985). Changes of hepatitis B virus DNA in liver and serum caused by recombinant leukocyte interferon treatment: analysis of intrahepatic replicative hepatitis B virus DNA. *Hepatology*, **5**, 728–734.

Yokota, T., Konno, K., Chonan, E. *et al.* (1990). Comparative activities of several nucleoside analogs against duck hepatitis B virus *in vitro*. *Antimicrob. Agents Chemother.*, **34**, 1326–1330.

Zoulim, F., Dannaoui, E., Borel, C. *et al.* (1996). 2′,3′-dideoxy-beta-L-5-fluorocytidine inhibits duck hepatitis B virus reverse transcription and suppresses viral DNA synthesis in hepatocytes, both *in vitro* and *in vivo*. *Antimicrob. Agents Chemother.*, **40**, 448–453.

Chapter 124

The Woodchuck Model of Hepatitis B Virus Infection

B. C. Tennant

Background

The hepatitis B virus (HBV) causes one of the most important infectious diseases of humankind. More than 350 million people worldwide are chronically infected with HBV. Although safe and effective HBV vaccines are now available and have been responsible for remarkable reductions in the rate of new infections, individuals with established chronic HBV infection are at high risk of developing life-threatening liver disease including chronic hepatitis, hepatic cirrhosis, and primary hepatocellular carcinoma (HCC). For such individuals, new and improved methods of therapy are desperately needed.

HBV infects only humans and other great apes and the chimpanzee is the only experimental animal species susceptible to HBV. Chimpanzees have been essential for the pre-clinical testing of candidate HBV vaccines for safety and efficacy. Their use in drug discovery and development has been restricted, however, because of cost, limited availability, and because the chimpanzee is an endangered species.

Natural history of woodchuck hepatitis virus infection

A mammalian virus closely related to HBV, the woodchuck hepatitis virus (WHV), was first described in the Eastern woodchuck by Summers *et al.* (1978) in a colony of woodchucks maintained at the Penrose Research Laboratory of the Philadelphia Zoological Gardens. For 18 years, colony woodchucks had experienced a high rate of chronic hepatitis, and 23% had been found at autopsy to have HCC. In 15% of the woodchucks, the serum contained DNA polymerase activity, and it was possible to demonstrate in such sera the presence of spherical and filamentous particles that resembled HBV. Virion DNA was similar in size and structure to HBV DNA, and it was concluded that WHV was a member of the same unique class of viruses to which HBV belonged (Summers *et al.*, 1978). WHV has been classified as a member of the family Hepadnaviridae, a group of *hepatotropic, DNA* viruses of which HBV is considered the prototype (Robinson 1980; Gust *et al.*, 1982; Melnick, 1982; Robinson *et al.*, 1982; Schodel *et al.*, 1989; Table 124.1).

The replication of WHV appears to be identical to that of HBV and other Hepadnaviridae and involves entry of the virion into the hepatocyte, removal of the surface envelope, and translocation of the nucleocapsid to the nucleus (Tiollais *et al.*, 1985; Ganem and Varmus, 1987). Following completion of plus-strand synthesis, viral DNA is converted to covalently closed circular DNA (cccDNA) which serves as the template for synthesis of viral messenger RNA and of the RNA pre-genome (Seeger *et al.*, 1986; Summers, 1987). In the cytoplasm, pre-genomic RNA and viral polymerase are packaged within core particles where synthesis of minus-strand viral DNA is completed by reverse transcription with concomitant degradation of RNA. Minus-strand DNA serves as the template for synthesis of the characteristically incomplete plus-strand (Ganem and Varmus, 1987). Cyclic transition of DNA to RNA during the life cycle of hepadnaviruses is similar to that of retroviruses (Ganem and Varmus, 1987; Summers, 1987; Schroder and Zentgraf, 1990). While integration of proviral DNA into the host genome is an essential step in retroviral replication, integration is not necessary for replication of hepadnaviruses. Hepadnavirus replicative intermediates may, however, be integrated into host genomic DNA and have a role in hepatocarcinogenesis (Schroder and Zentgraf, 1990).

Table 124.1 Classification of Hepadnavirus*

Family	*Hepadnaviridae*
Genus	*Orthohepadnavirus*
Type species	Hepatitis B virus (HBV)
Other members	
	Woodchuck hepatitis virus (WHV)
	California ground squirrel hepatitis virus (GSHV)
	Arctic ground squirrel hepatitis virus (AGSHV)
Genus	*Avihepadnavirus*
Type species	Duck hepatitis B virus (DHBV)
Other member	Heron hepatitis B virus (HHBV)

* Classification of the International Committee for Taxonomy of Viruses.

Handbook of Animal Models of Infection
ISBN 0-12-775390-7

The natural range of the woodchuck is from northern Georgia, Alabama and Mississippi on the south, to Quebec and Labrador on the north, west to Oklahoma, Kansas, Nebraska, and North and South Dakota, across Canada to British Columbia and the Yukon Territory, and includes a small area of south-eastern Alaska. A comprehensive study of the prevalence of WHV infection in this large population has not been completed. The woodchucks from which WHV was originally isolated by Summers *et al.* (1978) originated in Pennsylvania. Tyler *et al.* (1981) subsequently conducted a serologic study of woodchucks in Pennsylvania, New Jersey and Maryland. They found that 23% of adult woodchucks tested were positive for WHsAg, and 36% were positive for anti-WHs antibody, an overall rate of WHV infection of 59%. Similar WHV seroprevalence data from the mid-Atlantic states including Delaware have been reported by Wong *et al.* (1982).

In contrast to the high rates of WHV infection in the mid-Atlantic states, WHV infection in New York state and New England is unusual. In New York, the rate of WHV infection, based on detection of anti-WHs antibody, was estimated to be approximately 2% (Wong *et al.*, 1982). Our work suggests that the rate of persistent, WHsAg-positive infection is no greater than 0.2%. Although the number of woodchucks tested is limited, no evidence of WHV infection has been found in Vermont, Massachusetts, or Iowa (Young and Sims, 1979; Summers, 1981; Lutwick *et al.*, 1982). Woodchucks from the low-prevalence area of Central New York have been valuable for the development of breeding colonies to produce laboratory-born and reared experimental animals.

The laboratory woodchuck

The woodchuck has been developed as a laboratory animal model for HBV research. Woodchucks trapped in their native habitat with chronic WHV infection were found to have chronic hepatitis, and many ultimately developed HCC. Initial experimental work, however, showed that while acute WHV infection could be produced in susceptible adult woodchucks, the infection was characteristically self-limited. This suggested that persistent WHV infection in the wild woodchuck population was the result of WHV infection early in life, similar to the situation in humans. If this were true, experimental studies of chronic WHV infection and HCC would require inoculation of WHV during the neonatal period. For this, methods were required for breeding and maintaining woodchucks under laboratory conditions.

Laboratory rearing of woodchucks was also required to control several diseases of wild woodchucks, including infection with *Ackertia marmota* and *Capillaria* sp., both of which were associated with liver disease. Rabies is also an important infection of wildlife, including woodchucks in the Eastern United States. The raccoon is a species of major importance in the current epidemic. Raccoons share burrows with woodchucks and may be a source of rabies in wild woodchucks.

Experimental WHV infection

A woodchuck colony was established at Cornell University in 1979. It has been possible to produce woodchucks in the laboratory, to infect neonatal woodchucks successfully with WHV, and to produce woodchucks with persistent WHV infection for studies of pathogenesis. Two principal factors influence the outcome of experimental neonatal WHV infection. One factor is age. High rates of chronic WHV infection (50–70%) develop in woodchucks infected during the first week of life. In woodchucks infected with WHV at 2 months of age or older, however, chronic carrier rates of 10% or less are characteristically observed.

Another factor determining outcome of neonatal infection has been the infectious dose of the WHV inoculum. In one study, 24 of 24 neonatal woodchucks inoculated with a standard dose of 100 μl of a 10^{-1} dilution of serum (5 million $WCID_{50}$ or 50% infectious dose for woodchucks) became infected based on development of anti-core antibody (anti-WHc). Twenty-three of 24 became detectably viremic, and 17 of the 24 became chronic carriers. At a dilution of 10^{-5}, the same volume of inoculum resulted in WHV infection in 21 of 23 woodchucks. Ten of 21 woodchucks inoculated at the low dose became viremic, but only 2 of 21 became chronic carriers. The difference in infectious dose altered the incubation period. At the high dose, the first serologic marker of WHV infection was detected at 2 months of age on average and at the low dose was slightly more than 3 months. Because of the influence of age on the rate of chronicity, this difference in incubation period may be a relevant factor in determining the relationship between infectious dose and chronicity.

Following neonatal inoculation with WHV, there was characteristically no clinical evidence of acute illness. After 1 year, chronic WHV carriers had mild histologic evidence of hepatitis. Thereafter, portal hepatitis became progressively more severe. In our studies, the median time to death in carriers was 29 months and the longest survivor lived for 56 months. The lifetime risk of developing HCC was virtually 100% in chronic WHV carriers. Woodchucks in which WHV infection was resolved developed anti-WHsAg antibody and had a significantly increased life expectancy compared to carriers. At 28 months of age, when 50% of the chronic carriers had died of HCC, 80% of woodchucks with resolved WHV infection were alive. Fifty percent of the recovered group survived for 60 months. The lifetime risk of HCC was 20% in woodchucks infected with WHV at birth and in which there was resolution of infection. In uninfected control woodchucks raised under similar laboratory conditions, survival at 28 months was 90% and was 60% at 60 months. HCC was not observed in uninfected control woodchucks.

Antiviral studies

Adult woodchucks with experimentally induced chronic WHV infection have been used for the preclinical evaluation of antiviral agents. Woodchucks are tested prior to initiation of therapy for serologic markers of WHV infection, and woodchucks positive for WHsAg, WHV DNA, and anti-WHc antibody are used. Serum WHV DNA concentration is measured before treatment, at regular intervals during 4-, 12-, or 24-week periods of treatment and for 12 weeks following drug withdrawal. Ultrasound-directed percutaneous liver biopsies are obtained before treatment, after treatment is ended, and during post-treatment follow-up. The biopsies are examined histologically by routine light microscopy and histochemically for WHc and WHs antigens. Aliquots of the biopsies are also analyzed quantitatively for WHV nucleic acids.

The woodchuck model has now been used to test 15 drugs for antiviral activity (Table 124.2). Of these, 13 were nucleoside analogs, and two were immune-response modifiers. The nucleoside analogs, except for acyclovir and AZT, were all active *in vitro* against HBV in the 2.2.15 cell system (Korba and Gerin, 1992). Drugs were ranked based on their *in vitro* selective indices (SI; Table 124.2). Acyclovir and AZT had no selectivity in 2.2.15 cells and had no antiviral effect in the woodchuck model. Adenine arabinoside 5′-monophosphate (ARA-AMP), which had moderate antiviral activity *in vitro*, had significant antiviral activity in woodchucks *in vivo* at the 15 mg/kg per day dose (Enriquez *et al.*, 1995). Most nucleoside analogs with intermediate antiviral activity *in vitro* against HBV had intermediate antiviral activity *in vivo*, except for 1,-2′deoxy-2′-fluoro-1-β-D-arabinofuranosyl-5-iodo-uracil (D-FIAU) which had potent antiviral activity in woodchucks but was highly toxic (see below). Lobucavir also had intermediate *in vitro* activity in the 2.2.15 cell system but at the dosage studied was a highly potent *in vivo* antiviral drug against WHV.

Carbocyclic 2′-deoxyguanosine (2′CDG), which had very potent antiviral activity in woodchucks with chronic WHV infection, was also remarkably toxic. Lamivudine, which had the second highest SI *in vitro*, was a potent antiviral drug in woodchucks and was without apparent toxicity at doses of 5 or 15 mg/kg per day for 28 days.

Extended treatment of chronic WHV carrier woodchucks with lamivudine has delayed the development of HCC and significantly extended survival (Peek *et al.*, 1997a). Cis-5-fluoro-1-[2-(hydroxymethyl)-1,3-oxathiolan-5-YL]-cytosine (FTC), which had the highest *in vitro* selective index, was also a remarkably potent drug and was without toxicity. With most nucleosides tested, including FTC and lamivudine, withdrawal of drug treatment was followed by prompt return of serum WHV DNA to pre-treatment levels, characteristically within 1–2 weeks after treatment ended. When D-FIAU was withdrawn after 4 weeks of treatment, return of serum WHV DNA to the pre-treatment value was delayed for more than 12 weeks.

A similar delay in WHV recrudescence of 13 weeks was observed following Lobucavir withdrawal. The most potent of the nucleoside analogs so far tested in woodchucks is 1-(2-fluoro-5-methyl-β-L-arabinofuranosyl)-uracil (L-FMAU). After 12 weeks of treatment, serum WHV DNA was undetectable in all treated woodchucks and remained undetectable for more than 28 weeks after treatment ended in 3 of 4 chronic WHV carriers (Peek *et al.*, 1997b). Interferon-α and thymosin α_1 (TA1) are immune-response modifiers that were tested in woodchucks. The interferon-α (rhuIFN α B/D) used in woodchucks had significant antiviral activity both *in vitro* (Table 124.2) and *in vivo* (Gangemi *et al.*, 1996). TA1 given at a dose of 900 μg/m^2 twice weekly by subcutaneous injection also produced a highly significant antiviral effect in chronic WHV carriers (Gerin *et al.*, 1992; Tennant *et al.*, 1993) but had no detectable *in vitro* activity. As with the nucleosides, serum WHV DNA returned to near pre-treatment levels within 12 weeks following withdrawal of either interferon-α or TA1.

Toxicity of antiviral drugs

The majority of antiviral drugs tested so far have been found to be without toxicity in the woodchuck after 4 or 12 weeks of treatment. 2′CDG caused acute toxicity at a dose as low as 0.06 mg/kg per day. With 1,-2′deoxy-2′-fluoro-1-β-D-arabinofuranosyl-5-ethyl-uracil (D-FEAU; Korba *et al.*, 1991) and D-FIAU (Tennant *et al.*, 1998) toxicity was not observed until 6–7 weeks of treatment (Table 124.2). The closely related D-FMAU has also been shown to have delayed toxicity in woodchucks (Fourel *et al.*, 1990; Peek *et al.*, 1997b).

The three pyrimidine drugs, D-FEAU, D-FIAU, and D-FMAU demonstrated a similar, delayed pattern of toxicity. D-FEAU did not appear to be toxic after either 10 days or 4 weeks of treatment, but after 12 weeks of continued treatment at 0.2 mg/kg per day, the drug was highly toxic and treatment was associated with significant mortality. In a preliminary trial of D-FIAU, using doses of 0.3 and 1.5 mg/kg per day given intraperitoneally for 4 weeks, no significant toxicity was observed. Signs of delayed toxicity were observed after 7–8 weeks of oral treatment (Tennant *et al.*, 1998). A similar pattern of delayed toxicity has also been observed with D-FMAU (Peek *et al.*, 1997b).

A clinical study of D-FIAU in HBV-infected patients initiated in 1993 was terminated because of severe clinical toxicity (McKenzie *et al.*, 1995). Prior to the trial, preclinical toxicological investigations of D-FIAU had been performed using conventional laboratory animal species (mice, rats, dogs, primates) with normal livers and the

Table 124.2 Comparative antiviral activity and toxicity of antiviral agents *in vitro* and *in vivo*

Drug	*In vitro*		*In vivo*				
	Rank	Selective index (2.2.15 cells)*	Dose (mg/kg per day)	Route	Treatment duration (days)	Toxicity†	Antiviral efficacy‡
FTC	1	9054	10	p.o.	28	–	+++
Lamivudine	2	8900	5	p.o.	28	–	++
			15	p.o.	28	–	+++
			5	p.o.	84	–	+++
L-FMAU	3	838	10	p.o.	84	–	++++
2′CDG	4	126	1.5	i.p.	7	++++	+
			0.15	i.p.	7.17	++++	Not tested
			0.075	i.p.	28	+++	+++
			0.06	p.o.	14	++	++
			0.019	p.o.	28	+	++
IFN-α	5	76	5 million IU/kg per 2 days	s.q.	84	–	+
Lobucavir	6	50	20	p.o.	84	+	++++
BW816U88	7	26	15	i.p.	84	–	+
BW145U87	8	18	15	i.p.	84	–	+
BW1592U89	8	18	15	i.p.	84	–	+
D-FIAU	9	9	0.3	i.p.	28	–	+/–
			1.5	i.p.	28	–	++
			1.5	p.o.	84	++++	++++
ARA-AMP	10	5	15	i.p.	28	–	++
Ribavirin	10	5	40	i.p.	28	+	+
D-FEAU	11	2.6	2.0, 0.2	i.p.	84	++++	+++
AZT	12	Negative	15	i.p.	28	–	–
Acyclovir	12	Negative		p.o.	28	–	–

Structures of drugs
FTC, Cis-5-fluoro-1-[2-(hydroxymethyl)-1,3-oxathiolan-5-YL]-cytosine
Lamivudine (3TC), (-)-β-L-2′,3′-dideoxy,3′-thiacytidine
L-FMAU, 1-(2-fluoro-5-methyl-β-L-arabinofuranosyl)-uracil
2′CDG, Carbocyclic 2′-deoxyguanosine
IFN-α, rhu interferon-α B/D, recombinant human interferon-α B/D
Lobucavir (BMS-180194): [IR-(1α, 2β, 3α)]-2-amino-9-[2,3-bis(hydroxymethyl)cyclobutyl]-1,9-dihydro-6H-purin-6-one
BW816U88, (+–)-Cis-4-(2-amino-6-(cyclopropylamino)-9H-purin-9-YL)-2-cyclopentene-1-methanol, a ddG prodrug
BW145U87, (2R,5S)-2-Amino-6-(cyclopropylamino)-9-(tetrahydro-5-(hydroxymethyl)-2-furyl)-9H-purine, a ddG prodrug
BW1592U89, (1S,4R)-4-(2-Amino-6-(cyclopropylamino)-9H-purin-9-YL)-2-cyclopentene-1-methanol, a ddG prodrug
D-FIAU, 1,-2′ Deoxy-2′-fluoro-1-β-D-arabinofuranosyl-5-iodo-uracil
ARA-AMP, Adenine arabinoside 5′-monophosphate
Ribavirin, 1-β-D-ribofuranosyl-1,2,4 triazole-3-carboxamide
D-FEAU, 1,-2′Deoxy-2′-fluoro-1-β-D-arabinofuranosyl-5-ethyl-uracil
AZT, 3′-Azido-3′-deoxythymidine, zidovudine

* The Selective Index (SI) is defined as the EC_{90} (the concentration of drug at which 90% of HBV DNA production was inhibited) divided by the CC_{50} (the concentration of drug at which cytotoxicity of 50% of 2.2.15 cells was observed).

† 2′CDG produced prompt multiorgan failure in a dose-dependent manner. Both FIAU and FEAU produced delayed toxicity not recognizable clinically or with biochemical tests until 7 weeks or more of antiviral drug therapy. With all three drugs, the actual mechanisms of toxicity are not known but with each there was evidence of multiorgan failure. Ribavirin toxicity was limited to reversible anemia. Uniformly, fatal responses were classified as (+ + +) or (+ + + +), the latter being the most prompt. Clinically significant but reversible toxicity was identified as (+) or (+ +) responses.

‡ Antiviral activity was defined as (+) if there was at least a 10-fold reduction in serum WHV DNA at the end of the treatment period compared to controls; (+ +) and (+ + +) responses were defined as 2- and 3-log reductions in serum WHV DNA compared to controls; a (+ + + +) response was defined as a greater than 3-log reduction of serum WHV DNA at the end of treatment compared to controls.

question was raised whether the severe hepatotoxicity associated with D-FIAU treatment in patients with chronic HBV infection might have been related to pre-existing chronic hepatitis. A study was designed using the woodchuck to investigate the possibility that chronic HBV infection might have determined the toxicity observed in patients. The study involved four groups of woodchucks. One group was WHV-negative and received placebo. Woodchucks of the second group were also WHV-negative and received D-FIAU at a dose of 1.5 mg/kg per day. The third group of WHV carrier woodchucks received placebo, and the fourth group of carriers received D-FIAU at the 1.5 mg/kg per day dose.

After 7 weeks of treatment, a reduction in mean body weight was observed in both FIAU-treated groups compared to placebo-treated controls that gained weight. From 7 weeks onward, there was a progressive reduction in food intake in the D-FIAU-treated groups compared to placebo control groups. D-FIAU treatment was associated with severe metabolic acidosis. After 8 weeks, serum bicarbonate levels began to fall gradually and then fell precipitously during the final days of life. The base deficit in the two D-FIAU-treated groups was significantly greater than that of the two groups not treated with D-FIAU. Acidosis was also associated with significant elevation in serum lactate—a finding that was also observed in human patients with D-FIAU toxicity (McKenzie *et al.*, 1995).

Additional evidence of liver disease in the D-FIAU-treated groups compared with the placebo-treated animals included decreases in the serum fibrinogen concentration and increases in prothrombin time (Tennant *et al.*, 1998).

The D-FIAU-treated animals developed fatty liver which was primarily microvesicular in type and, as such, was similar to that seen in humans (McKenzie *et al.*, 1995). Using an arbitrary scoring system to quantify the amount of steatosis present, it was found that significant amounts of fatty change were present only in the D-FIAU-treated animals. Half of the D-FIAU-treated woodchucks died after 12 weeks or less and the remaining woodchucks in the drug-treated groups died within 4 weeks after treatment ended. Survival was not influenced by the presence or absence of chronic WHV infection, suggesting that the woodchuck was uniquely susceptible to D-FIAU toxicity, similar to humans and different from that observed in mice and rats treated at much higher doses (Tennant *et al.*, 1998).

CK Chu and Y-C Cheng have synthesized and tested the L-isomers of FEAU, FIAU, and FMAU and have compared their cytotoxicity and anti-HBV activity *in vitro* to the D-isomers. All three L-isomers were much less cytotoxic than the D-isomers and L-FMAU was found to have significant *in vitro* antiviral activity against HBV (Ma *et al.*, 1996; Pai *et al.*, 1996). The pharmacokinetic (Witcher *et al.*, 1997) and pharmacodynamic (Peek *et al.*, 1997b) studies so far completed in the woodchuck suggest that L-FMAU is among the most important drug candidates available for development as therapy for chronic HBV infection.

Conclusions

The woodchuck model has played a valuable role in the investigation of the pathogenesis of hepadnaviral infections and hepatocarcinogenesis. It is now becoming valuable in the development of antiviral drugs and possibly may be useful in evaluating new forms of immunotherapy both for efficacy and possible toxicity. In particular, the woodchuck seems to be ideal for studying the impact of antiviral and immunotherapy on the outcome of hepadnavirus infection and on survival. The median life expectancy of the chronic WHV carrier woodchuck is 29 months and almost all develop HCC, providing an opportunity to determine the effects of new forms of therapeutic intervention in a relatively short period of time.

Acknowledgments

Part of the work described above was supported by Contract NO1-AI-35164 to BCT from the National Institute of Allergy and Infectious Diseases, NIH. The work represents an ongoing collaboration with Drs Paul Cote, Brent Korba and John Gerin of the Division of Molecular Virology and Immunology, Georgetown University School of Medicine, Rockville, MD.

References

Enriquez, P. M., Jung, C., Josephson, L., Tennant, B. C. (1995). Conjugation of adenine arabinoside 5′-monophosphate to arabinogalactan: synthesis, characterization, and antiviral activity. *Bioconjugate Chem.*, **6**, 195–202.

Fourel, I., Hantz, O., Watanabe, K. A. *et al.* (1990). Inhibitory effects of 2′-fluorinated arabinosyl-pyrimidine nucleosides on woodchuck hepatitis virus replication in chronically infected woodchucks. *Antimicrob. Agents Chemother.*, **34**, 473–475.

Ganem, D., Varmus, H. E. (1987). The molecular biology of the hepatitis B viruses. *Annu. Rev. Biochem.*, **56**, 651–693.

Gangemi, J. D., Korba, B. E., Tennant, B. C., Ueda, H., Gilbert, J. (1996). Antiviral and anti-proliferative activities of α interferons in experimental hepatitis B virus infections. *Antiviral Ther.*, **1** (suppl. 4), 64–70.

Gerin, J. L., Korba, B. E., Cote, P. J., Tennant, B. C. (1992). A preliminary report of a controlled study of thymosin alpha-1 in the woodchuck model of hepadnavirus infection. In *Innovations in Antiviral Development and the Detection of Virus Infection*. (eds Block, T. M., Jungkind, D., Crowell, R. L., Denison, M., Walsh, L. R.), pp. 121–123. Plenum Press, New York.

Gust, I. D., Burrell, C. J., Coulepis, A. G., Robinson, W. S., Zuckerman, A. J. (1982). Taxonomic classification of human hepatitis B virus. *Intervirology*, **25**, 14–29.

Korba, B. E., Gerin, J. L. (1992). Use of a standardized cell culture assay to assess activities of nucleoside analogs against hepatitis B virus replication. *Antiviral Res.*, **19**, 55–70.

Korba, B. E., Cote, P. J., Tennant, B. C., Gerin, J. L. (1991). Woodchuck hepatitis virus infection as a model for the development of antiviral therapies against HBV. In *Viral Hepatitis and Liver Disease*. (eds Hollinger, F. B., Lemon, S. M., Margolis, H.), pp. 663–665. Williams & Wilkins, Baltimore.

Lutwick, L. I., Hebert, M. B., Sywassink, J. M. (1982). Cross-reactions between the hepatitis B and woodchuck hepatitis viruses. In: *Viral Hepatitis*. (eds Szmuness, W., Alter, H. J., Maynard, J. E.), p. 711. Franklin Institute Press, Philadelphia.

Ma, T., Pai, S. B., Zhu, Y. L. *et al.* (1996). Structure-activity relationships of 1-(2-deoxy-2-fluoro-β-l-arabinofuranosyl) pyrimidine nucleosides as anti-hepatitis B virus agents. *J. Med. Chem.*, **39**, 2835–2843.

McKenzie, R., Fried, M. W., Sallie, R. *et al.* (1995). Hepatic failure and lactic acidosis due to fialuridine (FIAU), an investigational nucleoside analogue for chronic hepatitis B. *N. Engl. J. Med.*, **333**, 1099–1105.

Melnick, J. L. (1982). Classification of hepatitis A virus as enterovirus type 72 and hepatitis B virus as hepadnavirus type 1. *Intervirology*, **18**, 105–106.

Pai, S. B., Liu, S.-H., Zhu, Y.-L., Chu, C. K., Cheng, Y.-C. (1996). Inhibition of hepatitis B virus by a novel L-nucleoside, 2-fluoro-5-methyl-β-l-arabinofuranosyl uracil. *Antimicrob. Agents Chemother.*, **40**, 380–386.

Peek, S. F., Jacob, J. R., Tochkov, I. A. *et al.* (1997a). Substantial antiviral activity of 1-(2-fluoro-5-methyl-β-l-arabinofuranosyl) uracil (L-FMAU) in the woodchuck (*Marmota monax*) model of hepatitis B virus infection. *Hepatology*, **26**, 425a.

Peek, S. F., Tochkov, I. A., Erb, H. N. *et al.* (1997b). 3′-thiacytidine (3TC) delays development of hepatocellular carcinoma (HCC) in woodchucks with experimentally induced chronic woodchuck hepatitis virus (WHV) infection. Preliminary results of a lifetime study. *Hepatology*, **26**, 368A.

Robinson, W. S. (1980). Genetic variation among hepatitis B and related viruses. *Ann. N. Y. Acad. Sci.*, **354**, 371–378.

Robinson, W. S., Marion, P. L., Feitelson, M., Siddiqui, A. (1982). The hepadnavirus group: hepatitis B and related viruses. In *Viral Hepatitis*. (eds Szmuness, W., Alter, H. J., Maynard, J. E.), p. 57. Franklin Institute Press, Philadelphia.

Schodel, F., Sprengel, R., Weimer, T., Fernholz, D., Schneider, R., Will, H. (1989). Animal hepatitis B viruses. *Adv. Viral Oncol.*, **8**, 73–102.

Schroder, H. B., Zentgraf, H. (1990). Hepatitis B virus related hepatocellular carcinoma: chronicity of infection – the opening to different pathways of malignant transformation. *Biochem. Biophys. Acta.*, **1032**, 137–156.

Seeger, C., Ganem, D., Varmus, H. E. (1986). Biochemical and genetic evidence for the hepatitis B virus replication strategy. *Science*, **232**, 477–484.

Summers, J. (1981). Three recently described animal virus models for human hepatitis B. *Hepatology*, **1**, 179–183.

Summers, J. (1987). The replication cycle of hepatitis B viruses. *Cancer*, **61**, 1957–1962.

Summers, J., Smolec, J. M., Snyder, R. (1978). A virus similar to human hepatitis B virus associated with hepatitis and hepatoma in woodchucks. *Proc. Natl. Acad. Sci. USA*, **75**, 4533–4537.

Tennant, B. C., Korba, B. E., Baldwin, B. H. *et al.* (1993). Treatment of chronic woodchuck hepatitis virus infection with thymosin alpha-$_1$ (TA$_1$). *Antiviral Res.*, **20** (suppl. 1), 163.

Tennant, B. C., Baldwin, B. H., Graham, L. A. *et al.* (1998). Antiviral activity and toxicity of fialuridine in the woodchuck model of hepatitis B virus infection. *Hepatology*, **28**, 179–191.

Tiollais, P., Pourcel, C., Dejean, A. (1985). The hepatitis B virus. *Nature*, **317**, 489–495.

Tyler, G. V., Summers, J., Snyder, R. (1981). Woodchuck hepatitis virus in natural woodchuck population. *J. Wildlife Dis.*, **17**, 297–301.

Witcher, J. W., Boudinot, F. D., Baldwin, B. H. *et al.* (1997). Pharmacokinetics of 1-(2-fluoro-5-methyl-β-l-arabinofuranosyl) uracil in woodchucks. *Antimicrob. Agents Chemother.*, **41**, 2184–2187.

Wong, D. C., Shih, J. W. K., Purcell, R. H., Gerin, J. L., London, W. T. (1982). Natural and experimental infection of woodchucks with woodchuck hepatitis virus as measured by new specific assays for woodchuck surface antigen and antibody. *J. Clin. Microbiol.*, **15**, 484–490.

Young, R. A., Sims, E. A. H. (1979). The woodchuck, *Marmota monax*, as a laboratory animal. *Lab. Anim. Sci.*, **29**, 770.

Chapter 125

Animal Models of Papillomavirus Infections

N. D. Christensen and J. W. Kreider

Background of model systems

Human papillomaviruses (HPVs) include more than 80 types, some of which can induce neoplastic proliferation of human epithelial cells. The lesions can result in life-threatening disease due to mass obstruction (juvenile laryngeal papillomatosis), or increased risk of squamous cell carcinoma of human anogenital tissues and skin (reviewed in Schiffman *et al.*, 1993; zur Hausen, 1994). The study of HPV infections and their treatment in intact hosts provide unique challenges. Animal models of HPV infection are handicapped by the strong species and tissue restriction of HPV. In short, HPV infection does not result in viral replication or production of papillomas in laboratory animal tissues. The site-specificity of HPV infections (exclusively targeting epithelial tissues) requires an animal model system of equivalence in which natural and/or experimental papillomas can be induced. Animal tumor cell lines transfected with HPV gene products under the control of heterologous promoters are an inadequate substitute for HPV infection. We have used two model systems: first, the cutaneous-targeting cottontail rabbit papillomavirus (CRPV)/domestic rabbit model (Shope and Hurst, 1933); and second, the athymic mouse/HPV-11-infected human foreskin xenograft model (Kreider *et al.*, 1985, 1987a; Bonnez *et al.*, 1993) to test potential therapeutic agents for anti-papillomavirus activity. Both model systems have been established and validated by our laboratory as effective and appropriate models for assessing therapeutic treatment of HPV infections (Kreider *et al.*, 1990, 1992; Kreider and Pickel, 1993; Okabayashi *et al.*, 1993; Gangemi *et al.*, 1994; Kreider and Christensen, 1994).

Current treatments for papillomavirus infections are primarily extirpative, using scalpel, lazer, freezing, or toxic agents applied topically (trichloroacetic acid, phenol, sulfcyclic acid, popophyllin, 5-fluorouracil, acyclic nucleotides and podofilox). Recurrence rates are very high, usually greater than 50% (Auborn and Steinberg, 1990). In addition, subclinical infections often go undetected and untreated. These latter infections may be reactivated by a variety of poorly characterized events and agents such as environmental carcinogens and/or co-factors, hormones, wounding, immune suppression and other sexually transmitted disease agents to produce new active clinical disease. Thus, current treatments are unsatisfactory, and there is an urgent need to develop drugs with greater efficacy, as well as to develop animal model systems that demonstrate clinical correlations.

There are few alternative animal model systems (in addition to those discussed above) for testing therapeutic agents for HPV infections. Those that are potentially available include bovine papillomavirus (BPV) infection of cattle (Olson, 1963; Campo, 1987), canine oral papillomavirus (COPV) infection of dogs (Sundberg *et al.*, 1994; Suzich *et al.*, 1995), and hamster oral papillomavirus infection of hamsters (Iwasaki *et al.*, 1997). The first two models are considerably more expensive than rabbits, and have not been used for antiviral testing. We have also developed a second rabbit papillomavirus model using the rabbit oral papillomavirus (ROPV) which targets mucosal tissues (Parsons and Kidd, 1936; Christensen *et al.*, 1996a). In this chapter, we will describe our experience with the two rabbit papillomavirus model systems and the athymic mouse xenograft system (for the production and growth of HPV-11 and HPV-40 infections in human tissues) for testing antiviral compounds.

Animal species

Outbred New Zealand White (NZW) rabbits are routinely used for infection by CRPV and ROPV. We also have an inbred strain of rabbits (EIII/JC) that can be used for studies requiring a constant genotype. All rabbits are equally susceptible to infection by both viruses, and the immunity that develops in response to CRPV does not provide cross-protection to ROPV and vice versa (Parsons and Kidd, 1942). For studies using HPV-11 and HPV-40 infection of human foreskin fragments that are implanted subrenally, we use athymic and severe combined immunodeficiency (SCID) mice obtained from various vendors with equal success.

Preparation and housing of animals

Outbred and inbred domestic rabbits are housed individually in stainless-steel cages without further specialized housing. Inbred (EIII/JC) rabbits are housed in a separate

Handbook of Animal Models of Infection
ISBN 0-12-775390-7

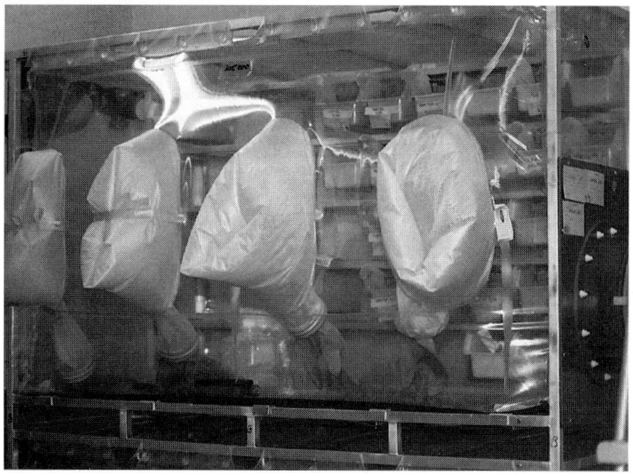

Figure 125.1 Flexible film isolator for housing athymic and severe combined immunodeficiency (SCID) mice. Each plastic isolator can house up to 32 cages of mice.

room where handling noises are minimized to prevent the audiogenic seizures to which these rabbits are prone. Special quarters including nest box are provided for breeding females. Daily veterinary care is provided to insure that the health and well-being of the animals are maintained.

Athymic and SCID mice are housed in specialized isolation facilities. Flexible film isolators are serviced by an air-handling system which delivers highly filtered air under pressure. The system derives air from non-animal usage areas, pulling first across a coarse filter, then pressurizing it by means of a high-speed blower, and finally passing it through high-efficiency particulate air (HEPA) filtration. Air enters the room via a supply duct and is routed by means of a flexible hosing system to individual isolators. Each isolator is provided with inlet and exhaust filters of triple-wrapped fiberglass media on a solid stainless-steel support frame. These filters not only prevent possible influx of air from the exhaust port but also maintain pressurization of the isolator for periods of several hours in the event of a failure in the air supply. Large flexible film isolators are equipped to house 32 cages of mice (Figure 125.1) and small flexible film isolators contain six cages along with associated working supplies or materials.

All supplies and research materials are sterilized by chemicals or steam prior to introduction to the isolator. Contaminated materials from the isolator are autoclaved prior to discard. A separate transfer isolator is used for surgical procedures on the nude mice. This unit interfaces with the receiving ports of isolators housing the nude mice.

Surgical procedures and grafting of xenografts

No invasive surgical procedures are conducted on the rabbits. Infection procedures require only superficial scarification and needle puncture under general anesthesia as described below.

Surgical procedures on athymic and SCID mice are conducted under Nembutal anesthesia (75 μg Nembutal per gram body weight). The skin of the mice is sterilized first with 70% ethanol. The kidneys are exposed by means of paravertebral, mid dorsal incisions beginning at the costovertebral angle. A small nick is made in the renal capsule, and the graft is inserted with the aid of an electron microscopy toothless forceps. Egress of the graft from the incision is prevented with mild pressure from a sterile cotton swab. The kidney is delivered into the abdomen, and the skin wound closed with stainless-steel clips.

After 3 months of growth, the mice are killed and the kidneys removed. The grafts are measured in three dimensions and the geometric mean diameter (GMD) for each xenograft is calculated. Microscopic sections stained with hematoxylin and eosin (H&E) are routinely prepared and the morphology of the graft evaluated. We also examine routinely a section of liver to detect the presence of mouse hepatitis virus, an extremely infectious and highly lethal infection which can completely destroy an athymic mouse colony. We have had no incidents of contamination of athymic mice for over 15 years. We experience excellent survival of the mice for their entire natural lifetime (> 1 year).

Preparation of infectious stocks of CRPV, ROPV, HPV-11 and HPV-40

The preparation of infectious stocks of various animal and especially of HPV has presented researchers with significant challenges. We have established an athymic mouse xenograft system (Kreider *et al.*, 1985) using initial stocks of infectious papillomavirus virions and tissues of the appropriate host for the production of our virus stocks. The details of the athymic mouse xenograft system will be described below under the infection protocols.

Infectious stocks of CRPV, ROPV, HPV-11 and HPV-40 were produced using the athymic mouse xenograft system (Kreider *et al.*, 1985). The initial starting material for each of these virus stocks was as follows: for CRPV, we obtained cutaneous papillomas from cottontail rabbits trapped in Kansas, USA (Kreider, 1980); for ROPV we obtained several small tongue papillomas from domestic NZW rabbits obtained from a local Pennsylvanian rabbit breeder (Christensen *et al.*, 1996a); for HPV-11, we obtained pooled condylomas from patients attending a Philadephia hospital (Kreider *et al.*, 1985); and for HPV-40, we obtained an anal condyloma from a transplant patient attending Hershey Medical Center (Christensen *et al.*, 1997). Initial infectious stocks of virions were incubated with the appropriate matching host tissue fragments for each virus type prior to implanting subrenally into athymic or SCID mice. For the production of CRPV we used fragments of cottontail rabbit cutaneous tissues incubated with CRPV, and for ROPV production, fragments of NZW rabbit oral mucosal tissue that were incubated with infectious ROPV. For production of infectious stocks of the two HPV types, fragments of human foreskin from neonatal circumcisions were infected with a suspension of HPV virions obtained from condylomas, and placed beneath the renal capsule of athymic mice. After 3–5 months, the grafts from each set of virus-infected xenografts were harvested separately, homogenized, and centrifuged at low speed. The supernatant was then used as the starting source of our stocks of infectious virions. This method provided us with high-titered stocks of HPV-11 (Kreider *et al.*, 1985, 1987a), HPV-40 (Christensen *et al.*, 1997), CRPV (Christensen and Kreider, 1990) and ROPV (Christensen *et al.*, 1996a). Viral stocks of each papillomavirus are stored in phosphate-buffered saline (PBS) at –70°C.

Infectivity procedures and establishment of papillomas

Rabbit cutaneous papillomas induced by CRPV

Rabbits are lightly anesthetized with a mixture of ketamine HCl (40 mg/kg) and xylazine (5 mg/kg), and their backs shaved with electric clippers. During this procedure the rabbits are immobilized on a stretcher-style rabbit board. Two sites on each side of the flanks of the mid-dorsum are scarified with a scalpel until abraded areas of approximately 2 × 2 cm are produced. The pattern of papilloma "targets" is illustrated in Figure 125.2.

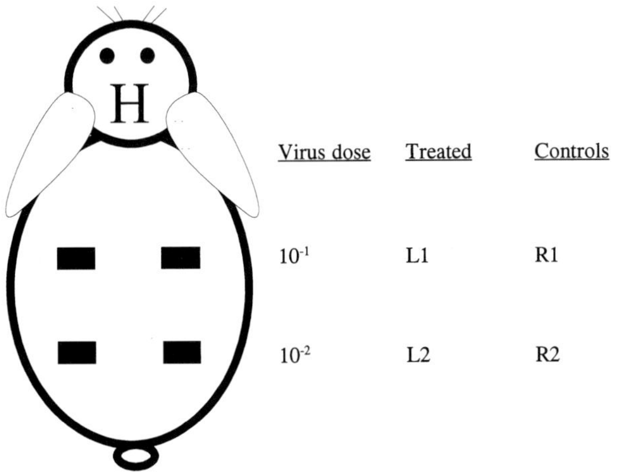

Figure 125.2 Diagram of sites scarified and infected with cottontail rabbit papillomavirus (CRPV). Four sites on the dorsum of rabbits are infected with two doses of CRPV extract. High-dose virus (10^{-1} dilution of CRPV extract) and low-dose virus (10^{-2} dilution of CRPV extract) are placed on to the scarified sites as indicated on the diagram. Topical application of antiviral compounds is usually begun 21 days after virus inoculation when papillomas are visible and in the active growth stage.

The view is of the dorsum of the rabbit, with the location of the head indicated (H), and papilloma challenge sites are represented by rectangles. The two anterior papillomas (L1, L = left and R1, R = right) on each side are induced with a dilution of 10^{-1} CRPV extract, whereas the two posterior papillomas (L2 and R2) are induced with a dilution of 10^{-2}. The use of four papillomas per rabbit reduces the impact of potential idiosyncratic events (trauma from handling, moving in the wire cage, chewing by the host) which might interfere with the growth of a single papilloma and thereby bias the data.

CRPV extract (10% w/v in PBS) is prepared by homogenization of cottontail papillomas (obtained from Pennsylvania cottontail xenografts as described above). Each scarified site receives 0.05 ml of papilloma extract, rubbed gently into the wound, and dried with a stream of warm air from a hair-dryer. The animals are observed, beginning at 2 weeks, for the development of papillomas. Papillomas are measured in three dimensions (length × width × height in mm) and the GMD is calculated. Rabbits are fitted with Elizabethan collars to prevent biting, chewing and scratching the papillomas. Severe toxicity and death can result if a rabbit gains access to a topical drug. "Noise" in the data due to chewing of papillomas may interfere with drug assessment even if the drug is given systemically. The prevention of chewing of the papillomas by these collars therefore is key to the effective assessment of antiviral compounds.

Rabbit oral papillomas induced by ROPV

Papillomas on the underside of the tongue of rabbits are generated using ROPV (Christensen *et al.*, 1996a). *Note: CRPV does not induce papillomas in rabbit mucosal tissues* (Kidd and Parsons, 1936). The rabbits are anesthetized using a mixture of ketamine HCl (40 mg/kg) and xylazine (5 mg/kg). Tongues are exposed using forceps, and several pinpricks (10–30) using a 25 G needle are applied over a 1 cm^2 region of mucosal surface. A suspension of ROPV prepared from ROPV-infected NZW rabbit tongue xenografts (Christensen *et al.*, 1996a) is pipetted on to the wounded area in a 20 µl volume. Infected regions are examined visually every 3 days, beginning at 2 weeks after virus inoculation. Papillomas reach a maximum size of 2 mm in diameter and 1 mm in height by Day 25–35. Viral infection is quantitated by counting the number and size of papillomas.

Infection of human foreskin fragments with HPV-11 and HPV-40

Human foreskins are obtained from routine neonatal circumcisions at area hospitals. The foreskins are held in cell culture medium (Dulbecco's minimal essential medium containing gentamicin) for no more than 48 hours prior to use. The foreskins are stretched over a rolled gauze pad and anchored with sterile hypodermic 18 G needles, pressed into an underlying corkboard. A sterile scalpel is used to shave a thin split-thickness graft of skin from the surface. The slice obtained is cut into fragments measuring 2 × 2 × 0.5 mm. The fragments are placed in 0.2 ml of HPV-11 (or HPV-40) extract which was produced in our laboratory (Kreider *et al.*, 1987a; Christensen *et al.*, 1997). After 1 hour incubation at 37°C, the tissues are grafted to the athymic mice.

Key parameters to monitor infection and response to treatment

Papilloma size

CRPV-induced rabbit papillomas are measured weekly beginning at day 14. GMDs are determined for each papilloma, and growth curves are plotted against time for each treatment group. Statistical comparisons between mean size of treated versus untreated papillomas in response to microbicidal agents delivered topically or intralesionally are made using groups of 5 rabbits, and Student's t-test. For drugs that are delivered systemically or subcutaneously we compare mean papilloma sizes for groups of rabbits that receive drugs versus rabbits receiving placebo. ROPV-induced papillomas are measured by size (diameter × height) and number on a weekly basis beginning 14 days after inoculation. Only a single size determination of subrenal xenografts (length × width × height in mm) of HPV-11 and HPV-40 infections is made at the termination of each experiment.

In situ immunohistochemistry and viral DNA detection

In situ hybridization to detect viral DNA and immunohistochemistry to detect viral capsid antigen are routinely used to semi-quantitate viral infection for ROPV, HPV-11 and HPV-40. The level of CRPV viral DNA in domestic rabbits is usually below detection levels using standard *in situ* hybridization procedures. Biotinylated DNA probes are prepared for each viral type, and the probes detected using streptavidin conjugated alkaline phosphatase and substrate (Christensen and Kreider, 1990; Christensen *et al.*, 1997). Papillomavirus capsid antigen is detected by immunohistochemistry using polyclonal antisera to the group specific antigen (Jenson *et al.*, 1980) of papillomaviruses.

Viral content or titer

ROPV-induced rabbit papillomas and HPV-11 and HPV-40 containing xenografts produce high titers of infectious virions. In contrast, CRPV-induced domestic rabbit papillomas are virion-poor. For the first three viral types, however, we are able to quantitate the virus titers obtained from treated and untreated papillomas and xenografts by infectivity assays. For ROPV titer determination, we infect the tongues of individual rabbits with serial 10-fold dilutions of virus extract and determine the extinction point (dilution at which no papillomas are induced as measured by visual inspection). For HPV-11 and HPV-40 titer determinations we use infection of human foreskins with serial 10-fold dilutions of each virus followed by implantation into athymic mice. Viral titers are determined as the dilution of virus at which no condylomatous transformation of human foreskin xenografts are observed. In addition, we use reverse transcriptase polymerase chain reaction (RT-PCR) detection of spliced viral transcripts in monolayer cultures of human epithelial cell lines infected with HPV-11 or -40 virions (Smith *et al.*, 1995). The viral titer is calculated as the dilution of virus that no longer produces detectable levels of viral spliced transcripts following RT-PCR with nested primers (Smith *et al.*, 1993, 1995; Christensen *et al.*, 1996b).

Drug toxicity

When agents are delivered systemically, we examine a variety of parameters to determine drug toxicity. The first endpoint consists of simple observations of the treated animals: abnormal behaviors such as lassitude, anorexia, and hair loss are some of the earliest signs of toxicity. Records are made daily of abnormal behavior. We also weigh rabbits routinely each week and plot weights as a function of the duration of treatments. For drugs which are known to have myelotoxic potential, or if pallor of gums and iris (in albino rabbits) is noted, we conduct blood counts including white blood cell numbers/mm^3, red blood cells/mm^3, and hemoglobin and hematocrit. Drugs which may be hepatotoxic are evaluated with serum bilirubin and serum glutamic-oxaloacetic transaminase. Drugs which are nephrotoxic are evaluated with blood urea nitrogen and creatine levels.

If rabbits die spontaneously, a necropsy is conducted, and bacterial cultures and H&E sections examined where indicated. Autopsies or acute deaths soon after drug administration include quick freezing of blood and selected viscera for potential drug concentration determinations.

Antimicrobial therapy

Treatment of CRPV-induced rabbit papillomas

Each rabbit is infected at four cutaneous sites on the dorsal region as shown in Figure 125.2. Each treatment group contains 5 rabbits per treatment. The choice of treatment protocol, selection of drug doses and testing regimes have been perfected by our laboratory over several years of drug test-

Table 125.1 Protocols for the treatment of cottontail rabbit papillomavirus (CRPV)-induced rabbit papillomas

Protocol	*Time course for treatment of CRPV-induced rabbit papillomas*
A Treatments begin at time of CRPV infection	↓___↓___↓___↓___↓___↓___↓___↓___↓___↓___↓___ 0 Day 21 Days → 90Days
B Treatments begin at Day 21 (papillomas appear)	____________↓___↓___↓___↓___↓___↓___↓___↓___ 0 Day 21 Days → 90 Days

Arrows show the administration of test agents.

ing, and offer the best course of treatment for testing antiviral compounds in the most cost-effective manner (Kreider *et al.*, 1990, 1992; Kreider and Pickel, 1993; Okabayashi *et al.*, 1993; Gangemi *et al.*, 1994; Kreider and Christensen, 1994). Our standard course of treatments for a single agent is to use a minimum of three doses of each agent in each protocol (Table 125.1).

In previous drug-testing programs we have delivered compounds in the following dosing regimes:

New compounds

1. Five times a week, twice a day topically, 5–8 weeks (protocol A); from week 3 to 11 (protocol B)
2. Five times a week, twice a day systemically, 5–8 weeks (protocol A); from week 3 to 11 (protocol B)
3. Once a day or once a week, intralesionally, from week 3 to 11 (protocol B).

Previously tested compounds

1. Five times a week (twice a day; once a day), 5–8 weeks (protocol A); from week 3 to 11 (protocol B)
2. Once a day, systemically, from week 3 to 11 (protocol B).

Each experiment includes an uninfected and untreated control group (group A), which receives either the drug solvent alone, PBS, or other appropriate placebo. The second group (group B) receives the highest treatment dose but will not be infected with virus. The third group (group C) is infected with virus but treated with placebo (as per group A). The fourth group (group D) receives one-tenth the optimal dose. Group E is given the optimal dose of the drug, predicted upon previous literature, and/or advice of the producer. Group F receives 10 times the optimal dose (in practice, this is often only two to three times the optimal dose based on solubility and

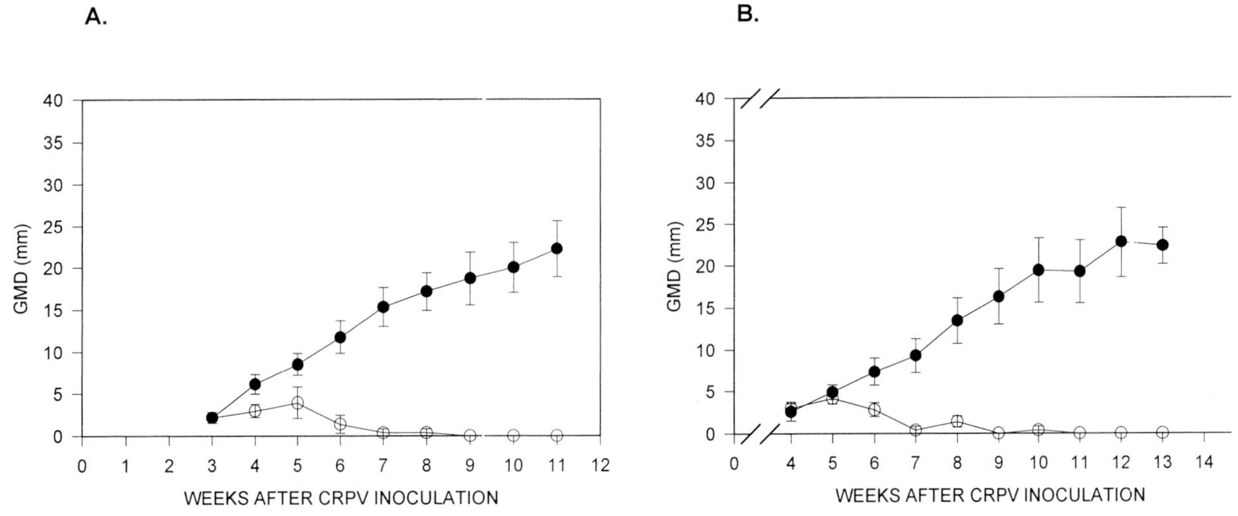

Figure 125.3 Topical antiviral treatment of cottontail rabbit papillomavirus (CRPV)-induced rabbit papillomas following treatment with acyclic nucleotides **(A)** PMEG and (B) cHPMPC. **(A)** Papillomas were treated with 0.1% PMEG topically, once per day, 5 days a week for 8 weeks beginning on week 3. **(B)** Papillomas were treated with 1.0% HPMPC twice a day, 5 days per week for 8 weeks beginning on week 4. Mean ± SEM (vertical bars) of treated (●) and untreated (○) papillomas (5 papillomas per group) are plotted against time in weeks after CRPV infection. Statistical significance is determined by comparing mean geometric mean diameter (GMD) values using Student's *t*-test.

Table 125.2 Protocols for the treatment of human papillomavirus (HPV)-11 xenografts

Protocol	*Time course for treatment of human xenografts*		
A Treatments begin at time of HPV-11 infection	↓___↓___↓___↓___↓___↓___ 0 Day	↓___↓___↓___ 60 Days	↓___↓___ → 90Days
B Treatments begin at Day 60 (latency period completed)	______________________ 0 Day	↓___↓___↓___ 60 Days	↓___↓___ → 90 Days

Arrows show the administration of test agents.

concentration determinants). If this highest dose extends into the known toxic range of the agent, the dose is reduced.

The above schedules allow us to construct a dose–response curve across this broad range. Our previous experiences (Kreider *et al.*, 1990, 1992; Kreider and Pickel, 1993; Okabayashi *et al.*, 1993; Gangemi *et al.*, 1994; Kreider and Christensen, 1994) have demonstrated that the above protocols have been effective for determining optimal drug activity. Protocol A provides the greatest sensitivity for a spectrum of agents of differing mechanisms of action, since treatments begin during the latent period (21 days) and continue during the growth phase of the papilloma. Protocol B is very analogous to the clinical treatment protocols in which treatments are usually begun on advanced lesions. In practice, protocol B is the most popular in our previous and current testing. The outcome of topical treatments of cutaneous rabbit papillomas with several antiviral compounds is shown in Figure 125.3.

Topical application of the acyclic nucleotides 9[(2-phosphonylmethoxy)ethyl]guanine (PMEG) (Figure 125.3A) and 1-[(S)-3-hydroxy-2-phosphonylmethoxypropyl]cytosine (HPMPC) (Figure 125.3B) result in complete eradication of rabbit papillomas.

Treatment of HPV-11 (and-40)-induced human xenografts in athymic mice

We use two main protocols for drug testing in the HPV-11 (and -40) athymic mouse xenograft system (Table 125.2).

The first (protocol A) provides for continuous administration of agents throughout the observation period of 90 days. We believe that this schedule offers the greatest opportunity for detecting effectiveness of a variety of agents which differ greatly in mechanism of action. This protocol is the initial screen. Protocol B is designed to be highly analogous to the clinical situation wherein patients usually present with established infections. Treatment is delayed until the latent period of infection (50–70 days; Kreider *et al.*, 1987a) has expired and the grafts have been transformed but the mass of papilloma is still small but rapidly expanding during the treatment period. Thus, the infection is established and the challenge is to alter the growth pattern of the proliferating cells.

Pitfalls (advantages/disadvantages) of the model

The CRPV domestic rabbit papillomavirus model system

Advantages

1. A major advantage of the CRPV/domestic rabbit system in our laboratory is that we have shown that its response to treatment with podofilox, 5-fluorouracil collagen matrix and Cidofovir (HPMPC) is very similar to clinical lesions, thus supporting its validity as a model system of HPV infections.
2. CRPV infection is a spontaneous, enzootic disease of wild cottontail rabbits. The pathogenesis of CRPV infection in laboratory-infected domestic rabbits closely parallels the spontaneous disease.
3. The hosts are large, outbred and the tumors are autochthonous.
4. The cutaneous location of tumors facilitates inspection, measurement, biopsy, and manipulation even at an early stage of development of papilloma or carcinoma.
5. Tumors are produced with 100% yield.
6. Tumor cell progression occurs at predictable stages in well-defined lesions. The time scale of progression of CRPV-induced lesions is similar to the time scale of human disease.
7. Carcinomatous transformation occurs with high yield.
8. Metastatic pattern is predictable and defined.
9. Spontaneous regression of benign lesions occurs with low (or high) frequency depending upon CRPV viral isolate (Rous and Beard, 1935; Salmon *et al.*, 1997). We have an isolate that produces low levels of spontaneous regressions; this is ideal for treatment analysis because there is no need for increased numbers of rabbits in order to obtain statistical significance.
10. *In vitro* demonstrations of CRPV-specific humoral and cell-mediated immunity have been achieved in our laboratory (Christensen and Kreider, 1990; Höpfl *et al.*, 1993, 1994, 1995).

Disadvantages

1. Infectious CRPV is not produced in most domestic rabbit papillomas. This is not a serious limitation in our laboratory, however, because we have developed a system for producing CRPV in cottontail rabbit tissues subrenally grafted in athymic mice (Christensen and Kreider, 1990). In addition, we have used recombinant DNA to produce virion-rich papillomas on cottontails (unpublished data).
2. Precise assays are not available for infectivity and transformation. Viral titrations are customarily expressed in terms of infectious dose 50 (ID_{50}) or dilution of extract that induces papillomas at 50% of challenged sites. Our stocks have an ID_{50} of 10^{-4}.
3. Papilloma cells are difficult to culture.
4. Cell types that are susceptible to CRPV infection are limited and confined to the epidermis of rabbits, hares and fetal rats (Kidd and Parsons, 1936; Greene, 1953; Kreider *et al.*, 1971).
5. Rabbits are outbred. This leads to increased variability in the system but is parallel to human papillomavirus infections. We have a colony of syngeneic rabbits (strain EIII/JC) for studies requiring a constant genotype.
6. Rabbits are expensive and a long latency is required for data yield, especially stage IV tumors. This has probably been the major factor accounting for the comparative unpopularity of the CRPV/rabbit model.

The athymic mouse xenograft system for HPV-11 and HPV-40 infections

Advantages

1. The xenograft system has been validated as a model system for testing anti-papillomavirus agents. These results have paralleled clinical trials with the same compounds.
2. Infected foreskin xenografts produce substantial quantities of infectious virion.
3. We have produced infectious stocks of a second HPV type, HPV-40. This overcomes a potential limitation in which we had produced infectious stocks of only one genital-targeting HPV type.
4. The stocks of infectious virions of HPV-11 and HPV-40 allow us to set up primary infections in an *in vitro* monolayer culture system for additional screening of antiviral compounds.

Disadvantages

1. Athymic mice lack a T-dependent immune response. This barrier cannot be overcome, however, since to do so would result in the prompt rejection of the human tissue graft.
2. Subrenal xenografts infected with HPV-11 and HPV-40 cannot be monitored weekly to assess treatment outcome.
3. Athymic mice are extremely vulnerable to murine pathogens, especially murine hepatitis virus. We have excellent barrier facilities at Hershey Medical Center, and have not experienced an outbreak of murine hepatitis infection.

Contributions of the models to infectious disease therapy

The CRPV/domestic rabbit and the HPV-11/athymic mouse xenograft system have proven to be very useful and relevant model systems to test anti-papillomavirus compounds. The athymic mouse xenograft system using HPV-11 infection of human foreskin fragments implanted subrenally has been developed by our laboratory and also used for antiviral screening (Kreider *et al.*, 1985, 1990; Kreider and Christensen, 1994). This latter model offers the advantage that it involves the interaction of a human papillomavirus with human tissue. The athymic mouse xenograft method also allows replication of HPV-11 and -40 to high levels, enabling for the first time serial passage and production of viral stocks (Kreider *et al.*, 1987a). Normal human tissues from a variety of sites have been transformed, thus reproducing in the laboratory the complete spectrum of human diseases induced by these viruses (Kreider *et al.*, 1987b; Christensen *et al.*, 1997).

We have tested a variety of different compounds in these two animal models of papillomavirus infection including acyclic nucleotides (PMEG, 9[2-phosphonylmethoxy) ethyl]adenine (PMEA), HPMPC, cyclic HPMPC (cHPMPC), podophyllin, podophyllotoxin, FIAU, 5-fluorouracil, matrigel 5-fluorouracil, and various immunomodulators for efficacy (Kreider *et al.*, 1990, 1992; Kreider and Pickel, 1993; Okabayashi *et al.*, 1993; Gangemi *et al.*, 1994; Kreider and Christensen, 1994). We have observed strong clinical correlations between the successful treatment of CRPV-induced rabbit papillomas and HPV-11 induced xenografts with treated condylomas of patients (Kreider *et al.*, 1992; Kreider and Pickel, 1993; Syed *et al.*, 1994; Swinehart *et al.*, 1997).

In conclusion, we have validated the CRPV/rabbit and the athymic mouse/HPV-11 human foreskin xenograft assay systems as effective and appropriate animal models for therapeutic treatment of HPV infections in humans.

Acknowledgments

These studies were supported by PHS grant AI85337 from the NIAID, NIH, and by the Jake Gittlen Memorial Golf

Tournament. We thank Martin Pickel and Patricia Welsh for expert technical assistance.

References

Auborn, K. J., Steinberg, B. M. (1990). *Papillomaviruses and Human Cancer* (ed Pfister, H.), pp. 203–224. CRC Press, Boca Raton.

Bonnez, W., Rose, R. C., Da Rin, C., Borkhuis, C., De Mesy Jensen, K. L., Reichman, R. C. (1993). Propagation of human papillomavirus type 11 in human xenografts using the severe combined immunodeficiency (SCID) mouse and comparison to the nude mouse model. *Virology*, **197**, 455–458.

Campo, M. S. (1987). Papillomas and cancer in cattle. *Cancer Surv.*, **6**, 39–53.

Christensen, N. D., Kreider, J. W. (1990). Antibody-mediated neutralization *in vivo* of infectious papillomaviruses. *J. Virol.*, **64**, 3151–3156.

Christensen, N. D., Cladel, N. M., Reed, C. A. *et al.* (1996a). Laboratory production of infectious stocks of rabbit oral papillomavirus. *J. Gen. Virol.*, **77**, 1793–1798.

Christensen, N. D., Reed, C. A., Cladel, N. M., Hall, K., Leiserowitz, G. S. (1996b). Monoclonal antibodies to HPV-6 L1 virus-like particles identify conformational and linear neutralizing epitopes on HPV-11 in addition to type-specific epitopes on HPV-6. *Virology*, **224**, 477–486.

Christensen, N. D., Koltun, W. A., Cladel, N. M. *et al.* (1997). Coinfection of human foreskin fragments with multiple human papillomavirus types (HPV-11, -40, -LVX82/MM7) produces regionally separate HPV infections within the same athymic mouse xenograft. *J. Virol.*, **71**, 7337–7344.

Gangemi, J. D., Pirisi, L., Angell, M., Kreider, J. W. (1994). HPV replication in experimental models: effects of interferon. *Antiviral Res.*, **24**, 175–190.

Greene, H. S. (1953). The induction of the Shope papilloma in homologous transplants of embryonic rat skin. *Cancer Res.*, **13**, 681–683.

Höpfl, R. M., Christensen, N. D., Angell, M. G., Kreider, J. W. (1993). Skin test to assess immunity against cottontail rabbit papillomavirus antigens in rabbits with progressing papillomas or after papilloma regression. *J. Invest. Dermatol.*, **101**, 227–231.

Höpfl, R., Christensen, N. D., Heim, K., Kreider, J. W. (1994). Skin test reactivity to papilloma cells is long lasting in domestic rabbits after regression of cottontail rabbit papillomavirus induced papillomas. In *Immunology of Human Papillomaviruses*. (ed Stanley, M. A.), pp. 259–265. Plenum Press, New York.

Hopfl, R., Christensen, N. D., Angell, M. G., Kreider, J. W. (1995). Leukocyte proliferation *in vitro* against cottontail rabbit papillomavirus in rabbits with persisting papillomas/cancer or after regression. *Arch. Dermatol. Res.*, **287**, 652–658.

Iwasaki, T., Maeda, H., Kameyama, Y., Moriyama, M., Kanai, S., Kurata, T. (1997). Presence of a novel hamster oral papillomavirus in dysplastic lesions of hamster lingual mucosa induced by application of dimethylbenzanthracene and excisional wounding: molecular cloning and complete nucleotide sequence. *J. Gen. Virol.*, **78**, 1087–1093.

Jenson, A. B., Rosenthal, J. D., Olson, C., Pass, F., Lancaster, W. D., Shah, K. V. (1980). Immunologic relatedness of papillomaviruses from different species. *J.N.C.I.*, **64**, 495–500.

Kidd, J. G., Parsons, R. J. (1936). Tissue affinity of Shope papilloma virus. *Proc. Soc. Exp. Biol. Med.*, **35**, 438–441.

Kreider, J. W. (1980). *Viruses in Naturally Occurring Cancers. Book A. Cold Spring Harbor Conference on Cell Proliferation*, vol. 7 (eds Essex, M., Todaro, G., zur Hausen, H.), pp. 283–299. Cold Spring Harbor Press, Cold Spring Harbor.

Kreider, J. W., Christensen, N. D. (1994). Model systems for the treatment of human papillomavirus infections in athymic mice. In *Proceedings of the First Workshop on Antiviral Claims for Topical Antiseptics* (eds Biswal, N., Dempsey, W. L., Lard, S. L., Trachewsky, D.), pp. 99–106. United States Government Printing Office, Washington, DC.

Kreider, J. W., Pickel, M. D. (1993). Influence of schedule and mode of administration on effectiveness of podofilox treatment of papillomas. *J. Invest. Dermatol.*, **101**, 614–618.

Kreider, J. W., Benjamin, S. V., Pruchnic, W. F., Strimlan, C. V. (1971). Immunologic mechanisms in the induction and regression of Shope papilloma virus-induced epidermal papillomas of rats. *J. Invest. Dermatol.*, **56**, 102–112.

Kreider, J. W., Howett, M. K., Wolfe, S. A. *et al.* (1985). Morphological transformation *in vivo* of human uterine cervix with papillomavirus from condylomata acuminata. *Nature*, **317**, 639–641.

Kreider, J. W., Howett, M. K., Leure-Dupree, A. E., Zaino, R. J., Weber, J. A. (1987a). Laboratory production in vivo of infectious human papillomavirus type 11. *J. Virol.*, **61**, 590–593.

Kreider, J. W., Howett, M. K., Stoler, M. H., Zaino, R. J., Welsh, P. (1987b). Susceptibility of various human tissues to transformation *in vivo* with human papillomavirus type 11. *Int. J. Cancer*, **39**, 459–465.

Kreider, J. W., Balogh, K., Olson, R. O., Martin, J. C. (1990). Treatment of latent rabbit and human papillomavirus infections with 9-(2-phosphonylmethoxy)ethylguanine (PMEG). *Antiviral Res.*, **14**, 51–58.

Kreider, J. W., Christensen, N. D., Christian, C. B., Pickel, M. D. (1992). Preclinical system for evaluating topical podofilox treatment of papillomas: dose–response and duration of growth prior to treatment. *J. Invest. Dermatol.*, **99**, 813–818.

Okabayashi, M., Pickel, M. D., Budgeon, L. R., Cladel, N. M., Kreider, J. W. (1993). Podofilox-induced regression of shope papillomas may be independent of host immunity. *J. Invest. Dermatol.*, **101**, 852–857.

Olson, C. (1963). Cutaneous papillomatosis in cattle and other animals. *Ann. N.Y. Acad. Sci.*, **108**, 1042–1056.

Parsons, R. J., Kidd, J. G. (1936). A virus causing oral papillomatosis in rabbits. *Proc. Soc. Exp. Biol. Med.*, **35**, 441–443.

Parsons, R. J., Kidd, J. G. (1942). Oral papillomatosis of rabbits: a virus disease. *J. Exp. Med.*, **77**, 233–250.

Rous, P., Beard, J. W. (1935). Comparison of tar tumors of rabbits and virus-induced tumors. *Proc. Soc. Exp. Biol. Med.*, **33**, 358–360.

Salmon, J., Ramoz, N., Cassonnet, P., Orth, G., Breitburd, F. (1997). A cottontail rabbit papillomavirus strain (CRPVb) with strikingly divergent E6 and E7 oncoproteins: an insight in the evolution of papillomaviruses. *Virology*, **235**, 228–234.

Schiffman, M. H., Bauer, H. M., Hoover, R. N. *et al.* (1993). Epidemiologic evidence showing that human papillomavirus infection causes most cervical intraepithelial neoplasia. *J.N.C.I.*, **85**, 958–964.

Shope, R. E., Hurst, E. W. (1933). Infectious papillomatosis of rabbits; with a note on the histopathology. *J. Exp. Med.*, **58**, 607–624.

Smith, L. H., Foster, C., Hitchcock, M. E., Isseroff, R. (1993). *In vitro* HPV-11 infection of human foreskin. *J. Invest. Dermatol.*, **101**, 292–295.

Smith, L. H., Foster, C., Hitchcock, M. E. *et al.* (1995). Titration of HPV-11 infectivity and antibody neutralization can be measured *in vitro*. *J. Invest. Dermatol.*, **105**, 438–444.

Sundberg, J. P., Smith, E. K., Herron, A. J., Jenson, A. B., Burk, R. D., Van Ranst, M. (1994). Involvement of canine oral papillomavirus in generalized oral and cutaneous verrucosis in a Chinese Shar Pei dog. *Vet. Pathol.*, **31**, 183–187.

Suzich, J. A., Ghim, S. J., Palmer-Hill, F. J. *et al.* (1995). Systemic immunization with papillomavirus L1 protein completely prevents the development of viral mucosal papillomas. *Proc. Natl. Acad. Sci. USA*, **92**, 11553–11557.

Swinehart, J. M., Sperling, M., Phillips, S. *et al.* (1997). Intralesional fluorouracil epinephrine injectable gel for treatment of condylomata acuminata: a phase 3 clinical study. *Arch. Dermatol.*, **133**, 67–74.

Syed, T. A., Lundin, S., Ahmad, S. A. (1994). Topical 0.3% and 0.5% podophyllotoxin cream for self-treatment of condylomata acuminata in women. A placebo-controlled, double-blind study. *Dermatology*, **189**, 142–145.

zur Hausen, H. (1994). Molecular pathogenesis of cancer of the cervix and its causation by specific human papillomavirus types. *Curr. Top. Microbiol. Immunol.*, **186**, 131–156.

Chapter 126

Adult Mouse Model for Rotavirus

R. L. Ward and M. M. McNeal

Background of human infection

Rotaviruses are the primary cause of severe infantile gastroenteritis worldwide. It is estimated that in the United States alone there are approximately 100 000 hospitalizations and 100 deaths due to rotavirus annually with costs of nearly 400 million dollars for hospital care alone (Matson and Estes, 1990; Jin *et al.*, 1996). On a world scale, rotaviruses are estimated to be responsible for nearly one million deaths each year (Institute of Medicine, 1986). Because treatment for rotavirus illness is presently limited to non-specific supportive measures such as oral or intravenous rehydration, the development of an effective rotavirus vaccine is a high priority.

Rotavirus transmission occurs by the fecal–oral route, which provides an efficient mechanism for universal exposure. Approximately 90% of children in both developed and developing countries experience at least one rotavirus infection by 3 years of age (Kapikian and Chanock, 1996). Rotavirus disease is primarily, if not solely, restricted to the gastrointestinal tract where symptoms typically include diarrhea and vomiting accompanied by fever, nausea, anorexia, cramping and malaise. Illnesses may be mild or produce severe dehydration. Symptoms resolve within several days in immunologically normal patients but can persist in immunodeficient individuals (Saulsbury *et al.*, 1980). Rotavirus disease is seasonal and occurs almost solely between October and May in temperate climates (Cook *et al.*, 1990; Haffejee, 1995). Even in tropical regions most rotavirus illnesses occur during the dry, cooler winter months. Although an understanding of the causes of seasonality would be invaluable in designing methods for prevention of rotavirus disease, there are no satisfactory explanations for this phenomenon. Also unexplained is the observation that the annual rotavirus epidemic in the United States begins in the south-west in the fall and gradually shifts to the north-eastern states where peak incidence is typically in the early spring (LeBaron *et al.*, 1990).

Rotavirus infection normally provides short-term immunity, at least against severe illnesses, but does not provide lifelong protection (Ward, 1996). Rotavirus infections occur in persons of all ages, even in neonates, but severe illness is usually found only after the first infection and when the infected child is at least several months of age (Kapikian and Chanock, 1996). Causes for reduced severity of symptoms during the first months of life and after the first years are being intensely investigated. Possibly, maternal antibody or other age-dependent factors play a role in the protection of the youngest subjects while active immunity is believed to provide at least partial protection after the initial rotavirus infection in subjects of all ages. The mechanisms of active immunity against rotavirus in humans are poorly understood. This has stimulated the development of animal models, the most versatile of which is the adult mouse model.

Background of model

Prior to the development of the adult mouse model, studies of active immunity were conducted primarily in calves and piglets. These models have restrictions that limit their usefulness. Mice had been used in many studies involving passive immunity but were not considered for active immunity studies because they are susceptible to rotavirus disease, i.e., diarrhea, for only their first 2 weeks of life. This provides insufficient time in which to immunize and develop adequate T- and B-cell responses to protect against rotavirus diarrhea following challenge.

Many of the limitations of this model were circumvented when protection against infection rather than disease was employed as the indicator of immunity. Initial experiments leading to the development of this model demonstrated that inbred BALB/c mice of all ages are susceptible to infection with unpassaged and even culture-adapted murine rotavirus (Ward *et al.*, 1990). Following oral challenge with approximately 1000 tissue-culture infectious virus (i.e., focus-forming units), nearly all mice shed large quantities of viral antigen in their stool beginning 2 days after challenge and lasting up to about 8 days. These mice also develop large serum and stool rotavirus antibody responses detectable by 1–2 weeks after inoculation. Once infected, mice are fully resistant to reinfection.

Based on these results, the adult mouse model was developed and has now been used in multiple laboratories. The

Handbook of Animal Models of Infection
ISBN 0-12-775390-7

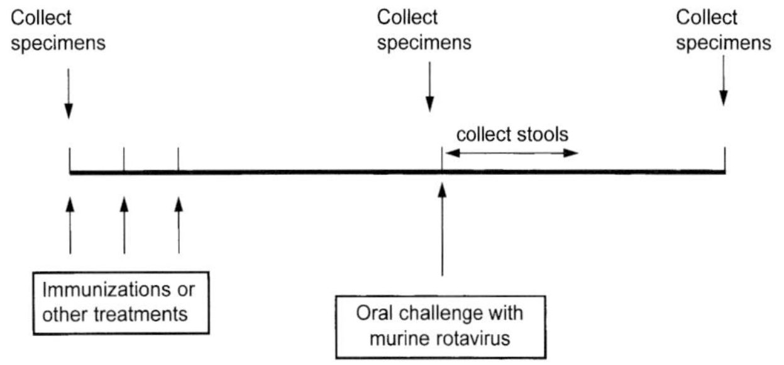

Figure 126.1 A generalized example of methods used in studies with the adult mouse model for rotavirus. Specimens collected are typically serum and stool that are used to measure rotavirus antibody titers, blood lymphocytes used to measure cytotoxic T lymphocyte activity, or lymphoid tissue used to measure specific immunological functions. Immunizations are typically with virus or viral components. Examples of other treatments are CD4 or CD8 cell depletion with monoclonal antibodies. Oral challenge is always with epizootic diarrhea of infant mice (EDIM) or other murine rotaviruses. Stools collected after challenge are measured for rotavirus antigen.

pertinent features of the model, as summarized by a generalized example (Figure 126.1), are that naive mice are first inoculated with experimental materials, then challenged with murine rotavirus to determine the effects of treatment on the quantity of viral antigen shed. Immunological effectors (e.g., T cells and rotavirus antibody) are also measured during the course of experimentation to determine immunological correlates of protection.

Animal species

Mice used in this model should have no pre-existing antibody to rotavirus. BALB/c mice have been utilized extensively due to their susceptibility to infection with murine rotavirus strains. Almost all naive BALB/c mice of any age will become infected and shed rotavirus when challenged with murine rotavirus. Other strains of mice, such as C57BL/6, are much more resistant to infection following oral inoculation with murine rotaviruses as adults. Genetically engineered knockout mice have also been successfully utilized for immunological studies on the mechanism of protection against rotavirus. Two B-cell-deficient mouse strains that make no antibody, μMt and J_HD, have been used most extensively with this model (Franco and Greenberg, 1995; McNeal *et al.*, 1995, 1997). Utilization of any mouse strain in this model system requires that its susceptibility to the challenge virus be clearly established. The ages of mice employed in this model can vary; typically, the first treatments are at 6 weeks of age or older when their immune systems are fully mature. Both sexes of mice have been employed but for practical consideration female mice are preferred.

Preparation of animals

Because rotavirus can spread from one animal to the next, even between open cages, mice are housed in sterile static microisolator cages. Cages, food and water are, therefore, sterilized as well as supplies or instruments employed in experimentation. Cages are opened only in a biosafety cabinet where mice are handled by individual groups using sterile techniques. When anesthesia is needed for immunization or collection of serum samples, animals are typically sedated using methoxyflurane (Metofane, Mallinckrodt Veterinary, Inc.) as directed by the manufacturer.

Prior to the start of an experiment, mice are identified, usually by ear-tagging. Serum samples are obtained before immunization and at various times throughout the experiment by bleeding from the retro-orbital capillary plexus with a hematocrit capillary tube under light anesthesia. Approximately 50 μl of sera can readily be obtained from each animal. Stool samples are also obtained at various time points to determine antibody responses and the presence of viral antigen. One or more stool pellets from each animal are placed in a buffered solution (0.5 ml) and kept frozen until analyzed. Before analysis, the stool is homogenized, either manually or by vortexing, and most particulate matter is removed by low-speed centrifugation.

To determine which immune mechanisms are involved in protection, components of the immune system can be depleted by inoculation with specific monoclonal antibodies (mAbs). CD8 and CD4 cells have been depleted by intraperitoneal inoculation with mAbs 2.43 and GK1.5, respectively. Typically, 1 mg of mAb is injected once a day for 4 days prior to immunization, and depletion is maintained by twice-weekly injections. If other immune cells or specific cytokines are to be investigated, mAbs to many are available commercially or from the American Type Culture Collection. To measure the extent of depletion, immune cells are isolated from sites such as spleen, mesenteric lymph nodes, lamina propria or intraepithelial lymphocytes. These are reacted with commercially available, fluorescently tagged mAbs against cell surface components different from those recognized by the mAbs used for depletion (McNeal *et al.*, 1995). The cells are then separated for identification by flow cytometry (fluorescence-activated cell sorter, or FACS).

Challenge virus

The murine epizootic diarrhea of infant mice (EDIM) strain of rotavirus has been used exclusively as the challenge virus in our laboratory, while this and other murine rotavirus strains have been utilized in other laboratories (Burns *et al.*, 1995). The EDIM virus was originally obtained from M. Collins (Microbiological Associates, Bethesda, MD) and used to infect orally neonatal BALB/c mice. Stools were collected from pups with diarrhea and stored frozen in buffer. This material was treated with 1,1,2-trichloro-1,2,2-trifluoroethane (freon) using two-phase separation to remove particulates. The aqueous phase was collected after low-speed centrifugation, and virus was pelleted by high-speed centrifugation. Virus was then resuspended in tissue culture medium and stored in aliquots at –70°C. This preparation represents the wild type (wt) EDIM virus.

The EDIM virus was also adapted to grow in tissue culture by multiple passages in MA104 cells (Ward *et al.*, 1990). A large stock of the ninth passage (p9) was prepared and also stored in aliquots of –70°C. This preparation has been used as the challenge virus in most of our studies. Both wt and p9 preparations of EDIM are titered by a fluorescent focus assay (FFA) in MA104 cells to measure infectivity in cell culture (Ward *et al.*, 1984). The 50% infectious dose (ID_{50}) *in vivo* was also determined for the wt and p9 preparations using BALB/c mice. During experimentation, mice are routinely challenged with approximately 1000 ID_{50}. The wt EDIM has been employed as the challenge virus only when the mouse strain may be less susceptible to the culture adapted virus, such as μMt B-cell-deficient mice that have a C57BL/6 background.

Vaccine materials

A number of different materials have been tested with this model to determine their effectiveness as possible vaccine candidates. Live homologous and heterologous viruses were initially used following growth in MA104 cells by employing standard procedures. For some experiments, the virus was concentrated or extensively purified prior to use as a vaccine. The steps used in purification are as follows: pelleting by ultracentrifugation, resuspension in buffer, extraction with freon, pelleting on to a 1.4 g/ml cesium chloride (CsCl) cushion and banding of virus particles during CsCl gradient centrifugation. Infectious triple-layered virus bands at a density of 1.36 g/ml in CsCl and doubled-layered particles lacking outer capsid (neutralization) proteins, VP4 and VP7, band at 1.38 g/ml. Both purified particles are collected and dialyzed against phosphate-buffered saline (PBS) containing 10 mmol/l $CaCl_2$, and 20% glycerol before being stored at –70°C. Protein concentrations are determined and virus is titered by FFA. In addition to purified homologous and heterologous viruses, double-layered particles lacking VP4 and VP7 have also been used to immunize mice. This permits analysis of the effects of immunization with particles lacking neutralization proteins.

Inactivated purified viral particles are also being used for vaccinating mice. We inactivate virus in 60 mm Petri dishes after addition of psoralen (4′-aminomethyltroxsalen HCl, Lee Biomolecular Research, Inc.) to a concentration of 40 μg/ml using long-wavelength ultraviolet light (345 nm). For this, the dishes are placed on ice on a rotating platform and irradiated (40 minutes) at a distance of 10 cm. The viral titer of the preparation is checked by a FFA to insure complete inactivation.

With the increased development of novel vaccine strategies, a wide range of materials such as expressed viral proteins and virus-like particles, naked DNA and microencaspulated viral proteins and nucleic acids are being tested as vaccines in this model.

Immunization process

Since rotavirus is a disease of the intestinal tract, vaccine candidates so far evaluated in humans have all been delivered orally. Live and inactivated rotaviruses have also been administered by this route in the adult mouse model. We have inoculated mice by mouth using a pipet containing small volumes of the vaccine preparation. Oral gavage with a 22 G × 1 inch gauge blunt-ended needle is used when the volume of inoculum is larger than 50 μl. The other mucosal route which has recently been explored is intranasal (i.n.) immunization. Vaccine is given by instilling 5–10 μl droplets to each nare while mice are under light anesthesia. Maximum volumes of 100 μl can be administered if given slowly so as not to block respiration.

Parenteral routes of immunization have also been used for different vaccine materials. Intramuscular (i.m.) injection into the hind leg muscle and intraperitoneal (i.p.) inoculation can be used with whole virus, viral particles or viral proteins. DNA vaccines have also been administered using a variety of routes (i.m., i.p., subcutaneous) but the best results, based on antibody responses to expressed rotavirus proteins, have been obtained with the gene gun or Accell particle delivery device (Geniva), used as described by the manufacturer (Herrmann *et al.*, 1996; Choi *et al.*, 1997).

The number of doses and amount of inoculum administered are based on the vaccine itself. We typically use 10–20 μg of protein for i.m., i.p. or i.n. immunizations. One or two doses are administered, usually separated by 2–4 weeks. Mice are then challenged 2 weeks or longer after the last immunization.

Challenge process

Mice in our laboratory are orally challenged with approximately 1000 ID_{50} of wt EDIM or p9 of tissue culture-adapted EDIM. Other laboratories use either lower or higher quantities of inoculum. Mice are fed the virus from a pipet in a volume of 20–50 ul. Oral gavage, as described above, can also be used to deliver the challenge virus.

Key parameters to monitor response to the vaccine

The most important response to the vaccine in this model is protection against murine rotavirus infection. This is determined by the shedding of rotavirus stool following challenge and by significant (i.e., ≥ fourfold) rises in serum rotavirus immunoglobulin A (IgA). Stools are collected for 7–10 days after challenge and analyzed for rotavirus antigen by an enzyme-linked immunosorbent assay (ELISA). In this procedure, ELISA plates are coated with either normal rabbit sera (negative wells) or rabbit IgG to rotavirus (positive wells). After washing, as is done between each step with PBS containing 0.05% Tween 20, a 10% stool suspension is added to duplicate positive and duplicate negative wells. Following incubation (1 hour at 37°C) on a rotating platform, wells are blocked with normal goat serum (Vector Laboratories, Inc.) in 5% non-fat dry milk. Hyperimmune guinea-pig serum to rotavirus along with normal rabbit serum is then added (30 minutes, room temperature or r.t.). Biotinylated goat anti-guinea-pig IgG (Vector Laboratories, Inc.) with normal rabbit serum is next added followed by peroxidase-conjugated avidin-biotin (each step: 30 minutes, r.t.). Substrate (*o*-pheylenediamine with H_2O_2) in citric acid phosphate buffer is added (15 minutes, r.t.), and the reaction is stopped with 1 mol/l H_2SO_4. Finally, the absorbency at wavelength 490 nm (A_{490}) is measured and the net optical densities determined. The stool is considered positive for rotavirus if the average A_{490} for the positive wells is more than twice the average of the negative wells and ≥0.15. The absorbance values for each animal or each group can be graphed to determine a shedding curve (Figure 126.2). Protection from shedding is determined statistically by comparison of the shedding curves for an unimmunized control group with shedding curves of groups receiving experimental vaccines.

The other major parameters to evaluate the effects of the experimental vaccine are immunological responses. Serum and stool specimens collected prior to vaccination, prior to challenge, and after challenge are typically measured for rotavirus IgG and IgA by an ELISA method. Briefly, ELISA plates are coated overnight at 4°C with anti-rotavirus rabbit IgG. EDIM lysate or mock-infected cell lysate are each added to duplicate wells followed by serial twofold dilutions of pooled sera from EDIM-infected mice (assigned concentrations of 160 000 or 10 000 units/ml of rotavirus IgG or IgA, respectively) to generate a standard curve. Mouse sera or stool suspensions to be tested are also added to duplicate wells of each lysate. This is followed by sequential addition of biotin-conjugated goat anti-mouse IgG or IgA, peroxidase-conjugated avidin-biotin, and *o*-phenylenediamine substrate. Color development is again

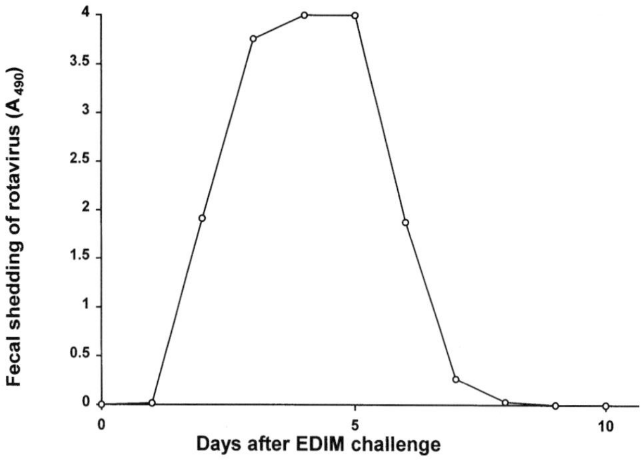

Figure 126.2 Pattern of fecal shedding of rotavirus antigen after challenge with epizootic diarrhea of infant mice (EDIM) as determined by enzyme-linked immunosorbent assay. Values represent the average absorbency at wavelength 490 nm (A_{490}) values for a group of 8 naive BALB/c mice challenged at 3 months of age.

stopped after 15 minutes with 1 mol/l H_2SO_4 and the A_{490} is measured. Titers of rotavirus IgG or IgA, expressed as units/ml, are determined from the standard curve generated by subtraction of the average A_{490} values of the duplicate cell lysate wells from the average of the wells coated with EDIM lysate. Serum-neutralizing antibody titers to EDIM virus are also routinely measured in these specimens using an antigen reduction neutralization assay (Knowlton *et al.*, 1991).

There are other immunological parameters that can be measured in this model depending on the focus of the experiment. For example cytotoxic T lymphocyte (CTL) activities have been measured following gene gun immunization as possible correlates of protection (Herrmann *et al.*, 1996). It is anticipated that other immunological effectors, including an array of cytokines, will be analyzed in future studies for their involvement in immunity against rotavirus.

Pitfalls (advantages/disadvantages) of the model

The most obvious limitation of the model is that it measures protection against infection rather than disease. The goal of most vaccines is to prevent symptomatic infection by the micro-organisms against which the vaccines are developed. Prevention of asymptomatic infection is often not necessary and may be undesirable since such infections should boost immunity. Even so, infection must occur to develop rotavirus disease, so immune responses that prevent infection will also prevent disease. Presumably, the same mechanisms of immunity will prevent both symptomatic and asymptomatic rotavirus infections. Therefore, the inability of this model to measure protection against disease should not limit its usefulness.

Another possible limitation of the model is its applicability to human disease — a potential problem with any model. Animals are most susceptible to rotavirus disease during their first days or weeks of life, depending on the species. In humans, rotavirus disease is most prevalent after 6 months of age (Kapikian and Chanock, 1996). Reasons for these differences are not fully explained, but they limit the applicability of the model. The pathogenesis of rotavirus in mice also appears to be different from other animals, such as piglets and calves, where extensive destruction of intestinal villi has been repeatedly observed following rotavirus infection (Greenberg *et al.*, 1994). Comparable damage has also been found in humans with rotavirus disease. In contrast, rotavirus infection in mice causes little, if any, discernible intestinal damage. Therefore, the mechanisms of diarrheal induction by rotavirus may be quite different in mice.

The advantages of this model reside in its simplicity, the availability of inbred and immunologically impaired mouse strains, and the extensive knowledge-base established for the murine immune system. These three factors combine to allow large quantities of detailed information to be compiled at minimal costs. With this model, it should not only be possible to obtain a complete understanding of the mechanisms by which immunity to rotavirus is developed, but also readily to test the effectiveness of novel vaccination strategies.

Contributions of the model to infectious disease therapy

Studies on the mechanisms of immunity to rotavirus

Initial studies with the model indicated that neutralizing intestinal antibody titers did not correlate with protection (Ward *et al.*, 1992). Subsequently, it was found that titers of both stool and serum rotavirus IgA correlated with protection (Feng *et al.*, 1994; McNeal *et al.*, 1994). Even non-neutralizing IgA against the VP6 protein of rotavirus produced following subcutaneous transplantation of hybridoma cells into the backs of mice protected against rotavirus infection (Burns *et al.*, 1996). Further experimentation using B-cell-deficient mice that were unable to produce antibody revealed that protection against rotavirus infection in previously immunized mice is primarily dependent on antibody (Franco and Greenberg, 1995; McNeal *et al.*, 1995). However, resolution of rotavirus infection is not antibody-dependent in these mice but is due primarily to CD8 cells, presumably through their CTL activities. That is, these mice were able to resolve their initial infection, but in the absence of CD8 cells (eliminated by inoculation of anti-CD8 cell monoclonal antibodies), they chronically shed large quantities of rotavirus. In normal mice depleted of CD8 cells, there is merely a delay in resolution of infection which corresponds to the time when antibody production is initiated and, once shedding is resolved in these mice, they remain fully protected against rotavirus infection (McNeal *et al.*, 1997). CD4 cells also play a role in rotavirus immunity because resolution of rotavirus infection required at least several weeks to be complete in immunologically normal mice depleted of this T-cell population. Since antibody production was largely suppressed in these animals, it is possible that complete resolution of infection, as well as protection against subsequent infection, relied on CD4 cell-dependent antibody production. The mechanisms by which rotavirus antibody provides immunity remain to be determined. If these results are directly applicable to human rotavirus infections, they should provide invaluable information for designing future rotavirus vaccines.

Development of novel vaccination strategies

Within a few years of the discovery of human rotavirus, a live, orally deliverable vaccine candidate was already being evaluated in clinical trials (Midthun and Kapikian, 1996). Although this vaccine candidate was a bovine rotavirus (RIT 4237) that was serotypically unrelated to the circulating human rotaviruses, it provided excellent protection in the initial trials in Finland. However, because it provided limited protection in developing countries, its evaluation was discontinued. All subsequent vaccine candidates that have reached clinical trials have also been live rotaviruses that are delivered orally. The emphasis on these types of vaccines for rotavirus is based on the findings that natural rotavirus infection can provide excellent protection against subsequent rotavirus disease, and rotavirus infection may be solely restricted to the intestinal tract. Thus, these vaccines were designed to stimulate the intestinal immunity found after natural rotavirus infection. Parenteral vaccines were generally ignored because initial studies in lambs and mice suggested that the presence of intestinal but not circulating rotavirus antibody correlated with protection (Snodgrass and Wells, 1976; Offit and Clark, 1985). However, in 1992 it was shown, through the utilization of the adult mouse model, that parenteral immunization may also be effective against rotavirus infection (McNeal *et al.*, 1992). The use of this model has expanded into several laboratories where the potential effectiveness of parenteral immunization by a variety of routes is being evaluated by employing novel strategies, including the use of different immunogens and adjuvants. In addition to parenteral immunization, vaccination by non-oral mucosal routes, particularly intranasal, has recently been found to provide excellent protection against rotavirus using the adult mouse model (unpublished results). Therefore, it is anticipated that the results found with this experimental model will result in second-generation vaccine candidates to supplement or possibly supplant the first generation of live oral vaccines.

References

Burns, J. W., Krishnaney, A. A., Vo, P. T., Rouse, R. V., Anderson, L. J., Greenberg, H. B. (1995). Analyses of homologous rotavirus infection in the mouse model. *Virology*, **207**, 143–153.

Burns, J. W., Siadat-Pajouh, M., Krishnaney, A. A., Greenberg, H. B. (1996). Protective effect of rotavirus VP6-specific IgA monoclonal antibodies that lack neutralizing activity. *Science*, **272**, 104–107.

Choi, A. H., Knowlton, D. R., McNeal, M. M., Ward, R. L. (1997). Particle bombardment-mediated DNA vaccination with rotavirus VP6 induces high levels of serum rotavirus IgG but fails to protect mice against challenge. *Virology*, **232**, 129–138.

Cook, S. M., Glass, R. I., LeBaron, C. W. *et al.* (1990). Global seasonality of rotavirus infections. *Bull. WHO*, **68**, 171–177.

Feng, N., Burns, J. W., Bracy, L., Greenberg, H. B. (1994). Comparison of mucosal and systemic humoral immune responses and subsequent protection in mice orally inoculated with a homologous or a heterologous rotavirus. *J. Virol.*, **68**, 7766–7773.

Franco, M. A., Greenberg, H. B. (1995). Role of B cells and cytotoxic T lymphocytes in clearance of an immunity to rotavirus infection in mice. *J. Virol.*, **69**, 7800–7806.

Greenberg, H. B., Clark, H. F., Offit, P. A. (1994). Rotavirus pathology and pathophysiology. *Curr. Topics Microbiol. Immunol.*, **185**, 255–283.

Haffejee, I. E. (1995). The epidemiology of rotavirus infections: a global perspective. *J. Pediatr. Gastroenterol. Nutr.*, **20**, 275–286.

Herrmann, J. E., Chen, S. C., Fynan, E. F. *et al.*, (1996). Protection against rotavirus infections by DNA vaccination. *J. Infect. Dis.*, **174**, S93–S97.

Institute of Medicine (1986). New Vaccine Development: Establishing Priorities. Diseases of Importance in Developing Countries. In *The prospects for immunizing against rotavirus*, vol. II, pp. 308–318. National Academy Press, Washington, DC.

Jin, S., Kilgore, P. E., Holman, R. C., Clarke, M. J., Gangarosa, E. J., Glass, R. I. (1996). Trends in hospitalizations for diarrhea in United States children from 1979 through 1992: estimates of the morbidity associated with rotavirus. *Pediatr. Infect. Dis. J.*, **15**, 397–404.

Kapikian, A. Z., Chanock, R. M. (1996). Rotaviruses. In *Fields Virology*, 3rd edn (eds Fields, B. N., Knipe, D. M.), pp. 1657–1708. Lippincott–Raven Publishers, New York, NY.

Knowlton, D. R., Spector, D. M., Ward, R. L. (1991). Development of an improved method for measuring neutralizing antibody to rotavirus. *J. Virol. Methods*, **33**, 127–134.

LeBaron, C. W., Lew, J., Glass, R. I., Weber, J. M., Ruiz-Palacios, G. M. (1990). Annual rotavirus epidemic patterns in North America. *J.A.M.A.*, **264**, 983–988.

Matson, D. O., Estes, M. K. (1990). Impact of rotavirus infection at a large pediatric hospital. *J. Infect. Dis.*, **162**, 598–604.

McNeal, M. M., Sheridan, J. F., Ward, R. L. (1992). Active protection against rotavirus infection of mice following intraperitoneal immunization. *Virology*, **191**, 150–157.

McNeal, M. M., Broome, R. L., Ward, R. L. (1994). Active immunity against rotavirus infection in mice is correlated with viral replication and titers of serum rotavirus IgA following vaccination. *Virology*, **204**, 642–650.

McNeal, M. M., Barone, K. S., Rae, M. N., Ward, R. L. (1995). Effector functions of antibody and CD8+ cells in resolution of rotavirus infection and protection against reinfection in mice. *Virology*, **214**, 387–397.

McNeal, M. M., Rae, M. N., Ward, R. L. (1997). Evidence that resolution of rotavirus infection in mice is due to both CD4 and CD8 cell-dependent activities. *J. Virol.*, **71**, 8735–8742.

Midthun, K., Kapikian, A. Z. (1996). Rotavirus vaccines: an overview. *Clin. Microbiol. Rev.*, **9**, 423–434.

Offit, P. A., Clark, H. F. (1985). Protection against rotavirus-induced gastroenteritis in a murine model by passively acquired gastrointestinal but not circulating antibodies. *J. Virol.*, **54**, 58–64.

Saulsbury, F. T., Winkelstein, J. A., Yolken, R. H. (1980). Chronic rotavirus infection in immunodeficiency. *J. Pediatr.*, **97**, 61–65.

Snodgrass, D. R., Wells, P. W. (1976). Rotavirus infection in lambs: studies on passive protection. *Arch. Virol.*, **52**, 201–205.

Ward, R. L. (1996). Mechanisms of protection against rotavirus in humans and mice. *J. Infect. Dis.*, **174**, S51–S58.

Ward, R. L., Knowlton, D. R., Pierce, M. J. (1984). Efficiency of human rotavirus propagation in cell culture. *J. Clin. Microbiol.*, **19**, 748–753.

Ward, R. L., McNeal, M. M., Sheridan, J. F. (1990). Development of an adult mouse model for studies on protection against rotavirus. *J. Virol.*, **64**, 5070–5075.

Ward, R. L., McNeal, M. M., Sheridan, J. F. (1992). Evidence that active protection following oral immunization of mice with live rotavirus is not dependent on neutralizing antibody. *Virology*, **188**, 57–66.

Chapter 127

Animal Models for Lentivirus Infections—Feline Immunodeficiency Virus

M. J. Burkhard and E. A. Hoover

Background of human infection

Animal models have their greatest value in elucidating the early pathogenesis, protective immune responses, and intervention strategies in human immunodeficiency virus (HIV) infection. Several animal models have been described. While HIV can productively infect chimpanzees, these animals typically remain asymptomatic, connoting differences in the host–HIV relationship between the two species.

Simian immunodeficiency virus (SIV) infection of Asian macaques is a valuable HIV model in which immunodeficiency disease can be induced. Paradoxically, SIV does not cause clinical disease in its natural west African simian hosts. Feline immunodeficiency virus (FIV) is a naturally occurring feline lentivirus analog of HIV and SIV. FIV causes immunodeficiency disease in its natural host and represents a non-primate model which is useful to study certain aspects of HIV infection not addressable in humans.

Background of model

FIV was discovered in 1986 by Pedersen and colleagues (Pedersen *et al.*, 1987) and has since been shown to be the feline counterpart of HIV. Although genetic analysis has shown FIV to be distinct from other lentiviruses, FIV and HIV share many genomic, structural, and biochemical characteristics. FIV infection in cats results in virus kinetics, pathogenesis, and immune responses similar to those seen in HIV infection of humans. FIV infection is characterized by three stages of disease: first, a primary transient illness with rapid virus replication; second, an asymptomatic period marked by early initiation of an antiviral immune response concurrent with down-modulation of viral replication but accompanied by gradual and progressive immune impairment; and third, a final stage in which viral load increases concomitant with the onset of clinical acquired immunodeficiency syndrome (AIDS)-like syndromes (Yamamoto *et al.*, 1988; Tompkins *et al.*, 1991; Pedersen, 1993).

Similar to HIV, FIV is a lymphotropic lentivirus infecting primarily CD4+ T cells in the early stages, with increasing viral burden in CD8+ T cells, macrophages, and B cells as disease progresses (Beebe *et al.*, 1992; English *et al.*, 1993; Dean *et al.*, 1996). Despite similar cell tropism, there is no evidence that FIV uses the CD4 molecule as a receptor or coreceptor (Willett *et al.*, 1997b). Alternate receptors such as the CD9 receptor and chemokine receptors have been suggested as means of viral entry. Anti-CD9 antibodies block productive infection of primary T cells by interfering with the virus life cycle post-entry into the cell, but the expression of CD9 is not sufficient to confer susceptibility to FIV infection. More recent work suggests that, as with HIV, it is likely that FIV uses chemokine receptors for cell entry (Willett *et al.*, 1997b).

FIV is neurotropic; virus enters the central nervous system early in the course of infection (Dow *et al.*, 1990). Associated neurologic symptoms include sleep disorders, behavioral changes, and altered visual and auditory evoked potential changes (Phillips *et al.*, 1994, 1996; Prospéro-García *et al.*, 1994). Lesions are principally cortical neuron loss and gliosis and minimal perivascular mononuclear cell infiltration affecting the brainstem and basal nuclei (Dow *et al.*, 1990; Phillips *et al.*, 1994, 1996).

Animal species

Domestic cats are the natural host for FIV.

Preparation of animals

No specialized treatment is required. Cats free of other diseases and pathogens can simplify interpretation of data and disease responses. In particular, pre-existent infection with feline leukemia virus (FeLV) would make interpretation of FIV infection especially complex since synergistic acceleration of FIV replication, CD4 T-cell depletion, and clinical disease have been documented (Pedersen *et al.*, 1990). As with HIV, SIV and all known retroviruses, susceptibility to FIV is inversely related to age; younger animals, especially neonates, have more rapid disease progression than adults.

Handbook of Animal Models of Infection
ISBN 0-12-775390-7

Storage, preparation of inocula

Based on envelope gene amino acid sequence, FIV can be classified into at least five subtypes or clades (Sodora *et al.*, 1994; Bachmann *et al.*, 1997). Such strain differences are probably responsible for the varied tissue tropism, replication patterns, and associated clinical signs. FIV strains that infect both macrophages and lymphocytes appear more likely to cross the placenta and mucosal surfaces or enter into the nervous system than those that predominantly infect lymphocytes (Dow *et al.*, 1990; O'Neil *et al.*, 1995; Burkhard *et al.*, 1997). Experimental studies have suggested that, because the virus burden is greatest and immune system selection least during the acute stage of infection, virus variants transmitted from acutely infected cats may accelerate the disease in newly infected cats (Diehl *et al.*, 1996).

Inoculation dose varies immensely in experimental transmission studies due to the influence of route, inoculum source, and study design (i.e., primary infection versus vaccine immunity challenge). Inocula used include whole blood, plasma, infected cells, and tissue culture supernatant. Heparin or ethylenediaminetetraacetic acid (EDTA) anticoagulated fresh whole blood from a FIV-infected cat can be used as a virus inoculum source. Plasma is harvested from the above or from Alsevier's anticoagulated whole blood by centrifugation and stored at –70°C or in liquid nitrogen until used. Harvesting of fresh peripheral blood mononuclear cells (PBMC) from anticoagulated whole blood is accomplished by density gradient centrifugation. Tissue culture-adapted viruses are frequently grown in adherent cell lines such as Crandell feline kidney (CrFK) cells. However, isolates demonstrating CrFK tropism appear less capable of establishing a virus burden *in vivo* versus virus propagated in PBMC.

Infection process

FIV is infectious via intravenous, intraperitoneal, intradermal, or subcutaneous injection (Pedersen, 1993) as well as by atraumatic instillation on to the oral, vaginal, or rectal mucosa (Bishop *et al.*, 1996; Burkhard *et al.*, 1997). Inocula have included whole blood, plasma, infected cells, and tissue culture supernatant. Either cell or cell-free inocula can establish infection transmucosally. Infection across the oral mucosa requires significantly more virus (10 ×) than other routes of mucosal exposure (Moench *et al.*, 1993). FIV has been transmitted to cats by artificial insemination of fresh, but not cryopreserved, semen (Jordan *et al.*, 1996). At least some strains of FIV are transmissible from infected queen to kitten, via both prenatal and postnatal pathways including *in utero*, milk/colostral, and intrapartum routes. Vertical transmission is probably most assured when pregnant queens are in the acute-phase infection, during which maximal viremia occurs (O'Neil *et al.*, 1995). However, queens chronically infected with at least one subtype B FIV isolate prior to pregnancy transmitted virus to >50% of kittens prenatally (O'Neil *et al.*, 1996). Low maternal CD4 T-cell count is associated with an increased rate of mother-to-kitten transmission.

Key parameters to monitor infection

Virus detection and quantification

Detection methods for FIV parallel those for HIV. Plasma viral RNA, as measured by quantitative competitive reverse transcriptase-polymerase chain reaction (QC-PCR), is the most relevant means to monitor active virus burden (Diehl *et al.*, 1996). Infectious virus titer in plasma, as determined by coculture of plasma with naive feline PBMC, is a less sensitive measure of cell-free viremia and does not always correlate with RNA PCR titer, potentially due to interference by antibody in plasma (Diehl *et al.*, 1996). During the asymptomatic period of chronic FIV infection, plasma viremia, whether measured by RNA PCR or PBMC coculture, is often low or undetectable.

Cell-associated infectious virus titers are determined by coculture of serial dilutions of PBMC or tissue cells (e.g., lymph node) from infected cats with naive PBMC. For some cell culture-adapted virus isolates, coculture with indicator T-cell lines can be substituted for PBMC (Novotney *et al.*, 1990; English *et al.*, 1991, 1993). For unselected clinical isolates, however, PBMC indicator cells are required. Viral replication is usually detected by FIV p26 Gag capture enzyme-linked immunosorbent assay (ELISA) of culture supernatants (Dreitz *et al.*, 1995). Serial dilution DNA PCR may also be used to quantitate provirus in either PBMC or tissue cells (Burkhard *et al.*, 1997). *In situ* hybridization, especially those employing riboprobes and immunohistochemistry, can help to identify the specific FIV-infected cell phenotypes in tissues (Beebe *et al.*, 1992). Strain-specific probes can increase the sensitivity of such assays.

CD4 cytopenia

One of the hallmarks of FIV infection, as with HIV infection, is the loss of CD4+ T lymphocytes (CD4 cells). CD4 cytopenia typically occurs late in the course of disease, but can occur within weeks of infection in accelerated disease models and in neonates. Less than 200 CD4 cells/μl of blood is considered consistent with AIDS; however, decreased numbers of CD4 cells alone are not responsible for the terminal immune dysfunction. CD4 cell counts less than 50 cells/μl have been seen without obvious clinical immunodeficiency.

Inversion of the CD4/CD8 ratio is seen earlier than CD4 cytopenia because of the combined effect of a decrease in CD4 lymphocytes and an increase in CD8 lymphocytes.

Antibody

Antibodies directed against FIV can be detected in serum or plasma by ELISA or western blot as early as 2 weeks post-infection, and typically persist throughout the disease (Yamamoto *et al.*, 1988; O'Neil *et al.*, 1995). Antibody responses to Env precede those directed against Gag. Cats without detectable antibody production generally progress rapidly to terminal immunodeficiency (Diehl *et al.*, 1995; O'Neil *et al.*, 1995; Burkhard *et al.*, 1997). A polyclonal gammopathy directed against non-viral proteins may be seen in FIV-infected cats as early as 6 weeks post-infection (Flynn *et al.*, 1994).

FIV contains an immunodominant region which shares both structural and functional homology with the V3 domain of HIV (Pancino *et al.*, 1993, 1995a,b). The V3 hypervariable region is an important target for virus-neutralizing antibody, although the role of virus-neutralizing antibody in FIV has not been clearly established. The presence of neutralizing antibody does not correlate with virus clearance or disease progression in FIV-infected cats (Yamamoto *et al*, 1991; Tozzini *et al*, 1992) but may show association with protection from challenge following vaccination by whole inactivated virus vaccines (Hosie *et al.*, 1995).

Cytotoxic T lymphocyte activity

Cytotoxic T lymphocyte (CTL) activity has been detected in acutely and chronically FIV-infected cats and in vaccinated cats. Methods to detect FIV-specific CTL have utilized both fresh and *in vitro* restimulated effector cells and autologous cultured skin fibroblast and lymphocyte targets (Song *et al.*, 1992; Flynn *et al.*, 1995; Beatty *et al.*, 1996). CTL directed against Gag, Pol, and Env epitopes have been detected in infected animals (Flynn *et al.*, 1996). Their relationship to control of viral infection remains uncertain; however, vaccine-induced protection against FIV challenge has been associated with CTL directed against Env (Hosie and Flynn, 1996).

Lymphocyte function and cytokine production

Immune dysfunction precedes the decline in CD4 cell counts. Both mitogen and antigen-induced lymphoproliferation are impaired or altered in FIV infection (Pedersen, 1993). Although the presence of feline T-helper subtypes has not been well-characterized, feline cytokine production as quantitated by RT-PCR is altered in FIV infection (Rottman *et al.*, 1995). Preliminary data from several groups suggest that, similar to HIV, in FIV-infected cats there is a shift away from T_h1 cytokine production.

Thymic atrophy

Thymic atrophy has been seen following *in utero*, neonatal, or juvenile infection with FIV (Woo *et al.*, 1997). Alterations in thymic structure, function, and cytokine production appear similar to those seen in HIV-infected children.

Hematologic abnormalities

Neutropenia, lymphopenia, and anemia are typical of FIV infection in cats (Callanan *et al.*, 1992; Sparkes *et al.*, 1993).

Antiviral therapy

Because of the similarities between FIV and HIV infections, FIV can be useful in testing of antiviral drugs directed toward targets shared by the two viruses. Two such targets are reverse transcriptase and protease.

Reverse transcriptase inhibitors

Although 3′azido-2′,3′-deoxythymidine (AZT) inhibits the replication of FIV *in vitro* at fairly low concentration (0.2 μg/ml) it is not as effective *in vivo* as the *in vitro* sensitivity might imply. At doses ranging from 0.2 to 50 mg/kg administered subcutaneously, AZT delays but does not prevent infection, even in cats treated prophylactically (Smyth and Bennett, 1994). However, virus burdens are lowered by AZT treatment and CD4 and CD8 cell counts are higher in comparison with those in untreated cats (Hayes *et al*, 1993). In cats, as in humans, AZT produces dose-related reversible macrocytic anemia (Hayes *et al.*, 1993; Smyth and Bennett, 1994). FIV mutants resistant to AZT have been detected by treatment of chronically infected cells with AZT *in vitro* followed by mapping of the amino acid mutations mediating resistance (Barlough *et al.*, 1993; North *et al.*, 1989a,b). The same mutations have yet to be produced by *in vivo* therapy but no studies of long-term AZT therapy have been conducted in the FIV model.

Acyclic nucleoside phosphonate derivatives have shown promise as therapeutic agents in the FIV model. 9-(2-phosphonylmethoxyethyl) adenine (PMEA) potently inhibits FIV replication *in vitro* and, like AZT, delays onset of virus infection and reduces virus burden, yet does not prevent infection when administered at the time of virus infection (Egberink *et al.*, 1990; Myles *et al.*, 1997). PMEA has been administered subcutaneously (3–6 mg/kg for 7 weeks) with reduced signs of toxicity when compared to AZT (Philpott *et al.*, 1992).

More recently, a similar acyclic nucleoside phosphonate, 9-(2-phosphonomethoxypropyl) adenine (PMPA) has been shown to block FIV infection and T-cell depletion completely when given parenterally at the time of virus exposure (30 mg/kg per day; Myles *et al.*, 1997). Orally administered PMPA provided partial protection. PMPA is unique in its very low toxicity combined with potent efficacy *in vivo*, and is currently being evaluated in animals with established infection. The same drug has been shown to be effective against acute and chronic SIV infection (Tsai *et al.*, 1997).

High-dose combination of AZT plus lamivudine (3TC) have also been shown to prevent FIV infection in a small number of cats treated with 75 mg/kg per day orally (Arai *et al.*, 1997).

Protease inhibitors

The FIV protease is structurally similar to that of HIV; however the HIV protease inhibitors currently in clinical trials do not inhibit the FIV protease (Laco *et al.*, 1997).

Vaccine strategies

The FIV model provides the opportunity to identify lentivirus vaccine strategies which induce protective immunity. FIV infection is unique in demonstrating successful vaccination against homologous virus challenge through the use of whole inactivated virus (WIV) immunization (Yamamoto *et al.*, 1991; Hosie *et al.*, 1995). It has been suggested that the gp120 Env may play a critical role in WIV vaccine protection (Willett *et al.*, 1997a). As with HIV and SIV, investigation of FIV subunit, peptide, and DNA vaccine approaches has not yielded protective immunity but has allowed dissection of the vaccine-associated immune responses (Willett *et al.*, 1997a).

Key parameters to monitor response to treatment

The key parameters to monitor response to treatment are similar to those used to monitor infection. Quantification of viremia is the most discriminating analysis, but surrogate markers of disease such as CD4 lymphocyte counts also provide information regarding treatment efficacy. It is critical to evaluate both blood and tissue sources for virus post-treatment as substantial reservoirs of tissue virus can exist in the face of minimal blood virus.

Advantages/disadvantages of the model

General strengths of the FIV system include a natural pathogen in an outbred host, similar virus genomic structure, function, target cells, tissue tropism, host–virus relationship, and clinical disease. The FIV system can be valuable for testing proof of principle for experimental intervention or treatment strategies and to address issues in early pathogenesis requiring invasive and interventive protocols, e.g. mucosal crossing, early target cells, and virus/cell trafficking. Promising strategies can also be tested in naturally infected animal populations. The number of animals possible per group can be expanded without expanding costs relative to primate model alternatives.

Despite many similarities between HIV and FIV, there are significant limitations which need to be considered if the FIV system is to be useful. FIV and HIV have very similar tissue tropism; however, FIV has not been shown to require CD4 as well as a chemokine coreceptor for cell infection (Willett *et al.*, 1997b). While it is likely that FIV uses a chemokine receptor for viral entry, that receptor has yet to be identified for clinical isolates (Willett *et al.*, 1997b). Some HIV genes/gene products, such as *nef*, *vpu*, and *vpr*, cannot be evaluated in the FIV model since these genes have no counterparts in the simpler FIV genome. Immunoreagents specific for HIV, e.g., receptor or gene product agonists/inactivators, may not be assessable if they rely on unique sequence specificity versus more broad activities, e.g., reverse transcriptase inhibitors. Finally, for obvious reasons, development of viral and immunologic reagents needed for assays and therapeutics lags behind those related to HIV, posing limitations to questions addressable.

Contributions of the model to infectious disease therapy

The FIV system provides the opportunity to study the pathogenesis and intervention of lentivirus infection in a natural host. Aspects of HIV and other lentivirus infections which can now be modeled in FIV infection include:

1. Early pathogenesis of mucosal infection and immunity
2. Neurotropism and neuropathogenesis
3. Mechanisms of vertical transmission and its intervention
4. Efficacy of immune-mediated and antiviral therapies
5. Efficacy and immune correlates of experimental vaccines.

Critical questions to be addressed

The mechanism of FIV entry, cell receptors, and early target cells need to be identified to utilize further the FIV model for pathogenesis, intervention, and vaccine strategies. Understanding the role of cellular immune responses and virus reservoirs in virus down-modulation versus true viral clearance is necessary to identify the correlates of vaccine protection. The role of cytokines, T-helper lymphocyte responses, and chemokines needs to be addressed to understand the role of the immune system both in disease protection and disease progression. The potential impact of low-level mucosal virus exposure in generating local protective immunity has direct relevance to HIV transmission worldwide.

References

Arai, M., Dunn, B. M., Yamamoto, J. K. (1997). AZT/3TC therapy in cats: its potential use in combination with FIV protease inhibitor. *Second International Feline Immunology Workshop, University of California, Davis, CA* Abstract, 9.

Bachmann, M. H., Sodora, D. L., Mathiason-Dubard, C. *et al.* (1997). Genetic diversity of feline immunodeficiency virus: dual infection, recombination and distinct evolutionary rates between envelope gene subtypes. *J. Virol.*, **71**, 4241–4253.

Barlough, J., North, T., Oxford, C. *et al.* (1993). Feline immunodeficiency virus infection of cats as a model to test the effect of certain *in vitro* selection pressures on the infectivity and virulence of resultant lentivirus variants. *Antiviral Res.*, **22**, 259–272.

Beatty, J. A., Willett, B. J., Gault, E. A., Jarrett, O. (1996). A longitudinal study of feline immunodeficiency virus (FIV)-specific cytotoxic T-lymphocytes (CTL) in experimentally infected cats using antigen-specific induction. *J. Virol.*, **70**, 6199–6206.

Beebe, A. M., Gluckstern, T. G., George, J., Pedersen, N. C., Dandekar, S. (1992). Detection of feline immunodeficiency virus infection in bone marrow of cats. *Vet. Immunol. Immunopathol.*, **35**, 37–49.

Bishop, S. A., Stokes, C. R., Gruffydd-Jones, T. J., Whiting, C. V., Harbour, D. A. (1996). Vaginal and rectal infection of cats with feline immunodeficiency virus. *Vet. Microbiol.*, **51**, 217–227.

Burkhard, M. J., Obert, L. A., O'Neil, L. L, Diehl, L. J., Hoover, E. A. (1997). Mucosal transmission of cell-associated and cell-free feline immunodeficiency virus. *AIDS Res. Hum. Retroviruses*, **13**, 347–355.

Callanan, J. J., Thompson, H., Toth, S. R. *et al.* (1992). Clinical and pathological findings in feline immunodeficiency virus experimental infection. *Vet. Immunol. Immunopathol.*, **35**, 3–13.

Dean, G. A, Reubel, G. H., Moore, P. F, Pedersen, N. C. (1996). Proviral burden and infection kinetics of feline immunodeficiency virus in lymphocyte subsets of blood and lymph node. *J. Virol.*, **70**, 5165–5169.

Diehl, L. J., Mathiason-Dubard, C. K., O'Neil, L. L. *et al.* (1995). Induction of accelerated felines immunodeficiency virus disease by acute-phase virus passage. *J. Virol.*, **69**, 6149–6157.

Diehl, L. J., Mathiason-Dubard, C. K., O'Neil, L. L., Hoover, E. A. (1996). Plasma viral RNA load predicts disease progression in accelerated feline immunodeficiency virus infection. *J. Virol.*, **70**, 2503–2507.

Dow, S. W., Poss, M. L., Hoover, E. A. (1990). Feline immunodeficiency virus: a neurotropic lentivirus. *J. AIDS*, **3**, 658–688.

Dreitz, M. J., Dow, S. W., Fiscus, S. A., Hoover, E. A. (1995). Development of monoclonal antibodies and capture immunoassays for feline immunodeficiency virus. *Am. J. Vet. Res.*, **56**, 764–768.

Egberink, H., Borst, M., Niphuis, H. *et al.* (1990). Suppression of feline immunodeficiency virus infection *in vivo* by 9-(2-phosphonomethoxyethyl)adenine. *Proc. Natl Acad. Sci. USA*, **87**, 3087–3091.

English, R. V., Johnson, C. M., Gebhard, D. H., Tompkins, M. B. (1993). *In vivo* lymphocyte tropism of feline immunodeficiency virus. *J. Virol.*, **67**, 5175–5186.

Flynn, J. N., Cannon, C. A., Lawrence, C. E., Jarrett, O. (1994). Polyclonal B-cell activation in cats infected with feline immunodeficiency virus. *Immunology*, **81**, 626–630.

Flynn, J. N., Beatty, J. A., Cannon, C. A. *et al.* (1995). Involvement of gag- and env-specific cytotoxic T lymphocytes in protective immunity feline immunodeficiency virus. *AIDS Res. Hum. Retroviruses*, **11**, 1107–1113.

Flynn, J. N., Keating, P., Hosie, M. J. *et al.* (1996). Env-specific CTL predominate in cats protected from feline immunodeficiency virus by vaccination. *J. Immunol.*, **157**, 3658–3665.

Hayes, K. A, Lafrado, L. J., Erickson, J. G., Marr, J. M., Mathes, L. E. (1993). Prophylactic ZDV therapy prevents early viremia and lymphocyte decline but not primary infection in feline immunodeficiency virus-inoculated cats. *J. AIDS*, **6**, 127–134.

Hosie, M. J., Flynn, J. N. (1996). Feline immunodeficiency virus vaccination: characterization of the immune correlates of protection. *J. Virol.*, **70**, 7561–7568.

Hosie, M. J., Osborne, R., Yamamoto, J. K., Neil, J. C., Jarrett, O. (1995). Protection against homologous but not heterologous challenge induced by inactivated feline immunodeficiency virus vaccines. *J. Virol.*, **69**, 1253–1255.

Jordan, H. L., Howard, J., Sellon, R. K., Wildt, D. E., Tompkins, W. A., Kennedy-Stoskopf, S. (1996). Transmission of feline immunodeficiency virus in domestic cats via artificial insemination. *J. Virol.*, **70**, 8224–8228.

Laco, G. S., Fitzgerald, M. C., Morris, G. M., Olson, A. J., Kent, S. B., Elder, J. H. (1997). Molecular analysis of the feline immunodeficiency virus protease: generation of a novel form of the protease by autoproteolysis and construction of cleavage-resistant proteases. *J. Virol.*, **71**, 5505–5511.

Moench, T. R, Whaley, K. J, Mandrell, T. D., Bishop, B. D., Witt, C. J., Cone, R. A. (1993). The cat/feline immunodeficiency virus model for transmucosal transmission of AIDS: nonoxynol-9 contraceptive jelly blocks transmission by an infected cell inoculum. *AIDS*, **7**, 797–802.

Myles, M. H, Ebner, J. P., Bischofberger, N., Hoover, E. A. Efficacy of 9-(2-phosphonylmethoxypropyl) adenine for therapy of acute feline immunodeficiency virus infection. *In preparation*, 1997.

North, T. W., Cronn, R. C., Remington, K. M., Tandberg, R. T., Judd, R. C. (1989a). Characterization of reverse transcriptase from feline immunodeficiency virus. *Am. Soc. Biochem. Mol. Biol.*, **265**, 5121–5127.

North, T. W., North, G. L. T., Pedersen, N. C. (1989b). Feline immunodeficiency virus, a model for reverse transcriptase-targeted chemotherapy for acquired immune deficiency syndrome. *Antimicrob. Agents Chemother.*, **33**, 915–919.

Novotney, C., English, R. V., Housman, J. *et al.* (1990). Lymphocyte population changes in cats naturally infected with feline immunodeficiency virus. *AIDS*, **4**, 1213–1218.

O'Neil, L. L., Burkhard, M. J., Diehl, L. J., Hoover, E. A. (1995). Vertical transmission of feline immunodeficiency virus. *AIDS Res. Hum. Retroviruses*, **11**, 171–182.

O'Neil, L. L., Burkhard, M. J., Hoover, E. A. (1996). Frequent perinatal transmission of feline immunodeficiency virus by chronically infected cats. *J. Virol.,* **70**, 2894–2901.

Pancino, G., Fossati, I., Chappey, C. *et al.* (1993). Structure and variations of feline immunodeficiency virus envelope glycoproteins. *Virology*, **192**, 659–662.

Pancino, G., Camoin, L., Sonigo, P. (1995a). Structural analysis of the principal immunodominant domain of the feline immunodeficiency virus transmembrane glycoprotein. *J. Virol.*, **69**, 2110–2118.

Pancino, G., Castelot, S., Sonigo, P. (1995b). Differences in feline immunodeficiency virus host cell range correlate with envelope fusogenic properties. *Virology*, **206**, 796–806.

Pedersen, N. C. (1993). The feline immunodeficiency virus. In *The Retroviridae*, vol 2. (ed Levy, J. A.), pp. 181–228. Plenum, New York.

Pedersen, N. C., Ho, E. W., Brown, M. L., Yamamoto, J. K. (1987). Isolation of a T-lymphotropic virus from domestic cats with an immunodeficiency-like syndrome. *Science*, **235**, 790–793.

Pedersen, N. C., Torten, M., Rideout, B. *et al.* (1990). Feline leukemia virus infection as a potentiating cofactor for the primary and secondary stages of experimentally induced feline immunodeficiency virus infection. *J. Virol.*, **64**, 598–606.

Phillips, T., Prospero-Garcia, O., Puaoi, D. *et al.* (1994). Neurological abnormalities associated with feline immunodeficiency virus infection. *J. Gen. Virol.*, **75**, 979–987.

Phillips, T. R., Prospero-Garcia, O., Wheeler, D. W. *et al.* (1996). FIV induced neurologic abnormalities. *Third International Feline Retrovirus Research Symposium* Abstract, 30.

Philpott, M. S., Ebner, J. P., Hoover, E. A. (1992). Evaluation of 9-(2-phosphonylmethoxyethyl) adenine therapy for feline immunodeficiency virus using a quantitative polymerase chain reaction. *Vet. Immunol. Immunopathol.*, **35**, 155–166.

Prospéro-García, O., Herold, N., Phillips, T. R., Elder, J. H., Bloom, F. E., Henriksen, S. J. (1994). Sleep patterns are disturbed in cats infected with feline immunodeficiency virus. *Proc. Natl Acad. Sci. USA*, **91**, 12947–12951.

Rottman, J. B., Freeman, E. B., Tonkonogy, S., Tompkins, M. B. (1995). A reverse transcription-polymerase chain reaction technique to detect feline cytokine genes. *Vet. Immunol. Immunopathol.*, **45**, 1–18.

Smyth, N. R., Bennett, M. (1994). Effect of 3′azido-2′,3′-deoxythymidine (AZT) on experimental feline immunodeficiency virus infection in domestic cats. *Res. Vet. Sci.*, **57**, 220–224.

Sodora, D. L., Shpaer, E. G., Kitchell, B. E. *et al.* (1994). Identification of three feline immunodeficiency virus (FIV) *env* gene subtypes and comparison of the FIV and human immunodeficiency virus type 1 evolutionary patterns. *J. Virol.*, **68**, 2230–2238.

Song, W., Collisson, E. W., Billingsley, P. M., Brown, W. C. (1992). Induction of feline immunodeficiency virus-specific cytolytic T-cell responses from experimentally infected cats. *J. Virol.*, **66**, 5409–5417.

Sparkes, A. H., Hopper, C. D., Millard, W. G., Gruffydd-Jones, T. J., Harbour, D. A. (1993). Feline immunodeficiency virus infection. Clinicopathologic findings in 90 naturally occurring cases. *J. Vet. Intern. Med.*, **7**, 85–90.

Tompkins, M. B., Nelson, P. D., English, R. V., Novotney, C. (1991). Early events in the immunopathogenesis of feline retrovirus infections. *J. Am. Vet. Med. Assoc.*, **199**, 1311–1315.

Tozzini, F., Matteucci, D., Bandecchi, P. *et al.* (1992). Simple *in vitro* methods for titrating feline immunodeficiency virus (FIV) and FIV neutralizing antibodies. *J. Virol. Methods*, **37**, 241–252.

Tsai, C. C., Follis, K. E., Beck, T. W., Sabo, A., Bischofberger, N., Dailey, P. J. (1997). Effects of (R)-9-(2-phosphonylmethoxypropyl)adenine monotherapy on chronic SIV infection in macaques. *AIDS Res. Hum. Retroviruses*, **13**, 707–712.

Willett, B. J., Flynn, N. J., Hosie, M. J. (1997a). FIV infection of the domestic cat: an animal model for AIDS. *Immunol. Today*, **18**, 182–189.

Willett, B. J., Picard, L., Clapham, P. R. (1997b). Shared usage of the chemokine receptor CXCR4 by the feline and human immunodeficiency viruses. *J. Virol.*, **71**, 6407–6415.

Woo, J. C., Dean, G. A., Moore, P. F. (1997). Immunopathologic changes in the thymus during the acute stage of experimentally induced feline immunodeficiency virus infection in juvenile cats. *J. Virol.*, **71**, 8632–8641.

Yamamoto, J. K., Sparger, E., Ho, E. W. *et al.* (1988). The pathogenesis of experimentally induced feline immunodeficiency virus (FIV) infection in cats. *Am. J. Vet. Res.*, **49**, 1246–1258.

Yamamoto, J. K., Okuda, T., Ackley, C. D. *et al.* (1991). Experimental vaccine protection against feline immunodeficiency virus infection. *AIDS Res. Hum. Retroviruses*, **7**, 911–922.

Chapter 128

Animal Models of HIV Infection: SIV Infection of Macaques

K. K. A. Van Rompay and N. L. Aguirre

Background of human infection

The World Health Organization and UN-AIDS estimate that as of December 1997, at least 40 million people are infected with or have died from human immunodeficiency virus (HIV) infection worldwide. The principal modes of HIV transmission are sexual contact, parenteral exposure to blood and mother-to-child transmission during the perinatal period. Two subtypes of HIV have been described: HIV type 1 and 2.

The primary target cells for HIV replication are CD4+ cells, in particular T-helper lymphocytes and macrophages. Because of the pivotal role of these cells in the ontogeny of acquired immune responses, the progressive depletion of CD4+ T cells during HIV infection ultimately leads to the development of immunodeficiency and the occurrence of opportunistic infections and malignancies (reviewed by Levy, 1997). Newborns infected perinatally with HIV often develop immunodeficiency sooner than HIV-infected adults (reviewed by Oxtoby, 1994).

During recent years, the development of novel anti-HIV drugs has led to major improvements in the clinical management of HIV-infected people in the developed countries. There is, however, much room for improvement, as not everyone experiences these "miraculous" effects. In addition, the majority of HIV-infected people in the world, living in the developing countries, have not benefited from these recent advances. Accordingly, the challenge for us now is to find and identify better antiviral drugs.

Due to the nature of the HIV disease course, and the logistics and financial and ethical aspects of human clinical trials, it is becoming increasingly complicated to test the efficacy of novel anti-HIV drugs, because it is considered unethical to treat "control" groups with anything less than the currently best available treatment (Cohen, 1997a,b; Lange, 1997). Animal models that allow rapid evaluation of the efficacy of anti-HIV compounds are very useful to sort out those drugs which are most promising to enter human clinical trials, from those drugs that should probably be discarded. While murine and feline models (described elsewhere in this book) are very appropriate for initial screening, further testing can be done in non-human primate models that better resemble HIV infection and acquired immunodeficiency syndrome (AIDS), such as simian immunodeficiency virus (SIV) infection of macaques.

Background of animal model

SIV was originally isolated from rhesus macaques in captivity that showed immunosuppression and lymphomas, and was called SIVmac. Since then, a number of SIV strains have also been isolated from other primate hosts. To date, five groups of primate lentiviruses have been identified (reviewed by Desrosiers, 1990; Hu *et al.*, 1993):

1. SIVagm (African green monkey)
2. SIVsmm (sooty mangabey)/SIVmac (macaque)/SIVmne (nemestrina)/SIVstm (stumptail)/HIV-2
3. SIVmnd (mandrill)
4. SIVsyk (sykes monkey)
5. HIV-1/SIVcpz (chimpanzee).

The SIVsmm/SIVmac/SIVmne group has been used most extensively. These viruses are closely related to HIV-2, and are more distantly related to HIV-1 (80–90% and 50–60% amino acid homology in polymerase gene, respectively).

The sooty mangabey, African green monkey, mandrill and chimpanzee isolates have all been derived from African non-human primate species, and are found in animals in their natural habitat; none has yet been shown to induce disease in its presumed natural host. Macaques, including rhesus macaques, are Asian Old World primates; evidence suggests that macaques are not naturally infected with SIV but that some captive macaques have acquired their SIV (SIVmac), which is genetically similar to SIVsmm, from sooty mangabey monkeys while in captivity at one or more of the Regional Primate Research Centers (reviewed by Gardner, 1996).

SIVmac and its closely related SIVsmm and SIVmne, upon experimental inoculation of a variety of Asian macaque species, induce a disease which very closely resembles human HIV infection and AIDS (Letvin and King,

Handbook of Animal Models of Infection
ISBN 0-12-775390-7

1990). Compared to HIV infection of humans, SIV infection in macaques has an accelerated time course, as a much greater proportion of infected monkeys die from the infection within a comparatively short period of time (3–30 months compared to 7–10 years for humans). SIVsmmPBj14 is a unique case: instead of causing chronic, progressive immunodeficiency, this variant of SIVsmm induces a very acute clinical disease (characterized by severe diarrhea) in pigtail macaques within 2 weeks following inoculation (Fultz *et al.*, 1989; Israel *et al.*, 1993); evidence suggests that the pathogenesis is due to extensive T-lymphocyte activation induced by a particular *nef* allele in the virus (Du *et al.*, 1995).

Animal species

Most antiviral drug studies use rhesus macaques (*Macaca mulatta*), or cynomolgus macaques (*M. fascicularis*). Although juvenile and adult macaques are suitable for short-term treatment aimed at preventing or modulating infection, the use of newborn macaques is more practical to evaluate reliably the efficacy of prolonged drug therapy on disease progression (Van Rompay *et al.*, 1994).

Preparation of animals

Animal housing and laboratory facilities need to comply with biosafety level 2 requirements for containment of infectious biohazardous agents. Animals need to be simian type D retrovirus-free and SIV-seronegative.

Details of experimental procedures

For virus inoculation and sample collection, animals are usually immobilized with ketamine HCl (Parke-Davis, Morris Plains, NJ) 10–15 mg/kg injected intramuscularly.

Storage, preparation of inocula

Aliquots of SIV virus stocks are kept frozen (–70 to –130°C), and thawed immediately prior to usage. The use of virulent instead of attenuated SIV isolates is recommended, as the demonstration of antiviral effects in a "worst-case scenario" is a more powerful indicator of the true potency of an antiviral compound. The importance of using a minimal virus inoculum dose (which better mimicks human exposure to HIV) has been demonstrated for chemoprophylaxis studies (Böttiger *et al.*, 1992a)

Infection process

In most macaque studies, SIV was inoculated intravenously. Because most human infections occur following exposure to HIV at mucosal sites, recent macaque studies have used mucosal routes (oral, rectal) of virus inoculation (Van Rompay *et al.*, 1996, 1998a,b; Böttiger *et al.*, 1997). As expected, a higher inoculum dose is required for mucosal than for intravenous SIV inoculation. Intravaginal SIV inoculation has been used to test the efficacy of topical application of microbicides or antiviral drugs against sexual transmission (Miller, 1994; Miller *et al.*, 1996).

Key parameters to monitor infection and response to treatment

A major advantage of SIV infection of macaques is that the efficacy of antiviral drugs can be evaluated thoroughly, not only by measuring reduction in virus levels, but also by monitoring disease progression. In addition, the laboratory and clinical parameters which are used to monitor the course of SIV infection in macaques are very similar or identical to those which are used for HIV infection of humans (Table 128.1).

Depending on the experimental design, antiviral drug studies in macaques aim to prevent infection (chemoprophylaxis) or to evaluate the efficacy of antiviral drug treatment during early or late stages of SIV infection (chemotherapy).

Chemoprophylaxis

The chemoprophylactic efficacy of an antiviral compound is determined by its ability to prevent SIV infection following inoculation with a virus dose sufficient to cause persistent infection in untreated control macaques. Protection against SIV infection is defined by the inability to detect virus as well as antiviral immune responses. Depending on the study design, the sensitivity of the assays used, and the frequency of sampling, sometimes transient detection of virus or antiviral immune responses can be found during the early time period following virus inoculation, but animals become negative by all criteria later on and remain healthy (Van Rompay *et al.*, 1998a). Because of such cases of possible drug-induced transient or abortive infection, chemoprophylactic success may have to be redefined as "prevention of establishment of persistent infection".

Chemotherapy

If early drug administration does not prevent establishment of persistent SIV infection, or if drug treatment is not started until later stages of SIV infection, antiviral drug

Table 128.1 Key parameters to monitor response to treatment

Viral parameters	
Virus levels in blood and lymphoid tissues (quantitative virus isolation, PCR, RT-PCR, bDNA . . .)	
Antiviral drug resistance (phenotypic and genotypic)	
Immune markers	
Antibody responses, cytotoxic T lymphocyte responses	
T lymphocyte phenotyping (CD4+, CD8+ . . .)	
Cytokine and chemokine levels	
Clinical parameters	
Weight gain	
Clinical disease (opportunistic infections, survival)	

PCR, Polymerase chain reaction; RT-PCR, reverse transcriptase polymerase chain reaction; bDNA, branched DNA.

efficacy can be evaluated by a number of laboratory and clinical parameters (Table 128.1).

Important indicators of antiviral drug efficacy are virus levels in blood and lymphoid tissues (measured using quantitative virus isolation, viral RNA and proviral DNA techniques). The significance of any of these parameters is straightforward: an antiviral drug is expected to inhibit virus replication, and therefore reduce virus levels.

Occasionally, the rate at which macaques mount antiviral antibody responses is used as an indicator of the therapeutic efficacy of antiviral compounds. This parameter, however, has to be used with extra caution, because the rate of antiviral antibody production during SIV infection is probably affected by two opposing forces which are both correlated with the amount of virus replication: high virus replication will result in sufficient antigen presentation to the immune system, but the virus-induced immunosuppression can dampen the development of antibody responses. Accordingly, the effect of beneficial antiviral drug therapy on the development of antiviral immune responses can vary significantly, depending on the virulence of the SIV inoculum, the potency of the antiviral drug, and the timing of drug administration. For instance, chemotherapy during the early stages of infection can reduce the initial viremia and therefore, by preventing immunosuppression, enhance antiviral antibody responses (Van Rompay *et al.*, 1995, 1996). In contrast, in a few studies, highly active antiviral drug treatment was started very early so that, although infection was not prevented, viral antigen levels during the drug treatment period were probably not sufficiently high to induce strong antiviral antibody responses; in these studies, antibody responses were delayed until drug treatment was withdrawn (Balzarini *et al.*, 1991; Tsai *et al.*, 1994; Van Rompay *et al.*, 1998a). In both scenarios, a beneficial outcome on disease progression was seen, regardless of whether antibody responses were accelerated or delayed. Thus, the kinetics of antiviral antibody responses must always be interpreted in correlation with virus levels and clinical disease outcome.

Similar to HIV infection of humans, SIV infection of macaques induces changes in T-lymphocyte subsets, in particular a decline in CD4+ T cells. The beneficial effects of antiviral chemotherapy on stabilizing T-lymphocyte subsets during SIV infection have been demonstrated in most studies using flow cytometry techniques.

Although reduced virus levels and stabilization of host immune parameters are indicative of the efficacy of chemotherapy, the ultimate goal of antiviral therapy for HIV-infected patients is to prolong disease-free survival indefinitely. Monitoring clinical disease during drug treatment takes into account possible side-effects of antiviral drugs. In addition, it is the most appropriate way to evaluate certain immunomodulatory compounds which have no direct antiviral properties, but which still may improve the clinical outcome for HIV-infected patients. In order to determine the effects of drugs on disease progression rapidly, juvenile/adult macaques are problematic because of the variable time course of SIV disease progression (the asymptomatic period can range from months to years). In contrast, the disease course in newborn macaques following infection with highly virulent SIV isolates is much more rapid and uniform, as most animals develop disease within 2–4 months; therefore, this animal model of pediatric AIDS, which presents a "worst-case scenario", is very useful for testing novel antiviral drugs, because efficacy can be assessed by monitoring clinical disease progression as well as viral and immune parameters, using a limited number of animals and in a relatively short period of time (Van Rompay *et al.*, 1994, 1995, 1996).

Antimicrobial therapy

A number of antiviral compounds have been evaluated in the SIV macaque model. A summary of representative studies is given in Tables 128.2 and 128.3.

A few studies have demonstrated the potential of topical application of vaginal microbicides or 9-[2-(phosphonomethoxy)propyl]adenine (PMPA) gel to protect adult female macaques against intravaginal SIV transmission (Miller, 1994; Miller *et al.*, 1996). The majority of macaque studies used systemic (oral or parenteral) drug administration.

Prevention or reduction of acute SIV viremia

Most studies in non-human primates have focused on testing whether short-term drug administration starting around the time of virus inoculation was effective in preventing infection or delaying viremia (Table 128.2). Although the first studies, which mostly used zidovudine (AZT), were not highly effective in preventing infection, a likely reason for this was the high dose of virus used in these experiments; more recent studies have now shown that

Table 128.2 Examples of antiviral drug treatment during early simian immunodeficiency virus (SIV) infection

Reference	*Virus*	*Inoculum size and route*	*Start of treatment relative to virus inoculation*	*Drug, daily dose, route, duration*	*Outcome*
McClure *et al.*, 1990	SIVsmmPBJ14	20 $TCID_{50}$, i.v.	1–72 hr after	AZT 3 × 33 mg/kg s.q. for 14 days	Early treatment prevented acute disease
McClure *et al.*, 1990	SIVsmm	150 $TCID_{50}$, i.v.	24 hr before	CS-87, AZT, D4T, FDT 3 × 33 mg/kg s.q. for 14 days	No protection of infection
Balzarini *et al.*, 1991	SIVmacBK28	i.v.	1 day prior	PMEA, 2 × 5 or 2 × 10 mg/kg i.m. for 29 days	Delayed seroconversion
Böttiger *et al.*, 1992a	SIVsm/HIV-2	2–10 AID_{50}, i.v.	8 hr prior	FLT, 3 × 5 mg/kg, s.q. for 10 days	Half of animals protected
Van Rompay *et al.*, 1992	SIVmac251	1–10 AID_{50}, i.v.	2 hr prior	AZT, 3 × 50 mg/kg oral for 6 weeks	Protection against infection
Martin *et al.*, 1993	SIV deltaB670	10 AID_{50}, i.v.	1–72 hr after	AZT, 4 × 25 mg/kg, s.q. for 28 days	Reduced viremia, antigenemia
Le Grand *et al.*, 1994	SIVmac251	4 AID_{50}, i.v.	7 days prior	AZT, 3 × 15 mg/kg, s.q. for 35 days	Reduced viremia
Tsai *et al.*, 1994	SIVmne	10^3 $TCID_{50}$, i.v.	48 hr prior	PMEA, 10 or 20 mg/kg per day, s.q. for 28 days	Protection proportional to dosage regimen
Van Rompay *et al.*, 1995	SIVmac251	10–100 AID_{50}, i.v.	From 2 hr before to 7 days after	AZT, 3 × 25 mg/kg oral for > 15 months	Reduced viremia, enhanced antiviral immune responses, delayed disease
Rausch *et al.*, 1995	SIVdeltaB670	10 AID_{50}, i.v.	2 hr before	AZT, continuous or intermittent infusion	Reduced antigenemia, delayed disease
Tsai *et al.*, 1995a	SIVmne	10 AID_{50}, i.v.	From 48 hr before to 24 hr after	PMPA 20 or 30 mg/kg s.q. for 4 weeks	Complete protection
Überla *et al.*, 1995	RT-SHIV	200–3000 $TCID_{50}$, i.v.	8 hr before	AZT 3 × 15 mg/kg or LY300046-HCl 3 × 5 mg/kg s.q. for 5 days	LY300046-HCl but not AZT delayed antigenemia
Joag *et al.*, 1997	SIVmac239	2000 $TCID_{50}$, i.v.	1 day before	PMEA, 20 mg/kg, s.q. for 30 days	Reduced viremia, delayed disease
Böttiger *et al.*, 1997	SIVsm/HIV-2	10–1000 AID_{50}, i.v. or intrarectal	From 8 hr before to 6 days after	BEA-005, 3 × (0.05–10) mg s.q. for 1–10 days	Depending on regimen, and virus inoculum: protection or delayed viremia
Van Rompay *et al.*, 1998b	SIVmac251	10^5 $TCID_{50}$ oral	4 hr before	2 doses of PMPA, 30 mg/kg s.q., at –4 hr and +20 hr	Protection

$TCID_{50}$, Tissue culture infectious doses 50%; AID_{50}, animal infectious dose 50%; i.v. intravenously; s.q. subcutaneously; AZT, zidovudine; PMPA, 9-[2-(phosphonomethoxy)propyl]adenine; PMEA, 9-[2-(phosphonomethoxy)ethyl]adenine; BEA-005, 2′,3′-dideoxy-3′-hydroxymethyl cytidine.

Table 128.3 Examples of antiviral drug therapy of chronic simian immunodeficiency virus (SIV) infection

Reference	*Virus*	*Start of treatment relative to virus inoculation*	*Drug, daily dose, route, duration*	*Outcome*
Böttiger *et al.*, 1992b	SIVsm	111–174 days after	AZT, 20 mg/kg s.q. for 9 weeks; FLT for 4 or 9 weeks	No reduction in viral antigen levels in lymph nodes
Van Rompay *et al.*, 1992	SIVmac251	6 weeks after	AZT, 3 × 50 mg/kg oral for 10 weeks	No reduction of infectious virus levels in peripheral blood
Tsai *et al.*, 1995b	SIVmne	Chronic infection	PMEA, 20–30 mg/kg s.q. for 4–8 weeks	Reduced virus levels
Van Rompay *et al.*, 1996	SIVmac251	3 weeks after	PMPA 30 mg/kg, s.q. for >1 year	Reduced virus levels, enhanced immune responses, prolonged survival
Tsai *et al.*, 1997	SIVmne	> 19 weeks after	PMPA 30 or 75 mg/kg, s.q. for 28 days	Strong reduction in virus levels

s.q., Subcutaneously; AZT, zidovudine; FLT, 3′-fluorothymidine; PMEA, 9-[2-(phosphonomethoxy)ethyl]adenine; PMPA, 9-[2-(phosphonomethoxy)propyl]adenine.

antiviral drug administration starting prior to, or at the time of, virus inoculation could prevent infection when a minimal dose of SIV was used for the inoculation (reviewed by Black, 1997). Two compounds were highly effective in preventing infection when treatment was started after virus inoculation: PMPA and 2′,3′-dideoxy-3′-hydroxymethyl cytidine (BEA-005; Tsai *et al.*, 1995a; Böttiger *et al.*, 1997; Van Rompay *et al.*, 1998a).

When infection was not prevented, early drug treatment reduced or delayed the peak of rapid virus replication that occurs during the first weeks of infection (Table 128.2). Reduction in early virus replication resulted in delayed disease progression (Rausch *et al.*, 1995; Van Rompay *et al.*, 1995, 1996; Joag *et al.*, 1997).

Therapy of chronic SIV infection

Few non-human primate studies have investigated the effects of antiretroviral drug therapy on established chronic SIV infection (i.e., after the acute viremia stage; Table 128.3). Initial studies with AZT failed to demonstrate therapeutic effects for macaques once SIV infection was established (Böttiger *et al.*, 1992b; Van Rompay *et al.*, 1992); this was partially due to the relatively weak efficacy of AZT monotherapy to inhibit SIV replication, but also to the lack of sensitive assays to measure virus replication during the asymptomatic stage. During recent years, the discovery of more potent antiviral drugs and the development of more sensitive assays to monitor virus levels during the asymptomatic stage (such as quantitation of viral RNA in plasma) have demonstrated the usefulness of macaques with chronic infection to test the efficacy of novel antiviral drugs, such as 9-[2-(phosphonomethoxy)ethyl]adenine (PMEA) and PMPA (Tsai *et al.*, 1995b, 1997; Van Rompay *et al.*, 1996; Nowak *et al.*, 1997). Studies with infant macaques have also studied the emergence of drug-resistant SIV mutants during prolonged drug treatment (Van Rompay *et al.*, 1995, 1996).

Pitfalls (advantages/disadvantages) of the model

SIV infection of macaques is generally believed to be the best animal model for HIV infection. Of all members of the animal kingdom, non-human primates are phylogenetically the closest to humans, and have a very similar physiology (including drug pharmacokinetics and metabolism, fetal and neonatal development, and immunology). In addition, SIV is closely related to HIV, and causes a disease in macaques which very closely resembles human AIDS. We can use the same parameters to monitor infection and disease progression as those which are used for HIV-infected patients. Although the high costs of purchasing and housing macaques limit the number of animals per study, well-designed studies with relatively small animal numbers allow recognition of statistically and biologically significant differences. All these factors allow a more reliable extrapolation of results obtained in non-human primate models towards clinical applications for the human population.

The main disadvantage is that SIV, although closely related, is not identical to HIV-1. For antiviral drug studies, it is noteworthy that the polymerase region of SIV has about 60% amino acid homology to that of HIV-1 (Myers *et al.*, 1991). Although SIV is susceptible to many of the same inhibitors of reverse transcriptase (RT) and protease enzymes at approximately the same *in vitro* concentrations as HIV-1, an exception is the non-nucleoside inhibitors (such as nevirapine) which are specific for HIV-1 and do not inhibit replication of HIV-2 and SIV (reviewed by De Clercq, 1995). To overcome this problem, SIV/HIV-1 chimeric viruses that contain the HIV-1 RT into an SIV background and that are infectious for macaques have been constructed (Überla *et al.*, 1995).

Contributions of the model to infectious disease therapy

Because SIV infection of macaques closely resembles HIV infection of humans, this animal model continues to be extremely valuable in gaining a better understanding of transmission and disease pathogenesis, and for vaccine development. In addition, because their physiology is so similar to that of humans, the macaque model has been very useful for studying toxicity and pharmacokinetics of antiviral drugs, including transplacental drug transfer.

By using different study designs (i.e., by manipulating variables such as initiation of drug treatment relative to virus inoculation, the duration of treatment, the age of the animals, the virulence and drug susceptibility of the virus inoculum), investigators have demonstrated that macaque models are well-suited to address very specific questions which are directly relevant to drug treatment of human HIV infection. Macaque studies have provided us with important insights regarding pre- and post-exposure prophylaxis, the effects of early drug intervention, and the emergence and clinical implications of drug-resistant viral variants.

The observation that a 6-week AZT treatment regimen prevented SIV infection of infant macaques (Van Rompay *et al.*, 1992) has preceded the demonstration that AZT can prevent infection of humans following exposure to HIV in two clinical settings. First of all, AIDS Clinical Trials Group Protocol 076 demonstrated that AZT administration to HIV-infected pregnant women beginning at 14–34 weeks of gestation, and continuing to their newborns during the first 6 weeks of life, reduces the rate of vertical HIV transmission by two-thirds; implementation of these results has been a major advance in the combat against pediatric

HIV infection (Connor *et al.*, 1994). More recently, it has also been demonstrated that AZT administration to health care workers following percutaneous HIV exposure by needlestick accidents reduces the risk of infection by 80% (MMWR, 1995, 1996). The development of antiviral drugs with even higher chemoprophylactic potential in non-human primates, such as PMPA (Tsai *et al.*, 1995a; Van Rompay *et al.*, 1998a,b), raises the hope that in the future more effective and shorter intervention regimens may become available to reduce vertical HIV transmission drastically, especially for the developing countries.

As described earlier, antiviral drug studies in macaques demonstrated that even when SIV infection was not prevented, early drug treatment during the first weeks of infection reduced the peak of rapid virus replication and resulted in enhanced antiviral immune responses and delayed disease progression. In recent years, this observation has been confirmed in HIV-infected people, as there is increasing evidence that early drug intervention during the initial stages of infection offers strong benefits in terms of long-term suppression of viral replication, enhanced antiviral immune responses, and delayed disease progression (Kinloch-de Loës *et al.*, 1995; Lafeuillade *et al.*, 1997; Rosenberg *et al.*, 1997).

In recent years, SIV infection of macaques has also been used to gain a better understanding of the clinical implications of drug-resistant viral mutants. Although the emergence of drug-resistant HIV mutants is a sign of incomplete suppression of virus replication, it is unclear whether drug-resistant mutants are fully virulent and whether their emergence contributes to reduced efficacy or failure of drug therapy. Although many studies have focused on determining the *in vitro* growth kinetics of drug-resistant HIV mutants, studies in the SIV-macaque model have demonstrated repeatedly that the correlation between viral *in vitro* growth properties and *in vivo* virulence is often very weak (Kestler *et al.*, 1991; Lohman *et al.*, 1994). An animal model allows an approach which is impossible and unethical to do in humans, but which is the most direct way to study the *in vivo* virulence of drug-resistant mutants—inoculation of animals with drug-resistant viral mutants. In recent years, several studies with infant macaques have investigated the emergence of drug-resistant SIV mutants during prolonged therapy with AZT or PMPA (Van Rompay *et al.*, 1995, 1996); the clinical implications were further investigated by inoculation of these drug-resistant viruses into naive newborn macaques (Van Rompay *et al.*, 1997, 1999).

Accordingly, this animal model of AIDS provides a system to study first, whether drug resistance mutations affect viral virulence, and second, if drug resistance can be responsible for loss or reduction of the efficacy of drug treatment.

The important role that the SIV animal model can play in the clinical development of anti-HIV drugs is exemplified by PMEA and PMPA. These compounds are not highly potent inhibitors of HIV and SIV replication *in vitro*; however, when PMEA and PMPA were tested in macaques, a surprisingly high antiviral activity was observed (Tsai *et al.*, 1995a,b, 1997; Van Rompay *et al.*, 1996). Results of these macaque studies played a major role in moving these drugs into human clinical trials.

In conclusion, the optimization of the SIV-macaque model for antiviral drug testing, coupled with a better understanding of HIV disease pathogenesis, is leading to a growing recognition of the importance of results obtained in this animal model. It can be expected that, in coming years, the value of these non-human primate models will continue to grow as excellent and very adaptable tools to rapidly gather data that provide a solid scientific basis to guide human clinical trials.

References

Balzarini, J., Naesens, L., Slachmuylders, J. *et al.* (1991). 9-(2-phosphonylmethoxyethyl)adenine (PMEA) effectively inhibits retrovirus replication in vitro and simian immunodeficiency virus infection in rhesus monkeys. *AIDS*, **5**, 21–28.

Black, R. J. (1997). Animal studies of prophylaxis. *Am. J. Med.*, **102**, 39–43.

Böttiger, D., Vrang, L., Öberg, B. (1992a). Influence of the infectious dose of simian immunodeficiency virus on the acute infection in cynomolgus monkeys and on the effect of treatment with 3′-fluorothymidine. *Antivir. Chem. Chemother.*, **3**, 267–271.

Böttiger, D., Ståhle, L., Li, S.-L., Öberg, B. (1992b). Long-term tolerance and efficacy of 3′-azido-thymidine and 3′-fluorothymidine treatment of asymptomatic monkeys infected with simian immunodeficiency virus. *Antimicrob. Agents Chemother.*, **36**, 1770–1772.

Böttiger, D., Johansson, N. G., Samuelsson, B. *et al.* (1997). Prevention of simian immunodeficiency virus, SIVsm, or HIV-2 infection in cynomolgus monkeys by pre- and postexposure administration of BEA-005. *AIDS*, **11**, 157–162.

Cohen, J. (1997a). AIDS trials ethics questioned. *Science*, **276**, 520–523.

Cohen, J. (1997b). Ethics of AZT studies in poorer countries attacked. *Science*, **276**, 1022.

Connor, E. M., Sperling, R. S., Gelber, R. *et al.* (1994). Reduction of maternal–infant transmission of human immunodeficiency virus type 1 with zidovudine treatment. *N. Engl. J. Med.*, **331**, 1173–1180.

De Clercq, E. (1995). Antiviral therapy for human immunodeficiency virus infections. *Clin. Microbiol. Rev.*, **8**, 200–239.

Desrosiers, R. C. (1990). The simian immunodeficiency viruses. *Annu. Rev. Immunol.*, **8**, 557–578.

Du, Z. J., Lang, S. M., Sasseville, V. G. *et al.* (1995). Identification of a nef allele that causes lymphocyte activation and acute disease in macaque monkeys. *Cell*, **82**, 665–674.

Fultz, P. N., McClure, H. M., Anderson, D. C., Switzer, W. M. (1989). Identification and biologic characterization of an acutely lethal variant of simian immunodeficiency virus from sooty mangabeys (SIV/SMM). *AIDS Res. Hum. Retroviruses*, **5**, 397–409.

Gardner, M. B. (1996). The history of simian AIDS. *J. Med. Primatol.*, **25**, 148–157.

Hu, S.-L., Haigwood, N. L., Morton, W. R. (1993). Non-human primate models for AIDS research. In *HIV Molecular*

Organization, Pathogenicity and Treatment (eds Morrow, W. J. W. Haigwood, N. L.), pp. 294–327. Elsevier, Amsterdam.

Israel, Z. R., Dean, G. A., Maul, D. H. *et al.* (1993). Early pathogenesis of disease caused by SIVsmmPBj14 molecular clone 1.9 in macaques. *AIDS Res. Hum. Retroviruses*, **9**, 277–286.

Joag, S. V., Li, Z., Foresman, L. *et al.* (1997). Early treatment with 9-(2-phosphonylmethoxyethyl)adenine reduces virus burdens for a prolonged period in SIV-infected rhesus macaques. *AIDS Res. Hum. Retroviruses*, **13**, 241–246.

Kestler, H. W., III, Ringler, D. J., Mori, K. *et al.* (1991). Importance of the *nef* gene for maintenance of high virus loads and for development of AIDS. *Cell*, **65**, 651–662.

Kinloch-de Loës, S., Hirschel, B. J., Hoen, B. *et al.* (1995). A controlled trial of zidovudine in primary human immunodeficiency virus infection. *N. Engl. J. Med.*, **333**, 408–413.

Lafeuillade, A., Poggi, C., Tamalet, C., Profizi, N., Tourres, C., Costes, O. (1997). Effects of a combination of zidovudine, didanosine, and lamivudine on primary human immunodeficiency virus type 1 infection. *J. Infect. Dis.*, **175**, 1051–1055.

Lange, J. M. A. (1997). Current problems and the future of antiretroviral drug trials. *Science*, **276**, 548–550.

Le Grand, R., Clayette, P., Noack, O. *et al.* (1994). An animal model for antilentiviral therapy: effect of zidovudine on viral load during acute infection after exposure of macaques to simian immunodeficiency virus. *AIDS Res. Hum. Retroviruses*, **10**, 1279–1287.

Letvin, N. L., King, N. W. (1990). Immunologic and pathologic manifestations of the infection of rhesus monkeys with simian immunodeficiency virus of macaques. *J. AIDS*, **3**, 1023–1040.

Levy, J. (1997). *HIV and the Pathogenesis of AIDS*, 2nd edn. American Society for Microbiology Press, Washington, DC.

Lohman, B. L., McChesney, M. B., Miller, C. J. *et al.* (1994). A partially attenuated simian immunodeficiency virus induces host immunity that correlates with resistance to pathogenic virus challenge. *J. Virol.*, **68**, 7021–7029.

Martin, L. N., Murphey-Corb, M., Soike, K. F., Davison-Fairburn, B., Baskin, G. B. (1993). Effects of initiation of 3′-azido-3′-deoxythymidine treatment at different times after infection of rhesus monkeys with simian immunodeficiency virus. *J. Infect. Dis.*, **168**, 825–835.

McClure, H. M., Anderson, D. C., Ansari, A. A., Fultz, P. N., Klumpp, S. A., Schinazi, R. F. (1990). Nonhuman primate models for evaluation of AIDS therapy. *Ann. N.Y. Acad. Sci.*, **616**, 287–298.

Miller, C. J. (1994). Use of the SIV/rhesus macaque system to test virucides designed to prevent the sexual transmission of HIV. In *Barrier Contraceptives: Current Status and Future Prospects* (eds Mauck, C. K. *et al.*), pp. 213–223. Wiley-Liss, New York.

Miller, C., Rosenberg, Z., Bischofberger, N. (1996). Use of topical PMPA to prevent vaginal transmission of SIV. Ninth International Conference on Antiviral Research, Fukushima, May 19–24 (late breaker abstract).

M. M. W. R. (1995). Case-control study of HIV seroconversion in health-care workers after percutaneous exposure to HIV-infected blood — France, United Kingdom, and United States, January 1988–August 1994. *M.M.W.R.*, **44**, 929–933.

M. M. W. R. (1996). Update: provisional Public Health Service recommendations for chemoprophylaxis after occupational exposure to HIV. *M.M.W.R.*, **45**, 468–472.

Myers, G., Berzofsky, J. A., Korber, B., Smith, R. F., Pavlakis, G. N. (1991). *Human Retroviruses and AIDS. A Compilation and Analysis of Nucleic Acid and Amino Acid Sequences*. Los Alamos National Laboratory, Los Alamos, New Mexico.

Nowak, M. A., Lloyd, A. L., Vasquez, G. M. *et al.* (1997). Viral dynamics of primary viremia and antiretroviral therapy in simian immunodeficiency virus infection. *J. Virol.*, **71**, 7518–7525.

Oxtoby, M. J. (1994). Vertically acquired HIV infection in the United States. In *Pediatric AIDS. The Challenge of HIV Infection in Infants, Children and Adolescents* (eds Pizzo, P. A., Wilfert, C. M.), pp. 3–20. Williams & Wilkins, Baltimore, MD.

Rausch, D. M., Heyes, M. P., Murray, E. A., Eiden, L. E. (1995). Zidovudine treatment prolongs survival and decreases virus load in the central nervous system of rhesus macaques infected perinatally with simian immunodeficiency virus. *J. Infect. Dis.*, **172**, 59–69.

Rosenberg, E. S., Billingsley, J. M., Caliendo, A. M. *et al.* (1997). Vigorous HIV-1-specific CD4+ T cell responses associated with control of viremia. *Science*, **278**, 1447–1450.

Tsai, C.-C., Follis, K. E., Sabo, A. *et al.* (1994). Preexposure prophylaxis with 9-(-2-phosphonylmethoxyethyl)adenine against simian immunodeficiency virus infection in macaques. *J. Infect. Dis.*, **169**, 260–266.

Tsai, C.-C., Follis, K. E., Beck, T. W. *et al.* (1995a). Prevention of simian immunodeficiency virus infection in macaques by 9-(2-phosphonylmethoxypropyl)adenine (PMPA). *Science*, **270**, 1197–1199.

Tsai, C.-C., Follis, K. E., Sabo, A., Grant, R., Bischofberger, N. (1995b). Efficacy of 9-(2-phosphonylmethoxyethyl)adenine treatment against chronic simian immunodeficiency virus infection in macaques. *J. Infect. Dis.*, **171**, 1338–1343.

Tsai, C.-C., Follis, K. E., Beck, T. W., Sabo, A., Bischofberger, N., Dailey, P. J. (1997). Effects of (R)-9-(2-phosphonylmethoxypropyl)adenine monotherapy on chronic SIV infection. *AIDS Res. Hum. Retroviruses*, **13**, 707–712.

Überla, K., Stahl-Hennig, C., Böttiger, D. *et al.* (1995). Animal model for the therapy of acquired immunodeficiency syndrome with reverse transcriptase inhibitors. *Proc. Natl Acad. Sci. USA*, **92**, 8210–8214.

Van Rompay, K. K. A., Marthas, M. L., Ramos, R. A. *et al.* (1992). Simian immunodeficiency virus (SIV) infection of infant rhesus macaques as a model to test antiretroviral drug prophylaxis and therapy: oral 3′-azido-3′-deoxythymidine prevents SIV infection. *Antimicrob. Agents Chemother.*, **36**, 2381–2386.

Van Rompay, K. K. A., Otsyula, M., Marthas, M. L., Pedersen, N. C. (1994). Simian immunodeficiency virus infection of newborn and infant rhesus macaques: an animal model for testing antiretroviral drugs. *Int. Antiviral News*, **2**, 5–6.

Van Rompay, K. K. A., Otsyula, M. G., Marthas, M. L., Miller, C. J., McChesney, M. B., Pedersen, N. C. (1995). Immediate zidovudine treatment protects simian immunodeficiency virus-infected newborn macaques against rapid onset of AIDS. *Antimicrob. Agents Chemother.*, **39**, 125–131.

Van Rompay, K. K. A., Cherrington, J. M., Marthas, M. L. *et al.* (1996). 9-[2-(Phosphonomethoxy)propyl]adenine therapy of established simian immunodeficiency virus infection in infant rhesus macaques. *Antimicrob. Agents Chemother.*, **40**, 2586–2591.

Van Rompay, K. K. A., Greenier, J. L., Marthas, M. L. *et al.* (1997). A zidovudine-resistant simian immunodeficiency virus mutant with a Q151M mutation in reverse transcriptase causes AIDS in newborn macaques. *Antimicrob. Agents Chemother.*, **41**, 278–283.

Van Rompay, K. K. A., Cherrington, J. M., Marthas, M. L. *et al.* (1999). 9-[2-(Phosphonomethoxy)propyl]adenine (PMPA) therapy prolongs survival of infant macaques inoculated with simian immunodeficiency virus with reduced susceptibility to PMPA. *Antimicrob. Agents Chemother.*, **43**, in press.

Van Rompay, K. K. A., Marthas, M. L., Lifson, J. D. *et al.* (1998a). Administration of 9-[2-(phosphonomethoxy)-propyl]adenine (PMPA) for prevention of perinatal simian immunodeficiency virus infection in rhesus macaques. *AIDS Res. Hum. Retroviruses*, **14**, 761–763.

Van Rompay, K. K. A., Berardi, C. J., Aguirre, N. L. *et al.* (1998b). Two doses of 9-[2-(phosphonomethoxy)-propyl]adenine (PMPA) protect newborn macaques against oral simian immunodeficiency virus infection. *AIDS*, **12**, F79–F83.

Chapter 129

The SCID-hu Thy/Liv Mouse: an Animal Model for HIV-1 Infection

C. A. Stoddart

Background of human infection

Estimates for 1997 by The Joint United Nations Programme on HIV/AIDS (UN-AIDS) and the World Health Organization indicate that over 30 million people are infected with human immunodeficiency virus type 1 (HIV) worldwide. Better prevention and potent new antiretroviral drugs have limited the spread of HIV-1 and reduced mortality from acquired immunodeficiency syndrome (AIDS) in the United States and western Europe, but HIV-1 continues to spread in the developing world, where more than 90% of HIV-1-infected people live. An estimated 5.8 million people acquired HIV-1 infection and 2.3 million people died of AIDS in 1997.

After entry into the body and probable interaction with a dendritic cell, HIV-1 is carried to a CD4+ T cell in a lymphoid organ where it establishes infection and disseminates progeny virus throughout the body (see Fauci, 1996 and Cohen *et al.*, 1997, for recent reviews). Viral replication is partially controlled by robust immune responses, but the virus escapes from immune control and produces a chronic, persistent infection that leads to fatal disease in about 10 years. In the absence of antiretroviral therapy, tremendous viral turnover rates of $\sim 10^{10}$ virions per day and continual loss of infected CD4+ T cells result in the eventual destruction of the immune system and death from wasting and opportunistic infections.

Although the pathogenesis of HIV-1-induced disease is complex and multifactorial (Fauci, 1996), recent advances in our understanding of the role of host factors, viral coreceptors, and cellular tropism in host–HIV-1 interactions suggest new strategies for prevention and therapy. The discovery and identification of HIV-1-suppressing cytokines and inhibitors of coreceptor-mediated HIV-1 entry may lead to promising and novel antiviral agents that could first be evaluated in animal models of HIV-1 disease (McCune, 1997; Moore, 1997).

Background of model

The SCID-hu mouse was designed to provide a small animal model for studying human hematopoiesis (McCune *et al.*, 1988), facilitating examination of the pathogenesis of HIV-1 in human hematolymphoid organs (Namikawa *et al.*, 1988; Kaneshima *et al.*, 1991; McCune *et al.*, 1991), and evaluation of agents possessing anti-HIV-1 activity *in vitro* (McCune *et al.*, 1990; Rabin *et al.*, 1996). In this model, severe combined immunodeficient (SCID) mice are implanted with a variety of human fetal organs, including bone, liver, thymus, lymph node, and spleen. In defined instances, the fetal implants become tolerant of the mouse environment and, reciprocally, growth of the human tissue is permitted by the immunocompromised status of the recipient SCID mouse (McCune, 1996, 1997).

The SCID-hu Thy/Liv construct is the result of implantation of interactive human lymphoid organs (fragments of human fetal thymus and liver) beneath the mouse kidney capsule. In a highly reproducible manner, the human fetal organs fuse, vascularize, and grow, eventually reaching a total mass of 10^7–10^8 human cells in 80–90% of recipient mice. A stable organ termed 'Thy/Liv' is thus established, and it successfully reproduces the differentiation, proliferation, and function of human hematopoietic progenitor cells (derived from the fetal liver) within the context of the human thymus (Namikawa *et al.*, 1990; McCune, 1996). The Thy/Liv implant possesses histologically normal cortical and medullary compartments that are capable of long-term multilineage human hematopoiesis (Namikawa *et al.*, 1990; Krowka *et al.*, 1991), including the generation of large numbers of CD4-bearing thymocytes which can serve as target cells for HIV-1 infection and replication. Importantly for a model of antiviral chemotherapy, it is possible to construct large numbers of SCID-hu Thy/Liv mice (typically 45–60) with tissue from the same fetal donor, and the Thy/Liv organ itself is amenable to experimental manipulation (e.g., infection with HIV-1).

Cells in the Thy/Liv organ are permissive to infection by primary, but not by tissue culture-adapted, isolates after direct inoculation of HIV-1 (Kaneshima *et al.*, 1991; Su *et al.*, 1997). Infection with some isolates leads to thymocyte depletion and thymic involution within 3–5 weeks (Aldrovandi *et al.*, 1993; Bonyhadi *et al.*, 1993; Stanley *et al.*, 1993; Kaneshima *et al.*, 1994). Both T cells and myeloid cells are targets of viral replication, and infection is accompanied by an absolute decrease in the number of double-positive

Handbook of Animal Models of Infection
ISBN 0-12-775390-7

(CD4+CD8+) immature cortical cells and single-positive (CD4+ or CD8+) mature medullary cells. The loss of CD4+ cells is typically greater than that of CD8+ cells, resulting in an inversion of the CD4/CD8 ratio. Viral replication and thymocyte depletion are inhibited by treating the mice with the nucleoside analogs zidovudine (AZT) and didanosine (ddI), with the non-nucleoside reverse transcriptase inhibitor nevirapine, and with a bicyclam inhibitor of viral entry (McCune *et al.*, 1990; Datema *et al.*, 1996; Rabin *et al.*, 1996).

Animal species

Male inbred homozygous *scid/scid* (CB-17/IcrTac-*scid*fDF) mice are used. The autosomal, recessive *scid* mutation arose spontaneously in CB-17 mice (Bosma *et al.*, 1983), and homozygous mice fail to develop immunocompetent T or B lymphocytes because of a defect in VDJ somatic recombination (Malynn *et al.*, 1988). Implantation of human fetal tissues is performed when the mice are 8–12 weeks of age. Male SCID mice are used because they have larger kidneys, less peritoneal fat, produce less scar tissue after surgical manipulation, and thus have more accessible Thy/Liv implants than do female SCID mice.

Preparation of animals

Mice are housed under specific pathogen-free conditions in mouse cages equipped with filter-top isolator bonnets. Cages, bedding, and water bottles are sterilized by autoclaving, and drinking water is maintained at pH 2.5–2.9 by addition of 12 N HCl. Mice are fed *ad libitum* with γ-irradiated low-fat mouse chow, and food pellets containing trimethoprim-sulfamethoxazole (SCID's MD's, Bio-Serv, Frenchtown, NJ) are added to the food bins to prevent opportunistic infection with *Pneumocystis carinii*. Each cage rack contains one cage of immunocompetent DBA/2 sentinel mice that receives soiled bedding from the SCID mouse cages at each cage cleaning. Sentinel mice are screened every 6 weeks by serology for a panel of murine pathogens, including murine hepatitis virus, Sendai virus, pneumonia virus of mice, *Mycoplasma pulmonis*, parvovirus, and Theiler's murine encephalitis virus (GD-VII). Lungs from 2 or 3 euthanized SCID mice are examined histologically for evidence of *P. carinii* every 3 months. Mice are rested for a minimum of 1 week after arrival before surgical manipulation or drug dosing.

Details of surgery

Overview

The left kidneys of anesthetized SCID mice are surgically exposed, and 1–2 mm^3 fragments of human fetal thymus and liver are implanted beneath the kidney capsule to create SCID-hu Thy/Liv mice. Cohorts of 45–60 mice are produced with tissue from the same fetal donor.

Materials required

Materials required include cell culture medium (RPMI 1640 containing 10% fetal bovine serum, L-glutamine, and antibiotics), dissection microscope, 102 mm microdissecting straight scissors, 127 mm Semken serrated forceps, 102 mm microdissecting (serrated, straight, extra-delicate) forceps, Vannas straight spring forceps, anesthetic (described below), rodent ear punch, electric hair clippers, 70% ethanol solution, povidone-iodine skin disinfectant, isothermal heating pads (DeltaPhase, Braintree Scientific), 18 G trocar (catalog #7926, Popper and Sons, New York), high-vacuum grease (Dow-Corning), 110 mm pattern #5 Dumont straight tweezers, 114 mm Moloney serrated curved forceps, suture material (Ethibond 4.0 non-dissolving polyester), Olsen-Hegar (serrated jaws, extra-delicate) needle holder, and Autoclip surgical stapler and 9 mm metal wound clips. Except for trocars, all surgical instruments are purchased from Roboz Surgical, Rockville, MD.

Preparation of tissue for implantation

Fresh, intact human fetal thymus and liver are acquired from products of conception aged 16–24 gestational weeks. All procedures are performed with approval and oversight by institutional human subjects and animal review boards. The tissues are kept in cold medium and minced under a dissection microscope into 1–2 mm^3 pieces. Microdissecting scissors and 127 mm Semken forceps are used for gross dissection of tissue, followed by fine dissection with 102 mm microdissecting forceps and Vannas scissors into appropriately sized fragments. Liver fragments are prepared from undamaged core sections, and thymus fragments that contain thymic lobules (not just connective tissue) are selected by viewing under the dissection microscope. One piece of thymus and two pieces of liver are prepared for each mouse to be implanted, and the coimplanted tissues are obtained from the same fetal donor.

Anesthesia

Mice are anesthetized with 100 mg/kg ketamine plus 8 mg/kg xylazine by intraperitoneal injection. The anesthetic solution contains 10 mg/ml ketamine (1 : 10 dilution) and 0.8 mg/ml xylazine (1:25 dilution) in phosphate-buffered saline (PBS) and is administered to the mice at 10 μl per gram body weight. Mice that recover prematurely from anesthesia are given methoxyflurane (Metofane) by brief (30 second) inhalation.

Surgical procedure

Anesthetized mice are numbered by ear-punching, and the left side is shaved and disinfected with 70% ethanol followed by povidone-iodine solution. Mice are kept warm on heating pads and their eyes are kept moistened with PBS. The mouse is placed on its right side and a 10 mm lateral incision is made through the skin of the left side with microdissecting scissors. An 8 mm incision through the abdominal wall is then made immediately posterior to the vein in the peritoneal fascia that lies over the spleen. Gentle pressure is applied around the incision so that kidney emerges and rests outside the abdomen, and the kidney is kept moistened with medium to facilitate the sliding motion of surgical instruments and tissue. A greased trocar needle is loaded with a sandwich of tissue in the order: liver, thymus, liver. The kidney capsule is pricked at the caudal end with Dumont #5 pattern forceps, and the loaded trocar needle is inserted, bevel down, through the pricked hole and slid along the dorsal arch of the kidney to a point one-third of the way from the rostral end (Figure 129.1).

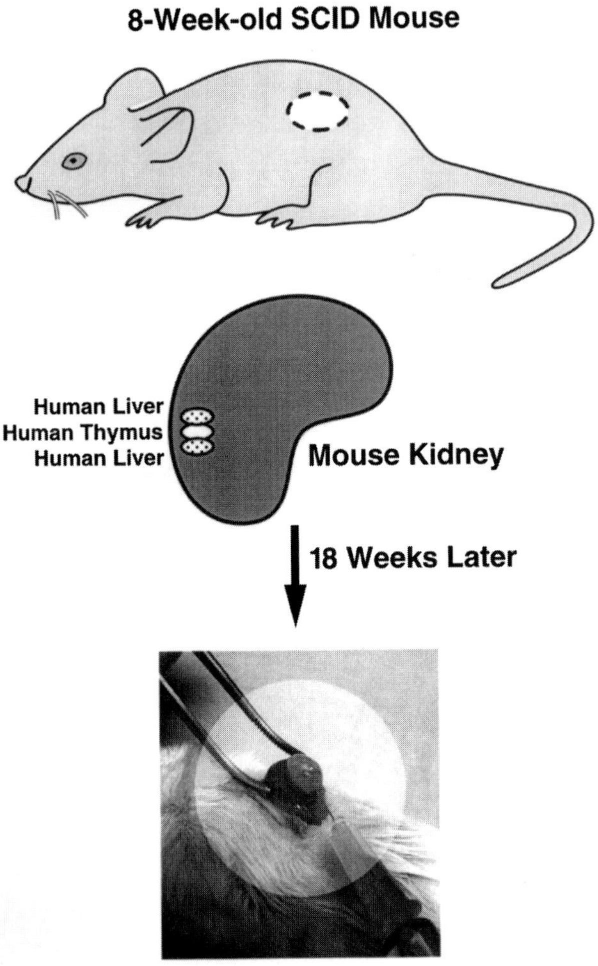

Figure 129.1 Construction of the severe combined immunodeficiency (SCID-hu) Thy/Liv mouse and inoculation of human immunodeficiency virus type 1 (HIV-1) into an established Thy/Liv organ by direct injection 18 weeks after implantation.

The plunger (coated with high-vacuum grease to maintain a proper seal) is slowly depressed while the trocar is withdrawn, pushing the tissue fragments out and on to the surface of the kidney underneath the capsule. If the implanted fragments need to be grouped or repositioned, they are pushed by sliding the moistened trocar over the kidney capsule. The kidney is moistened with medium and slipped back into the abdomen by pulling the abdominal wall over the kidney with 114 mm Moloney curved forceps. The abdominal wall is approximated with a single suture, and the skin is closed with one 9 mm metal clip. Mice are placed in a clean cage and monitored until recovery from anesthesia.

Implant assessment

The size and quality of the Thy/Liv implants are determined 12–14 weeks after implantation in 20% of randomly chosen mice per cohort. Mice are anesthetized and their kidneys are exposed as described above. Implants should be convex in shape, white or cream-colored, vascularized, and have no evidence of bacterial infection. If >80% of the examined implants are $\geq 30\ mm^3$ (Figure 129.2), the entire cohort is deemed acceptable and used for HIV-1 inoculation.

Storage, preparation of inocula

All procedures using infectious HIV-1 are carried out in a Biosafety Level 3 (BL3) facility or in a restricted animal barrier facility under BL3 guidelines. Seed virus from HIV-1 molecular clones is prepared by electroporation of 25 μg of HIV-1 DNA per 5×10^6 fresh phytohemagglutinin (PHA)-activated peripheral blood mononuclear cells (PBMC) at 960 μF and 280 V. Clinical HIV-1 isolates are derived from HIV-1-seropositive patients, and seed virus is obtained by cocultivation of patient-derived PHA-activated PBMC with PHA-activated PBMC from normal uninfected donors. Working stocks of HIV-1 for implant inoculation are kept at low passage and are prepared by inoculating 10^8 fresh PHA-activated PBMC with 2×10^5 50% tissue culture infectious doses ($TCID_{50}$) of seed virus in 5 ml of complete medium containing 5 μg polybrene per ml. After 2 hours at 37°C, the cells are diluted to a density of $2–3 \times 10^6$ per ml in complete medium containing 50 U interleukin-2 (from human lymphocytes, Boehringer-Mannheim) per ml.

On day 3, the cells are counted, and an equal number of uninfected PHA-activated PBMC are added for a final cell concentration of 1.5×10^6 per ml. Virus-containing supernatants are collected daily on days 4–7, and fresh cells are added after each collection, as described above for day 3. Alternatively, working stocks of HIV-1 molecular clones can be obtained by calcium phosphate transfection of 293T cells (15–30 μg HIV-1 DNA per 5×10^6 cells) with supernatants harvested on days 2 and 3. Virus-containing supernatants are aliquoted and stored in liquid nitrogen, and representative aliquots are analyzed for infectious virus titer by limiting dilution ($TCID_{50}$) assay in PHA-activated PBMC. Working stocks should have a titer of $\geq 10^{4.6}$ $TCID_{50}$ per ml so that each implant can be inoculated with at least 1000–2000 $TCID_{50}$ in a volume of 25–50 μl.

Infection process

Implants are inoculated 16–24 weeks after tissue implantation. Procedures are performed in a restricted animal barrier facility under BL3 guidelines (i.e., with two pairs of gloves, mask, eye-covering, gown, and booties). To maximize visual and manual access, teams of three operators work side-by-side on a bench top. One operator shaves the mouse and exposes the left kidney carrying the Thy/Liv implant as described above. The second operator then gently immobilizes the kidney with forceps and marks an opening in the kidney capsule over the implant with India ink injected with a 1 cc tuberculin syringe and 30 G × ½ inch sharp needle (Figure 129.1). Using the ink mark as a guide, the Thy/Liv implant is injected with 25–50 μl of stock virus (1000–2000 $TCID_{50}$) in 1–3 places with a 250 μl Hamilton glass syringe and 30 G × ½ inch blunt needle. A third operator closes the incision as described above.

Viral inoculum-containing vials are kept frozen on dry ice until several mice are anesthetized and their implants exposed. Implants are inoculated within 30 minutes of thawing the vial, after which another vial is thawed and used.

Key parameters to monitor infection

Implant processing

At various times after HIV-1 inoculation (but typically after 12 days), mice are euthanized by CO_2 inhalation followed by cervical dislocation, and the Thy/Liv implants are surgically excised and transferred into 6-well tissue culture plates containing cold PBS with 2% fetal bovine serum (FBS). A single cell suspension is made by placing the implant into a sterile nylon mesh bag, submerging the bag in PBS/2% FBS in a 60 mm tissue culture dish, and gently grinding the tissue between the nylon layers with forceps. The cells are counted with a Coulter counter, and appropriate numbers of cells are aliquoted for each assay. For p24 ELISA, pellets of 2.5×10^6 cells are resuspended in 400 μl of p24 lysis buffer (1% Triton X-100, 0.5% sodium deoxycholate, 5 mmol/l ethylenediaminetetraacetic acid (EDTA), 25 mmol/l Tris Cl, 250 mmol/l NaCl and 1% aprotinin), rotated overnight at 4°C, and stored at –20°C. For quantitation of infectious HIV-1, cells are serially diluted in fivefold increments and added to PHA-activated PBMC in 96-well plates. For DNA polymerase chain reaction (PCR), pellets containing 1×10^3 cells are stored at –80°C. For RNA quantitation by branched DNA (bDNA) assay, dry pellets of $2–5 \times 10^6$ cells are frozen and stored at –80°C. For flow cytometry analysis, 10^6 cells per well are placed in a 96-well plate for staining and fixation, and analyzed on the same day.

p24 ELISA assay

Samples are lysed in p24 lysing buffer and transferred into HIV p24 antibody-coated microplates (Dupont) for quantitative enzyme-linked immunosorbent assay (ELISA). A standard curve is generated, and results are calculated as pg p24 per 10^6 cells.

Quantitative microculture assay

Serially diluted thymocytes are added to 96-well plates containing 10^5 PHA-activated PBMC per well and incubated at 37°C. A range of 32–100 000 thymocytes per well are cocultivated in duplicate with the PBMC. After 7 days, cell pellets are lysed and assayed for p24, and wells containing detectable p24 (≥30 pg/ml) are scored positive for HIV-1 infection. Implant HIV-1 titers are expressed as $\log_{10}$ $TCID_{50}$ per 10^6 implant thymocytes, and the $\log_{10}$ values are used for calculation of geometric means. The limit of detection is 1.0 $\log_{10}$ $TCID_{50}$ per 10^6 cells.

DNA PCR

Infection of implant thymocytes is assessed by PCR amplification with primers specific for the conserved U5-*gag* region of the HIV-1 genome (SK145: AGTGGGGGGA-CATCAAGCAGCCATGCAAAT, SK431: TGCTAT-GTCAGTTCCCCTTGGTTCTCT; Gibco BRL). Amplification of human β-globin DNA is performed as a control for the presence of human DNA. Cell pellets containing 1×10^3 cells are lysed in 100 μl of 1% NP40, 1% Tween-20, 2.5 mmol/l $MgCl_2$, 5 mmol/l KCl and 10 μg proteinase K per ml. Samples are vortexed, microfuged, incubated at 60°C for 1 hour, heated at 95°C for 10 minutes to inactivate the proteinase K, vortexed, microfuged, and assayed by a standard PCR assay as previously described (McCune *et al.*, 1990). Amplification of 10 μl lysate (corresponding to 100 cell equivalents) is continued for 40 cycles (cycle 1: 6 minutes at 95°C, 2 minutes at 60°C, and 1.5 minutes at 72°C; cycles 2–5: 2 minutes at 95°C, 2 minutes at 60°C, and 1.5 minutes at 72°C; cycles 6–40: 1 minute at 95°C, 1 minute at 60°C, and 1.5 minutes at 72°C; final extension: 10 minutes at 72°C and 8 hours at 27°C). Amplified products are subjected to electrophoresis in a 2%

agarose gel with molecular weight standards (BstE II-digested lambda DNA). Gels contained 0.5 μg ethidium bromide per ml, and images are acquired on an Eagle Eye II gel imager (Stratagene). Control samples A, B, and C are prepared by mixing HIV-1-positive ACH-2 cells with HIV-1-negative implant thymocytes at concentrations of 10%, 2%, and 0%, respectively. Control A should yield a dark HIV-1 DNA band (scored +), control B a light HIV-1 band (+/–), and control C no HIV-1 band (–). The limit of HIV-1 DNA detection is therefore one to two HIV-1-positive cells per 100 input cells. The H_2O control sample (–) should yield neither human β-globin nor HIV-1 DNA bands.

Viral RNA quantitation by bDNA assay

Cells are disrupted with sterile disposable pestles and a cordless motor grinder (Kontes, Vineland, NJ) in 500 μl of 8 mol/l guanidine HCl with 0.5% sodium *N*-lauroylsarcosine. The RNA is extracted by adding 500 μl of 100% EtOH containing 20 μg polyadenylic acid (Sigma) per ml, and each sample is vortexed thoroughly and pelleted at 12,000 *g* for 20 minutes at 4°C. Supernatants are aspirated to remove DNA, and RNA pellets are washed with 500 μl of 70% EtOH, placed on dry ice, and digested with reagents supplied by the manufacturer (Quantiplex HIV-1 RNA assay 2.0, Chiron Corporation, Emeryville, CA). Implant HIV-1 RNA load is expressed as copies per 10^6 implant thymocytes, and the $\log_{10}$ values are used for calculation of geometric means. The limit of detection is typically 10^3 RNA copies per 10^6 cells.

FACS analysis for thymocyte depletion

Pellets containing 10^6 cells are resuspended in 50 μl of a monoclonal antibody cocktail containing fluorescein isothiocyanate-conjugated anti-CD4, phycoerythrin-conjugated anti-CD8, and tricolor-conjugated anti-CD3. Cells from one implant are also stained with conjugated, isotype-matched antibodies to control for non-specific antibody binding. Cells are incubated for 30 minutes in the dark, washed three times with PBS/2% FBS, resuspended in 200 μl of PBS/2% FBS containing 1% paraformaldehyde in 1.5 ml FACS tubes, and analyzed on a FACScan (Beckton Dickinson). After collecting 10 000 events, percentages of marker-positive (CD4+, CD8+, CD4+CD8+, and CD3–CD4+ CD8–) thymocytes in the implant samples are determined by gating on a live lymphoid cell population identified by forward- and side-scatter characteristics.

Kinetics of viral replication

Intrathymic injection of 1000–2000 $TCID_{50}$ of HIV-1 NL4-3 and various clinical isolates leads to productive infection of virtually all inoculated implants, and time-dependent increases in implant viral load are observed (Rabin *et al.*, 1996). Proviral DNA is first detectable in implant thymocytes by PCR at 4 days postinoculation (p.i.) and 100% of infected implants are positive by 11–14 days p.i. Infectious virus and viral RMA can first be detected at 7 days p.i. and p24 antigen becomes measurable by ELISA at 10 days p.i. Although the time course of the infection depends on the virus isolate, input titer, and fetal tissue donor, implant viral load parameters (p24, HIV-1 RNA, and infectivity titer) generally peak by 21 days p.i. and decrease thereafter. In this model, HIV-1 cannot be reliably recovered from the mouse peripheral blood or organs.

At 12–14 days p.i., the percentage of CD4+CD8+ immature cortical thymocytes (which is typically 80–90% of total implant thymocytes) begins to decrease and becomes <5% by 28–35 days p.i. as HIV-1-mediated thymocyte depletion occurs. At the same time, the CD4/CD8 ratio decreases from approximately 2.0 to <1.0. In the case of HIV-1 NL4-3, there is also a time-dependent decrease in the percentage of CD3–CD4+CD8– intrathymic T-progenitor cells (Su *et al.*, 1996). This phenomenon dictates that endpoint analyses for antiviral evaluations be performed by 12–14 days after virus inoculation, before significant cellular depletion is evident. At this time point, mean p24 values will range from 50 to 1700 pg per 10^6 cells depending on the donor tissue used.

Antiretroviral therapy

Large SCID-hu cohorts of 45–60 mice made with tissue from the same fetal donor permit antiviral evaluations that involve six to seven dosing groups of 5–8 mice each. In a typical experiment there will be three groups of mice treated with different doses (in half-log increments, e.g. 10, 30, and 100 mg/kg per day) of the test agent, one untreated group, one group treated with a positive antiretroviral control drug (typically ddI), and one untreated group that is mock-inoculated with medium alone. An additional group of mock-inoculated, drug-treated mice can be used to assess potential toxicity of the test agent on the Thy/Liv implant in the absence of virus-induced effects.

Treatment is usually initiated 24 hours before HIV-1 inoculation and continues daily until termination and implant collection 12 days later (Figure 129.2). Test agents can be administered intraperitoneally, intravenously, subcutaneously, orally, or by mini-osmotic pump with a dosing schedule determined by prior pharmacokinetic analyses in rodents. In experiments requiring twice daily dosing, mice are treated at 7 : 00 a.m. and 6 : 00 p.m. Intraperitoneal injections are performed with a 26 G × ½ in needle, subcutaneous injections are made at the back of the neck with a 26 G × ⅜-in intradermal needle, and oral gavage is performed with an 18 G × 3-in curved feeding needle; volumes of 100–300 μl are administered per injection. For continuous subcutaneous administration, the test agent is loaded

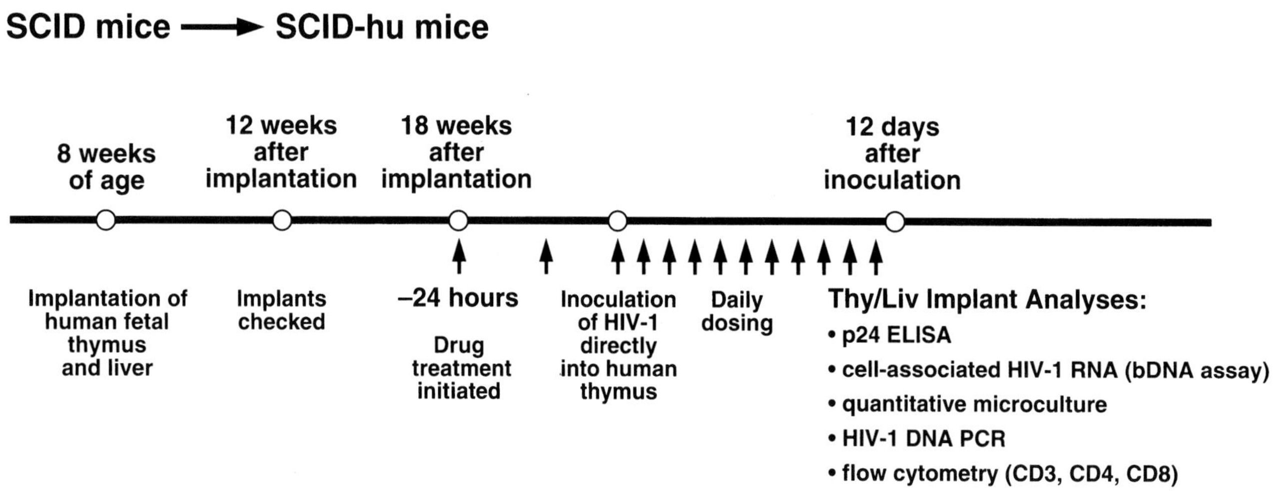

Figure 129.2 Standard experimental schedule for evaluation of antiretroviral agents in the human immunodeficiency virus (HIV-1)-infected severe combined immunodeficiency (SCID-hu) Thy/Liv mouse. ELISA, Enzyme-linked immunosorbent assay; bDNA, branched DNA; PCR, polymerase chain reaction.

into Alzet mini-osmotic pumps (model #2002 or #2004, Alza, Palo Alto, CA) according to the manufacturer's instructions. Drug-loaded pumps are incubated in 0.86% NaCl at 37°C for >40 hours before subcutaneous insertion.

Key parameters to monitor response to treatment

Each Thy/Liv implant is assessed for the following endpoint parameters: p24 (pg per 10^6 cells), HIV-1 RNA (copies per 10^6 cells), HIV-1 titer ($TCID_{50}$ per 10^6 cells), proviral DNA in 100 thymocytes (+, +/–, or –), implant cell yield, representation of various thymocyte subpopulations (CD3–CD4+CD8–, CD4+CD8+, CD4+, and CD8+) as a percentage of total live thymocytes, and the CD4/CD8 ratio. Data for mice in each group are compared to those for untreated infected mice at the same termination time point by nonparametric analysis (Mann–Whitney U-test, StatView 4.1, Abacus Concepts, Berkeley, CA), and P values ≤0.05 are considered statistically significant.

The SCID-hu Thy/Liv model has been validated in extensive and repeated dose–response studies with the nucleoside analogs AZT and ddI (Rabin *et al.*, 1996), the non-nucleoside reverse transcriptase inhibitor nevirapine (Rabin *et al.*, 1996), and a bicyclam that inhibits CXCR4-mediated viral entry (Datema *et al.*, 1996; Schols *et al.*, 1997). All four anti-HIV-1 agents inhibit implant p24 production in a dose-dependent manner and can reduce p24 levels to almost undetectable levels at the higher doses (see data from representative experiments in Figure 129.3). Statistically significant inhibition of viral replication is also made manifest by the protection of CD4+CD8+ thymocytes from depletion in ddI-treated mice 4 weeks after inoculation with NL4-3 (Figure 129.4).

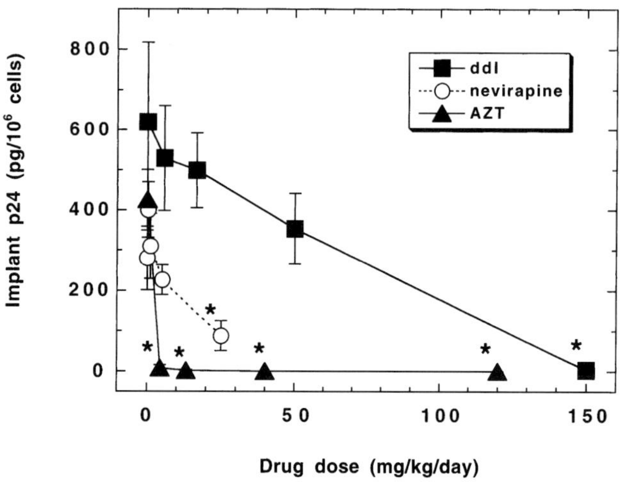

Figure 129.3 Dose-dependent antiretroviral activity of zidovudine (AZT), didanosine (ddI), and nevirapine in human immunodeficiency virus type 1 (HIV-1) NL4-3-infected severe combined immunodeficiency (SCID-hu) Thy/Liv mice (Rabin et al., 1996). Drugs were administered twice daily (AZT and nevirapine by oral gavage; ddI by intraperitoneal injection) beginning 24 hours before virus inoculation, and Thy/Liv implants were collected 14 days after inoculation. Statistically significant differences in p24 production in treated mice versus untreated mice were determined by the Mann–Whitney U-test, and P values ≤0.05 are indicated by an asterisk.

Pitfalls (advantages/disadvantages) of the model

The key advantage of the SCID-hu Thy/Liv model is that the Thy/Liv implant is an intact human organ system that recapitulates the full range of cellular differentiation and function of human hematopoiesis in an easily manipulated

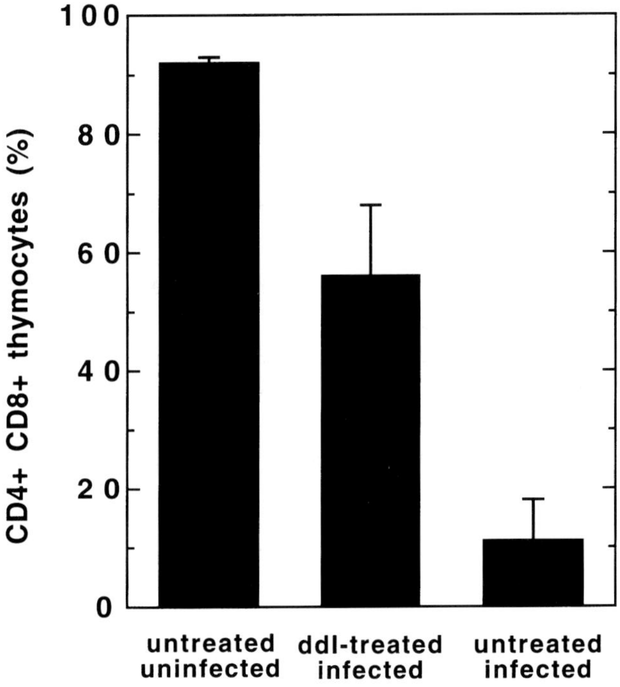

Figure 129.4 Protection of Thy/Liv implants from human immunodeficiency virus type 1 (HIV-1) NL4-3-induced depletion in mice treated with 50 mg of ddI per kg/day (Rabin et al., 1996).

small animal model (McCune, 1997). The implants contain primary human cells in their native physiological state and support hematopoiesis and T-cell generation for long periods of time (>11 months). Moreover, cohorts containing 45–60 mice can be produced with tissues from the same donor and this number permits statistical comparisons among six to seven dosing groups of 5–8 mice each.

Although the Thy/Liv implants mimic early human thymic development, they may not reflect aspects of the adult thymic microenvironment. One important limitation is that the Thy/Liv model lacks a human peripheral immune system and so does not support lymphocyte trafficking, homing, and antigen presentation. The human immune response to HIV-1 infection cannot, therefore, be studied in this model. Mouse cells may migrate into the Thy/Liv implants and perturb human thymocyte maturation. The model requires access to and use of human fetal tissue and considerable surgical skills for implantation of this tissue into SCID mice; each mouse must be "handmade." Finally, both engrafted and unengrafted SCID mice must be kept under specific pathogen-free conditions, and they must be monitored closely for signs of opportunistic infection. As a model for evaluation of antiviral therapies, the model is a severe test of antiviral efficacy because virus of high titer is injected directly into the organ and permitted access to large numbers of susceptible target cells for the initial round of viral replication. In addition, access of a drug to the Thy/Liv implant will be influenced by the pharmocokinetic properties of the mouse, which may differ substantially from that of humans for particular test compounds.

Contributions of the model to infectious disease therapy

The SCID-hu mouse was pivotal in the identification of a candidate human hematopoietic stem cell and has been used to define events in normal human hematopoiesis (McCune, 1997). As a useful model of HIV-1 disease, it has been used to evaluate differential HIV-1 pathology and cell tropism of tissue culture-adapted versus primary isolates (Bonyhadi *et al.*, 1993; Su *et al.*, 1997; Berkowitz *et al.*, 1998) and the role of accessory genes in *in vivo* replication (Jamieson *et al.*, 1994; Su *et al.*, 1997).

The SCID-hu Thy/Liv construct is a reliable small animal model for the preclinical evaluation of anti-HIV-1 agents *in vivo*. It is the first animal model in which the activity of such agents can be tested in the context of an HIV-1-infected human target organ (McCune *et al.*, 1990; Rabin *et al.*, 1996). Given the number of animals (n = 5–8) in a given dosing group, it is possible to assign effective dose ranges of antiviral agents with statistical confidence. Among animal models for HIV-1, this is unique: non-human primate models cannot provide dosing groups of such size and, hence, cannot achieve the same statistical power; additionally, some of the compounds studied (e.g., nevirapine) are not active against retroviruses that replicate in other species (e.g., murine leukemia virus, feline immunodeficiency virus, and simian immunodeficiency virus). It is also notable that the current SCID-hu model can be used to explore alternative treatment regimens such as variable dosing routes, and dosing intervals. Additionally, the large cohort sizes and the reproducibility of the model permit statistical evaluation of combination therapies against HIV-1.

Use of the model over the last several years has largely been devoted to the validation of licensed antiviral agents in current clinical use (i.e., AZT, ddI, and nevirapine); it is now poised for evaluation of new agents in preclinical development. Data obtained in these mice can guide selection of effective new therapies for the treatment of HIV-1 infection and disease.

References

Aldrovandi, G. M., Feuer, G., Gao, L. *et al.* (1993). The SCID-hu mouse as a model for HIV-1 infection. *Nature*, **363**, 732–736.

Berkowitz, R. D., Alexander, S., Bare, C. *et al.* (1998) CCR-5 and CXCR4-Utilizing Strains of Human Immunodeficiency Virus Type 1 Exhibit Differential Tropism and Pathogenesis In Vivo. *J. Virol.*, **72**, 10108–10117.

Bonyhadi, M. L., Rabin, L., Salimi, S. *et al.* (1993). HIV induces thymus depletion *in vivo*. *Nature*, **363**, 728–732.

Bosma, G. C., Custer, R. P., Bosma, M. J. (1983). A severe combined immunodeficiency mutation in the mouse. *Nature*, **301**, 527–530.

Cohen, O. J., Kinter, A., Fauci, A. S. (1997). Host factors in the pathogenesis of HIV disease. *Immunol. Rev.*, **159**, 31–48.

Datema, R., Rabin, L., Hincenbergs, M. *et al.* (1996). Antiviral efficacy *in vivo* of the anti-human immunodeficiency virus bicyclam SDZ SID 791 (JM 3100), an inhibitor of infectious cell entry. *Antimicrob. Agents Chemother.*, **40**, 750–754.

Fauci, A. S. (1996). Host factors and the pathogenesis of HIV-induced disease. *Nature*, **384**, 529–534.

Jamieson, B. D., Aldrovandi, G. M., Planelles, V. *et al.* (1994). Requirement of human immunodeficiency virus type 1 nef for *in vivo* replication and pathogenicity. *J. Virol.*, **68**, 3478–3485.

Jamieson, B. D., Pang, S., Aldrovandi, G. M., Zha, J., Zack, J. A. (1995). *In vivo* pathogenic properties of two clonal human immunodeficiency virus type 1 isolates. *J. Virol.*, **69**, 6259–6264.

Kaneshima, H., Shih, C. C., Namikawa, R. *et al.* (1991). Human immunodeficiency virus infection of human lymph nodes in the SCID-hu mouse. *Proc. Natl Acad. Sci. USA*, **88**, 4523–4527.

Kaneshima, H., Su, L., Bonyhadi, M. L., Connor, R. I., Ho, D. D., McCune, J. M. (1994). Rapid-high, syncytium-inducing isolates of human immunodeficiency virus type 1 induce cytopathicity in the human thymus of the SCID-hu mouse. *J. Virol.*, **68**, 8188–8192.

Kitchen, S. G., Zack, J. A. (1997). CXCR4 expression during lymphopoiesis: implications for human immunodeficiency virus type 1 infection of the thymus. *J. Virol.*, **71**, 6928–6934.

Krowka, J. F., Sarin, S., Namikawa, R., McCune, J. M., Kaneshima, H. (1991). Human T cells in the SCID-hu mouse are phenotypically normal and functionally competent. *J. Immunol.*, **146**, 3751–3756.

Malynn, B. A., Blackwell, T. K., Fulop, G. M. *et al.* (1988). The scid defect affects the final step of the immunoglobulin VDJ recombinase mechanism. *Cell*, **54**, 453–460.

McCune, J. M. (1996). Development and applications of the SCID-hu mouse model. *Semin. Immunol.*, **8**, 187–196.

McCune, J. M. (1997). Animal models of HIV-1 disease. *Science*, **278**, 2141–2142.

McCune, J. M., Namikawa, R., Kaneshima, H., Shultz, L. D., Lieberman, M., Weissman, I. L. (1988). The SCID-hu mouse: murine model for the analysis of human hematolymphoid differentiation and function. *Science*, **241**, 1632–1639.

McCune, J. M., Namikawa, R., Shih, C. C., Rabin, L., Kaneshima, H. (1990). Suppression of HIV infection in AZT-treated SCID-hu mice. *Science*, **247**, 564–566.

McCune, J., Kaneshima, H., Krowka, J. *et al.* (1991). The SCID-hu mouse: a small animal model for HIV infection and pathogenesis. *Annu. Rev. Immunol.*, **9**, 399–429.

Moore, J. P. (1997). Coreceptors: implications for HIV pathogenesis and therapy. *Science*, **276**, 51–52.

Namikawa, R., Kaneshima, H., Lieberman, M., Weissman, I. L., McCune, J. M. (1988). Infection of the SCID-hu mouse by HIV-1. *Science*, **242**, 1684–1686.

Namikawa, R., Weilbaecher, K. N., Kaneshima, H., Yee, E. J., McCune, J. M. (1990). Long-term human hematopoiesis in the SCID-hu mouse. *J. Exp. Med.*, **172**, 1055–1063.

Rabin, L., Hincenbergs, M., Moreno, M. B. *et al.* (1996). Use of standardized SCID-hu Thy/Liv mouse model for preclinical efficacy testing of anti-human immunodeficiency virus type 1 compounds. *Antimicrob. Agents Chemother.*, **40**, 755–762.

Schols, D., Struyf, S., Van Damme, J., Este, J. A., Henson, G., De Clercq, E. (1997). Inhibition of T-tropic HIV strains by selective antagonization of the chemokine receptor CXCR4. *J. Exp. Med.*, **186**, 1383–1388.

Stanley, S. K., McCune, J. M., Kaneshima, H. *et al.* (1993). Human immunodeficiency virus infection of the human thymus and disruption of the thymic microenvironment in the SCID-hu mouse. *J. Exp. Med.*, **178**, 1151–1163.

Su, L., Kaneshima, H., Bonyhadi, M. *et al.* (1995). HIV-1-induced thymocyte depletion is associated with indirect cytopathogenicity and infection of progenitor cells *in vivo*. *Immunity*, **2**, 25–36.

Su, L., Kaneshima, H., Bonyhadi, M. L. *et al.* (1997). Identification of HIV-1 determinants for replication *in vivo*. *Virology*, **227**, 45–52.

Chapter 130

Animal Models for HIV Infection: hu-PBL-SCID Mice

D. E. Mosier

Idealized animal models for HIV-1 infection — the context for available animal models

Human immunodeficiency virus type 1 (HIV-1) is the retrovirus responsible for acquired immunodeficiency syndrome (AIDS) in humans. HIV-1 was discovered in 1983 (Barre-Sinoussi *et al.*, 1983) and is now estimated to infect 30 million people worldwide. HIV-1 is readily infectious only for humans and selected higher primates (Fultz *et al.*, 1986; Fultz, 1991); transient or latent infection has also been reported in pigtailed macaques and rabbits (Kulaga *et al.*, 1989). There is universal agreement that an animal model for HIV-1 infection would accelerate progress in AIDS research, particularly in the pathogenesis of immune deficiency and the development of vaccines to prevent HIV-1 infection. The ideal animal model would mimic HIV-1 infection of humans as closely as possible.

HIV-1 infection is transmitted to humans by blood products or sexual contact. HIV-1 varies in the cell types it infects, and this cell tropism is primarily determined by the cellular receptors used for virus entry (Berger, 1997). Macrophage-tropic (M-tropic) HIV-1 infects both normal T cells and macrophages, but cannot infect established T-cell lines or T-cell tumors. M-tropic virus uses human CD4 and the chemokine receptor CCR5 for virus binding and entry. T-cell tropic HIV-1 infects primary T cells and T-cell lines, but cannot infect macrophages. T-tropic virus uses CD4 and the chemokine receptor CXCR4 for binding and entry. Transitional viruses exist that can infect both macrophages and T-cell lines (M/T- or dual-tropic), and these viruses can use either CCR5 or CXCR4 as co-receptors for entry. Primary infection of humans is almost always due to transmission of M-tropic virus (Zhu *et al.*, 1993), and evolution to M/T-tropic and T-tropic viruses occurs late in disease in about 50% of patients (Tersmette *et al.*, 1989). The characteristic feature of HIV-1 infection — loss of CD4 T cells — is accelerated when the predominant virus population switches from M-tropic to M/T- or T-tropic (Bozzette *et al.*, 1993; Connor *et al.*, 1993).

Primary infection with HIV-1 results in widespread dissemination of virus and high plasma levels of viral RNA, which subsequently decline to low levels concomitant with the onset of cellular immunity to the virus (Safrit *et al.*, 1994). In most patients, virus infection persists for many years before CD4 T-cell levels decline to less than 10% of normal, at which time the individual is at high risk for the opportunistic infections and malignancies that define AIDS.

Persistent infection with HIV-1 results in a chronically activated immune system, with CD8 T cells, B cells, and natural killer cells all showing signs of activation (Lane *et al.*, 1985; Giorgi and Detels, 1989). HIV-1 thus establishes a chronic infection that resists or subverts a strong host immune response, and ultimately destroys both primary and secondary lymphoid tissue (Pantaleo *et al.*, 1993; Stanley *et al.*, 1993).

The ideal animal model thus would be more susceptible to infection with M-tropic HIV-1 than T-tropic HIV-1, would establish a vigorous primary infection which is partly controlled by antiviral immunity, and then would result in a chronic infection with slow loss of CD4 T cells and eventual onset of AIDS-defining infections or tumors. No currently available animal model fulfills all of these criteria, although simian immunodeficiency virus type 1 (SIV-1) infection of macaques comes relatively close, and infection of primates with some recombinant SHIV (SIV/HIV hybrids) viruses (Reimann *et al.*, 1996; Stephens *et al.*, 1997) produces a disease course similar to that observed in humans. At the time that SCID mouse–human chimeric models for HIV-1 infection were developed (McCune *et al.*, 1988; Mosier *et al.*, 1988), only humans and chimpanzees were known to be susceptible to HIV-1 infection and there was a clear need for an animal model in which HIV-1 could be studied. The absence of an active immune response to HIV-1 infection in hu-PBL-SCID and SCID-hu thy/liv mice (see below) is a limitation for modeling the natural course of disease; instead, these animals develop an unopposed primary infection.

Handbook of Animal Models of Infection
ISBN 0-12-775390-7

Current animal models employing human cells or tissue transplanted to SCID mice

The hu-PBL-SCID model

Our laboratory has pioneered the use of severe combined immunodeficient mice (Bosma *et al.*, 1983; Custer *et al.*, 1985) reconstituted with human peripheral blood leukocytes (termed hu-PBL-SCID mice) as a small animal model for HIV infection (Mosier *et al.*, 1988, 1989, 1991, 1993a,b; Mosier, 1990, 1995, 1996a–d; Torbett *et al.*, 1991; Parren *et al.*, 1995; Gulizia *et al.*, 1996, 1997; Picchio *et al.*, 1997). A related but different small animal model for HIV infection has been created by grafting SCID mice with fetal human thymus and liver (McCune *et al.*, 1988, 1990a,b; Namikawa *et al.*, 1988; Aldrovandi *et al.*, 1993; Bonyadi *et al.*, 1993; Kollmann, *et al.*, 1994a,b; Goldstein *et al.*, 1996; Jamieson *et al.*, 1996). These animals are called either SCID-hu or SCID-hu thy/liv mice. The methodology and advantages/disadvantages of the hu-PBL-SCID model will be considered here, and a comparison with the SCID-hu thy/liv model discussed later.

The importance of the SCID mouse recipient of human grafts

SCID mice lack mature T and B lymphocytes and have increased sensitivity to ionizing irradiation because of an inherited mutation in the enzyme DNA-dependent protein kinase (Blunt *et al.*, 1995). The innate immune system, including natural killer (NK) cells, is intact in SCID mice (Dorshkind *et al.*, 1985). The SCID mutation was originally identified in the CB-17 Igh congenic line (Bosma *et al.*, 1983) and has now been crossed to several other mouse strains (Nonoyama *et al.*, 1993; Greiner *et al.*, 1995; Shultz *et al.*, 1995). NOD.SCID mice (NOD.SCID mice are the SCID mutation backcrossed onto the nonobese diabetic (NOD) genetic background) have been reported to be better recipients of human spleen cells than SCID mice (Greiner *et al.*, 1995), although our studies have not found a significant difference between NOD.SCID and SCID recipients of human PBL grafts. We have found that CB-17 SCID mice reared in a specific pathogen-free environment are good recipients for human cell engraftment.

We produce our own mice in a closed breeding colony because we have repeatedly found that commercial sources of SCID mice are more difficult to engraft with human cells. We screen all our SCID mice for the production of mouse immunoglobulin, a marker for 'leakiness' which appears to reflect the pathogen-dependent rescue of very rare B cells (Carroll and Bosma, 1988; Mosier *et al.*, 1993c; Riggs *et al.*, 1994). Mice with levels of mouse immunoglobulin M (IgM) greater than 10 μg per ml are discarded. With these precautions, the rate of successful PBL engraftment following injection of 20×10^6 cells is greater than 95%. Some investigators have found better PBL engraftment following low-dose irradiation and anti-asialo GM1 antibody treatment to reduce NK cell levels (Sandhu *et al.*, 1996; Albert *et al.*, 1997); we have not found this to be necessary with our SCID mouse colony.

One might think that other severely immunodeficient mice, such as RAG-1 or RAG-2 knockout strains (Oettinger *et al.*, 1990; Martin *et al.*, 1994), would be as good as recipients for human cells as SCID mice. This has not been our experience. The biology that permits successful establishment of human xenografts is still poorly understood. It is possible that some degree of xenostimulation leading to a mild, chronic graft-versus-host response is optimal for persistent human cell engraftment (Murphy *et al.*, 1992), and that CB-17 SCID mice are best at providing this stimulation. SCID.beige mice have some defects in NK cell function and may be slightly better as recipients of human cells (Mosier *et al.*, 1993c; Zhang *et al.*, 1996). However, the beige mutation also affects neutrophil function, and we find SCID.beige mice more difficult to maintain than SCID mice because of their enhanced sensitivity to bacterial infections.

We house all SCID mice before and after human PBL grafting in sterile microisolator cages with irradiated food, water, and bedding. The animals are kept within a Biosafety Level 3 (BSL-3) facility so that they do not have to be moved prior to infection with HIV-1. National Institutes of Health (NIH) and Centers for Disease Control guidelines for HIV-1 infection in this animal model recommend BSL-3 practices in a BSL-2-level laboratory (Milman, 1990).

Sources and preparation of human cells for engraftment

We have routinely used adult peripheral blood mononuclear cells purified from whole venous blood for generation of hu-PBL-SCID mice (Mosier *et al.*, 1988, 1991). Mononuclear cells are separated by density using centrifugation over Ficoll-Hypaque or a lymphocyte separation media. Donors are consenting adults who have been prescreened for infection with HIV-1, hepatitis B, hepatitis C, and Epstein–Barr virus (EBV). Most donors are EBV-seropositive, but certain long-term experiments may benefit from using EBV-negative donors since many EBV-positive donors will produce a lymphoproliferative syndrome in hu-PBL-SCID mice (Picchio *et al.*, 1992; Rochford and Mosier, 1994). Cord blood cells will also engraft in SCID mice, but do so much better in neonatal SCID recipients than in adult mice (Reinhardt *et al.*, 1994). Human spleens and tonsils have also been used as sources of cells for SCID mouse engraftment (Nadal *et al.*, 1992; Greiner *et al.*, 1995). Cryopreserved cells can be used for reconstituting SCID mice if they are carefully collected and frozen. Preservative-free 0.2% ethylenediaminetetraacetic acid (EDTA) should be used for anticoagulation. We routinely use $25–30 \times 10^6$ cryopreserved PBL per mouse for reconstitution to compensate for the slightly lower survival of the cells.

It is also possible to use PBL from HIV-1-infected

donors, which leads to the reactivation of virus replication in activated CD4 memory T cells (Boyle *et al.*, 1995; D. Mosier, unpublished results). This appears to occur even if the PBL donor has very low or undetectable viral RNA levels in plasma, and may prove to be an important method for recovering virus from small numbers of latently infected cells.

The PBL graft is established by the intraperitoneal injection of 20×10^6 human cells. One can use higher cell numbers, but the risk of significant graft-versus-host disease and EBV-associated lymphoproliferative disease increases with increasing cell numbers. From 20 to 25 SCID mice can be grafted with PBL recovered from a single unit of blood when 20×10^6 human cells per mouse are used, which is important in establishing meaningful numbers of mice per experimental treatment or manipulation. We find that a group size of 5 mice is sufficient for statistical significance in most experiments.

Monitoring the function of engrafted human cells

The functional status of human cells engrafted in SCID mice has recently been reviewed (Mosier, 1996a,c,d). Engraftment of human T cells predominates, and activation of T cells leads to spontaneous B-cell stimulation and human Ig production. Monitoring of plasma concentrations of human IgG thus provides an indirect measure of the level of human cell engraftment. Human IgG levels do not reach peak levels until 6 weeks or more after PBL engraftment, but should be detectable within 7–10 days of injection of PBL. Human cells can be directly counted by two- (or more) color flow cytometry. Typically we recover $1–2 \times 10^6$ human cells by peritoneal lavage following the injection of 5–10 ml of prewarmed tissue culture medium into the peritoneal cavity. The abdomen of the anesthetized mouse is then gently massaged and the medium withdrawn into the same syringe used for injection. Human cells are also found in the local lymph nodes draining the peritoneal cavity and in the spleen. Peripheral blood and distant lymph nodes have few human cells. Mouse and human cells are distinguished by staining with antibodies binding mouse class I major histocompatibility complex molecules (H-$2K^d$) and human CD45. Human cells can then be gated by size (forward versus side light scatter), and analyzed for expression of CD3, CD4, CD8, CD19, and other human markers. Human T cells expand following PBL engraftment of SCID mice, whereas other cells found in peripheral blood decline with time. Because the percentage of human cells differs between different sites of engraftment and between mice, we have found it convenient to standardize the frequency of CD4 T cells recovered after HIV-1 infection by expressing them as a percentage of recovered CD3 T cells.

Additional assays possible in hu-PBL-SCID mice include measurement of specific antibody responses to recall or novel antigens, and measurement of antigen-specific T-cell proliferative or cytolytic responses. In general, these adaptive immune responses are weaker in hu-PBL-SCID mice than in immunized or infected humans (Markham and Donnenberg, 1992; Mosier, 1996b; Albert *et al.*, 1997).

Preparation of HIV-1 virus stocks

HIV-1 isolates can be obtained directly from patients or from laboratory isolates grown in cell lines. The NIH AIDS Repository maintains many common HIV-1 strains. Prolonged passage of HIV-1 in T-cell lines tends to select for viruses that use CXCR4 for entry, and many laboratory stocks are poorly representative of patient isolates. We prepare infectious virus pools by propagation of virus in human peripheral blood mononuclear cell (PBMC) cultures that have been activated by the addition of phytohemagglutinin (PHA) and interleukin-2. Virus-containing medium is collected weekly, and endpoint titration is performed to determine tissue culture infectious doses (TCID), and aliquots of virus frozen in liquid nitrogen until use for infection. Patient isolates are prepared after as short a culture period as possible, usually 7–12 days. We do not use HIV-1 that has been grown in human T-cell lines such as H9, CEM, or Jurkat. There is some indication that M-tropic virus grown in purified CD4 T cells is different than virus grown in activated macrophages (Willey *et al.*, 1996), so virus produced in whole PBMC cultures may be more infectious than the same virus grown in purified CD4 T cells.

Safely infecting mice with HIV-1 isolates

We have routinely infected hu-PBL-SCID mice by intraperitoneal injection of 1–1000 TCID of HIV-1 at 2 weeks after PBL reconstitution. Mice are anesthetized with Metofane prior to injection, and injected using a 1 ml tuberculin syringe with a 25 G needle. The injection procedure is performed in a class II biological safety cabinet. Only individuals who have mastered the intraperitoneal injection technique are allowed to infect mice. Subsequent to infection, mice are handled with 12 inch forceps. A biosafety training and monitoring program is mandatory for all individuals working with HIV-1-infected hu-PBL-SCID mice.

Monitoring the course and consequences of HIV-1 infection

HIV-1 infection of hu-PBL-SCID mice can be followed by several techniques. Unfortunately, the best technique, serial determinations of plasma virus RNA copy number using quantitative polymerase chain reaction (PCR) techniques (e.g., Roche Amplicor Monitor Assay), is relatively expensive. This assay is sensitive down to 200 copies/ml of plasma, and infected mice often show RNA copy number in excess of 10^6/ml. Plasma p24 antigenemia can be followed serially using commercial p24 enzyme-linked immunosorbent assay (ELISA) kits, but these assays are less sensitive (~10 pg p24/ml) and only mice with high viral loads will show detectable p24 in plasma. Virus can be recovered from sites of human cell engraftment by co-culturing the recovered human cells with human T-cell blasts generated by PHA (2 µg/ml) and interleukin-2 (10 units/ml) stimulation for 2–3 days. These co-cultures can be carried out for up to 4 weeks, and are capable of detecting fewer than 10 infected cells in a hu-PBL-SCID mouse. This procedure can be made more quantitative by serial dilution of input hu-PBL-SCID mouse cells and calculating the frequency of cells capable of giving rise to an HIV-1 positive co-culture (Koup *et al.*, 1994). One may also perform quantitative PCR assays for proviral sequences (Mosier *et al.*, 1991, 1993b), which gives an estimate of the number of infected cells if one assumes a single integration event per infected cell. Finally, recovered tissue from hu-PBL-SCID mice can be processed for *in situ* hybridization for HIV-1 RNA, which can provide direct information about the frequency and distribution of infected cells. *In situ* PCR has been performed on human lymph node biopsies (Peng *et al.*, 1995; Haase *et al.*, 1996), but no one has applied it to samples from hu-PBL-SCID mice to date. A variant of the hu-PBL-SCID model has been created by injecting infected human monocytes into the brains of SCID mice to create a model for HIV-1 encephalitis (Persidsky and Gendelman, 1997).

The important contribution of HIV-1 isolates to the observed course of infection

As noted above, HIV-1 isolates differ in their cell tropism, cytopathic effects, and use of chemokine co-receptors for virus entry. These properties have a substantial effect on the course of HIV-1 infection in hu-PBL-SCID mice. Infection with M-tropic isolates that use CCR5 as an entry co-factor results in a persistent infection with sustained high viral RNA levels. Infection with T-tropic isolates that use CXCR4 as an entry co-factor results in a shorter course of viremia with lower peak viral RNA copy number. Infection with dual-tropic isolates that can use either CCR5 or CXCR4 for virus entry results in a very short duration of infection which is terminated by deletion of all CD4 target cells for infection. The choice of virus isolate used for infection thus has a major impact on the level of virus replication and the kinetics of CD4 T-cell depletion. These differences are illustrated in Figure 130.1. The status of the CCR5 genotype of the PBL donor can also influence the kinetics of virus infection (Picchio *et al.*, 1997).

Contributions to antiviral therapy

Hu-PBL-SCID mice have been used to test the efficacy of a number of candidate antiviral compounds (Koup *et al.*, 1993; Ussery *et al.*, 1995) as well as immunologic approaches to controlling HIV-1 infection (Mosier *et al.*, 1993a; Safrit *et al.*, 1993; van Kuyk *et al.*, 1994; Gauduin *et al.*, 1995; Parren *et al.*, 1995; Mosier, 1996b). Although promising results have emerged, published reports of drug efficacy are relatively rare. Some of the results are proprietary, but a more common problem has been the fast-track development of

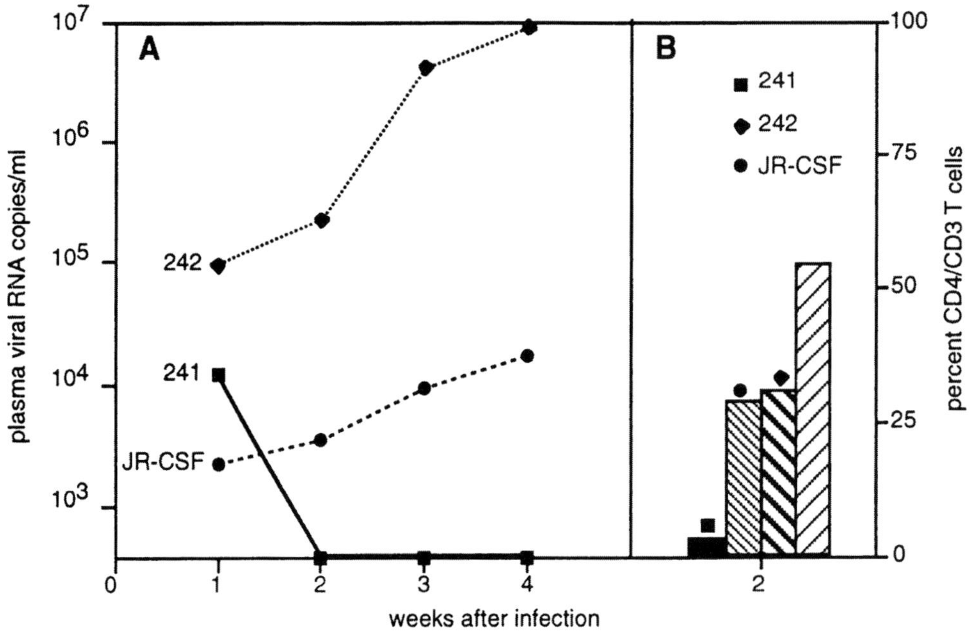

Figure 130.1 **(A)** Plasma human immunodeficiency virus type 1 (HIV-1) RNA levels at 1–4 weeks after infection of hu-PBL-SCID mice with three different isolates of HIV-1. Strains 242 and JR-CSF are macrophage-tropic and use CCR5 as a co-factor for viral entry, while strain 241, which differs from 242 by only a single amino acid in V3 of gp120, is dual-tropic and can use either CCR5 or CXCR4 for virus entry (Chesebro *et al.*, 1996; Speck *et al.*, 1997). **(B)** Mean recovery of CD4 T cells in each group of mice at 2 weeks after infection. The cross-hatched column to the right of panel B is the mean CD4 T-cell recovery in uninfected hu-PBL-SCID mice.

antivirals for HIV-1 that has no requirement for animal testing for efficacy, and a large population of patients eager to participate in drug trials. Current studies in our laboratory support the utility of the hu-PBL-SCID model in developing novel approaches to HIV-1 therapy. It has proven particularly useful to perform daily monitoring of plasma virus RNA levels following the introduction of candidate antivirals.

Advantages and disadvantages of the hu-PBL-SCID model

Hu-PBL-SCID mice are highly susceptible to infection by all isolates of HIV-1 and HIV-2 tested to date, and produce uniform patterns of virus replication with a given isolate. The model is relatively easy to establish compared to comparable animal models for lentivirus infections, but it does require biocontainment facilities for HIV-1-infected mice. The human graft consists mainly of activated, memory T cells which are a highly permissive target for HIV-1 replication. The course of infection can be varied by choosing the appropriate virus isolate. Virus replication takes place both in cells within the peritoneal cavity and in reconstituted lymphoid tissue.

There are two major limitations of the model. The human xenograft does not have a source for T-cell replacement, so infection is often terminated by exhaustion of target cells. There is no immune response to the virus, so the infection is unopposed. Virus often reappears when antiviral treatments are halted, but this might not happen in an intact host where the immune response could effectively deal with small residual viral loads. Elements of the mouse innate immune system may contribute to antiviral activity, but it is unclear whether this activity differs from that seen in naturally infected humans.

The SCID-hu thy/liv model

This animal model differs from the hu-PBL-SCID model in that the human graft consists of fragments of fetal thymus and liver placed surgically in close proximity under the renal capsule. The model has been extensively reviewed elsewhere (McCune, 1990; Aldrovandi *et al.*, 1993; Goldstein *et al.*, 1996; Rabin *et al.*, 1996; Bristol *et al.*, 1997). The advantage of this model is that T-cell development takes place in the fetal thymus graft and continues for several months, making use of lymphoid progenitors contained in the fetal liver. Direct injection of HIV-1 into the thymus graft establishes a persistent infection that leads to the depletion of developing T cells in the thymus (Aldrovandi *et al.*, 1993; Bonyadi *et al.*, 1993; Kaneshima *et al.*, 1994; Jamieson *et al.*, 1996). Both CD4 and CD8 T cells become depleted, apparently because of infection of the common CD4/8 double-positive precursor (Kitchen *et al.*, 1997). This model is highly relevant to infection of the thymus in maternally transmitted infection of newborns. HIV-1 infection is of longer duration in the thymus graft of SCID-hu mice than in the PBL graft of hu-PBL-SCID mice (Jamieson *et al.*, 1996).

The SCID-hu thy/liv model has been successfully used to screen antivirals for activity against HIV-1 infection (McCune *et al.*, 1990c; Shih *et al.*, 1991; Rabin *et al.*, 1996; Pettoello-Montavani *et al.*, 1997). It has many of the same advantages and disadvantages as the hu-PBL-SCID model, but it differs in that most HIV-1 replication takes place within an organized lymphoid organ. It is more cumbersome to place fetal tissue fragments surgically under the kidney capsule than to inject PBL intraperitoneally, and the pace of screening experiments is faster with the hu-PBL-SCID model.

References

Albert, S., McKerlie, C., Pester, A. *et al.* (1997). Time-dependent induction of protective anti-influenza immune responses in human peripheral blood lymphocyte/SCID mice. *J. Immunol.*, **159**, 1393–1403.

Aldrovandi, G., Feuer, G., Gao, L. *et al.* (1993). The SCID-hu mouse as a model for HIV-1 infection. *Nature*, **363**, 732–736.

Barre-Sinoussi, F., Chermann, J. C., Rey, F. *et al.* (1983). Isolation of a T-lymphotropic retrovirus from a patient at risk for acquired immune deficiency syndrome (AIDS). *Science*, **220**, 868–871.

Berger, E. (1997). HIV entry and tropism: the chemokine receptor connection. *AIDS*, **11**, S3–S16.

Blunt, T., Finnie, N. J., Taccioli, G. E. *et al.* (1995). Defective DNA-dependent protein kinase activity is linked to V(D)J recombination and DNA repair defects associated with the murine scid mutation. *Cell*, **80**, 813–823.

Bonyadi, M., Rabin, L., Salimi, S. *et al.* (1993). HIV induces thymus depletion *in vivo*. *Nature*, **363**, 728–732.

Bosma, G. C., Custer, R. P., Bosma, M. J. (1983). A severe combined immunodeficiency mutation in the mouse. *Nature*, **301**, 527–530.

Boyle, M. J., Connors, M., Flanigan, M. E. *et al.* (1995). The human HIV/peripheral blood lymphocyte (PBL)-SCID mouse. A modified human PBL-SCID model for the study of HIV pathogenesis and therapy. *J. Immunol.*, **154**, 6612–6623.

Bozzette, S., McCutchan, J., Spector, S. *et al.* (1993). A cross-sectional comparison of persons with syncytium- and non-syncytium-inducing human immunodeficiency virus. *J. Infect. Dis.*, **168**, 1374–1379.

Bristol, G., Gao, L., Zack, J. (1997). Preparation and maintenance of SCID-hu mice for HIV research. *Methods*, **12**, 343–347.

Carroll, A. M., Bosma, M. J. (1988). Detection and characterization of functional T cells in mice with severe combined immune deficiency. *Eur. J. Immunol.*, **18**, 1965.

Chesebro, B., Wehrly, K., Nishio, J. *et al.* (1996). Mapping of independent V3 envelope determinants of human immunodeficiency virus type 1 macrophage tropism and syncytial formation in lymphocytes. *J. Virol.*, **70**, 9055–9059.

Connor, R., Mohri, H., Cao, Y. *et al.* (1993). Increased viral burden and cytopathicity correlate temporally with CD4+ T lymphocyte decline and clinical progression in human immunodeficiency virus type-1 infected individuals. *J. Virol.*, **67**, 1772–1777.

Custer, R. P., Bosma, G. C., Bosma, M. J. (1985). Severe combined immunodeficiency (SCID) in the mouse: pathology, reconstitution, neoplasms. *Am. J. Pathol.*, **120**, 464.

Dorshkind, K., Pollack, S. B., Bosma, M. J. *et al.* (1985). Natural killer (NK) cells are present in mice with severe combined immunodeficiency (SCID). *J. Immunol.*, **134**, 3798–3801.

Fultz, P. N. (1991). Human immunodeficiency virus infection of chimpanzees: an animal model for asymptomatic HIV carriers and vaccine efficacy. *AIDS Res. Rev.*, **1**, 207.

Fultz, P. N., McClure, H. M., Swenson, R. B. *et al.* (1986). Persistent infection of chimpanzees with human T-lymphotropic virus type III/lymphadenopathy-associated virus: a potential model for acquired immunodeficiency syndrome. *J. Virol.*, **58**, 116–124.

Gauduin, M. C., Safrit, J. T., Weir, R. *et al.* (1995). Pre- and post-exposure protection against human immunodeficiency virus type 1 infection mediated by a monoclonal antibody. *J. Infect. Dis.*, **171**, 1203–1209.

Giorgi, J. V., Detels, R. (1989). T-cell subset alterations in HIV-infected homosexual men: NIAID multicenter AIDS cohort study. *Clin. Immunol. Immunopathol.*, **52**, 10–18.

Goldstein, H., Pettoello-Mantovani, M., Katopodis, N. F. *et al.* (1996). SCID-hu mice: a model for studying disseminated HIV infection. *Semin. Immunol.*, **8**, 223–231.

Greiner, D. L., Shultz, L. D., Yates, J. *et al.* (1995). Improved engraftment of human spleen cells in NOD/LtSz-scid/scid mice as compared with C.B-17-scid/scid mice. *Am. J. Pathol.*, **146**, 888–902.

Gulizia, R., Levy, J., Mosier, D. (1996). The envelope gp120 gene of human immunodeficiency virus type 1 determines the rate of CD4-positive T cell depletion in SCID mice engrafted with human peripheral blood leukocytes. *J. Virol.*, **70**, 4184–4187.

Gulizia, R. J., Collman, R. G., Levy, J. A. *et al.* (1997). Deletion of nef slows but does not prevent CD4-positive T-cell depletion in human immunodeficiency virus type 1-infected human-PBL-SCID mice. *J. Virol.*, **71**, 4161–4164.

Haase, A., Henry, K., Zupancic, M. *et al.* (1996). Quantitative image analysis of HIV-1 infection in lymphoid tissue. *Science*, **274**, 985–989.

Jamieson, B. D., Aldrovandi, G. M., Zack, J. A. (1996). The SCID-hu mouse: an *in-vivo* model for HIV-1 pathogenesis and stem cell gene therapy for AIDS. *Semin. Immunol.*, **8**, 215–221.

Kaneshima, H., Su, L., Bonyhadi, M. L. *et al.* (1994). Rapid-high, syncytium-inducing isolates of human immunodeficiency virus type 1 induce cytopathicity in the human thymus of the SCID-hu mouse. *J. Virol.*, **68**, 8188–8192.

Kitchen, S., Uittenbogaart, C., Zack, J. (1997). Mechanism of human immunodeficiency virus type 1 localization in CD4-negative thymocytes: differentiation from a CD4-positive precursor allows productive infection. *J. Virol.*, **71**, 5713–5722.

Kollmann, T. R., Kim, A., Zhuang, X. *et al.* (1994a). Reconstitution of SCID mice with human lymphoid and myeloid cells after transplantation with human fetal bone marrow without the requirement for exogenous human cytokines. *Proc. Natl Acad. Sci. USA*, **91**, 8032–8036.

Kollmann, T. R., Pettoello-Mantovani, M., Zhuang, X. *et al.* (1994b). Disseminated human immunodeficiency virus 1 (HIV-1) infection in SCID-hu mice after peripheral inoculation with HIV-1. *J. Exp. Med.*, **179**, 513–522.

Koup, R. A., Brewster, F., Grob, P. *et al.* (1993). Nevirapine synergistically inhibits HIV-1 replication in combination with zidovudine, interferon or CD4 immunoadhesin. *Aids*, **7**, 1181–1184.

Koup, R. A., Hesselton, R. M., Safrit, J. T. *et al.* (1994). Quantitative assessment of human immunodeficiency virus type 1 replication in human xenografts of acutely infected Hu-PBL-SCID mice. *AIDS Res. Hum. Retroviruses*, **10**, 279–284.

Kulaga, H., Folks, T., Rutledge, R. *et al.* (1989). Infection of rabbits with human immunodeficiency virus 1. *J. Exp. Med.*, **169**, 321–326.

Lane, H. C., Depper, J. M., Greene, W. C. *et al.* (1985). Qualitative analysis of immune function in patients with the acquired immunodeficiency syndrome. Evidence for a selective defect in soluble antigen recognition. *N. Engl. J. Med.*, **313**, 79–84.

Markham, R. B., Donnenberg, A. D. (1992). Effect of donor and recipient immunization protocols on primary and secondary human antibody responses in SCID mice reconstituted with human peripheral blood mononuclear cells. *Infect. Immun.*, **60**, 2305–2308.

Martin, A., Valentine, M., Unger, P. *et al.* (1994). Engraftment of human lymphocytes and thyroid tissue into scid and rag2-deficient mice: absent progression of lymphocytic infiltration. *J. Clin. Endocrinol. Metab.*, **79**, 716–723.

McCune, J. M. (1990). The rational design of animal models for HIV infection. *Semin. Virol.*, **1**, 229–235.

McCune, J. M., Namikawa, R., Kaneshima, H. *et al.* (1988). The SCID-hu mouse: murine model for the analysis of human hematolymphoid differentiation and function. *Science*, **241**, 1632–1639.

McCune, J. M., Kaneshima, H., Rabin, L. *et al.* (1990a). Preclinical evaluation of antiviral compounds in the SCID-hu mouse. *Ann. N. Y. Acad. Sci.*, **616**, 281–286.

McCune, J. M. (1990b). The rational design of animal models for HIV infection. *Sem. Virol.*, **1**, 229–235.

McCune, J. M., Namikawa, R., Shih, C.-C. *et al.* (1990c). Suppression of HIV infection in AZT-treated SCID-hu mice. *Science*, **247**, 564–566.

Milman, G. (1990). HIV research in the SCID mouse: biosafety considerations. *Science*, **250**, 1152.

Mosier, D. E. (1990). Immunodeficient mice xenografted with human lymphoid cells: new models for *in vivo* studies of human immunobiology and infectious diseases. *J. Clin. Immunol.*, **10**, 185–191.

Mosier, D. E. (1995). Distinct rate and patterns of human CD4+ T-cell depletion in hu-PBL-SCID mice infected with different isolates of the human immunodeficiency virus. *J. Clin. Immunol.*, **15**, 130S–133S.

Mosier, D. (1996a). Viral pathogenesis in hu-PBL-SCID mice. *Semin. Immunol.*, **8**, 255–262.

Mosier, D. E. (1996b). Evaluation of protective immunity to HIV-1 in human PBL-SCID mice. *Antibiot. Chemother.*, **48**, 125–130.

Mosier, D. E. (1996c). Human immunodeficiency virus infection of human cells transplanted to severe combined immunodeficient mice. *Adv. Immunol.*, **63**, 79–125.

Mosier, D. E. (1996d). Small animal models for acquired immune deficiency syndrome (AIDS) research. *Lab. Anim. Sci.*, **46**, 257–265.

Mosier, D. E., Gulizia, R. J., Baird, S. M. *et al.* (1988). Transfer of a functional human immune system to mice with severe combined immunodeficiency. *Nature (Lond.)*, **335**, 256–259.

Mosier, D. E., Gulizia, R. J., Baird, S. M. *et al.* (1989). Studies of HIV infection and the development of Epstein–Barr virus-related B cell lymphomas following transfer of human lymphocytes to mice with severe combined immunodeficiency. In *Current Topics in Microbiology and Immunology*, vol. 152 (eds Bosma, M. J., Phillips, R. A., Schuler, W.), pp. 195–199. Springer-Verlag, Berlin-Heidelberg.

Mosier, D. E., Gulizia, R. J., Baird, S. M. *et al.* (1991). Human immunodeficiency virus infection of human-PBL-SCID mice. *Science*, **251**, 791–794.

Mosier, D., Gulizia, R., MacIsaac, P. *et al.* (1993a). Resistance to human immunodeficiency virus 1 infection of SCID mice reconstituted with peripheral blood leukocytes for donors vaccinated with vaccinia gp160 and recombinant gp160. *Proc. Natl Acad. Sci. USA*, **90**, 2443–2447.

Mosier, D., Gulizia, R., MacIsaac, P. *et al.* (1993b). Rapid loss of CD4+ T cells in human-PBL-SCID mice by noncytopathic HIV isolates. *Science*, **260**, 689–692.

Mosier, D. E., Stell, K. L., Gulizia, R. J. *et al.* (1993c). Homozygous scid/scid;beige/beige mice have low levels of spontaneous or neonatal T cell-induced B cell generation. *J. Exp. Med.*, **177**, 191–194.

Murphy, W., Bennett, M., Anver, M. *et al.* (1992). Human–mouse lymphoid chimeras: host vs. graft and graft vs. host reactions. *Eur. J. Immunol.*, **22**, 1421–1427.

Nadal, D., Albini, B., Schläpfer, E. *et al.* (1992). Role of Epstein–Barr virus and interleukin 6 in the development of lymphomas of human origin in SCID mice engrafted with human tonsillar mononuclear cells. *J. Gen. Virol.*, **73**, 113–121.

Namikawa, R., Kaneshima, H., Lieberman, M. *et al.* (1988). Infection of the SCID-hu mouse by HIV-1. *Science*, **242**, 1684–1686.

Nonoyama, S., Smith, F. O., Ochs, H. D. (1993). Specific antibody production to a recall or a neoantigen by SCID mice reconstituted with human peripheral blood lymphocytes. *J. Immunol.*, **151**, 3894–3901.

Oettinger, M. A., Schatz, D. G., Gorka, C. *et al.* (1990). RAG-1 and RAG-2, adjacent genes that synergistically activate V(D)J recombination. *Science*, **248**, 1517–1523.

Pantaleo, G., Graziosi, C., Demerest, J. *et al.* (1993). HIV infection is active and progressive in lymphoid tissue during the clinical latent stage of disease. *Nature*, **362**, 355–358.

Parren, P., Ditzel, H., Gulizia, R. *et al.* (1995). Protection against HIV-1 infection in hu-PBL-SCID mice by passive immunization with a neutralizing human monoclonal antibody against the gp120 CD4-binding site. *AIDS*, **9**, 1–6.

Peng, H., Reinhart, T. A., Retzel, E. F. *et al.* (1995). Single cell transcript analysis of human immunodeficiency virus gene expression in the transition from latent to productive infection. *Virology*, **206**, 16–27.

Persidsky, Y., Gendelman, H. (1997). Development of laboratory and animal model systems for HIV-1 encephalitis and its associated dementia. *J. Leuk. Biol.*, **62**, 100–106.

Pettoello-Montavani, M., Kollman, T., Raker, C. *et al.* (1997). Saquinavir-mediated inhibition of human immunodeficiency virus (HIV) infection in SCID mice implanted with human fetal thymus and liver tissue: an *in vivo* model for evaluating the effect of drug therapy on HIV infection in lymphoid tissue. *Antimicrob. Agents Chemother.*, **41**, 1880–1887.

Picchio, G. R., Kobayashi, R., Kirven, M. *et al.* (1992). Heterogeneity among Epstein–Barr virus-seropositive donors in the generation of immunoblastic B-cell lymphomas in SCID mice receiving human peripheral blood leukocyte grafts. *Cancer Res.*, **52**, 2468–2477.

Picchio, G. R., Gulizia, R. J., Mosier, D. E. (1997). Chemokine receptor CCR5 genotype influences the kinetics of human immunodeficiency virus type 1 infection in human PBL-SCID mice. *J. Virol.*, **71**, 7124–7127.

Rabin, L., Hincenbergs, M., Moreno, M. *et al.* (1996). Use of standardized SCID-hu Thy/Liv mouse model for preclinical efficacy testing of anti-human immunodeficiency virus type 1 compounds. *Antimicrob. Agents Chemother.*, **40**, 755–762.

Reimann, K., Li, J., Voss, G. *et al.* (1996). An *env* gene derived from a primary human immunodeficiency virus type 1 isolate confers high *in vivo* replicative capacity to a chimeric simian/human immunodeficiency virus in rhesus monkeys. *J. Virol.*, **70**, 3198–3206.

Reinhardt, B., Torbett, B. E., Gulizia, R. J. *et al.* (1994). Human immunodeficiency virus type 1 infection of neonatal severe combined immunodeficient mice xenografted with human cord blood cells. *AIDS Res. Hum. Retroviruses*, **10**, 131–141.

Riggs, J. E., Feeney, A. J., Kirven, M. *et al.* (1994). VH11 bias and normal V-D-J junctions in SCID B lymphocytes rescued by neonatal T cell transfer. *Mol. Immunol.*, **31**, 783–791.

Rochford, R., Mosier, D. (1994). Immunobiology of Epstein–Barr virus-associated lymphomas. *Clin. Immunol. Immunopathol.*, **71**, 256–259.

Safrit, J. T., Fung, M. S., Andrews, C. A. *et al.* (1993). hu-PBL-SCID mice can be protected from HIV-1 infection by passive transfer of monoclonal antibody to the principal neutralizing determinant of envelope gp120. *Aids*, **7**, 15–21.

Safrit, J., Andrews, C., Zhu, T. *et al.* (1994). Characterization of human immunodeficiency virus type 1-specific cytotoxic T lymphocyte clones isolated during acute seroconversion: recognition of autologous virus sequences within a conserved immunodominant epitope. *J. Exp. Med.*, **179**, 463–472.

Sandhu, J. S., Boynton, E., Gorczynski, R. *et al.* (1996). The use of SCID mice in biotechnology and as a model for human disease. *Crit. Rev. Biotechnol.*, **16**, 95–118.

Shih, C.-C., Kaneshima, H., Rabin, L. *et al.* (1991). Post-exposure prophylaxis with AZT suppresses HIV-1 infection in SCID-hu mice in a time-dependent manner. *J. Infect. Dis.*, **163**, 625–627.

Shultz, L. D., Schweitzer, P. A., Christianson, S. W. *et al.* (1995). Multiple defects in innate and adaptive immunologic function in NOD/LtSz-scid mice. *J. Immunol.*, **154**, 180–191.

Speck, R., Wehrly, K., Platt, E. *et al.* (1997). Selective employment of chemokine receptors as HIV-1 coreceptors determined by individual amino acids within the envelope V3 loop. *J. Virol.*, **71**, 7136–7139.

Stanley, S., McCune, J., Kaneshima, H. *et al.* (1993). Human immunodeficiency virus infection of the human thymus and disruption of the thymic microenvironment of the SCID-hu mouse. *J. Exp. Med.*, **178**, 1151–1163.

Stephens, E., Mukherjee, S., Sahni, M. *et al.* (1997). A cell-free stock of simian-human immunodeficiency virus that causes AIDS in pig-tailed macaques has a limited number of amino acid substitutions in both SIVmac and HIV-1 regions of the genome and has altered cytotropism. *Virology*, **231**, 313–321.

Tersmette, M., Lange, J. M., de Goede, R. E. *et al.* (1989). Association between biological properties of human immunodeficiency virus variants and risk for AIDS and AIDS mortality. *Lancet*, **1**, 983–985.

Torbett, B. E., Picchio, G., Mosier, D. E. (1991). hu-PBL-SCID mice: a model for human immune function, AIDS, and lymphomagenesis. *Immunol. Rev.*, **124**, 139–164.

Ussery, M., Broud, D., Wood, O. *et al.* (1995). Antiviral agents reduce viral load and delay the detection of infected cells *in vivo* in the HIV-infected hu-PBMC SCID mouse model. *Antiviral Res.*, **26**, A238.

van Kuyk, R., Torbett, B. E., Gulizia, R. J. *et al.* (1994). Cloned human CD8+ cytotoxic T lymphocytes protect human peripheral blood leukocyte-severe combined immunodeficient mice from HIV-1 infection by an HLA-unrestricted mechanism. *J. Immunol.*, **153**, 4826–4833.

Willey, R., Shibata, R., Freed, R. *et al.* (1996). Differential glycosylation, virion incorporation, and sensitivity to neutralizing antibodies of human immunodeficiency virus type I envelope produced from infected primary T-lymphocyte and macrophage cultures. *J. Virol.*, **70**, 6431–6436.

Zhang, C., Cui, Y., Houston, S. *et al.* (1996). Protective immunity to HIV-1 in SCID/beige mice reconstituted with peripheral blood lymphocytes of exposed but uninfected individuals. *Proc. Natl Acad. Sci. USA*, **93**, 14720–14725.

Zhu, T., Mo, H., Wang, N. *et al.* (1993). Genotypic and phenotypic characterization of HIV-1 patients with primary infection. *Science*, **261**, 1179–1181.

Chapter 131

Chimpanzee Model of HIV-1 Infection

P. N. Fultz

Background of human infection

Although human immunodeficiency virus type 1 (HIV-1) was not isolated until 1983, the disease it induces, acquired immunodeficiency syndrome (AIDS), was described first in 1981 (Gottlieb *et al.*, 1981). Since that time HIV-1 has spread worldwide and continues to be a major infectious disease problem, especially in India, Asia, Africa and developing nations (World Health Organization, 1993). Because of the morbidity and mortality associated with HIV-1 infection, the identification of relevant animal models to study the pathogenesis of HIV-1 and to develop therapies and vaccines was a priority early in the epidemic. Furthermore, since the virus is transmitted efficiently after either parenteral or mucosal exposure, including by transfusion or intravenous (i.v.) drug abuse, between sexual partners, or from mother to child (Levy, 1993), it was important that these modes of transmission be a feature of any model system.

Background of model

Of attempts to infect several species of small animals and non-human primates with HIV-1, only infection of chimpanzees (*Pan troglodytes*) and gibbon apes (*Hylobates lar*) was successful (Alter *et al.*, 1984; Francis *et al.*, 1984; Gajdusek *et al.*, 1985; Morrow *et al.*, 1987; Lusso *et al.*, 1988). Although it was reported that pig-tailed macaques (*Macaca nemestrina*) could be infected by HIV-1 (Agy *et al.*, 1992; Frumkin *et al.*, 1993), other groups achieved only transient infection in this species (Gartner *et al.*, 1994; Otten *et al.*, 1994). Thus, because the macaque HIV-1 model could not be reproduced and sufficient numbers of gibbon apes were not available to characterize that system, infection of chimpanzees with HIV-1 was developed as the model of choice.

Before the etiologic agent of AIDS was identified, blood plasma and tissue samples from humans with the disease were inoculated into chimpanzees in attempts to amplify the organism and elicit disease (Alter *et al.*, 1984; Gajdusek *et al.*, 1985). During this same period, the first isolate of HIV-1, called lymphadenopathy-associated virus (LAV-1) (Barré-Sinoussi *et al.*, 1983), was injected into chimpanzees in the form of cell-free virus and autologous peripheral blood mononuclear cells (PBMC) infected with LAV-1 *in vitro* (Francis *et al.*, 1984; Gajdusek *et al.*, 1985). Although these investigators were successful in establishing persistent infections, there was no evidence of immunodeficiency disease. Subsequently, it was demonstrated that chimpanzees could be reproducibly infected with the LAV-1 and IIIB strains (both of which were later shown to be derived from the same LAI isolate; Wain-Hobson *et al.*, 1991) and that virus could be isolated routinely from their PBMC for months and years after initial infection (Fultz *et al.*, 1986b; Nara *et al.*, 1987).

The natural history of HIV-1 infection in chimpanzees paralleled that observed in HIV-1-infected humans with respect to virologic parameters and induction of HIV-1-specific immunity (Goudsmit *et al.*, 1987; Fultz *et al.*, 1989; Fauci, 1993; Fultz, 1993, 1997; Levy, 1993; Heeney, 1995). However, in chimpanzees there was only one report of transient disease which appeared to be directly related to HIV-1 infection (Fultz *et al.*, 1991). Several possible explanations for lack of disease in HIV-1-infected chimpanzees were proposed, including failure of HIV-1 to infect chimpanzee macrophages and to induce a state of chronic T-cell activation and dysfunction (Nara *et al.*, 1989; Heeney *et al.*, 1993; Schuitemaker *et al.*, 1993; Gougeon *et al.*, 1997). With the recent focus on apoptosis in HIV disease, several investigators reported failure to detect aberrant apoptosis in HIV-1-infected chimpanzees and invoked these results as additional evidence that chimpanzees might be inherently resistant to disease (Heeney, 1995). Because most chimpanzees were infected with the closely related LAV-1 and IIIB strains, it was also possible that other strains might be more pathogenic. However, it was generally accepted that HIV-1 does not cause disease in chimpanzees, which led some investigators to question the validity of the model, especially for testing candidate HIV-1 vaccines.

Recently, this conclusion was shown to be wrong when a chimpanzee (C-499) infected with two HIV-1 strains for about 10 years (Fultz *et al.*, 1987b) died of immunodeficiency disease associated with loss of CD4+ T cells (Novembre *et al.*, 1997). This chimpanzee was the same one that previously had developed transient immunodeficiency

Handbook of Animal Models of Infection
ISBN 0-12-775390-7

(Fultz *et al.*, 1991). That chimpanzees are not inherently resistant to HIV-1-induced disease is supported by the detection of persistently high levels of viremia associated with declines in CD4+ T cells in chimpanzees either transfused with blood from C-499 or inoculated i.v. with cell-free virus isolated from C-499's PBMC (Novembre *et al.*, 1997; Villinger *et al.*, 1997; Davis *et al.*, 1998). Thus, the existence of a strain of HIV-1 pathogenic for chimpanzees validates the model and will facilitate testing of novel therapeutic approaches and candidate vaccines not only for protection against infection but also for effects on viral burdens and prevention of disease progression.

Animal species

Outbred chimpanzees (*P. troglodytes*) of any age and sex can be infected with HIV-1 i.v., mucosally and perinatally (Fultz *et al.*, 1986a,b; Eichberg *et al.*, 1988; Girard *et al.*, 1992). Early after the model was established, it was recommended that adult animals previously employed in experiments with one or more of the hepatitis viruses be used. The rationale for this decision was based on the limited numbers and costs associated with the use of chimpanzees in infectious disease research and the consideration that HIV-1 infection potentially could induce terminal disease. After more than 12 years of follow-up of some HIV-1-infected chimpanzees, a history of hepatitis infection, including being a chronic carrier, does not appear to influence the course of HIV-1 infection. Most investigators have adhered to this practice to maximize the use of these valuable animals.

Preparation of animals

HIV-1-infected chimpanzees are housed in animal Biosafety Level 2 facilities with restricted access. These facilities should have anterooms which contain a sink and appropriate supplies and equipment for disposing of infectious waste. Anterooms are also required so that personnel can put on protective clothing, masks and gloves. Animals can be housed singly, in pairs or in larger social groups, depending on the experimental protocol and the compatibility of individual animals. HIV-1 does not appear to be transmitted easily between chimpanzees that are housed together and do not engage in sex (Fultz *et al.*, 1987a). Appropriate environmental enrichment should be provided.

Storage, preparation of inocula

Cell-free HIV-1 inoculum

In the majority of studies in the HIV-1 chimpanzee model, cell-free virus was used as the inoculum. Cell-associated virus and transfusion of whole blood from an infected to an uninfected animal have also been employed (Fultz *et al.*, 1986b, 1992b). Initially, some HIV-1 stocks were prepared from supernatants of infected human T-cell lines, but the demonstration that HIV-1 isolates passaged solely in primary human lymphocytes had different biologic properties from those grown in continuous T-cell lines led to the preparation of most HIV-1 stocks from culture supernatants of infected human or chimpanzee PBMC. A representative method is described here.

PBMC are separated from whole blood or buffy-coat preparations by Ficoll-diatrizoate gradient centrifugation, the cells are cultured at ~1×10^6 cells/ml in RPMI 1640 medium containing 15% fetal bovine serum (FBS) and antibiotics. After activation for 2–3 days by the addition of phytohemagglutinin (1 μg/ml), PBMC are washed and resuspended in RPMI 1640 containing 10% FBS, 10 units interleukin-2/ml and either diethylaminoethyl (DEAE) dextran or polybrene to facilitate virus entry into cells. When virus production peaks, as measured by reverse transcriptase (RT) activity or p24*gag* antigen in culture supernatants, cells are removed from the culture fluid by low-speed centrifugation, and the supernatants are filtered through 0.2 or 0.45 μm filters, aliquoted into cryovials (0.5–1 ml/vial), and stored in liquid nitrogen vapor. Although some investigators store virus stocks at –70 to –85°C, this practice is not advised because, during long-term storage, the infectious titer can decrease. Titers of the virus stocks are determined by limiting dilution and infection of indicator cells (normal human or chimpanzee PBMC or T-cell lines) and are usually expressed as 50% tissue culture infectious doses ($TCID_{50}$), which are calculated using an appropriate statistical method, such as Reed-Muench or Spearman-Karber.

HIV-1-infected cell inoculum

HIV-1-infected PBMC used as the inoculum in some pathogenesis and vaccine challenge protocols consisted of PBMC obtained from chimpanzees 2–4 months after initial HIV-1 infection when numbers of infectious cells in blood are high (Fultz *et al.*, 1986b, 1992b). After separation of PBMC from heparinized whole blood, the cells are washed, resuspended in freezing medium (RPMI 1640 containing antibiotics, 25% FBS and 10% dimethylsulfoxide), aliquoted at $1–2 \times 10^7$ PBMC/ml per vial, and maintained in liquid nitrogen vapor. Before use, cryopreserved cell-free HIV-1 and infected PBMC should be thawed rapidly in a 37°C water bath. After washing PBMC in phosphate-buffered saline (PBS) or medium, the cells are resuspended in medium, counted in a hemacytometer, and the viability is determined by trypan blue dye exclusion. Limiting dilution analysis, similar to that used for cell-free HIV-1 stocks, is also used to determine the number of infectious cells in PBMC. The results are often expressed as the minimum number of PBMC required to transfer infection to indicator cells (Fultz *et al.*, 1992a; Girard *et al.*, 1996).

Infection process

Parenteral

If a particular strain of HIV-1 can infect chimpanzees, successful infections are generally established in 100% of animals inoculated i.v. with at least 10 $TCID_{50}$ (Arthur *et al.*, 1989; Barré-Sinoussi *et al.*, 1997). Before any procedure, including virus inoculation and routine blood collection, chimpanzees are anesthetized by intramuscular (i.m.) injection of ketamine hydrochloride at a concentration of 10–15 mg/kg of body weight. Most animals are inoculated i.v. into the cephalic or saphenous vein with the desired amount of virus or virus-infected cells in a volume of 1 ml. Dilutions of the HIV-1 stock should be made in RPMI 1640 or PBS without addition of serum to limit exposure of the animals to serum proteins. This precaution is important in the event the animals are inoculated with virus more than once, such as in superinfection experiments or during rechallenge in vaccine protection experiments. Re-exposure to bovine proteins potentially could induce anaphylactic shock.

Mucosal

A model for establishing HIV-1 infection by a genital mucosal route was developed by Girard *et al.* (1992). Both cell-free and cell-associated $HIV\text{-}1_{LAI/IIIB}$ were used to establish infection in adult female chimpanzees by depositing virus in a total volume of approximately 0.25 ml in the cervical os. The addition of 25% normal human seminal plasma to the inoculum appeared to increase the frequency of infection, presumably by maintaining the viability of virus or infected PBMC before cervical deposition and, perhaps, by helping to buffer the acidic cervicovaginal environment.

Inoculation via the cervical os requires the use of a sterile speculum and a colposcope for direct observation to insure that the inoculum is deposited 1–2 mm into the endocervical canal without bleeding or other trauma. During virus inoculation, the animals are placed in a ventral decubitus position, and after inoculation, they are left with their hindquarters elevated for about 30 minutes. Unlike i.v. inoculation, not all chimpanzees exposed cervically to relatively high doses (generally at least 1000 $TCID_{50}$) of various HIV-1 strains become infected, and some animals appear to be resistant to infection by this route. In addition to exposure via the cervix, one adult chimpanzee became infected after swabbing its vaginal wall with cell-free HIV-1 (Fultz *et al.*, 1986a), but this method has been used infrequently. Attempts to develop a model of rectal infection using adult male chimpanzees appears less reproducible than cervical inoculation. Only one of five chimpanzees exposed by this route to high doses of cell-free LAI-derived strains has become infected (Fultz, unpublished).

Key parameters to monitor infection

The natural history of HIV-1 in chimpanzees infected with LAI-derived strains mirrors that of HIV-1 infection of humans and is characterized by initial high viral burdens that decrease as HIV-1-specific immune responses become evident, followed by a long clinically asymptomatic phase (Fultz, 1993). Thus, infection can be confirmed by detection of virus and HIV-1-specific immune responses. Although the majority of chimpanzees are infected with $HIV\text{-}1_{LAI/IIIB}$, several other strains, including genetically diverse strains from other clades/subtypes, have been used to establish infections (Table 131.1). Until recently, the HIV-1 strains used for experimental infection of chimpanzees have induced minimal signs of disease in this species. However, as discussed above, one long-term infected chimpanzee has died of AIDS (Novembre *et al.*, 1997), and there are reports of other long-term HIV-1-infected animals with progressive loss of CD4+ lymphocytes. Furthermore, strains of HIV-1 from the animal that died induce manifestations of disease and are being used to evaluate progression of disease in infected chimpanzees.

Table 131.1 Human immunodeficiency virus type 1 (HIV-1) strains shown to establish persistent infections in chimpanzees

Strain	*Subtype*	*Route*	*References*
LAI/LAV-1b	B	i.v.	Francis *et al.*, 1984
		i.v.	Fultz *et al.*, 1986b
		Rectal	Fultz *et al.*, 1999
LAI/IIIB	B	i.v.	Gajdusek *et al.*, 1985
		i.v.	Nara *et al.*, 1987
		Cervical	Girard *et al.*, 1992
LAI/JC	B	i.v	Novembre *et al.*, 1997
LAI/JC499	B	i.v./cervical	Davis *et al.*, 1998
SF2	B	i.v.	Fultz *et al.*, 1987b
		i.v.	Morrow *et al.*, 1989
DH12	B	i.v.	Shibata *et al.*, 1995
		Cervical	Girard *et al.*, 1998
Merck 5016	B	i.v.	Conley *et al.*, 1996
90CR402	E	i.v./cervical	Barré-Sinoussi *et al.*, 1997
92UG029	A	i.v.	Fultz and Girard, unpublished
92UG024	D	i.v.	Fultz, unpublished

Virological

The most frequently used method to confirm that a chimpanzee is infected with HIV-1 is isolation of virus from PBMC by cocultivation of an animal's PBMC with mitogen-activated normal human PBMC or a susceptible human CD4+ T-cell line that will amplify the virus (Fultz *et al.*, 1986b; Nara *et al.*, 1987). Briefly, depending on the size of the culture vessel, from 1×10^6 to 1×10^7 chimpanzee PBMC are cocultured with an equal number of human PBMC at 1×10^6 cells/ml. Production of HIV-1 in the culture supernatant can be monitored by enzymatic assay for cell-free RT activity or by enzyme immunoassay (EIA) for p24*gag* antigen. Since CD8+ T lymphocytes can suppress HIV-1 replication, removal of these cells from chimpanzee PBMC by positive selection with immunomagnetic beads, for example, and coculturing only the CD4+ T cells often enhances virus detection (Castro *et al.*, 1991). In addition to isolation of HIV-1, serum or plasma samples can be tested directly for p24*gag* antigen by EIA. However, false-negative results can be obtained if an animal has developed a virus-specific antibody response which can interfere with the assay. More recent techniques to measure virion RNA in plasma samples, such as quantitative reverse transcriptase polymerase chain reaction (RT-PCR), branched DNA (bDNA) signal amplification and nucleic acid sequence-based amplification (NASBA), are significantly more sensitive than EIA. Quantitative RT-PCR and the bDNA assay require ethylenediaminetetraacetic acid (EDTA)- or citrate-treated plasma for optimal results, whereas NASBA is equally sensitive using EDTA-, citrate- or heparin-treated plasma. The above methods can also be used to detect and quantify HIV-1 in bone marrow aspirates obtained from the iliac crest or other tissues or fluids such as cerebrospinal fluid or vaginal washes.

In some instances, especially following HIV-1 exposure via a mucosal surface, infection can be established in an animal, but HIV-1 replication is limited to such an extent that virus cannot be isolated from PBMC or detected in plasma. Culturing of single-cell suspensions of lymph node tissue obtained by biopsy is sometimes, but not always, more sensitive. Failure to detect HIV-1 by coculture implies that there may be less than one cell in 1×10^6 to 1×10^7 cells, depending on the number of cells originally cultured. When all cocultures are negative for virus, it is sometimes possible to detect HIV-1 proviral DNA in lymphocytes using a sensitive nested PCR assay. However, since PCR assays routinely use only 1 μg of DNA per reaction, which is equivalent to about 1.5×10^5 cells, it is often necessary to perform multiple nested PCR assays in parallel. For example, approximately 10 independent reactions would be required to detect one provirus in 1.5×10^6 cells. This consideration is often ignored by investigators attempting to verify whether infection has been established, particularly in vaccine challenge studies.

Immunological

Immunologically, the most definitive determination of whether a chimpanzee has been infected with HIV-1 after inoculation is the development of serum antibodies to the virus. Antibody titers are generally determined using one of several commercial EIA kits. Twofold serial dilutions of serum are tested, and the titer is defined as the highest dilution of serum that results in an optical density reading above the cut-off value recommended by the manufacturer. Serum samples can also be tested for the presence of antibodies to specific HIV-1 proteins by immunoblot assay using commercially available kits or radioimmunoprecipitation of HIV-1 antigens from radiolabeled lysates of cells infected with HIV-1, followed by resolution of the proteins on polyacrylamide gels and autoradiography. It is possible, however, for an animal to be infected with HIV-1, but not seroconvert. This phenomenon has been reported in the simian immunodeficiency virus (SIV) macaque model after rectal or vaginal exposure (Pauza *et al.*, 1993; Miller *et al.*, 1994). We detected transient viremia with isolation of virus from PBMC during the first month after cervical HIV-1 exposure of a chimpanzee, but the animal never seroconverted (Girard *et al.*, 1998).

Minor surgical procedures

Minor surgical procedures are performed routinely on HIV-1-infected chimpanzees, the most frequent being biopsies of superficial lymph nodes. Other procedures include rectal or cervical biopsies and bone marrow aspirates. For such procedures, the dose of ketamine can be decreased about 50% (8 mg/kg) and combined with xylazine (Rompun, 0.6–1.0 mg/kg) for deeper anesthesia. Before any invasive procedure, with the exception of mucosal tissue biopsies, the site is shaved and cleaned with Betadine. Sterile surgical techniques are observed for lymph node biopsies. Animals should be monitored for adverse effects, such as bacterial infections, for about 10 days after these procedures and be given analgesics, if indicated.

Antiviral therapy

Because most chimpanzees infected with HIV-1 have not exhibited signs of disease, there have been no studies conducted in this model using clinical endpoints to evaluate therapies. In addition, because many investigators have assumed erroneously that all HIV-1-infected chimpanzees have extremely low viral burdens, antiviral therapies designed to lower viral burdens have not been attempted. Two prophylaxis studies to determine whether non-nucleoside RT inhibitors could prevent cell-free HIV-1 infection in chimpanzees have been reported (Schleif *et al.*, 1994; Grob *et al.*, 1997). In the study by Schleif *et al.* (1994), two chimpanzees were given the pyridinone derivative L-696,229 every 6–8 hours daily for 6 weeks before and then for either 20 or 30 weeks after HIV-$1_{LAI/IIIB}$ inoculation. The chimpanzees were trained to take the drug orally in a sweetened orange-flavored liquid. Although both animals became infected, both seroconversion and virus isolation were delayed by 8 and 36 weeks, compared to the control animal. In the second study (Grob *et al.*, 1997), nevirapine was administered to three chimpanzees by gavage 6–36 hours before and then daily for either 10 or 20 days after i.v. inoculation of HIV-$1_{LAI/IIIB}$. None of the three treated animals seroconverted, and virus was not isolated from PBMC or detected in plasma by bDNA assay during 23 months of follow-up. However, proviral DNA was detected sporadically by PCR in PBMC from all three chimpanzees. In a third study, a chimeric protein consisting of the two NH_2-terminal domains of the CD4 molecule covalently linked to the Fc portion of a human IgG1 antibody, called an immunoadhesin, was shown to prevent infection of chimpanzees when the immunoadhesin was administered 8 hours before and for 9 weeks after HIV-1 inoculation (Ward *et al.*, 1991). These three studies demonstrate the feasibility of testing antiviral therapies in HIV-1-infected chimpanzees. Furthermore, with the identification of HIV-1-infected chimpanzees with persistent plasma viremia greater than 10^4 RNA copies/ml (Novembre *et al.*, 1997; Davis *et al.*, 1998; Fultz, unpublished) and more sensitive assays, such as quantitative RT-PCR, the potential for testing novel therapies for the ability to effect a decrease in viral burden exists.

A different approach for treatment of HIV-1-infected humans has involved attempts to enhance ongoing anti-HIV immune responses in infected individuals by immunization with HIV-1 antigens (Mascola *et al.*, 1996). Two such studies were done in the HIV-1 chimpanzee model, both of which resulted in increases in HIV-1-specific antibody and proliferative responses. Gibbs *et al.* (1991) administered gp120-depleted, fixed HIV-1 virions to two chimpanzees that had been infected for about 4 years, whereas Fultz *et al.* (1992a) repeatedly inoculated an HIV-1-infected chimpanzee with both HIV-related and unrelated antigens or adjuvants. In this latter study, although specific immune responses increased, each inoculation was followed by an immediate, transient increase in viral replication, which is consistent with HIV's requirement for activated lymphocytes in order to replicate. Subsequent studies showed that immunization of HIV-infected humans also induced transient increases in expression of virus (Staprans *et al.*, 1995; Stanley *et al.*, 1996), demonstrating that the HIV-1 chimpanzee model was predictive of HIV-1 infection in humans.

Prophylactic vaccination

Passive immunization

The most informative studies to date in the HIV-1 chimpanzee model have been those to evaluate immune-mediated prophylaxis, both active and passive, against HIV-1 infection (Table 131.2). Neutralizing anti-HIV-1 immunoglobulins (HIVIG), purified from serum of HIV-1-infected humans, were administered i.v. to chimpanzees 24 hours before they were challenged i.v. with HIV-1. Only 1 of 6 animals was protected from establishment of a productive infection during the period of observation (Prince *et al.*, 1988, 1991); this chimpanzee had received the highest dose of HIVIG and was challenged with a low dose of virus. The study suggested, however, that neutralizing antibodies to HIV-1 may be protective.

To test this possibility more directly, Emini *et al.* (1990a) mixed and incubated aliquots of HIV-1 with either neutralizing IgG from an infected chimpanzee or a purified murine monoclonal antibody (0.5 β) *in vitro* and then inoculated the mixture i.v. into 1 chimpanzee each. (The 0.5 β monoclonal antibody recognizes a major neutralizing epitope in the V3 loop of the external HIV-1 envelope glycoprotein, gp120.) The chimpanzee IgG prevented infection, and the monoclonal antibody delayed onset of infection. In a follow-up study, a chimeric mouse–human monoclonal

Table 131.2 Prophylactic passive immunization of chimpanzees

Antibody	*Specificity*	*Challenge virus**	*Protected†*	*References*
Human IgG	HIV-1	LAI/IIIB	0/4	Prince *et al.*, 1988
Human IgG	HIV-1	LAI/IIIB	0/1‡	Emini *et al.*, 1990a
Monoclonal 0.5β	IIIB-V3	LAI/IIIB	0/1‡	Emini *et al.*, 1990a
Chimp IgG§	LAI/IIIB	LAI/IIIB	0/1‡	Emini *et al.*, 1990a
Human IgG	HIV-1	LAI/IIIB	1/2¶	Prince *et al.*, 1991
Monoclonal CB1**	IIIB-V3	LAI/IIIB	2/2	Emini *et al.*, 1992
Monoclonal 2F5	gp41 epitope	5016	0/2	Conley *et al.*, 1996
Chimeric CD4-IgG(Fc)	gp120	LAI/IIIB	2/2	Ward *et al.*, 1991

* All challenges were intravenous.
† Number protected/number treated and challenged.
‡ Antibodies and virus were incubated *in vitro*, and the mixture was used as the challenge.
§ Neutralizing IgG from an HIV-1-infected chimpanzee.
¶ Protected animal received a higher dose of IgG and lower virus challenge dose.
** Chimeric mouse–human IgG1 with 0.5β specificity.

antibody with the variable region of 0.5 β was infused into 1 chimpanzee 24 hours before and into another animal 24 hours after inoculation of HIV-1 (Emini *et al.*, 1992); both animals were protected. All of these animals were inoculated with cell-free HIV-1, suggesting that neutralizing antibodies play a role in protection against cell-free virus. Furthermore, these studies suggest that passive transfer of HIVIG or a cocktail of monoclonal antibodies administered to health-care workers after needlestick accidents or to children of HIV-1-infected mothers may be prophylactic.

Active immunization

Effective therapies for HIV-infected humans are expensive and not readily available worldwide, especially in underdeveloped nations; therefore, an affordable, easily administered vaccine is needed to stop the epidemic spread of HIV. Several strategies to develop an effective vaccine have been tested in the HIV-1 chimpanzee model, with mixed results (Table 131.3). Although several vaccine prototypes have elicited protective immune responses, most of the live virus challenges were done by optimizing the conditions and time of challenge. The majority of studies used vaccines made from antigens of the same HIV-1 strain that was later used to challenge the animals. Such homologous vaccinations and exposures will not occur in life because the diversity of HIV-1 strains is too great. Multiple major subtypes are circulating throughout the world, and within each subtype, diversity can approach 20% in the envelope proteins. Furthermore, the diversity between strains of different subtypes can approach 40% (Myers *et al.*, 1994). Thus, any effective vaccine must elicit a broadly cross-reactive immune response.

Table 131.3 Prophylactic immunization of chimpanzees against human immunodeficiency virus type 1 (HIV-1) infection

	HIV-1 strain			
Immunogen	*Vaccine*	*Challenge**	*Protected†*	*References*
Whole inactivated virus				
HIV-1	LAI/IIIB	LAI/LAV	0/3	Niedrig *et al.*, 1993
HIV-1 + gp160 + V3	LAI/LAV	LAI/IIIB	1/1	Girard *et al.*, 1991
Purified proteins				
rgp120 Env	LAI/IIIB	LAI/IIIB	0/2	Arthur *et al.*, 1989
rgp120 Env	LAI/IIIB	LAI/IIIB	0/2	Berman *et al.*, 1988
rgp120 Env	LAI/IIIB	LAI/IIIB	2/2	Berman *et al.*, 1990
rgp160 Env	LAI/IIIB	LAI/IIIB	0/2	Berman *et al.*, 1990
p55 Gag	LAI/IIIB	LAI/IIIB	0/1	Emini *et al.*, 1990b
rgp160 + p18 Gag + V3	LAI/LAV	LAI/IIIB	1/1	Girard *et al.*, 1991
rgp160 Env + V3 +/– p18 Gag	LAI/LAV	cells-IIIB	3/3	Fultz *et al.*, 1992b
rgp120 Env	LAI/IIIB	LAI/IIIB	0/2	Bruck *et al.*, 1994
rgp160 Env	LAI/IIIB	LAI/IIIB	1/2	Bruck *et al.*, 1994
rgp160 + V3	LAI + MN	SF2	2/2	Girard *et al.*, 1995
rgp160 + V3 + Nef	LAI + LAV	SF2	1/1	Girard *et al.*, 1995
rgp160 + V3	LAI + MN	90CR402	0/2	Girard *et al.*, 1996
rgp120 Env	MN	SF2	3/3	Berman *et al.*, 1996
Live recombinant viruses				
vaccinia-gp160 Env	LAI/LAV	LAI/LAV	0/2	Hu *et al.*, 1987
vaccinia-gp160 Env + V3 + Nef + p18 Gag	LAI/LAV	LAI/IIIB	0/1	Girard *et al.*, 1991
vaccinia-gp160 Env	LAI/LAV	LAI/LAV	0/4	Pincus *et al.*, 1994
canarypox-gp160 Env +rgp160 Env	LAI + MN	SF2	0/2	Girard *et al.*, 1995
vaccinia-gp160+V3 +Nef + rgp160 Env	LAI + MN + ELi	SF2	1/1	Girard *et al.*, 1995
canarypox-gp120/TM –Gag-Pro	LAI/LAV	cells-IIIB	1/2	Girard *et al.*, 1997
canarypox-gp120/TM –Gag-Pro	MN + LAI	DH12	0/2	Girard *et al.*, 1997
adeno-gp160 + rgp120 Env	MN + SF2	SF2	4/4	Lubeck *et al.*, 1997
DNA-gp160-Gag/Pol	MN + IIIB	SF2	1/3	Boyer *et al.*, 1997

* All virus challenges were intravenous.
† Number protected/number immunized and challenged.

In addition to the homologous challenges in which the immunogens were derived from the challenge virus, two types of heterologous challenges have been reported. The first of these involved immunization of animals with Env immunogens from one subtype B strain, either LAI/IIIB or MN, and challenging them with a second, unrelated subtype B strain, SF2 (Girard *et al.*, 1995; Berman *et al.*, 1996). These two strains differ by about 10% at the nucleotide level in the C2-V5 region of the *env* gene. Of a total of 7 immunized chimpanzees, 5 animals appeared to be protected from an i.v. challenge. However, the caveat in these experiments is that the SF2 strain infects chimpanzees poorly, and it is possible that the pre-existing immunity sequestered the virus in lymph nodes or lowered viral burdens to undetectable levels. In a more rigorous challenge, 2 of the animals apparently protected from infection with HIV-1_{SF2} were subsequently boosted and then challenged with a subtype E strain, 90CR402 (Girard *et al.*, 1996). Both animals became infected with HIV-$1_{90CR402}$. These studies illustrate the difficulties involved in developing an efficacious vaccine against HIV-1 and indicate that there will continue to be a need for the HIV-1 chimpanzee model in this area of research.

Table 131.4 Advantages and disadvantages of the human immunodeficiency virus type 1 (HIV-1) chimpanzee model

Advantages
Chimpanzees are the only animals infectable with HIV-1
Chimpanzees are 98% genetically identical to humans and have comparable life spans
Chimpanzees can be infected by multiple HIV-1 strains from different clades
Infection established with cell-free and cell-associated HIV-1
Infection established by parenteral and mucosal routes
High viral burdens established and maintained by some HIV-1 strains
Immunodeficiency and other sequelae of HIV-1-induced disease can develop
Disadvantages
High cost of maintenance
Limited numbers available
Experimental groups are small and preclude statistical analysis

Advantages and disadvantages of the model

Evaluation of HIV-1 infection of chimpanzees has several advantages over all other lentivirus non-human primate models (Table 131.4):

1. Chimpanzees are the only animal species that can be infected reproducibly with HIV-1.
2. Chimpanzees are genetically the closest relatives to humans.
3. Infection can be established by mucosal and i.v. routes with low doses of cell-free and cell-associated HIV-1.
4. Chimpanzees can be infected with multiple HIV-1 strains representing different subtypes, that is, a surrogate virus such as SIV is not required.
5. Immunodeficiency disease can result from HIV-1 infection.

Although primary HIV-1 isolates representing clades A, B, D and E have been shown to establish chronic infections in chimpanzees (Table 131.1), with a few exceptions, viral burdens in the animals are generally lower than those attained by strains derived from the LAI isolate (Shibata *et al.*, 1995; Conley *et al.*, 1996; Barré-Sinoussi *et al.*, 1997; Davis *et al.*, 1998; Fultz, unpublished data). (Primary isolates are defined as those viruses isolated directly from PBMC and passaged a minimal number of times *only* in PBMC from normal donors.) In addition, it appears that only T-cell line-tropic, syncytium-inducing (SI) and not non-syncytium-inducing (NSI) HIV-1 isolates replicate in chimpanzee PBMC *in vitro* and, presumably, can establish infections *in vivo*. This finding is of concern to some investigators because NSI strains appear to be transmitted between humans more frequently than SI strains. Vaccine-elicited protection might vary depending on the biological properties of a particular HIV-1 strain.

There are several disadvantages associated with the use of chimpanzees; these apply not only to their use as a model for HIV-1 infections, but also in most unrelated studies. The disadvantages include the limited number of chimpanzees available for research and the cost of maintaining these animals, generally for their lifetimes, which can extend to more than 40 years in captivity. Because the number of chimpanzees per group in any experimental protocol is low, varying from 1 to 3 animals, statistical significance cannot be determined. Thus, if the results for a control animal differ from that of animals in the experimental groups, one can only conclude that a therapy or vaccine regimen potentially could or could not be effective.

Contributions of the model to infectious disease therapy

The most informative use of the HIV-1 chimpanzee model has been in studies to evaluate the protective efficacy of candidate vaccines, which have included purified recombinant proteins and peptide antigens; recombinant live vaccinia, canarypox and adenoviruses expressing HIV-1 antigens; and DNA vectors encoding HIV-1 antigens (see Table 131.3). Because of the genetic diversity of HIV-1, chimpanzees will continue to be of value as additional novel vaccine approaches are developed and require testing. To date, the HIV-1 chimpanzee model has been used sparingly to test new antiviral therapies; however, this may change now that there exist strains of HIV-1 that induce loss and dysfunction of CD4+ T cells within weeks of inoculation of chimpanzees (Novembre *et al.*, 1997; Davis *et al.*, 1998). The use of these strains to infect chimpanzees will also provide animals with which to test novel therapies. In addition, the sensitive techniques that are now available to detect HIV-1 make it possible to use the more than 150 HIV-1-infected chimpanzees currently residing in primate facilities in therapy studies. Thus, just as chimpanzees played a critical role in the development of a vaccine against hepatitis B virus, it is likely that their use to identify an efficacious vaccine against HIV-1 will be just as important.

Acknowledgments

Unpublished results presented in this chapter were funded, in part, by National Institutes of Health grant AI28147 and by the French ANRS.

References

Agy, M. B., Frumkin, L. R., Corey, L. *et al.* (1992). Infection of *Macaca nemestrina* by human immunodeficiency virus type-1. *Science*, **257**, 103–106.

Alter, H. J., Eichberg, J. W., Masur, H. *et al.* (1984). Transmission of HTLV-III infection from human plasma to chimpanzees: an animal model for AIDS. *Science*, **226**, 549–552.

Arthur, L. O., Bess, J. W., Waters, D. J. *et al.* (1989). Challenge of chimpanzees (*Pan troglodytes*) immunized with human immunodeficiency virus envelope glycoprotein gp120. *J. Virol.*, **63**, 5046–5053.

Barré-Sinoussi, F., Chermann, J. C., Rey, F. *et al.* (1983). Isolation of a T-lymphotropic retrovirus from a patient at risk for acquired immune deficiency syndrome (AIDS). *Science*, **220**, 868–871.

Barré-Sinoussi, F., Georges-Courbot, M.-C., Fultz, P. N. *et al.* (1997). Characterization and titration of an HIV type 1 subtype E chimpanzee challenge stock. *AIDS Res. Hum. Retroviruses*, **13**, 583–591.

Berman, P. W., Groopman, J. E., Gregory, T. *et al.* (1988). Human immunodeficiency virus type 1 challenge of chimpanzees immunized with recombinant envelope glycoprotein gp120. *Proc. Natl Acad. Sci. USA*, **85**, 5200–5204.

Berman, P. W., Gregory, T. J., Riddle, L. *et al.* (1990). Protection of chimpanzees from infection by HIV-1 after vaccination with recombinant glycoprotein gp120 but not gp160. *Nature*, **345**, 622–625.

Berman, P. W., Murthy, K. K., Wrin, T. *et al.* (1996). Protection of MN-rgp120-immunized chimpanzees from heterologous infection with a primary isolate of human immunodeficiency virus type 1. *J. Infect. Dis.*, **173**, 52–59.

Boyer, J. D., Ugen, K. E., Wang, B. *et al.* (1997). Protection of chimpanzees from high-dose heterologous HIV-1 challenge by DNA vaccination. *Nature Med.*, **3**, 526–532.

Bruck, C., Thiriart, C., Fabry, L. *et al.* (1994). HIV-1 envelope-elicited neutralizing antibody titres correlate with protection and virus load in chimpanzees. *Vaccine*, **12**, 1141–1148.

Castro, B. A., Walker, C. M., Eichberg, J. W., Levy, J. A. (1991). Suppression of human immunodeficiency virus replication by $CD8^+$ cells from infected and uninfected chimpanzees. *Cell. Immunol.*, **132**, 246–255.

Conley, A. J., Kessler, J. A., Boots, L. J. *et al.* (1996). The consequence of passive administration of an anti-human immunodeficiency virus type 1 neutralizing monoclonal antibody before challenge of chimpanzees with a primary virus isolate. *J. Virol.*, **70**, 6751–6758.

Davis, I. C., Girard, M., Fultz, P. N. (1998). Loss of CD4+ T cells in HIV-1-infected chimpanzees is associated with increased lymphocyte apoptosis. *J. Virol.*, **72**, 4623–4632.

Eichberg, J. W., Lee, D. R., Aan, J. S. *et al.* (1988). *In utero* infection of an infant chimpanzee with HIV. *N. Engl. J. Med.*, **319**, 722–723.

Emini, E. A., Nara, P. L., Schleif, W. A. *et al.* (1990a). Antibody-mediated *in vitro* neutralization of human immunodeficiency virus type 1 abolishes infectivity for chimpanzees. *J. Virol.*, **64**, 3674–3678.

Emini, E. A., Schleif, W. A., Quintero, J. C. *et al.* (1990b). Yeast-expressed p55 precursor core protein of human immunodeficiency virus type 1 does not elicit protective immunity in chimpanzees. *AIDS Res. Hum. Retroviruses*, **6**, 1247–1250.

Emini, E. A., Schleif, W. A., Nunberg, J. H. *et al.* (1992). Prevention of HIV-1 infection in chimpanzees by gp120 V3 domain-specific monoclonal antibody. *Nature*, **355**, 728–730.

Fauci, A. S. (1993). Multifactorial nature of human immunodeficiency virus disease: implications for therapy. *Science*, **262**, 1011–1018.

Francis, D. P., Feorino, P. M., Broderson, J. R. *et al.* (1984). Infection of chimpanzees with lymphadenopathy-associated virus. *Lancet* **ii**, 1276–1277.

Frumkin, L. R., Agy, M. B., Coombs, R. W. *et al.* (1993). Acute infection of *Macaca nemestrina* by human immunodeficiency virus type 1. *Virology*, **195**, 422–431.

Fultz, P. N. (1993). Nonhuman primate models for AIDS. *Clin. Infect. Dis.*, **17** (suppl. 1), S230–S235.

Fultz, P. N. (1997). Animal models for human immunodeficiency virus infection and disease. In: *AIDS and Other Manifestations of HIV Infection*, 3rd edn (ed Wormser, G. P.), pp. 201–215. Lippincott-Raven, New York.

Fultz, P. N., McClure, H. M., Daugharty, H. *et al.* (1986a). Vaginal transmission of human immunodeficiency virus (HIV) to a chimpanzee. *J. Infect. Dis.*, **154**, 896–900.

Fultz, P. N., McClure, H. M., Swenson, R. B. *et al.* (1986b). Persistent infection of chimpanzees with human T-lymphotropic virus type III/lymphadenopathy-associated virus: a potential model for acquired immunodeficiency syndrome. *J. Virol.*, **58**, 116–124.

Fultz, P. N., Greene, C., Switzer, W., Swenson, B., Anderson, D., McClure, H. M. (1987a). Lack of transmission of human immunodeficiency virus from infected to uninfected chimpanzees. *J. Med. Primatol.*, **16**, 341–347.

Fultz, P. N., Srinivasan, A., Greene, C. R., Butler, D., Swenson, R. B., McClure, H. M. (1987b). Superinfection of a chimpanzee with a second strain of human immunodeficiency virus. *J. Virol.*, **61**, 4026–4029.

Fultz, P. N., McClure, H. M., Swenson, R. B., Anderson, D. C. (1989). HIV infection of chimpanzees as a model for testing chemotherapeutics. *Intervirology*, **30**, 51–58.

Fultz, P. N., Siegel, R. L., Brodie, A. *et al.* (1991). Prolonged CD4+ lymphocytopenia and thrombocytopenia in a chimpanzee persistently infected with HIV-1. *J. Infect. Dis.*, **163**, 441–447.

Fultz, P. N., Gluckman, J.-C., Muchmore, E., Girard, M. (1992a). Transient increases in numbers of infectious cells in an HIV-infected chimpanzee following immune stimulation. *AIDS Res. Hum. Retroviruses*, **8**, 313–317.

Fultz, P. N., Nara, P., Barré-Sinoussi, F. *et al.* (1992b). Vaccine protection of chimpanzees against challenge with HIV-1-infected peripheral blood mononuclear cells. *Science*, **256**, 1687–1690.

Fultz, P. N., Wei, Q., Yue, L. (1999). Rectal transmission of human immunodeficiency virus type 1 to chimpanzees. *J. Infect. Dis.*, **179**(53), in press.

Gajdusek, D. C., Gibbs, C. J., Rodgers-Johnson, P. *et al.* (1985). Infection of chimpanzees by human T-lymphotropic retroviruses in brain and other tissues from AIDS patients. *Lancet*, **1**, 55–56.

Gartner, S., Liu, Y., Polonis, V. *et al.* (1994). Adaptation of HIV-1 to pigtailed macaques. *J. Med. Primatol.*, **23**, 155–163.

Gibbs, Jr., C. J., Peters, R., Gravell, M. *et al.* (1991). Observations after human immunodeficiency virus immunization and challenge of human immunodeficiency virus seropositive and seronegative chimpanzees. *Proc. Natl. Acad. Sci. USA*, **88**, 3348–3352.

Girard, M., Kieny, M.-P., Pinter, A. *et al.* (1991). Immunization of chimpanzees confers protection against challenge with human immunodeficiency virus. *Proc. Natl. Acad. Sci. USA*, **88**, 542–546.

Girard, M., Mahoney, J., Rimsky, L. *et al.* (1992). HIV-1 genital infection: a chimpanzee model. In *Retroviruses of Human AIDS and Related Animal Diseases* (eds Girard, M., Valette, L.), pp. 75–79. Fondation Merieux, Lyon, France.

Girard, M., Meignier, B., Barré-Sinoussi, F. *et al.* (1995). Vaccine-induced protection of chimpanzees against infection by a heterologous human immunodeficiency virus type 1. *J. Virol.*, **69**, 6239–6248.

Girard, M., Yue, L., Barré-Sinousi, F. *et al.* (1996). Failure of a human immunodeficiency virus type 1 (HIV-1) subtype B-derived vaccine to prevent infection of chimpanzees by an HIV-1 subtype E strain. *J. Virol.*, **70**, 8229–8233.

Girard, M., van der Ryst, E., Barré-Sinoussi, F. *et al.* (1997). Challenge of chimpanzees immunized with a recombinant canarypox-HIV-1 virus. *Virology*, **232**, 98–104.

Girard, M., Mahoney, J., Wei, Q. *et al.* (1998). Genital infection of female chimpanzees with human immunodeficiency virus type 1. *AIDS Res. Hum. Retroviruses*, **14**, 1357–1367.

Gottlieb, M. S., Schroff, R., Schanker, H. M. *et al.* (1981). *Pneumocystis carinii* pneumonia and mucosal candidiasis in previously healthy homosexual men: evidence of a new acquired cellular immunodeficiency. *N. Engl. J. Med.*, **305**, 1425–1431.

Goudsmit, J., Smit, L., Krone, W. J. A. *et al.* (1987). IgG response to human immunodeficiency virus in experimentally infected chimpanzees mimics the IgG response in humans. *J. Infect. Dis.*, **155**, 327–331.

Gougeon, M.-L., Lecoeur, H., Boudet, F. *et al.* (1997). Lack of chronic immune activation in HIV-infected chimpanzees correlates with the resistance of T cells to Fas/Apo-1 (CD95)-induced apoptosis and preservation of a T helper 1 phenotype. *J. Immunol.*, **158**, 2964–2976.

Grob, P. M., Cao, Y., Muchmore, E. *et al.* (1997). Prophylaxis against HIV-1 infection in chimpanzees by nevirapine, a non-nucleoside inhibitor of reverse transcriptase. *Nature Med.*, **3**, 665–670.

Heeney, J. L. (1995). AIDS: a disease of impaired Th-cell renewal? *Immunol. Today*, **16**, 515–520.

Heeney, J., Jonker, R., Koornstra, W. *et al.* (1993). The resistance of HIV-infected chimpanzees to progression to AIDS correlates with absence of HIV-related T-cell dysfunction. *J. Med. Primatol.*, **22**, 194–200.

Hu, S.-L., Fultz, P. N., McClure, H. M. *et al.* (1987). Effect of immunization with a vaccinia-HIV *env* recombinant on HIV infection of chimpanzees. *Nature*, **328**, 721–723.

Levy, J. A. (1993). Pathogenesis of human immunodeficiency virus infection. *Microbiol. Rev.*, **57**, 183–289.

Lubeck, M. D., Natuk, R., Myagkikh, M. *et al.* (1997). Long-term protection of chimpanzees against high-dose HIV-1 challenge induced by immunization. *Nature Med.*, **3**, 651–658.

Lusso, P., Markham, P. D., Ranki, A. *et al.* (1988). Cell-mediated immune response toward viral envelope and core antigens in gibbon apes (*Hylobates lar*) chronically infected with human immunodeficiency virus-1. *J. Immunol.*, **141**, 2467–2473.

Mascola, J. R., Snyder, S. W., Weislow, O. S. *et al.* (1996). Immunization with envelope subunit vaccine products elicits neutralizing antibodies against laboratory-adapted but not primary isolates of human immunodeficiency virus type 1. *J. Infect. Dis.*, **173**, 340–348.

Miller, C. J., Marthas, M., Torten, J. *et al.* (1994). Intravaginal inoculation of rhesus macaques with cell-free simian immunodeficiency virus results in persistent or transient viremia. *J. Virol.*, **68**, 6391–6400.

Morrow, W. J. W., Wharton, M., Lau, D., Levy, J. A. (1987). Small animals are not susceptible to human immunodeficiency virus infection. *J. Gen. Virol.*, **68**, 2253–2257.

Morrow, W. J. W., Homsy, J., Eichberg, J. W. *et al.* (1989). Long-term observation of baboons, rhesus monkeys, and chimpanzees inoculated with HIV and given periodic immuno-suppressive treatment. *AIDS Res. Hum. Retroviruses*, **5**, 233–245.

Myers, G., Korber, B., Wain-Hobson, S., Jeang, K.-T., Henderson, L. E., Pavlakis, G. N. (1994). *Human Retroviruses and AIDS*. Los Alamos National Laboratory, Los Alamos, NM.

Nara, P. L., Robey, W. G., Arthur, L. O. *et al.* (1987). Persistent infection of chimpanzees with human immunodeficiency virus: serological responses and properties of reisolated viruses. *J. Virol.*, **61**, 3173–3180.

Nara, P., Hatch, W., Kessler, J., Kelliher, J., Carter, S. (1989). The biology of human immunodeficiency virus-1 IIIB infection in the chimpanzee: *in vivo* and *in vitro* correlations. *J. Med. Primatol.*, **18**, 343–355.

Niedrig, M., Gregersen, J.-P., Fultz, P. N., Broker, M., Mehdi, S., Hilfenhaus, J. (1993). Immune response of chimpanzees after immunization with the inactivated whole immunodeficiency virus (HIV-1), three different adjuvants and challenge. *Vaccine*, **11**, 67–74.

Novembre, F. J., Saucier, M., Anderson, D. C. *et al.* (1997). Development of AIDS in a chimpanzee infected with human immunodeficiency virus type 1. *J. Virol.*, **71**, 4086–4091.

Otten, R. A., Brown, B. G., Simon, M. *et al.* (1994). Differential replication and pathogenic effects of HIV-1 and HIV-2 in *Macaca nemestrina*. *AIDS*, **8**, 297–306.

Pauza, C. D., Emau, P., Salvato, M. S. *et al.* (1993). Pathogenesis of SIVmac251 after atraumatic inoculation of the rectal mucosa in rhesus monkeys. *J. Med. Primatol.*, **22**, 154–161.

Pincus, S. H., Messer, K. G., Hu, S.-L. (1994). Effect of nonprotective vaccination on antibody response to subsequent human immunodeficiency virus infection. *J. Clin. Invest.*, **93**, 140–146.

Prince, A. M., Horowitz, B., Baker, L. *et al.* (1988). Failure of a human immunodeficiency virus (HIV) immune globulin to protect chimpanzees against experimental challenge with HIV. *Proc. Natl Acad. Sci. USA*, **85**, 6944–6948.

Prince, A. M., Reesink, H., Pascual, D. *et al.* (1991). Prevention of HIV infection by passive immunization with HIV immunoglobulin. *AIDS Res. Hum. Retroviruses*, **7**, 971–973.

Schleif, W. A., Murthy, K. K., Sardana, V. V. *et al.* (1994). Attempted prophylaxis of human immunodeficiency virus type 1 infection in chimpanzees with a nonnucleoside reverse transcriptase inhibitor. *AIDS Res. Hum. Retroviruses*, **10**, 107–110.

Schuitemaker, H., Meyaard, L., Kootstra, N. A. *et al.* (1993). Lack of T cell dysfunction and programmed cell death in human immunodeficiency virus type 1-infected chimpanzees correlates with absence of monocytotropic variants. *J. Infect. Dis.*, **168**, 1140–1147.

Shibata, R., Hoggan, M. D., Broscius, C. *et al.* (1995). Isolation and characterization of a syncytium-inducing, macrophage/T-cell line-tropic human immunodeficiency virus type 1 isolate that readily infects chimpanzee cells *in vitro* and *in vivo*. *J. Virol.*, **69**, 4453–4462.

Stanley, S. K., Ostrowski, M. A., Justement, J. S. *et al.* (1996). Effect of immunization with a common recall antigen on viral expression in patients infected with human immunodeficiency virus type 1. *N. Engl. J. Med.*, **334**, 1222–1230.

Staprans, S. I., Hamilton, B. L., Follansbee, S. E. *et al.* (1995). Activation of virus replication after vaccination of HIV-1-infected individuals. *J. Exp. Med.*, **182**, 1727–1737.

Villinger, F., Brar, S. S., Brice, G. T. *et al.* (1997). Immune and hematopoietic parameters in HIV-1-infected chimpanzees during clinical progression towards AIDS. *J. Med. Primatol.*, **26**, 11–18.

Wain-Hobson, S., Vartanian, J.-P., Henry, M. *et al.* (1991). LAV revisited: origins of the early HIV-1 isolates from Institut Pasteur. *Science*, **252**, 961–965.

Ward, R. H. R., Capon, D. J., Jett, C. M. *et al.* (1991). Prevention of HIV-1 IIIB infection in chimpanzees by CD4 immunoadhesin. *Nature*, **352**, 434–436.

World Health Organization (1993). *The HIV/AIDS Pandemic. 1993 Overview*. World Health Organization, Geneva.

Index

ISBN 0-12-775390-7